Cytokines

in Health and Disease

Cytokines

in Health and Disease

Second Edition, Revised and Expanded

edited by

Daniel G. Remick

University of Michigan
Ann Arbor, Michigan

Jon S. Friedland

Royal Postgraduate Medical School
Hammersmith Hospital
London, England

MARCEL DEKKER

MARCEL DEKKER, INC. NEW YORK • BASEL • HONG KONG

Library of Congress Cataloging-in-Publication Data

Cytokines in health and disease / edited by Daniel G. Remick, Jon S. Friedland. — 2nd ed., rev. and expanded.
p. cm.
Includes bibliographical references and index.
ISBN 0-8247-9823-6 (hardcover : alk. paper)
1. Cytokines. I. Remick, Daniel G. II. Friedland, Jon S.
[DNLM: 1. Cyokines—physiology. 2. Immunity, Cellular. QW 568 C9961 1997]
QR185.8.C95C96 1997
616.07'9—dc21
DNLM/DLC
for Library of Congress

97-12544
CIP

The publisher offers discounts on this book when ordered in bulk quantities. For more information, write to Special Sales/Professional Marketing at the address below.

This book is printed on acid-free paper.

MARCEL DEKKER, INC.
270 Madison Avenue, New York, New York 10016
http://www.dekker.com

Current printing (last digit):
10 9 8 7 6 5 4 3 2 1

PRINTED IN THE UNITED STATES OF AMERICA

To our wives,
Robin and Jo

Preface

Since the first edition of this book was published 5 years ago, interest in the field of cytokines has continued to grow exponentially. The word cytokine has moved into common parlance both in the laboratory and in the clinic. The molecular and immunological understanding of these protein mediators has been expanding quickly and, almost as rapidly, new knowledge is being applied first in animal models and then in the clinic. However, the scientist working on interferons and struggling to keep up with the literature in that field may know little about interleukin-10. Similarly, the clinician interested in infectious diseases may be unable to take advantage of developments in oncology published in different specialty journals. More than ever, there is a need to provide a single authoritative source of information on cytokines for the busy basic scientist and the practicing physician. This is the principal aim of our book.

We believe this new edition has been improved in many ways that will benefit our readers. At a glance, it is obvious that the book has been expanded to provide an even more useful outline of the information now available. However, the editors have kept to a single volume so that *Cytokines in Health and Disease* remains a functional reference work. In addition, the book has been given a more clearly defined structure. The book is divided into four parts: biochemistry and function of the principal cytokines; cytokines and inflammatory cells; cytokines in organs/systems; and cytokines in infectious diseases. There is an improved index and a helpful glossary of terms for those less familiar with the field of cytokines.

This edition has taken on a more international flavor, with the addition of Jon Friedland as a European editor, and there are many more contributors from outside the United States. Although every effort has been made to make the book up to date, the speed of progress in fields such as HIV infection, for example, means that new developments inevitably occur while the book is in press. However, the book is intended to allow the reader to obtain a solid foundation in cytokine biology and its application as well as an understanding of the direction in which different areas are moving. Most chapters contain a final section in which the author attempts to look into the future and predict what might lie ahead.

We are delighted that the contributors to this volume include many leaders in the field of cytokine biology. We hope that this volume, like the last, will be a useful reference work for scientists and clinicians around the world. However, the last word has not been written and suggestions for future editions are always welcome.

Daniel G. Remick
Jon S. Friedland

Contents

Contributors

Rasul Abdolrasulnia Vanderbilt University and VA Medical Center, Nashville, Tennessee

Pierette M. Appasamy, Ph.D. Pittsburgh Cancer Institute, University of Pittsburgh School of Medicine, Pittsburgh, Pennsylvania

Frances R. Balkwill, Ph.D. Biological Therapies Laboratory, Imperial Cancer Research Fund, London, England

Michele Barry, Ph.D. Department of Biochemistry, University of Alberta, Edmonton, Alberta, Canada

Andrew Bean DNAX Research Institute, Palo Alto, California

Etty N. Benveniste, Ph.D. Department of Cell Biology, University of Alabama at Birmingham, Birmingham, Alabama

Jonathan S. Bromberg, M.D., Ph.D. Department of Surgery and Microbiology and Immunology, University of Michigan, Ann Arbor, Michigan

Frances Burke Biological Therapies Laboratory, Imperial Cancer Research Fund, London, England

Federico Bussolino Università di Torino, Torino, Italy

Michael Caligiuri, M.D. Division of Medicine, Roswell Park Cancer Institute, Buffalo, New York

William Carson Division of Medicine, Roswell Park Cancer Institute, Buffalo, New York

Stephen W. Chensue University of Michigan Medical School, Ann Arbor, Michigan

Gerard Cox, M.B., B.A.O., B.Ch. St. Joseph's Hospital and McMaster University, Hamilton, Ontario, Canada

Charles J. Czuprynski, Ph.D. School of Veterinary Medicine, University of Wisconsin-Madison, Madison, Wisconsin

Elisabetta Dejana, Ph.D. Istituto di Ricerche Farmacologiche "Mario Negri," Milan, Italy

Deborah M. DeMarco, M.D. Department of Rheumatology, University of Massachusetts Medical Center, Worcester, Massachusetts

Anna Mae Diehl, M.D. Johns Hopkins University School of Medicine, Baltimore, Maryland

Xunxiang Du, M.D. Section of Pediatric Hematology/Oncology, James Whitcomb Riley Hospital for Children, and Herman B. Wells Center for Pediatric Research, Indiana University School of Medicine, Indianapolis, Indiana

Denise A. Faherty, Ph.D. Department of Medical Affairs, Hoffmann-La Roche, Inc., Nutley, New Jersey

Joseph C. Fantone, M.D. Department of Pathology, University of Michigan Medical School, Ann Arbor, Michigan

Nancy L. Farner Department of Human Oncology, University of Wisconsin and University of Wisconsin Comprehensive Cancer Center, Madison, Wisconsin

Kenneth R. Feingold, M.D. Department of Medicine, University of California, San Francisco, California

David P. Fivenson, M.D. Henry Ford Health System, Detroit, Michigan

Ragini Fredrich, M.D. Cedars-Sinai Medical Center, Los Angeles, California

Jon S. Friedland, M.A., Ph.D., M.R.C.P. Infectious Diseases, Royal Postgraduate Medical School, Hammersmith Hospital, London, England

Maurice K. Gately, M.D., Ph.D. Department of Inflammation/Autoimmune Diseases, Hoffmann-La Roche, Inc., Nutley, New Jersey

Jack Gauldie, Ph.D. Department of Pathology, McMaster University, Hamilton, Ontario, Canada

Frank Gonzalez-Crussi, M.D., F.R.C.P.(C) Department of Pathology, Children's Memorial Hospital, Northwestern University Medical School, Chicago, Illinois

Georges E. Grau, M.D. Department of Anesthesiology, Pharmacology, and Surgical Intensive Care, University Hospital, Geneva, Switzerland

Carl Grunfeld, M.D., Ph.D. Department of Medicine, University of California, San Francisco, California

Mary Haak-Frendscho, Ph.D. School of Veterinary Medicine, University of Wisconsin-Madison, and Promega Corporation, Madison, Wisconsin

Jacquelyn A. Hank Department of Pediatrics, University of Wisconsin and University of Wisconsin Comprehensive Cancer Center, Madison, Wisconsin

Göran K. Hansson, M.D., Ph.D. Department of Medicine, King Gustaf V Research Institute, Karolinska Institute, Stockholm, Sweden

Daniel Hill, M.D. Division of Digestive Diseases and Nutrition, University of Kentucky Medical Center, Lexington, Kentucky

Wolfgang Holter, M.D. Children's Cancer Research Institute, St. Anna's Childrens Hospital, Vienna, Austria

Wei Hsueh, M.D., Ph.D. Department of Pathology, Children's Memorial Hospital, Northwestern University Medical School, Chicago, Illinois

Pamela Hunt Amgen, Inc., Thousand Oaks, California

Martino Introna Istituto di Ricerche Farmacologiche "Mario Negri," Milan, Italy

Stanley C. Jordan, M.D. Pediatric Nephrology, Cedars-Sinai Medical Center, Los Angeles, California

Soverin Karmiol, M.D. Department of Pathology, University of Michigan Medical School, Ann Arbor, Michigan

Sylvie C. Kossodo, Ph.D. Department of Anesthesiology, Pharmacology, and Surgical Intensive Care, University Hospital, Geneva, Switzerland

Steven L. Kunkel, Ph.D. Department of Pathology, University of Michigan Medical School, Ann Arbor, Michigan

Nicholas W. Lukacs University of Michigan Medical School, Ann Arbor, Michigan

Siddhartha Mahanty, M.B.B.S., M.P.H. National Institute of Allergy and Montreal General Hospital, Montreal, Quebec, Canada

Alberto Mantovani, M.D. Istituto di Ricerche Farmacologiche "Mario Negri," Milan, Italy

Craig McClain, M.D. Director, Division of Digestive Diseases and Nutrition, University of Kentucky Medical Center, Lexington, Kentucky

Grant McFadden, Ph.D. Department of Biochemistry, University of Alberta, Edmonton, Alberta, Canada

Ian K. McNiece, Ph.D. Department of Developmental Hematology, Amgen, Inc., Thousand Oaks, California

Riaz A. Memon, M. D. Department of Medicine, University of California, San Francisco, California

Linde Meyaard Central Laboratory of the Netherlands Red Cross Blood Transfusion Service and the Laboratory of Experimental and Clinical Immunology of the University of Amsterdam, Amsterdam, The Netherlands

Frank Miedema Central Laboratory of the Netherlands Red Cross Blood Transfusion Service and the Laboratory of Experimental and Clinical Immunology of the University of Amsterdam, Amsterdam, The Netherlands

Isabelle Millet, Ph.D. Section of Endocrinology, Department of Internal Medicine, Yale University School of Medicine, New Haven, Connecticut

Adrian J. Minty, Ph.D. Sanofi Recherche, Labège, France

Kevin W. Moore DNAX Research Institute, Palo Alto, California

Brian J. Nickoloff, M.D., Ph.D. Department of Pathology, and Director, Skin Disease Research Laboratories, Cardinal Bernardin Cancer Center, Loyola University Medical Center, Maywood, Illinois

Thomas Nutman, M.D., Ph.D. Helminth Immunology Section, Laboratory of Parasitic Diseases, National Institute of Allergy and Infectious Disease, National Institutes of Health, Bethesda, Maryland

Edward J. Pearce College of Veterinary Medicine, Cornell University, Ithaca, New York

Sem Hin Phan, M.D., Ph.D. Department of Pathology, University of Michigan Medical School, Ann Arbor, Michigan

Fiona Powrie Nuffield Department of Surgery, University of Oxford, John Radcliffe Hospital, Oxford, England

Jeffrey D. Punch, M.D. Department of Surgery, University of Michigan, Ann Arbor, Michigan

Daniel G. Remick, M.D. Department of Pathology, University of Michigan, Ann Arbor, Michigan

Jean-Christophe Renauld, M.D. Ludwig Institute for Cancer Research, Catholic University of Louvain, Brussels, Belgium

Nancy H. Ruddle, Ph.D. Department of Epidemiology and Public Health, Yale University School of Medicine, New Haven, Connecticut

John A. Schmidt, M.D. Inflammation Research, Department of Immunology and Inflammation, Merck Research Laboratories, Rahway, New Jersey

Hanneke Schuitemaker Central Laboratory of the Netherlands Red Cross Blood Transfusion Service and the Laboratory of Experimental and Clinical Immunology of the University of Amsterdam, Amsterdam, The Netherlands

Ganes C. Sen, Ph.D. Department of Molecular Biology, The Cleveland Clinic Foundation Research Institute, Cleveland, Ohio

Steven Shedlofsky, M.D. Division of Digestive Diseases and Nutrition, University of Kentucky Medical Center, Lexington, Kentucky

Virginia L. Shepherd, Ph.D. Department of Medicine and Biochemistry, Vanderbilt University and VA Medical Center, Nashville, Tennessee

Alan Sher National Institute of Allergy and Infectious Disease, National Institutes of Health, Bethesda, Maryland

William P. Sheridan, M.B., B.S., F.R.A.C.P. Amgen, Inc., Thousand Oaks, and UCLA School of Medicine, Los Angeles, California

Scott Simonet, Ph.D. Amgen, Inc., Thousand Oaks, California

Paul M. Sondel, M.D., Ph.D. Department of Pediatrics, Human Oncology, and Genetics, University of Wisconsin and University of Wisconsin Comprehensive Cancer Center, Madison, Wisconsin

Sigurd Steinshamn, M.D. Section of Pulmonary Medicine, Department of Medicine, University of Trondheim, Trondheim, Norway

Veronika Stemme, M.D., Ph.D. Department of Medicine, King Gustaf V Research Institute, Karolinska Institute, Stockholm, Sweden

Robert M. Strieter, M.D. University of Michigan Medical School, Ann Arbor, Michigan

Xiaoming Sun, M.D. Department of Pathology, Children's Memorial Hospital, Northwestern University Medical School, Chicago, Illinois

Xiaodi Tan, M.D. Department of Pathology, Children's Memorial Hospital, Northwestern University Medical School, Chicago, Illinois

Michael J. Tocci, Ph.D. Inflammation Research, Department of Immunology and Inflammation, Merck Research Laboratories, Rahway, New Jersey

Kevin J. Tracey, M.D. Laboratory of Biomedical Science, North Shore University Hospital, and Picower Institute of Medical Research, Manhasset, New York

Hidekazu Tsukamoto, Ph.D., D.V.M. USC Center for Liver Diseases, University of Southern California, Los Angeles, California

Thomas R. Ulich, M.D. Preclinical Development, Amgen, Inc., Thousand Oaks, California

Anders Waage, M.D. Section of Hematology, Department of Medicine, University of Trondheim, Trondheim, Norway

Jeffrey S. Warren, M.D. Department of Pathology, University of Michigan Medical School, Ann Arbor, Michigan

Mark D. Wewers, M.D. Department of Internal Medicine, The Ohio State University, Columbus, Ohio

David A. Williams, M.D. Section of Pediatric Hematology/Oncology, James Whitcomb Riley Hospital for Children; Herman B. Wells Center for Pediatric Research, Indiana University School of Medicine; Howard Hughes Medical Institute, Indianapolis, Indiana

Chang You Wu, M.D., Ph.D. Department of Inflammation/Autoimmune Diseases, Hoffmann-La Roche, Inc., Nutley, New Jersey

Tasuku Yokota, M.D. Department of Medicine, King Gustaf V Research Institute, Karolinska Institute, Stockholm, Sweden

Robert B. Zurier, M.D. Department of Medicine, University of Massachusetts Medical Center, Worcester, Massachusetts

Abbreviations

ACC	Acetyl CoA carboxylase
ACTH	Adrenocorticotrophic hormone
AD	Atopic dermatitis
Ad2-MLTF	Adenovirus 2 major late promoter transcription factor
ADCC	Antibody-dependent cell-mediated immunity
AH	Alcoholic hepatitis
AIDS	Acquired immunodeficiency syndrome
AIH	Autoimmune hepatitis
ALD	Alcoholic liver disease
ALL	Acute lymphoblastic leukemia
APC	Antigen-presenting cell
APP	Acute-phase protein
ASFV	African swine fever virus
BBB	Blood–brain barrier
BCGF II	B-cell growth factor II
BMMC	Bone-marrow-derived mast-cell lines
BMT	Bone-marrow transplant
C5a	5th Component of complement cascade
CCHP	Complement control protein homolog
CED	*Caenorhabitis elegans* death
CFU	Colony-forming units
CIA	Collagen-induced arthritis
CMI	Cell-mediated immunity
CML	Chronic myelogenous leukemia
CMV	Cytomegalovirus
CNS	Central nervous system
CNTF	Ciliary neurotropic factor
CRF	Corticotrophin releasing factor
CRM	Cytokine response modifer
CSA	Cyclosporin A
CSF	Cerebrospinal fluid
CTLA	Cytotoxic T-lymphocyte-associated antigen
CVU	Chronic venous ulcer
DDC	Dermal dendritic cell
DETC	Dendritic epidermal T cells
DLN	Draining lymph nodes
EAE	Experimental autoimmune encephalitis
EBV	Epstein-Barr virus
EC	Endothelial cell
ECM	Extracellular matrix
EGF	Epidermal growth factor
ELAM	Endothelial leukocyte adhesion molecule
EMS	Eosinophilia myalgia syndrome

ENA-78	Epithelial neutrophil activating protein-78
ENL	Erythema nodosum leprosum
ER	Endoplasmic reticulum
ESR	Erythrocyte sedimentation rate
FGF	Fibroblast growth factor
FHF	Fulminant hepatic failure
FL	Fetal liver
fMLP	N-formylmethionyl peptides
G-CSF	Granulocyte colony-stimulating factor
GAS	Gamma-interferon activated sequence
GCP-2	Granulocyte chemotactic protein-2
GLA	Gamma linolenic acid
GM-CSF	Granulocyte-macrophage colony-stimulating factor
GR	Growth response
GRE	Glucocorticoid responsive element
Gro-α,β,γ	Growth-related oncogene alpha, beta, gamma
GSH	Glutathione
GvHD	Graft vs. host disease
HA	Hyaluronic acid
HIV	Human immunodeficiency virus
HMG	High mobility group antigen
HPP	High proliferative potential
HSP	Heat shock protein
HTLV-1	Human T lymphocytotrophic virus 1
HUVEC	Human umbilical vein endothelial cells
HVS	Herpesvirus saimiri
ICAM-1	Intercellular adhesion molecule-1
ICE	IL-1 converting enzyme
IFN	Interferon
IGF	Insulin-like growth factor
IL	Interleukin
IL-1R	IL-1 receptor
ip	Intraperitoneal
IP-10	gamma-interferon-inducible protein 10
IRF-1	Interferon-γ regulatory factor-1
ISRE	Interferon-stimulated response elements
iv	Intravenous
IVIG	Intravenous immunoglobulin
JAK	Janus-associated kinases
KB	Ketone body
KGF	Keratinocyte growth factor
LAK	Lymphokine activated killer cells
LAM	Lipoarabinomannan
LBP	Lipopolysaccharide binding protein
LDL	Low-density lipoprotein
LGL	Large granular lymphocytes
LIF	Leukemia inhibitory factor
LPS	Lipopolysaccharide
LRE	LPS responsive elements
LTB4	Leukotriene B4
M-CSF	Macrophage colony-stimulating factor
MAC	Mycobacterium avium intracellulare complex
MBP	Myelin basic protein
MCP	Monocyte chemoattractant protein
MEA	Mast-cell growth-enhancing activity
MHC	Major histocompatibility complex

MIF	Macrophage inhibitory factor
MIG	Monokine-induced by gamma interferon
MIP-1α,β	Macrophage inflammatory protein-1 alpha, beta
MLC	Mixed lymphocyte reaction
MLTF	Major late promotor transcription factor
MMP	Matrix metalloproteinase
MPC	Mononuclear phagocytic cells
MS	Multiple sclerosis
MSR	Macrophage scavenger receptor
MST	Mean survival time
MSU	Monosodium urate
MTD	Maximum tolerated dose
NANB	NonA, nonB
NAP-2	Neutrophil activating factor-2
NF	Nuclear factor
NF-AT	Nuclear factor of activated T cells
NF-κB	Nuclear factor-kappa B
NK	Natural killer
NO	Nitric oxide
NSAID	Nonsteroidal antiinflammatory agents
OA	Osteoarthritis
OAF	Osteoclast activating factor
OLT	Orthotopic liver transplant
OND	Other neurological diseases
PA	Plasminogen activator
PAF	Platelet activating factor
PAI	Plasminogen activator inhibitor
PBC	Primary biliary cirrhosis
PBMC	Peripheral blood mononuclear cells
PDGF	Platelet-derived growth factor
PF4	Platelet factor 4
PG	Prostaglandin
PH	Partial hepatectomy
PK	Protein kinase
PLP	Proteolipid protein
PMN	Polymorphonuclear leukocytes
PSF	Primary sclerosing cholangitis
PTK	Protein tyrosine kinases
RA	Rheumatoid arthritis
RANTES	Receptor-activated normal T cell expressed and secreted
RNI	Reactive nitrogen intermediates
ROI	Reactive oxygen intermediates
sc	Subcutaneous
SCF	Stem-cell factor
SCID	Severe combined immunodeficiency
SEA	Schistosome egg antigen
SEB	Staphylococcal enterotoxin B
SFV	Shope fibroma virus
SIRS	Systemic inflammatory response syndrome
SLE	Systemic lupus erythematosus
SMC	Smooth muscle cells
SPF	Specific pathogen-free
SRE	Serum response elements
SRTF	Sterol regulated transcription factor
STAT	Signal transducer and activator of transcription
TG	Triglyceride

TGF-β	Transforming growth factor beta
TH1 / 2	T helper type 1 / 2
TIL	Tumor-infiltrating lymphocytes
TNF	Tumor necrosis factor
tPA	Tissue-type plasminogen activator
TRF	T-cell replacing factor
TSF	T-cell stimulating factor
UIS	Upstream inducible sequence
UV	Ultraviolet
VCAM-1	Vascular cell adhesion molecule-1
VCP	Virus complement control protein
VEGF	Vascular endothelial growth factor
VGF	Vaccinia growth factor
VLDL	Very low density lipoprotein

1
Interleukin-1: Structure and Function

Michael J. Tocci and John A. Schmidt
Merck Research Laboratories, Rahway, New Jersey

INTRODUCTION

Cells involved in inflammatory responses are regulated in part by the actions of cytokines from the interleukin-1 (IL-1) gene family. Along with a myriad of other cytokines, the members of the IL-1 family participate in a complex network of signaling molecules that act in a hormonelike fashion (1). The three known constituents of the IL-1 family, IL-1α, IL-1β, and IL-1 receptor antagonist (IL-1RA), are structurally related to one another and bind with similar affinities to IL-1 receptors (IL-1R) on cells (2,3). IL-1α and IL-1β are potent agonists that elicit broad-ranging biological responses in a large number of cell types, whereas IL-1RA blocks the biological responses of the two agonists by competing for key binding sites on cell surface IL-1R (4). Nearly all of the cell types that produce IL-1α and IL-1β also produce IL-1RA. The uniqueness of this system is striking in that IL-1RA is the only known example of a naturally occurring protein that acts as a specific receptor antagonist of a cytokine or hormonelike molecule.

The three members of the IL-1 family differ in several significant ways. IL-1α through an ill-defined mechanism remains primarily cell-associated and is found largely in the cytosol and on the plasma membrane of cells (2). In contrast, IL-1β is proteolytically processed and released from cells by an unusual route that does not involve a classical secretory leader (5,6). Perhaps to accommodate these differences in compartmentalization, IL-1RA has evolved a strategy to generate both secreted and intracellular forms (4). IL-1 gene family members are regulated at the transcriptional, translational, and posttranslational levels (2). The genes are often co-induced and expressed at different levels depending upon cell type and stimulus. Once expressed, the members of the IL-1 gene family appear to work in concert to control crucial inflammatory and host defense responses, thus the balance between agonists and antagonists in the IL-1 system is likely to have profound effects on the pathogenesis of inflammatory diseases (4,7).

IL-1 has been implicated as a major inflammatory mediator in a variety of pathological situations such as rheumatoid arthritis, inflammatory bowel disease (IBD), Alzheimer's disease, and diabetes. The exact role and importance of each member of the IL-1 family in normal physiology and pathophysiology are not yet clearly defined due to redundancy with other cytokines. For example, tumor necrosis factor (TNF) and IL-1 elicit similar biological responses in many cell types yet are structurally unrelated and bind separate receptors (2,8–10). This finding suggests that the receptors for these cytokines share some components of a common signaling pathway. In some instances cells respond additively to co-stimulation with IL-1 and TNF, while in others dramatic synergy is achieved. Thus it is quite likely that a secondary signal transduction pathway that amplifies or prolongs the actions of these cytokines is activated in some cell types. The discovery of IL-1RA, the use of blocking agents such as IL-1–neutralizing antibodies and recombinant soluble IL-1R molecules, as well as the generation of gene knock-out mice in the IL-1 system have begun to clarify the biological functions of IL-1. Recent studies indicate that the ratio of IL-1 to IL-1RA varies among patients and correlates with disease severity. The characterization of the IL-1 molecules has already greatly expanded our fundamental understanding of how organisms respond to infection, injury, and self-imposed autoimmune disorders. In this chapter we will provide an overview of the biological and biochemical aspects of the IL-1 system.

HISTORICAL BACKGROUND

In 1979 the term IL-1 was adopted to describe a small group of proteins that were discovered independently and named for a series of discrete biological activities. The activities associated with endogenous pyrogen (EP), leukocyte endogenous mediator (LEM), lymphocyte-activating factor (LAF), B-cell–activating factor (BAF), osteoclast-activating factor (OAF), epidermal cell-derived thymocyte-activating factor (ETAF), hemopoietin-1 (HP-1), and mononuclear cell factor (MCF) would eventually be attributed to two distinct forms of IL-1 that were virtually indistinguishable in their major biological activities. The research on EP began in the latter part of the 1940s. Other activities such as those of LEM, MCF, and LAF were subsequently identified during the 1960s and continued up to the late 1970s, when various groups were able to show that several of these activities co-purified with two proteins. The two proteins possessed molecular masses of approximately 18,000 and 38,000 Da, and isoelectric points (pI) of 7.0 and 5.6, respectively (reviewed in Refs. 11,12). In 1984 two different cDNAs encoding IL-1 were cloned from murine and human cells (5,6,13). One cDNA, designated IL-1α, encoded the acidic (pI 5.5) protein, while the other, termed IL-1β, encoded the neutral (pI 7.0) protein. Both proteins had predicted molecular masses of 31 kDa. Moreover, amino acid sequencing of purified proteins showed that the COOH-terminal 17.5 kDa portions of both IL-1α and IL-1β were present in the culture fluids of activated monocytes (14–16). Thus, two related proteins were associated with these different biological activities (2,6,11). It was soon determined that the 31 kDa form of IL-1α possessed biological activity, whereas the 31 kDa precursor of IL-1β required processing to generate the biologically active COOH-terminal 17.5 kDa molecule (17). With these accomplishments, more than 30 years of research in fever induction, inflammation, and immune cell activation suddenly converged on IL-1 as the master mediator.

Inhibitors of IL-1 activity were identified in biological fluids from a variety of sources. The investigation of these activities eventually led to the discovery of IL-1RA and the subsequent finding based on molecular cloning that this molecule was itself a member of the IL-1 gene family (2,4). This discovery remains the only known case of a naturally occurring protein receptor antagonist.

SOURCES AND STIMULI OF IL-1 PRODUCTION

Activated T cells, cytokines (IL-2, GM-CSF, TNF, TNFα, β, and γ, and IL-1), thrombin, C5a, steroid hormones, Bradykinin, bacterial cell wall components, bacterial toxins (LPS, lipid A, toxic shock strain of *Staphylococcus aureus*), viruses (poxviruses, cytomegalovirus), mitogens, immune complexes, Fc fragments of immunoglobulin, muramyl dipeptides, particulates (urate crystals, silica, zymosan, carrageenan), ultraviolet light, ionophores, phorbol esters, turpentine, retinoic acid, and adherence to plastic can all induce the synthesis of IL-1 (reviewed in Refs. 11,12). In certain hormone-responsive tissues, low levels of progesterone and adrenocorticotropic hormone (ACTH) appear to upregulate IL-1 gene expression (18). Certain stimuli differentially affect the half-life of IL-1 mRNAs and their translational efficiency, thus the amount of IL-1 protein does not always correlate with mRNA levels (19–22). In LPS-stimulated monocytes, for example,

Table 1 Cellular Sources of IL-1 and IL-1RA

Cell type	IL-1α	IL-1β	sIL-1RA[a]	icIL-1RA[b]
Monocytes	++++	++++	++[c]	+[e]
Alveolar macrophages	±	++	+++[d]	++[e]
Synovial macrophages	±	++	+++[d]	++[e]
Langerhans cells	+	+	u	u
Microglia	++[c]	++[c]	+	u
Astrocytes	++[c]	++[c]	+	u
Mesangial cells	+[d]	+	u	u
B cells	+[c]	+[c]	+[c]	u
T_H cells	+[c]	+[c]	+[c]	u
Neutrophils	±	±	+[c]	+[c]
Eosinophils	+	—	u	u
Mast cells	+	+	u	u
Platelets	+	+	u	u
Keratinocytes	+++[d]	+	—	+++[d]
Endothelial cells	+	+	—	—
Chondrocytes	++	++	++	u
Synovial fibroblasts	+[d]	+[a]	+[a]	+[c–e]
Dermal fibroblasts	+[d]	+[d]	+	+[e]
Uterine stromal cells	u	+	+	u

Expression levels: very high, (++++); high, (+++); moderate, (++); low, (+); very low, (±), undetectable, (—); unknown, (u).
[a]sIL-1RA = secreted IL-1RA.
[b]icIL-1RA = intracellular IL-1RA.
[c]Inducible expression.
[d]Constitutive expression.
[e]Express type I and type II icIL-1RA.

IL-1β mRNA is 10- to 20-fold more abundant than that of IL-1α, however, under these conditions IL-1α mRNA has a slightly longer half-life and is translated more efficiently than IL-1β mRNA (23,24), resulting in the production of nearly equivalent levels of IL-1α and IL-1β. Alternatively, some stimuli induce IL-1β transcription but not translation, and still others promote translation but not secretion (24,25). These findings provide a glimpse of the complex control mechanisms cells have evolved to regulate IL-1 expression.

A large number of stimuli trigger IL-1RA production. IL-1, IL-4, and IL-10 are weak inducers of IL-1RA and enhance its production by LPS, GM-CSF, IL-3, and TNF. Adherence induces moderate levels of IL-1RA production, as does TNF. Adherent IgG or IgG complexes are strong inducers of IL-1RA and presumably activate cells by triggering cell-surface Fc receptors. In human monocytes, certain stimuli, such as adherent IgG, selectively stimulate the production of IL-1RA, whereas others such as LPS induce both IL-1β and IL-1RA production (4).

Many cell types produce members of the IL-1 family. Table 1 summarizes the relative production of IL-1 family members by various cell types.

Monocytes, Macrophages, and Kupffer and Langerhans Cells

Peripheral blood monocytes are the major producers of IL-1 on a per cell basis. The transcripts for IL-1α and IL-1β constitute over 5% of the total cellular polyadenylated RNA in LPS-activated monocytes (5,6,26). In situ hybridization expriments show that greater than 95% of LPS-activated monocytes simultaneously produce IL-1α and IL-1β mRNA, whereas only 40% produce IL-1RA (25,26). As monocytes mature into macrophages, they progressively lose the capacity to produce IL-1. Specialized tissue macrophages such as alveolar macrophages, peritoneal macrophages, Kupffer cells in liver, and synovial macrophages have all been shown to produce IL-1. These cells, however, express substantially (~ 10-fold) lower levels of IL-1α and IL-1β mRNA than monocytes. Synovial tissue macrophages, for instance, contain IL-1β transcripts, but lack detectable levels of IL-1α mRNA and protein (27–32). Myelogenous leukemia (ML) cells from patients with chronic (CML) and acute (AML) disease constitutively produce IL-1β as an autocrine growth factor (33–35). IL-1RA inhibits the proliferation of CML and AML cells in vitro (36). Finally, dendritic cells, which function as key antigen presenting cells in tissues, synthesize IL-1, most likely as a co-stimulatory molecule and growth factor (37).

Langerhans cells in the epidermis have also been reported to make IL-1 and differentiate into dendritic cells in response to IL-1 (38,39). IL-1β accumulates rapidly in Langerhans cells in response to contact allergens and is therefore believed to play a role in contact hypersensitivity (40,41).

IL-1RA is produced by monocytes, as well as alveolar, synovial, and peritoneal macrophages. Kupffer cells of the liver also produce IL-1RA. Compared to peripheral blood monocytes, macrophages produce considerably more IL-1RA (42). Alveolar and synovial macrophages produce both secreted and intracellular forms of IL-1RA (4). Myelomonocytic cells express both alternatively spliced isoforms of icIL-1RA (43) (Fig. 1).

Several stimuli (e.g., LPS) co-induce transcription of

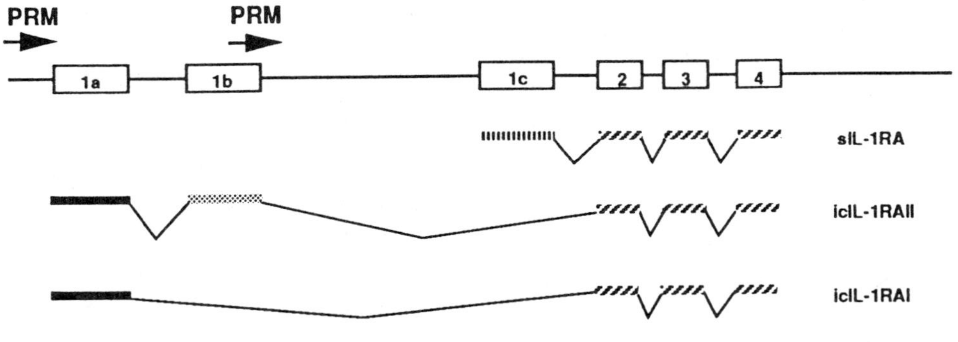

Figure 1 Three forms of IL-1RA are generated by alternative promoter usage and mRNA splicing. The top line represents the human IL-1RA gene (*IL1RN*). The exons are boxed and numbered. One promoter regulatory region, which is located upstream of exon 1a, is used to generate the transcripts that encode the two forms of icIL-1RA, while the second promoter, which is located in exon 1b, is used to transcribe the mRNA for sIL-1RA. The lower three line diagrams represent the alternatively spliced forms of IL-1RA. The combination of separate promoters and alternate first exon usage makes possible the production of three proteins that differ in their N-terminal sequences but are otherwise identical in the C-terminal receptor-binding domain.

all three IL-1 family member genes. In general, the expression of IL-1RA lags slightly behind that of IL-1α and IL-1β, peaking 8–16 h postinduction. Steady-state mRNA levels for IL-1α and IL-1β generally achieve maximum levels between 4 and 8 h postinduction. IL-1RA and IL-1β are regulated differently in human monocytes (28). After LPS stimulation, the amounts of IL-1β and IL-1RA produced are roughly equivalent, as are the mRNA half-lives (~ 2–4 h). In contrast, IL-1RA transcription is prolonged and mRNA half-life extended more than fivefold by induction with adherent IgG (4).

Lymphocytes

Both normal human B cells as well as Epstein-Barr virus–transformed B-lymphoblastoid cell lines produce IL-1α and IL-1β (44,45). In addition, cloned T-cell lines, helper T lymphocytes, and NK cells produce both cytokines, most likely for use as co-stimulatory factors in T-cell activation and differentiation (46). Thymic epithelial cells reportedly produce both IL-1s during T-cell differentiation (47,48). Lymphocytes also produce small amounts of IL-1RA.

Neutrophils

Polymorphonuclear neutrophils (PMNs) appear to contain small amounts of stored IL-1β and both glycosylated sIL-1RA and nonglycosylated icIL-1RA. PMNs produce less than 1% of the IL-1RA generated by monocytes or macrophages. Mature neutrophils, however, do not contain detectable levels of IL-1β or IL-1RA mRNA. The proteins appear to be transcribed and translated prior to full differentiation and are stored within the mature cell for later use. Several stimuli such as LPS, IL-4, TNF, and GM-CSF promote the release of glycosylated sIL-RA from PMNs (4,49).

Eosinophils express small amounts of IL-1α, whereas mast cells can produce both IL-1α and IL-1β upon activation. Platelets are unusual in that they contain IL-1β but do not appear to synthesize it.

Astrocytes, Microglia, and Neurons

Two endogenous cell types produce IL-1α and IL-1β in the CNS: microglia and astrocytes (50). Both cell types secrete IL-1β in response to incubation with LPS. In addition, glial cells such as Müller cells in the retina produce IL-1 upon stimulation with LPS or IFN-γ. In acute-phase brain injuries, microglia are the first cells to accumulate at sites of trauma and are most likely the primary source of IL-1. The IL-1 released by microglia stimulate the proliferation of nearby astrocytes. The activated astrocytes also produce IL-1 providing a potential autocrine mechanism for the growth of these cells and the generation of additional immune and inflammatory mediators. IL-1 has been shown to modulate gene expression in astrocytes being a potent inducer of cytokine (IL-6, TNF, GM-CSF, G-CSF) and adhesion molecule (ICAM-1) expression. In astrocyte-derived cell lines the production of IL-1β can be blocked by incubation with potent inhibitors of IL-1β–converting enzyme (ICE).

Immunohistochemical staining of human brain tissue indicates that neurons in the hypothalamus produce IL-1β. In rat brains IL-1β staining was also detected in the hippocampus and the olfactory tubercle region (51–53). The functional consequences of IL-1 expression in neurons have not yet been determined.

Little is known about the expression of IL-1RA in the CNS. In situ hybridization studies indicate that IL-1RA and IL-1 mRNA co-localize in the same regions of the brain, specifically the paraventricular nucleus of the hypothalamus, the hippocampus, and the cerebellum (54).

Keratinocytes, Fibroblasts, and Chondrocytes

Keratinocytes constitutively produce IL-1α. The level of IL-1α mRNA in human keratinocytes is equal to or greater than that of IL-1β (55–59). Both IL-1α, the predominant form of IL-1 in keratinocytes, and IL-1β, consisting entirely of the 31 kDa precursor (pIL-1β), remain cell-associated. The keratinocyte is the only cell type known that produces but does not process and secrete IL-1β. Interestingly, these cells also lack the processing enzyme for pIL-1β, ICE (60,61). Keratinocytes also constitutively produce large amounts of the alternatively spliced intracellular form of IL-1RA (4,62,63). Both type I and type II isoforms of icIL-1RA are present in keratinocytes (43). As expected this material remains primarily intracellular and does not appear to be actively secreted from the cells. TNF is the only one of several cytokines examined that enhances the production of icIL-1RA in keratinocytes. Thus, keratinocytes appear to produce large intracellular stores of IL-1 and IL-1RA, perhaps as a first line of defense in the event of traumatic injury to the skin.

Very small amounts of IL-1α and IL-1β mRNA have been detected in fibroblasts using highly sensitive techniques such as PCR and S1 nuclease protection. Dermal and synovial fibroblasts produce both the soluble and intracellular forms of IL-1RA. Recent findings indicate that fibroblasts contain both type I and type II icIL-1RA (43). In dermal fibroblasts icIL-1RA is constitutively produced at low levels and can be induced to higher levels upon incubation with PMA. In contrast, sIL-1RA is not expressed in resting fibroblasts and is produced only at low levels after PMA treatment (4).

Chondrocytes are an important source of IL-1 in articular cartilage. In culture these cells become activated and produce IL-1α, IL-1β, and IL-1RA constitutively (64).

Trophoblasts, Endometrium, and Uterine Stroma

IL-1β mRNA and immunoreactive protein have been detected in human and rodent uterine tissue during normal cycling of reproductive tissue and pregnancy. Expression, which appears to be under steroid hormone control, occurs in endometrium, specifically endometrial macrophages and endothelial cells, during the late secretory phase of uterine cycling. IL-1RA has also been detected by antibody staining in endometrial tissue. Immunohistochemical analysis indicates that both IL-1α and IL-1β are produced by human preimplantation embryos and placenta, in stromal cells, villous syncytiotrophoblasts, and extravillous trophoblasts (65). IL-1α and IL-1β have been detected in amniotic fluid, increasing at parturition or as a result of intrauterine infection. Various cytokines including IL-1 appear to be produced by ovarian tissue affecting the growth and differentiation of granulosa cells. The source of IL-1 is most likely resident ovarian macrophages (18). These findings suggest that IL-1 plays a role in normal gestation and premature births.

Endothelial and Other Cell Types

Endothelial cells are one of the only cell types that produce IL-1 but do not make IL-1RA. IL-1β is the major molecular species expressed by endothelial cells. Expression is induced by LPS, IL-1, and thrombin. LPS, IL-1, TNF, and GM-CSF do not induce detectable IL-1RA mRNA in these cells. Endothelial cells produce IL-6, IL-8, monocyte chemoattractant protein (MCP), and cell surface adhesion molecules (e.g., ICAM) in response to IL-1 as well as LPS. In models of IBD, IL-1α is present in gut epithelium, probably being generated by tissue macrophages (66). Certain renal mesangial cell lines constitutively produce IL-1α but very little if any IL-1β (67).

BIOCHEMICAL CHARACTERISTICS

IL-1 cDNA Sequences

The mammalian species of IL-1 have been the most well studied. Complementary DNAs (cDNA) for IL-1α and IL-1β have been cloned from mouse, rat, rabbit, sheep, bovine, and human sources (5,6,13,68–71). In addition, IL-1 and IL-1–like molecules have been described in a variety of lower species (72,73) including invertebrates (mussels, stafish, tunicates), fishes (carp, catfish), amphibians (African toads, American newts), and birds (chickens). IL-1RA cDNAs have been cloned from mouse, rat, rabbit, and human cells (43,74–79). Moreover, biological activities similar to IL-1RA have been identified in invertebrates such as tunicates (4,72). This phylogenetic pattern indicates that the members of the IL-1 family evolved before antibodies or specialized lymphocytes (39,72,73).

IL-1α mRNAs range from 2.0 to 2.2 kb in length and encode polypeptides of 270–271 amino acids. IL-1β mRNAs are approximately 1.7 kb in length and give rise to slightly smaller precursor proteins of 267–271 amino acids. Mature IL-1β is approximately 152 amino acids (17.5 kDa) in length. Both mRNAs possess large 3′ untranslated regions (3′UTR) containing AU-rich sequences that have been shown to regulate mRNA stability and translation (80–82). 3′UTR sequences such as AAUAAAUA have recently been shown to be recognized by sequence-specific RNA-binding proteins. These proteins are believed to regulate mRNAs at various levels, including half-life, cytoplasmic transport, and translation by ribosomes.

Three mRNAs for IL-1RA have been identified by cDNA cloning (see Fig. 1). A 1.8 kb transcript encodes the sIL-1RA glycoprotein consisting of a 177-amino-acid precursor protein. The 25-amino-acid signal peptide is removed, generating a 152-amino-acid mature sIL-1RA. A slightly larger 1.9 kb transcript encodes icIL-1RA. This variant of IL-1RA is a 155-amino-acid nonglycosylated protein that lacks a hydrophobic secretory leader (4). More recently a second alternately spliced mRNA has been identified that encodes another isoform of icIL-1RA, termed type II (43). The type II transcript (1.9 kb) contains a 63 bp sequence that is inserted in frame three codons downstream of the initiation codon of icIL-1RA. Type II icIL-1RA (25 kDa) differs from type I (22 kDa) by the insertion of 21 amino acids yielding a protein of 176 amino acids. The insertion generates a leaderless N-terminus that is rich in Gly and Glu residues. The biological activity of type II icIL-1RA is indistinguishable from that of type I.

The level of amino acid identity between individual members of the IL-1 family is low. The mature form of IL-1α shares only 22% sequence similarity with mature IL-1β and 18% with IL-1RA, while IL-1β displays 26% sequence similarity with IL-1RA and therefore is more closely related to IL-1RA than is IL-1α. The sequence similarity across mammalian species for an individual family member is high, ranging from 60 to 80%. Thus, amino acid sequences for a given family member retain significantly higher identities across species boundaries than the three family members within a single species.

IL-1 Genomic Sequences

IL-1α, IL-1β, and IL-1RA are encoded by separate genes, designated *IL1A*, *IL1B*, and *IL1RN*, respectively. All three genes are clustered on the long arm of human chromosome 2 in a region (q13-q21) that spans more than 430 kilobases (83). The *IL1A* and *IL1B* genes have 7 exons each that encode similar regions of the respective proteins. The human *IL1A* and *IL1B* genes are approximately 10.2 and 7.5 kb in size, respectively. Sequence analysis suggests that the three genes arose from a common ancestral gene more than 350 million years ago. The presence of a primordial poly A tail near the 3′ end of the human and murine *IL1B* genes suggests that it may have been generated from an ancestral *IL1A* sequence by a reverse transcription gene insertion event (11,84–86).

The *IL1RN* gene appears to have evolved before the IL-1α/IL-1β divergence, indicating that the evolution of the receptor antagonist was an important early event in regulating IL-1 activity (4,87). A single gene gives rise to both sIL-1RA and icIL-1RA (Fig. 1). The two proteins are identical except for the first exon. In sIL-1RA the first exon encodes the hydrophobic secretory leader, while in icIL-1RA the signal sequence is replaced by a short stretch of nonhydrophobic residues. This feat is accomplished by generating two alternatively spliced mRNA transcripts, each of which is controlled by a distinct promoter regulatory region. The *IL1RN* gene also contains 8 exons, spanning approximately 17 kb of genomic sequence (4,43,87). To generate icIL-1RA, the gene uses a 5′ promoter region and first exon that give rise to a 152-amino-acid protein lacking the secretory leader. The promoter region for sIL-1RA is 9.6 kb downstream of the first promoter regulatory region and is actually contained within the coding region of the first exon for icIL-1RA. The 63 bp sequence insertion for type II icIL-1RA is encoded by a newly discovered exon that lies between the first and second exon of icIL-1RA. This new exon is approximately 2 kb downstream of the first icIL-1RA–specific exon. The sIL-1RA coding sequences encompass a 6.4 kb region of genomic DNA and uses an alternative first exon that encodes the signal peptide.

IL-1 Gene Expression

The three IL-1 gene family members are regulated transcriptionally by a variety of environmental signals, which include bacterial cell wall components, cytokines, immune stimuli, and inflammatory mediators. Transcription is induced rapidly in response to such stimuli and does not require de novo protein synthesis. Thus, IL-1 family members are designated as immediate early response genes. The identification and characterization of genomic regulatory sequences that control transcriptional activation of the IL-1 gene family members have only just begun and there is still much to be learned about their regulation.

IL1B Gene Transcription

At present most is known about the factors that regulate *IL1B* transcription (88). DNA sequences at the 5′ end of the *IL1B* gene have been shown to control its expression. This region has been divided into two functional domains: the proximal promoter sequences encompassing nucleotides +12 to –131 and the upstream inducible sequences (UIS) spanning nucleotides –2729 to –3134. The proximal promoter appears to control cell specificity, while the UIS acts as an inducible enhancer region (89). The IL-1 UIS is separated by nearly 2500 base pairs (bp) from the proximal promoter region. Most of the work to characterize the DNA sequences and factors regulating *IL1B* gene transcription have been performed in cell lines of monocyte lineage such as U937 or THP.1.

In monocytic cells, transcription of the *IL1B* gene is dependent upon the presence of a hematopoietic specific transcription factor Spi-1 (NF-βA), also known as PU.1, which binds to proximal promoter sequences between –39 and –50, and –97 and –115 bp. Expression of this Ets domain protein family member is tissue specific and developmentally regulated, however, once expressed, Spi-1 is constitutively active (90). Spi-1 reportedly interacts with the TATA box-binding protein (TBP) and may be involved in recruiting other factors associated with the transcription initiation complex to the promoter. Mutational analysis of the proximal promoter region indicate that an NF-IL-6 site located between –83 and –91 bp cooperates with Spi-1 in the regulation of the *IL1B* gene. A Spi-1 responsive element has also been identified near the transcription start site (–30 to –202) of the *IL1RN* gene (88).

Two ubiquitous transcription factors NF-IL6 and CREB bind DNA sequences in the UIS enhancer region of the *IL1B* gene and regulate its expression. The sequence between –2729 and –2782 bp responds to LPS or LPS and cAMP and is believed to bind a heterodimeric factor consisting of NF-IL6 and CREB (91). Both of these factors are sequestered in an inactive form in resting cells until activated by external signals.

A type II gamma-interferon activation sequence (GAS) element has also been identified in the *IL1B* UIS between nucleotides –2850 and –2910. This element binds a protein that is antigenically related to the Stat1 transcription factor. Stat family members are also present in resting cells and are activated by phosphorylation through Jak tyrosine kinases upon receptor triggering. Once activated, Stat proteins are transported to the nucleus, whereupon they bind DNA as obligate dimers.

In the murine *IL1B* enhancer a novel *cis*-acting element that appears to be required for LPS inducibility has been identified between bases –2271 and 2281. This element binds an 85–90 kDa protein termed IL1β-UNF1 (92). It is not yet known if there is a human counterpart for this protein.

NFκB, a member of the *rel* DNA-binding protein gene family, is another important transcription factor that is present in resting cells and activated by stimuli that induce IL-1 expression. The heterodimeric complex of p65 and p50 is sequestered in the cytosol by the inhibitor IκB. Compounds that inhibit NFκB activation block LPS-induced *IL1B* transcription. NFκB appears to interact with sequences at two sites –297 and –2757 bp in the *IL1B* 5′ regulatory region. However, both of these sites appear to be required for achieving maximal transcription levels and not the induction of transcription. One possible explanation for this apparent dichotomy is that NFκB can associate with NF-IL6 in the absence of an NFκB DNA-binding site. This association increases the transactivation function of NF-IL6 when it is bound to its DNA recognition element and could thus explain the requirement for NFκB in LPS-induced *IL1B* transcription (93–96).

The fos•jun heterodimer AP-1 has also been implicated in *IL1B* expression (97–99). However, AP-1 does not seem to be involved in the induction of transcription but rather appears to help sustain *IL1B* gene transcription in response to certain stimuli such as PMA (87,96).

Dexamethasone inhibits IL-1 gene transcription as well as translation. Two sites in the *IL1B* UIS which bind NF-IL6 are sensitive to 5×10^{-7} M dexamethasone (88,89,97,100). This finding agrees with studies showing that glucocorticoid receptors can associate directly with NF-IL6 and suggests that inhibition of *IL1B* transcription by dexamethasone occurs through this heteromeric protein-protein interaction (88).

IL1A Gene Transcription

Much less is known about the regulation of *IL1A* gene transcription. The *IL1A* promoter region has a poorly defined TATAA element (TTACAAA) suggesting that other factors such as SP-1 may be required along with the TBP to initiate *IL1A* transcription (101). Sequence elements homologous to SP-1– and AP-1–binding sites are located at –21 through –25 and –5 through –12 bp, respectively. Initial transfection studies with *IL1A* gene 5′ upstream flanking sequences indicate that potential positive regulatory elements are located at –47 to –70 and –70 to –103 bp (86,101). These sequence elements correspond to potential binding sites for NFκB and the adenovirus-2 major late promoter transcription factor (Ad2-MLTF), respectively, although the exact nature of the factors that bind to these sites remain to be determined. Both the NFκB and MLTF elements are active in COS cells and cell lines that constitutively produce IL-1α, such as the human osteosarcoma cell line, MG-63, and the human esophageal carcinoma cell line, EC-G1, indicating that they function in constitutive expression. A sequence homologous to a glucocorticoid responsive element (GRE) is also present at –396 bp but has not yet been shown to respond to dexamethasone.

The genomic architecture of the *IL1A* gene also appears to differ from that of the *IL1B* gene in that sequences contained within exon 1, intron 1, and part of exon 2 (+1 to +731 bp) corresponding to the 5′ UTR of the IL-1α mRNA appear to function in retinoic acid, phorbol ester (PMA), and LPS-inducible transcription. Reporter gene constructs containing these sequences are 2- to 20-fold more active than constructs lacking them when transfected into the monocytic cell lines RAW264.7 and THP.1, respectively (88,102). These sequences contain putative binding sites for NF-IL6 and NFκB.

IL1RN Gene Transcription

Recent studies indicate that transcription of the *IL1RN* gene involves the cell-specific regulation of two different promoter-enhancer regions; one controlling expression of *sIL1RN* and the other *icIL1RN* (Fig. 1). The promoter region for *sIL1RN* lies within the first intron of *icIL1RN*. The sequence of this region contains a well-defined TATAA element as well as consensus sequences for NFκB, AP-1, CREB, and Spi-1 (4,103,104). Three LPS-responsive elements (LRE) have been mapped to the proximal promoter region of *sIL1RN* at –84 to –93 (LRE1), –148 to –200 (LRE2), and –200 to –250 (LRE3). LRE1 has been identified as a functional NFκB-binding site that is capable of binding p50 homodimers and p50•p65 heterodimers. LRE2 and LRE3 display cooperativity as neither element is functional alone. The factors that associate with these sites have not yet been identified (104). Sequences controlling *icIL1RN* gene transcription are located approximately 9.5 kbp further upstream of the *sIL1RN* promoter region. The sequence of this region lacks defined TATAA or CAAT elements indicating that factors used in TATA-less gene transcription or those that cooperate with TBP such as SP-1 or Spi-1 may be involved in transcriptional initiation of the *icIL1RN* gene.

IL-1 Synthesis

The mRNAs for IL-1α and IL-1β are translated into 31 kDa polypeptides on free polysomes (22–24). The 31 kDa forms of IL-1α and IL-1β are found in the cytosol and can be detected on the plasma membrane of activated cells by immunocytochemical techniques (105–109). The 31 kDa form of IL-1α is biologically active and appears to remain associated with the intact cell. Although IL-1α is not specifically processed, the 17.5 kDa COOH-terminal portion of IL-1α, which retains full biological activity, can be released from cells by the action of the cysteine proteinase calpain or certain other "bystander" proteases that are present in biological fluids (110). The identifications of Leu^{119} and Ser^{121} at N-terminus of 17.5 kDa IL-1α isolated from cultured LPS-stimulated monocytes are consistent with this hypothesis (14). IL-1α is released when cell integrity is compromised such as in necrosis, apoptosis, or cell permeation (111,112).

Alternatively, IL-1β is synthesized as an inactive 31 kDa precursor protein (pIL-1β) that is processed to its active form upon release from the cell by the specific proteinase ICE (113,114). The association of IL-1α and IL-1β with the plasma membrane is unusual as neither protein contains a standard signal peptide sequence. Targeting to the plasma membrane may be achieved through posttranslational modification, such as myristolylation. IL-1α and IL-1β are myristoylated on internal Lys residues by a unique myristyltransferase enzyme and presumably targeted to the plasma membrane through this unusual posttranslational modification (67,108). This mechanism for targeting IL-1β to the plasma membrane remains controversial at this time.

IL-1RA can be synthesized as a secreted protein that contains a classical hydrophobic leader sequence and is exported via the endoplasmic reticulum and Golgi apparatus, or two intracellular forms that lack the secretory leaders. The latter forms are generated from the same gene as a result of alternative exon usage. The intracellular isoforms of IL-1RA are not glycosylated.

Interleukin-1β Converting Enzyme

Interleukin-1β converting enzyme (ICE; EC3.4.22.36) is a novel thiol proteinase that cleaves the inactive 31 kDa precursor of IL-1β (pIL-1β) to its biologically active, 17 kDa mature form (113–115). The enzyme exhibits an unusual substrate specificity, cleaving on the C-terminal side of Asp at two highly conserved sites in pIL-1β, and requires Asp in the P_1 position for cleavage of both macromolecular (pIL-1β) and peptide substrates (116,117). The minimum substrate recognized by the enzyme consists of four residues Ac-Tyr-Val-Ala-Asp to the left of the cleavage site and methylamine to the right. This tetrapeptide ($k_{cat}/K_m = 4 \times 10^5\ M^{-1}s^{-1}$) is cleaved comparably to pIL-1β. Several classes of potent, specific reversible and irreversible inhibitors of the active enzyme (aldehydes, nitriles, ketones, acyloxymethylketones) have been developed based on this tetrapeptide sequence. These inhibitors prevent the processing of pIL-1β in human monocytes in vitro (L-709,049, $IC_{50} \sim 1\ \mu M$) and murine macrophages in vivo (L-709,049, $ED_{50} \sim 1$ mg/kg, IP) in a dose-dependent manner (60,115,118–122).

In addition to the synthetic inhibitors, a naturally occurring inhibitor of ICE has been identified. Poxviruses have evolved several genes to evade the actions of various cytokines and the host's immune response. One of these, the cytokine response modifier A (*crmA*) gene, encodes a member of the serpin gene family that has been shown to be an extremely potent inhibitor of ICE ($K_i \leq 7$ pM) (123,124). *CrmA* is the only serpin family member that has been shown to inhibit a cysteine proteinase, implying that ICE possesses substrate-binding features in common with serine proteinases. The interaction of *crmA* with ICE is unusual in that the enzyme-serpin complex dissociates in the presence of SDS (124).

The human gene (*IL1BC*), consisting of 10 exons, spans approximately 10.2 kb of DNA on chromosome 11 band position q22.2-22.3 (125). A single transcription start site has been identified 33 bp upstream of the initiator Met codon. Functional regulatory sequences in the 5′ flanking region of the *IL1BC* gene have not yet been identified, although consensus sequences for NFκB, AP-1, AP-3, and NFIL-6 are present within 600 bp of the transcription start site. The human and murine genes lack TATA elements but contain completely conserved 14 bp sequences that span the transcription start site of both genes, suggesting that this sequence is involved in regulating their transcription (125,126).

ICE is constitutively expressed as a 45 kDa proenzyme (p45) that is activated autoproteolytically by cleavages at as many as four Asp-X sites (where X is Ser^{104}, Asp^{119}, Ser^{298}, and Ala^{317}) generating two subunits of 19.8 (p20) and 10.2 (p10) kDa (118,127,128). Both subunits are required for enzymatic activity and are present in a 1:1 ratio. Cys^{285} in p20 has been identified as the catalytic thiol. ICE sequences that have been cloned from mouse and rat cells exhibit over 90% sequence identity to one another and nearly 68% identity to human ICE. The p10 subunit is the most highly conserved, possessing over 80% amino acid identity between species. Three of the four Asp-X cleavage sites that flank the p20 and p10 subunits are entirely conserved, as is a stretch of more than 19 residues encompassing the active site Cys^{285} (60,61,118,119,129).

The primary ICE transcript is alternatively spliced, giving rise to five different mRNA species and potentially five different isoforms of the enzyme (α-ε). The mRNAs for ICE δ, γ, and ε are generated in part by an unusual splicing event that occurs within the coding sequence of exon 2 (130). In these three mRNA species, an atypical splice donor site in the codon for Gly^{20} is used instead of the downstream splice site at the end of exon 2. Four of the isoforms—α, β, δ, and γ—contain the catalytic thiol residue Cys^{285}. Neither ICE_δ nor ICE_γ appears to possess autocatalytic activity, even though they both possess the conserved catalytic thiol residue. ICE_ε encodes a variant of p10 that is capable of complexing with the p20•p10 form of ICE (ICE_α). While the functional significance of each isoform is not yet known, it is now possible to ask if these variants are involved in regulating ICE activity or altering its substrate specificity.

IL-1β Processing and Secretion

The lack of any detectable intracellular mIL-1β in activated monocytes strongly suggests that IL-1β processing and secretion are concordant events (25,131). Several observations indicate that mIL-1β is released from cells via a nonclassical secretory pathway (106,132,133). Unlike most secreted proteins, neither pIL-1β nor mIL-1β contains classical hydrophobic transmembrane signal sequences. Immunoelectron-microscopic studies have shown that IL-1β is localized in the cytosolic ground substance and appears to be excluded from the endoplasmic reticulum (ER), Golgi apparatus, and secretory vesicles (105–107). The idea that pIL-1β does not pass through the Golgi complex is further supported by the fact that mIL-1β is not glycosylated, even though it contains several glycosylation sites. Finally, transfected fibroblasts efficiently express, but do not process or secrete pIL-1β or mIL-1β (133,134). More recent immunolocalization studies indicate that ICE and pIL-1β are present on the plasma membrane of activated human monocytes during secretion, further supporting the idea that ICE is involved in both IL-1β processing and transport (109).

The yeast-mating factor system may be a model for the secretion of eukaryotic proteins, such as IL-1, that lack conventional signal peptides across the plasma membrane. In *Saccharomyces cerevisiae* a-factor is palmitoylated at its COOH-terminus, processed by a specific proteinase and transported across the plasma membrane by protein(s) that constitute a transmembrane pore (135). Similarly, IL-1β is synthesized as an inactive precursor that must be processed to an active form by a specific proteinase, ICE. Immunolocalization studies using antibodies to the prodomain of IL-1β indicate that the precursor is targeted to the membrane of cells where processing and secretion take place (109). The targeting mechanism by which this occurs is unknown. Under normal conditions processing and secretion of IL-1β appear to be closely coupled since mature IL-1β is efficiently secreted and is not found inside cells (25,118). The cleaved N-terminal propeptide domain of the IL-1β precursor can be found, along with mature IL-1β, extracellularly (109,136). In human peripheral whole blood monocytes treated with the ICE inhibitor Ac-Tyr-Val-Ala-Asp-CHO, the 31 kDa IL-1β precursor is secreted intact, indicating that processing can be uncoupled from secretion (118).

Recent cDNA co-transfection experiments indicate that ICE not only processes pIL-1β but is also required for the export of IL-1β from cells (137,138). These studies show that, as in human monocytes, insect *Sf9* cells processed pIL-1β and secreted mIL-1β only when ICE was co-expressed. Moreover, like monocytes, mIL-1β is not detected within *Sf9* cells, suggesting that processing occurs at the plasma membrane and is concomitant with secretion. The co-expression of *crmA* in *Sf9* cells blocks IL-1β processing and secretion and prevents ICE proenzyme activation. *CrmA* appears to act only as an intracellular inhibitor of ICE in this system and in human monocytes (109,137), further supporting the idea that processing by ICE does not occur on the outside of the cell. The localization of active ICE to the plasma membrane of human monocytes coupled with the reconstitution of processing and secretion by co-expression of ICE in insect cells suggest that ICE is part of a transmembrane pore complex that is required for IL-1β secretion (109,137).

The Structure of ICE

The three-dimensional structure of ICE (p20•p10) complexed with tetrapeptide inhibitors has been determined by x-ray diffraction to 2.0 Å resolution. The crystal structure of the enzyme confirms the intimate association of the p20 and p10 subunits and, in fact, suggests that the active enzyme may actually exist as a homodimer of $[p20•p10]_2$. Residues at the homodimer interface are highly conserved, and mutation of one of these (His^{322}) inactivates the enzyme. This latter observation suggests that compounds that prevent dimer association may be useful as nonactive site-directed inhibitors. The key residues that constitute the active site of ICE are contributed by both subunits. In ICE Cys^{285} and His^{237} comprise the catalytic dyad that is found in all known cysteine proteinases. The Asp residue of substrates and inhibitors binds in the deepest pocket (S_1 site) of the active site and is sandwiched between Arg^{179} in p20 and Arg^{341} in p10. The distances between the subunits in the crystal structure further suggest that the p10 subunit from one p45 molecule interacts with the p20 from the other p45 in the homodimer (138–140). This arrangement of subunits in the $(p20•p10)_2$ homodimeric complex implies that activation of the proenzyme occurs via an intermolecular cleavage mechanism.

The ICE Gene Family

When it was first determined, the sequence of ICE was unique, making it the first member of an entirely new cysteine proteinase class. Recent discoveries have identified several new ICE-related cDNA sequences. The first of these is a gene from *Caenorhabitis elegans*, *ced-3*, (*C. elegans* death), which is required for programmed cell death and exhibits 23% sequence identity with human ICE (141). This finding along with studies showing that ectopic overexpression of an ICE-*lacZ* fusion protein induces apoptosis has led to the hypothesis that ICE is also involved in programmed cell death (142). The murine *nedd-2* gene, its human counterpart *Ich-1* (ICE/ced-3 homolog-1), MCH, and *CPP32*β are now also thought to be involved in apoptotic cell death (143–148). Two other cDNAs encoding putative ICE-related thiol proteinases have also been cloned: *ICE-II* (also known as TX) and *ICE-III* (149,150). All of the ICE homologs contain Asp-X sites that define regions within the predicted proteins with significant identity to both the p20 and p10 subunits of ICE. The sequence Lys-Pro-Lys-X_4-Gln-Ala-Cys-Arg-Gly encompassing the active site Cys of ICE is entirely conserved in all of the homologs, supporting the idea that these sequences also encode novel thiol proteinases. Nearly all of the other key residues that are required for catalysis are also conserved. Sequence comparisons indicate that *CPP32*β, *Ich-1*, and *nedd-2* are more closely related to *ced-3* than to ICE. These observations, along with the finding that *crmA* can inhibit apoptosis in cultured dorsal root ganglion neurons, implied that ICE or, more likely, ICE-related genes are involved in programmed cell death in mammals (151). The discovery that macrophages and cytotoxic T lymphocytes (CTL) from ICE-deficient mice undergo apoptosis suggests that other ICE-like proteases may substitute for ICE or that ICE does not play a critical role in apoptotic cell death. The potency or specificity of *crmA* toward many of these ICE-related gene products is not known at this time, although recent studies indicate that *CPP32*β like ICE is also inhibited by *crmA* (148,151,152).

Recently, *CPP32*β has been shown to cleave the enzyme polyadenosine ribose polymerase (PARP), a marker of apoptosis at the sequence Asp-Glu-Val-Asp. *CPP32*β does not cleave pIL-1β, nor does ICE cleave PARP. Moreover, inhibitors based on the ICE cleavage site in pIL-1β (AcTyr-Val-Ala-Asp-CHO) do not inhibit *CPP32*β, indicating that these enzymes may be inhibited selectively (147,148). Modeling of the active site of CPP32β based on the x-ray structure of ICE lend credence to this idea. The model shows that the Asp subsite (S_1) binding pocket is highly conserved among all of these enzymes. However, the residues in ICE that constitute the S_4-binding site and which are critical in binding the P_4 Tyr residue are not conserved in *CPP32*β or the other newly discovered ICE homologs. Table 2 summarizes the known ICE family members and their purported biological roles.

IL-1 RECEPTORS

Three related cell surface proteins are involved in IL-1 binding and signaling. These molecules, designated as the type 1 IL-1R, the type 2 IL-1R, and the newly discovered IL-1R associated protein (IL-1RAP) as well as the 2F1 and T1/ST2 genes, constitute the major members of the IL-1R gene family (3,153,154). The IL-1R genes are members of the much larger immunoglobulin supergene family, by virtue of the fact that the extracellular portion of each is composed of three Ig-like loop domains. The five genes display approximately 26–28% sequence identity in the extracellular domains. The IL-1R genes are located in the same region of human chromosome 2 as their ligands. The clustering of the IL-1R genes on chromosome 2 as well as sequence homologies among them suggest that they arose

Table 2 ICE/ced3 Family Members

Sequence	Function	Substrates	Refs.
ICE (mouse)	Proteolytic processing	IL-1β, ICE, CPP32β	61,119
ICE (rat)	Proteolytic processing	IL-1β, ICE	129
ICEα (human)	Proteolytic processing	IL-1β, ICE	60,118
ICEβ (human)	Proteolytic processing	IL-1β, ICE	130
ICEγ (human)	Proteolytic processing	IL-1β, ICE	
ICEδ (human)	Inhibitory[c]	NA[a]	
ICEε (human)	Inhibitory[c]	NA[a]	
ced-3	Apoptotic	Unknown	141
nedd-2 (mouse)	Apoptotic	Unknown	143
Ich-1_S (human)	Antiapoptotic	NA	145
Ich-1_L (human)	Apoptotic	Unknown	
CPP32 (human)	Apoptotic	PARP, SRTF[b]	144,147, 148
MCH (human)	Apoptotic	Lamin	146
ICEII (human)	Apoptotic[c]	Unknown	149,150
ICEIII (human)			149

[a]NA = not applicable.
[b]SRTF = sterol-regulated transcription factor.
[c]Proposed function.

by gene duplication from a common ancestral gene. Moreover, another gene termed the T1/ST2/*Fit-1* receptor, which exhibits 26% sequence similarity with the extracellular domains of the IL-1R, has been identified in this same region of chromosome 2 (3,155). The intron-exon structure of this receptor is nearly identical to that of the type 1 IL-1R. The T1/ST2/*Fit-1* receptor does not appear to be involved in IL-1 binding or signaling (3,155,156). The 2F1 receptor like IL-1RAP does not bind any of the IL-1 family ligands but appears to be capable of transducing signals similar to those of the type 1 receptor (154).

The type 1 IL-1R is an 85 kDa glycoprotein (p85) consisting of 552 amino acids. Virtually all cells express type 1 IL-1R. Structurally, the type I receptor is further divided into a 319-residue extracellular ligand-binding domain, a 20-residue transmembrane-spanning region, and a 213-residue cytoplasmic tail (157). The type 1 IL-1R transmits a signal triggering the cellular response to IL-1 (158). The number of type 1 IL-1R on cells is very low, ranging from as few as 10 to as many as 30,000. Interferon-γ, IL-1, IL-2, IL-4, PDGF, vitamin D_3, PGE_2, and PMA can increase the number of type 1 receptors on cells. Membrane-associated type 1 IL-1R as well as a soluble recombinant version of the receptor (sIL-1R type 1) bind all three members of the IL-1 family with high affinities. The type 1 IL-1R binds all three members of the IL-1 family, but has the greatest affinity for IL-1RA (Table 3) (3). From a biological perspective, the affinity of IL-1RA for the type 1 IL-1R is so great as to be considered virtually irreversible.

The type 2 IL-1R is a 68 kDa glycoprotein (p68). The 386-amino-acid protein possesses a 332-residue extracellular domain, a 26-residue transmembrane region and a short 29-residue cytoplasmic tail (159–161). Type 2 receptors do not transduce a signal upon IL-1 binding, instead the type 2 IL-1R is shed from the cell surface in a soluble form (159). The mechanism by which the receptor is shed is not yet known, although a membrane-associated protease most likely performs this task. The sIL-1R type 2 is found circulating in the blood of healthy individuals. Both the membrane-associated and soluble forms of this receptor display a distinct preference for IL-1β over the other IL-1 family members (3). A variety of inflammatory mediators and growth factors, such as IL-1, IL-4, PGE_2, corticosteroids, and hematopoietic growth factors increase the cell surface expression of type 2 receptors. Based on these findings the type 2 sIL-1R is believed to act as a decoy receptor, functioning as a constitutive natural antagonist to sequester the ligand (162). Thus, the receptor provides another means of regulating the activity of IL-1β. This same strategy is exploited by poxviruses to circumvent the host's inflammatory response. Vaccinia and cowpox viruses encode a homolog of the type 2 sIL-1R, the B15R gene product, that binds IL-1 (163,164).

An IL-1R–associated protein (IL-1RAP) with limited homology to the type 1 IL-1R has been cloned. Like the type 1 and type 2 IL-1Rs, IL-1RAP is also a member of the Ig superfamily. IL-1RAP co-precipitates with the type 1 IL-1R upon agonist binding but not upon binding of IL-1RA. This protein has not been shown to bind IL-1 directly, but appears to increase the affinity of the type 1 receptor for IL-1α and IL-1β by approximately threefold.

A fourth member of the IL-1R gene family, termed 2F1, has recently been cloned using degenerate PCR primers derived from conserved regions of the three known IL-1R family members (154). 2F1 is approximately 30% identical in its amino acid sequence to the type 1 IL-1R and IL-1RAP. Like IL-1RAP, 2F1 alone does not bind IL-1α, IL-1β, or IL-1RA. However, IL-1–stimulated COS cells transfected with chimeric molecules containing the extracellular domain of the type 1 IL-1R fused to the cytoplasmic tail of 2F1 are capable of inducing NFκB translocation to the nucleus and transcriptional activation of the IL-8 promoter demonstrating that 2F1 mediates responses similar to that of the type 1 IL-1R. At this point it is not known if 2F1 represents the potential γ-chain of the IL-1R complex or binds an as yet unidentified ligand of its own.

Table 3 Binding Affinities[a] of IL-1 Family Member Ligands for Recombinant Human IL-1 Receptors

IL-1R[b]	IL-1α	IL-1β	IL-1RA	Ref.
Type 1 membrane	2.1×10^9 M^{-1}	1.0×10^9 M^{-1}	1.0×10^{10} M^{-1}	157
Type 1 soluble[c]	3.0×10^7 M^{-1}	2.0×10^8 M^{-1}	5.0×10^8 M^{-1}	3
Type 2 membrane	1.5×10^9 M^{-1}	1.0×10^{10} M^{-1}	2.5×10^7 M^{-1}	160
Type 2 soluble	1.0×10^7 M^{-1}	3.0×10^8 M^{-1}	1.0×10^7 M^{-1}	3,159

[a]K_a.
[b]Recombinant version of receptors.
[c]Not shown to occur naturally.

Low picomolar concentrations of IL-1 are capable of eliciting a broad range of biological responses. This observation is inconsistent with the experimentally determined dissociation constants (K_d) for ligand binding to type 1 IL-1R, which range from 10 nM to 1 pM (15). Most cells have very few type 1 IL-1R on their surface, usually ranging from 10–100 to several thousand per cell (2,3,12,159). It also appears that there are a significant number of spare type 1 IL-1R present on cells since only 2–3% of the receptors need to be occupied in order to generate a biological response. Thus, the response to IL-1 is mediated by a small number of high-affinity receptors. The reason for this apparent discrepancy in the ability of IL-1 to trigger a biological response and the K_d is still unresolved. One possible explanation for this effect could be the nature of the type 1 IL-1R signaling mechanism, such as a kinase cascade. It is also possible that additional accessory chains such as IL-1RAP, which have yet to be identified, are associated with type 1 IL-1R binding and signaling events (153). Table 3 summarizes the interactions of the various IL-1 and IL-1R family members.

IL-1 Signal Transduction

Like all cytokine hormones, IL-1 acts as a switch turning specific genes on or off, thereby enabling the cell to respond appropriately to environmental stresses. Signals from outside the cell are transmitted across the plasma membrane by cell surface receptors to cytosolic signaling molecules and eventually to the nucleus where cytokine- or hormone-responsive target genes reside. IL-1 acts as a co-mitogen on certain target cells, but its major activity is by far proinflammatory. IL-1RA acts as a negative regulator of this system, serving to dampen or turn down the IL-1 response.

The molecular mechanism of IL-1 signaling has not yet been fully elucidated. Over the years numerous studies have produced conflicting data, suggesting that IL-1 signal transduction occurs via a variety of mechanisms. Recent studies indicate that a serine-threonine protein kinase is associated with the type 1 IL-1R. This protein kinase co-precipitates with the type 1 IL-1R from murine T-helper cell lines D10N and EL-4 and human fibroblasts. Treatment of the cells with IL-1α or IL-1β rapidly activates this kinase in a dose-dependent fashion, whereas excess IL-1RA competitively inhibits its activation (165,166). This kinase can use myelin basic protein (MBP) or histone H1 as substrates, but not heat-shock protein (hsp) 27 or casein.

In mammalian cells, LPS or IL-1 treatment activates a protein kinase cascade that leads to altered gene expression (167). IL-1α treatment activates p54α MAP kinase in vivo (168). A 42 kDa MAP kinase family member termed p38 is phosphorylated on Tyr residues in response to LPS or IL-1 (169–174). An as yet unidentified upstream activator kinase appears to phosphorylate p38 upon Tyr^{182} and Thr^{180}, which in turn phosphorylates a 50 kDa protein kinase (p50) on Thr and Ser residues. The p50 kinase appears to be MAPKAPK-2 (165–167). Immunohistochemical staining indicates that p38 is found in the cytoplasm and the nucleus of activated cells (172). All three kinases are inactivated by protein phosphatase 2B. In addition, p38 is inactivated by protein tyrosine phosphatase 1B. Inhibitors of this signaling pathway prevent the production of IL-1, TNF, and other cytokines. The production of IL-1α, IL-1β, and TNF by LPS-stimulated monocytes is inhibited by a series of compounds from the pyridinyl imidazole class, which block the activity of p38 MAP kinase (173). These compounds also block IL-1R–mediated signaling in human fibroblasts.

In human monocytes LPS can also activate alternate MAP kinase pathways. LPS triggers Ras GTP•GDP exchange and Raf-1 activation leading to the subsequent phosphorylation of MEK and MAP kinases and the transcriptional activator Elk-1 (175,176). Activation of the Ras•Raf-1 pathway with PMA alone leads to low-level cytokine production, indicating that LPS activates additional signaling events that augment the cells response to its environment (175). In IL-1–treated cells the ETS-domain transcription factor Elk-1 is also activated. Upon phosphorylation by JNK family MAP kinases, Elk-1 binds to serum response elements (SRE) in the promoters of target genes, activating transcription through increased DNA binding and the formation of a ternary complex with other serum response factors. The activation of the Raf-1 MAP kinase pathway alone does not appear to be sufficient to promote high-level IL-1 expression. The heterodimeric transcription factor AP-1, which is composed of fos and jun proteins, is activated by IL-1, presumably through the JNK pathway in a variety of cells, including T cells and synovial fibroblasts (98,99). Thus, IL-1 may activate different MAP kinase pathways in various cell types to coordinate different biological responses to environmental stimuli (Fig. 2) (177).

Presently, there is only indirect circumstantial evidence that JAK kinases are activated by LPS or IL-1R binding. However, the possibility that a Stat family member binds to sequence elements in the IL-1β promoter regulatory region suggests that JAK kinases could play a role in IL-1β expression (88).

The role of the second messenger cyclic AMP (cAMP) in IL-1 signaling is controversial. Elevated cAMP levels and inhibitors of cAMP-dependent protein kinase A (PKA) have not been demonstrated in all cells that respond to IL-1. It is possible that cAMP acts as a co-stimulus in some cell types but not others, serving to augment the primary signal transduction pathway. Alternatively, cAMP levels may modulate the activity of serine/threonine protein phosphatases, particularly PP1, through a PKA feedback loop (178).

The importance of the short-lived lipid-mediator ceramide in IL-1R signaling is also uncertain. Ceramide is produced from the phospholipid sphingomyelin by a membrane-associated enzyme sphingomyelinase. Intracellular ceramide activates protein kinase and phosphatase activities as well as the synthesis of diacylglycerol. The cytosolic serine-threonine phosphatase PP2A appears to be activated by ceramide in glioma cells (179,180). In addition, ceramide activates a kinase in HL60 myeloid cells that has been identified as p42 MAP kinase (181,182). This finding links ceramide to the MAP kinase activation pathway described above. Thus it is possible that ceramide may act primarily as a secondary signal to augment the MAP kinase activation pathway. Several unresolved questions exist concerning the physiological relevance of ceramide in IL-1 signal transduction. These questions include the timing of the ceramide response relative to IL-1R engagement and the levels of ceramide required to generate a biological response in cells.

Structure of IL-1

The three-dimensional structure of recombinant human mature IL-1β has been determined by x-ray crystallography at a resolution of 2.0 Å and an R-factor value of 17.2% (183,184). The structure reveals a compact spherical protein of the all β-type, confirming predictions from earlier hydrodynamic and circular dichroism studies. Overall the protein forms a distorted tetrahedron consisting of 12 β-strands. Each face is a triangular unit containing three antiparallel β-strands connected by loops. The edges of the tetrahedron where the faces meet are formed by two antiparallel β-pleated sheets, one from each triangular unit. Hydrogen bonding patterns occur at the vertices and along the edges of each symmetrical unit. The NH_2 and COOH termini of the protein are exposed to solvent, lying in close proximity to one another at one of the vertices. This structure is referred to as a β-barrel. The vast majority of residues that are conserved between species contribute to the internal hydrophobic core of the molecule or stabilize local structural motifs.

Three-dimensional structures of the mature forms of human IL-1α (185) and IL-1RA (186) have also been determined and are nearly identical to that of IL-1β.

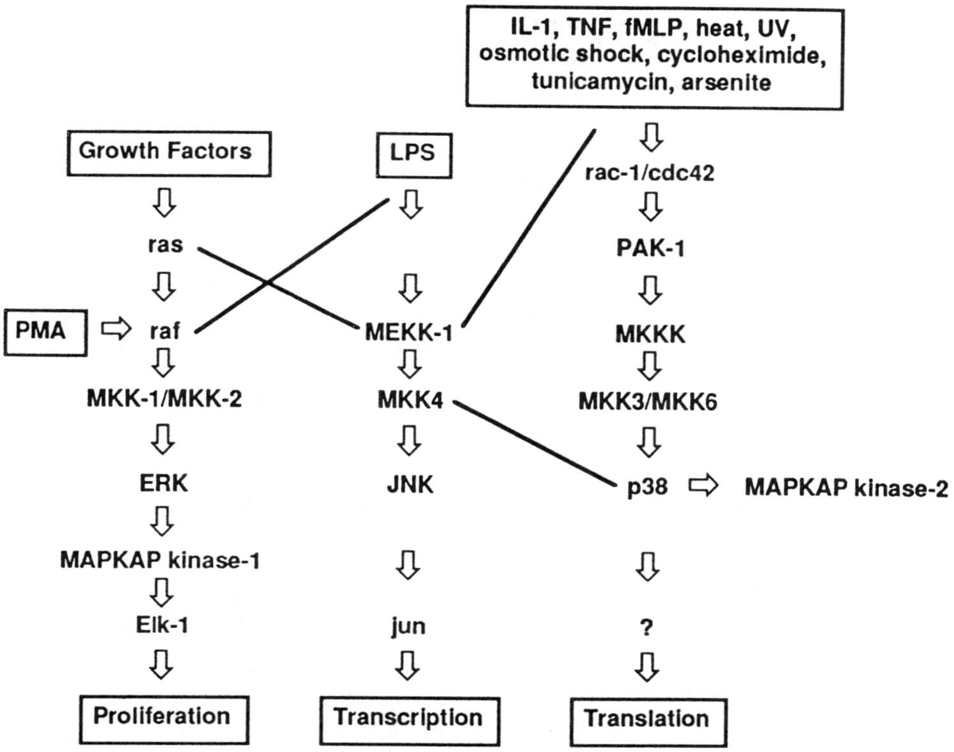

Figure 2 Diagrammatic representation of proposed protein kinase signaling cascades involved in LPS and IL-1 signal transduction. Solid lines indicate potential cross-talk points between pathways.

Despite limited amino acid sequence homology (~18–26%) the three structures are nearly superimposable. The major distinction between the structures, besides obvious amino acid side-chain differences, occurs in the loop regions connecting the β-strands. The total solvent-accessible surface area is also similar for each of the three proteins—approximately 7500 $Å^2$. These results help explain the biological activities and similar potencies of IL-1α and IL-1β on cells.

The structure of pIL-1β has not yet been reported. Initial studies examining the sensitivity of recombinant human pIL-1β from *E. coli* to proteases and alkylating agents suggest that the precursor is in a more open conformation (187). The fibroblast growth factors (FGF) are distantly related to the IL-1 gene family. Both acidic and basic FGF possess limited (~12–15%) amino acid homologies with IL-1β and IL-1α (188,189). Even more striking is the finding that the three-dimensional structure of bFGF can be superimposed on that of the IL-1 family members. As with the IL-1 family members, the major structural differences are in the amino acid side chains and the size of the loop regions that connect the β-strands (190).

Receptor–Ligand Interactions

IL-1RA, like IL-1α and IL-1β, has been shown to bind to the type 1 IL-1R on fibroblasts, chondrocytes, endothelial cells, and T cells. Unlike IL-1α and IL-1β, IL-1RA does not transduce a signal when bound to the type 1 receptor. Equilibrium binding studies indicate that the K_d of IL-1RA for the type I IL-1R is approximately 150–220 pM. The receptor association and dissociation rates for IL-1RA are slower than that of IL-1α or IL-1β (4,191). IL-1RA binds weakly to the membrane-anchored type 2 IL-1R but, like IL-1α, fails to bind to its soluble form. Thus, the type 2 sIL-1R may act as a specific inhibitor of circulating IL-1β by lowering its effective concentration in biological fluids and increasing the level of IL-1RA relative to mature active IL-1β.

Site-directed mutagenesis has been used to map the residues on the surface of IL-1β and IL-1RA that are involved in type I receptor binding (192–196). The results have revealed two discrete receptor-binding sites on IL-1β (sites A and B), only one of which is retained in IL-1RA (site A). These two sites are separated by approximately 25 $Å^2$ in IL-1β. Modeling studies suggest that IL-1α also contains two binding sites, which are similar to those in IL-1β. Site A is located on one side of the β-barrel, encompassing approximately 330 $Å^2$. In IL-1β, site A consists of Arg^{11}, Gln^{15}, His^{30}, and Gln^{32} (192,193). In IL-1RA there appear to be only five residues—Trp^{16}, Gln^{20}, Tyr^{34}, Gln^{36}, and Tyr^{147}—that are important for binding, accounting for approximately 53 kJ/mol of binding energy. This is in close agreement with experimentally determined free energy of binding of IL-1RA (52 kJ/mol) (196). Since the mutagenesis studies suggest that IL-1RA contains only a single binding site, this site is believed to be responsible for most of its binding energy, and thus it appears that IL-1RA may generate more binding energy from this site, the equivalent of site A, than do IL-1α or IL-1β. It appears that this site is functionally and structurally equivalent across the three IL-1 family members.

The B site is approximately 300 $Å^2$ in IL-1α and 1100 $Å^2$ in IL-1β. This site is positioned at the open end of the β-barrel and corresponds structurally to an exposed loop. In IL-1β the B site includes residues Arg^4, Lys^6, Phe^{46}, Ile^{56}, Lys^{92}, Lys^{93}, Lys^{103}, and Glu^{105} (191,193). The corresponding residues in IL-1α are Arg^{16}, Ile^{18}, Asp^{64}, Asp^{65}, Ile^{68}, Lys^{100}, Asn^{108}, Trp^{113}, Gly^{117}, and Gln^{136} (195,196). The exposed loop encompassing the B-binding site is one of the major structural differences between IL-1RA and both IL-1α and IL-1β (196). Thus, by mutagenesis it appears that both agonists contain two type 1 receptor-binding sites, while the antagonist contains only one.

The crystal structure of the type 1 IL-1R complexed with IL-1RA has recently been solved to 2.7 $Å^2$ resolution providing the first picture of the way the receptor interacts with one of the ligands in the IL-1 system (197). Structurally, the bound and unbound ligands are nearly identical, indicating that IL-1RA does not appear to undergo any major conformational changes upon binding to the type 1 receptor. The receptor forms a 1:1 complex with the ligand in which approximately 1700 $Å^2$ of IL-1RA surface area are buried. Thus, over 20% of the total surface area of IL-1RA is buried through interaction with the receptor. IL-1RA binds within a fold created by the three Ig-like domains of the receptor. This is fundamentally different from the human growth hormone receptor and the IFN-γR, which bind to the outside surfaces of folds on their respective receptors. Each of the three immunoglobulinlike (Ig) domains of the receptor contact IL-1RA (Fig. 3). The first two Ig domains interact with one another through an extensive network of interdomain contacts, while the third is separated from the first two by a flexible hinge region. Ig domain 2 interacts with the region in IL-1RA that corresponds to site A, which, as described above, has been shown through mutagenesis studies to be important for binding. The third Ig domain contacts a loop that is analogous to the proposed receptor-triggering region in IL-1α and IL-1β, but which is structurally distinct in IL-1RA. While a far greater number of residues contact the receptor than were identified by mutagenesis, it appears that all of the residues identified by mutagenesis as being important in binding to the receptor are confirmed by the

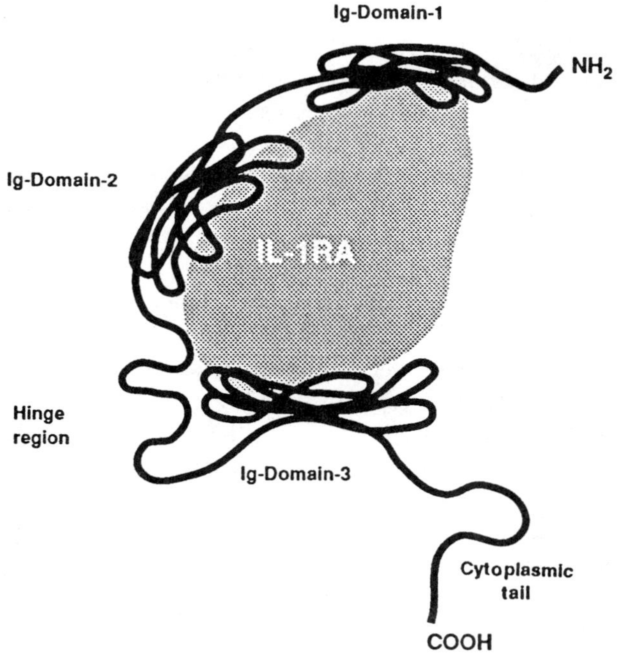

Figure 3 Schematic representation of IL-1RA bound to the extracellular domain of the type 1 IL-1R showing extensive interactions with the three Ig-like domains in the receptor and the flexible hinge region separating Ig-like domains 1 and 2 from 3.

structural data. Thus, it appears that a limited number of key residues within each binding site contribute the majority of the binding energy.

MAJOR BIOLOGICAL ACTIVITIES

IL-1 exhibits an extraordinary range of biological activities. Both isoforms of IL-1 possess co-mitogenic properties, recruit cells to sites of inflammation, and stimulate the production of proinflammatory mediators. IL-1 can act locally or systemically. The biological effects of IL-1α and IL-1β result from their ability to modulate gene expression in target cells. The expression of a large number of cytokines, chemokines, cytokine-related genes (i.e., adhesion and cytokine receptors), proteinases, and enzymes that produce potent low molecular weight inflammatory mediators are induced in response to IL-1 treatment. One of the key factors in regulating the biological response to IL-1 is IL-1RA. The levels of IL-1RA detected at sites of inflammation do not appear to be high enough to totally block IL-1R signaling, thus physiologically IL-1RA may serve to downregulate the response to IL-1 rather than inhibit it entirely.

Evidence that IL-1 is a key mediator in a variety of physiological and pathological processes has been obtained by administration of recombinant IL-1, macromolecules that act as receptor antagonists, i.e., IL-1RA, sIL-1R, and anti–IL-1R monoclonal antibodies (mAb) (4,198–201), or inhibitors of IL-1 production such as ICE inhibitors, or p38 kinase inhibitors (121,122,173). These agents have been shown to reduce lethality in rodent, rabbit, and primate models of severe endotoxic shock; prolong graft survival in murine models of graft rejection; reduce leukocyte infiltration and tissue damage in a rabbit model of IBD; and reduce joint swelling and histological scores in a rat model of streptococcal cell wall–induced arthritis (2). Additional evidence that IL-1 plays a key role in inflammatory disease has been obtained in patients with Lyme's arthritis, where the duration of clinical symptoms correlates positively with the ratio of IL-1β: IL-1RA in synovial fluid (202). In the murine model of collagen-II–induced arthritis, neutralizing antibodies to IL-1β, but not IL-1α, have been shown to significantly reduce joint pathology scores and markers of inflammation such as CRP and SAA when given at the time of onset or after initiation of disease. Neutralizing antibodies to TNF were less effective in reducing these parameters. In addition, IL-1β–neutralizing antibodies block the temperature increase and corresponding rise in IL-6 production in mice challenged with turpentine (203–205). Most recently, studies using mice containing either a disrupted IL-1β (*il1b*) gene or a disrupted ICE (*il1bc*) gene have confirmed several of the key studies performed with biological and chemical agents (126,206–211). Reports indicate that *il1b* knock-out mice fail to develop fever and anorexia and do not produce IL-6 in response to turpentine injection (207,210). *il1bc* knock-out mice are resistant to lethal LPS challenge but *il1b* are not (207–209).

Circumventing Host Defense: Viral Inhibitors of IL-1 Activity

Members of the poxvirus group have evolved two separate strategies to specifically circumvent the action of IL-1β (123,124,163,164). Cowpox virus encodes the *B15R* gene, a secreted homolog of the type 2 IL-1R, and *crmA* a novel cysteine containing serpin inhibitor of ICE. It is interesting that both of these strategies target the action of IL-1β and confer significant advantages on the pathogenicity of the virus. The B15R gene exhibits 30% sequence similarity with the type 2 receptor. Viruses containing this gene have a selective advantage over those that do not due to their ability to limit the host's inflammatory response (163,164).

The *crmA* gene product is an extremely potent inhibitor of ICE ($K_i < 7$ pM) (123,124). This unusual viral serpin homolog inhibits the activity of p20•p10 ICE, prevents ICE zymogen activation, and blocks processing and secretion of IL-1β in ICE-transfected cells (117,123,124). *CrmA* contains the dipeptide Cys-Asp in its putative serpin stress loop, making it unique with respect to all other known serpin family members. This sequence presumably confers the unusual specificity of *crmA* for ICE and other Asp-specific ICE-related proteases. Most recently *crmA* has been shown to be a less potent inhibitor of Apopain (*CPP32β*). The expression of *crmA* in cells inhibits apoptosis (151,152). In this regard *crmA* may serve not only to inhibit IL-1β but also block apoptosis conferring two distinct advantages on replication of the virus (148,151,152). The evolutionary strategy employed by the poxviruses confirms the central role of IL-1β as a major mediator of the systemic inflammatory host response.

Gene Knock-Out Mice

Gene-targeting techniques have been used to generate mice that are deficient in specific members of the IL-1 system. Homozygous single null mutation mice have been generated for *il1b*, *il1bc*, type I *il1r*, and type II *il1r*, whereas efforts to generate *il1a*- and *il1rn*-deficient mice are nearing completion. Phenotypic analyses of these mice have begun to generate interesting data on the biological functions of these genes and the redundancy in the IL-1 system, confirming previous studies that used IL-1RA or neutralizing antibodies to block the actions of IL-1.

The *il1b* gene (206) was the first member of the IL-1 family to be targeted by homologous recombination in murine embryonic stem cells. As expected, *il1b* null mice lack IL-1β but produce normal amounts of IL-1α and TNF when challenged with LPS or *Proprionibacterium acnes*. IL-1β–deficient mice are fertile and show no developmental abnormalities. These mice remain healthy under specific pathogen–free (SPF) conditions, and while they do not exhibit increased susceptibility to infection with the intracellular pathogen, *Listeria monocytogenes* (207), they do exhibit a slight increase in susceptibility and more rapid mortality to influenza virus infection (210). In contrast to previous studies using IL-1β–neutralizing antibodies (40,41), *il1b* knock-out mice respond normally in models of contact or delayed-type hypersensitivity (204).

In a mouse model of turpentine-induced sterile abscess formation, IL-1β-neutralizing antibodies have been shown to block IL-6 production and the febrile response (204). In this same model IL-1β–deficient mice fail to develop the fever or anorexia associated with the turpentine injection and do not produce IL-6 or acute-phase response proteins (207). These studies confirm the central role of IL-1β in IL-6 and fever induction. In contrast, IL-1β–deficient mice are not resistant to lethal LPS challenge. Interestingly, unlike the turpentine challenge, the magnitude of the fever is reduced by only 50% in IL-1β–deficient mice upon LPS administration, and there is no apparent difference in IL-6 or acute-phase protein synthesis as compared to wild-type littermates (207). ICE-deficient animals survive the lethal LPS challenge despite the acute inflammatory response. Preliminary results indicate that type 1 *il1r* gene knock-out mice do not develop fevers upon injection with turpentine providing confirmation of this phenotype.

B10.RIII mice are highly susceptible to collagen-induced arthritis (CIA). IL-1β–neutralizing antibodies block the onset as well as the progression of established CIA in B10.RIII mice (200,201). Recent studies show that IL-1β–deficient mice backcrossed (N3) onto the B10.RIII background fail to develop CIA, confirming the earlier neutralizing antibody studies (211).

il1bc null mice generated by gene-targeting experiments have confirmed the role of ICE in IL-1β processing (126,128,209). These mice lack the 45 kDa ICE proenzyme. Cells from homozygous ICE knock-out mice contain normal intracellular levels pIL-1β when activated and the level of mature IL-1β is reduced by more than 95%. However, a small amount of biologically active mature IL-1β is still generated in these animals. This material represents less than 1% of the mature IL-1β produced by normal mice, and its production is not inhibited by a potent ICE inhibitor. In heterozygous knock-out mice the level of IL-1β is 50% of that found in wild-type mice, indicating that mature IL-1β production is determined by the amount of ICE. Surprisingly, IL-1α levels in these mice are reduced by > 90% after LPS challenge. This observation suggests that IL-1α and IL-1β share some common components of a novel secretory pathway. In stark contrast to IL-1β knock-out mice, ICE-deficient mice develop normal fever responses in response to LPS or turpentine but are resistant to lethal LPS challenge. As stated earlier, CTL-mediated apoptosis is largely unaffected in ICE-deficient mice. There is, however, a modest effect on Fas-ligand–mediated apoptosis in these mice. In cells from normal mice this process is inhibitable by *CrmA* and to some extent by a tetrapeptide ICE inhibitor (152,208,209,212,213). Table 4 summarizes the phenotypes of *Il1b* and *IL1bc* knock-out mice.

Immune and Hematopoietic System Interactions

Once termed LAF, IL-1 is now believed to function primarily as a co-mitogen for B and T lymphocytes rather than as a potent primary stimulus (12). Of the two subpopulations of T cells, T_H1 and T_H2, only T_H2 cells, which constitute a very small proportion of the total T-cell

Table 4 Challenge Phenotypes of IL-1β and ICE-Deficient Mice

Challenge	IL-1β KO	ICE KO	WT[a]	Ref.
Fever (ΔT_b)[b]				207
LPS	+	++	++	
Turpentine	–	++++	++++	
APP[c] induction				
LPS	+++	N.D.[d]	+++	207,210
Turpentine	–	N.D.	+++	
Mortality,				
LPS	100%	< 10%	100%	208,209
Weight loss,				
turpentine	none	20%	20%	207
CIA[e]	–	+	+	211
Apoptosis	+	+	+	207–209

[a]WT = wild type.
[b]ΔT_b = change in body temperature.
[c]APP = acute phase protein.
[d]N.D. = not done.
[e]CIA = collagen-induced arthritis.

population, require IL-1. These cells are largely of the T-helper class. T_H2 cells bear IL-1R and appear to proliferate and respond to IL-4 more vigorously in the presence of IL-1 (214,215). Thus the vast majority of T cells do not appear to respond to IL-1. The role of IL-1 production or ICE in T-cell–mediated cell killing and apoptosis is under intense investigation. However, T cells from *il1bc* gene knock-out mice efficiently kill target cells and are not defective in apoptosis suggesting that other ICE-like proteases are involved in this process (208,209).

Like macrophages and dendritic cells, B cells can function as antigen-presenting cells. B cells produce mainly membrane-associated IL-1, most likely IL-1α. However, both in vitro and in vivo experiments indicate that B-cell responses are unaffected by treatment with high doses of IL-1RA. Neither primary nor recall antigen responses are affected by treatment with IL-1RA. IL-1RA also does not appear to significantly affect immunoglobulin production. Thus it appears that IL-1 does not perform a critical role in B-cell function (216,217).

The discovery that HP-1 was identical to IL-1α provided the first indication that IL-1 could function as a hematopoietic growth factor (218,219). This property of IL-1 is largely due to its ability to induce or enhance GM-CSF, G-CSF, and M-CSF production. IL-1 along with various other cytokines promotes differentiation of specific lineage and multilineage progenitor cells. Injection of IL-1 into mice increases circulating CSF, protects stem cells, and accelerates the return of granulocytes after irradiation or cytotoxic drug treatments (220,221). In summary, IL-1 appears to enhance hematopoiesis in concert with other cytokines by increasing the levels of colony-stimulating factors. In this capacity IL-1 may be useful as adjunct therapy in cancer patients.

Neuroendocrine and Neuroimmune System Interactions

IL-1 stimulates the hypothalamic-pituitary-adrenal axis to release bioactive neuropeptides into the hypothalamic portal and systemic circulation. Increased levels of corticotrophin-releasing factor (CRF), adrenocorticotropin-releasing hormone (ACTH), endorphins, vasopressin, and somatostatin are detected after IL-1 administration. The induction of glucocorticoids by IL-1 in vivo appears to be due to the release of ACTH from the pituitary or CRF from the hypothalamus and not a direct effect on the adrenocortex (222–227). Products of the cyclooxygenase pathway appear to mediate IL-1–induced ACTH release (228). Since glucocorticoids decrease IL-1 production a negative feedback loop between the immune and neuroendocrine systems may serve to modulate the production of IL-1 via CRF, ACTH, and hydrocortisol (227,229,230). In fact, adrenalectomized mice are more sensitive to lethal doses of IL-1 indicating that corticosteroids play a protective role in minimizing the effects of IL-1 (231).

Under stressful conditions IL-1 appears to act directly on neurons. Signals received from peripheral nerves are transmitted to terminal neurons, which in turn produce proinflammatory substances. In rats this pathway results in the release of norepinephrine, dopamine, and serotonin from the hypothalamus (51,52,232).

Early on in its history IL-1 was identified as an extremely potent pyrogen (9,10). Systemic injection of IL-1 into humans or experimental animals (mice, rats, rabbits) results in the release of IL-6 as well as numerous neuropeptides and the rapid onset of fever, sleep, anorexia, and malaise (10,204,233). The level of IL-6 in the circulation and the cerebral spinal fluid has been shown to correlate with the magnitude of the fever response. The effects of systemic IL-1 administration on the hypothalamus indicates that IL-1 signals across the blood-brain barrier. It is unlikely that significant amounts of IL-1 cross the blood brain barrier since IL-1–neutralizing antibodies given intravenously block fever induction (203,204). One plausible explanation for these findings is that IL-1 activates hypothalamic neurons in the organum vasculosum of the lamina terminalis. Here the blood-brain barrier is loosely organized, consisting of specialized endothelial cells that can respond to IL-1 by producing arachidonic acid metabolites and cyclic AMP. Thus it is possible that glial cells become activated via an IL-1–dependent mechanism to produce more IL-1. Alternatively, dendritic projections inserted between the endothelial cell junctions might allow hypothalamic neurons to respond directly to IL-1 (10). Recent studies show that IL-1β–neutralizing antibodies block endotoxin-induced fever and that *il1b* knock-out mice are completely resistant to turpentine-induced fevers (203,207). Interestingly, *il1b* knock-out mice do not develop anorexia upon turpentine injection (207). The anorexia is mediated largely through prostaglandins.

IL-1 produced by microglia and astrocytes is also believed to play an inflammatory role in certain neurodegenerative diseases such as multiple sclerosis and Alzheimer's disease. The degenerative lesions associated with these chronic autoimmune diseases contain immune and inflammatory cells capable of producing and responding to IL-1 and are therefore believed to play a role in disease progression. Experiments employing neutralizing antibodies or gene knock-out mice will help address these issues.

Muscular-Skeletal System: Effect on Bone and Cartilage

IL-1 has significant effects on bone and connective tissue cells, implying that it plays a role in connective tissue remodeling (8,10,64). From a historical perspective, IL-1 was shown to be equivalent to two factors that targeted cells from bone and cartilage, OAF and MCF (10). Subsequently, in vivo experiments have shown that systemic administration of IL-1 stimulates bone resorption and that the injection of IL-1 directly into joints mimics the pathological changes seen in inflammatory arthritis (234,235). In inflammatory Lyme's arthritis the severity and duration of the disease correlates with the level of IL-1β in synovial fluid, and in patients with a high IL-1RA:IL-1β ratio disease severity and duration were decreased (202). This particular finding points out the probable physiological role of IL-1RA in modulating the IL-1 response. Such studies suggest a major role for IL-1 in destructive joint diseases (236). In murine collagen-induced arthritis (CIA) neutralizing antibodies to IL-1β but not IL-1α block the onset as well as the progression of already initiated disease (200,201). Moreover, recent studies indicate that IL-1β knock-out mice fail to develop CIA, confirming the importance of this single cytokine in this model of inflammatory arthritis (211).

Both forms of IL-1 are potent activators of connective tissue cells (8,12,64). Although now recognized as a relatively poor mitogen, IL-1 can induce the proliferation of synovial fibroblasts as well as osteoclasts. In rheumatoid arthritis the co-mitogenic property of IL-1 is believed to contribute to the expansion of the synovial membrane into the enlarged pannus. As mentioned earlier, IL-1 induces the production of inflammatory mediators (e.g., PGE_2, plasminogen activator) and the induction of specific proteinases (e.g., collagenase, stromelysin) in both chondrocytes and synovial fibroblasts (236,237). IL-1 also inhibits the synthesis of one of the major components of cartilage, proteoglycans (236). Thus IL-1 affects several metabolic processes that are responsible for maintaining the integrity of bone and cartilage.

Functions in Skin

Keratinocytes and dermal fibroblasts also produce IL-1 (12). In keratinocytes, expression of IL-1α mRNA is induced by stresses such as ultraviolet light (55–59). One possible function of IL-1 in skin is to recruit neutrophils to sites of skin damage, particularly in inflammatory skin diseases such as psoriasis. However, at this time there is no experimental data to support this notion.

SYSTEMIC INFLAMMATION AND SEPTIC SHOCK

Hepatic System: Acute Phase Protein Response

Early studies suggested that IL-1 regulated acute-phase protein synthesis by the liver. Later studies have shown that IL-6 is also important in stimulating acute-phase protein production. Induction of IL-6 by IL-1 partly explains IL-1's role in acute-phase protein production; however, it has subsequently been established that IL-1 alone increases the transcription of some acute phase liver proteins and decreases transcription, independent of mRNA, half-life of other hepatic proteins. In HepG2 cells, for instance, IL-1 and IL-6 regulate separate but overlapping sets of acute phase proteins, providing a possible explanation for these observations (238). This phenomenon can be seen in *il1b* gene knock-out mice where turpentine-challenge fails to induce acute-phase protein synthesis by the liver, whereas LPS-challenged mice exhibit a normal acute-phase response (207). Serum amyloid A, fibrinogen, factor B, metallothioneins, serum clotting factors, components of the complement system, and IL-6 are induced by IL-1. The liver is also one of the primary sources of type 2 nitric oxide synthase (NOS2), which is induced in hepatocytes by a combination of cytokines including IL-1. In contrast, IL-1 decreases albumin, cytochromes, transferrin, and lipoprotein lipase transcription (239).

Effects on Vascular Endothelium and Smooth Muscle

TNF and IL-1 play critical roles in severe systemic inflammation. Septic shock is induced by a variety of toxin-producing microorganisms. Infusion of IL-1 mimics the systemic inflammation seen in septic shock: hypotension, acidosis, decreased blood flow, and organ failure often followed by death. TNF significantly augments this response. IL-1–induced hypotension appears to be triggered by the release of nitric oxide, prostaglandins, and platelet-activating factor. In animal models, blocking either IL-1 or TNF reduced mortality (240–248). IL-1RA is effective in reducing mortality in animal models of septic shock but less so in human patients. In patients with sepsis recombinant human IL-1RA increased survival in those individuals with organ dysfunction or a greater than 24% predicted risk of mortality (245). *il1bc* gene knock-out mice are unable to generate mature IL-1β. These mice also have reduced levels of IL-1α and TNF upon LPS challenge and are resistant to LPS-induced lethal shock (208,209). However, neither *tnf* nor *il1b* gene knock-out mice show significant resistance to LPS-induced mortality. These mice do not produce TNF or IL-1β, respectively, but produce normal levels of other cytokines, indicating that multiple cytokines contribute to the severe systemic inflammation that occurs during sepsis.

Infusion of IL-1 results in increased prostanoid production, hypotension, decreased vascular resistance, and increased heart rate and cardiac output (240). Overall the effects of IL-1 on the various components of the vascular system are consistent with clinical conditions such as disseminated intravascular coagulation, vasculitis, and hypotension due to sepsis (Table 5). In the vascular system, smooth muscle and endothelial cells synthesize and respond to IL-1. IL-1 plays an important role in vasorelaxation. Vascular smooth muscle cell contraction is inhibited by IL-1 through a prostaglandin-independent process (249). This effect is due to the induction of NOS2 activity in these cells and the subsequent synthesis of nitric oxide. These cells also produce PGE_2 and PGI_2 in response to IL-1 (250).

Upon IL-1 treatment endothelial cells increase their adherent properties, promoting leukocyte–endothelial cell interactions. IL-1 increases adherence of neutrophils, monocytes, and B and T lymphocytes to endothelial cells (251–253). This is due to the induction of adhesion receptors, in particular, ICAM-1 and E-selectin on the cell surface (12,254). IL-1 increases neutrophil adherence to endothelial cells and induces a host of chemoattractants. The neutrophil chemoattractant properties of IL-1 are due to its ability to promote the expression of a variety of chemokines, several of which act as neutrophil attractants and activators. Endothelial cells also produce a variety of factors, such as platelet-activating factor, tissue factor, PGE_2, PGI_2, and PAI-1, that promote coagulant activity when activated by IL-1 (250,255,256). Procoagulant activity is further enhanced by decreasing tissue plasminogen activator expression.

Gastrointestinal and Endocrine System Effects

IL-1 has been implicated as a proinflammatory mediator in IBD and Crohn's disease. IL-1α expression has been demonstrated in inflamed intestinal tissue from animal models with chemically induced colitis. Elevated levels of PGE_2 and LTB_4 were also present. Administration of

Table 5 Effect of IL-1 on Selected Organs and Tissues

Organ/Tissue	Effect on gene expression	Response
Lymphoid	NOS2, COX2, LTB_4 synthase, phospholipase A_2, IL-1, IL-2, IL-2R, IL-3, IL-6, IL-8, other chemokines, protooncogenes, T_H-cell activation PAF, PAI, plasminogen activator induced	Immune and inflammatory responses
CNS	ACTH, endorphins, somatostatin, vasopressin, IL-6/LIF/CNTF induced	Fever, anorexia, pain, sleep
Hematopoietic	G-CSF, M-CSF, GM-CSF induced	Hematopoiesis
Musculoskeletal	Proteoglycan synthesis inhibited, collagen types I, III, & IV synthesis induced, type II collagen synthesis inhibited, matrix metalloproteinases chemokines and IL-6 induced	Bone and cartilage remodeling, joint destruction
Hepatic	IL-6, acute-phase proteins (SAA, C-reactive protein, lactoprotein, albumin, transferrin, α1-acid glycoprotein, haptoglobin, lipase, cytochromes), NOS2, fibrinogen, factor B, clotting factors induced	Acute-phase protein response
Vascular	IL-8, other chemokines, ICAM, PAF, PAI, PGE_2 induced, tissue factor	Leukocyte trafficking, vascular relaxation, hyper-coagulation

IL-1RA reduces colonic inflammation in these models and decreases PGE_2 and LTB_4, but does not lower IL-1α levels (79,257).

Insulin-dependent diabetes mellitus (IDDM) is an autoimmune disorder affecting the pancreas. In IDDM the insulin-producing β-cells of the islets of Langerhans are gradually destroyed, resulting in insulin insufficiency. Prolonged incubation with IL-1 is cytotoxic to isolated β-cells and intact islets (258,259). Toxicity appears to be mediated by the induction of NOS2 (260–262).

Reproductive System Interactions

A complex cytokine network participates in the cellular communication events required for tissue remodeling during mammalian reproduction. In this regard various cytokines, including IL-1, are produced by uterine epithelium, infiltrating leukocytes, and placental tissues and aid in the communication between maternal and embryonic cells (41). Both *il1b* and *ilbc* knock-out mice are fertile and have normal gestation periods and litter sizes, indicating that either IL-1β is not required for these processes or that other cytokines compensate for its absence in this complex communication network (207–209). Elevated IL-1 levels have been detected prior to normal parturition and in premature labor due to intrauterine infections. In mice IL-1RA prevents premature parturition induced by IL-1 (18,65,263). High IL-1RA levels have been detected in amniotic fluid of women during normal pregnancy, suggesting a possible protective role in pregnancy.

CONCLUDING REMARKS

The members of the IL-1 gene family play an integral part in regulating the host inflammatory responses. The interactions between the individual family members as well as with other cytokines and biological mediators are complex due to redundancy of function, specificity at the cellular level, and intricate control over expression. Despite these complexities, studies over the past 10 years have confirmed much of the pioneering work of those who initially defined the activities of IL-1 (9,10) and have demonstrated that IL-1 sits atop a proinflammatory cascade in key pathophysiological responses. Most recently, studies have shown that blockade of IL-1, or even a single member of the IL-1 family IL-1β, interrupts this network of cytokines and blocks host inflammatory responses. These findings hold great promise for the potential to discover and develop therapeutic agents that will truly modify the course of acute and chronic inflammatory diseases.

REFERENCES

1. Smith EM. Hormonal activities of cytokines. In: Blalock JE, ed. Neuroimmunoendocrinology. 2nd rev. ed. Basel: Karger, 1992:154–169.
2. Dinarello CA. The IL-1 family: 10 years of discovery. FASEB J 1994; 8:1314–1325.
3. Sims JE and Dower SK. Interleukin-1 receptors. Eur Cytokine Netw 1994; 5:539–546.
4. Arend WP. IL-1 receptor antagonist. Adv Immunol 1993; 54:167–227.
5. Auron PE, Webb AC, Rosenwasser LJ, et al. Nucleotide sequence of human monocyte interleukin-1 precursor. Proc Natl Acad Sci USA 1984; 81:7907–7912.
6. March CJ, Mosley B, Larsen A, et al. Cloning, sequence and expression of two distinct human interleukin-1 cDNAs. Nature 1985; 315:641–645.
7. Gershenwald JE, Fong YM, Fahey TJ, et al. Interleukin 1 receptor blockade attenuates the host inflammatory response. Proc Natl Acad Sci USA 1990; 87:4966–4973.
8. Rupp EA, Cameron PM, Ranawat CS, et al. Specific bioactivities of monocyte-derived interleukin 1α and interleukin 1β are similar to each other on cultured murine thymocytes and on cultured human connective tissue cells. J Clin Invest 1986; 78:836–839.
9. Dinarello CA. Interleukin-1 and its biologically related cytokines. Adv Immunol 1989; 44:153–205.
10. Dinarello CA. The biological properties of interleukin-1. Eur Cytokine Netw 1994; 5:517–531.
11. Dinarello CA. Interleukin-1. Reviews of infectious diseases. 1984; 6:51–95.
12. Schmidt JA, Tocci MJ. Interleukin-1. In: Sporn M, Roberts A, eds. The Handbook of Experimental Pharmacology, Part I, Peptide Growth Factors and Their Receptors. Berlin: Springer-Verlag, 1990:473–521.
13. Lomedico PT, Gubler U, Hellman CP, et al. Cloning and expression of murine interleukin-1 cDNA in E. coli. Nature 1984; 312:458–463.
14. Cameron PM, Limjuco GA, Chin J, et al. Purification to homogeneity and amino acid sequence analysis of two anionic species of human interleukin-1. J Exp Med 1985; 164:237–250.
15. Chin J, Rupp E, Cameron PM, MacNaul K, et al. Identification of high affinity receptors for interleukin-1α and interleukin-1β on cultured human rheumatoid synovial cells. J Clin Invest 1988; 82:420–426.
16. Van Damme J, DeLey M, Opdenakker G, et al. Homogeneous interferon inducing factor is related to endogenous pyrogen and interleukin-1. Nature 1985; 314:266–268.
17. Mosley B, Urdal DL, Prickett KS, et al. The interleukin-1 receptor binds the human interleukin-1α precursor but not the interleukin-1β precursor. J Biol Chem 1987; 262:2941–2946.
18. Adashi EY. The potential role of cytokines to ovarian physiology: the emerging role of resident ovarian cells of the white blood cell series. Endocrine Rev 1990; 11:454–464.
19. Kaspar RL, Gerke L. Peripheral blood mononuclear cells stimulated with C5a or lipopolysaccharide to synthesize equivalent levels of IL-1β mRNA show unequal IL-1β protein accumulation but similar polysome profiles. J Immunol 1994; 153:277–286.
20. Schindler R, Gelfand JA, Dinarello CA. Recombinant C5a stimulates transcription rather than translation of IL-1 and TNF; cytokine synthesis induced by LPS, IL-1 or PMA. Blood 1990; 76:1631–1638.
21. Schindler R, Clark BD, Dinarello CA. Dissociation between interleukin-1β mRNA and protein synthesis in human periph-

eral blood mononuclear cells. J Biol Chem 1990; 265:10232–10237.
22. Fenton MJ, Vermeullen MW, Clark BD, et al. Human pro-IL-1 beta gene expression in monocytic cells is regulated by two distinct pathways. J Immunol 1988; 140:2267–2273.
23. Fenton MJ, Clark BD, et al. Transcriptional regulation of the human prointerleukin 1β gene. J Immunol 1987; 138:3972–3979.
24. Fenton MJ. Transcriptional and posttranscriptional regulation of interleukin-1 gene expression. Int J Immunol 1992; 14:401–411.
25. Chin J, Kostura MJ. Dissociation of IL-1β synthesis and secretion in human blood monocytes stimulated with bacterial cell wall products. J Immunol 1993; 151:5574–5581.
26. Tocci MJ, Hutchinson NI, Cameron PM, et al. Expression in Escherichia coli and characterization of recombinant human interleukin-1β: comparison with native IL-1β. J Immunol 1987; 138:1109–1114.
27. MacNaul K, Hutchinson NI, Parsons J, et al. Analysis of human interleukin-1 and tumor necrosis factor-α gene expression in normal monocytes and rheumatoid synoviocytes by in situ hybridization. J Immunol 1990; 145:4154–4166.
28. Arend WP, Smith MF, Janson RW, Joslin FG. IL-1 receptor antagonist and IL-1β production in human monocytes are regulated differently. J Immunol 1991; 147:1530–1536.
29. Elias JA, Schreiber AD, Gustilo K, et al. Differential interleukin 1 elaboration by unfractionated and density fractionated human alveolar macrophages and blood monocytes: relationship to cell maturity. J Immunol 1985; 135:3198–3204.
30. Wewers MD, Saltini C, Sellers S, et al. Evaluation of alveolar macrophages in normals and individuals with active pulmonary sarcoidosis for the spontaneous expression of the interleukin-1β gene. Cell Immunol 1987; 107:479–488.
31. Bernaudin JF, Yamauchi K, Wewers MD, et al. Dissimilar expression of the Interleukin-1β gene among populations of activated human mononuclear phagocytes. J Immunol 1988; 140:3822–3829.
32. Burchett DK, Weaver WM, Westall JA, et al. Regulation of tumor necrosis factor/cachectin and IL-1 secretion in human mononuclear phagocytes. J Immunol 1988; 140:3473–3481.
33. Rambaldi A, Torcia M, Bettoni S, et al. Modulation of cell proliferation and cytokine production in acute myeloblastic leukemia by interleukin-1 receptor antagonist and lack of its expression by leukemic cells. Blood 1991; 78:3248–3253.
34. Estrov Z, Kurzrock R, Wetzler M, et al. Suppression of chronic myelogenous leukemia colony growth by interleukin-1 (IL-1) receptor antagonist and soluble IL-1 receptors: A novel application for inhibitors of IL-1 activity. Blood 1991; 78:1476–1484.
35. Krakauer T, Oppenheim JJ. Interleukin-1 production by a human acute monocytic leukemia cell line. Cell Immunol 1983; 80:223–229.
36. Dinarello CS, Thompson RC. Blocking IL-1: interleukin-1 receptor antagonism in vivo and in vitro. Immunol Today 1991; 12:404–410.
37. Duff GW. Arthritis and interleukins. Br J Rheumatol 1988; 27:2–5.
38. Sauder DN, Mounessa NL, Katz SI, Dinarello CA, Gallin JI. Chemotactic cytokines: the role of leukocyte pyrogen and epidermal cell thymocyte-activation factor in neutrophil chemotaxis. J Immunol 1984; 132:828–832.
39. di Giovine FS, Duff GW. Interleukin 1: the first interleukin. Immunol Today 1990; 11:13–20.
40. Enk AH, Katz SI. Early molecular events in the induction phase of contact sensitivity. Proc Natl Acad Sci USA 1991; 89:1398–1402.
41. Enk AH, Angeloni VL, Udey MC, Katz SI. An essential role for Langerhans cell-derived IL-1β in the initiation of primary immune responses in skin. J Immunol 1993; 150:3698–3704.
42. Nicod LP, Galve-de-Rochemonteix B, Dayer JM. Modulation of IL-1 receptor antagonist and TNF-soluble receptors produced by alveolar macrophages and blood monocytes. Ann NY Acad Sci 1994; 72:5323–5330.
43. Muzio M, Polentarutti N, Sironi M, et al. Cloning and characterization of a new isoform of the interleukin-1 receptor antagonist. J Exp Med 1995; 182:623–628.
44. Acres RB, Larsen A, Gillis S, Conlon PJ. Production of IL-1α and IL-1β by clones of EBV transformed B cells. Mol Immunol 1987; 24:479–485.
45. Pistoia V, Cozzolino F, Rubartelli A, et al. In vitro production of interleukin 1 by normal and malignant human B lymphocytes. J Immunol 1986; 136:1688–1692.
46. Acres RB, Larsen A, Conlon PJ. IL-1 expression in a clone of human T cells. J Immunol 1987; 138:2132–2136.
47. Le JM, Weinstein D, Gubler U, Vilcek J. Induction of membrane-associated interleukin 1 by tumor necrosis factor in human fibroblasts. J Immunol 1987; 138:2137–2142.
48. Le PT, Tuck DT, Dinarello CA, Haynes BF, Singer KH. Human thymic epithelial cells produce interleukin 1. J Immunol 1987; 138:2520–2526.
49. Lindemann A, Riedel D, Oster W, et al. Granulocyte/Macrophage colony-stimulating factor induces interleukin 1 production by human polymorphonuclear neutrophils. J Immunol 1988; 140:837–839.
50. Benveniste EN. Hormonal activities of cytokines. In Blalock JE, ed. Neuroimmunoendocrinology. 2nd rev. ed. Karger: Basel, 1992:106–142.
51. Breder CD, Dinarello CA, Saper CB. Interleukin-1 immunoreactive innervation of the human hypothalamus. Science 1988; 240:321–324.
52. Lechan RM, Toni R, Clark BD, et al. Immunoreactive interleukin-1β localization in the rat forebrain. Brain Res 1990; 514:135–141.
53. Merrill JE. Tumor necrosis factor alpha, interleukin-1 and related cytokines in brain development: normal and pathological. Dev Neurosci 1992; 14:1–10.
54. Licinio J, Wong ML, Gold PW. Localization of interleukin-1 receptor antagonist mRNA in rat brain. Endocrinology 1991; 129:562–564.
55. Luger TA, Stadler BM, Luger BM, et al. Murine epidermal cell-derived thymocyte activating factor resembles murine interleukin 1. J Immunol 1982; 125:2147–2152.
56. Kupper TS, Ballard DW, Chua AO, et al. Human keratinocytes contain mRNA indistinguishable from monocyte interleukin-1α and β mRNA. Keratinocyte epidermal cell-derived thymocyte activating factor is identical to interleukin-1. J Exp Med 1986; 164:2095–2100.
57. Kupper TS, Chua AO, Flood P, et al. Interleukin-1 gene expression in cultured human keratinocytes is augmented by ultraviolet light. J Clin Invest 1988; 80:430–436.
58. Kupper TS, Lee F, Birchall N, et al. Interleukin-1 binds to specific receptors on human keratinocytes and induces granulocyte macrophage colony stimulating factor mRNA and protein. A potential autocrine role for interleukin-1 in epidermis. J Clin Invest 1988; 82:1787–1792.
59. Blanton R, Kupper TJ, McDougall JK, Dower S. Regulation of interleukin-1 and its receptor in human keratinocytes. Proc Natl Acad Sci USA 1989; 86:1273–1277.
60. Cerretti DP, Kozlosky CJ, Mosley B, et al. Molecular cloning

of the interleukin-1β converting enzyme. Science 1992; 256:97–100.
61. Nett MA, Cerretti DP, Berson DR, et al. Molecular cloning of the murine IL-1β converting enzyme cDNA. J Immunol 1992; 149:3254–3259.
62. Gruaz-Chatellard D, Baumberger C, Saurat JH, et al. Interleukin 1 receptor antagonist in human epidermis and cultured keratinocytes. FEBS Lett 1991; 294:137–140.
63. Bigler CF, Norris DA, Weston WL, Arend WP. Interleukin-1 receptor antagonist production by human keratinocytes. J Invest Dermatol 1992; 98:38–44.
64. Ollivierre F, Gubler U, Towle CA, et al. Expression of IL-1 genes in bovine chondrocytes: a mechanism for autocrine control of cartilage matrix degradation. Biochem Biophys Res Commun 1986; 141:904–911.
65. Robertson SA, Seamark RF, Guilbert LJ, Wegmann TG. The role of cytokines in gestation. Crit Rev Immunol 1994; 14:239–292.
66. Cominelli F, Nast CC, Clark BD, et al. Interleukin 1 (IL-1) Gene expression, syntheses and effect of specific IL-1 receptor blockade in rabbit immune complex colitis. J Clin Invest 1990; 86:972–980.
67. Bursten SL, Locksley RM, Ryan JL, Lovett DH. Acylation of monocyte and glomerular mesangial cell proteins: myristyl acylation of the interleukin-1 precursors. J Clin Invest 1988; 82:1479–1487.
68. Gray PW, Glaister D, Chen E, Goeddel DV, Pennica D. Two interleukin 1 genes in the mouse: cloning and expression of the cDNA for murine interleukin 1β. J Immunol 1986; 137:3644–3648.
69. Furutani Y, Notake M, Yamyoshi M, et al. Cloning and characterization of the cDNAs for human and rabbit interleukin-1 precursor. Nucleic Acid Res 1986; 13:5869–5874.
70. Mori S, Goto F, Goto K, et al. Cloning and sequence analysis of a cDNA for lymphocyte proliferation potentiating factor of rabbit polymorphonuclear leukocytes: identification as rabbit interleukin-1β. Biochem Biophys Res Commun 1988; 150:1237–1243.
71. Maliszewski CR, Baker PE, Schoenborn MA, et al. Cloning, sequence and expression of bovine interleukin 1α and interleukin 1β complementary DNAs. Mol Immunol 1988; 25:429–437.
72. Secombes CJ. The phylogeny of cytokines. In: Thomson AW, ed. The Cytokine Handbook. 2nd ed. New York: Academic Press, 1994:573–576.
73. Hughes AL. Evolution of the IL-1 gene family in mammals. J Mol Evol 1994; 39:6–12.
74. Goto F, Goto K, Miyata T, et al. Interleukin-1 receptor antagonist in inflammatory exudate cells of rabbits. Production, purification and determination of primary structure. Immunology 1992; 77:235–244.
75. Shuck M, Eessalu TE, Tracey DE, Bienkowski MJ. Cloning, heterologous expression and characterization of murine interleukin 1 receptor antagonist protein. Eur J Immunol 1991; 21:2775–2780.
76. Zahedi K, Seldin MF, Rits M, Ezekowitz AB, Whitehead AS. Mouse IL-1 receptor antagonist protein. Molecular characterization, gene mapping, and expression of mRNA in vitro and in vivo. J Immunol 1991; 146:4228–4233.
77. Matsushime H, Roussel MF, Matsushima K, et al. Cloning and expression of murine interleukin-1 receptor antagonist in macrophages stimulated by colony-stimulating factor 1. Blood 1991; 78:616–623.
78. Eisenberg SP, Evans RJ, Arend WP, et al. Primary structure and functional expression from complementary DNA of a human interleukin-1 receptor antagonist. Nature 1990; 343:341–346.
79. Cominelli F, Bortolami M, Pizarro TT, et al. Rabbit interleukin-1 receptor antagonist: cloning, expression, functional characterization and regulation during intestinal inflammation. J Biol Chem 1994; 269:6962–6971.
80. Caput D, Beutler B, Hartog K, Thayer R, et al. Identification of a common nucleotide sequence in the 3′ untranslated region of mRNA molecules specifying inflammatory mediators. Proc Natl Acad Sci USA 1986; 83:1670–1674.
81. Cosman D. Control of mRNA stability. Immunol Today 1987; 8:16–17.
82. Decker CJ, Parker R. Diversity in the cytoplasmic functions for the 3′ untranslated region of eukaryotic transcripts. Curr Opin Cell Biol 1995; 7:386–392.
83. Nicklin MJH, Weith A, Duff GW. A physical map of the region encompassing the human interleukin-1α, interleukin-1β, and interleukin-1 receptor antagonist genes. Genomics 1994; 19:382–384.
84. Clark BW, Collins KL, Gandy MS, Webb AC, Auron PE. Genomic sequences for prointerleukin-1 beta: possible evolution from a reverse transcribed prointerleukin-1 alpha gene. Nucleic Acid Res 1986; 14:7897–7914.
85. Bensi G, Raugei G, Pall E, et al. Human interleukin-1 beta gene. Gene 1987; 52:95–101.
86. Furutani Y, Notake M, Fukui T, et al. Complete nucleotide sequence of the gene for human interleukin-1 alpha. Nucleic Acid Res 1986; 14:3167–3179.
87. Eisenberg SP, Brewer MT, Verderber E, et al. IL-1 receptor antagonist is a member of the IL-1 gene family. Proc Natl Acad Sci USA 1991; 88:5232–5236.
88. Auron PE, Webb AC. IL-1: a gene expression system regulated at multiple levels. Eur Cytokine Netw 1994; 5:573–594.
89. Shirikawa F, Saito K, Bonagura CA, et al. The human prointerleukin-1β gene requires DNA sequences both proximal and distal to the transcription start site for tissue-specific induction. Mol Cell Biol 1993; 13:1332–1344.
90. Kominato Y, Galson DL, Waterman WR, et al. Monocyte expression of the human prointerleukin-1β gene (*IL1B*) is dependent upon promoter sequences which bind the hematopoietic transcription factor Spi-1/PU.1. Mol Cell Biol 1995; 15:58–68.
91. Tsukuda J, Saito K, Waterman WR, Webb AC, Auron PE. Transcription factors NF-IL6 and CREB recognize a common essential site in the human prointerleukin-1β gene. Mol Cell Biol 1993; 14:7285–7297.
92. Godambe SA, Chaplin DD, Takova T, et al. A novel *cis*-acting element required for lipopolysaccharide-induced transcription of the murine interleukin-1β gene. Mol Cell Biol 1995; 15:112–119.
93. Gogswell JP, Godlevski MM, Wisely GB, et al. NF-κB regulates IL-1β transcription through a consensus NF-κB site and a non-consensus CRE-like site. J Immunol 1994; 153:712–721.
94. Gray JG, Chandra G, Clay WC, et al. A CRE/ATF site in the upstream regulatory sequences of the human interleukin-1β gene is necessary for induction in U937 and THP-1 monocytic cell lines. Mol Cell Biol 1993; 13:6678–6689.
95. Hiscott J, Maois J, Gaofoulis J, et al. Characterization of a functional NF-κB in the human interleukin-1β promoter: evidence for a positive autoregulatory loop. Mol Cell Biol 1993; 13:6231–6240.
96. LeClair KP, Blanar MA, Sharp PA. The p50 subunit of NF-κB associates with the NF-IL6 transcription factor. Proc Natl Acad Sci USA 1992; 89:8145–8190.
97. Nishida T, Takano M, Nishio Y, et al. The transcription of the

interleukin-1 beta gene is induced with PMA and inhibited with dexamethasone in U937 cells. Biochem Biophys Res Commun 1988; 156:269–277.
98. Muegge K, Williams TM, Kant J, et al. Interleukin-1 co-stimulatory activity on the interleukin-2 promoter via AP-1. Science 1989; 246:249–251.
99. Muegge K, Vila M, Gusella GL, et al. IL-1 induction of the c-jun promoter. Proc Natl Acad Sci USA 1993; 90:7054–7059.
100. Nishio Y, Isshiki H, Kishimoto T, Akira S. A nuclear factor for interleukin-6 expression (NF-IL6) and the glucocorticoid receptor synergistically activate transcription of the rat a1-acid glycoprotein gene via direct protein-protein interaction. Mol Cell Biol 1993; 13:1854–1862.
101. Furutani Y. Molecular studies on interleukin-1α. Eur Cytokine Netw 1994; 5:533–538.
102. Mora M, Carinci V, Bens G, et al. Differential expression of the interleukin-1 alpha and beta genes. Prog Clin Biol Res 1990; 349:205–211.
103. Smith MF, Eidlen D, Brewer MT, et al. The human IL-1 receptor antagonist promoter: cell type-specific activity and identification of regulatory regions. J Immunol 1992; 149:2000–2008.
104. Smith MF, Eidlen D, Arend WP, Gutierrez-Hartmann A. LPS-induced expression of the human IL-1 receptor antagonist gene is controlled by multiple interacting promoter elements. J Immunol 1994; 153:3584–3593.
105. Bayne EK, Rupp EA, Lumjuco GA, et al. Immunocytochemical detection of interleukin-1 within stimulated human monocytes. J Exp Med 1986; 163:1267–1273.
106. Bakouche O, Brown DC, Lachman LB. Subcellular localization of human monocyte interleukin-1: evidence for an inactive precursor molecule and possible mechanism for IL-1 release. J Immunol 1987; 138:4249–4255.
107. Singer II, Scott S, Hall GL, et al. Interleukin-1β is localized in the cytoplasmic ground substance but is largely absent from the golgi apparatus and plasma membranes of stimulated human monocytes. J Exp Med 1988; 167:389–399.
108. Stevenson FT, Torrano F, Locksley RM, Lovett DH. Interleukin-1: the patterns of translation and intracellular distribution support alternative secretory mechanisms. J Cell Physiol 1992; 152:223–229.
109. Singer II, Scott S, Chin J, et al. Interleukin-1β converting enzyme is localized on the external cell-surface membranes and in the cytoplasmic ground substance of activated human monocytes by immuno-electronmicroscopy. J Exp Med 1995; 182:1447–1459.
110. Kobayashi Y, Yamamoto K, Saiso T, et al. Identification of calcium-activated neutral protease as a processing enzyme of human interleukin-1α. Proc Natl Acad Sci USA 1990; 87:5548–5552.
111. Hogquist KA, Nett MA, Unanue ER, Chaplin DD. Interleukin-1 is processed and released during apoptosis. Proc Natl Acad Sci USA 1991; 88:8485–8489.
112. Hogquist KA, Nett MA, Sheehan KC, et al. Release of IL-1 from mononuclear phagocytes. J Immunol 1991; 146:2181–2187.
113. Black RA, Kronheim SR, Sleath PR. Activation of interleukin-1β by a co-induced protease. FEBS Lett 1989; 247:386–390.
114. Kostura MJ, Tocci MJ, Limjuco G, et al. Identification of a monocyte specific pre-interleukin-1β convertase activity. Proc Natl Acad Sci USA 1989; 86:5227–5231.
115. Thornberry NA. Interleukin-1β converting enzyme. In: Barrett A, ed. Proteolytic Enzymes, Part D, Methods in Enzymology. Vol. 230. New York: Academic Press.
116. Sleath PR, Hendrickson RC, Kronheim SR, et al. Substrate specificity of the protease that processes human interleukin-1β. J Biol Chem 1990; 265:14526–14528.
117. Howard AD, Kostura MJ, Thornberry NA, et al. IL-1β converting enzyme requires aspartic acid residues for processing of the IL-1β precursor at two distinct sites and does not cleave 31-kDa IL-1α. J Immunol 1991; 146:2964.
118. Thornberry NA, Bull HG, Calaycay J, et al. A novel heterodimeric cysteine protease is required for interleukin-1β processing in monocytes. Nature 1992; 356:768–774.
119. Molineaux SM, Casano FJ, Rolando AM, et al. Interleukin-1β (IL-1β) processing in murine macrophages requires a structurally conserved homologue of human IL-1β converting enzyme. Proc Natl Acad Sci USA 1993; 90:1809–1813.
120. Thornberry NA, Peterson EP, Zhao J, et al. Inactivation of interleukin-1β converting enzyme by peptide acyloxy-methylketones. Biochemistry 1994; 33:3934–3939.
121. Miller BE, Krasney PA, Gauvin DM, et al. Inhibition of mature IL-1β production in murine macrophages and a murine model of inflammation by WIN 67694, an inhibitor of IL-1β converting enzyme. J Immunol 1995; 154:1331–1338.
122. Fletcher DS, Agarwal L, Chapman K, et al. A synthetic inhibitor of interleukin-1β converting enzyme prevents endotoxin-induced IL-1β production in vitro and in vivo. J Interferon Cytokine Res 1995; 15:243–248.
123. Ray CA, Black RA, Kronheim SR, et al. Viral inhibition of inflammation: cowpox virus encodes an inhibitor of the interleukin-1β converting enzyme. Cell 1992; 69:597–604.
124. Komiyama T, Ray CA, Pickup DJ, et al. Inhibition of interleukin-1β converting enzyme by the cowpox virus serpin *crmA*: an example of cross-class inhibition. J Biol Chem 1994; 269:19331–19335.
125. Cerretti DP, Tamaralayne Hollingsworth L, Kozlowsky CJ, et al. Molecular characterization of the gene for human interleukin-1β converting enzyme (IL1BC). Genomics 1994; 20:468–473.
126. Casano FJ, Rolando AM, Mudgett JS, Molineaux SM. The structure and complete nucleotide sequence of the murine gene encoding interleukin-1β converting enzyme. (ICE). Genomics 1994; 20:474–481.
127. Miller DK, Ayala JA, Egger LA, et al. Purification and characterization of active human interleukin-1β converting enzyme from THP.1 monocytic cells. J Biol Chem 1993; 268:18062–18073.
128. Ayala J, Yamin TT, Egger LA, et al. Interleukin-1β converting enzyme is present in monocytes as an inactive precursor. J Immunol 1994; 153:2592–2597.
129. Keane KM, Giegel DA, Lipinski WJ, et al. Cloning, tissue expression and regulation of rat interleukin-1β converting enzyme. Cytokine 1995; 7:105–110.
130. Alnemri ES, Fernandes-Alnemri T, Litwack G. Cloning and expression of four novel isoforms of human interleukin-1β converting enzyme with different apoptotic activities. J Biol Chem 1995; 270:4312–4317.
131. Hazuda DJ, Lee JC, Young PR. The kinetics of interleukin-1 secretion from activated monocytes. Differences between interleukin-1α and interleukin-1β. J Biol Chem 1988; 263:8473–8479.
132. Rubartelli A, Cozzalino F, Talo M, Sitia R. A novel secretory pathway for interleukin-1 beta, a protein lacking a signal sequence. EMBO J 1990; 9:1503–1510.
133. Kuchler K. Unusual routes of protein secretion: the easy way out. Trends Cell Biol 1993; 3:421–423.
134. Young P, Hazuda DJ, Simon PL. Human interleukin-1β is not secreted from hamster fibroblasts when expressed con-

stitutively from a transfected cDNA. J Cell Biol 1988; 107:447–456.
135. McGrath JP, Varshavsky A. The yeast STE6 gene encodes a homologue of the mammalian multidrug resistance P-glycoprotein. Nature 1989; 340:400–403.
136. Higgins GC, Foster JL, Postlethwaite AE. Interleukin-1β propeptide is detected intracellularly and extracellularly when human monocytes are stimulated with LPS in vitro. J Exp Med 1994; 180:607–614.
137. Howard AD, Palyha OC, Griffin PR, et al. Human IL-1β processing and secretion in recombinant baculovirus-infected Sf9 cells is blocked by the cowpox virus serpin *crmA*. J Immunol 1995; 154:2321–2332.
138. Wilson KP, Black JA, Thomson JA, et al. Structure and mechanism of interleukin-1β converting enzyme. Nature 1995; 370:270–275.
139. Walker N, Talanian RV, Brady KD, et al. Crystal structure of the cysteine protease interleukin-1β-converting enzyme: a $(p20/p10)_2$ homodimer. Cell 1995; 78:343–352.
140. Gu Y, Wu J, Faucheu C, et al. Interleukin-1β converting enzyme requires oligomerization for activity of processed forms in vivo. EMBO J 1995; 14:1923–1931.
141. Yuan J, Shaham S, Ledoux S, Ellis HM, Horovitz HR. The *C. elegans* cell death gene *ced-3* encodes a protein similar to mammalian interleukin-1β converting enzyme. Cell 1993; 75:641–652.
142. Miura M, Zhu H, Rotello R, Hartwieg E, Yuan J. Induction of apoptosis in fibroblasts by IL-1β converting enzyme, a mammalian homolog of the *C. elegans* cell death gene *ced-3*. Cell 1993; 75:653–658.
143. Kumar S, Kinoshita M, Noda M, et al. Induction of apoptosis by the mouse *nedd2* gene, which encodes a protein similar to the products of the *Caenorahabitis elegans* cell death protein *Ced-3* and the mammalian interleukin-1β converting enzyme. Genes Dev 1994; 8:1613–1620.
144. Fernandes-Alnemri T, Litwack G, Alnemri ES. CPP32, a novel human apoptotic protein with homology to *Caenorahabitis elegans* cell death protein *ced-3* and mammalian interleukin-1β converting enzyme. J Biol Chem 1994; 269:30761–30766.
145. Wang L, Miura M, Bergeron L, Yuan J. *Ich-1*, an ICE/*ced-3*-related gene encodes both positive and negative regulators of programmed cell death. Cell 1994; 78:739–746.
146. Earnshaw S, Littwack G, et al. Identification of MCH2, a new member of the ICE/ced3 cysteine proteinase family.
147. Nicholson DW, Ali A, Thornberry N, et al. Identification and inhibition of the ICE/CED-3 protease necessary for mammalian apoptosis. Nature 1995; 376:37–43.
148. Tewari M, Quan TL, O'Rourke K, et al. Yama/CPP32β, a mammalian homolog of CED-3, is a CrmA-inhibitable protease that cleaves the death substrate poly(ADP-ribose) polymerase. Cell 1995; 81:801–809.
149. Munday NA, Vaillancourt JP, Ali A, et al. Molecular cloning and pro-apoptotic activity of ICE_{rel}II and ICE_{rel}III, members of the ICE/CED-3 family of cysteine proteinases. J Biol Chem 1995; 270:15870–15876.
150. Faucheau C, Diu A, Chan AWE, et al. A novel human protease similar to the interleukin-1β converting enzyme induces apoptosis in transfected cells. EMBO J 1995; 14:1914–1922.
151. Gagliardini V, Fernandez P-A, Lee RKK, et al. Prevention of vertebrate cell death by the *crmA* gene. Science 1994; 263:826–828.
152. Tewari M, Dixit VM. Fas and tumor necrosis factor-induced apoptosis is inhibited by the pox virus *crmA* gene product. J Biol Chem 1995; 270:3255–3260.
153. Greenfeder SA, Nunes P, Kwee L, Labow M, Chizzonite RA, Ju G. Molecular cloning and characterization of a second subunit of the interleukin-1 receptor complex. J Biol Chem 1995; 270:13757–13765.
154. Parnet P, Garka KE, Bonnert TP, Dower SK, Sims JE. 2F1: a novel receptor similar to the type 1 Interleukin-1 receptor and T1/ST2. AACR 1995;
155. Bergers G, Reikerstorfer A, Braselmann S, Graninger P, Busslinger M. Alternative promoter usage of the Fos-responsive gene *Fit-1* generates mRNA isoforms coding for either secreted or membrane bound proteins related to the IL-1 receptor. EMBO J 1994; 13:1176–1188.
156. Tominga S, Jenkins NA, Gilbert DJ, Copeland NG, Tetsuka T. Molecular cloning of the murine ST2 gene. Characterization and chromosomal mapping. Biochim Biophys Acta 1991; 1090:1–7.
157. Sims JE, March CJ, Cosman D, et al. cDNA expression cloning of the IL-1 receptor, a member of the immunoglobulin superfamily. Science 1988; 241:585–589.
158. Sims JE, Gayke MA, Slack JL, et al. Interleukin-1 signaling occurs exclusively via the type I receptor. Proc Natl Acad Sci USA 1993; 90:6155–6159.
159. Sims JE, Giri JG, Dower SK. The two interleukin-1 receptors play different roles in IL-1 activities. Clin Immunol Immunopathol 1994; 72:9–14.
160. McMahon CJ, Slack JL, Mosley B, et al. A novel IL-1 receptor cloned form B cells by mammalian expression is expressed in many cell types. EMBO J 1991; 10:2821–2832.
161. Symons JA, Eastgate JA, Duff GW. Purification and characterization of a novel soluble receptor for interleukin-1. J Exp Med 1991; 174:1251–1254.
162. Colotta F, Re F, Muzio M, et al. Interleukin-1 type II receptor: a decoy target for IL-1 that is regulated by IL-4. Science 1994; 261:472–475.
163. Spriggs MK, Hruby DE, Maliszewski CR, et al. Vaccinia and cowpox viruses encode a novel secreted interleukin-1 binding protein. Cell 1992; 71:145–152.
164. Alcami A, Smith GL. A soluble receptor for interleukin-1β encoded by vaccinia varus: a novel mechanism of virus modulation of the host response to infection. Cell 1992; 71:153–167.
165. Martin M, Fleur Bol G, Eriksson A, et al. Interleukin-1 induced activation of a protein kinase co-precipitating with the type I interleukin-1 receptor in T cells. Eur J Immunol 1994; 24:1566–1571.
166. Croston GE, Cao Z, Goedell DV. NFκB activation by interleukin-1 (IL-1) requires an IL-1 receptor-associated protein kinase activity. J Biol Chem 1995; 270:16514–16517.
167. Matushima K, Kobayshi Y, Copeland TD, et al. Phosphorylation of a cytosolic 65 kDa protein induced by interleukin-1 in glucocorticoid pretreated normal human peripheral blood mononuclear leukocytes. J Immunol 1987; 139:3367–3374.
168. Kracht M, Truong O, Totty NF, et al. Interleukin-1α activates two forms of p54α mitogen-activated protein kinase in rabbit liver. J Exp Med 1994; 180:2017–2025.
169. Freshney NW, Rawlinson L, Guesdon F, et al. Interleukin-1 activates a novel protein kinase cascade that results in phosphorylation of Hsp27. Cell 1994; 78:1039–1049.
170. Rouse J, Cohen P, Trigon S, et al. A novel kinase cascade triggered by stress and heat shock that stimulates MAPKAP kinase-2 and phosphorylation of the small heat shock protein. Cell 1994; 78:1027–1037.
171. Han J, Lee JD, Bibbs L, Ulevitch RL. A MAP kinase targeted by endotoxin and hyperosmolarity in mammalian cells. Science 1994; 265:808–811.
172. Raingeaud J, Gupta S, Rogers JS, et al. Pro-inflammatory

cytokines and environmental stress cause p38 mitogen activated protein kinase activation by dual phosphorylation on tyrosine and threonine. J Biol Chem 1995; 270:7420–7246.
173. Lee JC, Laydon JT, McDonnell PC, et al. A protein kinase involved in the regulation of inflammatory cytokine biosynthesis. Nature 1994; 372:739–746.
174. Zhang S, Han J, Sells MA, et al. Rho family GTPases regulate p38 mitogen-activated protein kinase through the downstream mediator Pak-1. J Biol Chem 1995; 270:23934–23936.
175. Hambleton J, McMahon M, DeFranco AL. Activation of Raf-1 and mitogen-activated protein kinase in murine macrophages partially mimics lipopolysaccharide-induced signaling events. J Exp Med 1995; 182:147–154.
176. Reimann T, Buscher R, Hipskind S, et al. Lipopolysaccharide induces activation of the Raf-1/MAP kinase pathway: a putative role in the induction of the IL-1β and the TNFα genes. J Immunol 1994; 153:5740–5749.
177. Whitmarsh AJ, Shore P, Sharrocks AD, Davis RJ. Integration of MAP kinase signal transduction pathways at the serum response element. Science 1995; 269:403–407.
178. Brooks JW, Mizel SB. Interleukin-1 signal transduction. Eur Cytokine Netw 1994; 5:547–561.
179. Dobrowsky RT, Hannun YA. Ceramide stimulates a cytosolic protein phosphatase. J Biol Chem 1992; 267:5048–5053.
180. Dobrowsky RT, Kamibayashi C, Mumby MC, Hannun YA. Ceramide activates heterotrimeric protein phosphatase 2A. J Biol Chem 1993; 268:15523–15529.
181. Mathias S, Dressler KA, Kolesnick RN, Golde DW. Characterization of a ceramide activated protein kinase: stimulation by TNFα. Proc Natl Acad Sci USA 1991; 88:10009–10012.
182. Raines MA, Kolesnick RN, Golde DW. Sphingomyelinase and ceramide activate mitogen-activated protein kinase in myeloid HL-60 cells. J Biol Chem 1993; 268:14572–14580.
183. Priestle JP, Schar HP, Grutter MG, et al. Crystallographic refinement of interleukin-1 beta at 2.0 Å resolution. Proc Natl Acad Sci USA 1988; 86:9667–9671.
184. Clore GM, Wingfield PT, Gronenborn AM. High-resolution three-dimensional structure of interleukin-1β in solution by three- and four-dimensional nuclear magnetic resonance spectroscopy. Biochemistry 1991; 30:2315–2323.
185. Graves BJ, Hatada MH, Hendrickson WA, et al. Structure of interleukin-1α at 2.7-Å resolution. Biochemistry 1990; 29:2679–2684.
186. Vigers GP, Caffes P, Evans RJ, et al. X-ray structure of interleukin-1 receptor antagonist at 2.0-Å resolution. J Biol Chem 1994; 269:12874–12879.
187. Hazuda D, Webb RL, Simon P, Young P. Purification and characterization of the human recombinant precursor interleukin-1β. J Biol Chem 1989; 264:1689–1693.
188. Gimenez-Gomez G, Todkey J, Bennet C, et al. Brain-derived fibroblast growth factor: complete amino acid sequence and homologies. Science 1985; 230:1385–1388.
189. Thomas KA, Rios-Candelore M, Gimenez-Gomez G, et al. Pure brain-derived fibroblast growth factor is a potent vascular endothelial cell mitogen with sequence homology to interleukin-1. Proc Natl Acad Sci USA 1985; 82:6409–6413.
190. Murzin AG, Lesk AM, Chotia C. Beta-trefoil fold. Patterns of structure and sequence in the Kunitz inhibitors interleukin-1 beta and -1 alpha and fibroblast growth factor. J Mol Biol 1992; 22:531–543.
191. Dripps DJ, Brandhuber BJ, Thompson RC, Eisenberg SP. Interleukin-1 (IL-1) receptor antagonist binds to the 80-kDa IL-1 receptor but does not initiate IL-1 signal transduction. Biol Chem 1991; 266:10331–10336.
192. MacDonald HR, Wingfield P, Schmeissner U, et al. Point mutations of human interleukin-1 with decreased receptor binding affinity. FEBS Lett 1986; 209:295–298.
193. Gerke L, Jobling SA, Paik LS, et al. A point mutation uncouples human interleukin-1 beta biological activity and receptor binding. J Biol Chem 1990; 265:5922–5925.
194. Labriola-Tompkins E, Chandran C, Kaffka KL, et al. Identification of the discontinuous binding site in human interleukin 1β for the type I interleukin 1 receptor. Proc Natl Acad Sci USA 1991; 88:11182–11186.
195. Labriola-Tompkins E, Chandran C, Varnell TA, et al. Structure-function analysis of human IL-1α: identification of residues required for binding to the human type I IL-1 receptor. Protein Eng 1993; 6:535–539.
196. Evans RJ, Bray J, Childs JD, et al. Mapping receptor binding sites in IL-1 receptor antagonist and IL-1β by site directed mutagenesis. J Biol Chem 1995; 270:11477–11483.
197. Schreuder H. Crystal structure of the interleukin-1 receptor-antagonist complex. Cytokine 1995; 7:599.
198. Dinarello CA, Thompson RC. Blocking IL-1: interleukin 1 receptor antagonist in vivo and in vitro. Immunol Today 1991; 12:404–410.
199. Dayer J-M, Burger D. Interleukin-1, tumor necrosis factor and their specific inhibitors. Eur Cytokine Netw 1994; 5:563–571.
200. Geiger T, Towbin H, Consenti-Vargas A, et al. Neutralization of interleukin-1β activity in vivo with a monoclonal antibody alleviates collagen-induced arthritis in DBA/1 mice and prevents the associated acute-phase response. Clin Exp Rheumatol 1993; 11:515–522.
201. Van den Berg WB, Joosten LAB, Helsen M, Van de Loo FAJ. Amelioration of established murine collagen-induced arthritis with anti-IL-1 treatment. Clin Exp Immunol 1994; 95:237–243.
202. Arend WP. Interleukins and arthritis: IL-1 antagonism in inflammatory arthritis. Lancet 1993; 341:155–156.
203. LeMay LG, Otterness IG, Vander AJ, Kluger MJ. In vivo evidence that the rise in plasma IL-6 following injection of a fever-inducing dose of LPS is mediated by IL-1 beta. Cytokine 1990; 2:199–209.
204. Kluger MJ. Fever: role of pyrogens and cryogens. Physiol Rev 1991; 71:93–127.
205. Klir JJ, McClellan JL, Kluger MJ. Interleukin-1 beta causes the increase in anterior hypothalamic interleukin-6 during LPS-induced fever in rats. Am J Physiol 1994; 266:R1845–1848.
206. Telford JL, Macchia G, Massone A, et al. Murine interleukin-1β gene: structure and evolution. Nucleic Acid Res 1986; 14:9955–9963.
207. Zheng H, Fletcher D, Kozak W, et al. Resistance to fever induction and impaired acute phase response in Interleukin-1β deficient mice. Immunity 1995; 3:9–19.
208. Li P, Allen H, Banerjee S, et al. Mice deficient in IL-1β converting enzyme are defective in production of mature IL-1β and resistant to endotoxic shock. Cell 1995; 80:401–411.
209. Kuida K, Lippke JA, Ku G, et al. Altered cytokine export and apoptosis in mice deficient in interleukin-1β converting enzyme. Science 1995; 267:2000–2003.
210. Kozak W, Zheng H, Conn CA, et al. Thermal and behavioral effects of lipopolysaccharide and influenza in interleukin-1β deficient mice. Am J Physiol 1995;
211. Christen A, Mudgett JM, Shen F, Kostura M, Visco D. Lack of collagen induced arthritis in IL-1β knock mice. Abstracts of the Orthopedic Research Association (in press).
212. Enari M, Hug H, Nagata S. Involvement of an ICE-like protease in Fas-mediated apoptosis. Nature 1995; 375:78–81.
213. Los M, Van de Craen M, Penning LC, et al. Requirement of an

ICE/CED-3 protease for Fas/APO-1-mediated apoptosis. Nature 1995; 375:81–83.

214. Kupper T, Horowitz M, Lee F, et al. Autocrine growth of T cells independent of IL-2: identification of IL-4 (IL-4, BSF-1) as an autocrine growth factor for a cloned antigen specific helper T cell. J Immunol 1987; 138:4280–4288.
215. Lichtman AH, Chin J, Schmidt JA, Abbas AK. Role of interleukin 1 in the activation of T lymphocytes. Proc Natl Acad Sci USA 1988; 85:9699–9703.
216. Nicod LP, Habre FE, Dayer JM. Natural and recombinant interleukin-1 receptor antagonist does not inhibit human T-cell proliferation induced by mitogens, soluble-antigens or allogeneic determinants. Cytokine 1992; 4:29–35.
217. Havell EA, Moldawer LL, Helfgott D, Kilian PL, Sehgal PB. Type 1 IL-1 receptor blockade exacerbates murine listeriosis. J Immunol 1992; 148:1486–1492.
218. Mochizuki DY, Eisenman JR, Conlon PJ, et al. Interleukin 1 regulates hematopoietic activity, a role previously ascribed to hemopoietin 1. Cell Biol 1987; 84:5267–5271.
219. Moore MAS, Warren DJ. Synergy of interleukin 1 and granulocyte colony-stimulation factor: in vivo stimulation of stem-cell recovery and hemopoietin 1. Cell Biol 1987; 84:7134–7138.
220. Oppenheim JJ, Neta R, Tiberghein P, et al. Interleukin-1 enhances survival of lethally irradiated mice treated with allogeneic bone marrow cells. Blood 1989; 74:2257–2263.
221. Schwartz GN, MacVittie TJ, Vigneulle RM, et al. Enhanced hematopoietic recovery in irradiated mice pretreated with interleukin-1 (IL-1). Immunopharmacol Immunotoxicol 1987; 9:371–377.
222. Woloski BMRNJ, Smith EM, Meyer WJ III, et al. Corticotropin-releasing activity of monokines. Science 1985; 230:1035–1037.
223. Sapolsky R, Rivier C, Jamamoto G, et al. Interleukin 1 stimulates the secretion of hypothalamic corticotropin-releasing factor. Science 1987; 238:522–524.
224. Berkenbosch F, Van Oers J, Del Ray A, Tilders F, Besedovsky H. Corticotropin-releasing factor-producing neurons in the rat activated by interleukin 1. Science 1987; 238:524–526.
225. Bernton EW, Beach JE, Holaday JW, et al. Release of multiple hormones by a direct action of interleukin-1 on pituitary cells. Science 1987; 238:519–521.
226. Besedovsky H, Del Ray A, Sorkin E, Dinarello CA. Immunoregulatory feedback between interleukin-1 and glucocorticoid hormones. Science 1986; 233:652–654.
227. Chrousos GP. The hypothalamic pituitary-adrenal axis and immune mediated inflammation. N Engl J Med 1995; 332:1351–1361.
228. Revhaug A, Michie HR, Nanson JM, et al. Inhibition of cyclo-oxygenase attenuates the metabolic response to endotoxin in humans. Arch Surg 1988; 123:162–169.
229. Lumpkin MD. The regulation of ACTH secretion by IL-1. Science 1987; 238:452–454.
230. Del Rey A, Besedovsky H, Sorkin E, Dinarello CA. Interleukin-1 and glucocorticoid hormones integrate an immunoregulatory feedback circuit. Ann NY Acad Sci 1987; 496:85–91.
231. Bertini R, Bianchi M, Ghezzi P. Adrenalectomy sensitizes mice to the lethal effects of interleukin 1 and tumor necrosis factor. J Exp Med 1988; 167:1708–1717.
232. Shintani F, Nakaki T, Kanba S, et al. Involvement of interleukin-1 in immobilization stress-induced increase in plasma adrenocorticotropic hormone and in the release of hypothalamic monoamines in the rat. J Neurosci 1995; 15:1961–1970.
233. Mrosovsky N, Molony LA, Conn CA, Kluger MJ. Anorexic effects of interleukin-1 in the rat. Am J Physiol 1989; 257:R1315.
234. Feige U, Karbowski A, Rordorf-Adam C, Pataki A. Arthritis induced by continuous infusion of hr-interleukin-1α into the rabbit knee-joint. Int J Tiss Reac 1989; 11:225–232.
235. Konig A, Muhlbauer RC, Fleisch H. Tumor necrosis factor and interleukin-1 stimulate bone resorption in vivo as measured by [3-H] tetracycline excretion from prelabeled mice. J Bone Miner Res 1988; 3:621–628.
236. Krane SM, Conca W, Stephenson ML, Amento EP, Goldring MB. Mechanisms of matrix degradation in rheumatoid arthritis. Ann NY Acad Sci 1990; 580:340–352.
237. Dayer JM, de Rochemoteix B, Burrus B, et al. Human recombinant interleukin 1 stimulates collagenase and prostaglandin E2 production by human synovial cells. J Clin Invest 1986; 77:645–651.
238. Bauman H, Richards C, Gauldie J. Interaction among hepatocyte-stimulating factors, interleukin 1, and glucocorticoids for regulation of acute phase plasma proteins in human hepatoma (HepG2) cells. J Immunol 1987; 139:4122–4128.
239. Kushner I, Mackiewicz A. The acute phase response: an overview. In: Mackiewicz A, Kushner I, Bauman H, eds. Acute Phase Proteins. Boca Raton, FL: CRC Press, 1993:255–271.
240. Okusawa S, Gelfand JA, Ikejima T, et al. Interleukin 1 induces a shock-like state in rabbits. Synergism with tumor necrosis factor and the effect of cyclooxygenase inhibition. J Clin Invest 1988; 81:1162–1172.
241. Beutler B, Milsark IW, Cerami A. Passive immunization against cachectin/tumor necrosis factor protects mice from lethal effect of endotoxin. Science 1985; 229:869.
242. Tracey KJ, Fong Y, Hesse DG, et al. Anticachectin/TNF monoclonal antibodies prevent septic shock during lethal bacteremia. Nature 1987; 330:662–666.
243. Alexander HR, Doherty GM, Buresh CM, et al. A recombinant human receptor antagonist to interleukin 1 improves survival after lethal endotoxemia in mice. J Exp Med 1991; 173:1029–1040.
244. Ohlsson K, Bjork P, Bergenfeldt M, et al. Interleukin-1 receptor antagonist reduces mortality from endotoxin shock. Nature 1990; 348:550–553.
245. Fisher CJ, Dhainhaut J-F, Opal SM, et al. Recombinant human interleukin-1 receptor antagonist in the treatment of patients with sepsis syndrome: results from a randomized double-blind, placebo-controlled trial. JAMA 1994; 271:1836–1843.
246. Wakabayashi G, Gelfand JA, Burke JF, et al. A specific receptor antagonist for interleukin-1 prevents *Escherichia coli*-induced shock in rabbits. FASEB J 1991; 5:338–344.
247. Fischer E, Marano MA, van Zee KJ, et al. IL-1 receptor blockade improves survival and hemodynamic performance in E. coli septic shock, but fails to alter host responses to sublethal endotoxemia. J Clin Invest 1992; 89:1551–1562.
248. Fischer E, Marano MA, Barber AE, et al. Comparison between the effects of interleukin-1α administration and sublethal endotoxemia in primates. Am J Physiol 1991; 261:R442.
249. Beasley D, Cohen RA, Levinsky NG. Interleukin 1 inhibits contraction of vascular smooth muscle. J Clin Invest 1989; 83:331–338.
250. Dejana E, Breviario F, Erroi A, et al. Modulation of endothelial cell functions by different molecular species of interleukin 1. Blood 1987; 69:695–699.
251. Bevilacqua MP, Pober JS, Majeau GR, et al. Interleukin 1 (IL-1) induces biosynthesis and cell surface expression of procoagulant activity in human vascular endothelial cells. J Exp Med 1984; 160:618–623.
252. Breviario F, Bertocchi F, Dejana E, Bussolino F. IL-1 induced adhesion of polymorphonuclear leukocytes to cultured human endothelial cells. J Immunol 1988; 141:3391–3397.

253. Cavendar DE, Haskard DO, Joseph B, Ziff M. Interleukin 1 increases the binding of human B and T lymphocytes to endothelial cell monolayers. J Immunol 1986; 136:203–207.

254. Pober JS, Bevilacqua MP, Mendrick DL, et al. Two distinct monokines, interleukin 1 and tumor necrosis factor, each independently induce biosynthesis and transient expression of the same antigen on the surface of cultured human vascular endothelial cells. J Immunol 1986; 136:1680–1687.

255. Rossi V, Breviario F, Ghezzi P, et al. Prostacyclin synthesis induced in vascular cells by interleukin-1. Science 1985; 229:174–176.

256. Schleff RR, Bevilacqua MP, Sawdey M, et al. Cytokine activation of vascular endothelium. Effects of tissue-type plasminogen activator and type 1 plasminogen activator inhibitor. J Biol Chem 1988; 263:5797–5803.

257. Thomas TK, Will PC, Srivastava A, et al. Evaluation of an interleukin-1 receptor antagonist in the rat acetic acid-induced colitis model. Agents Actions 1991; 34:187–190.

258. Mandrup-Poulsen T, Bendtzen K, Dinarello CA, Nerup J. Human tumor necrosis factor potentiates human interleukin 1-mediated rat pancreatic β-cell cytotoxicity. J Immunol 1987; 139:4077–4082.

259. Mandrup-Poulsen T, Zumsteg U, Reimers J, et al. Involvement of interleukin-1 and interleukin-1 receptor antagonist in pancreatic β-cell destruction in insulin-dependent diabetes mellitus. Cytokine 1993; 5:185–192.

260. Moncada S, Higgs A. The L-arginine-nitric oxide pathway N. Eng J Med 1993; 329:2002–2012.

261. Corbett JA, Sweetland MA, Wang JL, Lancaster JR, McDaniel ML. Nitric oxide mediates cytokine-induced inhibition of insulin secretion by human islets of Langerhans. Proc Natl Acad Sci USA 1993; 90:1731–1735.

262. Nathan C, Xie Q-W. Nitric oxide synthases: roles, tolls, and controls. Cell 1994; 78:915–918.

263. Romero R, Mazor M, Brandt F, et al. Interleukin-1α and interleukin-1β in preterm and term human parturition. Am J Reprod Immunol 1992; 27:117–122.

2

Interleukin-2: Molecular and Clinical Aspects

Nancy L. Farner, Jacquelyn A. Hank, and Paul M. Sondel
University of Wisconsin and University of Wisconsin Comprehensive Cancer Center, Madison, Wisconsin

HISTORY

Interleukin-2 (IL-2) was initially described as a T-cell growth factor (TCGF) in 1976 when it was characterized as a component of conditioned culture medium from lectin-stimulated peripheral blood mononuclear cells (1–3). TCGF stimulated resting T cells to become activated and also supported long-term growth of activated human T cells. TCGF was subsequently purified and cloned and is now defined as a 15 kDa glycoprotein encoded by a single gene on human chromosome 4 (4,5). The entire gene spans 6 kb and includes four exons separated by one short and two long introns. IL-2 is the most extensively investigated cytokine to date. It is considered the primary growth factor of T cells, a potent modulator of T-cell and natural killer (NK)-cell function, and plays a major role in the immune response. With the central role of IL-2 as a modulator of immunological activity, IL-2 has been examined for its ability to stimulate antitumor responses, overcome immunodeficiency, and function as an adjuvant in conjunction with immunizations. In addition, molecular advances have raised the possibility of utilizing IL-2 for the treatment of several disease states thought to be influenced by IL-2 responses.

BIOCHEMISTRY

The human IL-2 molecule is composed of 133 amino acids, while its species counterparts range up to 149 amino acids (6,7). The posttranslational modifications of IL-2 include removal of the N-terminal signaling peptide, glycosylation, and formation of a disulfide bond. Size heterogeneity of IL-2 is attributed mostly to the extent of glycosylation. Glycosylation does not appear necessary for effective IL-2 function, whereas the disulfide bond formation between Cys 58 and Cys 105 is critical to IL-2 function (7). The resultant folding of IL-2 is such that two sets of α-helices (which compose 67% of the protein) are formed in parallel and are connected by two β-pleated sheets. Because of this physical formation, IL-2 is a member of the cytokine classification known as the short-chain four α-helical bundle group, all of which are monomers (8–10). IL-3, IL-4, IL-7, IL-9, IL-13, GM-CSF, M-CSF, SCF, and IFN-γ are also members of this group, whose structure is commonly known as "up-up-down-down," indicating the relative direction of each of the four α-helices from the C-terminus to the N-terminus (Fig. 1). The long-chain four α-helical bundle group is usually composed of dimers with longer α-helical structures and includes IL-6, IL-10, and erythropoietin.

IL-2 has been cloned from several species, and the sequences are quite similar with homology ranging from 65–70% (Fig. 2) (5,8). However, mouse IL-2 contains a 12-residue polyglutamine segment insertion near the N-terminus that once suggested a unique functional distinction between human and mouse IL-2. This is no longer thought to be the case (11,12). Even so, mouse and human IL-2 do not act equally on their species counterparts: human IL-2 can effectively stimulate mouse T-cell proliferation, while mouse IL-2 does not effectively stimulate human T-cell proliferation (13).

The components that control IL-2 production have been localized to a transcriptional enhancer region 5′ of the IL-2 gene (4,14–17). Within this region, binding sites for NFAT-1, NF-κB, and AP-1 have been identified. Activation of the IL-2 gene is dependent on NFAT-1 activation,

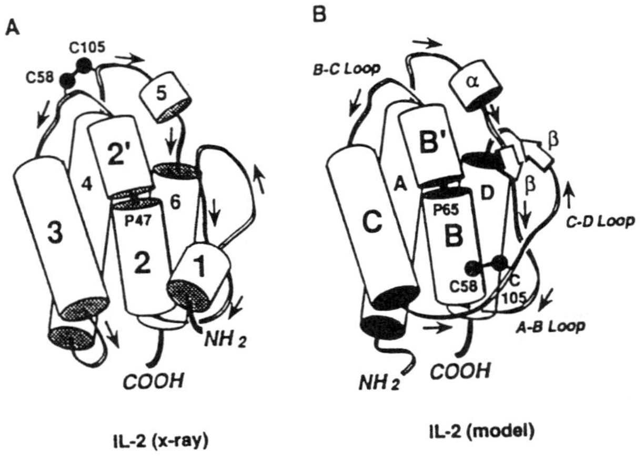

Figure 1 Two models of IL-2 structure. The model in panel A was deduced from x-ray crystallography. The model in panel B was based on x-ray scaffold data and ligand binding data. The second model depicts the structure as similar to other members of the cytokine family with four α-helices. Reprinted with permission from Ref. 10. Copyright 1992 American Association for the Advancement of Science.

which occurs approximately 10 min before IL-2 mRNA is detectable. This lag occurs supposedly while the NFAT-1 complex is being formed. Following IL-2 gene activation, stabilization of the IL-2 mRNA occurs, which is mediated through the AU-rich sequence in the 3′ untranslated region.

PRINCIPAL CELLULAR SOURCES AND BIOLOGICAL ACTIVITIES

Production and secretion of IL-2 is conducted primarily by mature T cells within 4–12 h following activation by antigen or mitogen (1,18). Whether IL-2 acts on a mature T cell in a paracrine or autocrine fashion, the response to IL-2 by T cells expressing IL-2 receptors is to undergo proliferation and further activation. IL-2 receptor expression is induced on helper and cytotoxic T cells that have recognized antigen. Thus, the IL-2–induced proliferation preferentially expands those T-cell clones able to recognize specific antigen, amplifying the number of cells involved in the antigen-specific immune response. Hence, the magnitude of the immune responses instigated by T cells is determined by IL-2. In addition, failure to respond to IL-2 or produce adequate quantities of IL-2 may result in T-cell anergy (19). Immune responses mediated by cytotoxic T cells are turned off by reducing the amount of IL-2 produced, which naturally occurs once the antigen stimulus is cleared.

IL-2 is also important in B-cell activity as it is a B-cell growth factor and also stimulates antibody synthesis (12,13,20). However, IL-2 is not considered a cytokine that can induce isotype switching in mature B-cell immunoglobulin production.

Natural killer cells are a subset of lymphocytes that are also cytotoxic, but are distinct from T cells in that they are not triggered by a specific antigen and do not require prior sensitization to their target. Their ability to proliferate and perform cytotoxic activities is influenced strongly by IL-2. Following activation by IL-2, NK cells become large and more granular, and have been designated lymphokine-activated killer (LAK) cells, as they are capable of lysing fresh tumor cells or virally infected cells (21,22).

Monocyte function can also be enhanced by IL-2, although these activities do not include proliferation and expansion, which are central events in T-, B-, and NK-cell responses to IL-2 (23,24). Instead, monocyte differentiation into activated macrophages or phagocytes, which

```
HUMAN    APTSSST--- ---------- -KKTQLQLEH LLLDLQMILN GINNYKNPKL TRMLTFKFYM PK-KATELKH LQCLE   60
GIBBON   APTSSST--- ---------- -KKTQLQLEH LLLDLQMILN GINNYKNPKL TRMLTFKFYM PK-KATELKH LQCLE   60
SHEEP    APTSSST--- ---------- -GNTMKEVKS LLLDLQLILE KVKNPENLKL SRMLTFNFYM PKVNATELKH LKCLL   61
COW      APTSSST--- ---------- -GNTMKEVKS LLLDLQLILE KVKNPENLKL SRMLTFDFYV PKVNATELKH LKCLL   61
PIG      APTSSST--- ---------- -KNTKKQLEP LLLDLQLLLK EVKNYENADL SRMLTFKFYM PK-QATELKH LQCLV   60
MOUSE    APTSSSTSSS TAEAQQQQQQ QQQQQQHLEQ LLMDLQELLS RMENYRNLKL PRMLTFKFYL PK-QATELKD LQCIE   74
RAT      APTSSPA--- ---------- -KETQQHLEQ LLLDLQVLLR GIDNYKNLKL PMMLTFKFYL PK-QATELKH LQCLE   60

HUMAN    EELKPLEEVL NLAQSKNFHL R-PRDLISNI NVIVLELKGS ETTFMCEYAD ETATIVEFLN RWITFCQSII STLT-  133
GIBBON   EELKPLEEVL NLAQSKNFHL R-PRDLISNI NVIVLELKGS ETTFMCEYAD ETATIVEFLN RWITFCQSII STLT-  133
SHEEP    EELKLLEEVL ILAPSKNLNT REIKDSMDNI KRIVLELQGS ETRFTCEYDD ATVKAVEFLN KWITFCQSIY STMT-  135
COW      EELKLLEEVL NLAPSKNLNP REIKDSMDNI KRIVLELQGS ETRFTCEYDD ATVNAVEFLN KWITFCQSIY STMT-  135
PIG      EELKALEGVL NLGQSKNSDS ANIKESMNNI NVTVLELKGS ETSFKCEYDD ETVTAVEFLN KWITFCQSIY STLT-  134
MOUSE    DELGPLRHVL DLTQSKSFQL EDAENFISNI RVTVVKLKGS DNTFECQFDD ESATVVDFLR RWIAFCQSII STSPQ  149
RAT      NELGALQRVL DLTQSKSFHL EDAGNFISNI RVTVVKLKGS ENKFECQFDD EPATVVEFLR RWIAICQSII STMTQ  135
```

Figure 2 Conservation of the amino acid sequence of IL-2 in seven species. Boxed regions indicate deduced amino acid sequences with perfect conservation among the species. (From Ref. 8, by permission of Oxford University Press.)

engulf and destroy foreign cells by digesting them from within, is the prominent IL-2 response by monocytes. IL-2 induces monocytes to secrete IL-1, increases monocyte-mediated cytotoxicity, induces differentiation of macrophage precursors, and upregulates expression of other growth factor receptors on precursors and mature monocytes.

As IL-2 influences the activities of T and B lymphocytes, NK cells, and monocytes, which are key components of the immune system (see Table 1 for summary), it follows that lack of this cytokine may result in immunological defects or immunodeficiencies. Although several types of immunodeficiencies in humans and mice have been described, recent studies suggest that individuals lacking the ability to produce IL-2 are not profoundly immunodeficient (25,26). For example, a mouse strain was developed that does not produce IL-2 due to "knock-out" of the IL-2 gene (25). These mice develop a fairly intact cellular immune system as thymic development and T-cell differentiation and activity is normal. However, efficient activation of their immune response requires exogenous IL-2 and an autoimmune state of ulcerative colitis is notable. Although IL-2 may still be necessary to support a fully normal immune response network, in the absence of IL-2, other mechanisms may be at work to allow the development of an intact immune system. These mechanisms include responses driven by other cytokines (such as IL-4, IL-7, IL-9, IL-13, and IL-15) that can interact with receptors expressed on the immune cells, including IL-2 receptor components.

Table 1 Immune Cell Responses to IL-2

Cell type	Response
T lymphocytes	Clonal expansion
	Cytotoxic activity
	Produce and secrete IL-2
	Soluble TAC release
B lymphocytes	Proliferation
	Antibody production
Monocytes	Differentiation into macrophages
	IL-1 secretion
	Cytotoxic activity by macrophages and monocytes
	Growth factor receptor expression up-regulation
Natural killer cells	Proliferation
	Cytotoxic activity
	Induction of c-myc mRNA
	IFN production and secretion

THE IL-2 RECEPTOR

The responses induced by IL-2 on T, B, and NK cells and monocytes are initiated by IL-2 binding to the IL-2 receptor (IL-2R). Three forms of the IL-2R exist, each with a distinct binding affinity for IL-2: the low-, intermediate-, and high-affinity forms. These receptors are composed through the combination of up to three distinct receptor subunits (IL-2Rα,β, and γ) (27–30). The low-affinity IL-2R ($K_d = 2–20$ nM) is composed of the IL-2Rα subunit, initially called the T-cell activation (TAC) marker. This 55 kDa protein is readily detectable on activated T cells and has the CD designation of CD25 (31–33). IL-2 binding to IL-2Rα alone does not induce a detectable biological response due in part to the lack of intracellular docking components for kinase activity (4,32,34–37).

IL-2Rβ (CD122) alone cannot bind IL-2, and IL-2Rγ binds IL-2 with an affinity that is even lower than IL-2Rα. The IL-2Rβ and IL-2Rγ components in combination create the intermediate-affinity IL-2R ($K_d = 0.5–2$ nM), which is expressed predominantly on NK cells and monocytes (38–40). The high-affinity IL-2R ($K_d = 2–20$ pM) is composed of all three subunits together and is detected predominantly on activated T cells along with the low-affinity IL-2 receptor. The intermediate- and high-affinity IL-2Rs can mediate responses to IL-2, although a higher concentration of IL-2 is required to trigger a response from cells expressing intermediate-affinity complexes. Both IL-2Rβ and IL-2Rγ are required in intermediate- and high-affinity forms of the IL-2R complex for appropriate cell signaling to occur (41,42). Figure 3 illustrates the configuration of the three receptor types and mature immune cells that express each form.

Although IL-2 does not appear to be critical to cellular immune function (as evidenced in the IL-2 knock-out mice), the IL-2Rγ chain is a critical element. Alterations in the gene for IL-2Rγ cause X-linked severe combined immunodeficiency (XSCID), which is manifested by a profound absence of both cell-mediated and humoral immunity in males who carry the defective IL-2Rγ gene in their single copy of the X chromosome (43–45). Mutations in IL-2Rγ cause profound immunological defects because IL-2Rγ is a shared receptor component of IL-2, IL-4 (critical for B-cell development), and IL-7 (critical for T-cell development) (46–48). IL-2Rγ has thus been renamed the common γ chain and is designated γ_c.

Since the cause of XSCID is now apparent, several methods are being explored to screen for carriers of XSCID and identify male children who have a defective γ_c gene so that early treatment can be engaged (45). However, it should be noted that up to 40% of all male children diagnosed with XSCID have spontaneous muta-

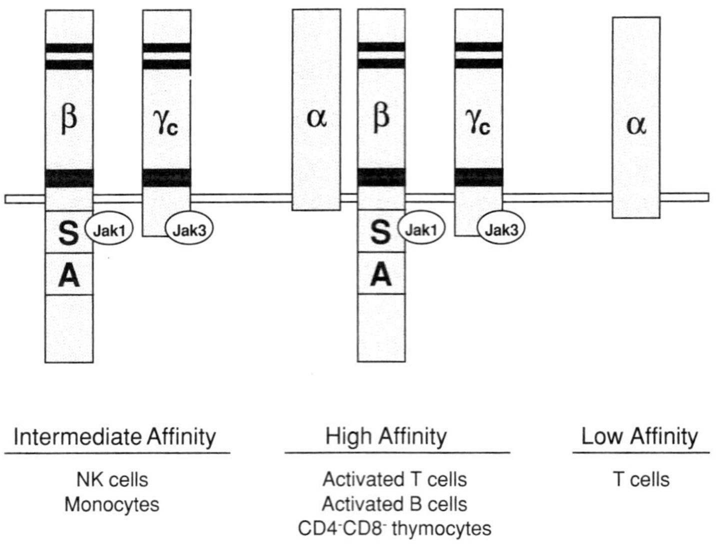

Figure 3 IL-2 receptor forms and immune cells that express them. Individual subunits that compose the three forms of IL-2 receptors are as indicated. JAK1 is a JAK family kinase that associates with the S (serine-rich) region on IL-2Rβ and is a primary kinase associated with IL-2 signaling. JAK3 is a JAK family kinase that associates with IL-2Rγ and is required for IL-2 signaling. The A (acidic) region is also required for proper signaling to occur. Immune cell subsets that express each IL-2 receptor form are listed below the schematics.

tions in the $γ_c$ gene not inherited from their mothers or express other autosomal genetic defects that are misdiagnosed as XSCID. In these cases, screening methods that target identification of possible carriers may not be useful, raising the necessity for further defining defects that could lead to XSCID-like characteristics.

Other candidate genes whose defects could lead to an immunodeficiency mimicking XSCID may include gene products involved in downstream mechanisms of $γ_c$ engagement. JAK3, a janus family kinase, is known to be a critical component of the signaling pathways through $γ_c$, including the $γ_c$ responses to IL-2, IL-4, IL-7, and other cytokines (49). Hence, alterations in JAK3 that cause disablement may also lead to profound immunodeficiency as several cytokines depend on JAK3 to relay signals. Whether there are other gene products that interact with $γ_c$ that could cause an XSCID-like state has yet to be determined.

CLINICAL ASPECTS OF IL-2

Tolerance and Autoimmunity

Given the complexity of the immune system, it is well appreciated that normal activities can easily go awry. Autoimmunity is the action of the immune system to attack "self," a phenomenon triggered by the breakdown of immune regulation (50). One important component of immune regulation is self-tolerance in which self-reactive cellular components of the immune system are either eliminated (clonal selection) or prevented from responding to self antigens (anergy). If these tolerance mechanisms are overcome, self-reactive T cells can become cytotoxic to "self" and cause autoimmunity.

Interleukin-2, because of its influence on T-cell activation, is thought to be capable of reversing self-tolerance or nonresponsiveness in some animal models, potentially resulting in generation of activated self-reactive T cells capable of eliciting autoimmunity (51,52). This hypothesis has been tested in mice that were infected with retrovirus containing the IL-2 gene (53). Overproduction of IL-2 in the microenvironment led to local inflammation in the tissue and the release of self-reactive T cells capable of inducing an autoimmune state.

Immunodeficiency

Although IL-2 knock-out mice suggest that IL-2 is not an essential component in the development of a competent immune response, there have been reports of human immunodeficiency resulting from low or no production of IL-2 (54). In one case, the affected patient was incapable of producing IL-2, although her T-cell number and distribution were fairly normal. Therapy with recombinant IL-2 (rIL-2) effectively restored her immune function. However, in this case and several others, it is unclear if the IL-2 defect is caused by a mutation in the IL-2 gene alone or in regulatory factor(s) that may control transcription of the IL-2 gene, along with other cytokine genes.

Infectious Diseases

Pathogenic microorganisms avoid clearance by eluding both specific cellular and humoral immune responses using a variety of mechanisms. Some of these pathways toward avoidance of immune attack have been defined and are providing leads for the development of treatment for infectious diseases. The following examples typify the role IL-2 may play in the development of systemic immunity against pathogenic microorganisms and how it may prove to be useful in creating more effective therapies.

Parasitic

The parasite that causes leishmaniasis, *leishmania major,* resides in macrophages of its human host. Murine models of leshmaniasis indicate that the antigens shed by *L. major* are presented by macrophages that specifically activate a $CD4^+$ helper T-cell subpopulation (Th2 cells), in which susceptible mouse strains secrete IL-4, which prevents the macrophage from lysing the parasite within it (55). One strategy to combat human *L. major* infections would be to block IL-4 binding with an anti–IL-4 antibody. In murine models, this has led to the clearance of the infections as Th1 $CD4^+$ helper T-cells expand in the absence of IL-4 secreted by Th2 cells, produce IFN-γ, and activate the host macrophage to lyse the parasite (55). It is thought that since IL-2 supports expansion of the Th1 subset, long-term treatment with this cytokine may prevent *L. major* infection or may help clear the infection more quickly. This concept still requires experimental testing in vivo.

Bacterial

Since clearance of parasitic infections is possible with the use of IL-2, certain bacterial infections may also be susceptible to clearance using IL-2 as a therapeutic reagent. Kaplan et al. reported that patients with lepromatous leprosy given low-dose rIL-2 as dermal injections for 8 days had reductions in their total body burden of *Mycobacterium leprae* (56). The immune mechanisms that are affected by delivery of rIL-2 and subsequently result in the clearance of the infection probably include monocyte activation, promotion of antibacterial antibody secretion from B cells, and expansion of Th1 cells, which support both mechanisms. Using IL-2 in this manner may help clear other facultative intracellular pathogen infections as well.

Viral

Complex biochemical and cellular pathways are involved in the pathological mechanisms of viral infections (57). In many instances, multiple immune interactions are involved in control of viral infection. Adequate viral antigen presentation to helper and cytotoxic T cells is an important component of an effective immune response able to alleviate the infection. Ideally, important viral antigens are processed and presented on the surface of infected cells and nearby macrophages soon after viral infection. If this occurs, the response by helper and cytotoxic T cells can be effective, allowing clearance of virally infected cells within 5–7 days. In contrast, the clearance rate by cytotoxic T cells is much lower if the viral antigen(s) are not detected due to low-level presentation or if they are ubiquitous due to widespread dissemination of virally infected tissues prior to an effective T-cell response. As such, IL-2 plays a significant role in triggering activities by helper and cytotoxic T cells and NK cells. The presence of these IL-2–activated effector cells may hasten an effective response to viral infection.

This mechanism was tested in mice infected with a vaccinia virus (VV) genetically altered to produce IL-2 (58). Mice infected with this VV variant had rapid clearance of the infection, whereas mice infected with the wild-type VV developed a chronic infection. In this setting, IL-2 may be working through more than one pathway, including expansion of VV reactive cytotoxic effector cells, activation of NK cells directly, and secondary cytokine secretion of IFN-γ, which directly inhibits viral replication, enhances antigen presentation, and activates macrophages and NK cells.

The human immunodeficiency virus (HIV) is a retrovirus that infects $CD4^+$ T-helper cells, the cells necessary to mediate cellular and humoral responses to microbes. This virus depletes T cells by undergoing the lytic cycle (and hence destroying the cell), causing other cytotoxic immune cells to lyse the infected T cell itself or triggering apoptosis (59).

The role IL-2 plays in HIV infection and the development of acquired immunodeficiency syndrome (AIDS) is complex. Some evidence suggests that HIV infection skews the Th1:Th2 ratios in vast favor of Th2 cells. This may explain why HIV-infected individuals are more susceptible to intracellular microbes (as Th2 cells do not activate macrophages). Also, IL-2 production by the Th1 cells may be unfavorable for the HIV infection. Recent studies have reported that clinical treatment with low-dose IL-2 increases $CD4^+$ cell counts in HIV^+ patients and, in some cases, enhances NK/LAK activity (60–62). However, it should also be noted that the same mechanisms that promote T-cell activation may enhance the opportunity for HIV infection or replication in HIV-responsive $CD4^+$ cells (which could include exposure to IL-2). Further clarification of the role IL-2 plays in HIV infection may provide other avenues for possible antiviral treatment.

Allergy

Hypersensitivity to antigen commonly leads to allergies. The cascade of events resulting in allergy is complex, centering on the release of IgE from B cells, which binds to mast cells via their Fc receptor (63). Once the antibody molecule encounters its specific antigen, the mast cell is triggered, causing degranulation of molecules such as histamine. Since IL-4 induces B cells to undergo immunoglobulin class switching and produce IgE, it follows that the Th2 $CD4^+$ helper T-cells are predominant in allergic responses. Indeed, in allergic individuals, most $CD4^+$ T cells are of the Th2 subset, whereas nonallergic individuals contain predominantly Th1 (IL-2 secreting) $CD4^+$ T cells (64,65). Whether this is a primary or secondary response to the allergen remains unclear.

Therapeutic intervention of allergic response using IL-2 remains highly speculative. One approach would be to skew the Th1:Th2 ratio toward Th1 with rIL-2, preventing Th2 expansion and IL-4 secretion. This could decrease the quantity of IgE produced by B cells and reduce mast cell triggering. The second approach would involve causing anergy in the $CD4^+$ T cells, which are involved in the generation of an allergic state. Long-term anergy has been obtained experimentally in mice given antigenic peptides derived from allergens in the absence of secondary signals such as IL-2 (66,67). Th2 cells are also rendered anergic with this method, although the mechanism is unclear. An agent that transiently neutralizes endogenous IL-2 may prove useful here. Clearly, the role of IL-2 in allergic responses requires further preclinical evaluation prior to pursuing clinical testing in allergy treatment.

Cancer Therapy

Currently, the most heavily investigated therapeutic use of IL-2 is as an antitumor reagent (68–70). The antitumor activity of IL-2 was initially acknowledged following a variety of experiments performed in animal models in which IL-2 prevented growth of induced or transplanted tumors (71–73). Depending on the animal model and its experimental parameters, evidence for antitumor activity has been carefully characterized for T cells, NK cells, monocytes, antibodies produced by B cells, and growth factors/cytokines produced by all of these cell types. Understanding the manner in which each cell type mentioned elicits an antitumor response is pivotal in our ability to efficiently harness these cells to enhance antitumor responses by the immune system.

For example, patients receiving IL-2 alone have heightened immune activation, which includes a high percentage of activated NK cells (74–78). These NK cells mediate high cytotoxic activity in vitro, and hence could be considered a potential antitumor effector population. This led to one experimental protocol in which patients' lymphocytes were grown in the presence of IL-2 in vitro to produce large numbers of activated NK cells, which were then given back to the patients with additional IL-2 (79–83). However, further studies indicated that these activated NK cells, whether generated in patients receiving systemic IL-2 or generated by IL-2 activation in vitro, were responsible for a large portion of the nonspecific toxicities seen with IL-2 (84,85). These include fever, chills, and multiple organ toxicities, which result from the capillary leakage associated with endothelial damage by IL-2–activated NK cells. Nevertheless, protocols using IL-2 are providing clinical benefit for at least some patients with renal cell cancer and malignant melanoma (86,87). IL-2 has thus become an approved treatment for renal cell cancer and is continuing to be tested for potential benefit in several oncological settings.

While the antitumor benefit of IL-2 is realized in only a minority of patients receiving IL-2, its dose-dependent toxicities are seen in nearly all patients treated with IL-2 (88,89). An alternative approach would be to provide tumor specificity to the IL-2–activated NK cells. This could be accomplished with the use of a tumor-specific monoclonal antibody as a mechanism that allows activated Fc receptor positive effector cells to target tumors (see Fig. 4). With the use of a combination of IL-2 and antitumor antibodies (70,77,89,90), the NK cells (now activated by IL-2) bind to the antibody-coated tumor cells via their Fc receptors. This approach might also allow for enhanced monocyte and granulocyte interactions with the antibody-coated tumor through their Fc receptors.

Another approach is to generate fusion proteins in which the antitumor antibody is attached directly to IL-2 molecules at the constant region of the antibody (91–94). Hence, the antibody binds the tumor and the NK cells bind to the IL-2 via IL-2 receptors. This approach would also allow for enhanced monocyte and T-cell interactions with the tumor as these subsets also express IL-2 receptors. An additional benefit of the IL-2 fusion protein is that IL-2 concentrations will be elevated in the microenvironment surrounding the tumor, which would theoretically heighten lymphocyte activation in the tumor region. This may activate specific T-cell immunity to tumor antigens as the IL-2 in the tumor vicinity may augment T-cell immunity.

Another approach used to enhance T-cell immunity is

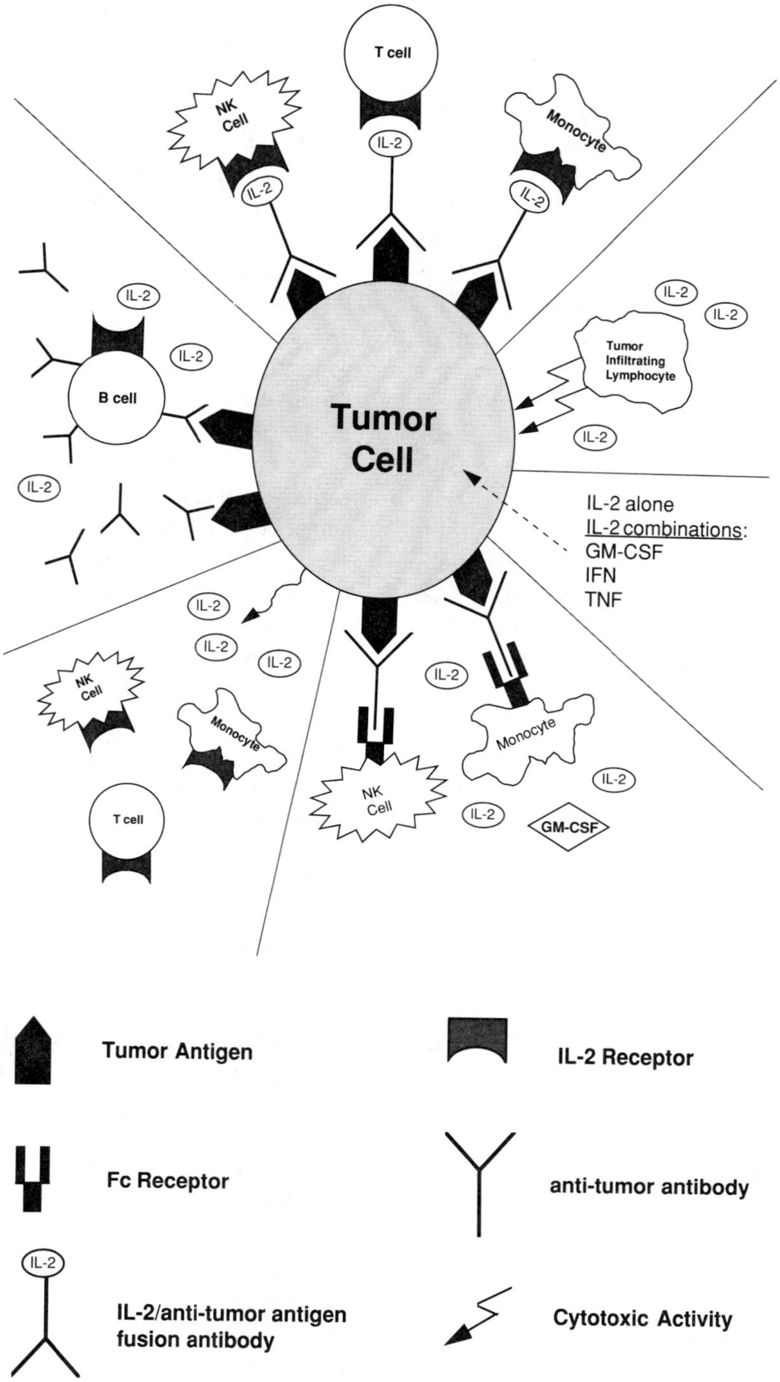

Figure 4 Illustration of antitumor mechanisms that are based in part on the role IL-2 plays in the immune response. From top panel going clockwise: IL-2/antitumor antibody fusion protein binds to tumor, and effector cells are brought into the site via their IL-2 receptors, Tumor infiltrating lymphocytes (TIL) harvested from tumors and expanded in vivo, are given back to the patient with the malignancy along with exogenous IL-2, IL-2 given systemically or in combination with other cytokines to activate effector cells systemically, antitumor antibody binds to the tumor and exogenous IL-2 or GM-CSF is given to activate effector cells that are brought into the site via their Fc receptors, IL-2 secreted by the tumor itself attracts effector cells, which utilize the IL-2 to become active and initiate an antitumor response, B cells whose antibody recognize the tumor will secrete more antibody near the tumor site with support from exogenous IL-2 and can bring effector cells into the tumor site, which bind the tumor-bound antibody generated by the B cells via their Fc receptors.

to introduce IL-2 directly at the tumor site. The gene for IL-2 has been placed into tumor cells directly in some models anticipating that IL-2 expression at the tumor site will enhance a tumor-reactive T-cell immune response that will become systemic and destroy tumor cells at distant sites (95). IL-1, IL-4, IL-6, IL-7, IL-12, GM-CSF, IFN-γ, and TNF have also been utilized in this cytokine gene therapy approach, and all have demonstrated utility in rodent models as evidenced by decreased tumorigenicity in tumors secreting these growth factors (96,97).

This method may also be successful in the treatment of leukemia. It has been demonstrated that IL-2–secreting leukemic cells are more antigenic and are also less likely to be tumorigenic than their nonsecreting parental cells (98). A current focus in this gene therapy research is to increase the efficiency in which the tumors receive the IL-2 gene and secrete the cytokine (99,100). However, an important point to consider is that some tumor cells of nonimmunological origin express IL-2 receptor components, including lung and renal cell cancers. This raises the possibility that IL-2 not only influences the activities on nonimmune cells, but that antitumor affects in response to cytokines such as IL-2 may be due to direct action of the cytokine on the tumor (101). Increasing the efficiency of IL-2 gene delivery at the tumor site may enhance these phenomena.

In this regard, since NK cells mediate somewhat nonspecific cytotoxicity, and since other effector cell types can play an important role in antitumor activity, combination therapy of IL-2 with other cytokines is being investigated. IL-2 has been combined with IFN-α, IFN-γ, and IL-4. Another treatment currently under investigation is the combination of IL-2 and GM-CSF, which could potentially activate T and NK cells with IL-2 as well as monocytes and neutrophils with GM-CSF. Clarification of appropriate dosing schedules and mechanisms of antitumor activity by this approach will clearly enhance the potential of such treatment.

CONCLUSIONS AND FUTURE DIRECTIONS

Although IL-2 is one of the most extensively studied cytokines to date, several aspects of its activities have yet to be defined; for example, the role of IL-2 in the development of each immune cell lineage. The ability of IL-2–deficient mice to develop a largely intact immune system has raised questions regarding the central role of IL-2 versus the IL-2 receptor complex in immune function and development. The development of an autoimmune phenomenon resembling ulcerative colitis in IL-2–deficient mice also points to the area of tissue immunity and the efforts that should be made to understand its complexity (102). A clear understanding of when IL-2 is critical for a particular function and when it is not will be an important focus for the future. With the role of the γ_c chain expanding into the receptors for IL-4, IL-7, and IL-15, whether any of these cytokines can support functions once thought to be mediated solely by IL-2 is a critical area of investigation (103–105). Clearly IL-15 shares several functions with IL-2, which may explain why IL-2–deficient mice can develop an intact immune system. LAK cells can be generated with IL-7 or IL-15 in the absence of IL-2, again raising the question of which responses specifically require IL-2 (104).

Another potentially critical area is the location of contact points between IL-2 and each of its receptor components. These studies are important for several reasons. First, if contact residues can be established, it may be possible to offer alternative amino acid sequences at the probable contact sites, which may either increase or decrease the affinity of IL-2 for its receptor. If the affinity of IL-2 for IL-2R binding can be increased, it may be possible to generate a response from T cells with lower concentrations of specifically mutated IL-2 with the added benefit of faster kinetics than with wild-type (wt) IL-2. Alternatively, if the affinity of IL-2 can be decreased, some forms of nonspecific toxicity may be alleviated or at least downregulated because this mutated IL-2 form might not be capable of interacting as strongly as wt IL-2. This approach can be taken one step further by identifying IL-2 residues that contact each subunit individually. An IL-2 variant generated by Heaton et al. has been characterized that does not bind to the IL-2Rα chain and causes less nonspecific toxicity than wt IL-2 (106). This approach may also apply to molecules with tighter or weaker binding of γ_c, which could cause different signaling pathways to be activated in comparison to wt IL-2.

Another aspect with clinical relevance relates to IL-2 stability. Since the half-life of IL-2 following intravenous bolus injection is approximately 45 minutes, continuous infusion of systemic IL-2 has been an effective method to maintain steady-state IL-2 levels in vivo. However, with the advent of gene therapy based upon using the body's own cells to produce IL-2 and release it from within, clarifying the control mechanisms of IL-2 production, which include mRNA production, stabilization, and processing, will heighten the effectiveness of this alternative approach to systemic IL-2 infusion.

In several respects, information gleaned from studying IL-2 has been applicable in some capacity to other cytokine systems. One final question yet to be answered is whether multichain cytokine receptor complexes are preformed, or if each component is free-standing prior to ligand binding. Since IL-2 and its receptor complex are well characterized, insight into this question will likely be gained using IL-2 and its receptor as a model. Such studies will clarify the way we look at cellular dynamics and the way cytokines can and do engage the cell. They have already brought advancements in the ways IL-2 is used therapeutically. Even though the elasticity of the immune system has provided many other cytokine systems that can support some immune functions in the absence of IL-2, this cytokine and its receptor complex play pivotal roles in immune response and development. Hence, IL-2 will remain one of the stronger reagents available to clarify cytokine/cytokine receptor biology and function.

ACKNOWLEDGMENTS

This work was supported by grants from the American Cancer Society, IM-678, NIH CA-32685, p30-CA14520, CM-87290, RR-03186.

REFERENCES

1. Smith KA. Interleukin-2: inception, impact, and implications. Science 1988; 240:1169–1176.
2. Morgan DA, Ruscetti FW, Gallo RC. Selective in vitro growth of T lymphocytes from normal human bone marrows. Science 1976; 193:1007–1008.
3. Gillis S, Smith KA. Long term culture of tumour-specific cytotoxic T cells [letter]. Nature 1977; 268:154–156.
4. Siegel JP, Sharon M, Smith PL, Leonard WJ. The IL-2 receptor beta chain (p70): role in mediating signals for LAK, NK and proliferative activities. Science 1987; 238:75–78.
5. Arai K, Lee F, Miyajima A, Miyatake S, Arai N, Yokota T. Cytokines: coordinators of immune and inflammatory responses. Ann Rev Biochem 1990; 59:783–836.
6. Kuziel WA, Greene WC. Interleukin-2 and the interleukin-2 receptor: new insights into structure and function. J Invest Dermatol 1990; 94(6 Suppl):275–325.
7. Robb RJ, Kitny RM, Chowdhry V. Purification and partial sequence analysis of human T cell growth factor. PNAS USA 1983; 80:5990–5994.
8. Goldsmith MA, Greene WC. Interleukin-2. In: Nicola NA, ed. Guidebook to Cytokines and Their Receptors. New York: Oxford University Press, 1994:27–30.
9. Brandhuber BJ, Bopone T, Kenney WC, McKay DB. Three dimensional structure of interleukin 2. Science 1987; 238:1707–1709.
10. Bazan JF. Unraveling the structure of IL-2. Science 1992; 257:410–413.
11. Kashima N, Nishi Takaoka C, Fujita T, Taki S, Yamada G. Unique structure of murine interleukin 2 as deduced from cloned cDNAs. Nature 1985; 313:402–404.
12. Zurawski SM, Mosmann TR, Benedik M, Zurawski G. Alterations in the amino terminal third of mouse interleukin 2: effects on biological activity and immunoreactivity. J Immunol 1986; 137:3354–3360.
13. Mosman T, Yokota T, Kastelein R, Zurawski SM, Arai N. Species specificity of T cell stimulating activities of IL-2 and BSF1 (IL-4) comparison of normal and recombinant mouse and human IL-2 and BSF1 (IL-4). J Immunol 1987; 138:1813–1816.
14. Crabtree GR. Contingent genetic regulatory events in T lymphocyte activation. Science 1989; 243:355–361.
15. Ullman KS, Northrop JP, Verweij CL, Crabtree GR. Transmission of signals from the T lymphocyte antigen receptor to the genes responsible for cell proliferation and immune function: the missing link. Ann Rev Immunol 1990; 8:421–452.
16. Riegel JS, Corthesy B, Flanagan WM, Crabtree GR. Regulation of the interleukin-2 gene. Chem Immunol 1992; 51:266–298.
17. Umlauf SW, Beverly B, Kang SM, Brorson K, Tran AC, Schwartz RH. Molecular regulation of the IL-2 gene: rheostatic control of the immune system. Immun Rev 1993; 133:177–197.
18. Smith KA. T cell growth factor. Immun Rev 1980; 51:337–357.
19. Johnson JG, Jenkins MK. The role of anergy in peripheral T cell responsiveness. Life Sci 1994; 55:1767–1780.
20. Aversa G, Cocks BG, Punnonen J, Carballido JM, DeVries JE. Contact mediated signals and cytokines involved in B cell activation and isotype switching in pre B and mature B cells. Res Immunol 1994; 145:222–226.
21. Trinchieri G, Perussia B. Immune interferon: a pleiotropic lymphokine with multiple effects. Immunol Today 1985; 6:131–136.
22. Whiteside TL, Herberman RB. The role of natural killer cells in human disease. Clin Immunol Immunopathol 1989; 53:1–23.
23. Moretta L, Ciccone E, Mingari MC, Biassoni R, Moretta A. Human natural killer cells: origin, specificity, and receptors. Adv Immunol 1994; 55:341–380.
24. Espinoza-Delgado I, Bosco MC, Musso T, Gusella GL, Longo DL, Varesio L. Interleukin-2 and human monocyte activation. J Leukocyte Biol 1995; 57:13–19.
25. Schorle H, Holtschke T, Hunig T, Schimpl A, Horak I. Development and function of T cells in mice rendered interleukin-2 deficient by gene targeting. Nature 1991; 352:621–624.
26. Kindig T, Schorle H, Backman M, Hengartner H, Zinkernagel R, Horak I. Immune responses in IL-2 deficient mice. Science 1993; 262:1059–1061.
27. Robb RJ, Smith KA. Heterogeneity of human T cell growth factor (TCGF) due to variable glycosylation. Mol Immunol 1991; 18:1087–1094.
28. Robb RJ, Munck A, Smith KA. T cell growth factor receptors: quantitations, specificity, and biological relevance. J Exp Med 1981; 154:1455–1474.
29. Minami Y, Kono T, Miyazaki T, Taniguchi T. The IL-2 receptor complex: Its structure, function, and target genes. Ann Rev Immunol 1993; 11:245–267.
30. Waldmann TA. The interleukin-2 receptor. J Biol Chem 1991; 266:2681–2684.
31. Leonard WJ, Depper JM, Crabtree GR, Rudikoff S, Pumphrey J, Robb RJ, Kronke M, Svetlik PB, Peffer NJ, Waldmann TA, Greene WC. Molecular cloning and expression of cDNAs for the human interleukin-2 receptor. Nature 1984; 311:626–631.
32. Nikaido T, Shimizu A, Ishaida N, Sabe H, Teshigawara K, Maida M, Uchiyama T, Yodoi J, Honjo T. Molecular cloning of cDNA encoding human interleukin-2 receptor. Nature 1984; 311:631–635.
33. Cosman D, Cerretti DP, Larsen A, Park LS, March C, Dower S, Gillis S, Urdal D. Cloning, sequence, and expression of human interleukin-2 receptor. Nature 1984; 312:768–771.
34. Robb RJ. Conversion of low affinity interleukin-2 receptors to

high affinity state following fusion of cell membranes. PNAS USA 1986; 83:3992–3996.
35. Robb RJ, Greene WC, Rusk CM. Low and high affinity cellular receptors for interleukin-2: implications for the level of Tac antigen. J Exp Med 1984; 160:1126–1146.
36. Tsudo M, Goldman CK, Bongiovanni KF, Chan WC, Winton EF, Yagita M, Grimm EA, Waldmann TA. The p75 peptide is the receptor for interleukin-2 expressed on large granular lymphocytes and is responsible for interleukin-2 activation of these cells. PNAS USA 1987; 84:5394–5398.
37. Sabe H, Kondo S, Shimizu A, Tagaya Y, Yodoi J. Properties of human interleukin 2 receptor expressed on non-lymphoid cells by cDNA transfection. Mol Biol Med 1984; 2:379–388.
38. Takeshita T, Asao H, Ohtani K, Ishii N, Kumaki S, Tanaka N, Munakata H, Nakamura M, Sugamura K. Cloning of the γ chain of the human IL-2 receptor. Science 1992; 257:379–382.
39. Takeshita T, Ohtani K, Asao H, Kumaki S, Nakamura M, Sugamura K. An associated molecule, p64, with IL-2 receptor β chain. Its possible involvement in the formation of the functional intermediate affinity IL-2 receptor complex. J Immunol 1992; 148:2154–2158.
40. Voss SD, Sondel PM, Robb RJ. Characterization of the IL-2 receptors (IL-2R) expressed on human NK cells activated in vivo by IL-2: association of the p64 γ chain with the IL-2Rβ chain in functional intermediate affinity IL-2 receptors. J Exp Med 1992; 176:531–541.
41. Nakamura Y, Russel SM, Mess SA, Friedmann M, Erdos M, Francois C, Jacques Y, Adelstein S, Leonard WJ. Heterodimerization of the IL-2 receptor beta- and gamma-chain cytoplasmic domains is required for signaling. Nature 1994; 369:330–333.
42. Nelson BH, Lord JD, Greenberg PD. Cytoplasmic domains of the interleukin-2 receptor beta and gamma chains mediate the signal for T cell proliferation. Nature 1994; 369:333–336.
43. Voss SD, Hong R, Sondel PM. Severe combined immune deficiency, interleukin-2 and the IL-2 receptor: Experiments of nature continue to point the way. Blood 1994; 83:626–635.
44. Puck JM. X-linked immunodeficiencies. Adv Hum Genet 1993; 21:107–144.
45. Farner NF, Voss SD, PM Sondel. XSCID and the γ_c receptor component: prospects for molecular diagnosis. J Clin Lab Immunol 1995; 2:518–523.
46. Russell SM, Keegan AD, Harada N, Nakamura Y, Noguchi M, Leland P, Friedmann NC, Miyajima A, Puri RK, Paul WE, Leonard WJ. Interleukin-2 receptor γ chain: a functional component of the interleukin-4 receptor. Science 1993; 262:1880–1883.
47. Kondo M, Takeshita T, Ishii N, Nakamura M, Watanabe S, Arai K, Sugamura K. Sharing of the interleukin-2 (IL-2) receptor γ chain between receptors for IL-2 and IL-4. Science 1993; 262:1874–1876.
48. Noguchi M, Nakamura Y, Russell SM, Ziegler SF, Tsang M, Cao X, Leonard WJ. Interleukin-2 receptor γ chain: a functional component of the interleukin-7 receptor. Science 1993; 262:1877–1880.
49. Ihle JN. The janus kinase family and signaling through members of the cytokine receptor superfamily. PSEBM 1994; 206:268–272.
50. Schwartz RS. Autoimmunity and autoimmune disease. In: WE Paul, ed. Fundamental Immunology. New York: Raven Press Ltd., 1993:1033–1097.
51. Beverly B, Kang SM, Lenardo MJ, Schwartz RH. Reversal of in vitro T cell clonal anergy by IL-2 stimulation. Int Immunol 1994; 4:661–671.
52. Esery G, Feldmann M, Lamb JR. Interleukin-2 can prevent and reverse antigen-induced unresponsiveness in cloned human T lymphocytes. Immunology 1988; 64:413–417.
53. Heath WR, Allison J, Hoffmann MW, Schonrich G, Hammerling G, Arnold B, Miller JFAP. Autoimmune diabetes as a consequence of locally produced interleukin-2. Nature 1992; 359:547–549.
54. Pahwa R, Chatila T, Pahwa S, Paradise C, Day NK, Geha R, Schwartz SA, Slade H, Oyaizu N, Good RA. Recombinant Interleukin-2 therapy in severe combined immunodeficiency disease. PNAS USA 1989; 86:5069–5073.
55. Paul WE, Seder RA. Lymphocyte responses and cytokines. Cell 1994; 76:241–251.
56. Kaplan G, Britton WJ, Hancock GE, Theuvenet WJ, Smith KA, Job CK, Roche PW, Molloy A, Burkhardt R, Barker J, Pradhan HM, Cohn ZA. The systemic influence of recombinant interleukin-2 on the manifestations of lepratomous leprosy. J Exp Med 1991; 173:993–1006.
57. Biron CA. Cytokines in the generation of immune responses to and resolution of virus infection. Curr Opin Immunol 1994; 6:530–538.
58. Flexner C, Hugin A, Moss B. Prevention of vaccinia virus infection in immunodeficient mice by vector-directed IL-2 expression. Nature 1987; 330:259–262.
59. Rosenberg ZF, Fauci AS. The immunopathogenesis of HIV infection. Adv Immunol 1989; 47:377–431.
60. Wood R, Montoya JG, Kindu SK, Schwartz DH, Merigan TC. Safety and efficacy of polyethylene glycol-modified interleukin-2 and zidovudine in human immunodeficiency virus type 1 infection: a phase I/II study. J Infect Dis 1993; 167:519–525.
61. Teppler H, Kaplan G, Smith KA, Montana AL, Meyn P, Cohn ZA. Prolonged immunostimulatory effect of low-dose polyethylene glycol interleukin-2 in patients with human immunodeficiency type 1 infection. J Exp Med 1993; 177:483–492.
62. Teppler H, Kaplan G, Smith KA, Cameron P, Montana A, Meyn P, Cohn Z. Efficacy of low doses of the polyethylene glycol derivative of interleukin-2 in modulating the immune response of patients with human immunodeficiency virus type 1 infection. J Infect Dis 1993; 167:291–298.
63. Plaut M, Zimmerman EM. Allergy and mechanisms of hypersensitivity. In: WE Paul, ed. Fundamental Immunology. 3rd ed. New York: Raven Press Ltd., 1993:1399–1425.
64. Kapsenberg ML, Wierenga EA, Bos JD, Jansen HM. Functional subsets of allergen reactive human CD4+ T cells. Immunol Today 1991; 12:392–395.
65. Romagnani S. Human TH1 and TH2 subsets: doubt no more. Immunol Today 1991; 12:256–257.
66. O'Hehir RE, Garman RD, Greenstein JL, Lam JR. The specificity and regulation of T cell responsiveness to allergens. Adv Immunol 1991; 9:67–95.
67. Jenkins MK. The role of cell division in the induction of clonal anergy. Immunol Today 1992; 13:69–73.
68. Sondel PM, Mackway-Girardi A, eds. Oncology Overview: Selected Abstracts on Therapeutic Use of Interleukin-2. U.S. Department of Health Service, 1991:vii–xiv.
69. Sondel PM. Interleukin-2. In: DW Galvani and JC Crowley, eds. Cytokine Therapy. Cambridge: Cambridge University Press, 1992:78–90.
70. Voss SD, Sondel PM. The clinical biology and mechanisms of action of systemically administered IL-2. In: MS Mitchel, ed. Biological Approaches to Cancer Treatment: Biomodulation. New York: McGraw-Hill, 1993:411–439.
71. Smith KA, Lachmann LB, Oppenheim JJ, Favata MF. The functional relationship of the interleukins. J Exp Med 1980; 151:1551–1556.
72. Cheever MA, Thompson DB, Klarnet JR, Greenberg PD.

Antigen driven long term cultured T cells proliferate in vivo, distribute widely, mediate specific tumor therapy and persist long term as functional memory T cells. J Exp Med 1986; 163:1100–1112.
73. Mule JJ, Yan JC, Lafreniere R, Shu S, Rosenberg SA. Identification of cellular mechanisms operation in vivo during the regression of established pulmonary metastases by the systemic administration of high dose recombinant interleukin 2. J Immunol 1987; 139:285–295.
74. Phillips JH, Lanier LL. Dissection of the lymphokine activated killer phenomenon. Relative contribution of peripheral blood natural killer cells and T lymphocytes to cytosis. J Exp Med 1986; 164:814–825.
75. Sosman JA, Kohler PC, Hank JA, Moore KH, Bechhofer R, Storer B, Sondel PM. Repetitive weekly cycles of recombinant human interleukin-2: responses of renal carcinoma with acceptable toxicity. JNCI 1988; 80:60–63.
76. Goldstein D, Sosman JA, Hank JA, Weil-Hillman G, Moore KH, Borchert A, Bechhofer R, Storer B, Kohler PC, Levitt D, Sondel PM. Repetitive weekly cycles of interleukin-2: effect of outpatient treatment with a lower dose of interleukin 2 on non-major histocompatibility complex-restricted killer activity. Cancer Res 1989; 49:6832–6839.
77. Weil-Hillman G, Fisch P, Prieve AF, Sosman JA, Hank JA, Sondel PM. Lymphokine activated killer activity induced by in vivo interleukin 2 therapy: predominant role for lymphocytes with increased expression of CD2 and Leu19 antigens but negative expression of CD16 antigens. Cancer Res 1989; 49:3680–3688.
78. Sosman JA, Hank JA, Moore KA, Borchert A, Schell K, Kohler PC, Goldstein D, Bechhofer R, Storer B, Albertini MR, Leung PE, Levitt D, Sondel PM. Prolonged interleukin 2 (IL-2) treatment can augment immune activation without enhancing antitumor activity in renal cell carcinoma. Cancer Invest 1991; 9:35–48.
79. Sosman JA, Kohler PC, Hank JA, Moore KH, Bechhofer R, Storer B, Sondel PM. Repetitive weekly cycles of interleukin 2. II. Clinical and immunobiologic effects of dose, schedule, and addition of indomethacin. JNCI 1988; 80:1451–1461.
80. West WH, Tauer KW, Yanneli JR, Marshall GD, Orr DW, Thurman GB, Oldham RK. Constant infusion recombinant interleukin 2 in adoptive immunotherapy of cancer. NEJM 1987; 316:898–905.
81. Paciucci PA, Holland JF, Ryder JS, Konefal RG, Bekesi GJ, Odchimar R, Gordon R. Immunotherapy with interleukin 2 by constant infusion with and without adoptive cell transfer and with weekly doxorubicin. Cancer Treat Rev 1989; 16:67–81.
82. Eberlein TJ, Schoof DD, Jung S, Davidson D, Gramolini B, McGrath K, Massaro A, Wilson RE. A new regimen of interleukin 2 and lymphokine activated killer cells: efficacy without significant toxicity. Arch Int Med 1988; 48:2571–2576.
83. Rosenberg SA, Lotze MT, Muul LM, Leitman S, Chang AE, Ettinghausen SE, Matory YL, Skibber JM, Shiloni E, Vetto JT, Seipp CA, Simpson C, Reichert CM. Observations on the systemic administration of autologous lymphokine activated killer cells and recombinant interleukin 2 to patients with metastatic cancer. NEJM 1985; 313:1485–1492.
84. Sondel PM, Hank JA, Kohler PC, Chen BP, Minkoff DZ, Molenda JA. Destruction of autologous human lymphocytes by interleukin-2 activated cytotoxic cells. J Immunol 1986; 137:502–511.
85. Albertini MA, Sosman JA, Hank JA, Moore KH, Borchert A, Schell K, Kohler PC, Bechhofer R, Storer B, Sondel PM. The influence of autologous lymphokine activated killer cell infusions on the toxicity and antitumour effect of repetitive cycles of interleukin 2. Cancer 1990; 66:2457–2464.
86. Sznol M, Parkinson DR. Clinical applications of IL-2. Oncology 1994; 8:61–71.
87. Rosenberg SA, Lotze MT, Muul LM, Chang AE, Avis FP, Leitman S, Linhan WM, Robertson CN, Lee RE, Rubin JT, Seipp CA, Simpson C, White DE. A progress report on the treatment of 157 patients with advanced cancer using lymphokine activated killer cells and interleukin-2 or high dose interleukin 2 alone. NEJM 1987; 316:889–897.
88. Dutcher JP, Creekmore SP, Weiss GR, Margolin KA, Markoweitz AB, Roper M, Parkinson DR, Ciobanu N, Fisher RI, Boldt DH, Dorshow JH, Rayner AA, Hawkins M, Atkins M. A phase II study of interleukin 2 and lymphokine activated killer cells in patients with metastatic malignant melanoma. J Clin Oncol 1989; 7:477–485.
89. Sondel PM, Kohler PC, Hank JA, Moore KH, Rosenthal NS, Sosman JA, Bechhofer R, Storer B. Clinical and immunological effects of recombinant interleukin 2 given by receptive weekly cycles to patients with cancer. Cancer Res 1988; 48:2561–2567.
90. Hank JA, Robinson RR, Surfus J, Mueller BM, Reisfeld RA, Cheung NK, Sondel PM. Augmentation of antibody dependent cell mediated cytotoxicity following in vivo therapy with recombinant interleukin 2. Cancer Res 1990; 50:5234–5239.
91. Iwanuma Y, Kato K, Yagita H, Okumura K. Induction of tumor-specific cytotoxic T lymphocytes and natural killer cells by tumor cells transfected with the interleukin-2 gene. Cancer Immunol Immunother 1995; 40:17–23.
92. Schmidt Wolf GD, Schmidt Wolf IGH. Cytokines and gene therapy. Immunol Today 1995; 16:173–175.
93. Tepper RI, Mule JJ. Experimental and clinical studies of cytokine gene-modified tumor cells. Hum Gene Ther 1994; 5:153–164.
94. Cignetti A, Guarni A, Carbone A, Forni M, Cronin K, Fopmi G, Gansbacher B, Foa R. Transduction of the IL-2 gene into human acute leukemia cells: induction of tumor rejection without modifying cell proliferation and IL-2 receptor expression. JNCI 1994; 86:785–791.
95. Fell HP, Gayle MA, Grosmaire L, Ledbetter JA. Genetic construction and characterization of a fusion protein consisting of a chimeric F(ab′) with specificity for carcinomas and human IL-2. J Immunol 1991; 146:2446–2452.
96. Landolfi NF. A chimeric IL-2/lg Molecule possesses the functional activity of both proteins. J Immunol 1991; 146:915–919.
97. Gillies SD, Reilly EB, Lo KM, Reisfeld RA. Antibody targeted interleukin 2 stimulates T cell killing of autologous tumor cells. PNAS USA 1992; 89:1428–1432.
98. Sabzevari H, Gillies SD, Mueller BM, Pancook JD, Reisfeld RA. A recombinant antibody interleukin 2 fusion protein suppresses growth of hepatic human neuroblastoma metastases in severe combined immunodeficient mice. PNAS USA 1994; 91:9296–9630.
99. Sun WH, Burkholder JK, Sun J, Culp J, Turner J, Lu XG, Pugh TD, Ershelr WB, Yang NS. In vivo cytokine gene transfer by gene gun reduces tumor growth in mice. PNAS 1995; 92:2889–2893.
100. Cheng L, Ziegelhoffer PR, Yang NS. In vivo promoter activity and transgene expression in mammalian somatic tissues evaluated by using particle bombardment. PNAS USA 1993; 90:4455–4459.
101. Weidmann E, Sacchi M, Plaisance S, Heo SD, Yasumura S, Lin W, Johnson JT, Herberman RB, Azzarone B, Whiteside TL. Receptors for Interleukin-2 on human squamous cell carci-

noma cell lines and tumor in situ. Cancer Research 1992; 52:5963–5970.

102. Sadlock B, Merz H, Schorle H, Schimp A, Feler A, Horck I. Ulcerative Colitis like disease in mice with a disrupted IL-2 gene. Cell 1993; 75:253–261.

103. Grabstein KH, Eisenman J, Shanebeck K, Rauch C, Srinivasan S, Fung V, Beers C, Richardson J, Schoenborn MA, Ahdieh M, Johnson L, Alderson MR, Watson JD, Anderson DM, Giri JG. Cloning of a T cell growth factor that interacts with the β chain of the interleukin-2 receptor. Science 1994; 264:965–968.

104. Carson WE, Giri JG, Lindemann MJ, Linett ML, Ahdieh M, Paxton R, Anderson D, Eisenmann J, Grabstein D, Caligiuri MA. Interleukin (IL)-15 is a novel cytokine that activates human natural killer cells via components of the IL-2 receptor. J Exp Med 1994; 180:1395–1403.

105. Giri JG, Ahdieh M, Eisenbman J, Shanebeck K, Grabstein K, Kumaki S, Namen A, Park LS, Cosman D, Anderson D. Utilization of the β and γ chains of the IL-2 receptor by the novel cytokine, IL-15. EMBO 1994; 13:2822–2830.

106. Heaton KM, Ju G, Grimm EA. Human interleukin 2 analogues that preferentially bind the intermediate affinity interleukin 2 receptor lead to reduced secondary cytokine secretion: Implication for the use of these interleukin 2 analogues in cancer immunotherapy. Cancer Res 1993; 53:2597–2602.

3
Interleukin-3 and the Colony-Stimulating Factors

Ian K. McNiece
Amgen, Inc., Thousand Oaks, California

INTRODUCTION

High-dose chemotherapy for the treatment of solid tumors requires cellular support to overcome pancytopenia. Both bone marrow cells (BMC) and peripheral blood progenitor cells (PBPC) are cellular sources that provide recovery of hematopoietic cells. In particular, PBPC mobilized by growth factors have been shown to provide very rapid recovery of neutrophils and platelets resulting in marked reductions in costs of associated treatments (1,2). Also, growth factors stimulate progenitor cells present in these cellular grafts to provide enhanced recovery of mature cells (3). It is the aim of this review to describe the present use of hematopoietic growth factors in cellular therapy for support of high-dose chemotherapy. Other experimental uses of growth factors such as ex vivo expansion and gene therapy will also be discussed and their clinical potential outlined.

COLONY-STIMULATING FACTORS

There are four colony-stimulating factors (CSFs), which have been named for the colony morphology that they stimulate: (1) macrophage CSF (M-CSF or CSF-1) stimulates the formation of macrophage colonies (4), (2) granulocyte CSF (G-CSF) stimulates the formation of neutrophilic granulocyte colonies (5), (3) granulocyte-macrophage CSF (GM-CSF) stimulates the formation of mixed neutrophilic granulocyte and macrophage colonies (6), and (4) multi-CSF or interleukin-3 (IL-3) stimulates the formation of mixed colonies of neutrophilic granulocytes, macrophages, megakaryocytes, and erythrocytes (7,8). The human analog of each mouse CSF has been identified and recombinantly cloned (9–13). A summary of the cell types stimulated by these molecules is presented in Table 1.

Macrophage Colony-Stimulating Factor

The stimulatory actions of CSF-1 were originally described for the mononuclear phagocyte series (14). CSF-1 was shown to stimulate the proliferation of monocytes, macrophages, and their committed bone marrow progenitors (macrophage colony-forming cells, M-CFC) (4,14). More recent studies have demonstrated synergistic effects of CSF-1 on primitive hematopoietic progenitor cells in

Table 1 Lineage Potential of CSFs

Growth factor	Progenitor cell	Lineage[a]
CSF-1	GM-CFC, M-CFC, Mk-CFC, HPP-CFC[b]	Mac, Meg
G-CSF	GM-CFC, G-CFC, HPP-CFC[b], Mk-CFC[b], CFU-blast[b]	Gran, Meg
GM-CSF	GM-CFC, M-CFC, G-CFC, Mk-CFC, HPPCFC[b], CFU-GEMM[b], BFU-Mk, BFU-E[b], Eo-CFC	Gran, Mac, Meg, Eryth, Eos
IL-3	GM-CFC, M-CFC, G-CFC, Mk-CFC, HPP-CFC[b], CFU-GEMM[b], BFU-Mk, BFU-E[b], CFU-blast, CFU-S, Mast-CFC	Gran, Mac, Meg, Eryth, Mast

[a]Gran, granulocytes; Mac, macrophages; Meg, megakaryocytes; Eryth, erythrocytes; Eos, eosinophils; Mast, mast cells.
[b]The CSFs alone do not stimulate these progenitor cells; however, they do act in combination with one or more growth factors to stimulate the progenitor cells.

combination with other hematopoietic growth factors, such as IL-3, GM-CSF, G-CSF, IL-1α, and IL-6 (15–20). CSF-1 has also been shown to stimulate mRNA expression of platelet factor 4 in megakaryocytes, suggesting a possible role in platelet production (21). CSF-1 affects the function of mature monocytes and macrophages, enhances the production of IL-1, tumor necrosis factor (TNF), G-CSF, and interferon, and potentiates the release of prostaglandins and biocidal oxygen metabolites (9,22,23).

CSF-1 is produced by mesenchymal cells, including stromal cells of the bone marrow, fibroblasts, and endothelial cells (4,24,25). Peripheral blood monocytes and macrophages can be induced to express CSF-1 by phorbol esters, GM-CSF, interferon-γ, or TNFα (26–29). In addition to its role in hematopoiesis, CFS-1 synthesized by normal uterine glandular epithelial cells in response to ovarian hormones during pregnancy appears to stimulate receptor-bearing trophoblasts of the adjacent placenta (30–32), suggesting a role of CSF-1 in placental development.

In semisolid agar culture of mouse bone marrow cells, both mouse- and human-derived CSF-1 stimulate the formation of colonies of mature macrophages. Mouse-derived CSF-1 has no activity on human bone marrow progenitors or human monocytes or macrophages. Surprisingly, human CFS-1 also has very little effect on human bone marrow progenitors, stimulating only cluster formation (colonies containing less than 50 cells) in semisolid agar cultures of human bone marrow cells (9). This finding complicates our understanding of the role of CSF-1 in humans.

Native human and mouse CSF-1 are heavily glycosylated dimer proteins of 45,000–70,000 daltons. The unglycosylated subunits of murine and human urinary CSF-1 are reported to have a size of 14,500 daltons (24), whereas the human cDNA predicts a polypeptide of 26,000 daltons. Removal of the carbohydrate moieties from CSF-1 does not affect its biological activity; however, reduction in the disulfide bonds abrogates all activity (24,33). At least two differentially spliced mRNAs for human CSF-1 are expressed from a common gene, and each encodes a different CSF-1 precursor. Additional processing and glycosylation result in at least two forms: a 70,000–90,000 dalton glycoprotein consisting of a 35,000–45,000 dalton glycoprotein consisting of a dimer of a 20,000–25,000 dalton subunit. The two resulting proteins are 554 and 256 amino acids in length, respectively. In mouse cells only a 552-amino-acid form has been found thus far (34,35).

CSF-1 has been shown to bind to a single class of high-affinity cell surface receptors expressed primarily on mature mononuclear phagocytes and their immature bone marrow precursors (36–39). Sherr and colleagues (40) demonstrated that the CSF-1 receptor is the product of the c-*fms* protooncogene. Binding of CSF-1 to its receptor activates the receptor kinase and leads to autophosphorylation of the receptor on tyrosine (40–42), internalization and degradation of the receptor-ligand complexes (37,43,44), and phosphorylation of a series of cellular proteins (44). CSF-1 receptors can be trans-downmodulated by IL-3, GM-CSF, and CSF-1 at 37°C but not by G-CSF (45).

Granulocyte-Macrophage Colony-Stimulating Factor

GM-CSF was originally described by its ability to stimulate the proliferation and differentiation of progenitor cells for neutrophilic granulocyte and macrophage lineages (46). Subsequent studies demonstrated the ability of GM-CSF to also stimulate progenitor cells, giving rise to eosinophils and megakaryocytes (47,48). GM-CSF also shows additive and synergistic effects with many other growth factors. In combination with erythropoietin (Epo), human or mouse GM-CSF stimulates early erythroid progenitor cells (BFU-E) and with CSF-1, IL-3, or G-CSF stimulates early myeloid progenitors [high proliferative potential CFC, HPP-CFC (19,44–51)]. GM-CSF has additive effects, with IL-3 in stimulating the development of maximal numbers of eosinophil, erythroid, and megakaryocyte colonies (48,52,53). In human cultures, GM-CSF synergizes with CSF-1 or G-CSF to stimulate increased colony numbers of macrophages or granulocytes, respectively (54,55).

Mouse GM-CSF stimulates only mouse bone marrow cells, not human bone marrow cells; human GM-CSF stimulates only human bone marrow cells, not mouse bone marrow cells. In addition to the stimulatory effects of GM-CSF on a wide range of progenitor cells, it also stimulates the functional activity of mature cells, inhibits the migration of neutrophils, is chemotactic for neutrophils and monocytes, and enhances the ability of neutrophils to kill antibody-coated tumor cells and phagocytose bacteria, parasites, and yeast (56–62). GM-CSF has similar effects on monocytes and macrophages, stimulating the antibody-dependent cytotoxicity of macrophages and the killing of parasites (63,64).

GM-CSF is produced by a range of cell types, including T cells, endothelial cells, and fibroblasts (65–67). Keratinocytes and the Hodgkin's tumor cell line L428KSA have also been found to produce GM-CSF (68,69). IL-1, TNFα, and lipopolysaccharide (LPS) have all been shown to induce the expression of GM-CSF in endothelial and macrophage cells (70–74).

Mouse GM-CSF was originally detected and isolated from medium conditioned by lung tissue from mice previously injected with endotoxin (6,75,76). The mouse GM-CSF was purified from lung tissue and the cDNA for

GM-CSF isolated from lung mRNA (77,78). In 1985, two groups isolated cDNA clones encoding human GM-CSF using functional expression in mammalian cells: Wong and colleagues isolated a cDNA library from the human T-lymphoblast cell line Mo (12), whereas Lee and colleagues isolated a cDNA library from a concanavalin A (ConA)–stimulated human helper T-cell clone T7 (13).

Although natural GM-CSF is glycosylated, *Escherichia coli*–derived GM-CSF, which is nonglycosylated, has high levels of biological activity (79). Mature mouse GM-CSF consists of 124 amino acids encoding a 14,000 dalton protein; the human GM-CSF contains an additional three amino acids between residues 21 and 22 (12,78). The mouse and human GM-CSF have only approximately 56% amino acid sequence homology (12), consistent with the lack of activity of human GM-CSF on mouse bone marrow.

Equilibrium binding studies have demonstrated low numbers of high-affinity binding sites on myeloid leukemic cells and on normal monocytes and neutrophils (80). GM-CSF receptors have also been demonstrated on nonhematopoietic tissues, including small-cell lung carcinoma and malignant neural tissue (81). Cross-linking studies suggest that the GM-CSF receptor is an 84 kDa polypeptide (81). GM-CSF induces a rapid internalization of its receptor followed by slow degradation of GM-CSF and receptor recycling. After binding to its receptor, GM-CSF stimulates the serine-specific phosphorylation of a 68 kDa cytosolic protein and a 68 kDa membrane protein as well as the tyrosine-specific phosphorylation of a 90 kDa membrane protein (82).

Granulocyte Colony-Stimulating Factor

Mouse G-CSF was first described as a factor in the serum of endotoxin-injected mice that induced terminal differentiation of the mouse myelomonoytic leukemic cell line WEHI-3 D^+ (83). G-CSF was soon purified from media conditioned by lungs from endotoxin-treated mice (84). Human G-CSF was subsequently purified from the bladder carcinoma cell line 5637 (83). The amino acid sequence was determined for both the human and mouse G-CSF by direct sequencing of the purified protein (86,87). The mouse G-CSF consists of 178 amino acids (molecular weight 19,000), and the human consists of 174 amino acids (molecular weight 18,600). Both mouse and human G-CSF are glycoproteins that have two disulfide bonds that stabilize the molecules (88,89).

Almost all murine tissues produce G-CSF when cultured in vitro (90). Endotoxin or the products of endotoxin-induced cells have been shown to stimulate macrophages, endothelial cells, and fibroblasts to increase G-CSF production (91–93). In semisolid agar culture of mouse bone marrow cells, both human and mouse G-CSF stimulate colony formation of predominantly (90%) neutrophilic granulocytes. Human bone marrow cells are similarly stimulated by human and mouse G-CSF to form neutrophil colonies (83,85). Recombinant human G-CSF has been shown to synergize with multiple factors, including CSF-1, IL-3, GM-CSF, and IL-1α, to form mature macrophage cells (19,94,95). G-CSF has not been reported to affect megakaryocyte formation alone but has been shown to augment the megakaryocyte colony-stimulating ability of IL-3 (96). Similarly, blast colony formation stimulated by IL-3 is augmented by G-CSF, but G-CSF alone has no stimulatory effect on blast cell formation (97). In preliminary data, G-CSF has been implicated in the generation of mouse pre-B cells; however, this effect may be mediated through accessory cells (98). G-CSF is a unique CSF: both mouse and human G-CSF stimulate mouse and human cells, suggesting strong conversation of the active regions of these molecules (99). Mouse and human G-CSF stimulate both bone marrow–derived and peritoneal exudate mouse neutrophils in their ability to demonstrate antibody-dependent cell-mediated cytotoxicity against coated mouse thymoma cells (58). G-CSF has also been shown to prime human neutrophils for enhanced responsiveness to other agents, such as ionophore or chemoattractants (100).

Binding studies reveal one class of high-affinity receptors on normal and leukemic myeloid cells, and cross-linking studies indicate that the G-CSF receptor is a single subunit protein of molecular weight 150 kDa (101). The cDNA cloning of the G-CSF receptor was recently achieved by Fukunaga and colleagues (102) using expression cloning techniques. The G-CSF receptor is an 812-amino-acid polypeptide with a molecular weight of 90,814 daltons and contains a single transmembrane domain. The extracellular domain consists of 601 amino acids, with 220 amino acids closely resembling the rat prolactin receptor. The cytoplasmic domain shows a significant similarity with parts of the cytoplasmic domain of the murine IL-4 receptor (102).

Interleukin-3 or Multi-CSF

Initially IL-3 was defined as a factor inducing the enzyme 20α-hydroxysteroid dehydrogenase, which was believed to be a unique marker for T-cell maturation, and so IL-3 was thought to act in early T-cell differentiation (103,104). IL-3 was subsequently purified to homogeneity in the early 1980s by Ihle and colleagues (105), and the purified protein was shown to be identical to a number of molecules reported to have vastly different biological actions (7). IL-3 was shown to be identical to burst-promoting activity

(BPA) (106), stem cell activating factor (107), P-cell stimulating factor (108), mast cell growth factor (109), synergistic factor (17), and multi-CSF (8). In vitro IL-3 stimulates (1) granulocyte and/or macrophage colony formation, (2) erythroid burst-forming cells, (3) CFC that contain granulocyte, macrophages, erythroid cells, and megakaryocytes, (4) spleen colony–forming cell (CFU-S) generation, (5) eosinophil colony formation, (6) mast cell colony formation, and (7) the proliferation of factor-dependent mast cell lines (7,110).

The cDNA encoding mouse IL-3 was isolated from WEHI-3–derived mRNA by Fung and colleagues in 1984, utilizing functional expression in oocytes and the IL-3–dependent cell line 32Dcl-23 to detect biological activity (111). Recombinant IL-3 has been found to have biological and biochemical properties identical to those of the native IL-3 protein (110,112).

The mouse IL-3 cDNA predicts a mRNA of 851 bases that encodes a polypeptide of 166 amino acids (113). The mature mouse IL-3 has a molecular weight of approximately 16,000 daltons, but native IL-3 has a molecular weight of 28,000 daltons, probably due to glycosylation (114).

In addition to the many activities of IL-3 in various mature cell types, it also synergizes with a number of other growth factors to stimulate primitive progenitor cells. In combination with erythropoietin, IL-3 stimulates early erythroid progenitors (7). The combination of IL-3 with G-CSF, CSF-1, GM-CSF, and IL-1 plus CSF-1 stimulates early myeloid progenitors termed high proliferative potential colony-forming cells (16,17,19,50). G-CSF augments the megakaryocyte-stimulating activity of IL-3, as does megakaryocyte potentiator (92,115). Ogawa and colleagues have demonstrated that IL-3 synergizes with G-CSF, IL-1, and IL-6 to stimulate primitive cells termed blast cells (93,116).

Identification of human IL-3 was complicated by the low level of homology between the mouse and human amino acid sequences [only 23% (11)]. Using mRNA from a gibbon T-cell line the gibbon IL-3 cDNA was isolated, and using the cDNA as a probe the human IL-3 gene was isolated from a human genomic library (11). The homology between the mouse and human IL-3 is very low and is reflected in the lack of cross-reactivity between the two molecules, with human IL-3 stimulating only human cells and mouse IL-3 stimulating only mouse cells.

The predominant source of IL-3 appears to originate from activated T cells (117); however, other reports suggest that epidermal cells, yolk sac, and astrocytes may produce IL-3 (118–120). The best known source of IL-3 is the WEHI-3B cell line, which produces IL-3 constitutively. WEHI-3B is a myelomonocytic leukemic cell line that was induced in Balb/c mice and was the source used by Ihle and colleagues (105) to purify mouse IL-3. More recent studies utilizing polymerase chain reaction (PCR) technology have detected transcripts for IL-3 in cDNA prepared from total cytoplasmic RNA from normal human long-term bone marrow cultures in the absence of mature T cells (121).

The mouse IL-3 receptor was isolated by Krystal and colleagues (122) using the IL-3–dependent cell line B6SutA1. The IL-3 receptor is proposed to be a 140 kDa glycoprotein that is cleaved to a 70 kDa surface protein by a receptor-associated protease.

BIOLOGICAL AND CLINICAL ROLES

High-Dose Chemotherapy and Colony-Stimulating Factors

The major dose-limiting effect of high-dose chemotherapy is extensive myelotoxicity. Bone marrow transplantation (BMT) has made possible the escalation of the doses of chemotherapy drugs; however, significant periods of neutropenia and thrombocytopenia still occur. The use of CSFs following chemotherapy with or without BMT has led to significant decreases in the periods of neutropenia.

A number of studies have shown that the administration of rhG-CSF to patients following chemotherapy leads to decreased severity and duration of neutropenia (1,123–125). The use of rhG-CSF reduces the incidence of febrile neutropenic events, the duration of antibiotic use, the length of hospital stay and decreased severity of mucositis (124–127). rhGM-CSF has also been administered to patients following chemotherapy. Patients receiving rhGM-CSF experienced earlier recoveries in neutrophil counts, which resulted in shorter periods of neutropenia between cycles of chemotherapy (128). However, no difference was seen in the periods of fever or length of hospital stay. A study in children has suggested better tolerability of rhGM-CSF with dose-related reduction in the number of hospitalizations and days of neutropenic fever (129).

In recent years, CSFs have been shown to mobilize progenitor cells from the bone marrow to the peripheral blood. The mobilization of PBPC in patients treated with rhG-CSF was first described by Sheridan and colleagues (1). As these investigators predicted, rapid recovery of neutrophils occurred when PBPC were transplanted following high-dose chemotherapy. What was unexpected was a more rapid recovery of platelets when compared to the recovery following the use of bone marrow. Many studies have subsequently confirmed these findings, and mobilized PBPC are rapidly becoming the preferred cellular source to support high-dose chemotherapy in the autologous setting for solid tumors. The mechanism by which PBPC are mobilized is not understood. GM-CSF also mobilizes PBPC but is less effective than G-CSF (130,131). Both G-CSF and GM-CSF are effective in mobilizing PBPC following chemotherapy, e.g., cyclophosphamide at doses of approximately 4–8 g/m^2. IL-3 alone or in combination with other CSFs has been shown to be effective in mobilization of PBPC (132). In sequential or simultaneous combination with G-CSF, IL-3 stimulates mobilization of PBPC.

The use of mobilized PBPC for allogeneic transplantation has been reported by several groups (135,136). In particular, rhG-CSF has been used to mobilize PBPC in normal donors. An important consideration in the use of PBPC for allogeneic transplant is the increased number of T cells compared bone marrow. PBPC harvests may contain up to 10 times the incidence of T cells as conventional marrow grafts (137,138).

Expansion of Hematopoietic Progenitor Cells—Use of Combination Therapy

Several studies have described the ability of CSFs to synergize with other growth factors to expand the number of mature cells, mature progenitor cells, or primitive progenitor cells. In liquid culture of purified mouse bone marrow cells ($Lin^-/Sca\text{-}1^+$), SCF plus IL-6 was shown to stimulate the generation of primitive progenitor cells HPP-CFC, while SCF plus IL-3 generated mature progenitor cells or GM-CFC (139). Similar studies using human bone marrow cells or peripheral blood progenitor cells (PBPC) have demonstrated the ability of growth factor cocktails to expand mature cells approximately 100-fold and GM-CFC 20-fold or higher (140). In all combinations of growth factors evaluated, the presence of SCF is essential for optimal expansion.

The potential clinical uses of ex vivo expanded cells include the elimination of thrombocytopenia and neutropenia, tumor purging, and the improvement of suboptimal bone marrow or leukapheresis harvests. The clinical development of ex vivo expansion involves a number of regulatory issues. These issues include a procedure that is reproducible while maintaining sterility. In order to address these issues, McNiece et al. have identified a procedure that can result in at least 20-fold expansion of GM-CFC in large-scale cultures (141). CD34-expressing (CD34+) cells from a leukapheresis harvest are isolated and cultured in 3 × 1 L Teflon culture bags at 1.0×10^4 $CD34^+$ cells/ml in defined media (DM, Amgen, Inc.). The DM contains no fetal calf serum or other animal products. Comparison of this media to a commercially available medium resulted in greater expansion with DM than Ex Vivo 15 medium from Biowhittaker Laboratories Inc. The addition of autologous plasma to either DM or Ex Vivo 15 results in increased expansion; however, the DM in the absence of autologous plasma results in equivalent expansion to the Ex Vivo 15 plus autologous plasma. Different lots of autologous plasma have been found to give highly variable expansion and in some instances inhibit expansion. Therefore, the development of the DM provides a highly reproducible culture media and eliminates the variability introduced with media requiring the addition of autologous plasma (141).

Maximal expansion in the DM has been obtained with the combination of SCF/IL-3/IL-6/G-CSF, all added at 100 ng/ml (141). The use of culture bags enables the transfer of cells for washing in a closed system with sterile transfer sets. Under these conditions maximal expansion of GM-CFC is obtained at day 10 of culture without refeeding or further addition of factors. Following the expansion culture, the cells are washed free of growth factors and media on a cell washer (COBE, Inc.) with minimal cell loss. A comprehensive list of quality assurance and quality control (QA/QC) will be necessary for validation of the final cell product. As the parameters for optimal ex vivo expansion have been defined, what remains to be accomplished is to perform the clinical studies to determine the potential of this technology.

Gene Therapy Procedures

The current technology employed for the transfer of genes into hematopoietic cells is based upon retroviral transduction into stem cells. Retroviruses will not transfect quiescent cells, so growth factors are used to stimulate quiescent and slowly cycling cells into active cell cycle. Several studies in animal models have demonstrated efficient transduction.

It has been shown that primitive murine multipotential colony-forming unit spleen cells (CFU-S) can be expanded up to threefold by culturing for 6 days with SCF, IL3, and IL6 (142). A more remarkable expansion of four- to eightfold was seen when murine bone marrow cells were enriched for CFU-S via elutriation and lineage subtraction prior to culture. Combinations of SCF and IL-3 or IL-3 and IL-6 maintain CFU-S number, but are not as effective at expansion (142,143). The long-term repopulating ability of murine bone marrow cells measured in competitive repopulation experiments has also been shown to be enhanced severalfold by in vitro culture in SCF, IL-3, and IL-6 (142,143). IL-3 and IL-6 alone maintain but do not expand cells with repopulating ability (144). $CD34^+$ cells isolated from rhesus bone marrow expand severalfold in 4 days of culture with SCF, IL-3, and IL-6 (145). Autologous transplantation with these cultured cells after total body irradiation (TBI) has produced rapid trilineage reconstitution.

There is extensive evidence in murine models that retroviral vectors can introduce foreign genes at high efficiency into hematopoietic progenitor cells and stem cells (143,146–148). A number of variables appear to be critical to effect efficient gene transfer into primitive progenitor cells and stem cells, such as 5-fluorouracil (5-FU) pretreatment of the donor animal prior to marrow harvest, use of very high-titer viruses, and the inclusion of hematopoietic growth factors in the media during transduction (143,144,149,150). Culture in the presence of growth factors for at least 48 h appears to be necessary for efficient retroviral gene transfer into murine stem cells (143,148). This is presumably due to the increased likelihood of proviral integration in actively cycling cells. Pretreatment of the donor with 5-FU and in vitro culture in IL-3 and IL-6 produces long-term engraftment with marked cells in up to 90% of recipient mice with up to 20% of circulating cells positive for the provirus (143,144).

Experience in larger animals is less extensive. Two groups working with dogs have shown gene transfer in 0.1–10% of circulating cells or progenitor cells of multiple lineages up to 2 years posttransplantation (151,152). Early primate studies using short exposure to low-titer viral supernatant without addition of growth factors could not demonstrate transduction of long-term repopulating cells despite transient circulation of marked cells presumably derived from successfully transduced committed progenitor cells (153). Incorporation of the modifications found to increase gene transfer efficiency in the murine models has also improved the results in primates. Six out of 6 primates transplanted with 5-FU–primed marrow transduced in coculture with a high-titer but helper-contaminated producer line in the presence of IL-3 and IL-6 were positive for proviral DNA in hematopoietic tissue at 6–12 months (149). Three out of 7 animals transplanted with CD34-enriched transduced marrow had the marker gene detectable in marrow, peripheral blood, purified granulocytes and T cells at levels of 1–10% for greater than 100 days posttransplant. However, the presence of high-level helper virus contamination of the producer cell line used in these experiments and chronic viremia in these animals makes interpretation of these results difficult (149). Three of these animals subsequently developed T-cell lymphomas related to the helper virus contamination.

Preclinical studies designed to directly assess the feasibility and efficiency of gene transfer to human stem cells or long-term repopulating cells are not possible. Various surrogates have been proposed, including the long-term bone marrow culture initiating cell assay (LTBMCIC) (154); however, it is not known if this assay really predicts the in vivo behavior of human stem cells. Human LTBMCIC have been successfully marked by retroviruses (155–157). Several different laboratories have shown that up to 40% of colony-forming unit-cells (GM-CFC) cultured out of 5- to 8-week long-term bone marrow cultures initiated with retrovirally transduced marrow are marked with the proviral genome. All of these studies have found that culture for at least 3 days with either hematopoietic growth factors or autologous stroma is necessary for efficient transduction. The transduction process had no effect on the number of total LTBMCIC present or on clonagenic progenitor cell production by these cultures. These published studies used higher-titer viruses. Of note in all of these studies is that the transduction efficiency of GM-CFC was nearly identical to the transduction efficiency of LTBMCIC.

To date, results from clinical studies have been disappointing, and it is obvious that optimization of methodology is still required for effective gene transfer into human hematopoietic cells.

REFERENCES

1. Sheridan WP, Begley CG, Juttner CA, Szer J, To LB, Maher D, McGrath KM, Morstyn G, Fox RM. Effect of peripheral-blood progenitor cells mobilized by filgrastim (G-CSF) on platelet recovery after high-dose chemotherapy. Lancet 1992; 339:6450.
2. Henon PR. Autologous blood stem-cell versus bone marrow transplantation: comparison of cost-effectiveness and of clinical benefits. In: Levitt D, Mertelsmann R, eds. New York: Marcel Dekker Inc., 1995:421–434.
3. Cebon J, Layton JE, Maher D, Morstyn G. Endogenous hematopoietic growth factors in neutropenia and infection. Br J Haematol 1994; 86:265–274.
4. Stanley ER, Heard PM. Factors regulating macrophage production and growth. Purification and some properties of the colony stimulating factor from medium conditioned by mouse L cells. J Biol Chem 1977; 252:4305.
5. Nicola NA, Metcalf D. Analysis of purified fetal liver haemopoietic progenitor cells in liquid culture. J Cell Physiol 1982; 112:257.
6. Burgess AW, Camakaris J, Metcalf D. Purification and properties of colony-stimulating factor from mouse lung-conditioned medium. J Biol Chem 1977; 252:1998.
7. Ihle JN, Keller J, Oroszlan S, et al. Biological properties of homogeneous interleukin 3. J Immunol 1983; 131:282.
8. Cutler RL, Metcalf D, Nicola NA, Johnson GR. Purification of a multipotential colony-stimulating factor from pokeweed mitogen-stimulated mouse spleen cell conditioned medium. J Biol Chem 1985; 260:6579.
9. Ralph P, Warren MK, Ladner MD, Kawasaki ES, Boosman A, White TJ. Molecular and biological properties of human growth factor, CFS-1. Cold Spring Harbor Symp Quant Biol 1986; 51:679.
10. Souza LM, Boone TC, Gabrilove J, et al. Recombinant granulocyte colony-stimulating factor: effects on normal and leukemic myeloid cells. Science 1986; 232:61.
11. Yang YC, Ciarletta AB, Temple PA, et al. Human IL-3 (multi-CFS): identification by expressing cloning of a novel hematopoietic growth factor related to murine IL-3. Cell 1986; 47:3.

12. Wong GG, Witek JS, Temple PA, et al. Human GM-CSF: molecular cloning of the complementary DNA and purification of the natural and recombinant proteins. Science 1985; 228:810.
13. Lee F, Yokota T, Otsuka L, et al. Isolation of a cDNA for a human granulocyte-macrophage colony stimulating factor by functional expression in mammalian cells. Proc Natl Acad Sci USA 1985; 82:4360.
14. Stanely ER, Guilbert LJ, Tushinski RJ, Bartelmez SH. CSF-1: a mononuclear phagocyte lineage-specific hemopoietic growth factor. J Cell Biochem 1983; 21:151.
15. Bradley TR, Hodgson GS. Detection of primitive macrophage progenitor cells in mouse bone marrow. Blood 1979; 54:1446.
16. Bartelmez SH, Stanely ER. Synergism between hemopoietic growth factors (HGFs) detected by their effects on cells bearing receptors for a lineage specific HGF: assay of hemopoietin-1. J Cell Physiol 1984; 13:329.
17. McNiece IK, Kriegler AB, Hapel AJ, et al. Recombinant interleukin-3 exhibits synergistic factor activity. Cell Biol Int Rep 1984; 8:812.
18. McNiece IK, Kriegler AB, Bradley TR, Hodgson GS. Subpopulations of mouse bone marrow high-prolifertive-potential colony-forming cells (HPP-CFC). Exp Hematol 1986; 14:856.
19. McNiece IK, Stewart FM, Deacon DM, Quesenberry PJ. Synergistic interactions between hematopoietic growth factors as detected by in vitro mouse bone marrow colony formation. Exp Hematol 1988; 16:383.
20. Rennick D, Jackson J, Yang G, Wideman J, Lee F, Hudak S. Interleukin-6 interacts with interleukin-4 and other hematopoietic growth factors to selectively enhance the growth of megakaryocytic, erythroid, myeloid, and multipotential progenitor cells. Blood 1989; 73:1828.
21. Hunt P, personal communication, 1990.
22. Metcalf D, Nicola NA. Synthesis by mouse peritoneal cells of G-CSF, the differentiation inducer for myeloid leukemia cells: stimulation by endotoxin, M-CSF, and multi-CSF. Leuk Res 1985; 9:35.
23. Warren MK, Ralph P. Macrophage growth factor CSF-1 stimulates human monocyte production of interferon, tumor necrosis factor, and colony stimulating activity. J Immunol 1986; 137:2281.
24. Das SK, Stanley ER. Structure-function studies of a colony stimulating factor (CSF-1). J Biol Chem 1982; 257:13679.
25. Tushinski RJ, Oliver IT, Guilbert IJ, Tynan PW, Warner JR, Stanely ER. Survival of mononuclear phagocytes depends on a lineage-specific growth factor that the differentiated cells selectively destroy. Cell 1982; 28:71.
26. Horiguchi J, Warren MK, Kufe D. Expression of the macrophage-specific colony-stimulating factor in human monocytes treated with granulocyte-macrophage colony-stimulating factor. Blood 1987; 69:1259.
27. Horiguchi J, Warren MK, Ralph P, Kufe D. Expression of the macrophage specific colony stimulating factor (CSF-1) during human monocytic differentiation. Biochem Biophys Res Commun 1986; 141:924.
28. Rambaldi A, Young DC, Griffin JD. Expression of the M-CFS (CSF-1) gene by human monocytes. Blood 1987; 69:1409.
29. Oster WA, Lindermann A, Horn S, Mertelsmann R, Herrmann F. Tumor necrosis factor (TNF)-alpha but not TNF-beta induces secretion of colony-stimulating factor for macrophages (CSF-1) by human monocytes. Blood 1987; 70:1700.
30. Rettenmeir CW, Roussel MF, Ashmun RA, Ralph P, Price K, Sherr CJ. Synthesis of membrane-bound colony-stimulating factor-1 (CSF-1) and downmodulation of CSF-1 receptors in NIH 3T3 cells transformed by cotransfection of the human CSF-1 and c-fms (CSF-1 receptor) genes. Mol Cell Biol 1987; 7:2378.
31. Bartocci A, Pollard JW, Stanely ER. Regulation of colony-stimulating factor 1 during pregnancy. J Exp Med 1986; 164:956.
32. Pollard JW, Bartocci R, Arceci R, Orlofsky A, Ladner MB, Stanely ER. Apparent role of the macrophage growth factor, CSF-1, in placental development. Nature 1987; 330:484.
33. Das SK, Stanely ER, Guilbert LJ, Forman LW. Human colony-stimulating factor (CSF-1) radioimmunoassay: resolution of three subclasses of human colony-stimulating factors. Blood 1981; 58:630.
34. Wong GG, Temple PA, Leary AC, et al. Human CFS-1: molecular cloning and expression of 4kb cDNA encoding the human urinary protein. Science 1987; 235:1504.
35. Kawasaki ES, Ladner MB, Wang AM, et al. Molecular cloning of a complimentary DNA encoding human macrophage specific colony stimulating factor (CSF-1). Science 1985; 230:291.
36. Guilbert LJ, Stanely ER. Specific interaction of murine colony-stimulating factor with mononuclear phagocyte cells. J Cell Biol 1980; 85:153.
37. Guilbert LJ, Stanely ER. The interaction of ^{125}I-colony-stimulating factor-1 with bone marrow-derived macrophages. J Biol Chem 1986; 261:4024.
38. Byrne PV, Guilbert LJ, Stanely ER. Distribution of cells bearing receptors for a colony-stimulating factor (CSF-1) in murine tissues. J Cell Biol 1981; 91:848.
39. Morgan CJ, Stanely ER. Chemical crosslinking of the mononuclear phagocytic specific growth factor CSF-1 to its receptor at the cell surface. Biochem Biophys Res Commun 1984; 119:35.
40. Sherr CJ, Rettenmeir CW, Sacca R, Roussel MF, Look AT, Stanely ER. The c-fms proto-oncogene product is related to the receptor for the mononuclear phagocyte growth factor, CSF-1. Cell 1985; 41:665.
41. Sacca R, Stanely ER, Sherr CJ, Rettenmeir CW. Specific binding of the mononuclear phagocyte colony-stimulating factor CSF-1 to the product of the v-fms oncogene. Proc Natl Acad Sci USA 1986; 83:3331.
42. Yeung YG, Jubinsky PT, Sengupta A, Yeung DCY, Stanely ER. Purification of the colony-stimulating factor 1 receptor and demonstration of its tyrosine kinase activity. Proc Natl Acad Sci USA 1987; 84:1268.
43. Wheeler EF, Rettenmeir CW, Look AT, Sherr CJ. The v-fms oncogene induces factor independence and tumorigenicity in CSF-1 dependent macrophage cell line. Nature 1986; 324:377.
44. Downing JR, Rettenmeir CW, Sherr CJ. Ligand-induced tyrosine kinase activity of the colony stimulating factor-1 receptor in a murine macrophage cell line. Mol Cell Biol 1988; 8:1795.
45. Walker F, Nicola NA, Metcalf D. Hierarchical downmodulation of hemopoietic growth factor receptors. Cell 1985; 43:269.
46. Metcalf D. Hemopoietic colonies. In: In Vitro Cloning of Normal and Leukemic Cells. Berlin: Springer-Verlag, 1977:
47. Metcalf D, Burgess AW, Johnson GR, et al. In vitro actions on hemopoietic cells of recombinant murine GM-CSF purified after production in E. coli: comparison with purified native GM-CSF. J Cell Physiol 1986; 128:421.
48. Robinson BE, McGrath HE, Quesenberry PJ. Recombinant murine granulocyte macrophage colony-stimulating factor has megakaryocyte colony-stimulating factor, has megakaryocyte colony-stimulating activity and augments megakaryocyte colony stimulation by interleukin 3. J Clin Invest 1987; 79:1648.
49. Metcalf D, Johnson GR, Burgess AW. Direct stimulation by

purified GM-CSF of the proliferation of multipotential and erythroid precursor cells. Blood 1980; 55:138.
50. Migliaccio AR, Bruno M, Migliaccio G. Evidence for direct action of human bio-synthetic (recombinant) GM-CSF on erythroid progenitors in serum-free culture. Blood 1987; 70:1867.
51. McNiece IK, Stewart FM, Deacon DM, et al. Detection of human CFC with a high proliferative potential. Blood 1989; 74:609.
52. Mazur EM, Cohne JL, Wong GG, Clark SC. Modest stimulating effect of recombinant human GM-CSF on colony growth from peripheral blood human megakaryocyte progenitor cells. Exp Hematol 1987; 15:1128.
53. Messner HA, Yamasaki K, Jamal N, et al. Growth of human hemopoietic colonies in response to recombinant gibbon interleukin 3: comparison with human recombinant granulocyte and granulocyte-macrophage colony-stimulating factor. Proc Natl Acad Sci USA 1987; 84:6765.
54. Caracciolo D, Shirshat N, Wong GG, Lange B, Clark S, Rovera G. Recombinant human macrophage colony-stimulating factor (M-CSF) requires subliminal concentrations of granulocyte/macrophage (GM)-CSF for optimal stimulation of human macrophage colony formation in vitro. J Exp Med 1987; 166:1851.
55. McNiece I, Andrews R, Stewart M, Clark S, Boone T, Quesenberry P. Action of interleukin-3, G-CSF, and GM-CSF on highly enriched human hematopoietic progenitor cells: synergistic interaction of GM-CSF plus G-CSF. Blood 1989; 41:110.
56. Burgess AW, Begley CG, Johnson GR, et al. Purification and properties of bacterially synthesized human granulocyte-macrophage colony stimulating factor. Blood 1987; 69:43.
57. Fleischmann J, Golde DW, Weisbart RH, Gasson JC. Granulocyte-macrophage colony-stimulating factor enhances phagocytosis of bacteria by human neutrophils. Blood 1986; 68:708.
58. Lopez AF, Nicola NA, Burgess AW, et al. Activation of granulocyte cytotoxic function by purified mouse colony-stimulating factors. J Immunol 1983; 131:2983.
59. Vadas MA, Nicola NA, Metcalf D. Activation of anti-body-dependent cell-mediated cytotoxicity of human and eosinophils by separate colony stimulating factors. J Immunol 1983; 130:795.
60. Villalta F, Kierszenbaum J. Effects of human colony-stimulating factor on the uptake and destruction of a pathogenic parasite (Trypanosoma cruzi) by human neutrophils. J Immunol 1986; 137:1703.
61. Weisbart RM, Golde DW, Clark SC, Wong GG, Gasson JC. Human granulocyte-macrophage colony-stimulating factor is a neutrophil activator. Nature 1985; 314:361.
62. Williamson DJ, Brown TC. Enhancement of neutrophil-mediated phagocytosis by human granulocyte-macrophage colony-stimulating factor demonstrated using a novel mathematical mode. Immunol Cell Biol 1987; 65:329.
63. Grabstein KH, Urdal DL, Tushinski RJ, et al. Induction of macrophage tumoricidal activity by granulocyte-macrophage colony-stimulating factor. Science 1986; 232:506.
64. Handman E, Burgess AW. Stimulation by granulocyte-macrophage colony stimulating factor of Leishmania tropica killing by macrophages. J Immunol 1979; 122:1134.
65. Kelso A, Gough N. Expression of hemopoietic growth factor genes in murine T lymphocytes. In: Webb DR, Goeddel DV, eds. The Lymphokines. Vol. 3. New York: Academic Press, 1987:209.
66. Koury MJ, Pragnell IB. Retroviruses induce granulocyte-macrophage colony-stimulating activity in fibroblasts. Nature 1982; 299:638.
67. Quesenberry PJ, Gimbrone MA. Vascular endothelium as a regulator of granulopoiesis. Production of colony-stimulating activity by cultured human endothelial cells. Blood 1980; 56:1060.
68. Chodakewitz JA, Kupper TS, Coleman DL. Keratinocyte-derived granulocyte/macrophage colony-stimulating factor induces DNA synthesis by peritoneal macrophages. J Immunol 1988; 140:832.
69. Byrne PV, Heit WF, March CJ. Human granulocyte-macrophage colony-stimulating factor purified from a Hodgkin's tumor cell line. Biochim Biophys Acta 1986; 874:266.
70. Broudy VC, Kaushansky K, Segal GM, Harlan JM, Adamson JW. Tumor necrosis factor type alpha stimulates human endothelial cells to produce granulocyte/macrophage colony stimulating factor. Proc Natl Acad Sci USA 1986; 83:7467.
71. Munker R, Gasson J, Ogawa M, Koeffler HP. Recombinant human TNF induces production of granulocyte-monocyte colony-stimulating factor. Nature 1986; 323:79.
72. Seelentag WK, Mermod J-J, Montesano R, Vassali P. Additive effects of interleukin 1 and tumor necrosis factor-alpha on the accumulation of the three granulocyte and macrophage colony-stimulating factor mRNA's in human endothelial cells. EMBO J 1987; 6:2261.
73. Seiff CA, Tsai S, Faller DV. Interleukin-1 induces cultured human endothelial cell production of granulocyte-macrophage colony stimulating factor. J Clin Invest 1987; 79:48.
74. Zsebo KM, Yuscenkoff VN, Schiffer S, et al. Vascular endothelial cells and granulopoiesis interleukin-1 stimulates release of G-CSF and GM-CSF. Blood 1988; 71:99.
75. Sheridan JW, Metcalf D. Purification of mouse lung-conditioned medium colony stimulating factor (CSF). Proc Soc Exp Biol Med 1974; 146:218.
76. Sheridan JW, Metcalf D, Stanely ER. Further studies on the factor in lung-conditioned medium stimulating granulocyte and monocyte colony formation in vitro. J Cell Physiol 1974; 84:147.
77. Sparrow LG, Metcalf D, Hunkapiller MW, Hood LE, Burgess AW. Purification and partial amino acid sequence of asialo murine granulocyte-macrophage colony stimulating factor. Proc Natl Acad Sci USA 1985; 82:292.
78. Gough NM, Gough J, Metcalf D, et al. Molecular cloning of cDNA encoding a murine haemopoietic growth regulator, granulocyte-macrophage colony stimulating factor. Nature 1984; 309:763.
79. DeLarmarter JF, Mermod J-J, Liang C-M, Eliason JF, Thatcher DR. Recombinant murine GM-CSF from E. coli has biological activity and is neutralized by a specific antiserum. EMBO J 1985; 4:2575.
80. DiPersio J, Billing P, Kaufman S, Eghtesady P, Williams RE, Gasson JC. Characterization of the human granulocyte-macrophage colony-stimulating factor receptor. J Biol Chem 1988; 263:1834.
81. Baldwin GC, Gasson JC, Kaufman SE, et al. Nonhematopoietic tumor cells express functional GM-CSF receptors. Blood 1989; 73:1033.
82. Sorenson PH, Mui AL-F, Murphy SC, Krystal G. Interleukin-3, GM-CSF, and TPA induce distinct phosphorylation events in an interleukin 3-dependent multipotential cell line. Blood 1989; 73:406.
83. Burgess AW, Metcalf D. Characterization of a serum factor stimulating the differentiation of myelomonocytic leukemic cells. Int J Cancer 1980; 39:647.
84. Nicola NA, Metcalf D, Matsumoto M, Johnson GR. Purification of a factor inducing differentiation in a murine myelomonocytic leukemia cells. Identification as granulocyte colony-stimulating factor. J Biol Chem 1983; 258:9017.

85. Welte K, Platzer E, Lu L, et al. Purification and biochemical characterization of human pluripotent hematopoietic colony-stimulating factor. Proc Natl Acad Sci USA 1985; 82:1526.
86. Simpson RJ, Nice EC, Nicola NA. Structural studies on the murine granulocyte colony-stimulating factor. Biol Chem Hope Seyler 1987; 368:1327.
87. Nagata S, Tsuchiya M, Asano S, et al. Molecular cloning and expression of cDNA for human granulocyte colony-stimulating factor. Nature 1986; 319:415.
88. Nicola NA, Metcalf D. Differentiation induction in leukemic cells by normal growth regulators: molecular and binding properties of purified granulocyte colony-stimulating factor. In: Bishop JM, Rowley JD, Greaves M, eds. Genes and Cancer. New York: Alan R. Liss, 1984:591.
89. Nicola NA. Granulocyte-colony-stimulating factor. Methods Enzymol 1987; 116:600.
90. Nicola NA, Metcalf D. Biochemical properties of differentiation factors for murine myelomonocytic leukemic cells in organ conditioned media. Separation from colony-stimulating factors. J Cell Physiol 1981; 112:257.
91. Vadas MA, Lopez AF. Regulation of granulocyte function by colony-stimulating factors and monoclonal antibodies. Lymphokines 1985; 12:179.
92. Koeffer HP, Gasson J, Raynard J, Souza L, Shepard M, Munker R. Recombinant human TNF alpha stimulates production of granulocyte colony-stimulating factor. Blood 1987; 70:55.
93. Rennick D, Yang G, Gemmell L, Lee F. Control of hemopoiesis by a bone marrow stromal cell clone: lipopolysaccharide and interleukin-1-inducible production of colony-stimulating factors GM-CSF and G-CSF. Blood 1987; 69:682.
94. McNiece IK, Kriegler AB, Quesenberry PJ. Studies on the myeloid synergistic factor from 5637: comparison with interleukin-1 alpha. Blood 1989; 73:919.
95. Moore MAS, Warren DJ. Synergy of interleukin 1 and granulocyte colony-stimulating factor: in vivo stimulation of stem-cell recovery and hematopoietic regeneration following 5-fluorouracil treatment of mice. Proc Natl Acad Sci USA 1987; 84:7134.
96. McNiece IK, McGrath HE, Quesenberry PJ. Granulocyte colony stimulating factor augments in vitro megakaryocyte colony formation by interleukin-3. Exp Hematol 1988; 16:807.
97. Ikebuchi K, Clark S, Ihle JN, Souza LM, Ogawa M. Granulocyte colony-stimulating factor enhances interleukin-3 dependent proliferation of multipotential hemopoietic progenitors. Proc Natl Acad Sci USA 1988; 85:3445.
98. Woodward TA, McNiece IK, Witte PL, et al. Further studies on growth factor production by the TC-1 stromal cell line: Pre-B stimulating activity. Blood 1990; 75:2130.
99. Nicola NA, Begley CG, Metcalf D. Identification of the human analogue of a regulator that induces differentiation in murine leukaemic cells. Nature 1985; 314:625.
100. Avalos BR, Hedzat C, Baldwin GC, Golde DW, Gasson JC, DiPersio JF. Biological activities of human G-CSF and characterization of the human G-CSF receptor. Blood 1987; 70(suppl 1):165a.
101. Avalos BR, Gasson JC, Connors LG, Golde DW, Souza LM, Slamn DJ. Molecular characterization of the human G-CSF receptor. J Cell Biochem Suppl 1989; 13C:22.
102. Fukunaga R, Ishizaka-Ikeda E, Seto Y, Nagata S. Expression cloning of a receptor for murine granulocyte colony-stimulating factor. Cell 1990; 61:341.
103. Ihle JN, Pepersack L, Rebar L. Regulation of T-cell differentiation: in vitro induction of 20 alpha-hydroxysteroid dehydrogenase in splenic lymphocytes from athymic mice by a unique lymphokine. J Immunol 1981; 126:2184.
104. Ihle JN, Rebar L, Kellar J, Lee JC, Hapel AJ. Interleukin 3: possible roles in the regulation of lymphocyte differentiation and growth. Immunol Rev 1982; 63:5.
105. Ihle JN, Keller J, Henderson L, Klein F, Palaszynski E. Procedures for the purification of interleukin 3 to homogeneity. J Immunol 1982; 129:2431.
106. Iscove NN, Roitsch CA, Williams N, Guilbert LJ. Molecules stimulating early red cell, granulocyte, macrophage and megakaryocyte precursors in culture: similarity in size, hydrophobicity and charge. J Cell Physiol Suppl 1982; 1:65.
107. Wagemaker G, Merchav S, Burger H. Multiplication of pluripotent hemopoietic stem cells induced by purified stem cell activating factor (SAF) in serum free suspension cultures. Exp Hematol 1982; 10(suppl 11):162.
108. Clark-Lewis I, Kent SBH, Schrader JW. Purification to apparent homogeneity of a factor stimulating the growth of multiple lineages of hemopoietic cells. J Biol Chem 1984; 259:7488.
109. Yung YP, Eger R, Tertian G, Moore MAS. Long-term in vitro culture of murine mast cells. II. Purification of a mast cell growth factor and its dissociation from TCGF. J Immunol 1981; 127:794.
110. Hapel AJ, Fung MC, Johnson RM, Young IG, Johnson GR, Metcalf D. Biological properties of molecularly cloned and expressed murine interleukin 3. Blood 1985; 65:1453.
111. Fung MC, Hapel AJ, Ymer S, et al. Molecular cloning of cDNA for murine interleukin-3. Nature 1984; 307:233.
112. Rennick DM, Lee FD, Yokota T, Arai K-I, Cantor H, Nabel GJ. A cloned MCGF cDNA encodes a multilineage hematopoietic growth factor: multiple activities of interleukin 3. J Immunol 1985; 134:910.
113. Lee F, Abrams J, Arai K, et al. The expression and characterization of recombinant mouse IL-3. In: Schrader JW, ed. Lymphokines. Vol. 15. San Diego: Academic Press, 1988:163.
114. Ihle JN. Biochemical and biological properties of interleukin-3: a lymphokine mediating the differentiation of a lineage of cells that includes prothymocytes and mastlike cells. Contemp Top Mol Immunol 1984; 10:93.
115. Oon SH, Williams N. Biochemical characterization of an in vitro murine megakaryocyte growth activity: megakaryocyte potentiator. Leuk Res 1986; 10:403.
116. Ikebuchi K, Wong GG, Clark SC, Ihle JN, Hirai Y, Ogawa M. Interleukin 6 enhancement of interleukin 3-dependent proliferation of multipotential hemopoietic progenitors. Proc Natl Acad Sci USA 1987; 84:9035.
117. Kelso A, Metcalf D. Clonal heterogeneity in colony stimulating factor production by murine T-lymphocytes. J Cell Physiol 1985; 123:101.
118. Luger TA, Wirth U, Kock A. Epidermal cells synthesis a cytokine with interleukin-3-like properties. J Immunol 1985; 134:915.
119. Labastie MC, Thiery JP, Le Douarin NM. Mouse yolk sac and intra embryonic tissues produce factors able to elicit differentiation of erythroid burst-forming units and colony-forming units respectively. Proc Natl Acad Sci USA 1984; 81:1453.
120. Frei K, Bodmer S, Scherdel C, Fontana A. Astrocyte-derived interleukin-3 as a growth factor for microglia cells and peritoneal macrophages. J Immunol 1986; 137:3521.
121. Barge AJ, Johnson GA, Simmons P, Torok-Storb B. Interleukin 3 transcript, bio-activity and protein is detectable in normal bone marrow stroma in the absence of mature T-lymphocytes. Blood 1989; 74(suppl 1):115a.
122. Murthy SC, Mui AL-F, Krystal G. Characterization of the interleukin 3 receptor. Exp Hematol 1990; 18:11.

123. Rapoport AP, Abboud CN, DiPersio JF. Granulocyte-macrophage colony-stimulating factor (GM-CSF): receptor biology, signal transduction, and neutrophil activation. Blood Rev 1992; 6:43–57.
124. Morstyn G, Cebon J, Layton J, et al. Cytokines in infections and as anticancer agents. Ann Hematol 1991; 29a.
125. Kawakami M, Tsutsumi H, Kumakawa T, et al. Levels of serum granulocyte colony-stimulating factor in patients with infections. Blood 1990; 76:1962–1964.
126. Demetri GD, Griffin JD. Granulocyte colony-stimulating factor and its receptor. Blood 1991; 78:2791–2802.
127. Kojima S, Fukuda M, Miyajima Y, Matsuyama T, Horibe K. Treatment of aplastic anemia in children with recombinant human granulocyte colony-stimulating factor. Blood 1991; 77:937–941.
128. Kaplan LD, Kahn JO, Crowe S, Northfelt D, Neville P, Grossberg H, Abrams DI, Tracey J, Mills J, Volberding PA. Clinical and virologic effects of recombinant human granulocyte-macrophage colony-stimulating factor in patients receiving chemotherapy for human immunodeficiency virus-associated non-Hodgkin's lymphoma: results of a randomized trial. J Clin Oncol 1991; 9:929–940.
129. Furman WL, Fairclough DL, Huhn RD, Pratt CB, Stute N, Petros WP, Evans WE, Bowman LC, Douglass EC, Santana VM, Meyer WH, Crist WM. Therapeutic effects and pharmacokinetics of recombinant human granulocyte-macrophage colony-stimulating factor in childhood cancer patients receiving myelosuppressive chemotherapy. J Clin Oncol 1991; 9:1022–1028.
130. Willerford DM, Preffer F, Mcafee S, Scavone M, Spitzer TR. Mobilization of early and late hematopoietic precursors using GM-CSF, G-CSF, or combined GM-CSF and G-CSF following chemotherapy. Blood 1994; 84:A90.
131. Spitzer G, Mathews M, Dunphy F, Bowers C, Petruska P, Velasquez W, et al. Are growth-factor (GF) combinations superior to G-CSF alone for mobilization of peripheral-blood stem-cells (PBSC)—a randomized study comparing G-CSF to G-CSF with GM-CSF for PBSC mobilization. Blood 1994; 84:A107.
132. Scheding S, Brugger W, Mertelsmann R, Kanz L. Peripheral-blood stem-cells—in vivo biology and therapeutic potential. Stem Cells 1994; 12(suppl 1):203.
133. Bolwell B, Rosenfeld C, Lefever A, List A, Andrews F, Schuster M, et al. A phase-II trial of G-CSF with or without SDZ ILE-964 (IL-3) for the mobilization of peripheral-blood progenitor cells (PBPC) for autologous progenitor-cell transplantation. Blood 1994; 84:A106.
134. Huhn RD, Yurkow EJ, Tushinski R, Hoffman R, Clarke L, Sheay W, et al. Priming with recombinant human interleukin-3 (rhIL-3) enhances the mobilization of blood progenitor cells (PBPC) by rh-granulocyte colony-stimulating factor (rhG-CSF) in normal volunteers. Blood 1994; 84:A541.
135. Kessinger A, Smith DM, Stranford SE, Landmark JD, Dooley DC, Law P, et al. Allogeneic transplantation of blood-derived T-cell depleted hemopoietic stem cells after myeloablative treatment in a patient with acute lymphoblastic leukemia. Bone Marrow Transplant 1989; 4:643.
136. Schmitz N, Linch DC, Dreger P, Boogaerts MA, Goldstone AH, Ferrant A, et al. A randomized phase III study of filgrastim-mobilized peripheral blood progenitor cell transplantation (PBPCT) in comparison with autologous bone marrow transplantation (ABMT) in patients with Hodgkins disease (HD) and non-Hodgkins lymphoma (NHL). Blood 1994; 84:204a.
137. Weaver CH, Buckner CD, Longin K, Appelbaum FR, Rowley S, Lilleby K, et al. Syngeneic transplantation with peripheral blood mononuclear cells collected after administration of recombinant human granulocyte colony-stimulating factor. Blood 1993; 82:1981.
138. Dreger P, Viehmann K, Steinmann J, Eckstein V, Muller-ruchholtz W, Loffler H. G-CSF-mobilized peripheral-blood progenitor cells for allogeneic transplantation—comparison of T-cell depletion strategies using different CD34(+) selection systems or CAMPATH-1. Exp Hematol 1994; 23:147.
139. Williams N, Bertoncello I, Kavnoudias H, Zsebo K, McNiece I. Recombinant rat stem cell factor stimulates the amplification and differentiation of fractionated mouse stem cell populations. Blood 1992; 79:58.
140. Haylock DN, To LB, Dowse TL, Juttner CA, Simmons PJ. Ex vivo expansion and maturation of peripheral blood CD34+ cells into the myeloid lineage. Blood 1992; 80:1405.
141. Shieh J-H, Chen Y-F, Briddell R, Stoney G, McNiece IK. High purity of blast cells in CD34 selected populations are essential for optimal ex vivo expansion of human GM-CFC. Exp Hematol 1994; 22(8):756 (abstract).
142. Bodine DM, Orlic D, Birkett NC, Seidel NE, Zsebo KM. Stem cell factor increases colony-forming unit-spleen number in vitro in synergy with interleukin-6, and in vivo in SI/SI-d mice as a single factor. Blood 1992; 79:913.
143. Bodine DM, Karlsson S, Nienhuis AW. Combination of interleukins 3 and 6 preserves stem cell function in culture and enhances retrovirus-mediated gene transfer into hematopoietic stem cells. Proc Natl Acad Sci USA 1989; 86:8897.
144. Bodine DM, McDonough KT, Seidel NE, Nienhuis AW. Survival and retrovirus infection of murine hematopoietic stem cells in vitro: effects of 5-FU and method of infection. Exp Hematol 1991; 19:206.
145. Donahue RE, Kessler SW, Bodine DB, Agricola BA, Byrne ER, Metzger ME, McDonough KT, Bacher JD, Zsebo KM, Nienhuis AW. Gene transfer into primate immunoselected CD34+ cells: in vivo evidence that CD34 cells are responsible for both short- and long-term reconstitution of lymphoid and myeloid populations. Blood 1991; 78:79a.
146. Nienhuis AW, McDonough KT, Bodine DM. Gene transfer into hematopoietic stem cells. Cancer 1991; 67:2700.
147. Karlson S. Treatment of genetic defects in hematopoietic cell function by gene transfer. Blood 1991; 78:2481.
148. Lim B, Williams DA, Orkin SH. Retrovirus-mediated gene transfer of human adenosine deaminase: expression of functional enzyme in murine hematopoietic stem cells in vivo. Mol Cell Biol 1987; 7:3459.
149. Bodine DM, McDonough KT, Brandt SJ, Ney PA, Aricola B, Byrne E, Nienhuis AW. Development of a high-titer retrovirus producer cell line capable of gene transfer into rhesus monkey hematopoietic stem cells. Proc Natl Acad Sci USA 1990; 87:3738.
150. Wieder R, Cornetta K, Kessler SW, Anderson WF. Increased efficiency of retroviral-mediated gene transfer and gene expression in primate bone marrow after 5-fluorouracil-induced hematopoietic suppression and recovery. Blood 1991; 77:448.
151. Schuening FG, Kawahara K, Miller AD, To R, Goehle S, Stewart D, Mullally K, Fisher L, Graham TC, Appelbaum FR, Hackman R, Osborne WRA, Storb R. Retrovirus-mediated gene transduction into long-term repopulating marrow cells of dogs. Blood 1991; 78:2568.
152. Carter RF, Abrams-Ogg ACG, Dick JE, Kruth SA, Valli VE, Kamel-Reid S, Dube ID. Autologous transplantation of canine long-term marrow culture cell genetically marked by retroviral vectors. Blood 1992; 79:356.
153. Kantoff PW, Gillio AP, McLachlin JR, Bordignnon C, Eglitis

MA, Kernan NA, Moen RC, Kohn DB, Yu SF, Karson E, Karlsson S, Zweibel J, Gilboa E, Blaese RM, Nienhuis AO, Reilly RJ, Anderson WF. Expression of human deaminase in nonhuman primates after retrovirus-mediated gene transfer. J Exp Med 1987; 166:219.

154. Sutherland HJ, Lansdorp PM, Henkelman DH, Eaves AC, Eaves CJ. Functional characterization of individual human hematopoietic stem cells cultured at limiting dilution on supportive marrow stromal layers. Proc Natl Acad Sci USA 1990; 87:3584–3588.

155. Hughes PFD, Eaves CJ, Hogge DE, Humphries RK. High efficiency gene transfer to human hematopoietic cells maintained in long-term marrow culture. Blood 1989; 74:1915.

156. Fraser C, Eaves CJ, Szilvassy SJ, Humphries RK. Expansion in vitro of retrovirally marked totipotent hematopoietic stem cells. Blood 1990; 76:1071.

157. Nolta JA, Kohn DB. Comparison of the effects of growth factors on retroviral vector-mediated gene transfer and the proliferative status of human hematopoietic progenitor cells. Human Gene Ther 1990; 1:257.

4

Interleukin 4: Structure and Function

Wolfgang Holter

Children's Cancer Research Institute, St. Anna's Children's Hospital, Vienna, Austria

INTRODUCTION AND HISTORY OF INTERLEUKIN-4

Interleukin 4 (IL-4) was characterized functionally as early as 1982 as a T-cell–derived B-cell growth factor different from IL-2 (1). Molecular cloning of the cytokine was reported in 1986 (2–4). Initially, IL-4 was recognized as a B-cell growth factor and as a switch factor for IgE production. During the following years much evidence has accumulated regarding the effects of IL-4 on B cells, T cells, monocytes, mast cells, endothelial cells, and fibroblasts (reviewed in Ref. 5). Similarly, the regulation of its production was studied intensely. Most importantly, however, with the definition of IL-4 it also became clear that T-cell clones can be grouped into Th1-type clones secreting IL-2 and IFN-γ and Th2-type clones secreting IL-4 and IL-5, respectively (reviewed in Ref. 6). The importance of the Th1 and Th2 dichotomy for the regulation of allergic reactions and also for the fight against infectious disease soon became clear. Thus, whether an organism responds with a Th1- or a Th2-type immune response to an infection such as leprosy, leishmaniasis, or HIV will determine its success or failure in fighting the invading pathogen in the first two cases and may be responsible for some of the pathophysiology seen during the HIV infection (7). Finally, evidence is accumulating that whether an IL-4–dominated Th2-type or an IFN-γ–dominated Th1-type immune response is developed may also determine the outcome of the immune response to tumor cells and to allogeneic cells during graft-versus-host reactions. Taken together, the discovery of IL-4 has made it possible to put the old classification of immune responses as humoral or cellular dominated on a new basis, both conceptually as well as in terms of understanding the underlying mechanisms of immune factors.

THE STRUCTURE OF IL-4

Protein Structure

Murine and human IL-4 are glycoproteins with a molecular weight of 14,000–19,000 daltons, depending on the degree of glycosylation (2,8). Recombinant deglycosylated forms are fully biologically active. IL-4 is a globular and highly compact molecule with a hydrophobic core. It consists of a left-handed four–α-helix bundle with two overhand connections including a two-stranded antiparallel β-sheet (9). The overall structure is very similar to that of human growth hormone and GM-CSF, although the primary sequence between these molecules has little similarity, apart from an approximate 30% homology in the core of the molecule, including parts of the four helices (10,11). Correspondingly, the receptors for all three growth factors belong to the same family of hematopoietic growth factor receptors, each possessing tandem repeats of a fibronectin type III–like domain.

The binding between IL-4 and its receptor occurs at a 1:1 stoichiometry (12) and involves residues around amino acid (AA) 9 in helix A and AA 88 in helix C. MAbs against these epitopes as well as mutations of the respective AAs prevent binding to the receptor. A further epitope on helix D located at position 121–124 seems to be involved in signaling as mutations of these AAs and mAbs against helix D inhibit IL-4 function but not binding (13,14).

The IL-4 Gene

The cDNAs for both human and murine IL-4 have been cloned (2–4). The gene is found on mouse chromosome 11 and on the human chromosome 5 (15,16). The IL-4 gene is a member of a related family of lymphokine genes, which also includes those coding for IL-3, IL-5, GM-CSF, and IL-13. All of these genes are located within relatively close chromosomal distance (17).

The human IL-4 gene consists of four exons and three introns spanning 10 kb (18). It codes for a protein of 153 amino acid residues including a putative signal peptide. In the 5′ region several sequences have been identified,which show homology to a variable extent with areas in the 5′-flanking region of the IL-2, IL-3, IL-5, and GM-CSF genes (18). Recent studies describe the identification of different regulatory sequences in the IL-4 promoter region. An 11-base-pair DNA motif identified by Abe et al. is located within nucleotide positions -79 to -69 relative to the transcription start site, confers responsiveness to PMA/A23187 stimulation of an indicator plasmid, and is not homologous to sequences of the IL-2 promoter (19). Li-Weber et al. reported on the identification of two negative regulatory regions (NRE-1 and NRE-2) and one positive regulatory element (PRE-1) further upstream in the IL-4 promoter (20,21). Both groups also identified DNA-binding proteins specific for their respective regulatory sequences.

Using transgenic mice it was determined that 3kb of murine IL-4 5′ sequence is sufficient to confer in vivo tissue specific and inducible expression (22). Within the promoter five repeated elements sharing a common consensus ATTTTCCNNT sequence were located between -40 and -250. Each of these elements was shown to interact with the T-cell–specific factor NF, possibly related to NF-AT, previously implicated in the regulation of the IL-2 gene (23). Interestingly, a multimer of a sequence designated CS1 and related to this common consensus sequence linked to a heterologous promoter could function in an IL-4–producing Th2-type clone but not in a non–Il-4-producing Th1-type clone (24), indicating that the regulation of selective cytokine gene transcription in Th1 versus Th2 clones occurs at the level of transcription factors. A further regulatory element located at position -114 to -107 in the IL-4 gene is homologous to the Y box found in all major histocompatibility complex (MHC) class II gene promoters (23).

Apart from transcriptional control, the amount of cellular IL-4 mRNA may be regulated also posttranscriptionally as it contains an AU rich region similar to that found in the mRNAs of other lymphokines and protooncogenes where it mediates mRNA degradation (25). Finally, an alternatively spliced transcript lacking exon 2 has been found in PHA-stimulated human mononuclear cells (26).

CELLULAR SOURCES

IL-4 production seems to be very restricted and under strong negative control. It has been found to occur in thymocytes, mature T cells, certain malignant T cells, mast cells and basophils, and, occasionally, in transformed B cells.

IL-4 Production in the Thymus

In the murine system IL-4 production by thymocytes has been investigated and linked to certain stages of T-cell thymic development. Whereas $CD4^+CD8^-HSA^{high}$ thymocytes produced only IL-2, their $CD4^+CD8^-HSA^{low}$ progeny produced high amounts of IL-4, IL-5, IL-10, and IFN-γ, compatible with a model postulating a thymocyte activation step as a basis for thymic selection (27), because the acquisition of competence to produce these cytokines has been linked to an activation and differentiation step (28) (see below). However, the mature virgin T cells leaving the thymus and populating the periphery again have a very restricted cytokine pattern producing only IL-2, indicating that the extended cytokine-producing phenotype is reversible.

IL-4 Production by Freshly Isolated T Cells

Ehlers and Smith showed that human neonatal T cells cannot synthesize IL-4–specific mRNA as detectable by PCR, yet such RNA can be detected after an in vitro stimulation course with a CD3 mAB followed by IL-2 (28). This finding was in line with the notion that IL-4 production is not the property of virgin T cells, but that of differentiated memory and/or effector cells (29), and indicates that the capacity to produce IL-4 is acquired during an IL-2–dependent differentiation step. Also in line with this model are several reports that describe only low IL-4 production by freshly isolated T cells of adult cell donors and an enhancement following an in vitro or an in vivo polyclonal activation course (30–34). The frequency of IL-4–producing cells increased substantially under such conditions. The culture conditions optimal for the generation of IL-4–producing cells in these studies were found to be continued engagement of the T-cell receptor (34) or culture in the presence of IL-4 (35–37). The addition of IFN-γ to the culture medium seems to inhibit the generation of IL-4–producing cells (35,38).

In the mouse a subpopulation of $CD4^+$ lymphocytes has been identified, which seems to be capable of producing IL-4 without a prior IL-4–dependent differentiation step. These cells have the phenotype $CD4^{+}NK1.1^{+}CD44^{bright}CD45RB^{dull}$ and could act as a first regulatory T-cell subset responsible, if activated, for the development of a subsequent Th2-type immune reaction (39).

Apart from the role that IL-2 apparently plays in the growth and differentiation of precursor cells, which develop into IL-4–producing cells (28,40,41), there is evidence that IL-2 is also necessary for the actual production phase of IL-4 following CD3 mAb stimulation. MAbs against IL-2 and the IL-2R interfered with IL-4 production in freshly isolated T cells as well as in the malignant cell line HUT78, the only malignant T-cell line described to produce high amounts of IL-4 (42,43).

Most of the studies reported in the literature used calcium ionophore plus PMA, plant lectins, and CD3 or CD2 mAbs as stimuli (29,31,44–49). Yet, there is evidence that the nature of the stimulating agent is very important for the amount of IL-4 that is induced.

An efficient route of IL-4 induction in adult human T cells was found to be stimulation by mAbs recognizing the T-cell surface molecules CD2 plus CD28 (42,50). In this system, IL-4 is also produced by memory cells. In mAb stimulation studies, CD28 co-stimulation increased IL-4 production approximately threefold, i.e., much less than IL-2 production (42). Similarly, co-stimulation with B7-1, a natural ligand binding to CD28, modestly increased IL-4 production by increasing the frequency of IL-4–producing cells in a model system employing a Th2 clone (51).

In freshly isolated T cells the protein kinase C activator PMA was found to inhibit IL-4 production induced by CD3 plus CD28 or CD2 plus CD28 stimulation, although it enhanced IL-2 production (50), whereas in preactivated cells, also PMA co-stimulated IL-4 production (52). The reason for this differential regulation of the IL-4 promoter by PMA has not been defined on the molecular level. In independent experiments, PMA-enhanced Con A induced IL-4 mRNA and protein synthesis in T cells. In this study, PMA was found to both augment the transcription rate of IL-4 and to increase the stability of the mRNA from a half-life of 90 min to one of more than 3 h (53).

Typically IL-4 production is a function of CD4+ T cells. However, it has been reported that CD8+ T cells can also be induced to secrete IL-4 following an appropriate in vitro preculture course (54). In the majority of cases IL-4–producing CD8+ T cell lines have a non-cytotoxic "suppresser" or helper phenotype (55–57), although IL-4 production by cytotoxic CD8+ T cells has been described (58).

With regard to the TCR phenotype, the majority of the IL-4–producing cells apparently carry the TCR-$\alpha\beta$. Recently, $\gamma\delta$ T-cell clones obtained from the synovial fluid and the peripheral blood of an arthritis patient were also found to produce varying amounts of IL-4 (59). Importantly, $\gamma\delta$+ T cells staining intracellularly for IL-4 were found in vivo early after infection with *Nippostrongylus brasiliensis,* which suggests an important role for these cells in establishing an appropriate primary immune response during infection (60).

Regulation of IL-4 Production by Drugs and Cytokines

As outlined above in the context of cell differentiation, IL-2 has a prominent role indirectly and directly in the regulation of IL-4–producing cells. Hence, all agents affecting IL-2 production will also influence IL-4 production. Surprisingly, corticosteroids, while suppressing IL-2 production, were reported to leave the production of IL-4 unimpaired in a murine model (61). In another study, however, it was reported that hydrocortisone could suppress the production of IL-4 by PHA plus PMA-stimulated human cells (45). Agents that increase the production of cAMP, including PGE_2, were reported to leave IL-4 production unaffected or even to enhance it, while inhibiting IL-2 production (62,64). One explanation for the in-part contradictory literature was provided by showing that, in situations where IL-2 production is limited, PGE_2 suppresses both IL-2 and (indirectly) IL-4 production, whereas under stimulatory conditions where IL-2 is produced in abundant quantities, IL-4 production becomes relatively insensitive to IL-2–inhibiting agents (65).

Concerning cytokines, IL-7 was shown to augment IL-4 production in ConA-stimulated human T cells (66). IL-7 itself did not induce IL-4 transcription but led to an increase in the half-life of ConA-induced IL-4 mRNA. One cytokine found to inhibit IL-4 production is TGF-β_1. This molecule has a variety of regulatory effects on cells of the immune system and is also produced by activated immune cells (67,68). Whereas T-cell growth is inhibited by TGF-β_1 through a nontoxic mechanism, IL-2 production is unaffected (69). Interestingly, IL-4 as well as IFN-γ production can be suppressed by TGF-β_1 in vitro (70). TGF-β_1 seems to inhibit signal transduction through the IL-2 receptor (70,71), a finding that might provide a mechanistic explanation for inhibition of IL-2–dependent cytokines such as IL-4.

IL-4 Production by T-Helper Subsets

Long-term cultured T-cell clones can be grouped according to their cytokine secretion pattern into Th1, Th2, and Th0 clones, of which only those with a Th2 or Th0 phenotype produce IL-4 (6). In addition to IL-4, Th2 clones typically also secrete IL-3, IL-5, IL-10, and GM-CSF, but not IL-2, IFN-γ, or TNG-β. These latter cytokines along with others are produced by Th1 and Th0 clones. Although this concept of T-cell differentiation was originally proposed for murine T-cell clones with IL-4 the prototypic cytokine for Th2 clones, the description of a number of human T-cell clones with the Th1 or the Th2 phenotype has extended it to the human system as well (72–75). On the other hand, several reports describe T-cell clones producing an undifferentiated spectrum of lymphokines (46,76,77). The existence of these clones has raised the question of whether the Th1/Th2 concept can be applied only to special cases of late T-cell differentiation or whether it is of relevance also for the majority of T lymphocytes involved in an in vivo immune response. Yet, several disease models both in mouse and humans suggest the existence of polarized immune responses, which as a whole may be classified with the help of the Th1/Th2 subtype concept (78–81).

The demonstration of functional differences between helper T-cell clones raises the question as to which signals lead to the generation of which phenotype. Clear evidence for a role of cytokines in this respect is provided by several polyclonal in vitro and in vivo models (35–37,82) as well as by a report on T-cell cloning by Maggi et al. (83). This latter study showed that a 5-day culture of PPD-stimulated human T cells in the presence of IL-4 prior to T-cell cloning strongly selected for the subsequent generation of Th2 T-cell clones, whereas in the absence of IL-4 Th1 T-cell clones were obtained. On the other hand, the addition of an anti–IL-4 antibody or of IFN-γ to allergen-stimulated mononuclear cells from an atopic patient biased the subsequently obtained T-cell clone repertoire toward an Th1-like phenotype. A similar decisive role for IL-4 for the generation of Th2 cells and for the suppression of Th1 cells has been shown in vitro using a murine antigen-specific transgenic TCR system (84).

It is also becoming clear from a number of recent studies that, in vivo, the nature of the stimulating antigen is important for the selection of the responding T-cell subset. Most T-cell clones specific for *Dermatophagoides pteronyssimus* or for *Toxocara canis* excretory-secretory antigen generated from atopic patients or healthy individuals secrete high levels of IL-4 and IL-5, whereas T-cell clones specific for tetanus toxoid or *Mycobacterium* secrete IFN-γ and IL-2 but no IL-4 (72,73). It is possible that the physicochemical properties of the stimulating antigen might be responsible for the selection of one T-cell phenotype versus the other (85,86). Probably of greatest practical importance, however, is the fact that the route of immunization and/or the nature of the antigen selects the antigen-presenting cell, which in turn influences T-cell differentiation by secreting appropriate cytokines.

Prominent among such APC-derived cytokines is IL-12. Interleukin-12 is produced primarily by monocytes and macrophages stimulated by bacteria or intracellular parasites (87–89). IL-12 induces IFN-γ production in T cells and NK cells. Such indirect as well as IFN-γ–independent mechanisms result in Th1 cell differentiation and suppression of IL-4–producing Th2 cells (88,90,91). When cells that are poor producers of IL-12, such as keratinocytes, present antigen to T cells, the resulting Th-cell development will be biased toward IL-4 production (92). Once fully differentiated Th-cell clones are established, they may be stimulated differently by different antigen-presenting cells, with B cells stimulating efficiently established Th2 clones and adherent cells activating Th1 clones (93).

Cytokine production depends on the genetic background of the host. Following infection with *Leishmania major,* BALB/c mice develop a Th2-type immune response with IL-4 mRNA-containing cells in the draining lymph node and in the spleen and suffer from progressive disease, whereas C57BL/6 mice, which are able to control the infection, develop a Th1-type response with strong expression of IFN-γ mRNA but no expression of IL-4 mRNA in either organ (94). In another model, mice with different MHC class II haplotypes respond differently to the same antigen with regard to IL-4 and IFN-γ production (95).

An important question with regard to Th-cell differentiation is whether Th1 and Th2 cells develop from a common precursor cell or whether there are separate precursor cells for each subset. In the latter case, cytokines would result in selective outgrowth of a specific subset. By now, evidence from at least two different experimental systems suggests that the first model is correct. When single cells were cloned and the resulting cell clone split and cultured under different conditions, the resulting subclones differed with regard to their potential to produce IL-4 (96). In a transgenic suicide model, in which IL-4–producing cells can be eliminated by feeding the drug ganciclovir, such treatment also destroys IFN-γ production, indicating that a common precursor cell exists that can produce IL-4 (97).

The existence of functionally differentiated helper subtypes leads to the question of molecular differences between such cells. Several lines of evidence show differences in signal transduction between Th1 and Th2 clones. Elevation of cytosolic free Ca^{2+} and the generation of inositol phosphates in response to ConA or to a T-cell receptor antibody was demonstrated in Th1 clones but not in Th2 clones in one study (98). In a subsequent investigation, a slow but sustained increase in Ca^{2+} levels was demonstrated in Th2 cells in the absence of tyrosine phosphorylation or PIP2 breakdown, a pattern different from that observed concomitantly in a Th1-cell clone

(99). CD28 co-stimulation induces IL-2 production in Th1 clones but not in Th2 cells. In the latter cells, however, the responsiveness to IL-4 is induced following CD28 triggering (100).

A clear functional difference between Th1 (IL-4–nonproducing) and Th2 (IL-4–producing) clones has also been found in terms of anergy induction. Early experiments had shown that a CD3 antibody induced unresponsiveness to IL-2 in Th1 clones, but not in Th2 cells (101). Whereas induction of anergy in Th1 cells lead to abrogation of a subsequent proliferative response to Ag plus APC, cell proliferation of Th2 cells remained unimpaired under such conditions (102). In another clone producing both IL-2 and IL-4 cells, anergy induction led to abrogation of a proliferative response to IL-4 and to abrogation of IL-2 production, but not of IL-4 production (103). During the differentiation phase both sets of clones seemed to be differently sensitive to anergy induction. Culture of small resting T cells in the presence of a CD3 mAb with or without IL-2 led to unresponsiveness to subsequent restimulation, whereas the presence of IL-4 during the primary culture prevented anergy induction and allowed cells to differentiate into IL-4–producing cells (104). These phenomena may provide the basis for tolerance induction in vivo.

IL-4 Production by Non–T Cells

In addition to T cells, recent reports show that IL-4 can be produced by murine mast cell lines as well as by freshly isolated murine non-T, non-B cells (105,106). Among isolated spleen cells, IL-4–producing cells lack CD3, CD4, CD8, Thy-1, B220, HLA Class II, and express Fc_ε receptors (107). Synthesis of IL-4 in these non–T cells can be induced by calcium ionophore or by cross-linking Fc_ε or Fc_γ receptors and is upregulated by IL-3. Splenic $Fc_\varepsilon R^+$ cells contain all the IL-4–producing cells and acquire the characteristics of basophils upon culture in IL-3 (108). Interestingly, the frequency of IL-4–producing cells among $Fc_\varepsilon R^+$ cells is 1 in 20 to 1 in 40 in naive animals and increases to 1 in 5 animals infected with *N. brasiliensis* or injected with anti-IgD (109). Human bone marrow–derived $Fc_\varepsilon R^+$ cells, which lack T- and B-cell markers, have also been reported to produce IL-4 in response to Fc receptor cross-linking (110). Finally, peripheral blood human basophils released IL-4 following a priming culture in IL-3 and cross-linking of their IgE receptor (111). B-cell IL-4 production has been described in malignant cell lines but not in normal, untransformed B cells (112,113).

BIOLOGICAL ACTIVITIES

B-Cell Function

IL-4 is an important cofactor for mature B-cell growth, which was in fact the characteristic feature for its initial isolation (1). Long-term B-cell cultures can be grown in vitro in response to IL-4 plus CD40 stimulation (114). Alone or together with co-stimulatory signals, IL-4 rescues B cells from apoptosis, thereby taking part in shaping the repertoire of immunoglobulin production by selecting B-cell clones for further differentiation and survival (115,116). Consistent with this was the demonstration of IL-4 mRNA in T cells isolated from germinal centers, the site where maturation of the antibody response occurs (117). Apart from co-stimulating the growth of mature B cells, it is likely that IL-4 also supports the development of immature B cells from bone marrow precursors (118,119). With regard to Ig production, IL-4 supports IgE and IgG1 in murine and IgE and IgG4 production in human cells in response to signals received through CD40 (120–123). The importance of IL-4 for immunoglobulin production was best demonstrated in the IL-4 knock-out mouse, in which no basal IgE or IgE is produced in response to parasite infection (124). Apart from its promoting effects on cell growth and Ig production, IL-4 enhances the expression of HLA II antigens, IgM, CD23, and Il-4 receptors on the surface of B cells (125–128). Furthermore, IL-4 induces B7-1 and B7-2 on small resting B cells, which function as important T-cell co-stimulatory molecules preventing anergy induction and inducing cytokine synthesis (129,130). This finding serves as an example of the interdependence of cell-bound and soluble signals between T and B cells during development of an immune response.

T-Cell Function

During thymic development IL-4 suppresses the frequency of $CD4^+CD8^+$ cells both in organ culture as well as in vivo in IL-4–overexpressing transgenic animals (131,132). In IL-4 transgenic mice, T-cell development seems to be profoundly disturbed, resulting in peripheral lymphopenia (132). In mature T cells, IL-4 is co-stimulatory under certain conditions and supports the differentiation of $CD4^+$ into Th2-type cells (see above) and simultaneously suppresses the development of $CD4^+$ Th1-cell clones with cytolytic potential (133). With regard to $CD8^+$ T-cell differentiation, IL-4 co-stimulates the development of $CD8^+$ cells together with IL-2 or IL-7 into antigen-specific cytotoxic cells while suppressing the IL-2–dependent generation of antigen-nonspecific lymphokine activated killer cells (134,135). Finally, IL-4 strongly co-stimulates IL-9 production by CD3 plus CD28-stimulated murine T cells in the presence of IL-2 and TGF-β (136).

Monocyte Function

On monocytes IL-4 acts in an anti-inflammatory manner. IL-4 suppresses the activation-induced synthesis of a number of monocytic cytokines, including IL-1, IL-6, IL-10, Il-12, G-CSF, M-CSF, IFN-β, and TNF-α (137–143). At the same time, the production of the IL-1 receptor antagonist (IL-1ra) is enhanced by IL-4 (144, 145). The latter effect can be observed in vivo, since serum IL-1ra levels have been found increased in tumor patients treated with IL-4 (146). However, the exact timing of monocyte exposure to IL-4 relative to activating stimuli may be very important. Only simultaneous incubation of monocytes with IL-4 and LPS or *Staphylococcus aureus* lead to inhibition of TNF-α and IL-12 production, whereas preincubation (priming) of the cells with IL-4 resulted in enhancement of cytokine production in this system (143). IL-4 also suppresses LPS-induced PGE_2 and nitrogen oxide production and the TGF-β–induced expression of CD16 in monocytes (137,147–149). Inhibition of monocyte generation of reactive oxygen metabolites by IL-4 results in decreased killing of microorganisms such as *Schistosoma mansoni* and *Leishmania donovani* (147,150). Finally, IL-4 enhances apoptosis and suppresses the growth of stimulated monocytes and supports the generation of dendritic cells from peripheral blood mononuclear cells (151–153).

Effects on Bone Marrow Cells

IL-4 together with other hematopoietic growth factors supports the growth of different lineages of bone marrow progenitor cells in soft agar cultures (154,155), with the exception of megakaryocytic progenitor cells, which are strongly inhibited (156). In stroma-cell–supported bone marrow cultures, IL-4 seems to inhibit lymphoid progenitor cell growth while promoting myeloid progenitors of immature phenotype (157–159). In stroma cells IL-4 inhibits IL-1 and GM-CSF production, thereby indirectly influencing progenitor cell differentiation (159), whereas it induces an enhanced expression of VCAM-1, resulting in an increased adhesion of $CD34^{pos}$ progenitor cells (160).

Effects on Fibroblasts and Endothelial Cells

Fibroblasts express IL-4 receptors, proliferate, and produce collagen and fibronectin in response to IL-4 (161–163). In addition, the production of inflammatory cytokines such as RANTES and IL-8 by fibroblasts is regulated by IL-4 (164). RANTES synthesis is also inhibited in endothelial cells by IL-4 (165). With regard to the expression of adhesion molecules regulating leukocyte migration and tissue infiltration, IL-4 upregulates VCAM-1 expression in large-vessel endothelial cells, but not in microvascular endothelial cells (166). IL-4 increases the adhesion of basophils and eosinophils, but not of neutrophils, to endothelial cells (167).

IL-4 in Infectious Diseases

Bacterial, Protozoal, and Parasitic Infections

IL-4 is a prototypic cytokine inducing humoral immune responses with the regulated synthesis of specific antibodies of high affinity and appropriate subclass. IL-4–producing Th2 cells are essential to fight extracellular parasites too large to be engulfed and killed intracellularly by macrophages. In these situations IL-4 leads to the preferential production of IgE, of IgG1 in mouse or IgG4 in humans, development of eosinophils through supporting the generation of IL-5–producing cells, and the inhibition of Th1-supported DTH responses. Early and strong Th2 responses are associated with protective and lasting immunity to parasites such as *Heligmosomoides polygyrus* (168,169).

The successful fight against many intracellular infectious agents, however, primarily depends on the efficient generation of a Th1-dominated immune response as exemplified in the disease model *Leishmania donovani* (78,94,170). In this disease, resistant animals mount a successful immune response of a Th1 type, whereas susceptible animals show an IL-4–dominated Th2-type response and succumb to the infection. Correspondingly, susceptible animals can be cured by intervention with an anti–IL-4 mAb or Th1-inducing cytokines such as IL-12 (171–173). Interestingly, however, in mice resistant to *L. donovani* one can demonstrate the presence of an antigen-specific Th2-type response together with the dominating Th1-type response. The latter appears primarily responsible for elimination of the infection, whereas the former seems to prevent autoimmune complications of the anti-*Leishmania* Th1 response. The evidence for such a model comes from experiments in which the transfer of the antigen-specific Th1 response to a syngenic animal resulted in a severe colitis abrogated by cotransfer of Th2 type cells (174). It seems that although a certain polarization of the immune response is important for fighting infectious agents, a finely tuned balance of regulatory T-cell cytokines including IL-4 is necessary to prevent harm to the host at the same time.

Fungal Infections

As in protozoal infections, murine candidiasis is associated with an IL-4–dominated Th2 response in susceptible animals, whereas successful fight against the fungus is Th1 dominated and associated with IL-12 production (175). In accordance with this, anti–Il-4 treatment cures mice infected with *Candida albicans* and decreases IL-4 production in vivo while elevating IFN-γ production (176).

Viral Infections

Protective immunity against viruses is often of a Th1 type, but sometimes viral infection leads to an IL-4–dominated Th2 cytokine pattern in the host. Measles virus infection results in elevated serum IL-4 levels and high IL-4 production in vitro. This is associated with a strong antibody response against the virus and cutaneous anergy lasting for weeks following infection (177). In RSV infection, immunization with inactivated virus leads to a Th2-type response, and treatment with anti–IL-4 mAb during vaccination but not during challenge with live virus reduces clinical illness following infection (178).

Recently, the type of immune response elicited by HIV infection and whether a Th2-type response is advantageous or detrimental in the progression to AIDS have attracted considerable interest. Studying peripheral blood T cells, evidence has been presented that during HIV infection a gradual shift from a Th1-type cytokine pattern to a Th2-type one with elevated levels of IL-4 and IL-10 occurs, along with a loss of cutaneous reactivity, appearance of hyperimmunglobulinemia, and loss of specific peripheral blood T-cell reactivity (7,179,180). However, a general shift to Th2 cytokine production during progression of AIDS has been questioned because no general elevation of Th2 cytokine mRNA expression was found in lymph nodes of HIV patients, the principal anatomical site of virus replication (181). A third group has reported a somewhat in-between result, with a shift from a Th1- to a Th0-type response during progression of the disease (182). Most interestingly, in the latter study, HIV replication was observed in Th2 and Th0 clones from infected individuals, but not in Th1 T-cell clones (182). This finding provides an attractive basis for the hypothesis that a strong Th1-type response, as found in exposed but seronegative persons, is of advantage for infected individuals (7). In line with the evidence obtained from human patients, it was found that mice deficient in IL-4 are partially resistant to the progression of MAIDS, a murine retroviral model disease for HIV infection (183). Altogether, the evidence strongly suggests that IL-4 production in human viral diseases is disadvantageous for the host.

THE IL-4 RECEPTOR

IL-4 binds to its target cells through specific high-affinity receptors. The genes for both the human and the murine receptors have been cloned and localized to chromosome 16 (p11.2-p12.1) and chromosome 7, respectively (184–187). In accordance with the wide spectrum of IL-4–responsive cells (see above), the receptors are expressed on freshly isolated T and B cells, monocytes, fibroblasts, mast cells, stroma cell lines, endothelial cells, and a large variety of different cancer cell lines. Expression of IL-4R is rapidly upregulated in lymphocytes following activation or incubation of the cells in IL-4 (127,188).

The IL-4 receptor is a member of the hematopoietin receptor superfamily with the main IL-4–binding receptor chain a 140 kDa glycoprotein containing the characteristic WSXWS motif. It is 825 amino acids in length, the major part (553 AA) of the molecule being intracellular (186). The intracellular part of the molecule contains no consensus sequences characteristic of serine/threonine or tyrosine kinases. However, mutation studies have identified two regions indispensable for signaling in the pre–B-cell line Ba/F3 and the myeloid progenitor cell line 32D between AA 353–390 and 487–605 (189).

Evidence from different experimental systems suggests that the IL-4–binding receptor chain may not be the only component of the IL-4R complex, similar to other multichain receptor complexes of the hematopoietin receptor superfamily. Cross-linking studies using ^{125}I-labeled IL-4 have demonstrated, in addition to the 140 kDa IL-4–binding receptor chain, the presence of smaller molecules, some but not all of which may represent degradation products of the 140 kDa chain (190). Cross-linking studies suggested that the 75 kDa IL-25R γ-chain might be a component of the IL-4R (191). Moreover, transfection of the IL-2R γ-chain into COS-7 cells transfected with the IL-4R chain increased their binding affinity and number of binding sites for IL-4 (191). On the other hand, B cells from patients with X-linked severe combined immunodeficiency, which is caused by a defect of the IL-2R γ-chain, proliferate in response to IL-4, indicating that the IL-2R γ-chain is not an absolute prerequisite for the function of the IL-4R (192). The B9 plasmacytoma cell line does not express the IL-2R γ-chain, yet proliferates in response to IL-4. Following transfection of the γ-chain, however, transfected cells are 10-fold more sensitive to IL-4, indicating that the transfected γ-chain can participate in the IL-4 receptor and increase its affinity for IL-4 (193). It is therefore likely that two types of IL-4R exist. These receptors may differ in affinity as well as signal transduction, allowing highly differentiated cellular responses to the cytokine to occur in a regulated manner (193,194).

The signaling pathway of the IL-4R is currently under intensive study. Early biochemical signals detected include tyrosine phosphorylation of a number of target proteins including the IL-4R–binding chain (195) and Jak kinases

(196). In myeloid cells, an important component of the IL-4R–signaling pathway appears to be the 170 kDa 4PS, which is closely related to the IRS protein, earlier found in the insulin and insulinlike growth factor 1 receptor pathway. IL-4 results in the phosphorylation of 4PS, which in turn may mediate the association of the IL-4R to further signaling proteins such as PI 3-kinase (reviewed in Ref. 189). Finally, activation in part through phosphorylation of a number of DNA-binding proteins occurs, such as Stat (197), IL-4 NAF (198), or STF-IL-4 binding to the IFN target sequence GAS (199), thereby mediating activation or inhibition of the transcription of IL-4 target genes.

In addition to the cell-bound receptor, a splice variant resulting in a soluble 40 kDa IL-4R exists (189,200). Because the dissociation rate of IL-4 from the soluble receptor is much faster than that from the cell-bound receptor, it has been suggested that the soluble molecule might function as an IL-4 transporter protein (201). Interestingly, exogenously applied soluble IL-4 receptors could block the effect of endogenously produced IL-4 but at certain concentrations potentiated the effect of exogenously supplied IL-4 (202). These results also suggest that the naturally occurring soluble receptor functions as a biologically important regulatory molecule of IL-4.

REFERENCES

1. Howard M, Farrar J, Hilfiker M, Johnson B, Takatsu K, Hamaoka T, Paul WE. Identification of a T-cell derived B-cell growth factor distinct from interleukin 2. J Exp Med 1982; 155:914–923.
2. Yokota T, Otsuka T, Mosmann T, Banchereau J, de France T, Blanchard D, de Vries JE, Lee F, Arai K-I. Isolation and characterization of a human interleukin cDNA clone, homologous to mouse B-cell stimulatory factor 1, that expresses B-cell- and T-cell-stimulating activities. Proc Natl Acad Sci USA 1986; 83:5894–5898.
3. Noma Y, Sideras T, Naito T, Bergstedt-Lindqvist S, Azuma C, Severinson E, Tanabe T, Kinashi T, Matsuda F, Yaoita T, Honjo T. Cloning of cDNA encoding the murine IgG1 induction factor by a novel strategy using SP6 promoter. Nature 1986; 319:640–646.
4. Lee F, Yokota T, Otsuka T, Meyerson P, Villaret D, Coffman R, Mosmann T, Rennick D, Roehm N, Smith C, Zlotnick A, Arai K. Isolation and characterization of a mouse interleukin cDNA clone that expresses B-cell stimulatory factor 1 activities and T-cell-and mast-cell-stimulating activities. Proc Natl Acad Sci USA 1986; 83:2061–2065.
5. Paul WE. Interleukin-4: a prototypic immunoregulatory lymphokine. Blood 1991; 77:1859–1870.
6. Mosmann TR, Coffman RL. TH1 and TH2 cells: different patterns of lymphokine secretion lead to different functional properties. Ann Rev Immunol 1989; 7:145–173.
7. Clerici M, Shearer GM. A Th1-Th2 switch is a critical step in the etiology of HIV infection. Immunol Today 1993; 14:107–111.
8. Grabstein K, Eisenman J, Mochizuki D, Shanebeck K, Conlon P, Hopp T, March C, Gillis S. Purification to homogeneity of B cell stimulating factor. J Exp Med 1986; 163:1405–1414.
9. Walter MR, Cook WJ, Zhao BG, Cameron RPJ, Ealick SE, Walter RLJ, Reichert P, Nagabhushan TL, Trotta PP, Bugg CE. Crystal structure of recombinant human interleukin-4. J Biol Chem 1992; 267:20371–20376.
10. Wlodaver A, Pavlovsky A, Gustchina A. Crystal structure of human recombinant interleukin-4 at 2.25 A resolution. FEBS Lett 1992; 309:59–64.
11. Powers R, Garrett DS, March CJ, Frieden EA, Gronenborn AM, Clore GM. Three-dimensional solution structure of human interleukin-4 by multidimensional heteronuclear magnetic resonance spectroscopy. Science 1992; 256:1673–1677.
12. Hoffman RC, Schalk-Hihi C, Castner BJ, Gibson MG, Rasmussen BD, Zdanov A, Gustchina A, March CJ, Wlodawer A. Stoichiometry of the complex of human interleukin-4 with its receptor. FEBS Lett 1994; 347:17–21.
13. Kruse N, Shen BJ, Arnold S, Tony HP, Muller T, Sebald W. Two distinct functional sites of human interleukin 4 are identified by variants impaired in either receptor binding or receptor activation. EMBO J 1993; 12:5121–5129.
14. Ramanathan L, Ingram R, Sullivan L, Greenberg R, Reim R, Trotta PP, Le HV. Immunochemical mapping of domains in human interleukin 4 recognized by neutralizing monoclonal antibodies. Biochemistry 1993; 32:3549–3556.
15. D'Eustachio P, Brown M, Watson C, Paul WE. The IL-4 gene maps to chromosome 11, near the gene encoding IL-3. J Immunol 1988; 141:3067–3071.
16. Le Beau MM, Lemons RS, Espinosa R3, Larson RA, Arai N, Rowley JD. Interleukin-4 and interleukin-5 map to human chromosome 5 in a region encoding growth factors and receptors and are deleted in myeloid leukemias with a del(5q). Blood 1989; 73:647–650.
17. Zurawski G, de Vries JE. Interleukin 13, an interleukin 4-like cytokine that acts on monocytes and B cells, but not on T cells. Immunol Today 1994; 15:19–26.
18. Arai N, Nomura D, Villaret C, de Waal Malefijt R, Seiki M, Yoshida M, Minoshima S, Fukuyama R, Maekawa M, Kudoh J, Shimizu N, Yokota K, Abe E, Yokota T, Takebe T, Arai K. Complete nucleotide sequence of the chromosomal gene for human IL-4 and its expression. J Immunol 1989; 142:274–282.
19. Abe E, De Waal Malefyt R, Matsuda I, Arai K-I, Arai N. An 11-base-pair DNA sequence motif apparently unique to the human interleukin 4 gene confers responsiveness to T-cell activation signals. Proc Natl Acad Sci USA 1992; 89:2864–2868.
20. Li-Weber M, Eder A, Czepa-Krafft H, Krammer PH. T cell-specific negative regulation of transcription of the human cytokine IL-4. J Immunol 1992; 148:1913–1918.
21. Li-Weber M, Krafft H, Krammer PH. A novel enhancer element in the human IL-4 promoter is suppressed by a position-independent silencer. J Immunol 1993; 151:1371–1382.
22. Todd MD, Grusby MJ, Lederer JA, Lacy E, Lichtman AH, Glimcher LH. Transcription of the interleukin 4 gene is regulated by multiple promoter elements. J Exp Med 1993; 177:1663.
23. Szabo SJ, Gold JS, Murphy TL, Murphy KM. Identification of cis-acting regulatory elements controlling interleukin-4 gene expression in T cells: roles for NF-Y and NF-ATc. Mol Cell Biol 1993; 13:4793–4805.
24. Bruhn KW, Nelms K, Boulay JL, Paul WE, Lenardo MJ. Molecular dissection of the mouse interleukin-4 promoter. Proc Natl Acad Sci USA 1993; 900:9707–9711.
25. Shaw G, Kamen R. A conserved AU sequence from the 3′

untranslated region of GM-CSF mRNA mediates selective mRNA degradation. Cell 1986; 46:659–667.
26. Sorg RV, Enczmann J, Sorg UR, Schneider EM, Wernet P. Identification of an alternatively spliced transcript of human interleukin-4 lacking the sequence encoded by exon 2. Exp Hematol 1993; 21:560–563.
27. Bendelac A, Matzinger P, Seder RA, Paul WE, Schwartz RH. Activation events during thymic selection. J Exp Med 1992; 175:731–742.
28. Ehlers S, Smith KA. Differentiation of T cell lymphokine gene expression: the in vitro acquisition of T cell memory. J Exp Med 1991; 173:25–36.
29. Lewis DB, Prickett KS, Larsen A, Grabstein K, Weaver M, Wilson CB. Restricted production of interleukin 4 by activated human T cells. Proc Natl Acad Sci USA 1988; 85:9743–9747.
30. Powers G, Abbas AK, Miller RA. Frequencies of IL-2- and IL-4-secreting T cells in naive and antigen-stimulating lymphocyte populations. J Immunol 1988; 140:3352–3357.
31. Bettens F, Walker C, Gauchat J-F, Wyss T, Pichler WJ. Lymphokine gene expression related to CD4 T cell subset (CD45R/CDw29) phenotype conversion. Eur J Immunol 1989; 19:1569–1574.
32. Seder RA, Le Gros G, Ben-Sasson SZ, Urban Jr. J, Finkelman FD, Paul WE. Increased frequency of interleukin 4-producing T cells as a result of polyclonal priming. Use of a single-cell assay to detect interleukin 4-producing cells. Eur J Immunol 1991; 21:1241–1247.
33. Röcken M, Müller KM, Saurat J-H, Hauser C. Lectin-mediated induction of IL-4-producing $CD4^+$ T cells. J Immunol 1991; 146:577–584.
34. Röcken M, Müller KM, Saurat J-H, Müller I, Louis JA, Cerottini J-C, Hauser C. Central role for TCR/CD3 ligation in the differentiation of $CD4^+$ T cells toward a Thl or Th2 functional phenotype. J Immunol 1992; 148:47–54.
35. Betz M, Fox BS. Regulation and development of cytochrome c-specific IL-4-producing T cells. J Immunol 1990; 145:1046–1052.
36. Schmitt E, Van Brandwijk R, Fischer H-G, Rüde E. Establishment of different T cell sublines using either interleukin 2 or interleukin 4 as growth factors. Eur J Immunol 1990; 20:1709–1715.
37. Swain SL, Weinberg AD, English M, Huston G. IL-4 directs the development of Th2-like helper effectors. J Immunol 1990; 145:3796–3806.
38. Gajewski TF, Joyce J, Fitch FW. Antiproliferative effect of IFN-γ in immune regulation. III. Differential selection of T_H1 and T_H2 murine helper T lymphocyte clones using recombinant IL-2 and recombinant IFN-γ. J Immunol 1989; 143:15–22.
39. Yoshimoto T, Paul WE. CD4pos, NK1.1pos T cells promptly produce interleukin 4 in response to in vivo challenge with anti-CD3. J Exp Med 1994; 179:1285–1295.
40. Ben-Sasson SZ, Le Gros G, Conrad DH, Finkelman FD, Paul WE. IL-4 production by T cells from naive donors: IL-2 is required for IL-4 production. J Immunol 1990; 145:1127–1136.
41. Le Gros G, Ben-Sasson SZ, Seder R, Finkelman FD, Paul WE. Generation of interleukin 4 (IL-4)-producing cells in vivo and in vitro: IL-2 and IL-4 are required for in vitro generation of IL-4 producing cells. J Exp Med 1990; 172:921–929.
42. Holter W, Majdic O, Kalthoff FS, Knapp W. Regulation of interleukin-4 production in human mononuclear cells. Eur J Immunol 1992; 22:2765–2767.
43. Holter W, Kalthoff FS, Majdic O, Knapp W. Interleukin-4 production by the malignant cell line HUT78. Lymphok Cytok Res 1993; 12:33–37.
44. Andersson U, Andersson J, Lindfors A, Wagner K, Möller G, Heusser CH. Simultaneous production of interleukin 2, interleukin 4 and interferon-γ by activated human blood lymphocytes. Eur J Immunol 1990; 20:1591–1596.
45. Wu CY, Fargeas C, Nakajima T, Delespesse G. Glucocorticoids suppress the production of interleukin 4 by human lymphocytes. Eur J Immunol 1991; 21:2645–2647.
46. Salmon M, Kitas GD, Bacon PA. Production of lymphokine mRNA by $CD45R^+$ and $CD45R^-$ helper T cells from human peripheral blood and by human $CD4^+$ T cell clones. J Immunol 1989; 143:907–912.
47. Carding SR, West J, Woods A, Bottomly K. Differential activation of cytokine genes in normal CD4-bearing T cells is stimulus dependent. Eur J Immunol 1989; 19:231–238.
48. Flamand V, Abramowicz D, Goldman M, Biernaux C, Huez G, Urbain J, Moser M, Oberdan L. Anti-CD3 antibodies induce T cells from unprimed animals to secrete IL-4 both in vitro and in vivo. J Immunol 1990; 144:2875–2882.
49. Valentin H, Groux H, Gelin C, Chretien I, Bernard A. Modulation of lymphokine release and cytolytic activities by activating peripheral blood lymphocytes via CD2. J Immunol 1990; 144:875–882.
50. Van der Pouw-Kraan T, Van Kooten C, Rensink I, Aarden L. Interleukin (IL)-4 production by human T cells: Differential regulation of IL-4 vs. IL-2 production. Eur J Immunol 1992; 22:1237–1241.
51. Bucy RP, Panoskaltis-Mortari A, Huang GQ, Li J, Karr L, Ross M, Russell JH, Murphy KM, Weaver CT. Heterogeneity of single cell cytokine gene expression in clonal T cell populations. J Exp Med 1994; 180:1251–1261.
52. Van der Pouw Kraan T, de Jong R, Aarden L. Development of human Thl and Th2 cytokine responses: the cytokine production profile of T cells is dictated by the primary in vitro stimulus. Eur J Immunol 1993; 23:1–5.
53. Dokter WH, Esselink MT, Sierdsema SJ, Halie MR, Vellenga E. Transcriptional and postranscriptional regulation of the interleukin-4 and interleukin-3 genes in human T cells. Blood 1993; 81:35–40.
54. Seder RA, Boulay J-L, Finkelman F, Barbier S, Ben-Sasson SZ, Le Gros G, Paul WE. $CD8^+$ T cells can be primed in vitro to produce IL-4. J Immunol 1992; 148:1652–1656.
55. Salgame P, Abrams JS, Clayberger C, Goldstein H, Convit J, Modlin RL, Bloom BR. Differing lymphokine profiles of functional subsets of human CD4 and CD8 T cell clones. Science 1991; 254:279–282.
56. Le Gros G, Erard F. Non-cytotoxic, IL-4, IL-5, IL-10 producing CD8+ T cells: their activation and effector functions. Curr Opin Immunol 1994; 6:453–457.
57. Grubeck-Loebenstein B, Trieb K, Sztankay A, Holter W, Anderl H, Wick G. Retrobulbar T cells from patients with Graves' ophthalmopathy are CD8+ and specifically recognize autologous fibroblasts. J Clin Invest 1994; 93:2738–2743.
58. Maccalli C, Mortarini R, Parmiani G, Anichini A. Multiple sub-sets of CD4+ and CD8+ cytotoxic T-cell clones directed to autologous human melanoma identified by cytokine profiles. Int J Cancer 1994; 5:56–62.
59. Chomarat P, Kjeldsen-Kragh J, Quayle AJ, Natvig JB, Miossec P. Different cytokine production profiles of gamma delta T cell clones: relation to inflammatory arthritis. Eur J Immunol 1994; 24:2087–2091.
60. Ferrick DA, Schrenzel MD, Mulvania T, Hsieh B, Ferlin WG, Lepper H. Differential production of interferon-gamma and interleukin-4 in response to Thl- and Th2 stimulating pathogens by gamma delta T cells in vivo. Nature 1995; 373:255–257.
61. Daynes RA, Araneo BA. Contrasting effects of gluco-

corticoids on the capacity of T cells to produce the growth factors interleukin 2 and interleukin 4. Eur J Immunol 1989; 19:2319–2325.

62. Betz M, Fox BS. Prostaglandin E_2 inhibits production of Th1 lymphokines but not of Th2 lymphokines. J Immunol 1991; 146:108–113.

63. Muñoz E, Zubiaga AM, Merrow M, Sauter NP, Huber BT. Cholera toxin discriminates between T helper 1 and 2 cells in T cell receptor-mediated activation: Role of cAMP in T cell proliferation. J Exp Med 1990; 172:95–103.

64. Novak, TJ, Rothenberg EV. cAMP inhibits induction of interleukin 2 but not of interleukin 4 in T cells. Proc Natl Acad Sci USA 1990; 87:9353–9357.

65. Hilkens CM, Vermeulen H, van Neerven RJ, Snijdewint FG, Wierenga EA, Kapsenberg ML. Differential modulation of T helper type 1 (Th1) and T helper type 2 (Th2) cytokine secretion by prostaglandin E2 critically depends on interleukin-2. Eur J Immunol 1995; 25:59–63.

66. Dokter WH, Sierdsema SJ, Esselink MT, Halie MR, Vellenga E. Interleukin-4 mRNA and protein in activated human T cells are enhanced by interleukin-7. Exp Hematol 1994; 22:74–79.

67. Kehrl JH, Wakefield LM, Roberts AB, Jakowlew S, Alvarez-Mon M, Derynck R, Sporn MB, Fauci AS. Production of transforming growth factor β by human T lymphocytes and its potential role in the regulation of T cell growth. J Exp Med 1986; 163:1037–1050.

68. Assoian RK, Fleurdelys BE, Stevenson HC, Miller PJ, Madtes DK, Raines EW, Ross R, Sporn MB. Expression and secretion of type β transforming growth factor by activated human macrophages. Proc Natl Acad Sci USA 1987; 84:6020–6024.

69. Wahl SM, Hunt DA, Wong HL, Dougherty S, McCartney-Francis N, Wahl LM, Ellingsworth L, Schmidt JA, Hall G, Roberts AB, Sporn MB. Transforming growth factor-β is a potent immunosuppressive agent that inhibits IL-1-dependent lymphocyte proliferation. J Immunol 1988; 140:3026–3032.

70. Holter W, Kalthoff FS, Pickl WF, Ebener C, Majdic O, Kraft D, Knapp W. Transforming growth factor-β inhibits IL-4 and IFN-γ production by stimulated human T cells. Int Immunol 1994; 6:469–475.

71. Ortaldo JR, Mason AT, O'Shea JJ, Smyth MJ, Falk LA, Kennedy ICS, Longo DL, Ruscetti FW. Mechanistic studies of transforming growth factor-β inhibition of IL-2-dependent activation of $CD3^-$ large granular lymphocyte functions. Regulation of IL-2Rβ (p75) signal transduction. J Immunol 1991; 146:3791–3798.

72. Wierenga EA, Snoek M, De Groot C, Chrétien I, Bos JD, Jansen HM, Kapsenberg ML. Evidence for compartmentalization of functional subsets of $CD4^+$ T lymphocytes in atopic patients. J Immunol 1990; 144:4651–4656.

73. Del Prete G, De Carli M, Mastromauro C, Biagiotti R, Macchia D, Falagiani P, Ricci M, Romagnani S. Purified protein derivative of *Mycobacterium tuberculosis* and excretory-secretory antigen(s) of *Toxocara canis* expand in vitro human T cells with stable and opposite (type 1 T helper or type 2 T helper) profile of cytokine production. J Clin Invest 1991; 88:346–350.

74. Kapsenberg ML, Wierenga EA, Bos JD, Jansen HM. Functional subsets of allergen-reactive human $CD4^+$ T cells. Immunol Today 1991; 12:392–395.

75. Romagnani S. Human T_H1 and T_H2 subsets: Doubt no more. Immunol Today 1991; 12:256–257.

76. Paliard X, de Waal Malefijt R, Yssel H, Blanchard D, Chrétien I, Abrams J, de Vries J, Spits H. Simultaneous production of IL-2, IL-4, and IFN-γ by activated human $CD4^+$ and $CD8^+$ T cell clones. J Immunol 1988; 141:849–855.

77. Yasukawa M, Inatsuki A, Horiuchi T, Kobayashi Y. Functional heterogeneity among herpes simplex virus-specific human $CD4^+$ T cells. J Immunol 1991; 146:1341–1347.

78. Heinzel FP, Sadick MD, Holaday BJ, Coffman RL, Locksley RM. Reciprocal expression of interferon γ or interleukin 4 during the resolution or progression of murine leishmaniasis. J Exp Med 1989; 169:59–72.

79. Haanen JBAG, de Waal Malefijt R, Res PCM, Kraakman EM, Ottenhoff THM, de Vries RRP, Spits H. Selection of a human T helper type 1-like T cell subset by mycobacteria. J Exp Med 1991; 174:583–592.

80. Yamamura M, Uyemura K, Deans RJ, Weinberg K, Rea TH, Bloom BR, Modlin RL. Defining protective responses to pathogens: cytokine profiles in leprosy lesions. Science 1991; 254:277–279.

81. Robinson DS, Hamid Q, Ying S, Tsicopoulos A, Barkans J, Bentley AM, Corrigan C, Durham SR, Kay AB. Predominant T_{H2}-like bronchoalveolar T-lymphocyte population in atopic asthma. N Engl J Med 1992; 326:298–304.

82. Scott P. IFN-γ modulates the early development of Th1 and Th2 responses in a murine model of cutaneous leishmaniasis. J Immunol 1991; 147:3149–3155.

83. Maggi E, Parronchi P, Manetti M, Simonelli C, Piccinni M-P, Rugiu FS, De Carli M, Ricci M, Romagnani A. Reciprocal regulatory effects of IFN-γ and IL-4 on the in vitro development of human Th1 and Th2 clones. J Immunol 1992; 1482142–2147.

84. Seder RA, Paul WE, Davis MM, Fazekas-de-St.-Groth B. The presence of interleukin 4 during in vitro priming determines the lymphokine producing potential of CD4+ T cells from T cell receptor transgenic mice. J Exp Med 1992; 176:1091–1098.

85. Baum CG, Szabo P, Siskind GW, Becker CG, Firpo A, Clarick CJ, Francus T. Cellular control of IgE induction by a polyphenol-rich compound: preferential activation of Th2 cells. J Immunol 1990; 145:779–784.

86. Liew FY, Millott SM, Schmidt JA. A repetitive peptide of *Leishmania* can activate T helper type 2 cells and enhance disease progression. J Exp Med 1990; 172:1359–1365.

87. D'Andrea A, Rengaraju M, Valiante NM, Chehimi J, Kubin M, Aste M, Chan SH, Kobayashi M, Young D, Nickbarg E, Chizzonite R, Wolf SF, Trinchieri G. Production of natural killer cell stimulatory factor (interleukin 12) by peripheral blood mononuclear cells. J Exp Med 1992; 176:1387–1398.

88. Hsieh C-S, Macatonia SE, Tripp CS, Wolf SF, O'Garra A, Murphy KM. Development of Th1 CD4+ T cells through IL-12 produced by *Listeria*-induced macrophages. Science 1993; 260:547–549.

89. Grazzinelli RT, Hieny S, Wynn TA, Wolf S, Sher A. Interleukin 12 is required for the T-lymphocyte-independent induction of interferon gamma by an intracellular parasite and induces resistance in T-cell-deficient hosts. Proc Natl Acad Sci USA 1993; 90:6115–6119.

90. Manetti R, Parronchi P, Giudizi MG, Piccinni MP, Maggi E, Trinchieri G, Romagnani S. Natural killer cell stimulatory factor (interleukin-12 [IL-12]) induces T helper type 1 (Th1)-specific immune responses and inhibits the development of IL-4-producing Th cells. J Exp Med 1993; 177:1199–1204.

91. Wang Z, Zheng S, Corry DB, Dalton DK, Seder RA, Reiner SL, Locksley RM. Interferon-gamma-independent effects of interleukin-12 administered during acute or established infection with Leishmania major. Proc Natl Acad Sci USA 1994; 91:12932–12936.

92. Goodman RE, Nestle F, Naidu YM, Green JM, Thompson CB, Nickoloff BJ, Turka LA. Keratinocyte-derived T cell co-stimulation induces preferential production of IL-2 and IL-4 but not IFN-gamma. J Immunol 1994; 152:5189–5198.

93. Gajewski TF, Pinnas M, Wong T, Fitch FW. Murine Th1 and

Th2 clones proliferate optimally in response to distinct antigen-presenting cell populations. J Immunol 1991; 146:1750–1758.
94. Heinzel FP, Sadick MD, Mutha SS, Locksley RM. Production of interferon γ, interleukin 2, interleukin 4, and interleukin 10 by CD4+ lymphocytes *in vivo* during healing and progressive murine leishmaniasis. Proc Natl Acad Sci USA 1991; 88:7011–7015.
95. Murray JS, Madri J, Tite J, Carding SR, Bottomly K. MHC control of CD4+ T cell subset activation. J Exp Med 1989; 170:2135–2140.
96. Rocken M, Saurat JH, Hauser C. A common precursor for CD4+ T cells producing IL-2 or IL-4. J Immunol 1992; 148:1031–1036.
97. Kamogawa Y, Minasi LA, Carding SR, Bottomly K, Flavell RA. The relationship of IL-4- and IFN gamma-producing T cells studied by lineage ablation of IL-4-producing cells. Cell 1993; 75:985–995.
98. Gajewski TF, Schell SR, Fitch FW. Evidence implicating utilization of different T cell receptor-associated signaling pathways by T_H1 and T_H2 clones. J Immunol 1990; 144:4110–4120.
99. Tamura T, Yanagida T, Nariuchi H. Difference in signal transduction pathway for IL-2 and IL-4 production in T helper 1 and T helper 2 cell clones in response to anti-CD3. J Immunol 1993; 151:6051–6061.
100. McArthur JG, Raulet DH. CD28-induced costimulation of T helper type 2 cells mediated by induction of responsiveness to interleukin 4. J Exp Med 1993; 178:1645–1653.
101. Williams ME, Lichtman AH, Abbas AK. Anti-CD3 antibody induces unresponsiveness to IL-2 in Th1 clones but not in Th2 clones. J Immunol 1990; 144:1208–1214.
102. Gilbert KM, Hoang KD, Weigle WO. Th1 and Th2 clones differ in their response to a tolerogenic signal. J Immunol 1990; 144:2063–2071.
103. Mueller DL, Chiodetti L, Bacon PA, Schwartz RH. Clonal anergy blocks the response to IL-4, as well as the production of IL-2, in dual-producing T helper cell clones. J Immunol 1991; 147:4118–4125.
104. Williams ME, Shea CM, Lichtman AH, Abbas AK. Antigen receptor-mediated anergy in resting T lymphocytes and T cell clones. Correlation with lymphokine secretion patterns. J Immunol 1992; 149:1921–1926.
105. Plaut M, Pierce JH, Watson CJ, Hanley-Hyde J, Nordan RP, Paul WE. Mast cell lines produce lymphokines in response to cross-linkage of FcεRI or to calcium ionophores. Nature 1989; 339:64–67.
106. Le Gros G, Ben-Sasson SZ, Conrad DH, Clark-Lewis I, Finkelman FD, Plaut M, Paul WE. IL-3 promotes production of IL-4 by splenic non-B, non-T cells in response to Fc receptor cross-linkage. J Immunol 1990; 145:2500–2506.
107. Ben-Sasson SZ, Le Gros G, Conrad DH, Finkelman FD, Paul WE. Cross-linking Fc receptors stimulate splenic non-B, non-T cells to secrete interleukin 4 and other lymphokines. Proc Natl Acad Sci USA 1990; 87:1421–1425.
108. Seder RA, Paul WE, Dvorak AM, Sharkis SJ, Kagey-Sobotka A, Niv Y, Finkelman FD, Barbieri SA, Galli SJ, Plaut M. Mouse splenic and bone marrow cell populations that express high-affinity $Fc_ε$ receptors and produce interleukin 4 are highly enriched in basophils. Proc Natl Acad Sci USA 1991; 88:2835–2839.
109. Seder RA, Plaut M, Barbieri S, Urban Jr. J, Finkelman FD, Paul WE. Purified FcεR+ bone marrow and splenic non-B, non-T cells are highly enriched in the capacity to produce IL-4 in response to immobilized IgE, IgG2a, or ionomycin. J Immunol 1991; 147:903–909.
110. Piccinni M-P, Macchia D, Parronchi P, Giudizi MG, Bani D, Alterini R, Grossi A, Ricci M, Maggi E, Romagnani S. Human bone marrow non-B, non-T cells produce interleukin 4 in response to cross-linkage of Fcε and Fcγ receptors. Proc Natl Acad Sci USA 1991; 88:8656–8660.
111. Brunner T, Heusser CH, Dahinden CA. Human peripheral blood basophils primed by interleukin 3 (IL-3) produce IL-4 in response to immunoglobulin E receptor stimulation. J Exp Med 1993; 177:605–611.
112. Rao TD, Maghazachi AA, Faria AM, Basch RS, Phillips-Quagliata JM. T560: an (H-2b x H-2a) F1 hybrid, phosphorylcholine (PC)-binding, murine B cell lymphoma that bears receptors for IgA and IgG, presents antigen and secretes IL-4. Int Immunol 1992; 4:107–118.
113. Louie SW, Ramirez LM, Krieg AM, Maliszewski CR, Bishop GA. Endogenous secretion of IL-4 maintains growth and Thy-1 expression of a transformed B cell clone. J Immunol 1993; 150:399–406.
114. Bancherau J, de-Paoli P, Valle A, Garcia E, Rousset F. Long-term human B cell lines dependent on interleukin-4 and antibody to CD40. Science 1991; 251:70–72.
115. Mangeney M, Richard Y, Coulaud D, Tursz T, Wiels J. CD77: an antigen of germinal center B cells entering apoptosis. Eur J Immunol 1991; 21:1131–1140.
116. Illera VA, Perandones CE, Stunz LL, Mower DAJ, Ashman RF. Apoptosis in splenic B lymphocytes. Regulation by protein kinase C and IL-4. J Immunol 1993; 151:2965–2973.
117. Butch AW, Chung GH, Hoffmann JW, Nahm MH. Cytokine expression by germinal center cells. J Immunol 1993; 150:39–47.
118. Peschel C, Green I, Paul WE. Preferential proliferation of immature B lineage cells in long-term stroma cell-dependent cultures with IL-4. J Immunol 1989; 142:1558–1568.
119. Punnonen J, Aversa G, de Vries JE. Human pre-B cells differentiate into Ig-secreting plasma cells in the presence of interleukin-4 and activated CD4+ T cells or their membrane. Blood 1993; 82:2781–2789.
120. Stavnezer J, Radcliffe G, Lin YC, Nietupski J, Bergreu L, Sitia R, Severinson E. Immunoglobulin heavy-chain switching may be directed by prior induction of transcripts from constant-region genes. Proc Natl Acad Sci USA 1988; 85:7704–7708.
121. Jabara HH, Fu SM, Geha RS, Vercelli D. CD40 and IgE: synergism between anti-CD40 monoclonal antibody and interleukin 4 in the induction of IgE synthesis by highly purified human B cells. J Exp Med 1990; 172:1861–1864.
122. Gascan H, Gauchat JF, Roncarolo MG, Yssel H, Spits H, de Vries JE. Human B cell clones can be induced to proliferate and to switch to IgE and IgG4 synthesis by interleukin 4 and a signal provided by activated CD4+ T cell clones. J Exp Med 1991; 173:747–750.
123. Armitage RJ, Fanslow WC, Strockbine L, Sato TA, Clifford KN, Macduff BM, Anderson DM, Gimpel SD, Davis-Smith T, Maliszewski RC, Clark EA, Smith CA, Grabstein KH, Cosman D, Spriggs MK. Molecular and biological characterization of a murine ligand for CD40. Nature 1992; 357:80–82.
124. Kühn R, Rajewski K, Müller W. Generation and analysis of interleukin-4 deficient mice. Science 1991; 254:707–710.
125. De France T, Aubry JP, Rousset F, Vanbervliet B, Bonnefoy JY, Arai N, Takebe Y, Yokota T, Lee F, Arai K-I, de Vries JE, Banchereau J. Human recombinant interleukin 4 induces Fcε receptors (CD23) on normal human B lymphocytes. J Exp Med 1987; 165:1459–1467.
126. Rousset F, de Waal Malefijt R, Slierendregt B, Aubry J-P, Bonnefoy J-Y, Defrance T, Banchereau J, de Vries JE. Regulation of Fc receptor for IgE (CD23) and class II MHC antigen

expression on Burkitt's lymphoma cell lines by human IL-4 and IFN-γ. J Immunol 1988; 140:2625–2632.
127. Renz H, Domenico J, Gelfand EW. IL-4-dependent up-regulation of IL-4 receptor expression in murine T and B cells. J Immunol 1991; 146:3049–3055.
128. Law CL, Armitage RJ, Villablanca JG, LeBien TW. Expression of interleukin-4 receptors on early human B-lineage cells. Blood 1991; 78:703–710.
129. Stack RM, Lenschow DJ, Gray GS, Bluestone JA, Fitch FW. IL-4 treatment of small splenic B cells induces costimulatory molecules B7-1 and B7-2. J Immunol 1994; 152:5723–5733.
130. Valle A, Aubry JP, Durand I, Bancherau J. IL-4 and IL-2 upregulate the expression of antigen B7, the B cell counterstructure to T cell CD28: an amplification mechanism for T-B cell interactions. Int Immunol 1991; 3:229–235.
131. Plum J, De Smedt M, Leclercq G, Tison B. Inhibitory effect of murine recombinant IL-4 on thymocyte development in fetal thymus organ cultures. J Immunol 1990; 145:1066–1073.
132. Tepper RI, Levinson DA, Stange BZ, Campos-Torres J, Abbas AK, Leder P. IL-4 induces allergic-like inflammatory disease and alters T cell development in transgenic mice. Cell 1990; 62:457–467.
133. Parronchi P, De Carli M, Manetti R, Simonelli C, Sampognaro S, Piccinni MP, Macchia D, Maggi E, Del Prete G, Romagnani S. IL-4 and IFN (α and γ) exert opposite regulatory effects on the development of cytolytic potential by Th1 or Th2 human T cell clones. J Immunol 1992; 149:2977–2983.
134. Selvan RS, Nagarkatti PS, Nagarkatti M. Role of IL-2, IL-4 and IL-6 in the growth and differentiation of tumor-specific CD4+ T helper and CD8+ T cytotoxic cells. J Immunol 1990; 45:1096–1104.
135. Stotter H, Custer MC, Bolton ES, Guedez L, Lotze MT. IL-7 induces human lymphokine-activated killer cell activity and is regulated by IL-4. J Immunol 1991; 146:150–155.
136. Schmitt E, Germann T, Goedert S, Hoehn P, Huels C, Koelsch S, Kuhn R, Muller W, Palm N, Rude E. IL-9 production of naive CD4+ T cells depends on IL-2 is synergistically enhanced by a combination of TGF-beta and IL-4, and is inhibited by IFN-gamma. J Immunol 1994; 153:3989–3996.
137. Hart PH, Vitti GF, Burgess DR, Whitty GA, Piccoli DS, Hamilton JA. Potential antiinflammatory effects of interleukin 4: suppression of human monocyte tumor nercrosis factor α, interleukin 1, and prostaglandin E_2. Proc Natl Acad Sci USA 1989; 86:3803–3807.
138. Donnelly RP, Fenton MJ, Finbloom DA, Gerrard TL. Differential regulation of IL-1 production in human monocytes by IFN-γ and IL-4. J Immunol 1990; 145:569–575.
139. de Waal Malefyt R, Abrams J, Bennett B, Figdor CG, de Vries JE. Interleukin 10 (IL-10) inhibits cytokine synthesis by human monocytes: an autoregulatory role of IL-10 produced by monocytes. J Exp Med 1991; 174:1209–1220.
140. Vellenga E, Dokter W, de Wolf JT, van de Vinne B, Esselink MT, Halie MR. Interleukin-4 prevents the induction of G-CSF mRNA in human adherent monocytes in response to endotoxin and IL-1 stimulation. Br J Haematol 1991; 79:22–26.
141. Nickolaus P, Zawatzky R. Inhibition by interleukin-4 of constitutive beta interferon synthesis in mouse macrophages. J Virol 1994; 68:6763–6766.
142. Gruber MF, Williams CC, Gerrard TL. Macrophage-colony-stimulating factor expression by anti-CD45 stimulated human monocytes is transcriptionally up-regulated by IL-1 beta and inhibited by IL-4 and IL-10. J Immunol 1994; 152:1354–1361.
143. D'Andrea A, Ma X, Aste-Amezaga M, Paganin C, Trinchieri G. Stimulatory and inhibitory effects of interleukin (IL)-4 and IL-13 on the production of cytokines by human peripheral blood mononuclear cells: priming for IL-12 and tumor necrosis factor alpha production. J Exp Med 1995; 181:537–546.
144. Vannier E, Miller LC, Dinarello CA. Coordinated antiinflammatory effects of interleukin 4: interleukin 4 suppresses interleukin 1 production but up-regulates gene expression and synthesis of interleukin 1 receptor antagonist. Proc Natl Acad Sci USA 1992; 89:4076–4080.
145. Fenton MJ, Buras JA, Donnelly RP. IL-4 reciprocally regulated IL-1 and IL-1 receptor antagonist expression in human monocytes. J Immunol 1992; 149:1283–1288.
146. Wong HL, Costa GL, Lotze MT, Wahl SM. Interleukin (IL) 4 differentially regulates monocyte IL-1 family gene expression and synthesis in vitro and in vivo. J Exp Med 1993; 177:775–781.
147. Oswald IP, Gazzinelli RT, Sher A, James SL. IL-10 synergizes with IL-4 and transforming growth factor-beta to inhibit macrophage cytotoxic activity. J Immunol 1992; 148:3578–3582.
148. Niiro H, Otsuka T, Kuga S, Nemoto Y, Abe M, Hara N, Nakano T, Ogo T, Niho T. IL-10 inhibits prostaglandin E2 production by lipopolysaccharide-stimulated monocytes. Int Immunol 1994; 6:661–664.
149. Wong HL, Welch GR, Brandes ME, Wahl SM. IL-4 antagonizes induction of FcγRIII (CD16) expression by transforming growth factor-β on human monocytes. J Immunol 1991; 147:1843–1848.
150. Ho JL, He SH, Rios MJ, Wick EA. Interleukin-4 inhibits human macrophage activation by tumor necrosis factor, granulocyte-monocyte colony-stimulating factor, and interleukin-3 for antileishmanial activity and oxidative burst capacity. J Infect Dis 1992; 165:344–351.
151. Mangan DF, Robertson B, Wahl SM. IL-4 enhances programmed cell death (apoptosis) in stimulated human monocytes. J Immunol 1992; 148:1812–1816.
152. Romani N, Gruner S, Brang D, Kampgen E, Lenz A, Trockenbacher B, Konwalinka G, Fritsch PO, Steinman RM, Schuler G. Proliferating dendritic cell progenitors in human blood. J Exp Med 1994; 180:83–93.
153. Sallusto F, Lanzavecchia A. Efficient presentation of soluble antigen by cultured human dendritic cells is maintained by granulocyte/macrophage colony-stimulating factor plus interleukin 4 and downregulated by tumor necrosis factor alpha. J Exp Med 1994; 179:1109–1118.
154. Peschel C, Paul WE, Ohara J, Green I. Effects of B-cell stimulatory factor-1/interleukin-4 on hematopoietic progenitor cells. Blood 1987; 70:254–263.
155. Musashi M, Clark SC, Sudo T, Urdal DL, Ogawa M. Synergistic interactions between interleukin-11 and interleukin-4 in support of proliferation of primitive hematopoietic progenitors of mice. Blood 1991; 78:1448–1451.
156. Sonoda Y, Kuzuyama Y, Tanaka S, Yokota S, Maekawa T, Clark SC, Abe T. Human interleukin-4 inhibits proliferation of megakaryocyte progenitor cells in culture. Blood 1993; 81:624–630.
157. Renard N, Duvert V, Banchereau J, Saeland S. Interleukin-13 inhibits the proliferation of normal and leukemic human B-cel precursors. Blood 1994; 84:2253–2260.
158. Ryan DH, Nuccie BL, Ritterman I, Liesveld JL, Abboud CN. Cytokine regulation of early human lymphopoiesis. J Immunol 1994; 152:5250–5258.
159. Keller U, Aman JJ, Derigs G, Huber C, Peschel C. Human interleukin-4 enhances stromal cell-dependent hematopoiesis: costimulation with stem cell factor. Blood 1994; 84:2189–2196.
160. Simmons PJ, Masinovsky B, Longenecker BM, Berenson R, Torok-Storb B, Gallatin WM. Vascular cell adhesion molecule-

1 expressed by bone marrow stromal cells mediates the binding of hematopoietic progenitor cells. Blood 1992; 80:388–395.
161. Lowenthal JW, Castle BE, Christiansen J, Schreurs J, Rennick D, Arai N, Hoy P, Takebe Y, Howard M. Expression of high affinity receptors for murine interleukin 4 (BSF-1) on hemopoietic and nonhemopoietic cells. J Immunol 1988; 140:456–464.
162. Postlethwaite AE, Holness MA, Katai H, Raghow R. Human fibroblasts synthesize elevated levels of extracellular matrix proteins in response to interleukin 4. J Clin Invest 1992; 90:1479–1485.
163. Sempowski GD, Beckmann MP, Derdak S, Phipps RP. Subsets of murine lung fibroblasts express membrane-bound and soluble IL-4 receptors. Role of IL-4 in enhancing fibroblast proliferation and collagen synthesis. J Immunol 1994; 152:3606–3614.
164. Rathanaswami P, Hachicha M, Sadick M, Schall TJ, McColl SR. Expression of the cytokine RANTES in human rheumatoid synovial fibroblasts. Differential regulation of RANTES and interleukin-8 genes by inflammatory cytokines. J Biol Chem 1993; 268:5834–5839.
165. Marfaing-Koka A, Devergne O, Gorgone G, Portier A, Schall TJ, Galanaud P, Emilie D. Regulation of the production of the RANTES chemokine by endothelial cells. Synergistic induction by IFN-gamma plus TNF-alpha and inhibition by IL-4 and IL-13. J Immunol 1995; 154:1870–1878.
166. Swerlick RA, Lee KH, Li LJ, Sepp NT, Caughman SW, Lawley TJ. Regulation of vascular cell adhesion molecule 1 on human dermal incrovascular endothelial cells. J Immunol 1992; 149:698–705.
167. Schleimer RP, Sterbinsky SA, Kaiser J, Bickel CA, Klunk DA, Tomioka K, Newman W, Luscinskas FW, Gimbrone Jr MA, McIntyre BW, Bochner BS. IL-4 induces adherence of human eosinophils and basophils but not neutrophils to endothelium. J Immunol 1992; 148:1086–1092.
168. Urban Jr JF, Katona IM, Paul WE, Finkelman FD. Interleukin 4 is important in protective immunity to a gastrointestinal nematode infection in mice. Proc Natl Acad Sci USA 1991; 88:5513–5517.
169. Wahid FN, Behnke JM, Grencis RK, Else KJ, Ben-Smith AW. Immunological relationships during primary infection with Heligmosomoides polygyrus: Th2 cytokines and primary response phenotype. Parasitology 1994; 108:461–471.
170. Reiner SL, Locksley RM. The regulation of immunity to leishmania major. Annu Rev Immunol 1995; 13:151–177.
171. Sadick MD, Heinzel FP, Holaday BJ, Pu RT, Dawkins RS, Locksley RM, Cure of murine leishmaniasis with anti-interleukin 4 monoclonal antibody. J Exp Med 1990; 171:115–127.
172. Heinzel FP, Schoenhaut DS, Rerko RM, Rosser LE, Gately MK. Recombinant interleukin 12 cures mice infected with *Leishmania major.* J Exp Med 1993; 177:1505–1509.
173. Sypek JP, Chung CL, Mayor SEH, Subramanyam JM, Goldman SJ, Sieburth DS, Wolf SF, Schaub RG. Resolution of cutaneous leishmaniasis: interleukin 12 initiates a protective T helper type 1 immune response. J Exp Med 1993; 177:1797–1802.
174. Powrie F, Correa-Oliveira R, Mauze S, Coffman RL. Regulatory interactions between CD45RBhigh and CD45RBlow CD4+ T cells are important for the balance between protective and pathogenic cell-mediated immunity. J Exp Med 1994; 179:589–600.
175. Romani L, Mencacci A, Tonnetti L, Spaccapelo R, Cencei E, Wolf S, Puccetti P, Bistoni F. Interleukin-12 but not interferon-gamma production correlates with induction of T helper type-1 phenotype in murine candidiasis. Eur J Immunol 1994; 24:909–915.
176. Romani L, Mencacci A, Grohmann U, Mocci S, Mosci P, Puccetti P, Bistoni F. Neutralizing antibody to interleukin 4 induces systemic protection and T helper type 1-associated immunity in murine candidiasis. J Exp Med 1992; 176:19–25.
177. Griffin DE, Ward BJ. Differential CD4 T cell activation in measles. J Infect Disease 1993; 168:275–281.
178. Tang YW, Graham BS. Anti-IL-4 treatment at immunization modulates cytokine expression, reduces illness, and increases cytotoxic T lymphocyte activity in mice challenged with respiratory syncytial virus. J Clin Invest 1995; 94:1953–1958.
179. Clerici M, Haki FT, Venzon DJ, Blatt S, Hendrix CW, Wynn TA, Shearer GM. Changes in interleukin-2 and interleukin-4 production in asymptomatic, human immunodeficiency virus-seropositive individuals. J Clin Invest 1993; 91:759–765.
180. Meyaard L, Otto SA, Keet IP, van Lier RA, Miedema F. Changes in cytokine secretion patterns of CD4+ T-cell clones in human immunodeficiency virus infection. Blood 1994; 84:4262–4268.
181. Graziosi C, Pantaleo G, Gantt KR, Fortin JP, Demarest JF, Cohen OJ, Sekaly RP, Fauci AS. Lack of evidence for the dichotomy of Th1 and Th2 predominance in HIV-infected individuals. Science 1994; 265:248–252.
182. Maggi E, Mazzetti M, Ravina A, Annunziato F, de Carli M, Piccinni MP, Manetti R, Carbonari M, Pesce AM, del Prete G, Romagnani S. Ability of HIV to promote a Th1 to Th0 shift and to replicate preferentially in Th2 and Th0 cells. Science 1994; 265:244–248.
183. Kanagawa O, Vaupel BA, Gayama S, Koehler G, Kopf M. Resistance of mice deficient in IL-4 to retrovirus-induced immunodeficiency syndrome (MAIDS). Science 1993; 262:240–242.
184. Mosely B, Beckmann MP, March CJ, Idzerda RL, Gimpel SD, VandenBos T, Friend D, Alpert A, Anderson D, Jackson J, Wignall J, Smith C, Gallis B, Sims JE, Urdal DL, Widmer MB, Cosman D, Park LS. The murine interleukin-4 receptor: molecular cloning and characterization of secreted and membrane forms. Cell 1989; 59:335–348.
185. Harada N, Castle BE, Gorman DM, Itoh N, Schreurs J, Barrett RL, Howard M, Miyajima A. Expression cloning of a cDNA encoding the murine interleukin 4 receptor based on ligand binding. Proc Natl Acad Sci USA 1990; 87:857–861.
186. Idzerda RL, March CJ, Mosely B, Lyman SD, VandenBos T, Gimpel SD, Din WS, Grabstein KH, Widmer MB, Park LS, Cosman D, Beckmann MP. Human interleukin 4 receptor confers biological responsiveness and defines a novel receptor superfamily. J Exp Med 1990; 171:861–873.
187. Pritchard MA, Baker E, Whitmore SA, Sutherland GR, Idzerda RL, Park LS, Cosman D, Jenkins NA, Gilbert DJ, Copeland NG. The interleukin-4 receptor gene (IL4R) maps to 16p11.2-16p12.1 in human and to the distal region of mouse chromosome 7. Genomics 1991; 10:801–806.
188. Ohara J, Paul WE. Up-regulation of interleukin 4/B-cell stimulatory factor 1 receptor expression. Proc Natl Acad Sci USA 1988; 85:8221–8225.
189. Keegan AD, Nelms K, Wang LM, Pierce JH, Paul WE. Interleukin 4 receptor: signaling mechanisms. Immunol Today 1994; 15:423–432.
190. Keegan AD, Beckmann MP, Park LS, Paul WE. The IL-4 receptor: biochemical characterization of IL-4-binding molecules in a T cell line expressing large numbers of receptors. J Immunol 1991; 146:2272–2279.
191. Russell SM, Keegan AD, Harada N, Nakamura Y, Noguchi M, Leland P, Friedmann MC, Miyajima A, Puri RK, Paul WE.

Interleukin-2 receptor gamma chain: a functional component of the interleukin-4 receptor. Science 1993; 262:1880–1883.

192. Matthews DJ, Clark PA, Herbert J, Morgan G, Armitage RJ, Kinnon C, Minty A, Grabstein KH, Caput D, Ferrara P, Callard R. Function of the interleukin-2 (IL-2) receptor gamma-chain in biologic responses of X-linked severe combined immunodeficient B cells to IL-2, IL-4, IL-13, and IL-15. Blood 1995; 85:38–42.

193. He Y-W, Malek TR. The IL-2 receptor γ_c chain does not function as a subunit shared by the IL-4 and IL-13 receptors. J Immunol 1995; 155:9–12.

194. Rigley KP, Thurstan SM, Callard RE. Independent regulation of interleukin 4 (IL-4)-induced expression of human B cell surface CD23 and IgM: functional evidence for two IL-4 receptors. Int Immunol 1991; 3:197–203.

195. Izuhara K, Harada N. Interleukin-4 (IL-4) induces protein tyrosine phosphorylation of the IL-4 receptor and association of phosphatidylinositol 3-kinase to the IL-4 receptor in a mouse T cell line, HT2. J Biol Chem 1993; 268:13097–13102.

196. Witthuhn BA, Silvennoinen O, Miura O, Lai KS, Cwik C, Liu ET, Ihle JN. Involvement of the Jak-3 Janus kinase in signalling by interleukins 2 and 4 in lymphoid and myeloid cells. Nature 1994; 370:153–157.

197. Hou J, Schindler U, Henzel WJ, Ho TC, Brasseur M, McKnight SL. An interleukin-4-induced transcription factor: IL-4 Stat. Science 1994; 265:1701–1706.

198. Kotanides H, Reich NC. Requirement of tyrosine phosphorylation for rapid activation of a DNA binding factor by IL-4. Science 1993; 262:1265–1267.

199. Schindler C, Kashleva H, Pernis A, Pine R, Rothman P. STF-IL-4: a novel IL-4-induced signal transducing factor. EMBO J 1994; 13:1350–1356.

200. Fernandez-Botran R, Vitetta ES. A soluble, high affinity, interleukin-4-binding protein is present in the biological fluids of mice. Proc Natl Acad Sci USA 1990; 87:4202–4206.

201. Fernandez-Botran R, Vitetta ES. Evidence that natural murine soluble interleukin 4 receptors may act as transport proteins. J Exp Med 1991; 174:673–681.

202. Sato TA, Widmer MB, Finkelman FD, Madani H, Jacobs CA, Grabstein KH, Maliszewski CR. Recombinant soluble murine IL-4 receptor can inhibit or enhance IgE responses in vivo. J Immunol 1993; 150:2717–2723.

5
Interleukin-5 and Its Receptor

Siddhartha Mahanty
National Institute of Allergy and Montreal General Hospital, Montreal, Quebec, Canada

Thomas Nutman
National Institute of Allergy and Infectious Diseases, National Institutes of Health, Bethesda, Maryland

INTRODUCTION

Interleukin-5 (IL-5) is a member of the cytokine family of polypeptides and glycoproteins that regulate cellular and humoral immune responses. These molecules mediate the interaction of T cells, B cells, and other effector cells of the immune system. The majority of cytokines have a broad range of activity on a variety of target cells, but some cytokines appear to be relatively restricted in the range of target cells. IL-5 is one such cytokine that acts on selected lineages of cells (1). Although this cytokine was initially described as a growth factor for a murine B-cell lymphoma (2), its association with eosinophil differentiation (3) has made IL-5 of special interest in the study of parasitic infections and allergic disorders.

An interesting feature of the biological activity of IL-5 is its overlap of function with that of IL-3 and granulocyte-macrophage colony-stimulating factor (GM-CSF). Although the action and biological activity of these cytokines overlap substantially, there are cytokine-specific actions that suggest that separate intracellular signaling mechanisms are also activated, in addition to common pathways of cell activation (4). Experiments done by several groups have successfully elucidated the molecular basis for this functional overlap (reviewed in Ref. 5).

In this review, we summarize current knowledge of the structure and function of IL-5, regulation of the gene encoding this cytokine, the interaction of the molecule with its receptor [the IL-5 receptor (IL-5R)], the signals transduced by this interleukin, and IL-5's biological role in selected disease states. Finally, therapeutic strategies that could involve the use of IL-5 or IL-5 blockade—either by specific anti-IL-5 antibodies or soluble receptors—will be discussed.

HISTORICAL BACKGROUND

The observation that eosinophilia could occur in the absence of changes in populations of other leukocytes suggested not only that there was a control mechanism specific for eosinophils but also that there was a T-cell dependence of this control. Consequently, other studies examining the role of T cells in the growth and functional activation of B cells revealed that T-cell supernatants had the capacity to induce B-cell proliferation and antibody secretion (6). This capacity was attributed to a differentiation factor designated T-cell–replacing factor (TRF) (2,7). B-cell growth–stimulating activity in T-cell supernatants was attributed to another factor, designated B-cell growth factor (BCGFII)—a molecule that was subsequently found to be identical to TRF (8). An eosinophil differentiation factor (EDF) was concurrently reported in supernatants of murine spleen cells (9). The discovery that EDF and TRF activity copurified (10)—followed by the demonstration by cDNA cloning and sequence analysis (11,12) that the factors responsible for these two activities were identical—established IL-5 to be primarily a T-cell–derived cytokine with multiple biological actions on murine B cells and murine and human eosinophil progenitors (3).

BIOCHEMISTRY AND STRUCTURE

Organization of the IL-5 Gene

The amino acid sequence deduced from the nucleotide sequence of the entire IL-5 cDNA indicated that murine IL-5 consisted of 133 amino acids including a hydrophobic signal sequence of 20 amino acids (12). The primary nucleic acid sequence failed to show homology to any of the other cytokines then identified except to short sequences of murine IL-3, GM-CSF, and interferon (IFN)-γ (12). The human IL-5 gene was isolated using the murine cDNA as a probe (11). Chromosomal localization mapped the murine gene to a 230 kb region on chromosome 11 and the corresponding human gene on chromosome 5 (5q23.3-31) (13–15). The genes for IL-4, IL-3, and GM-CSF are located on the same chromosomes, and, in the case of the human genes, on the same arm of chromosome 5 (16). Deletion of the IL-5 gene has been reported with the $5q^-$ syndrome (14), some myeloid leukemias (15), eosinophilic leukemias (17), and certain immunodeficiency states (e.g., severe combined immunodeficiency, SCID) (18).

The organization of the IL-5 gene in the three species studied to date shows that there are four exons and three introns in IL-5 genes (19–21). Conserved TATA-like and cytokine consensus sequences are found 26 and 73 bp, respectively, upstream of the transcription initiation sites (20,22). The exon-intron organization and location of cysteine codons of IL-5 genes resemble those of GM-CSF, IL-2, and IL-4 (and are quite distinct from the G-CSF gene), suggesting an overall similarity between the structure of the IL-5 gene and those of these other cytokines (3,23). There also appear to be regulatory elements 5′ to the coding region of the IL-5 and GM-CSF genes in humans and the mouse that are identical. Furthermore, clustering of the IL-5 gene close to the genes for GM-CSF, IL-4, and IL-3 on chromosome 5 in humans and chromosome 11 in the mouse have been interpreted to indicate that these genes may have been derived by duplications from a common ancestral gene.

Regulation of Gene Expression

Peripheral blood mononuclear cells from normal individuals rarely show detectable IL-5 mRNA without mitogen stimulation (Ref. 24 and S. Mahanty, unpublished). Under conditions associated with a marked expansion of Th2-cell populations, such as in individuals with helminth infections, IL-4 and IL-5 production are generally regulated coordinately (25); however, in vitro stimulation of T cells directly by cytokines (in contrast to T-cell receptor–mediated stimulation by antigens) can result in differential secretion of IL-4 or IL-5 (26). In this and other in vitro systems (27), IL-2 is able to selectively induce expression of mRNA for IL-5 but not for IL-4.

While constitutive expression of mRNA for IL-5 has been demonstrated in some murine T-cell hybrid lines (28), stimulation with T-cell mitogens, phorbol myristate acetate (PMA) plus a calcium ionophore (ionomycin) or antigen markedly augments expression of IL-5 mRNA (29), particularly from T-cell clones that have the characteristics of Th2 cells (30,31).

The effect of other cytokines on the expression of IL-5 mRNA is an area of active research at present, particularly with the increasing therapeutic use of cytokines such as IL-2 and IFN-γ. Passive administration of IL-2 to mice can induce IL-5 mRNA expression in T cells (32). Selective increases in levels of IL-5 in the serum of patients infused with IL-2 have also been reported (33–35), suggesting that IL-2 can augment IL-5 production either directly or through another intermediary. The discovery of nuclear factors that interact with elements shared by cytokine genes (36) suggests one possible mechanism by which other cytokines could regulate the expression of the IL-5 gene. Further evidence for such a control mechanism for induction of IL-5 and GM-CSF gene transcription comes from reports that an identical 10-base sequence occurs upstream of the TATA box of the IL-5 and GM-CSF genes in both humans and mice (13); similar sequences are also present in other cytokine genes, including the human GM-CSF gene (37).

Protein Structure of IL-5

IL-5 protein was initially purified from murine T-cell supernatants. The eosinophil differentiation activity of supernatant from one T-cell clone was found to have an M_r of 45,000 daltons and was therefore clearly distinct from IL-3 or GM-CSF (9,10). Both human and murine IL-5 have been characterized by cloning, sequencing, and expression (38). Table 1 compares the biochemical, molecular, and functional properties of human, murine, and rat IL-5. It should be noted that while cross-reactivity in terms of bioassays and immunoassays has been observed between murine and human IL-5, the human cytokine is approximately 100-fold less active on mouse target cells than on human bone marrow cells, while mouse IL-5 is 20-fold less active on human than on homologous bone marrow (39).

Biochemical characterization of the IL-5 molecule has been facilitated by the production of murine and human recombinant IL-5 (rIL-5) in eukaryotic systems (40,41). When recombinant murine (mrIL-5) and human IL-5 (hrIL-5) were produced in Chinese hamster ovary cells, the cDNA for mrIL-5 was found to encode a mature peptide of 113 amino acids (M_r 45–60 kDa), dropping to

Table 1 Molecular and Biochemical Features of Murine, Rat, and Human IL-5

	Human	Mouse	Rat
Molecular structure M_r (daltons)	30,000–40,000	40,000–50,000	40,000–50,000
No. of amino acids			
Total	134	133	133
Mature product	115	113	
Predicted size of peptide (daltons)	13,149	13,111	
Active form	Homodimer	Homodimer	Homodimer
N-Glycosylation sites	2	3	?
Genomic structure	4 exons 3 introns	4 exons 3 introns	4 exons 3 introns
Chromosomal location	5q23-31	11	?
Homology with human IL-5	—	67%	62%
Copy number	1	1	1
Cell sources	T cells EBV-transformed B cells Eosinophils Reed-Sternberg cells	T cells Mast cells	
Functional cross-reactivity	Rat, hamster, multi-mammate rat, gerbil, ferret, vole	Rat, hamster, multi-mammate rat, gerbil, ferret, vole	

an M_r 20–30 kDa under reducing conditions (38,42). The mRNA for hrIL-5 encodes for a 115-amino-acid peptide that is 70% homologous in protein sequence to the murine peptide. The active form of IL-5 exists as an antiparallel dimer, unusual among T-cell–derived cytokines, which are linked by two pairs of cysteines, and contain at least two glycosylation sites (40,43); glycosylation, however, does not appear to be necessary for biological function (44).

IL-5 Receptors

IL-5 has multiple biological activities that appear to be exerted by interactions with cell surface receptors. These receptors have been shown to be heterogeneous in terms of their affinity and specificity and are low in numbers per cell (typically 100–1000 per cell), features that initially hindered their purification (4). The availability of highly purified rIL-5 and anti-IL-5 monoclonal antibodies, however, has made it possible to examine binding sites for IL-5. As the cDNAs for these receptors have been cloned by direct expression strategies, it has been possible to define their affinities and specificities (4,43).

Binding kinetics with ^{35}S-labeled or radioiodinated murine rIL-5 have revealed two classes of IL-5–binding sites. Approximately 10–15% of these receptors have high-affinity binding (dissociation constant, Kd, of ~150 pM), while the remainder have much lower affinity (Kd ~20 nM) (42,45). Lipopolysaccharide (LPS) has been shown to upregulate high-affinity IL-5R on murine B cells, and both IL-2 and IL-5 downregulate the same receptor (45). As would be expected, the density of high-affinity IL-5R correlates with responsiveness to IL-5. The concentration of IL-5 that is required for interaction with the two receptors (high and low affinity) indicates that the biological activity of IL-5 occurs only on engagement of the high-affinity receptor (reviewed in Ref. 43).

High-affinity IL-5R for both humans and mice have been cloned (46,47). Recent evidence indicates that the β subunit of IL-5R is shared with other cytokine receptors, in particular with the receptors for GM-CSF and IL-3 (which has a homologous subunit, AIC2A). The tertiary structure of the amino terminal of this subunit is felt to determine the high-affinity binding to these receptors (48,49). These findings may explain the observations that these cytokines can compete with one another for receptor occupancy (50,51). Thus, it appears that the receptors for IL-3, IL-5, and GM-CSF are each composed of two polypeptide chains, designated α and β. Each α chain binds only to its cognate ligand but is able to transmit a signal into the cell only in the presence of the β chain, an interaction that (for IL-5, at least) probably occurs in a proline-rich intracellular domain (5). Although the same β chain is a component of the receptors for all three cytokines, there does not appear to be significant contact of the IL-5 molecule with the β chain; however, the inability of α-chain binding alone to transmit signals for proliferation or differentiation in transfection experiments (46) and the remarkable absence of eosinophilia in IL-5Rα transgenic mice (52,53) suggest that the β chain participates in signal transduction after binding of IL-5 with the receptor complex.

In the mouse, a variety of cell types have been studied for their capacity to bind specifically radiolabeled IL-5; among the primary cells studied, only LPS-activated B cells have been found to express the high-affinity IL-5R (42,45,54). In humans, IL-5R has also been shown to be present on eosinophils and leukemic cell lines (55–57). On eosinophils, the receptor has been found to have higher affinity for IL-5 on activated, hypodense eosinophils (56) than on normodense, "resting" eosinophils. Interestingly, human eosinophils may only have one class of high-affinity IL-5R, with an M_r of ~50–60 kDa based on cross-linking studies and Scatchard analysis (5,57). Although murine IL-5 is able to bind human IL-5R quite efficiently, it remains to be shown whether the human IL-5 receptor system is similar to that described in the mouse.

The IL-5 Signaling Cascade

Recent studies comparing and contrasting cell activation by IL-5 with that induced by other cytokines whose signal transduction pathways have been elucidated (reviewed in Ref. 58) have extended our knowledge of the intracellular events that follow binding of IL-5R by IL-5. Receptors belonging to the cytokine receptor superfamily generally fall into two categories: those with and those without intrinsic protein tyrosine kinase activity. Receptors that lack tyrosine kinase activity appear to use intracellular kinases in common pathways of activation and transcription of a variety of genes. The Janus kinases (JAK) are a family of related peptides shown recently to be activated by several cytokines (IL-2, IL-3, IL-6, IFN-γ, etc.) upon binding with their respective receptors (59). These kinases, in turn, transduce cytokine-generated signals to the nucleus to initiate transcription of genes via a family of related transcription factors called signal transducers and activators of transcription (STATs) (60). Similar signal-transduction cascades are set off by many cytokines and growth factors [IL-6, ciliary neural growth factor (CNTF)], some of which also have a common gp130-signaling subunit. The membrane-proximal region of this subunit is required for signal transduction and has been shown to associate with JAK1, JAK2, and TYK2 (61,62). JAK activation can occur by one of at least three mechanisms: homodimerically, with a single receptor chain associating with a single JAK; heterodimerically, where a single receptor chain associates with multiple JAKs; or individual receptor chains may associate with specific JAKs, resulting in obligate heterodimeric JAK complexes (59).

The intracellular signal transduction element of IL-5Rαβ lacks intrinsic kinase activity, as do those of several other cytokine receptor molecules (e.g., IL-2, IL-4, and IFN-γ) (58,59); however, in IL-5–dependent murine cell lines, there is rapid phosphorylation of several serine and tyrosine residues on the 60 kDa subunit (α chain) of IL-5R after engagement of the receptor by IL-5 and IL-3 (reviewed in Ref. 63). Four proteins (p140, p92, p53, and p45) have been reported to undergo tyrosine phosphorylation within 5 minutes of exposure to IL-5 (64), suggesting that tyrosine kinases are involved early in the signal-transducing pathways of IL-5. Similar patterns of protein phosphorylation are seen upon stimulation with IL-3 and GM-CSF (63). The tyrosine and serine kinases activated by IL-5 appear to induce phosphorylation and activation of the IL-2Rβ chain in a manner similar to IL-2 itself, making the cell more responsive to the effect of IL-2 (53).

Although the cytoplasmic domains of the β chain of the IL-3, GM-CSF, and IL-5 receptors are relatively short, they are required for signal transduction, because truncation of their cytoplasmic domains leads to loss of cellular responsiveness (65–67). Ligand-induced heterodimerization of the IL-5 receptor (as for GM-CSF and IL-3) appears to be important for bringing the cytoplasmic domains of the two components together (58). Ligand-bound receptor complexes are capable of binding JAK2 in the membrane-proximal region (68).

According to the model of JAK-STAT–dependent IL-5 signal transduction, binding of IL-5 to IL-5R induces dimerization of receptor components and activation of the closely associated JAKs (58). In signal transduction cascades initiated by other cytokines (IL-2, IL-3, and GM-CSF), JAKs have been shown to physically associate with membrane-proximal regions of the cytokine-receptor complex that, upon dimerization, would be brought into close proximity, leading to tyrosine phosphorylation and activation of the JAKs. The activated JAKs can then phosphorylate the receptors as well as cytoplasmic proteins belonging to the STAT family of transcription factors (60,69,70). Activated STATS then translocate to the nucleus and bind IFN-γ–activating sequences (GAS) (59,70,71) in the upstream regulatory regions of the genes encoding IL-5–inducible proteins (Fig. 1). Evidence for activation of the JAK-STAT pathway of signal transduction has been found in murine IL-5–dependent cell lines (72) and in human eosinophils (71,73).

Apart from the JAK-STAT pathway, IL-5 has also been found to induce tyrosine phosphorylation and activation of a number of other kinases belonging to the *ras* pathway (74). Two species of *lyn* kinases are phosphorylated and activated rapidly (within 1 min) in human eosinophils exposed to IL-5 (74,75). These proteins are closely associated with the IL-5 β chain. Involvement

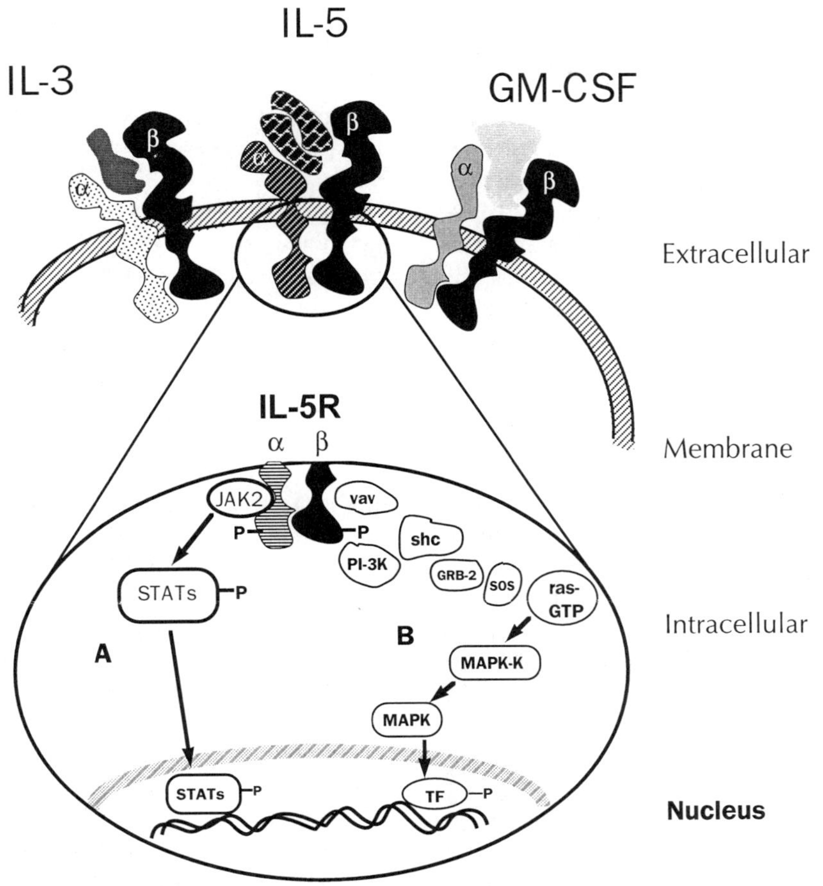

Figure 1 Schematic representation of signal transduction pathways activated by IL-5 receptor binding. The initial event is ligand-dependent aggregation of receptors, bringing JAKs in close proximity to each other and to other receptors. Two major pathways activated by cytokine binding depicted in this schematic diagram are the JAK-STAT pathway (A) and the *ras* pathway (B). The membrane-proximal region of the β chain is required for activation of the JAK-STAT pathway, via JAK2 (for IL-5), and JAKs do not necessarily interact with the α subunit, as shown in this illustration. Activation of the *ras* pathway through SHC binding and phosphorylation requires the membrane-distal region, whereas *vav* phosphorylation requires the membrane-proximal region of the β chain. The β chains of IL-5 and GM-CSF are identical and have high similarity (at the level of amino acid sequence) to that for IL-3. It is likely that similar or identical signal transduction pathways are activated by the cytokines. JAK, Janus kinase; STAT, signal transducer and activator of transcription; PI-3K, phosphoinositol-3 kinase; *vav*, p95vav proto-oncogene product; *lyn*, tyrosine kinase of the *src* family of proto-oncogenes; MAPK, mitogen-activated protein kinase; MAPK-K, MAPK kinase; TF, transcription factor (unidentified). Arrows indicate phosphorylation of tyrosine residues.

of the *ras* pathway in these cells was demonstrated by increased binding of ^{32}P-GTP to *ras* after stimulation with IL-5 (74). Compelling evidence for the involvement of this pathway in murine B cells also comes from the xid mouse, B cells of which are unresponsive to IL-5 and have been demonstrated to have defects in the activation of *btk* following IL-5 stimulation (75–77). Figure 1 depicts a proposed scheme for signal transduction after ligand binding of the IL-5R complex based on current knowledge of intracellular phosphorylation events immediately following IL-5 stimulation of responsive cells.

Detection of IL-5

An assay developed for the detection of eosinophil-differentiating activity of IL-5 relied on its ability to generate eosinophil colonies in murine bone marrow cultures either on semisolid agar (78,79) or in liquid suspension (80). This assay has been adapted to microplates to conserve reagents and simplify analysis; however, it is of limited usefulness because it also detects other eosinophilopoietic cytokines, e.g., IL-3 and GM-CSF (3). The ability of IL-5 to prolong the survival of eosinophils in vitro has been used (in conjunction with antibodies to GM-CSF and IL-3) in eosinophil survival assays to detect the presence of IL-5 in serum and cell supernatants (81–84).

The availability of monoclonal antibodies specific for cloned rIL-5 resulted in the development of more sensitive immunoassays that made it possible to detect concentrations of IL-5 in the picogram range (85). The use of ELISA for the detection of immunoreactive IL-5 has made it possible to study the production of IL-5 by various cell populations in humans and mice. Long-term stability and lack of significant consumption by cells in vitro—a problem that diminishes the usefulness of in vitro secretion studies of some other cytokines (such as IL-2 and IL-4)—make IL-5 detection by ELISA an accepted method for assaying IL-5 production. A technique using nitrocellulose-bottomed multiple wells, derived from the older plaque-forming assay, has also resulted in a sensitive method (ELISPOT or filter immunoplaque assay) for the detection of single cytokine-secreting cells (86 and reviewed in Ref. 87). This technique has been successfully applied to IL-5–secreting cells to demonstrate expansion of these populations in humans with eosinophilia (25) and in the mouse (88,89).

Immunofluorescence techniques have been applied successfully to detect IL-5 in the cytoplasm of individual cells (90,91), and polymerase chain reaction has also been used to detect mRNA for IL-5 in as few as 10–100 cytokine-secreting cells (92,93). Anatomical localization of IL-5–secreting cells has been demonstrated by in situ hybridization techniques (94–96). Flow cytometry has been used to determine the frequency of IL-5–secreting cells by intracellular staining with fluorescein-conjugated monoclonal anti–hIL-5 antibodies (97).

SOURCES OF IL-5 AND DISTRIBUTION OF IL-5–SECRETING CELLS

Activated T-helper cell populations ($CD4^+$, $CD45R0^+$ cells) are the main source of IL-5, but secretion of IL-5 from several other cell types has been reported in humans and in mice. Other T-cell subpopulations that can produce IL-5 include (in the mouse) $CD3^+$ $\gamma\delta$ T-cell receptor $(TCR)^+$ T cells in the intestinal mucosa (88) and $CD8^+$ cells (98). Non–T-cell populations have also been demonstrated to be capable of secreting IL-5; e.g., Epstein-Barr virus (EBV)–transformed B cells from humans secrete IL-5 in supernatants (99), and mRNA for IL-5 has been demonstrated by in situ hybridization in eosinophils found in human intestinal mucosa (100). Reed-Sternberg cells from individuals with Hodgkin's disease accompanied by eosinophilia have also been found to express mRNA for IL-5 and may be involved in the generation of eosinophilia associated with this disorder (101).

IL-5 mRNA has also been demonstrated in sites of allergic reactions, such as in the airway mucosa in individuals with asthma and seasonal allergic rhinitis. Although most (> 70%) of the IL-5 $mRNA^+$ cells in nasal and tracheal mucosa and bronchoalveolar lavage fluid in allergic individuals are T cells (102,103), mast cells and eosinophils have also been shown to express mRNA for IL-5 using in situ mRNA detection (104–106). IL-5 mRNA expression in mast cells appears to be restricted to the $tryptase^+$ $chymase^+$ subpopulation (107) and is upregulated by cross-linkage of high-affinity Fcε receptors (FcRI) (108).

BIOLOGICAL ACTIVITY OF IL-5

IL-5 induces marked proliferation of eosinophil precursors and eosinophil activation and survival in humans and mice (reviewed in Refs. 63 and 109); however, in other respects, the biological activities of IL-5 have been found to differ in the mouse and human. For example, in humans there has been no demonstration of capacity to induce either proliferation or cofactor activity for immunoglobulin (Ig) secretion using a wide range of assay systems (110). In marked contrast, murine IL-5 appears to induce both proliferation and Ig secretion by B cells both in normal, activated cells and in leukemic cell lines of B-cell origin (reviewed in Ref. 5). Table 2 compares and contrasts the reported biological activities of human and murine IL-5.

Effect of IL-5 on Eosinophils and Eosinophilopoiesis

The eosinophilopoietic action of IL-5 was first demonstrated in mouse bone marrow cultures and was found to be cell lineage specific (9). Production of eosinophils in this system is markedly enhanced in parasite-infected mice, and anti-IL-5 administration in vivo to parasitized mice completely abolishes the eosinophilia, suggesting that IL-5 is the main factor responsible for eosinophilia in this model (111). The effect of IL-5 on human eosinophils is very similar to that seen in mice: IL-5 prolongs eosinophil survival in vitro (112,113), and eosinophilic colonies are preferentially expanded by IL-5 from human bone marrow in culture (114). Studies of the kinetics of IL-5–mediated generation of eosinophils from bone marrow have concluded that IL-5 is only active at a late stage of eosinophil differentiation (32,109) and that preincubation of murine bone marrow cells with IL-3, GM-CSF, or M-CSF induced a marked increase in eosinophil precursor cells that are IL-5 responsive (5,115). Similar effects of

Table 2 Targets and Biological Activity of Human and Murine IL-5

Target cells	Species: Human	Species: Mouse
Eosinophils	Differentiation of precursor cells; activation, release of granule proteins	Differentiation of precursor cells; activation, release of granule proteins
B cells	IgA and IgM secretion by peripheral B cells	Differentiation; Ig secretion; enhancement of IL-4–directed switching to IgG1 and IgE
T cells	Induction of IL-2 receptor expression	Induction of IL-2 receptor expression
Basophils	Proliferation; histamine release; secretion of leukotriene C4	Not reported

IL-5 have been described with human bone marrow cultures in vitro (116).

Apart from increased survival and generation of eosinophils, IL-5 has been reported to have several other direct effects on eosinophil function. IL-5 has been shown to enhance adhesion of human eosinophils to plasma-coated glass, a phenomenon probably mediated by augmented expression of integrin molecules of the CD11/18 family (117 and reviewed in Ref. 118). IL-5 also enhances eosinophil-related cytotoxicity and the respiratory burst in eosinophils (119). In states of hypereosinophilia, such as the idiopathic hypereosinophilic syndrome and the eosinophilia-myalgia syndrome, circulating eosinophils have a higher proportion of hypodense ("activated") vs. normodense (inactivated) eosinophils, which have been shown to correlate with IL-5 in serum (81,120). Similar rises in serum IL-5 levels are observed in helminth-infected individuals who develop posttreatment eosinophilia (121), in individuals receiving IL-2/lymphokine-activated killer (LAK) cells (122), and in the episodic angioedema with eosinophilia syndrome (82).

Regulation of B-Cell Proliferation and Antibody Production

The effect of IL-5 on human B cells remains controversial. Although it has been shown to induce IgA and IgM production by peripheral blood B cells in vitro (123) and to augment Ig production by *Staphylococcus aureus* Cowan I- or pokeweed mitogen–stimulated human B-cell blasts in the presence of IL-2, hIL-5 has been shown to have no detectable effect on B-cell function as determined by a variety of standard B-cell functional assays (110).

In the mouse, the ability of IL-5 to induce differentiation and Ig secretion by peritoneal B cells and B-cell lines was the first reported activity of IL-5 (2,7). Investigation of the action of IL-5 on murine B cells at the molecular level has revealed that IL-5 enhances the IL-4–directed isotype switching to IgG1 and IgE (in the mouse) (124,125), regulates Ig heavy-chain gene transcription (126), and induces transcription of Ig α-chain mRNA (127). IL-5 thus induces polyclonal IgM production by murine B cells as well as an increase in the levels of mRNA of secreted forms of the μ chain. Using antigen-specific transfected cell lines, a region between –250 and –125 bp lying 5′ of the Ig transcription site has been identified to be necessary for this action of IL-5 (128). Increased IgA synthesis in committed cells in the absence of proliferation has been observed (129); however, IL-5 only acts on $sIgA^+$ (not $sIgA^-$) cells to induce IgA production (130). Thus, IL-5 can only stimulate IgA from cells already committed to become IgA-secreting cells (63).

Recent in vitro studies of the effect of mrIL-5 on B-cell development have revealed that IL-5 preferentially acts on $Ly1^+$ ($CD5^+$) B cells from stromal bone marrow cells to induce sIgM expression (131,132). These B cells are found mainly in the peritoneal cavity in the mouse and are thought to be important in initial antibody responses to carbohydrate and glycoprotein antigens.

Other Effects of IL-5

Apart from their effect on eosinophil generation and activation, a variety of other effects—related primarily to allergic phenomena—have been attributed to IL-5. rIL-5 directly induces release of histamine and generation of leukotriene C4 from human basophils in vitro (51,133,134), and IL-5 appears to be a basophilopoietic cytokine in humans (135). IL-5 production by specialized subpopulations of T lymphocytes has been elegantly demonstrated using Northern analysis and in situ hybridization both in mesenteric lymph nodes and in the lamina propria in humans (136,137) as well as in bronchoalveolar lavage cells in individuals with asthma (95,96). The central role of eosinophils as an important proinflammatory cell in asthma is becoming widely accepted (138), and the involvement of IL-5 in recruitment and activation of eosinophils in asthma is the focus of current and likely future research.

Pathophysiology of IL-5 Transgenic Mice

Transgenic mice carrying the murine IL-5 gene with promoters for constitutive expression of the cytokine have been used to investigate the role of IL-5 in eosinophil development and B-cell function. Two such models of transgenic IL-5 production have been studied (139,140); both sets of transgenic mice exhibited elevated serum IL-5 levels and eosinophilia. A prominent feature of IL-5 transgenic mice is a marked increase in the numbers of circulating lymphocytes and eosinophils. The increase in eosinophil populations is 75-fold compared with age-matched normal mice (140), but cell cycle analysis of these eosinophils suggests that there is no increase in the proliferative capacity of eosinophils in these animals (141). Tissue infiltration with eosinophils is apparent, and the spleen is typically enlarged approximately five times (in the number of cells). Interestingly, the expanded population of eosinophils in these animals appears to be a resting population that needs to be activated to release granule contents.

Other abnormalities noted in these mice include expansion of a distinctive population of B cells with a high density of surface IgM and the IL-5 receptor (IL-5R) in the bone marrow and peritoneum. The significance and function of this population is not clearly understood, but this cell population has been postulated to be involved in the generation of autoantibodies (5,63).

The IL-5 transgenic mouse may serve as a model for diseases characterized by eosinophilia such as idiopathic hypereosinophilic syndrome, myositis, and fasciitis, although the fact that the eosinophils are in the resting state in these animals (vs. increased proportions of activated eosinophils that are commonly seen in these conditions) must be kept in mind.

POTENTIAL THERAPEUTIC ROLES FOR IL-5 AND ANTIBODIES TO IL-5

In general, increased IL-5 production is observed in situations where there is an expansion of Th2 cell populations. In experimental helminth infections of the mouse, IL-5 does not appear to have an important role in the induction of protective immunity as evidenced by the lack of protection afforded to mice treated with anti–IL-5 at doses that abolish eosinophilia (111,142,143); however, eosinophils have been demonstrated to have helminthotoxic activity against *Schistosoma mansoni* and *Schistosoma hematobium* both in vitro and in reinfection surveys of children (reviewed in Ref. 1). Similarly, eosinophilic infiltrates correlate with a better prognosis in some cancers (e.g., colonic, gastric, and lung carcinoma), suggesting a potential role in cancer and graft rejection. Local induction of eosinophilia could be envisaged to be beneficial in these conditions, and, therefore, IL-5 could have a role in immunotherapy for appropriately selected individuals.

The association of eosinophilia and neoplasia has been noted for many years (144), and its presence has been associated with a favorable prognosis in some cancers, notably carcinomas of the urinary bladder (145), uterine cervix (146), and head and neck (147). Although local production of eosinophilopoietic cytokines (GM-CSF, IL-3, and IL-5) by tumor cells has not been investigated extensively, a study of bone marrow and blood eosinophilia in two patients with solid malignancies found that tumor cells secreted IL-3 and GM-CSF in vitro, measured by a sensitive bioassay and neutralizable by antibodies to GM-CSF and IL-3; however, in these experiments, no IL-5 was measured in the supernatants (148). Eosinophilia has also been reported to accompany Hodgkin's lymphoma (149) and some cases of nonlymphocytic leukemias. In Hodgkin's lymphoma, Reed-Sternberg cells have been demonstrated to be capable of secreting IL-5 (101), and in some leukemia-associated eosinophilia, chromosomal translocations involving chromosome 5—the location of the IL-5 gene (150,151)—have been observed, although there has been no demonstration of elevated IL-5 in these cases.

The role of eosinophil-derived toxic products in necrosis of tumor tissue (152) and correlation of eosinophilia with clinical response in patients treated with IL-2 (153) raise interesting possibilities for the use of IL-5 as a therapeutic modality in treatment for certain tumors. Thus, in cases of carcinoma where eosinophilia appears to be associated with better prognosis or involved in the inflammatory response to tumor cells and where it can be demonstrated that the eosinophilia is primarily mediated by IL-5, treatment with exogenous IL-5 may be added to the chemotherapeutic regimes available.

Alternatively, there is compelling evidence that an excess of eosinophils and their products can be involved in the pathogenesis of several diseases. There is increasing evidence that eosinophils may be a major factor in the development of respiratory malfunction of asthma, as seen by the ability of eosinophil products such as major basic protein, eosinophilic cationic protein, and eosinophil peroxidase to cause direct cytotoxicity to lung epithelium in vitro and deposition of these products in damaged lung tissue from asthmatic patients (109,118). In the idiopathic hypereosinophilic syndrome and allergic granulomatosis (Churg-Strauss syndrome), deposits of eosinophil products have been demonstrated in the heart, kidney, and neural tissue. Prolonged peripheral eosinophilia with extremely elevated eosinophil counts from any cause can potentially result in end organ damage to the heart (endomyocardial fibrosis) or nerves (peripheral neuropathy). Whether the factors regulating eosinophilia locally are understood sufficiently well to use local targeting of IL-5 by humanized anti–IL-5 antibodies remains an open question.

SUMMARY

IL-5 is a cytokine derived primarily from T cells that has multiple regulatory functions on eosinophils and (in the mouse) on antibody-secreting B cells. Many of these functions overlap with those of IL-3 and GM-CSF. A complex network of cytokines appears to regulate the transcription and translation of the gene for IL-5. Molecular events that follow engagement of the IL-5 receptor have been elucidated in murine and human IL-5–responsive cells. The functional overlap of IL-5 with IL-3 and GM-CSF can be explained on the basis of a common receptor molecule and intracellular portions of the receptors resulting in the activation of similar, if not identical, intracellular events after receptor-ligand binding. Abnormally high levels of this cytokine are associated with infections with tissue-dwelling parasites and a diverse group of hypereosinophilic conditions of no known etiology. Our understanding of the biological role of IL-5 in the regulation of Ig production and the development of immunity to parasites is far from complete, but basic knowledge of its action at the cellular level is accumulating and will be critical for the intelligent application of immunotherapy with IL-5 or antibodies to IL-5 in infectious, neoplastic, and other diseases.

REFERENCES

1. Sanderson CJ. The biological role of interleukin 5. Int J Cell Cloning 1990; 1:147–153.
2. Swain S, Dutton RW. Production of a B cell growth-promoting activity, DL(BCGF), from a cloned T cell line and its assay on the BCL1, B cell tumor. J Exp Med 1982; 156:1821–1834.
3. Sanderson CJ. Eosinophil differentiation factor (interleukin-5). Immunol Ser 1990; 49:231–256.
4. Goodall GJ, Bagley CJ, Vadas MA, Lopez AF. A model for the interaction of GM-CSF, IL-3 and IL-5 receptors with their ligands. Growth Factors 1993; 8:87–97.
5. Takatsu K, Takaki S, Hitoshi Y. Interleukin-5 and its receptor system: implications in the immune system and inflammation. Adv Immunol 1994; 57:145–190.
6. Schimpl A, Wecker E. Replacement of T-cell function by a T cell product. Nature 1972; 237:15–17.
7. Takatsu K, Tominaga A, Hamaoka T. Antigen-induced T cell-replacing factor (TRF). I. Functional characterization of a TRF-producing helper T cell subset and genetic studies of TR production. J Immunol 1980; 124:2414–2422.
8. Harada N, Kikuchi Y, Tominaga A, Takaki S, Takatsu K. BCGFII activity on activated B cells of purified murine T cell replacing factor (TRF) from a T cell hybridoma (B151K12). J Immunol 1985; 134:3944–3951.
9. Sanderson CJ, Warren DJ, Strath M. Identification of a cytokine that stimulates eosinophil differentiation in vitro. Its relationship to IL-3, and functional properties of eosinophils produced in cultures. J Exp Med 1985; 162:60–74.
10. Sanderson CJ, Campbell HD, Young IG. Molecular and cellular biology of eosinophil differentiating factor (interleukin-5) and its effect on human and mouse B cells. Immunol Rev 1988; 102:29–50.
11. Azuma C, Tanabe T, Konishi M, et al. Cloning of the cDNA for human T-cell replacing factor (interleukin-5) and comparison with the murine homologue. Nucl Acids Res 1986; 14:9149–9158.
12. Kinashi TN, Harada E, Severinson E, et al. Cloning of a complementary DNA encoding T cell replacing factor and identity with B cell growth factor II. Nature 1986; 324:70–73.
13. Campbell HD, Tucker WQJ, Hort Y, et al. Molecular cloning, nucleotide sequence, and expression of the gene encoding human eosinophil differentiation factor. Proc Natl Acad Sci USA 1987; 84:6629–6633.
14. Sutherland GR, Baker E, Callen DF, et al. Interleukin-5 is at 5q31 and is deleted in the 5q- syndrome. Blood 1988; 71:1150–1152.
15. leBeau MM, Lemons RS, Espinosa R, et al. Interleukin-4 and interleukin-5 map to human chromosome 5 in a region encoding growth factors and receptors and are deleted in myeloid leukemias with a del(5q). Blood 1989; 73:647–650.
16. Yang Y-C, Kovacic S, Kriz R, et al. The human genes for GM-CSF and IL-3 are closely linked in tandem on chromosome 5. Blood 1988; 71:958–961.
17. Yates P, Potter MN. Eosinophilic leukaemia with an abnormality of 5q31, the site of the IL-5 gene. Clin Lab Haematol 1991; 13:211–215.
18. Geha RS, Castigli E, Chatila T. Novel immune deficiencies: defective transcription of lymphokine genes. Clin Immunol Immunopathol 1991; 61:S16–S20.
19. Campbell HD, Sanderson CJ, Wang Y, et al. Isolation, structure and expression of cDNA and genomic clones for murine eosinophil differentiation factor. Comparison with other eosinophilopoietic lymphokines and identity with interleukin-5. Eur J Biochem 1988; 174:345–352.
20. Tanabe T, Konishi M, Mizuta T, Noma T, Honjo T. Molecular cloning and structure of the human interleukin-5 gene. J Biol Chem 1987; 262:16580–16584.
21. Mizuta TR, Tanabe T, Nakakubo H, Noma T, Honjo T. Molecular cloning and structure of the mouse interleukin-5 gene. Growth Factors 1989; 1:51–57.
22. Taga T, Kishimoto T. Role of a two-chain IL-6 receptor system in immune and hematopoietic cell regulation. Crit Rev Immunol 1992; 11:265–280.
23. Takatsu K. Interleukin 5 (IL-5) and its receptor. Microbiol Immunol 1991; 35:593–606.
24. Limaye AP, Abrams JS, Silver JE, Ottesen EA, Nutman TB. Regulation of parasite-induced eosinophilia: selectively increased interleukin 5 production in helminth-infected patients. J Exp Med 1990; 172:399–402.
25. Mahanty S, Abrams JS, King CL, Limaye AP, Nutman TB. Parallel regulation of IL-4 and IL-5 in human helminth infections. J Immunol 1992; 148:3567–3571.
26. Bohjanen PR, Okajima M, Hodes RJ. Differential regulation

of interleukin 4 and interleukin 5 gene expression: a comparison of T-cell gene induction by anti-CD3 antibody or by exogenous lymphokines. Proc Natl Acad Sci USA 1990; 87:5283–5287.

27. Steel C, Nutman TB. Regulation of IL-5 in onchocerciasis. A critical role for IL-2. J Immunol 1993; 150:5511–5518.
28. Tominaga A, Matsumoto M, Harada M, et al. Molecular properties and regulation of mRNA for T cell-replacing factor/IL-5. J Immunol 1988; 140:1175–1181.
29. Takatsu K, Tominaga A. Interleukin 5 and its receptor. Prog Growth Factor Res 1991; 3:87–102.
30. Bass H, Mosmann T, Strober S. Evidence for mouse Th1- and Th2-like helper T cells in vivo. Selective reduction of Th1-like cells after total lymphoid irradiation. J Exp Med 1989; 170:1495–1511.
31. Street NE, Schumacher JH, Fong TA, et al. Heterogeneity of mouse helper T cells. Evidence from bulk cultures and limiting dilution cloning for precursors of Th1 and Th2 cells. J Immunol 1990; 144:1629–1639.
32. Yamaguchi Y, Suda T, Suda J, et al. Purified IL-5 supports the terminal differentiation and proliferation of murine eosinophil precursors. J Exp Med 1988; 167:43–56.
33. Enokihara H, Kajitani H, Nagashima S, et al. Interleukin 5 activity in sera from patients with eosinophilia. Br J Haematol 1990; 75:458–462.
34. Macdonald D, Gordon AA, Kajitani H, Enokihara H, Barrett AJ. Interleukin-2 treatment-associated eosinophilia is mediated by interleukin-5 production. Br J Haematol 1990; 76:168–173.
35. Schaafsma MR, Falkenburg JH, Landegent JE, et al. In vivo production of interleukin-5, granulocyte-macrophage colony-stimulating factor, macrophages colony-stimulating factor, and interleukin-6 during intravenous administration of high-dose interleukin-2 in cancer patients. Blood 1991; 78:1981–1987.
36. Miyatake S, Shlomai J, Arai K, Arai N. Characterization of the mouse granulocyte-macrophage colony-stimulating factor (GM-CSF) gene promoter: nuclear factors that interact with an element shared by three lymphokine genes—those for GM-CSF, interleukin-4 (IL-4), and IL-5. Mol Cell Biol 1991; 11:5894–5901.
37. Nagata SM, Tsuchiya M, Asano S, et al. The chromosomal gene structure and two mRNAs for human granulocyte colony-stimulating factor. EMBO J 1986; 5:575–581.
38. Takatsu K, Tominaga A, Harada N, et al. T cell-replacing factor (TRF)/interleukin 5 (IL-5): molecular and functional properties. Immunol Rev 1988; 102:107–135.
39. Takatsu K, Yamaguchi N, Hitoshi Y, et al. Signal transduction through interleukin-5 receptors. Cold Spring Harb Symp Quant Biol 1989; 2:745–751.
40. Minamitake Y, Kodama S, Katayama T, et al. Structure of recombinant human interleukin 5 produced by Chinese hamster ovary cells. J Biochem (Tokyo) 1990; 107:292–297.
41. Tavernier J, Devos R, Cornelis S, et al. A human high affinity interleukin-5 receptor (IL5R) is composed of an IL5-specific α chain and a β chain shared with the receptor for GM-CSF. Cell 1991; 66:1175–1184.
42. Mita S, Harada N, Naomi S, et al. Receptors for T cell-replacing factor/interleukin 5: specificity, quantitation and its implication. J Exp Med 1988; 168:863–878.
43. Tominaga A, Takaki A, Hitoshi Y, Takatsu K. Role of the interleukin 5 receptor system in hematopoiesis: molecular basis for overlapping function of cytokines. Bioessays 1992; 14:527–533.
44. Kunimoto DY, Allison KC, Watson C, et al. High-level production of murine interleukin-5 (IL-5) utilizing recombinant baculovirus expression. Purification of the rIL-5 and its use in assessing the biologic role of IL-5 glycosylation. Cytokine 1991; 3:224–230.
45. Mita S, Tominaga A, Hitoshi Y, et al. Characterization of high affinity receptors for interleukin 5 on interleukin 5-dependent cell lines. Proc Natl Acad Sci USA 1989; 86:2311–2315.
46. Takaki S, Tominaga A, Hitoshi Y, et al. Molecular cloning and expression of the murine interleukin-5 receptor. EMBO J 1990; 9:4367–4374.
47. Takaki S, Mita S, Kitamura T, et al. Identification of the second subunit of the murine interleukin-5 receptor: interleukin-3 receptor-like protein, AIC2B is a component of the high affinity interleukin-5 receptor. EMBO J 1991; 10:2833–2838.
48. Devos R, Plaetinck G, vanderHeyden J, et al. Molecular basis of a high affinity murine interleukin-5 receptor. EMBO J 1991; 10:2133–2137.
49. Shanafelt AB, Miyajima A, Kitamura T, Kastelein RA. The amino-terminal helix of GM-CSF and IL-5 governs high affinity binding to their receptors. EMBO J 1991; 10:4105–4112.
50. Lopez AF, Eglinton JM, Lyons AB, et al. Human interleukin-3 inhibits the binding of granulocyte-macrophage colony-stimulating factor and interleukin-5 to basophils and strongly enhances their functional activity. J Cell Physiol 1990; 145:69–77.
51. Lopez AF, Vadas MA, Woodcock JM, et al. Interleukin-5, interleukin-3, and granulocyte-macrophage colony-stimulating factor cross-compete for binding to cell surface receptors on human eosinophils. J Biol Chem 1991; 266:24741–24747.
52. Koike M, Kikuchi Y, Tominaga A, et al. Establishment and analysis of transgenic mice expressing IL-5 receptor α chain and functional analysis of their B cells (abstr). J Immunol 1993; 150:206A.
53. Koike M, Takatsu K. IL-5 and its receptor: which role do they play in the immune response? Int Arch Allergy Immunol 1994; 104:1–9.
54. Banchereau J, Galizzi JP. Receptors for interleukin 4, interleukin 5 and interleukin 6. Ann Ist Super Sanita 1990; 26:453–467.
55. Plaetinck G, vanderHeyden J, Tavernier J, et al. Characterization of interleukin 5 receptors on eosinophilic sublines from human promyelocytic leukemia (HL-60) cells. J Exp Med 1990; 172:683–691.
56. Chihara J, Plumas J, Gruart V, et al. Characterization of a receptor for interleukin 5 on human eosinophils: variable expression and induction by granulocyte/macrophage colony-stimulating factor. J Exp Med 1990; 172:1347–1351.
57. Migita M, Yamaguchi N, Mita S, et al. Characterization of the human IL-5 receptors on eosinophils. Cell Immunol 1991; 133:484–497.
58. Taga T, Kishimoto T. Signaling mechanisms through receptors that share signal transducing receptor components. Curr Opin Immunol 1995; 7:17–23.
59. Ihle JN, Kerr IM. Jaks and Stats in signaling by the cytokine receptor superfamily. Trends Genet 1995; 11:69–74.
60. Rothman P, Kreider B, Azam M, et al. Cytokines and growth factor signal through tyrosine phosphorylation of a family of related transcription factors. Immunity 1994; 1:457–468.
61. Narazaki M, Witthuhn BA, Yoshida K, et al. Activation of JAK2 kinase mediated by the interleukin 6 signal transducer gp130. Proc Natl Acad Sci USA 1994; 91:2285–2289.
62. Stahl N, Boulton TG, Farruggella T, et al. Association and activation of Jak-Tyk kinases by CNTF-LIF-OSM-IL-6 β receptor components. Science 1994; 263:92–95.
63. Takatsu K. Interleukin-5. Curr Opin Immunol 1992; 4:299–306.
64. Murata Y, Yamaguchi N, Hitoshi Y, Tominaga A, Takatsu K.

Interleukin 5 and interleukin 3 induce serine and tyrosine phosphorylations of several cellular proteins in an interleukin 5-dependent cell line. Biochem Biophys Res Commun 1990; 173:1102–1108.

65. Takaki S, Kanazawa H, Shiiba M, Takatsu K. A critical cytoplasmic domain of the interleukin-5 (IL-5) receptor chain β and its function in IL-5-mediated growth signal transduction. Mol Cell Biol 1994; 14:7404–7413.
66. Polotskaya A, Zhao Y, Lilly MB, Kraft AS. Mapping the intracytoplasmic regions of the granulocyte-macrophage colony-stimulating factor receptor for cell growth regulation. J Biol Chem 1994; 269:14607–14613.
67. Weiss M, Yokoyama C, Shikama Y, et al. Human granulocyte-macrophage colony-stimulating factor receptor signal transduction requires the proximal cytoplasmic domains of the α and β subunits. Blood 1993; 82:3298–3306.
68. Quelle FW, Sato N, Witthuhn BA, et al. JAK2 associates with the β chain of the receptor for granulocyte-macrophage colony-stimulating factor, and its activation requires the membrane-proximal region. Mol Cell Biol 1994; 14:4335–4341.
69. Silvennoinen O, Witthuhn B, Quelle FW, et al. Structure of the JAK2 protein kinase and its role in IL-3 signal transduction. Proc Natl Acad Sci USA 1993; 90:8429–8433.
70. Darnell JE, Kerr IM, Stark GR. Jak-STAT pathways and transcriptional activation in response to IFNs and other extracellular signaling proteins. Science 1994; 264:1415–1421.
71. Pazdrak K, Stafford S, Alam R. The activation of the Jak-STAT 1 signaling pathway by IL-5 in eosinophils. J Immunol 1995; 155:397–402.
72. Sato S, Katagiri T, Takaki S, et al. IL-5 receptor-mediated tyrosine phosphorylation of SH2/SH3-containing proteins and activation of Bruton's tyrosine and Janus 2 kinases. J Exp Med 1994; 180:2101–2011.
73. vanderBruggen T, Caldenhoven E, Kanters D, et al. Interleukin-5 signaling in human eosinophils involves JAK2 tyrosine kinase and Stat1 α. Blood 1995; 85:1442–1448.
74. Pazdrak K, Schreiber D, Forsythe P, Justement L, Alam R. The intracellular signal transduction mechanism of interleukin 5 in eosinophils: the involvement of *lyn* tyrosine kinase and the *ras-raf*-1-MEK-microtubule-associated protein kinase pathway. J Exp Med 1995; 181:1827–1834.
75. Koike M, Kikuchi Y, Tominaga A, et al. Defective IL-5-receptor-mediated signaling in B cells of X-linked immunodeficient mice. Int Immunol 1995; 7:21–30.
76. Appleby MW, Kerner JD, Chien S, et al. Involvement of p59*fyn* in interleukin-5 receptor signaling. J Exp Med 1995; 182:811–820.
77. Li T, Tsukada S, Satterthwaite A, et al. Activation of Bruton's tyrosine kinase (BTK) by a point mutation in its pleckstrin homology (PH) domain. Immunity 1995; 2:451–460.
78. Metcalf D, Parker JW, Chester EM, Kincade PW. Formation of eosinophilic-like granulocyte colonies by mouse bone marrow cells in vitro. J Cell Physiol 1974; 84:275–290.
79. Clutterbuck EJ, Sanderson CJ. Regulation of human eosinophil precursor production by cytokines: a comparison of recombinant human interleukin-1 (rhIL-1), rhIL-3, rhIL-5, rhIL-6, and rh granulocyte-macrophage colony-stimulating factor. Blood 1990; 75:1774–1779.
80. Strath M, Sanderson CJ. The production and functional properties of eosinophils from bone marrow cultures. J Cell Sci 1985; 74:207–217.
81. Owen WF, Rothenberg ME, Petersen J, et al. Interleukin 5 and phenotypically altered eosinophils in the blood of patients with the idiopathic hypereosinophilic syndrome. J Exp Med 1989; 170:343–348.
82. Butterfield JH, Leiferman KM, Abrams J, et al. Elevated serum levels of interleukin-5 in patients with the syndrome of episodic angioedema and eosinophilia. Blood 1992; 79:688–692.
83. Ohnishi T, Sur S, Collins DS, et al. Eosinophil survival activity identified as interleukin-5 is associated with eosinophil recruitment and degranulation and lung injury twenty-four hours after segmental antigen lung challenge. J Allergy Clin Immunol 1993; 92:607–615.
84. Ohnishi T, Kita H, Weiler D, et al. IL-5 is the predominant eosinophil-active cytokine in the antigen-induced pulmonary late-phase reaction. Am Rev Respir Dis 1993; 147:901–907.
85. Fukuda Y, Hashino J, Haruyama M, et al. A sandwich enzyme-linked immunosorbent assay for human interleukin-5. J Immunol Methods 1991; 143:89–94.
86. Czerkinsky C, Nilsson L-A, Nygren H, Ouchterlony O, Tarkowski A. A solid-phase enzyme-linked immunospot (ELISPOT) assay for enumeration of specific antibody-secreting cells. J Immunol Methods 1983; 65:109–114.
87. Czerkinsky C, Andersson G, Ferrua B, et al. Detection of human cytokine-secreting cells in distinct anatomical compartments. Immunol Rev 1991; 119:5–21.
88. Taguchi T, McGhee JR, Coffman RL, et al. Analysis of Th1 and Th2 cells in murine gut-associated tissues. Frequencies of $CD4^+$ and $CD8^+$ T cells that secrete IFN-γ and IL-5. J Immunol 1990; 145:68–77.
89. Kaye PM, Curry AJ, Blackwell JM. Differential production of Th1- and Th2-derived cytokines does not determine the genetically controlled or vaccine-induced rate of cure in murine visceral leishmaniasis. J Immunol 1991; 146:2763–2770.
90. Sander B, Cardell S, Möller G, Möller E. Differential regulation of lymphokine production in mitogen-stimulated murine spleen cells. Eur J Immunol 1991; 21:1887–1892.
91. Kagnoff MF, LY DG, McCaffery JM, Schoenbeck S. Detection by immunofluorescence of lymphokine-producing T cells. Immunol Res 1991; 10:255–257.
92. Yamamura M, Uyemura K, Deans RJ, et al. Defining protective responses to pathogens: cytokine profiles in leprosy lesions. Science 1991; 254:277–279.
93. vanSeventer GA, Newman W, Shimizu Y, et al. Analysis of T cell stimulation by superantigen plus major histocompatibility complex class II molecules or by CD3 monoclonal antibody: costimulation by purified adhesion ligands VCAM-1, ICAM-1, but not ELAM-1. J Exp Med 1991; 174:901–913.
94. Kay AB, Ying S, Varney V, et al. Messenger RNA expression of the cytokine gene cluster, interleukin 3 (IL-3), IL-4, IL-5, and granulocyte/macrophage colony-stimulating factor, in allergen-induced late-phase cutaneous reactions in atopic subjects. J Exp Med 1991; 173:775–778.
95. Hamid Q, Azzawi M, Ying S, et al. Interleukin-5 mRNA in mucosal bronchial biopsies from asthmatic subjects. Int Arch Allergy Appl Immunol 1991; 94:169–170.
96. Hamid Q, Azzawi M, Ying S, et al. Expression of mRNA for interleukin-5 in mucosal bronchial biopsies from asthma. J Clin Invest 1991; 87:1541–1546.
97. Elson LH, Nutman TB, Metcalfe DD, Prussin C. Flow cytometric analysis for cytokine production identifies T helper 1, T helper 2, and T helper 0 cells within the human $CD4^+CD27^-$ lymphocyte subpopulation. J Immunol 1995; 154:4294–4301.
98. Cardell S, Sander B, Möller G. Helper interleukins are produced by both CD4 and CD8 splenic T cells after mitogen stimulation. Eur J Immunol 1991; 21:2495–2500.
99. Paul CC, Keller JR, Armpriester JM, Baumann MA. Epstein-Barr virus transformed B lymphocytes produce interleukin-5. Blood 1990; 75:1400–1403.
100. Desreumaux P, Janin A, Colombel JF, et al. Interleukin 5

messenger RNA expression by eosinophils in the intestinal mucosa of patients with coeliac disease. J Exp Med 1992; 175:293–296.
101. Samoszuk M, Nansen L. Detection of interleukin-5 messenger RNA in Reed-Sternberg cells of Hodgkin's disease with eosinophilia. Blood 1990; 75:13–16.
102. Ying S, Durham SR, Barkans J, et al. T cells are the principal source of interleukin-5 mRNA in allergen-induced rhinitis. Am J Respir Cell Mol Biol 1993; 9:356–360.
103. Krishnaswamy G, Liu MC, Su SN, et al. Analysis of cytokine transcripts in the bronchoalveolar lavage cells of patients with asthma. Am J Respir Cell Mol Biol 1993; 9:279–286.
104. Kay AB, Ying S, Durham SR. Phenotype of cells positive for interleukin-4 and interleukin-5 mRNA in allergic tissue reactions. Int Arch Allergy Immunol 1995; 107:208–210.
105. Bradding P, Roberts JA, Britten KM, et al. Interleukin-4, -5, and -6 and tumor necrosis factor-α in normal and asthmatic airways: evidence for the human mast cell as a source of these cytokines. Am J Respir Cell Mol Biol 1994; 10:471–480.
106. Bradding P, Feather IH, Wilson S, et al. Immunolocalization of cytokines in the nasal mucosa of normal and perennial rhinitic subjects. The mast cell as a source of IL-4, IL-5, and IL-6 in human allergic mucosal inflammation. J Immunol 1993; 151:3853–3865.
107. Bradding P, Okayama Y, Howarth PH, Church MK, Holgate ST. Heterogeneity of human mast cells based on cytokine content. J Immunol 1995; 155:297–307.
108. Okayama Y, Petit-Frere C, Kassel O, et al. IgE-dependent expression of mRNA for IL-4 and IL-5 in human lung mast cells. J Immunol 1995; 155:1796–1808.
109. Gleich GJ, Adolphson CR, Leiferman KM. The biology of the eosinophilic leukocyte. Annu Rev Med 1993; 44:85–101.
110. Clutterbuck EJ, Shields J, Gordon J, et al. Recombinant human interleukin-5 is an eosinophil differentiation factor but has no activity in standard human B cell growth factor assays. Eur J Immunol 1987; 17:1743–1750.
111. Rennick DM, Thompson SL, Coffman RL, et al. In vivo administration of antibody to interleukin-5 inhibits increased generation of eosinophils and their progenitors in bone marrow of parasitized mice. Blood 1990; 76:312–316.
112. Her E, Frazer J, Austen KF, Owen WJ. Eosinophil hematopoietins antagonize the programmed cell death of eosinophils. Cytokine and glucocorticoid effects on eosinophils maintained by endothelial cell-conditioned medium. J Clin Invest 1991; 88:1982–1987.
113. Tai PC, Sun L, Spry CJ. Effects of IL-5, granulocyte/macrophage colony-stimulating factor (GM-CSF) and IL-3 on the survival of human blood eosinophils in vitro. Clin Exp Immunol 1991; 85:312–316.
114. Lu L, Lin ZH, Shen RN, et al. Influence of interleukins 3, 5 and 6 on the growth of eosinophil progenitors in highly enriched human bone marrow in the absence of serum. Exp Hematol 1990; 18:1180–1186.
115. Warren DJ, Sanderson CJ. Production of a T cell hybrid producing a lymphokine stimulating eosinophil differentiation. Immunology 1985; 54:615–623.
116. Lu P, Zhou X, Chen SJ, et al. CTLA-4 ligands are required to induce an in vivo interleukin 4 response to a gastrointestinal nematode parasite. J Exp Med 1994; 180:693–698.
117. Walsh GM, Wardlaw AJ, Hartnell A, Sanderson CJ, Kay AB. Interleukin-5 enhances the in vitro adhesion of human eosinophils, but not neutrophils, in a leukocyte integrin (CD11/18)-dependent manner. Int Arch Allergy Appl Immunol 1991; 94:174–178.
118. Weller PF. Cytokine regulation of eosinophil function. Clin Immunol Immunopathol 1992; 62:S55–9.
119. Lopez AF, Sanderson CJ, Gamble JR, et al. Recombinant human interleukin-5 is a selective activator of human eosinophil function. J Exp Med 1988; 167:219–224.
120. Owen WJ, Petersen J, Sheff DM, et al. Hypodense eosinophils and interleukin 5 activity in the blood of patients with the eosinophilia-myalgia syndrome. Proc Natl Acad Sci USA 1990; 87:8647–8651.
121. Limaye AP, Abrams JS, Silver JE, et al. Interleukin-5 and the posttreatment eosinophilia in patients with onchocerciasis. J Clin Invest 1991; 88:1418–1421.
122. van Haelst, Pisani C, Kovach JS, Kita H, et al. Administration of interleukin-2 (IL-2) results in increased plasma concentrations of IL-5 and eosinophilia in patients with cancer. Blood 1991; 78:1538–1544.
123. Yokota T, Coffman RL, Hagiawara H, et al. Isolation and characterization of lymphokine cDNA clones encoding mouse and human IgA-enhancing factor and eosinophil colony-stimulating factor activities: relationship to interleukin 5. Proc Natl Acad Sci USA 1987; 84:7388–7392.
124. Purkerson JM, Isakson PC. Isotype switching in anti-immunoglobulin-activated B lymphoblasts: differential requirements for interleukin 4 and other lymphokines to elicit membrane vs. secreted IgG1. Eur J Immunol 1991; 21:707–714.
125. Purkerson JM, Isakson PC. Interleukin 5 (IL-5) provided a signal that is required in addition to IL-4 for isotype switching to immunoglobulin (Ig) G1 and IgE. J Exp Med 1992; 175:973–982.
126. Yuan D, Dang T, Sanderson C. Regulation of Ig H chain gene transcription by IL-5. J Immunol 1990; 145:3491–3496.
127. Webb CF, Das C, Coffman RL, Tucker PW. Induction of immunoglobulin M mRNA in a B cell transfectant stimulated with interleukin-5 and a T-dependent antigen. J Immunol 1989; 143:3934–3939.
128. Webb CF, Das C, Eaton S, Calame K, Tucker PW. Novel protein-DNA interactions associated with increased immunoglobulin transcription in response to antigen plus interleukin-5. Mol Cell Biol 1991; 11:5197–5205.
129. Sonoda E, Matsumoto R, Hitoshi Y, et al. Transforming growth factor β induces IgA production and acts additively with interleukin 5 for IgA production. J Exp Med 1989; 170:1415–1420.
130. Sonoda E, Hitoshi Y, Yamaguchi N, et al. Differential regulation of IgA production by TGF-β and IL-5: TGF-β induces surface IgA-positive cells bearing IL-5 receptor, whereas IL-5 promotes their survival and maturation into IgA-secreting cells. Cell Immunol 1992; 140:158–172.
131. Katoh S, Tominaga A, Migita M, Kuda A, Takatsu K. Conversion of normal Ly-1-positive B lineage cells into Ly-1-positive macrophages in long term-bone marrow cultures. Dev Immunol 1990; 1:113–125.
132. Katoh S, Bendig MM, Kanai Y, et al. Maintenance of CD5$^+$ B cells at an early developmental stage by interleukin-5: evidence from immunoglobulin gene usage in interleukin-5 transgenic mice. DNA Cell Biol 1993; 12:481–491.
133. Bischoff SC, Brunner T, De Weck AL, Dahinden CA. Interleukin 5 modifies histamine release and leukotriene generation by human basophils in response to diverse agonists. J Exp Med 1990; 172:1577–1582.
134. Hirai K, Yamaguchi M, Misaki Y, et al. Enhancement of human basophil histamine release by interleukin 5. J Exp Med 1990; 172:1525–1528.
135. Denburg JA, Silver JE, Abrams JS. Interleukin-5 is a human basophilopoietin: induction of histamine content and baso-

philic differentiation of HL-60 cells and of peripheral blood basophil-eosinophil progenitors. Blood 1991; 77:1462–1468.

136. James SP, Kwan WC, Sneller MC. T cells in the inductive and effector components of the intestinal mucosal immune system of non-human primates differ in lymphokine mRNA expression, lymphokine utilization, and regulatory function. J Immunol 1990; 144:1251–1256.

137. James SP, Mullin GE, Kanof ME, Zeitz M. Role of lymphokines in immunoregulatory function of mucosal T cells in humans and non-human primates. Immunol Res 1991; 10:230–238.

138. Kay AB. T cells, cytokines and asthma. J R Coll Phys Lond 1994; 28:325–331.

139. Dent LA, Strath M, Mellor AL, Sanderson CJ. Eosinophilia in transgenic mice expressing interleukin 5. J Exp Med 1990; 172:1425–1431.

140. Tominaga A, Takaki S, Koyama N, et al. Transgenic mice expressing a B cell growth and differentiation factor gene (interleukin 5) develop eosinophilia and autoantibody production. J Exp Med 1991; 173:429–437.

141. Hitoshi Y, Yamaguchi N, Korenaga M, et al. In vivo administration of antibody to murine IL-5 receptor inhibits eosinophilia of IL-5 transgenic mice. Int Immunol 1991; 3:135–139.

142. Pearce EJ, Sher A. Immunity to helminths. Curr Opin Immunol 1989; 2:375–379.

143. Sher A, Coffman RL, Hieny S, Cheever AW. Ablation of eosinophil and IgE responses with anti-IL-5 or anti-IL-4 antibodies fails to affect immunity against *Schistosoma mansoni* in the mouse. J Immunol 1990; 145:3911–3916.

144. Lowe D, Jorizzo J, Hutt MSR. Tumour-associated eosinophilia: a review. J Clin Pathol 1981; 34:1343–8

145. Lowe W, Fletcher CDM, Gower RL. Tumor associated eosinophils in the bladder. J Clin Pathol 1984; 37:500–502.

146. Kapp DS, Livolsi VA. Intense eosinophilic stromal infiltrate in carcinoma of the uterine cervix. Gynecol Oncol 1983; 16:19–30.

147. Goldsmith MM, Gresson DH, Askin FB. The prognostic significance of stromal eosinophilia in head and neck cancer. Otolaryngol Head Neck Surg 1987; 96:319–324.

148. Stefanini M, Claustro JC, Motos RA, Bendigo LL. Blood and bone marrow eosinophilia in malignant tumors. Role and nature of blood and tissue eosinophil colony-stimulating factor(s) in two patients. Cancer 1991; 68:543–548.

149. Desenne JJ, Acquatella G, Stern R, et al. Blood eosinophilia in Hodgkin's disease. A follow-up of 25 cases in Venezuela. Cancer 1992; 69:1248–1253.

150. Srivastava A, Boswell HS, Heerema NA, et al. KRAS2 oncogene overexpression in myelodysplastic syndrome with translocation 5;12. Cancer Genet Cytogenet 1988; 35:61–71.

151. Berkowicz M, Rosner E, Rechavi G, et al. Atypical chronic myelomonocytic leukemia with eosinophilia and translocation (5;12). A new association? Cancer Genet Cytogenet 1991; 51:277–278.

152. Huland E, Huland H. Tumor-associated eosinophilia in interleukin-2-treated patients: evidence of toxic eosinophil degranulation on bladder cancer cells. Cancer Res Clin Oncol 1992; 118:463–467.

153. Arinaga S, Karimine N, Takamuku K, et al. Correlation of eosinophilia with clinical response in patients with advanced carcinoma treated with low-dose recombinant interleukin-2 and mitomycin C. J Immunol 1990; 144:4898–4905.

6
Interleukin-6

Gerard Cox
St. Joseph's Hospital and McMaster University, Hamilton, Ontario, Canada

Jack Gauldie
McMaster University, Hamilton, Ontario, Canada

INTRODUCTION

Interleukin-6 (IL-6) is the name given to a highly pleotropic cytokine (1), whose history and discovery typifies the complex nature of cell-derived factors and the limitations of the systems with which we have to study their characteristics. In retrospect (it's amazing we feel comfortable looking retrospectively after only a few years!) this cytokine was discovered many times during the 1980s. It is an excellent example of factor discovery as summarized by Hotta and Baird (2) in their article on transforming growth factor β (TGF-β). That so many groups were unknowingly working on the same molecule at the same time and came to the realization within a short period of time allows us to paraphrase Winston Churchill "never in the history of cytokine biology has so much been known about a molecule by so many, but so few realized it."

The earliest record of cloning the molecule belongs to Weissenbach et al. (3), who identified the activity as an interferon (interferon-β_2) (4), while another group (5) also described an IFN-β_2 message that proved to have the same activity (6). These activities were derived from cytokine of poly I:C stimulated fibroblasts.

Similar cellular approaches led Content et al. (7) to describe a 26 kDa protein of unknown function, which proved to be the same as IFN-β_2 (8). T-cell–derived factors involved in B-cell development and immunoglobulin synthesis were described by Teranishi et al. (9), and cloning (10) proved it to be the same molecule (11). At the same time, two groups were working on a human factor that modified the proliferation of plasmacytoma/hybridoma cells, one derived from fibroblasts (12,13) and a second from monocytes (14,15), both of which were shown to be IL-6. Similar studies had been completed in the mouse with the cytokine IL-HPI (16,17), and a further activity PCT-GF (18,19) was shown to be identical to IL-HPI and homologous to human HPGF/HGF, or IL-6. Murine IL-6 was also cloned as a hemopoietic factor from stromal cells (20). Possibly the earliest description of IL-6 activity was the report of a stem cell differentiation factor, MGI-2A (21), which was later shown to be IL-6 (22). The intriguing aspect here is that Sachs (MGI-2A) and Ravel (IFN-β_2) unknowingly worked on the molecule within the same institute for many years before they were able to show identity, a testament to the difficulties of working in cell systems in the absence of cloning. The importance of IL-6 as part of the body's immediate defense or stress response came with the search for the cytokine that regulated the hepatic response in inflammation. Termed hepatocyte stimulating factor (HSF) (23,24), its identification as IL-6 indicated a most important function of IL-6 (25)—that of regulating the acute-phase response.

Thus, the discovery that such diverse activities were resident in one molecule brought a large amount of data together within a short period of time. The molecule was cloned by numerous groups, and recombinant material was readily available. From the first realization of shared functions to the first summary conference (26), only 2 years elapsed. More recently, it has become evident that

IL-6 is the prototype of a family of cytokines (27). These are helical cytokines that share a common signal-transducing component (gp130) in their receptors (28). The group includes leukemia inhibitory factor (LIF), oncostatin M, ciliary neurotrophic factor (CNTF), IL-11, and IL-6. Not surprisingly, these agents share a number of actions—an example of redundancy in cytokine biology—and participate in a wide variety of cellular processes—an example of pleiotropy in cytokine biology (29–31). Many of the discoveries of the biology of IL-6 have served as paradigms for the examination of the biology of both related and unrelated cytokines (32,33).

BIOCHEMISTRY AND MOLECULAR BIOLOGY

IL-6 Gene Structure

The IL-6 gene has been cloned from human, mouse, and rat, and are of approximately 5, 7, and 10 kb in length, respectively (34–36). IL-6 is a single-copy gene that is located on chromosome 7p21 for human (37,38) and on chromosome 5 in the mouse (39). In all three species, there are five exons and four interferons, and the exon/intron boundaries are conserved between the three species (36). The gene structure is homologous to murine and human G-CSF (10) and to the chicken myelomonocytic growth factor (40).

The cDNAs of the three species share reasonable homology (rat:mouse 92%; mouse:human 65%; rat:human 68%), with a different comparison with the predicted amino acid sequences (rat:mouse 93%; mouse:human 42%; rat:human 58%) (36).

The mRNA transcripts appear to initiate from several cap sites in the human, possibly in different cell types (34), while the rat has two possible cap sites, only one of which has been proven to be used (36). The human has two species of mRNA, each approximately 1.3 kb (10,41,42), using two different polyadenylation signals 78 bp apart (Gauldie J., Northemann W., and Fey G.H., unpublished). The two species are equally abundant in human fibroblasts and macrophages. The mouse has the same two polyadenylation signals, but to date only one species of mRNA has been reported (32). The rat also has two mRNA species: one at 1.2–1.3 kb and another, occurring at about 20% abundance of the other in both fibroblasts and macrophages, with a size of 2.4 kb, using the second polyadenylation site (36). This increased size is caused by the insertion into the gene, of a 730 bp L1 repetitive sequence element (Fig. 1). Since the human transcripts differ in their content of AUUUA sequences, presumably coding for message stability (43)—the longer of the human transcripts having six such sequences and the shorter four—the two messages may thus have differential stability under different conditions of stimulation, perhaps performing fine-tuning on the regulation of gene expression.

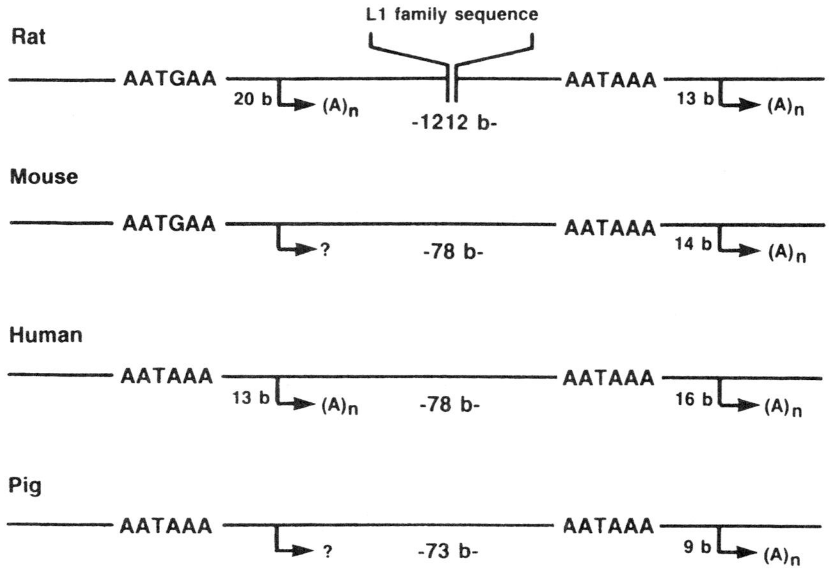

Figure 1 Comparison of the polyadenylation sites used in rat, mouse, human, and porcine IL-6 genes. The AATGAA polyadenylation signal for the rodent and AATAAA signal for human and porcine are shown adjacent to the polyadenylation site $(A)_n$. The position of the L1 family sequence insert (730 bp) is shown, and the distance between the two potential poly(A) sites in each gene is marked. Only the human and rat are confirmed for both RNA species.

Protein Structure

The open reading frame in the gene codes for 211-amino-acid protein in mouse and rat and 212 in the human. Depending on the site of cleavage of the signal peptide, this yields a mature protein of 183–185 amino acids. Human IL-6 has two N-glycosylation sites and several O-glycosylation sites and expresses glycosyl and phosphoryl microheterogeneity (44–46) with an apparent molecular weight of 23–30 kDa (47–49) and reviewed by Sehgal (50). Neither rat nor mouse has N-glycosylation sites, and purification of murine IL-6 from P388D$_1$ cells indicated the lack of glycosyl residues in the natural protein (51). Figure 2 shows a homologous comparison of human rat and mouse IL-6 proteins. All proteins have four cysteine residues at conserved positions in the chain, and the intrachain disulfide bonds are identical in mouse and human (ap45-ap51 and ap74-ap84) (52). This bridging is similar for IL-6 and G-CSF and is important for proper folding of the molecule (53). Rat IL-6 has an additional free cysteine at position 103, possibly able to form covalent structures with other proteins. Given the importance

```
  1 MKFLSARDFQPVAF.LGLMLLTATAFPTSQVRRGDFTEDTTHNRPVYTTS 49  R
  1 MKFLSARDFHPVAF.LGLMLVTTTAFPTSQVRRGDFTEDTTPNRPVYTTS 49  M
  1 MNSFSTSAFGPVAFSLGLLLVLPAAFPAPVPPGEDSKDVAAPHRQPLTSS 50  H
  1 MNSLSTSAFSPVAFSLGLLLVMATAFPTPGRLEEDAKGDATSDKMLFTSP 50  P

 50 Q.VGGLITYVLREILEMRKELCNGNSDCMNSDDALSENNLKLPEIQRNDG 98  R
 50 Q.VGGLITHVLWEIVEMRKELCNGNSDCMNNDDALAENNLKLPEIQRNDG 98  M
 51 ERIDKQIRYILDGISALRKETCNKSNMCESSKEALAENNLNLPKMAEKDG 100 H
 51 DKTEELIKYILGKISAMRKEMCEKYEKCENSKEVLAENNLNLPKMAEKDG 100 P

 99 CFQTGYNQEICLLKICSGLLEFRFYLEFVKNNLQDNKKDKARVIQSNTET 148 R
 99 CYQTGYNQEICLLKISSGLLEYHSYLEYMKNNLKDNKKDKARVLQRDTET 148 M
101 CFQSGFNEETCLVKIITGLLEFEVYLEYLQNRFESS.EEQARAVQMSTKV 149 H
101 CFQSGFNQETCLMRITTGLVEFQIYLDYLQKEYESN.KGNVEAVQISTKA 149 P

149 LVHIFKQEIKDSYKIVLPTPTSNALLMEKLESQKEWLRTKTIQLILKALE 198 R
149 LIHIFNQEVKDLHKIVLPTPISNALLTDKLESQKEWLRTKTIQFILKSLE 198 M
150 LIQFLQKKAKNLDAITTPDPTTNASLLTKLQAQNQWLQDMTTHLILRSFK 199 H
150 LIQTLRQKGKNPDKATTPNPTTNAGLLDKLQSQNEWMKNTKIILILRSLE 199 P

199 EFLKVTMRSTRQT  211 R
199 EFLKVTLRSTRQT  211 M
200 EFLQSSLRALRQM  212 H
200 DFLQFSLRAIRIM  212 P
```

Figure 2 Comparison of rat (R), mouse (M), human (H), and porcine (P) IL-6 amino acid sequences showing homology between the species (dashes). The cysteine residues are marked with boxes. The rat has an additional cysteine at position 114. Gaps (.) are introduced to produce optimal alignments.

attributed to disulfide bridging, it was quite surprising that Jambou et al. found significant biological activity with a synthetic cysteine-free form of IL-6 (54).

The N-terminus of the molecule is not necessary for bioactivity. Residues 1-28 can be removed and the human molecule maintains activity in the hybridoma bioassay (55), while removal of two residues from the C-terminus causes marked loss of activity (56). It is likely that the C- and N-termini are held in proximity in the intact molecule, as shown by studies using epitope-specific monoclonal antibodies (41). IL-6 is one of a group of cytokines that are predicted to share a common tertiary folding structure (57). This is similar to the four α-helix structure of growth hormone. The molecule is predicted to comprise four amphipathic helices that pack into an antiparallel bundle so that hydrophobic elements are buried. The majority of the homology of IL-6 with other members of the group occurs in the areas of these helices, while the loops that connect them exhibit differences between cytokines. The disulfide bridges between the cysteine residues would be expected to impart stability to the final folded structure.

Gene Regulation

Expression of the gene is stimulated by a wide range of mediators, including a number of cytokines and growth factors, bacterial endotoxins, and neuropeptides. The most prominent stimulus in fibroblasts is IL-1, with PDGF and FGF also causing upregulation. In monocytes/macrophages, lipopolysaccharide (LPS) is the most potent stimulator, with phorbol ester (TPA) also being prominent. In macrophages, there is a rapid increase in IL-6 mRNA after LPS addition, with raised levels detected at 1–2 h and maximum accumulation occuring at 4–6 h (36). Inhibition of stimulated expression occurs with corticosteroid treatment. Pretreatment with 10^{-8}–10^{-6} M dexamethasone shuts off IL-6 production in vitro in most cells, and IL-1 or LPS is then not able to upregulate the expression (58).

The regulation of IL-6 gene expression occurs through at least three different pathways of signal transduction, including cyclic AMP (59), calcium activation (60), and diacylglycerol (61). Regulation is achieved through the coordination of at least three *cis*-acting elements in the 5′ region of the gene. The region at -113 to -225 codes for a serum-responsive element (SRE) similar to that seen in c-fos oncogene (62). The second region at -145 to -158 has been called the NFIL-6 region (63), and the nuclear factor (NF-IL-6) has been shown to have high homology to C/EPB, another important linear specific nuclear factor (64). A third site, termed NFKB-like, at -63 to -73 has proven to be necessary for induction (65,66).

With respect to negative regulation by corticosteroids, there are several potential sites of action in the 5′ regulatory region of the gene. There are two homologous AP-1 sites, but neither has been found to be functional for induction. These sites may be important for downregulation, as seen for the collagenase gene stimulated by IL-1 treatment of fibroblasts (67–69). Ray et al. described a steroid-sensitive negative regulatory region that appears to overlap the TATA box and the Inr element (70). Later, Ray and Prefontaine showed that activation of the IL-6 promoter by a combination of NF–IL-6 and the p65 subunit of NF-κB could be inhibited by corticosteroid (71). They also showed that direct interaction between NF-κB and the glucocorticoid receptor was partially responsible for this effect. Zitnik et al. found that inhibition of mRNA transcription explained only part of the response in human lung fibroblasts and that selective destabilization of IL-6 mRNA also occurred with steroid treatment (72).

CELLULAR SOURCES OF IL-6

Cells of Origin

Most, if not all, nucleated cells have been shown to express and synthesize IL-6 in vitro. The most prominent source appears to be stimulated monocyte/macrophages, and cytokine (IL-1 or TNF) stimulated the stromal cells-fibroblasts and epithelial cells as well as endothelial cells. In addition, T- and B-cell lines can express IL-6, as can mast cell lines, numerous tumor cell lines, and even polymorphonuclear leukocytes (25,32,73,75). Within the central nevous system, IL-6 can be expressed by astrocytes, microglial cells, and folliculostellate cells in vitro (76,77).

While many cells can be shown to secrete IL-6 in vitro, examination of normal and inflamed tissue by Northern gel analysis and in situ hybridization reveals that only a few cells are active in pathological conditions. Thus, although synoviocytes and chondrocytes can release high levels of IL-6 upon IL-1 or TNF stimulation in vitro, and high levels of IL-6 are found in the synovial fluid of a rheumatoid joint, only the synovial fibroblasts were seen to be active by in situ examination (78,79).

Studies in both humans and animals have shown an increase in circulating levels of IL-6 with age (80). Lymphoid cells of aged, but not adult, mice spontaneously produce IL-6 in vitro. Administration of the steroid hormone dihydroepiandrosterone (DHEA) reduced plasma levels of IL-6 (80). The increases in serum immunoglobulins and amyloid P that were associated with elevated IL-6 levels were also reduced with DHEA treatment, indicating the IL-6 was biologically active and perhaps contributing to the consequences of aging.

In an animal model of urinary tract infection, Rugo et al. found significant elevation of IL-6 in the urine but not in the plasma, indicating compartmentalization of the cytokine response (81). By immunohistochemistry, the mesangial cells were the main cell source of IL-6 and not the infiltrating inflammatory cells.

IL-6 in Plasma: Action as a Hormone

Among the various cytokines that have been described, IL-6 shows signs of being involved as an autocrine, paracrine, and exocrine inflammatory hormone. Its activity as a growth factor for plasmacytomas demonstrates its autocrine capability (82), as does its role in renal carcinoma (83) and the proliferation of mesangial cells (84). This interpretation has been challenged (85,86), and a paracrine role has been suggested. Paracrine activity is also suggested by data showing Kupffer's cell activation and IL-6 release with subsequent hepatocyte response (see later) in the generation of acute-phase proteins in LPS-treated rats (87).

The exocrine function of IL-6 can be shown in a system in which a sterile abscess induced subcutaneously causes an increase of IL-6 in the plasma and response of the hepatocyte with acute-phase protein synthesis. Primary cultures of hepatocytes can be shown to make IL-6 in vitro (88), but at no time during the challenge does the hepatocyte show IL-6 message or protein. Thus the signal must be derived from the distal influenced tissue and travel through the circulation to elicit the metabolic change in the hepatocyte (89).

Under experimental conditions in animals or associated with trauma or infection in humans, IL-6 levels are shown to rise in plasma shortly after challenge (90–93). There is a temporal lag compared to levels of TNF, and thus it is likely that macrophage/monocytes are the primary cell of activation, releasing their many cytokines, but predominantly IL-1 and TNF. These early alarm cytokines can recruit adjacent stromal cells (fibroblast, smooth muscle cells) and endothelial cells to release high levels of IL-6 and other mediators (94). Such a sequence was shown in mice treated with IL-1 and showing a subsequent IL-6 response (95). This augmentation of signal is important in the generation of raised plasma levels suitable to cause systemic changes. Figure 3 outlines a scheme involving the sequential activation and cytokine cascade leading to raised IL-6 levels in inflammation. Since IL-6 can induce cortisol levels (see below) and cortisol is required for the hepatic acute-phase response, IL-6 play an inductive role in providing a second required signal to the liver (48,96). At the same time, cortisol is a potent inhibitor of IL-6 gene expression (58,70–72) and can subsequently reduce the levels of IL-6. Obviously, the timing and intensity of these hormone cascades are carefully controlled to elicit the appropriate effect needed in the correct temporal sequence to cope with the infective or traumatic challenge.

IL-6 RECEPTORS

As would be expected from the pleotropic nature of its action, many diverse cells have IL-6 receptors detected through binding studies (97–99). With the cloning of the human IL-6–binding receptor (100), the novel characteristics of IL-6–mediated signal transduction became apparent. Since then other cytokines have been shown to utilize similar receptor structure and signaling mechanisms (28). The 80 kDa receptor, a member of the Ig supergene family, is specific for IL-6 and is present on monocytes, hepatocytes, and T and B cells. This can be shed from the surface of cells (perhaps due to specific proteases) (101) and found in a soluble form in biological fluids and cell supernatants. IL-6 will bind to either the soluble or the membrane-associated form. This complex then associated with a second molecule, a transmembrane 130 kDa glycoprotein (102), which is required for signal transduction. gp130 is shared with the other helical cytokines—oncostatin M, LIF, CNTF, IL-11 (28)—and cannot bind IL-6 (or the other cytokines) directly. This two-chain receptor model helps explain the low- and high-affinity receptors detected on various cells (103) and is structurally similar to many cytokine receptors (94,104). Association of IL-6 with the 80 kDa receptor gives rise to a low-affinity interaction. This is converted to a high-affinity interaction by the association with monomeric or dimeric gp130.

The regulation of expression of the receptor has been studied in vitro in the human. Bauer et al. (105) have reported that in monocytes, LPS and IL-6 decrease the expression of IL-6R, as does corticosteroid. Hepatocytes seem to behave in the opposite fashion with IL-6 and IL-1 as well as corticosteroid, causing increased expression of the receptor (106). Similar positive regulation of IL-6R by glucocorticoid occurs with epithelial cells (107). The homologous rat 80 kDa receptor was cloned (108), and we used this probe to examine the behavior of the receptor in vivo. The role of corticosteroid as an upregulator does not appear to be as dramatic in vivo, and during inflammation the receptor expression on hepatocytes appears to be both up- and downregulated depending on the kinetics of inflamation (242,243). This area requires considerable work to clarify the regulation of the ligand-binding receptor.

Both chains of the IL-6 receptor belong to the receptor family that is characterized by a WSXWS motif and a region of approximately 200 amino acids that includes four

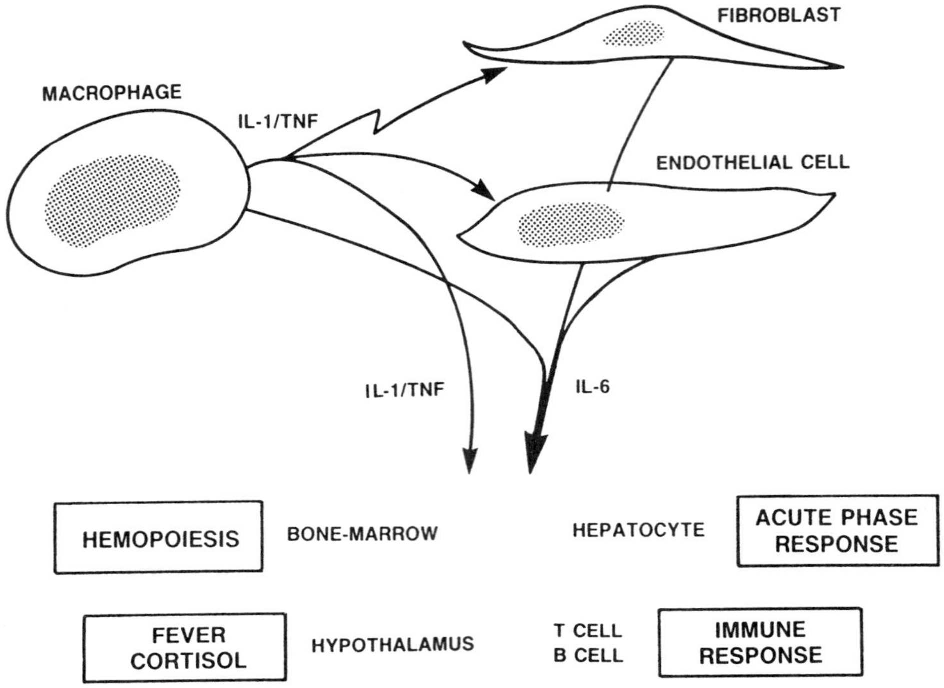

Figure 3 Sequence of cytokine cascades that result in the raised IL-6 levels and the multiple effects of exposure of cells and tissue to the various signals. The initial release of IL-1 and TNF result in the activation of adjacent tissue cells and further release of IL-6.

cysteines (109). This family includes receptors for interleukins (IL-2–IL-5), colony-stimulating factors (GM- and G-CSF), and growth factors (growth hormone, erythropoietin, prolactin) (110). Given this degree of relatedness among receptors, it is important to try to understand how specific cellular responses are mediated. This is dictated at one level by the cell phenotype and at the subcellular level by distinct subregions of the cytoplasmic domains of the receptor utilizing different signaling mediators (110). In the case of IL-6 signal transduction, the Janus kinase family of tyrosine kinases (Jak) are integrally involved (111). Activation of IL-6 receptor-associated Jak leads to the recruitment and tyrosine phosphorylation of a signaling protein of the STAT (Signal Transducer and Activation for Transcription) family. Activation of STAT proteins leads to dimerization and increased DNA-binding activity. Specifically, IL-6 activates STAT-3, which is also known as APRF (acute-phase response factor) via Jak-1 (28).

FUNCTIONS OF INTERLEUKIN-6

B-Cell Function

One of the earliest identified activities of IL-6 was as a T-cell–derived signal that induced the final maturation of B lymphocytes to plasma cells (112,113). Using pokeweed mitogen (PWM)–stimulated peripheral blood mononuclear cells, recombinant human IL-6 (rhIL-6) caused 3- to 10-fold increases in IgG, IgM, and IgA antibody production in a dose-dependent fashion (114). Evidence that IL-6 could act directly on activated B cells includes the detection of IL-6 receptor on the surface of activated B cells (though not on resting B cells, which are not stimulated to produce Ig by IL-6) and experiments showing the requirement for IL-6 when purified B cells are cultured in the presence of irradiated T cells and mitogen.

That IL-6 is in fact an essential factor for antibody production is suggested by the finding that anti–IL-6 diminished antibody production by 90% in PWM-stimulated peripheral blood lymphocytes (PBL) (114). Because this effect was present when antibody was added even after 4 days of culture (8 days total), IL-6 likely acts a later stage of B-cell differentiation.

The effect of IL-6 on immune functions in vivo has been studied using mice. Both the primary and the secondary antibody responses to sheep red blood cells were enhanced by IL-6, modifying both the intensity and the kinetics of antibody production (115). Experiments using transgenic mice have shown dramatic effects of excess IL-6 with some variation in the outcome depending on the different experimental methods used. These include the development of plasmacytomas and myeloma and high levels of IgG1 (as well as mesangio-proliferative glomerulonephritis and megakaryocytosis) (116,117). Brandt et al., using an IL-6 retrovirus construct in mice, induced a syndrome that resembled Castleman's disease (118), while Hawley et al., using different promoters, caused a rapidly fatal myeloproliferative disorder (119). In the latter system there was little evidence of B-cell or plasma-cell expansion. The in vivo consequences of IL-6 deficiency have been studied using IL-6 knock-out mice. These develop (almost) normally and have normal levels of B cells, indicating that IL-6 is not essential for B-cell development (120). Also, while basal levels of circulating immunoglobulins were normal after challenge with vesicular stomatitis virus (VSV), the IgM response was normal and the IgG response was modestly reduced (124). In contrast, IL-6–deficient mice were unable to mount an IgA response to local challenge, and this was corrected by replacing IL-6 (120). These data support an important role for IL-6 in the generation of local responses rather than the development of B cells and systemic responses.

B-Cell Growth

Evidence for the role of IL-6–mediating stimulation of proliferation of B cells is mixed. In the work referred to above by Muraguchi, the addition of anti–IL-6 did not alter mitogen-induced proliferation of the B cells (114). Alderson and Pike used a single hapten-specific murine B-cell line to show enhancement of antibody production without proliferation in response to human IL-6 (121). On the other hand, one of the earliest described for IL-6 was that of a B-cell growth factor (14), which is a potent growth factor for myeloma and plasmacytoma cells (18,82,86).

Work with Epstein-Barr virus (EBV)–transformed human B-cell lines demonstrated stimulation of Ig production by IL-6 but did not support a role for the molecule in growth promotion. Taga et al. demonstrated IL-6 receptors on EBV-transformed B cells (98), and IL-6 promoted the proliferation of these lines. Yokoi et al. showed that such lines expressed message RNA for IL-6 and produced the protein (122). Thus IL-6 could act as an autocrine growth factor. It appears that the effect is present mainly when cells are cultured at low density, which may be why the early work did not detect this activity.

That IL-6 may play a role in the development of human B-cell malignancies is suggested by the finding of IL-6 receptors on freshly isolated human multiple myeloma cells (82). These cells also produce IL-6, respond to exogenous IL-6 by increased proliferation, and show inhibition of growth by anti–IL-6, all of which suggests that IL-6 can act as an autocrine factor for human myeloma cells. Whether IL-6 functions as an autocrine or a paracrine hormone in this regard (85), its activity might depend on the expression of transferring receptors by the myeloma cells (123). From the in vivo studies of excess or deficient IL-6, it is clear that it can promote overproliferation of B cells but is not required for normal development of these cells (116–120,124).

T-Cell Function

In contrast to B cells, both resting and activated T cells express IL-6R, and a variety of effects on T cells at different levels of differentiation have been described. IL-6 stimulates the proliferation of both thymocytes and peripheral T cells, in response to mitogen (125,126). In addition, IL-6 promotes differentiation and proliferation (127,128) as well as augmenting the activity of natural killer (NK) cells (129). Many of these actions may be enhanced through synergy with IL-1 (130) and appear to be mediated via IL-2 and the IL-2R (42). Induction and enhancement of IL-2R expression in T cells by IL-6 has been shown (125,131). Addition of antibody to the p55 chain of IL-2R partially abrogated proliferation of thymocytes and T cells in response to IL-6 (124,125), and anti–IL-2R blocked the effect of IL-6 on NK cell activity (132). A second IL-2–independent pathway of stimulation of thymocyte and T-cell proliferation is suggested, since anti–IL-2R failed to inhibit IL-6 responses under different experimental conditions of low phytohemagglutinin (PHA) concentrations (31,33).

In a murine system to study delayed-type hypersensitivity (DTH) responses, IL-6, in addition to TNF-α, was shown to enhance DTH in a dose-dependent manner. The augmentation required the presence of T cells in the inoculum, again suggesting it might be mediated by IL-2 (134). Interestingly, Scott et al. failed to show significant IL-6 effects in a system examining T-cell "retention," which comprises survival and proliferation, in the presence of rheumatoid synoviocyte culture medium (135). One might have expected to see a role for IL-6 in this system, given its effects on T-cell proliferation. This example highlights the difficulties of extrapolating from in vitro data to human diseases.

IL-6–deficient mice show modest reduction in the numbers of thymocytes and peripheral T cells, to 60–80% of controls (124). However, the cells did have normal patterns of expression of surface markers. T-cell responses to virus infections were variably affected in that cytotoxic T-cell (CTL) function against lymphocytic choriomeningitis virus (LCV) and attenuated vaccinia virus was normal but was 3- to 10-fold reduced against vaccinia virus. Also, in these mice the T-cell–dependent antibody response to

VSV is reduced, as is macrophage killing of *Listeria monocytogenes*, which depends on appropriate T-cell activation. These data support an important role for IL-6 in the generation of certain T-cell responses and a modest role in T-cell development. After intraperitoneal infection with an adenovirus vector expressing rodent IL-6 cDNA, mice showed marked splenomegaly and accumulation of lymphoid aggregates in the liver, which resolved as IL-6 levels declined (136). Both B- and T-cell compartments were found to be enlarged. When such infection and expression was targeted to the bronchial epithelium, there was striking accumulation of T cells in the airway lumen and massive lumphocytic infiltration of the parenchyma (137,138). Only minor B-cell accumulation was noted, and again, the changes subsided after 10 days, subsequent to declining IL-6 levels.

IL-6 as a Hemopoietic Factor

Support for a role of IL-6 as a hemopoietic growth factor includes the finding that bone marrow stromal cells are potent sources of this cytokine (20). The myeloid growth factor known as macrophage and granulocyte inducer type 2 (MGI-2) or differentiation factor was shown to be IL-6 by Shabo et al. (22). This murine factor induces monocytic and granulocytic differentiation of normal myeloid precursor cells and some clones of murine leukemic cells. Chiu and others showed that murine IL-6 is both a growth factor for hemopoietic progenitor and leukemic cells and a differentiation factor, inducing dose-dependent monocytic development from the M1 and WEHI-3B myeloid leukemic cell lines (139). In addition to morphological evidence of differentiation, the cells had increased expression of FcγRII mRNA and enhanced phagocytic activity within 6 hours. While exhibiting differentiation, the M1 cells had markedly diminished rates of proliferation and subsequent loss of viability. Whether this represents true antiproliferative activity of IL-6 or occurs as a result of differentiation is unclear. IL-6 can interact with IL-1, IL-3, IL-4, and the CSFs to promote colony formation by multipotential hemopoietic progenitors (140–145). Human IL-6 also stimulated murine hemopoietic colony formation (146), and Suda et al. showed that human progenitors could similarly be stimulated to proliferate and differentiate in response to IL-6 (147).

Ishibashi et al. reported the effect of IL-6 on megakaryocyte maturation and, later, that administration of rhIL-6 to mice induced elevated platelet levels and increased megakaryocyte cell diameter (148,149). Hill et al. noted that treatment of mice with rhIL-6 (8 injections s.c. over 54 hours) resulted in elevated megakaryocyte counts and platelet levels in a dose-dependent fashion (150). No change in hematocrit and only a modest increase in white cell count with the highest dose of IL-6 were detected in this study.

The recent increase in interest in the use of colony-stimulating factors in the recovery from leukopenia associated with chemotherapy and radiotherapy has stimulated study of the effects of IL-6. In mice, Okano et al. reported that following bone marrow transplantation, administration of IL-6 resulted in increases in white blood cell counts both in marrow and in the circulation (151). However, they found no effect on red blood cell or platelet counts. Furthermore, survival rates following transplantation could be increased from 20 to 90% when the transplanted material was preincubated with IL-6 and IL-3.

As might have been expected from the above experience, transgenic mice with excess IL-6 show increases in megakaryocytes and platelet counts (116,117). However, with the exception of the experience of Hawley et al. (119), other levels of white blood cells and red cells appear to be normal in the presence of excess IL-6 or in its absence (116–118). Thus, it would appear that the functional redundancy among cytokines can compensate for IL-6 deficiency, at least in regard to leukocyte and red blood cell development.

IL-6 AND THE ACUTE-PHASE REACTION

Injury or infection induces the acute-phase response—a coordinated homeostatic reaction, which includes fever, leukocytosis, and changes in the circulating levels of particular proteins, the acute-phase proteins (APP). IL-6 is the cytokine primarily responsible for regulation of production of the acute-phase proteins by the liver. This, and regulation of B-cell function, are likely the most important physiological roles of IL-6. The majority of these acute-phase reactants are glycoproteins, which play a variety of roles in the homeostatic response to injury. From Table 1, we see that the proteins can both mediate the process and be consumed by it (reviewed in Refs. 96,152,153).

In 1983 Ritchie and Fuller described a monocyte-derived acute-phase regulatory protein and termed it hepatocyte-stimulating factor (HSF) (23). This molecule required the presence of corticosteroid for maximal activity. Others, including our group, confirmed that monocyte-conditioned medium could regulate most of the APP genes in isolated rat hepatocytes and hepatoma cells in a coordinated manner similar to that seen in vivo. We showed that HSF was identical to the previously cloned molecule interferon-β2 (B-cell stimulating factor), now known as interleukin-6 (25). Prior to the recognition of HSF/IL-6 and development of anti–IL-6 and rIL-6, much of the regulation of hepatocyte function was falsely attributed to inter-

Table 1 Acute-Phase Proteins

Protein	Symbol	Species	Function
		Major changes	
C-reactive protein	CRP	Human, rabbit	Opsonin
Serum amyloid A protein	SAA	Human, mouse, rat	Apolipoprotein
Cysteine proteinase inhibitor	CPI	Rat	Antiproteinase
α_2-Macroglobulin	α2M	Rat	Antiproteinase
α_1-Acid glycoprotein	α1AGP	Most species	Transport
		Medium changes	
Fibrinogen	FBG	Most species	Coagulation
α_1-Proteinase inhibitor	α1Pi	Human, rat, mouse	Antiproteinase
α_1-Antichymotrypsin	α1ACHY	Human, rat, mouse	Antiproteinase
Haptoglobin	HPT	Most species	Binds and removes hemoglobin
Hemopexin	HPX	Most species	Binds heme
Ceruloplasmin	CER	Mouse	O_2 scavenger, transport
Complement C3	C3	Human, mouse	Opsonin
Serum amyloid P	SAP	Human, mouse	(?)
		Negative changes	
Albumin	Alb	Most species	(?)
Transferrin	Tf	Most species	Transport

leukin-1 (IL-1) and tumor necrosis factor (TNF). The evidence was derived from studies involving the in vivo administration of IL-1 or TNF and because these cytokines are potent inducers of IL-6, the observed activities were incorrectly attributed to IL-1 and TNF. The APP can be divided into two groups, those regulated primarily by IL-1–type cytokines and those regulated mainly by IL-6–type cytokines (see Table 2). Other mediators such as TGF-β, insulin and insulinlike growth factor, and hepatocyte growth factor can modulate the response of the liver cells to IL-1 and IL-6 by attenuating the induction of APP genes or altering IL-6 receptor expression (153).

There is in vivo evidence that supports the hypothesis that IL-6 is the principal regulator of APP production.

Table 2 Acute-Phase Protein Genes Regulated by Inflammatory Cytokines

Induced by IL-1 and IL-6	Induced by IL-6 only
C-reactive protein (human)	Cysteine proteinase inhibitor
α_1-Acid glycoprotein	α_2-Macroglobulin
C3	Fibrinogen
Factor B	α_1-Proteinase inhibitor
SAA	Haptoglobin (human)
SAP	α_1-Antichymotrypsin
Haptoglobin (rat)	Ceruloplasmin
Hemopexin (rat)	C1 esterase inhibitor

Human rIL-6 given i.p to rats mimics the AP response seen following turpentine, but it occurs sooner and requires coadministration of corticosteroid (dexamethasone) (154,155), which may induce IL-6R. Also, the response seen in vivo parallels the in vitro response of rat hepatoma cells to IL-6. In burn patients the increase in IL-6 preceded elevations in C-reactive protein levels and correlated directly with the febrile response (93). In other clinical situations elevated IL-6 levels have been detected early in the course of infectious diseases (156,162,163). When rTNF was given to patients with malignancies, there was a prompt (5-minute) increase in IL-6, peaking at 3 hours, followed at 4 h by elevated cortisol levels; at 8 h increased C-reactive protein was found (90).

As mentioned above, other elements of the response to injury include fever and leukocytosis. While endogenous pyrogen (interleukin-1) is well known to cause fever, rhIL-6 given i.v. to rabbits induced fever in a dose-dependent manner, detectable at 15 min and peaking at 1 hour. Furthermore, Rothwell et al. have shown that antibody to rat IL-6 inhibits fever induced by systemic LPS treatment (158). As yet there is no concrete evidence that IL-6 contributes to the neutrophilia of early inflammation, although it does enhance proliferation of hemopoietic precursors (see above) and may activate mature neutrophils (see below). Interleukin-6 stimulates ACTH release via the release of corticotrophin-releasing hormone (159) and subsequently corticosteroid production, and thus may indirectly lead to neutrophilia via corticosteroid-induced demargination and increased survival of neutrophils (160).

Temporary overexpression of IL-6 has been achieved by intraperitoneal infection of mice with a recombinant adenovirus vector expressing rodent IL-6 cDNA (136). Infected animals showed a fourfold increase in serum APP (haptoglobin and α_1-acid glycoprotein) levels during the period of elevated IL-6 levels.

The generation of the APR in response to different stimuli has been examined in IL-6–deficient mice (30,124). These mice responded normally to exogenous IL-6, indicating that it was not essential in hepatic development. The response to systemic LPS was only slightly reduced in IL-6–deficient mice, while the response to subcutaneously injected turpentine (causing sterile tissue damage) and intravenous bacterial infection with *Listeria monocytogenes* was markedly attenuated (124), indicating that LPS caused release of numerous cytokines, some of which could mimic the action of IL-6, while turpentine inflammation involves primarily IL-6 as the eliciting cytokine. The APP studied in these experiments were all from the group that are regulated by IL-6– and IL-1–type cytokines, which provides even stronger evidence that IL-6 is central to the regulation of the APR in vivo.

IL-6 AND OTHER ORGANS/SYSTEMS

The Nervous System

The effects of IL-6 on the neoplastic chromaffin cell line, PC12, have been studied. IL-6 maintained viability and induced differentiation accompanied by chemical and morphological changes, similar to that induced by nerve growth factor (NGF) (164). The finding of specific IL-6 receptors, distinct from NGF receptors, suggests that IL-6 acts directly on the PC12 cells. This is relevant in light of the finding that IL-6 can induce NGF secretion by astrocytes (132).

Stimulation of astrocytoma or glioblastoma cells with IL-1 induced expression of IL-6 mRNA. IL-6 was produced in a model of CNS viral infection in mice, and later microglial cells and astrocytes were shown to produce IL-6 following infection with virus (76,156). This potential for local IL-6 production by central nervous system cells provides a mechanism for the intrathecal production of antibody in inflammatory CNS diseases such as multiple sclerosis. However, while patients with MS have been shown to have elevated levels of IL-6, Frei et al. found no correlation between IL-6 levels and disease activity (165). Also recombinant IL-6 suppressed demyelination in an animal model of multiple sclerosis (166).

Another potential consequence of excess locally produced IL-6 is the generation of amyloid proteins (167). In Alzheimer's disease there are amyloid plaques present in the cerebral cortex, which are thought to derive from α_1-antichymotrypsin and amyloid protein precursor. Local production of these proteins may be regulated by IL-6, and IL-6 immunostaining has been demonstrated in plaques from patients with Alzheimer's disease (168,169).

One effect of IL-6 on the CNS is on the release of ACTH. Administration of antibody to corticotropin-releasing factor abolishes the effect, implying that IL-6 acts on the hypothalamus to release CRF, which in turn acts on the anterior pituitary to stimulate ACTH release (159).

Bone Metabolism

IL-6 may play an important role in regulating bone formation and resorption. Osteoblasts express IL-6 mRNA and produce IL-6, particularly in response to bone-resorbing agents, like IL-1, TNF, and LPS (170). This expression of IL-6 can be inhibited by estrogen treatment (171). IL-6 can stimulate the number, through increasing the proliferation of hematopoietic progenitors from which osteoclasts are derived, and the activity of osteoclasts (172). Furthermore, the administration of anti–IL-6 antibodies can prevent the increase in osteoclast precursors that occurs following estrogen depletion (173). IL-6–deficient mice have normal bone mass and do not show any increase in bone turnover or reduction in bone mass as a result of estrogen deficiency (174). That IL-6 is required for bone loss caused by estrogen withdrawal is of particular interest given that IL-6 levels in both males and females appear to rise with age (80).

In Paget's disease there is increased turnover of bone and elevated levels of IL-6 have been detected in the serum and bone of patients (175). Furthermore, antibody to IL-6 prevented development of typical osteoclastlike cells in marrow cultures treated with patient-derived medium.

Glomerulonephritis

In the subtype of glomerulonephritis known as mesangial proliferative glomerulonephritis, excess IL-6 is produced by the mesangial cells, which acts an autocrine growth factor (176). Thus, IL-6 is detectable in the urine during this disease but not in other forms of glomerulonephritis (177). Certain strains of transgenic mice that overexpress IL-6 develop mesangial proliferative glomerulonephritis, suggesting a central role for this cytokine in this disease (116,117). However, raised urinary IL-6 levels are not specific for glomerular disease since they are also raised during infection, and again the IL-6 was found to be produced by the mesangial cells (81).

Connective Tissue and Joint Diseases

Raised levels of IL-6 have been detected in the body fluids of patients with a variety of systemic autoimmune diseases, for example, the synovial fluid of rheumatoid joints (178) and the cerebrospinal fluid of patients with systemic lupus erythematosis (SLE-associated CNS disease) (179). Levels of IL-6 may fluctuate with the disease activity and therapy (180,181). Given the role of IL-6 in regulating acute-phase protein production and antibody production, it is possible that IL-6 acts both as an indicator of inflammation as well as an important mediator in the generation of disease effects. The detection of IL-6 in synovial fluid (182) or serum (183) and its correlation with other measures of disease activity support this idea. In ankylosing spondylitis a close correlation has also been found between serum IL-6 levels and disease activity and severity (184). In progressive systemic sclerosis, complexes of IL-6 antibody with IL-6 that retain activity have been found in the serum of some patients (185). Furthermore, alveolar macrophages from patients with systemic sclerosis that include lung involvement show abnormal secretion of IL-6 in vitro (186). Similarly experiments using animal models of arthritis induced by various agents have found correlations between IL-6 levels and disease activity (187,188).

However, the role of IL-6 in the generation of joint inflammation remains unclear, since none of the animal models involving excess production of IL-6 or continuous activation of the IL-6 receptor have developed arthritis (116–120,189). Indeed it is possible that IL-6 acts as an anti-inflammatory agent since administration of IL-6 suppressed arthritis in one model (190) and increasing IL-6 serum levels have been found in patients that have clinical improvement (191).

Endocrine System

In recent years there has been considerable interest in the potential role of IL-6 in regulating various elements of the endocrine system. Preovulatory ovicles have been shown to synthesize IL-6, and it is proposed that IL-6 plays a role in regulating ovulation (192,193). The mesenchymal cells in the pili of the placenta produce increased amounts of IL-6 following stimulation by IL-1 (194,195). Towards the end of pregnancy detection of increased levels of IL-6 in cervical secretions have been shown to correlate with increased incidence of preterm labor (196,197). The detection of high levels of IL-6 in the amniotic fluid may be a strong indicator of intrauterine infection in the absence of premature rupture of the membranes (198,199). In mothers with preeclampsia, elevated plasma levels of IL-6 have been found, perhaps indicating endothelial cell activation as part of the pathogenesis of the condition (200).

IL-6 mRNA is found in adrenal gland tissue, and a possible role of IL-6 in the autocrine or paracrine regulation of adrenal gland secretion is proposed (201). As mentioned previously, IL-6 can stimulate ACTH production from the anterior pituitary (159,202) and IL-6 may regulate posterior pituitary function and in particular secretion of vasopressin (203). Also, human pituitary adenoma tissue has been shown to express mRNA for IL-6 and to produce active protein (204).

IL-6 may interact in a number of ways with thyroid hormone biology. Exogenous IL-6 can alter thyroid metabolism in humans (205). In children with acute respiratory failure there is a correlation between IL-6 levels and changes in circulating thyroid hormone levels (206). In subjects with amiodarone-induced thyrotoxicosis, elevated levels of serum IL-6 have been detected (207). Also, in some conditions, IL-6 may be a marker of diseases associated with destruction of the thyroid gland (208). The relevance of these observations to normal or disease is unclear—it is possible that IL-6 levels in these conditions reflect activation of the immune/inflammatory system rather than a specific role of IL-6 in disease pathogenesis.

In the case of insulin production, IL-6 may play a role in human islet function (209). IL-6 may also be involved in the development of diabetes. Administration of cyclophosphamide to NOD/Wehi mice induces autoimmune diabetes mellitus, which is associated with increasing levels of serum IL-6. Campbell et al. found that treatment with either IL-6 antibody or anti–interferon gamma antibody significantly reduced the incidence of diabetes (210). One possible explanation is that IL-6 was involved in the generation of the auto-antibodies that destroyed the pancreatic islet cells.

Neoplasia

Given its strong growth-promoting effects on B cells, the potential role of IL-6 in regulating the growth of B-cell tumors was investigated early (82,122). An autocrine role of IL-6 was mentioned above for myeloma tumors (82,211). In addition to promoting growth, IL-6 can contribute to tumor cell accumulation by inhibiting apoptosis, that is, cell death, induced by dexamethasone (212). IL-6 may also be responsible for mediating the systemic effects of neoplasms, as Klein et al. found that administration of IL-6 antibodies caused significant improvement in a patient with a primary plasma cell leukemia (213).

Increased expression of IL-6 has been found in both benign hypertrophy and malignancy of the prostate (214). Also, IL-6 production was documented by cells of a malignant paraganglioma (215) as well as by pituitary adenomas (204). In contrast, IL-6 production by meningioma cells was felt to be growth inhibitory (216), consistent with prior experience that showed inhibition of tumor growth in mice by IL-6 (217). One of the first descriptions of IL-6 contributing of the manifestations of neoplasia was in cardiac myxoma (218). Cells of this tumor have been shown to produce IL-6, and IL-6 has been implicated in producing fever as well as excess gammaglobulins seen with this tumor in vivo (10,49,218). IL-6 may mediate some of the systemic effects of other solid tumors (219,220). In patients with epithelial ovarian cancers, serum as well as ascitic fluid levels of IL-6 may have prognostic significance, particularly in regard to tumor burden (221,222). Because of the potential anti-inflammatory as well as antitumor effects of IL-6, a preliminary study in humans is currently underway (213). Whatever the potential role of IL-6 in the generation of solid tumors, the animal models of excess production of IL-6 have not developed cancers (116–120). However, animals with ongoing activation of the IL-6 receptor gp130 subunit showed cardiac hypertrophy but no other evidence of neoplasia (189).

Lung Disease

Elevation of local as well as systemic levels of IL-6 has been found in sepsis involving the respiratory tract (224,225). In the systemic immune response syndrome that is frequently triggered by sepsis and also involves the lung through the development of adult respiratory distress syndrome (ARDS), high levels of circulating IL-6 have been found during the prodrome of the illness (225,226).

IL-6 can be produced by a variety of cells present in lung tissue including type 2 epithelial cells that line the alveoi (227). Excess local production of IL-6 in the airways of mice resulted in a local lymphocytic infiltrate as well as reduced responsiveness to inhaled methacholine (137,138). In human asthma, peripheral blood cells as well as alveolar macrophages have been shown to produce greater amounts of IL-6 than control cells, and this production is less sensitive to inhibition by IL-4 (228,229). Excess local production of IL-6 has also been found in systemic sclerosis and in sarcoidosis (186,203). However, elevated systemic levels of IL-6 are not associated with any distinct pulmonary pathology (116–120).

IL-6 as an Anti-Inflammatory Mediator

Interestingly, IL-6 may modulate the production of other cytokines by monocytes. Schindler et al. showed that IL-6 inhibited both LPS- and PHA-induced IL-1B and TNF production by human PBMC (231). Furthermore, they report that the control appeared to be mediated at the level of transcription. In a system employing GM-CSF primed human U937 histiocytic lymphoma cells, Aderka et al. found that treatment with IL-6 prevented subsequent TNF production in response to LPS and also showed that IL-6 was effective in reducing TNF production in vivo in mice (232). In a model of acute lung inflammation induced by LPS, which is known to depend on production of TNF, Ulich et al. found that coadministration of IL-6 significantly inhibited neutrophil influx (233). As mentioned above, IL-6 has been shown to diminish demyelination in a model of multiple sclerosis as well as to exhibit certain antitumor effects (166,217). Administration of IL-6 inhibited delayed-type hypersensitivity reactions as well as adjuvant-induced arthritis in an animal model (190). Since administration of IL-6 induces IL-1 receptor antagonist as well as soluble TNF-receptor (234) and these mediators would diminish macrophage-mediated inflammatory responses, this is one likely mechanism of the anti-inflammatory effects of IL-6.

Miscellaneous

There is evidence that IL-6 can modify the behavior of mature granulocytes. While rIL-6 could not directly stimulate the respiratory burst of human neutrophils or monocytes, it did prime these cells for enhanced generation of oxygen radicals following stimulation with formyl-methionylleucyl-phenylalanine (235). In addition, IL-6 was found to be a secretogogue for human neutrophils, causing release of both primary and secretory granule contents (236).

The growth and differentiation of epithelial cells may also be regulated by IL-6. In psoriasis there is an abnormal proliferation of skin cells leading to thickening. Elevated IL-6 levels have been detected in the serum, and cells in active psoriatic plaques have stained strongly for IL-6 mRNA and protein and for IL-6R (237,238). In the mouse model using adenovirus infection to achieve overexpression of IL-6, we noted moderate degrees of bronchial epithelial cell hyperplasia during the period of IL-6 excess (137).

Patients with AIDS have increased serum levels of IL-6 (239). In vitro work shows that HIV infection or elements of the HIV envelope can induce IL-6 production by T-cell clones (240). Monocytes-macrophages from the peripheral blood of patients spontaneously produce more IL-6 than normals (239). The addition of antisense oligonucleotides to IL-6 can inhibit the proliferation of Kaposi's sarcoma cells, suggesting it is acting as an autocrine growth factor similar to its role in myeloma cell growth (241).

REFERENCES

1. Paul WE. Pleiotropy and redundancy: T cell derived lymphokines in the immune response. Cell 1989; 57:521–524.
2. Hotta M, Baird A. Differential effects of transforming growth factor type beta on the growth and function of adrenocortical cells in vitro. Proc Natl Acad Sci USA 1986; 83:7795–7799.
3. Weissenbach J, Chernajovsky Y, Zeevi M, et al. Two interferon mRNAs in human fibroblasts: in vitro translation and Escherichia coli cloning studies, Proc Natl Acad Sci USA 1980; 77:7152–7156.
4. Zilberstein A, Ruggieri R, Korn JH, Revel M. Structure and expression of cDNA and genes for human interferon-β2, a distinct species inducible by growth-stimulatory cytokines. EMBO J 1986; 5:2529–2537.
5. Sehgal PB, Sagar AD. Heterogeneity of poly(I)-poly(C)-induced human fibroblast interferon mRNA species. Nature 1980; 288:95–97.
6. May LT, Helfgott DC, Sehgal PB. Anti-β-interferon antibodies inhibit the increased expression of HLA-B7 mRNA in tumor necrosis factor treated human fibroblasts: structural studies of the β2 interferon involved. Proc Natl Acad Sci USA 1986; 83:8957–8961.
7. Content J, de Wit L, Pierard D, Derynck R, de Clercq E, Fiers W. Secretory proteins induced in human fibroblasts under conditions used for the production of interferon B. Proc Natl Acad Sci USA 1982; 79:2768–2772.
8. Haegeman G, Content J, Volckaert G, Derynck R, Tavernier J, Fiers W. Structural analysis of the sequence coding for an inducible 26-kDa protein in human fibroblasts. Eur J Biochem 1986; 159:625–632.
9. Teranishi T, Hirano T, Naomichi A, Onoue K. Human helper T cell factor(s) (ThF). II. Induction of IgG production in B lymphoblastoid cell lines and identification of T cell-replacing factor-(TRF) like factor(s). J Immunol 1982; 128:193–198.
10. Hirano T, Yasukawa K, Harada H, et al. Complementary DNA for a novel human interleukin (BSF-2) that induces B lymphocytes to produce immunoglobulin. Nature 1986; 324:73–76.
11. Billiau A. BSF-2 is not just a differentiation factor. Nature 1986; 324:415.
12. van Damme J, Cayphas S, Opdenakker G, Billiau A, Van Snick J. Interleukin 1 and poly(rI)-poly(rC) induce proliferation of a hybridoma growth factor by human fibroblasts. Eur J Immunol 1987; 17:1–7.
13. van Damme J, Opdenakker G, Simpson RJ, et al. Identification of the human 26-kD protein, interferon β2, as a cell hybridoma/plasmacytoma growth factor induced by interleukin-1 and tumor necrosis factor. J Exp Med 1987; 165:914–919.
14. Aarden LA, Landsdorp PM, De Groot ER, Evers RF, Pannelkoek H, Aarden LA. Molecular cloning and expression of hybridoma growth factor in E. coli. J Immunol 1987; 139:4116–4121.
15. Brakenhoff JPJ, De Groot ER, Evers RF, Pannekoek H, Aarden LA. Molecular cloning and expression of hybridoma growth factor in E. coli. J Immunol 1987; 139:4116–4121.
16. van Snick J, Cayphas S, Vink A, Uttenhoye C, Coulie P, Simpson R. Purification and NH_2-terminal amino acid sequence of a new T cell-derived lymphokine with growth factor activity for B cell hybridomas. Proc Natl Acad Sci USA 1986; 83:9679–9683.
17. van Snick J, Cayphas S, Szikora J-P, et al. cDNA cloning of murine interleukin-HP1; homology with human interleukin-6. Eur J Immunol 1988; 18:193–198.
18. Nordan R, Potter M. A macrophage-derived factor required by plasmacytomas for survival and proliferation in vitro. Science 1986; 233:566–569.
19. Nordan R, Pumphrey J, Rudikoff S. Purification and NH_2-terminal amino acid sequence of a plasmacytoma growth factor derived from the murine macrophage cell line, P388DI. J Immunol 1987; 139:813–817.
20. Chiu CP, Moulds C, Coffman RL, Rennick D, Lee F. Multiple biological activities are expressed by a mouse interleukin-6 cDNA clone isolated from bone marrow stromal cells. Proc Natl Acad Sci USA 1988; 85:7099–7103.
21. Fibach E, Sachs L. Control of normal differentiation of myeloid leukemic cells. XI. Induction of a specific requirement for cell viability and growth during the differentiation of myeloid leukemic cells. J Cell Physiol 1976; 89:259–266.
22. Shabo Y, Lotem J, Rubinstein M, et al. The myeloid blood cell differentiation inducting protein MGI-2A is interleukin-6. Blood 1988; 72:2070–2073.
23. Ritchie DG, Fuller GM. Hepatocyte-stimulating factor: a monocyte-derived acute-phase regulatory protein. Ann NY Acad Sci 1983; 408:490–502.
24. Koj A, Gauldie J, Regoeczi E, Sauder DN, Sweeney GD. The acute-phase response of cultured rat hepatocytes. System characterization and the effect of human cytokines. Biochem J 1984; 224:505–514.
25. Gauldie J, Richards C, Harnish D, Lansdorp P, Baumann H. Interferon β2/B-cell stimulatory factor type 2 shares identity with monocyte-derived hepatocyte-stimulating factor and regulates the major acute phase protein response in liver cells. Proc Natl Acad Sci USA 1987; 84:7251–7255.
26. Sehgal PB, Grieninger G, Tosato G, eds. Regulation of the acute phase and immune responses: interleukin-6. Ann NY Acad Sci 1987; 557:583.
27. Rose TM, Bruce AG. Oncostatin M is a member of a cytokine family that includes leukemia-inhibitory factor, granulocyte colony-stimulating factor, and interleukin 6. Proc Natl Acad Sci USA 1991; 88:8641–8645.
28. Hirano T, Matsuda T, Nakajima K. Signal transduction through gp 130 that is shared among the receptors for the interleukin 6 related cytokine subfamily. Stem Cells 1994; 12:262–277.
29. Le J, Vilcek J. Interleukin 6: a multifunctional cytokine regulating immune reactions and the acute phase protein response. Lab Invest 1989; 61:588–602.
30. van Snick J. Interleukin-6: an overview. Annu Rev Immunol 1990; 8:253–278.
31. Lotz M. Interleukin-6. Cancer Invest 1993; 11:732–742.
32. Akira S, Hirano T, Taga T, Kishimoto T. Biology of multifunctional cytokines: IL 6 and related molecules (IL 1 and TNF). FASEB J 1990; 4:2860–2867.
33. Kishimoto T, Akira S, Taga T. Interleukin-6 and its receptor: a paradigm for cytokines. Science 1992; 258:593–597.
34. Yasukawa K, Hirano T, Watanabe Y, et al. Structure and expression of human B cell stimulatory factor-2 (BSF2/IL-6) gene. EMBO J 1987; 6:2939–2945.
35. Tanabe O, Akira S, Kamiya T, Wong GG, Hirano T, Kishimoto T. Genomic structure of the murine IL-6 gene: high degree conservation of potential regulatory sequences between mouse and human. J Immunol 1988; 141:3875–3881.
36. Northemann W, Braciak TA, Hatton M, Lee F, Fey GH. Structure of the rat interleukin 6 gene and its expression in macrophage-derived cells. J Biol Chem 1989; 264:16072–16082.
37. Sehgal PB, Zilberstein A, Ruggieri R, et al. Human chromosome 7 carries the β2 interferon gene. Proc Natl Acad Sci USA 1986; 83:5219–5222.
38. Bowcock AM, Kidd JR, Lathrop M, et al. The human "β2

interferon/hepatocyte stimulating factor/interleukin-6" gene: DNA polymorphism studies and localization to chromosome 7p21. Genomics 1988; 3:8–16.
39. Mock BA, Nordan RP, Justice MG, et al. The murine IL-6 gene maps to the proximal region of chromosome 5. J Immunol 1989; 142:1372–1376.
40. Leutz A, Damm K, Sterneck E, et al. Molecular cloning of the chicken myelomonocytic growth factor (cMGF) reveals relationship to interleukin 6 and granulocyte colony stimulating factor. EMBO J 1989; 8:175–181.
41. Brakenhoff JPJ, Hart M, de Groot ER, di Padova F, Aarden LA. Structure-function analysis of human IL-6. Epitope mapping of neutralizing monoclonal antibodies with amino- and carboxyl-terminal deletion mutants. J Immunol 1990; 145:461–468.
42. Garman RD, Jacobs AJ, Clark SC, Raulet DH. B-cell stimulatory factor 2 (β2 interferon) functions as a second signal for interleukin 2 production by mature murine T cells. Proc Natl Acad Sci USA 1987; 84:7629–7633.
43. Shaw G, Kamen R. A conserved AU sequence from the 3′ untranslated region of GM-CSF mRNA mediates selective mRNA degradation. Cell 1986; 46:659–667.
44. May LT, Ghrayeb J, Santhanam U, et al. Synthesis and secretion of multiple forms of "β2-interferon/B cell differentiation factor 2/hepatocyte stimulating factor" by human fibroblasts and monocytes. J Biol Chem 1988; 262:7760–7766.
45. May LT, Santhanam U, Tatter SB, Bhardwaj N, Ghrayeb J, Sehgal PB. Phosphorylation of secreted forms of human β2-interferon/hepatocyte stimulating factor/interleukin-6. Biochem Biophys Res Comm 1988; 152:1144–1150.
46. Helfgott DC, Tatter SB, Santhanam U, et al. Multiple forms of INF-β1/IL-6 in serum and body fluids during acute bacterial infection. J Immunol 1989; 142:948–953.
47. Baumann H, Onorato V, Gauldie J, Jahreis GP. Distinct sets of acute phase plasma proteins are stimulated by separate hepatocyte stimulating factors and monokines in rat hepatoma cells. J Biol Chem 1987; 262:9756–9768.
48. Baumann H, Richards C, Gauldie J. Interactions among hepatocyte stimulating factors, interleukin-1, and glucocorticoids for regulation of acute phase plasma proteins in human hepatoma (HepG2) cells. J Immunol 1987; 139:4122–4128.
49. Hirano T, Taga T, Yasukawa K, et al. Human B cell differentiation factor defined by an anti-peptide antibody and its possible role in autoantibody production. Proc Natl Acad Sci USA 1987; 84:228–231.
50. Sehgal PB. Interleukin-6: a regulator of plasma protein gene expression in hepatic and non-hepatic tissues. Mol Biol Med 1990; 7:117–130.
51. Fuller GM, Grenett HE. The structure and function of the mouse hepatocyte stimulating factor. Ann NY Acad Sci 1989; 557:31–45.
52. Simpson RJ, Moritz RL, Van Roost E, Van Snick J. Characterization of a recombinant murine interleukin-6: assignment of disulfide bonds. Biochem Biophys Res Commun 1988; 157:364–372.
53. Clogston CL, Boone TC, Crandall BC, Mendiaz EA, Lu HS. Disulfide structures of human interleukin-6 are similar to those of human granulocyte colony stimulating factor. Arch Biochem Biophys 989; 272:144–151.
54. Jambou RC, Snouwaert JN, Bishop GA, Stebbins JR, Frelinger JA, Fowlkes DM. High-level expression of a bioengineered, cysteine-free hepatocyte-stimulating factor (interleukin 6)-like protein. Proc Natl Acad Sci USA 1988; 85:9426–9430.
55. Brakenhoff PJ, Hart M, Aarden LA. Analysis of human IL-6 mutants expressed in Escherichia coli. Biologic activities are not affected by deletion of amino acids 1-28. J Immunol 1989; 143:1175–1182.
56. Kruttgen A, Rose-John S, Moller C, et al. Structure-function analysis of human IL-6: evidence for the involvement of the carboxy-terminus in function. FEBS Lett 1990; 262:323–326.
57. Bazan JF. Neuropoietic cytokines in the hematopoietic fold. Neuron 1991; 7:197–208.
58. Kohase M, Henrickson-Destefano D, Sehgal PB, Vilcek J. Dexamethasone inhibits feedback regulation of the mitogenic activity of TNF, IL-1 and EGF in human fibroblasts. J Cell Physiol 1987; 132:271–278.
59. Zhang Y, Lin Y, Vilcek J. Synthesis of interleukin-6 (interferon-β2/B cell stimulatory factor 2) in human fibroblasts is triggered by an increase in intracellular cyclic AMP. J Biol Chem 1988; 263:6117–6182.
60. Sehgal PB, Walther Z, May LT. Rapid enhancement of β2-interferon/B cell differentiation factor BSF-2 gene expression in human fibroblasts by diacylglycerols and the calcium ionophore A23187. Proc Natl Acad Sci USA 1987; 84:3663–3667.
61. Ray A, Tatter SB, May LT, Sehgal PB. Activation of the human "β2-interferon/hepatocyte stimulating factor/interleukin-6" promoter by cytokines, viruses and second messengers. Proc Natl Acad Sci USA 1988; 85:6701–6705.
62. Ray A, Sassone-Corsi P, Sehgal PB. A multiple cytokine- and second messenger-responsive element in the enhancer of the human interleukin-6 gene: similarities with c-fos gene regulation. Mol Cell Biol 1989; 9:5537–5547.
63. Isshiki H, Akira S, Tanabe O, et al. Constitutive and interleukin-1 (IL-1)-inducible factors interact with the IL-1-responsive element in the IL-6 gene. Mol Cell Biol 1990; 10:2757–2764.
64. Akira S, Isshiki H, Sugita T, et al. A nuclear factor for IL-6 expression (NF-IL-6) is a member of a C/EBP family. EMBO J 1990; 9:1897–1906.
65. Shimizu H, Mitomo K, Watanabe T, Okamoto S, Yamamoto K-I. Involvement of MF-kappa B-like transcription factor in the activation of the interleukin-6 gene by inflammatory lymphokines. Mol Cell Biol 1990; 10:561–568.
66. Libermann TA, Baltimore D. Activation of interleukin-6 gene expression through the NF-kappa β transcription factor. Mol Cell Biol 1990; 10:2327–2334.
67. Schule R, Rangarajan P, Kliewer S, et al. Functional antagonism between oncoprotein c-Jun and the glucocorticoid receptor. Cell 1990; 62:1217–1226.
68. Yang-Yen HF, Chambard J-C, Sun Y-L, et al. Transcriptional interference between c-Jun and the glucocorticoid receptor: mutual inhibition of DNA binding due to direct protein-protein interaction. Cell 1990; 62:1205–1215.
69. Jonat C, Rahmsdorf HJ, Park K-K, et al. Antitumor promotion and antiinflammation: down-modulation of AP-1 (Fos/Jun) activity by glucocorticoid hormone. Cell 1990; 62:1189–1204.
70. Ray A, LaForge KS, Sehgal PB. On the mechanisms for efficient repression of the IL-6 promoter by glucocorticoids: enhancement of TATA box, and RNA start site (Inr motif) occlusion. Mol Cell Biol 1990; 10:5736–5746.
71. Ray A, Prefontaine KE. Physical association and functional antagonism between the p65 subunit of transcription factor Nf-κB and the glucocorticoid receptor. Proc Natl Acad Sci USA 1994; 91:752–756.
72. Zitnik RJ, Narda L, Elias W, Elias JA. Glucocorticoid inhibition of interleukin-1-induced interleukin-6 production of human lung fibroblasts: evidence for transcriptional and post-transcriptional regulatory mechanisms. Am J Resp Cell Mol Biol 1994; 10:643–650.
73. Kato K, Yokoi T, Takano N, et al. Detection by in situ hybridization and phenotypic characterization of cells express-

ing IL-6 mRNA in human stimulated blood. J Immunol 1990; 144:1317–1322.

74. Plaut M, Pierce JH, Watson CJ, Hanley-Hyde J, Nordan RP, Paul WE. Mast cell lines produce lymphokines in response to cross-linkage of FεRI or to calcium ionophores. Nature 1989; 339:64–67.
75. Cicco NA, Lindemann A, Content J, et al. Inducible production of interleukin-6 by human polymorphonuclear neutrophils: role of granulocyte-macrophage colony-stimulating factor and tumor necrosis factor-alpha. Blood 1990; 75:2049–2052.
76. Frei K, Malipiero UV, Leist TP, Zinkernagel RM, Schwab ME, Fontana A. On the cellular source and function of interleukin 6 produced in the central nevous system in viral diseases. Eur J Immunol 1989; 19:689–694.
77. Vankelcomm H, Carmeliet P, van Damme J, Billiau A, Denef C. Production of interleukin-6 by folliculo-stellate cells of the anterior pituitary gland in a histiotypic cell aggregate culture system. Neuroendocrinology 1989; 49:102–106.
78. Firestein GS, Alvaro-Garcia JM, Maki R. Quantitative analysis of cytokine gene expression in rheumatoid arthritis. J Immunol 1990; 144:3347–3353.
79. Guerne P-A, Zuraw BL, Vaughan JH, Carson DA, Lotz M. Synovium as a source of interleukin-6 in vitro. Contribution to local and systemic manifestations of arthritis. J Clin Invest 1989; 83:585–592.
80. Daynes RA, Araneo BA, Ershler WB, Maloney C, Li G, Ryu S. Altered regulation of IL-6 production with normal aging. J Immunol 1993; 150:5219–5220.
81. Rugo HS, O'Hanley P, Bishop AG, Pearce MK, Abrams JS, Howard M, O'Garra A. Local cytokine production in a murine model of Escherichia coli pyelonephritis. J Clin Invest 1992; 89:1032–1039.
82. Kawano M, Hirano T, Matsuda T, et al. Autocrine generation and requirement of BSF-2/IL-6 for human multiple myelomas. Nature 1988; 332:83–85.
83. Miki S, Iwano M, Miki Y, et al. IL-6 functions as an in vitro autocrine growth factor in renal carcinomas. FEBS Lett 1989; 250:607–610.
84. Reuf C, Budde K, Lacy J, et al. Interleukin 6 is an autocrine growth factor for mesangial cells. Kidney Int 1990; 38:249–257.
85. Klein B, Zhang XG, Jourdan M, et al. Paracrine rather than autocrine regulation of myeloma-cell growth and differentiation by interleukin-6. Blood 1989; 73:517–526.
86. Klein B, Zhang XG, Jourdan M, et al. Interleukin-6 is the central tumor growth factor in vitro and in vivo in multiple myeloma. Eur Cytokine Net 1990; 1:193–201.
87. Podor TJ, Hirsh J, Gelehrter T, et al. Type 1 plasminogen activator inhibitor (PAI1) is not an acute phase reactant in rats: plasma PAI1 expression is independent of hepatocyte synthesis and elevations in plasma interleukin-6. J Immunol 1993; 150:225–235.
88. Baffet G, Braciak T, Gauldie J, Fey GH, Northeman W. Autocrine activity of IL-6 secreted by hepatocarcinoma cell lines. Mol Biol Med 1991; (8):141–156.
89. Gauldie J, Northeman W, Fey GH. IL-6 functions as an exocrine hormone in inflammation: hepatocytes undergoing acute phase responses require exogenous IL-6. J Immunol 1990; 144:3804–3808.
90. Jablons DM, Mule JJ, McIntosh JK, et al. IL-6/IFN-β-2 as a circulating hormone. Induction by cytokine administration in humans. J Immunol 1989; 142:1542–1547.
91. Brouckaert P, Spriggs DR, Demetri G, Kufe DW, Fiers W. Circulating interleukin-6 during a continuous infusion of tumor necrosis factor and interferon-γ. J Exp Med 1989; 169:2257–2262.
92. Hack CE, de Groot ER, Felt-Bersma RJF, et al. Increased plasma levels of interleukin 6 in sepsis. Blood 1990; 75:1824–1831.
93. Nijsten MWN, de Groot ER, ten Duis HJ, Hack CE, Aarden LA. Serum levels of interleukin-6 on acute phase responses (letter). Lancet 1987; 2:921.
94. Arai K, Lee F, Miyajima A, Miyatake S, Arai N, Yokota T. Cytokines: coordinators of immune and inflammatory responses. Annu Rev Biochem 1990; 59:783–836.
95. Libert C, Brouckaert P, Shaw A, Fiers W. Induction of IL-6 by human and murine recombinant IL-1 in mice. Eur J Immunol 1990; 20:691–695.
96. Gauldie J, Baumann H. Cytokines and acute phase protein expression. In: Kimball ES, ed. Cytokines in Inflammation. Boca Raton, FL: CRC Press Inc., 1991:275–305.
97. Kishimoto T, Taga T, Matsuda T, Hirano T. Interleukin-6 and its receptor in immune regulation. Immunology 1989; 7:633–640.
98. Taga T, Kawanishi K, Hardy RR, Hirano T, Kishimoto T. Receptors for the B cell stimulatory factor 2 (BSF-2): quantitation, specificity, distribution and regulation of the expression. J Exp Med 1987; 166:967–981.
99. Coulie PG, Vanhecke A, van Damme J, et al. High-affinity binding sites for human 26-kDa protein (interleukin 6, B cell stimulatory factor-2, human hybridoma plasmacytoma growth factor, interferon β2), different from those of type I interferon (α,β) on lymphoblastoid cells. Eur J Immunol 1987; 17:1435–1440.
100. Yamasaki K, Taga T, Hirata Y, et al. Cloning and expression of the human interleukin-6 (BSF-2/IFNβ2) receptor. Science 1988; 241:825–882.
101. Mullberg J, Schooltink H, Stoyan T, Gunther M, Graeve L, Buse G, Mackiewicz A, Heinrich PC, Rose-John S. The soluble interleukin-6 receptor is generated by shedding. Eur J Immunol 1993; 23:474–480.
102. Taga T, Hibi M, Hirata Y, et al. Interleukin 6 (IL-6) triggers the association of its receptor (IL-6-R) with a possible signal transducer, gp 130. Cell 1989; 58:573–581.
103. Coulie PG, Stevens M, van Snick J. High- and low-affinity receptors for murine interleukin 6. Distinct distribution on B and T cells. Eur J Immunol 1989; 19:2107–2114.
104. Taga T, Kishimoto T. Role of a chain IL-6 receptor system in immune and hempoietic cell regulation. Crit Rev Immunol 1992; 11(5):265–280.
105. Bauer J, Bauer TM, Kalb T, et al. Regulation of interleukin 6 receptor expression in human monocytes and monocyte-derived macrophages. Comparison with the expression in human hepatocytes. J Exp Med 1989; 170:1537–1549.
106. Rose-John S, Schooltink H, Lenz D, et al. Studies on the structure and regulation of the human hepatic interleukin-6 receptor. Eur J Biochem 1990; 190:79–83.
107. Snyers L, DeWitt L, Content J. Glucocorticoid up-regulation of high affinity IL-6 receptors on human epithelial cells. Proc Natl Acad Sci USA 1990; 87:2838–2842.
108. Baumann M, Baumann H, Fey GH. Molecular cloning, characterization and functional expression of the rat liver interleukin 6 receptor. J Biol Chem 1990; 265:19853–19862.
109. Bazan JF. Haemopoietic receptors and helical cytokines. Immunol Today 1990; 11:350–354.
110. Kishimoto T, Taga T, Akira S. Cytokine signal transduction. Cell 1994; 76:253–262.
111. Darnell JE, Jr., Kerr IM, Stark GR. Jak-STAT pathways and

transcriptional activation in response to IFNs and other extracellular signaling proteins. Science 1994; 264:1415–1420.

112. Muraguchi A, Kishimoto T, Miki T, et al. T cell-replacing factor (TRF)-induced IgG secretion in human B blastoid cell line and demonstration of acceptors for TFR. J Immunol 1981; 127:412–416.
113. Hirano T, Taga T, Nakano N, et al. Purification to homogeneity and characterization of human B cell differentiation factor (BCDF or BSFp-2). Proc Natl Acad Sci USA 1985; 82:5490–5494.
114. Muraguchi A, Hirano T, Tang B, et al. The essential role of B-cell stimulatory factor (BSF-2/IL-6) for the terminal differentiation of B cells. J Exp Med 1988; 167:332.
115. Takatsuki F, Okano A, Suzuki C, et al. Human recombinant interleukin 6/B cell stimulatory factor 2 (IL-6/BSF-2) augments murine antigen-specific antibody responses in vitro and in vivo. J Immunol 1988; 141:3072–3077.
116. Suematsu S, Matsuda T, Aozasa K, Akira S, Nakano N, Ohno S, Miyazaki J, Yamamura K, Hirano T, Kishmoto T. IgG1 plasmacytosis in interleukin 6 transgenic mice. Proc Natl Acad Sci USA 1989; 86:7547–7551.
117. Suematsu S, Matsusaka T, Matsuda T, Ohno S, Miyazaki J, Yamamura K, Hirano T, Kishimoto T. Generation of plasmacytomas with the chromasomal translocation t(12;15) in interleukin 6 transgenic mice. Proc Natl Acad Sci USA 1992; 89:232–235.
118. Brandt SJ, Bodine DM, Dunbar CE, Neinhuis AW. Dysregulated interleukin 6 expression produces a syndrome resembling Castleman's disease in mice. J Clin Invest 1990; 86:592–599.
119. Hawley RG, Fong AZC, Burns BF, Hawley TS. Transplantable myeloproliferative disease induced in mice by an interleukin 6 retrovirus. J Exp Med 1992; 176:1149–1163.
120. Ramsay AJ, Husband AJ, Ramshaw IA, Bao S, Matthaei KI, Koehler G, Kopf M. The Role of Interleukin-6 in mucosal IgA antibody response in vivo. Science 1994; 264:561–563.
121. Alderson AB, Pike BL. Recombinant human interleukin-6 (B cell stimulating factor 2) enhances immunoglobulin secretion by single murine hapten-specific B cells in the absence of cell division. Int Immunol 1989; 1:20–28.
122. Yokoi T, Miyawaki T, Yachie A, Kato K, Kasahara Y, Taniguchi N. Epstein-Barr virus immortalized B cells produce IL-6 as an autocrine growth factor. Immunology 1990; 70:100–105.
123. Neckers LM, Nordan RP. Regulation of murine plasmacytoma transferrin receptor expression and G_1 traversal by plasmacytoma cell growth factor. J Cell Physiol 1988; 135:495–501.
124. Kopf M, Baumann H, Freer G, Freudenberg M, Lamers M, Kishimoto T, Zinkernagel R, Bluethmann H, Kohler G. Impaired immune and acute-phase responses in interleukin-6-deficient mice. Nature 1994; 368:339–342.
125. Le J, Fredrickson R, Reis LFL, et al. Interleukin 2-independent and interleukin 2 dependent pathways of regulation of thymocyte function by interleukin 6. Proc Natl Acad Sci USA 1988; 85:8643–8647.
126. Helle M, Boeije L, Aarden LA. IL-6 is an intermediate in IL-1 induced thymocyte proliferation. J Immunol 1989; 142:4334–4338.
127. Takai Y, Wong GC, Clark SC, Burakoff SJ, Herrman SH. B cell stimulatory factor-2 is involved in the differentiation of cytotoxic T lymphocytes. J Immunol 1988; 140:508–512.
128. Okada M, Kitihara M, Kishimoto S, Matsuda T, Hirano T, Kishimoto T. IL-6/BSF-2 functions as a killer helper factor in the in vitro induction of cytotoxic T cells. J Immunol 1988; 141:1543–1549.
129. Luger TA, Krutman J, Kirnbauer R, et al. IFN-B2/IL-6 augments the activity of human natural killer cells. J Immunol 1989; 143:1206–1209.
130. Mizutani H, May LT, Sehgal PB, Kupper TS. Synergistic interactions of IL-1 and IL-6 in T-cell activation. J Immunol 1989; 143:896–901.
131. Noma T, Mizuta T, Rosen A, Hirano T, Kishimoto T, Honjo T. Enhancement of the interleukin 2 receptor expression on T cells by multiple B-lymphotropic lymphokines. Immunol Lett 1987; 15:249–253.
132. Kishimoto T. The biology of interleukin-6. Blood 1989; 74:1–10.
133. Houssiau F, Coulie PG, Olive D, van Snick J. Synergistic activation of human T cells by interleukin 6. Eur J Immunol 1988; 18:653–656.
134. Jayarman S, Martin CA, Dorf ME. Enhancement of in vivo cell-mediated immune responses by three distinct cytokines. J Immunol 1990; 144:942–951.
135. Scott S, Pandolfi F, Kurnick JT. Fibroblasts mediate T cell survival: a proposed mechanism for retention of primed T cells. J Exp Med 1990; 172:1873–1876.
136. Braciak TA, Mittal SK, Graham FL, Richards CD, Gauldie J. Construction of recombinant human type 5 adenoviruses expressing rodent IL-6 genes. J Immunol 1993; 51:5145–5153.
137. Xing Z, Braciak T, Jordana M, Croitoru K, Graham FL, Gauldie J. Adenovirus-mediated cytokine gene transfer at tissue sites. J Immunol 1994; 153:4059–4069.
138. DiCosomo BF, Geba GP, Picarella D, Elias JA, Rankin JA, Stripp BA, Whitsett JA, Flavell RA. Airway epithelial cell expression of interleukin-6 in transgenic mice. Uncoupling of airway inflammation and bronchial hyperreactivity. J Clin Invest 1994; 94:2028–2035.
139. Chiu CP, Lee F. IL-6 is a differentiation factor for M1 and Wehi-3B myeloid leukemic cells. J Immunol 1989; 142:1909–1915.
140. Bot FJ, Eijk L, Broeders L, Aarden LA, Lowenberg B. Interleukin-6 synergizes with M-CSF in the formation of macrophage colonies from purified human marrow progenitor cells. Blood 1989; 73:435–437.
141. Caracciolo D, Clark SC, Rovera G. Human interleukin 6 supports granulocytic differentiation of hematopoietic progenitor cells and acts synergistically with GM-CSF. Blood 1988; 72:823–826.
142. Ikebuchi K, Wong GC, Clark SC, Ihle JN, Hirai Y, Ogawa M. Interleukin 6 enhancement of interleukin 3-dependent proliferation of multipotential hemopoietic precursors. Proc Natl Acad Sci USA 1987; 84:9035–9039.
143. Ikebuchi K, Ihle JN, Hirai Y, Wong GC, Clark SC, Ogawa M. Synergistic factors for stem cell proliferation: further studies of the target stem cells and the mechanism of stimulation by interleukin-1, interleukin-6, and granulocyte colony-stimulating factor. Blood 1988; 72:2007–2014.
144. Leary AG, Ikebuchi K, Hirai Y, et al. Synergism between interleukin-6 and interleukin-3 in supporting proliferation of human haematopoietic stem cells: comparison with interleukin-1α. Blood 1988; 71:1759–1763.
145. Rennick D, Jackson J, Yang G, Wideman J, Lee F, Hudak S. Interleukin 6 interacts with interleukin 4 and other hemopoietic growth factors to selectively enhance the growth of megakaryocytic, erythroid, myeloid, and multipotential progenitor cells. Blood 1989; 73:1838–1835.
146. Wong GC, Witek-Giannotti JS, Temple PA, et al. Stimulation of murine hemopoietic colony formation by human IL-6. J Immunol 1988; 140:3040.
147. Suda T, Yamaguchi Y, Suca J, Miura Y, Okano A, Akiyama Y.

Effect of interleukin 6 (IL-6) on the differentiation and proliferation of murine and human hemopoietic progenitors. Exp Hematol 1988; 16:891.
148. Ishibashi T, Kimura H, Shikama T, et al. Interleukin-6 is a potent thrombopoietic factor in vivo in mice. Blood 1989; 74:1241–1244.
149. Ishibashi T, Kimura H, Uchida T, Kariyone S, Friese P, Burstein SA. Human interleukin-6 is a direct promoter of maturation of megakaryocytes in vitro. Proc Natl Acad Sci USA 1988; 86:5953–5957.
150. Hill RJ, Warren MK, Levin J. Stimulation of thrombopoiesis in mice by humanre combinant interleukin 6. J Clin Invest 1990; 85:1242–1247.
151. Okano A, Suzuki C, Takatsuki F, et al. Effects of interleukin 6 on haematopoiesis in bone-marrow transplanted mice. Transplantation 1989; 47:738–740.
152. Koj A. Definition and classification of acute phase proteins in the acute phase response to injury and infection. In: Gordon AH, Koj A, eds. The Acute Phase Response to Injury and Infection. Amsterdam: Elsevier, 1985:139–232.
153. Baumann H, Gauldie J. The acute phase response. Immunol Today 1994; 15:74–80.
154. Geiger T, Andus T, Klapproth J, Hirano T, Kishimoto T, Heinrich PC. Induction of rat acute-phase proteins by interleukin-6 in vivo. Eur J Immunol 1989; 142:808–821.
155. Marinkovic S, Jahreis GP, Wong GG, Baumann H. IL-6 modulates the synthesis of a specific set of acute phase plasma proteins in vivo. J Immunol 1989; 142:808–812.
156. Frei K, Leist TP, Meager A, et al. Production of B-cell stimulatory factor-2 and gamma in the central nervous system during viral meningitis and encephalitis. J Exp Med 1988; 168:449–453.
157. Helle M, Brakenhoff J, De Groot E, Aarden L. Interleukin-6 is involved in interleukin-1-induced activities. Eur J Immunol 1988; 18:957–959.
158. Rothwell NJ, Busbridge NJ, Lefeuvre RA, Hardwick AJ, Gauldie J, Hopkins SJ. Interleukin-6 is a centrally acting endogenous pyrogen. Can J Phys Pharmac 1991; 69:1465–1469.
159. Naitoh Y, Fukata J, Tominaga T, et al. Interleukin 6 stimulates the secretion of adrenocorticotrophic hormone in conscious, freely moving rats. Biochem Biophys Res Commun 1988; 155:1459–1463.
160. Cox G. Glucocorticoid treatment inhibits apoptosis in human neutrophils. J Immunol 1995; 154:4719–4725.
161. Fattori E, Cappelletti M, Costa P, et al. Defective inflammatory response in interleukin 6-deficient mice. J Exp Med 1994; 180:1243–1250.
162. Houssiau FA, Bukasa K, Sindic CJM, Van Damme J, Van Snick J. Elevated levels of the 26K human hybridoma growth factor (interleukin 6) in cerebrospinal fluid of patients with acute infection of the central nevous system. Clin Exp Immunol 1988; 71:320–332.
163. Steinmetz HT, Herbertz A, Bertram M, Diehl V. Increase in interleukin-6 serum level preceding fever in granulocytopenia and correlation with death from sepsis. J Infect Dis 1995; 171:225–228.
164. Satoh T, Nakamura S, Taga T, et al. Induction of neural differentiation in PC12 cells by B cell stimulatory factor/interleukin 6. Mol Cell Biol 1988; 8:3546.
165. Frei K, Fredrikson S, Fontana A, Link H. Interleukin-6 is elevated in plasma in multiple sclerosis. J Neuroimmunol 1991; 31:147–153.
166. Rodriguez M, Pavelko KD, McKinney CW, Leibowitz JL. Recombinant human interleukin-6 suppresses demyelination in a viral model of multiple sclerosis. J Immunol 1994; 153:3811–3821.
167. Vandenabeele P, Fiers W. Is amyloidogenesis during Alzheimer's disease due to an IL-1-/IL-6-mediated 'acute phase response' in the brain? Immunol Today 1991; 12:217–219.
168. Bauer J, Strauss S, Schreiter-Greasser U, Ganter U, Schlegel P, Witt I, Yolk B, Berger M. Interleukin-6 and alpha-2-macroglobulin indicated an acute-phase state in Alzheimer's disease cortices. FEBS Lett 1991; 285:111–114.
169. Strauss S, Bauer J, Ganter U, Jonas U, Berger M, Volk B. Detection of interleukin-6 and alpha 2-macroglobulin immunoreactivity in cortex and hippocampus of Alzheimer's disease patients. Lab Invest 1992; 66:223–230.
170. Ishimi Y, Miyaura C, He Jin C, et al. IL-6 is produced by osteoblasts and induces bone resorption. J Immunol 1990; 145:3297–3303.
171. Girasole G, Jilka RL, Passeri G, et al. 17-Beta-Estradiol inhibits IL-6 production by bone marrow derived stromal cells and osteoblasts in vitro: a potential mechanism for antiosteoporotic effect of estrogen. J Clin Invest 1992; 89:883–891.
172. Michalevicz R, Lifshitz D, Revel M. Inteferon Beta2/IL-6 and IL-3 synergize in stimulating proliferation of human early hematopoietic progenitor cells. Scanning Microsc 1989; 3:1143–1150.
173. Jilka RL, Hangoc G, Girasole G, et al. Increased osteoclast development after oestrogen loss: mediation by interleukin-6. Science 1992; 257:88–91.
174. Poli V, Balaena R, Fattori E, et al. Interleukin-6 deficient mice are protected from bone loss caused by estrogen deficiency. EMBO 1994; 13:1189–1194.
175. Roodman GD, Kurihara N, Ohsaki Y, Kukita A, Hosking D, Demulda A, Smith JF, Singer FR. Interleukin 6. A potential autocrine/paracrine factor in Paget's disease of bone. J. Clin Invest 1992; 89:46–52.
176. Harii Y, Muraguchi A, Iwano M, et al. Involvement of IL-6 is mesangial proliferative glomerulonephritis. J Immunol 1989; 143:3949–3955.
177. Gordon C, Richards N, Howie AJ, Richardson K, Michael J, Adu D, Emery P. Urinary IL-6: a marker for messangial proliferation glomerulonephritis? Clin Exp Immunol 1991; 86:145–149.
178. Hirano T, Matsuda T, Turner M, et al. Excessive production of interleukin-6/B cell stimulatory factor-2 in rheumatoid arthritis. Eur J Immunol 1988; 18:1797–1801.
179. Hirohata S, Miyamoto T. Elevated levels of interleukin-6 in cerebrospinal fluid from patients with systemic lupus erythematosus and central nervous system involvement. Arthritis Rheum 1990; 33:644–649.
180. Tanaka Y, Saito K, Shirakawa F, et al. Production of B cell stimulating factors in patients with systemic lupus erythematosis. J Immunol 1988; 141:3043–3049.
181. Linker-Israeli M, Deans RJ, Wallace DJ, Prehn J, Ozeri-Chen T, Klinenberg JR. Elevated levels of endogenous IL-6 in systemic lupus erythematosus. A putative role in pathogenesis. J Immunol 1991; 147:117–123.
182. Miltenburg AM, van Laar JM, de Kuiper R, Daha MR, Breedveld FC. Interleukin-6 activity in paired samples of synovial fluid. Correlation of synovial fluid interleukin-6 levels with clinical and laboratory parameters of inflammation. Br J Rheumatol 1991; 30:186–189.
183. de Benedetti F, Massa M, Robbioni, Ravelli A, Burgio GR, Martini A. Correlation of serum interleukin-6 levels with joint involvement and thrombocytosis in systemic juvenile rheumatoid arthritis. Arthritis Rheum 1991; 34:1158–1163.
184. Gratacos J, Collado A, Filella X, et al. Serum cytokines in

ankylosing spondylitis: close correlation between serum IL-6 and disease activity and severity. Br J Rheumatol 1994; 33:927–931.
185. Takemura H, Suzuki H, Yoshizaki K, Otgata A, et al. IL-6-anti-IL-6 autoantibody complexes with IL-6 activity in sera from patients with systemic sclerosis. J Immunol 1994; 152:485–495.
186. Crestani B, Seta N, de Brandt M, et al. Interleukin-6 secretion by monocytes and alveolar macrophages in systemic sclerosis with lung involvement. Am J Respir Crit Care Med 1994; 145:1260–1265.
187. Takai Y, Seki N, Senoh H, et al. Enhanced production of IL-6 in mice with type II collagen-induced arthritis. Arthritis Rheum 1989; 32:594–600.
188. Leisten JC, Gaarde WA, Scholtz W. IL-6 levels correlate with footpad swelling in adjuvant-induced arthritis Lewis rats treated with cyclosporin A or indomethacin. Clin Immunol Immunopathol 1990; 56:108–115.
189. Hirota H, Yoshida K, Kishimoto T, Taga T. Continuous activation of gp130, a signal-transducing receptor component for interleukin-6-related cytokines, causes myocardial hypertrophy. Proc Natl Acad Sci 1995; 92:4862–4866.
190. Mihara M, Ikuta M, Koishihara Y, Ohsugi Y. Interleukin 6 inhibits delayed-type hypersensitivity and the development of adjuvant arthritis. Eur J Immunol 1991; 21:2327–2331.
191. Wood NC, Symons JA, Dickens E, Duff GW. In situ hybridization of IL-6 in rheumatoid arthritis. Clin Exp Immunol 1992; 87:183–189.
192. Machelon V, Emilie D, Lefevre A, Nome F, Durand-Gasselin I, Testart J. Interleukin-6 biosynthesis in human preovulatory follicles: some of its potential roles at ovulation. J Clin Endocr Metab 1994; 79:633–642.
193. De M, Sanford T, Wood GW. Interleukin-1, interleukin-6, and tumor necrosis factor alpha are produced in the mouse uterus during the estrous cycle and are induced by estrogen and progesterone. Dev Biol 1992; 151:297–305.
194. Tabibzadeh SS, Santhanam U, Sehgal PB, May LT. Cytokine-induced production of IFN-B2/IL-6 by freshly explanted human endometrial stromal cells. J Immunol 1989; 142:3134–3139.
195. Kauma SW, Turner TT, Harty JR. IL-1 beta stimulates IL-6 production in placental villous core mesenchymal cells. Endocrinology 1994; 134:457–460.
196. Inglis SR, Jeremias J, Kuno K, et al. Detection of tumor necrosis factor alpha, interleukin-6, and fetal fibronectin in the lower genital tract during pregnancy: relation to outcome. Am J Obstet Gynecol 1994; 171:5–10.
197. Lockwood CJ, Ghidini A, Wein R, Lapinski R, Casal D, Berkowitz RL. Increased IL-6 concentrations in cervical secretions are associated with preterm delivery. Am J Obstet Gynecol 1994; 171:1097–1102.
198. Romero R, Avila C, Santhanam U, Sehgal PB. Amniotic fluid interleukin 6 in preterm labor. Association with infection. J Clin Invest 1990; 85:1392–1400.
199. Coultrip LL, Lien JM, Gomez R, Kapernick P, Khoury A, Grossman JH. The value of amniotic fluid IL-6 determination in patients with preterm labor and intact membranes in the detection of microbial invasion of the amniotic cavity. Am J Obstet Gynecol 1994; 171:901–911.
200. Greer IA, Lyall F, Perara T, Boswell F, Macara LM. Increased concentrations of cytokines interleukin-6 and interleukin-1 receptor antagonist in plasma of women with preeclampsia: a mechanism for endothelial dysfunction? Obstet Gynecol 1994; 84:937–940.
201. Gonzalez-Hernandez JA, Bornstein SR, Ehrhart-Bornstein M, Spath-Schwalbe E, Jirikowski G, Scherbaum WA. Interleukin-6 messenger ribonucleic acid expression in human adrenal gland in vivo: new clue to a paracrine or autocrine regulation of adrenal function. J Clin Endocrinol Metab 1994; 79:1492–1497.
202. Spath-Schwalbe E, Born J, Schrezenmeier H, Bornstein SR, Stromeyer P, Dreschsler S, Fehm HL, Porzsolt F. Interleukin-6 stimulates the hypothalamus-pituitary-adrenocortical axis in man. J Clin Endocrinol Metab 1994; 79:1212–1214.
203. Mastorakos G, Weber JS, Magiakou MA, Gunn H, Chrousos GP. Hypothalamic-pituitary-adrenal axis activation and stimulation of systemic vasopressin secretion by recombinant interleukin-6 in humans: potential implications for the syndrome of inappropriate vasopressin secretion. J Clin Endocrinol Metab 1994; 79:934–939.
204. Jones TM, Daniels M, James RA, et al. Production of bioactive and immunoreactive IL-6 and expression of IL-6 mRNA by human pituitary adenomas. J Clin Endoc Metab 1994; 78:1492–1497.
205. Southard JM, vander Poll T, Endert E, et al. Effects of acute and chronic IL-6 administration on thyroid hormone metabolism in humans. J Clin Endoc Metab 1994; 79:1342–1346.
206. Hashimoto H, Igarashi N, Yachie A, Miyawaki T, Sato T. The relationship between serum levels of interleukin-6 and thyroid hormone in children with acute respiratory infection. J Clin Endocrinol Metab 1994; 78:288–291.
207. Bartalena L, Grasso L, Brogioni S, Aghini-Lombardi F, Braverman LE, Martino E. Serum interleukin-6 in amiodarone-induced throtoxicosis. J Clin Endocrinol Metab 1994; 78:423–427.
208. Bartalena L, Brogioni S, Grasso L, et al. Interleukin-6: a marker of thyroid destructive processes? J Clin Endoc Metab 1994; 79:1424–1427.
209. Eizerik DL, Sandler S, Welsh ?, et al. Cytokines suppress human islet function irrespective of their effects on nitric oxide generation. J Clin Invest 1994; 93:1968–1974.
210. Campbell IL, Kay TW, Oxbrow L, Harrison LC. Essential role for interferon-gamma and interleukin-6 in autoimmune insulin-dependent diabetes in NOD/Wehi mice. J Clin Invest 1991; 87:739–742.
211. Zhang XG, Klein B, Bataille R. Interleukin-6 is a potent myeloma-cell growth factor in patients with aggressive multiple myeloma. Blood 1989; 74:11–13.
212. Hardin J, MacLeod S, Grigorieval I, et al. Interleukin-6 prevents dexamethasone-induced myeloma cell death. Blood 1994; 84:2759–2766.
213. Klein B, Wijdenes J, Zhang XG, et al. Murine anti-interleukin-6 monoclonal antibody therapy for a patient with plasma cell leukemia. Blood 1991; 78:1198–1204.
214. Siegsmund MJ, Yamazaki H, Pastan I. Interleukin-6 receptor mRNA in prostate carcinomas and benign prostate hyperplasia. J Urology 1994; 151:1396–1399.
215. Omura M, Sato T, Cho R, Lizuka T, et al. A patient with malignant paraganglioma that simultaneously produces adrenocorticotropic hormone and interleukin-6. Cancer 1994; 74:1634–1639.
216. Todo T, Adams EF, Rafferty B, Fahlbusch R, Dingerman T, Werner H. Secretion of interleukin-6 by human meningioma cells: possible autocrine inhibitory regulation of neoplastic cell growth. J Neurosurgery 1994; 81:394–401.
217. Mulé JJ, McIntosh JK, Jablons DM, Rosenberg SA. Antitumor activity of recombinant interleukin 6 in mice. J Exp Med 1990; 171:629–636.
218. Graham SL, Sellers AL. Atrial myxoma with multiple myeloma. Arch Intern Med 1979; 139:116–117.
219. Tatsuno I, Nishikawa T, Sasano H, Shizawa S, Iwase H, Satoh

S. Interleukin-6-producing gastric carcinoma with fever, hypergammaglobulinemia, and plasmacytosis in bone marrow. Gastroenterology 1994; 107:543–547.
220. Miki S, Iwano M, Miki Y, Yamamoto M, Tang B, Yokokawa K, Sonoda T, Hirano T, Kishimoto T. Interleukin-6 (IL-6) functions as an in vitro autocrine growth factor in renal cell carcinomas. FEBS Lett 1989; 250:607–610.
221. Berek JS, Chung C, Kaldi K, Watson JM, Knox RM, Martinez-Maza O. Serum interleukin-6 levels correlated with disease status in patients with epithilial ovarian cancer. Am J Obstet Gynecol 1991; 164:1038–1042.
222. Plante M, Rubin SC, Wong GY, Federici MG, Finstad CL, Gastl GA. Interleukin-6 levels in serum and ascites as a prognostic factor in patients with epithelial ovarian cancer. Cancer 1994; 74:1021–1025.
223. van Gameran MM, Willemse PH, Mulder NH, et al. Effects of recombinant IL-6 in cancer patients: a phase I-II study. Blood 1994; 84:1434–1441.
224. Dehoux MS, Boutten A, Ostinelli J, Seta N, et al. Compartmentalized cytokine production within the human lung in unilateral pneumonia. Am J Respir Crit Care Med 1994; 150:710–716.
225. Moussa K, Mitchie HJ, Cree IA, et al. Phagocyte function and cytokine production in community acquired pneumonia. Thorax 1994; 49:107–111.
226. Donnelly TJ, Meade P, Jagels M, et al. Cytokine, complement and endotoxin profiles associated with development of adult respiratory distress syndrome after severe injury. Crit Care Med 1994; 22:768–776.
227. Crestani B, Cornillet P, Dehoux M, Rolland C, Guenounou M, Aubier M. Alveolar type II cells produce interleukin-6 in vitro and in vivo. Regulation by alveolar macrophage secretory products. J Clin Invest 1994; 94:731–740.
228. Chanez P, Vignola AM, Paul-Eugene N, et al. Modulation by interleukin-4 of cytokine release from mononuclear phagocytes in asthma. J All Clin Immunol 1994; 94:997–1005.
229. Gosset P, Tsicopuoulos A, Wallaert B, et al. Increased secretion of tumor necrosis factor alpha and interleukin-6 by alveolar macrophages consecutive to the development of the late asthmatic reaction. Clin Immunol 1991; 88:561–571.
230. Koichi S, Yasutaka I, Katsutoshi T, Toshihide S, Masahiko Y, Munehiko M. Significance of interleukin-6 in patients with sarcoidosis. Chest 1994; 106:156–160.
231. Schindler R, Mancilla J, Ghorbani R, Clark SC, Dinarello CA. Correlations and interactions in the production of interleukin-6 (IL-6), IL-1, and tumor necrosis factor (TNF) in human blood mononuclear cells: IL-6 suppresses IL-1 and TNF. Blood 1990; 75:40–47.
232. Aderka D, Le J, Vilcek J. IL-6 inhibits liposaccharide-induced tumor necrosis factor production in cultured human monocytes, U937 cells, and in mice. J Immunol 1989; 143:3517–3523.
233. Ulich TR, Yin S, Guo K, Remick D, del Castillo J. Intratracheal injection of endotoxin and cytokines. II. Interleukin-6 and transforming growth factor beta inhibit acute inflammation. Am J Pathol 1991; 138:1097–1101.
234. Tilg H, Trehn E, Atkins MB, Dinarello CA, Mier JW. Interleukin-6 as an anti-inflammatory cytokine: induction of circulating IL-1 receptor antagonist and soluble tumor necrosis factor receptor p55. Blood 1994; 83:113–118.
235. Kharmazmi A, Nielsen H, Rechnitzer C, Bendtzen K. Interleukin-6 primes neutrophil and monocyte oxidative burst response. Immunol Lett 1989; 21:177–184.
236. Borish L, Rosenbaum R, Albury L, Clark S. Activation of neutrophils by recombinant interleukin-6. Cell Immunol 1989; 121:280–289.
237. Grossman RM, Krueger J, Yourish D, et al. Interleukin-6 is expressed in high levels in psoriatic skin and stimulates proliferation of cultured human keratinocytes. Proc Natl Acad Sci 1989; 86:6367–6371.
238. Ohta Y, Katayama I, Funato T, et al. In situ expression of mRNA of IL-1 and IL-6 in psoriasis: IL-6 involved in formation of psoriatic lesions. Arch Dermatol Res 1991; 283:351–356.
239. Breen EC, Rezai AR, Nakajima K, Beall GN, Mitsuyasu RT, Hirano T, Kishimoto T, Martinez-Maza O. Infection with HIV is associated with elevated IL-6 levels and production. J Immunol 1990; 144:480–484.
240. Villiger PM, Cronin MT, Amenomori T, Wachsman W, Lotz M. IL-6 production by human T lymphocytes. Expression in HTLV-1 infected but not in normal T cells. J Immunol 1991; 146:550–559.
241. Miles SA, Rezai AR, Salazar-Gonzalez JF, Meyden MV, Stevens RH, Logan DM, Mitsuyasu RT, Taga T, Hirona T, Kishimoto T, Martinez-Maza O. AIDS Kaposi sarcoma-derived cells produce and respond to interleukin 6. Proc Natl Acad Sci USA 1990; 87:4068–4072.
242. Geisterfer M, Richards C, Baumann H, Fey G, Giogune D, Gauldie J. Regulation of IL-6 receptor in acute inflammation in vivo. Cytokine 1993; 5:1–7.
243. Geisterfer M, Richards C, Gauldie J. Cytokines, Oncostatin M and Interleukin-1, regulate the expression of the IL-6 receptor (gp80,gp130). Cytokine 1995; 7:503–509.

7
Interleukin-7 and Lymphopoiesis

Pierette M. Appasamy
University of Pittsburgh School of Medicine, Pittsburgh, Pennsylvania

INTRODUCTION

This chapter describes the properties of interleukin 7 (IL-7), a cytokine that is most well known for its effects on early B and T lymphopoiesis. The actions of IL-7, however, are not limited to lymphopoiesis. Rather, IL-7 can now be regarded as a multilineage cytokine, since it is a cofactor for myelopoiesis as well as an activation factor for mature monocytes and macrophages, natural killer (NK) cells, and cytotoxic T lymphocytes (CTLs). IL-7 also stimulates lymphokine-activated killer cell (LAK) activity. A great number of in vivo studies in mice predict that IL-7 may be of clinical utility in antitumor therapy and acceleration of lymphoid and myeloid recovery in leukopenic patients, especially in combination with bone marrow or peripheral blood stem cell transplants. Antimicrobial effects of IL-7 in mice have been reported, and, while less well studied, they appear to hold some promise for clinical acceleration of these responses. As IL-7 was not covered in the first edition of this book, I have attempted to include as many of the most significant studies as possible, but regret that not all references could be included given space limitations. I have reviewed elsewhere much of the literature described herein, with particular emphasis on the antitumor properties of IL-7 (1,2). Other excellent reviews of the functions of IL-7 have been published, primarily concentrating on the role of IL-7 in B-cell generation (3–7).

HISTORICAL PERSPECTIVE

The identification of IL-7, originally named lymphopoietin-1, by a group of investigators at the Immunex Research and Development Corporation in 1988 (8) was the culmination of a long search for factors that regulate early steps in B-lineage development (Table 1). Prior to the description of IL-7 as one of those factors, it was known that murine bone marrow stromal cells from long-term Whitlock-Witte cultures (9) were a source of soluble factors that supported in vitro growth of B-cell progenitors, and this resulted in the development of a rapid and specific bioassay to detect precursor B-cell growth factor activity (8). A stable stromal cell line from Whitlock-Witte cultures that secreted enhanced levels of B-cell precursor growth factor activity was established by transforming primary stromal cell cultures with a plasmid vector encoding SV40 transforming sequences, and this cell line (IxN/A6) was used to isolate and characterize IL-7. The cDNA encoding murine IL-7 was cloned shortly thereafter from the IxN/A6 cell line (10), and the human cDNA was cloned in 1989 using the murine cDNA as a probe (11).

After the identification and cloning of IL-7, progress was rapidly made in the characterization of the murine IL-7 receptor (IL-7R) and its distribution on lymphocytes (12) (Table 1). Murine and human IL-7Rα cDNAs were subsequently cloned, at which time it was recognized that there were both membrane and soluble forms of the IL-7R and that the IL-7Rα was a member of a then newly recognized receptor family, the hematopoietin/cytokine receptor superfamily (13).

Soon after IL-7 was identified as a growth factor for B-cell precursors, it was found to directly stimulate proliferation of fetal and adult murine thymocytes, including $CD4^-CD8^-CD3^-$ cells, i.e., developmentally early thymocytes (14,15). Together, these data clearly established IL-7 as a growth factor for B- and T-cell precursors.

Table 1 Major Events in the Development of IL-7

	Year	Ref.
Purification of IL-7 and identification as a growth factor for B-cell progenitors	1988	8
Molecular cloning of murine and human IL-7	1988, 1989	10,11
Identification of IL-7 as a thymocyte growth factor	1989	14,15
Characterization and cloning of the IL-7 receptor/recognition of the IL-7R as a member of the hematopoietin receptor superfamily	1990	12,13
Identification of IL-7 as an activator/enhancer of LAK and CTL activity	1990, 1991	19–23
Enhancement of antitumor responses with IL-7	1991 and later	24–31
Identification of the in vivo ability of IL-7 to enhance lymphoid regeneration	1991	32
Identification of IL-2Rγc as a component of IL-7R	1993, 1994	36,37
Generation of IL-7 transgenic mice	1991 and later	40–43
Generation of IL-7R, IL-7 knock-out mice	1994, 1995	44,45

However, unlike the B-cell lineage in which IL-7 does not appear to influence mature B cells, IL-7 was shown to activate mature, peripheral T lineage cells (16–18) including CTLs (19,20), and stimulate T-cell LAK activity (21,22). LAK activity associated with an NK-cell phenotype has also been demonstrated (23) (Table 1).

Interest in the effects of IL-7 in enhancing antitumor responses rapidly grew following the demonstration of its effects on CTL and LAK activity. IL-7 was subsequently shown to enhance the in vitro generation of antitumor CTLs with immunotherapeutic efficacy in vivo (24,25). Thereafter, a number of studies were published describing the in vivo effects of tumor cells transfected with the IL-7 gene in reducing tumorigenicity (26–31). The in vivo ability of IL-7 to enhance lymphoid regeneration was recognized as early as 1991 (32) and has since been the subject of a number of studies (33–35) (Table 1).

One of the most exciting recent developments in the IL-7 field has been the identification of the common chain of the IL-2 receptor (γc) as a functional component of the IL-7R (36,37). More recently, signaling through the IL-7Rα/γc complex has been shown to involve activation of members of the Janus family of tyrosine kinases and Stat signal–transducing molecules (38,39).

Many of the expected in vivo effects of IL-7 were confirmed in 1991, and later, by the development of IL-7–transgenic mice (40–43). Many of these studies also illustrated that overproduction of IL-7 can be related to the development of lymphoproliferative diseases including lymphomas. Mice genetically deficient in IL-7Rα or IL-7 were only recently generated (44,45); however, these provide arguably the most important tools in defining the critical role of IL-7 in regulating precursor B- and T-cell development in vivo (Table 1).

BIOCHEMISTRY AND MOLECULAR BIOLOGY OF IL-7

The IL-7 protein was purified from the IxN/A6 murine bone marrow stromal cell line immortalized by transfection with pSV3*neo* and shown to have a M_r of 25×10^3 (8). Biochemical purification indicated that IL-7 was stable to extremes of pH (2.1–8.0) and acetonitrile or *n*-propanol and in the presence of SDS or heat. However, reduction of the protein by the addition of 2-mercaptoethanol destroyed all biological activity, indicating that disulfide bonds are important in IL-7 activity. A cDNA encoding murine IL-7 was isolated by transfection of an IxN/A6 cDNA library into COS-7 cells, which were screened for IL-7 activity (10). A single clone of 1.6 kb with a major open reading frame of 462 bp was identified, the sequence of which predicted an unmodified protein of M_r 14,900. The cDNA contains two potential glycosylation sites contributing to the higher M_r of the native protein and six cysteine residues, at least some of which are important in the function of IL-7 (10). The human cDNA encoding IL-7 was cloned by nucleic acid hybridization with the murine cDNA, from a human liver adenocarcinoma cell line (11). The human and murine IL-7 amino acid sequences are highly homologous (60%). However, the human IL-7 cDNA encodes a protein with an additional 19 amino acids (positions 96–114), although these do not appear to be important for bioactivity (11). Human recombinant IL-7 is active on murine and human B-cell progenitors. In contrast, murine recombinant IL-7 does not stimulate growth of human B-cell progenitors. IL-7 transcripts appear to be differentially spliced and polyadenylated, since several different RNA species are detected on Northern blots (10). IL-7 is predicted to form a four–α-helix bundle structure common to several hematopoietins and has been classified as being a member of the "small hematopoietin" subclass, which also includes IL-9 (46).

The human IL-7 gene is a single-copy gene located on chromosome 8q12-13 (47) and consists of six exons (48). The murine IL-7 gene has only five exons; human IL-7 exon 5 encodes the 19-amino-acid insert absent from murine IL-7 (48). Interestingly, the localization of the human IL-7 gene to 8q12-13 establishes IL-7 in proximity to the gene encoding p53/p56^{lyn}, a member of the Src family of tyrosine kinases, at 8q11; IL-7 stimulates p53/p56^{lyn}, kinase activity (49,50). A second gene, HYRC, that may be involved in immunoglobulin V(D)J recombination is also found at 8q11 (50). Therefore, at least three genes that appear to have some role in early lymphoid development occur at 8q11-13, which has been postulated to constitute a lymphoid-signaling cluster (50). The murine and human IL-7 genes lack traditional TATA sequences but include potential binding sites for other known DNA-binding regulatory proteins (48), including an interferon (IFN)-stimulated response element (ISRE) followed by an initiator sequence (51). IFN-γ was recently shown to regulate IL-7 gene expression in keratinocytes (51). However, the mechanisms by which expression of the IL-7 gene is regulated in bone marrow and thymic stroma remain to be determined.

PRINCIPAL CELLULAR SOURCES OF IL-7 AND STIMULI FOR IL-7 PRODUCTION

IL-7 is considered to be expressed constitutively in the stroma of generative lymphoid organs (i.e., bone marrow and thymus), which is consistent with the principal role of IL-7 as a growth factor for precursor B cells and immature thymocytes (Table 2). However, IL-7 mRNA or protein can be found in some secondary lymphoid tissues (i.e., spleen) and probably plays a role in regulating T-cell functions and perhaps NK-cell and monocyte/macrophage activities in these locations. IL-7 production by murine long-term bone marrow cultures declines with age, correlating with decreased pre–B-cell numbers (52). IL-7 mRNA expression is distributed throughout the thymus, although the subcapsular and cortical regions and the epithelial outer layer of Hassall's corpuscles are enriched in IL-7 transcripts (53). IL-7 mRNA and protein levels are highest in the murine thymus early in fetal development and decrease as development progresses into the postnatal period (54–56). This may be related to the relative abundance of immature thymocytes early in thymic development compared to later, as immature thymocytes are the major targets of IL-7–induced proliferation in the thymus (14,15). IL-7 is not considered normally to be a product of lymphocytes,

Table 2 Cells and Tissues That Produce IL-7

IL-7 producer	Protein	mRNA	Ref.
Bone marrow stromal cells	+	+	8,74,189
Fetal liver stromal cells	n.d.	+(m)	76
Spleen	n.d.	+(m,h)	10,11
Fetal and adult thymus	+	+(m,h)	10,11,53,55,190
Thymic stromal cells	+	+(m)	93,190
Hassall's corpuscles	n.d.	+(h)	53
Kidney	n.d.	+(m)	10
Keratinocytes	+	+(m,h)	57,58
Intestine	+(h)	+(m,h)	[a]
Uterus	n.d.	+(m)	[a]
Brain	n.d.	+(m)	[a]
Adult liver	n.d.	+(r)	[a]
B-cell chronic lymphocyte leukemia	+	+(h)	191,192
EBV$^+$ B-cell lines	+	+(h)	66
Burkitt's lymphoma	+	+(h)	66
Bladder carcinoma	n.d.	+(h)	193
Colon carcinoma	+(h)	+(h)	194
Melanoma cells	n.d.	+(h)	195
Hepatoma cell line	n.d.	+(h)	11
Inflammatory malignant fibrous histocytoma	+(h)	n.d.	196

n.d., Not done; m, mouse; h, human; r, rat.
[a]From P. M. Appasamy and T. W. Kenniston, unpublished.

however, production by transformed B cells has been documented, and some Epstein-Barr virus positive (EBV^+) B-cell lines constitutively secrete IL-7 (Table 2).

IL-7 is expressed in some nonlymphoid tissues including the kidney, skin, intestine, uterus, brain, and adult liver (Table 2). The role of IL-7 in nonlymphoid tissues is unknown, however, T lymphocytes are found in many of these tissues and IL-7 may function in these locations as a growth or viability factor for tissue T cells. IL-7 produced in certain peripheral tissues (e.g., skin, intestinal epithelium) may not only be a growth/viability factor for T cells (57–59) but may also be involved in extrathymic development of T cells. For instance, Kenai et al. (60) showed that the appearance of extrathymically differentiated T cells in athymic nude mice was blocked by anti–IL-7 monoclonal antibody treatment.

A number of lymphoid and nonlymphoid tumor cells and cell lines also produce IL-7 mRNA (Table 2). It has been proposed that the expression of IL-7R and production of IL-7 by tumor cells are related to a role for IL-7 as an autocrine growth factor for certain tumors. Alternatively, IL-7 may play a role in neoplastic transformation. Transfection of pre–B cells with a retroviral construct containing IL-7, alone (61) or with V-Ha-*ras* (62), can induce tumorigenicity. In addition, lymphoid tumors have been shown to develop in several of the IL-7–transgenic strains (41,43,63). Interestingly, serum IL-7 levels are elevated in a significant percentage of patients with untreated Hodgkin's lymphoma (64), suggesting a correlation between the development or growth of this B-cell lymphoma and IL-7 production. However, overexpression of IL-7 is not necessary for pre–B-cell transformation (65), and some B-cell tumors that secrete IL-7 do not express IL-7R (66).

MAJOR BIOLOGICAL ACTIVITIES OF IL-7

The biological activities of IL-7 have been defined primarily using in vitro assays of pre–B-cell growth, thymocyte or peripheral T-cell proliferation, or LAK activity. The in vivo effects of administration of exogenous IL-7 or overproduction of IL-7 in IL-7–transgenic mice have confirmed the in vitro findings, and in some cases, identified additional effects of IL-7, e.g., the ability of IL-7 to influence myeloid progenitor development. The recently described IL-7Rα and IL-7 "knock-out" mice have provided the strongest in vivo data that clearly establish IL-7 as an essential growth factor for B-cell precursors and thymocytes. The following discussion highlights the major effects of IL-7 on various hematolymphoid lineages determined from these in vitro and in vivo studies.

IL-7 as a Major Growth Factor for Late Pro–B and Early Pre–B Cells

The role of IL-7 in lymphoid development is most well established for B-lineage cells. For the purpose of this chapter, we have used the nomenclature of Hardy and colleagues (67) to define early stages of differentiation of murine B-cell precursors. These cells, found in the bone marrow or fetal liver, are divided into (1) prepro–B, (2) early pro–B, (3) late pro–B, and (4) pre–B cells (67,68), based upon expression of antigens including HSA, CD43, and B220, but not surface IgM (sIgM) (Fig. 1). Prepro–B cells do not respond to IL-7, but require stromal cell contact and possibly other stromal cell factors (68) (Fig. 1). Early pro–B cells require both IL-7 and stromal cell molecules, whereas late pro–B cells can proliferate in response to IL-7 alone. IL-7 alone stimulates the in vitro growth of early pre–B cells but probably not late pre–B

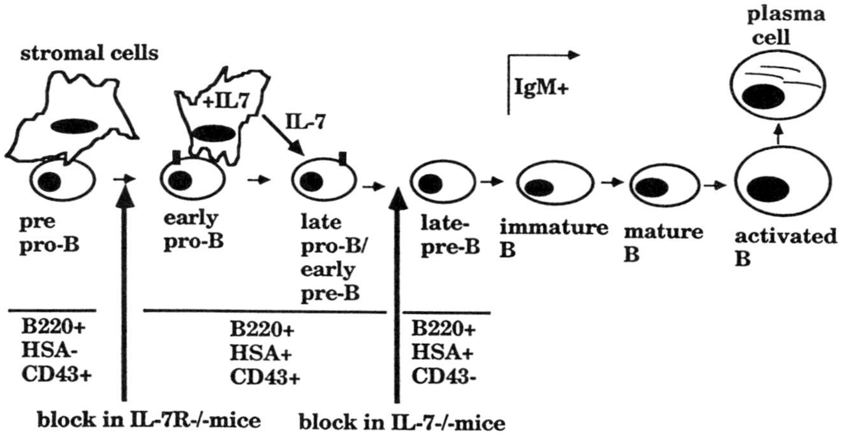

Figure 1 Role of IL-7 in B-lineage development. The hatched bar represents the presence of IL-7 receptors.

cells (68) (Fig. 2). Cytoplasmic expression of a functional Igμ protein may be necessary to drive differentiation from the stromal cell/IL-7–dependent stage to the IL-7–dependent stage (69). Other stromal cell factors are probably involved in the differentiation of pre–B cells to immature B cells (70). In turn, rearrangement of κ light chains (71) and expression of surface IgM (12,69) are correlated with the downregulation of IL-7R expression and IL-7 responsiveness. Mature B cells are, therefore, IL-7 unresponsive; however, one report suggests that anti-IgM–activated mature B cells can proliferate in response to IL-7 (72). The reader should be aware that in much of the earlier literature describing the effects of IL-7, pre–B cells are not subdivided into early and late pre–B cells; in those studies, IL-7 alone is sufficient for growth of pre–B cells that are defined as $B220^+sIgM^-$ (10,67,68,70,73–77). Although it is difficult to directly compare the stages of murine and human early B-cell development since the antigens used to define various stages appear to be species-specific, human pro–B but not pre–B cells have been shown to be responsive to IL-7 (77,78).

The in vitro studies of the effects of IL-7 on B-lineage development have been confirmed by studies in which exogenous IL-7 was administered (500 ng., s.c., twice daily) to mice with cyclophosphamide-induced lymphopenia (32). IL-7 administration accelerated regeneration of splenic and lymph node cellularity and B-cell progenitor numbers were increased in the spleen and bone marrow, followed by an increase in mature B-cell numbers (32). IL-7 administration to normal, nonlymphopenic mice also results in an increase in B-cell progenitors in the bone marrow, followed by an increase in immature and mature B cells in the spleen and lymph nodes (33). IL-7–transgenic mice have an increase in the number of B-cell precursors in the bone marrow and mature B cells in the periphery (40–42). In addition, some of these mice display a lymphoproliferative disorder in which B-lineage tumors develop (43), suggesting that dysregulation of IL-7 expression could be involved in tumorigenesis. The most definitive data demonstrating a requirement for IL-7 in B-cell precursor development comes from mice in which the IL-7Rα (44) or IL-7 (45) genes have been disrupted by homologous recombination. These mice have markedly reduced numbers of mature B lymphocytes. Late pre–B cells ($B220^+HSA^+CD43^-$) were absent from the bone marrow when the IL-7 or IL-7Rα gene was disrupted (Fig. 1). In addition, the $B220^+HSA^+CD43^+$ population also appeared to be diminished in IL-7Rα gene knockout mice, suggesting that the block in these mice is earlier than the block in $IL\text{-}7^{-/-}$ mice (Fig. 1). The difference in the phenotypes of these two knock-out mice has been proposed to be due to the ability of other cytokines to bind to the IL-7Rα chain that may be required for B-lineage development (44,45). The second chain that constitutes the high-affinity IL-7R, γ_c, has also been knocked out, and these mice have significantly reduced numbers of B lymphocytes, as well as T cells (79,80) and late pre–B cells (79), suggesting a block at the IL-7–dependent stages.

Studies in which IL-7–stimulated responses are affected by administration of anti–IL-7 or anti–IL-7R anti-

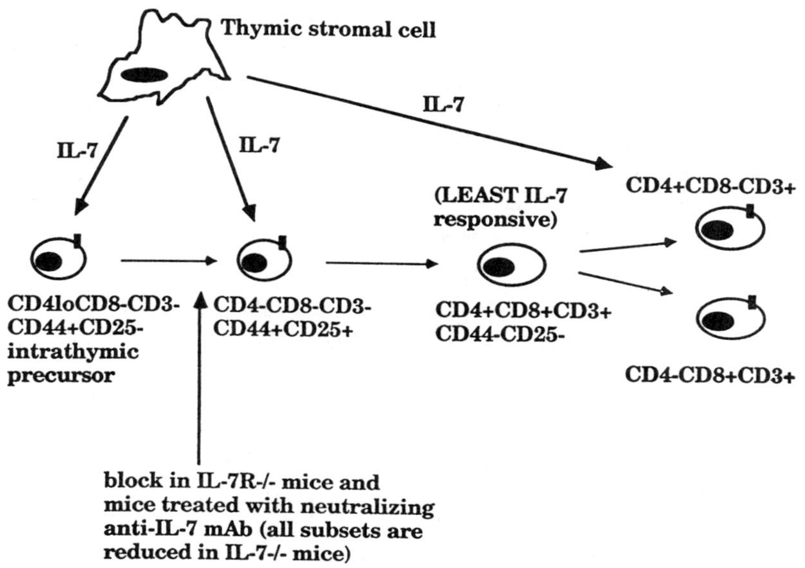

Figure 2 Role of IL-7 in T-lineage development in the thymus. The hatched bar represents the presence of IL-7 receptors.

bodies or in IL-7R knock-out mice also provide useful information about the in vivo functions of IL-7 on the B-cell lineage. Grabstein et al. (81) injected monoclonal anti–IL-7 antibodies into mice (3 mg every third day for 10 days) and found a reduction in the number of immature, pre–B, and pro–B cells in the bone marrow. Prepro–B cells survived in these mice, consistent with the in vitro data that they probably do not require IL-7 for their growth. In utero treatment with the anti–IL-7 antibody also reduced the number of B-cell precursors in the spleen (81). Continuous injection of an antibody to the cloned IL-7R into adult, neonatal, or fetal mice also caused a reduction in B-cell progenitors and mature B cells (82).

Several cytokines act in concert with IL-7 to regulate B-cell precursor development. Stem cell factor (SCF)/*c-kit* ligand, a growth factor for erythroid and myeloid progenitor cells, can synergize with IL-7 in vitro in stimulating B-cell precursor growth (83–85). IL-1α (86), TGF-β (73), and IFN-γ (87) inhibit IL-7–induced B-cell precursor growth. The additional bone marrow stromal cell factors that are probably essential for the differentiation of B-cell progenitors have yet to be identified. There are a number of genes upregulated by IL-7 that may be involved in B-lineage development. These include *N-myc* and *c-myc* (88), precursor lymphocyte-specific regulatory light chain (PLRLC), a regulatory myosin light chain (89), CD19, a cell adhesion molecule with Ig domains (90), and BP-1/6C3, an aminopeptidase (91).

IL-7 as a Growth and Possible Maturation Factor for Immature Thymocytes and γδ T Cells

T-cell development in the thymus has been subdivided into several sequential stages based upon expression of the cell surface antigens CD4, CD8, CD3, CD44, and CD25 (IL-2Rα) (92). $CD4^{lo}CD8^-CD3^-CD44^+CD25^-$ cells are the earliest precursors in the thymus, and these cells give rise to $CD4^-CD8^-CD3^-CD44^+CD25^+$ cells, which rearrange TCR β (Fig. 2). At the next stage of development, CD4, CD8, and CD3 begin to be expressed and CD25 is lost. When cell surface TCR is expressed, thymocytes can undergo positive and negative selection and diverge as $CD4^+CD8^-CD3^+$ or $CD4^-CD8^+CD3^+$ thymocyte subsets, which can populate the peripheral lymphoid tissues. As mentioned above, IL-7 is produced in the thymus, apparently by epithelial stromal cells (93) and primarily, although not exclusively, in the fetal period (54–56). Unseparated adult and fetal thymocytes proliferate in response to IL-7 in the absence of co-mitogens (14,15), and antibodies to IL-7 block proliferation in fetal thymic organ cultures (55,94). $CD4^-CD8^-$ immature thymocytes (fetal and adult) proliferate particularly well in response to IL-7 (14,15,94,95), while $CD4^+CD8^+$ thymocytes respond poorly to IL-7 (Fig. 2) (95). The responsiveness to IL-7 correlates with the expression of IL-7Rα primarily by $CD4^-CD8^-$, $CD4^+CD8^-$, and $CD4^-CD8^+$, but not $CD4^+CD8^+$ thymocytes (82).

There is a substantial body of evidence to support the notion that IL-7 preferentially stimulates growth and/ or development of γδ TCR^+ compared to αβ TCR^+ thymocytes, particularly $CD4^-CD8^-$ γδ TCR^+ cells (53,94,96–101). This may be related to differences in the expression of IL-7Rα by γδ TCR^+ versus αβ TCR^+ cells (102), differences in the activation requirements of γδ TCR^+ versus αβ TCR^+ cells, or perhaps to a role of IL-7 in regulating differentiation of γδ cells. As mentioned above, IL-7 is produced at relatively high levels in the fetal versus the adult thymus, and fetal thymocytes, which are primarily γδ TCR^+ cells, grow particularly well in response to IL-7. αβ TCR^+ thymocytes do not begin to populate the thymus in large numbers until just before birth (103). IL-7 is also a growth factor for γδ TCR^+ dendritic epidermal T cells (DETC) in the skin (58,59).

The data described above establish IL-7 as a growth factor for thymocytes, particularly immature $CD4^-CD8^-$ cells in the fetal thymus, and for γδ TCR^+ thymocytes. It has been much more difficult to establish a direct role for IL-7 in promoting or regulating differentiation of thymocytes. Some of the data suggesting a role for IL-7 in differentiation could also be interpreted as indicating a role for IL-7 in maintaining viability of thymocyte populations and allowing programs of differentiation to proceed, perhaps under the influence of other cytokines. For example, culture of human triple negative, TCR^- thymocytes with IL-7, or IL-2, results in the appearance of $CD3^+$, TCR $\gamma\delta^+$ cells (104). $CD3^-CD4^+CD8\alpha^+\beta^-$ thymocytes also can be identified following a culture of $CD4^-CD8^-CD3^-$ thymocytes with IL-7 (105). IL-7 stimulates proliferation and development of $CD4^+CD8^+$ human fetal thymocytes into $CD4^+CD8^-$ cells (106). Phorbol ester and IL-7 promote the generation of γδ TCR^+ cells from murine fetal thymocytes in vitro (100).

A number of observations support the notion that IL-7 may stimulate differentiation of a pre–T-cell population in the murine fetal liver (FL) or bone marrow, perhaps mimicking the events that normally take place in the fetal thymus. For instance, we have shown that IL-7 stimulates expression of mRNA encoding rearranged TCR γ genes (Vγ4-Jγ1-Cγ1 and Vγ6-Jγ1-Cγ1) in cultures of FL cells (107). The TCR γ transcripts expressed in response to IL-7 in the FL include mature and potentially functional recombinations of TCR γ V-J-C, and these rearrangements appear to have taken place in vitro, presumably in response to IL-7

(108). Culture of FL cells with IL-7 also results in the appearance of a population of Thy-1^+ $CD44^+$ cells, a phenotype similar to that seen in the early fetal thymus (109). We have recently also found that TCR γ genes are expressed at lower levels in the thymus of fetal or adult IL-7R$\alpha^{-/-}$ mice compared to age-matched controls, suggesting a role for IL-7 in regulating rearrangement and/or expression of TCR γ genes (P. Appasamy and N. Perumal, unpublished). A role for IL-7 in regulating TCR β rearrangement has also been suggested since IL-7, but not other cytokines or phorbol ester, allows TCR β rearrangement to proceed in short-term cultures of fetal thymocytes (110). In addition, putative FL or bone marrow lymphoid progenitors can differentiate into cells resembling early thymocytes ($CD4^+CD8^-CD3^-$ cells) when cultured in the presence of IL-7, a thymic epithelial cell line and supernatants from a FL stromal cell line (111).

Several in vivo studies have confirmed the above-described in vitro findings that IL-7 is a growth factor for immature thymocytes and $\gamma\delta$ T cells. Overexpression of IL-7 in transgenic mice in which IL-7 expression was driven by Ig heavy-chain core enhancer elements and a κ light-chain enhancer and promoter results in an increase in thymocyte and peripheral T-cell numbers, without affecting thymocyte or peripheral CD4/CD8 subsets (40). The thymic cortex was greatly reduced in size and $CD4^+CD8^+$ thymocytes were essentially absent in IL-7 transgenic mice in which the IL-7 cDNA was expressed from an IgH chain promoter and enhancer (41). These mice also developed a progressive alopecia and lymphoid infiltration in the dermis, composed mostly of TCR $\alpha\beta^+$ Thy1^+ T cells, but also including TCR $\gamma\delta^+$ T cells (42). IL-7–transgenic mice in which IL-7 expression was driven by the SRα promoter also developed alopecia, dermatitis, and dermal lymphoid infiltrates of TCR $\gamma\delta^+$ T cells. These mice had increased numbers of TCR $\gamma\delta^+$ T cells in lymphoid tissues. The effects of IL-7 on lymphocytes in the skin are not surprising, since IL-7 can stimulate growth of TCR $\gamma\delta^+$ DETC (58,59) and Sézary lymphoma cells (112). In a recent study, the IL-7 transgene expressed under the control of the Ig promoter/enhancer was introduced into nude mice and shown to overcome the marked T-cell deficiency normally seen in nude mice, presumably by stimulating intrathymic maturation (113).

As with B-cell development, the strongest data supporting a requirement for IL-7 in early thymocyte development comes from mice in which the IL-7Rα (44) or IL-7 (45) genes have been disrupted by homologous recombination. IL-7R$\alpha^{-/-}$ mice display a marked but variable reduction in thymocyte cellularity, ranging from 0.01 to 10% of control thymocyte cellularity (44). The distribution of CD4 and CD8 subsets in the IL-7R$\alpha^{-/-}$ thymus is normal in mice in which thymocyte cellularity is at least 1% of control mice. However, when the cellularity is less than 1% of controls, the thymus is composed primarily of $CD4^-CD8^-$ cells, i.e., immature thymocytes. The source of variability in thymocyte numbers is unknown. Thymocyte development in the IL-7R$\alpha^{-/-}$ mice is blocked at a very early developmental stage, since the $CD4^-CD8^-$ thymocytes, which are the predominant population in the IL-7R$\alpha^{-/-}$ mice with < 1% thymic cellularity, were composed mainly of $CD25^-$ cells (Fig. 2). These data suggest that the IL-7Rα is required for proliferation and possibly differentiation of immature thymocytes. Similarly, in IL-$7^{-/-}$ mice, thymic cellularity was substantially reduced, although the distribution of CD4/CD8 subsets in the thymus was normal, demonstrating a requirement for IL-7 in thymocyte proliferation (45). In experiments in which mAb to IL-7Rα was injected into adult mice, there was a substantial decrease in all populations of thymocytes (82). More recently, administration of neutralizing mAb to IL-7 was shown to result in a profound reduction in thymic cellularity, and expansion of $CD4^-CD8^-CD3^-$ cells was blocked at the transition to the $CD44^+CD25^+$ stage (Fig. 2) (81,114).

As with B-cell precursors, additional cytokines can influence the effects of IL-7 on thymocytes. Although IL-7–induced proliferation of thymocytes does not require IL-2, IL-4, IL-6, IL-1α, or IL-1β (14,115), IL-7 can synergize with IL-2 (95) or with SCF (116,117) to promote thymocyte growth. TNF-α can enhance thymocyte proliferation stimulated by IL-7 (118). IL-7–driven thymocyte proliferation can be inhibited by TGF-β1 or β2 (119). IL-1–induced thymocyte growth has been shown to be IL-7 dependent (120), and GM-CSF produced by thymic accessory cells in response to IL-1 participates in the IL-7–induced expansion of $CD4^-CD8^-$ thymocytes (121).

IL-7 as a Co-Mitogen for Peripheral T Lymphocytes

Although it is clear that the predominant effects of IL-7 on the T-cell lineage are on proliferation and possibly differentiation of thymocytes, IL-7 can also stimulate proliferation of mature, peripheral T lymphocytes. For example, IL-7 can function in vitro as a potent costimulator, with ConA, of mature murine splenic T-cell proliferation ($CD4^+$ and $CD8^+$ cells); these effects have been shown to have an IL-2–independent component (17,18). Human peripheral blood T cells proliferate in response to IL-7 when stimulated with anti-CD3 or lectin in short-term cultures (122). These effects are partially IL-2–dependent; in long-term cultures, IL-7 alone was sufficient to promote growth, independently of other cytokines (122). Human memory T lymphocytes respond better to IL-7 than naive cells and do not require co-mitogens or IL-2 (16). IL-7 augments the cytotoxicity of human peripheral blood $CD4^-CD8^+$ CTLs stimulated with ConA and IL-2, or antigen (19). In fact, endogenous IL-7 may be involved in the development of CTL cytotoxicity since neutralizing antiserum to IL-7 almost completely blocks alloantigen-stimulated CTL activity (19). There is an IL-2–dependent and an IL-2–independent component of the IL-7–induced generation of CTL (20,22,123–125). IL-7 can also support the generation of $CD4^-CD8^+$ CTL from murine thymocytes (123). The ability of IL-7 to enhance CTL generation is approximately equal to the effects of IL-2. However, maximal IL-7–induced cytolytic activity is delayed when compared with IL-2–induced activity (22,124). Recently, IL-7 was shown to synergize with IL-12 in the enhancing the proliferation and cytotoxic activity of human $CD8^+$ CTLs (126). In these studies, IL-7 upregulated expression of a subunit of the IL-12 receptor.

Tumor-infiltrating lymphocytes (TILs) isolated from human tumors expanded in vitro with IL-2 have cytolytic activity that is both specific and nonspecific for autologous tumor (127). IL-7 can enhance IL-2–induced proliferation of $CD4^+$ and $CD8^+$ lymphocytes, generation of LAK activity, and IFN-γ secretion from TILs derived from renal cell carcinoma (128) and can synergize with IL-2 to induce greater proliferation than with either cytokine alone (129). In these studies, the IL-7–induced proliferation was partially IL-2 dependent and IL-7 induced expression of IL-2Rα. IL-7, in conjunction with low-dose IL-2, can be used to expand TILs isolated from human melanoma (130).

NK cells and some subsets of T cells have an inherent capacity to lyse certain tumor target cells without MHC restriction or prior sensitization. LAK activity or activation of these cells with cytokines including IL-2, IL-4, IL-7, IL-10, IL-12, and IL-15 induces a broad capacity for lysis of autologous tumors and cultured tumor cell lines (M. Lotze, personal communication) (131). IL-7 promotes the generation and viability of murine LAK cells, which lyse NK-resistant tumor cell lines or freshly harvested tumor cells, from resting lymph node and splenic lymphocytes (21). IL-7–induced LAK activity of murine lymph node cells is MHC-unrestricted and associated with T cells (21) IL-7 can also stimulate IL-2–independedept development of LAK cells from cultures of human peripheral blood cells, but is much less effective than IL-2 and is mediated largely by T cells (22). IL-7 can also stimulate NK-cell LAK activity from human peripheral blood (23). Peripheral blood NK cells generate high levels of LAK activity in response to IL-7, in the absence of IL-2, although of a lesser magnitude than that generated with IL-2 (132,133). IL-7 can induce LAK activity from human peripheral blood cells obtained after autologous or syngeneic bone marrow transplant (134). This LAK activity was comparable quantitatively to IL-2–induced LAK and was particularly strong if patients had received IL-2 therapy.

IL-7 and NK-Cell Activation

NK cells are subset of lymphocytes generally defined as large granular lymphocytes that do not rearrange TCR genes but display NK-associated antigens and mediate MHC-unrestricted lysis of target cells (135). IL-7 induces expression of TNF receptors and high-affinity IL-2R on NK cells and production of low levels of TNF but not IL-2 or IL-6 (132,136). IL-7 also stimulates production of GM-CSF by NK cells (136). IL-7 is nearly as effective as IL-2 in stimulating proliferation of NK cells, and the effects of IL-7 on growth can be abrogated with anti–TNF-α antibodies (133). Human $CD56^{dim}$ NK cells proliferate much less well than $CD56^{bright}$ NK cells in response to IL-7 (137). As described above, IL-7 stimulates LAK activity from mature NK cells (132,133). IL-7 does not stimulate the growth or maturation of murine bone marrow NK precursors, but, instead, appears to maintain the viability of these cells and their subsequent responsiveness to IL-2 (138). Thus, the major effect of IL-7 on the NK cell lineage appears to be the stimulation of LAK activity by mature NK cells.

IL-7, Monocytes/Macrophages, and Committed Progenitors

The most well characterized effects of IL-7 are clearly those affecting early stages of development of B- and T-lymphoid lineages. However, IL-7 has also recently been shown to influence the activity of mature monocytes/macrophages as well as developing myeloid cells (Fig. 3). Macrophage tumor cell lines express IL-7Rα (12). Human peripheral blood monocytes produce a number of cytokines, including IL-6, TNF-α, IL-1α and IL-1β (139), macrophage inflammatory protein (MIP)-1β (140), and IL-8 (141) in response to IL-7. IL-7 can stimulate monocyte/macrophage tumoricidal activity against human melanoma cells (139). IL-7 also enhances in vitro murine macrophage antimicrobial activity against *Leishmania major* (142). Recently, IL-7 was shown to enhance proliferation of bone marrow macrophages induced by colony-stimulating factors in vitro (143).

The first suggestion that IL-7 might also affect development of myeloid cells came from studies in which human recombinant IL-7 administered to mice stimulated splenic extramedullary myelopoiesis (34,144). In those studies, IL-7 increased the number of immature granulo-

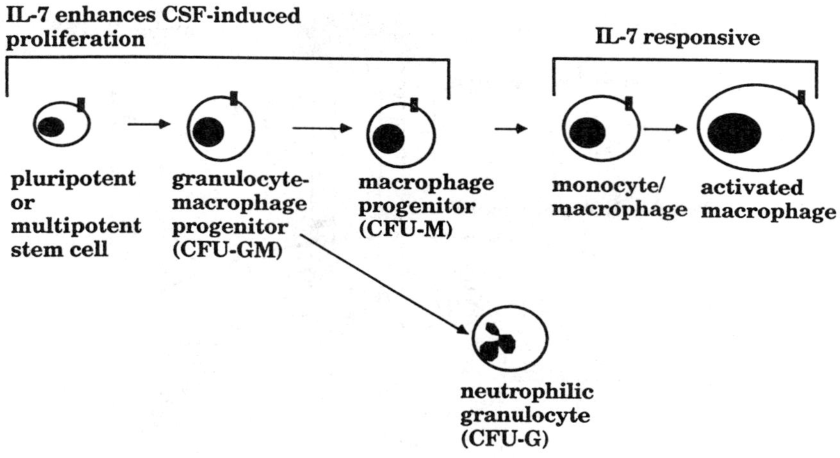

Figure 3 Role of IL-7 in myelopoiesis. The hatched bar represents the presence of IL-7 receptors.

cytes in normal and irradiated mice, or mice treated with cytotoxic agents, without affecting the number of mature granulocytes. In other studies, IL-7 administration stimulated the export of myeloid progenitors from the bone marrow into the spleen and liver (145). IL-7 can enhance IL-3–, CSF-1–, or GM-CSF–induced myeloid colony formation from highly enriched murine primitive hematopoietic precursors in vitro (Lin^-Sca^+ bone marrow cells) (146). IL-7 can also synergize with SCF to promote myeloid development from these precursors (147). More recently, IL-7 was shown to enhance CSF-1–, GM-CSF–, G-CSF–, and IL-3–stimulated proliferation of purified murine macrophage or granulocyte/macrophage progenitors in vitro but not progenitors committed to granulocyte differentiation (Fig. 3) (143). IL-7 can also synergize with CSF-1, GM-CSF, G-CSF, IL-3, and SCF in stimulation of myelopoiesis of purified human $CD34^+$ progenitors in vitro (148). Administration of IL-7 to lethally irradiated mice can mobilize long-term reconstituting hematopoietic stem cells to the peripheral blood (145). Although these data support a role for IL-7 in regulating myelopoiesis in conjunction with other cytokines, IL-7 is probably not required for baseline myelopoiesis since $IL\text{-}7^{-/-}$ mice have normal ratios and absolute numbers of monocytes and granulocytes in their peripheral blood (45). Similarly, bone marrow cellularity and CFU-GM potential are unaltered in $IL\text{-}7R\alpha^{-/-}$ mice (44).

There are no reported effects of IL-7 on erythroid lineage cells. IL-7 can stimulate the in vitro formation of eosinophil precursors and more mature eosinophils, apparently as a result of induction of IL-5 expression (149). Administration of rhuIL-7 has no reported effect on the number of blood platelets or bone marrow megakaryocytes in normal or thrombocytopenic mice (34). However, IL-7 administration can increase the number of megakaryocytes in the spleens of normal and myelosuppressed mice (34).

THE IL-7 RECEPTOR

Molecular and Structural Characterization of IL-7R

Cellular distribution studies have demonstrated the presence of IL-7–binding sites on pre–B cells, thymocytes, some T-cell lines, and bone marrow–derived macrophages, but not on mature B cells, which is consistent with the targets of IL-7 action (12). The IL-7R was subsequently found to include two subunits, IL-7Rα and the common γ chain, γ_c (Fig. 4); γ_c is also a component of the receptors for IL-2, IL-4, IL-9, and IL-15 (36,37,150–153). IL-7Rα and γ_c form a heterodimer, and both are required for high-affinity binding of IL-7. The IL-7Rα, designated as CDw127 (154), was initially characterized on a murine IL-7–dependent cloned pre–B-cell line (IxN/2b), as a 75–79 kDa glycoprotein conferring rapid and saturable IL-7 binding, with intermediate high ($K_d \sim 100$ pM)- and low-affinity ($K_d \sim 1$ nM) binding (12). Six extracellular and four intracellular cysteine residues are present in human and murine IL-7Rα, and each encodes a protein of 439 amino acids (13). There is also a human mRNA species encoding an IL-7Rα protein with a truncated cytoplasmic region and a species encoding a secreted form of the IL-7R, which binds IL-7 in solution (Fig. 4) (13) and may function in clearance of circulating cytokines (155). The IL-7R may contain an as-yet-unidentified subunit distinct from IL-7Rα and γ_c since IL-7 binds at low affinity to COS cells in the absence of these subunits (13,36).

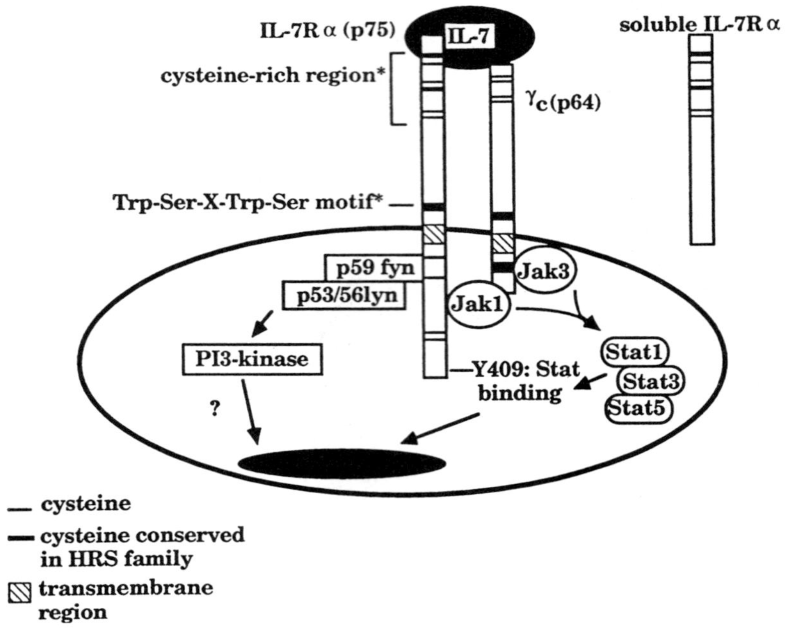

Figure 4 Model of IL-7/IL-7R interaction and signaling.* These elements are conserved between members of the hematopoietin receptor superfamily.

The human IL-7Rα gene has been identified on chromosome 5p13 (156). A number of other cytokine receptor genes have been localized to this area of chromosome 5, including receptors for LIF, growth hormone, and prolactin (157), suggesting that they derive from a common ancestral gene by gene-duplication events. The sequence of the 5′ flanking region of the murine IL-7Rα contains TATA and CAATT sequences, several potential glucocorticoid receptor-binding sites, potential AP-1 and AP-2 sites, and a sequence resembling an interferon (IFN) response element (IRE) (158). The IRE-like sequence is capable of specifically binding in vitro to two IFN regulatory factors (IRF-1 and IRF-2). This sequence also confers IFN responsiveness to a reporter gene when transfected into murine L929 cells. These data suggest that the IL-7Rα may be regulated by IFNs. Overexpression of human IRF-1 in transgenic mice results in depletion of B cells in lymphoid tissues (159).

The second subunit of the IL-7R, γ_c (p64), as mentioned above, associates with the IL-7Rα (Fig. 4) as well as receptors for IL-2, IL-4, IL-9, and IL-15. The γ_c chain appears to be constitutively expressed (160). Although γ_c alone does not bind IL-7, the association of IL-7Rα and γ_c results in the formation of a high-affinity IL-7R complex (Kd 35–40 pM) (36,37). The presence of γ_c is required for internalization of IL-7 by the high-affinity receptor complex (36). X-linked severe combined immunodeficiency (XSCID) in humans, in which there is a profound and life-threatening diminution in cell-mediated and humoral immunity, was recently attributed to mutations of the γ_c (161). The nearly complete absence of mature lymphocytes (NK, B, and T cells) in XSCID appears to be due to the loss of the activities of all five cytokine systems which utilize the γ_c and are required for lymphocyte development. IL-7Rα and γ_c are members of the hematopoietin/cytokine receptor superfamily, which is characterized by a pair of conserved extracellular cytseine residues, a Trp-Ser-X-Trp-Ser motif in the extracellular domain, and the absence of intrinsic tyrosine kinase activity (13) (Fig. 4). Other members of this superfamily are receptors for IL-2β, IL-4, IL-5, IL-6, IL-9, IL-12, IL-15, GM-CSF, G-CSF, erythropoietin, leukemia inhibitor factor (LIF), ciliary neurotrophic factor (CNTF), prolactin, and growth hormone (162).

Signaling Through IL-7R

As with other members of the hematopoietin/cytokine receptor superfamily, IL-7Rα and γ_c do not contain intrinsic kinase catalytic domains (13). The γ_c chain has a *src* homology 2 (SH2) domain in its cytoplasmic tail, which may play a role in mediating protein-protein interactions that can regulate signaling processes. Initially, IL-7 was shown to stimulate inositol phospholipid (IP) hydrolysis

(106,163,164), however, others have since shown that IL-7 does not induce calcium fluxes, changes in IP levels, or activation of phosphatidylinositol (PI)-specific phospholipase C-γ1 (165). Protein kinase C (PKC) activation is probably not utilized in the IL-7–signaling pathway (166).

IL-7 stimulates tyrosine phosphorylation of several cellular proteins in B- and T-lineage cells (49,106,163–165,167). Specific members of the nonreceptor *src* family of tyrosine kinases are activated by IL-7. The activity of the *src*-like tyrosine kinase $p59^{fyn}$, but not $p62^{yes}$, is stimulated by IL-7 in murine pre–B cells (168). IL-7 stimulates PI3-kinase activity, which may be a consequence of recruitment of the *src*-like tyrosine kinases and may be involved in the regulation of cell proliferation (168–170). IL-7Rα associates with $p59^{fyn}$ through a 95-amino-acid segment of the cytoplasmic tail of the IL-7Rα (Fig. 4) (168). $p59^{fyn}$ and $p53/56^{lyn}$ activity are stimulated by IL-7 in human pre–B cells (49). Another protein tyrosine kinase, *pim-1,* may be involved in the responsiveness of pre–B cells to IL-7, since IL-7–induced pre–B-cell expansion is reduced in *pim-1*–deficient mice (171).

More recently, IL-7 was shown to activate two members of the Janus (Jak) family of nonreceptor tyrosine kinases, Jak1 and Jak3 (38,39,152), which are also activated by other cytokines sharing the γ_c, including IL-2, IL-4, and IL-9. The primary role of γ_c may be in the recruitment of Jak 3 (39). The Jak tyrosine kinases may provide a direct signal transduction pathway to the nucleus by phosphorylation and activation of signal transducers and activators of transcription (Stats). Activated Stat proteins move to the nucleus and function as transcriptional activators upon binding to specific DNA elements (reviewed in Ref. 172). IL-7 can activate Stat 1, Stat 3, and Stat 5 (38,173). These same Stat proteins are induced by IL-7, IL-2, and IL-15, but not IL-4 (173), which may provide an additional molecular basis for the redundancy in certain responses elicited by these cytokines. At least one region of the cytoplasmic domain of IL-7Rα that may be important for binding/activation of Stat proteins has been mapped, to an area spanning tyrosine 409 at the carboxy terminus of the IL-7R (173) (Fig. 4). Similar motifs for binding of Stats are present in the IL-2Rβ.

That two distinct tyrosine kinase pathways have been shown to be activated in response to IL-7 suggests that the IL-7R has the capacity to stimulate multiple-signaling pathways. Multiple-signaling pathways may exist to allow distinct responses to IL-7 in different cell types or at particular stages of differentiation or to target distinct metabolic or physiological programs. For example, activated T cells express an altered IL-7R structure (174), which could be related to signaling events in activated versus unactivated cells. The pattern of Stats activated in PBLs in response to IL-7 differs depending upon the activation state of the responding cells (173). The concept of recruitment of multiple-signaling molecules appears to be shared by other cytokine and growth factor receptors (175).

POTENTIAL CLINICAL UTILITY OF IL-7

Based upon the in vitro and in vivo biological properties of IL-7 described above, it was recognized that IL-7 might have therapeutic value in enhancing antitumor responses, by activation of lymphocytes and macrophages; in promoting lymphoid and myeloid recovery from primary immunodeficiency or immunodeficiency secondary to chemotherapy, radiotherapy, or to other diseases; or in enhancing antimicrobial responses (Table 3). Several studies have been published in which the potential utility of IL-7 in these therapies has been tested in mice. The pharmacokinetics of soluble and liposome-formulated IL-7 administration in guinea pigs have recently been described (176). Unfortunately, as with most cytokines, high doses of IL-7 have been shown to produce some toxic effects in mice (177).

Table 3 Potential Clinical Applications of IL-7 in Therapy of Human Disease

Application	Cellular mechanisms
Promotion of antitumor responses by enhancing adoptive immunotherapy, by direct administration of IL-7, or by transfection of IL-7 gene	Induction of LAK, CTL, and macrophage activity, alone or in combination with other cytokines
Promotion of lymphopoietic and myelopoietic recovery in immunodeficiency states	Enhanced generation of lymphoid myeloid cells
Promotion of engraftment for bone marrow/stem cell transplant	Accelerated generation of lymphoid and myeloid lineages and activation of mature T cells
Enhancement of antimicrobial responses	Activation of T cells and macrophages

Antitumor Activities of IL-7

There have been several studies suggesting that IL-7 can stimulate growth or activation of antitumor lymphocytes in vitro, followed by the addition of the lymphocytes back into tumor-bearing mice (adoptive immunotherapy). Addition of IL-7 to in vitro culture of lymphocytes from draining lymph nodes (DLN) of mice injected with syngeneic fibrosarcomas enhances antitumor responsiveness when these CTL are infused in fibrosarcoma-bearing mice, and IL-7 was found to be more effective than IL-2 or IL-4 (25). Interestingly, DLN CTL can be grown in IL-7–supplemented medium for much longer periods (22 months) than in IL-2 (5–6 weeks) in the absence of additional stimulation and still retain their ability to specifically reject tumors in vivo (178). CTLs generated from DLN lymphocytes cultured with IL-7 were as effective as IL-2–generated CTL in adoptive immunotherapy against syngeneic fibrosarcoma pulmonary metastases (24). These effects were later shown to be partially IL-2 independent in vitro, and IL-7 was more effective than IL-2 in vitro in generating specific immunity (124).

Several studies in which recombinant IL-7 was directly administered to tumor-bearing mice have been described, and these generally point to a limited role for IL-7 alone; rather, IL-7 appears to be most effective in enhancing the effects of other chemical or biological therapies. IL-7 administration (10 μg/day for 10 days, i.p.) to SCID mice can enhance the antitumor (human colon carcinoma xenograft) effects of adoptively transferred human peripheral blood T cells given prior to IL-7 administration (179). These effects were shown to be partially due to increased production of IFN-γ. IL-7 alone had no effect on survival of tumor-bearing recipients (179). In these studies, IL-7 appeared to promote engraftment of the human T cells. Administration of IL-7 (10 μg/injection, twice daily for 7 days, i.p.) also caused a 75% reduction in pulmonary metastases of the renal adenocarcinoma Renca (177). These effects were due to generation of tumor-specific $CD8^+$ CTLs. The ability of IL-7 to augment survival of mice with metastatic breast cancer has been examined (180) using chemotherapy and bone marrow transplant (BMT) analogous to a model used in treatment of human breast cancer. IL-7 (5 μg twice daily for 20 days, i.v.) had an additive therapeutic effect and prolonged survival relative to chemotherapy and BMT alone. These effects were primarily due to more rapid lymphoid reconstitution with IL-7 treatment, although IL-7 alone had no therapeutic activity. IL-7 (5 ng/injection, twice a day), in combination with local hyperthermia, has been shown to reduce the size of primary B16 melanoma lesions, as well as lung metastases such that survival was significantly prolonged; IL-7 alone decreased tumor size and metastatic lesions but did not prolong survival (181). These effects appeared to be a result of an abrogation in the tumor-induced decrease in T and NK cell numbers and an increase in IL-1α, IL-6, TNF-α, and IFN-γ levels. In our own studies, we found that IL-7 administration (0.5 μg/day twice daily for 4 days, i.p.) to mice, injected 3 days before with B16 melanoma, also caused a significant reduction in the number of metastatic lung nodules (P. Appasamy and R. Goldfarb, unpublished).

Several investigators have used IL-7 gene–transfected tumor cells to evaluate the ability of IL-7 to enhance antitumor responses in vivo. These studies are of interest because of the potential use of gene-transfected cells as antitumor vaccines in human cancers. In each of these studies, tumors were effectively rejected by host mice while their in vitro growth rate was unaffected by IL-7 transfection (26–31,182,183). Presumably, a higher concentration of IL-7 at the tumor site as a result of transfection results in more effective antitumor responses than when IL-7 is injected. The antitumor responsive cells were, in most cases, T lymphocytes (26–30); however, macrophages (26), eosinophil, and basophils (28) were also often found at the site of tumor rejection. In addition, transfection of the IL-7 gene into a murine fibrosarcoma can decrease tumor production of TGF-β1, an immunosuppressive molecule (184). Clinical trials utilizing IL-7–transfected tumor cells have now been proposed by two groups, one in Berlin (185) and one at UCLA (J. Economou, personal communication).

Lymphoid and Myeloid Regeneration in Lymphopenic and Leukopenic Mice

One of the most promising areas of study on the clinical applications of IL-7 is the utility of IL-7 administration to accelerate lymphoid regeneration in patients with lymphopenia as a consequence of primary or secondary B- or T-cell immunodeficiencies (e.g., AIDS, SCID, DiGeorge syndrome, infectious disease) or subsequent to chemotherapy. With the more recent demonstration that IL-7 can accelerate myelopoiesis, IL-7 may be effective in regeneration of this compartment of white blood cells as well. Acceleration of lymphoid regeneration (B- and T-cell compartments) has been demonstrated in mice with cyclophosphamide-induced lymphopenia (32). IL-7 has been shown to accelerate recovery of peripheral blood lymphocytes in mice made leukopenic by administration of cyclophosphamide or 5-fluorouracil (144). Administration of IL-7 to mice made leukopenic by irradiation promotes a rapid increase in total leukocyte numbers and numbers of lymphocytes (34). In these studies, IL-7 also increased the numbers of some immature myeloid cells. Following a split-dose chemotherapy protocol of cyclophosphamide, cisplatin, and nitrosourea for treatment of metastatic breast cancer in mice, treatment with IL-7 and BMT could significantly prolong survival beyond BMT and chemotherapy alone, presumably due to enhance T-cell reconstitution (180). In an interesting recent study, administration of recombinant human IL-7 to mice was shown to mobilize long-term reconstituting peripheral blood stem cells (145). In these studies, survival was significantly increased among lethally irradiated mice that received PBL from rhIL-7–treated donors, compared to mice that received normal PBL. These experiments illustrate that IL-7 may be useful in mobilizing hematopoietic stem cells to the peripheral blood for use in stem cell transplantation or gene modification. Since IL-7 can stimulate proliferation of donor T cells in MHC-matched allogeneic bone marrow transplants and is produced in the spleen during BMT, it has been proposed to play an in vivo role in regulating T-cell reconstitution and activation during BMT (80).

Antimicrobial Properties of IL-7

Not unexpectedly, IL-7 has been shown to have antimicrobial properties, primarily due to T-cell, NK-cell, and macrophage activation. IL-7 injection can prolong survival time and reduce virus levels in mice infected with the polycythemia-inducing strain of Friend virus complex (FVC) (186). IL-7 was effective in restoring FVC-induced decreases in NK-cell activity and production of IL-6 and IFN-γ. The severity and course of primary herpes simplex virus (HSV)-2 infection in guinea pigs can be reduced by administration of IL-7 in association with HSV antigen (187). IL-7 enhances the antimicrobial activity of murine macrophages against *Leishmania major* (142). Recently, liposome-formulated IL-7 was shown to enhance antibody and CTL responses to recombinant HIV envelope protein in mice (188), suggesting that concurrent administration of IL-7 may enhance vaccination to this and other proteins.

CONCLUSIONS

There are now more than 500 published references evaluating the biological properties of IL-7. The role of IL-7 in stimulating B-cell precursor and thymocyte growth is quite well defined. Clinical trials in humans have yet to be performed on the utility of IL-7 in enhancing antitumor responses or accelerating regeneration of lymphoid or myeloid lineages, which may be partly due to limited perception of the utility by the controlling company. However, RAC and FDA approval has recently been obtained for a clinical trial of IL-7–transduced melanoma vaccines (J. Economou, personal communication). Another clinical study in which the IL-7 gene will be transfected into lymphocytes or tumor cells for therapy of human metastatic colon carcinoma, renal cell carcinoma, melanoma, or lymphoma has been proposed (185). There are still a number of areas in the field of IL-7 biology that remain unexplored. For example, relatively little is known about the molecular mechanisms by which IL-7 functions as a lymphopoietic growth factor, and this is certainly an area for extensive future investigation. Regulation of the expression of IL-7 and IL-7Rα remains virtually unstudied. Whether IL-7 truly regulates T-lineage differentiation or is restricted to effects on growth of thymocytes remains to be clarified. The mechanisms by which IL-7 is a particularly good growth factor for $\gamma\delta$ TCR^+ compared to $\alpha\beta$ TCR^+ T cells are only beginning to be understood. The results of investigation into each of these areas of IL-7 biology are expected to contribute to the understanding of hematolymphopoiesis in normal and diseased states.

ACKNOWLEDGMENTS

I thank Drs. David Tweardy, William Chambers, Narayanan Perumal, Markus Maeurer, and Michael Lotze for their help during the preparation of this manuscript, and Dr. James Economou for sharing information on the status of his clinical trial of IL-7–transduced melanoma vaccines. This work was supported in part by Grant AI31515 from the National Institutes of Health.

REFERENCES

1. Appasamy PM. Interleukin-7 and lymphopoiesis: biologic and clinical implications. In: Kurzrock R, Talpaz M, eds. Cancer Treatment and Research. Boston: Kluwer Academic Publishers, 1995:235–260.
2. Appasamy PM. Interleukin-7: biology and potential clinical applications. Cancer Invest 1993; 11:487–499.
3. Goodwin RG, Namen AE. The cloning and characterization of interleukin-7. In: Cruse JM, Lewis RE Jr, eds. The Year in Immunology. Basel: Karger, 1990:127–139.
4. Widmer MB, Morrissey PJ, Goodwin RG, Grabstein KH, Park LS, Watson JD, Kincade PW, Colon PJ, Namen AE. Lymphopoiesis and IL-7. Int J Cell Cloning 1990; 8(suppl 1):168–172.
5. Park LS, Morrissey PJ, Davison B, Grabstein K. The role of IL-7 and its receptor in B-cell ontogeny. In: Gupta S, Waldmann TA, eds. Mechanisms of Lymphocyte Activation and Immune Regulation IV: Cellular Communications. New York: Plenum Press, 1992:125–129.
6. Costello R, Duffaud F. The pleiotropic effects of interleukin 7 and their pathologic and therapeutic implications. Eur J Med 1992; 1:119–121.
7. Costello R, Imbert J, Olive D. Interleukin-7, a major T-lymphocyte cytokine. Eur Cytokine Netw 1993; 4:253–262.
8. Namen AE, Schmierer AE, March CJ, Overell RW, Park LS, Urdal DL, Mochizuki DY. B cell precursor growth-promoting activity. Purification and characterization of a growth factor

active on lymphocyte precursors. J Exp Med 1988; 167:988–1002.
9. Whitlock CA, Witte ON. Long term culture of B lymphocytes and their precursors from murine bone marrow. Proc Natl Acad Sci USA 1982; 79:3608–3612.
10. Namen AE, Lupton S, Hjerrild K, Wignal J, Mochizuki DY, Schmierer A, Mosley B, March CJ, Urdal D, Gillis S. Stimulation of B-cell progenitors by cloned murine interleukin-7. Nature 1988; 333:571–573.
11. Goodwin RG, Lupton S, Schmierer A, Hjerrild KJ, Jerzy R, Clevenger W, Gillis S, Cosman D, Namen AE. Human interleukin 7: molecular cloning and growth factor activity on human and murine B-lineage cells. Proc Natl Acad Sci USA 1989; 86:302–306.
12. Park LS, Friend DJ, Schmierer AE, Dower SK, Namen AE. Murine interleukin 7 (IL-7) receptor. Characterization on an IL-7-dependent cell line. J Exp Med 1990; 171:1073–1089.
13. Goodwin RG, Friend D, Ziegler SF, Jerzy R, Falk BA, Gimpel S, Cosman D, Dower SK, March CJ, Namen AE, et al. Cloning of the human and murine interleukin-7 receptors: demonstration of a soluble form and homology to a new receptor superfamily. Cell 1990; 60:941–951.
14. Conlon PJ, Morrissey PJ, Nordan RP, Grabstein KH, Prickett KS, Reed SG, Goodwin R, Cosman D, Namen AE. Murine thymocytes proliferate in direct response to interleukin-7. Blood 1989; 74:1368–1373.
15. Watson JD, Morrissey PJ, Namen AE, Conlon PJ, Widmer MB. Effect of IL-7 on the growth of fetal thymocytes in culture. J Immunol 1989; 143:1215–1222.
16. Welch PA, Namen AE, Goodwin RG, Armitage R, Cooper MD. Human IL-7: a novel T cell growth factor. J Immunol 1989; 143:3562–3567.
17. Grabstein KH, Namen AE, Shanebeck K, Voice RF, Reed SG, Widmer MB. Regulation of T cell proliferation by IL-7. J Immunol 1990; 144:3015–3020.
18. Morrissey PJ, Goodwin RG, Nordan RP, Anderson D, Grabstein KH, Cosman D, Sims J, Lupton S, Acres B, Reed SG. Recombinant interleukin 7, pre-B cell growth factor, has costimulatory activity on purified mature T cells. J Exp Med 1989; 169:707–716.
19. Hickman CJ, Crim JA, Mostowski HS, Siegel JP. Regulation of human cytotoxic T lymphocyte development by IL-7. J Immunol 1990; 145:2415–2420.
20. Smyth MJ, Norihisa Y, Gerard JR, Young HA, Ortaldo JR. IL-7 regulation of cytotoxic lymphocytes: pore-forming protein gene expression, interferon-gamma production, and cytotoxicity of human peripheral blood lymphocytes subsets. Cell Immunol 1991; 138:390–403.
21. Lynch DH, Miller RE. Induction of murine lymphokine-activated killer cells by recombinant IL-7. J Immunol 1990; 145:1983–1990.
22. Alderson MR, Sassenfeld HM, Widmer MB. Interleukin 7 enhances cytolytic T lymphocyte generation and induces lymphokine-activated killer cells from human peripheral blood. J Exp Med 1990; 172:577–587.
23. Stotter H, Custer MC, Bolton ES, Guedez L, Lotze MT. IL-7 induces human lymphokine-activated killer cell activity and is regulated by IL-4. J Immunol 1991; 146:150–155.
24. Jicha DL, Mule JJ, Rosenberg SA. Interleukin 7 generates antitumor cytotoxic T lymphocytes against murine sarcomas with efficacy in cellular adoptive immunotherapy. J Exp Med 1991; 174:1511–1515.
25. Lynch DH, Namen AE, Miller RE. In vivo evaluation of the effects of interleukins 2, 4 and 7 on enhancing the immunotherapeutic efficacy of anti-tumor cytotoxic T lymphocytes. Eur J Immunol 1991; 21:2977–2985.
26. Hock H, Dorsch M, Diamantstein T, Blankenstein T. Interleukin 7 induces $CD4^+$ T cell-dependent tumor rejection. J Exp Med 1991; 174:1291–1298.
27. Aoki T, Tashiro K, Miyatake S, Kinashi T, Nakano T, Oda Y, Kikuchi H, Honjo T. Expression of murine interleukin 7 in a murine glioma cell line results in reduced tumorigenicity in vivo. Proc Natl Acad Sci USA 1992; 89:3850–3854.
28. McBride WH, Thacker JD, Comora S, Economou JS, Kelley D, Hogge D, Dubinett SM, Dougherty GJ. Genetic modification of a murine fibrosarcoma to produce interleukin 7 stimulates host cell infiltration and tumor immunity. Cancer Res 1992; 52:3931–3937.
29. Miller AR, McBride WH, Dubinett SM, Dougherty GJ, Thacker JD, Shau H, Kohn DB, Moen RC, Walker MJ, Chiu R, et al. Transduction of human melanoma cell lines with the human interleukin-7 gene using retroviral-mediated gene transfer: comparison of immunologic properties with interleukin-2. Blood 1993; 82:3686–3694.
30. Hock H, Dorsch M, Kunzendorf U, Qin Z, Diamantstein T, Blankenstein T. Mechanisms of rejection induced by tumor cell-targeted gene transfer of interleukin 2, interleukin 4, interleukin 7, tumor necrosis factor, or interferon gamma. Proc Natl Acad Sci USA 1993; 90:2774–2778.
31. Hock H, Dorsch M, Kunzendorf U, Uberla K, Qin Z, Diamantstein T, Blankenstein T. Vaccinations with tumor cells genetically engineered to produce different cytokines: effectivity not superior to a classical adjuvant. Cancer Res 1993; 53:714–716.
32. Morrissey PJ, Conlon P, Braddy S, Williams DE, Namen AE, Mochizuki DY. Administration of IL-7 to mice with cyclophosphamide-induced lymphopenia accelerates lymphocyte repopulation. J Immunol 1991; 146:1547–1552.
33. Morrissey PJ, Conlon P, Charrier K, Braddy S, Alpert A, Williams D, Namen AE, Mochizuki D. Administration of IL-7 to normal mice stimulates B-lymphopoiesis and peripheral lymphadenopathy. J Immunol 1991; 147:561–568.
34. Faltynek CR, Wang S, Miller D, Young E, Tiberio L, Kross K, Kelley M, Kloszewski E. Administration of human recombinant IL-7 to normal and irradiated mice increases the numbers of lymphocytes and some immature cells of the myeloid lineage. J Immunol 1992; 149:1276–1282.
35. Fraser CC, Thacker JD, Hogge DE, Fatur Saunders D, Takei F, Humphries RK. Alterations in lymphopoiesis after hematopoietic reconstitution with IL-7 virus-infected bone marrow. J Immunol 1993; 151:2409–2418.
36. Noguchi M, Nakamura Y, Russell SM, Ziegler SF, Tsang M, Cao X, Leonard WJ. Interleukin-2 receptor gamma chain: a functional component of the interleukin-7 receptor [see comments]. Science 1993; 262:1877–1880.
37. Kondo M, Takeshita T, Higuchi M, Nakamura M, Sudo T, Nishikawa S-I, Sugamura K. Functional participation of the IL-2 receptor γ chain in IL-7 receptor complexes. Science 1994; 263:1453–1454.
38. Zeng YX, Takahashi H, Shibata M, Hirokawa K. JAK3 Janus kinase is involved in interleukin 7 signal pathway. FEBS Lett 1994; 353:289–293.
39. Musso T, Johnston JA, Linnekin D, Varesio L, Rowe TK, O'Shea D, McVicar DW. Regulation of JAK3 expression in human monocytes: phosphorylation in response to interleukins 2, 4, and 7. J Exp Med 1995; 181:1425–1431.
40. Samaridis J, Casorati G, Traunecker A, Iglesias A, Gutierrez JC, Muller U, Palacios R. Development of lymphocytes in interleukin 7-transgenic mice. Eur J Immunol 1991; 21:453–460.

41. Rich BE, Campos Torres J, Tepper RI, Moreadith RW, Leder P. Cutaneous lymphoproliferation and lymphomas in interleukin 7 transgenic mice. J Exp Med 1993; 177:305–316.
42. Uehira M, Matsuda H, Hikita I, Sakata T, Fujiwara H, Nishimoto H. The development of dermatitis infiltrated by gamma delta T cells in IL-7 transgenic mice. Int Immunol 1993; 5:1619–1627.
43. Fisher AG, Burdet C, LeMeur M, Haasner D, Gerber P, Ceredig R. Lymphoproliferative disorders in an IL-7 transgenic mouse line. Leukemia 1993; 7(suppl 2):S66–S68.
44. Peschon JJ, Morrissey PJ, Grabstein KH, Ramsdell FJ, Maraskovsky E, Gliniak BC, Park LS, Ziegler SF, Williams DE, Ware CB, Meyer JD, Davison BL. Early lymphocyte expansion is severely impaired in interleukin-7 receptor deficient mice. J Exp Med 1994; 180:1955–1960.
45. von Freeden-Jeffry U, Vieira P, Lucian LA, McNeil T, Burdach SEG, Murray R. Lymphopenia in interleukin (IL)-7 gene-deleted mice identifies IL-7 as a nonredundant cytokine. J Exp Med 1995; 181:1519–1526.
46. Boulay J, Paul WE. Hematopoietin sub-family classification based on size, gene organization and sequence homology. Current Biol 1993; 3:573–581.
47. Sutherland GR, Baker E, Fernandez KE, Callen DF, Goodwin RG, Lupton S, Namen AE, Shannon MF, Vadas MA. The gene for human interleukin 7 (IL7) is at 8q12-13. Hum Genet 1989; 82:371–372.
48. Lupton SD, Gimpel S, Jerzy R, Brunton LL, Hjerrild KA, Cosman D, Goodwin RG. Characterization of the human and murine IL-7 genes. J Immunol 1990; 144:3592–3601.
49. Seckinger P, Fougereau M. Activation of src family kinases in human pre-B cells by IL-7. J Immunol 1994; 153:97–109.
50. Corey SJ, Shapiro DN. Localization of the human gene for Src-related protein tyrosine kinase LYN to chromosome 8q11-12: a lymphoid signaling cluster? Leukemia 1994; 8:1914–1917.
51. Ariizumi K, Meng Y, Bergstresser PR, Takashima A. IFN-γ-dependent IL-7 gene regulation in keratinocytes. J Immunol 1995; 154:6031–6039.
52. Updyke LW, Cocke KS, Wierda D. Age-related changes in production of interleukin-7 (IL-7) by murine long-term bone marrow cultures (LTBMC). Mech Ageing Dev 1993; 69:109–117.
53. He W, Zhang Y, Deng Y, Kabelitz D. Induction of TCT-γδ expression on triple negative ($CD3^-4^-8^-$) human thymocytes. J Immunol 1995; 154:3726–3731.
54. Montgomery RA, Dallman MJ. Analysis of cytokine gene expression during fetal thymic ontogeny using the polymerase chain reaction. J Immunol 1991; 147:554–560.
55. Wiles MV, Ruiz P, Imhof BA. Interleukin-7 expression during mouse thymus development. Eur J Immunol 1992; 22:1037–1042.
56. Moore NC, Anderson G, Smith CA, Owen JJT, Jenkinson EJ. Analysis of cytokine gene expression in subpopulations of freshly isolated thymocytes and thymic stromal cells using semiquantitative polymerase chain reaction. Eur J Immunol 1993; 23:922–927.
57. Heufler C, Topar G, Grasseger A, Stanzl U, Koch F, Romani N, Namen AE, Schuler G. Interleukin 7 is produced by murine and human keratinocytes. J Exp Med 1993; 178:1109–1114.
58. Matsue H, Bergstresser PR, Takashima A. Keratinocyte-derived IL-7 serves as a growth factor for dendritic epidermal T cells in mice. J Immunol 1993; 151:6012–6019.
59. Matsue H, Bergstresser PR, Takashima A. Reciprocal cytokine-mediated cellular interactions in mouse epidermis: promotion of gamma delta T-cell growth by IL-7 and TNF alpha and inhibition of keratinocyte growth by gamma IFN. J Invest Dermatol 1993; 101:543–548.
60. Kenai H, Matsuzaki G, Nakamura T, Yoshikai Y, Nomoto K. Thymus-derived cytokine(s) including interleukin-7 induce increase of T cell receptor alpha/beta+ CD4- CD8- T cells which are extrathymically differentiated in athymic nude mice. Eur J Immunol 1993; 23:1818–1825.
61. Overell RW, Clark L, Lynch D, Jerzy R, Schmierer A, Weisser KE, Namen AE, Goodwin RG. Interleukin-7 retroviruses transform pre-B cells by an autocrine mechanism not evident in Abelson murine leukemia virus transformants. Mol Cell Biol 1991; 11:1590–1597.
62. Chen SC, Redenius D, Young JC, Schwartz RC. Synergy of IL-7 and v-Ha-ras in the in vitro neoplastic progression of murine pre-B cells. Oncogene 1993; 8:2119–2125.
63. Fisher AG, Burdet C, Bunce C, Merkenschlager M, Ceredig RH. Lymphoproliferative disorders in IL-7 transgenic mice: expansion of immature B cells which retain macrophage potential. Int Immunol 1995; 7:415–423.
64. Trumper L, Jung W, Dahl G, Diehl V, Pfeundschuh M. Interleukin-7, interleukin-8, soluble TNF receptor, and p53 protein-levels are elevated in the serum of patients with Hodgkins disease. Ann Oncol 1994; 5:93–96.
65. Young JC, Gishizky ML, Witte ON. Hyperexpression of interleukin-7 is not necessary or sufficient for transformation of a pre-B lymphoid cell line. Mol Cell Biol 1991; 11:854–863.
66. Benjamin D, Sharma V, Knobloch TJ, Armitage RJ, Dayton MA, Goodwin RG. Human B cell lines constitutively secrete IL-7 and express IL7 receptors. J Immunol 1994; 152:4749–4757.
67. Li Y, Hayakawa K, Hardy RR. The regulated expression of B lineage associated genes during B cell differentiation in bone marrow and fetal liver. J Exp Med 1993; 178:951–960.
68. Hardy RR, Carmack CE, Shinton SA, Kemp JD, Hayakawa K. Resolution and characterization of pro-B and pre-pro-B cell stages in normal mouse bone marrow. J Exp Med 1991; 173:1213–1225.
69. Era T, Ogawa M, Nishikawa S, Okamoto M, Honjo T, Akagi K, Miyazaki J, Yamamura K. Differentiation of growth signal requirement of B lymphocyte precursor is directed by expression of immunoglobulin. EMBO J 1991; 10:337–342.
70. Cumano A, Dorshkind K, Gillis S, Paige CJ. The influence of S17 stromal cells and interleukin 7 on B cell development. Eur J Immunol 1990; 20:2183–2189.
71. Henderson AJ, Narayanan R, Collins L, Dorshkind K. Status of kappa L chain gene rearrangements and c-kit and IL-7 receptor expression in stromal cell-dependent pre-B cells. J Immunol 1992; 149:1973–1979.
72. Joshi PC, Choi YS. Human interleukin 7 is a B cell growth factor for activated B cells. Eur J Immunol 1991; 21:681–686.
73. Lee G, Namen AE, Gillis S, Ellingsworth LR, Kincade PW. Normal B cell precursors responsive to recombinant murine IL-7 and inhibition of IL-7 activity by transforming growth factor-beta. J Immunol 1989; 142:3875–3883.
74. Sudo T, Ito M, Ogawa Y, Iizuka M, Kodama H, Kunisada T, Hayashi S, Ogawa M, Sakai K, Nishikawa S. Interleukin 7 production and function in stromal cell-dependent B cell development. J Exp Med 1989; 170:333–338.
75. Hayashi S, Kunisada T, Ogawa M, Sudo T, Kodama H, Suda T, Nishikawa S. Stepwise progression of B lineage differentiation supported by interleukin 7 and other stromal cell molecules. J Exp Med 1990; 171:1683–1695.
76. Gunji Y, Sudo T, Suda J, Yamaguchi Y, Nakauchi H, Nishikawa S, Yanai N, Obinata M, Yanagisawa M, Miura Y, et al. Support

of early B-cell differentiation in mouse fetal liver by stromal cells and interleukin-7. Blood 1991; 77:2612–2617.

77. Ryan DH, Nuccie BL, Ritterman I, Liesveld JL, Abboud CN. Cytokine regulation of early human lymphopoiesis. J Immunol 1994; 152:5248–5249.
78. Dittel BN, LeBien TW. The growth response to IL-7 during normal human B cell ontogeny is restricted to B-lineage cells expressing CD34. J Immunol 1995; 154:58–67.
79. DiSanto JP, Muller W, Guy-Grand D, Fischer A, Rajewsky K. Lymphoid development in mice with a targeted deletion of the interleukin 2 receptor γ chain. Proc Natl Acad Sci USA 1995; 92:377–381.
80. Cao X, Shores EW, Hu-Li J, Anver MR, Kelsall BL, Russell SM, Drago J, Noguchi M, Grinberg A, Bloom ET, Paul WE, Katz SI, Love PE, Leonard WJ. Defective lymphoid development in mice lacking expression of the common cytokine receptor γ chain. Immunity 1995; 2:223–238.
81. Grabstein KH, Waldschmidt TJ, Finkelman FD, Hess BW, Alpert AR, Boiani NE, Namen AE, Morrissey PJ. Inhibition of murine B and T lymphopoiesis in vivo by an anti-interleukin 7 monoclonal antibody. J Exp Med 1993; 178:257–264.
82. Sudo T, Nishikawa S, Ohno N, Akiyama N, Tamakoshi M, Yoshida H. Expression and function of the interleukin 7 receptor in murine lymphocytes. Proc Natl Acad Sci USA 1993; 90:9125–9129.
83. McNiece IK, Langley KE, Zsebo KM. The role of recombinant stem cell factor in early B cell development. Synergistic interaction with IL-7. J Immunol 1991; 146:3785–3790.
84. Funk PE, Varas A, Witte PL. Activity of stem cell factor and IL-7 in combination on normal bone marrow B lineage cells. J Immunol 1993; 150:748–752.
85. Billips LG, Petitte D, Dorshkind K, Narayanan R, Chiu CP, Landreth KS. Differential roles of stromal cells, interleukin-7, and kit-ligand in the regulation of B lymphopoiesis. Blood 1992; 79:1185–1192.
86. Suda T, Okada S, Suda J, Miura Y, Ito M, Sudo T, Hayashi S, Nishikawa S, Nakauchi H. A stimulatory effect of recombinant murine interleukin-7 (IL-7) on B-cell colony formation and an inhibitory effect of IL-1 alpha. Blood 1989; 74:1936–1941.
87. Garvy BA, Riley RL. INF-γ abrogates IL-7 dependent proliferation in pre-B cells, conciding with onset of apoptosis. Immunology 1994; 81:381–388.
88. Morrow MA, Lee G, Gillis S, Yancopoulos GD, Alt FW. Interleukin-7 induces N-myc and c-myc expression in normal precursor B lymphocytes. Genes Dev 1992; 6:61–70.
89. Oltz EM, Yancopoulos GD, Morrow MA, Rolink A, Lee G, Wong F, Kaplan K, Gillis S, Melchers F, Alt FW. A novel regulatory myosin light chain gene distinguishes pre-B cell subsets and is IL-7 inducible. EMBO J 1992; 11:2759–2767.
90. Wolf ML, Weng WK, Stieglbauer KT, Shah N, LeBien TW. Functional effect of IL-7-enhanced CD19 expression on human B cell precursors. J Immunol 1993; 151:138–148.
91. Sherwood PJ, Weissman IL. The growth factor IL-7 induces expression of a transformation-associated antigen in normal pre-B cells. Int Immunol 1990; 2:399–406.
92. Shortman K. Cellular aspects of early T-cell development. Curr Opin Immunol 1992; 4:140–146.
93. Gutierrez JC, Palacios R. Heterogeneity of thymic epithelial cells in promoting T-lymphocyte differentiation in vivo. Proc Natl Acad Sci USA 1991; 88:642–646.
94. Plum J, De Smedt M, Leclercq G. Exogenous IL-7 promotes the growth of CD3-CD4-CD8-CD44+CD25+/– precursor cells and blocks the differentiation pathway of TCR-alpha beta cells in fetal thymus organ culture. J Immunol 1993; 150:2706–2716.
95. Okazaki H, Ito M, Sudo T, Hattori M, Kano S, Katsura Y, Minato N. IL-7 promotes thymocyte proliferation and maintains immunocompetent thymocytes bearing alpha beta or gamma delta T-cell receptors in vitro: synergism with IL-2. J Immunol 1989; 143:2917–2922.
96. Leclercq G, De Smedt M, Plum J. Cytokine production and responsiveness of fetal T-cell receptor V gamma 3 thymocytes. Scand J Immunol 1992; 36:833–841.
97. Leclercq G, De Smedt M, Plum J. Presence of CD8 alpha-CD8 beta-positive TcR gamma/delta thymocytes in the fetal murine thymus and their in vitro expansion with interleukin-7. Eur J Immunol 1992; 22:2189–2193.
98. Lynch F, Shevach EM. Activation requirements of newborn thymic gamma delta T cells. J Immunol 1992; 149:2307–2314.
99. He W, Kabelitz D. Differential effects of interleukin-7 and interleukin-2 on T-cell receptor gamma delta-expressing cells within CD4-CD8-postnatal human thymocytes. Int Arch Allergy Immunol 1993; 102:321–326.
100. Watanabe Y, Sudo T, Minato N, Ohnishi A, Katsura Y. Interleukin 7 preferentially supports the growth of gamma delta T cell receptor-bearing T cells from fetal thymocytes in vitro. Int Immunol 1991; 3:1067–1075.
101. Tomana M, Ideyama S, Iwai K, Gyotoku J, Germeraad WT, Muramatsu S, Katsura Y. Involvement of IL-7 in the development of gamma delta T cells in the thymus. Thymus 1993; 21:141–157.
102. Heinrich DW, Ziegler HK. αβ and γσ T cells from listeria monocytogenes (LM) immunized mice differ in their IL-7 receptor expression (abstr). FASEB J 1995; A1062.
103. Balasubramaniam V, Appasamy PM. Development of murine pre-T cells into γδ T-cell receptor bearing cells. Dev Comp Immunol 1994; 18:179–191.
104. Groh V, Fabbi M, Strominger JL. Maturation or differentiation of human thymocyte precursors in vitro? Proc Natl Acad Sci USA 1990; 87:5973–5977.
105. Hori T, Cupp J, Wrighton N, Lee F, Spits H. Identification of a novel human thymocyte subset with a phenotype of CD3-CD4+ CD8 alpha + beta-1. Possible progeny of the CD3- CD4-CD8- subset. J Immunol 1991; 146:4078–4084.
106. Uckun FM, Tuel Ahlgren L, Obuz V, Smith R, Dibirdik I, Hanson M, Langlie MC, Ledbetter JA. Interleukin 7 receptor engagement stimulates tyrosine phosphorylation, inositol phospholipid turnover, proliferation, and selective differentiation to the CD4 lineage by human fetal thymocytes. Proc Natl Acad Sci USA 1991; 88:6323–6327.
107. Appasamy PM, Kenniston TW, Jr., Weng Y, Holt EC, Kost J, Chambers WH. Interleukin 7-induced expression of specific T cell receptor gamma variable region genes in murine fetal liver cultures. J Exp Med 1993; 178:2201–2206.
108. Appasamy PM, Weng Y, Kenniston TW, Jr., DeLeo AB, Tang L. Expression of diverse and functional TCR γ and Ig heavy chain transcripts in fetal liver cells cultured with IL-7. Mol Immunol 1995; 32:805–817.
109. Appasamy PM. IL-7 induced T cell receptor-gamma gene expression by pre-T cells in murine fetal liver cultures. J Immunol 1992; 149:1649–1656.
110. Muegge K, Vila MP, Durum SK. Interleukin-7: a cofactor for V(D)J rearrangement of the T cell receptor beta gene. Science 1993; 261:93–95.
111. Golunski E, Palacios R. Fetal liver and bone marrow JORO 75+ lymphocyte progenitors are precursors of CD4+8- TCR/CD3- early thymocytes. J Exp Med 1994; 179:721–725.
112. Dalloul A, Laroche L, Bagot M, Mossalayi MD, Fourcade C, Thacker DJ, Hogge DE, Merle Beral H, Debre P, Schmitt C.

Interleukin-7 is a growth factor for Sezary lymphoma cells. J Clin Invest 1992; 90:1054–1060.
113. Rich BE, Leder P. Transgenic expression of interleukin 7 restores T cell populations in nude mice. J Exp Med 1995; 181:1223–1228.
114. Bhatia SK, Tygrett LT, Grabstein KH, Waldschmidt TJ. The effect of in vivo IL-7 deprivation on T cell maturation. J Exp Med 1995; 181:1399–1409.
115. Everson MP, Eldridge JH, Koopman WJ. Synergism of interleukin 7 with the thymocyte growth factors interleukin 2, interleukin 6, and tumor necrosis factor alpha in the induction of thymocyte proliferation. Cell Immunol 1990; 127:470–482.
116. Godfrey DI, Zlotnik A, Suda T. Phenotypic and functional characterization of c-kit expression during intrathymic T cell development. J Immunol 1992; 149:2281–2285.
117. Morrissey PJ, McKenna H, Widmer MB, Braddy S, Voice R, Charrier K, Williams DE, Watson JD. Steel factor (c-kit ligand) stimulates the in vitro growth of immature CD3-/CD4-/CD8- thymocytes: synergy with IL-7. Cell Immunol 1994; 157:118–131.
118. Suda T, Murray R, Guidos C, Zlotnik A. Growth-promoting activity of IL-1 alpha, IL-6, and tumor necrosis factor-alpha in combination with IL-2, IL-4, or IL-7 on murine thymocytes. Differential effects on CD4/CD8 subsets and on CD3+/CD3- double-negative thymocytes. J Immunol 1990; 144:3039–3045.
119. Chantry D, Turner M, Feldmann M. Interleukin 7 (murine pre-B cell growth factor/lymphopoietin 1) stimulates thymocyte growth: regulation by transforming growth factor beta. Eur J Immunol 1989; 19:783–786.
120. Herbelin A, Machavoine F, Schneider E, Papiernik M, Dy M. IL-7 is requisite for IL-1-induced thymocyte proliferation. Involvement of IL-7 in the synergistic effects of granulocyte-macrophage colony-stimulating factor or tumor necrosis factor with IL-1. J Immunol 1992; 148:99–105.
121. Herbelin A, Machavoine F, Vicari A, Schneider E, Papiernik M, Ziltener H, Penit C, Dy M. Endogenous granulocyte-macrophage colony-stimulating factor is involved in IL-1- and IL-7-induced murine thymocyte proliferation. J Immunol 1994; 153:1973–1981.
122. Armitage RJ, Namen AE, Sassenfeld HM, Grabstein KH. Regulation of human T cell proliferation by IL-7. J Immunol 1990; 144:938–941.
123. Bertagnolli M, Herrmann S. IL-7 supports the generation of cytotoxic T lymphocytes from thymocytes. Multiple lymphokines required for proliferation and cytotoxicity. J Immunol 1990; 145:1706–1712.
124. Jicha DL, Schwarz S, Mule JJ, Rosenberg SA. Interleukin-7 mediates the generation and expansion of murine allosensitized and antitumor CTL. Cell Immunol 1992; 141:71–83.
125. Kos FJ, Mullbacher A. IL-2-independent activity of IL-7 in the generation of secondary antigen-specific cytotoxic T cell responses in vitro. J Immunol 1993; 150:387–393.
126. Mehrotra PT, Grant AJ, Siegel JP. Synergistic effects of IL-7 and IL-12 on human T cell activation. J Immunol 1995; 154:5093–5102.
127. Whiteside T, Miescher S, Hurlimann J, Moretta L, Von Fliedner V. Clonal analysis and in situ characterization of lymphocytes infiltrating human breast carcinomas. Cancer Immunol Immunother 1986; 23:169–178.
128. Sica D, Rayman P, Stanley J, Edinger M, Tubbs RR, Klein E, Bukowski R, Finke JH. Interleukin 7 enhances the proliferation and effector function of tumor-infiltrating lymphocytes from renal-cell carcinoma. Int J Cancer 1993; 53:941–947.
129. Ditonno P, Tso CL, Sakata T, deKernion JB, Belldegrun A. Regulatory effects of interleukin-7 on renal tumor infiltrating lymphocytes. Urol Res 1992; 20:205–210.
130. Lotze MT, Zeh HJ, III, Elder EM, Cai Q, Pippin BA, Rosenstein MM, Whiteside TL, Herberman R. Use of T-cell growth factors (interleukins 2, 4, 7, 10, and 12) in the evaluation of T-cell reactivity to melanoma. J Immunother 1992; 12:212–217.
131. Lotze MT, Grimm EA, Mazumder A, Strausser JL, Rosenberg SA. Lysis of fresh and cultured autologous tumor by human lymphocytes cultured in T cell growth factor. Cancer Res 1981; 41:4420–4425.
132. Naume B, Espevik T. Effects of IL-7 and IL-2 on highly enriched CD56+ natural killer cells. A comparative study. J Immunol 1991; 147:2208–2214.
133. Naume B, Gately M, Espevik T. A comparative study of IL-12 (cytotoxic lymphocyte maturation factor)-, IL-2-, and IL-7-induced effects on immunomagnetically purified CD56+ NK cells. J Immunol 1992; 148:2429–2436.
134. Pavletic Z, Benyunes MC, Thompson JA, Lindgren CG, Massumoto C, Alderson MR, Buckner CD, Fefer A. Induction by interleukin-7 of lymphokine-activated killer activity in lymphocytes from autologous and syngeneic marrow transplant recipients before and after systemic interleukin-2 therapy. Exp Hematol 1993; 21:1371–1378.
135. Lotzova E. Definition and functions of natural killer cells. Nat Immun 1993; 12:169–176.
136. Naume B, Johnsen AC, Espevik T, Sundan A. Gene expression and secretion of cytokines and cytokine receptors from highly purified CD56+ natural killer cells stimulated with interleukin-2, interleukin-7 and interleukin-12. Eur J Immunol 1993; 23:1831–1838.
137. Robertson MJ, Manley TJ, Donahue C, Levine H, Ritz J. Costimulatory signals are required for optimal proliferation of human natural killer cells. J Immunol 1993; 150:1705–1714.
138. Pollack SB, Tsuji J. Effects of rIL-7 on murine bone marrow NK precursor cells. Cell Immunol 1993; 151:1–11.
139. Alderson MR, Tough TW, Ziegler SF, Grabstein KH. Interleukin 7 induces cytokine secretion and tumoricidal activity by human peripheral blood monocytes. J Exp Med 1991; 173:923–930.
140. Ziegler SF, Tough TW, Franklin TL, Armitage RJ, Alderson MR. Induction of macrophage inflammatory protein-1 beta gene expression in human monocytes by lipopolysaccharide and IL-7. J Immunol 1991; 147:2234–2239.
141. Standiford TJ, Strieter RM, Allen RM, Burdick MD, Kunkel SL. IL-7 up-regulates the expression of IL-8 from resting and stimulated human blood monocytes. J Immunol 1992; 149:2035–2039.
142. Gessner A, Vieth M, Will A, Schroppel K, Rollinghoff M. Interleukin-7 enhances antimicrobial activity against Leishmania major in murine macrophages. Infect Immun 1993; 61:4008–4012.
143. Jacobsen FW, Veiby OP, Jacobsen Sten EW. IL-7 stimulates CSF-induced proliferation of murine bone marrow macrophages and Mac-1^+ myeloid progenitors in vitro. J Immunol 1994; 153:270–276.
144. Damia G, Komschlies KL, Faltynek CR, Ruscetti FW, Wiltrout RH. Administration of recombinant human interleukin-7 alters the frequency and number of myeloid progenitor cells in the bone marrow and spleen of mice. Blood 1992; 79:1121–1129.
145. Grzegorzewski K, Komschlies KL, Mori M, Kaneda K, Usui N, Faltynek CR, Keller JR, Ruscetti FW, Wiltrout RH. Administration of recombinant human interleukin-7 to mice induces the exportation of myeloid progenitor cells from the bone marrow to peripheral sites. Blood 1994; 83:377–385.

146. Jacobsen FW, Veiby OP, Skjonsberg C, Jacobsen SE. Novel role of interleukin 7 in myelopoiesis: stimulation of primitive murine hematopoietic progenitor cells. J Exp Med 1993; 178:1777–1782.
147. Fahlman C, Blomhoff HK, Veiby OP, McNiece IK, Jacobsen SE. Stem cell factor and interleukin-7 synergize to enhance early myelopoiesis in vitro. Blood 1994; 84:1450–1456.
148. Jacobsen FW, Rusten LS, Jacobsen SE. Direct synergistic effects of interleukin-7 on in vitro myelopoiesis of human CD34+ bone marrow progenitors. Blood 1994; 84:775–779.
149. Vellenga E, Esselink MT, Straaten J, Stulp BK, De Wolf JT, Brons R, Giannotti J, Smit JW, Halie MR. The supportive effects of IL-7 on eosinophil progenitors from human bone marrow cells can be blocked by anti-IL-5. J Immunol 1992; 149:2992–2995.
150. Kondo M, Takeshita T, Ishii N, Nakamura M, Watanabe S, Arai K, Sugamura K. Sharing of the interleukin-2 (IL-2) receptor gamma chain between receptors for IL-2 and IL-4. Science 1993; 262:1874–1877.
151. Russell SM, Keegan AD, Harada N, Nakamura Y, Noguchi M, Leland P, Friedmann MC, Miyajima A, Puri RK, Paul WE, Leonard WJ. Interleukin-2 receptor gamma chain: a functional component of the interleukin-4 receptor. Science 1993; 262:1880–1883.
152. Russell SM, Johnson JA, Noguchi M, Kawamura M, Bacon CM, Friedmann M, Berg M, McVicar DW, Witthuhn BA, Silvennoinen O, et al. Interaction of IL-2R beta and gamma c chains with Jak1 and Jak3: implications for XSCID and SCID. Science 1994; 266:1042–1045.
153. Giri JG, Ahdieh M, Eisenman J, Shanebeck K, Grabstein K, Kumaki S, Namen A, Park LS, Cosman D, Anderson D. Utilization of the beta and gamma chains of the IL-2 receptor by the novel cytokine IL-15. EMBO J 1994; 13:2822–2830.
154. Schlossman SF, Boumsell L, Gilks W, Harlan JM, Kishimoto T, Morimoto C, Ritz J, Shaw S, Silverstein RL, Springer TA, Tedder TF, Todd RF. CD antigens 1993. J Immunol 1994; 152:1–2.
155. Aggarwal BB, Pocsik E. Cytokines: From clone to clinic. Arch Biochem Biophys 1992; 292:335–359.
156. Lynch M, Baker E, Park LS, Sutherland GR, Goodwin RG. The interleukin-7 receptor gene is at 5p13. Hum Genet 1992; 89:566–568.
157. Gearing DP, Druck T, Huebner K, Overhauser J, Gilbert DJ, Copeland NG, Jenkins NA. The leukemia inhibitory factor receptor (LIFR) gene is located within a cluster of cytokine receptor loci on mouse chromosome 15 and human chromosome 5p12-p13. Genomics 1993; 18:148–150.
158. Pleiman CM, Gimpel SD, Park LS, Harada H, Taniguchi T, Ziegler SF. Organization of the murine and human interleukin-7 receptor genes: two mRNAs generated by differential splicing and presence of a type I-interferon-inducible promoter. Mol Cell Biol 1991; 11:3052–3059.
159. Yamada G, Ogawa M, Akagi K, Miyamoto H, Nakano N, Itoh S, Miyazaki J, Nishikawa S, Yamamura K, Taniguchi T. Specific depletion of the B-cell population induced by aberrant expression of human interferon regulatory factor 1 gene in transgenic mice. Proc Natl Acad Sci USA 1991; 88:532–536.
160. Taniguchi T, Minami Y. The IL-2/IL-2 receptor system: a current overview. Cell 1993; 73:5–8.
161. Leonard WJ, Noguchi M, Russell SM, McBride OW. The molecular basis of X-linked severe combined immunodeficiency: the role of the interleukin-2 receptor gamma chain as a common gamma chain, gamma c [Review]. Immunol Rev 1994; 138:61–86.
162. Cosman D. The hematopoietin receptor superfamily. Cytokine 1993; 5:95–106.
163. Dibirdik I, Langlie MC, Ledbetter JA, Tuel Ahlgren L, Obuz V, Waddick KG, Gajl Peczalska K, Schieven GL, Uckun FM. Engagement of interleukin-7 receptor stimulates tyrosine phosphorylation, phosphoinositide turnover, and clonal proliferation of human T-lineage acute lymphoblastic leukemia cells. Blood 1991; 78:564–570.
164. Uckun FM, Dibirdik I, Smith R, Tuel Ahlgren L, Chandan Langlie M, Schieven GL, Waddick KG, Hanson M, Ledbetter JA. Interleukin 7 receptor ligation stimulates tyrosine phosphorylation, inositol phospholipid turnover, and clonal proliferation of human B-cell precursors. Proc Natl Acad Sci USA 1991; 88:3589–3593.
165. Roifman CM, Wang GX, Freedman M, Pan ZQ. IL-7 receptor mediates tyrosine phosphorylation but does not activate the phosphatidylinositol-phospholipase C-gamma 1 pathway. J Immunol 1992; 148:1136–1142.
166. Seckinger P, Milili M, Schiff C, Fougereau M. Interleukin-7 regulates c-myc expression in murine T cells and thymocytes: a role for tyrosine kinase(s) and calcium mobilization. Eur J Immunol 1994; 24:716–722.
167. Yip Schneider MT, Horie M, Broxmeyer HE. Characterization of interleukin-7-induced changes in tyrosine phosphorylation and c-myc gene expression in normal human T cells. Exp Hematol 1993; 21:1648–1656.
168. Venkitaraman AR, Cowling RJ. Interleukin 7 receptor functions by recruiting the tyrosine kinase p59fyn through a segment of its cytoplasmic tail. Proc Natl Acad Sci USA 1992; 89:12083–12087.
169. Dadi HK, Ke S, Roifman CM. Interleukin 7 receptor mediates the activation of phosphatidylinositol-3 kinase in human B-cell precursors. Biochem Biophys Res Commun 1993; 192:459–464.
170. Dadi HK, Roifman CM. Activation of phosphatidylinositol-3 kinase by ligation of the interleukin-7 receptor on human thymocytes. J Clin Invest 1993; 92:1559–1563.
171. Domen J, van der Lugt NM, Acton D, Laird PW, Linders K, Berns A. Pim-1 levels determine the size of early B lymphoid compartments in bone marrow. J Exp Med 1993; 178:1665–1673.
172. Darnell JE, Jr., Kerr IM, Stark GR. Jak-STAT pathways and transcriptional activation in response to IFNs and other extracellular signaling proteins. Science 1994; 264:1415–1420.
173. Lin J-X, Migone T-S, Tsang M, Friedmann M, Weatherbee JA, Zhou L, Yamauchi A, Bloom ET, Mietz J, John S, Leonard WJ. The role of shared receptor motifs and common stat proteins in the generation of cytokine pleiotropy and redundancy by IL-2, IL-4, IL-7, IL-13 and IL-15. Immunity 1995; 2:331–339.
174. Foxwell BM, Taylor Fishwick DA, Simon JL, Page TH, Londei M. Activation induced changes in expression and structure of the IL-7 receptor on human T cells. Int Immunol 1992; 4:277–282.
175. Taniguchi T. Cytokine signaling through nonreceptor protein tyrosine kinases. Science 1995; 268:251–255.
176. Bui T, Faltynek C, Ho RJ. Differential disposition of soluble and liposome-formulated human recombinant interleukin-7: effects on blood lymphocyte population in guinea pigs. Pharm Res 1994; 11:633–641.
177. Komschlies KL, Gregorio TA, Gruys ME, Back TC, Faltynek CR, Wiltrout RH. Administration of recombinant human IL-7 to mice alters the composition of B-lineage cells and T cell subsets, enhances T cell function, and induces regression of established metastases. J Immunol 1994; 152:5776–5784.
178. Lynch DH, Miller RE. Interleukin 7 promotes long-term in

vitro growth of antitumor cytotoxic T lymphocytes with immunotherapeutic efficacy in vivo. J Exp Med 1994; 179:31–42.
179. Murphy WJ, Back TC, Conlon KC, Komschlies KL, Ortaldo JR, Sayers TJ, Wiltrout RH, Longo DL. Antitumor effects of interleukin-7 and adoptive immunotherapy on human colon carcinoma xenografts. J Clin Invest 1993; 92:1918–1924.
180. Talmadge JE, Jackson JD, Kelsey L, Borgeson CD, Faltynek C, Perry GA. T-cell reconstitution by molecular, phenotypic, and functional analysis in the thymus, bone marrow, spleen, and blood following split-dose polychemotherapy and therapeutic activity for metastatic breast cancer in mice. J Immunother 1993; 14:258–268.
181. Wu B, Shen RN, Wang WX, Broxmeyer HE, Lu L. Antitumor effect of interleukin 7 in combination with local hyperthermia in mice bearing B16a melanoma cells. Stem Cells 1993; 11:412–421.
182. Tepper RI, Mule JJ. Experimental and clinical studies of cytokine gene-modified tumor cells. Human Gene Ther 1994; 5:153–164.
183. Allione A, Consalvo M, Nanni P, Lollini PL, Cavallo F, Giovarelli M, Forni M, Gulino A, Colombo MP, Dellabona P, et al. Immunizing and curative potential of replicating and nonreplicating murine mammary adenocarcinoma cells engineered with interleukin (IL)-2, IL-4, IL-6, IL-7, IL-10, tumor necrosis factor alpha, granulocyte-macrophage colony-stimulating factor, and gamma-interferon gene or admixed with conventional adjuvants. Cancer Res 1994; 54:6022–6026.
184. Dubinett SM, Huang M, Dhanani S, Economou JS, Wang J, Lee P, Sharma S, Dougherty GJ, McBride WH. Downregulation of murine fibrosarcoma transforming growth factor-β1 expression by interleukin 7. J Natl Cancer Inst 1995; 87:593–597.
185. Schmidt-Wolf IGH, Huhn D, Neubauer A, Wittig B. Interleukin-7 gene transfer in patients with metastatic colon carcinoma, renal cell carcinoma, melanoma, or with lymphoma. Human Gene Therapy 1994; 5:1161–1168.
186. Lu L, Zhou Z, Wu B, Xiao M, Shen RN, Williams DE, Kim YJ, Kwon BS, Ruscetti S, Broxmeyer HE. Influence of recombinant human interleukin (IL)-7 on disease progression in mice infected with Friend virus complex. Int J Cancer 1992; 52:261–265.
187. Bui T, Faltynek CR, Ho RJY. Biological response of recombinant interleukin-7 on herpes-simplex virus-infection in guinea-pigs (abstr). Vaccine 1994; 12(7):646–652.
188. Bui T, Dykers T, Hu SL, Faltynek CR, Ho RJ. Effect of MTP-PE liposomes and interleukin-7 on induction of antibody and cell-mediated immune responses to a recombinant HIV-envelope protein. AIDS 1994; 7:799–806.
189. Witte PL, Frantsve LM, Hergott M, Rahbe SM. Cytokine production and heterogeneity of primary stromal cells that support B lymphopoiesis. Eur J Immunol 1993; 23:1809–1817.
190. Sakata T, Iwagami S, Tsuruta Y, Teraoka H, Tatsumi Y, Kita Y, Nishikawa S, Takai Y, Fujiwara H. Constitutive expression of interleukin-7 mRNA and production of IL-7 by a cloned murine thymic stromal cell line. J Leukoc Biol 1990; 48:205–212.
191. Frishman J, Long B, Knospe W, Gregory S, Plate J. Genes for interleukin 7 are transcribed in leukemic cell subsets of individuals with chronic lymphocytic leukemia. J Exp Med 1993; 177:955–964.
192. Plate JM, Knospe WH, Harris JE, Gregory SA. Normal and aberrant expression of cytokines in neoplastic cells from chronic lymphocytic leukemias. Hum Immunol 1993; 36:249–258.
193. Kaashoek JGJ, Mout R, Faulkenburg JHF, Willemze R, Fibbe WE, Landegent JE. Cytokine production by the bladder carcinoma cell line 5637: rapid analysis of mRNA expression levels using a cDNA-PCR procedure. Lymphokine Cytokine Res 1991; 10:231.
194. Maeurer MG, Walter W, Zitvogel L, Storkus WJ, Lotze MT. The role of Interleukin-7 in the expansion of colon cancer-specific and MHC Class II restricted tumor infiltrating lymphocytes (TIL) in vitro and in vivo. Cancer Gene Therapy 1996 (in press).
195. Mattei S, Colombo MP, Melani C, Silvani A, Parmiani G, Herlyn M. Expression of cytokine/growth factors and their receptors in human melanoma and melanocytes. Int J Cancer 1994; 56:853–857.
196. Melhem MF, Meisler AI, Saito R, Finley GG, Hockman HR, Koski RA. Cytokines in inflammatory malignant fibrous histiocytoma presenting with leukemoid reaction. Blood 1993; 82:2038–2044.

8
Chemokines and the Inflammatory Response

Steven L. Kunkel, Nicholas W. Lukacs, Stephen W. Chensue, and Robert M. Strieter
University of Michigan Medical School, Ann Arbor, Michigan

INTRODUCTION

In the most basic terms, inflammation constitutes the host's reaction to a number of direct and indirect insults, including soft tissue trauma, cancer, infections, and autoimmune disease. The ability of the host to respond to the above disorders is often characterized as an acute or chronic inflammatory reaction, depending upon the longevity of the response. The length of any disease is likely controlled by the etiology of the disorder and the subsequent production of mediators that activate and elicit specific subpopulations of leukocytes. The actual mechanisms that control the elicitation of specific leukocyte populations, which ultimately dictates the initiation, maintenance, and repair process of acute and chronic inflammation, are currently under study. While various aspects of leukocyte movement remain an enigma, it is becoming increasingly clear that both peripheral blood leukocytes and tissue-associated leukocytes, in combination with endothelial cells and resident structural cells, are key participants in maintaining chemotactic factor–generating cytokine cascades. The end result of these cell interactions is the mobilization of subpopulations of leukocytes to an inflamed area. Thus, one of the most important host responses to tissue injury is the concerted effort of various cells to maximize the recruitment process in order for the appropriate leukocyte population to participate in an inflammatory response.

Like all processes associated with the immune/inflammatory response, chemotaxis is controlled by both positive and negative signals, which are expressed in a coordinated manner. Many of the initial interactions that are key to the elicitation of individual inflammatory cells occur in the vascular wall of blood vessels in an area of tissue injury. In this microenvironment, activated endothelial cells first physically interact with peripheral blood leukocytes in a sequence of events that involves an initial rolling of leukocytes on activated endothelium, activation of these leukocytes, firm attachment of the leukocytes via adhesion molecules in a manner that binds certain leukocytes to the wall of the endothelium, and finally transendothelial cell migration (1,2). The attachment of leukocytes to the endothelium is mediated, in part, by the production of early response cytokines, such as tumor necrosis factor–alpha (TNF) and interleukin-1 (IL-1), as these mediators can induce the expression of adhesion molecules on both endothelial cells and leukocytes (3,4).

The major cytokine-induced adherence proteins that are expressed on the surface of endothelial cells are designated intercellular adhesion molecule-1 (ICAM-1), endothelial leukocyte adhesion molecule (E-selectin), and vascular cell adhesion molecule (VCAM-1). Although de novo transcription and translation is required for IL-1 and/or TNF to induce the expression of ICAM, E-selectin, or VCAM-1, this process is quite rapid and occurs within a few hours of stimulation (2,3). Interestingly, the expression of adhesion molecules is not restricted to the surface of the endothelium, as fibroblasts and epithelial cells can also express adhesion molecules and physically interact with leukocytes in a manner that would promote cell-directed movement. The above adhesion proteins appear to serve as the natural ligand for additional polypeptides that are expressed on the surface of leukocytes. One group of leukocyte-derived adhesion proteins belongs to the CD11/CD18 family of surface proteins and includes CD11a/CD18, CD11b/CD18, and CD11c/CD18 (2). The interaction between adhesion molecules on endothelial cells and their ligands on leukocytes allow for a high degree of binding for certain subpopulations of peripheral blood leukocytes to a restricted area of activated endothelium.

The adherence phenomenon is a unique interaction in that the binding must be firm to allow significant attachment of the inflammatory cells to the endothelium, yet be regulated by some mechanism(s) that allows the adherence process to be reversible. This aspect of an inflammatory response is somewhat paradoxical in that the initial adhesion of leukocytes to the endothelium is a strict requirement for the recruitment process, however, this tethering cannot be permanent. The localization of leukocytes to an area of activated endothelium is important in delivering these cells to the general region of tissue injury and inflammation, but if the remaining cascade of chemotactic factors is not expressed, these cells will rejoin the circulating pool and the recruitment process will be aborted. Therefore, the subsequent steps in the chemotactic response are as critical as the initial ones in ensuring the successful delivery of leukocytes to an area of inflammation. The reversible nature of the adhesion molecules is a fundamental process of cell elicitation, as different populations of peripheral blood leukocyte must be able to respond to the next step in chemotaxis, the movement of inflammatory cells in response to chemotactic signals.

HISTORY OF THE CHEMOKINES

Historically, the pathology of various disorders and diseases has been characterized, in part, by the microscopic identification of specific leukocyte populations recruited to an area of inflammation. Early pathologists recognized that certain diseases were associated with specific populations of leukocytes, some of which were observed in association with the vascular wall and others of which were found in the involved tissue. However, the mechanisms responsible for the accumulation of the leukocytes were not known. Initial investigations directed at understanding leukocyte-directed migration to a known stimulus were performed using dead microorganisms or products of human leukocytes. These first descriptive studies were important in providing the groundwork for a number of subsequent investigations demonstrating that bacteria products, such as *N*-formylmethionyl peptides or synthetic *N*-formylmethionyl derivatives, were potent inducers of leukocyte activation and chemotaxis (5).

Subsequent investigations identified that both bacterial-derived peptides and host-derived polypeptides, like the split product derived from the fifth component of complement (C5a), were highly chemotactic for leukocytes (6). These seminal observations provided some of the first data about the mechanism for the recruitment of leukocytes during infectious processes (6,7). However, these studies did not address the specificity associated with leukocyte recruitment during inflammation. The identification of other chemotactic factors such as platelet-activating factor, leukotriene B_4, and split products of fibronectin only complicated the issue of leukocyte movement, since these chemotactic agents also lack any degree of specificity for the recruitment of leukocyte subpopulations (8). An important aspect of these initial investigations was identifying the redundancy in the number and types of factors that could induce leukocyte chemotaxis. The redundant nature of chemotactic factors stresses the importance of cell recruitment during an infectious/inflammatory process, as a number of systems are in place to ensure the successful elicitation of leukocytes to participate in host defense.

However, the fact that multiple chemotactic agents possess little in the way of cell specificity was still at odds with the pathology/cytology of certain acute and chronic diseases, which are characterized by specific leukocyte populations. For example, many acute inflammatory reactions with a bacterial etiology are often dominated by the presence of neutrophils, while many chronic immune reactions are characterized by the presence of specific mononuculear leukocyte populations. Insight into this longstanding enigma has been recently provided by discoveries demonstrating that family members of a supergene cluster of chemotactic cytokines, called chemokines, possess a relatively high degree of specificity for the elicitation of leukocyte subpopulations (9–12). These chemokines belong to two groups of related polypeptide families, identified by the location of two cysteine amino acids near the amino terminus that comprised the primary amino acid structure (Table 1). In one supergene family, the two amino-terminal cysteines are separated by an nonconserved amino acid in a C-X-C motif, and in the other

Table 1 Alignment of Amino-Terminal Amino Acids in C-C and C-X-C Chemokines Showing the Location of the Two Amino-Terminal Cysteines

	C-C chemokine alignment
MCP-1	QPDAINAPVT**CC**YNFTNRK
MCP-2	QPDSVSIPIT**CC**FNNVINRKI
MCP-3	QPVGINTSTT**CC**YRFINKKI
RANTES	SPYSSDTTP**CC**FAYIARPL
MIP-1α	SLAADTPTA**CC**SYTSRQI
	C-X-C chemokine alignment
IL-8	EGAVLPRSAKELR**C**Q**C**
ENA-78	AGPAAAVLRELR**C**V**C**
GRO-α	ASVATELR**C**Q**C**
IP-10	VPLSRTVR**C**T**C**

family the amino-terminal cysteines are in juxtaposition in a C-C motif. The discovery of two related chemokine families of polypeptide mediators of inflammation has proven to be important, as they provide a potential mechanism for the selective recruitment of peripheral blood leukocytes to a specific site of inflammation.

BIOCHEMISTRY OF THE CHEMOKINES

Accumulating evidence supports the concept that members of the chemokine supergene families have proinflammatory and reparative activities. In their monomeric form, these chemokines are basic, heparin-binding proteins weighing less than 10,000 daltons. The C-X-C chemokines appear to have specificity for the elicitation of neutrophils, are all clustered on human chromosome 4, and possess approximately 20–55% homology in their primary structure with other family members. Interest in this area is exemplified by investigations that have identified 12 different C-X-C chemokines, including platelet factor-4, platelet basic protein, connective tissue activating protein III, β-thromboglobulin, neutrophil-activating factor-2, interleukin-8, growth-related oncogene α, β, and γ, γ-interferon–inducible protein (IP-10), epithelial neutrophil-activating protein-78, monokine induced by γ-interferon (MIG), and granulocyte chemotactic protein-2 (GCP-2) (13–20). Connective tissue–activating protein-III, β-thromboglobulin, and neutrophil-activating protein-2 are all *N*-terminal truncations of platelet basic protein. These cleavage products are formed when platelet basic protein is released from platelet α granules and is proteolytically digested by leukocyte-derived proteases.

Recent investigations have demonstrated that an important feature in the primary structure of C-X-C chemokines may account for the neutrophil chemotactic and activating properties of these mediators (Table 2). These studies identified three critical amino acid residues immediately preceding the first *N*-terminal cysteine, which are important in binding to neutrophil receptors. These amino acids are Glu-Leu-Arg or the ELR motif, which is absent in certain members of the C-X-C chemokine family (21,22). In particular, platelet factor-4, IP-10, and MIG all lack the ELR motif and do not possess neutrophil-activating properties. However, when the ELR motif was synthetically introduced into platelet factor-4, this polypeptide gained chemotactic activity. Therefore, certain members of the C-X-C supergene family may have different biological activities. Platelet factor-4 (a non-ELR–containing C-X-C chemokine) was one of the first members of this family to be described. This factor was originally identified for its ability to bind heparin, leading to the inactivation of the anticoagulation function of heparin. Interleukin-8 is one of the most studied C-X-C chemokines and, as described below, is produced by an array of different immune and nonimmune cells.

A family of polypeptide mediators that is related to the C-X-C chemokines is the C-C chemokine family (Table 2). This group is defined by the location of the first two N-terminal cysteines being found in juxtaposition to one another. In general terms, the C-C chemokine family has relative specificity for the elicitation of mononuclear cells. The C-C chemokine family includes macrophage inflammatory protein-1 (MIP-1α) and β (MIP-1β), monocyte chemoattractant protein 1 (MCP-1), 2 (MCP-2), and 3 (MCP-3), I-309, and RANTES (23–25). The C-C chemokine family members are low molecular weight polypeptides with a basic isoelectric point that bind heparin. The ability to bind heparin was an important biochemical property used in purifying many of the C-C chemokines. In addition, the heparin-binding activity is likely related to key biological events in vivo. Like the C-X-C family, certain members of the C-C family have been identified as deriving from a number of cellular sources, including a variety of tumor cell lines.

Table 2 The C-X-C Chemokine Family

C-X-C chemokines containing the ELR amino acid motif:
Interleukin-8 (IL-8)
Growth-regulated oncogene-α (GRO-α)
Growth-regulated oncogene-β (GRO-β)
Growth-regulated oncogene-γ (GRO-γ)
Epithelial neutrophil-activating protein (ENA-78)
Neutrophil-activating protein-2 (NAP-2)
Granulocyte chemotactic protein-2 (GCP-2)
C-X-C chemokines that lack the ELR amino acid motif
Monokine induced by γ-interferon (MIG)
Platelet factor 4 (PF4)
γ-Interferon inducible protein-10 (IP-10)

CELLULAR SOURCES OF CHEMOKINES

One of the key discoveries concerning the chemokine family is that these polypeptides are products of both inflammatory and noninflammatory cells. Investigative studies have identified that monocytes/macrophages and lymphocytes are not the only source of these chemotactic agents, as stromal cells, epithelial cells, endothelial cells, hepatocytes, and even neutrophils can generate significant chemokine levels (26–35). The expression of various chemokines by noninflammatory cells raises intriguing questions regarding the role of normal resident cells in the initiation and maintenance of an inflammatory response. For example, fibroblasts, smooth muscle cells, epithelial cells, and endothelial cells may serve as important effector cells during the recruitment phase of inflammation. Recent studies have demonstrated that normal noninflammatory resident cells can support inflammation via the induction of appropriate chemokines. The synthesis of C-X-C or C-C family members by these noninflammatory cells is dependent upon a cytokine network or cascade, whereby a proximal cytokine can influence the expression of a more distal cytokine (36–38). At the tissue level, the expression of a given chemokine by a noninflammatory cell is likely dictated by a communication loop with resident, tissue macrophages. These latter cells appear to be the first to respond to an initial insult with the generation of early-response cytokines, such as TNF or IL-1. These early-response cytokines stimulate the production of distal chemokines from surrounding resident stromal or parenchymal cells. Interestingly, many normal noninflammatory, resident cells are usually not susceptible to an exogenous challenge, such as lipopolysaccharide (LPS), for the generation of chemokines. Thus, the activation and subsequent expression of IL-1 or TNF by local tissue macrophages can magnify the response needed to initiate the recruitment process. This response may be caused in disease states by mononuclear phagocytes responding to an initial challenge, which can generate early-response mediators that network with chemokine-generating resident noninflammatory cells.

This scenario can occur in various tissues, like the lung, where either alveolar or interstitial macrophages can respond to an initial insult with the expression of the early-response cytokines IL-1 and TNF. These cytokines can then activate resident lung endothelial cells, epithelial cells, and fibroblasts, resulting in the production of chemotactic cytokines. This network has been demonstrated experimentally using an in vitro co-culture system of human alveolar macrophages and human lung fibroblasts (36). The addition of LPS to this two-cell system results in the rapid expression of IL-1 and TNF by the macrophages and the delayed expression of IL-8 by the pulmonary fibroblasts (Fig. 1). The production of fibroblast-derived IL-8 in this co-culture system was dramatically reduced upon the addition of neutralizing antibodies to either IL-1 or TNF at the time of LPS challenge. Additional studies using a co-culture system have identified MCP-1, a C-C family member, as a chemokine that may be expressed in this network fashion.

The above system of chemokine-producing cytokine cascades is not unique to the lung, as experiments with Kupffer cells and hepatocytes have generated similar data from the liver. In initial studies, cultured human hepatocytes challenged with either IL-1 or TNF could express significant levels of steady state IL-8 mRNA (39), again demonstrating the key role that these early-response cytokines appear to play during the initiation of inflammation. Furthermore, conditioned media from LPS-treated human Kupffer cells overlaid on human hepatocytes demonstrated an increase in the expression of steady-state IL-8 mRNA from cultured hepatocytes. Interestingly, the addition of neutralizing antibodies to either IL-1 or TNF resulted in a diminution of IL-8 mRNA expression, while the addition of neutralizing antibodies to both IL-1 and TNF reduced the pattern of steady-state IL-8 mRNA to near basal levels. Thus, the cytokine network between fixed, resident leukocytes and normal tissue cells is paramount in the successful chemokine-dependent elicitation of leukocytes to an area of inflammation.

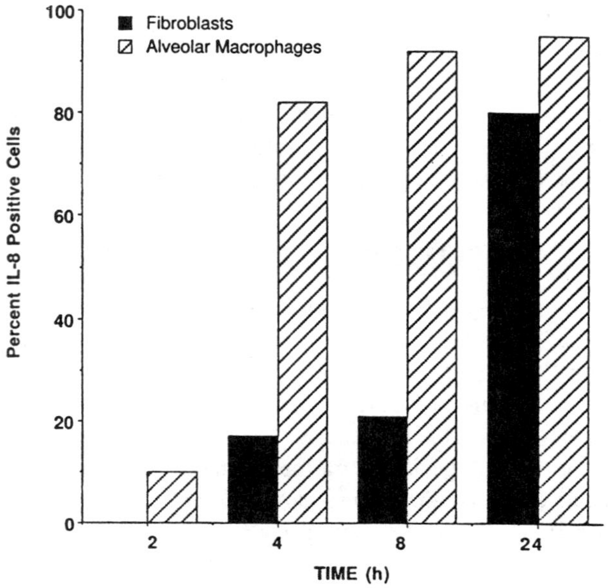

Figure 1 Time-dependent expression of IL-8 antigen by human lung fibroblasts cocultured with human alveolar macrophages. At time 0 the coculture was challenged with LPS, and IL-8–positive cells were identified by immunohistolocalization.

Fibroblasts, epithelial cells, hepatocytes, and endothelial cells can all produce high levels of IL-8 when appropriately stimulated. In addition, the neutrophil has also been identified as an important source of various cytokines. Moreover, investigations have identified the neutrophil as a rich source of a number of C-X-C and C-C chemokines. In initial studies, neutrophils were shown to be responsive to either LPS or a phagocytic stimulus with the subsequent production of IL-8 (40,41), however, it is now known that the neutrophil can generate IL-8 in response to a number of specific challenges. For example, when neutrophils adhere, IL-8 mRNA is rapidly expressed followed by the production of biologically active IL-8. As shown in Figure 2, isolated adherent human neutrophils

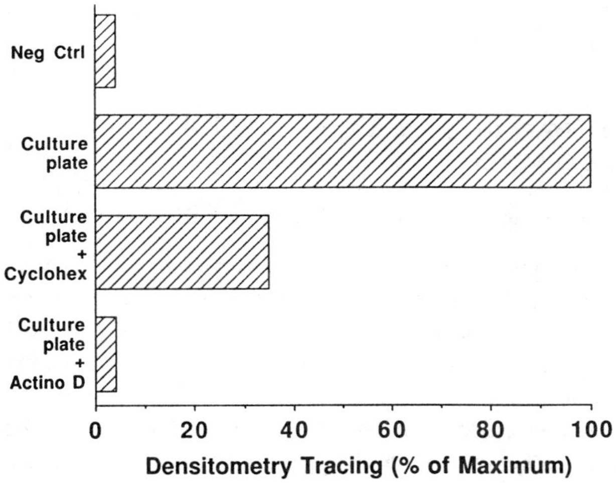

Figure 2 Expression of steady-state levels of IL-8 mRNA as determined by laser densitometry of unstimulated, freshly isolated neutrophils (Neg Ctrl), neutrophils adherent to plastic tissue culture plates, adherent neutrophils in the presence of cycloheximide, and adherent neutrophils in the presence of actinomycin D.

cultured on plastic plates expressed significant levels of steady-state IL-8 mRNA by 2 h, as compared to nonadherent neutrophils cultured in teflon chambers. The levels of adherence-induced neutrophil-derived IL-8 mRNA appeared to be under the regulation of an intracellular regulatory protein, since the addition of cycloheximide to the cultured neutrophils resulted in a "superinduction" (threefold) of steady-state levels of IL-8 mRNA. Furthermore, the addition of actinomycin D to the neutrophils at the beginning of the adherence step completely blocked the increase in IL-8 mRNA expression (40). These initial studies demonstrated that human neutrophils not only undergo transcription for IL-8 during the adherence process, but that this process is likely regulated in a complex manner. The production of a chemotactic factor by adherent neutrophils is significant, since the adherence and chemotactic responses are two fundamental processes required for the active participation of this early responding cell during inflammation. These observations support the concept that the adherence response, which initially localizes the leukocyte to an area of inflammation, may also be the signal to further activate neutrophils for the production of a chemotactic cytokine. Thus, the activation of specific receptors involved in binding the leukocyte to the endothelial cell may also initiate biosynthetic pathways for the production of appropriate inflammatory mediators to maintain the inflammatory response. The production of a neutrophil chemotactic factor by the neutrophil itself is not without precedence, as leukotriene B and platelet activating factor are potent neutrophil-derived chemotactic factors.

In addition to the adherence phenomenon, other stimuli are important in inducing the expression of IL-8 mRNA and biologically active protein from the neutrophil. Studies have shown that neutrophil-derived IL-8 is generated in both a time- and dose-related manner when challenged with either LPS, TNF, or IL-1. Significant levels of IL-8 protein were first observed 4 h post-stimulation with either LPS, IL-1, or TNF. The production of IL-8 appeared to peak at about 24 h after the addition of the stimulating factors. The synthesis of neutrophil-derived IL-8 was also produced in a concentration-dependent manner. The limiting concentrations of LPS, TNF, and IL-1 for the induction of significant levels of neutrophil-derived IL-8 were 1 ng/ml, 20 pg/ml, and 20 pg/ml, respectively (40). The half-maximum concentration for each of these agents needed to cause the synthesis of neutrophil-derived IL-8 was 10 ng/ml, 2 ng/ml, and 2 ng/ml, respectively, for LPS, IL-1, and TNF. Further investigations demonstrated that steady-state levels of IL-8 mRNA were also generated in a time- and dose-dependent manner. Neutrophils stimulated with either LPS, TNF, or IL-1 rapidly expressed IL-8 mRNA. By 1 h postchallenge the levels of IL-8 mRNA peaked and then declined to undetectable levels by 24 h. This kinetic pattern of neutrophil-derived IL-8 mRNA is in contrast to the pattern of expression found in other IL-8–producing cells. Many noninflammatory cells demonstrate protracted expression kinetics, as demonstrated by elevated levels of IL-8 mRNA 24–48 h after stimulation. The rapid pattern of IL-8 mRNA expression may reflect the role of the neutrophil in acute inflammation and stresses the importance of the early appearance of this cell in inflamed tissue.

Many classical studies have shown that either C5a, fMLP, or LTB_4 are potent neutrophil-activating compounds (5,6,8). In addition to inducing a chemotactic response, these three agents can activate neutrophils to release proteolytic enzymes and reactive oxygen species. In addition, these neutrophil-activating factors have also been shown to increase the expression of LPS-dependent neutrophil-derived IL-8 mRNA and the production of antigenic IL-8. While C5a, fMLP, and LTB_4 were not found to have a significant direct agonistic effect for IL-8 expression, the addition of LPS plus one of these agents increased IL-8 generation 100–250% above LPS alone (40). The above studies suggest that a synergistic relationship may exist between LPS and known classical chemotactic agents with the subsequent production of IL-8 by neutrophils.

Interleukin-8 is not the only chemokine family member produced by human neutrophils, as recent investigations have demonstrated that MIP-1α and MIP-1β are also generated by activated neutrophils (42). While MIP-1 was originally identified as a product of LPS-stimulated monocytes and macrophages, it is now known that this chemokine member can be generated by other cells. Isolated human neutrophils treated with graded doses of LPS demonstrated a time-dependent increase in the production of steady-state levels of MIP-1, which peaked at 24 h poststimulation. Interestingly, a major neutrophil-activating cytokine, GM-CSF, did not directly increase gene expression for neutrophil-derived MIP-1, but did serve a synergistic role in combination with LPS. The addition of 10 U/ml of GM-CSF to LPS-stimulated neutrophils resulted in a twofold increase in the expression of MIP-1 mRNA from neutrophils. The molecular mechanism(s) that regulate the effect of GM-CSF plus LPS appear to be complex, as the addition of cycloheximide, an inhibitor of protein synthesis, can cause a superinduction of MIP-1 mRNA. The assessment of antigenic MIP-1 production from LPS- or LPS plus GM-CSF–treated neutrophils also demonstrated an increase in the translation of MIP-1 polypeptide. Recent information has suggested that MIP-1 is a potent chemoattractant for monocytes and that its activity to move neutrophils is somewhat limited. Thus, it was of interest to determine whether the chemotactic activity for monocytes associated with the conditioned media recovered from LPS- and LPS plus GM-CSF–treated neutrophils was attributable to MIP-1. Using modified Boyden chambers, supernatant from stimulated neutrophils did indeed contain chemotactic activity for monocytes, which was suppressed by approximately 45% with rabbit antihuman MIP-1 antisera (42).

The production of a neutrophil-derived, monocyte chemotactic factor that belongs to the C-C chemokine supergene family is interesting, as this mediator may be involved in the induction of sequential "waves" of blood-borne leukocytes recruited during the evolution of an inflammatory reaction. Scientists have known for over a decade that many inflammatory processes involve characteristic populations of leukocytes, which target an area of injury. Classically, the neutrophil is one of the first blood-borne leukocytes to arrive at a site of inflammation (30). It is likely that this early response cell is responding to specific adherence and chemotactic factors, which guide the neutrophil to the site of reactivity. The neutrophil can then actively participate in the acute inflammatory reaction and, if necessary, begin the process that will aid in the recruitment of mononuclear phagocytic cells. This latter process will likely involve the production of neutrophil-derived chemotactic factors that possess specificity for the movement of mononuclear cells. Although neutrophils may not be as biosynthetically active as other immune/inflammatory cells, the fact that they comprise 70% of the circulating pool of leukocytes may compensate for this deficiency.

CHEMOKINES IN ANIMAL MODELS OF DISEASE

In Vivo and In Vitro Effects of Chemokines

Initial experimental animal studies assessing the in vivo activity of human IL-8 demonstrated that a single 100-μg intravenous injection into rats or rabbits caused marked neutrophilia, which was observed 15 min after administration and peaked by 1–2 h (43,44). Histological examination revealed both leukostasis and congestion in the microvasculature of the lung in the IL-8–treated animals. Interestingly, repeated intravenous administration of IL-8 resulted in significant pathology, which was restricted only to the lungs. These pathological changes included the formation of neutrophil aggregates in the small to medium-sized pulmonary vessels, diffuse septal and intraalveolar edema, and foci of hemorrhage (43). Animals that received five daily injections (5 nmol each) of IL-8 showed a substantial broadening of the alveolar septa, which contained fibroblasts and a mixed leukocytic infiltrate (43). This information supported earlier studies demonstrating that the intradermal administration of IL-8 in experimental animals can induce leukocyte accumulation in the skin (45). Furthermore, it provided interesting information concerning the lung as an especially sensitive target for the pathological effects of IL-8 and other C-X-C chemokines.

The above data demonstrate a significant interest in understanding the contribution of C-X-C chemokines to the evolving pathology of acute neutrophil-mediated inflammatory responses. Although the exact rodent homolog of human IL-8 has not been identified (46), experimental studies have shown that murine macrophage inflammatory protein-2 (MIP-2), a murine chemokine functionally equivalent to human IL-8, is a major LPS-inducible chemokine (47,48). This was first demonstrated during the original isolation studies, as murine MIP-2 was purified from the conditioned media of LPS-stimulated mouse RAW 264.7 cells, a macrophage cell line, hence the name macrophage inflammatory protein-2 (47). While murine MIP-2 does have functional and some structural homology to human IL-8, including an E-L-R-C-X-C motif near the amino terminus, this murine chemokine also has homology to human GRO-β and GRO-γ (also known as human MIP-2α and MIP-2β) (49). Other closely related rodent chemokines to murine MIP-2 include mouse and rat GRO-α (rat CINC) and rat MIP-2 (50,51).

Investigations have shown murine MIP-2 is highly chemotactic for neutrophils and elicits a localized neutrophil infiltrate when administered in vivo. Initial in vitro studies demonstrated that MIP-2 was significantly chemotactic for human neutrophils at 10 ng/ml and was more chemotactic for neutrophils than fMLP on a nanomolar basis (47). In vivo investigations demonstrated that a subcutaneous or intracisternal challenge of MIP-2 to mice or rabbits, respectively, induced significant and rapid neutrophilic accumulation (52–4). In the former study, the in vivo administration of 100 ng of MIP-2 into the footpad of a C3H/HeJ (LPS-resistant) mouse induced the elicitation of neutrophils that was not dependent upon LPS. Further analyses of rat alveolar macrophages challenged in vitro with LPS demonstrated a marked increase in steady-state MIP-2 mRNA levels within 30 min postchallenge (52,53). The levels of MIP-2 mRNA peakedl at 1 h and remained elevated for the next 4–6 h. Similar MIP-2 mRNA kinetic data were obtained from alveolar macrophages recovered from animals intratracheally challenged with LPS. These studies were limited in that the only parameter of MIP-2 expression measured was steady-state levels of mRNA, as neither the production of antigenic MIP-2 nor biological activity was assessed.

Chemokines in Models of Acute Lung Injury

In a rabbit model of lung reperfusion injury, antigenic IL-8 was observed both in the BAL fluid and in lung tissue homogenates after 2 h of ischemia followed by 2 h of reperfusion (55). The levels of IL-8 correlated with neutrophil infiltration and injury of the pulmonary parenchyma. The administration of neutralizing IL-8 antibodies prevented neutrophil infiltration and lung injury, suggesting a causal role for IL-8 in this model. Furthermore, the initiation of endotoxin-induced pleural inflammation in the rabbit was associated with a significant increase in IL-8 expression in the pleural space, which was associated with a profound influx of neutrophils (56). Passive immunization of rabbits with neutralizing antibodies to rabbit IL-8 resulted in a marked reduction in neutrophil influx into the pleural space (56). This study was important in that species-specific reagents were used to confirm the role of IL-8 in an experimental model of acute inflammation, since previous reports had demonstrated some efficacy of antibodies to human IL-8 in a rodent model of acute lung injury (57). In addition, these findings support previous studies that the mesothelium is a likely source of IL-8 in human diseases.

Chemokines in Experimental Endotoxemia

In sepsis, or endotoxemia, the release of LPS during bacterial contamination/infection of the host results in the sequential release of the early-response cytokines TNF and IL-1. Subsequently, TNF and IL-1 production can lead to the systemic activation and accumulation of neutrophils in several organs. One of the best characterized models of systemic inflammation is the mouse endotoxemia model, usually induced by a bolus intraperitoneal injection of LPS. This model is not technically difficult and has outlined many basic mechanisms of cytokine activation pathways that affect multiple organ systems, including liver, lung, kidney, heart, and intestinal tract. The detection of chemokines and cytokines induced during the LPS response can be assayed from the serum or in aqueous extracts of organ homogenates. During endotoxemia there appears to be an initial early peak in TNF and IL-1 production, followed by the accumulation of neutrophils in multiple organ systems. The examination of neutrophil-specific chemokines has identified MIP-2, a functional homolog of IL-8, as a potential mediator of neutrophil accumulation leading to tissue injury in this model. The in vivo induction of MIP-2 occurs very quickly after LPS injection, with detectable protein in the serum by 30 min post–LPS injection. MIP-2 does not appear to be induced by TNF, but directly activated by the LPS, as in vivo neutralization of TNF does not significantly inhibit early MIP-2 production. The primary source of MIP-2 during endotoxemia appears to be monocytes and macrophages, as indicated by immunohistochemical localization. Interestingly, epithelial cell populations within the lung have also been identified as a source of MIP-2 by in situ hybridization, indicating that multiple cellular sources likely contribute to MIP-2 levels. Neutralization of MIP-2 in vivo by passive immunization with rabbit antimouse MIP-2 polyclonal serum during LPS-mediated sepsis decreased neutrophil accumulation within the lung, as determined by myeloperoxidase levels, but had a minimal effect on LPS-induced mortality. However, neutralization of MIP-2, along with inhibition of TNF, synergistically decreased mortality in this model, suggesting that multiple mechanisms of injury were responsible for mortality.

C-C chemokines are also expressed in models of endotoxemia. For example, MIP-1α and RANTES appear to be generated in significant levels within the lung after LPS challenge. These members of the C-C chemokine family have chemotactic activity for monocytes, lymphocytes, and eosinophils. During endotoxemia, both RANTES and MIP-1α appear to mediate mononuclear phagocytic cell recruitment into the lungs. Passive immunization of mice with anti–MIP-1α antibodies significantly reduced inflammation, monocyte accumulation, and mortality associated with the host's response to LPS. Interestingly, passive neutralization of RANTES appeared to significantly inhibit macrophage but not lymphocyte or neutrophil influx during LPS challenge. In addition, neutralization of TNF in the endotoxin model significantly reduced RANTES mRNA expression and protein production within the lung. Taken together, it appears that multiple chemokines are operative during endotoxemia, leading to both neutrophil and monocyte infiltration. The results from these studies may give significant insight into the development of pharmacological agents to inhibit the production, release, or activity of chemokines during inflammation. This may be especially pertinent in endotoxemia or sepsis-related disorders, as it appears that targeting early-response cytokines, such as IL-1 and TNF, has not been effective in these patient populations.

Chemokines in Experimental Allergic Airway Inflammation

The production of chemokines during asthmatic responses may specifically contribute to the influx of eosinophils, a major effector cell inducing airway injury and subsequent late-phase reactivity. A number of C-C chemokines appear to preferentially recruit human eosinophils in vitro, including RANTES, MIP-1α, and MCP-3. To determine the relative contribution of the eosinophil-specific chemokines, a murine model of allergic airway inflammation utilizing schistosome parasite soluble egg antigen (SEA) has been developed. The ability of SEA to induce a TH_2-type (IL-4, IL-5) response, with high serum IgE (~10^5 titer) and hypereosinophils, was a key aspect of the development of this model. Localization of the soluble SEA antigen to the airway 21 days after the primary challenge induces an airway-localized TH_2-type response characterized by intense peribronchial leukocyte accumulation, airway eosinophilia, antigen-specific histamine release, and airway hyperreactivity. These responses appear to be similar to those observed in patient populations with allergic asthma. In this model the eosinophilic influx appears to be regulated by both IL-4 and TNF, as neutralization of these cytokines in vivo significantly attenuates the eosinophil recruitment into the lung and airway. Although IL-4 and TNF are not directly chemotactic for leukocytes, they have been reported to induce various C-C and C-X-C chemokines. Of particular interest has been the induction of C-C chemokines by these cytokines, which promote eosinophil chemotaxis. During the development of the eosinophilic response, MIP-1α and RANTES levels are elevated in the lung homogenates and BAL, peaking between 8 and 48 h post–airway allergen challenge. In this model, MIP-1α appears to be produced by macrophages and airway epithelial cells, while RANTES appears to be produced primarily by epithelial cells within the lung. In addition, the production of both MIP-1α and RANTES was significantly decreased in lung homogenates when either IL-4 or TNF was neutralized in vivo. Furthermore, the neutralization of either chemokine resulted in a significant reduction in airway eosinophilia, indicating that multiple chemokines may contribute to the recruitment of eosinophils. Interestingly, passive immunization with anti–MCP-1 did not reduce the influx of eosinophils into the airway, thus demonstrating chemokine specificity for the response. The use of this animal model, which has many of the attributes of human asthma, is important in defining the specific role that C-C chemokines play in allergic airway inflammation.

Chemokines in Experimental Rheumatoid Arthritis

Rheumatoid arthritis (RA) is an autoimmune disease characterized by the recruitment of various leukocyte subpopulations into an inflammatory pannus and synovial space. The chronic nature of this disease results in joint injury with cartilage and bone erosion. Type II collagen-induced arthritis (CIA) in the mouse has proven to be a useful model of RA, as it has many of the cellular and humoral immune characteristics of human RA. Interestingly, joint inflammation found in RA is characterized by both neutrophil and mononuclear cells, suggesting that both C-C and C-X-C chemokines may play prominent roles in the development of RA. The CIA model of RA has been very useful in describing the pathophysiological roles of various cytokines during disease development. In human RA a number of chemokines have been identified, including the C-X-C chemokines IL-8 and ENA-78, as well as C-C chemokines MIP-1α/β and MCP-1 (58,59). The CIA model has been useful in determining the relative contribution, temporal production patterns, as well as the regulation of chemokines during joint inflammation. Measurement of chemokines at different time points during development of CIA has demonstrated a correlation of MIP-1α, MCP-1, and MIP-2 (murine IL-8 homolog) with the onset and severity of clinical disease. Furthermore, depletion of either MIP-2 or MIP-1α by passive immunization significantly decreased both leukocyte influx and clinical severity and delayed the onset of joint inflammation. These data indicate that disruption of either the neutrophil or mononuclear recruitment could significantly alter the severity of the inflammation. Thus, both cell types appear to contribute to the joint injury. Interestingly, the production of chemokines (MIP-2 and MIP-1α) and the intensity of the leukocyte influx was regulated by IL-10 production within the joint, as passive neutralization with anti–IL-10 antibodies increased chemokine levels and joint inflammation. The source of IL-10 during the development of CIA appears to be synovial cells within the pannus itself. The above studies suggest that chronic inflammation may be controlled by the appropriate use of immunomodulators, such as IL-10, which may control the production of chemokines and reduce leukocyte recruitment.

Chemokines in Experimental Granuloma Formation

Chronic inflammatory responses have been characterized by sustained infiltration of mononuclear cells leading to tissue injury and organ dysfunction. The clinical manifestations of granulomatous diseases are usually the consequence of an intense inflammatory reaction involving both immune and nonimmune cells. These interactions lead to the initiation and maintenance of the immune response and often dictate the subsequent tissue destruction and end-stage disease associated with granulomatous inflammation. The use of a chronic model of granulomatous inflammation induced via the embolization of *S. mansoni* parasite eggs into the lungs of naive and sensitized mice has provided a convenient model to study granulomatous delayed-type hypersensitivity responses. The primary granulomatous response induced in naive mice develop a TH_1 immune reaction that is γ-interferon mediated. The cellular response in this model peaks at 16 days postchallenge. In addition, a secondary granuloma can be initiated in SEA-sensitized mice, and this granuloma is mediated by a TH_2 immune reaction and IL-4–dependent responses. The TH_2 response represents an accelerated granulomatous reaction, peaking at 4 days and resolving by 16 days postchallenge. The primary granuloma is characterized by mainly mononuclear cell accumulation, whereas the secondary granuloma formation is characterized by macrophages, eosinophils, and fibroblasts. These two distinct phases of granuloma formation provide a unique opportunity to examine the differential role of chemokines during the progression of inflammation in lesions with distinct cytokine profiles.

Initial studies demonstrated a significant induction of MIP-1α mRNA in the primary and secondary lesions which corresponded to the cell-recruitment patterns of granuloma formation (60). The production of MIP-1α protein from isolated granulomas demonstrated a progressive increase during primary lesion development, while TH_2 lesions had accelerated production of MIP-1α following granuloma formation. In vivo neutralization of MIP-1α during primary and secondary granuloma formation demonstrated differential results. During primary granuloma formation, the size of the lesion was reduced by approximately 60%, whereas secondary granuloma formation was reduced only 15–20%. Immunohistochemical localization of MIP-1α identified mononuclear phagocytes as the primary source of this C-C chemokine. These results suggest that MIP-1α appears to play a greater role in leukocyte recruitment during primary, TH_1-mediated, than in secondary, TH_2-mediated granuloma formation.

In addition, the C-C chemokine monocyte chemotactic protein-1 (MCP-1) was also assessed for its production pattern during granuloma development. MCP-1 is chemotactic for mononuclear cells and can be induced in a number of cell populations by IL-1 and TNF. Interestingly, IL-4, a TH_2 cytokine, has been shown to induce MCP-1 from endothelial cells, as well as smooth muscle cell populations. The expression of MCP-1 mRNA in granulomatous lungs was much stronger in lungs with TH_2, as compared to TH_1 granulomas. The production of MCP-1 in isolated granulomas demonstrated high constitutive levels of MCP-1 in TH_2 lesions, which correlated with the size of the TH_2-mediated response. Interestingly, when MCP-1 was neutralized in vivo, there was no evidence of diminished formation of TH_1 lesions, but a significant decrease was observed in TH_2 granulomas. The predominant cellular expression of MCP-1 appeared to be within vascular smooth muscle cell populations within the TH_2 lesions. These data contradict the MIP-1 neutralization data and suggest that specific chemokines are operative at certain stages of disease, possibly dependent upon the local expression of supporting cytokine profiles (TH_1- vs. TH_2-type cytokines).

SUMMARY AND FUTURE DIRECTIONS

A number of investigations support the role of various chemokines as key mediators of leukocyte elicitation. These chemotactic cytokines belong to unique supergene families and possess potent recruiting activity for leukocyte subpopulations. The relatively specific bioactivity of these mediators likely contributes to the complex gatekeeping roles of cell-derived cytokines during leukocyte elicitation. Thus, C-X-C chemokines appear to be involved in the recruitment of neutrophils, while C-C chemokines appear to be involved in the movement of mononuclear cells. Recent studies have “reduced to practice” the role of specific chemokines, as in vivo investigations utilizing neutralizing antibodies against individual chemokines have identified the contribution of chemokines to the evolution of animal models of disease. Interest in determining the role of both C-C chemokines and C-X-C chemokines during the initiation and maintenance of inflammation is not restricted to academia, as both pharmaceutical and biotechnology companies have set their sights on chemokines as therapeutic targets that may bring about a new generation of anti-inflammatory agents.

ACKNOWLEDGMENTS

These studies were supported in part by National Institutes of Health grants HL50057, P50HL46487, HL319634, and HL35276.

REFERENCES

1. Springer TA. Traffic signals for lymphocyte recirculation and leukocyte emigration: the multistep paradigm. Cell 1994; 76:301–314.
2. Carlos TM, Harlan JM. Leukocyte-endothelial adhesion molecules. Blood 1994; 84:2068–2101.
3. Pohlman TH, Stanness KA, Beatty PG, Ochs HD, Harlan JM. An endothelial cell surface factor(s) induced in vitro by LPS, IL-1, and TNF-alpha increases neutrophil adherences by a CDW 18-dependent mechanism. J Immunol 1986; 136:4548–4554.
4. Dustin ML, R. Rothlein R, Bhan AK, Dinarello CA, Springer TA. Induction by IL-1 and interferon-γ: tissue distribution, biochemistry, and function of a natural adherence molecule (ICAM-1). J Immunol 1986; 137:1245–1254.
5. Shiffman E, Corcoran BA, Wahl SM. N-formylmethionyl peptides as chemoattractants for leukocytes. Proc Natl Acad Sci USA 1975; 72:1059–1062.
6. Fernandez HN, Henson PM, Otani A, Hugli TE. Chemotactic response to human C3a and C5a anaphylatoxins. I. Evaluation of C3a and C5a leukotaxis in vitro and under stimulated in vivo conditions. J Immunol 1978; 120:109–115.
7. Ward PA, Newman LJ. A neutrophil chemotactic factor from human C′5. J Immunol 1969; 102:93–99.
8. Ford-Hutchinson AW, Bray MA, Doig MV, Smith MJ. Leukotriene B4 a potent chemokinetic and aggregating substance released from polymorphonuclear leukocytes. Nature 1980; 286:262–265.
9. Baggiolini M, Walts A, Kunkel SL. NAP-1/IL-8, a novel cytokine that activates neutrophils. J Clin Invest 1989; 57:2837–2841.
10. Leonard EJ, Yoshimura T. Human monocyte chemoattractant protein-1 (MCP-1). Immunol Today 1990; 11:97–100.
11. Matsushima K, Morishita K, Yoshimura T, et al. Molecular cloning of a human monocyte-derived neutrophil chemotactic factor (MDNCF) and the induction of MDNCF mRNA by interleukin-1 and tumor necrosis factor. J Exp Med 1988; 167:1883–1893.
12. Matsushima K, Oppenheim JJ. Interleukin-8 and MCAF: novel inflammatory cytokines inducible by IL-1 and TNF. Cytokine 1990; 1:2–13.
13. Oppenheim JJ, Zachariae COC, Mukaida N, Matsushima K. Properties of the novel proinflammatory supergene "intercrine" cytokine family. Ann Rev Immunol 1991; 9:617–648.
14. Farber JM. HuMIG: a new member of the chemokine family of cytokines. Biochem Biophys Res Comm 1993; 192:223–230.
15. Walz A, Burgener R, Car B, Baggiolini M, Kunkel SL, Strieter RM. Structure and neutrophil-activating properties of a novel inflammatory peptide (ENA-78) with homology to interleukin-8. J Exp Med 1991; 74:1355–1362.
16. Proost P, De Wolf-Peeters C, Conings R, Opdenakker G, Billiau A, Van Damme J. Identification of a novel granulocyte chemotactic protein (GCP-1) from human tumor cells: in vitro and in vivo comparison with natural forms of GROa, IP-10, and IL-8. J Immunol 1993; 150:1000–1010.
17. Deutsch E, Kain W. Studies on platelet factor 4. In: Jonson SA, Monto RW, Rebuck JW, Horn RC, eds. Blood Platelets. Boston: Little, Brown, 1961:337–345.
18. Kaplan G, Luster AD, Hancock G, Cohn Z. The expression of a gamma-interferon-induced protein (IP-10) in delayed immune responses in human skin. J Exp Med 1987; 166:1098–1108.
19. Ansiowicz A, Zajchowski D, Stenman G, Sager R. Functional diversity of gro gene expression in human fibroblasts and mammary epithelial cells. Proc Natl Acad Sci USA 1988; 85:9645–9649.
20. Ansiowicz A, Bardwell L, Sager R. Constitutive overexpression of a growth-regulated gene in transformed Chinese hamster and human cells. Proc Natl Acad Sci USA 1987; 84:7188–7192.
21. Hebert CA, Vitangcol RV, Baker JB. Scanning mutagenesis of interleukin-8 identifies a cluster of residues required for receptor binding. J Biol Chem 1991; 266:18989–94.
22. Clark-Lewis I, Dewald B, Geiser T, Moser B, Baggiolini M. Platelet factor 4 binds to interleukin 8 receptors and activates neutrophils when its N terminus is modified with Glu-Leu-Arg. Proc Natl Acad Sci USA 1993; 90:3574–3577.
23. Davatelis GP, Tekamp-Olson SD, Wolpe SD, Herman K, Luedke C, Gallegos C, Coit D, Merryweather J, Cerami A. Cloning and characterization of a cDNA for murine macrophage inflammatory protein (MIP) a novel monokine with inflammatory and chemokinetic properties. J Exp Med 1988; 167:1939–1944.
24. Yoshimura T, Yuhki N, Moore S, Appella E, Lerman M, Leonard EJ. Human monocyte chemoattractant protein-1 (MCP-1). Full-length cDNA cloning, expression of mitogen-stimulated blood mononuclear leukocytes and sequence similarity to mouse competence gene JE. FEBS Lett 1989; 244:487–493.
25. Rollins BJ, Morrison ED, Stiles CD. Cloning and expression of JE, a gene inducible by platelet-derived growth factor and whose product has cytokine-like properties. Proc Natl Acad Sci USA 1988; 85:3738–3749.
26. Strieter RM, Kunkel SL, Showell HJ, Marks RM. Monokine-induced gene expression of human endothelial cell-derived neutrophil chemotactic factor. Biochem Biophys Res Commun 1988; 156:1340–1345.
27. Strieter RM, Kunkel SL, Showell H, Remick DG, Phan SH, Ward PA, Marks RM. Endothelial cell gene expression of a neutrophil chemotactic factor by TNF-α, LPS, and IL-1β. Science 1989; 243:1467–1469.
28. Strieter RM, Phan SH, Showell HJ, Remick DG, Lynch JP, Genard M, Raiford C, Eskandari M, Marks RM, Kunkel SL. Monokine-induced neutrophil chemotactic factor gene expression in human fibroblasts. J Biol Chem 1989; 264:10621–10626.
29. Thornton AJ, Strieter RM, Lindley I, Baggiolini M, Kunkel SL. Cytokine-induced gene expression of a neutrophil chemotactic factor/interleukin-8 by human hepatocytes. J Immunol 1990; 144:2609–2613.
30. Elner VM, Strieter RM, Elner SG, Baggiolini M, Lindley I, Kunkel SL. Neutrophil chemotactic factor (IL-8) gene expression by cytokine-treated retinal pigment epithelial cells. Am J Path 1990; 136:745–750.
31. Strieter RM, Chensue SW, Basha MA, Standiford TJ, Lynch JP, Kunkel SL. Human alveolar macrophage gene expression of interleukin-8 by TNF-α, LPS and IL-1β. Am J Respir Cell Mol Biol 1990; 2:321–326.
32. Brown Z, Strieter RM, Chensue SW, Ceska P, Lindley I, Nield GH, Kunkel SL, Westwick J. Cytokine activated human mesangial cells generate the neutrophil chemoattractant—interleukin 8. Kidney Int 1991; 40:86–90.
33. Hotta K, Hayashi K, Ishikawa J, Tagawa M, Hashimoto K, Mizuno S, Suzuki K. Coding region structure of interleukin-8 gene of human lung giant cell carcinoma LU65C cells that produce LUCT/interleukin-8: homogeneity in interleukin-8 genes. Immunol Lett 1990; 24:165–169.

34. VanMeir E, Ceska M, Effenberger F, Walz A, Grouzmann E, Desbaillets I, Frei D, Fontana A, deTribolet N. Interleukin-8 is produced in neoplastic and infectious diseases of the human central nervous system. Cancer Res 1992; 52:4297–4305.
35. Abruzzo LV, Thornton AJ, Liebert M, Grossman HB, Evanoff H, Westwick J, Strieter RM, Kunkel SL. Cytokine-induced gene expression of interleukin-8 in human transitional cell carcinomas and renal cell carcinomas. Am J Pathol 1992; 140:365–373.
36. Rolfe MW, Kunkel SL, Standiford TJ, et al. Pulmonary fibroblasts expression of interleukin-8: a model for alveolar macrophage derived cytokine networking. Am J Respir Cell Mol Biol 1991; 5:579–585.
37. Rolfe MW, Kunkel SL, Standiford TJ, et al. Expression and regulation of human pulmonary fibroblast-derived monocyte chemotactic peptide-1. Am J Physiol 1992; 263:536–545.
38. Standiford TJ, Kunkel SL, Basha MA, et al. Interleukin-8 expression by pulmonary epithelial cells: A model for cytokine networks in the lung. J Clin Invest 1990; 86:1945–1953.
39. Thornton AJ, Ham J, Kunkel SL. Kupffer cell-derived cytokines induce the synthesis of a leukocyte chemotactic peptide, interleukin-8, in human hepatoma and primary hepatocyte cultures. Hepatol 1991; 14:1112–1122.
40. Strieter RM, Kasahara K, Allen RM, Standiford TJ, Rolfe MW, Becker FS, Chensue SW, Kunkel SL. Cytokine-induced neutrophil-derived interleukin-8. Am J Pathol 1992; 141:397–407.
41. Kasama T, Strieter RM, Lukacs NW, Lincoln PM, Burdick MD, Kunkel SL. Interferon gamma modulates the expression of neutrophil-derived chemokines. J Invest Med 1995; 43:58–67.
42. Kasama T, Strieter RM, Standiford TJ, Burdick MD, Kunkel SL. Expression and regulation of human neutrophil-derived macrophage inflammatory protein 1 alpha. J Exp Med 1993; 178:63–71.
43. Zwahlen R, Walz A, Rot A. In vitro and in vivo activity and pathophysiology of human interleukin-8 and related peptides. Inter Rev Exp Pathol 1993; 34B:27–42.
44. Coldtiz I, Zwahlen R, Dewald B, Baggiolini M. In vivo inflammatory activity of neutrophil-activating factor, a novel chemotactic peptide derived from monocytes. Am J Pathol 1989; 134:755–764.
45. Rampart M, Van Damme J, Zonnekeyn L, Herman AG. Granulocyte chemotactic protein-interleukin-8 induces plasma leakage and neutrophil accumulation in rabbit skin. Am J Pathol 1989; 1345:21–30.
46. Yoshimura T. The gene for neutrophil attractant protein-1 (NAP-1) is highly conserved in guinea pig but not in rat or mouse. J Immunol 1993; 151:6225–6236.
47. Wolpe SD, Sherry B, Juers D, Davatelis G, Yurt RW, Cerami A. Identification and characterization of macrophage inflammatory protein-2. Proc Natl Acad Sci USA 1990; 272:911–927.
48. Sherry B, Horii Y, Manogue KR, Widmer M, Cerami A. Macrophage inflammatory proteins 1 and 2: an overview. Cytokine 1992; 4:127–130.
49. Tekamp-Olson P, Gallegos C, Bauer D, McClain J, Sherry B, Fabre M, van Deventer S, Cerami A. Cloning and characterization of cDNAs for murine macrophage inflammatory protein 2 and its human homologues. J Exp Med 1990; 272:911–927.
50. Zagorski J, DeLarco JE. Rat CINC (cytokine-iunduced neutrophil chemoattractant) is the homologue of the human GRO protein but is encoded by a single gene. Biochem Biophys Res Comm 1993; 190:104–110.
51. Oquendo P, Alberta J, Wen D, Graycar J, Derynck R, Stiles C. The platelet-derived growth factor inducible KC gene encodes a secretory protein related to platelet alpha-granule proteins. J Biol Chem 1989; 264:4133–4137.
52. Xing Z, Jordana M, Kirpalani H, Driscoll KE, Schall TJ, Gauldie J. Cytokine expression by neutrophils and macrophages in vivo: Endotoxin induces TNF, MIP-2, IL-1, IL-6, but not RANTES or TGF mRNA expression in acute lung inflammation. Am J Respir Cell Mol Biol 1994; 10:148–153.
53. Huang S, Paulauskis JD, Godleski JJ, Kobzik L. Expression of macrophage inflammatory protein 2 and KC mRNA in pulmonary inflammation. Am J Pathol 1992; 141:981–988.
54. Driscoll KE, Hassenbein DG, Carter J, Poynter J, Asquith TN, Grant RA, Whitten J, Purdon MP, Takigiku R. Macrophage inflammatory proteins 1 and 2: expression by rat alveolar macrophages, fibroblasts, and epithelial cells and in rat lungs after mineral dust exposure. Am J Respir Cell Mol Biol 1993; 8:311–318.
55. Sekido N, Mukaida N, Harada A, Nakanishi I, Watanabe Y, Matsushima K. Prevention of lung reperfusion injury in rabbits by a monoclonal antibody against interleukin-8. Nature 1994; 365:654–657.
56. Broaddus VC, Boylan AM, Hoeffel JM, Kim KJ, Sadick M, Chuntharapai A, Hebert CA. Neutralization of IL-8 neutrophil influx in a rabbit model of endotoxin-induced pleuristy. J Immunol 1994; 152:2960–2967.
57. Mulligan MS, Jones ML, Bolanowski MA, Baganoff MP, Deppeler CL, Meyers DM, Ryan US, Ward PA. Inhibition of lung inflammatory reactions in rats by an anti-human IL-8 antibody. J Immunol 1993; 150:5585–5592.
58. Koch AE, Kunkel SL, Harlow LA, Mazarakis DD, Haines GK, Burdick MD, Pope RM, Strieter RM. Macrophage inflammatory protein-1 alpha. A novel chemotactic cytokine for macrophages in rheumatoid arthritis. J Clin Invest 1994; 93:921–928.
59. Koch AE, Kunkel SL, Harlow LA, Johnson B, Evanoff HL, Haines GL, Burdick MD, Pope RM, Strieter RM. 1992. Enhanced production of monocyte chemoattractant protein-1 in rheumatoid arthritis. J Clin Invest 1992; 90:772–779.
60. Lukacs NW, Kunkel SL, Strieter RM, Warmington K, Chensue SW. 1993. The role of macrophage inflammatory protein 1 alpha in Schistosoma mansoni egg-induced granulomatous inflammation. J Exp Med 1993; 177:1551–1559.

9
Interleukin-9

Jean-Christophe Renauld
Ludwig Institute for Cancer Research, Catholic University of Louvain, Brussels, Belgium

INTRODUCTION

Interleukin-9 (IL-9) was first identified in 1988 as a growth-promoting factor that allowed for the antigen-independent proliferation of murine T-helper cell clones (1). Stable T-cell lines could be derived in the absence of feeder cells and antigen, provided supernatant from activated TH2 cells were added to the culture. The growth factor present in such supernatants was further purified and designated P40, and its corresponding cDNA was cloned (2).

Independently, Hültner and coworkers observed that the proliferation of mast cell lines in response to IL-3 or IL-4 could be enhanced by a factor produced by activated splenocytes (3,4). This activity was partially purified and called mast cell growth–enhancing activity (MEA). Incidentally, high levels of MEA were also found in the supernatant from a murine M1s-reactive T-cell line derived by Schmitt and colleagues, who had observed that these cells produced a T-cell growth factor distinct from IL-2 and IL-4, which was called TCGF-III (5). The molecular cloning of a murine P40 cDNA and the availability of recombinant protein led to the demonstration that the same factor, now termed IL-9, was responsible for all these biological activities (6).

Human IL-9 cDNA was identified by expression cloning of a factor stimulating the proliferation of a human megakaryoblastic leukaemia (7) and by cross-hybridization with a mouse probe (8). Since then, the spectrum of its recognized biological activity has expanded significantly to include direct effects on normal hemopoietic progenitors (9), human T cells (10), B cells (11,12), fetal thymocytes (13), thymic lymphomas (14), and murine neuronal cell lines (15).

STRUCTURE AND MOLECULAR BIOLOGY OF IL-9

The purification of mouse IL-9 from the supernatants of activated helper T cells was made possible by the use of stable factor-dependent T-cell lines, derived from normal antigen-dependent clones. The purified protein, originally designated P40 on the basis of its apparent size in gel filtration, was characterized by an elevated pI (~10) and a high level of glycosylation (1). The amino acid sequence has been obtained by the complete sequencing of the purified protein and by the characterization of a full-length cDNA clone (2,16).

Natural human IL-9 was never purified, but its cDNA was cloned independently by expression cloning of a factor stimulating the growth of a human megakaryoblastic leukemia (7) and by cross-hybridization with the mouse gene (8). Both human and murine protein sequences contain 144 residues with a signal peptide of 18 amino acids. The overall identity reached 69% at the nucleotide level and 55% at the protein level. In accordance with the heavy glycosylation observed with the natural murine protein, four potential N-linked glycosylation sites were noticed in both sequences. This glycosylation is probably responsible for the discrepancy observed between the predicted relative molecular mass (14,15) and the M_r measured for native IL-9. The sequence is also characterized by the presence of 10 cysteines, which are perfectly matched in both mature proteins and a strong predominance of cationic residues, which explains the elevated pI (~10) measured with purified natural IL-9.

The human IL-9 gene is a single-copy gene and was mapped on chromosome 5 in the 5q31->q35 region (17).

Interestingly, this region also contains various growth factor and growth factor receptor genes, such as IL-3, IL-4, IL-5, IL-13, GM-CSF, and CSF-1R, and has been shown to be deleted in a series of hematological disorders (18). In addition, a linkage has been recently described between markers in this region and a gene controlling total serum IgE concentration (19). Radiation hybrid mapping analysis has located the IL-9 gene between the IL-3 and the EGR-1 (early growth response-1) genes (20,21). However, in the mouse, the IL-9 gene does not seem to be linked to the same gene cluster as it has been localized on mouse chromosome 13 (22) while the IL-3, IL-4, IL-5, and GM-CSF genes are located on chromosome 11.

A similar structure is shared by the human and murine IL-9 genes with 5 exons and 4 introns stretching over about 4 kb (23). The 5 exons are identical in size for both species and show homology levels ranging from 56 to 74%. In contrast, no significant sequence homology was found in the introns, although their size is roughly conserved. However, 3′ and 5′ flanking regions show a high level of identity supporting a possible involvement of these sequences in the transcription of posttranscriptional regulation of IL-9 expression. Particularly, numerous ATTTA motifs were found in the 3′ untranslated region of both genes. These sequences, frequently present in cytokine mRNAs, are thought to be involved in the modulation of the mRNA stability and in the regulation of translation (24).

The promoter of the IL-9 gene contains a TATA box sequence and potential recognition sites for several TPA-inducible transcription factors such as AP-1, which could provide a structural basis for the induction of IL-9 expression by phorbol esters. A consensus sequence for IRF-1 (interferon regulatory factor-1) was also identified in both promoters, but its physiological relevance remains more elusive (23). Kelleher and coworkers further identified other consensus sequences in the 5′ flanking region of the human gene (e.g., SP1, NF-κB, octamer, AP-3, AP-5, glucocorticoid responsive element, cAMP response element) and suggested that the NF-κB site and cAMP response element could be involved in the constitutive expression of IL-9 by HTLV-1–transformed T cells (25). However, functional analyses using deletion or substitution mutants of the IL-9 promoter are still lacking to clarify the regulatory mechanisms underlying IL-9 expression.

CELLULAR SOURCES AND STIMULI FOR IL-9 PRODUCTION

So far, IL-9 expression seems to be restricted to activated T cells. In the mouse, IL-9 is preferentially produced by TH2 clones in vitro (26,27), while in vivo, IL-9 production is also associated with TH2-like T-cell responses such as anti-IgD–mediated polyclonal activation (28), helminth infections (29), and *Leishmania major* infection of susceptible but not resistant mice (27). In the human, the regulation of IL-9 expression has been studied in vitro using freshly isolated peripheral blood mononuclear cells (PBMC). While no IL-9 message could be detected in these cells in the absence of any stimulation or after activation of B cells or monocytes, T-cell mitogens such as phytohemagglutinin (PHA) or anti-CD3 monoclonal antibodies (mAb) induced a substantial IL-9 expression that was further enhanced by addition of phorbol myristate acetate (PMA). Sorting experiments confirmed that IL-9 was preferentially produced by T-cell–enriched lymphocyte populations and, more specifically, by CD4+ T cells (23). Similar IL-9 expression has also been observed after stimulation through CD3 and CD28. In this system, IL-9 expression was restricted to the CD4+ CD45RO+ T-cell subset (so-called memory T cells) (30).

When peripheral T cells are activated by lectins or other mitogens, IL-9 mRNA expression appears in the late stages of T-cell activation, with a peak at 28 h. Further experiments identified IL-2 as a key mediator for IL-9 induction in T cells since anti–IL-2R mAb blocked IL-9 expression after stimulation with PMA and anti-CD3 mAb (31). More recently, IL-4 and IL-10, whose expression is also dependent on IL-2, were found to be required for IL-9 expression by peripheral human T cells. This expression thus results from a cascade of cytokines with IL-2 being required for IL-4 production, a combination of IL-2 and IL-4 for IL-10 production, and a combination of IL-4 and IL-10 for IL-9 production (30). Interestingly, CD45RA+ T cells were capable of producing IL-9, provided that both exogenous IL-4 and IL-10 were added to the cultures.

The central role played by IL-2 for IL-9 expression in freshly isolated human T cells was also confirmed in the mouse, as IL-2 was capable of inducing IL-9 in fresh murine splenocytes co-stimulated with PMA (P. Monteyne, personal communication), and IL-9 production is significantly reduced in IL-2–deficient mice (32). However, other regulatory mechanisms could be involved in IL-9 expression by some T-cell lines and tumors. It was indeed reported that IL-1, and not IL-2, serves as a secondary signal for IL-9 expression in murine T-cell lines (33) and that HTLV-I–transformed T cells produce IL-9 constitutively (7,25). TGF-β was also shown to be a potent inducer of IL-9 production by murine peripheral T cells (32). Recent observations indicating that IL-9 is produced in vivo in fetal thymus and in the central nervous system further suggest that various other regulatory mechanisms could take place in different cell types in vivo.

BIOLOGICAL ACTIVITIES OF IL-9

Mast Cells

Using IL-3–dependent bone marrow–derived mast cell lines (BMMC), Hültner and colleagues observed that spleen cell–conditioned medium contained a factor that was able to promote, in synergy with IL-3, the proliferation of mast cells (3–6). This factor, originally designated mast cell growth–enhancing activity (MEA) was eventually identified as IL-9, and the proliferative activity of IL-9, alone or in synergy with IL-3, was confirmed with most permanent mast cell lines such as MC-6, H7, and MC-9 (34, and unpublished data from our laboratory).

The effect of IL-9 on BMMC varies according to the length of in vitro culture. Early after derivation of primary BMMC from bone-marrow progenitors, IL-9 alone is not sufficient to sustain mast cell growth, but increases the survival of the cells and synergistically enhances the proliferation induced by IL-3 or steel factor. After more time in vitro, when stable mast cell lines are obtained, IL-9 alone becomes capable of inducing their proliferation, without any need for additional factors.

In addition to this growth-promoting activity, IL-9 also appears to be a potent regulator of mast cell effector molecules. It was indeed shown to induce IL-6 secretion by mast cell lines (6,35). Moreover, IL-9 seems to play a key role in the regulation of the expression of mast cell proteases. Stimulation of BMMC by IL-9 indeed induces the expression of transcripts encoding mMCP-1, mMCP-2, and mMCP-4, thereby suggesting that IL-9 could be involved in the differentiation of these cells. By contrast, IL-4 and IL-3 seem to suppress the differentiation of BMMC, since they inhibit the IL-9–induced expression of mast cell proteases (36). In addition, the recent identification of genes specifically induced by IL-9 has shown that other protease genes belonging to the granzyme family, such as granzyme B, are produced by mast cells in response to IL-9. Moreover, expression of the α chain of the high-affinity IgE receptor is also upregulated by IL-9, again indicating that, beyond its activity on the proliferation, this factor could be an important mediator of mast cell differentiation (37).

In vivo, infections of mice by helminth parasites typically induce an immune response characterized by a strong IgE production and a massive mucosal mastocytosis resulting from TH2 cytokine production. IL-9 expression by mesenteric lymph nodes is indeed induced by infection with *Trichinella spiralis,* in a similar fashion as other TH2 cytokines such as IL-3, IL-4, and IL-5 (38), and in the case of *Trichuris muris* infection, resistance was found to correlate with the production of IL-9 in mesenteric lymph nodes (39,40).

The immune responses induced by *Heligmosomoides polygyrus* or *Nippostrongylus brasiliensis,* two gastrointestinal nematode parasites, are characterized by a potent induction of TH2 cytokine expression but with considerable differences between the kinetics and level of induction of these factors (29,41). During *H. polygyrus* infection, the patterns of secretion of IL-3 and IL-9 can clearly distinguish mice with fast or slow responder phenotypes. In slow-responder mice, the initial IL-3 and IL-9 production declines to background levels by week 6 postinfection. By contrast, fast responders continue to secrete high amounts of these cytokines, thereby raising the hypothesis that IL-3 and IL-9 facilitate expulsion of adult worms (42).

The protective role of TH2 activation during parasite infections is further supported by the deleterious effect of IL-12, a cytokine known to promote TH1 responses. IL-12 injection during a primary *Nippostrongylus brasiliensis* infection indeed completely abolished the IL-4, IL-5, and IL-9 expression induced by the parasite as well as the intestinal mastocytosis, blood eosinophilia, and IgE production. Most importantly, IL-12 treatment resulted in an increased parasite survival and egg production during primary infections, thereby confirming the protective role of the TH2 response in this system (43).

The implication of IL-3 and IL-4 in *Nippostrongylus*-induced mastocytosis was demonstrated by the observation that either anti–IL-3 or anti–IL-4 antibodies partially inhibit the mucosal mast cell response, while the combination of both anti–IL-3 and anti–IL-4 antibodies blocked most of the mastocytosis (44). Although an anti–IL-9 antiserum by itself did not influence mast cell hyperplasia, addition of anti–IL-9 to anti–IL-3 and anti–IL-4 antibodies increased the suppression of the mast cell response (F. Finkelman, personal communication), demonstrating that IL-9 may contribute to the development of mucosal mast cell hyperplasia induced by worm infections, particularly under conditions where IL-3 and IL-4 are limiting. Recent observations in IL-9 transgenic mice have confirmed the activity of IL-9 in mast cell responses, showing that IL-9 overexpression results in a massive intestinal mastocytosis and increased circulating levels of mast cell–specific proteases (J.-C. Renauld et al., unpublished observations).

T Cells

Initial experiments in a murine system indicated that the activity of IL-9 for T cells was apparently restricted to some T-helper-cell clones (1). Noticeably, freshly isolated T cells never responded to this cytokine. However, it appeared that the sensitivity to the growth-promoting activity of IL-9 is not a characteristic of a particular T-cell subpopulation but can be gradually acquired by long-term in vitro culture. Interestingly, this process is reminiscent of—and eventually leads to—some kind of tumoral transformation.

T-helper clones can be derived from lymph nodes of mice immunized with proteins and maintained in vitro by biweekly stimulations with antigen and syngeneic antigen-presenting cells. After stimulation, the T cells go through a stage of blastic activation and proliferate intensively before undergoing a considerable size reduction. A quiescent period of 1 or 2 weeks is then needed before restimulation of the cells, otherwise no stable T-cell clones can be maintained in culture.

In this experimental system, proliferative responses to cytokines can be evaluated at the end of the quiescent period. Early T-cell clones, like freshly isolated T cells, were not responsive to IL-9, although significant proliferations were induced by IL-2. After a few passages in vitro, T-helper clones responded to IL-9 by blast formation, IL-6 secretion, and enhanced survival but no significant proliferation. After some more weeks in culture, short-term proliferations could be observed, but only in synergy with another factor like IL-4 or IL-3. After months of culture, IL-9 alone induced significant proliferations.

During this progressive transition, major changes appeared in the phenotype of the corresponding cultures. Cells remained blastic during the complete culture period and required more frequent passages. At this stage, cultures could be maintained by addition of IL-9 and permanent IL-9–dependent cell lines could be derived. An increase in cell size was noticed, as well as an accelerated growth rate. Progressively, most T-cell markers such as Thy 1, CD4, CD3, and TCR expression were lost.

This progressive deregulation undergone by T cells during acquisition of IL-9 responsiveness raised the possibility that some form of tumorigenic transformation occurred in these cell lines. Transfection of an IL-9–dependent T-cell line with the IL-9 cDNA results in transfectant cells secreting IL-9 and growing autonomously in vitro. Most interestingly, when these cells were injected into syngeneic mice, a very high tumor incidence was observed and animals died in 3–4 months as a result of widespread lymphoma development (45). In addition, factor-independent cells were sporadically generated during in vitro cultures and were similarly found to form tumors in vivo. Similar observations have been reported in a rat model, using an IL-2–dependent T-cell lymphoma that, after infection with murine polytropic retroviruses, became IL-2 independent by induction of an autocrine loop involving IL-9 and its receptor (46).

The significance of these observations is supported by the analysis of IL-9 transgenic mice that constitutively express high levels of this cytokine, since 5–10% of these mice spontaneously developed lymphoblastic lymphomas (47). Interestingly, no preneoplastic T-cell hyperplasia has ever been observed in these mice, thereby confirming the lack of activity of IL-9 on normal resting T cells in vitro and contrasting with other transgenic models such as the IL-7 transgenic mice, in which a proliferation of normal T cells in the skin preceeds the onset of lymphomas (48). Moreover, the IL-9 transgenic mice were highly susceptible to chemical mutagenesis as all transgenic animals developed such tumors after injection of doses of a mutagen (*N*-methyl-*N*-nitroso-urea) that were totally innocuous in control mice. Similarly, these transgenic mice exhibit a high sensitivity to the tumorigenic effect of gamma irradiation (49).

The growth-promoting activity of IL-9 for T-cell tumors was also investigated in vitro using other models of thymic lymphomas generated in normal mice. In these experiments, IL-9 was also found to significantly stimulate the in vitro proliferation of primary lymphomas induced either by chemical mutagenesis in DBA/2 mice or by x-ray radiation in B6 mice (14). However, the major role for IL-9 in T-cell lymphoma development could extend beyond its effect on cell proliferation. IL-9 has indeed also been found to protect such tumor cells against dexamethasone-induced apoptosis even for cell lines whose in vitro proliferation is completely independent on any cytokine. Moreover, with some cell lines, IL-9 seems more potent in protecting cells against apoptosis rather than in inducing proliferation. By contrast, IL-2 seems more efficient as a proliferation inducer (50). This activity could therefore be particularly relevant for the oncogenic potential of IL-9 in vivo.

In the human, a link between dysregulated IL-9 production and lymphoid malignancies has been initially suggested by the observation that lymph nodes from patients with Hodgkin's and large-cell anaplastic lymphomas constitutively produce IL-9 (51). Constitutive IL-9 expression was also detected in HTLV-1–transformed T cells (25) and in Hodgkin's cell lines (51–53). Moreover, an in vitro autocrine loop involving IL-9 has been observed for one of these Hodgkin's cell lines (52), thereby suggesting a potential involvement of IL-9 in the pathology of this disease.

A different, perhaps more physiological, aspect of the T-cell growth factor activity of IL-9 has been unraveled using human peripheral T cells (54). In line with previous reports in the murine model, IL-9 did not induce any proliferation of freshly isolated T cells, either alone or in synergy with other cytokines or T-cell co-stimuli. By contrast, significant proliferations could be induced by IL-9 when PBMC were activated for only 10 days with PHA, IL-2, and irradiated allogeneic feeder cells, thereby indicating that responses to IL-9 require previous activation. Similar results were obtained by Lehrnbecher et al. (55) using T cells activated with PHA for 3 or 7 days. When human T-cell clones were derived from established PHA-stimulated T-cell lines, it was found that most of these clones proliferated in response to IL-9, irrespective of their CD4 or CD8 phenotype (54).

The results obtained with human T cells differ in several ways from those initially reported in the murine model. As described earlier, murine T cells require months of in vitro culture before responding to IL-9. Moreover, when murine T cells become responsive to IL-9, factor-dependent lines are readily derived that grow independently of their antigen and feeder cells, while no such long-term proliferation has yet been obtained in the human system. However, these discrepancies could be simply related to the differences in the experimental systems used and, in particular, to the differences in the stimulation procedures. Human IL-9–responsive T cells are strongly activated with PHA, IL-2, and feeder cells on a weekly basis, and this strong stimulation was found to be crucial for IL-9 responses. By contrast, a much less potent, antigen-specific, stimulation is used in the murine model. This difference could explain why freshly raised murine T-cell clones do not respond to IL-9. On the other hand, the acquisition of IL-9 responsiveness by murine T cells is associated to a cryptic in vitro transformation process. The fact that no permanent IL-9–dependent human T-cell lines or clones could be derived so far suggests that such an in vitro transformation process does not take place under the conditions used in the human model. Taken together, these results indicate that different mechanisms may render T cells responsive to IL-9: potent activation of freshly isolated cells, as observed in the human model, or progressive dysregulation resulting from long-term in vitro culture, as observed in the mouse model.

Other Activities of IL-9

The involvement of IL-9 in mast-cell activation and proliferation as well as IL-9 production during parasite infections and TH2 activation suggest a potential role of this factor in Ig-mediated responses. Analyzing the ability of IL-9 to modulate immunoglobulin production by mouse B cells indicated that IL-9 acts in synergy with suboptimal doses of IL-4 for IgE and IgG1 production by LPS-activated semi-purified B cells but does not induce any IgE or IgG1 production in the absence of IL-4 (11). The influence of IL-9 on the IL-4–induced IgG1 production correlated with an increase in the number of IgG1-secreting cells. By contrast, IL-9 did not affect the IL-4–induced CD23 expression by LPS-activated B cells, indicating that its activity is not a simple upregulation of the IL-4 responsiveness by the B cells. Moreover, these experiments have not ruled out the possibility that the effect of IL-9 observed on murine B cells would be mediated by accessory cells.

In the human, very similar observations have been reported with semipurified peripheral B cells (12). In this experimental system, IL-9 cooperated with IL-4 for IgE and IgG but not for IgM production. Moreover, IL-9 also potentiated the IL-4–induced IgE production by sorted CD20+ cells upon costimulation by irradiated EL4 murine T cells, thereby suggesting a direct activity on B cells. However, IL-9 does not potentiate the IL-4–induced activation when B cells are stimulated by anti-CD40 antibodies.

Although the activity of IL-9 on Ig production in vitro is rather modest, its potential in vivo implications are strongly supported by preliminary observations made in IL-9 transgenic mice. These mice indeed show a clear increase in serum immunoglobulin titers, irrespectively of the Ig class or subclass. In addition, antigen-specific antibody responses were also enhanced in IL-9 transgenic mice as compared with normal animals (J.-C. Renauld et al., unpublished data).

The first evidence for an involvement of IL-9 in the hematopoietic system was provided by the identification and cloning of the human protein as a growth factor for the megakaryoblastic leukemia Mo7E, a cell line displaying early markers of differentiation, such as CD33 and CD34, and markers for bipotent erythro-megakaryoblastic hematopoietic precursors (7,56). Although IL-9 did not seem to be active on normal megakaryoblastic precursors, it was found to support the clonogenic maturation of erythroid progenitors in the presence of erythropoietin (9). This activity was confirmed by several groups and reproducibly observed with highly purified progenitors after sorting for CD34+ cells and T-cell depletion (57–60). In the mouse, a similar erythroid burst–promoting activity has been described but appeared to be dependent on the presence of T cells (33).

By contrast, granulocyte or macrophage colony formation (CFU-GM, CFU-G, or CFU-M) was usually not affected by IL-9. However, Schaafsma and colleagues observed that IL-9 also promoted some granulocytic as well as monocytic colony (CFU-GM) growth from CD34+ CD2– progenitors in two out of eight different bone marrow donors (59). However, experiments comparing the effects of IL-9 on fetal and adult progenitors have shown that IL-9 is more effective on fetal cells of the erythroid lineage and that addition of IL-9 to cultures of fetal progenitors induced maturation of CFU-mix and CFU-GM (61). The observation that murine day 15 fetal thymocytes (13), but not adult thymocytes (62), respond synergistically to IL-9 and IL-2 further supports the hypothesis that the spectrum of activity of IL-9 is larger on fetal progenitors. The mechanisms responsible for this wider spectrum of action and the physiological role of IL-9 in the regulation of hematopoiesis remain to be determined.

A role for hematopoietic cytokines in the differentiation of neuronal cells has been suggested by studies on immortalized murine embryonic hippocampal progenitor cell lines, which show little evidence of morphological maturation. In combination with bFGF and TGF-α, IL-9 enhanced neurite outgrowth as well as other morphological modifications and conferred some electrical excitability to these cells (15). In vivo, further investigations will have to address the relevance of such observations for normal brain development.

THE IL-9 RECEPTOR

The existence of specific high-affinity receptors for IL-9 (Kd ~ 100 pM) was demonstrated on a variety of mouse hemopoietic cells including T cells, mast cells, and macrophages (63). The murine IL-9 receptor, identified by cDNA expression cloning (64), contains 468 amino acids including an extracellular domain composed of 233 acids. The position of four cysteine residues, as well as the presence of a WSEWS motif located a few amino acids upstream from the transmembrane region, indicate that the IL-9 receptor is a member of the hematopoietin receptor superfamily (65). The human IL-9 receptor cDNA, isolated by cross-hybridization with the mouse probe, encodes a 522-amino-acid protein with a 53% identity with the mouse IL-9 receptor. The extracellular region is particularly conserved with 67% identity, while the cytoplasmic domain is significantly larger in the human receptor (231 vs. 177 residues) (64).

In the mouse, the IL-9R gene is a single-copy gene composed of 9 exons and 8 introns and shares many characteristics with other genes encoding cytokine receptors (J.-C. Renauld et al., unpublished data). By contrast, the human genome contains at least four IL-9R pseudogenes with ~90% homology with the IL-9R gene, which is located in the subtelomeric region of chromosomes X and Y (66,67).

Accumulating data indicate that most cytokine receptors consist of at least two chains: one chain that is specific for the ligand and another that can be shared by several cytokine receptors (68). Recently, the γ chain of the IL-2 receptor has been shown to be involved in the IL-9 receptor since an antibody directed against this molecule completely inhibits the activity of IL-9 (69). As this chain has also been shown to be involved in the IL-7- and IL-4 receptors, it could explain the overlapping activities observed for these T-cell growth factors.

The mechanisms involved in signal transduction through the IL-9 receptor are still poorly understood. IL-9 stimulation of human megakaryoblastic leukemia Mo7E induced at least four unidentified tyrosine phosphorylated bands in Western blot analysis with antiphosphotyrosine antibodies. In contrast, IL-9 did not induce or enhance the phosphorylation of the serine-threonine kinases Raf-1 or MAP, an event frequently observed upon stimulation with cytokines (70). The cytoplasmic domain of the IL-9 receptor does not contain any motif characteristic of a potential tyrosine- or serine/threonine-kinase activity. However, a high percentage of serine and proline residues was found in this region, as already observed for most cytokine receptors. Moreover, an intriguing stretch of nine successive serines was observed in the human receptor. Finally, some sequence homology with other receptors was noticed proximally to the transmembrane domain, in particular a Pro-X-Pro sequence preceded by a cluster of hydrophobic residues that partially fits the "box 1" consensus sequence shared by many cytokine receptors (IL-4R, IL-7R, IL-3R, EPOR, IL-2Rβ, G-CSFR, . . .) (71). Downstream from this motif, a striking homology was observed with the β chain of the IL-2 receptor and with the erythropoietin receptor. As a result, for the first 33 amino acids of the cytoplasmic domain, a 40% identity was noticed between the human IL-9R and IL-2Rβ and a 27% identity with the erythropoietin receptor. This region could be involved in the activation of Jak kinases by these cytokine receptors. Immunoprecipitation experiments have indeed shown that IL-9, like other cytokines such as IL-2, induces Jak1 and Jak3 phosphorylation (72,73).

CONCLUSION

Although cytokines appear to be involved in a network characterized by an ever-growing complexity, recent advances, particularly in the field of their receptors, point to the existence of superfamilies with common structural features. In this respect, IL-9 seems to be structurally related to IL-2, IL-4, and IL-7 and shares with these factors a common receptor chain (IL-2Rγ or γc). These observations can at least partially explain the redundancies found in the biological activities of these cytokines, such as the fact that they are four major T-cell growth factors.

However, IL-9 has some definite specificities such as the preferential activity for preactivated or transformed T cells. In addition, IL-9 seems to be a key regulator of mast-cell growth and differentiation. Interestingly, recent observations have shown that this cytokine induced expression of a similar set of genes by mast cells and T cells, including mast-cell–specific proteases and the α chain of the high-affinity IgE receptor (FcεRIα) (37). Although the functional consequences of these observations are still to be defined, they definitely point to the important relationship between mast cells and T cells and underlie the potential involvement of IL-9 in physiological situations implicating these cell types.

REFERENCES

1. Uyttenhove C, Simpson RJ, Van Snick J. Functional and structural characterization of P40, a mouse glycoprotein with T-cell growth factor activity. Proc Natl Acad Sci USA 1988; 85:6934–6938.
2. Van Snick J, Goethals A, Renauld J-C, Van Roost E, Uyttenhove C, Rubira MR, Moritz RL, Simpson RJ. Cloning and characterization of a cDNA for a new mouse T cell growth factor (P40). J Exp Med 1989; 169:363–368.
3. Hültner L, Moeller J, Schmitt E, Jäger G, Reisbach G, Ring J, Dörmer P. Thiol-sensitive mast cell lines derived from mouse bone marrow respond to a mast cell growth-enhancing activity different from both IL-3 and IL-4. J Immunol 1989; 142:3440–3446.
4. Moeller J, Hültner L, Schmitt E, Dörmer P. Partial purification of a mast cell growth-enhancing activity and its separation from IL-3 and IL-4. J Immunol 1989; 142:3447–3451.
5. Moeller J, Hültner L, Schmitt E, Breuer M, Dörmer P. Purification of MEA, a mast cell growth-enhancing activity, to apparent homogeneity and its partial amino acid sequencing. J Immunol 1990; 144:4231–4234.
6. Hültner L, Druez C, Moeller J, Uyttenhove C, Schmitt E, Rüde E, Dörmer P, Van Snick J. Mast cell growth-enhancing activity (MEA) is structurally related and functionally identical to the novel mouse T cell growth factor P40/TCGFIII (interleukin 9). Eur J Immunol 1990; 20:1413–1416.
7. Yang Y, Ricciardi S, Ciarletta A, Calvetti J, Kelleher K, Clark SC. Expression cloning of a cDNA encoding a novel human hematopoietic growth factor: human homologue of murine T-cell growth factor P40. Blood 1989; 74:1880–1884.
8. Renauld J-C, Goethals A, Houssiau F, Van Roost E, Van Snick J. Cloning and expression of a cDNA for the human homolog of mouse T cell and mast cell growth factor P40. Cytokine 1990; 2:9–12.
9. Donahue RE, Yang YC, Clark SC. Human P40 T cell growth factor (interleukin-9) supports erythroid colony formation. Blood 1990; 75:2271–2275.
10. Houssiau F, Renauld J-C, Stevens M, Lehmann F, Coulie PG, Van Snick J. Human T cell lines and clones respond to IL-9. J Immunol 1993; 150:2634–2640.
11. Petit-Frère C, Dugas B, Braquet P, Mencia-Huerta JM. Interleukin-9 potentiates the interleukin-4-induced IgE and IgG1 release from murine B lymphocytes. Immunology 1993; 79:146–151.
12. Dugas B, Renauld J-C, Pene J, Bonnefoy JY, Petit-Frère C, Braquet P, Bousquet J, Van Snick J, Mencia-Huerta JM. Interleukin-9 potentiates the interleukin-4-induced immunoglobulin (IgG, IgM and IgE) production by normal human B lymphocytes. Eur J Immunol 1993; 23:1687–1692.
13. Suda T, Murray R, Fischer M, Tokota T, Zlotnik A. Tumor necrosis factor-α and P40 induce day 15 murine fetal thymocyte proliferation in combination with IL-2. J Immunol 1990; 144:1783–1787.
14. Vink A, Renauld J-C, Warnier G, Van Snick J. Interleukin 9 stimulates in vitro growth of mouse thymic lymphomas. Eur J Immunol 1993; 23:1134–1138.
15. Mehler MF, Rozental R, Dougherty M, Spray DC, Kessler JA. Cytokine regulation of neuronal differentiation of hippocampal progenitor cells. Nature 1993; 362:62–65.
16. Simpson RJ, Moritz RL, Gorman JJ, Van Snick J. Complete amino acid sequence of a new murine T-cell growth factor P40. Eur J Biochem 1989; 183:715–722.
17. Modi WS, Pollock DD, Mock BA, Banner C, Renauld J-C, Van Snick J. Regional localization of the human glutaminase (GLS) and interleukin-9 (IL9) genes by in situ hybridization. Cytogenet Cell Genet 1991; 57:114–116.
18. Sokal G, Michaux JL, Van Den Berghe H, Cordier A, Rodhain J, Ferrant A, Moriau M, De Bruyère M, Sonnet J. A new hematologic syndrome with a distinct karyotype: the 5q-chromosome. Blood 1975; 46:519.
19. Marsh DG, Neely JD, Breazeale DR, Ghosh B, Freidhoff LR, Ehrlich-Kautzky E, Schou C, Krishnaswany G, Beaty TH. Linkage analysis of IL-4 and other chromosome 5q31.1 markers and total serum immunoglobulin E concentrations. Science 1994; 1152–1156.
20. Warrington J, Bailey S, Armstrong E, Aprelikova O, Alitalo K, Dolganov G, Wilcox A, Sikela J, Wolfe S, Lovett M, Vasmuth J. A radiation hybrid map of 18 growth factor, growth factor receptor, hormone receptor, or neurotransmitter receptor genes on the distal region of the long arm of chromosome 5. Genomics 1992; 13:803–808.
21. Saltman DL, Dolganov GM, Warrington JA, Wasmuth JJ, Lovett M. A physical map of 15 loci on human chromosome 5q23-q33 by two-color fluorescence in situ hybridization. Genomics 1993; 16:726–732.
22. Mock BA, Krall M, Kozak CA, Nesbitt MN, McBride OW, Renauld J-C, Van Snick J. IL9 maps to mouse chromosome 13 and human chromosome 5. Immunogenetics 1990; 31:265–270.
23. Renauld J-C, Goethals A, Houssiau F, Merz H, Van Roost E, Van Snick J. Human P40/IL-9: expression in activated CD4+ T cells, genomic organization, and comparison with the mouse gene. J Immunol 1990; 144:4235–4241.
24. Kruys V, Marinx O, Shaw G, Deschamps J, Huez G. Translational blockade imposed by cytokine-derived UA-rich sequences. Science 1989; 245:852.
25. Kelleher K, Bean K, Clark SC, Leung WY, Yang-Feng TL, Chen JW, Lin PF, Luo W, Yang YC. Human Interleukin-9: genomic sequences, chromosomal location, and sequences essential for its expression in human T cell leukemia virus (HTLV)-I-transformed human T cells. Blood 1991; 77:1436–1441.
26. Schmitt E, van Brandwijk R, Van Snick J, Siebold B, Rüde E. TCGFIII/P40 is produced by naive murine CD4+ T cells but is not a general T cell growth factor. Eur J Immunol 1989; 19:2167–2170.
27. Gessner A, Blum H, Röllinghoff M. Differential regulation of IL-9 expression after infection with *Leishmania major* in susceptible and resistant mice. Immunobiology 1993; 189:419–435.
28. Svetic A, Finkelman FD, Jian YC, Dieffenbach CW, Scott DE, McCarthy KF, Steinberg AD, Gause WC. Cytokine gene expression after in vivo primary immunization with goat antibody to mouse IgD antibody. J Immunol 1991; 147:2391–2397.
29. Svetic A, Madden KB, Zhou XD, Lu P, Katona IM, Finkelman FD, Urban JF, Gause WC. A primary intestinal helminthic infection rapidly induces a gut-associated elevation of TH2-associated cytokines and IL-3. J Immunol 1993; 150:3434–3441.
30. Houssiau F, Schandené L, Stevens M, Cambiaso C, Goldman M, Van Snick J, Renauld J-C. A cascade of cytokines is responsible for IL-9 expression in human T cells. Involvement of IL-2, IL-4 and IL-10. J Immunol 1995; 154:2624–2630.
31. Houssiau F, Renauld J-C, Fibbe W, Van Snick J. IL-2 dependence of IL-9 expression in human T lymphocytes. J Immunol 1992; 148:3147–3151.

32. Schmitt E, Germann T, Goedert S, Hoehn P, Huels C, Koelsch S, Kuhn R, Muller W, Palm N, Rüde E. IL-9 production of naive CD4+ T cells depends on IL-2, is synergistically enhanced by a combination of TGF-beta and IL-4, and is inhibited by IFN-gamma. J Immunol 1994; 153:3989–3996.
33. Schmitt E, Beuscher HU, Huels C, Monteyne P, van Brandwijk R, Van Snick J, Rüde E. IL-1 serves as a secondary signal for IL-9 expression. J Immunol 1991; 147:3848–3854.
34. Williams DE, Morrissey PJ, Mochizuki DY, de Vries P, Anderson D, Cosman D, Boswell HS, Cooper S, Grabstein KH, Broxmeyer HE. T cell growth factor P40 promotes the proliferation of myeloid cell lines and enhances erythroid burst formation by normal murine bone marrow cells in vitro. Blood 1990; 76:906–911.
35. Hültner L, Moeller J. Mast cell growth-enhancing activity (MEA) stimulates interleukin 6 production in a mouse bone marrow-derived mast cell line and a malignant subline. Exp Hematol 1990; 18:873–877.
36. Eklund KK, Ghildyal N, Austen KF, Stevens RL. Induction by IL-9 and suppression by IL-3 and IL-4 of the levels of chromosome 14-derived transcripts that encode late-expressed mouse mast cell proteases. J Immunol 1993; 151:4266–4273.
37. Louahed J, Kermouni A, Van Snick J, Renauld J-C. IL-9 induces expression of granzymes and high affinity IgE receptor in murine T helper clones. J Immunol 1995; 154:5061–5070.
38. Grencis RK, Hültner L, Else KJ. Host protective immunity to Trichinella spiralis in mice: activation of Th cell subsets and lymphokine secretion in mice expressing different response phenotypes. Immunology 1991; 74:329–332.
39. Else K, Hültner L, Grencis R. Cellular immune responses to the murine nematode parasite Trichuris muris. II Differential induction of Th cell subsets in resistant mice versus susceptible mice. Immunology 1992; 75:232–237.
40. Else K, Hültner L, Grencis R. Modulation of cytokine production and response phenotypes in murine trichuriasis. Parasite Immunology 1992; 14:441–449.
41. Urban JF, Madden KB, Svetic A, Cheever A, Trotta PP, Gause WC, Katona IM, Finkelman FD. The importance of TH2 cytokines in protective immunity to nematodes. Immunol Rev 1992; 127:205–220.
42. Wahid FN, Behnke JM, Grencis RK, Else KJ, Ben-Smith AW. Immunological relationships during primary infection with Heligmosomoides polygyrus: Th2 cytokines and primary response phenotype. Parasitology 1994; 108:461–471.
43. Finkelman FD, Madden KB, Cheever AW, Katona IM, Morris SC, Gately MK, Hubbard BR, Gause WC, Urban JF. Effects of Interleukin 12 on immune responses and host protection in mice infected with intestinal nematode parasites. J Exp Med 1994; 179:1563–1572.
44. Madden K, Urban K, Ziltener H, Schrader J, Finkelman F, Katona I. Antibodies to IL3 and IL4 suppress Helminth-induced intestinal mastocytosis. J Immunol 1991; 147:1387–1391.
45. Uyttenhove C, Druez C, Renauld J-C, Herin M, Noel H, Van Snick J. Autonomous growth and tumorigenicity induced by P40/interleukin 9 cDNA transfection of a mouse P40-dependent T cell line. J Exp Med 1991; 173:519–522.
46. Flubacher MM, Bear SE, Tsichlis PN. Replacement of interleukin-2 (IL-2)-generated mitogenic signals by a mink cell focus-forming (MCF) or xenotropic virus-induced IL-9-dependent autocrine loop: implications for MCF virus-induced leukemogenesis. J Virol 1994; 68:7709–7716.
47. Renauld J-C, van der Lugt N, Vink A, van Roon M, Godfraind C, Warnier G, Merz H, Feller A, Berns A, Van Snick J. Thymic lymphomas in interleukin 9 transgenic mice. Oncogene 1994; 9:1327–1332.
48. Rich B, Campos-Torres J, Tepper R, Moreadith R, Leder, P. Cutaneous lympoproliferation and lymphomas in IL-7 transgenic mice. J Exp Med 1993; 177:305–317.
49. Renauld J-C, Kermouni A, Vink A, Louahed J, Van Snick J. Interleukin-9 and its receptor: involvement in mast cell differentiation and T cell oncogenesis. J Leukoc Biol 1995; 57:353–360.
50. Renauld J-C, Vink A, Louahed J, Van Snick J. Interleukin-9 is a major anti-apoptotic factor for thymic lymphomas. Blood 1995; 85:1300–1305.
51. Merz H, Houssiau F, Orscheschek K, Renauld J-C, Fliedner A, Herin M, Noël H, Kadin M, Mueller-Hermelink KH, Van Snick J, Feller AC. Interleukin-9 expression in human malignant lymphomas: unique association with Hodgkin's disease and large cell anaplastic lymphoma. Blood 1991; 78:1311–1317.
52. Gruss HJ, Brach MA, Drexler HG, Bross KJ, Herrman F. Interleukin 9 is expressed by primary and cultured Hodgkin and Reed-Sternberg cells. Cancer Res 1992; 52:1026–1031.
53. Trümper LH, Brady G, Bagg A, Gray D, Loke SL, Griesser H, Wagman R, Graziel R, Gascoyne RD, Vicini S, Iscove NN, Cossman J, Mak TW. Single cell analysis of Hodgkin and Reed-Sternberg cells: molecular heterogeneity of gene expression and p53 mutations. Blood 1993; 81:3097–3115.
54. Houssiau F, Renauld J-C, Stevens M, Lehmann F, Coulie PG, Van Snick J. Human T cell lines and clones respond to IL-9. J Immunol 1993; 150:2634–2640.
55. Lehrnbecher T, Poot M, Orscheschek K, Sebald W, Feller AC, Merz H. Interleukin 7 as interleukin 9 drives phytohemagglutinin-activated T cells through several cell cycles; no synergism between interleukin 7, interleukin 9 and interleukin 4. Cytokine 1994; 6:279–284.
56. Hendrie P, Miyazawa K, Yang YC, Langefeld C, Broxmeyer H. Mast cell growth factor (c-kit ligand) enhances cytokine stimulation and proliferation of the human factor-dependent cell line, Mo7E. Exp Hematol 1991; 19:1031–1037.
57. Birner A, Hültner L, Mergenthaler HG, Van Snick J, Dörmer P. Recombinant murine Interleukin 9 enhances the erythropoietin-dependent colony formation of human BFU-E. Exp Hematol 1992; 20:541–545.
58. Sonoda Y, Maekawa T, Kuzuyama Y, Clark SC, Abe T. Human interleukin-9 supports formation of a subpopulation of erythroid bursts that are responsive to interleukin-3. Am J Hematol 1992; 41:84–91.
59. Schaafsma MR, Falkenburg JH, Duinkerken N, Van-Snick J, Landegent JE, Willemze R, Fibbe WE. Interleukin-9 stimulates the proliferation of enriched human erythroid progenitor cells: additive effect with GM-CSF. Ann Hematol 1993; 66:45–49.
60. Lu L, Leemhuis T, Srour E, Yang YC. Human Interleukin 9 stimulates proliferation of $CD34^{+++}DR^{+}CD33^{-}$ erythroid progenitors in normal human bone marrow in the absence of serum. Exp Hematol 1992; 20:418–424.
61. Holbrook ST, Ohis RK, Schribler KR, Yang YC, Christensen RD. Effect of interleukin-9 on clonogenic maturation and cell-cycle status of foetal and adult hematopoietic progenitors. Blood 1991; 77:2129–2134.
62. Suda T, Murray R, Guidos C, Zlotnik A. Growth-promoting activity of IL-1α, IL-6, and Tumor Necrosis Factor-α in combination with IL-2, IL-4, or IL-7 on murine thymocytes. J Immunol 1990; 144:3039–3045.
63. Druez C, Coulie P, Uyttenhove C, Van Snick J. Functional and biochemical characterization of mouse P40/IL-9 receptors. J Immunol 1990; 145:2494–2499.
64. Renauld J-C, Druez C, Kermouni A, Houssiau F, Uyttenhove C, Van Roost E, Van Snick J. Expression cloning of the murine and

human interleukin 9 receptor cDNAs. Proc Natl Acad Sci USA 1992; 89:5690–5694.

65. Bazan F. Structural design and molecular evolution of a cytokine receptor superfamily. Proc Natl Acad Sci USA 1990; 87:6934.

66. Kermouni A, Van Roost E, Arden KC, Vermeesch JR, Weiss S, Godelaine D, Flint J, Lurquin C, Szikora JP, Higgs DR, Marynen P, Renauld J-C. The IL-9 receptor gene: genomic structure, chromosomal localization in the pseudoautosomal region of the long arm of the sex chromosomes and identification of IL-9R pseudogenes at 9qter, 10pter, 16pter and 18pter. Genomics 1995; 29:371–382.

67. Chang MS, Engel G, Benedict C, Basu R, McNimch J. Isolation and characterization of the human interleukin-9 receptor gene. Blood 1994; 83:253–262.

68. Kishimoto T, Taga T, Akira S. Cytokine signal transduction. Cell 1994; 76:3199–3205.

69. Kimura Y, Takeshita T, Kondo M, Ishii N, Nakamura M, Van Snick J, Sugamura K. Sharing of the IL-2 receptor γ chain with the functional IL-9 receptor complex. Int Immunol 1995; 7:115–120.

70. Miyazawa K, Hendrie P, Kim YJ, Mantel C, Yang YC, Se Kwon B, Broxmeyer H. Recombinant human Interleukin-9 induces protein tyrosine phosphorylation and synergizes with steel factor to stimulate proliferation of the human factor-dependent cell line, Mo7E. Blood 1992; 80:1685–1692.

71. Murakami M, Narazaki M, Hibi M, Yawata H, Yasukawa K, Hamaguchi M, Taga T, Kishimoto T. Critical cytoplasmic region of the interleukin 6 signal transducer gp130 is conserved in the cytokine receptor family. Proc Natl Sci USA 1991; 88:11349–11353.

72. Yin T, Lik-Shing Tsang M, Yang Y-C. Jak1 kinase forms complexes with interleukin-4 receptor and 4PS/insulin receptor substrate-1-like protein and is activated by interleukin-4 and interleukin-9 in T lymphocytes. J Biol Chem 1994; 269:26614–26617.

73. Russel SM, Johnston JA, Noguchi M, Kawamura M, Bacon CM, Friedman M, Berg M, McVicar DW, Witthuhn BA, Silvennonien O, Goldman AS, Schmalstieg FC, Ihle HM, O'Shea JJ, Leonard WJ. Interaction of IL-2Rβ and γc chains with Jak1 and Jak3: implications for XSCID and XCID. Science 1994; 266:1042.

10
Interleukin-10

Fiona Powrie
University of Oxford, John Radcliffe Hospital, Oxford, England

Andrew Bean and Kevin W. Moore
DNAX Research Institute, Palo Alto, California

INTRODUCTION AND HISTORICAL PERSPECTIVE

Interleukin-10 has several powerful and important regulatory activities in lymphoid and myeloid cells. IL-10 suppresses activation of macrophages and dendritic cells, inhibiting their abilities to both secrete cytokines and function as accessory cells for T-cell and natural-killer (NK)-cell stimulation. IL-10 also participates in stimulating proliferation and differentiation of B cells, mast cells, and mature and immature T cells. Two herpesviruses [Epstein-Barr virus (EBV) and equine herpesvirus type II (EHV)] are known to encode their own homologs of the IL-10 gene; the EBV product (BCRF1/viral IL-10, vIL-10) shares a subset of IL-10's activities, which may be important in the host-virus relationship. This chapter is an overview of the structure and function of IL-10 and information available about its receptor. Detailed reviews have been published recently elsewhere (1–3).

IL-10 was initially characterized in 1988–1989 at DNAX Research Institute as the "cytokine synthesis inhibitory factor" (CSIF) by Mosmann and colleagues in their studies of the regulation of mouse T-helper-cell subset (TH_1 and TH_2) functions. An activity produced by mouse TH_2 clones was identified that inhibited cytokine production by activated TH_1 clones (4). cDNA clones encoding both mouse and human IL-10 (mIL-10, hIL-10) were reported a short while later, and their relationship to an EBV gene was discovered (5,6). It was soon established that IL-10's inhibition of cytokine synthesis by both T cells (7,8) and NK cells (9) was mediated via its action on the macrophage/monocyte (but not B-cell) antigen-presenting or accessory cell. Because of its ability to inhibit activation of cells of the macrophage/monocyte and dendritic cell lineages, IL-10 was termed a "macrophage-deactivating factor" (2,10).

With the availability of recombinant IL-10 for in vitro and in vivo experimentation, progress was rapid, leading to identification of additional activities of the cytokine on mast cells, T cells, and B cells (11–15) and to in vivo confirmation of many of the activities discovered in vitro, as will be discussed below. Particularly illuminating have been studies of IL-10–deficient mice, which develop severe, chronic inflammatory bowel disease and exhibit other unregulated inflammatory responses (16–18). This latter work has suggested that a principal role of IL-10 in vivo is to limit inflammatory responses.

Investigations of the functions of cellular IL-10 (cIL-10) and vIL-10 led to the characterization and cloning of IL-10 receptors and their relationship to the interferon receptor family (19–21), as well as the beginnings of elucidation of the mechanisms of IL-10 signal transduction (22–25). Because of its inhibitory effects on macrophage activation, IL-10 offers an excellent opportunity for studying "cross-talk" between stimulatory and inhibitory cytokines at the level of signal transduction, an aspect of cytokine biology that has so far been largely neglected.

BIOCHEMISTRY AND MOLECULAR BIOLOGY OF IL-10

Cellular IL-10

The mIL-10, rat IL-10, and hIL-10 amino acid sequences were obtained from cDNA clone sequences (5,6,26). A recent version of the Genbank database also contains entries for sheep, pig, red deer, pigtailed macaque, and mangabey and rhesus monkey IL-10 cDNA sequences. The mIL-10 and hIL-10 cDNA clones encode 178-amino-acid open reading frames, which are 73% identical. These sequences reveal IL-10 to be a member of the α-helical bundle cytokine superfamily (27).

Recombinant hIL-10 is an 18 kDa polypeptide, which is not glycosylated. In contrast, both recombinant and T-cell–derived mIL-10 are produced as a mixture of 17, 19, and 21 kDa polypeptides due to heterogeneous N-glycosylation at a site near the N-terminus (5,28). hIL-10 is active on both mouse and human cells, but mIL-10 is not active on human cells. At least one monoclonal antibody cross-reacts with mIL-10 and hIL-10 (21).

Early evidence suggested that mIL-10 and hIL-10 were expressed as acid-sensitive noncovalent homodimers (4,19,29). This has been borne out recently by an x-ray structure of hIL-10 (30), which was interpreted as a dimer of two interpenetrating polypeptide chains. It is not conclusively established whether the monomeric cytokine is biologically active. Both mIL-10 and hIL-10 contain two intrachain disulfide bonds, the former having a fifth cysteine, which remains unpaired. Reduction of the disulfide bonds abolished the cytokine's activity (29).

The mIL-10 gene is composed of five exons arrayed over ~ 5.5 kb of genomic DNA, and the m,hIL-10 genes are located on mouse and human chromosome 1 (31). The human IL-10 gene has a similar structure (R. de Waal Malefyt, unpublished). Transcription of the IL-10 gene appears to generate primarily a single mRNA of ~ 1.4 kb (mIL-10) or ~ 2 kb (hIL-10), although a shorter mRNA was reported in a mouse T-helper 2 (TH_2) clone (5,6). Several possible transcriptional regulatory sequence elements were identified in the mIL-10 gene, which together resembled those found in IL-6 genes (31), and as described later, the pattern of IL-10 expression is not unlike that of IL-6. Comparatively little is known about regulation of IL-10 transcription. However, mIL-10 production stimulated by LPS in vivo was not inhibited by cyclosporin, while anti-CD3 stimulated IL-10 expression was (32), suggesting that stimulation of T-cell IL-10 production (anti-CD3) could be inhibited by cyclosporin, whereas monocyte IL-10 expression (LPS) could not. These data suggested different cell type–specific regulation mechanisms of IL-10 expression.

Viral IL-10 Homologs

The EBV and EHV2 genomes contain viral homologs of IL-10 (5,6,33) that lack introns. Homology to the mammalian cytokines is restricted mainly to the mature protein and is far less marked or nonexistent in the leader sequence or flanking untranslated regions. The hIL-10 and EBV (vIL-10) mature protein sequences are 84% identical, with most differences found in the *N*-terminal 20 amino acids. Not surprisingly, most anti–hIL-10 antibodies cross-react with vIL-10 (1,34). No equine IL-10 sequence is available for comparison, but the EHV2 IL-10 homolog also differs most from the mouse, rat, and human proteins in the *N*-terminal region (33). vIL-10 is a 17 kDa nonglycosylated polypeptide, which expresses some, but not all of the activities of IL-10, although it is 3- to 10-fold less potent than the cellular cytokine in those systems where it is active (2,6,11,12,35).

CELLULAR SOURCES OF IL-10

IL-10 was initially identified as a product of antigen-stimulated murine TH_2 T-cell clones (TCC), which was not expressed by similarly activated Th1 TCC (4). Murine IL-10 has since been shown to be produced by peripheral CD4+ TH_2 populations in a number of situations, particularly during chronic infectious disease states (3,36), as well as by CD8+ T cells, which exhibit a TH_2-type pattern of cytokine secretion (37–39). In murine systems IL-10 production often correlates with production of other TH_2 cytokines, such as IL-4, but in humans IL-10 is produced by both TH_1- and TH_2-type CD4+ TCC (40). In addition to peripheral T-cell populations, IL-10 is produced by activated fetal thymocytes (12). IL-10 is also secreted upon activation by a vast array of other cell types including activated human monocytes (34) and B cells (especially those infected with EBV) (41,42), activated murine macrophages (43), and splenic and peritoneal B cells (44). Among B cells, IL-10 production was significantly enriched within the CD5+ subset (45). IL-10 mRNA has also been detected in mouse mast cell lines (5). Several nonhematopoietic sources of IL-10 have been identified including mouse keratinocytes (46,47) and melanoma cell lines (48).

BIOLOGICAL ACTIVITIES

IL-10 In Vitro

CSIF Activity

Inhibition of TH_1 TCC proliferation and cytokine synthesis was the activity used to originally characterize IL-10.

While IL-10 inhibited production of all TH_1 cytokines measured to some extent, its effect was most pronounced on IL-3 and IFN-γ synthesis, both of which are synthesized late after T-cell activation (4). IL-10 has since been shown to inhibit proliferation and cytokine synthesis by normal populations of mouse and human T cells responding to antigen, mitogen, or alloantigen (8,49–52) as well as production of IFN-γ by NK cells (9). The effects on T-cell proliferation can in some systems be overcome by addition of IL-2, suggesting that IL-10 may inhibit T-cell proliferation indirectly via inhibition of IL-2 production (53,54).

Using mouse Th1 TCC as targets, Mosmann and colleagues showed that IL-10 mediated its effects on cytokine synthesis indirectly: IFN-γ production was not inhibited when T-cell clones were stimulated with anti-CD3 or concanavalin A in the absence of APC (4). It was later shown that the CSIF activity of IL-10 was dependent on the type of APC, since IL-10 inhibited IFN-γ production by TCC when macrophages, but not B cells, were used as APC (7,8). Inhibition of IFN-γ production by human NK cells was also mediated via an effect on macrophages (9,55).

Inhibition of the induction of MHC class II expression is thought to play a significant part in the inhibitory effects of IL-10 on the antigen-specific activation of human TCC. IL-10 inhibited spontaneous and IL-4–, IL-13–, or IFN-γ–induced MHC class II expression by cultured human monocytes (34,56). Such an effect has been less readily observed using mouse cells where IL-10 reportedly had no effect on IFN-γ–induced MHC class II levels on a macrophage line or resident peritoneal macrophages (7). While inhibition of MHC class II expression correlates with the CSIF activity of IL-10 in some systems, the fact that IL-10 inhibits T-cell activation that is independent of antigen presentation and processing suggests additional mechanisms.

It is now apparent that the CSIF activity of IL-10 primarily involves inhibition of the costimulatory properties of macrophages for T and NK cell activation. IL-10 inhibits IFN-γ–induced production of a number of monokines, several of which, including TNF-α, IL-1, and IL-12, have been identified as costimulators of T- and NK-cell activation (10,34,43). IL-12 may be particularly important in this regard as it is a critical cofactor for IFN-γ production by T and NK cells (57–60), as well as for the development of TH_1 responses in mouse and human systems (61,62). In addition to inhibiting expression of soluble costimulators of T-cell activation, in mouse and human systems IL-10 inhibits activation-induced expression of the membrane-bound costimulators B7.1 (CD80) (53,63) and B7.2 (CD86) (64), which interact with CTLA4/CD28 on T cells. Thus it seems likely that the inhibition of accessory cell–derived costimulatory signals, such as IL-12, CD80, and CD86, significantly explains the CSIF activity of IL-10.

Effects on Macrophages

In addition to their APC function, macrophages also mediate key effector functions, critical for the host cell-mediated immune response against a variety of invading microorganisms. These include recruitment of immune cells into sites of inflammation, as well as activation of killing mechanisms effective against intracellular and extracellular pathogens, including bacteria, protozoa, and fungi. Many of the functions of macrophages are mediated by the vast array of cytokines and chemokines produced upon activation. IL-10 inhibits synthesis of many of the cytokines and chemokines normally induced in human and murine macrophages by stimulation with LPS or LPS plus IFN-γ. These include IL-1α, IL-1β, GMCSF, TNF, IL-6, IL-8, MIP-1α, IL-10, and IL-12 (10,34,43,55,65). IL-10 has also been shown to inhibit macrophage microbicidal activity by inhibiting IFN-γ–induced nitric oxide synthase and the production of reactive nitrogen (NO) and reactive oxygen (H_2O_2) intermediates, which are involved in macrophage killing mechanisms (66,67). These effects may be indirect, for example, inhibition of NO production by IL-10 is due in part to the blocking of endogenous production of TNF-α, which is required as a cofactor for NO production (68). The inhibition of a number of inflammatory functions of macrophages has led to the description of IL-10 as a macrophage-deactivating factor (10) and undoubtedly plays a major part in the potent anti-inflammatory properties of IL-10 in vivo (see below). In contrast to its effect on the production of most monokines, IL-10 induces production of the IL-1 receptor antagonist (IL-1ra) by human monocytes (34).

IL-10 is synthesized late by activated human monocytes and has been shown to inhibit its own production in an autocrine fashion (34). Such a negative feedback loop may provide an important mechanism to prevent prolonged immune suppression.

Effects on Dendritic Cells

Dendritic cells (DC) are APC which are uniquely potent in the induction of primary T-cell responses (69). IL-10 has been shown to inhibit antigen-, alloantigen-, and mitogen-induced IFN-γ synthesis, but not proliferation, of mouse T cells stimulated in the presence of DC (51,53). Similar results were found with human T cells, however, in these studies proliferation was also inhibited (70,71). Differences may reflect the different sources of DC used in the murine and human studies or the levels of IL-2 induced in the various culture systems. Evidence suggests that IL-10 inhibits the APC function of DC in a similar way to that described for macrophages, via inhibition of the upregulation of the costimulatory molecules CD80 and CD86, and of IL-12 production (53,64,72). That IL-10 can inhibit IFN-γ production in primary responses suggests that the presence of IL-10 may inhibit the development of TH_1 cells; however, in contrast to IL-4, IL-10 has not been shown to induce development of TH_2 cells (73).

Direct Effects on T Cells

Many of the inhibitory effects of IL-10 on T-cell function are mediated indirectly via APC. However, IL-10 does partially inhibit the proliferation of human T-cell clones induced by anti-CD3 or anti-CD2 mAb in the absence of added accessory cells. The mechanism involved inhibition of IL-2 production as the effect was overcome by addition of IL-2 (15,54). In contrast, IL-10 has positive effects on the proliferation of a subset of thymocytes in the mouse, acting as a cofactor, together with IL-2 and IL-4, for their growth (12). In addition, IL-10 in combination with IL-2 stimulated proliferation and differentiation of CD8+ cytotoxic effector cells in the mouse (74).

Effects on Hematopoiesis

IL-10 may contribute to hematopoiesis in the mouse. Although unable to act as a primary growth factor, IL-10 was able to costimulate the growth of many progenitor cell types, including primitive stem cells, megakaryocytes, mast cell precursors, and fetal bone marrow B-cell precursors (13,75,76).

Effects on B Cells

In contrast to macrophages and DC, B cells are refractory to IL-10–induced inhibition of APC function. Indeed, IL-10 mediates a number of stimulatory activities on B cells. In the mouse system, IL-10 enhanced the viability of splenic B cells in culture and increased their expression of class II MHC molecules (11). More extensive studies have been reported in the human system, where IL-10 is a cofactor for B-cell growth and differentiation (14,77,78). IL-10 stimulated enhanced proliferation and the secretion of IgM, IgA, IgG1, and IgG3 by anti-CD40–stimulated resting B cells, suggesting that, in combination with stimulation through CD40, IL-10 may be a switch factor for IgG1 and IgG3 isotypes (79).

Viral IL-10

Activities of Viral IL-10

vIL-10/BCRF1 exhibits many of the properties of hIL-10, including CSIF and macrophage-deactivating factor activity on mouse and human cells (6,8,9,34,35,80), and the known activities of hIL-10 on human B cells (14,77). Like mIL-10, the viral cytokine enhanced viability of mouse B cells in culture (11).

However, vIL-10 appears to lack several of the activities of mIL-10 and hIL-10 on mouse cells. vIL-10 neither effectively enhanced class II MHC expression on mouse B cells (11) nor costimulated mouse thymocyte or mast cell proliferation (6,12). In these systems, we could not demonstrate antagonism of hIL-10 activity by vIL-10 (1), and as noted below, vIL-10 has at least 100- to 1000-fold lower affinity than hIL-10 for the cloned IL-10 receptor (IL-10R) (20,21). These differences have been described so far only on mouse cells, and analogous differences on human cells have not yet been reported. However, the lack of vIL-10 activity in certain assays suggested the existence of different IL-10 receptor and/or signal transduction complexes on vIL-10–responsive and –nonresponsive cells and that vIL-10 has conserved only a subset of activities of the cellular cytokine. These in vitro data recently received striking confirmation in vivo in experiments in which immunogenic and allogeneic mouse tumor cells were infected with recombinant retroviruses containing IL-10 or vIL-10 cDNAs (81). Rejection of mIL-10–expressing tumors was significantly accelerated compared to controls, while either vIL-10–expressing cells were not rejected at all or the rejection kinetics were greatly retarded.

vIL-10 and the EBV Life Cycle

What advantage might have been conferred on the ancestor of EBV when it gained its own IL-10 gene? vIL-10 is expressed in the late phase of the lytic cycle (82), and vIL-10's ability to inhibit activation of virus-specific T and NK cells might be crucial at this time, when viral proteins and infectious virus are produced and fresh B cells become infected. Moreover, after chronic, latent infection is established, the ability of vIL-10 to enhance antiviral cytotoxic activity (83) might help maintain a balance between latency and lytic infection.

Transformation of human B cells by EBV involves a number of viral gene products, but the activities of vIL-10 on human B cells suggest a potential contribution to this mechanism. By virtue of its B-cell–stimulating activities, vIL-10 could enhance both the numbers and susceptibility to infection/transformation of B lymphocytes: B-cell transformation by EBV induced production of hIL-10, and outgrowth of EBV-transformed cells in vitro was delayed by the presence of an antiserum, which blocked hIL-10 (and presumably also vIL-10) activity (42). However, the exact role of vIL-10 in this process has been difficult to assess. One group (84) reported that vIL-10 was expressed as early as 4 h after viral infection, with hIL-10 transcription following at 12–18 h. Antisense vIL-10 (but not control vIL-10 sense or hIL-10 antisense) oligonucleotides blocked B-cell transformation and EBV-induced Ig secretion in their experiments. The inhibitory effects of antisense vIL-10 on transformation were reversed by addition of exogenous hIL-10 to the cultures. Taken together, these results implied roles for both vIL-10 and infection-induced hIL-10 in transformation of B cells by EBV. In contrast, another report showed that vIL-10–deleted EBV were not impaired in their ability to transform B cells in vitro (85). However, cells infected with the vIL-10^- EBV were, unlike wild-type virus-infected cells, unable to block IFN-γ production by autologous human peripheral blood cells. Despite the different findings, it is likely that some combination of the known activities of vIL-10 likely confers a number of adaptive advantages on EBV in its interaction with the host immune system. It is thus possible that acquisition of the vIL-10 gene by an ancestral virus might have been an important step in the evolution of EBV from pathogenic organism to a benign viral parasite.

IL-10 In Vivo

Consistent with its inhibitory effects on a number of in vitro assays of cell-mediated immunity, IL-10 has been shown to be a potent inhibitor of inflammatory responses in vivo. Studies in IL-10–deficient mice or mice treated with anti–IL-10 mAb have identified this cytokine as a key molecule produced by the host to limit the severity of inflammatory responses, especially those mediated by the mononuclear phagocytic cell lineage.

Cutaneous Inflammation

IL-10 inhibited several models of cutaneous inflammation, including delayed-type hypersensitivity (DTH) and contact hypersensitivity (CHS), which are classical experimental examples of inflammatory cell-mediated immunity. DTH and CHS are complex immune reactions, which are T-cell dependent and require prior sensitization. DTH has been shown to be mediated by TH_1 cells (86). Injection of rIL-10 inhibited the sensitization phase of DTH (46) as well as the elicitation of CHS (87) and DTH in sensitized mice (88,89) indicating that IL-10 is capable of inhibiting both the priming and effector function of TH_1 responses in vivo. Endogenous IL-10 plays an important role in regulating these inflammatory responses as IL-10–deficient mice exhibited enhanced and prolonged CHS. This was particularly evident in response to the chemical irritant croton oil. IL-10–deficient mice mounted an enhanced response and developed severe tissue damage, which did not occur in control mice (17). The relevant source of IL-10 in vivo and the target(s) of its action are not known, although there is evidence in some systems that a reduction in TNF-α production may be an important factor (17).

Endotoxin- and Enterotoxin-Induced Shock

Infusion of mice with the bacterial cell wall product lipopolysaccharide (LPS) induces a state of endotoxic shock, which is used as a model of shock induced by infection with gram-negative bacteria. Induction from macrophages of a number of inflammatory mediators, particularly, TNF-α, is thought to account for the toxicity of LPS (90,91). Injection of rIL-10 prior to LPS sensitization protected mice from a lethal challenge of LPS (92,93). Host-derived IL-10 limits the sensitivity of mice to LPS, as a 20-fold lower dose was required to induce death in IL-10–deficient mice than in normal controls (18). IL-10 also protected mice from lethal staphyloccocal enterotoxin B–induced shock, indicating that IL-10 can inhibit superantigen-induced T-cell activation in vivo, as reported in vitro (94).

Regulation of Intestinal Immune Responses

The gastrointestinal tract is one of the major sources of antigenic stimulation to the host, yet intestinal immune responses are notable for the lack of cell-mediated immune responses to these antigens. IL-10 appears to play an important role in this natural immunoregulatory mechanism, because IL-10–deficient mice developed severe inflammation in the large intestine characterized by leukocytic infiltrates, goblet cell depletion, and epithelial cell hyperplasia (16). Disease did not appear to be due to an infectious agent, rather a dysregulated inflammatory response to components of the enteric flora. In addition, administration of rIL-10 completely prevented the development of a TH_1-mediated colitis induced in severe combined immunodeficient (SCID) mice by transfer of a subset of normal CD4+ T cells (95). These results illustrate the importance of IL-10 in regulation of inflammatory responses in the colon and suggest its therapeutic utility for the treatment of IBD in humans.

Autoimmunity and Tolerance

Administration of rIL-10 inhibited the onset of diabetes in a murine model of insulin-dependent diabetes, which develops spontaneosly in nonobese diabetic (NOD) mice (96). Mice were still significantly protected from disease 1 month after cessation of treatment, suggesting that IL-10 may have induced a state of tolerance in the diabetes-prone mice. In contrast, the incidence of diabetes was enhanced in transgenic NOD mice expressing IL-10 under control of the rat insulin promoter (97). IL-10 has been shown to be chemotactic in vivo (98), and the different results of the two studies may reflect exposure to high levels of IL-10 at different times during development of the autoimmune response. IL-10 has also been shown to play a protective role in experimental allergic encephalomyelitis (EAE),

TH_1-mediated autoimmune disease induced in rats and mice by immunization with guinea pig myelin basic protein in adjuvant. Administration of rIL-10 prevented disease development in rats (99), and increases in IL-10 mRNA correlated with recovery from disease (100), suggesting that endogenous IL-10 may play a role in the natural recovery phase of the disease.

There is evidence that IL-10 may be involved in transplantation tolerance. SCID patients transplanted with allogeneic fetal hematopoietic stem cells developed a chimerism in which the T cells were of donor and the B cells and monocytes of host origin. Despite the presence of antihost T cells, tolerance toward host and donor was maintained, and these patients showed no signs of graft-versus-host disease (GVHD). Unusually high levels of IL-10 were produced in vitro by donor antihost T cells and monocytes, suggesting that IL-10 may be involved in this mechanism of tolerance in vivo (101).

The role of IL-10 in the immune suppression associated with pregnancy has recently been investigated. In the mouse, the CBA × DBA/2 mating combination is associated with high rates of spontaneous fetal resorption. Placentae from these mice were shown to be deficient in IL-4 and IL-10 production compared with mating combinations exhibiting a low rate of fetal resorption. The high rate of fetal resorption was completely reversed by administration of rIL-10. Furthermore, agents known to counteract fetal resorption enhanced placental IL-4 and IL-10 production (102). Based on these results it was postulated that placentally produced anti-inflammatory cytokines, such as IL-10, play a vital role in the survival of the fetal allograft, probably by counteracting deleterious inflammatory cytokines.

Given the potent activities of IL-10 as a B-cell growth and differentiation factor, IL-10 may be expected to be deleterious in autoimmune diseases involving B-cell hyperreactivity, such as systemic lupus erythematosus. Evidence from a murine model of SLE, which develops spontaneously in NZB/W F1 mice, supports this notion. Treatment of these mice from birth with anti–IL-10 delayed the onset of autoimmunity, whereas injection of rIL-10 accelerated disease (103).

Infectious Disease

A number of infectious diseases such as lepromatous leprosy, filiriasis, visceral leishmaniasis, and acquired immunodeficiency syndrome (AIDS) are characterized by a lack of cell-mediated immunity to the relevant pathogen. There is some evidence that overproduction of IL-10 may be involved in the immunosuppression associated with these diseases, as in some cases antigen-induced proliferative responses in vitro could be restored by addition of anti–IL-10 to cultures of peripheral blood lymphocytes from patients (104–107). A similar immunosuppressive role for IL-10 has been demonstrated in a number of experimental parasitic infections in mice (108). Neutralization of IL-10 in vitro led to enhanced antigen specific IFN-γ production by CD4+ T cells from mice infected with *Leishmania major* (52) or *Schistosoma mansoni* (109), and injection of anti–IL-10 mAb prevented acute disease and death in mice normally susceptible to infection with *Trypanosoma cruzi* (110). The host antimicrobial response is heavily dependent on IFN-γ–induced defense mechanisms, especially induction of macrophage microbicidal activity. Based on in vitro studies, it seems likely that IL-10 acts in vivo to antagonize protective cell-mediated immunity at two levels: by inhibiting production of IFN-γ by T cells and NK cells and by antagonizing IFN-γ–induced macrophage activation.

Therapeutic Utility of IL-10

The wide range of biological activities of IL-10 in vitro and in vivo suggest numerous clinical applications. The potent anti-inflammatory properties of IL-10 discussed above make IL-10 a candidate for the treatment of a number of autoimmune diseases with a Th1/macrophage-mediated pathogenesis, including GVHD, sepsis, diabetes, IBD, multiple sclerosis, psoriasis, and rheumatoid arthritis. Conversely, antagonists of IL-10, including anti–IL-10 mAb, anti-IL-10 receptor mAb, or receptor antagonists may reverse the suppression of cell-mediated immunity and promote activation of host defense mechanisms in a number of infectious diseases, including lepromatous leprosy, visceral leishmaniasis, and human immunodeficiency virus (HIV) infection. In addition, IL-10 may be useful for the treatment of B-CLL, given its reported ability to induce apoptosis of B-CLL cells in vitro (111).

Initial studies with IL-10 in patients appear promising in this regard. In a phase I study in healthy humans, IL-10 at doses as high as 25 μg/kg caused no adverse symptoms. Administered IL-10 inhibited cytokine production: whole blood production of IL-6 in response to stimulation with LPS in vitro was significantly reduced in IL-10–treated patients compared with placebo-treated controls (112). In another study, topical administration of IL-10 to patients with ulcerative colitis led to substantial inhibition of in vitro–induced IL-1 and TNF-α production by peripheral blood monocytes and by lamina propria mononuclear cells, indicating that in vivo administration of IL-10 in humans is capable of inhibiting the local and systemic production of proinflammatory cytokines (113).

IL-10 RECEPTORS

The biological effects of IL-10 are mediated by interaction with cell-surface receptors. The ligand-binding subunit of mouse and human receptors for IL-10 (mIL-10R; hIL-10R) has been identified, characterized, and cloned (19–21). Binding affinity was high, with Kd in the 35–250 pM range. Cross-linking studies with ^{125}I-hIL-10 indicated a molecular size of 90–120 kDa for hIL-10R, although a number of ^{125}I-labeled bands were detected in one study (19), suggesting proteolytic degradation or the presence of multiple polypeptides cross-linked to labeled IL-10.

IL-10R cDNA clones were isolated from mouse mast cell and macrophage lines (20) and from a human B-cell line (21), cell types that respond to m,hIL-10 (only macrophages and B cells respond to vIL-10). Despite differences in the m,hIL-10/vIL-10 responses of the parent cells, IL-10R encoded by the mast cell and macrophage-derived cDNA clones were identical. hIL-10R cDNA clones exhibited 70% DNA sequence homology to the mIL-10R cDNA. The mIL-10R and hIL-10R cDNAs encode open reading frames of 576 and 578 amino acids, respectively. The predicted hIL-10R and mIL-10R amino acid sequences are 60% identical. hIL-10 bound to m,hIL-10R, and mIL-10 bound only to mIL-10R, as expected from the known species specificity of IL-10 (1). mIL-10R and hIL-10R mRNA was detected in all IL-10–responsive cells examined and in most hemopoietic cell lines, but not in the few nonhemopoietic cells tested. A recombinant soluble hIL-10R has been described that exhibits a degree of antagonist activity in vitro (114). Evidence suggested that the hIL-10/shIL-10R complex was multimeric, consisting of two ligand dimers bound to up to four shIL-10R molecules.

Recombinant IL-10R function was studied utilizing a mouse pro-B cell line (Ba/F3) transfected with recombinant IL-10R cDNAs (20,21). m,hIL-10R bound ^{125}I-hIL-10 with high affinity, however, affinity for vIL-10 was at least 100- to 1000-fold lower (2,20–22). The mIL-10R– and hIL-10R–expressing Ba/F3 cell lines both exhibited a short-term proliferative response to IL-10 (20–22). In addition to proliferation, mIL-10R$^+$ Ba/F3 transfectants newly expressed a number of cell-surface antigens when cultured in IL-10. Site-directed mutagenesis and deletion analysis of the cytoplasmic domain of mIL-10R showed that different regions of the cytoplasmic domain are responsible for the various biological responses (22).

mIL-10R and hIL-10R are structurally related to the interferon receptor (IFNR) family. This structural relationship is consistent with the resemblance of the IL-10 structure to IFNγ (30) and is of special interest because IL-10 and IFNγ antagonize each others' functions and production in several systems (1,115), yet have some similar effects, such as induction of FcγR expression. These findings suggest the possibility of interaction between the signaling pathways of IL-10R and IFNR. Because of the multisubunit nature of IFN-γR and IFN-αβR, as well as the existence of vIL-10–responsive and nonresponsive cells, it is likely that IL-10R will also be a complex of the known subunit and at least one additional component. The possibility that IL-10R might share a second receptor subunit with an IFNR seems unlikely at present because neither mIFN-γ nor mIFN-α completed for mIL-10 binding to mIL-10R (20), and up to a 100-fold excess of hIL-10 did not complete for binding to hIFN-γR on a human monocyte line, U937 (21).

An additional possibility is that the IL-10 and IFN signal transduction pathways may share signaling molecules or that the IL-10/IL-10R interaction could directly antagonize the IFNR signal transduction pathway, perhaps by interacting with or sequestering one or more of its component proteins. Like IFN, IL-10 activates STAT family transcription factors (23–25,116,117), although activation of STAT1α and STAT3 by IL-10 is different from the activation of STAT1α only by IFN-γ, and STAT1αβ and STAT2 by IFN-α. Binding of IL-10 to IL-10R also induces tyrosine phosphorylation of the Janus kinases Jak1 and Tyk2 (22,25). At present little else is known about IL-10 signal transduction, except that activation of the raf/ras/MAP kinase cascade likely does not occur in response to IL-10 and may be inhibited in some cases (118).

REFERENCES

1. Moore KW, O'Garra A, de Waal Malefyt R, et al. Interleukin-10. Ann Rev Immunol 1993; 11:165–190.
2. Ho AS-Y, Moore KW. Interleukin-10 and its receptor. Ther Immunol 1994; 1:173–185.
3. Mosmann T. Properties and functions of interleukin-10. Adv Immunol 1994; 56:1–26.
4. Fiorentino DF, Bond MW, Mosmann TR. Two types of mouse helper T cell. IV. Th2 clones secrete a factor that inhibits cytokine production by Th1 clones. J Exp Med 1989; 170:2081–2095.
5. Moore KW, Vieira P, Fiorentino DF, et al. Homology of cytokine synthesis inhibitory factor (IL-10) to the Epstein Barr virus gene BCRFI. Science 1990; 248:1230–1234.
6. Vieira P, de Waal-Malefyt R, Dang M-N, et al. Isolation and expression of human cytokine synthesis inhibitory factor (CSIF/IL10) cDNA clones: homology to Epstein-Barr virus open reading frame BCRFI. Proc Natl Acad Sci USA 1991; 88:1172–1176.
7. Fiorentino DF, Zlotnik A, Vieira P, et al. IL-10 acts on the antigen-presenting cell to inhibit cytokine production by Th1 cells. J Immunol 1991; 146:3444–3451.
8. de Waal Malefyt R, Haanen J, Spits H, et al. IL-10 and viral IL-10 strongly reduce antigen-specific human T cell proliferation by diminishing the antigen-presenting capacity of monocytes via downregulation of class II MHC expression. J Exp Med 1991; 174:915–924.

9. Hsu D-H, Moore KW, Spits H. Differential effects of interleukin-4 and -10 on interleukin-2-induced interferon-γ synthesis and lymphokine-activated killer activity. Int Immunol 1992; 4:563–569.
10. Bogdan C, Vodovotz Y, Nathan C. Macrophage deactivation by interleukin 10. J Exp Med 1991; 174:1549–1555.
11. Go NF, Castle BE, Barrett R, et al. Interleukin 10 (IL-10), a novel B cell stimulatory factor: unresponsiveness of X chromosome-linked immunodeficiency B cells. J Exp Med 1990; 172:1625–1631.
12. MacNeil I, Suda T, Moore KW, et al. IL-10: a novel cytokine growth cofactor for mature and immature T cells. J Immunol 1990; 145:4167–4173.
13. Thompson-Snipes L, Dhar V, Bond MW, et al. Interleukin-10: a novel stimulatory factor for mast cells and their progenitors. J Exp Med 1991: 173:507–510.
14. Rousset F, Garcia E, Defrance T, et al. IL-10 is a potent growth and differentiation factor for activated human B lymphocytes. Proc Natl Acad Sci USA 1992; 89:1890–1893.
15. de Waal Malefyt R, Yssel H, de Vries JE. Direct effects of IL-10 on subsets of human CD4+ T cell clones and resting T cells. J Immunol 1993; 150:4754–4765.
16. Kuhn R, Lohler J, Rennick D, et al. Interleukin-10 deficient mice develop chronic enterocholitis. Cell 1993; 75:263–274.
17. Berg DJ, Leach MW, Kuhn KR, et al. Interleukin 10 but not interleukin 4 is a natural suppressant of cutaneous inflammatory responses. J Exp Med 1995; 182:99–108.
18. Berg DJ, Kuhn R, Rajewsky K, et al. Interleukin-10 is a central regulator of the response to LPS in murine models of endotoxic shock and the Schwartzman reaction but not endotoxin tolerance. J Clin Invest 1995; 96:2339–2347.
19. Tan JC, Indelicato S, Narula SK, et al. Characterization of interleukin-10 receptors on human and mouse cells. J Biol Chem 1993; 268:21053–21059.
20. Ho AS-Y, Liu Y, Khan TA, et al. A receptor for interleukin-10 is related to interferon receptors. Proc Natl Acad Sci USA 1993; 90:11267–11271.
21. Liu Y, Wei SH-Y, Ho AS-Y, et al. Expression cloning and characterization of a human interleukin-10 receptor. J Immunol 1994; 152:1821–1829.
22. Ho AS-Y, Wei SH-Y, Mui AL-F, et al. Functional regions of the mouse IL-10 receptor cytoplasmic domain. Mol Cell Biol 1995; 15:5043–5053.
23. Weber-Nordt RM, Meraz MA, Schreiber RD. LPS-dependent induction of IL-10 receptor expression on murine fibroblasts. J Immunol 1994; 153:3734–3744.
24. Weber-Nordt RM, Riley JK, Greenlund AC, et al. Stat3 recruitment by two distinct ligand-induced tyrosine phosphorylated docking sites in the IL-10 receptor intracellular domain.
25. Finbloom DS, Winestock KD. IL-10 induces the tyrosine phosphorylation of tyk2 and Jak1 and the differential assembly of STAT1α and STAT3 complexes in human T cells and monocytes. J Immunol 1995; 155:1079–1090.
26. Goodman RE, Oblak J, Bell RG. Synthesis and characterization of rat interleukin-10 (IL-10) cDNA clones from the RNA of cultures OX8-OX22- thoracic duct T cells. Biochem Biophys Res Commun 1992; 189:1–7.
27. Sprang SR, Bazan JF. Cytokine structural taxonomy and mechanisms of receptor engagement. Curr Opin Struct Biol 1993; 3:815–827.
28. Mosmann TR, Schumacher J, Fiorentino DF, et al. Isolation of monoclonal antibodies specific for IL4, IL5, IL6, and a new Th2-specific cytokine (IL-10), cytokine synthesis inhibitory factor, by using a solid phase radioimmunoadsorbent assay. J. Immunol 1990; 145:2938–2945.
29. Windsor WT, Syto R, Tsarbopoulos A, et al. Disulfide bond assignments and secondary structure analysis of human and murine interleukin 10. Biochemistry 1993; 32:8807–8815.
30. Zdanov A, Schalk-Hihi C, Gustchina A, et al. Crystal structure of interleukin-10 reveals the functional dimer with an unexpected topological similarity to interferon γ. Structure 1995; 3:591–601.
31. Kim JM, Brannan CI, Copeland NG, et al. Structure of the mouse interleukin-10 gene and chromosomal localization of the mouse and human genes. J Immunol 1992; 148:3618–3623.
32. Durez P, Abramowicz D, Gerard C, et al. In vivo induction of interleukin-10 by anti-CD3 monoclonal antibody or bacterial lipopolysaccharide: differential modulation by cyclosporin A. J Exp Med 1993; 177:551–555.
33. Rode H-J, Janssen W, Rosen-Wolff A, et al. The genome of equine herpesvirus type 2 harbors an interleukin-10 (IL-10)-like gene. Virus Genes 1993; 7:111–116.
34. de Waal Malefyt R, Abrams J, Bennett B, et al. IL-10 inhibits cytokine synthesis by human monocytes: an autoregulatory role of IL-10 produced by monocytes. J Exp Med 1991; 174:1209–1220.
35. Hsu D-H, de Waal Malefyt R, Fiorentino DF, et al. Expression of IL-10 activity by Epstein-Barr Virus Protein BCRFI. Science 1990; 250:830–832.
36. Sher A, Gazzinelli RT, Oswald IP, et al. Role of T-cell derived cytokines in the downregulation of immune responses in parasitic and retroviral infection. Immunol Rev 1992; 127:183–204.
37. Salgame P, Yamamura M, Bloom BR, et al. Evidence for functional subsets of CD4+ and CD8+ T cells in human disease: lymphokine patterns in leprosy. Chem Immunol 1992; 54:44–59.
38. Sad S, Marcotte R, Mosmann TR. Cytokine-induced differentiation of precursor mouse CD8+ T cells into cytotoxic CD8+ T cells secreting Th1 or Th2 cytokines. Immunity 1995; 2:271–279.
39. Le Gros G, Erard F. Non-cytotoxic, IL-4, IL-5, IL-10 producing CD8+ T cells: their activation and effector functions. Curr Opin Immunol 1994; 6:453–457.
40. Yssel H, De Waal Malefyt R, Roncarolo MG, et al. IL-10 is produced by subsets of human CD4+ T cell clones and peripheral blood T cells. J Immunol 1992; 149:2378–2384.
41. Benjamin D, Knoblach TJ, Dayton MA. Human B-cell interleukin 10: B cell lines derived from patients with AIDS and Burkitt's lymphoma constitutively secrete large quantities of interleukin 10. Blood 1992; 80:1289–1298.
42. Burdin N, Perrone C, Banchereau J, et al. Epstein-Barr virus transformation induces B lymphocytes to produce human interleukin-10. J Exp Med 1993; 177:295–304.
43. Fiorentino DF, Zlotnik A, Mosmann TR, et al. IL-10 inhibits cytokine production by activated macrophages. J Immunol 1991; 147:3815–3822.
44. O'Garra A, Stapleton G, Dhar V, et al. Production of cytokines by mouse B cells: B lymphomas and normal B cells produce interleukin 10. Int Immunol 1990; 2:821–832.
45. O'Garra A, Chang R, Go N, et al. Ly-1 B (B-1) cells are the main source of B cell-derived interleukin 10. Eur J Immunol 1992; 22:711–717.
46. Rivas JM, Ullrich SE. Systemic suppression of delayed-type hypersensitivity by supernatants from UV-irradiated keratinocytes. An essential role for keratinocyte-derived IL-10. J Immunol 1992; 149:3865–3871.
47. Enk AH, Katz SI. Identification and induction of keratinocyte-derived IL-10. J Immunol 1992; 149:92–95.
48. Chen Q, Daniel V, Maher DW, et al. Production of IL-10 by

melanoma cells: examination of its role in immunosuppression mediated by melanoma. Int J Cancer 1994; 56:755–760.
49. Del Prete G, De Carli M, Almerigogna F, et al. Human IL-10 is produced by both type 1 helper (Th1) and type 2 helper (Th2) T cell clones and inhibits their antigen-specific proliferation and cytokine production. J Immunol 1993; 150:353–360.
50. Bejarano MT, de Waal Malefyt R, Abrams JS, et al. Interleukin 10 inhibits allogeneic proliferative and cytotoxic T cell responses generated in primary mixed lymphocyte cultures. Int Immunol 1992; 4:1389–1397.
51. Macatonia SE, Doherty TM, Knight SC, et al. Differential effect of IL-10 on dendritic cell-induced T cell proliferation and IFN-gamma production. J Immunol 1993; 150:3755–3765.
52. Powrie F, Coffman RL. Inhibition of cell-mediated immunity by IL4 and IL10. Res Immunol 1993; 144:639–643.
53. Ding L, Shevach EM. IL-10 inhibits mitogen-induced T cell proliferation by selectively inhibiting macrophage costimulatory function. J Immunol 1992; 148:3133–3139.
54. Taga K, Tosato G. IL-10 inhibits human T cell proliferation and IL-2 production. J Immunol 1992; 148:1143–1148.
55. D'Andrea A, Aste-Amezaga M, Valiante NM, et al. Interleukin 10 (IL-10) inhibits human lymphocyte interferon gamma production by suppressing natural killer cell stimulatory factor/ IL-12 synthesis in accessory cells. J Exp Med 1993; 178:1041–1048.
56. de Waal Malefyt R, Figdor CG, Huijbens R, et al. Effects of IL-13 on phenotype, cytokine production, and cytotoxic function of human monocytes. Comparison with IL-4 and modulation by IFN-gamma or IL-10. J Immunol 1993; 151:6370–6381.
57. Kobayashi M, Fitz L, Ryan M, et al. Identification and purification of natural killer cell stimulatory factor (NKSF), a cytokine with multiple biologic effects on human lymphocytes. J Exp Med 1989; 170:827–845.
58. Chan SH, Perussia B, Gupta JW, et al. Induction of interferon gamma production by natural killer cell stimulatory factor: characterization of the responder cells and synergy with other inducers. J Exp Med 1991; 173:869–879.
59. Murphy EE, Terres G, Macatonia SE, et al. B7 and interleukin 12 cooperate for proliferation and interferon gamma production by mouse T helper clones that are unresponsive to B7 costimulation. J Exp Med 1994; 180:223–231.
60. Kubin M, Kamoun M, Trinchieri G. Interleukin 12 synergizes with B7/CD28 interaction in inducing efficient proliferation and cytokine production of human T cells. J Exp Med 1994; 180:211–222.
61. Hsieh CS, Macatonia SE, Tripp CS, et al. Development of TH1 CD4+ T cells through IL-12 produced by Listeria-induced macrophages. Science 1993; 260:547–549.
62. Manetti R, Gerosa F, Giudizi MG, et al. Interleukin 12 induces stable priming for interferon gamma (IFN-gamma) production during differentiation of human T helper (Th) cells and transient IFN-gamma production in established Th2 cell clones. J Exp Med 1994; 179:1293–1283.
63. Willems F, Marchant A, Delville JP, et al. Interleukin-10 inhibits B7 and intercellular adhesion molecule-1 expression on human monocytes. Eur J Immunol 1994; 24:1007–1009.
64. Kawamura T, Furue M. Comparative analysis of B7-1 and B7-2 expression in Langerhans cells: differential regulation by T helper type 1 and T helper type 2 cytokines. Eur J Immunol 1995; 25:1913–1917.
65. Kubin M, Chow JM, Trinchieri G. Differential regulation of interleukin-12 (IL-12), tumor necrosis factor alpha, and IL-1 beta production in human myeloid leukemia cell lines and peripheral blood mononuclear cells. Blood 1994; 83:1847–1855.
66. Gazzinelli RT, Oswald IP, James SL, et al. IL-10 inhibits parasite killing and nitrogen oxide production by IFN-gamma-activated macrophages. J Immunol 1992; 148:1792–1796.
67. Cunha FQ, Moncada S, Liew FY. Interleukin-10 (IL-10) inhibits the induction of nitric oxide synthase by interferon-gamma in murine macrophages. Biochem Biophys Res Commun 1992; 182:1155–1159.
68. Oswald IP, Wynn TA, Sher A, et al. Interleukin 10 inhibits macrophage microbicidal activity by blocking the endogenous production of tumor necrosis factor alpha required as a costimulatory factor for interferon gamma-induced activation. Proc Natl Acad Sci USA 1992; 89:8676–8680.
69. Steinman RM. The dendritic cell system and its role in immunogenicity. Annu Rev Immunol 1991; 9:271–296.
70. Enk AH, Angeloni VL, Udey MC, et al. Inhibition of Langerhans cell antigen-presenting function by IL-10. A role for IL-10 in induction of tolerance. J Immunol 1993; 151:2390–2398.
71. Caux C, Massacrier C, Vanbervliet B, et al. Interleukin 10 inhibits T cell alloreaction induced by human dendritic cells. Int Immunol 1994; 6:1177–1185.
72. Macatonia SE, Hosken NA, Litton M, et al. Dendritic cells produce IL-12 and direct the development of Th1 cells from naive CD4+ T cells. J Immunol 1995; 154:5071–5079.
73. Hsieh CS, Heimberger AB, Gold JS, et al. Differential regulation of T helper phenotype development by interleukins 4 and 10 in an alpha beta T-cell-receptor transgenic system. Proc Natl Acad Sci USA 1992; 89:6065–6069.
74. Chen W-F, Zlotnik A. Interleukin 10: a novel cytotoxic T cell differentiation factor. J Immunol 1991; 147:528–534.
75. Rennick D, Hunte B, Dang W, et al. Interleukin-10 promotes the growth of megakaryocyte, mast cell, and multilineage colonies: analysis with committed progenitors and Thy1loSca1+ stem cells. Exp Hematol 1994; 22:136–141.
76. Saeland S, Duvert V, Moreau I, et al. Human B cell precursors proliferate and express CD23 after CD40 ligation. J Exp Med 1993; 178:113–120.
77. Defrance T, Vanbervliet B, Briere F, et al. Interleukin 10 and transforming growth factor β cooperate to induce anti-CD40-activated naive human B cells to secrete immunoglobulin A. J Exp Med 1992; 175:671–682.
78. Nonoyama S, Hollenbaugh D, Aruffo A, et al. B cell activation via CD40 is required for specific antibody production by antigen-stimulated human B cells. J Exp Med 1993; 178:1097–1102.
79. Briere F, Servet-Delprat C, Bridon JM, et al. Human interleukin 10 induces naive surface immunoglobulin D+ (sIgD+) B cells to secrete IgG1 and IgG3. J Exp Med 1994; 179:757–762.
80. Niiro H, Otsuka T, Abe M, et al. Epstein-Barr virus BCRF1 gene product (viral interleukin 10) inhibits superoxide anion production by human monocytes. Lymphokine Cytokine Res 1992; 11:209–214.
81. Suzuki T, Tahara H, Narula S, et al. Viral interleukin 10 (IL-10), the human herpes virus 4 cellular IL-10 homologue, induces local anergy to allogeneic and syngeneic tumors. J Exp Med 1995; 182:477–486.
82. Hudson GS, Bankier AT, Satchwell SC, et al. The short unique region of the B95-8 Epstein-Barr virus genome. Virology 1985; 147:81–98.
83. Stewart JP, Rooney CM. The interleukin-10 homolog encoded by Epstein-Barr virus enhances the reactivation of virus-specific cytotoxic T cell and HLA-unrestricted killer responses. Virology 1992; 191:773–782.
84. Miyazaki I, Cheung RK, Dosch H-M. Viral interleukin 10 is critical for the induction of B cell growth transformation by Epstein-Barr virus. J Exp Med 1993; 178:439–447.

85. Swaminathan S, Hesselton R, Sullivan J, et al. Epstein-Barr virus recombinants with specifically mutated BCRF1 genes. J Virol 1993; 67:7406–7413.
86. Cher DJ, Mormann TR. Two types of murine helper T cell clone. II. Delayed-type hypersensitivity is mediated by TH1 clones. J Immunol 1987; 138:3688–3694.
87. Ferguson TA, Dube P, Griffith TS. Regulation of contact hypersensitivity by interleukin 10. J Exp Med 1994; 179:1597–1604.
88. Powrie F, Menon S, Coffman RL. Interleukin-4 and interleukin-10 synergize to inhibit cell-mediated immunity in vivo. Eur J Immunol 1993; 23:2223–2229.
89. Li L, Elliott JF, Mosmann TR. IL-10 inhibits cytokine production, vascular leakage, and swelling during T helper 1 cell-induced delayed-type hypersensitivity. J Immunol 1994; 153:3967–3978.
90. Beutler B, Milsark IW, Cerami AC. Passive immunization against cachectin/tumor necrosis factor protects mice from lethal effect of endotoxin. Science 1985; 229:869–871.
91. Tracey KJ, Fong Y, Hesse DG, et al. Anti-cachectin/TNF monoclonal antibodies prevent septic shock during lethal bacteraemia. Nature 1987; 330:662–664.
92. Howard M, Muchamuel T, Andrade S, et al. Interleukin 10 protects mice from lethal endotoxemia. J Exp Med 1993; 177:1205–1208.
93. Gerard C, Bruyns C, Marchant A, et al. Interleukin 10 reduces the release of tumor necrosis factor and prevents lethality in experimental endotoxemia. J Exp Med 1993; 177:547–550.
94. Bean AG, Freiberg RA, Andrade S, et al. Interleukin 10 protects mice against staphylococcal enterotoxin B-induced lethal shock. Infect Immun 1993; 61:4937–4939.
95. Powrie F, Leach MW, Mauze S, et al. Inhibition of Th1 responses prevents inflammatory bowel disease in scid mice reconstituted with CD45RBhi CD4+ T cells. Immunity 1994; 1:553–562.
96. Pennline KJ, Roque-Gaffney E, Monahan M. Recombinant human IL-10 prevents the onset of diabetes in the nonobese diabetic (NOD) mouse. Clin Immunol Immunopathol 1994; 71:169–175.
97. Wogensen L, Lee MS, Sarvetnick N. Production of interleukin 10 by islet cells accelerates immune-mediated destruction of beta cells in nonobese diabetic mice. J Exp Med 1994; 179:1379–1384.
98. Wogensen L, Huang X, Sarvetnick N. Leukocyte extravasation into the pancreatic tissue in transgenic mice expressing interleukin 10 in the islets of Langerhans. J Exp Med 1993; 178:175–185.
99. Rott O, Fleischer B, Cash E. Interleukin-10 prevents experimental allergic encephalomyelitis in rats. Eur J Immunol 1994; 24:1434–1440.
100. Kennedy MK, Torrance DS, Picha KS, et al. Analysis of cytokine mRNA expression in the central nervous system of mice with experimental autoimmune encephalomyelitis reveals that IL-10 mRNA expression correlates with recovery. J Immunol 1992; 149:2496–2505.
101. Bacchetta R, Bigler M, Touraine JL, et al. High levels of interleukin 10 production in vivo are associated with tolerance in SCID patients transplanted with HLA mismatched hematopoietic stem cells. J Exp Med 1994; 179:493–502.
102. Chaouat G, Assal Meliani A, Martal J, et al. IL-10 prevents naturally occurring fetal loss in the CBA × DBA/2 mating combination, and local defect in IL-10 production in this abortion-prone combination is corrected by in vivo injection of IFN-tau. J Immunol 1995; 154:4261–4268.
103. Ishida H, Muchamuel T, Sakaguchi S, et al. Continuous administration of anti-interleukin 10 antibodies delays onset of autoimmunity in NZB/W F1 mice. J Exp Med 1994; 179:305–310.
104. Sieling PA, Abrams JS, Yamamura M, et al. Immunosuppressive roles for IL-10 and IL-4 in human infection. In vitro modulation of T cell responses in leprosy. J Immunol 1993; 150:5501–5510.
105. King CL, Mahanty S, Kumaraswami V, et al. Cytokine control of parasite-specific anergy in human lymphatic filariasis. Preferential induction of a regulatory T helper type 2 lymphocyte subset. J Clin Invest 1993; 92:1667–1673.
106. Carvalho EM, Bacellar O, Brownell C, et al. Restoration of IFN-gamma production and lymphocyte proliferation in visceral leishmaniasis. J Immunol 1994; 152:5949–5956.
107. Clerici M, Wynn TA, Berzofsky JA, et al. Role of interleukin-10 in T helper cell dysfunction in asymptomatic individuals infected with the human immunodeficiency virus. J Clin Invest 1994; 93:768–775.
108. Sher A, Coffman RL. Regulation of immunity to parasites by T cells and T cell-derived cytokines. Annu Rev Immunol 1992; 10:385–409.
109. Sher A, Fiorentino D, Caspar P, et al. Production of IL-10 by CD4+ T lymphocytes correlates with down-regulation of Th1 cytokine synthesis in helminth infection. J Immunol 1991; 147:2713–2716.
110. Reed SG, Brownell CE, Russo DM, et al. IL-10 mediates susceptibility to Trypanosoma cruzi infection. J Immunol 1994; 153:3135–3140.
111. Fluckiger AC, Durand I, Banchereau J. Interleukin 10 induces apoptotic cell death of B-chronic lymphocytic leukemia cells. J Exp Med 1994; 179:91–99.
112. Chernoff AE, Granowitz EV, Shapiro L, et al. A randomized, controlled trial of IL-10 in humans. Inhibition of inflammatory cytokine production and immune responses. J Immunol 1995; 154:5492–5499.
113. Schreiber S, Heinig T, Thiele H-G, et al. Immunoregulatory role of interleukin 10 in patients with inflammatory bowel disease. Gastroenterology 1995; 108:1434–1444.
114. Tan JC, Braun S, Rong H, et al. Characterization of recombinant extracellular domain of human interleukin-10 receptor. J Biol Chem 1995; 270:12906–12911.
115. Chomarat P, Rissoan M-C, Banchereau J, et al. Interferon gamma inhibits interleukin 10 production by monocytes. J Exp Med 1993; 177:523–527.
116. Larner AC, David M, Feldman GM, et al. Tyrosine phosphorylation of DNA binding proteins by multiple cytokines. Science 1993; 261:1730–1733.
117. Lehmann J, Seegert D, Strehlow I, et al. IL-10-induced factors belonging to the p91 family of proteins bind to IFN-γ-responsive promoter elements. J Immunol 1994; 153:165–172.
118. Geng Y, Gulbins E, Altman A, et al. Monocyte deactivation by interleukin 10 via inhibition of tyrosine kinase activity and the Ras signaling pathway. Proc Natl Acad Sci USA 1994; 91:8602–8606.

11
Interleukin-11

Xunxiang Du
James Whitcomb Riley Hospital for Children and Indiana University School of Medicine, Indianapolis, Indiana

David A. Williams
James Whitcomb Riley Hospital for Children, Indiana University School of Medicine, and Howard Hughes Medical Institute, Indianapolis, Indiana

INTRODUCTION AND HISTORICAL PERSPECTIVE

The proliferation and differentiation of hematopoietic stem cells is dependent on the interactions of hematopoietic cells and cells and proteins within the hematopoietic microenviroment. Stromal cells play an important role in this hematopoietic microenviroment by producing positive and negative growth factors [including interleukin (IL)-11] and extracellular matrix (ECM) for stabilization of growth factors in a high local concentration and adhesion sites for hematopoietic cells. Stromal cells are a heterogeneous cell population consisting of multiple cell types, including macrophages, endothelial cells, fibroblasts (adventitial reticular cells), and adipocytes. Immortalized stromal cell lines provide simplified systems for analysis of the interactions between stromal and hematopoietic cells.

One such immortalized bone marrow stromal cell line, PU-34, was derived from a primate (*Macaca fascicularis*) bone marrow fibroblast cell (1). Conditioned medium from PU-34 cells can maintain multilineage hematopoiesis in vitro for several weeks. PU-34 produces IL-6, IL-7, granulocyte/macrophage–colony-stimulating factor (GM-CSF), macrophage (M)-CSF, granulocyte (G)-CSF, and leukocyte inhibitory factor (LIF). In addition, conditioned medium from IL-1α–induced PU-34 cells can stimulate the proliferation of an IL-6–dependent plasmacytoma cell line, T1165. This IL-6–like activity cannot be neutralized by IL-6 antibody, and this activity became the basis for functional expression cloning of primate IL-11 in COS-1 cells using a cDNA library generated from PU-34 cells (2). The resulting cDNA contained a single long reading frame of 597 nucleotides encoding a predicted 199-amino-acid polypeptide with a 21-amino-acid leader sequence (Fig. 1). The human IL-11 cDNA was subsequently isolated from a human fetal lung fibroblast cell line (MRC 5) stimulated with phorbol 12-myristate 13-acetate (PMA) and IL-1α by using the primate IL-11 cDNA as a probe. The primate and human IL-11–coding regions share 97% homology at the nucleotide level (1,2).

A year later in 1991, the adipogenesis inhibitory factor was identified in a human bone marrow–derived stromal cell line (KM-102) after stimulation of the cells with PMA and calcium ionophore (A23187). The cDNA of AGIF was cloned by a combination of oligonucleotide probe screening and expression cloning in COS-1 cells and was found to be identical to IL-11 (3,4).

Recently, a murine IL-11 cDNA was expression cloned using a cDNA library generated from a lipopolysaccharide (LPS)-induced murine fetal thymic cell line (T2). The murine IL-11 cDNA shares 86% homology with human IL-11 at the nucleotide level and predicts a protein with 88% homology at amino acid level (5).

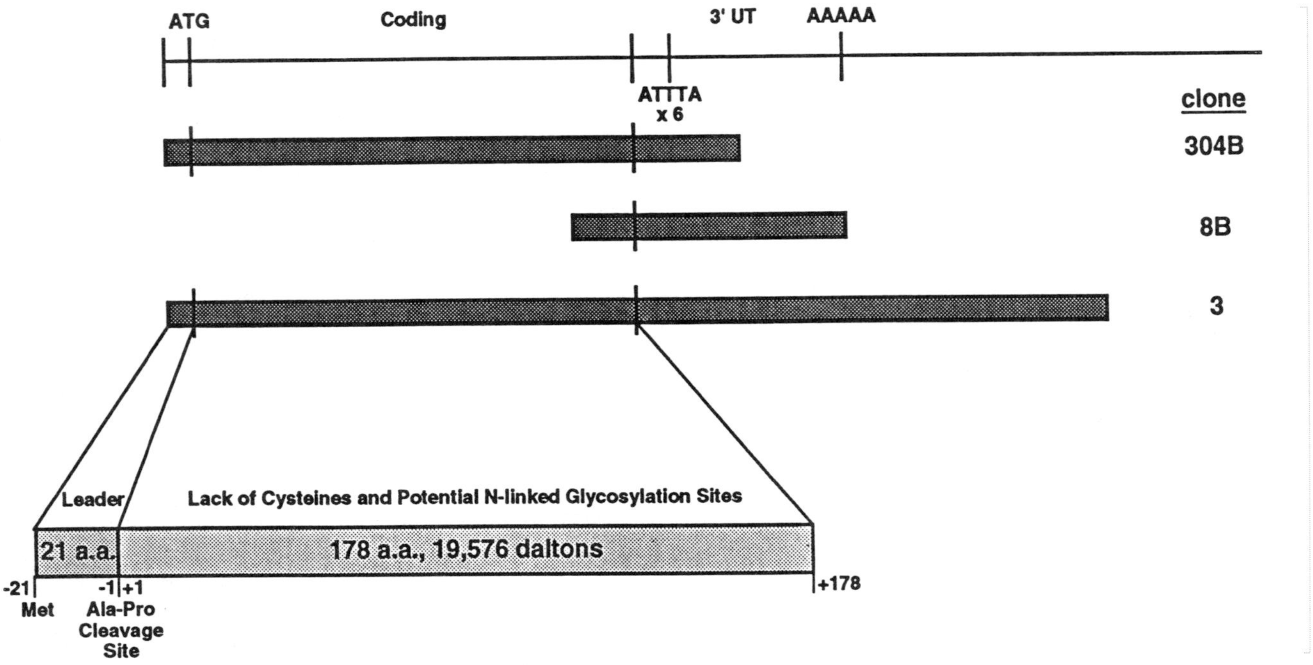

Figure 1 Schematic representation of overlapping IL-11 cDNA clones derived from PU-34 cDNA library and predicted protein encoded by these cDNAs.

BIOCHEMISTRY AND GENE STRUCTURE OF IL-11

The human IL-11 genomic sequence was characterized by McKinley et al. in 1992 (6). The gene spans 7 kb, consists of 5 exons and 4 introns, and was mapped to chromosome 19 at band 19q13.3-q13.4. This region of chromosome 19 contains several zinc finger genes. The 5′-flanking region of the human IL-11 gene contains several DNA motifs involved in transcriptional control, such as CTF/NF1, SP-1, several copies of interferon-inducible elements, and an AP-1 site related to transcriptional regulation by c-jun/c-fos complexes after PMA treatment. A "TATA" box-like sequence, TATATAA, is located about 180 nucleotides upstream from the translation initiation codon ATG (6). A 10-bp promoter sequence (5′ GGTGAGTCAG 3′) in this region that contains an AP-1 site (TGAGTCA) may be an important regulatory element in IL-11 gene expression in a variety of cells. This 10-bp promoter region and JunD/AP-1 complexes are responsible for the basal level transcription of IL-11 gene in bone marrow fibroblast (PU-34) cells (7). The 5′-flanking region also shares significant sequence similarities with the 5′-regulatory regions of other cytokine genes, such as IL-2, -3, -6, and -9 and GM-CSF. The 3′ untranslated region of the IL-11 gene contains multiple ATTTA repeats common to several cytokine genes and *Alu* repeats (6). The ATTTA repeats have been shown to play a role in mRNA destabilization (8). In addition, a DNA sequence (ACATGGCAAAACCC) that has 71% similarity with the IL-1–responsive element of the IL-6 gene is located in the 3′-noncoding region of IL-11. This element is essential for the IL-1–induced expression of IL-6 gene in certain cell types (9) but is not functional in IL-11 transcriptional regulation in PU-34 cells (7). There are two polyadenylation sites located at nucleotide positions 6762 and 5591 corresponding to the 2.5- and 1.5-kb IL-11 transcripts expressed in several IL-1–induced cell lines (2,6). The characterization of murine IL-11 genomic DNA is currently underway, but our laboratory has found significant conservation in genomic structure and putative transcriptional control sequences 5′ to the coding sequences (M. McAndrews-Hill, E. Everett, and D. A. Williams, unpublished data).

The human IL-11 cDNA encodes an IL-11 protein precursor of 199 amino acids including a 21-amino-acid leader sequence. The mature form of human IL-11 is 178 amino acids with the preprotein cleaved between alanine (#–1) and proline (#+1) (Fig. 1). The theoretical molecular weights of recombinant human and murine IL-11 are 19144 daltons (4) and 19154 daltons, respectively (C. Wood, personal communication). However, both human and murine IL-11 secreted from transfected COS cells migrate at approximately 23 kDa on SDS-polyacrylamide gel electrophoresis (PAGE). The slower mobility of IL-11 on SDS-PAGE gels might be due to the strongly basic character of IL-11 (calculated pI = 11.3). Human IL-11 is proline-rich, containing 12% prolines. Unlike many other cytokines, IL-11 has no cysteine residues or potential N-linked glycosylation sites. Human and primate mature IL-11 share 94% identity in the amino acid sequence (3,4).

Based on secondary structure predictions and the location of exon-intron boundaries in the human IL-11 genomic structure, Czupryn et al. proposed a four-helix bundle topology as a structural model for rhIL-11. According to this model, methionine residue (Met^{58}) and lysines (Lys^{41} and Lys^{98}) are located on the surface of the protein. Chemical modifications (alkylation or site-directed mutagenesis) of Met^{58} resulted in a 25-fold decrease in in vitro bioactivity of rhIL-11. Chemical modification of Lys^{41} and Lys^{98} resulted in a threefold decrease in bioactivity. The carboxyl-terminal portion of several cytokines, including hGH, IL-4, and IL-6, has been shown to be critical for their activity and receptor binding. rhIL-11 lacking four carboxyl-terminal residues had 25-fold lower bioactivity, and elimination of 8 or more residues completely abolished activity. These data suggest that the rhIL-11 molecule contains two receptor-binding sites. Met^{58} and the carboxyl terminus of rhIL-11 are involved in the primary receptor-binding site (site I), whereas Lys^{41} and Lys^{98} as well as positively charged arginine residues, which are found on the exposed face of helix C, may be involved in receptor binding site II (10).

CELLULAR SOURCES

Studies done in vitro have shown that the IL-11 gene is expressed in a variety of cells of mesenchymal origin (Table 1). Several types of cells can constitutively express detectable IL-11 mRNA, such as primate bone marrow fibroblasts (PU-34 cells) (1,2,7) and the human bone marrow stromal cell line, KM-102 (3); human trophoblast cell line TPA30-1 (1); bronchial and alveolar epithelial cell lines (9HTE and A549) (11); normal human epidermis (12); the human thyroid carcinoma cell line, NIM-1 (13); and human osteosarcoma cell lines, SAOS (14) and SaOS-2 cells (15) (see Table 1). IL-11 mRNA is also abundant in normal murine testis and detectable in murine hippocampal neuronal cells (granular layer dentate gyrus and pyramidal cellular layer of hippocampus) as well as in motor neurons and sympathetical neuron of spinal cord (15a).

Although the basal level of gene expression is low, IL-11 gene expression can be induced and/or upregulated by several inflammatory cytokines and agonists as well as hormones. IL-1α, TGF-β, and PMA are the best studied inducers of IL-11 gene expression. IL-1α and TGF-β, either alone or synergistically, induce or increase IL-11 gene expression in a variety of tissues (Table 1). In addition, parathyroid hormone (PTH) and parathyroid hormone–related protein (PTHrP) can stimulate IL-11 mRNA accumulation and protein production in a dose- and time-dependent fashion in osteoblastlike osteosarcoma cell lines (14–16). Histamine (17) and asthma-associated viruses such as respiratory syncytial virus (RSV), parainfluenza virus type 3, and rhinovirus can also stimulate respiratory epithelial and fibroblast IL-11 gene expression and protein production (11,18).

Signaling pathways and regulatory levels involved in induction of IL-11 expression are different in various cell types. In bone marrow fibroblast cells, the stimulation of IL-11 expression is largely at the posttranscriptional level, since IL-1α induction does not change the rate of transcription of the IL-11 gene but significantly increases IL-11 mRNA half-life (7). In respiratory epithelial and fibroblast cells, stimulation of IL-11 gene expression by IL-1α and TGF-β1 and TGF-β2 is likely transcriptionally regulated

Table 1 Cells and Cell Lines Expressing IL-11

Cell type	Cell line name	Inducers	Secondary message	Ref.
Bone marrow fibroblast	PU34, KM102	PMA, IL-1α, TPA		2,3,15,20
Fetal lung fibroblast	MRC 5	PMA, IL-1α		2
Lung fibroblast	CCL 202	IL-1α, TGF-β1, TGF-β2, RSV	Calcium/Calmodulin	19,103
Bronchial and alveolar epithelial	9HTE, A549	Rhinovirus, parainfluenza type 3 histamine	H_1 receptor	11,17,18
Connective tissue	Chondrocyte, synoviocyte	IL-1β, TGF-β_1, PMA	PKC	21
Vein endothelial cells		PMA, IL-1α, TGF-β_1, β_2		104
Epidermis	Keratinocytes			12
Trophoblast cells	TPA 30-1	PMA, IL-1α		2
Osteosarcoma	SAOS, MG-63, HOBIT, SAKA	IL-1α, TGF-β_1, PTH, PTHrP	cAMP, PKC	14–16
Thyroid carcinoma	NIM-1			13

since the half-life of IL-11 mRNA is unchanged and the relative rate of IL-11 nuclear transcription is increased (11,19). A tyrosine kinase inhibitor, genistein, can block IL-1α induction, while the serine and threonine kinase inhibitor H7 [which is able to block protein kinase C (PKC)–mediated gene expression in PU-34 cells] has little effect on IL-1α induction. These studies suggest that tyrosine kinase is involved in the induction of IL-11 mRNA by IL-1α stimulation and that this induction is independent of PKC (7) in bone marrow stromal, PU-34 cells. IL-1α– or 12-O-tetradecanoyl-phorbol-13 acetate (TPA)–induced IL-11 gene expression in PU-34 cells can also be abolished by heparin, which has been shown to trans-repress AP-1–mediated gene transcription, through its competition for mRNA-binding proteins and mRNA destabilization (20).

PKC-mediated signaling events are probably involved in the induction of IL-11 in connective tissues and osteosarcoma cell lines (15,21). Induction of IL-11 mRNA by the protein synthesis inhibitor cyclohexamide suggests that transcription of IL-11 is negatively regulated by protein(s) with short half-lives in connective tissues (21). IL-1α and TGF-β1 or TGF-β2 synergistically stimulate lung fibroblast IL-11 mRNA expression via a pathway that is to a great extent calmodulin-dependent. Calmodulin antagonists can significantly decrease the inductive effects of the combination of IL-1α and TGF-β on IL-11 gene expression. Unlike the effects seen in human articular chondrocytes and synoviocytes, increased intracellular calcium and inhibition of Na^+/H^+ pump activity can induce IL-11 mRNA accumulation in lung fibroblast cells. Although activation of PKC can also induce IL-11 gene expression in lung fibroblast cells, the stimulation of IL-11 gene expression by IL-1α and TGF-β in this type of cells is largely via a PKC-independent pathway (19), which is similar to PU-34 cells (7). The synergy of histamine and TGF-β1 on IL-11 mRNA accumulation and protein production in human lung fibroblasts cannot be altered by H_2 receptor agonists and antagonists. However, the synergy can be abrogated by H_1 antagonists and calmodulin antagonists and reproduced by H_1 agonists. These data suggest that the interaction of histamine and TGF-β1 in induction of IL-11 is, to a great extent, at the transcriptional level, which is dependent on H_1 receptors and a calcium/calmodulin-dependent activation pathway (17).

THE IL-11 RECEPTOR

Yin et al. demonstrated the existence of a single class of high-affinity IL-11 receptor with a Kd of 3.49×10^{-9} M and 5000 specific binding sites/cell on 3T3-L1 cells. Cross-linking studies indicated that the IL-11 receptor consists of a single polypeptide chain of 151 kDa (22). Recently, Hilton et al. described the cloning of the cDNA encoding a murine IL-11 receptor α chain (23). The cloning strategy utilized oligonucleotides designed to hybridize to the WSXWS motif conserved in receptors of the hematopoietin receptor superfamily. The mRNA of IL-11 receptor α chain is detectable in several murine cell lines, such as 3T3-L1 cells, stromal cell line BAd, embryonic carcinoma cell line PC13, and factor-dependent hematopoietic cell lines FDCP-1 and D35. A wide range of primary tissues express IL-11 receptor α-chain mRNA, including hematopoietic tissues (bone marrow, spleen, and thymus), liver, brain, heart, kidney, muscle, and salivary gland. The extracellular region of the IL-11 receptor α-chain shares sequence similarity to the α chains of IL-6 and ciliary neurotropic factor (CNTF) receptors (24 and 22% amino acid identity, respectively). The binding of IL-11 to the receptor α chain occurs at low affinity (Kd = 10 nM) and, although necessary, is not sufficient for signal transduction. The generation of a high-affinity IL-11 receptor capable of generating a biological signal leading to proliferation and differentiation responses to IL-11 requires co-expression of the receptor α chain and gp130 (23). By amplifying human cDNA libraries with oligonucleotide primers, which cover the nucleotide sequences encoding the WSXWS motifs, Chérel et al. (24) recently cloned two isoforms of human IL-11 receptor α chain. Both isoforms of human IL-11 receptor are identical in their extracellular and transmembrane domains, but differ in their cytoplasma domain. One isoform of human IL-11 receptor, like the human IL-6 receptor and the murine IL-11, has a short intracytoplamic domain. The other isoform, like the human CNTF receptor, lacks a cytoplamic domain. Preliminary study has shown that human IL-11 receptor gene is expressed in myeloid (K562), megakaryocytic (Mo7E), and erythroid (TF1) leukemia cell lines as well as the osteosarcoma cell lines (MG-63 and Saos-2) (24).

gp130 is the common subunit of the IL-6, Oncostatin M (OSM), LIF, and CNTF as well as IL-11 receptors (25–28). Binding of IL-11, IL-6, and related cytokines to specific cell-surface receptors induces receptor dimerization, tyrosine phosphorylation (22,29), and activation of gp130 (25,27,30) followed by activation and protein tyrosine and serine/threonine phosphorylation of gp130-associated protein tyrosine kinases and H7 (a serine/threonine kinase inhibitor)-sensitive kinases (31,32). IL-11–induced protein tyrosine phosphorylation is transient and PKC independent (29). IL-11–induced phosphorylation of H7-sensitive kinases is also PKC independent and cAMP, cGMP, and calcium/calmodulin independent (30). These phosphorylated protein kinases include JAK-TYK kinases (31,33) and mitogen-activated protein (MAP) kinases as well as ribosomal S6 protein ($pp90^{rsk}$) kinases (32). Phosphorylation of gp130-associated protein kinases leads to early-response gene expression (such as tis8, tis11, tis21, and junB) depending on the cell type examined (25,30,32,34). It is interesting that IL-11 and IL-6 can trigger the same primary response gene expression in all the cell lines tested, further supporting that IL-11 and IL-6 may share common signal transduction pathways (30). Among these protein kinases, $pp90^{rsk}$ is one of the H7-sensitive kinases critical for the activation of early response genes by IL-11, IL-6, LIF, and OSM in 3T3-L1 cells (32).

Acute-phase response factor (APRF) is antigenically and functionally related to members of the signal transducer and activator of transcription (STAT) family. IL-11 and other cytokines sharing the signal transducing subunit gp130 can activate APRF by tyrosine phosphorylation in variety of cell types. This fact indicates a central role of this transcription factor in gp130-mediated signaling (35).

BIOLOGICAL ACTIVITY

Hematopoiesis

Megakaryocytopoiesis and Thrombopoiesis

As a single agent, IL-11 has no influence on megakaryocyte (MK) colony formation in serum-free methylcellulose cultures (36), but can stimulate the proliferation of the human MK cell lines (CMK and MegJ) in a dose-dependent fashion (37). In combination with IL-3, IL-11 stimulates the formation of both colony-forming units–MK (CFU-MK) and burst-forming units–MK (BFU-MK) colonies (38). Several investigators demonstrated that IL-11 alone stimulates acetylcholinesterase (AchE) production, which is a general measure of MK mass, reflecting both the number and the size of MK, in murine liquid serum-free cultures. This effect is similar to IL-6 and LIF. IL-11 also increases the size of both MK colonies and cells (36,39). Either alone or with IL-3, IL-11 increases the percentage of high ploidy megakaryocytes derived from both murine and human bone marrow cells (39). These in vitro data suggest that, like IL-6 and LIF, IL-11 is a MK potentiator that affects the immediate precursors of polyploid MKs. Recently, Kaushansky et al. (40) showed that neutralization of thrombopoietin (Tpo, Mpl ligand) activity eliminates MK formation in response to IL-11, IL-6, and SCF. These data suggest that the effects of IL-11, IL-6, and SCF on megakaryocytopoiesis and thrombopoiesis may be mediated, at least in part, via Tpo.

In vivo treatment of IL-11 results in marked stimulation of megakaryocytopoiesis in both small rodents and nonhuman primates (41–44). In normal mice, IL-11 administration increases megakaryocyte progenitors, stimulates megakaryocyte endoreplication, and increases peripheral platelet counts. Similar results have been reported in rats treated with IL-11, with a dose-dependent increase in platelet counts and increased bone marrow megakaryocyte size and ploidy. In nonhuman primates, IL-11 treatment increases circulating platelet concentrations by enhancing the size and maturation rate of bone marrow megakaryocytes. IL-11 administration followed by chemotherapy in women with breast cancer significantly increased bone marrow megakaryocyte and CFU-MK number. IL-11 treatment in these patients is also associated with higher ploidy of megakaryocytes (45–47). These data indicate that IL-11 promotes multiple stages of megakaryopoiesis and thrombopoiesis both in vitro and in vivo in small rodents, nonhuman primates, and humans, including both the production and maturation of megakaryocytes.

Hematopoietic Stem and Multipotential Progenitor Cells

IL-11 alone has little effect on murine blast colony formation, CFU-granulocyte, erythrocyte, macrophage, megakaryocyte (CFU-GEMM) colony formation, or colony formation from a variety of purified stem cell populations, such as human $CD34^+/HLA\text{-}DR^-$, $CD34^+/CD33^-/DR^-$ (48), and $CD34^+/DR^-/CD15^-$ cells (49) and $Lin^-/Sca\ 1^+$ day 2 post 5-fluorouracil (FU_{2d}) mouse bone marrow cells (50) in in vitro cultures. Addition of IL-11 to murine stromal-dependent long-term born marrow cultures (LTBMC) has no significant effect on total production of high proliferating potential–colony-forming cells (HPP-CFC), although IL-11 "mobilized" HPP-CFC from the adherent layer to nonadherent layer in these cultures (51). In both human and murine stromal-dependent LTBMC, IL-11 addition decreases the size of stem cell compartment, decreases the frequency of LTC-IC in human LTBMC, and decreases the frequency of long-term reconstituting stem cells in murine LTBMC. In contrast to the effects of IL-11 on stem cell compartments in these stromal-dependent LTBMC, IL-11 addition significantly increases nonadherent cellularity, progenitor production (CFU-GM, CFU-MK, BFU-E, and CFU-Mix), and day 12 colony-forming unit-spleen ($CFU\text{-}S_{12}$) production (51). IL-11 also acts synergistically with other early growth factors such as IL-3 (52–54), IL-4 (52,53,55), IL-7, IL-12 (56–58), IL-13 (59), SCF (48), flt3 ligand (60), and G-CSF (48) to stimulate the proliferation and expansion of primitive stem cells (long-term culture-initiating cells, or LTC-IC, and HPP-CFC) and multipotential as well as committed progenitor cells in suspension and semisolid culture systems (52–54). This affect appears to be due to the stimulation of the quiescent

(G_0) population of the cells into active cell cycle (54). These data suggest that IL-11 in combination with other cytokines present in hematopoietic microenvironment or in culture systems may increase commitment of primitive stem cells into the multilineage progenitor compartment.

Myelopoiesis

When added to normal murine bone marrow cells in vitro in serum-containing cultures, IL-11 alone or in combination with IL-3 supports colony formation predominantly of the granulocyte/macrophage type in a dose-dependent manner and supports CFU-GEMM colony formation from FU_{2d} bone marrow. IL-11 stimulates CFU-GM colony formation from human unfractioned bone marrow cells and from $CD34^+/DR^+$ cells and has additive effects on CFU-GM colony formation from $CD34^+/DR^+$ cells in combination with IL-3, SCF, M-CSF, and GM-CSF. As a single agent, IL-11 induces cycling of human fetal CFU-Mix, CFU-GM, and BFU-E. In addition to proliferative effects on myeloid progenitors, IL-11 also modulates the differentiation and maturation of myeloid progenitor cells. IL-11 in combination with SCF stimulates myeloid colony formation from murine $Lin^-/Sca\ 1^+$ bone marrow cells, resulting in colonies composed mostly of granulocytes and myeloid blasts. The combination of these two cytokines with IL-13 or IL-4 can reduce the proportion of granulocytes and blasts in myeloid colonies, with a concomitant increase in macrophages (59).

Administration of IL-11 in vivo enhances the cycling rates of BFU-E and CFU-GEMM and increases the absolute number of CFU-GM, BFU-E, and CFU-GEMM in both bone marrow and spleen of normal mice (61). IL-11 has no effects on peripheral leukocyte counts when administered to normal mice and rats (41,42). Similar to these effects in rodents, IL-11 treatment resulted in no changes in peripheral leukocyte counts in nonhuman primates (cynomolgus monkeys) (43). IL-11 has a synergistic effect with G-CSF on hematopoietic progenitor mobilization (62).

Erythropoiesis

IL-11 can stimulate the in vitro growth of macroscopic erythroblast colonies containing a high frequency of CFU-E when replated into secondary cultures in combination with SCF and erythropoietin (Epo). In addition, IL-11 synergizes with IL-3 in serum-containing cultures to stimulate BFU-E growth in a dose-dependent fashion, even in the absence exogenous Epo (63). IL-11 with Epo stimulates formation of BFU-E derived from human $CD34^+CD33^-DR^-$ cells (64). IL-11 alone can support the maturation of less primitive CFU-E from fetal liver or bone marrow cells and from the cells replated at low density from macroscopic erythroblast colonies (63). Thus, at least in vitro, IL-11 may act both on primitive erythroid progenitor cells and affect terminal differentiation of erythroid precusors.

IL-11 administration results in no significant effects on erythrocyte counts and hematocrits in normal mice and rats. IL-11 administration to nonhuman primates (cynomolgus monkeys) has been associated with a moderate, transient anemia. Recovery of red blood cell counts and hemoglobin concentration in these animals was seen by day 28 of treatment. Reticulocytes were significantly increased after 28 days of IL-11 treatment in these animals.

Lymphopoiesis

IL-11 in combination with SCF or IL-4 effectively supports the generation of B cells in the primary colonies from marrow cells of 5-FU–treated mice, an effect similar to IL-6 (58,65,66). The combination of IL-11, SCF, and IL-7 can support pre–B-cell development from $AA4.1^+/B220^-/Mac\text{-}1^-/Ly6A^+$ day 12 fetal liver cells (67). IL-11 stimulates the generation of mouse spleen sheep red blood cell (SRBC)–specific plaque-forming cells in a dose-dependent manner in vitro (2,68). Depletion of $Thy\text{-}1.2^+$ or $L3T4^+(CD4)$ cells dramatically decrease this IL-11–induced enhancement of SRBC-specific plaque-forming cells and suggest that $L3T4(CD4)^+$ T cells, but not Lyt-2 $(CD8)^+$ T cells are required for IL-11– stimulated antibody responses. Similarly, Anderson and his colleagues have demonstrated that IL-11 or IL-6 addition with pokeweed mitogen to a mixture of human B cells, $CD34^+/45RA^-$ T cells and monocytes leads to increased DNA synthesis and Ig secretion by B cells. Anti–IL-6 antibody only partially blocks IL-11–related increases in Ig secretion. Thus, it appears that IL-11 promotes the differentiation of B lymphocytes by selective stimulation of $CD34^+/45RA^-$ T cells and monocytes, in a manner similar to IL-6 (69). IL-11 administration in vivo also results in a dose-dependent increase in the number of spleen SRBC-specific plaque-forming cells as well as serum SRBC-specific antibody titer in both the primary and secondary immune responses (68).

Effects on Blast-Cell Lines and Leukemic Cells

IL-11 stimulates the proliferation of several murine plasmacytoma cell lines, including T1165, T10, B9, and B9-Ty1, in a dose-dependent manner, and the IL-11–induced stimulation of these cells cannot be neutralized by anti–IL-6 antibody (2,70,71). In contrast to these mitogenic effects of IL-11 on murine plasmacytoma cell lines, exogenous IL-11 does not increase DNA synthesis or Ig secretion of several human myeloma cell lines examined or freshly isolated human myeloma cells (72). Moreover, there are no increases in IL-11 mRNA expression, protein production, or expression of IL-11–binding protein on the cell surface of human myeloma cell lines or freshly isolated human myeloma cells in these studies. These data suggest that IL-11 does not play a role, either in autocrine or in paracrine fashion, in the growth of human myeloma cells or derived cell lines (72). In contrast, IL-11 in combination with IL-3, GM-CSF, or SCF does support colony formation from myeloid leukemia cell lines and triggers acute myeloblasts from patients with acute myeloblastic leukemia into S-phase. Several acute pre–B-lymphocytic leukemia and acute T-lymphocytic leukemia cell lines fail to respond to IL-11 (73).

Nonhematopoiesis

Adipogenesis

IL-11 directly inhibits the process of adipogenesis in murine 3T3-L1 cells and decreases lipoprotein lipase (LPL) mRNA expression, protein synthesis, and enzyme activity in these cells in a dose-dependent fashion (74). IL-11 can also inhibit adipocyte differentiation induced by insulin and dexamethasone in 3T3-L1 cells. IL-11 inhibits adipose accumulation in H-1/A cells, a cell line derived from the adherent cell layer of a 14-week-old murine LTBMC (4). Moreover, IL-11 markedly inhibits adipose accumulation in the adherent layer of primary human LTBMC, apparently due to a block in the differentiation of preadipocytes (75).

Acute-Phase Responses

Similar to IL-6 and LIF, IL-11 stimulates production of acute-phase reactants both in vitro and in vivo. IL-11 stimulates the production of type 1 (α_1-acid glycoprotein, complement component C3, haptoglobin, and hemopexin) and type 2 (thiostatin and fibrinogen) acute-phase proteins in a dose-dependent fashion in rat hepatoma (H-35) cells and rat hepatocytes (76). In human hepatoma HepG2 cells, IL-11 increases the mRNA level of microsomal heme oxygenase, the rate-limiting enzyme in heme catabolism, in a dose- and time-dependent fashion. This response is similar to the response observed with IL-6, whereas IL-11 has a significantly weaker effect than IL-6 on the induction of haptoglobin mRNA, a major acute-phase protein (77).

Osteoclastogenesis

As mentioned above, IL-11 mRNA is expressed in osteosarcoma cells and message levels are induced in normal osteoblast cells by IL-1. Gene expression in these cells can also be upregulated by TGF-β and parathyroid hormone (PTH) (14,16). IL-11 in combination with 1,25-dihydroxyvitamin $D_3[1,25(OH)_2D_3]$ and PTH has been shown to stimulate osteoclast development and bone resorption in bone marrow cultures and in cocultures of bone marrow with calvaria cells (78). IL-11 inhibits bone nodule formation in rat calvaria cell cultures in a dose-dependent fashion (79).

The bone-resorptive activity of IL-11 can be inhibited by calcitonin. Neutralizing antibody to IL-11 can partially negate the bone-resorptive effects of PTH. IL-11 antibody can also block IL-1, TNF, and $1,25(OH)_2D_3$-induced osteoclast development (78,80,81). IL-11–stimulated osteoclast formation can be suppressed in a dose-dependent fashion by the cyclooxygenase inhibitor indomethacin. These data suggest that IL-11 plays a modulating role in osteoclastogenesis and bone resorption but via the direct actions of other cytokines.

Embryogenesis

As previously mentioned, IL-11 is expressed in several cell lines derived from fetal tissues and may be involved, based on in vitro assays, in embryogenesis. Thus, IL-11 can stimulate differentiation and maturation of embryonic hippocampal progenitor cells (82) and can enhance hematopoietic differentiation of embryonic stem cells in the later stages of embryoid body formation (day 10 to day 14 of differentiation in vitro) (83). IL-11 increases the expression of both α- and β-human chorionic gonadotropin (hCG) and the expression of urokinase-type plasminogen activator in trophoblast and cytotrophoblast cells. IL-11 expression can be induced in placental fibroblasts by IL-1α and PTHrP. However, it needs to be further elucidated whether IL-11 is involved in trophoblast differentiation, placental implantation, and maintenance of pregnancy (84).

Neurogenesis

IL-11 stimulates the proliferation of a rat hippocampal neuronal progenitor cell line, H19-7 cells, in a dose-dependent fashion (Du and Williams, unpublished data). Mehler and colleagues have provided evidence that IL-11 and several other hematopoietic growth factors, including IL-5, IL-7, and IL-9, are survival and/or differentiation factors for murine hippocampal progenitors (82). IL-11 promotes neuronal differentiation of an immortalized neural stem and progenitor cell line (MK31) from dissociated embryonic (day 17) mouse hippocampus. IL-11 alone or in combination with TGF-α enhances cellular polarity, neurite outgrowth, nuclear enlargement, and complexity of MK31 cells. Moreover, IL-11 alone or with TGF-α promotes cellular expression of the phosphorylated form of the high-molecular-weight mature neurofilament protein (NFH-P), with reduced expression of nestin and NF66. IL-11 alone can induce inward currents and propagation of immature action potentials in MK31 cells (82). Taken together the data would suggest that IL-11 induces intermediate stages of neuronal maturation in the mammalian brain. Fann and Patterson have demonstrated that IL-11 and other cytokines using gp130 can induce preprotachykinin mRNA (which produces the substance P) expression in cultured sympathetic neurons (85). It is intriguing to speculate that the mechanisms regulating the proliferation and differentiation of neural and hematopoietic cells are similar.

Extracellular Matrix Metabolism and Other Effects

Metalloproteinases and their specific inhibitors play an important role in extracellular matrix (ECM) protein degradation. IL-11 can mediate metabolism of ECM in connective tissue chondrocytes and synoviocytes by inducing the synthesis of the tissue inhibitor of metalloproteinase 1 (21). Through this effect, IL-11 potentially balances IL-1β–induced ECM breakdown and may have a protective effect on connective tissues. In hepatocytes, IL-11 also induces the expression of the tissue inhibitor of metalloproteinase-1, which may be involved in the pathogenesis of liver fibrosis and cirrhosis (86).

A number of inflammatory cytokines, including IL-1, IL-6, tumor necrosis factor (TNF), and macrophage inflammatory protein-1 (MIP-1), act in the brain to produce fever. Lopez-Valpuesta and Myers (87) found that IL-11 induces fever when the cytokine is directly delivered to the thermosensitive and pyrogen-reactive region of the hypothalamus (anterior hypothalamic preoptic area, AH/POA). The pharmacological blockade by indomethacin of the febrile response to IL-11 suggests that prostaglandin E series may serve as mediator of IL-11–induced fever (87). In small intestine, IL-11 can enhance gastrointestinal absorption of iron (88). IL-11 protein can be detected in epidermal and adnexal keratinocytes in normal skin as well as in multinucleated giant cells (MNGC) in lesions of cutaneous sarcoidosis and in vitro. IL-11 promotes the formation of MNGC. It seems possible that local production of IL-11 may play a role in granulomatous inflammation of the skin (12).

Preclinical Studies

Syngeneic Bone Marrow Transplant Models

In in vivo preclinical animal models, our laboratory and other investigators have shown that IL-11 stimulates multilineage hematopoietic reconstitution and enhances antigen-specific responses after the administration of multiple cytoablative therapies. In contrast to the effects of IL-11 administration in normal mice and in some preclinical models, IL-11 significantly increases peripheral total leukocyte counts in bone marrow transplant (BMT) mice and in sublethally irradiated (nontransplanted) mice (89). However, in these models, the major effect was a dose-dependent increase in peripheral platelet counts. The increased total leukocytes reflects mainly an acceleration of peripheral neutrophil recovery (89), which leads to significantly decreased mortality and morbidity from lethal exogenous infection with *Pseudomonas aeruginosa* (90). IL-11–treated BMT mice demonstrated a shortened period of neutropenia (by at least 7 days) compared with vehicle-treated mice (89). Acceleration of recoveries of peripheral neutrophils and platelets are also seen in mice immunosuppressed by cyclophosphamide treatment (68). BMT mice treated with the combination of IL-11 and SCF show additive increases in total leukocyte counts and platelet counts as well as hematocrits (instead of the mildly decreased hematocrits seen in BMT mice treated with IL-11 alone) (91). The combination of these two cytokines shortens the period of cytopenia of all lineages and is well tolerated post-BMT. In no other cytokine combinations have such profound effects on neutrophil and platelet lineages in vivo been seen. In contrast to SCF, which increases splenic hematopoiesis to a large extent, IL-11 appears to stimulate medullary hematopoiesis (91).

IL-11 administration is associated with increases in bone marrow and spleen cellularity and megakaryocyte numbers, and CFU-GM, CFU-Mix, and CFU-MK content in BMT mice. Although IL-11 treatment is associated with mild anemia, BFU-E content in both bone marrow and spleen are not decreased. In contrast to the effect of IL-11 on the number of CFU-S_{12} in murine LTBMC in vitro (51) (see above), IL-11 administration in vivo has little effect on CFU-S_{12} content in either bone marrow or spleen of BMT mice (89,91).

Lethally irradiated mice transplanted with syngeneic bone marrow cells infected with a retrovirus expressing the human IL-11 cDNA show no changes in peripheral leukocyte counts despite a greater than 20-fold increase in myeloid progenitor content in spleen. These mice also demonstrate systemic effects of chronic IL-11 exposure, including loss of body weight (due to loss of fat) accompanied by a hyperactive state, thymic atrophy, inflammation of eye lids, and increased acute-phase reactants (haptoglobin level) (92).

Chemotherapy Models

In a murine model of cumulative severe myelosuppression seen in chloroethylnitrosourea-containing chemotherapy regimens, IL-11 markedly diminishes 1,3-bis(2-chloroethyl)-1-nitrosourea (BCNU)–induced pancytopenia and reduces chemotherapy-related mortality. These data suggest that IL-11 might be useful in dose intensification in the treatment of tumors that are nitrosourea sensitive (93). In addition to acceleration of the recovery of both peripheral neutrophils and platelets in mice immunosuppressed by cyclophosphamide treatment as mentioned above (68), IL-11 administration augments the number of spleen SRBC-specific plaque-forming cells as well as serum SRBC-specific antibody titer. These data are consistent with the effects of IL-11 on immune cells in vitro.

Combined Modality Models

In mice treated with a combination of 5-FU (at 150 mg/kg) and radiation therapy (sublethal irradiation, 600–700 cGy TBI/mouse), IL-11 administration markedly decreases morbidity and mortality due to sepsis by endogenous infection of gut organisms (94). The increased survival is associated with a rapid repair of small intestinal villus structures. IL-11 administration is associated with increases in the mitotic index of cells in the small intestinal crypt, the location of villous stem and progenitor cells. The frequency of staining of these cells with a monoclonal antibody against proliferating cell nuclear antigen (PCNA) is also increased. More recently, our laboratory has demonstrated decreased apoptosis of villous cells in this same model (95). This effect on chemotherapy- and radiation therapy–induced gut mucosal damage may prove to be important in clinical use in cancer and bone marrow transplant protocols in the future.

In contrast to the above combined-modality model, in which IL-11 has no significant effect on peripheral blood cell counts with only mild increase in platelet counts at the first 10 days postirradiation, administration of IL-11 to mice receiving 500 cGy sublethal irradiation followed by a single injection of the chemotherapeutic drug carboplatin (1.2 mg/mouse) results in the stimulation of multilineage hematopoiesis in bone marrow and spleen. IL-11 blunted the platelet and hematocrit nadirs and shortened the period of thrombocytopenia and anemia in this mode (96).

Inflammatory Disease and Sepsis Models

Oral mucositis is one of the most troublesome complications of cancer therapy, and mucosal damage can lead to significant morbidity and mortality in chemointensive protocols. In a hamster model of oral mucositis (5-FU 60 mg/kg, ip at day 0 and day 2, and left cheek pouch mucosa irradiation at day 4), IL-11 decreases the extent and severity of the oral lesions and increases animal survival. Acute colitis in rats due to chemical damage (acetic acid) and chronic inflammatory bowel disease in HLA-B27 transgenic rats are improved at both the gross and microscopic level by administration of IL-11 (97).

IL-11 administration also improves survival and enhances platelet recovery following experimental group B streptococcal (GBS) sepsis in neonatal rats. The significant increase of IL-11 mRNA in bone marrow of GBS-infected neonatal rats suggests a potential role of endogenous IL-11 in the pathophysiological response of neonatal animals with bacterial sepsis and its associated thrombocytopenia (98).

Human Studies and Clinical Trials

Given the multiple preclinical studies demonstrating thrombopoietic effects of IL-11, it is not surprising that initial clinical trials demonstrate stimulation of platelet production in humans treated with IL-11 (45–47). IL-11 also increases plasma von Willebrand's factor and fibrinogen concentrations in normal human subjects (99). These data suggest that IL-11 may augment hemostasis in thrombocytopenic patients (45–47). Recently, Suen et al. found that the circulating IL-11 levels in BMT/peripheral stem cell transplant (PSCT) patients are significantly elevated until 4 months postplatelet engraftment. In normal adults and bone marrow transplant (BMT) patients before myeloablative therapies, IL-11 levels are undetectable (< 50 pg/ml). The elevation of endogenous IL-11 levels may compensate for residual damage to the microenvironment and promote hematopoietic recovery, and IL-11 may play a physiological role during acute thrombocytopenia associated with myeloablative therapy and BMT (100).

Currently, IL-11 is being evaluated in several clinical trials including in childhood cancers. Women with advanced-stage breast cancer (stage IIIb or IVBC) undergoing high-dose chemotherapy (cyclophosphamide and doxorubicin) were treated with increasing doses of IL-11 (10, 25, 50, 75, and 100 mg/kg/day) both before therapy and after each of four cycles of combined chemotherapy (45–47). In these studies, IL-11 at doses of 25–50 mg/kg/day was found to stimulate platelet production and maturation as evidenced by increased bone marrow CFU-MK, megakaryocyte number, and higher megakaryocyte ploidy. IL-11 treatment is also associated with increased bone marrow cellularity as well as the number and cycling status (by PCNA staining) of immature erythroid and myeloid precusors. IL-11 administration is associated with less severe thrombocytopenia after high-dose chemotherapy in women with advanced stages of breast cancers (45–47). IL-11 treatment is well tolerated at doses of 10–50 mg/kg/day. A reversible and non–dose-related decrease in hemoglobin concentration (mean 20%) was observed as early as day 3 of treatment. Hemoglobin concentration returned to normal after completion of treatment. This mild anemia may be dilutional because IL-11 increases plasma volume and decreases urine sodium excretion in normal human subjects (101). IL-11 administration increased the plasma concentration of acute-phase reactants, including C-reactive protein, fibrinogen, and haptoglobin at all doses. Other side effects include reversible edema, fatigue, and myalgias in all patients treated with 75 mg/kg/day IL-11. In breast cancer patients receiving high-dose cyclophosphamide, BCNU, and thiotepa followed by ABMT, the combination of IL-11 with G-CSF can accelerate peripheral neutrophil and platelet recoveries and is well tolerated in doses up to 50 μg/kg/day (102).

In IL-11 phase II clinical trial, patients who had solid tumors or lymphomas and received at least one platelet transfusion during their chemotherapy were given an additional cycle of the same chemotherapy without dose reduction and were randomized to receive rhIL-11 (50 or 25 mg/kg) or placebo. The patients treated with rhIL-11 in this phase II study were less likely to require platelet transfusions than the patients receiving placebo. For the patients treated with IL-11 at 50 and 25 mg/kg, 30 and 18% required no platelet transfusions, compared to the patients treated with placebo, only 4% of whom required no platelet transfusions (105). IL-11 was generally well tolerated and there were no severe side effects similar to the results seen in phase I clinical trial. Taken together, these data indicate that IL-11 is a very promising cytokine for the treatment of thrombocytopenia resulting from cytoablative cancer therapies.

ACKNOWLEDGMENTS

This work is supported by NIDDK R01 DK 48605 (DAW and XD) and RMA 95-6 (XD). Dr. Williams receives payments from Children's Hospital, Boston, MA, based on certain milestones set forth in an interleukin-11 agreement between Genetics Institute, Cambridge, MA, and Children's Hospital. We thank Karen Roller for preparation of the manuscript.

REFERENCES

1. Paul SR, Yang Y-C, Donahue RE, Goldring S, Williams DA. Stromal cell-associated hematopoiesis: immortalization and characterization of a primate bone marrow-derived stromal cell line. Blood 1991; 77:1723–1733.
2. Paul SR, Bennett F, Calvetti JA, et al. Molecular cloning of a cDNA encoding interleukin 11, a stromal cell-derived lymphopoietic and hematopoietic cytokine. Proc Natl Acad Sci USA 1990; 87:7512–7516.
3. Kawashima I, Ohsumi J, Mita-Honjo K, et al. Molecular cloning of cDNA encoding adipogenesis inhibitory factor and identity with interleukin-11. FEBS 1991; 283:199–202.
4. Ohsumi J, Miyadai I, Ishikawa-Ohsumi H, Sakakibara S, Mita-Honjo K, Takiguchi T. Adipogenesis inhibitory factor. A novel inhibitory regulator of adipose conversion in bone marrow. FEBS 1991; 288:13–16.
5. Morris J, Neben S, Morris G, et al. Molecular cloning of the murine interleukin-11 cDNA (abstr). Exp Hematol 1994; 22:737.
6. McKinley D, Wu Q, Yang-Feng T, Yang Y-C. Genomic sequence and chromosomal location of human interleukin (IL)-11 gene. Genomics 1992; 13:814–819.
7. Yang L, Yang Y-C. Regulation of interleukin(IL)-11 gene expression in IL-1 induced primate bone marrow stromal cells. J Biol Chem 1994; 269:32732–32739.
8. Shaw G, Kamen R. A conserved AU sequence from the 3′ untranslated region of GM-CSF mRNA mediates selective mRNA degradation. Cell 1986; 46:659–667.
9. Isshiki H, Akira S, Tanabe O, et al. Constitutive and interleukin-1 (IL-1)-inducible factors interact with the IL-1-responsive element in the IL-6 core. Mol Cell Biol 1990; 10:2757–2764.
10. Czupryn MJ, McCoy JM, Scoble HA. Structure-function relationships in human interleukin-11. J Biol Chem 1995; 270:978–985.
11. Elias JA, Zheng T, Einarsson O, et al. Epithelial interleukin-11: regulation by cytokines, respiratory syncytial virus, and retinoic acid. J Biol Chem 1994; 269:22261–22268.
12. Longley J, Rochester C, Tyrrell L, Einarsson O, Elias J. Interleukin-11 (IL-11) in normal skin and sarcoidosis (abstr). J Invest Dermatol 1993; 100:524.
13. Kawashima I, Takiguchi Y. Interleukin-11: a novel stroma-derived cytokine. Prog Growth Factor Res 1992; 4:191–206.
14. Horowitz M, Fields A, Zamparo J, Heyden D, Tang W, Elias J. IL-11 secretion and its regulation in bone cells (abstr). J Bone Min Res 1993; 8:S143.
15. Elias JA, Tang W, Horowitz MC. Cytokine and hormonal stimulation of human osteosarcoma interleukin-11 production. Endocrinology 1995; 136:489–498.
15a. Du XX, Everett ET, Wang G, Lee W-H, Yang Z, Williams DA. Murine interleukin-11 is expressed at high levels in the hippocampus and expression is developmentally regulated in the testis. J Cell Physiol 1996; 168:362–372.
16. Passeri G, Bellido T, Girasole G, Tkaczyk A, Manolagas SC, Jilka RL. Transforming growth factor-β(TGFβ) and interleukin-1 (IL-1) induce the interleukin-11 (IL-11) mRNA in both bone marrow-derived stromal cells and osteoblasts from humans (abstr). J Bone Min Res 1993; 8:S162.
17. Zheng T, Nathanson MH, Elias JA. Histamine augments cytokine-stimulated IL-11 production by human lung fibroblasts. J Immunol 1994; 153:4742–4752.
18. Einarsson O, Geba GP, Panuska JR, Zhu Z, Landry M, Elias JA. Asthma-associated viruses specifically induce lung stromal cells to produce interleukin-11, a mediator of airways hyperreactivity. CHEST 1995; 107(suppl):132s–133s.
19. Elias JA, Zheng T, Whiting NL, et al. IL-1 and transforming growth factor-β regulation of fibroblast-derived IL-11. J Immunol 1994; 152:2421–2429.
20. Yang L, Yang Y-C. Heparin inhibits the expression of interleukin-11 and granulocyte-macrophage colony-stimulating factor in primate bone marrow stromal fibroblasts through mRNA destabilization. Blood 1995; 86:2526–2533.
21. Maier R, Ganu V, Lotz M. Interleukin-11, an inducible cytokine in human articular chondrocytes and synoviocytes, stimulates the production of the tissue inhibitor of metalloproteinases. J Biol Chem 1993; 268:21527–21532.
22. Yin T, Miyazawa K, Yang YC. Characterization of interleukin-11 receptor and protein tyrosine phosphorylation induced by interleukin-11 in mouse 3T3-L1 cells. J Biol Chem 1992; 267:8347–8351.
23. Hilton DJ, Hilton AA, Raicevic A, et al. Cloning of a murine IL-11 receptor α-chain; requirement for gp130 for high affinity binding and signal transduction. EMBO J 1994; 13:4765–4775.
24. Chérel M, Sorel M, Lebeau B, et al. Molecular cloning of two isoforms of a receptor for the human hematopoietic cytokine interleukin-11. Blood 1995; 86:2534–2540.

25. Yin T, Taga T, Tsang ML-S, Yasukawa K, Kishimoto T, Yang Y-C. Involvement of IL-6 signal transducer gp130 in IL-11-mediated signal transduction. J Immunol 1993; 151:1–7.
26. Nishimoto N, Ogata A, Shima Y, et al. Oncostatin M, leukemia inhibitory factor, and interleukin 6 induce the proliferation of human plasmacytoma cells via the common signal transducer, GP130. J Exp Med 1994; 179:1343–1347.
27. Fourcin M, Chevalier S, Lebrun J-J, et al. Involvement of gp130/interleukin-6 receptor transducing component in interleukin-11 receptor. Eur J Immunol 1994; 24:277–280.
28. Zhang X-G, Gu J-J, Lu Z-Y, et al. Ciliary neurotropic factor, interleukin 11, leukemia inhibitory factor, and oncostatin M are growth factors for human myeloma cell lines using the interleukin 6 signal transducer GP130. J Exp Med 1994; 177:1337–1342.
29. Yang Y-C, Yin T. Interleukin-11 and its receptor. Biofactors 1992; 4:1:15–21.
30. Yin T, Yang Y-C. Interleukin-11 mediated signal transduction pathways: comparison with those of interleukin-6. In: Moody TW, ed. Growth Factors, Peptides, and Receptors. New York: Plenum Press, 1993:323–333.
31. Yin T, Yasukawa K, Taga T, Kishimoto T, Yang Y-C. Identification of a 130-kilodalton tyrosine-phosphorylated protein induced by interleukin-11 as JAK2 tyrosine kinase, which associates with gp130 signal transducer. Exp Hematol 1994; 22:467–472.
32. Yin T, Yang Y-C. Mitogen-activated protein kinases and ribosomal S6 protein kinases are involved in signaling pathways shared by interleukin-11, interleukin-6, leukemia inhibitory factor, and oncostatin M in mouse 3T3-L1 cells. J Biol Chem 1994; 269:3731–3738.
33. Berger LC, Hawley TS, Lust JS, Goldman SJ, Hawley RG. Tyrosine phosphorylation of JAK-TYK kinases in malignant plasma cell lines growth-stimulated by interleukins 6 and 11. Biochem Biophys Res Comm 1994; 202:596–605.
34. Yin T, Yang Y-C. Protein tyrosine phosphorylation and activation of primary response genes by interleukin 11 in B9-TY1 cells. Cell Growth Diff 1993; 4:603–609.
35. Wegenka UM, Lutticken C, Buschmann J, et al. The interleukin-6-activated acute-phase response factor is antigenically and functionally related to members of the signal transducer and activator of transcription (STAT) family. Mol Cell Biol 1994; 14:3186–3196.
36. Yonemura Y, Kawakita M, Masuda T, Fujimoto K, Kato K, Takatsuki K. Synergistic effects of interleukin 3 and interleukin 11 on murine megakaryopoiesis in serum-free culture. Exp Hematol 1992; 20:1011–1016.
37. Kobayashi S, Teramura M, Sugawara I, Oshimi K, Mizoguchi H. Interleukin-11 acts as an autocrine growth factor for human megakaryoblastic cell lines. Blood 1993; 81:889–893.
38. Bruno E, Briddell RA, Cooper RJ, Hoffman R. Effects of recombinant interleukin-11 on human megakaryocyte progenitor cells. Exp Hematol 1991; 19:378–381.
39. Burstein SA, Mei RL, Henthorn J, Friese P, Turner K. Leukemia inhibitory factor and interleukin-11 promote maturation of murine and human megakaryocytes in vitro. J Cell Physiol 1992; 153:305–312.
40. Kaushansky K, Broudy VC, Lin N, et al. Thrombopoietin, the Mpl ligand, is essential for full megakaryocyte development. Proc Natl Acad Sci USA 1995; 92:3234–3238.
41. Yonemura Y, Kawakita M, Masuda T, Fujimoto K, Takasuki K. Effect of recombinant human interleukin-11 on rat megakaryopoiesis and thrombopoiesis in vivo: comparative study with interleukin-6. Br J Haematol 1993; 84:16–23.
42. Neben TY, Loebelenz J, Hayes L, et al. Recombinant human interleukin-11 stimulates megakaryocytopoesis and increases peripheral platelets in normal and splenectomized mice. Blood 1993; 81:901–908.
43. Mason L, Timony G, Perkin C, et al. Human recombinant interleukin-11 promotes maturation of bone marrow megakaryocytes in non-human primates: an electron microscopic study (abstr). Blood 1993; 82:69a.
44. Bree A, Schlerman F, Timony G, McCarthy K, Stoudimire J. Pharmacokinetics and thrombopoietic effects of recombinant human interleukin-11 (rhIL-11) in nonhuman primates and rodents (abstr). Blood 1991; 78:132a.
45. Gordon MS, Battiato L, Hoffman R, et al. Subcutaneously (SC) administered recombinant human interleukin-11 (neumega™ rhIL-11 growth factor; rhIL-11) prevents thrombocytopenia following chemotherapy (CT) with cyclophosphamide (C) and doxorubicin (A) in women with breast cancer (abstr). Blood 1993; 82:318a.
46. Gordon MS, Sledge GW, Battiato L, et al. The in vivo effects of subcutaneously (SC) administered recombinant human interleukin-11 (neumega™ rhIL-11 growth factor; rhIL-11) in women with breast cancer (BC) (abstr). Blood 1993; 82:498a.
47. Orazi A, Cooper R, Tong J, et al. Recombinant human interleukin-11 (neumega™ rhIL-11 growth factor: rhIL-11) has multiple profound effects on human hematopoiesis (abstr). Blood 1993; 82:369a.
48. Lemoli RM, Rogli M, Fortuna A, et al. Interleukin-11 stimulates the proliferation of human hematopoietic $CD34^+$ and $CD34^+CD33^-DR^-$ cells and synergizes with stem cell factor, interleukin-3, and granulocyte-macrophage colony-stimulating factor. Exp Hematol 1993; 21:1668–1672.
49. Brandt JE, Hoffman R. Interleukin-11 plays a pivotal role in the in vitro expansion of human hematopoietic progenitor and stem cells (abstr). Blood 1993; 82:368a.
50. Neben S, Donaldson D, Sieff C, et al. Synergistic effects of interleukin-11 with other growth factors on the expansion of murine hematopoietic progenitors and maintenance of stem cells in liquid culture. Exp Hematol 1994; 22:353–359.
51. Du XX, Scott D, Yang ZX, Cooper R, Xiao XL, Williams DA. Interleukin-11 stimulates multilineage progenitors, but not stem cells, in murine and human long-term marrow cultures. Blood 1995; 86:128–134.
52. Tsuji K, Lyman SD, Sudo T, Clark SC, Ogawa M. Enhancement of murine hematopoiesis by synergistic interactions between steel factor (ligand for c-kit), interleukin-11, and other early acting factors in culture. Blood 1992; 11:2855–2860.
53. Musashi M, Yang Y-C, Paul SR, Clark SC, Sudo T, Ogawa M. Direct and synergistic effects of interleukin 11 on murine hemopoiesis in culture. Proc Natl Acad Sci USA 1991; 88:765–769.
54. Leary AG, Zeng HQ, Clark SC, Ogawa M. Growth factor requirements for survival in G_0 and entry into the cell cycle of primitive human hemopoietic progenitors. Proc Natl Acad Sci USA 1992; 89:4013–4017.
55. Musashi M, Clark SC, Sudo T, Urdal DL, Ogawa M. Synergistic interactions between interleukin-11 and interleukin-4 in support of proliferation of primitive hematopoietic progenitors of mice. Blood 1991; 78:1448–1451.
56. Ploemacher RE, van Soest PL, Boudewijn A, Neben S. Interleukin-12 enhances interleukin-3 dependent multilineage hematopoietic colony formation stimulated by interleukin-11 or steel factor. Leukemia 1993; 7:1374–1380.
57. Ploemacher RE, van Soest PL, Voorwinden H, Boudewijn A. Interleukin-12 synergizes with interleukin-3 and steel factor to

enhance recovery of murine hematopoietic stem cells in liquid culture. Leukemia 1993; 7:1381–1388.
58. Hirayama F, Katayama N, Neben S, et al. Synergistic interaction between interleukin-12 and steel factor in support of proliferation of murine lymphohematopoietic progenitors in culture. Blood 1994; 83:92–98.
59. Jacobsen SEW, Okkenhaug C, Veiby OP, Caput D, Ferrara P, Minty A. Interleukin 13: novel role in direct regulation of proliferation and differentiation of primitive hematopoietic progenitor cells. J Exp Med 1994; 180:75–82.
60. Jacobsen SEW, Okkenhaug C, Myklebust J, Veiby OP, Lyman SD. The FLT3 ligand potently and directly stimulates the growth and expansion of primitive murine bone marrow progenitor cells in vitro: synergistic interactions with interleukin (IL) 11, IL-12, and other hematopoietic growth factors. J Exp Med 1995; 181:1357–1363.
61. Hangoc G, Yin T, Cooper S, Schendel P, Yang Y-C, Broxmeyer HE. In vivo effects of recombinant interleukin-11 on myelopoiesis in mice. Blood 1993; 81:965–972.
62. Quinto CM, Leonard JP, Kozitza MK, Goldman SJ. Synergistic interactions between rhIL-11 and G-CSF in the mobilization of hematopoietic progenitors (abstr). Blood 1993; 82:369a.
63. Quesniaux VFJ, Clark SC, Turner K, Fagg B. Interleukin-11 stimulates multiple phases of erythropoiesis in vitro. Blood 1992; 80:1218–1223.
64. Yee NS, Langen H, Besmer P. Mechanism of kit ligand, phorbol ester, and calcium-induced down-regulation of c-kit receptors in mast cells. J Biol Chem 1993; 268:14189–14201.
65. Hirayama F, Shih J-P, Awgulewitsch A, Warr GW, Clark SC, Ogawa M. Clonal proliferation of murine lymphohemopoietic progenitors in culture. Proc Natl Acad Sci USA 1992; 89:5907–5911.
66. Hirayama F, Clark SC, Ogawa M. Negative regulation of early B lymphopoiesis by interleukin 3 and interleukin 1α. Proc Natl Acad Sci USA 1994; 91:469–473.
67. Le BL, Cumano A, Iscove NN, Paige CJ. Stromal cell independent growth of bipotent B cell-macrophage precursors from murine fetal liver. Int Immunol 1994; 6:401–407.
68. Yin T, Schendel P, Yu-Chung Y. Enhancement of in vitro and in vivo antigen-specific antibody responses by interleukin-11. J Exp Med 1992; 175:211–216.
69. Anderson KC, Morimoto C, Paul SR, et al. Interleukin-11 promotes accessory cell-dependent B-cell differentiation in humans. Blood 1992; 80:2797–2804.
70. Burger R, Gramatzki M. Responsiveness of the interleukin (IL)-6-dependent cell line B9 to IL-11. J Immunol Methods 1993; 158:147–148.
71. Lu Z-Y, Zhang X-G, Gu Z-J, et al. A highly sensitive quantitative bioassay for human interleukin-11. J Immunol Methods 1994; 173:19–26.
72. Zanjani ED, Pallavicini MG, Ascensao JL, et al. Engraftment and long-term expression of human fetal hematopoietic stem cells in sheep following transplantation in utero. J Clin Invest 1992; 89:1178–1188.
73. Hu JP, Cesano A, Santoli D, Clark SC, Hoang T. Effects of interleukin-11 on the proliferation and cell cycle status of myeloid leukemic cells. Blood 1993; 81:1586–1592.
74. Ohsumi J, Miyadai K, Kawashima I, Sakakibara S, Yamaguchi J, Itoh Y. Regulation of lipoprotein lipase synthesis in 3T3-L1 adipocytes by interleukin-11/adipogenesis inhibitory factor. Biochem Mol Biol Int 1994; 32:705–712.
75. Keller DC, Du XX, Srour EF, Hoffman R, Williams DA. Interleukin-11 inhibits adipogenesis and stimulates myelopoiesis in long term marrow cultures. Blood 1993; 82:1428–1435.
76. Baumann H, Schendel P. Interleukin-11 regulates the hepatic expression of the same plasma protein genes as interleukin-6. J Biol Chem 1991; 266:20424–20427.
77. Fukuda Y, Sassa S. Effect of interleukin-11 on the levels of mRNAs encoding heme oxygenase and haptoglobin in human HepG2 hepatoma cells. Biochem Biophys Res Comm 1993; 193:297–302.
78. Girasole G, Passeri G, Jilka RL, Manolagas SC. Interleukin-11: a new cytokine critical for osteoclast development. J Clin Invest 1994; 93:1516–1524.
79. Hughes FJ, Howells GL. Interleukin-11 inhibits bone formation in vitro. Calcif Tissue Int 1993; 53:362–364.
80. Girasole G, Passeri G, Knutson S, Jilka RL, Manolagas SC. A distinct and hierarchically central role of interleukin-11 among other cytokines in osteoclast development (abstr). J Bone Min Res 1993; 8:S117.
81. Glowacki J, Shevde N, Greenberger JS. Distinct roles of interleukins-1, -6, and -11 in osteoclastic differentiation of hematopoietic progenitors (abstr). J Bone Min Res 1993; 8:S164.
82. Mehler MF, Rozental R, Dougherty M, Spray DC, Kessler JA. Cytokine regulation of neuronal differentiation of hippocampal progenitor cells. Nature 1993; 362:62–65.
83. Keller G, Kennedy M, Carlsson L. Hematopoietic commitment of embryonic stem (ES) cells in culture (abstr). Exp Hematol 1994; 22:773.
84. Ireland RC, Rubin LP, Iovene C, et al. A potential role for interleukin-11 in trophoblast differentiation and placentation (abstr). Blood 1994; 84:276a.
85. Fann M-J, Patterson PH. Neuropoietic cytokines and activin A differentially regulate the phenotype of cultured sympathetic neurons. Proc Natl Acad Sci USA 1994; 91:43–47.
86. Roeb E, Graeve L, Hoffmann R, Decker K, Edwards DR, Heinrich PR. Regulation of tissue inhibitor of metalloproteinases-1 gene expression by cytokines and dexamethasone in rat hepatocyte primary cultures. Hepatol 1993; 18:1437–1442.
87. Lopez-Valpuesta FJ, Myers RD. Fever produced by interleukin-11 (IL-11) injected into the anterior hypothalamic pre-optic area of the rat is antagonized by indomethacin. Neuropharmacology 1994; 33:989–994.
88. Baynes RD, Cook JD. Interleukin-11 enhances gastrointestinal absorption of iron (abstr). Blood 1994; 84:276a.
89. Du XX, Neben T, Goldman S, Williams DA. Effects of recombinant human interleukin-11 on hematopoietic reconstitution in transplant mice: acceleration of recovery of peripheral blood neutrophils and platelets. Blood 1993; 81:27–34.
90. Du XX, Keller D, Goldman S, Williams DA. Functional effects of interleukin-11 treatment in vivo following bone marrow transplantation (BMT) and combined modality therapy in mice. Exp Hematol 1992; 20:768.
91. Kurtz A, Hartl W, Jelkmann W, Zapf J, Bauer C. Activity in fetal bovine serum that stimulates erythroid colony formation in fetal mouse livers is insulinlike growth factor 1. J Clin Invest 1985; 76:1643–1648.
92. Hawley RG, Fong AZC, Ngan BY, de Lanux VM, Clark SC, Hawley TS. Progenitor cell hyperplasia with rare development of myeloid leukemia in interleukin 11 bone marrow chimeras. J Exp Med 1993; 178:1175–1188.
93. Maze R, Moritz T, Williams DA. Increased survival and multilineage hematopoietic protection from delayed and severe myelosuppressive effects of a nitrosourea with recombinant interleukin-11. Cancer Res 1994; 54:4947–4951.
94. Du XX, Doerschuk CM, Orazi A, Williams DA. Bone marrow stromal-derived growth factor, interleukin-11, stimulates re-

covery of small intestinal mucosal cells after cytoablative therapy. Blood 1994; 83:33–37.

95. Orazi A, Du XX, Yang Z, Kashai M, Williams DA. Interleukin-11 prevents apoptosis and accelerates recovery of small intestinal mucosa in mice treated with combined chemotherapy and radiation. Lab Invest 75:33–42.

96. Leonard JP, Quinto CM, Kozitza MK, Neben TY, Goldman SJ. Recombinant human interleukin-11 stimulates multilineage hematopoietic recovery in mice after a myelosuppressive regimen of sublethal irradiation and carboplatin. Blood 1994; 83:1499–1506.

97. Keith JCJr, Albert L, Sonis ST, Pfeiffer CJ, Schaub RG. IL-11, a pleiotropic cytokine: exciting new effects of IL-11 on gastrointestional mucosal biology. Stem Cells 1994; 12:79–90.

98. Chang M, Williams A, Ishizawa L, Knoppel A, Van de ven C, Cairo MS. Role of interleukin-11 (IL-11) during experimental group B streptococcal (GBS) sepsis in neonatal rats: prophylactic use of IL-11 improves survival and enhances platelet recovery (abstr). Blood 1994; 84:477a.

99. Kaye JA, Loewy J, Blume J, et al. Recombinant human interleukin eleven (neumega™ rhIL-11 growth factor) increases plasma von Willebrand factor and fibrinogen concentrations in normal human subjects (abstr). Blood 1994; 84:276a.

100. Suen Y, Chang M, Killen R, Cameron K, Rosenthal J, Cairo M. Physiological role of endogenous IL-11 during thrombocytopenia post myeloablative therapy (MAT) and BMT (abstr). Blood 1994; 84:91a.

101. Ault KA, Mitchell J, Knowles C, et al. Recombinant human interleukin eleven (neumega™ rhIL-11 growth factor) increases plasma volume and decreases urine sodium excretion in normal human subjects (abstr). Blood 1994; 84:276a.

102. Champlin RE, Mehra R, Kaye JA, et al. Recombinant human interleukin eleven (rhIL-11) following autologous BMT for breast cancer (abstr). Blood 1994; 84:395a.

103. McCredie KB, Hersh EM, Freireich EJ. Cells capable of colony formation in the peripheral blood of man. Science 1971; 171:293–294.

104. Suen Y, Chang M, Lee SM, Buzby JS, Cairo MS. Regulation of interleukin-11 protein and mRNA expression in neonatal and adult fibroblasts and endothelial cells. Blood 1994; 84:4125–4134.

105. Tepler I, Elias L, Smith I, Hussein M, Rosen G, Chang AY-C, Moore JO, Gordon MS, Kuca B, Beach KJ, Loewy JW, Garnick MB, Kaye JA. A randomized placebo-controlled trial of recombinant interleukin-11 in cancer patients with severe thrombocytopenia due to chemotherapy. Blood 1996; 87:3607.

12

Interleukin-12: A Heterodimeric Cytokine That Promotes Cell-Mediated Immunity

Maurice K. Gately, Chang You Wu, and Denise A. Faherty
Hoffmann-La Roche, Inc., Nutley, New Jersey

INTRODUCTION/HISTORY OF IL-12

Interleukin-12 (IL-12) is a heterodimeric cytokine which exerts important immunoregulatory effects on T and natural killer (NK) cells. Human IL-12 was discovered independently by investigators at Hoffmann-La Roche, Inc. and by Trinchieri and colleagues at the Wistar Institute in collaboration with investigators at Genetics Institute. Trinchieri et al. initially identified IL-12 on the basis of its ability to enhance NK lytic activity and to induce interferon-γ (IFN-γ) production by NK and T cells and called this cytokine natural killer cell stimulatory factor (NKSF) (1), whereas the Roche group identified IL-12 on the basis of its ability to synergize with IL-2 in inducing lymphokine-activated killer (LAK) cells and cytolytic T-lymphocyte (CTL) responses and thus called this factor cytotoxic lymphocyte maturation factor (CLMF) (2,3). Cloning and sequencing of the cDNAs encoding NKSF and CLMF showed these two factors to be identical, and designation of this factor as "interleukin-12" was proposed (4,5). Mouse IL-12 was initially described by Germann et al. (6–8) as a cytokine that enhances IL-2–induced proliferation of mouse type 1 T-helper (Th1) cells and induced IFN-γ production by mouse Th1 and NK cells. Germann et al. called this factor T-cell–stimulating factor (TSF). TSF itself was never isolated or cloned; however, subsequent to the cloning and expression of mouse IL-12 using human IL-12 cDNA clones as probes (9), it was shown that TSF and mouse IL-12 were identical (10).

Like other cytokines, IL-12 is pleiotropic in its activities. In addition to its ability to enhance the lytic activity of NK/LAK cells and CTL, to induce the secretion of IFN-γ by both T and NK cells, and to stimulate the proliferation of activated T and NK cells, IL-12 has been shown to play a unique and critical role in facilitating cell-mediated immunity by promoting the maturation and activation of TH_1 cells (11–13). As will be discussed below, endogenous IL-12 likely plays an important role in the host defense against predominantly intracellular microbial pathogens. In addition, recombinant IL-12 (rIL-12) has been shown to have striking therapeutic effects in a number of mouse tumor models (14,15) and mouse models of infectious diseases (16–21). Clinical trials investigating the potential therapeutic effects of rIL-12 in human cancer patients and HIV-infected patients have recently been initiated. In this manuscript, we will focus on the structure of IL-12 and its receptor, the cellular sources of IL-12 and factors regulating its biosynthesis, the biological activities of IL-12, and the role of IL-12 in normal and pathological immune responses. Additional information can be obtained from several recently published reviews (22–25).

STRUCTURE OF IL-12

The protein structure of IL-12 is unique among the interleukins in that it is a heterodimer composed of two disulfide-bonded subunits with approximate molecular masses of 40 and 35 kDa, respectively (1,3,26). The heavy chain (designated p40) of human IL-12 is composed of 306 amino acids and contains 10 cysteine residues and 4 potential N-linked glycosylation sites, whereas the light chain (designated p35) consists of 197 amino acids and possesses 7 cysteine residues and 3 potential N-linked glycosylation sites (4,5). Chemical and enzymatic deglycosylation of the human IL-12 heterodimer demonstrated that the p40 and p35 subunits contain approximately 10 and 20% carbohydrate, respectively (26). The two subunits of human IL-12 are structurally unrelated to one another and are encoded by genes located on separate chromosomes (27). The gene for the human p40 subunit is located on chromosome 5 (5q31-q33) in a region containing genes encoding receptors of various types including the macrophage colony-stimulating factor receptor (25,27). It is noteworthy that, as discussed further below, the IL-12 p40 subunit itself is structurally a member of the hematopoietin/cytokine receptor family (9,28). The gene for the human IL-12 p35 subunit is located on chromosome 3 (3p12-3q13.2) (27).

The cDNAs for mouse IL-12 have also been cloned and expressed, and the structure of mouse IL-12 is similar to that of human IL-12 (9). The predicted amino acid sequence of the mouse IL-12 p40 subunit is 70% identical to human IL-12 p40, whereas the human and mouse p35 subunits share 60% amino acid sequence identity (9). Mouse IL-12 is fully active on human lymphocytes, but human IL-12 is inactive on mouse lymphocytes. Differences in the p35 subunits of human and mouse IL-12 are critical for this species specificity (9). Structure-function analysis of the mouse p35 subunit identified five amino acid residues at three discontinuous sites, which contributed greatly to the species specificity observed for mouse and human IL-12 (29). Recently, the cDNA encoding the IL-12 p35 subunit of feline IL-12 was also cloned, and feline p35 was shown to be structurally similar to human and mouse p35, displaying three leucine zipper motifs (30).

There is no sequence homology between the IL-12 p35 and p40 subunits. However, the IL-12 p35 subunit displays a "distant but significant relationship" to IL-6, granulocyte colony-stimulating factor, and chicken myelomonocytic growth factor (31). In contrast, although having no homology to any other cytokines, the sequence of the IL-12 p40 subunit is related to the extracellular domain of the IL-6 receptor α subunit (IL-6Rα) and to the ciliary neurotrophic factor receptor (9,28). Thus, IL-12 is structurally analogous to a secreted, disulfide-linked complex of a cytokine with a cytokine receptor. The structural relationships of IL-12 p35 and p40 to IL-6 and IL-6Rα, respectively, led to speculation that (1) IL-12 p40 might exist in membrane-bound form and (2) the IL-12 receptor (IL-12R) might be related to gp130, the signal-transducing component of the IL-6R (28). At present, there is no evidence for the existence of membrane-associated forms of IL-12 or either of its subunits. Nonetheless, as will be discussed below, at least one component of the IL-12R is structurally related to gp130 (32).

Disulfide-linked IL-12 heterodimer is necessary for biological activity since neither IL-12 subunit alone displays significant biological activity over the range of concentrations at which the IL-12 heterodimer is active (4,5). However, it has been reported that mixtures of soluble recombinant p40 and p35 can mediate IL-12 activity, but at concentrations several logs higher than that at which the IL-12 heterodimer is active (33). Culture supernates containing mouse IL-12 p40 were shown to inhibit specifically the biological activity of mouse IL-12 heterodimer in several in vitro assays (34), and a disulfide-linked homodimeric form of IL-12 p40 has been shown to act as an IL-12 antagonist in both human (35) and murine (36) systems. Competitive binding assays performed under high-affinity binding conditions suggested that mouse p40 homodimer binds to the mouse IL-12R with an affinity similar to that of mouse IL-12 heterodimer, whereas human IL-12 p40 homodimer appears to bind to the human IL-12R with an affinity that is 5- to 10-fold lower than that of human IL-12. Likewise, human p40 homodimer is approximately 10-fold less potent than mouse p40 homodimer in its ability to inhibit IL-12 bioactivity (35,36). The monomeric forms of both human and mouse p40 displayed little, if any, ability to block IL-12 binding to its receptor or to inhibit IL-12 bioactivity. It is conceivable that IL-12 p40 homodimer might act as a physiological regulator of IL-12 activity; however, this has not yet been demonstrated.

PRODUCTION OF IL-12

Human IL-12 was originally purified from the culture supernates of Epstein-Barr virus (EBV)–transformed B-cell lines activated with phorbol ester (1) or phorbol ester plus calcium ionophore (3). Screening of many different human cell lines for IL-12 production showed that most EBV-transformed human B-cell lines constitutively produced low levels of IL-12 p40 and that the production of IL-12 heterodimer could be induced by stimulation with phorbol diester (37). A number of T-cell lines, and solid tumor–derived cell lines did not produce IL-12 either constitutively or upon stimulation with phorbol ester (37); however, myeloid cell lines, THP-1, ML-3, and HL-60, produced low levels of IL-12 when stimulated with endotoxin lipopolysaccharide (LPS) (38). In all cell lines producing IL-12, the p40 subunit was secreted in large excess over the heterodimer, but secretion of free p35 was not observed (26,37,38). Constitutive expression of p35 mRNA was seen in a variety of cell lines of both hematopoietic and nonhematopoietic origin, whereas expression of p40 mRNA was found less frequently and correlated with ability to produce IL-12 (37,39). Nonetheless, in the

RPMI-8866 and NC-37 B-cell lines from which human IL-12 was originally isolated, expression of p40 and p35 mRNA was coordinately upregulated following activation by phorbol ester (4,5).

In contrast to the results of cell line studies in which B-cell lines were the prominent producers of IL-12, monocyte/ macrophages appear to constitute the primary normal cellular source of IL-12 in both human (37) and murine (12,18,40) systems. In a number of studies performed in vitro, both human monocytes and mouse macrophages have been shown to produce IL-12 in response to a variety of bacteria, mycobacteria, protozoan parasites, and various substances derived from these microorganisms, including endotoxin and several bacterial superantigens (12,18,37,40–46). Unlike B-cell lines, normal human monocytes did not produce IL-12 in response to phorbol diester (37). Recently, human monocytes were shown to release IL-12 as a result of culture with antigen-activated T cells, and IL-12 production in these cultures was suppressed by the addition of soluble CD40 (47). In addition, IL-12 production by human monocytes could be induced by culture with anti-CD40 monoclonal antibody (mAb) immobilized on irradiated, F_c receptor (CD32)–transfected L cells (47). Likewise, we have found that production of IL-12 p40 subunit by anti-CD3–activated human peripheral blood mononuclear cells (PBMC) can be inhibited by mAb to human CD40 ligand (C. Wu and M. Gately, unpublished results). These data indicate that activated T cells can induce IL-12 production by monocytes via the interaction of T-cell–associated CD40 ligand with CD40 on monocytes and provide a mechanism whereby IL-12 production may be elicited in the course of virtually any T-cell–dependent immune response.

In addition to monocytes and macrophages, other cell types have also been reported to produce IL-12 in vitro. *Staphylococcus aureus* Cowan I strain (SAC) elicited IL-12 production by monocyte-depleted human peripheral blood lymphocytes, most probably from B cells (37). On the other hand, purified human T and NK cells did not produce detectable amounts of IL-12 or IL-12 p40 mRNA when activated with phorbol diester, IL-2 and/or anti-CD3, or anti-CD16 antibodies (37). Incubation of human polymorphonuclear leukocytes with LPS led to secretion of IL-12 (48), as did culture of primary human keratinocyte cell lines with phorbol diester or with trinitobenzenesulfonate, an analog of the contact allergen trinitrochlorobenzene to which the keratinocyte donors had been sensitized (49,50). Mouse bone marrow mast cells cultured in mast cell growth factor or IL-3 exhibited properties similar to connective tissue mast cells or mucosal mast cells, respectively. Mast cell growth factor–induced mast cells expressed IL-12 p40 and p35 mRNA but not IL-4 mRNA, whereas IL-3–induced mast cells expressed IL-4 but not IL-12 transcripts (51). Whether or not dendritic cells can produce IL-12 has been controversial. Early results suggested that mouse dendritic cells did not produce IL-12 (52); however, more recent studies raise the possibility that dendritic cells may be capable of producing IL-12 in small amounts (25,53). Expression of both IL-12 p35 and p40 mRNAs was detected in mouse thymus and in several thymic stromal cell lines (9,54).

Upregulation of IL-12 mRNA and/or production of IL-12 protein in vivo has been demonstrated in mice following infection with various micro-organisms (41,43,55–58), administration of endotoxin (59–61), or administration of rIL-12 itself (19). Likewise, IL-12 protein has been demonstrated in the pleural fluid of human patients suffering from tuberculous pleuritis (42) and by immunohistochemistry in lesions from patients with tuberculoid leprosy (44). Production of IL-12 in vivo as well as in vitro has invariably been accompanied by production of excess p40 subunit (37,38,42,45,59,60). Whether this excess p40 contains inhibitory p40 homodimer that could act as a physiological regulator of IL-12 activity is at present unknown. In situ hybridization studies on spleens from normal mice or injected with LPS or staphylococcal enterotoxin B (SEB) showed that expression of p40 mRNA was strongly upregulated after injection of LPS or SEB, whereas p35 mRNA expression was increased only slightly (61). Importantly, in both spleens from normal mice and spleens from mice that had received LPS or SEB, the p35 and p40 mRNAs were largely localized to different areas of the spleen and hence were produced, for the most part, by different cells (61). Since both p40 and p35 mRNA must be expressed in the same cell for production of bioactive IL-12 heterodimer (4,5), these results illustrate the potential problems inherent in interpreting changes in p40 or p35 mRNA levels in whole tissues or mixed cell populations as indicative of changes in IL-12 production. Measurements of IL-12 protein are essential to confirm any conclusions in this regard.

IL-12 production can be either positively or negatively regulated by several cytokines. IFN-γ by itself induced little or no IL-12 production, but can strongly potentiate the secretion of both IL-12 heterodimer and the IL-12 p40 chain from human monocytes, granulocytes, and myeloid cell lines and from mouse macrophages stimulated with LPS, SAC, or heat-killed *Listeria monocytogenes* (38,48, and M. Gately, unpublished results). The potentiating effect of IFN-γ on IL-12 production by LPS-stimulated granulocytes was due both to enhancement of the LPS-induced accumulation of p40 mRNA and to a directly induced increase in p35 mRNA expression (48). Suppression of IL-12 production can be caused by the Th2-associated cytokines IL-4 (44,62,63), IL-10 (12,38,40,44,46,48,63,64), and IL-13 (62,63,65) and by transforming growth factor (TGF)-β (63). However, the regulatory effects of IL-4 and IL-13 on IL-12 production are complex. Addition of these cytokines to cultures of human PBMC simultaneously with LPS or SAC led to inhibition of IL-12 production, but preincubation of PBMC with IL-4 or IL-13 for ≥ 20 h prior to addition of LPS or SAC resulted in enhanced secretion of IL-12 (63). Prostaglandin E_2 (PGE_2), a potent factor in promoting Th2 responses, inhibited LPS-induced IL-12 production by human monocytes (66). The inhibitory effect of PGE_2 was independent of IL-10 and appeared to be mediated via an increase in intracellular cAMP, since other cAMP-increasing agents similarly inhibited the production of IL-12 (66). Interestingly, PBMC from HIV-infected individuals, as compared to normal human PBMC, produced diminished amounts of IL-12 in response to SAC in vitro, whereas production of tumor necrosis factor (TNF)-α, IL-1β, IL-6, and IL-10 was not depressed (67). This effect appeared to be directly related to HIV infection of monocytes and was independent of IL-10 (67). These results indicate that IL-12 production by monocytes can be regulated independent of changes in the secretion of other monokines.

BIOLOGICAL ACTIVITIES OF IL-12

Regulation of TH_1/TH_2 Responses by IL-12

Helper T cells can be divided into two subsets: TH_1 cells, which secrete IL-2, IFN-γ, and TNF-β, and TH_2 cells, which produce IL-4, IL-5, IL-6, IL-10, and IL-13 (68,69). TH_1 cells preferentially induce cell-mediated immune responses, while TH_2 cells facilitate humoral immunity (68,69). Mounting evidence suggests that IL-12 plays a unique role in promoting the maturation and activation of TH_1 cells (10–13, 70–75), whereas IL-4 is critical for inducing TH_2 cell responses (69).

In an in vitro mouse model using ovalbumin-specific αβ T-cell receptor transgenic $CD4^+$ T cells, Hsieh et al. (12) demonstrated that heat-killed *L. monocytogenes* preferentially induced the development of TH_1 cells from naive mouse T cells by inducing production of IL-12 by macrophages. Mouse rIL-12, but none of several other cytokines tested, could substitute for *Listeria*-activated macrophages in this model. Addition of IL-10 inhibited Th1 development mediated by *Listeria*-stimulated macrophages but not that induced by added rIL-12, suggesting that the inhibitory effect of IL-10 was due to its ability to inhibit IL-12 production by macrophages (12). In this model, the ability of IL-12 to direct TH_1 development was inhibited by anti–IFN-γ, although addition of IFN-γ to primary cultures was not sufficient by itself to support a TH_1 response (52). When IL-4, which caused the development of TH_2 cells in this model, was added simultaneous to IL-12, the effects of IL-4 were dominant over those of IL-12 (12). In contrast to the marked increases in IFN-γ production it induced, IL-12 caused no change in the amount of IL-2 secreted by $CD4^+$ T cells in this model (12), suggesting that the production of IL-2 by TH_1 cells may be regulated differently from the secretion of IFN-γ. In additional studies in other in vitro models of mouse TH_1 cell development, it was shown that IL-12 could exert its effects directly on naive $CD4^+$ T cells (71,72). IFN-γ was a necessary costimulus of IL-12–dependent TH_1 development in most (71,73), but not all (71), models, and induction of TH_1 responses by IL-12 could be inhibited by TGF-β (73), as well as by IL-4 (72,73). $CD8^+$ CTL, like $CD4^+$ helper T cells, can secrete a variety of cytokines and can be divided into type 1 and type 2 subsets based on their ability to produce IFN-γ or IL-4 and IL-5, respectively (76,77). Similar to the results for $CD4^+$ T-helper cells, IL-12 was found to promote the development of type 1 $CD8^+$ CTL, whereas IL-4 induced the development of type 2 $CD8^+$ CTL (76,77).

In addition to promoting the commitment of naive T cells to the TH_1 pathway, IL-12 has been shown to serve as an important costimulus for the activation of fully differentiated mouse TH_1 cells (13,74). IL-12 was shown to synergize with B7 as costimuli in facilitating anti-CD3–induced proliferation and IFN-γ secretion by cloned Th1 cells (13). More importantly, antigen-specific proliferation and IFN-γ secretion by cloned TH_1 cells were substantially inhibited by anti–IL-12 and nearly abolished by the combination of anti–IL-12 plus CTLA4-Ig, a specific inhibitor of B7-CD28 interactions (13,74). The ability of IL-10 to inhibit antigen-induced proliferation and IFN-γ production by TH_1 clones in the presence of splenic antigen-presenting cells (APC) could be largely reversed by addition of IL-12, suggesting that the inhibitory effect of IL-10 was largely due to downregulation of IL-12 production by splenic APC (13,74). In contrast to its ability to act as a stimulus or costimulus to elicit a number of biological responses by cloned mouse TH_1 cells, IL-12 appeared to lack biological activity on mouse TH_2 cells (10,74,78). Recent data suggest that the unresponsiveness of cloned mouse TH_2 cells to IL-12 is not due to lack of high-affinity IL-12R but rather due to extinction of IL-12 signaling which can occur within 3 days after activation of naive T cells in the presence of IL-4 (79). It should be noted that this loss of responsiveness to IL-12 has not been observed in cloned human TH_2 cells (see below).

Consistent with the in vitro results described above, IL-12 has been shown to promote TH_1-type cytokine responses while suppressing the development of TH_2-type responses in mice in vivo following immunization with protein antigens (80,81) or challenge with various infectious agents (16,17,19,43,55,82–84). For maximum suppression of a TH_2-type response, IL-12 must be administered early before such a response has become established (82,83). When IL-12 was administered for the first time during a secondary immune response, IFN-γ expression was still enhanced, but little or no suppression of IL-4 expression was observed (82). The role of endogenous IL-12 in promoting TH_1 responses is also supported by studies in IL-12 knock-out mice in which it was shown that lymph node cells from mice immunized with a protein antigen proliferated normally in response to antigen restimulation in vitro, but secretion of IFN-γ was reduced by ~ 90% (85).

IL-12 has also been shown in vitro to promote the development of human TH_1 cells while inhibiting the generation of human TH_2 cells (11,70,75). However, in contrast to mouse TH_2 cells, which appear to lose responsiveness to IL-12 during their development (10,74,78,79), activation of cloned human TH_2 cells in the presence of IL-12 induced or enhanced secretion of IFN-γ (75,86). IL-12 did not significantly alter the amount of IL-4 secreted by activated human TH_2 clones (75,86). The inducing effect of IL-12 on IFN-γ production by human TH_2 clones was transient, since subsequent culture in the absence of IL-12 resulted in loss of the ability to produce IFN-γ in response to activation (75,86). Additional evidence suggests that IL-12 may regulate the balance between TH_1-type (type 1) and TH_2-type (type 2) cytokine production in humans in vivo. Human patients having the tuberculoid form of leprosy restrict the growth of *M. leprae* and express type 1 cytokines in their lesions, whereas patients with lepromatous leprosy express type 2 cytokines and develop disseminated infection (44). It was found that expression of IL-12 heterodimer, as measured by immunohistochemistry, and of IL-12 p40 mRNA was approximately 10-fold greater in lesions of tuberculoid leprosy patients than in lepromatous lesions (44). Decreased production of IL-12 also appeared to be correlated with aberrant production of type 2 cytokines by PBMC from patients having an advanced form of cutaneous T-cell lymphoma, Sezary syndrome (87).

Exceptions to the generalization that IL-12 promotes TH_1-type responses while suppressing TH_2-type responses have been reported. In an in vitro system, activation of naive mouse $CD4^+$ T cells by immobilized anti-CD3 in the presence of high concentrations of IL-4 plus IL-12 led to the development of increased TH_2-cell activity (as measured by IL-4 production in secondary cultures), as compared to the TH_2-cell activity generated when T cells were activated in the presence of IL-4 alone (72). Thus, in this model IL-12 appeared to enhance IL-4–dependent differentiation of TH_2 cells; nonetheless, in the absence of IL-4, IL-12 induced a TH_1 response (72). Administration of IL-12 to BALB/c mice infected with *Leishmania major* resulted in increased expression of IFN-γ mRNA in draining lymph nodes and decreased expression of IL-4 mRNA if IL-12 treatment was begun at the time of infection. However, if IL-12 treatment was begun on day 14 when the TH2 response to infection was well established, IL-12 caused enhanced expression of IL-4, as well as IFN-γ, mRNA (88). The effects of IL-12 on IL-4 mRNA expression in this model were similar in normal and IFN-γ knock-out mice, indicating that IL-12 could regulate IL-4 expression independent of IFN-γ (88). Different results were obtained when the effects of IL-12 on schistosome egg–induced TH_2 responses and granuloma formation were compared in normal and IFN-γ knock-out mice. Rather than suppressing granuloma formation and eosinophilia as had been observed in wild-type mice (19), IL-12 exacerbated TH_2-dependent granuloma formation and failed to reduce peak tissue eosinophilia in the IFN-γ knock-out mice (89). IL-12 also caused substantial elevation of serum IgE levels in the IFN-γ knock-out mice following egg injection. In these studies IL-12 treatment was begun at the time of egg injection. Hence, in this model the IL-12–induced suppression of schistosome egg–induced TH_2 responses observed in wild-type mice was clearly IFN-γ dependent. IL-12 was also shown to exert differing effects on the development of naive human $CD4^+$ T cells into TH_1 and TH_2 cells, depending upon whether neonatal or adult human T cells were used (90). Priming of naive cells from human adults in the presence of IL-12 led to the development of TH_1 cells, whereas priming of neonatal human T cells in the presence of IL-12 resulted in a cell population that produced high levels of both IFN-γ and IL-4 following secondary stimulation (90). Preincubation of unstimulated neonatal T cells with IL-12 before priming resulted in the development of TH_1-like cells after priming in the presence of IL-12 (70,90). Preincubation of unstimulated neonatal T cells with IL-4 prior to activation by phorbol ester and ionomycin resulted in production of IFN-γ and IL-2, but not IL-4, following activation. In contrast, preincubation with IL-4 together with IL-12 led to production of IL-4, in addition to IL-2 and IFN-γ, in response to phorbol ester and ionomycin (91). In the aggregate, these studies indicate that although IL-12 most commonly promotes the development and activation of TH_1 cells and suppresses TH_2 responses, under some circumstances IL-12 may enhance TH_2-like responses, depending upon the cytokine milieu in which IL-12 acts and the maturational state of the T cells.

Induction of Cytokine Secretion by IL-12

Consistent with its actions in promoting TH_1 responses, IL-12 has been shown in vitro to be a potent inducer of IFN-γ secretion by both resting and activated T and NK cells in both human (1,5,64,70,92–98) and murine (8–10,18,40) systems. This property of IL-12 is central to many of its biological activities in vivo (see below). IL-12–induced IFN-γ production by resting human peripheral blood lymphocytes or neonatal T cells required the presence of $HLA\text{-}DR^+$ accessory cells (70,92), whereas activated human T cells incubated with IL-12 appeared to produce IFN-γ independent of accessory cells (92,93). In eliciting the production of IFN-γ by T and/or NK cells, IL-12 has been shown to synergize with a variety of cytokines and other lymphocyte-activating agents including TNF-α (18,40,70), IL-1 (64,70), IL-2 (1,10,40,92,93,96), IL-15 (99), B7/CD28 interactions (13,100), soluble CD23 (101), anti-CD3 (92,96), phytohemagglutinin (PHA) (92), and phorbol esters (92,93,96). On the other hand, IL-12–induced IFN-γ production has been reported to be downregulated by IL-4 (95), IL-10 (40,64), and TGF-β (70,102). Analysis of the molecular mechanism by which IL-12 regulates IFN-γ synthesis revealed that in PHA-activated human T lymphoblasts and in a human T-leukemia cell line, IL-12 increased transcription of the IFN-γ gene (93). However, synergy of IL-12 with IL-2 or PHA in inducing IFN-γ production was due, at least in part, to stabilization of cytoplasmic IFN-γ mRNA (93,98).

Administration of rIL-12 to mice has been shown to induce the production of IFN-γ in vivo (19,82,88,103–106). Daily administration of mouse rIL-12 to normal mice resulted in elevated serum IFN-γ levels that first became detectable 24 h after the initial injection of IL-12 and continued to increase for the subsequent 5 days (103). However, if rIL-12 administration (1 μg/day, 5 days/week) was continued for 3 more weeks, serum IFN-γ levels declined in the middle of the second week to a low level, which was sustained through the third and fourth weeks (106). This decline did not appear to result from antibody production against mouse rIL-12 (106). The mechanism of this downregulation of IL-12–induced IFN-γ production is unclear, but it could result from the induction of inhibitory cytokines such as IL-10 (104) or TGF-β.

The production of IFN-γ in response to administered rIL-12 has been shown to be essential to the therapeutic effects of IL-12 in a number of mouse tumor models and models of infectious diseases (15,16,18,19,21,105,107,108). However, in those models in which rIFN-γ was tested for therapeutic activity, it did not display therapeutic effects comparable to those seen with rIL-12 (16,105,108). This may indicate that the therapeutic effects of rIL-12 involved additional biological activities that were not mimicked by rIFN-γ or that continuous local production of IFN-γ can lead to therapeutic effects not achievable by intermittent systemic administration of large amounts of rIFN-γ, which has a short serum half-life (109). Recombinant IFN-γ, when given to mice by continuous administration, could reproduce the antiangiogenic effects caused rIL-12 via an IFN-γ-dependent mechanism (110).

IL-12 has been shown to induce the production of several other cytokines in addition to IFN-γ. IL-12 was shown to elicit the secretion of low amounts of TNF-α from alloactivated (111) or resting human NK cells (96,112) or from resting human T cells (96), but more TNF-α was secreted by IL-2–induced than by IL-12–induced NK or T cells (96,112). A fivefold increase in splenic mRNA for TNF-α was observed in mice injected with rIL-12 (104), and low levels of TNF-α have been detected in the serum of rIL-12–treated mice (15,113). Administration of rIL-12 to mice also resulted in large increases in expression of IL-10 mRNA in the spleen and/or lymph nodes (82,88,104), and the number of splenic IL-10–secreting cells was greatly increased (104). Upregulation of IL-10 production by IL-12 appeared to be independent of IFN-γ (88,104) and was observed in SCID mice deficient in T and B cells (88). Given the ability of IL-10 to inhibit the production and, to a lesser extent, the action of IL-12 (38,40,64), the induction of IL-10 production by IL-12 may represent a negative feedback mechanism by which IL-12 limits its own effects. IL-12 was found to elicit the secretion of small amounts of GM-CSF in vitro from human NK and T cells and to enhance production of GM-CSF induced by other stimuli, such as anti-CD3 or phorbol ester (94,96). IL-12 was also reported to induce the secretion of small amounts of IL-8 by purified human NK cells (94) and of IL-2 by PHA-activated human lymphoblasts (111); however, no increased IL-2 secretion was detected from human NK cells (112), human neonatal $CD4^+$ T cells (70), or mouse T cells (12) stimulated with IL-12. Interestingly, administration of rIL-12 to mice injected with schistosome eggs resulted in increased expression of IL-12 p40 mRNA in the lungs (19), suggesting that IL-12 may be capable of enhancing its own production or, alternatively, that it induces the production of IL-12 p40 homodimer, an IL-12 antagonist that could limit the action of IL-12 (35,36).

Enhancement of Cytolytic Lymphocyte Activity by IL-12

The ability of IL-12 to enhance cytolytic lymphocyte activity was one of the earliest known biological activities of IL-12 and is reflected in its original names: natural killer cell–stimulatory factor (1) and cytotoxic lymphocyte maturation factor (3). IL-12 was shown to enhance NK lytic activity in overnight cultures (1,114,115) and LAK activity in 3- to 7-day cultures (3,112,116). In the longer-term cultures, IL-12–activated effector cells were shown to be predominantly $CD56^+$ NK cells (116). IL-12 was shown to enhance NK/LAK activity independent of IL-2 (112,114,116,117), IFN-α (118), IFN-β (118), and IFN-γ (112,116–118). Activation of LAK cell activity by IL-12 was partially inhibited by anti-TNF-α (112,116), but IL-12–induced enhancement of NK activity in overnight cultures appeared to occur independent of TNF-α (117,118). Use of IL-12 in combination with IL-2 led to additive, supra-additive, or inhibitory effects, depending on the concentrations of the two cytokines and the length of the culture (3,114–117,119,120). IL-12 was effective at much lower concentrations than IL-2 in activating NK cells in overnight cultures (1,114) or at concentrations similar to those of IL-2 in longer-term cultures (116). In both cases, the maximum lytic activity induced by optimal concentrations of IL-12 was about 20–50% of the maximum IL-2–induced lytic activity (1,112,114–117). The enhancement of NK/LAK activity by IL-12 appears to be due in part to enhanced expression of perforin and granzymes (96,117,121), increased formation of cytoplasmic granules (117,118), and enhanced stimulus-dependent granule exocytosis (122). In addition, IL-12 has been shown to upregulate NK cell expression of a variety of adhesion molecules and cytokine receptors, including CD2, CD11a, CD54, CD56, the 75 kDa TNF receptor, and the receptors for IL-2 (α and β subunits), IL-4, and IL-12 (112,114,123,124). Purified human NK cells that were cultured for 3–6 days with IL-12 showed increased binding to target cells (124), whereas overnight incubation of unseparated PBMC with IL-12 did not lead to a significant increase in the number of target-binding cells (117). Mouse rIL-12 increased the lytic activity of mouse NK cells in overnight cultures (9), and enhanced NK/LAK activity was observed in the spleen, liver, lungs, and peritoneal cavity of mice treated with mouse rIL-12 (103). Likewise, PBMC from squirrel monkeys given human rIL-12 displayed increased LAK activity and lectin-dependent cell-mediated cytotoxicity (125).

In addition to its effects on NK/LAK activity, IL-12 can also facilitate the induction of specific human CTL responses (116,126). In mixed leukocyte cultures (MLCs) using irradiated allogeneic PBMC as stimulator cells, IL-12 enhanced both proliferation and CTL generation (126). If the responder and stimulator cells differed at the HLA class I locus but were HLA class II identical, the effects of added IL-12 were even more dramatic; little proliferation and no CTL generation were observed in the absence of added cytokine (126). IL-12–induced CTL responses in these cultures were associated with upregulation of perforin and granzyme B mRNA expression. The potentiating effect of IL-12 on CTL generation in this system appeared to be IL-2 independent since it was not inhibited by a neutralizing anti–IL-2Rα mAb (126). Similarly, IL-12 was reported to enhance, via an IL-2–independent mechanism, the lytic activity of purified human $CD8^+$ T cells stimulated by immobilized anti-CD3 (127). IL-12 was also shown to potentiate specific human CTL responses to irradiated allogeneic melanoma cells in vitro (116). Both IL-12 and IL-2 were active in this assay, but the concentration of IL-12 necessary to induce a half-maximal response was approximately 10-fold less than the concentration of IL-2 (116). However, in contrast to the results described above, the IL-12–mediated induction of CTL responses to allogeneic melanoma cells appeared to require in situ production of IL-2 since anti–IL-2 antibody inhibited CTL generation by more than 90% (116). Interestingly, in allogeneic human MLCs, neutralizing anti–IL-12 monoclonal antibodies partially inhibited CTL generation against target cells that differed from the responders at both HLA class I and class II loci (126). Likewise, anti–IL-12 antibodies have been reported to partially inhibit human alloantigen-induced proliferative responses (126,128). Thus, endogenous IL-12 produced in allogeneic human MLCs appears to play an important role in facilitating both the proliferative and cytotoxic responses induced by alloantigen.

Mouse IL-12 was reported to enhance allogeneic mouse CTL responses in vitro by a mechanism that was independent of IL-2 or IFN-γ and was not associated with increased proliferation (129). Unlike the results in human MLCs described above, neutralizing antibodies to mouse IL-12 had no inhibitory effect on the allogeneic CTL responses generated in mouse MLCs (D. Faherty, unpublished results). Administration of rIL-12 to mice can lead to increased CTL responses in vivo. When mice immunized with allogeneic splenocytes in the footpads were given daily injections of IL-12 starting on the day of immunization, primary CTL responses in the draining lymph nodes were increased by an average of 14-fold (103). The kinetics of the CTL response was not altered by IL-12. Maximum CTL responses were observed in mice given 1 μg IL-12 per day, and at doses of IL-12 < 100 ng/day no significant increase in CTL activity was observed (103). IL-12 also increased CTL responses in mice immunized with a p53 nonamer peptide vaccine (130). In these studies mice were immunized with the peptide in incomplete Freund's adjuvant on days 0 and 7, and splenic CTL activity was measured on day 14. Four daily doses of IL-12 were administered together with each immunization. In this study maximum CTL responses were observed in mice receiving 1 ng/day IL-12, and higher doses of IL-12 had a suppressive effect (130). The large difference in the optimum IL-12 dose for enhancing in vivo CTL responses in the studies of Gately et al. (103) as compared to Noguchi et al. (130) could reflect differences in the nature of the antigen used (allogeneic splenocytes versus peptide in incomplete Freund's adjuvant), the dosing regimen (primary versus secondary CTL response), and/or the tissue from which the CTL were harvested (lymph node versus spleen).

Effects of IL-12 Lymphocyte Proliferation

Both T and NK cells have been shown to proliferate in response to IL-12 in various short term in vitro assays, and this has been frequently used as a means to quantitate IL-12 bioactivity (131). IL-12 stimulates minimal proliferation by resting PBMC but can induce proliferation of T and NK cells that have been activated by a number of stimuli (1,3–5,10,74,111,112,114,119,132–135), many or all of which upregulate expression of high affinity IL-12R (137). The growth-promoting effects of IL-12 on activated T cells have been demonstrated using T-cell clones, CTL lines, T lymphoblasts derived from PBMC, and tumor-infiltrating-lymphocytes (TILs) (10,74,78,111,119,132–134). IL-12 stimulates the proliferation of both $CD4^+$ and $CD8^+$ TCR-$\alpha\beta^+$ T cells or T-cell clones, and TCR-$\gamma\delta$ lymphoblasts (111,132). The proliferation induced by IL-12 is independent of IL-2, since antibodies to IL-2 or the IL-2 receptor do not inhibit the proliferation of PHA-activated human lymphoblasts, CTL lines, or TILs (3,132–134).

The ability of T cells to proliferate in response to IL-12 is dependent upon their state of activation, since IL-12 caused little or no growth of either resting human T cells (1,119,132) or anergized mouse T cells (135). Likewise, alloantigen- or anti-CD3–activated human T cells, when cultured in the absence of the activating agent for 1–2 weeks, lost their ability to proliferate in response to IL-12 alone (133). IL-12 synergized with IL-4 to induce proliferation of resting mouse TH_1 clones, but neither cytokine alone induced significant proliferation (74). IL-12 also synergized with IL-2, IL-7, and IL-15 to promote proliferation of resting TH_1 clones, but no such synergy was noted with either TH_0 or TH_2 clones (74). IL-12 and IL-4 likewise caused synergistic proliferation of isolated human NK cells (123), and this synergy may have been due, at least in part, to upregulation of IL-12R expression by IL-4 and of IL-4R expression by IL-12 (123).

Although both IL-12 and IL-2 can induce proliferation of activated T and NK cells, the magnitude of the response differs. The maximum proliferation induced by IL-12 was $\leq 50\%$ of the maximum proliferation induced by IL-2 (114,119,132). However, significantly less IL-12 than IL-2 was needed to achieve half-maximal proliferation (119,132). When lectin-activated human T cells were cultured with suboptimal amounts of IL-2 and IL-12 in combination, additive proliferation was observed, up to the maximum proliferation induced by IL-2 alone (111,114,132). In contrast, when IL-2–activated human NK cells or TCR-$\gamma\delta^+$ lymphoblasts were cultured with IL-12 in combination with intermediate to high doses of IL-2, IL-12 reduced the IL-2–induced proliferative response (111,114) by a TNF-dependent mechanism (111). A similar inhibitory effect was observed when IL-12 was added to cultures of anti-CD3–activated human $CD8^+$ T cells incubated with high concentrations of IL-2, whereas IL-12 synergized with low concentrations of IL-2 to induce these cells to proliferate (127). High levels of human T-lymphoblast proliferation, comparable to or greater than those observed with IL-2, were obtained by combining IL-12 with an anti-CD28 antibody (100). Synergy between IL-12 and B7/CD28 interactions in stimulating the proliferation of mouse TH_1 clones was also reported (13).

IL-12 was recently reported to cause modest proliferation of SAC-activated human peripheral blood B lymphocytes cultured together with IL-2 (136). These results contrast with the results of an earlier study (137) in which IL-12 caused, at best, minimal proliferation of SAC-activated tonsillar B cells. The concentration of IL-12 used in the experiments of Jelinek and Braaten (136) was 20 ng/ml, which is approximately 20-fold higher than the concentration of IL-12 sufficient to cause maximum proliferation of activated T and NK cells and fivefold greater than the maximum concentration of IL-12 tested in the earlier report (137). This potential discrepancy raises the question of whether the observed effects of IL-12 on B cells (136) were mediated through a different receptor than the high affinity receptor found on activated T and NK cells (137).

Effects of IL-12 on Antibody Responses

Initial studies on the regulation of antibody responses by IL-12 focused on the ability of IL-12 to suppress IgE responses. IL-12 was shown to markedly inhibit IL-4-induced IgE production by human PBMC in vitro (95). IL-4– or pokeweed mitogen–stimulated synthesis of IgG, IgM, or IgA was unaffected by IL-12 (95). In these studies, IL-12 appeared to inhibit IgE synthesis both by inducing the production of IFN-γ, a known inhibitor of IgE synthesis, and by a novel mechanism independent of IFN-γ. The latter mechanism appeared to be indirect and to require the presence of T cells, because IL-12 failed to block IgE synthesis by highly purified B cells costimulated with IL-4 and anti-CD40 mAb (95) or by B cells cultured with monocytes in the presence of IL-4 and hydrocortisone (138). In accordance with these observations, human tonsillar B cells activated by culture with anti-IgM conjugated to dextran ± IL-2 or with SAC + IL-2 failed to bind measurable amounts of IL-12 (137). In contrast, when highly purified human peripheral blood B cells were cultured with SAC plus IL-2, addition of IL-12 significantly enhanced the secretion of IgG, IgM, and IgA (136). In this study the enhancing effect of IL-12 required the presence of both SAC and IL-2. The reason for the seeming discrepancy between the results of Jelinek and Braaten (136) and those of the earlier studies (137,138) is at present unclear. Possible explanations include differences in the source of the B cells and the culture conditions, the presence of a very small number of contaminating T and/or NK cells in the purified peripheral blood B-cell populations (136) through which the enhancing effect of IL-12 on immunoglobulin production was mediated, or the presence of intermediate- or low-affinity IL-12R on activated B cells which were not detected in the earlier study (137) and through which the effects of IL-12 on B cells are mediated. Consistent with the latter possibility, we have recently observed that expression of the β subunit of the IL-12R, which by itself can bind IL-12 with low affinity (32), is upregulated on activated human tonsillar and peripheral blood B cells (C. Y. Wu and M. Gately, unpublished results).

In vivo, IL-12 can either enhance or inhibit humoral immunity, depending on the Ig isotype and the stimulus to antibody formation. When mice that had been given goat antimouse IgD antibody (GaMδ) to elicit polyclonal Ig responses were treated with IL-12, IgG1 and IgE responses were reduced by >98% and IgG2a and IgG3 responses were decreased to a lesser extent (104). These effects were correlated with decreased expression of IL-3 and IL-4 mRNA and increased IFN-γ mRNA in the spleens of IL-12–treated mice (104). Administration of neutralizing anti–IFN-γ antibody to mice treated with GaMδ + IL-12 largely reversed the IL-12–mediated suppression of IgG1 synthesis but not of IgE synthesis and led to enhancement of IgG2a and IgG3 responses by IL-12 (104). Consistent with these results, IL-12 inhibited GaMδ-induced IgE responses in IFN-γ knock-out mice (88); however, the magnitude of the inhibition in the IFN-γ knock-out mice was not as great as that seen previously in wild-type mice (104). Administration of IL-12 at the time of a primary infection was found to inhibit IgE responses in helminth-infected mice, but IL-12 was not inhibitory if given for the first time during a secondary infection (82). In this model, inhibition of IgE synthesis by IL-12 was IFN-γ dependent (82). IL-12 was also found to prevent the elevated serum IgE levels associated with chronic graft-versus-host disease in mice, and this inhibitory effect was partially reversed by simultaneous administration of anti–IFN-γ mAb (139). In contrast to all of the previous studies in which IL-12 suppressed IgE responses, in IFN-γ knock-out mice injected with schistosome eggs, administration of IL-12 led to substantially increased IgE levels (89).

Additional studies examined the effects of IL-12 on IgG responses to protein antigens in vivo and showed that IL-12 can have dramatic effects in increasing the production of some IgG isotypes. When mice were immunized a single time with the hapten-protein conjugate TNP-KLH, IL-12 strongly inhibited anti-TNP IgG1 production but caused a modest increase in IgG2a production and had no effect on IgG3 responses (80). These results are consistent with observations that IgG1 antibody responses are associated with TH_2 cell activation, whereas IgG2a antibody responses are associated with the activation of Th1 cells (140). However, administration of IL-12 to mice immunized with the bee venom allergen PLA_2 or KLH adsorbed to alum resulted in enhancement of the production of antigen-specific antibodies of the IgG2a, IgG2b, and IgG3 subclasses by 10- to 1000-fold, slight (two- to fivefold) enhancement of antigen-specific IgG1 antibody levels, and suppression of antigen-specific IgE synthesis (141). In this study, mice received three to six biweekly immunizations, and five daily doses of IL-12 were given with each immunization. The enhancement of IgG2a responses by IL-12 could be largely inhibited by administration of anti–IFN-γ (141). These results suggest that it may be possible to use IL-12 as an adjuvant to enhance humoral immune responses to protein antigens adsorbed to alum.

Effects of IL-12 on Hematopoiesis and Thymopoiesis

In mouse models in vitro, IL-12 can synergize with stem cell factor (SCF) and, to a lesser extent, with some of the other colony-stimulating factors and with flt3 ligand to induce the proliferation of hematopoietic stem cells (142–146). IL-12 alone was inactive. Synergy between IL-12 and SCF was observed in single-cell cloning assays, indicating that IL-12 can exert an effect directly on stem cells (142,143). IL-12 in combination with several other colony-stimulating factors increased colony formation of high proliferative potential (HPP) colony-forming cells, indicating that it could act on a primitive stem cell population (142). In the presence of SCF, IL-12 could support the formation of both myeloid colonies (142,143) and colonies that yielded pre–B-cell colonies when replated in secondary culture containing SCF and IL-7 (143). In the presence of IL-3 + IL-11 or IL-3 + SCF, IL-12 could also enhance the development of megakaryocytes, erythrocytes, and mast cells, but not eosinophils, via an IFN-γ–

independent mechanism (144). Likewise, in studies using enriched hematopoietic progenitor cells from human peripheral blood and bone marrow, IL-12 synergized with a combination of SCF + IL-3 to induce the formation of mixed, myeloid, and erythroid colonies (147). However, if human hematopoietic progenitor cells were cultured together with human NK cells, addition of IL-12 suppressed formation of hematopoietic colonies supported by IL-3 and GM-CSF (147). The mechanism of suppression involved IL-12–induced secretion of IFN-γ and TNF-α, two cytokines known to have inhibitory effects on hematopoiesis.

Studies in mice have shown that IL-12 can influence hematopoiesis in vivo (148,149). Administration of 1 μg rIL-12 to mice daily for 7 days resulted in marked suppression of hematopoiesis in the bone marrow but enhanced hematopoiesis in the spleen (148,149). Measurement of hematopoietic colony-forming cells in the bone marrow and spleen of mice treated with rIL-12 in this fashion revealed a decrease in CFU-GM (granulocyte/macrophage progenitors), CFU-e (late erythroid progenitors), BFU-e (early erythroid progenitors), and HPP colony-forming cells in the bone marrow, whereas CFU-GM, BFU-e, and HPP colony-forming cells were dramatically increased (CFU-e were not measured) in the spleen (148,149). These changes were reversible following cessation of IL-12 treatment (148). Hence, administering IL-12 to mice caused a reversible myelosuppression accompanied by marked extramedullary hematopoiesis. In contrast to the bone marrow hypoplasia observed in mice given daily injections of mouse rIL-12 for 7 days (148,149), squirrel monkeys that received daily injections of human rIL-12 for 2 weeks displayed a dose-dependent trilineage hyperplasia of the bone marrow (125). Also, administration of rIL-12 to mice prior to lethal irradiation protected mice from death and resulted in a fivefold increase in the number of bone marrow cells 6 days after irradiation (150). This radioprotective effect of IL-12 on hematopoietic cells was independent of IFN-γ but required both IL-1 and SCF (150). Thus, it is clear that administration of rIL-12 can have dramatic effects on hematopoiesis in vivo; however, it is unclear whether endogenous IL-12 plays an important role in the regulation of hematopoiesis. IL-12 knock-out mice did not display any major hematological abnormalities (85); however, subtle alterations have not been excluded.

IL-12 may also influence thymopoiesis. Addition of IL-12 to fetal thymic organ cultures led to a significant reduction in cell number (54). Analysis of the thymocytes recovered from these cultures showed that most thymocyte subsets were decreased in number, but there was an increase in the number of $\alpha\beta TCR^+CD4^-CD8^+$ thymocytes. IL-12 in combination with SCF or with IL-2 + IL-4 stimulated the proliferation of some subsets of isolated thymocytes (54). Both p35 and p40 IL-12 mRNAs were detected in fetal (54) and adult (9,54) thymus, and two thymic epithelial cell lines expressed both transcripts (54). These results suggest that IL-12 may influence T-cell development in the thymus; however, it is not known whether endogenous IL-12 plays a significant role in regulating thymopoiesis in vivo.

Effects of IL-12 on Programmed Cell Death

Activation of T cells via their antigen-specific receptor can induce either proliferation or programmed cell death, also called apoptosis. IL-12 has recently been reported to inhibit apoptosis induced by culture of anti-CD3–activated human peripheral blood T cells with pokeweed mitogen (PWM) + staphylococcal enterotoxin B (SEB) (151) or by cross-linking of CD4 on cloned human TH_1 cells followed by activation with immobilized anti–T-cell receptor antibody (152). IL-12 suppressed apoptosis of both activated $CD4^+$ and activated $CD8^+$ T cells (151). Unlike normal human peripheral blood T cells, peripheral blood T cells from HIV-infected patients underwent apoptosis when cultured with PWM/SEB without prior in vitro activation, and IL-12, like IFN-γ and IL-2, could also inhibit apoptosis of these cells (151). Interestingly, anti–IL-12 antibody dramatically increased PWM/SEB-induced apoptosis in cultures of HIV^+ PBMC from patients early in the disease process, whereas anti–IFN-γ was without effect (151). In contrast to these results in which IL-12 prevented apoptosis, preincubation of human peripheral blood NK cells with either IL-2 or IL-12 primed them so that they underwent apoptosis in response to subsequent cross-linking of CD16 in the absence of added cytokines (153). Several other cytokines, including IFN-α and IFN-γ, were ineffective in priming NK cells for subsequent apoptosis (153). The discrepancy between the enhancing effects of apoptosis in this study (153) and the inhibitory effects of IL-12 in the previous studies (151,152) could reflect differences in the cell type examined or differences in the timing of exposure to IL-12 relative to the timing of addition of the apoptosis-inducing stimulus.

Role of Endogenous IL-12 in the Normal Host Immune Response to Infectious Agents

IL-12 can enhance the host immune response to intracellular pathogens both by promoting TH_1-dependent cell-mediated immunity (16,17,81) and by enhancing T-cell–independent innate resistance mediated by NK cells and macrophages (18,40). Studies using neutralizing anti–IL-12 antibodies have indicated that IL-12 produced endogenously in response to infection plays an important role in the host defense against intracellular pathogens. Administration of anti-IL-12 to *Leishmania major*–infected but normally resistant C57BL/6 or C3H mice exacerbated infection (17,154,155), and similar results were seen in a mouse model of candidiasis (84). Likewise, giving anti–IL-12 to mice injected with *Schistosoma mansoni* eggs led to a dramatic increase in the size of egg-induced pulmonary granulomas (19). In each of these models, the effects of anti–IL-12 correlated with a shift from a predominantly TH_1-type cytokine response to a TH_2-type response. In either normal or SCID mice infected with a sublethal dose of *Listeria monocytogenes*, injection of anti–IL-12 resulted in increased *Listeria* burden in the spleen, decreased macrophage I-A expression, and death of the mice; administration of rIFN-γ together with the anti–IL-12 reversed all of these effects (156). Administering anti–IL-12 to normal (43,157) or SCID (43,158) mice infected with *Toxoplasma gondii* was shown to decrease survival if the antibody was given at the time of infection; if antibody treatment was not started until after the infection had become established, no effect was observed (43,157). Anti–IL-12 treatment of mice was also shown to result in decreased resistance to infection by *Mycobacterium tuberculosis* (159) or *Brucella abortus* (160). Thus, endogenous IL-12 appears to play an essential role in the host defense against a variety of intracellular bacterial, fungal, and protozoan pathogens in mouse models. Consistent with this, IL-12 levels were significantly higher in pleural fluids from patients with tuberculous pleuritis than in serum from the same patients or in pleural effusions from patients with malignancies (42). In addition, anti–IL-12 antibody partially suppressed the proliferation of pleural fluid mononuclear cells to *M. tuberculosis* in vitro (42). These results suggest that IL-12 may play a role in the immune response to *M. tuberculosis* infection in humans. IL-12 may also play a role in the host defense against *M. leprae* in humans. Immunohistochemical analyses showed that IL-12–producing cells were prominent in the lesions of patients with tuberculoid leprosy, in whom mycobacterial growth is restricted, but rare in lesions of lepromatous leprosy patients, who develop disseminated infection (44). Hence, an important physiological role of IL-12 appears to be to contribute to host resistance to infection with intracellular bacterial, fungal, or protozoan pathogens. What role, if any, IL-12 plays in the host defense against viral infections remains to be defined.

Role of IL-12 in Immunopathology

Recent data suggest that IL-12 may play a role in the pathogenesis of endotoxin-induced shock and mortality (59,60,161). The generalized Schwartzman reaction is a model of endotoxin-induced shock in which mice are primed with a low dose of LPS in the footpad followed 1 day later by a lethal challenge dose of LPS given intravenously. In this model, administration of antibodies to IL-12 or IFN-γ prior to the priming dose of LPS could prevent mortality (161). A small amount (10 ng) of IL-12 could substitute for LPS in priming mice for a lethal response to the LPS challenge, and the priming effect of IL-12 could be blocked by treatment with anti-IFN-γ (161). Administration of anti-IL-12 antibodies given at challenge did not prevent mortality, and IL-12 could not substitute for the challenge dose of LPS (161). The results suggest that in this model the priming dose of LPS induces production of IL-12, which in turn stimulates the secretion of IFN-γ, resulting in priming of macrophages and other cell types for a lethal response to the LPS challenge. In a second model in which mice were sensitized with BCG and challenged with a low dose (1 μg) of LPS, antibodies against IL-12 could suppress LPS-induced IFN-γ production and mortality (60). In this model, TNF-α and additional unidentified cofactor(s) appeared to cooperate with IL-12 to induce maximum IFN-γ production (60). Likewise, in a third study administration of anti–IL-12 antibodies caused a significant reduction in the levels of serum IFN-γ observed following i.p. injection of a high dose (100–300 μg) of LPS (59). Thus, it appears that IL-12 may play an essential role in the pathogenesis of endotoxin-induced shock.

IL-12 also appears to play an important role in the pathogenesis of some autoimmune diseases. In an adoptive transfer model of experimental autoimmune encephalomyelitis (EAE), administration of anti–IL-12 antibodies for 6 days after transfer of proteolipid protein-primed lymph node cells delayed the onset of disease, but the mice ultimately developed paralysis similar in severity to that seen in control mice. However, if anti–IL-12 treatment was extended to day 12, not only was the onset of symptoms delayed, but also the severity of disease was dramatically reduced (162). Likewise, administration of a neutralizing anti–IL-12 mAb resulted in therapeutic effects in a mouse model of granulomatous colitis (163). In this model, disease was induced in BALB/c mice by rectal administration of low doses of trinitrobenzene sulfonic acid (TNBS) and was characterized by a granulomatous, transmural colitis with severe diarrhea and weight loss which appeared within the first week after TNBS administration and was maximum after 2 weeks. Treatment with anti–IL-12 beginning either early (day 5) or late (day 20) in the course of the disease strikingly improved the clinical and histological scores, frequently abrogating established colitis completely (163). Thus, in these studies IL-12 was found to play an essential role in the pathogenesis of EAE and of an autoimmune granulomatous colitis in mice. In addition, the administration of rIL-12 has been shown to accelerate the onset of spontaneous, insulin-dependent diabetes mellitus in NOD mice (164), to increase the incidence and severity of collagen-induced arthritis in DBA/1 mice immunized with type II collagen in incomplete Freund's adjuvant (165), and to accelerate the production of anti-dsDNA antibodies and the appearance of renal disease in NZBWF1 mice, which spontaneously develop a lupuslike syndrome (D. Faherty and U. Sarmiento, unpublished results). Studies examining the effects of blocking endogenous IL-12 production in these models are being pursued.

STRUCTURE AND PROPERTIES OF THE IL-12 RECEPTOR

The biological activities of IL-12 are mediated via specific, high-affinity receptors. Analysis of IL-12R expression by flow cytometry revealed that IL-12R were present on resting human NK cells (123), but were not detectable on resting human T cells (137), B cells (137), or monocytes (M. Gately, unpublished results). Expression of IL-12R was upregulated on mitogen- or alloantigen-activated T cells, including $CD4^+$ and $CD8^+$ T cells, (137) and on IL-2– or IL-12–activated NK cells (123,137), but IL-12R were not detected on B cells activated with anti-IgM conjugated to dextran ± IL-2 or with SAC + IL-2 (137). Initial binding assays using radiolabeled IL-12 showed that PHA-activated human lymphoblasts expressed 1000–9000 IL-12 binding sites per cell with an apparent equilibrium dissociation constant of 100–600 pM (166). A more detailed analysis of IL-12–binding sites using ^{125}I-labeled IL-12 with a lower specific activity showed that PHA-activated human lymphoblasts displayed three binding sites with apparent affinities of 5–20 pM, 50–200 pM, and 2–6 nM, respectively (32), suggesting the existence of different forms or different chains of the IL-12R. Affinity cross-linking and immunoprecipitation studies with anti–IL-12 antibodies identified one protein of approximately 110 kDa on human PHA-activated lymphoblasts that was cross-linked to IL-12 (166). In addition, an 85 kDa protein co-precipitated with the complex of IL-12 cross-linked to the 110 kDa protein, but it is not clear whether the association of the 85 kDa protein with the cross-linked complex was specific or nonspecific (166). Recently, a cDNA encoding a human IL-12R component was cloned by means of expression cloning using an anti–IL-12 mAb (32). This subunit is a 662-amino-acid type I transmembrane protein with an extracellular domain of 516 amino acids and a cytoplasmic domain of 91 amino acids (32). The subunit is a member of the hemopoietin receptor superfamily and most closely related over its entire length to gp130, the signal-transducing component of the IL-6R complex, and to the receptors for G-CSF and leukemia inhibitory factor. This cloned IL-12R subunit was designated IL-12Rβ (32). COS cells transfected with IL-12Rβ chain cDNA bound both human and mouse IL-12 with a single affinity of 2–5 nM, similar to the low-affinity IL-12–binding sites on PHA-activated human PBMC (32). This result suggests that one or more additional subunits are required to reconstitute the high-affinity IL-12R. Cell-surface labeling and immunoprecipitation experiments indicated that transfected COS cells expressed the IL-12Rβ chain both as a 100 kDa monomer and as a disulfide-linked dimer/oligomer, which formed independent of IL-12 binding (32). Only the dimer/oligomer form of IL-12Rβ mediated low-affinity binding of IL-12 (32). Polyclonal and monoclonal antibodies against the human IL-12Rβ subunit specifically inhibited IL-12–induced proliferation of PHA-activated human PBMC, IL-12–induced secretion of IFN-γ by resting human PBMC, and development of LAK activity in response to IL-12 (32; C. Wu and M. Gately, unpublished results). These data suggest that the IL-12Rβ chain is an essential component of the functional IL-12R on both T and NK cells. Studies using flow cytometry showed that IL-12Rβ chain was expressed on several human T, NK, and B-cell lines but not on nonlymphohematopoietic cell lines (C. Wu and M. Gately, unpublished results). IL-12Rβ expression was upregulated not only on activated human T and NK cells, but also on activated human tonsillar and peripheral blood B lymphocytes, which nonetheless did not display detectable IL-12 binding (C. Wu and M. Gately, unpublished results). These results suggest that activated B cells lack other component(s) of the IL-12R required to bind IL-12 with high affinity. The expression of IL-12Rβ chains on human PBMC can be positively or negatively regulated in vitro by cytokines. IL-2, IL-7, and IL-15 upregulated IL-12Rβ expression, whereas IL-4, IL-10, TGF-β, and PGE_2 were found to inhibit the expression of IL-12Rβ chain on anti-CD3– or IL-2–stimulated human PBMC (C. Wu and M. Gately, unpublished results). The subunit of the mouse of IL-12R homologous to human IL-12Rβ has recently been identified and cloned (167). When expressed in mouse Ba/F3 cells, mouse IL-12Rβ bound mouse IL-12 with affinities of 50 and 470 pM, corresponding to the affinities of the medium- and high-affinity IL-12–binding sites previously identified on mouse ConA-activated lymphoblasts (167). However, mouse IL-12Rβ–transfected Ba/F3 cells did not appear to transduce a signal in response to mouse IL-12, suggesting that an additional subunit(s) is required to generate the functional mouse IL-12R complex (167).

Elucidation of the IL-12 signal transduction pathway is at an early stage. Stimulation of PHA-activated human peripheral blood T cells with IL-12 resulted in the rapid tyrosine phosphorylation of several proteins and activation of a 44 kDa isoform of the mitogen-activated protein (MAP) kinase family (168). In a separate study, incubation of PHA-activated human peripheral blood T cells with IL-12 was shown to induce tyrosine phosphorylation of two members of the Janus family of protein tyrosine kinases, Tyk2 and Jak2, but not Jak1 or Jak3 (169). IL-12–induced tyrosine phosphorylation of Tyk2 and Jak2 was also demonstrated in the human NK cell line NK3.3 (169). Likewise, IL-12 was also shown to stimulate tyrosine phosphorylation of Jak2, Stat3, and Stat4 in mouse TH_1 cells but not in mouse Th2 cells (79,170).

SUMMARY

IL-12 is a heterodimeric cytokine that is produced primarily by macrophages and plays a central role in promoting TH_1-type responses and, hence, cell-mediated immunity. In addition to its role in stimulating the development of naive T cells into TH_1 cells and acting as a costimulus to the activation of already differentiated TH_1 cells, IL-12 induces IFN-γ secretion by both resting and activated T and NK cells, enhances the lytic activity of NK/LAK cells, and facilitates specific CTL responses. IL-12 has been found to display all of these activities in vivo, as well as in vitro. IL-12 can substantially enhance the production of specific IgG antibodies of the isotypes associated with TH_1 responses (IgG2a, IgG2b, and IgG3 in mice). On the other hand, IL-12 suppresses IgE antibody responses. Studies with neutralizing anti–IL-12 antibodies have suggested that endogenous IL-12 plays an important role in the normal host defense against infection by a variety of intracellular bacterial, fungal, and protozoan pathogens. However, IL-12 appears also to play a central role in the pathogenesis of some forms of immunopathology, including endotoxin-induced shock and some autoimmune diseases associated with aberrant TH_1 activity. Recombinant IL-12 has been shown to have therapeutic activity in a variety of mouse tumor and infectious disease models and is being evaluated in clinical trials in human cancer patients. IL-12 antagonists may also have therapeutic potential in the treatment of some autoimmune disorders.

REFERENCES

1. Kobayashi M, Fitz L, Ryan M, et al. Identification and purification of natural killer cell stimulatory factor (NKSF), a cytokine with multiple biological effects on human lymphocytes. J Exp Med 1989; 170:827–845.
2. Wong HL, Wilson DE, Jenson JC, et al. Characterization of a factor(s) which synergizes with recombinant interleukin-2 in promoting allogeneic human cytolytic T lymphocyte responses in vitro. Cell Immunol 1988; 111:39–54.
3. Stern AS, Podlaski FJ, Hulmes JD, et al. Purification to homogeneity and partial characterization of cytotoxic lymphocyte maturation factor from human B-lymphoblastoid cells. Proc Natl Acad Sci USA 1990; 87:6808–6812.
4. Gubler U, Chua AO, Schoenhaut DS, et al. Coexpression of two distinct genes is required to generate secreted, bioactive cytotoxic lymphocyte maturation factor. Proc Natl Acad Sci USA 1991; 88:4143–4147.
5. Wolf SF, Temple PA, Kobayashi M, et al. Cloning of cDNA for natural killer cell stimulatory factor, a heterodimeric cytokine with multiple biologic effects on T and natural killer cells. J Immunol 1991; 146:3074–3081.
6. Germann T, Schmitt E, Huhn G, et al. Involvement of soluble mediator(s) different from interleukin (IL)1 in the antigen-induced IL 2 receptor expression and proliferation of L3T4$^+$ (CD4$^+$) T lymphocytes. Eur J Immunol 1988; 18:459–465.
7. Germann T, Partenheimer A, Rüde E. Requirements for the growth of T_H1 lymphocyte clones. Eur J Immunol 1990; 20:2035–2040.
8. Germann T, Jin S-C, Mattner F, et al. Components of an antigen-/T cell receptor-independent pathway of lymphokine production. Eur J Immunol 1991; 21:1857–1861.
9. Schoenhaut DS, Chua AO, Wolitzky AG, et al. Cloning and expression of murine IL-12. J Immunol 1992; 148:3433–3440.
10. Germann T, Gately MK, Schoenhaut DS, et al. Interleukin-12/T cell stimulating factor, a cytokine with multiple effects on T helper type 1 (T_h1) but not on T_h2 cells. Eur J Immunol 1993; 23:1762–1770.
11. Manetti R, Parronchi P, Giudizi MG, et al. Natural killer cell stimulatory factor (interleukin 12 [IL-12]) induces T helper type 1 (Th1)-specific immune responses and inhibits the development of IL-4-producing Th cells. J Exp Med 1993; 177:1199–1204.
12. Hsieh C-S, Macatonia SE, Tripp CS, et al. Development of T_H1 CD4+ T cells through IL-12 produced by Listeria-induced macrophages. Science 1993; 260:547–549.
13. Murphy EE, Terres G, Macatonia SE, et al. B7 and interleukin 12 cooperate for proliferation and interferon γ production by mouse T helper clones that are unresponsive to B7 costimulation. J Exp Med 1994; 180:223–231.
14. Brunda MJ, Luistro L, Warrier R, et al. Antitumor and antimetastatic activity of interleukin-12 against murine tumors. J Exp Med 1993; 178:1223–1230.
15. Nastala CL, Edington HD, McKinney TG, et al. Recombinant IL-12 administration induces tumor regression in association with IFN-γ production. J Immunol 1994; 153:1697–1706.
16. Heinzel FP, Schoenhaut DS, Rerko RM, et al. Recombinant interleukin 12 cures mice infected with Leishmania major. J Exp Med 1993; 177:1505–1509.
17. Sypek JP, Chung CL, Mayor SEH, et al. Resolution of cutaneous leishmaniasis: interleukin 12 initiates a protective T helper type 1 immune response. J Exp Med 1993; 177:1797–1802.
18. Gazzinelli RT, Heiny S, Wynn TA, et al. Interleukin 12 is required for the T-lymphocyte-independent induction of

interferon γ by an intracellular parasite and induces resistance in T-cell-deficient hosts. Proc Natl Acad Sci USA 1993; 90:6115–6119.
19. Wynn TA, Eltoum I, Oswald IP, et al. Endogenous interleukin 12 (IL-12) regulates granuloma formation induced by eggs of Schistosoma mansoni and exogenous IL-12 both inhibits and prophylactically immunizes against egg pathology. J Exp Med 1994; 179:1551–1561.
20. Clemons KV, Brummer E, Stevens DA. Cytokine treatment of central nervous system infection: efficacy of interleukin-12 alone and synergy with conventional antifungal therapy in experimental cryptococcosis. Antimicrob Agents Chemother 1994; 38:460–464.
21. Gazzinelli RT, Giese NA, Morse HC III. In vivo treatment with interleukin 12 protects mice from immune abnormalities observed during murine acquired immunodeficiency syndrome (MAIDS). J Exp Med 1994; 180:2199–2208.
22. Brunda MJ. Interleukin-12. J Leukocyte Biol 1994; 55:280–288.
23. Brunda MJ, Gately MK. Antitumor activity of interleukin-12. Clin Immunol Immunopath 1994; 71:253–255.
24. Gately MK, Gubler U, Brunda MJ, et al. Interleukin-12: a cytokine with therapeutic potential in oncology and infectious diseases. Ther Immunol 1994; 1:187–196.
25. Trinchieri G. Interleukin-12: a cytokine produced by antigen-presenting cells with immunoregulatory functions in the generation of T-helper cells type 1 and cytotoxic lymphocytes. Blood 1994; 84:4008–4027.
26. Podlaski FJ, Nanduri VB, Hulmes JD, et al. Molecular characterization of interleukin 12. Arch Biochem Biophys 1992; 294:230–237.
27. Sieburth D, Jabs EW, Warrington JA, et al. Assignment of NKSF/IL 12, a unique cytokine composed of two unrelated subunits, to chromosomes 3 and 5. Genomics 1992; 14:59–62.
28. Gearing DP, Cosman D. Homology of the p40 subunit of natural killer cell stimulatory factor (NKSF) with the extracellular domain of the interleukin-6 receptor. Cell 1991; 66:9–10.
29. Zou JJ, Schoenhaut DS, Carvajal DM, et al. Structure-function analysis of the p35 subunit of mouse interleukin 12. J Biol Chem 1995; 270:5864–5871.
30. Bush K, Day NK, Kraus LA, et al. Molecular cloning of feline interleukin 12 p35 reveals the conservation of leucine-zipper motifs present in human and murine IL-12 p35. Mol Immunol 1994; 31:1373–1374.
31. Merberg DM, Wolf SE, Clark SC. Sequence similarity between NKSF and the IL-6/G-CSF family. Immunol Today 1992; 13:77–78.
32. Chua AO, Chizzonite R, Desai BB, et al. Expression cloning of a human IL-12 receptor component: a new member of the cytokine receptor superfamily with strong homology to gp130. J Immunol 1994; 153:128–136.
33. Trinchieri G. Interleukin-12 and its role in the generation of T_H1 cells. Immunol Today 1993; 14:335–338.
34. Mattner F, Fischer S, Guckes S, et al. The interleukin-12 subunit p40 specifically inhibits effects of the interleukin-12 heterodimer. Eur J Immunol 1993; 23:2202–2208.
35. Ling P, Gately MK, Gubler U, et al. Human IL-12 p40 homodimer binds to the IL-12 receptor but does not mediate biologic activity. J Immunol 1995; 154:116–127.
36. Gillessen S, Carvajal D, Ling P, et al. Mouse interleukin-12 (IL-12) p40 homodimer: a potent IL-12 antagonist. Eur J Immunol 1995; 25:200–206.
37. D'Andrea A, Rengaraju M, Valiante NM, et al. Production of natural killer cell stimulatory factor (interleukin-12) by peripheral blood mononuclear cells. J Exp Med 1992; 176:1387–1398.
38. Kubin M, Chow JM, Trinchieri G. Differential regulation of interleukin-12 (IL-12), tumor necrosis factor α, and IL-1β production in human myeloid leukemia cell lines and peripheral blood mononuclear cells. Blood 1994; 83:1847–1855.
39. Wolf S, Seiburth D, Perussia B, et al. Cell sources of natural killer cell stimulatory factor (NKSF/IL-12) transcripts and subunit expression. FASEB J 1992; 6:A1335.
40. Tripp CS, Wolf SF, Unanue ER. Interleukin 12 and tumor necrosis factor α are costimulators of interferon γ production by natural killer cells in severe combined immunodeficiency mice with listeriosis, and interleukin 10 is a physiologic antagonist. Proc Natl Acad Sci USA 1993; 90:3725–3729.
41. Reiner SL, Zheng S, Wang Z-E, et al. Leishmania promastigotes evade interleukin 12 (IL-12) induction by macrophages and stimulate a broad range of cytokines from $CD4^+$ T cells during initiation of infection. J Exp Med 1994; 179:447–456.
42. Zhang M, Gately MK, Wang E, et al. Interleukin 12 at the site of disease in tuberculosis. J Clin Invest 1994; 93:1733–1739.
43. Gazzinelli R, Wysocka M, Hayashi S, et al. Parasite-induced IL-12 stimulates early IFN-γ synthesis and resistance during acute infection with Toxoplasma gondii. J Immunol 1994; 153:2533–2543.
44. Sieling PA, Wang X-H, Gately MK, et al. IL-12 regulates T helper type 1 cytokine responses in human infectious disease. J Immunol 1994; 153:3639–3647.
45. Leung DYM, Gately M, Trumble A, et al. Bacterial superantigens induce T cell expression of the skin-selective homing receptor, the cutaneous lymphocyte-associated antigen, via stimulation of interleukin 12 production. J Exp Med 1995; 181:747–753.
46. Skeiky YAW, Guderian JA, Benson DR, et al. A recombinant Leishmania antigen that stimulates human peripheral blood mononuclear cells to express a Th1-type cytokine profile and to produce interleukin 12. J Exp Med 1995; 181:1527–1537.
47. Shu U, Kiniwa M, Wu CY, et al. Activated T cells induce interleukin-12 production by monocytes via CD40-CD40 ligand interaction. Eur J Immunol 1995; 25:1125–1128.
48. Cassatella MA, Meda L, Gasperini S, et al. Interleukin-12 production by human polymorphonuclear leukocytes. Eur J Immunol 1995; 25:1–5.
49. Müller G, Saloga J, Germann T, et al. Identification and induction of human keratinocyte-derived IL-12. J Clin Invest 1994; 94:1799–1805.
50. Aragane Y, Riemann H, Bhardwaj RS, et al. IL-12 is expressed and released by human keratinocytes and epidermoid carcinoma cell lines. J Immunol 1994; 153:5366–5372.
51. Smith TJ, Ducharme LA, Weis JH. Preferential expression of interleukin-12 or interleukin-4 by murine bone marrow mast cells derived in mast cell growth factor or interleukin-3. Eur J Immunol 1994; 24:822–826.
52. Macatonia SE, Hsieh C-S, Murphy KM, et al. Dendritic cells and macrophages are required for Th1 development of CD4+ T cells from αβ TCR transgenic mice: IL-12 substitution for macrophages to stimulate IFN-γ production is IFN-γ-dependent. Int Immunol 1993; 5:1119–1128.
53. Kanangat S, Nair S, Babu JS, et al. Expression of cytokine mRNA in murine splenic dendritic cells and better induction of T cell-derived cytokines by dendritic cells than by macrophages during in vitro costimulation assay using specific antigens. J Leuk Biol 1995; 57:310–316.
54. Godfrey DI, Kennedy J, Gately MK, et al. IL-12 influences

intrathymic T cell development. J Immunol 1994; 152:2729–2735.
55. Romani L, Mencacci A, Tonnetti L, et al. Interleukin-12 but not interferon-γ production correlates with induction of T helper type-1 phenotype in murine candidiasis. Eur J Immunol 1994; 24:909–915.
56. Borst KL, Clements JD. In vivo induction of interleukin-12 mRNA expression after oral immunization with Salmonella dublin or the B subunit of Escherichia coli heat-labile enterotoxin. Infect Immun 1995; 63:1076–1083.
57. Cooper AM, Roberts AD, Rhoades ER, et al. The role of interleukin-12 in acquired immunity to Mycobacterium tuberculosis infection. Immunology 1995; 84:423–432.
58. Coutellier J-P, van Broeck J, Wolf SF. Interleukin-12 gene expression after viral infection in the mouse. J Virol 1995; 69:1955–1958.
59. Heinzel FP, Rerko DM, Ling P, et al. Interleukin 12 is produced in vivo during endotoxemia and stimulates synthesis of gamma interferon. Infect Immun 1994; 62:4244–4249.
60. Wysocka M, Kubin M, Vieira LQ, et al. Interleukin-12 is required for interferon-γ production and lethality in lipopolysaccharide-induced shock in mice. Eur J Immunol 1995; 25:672–676.
61. Bette M, Jin S-C, Germann T, et al. Differential expression of mRNA encoding interleukin-12 p35 and p40 subunits in situ. Eur J Immunol 1994; 24:2435–2440.
62. Malefyt RdW, Figdor CG, Huijbens R, et al. Effects of IL-13 on phenotype, cytokine production, and cytotoxic function of human monocytes. Comparison with IL-4 and modulation by IFN-γ or IL-10. J Immunol 1993; 151:6370–6381.
63. D'Andrea A, Ma X, Aste-Amezaga M, et al. Stimulatory and inhibitory effects of interleukin (IL)-4 and IL-13 on the production of cytokines by human peripheral blood mononuclear cells: priming for IL-12 and tumor necrosis factor α production. J Exp Med 1995; 181:537–546.
64. D'Andrea A, Aste-Amezaga M, Valiante NM, et al. Interleukin 10 (IL-10) inhibits human lymphocyte interferon γ production by suppressing natural killer cell stimulatory factor/IL-12 synthesis in accessory cells. J Exp Med 1993; 178:1041–1048.
65. Doherty TM Kastelein R, Menon S, et al. Modulation of murine macrophage function by IL-13. J Immunol 1993; 151:7151–7160.
66. van der Pouw Kraan TCTM, Boeije LCM, Smeenk RJT, et al. Prostaglandin-E_2 is a potent inhibitor of human interleukin 12 production. J Exp Med 1995; 181:775–779.
67. Chehimi J, Starr SE, Frank I, et al. Impaired interleukin 12 production in human immunodeficiency virus-infected patients. J Exp Med 1994; 179:1361–1366.
68. Scott P. IL-12: initiation cytokine for cell-mediated immunity. Science 1993; 260:496–497.
69. Paul WE, Seder RA. Lymphocyte responses and cytokines. Cell 1994; 76:241–251.
70. Wu C-Y, Demeure C, Kiniwa M, et al. IL-12 induces the production of IFN-γ by neonatal human CD4 T cells. J Immunol 1993; 151:1938–1949.
71. Seder RA, Gazzinelli R, Sher A, et al. Interleukin 12 acts directly on CD4+ T cells to enhance priming for interferon γ production and diminishes interleukin 4 inhibition of such priming. Proc Natl Acad Sci USA 1993; 90:10188–10192.
72. Schmitt E, Hoehn P, Germann T, et al. Differential effects of interleukin-12 on the development of naive mouse CD4+ T cells. Eur J Immunol 1994; 24:343–347.
73. Schmitt E, Hoehn P, Huels C, et al. T helper type 1 development of naive CD4+ T cells requires the coordinate action of interleukin-12 and interferon-γ and is inhibited by transforming growth factor-β. Eur J Immunol 1994; 24:793–798.
74. Kennedy MK, Picha KS Shanebeck KD, et al. Interleukin-12 regulates the proliferation of Th1, but not Th2 or Th0, clones. Eur J Immunol 1994; 24:2271–2278.
75. Manetti R, Gerosa F, Giudizi MG, et al. Interleukin 12 induces stable priming for interferon γ (IFN-γ) production during differentiation of human T helper (Th) cells and transient IFN-γ production in established Th2 cell clones. J Exp Med 1994; 179:1273–1283.
76. Croft M, Carter L, Swain SL, et al. Generation of polarized antigen-specific CD8 effector populations: reciprocal action of interleukin (IL)-4 and IL-12 in promoting type 2 versus type 1 cytokine profiles. J Exp Med 1994; 180:1715–1728.
77. Sad S, Marcotte R, Mosmann TR. Cytokine-induced differentiation of precursor mouse $CD8^+$ T cells into cytotoxic $CD8^+$ T cells secreting Th1 or Th2 cytokines. Immunity 1995; 2:271–279.
78. Yanagida T, Kato T, Igarashi O, et al. Second signal activity of IL-12 on the proliferation and IL-2R expression of T helper cell-1 clone. J Immunol 1994; 152:4919–4928.
79. Szabo SJ, Jacobson NG, Dighe AS, et al. Developmental commitment to the Th2 lineage by extinction of IL-12 signaling. Immunity 1995; 2:665–675.
80. McKnight AJ, Zimmer GJ, Fogelman I, et al. Effects of IL-12 on helper T cell-dependent immune responses in vivo. J Immunol 1994; 152:2172–2179.
81. Afonso LCC, Scharton TM, Vieira LQ, et al. The adjuvant effect of interleukin-12 in a vaccine against Leishmania major. Science 1994; 263:235–237.
82. Finkelman FD, Madden KB, Cheever AW, et al. Effects of interleukin 12 on immune responses and host protection in mice infected with intestinal nematode parasites. J Exp Med 1994; 179:1563–1572.
83. Oswald IP, Caspar P, Jankovic D, et al. IL-12 inhibits Th2 cytokine responses induced by eggs of Schistosoma mansoni. J Immunol 1994; 153:1707–1713.
84. Romani L, Mencacci A, Tonnetti L, et al. IL-12 is both required and prognostic in vivo for T helper type 1 differentiation in murine candidiasis. J Immunol 1994; 153:5167–5175.
85. Warrier R, Carvajal D, Wu C, et al. Biological characterization of IL-12 p40 knockout mice. FASEB J 1995; 9:A241.
86. Yssel H, Fasler S, deVries JE, et al. IL-12 transiently induces IFN-γ transcription and protein synthesis in human CD4+ allergen-specific T_h2 T cell clones. Int Immunol 1994; 6:1091–1096.
87. Rook AH, Kubin M, Cassin M, et al. IL-12 reverses cytokine and immune abnormalities in Sezary syndrome. J Immunol 1995; 154:1491–1498.
88. Wang Z-E, Zheng S, Corry DB, et al. Interferon γ-independent effects of interleukin 12 administered during acute or established infection due to Leishmania major. Proc Natl Acad Sci USA 1994; 91:12932–12936.
89. Wynn TA, Jankovic D, Hieny S, et al. IL-12 exacerbates rather than suppresses T helper 1-dependent pathology in the absence of endogenous IFN-γ. J Immunol 1995; 154:3999–4009.
90. Shu U, Demeure CE, Byun D-G, et al. Interleukin 12 exerts a differential effect on the maturation of neonatal and adult human $CD45RO^-$ CD4 T cells. J Clin Invest 1994; 94:1352–1358.
91. Wu CY, Demeure CE, Gately M, et al. In vitro maturation of human neonatal CD4 T lymphocytes I. Induction of IL-4-producing cells after long-term culture in the presence of IL-4 plus either IL-2 or IL-12. J Immunol 1994; 152:1141–1153.

92. Chan SH, Perussia B, Gupta JW, et al. Induction of interferon γ production by natural killer cell stimulatory factor: characterization of the responder cells and synergy with other inducers. J Exp Med 1991; 173:869–879.
93. Chan SH, Kobayashi M, Santoli D, et al. Mechanisms of IFN-γ induction by natural killer cell stimulatory factor (NKSF/IL-12). Role of transcription and mRNA stability in the synergistic interaction between NKSF and IL-2. J Immunol 1992; 148:92–98.
94. Naume B, Johnsen A-C, Espevik T, et al. Gene expression and secretion of cytokines and cytokine receptors from highly purified CD56+ natural killer cells stimulated with interleukin-2, interleukin-7 and interleukin-12. Eur J Immunol 1993; 23:1831–1838.
95. Kiniwa M, Gately M, Gubler U, et al. Recombinant interleukin-12 suppresses the synthesis of IgE by interleukin-4 stimulated human lymphocytes. J Clin Invest 1992; 90:262–266.
96. Aste-Amezaga M, D'Andrea A, Kubin M, et al. Cooperation of natural killer cell stimulatory factor/interleukin-12 with other stimuli in the induction of cytokines and cytotoxic cell-associated molecules in human T and NK cells. Cell Immunol 1994; 156:480–492.
97. Alzona M, Jäck H-M, Fisher RI, et al. IL-12 activated IFN-γ production through the preferential activation of $CD30^+$ T cells. J Immunol 1995; 154:9–16.
98. Nagy E, Buhlmann JE, Henics T, et al. Selective modulation of IFN-γ mRNA stability by IL-12/NKSF. Cell Immunol 1994; 159:140–151.
99. Carson WE, Giri JG, Lindemann MK, et al. Interleukin (IL) 15 is a novel cytokine that activates human natural killer cells via components of the IL-2 receptor. J Exp Med 1994; 180:1395–1403.
100. Kubin M, Kamoun M, Trinchieri G. Interleukin 12 synergizes with B7/CD28 interaction in inducing efficient proliferation and cytokine production of human T cells. J Exp Med 1994; 180:211–222.
101. Armant M, Ishihara H, Rubio M, et al. Regulation of cytokine production by soluble CD23: costimulation of interferon γ secretion and triggering of tumor necrosis factor α release. J Exp Med 1994; 180:1005–1011.
102. Hunter CA, Bermudez L, Beernick H, et al. Transforming growth factor-β inhibits interleukin-12-induced production of interferon-γ by natural killer cells: a role for transforming growth factor-β in the regulation of T cell-independent resistance to Toxoplasma gondii. Eur J Immunol 1995; 25:994–1000.
103. Gately MK, Warrier RR, Honasoge S, et al. Administration of recombinant IL-12 to normal mice enhances cytolytic lymphocyte activity and induces production of IFN-γ in vivo. Int Immunol 1994; 6:157–167.
104. Morris SC, Madden KB, Adamovicz JJ, et al. Effects of IL-12 on in vivo cytokine gene expression and Ig isotype selection. J Immunol 1994; 152:1047–1056.
105. Brunda MJ, Luistro L, Hendrzak JA, et al. Role of interferon-γ in mediating the antitumor efficacy of interleukin-12. J Immunother 1995; 17:71–77.
106. Gately MK, Brunda MJ. Interleukin-12: a pivotal regulator of cell-mediated immunity. In: Kurzrock R, Talpaz M, eds. Cytokines: Interleukins and Their Receptors. Norwell, MA: Kluwer Academic Publishers, 1995:341–366.
107. Sedegah M, Finkelman F, Hoffman SL. Interleukin 12 induction of interferon γ-dependent protection against malaria. Proc Natl Acad Sci USA 1994; 91:10700–10702.
108. Murray HW, Hariprashad J. Interleukin 12 is effective treatment for an established systemic intracellular infection: experimental visceral leishmaniasis. J Exp Med 1995; 181:387–391.
109. Rutenfranz I, Kirchner H. Pharmacokinetics of recombinant murine interferon-gamma in mice. J Interferon Res 1988; 8:573–580.
110. Voest EE, Kenyon BM, O'Reilly MS, et al. Inhibition of angiogenesis in vivo by interleukin 12. J Natl Cancer Inst 1995; 87:581–586.
111. Perussia B, Chan SH, D'Andrea A, et al. Natural killer (NK) cell stimulatory factor or IL-12 has differential effects on the proliferation of TCR-αβ+, TCR-γδ+ T lymphocytes, and NK cells. J Immunol 1992; 149:3495–3502.
112. Naume B, Gately M, Espevik T. A comparative study of IL-12 (cytotoxic lymphocyte maturation factor)-, IL-2-, and IL-7-induced effects on immunomagnetically purified CD56+ NK cells. J Immunol 1992; 148:2429–2436.
113. Orange JS, Wolf SF, Biron CA. Effects of IL-12 on the response and susceptibility to experimental viral infections. J Immunol 1994; 152:1253–1264.
114. Robertson MJ, Soiffer RJ, Wolf SF, et al. Responses of human natural killer (NK) cells to NK cell stimulatory factor (NKSF): cytolytic activity and proliferation of NK cells are differentially regulated by NKSF. J Exp Med 1992; 175:779–788.
115. Chehimi J, Starr SE, Frank I, et al. Natural killer (NK) cell stimulatory factor increases the cytotoxic activity of NK cells from both healthy donors and human immunodeficiency virus-infected patients. J Exp Med 1992; 175:789–796.
116. Gately MK, Wolitzky AG, Quinn PM, et al. Regulation of human cytolytic lymphocyte responses by interleukin-12. Cell Immunol 1992; 143:127–142.
117. Cesano A, Visonneau S, Clark SC, et al. Cellular and molecular mechanisms of activation of MHC nonrestricted cytotoxic cells by IL-12. J Immunol 1993; 151:2943–2957.
118. Chehimi J, Valiante NM, D'Andrea A, et al. Enhancing effect of natural killer cell stimulatory factor (NKSF/interleukin-12) on cell-mediated cytotoxicity against tumor-derived and virus-infected cells. Eur J Immunol 1993; 23:1826–1830.
119. Zeh HJ III, Hind S, Storkus WJ, et al. Interleukin-12 promotes the proliferation and cytolytic maturation of immune effectors: implications for the immunotherapy of cancer. J Immunother 1993; 14:155–161.
120. Rossi AR, Pericle F, Rashleigh S, et al. Lysis of neuroblastoma cell lines by human natural killer cells activated by interleukin-2 and interleukin-12. Blood 1994; 83:1323–1328.
121. Salcedo TW, Azzoni L, Wolf SF, et al. Modulation of perforin and granzyme messenger RNA expression in human natural killer cells. J Immunol 1993; 151:2511–2520.
122. Bonnema JD, Rivlin KA, Ting AT, et al. Cytokine-enhanced NK cell-mediated cytotoxicity. Positive modulatory effects of IL-2 and IL-12 on stimulus-dependent granule exocytosis. J Immunol 1994; 152:2098–2104.
123. Naume B, Gately MK, Desai BB, et al. Synergistic effects of interleukin 4 and interleukin 12 on NK cell proliferation. Cytokine 1993; 5:38–46.
124. Rabinowich H, Herberman RB, Whiteside TL. Differential effects of IL12 and IL2 on expression and function of cellular adhesion molecules on purified human natural killer cells. Cell Immunol 1993; 152:481–498.
125. Sarmiento UM, Riley JH, Knaack PA, et al. Biologic effects of recombinant human interleukin-12 in squirrel monkeys (Sciureus saimiri). Lab Invest 1994; 71:862–873.
126. Chouaib S, Chehimi J, Bani L, et al. Interleukin 12 induces the differentiation of major histocompatibility complex class I-

primed cytotoxic T-lymphocyte precursors into allospecific cytotoxic effectors. Proc Natl Acad Sci USA 1994; 91:12659–12663.
127. Mehrotra PT, Wu D, Crim JA, et al. Effects of IL-12 on the generation of cytotoxic activity in human CD8+ T lymphocytes. J Immunol 1993; 151:2444–2452.
128. Clerici M, Lucey DR, Berzofsky JA, et al. Restoration of HIV-specific cell-mediated immune responses by interleukin-12 in vitro. Science 1993; 262:1721–1724.
129. Bloom ET, Horvath JA. Cellular and molecular mechanisms of the IL-12-induced increase in allospecific murine cytolytic T cell activity. Implications for the age-related decline in CTL. J Immunol 1994; 152:4242–4254.
130. Noguchi Y, Richards EC, Chen Y-T, et al. Influence of interleukin 12 on p53 peptide vaccination against established Meth A sarcoma. Proc Natl Acad Sci USA 1995; 92:2219–2223.
131. Gately MK, Chizzonite R. Measurement of human and mouse interleukin 12. In: Coligan JE, Kruisbeek AM, Margulies DH, et al., eds. Current Protocols in Immunology, vol. 1. New York: John Wiley & Sons, Inc. 1992: 6.16.1–6.16.8.
132. Gately MK, Desai BB, Wolitzky AG, et al. Regulation of human lymphocyte proliferation by a heterodimeric cytokine, IL-12 (cytotoxic lymphocyte maturation factor). J Immunol 1991; 147:874–882.
133. Bertagnolli MM, Lin B-Y, Young D, et al. IL-12 augments antigen-dependent proliferation of activated T lymphocytes. J Immunol 1992; 149:3778–3783.
134. Andrews JVR, Schoof DD, Bertagnolli MM, et al. Immunomodulatory effects of interleukin-12 on human tumor-infiltrating lymphocytes. J Immunother 1993; 14:1–10.
135. Quill H, Bhandoola A, Trinchieri G, et al. Induction of interleukin 12 responsiveness is impaired in anergic T lymphocytes. J Exp Med 1994; 179:1065–1070.
136. Jelinek DF, Braaten JK. Role of IL-12 in human B lymphocyte proliferation and differentiation. J Immunol 1995; 154:1606–1613.
137. Desai BB, Quinn PM, Wolitzky AG, et al. IL-12 receptor. II. Distribution and regulation of receptor expression. J Immunol 1992; 148:3125–3132.
138. Vezzio N, Gately M, Kiniwa M, et al. IFN-γ-independent suppression of IgE synthesis by IL-12. J Immunol 1993; 150:61A.
139. Via CS, Rus V, Gately MK, et al. IL-12 stimulates the development of acute graft-versus-host disease in mice that normally would develop chronic, autoimmune graft-versus-host disease. J Immunol 1994; 153:4040–4047.
140. Finkelman FD, Holmes J, Katona IM, et al. Lymphokine control of in vivo immunoglobulin isotype selection. Annual Rev Immunol 1990; 8:303–333.
141. Germann T, Bongartz M, Dlugonska H, et al. Interleukin-12 profoundly up-regulates the synthesis of antigen-specific complement-fixing IgG2a, IgG2b and IgG3 antibody subclasses in vivo. Eur J Immunol 1995; 25:823–829.
142. Jacobsen SEW, Veiby OP, Smeland EB. Cytotoxic lymphocyte maturation factor (interleukin 12) is a synergistic growth factor for hematopoietic stem cells. J Exp Med 1993; 178:413–418.
143. Hirayama F, Katayama N, Neben S, et al. Synergistic interaction between interleukin-12 and steel factor in support of proliferation of murine lymphohematopoietic progenitors in culture. Blood 1994; 83:92–98.
144. Ploemacher RE, van Soest PL, Boudewijn A, et al. Interleukin-12 enhances interleukin-3 dependent multilineage hematopoietic colony formation stimulated by interleukin-11 or steel factor. Leukemia 1993; 7:1374–1380.
145. Ploemacher RE, van Soest PL, Voorwinden H, Boudewijn A. Interleukin-12 synergizes with interleukin-3 and steel factor to enhance recovery of murine hemopoietic stem cells in liquid culture. Leukemia 1993; 7:1381–1388.
146. Jacobsen SEW, Okkenhaug C, Myklebust J, et al. The FLT3 ligand potently and directly stimulates the growth and expansion of primitive murine bone marrow progenitor cells in vitro: synergistic interactions with interleukin (IL) 11, IL-12, and other hematopoietic growth factors. J Exp Med 1995; 181:1357–1363.
147. Bellone G, Trinchieri G. Dual stimulatory and inhibitory effect of NK cell stimulatory factor/IL-12 on human hematopoiesis. J Immunol 1994; 153:930–937.
148. Tare NS, Bowen S, Warrier RR, et al. Administration of recombinant interleukin-12 to mice suppresses hematopoiesis in the bone marrow but enhances hematopoiesis in the spleen. J Interferon Cytokine Res 1995; 15:377–383.
149. Jackson JD, Yan Y, Brunda MJ, et al. Interleukin-12 enhances peripheral hematopoiesis in vivo. Blood 1995; 85:2371–2376.
150. Neta R, Stiefel SM, Finkelman F, et al. IL-12 protects bone marrow from and sensitizes intestinal tract to ionizing radiation. J Immunol 1994; 153:4230–4237.
151. Clerici M, Sarin A, Coffman RL, et al. Type 1/type 2 cytokine modulation of T-cell programmed cell death as a model for human immunodeficiency virus pathogenesis. Proc Natl Acad Sci USA 1994; 91:11811–11815.
152. Radrizzani M, Accornero P, Amidei A, et al. IL-12 inhibits apoptosis induced in a human Th1 clone by gp 120/CD4 cross-linking and CD3/TCR activation or by IL-2 deprivation. Cell Immunol 1995; 161:14–21.
153. Ortaldo JR, Mason AT, O'Shea JJ. Receptor-induced death in human natural killer cells: involvement of CD16. J Exp Med 1995; 181:339–344.
154. Scharton-Kerston T, Afonso LCC, Wysocka M, et al. IL-12 is required for natural killer cell activation and subsequent T helper 1 cell development in experimental leishmaniasis. J Immunol 1995; 154:5320–5330.
155. Heinzel FP, Rerko RM, Ahmed F, et al. Endogenous interleukin-12 (IL-12) is required for control of Th2 cytokine responses capable of exacerbating leishmaniasis in normally resistant mice. J Immunol 1995; 155:730–739.
156. Tripp CS, Gately MK, Hakimi J, et al. Neutralization of IL-12 decreases resistance to Listeria in SCID and C.B-17 mice. Reversal by IFN-γ. J Immunol 1994; 152:1883–1887.
157. Khan IA, Matsuura T, Kasper LH. Interleukin-12 enhances murine survival against acute toxoplasmosis. Infect Immun 1994; 62:1639–1642.
158. Hunter CA, Subauste CS, van Cleave VH, et al. Production of gamma interferon by natural killer cells from Toxoplasma gondii-infected SCID mice: regulation by interleukin-10, interleukin-12, and tumor necrosis factor alpha. Infect Immun 1994; 62:2818–2824.
159. Cooper AM, Roberts AD, Rhoades ER, et al. The role of interleukin-12 in acquired immunity to Mycobacterium tuberculosis infection. Immunology 1995; 84:423–432.
160. Zhan Y, Cheers C. Endogenous interleukin-12 is involved in resistance to Brucella abortus infection. Infect Immun 1995; 63:1387–1390.
161. Ozmen L, Pericin M, Hakimi J, et al. Interleukin 12, interferon γ, and tumor necrosis factor α are the key cytokines of the generalized Shwartzman reaction. J Exp Med 1994; 180:907–915.
162. Leonard JP, Waldburger KE, Goldman SJ. Prevention of exper-

imental autoimmune encephalomyelitis by antibodies against interleukin 12. J Exp Med 1995; 181:381–386.
163. Neurath MF, Fuss I, Kelsall BL, et al. Antibodies to IL-12 abrogate established granulomatous colitis in mice. J Exp Med 1995; 182:1281–1290.
164. Trembleau S, Penna G, Bosi E, et al. Interleukin 12 administration induces T helper type 1 cells and accelerates autoimmune diabetes in NOD mice. J Exp Med 1995; 181:817–821.
165. Germann T, Szeliga J, Hess H, et al. Administration of IL12 in combination with type II collagen induces severe arthritis in DBA/1 mice. Proc Natl Acad Sci USA 1995; 92:4823–4827.
166. Chizzonite R, Truitt T, Desai BB, et al. IL-12 receptor. I. Characterization of the receptor on phytohemagglutinin-activated human lymphoblasts. J Immunol 1992; 148:3117–3124.
167. Chua AO, Wilkinson VL, Presky DH, et al. Cloning and characterization of a mouse IL-12 receptor beta component. J Immunol 1995; 155:4286–4294.
168. Pignata C, Sanghera JS, Cossette L, et al. Interleukin-12 induces tyrosine phosphorylation and activation of 44-kD mitogen-activated protein kinase in human T cells. Blood 1994; 83:184–190.
169. Bacon CM, McVicar DW, Ortaldo JR, et al. Interleukin-12 (IL-12) induces tyrosine phosphorylation of JAK2 and TYK2: differential use of Janus family tyrosine kinases by IL-2 and IL-12. J Exp Med 1995; 181:399–404.
170. Jacobson NG, Szabo SJ, Weber-Nordt RM, et al. Interleukin 12 signaling in T helper type 1 (Th1) cells involves tyrosine phosphorylation of signal transducer and activator of transcription (Stat)3 and Stat4. J Exp Med 1995; 181:1755–1762.

13
Interleukin-13

Adrian J. Minty
Sanofi Recherche, Labège, France

INTRODUCTION: HISTORY OF THE CYTOKINE

Although the first publications describing the biological activities for IL-13 appeared as recently as 1993, the very first paper concerning IL-13 was one of the pioneering studies by Mosmann and coworkers on TH_1 and TH_2 lymphocyte subsets in the mouse (1). They described a cDNA clone called P600, which encoded an mRNA expressed exclusively in the TH_2 subset. P600 was so named because the restriction enzyme Pst1 generated a fragment of 600 base pairs in the cDNA clone. No biological activities were attributed to the protein encoded by this cDNA. However, upon the discovery of the human equivalent, it was rapidly obvious that this was a cytokine possessing a number of important anti-inflammatory and immunomodulatory activities. In view of the production of this cytokine by T lymphocytes and its activities on B lymphocytes and monocytes, the term interleukin-13 was immediately proposed and adopted (2–4).

CHARACTERIZATION OF THE IL-13 PROTEIN AND GENE

The IL-13 Protein

The secreted "mature" form of the human IL-13 is a 112-amino-acid protein (Table 1, Fig. 1), which exists partially as an N-glycosylated protein. A variant lacking a glutamine residue (78) due to alternative splicing, but retaining the same biological activity, has been described (3). This variant thus corresponds to the amino-acid sequences of the mouse and rat proteins derived from the corresponding cDNAs, which lack this glutamine (Fig. 1). We have described two forms of IL-13 with Asp (I) or Gly (II) at position 61. Form II is the prevalent mRNA in peripheral blood lymphocytes from a number of donors.

Human IL-13 contains two disulfide bonds (28–56; 44–70) (J. C. Guillemot, personal communication). Theoretical structural models (7) based on structures with a single disulfide bond are thus unlikely to be correct. IL-13 shows limited homology with IL-4 in the first and fourth

Table 1 Physicochemical Properties of IL-13

	Human	Mouse
Amino acids		
Precursor	132 (146)[a]	131
Signal sequence	20 (34)[a]	18[b]
Mature	112[c]	111[b]
Molecular mass (kDa)		
Predicted	12.4	12.2
Mass spectrometry	12.4, 14.4[d]	
SDS-PAGE	9–10, 17[d]	
Potential N-glycosylation sites	4	3
Disulfide bonds	28-56 44-70	

[a]There are two possible methionine initiation codons. The most probable translation initiation site would give a polypeptide of 132 amino acids and a signal peptide of 20 amino acids.
[b]There has been no experimental determination of the cleavage site for the signal peptide. This is a prediction based on homology with the human protein.
[c]This is the experimentally determined cleavage site for IL-13 made in recombinant CHO cells.
[d]N-glycosylated species.

```
                              HELIX A

                      h  h   h   h     §      hh h h
IL-13 R   GPVRRSTSPPVALRELIEELSNITQDQKTSLCNSSMVWSVDLTAG-----
          *** ** * *  * ************* * *** ******** **
IL-13 M   GPVPRSVSLPLTLKELIEELSNITQDQ-TPLCNGSMVWSVDLAAG-----
          **** *      * ****** **** * *********  * **
IL-13 H   GPVPPST----ALRELIEELVNITQNQKAPLCNGSMVWSINLTAG-----
             *        * * *  *   *  **§ **   *  *   §
IL-4  H   NFVHGHKCD-ITLQEIIKTLNSLT-EQKT-LCTELTVTDIFAASKNTTEK
            * * **    * ***  ** *    *  ***  *     * *****
IL-4  M   TRSHIHGCDKNHLREIIGILNEVT-GEGTP-CTEMDVPNVLTATKNTTES
          *   ****    *****  ** **   *** **** ** ***** ****
IL-4  R   TGNGIHGCNDSPLREIINTLNQVT-EKGTP-CTEMFVPDVLTATRNTTEN
                      E9          (28/24)
                                     |
                    HELIX B     S--------S       HELIX C
                                |
          §h      hh  hh      (56/65)             h  h  hh  h
IL-13 R   -GFC---AALESLTNISS----C---------------NAIHRTQRILNGL
           ***    ** ******     *               *** ****** **
IL-13 M   -GFC---VALDSLTNISN----C---------------NAIYRTQRILHGL
             *    ** ** * *     *                **  *** * *
IL-13 H   -MYC---AALESLINVSG----C---------------SAIEKTQRMLSGF
             *      *           *                     * * *
IL-4  H   ETFCRAATVLRQFYSHHEKDTRCLGATAQQFHRHKQLIRFLKRLDRNLWGL
          * *** *** ** * * * **                  * ** *    *
IL-4  M   ELVCRASKVLRIFYLKHGK-TPCL-------KKNSSVLMELQRLFRAFRCL
          ** **** *** **        ***       *  * ** **  * *   *
IL-4  R   ELICRASRVLRKFYFPRDV-PPCL-------KNKSGVLGELRKLCRGVSGL
          (44/46)                                   R88
             |
            S-S                    HELIX D
             |
          (70/99)    §              h  hh  h   h
IL-13 R   -----CNQKASDV-ASSPP--DTKIEVAQFISKLLNYSKQLFRYGH
               ** **     ** *  ******* ** *** * ***** *
IL-I3 M   -----CNRKAPTT-VSSLP--DTKIEVAHFITKLLSYTKQLFRHGPF
               *  *      ***   ******* *   **   * *** * *
IL-13 H   -----CPHKVSAGQFSSLHVRDTKIEVAQFVKDLLLHLKKLFREGRFN
          §    *                *           *  **   **
IL-4  H   AGLNSC---------------PVKEANQSTLENFLERLKTIMREKYSKCSS
               **                 *     *  *** ** **   **
IL-4  M   DSSISC---------------TMNESKSTSLKDFLESLKSIMQMDYS
           *  **              * ***  * ***********    *
IL-4  R   NSLRSC---------------TVNESTLTTLKDFLESLKSILRGKYLQSCTSMS
                                                R121.Y124
```

Figure 1 Indicated in bold are hydrophobic residues (h) buried in the structural core of IL-4. Indicated in outline are amino acids E9, R88, R121, Y124 important in IL-4-receptor interactions (5) and cysteine residues involved in disulfide bonding. The alignment is based on that of Zurawski et al. (6), with minor adjustments, using the conservation of the Cys-Cys disulfide bridges (numbered x/y, for their positions in human IL-13 and human IL-4) and the conserved hydrophobic residues to align helices B and C. Multiple alignments are possible in this region (6,7). [Mouse IL-13 sequence from Brown et al. (78); Rat IL-13 sequence from Lakkis and Cruet (79)].

α-helical regions of IL-4 (2,6). The B and C helices of IL-4 show virtually no homology with IL-13, but various alignments can be made of the IL-13 and IL-4 sequences in these regions, which retain hydrophobic residues known to be buried in the structural core of the IL-4 molecule (6,7) (see Fig. 1). Thus the four-helical (up, up, down, down) model for IL-4 (7) may also apply to IL-13, but with considerably shortened B and C helices. Circular dichroism studies on mouse and human IL-13 (6; M. C. Bonnet, personal communication) show an absorption spectrum characteristic of a highly α-helical protein.

Human IL-13 acts on mouse, rat, and rabbit cells (8–10), although the affinity for mouse IL-13 receptor is considerably lower. Mouse and human IL-13 are equally active on human cells (3).

The IL-13 Gene

The human IL-13 gene was first shown to be adjacent to the IL-4 gene on the long arm of chromosome 5 (23–31) by Morgan et al. (11), who were seeking to identify coding regions on a yeast artificial chromosome containing a human DNA fragment including the IL-4 and IL-5 genes. The human IL-13 gene is at the centromere-proximal end of the lymphokine gene cluster, including IL-4, IL-5, IL-3 and GM-CSF (Fig. 2). The mouse IL-13 gene has been mapped to the central region of chromosome 11. Mouse and human IL-13 genes have been isolated and characterized by McKenzie et al. (12) and Smirnov et al. (13). They are both formed by the 4 exon:3 intron structure found in a number of cytokine genes (Fig. 2B). The intron positions are approximately the same as those in the IL-4 gene, with exon 1 containing helix A and exon 4 containing helix D (Fig. 1). The intron sizes are considerably smaller than those in the IL-4 genes (1055, 251, and 345 in the human IL-13 gene and 1258, 576, and 311 in the mouse IL-13 gene). Upstream regions of the mouse and human IL-13 genes are highly conserved (76% homology over 400 base pairs) but show strikingly little homology with the IL-4 gene (12), although some shared regulatory sequences are present (13).

The human IL-13 mRNA (2,3) is a 1280 nucleotides in length with a short 5′ noncoding region (56 nucleotides) and a longer 3′ noncoding region (850 nucleotides) containing four copies of the (A/T)ATTTA(A/T) repeat, including one copy of the nonamer TTATTTATT, implicated in cytokine mRNA instability. There are two potential ATG initiation codons, with the second being present in a nucleotide sequence context closer to the consensus sequence, CC(A/G)CCAUGG, for translation initiation. Utilization of the first ATG (not yet established) would mean an extremely short 5′ noncoding region (12 nucleotides) and a 146-amino-acid protein, which could conceivably be membrane-anchored.

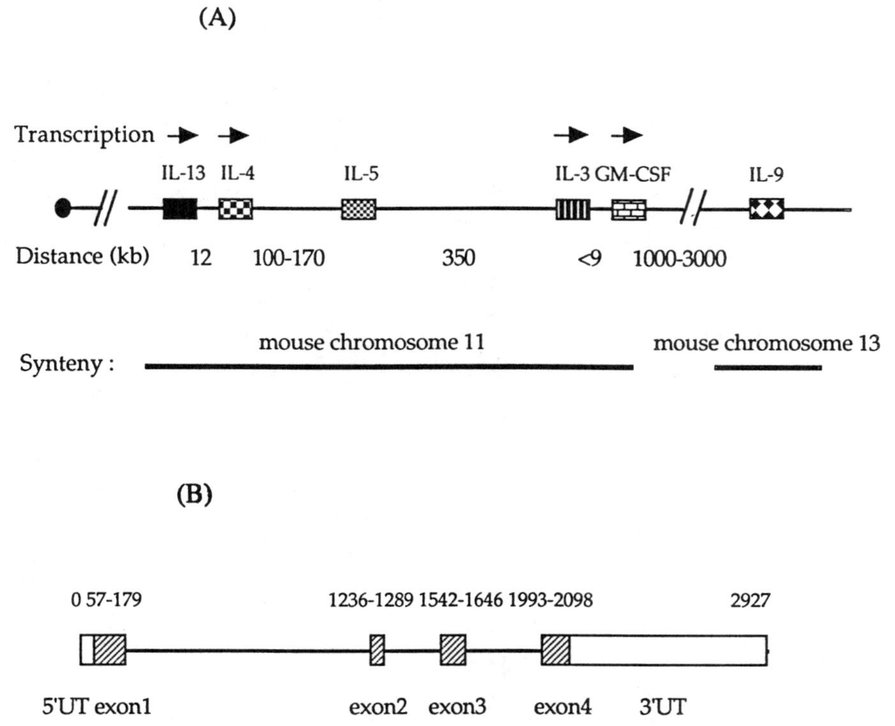

Figure 2 Chromosome location and structure of human IL-13 gene. (Source: Refs. 11–13.)

SOURCES AND INDUCERS OF IL-13

Human interleukin-13 was initially identified by differential screening of cDNA libraries from activated peripheral blood lymphocytes. It can be expressed by both CD4+ and CD8+ lymphocytes. Although in the mouse it is expressed almost exclusively by TH2 lymphocyte clones, in humans it can be expressed in both TH_1 and TH_2 lymphocyte clones (4; A. Minty and A. Bensussan, unpublished results). Initial experiments examining cytokine production by polyclonally activated populations of peripheral blood lymphocytes, which had been previously expanded in conditions favoring either a type I response (anti-CD3+IL-2+IL-12) or a type II response (anti-CD3+IL-4), indicate that IL-13 expression is primarily produced as part of a type II response (M. Bréban, A. Minty, S. Asselin, and D. Fradelizi, unpublished observations), as is the case in the mouse (14). However, IL-4 and IL-13 expression is clearly dissociated when naive human CD4+ 45RO- T lymphocytes are stimulated via the T-cell receptor, since effector cells develop which produce large amounts of IL-13, IL-5, and IFN-γ but no IL-4 (15).

Recently, IL-13 mRNA expression has been reported by B-lymphocyte cell lines (16) and by normal B lymphocytes after CD40 cross-linking and subsequent exposure to PMA and ionomycin (17). The latter B lymphocytes do not express detectable IL-4 mRNA. Both IL-13 and IL-4 can be expressed by activated mast cells (18), where they will act in concert with CD40 ligand expressed on these cells to induce IgE production on neighboring B lymphocytes. This IL-13 expression by B lymphocytes and mast cells may be important in the pathogenesis of atopy.

Optimal expression of IL-13 in T lymphocytes is obtained, as for other lymphokines, using a combination of two different inducing agents (PMA, PHA-P, anti-CD2, anti-CD28, anti-CD3, ionomycin). Cyclosporin A can suppress the T-lymphocyte production of IL-13 induced by PHA-P. The combination of PMA and a calcium ionophore gives good stimulation of mast cell and basophil cell lines (18).

RECEPTORS

The number of IL-13 and IL-4 receptors on a variety of cell types is given in Table 2. IL-13 binds with high affinity (Kd = 300 pM). It seems that cell types expressing IL-13 receptors are a subset of those expressing IL-4 receptors.

Table 2 Distribution of IL-13 Receptors on Different Cell Types

Cell type	Characteristics	IL-13r/cell	IL4r/cell
Renal carcinoma cells			
WS-RCC		2,000	2,300
PM-RCC		26,000	3,800
HL-RCC		150,000	
Other nonhematopoietic cells			
COS-3	Fibroblast	2,500	2,200
A431	Keratinocyte	205	315
Monocytes, premyeloid cells			
U937	Promyelomonocyte	420	700
THP-1	Monocyte	270	260
Peripheral blood	Monocyte	120	
TF-1	Premyeloid	130	
		250	750
HL60	Promyelomonocyte	UD	UD
B lymphocytes			
NALM-6	pre-B	UD	2,150
Daudi	Burkitt lymphoma	UD	450
Raji	Burkitt lymphoma	UD	2,300
DH	EBV-transformed	300	
T lymphocytes			
PHA-activated PBL		UD	700
H9	T cells	UD	
MLA-144	T lymphocyte	UD	350

UD = undetectable (< 50r/cell).
Source: Refs. 19–22.

In particular, IL-13 receptors appear to be absent from activated T lymphocytes, T-lymphocyte cell lines, and certain B-lymphocyte cell lines (6,16,20,22). Cross-competition experiments have shown that IL-4 can usually completely compete for IL-13 binding (except in renal carcinoma cell lines, which express large numbers of IL-13 receptors), whereas IL-13 can variably compete for IL-4 binding.

Since the cloned IL-4 receptor subunit (IL4R gp140) does not bind IL-13, these results were initially taken to indicate the presence of an additional shared component between IL-13 and IL-4 receptors (6). This conclusion was also drawn from studies showing that an IL-4 antagonist (Y124D) would interfere with IL-13 effects (23). Upon the discovery that the IL-2r γ chain could be part of the IL-4 receptor (24,25), it was postulated, and subsequently widely assumed, that this additional component would be the γ chain.

However, recent studies on the B lymphocytes of X-SCID patients (26) conclusively showed that the γ chain is not necessary for IL-13 and IL-4 effects on B-cell proliferation and Ig switching. In addition, IL-4/IL-13 effects and cross-competition can be seen in nonhematopoietic cell types not expressing γ chain (20). Furthermore, anti-γ chain antibodies can antagonize proliferation induced by IL-4, but not that induced by IL-13, in the MC/9 mast cell line, and transfection of the γ chain into the B9 plasmocytoma cell line increases proliferation only in response to IL-4 (27). The γ chain is thus not the shared component in IL-4 and IL-13 receptors.

A different model (Fig. 3) is suggested by experiments from many laboratories showing that antibodies against the IL-4 gp140 subunit will inhibit IL-13 responses (e.g., Refs. 28–30), indicating that this subunit is part of the IL-13 receptor. Although the extracellular region of the IL-4R gp140 will not form dimers in the presence of IL-4 (31), this subunit may thus form heterodimers with the IL-13–binding subunit. Such heterodimers are suggested by cross-linking experiments (22). In this model, the IL-4R gp140 will thus form heterodimers either with the γ chain or with the IL-13–binding chain. The IL-13–binding subunit has not yet been described, but cross-linking studies (20,22) show that it has a molecular weight of approximately 70 kDa.

BIOLOGICAL ACTIVITIES IN VITRO

Bioassay

We have developed a bioassay for mouse or human IL-13 that is relatively, though not completely, specific (30). This assay measures the proliferation of a subclone (B9.1.3) of the B9 plasmacytoma cell line developed by Aarden for measuring IL-6 bioactivity, isolated after continuous culture in IL-13. This subclone retains very high sensitivity to mouse or human IL-6 (pg/ml), but has acquired IL-13 responsiveness ($ED_{50} = 5$ ng/ml for human IL-13, 50 pg/ml for mouse IL-13). It has also acquired a proliferative response to mouse (but not human) IL-4, but at higher doses. In measuring human IL-13 in biological samples, controls using blocking anti–IL-6 antibodies must thus be employed.

The proliferative response of the human erythroleukemic TF1 cell line can also be used to assay mouse and human IL-13 (3). Although this assay is probably unsuitable for biological samples of human origin since this cell line responds to a bewildering variety of cytokines, it may be useful for measuring IL-13 in mouse samples, since other mouse cytokines such as IL-3, IL-4, and GM-CSF will not be active (18).

Overview of Biological Activities

IL-13 has been described by Zurawski and De Vries (4) as "an IL4-like cytokine that acts on monocytes and B cells, but not on T cells"—a generalization apparently borne out by binding studies showing cross-competition between IL-13 and IL-4 on a number of cell types and a lack of IL-13 receptors on T lymphocytes (6,20,22). Two observations suggest that T lymphocytes have the potential to express IL-13 receptors. Kohler et al. (32) originally observed that IL-13 will activate the transcription factor NF-IL-4 (IL-4Stat/Stat 6) in highly purified T lymphocytes, and Jinquan et al. (33) recently reported inhibition of chemokine-induced T-cell chemotaxis by IL-13. Recent work by Lin et al. (34) showed activation of Stat proteins by IL-13 in freshly isolated, but not in PHA-activated, peripheral blood lymphocytes. Thus, while the characteristics (purity, preparation method) of the lymphocyte populations used by Lin et al. are not discussed, it seems conceivable that IL-13 receptors may be present on unactivated T lymphocytes, but absent from activated T lymphocytes and T-cell lines.

IL-13–specific activities have been difficult to detect. IL-13 inhibition of proliferation in the absence of IL-4 effects has been observed on some acute lymphoblastic leukemia B-cell lines (29). IL-13–specific proliferation of the B9 plasmocytoma cell line has also been reported (4,30), but other B9 cell lines respond to both cytokines, albeit less favorably to IL-4 (27). Some of the IL-13 effects on mouse $Lin^{-}Sca1^{+}$ hematopoietic precursor cells appear more pronounced than those of IL-4 (35). However, it is difficult to exclude relatively trivial differences, for example, in cytokine stability, which will have important effects in long-term culture experiments. In addition, experiments

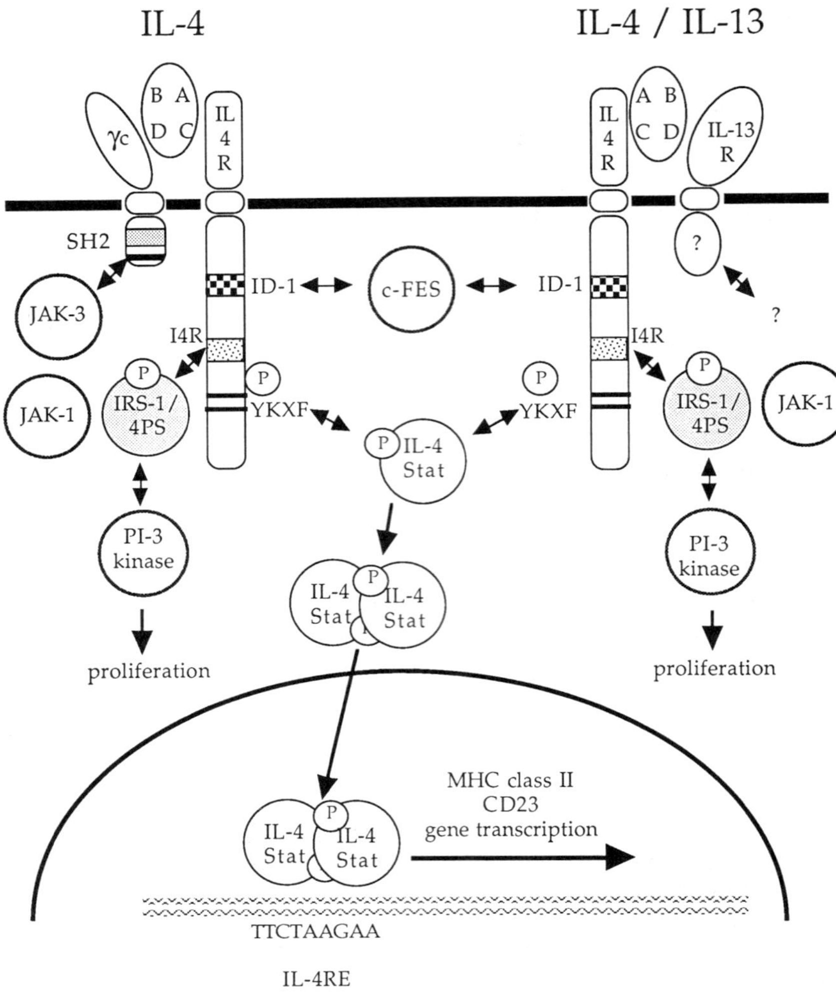

Figure 3 Signal transduction by IL-4 and IL-13 receptors. (For details, see Receptors and Signal Transduction sections.)

on nonhomogeneous populations can give apparently IL-13–specific responses, which result from a combination of a stimulatory activity (common to IL-13 and IL-4) on one cell type and an inhibitory activity specific to IL-4 on a second cell type. As an example, the IL-13–specific induction of IFN-γ in large granular lymphocyte (LGL) cultures, which we originally reported (2) may have resulted from IL-12 induction in monocytes in these cultures (common to IL-13 and IL-4) (36; A. Minty, unpublished results) acting in synergy with IL-2 to activate IFN-γ synthesis and concomitant inhibition of the IL-2–activating effect on LGL by IL-4 but not by IL-13.

IL-4–specific activities can be observed even on cell lines possessing IL-13 receptors, such as the A431 keratinocyte cell line (22,37). It is not clear whether the A431 cell line may express low levels of γ chain, accounting for a differential signaling pathway. Nor is it known yet whether IL-13–specific responses will be observed in cells possessing high numbers of IL-13 receptors, such as the renal carcinoma cell lines described by Obiri et al. (20), where IL-13 binding cannot be efficiently competed against by IL-4.

Signal Transduction

From the foregoing discussion and the finding that the IL-4R gp140 subunit is also part of the IL-13 receptor, it should be clear that, although IL-13 signal transduction has been little studied so far, much is already known from studies of IL-4. In support of this assumption, phosphorylation of the 4PS and NF-IL-4/IL-4Stat/Stat 6 proteins, considered as relatively specific IL-4 signal transducers, is also observed with IL-13 (32,34,38,77).

IL-4 signal transduction has recently been extensively

reviewed (39), and we will here consider only a few salient points. The IL-4 receptor gp140 subunit lacks intrinsic tyrosine kinase activity, as does its associated γ chain. Nevertheless, many cellular proteins are rapidly phosphorylated in response to IL-4, notably IRS-1 (insulin receptor substrate-1) and a related 170 kDa protein 4PS (IL-4-induced phosphotyrosine substrate) (Fig. 3). The γ chain has been reported to be necessary for the phosphorylation of IRS-1 (24), but is apparently not an absolute requirement. L cells, which lack γ chain, will nevertheless phosphorylate IRS-1 and/or 4PS in response to IL-4 or IL-13. However, introduction of the γ chain by transfection resulted in enhanced phosphorylation of IRS-1 and/or 4PS by IL-4, but not by IL-13 (77).

A model (39) for one type of IL-4 signal transduction involves tyrosine phosphorylation of tyr 497 within the I4R motif in the cytoplasmic region of the gp140 via kinases associated with the receptor (Fig. 3). These kinases may include JAK-1 and JAK-3 (40,41) and/or c-Fes-like protein, associated with gp140 (42). JAK-3 kinase has been shown to be associated with the 48 carboxy-terminal amino acids of the γ chain (43). Consistent with the idea that IL-13 does not use the γ chain is the lack of phosphorylation of JAK-3 by IL-13 (44).

The phosphorylated tyr 497 is a site for the "docking" of the IRS-1/4PS proteins (39). These proteins can then be phosphorylated by other receptor-associated kinases and serve as "docking proteins" for other signal-transducing molecules containing SH2 (Src homology) domains, including phosphoinositide-3 kinase. It is clear that multiple mechanisms of IL-4 signal transduction can occur, since no 4PS phosphorylation is observed in T-cell lines which proliferate with sub-picomolar quantities of IL-4 (39).

Another signaling pathway common to IL-4 and IL-13 is the IL-4 Stat/NF-IL-4/Stat 6 protein (32,45). This protein can directly interact with two domains in the carboxy-terminus of the gp140 molecule containing a phospho-YKXF motif via SH2 domains (45). Proximity to a receptor-associated kinase leads to phosphorylation of IL-4 Stat, after which the molecule dissociates from the receptor, dimerizes, and migrates to the nucleus, where it can activate, or repress, a number of genes including MHC class II genes, Fc receptors, and CD23, via an IL-4–responsive element (TTCTAAGAA) (45) (see Fig. 3). JAK kinases may also be involved in the tyrosine phosphorylation of IL-4 Stat, since cytokine-induced phosphorylation of other Stat proteins is mediated by JAK kinases. The

Table 3 Alternative Macrophage Activation by IL-13

	Cytokine treatment			
	IL-13 or IL-4	IL-10	IFN-γ	IFN-γ + IL-13
Production of proinflammatory cytokines and antagonists				
Inflammatory cytokines (treatment with LPS)	–	–	+	+/–
IL-1ra (treatment with LPS)	+	+	0/–	+
IL-12, TNF-α (treatment before LPS)	+	–	+	++
Cell surface antigens				
MHC class II / APC	+	–	+	++
Fc receptors / ADCC	–	+	+	+/–
CD11b / CD11c	+	0	0	+
CD14	–	0	0	–
Mannose receptors	+	+/0	–	ND
Other pro- and anti-inflammatory and procoagulant activities				
NO production	+/–[a]	–	+	–[a]
Respiratory burst	–/0[a]	–	+	ND
Procoagulant activity	–	–	+	–
15-lipoxygenase / 15-HETE	+	0	–	–
Protein tyrosine kinases				
Lsk	+	ND	–	–
Csk	0	ND	+	+

+ = Increase; – = decrease; 0 = no change. APC = Antigen-presenting cell function; ADCC = antibody-dependent cell-mediated cytotoxicity; ND = not done.
[a]NO production and respiratory burst are variably affected by IL-13 and IL-4 (see text).
Source: See text and Ref. 80.

complex interrelationship between the different phosphorylation events (IRS-1/4PS, c-Fes, IL-4Stat, JAK-1, JAK-3) is not yet established.

The simplest interpretation of IL-4 and IL-13 signaling based on the data so far available would be that there are two types of IL-4 receptor with both common and potentially different signaling pathways. These comprise the gp140 subunit dimerized with either the γ chain or the IL-13–binding subunit (Fig. 3). IL-13 will use only the second type of IL-4 receptor.

Effects on Monocytes and Macrophages

As discussed by Doyle et al. (8), IL-13 and IL-4 show an alternative form of macrophage activation, contrasting with the classical activation and deactivation observed with IFN-γ and IL-10 (Table 3). This alternative monocyte activation state can be classified as follows:

1. When IL-13 is given at the same time as bacterial lipopolysaccharide, it inhibits the production of proinflammatory cytokines (IL-1, TNF, IL-6, IL-8, MCP-1) and upregulates the natural inhibitors of IL-1 bioactivity and (IL-1ra and sIL-1rII) (2,46–48). It is for this reason that IL-13 was originally classified as an anti-inflammatory cytokine. In addition, IL-13 downregulates expression of CD14, the LPS receptor, on circulating monocytes (46), although it can upregulate this molecule on monocyte cell lines (P. Liauzun, personal communication). The notion of IL-13 as an anti-inflammatory cytokine now has to be somewhat modified to take into account the "priming" of TNF-α and IL-12 production by IL-13 treatment given before an inflammatory stimulus (LPS or SAC) (36). Even under these conditions, however, downregulation of other cytokines such as IL-1, IL-8, and IL-10 is seen, and we would argue that the overall activity of IL-13 under these conditions will nevertheless be anti-inflammatory, but also immunostimulatory.
2. IL-13 will also exert immunomodulatory activities via its upregulation of MHC class II molecules and antigen presentation (8,46, 49).
3. IL-13 exerts antiprocoagulant activities (downregulation of tissue factor) in common with other TH2 cytokines IL-4 and IL-10 (50).
4. IL-13 induces many morphological changes including homotypic aggregation, strong adherence, and the development of long cellular processes (46). It upregulates expression of the integrins CD11b, CD18, CD29, and CD40e (VLA-5). It also induces monocyte chemotaxis (10).
5. IL-13 inhibits antibody-dependent cell-mediated cytotoxicity, in part via downregulation of FcγRI, although these two events can be dissociated in the presence of IFN-γ or IL-10 where FγRI expression is maintained while ADCC is still inhibited by IL-13 (46,49).
6. IL-13 induces the enzyme 15-lipoxygenase, which catalyzes the C-15 hydroperoxidation of fatty acids (51). In particular, this leads to increased production of 15-(s)-hydroxyeicosatetraenoic acid (15-HETE) (52), which specifically antagonizes the functional responses to leukotrienes.
7. IL-13 will inhibit NO production by mouse peritoneal macrophages stimulated with LPS and IFN-γ (8) and by mouse macrophages derived from bone marrow by culture with GM-CSF (49). On the contrary, it increases NO production by mouse macrophages derived by culture with M-CSF (49). Similarly, IL-4 has been reported to upregulate NO production in human monocytes (53).
8. IL-13 does not affect PMA-induced superoxide production by mouse peritoneal macrophages (8) but inhibits PMA-induced superoxide production by human monocytes (54). It would thus appear that the effects of IL-4 and IL-13 on superoxide may be species-specific (55).

In many respects, IL-13 (as IL-4) would appear to be playing a protective role, since limitation of NO, superoxide, and inflammatory cytokine (e.g., TNF) synthesis will serve to protect against host tissue damage. IL-13 also increases mannose receptors (8), which may serve to mop up secreted enzymes bearing terminal mannosyl ligands (such as lysosomal hydrolases and plasminogen activators). Its effects on host defense are mixed: it decreases Fc receptors and ADCC, yet increases mannose receptors, which will promote phagocytosis of mannosylated parasites. As discussed above, the monocyte microbicidal activity through NO and superoxide production will be variably affected.

Effects on B Lymphocytes

IL-13 activities on B lymphocytes were among the first to be described and have been reviewed by Zurawski and De Vries (4) and Defrance et al. (56). IL-13 has a number of activities resembling IL-4, although it is often less potent and less efficient.

1. IL-13 induces proliferation of mature B cells in co-stimulation with anti-CD40 or anti-Ig (3,56–58).
2. In contrast, it inhibits the proliferation of normal pre-B cells in a culture system employing stromal

cells and IL-7 and their malignant counterparts in B-lineage acute lymphoblastic leukemia (ALL) (29). Although these effects are similar to those of IL-4, some ALL cell lines respond preferentially to IL-4, and some to IL-13. IL-4 also reduces viability of ALL cell lines in culture, whereas this is not observed with IL-13.
3. It inhibits the IL-2–induced proliferation and apoptosis of chronic lymphocytic leukemia B cells (59).
4. It stimulates Ig production (57,58) and Ig switching to IgE and IgG4 (23,56,57).
5. It upregulates CD72, MHC class II molecules, and CD23 (57).

Mouse B lymphocytes do not respond to IL-13 (4). The proportion of human B cells expressing IL-13 receptors is not yet clear. Binding studies on B-cell lines suggest that IL-13 receptors are present only on certain B cells (16,20,22). This would be consistent with upregulation of CD23 on only a subpopulation of cells (56,57). However, CD72 expression is increased by IL-13 on the majority of splenic B cells (57).

On tonsil-derived B lymphocytes, the proliferative effects of IL-13 are mainly on the IgD+ "naive" B-cell population (56). This may indicate a role for IL-13 in a primary immune response, as opposed to the effects of IL-2, which appear to be more on the IgD- "mature" B-cell population.

Although IL-13 appears to be only a minor regulator of IgG and IgM synthesis in adult tonsil–derived B cells compared to IL-2 or IL-10 (56), it significantly increases the production of IgG4 and IgE in costimulation with signaling from the CD40 ligand. IL-13 and IL-4 appear to be the only cytokines capable of directly inducing IgE production (56–58).

Effects on Other Hematopoietic Cells

IL-13 activities, similar to those of IL-4, have also been described on circulating human polymorphonuclear cells (48) and mast cells (60).

In the mouse, IL-13 has two principal effects on bone-marrow $Lin^{-}Scal^{+}$ primitive precursor hematopoietic cells, increasing their proliferation in synergy with other cytokines, especially SCF, and leading to formation of macrophage colonies, in the presence of cytokine combinations such as SCF + G-CSF and SCF + IL-11, which normally generate principally granulocyte colonies (35).

However, IL-13 acts on human CD34+ precursor cells from cord blood or from bone marrow to inhibit, rather than to promote, granulocyte/monocyte or monocyte colony formation (61,62). This may represent a species difference or a difference in maturity of the progenitor cell populations. The IL-13–induced increase in macrophage colonies from mouse bone marrow cells is seen with the more primitive Lin^{-}Scal+ cells but not with the more mature $Lin^{-}Scal^{-}$ mouse cells (35), whereas in causing the decrease of macrophage colony formation from human bone marrow cells, IL-13 appears to be acting on late-stage precursors (62).

IL-13 is also a potent stimulator of megakaryocyte formation from human cord blood CD34+ cells (61), particularly in combination with other megakaryocyte-stimulating factors such as IL-3.

Effects on Nonhematopoietic Cells

Accumulating data suggest that IL-13, like IL-4, has widespread pleiotropic activity. Other cell types directly affected by IL-13 include endothelial cells, keratinocytes, microglial cells, renal carcinoma cells, and HT-29 colon carcinoma cells. The effects of IL-13 on these cells will have repercussions on hematopoietic cells: for example, the upregulation of VCAM-1 on endothelial cells may have important consequences in leukocyte trafficking (63).

It is becoming clear that the IL-13 regulation of cytokine production by these nonhematopoietic cells is very different from its effects on circulating blood monocytes. For example, IL-6 and IL-8 production are inhibited by IL-13 in monocytes (2,46), whereas IL-6 is upregulated in endothelial cells, keratinocytes, and microglial cells (63,64,81) and IL-8 is upregulated in endothelial (52) and HT29 carcinoma cells (82). This illustrates the fact that local activity of IL-13 may be of critical significance in inflammatory situations.

IN VIVO EFFECTS AND IMPLICATION IN DISEASE STATES

Anti-Inflammatory Activities

The originally described anti-inflammatory activity on LPS-induced monokine synthesis has prompted trials of IL-13 in rodent inflammatory, autoimmune, disease models. In both models, collagen-induced arthritis (CIA) in mice and experimental autoimmune encephalomyelitis (EAE) in rats, a single injection of IL-13–producing CHO cells leads to a significant, long-lasting improvement in clinical status (9,83). Whether or not the in vivo anti-inflammatory effects of IL-13 are related to its inhibition in vitro of the production of monokines such as TNF remains to be established.

Antitumoral Activities

In similar experiments, co-injection of IL-13–producing CHO cells with HeLa cells prevents the HeLa cells from forming tumors in nude mice (65). Likewise, P815 mastocytoma cells genetically modified to produce IL-13 transiently form small tumors when injected into DBA/2 mice, which rapidly regress. Mice treated in this way can be subsequently reinjected with the parental P815 cells without undergoing tumor formation. The mechanism of the initial tumor rejection has not been determined, but monocytes and neutrophils are found to infiltrate the tumor (65) with little sign of the eosinophil infiltration seen in IL-4–treated tumors (66). This may be dictated in part by IL-13–influenced local chemokine production at the tumor site.

Antiviral Activities

Another potential clinical use for IL-13, which has so far received support only from in vitro experiments, is in the control of HIV-1 viral load in AIDS patients. Following the initial report of Montaner et al. (67) that IL-13 reduces virus production in cultured blood-derived human macrophages, this inhibition has also been observed in chronically infected monocyte cell lines (68) and in alveolar macrophages and cultures of total peripheral blood mononuclear cells from AIDS patients (69). Since the macrophage is probably a major reservoir of virus, this inhibitory effect of IL-13 may be clinically useful, even though IL-13 has no effect on virus production by T lymphocytes (67).

Immunomodulation by IL-13

A potential drawback to such a clinical use of IL-13 was the reported inhibitory effects of IL-13 on LPS-induced IL-12 p40 mRNA production (4–6), since IL-12 deficit has been implicated in the immunodeficiency associated with HIV infection (70,71). Such results have led to the suggestion that IL-13 may inhibit cell-mediated immunity (4).

However, IL-13 only weakly inhibits monocyte production of the biologically active IL-12 p70 heterodimer stimulated by LPS and does not inhibit or slightly increases its production by SAC-stimulated monocytes (36; our unpublished observations). Pretreatment with IL-13 before LPS or SAC stimulation actually increases IL-12 p70 production (36), thus contributing to the development of a type 1 lymphocyte response, instrumental in cellular immunity. In addition, IL-13 lacks the direct inhibitory effects of IL-4 on T-cell functions such as IFN-γ and perforin production and will participate in a humoral immune response via its activities on B lymphocytes. It may thus stimulate both arms of the immune response.

IL-13 and Allergy

IL-4 gene knock-out studies in mouse have shown that IL-4 is necessary for IgE production (72). In contrast, IL-13 knock-out mice (73) have apparently normal IgE responses, as might be expected from the lack of IL-13 activities on mouse B cells (4). However, in humans, a significant and variable part of the IgE-producing activity in activated T cells from atopic individuals is not inhibited by anti-IL-4 (74). The remaining activity may thus be IL-13. This would be consistent with genetic data identifying a locus regulating serum IgE levels in atopic humans within the region including or adjacent to the IL-4 gene (75).

Kimata et al. (76) recently reported that T cells from atopic dermatitis (AD) patients spontaneously produce IL-4, but not IL-13. However, they did not report the expression of these lymphokines from activated T cells. In contrast to the situation in AD patients, IL-13, but not IL-4, was spontaneously produced from T cells of patients with nephrotic syndrome (NS). It thus will be of interest to examine further IL-13 and IL-4 production in different allergic and nonallergic conditions where high IgE levels are observed.

Both IL-13 and IL-4 can also be expressed by activated mast cells (18), where they will act in concert with CD40 ligand expressed on these cells to induce IgE production on neighboring B lymphocytes. This IL-13 expression by mast cells may be important in the pathogenesis of atopy. The Y124D IL-4 mutant protein, which has been shown to inhibit both IL-4– and IL-13–induced IgE production (23), may be useful as a competitive antagonist in limiting allergic reactions.

SUMMARY AND CONCLUSION

IL-13 is a 112-amino-acid protein produced by activated CD4+ and CD8+ T lymphoctyes, CD40-ligand–activated B lymphocytes, and activated mast cells. It can be concomitantly expressed with IL-4 as part of a TH2-type response, but can also be expressed independently of IL-4. It shares

The following two recent articles describe the cloning of two distinct IL-13-binding subunits: Hilton DJ, Zhang J-G, Metcalf D, Alexander WS, Nicola NA, Willson TA. Cloning and characterization of a binding subunit of the interleukin-13 receptor that is also a component of the interleukin-4 receptor. Proc. Natl. Acad. Sci 1996; 93:497–501; and Caput D, Laurent P, Kayhad M, Lelias JM, Lefort S, Vita N, Ferrara P. Cloning and characterization of a specific interleukin (IL)-13 binding protein structurally related to the IL-5 receptor α chain. J Biol Chem 1996; 271:16921–16926.

receptor components and biological activities with IL-4 on most cell types, an exception being activated T lymphocytes, which apparently lack IL-13 receptors. Its anti-inflammatory and immunostimulatory activities may be useful in the treatment of autoimmune disease, malignancy, and HIV infection. Only a few years have passed since the initial reports of IL-13 biological activity—it is clear that surprises lie ahead. In particular, the long-awaited characterization of the IL-13 receptor should clarify many aspects of IL-13 biology.

NOTE ADDED IN PROOF

The following two recent articles describe the cloning of two distinct IL-13-binding subunits: Hilton DJ, Zhang J-G, Metcalf D, Alexander WS, Nicola NA, Willson TA. Cloning and characterisation of a binding subunit of the interleukin-13 receptor that is also a component of the interleukin-4 receptor. Proc Natl Acad Sci 1996; 93:497–501; and Caput D, Laurent P, Kayhad M, Lelias JM, Lefort S, Vita N, Ferrara P. Cloning and characterisation of a specific interleukin (IL)-13 binding protein structurally related to the IL-5 receptor α chain. J Biol Chem 1996; 271:16921–16926.

REFERENCES

1. Cherwinski HM, Schumacher JH, Brown KD, Mosmann TR. Two types of mouse helper T cell clone III. Further Differences in Lymphokine Synthesis between Th1 and Th2 clones revealed by RNA hybridization, functionally monospecific bioassays, and monoclonal antibodies. J Exp Med 1987; 166:1229–1244.
2. Minty A, Chalon P, Derocq J-M, Dumont X, Guillemot J-C, Kaghad M, Labit C, Leplatois P, Liauzun P, Miloux B, Minty C, Casellas P, Loison G, Lupker J, Shire D, Ferrara P, Caput D. Interleukin-13 is a new human lymphokine regulating inflammatory and immune responses. Nature 1993; 362:248–250.
3. McKenzie AN, Culpepper JA, de Waal Malefyt RB, Punnonen J, Aversa G, Sato A, Dang W, Cocks BG, Menon S, De Vries JE, Banchereau J, Zurawski G. Interleukin 13, a T-cell-derived cytokine that regulates human monocyte and B-cell function. Proc Natl Acad Sci USA 1993; 90:3735–3739.
4. Zurawski G, de Vries JE. Interleukin-13, an interleukin 4-like cytokine that acts on monocytes and B cells, but not on T cells. Immunol Today 1994; 15:19–26.
5. Kruse N, Shen BJ, Arnold S, Tony HP, Müller T, Sebald W. Two distinct functional sites of human interleukin 4 are identified by variants impaired in either receptor binding or receptor activation. EMBO J 1993; 12:5121–5129.
6. Zurawski SM, Vega F, Huyghe B, Zurawski G. Receptors for interleukin-13 and interleukin-4 are complex and share a novel component that functions in signal transduction. EMBO J 1993; 12:2663–2670.
7. Bamborough P, Duncan D, Richards WG. Predictive modelling of the 3-D structure of interleukin-13. Pro Eng 1994;7:1077–1082.
8. Doyle A, Herbein G, Montaner LJ, Minty AJ, Caput D, Ferrara P, Gordon S. Interleukin 13 alters the activation state of murine macrophages in vitro: comparison with interleukin-4 and interferon-g. Eur J Immunol 1994; 24:1441–1445.
9. Cash E, Minty A, Ferrara P, Caput D, Rott O. Macrophage-inactivating interleukin-13 suppresses experimental autoimmune encephalomyelitis in rats. J Immunol 1994; 153:4258–4267.
10. Magazin M, Guillemot JC, Vita N, Ferrara P. Interleukin-13 is a monocyte chemoattractant. Eur Cyt Network 1994; 397–400.
11. Morgan G, Dolganov GM, Robbins SE, Hinton LM, Lovett M. The selective isolation of novel cDNAs encoded by the regions surrounding the human interleukin 4 and 5 genes. Nucl Acids Res 1992; 20:5173–5179.
12. McKenzie ANJ, Li X, Largaespada DA, Sato A, Kaneda A, Zurawski SM, Doyle EL, Milatovich A, Francke U, Copeland NG, Jenkins NA, Zurawski G. Structural comparison and chromosomal localization of the human and mouse IL-13 genes. J Immunol 1993; 150:5436–5444.
13. Smirnov DV, Smirnova MG, Korobko VG, Frolova EI. Tandem arrangement of human genes for interleukin-4 and interleukin-13: resemblance in their organization. Gene 1995; 155:277–281.
14. Stumbles P, Mason D. Activation of CD4+ T cells in the presence of a nondepleting monoclonal antibody to CD4 induces a TH2-type response in vitro. J Exp Med 1995; 182:5–13.
15. Brinkmann V, Kristofic C. TCR-stimulated naive human CD4+ 45RO- T cells develop into effector cells that secrete IL-13, IL-5 and IFN-γ, but no IL-4, and help efficient IgE production by B cells. J Immunol 1995; 154:3078–3087.
16. Fior R, Vita N, Raphael M, Minty A, Maillot M-C, Crevon M-C, Caput D, Biberfeld P, Ferrara P, Galanaud P, Emilie D. Interleukin 13 gene expression by malignant and EBV-transformed human B lymphocytes. Eur Cytokine Netw 1994; 5:593–600.
17. Kindler V, Matthes T, Jeannin P, Zubler RH. Interleukin-2 secretion by human B lymphocytes occurs as a late event and requires additional stimulation after CD40 cross-linking. Eur J Immunol 1995; 25:1239–1243.
18. Burd PR, Thompson WC, Max EE, Mills FC. Activated mast cells produce interleukin 13. J Exp Med 1995; 181:1373–1380.
19. Obiri NI, Hilman GG, Haas GP, Sud S, Puri RK. Expression of high affinity interleukin-4 receptors on human renal cell carcinoma cells and inhibition of tumor cell growth in vitro by interleukin-4. J Clin Invest 1993; 91:88–93.
20. Obiri N, Debinski W, Leonard WJ, Puri RK. Receptor for interleukin 13. Interaction with interleukin 4 by a mechanism that does not involve the common γ chain shared by receptors for interleukins 2,4,7,9 and 15. J Biol Chem 1995; 270:8797–8804.
21. Park LS, Friend D, Sassenfeld HM, Urdal D. Characterization of the human B cell stimulatory factor 1 receptor. J Exp Med 1987; 166:476–488.
22. Vita N, Lefort S, Laurent P, Caput D, Ferrara P. The interleukin-13 receptor: characterization and comparison with the interleukin-4 receptor on several cell types. J Biol Chem 1995; 270:3512–3517.
23. Aversa G, Punnonen J, Cocks BG, de Waal Malefyt R, Vega F, Zurawski SM, Zurawski G, de Vries JE. An interleukin 4 (IL-4) mutant protein inhibits both IL-4 or IL-13-induced human immunoglobin G4 (IgG4) and IgE synthesis and B cell proliferation: support for a common component shared by IL-4 and IL-13 receptors. J Exp Med 1993; 178:2213–2218.
24. Russell SM, Keegan AD, Harada N, Nakamura Y, Noguchi M, Leland P, Friedman MC, Miyajimia A, Puri RK, Paul WE, Leonard WJ. Interleukin-2 receptor γ chain: a functional component of the interleukin-4 receptor. Science 1993; 262:1880–1883.
25. Kondo M, Takeshita T, Ishii N, Nakamura M, Watanabe S, Arai K, Sugamura K. Sharing of the interleukin-2 (IL-2) receptor γ

chain between receptors for IL-2 and IL-4. Science 1993; 262:1874–1883.
26. Matthews DJ, Clark PA, Herbert J, Morgan G, Armitage RJ, Kinnon C, Minty A, Grabstein KH, Caput D, Ferrara P, Callard R. Function of the interleukin-2 (IL-2) receptor gamma chain in biological responses of X-linked severe combined immunodeficient B cells to IL-2, IL-4, IL-13 and IL-15. Blood 1995; 85:38–42.
27. He Y -W, Malek TR. The IL-2 receptor γc chain does not function as a subunit shared by the IL-4 and IL-13 receptors. Implication for the structure of the IL-4 receptor. J Immunol 1995; 155:9–12.
28. Tony H-P, Shen BJ, Reusch P, Sebald W. Design of human interleukin-4 antagonists inhibiting interleukin-4-dependent and interleukin-13-dependent responses in T-cells and B-cells with high efficiency. Eur J Biochem 1994; 225:659–665.
29. Renard N, Duvert V, Banchereau J, Saeland S. Interleukin-13 inhibits the proliferation of normal and leukemic human B-cell precursors. Blood 1994; 84:2253–2260.
30. Labit-Le Bouteiller C, Astruc R, Minty A, Ferrara P, Lupker JH. Isolation of an IL-13-dependent subclone of the B9 cell line useful for the estimation of human IL-13 bioactivity. J Immunol Methods 1995: 181:29–36.
31. Hoffman RC, Schalk-Hihi C, Castner BJ, Gibson MG, Rasmussen BD, Zdanov A, Gustchina A, March CJ, Wlodawer A. Stoichiometry of the complex of human interleukin-4 with its receptors. FEBS Lett 1994; 347:17–21.
32. Kohler I, Alliger P, Minty A, Caput D, Ferrara P, Holl-Neugeauer B, Rank G, Rieber EP. Human interleukin-13 activates the interleukin-4-dependent transcription factor NF-IL4 sharing a DNA binding motif with an interferon-γ-induced nuclear binding factor. FEBS Lett 1994; 345:187–192.
33. Jinquan T, Deleuran B, Gesser B, Maare H, Deleuran M, Gronhoj Larsen C, Thestrup-Pedersen K. Regulation of human T lymphocyte chemotaxis in vitro by T cell-derived cytokines IL-2, IFN-γ, IL-4, IL-10 and IL-13. J Immunol 1995; 154:3742–3752.
34. Lin J-X, Migone T-S, Tsang M, Friedmann M, Weatherbee JA, Zhou L, Yamauchi A, Bloom ET, Mietz J, John S, Leonard WJ. The role of shared receptor motifs and common Stat proteins in the generation of cytokine pleiotropy and redundancy by IL-2, IL-4, IL-7, IL-13 and IL-15. Immunity 1995; 2:331–339.
35. Jacobsen SEW, Okkenhaug C, Veiby OP, Caput D, Ferrara P, Minty A. Interleukin-13: novel role in direct regulation of proliferation and differentiation of primitive hematopoietic progenitor cells. J Exp Med 1994; 180:75–82.
36. D'Andrea A, Ma X, Aste-Amezaga M, Paganin C, Trinchieri G. Stimulatory and inhibitory effects of interleukin (IL)-4 and IL-13 on the production of cytokines by human peripheral blood mononuclear cells: priming for IL-12 and tumor necrosis factor α production. J Exp Med 1995; 181:537–546.
37. Debenski W, Obiri NI, Pastan I, Puri RK. A novel chimeric protein composed of interleukin 13 and Pseudomonas exotoxin is highly cytotoxic to human carcinoma cells expressing receptors for interleukin 13 and interleukin 4. J Biol Chem 1995; 270:16775–16780.
38. Lefort S, Vita N, Reeb R, Caput D, Ferrara P. IL-13 and IL-4 share signal transduction elements as well as receptor components in TF-1 cells. FEBS Lett 1995; 366:122–126.
39. Keegan A, Nelms K, Wang LM, Pierce JH, Paul W. Interleukin-4 receptor: signalling mechanisms. Immunol Today 1994; 15:423–432.
40. Johnston JA, Kawamura M, Kirken RA, Chen Y-Q, Blake TB, Shibuya K, Ortaldo JR, McVicar DW, O'Shea JJ. Phosphorylation and activation of the Jak-3 Janus kinase in response to IL-2. Nature 1994; 370:151–157.
41. Witthuhn BA, Sivennoinen O, Miura O, Lai KS, Cwik C, Liu ET, Ihle JN. Involvement of the Jak-3 Janus kinase in signalling by interleukins 2 and 4 in lymphoid and myeloid cells. Nature 1994; 370:153–157.
42. Izuhara K, Feldman RA, Greer P, Harada N. Interaction of the c-fes protooncogene product with the interleukin-4 receptor. J Biol Chem 1994; 269:18623–18629.
43. Miyazaki T, Kawahara A, Fujii H, Nakagawa Y, Minami Y, Liu ZJ, Oshi I, Silvennoinen O, Witthuhn BA, Ihle JN, Taniguchi T. Functional association of Jak1 and Jak3 by selective association with IL-2 receptor subunits. Science 1994; 266:1045–1047.
44. Keegan AD, Johnston JA, Tortolani PJ, McReynolds LJ, Kinzer C, O'Shea JJ, Paul WE. Similarities and differences in signal transduction by interleukin 4 and interleukin 13: analysis of Janus kinase activation. Proc Natl Acad Sci USA 1995; 92:7681–7685.
45. Hou J, Schindler U, Henzel WJ, Ho TC, Brasseur M, McKnight SL. An interleukin-4-induced transcription factor: IL-4 Stat. Science 1994; 265:1701–1706.
46. de Waal Malefyt R, Figdor CG, Huijbens R, Mohan-Peterson S, Bennett B, Culpepper J, Dang W, Zurawski G, de Vries JE. Effects of IL-13 on phenotype, cytokine production, and cytotoxic function of human monocytes. J Immunol 1993; 151:6370–6381.
47. Muzio M, Re F, Sironi M, Polentarutti N, Minty A, Caput D, Ferrara P, Mantovani A, Colatta F. Interleukin-13 induces the production of interleukin-1 receptor antagonist (IL-1ra) and the expression of the mRNA for the intercellular (keratinocyte) form of IL-1ra in human myelomonocytic cells. Blood 1994; 83:1738–1743.
48. Colotta F, Re F, Muzio M, Bertini R, Polentarutti N, Minty A, Caput D, Ferrara P, Mantovani A. Interleukin-13 induces expression and release of interleukin-1 decoy receptor in human polymorphonuclear cells. J Biol Chem 1994; 269:12403–12406.
49. Doherty TM, Kastelein R, Menton S, Andrade S, Coffman RL. Modulation of murine macrophage function by IL-13. J Immunol 1993; 151:7151–7160.
50. Herbert J-M, Savi P, Laplace M-C, Lale A, Dol F, Dumas A, Labit C, Minty A. IL-4 and IL-13 exhibit comparable abilities to reduce pyrogen-induced expression of procoagulant activity in endothelial cells and monocytes. FEBS Lett 1993; 328:268–270.
51. Nassar GM, Morrow JD, Jackson Roberts L, Lakkis FG, Badr KF. Induction of 15-lipoxygenase by interleukin-13 in human blood monocytes. J Biol Chem 1994; 269:27631–27634.
52. Deleuran B, Iversen L, Deleuran M, Yssel H, Kragballe K, Stengaard-Pedersen K, Thestrup-Pedersen K. Interleukin 13 suppresses cytokine production and stimulates the production of 15-HETE in PBMC. Cytokine 1995; 7:319–324.
53. Kolb JP, Paul-Eugene N, Damais C, Yamaoka K, Drapier JC, Dugas B. Interleukin-4 stimulates cGMP production by IFN-γ activated human monocytes. Involvement of the nitric oxide pathway. J Biol Chem 1994; 269:9811–9816.
54. Sozzani P, Cambon C, Vita N, Séguelas M-H, Caput D, Ferrara P, Pipy B. Interleukin-13 inhibits protein kinase C-triggered respiratory burst in human monocytes: role of calcium and cyclic AMP. J Biol Chem 1995; 270:5084–5088.
55. Philips WA, Croatto M, Hamilton JA. The effect of interleukin-4 on the macrophage respiratory burst is species dependent. Biochem Biophys Res Comm 1992; 182:727–732.
56. Defrance T, Carayon P, Billian G, Guillemot J-C, Minty A, Caput D, Ferrara P. Interleukin 13 is a B cell stimulating factor. J Exp Med 1994; 179:135–143.
57. Punnonen J, Aversa G, Cocks BG, McKenzie AN, Menton S,

Zurawski G, de Waal Malefyt R, de Vries JE. Interleukin 13 induces interleukin 4-independent IgG4 and IgE synthesis and CD23 expression by human B cells. Proc Natl Acad Sci USA 1993; 90:3730–3734.
58. Punnonen J, de Vries J. IL-13 induces proliferation, Ig isotype switching and Ig synthesis by immature human fetal B cells. J Immunol 1994; 152:1094–1102.
59. Chaouchi N, Wallon C, Goujard C, Tertian G, Rudent A, Caput D, Ferrara P, Minty A, Vazquez A, Delfraissy J-F. IL-13 inhibits IL-2 induced proliferation and protects chronic lymphocytic leukemia B cells from in vitro apoptosis. Blood 1996; 87:1022–1029.
60. Nilsson G, Nilsson K. Effects of interleukin (IL)-13 on immediate-early response gene expression, phenotype and differentiation of human mast cells. Comparison with IL-4. Eur J Immunol 1995; 25:870–873.
61. Xi X, Schlegel N, Caen J, Minty A, Fournier S, Caput D, Ferrara P, Han ZC. Differential effects of recombinant human interleukin 13 on the in vitro growth of human haematopoietic progenitor cells. Br J Haematol 1995; 90:921–927.
62. Sakamoto O, Hashiyama M, Minty A, Ando M, Suda T. Interleukin-13 selectively suppresses the growth of human macrophage progenitors at the late stage. Blood 1995; 85:3487–3493.
63. Sironi M, Sciacci FL, Matteucci C, Conni M, Vecchi A, Bernasconi S, Minty A, Caput D, Ferrara P, Colotta F, Mantovani A. Regulation of endothelial and mesothelial cell function by interleukin-13: selective induction of vascular cell adhesion molecule-1 and amplification of interleukin-6 production. Blood; 84:1913–1921.
64. Derocq J-M, Segui M, Poinot-Chazel C, Minty A, Caput D, Ferrara P, Casellas P. Interleukin-13 stimulates interleukin-6 production by human keratinocytes. Similarity with interleukin-4. FEBS Lett 1994; 343:32–36.
65. Lebel-Binay S, Tulliez M, Quintin-Colonna F, Conjeaud H, Magazin M, Miloux B, Pecceu F, Ferrara P, Caput D, Fradelizi D. Experimental gene therapy of cancer using tumor cells engineered to secrete IL-13. 1995. Eur J Immunol 25:2340–2348.
66. Tepper RI, Coffman RL, Leder P. An eosinophil-dependent mechanism for the antitumor effect of interleukin-4. Science 1992; 257:548–551.
67. Montaner LJ, Doyle AG, Collin M, Herbein G, Illei P, James W, Minty A, Caput D, Ferrara P, Gordon S. Interleukin 13 inhibits human immunodeficiency virus type 1 production in primary blood-derived human macrophages in vitro. J Exp Med 1993; 178:743–747.
68. Mikovits JA, Meyers AM, Ortaldo JR, Minty A, Caput D, Ferrara P, Ruscetti FW. IL-4 and IL-13 have overlapping but distinct effects on HIV production in monocytes. J Leukocyte Biol 1994; 56:340–346.
69. Denis M, Ghadiran E. Interleukin-13 and interleukin-4 protect broncoalveolar macrophages from productive infection with human immunodeficiency virus type 1. AIDS Res Human Retroviruses 1994; 10:795–802.
70. Clerici M, Lucey DR, Berzofsky JA, Pinto LA, Wynn TA, Blatt SP, Dolan MJ, Hendrix CW, Wolff SF, Shearer GM. Restoration of HIV-specific cell-mediated immune responses by interleukin-12 in vitro. Science 1994; 262:1721–1724.
71. Chehimi J, Starr SE, Franck I, D'Andrea A, Ma X, MacGregor RR, Sennelier J, Trinchieri G. Impaired interleukin 12 production in immunodeficiency virus-infected patients. J Exp Med 1994; 179:1361–1366.
72. Kopf M, Le Gros G, Bachmann M, Lamers MC, Bluethmann H, Kohler G. Disruption of the murine IL 4 gene blocks TH-2 cytokine responses. Nature 1993; 362:245–248.
73. Krishna M, Murray R, Zurawski G, McKenzie ANJ. Study of infectious disease models in interleukin-13 deficient mice. Cytokine 1995; 7:637.
74. Zhang X, Polla B, Hauser C, Zubler RH. T cells from atopic individuals produce IgE-inducing activity incompletely blocked by anti-interleukin 4 antibody. Eur J Immunol 1992; 22:829–833.
75. Marsh DG, Neely JD, Breazeale DR, Ghosh B, Friedhoff LR, Ehrlich-Kautzy E, Schou C, Krishnaswamy G, Beatty TH. Linkage analysis of IL-4 and other chromosome 5q31.1 markers and total serum immunoglobulin E concentrations. Science 1994; 264:1152–1156.
76. Kimata H, Fujimoto M, Furusho K. Involvement of interleukin (IL)-13, but not IL-4, in spontaneous IgE and IgG4 production in nephrotic syndrome. Eur J Immunol 1995; 25:1497–1501.
77. Wang L-M, Michieli P, Lie W-R, Liu F, Lee C-C, Minty A, Sun X-J, Levine A, White MF, Pierce JH. The IRS-1-related 4PS substrate but not the IL-2Rγ chain is involved in IL-13-mediated signal transduction. Blood 1995; 86:4218–4227.
78. Brown KD, Zurawski SM, Mosmann TR, Zurawski G. A family of small inducible proteins secreted by leukocytes are members of a new superfamily that includes leukocyte and fibroblast-derived inflammatory agents, growth factors, and indicators of various activation processes. J Immunol 1989; 142:679–687.
79. Lakkis FG, Cruet EN. Cloning of rat interleukin-13 (IL-13) cDNA and analysis of IL-13 gene expression in experimental glomerulonephritis. Biochem Biophys Res Comm 1993; 197:612–618.
80. Musso T, Varesio L, Zhang X, Rowe TK, Ferrara P, Ortaldo JR, O'Shea JJ, McVicar DW. IL-4 and IL-13 induce Lsk, a Csk-like tyrosine kinase in human monocytes. J Exp Med 1994; 180:2383–2388.
81. Sebire G, Delfraissy JF, Demotesmainard J, Oreiteh A, Emilie D, Tavdieu M. Interleukin-13 and interleukin-4 act as interleukin-6 inducers in human microglial cells. Cytokine 1996; 8:636–641.
82. Kolios G, Robertson DAF, Jordan NJ, Minty A, Caput D, Ferrara P, Westnick J. Interleukin-8 production by the human colon epithelial cell line HT-29. Modulation by interleukin-13. Br J Pharmacol 119:351–359.
83. Bessis N, Boissier M-C, Ferrara P, Blankenstein T, Fradelizi D, Fournier C. Attenuation of collagen-induced arthritis in mice by treatment with vector cells engineered to secrete interleukin-13. Eur J Immunol 1996; 26:2399–2403.

14
The Interferons

Ganes C. Sen
The Cleveland Clinic Foundation Research Institute, Cleveland, Ohio

INTRODUCTION AND HISTORICAL PERSPECTIVE

Interferon (IFN) was the first discovered member of the family of secretory proteins known as cytokines, which have highly potent and diverse physiological functions both within and outside of the immune system. Interferon revealed itself as a substance that interferes with virus replication, hence its name (1). It was observed that influenza virus–infected chick embryo cells secrete a proteinaceous substance that induces resistance to infection by influenza and other viruses. Similar virus-induced IFNs (IFN-α/β) have since been found in all vertebrates except amphibians (2). A new class of IFN, IFN-γ, was later described as a virus-inhibitory protein produced by mitogen activated T lymphocytes (3). These early observations regarding the role of IFNs in the host defense against viral infection have been validated by later studies in mice, in which antibodies to IFN were administered (4) or the genes for the receptors for IFN-α/β and IFN-γ were disrupted (5). Although discovered as antiviral agents, IFNs have many additional biological activities (2). They protect the host against parasitic infection and certain tumors. Functioning of the various components of the immune system is profoundly affected by IFNs. In addition, they affect differentiation and proliferation of selective types of cells. IFNs are clinically used to combat viral infection as well as several specific types of cancer and other diseases such as multiple sclerosis.

Analysis of the IFN system as the first line of defense against viral infection has included studies on both IFN biosynthesis and IFN actions. Viral infection and other insults to a cell result in the induction of IFN synthesis. The newly synthesized IFN is secreted to the circulation, so that it can interact with neighboring uninfected cells and protect them from oncoming viral infection. The antiviral and all other cellular actions of IFNs are mediated by cell-surface receptors, which signal to the nucleus to transcriptionally activate a set of cellular genes, the IFN-inducible genes, whose products carry out the multiple effects of IFNs. Much information has been gathered in the recent past about the molecular mechanisms involved in IFN induction and IFN actions (6,7). Studies demonstrated that the pathways responsible for operating the two arms of the IFN systems, although distinct, share many common components and molecular strategies. These aspects of the IFN system will be discussed in greater detail in the following pages.

BIOCHEMISTRY AND MOLECULAR BIOLOGY

IFNs are a family of proteins encoded by different genes and sharing the common property of inhibiting virus replication. They were originally classified, by their sources, as leukocyte, fibroblast, and immune IFNs (2). The corresponding current nomenclature is IFN-α, IFN-β, and IFN-γ. IFN-α and IFN-β belong to the type I IFN superfamily, the other members of which are IFN-ω and IFN-τ. IFN-γ is the sole member of the type II IFN family (3). The members of the type I superfamily are structurally related, and they are encoded by a gene cluster located on the short arm of human chromosome 9. The genetic clustering and the structural homology suggest that the different type I IFN genes evolved from a single ancestral gene (2,8). In humans, there are at least 14 IFN-α subtypes, encoded by as many genes, but only one IFN-β species. The genes for type I IFN do not contain any intron. The single gene for IFN-γ, on the other hand, has three introns and is located on human chromosome 12. IFN-β is the most divergent members of the type I IFN family; it has 25–30% homology, at the amino acid level, with different members of the IFN-α subfamily (9). There is also significant homology in the regulatory regions of the IFN-α and IFN-β genes, so that they are often coordinately regulated.

All animal species have a large number of IFN-α subtypes. Most have only one IFN-β gene, except for the ungulates, which have five such genes (10,11). IFN-τ has been identified in cattle and sheep (12). It is produced in large quantities in the early preimplantation embryo and plays an essential role in successful completion of pregnancy. IFN-τ is structurally closest to IFN-ω (13). These type I IFNs do not have any structural homology with IFN-γ. Type I IFNs are active in monomeric forms (14,15). The mature IFN-α proteins have 165–166 residues, IFN-β has 166 amino acids, and IFN-ω and IFN-τ have 172 residues. All human IFN-α proteins have four conserved cysteines, which are thought to form two intramolecular disulfide bonds, one of which has been shown to be dispensable for biological activity (8,16). Cysteine residues are conserved in similar positions of IFN-α proteins of other species as well. Most IFN-α proteins are unglycosylated, but IFN-β is N-glycosylated and contains one intramolecular disulfide bond (17). The tertiary structure of IFN-β contains five α-helices as analyzed by NMR and x-ray crystallography (18,19). IFN-γ functions as a dimer (20). The mature glycosylated IFN-γ protein has 146 residues and does not contain any cysteine.

The biological significance of so many type I IFNs is unclear. They all compete for binding to the same receptor and appear to have similar biological activities (2,21). IFN-γ, in contrast, binds to a different receptor. Recent investigations have revealed shared components between the signal transduction pathways used by the type I and the type II IFNs (22). This may explain why many cellular genes are activated by both types of IFNs and consequently have partially overlapping biological activities. IFN-τ appears to have a special role in embryo implantation in some species (23,24). However, mutation of a subunit of the type I IFN receptor in mouse did not affect embryonic development, suggesting that type I IFNs are not required for embryogenesis in this species (5).

SOURCES OF INTERFERONS

Although low levels of endogenous IFNs can be detected in several tissues in vivo (25), IFN synthesis has to be induced by specific biological stimuli. Cells in culture do not synthesize IFN unless they are treated with an inducing agent. IFN-α or IFN-β synthesis can be induced in almost all cell types, the notable exception being embryonic carcinoma cells. Virus infection is the most efficient route of type I IFN induction (2). The potencies of different viruses, in this respect, vary over a wide range. Double-stranded (ds) RNA, a common viral product, is thought to be the inducing intermediate. In fact, when added to cell cultures exogenously, dsRNA induces type I IFN synthesis very efficiently. Infection with organisms other than viruses such as bacteria, protozoa, and mycoplasma can also induce IFN synthesis in mononuclear phagocytic cells. In addition, bacterial LPS and cytokines such as IL-1, IL-2, TNF, and CSF-1 can also trigger IFN synthesis in certain cell types. The natural inducer of IFN-τ in preimplantation embryos remains unknown. Type II IFN- (IFN-γ) synthesis is restricted to T lymphocytes and NK cells (2). The common natural inducer of IFN-γ synthesis is stimulation by antigens to which organisms have been presensitized. Nonspecific T-cell activators, such as PHA and ConA, can also induce IFN-γ synthesis in vitro. IL-2 and IL-12 can induce IFN-γ synthesis in T cells and NK cells (26–28). Other agents such as glucocorticoids, TGF-β, and IL-10 downregulate IFN-γ induction (29,30).

Type I IFN induction is regulated primarily at the level of transcription, although posttranscriptional regulation also contributes (31). The transcriptional induction involves both derepression of the genes and activation of preexisting transcription factors. The posttranscriptional regulation is achieved by stabilization of IFN mRNAs, which carry destabilization sequences in their 3′ untranslated regions. Inhibitors of protein synthesis cause superinduction of IFN mRNAs by blocking their rapid turnover and by blocking the synthesis of putative respressor proteins that are coinduced with IFNs and repress IFN mRNA transcription. IFN induction is also enhanced by pretreatment of the cells by a low dose of IFN, a phenomenon called priming. Recent findings that the synthesis of some of the transcription factors used for IFN induction is, in turn, induced by IFN suggests a rational basis of priming. IFN-γ synthesis is regulated primarily at the transcriptional level. Although the details are not known, a combination of positive and negative elements present in the 5′ flanking region and in the introns of the IFN-γ gene are thought to regulate IFN-γ synthesis (30,32).

The transcriptional regulation of IFN-α and IFN-β genes has been extensively studied at the molecular level (7,33). The underlying mechanism is complex and uses both positive and negative regulatory factors. The human IFN-β gene contains four positive regulatory elements PRD I–IV and one negative regulatory element, some of which structurally overlap (34,35). The PRD I site bears resemblance to the IFN-stimulated response element (ISRE) present in IFN-responsive genes. Several protein factors bind to both PRDI and ISRE. Such factors include multiple members of the IRF family, IRF-I, IRF-2, p48, IRF-3, ICSBP, and an unrelated zinc finger repressor protein, PRDIBFI (36–40). IRF-1 stimulates transcription, whereas IRF-2 represses it. IRF-1 synthesis is induced by many cytokines, including IFNs (41,42). The functional activity of IRF-1 protein, on the other hand, appears to be regulated by a posttranslational process induced by virus infection, but not by IFNs (43). In cells of mice lacking the IRF-1 gene, IFN-β and IFN-α induction by dsRNA is decreased, but induction by virus is normal (44). IRF-1 is, however, required for the induction of GBP and nitric oxide synthase genes by IFN-γ (45,46). IRF-1 also has a role in the regulation of B-cell growth and in cellular transformation (47). Overexpression of IRF-2 causes cellular transformation, which is blocked by concomitant overexpression of IRF-1 (48). The possible role of IRF-1 as a tumor suppressor is also suggested by the frequent deletion of its genetic locus in certain leukemias (49).

The PRD II site binds NK-κB, a common transcription factor, which is activated by dsRNA and by many other agents (50). Another protein HMG I (y) associates with NF-κB in binding PRD II. It also stimulates the binding of a virus-inducible complex containing ATF-2 to the PRD IV element (51). Thus, assembly of a multicomponent protein complex is necessary for activating transcription of the IFN-β gene. The IFN-α genes contain the PRD I–like elements, but not NF-κB–binding sites. Activation of IFN-α genes requires PRD I–binding proteins and additional factors that bind to the GAAATG sequence present in them (52).

BIOLOGICAL ACTIONS OF INTERFERONS

IFNs are pleiotropic biological response modifiers. The specific spectrum of actions is influenced by the nature of the cell type and the species of the IFN used. Their major function is in the host defense against viral and parasitic infection. All viruses are, however, not equally susceptible to IFN's antiviral effects, because many have evolved mechanisms to evade these effects (7,53,54). The physiological importance of the IFN system in host defense against natural viral infection has been convincingly demonstrated by injecting mice with antibodies to IFNs (4). In the recent past, several knockout mouse strains, which are missing either functional IFN genes or IFN-receptor genes, have been generated. The enhanced viral susceptibility of these strains underscores the importance of the IFN system in host defense (5,35,55–57).

IFNs also regulate cell growth and differentiation (53). The effect on cell growth is always negative in nature. The magnitude of the antigrowth effect depends on the cell types and the type of IFN used. It was noted early that certain tumor cells are exceptionally susceptible to the anticellular effects of IFN. These observations led to the clinical use of IFNs for treating specific types of human cancer. Currently, IFN-α is used to treat hairy cell leukemia and Kaposi's sarcoma (53). IFN-β is used to treat multiple sclerosis (58), and IFN-γ is effective in the treatment of chronic granulomatous diseases (59).

IFNs also have prominent effects on different components of the immune system. They activate monocyte/macrophages, cytotoxic T cells, and natural killer cells. They also modulate immunoglobulin synthesis by B cells. MHC class I and class II genes and F_c receptor gene are induced by IFNs, and the encoded proteins contribute to the potentiation of the immune system (2). Some of these effects are specific for IFN-γ: only IFN-γ induces MHC class II antigens and activates macrophages. One member of the type I IFN family, IFN-τ, or trophoblast IFN has a distinct effect in early pregnancy in cattle and sheep (12). In mice, however, gestation is normal even in the absence of functional type I IFN receptors through which IFN-τ signals (5).

INTERFERON RECEPTORS

IFNs bind to specific cell-surface receptors and elicit intracellular signals, which are transmitted to the IFN-inducible genes present in the nucleus, so that their transcription is augmented (6,60). Type I and type II IFNs bind to different receptors. Although IFN-αs and IFN-β compete for binding to the same receptors, there could be microheterogeneity among the cognate receptors. IFN-γ, on the other hand, does not compete with type I IFNs for receptor binding. Binding of IFNs to their receptors is, in general, species-specific. Human IFNs bind to receptors on human cells much more avidly than to those on mouse cells. This species specificity has been profitably exploited as a tool for studying the structures of IFN receptors. The species specificity is more pronounced for IFN-γ than for IFN-α. The abundance of receptors for type I IFNs is low, usually only a few thousand receptors are present per cell. The affinity for the ligand is on the order of 10^{-10} M.

Both type I and type II IFN receptors belong to the class II cytokine receptor family (61,62). They are composed of multiple transmembrane proteins whose extracellular domains bind ligands, and intracellular domains transmit the signal to other cytosolic proteins. The human type I receptor is encoded by genes located on chromosome 21 (63,64). The best-characterized IFN-α receptor has at least two chains. The cDNAs for both of these polypeptides have been cloned. These proteins do not have intrinsic tyrosine kinase activity, but their intracellular domains are associated with specific protein tyrosine kinases that are activated once ligands bind to the receptors (22,64–67).

IFN-γ receptor is a heterodimer of two transmembrane proteins, to one of which the ligand binds (68–71). Two different tyrosine kinases are bound to the intracellular domains of the two subunits. The two chains of the human IFN-γ receptor are encoded by two genes located on chromosome 6 and chromosome 21. Specific residues on both IFN-γRα and IFN-γRβ chains have been identified as the docking sites of the cognate JAK tyrosine kinases and the signal-transducing STAT 1 protein (72–74). Intracellular signaling is thought to require the dimerization of IFN-γ receptors upon binding of IFN-γ dimer (75).

SIGNAL TRANSDUCTION PATHWAYS

Investigation of the IFN signal transduction pathways has been extremely rewarding. These studies have defined a new mode of signal transduction called the JAK-STAT pathway (22,65,66). The same general principles have been found to be valid for many other cytokines and growth factors, although the specific proteins involved may be different. Thus, the IFN signaling pathway has become the paradigm for all cytokine signaling pathways, and more than a dozen cytokines and growth factors have already been shown to use similar pathways for transcriptional signaling (62,67). Receptors that use this signaling pathway lack intrinsic tyrosine kinase activity. Members of a specific family of protein tyrosine kinases called the JAKs are, however, bound to the intracellular domains of these receptors. The JAKs are activated upon ligand binding by the receptors, and they tyrosine-phosphorylate themselves and the receptors. The next step in signal transduction is JAK-mediated tyr-phosphorylation of the STAT (signal transducer and activator of transcription) proteins (76–78). Upon activation, these proteins translocate to the nucleus, where they form active transcription complexes by binding to specific *cis*-acting sequences present in the cognate genes. Thus, the signal from the receptors on the cell surface to the genes in the nucleus is physically carried by the STAT proteins.

Transcriptional induction of genes in response to IFN-α/β is rapid, but transient (79–81). It also does not require ongoing protein synthesis. The kinetics of gene induction by IFN-γ are usually slower, and the induction of some of these genes requires ongoing protein synthesis. Most IFN-α/β–inducible genes contain a *cis*-acting consensus sequence called ISRE (interferon-stimulated response element), which is the receiver of the signal (22). Double-stranded RNA, which is produced during virus infection, can also induce the transcription of many of these genes through ISRE. The signal transduction pathways used by IFN-α/β and dsRNA are, however, quite distinct (82). IFN-γ–inducible genes contain one or more of several distinct *cis*-elements (22,83). ISREs of some genes can also mediate IFN-γ response, although not all IFN-α/β–inducible genes are induced by IFN-γ (84). Another class of IFN-γ–inducible gene contains the IFN-γ activation site (GAS). Some genes, such as the GBP gene, contain both a GAS and an ISRE, and both elements are required for IFN-γ induction of these genes (22). Another set of IFN-γ–inducible genes is the MHC class II and the invariant chain genes. There are no ISRE and no GAS sequences in these genes. Instead, induction is mediated by a region encompassing the Z (or H), X, and Y box elements present in all of these genes (85).

The nature of the STAT proteins and the JAK kinases involved in IFN-α/β and IFN-γ signaling has been revealed by investigations using biochemical, molecular biological, and genetic approaches. The biochemical approach identified the crucial ISRE-binding protein complex whose formation parallels transcriptional induction. This factor, called ISGF-3, was purified. ISGF-3 is composed of three proteins: p113 (now called STAT 2), either p91 or p84 (now called STAT 1α or STAT 1β, respectively), and p48 (78). p48 is a member of the IRF-1 family of proteins, and it binds ISRE, albeit weakly (86). But once p48 complexes with STAT 1 and STAT 2, the resultant complex, ISGF 3, has high affinity for ISRE. Cloning and sequencing of p113 and p91 (87) revealed that they contain common structural motifs, which are hallmarks of substrates of tyrosine kinases (88). A complementary approach used somatic cell genetic analysis for dissecting the IFN-signaling pathways (22). Mutants of different components of these pathways were generated and classified by complementation analysis. The specific mutated gene in a given complementation group was identified by assaying for the restoration of full IFN response upon transfection of a complete cDNA or genomic expression library from a wild-type cell. This approach confirmed the participation of the IFN-α/β receptor, STAT 1, STAT 2, and p48 in the IFN-α/β–signaling pathway. In addition, it revealed that two members of the JAK family of protein tyrosine kinases, Jak 1 and Tyk 2, are also required for this signaling process (89,90). Similar analysis of the IFN-γ–signaling pathway revealed that Jak 1 and another member, Jak 2, but not Tyk 2, are required (91,92). The GAS-mediated IFN-γ–signaling pathway uses a homodimer of STAT 1α as the transcription factor (74,93). Consequently, neither STAT 2 nor p48 are required for this pathway. In contrast, p48 is required for IFN-γ signaling through ISRE. It should be apparent from the above discussion that type I and type II IFNs not only use the same strategy for transcriptional signaling, but also use some of the same proteins to accomplish it. This provides an opportunity for cross-talk between the two pathways, but it also calls for mechanisms maintaining specificity. The details of such mechanisms remain to be understood.

INTERFERON-INDUCIBLE PROTEINS

All biological actions of IFNs are mediated by IFN-induced proteins (6,7). In untreated cells they are usually absent or present in low quantities. Synthesis of many of these proteins is induced as a direct response to IFN treatment. Other proteins are induced later as a secondary response. Type I and type II IFNs induce distinct, but partially overlapping sets of proteins whose profiles depend on the specific cell type. To date, more than 50 such proteins have been identified. Unfortunately, the biochemical and the biological functions of many of these proteins remain to be determined. It is suspected that different members of this family are responsible for mediating different cellular functions of IFNs. In this vein, it has become clear that even the antiviral actions of IFNs are carried out by a number of IFN-induced proteins. The characteristics of some of these proteins are outlined below.

Both type I and type II IFNs induce the synthesis of a family of novel enzymes called 2-5 (A) synthetases (7,94,95). These enzymes synthesize 2′-5′–linked oligoadenylates [2-5 (A)] from ATP in the presence of dsRNA, which is needed as a mandatory cofactor. Their product, 2-5 (A), activates a latent ribonuclease, RNase L. Activated RNase L can cleave single-stranded RNAs in a sequence-independent fashion. The 2-5 (A) synthetase/RNase L system thus works as a concerted mechanism for mRNA degradation, whose activation requires two signals: IFNs for inducing the synthesis of the enzymes and dsRNA for activating 2-5 (A) synthetase. In an IFN-treated virus-infected cell, the latter activation is presumably done by viral dsRNA, which is produced as a by-product of replication of many viruses. Many lines of evidence indicate that IFN-induced inhibition of picornavirus replication is mediated by the 2-5 (A) synthetase pathway (96–99). This pathway, however, is not responsible for inhibiting the replication of other RNA viruses such as VSV. There are suggestions that it may contribute to the anticellular effects of IFNs.

The 2-5 (A) synthetases are a family of proteins. There are three size classes encoded by three sets of IFN-inducible genes (7). The smallest family members have molecular weights in the range of 40–46 kDa. Some of them are encoded by different genes, and some are alternative splicing products of the same gene (100). Different isozymes have different subcellular locations indicating different cellular functions. The medium-sized human 2-5 (A) synthetases are the products of alternatively spliced mRNAs of the same gene. The 69 kDa isozyme is myristoylated. The large synthetases have a molecular weight of about 100 kDa, and they are activated by a lower concentration of dsRNA. RNase L functions as a dimer, and the dimerization is brought about by 2-5 (A) (101). The RNase L protein has nine ankyrinlike repeat domains, a cysteine-rich domain, and a protein-kinase homology domain (97,102).

Another IFN-induced dsRNA-activated enzyme is a protein kinase, PKR (103). PKR is activated by autophosphorylation in response to dsRNA or polyanions such as heparin. Activated PKR can phosphorylate other substrates, only two of which have been identified: eIF-2, the translation initiation factor, and IκB, the inhibitor of the transcription factor NF-κB (104). Phosphorylation of eIF-2 leads to inhibition of protein synthesis, whereas phosphorylation of Iκb leads to transcriptional activation of genes. PKR is the prototype of a family of dsRNA-binding proteins, which share a common dsRNA-binding structural motif (105–107). The same motifs also mediate homo- and heterodimer formation between members of this family of proteins (108,109). Involvement of PKR in the antiviral effects of IFNs is suggested by the various inhibitors of this enzyme encoded or induced by different viruses (7). This enzyme has also been implicated in many other cellular functions such as cell growth, differentiation, and transcriptional signaling (110–112). Introduction of a mutated PKR into NIH/3T3 cells inhibits the functioning of the endogenous enzyme causing unregulated cell proliferation and tumorigenesis in mice injected with those cells (113,114). Similarly, overexpression of a cellular inhibitor of PKR, P58, also causes cellular transformation (115).

Another family of IFN-inducible proteins has an intrinsic GTPase activity (116). Different members of this family of proteins, called the Mx proteins, can inhibit the replication of influenza virus and VSV, a rhabdovirus (117). Some Mx proteins are nuclear, whereas others are cytoplasmic. Human MxA can bind cytoskeletal proteins and may be involved in vacuolar protein sorting (118). A different enzyme, nitric oxide synthase, is induced by IFN-γ in macrophages. This enzyme is responsible for inhibiting poxviruses and herpesviruses (119,120). IFN-γ also induces the synthesis of indoleamine 2,3-dioxygenase, an antiprotozoal enzyme (7). Many components of the immune system are induced by IFNs as well (2,121,122). These include MHC class I and class II proteins, proteasomes, and permeases.

ANTIVIRAL ACTIONS OF INTERFERONS

As indicated above, different IFN-induced proteins mediate IFN's actions against different viruses. Moreover, for activating some of the antiviral pathways, viral gene products such as dsRNA are needed in addition to the IFN-induced antiviral proteins. The points of attack are also different for different families of viruses. Viral penetration and uncoating, primary transcription, viral protein synthesis, viral genome replication, and viral assembly and release can all be inhibited by IFNs. The primary target depends on the virus, the cell type, and the type of IFN. And often more than one viral process is affected (7,54).

The role of the IFN system in protecting intact organisms against virus infection has been explored in three types of experimental models: natural mouse strains, which differ in their susceptibility to specific viruses (116), effects of administering IFN antibodies to virus-infected mice (4), and with targeted disruption of gene coding for IFNs (55), IFN receptors, (5,57) or IFN-induced proteins (46). The first type of investigation led to the discovery of the role of the Mx proteins against influenza virus. The second approach showed that IFN α/β protects mice against EMCV, HSV-1, SFV, MSV, MuLV, and polyoma virus. Similarly, IFN-γ protects them against LCMV and vaccinia viruses. The gene knock-out experiments have confirmed the role of IFN-α/β against VSV, SFV, vaccinia virus, and the role of IFN-γ against vaccinia virus.

To facilitate their own replication, many viruses have developed strategies to evade the IFN system (7,54). Several DNA viruses, such as adenovirus and hepatitis B virus, block IFN-signaling pathways, thus inhibiting the synthesis of any IFN-induced protein (123–126). In the case of adenoviruses, the responsible viral protein is E1A. In cells expressing E1A, IFN-α cannot activate ISGF3 (126). Other viruses counteract the functions of IFN-induced enzymes (7,54). Adenoviral VAI RNA (127), EB viral EBER RNA (128), HIV-1 TAR RNA (129), vaccinia E3L (130), K3L proteins (131), HIV-1 Tat protein (132), reovirus σ-3 protein (133), and influenza virus–induced cellular P58 protein (134) all block the activation or the action of PKR. Similarly, HSV-1 produces 2-5 (A) analogs, which block the action of RNase L (135). Myxoma viruses encode a secreted protein, which resembles the extracellular domain of IFN-γ receptors and binds IFN-γ, thus blocking its action (136). The existence of these viral defense mechanisms against the IFN system underscores the importance of co-evolution of these biological regulatory processes, so that both hosts and viruses can co-exist in nature.

CONCLUSIONS

Interferons were the first cytokines to be discovered. Since then, the list of these biological response modifiers has been growing steadily. They all share the common properties of high potency, strong specificity, pleiotropic functions, and the ability to cross-talk with other members of the family. Investigation of IFN's mechanism of action has revealed the existence of a novel type of transcriptional signaling pathway—the JAK-STAT pathway. Many other cytokines have now been shown to use similar pathways for signaling. Through IFN research, novel enzymes whose functions depend on dsRNA have been discovered. Finally, IFNs have proven to be of important clinical value in selective diseases.

In spite of such advances in knowledge, much remains to be learned about the IFN system. Why do so many different IFN species exist? What are the functions of many of the IFN-induced proteins? How is IFN biosynthesis physiologically regulated? Do IFNs have major roles in cell growth regulation? Why are certain tumors and not others sensitive to IFNs? Since there is so much overlap, how do IFNs and other cytokines maintain their signaling specificity? Given the vigor of research in this field, we can reasonably expect to find answers to many of these questions in the near future.

ACKNOWLEDGMENTS

I am grateful to Dorthy Herzberg for expert editorial assistance. This work was supported by the National Institutes of Health grants CA-62220 and CA-68782.

REFERENCES

1. Isaacs A, Lindenmann J. Virus interference. 1. The interferon. Proc R Soc Lond (Biol) 1957; 147:258–267.
2. DeMaeyer E, DeMaeyer-Guignard J. Interferons and Other Regulatory Cytokines, New York: John Wiley & Sons, 1988.
3. Wheelock EF. Interferon-like virus-inhibitor induced in human leukocytes by phytohemagglutinin. Science 1965; 149:310–311.
4. Gresser I. Role of interferon in resistance to viral infection in vivo. In: Vilcek J, DeMaeyer E, eds. Interferon Vol 2. Interferons and the Immune System. Amsterdam: Elsevier Science, 1984:221–247.
5. Müller U, Steinhoff U, Reis LFL, Hemmi S, Pavolovic J, Zinkernagel RM, Aguet M. Functional role of type I and type II interferons in antiviral defense. Science 1994; 264:1918–1921.
6. Sen GC, Lengyel P. The interferon system: a bird's eye view of its biochemistry. J Biol Chem 1992; 267:5017–5020.
7. Sen GC, Ransohoff RM. Interferon-induced antiviral actions and their regulation. Adv Virus Res 1993; 42:57–102.
8. Weissmann C, Weber H. The interferon genes. Prog Nucl Acid Res Mol Biol 1986; 33:251–300.
9. Taniguchi T, Mantei N, Schwarzstein M, Nagata S, Muramatsu M, Weissmann C. Human leukocyte and fibroblast interferons are structurally related. Nature 1980; 285:547–549.

10. Leung DW, Capon DJ, Goeddel DV. The structure and bacterial expression of three distinct bovine interferon-β genes. Biotechnology 1984; 2:458–464.
11. Wilson V, Jeffreys AJ, Barrie PA, et al. A comparison of vertebrate interferon gene families detected by hybridization with human interferon DNA. J Mol Biol 1983; 166:457–475.
12. Imakawa K, Anthony RV, Kazemi M, Marotti KR, Polites HG, Roberts RM. Interferon-like sequence of ovine trophoblast protein secreted by embryonic trophectoderm. Nature 1987; 330:377–379.
13. Stewart HJ, McCann SHE, Flint APF. Structure of an interferon-α2 gene expression in the bovine conceptus early in gestation. J Mol Endocrinol 1990; 4:275–282.
14. Chelbi-Alix MK, Thang MN. Cloned human interferons α: differential affinities for polyinosinic acid and relationship between molecular structure and species specificity. Biochem Biophys Res Commun 1987; 145:426–435.
15. Utsumi J, Yamazaki S, Kawaguchi K, Kimura S, Shimizu H. Stability of human interferon-β1: oligomeric human interferon-β1 is inactive but reactivated by monomerization. Biochem Biophys Acta 1989; 998:167–172.
16. Wetzel R. Assignment of the disulphide bonds of leukocyte interferon. Nature 1981; 289:606–607.
17. Knight E. Interferon: purification and initial characterization from human diploid cells. Proc Natl Acad Sci USA 1976; 73:520–523.
18. Utsumi J, Shimizu H. Human interferon β, protein structure and function. In: Baron S, Coppenhaver DH, Dianzani F, eds. Interferon: Principles and Medical Applications. Galveston: University of Texas Medical Branch, 1992:107–116.
19. Utsumi J, Yamazaki S, Hosoi K, et al. Characterization of *E. coli*-derived recombinant human interferon-β as compared with fibroblast human interferon-β. J. Biochem 1987; 101:1199–1208.
20. Yip YK, Barrowclough BS, Urban C, Vilcek J. Molecular weight of human γ interferon is similar to that of other human interferons. Science 1982; 215:411–413.
21. Pestka S, Langer JA, Zoon KC, Samuel CE. Interferons and their actions. Annu Rev Biochem 1987; 56:727–777.
22. Darnell Jr. JE, Kerr IM, Stark GR. Jak-STAT pathways and transcriptional activation in response to IFNs and other extracellular signaling proteins. Science 1994; 264:1415–1421.
23. Roberts RM, Imakawa K, Niwano Y, et al. Interferon production by the preimplantation sheep embryo. J Interferon Res 1989; 9:175–187.
24. Hansen TR, Kazemi M, Keisler DH, Malathy PV, Imakawa K, Roberts RM. Complex binding of the embryonic interferon, ovine trophoblast protein-1, to endometrial receptors. J Interferon Res 1989; 9:215–225.
25. Tovey MG, Streuli M, Gresser I, et al. Interferon messenger RNA is produced constitutively in the organs of normal individuals. Proc Natl Acad Sci USA 1987; 84:5038–5042.
26. Hardy KJ, Sawada T. Human γ interferon strongly upregulates its own gene expression in peripheral blood lymphocytes. J Exp Med 1989; 170:1021–1026.
27. Trinchieri G. Interleukin-12 and its role in the generation of T_H1 cells. Immunol Today 1993; 14:335–337.
28. Trinchieri G, Wysocka M, D'Andrea A, et al. Natural killer cell stimulatory factor (NKSF) or interleukin-12 is a key regulator of immune response and inflammation. Prog Growth Factor Res 1992; 4:355–368.
29. Moore KW, O'Garra A, de Waal Malefyt R, Vieira P, Mosmann TR. Interleukin 10. Annu Rev Immunol 1993; 11:165–190.
30. Young HA, Hardy KJ. Interferon-γ: producer cells, activation, stimuli, and molecular genetic regulation. Pharmacol Ther 1990; 45:137–151.
31. Maniatis T, Whittemore LA, Du W, et al. In: McKnight SL, Yamamoto K, eds. Transcriptional regulation. Part 2. Cold Spring Harbor, NY: Cold Spring Harbor Laboratory, 1992:1193–1220.
32. Chrivia JC, Wedrychowicz T, Young HA, Hardy KJ. A model of human cytokine regulation based on transfection of γ interferon gene fragments directly into isolated peripheral blood T lymphocytes. J Exp Med 1990; 172:661–664.
33. Leblanc J-F, Cohen L, Rodrigues M, Hiscott J. Synergism between distinct enhanson domains in viral induction of the human β interferon gene. Mol Cell Biol 1990; 10:3987–3993.
34. Goodbourn S, Maniatis T. Overlapping positive and negative regulatory domains of the human β-interferon gene. Proc Natl Acad Sci USA 1988; 85:1447–1451.
35. Nourbakhsh M, Hoffmann K, Hauser H. Interferon-β promoters contains a DNA element that acts as a position-independent silencer on the NF-κB site. EMBO J 1993; 12:451–459.
36. Harada H, Fujita T, Miyamoto M, et al. Structurally similar, but functionally distinct factors, IRF-1 and IRF-2, bind to the same regulatory elements of IFN and IFN-inducible genes. Cell 1989; 58:729–739.
37. Taniguchi T. Regulation of interferon-β gene: structure and function of cis-elements and trans-factors. J Interferon Res 1989; 9:633–640.
38. Au W-C, Moore PA, Lowther W, Juang Y-T, Pitha PM. Identification of a member of the interferon regulatory factor family that binds to the interferon-stimulated response element and activated expression of interferon-induced genes. Proc Natl Acad Sci USA 1995; 92:11657–11661.
39. Driggers P, Ennist D, Gleason S, et al. An interferon γ-regulated protein that binds the interferon-inducible enhancer element of major histocompatibility complex class 1 genes. Proc Natl Acad Sci USA 1991; 87:3743–3747.
40. Keller A, Maniatis T. Identification and characterization of a novel repressor of β-interferon gene expression. Genes Dev 1991; 5:868–879.
41. Harroch S, Revel M, Chebath J. Induction by interleukin-6 of interferon regulatory factor 1 (IRF-1) gene expression through the palindromic interferon response element pIRE and cell type-dependent control of IRF-1 binding to DNA. EMBO J 1994; 13:1942–1949.
42. Fujita T, Reis LFL, Watanabe N, Kimura Y, Taniguchi T, Vilcek J. Induction of the transcription factor IRF-1 and interferon-β mRNAs by cytokines and activators of second messenger pathways. Proc Natl Acad Sci USA 1989; 86:9936–9940.
43. Watanabe N, Sakakibara J, Novanessian A, Taniguchi T, Fujita T. Activation of IFN-β element by IRF-1 requires a post translational event in addition to IRF-1 synthesis. Nucl Acids Res 1991; 19:4421–4428.
44. Matsuyama T, Kimura T, Kitagawa M, et al. Targeted disruption of IRF-1 or IRF-2 results in abnormal type 1 IFN gene induction and lymphocyte development. Cell 1993; 75:83–97.
45. Kamijo R, Harada H, Matsuyama T, et al. Requirement for transcription factor IRF-1 in NO synthase induction in macrophages. Science 1994; 263:1612–1615.
46. Kimura T, Nakayama K, Penninger J, et al. Involvement of the IRF-1 transcription factor in antiviral responses to interferons. Science 1994; 264:1921–1924.
47. Yamada G, Ogawa M, Akagi K, et al. Specific depletion of the B-cell population induced by aberrant expression of human interferon regulatory factor 1 gene in transgenic mice. Proc Natl Acad Sci USA 1991; 88:532–536.
48. Harada H, Kitagawa M, Tanaka N, et al. Anti-oncogenic and

oncogenic potentials of interferon regulatory factors-1 and -2. Science 1993; 259:971–974.
49. Willman CL, Sever CE, Pallavicini MG, et al. Deletion of IRF-1, mapping to chromosome 5q31.1, in human leukemia and preleukemic myelodysplasia. Science 1993; 259:968–971.
50. Visvanathan KV, Goodbourn S. Double-stranded RNA activates binding of NF-κB to an inducible element in the human β-interferon promoter. EMBO J 1989; 8:1129–1138.
51. Du W, Thanos D, Maniatis T. Mechanisms of transcriptional synergism between distinct virus-inducible enhancer elements. Cell 1993; 74:887–898.
52. MacDonald N, Kuhl D, Maguire D, et al. Different pathways mediate virus inducibility of the human IFN-α and IFN-β genes. Cell 1990; 60:767–779.
53. Lengyel P. Tumor-suppressor genes: news about the interferon connection. Proc Natl Acad Sci USA 1993; 90:5893–5895.
54. Samuel C. Antiviral actions of interferon: interferon-regulated cellular proteins and their surprisingly selective antiviral activities. Virology 1991; 183:1–11.
55. Dalton DK, PItts-Meek S, Keshav S, Figari IS, Bradley A, Stewart TA. Multiple defects of immune cell function in mice with disrupted interferon-γ genes. Science 1993; 259:1739–1742.
56. Graham MB, Dalton DK, Giltinan D, Braciale VL, Stewart TA, Braciale TJ. Response to influenza infection in mice with a targeted disruption in the interferon γ gene. J Exp Med 1993; 178:1725–1732.
57. Huang S, Hendriks W, Althage A, et al. Immune response in mice that lack the interferon-γ receptor. Science 1993; 259:1742–1745.
58. Knobler RL, Greenstein JI, Johnson KP, et al. Systemic recombinant human interferon-β treatment of relapsing-remitting multiple sclerosis: pilot study analysis and six-year follow-up. J Interferon Res 1993; 13:333–340.
59. Sechler JM, Malech HL, White CJ, Gallin JI. Recombinant human interferon-γ reconstitutes defective phagocyte function in patients with chronic granulomatous disease of childhood. Proc Natl Acad Sci USA 1988; 85:4874–4878.
60. Farrar MA, Schreiber RD. The molecular cell biology of interferon-γ and its receptor. Ann Rev Immunol 1993; 11:571–611.
61. Bazan JF. Structural design and molecular evolution of a cytokine receptor superfamily. Proc Natl Acad Sci USA 1990; 87:6934–6938.
62. Taniguchi T. Cytokine signaling through non-receptor protein tyrosine kinases. Science 1995; 268:251–255.
63. Uze G, Lutfalla G, Gresser I. Genetic transfer of a functional human interferon α receptor into mouse cells. Cloning and expression of its cDNA. Cell 1990; 60:225–234.
64. Novick D, Cohen B, Rubinstein M. The human interferon α/β receptor: characterization and molecular cloning. Cell 1994; 77:391–400.
65. Pellegrini S, Schindler C. Early events in signaling by interferons. Trends Biochem Sci 1993; 18:338–342.
66. Schindler C, Darnell Jr. JE. Transcriptional responses to polypeptide ligands: the JAK-STAT pathway. Ann Rev Biochem 1995; 64:621–651.
67. Ihle JN. Cytokine receptor signaling. Nature 1995; 377:591–594.
68. Aguet M, Dembic Z, Merlin G. Molecular cloning and expression of the human interferon-γ receptor. Cell 1988; 55:273–280.
69. Soh J, Donnelly RJ, Kotenko S, et al. Identification and sequence of an accessory factor required for activation of the human interferon γ receptor. Cell 1994; 76:793–802.
70. Hemmi S, Bohni R, Stark G, DiMarco F, Aguet M. A novel member of the interferon receptor family complements functionality of the murine interferon γ receptor in human cells. Cell 1994; 76:803–810.
71. Gray PW, Leong S, Fennie E, et al. Cloning and expression of the cDNA for the murine interferon γ receptor. Proc Natl Acad Sci USA 1989; 86:8479–8501.
72. Cook JR, Jung V, Schwartz B, Wang P, Pestka S. Structural analysis of the human IFN-γ receptor: a small segment of the intracellular domain is specifically required for class I major histocompatibility complex antigen induction and antiviral activity. Proc Natl Acad Sci USA 1992; 89:11317–11321.
73. Farrar MA, Campbell JD, Schreiber RD. Identification of a functionally important sequence in the C terminus of the interferon-γ and its receptor. Proc Natl Acad Sci USA 1992; 89:11706–11710.
74. Greenlund AC, Farrar MA, Viviano BL, Schreiber RD. Ligand-induced IFNγ receptor tyrosine phosphorylation couples the receptor to its signal transduction system (p91). EMBO J 1994; 13:1591–1600.
75. Greenlund AC, Schreiber RD, Goeddel DV, Pennica D. Interferon-γ induces receptor dimerization in solution and on cells. J Biol Chem 1993; 268:18103–18110.
76. Schindler C, Shuai K, Prezioso VR, Darnell Jr. JE. Interferon-dependent tyrosine phosphorylation of a latent cytoplasmic transcription factor. Science 1992; 257:809–813.
77. Shuai K, Stark GR, Kerr IM, Darnell Jr. JE. A single phosphotyrosine residue of Stat91 required for gene activation by interferon-γ polypeptide signaling to the nucleus through tyrosine phosphorylation of Jak and Stat protein. Science 1993; 261:1744–1746.
78. Shuai K, Ziemiecki A, Wilks AF, et al. Polypeptide signaling to the nucleus through tyrosine phosphorylation of Jak and Stat protein. Nature 1993; 366:580–583.
79. Sen GC. Transcriptional regulation of interferon-inducible genes. In: Cohen PP, Foulke J, eds. Hormonal Regulation of Gene Transcription. Amsterdam: Elsevier, 1991:349–374.
80. Williams BR. Transcriptional regulation of interferon-stimulated genes. Eur J Biochem 1991; 200:1–11.
81. Revel M, Chebath J. Interferon-activated genes. Trends Biochem Sci 1986; 11:166–170.
82. Bandyopadhyay S, Leonard G, Bandyopadhyay T, Stark GR, Sen GC. Transcriptional induction by double-stranded RNA is mediated by interferon-stimulated response elements without activation of interferon-stimulated gene factor 3. J Biol Chem 1995; 270:19624–19629.
83. Ransohoff RM, Sen GC. Interferon γ signaling: current status and evidence for cross-talk with other interferons and cytokines. In: Baron S, Coppenhaver DH, Dianzani F, eds. Interferon: Principles and Medical Applications. Galveston, TX: University of Texas Medical Branch, 1992:175–191.
84. Reid LE, Brasnett AH, Gilbert GS, et al. A single DNA response element can confer inducibility by both α- and γ-interferons. Proc Natl Acad Sci USA 1989; 86:840–844.
85. Tsang SY, Nakanishi M, Peterlin BM. Mutational analysis of the DRA promoter: cis-acting sequences and trans-acting factors. Mol Cell Biol 1990; 10:711–719.
86. Veals SA, Schindler C, Leonard D, et al. Subunit of an α-interferon responsive transcription factor is related to interferon regulatory factor and Myb families of DNA-binding proteins. Mol Cell Biol 1992; 12:3315–3324.
87. Schindler C, Fu XY, Improta T, Aebersold R, Darnell Jr. JE. Proteins of transcription factor ISGF-3: one gene encodes the 91- and 84-kDa ISGF-3 proteins that are activated by interferon α. Proc Natl Acad Sci USA 1992; 89:7836–7839.
88. Fu XY. A transcription factor with SH2 and SH3 domains is

directly activated by an interferon α-induced cytoplasmic protein tyrosine kinase(s). Cell 1992; 70:323–335.
89. Velazquez L, Fellows M, Stark GR, Pellegrini S. A protein tyrosine kinase in the interferon α/β signaling pathway. Cell 1992; 70:313–322.
90. Muller M, Briscoe J, Laxton C, et al. The protein tyrosine kinase Jak1 complements defects in interferon-α/β and -γ signal transduction. Nature 1993; 366:129–135.
91. Watling D, Guschin D, Muller M, et al. Complementation by the protein tyrosine kinase Jak2 of a mutant cell line defective in the interferon-γ signal transduction pathway. Nature 1993; 366:166–170.
92. Silvennoinen O, Ihle JN, Schlessinger J, Levy DE. Interferon-induced nuclear signaling by Jak protein tyrosine kinases. Nature 1993; 366:583–585.
93. Shuai K, Horvath CM, Huang T, Gureshi SA, Cowburn D, Darnell Jr. JE. Interferon activation of the transcription factor Stat91 involves dimerization through SH2-phosphotyrosyl peptide interactions. Cell 1994; 76:821–828.
94. Williams BR, Silverman R. The 2-5 system: clinical and molecular aspects of the interferon regulated pathway. New York, Liss, 1985.
95. Ghosh SK, Kusari J, Bandyopadhyay S, Kumar R, Sen GC. Cloning, sequencing and expression of two murine 2′-5′-oligoadenylate synthetases: structure-function relationships. J Biol Chem 1991; 266:15293–15299.
96. Chebath J, Benech P, Revel M, Vigneron M. Constitutive expression of (2′-5′) oligo A synthetase confers resistance to picornavirus infection. Nature 1987; 330:587–588.
97. Hassel BA, Zhou A, Sotomayar C, Maran A, Silverman RH. A dominant negative mutant of 2-5A-dependent RNase suppresses antiproliferative and antiviral effects to interferon. EMBO J 1993; 12:3297–3304.
98. Gribaudo G, Lembo D, Cavallo G, Landolfo S, Lengyel P. Interferon action: binding of viral RNA to the 40 kDa 2′-5′-oligoadenylate synthetase in interferon-treated HeLa cells infected with encephalomyocarditis virus. J Virol 1991; 65:1748–1757.
99. Kumar R, Choubey D, Lengyel P, Sen GC. Studies on the role of the 2′-5′-oligoadenylate synthetase-RNase L Pathway in β interferon-mediated inhibition of encephalomyocarditis virus replication. J Virol 1988; 62:3175–3181.
100. Hovanessian A. RNA-activated enzymes: a specific protein kinase and 2′-5′ oligoadenylate synthetases. J Interferon Res 1991; 11:199–205.
101. Dong B, Silverman RH. 2-5A-dependent RNase molecules dimerize during activation by 2-5A. J Biol Chem 1995; 270:4133–4137.
102. Zhou A, Hassel BA, Silverman RH. Expression cloning of 2-5A-dependent RNase: a uniquely regulated mediator of interferon action. Cell 1993; 72:753–765.
103. Hovanessian A. The double-stranded RNA activated protein kinase induced by interferon. J Interferon Res 1989; 9:641–647.
104. Williams BRG. The role of the dsRNA-activated kinase, PRK, in signal transduction. Semin Virol 1995; 6:191–202.
105. Feng GS, Chong K, Kumar A, Williams BRG. Identification of double-stranded RNA-binding domains in the interferon-induced double-stranded RNA-activated p68 kinase. Proc Natl Acad Sci USA 1992; 89:5447–5451.
106. Meurs EF, Chong K, Galabru J, et al. Molecular cloning and characterization of the human double-stranded RNA-activated protein kinase induced by interferon. Cell 1990; 62:379–390.
107. Patel R, Sen G. Identification of the double-stranded RNA binding domain of the human interferon-inducible protein kinase. J Biol Chem 1992; 267:7671–7676.
108. Patel RC, Stanton P, McMillan NMJ, Williams BRG, Sen GC. The interferon-inducible double-stranded RNA-activated protein kinase, PKR, self-associates in vitro and in vivo. Proc Natl Acad Sci USA 1995; 92:8263–8287.
109. Cosentino GP, Venkatesan S, Serluca FC, Green SR, Mattews MB, Sonenberg N. Double-stranded RNA dependent protein kinase and TAR RNA-binding protein form homo and heterodimers in vivo. Proc Natl Acad Sci USA 1995; 92:9445–9449.
110. Petryshyn R, Chen J-J, London I. Detection of activated double-stranded RNA-dependent protein kinase in 3T3-F442A cells. Proc Natl Acad Sci USA 1988; 85:1427–1431.
111. Tiwari R, Kusari J, Kumar R, Sen G. Gene induction by interferons and double-stranded RNA: selective inhibition by 2-aminopurine. Mol Cell Biol 1988; 8:4289–4294.
112. Zinn K, Keller A, Wittemore L-A, Maniatis T. 2-aminopurine selectively inhibits the induction of β-interferon, c-fos, and c-myc gene expression. Science 1988; 240:210–213.
113. Koromilas AE, Roy S, Barber GN, Katze MG, Sonenberg N. Malignant transfection by a mutant of the IFN-inducible dsRNA-dependent protein. Science 1992; 257:1685–1689.
114. Meurs EF, Galabru J, Barber GN, Katze MG, Hovanessian AG. Tumor suppressor function of the interferon-induced double-stranded RNA-activated protein kinase. Proc Natl Acad Sci USA 1992; 90:232–236.
115. Barber GN, Thompson S, Lee TG, et al. The 58 kDa inhibitor of the interferon induced DS RNA activated protein kinase is a TPR protein with oncogenic properties. Proc Natl Acad Sci USA 1994; 91:4278–4282.
116. Staeheli P. Interferon-induced proteins and the antiviral state. Adv Virus Res 1990; 38:147–200.
117. Pavlovic J, Zurcher T, Haller O, Staeheli P. Resistance to influenza and vesicular stomatitis virus conferred by human MxA protein. J Virol 1990; 64:26–31.
118. Rothman J, Raymond C, Gilbert T, O'Hara P, Stevens T. A putative GTP binding protein homologous to interferon-inducible Mx proteins performs an essential function in yeast protein sorting. Cell 1990; 61:1063–1074.
119. Croen KD. Evidence for antiviral effect of nitric oxide. Inhibition of herpes simplex virus type 1 replication. J Clin Inv 1993; 91:2446–2452.
120. Karupiah G, Xie Q-W, Mark R, et al. Inhibition of viral replication by interferon-γ-induced nitric oxide synthetase. Science 1993; 261:1445–1448.
121. Kelly A, Powis S, Glynne R, Radley E, Beck S, Trowsdale J. Second proteasome-related gene in the human MHC class II region. Nature 1991; 353:667–668.
122. Snapper CM, Peschel C, Paul WE. IFN-γ stimulates IgG2a secretion by murine B cells stimulated by bacterial lipopolysaccharide. J Immunol 1988; 140:2121–2127.
123. Ackrill AM, Foster GR, Laxton CD, Flavell DM, Stark GR, Kerr IM. Inhibition of the cellular response to interferons by products of the adenovirus type 5 E1A oncogene. Nucl Acids Res 1991; 19:4387–4393.
124. Foster GR, Ackrill AM, Goldin RD, Kerr IM, Thomas HC, Stark GR. Expression of the terminal protein region of hepatitis B virus inhibits cellular responses to interferons α and γ and double-stranded RNA. Proc Natl Acad Sci USA 1991; 88:2888–2892.
125. Gutch MJ, Reich NC. Respression of the interferon signal transduction pathway by the adenovirus E1A oncogene. Proc Natl Acad Sci USA 1991; 88:7913–7917.
126. Kalvakolanu DVR, Bandyopadhyay S, Harter ML, Sen GC. Inhibition of interferon-inducible gene expression by adenovi-

ral E1A proteins: block in transcriptional complex formation. Proc Natl Acad Sci USA 1991; 88:7459–7463.

127. Mathews MB, Shenk T. Adenovirus-associated RNA and translation control. J Virol 1991; 65:5657–5662.

128. Clarke PA, Schwemmle M, Schinckinger J, Hilse K, Clemens MJ. Binding of Epstein-Barr virus small RNA EBER-1 to the double-stranded RNA-activated protein kinase DAI. Nucl Acids Res 1991; 19:243–248.

129. Gunnery S, Rice AP, Robertson HD, Mathews MB. Tat-responsive region RNA of human immunodeficiency virus 1 can prevent activation of the double-stranded-RNA-activated protein kinase. Proc Natl Acad Sci USA 1990; 87:8687–8691.

130. Chang H-W, Watson JC, Jacobs BL. The E3L gene of vaccinia virus encodes an inhibitor of the interferon-induced, double-stranded RNA-dependent protein kinase. Proc Natl Acad Sci USA 1992; 89:4825–4829.

131. Davies MV, Chang HW, Jacobs BL, Kaufman RJ. The E3L and K3L vaccinia virus gene products stimulate translation through inhibition of the double-stranded RNA-dependent protein kinase by different mechanisms. J Virol 1993; 67:1688–1692.

132. Roy S, Katze MG, Parkin NT, Edery I, Hovanessian AG, Sonenberg N. Control of the interferon-induced 68-kilodalton protein kinase by the HIV-1 tat gene product. Science 1990; 247:1216–1219.

133. Imani F, Jacobs BL. Inhibitory activity for the interferon-induced protein kinase is associated with the reovirus serotype 1 sigma 3 protein. Proc Natl Acad Sci USA 1988; 85:7887–7891.

134. Lee TG, Tomita J, Hovanessian AG, Katze MG. Purification and partial characterization of a cellular inhibitor of the interferon-induced protein kinase of Mr 68,000 from influenza virus-infected cells. Proc Natl Acad Sci USA 1990; 87:6208–6212.

135. Cayley PJ, Davies JA, McCullagh KG, Kerr IM. Activation of the ppp(A2′p)nA system in interferon-treated, herpes simplex virus-infected cells and evidence for novel inhibitors of the ppp(A2′p)nA-dependent RNase. Eur J Biochem 1984; 143:165–174.

136. Upton C, Mossman K, McFadden G. Encoding of a homolog of the IFN-γ receptor by myxoma virus. Science 1992; 258:1369–1372.

15
Interleukin-15

William Carson and Michael Caligiuri
Roswell Park Cancer Institute, Buffalo, New York

INTRODUCTION

Interleukin-15 (IL-15) is a novel cytokine, which binds to the β and γ subunits of the interleukin-2 receptor (IL-2R) and is able to activate cells that bear functional forms of this receptor. IL-15 may therefore be an important ligand in vivo for cells that express the IL-2R. In this chapter we will review what is currently known about IL-15 and consider its possible role in the immune response.

HISTORY AND BIOCHEMISTRY

IL-15 was recently cloned by Grabstein et al. (1). IL-15 exhibits no amino acid sequence homology with IL-2, yet is able to activate cells via components of the IL-2R. To clone this cytokine, Grabstein et al. screened conditioned media produced by a panel of cell lines for their ability to stimulate the proliferation of the IL-2–dependent murine cell line CTLL.2 in the presence of an anti-IL-2 monoclonal antibody (mAb). Using this strategy, a factor was identified in the supernatants of the simian kidney epithelial cell line CV-1 EBNA, which strongly stimulated CTLL proliferation. After several purification steps, a 14 kDa protein was isolated, which possessed the full activity of the conditioned media. Using the PCR reaction and degenerate oligonucleotide primers derived from a partial amino acid sequence of the purified protein, a full-length cDNA clone was obtained from a CV-1 EBNA cDNA plasmid library. Analysis of the amino acid sequence predicted by the cloned cDNA reveals a 162-amino-acid precursor polypeptide containing a 48-amino-acid leader sequence. This precursor form of IL-15 is cleaved at the experimentally determined NH_2-terminus to generate the mature protein, which is 114 amino acids in length. IL-15 exhibits no amino acid sequence homology to IL-2 or any other known members of the cytokine family. In fact, IL-15 is unrelated to any known protein in the GenBank or EMBV databases (1). However, analysis of the predicted amino acid sequence reveals strong helical moments at residues 1–17 and 94–112 as well as two disulfide cross-links, one of which is analogous to the single Cys^{42}-Cys^{88} disulfide bond present in IL-2. IL-2 is classified as a four-helix bundle cytokine, and the primary structure of IL-15 seems to predict a four-helix structure for this cytokine as well. These four helices are arranged in an up-up-down-down configuration and are connected by three loops. Human IL-15 was subsequently cloned from the IMTLH bone marrow stromal cell line, and comparison of the simian and human IL-15 cDNAs reveals 97% sequence homology within their coding regions. The open reading frame of the human IL-15 cDNA is 486 bp in length and is flanked by a 316 bp 5′ noncoding region and a 400 bp 3′ noncoding region.

Anderson et al. have analyzed the location and organization of the IL-15 gene in the mouse (2). Murine IL-15 exhibits 73% sequence identity at the nucleic acid and amino acid levels with human IL-15. The murine IL-15 gene locus is approximately 34 kb in length and contains 7 introns and 8 exons, of which exons 2, 4, 5, 6, 7, and 8 code for the mature protein. The last 3 introns delineate exons that encode the predicted helix-forming domains of IL-15. Using interspecific back-cross analysis, the IL-15 gene was localized to the central region of murine chromosome number 8. In humans, IL-15 is present as a single-copy gene on chromosome 4 at band q31 and appears to possess the same structure and organization as its murine counterpart. Using PCR amplification and a panel of rodent-human somatic cell hybrids, Mendrola and Huebner localized the IL-15 gene to chromosome 4q25-qter (J. Mendrola and K. Huebner, personal communication). This region of chromosome 4 also encodes the genes for IL-2, IL-8, epidermal growth factor, and fibroblast growth factor. Despite their relative proximities, IL-2 and IL-15 demonstrate no significant evolutionary homology, and there is no evidence to suggest that the gene for IL-15 is derived from the IL-2 locus (2). Several clinical disorders have been mapped to 4q25-qter, including Reiger syndrome, Williams-Beuren syndrome, and facioscapulohumeral muscular dystrophy. In order to assess the involvement of IL-15 in these diseases, a finer chromosomal localization will be required.

SOURCES OF IL-15

Analysis of human tissues reveals that IL-15 mRNA is abundantly expressed in skeletal muscle as well as placenta and, to a lesser degree, in heart, lung, liver, kidney, and pancreas (1). mRNA for IL-15 is also highly expressed by adherent peripheral blood mononuclear cells (primarily $CD14^+$ monocytes), which were stimulated with lipopolysaccharide (LPS) (2). This finding was confirmed at the protein level by Carson et al., who demonstrated the production of soluble IL-15 by purified monocytes within 5 h of LPS stimulation (3). In contrast, fresh PBMC express only low levels of IL-15 mRNA. T cells do not express any detectable IL-15 transcript in either the resting or activated state, but IL-15 transcript is expressed by several transformed T-cell lines (4). In fact, a lymphokine has been isolated from the conditioned medium of a human T-cell lymphotrophic virus I–induced adult T-cell leukemia that is a potent stimulator of T cells and NK cells. This factor has a molecular mass of between 12 and 16 kDa and shares many properties with IL-15. It is quite possible that this lymphokine (referred to as IL-T by its discoverers) is identical to IL-15 (5,6).

IL-15 AND IL-2 RECEPTORS

Binding of IL-15 to Components of the IL-2 Receptor

Grabstein's initial studies with IL-15 confirmed its ability to stimulate the proliferation of cells bearing functional IL-2R. Purified, recombinant (r) human IL-15 stimulated the proliferation of murine CTLL.2 to the same extent and with similar potency as IL-2 (1). Similar results were obtained with antigen-specific murine helper and cytotoxic T-cell clones (1). IL-15 also stimulated proliferation in human PHA-activated peripheral blood T cells and their derivative $CD4^+$ and $CD8^+$ subsets (1). IL-2 has been used to generate LAK cells from peripheral blood mononuclear cells (PBMC), and IL-2 is also critical to the induction of alloantigen-specific cytotoxic T lymphocytes (CTL) in mixed lymphocyte cultures (1). Grabstein cultured human PBMC in either IL-2 or IL-15 at concentrations ranging from 1 to 30 ng/ml and tested the resulting LAK cells against an NK-resistant cell line (Daudi). Interestingly, culture of PBMC in IL-15 gave rise to cytotoxic effectors that were slightly superior to those obtained in the presence of IL-2, especially when cytokine concentrations below 30 ng/ml were used. Similar results were obtained when CTL generated in a mixed lymphocyte reaction in the presence of IL-2 or IL-15 were tested for their cytolytic activity against allogeneic target cells.

IL-15 does not stimulate proliferation via the induction of IL-2 since a neutralizing antibody to IL-2 does not inhibit the proliferative response of PHA-activated PBMC to IL-15 (1). Likewise, anti–IL-2 Abs do not inhibit the generation of LAK cells or alloantigen-specific CTL by exogenous IL-15. In fact, experiments with blocking mAbs suggest the participation of the IL-2Rβ subunit but not the IL-2Rβ α subunit in IL-15 binding and function. Whereas an IL-2Rα mAb could not inhibit IL-15-induced proliferation of human T cells, antibodies to IL-2Rβ (e.g., Mik1β, TU11, and TU27) significantly inhibited all actions of IL-15 including the generation of LAK cells and CTLs and the proliferation of PHA-activated PBMC and murine CTLL.2 cells.

A variety of cell lines and a large number of antigen-dependent human and murine T-cell clones were subsequently tested for their ability to bind radiolabeled IL-15 (7). The murine T-cell lines S49 and CTLL.2 bound IL-15 as did the murine antigen-dependent T-cell clones AB8, 7B9, and RA8. The human antigen-dependent T-cell clone C.22 and EBV-immortalized human B cells also exhibited specific IL-15 binding. Freshly isolated NK cell and LPS-activated monocytes bound IL-15 as did human endothelial cells and several murine thymic epithelial and fetal liver stromal cell lines. All the above cells bound IL-15 with high affinity and an apparent Kd between 1×10^{-11} and 5×10^{-10} M. Importantly, simian and human IL-15 were able to bind to receptors on murine cells. At present, it is not clear whether murine IL-15 can bind to receptors on human cells.

The ability of IL-15 to activate murine and human lymphoid cells bearing functional IL-2R suggested that IL-15 was able to bind to components of this receptor. Grabstein et al. demonstrated that radiolabeled IL-15 will bind to PHA-activated PBMC in a specific manner (1). Moreover, unlabeled IL-15 will inhibit the binding of radiolabeled IL-2 to the YT cell line, which expresses the β and γ chains of the IL-2R but very low amounts of the IL-2Rα chain (1). In order to expand upon these data, Giri et al. conducted an elegant series of binding experiments utilizing cell types responsive to IL-2 and IL-15 as well as cell lines transfected with various components of the IL-2R (7). Since IL-15 had been shown to stimulate proliferation in CTLL.2 cells, Giri et al. initially analyzed the binding of radiolabeled IL-2 and IL-15 to this cell line. In contrast to IL-2, which exhibited both high- and low-affinity receptors for IL-2, IL-15 demonstrated only a single high-affinity receptor on CTLL.2 cells with an apparent Kd of 3.4×10^{-10} M and 1420 sites per cell. Giri obtained similar results with PHA-activated PBMC and further demonstrated that antibodies to the IL-2Rβ chain but not the IL-2Rα chain could inhibit the binding of radiolabeled IL-15 to activated PBMC.

The human NK-like cell line YTN17 expresses the IL-2Rβ chain at high surface density and also demonstrates high-affinity binding sites for IL-15, which can be blocked in the presence of neutralizing anti–IL-2Rβ mAb (7). These data suggested a specific interaction between the IL-2Rβ subunit and IL-15. To confirm this observation, the BAF/B03 murine pro-B-cell line was employed. The BAF/B03 cell line expresses the IL-2Rα and γ chains but not the IL-2Rβ chain and will proliferate in response to IL-3 but not IL-2 (8). Studies with this cell line revealed that IL-15 did not bind with any affinity, nor did it transduce a proliferative signal (7). However, after transfection with a mammalian expression vector containing the IL-2Rβ subunit, the BAF/B03 cell line acquired the ability to proliferate in response to IL-15 and also bound IL-15 with high affinity. In addition, the effect of IL-15 could be completely abolished in presence of an anti–IL-2Rβ mAb (7).

In order to determine whether the IL-2Rγ chain is a component of the IL-15–binding complex, COS-7 cells were transiently transfected with expression vectors for the IL-2Rβ and IL-2Rγ subunits, either alone or in combination (7). COS-7 cells did not bind IL-15 when transfected with either chain alone. However, a dramatic increase in binding of radiolabeled IL-15 was obtained when COS cells were co-transfected with both the IL-2Rβ and IL-2Rγ subunits. Evidence for a physical interaction between IL-15 and the IL-2Rβ and IL-2Rγ chains comes from cross-linking expriments in which radiolabeled ligand is covalently linked to cell surface receptors. Cross-linking of radiolabeled IL-15 to IL-2Rβγ–transfected COS cells led to the formation of a high molecular weight complex, which was consistent with the binding of IL-15 to surface IL-2Rβ and IL-2Rγ subunits. This complex was identical to that obtained with radiolabeled IL-2. Similar results were obtained when radiolabeled IL-15 was cross-linked to IL-2Rβγ–expressing YT cells.

It has been previously shown that IL-2 can induce the dimerization of IL-2Rβ and γ subunits that are expressed on the cell surface. In order to demonstrate the ability of IL-15 to mediate similar dimerization, Giri made use of soluble forms of the IL-2Rβ and γ subunits, which consisted of just the extracellular domain of the protein (7). When an anti–IL-2Rβ mAb was added to a mixture of soluble IL-2Rβ and γ subunits, only the IL-2Rβ chain could be precipitated. But the addition of IL-15 enabled this antibody to precipitate both receptor proteins, indicating that IL-15 interacts with the extracellular domains of both the IL-2Rβ and γ subunits.

The IL-15Rα Chain

Initial studies by Grabstein et al. suggested that IL-15 did not use the IL-2Rα chain for binding or signal transduction (1). Hence, a search was undertaken for another chain that might confer high-affinity binding for IL-15. Giri et al. have subsequently reported the characterization and molecular cloning of the murine IL-15Rα chain (9). This molecule is structurally related to the IL-2R α chain and was isolated from a murine helper T-cell clone known as D10. This cell line could proliferate in response to very low concentrations of IL-15 and had an EC_{50} for IL-15 which was 10-fold lower than that for IL-2. The D10 cell line also exhibited higher levels of IL-15 binding than did the IL-2–dependent CTLL.2 cell line and was subsequently found to overexpress the IL-15Rα chain. Analysis of the cloned cDNA revealed a Type I membrane protein with a predicted signal peptide of 32 amino acids, a 173-amino-acid extracellular domain, a 21-amino-acid transmembrane region, and a 37-amino-acid cytoplasmic tail. Comparison of the murine IL-2Rα and IL-15Rα proteins reveals the presence of a conserved protein binding motif known as the "sushi domain." The murine IL-2Rα protein has two such domains, whereas the IL-15Rα subunit has just one of these motifs in its extracellular region. The IL-15Rα also has a slightly longer cytoplasmic tail than does the IL-2Rα chain, but it does not exhibit any recognized catalytic or protein-binding domains. Expression of IL-15Rα transcript was detected in T-cell lines, B-cell lines, a macrophage line, a thymic stromal cell line, and a bone marrow stromal cell line. In addition, IL-15Rα mRNA was detected in a variety of normal tissues including spleen, heart, lung, and skeletal muscle. The expression of mRNA for IL-15Rα was highest within murine liver. IL-15Rα expression could be upregulated in T cells by treatment with anti-CD3 Ab, PHA, PMA, or IL-2. Similar results were obtained when monocytes were treated with IFN-γ.

Thus, the expression of the IL-15Rα chain appears to be augmented by activation signals in much the same manner as is the IL-2Rα subunit.

A role for IL-15Rα in IL-15 binding was confirmed through experiments with a cell line which constitutively expresses the IL-2Rγ chain (9). Transfection of this cell line with IL-15Rα and IL-2Rβ reconstituted high-affinity IL-15 binding as well as IL-15–induced proliferation. Analysis of the binding of IL-15 to the IL-15Rα chain revealed that this receptor subunit has an extremely high affinity for IL-15 ($Ka = 1.5 \times 10^{11}$). This finding is not predicted by our knowledge of the interaction between IL-2 and the IL-2Rα and points to the potential for significant differences in the biological roles of IL-2 and IL-15 in vivo. Despite the similarity of their extracellular domains, the isolated IL-2Rα and IL-15Rα protein subunits exhibit a 1000-fold difference in their affinity for their respective ligands. This finding, coupled with the widespread mRNA expression of IL-15Rα, implies that IL-15 may have multiple physiological roles. It will be of prime importance to determine whether the IL-15Rα subunit can itself transmit a signal in tissues, such as muscle, which appear to lack significant expression of the IL-2Rβ and IL-2Rγ chains. Alternatively, the IL-15Rα chain may serve as a mechanism for sequestration of IL-15 and/or a means by which IL-15 is presented to other cells IL-15Rα in association with IL-2Rβ and γ. A soluble form of the IL-15Rα has been produced which can inhibit CTLL proliferation to 10 ng/ml of IL-15 at a 1:1000 dilution (9). Thus, the soluble form of the IL-15Rα could be used to inhibit IL-15 binding in vivo, and experiments to evaluate the potential of this strategy are currently underway.

Signaling

Thus far, all the available evidence suggests that the binding of IL-15 to its receptor complex leads to a series of signaling events which are mediated by the IL-2Rβγ heterodimer and are therefore similar if not identical to those elicited by IL-2 (10). For virtually all cell types examined to date, the functional response to IL-15 is very similar to that of IL-2. However, some cells appear to respond preferentially to one cytokine or the other (1,7). Such a differential response might be explained in terms of cytokine-binding affinities, which in turn are governed by receptor density and composition.

The binding of IL-15 and IL-2 to their receptor(s) results in the activation of nonreceptor protein tyrosine kinases (PTKs) which are physically associated with the IL-2R (10). The Src-family PTKs Lck, Fyn, and Lyn associate with a cytoplasmic region of the IL-2Rβ chain known as the acidic region. It is believed that the binding of ligand induces IL-2R oligomerization, which in turn permits the associated PTKs to transphosphorylate each other. This event induces PTK enzymatic activity, and Lck subsequently phosphorylates the IL-2Rβ chain a Tyr^{355} and Tyr^{358}. Phosphorylation of these tyrosine residues may recruit other signaling molecules such as phosphatidylinositol-3 kinase. Activation of Lck is followed by the induction of the *c-fos* and *c-jun* genes, which are known to be critical targets for the proliferative signals generated by growth factor receptors. IL-2R activation may also initiate other signaling pathways and target molecules such as Raf-1 kinase, mitogen-activated protein kinase, p70 S6 kinase, $p34^{cdk2}$, $p34^{cdc2}$, $p27^{kip1}$, Shc, and glycosylphosphatidylinositol (10).

A previously unrecognized signal transduction pathway has been discovered through the study of transcriptional activation in response to interferon (IFN)-α and IFN-γ (10). This pathway results in rapid activation of IFN-responsive genes but does not utilize conventional second messenger pathways. Instead, a novel set of PTKs of the Janus kinase (Jak) family is involved in the transduction of the cytokine signal (10). Of prime interest is the finding that this system is operative in a variety of receptor-ligand systems in addition to the type I and II interferons (11). Receptors types implicated thus far include members of the cytokine receptor family (e.g., erythropoietin, growth hormone, prolactin, IL-2, IL-3, IL-4, IL-5, IL-6, GM-CSF, G-CSF, oncostatin M, leukemia inhibitory factor, ciliary neurotrophic factor) as well as growth factor receptors with intrinsic tyrosine kinase activity such as epidermal growth factor and platelet-derived growth factor. A number of Jak kinases have been identified, and it is apparent that different cytokine receptors selectively associate with distinct members of the Jak kinase family.

In the case of IL-15 and IL-2, the binding of either ligand induces oligomerization of the IL-2R and activation of prebound Jak kinases. The IL-2Rβ and γ subunits recruit Jak1 and Jak3, respectively (10). Jak3 requires the serine-rich, membrane-proximal domain of the IL-2Rβ subunit for activation, and its expression is limited to NK cells and activated T cells. The phosphorylation of specific tyrosine residues by activated Jak PTKs results in the creation of SH2-protein binding sites. These sites attract SH2-domain–bearing transducer molecules called STAT (signal transducer and activator of transcription) proteins, which are subsequently phosphorylated by the activated Jak PTKs kinase. STAT proteins are constitutively expressed within the cytoplasm of resting cells and exist there in an unphosphorylated state. Phosphorylation of STAT proteins results in their rapid nuclear translocation. Once in the nucleus, phosphorylated STAT proteins bind to STAT-specific DNA sequences and direct the transcription of selected genes. The above sequence of events is very rapid, with activated STAT proteins binding to DNA within minutes of receptor activation. Analysis of the promoter regions of the IFN-responsive genes has led to the identification of a number of STAT-binding sequences, which have high affinities for specific STAT proteins.

The STAT protein family has several members, each of which is involved in signaling for specific cytokines. IL-15 and IL-2 appear to utilize STAT5 and STAT3 (10). Lin et al. have performed a detailed analysis of STAT protein activation in peripheral blood lymphocytes in response to IL-15 and IL-2 (12). Activation of nuclear factors was evaluated using the electrophoretic mobility shift assay (EMSA) and a radiolabeled DNA probe corresponding to the STAT protein–binding site located within the promoter of the IFN-γ–activated gene, FcγRI. Analysis of resting and PHA-activated peripheral blood lymphocytes revealed that treatment of cells with IL-2 or IL-15 induced identical DNA-binding complexes (12).

The importance of the Jak-STAT signaling pathway in lymphocyte activation is underscored by the study of humans who have a loss of function mutations of the Jak3 gene. In this situation, a severe combined immunodeficient state is observed, which results in a complete lack of circulating T cells and NK cells, as well as severe impairment of B-cell function (13).

BIOLOGICAL ACTIVITY OF IL-15

Effects on T Cells

IL-15 was discovered through its ability to stimulate the proliferation of an IL-2–dependent murine T-cell line (1). As mentioned above, IL-15 can also stimulate the proliferation of PHA-activated peripheral blood T cells as well as their derivative $CD4^+$ and $CD8^+$subsets. In addition, IL-15 has been used to generate LAK cells from human PBMC, and it is also effective in stimulating allo-antigen specific CTLs in mixed lymphocyte culture (1).

Kennedy et al. have shown that IL-12 is an important costimulator of antigen-dependent proliferation in murine Th_1 clones and have recently demonstrated a role for IL-15 in this process (14). They have demonstrated that the ability of Th_1 clones to acquire responsiveness to IL-4 during antigen-specific activation is dependent upon activation via the T-cell receptor and co-stimulation with IL-12. They have also shown that resting Th_1 clones fail to proliferate in response to IL-12 or IL-4 alone but will respond to the combination of IL-12 and IL-4 with a suboptimal proliferative response. IL-12 was also found to synergize with IL-15 to induce the proliferation of resting Th_1 clones but not Th_0 or Th_2 clones (14).

IL-2 is a known chemoattractant for T cells, and Wilkinson and Liew have recently shown that IL-15 also exerts a chemoattractant effect upon T cells (15). They showed that both IL-15 and IL-2 will induce polarization of fresh and cultured human PBMC and found that the entire increase in polarized lymphocyte morphology could be attributed to the T-cell compartment. $CD4^+$ and $CD8^+$ cells appeared to polarize equally in response to IL-15 or IL-2. The effects of IL-15 on lymphocyte morphology could be strongly inhibited by neutralizing antihuman IL-15 antibodies or anti–IL-2Rβ antibodies, but blockade of the IL-2Rα receptor subunit had no effect upon IL-15–induced T-cell polarization. Lymphocytes cultured in 1 μg/ml of IL-15 exhibited significant active locomotion as compared to cells cultured in media alone, and collagen gels containing IL-15 exhibited greater invasion of lymphocytes than did control gels. It was also shown that T-cell locomotion could be enhanced by exposing the cells to positive concentration gradients of IL-15 or by simply increasing the absolute concentration of IL-15. These results imply that IL-15 can induce T-cell movement as well as accelerate this movement, and the authors hypothesized that accessory cells at inflammatory sites might attract T cells to these areas via their production of IL-15.

Effects on B Cells

Armitage et al. examined IL-15 for its ability to stimulate B cells in a series of in vitro activation assays and found that IL-15 is able to induce B-cell proliferation and differentiation in much the same manner as does IL-2 (16). In the absence of co-stimulatory signals, neither IL-2 nor IL-15 could induce significant B-cell proliferation. However, purified human tonsilar B cells could proliferate in response to IL-15 if first activated with phorbol ester or anti-IgM antibody. As might be expected from the results of earlier studies, an anti–IL-2Rβ mAb significantly inhibited the proliferation of activated B cells in response to IL-15, whereas an anti–IL-2Rα mAb had no effect. These data imply that IL-15 binds to the IL-2Rβ subunit of the IL-2/15R on B cells but not the IL-2Rα chain. It has been previously shown that co-stimulation of B cells with IL-2 and CD40 ligand (CD40L) results in significant immunoglobulin production (17). Similarly, IL-15 induces the secretion of IgM, IgG, and IgA from B cells when used in combination with CD40L. Importantly, the combination of IL-2 and IL-15 had an additive effect upon B-cell proliferation in this assay, which would imply that these two cytokines can act in concert to induce B-cell immunoglobulin production after their stimulation with CD40L. Finally, the authors demonstrated that IL-15 is able to partially replace IL-2 in the induction of antigen-specific antibody production in vitro. The assay employed makes use of sheep RBCs as the antigenic stimulus and is dependent upon T-cell help in the form of IL-2 and CD40L. IL-15 was not as potent as IL-2 in this assay, but IL-15 was the only cytokine capable of replacing IL-2 to any significant degree. These results demonstrate the ability of IL-15 to stimulate the differentiation and proliferation of human B cells, and it is conceivable that IL-15 may play a role in B-cell activation. As IL-2 is produced primarily by T cells, IL-15 derived from activated monocytes might prove to be an important costimulatory signal for the induction of immunoglobulin production in vivo.

B-cell responses to IL-15 and IL-2 were also studied in two patients with X-linked SCID (X-SCID) who demonstrated mutations of the IL-2Rγ chain (18,19). Analysis of genomic DNA revealed that one patient had a deletion of approximately 1.4 kb, which included the exons encoding for the transmembrane domain of the mature protein. The other patient exhibited a single base-pair mutation, which resulted in a frame shift and a premature stop codon at position 814. B cells from these X-SCID patients were costimulated with anti-IgM antibody and IL-2, IL-4, IL-13, or IL-15. B cells from a normal control responded strongly to IL-15 or IL-2, whereas X-SCID B cells did not proliferate in response to this treatment. In contrast, IL-13 and IL-4 were able to induce a strong proliferative response in B cells from both X-SCID patients and unaffected controls. In a similar experiment, it was shown that B cells from a normal patient secreted immunoglobulins of all classes in response to co-stimulation with CD40L and IL-15 or IL-2, whereas X-SCID B cells were insensitive to this treatment. These results imply that the common γ chain is absolutely necessary for B-cell activation in response to IL-15 and IL-2 and support the notion that the IL-2Rγ chain is vital for IL-15 and IL-2 receptor signaling. In a subsequent study, B cells from these X-SCID patients were transformed with the Epstein-Barr virus in order to generate B-lymphoblastoid cell lines. Analysis of these cell lines revealed a significant delay in IL-15 internalization as well as a marked decrease in the absolute amount of IL-15 that could be internalized over a given period of time (19).

Effects on NK Cells

NK cells are unique among lymphocytes in their constitutive expression of a functional IL-2 receptor (IL-2R), but it is important to note that IL-2 is produced solely and transiently by T cells after stimulation with specific antigen and is quickly bound to the heterotrimeric high-affinity T-cell IL-2Rαβγ after its secretion (20,21). The high-affinity IL-2R on activated T cells has 100-fold greater affinity for IL-2 than the intermediate-affinity IL-2Rβγ expressed on the majority of NK cells. The activated T cell uses the IL-2 it secretes in an autocrine fashion to promote cell differentiation. Thus, the availability of T-cell–derived IL-2 as a stimulatory cytokine for NK-cell function has yet to be conclusively demonstrated. IL-15 is produced by activated monocytes and may be an important ligand for the NK-cell IL-2R (3).

A comparison of the effects of rIL-2 and rIL-15 on functionally distinct populations of human NK cells reveals subtle differences in their proliferative actions (22). $CD56^{bright}$ NK cells comprise approximately 10% of circulating NK cells and constitutively express both a heterotrimeric high-affinity IL-2Rαβγ and an intermediate-affinity IL-2Rβγ (23). Binding of picomolar amounts of IL-2 to the high-affinity IL-2Rαβγ of $CD56^{bright}$ NK cells results in a brisk proliferative response, which can be inhibited by mAbs directed against either the IL-2Rα or the IL-2Rβ subunits of the receptor (Fig. 1A). Interestingly, IL-2 and IL-15 generate distinct dose-response curves in a $CD56^{bright}$ proliferation assay (Fig. 1B). Incubation of cells with picomolar concentrations of IL-2 (0.1–1 ng/ml) results in partial saturation of the high-affinity IL-2R and significant but suboptimal levels of proliferation. The high-affinity receptor becomes fully saturated as the concentration of IL-2 is raised to 10 ng/ml and a maximal proliferative response is then noted. However, as seen in Figure 1B, IL-15 does not induce a proliferative response in $CD56^{bright}$ NK cells until a concentration of 10 ng/ml has been achieved, at which point a maximal proliferative response is obtained. As a result, the dose-response curve for IL-15 in this assay is much steeper than that of IL-2. IL-2–induced proliferation of $CD56^{bright}$ NK cells can be inhibited by antibodies to either the α or β subunits of the IL-2R (Fig. 1A). In contrast, IL-15-induced proliferation of $CD56^{bright}$ NK cells is inhibited by anti–IL-2Rβ mAb, but not anti–IL-2Rα mAb (Fig. 1C). This finding has been confirmed and expanded upon using several different anti–IL-2Rα mAb. Thus, IL-15 need not bind to the IL-2Rα subunit in order to induce a proliferative response in $CD56^{bright}$ human NK cells.

$CD56^{dim}$ human NK cells express only the intermediate-affinity IL-2Rβγ, which requires nanomolar concentrations of IL-2 for activation and induces primarily a cytotoxic response in this cell population (23). Incubation of $CD56^{dim}$ NK cells in the presence of IL-15 results in LAK activity against the NK-resistant cell line COLO 205 in a dose-dependent fashion (22). Thus, at equal concentrations, IL-15 and IL-2 induce nearly identical levels of cytotoxicity in $CD56^{dim}$ NK cells (Fig. 2A). Like IL-2, IL-15–induced LAK activity can be completely abrogated in the presence of an anti–IL-2Rβ mAb (Fig. 2B). However, blockade of the α subunit of the IL-2R has no effect

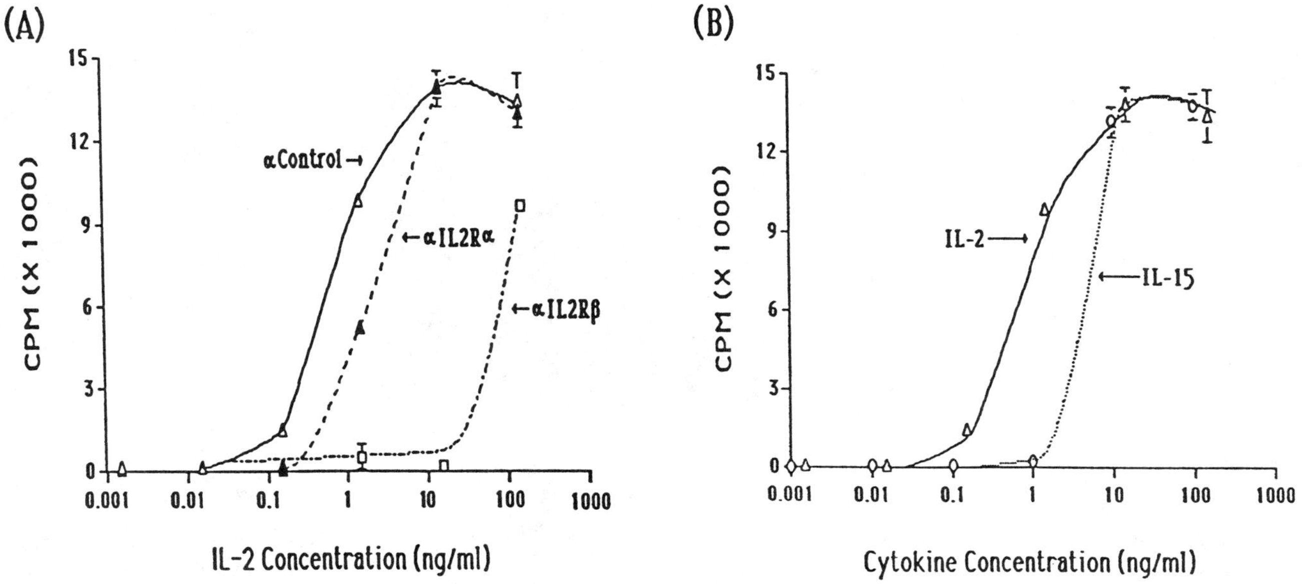

Figure 1 Proliferation of an NK-cell subset in the presence of IL-2 or IL-15. CD56[bright] NK cells were sorted from fresh peripheral blood lymphocytes and cultured in medium plus varying concentrations of recombinant IL-2 or IL-15 for 84 h and then assayed for methyl-[^{3}H]-thymidine incorporation during the final 12 h of incubation. Results represent the mean ± SE of triplicate wells. (A) Proliferation of CD56[bright] NK cells in the presence of varying concentrations of IL-2 plus anti–IL-2Rα, anti–IL-2Rβ, or isotype control mAb. (B) Proliferation of CD56[bright] NK cells in response to varying concentrations of IL-2 or IL-15. (C) Proliferation of CD56[bright] NK cells in the presence of varying concentrations of IL-15 plus anti–IL-2Rα, anti–IL-2Rβ, or isotype control mAb.

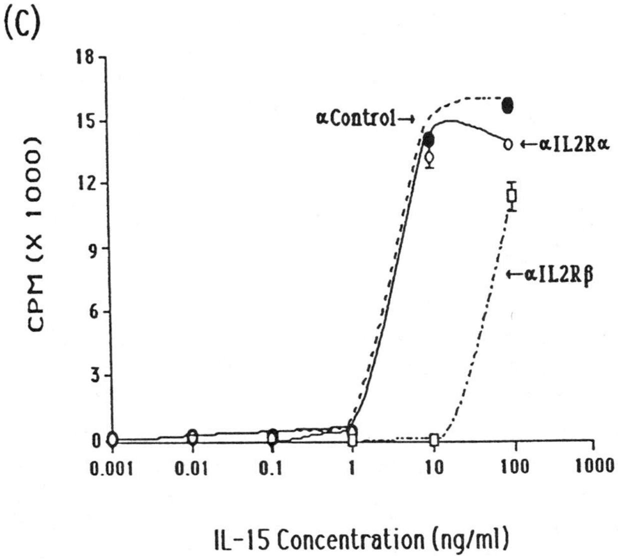

on the cytolytic function of this NK-cell subset in response to IL-2 or IL-15. The majority of NK cells also express a low-affinity receptor for the Fc portion of immunoglobulin (FcRγIII or CD16), which enables them to interact with antibody-coated target cells and mediate antibody-dependent cellular cytotoxicity (ADCC) (23). IL-15–activated CD56[dim] NK cells exhibit ADCC against the P815 murine mastocytoma cell line in the presence of a rabbit antimouse antiserum. A clear dose-response curve is noted, and the effects of IL-15 are identical to those of IL-2 (22).

Given that IL-15 is produced by activated monocytes, we wished to determine if IL-15 could act in a synergistic or additive manner with IL-12, a cytokine that is also stimulatory to NK cells and is also derived from monocytes. IL-12 stimulates lymphokine-activated killer (LAK) activity in human NK cells, and culture of NK cells with both IL-12 and IL-2 has an additive effect on NK cytolytic activity (23). In a cytotoxicity experiment using an NK-resistant target (COLO 205), we found that IL-2 and IL-15 stimulate similar levels of cytotoxicity in CD56[dim] NK cells in a standard 4-h ^{51}Cr-release assay (Fig. 2C) (22). At a concentration of 10 U/ml, IL-12 induced comparable levels of target cell lysis. IL-12 is known to induce LAK activity in human NK cells, and co-incubation of NK cells with both IL-2 and IL-12 can have an additive effect on NK cytotoxic activity (23). As can be seen in Fig. 2C, the combination IL-12 and IL-15 resulted in additive LAK activity that was nearly identical to that seen with IL-12 and IL-2. Interestingly, the combination of IL-2 and IL-15 did not result in an additive cytotoxic effect, which implies

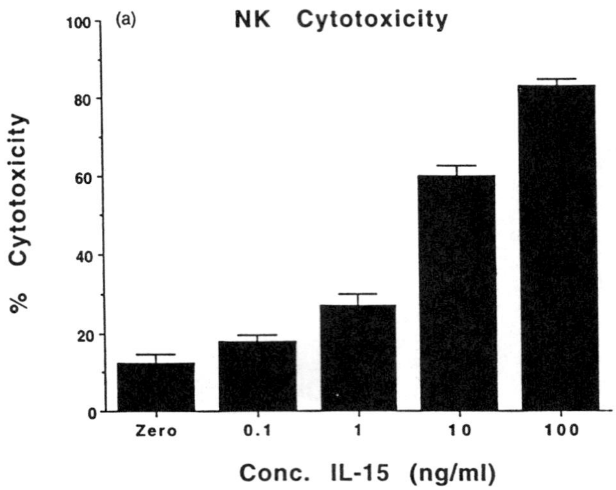

Figure 2 IL-15 induces LAK activity in CD56^{dim} NK cells. (a) Purified CD56^{dim} NK cells were incubated at increasing concentrations of IL-15 for 18 h and then tested against COLO 205 target cells in a standard 4 h ^{51}Cr-release assay. E:T ratio was 5:1. (b) IL-15–induced LAK activity of CD56^{dim} NK cells is abrogated by an anti–IL-2Rβ mAb. CD56^{dim} NK cells were cultured in medium and isotype control mAb (‾□), medium supplemented with isotype control mAb and then 1ng/ml IL-15 (‾■), or medium supplemented with anti–IL-2Rβ mAb and then 1 ng/ml IL-15 (‾□). E:T ratios of 2.5:1, 5:1, and 10:1 were used against COLO 205 target cells. (c) IL-15 or IL-2 synergize with IL-12 to enhance the cytotoxic activity of CD56^{dim} NK cells. CD56^{dim} NK cells were incubated in 1 ng/ml IL-2, 1 ng/ml IL-15, or 10 U/ml IL-12 alone or in combination for 24 h and then tested for cytotoxicity against COLO 205 target cells at an E:T ratio of 5:1. Results represent the mean ± SE of triplicate wells.

that the binding of these ligands to the IL-2Rβγ results in the transduction of similar (if not identical) signals. As might be predicted, the activity of all three cytokines together was no greater than the combination of IL-12 with either IL-15 or IL-2, further supporting the notion that IL-2 and IL-15 may be acting in a redundant fashion.

Another important effector function of the NK cell is the production of cytokines that can modulate the host immune response to infection. Specifically, NK cells can be stimulated in vitro and in vivo to produce relatively large quantities IFN-γ, TNF-α, and GM-CSF, all of which are strong activators of monocyte function (23). IL-2 is required to stimulate maximal NK cell secretion of these cytokines in several systems (see below). However, the production of IL-2 by T cells is temporally and spatially limited. The monocyte is presumed to take part in NK-cell activation via the secretion of cytokines such as IFN-α/β, IL-12, and TNF-α, which in turn induce the production of monocyte-stimulatory factors by the NK cell in what might be described as a paracrine feedback mechanism (24). However, the role of the monocytes in NK-cell function takes on new importance in light of its ability to produce IL-15 upon stimulation with factors such as LPS (1,3). We therefore investigated the production of IFN-γ, TNF-α, and GM-CSF by NK cells in response to IL-15, either alone or in combination with other monocyte-derived cytokines. NK cells were cultured under various conditions for 72 h, at which time supernatants were harvested and analyzed by ELISA. As shown in Figure 3A–C, the combination of IL-15 and IL-12 is a potent stimulus to the production of IFN-γ, TNF-α, and GM-CSF by human natural killer cells (22). We also found that TNF-α could combine synergistically with IL-15 to stimulate the secretion of GM-CSF. Interestingly, TGF-β, another important cytokine product of monocyte, is a strong inhibitor of the IFN-γ production of NK cells cultured with IL-15 and IL-12 (W. Carson, unpublished observations). It will obviously be important to determine what stimuli induce the simultaneous production of IL-15 with either IL-12 or TNF-α before one can propose the existence of a synergistic effect in vivo. Also, as evidenced by the results with TGF-β, inhibitory cytokines may exert a dominant effect upon NK cell cytokine secretion in vivo.

NK cells constitutively express a number of cytokine receptors, and upon binding specific ligand these receptors mediate enhanced effector functions and increased secretion of immunomodulatory cytokines (23). Importantly, NK cells secrete large amounts of IFN-γ in response to stimulation by monocyte-derived cytokines (22). IFN-γ production by NK cells is now known to be critical to the activation of host monocyte and the development of an effective innate immune response in the time period pre-

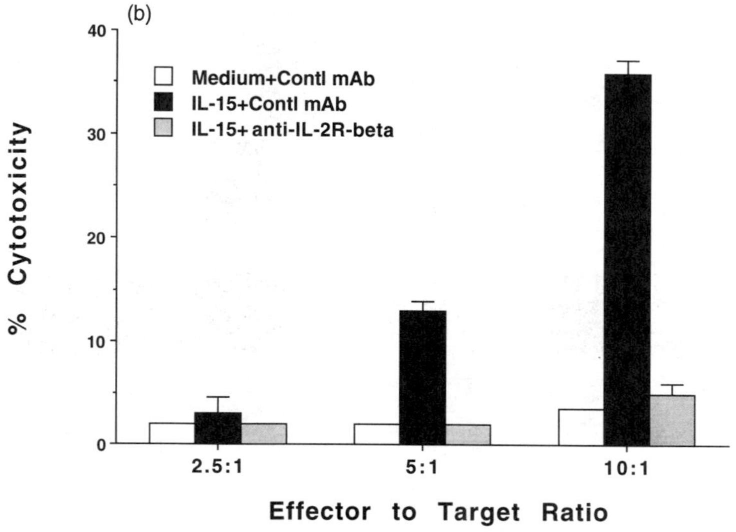

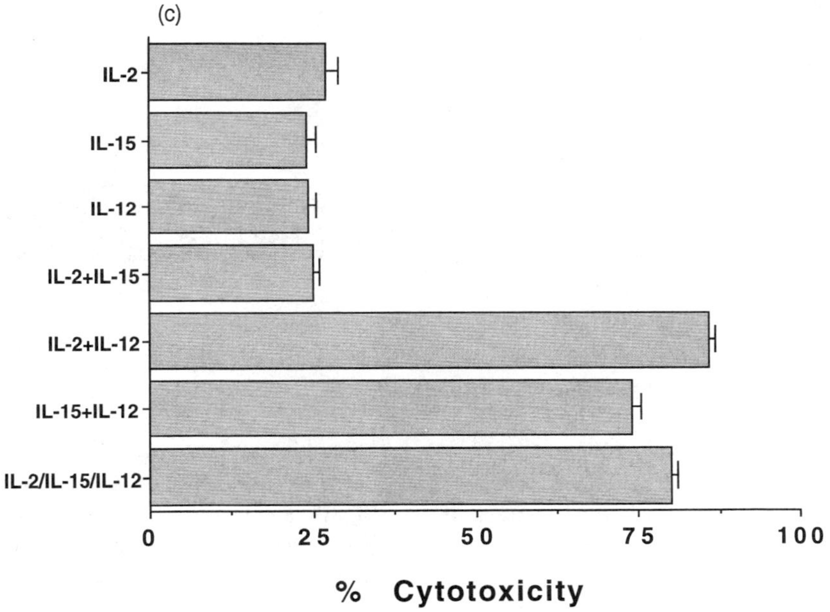

ceding the antigen-specific T-cell response (25). It has recently been shown in a number of in vitro and in vivo experimental systems that NK-cell production of IFN-γ is critical for the control of several intracellular pathogens (24,25). This phenomenon has been best examined by studying monocytes infected with *Listeria monocytogenes*, *Toxoplasma gondii*, and *Leishmania major* (reviewed in Ref. 25). In the case of infection with *L. monocytogenes* or *T. gondii*, the production of both TNF-α and IL-12 by infected monocytes has been shown to be critical in the induction of IFN-γ secretion by NK cells (26–28). This response was found to be markedly enhanced in the presence of exogenous IL-2, and at times was inefficient in the absence of IL-2 (26,29). While NK cells constitutively express a functional IL-2R (23), neither monocytes nor NK cells make IL-2, which suggests that either IL-2 produced by antigen-activated T cells or other cytokines (such as IL-15) may also be important for the production of IFN-γ by NK cells in vivo. Grabstein et al. have shown that fresh human PBMC express minimal IL-15 mRNA, while LPS-activated monocytes express abundant levels of IL-15 transcript (1). Neither resting nor activated $CD2^+$ lymphocytes (i.e., T cells and NK cells) show evidence of significant IL-15 transcript (1). We have investigated IL-15

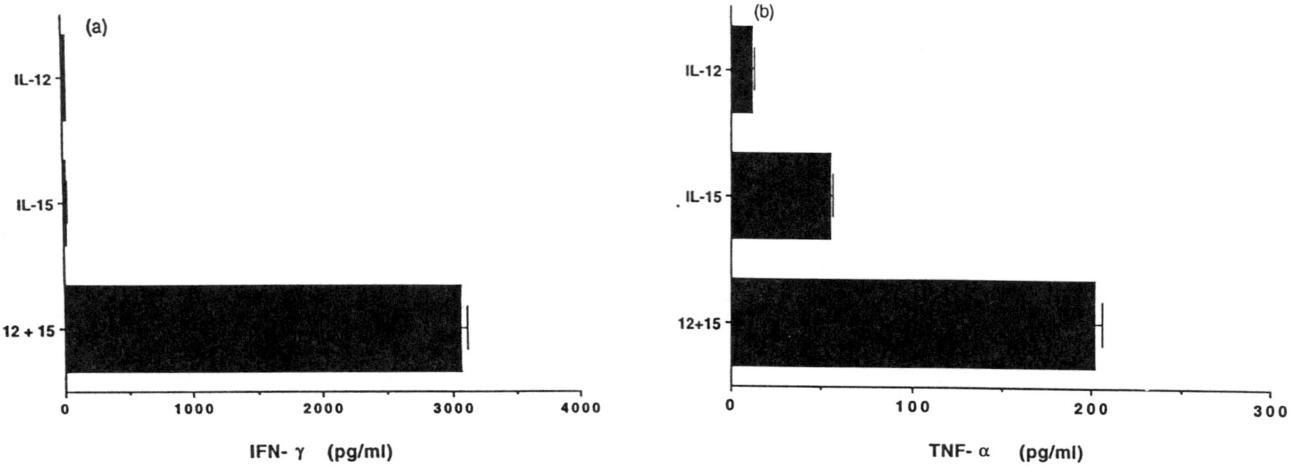

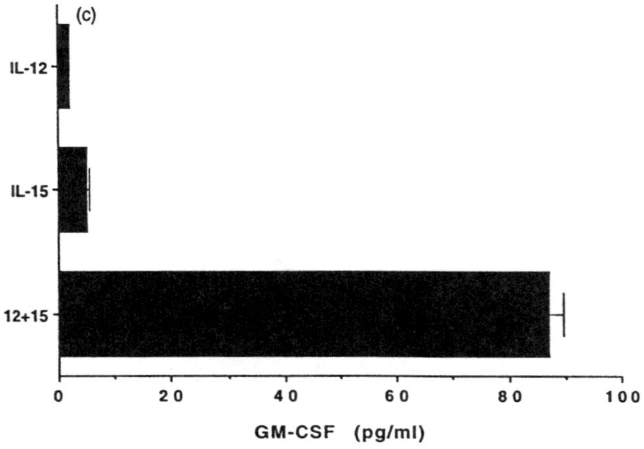

Figure 3 The combination of IL-15 and IL-12 stimulates $CD56^{dim}$ NK cell cytokine production. 1×10^5 $CD56^{dim}$ NK cells were plated in U-bottom wells in 200 μl of medium supplemented with IL-12 (10 U/m) and/or IL-15 (10 ng/ml). Cells were incubated for 72 h, at which time the culture supernatants were harvested and assayed for the presence of either IFN-γ, TNF-α, or GM-CSF using an ELISA. Cytokine production is measured in pg/ml. Results represent the mean of duplicate wells ± SE.

protein production in a variety of cell types using an antihuman IL-15 mAb. Analysis of human monocytes by immunohistochemical staining and Western blot analysis reveals IL-15 production within 5 h of LPS activation (3). Moreover, priming of monocytes with IFN-γ greatly enhances their ability to secrete IL-15 after stimulation with LPS (M. Marien and W. Carson, unpublished observations). Importantly, resting monocytes do not express IL-15 protein (3).

T cells and NK cells are the only known sources of IFN-γ (25), and studies with recombinant cytokines have shown that human NK cells require monocyte-derived products in order to produce IFN-γ (22). We therefore sought to determine if co-culture of LPS-activated monocytes and NK cells would result in IFN-γ production (3). Alone, neither monocytes nor NK cells produced detectable levels of IFN-γ in either the presence or absence of LPS (Fig. 4a). Co-culture of unstimulated monocytes with unstimulated NK cells did not induce IFN-γ production. However, when monocytes and NK cells were co-cultured in the presence of LPS, IFN-γ was easily detectable in culture supernatants at 72 h (Fig. 4a). Thus, LPS-activation of human monocytes induces IFN-γ production in NK cells. IFN-γ was detected in culture supernatants 18 h after the addition of LPS, peaked in concentration at 48 h, and underwent a significant decline by 96 h (3).

The kinetics and quantity of IL-15 production in this co-culture system were determined by ELISA and compared to IL-12 protein production in the same culture (3). Significant levels of IL-15 and IL-12 were produced by NK/mononcyte co-cultures within 24 h of LPS activation (Fig. 4b). We next investigated whether endogenous IL-15 production by LPS-activated human monocytes was contributing to IFN-γ secretion by human NK cells. Co-cultures of LPS-activated monocytes and NK cells were treated with neutralizing anti–IL-15 antiserum or a control antiserum. Culture in the presence of neutralizing anti–IL-15 antiserum resulted in a 77% decrease in IFN-γ production as compared to a control antiserum (Fig. 4c). IFN-γ production could also be partially inhibited in the presence of an anti–IL-2Rβ mAb (68% reduction), while an anti–IL-2–neutralizing antiserum had no significant effect (3). It is noteworthy that while endogenous IL-15 is required for optimal IFN-γ production, complete abrogation was never achieved with higher concentrations of anti–IL-15 antiserum or anti–IL-2Rβ mAb. This suggests that suboptimal amounts of IFN-γ can be produced by NK cells in the

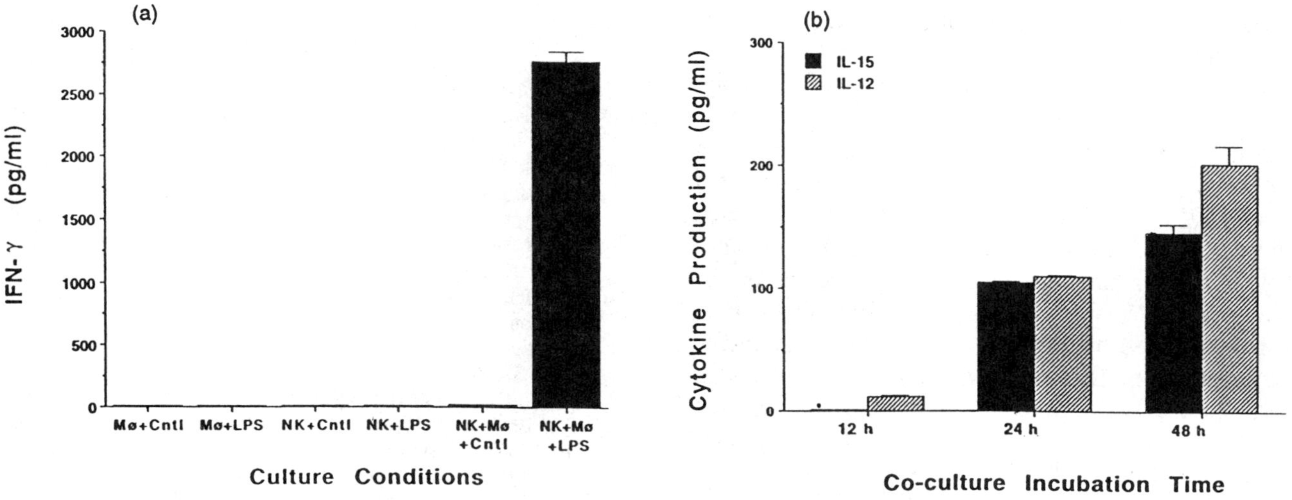

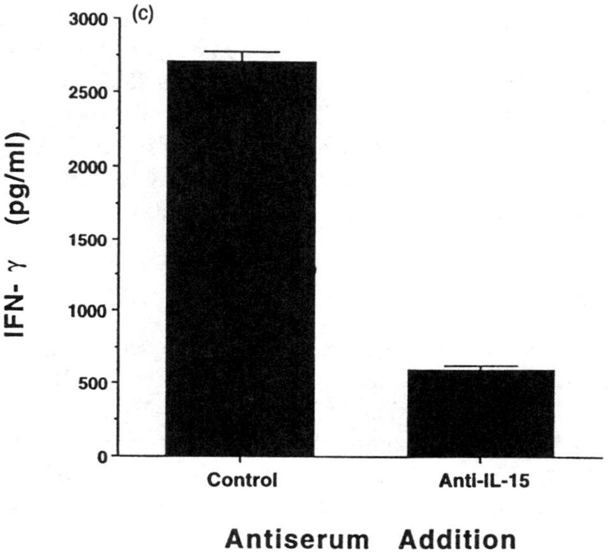

Figure 4 LPS stimulation induces IFN-γ production in co-cultures of human NK cells and monocytes. (a) NK cells and/or monocytes (Mø) were isolated from the fresh blood of normal donors and cultured at a density of 1×10^5 cells/well in 200 μl of 10% human AB serum, in the absence or presence of LPS (5 μg/ml). After 72 h of culture, supernatants from each culture condition were harvested and analyzed for IFN-γ protein production by ELISA. (b) Kinetics and quantitation of endogenous human IL-15 and IL-12 protein production. Supernatants from co-cultures of NK cells and LPS-activated monocytes were harvested at 12, 24, and 48 h and analyzed for human IL-15 and IL-12 production by ELISA (c). 1×10^5 NK cells and 1×10^5 monocytes were activated with LPS (5 μg/ml) in the presence of an anti–IL-15 antiserum or a control antiserum. After 72 h of culture, supernatants were harvested and analyzed for IFN-γ protein production by ELISA. Results represent the mean of duplicate wells ± SE.

presence of other monocyte-derived cytokines such as TNF-α and IL-12 (26–28). In fact, complete neutralization of IL-12 alone reproducibly abolished IFN-γ production in this co-culture system (3). This work suggests that IL-15 is an important monocyte-derived cytokine, which is required for optimal production of IFN-γ by human NK cells. As such, it is likely to have a significant role in the innate immune response to certain infections in vivo. Its production by activated monocytes provides additional insight into why NK cells constitutively express a functional IL-2R.

IL-15 IN DISEASE STATES

Gamero et al. have generated LAK cells from the PBMC of patients with malignant melanoma by culturing them in 10–100 ng/ml of IL-15 for 72 h (30). Specific lysis ranged from 5.6–29.6% when IL-15–generated LAK cells from 11 different patients were tested against a LAK-sensitive human melanoma cell line (FMEX) at an effector:target ratio of 25:1. It was also shown that LAK cells generated from patient PBMC were capable of mediating significant lysis of autologous melanoma tumor cells. Interestingly, IL-15 did not stimulate significant LAK activity from the PBMC of this patient population when used at concentrations below 10 ng/ml, and unstimulated PBMC exhibited no activity against FMEX or the standard LAK-sensitive cell lines K562 and Daudi. The LAK activity of these effectors could be inhibited in the presence of a blocking antibody to CD18 but not by antibodies to CD4/CD8 or MHC class I. This finding implies that the interaction of ICAM-1 on the target cell with CD11/CD18 on the surface of IL-15–induced LAK cells is critical to the cytolytic process. IL-15 was also capable of upregulating the transcription of the perforin gene in PBMC obtained from patients with malignant melanoma as well as in normal controls.

Seder et al. have examined the effects of IL-15 on PBMC and $CD4^+$ T cells obtained from patients infected with HIV (31). PBMC from HIV^+ patients were stimulated with polyclonal mitogens, tetanus toxoid, or HIV-specific antigens and then cultured with varying concentrations of IL-15. A dose-dependent response was observed, which could be inhibited by an anti–IL-2Rβ blocking antibody. Experiments with IL-2–neutralizing antibodies revealed that IL-15–induced proliferation was enhanced by the presence of endogenously-produced IL-2. Short-term cultures of PBMC or $CD4^+$ T cells did not respond to IL-15 or IL-2 with enhanced production of IL-2, IL-4, or IFN-γ. However, exogenous IL-12 induced significant production of both IL-2 and IFN-γ in these same cultures. IFN-γ production by mitogen-stimulated cultures was inhibited by neutralizing antibodies to IL-12 as well as IL-15. These experiments indicate that IL-15 may be an important component of the immune response in HIV-infected individuals and provide evidence that IL-15 may be useful as an immune modulator in this patient population.

CONCLUSIONS

IL-15 is a novel cytokine that exhibits no sequence homology with IL-2, yet is able to activate cells bearing functional IL-2R. The binding of IL-15 to the IL-2R appears to be mediated by the β and γ subunits. The newly discovered IL-15Rα chain is similar in structure to the α chain of the IL-2R but has extremely high affinity for its ligand and is expressed in a wide variety of tissues including several lymphoid cell lines. Thus far, it is believed that signaling in response to IL-15 involves activation of the same protein tyrosine kinases that have been implicated in IL-2 signaling. The actions of IL-15 on T cells, B cells, and NK cells are therefore very similar to those of IL-2, however, subtle differences do exist, which can be attributed to receptor density and composition. IL-15 is produced by activated monocytes, and its transcript is found in a variety of normal tissues including skeletal muscle, placenta, heart, lung, liver, and kidney. The wide expression of IL-15 and the IL-15Rα chain suggest that this cytokine may possess physiological functions not shared by IL-2.

REFERENCES

1. Grabstein KH, Eisenman J, Shanebeck K, Rauch C, Srinivasan S, Fung V, Beers C, Richardson J, Schoenborn MA, Ahdieh M, Johnson L, Alderson MR, Watson JD, Anderson DM, Giri JG. Cloning of a T cell growth factor that interacts with the β chain of the interleukin-2 receptor. Science 1994; 264:965–968.
2. Anderson DM, Johnson L, Glaccum MB, Copeland NG, Gilbert DJ, Jenkins NA, Valentine V, Kirstein MN, Shapiro DN, Morris SW, Grabstein K, Cosman D. Chromosomal assignment and genomic structure of IL-15. Genomics 1995; 25:701–706.
3. Carson WE, Ross ME, Baiocchi RA, Marien MJ, Boiani M, Grabstein K, Caligiuri MA. Endogenous production of interleukin 15 by activated human monocytes is critical for optimal production of interferon-γ by natural killer cells in vitro. J Clin Invest 1995; 96:2578–2582.
4. Giri JG, Anderson DM, Kumari S, Park LS, Grabstein K, Cosman D. IL-15, a novel T cell growth factor that shares activities and receptor components with IL-2. J Leukoc Biol 1995; 57:763–766.
5. Burton JD, Bamford RN, Peters C, Grant AJ, Kurys G, Goldman CK, Brennan J, Roessler E, Waldmann TA. A lymphokine, provisionally designated interleukin T and produced by a human adult T-cell leukemia line, stimulates T-cell proliferation and the induction of lymphokine-activated killer cells. Proc Natl Acad Sci USA 1994; 91:4935–4939.
6. Bamford RN, Grant AJ, Burton JD, Peters C, Kurys G, Goldman CK, Brennan J, Roessler E, Waldmann TA. The interleukin (IL) 2 receptor β chain is shared by IL-2 and a cytokine, provisionally designated IL-T, that stimulates T-cell proliferation and the induction of lymphokine-activated killer cells. Proc Natl Acad Sci USA 1994; 91:4940–4944.
7. Giri JG, Ahdieh M, Eisenman J, Shanebeck K, Grabstein K, Park LS, Cosman D, Anderson D. Utilization of the β and γ chains of the IL-2 receptor by the novel cytokine IL-15. EMBO J 1994; 13:2822–2830.
8. Hatakeyama M, Mori H, Doi T, Tanaguchi T. A restricted cytoplasmic region of the IL-2 receptor β chain is essential for growth signal transduction but not for ligand binding and internalization. Cell 1989; 59:837–845.
9. Giri JG, Kumaki S, Ahdieh M, Friend DJ, Loomis A, Shanebeck K, Dubose R, Cosman D, Park LS, Anderson DM. Identification and cloning of a novel IL-15 binding protein that is structurally related to the a chain of the IL-2 receptor. EMBO J 1995; 14:3654–3663.
10. Taniguchi T. Cytokine signaling through nonreceptor protein kinases. Science 1995; 268:251–255.
11. Darnell JE, Kerr IM, Stark GR. Jak-STAT pathways and transcriptional activation in response to IFNs and other extracellular signaling proteins. Science 1994; 264:1415–1421.
12. Lin J-X, Migone T-S, Tsang M, Freidmann M, Weatherbee JA, Zhou L, Yamauchi A, Bloom ET, Meitz J, John S, Leonard WJ. The role of shared receptor motifs and common Stat proteins in the generation of cytokine pleiotropy and redundancy by IL-2, IL-4, IL-7, IL-13, and IL-15. Immunity 1995; 2:331–339.
13. Macchi P, Villa A, Gillani S, Sacco MGA, Frattini A, Porta F, Ugazlo AG, Johnston JA, Candotti F, O'Shea JJ, Vezzoni P, Notarangelo LD. Mutations of Jak-3 gene in patients with autosomal severe combined immune deficiency (SCID). Nature 1995; 377:65–68.
14. Kennedy MK, Picha KS, Shanebeck KD, Anderson DM, Grabstein KH. Interleukin-12 regulates the proliferation of Th1, but not Th2 or Th0, clones. Eur J Immunol 1994; 24:2271–2278.
15. Wilkinson PC, Liew FY. Chemoattraction of human blood T lymphocytes by interleukin-15. J Exp Med 1995; 181:1255–1259.
16. Armitage RJ, Macduff BM, Eisenman J, Paxton R, Grabstein K. IL-15 has stimulatory activity for the induction of B cell proliferation and differentiation. J Immunol 1995; 154:483–490.
17. Armitage RJ, Macduff BM, Spriggs MK, Fanslow WC. Human B cell proliferation and Ig secretion induced by recombinant

CD40 ligand are modulated by soluble cytokines. J Immunol 1993; 150:3671–3680.

18. Kumaki S, Ochs HD, Timour M, Schooley K, Ahdieh M, Hill H, Sugamura K, Anderson D, Zhu Q, Cosman D, Giri J. Characterization of B-cell lines established from two X-linked severe combined immunodeficiency patients: Interleukin-15 binds to the B cells but is not internalized efficiently. Blood 1995; 86:1428–1436.
19. Matthews DJ, Clark PA, Herbert J, Morgan G, Armitage RJ, Kinnon C, Minty A, Grabstein K, Caput D, Ferrara P, Callard R. Function of the Interleukin-2 (IL-2) receptor g-chain in biologic responses of X-linked severe combined immunodeficient B cells to IL-2, IL-4, IL-13, and IL-15. Blood 1995; 85:38–42.
20. Smith KA. Lowest dose interleukin-2 immunotherapy. Blood 1993; 81:1414–1423.
21. Smith KA. Interleukin 2: inception, impact, and implications. Science 1988; 240:1169–1176.
22. Carson WE, Giri JG, Lindemann MJ, Linett ML, Ahdieh M, Anderson D, Eisenmann J, Grabstein K, Caligiuri MA. Interleukin-15 is a novel cytokine which activates human natural killer cells via components of the interleukin-2 receptor. J Exp Med 1994; 180:1395–1403.
23. Robertson RJ, Ritz J. Biology and clinical relevance of human natural killer cells. Blood 1990; 76:2421–2438.
24. Biron CA, Gazzinelli RT. Effects of IL-12 on immune responses to microbial infections: A key mediator in regulating disease outcome. Curr Opin Immunol 1995; 7:485–496.
25. Bancroft GJ. The role of natural killer cells in innate resistance to infection. Curr Opin Immunol 1993; 4:503–510.
26. Tripp CS, Wolf SF, Unanue ER. Interleukin 12 and tumor necrosis factor α are costimulators of interferon γ production by natural killer cells in severe combined immunodeficiency mice with listeriosis, and interleukin 10 is a physiologic antagonist. Proc Natl Acad Sci USA 1993; 90:3725–3729.
27. Gazzinelli RT, Hieny S, Wynn TA, Wolf S, Sher A. Interleukin 12 is required for the T-lymphocyte-independent induction of interferon γ by an intracellular parasite and induces resistance in T-cell-deficient hosts. Proc Natl Acad Sci USA 1993; 90:6115–6119.
28. Hunter CA, Subauste CS, Vancleave VH, Remington JS. Production of gamma interferon by natural killer cells from Toxoplasma gondii-infected SCID mice—regulation by interleukin-10, interleukin-12, and tumor necrosis factor alpha. Infect Immunity 1994; 62:2818–2824.
29. Sharma SD, Hofflin JM, Remington JS. In vivo recombinant interleukin-2 administration enhances survival against a lethal challenge with Toxoplasma gondii. J Immunol 1985; 135:4160–4163.
30. Gamero AM, Ussery D, Reintgen D, Puleo CA, Djeu JY. Interleukin 15 induction of lymphokine-activated killer cell function against autologous tumor cells in melanoma patient lymphocytes by a Cd18-dependent, perforin-related mechanism. Cancer Res 1995; 55:4988–4994.
31. Seder RA, Grabstein KH, Berzofshy JA, McDyer JF. Cytokine interactions in human immunodeficiency virus-infected individuals: roles of interleukin (IL)-2, IL-12, and IL-15. J Exp Med 1995; 182:1067–1078.

16

Tumor Necrosis Factor

Kevin J. Tracey
North Shore University Hospital and Picower Institute for Medical Research, Manhasset, New York

HISTORY

It was at the end of the nineteenth century that Dr. William Coley, a surgeon at New York's Memorial Hospital, first recognized the importance of spontaneous tumor regressions occurring in some of his surgically treated cancer patients who developed infections. These observations ultimately evolved into a search for a tumor-killing agent produced during the immunological response to the infection. Nearly 100 years later, tumor necrosis factor (TNF) was discovered in 1975 by Lloyd Old and his colleagues at the Memorial Sloan-Kettering Cancer Center and identified as a macrophage-derived factor capable of necrotizing tumors in mice (1). This discovery raised expectations that an immunological factor could be employed as a tumor-specific toxin in order to eliminate cancer, while leaving normal tissues unscathed. Literally across the street from Lloyd's laboratory in New York, Anthony Cerami and his colleagues at Rockefeller University were independently pursuing the identity of a macrophage-derived factor capable of causing the pathogenic manifestations of wasting that accompany chronic infections. Using a bioassay based on the inhibition of lipoprotein lipase, Bruce Beutler isolated and purified a 17 kDa, protein, which they termed "cachectin" (a term which, loosely translated from the Greek, means "I've got it bad") because of its putative role in causing the pathogenesis of infection-induced weight loss (2). It was soon realized that the amino-terminal sequence of murine cachectin was nearly homologous to the sequence of human TNF, and molecular cloning of cachectin and TNF confirmed their identity (3–7).

The identification of cachectin as TNF presaged the next newsworthy round in this history, stimulated by the discoveries that TNF was not selectively toxic to tumors, but rather it was sufficient to cause lethal shock and tissue injury in normal animals (8). The observations that an immunological factor, TNF, could kill normal host tissues raised doubt about the specificity of TNF against cancer. But it did provide hope for treating severe infection, in which time-tested therapies based solely on eliminating the bacteria failed to eradicate the high mortality rate. Proof of the concept that TNF is the pivotal mediator of lethality during infection came from the observation that application of monoclonal antibodies against TNF protected bacterially infected baboons from death, despite the presence of replicating bacteria in their bloodstream (9). Thus, for the first time it was realized that a macrophage product was both necessary and sufficient to cause all of the clinical manifestations of overwhelming infection. These and other studies prompted clinical trials in severely infected patients, with the hope that neutralizing a potentially pathogenic immune factor can improve survival results over and above antibiotics alone.

There has been a subsequent information explosion addressing the biology, activities, roles, and potential clinical uses of both TNF and anti-TNF. The tissue-injuring effects of TNF are now implicated in a spectrum of pathogenic inflammatory states, including cachexia, autoimmune diseases, organ transplant rejection, and infections. As reflected in this brief history and the remainder of this chapter, this cytokine can often be found near the center of the action in the host response to invasive threats, where it sometimes benefits, but at other times derails, the immunological defense strategy. Historical interest in TNF's contrasting activities is fueled even now by promising clinical results showing potential benefit from directly perfusing tumors with TNF to effect necrosis and from neutralizing the lethal toxicity of TNF produced during overwhelming infection. The volume of biological and molecular data contributed to the TNF literature by hundreds of laboratories cannot be comprehensively covered in a single chapter. Of necessity, therefore, this chapter is presented as an overview of these topics.

BIOCHEMISTRY

The gene for human TNF lies on the short arm of chromosome 6 near the HLA locus flanking the centromere. The coding regions, spanning approximately 3 kb of DNA, include four exons, but one of these (exon 4) accounts for more than 75% of the amino acid sequence encoded in the mature protein (6,7). The TNF gene family also includes lymphotoxin-α, the most closely related cytokine, which apparently arose by a duplication event. The other family members include ligands for CD27, CD30, CD40, OX40, 4-1BB, and Fas. For a time, lymphotoxin was also known as TNF-β, but recently two related forms (alpha and beta) of lymphotoxin have been identified. By consensus we now designate the three separate cytokines as TNF, lymphotoxin-α, and lymphotoxin-β. While these cytokines share some overlapping biological activities, important differences in the spectrum of biological activities have been identified, including distinct receptors and differential regulation of gene transcription and translation. The remainder of this chapter will focus on TNF (Table 1); the reader is referred to other chapters for descriptions of the lymphotoxins and other TNF family members.

Human TNF is initially translated as a prohormone composed of 233 amino acids, which is subsequently modified during posttranslational processing (4,7,10–12). Some of the translation product is degraded in the cell by an incompletely understood pathway (11,13). The prohormone translation product destined for excretion undergoes posttranslational modification by acylation with myristic acid (14). It has been suggested that this acylation step targets the nascent prohormone for insertion into membranes (14). This membrane-bound (26 kDa) form of TNF retains biological activity, because cytotoxicity to tumor cells mediated by cell-cell contact with macrophages expressing membrane TNF can be inhibited by neutralizing anti-TNF antibodies (15,16). During activation and induction of TNF secretion, the membrane form of TNF is cleaved by a serine protease, and three secreted monomers associate into a biologically active trimer (17). Recently developed TNF protease inhibitors inhibit the release of mature, secreted TNF from macrophages, indicating that this membrane-targeted secretory route is the predominate pathway for regulated TNF release (13).

Table 1 Biochemistry of hTNF

Size:	Initial translation product is 25 kD prohormone
	Processed by acylation with myristate and insertion into membrane
	Cleaved to secreted 17 kDa monomer (157 amino acids)
	Predominant biologically active form is trimer
Structure:	Monomer forms two antiparallel β-pleated sheets
	Three monomers associate by edge-to-face packing to form bell-shaped matrix
Gene:	Chromosome 6 flanking the centromere
	Juxtapositioned to the MHC
	Different allelic forms may mediate susceptibility to disease pathogenesis

When released from the cell surface membrane by proteolytic cleavage of the 26 kDa precursor, TNF is a 17 kDa monomer composed of 157 amino acids. Larger forms of the secreted protein have also been observed, suggesting that cleavage at additional sites can occur, but the 17 kDa form is believed to predominate in mammalian systems. Recent evidence suggests that the secreted 17 kDa monomer folds back onto itself, sandwichlike, into two β-pleated sheets (17,18). Three such sandwiched monomers form a bell-shaped trimer by association between the edges of one subunit and the face of an adjacent subunit (19). The exterior surface of the trimer is composed of polar residues, whereas the inner surfaces are lined by highly conserved hydrophobic domains. Critical sites for receptor signaling have been identified by studies of TNF muteins, peptide fragments, and antibodies directed against various sites of the TNF monomer (19,20). The base of the bell-shaped trimer binds to two or three membrane receptors. Although the N-terminus of the individual monomers can be exposed on the surface, it may not be required for receptor activation (19). Receptor activation presumably occurs by the clustering of two or three receptors around each TNF trimer, and this signaling can be prevented with antibodies that prevent recognition of the trimer by the receptors. Internalization of the ligand is not required for signal transduction.

PRINCIPAL CELL SOURCES AND BIOSYNTHESIS

Cells of the macrophage/monocyte lineage are the predominate source of TNF in vivo, but a number of other cell types are capable of producing biologically active TNF (Table 2). Recent in vivo studies suggest that much of the TNF appearing transiently in the bloodstream during acute

Table 2 Cell Sources and Factors That Stimulate TNF Production

Cell sources:	Macrophage/Monocytes (predominantly)
	Astrocytes
	Basophils
	Cardiac myocytes
	Eosinophils
	Langerhans cells
	Mast cells
	Microglial cells
	Keratinocytes
	Kupffer cells
	NK cells
	Polymorphonuclear leukocytes
	Tumor cells (e.g., epithelial, ovarian, glioblastoma multiforme, colorectal carcinoma)
Stimuli:	Bacterial endotoxin (LPS)
	Heat-killed bacteria
	Synthetic lipid A
	Toxin shock toxin-1
	Enterotoxin
	Mycobacterial cord factor
	TNF
	IL-1
	MIP-1α
	GM-CSF
	Phorbol esters
	Calcium ionophores
	Nitrites
	Hypoxia
	Oxygen radicals
	Irradiation
	Leukotrienes
	Mellitin
	C5a
	Fc receptor cross-linking
	Antibodies to LFA-3, CD44, CD45
	Viruses

endotoxemia can be accounted for by production and release from liver, presumably from macrophage-derived cells in residence there (21,22). Circulating monocytes, however, as well as fixed tissue macrophages found in other organs also contribute significant quantities of TNF during activation of the immune response to invasion. Detection of TNF in cultures of macrophages is commonly taken as evidence that the macrophage has been activated by exposure to a stimulating agent (23). Although TNF production has been studied most extensively in monocytes and macrophages, it is clear that the local production of TNF by nonhematopoietic cells can predominately influence the pathogenesis of a particular inflammatory state or disease. For instance, glial cells of the brain produce TNF (24–26), and the local effects of brain-derived TNF mediate a biological response that is quite distinct from TNF produced outside the central nervous system (27). TNF is also produced by cardiac myocytes (28), where its local effects can profoundly alter systemic physiology through altered myocardial cell function mediated via its local, paracrine activity (29). The local production of TNF in these two critical organs (heart and brain) serve to illustrate an important point: when considering the biology of TNF in vivo, the amount of TNF bioactivity present in local tissues, not serum, may greatly influence the systemic physiological response. Although the number of TNF-producing cell types is fairly large, the TNF gene has been silenced in a number of mammalian cells (30,31). TNF-secreting cells derived from human tumors have also been identified, where the biological effects of TNF have been observed to inhibit tumor cell growth in some cases, but to stimulate tumor cell growth in others (32).

The principal stimulus to TNF production in monocytes/macrophages is bacterial endotoxin (lipopolysaccharide, LPS) released from the cell wall of gram-negative bacteria. Although LPS is the most widely studied stimulating agent, TNF is also produced by exposure to a wide variety of other factors, both exogenous and endogenous. The exogenous group includes molecules associated with infectious pathogens such as enterotoxins, fungal wall proteins, viral antigens, and parasite components. Nonspecific chemical activators including phorbol esters and calcium ionophores are also effective triggers to TNF production. Endogenous proteins are also capable of eliciting TNF, and this list includes other cytokines (under some conditions MIP-1α, TNF itself, IL-1, and GM-CSF). TNF production can also be activated during the cellular response to severe stress induced by hypoxia, oxygen radicals, complement activation, and temperature shock. Thus, TNF production occurs as part of the conserved response to a wide range of cell stressors, presumably because it participates as a central mediator of the complex cytokine, hormonal, and cellular signaling cascade underlying the coordinated response to invasive or threatening stimuli.

The biosynthesis and release of TNF by activated cells is a carefully orchestrated and regulated cellular event. It has been teleologically reasoned that such tight-knit control is necessary in order to assure a timely mobilization of the TNF response to protect the host in times of need, but to simultaneously assure that toxic quantities of the potentially lethal cytokine are not produced indiscriminately. Minutes after exposure to LPS, human monocytes/macrophages obtained from peripheral blood release copious amounts of TNF. Even in the presence of persistent stimulation with LPS, however, monocyte production of newly synthesized TNF is extinguished within 3–4 h (33–36). After this robust response, monocytes become refractory or "tolerant" to repeated LPS exposure and fail to secrete TNF until after recovering for 3–7 days (11,37–39). The mechanism of the development of tolerance is not understood, but one possibility lies in the upregulation of glucocorticoid-binding sites on the cell membrane, which then inhibit subsequent TNF production by increasing cytoplasmic levels of IκB (40). The IL-1 receptor antagonist (IL-1ra) also prevents the development of LPS-induced tolerance, suggesting that IL-1 produced during the LPS response may mediate the tolerant state (42). Moreover, other anti-inflammatory cytokines (e.g., IL-10, IL-12, TGF-β), produced by macrophages during the activation response, may "feedback" on the macrophage, turning off TNF production and rendering it less responsive to subsequent LPS stimulation (43,44). Ultimately, a number of these signals may interact to shut down TNF production in a manner that is dependent upon the cellular milieu. For instance, monocytes cultured under serum-free conditions (in the absence of steroids) also shut off TNF production, suggesting that upregulation of the macrophage glucocorticoid receptor may have a more important role in blocking TNF production in vivo than in vitro. Consideration of the wide range of stimuli capable of eliciting TNF from macrophages and of the sudden downregulation of TNF that prevents its prolonged production suggests that mammals have evolved an integral cellular mechanism designed to spare the host from TNF toxicity.

Systemic TNF release occurs during exposure to LPS and other invasive stimuli. Serum TNF levels peak within 90–120 min after administration of intravenous LPS to mammals (from rodents to rabbits, dogs, pigs, cows, horses, baboons, monkeys, and humans) (9,45,46). Peak TNF levels stimulate the production of a cascade of other cytokines and hormones, including the glucose counterregulatory hormones epinephrine, glucagon, and glucocorticoids, and these hormones have been implicated in shutting off TNF production (47). Hypophysectomized and adrenalectomized animals, which fail to mount a circulating glucocorticoid response to LPS or TNF, develop progressive increases in LPS-stimulated circulating TNF, an observation that is cited as evidence that the "stress hormone response" helps to downmodulate the TNF response in a protective manner (48,49). At another level the cytokine MIF, which is also a pituitary hormone (50), can override the suppression of TNF by glucocorticoids, providing a mechanism enabling TNF production locally even in the face of systemic steroid excess. Interferon-γ, which is ubiquitous during the host inflammatory response, also overrides steroid-induced TNF suppression (51). Thus, the classical endocrine hormone system, and the immune-derived cytokine network, interact closely in vivo to regulate both the stimulation and suppression of TNF.

After a rapid distribution phase, circulating TNF is cleared from the circulation by binding to specific receptors in lungs, spleen, liver, skin, kidneys and other organs. A significant percentage is retained in the circulation by binding to proteolytic fragments of the TNF receptor (52–54). In addition to the quenching effect these receptor fragments have on neutralizing TNF, the TNF-binding proteins (TNF-BPs) can also serve as a reservoir of continued TNF release that extends TNF bioavailability in a "slow-release" mechanism (53–55). It is widely presumed that TNF has been conserved throughout mammalian evolution because of its biological activities that protect the host during infection, injury, and invasion, and that these protective actions are particularly effective in paracrine or local mechanisms. By this reasoning, very high TNF levels appearing in the bloodstream represent an unusual event in nature, associated only with extreme threat and or pathophysiological crisis (9).

Because of its biological significance, and its potential medicinal value as a drug target, the molecular regulation of monocyte TNF production has been studied in some detail. TNF production by activated macrophages is regulated at the levels of transcription, translation, and posttranslational processing. Transcription of the TNF gene increases on the order of approximately 20-fold in LPS activated monocytes (56). The promoter region is defined by a TATA box found 20 bp upstream from the transcription site; a "Y-box" control sequence is located 240 bp upstream from the TATA box (57). Experimental deletion of the κB enhancer and Y box from a promoter construct add to diminished reporter activity, suggesting that these sites can participate in the regulation of transcription (58–60). Short-lived repressor proteins have been implicated in preventing the transcription of TNF in quiescent, unstimulated macrophages (61). Once cells become activated, the transcription factor NF-κB translocates into the nucleus, where it binds to the enhancer sites in the TNF promoter to facilitate transcription (62). NF-κB appears to be necessary, but not sufficient to induce TNF production in human monocytes, because ultimately TNF protein production is also dependent upon the activities of other regulatory factors (e.g., kinases, proteases) downstream in the posttranscriptional processing stream (13,63–65). When two of the four NF-κB–binding sites are removed from an experimental promoter construct, the LPS sensitivity of the promoter is eliminated, suggesting that NF-κB is normally required for complete manifestation of the activation phenotype (66–68). The TNF promoter also appears to be under the control of other transcriptional elements including AP-1, AP-2, CREB, and a cyclosporine-responsive element (66,69).

Enhanced transcription alone does not account for the magnitude of the increase in TNF protein levels released by the activated macrophage. Time course studies of LPS-activated macrophages reveal that steady-state levels of TNF mRNA are induced 1 million-fold, remain elevated during the period of maximal cytokine synthesis (30 min–3 h), then decline during the subsequent 2 h as the macrophage becomes refractory or tolerant (70). High mRNA levels are accounted for in part by a stabilization of TNF mRNA resulting in slowed degradation (56). The factors that regulate mRNA stability remain uncertain, but one possibility is that a putative regulatory factor(s) interacts with the AU-rich consensus octamer present in the 3′-untranslated region of TNF and other cytokines (71,72). In agreement with this mechanism, the half-life of normally long-lived messages was shortened by insertion of an experimental construct bearing the AU-rich region (73). An additional level of control for TNF production, which enables the markedly increased levels of secreted TNF protein, occurs by enhancement of translational efficiency by 50-fold or more in LPS-activated macrophages (12,39,56,60). Both the 3′-untranslated regions and the 5′-promoter regions are required in order to produce this maximal stimulation of TNF protein production, perhaps as a result of a specific interaction between these structural regions (56,74,75).

A number of intracellular signals apparently participate in signaling altered gene expression in LPS-activated macrophages. It is clear that certain signaling molecules are proximal in the cascade of steps leading to mobilization of transcription factors and stimulation of TNF translation, but the identity of the cellular mechanisms controlling these individual regulatory steps is only partially understood. The LPS-induced signal pathways implicated in the mediation of transcriptional and posttranscriptional regulatory events of TNF biosynthesis include prostaglandins, leukotrienes, calcium, cyclic AMP, and protein tyrosine kinases. Nonspecific inhibitors of kinase activity prevent the biosynthesis of TNF (76). Evidence of the disparate regulation of transcription and translation is provided by the observations that the recently developed multivalent guanylhydrazone class of TNF inhibitors prevent TNF translation, but do not block the early translocation of NF-κB into the nucleus (33,70). The pyridinyl imidazole class of TNF suppressors specifically inhibits the mitogen-activated protein (MAP) kinase (p38 or CSBP) responsible for enhancing the translation of TNF (64,65). The glucocorticoid class of TNF inhibitors interferes more proximally in the LPS-signaling cascade by increasing cytoplasmic IκB and reducing NF-κB translocation (77). Studies of the regulation of TNF secretion using these and other classes of compounds that inhibit the transcription and/or translation of TNF (e.g., thalidomide, cyclosporine, pentoxifylline, and tyrophostins) verify that multiple independent but overlapping critical signals are necessary for competent upregulation of TNF (76,78–82).

Once the mature translation product is inserted into the membrane, TNF secretion is also regulated posttranslationally by the activity of a metalloproteinase that cleaves the prohormone from the membrane. The role of this enzyme activity has been defined in studies using an enzyme inhibitor to prevent macrophage TNF production (13). Although it is believed that this "TNF-converting enzyme" is active at the extracellular surface, it may also function intracellularly, perhaps acting upon nascent TNF following insertion into the Golgi apparatus or endoplasmic reticulum. Specific information about the cellular function and localization of the TNF-converting enzyme must await its eventual isolation, purification, and identification. In the meantime, the availability of these enzyme inhibitors has provided evidence that the majority of TNF produced during macrophage activation is released as secreted TNF, because they prevent the release of virtually all secreted TNF. Only a small fraction of intact 26 kDa TNF normally remains on the cell membrane of an activated macrophage, where it may retain biological activity in the mediation of cell-cell signaling (15,16).

In the final analysis, the multiple regulatory mechanisms underlying TNF production in humans are manifested as one of two broad categories of secretory phenotype. "High" or "low" human TNF producers have been identified in studies using peripheral blood mononuclear cells activated in response to LPS (83,84). The TNF gene contains a biallelic polymorphism within the 5′ DNA region that has been implicated in controlling transcription (58,67,84). The "TNF2" allele is strongly associated with the HLA-A1-B8-DR3-DQ2 haplotype (85), and HLA-DR3 and -DR4 have been correlated to relatively high rates of TNF production as compared to lower rates of TNF production by HLA-DR2 (83,84,86). It has been suggested that polymorphism in the TNF locus might therefore contribute to susceptibility or severity of diseases of autoimmunity, infection, inflammation, or malignancy (85–90). By this analysis, subjects homozygous for the TNF2 allele may be prone to overproduce TNF, and accordingly more apt to suffer the consequences of its toxicity during the response to an invasive illness. In agreement with this logic, a strong association between TNF2 homozygosity and death from cerebral malaria has been reported (91). Since the frequency of homozygosity for TNF2 was found to be 0.16 in that study, it is possible that this allele also confers some unknown biological advantage, perhaps against intracellular infections where TNF is known to confer protection, or perhaps by balancing the potentially deleterious effects of this allele during cerebral malaria. Since linkage disequilibrium frequently occurs in the MHC locus (where the TNF gene resides), an analysis of the role of allelic variants to disease phenotypes must consider the possibility that some other unrelated haplotype actually underlies the disease manifestation. In the malaria study (91), no other marker within the MHC was found to correlate with morbidity, suggesting that an overproduction of TNF was causal. It is anticipated that ongoing studies of TNF allelic differences, and the role of "high" and "low" responses, in a number of inflammatory, infectious, and autoimmune diseases will be enlightening to our understanding of the role of TNF in disease cure and pathology.

RECEPTORS

TNF mediates its effects in responding cells by interaction with two related membrane receptors termed TNF-R1 and TNF-R2 (or type I and type II). The molecular mass of the human receptor proteins is approximately 60 and 80 kDa, and of the murine forms approximately 55 and 75 kDa. The extracellular ligand-binding domains contain highly conserved areas between the two receptor types, but the cytoplasmic signaling components are distinct (92–94). Signaling occurs when two or three receptors commingle around a single TNF trimer (95,96). Neutralizing anti-TNF antibodies presumably interfere with this arrangement, whereas antibodies with TNF agonist activity can stimulate receptor signaling by bringing two or more receptors into close proximity (97,98). Two proteins that interact with the cytoplasmic signaling portion of the TNF receptors (TNF-receptor–associated protein, TRAF-1 and TRAF-2) have been identified and implicated as early signal-transducing proteins (99). The result of TNF receptor activation is the activation of secondary signal cascades involving prostaglandins, leukotrienes, calcium, ceramides, and tyrosine kinases. Different cell types manifest different biochemical responses to TNF, but in some immunocytes the result of this signal transduction is translocation of the transcription factor NF-κB into the nucleus (97,100), where it promotes the transcription of other proinflammatory cytokines. Much less is known about the cellular regulation of the putative enzyme responsible for cleaving the TNF receptors into soluble TNF-BPs (101,102), but serum levels of TNF-BPs are detected in normal subjects and become significantly increased during the mammalian response to endotoxin and other invasive stimuli (103–105). Since TNF-BPs are produced during activation of the cytokine network, they essentially function as cytokines themselves, whose presently defined role is best understood in terms of downmodulating the response to TNF.

BIOLOGICAL ACTIVITIES

TNF Effects in Models of Human Diseases

Acutely Lethal Shock and Tissue Injury

The demonstration that highly purified recombinant human TNF is sufficient to mediate all of the clinical manifestations of overwhelming infection, including lethal shock and tissue injury, provided the first evidence that a macrophage-derived factor is sufficient to trigger the pathophysiology of an infectious disease (8). These observations have subsequently been confirmed and extended to include larger mammals such as rabbit, dog, pig, monkey, baboon, horse, and cow (47,106–110). There is a clear species dependence for the toxic activity of rh-TNF in the mediation of shock and tissue injury, but clinical studies employing low doses of systemically administered TNF in humans show the same spectrum of potentially pathological effects including, fever, myalgia, headache, hypotension, vascular leakage syndrome, and accelerated protein degradation (111–113). Although the sudden release of toxic quantities of TNF into the circulation is decidedly unusual in nature, it occurs frequently in humans with overwhelming infection, trauma, burn injury, meningitis, and infections. It is therefore of interest to assess the organ-specific TNF toxicity in a variety of systems (Table 3).

Within minutes of acute elevation of serum TNF into the range observed during septic shock or infection, blood pressure begins to fall into the "shock" range (8,47). The cellular basis for the development of shock occurs in part from a depression of myocardial function associated with decreased contractility, decreased left ventricular ejection fraction, and diminished cardiac filling pressures (109,114). The myocardial suppressive effects may not be solely dependent upon circulating TNF, because cardiac myocytes also produce TNF in response to LPS, causing depressed contractility through direct negative inotropic actions of TNF (28,115). TNF-mediated cardiovascular depression is important in determining lethality from systemic or cardiac-mediated TNF toxicity, because neutralization of TNF cardiac toxicity and cardiovascular derangements with monoclonal antibodies or TNF-BPs prevents the onset of shock (9,29,116–118).

The vascular system itself is also sensitive to the toxic effects of acutely elevated TNF. This toxicity is manifested by the generalized vasodilation of resistance blood vessels due to the activity of endothelium-derived relaxing factor (EDRF) (119). The molecular identity of EDRF is nitric

Table 3 Effects of Acutely Elevated Serum TNF on Organ System Function

Cardiovascular:	Shock and decreased organ perfusion
	Depression of cardiac contractility
	Vascular leakage syndrome
	Systemic inflammatory response syndrome
	Lactic acidosis
Pulmonary:	Hypoxia
	Pulmonary edema
	Leukocyte sequestration
	Adult respiratory distress syndrome
Nervous:	Fever
	Anorexia
	Activation of hypothalamic-pituitary stress response (ACTH, GH, prolactin, MIF)
Endocrine:	Release of glucose counterregulatory hormones (epinephrine, norepinephrine, glucagon, glucocorticoids)
	Transient hyperglycemia followed by hypoglycemia
	Hyperaminoacidemia
Renal:	Hemorrhagic tubular necrosis
	Proteinuria
	Suppression of creatinine clearance
Musculoskeletal:	Myalgia
	Suppression of resting transmembrane potentials
	Stimulation of sodium uptake
	Increased proteolysis
	Futile cycling of glucose metabolism
Gastrointestinal and hepatic:	Hemorrhagic bowel necrosis
	Diarrhea
	Depressed motility and absorption
	Enhanced acute-phase protein secretion
	Suppression of albumin biosynthesis

oxide (NO), a short-lived biological mediator produced enzymatically by nitric oxide synthase (NOS). Macrophages, cardiac myocytes, endothelial cells, and other cells are induced to produce NO when stimulated by LPS, TNF, and other cytokines (120–123). Endothelial and myocardial toxicity from acutely elevated TNF levels can be mediated by the generation of toxic levels of NO and by direct cytotoxicity from TNF itself. Addition of TNF to cultured endothelium results in reorganization of cellular structural filaments and separation of the individual cells, and this mechanism may underlie the development of albumin and plasma leakage from the capillary beds during the response to sepsis and TNF toxicity (124–126). Endothelial cells are also activated by TNF to express adherence proteins on their cell surface, which directly mediate localization and recruitment of leukocytes to the area (127–129). TNF activation of adherent leukocytes can trigger the release of superoxide radicals, proteolytic enzymes, and NO, which then injure or disrupt endothelial integrity (130). Evidence that TNF toxicity to capillaries and endothelium contributes significantly to the development of systemic inflammatory response syndrome (SIRS) was also provided with the demonstration that monoclonal anti-TNF antibodies prevent the development of dehydration and vascular leakage (9).

Lethal quantities of TNF cause injury to multiple organs including the lungs, kidneys, bowel, and adrenal (8,9,47,106). The pathology of this multiple organ-failure syndrome has been attributed to the direct toxicity of TNF to normal tissues and to the effects of diminished blood flow mediated by TNF-induced shock. Examination of the lungs from TNF-infused animals reveals leukocyte sequestration, capillary leakage, activation of endothelium, and thrombosis of vasculature. Animals succumbing to acute TNF toxicity develop profound hypoxia and elevated airway resistance pressures similar to the human condition of adult respiratory distress syndrome (ARDS), a catastrophic complication of septic shock or other critical illnesses (e.g., pancreatitis). These observations have been extended by the development of animal models of ARDS based on administration of TNF (8,107,131–133).

Studies of acute TNF toxicity in normal mammals provide compelling evidence that this cytokine is sufficient to induce the manifestations of septic shock syndrome. To date TNF remains the only endogenous mediator yet identified capable of eliciting all the clinical manifestations of septic shock syndrome when administered in physiologically relevant quantities to normal mammals (134). It is likely that other mediators will be identified in this capacity, but the concept that a host product, and not an exogenous toxin, might activate lethal shock and tissue injury served to focus investigative efforts on methods to inactivate this toxicity in human disease.

Evidence that endogenously produced TNF is necessary for the clinical manifestations of septic shock syndrome was obtained by neutralizing TNF with monoclonal anti-TNF antibodies during overwhelming bacterial infection (9). Despite the presence of replicating bacteria in their bloodstream, baboons receiving neutralizing anti-TNF therapy were protected against all the manifestations of TNF toxicity including shock, tissue injury, hypoxia, renal failure, and vascular leakage syndrome (9), even when antibodies were administered after the onset of the infection (118). The protective effects of anti-TNF therapy have also been demonstrated in other models of shock induced by administration of very large quantities of LPS

(116,117). TNF-BPs and constructs of TNF-BP fusion proteins designed to have longer circulating half-lives are also effective in preventing the toxicity of TNF (55, 135,136). Elegant proof that TNF toxicity is crucial in the development of shock and tissue injury was recently obtained from transgenic mice rendered incapable of responding to TNF by "knock-out" of the TNF receptor (137). As predicted, these animals were protected against the lethality of exogenous LPS (137). Thus, TNF is necessary for development of shock and tissue injury during overwhelming infection with live bacteria or lethal doses of LPS.

Cytokine Cascade During TNF Toxicity

Although TNF is the only endogenous mediator to date that is both sufficient and necessary for the development of shock and lethal organ failure during overwhelming infection and septic shock, these diverse pathological effects are mediated by a cascade of secondary mediators activated by TNF. Evidence of the pivotal role of TNF in stimulating the secondary cytokine cascade during overwhelming infection was obtained in lethally infected baboons receiving anti-TNF antibodies (138). Neutralization of TNF prevented the development of secondary cytokine cascades with reductions in circulating levels of both IL-1 and IL-6 (138). This observation provides some insight into the mechanisms of TNF toxicity, because IL-1 (and other cytokines, e.g., IFN-γ) synergistically increase the toxicity of TNF to normal mammals (139,140). The list of secondary mediators induced by TNF as part of the "cytokine network" response to invasion or infection is expansive and continues to grow (Table 4). The scope of their associated biological activities accounts significantly for the generalized toxicity of TNF and enables this toxicity to be amplified and extended in a self-propagating cascade.

Table 4 Some Secondary Mediators Induced by Systemically Administered TNF

Cytokines:	IL-1
	IL-2
	IL-4
	IL-6
	IL-10
	IL-12
	IFN-γ
	TGF-β
	LIF
	MIF
Small molecules:	NO
	Oxygen free radicals
	Prostaglandins
	Leukotrienes
	Platelet-activating factor
Hormones:	Cortisol
	Epinephrine
	Norepinephrine
	Insulin
	Glucagon
Others:	TNF-BP
	IL-1ra
	Acute-phase proteins

Chronic TNF Toxicity

The physiological effects of chronic, repeated dosing of highly purified recombinant human TNF were first shown to cause anorexia, weight loss, and anemia (141). These manifestations of chronic TNF toxicity differed markedly from the dramatic effects of shock and tissue injury mediated by acute TNF toxicity. The pathophysiology of chronic TNF toxicity mediated the onset of cachexia, a state of wasting, weakness, and inanition that frequently complicates a number of chronic illnesses including cancer, AIDS, and chronic inflammatory diseases. The hallmark of the cachectic state is the persistent catabolism by the host of endogenous protein stores leading to the mobilization of skeletal muscle proteins and the accelerated excretion of whole-body nitrogen as urea (142,143). The persistent loss of nitrogen underlies the inevitable development of generalized weakness, immunosuppression, and respiratory failure.

A number of studies have confirmed that TNF is sufficient to cause cachexia in normal animals. TNF exposure causes increased resting energy expenditure, increased urinary urea nitrogen excretion, and depletion of carcass protein stores (27,141,142,144–149). As in the acute toxicity of TNF, some of these responses represent the direct effects of TNF receptor-ligand interaction, but others represent the activities of secondary mediators (Table 3), whose release is activated by TNF (150). A number of other cytokines that are activated by TNF are also sufficient to induce these parameters of the cachectic state in vivo including IL-1, IL-6, IFN-γ, and LIF. The activity of TNF in the central nervous system contributes to the mediation of the chronic cachectic state, because it can directly increase thermogenesis and suppress appetite by interacting with hypothalamic nuclei (151–154). Another important metabolic component of the cachectic state is the development of insulin resistance, and animals receiving exogenous TNF develop this metabolic defect as well (155). The molecular basis for TNF-induced insulin resistance remains largely unclarified, but insulin resistance has also been implicated in the development of severe obesity (156). This metabolic effect may underlie the observed association between increased TNF biosynthesis rates in adipocytes from genetically obese animals, who have a concomitant defect in their ability to suppress appetite (157). Thus, insulin resistance in the setting of diminished food intake occurring in the anorectic subject would predictably worsen the severity of nitrogen loss, but insulin resistance in the setting of excessive nutrient intake would predictably increase fat deposition.

Studies of the biological effects of chronic TNF exposure also provided dramatic evidence that the biological effects of TNF produced in vivo were dependent not solely on the levels circulating in blood, but also on the source of tissue-derived TNF. For instance, nude mice bearing genetically engineered tumors that constitutively secrete TNF after implantation into skeletal muscle develop profound cachexia characterized by weight loss and depletion of whole-body protein (27,144). But when the identical tumors were implanted stereotactically into the brain of nude mice, an entirely different metabolic effect developed (27). Brain tumor-bearing animals developed profound anorexia and grew hair, but had a relative sparing of whole-body nitrogen (27). This constellation of metabolic responses occurred only in the brain tumor group, despite the fact that serum TNF levels were comparable in both the hind limb and brain tumor animals. The high brain tissue levels of TNF in the brain tumor–bearing animals acted locally to trigger a distinct set of secondary signals, which did not occur in the animals with peripheral tumors. These observations must be kept in mind during an assessment of the biological activities of TNF in animal models, and when interpreting results of TNF levels in various clinical samples from different body compartments (e.g., bronchial lavage, joint fluid, and cerebrospinal fluid).

TNF in the Mediation of Human Disease

As noted in the introduction of this chapter, the volume of literature on the role of TNF as a disease mediator has swollen to the point that each of 12 or more separate diseases might warrant a complete chapter of its own. Herein is an abridged summary of the role of TNF in several widely studied diseases (see Table 5).

Table 5 TNF in Human Diseases

Septic shock and SIRS	Necessary and sufficient mediator in animal models of overwhelming infection (8,9) Intermittent serum TNF increases in septic shock patients (202)
Cancer cachexia:	Elevated production rates by PBMC of cachectic cancer patients (203,204) Sufficient mediator in animal models (27,141,144)
Rheumatoid arthritis:	Produced in inflamed joints (161–163) Sufficient to mediate experimental arthritis in animals (205)
AIDS:	Sufficient to activate latently infected virus (171,173,206) Sufficient mediator of cachexia in animal models (27,141,144)
Multiple sclerosis:	Produced in MS plaques (26) Necessary mediator in animal models (178)
Diabetes:	Enhanced production rates in diabetes (183) Cytotoxicity to islet cells (182)
Malaria:	Production correlates to disease severity (91,184,185) Necessary mediator in animal models (186)
Meningitis:	Production correlates to mortality (187,188)
ARDS:	Sufficient mediator in animal models (8,107,131–133) Correlation of TNF production to severity of disease (190)
Obesity:	Mediator of insulin resistance in animal models (155) Production increased in genetically obese animals (157)
Trauma:	Produced intermittently following burn injury (202) Necessary mediator of ischemia-reperfusion injury (124,207,208)
Asthma and chronic obstructive pulmonary disease:	TNF transgene develops pulmonary fibrosis (209) Increases airway responsiveness and sputum production (210)

Septic Shock and SIRS

Although one must be cautious when interpreting the biological meaning of transiently elevated serum TNF levels in septic shock without knowing simultaneously the levels in tissues for TNF and other cytokines (outlined above), there is now abundant evidence that TNF fulfills Koch's essential postulates as a causative, necessary, and sufficient mediator of septic shock syndrome during overwhelming infection (8,9). The results of recent clinical trials of passively inhibiting TNF with monoclonal antibodies showed improved survival in patients with severe septic shock (158). As predicted from the relatively short half-life of the antibody therapy (2–3 days), the maximal protective effects were observed within 24 h of receiving therapy. No additional antibody therapy was given, and the mortality curves for the remaining 30-day study are parallel, presumably reflecting the mortality attributable to other underlying disease. The definitive phase III trial of this therapeutic strategy is underway. Although animal studies suggest that TNF can contribute to the disease pathogenesis underlying SIRS (159), there is less available information from humans about the causal role of this cytokine.

Cachexia

Although TNF is capable of eliciting the metabolic manifestations of the cachexia syndrome in animals and humans and is detected in patients with cancer cachexia, demonstration of a causal role for TNF in this disease has been hampered by the finding that TNF is not unique in its ability to induce cachexia. Nonetheless, the results of animal studies suggest that a strong case can be made for neutralizing chronic TNF toxicity in cancer. TNF production is increased in macrophages derived from tumor bearing animals (150), and chronic exposure to elevated TNF levels mediates loss of skeletal muscle protein, anorexia, weight loss, anemia, and increased urinary nitrogen excretion (27,141,142,144,160).

Rheumatoid Arthritis

Local production of TNF in the inflamed joint has been implicated in the mediation of rheumatoid arthritis (161–164). The proinflammatory effects of TNF in the development of arthritis include the stimulation of other cytokines, NO, and superoxide radicals. TNF activation of collagenase and other proteases can contribute to the development of joint remodeling and tissue destruction (165–167). The results of a recent clinical trial indicate that anti-TNF antibodies can effectively inhibit the clinical score reflecting joint involvement and the levels of circulating acute phase proteins, giving an additional objective assay of antibody-mediated suppression of the inflammatory state. The pivotal causal role of TNF in this inflammatory disease was suggested by the prolonged period of remission (4–6 weeks) following antibody therapy.

AIDS

TNF has been implicated in the pathogenesis of AIDS cachexia, the activation of latent HIV infection, and the manifestations of organ toxicity in the HIV-infected patient. Studies of TNF production in monocytes from HIV-infected patients have found both enhanced and suppressed rates of production. Several groups have identified increased TNF production as a prognostic marker for disease severity (168–170). Addition of TNF to latently infected cells stimulates the activation of HIV viral expression through NF-κB translocation into the nucleus (171–173). The proinflammatory sequelae of TNF produced during the host response to secondary infections has also been implicated in the development of tissue injury to brain and lung (174–176). Evidence that the cytokine profile can influence the progression of HIV infection has further stimulated interest in developing therapeutics that modulate cytokine production (177).

Multiple Sclerosis

TNF has been implicated as a mediator of brain injury during acute exacerbations of MS, and anti-TNF antibodies suppress this injury (178). The biological basis of this deleterious effect has been attributed to the direct toxicity of TNF to oligodendrocytes and nervous tissue (179,180). TNF has been demonstrated in human MS plaques and in the serum and spinal fluid of patients during acute exacerbations (26,181). Small molecule inhibitors of TNF including glucocorticoids, pentoxifylline, and rolipram have been effective in preventing the onset of inflammation in animal models of EAE.

Diabetes

Along with IL-1 and NO, TNF has been implicated in the development of macrophage-dependent cytotoxicity of islets in murine models of insulinitis and diabetes (182). As discussed above, TNF administration induces a state of insulin resistance that is similar to the metabolic effects of type II diabetes in humans (155). An association between TNF production and the development of diabetes has also been described (183), raising important questions about the potential causality of this association.

Malaria

Early studies revealed a close correlation between serum TNF levels and outcome from malarial infection, and more recent data now indicate that the explanation for this correlation may have a basis in an underlying genetic predisposition of some individuals to produce larger quantities of TNF during infection (91,184,185). The principal mechanism of TNF-mediated lethality is through increased susceptibility to the complications of cerebral malaria, and anti-TNF antibodies prevent these cerebral complications in animal models (186).

Meningitis

Mortality from cerebral meningitis correlates to the levels of TNF detected in the serum and cerebrospinal fluid compartments (187,188). Animal studies indicate that TNF is involved in the development of the inflammatory response to intracerebral infection, but the proinflammatory mechanisms of TNF in this disease are incompletely characterized (189). By extrapolation from other examples of TNF-mediated tissue injury (8,27), it is likely that excessive local production of TNF in brain can injure tissue through activation of endothelium, leukocytes, and glial cells, by its direct cytotoxic effects on brain parenchyma (179), and by stimulating release of secondary mediators (NO, IL-1, IL-6, MIP-1α, and MIP-1β).

ARDS

TNF levels correlate to the severity of ARDS in clinical samples obtained from the circulation and bronchial washing of ARDS patients (190,191). When administered to animals via the systemic or intratracheal routes, TNF causes the pathological and physiological sequelae of ARDS (8,107,131–133). Antibodies against TNF prevent the manifestations of ARDS in baboons with overwhelming bacterial infection, indicating the potential for therapy.

FUTURE DIRECTIONS

Identification of TNF as a pathological mediator prompted the development of a number of experimental strategies to neutralize its undesired toxicity on an acute, chronic, and local basis. Ongoing clinical trials of anti-TNF therapies have already showed limited success in the treatment of septic shock, rheumatoid arthritis, and inflammatory bowel disease (158). There are a number of classes of small-molecule TNF inhibitors under development as well, including the multivalent guanylhydrazones (33,70), pyridinyl imidazoles (65), pentoxifylline (81), thalidomide (192), and rolipram. It is plausible that from a continued improvement in our understanding of the regulation of TNF biosynthesis at the molecular and tissue levels, these or other classes of therapeutics may be applied to the diseased host.

Recent advances in the local application of recombinant TNF to isolated cancer-bearing organs or extremities have met with success in causing tumor regressions (193–195), bringing back full-circle the story of applying endogenous protein to treat cancer. It is hoped that some cancers may yet be treated by applying TNF directly to the tumor, while simultaneously eliminating the systemic toxicity through either filtration of the effluent or pharmacological reduction of systemic TNF toxicity.

From the earliest studies of TNF toxicity in normal animals, our understanding of the biological potentially injurious as well as beneficial activities of this cytokine have expanded significantly. As for many hormones or drugs, some dose or quantity is always toxic to the host. The conservation of the TNF gene during evolution was clearly not achieved because of its toxicity, any more than insulin was conserved for its ability to cause lethal hypoglycemia. The beneficial effects of local, immunocyte-derived TNF in wound healing, controlled macrophage activation, developmental apoptosis, and an organized defense against intracellular organisms or other invaders cannot be overstated (102,196–200). It is plausible that from a better understanding of the protective role of TNF in host defense will come a strategy to improve or strengthen these defense mechanisms. As such, the principles for future development of TNF in clinical medicine have changed little: to eliminate undesired toxicity in some cases and to enhance its beneficial activities in others (201).

REFERENCES

1. Carswell EA, Old LJ, Kassel RL, Green S, Fiore N, Williamson B. An endotoxin-induced serum factor that causes necrosis of tumors. Proc Natl Acad Sci USA 1975; 72:3666–3670.
2. Beutler B, Mahoney J, Le Trang N, Pekala P, Cerami A. Purification of cachectin, a lipoprotein lipase-suppressing hormone secreted by endotoxin-induced RAW 264.7 cells. J Exp Med 1985; 161:984–995.
3. Beutler B, Greenwald D, Hulmes JD, et al. Identity of tumour necrosis factor and the macrophage-secreted factor cachectin. Nature 1985; 316:552–554.
4. Wang AM, Creasey AA, Ladner MB, et al. Molecular cloning of the complementary DNA for human tumor necrosis factor. Science 1985; 228:149–154.
5. Davis JM, Narachi MA, Alton K, Arakawa T. Structure of human tumor necrosis factor-a derived from recombinant DNA. Biochemistry 1987; 26:1322–1326.
6. Spies T, Blanck G, Bresnahan M, Sands J, Strominger JL. A new cluster of genes within the human major histocompatibility complex. Science 1989; 243:214–217.
7. Nedwin GE, Naylor SL, Sakaguch AY, et al. Human Lymphotoxin and TNF α genes: Structure, homology and chromosomal localization. Nucleic Acids Res 1985; 13:6361–6373.
8. Tracey KJ, Beutler B, Lowry SF, et al. Shock and tissue injury induced by recombinant human cachectin. Science 1986; 234:470–474.
9. Tracey KJ, Fong Y, Hesse DG, et al. Anti-cachectin/TNF monoclonal antibodies prevent septic shock during lethal bacteraemia. Nature 1987; 330:662–664.
10. Pennica D, Nedwin GE, Hayflick JS, et al. Human tumor necrosis factor: precursor structure, expression, and homology to lymphotoxin. Nature 1984; 312:724–729.
11. Jue D-M, Sherry B, Luedke C, Manogue KR, Cerami A. Processing of newly synthesized cachectin/tumor necrosis factor in endotoxin-stimulated macrophages. Biochemistry 1990; 29:8371–8377.
12. Beutler B, Krochin N, Milsark IW, Luedke C, Cerami A. Control of cachectin (tumor necrosis factor) synthesis: mechanisms of endotoxin resistance. Science 1986; 232:977–980.
13. McGeehan GM, Becherer JD, Bast Jr RC, et al. Regulation of tumor necrosis factor-α processing by a metalloproteinase inhibitor. Nature 1994; 370:558–561.
14. Stevenson FT, Burstein SL, Locksley RM, Lovett DH. Myristyl acylation of the TNF a precursor on specific lysine residues. J Exp Med 1992; 176:1053–1062.
15. Kriegler M, Perez C, DeFay K, Albert I, Lu SD. A novel form of TNF/cachectin is a cell surface cytotoxic transmembrane protein: ramifications for the complex physiology of TNF. Cell 1988; 53:45–53.
16. Perez C, Albert I, DeFay K, Zachariades N, Gooding L, Kriegler M. A nonsecretable cell surface mutant of tumor necrosis factor (TNF) kills by cell-to-cell contact. Cell 1990; 63:251–258.

17. Smith RA, Baglioni C. The active form of tumor necrosis factor is a trimer. J Biol Chem 1987; 262:6951–6954.
18. Jones EY, Stuart DI, Walker NP. Structure of tumour necrosis factor. Nature 1989; 338:225–228.
19. Sprang SR, Eck MJ. The 3-D structure of TNF. In: Beutler B, ed. Tumor Necrosis Factor: The Molecules and Their Emerging Role in Medicine. New York: Raven Press, 1992:11–32.
20. Lin LS. The design and activities of TNF muteins. In: Beutler B, ed. Tumor Necrosis Factor: The Molecules and Their Emerging Role in Medicine. New York: Raven Press, 1992:33–48.
21. Ciancio MJ, Hunt J, Jones SB, Filkins JP. Comparative and interactive in vivo effects of tumor necrosis factor alpha and endotoxin. Circ Shock 1991; 33:108–120.
22. Fong Y, Marano MA, Moldawer LL, et al. The acute splanchnic and peripheral tissue metabolic response to endotoxin in humans. J Clin Invest 1990; 85:1896–1904.
23. Auger MJ. Mononuclear phagocytes. BMJ 1989; 298:546–548.
24. Sharif SF, Hariri RJ, Chang VA, Barie PS, Wang RS, Ghajar JBG. Human astrocyte production of tumour necrosis factor-alpha, interleukin-1B, and interleukin-6 following exposure to lipopolysaccharide endotoxin. Neurol Res 1993; 15:xx–xx.
25. Sawada M, Kondo N, Suzumura A, Marunouchi T. Production of tumor necrosis factor-α by microglia and astrocytes in culture. Brain Res 1989; 491:394–397.
26. Selmaj K, Raine CS, Cannella B, Brosnan CF. Identification of lymphotoxin and tumor necrosis factor in multiple sclerosis lesions. J Clin Invest 1991; 87:949–954.
27. Tracey KJ, Morgello S, Koplin B, et al. Metabolic effects of cachectin/tumor necrosis factor are modified by site of production: cachectin/tumor necrosis factor-secreting tumor in skeletal muscle induces chronic cachexia, while implantation in brain induces predominately acute anorexia. J Clin Invest 1990; 86:2014–2024.
28. Kapadia S, Lee J, Torre-Amione G, Birdsall HH, Ma TS, Mann DL. Tumor necrosis factor-α gene and protein expression in adult feline myocardium after endotoxin administration. J Clin Invest 1995; 96:1042–1052.
29. Kapadia S, Torre-Amione G, Yokoyama T, Mann Douglas L. Soluble TNF binding proteins modulate the negative inotropic properties of TNF-α in vitro. Am J Physiol 1995; 268:H517–H525.
30. Kruys V, Thompson P, Beutler B. Extinction of the TNF locus, and of genes encoding the lipopolysaccharide signaling pathway. J Exp Med 1993; 177:1383–1390.
31. Giroir BP, Johnson JH, Brown T, Allen GL, Beutler B. The tissue distribution of tumor necrosis factor biosynthesis during endotoxemia. J Clin Invest 1993; 90:693–698.
32. Spriggs DR, Imamura K, Rodriguez C, Horiguchi J, Kufe DW. Induction of tumor necrosis factor expression and resistance in a human breast tumor cell line. Proc Natl Acad Sci USA 1987; 84:6563–6566.
33. Bianchi M, Bloom O, Raabe T, et al. Suppression of proinflammatory cytokines in monocytes by a tetravalent guanylhydrazone. J Exp Med 1996; 83:927–936.
34. Chensue SW, Terebuh PD, Remick DG, Scales WE, Kunkel SL. In vivo biologic and immunohistochemical analysis of interleukin 1 alpha, beta and tumor necrosis factor during experimental endotoxemia: kinetics, Kupffer cell expression, and glucocorticoid effects. Am J Pathol 1991; 138:395–402.
35. Dubravec DB, Spriggs DR, Mannick JA, Rodrick ML. Circulating human peripheral blood granulocytes synthesize and secrete tumor necrosis factor alpha. Proc Natl Acad Sci USA 1990; 87:6758–6761.
36. Beutler B, Tkacenko V, Milsark I, Krochin N, Cerami A. Effect of gamma interferon on cachectin expression by mononuclear phagocytes. Reversal of the lpsd (endotoxin resistance) phenotype. J Exp Med 1986; 164:1791–1796.
37. Zuckerman SH, Evans GF. Endotoxin tolerance: in vivo regulation of tumor necrosis factor and interleukin-1 synthesis is at the transcriptional level. Cell Immunol 1992; 140:513–519.
38. Han J, Beutler B, Huez G. Complex regulation of TNF mRNA turnover in lipopolysaccharide-activated macrophages. Biochim Biophys Acta 1991; 1090:22–28.
39. Brown HT, Beutler B. Endotoxin-responsive sequences control cachectin/tumor necrosis factor biosynthesis at the translational level. J Exp Med 1990; 171:465–475.
40. Salkowski CA, Vogel SN. Lipoplysaccharide increases glucocorticoid receptor expression in murine macrophages. J Immunol 1992; 149:4041–4047.
41. Tsujimoto M, Sugiyama M, Adachi H. Modulation of mitogenic activity of tumor necrosis factor by interferons and dexamethasone. Lymphokine Res 1989; 8:99–106.
42. Henricson BE, Neta R, Vogel SN. An interleukin-1 receptor antagonist blocks lipopolysaccharide-induced colony-stimulating factor production and early endotoxin tolerance. Infect Immun 1991; 59:1188–1191.
43. Frankenberger M, Pechumer H, Ziegler-Heitbrok HWL. Interleukin-10 is upregulated in LPS tolerance. J Inflamm 1995; 45:56–63.
44. Ziegler-Hietbrock HWL. Molecular mechanism in tolerance to lipopolysaccharide. J Inflamm 1995; 45:13–26.
45. Hesse DG, Tracey KJ, Fong Y, et al. Cytokine appearance in human endotoxemia and primate bacteremia. Surg Gynecol Obstet 1988; 166:147–153.
46. Michie HR, Manogue KR, Spriggs DR, et al. Detection of circulating tumor necrosis factor after endotoxin administration. N Engl J Med 1988; 318:1481–1486.
47. Tracey KJ, Lowry SF, Fahey III TF, et al. Cachectin/tumor necrosis factor induces lethal shock and stress hormone responses in the dog. Surg Gynecol Obstet 1987; 164:415–422.
48. Bertini R, Bianchi M, Ghezzi P. Adrenalectomy sensitizes mice to the lethal effects of interleukin-1 and tumor necrosis factor. J Exp Med 1988; 167:1708–1712.
49. Zuckerman SH, Shellhaas J, Butler LD. Differential regulation of lipopolysaccharide-induced interleukin 1 and tumor necrosis factor synthesis: effects of endogenous and exogenous glucocorticoids and the role of the pituitary-adrenal axis. Eur J Immunol 1989; 19:301–305.
50. Bernahagen J, Calandra T, Mitchell RA, et al. Macrophage migration inhibitory factor(MIF) is a pituitary-derived cytokine and potentiates lethal endotoxaemia. Nature 1993; 365:756–769.
51. Leudke CE, Cerami A. Interferon-gamma overcomes glucocorticoid suppression of cachectin/tumor necrosis factor biosynthesis by murine macrophages. J Clin Invest 1990; 86:1234–1240.
52. Lantz M, Malik S, Slevin ML, Olsson I. Infusion of tumor necrosis factor (TNF) causes an increase in circulating TNF-binding protein in humans. Cytokine 1990; 2:402–406.
53. Aderka D, Engelmann H, Maor Y, Brakebusch C, Wallach D. Stabilization of the bioactivity of tumor necrosis factor by its soluble receptors. J Exp Med 1992; 175:323–329.
54. Bemelmans MHA, Gouma DJ, Buurman WA. Influence of nephrectomy on tumor necrosis factor clearance in a murine model. J Immunol 1993; 150:2007–2017.
55. Mohler KM, Torrance DS, Smith CA, et al. Soluble tumor necrosis factor (TNF) receptors are effective therapeutic agents in lethal endotoxemia and function simultaneously as both TNF

carriers and TNF antagonists. J Immunol 1993; 151:1548–1561.
56. Beutler B, Han J, Kruys V, Giroir BP. Coordinate regulation of TNF biosynthesis at the levels of transcription and translation—patterns of TNF expression in vivo. In: Beutler B, ed. Tumor Necrosis Factors: The Molecules and Their Emerging Role in Medicine. New York: Raven Press, 1992:561–574.
57. Jongeneel CV. Transcriptional regulation of the tumor necrosis factor α gene. Immunobiology 1995; 193:210–216.
58. Sariban E, Imamura K, Luebbers R, Kufe DW. Transcriptional and posttranscriptional regulation of tumor necrosis factor gene expression in human monocytes. J Clin Invest 1988; 81:1506–1510.
59. Kruys V, Kemmer K, Shakhov A, Jongeneel V, Beutler B. Constitutive activity of the tumor necrosis factor promoter is cancelled by the 3′ untranslated region in nonmacrophage cell lines: a trans-dominant factor overcomes this suppressive effect. Proc Natl Acad Sci USA 1992; 89:673–677.
60. Jongeneel CV, Shakhov AN, Nedospasov SA, Cerottini JC. Molecular control of tissue-specific expression at the mouse TNF locus. Eur J Immunol 1989; 19:549–552.
61. Collart MA, Belin D, Vassalli JD, de Kossodo S, Vassalli P. Gamma interferon enhances macrophage transcription of the tumor necrosis factor/cachectin, interleukin 1, and urokinase genes, which are controlled by short-lived repressors. J Exp Med 1986; 164:2113–2118.
62. Kuprash DV, Udalova IA, Turetskaya RL, Rice NR, Nedospasov SA. Conserved kappaB element located downstream of the tumor necrosis factor α gene: distinct NF-kappaB binding pattern and enhancer activity in LPS activated murine macrophages. Oncogene 1995; 11:97–106.
63. Albrecht H, Schook LB, Jongeneel CV. Nuclear migration of NF-κB correlates with TNF-α mRNA accumulation. J Inflamm 1995; 45:64–71.
64. Raingeaud J, Gupta S, Rogers JS, et al. Pro-inflammatory cytokines and environmental stress causes p38 mitogen-activated protein kinase activation by dual phosphorylation on tyrosine and threonine. J Biochem 1995; 270(13):7420–7426.
65. Lee JC, Laydon JT, McDonnell PC, et al. A protein kinase involved in the regulation of inflammatory cytokine biosynthesis. Nature 1995; 372:739–746.
66. Trede NS, Tsytsykova AV, Chatila T, Goldfeld AE, Geha RS. Transcriptional activation of the human TNF-α promoter by superantigen in human monocytic cells: role of NF-kappaB. J Immunol 1995; 155:902–908.
67. Goldfeld AE, Strominger JL, Doyle C. Human tumor necrosis factor α gene regulation in phorbol ester stimulated T and B cell lines. J Exp Med 1991; 174:73–81.
68. Goldfeld AE, Doyle C, Maniatis T. Human tumor necrosis factor alpha gene regulation by virus and lipopolysaccharide. Proc Natl Acad Sci USA 1990; 87:9769–9773.
69. Goldfeld AE, McCaffrey PG, Strominger JL, Rao A. Identification of a novel cyclosporin-sensitive element in the human tumor necrosis factor α gene promoter. Proc Natl Acad Sci USA 1990; 87:9769–9773.
70. Cohen P, Dennis J, Nakshatri H, et al. Inhibition of monocyte tumor necrosis factor synthesis by a tetravalent guanylhydrazone occurs via translation regulation. Proc Natl Acad Sci USA 1995; 93:3967–3971.
71. Bohjanen PR, Petryniak B, June CH, Thompson CB, Lindsten T. An inducible cytoplasmic factor (AU-B) binds selectively to AUUUA multimers to the 3′ untranslated region of lymphokine mRNA. Mol Cell Biol 1991; II:3288–3295.
72. Vakalopoulou E, Schaack J, Shenk T. A 32-kilodalton protein binds to AU-rich domains in the 3′ untranslated regions of rapidly degraded mRNAs. Mol Cell Biol 1991; II:3355–3364.
73. Shaw G, Kamen R. A conserved AU sequence from the 3′-untranslated region of GM-CSF mRNA mediates selective mRNA degradation. Cell 1986; 46:659–667.
74. Han J, Huez G, Beutler B. Interactive effects of the tumor necrosis factor promoter and 3′-untranslated regions. J Immunol 1991; 146:1843–1848.
75. Beutler B, Brown T. A CAT reporter construct allows ultrasensitive estimation of TNF synthesis, and suggests that the TNF gene has been silenced in non-macrophage cell lines. J Clin Invest 1991; 87:1336–1344.
76. Novogrodsky A, Vanichkin A, Patya M, Gazit A, Osherov N, Levitzki A. Prevention of lipopolysaccharide-induced lethal toxicity by tyrosine kinase inhibitors. Science 1994; 264:1319–1322.
77. Scheinman RI, Cogswell PC, Lofquist AK, Baldwin Jr AS. Role of transcriptional activation of IkBα in mediation of immunosuppression by glucocorticoids. Science 1995; 270:283–286.
78. Lorenz JJ, Furdon PJ, Taylor JD, et al. A cyclic adenosine 3′, 5′-monophosphate signal is required for the induction of IL-1β by TNF-α in human monocytes. J Immunol 1995; 155:836–844.
79. Sinha B, Semmler J, Eisenhut T, Eigler A, Endres S. Enhanced tumor necrosis factor suppression and cyclic adenosine monophosphate accumulation by combination of phosphodiesterase inhibitors and prostanoids. Eur J Immunol 1995; 25:147–153.
80. Bradley JR, Johnson DR, Pober JS. Four different classes of inhibitors of receptor-mediated endocytosis decrease tumor necrosis factor-induced gene expression in human endothelial cells. J Immunol 1993; 150:5544–5555.
81. Remick DG, Nguyen DT, Eskandari MK, Strieter RM, Kunkel SL. Cyclosporine A inhibits TNF production without decreasing TNF mRNA levels. Biochem Biophys Res Commun 1989; 161:551–555.
82. Sampaio EP, Sarno EN, Galilly R, Cohn ZA, Kaplan G. Thalidomide selectively inhibits TNF-a production by stimulated human moncytes. J Exp Med 1991; 173:699–703.
83. Molvig J, Baek L, Christensen P, et al. Endotoxin-stimulated human monocyte secretion of interleukin 1, tumor necrosis factor alpha, and prostaglandin E_2 shows stable interindividual differences. Scand J Immunol 1988; 27:705–716.
84. Wilson AG, diGiovine FS, Duff GW. Genetics of tumour necrosis factor-α in autoimmune, infectious, and neoplastic diseases. J Inflamm 1995; 45:1–12.
85. Wilson AG, deVries N, Piciot F, de Giovine FS, van der Putte LBA, Duff GW. An allelic polymorphism within the human tumor necrosis factor α promoter region is strongly associated with HLA A1, B8, and DR 3 alleles. J Exp Med 1993; 177:557–560.
86. Jacob CO, Fronek Z, Lewis GD, Koo M, Hansen JA, McDevitt HO. Heritable major histocompatibility complex class II-associated differences in production of tumor necrosis factor α: relevance to genetic predisposition to systemic lupus erythematosus. Proc Natl Acad Sci USA 1990; 87:1233–1237.
87. Spies T, Morton CC, Nedospasov SA, Fiers W, Pious D, Strominger JL. Genes for the tumor necrosis factor -a and b are linked to the human major histocompatibility complex. Proc Natl Acad Sci USA 1986; 83:8699–8702.
88. Jacob CO, McDevitt HO. Tumour necrosis factor-alpha in murine autoimmune 'lupus' nephritis. Nature 1988; 331:356–358.
89. Jongenell CV, Briant L, Udalova IA, Sevin A, Nedospasov SA, Cambon-Thomsen A. Extensive genetic polymorphism in the

human tumor necrosis factor region and relation to extended HLA haplotypes. Proc Natl Acad Sci USA 1991; 88:9717–9721.

90. Wilson AG, Symons JA, McDowell TL, diGiovine FS, Duff GW. Effects of a tumor necrosis factor (TNFα) promoter base transition on transcriptional activity. Br J Rheumatol 1994; 33:89.
91. McGuire W, Hill AVS, Allsopp CEM, Greenwood BM, Kwiatkowski D. Variation in the TNF-α promoter region associated with susceptibility to cerebral malaria. Nature 1994; 371:508–511.
92. Brockhaus M, Schoenfeld H-J, Schlaeger E-J, Hunziker W, Lesslauer W, Loetscher H. Identification of two types of tumor necrosis factor receptors on human cell lines by monoclonal antibodies. Proc Natl Acad Sci USA 1990; 87:3127–3131.
93. Tartaglia LA, Ayres TM, Wong GHW, Goeddel DV. A novel domain within the 55 kd TNF receptor signals cell death. Cell 1993; 74:845–853.
94. Trefzer U, Brockhaus M, Loetscher H, et al. The 55-kD tumor necrosis factor receptor on human keratinocytes is regulated by tumor necrosis factor-alpha and by ultraviolet B radiation. J Clin Invest 1993; 92:462–470.
95. Banner DW, D'Arcy A, Janes W, et al. Crystal structure of the soluble human 55 kd TNF receptor-human TNF-b complex: implications for TNF receptor activation. Cell 1993; 73:431–445.
96. Stahl N, Yancopoulos GD. The alphas, betas, and kinases of cytokine receptor complexes. Cell 1993; 74:587–590.
97. Adam D, Kessler U, Krönke M. Cross-linking of the p55 tumor necrosis factor receptor cytoplasmic domain by a dimeric ligand induces nuclear factor-kappaB and mediates cell death. J Biol Chem 1995; 270:17482–17487.
98. Shalaby MR, Sundan A, Loetscher H, Brockhaus M, Lessaluer W, Espevik T. Binding and regulation of cellular functions by monoclonal antibodies against human tumor necrosis factor receptors. J Exp Med 1990; 172:1517–1520.
99. Rothe M, Sarma V, Dixit VM, Goeddel DV. TRAF-2-mediated activation of NF-κB by TNF receptor 2 and CD40. Science 1995; 269:1424–1426.
100. Hohmann H-P, Remy R, Poeschl B, Van Loon APGM. Tumor necrosis factors-alpha and -beta bind to the same two types of tumor necrosis factor receptors and maximally activate the transcription factor NF-kappa B at low receptor occupancy and within minutes after receptor binding. J Biol Chem 1990; 265:15183–15188.
101. Gullberg U, Lantz M, Lindvall L, Olsson I, Himmler A. Involvement of an Asn/Val cleavage site in the production of a soluble form of a human tumor necrosis factor (TNF) receptor. Site-directed mutagenesis of a putative cleavage site in the P55 TNF receptor chain. Eur J Cell Biol 1992; 58:307–312.
102. Decoster E, Vanhaesebroeck B, Vandenabeele P, Grooten J, Fiers W. Generation and biological characterization of membrane-bound, uncleavable murine tumor necrosis factor. J Biol Chem 1995; 270:18473–18478.
103. van Zee KJ, Kohno T, Fischer E, Rock CS, Moldawer LL, Lowry SF. Tumor necrosis factor soluble receptors circulate during experimental and clinical inflammation and can protect against excessive tumor necrosis factor-a in vitro and in vivo. Proc Natl Acad Sci USA 1992; 89:4845–4849.
104. Carpenter A, Evans TJ, Buurman WA, Bemelmans MH, Moyes D, Cohen J. Differences in the shedding of soluble TNF receptors between endotoxin-sensitive and endotoxin-resistant mice in response to lipopolysaccharide or live bacterial challenge. J Immunol 1995; 155:2005–2012.
105. Spinas GA, Bloesch D, Kaufmann M-T, Keller U, Dayer J-M. Induction of plasma inhibitors of interleukin 1 and TNF-alpha activity by endotoxin administration to normal humans. Am J Physiol Regul Integr Comp Physiol 1990; 259:R993–R997.
106. Remick DG, Kunkel RG, Larrick JW, Kunkel SL. Acute in vivo effects of human recombinant tumor necrosis factor. Lab Invest 1987; 56:583–590.
107. Colletti LM, Remick DG, Burtch GD, Kunkel SL, Strieter RM, Campbell Jr DA. Role of tumor necrosis factor-alpha in the pathophysiologic alterations after hepatic ischemia/reperfusion injury in the rat. J Clin Invest 1990; 85:1936–1943.
108. Redl H, Schlag G, Lamche H, et al. TNF- and LPS-induced changes of lung vascular permeability: Studies in unanesthetised sheep. Circ Shock 1990; 31:183–192.
109. Natanson C, Eichenholz PW, Danner RL, et al. Endotoxin and tumor necrosis factor challenge in dogs simulate the cardiovascular profile of human septic shock. J Exp Med 1989; 169:823–832.
110. Dinarello CA, Okusawa S, Gelfand JA. Interleukin-1 induces a shock-like state in rabbits: synergism with tumor necrosis factor and the effect of ibuprofen. Prog Clin Biol Res 1989; 299:203–215.
111. Abbruzzese JL, Levin B, Ajani JA, et al. A phase II trial of recombinant human interferon-gamma and recombinant tumor necrosis factor in patients with advanced gastrointestinal malignancies: results of a trial terminated by excessive toxicity. J Biol Response Mod 1990; 9:522–527.
112. Schiller JH, Storer BE, Witt PL, et al. Biological and clinical effects of intravenous tumor necrosis factor-alpha administered three times weekly. Cancer Res 1991; 51:1651–1658.
113. Starnes Jr HF, Warren RS, Jeevanandam M, et al. Tumor necrosis factor and the acute metabolic response to tissue injury in man. J Clin Invest 1988; 82:1321–1325.
114. Parrillo JE, Parker MM, Natanson C, et al. Septic shock in humans: advances in the undrestanding of pathogenesis, cardiovascular dysfunction, and therapy. Ann Intern Med 1990; 113:227–242.
115. Yokoyama T, Vaca L, Rossen R, Durante W, Hazarika P, Mann D. Cellular basis for the negative inotropic effects of tumor necrosis factor-alpha in the adult mammalian heart. J Clin Invest 1993; 92:2303–2312.
116. Beutler B, Milsark IW, Cerami AC. Passive immunization against cachectin/tumor necrosis factor protects mice from lethal effect of endotoxin. Science 1985; 229:869–871.
117. Mathison JC, Wilfson E, Ulevitch RJ. Participation of tumor necrosis factor in the mediation of gram-negative bacterial lipopolysaccharide-induced injury in rabbits. J Clin Invest 1988; 81:1925–1937.
118. Emerson TE, Lindsey DC, Jesmok GJ, Duerr ML, Fournel MA. Efficacy of monoclonal antibody against TNF-a in an endotoxemic baboon model. Circ Shock 1992; 38:75–84.
119. Kilbourn RG, Gross SS, Jubran A, et al. N^g-methyl-L-arginine inhibits tumor necrosis factor-induced hypotension: implications for the involvement of nitric oxide. Proc Natl Acad Sci USA 1990; 87:3629–3632.
120. Knowles RG, Merrett M, Salter M, Moncada S. Differential induction of brain, lung and liver nitric oxide synthase by endotoxin in the rat. Biochem J 1990; 270:833–836.
121. Bogle RG, Baydoun AR, Pearson JD, Moncada S, Mann GE. L-Arginine transport is increased in macrophages generating nitric oxide. Biochem J 1992; 284:15–18.
122. Nathan C. Nitric oxide as a secretory product of mammalian cells. FASEB J 1992; 6:3051–3064.
123. Maciejewski JP, Selleri C, Sato T, et al. Nitric oxide suppression of human hematopoiesis in vitro. Contribution to inhibi-

tory action of interferon-gamma and tumor necrosis factor-α. J Clin Invest 1995; 96:1085–1092.
124. Caty MG, Guice KS, Oldham KT, Remick DG, Kunkel SI. Evidence for tumor necrosis factor-induced pulmonary microvascular injury after intestinal ischemia-reperfusion injury. Ann Surg 1990; 212:694–700.
125. Abe Y, Sekiya S, Yamasita T, Sendo F. Vascular hyperpermeability induced by tumor necrosis factor and its augmentation by IL-1 and IFN-gamma is inhibited by selective depletion of neutrophils with a monoclonal antibody. J Immunol 1990; 145:2902–2907.
126. Stolpen AH, Guinan EC, Fiers W, Pober JS. Recombinant tumor necrosis factor and immune interferon act singly and in combination to reorganize human vascular endothelial cell monolayers. Am J Path 1986; 123:16–24.
127. Bevilacqua MP, Pober JS, Majeau GR, Fiers W, Cotran RS, Gimbrone Jr MA. Recombinant tumor necrosis factor induces procoagulant activity in cultured human vascular endothelium: characterization and comparison with the actions of interleukin 1. Proc Natl Acad Sci USA 1986; 83:4533–4537.
128. Bevilacqua MP, Stengelin S, Gimbrone MA Jr, Seed B. Endothelial leukocyte adhesion molecule 1: an inducible receptor for neutrophils related to complement regulatory proteins and lectins. Science 1989; 243:1160–1165.
129. Lasky LA. Selectins: interpreters of cell-specific carbohydrate information during inflammation. Science 1992; 258:964–969.
130. Slungaard A, Vercellotti GM, Walker G, Nelson RD, Jacob HS. Tumor necrosis factor-alpha/cachectin stimulates eosinophil oxidant production and toxicity towards human endothelium. J Exp Med 1990; 171:2025–2041.
131. Stephens KE, Ishizaka A, Larrick JW, Raffin TA. Tumor necrosis factor causes increased pulmonary permeability and edema. Am Rev Respir Dis 1988; 137:1364–1370.
132. Horvath CJ, Ferro TJ, Jesmok G, Malik AB. Recombinant tumor necrosis factor increases pulmonary vascular permeability independent of neutrophils. Proc Natl Acad Sci USA 1988; 85:9219–9223.
133. Ulich TR, Yin S, Guo K, Del Castillo J, Eisenberg SP, Thompson RC. The intratracheal administration of endotoxin and cytokines III. The interleukin-1 (IL-1) receptor antagonist inhibits endotoxin- and IL-1-induced acute inflammation. Am J Pathol 1991; 138:521–524.
134. Tracey KJ. The acute and chronic pathophysiological effects of TNF: Mediation of septic shock and wasting (cachexia). In: Beutler B, ed. Tumor Necrosis Factors: The Molecules and Their Emerging Role in Medicine. New York: Raven Press, 1992:255–273.
135. Peppel K, Poltorak A, Melhado I, Jirik F, Beutler B. Expression of a TNF inhibitorin transgenic mice. J Immunol 1993; 151:5699–5703.
136. Ashkenazi A, Marsters SA, Capon DJ. Protection against endotoxic shock by a tumor necrosis factor receptor immunoadhesin. Proc Natl Acad Sci USA 1991; 88:10535–10539.
137. Pfeffer K, Matsuyama T, Kundig TM, et al. Mice deficient for the 55 kd tumor necrosis factor receptor are resistant to endotoxin shock, yet succumb to L. monocytogenes infection. Cell 1993; 73:457–467.
138. Fong Y, Tracey KJ, Moldawer LL, et al. Antibodies to cachectin/TNF reduce interleukin-1-beta and interleukin-6 appearance during lethal bacteremia. J Exp Med 1989; 170:1627–1633.
139. Waage A, Espevik T. Interleukin 1 potentiates the lethal effect of tumor necrosis factor-alpha/cachectin in mice. J Exp Med 1988; 167:1987–1992.
140. Grau GE, Heremans H, Piguet PF, et al. Monoclonal antibody against interferon gamma can prevent experimental cerebral malaria and its associated overproduction of tumor necrosis factor. Proc Natl Acad Sci USA 1989; 86:5572–5574.
141. Tracey KJ, Wei H, Manogue KR, et al. Cachectin/tumor necrosis factor induces cachexia, anemia, and inflammation. J Exp Med 1988; 167:1211–1227.
142. Fukushima R, Saito H, Taniwaka K, et al. Different roles of IL-1 and TNF on hemodynamics and interorgan amino acid metabolism in awake dogs. Am J Physiol 1992; 2625:E275–E181.
143. Tracey KJ. Role of tumor necrosis factor in protein metabolism. Nestle Nutrition Workshop 1994; 33:229–242.
144. Oliff A, Defeo-Jones D, Boyer M, et al. Tumors secreting human TNF/cachectin induce cachexia in mice. Cell 1987; 50:555–563.
145. Costelli P, Carbo N, Tessitore L, et al. Tumor necrosis factor-α mediates changes in tissue protein turnover in a rat cancer cachexia model. J Clin Invest 1993; 92:2783–2789.
146. Milner E, Cioffi W, Mason A, Mcmanus W, Pruitt A. Accuracy of urinary urea nitrogen for predicting total urinary nitrogen in thermally injured patients. J Parenter Enteral Nutr 1993; 17:414–416.
147. Tracey KJ. TNF and other cytokines in the metabolism of septic shock and cachexia. Clin Nutr 1992; 11:1–11.
148. Gelin J, Moldawer LL, Loennroth C, Sherry B, Chizzonite R, Lundholm K. Role of endogenous tumor necrosis factor alpha and interleukin 1 for experimental tumor growth and the development of cancer cachexia. Cancer Res 1991; 51:415–421.
149. Teng MN, Park BH, Koeppen KHW, Tracey KJ, Fendly BM, Schreiber H. Long-term inhibition of tumor growth by tumor necrosis factor in the absence of cachexia or T-cell immunity. Proc Natl Acad Sci USA 1991; 88:3535–3539.
150. Stovroff MC, Fraker DL, Travis WD, Norton JA. Altered macrophage activity and tumor necrosis factor: Tumor necrosis and host cachexia. J Surg Res 1989; 46:462–469.
151. Strijbos PJ, Hardwick AJ, Relton JK, Carey F, Rothwell NJ. Inhibition of central actions of cytokines on fever and thermogenesis by lipocortin-1 involves CRF. Am J Physiol 1992; 263:E632–E636.
152. Dinarello CA, Cannon JG, Wolff SM, et al. Tumor necrosis factor (cachectin) is an endogenous pyrogen and induces production of interleukin 1. J Exp Med 1986; 163:1433–1450.
153. Morimoto A, Sakata Y, Watanabe T, Murakami N. Characteristics of fever and acute-phase response induced in rabbits by IL-1 and TNF. Am J Physiol 1989; 256:R35–R41.
154. Plata-Salaman CR, Oomura Y, Kai Y. Tumor necrosis factor and interleukin 1-beta: suppression of food intake by direct action in the central nervous system. Brain Res 1988; 448:106–114.
155. Lang CH, Dobrescu C, Bagby GJ. Tumor necrosis factor impairs insulin action on peripheral glucose. Endocrinology 1992; 130:43–52.
156. Clauser E, Leconte I, Auzan C. Molecular basis of insulin resistance. Horm Res 1992; 38:5–12.
157. Hotamisligil GS, Shargill NS, Spiegelman BM. Adipose expression of tumor necrosis factor—a direct role in obesity-linked insulin resistance. Science 1993; 259:87–91.
158. Abraham E, Wunderink R, Silverman H, et al. Efficacy and safety of monoclonal antibody to human tumor necrosis antibody to human tumor necrosis factor α in patients with sepsis syndrome. JAMA 1995; 273:934–941.
159. Bone RC. Toward an epidemiology and natural history of SIRS (systemic inflammatory response syndrome). JAMA 1992; 268:3452–3455.
160. Fong Y, Moldawer LL, Marano MA, et al. Cachectin/TNF or

IL-1 alpha induces cachexia with redistribution of body proteins. Am J Physiol 1989; 256:R659–R665.
161. Neale ML, Williams BD, Matthews N. Tumour necrosis factor activity in joint fluids from rheumatoid arthritis patients. Br J Rheumatol 1989; 28:104–108.
162. Bucala R, Ritchlin C, Winchester R, Cerami A. Constitutive production of inflammatory and mitogenic cytokines by rheumatoid synovial fibroblasts. J Exp Med 1991; 173:569–574.
163. Müssener Å, Klareskog L, Lorentzen JC, Kleinau S. TNF-α dominates cytokine mRNA expression in lymphoid tissues of rats developing collagen- and oil-induced arthritis. Scand J Immunol 1995; 42:128–134.
164. Brennan FM, Gibbons DL, Cope AP, Katsikis P, Maini RN, Feldmann M. TNF inhibitors are produced spontaneously by rheumatoid and osteoarthritic synovial joint cell cultures: evidence of feedback control of TNF action. Scand J Immunol 1995; 42:158–165.
165. Ito A, Sato T, Iga T, Mori Y. Tumor necrosis factor bifunctionally regulates matrix metalloproteinases and tissue inhibitor of metalloproteinases (TIMP) production by human fibroblasts. FEBS Lett 1990; 269:93–95.
166. Dayer JM, Beutler B, Cerami A. Cachectin/tumor necrosis factor stimulates collagenase and prostaglandin E2 production by human synovial cells and dermal fibroblasts. J Exp Med 1985; 162:2163–2168.
167. Yaron I, Meyer FA, Dayer JM, Bleiberg I, Yaron M. Some recombinant human cytokines stimulate glycosaminoglycan synthesis in human synovial fibroblast cultures and inhibit it in human articular cartilage cultures. Arthritis Rheum 1989; 32:173–180.
168. Hess G, Rossol S, Rossol R, Meyer zum Bueschenfelde K-H. Tumor necrosis factor and interferon as prognostic markers in human immunodeficiency virus (HIV) infection. Infection 1991; 19:S93–S97.
169. Lahdevirta J, Maury CP, Teppo AM, Repo H. Elevated levels of circulating cachectin/tumor necrosis factor in patients with acquired immunodeficiency syndrome. Am J Med 1988; 85:289–291.
170. Molina JM, Scadden DT, Byrn R, Dinarello CA, Groopman JE. Production of tumor necrosis factor alpha and interleukin 1 beta by monocytic cells infected with human immunodeficiency virus. J Clin Invest 1989; 84:733–737.
171. Griffin GE, Leung K, Folks TM, Kunkel S, Nabel GJ. Activation of HIV gene expression during monocyte differentiation by induction of NF-kappa B. Nature 1989; 339:70–73.
172. Folks TM, Clouse KA, Justement J, et al. Tumor necrosis factor alpha induces expression of human immunodeficiency virus in a chronically infected T-cell clone. Proc Natl Acad Sci USA 1989; 86:2365–2368.
173. Osborn L, Kunkel S, Nabel GJ. Tumor necrosis factor alpha and interleukin 1 stimulate the human immunodeficiency virus enhancer by activation of the nuclear factor kappa B. Proc Natl Acad Sci USA 1989; 86:2336–2340.
174. Porter SB, Sande MA. Toxoplasmosis of the central nervous system in the acquired immunodeficiency syndrome. N Engl J Med 1992; 327:1643–1648.
175. Koenig S, Gendelman HE, Orenstein JM, et al. Detection of AIDS virus in macrophages in brain tissue from AIDS patients with encephalopathy (abstr). Rep 1986; 1089–1903.
176. Huang ZB, Eden E. Effect of corticosteroids on IL1beta and TNFalpha release by alveolar macrophages from patients with AIDS and Pneumocystis carinii pneumonia. Chest 1993; 104:751–755.
177. Mosmann TR. Cytokine patterns during the progression to AIDS. Science 1994; 265:193–194.
178. Ruddle NH, Bergman CM, McGrath KM, et al. An antibody to lymphotoxin and tumor necrosis factor prevents transfer of experimental allergic encephalomyelitis. J Exp Med 1990; 172:1193–1200.
179. Simmons RD, Willenborg DO. Direct injection of cytokines into the spinal cord causes autoimmune encephalomyelitis-like inflammation. J Neurol Sci 1990; 100:37–42.
180. Powell MB, Steinman L. The role of lymphotoxin and TNF in demyelinating diseases of the CNS. In: Beutler B, ed. Tumor Necrosis Factors: The Molecules and Their Emerging Role in Medicine. New York: Raven Press, 1923:355–369.
181. Franciotta DM, Grimaldi LME, Martino GV, et al. Tumor necrosis factor in serum and cerebrospinal fluid of patients with multiple sclerosis. Ann Neurol 1989; 26:787–789.
182. Campbell IL, Iscaro AL, Harrison LC. IFN-gamma and tumor necrosis factor-alpha: cytotoxicity to murine islets of Langerhans. J Immunol 1988; 141:2325–2329.
183. Picot F, Briant L, Jongeneel CV, et al. Association of tumor necrosis factor (TNF) and class II major histocompatibility complex alleles with the secretion of TNF-α and TNF-β by human mononuclear cells: a possible link to insulin-dependent diabetes mellitus. Eur J Immunol 1993; 23:224–231.
184. Kwiatkowski D, Hill AVS, Sambou I, et al. TNF concentration in fatal cerebral, non-fatal cerebral, and uncomplicated Plasmodium falciparum malaria. Lancet 1990; 336:1201–1204.
185. Grau GE, Taylor TE, Molyneux ME, et al. Tumor necrosis factor and disease severity in children with falciparum malaria. N Engl J Med 1989; 320:1586–1591.
186. Grau GE, Fajardo LF, Piguet PF, Allet B, Lambert PH, Vassalli P. Tumor necrosis factor (cachectin) as an essential mediator in murine cerebral malaria. Science 1987; 237:1210–1212.
187. Waage A, Halstensen A, Shalaby R, Brandtzaeg P, Kierulf P, Espevik T. Local production of tumor necrosis factor-alpha, interleukin 1, and interleukin 6 in meningococcal meningitis. J Exp Med 1989; 170:1859–1967.
188. Marceline van Furth A, Seijmonsbergen EM, Langermans JAM, Groeneveld PHP, De Bel CE, Van Furth R. High levels of interleukin 10 and tumor necrosis factor α in cerebrospinal fluid during the onset of bacterial meningitis. Clin Infect Dis 1995; 21:220–222.
189. Saukkonen K, Sande S, Cioffe C, et al. The role of cytokines in the generation of inflammation and tissue damage in experimental gram-positive meningitis. J Exp Med 1990; 171:439–448.
190. Roten R, Markert M, Feihl F, Schaller M-D, Tagan M-C, Perret C. Plasma levels of tumor necrosis factor in the adult respiratory distress syndrome. Am Rev Respir Dis 1991; 143:590–592.
191. Fahey III TF, Tracey KJ, Cerami A. Tumor necrosis factor (cachectin) and the adult respiratory distress syndrome. In: Fishman AP, ed. Pulmonary Diseases and Disorders: New Directions. New York: McGraw-Hill, 1992; 175–183.
192. Moreira AL, Sampaio EP, Zmuidzinas A, Frindt P, Smith KA, Kaplan G. Thalidomide exerts its inhibitory action on TNF alpha by enhancing mRNA degradation. J Exp Med 1993; 177:1675–1680.
193. Frker DL, Alexander HR, Thom AK. Use of tumor necrosis factor is isolated hepatic perfusion. Circ Shock 1994; 44(1):45–50.
194. van Dullemen HM, van Deventer SJ, Hommes DW, et al. Treatment of Chron's disease with anti-tumor necrosis factor chimeric monoclonal antibody(cA2). Gastroenterology 1995; 109(1):129–135.
195. Eggimann P, Chiolero R, Chassot PG, Lienard D, Gerain J, Lejeune F. Systemic and hemodynamic effects of recombinant

tumor necrosis factor alpha in isolation perfusion of the limbs. Chest 1995; 107(4):1074–1082.
196. Havell EA. Production of tumor necrosis factor during murine listeriosis. J Immunol 1987; 139:4225–4231.
197. Titus RG, Sherry B, Cerami A. Tumor necrosis factor plays a protective role in experimental murine cutaneous leishmaniasis. J Exp Med 1989; 170:2097–2104.
198. Desiderio JV, Kiener PA, Lin PF, Warr GA. Protection of mice against Listeria monocytogenes infection by recombinant human tumor necrosis factor alpha. Infect Immun 1989; 57:1615–1617.
199. Fahey III TF, Sherry B, Tracey KJ, et al. Cytokine production in a model of wound healing: the appearance of MIP-1, MIP-2, Cachectin/TNF, and IL-1. Cytokine 1990; 2:92–99.
200. Chaplin CD, Hogquist KA. Interactions between TNF and interleukin-1. In: Beutler B, ed. Tumor Necrosis Factor: The Molecules and Their Emerging Role in Medicine. New York: Raven Press, 1992; 197–220.
201. Tracey KJ, Vlassara H, Cerami A. Cachectin/TNF (tumour necrosis factor). Lancet 1989; 1:1122–1126.
202. Marano MA, Fong Y, Moldawer LL, et al. Serum cachectin/TNF in critically ill patients with burns correlates with infection and mortality. Surg Gynecol Obstet 1990; 170:32–38.
203. Aderka D, Fisher S, Levo Y, Holtmann H, Hahn T, Wallach O. Cachectin/tumour-necrosis-factor production by cancer patients [letter]. Lancet 1985; 2:1190.
204. Balkwill F, Osborne R, Burke F, et al. Evidence for tumour necrosis factor/cachectin production in cancer. Lancet 1987; 2:1229–1232.
205. Henderson B, Pettipher ER. Arthritogenic actions of recombinant IL-1 and tumour necrosis factor alpha in the rabbit: evidence for synergistic interactions between cytokines in vivo. Clin Exp Immunol 1989; 75:306–310.
206. Folks TM, Justement J, Kinter A, Dinarello CA, Fauci AS. Cytokine-induced expression of HIV-1 in a chronically infected promonocyte cell line. Science 1987; 238:800–802.
207. Seekamp A, Warren JS, Remick DG, Till GO, Ward PA. Requirements for tumor necrosis factor-alpha and interleukin-1 in limb ischemia/reperfusion injury and associated lung injury. Am J Pathol 1993; 143:453–463.
208. Ghezzi P. TNF and the liver. In: Beutler B, ed. Tumor Necrosis Factor: The Molecules and Their Emerging Role in Medicine. New York: Raven Press, 1992:87–96.
209. Miyazaki Y, Araki K, Vesin C, et al. Expression of a tumor necrosis factor-α transgene in murine lung causes lymphocytic and fibrosing alveolitis. A mouse model of progressive pulmonary fibrosis. J Clin Invest 1995; 96:250–259.
210. Thomas PS, Yates DH, Barnes PJ. Tumor necrosis factor-α increases airway responsiveness and sputum neutrophilia in normal human subjects. Am J Respir Crit Care Med 1995; 152:76–80.

17
Lymphotoxin

Isabelle Millet and Nancy H. Ruddle
Yale University School of Medicine, New Haven, Connecticut

HISTORY

Lymphotoxin (LT), also known as LT-α or tumor necrosis factor-β (TNF-β), is a 25 kDa protein secreted predominately by activated lymphocytes and was the first of the TNFs to be described. At the time of its discovery, Ruddle and Waksman (1) called it cytotoxic factor and Granger and Williams (2) coined the term lymphotoxin, describing its production by lymphocytes and its toxicity to other cells. The factor was originally characterized by its toxicity to syngeneic embryo fibroblasts or L929 cells. Though its cytotoxic activity, which is predominantly due to its ability to induce apoptosis (3,4), is well established, it is now clear that LT is a pleiotropic cytokine with a multitude of activities that can also result in induction of proliferation or differentiation and activation of gene expression (reviewed in Refs. 5–7). The recent important revelation concerning LT is its crucial and unique role in lymphoid development (8,9). The realization that LT and tumor necrosis factor, a protein originally thought to be produced exclusively by activated macrophages, are 35% homologous (10) and share many activities, including killing L929 cells, inducing necrosis of the Meth A sarcoma in vivo (11), and activating polymorphonuclear leukocytes (12), prompted some to call lymphotoxin TNF-β and tumor necrosis factor TNF-α. In addition, the fact that the genes coding for these proteins are tightly linked as a gene family within the major histocompatibility complex (MHC) and that the proteins bind the same (p55 and p75) TNF receptors (TNFRs) reinforced the apparent logic of this nomenclature. However, in 1991, Browning and Ware and their colleagues began publishing a series of papers indicating that lymphotoxin is physically associated with another protein (p33) to form a heterotrimeric complex on the surface of activated T, B, and natural killer (NK) cells (13–15). The cloning of the gene for this protein revealed that it is structurally related and genetically linked to the LT-TNF complex (16). The arrangement of these genes within the human MHC is depicted in Figure 1. Because the p33 protein was found physically associated with lymphotoxin, it was termed LT-β, and the name of lymphotoxin was, confusingly enough, changed again and is called by some LT-α. In this chapter we will refer to lymphotoxin as LT, since that was its original name, because it is frequently found in its secreted form in the absence of LT-β, and because it plays an important role in lymph node development, a function that is absolutely distinct from TNF-α. We will call the p33 protein LT-β since as yet it has not been detected alone and no function for the protein has been identified. We will call tumor necrosis factor TNF-α. These three cytokines comprise an immediate gene family. The proteins are also highly homologous to and thus members of an extended family, which includes the ligands for CD40, CD27, CD30, fas, 4-1BB, and OX-40 (17). These ligands bind to a class of related receptors (see below).

BIOCHEMISTRY

The major secreted form of human LT is a glycoprotein that by SDS-PAGE has a molecular weight of 25 kDa (11,18.19). By gel filtration, its molecular weight is 60–70 kDa, due to complex formation. It has an isoelectric point of 5.8, is only slightly sensitive to trypsin, resistant to chymotrypsin, acid labile, and instable in detergents. It has been technically difficult to prepare large quantities of

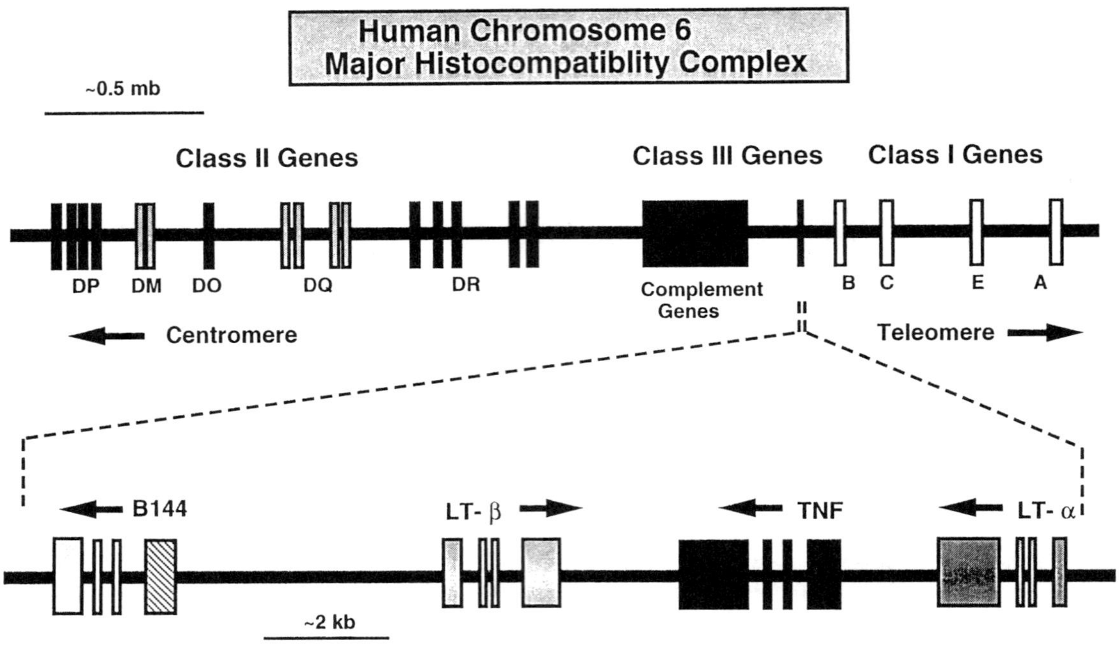

Figure 1 Genetic map of the tumor necrosis factor/lymphotoxin (TNF/LT) locus in the human MHC. (Reprinted with permission from Ref. 29.)

recombinant murine LT because of its insolubility; however, in vitro translation of murine LT cDNA results in two species of 17 and 34 kDa (10). The higher molecular weight form is presumably an oligomer. Both the in vitro translated murine material and recombinant *E. coli*–derived human material kill L929 cells, indicating that glycosylation is not necessary for function.

LT protein sequences are highly conserved among mammalian species (10,11,20–22). The cDNAs code for polypeptides of 205 amino acids (human), 204 amino acids (cow), 197 amino acids (rabbit), or 202 amino acids (mouse). There is some divergence in the length and amino acid sequences in the N-terminal signal of 34 residues (human), 33 residues (cow and mouse), or 26 residues (rabbit). The mature secreted proteins are highly homologous, with 75% identity in the 169 terminal amino acids. Nevertheless, antibodies against LTs derived from different species do not generally cross-react.

The molecular structure of the biologically active secreted form of LT is a homotrimer (23), which can bind to both (p55 and p75) TNFRs. LT crystallizes in the trimeric form (24) as does TNF-α (25). Analysis of human recombinant LT indicates the importance of surface loops near the base of each monomer. Mutation of Asp 50 or Tyr 108 results in loss of L929 killing activity (26). These regions of the LT molecule are comparable to regions of TNF-α, which are also crucial for function (27,28). The physical properties of LT were compared to those of other members of the extended TNF family in considerable detail in a recent review (29). The LT ligand-receptor complex has been crystallized at 2.85 Å resolution (30). This complex is composed of three soluble p55 TNFR molecules bound symmetrically to one LT trimer. These data and the above-noted mutational analysis indicate that the binding site of the LT homotrimer is at the junction between neighboring subunits.

LT also forms a heterotrimeric cell surface complex with LT-β, with the predominant form on T cells consisting of one LT molecule complexed with two LT-β molecules ($LT\alpha_1\beta_2$) (31). LT-β is a 25-26 kDa type II transmembrane protein, with a short cytoplasmic domain and an extensive stretch of hydrophobic amino acids comprising a membrane-anchoring domain. This organization is similar to that of TNF-α, which is also a type II transmembrane protein. However, it is not known whether LT-β or the LT-αβ complex can be clipped and released by a metalloproteinase as is TNF-α (32,33). It appears that LT-β requires LT for appropriate assembly and transport to the cell surface of lymphocytes (13,31). LT-β has one cysteine residue in the extracellular domain and two methionines in the last C-terminal domain (16). Human LT-β is composed of 244 amino acids, although the murine form contains an insert of 66 amino acids between the extracellular and transmembrane domains. The human and mouse sequences are 77% identical at the amino acid level if the insert is excluded from the analysis (34,35).

PRINCIPAL CELLULAR SOURCES

LT is predominately secreted by activated lymphocytes including T cells, B cells, and NK cells (6,36). T-cell sources include cytotoxic CD8 T cells and CD4 TH_1 inflammatory cells, but not TH_2 cells. Millet and Ruddle (37) demonstrated that LT gene transcription is generally not detectable unless T cells are activated. LT is induced in T cells by the antigen to which the cells are specifically sensitized (38,39), agents such as anti-CD3 that mimic activation though the T-cell receptors (36,40–42), and the *tax* gene of HTLV-1. LT is produced by normal human B cells after several types of activation, including mitogens (43), bacterially derived products such as lipopolysaccharide (LPS) (44), and anti-CD40 (45). LT mRNA has also been detected in LPS-activated, murine B cells (46). LT is produced constitutively by several malignantly transformed murine cell lines that represent early stages of B-cell development (47) and human myeloma cells (48).

Certain virus infections can induce LT from T or B cells. It is produced constitutively at a high level by HTLV-I–infected cell lines (49–52). A direct role for that virus is indicated by the observation that constitutive expression of LT occurs after infection of T-cell clones with HTLV-I (53). Furthermore, tax protein induces expression of LT mRNA in B cells (54) and transfection of HTLV-I *tax* gene in T cells activates LT gene promoter activity (55). LT is also induced by human immunodeficiency virus (HIV) *tat* III gene transfection of B cells (56) and is produced constitutively by most Epstein-Barr virus–transformed human B-cell lines.

Data are accumulating indicating that LT can be produced by cells of nonlymphoid lineages. These include melanocytes (57) and activated cells of the central nervous system, including activated astrocytes (58). LT is detectable in microglial cells and T cells in multiple sclerosis lesions (59). LT is almost definitely produced by a cell other than a mature T or B cell in the course of embryonic development. Mice genetically deficient in LT (LT knockout mice) that produce normal levels of LT-β and TNF-α have no lymph nodes or Peyer's patches and exhibit marked splenic disorganization with regard to T- and B-cell areas and germinal center development (8,9). Since mice deficient in mature T or B cells (*nu/nu* mice, scid mice, RAG knock-out mice) have lymph nodes, LT must be produced by another cell type. This could be an NK cell, a very primitive T or B cell, or a cell of an entirely different, nonlymphoid, lineage.

Molecular analysis of LT regulation has been carried out in both T and pre-B cells. Most studies suggest that LT is both transcriptionally and posttranscriptionally regulated in those cells. Primer extension and RNase protection analysis of murine LT mRNA derived from anti-CD3–activated T-cell clones and PD and PD31 pre-B cells have revealed two functional cap sites within 15 bases of each other (60). In all species analyzed, within 110 bp upstream of a TATA box are located consensus sites for SP1 transcription factor and near-consensus sites for AP-2. Several NF-κB–like sites are found in the LT genes of human, mouse, and rabbit. The human upstream NF-κB-like sequence, which is identical to the murine sequence, forms a complex after incubation with nuclear extracts from a variety of cells (PHA/PMA-activated Jurkat, C81-66-45, MT-2), which are known to contain active NF-κB (52). This site is a functional element in LT gene transcription in T cells (52,61) and is responsive to activation by HTLV-1 *tax* gene product (54,55).

Additional 5′ regulatory elements in LT regulation have been identified with CAT reporter constructs. Both positive and negative regulatory elements have been defined that appear to have some tissue specificity (52,60). One long poly (dA-dT) rich region binds a high-mobility group I nonhistone nuclear protein also known as HMGI(Y) and is crucial for LT production in pre-B cells (62). This protein does not act as a transcriptional factor per se, but rather stabilizes chromatin structure to allow interactions with other transcription factors. Intriguingly, it has recently been reported that HMGI(Y) mRNA is expressed during a very limited time frame in embryonic development. Its accumulation between days 10.5 and 16.5 postconception (63) is consistent with its putative regulation of LT at precisely the time that lymph nodes are organized. Histological studies in the rat have revealed the first morphological signs of popliteal lymph node development at the seventeenth day of gestation (64).

LT gene expression is also regulated at the posttranscriptional level. Jongeneel et al. (65) and Ferreri et al. (66) noted that, though activation of murine T-cell clones results in a 10-fold increase in LT transcription rate as determined by nuclear run-on analysis, in fact the overall rate of transcription is quite low compared to that of TNF-α. Nevertheless, high levels of LT mRNA accumulate under such activation conditions. This is due to the fact that LT mRNA has an unusually long half-life, particularly compared to that of TNF-α (37,67). This is probably because LT mRNA is relatively poor in AU-rich sequences compared to TNF-α (68). These areas have been implicated in mRNA destabilization (69). LT production is also regulated at the level of mRNA processing. Different activation schemes influence the cytoplasmic accumulation of differently spliced mRNA species. Weil et al. (70) determined that IL-2 induces nuclear accumulation of at least four species of LT mRNA. Two of these, a fully spliced form and an intron-3–retaining form, are exported to the cytoplasm with slightly different kinetics. These forms are also induced by anti-CD3 activation of murine T-cell clones (37).

The cell-surface LT-αβ complex has been detected on many human primary and cultured cells, including activated T and B cells, B lymphomas, IL-2–activated CD56+ NK cells (15), and LAK cells (71). Cell-surface LT, which is presumably in a LT-αβ complex, is also detectable on anti-CD40–activated human B cells (45). The LT-β gene is transcribed constitutively in many cells in which the LT gene is not active. These include resting T cells, resting B cells, and macrophages (37,72; I. Millet, unpublished) and in lymphoid cells in LT knock-out mice (I. Millet and N. H. Ruddle, unpublished). At least in activated T cells, transcription of the LT gene is rate limiting with regard to production of the complex (37). LT-β mRNA is also detected in the skin (35), though the cell of origin is not known. LT-β mRNA is also detectable in the brain in cortical areas and in Purkinje cells in the cerebellum (35). It is not known whether LT-β protein is actually present on the cell surface under these circumstances, as it has been reported that LT-β requires LT for appropriate assembly (13,31).

RECEPTORS

LT can bind to at least three different receptors. The secreted form binds and activates the two types of TNFRs, the 55 kDa form, also called p60, Type 1, or CD120a, and the higher molecular weight form called p75, p80, Type 2, or CD120b. Both TNFRs bind TNF-α and LT homotrimers and the less common cell surface form of LT ($LT\alpha_2\beta_1$). The predominant cell-surface complex ($LT\alpha_1\beta_2$) binds with high affinity to a unique receptor, which was initially identified as the TNFR-related protein (73) and then renamed LT-β receptor (LTβR), though it does not appear to bind cell surface LT-β in the absence of LT-α (31,74,75). It does not bind the $LT\alpha_2\beta_1$ heterotrimer. The two TNFRs and LTβR are members of, and homologous to, an ever-growing family of single-chain transmembrane proteins characterized by cysteine-rich repeats (76–78). Other members of this family include the low-affinity nerve growth factor receptor, OX 40, 4-1BB, CD40, CD27, fas, and products of viral open reading frames (17).

The p55 and p75 TNFRs appear to have different functions, though considerable controversy exists regarding their roles (79,80). As noted above, both p55 and p75 TNFRs bind LT and TNF-α and can carry out biological activities after interaction with their ligands. P55 TNFR appears to be more ubiquitously expressed, shows no species preference, and is the receptor whose binding induces killing of murine WEHI 164 cells (81). On the other hand, p75TNFR is expressed on a rather limited range of cells (82) and exhibits some degree of species preference, in that murine thymocytes respond to murine but not human TNF (81). It has been identified as a mediator of proliferation and has also been implicated in signaling through NF-κB (83). Its role in cytotoxicity should not be overlooked, however, as it has been demonstrated to play such a role in both tissue culture fibroblasts (80) and activated CD8 T cells (84).

The two types of TNFRs and LTβR are expressed on most nucleated cell types, but they are regulated differently, and their ratio on various cells and response to inductive signals differs. The genes for the TNFRs are located on different chromosomes; p55 TNFR is encoded by a gene on mouse chromosome 6 and human chromosome 12; p75 TNFR is encoded by a gene on mouse chromosome 4 and human chromosome 1 (summarized in Ref. 85). LTβR maps to murine chromosome 6, tightly linked to p55 TNFR (75). Its mRNA sequence is expressed in lymph nodes, thymic medullary cell lines, spleen, liver, heart, lung, kidney, and weakly in testis. Interestingly, it does not appear to be expressed by T cells or in the brain, despite the fact that its ligand is expressed in both of those locations.

Considerable information is emerging regarding signaling mechanisms and associated proteins for this family of receptors, which will be covered elsewhere in this volume and has recently been summarized (86). Briefly, several proteins have been identified. TRADD binds to the cytoplasmic death domain of p55 TNFR (87). Other proteins termed TRAF2 or TRAP (which may be the human homolog of TRAF2) bind p75 TNFR cytoplasmic regions (88,89). Other TRAF proteins whose function are not known have been identified (88). These proteins have been implicated in activation of NK-κB and cell death.

Mice deficient for the TNFRs have been produced through genetic engineering. The p55 TNFR knock-out mice appear phenotypically normal and have lymph nodes, but are highly susceptible to infection by *Listeria monocytogenes* and resistant to endotoxin shock (90). These data suggest that the p55 TNFR plays a crucial role in these two biological activities. It is most likely that TNF-α, a product of LPS-activated macrophages, is the mediator of endotoxin shock. LT has not been invoked in this syndrome. The mediator for *Listeria* resistance could be TNF-α and/or LT. The p75 TNFR knock-out mice also show some decreased sensitivity to endotoxin shock as well as normal T-cell development (91). Since both p55 and p75 TNFR knock-out mice have lymph nodes, as do the double-receptor knock-outs, LT acts in lymphoid development through a different receptor. This could be through the LT-$\alpha_1\beta_2$ complex binding LTβR, or LT may act alone in development through an as yet unidentified receptor.

MAJOR BIOLOGICAL ACTIVITIES

LT carries out many activities in tissue culture including killing several different tumor cell lines, inducing DNA fragmentation, activating fibroblast growth, and activating expression of several genes including those of the MHC and adhesion molecules (reviewed in Refs. 6,7,36). LT is a growth factor for activated B cell (43) and may act as an autocrine factor for B lymphomas (92) and some Epstein-Barr virus–infected B-cell lines (93). LT also acts as an osteoclast-activating factor (94). It stimulates replication of HIV (95). Its several other activities that have been defined in vitro have been extensively summarized (6,7,29) and compared with those of TNF-α.

LT and TNF-α carry out many of the same activities in vitro, though both qualitative and quantitative differences have been noted. The basis for occasional biological differences is not understood since both TNF receptors bind both ligands with high affinities. Dissociation constants ranging between 10^{-9} and 10^{-11} M have been reported. The diversity of LT's effects on a variety of different targets, from induction of proliferation to killing, reflects the variety of ways that target cells can respond to LT and suggests that different signaling mechanisms are used by individual cells in their response to that ligand.

At least one of the two molecular classes of TNFRs have been found on most nucleated cells. The number of receptors found on individual cells has been reported to range from less than 2,000 to more than 10,000. Though the presence of a receptor is required for a response to LT, the number of binding molecules expressed by a cell does not necessarily reflect its sensitivity to the ligand, presumably indicating the different biological functions of the receptors. It has also been reported that LT can form ion channels (96). Presumably this activity does not require receptor interaction.

The fact that antibodies to the p55 TNFR induce killing (97) suggests that a major part of LT's effects occurs through signal transduction mediated by one of the receptors. The recent isolation of several molecules associated with signaling is summarized above. It has been reported that TNF-α can activate an endosomal acid sphingomyelinase (98–100) generating ceramide, which activates a serine/threonine protein kinase, which then activates NF-κB. Since LT also binds p55 TNFR and enhances expression of many genes, at least in part, through activation of NF-κB (61,95) sphingomyelinase may be involved in LT's effects as well.

Both LT and TNF-α induce apoptosis or programmed cell death in some targets (3,4,61,95), as do many other members of the TNF family. Though this killing takes a long time, effects on the structure and integrity of target cell DNA are detectable considerably earlier. This phenomenon, also called "killing from within," is manifested as degradation of target cell DNA detectable as release of radiolabeled DNA, which can be visualized by gel electrophoresis in a laddered pattern, indicating degradation at regular intervals presumably internucleosomal spaces. Several mechanisms have been invoked to explain TNF effects on cellular DNA. These include generation of oxygen free radicals and activation or translocation to the nucleus of a cellular endonuclease. LT itself does not degrade DNA, but induces the process. Insight into the mechanism has emerged from a comparison of cell death molecules in *Caenorhabditis elegans* to those induced through the TNF receptors. Recently Duan et al. (101) demonstrated that ICE-LAP3, a homolog of *C. elegans* CED-3, is activated during TNF-induced apoptosis. The relationship of apoptosis to induction of gene expression is not clear. The conflicting effects of LT on target cells are similar to the effects of several other members of the TNFR family when ligand binding may result in proliferation or apoptosis (102), and in fact ceramide has been invoked in both activities (99).

LT plays a major role in lymphoid tissue development and organization and thus contributes to the overall function of the immune system. The revelation of this biological activity came from the analysis of mice made genetically deficient in LT production through the process of homologous recombination. The central observation in such mice is their lack of lymph nodes and Peyer's patches (8,9). Though the thymus appears normal, there is marked splenic disorganization with a loss of T- and B-cell areas and germinal centers and the accumulation of lymphocytes in the periportal areas of the liver and perivascularly in the lung (8,9). These latter defects may be secondary to the absence of the lymph nodes, since the splenic architecture, though not the lymph nodes, can be restored with infusions of normal bone marrow (103). It has been assumed by some authors (104) that LT exerts its effects in lymphoid development as an LTαβ complex because both p55 TNFR and p75 TNFR knock-out mice have normal lymph nodes. This invokes a role for another receptor, which could be LTβR. An alternative explanation is that there is another LT-specific receptor that has not yet been identified and is expressed for a limited time during development. The role (if any) of LT-β in lymphoid development will be revealed with the analysis of LT-β knock-out mice (Koni, L. et al., in preparation). The mechanism by which LT exerts its effects in development is not understood. There are several, non–mutually exclusive possibilities. LT may induce expression of endothelial adhesion molecules and induce the egress of lymphoid cells from the veins into the mesenchyme between those vessels and lymphoid vessels; LT could induce differentiation of dendritic cells, which then contribute to lymph node organization; it could act as a growth factor for lymphoid cells or induce expression of chemokines.

The *aly* mouse is a natural mutant that has no lymph nodes (105). It is not clear what LT's relationship is to this mutation. The *aly* mutant maps to mouse chromosome 11, which does not include the genes for any of the obvious candidates such as LT, LT-β, or their known receptors. It is possible that the *aly* mutation involves a gene for a signaling or adhesion molecule, which is normally induced or activated by LT (106).

In addition to its beneficial and unique role in development, LT, along with TNF-α, contributes to the pathogenesis of some autoimmune diseases. Their synergism with interferon and interleukin-1 in destroying β cells of the islets of Langerhans (107) suggests a role in insulin-dependent diabetes mellitus. LT's induction of myelin swelling, its cytoxicity against oligodendrocytes (108,109), and its production by cells of the central nervous systems (58) implicate it in certain demyelinating diseases, such as multiple sclerosis. Its presence in the plaques of multiple sclerosis patients also implicates it in the disease (59). Its role as an inflammatory mediator also suggests its involvement in such diseases with their prominent mononuclear infiltrates. LT production is positively correlated with T cells' ability to transfer experimental allergic encephalomyelitis, a murine model of an autoimmune demyelinating disease (110). Further evidence implicating LT as an effector molecule of that disease was derived from studies with a monoclonal antibody that neutralizes both LT and TNF-α and inhibits transfer of symptoms by T-cell clones (42). This process is due in part to the TNFs proinflammatory activities, particularly induction of adhesion molecules in the central nervous system (111). This induction of adhesion molecules results in infiltration of T and B cells and macrophages. In fact, populations of T cells representing several different Vβ T cell receptors are drawn into such lesions (112), allowing the possibility of determinant spreading and reactivity to new antigens at the local site.

LT's activity in vivo as a proinflammatory mediator and contributor to autoimmune damage is apparent from studies in which its production is artificially regulated in transgenic mice. Mice transgenic for a construct consisting of the rat insulin promoter II (RIP) driving LT express high levels of the transgene in β cells in the islets of Langerhans, the kidney, and skin (113). This inappropriate expression has dramatic consequences: the hair has a distinctly "spikey" appearance and there is a marked inflammatory infiltrate in the kidney and around the islets (113,114). This infiltrate consists of CD4+ T cells, CD8+ T cells, B220+ B cells, and dendritic cells (115). Presumably this a response to LT-induced elevated expression of adhesion molecules at these sites. Similar, but not identical patterns of inflammation are seen in the islets of mice transgenic for the RIP–TNF-α construct (114,116). Despite the very intense inflammation in and around the islets of such transgenic mice, no β cell destruction is seen. However, the infiltrates clearly contain activatible cells, since β-cell destruction occurs if there is a second signal such as transgene-directed (influenza virus) antigen in immunized mice (116) or islet-expressed costimulator molecule (117; S. Guerder, personal communication).

We have recently determined that the inflammation caused by the LT transgene has most of the characteristics of organized lymphoid tissue with regard to cellular composition (T cells, B cells, plasma cells, and antigen-presenting cells), delineated T- and B-cell areas, primary and secondary follicles, characteristic morphological and antigenic (ICAM-1, VCAM-1, MAdCAM-1, and PNAd) features of high endothelial venules, and the ability to respond to antigen and undergo Ig class switching when immunized with SRBC (115). The vascular changes, with the exception of PNAd, appear to be the direct consequence of transgene-derived LT expression as they were also observed in RIP-LT mice lacking mature T and B cells. These observations are consistent with the above-noted role of LT in lymph node development. They engender a unifying concept for chronic inflammation and lymphoid organogenesis in emphasizing that LT's role in these processes is pivotal and similar. The ability of LT to organize lymphoid tissue, which in the embryo is a physiological role, may in the inflammatory process in the adult be pathogenic. Establishment of an organized lymphoid structure at a local site may result in continual antigen presentation, determinant spreading, and exacerbation of clinical signs—clearly consequences of LT expression that are major contributors to pathology in autoimmune disease.

LT may also play a role in the pathogenesis of viral infections. As noted above, HTLV-I induces expression of LT. There are several sequelae of HTLV-I infection in which LT may be implicated. The presence of LT in the serum of patients with hypercalcemia (118), and the high constitutive LT production by several HTLV-I–infected cell lines (52) suggests that LT could contribute to the pathogenesis of clinical manifestations of infection with that virus such as hypercalcemia and myelopathy. Hypercalcemia could be facilitated through LT's activity as an osteoclast-activating factor, and HTLV-I–associated myelopathy through its role as a demyelinating and proinflammatory agent. LT may also contribute to the pathogenesis of AIDS. Several years ago, it was suggested (119) that HIV could induce LT and that the cytokine would induce T-cell death through T-cell suicide (apoptosis). This hypothesis was made before several facts were known concerning the diverse activities of both TNFs and the variety of pathologies associated with AIDS. Known LT activities that could contribute to HIV pathogenesis include cachexia, B-cell activation, T-cell death, oligodendrocyte killing, and inflammation. Proinflammatory activities of LT may contribute to the lesions of pulmonary hyperplasia seen in pediatric AIDS patients, which have characteristics of lymphoid organs (120,121). LT's ability to induce HIV activation and replication (95) in infected T cells and monocytes has been demonstrated. This occurs through LT and TNF-α's ability to activate NF-κB, which binds to the HIV promoter region. All of these facts add substantial support to the hypothesis. Further support derives from the observation that L929 killing activity is increased in supernatants of HIV-activated cells (50) and that transfection of the HIV-1 *tat* III gene results in induction of LT mRNA and protein from Raji cells (56). It is now clear that, as suggested in 1986, cytokines, including LT, do contribute to HIV infection symptoms and disease progression.

In conclusion, it is apparent that LT can be both beneficial and detrimental to the individual. That this is the case with regard not only to members of the TNF family, but to other cytokines as well, is just beginning to be appreciated and should engender caution in imposing therapeutic schemes to regulate their production or activity. On the one hand, LT's roles in viral pathogenesis and autoimmune disease can be considered as harmful, and protocols to eliminate or regulate its production or activity would appear at first glance to be appropriate. On the other hand, LT is beneficial in its unique ability to organize lymphoid tissue. The provision of structures that allow pathogen trapping, antigen presentation, T-cell activation, and immunoglobulin switching is clearly of major benefit. We are just beginning to understand the role of LT, the LT-αβ complex, and the various receptors in these processes. Judging from the startling results obtained with LT knock-out mice, the realization that LT and TNF-α are not redundant, and the recent emergence of LT-β, we can be sure that the future holds many surprises.

ACKNOWLEDGMENTS

This work was supported by NIH RO1 CA-16885 and ROI AI-34404 and National Multiple Sclerosis Society grant RG 2394-B. We are grateful to Cheryl Bergman, Alexander Kratz, Nick Ferreri, Antonio-Campos-Neto, Rosalba Sacca, Carolyn Cuff, and Matthew Hanson for helpful discussions and to Nina Paul and Regina Turetskaya for contributions to previous versions of the chapter.

REFERENCES

1. Ruddle NH, Waksman BH. Cytotoxicity mediated by soluble antigen and lymphocytes in delayed hypersensitivity. III. Analysis of mechanism. J Exp Med 1968: 128:1267.
2. Granger GA, Williams TW. Lymphocyte cytotoxicity in vitro: activation and release of a cytotoxic factor. Nature 1968; 218:1253.
3. Schmid DS, Tite JP, Ruddle NH. DNA fragmentation: manifestation of target cell destruction mediated by cytotoxic T cell lines, lymphotoxin-secreting helper T cell clones and cell-free lymphotoxin-containing supernatant. Proc Natl Acad Sci (USA) 1986; 83:1881–1886.
4. Schmid DS, Hornung R, McGrath KM, Ruddle NH. Target cell DNA fragmentation is mediated by lymphotoxin and tumor necrosis factor. Lymphokine Res 1987; 61:195–202.
5. Millet I, Ruddle NH. Tumor necrosis factor-β. In: Nicole N, ed. Guidebook to Cytokines and Their Receptors. Oxford: Oxford University Press, 1994:105–108.
6. Paul NL, Ruddle NH. Lymphotoxin. Ann Rev Immunol 1988; 6:407–438.
7. Ruddle N. Tumor necrosis factor-beta; lymphotoxin-alpha. In: Thomson A, ed. Cytokine Handbook. 2nd ed. London: Academic Press, 1994; 305–318.
8. De Togni PD, Goellner J, Ruddle NH, et al. Abnormal development of peripheral lymphoid organs in mice deficient in lymphotoxin. Science 1994; 264:703–707.
9. Banks TA, Rouse BT, Kerley MK, et al. Lymphotoxin-alpha-deficient mice. Effects on secondary lymphoid organ development and humoral immune responsiveness. J Immunol 1995; 144:1685–1683.
10. Li C-b, Gray PW, Lin PF, McGrath KM, Ruddle FH, Ruddle NH. Cloning and expression of murine lymphotoxin cDNA. J Immunol 1987; 138:4496.
11. Gray PW, Aggarwal BB, Benton CV, et al. Cloning and expression of the cDNA for human lymphotoxin: a lymphokine with tumor necrosis activity. Nature 1984; 312:721.
12. Shalaby MR, Aggarwal Bb, Rinderknecht E, Svedersky LP, Finkle BS, Palladino MA. Activation of human polymorphonuclear leukocyte functions by interferon-α and tumor necrosis factors. J Immunol 1985; 135:2069–2073.
13. Androlewicz MJ, Browning JL, Ware CF. Lymphotoxin is expressed as a heteromeric complex with a distinct 33-kDa glycoprotein on the surface of an activated human T cell hybridoma. J Biol Chem 1992; 267:2542–2547.
14. Browning JL, Androlewicz MJ, Ware CF. Lymphotoxin and an associated 33 kDa glycoprotein are expressed on the surface of an activated T cell hybridoma. J Immunol 1991; 147:1230–1237.
15. Ware CF, Crowe PD, Grayson MH, Androlewich MJ, Browning JL. Expression of surface lymphotoxin and tumor necrosis factor on activated T, B, and natural killer cells. J Immunol 1992; 149:3881–3888.
16. Browning JL, Ngam-ek A, Lawton P, et al. Lymphotoxin-beta, a novel member of the TNF family that forms a heteromeric complex with lymphotoxin on the cell surface. Cell 1993; 72:847–856.

17. Smith CA, Farrah T, Goodwin RG. The TNF receptor superfamily of cellular and viral proteins: activation, costimulation, and death. Cell 1994; 76:959–962.
18. Aggarwal BB, Moffat B, Harkins RN. Purification and characterization of lymphotoxin from human lymphoblastoid cell line 1788. In: Cohen JJ, ed. Interleukins, Lymphokines, and Cytokines. New York: Academic Press, Inc., 1983, p. 521.
19. Aggarwal BB, Henzel WJ, Moffat B, Kohr WJ, Harkins RN. Primary structure of human lymphotoxin derived from 1788 lymphoblastoid cell line. J Biol Chem 1985; 260:2334.
20. Goeddel DV, Aggarwal BB, Gray PW, et al. Tumor Necrosis Factors: Gene Structure and Biological Activities. Cold Spring Harbor, NY: Cold Spring Harbor Symposia on Quantitative Biology, 1986, p. 597.
21. Ito H, Shirau T, Yamamoto S, et al. Molecular cloning of the gene encoding rabbit tumor necrosis factor. DNA 1986; 5:157.
22. Shakhov AN, Kuprash DV, Azizov MM, Jongeneel CV, Nedospasov SA. Structural analysis of the rabbit TNF locus, containing the genes encoding TNF-β (lymphotoxin) and TNF-α (tumor necrosis factor). Gene 1990; 95:215–221.
23. Schoenfeld HJ, Poeschl B, Frey JR, et al. Efficient purification of recombinant human tumor necrosis factor β from Eschericia coli yields a biologically active protein with a trimeric structure that binds to both tumor necrosis factor receptors. J Biol Chem 1991; 266:3863–3869.
24. Eck MJ, Ultsch M, Rinderknecht ER, deVos AM, Sprang SR. The structure of human lymphotoxin (tumor necrosis factor-β) at 1.9-Å resolution. J Biol Chem 1992; 267:2119–2122.
25. Jones EY, Stuart DI, Walker NPC. Structure of tumour necrosis factor. Nature 1989; 338:225–228.
26. Goh CR, Loh S-C, Porter AG. Aspartic acid 50 and tyrosine 108 are essential for receptor binding and cytotoxic activity of tumor necrosis factor beta (lymphotoxin). Prot Eng 1991; 4:785–791.
27. Yamagishi JL, Kawashima H, Matsuo N, et al. Mutational analysis of structure-activity relationships in human tumor necrosis factor-alpha. Prot Eng 1990; 3:713–719.
28. Van Ostade Z, Tavernier J, Prange T, Fiers W. Localization of the active site of human tumor necrosis factor (hTNF) by mutational analysis. EMBO J 1991; 4:827–836.
29. Ware CF, VanArsdale TL, Crowe PD, Browning JL. The ligands and receptors of the lymphotoxin system. Curr Top Microbiol Immunol 1995; 175:218.
30. Banner DW, D'Arcy A, Jaanes W, et al. Crystal structure of the soluble human 55 kd TNF receptor-human TNFβ complex: implications for TNF receptor activation. Cell 1993; 73:431–445.
31. Browning JL, Douglas I, Ngam-ek A, et al. Characterization of surface lymphotoxin forms. J Immunol 1995; 154:33–36.
32. McGeehan GM, Becherer Jr JD, Bast RC, et al. Regulation of tumour necrosis factor-α processing by a metalloproteinase inhibitor. Nature 1994; 370:558–561.
33. Mohler KM, Sleath PR, Fitzner JN, et al. Protection against a lethal dose of endotoxin by an inhibitor of tumour necrosis factor processing. Nature 1994; 370:218–220.
34. Lawton P, Nelson J, Tizard R, Browning JL. Characterization of the mouse lymphotoxin-β gene. J Immunol 1995; 154:239–246.
35. Pokholok DK, Maroulakou IG, Kuprash DY, et al. Cloning and expression analysis of the murine lymphotoxin β gene. Proc Natl Acad Sci USA 1995; 92:674–678.
36. Turetskaya R, Fashena SJ, Paul NL, Ruddle NH. Genomic structure, induction, and production of TNF-β. In: Aggarwal BB, Vilcek J, eds. Tumor Necrosis Factors: Structure, Function and Mechanism of Action. New York: Marcel Dekker, Inc., 1992: 35–60.
37. Millet I, Ruddle NH. Differential regulation of lymphotoxin (LT), lymphotoxin-β (LT-β), and TNF-α in murine T cell clones activated through the TCR. J Immunol 1994; 152:4335–4346.
38. Conta BS, Powell MB, Ruddle NH. Production of lymphotoxin, IFN-γ and IFN-α, β by murine T cell lines and clones. J Immunol 1983; 130:2231.
39. Conta BS, Powell MB, Ruddle NH. Activation of Lyt 1+ and Lyt 2+ T cell clones lines: stimulation of proliferation, lymphokine production, and self destruction. J Immunol 1985; 134:2185–2190.
40. Sung S-S, Bjorndahl J, Wang CY, Kao HT, Fu SM. Production of tumor necrosis factor-cachectin by human T cell lines and peripheral blood T lymphocytes stimulated by phorbol myristate acetate and anti-CD3 antibody. J Exp Med 1988; 167:937–953.
41. Steffen M, Ottmann OG, Moore MAS. Simultaneous production of tumor necrosis factor-α and lymphotoxin by normal T cells after induction with IL-2 and anti-T3. J Immunol 1988; 140:2621–2624.
42. Ruddle NH, Bergman CM, McGrath KM, et al. An antibody to lymphotoxin and tumor necrosis factor prevents transfer of experimental allergic encephalomyelitis. J Exp Med 1990; 172:1193–1200.
43. Kehrl JH, Alvarez-Mon M, Delsing GA, Fauci AS. Lymphotoxin is an important T cell-derived growth factor for human B cells. Science 1987; 238:1144–1146.
44. Sung S-S, Jung LKL, Walters JA, Jeffes EWB, Granger G, Fu SM. Production of lymphotoxin by isolated human tonsillar B lymphocytes and B lymphocyte cell lines. J Clin Invest 1989; 84:236–243.
45. Worm M, Geha RS. CD40-mediated lymphotoxin a expression in human B cells is tyrosine kinase dependent. Eur J Immunol 1995; 25:2438–2444.
46. O'Garra A, Stapleton G, Dhar V, et al. Production of cytokines by mouse B cells: B lymphomas and normal B cells produce interleukin 10. Int Immunol 1990; 2:821–832.
47. Laskov R, Lancz G, Ruddle NH, et al. Production of tumor necrosis factor (TNF-α) and lymphotoxin (TNF-β) by murine pre B and B cells lymphomas. J Immunol 1990;144:4324–3430.
48. Garrett IR, Durie BGM, Nedwin GE, et al. Production of lymphotoxin, a bone-resorbing cytokine, by cultured human myeloma cells. N Engl J Med 1987; 317:527.
49. Tschachler E, Robert-Guroff M, Gallo RC, Reitz MS. Human T-lymphotropic virus I-infected T cells constitutively express lymphotoxin in vitro. Blood 1989; 73:194–201.
50. Ratner L, Polmar SH, Paul N, Ruddle N. Cytotoxic factors secreted by cells infected by human immunodeficiency virus type I. AIDS Res. Human Retroviruses 1987; 3:147–154.
51. Kronke M, Hensel G, Schluter C, Scheurich P, Schutze S, Pfizenmaier K. Tumor necrosis factor and lymphotoxin gene expression in human tumor cell lines. Cancer Res 1988; 48:5417–5421.
52. Paul NL, Lenardo MJ, Novak KD, Sarr T, Tang W-L, Ruddle NH. Lymphotoxin activation by human T-cell leukemia virus type I-infected cell lines: role for NF-κB. J Virol 1990; 64:5412–5419.
53. Tschachler E, Bohnlein E, Felzmann S, Reitz MSJ. Human T-lymphotropic virus type I tax regulates the expression of the human lymphotoxin gene. Blood 1993; 81:95–100.
54. Lindholm PF, Reid RL, Brady JN. Extracellular Tax1 protein stimulates tumor necrosis factor-β and immunoglobulin

kappa light chain expression in lymphoid cells. J Virol 1992; 66:1294–1302.

55. Paul NL, Millet I, Ruddle NH. The lymphotoxin (TNF-β) promoter is stimulated by HTLV-I tax activation. Cytokine 1993; 5:372–378.
56. Sastry KJ, Reddy RH, Pandito R, Totpol K, Aggarwal BB. HIV tat gene induces tumor necrosis factor-β (lymphotoxin) in a human B lymphoblastoid cell line. Cell 1990; 61:361.
57. Melani C, Silvani A, Parmiani G, Colombo MP. Lymphotoxin gene expression by melanocytes and melanoma cell lines and persistence of unspliced mRNA. FEBS Lett 1993; 335:114–118.
58. Lieberman AP, Pitha P, Shin HS, Shin ML. Production of tumor necrosis factor and other cytokines by astrocytes stimulated with lipopolysaccharide or a neurotropic virus. Proc Natl Acad Sci USA 1989; 86:6348–6352.
59. Selmaj K, Raine CS, Cannella B, Brosnan CF. Identification of lymphotoxin and tumor necrosis factor in multiple sclerosis lesions. J Clin Invest 1991; 87:949–954.
60. Fashena SJ, Tang W-L, Sarr T, Ruddle NH. The murine lymphotoxin gene promoter: characterization and negative regulation. J Immunol 1990; 145:177–183.
61. Messer GE, Weiss EH, Baeuerle PA. Tumor necrosis factor β (TNF-β) induces binding of the NF-κB transcription factor to a high-affinity κB element in the TNF-β promoter. Cytokine 1990; 2:389–397.
62. Fashena SJ, Reeves R, Ruddle NH. A poly (dA-dT) upstream activating sequence binds high mobility group I protein and contributes to lymphotoxin (tumor necrosis factor-β) gene regulation. Mol Cell Biol 1992; 12:894–903.
63. Zhou X, Benson KF, Ashar HR, Chada K. Mutation responsible for the mouse pygmy phenotype in the developmentally regulated factor HMGI-C. Nature 1995; 376:771–774.
64. Eikelenboom P, Nassy JJJ, Post J, Versteeg JCMB, Langevoort HL. The histogenesis of lymph nodes in rat and rabbit. Anat Rec 1978; 190:201–215.
65. Jongeneel CV, Shakhov AN, Nedospasov SA, Cerottini J-C. Molecular control of tissue-specific expression at the mouse TNF locus. Eur J Immunol 1989; 19:549–552.
66. Ferreri NR, Sarr T, Askenase PW, Ruddle NR. Molecular regulation of tumor necrosis factor-α and lymphotoxin production in T cells: inhibition by prostaglandin E2. J Biol Chem 1992; 267:9443–9449.
67. English K, Weaver WM, Wilson CB. Differential regulation of lymphotoxin and tumor necrosis factor genes in human T lymphocytes. J Biol Chem 1991; 266:7108–7113.
68. Caput DB, Beutler B, Hartog K, Thayer R, Brown-Shimer S, Cerami A. Identification of a common nucleotide sequence in the 3′ untranslated region mRNA molecules specifying inflammatory mediators. Proc Natl Acad Sci USA 1986; 83:1670–1674.
69. Shaw G, Kamen R. A conserved AU sequence from the 3′ untranslated region of GM-CSF mRNA mediates selective mRNA degradation. Cell 1986; 46:659.
70. Weil D, Brosset S, Dautry F. RNA processing is a limiting step for murine tumor necrosis factor β expression in response to interleukin-2. Mol Cell Biol 1990; 10:5865–5875.
71. Abe Y, Horiuchi A, Osuka Y, Kimura S, Granger GA, Gatanaga T. Studies on membrane-associated and soluble (secreted) lymphotoxin in human lymphokine-activated T-killer cells in vitro. Lymphokine Cytokine Res 1992; 11:115–121.
72. Boussiotis VA, Nadler LM, Strominger JL, Goldfield AE. Tumor necrosis factor a is an autocrine growth factor for normal human B cells. Proc Natl Acad Sci USA 1994; 91:7007–7011.
73. Baens M, Chaffanet M, Cassiman JJ, van den Berghe H, Maryen P. Construction and evaluation of a hnc DNA library of human 12p transcribed sequences derived from a somatic cell hybrid. Genomics 1993; 16:214–218.
74. Crowe PD, VanArsdale TL, Walter BN, et al. A lymphotoxin-beta specific receptor. Science 1994; 264:707–710.
75. Force WK, Walter BN, Hession C, et al. Mouse lymphotoxin-β receptor. J Immunol 1995; 155:5280–5288.
76. Loetscher H, Pan YCE, Lahm HW, et al. Molecular cloning and expression of the human 55 kd tumor necrosis factor receptor. Cell 1990; 61:351–359.
77. Schall TJ, Lewis M, Koder KJ, et al. Molecular cloning and expression of a receptor for human tumor necrosis factor. Cell 1990; 61:361–370.
78. Smith CA, Davis T, Anderson D, et al. A receptor for tumor necrosis factor defines an unusual family of cellular and viral proteins. Science 1990; 248:1019–1023.
79. Tartaglia LA, Rothe M, Hu Y-F, Goeddel DV. Tumor necrosis factor's cytotoxic activity is signaled by the p55 TNF receptor. Cell 1993; 73:213–216.
80. Heller RA, Song K, Fan N. Cytotoxicity by tumor necrosis factor is mediated by both p55 and p70 receptors. Cell 1993; 73:216.
81. Barrett K, Taylor-Fishwick DA, Cope AP, et al. Cloning, expression and cross-linking analysis of the murine p55 tumor necrosis factor receptor. Eur J Immunol 1991; 21:1649–1656.
82. Ware C, Crowe PD, Vanarsdale TL, et al. Tumor necrosis factor (TNF) receptor expression in T lymphocytes: differential regulation of the Type I TNF receptor during activation of resting and effector cells. J Immunol 1991; 147:4229–4238.
83. Hohmann H-P, Brockhaus M, Baeuerle PA, Remy R, Kolbeck R, VanLoon APGM. Expression of the types A and B tumor necrosis factor (TNF) receptors is independently regulated, and both receptors mediate activation of the transcription factor, NF-κB. J Biol Chem 1991; 265:22409–22417.
84. Zheng L, Fisher G, Miller RE, Peschon J, Lynch DH, Lenardo MJ. Induction of apoptosis in mature T cells by tumour necrosis factor. Nature 1995; 377:348–351.
85. Ruddle NH. Tumor necrosis factor (TNF-α) and lymphotoxin (TNF-β). Curr Opin Immunol 1992; 4:327–332.
86. Vandenabeele P, DeClercq W, Beyaert R, Fiers W. Two tumor necrosis factor receptors-Structure and function. Trends Cell Biol 1995; 5:392–399.
87. Hsu H, Xiong J, Goeddel D. The TNF receptor-1 associated protein TRADD signals cell death and NF-κB activation. Cell 1995; 81:495–504.
88. Rothe M, Wong SC, Henzel WJ, Goeddel DV. A novel family of putative signal transducers associated with the cytoplasmic domain of the 75 kDA tumor necrosis factor receptor. Cell 1994; 78:681–692.
89. Song HY, Donner DB. Association of a ring finger protein with the cytoplasmic domain of the human type-2 tumor necrosis factor receptor. Biochem J 1995; 309:825–829.
90. Pfeffer K, Matsuyama T, Kunding TM, et al. Mice deficient for the 55 kd tumor necrosis factor receptor are resistant to endotoxic shock, yet succumb to L. monocytogenes infection. Cell 1993; 73:457–467.
91. Erickson SL, Sauvage FJd, Kikly K, et al. Decreased sensitivity to tumour necrosis factor but normal T-cell development in TNF-receptor-2-deficient mice. Nature 1994; 372:560–563.
92. Seregina TM, Mekshenkov MI, Turetskaya RL, Nedospasov SA. An autocrine growth factor constitutively produced by a human lymphoblastoid B-cell line is serologically related to lymphotoxin (TNF-β). Mol Immunol 1989; 26:339–342.
93. Estrov Z, Kurzrock R, Pocsike, et al. Lymphotoxin is an

autocrine growth factor for Epstein-Barr virus-infected B cell lines. J Exp Med 1993; 177:763–774.
94. Bertolini DR, Nedwin GE, Bringman TS, Smith DD, Mundy TR. Stimulation of bone resorption and inhibition of bone formation in vitro by human tumor necrosis factor. Nature 1986; 319:516–518.
95. Folks TM, Clouse KA, Justement J, et al. Tumor necrosis factor-α expression of human immunodeficiency virus in a chronically infected T cell clones. Proc Natl Acad Sci USA 1989; 86:2365–2368.
96. Baldwin RL, Mirzabekov T, Kagan BL, Wisnieski BJ. Conformation and ion channel activity of lymphotoxin at neutral and low pH. J Immunol 1995; 154:790–798.
97. Espevik T, Brockhaus M, Loetscher H, Nonstad U, Shalaby R. Characterization of binding and biologic effects of monoclonal antibodies against a human tumor necrosis receptor. J Exp Med 1990; 171:415.
98. Schtuze S, Potthoff K, Machleidt T, Berkovic D, Wiegmann K, Kronke M. TNF activates NF-κB by phosphatidylcholine-specific phospholipase C-induced "acidic" sphingomyelin breakdown. Cell 1992; 71:765–776.
99. Kolesnick R, Fuks Z. Ceramide: a signal for apoptosis or mitogenesis? J Exp Med 1995; 181:1949–1952.
100. Kolesnick R, Golde DW. The sphingomyelin pathway in tumor necrosis factor and interleukin-1 signalling. Cell 1994; 77:325–328.
101. Duan H, Chinnaiyan AM, Hudson PL, Wing JP, He W-W, Dixit V. ICE-LAP3, a novel mammalian homolog of the Caenorhabditis elegans cell death protein CED-3 is activated during Fas- and tumour necrosis factor-induced apoptosis. J Biol Chem 1996; 271:1621–1625.
102. Rabizadeh S, Oh J, Zhong LT, et al. Induction of apoptosis by the low-affinity NGF receptor. Science 1993; 261:345–348.
103. Marianthasan S, Matsumoto M, Baranyay F, Nahm MH, Kanagawa O, Chaplin DD. Absence of lymph nodes in lymphotoxin-α (LTα)-deficient mice is due to abnormal organ development, not defective lymphocyte migration. J Inflamm 1995; 45:72–78.
104. Beutler B, van Huffel C. Unraveling function in the TNF ligand and receptor families. Science 1994; 264:667–668.
105. Miyawaki S, Nakamura Y, Suzuka H, et al. A new mutation, aly, that induces a generalized lack of lymph nodes accompanied by immunodeficiency in mice. Eur J Immunol 1994; 24:429–434.
106. Kuramoto T, Mashimo T, Koike R, et al. The alymphoplasia (aly) mutation co-segregates with the intercellular adhesion molecule-2 (ICAM-1) on mouse chromosome 11. Int Immunol 1994; 6:991.
107. Pukel C, Baquerizo H, Rabinovitch A. Destruction of rat islet cell monolayers by cytokines. Synergistic interactions of interferon-γ tumor necrosis factor, lymphotoxin, and interleukin 1. Diabetes 1988; 37:133–136.
108. Selmaj K, Raine CS. Tumor necrosis factor mediates myelin and oligodendrocyte damage in vitro. Ann Neurol 1988; 23:339–346.
109. Selmaj K, Raine CF, Farooq M, Norton WT, Brosnan CF. Cytokine cytotoxicity against oligodendrocytes. Apoptosis induced by lymphotoxin. J Immunol 1991; 147:1522–1529.
110. Powell MB, Mitchell D, Lederman J, et al. Lymphotoxin and tumor necrosis factor-alpha production by myelin basic protein specific clones correlates with encephalitogenicity. Int Immunol 1990; 2:539–544.
111. Barten D, Ruddle NH. Vascular cell adhesion molecule-1 modulation by tumor necrosis factor in experimental allergic encephalomyelitis. J Neuroimmunol 1993; 51:123–133.
112. Barten D, Clark RB, Ruddle NH. Presence of T cells with activated and memory phenotypes in inflammatory spinal cord lesions. J Immunol 1995; 155:5409–5418.
113. Picarella D, Kratz A, Li C-b, Ruddle NH, Flavell RA. Insulitis in transgenic mice expressing TNF-β (lymphotoxin) in the pancreas. Proc Natl Acad Sci (USA) 1992; 89:10036–10040.
114. Picarella DE, Kratz A, Li C-b, Ruddle NH, Flavell RA. Transgenic TNF-α production in islets leads to insulitis, not diabetes: distinct patterns of inflammation in TNF-α and TNF-β transgenic mice. J Immunol 1993; 149:4136–4150.
115. Kratz A, Campos-Neto A, Hanson MS, Ruddle NH. Chronic inflammation caused by lymphotoxin is lymphoid neogenesis. J Exp Med 1996; 183:1461–1472.
116. Higuchi Y, Herrera P, Muniesa P, et al. Expression of a tumor necrosis factor alpha transgene in murine pancreatic beta cells results in severe and permanent insulitis without evolution towards diabetes. J Exp Med 1992; 176:1719–1731.
117. Guerder S, Picarella D, Linsley PS, Flavell RA. Costimulator B7-1 confers antigen-presenting cell function to parenchymal tissue and in conjunction with tumor necrosis factor alpha leads to autoimmunity in transgenic mice. Proc Natl Acad Sci USA 1994; 91:5138–5142.
118. Ishibashi K, Ishitsuka K, Chumon Y, et al. Tumor necrosis factor-β in the serum of adult T cell leukemia with hypercalcemia. Blood 1991; 77:2451–2455.
119. Ruddle NH. Lymphotoxin production in AIDS. Immunol Today 1986; 7:8–9.
120. Joshi VV. Pathologic findings associated with HIV infection in children. In: Pizzo PA, Wilfert CM, eds. Pediatric AIDS. The Challenge of HIV Infection in Infants, Children, and Adolescents. Baltimore, MD: Williams & Wilkins, 1991:113–119.
121. Connor EM, Marquis J, Oleske JM. Lymphoid interstitial pneumonitis. In: Pizzo PA, Wilfert CM, eds. Pediatric AIDS. The Challenge of HIV Infection in Infants, Children, and Adolescents. Baltimore, MD: Williams & Wilkins, 1991:343–354.

18
Virokines and Viroceptors

Michele Barry and Grant McFadden
University of Alberta, Edmonton, Alberta, Canada

INTRODUCTION

The vertebrate immune system has evolved under the influence of powerful selection pressures exerted by a multiplicity of infectious agents, and viruses have played a major role in shaping the effector mechanisms of the humoral and cellular immune response pathways. Since only those viruses that have adapted to the rigors imposed by these same immune pathways have survived in outbred host populations to the present day, it should not be surprising that the study of viral survival tactics can provide important clues to the innermost workings of the immune system itself. In most cases, the smaller viruses with limited coding capacity have survived by recognizing weaknesses or gaps in the immune repertoire and exploiting these to permit virus replication and host-to-host spread. However, the larger DNA viruses, particularly the poxviruses, herpesviruses, and adenoviruses, have also evolved strategies which include virus-encoded proteins that function by interacting directly with molecules of the host immune defense arsenal (reviewed in Refs. 1–6). In this chapter, we review only those virus proteins that either mimic the activities of host cytokines/growth factors and related soluble regulators ("virokines") or else are homologs of cellular cytokine/growth factor receptors ("viroceptors"). Virtually every important immune pathway discovered thus far has also been found to be a potential target for inhibitory proteins encoded by one or more viruses, and it is likely that as more cytokines with antiviral activity are discovered, more examples of counteractive viral anticytokine proteins will be revealed.

IMMUNE SUBVERSION BY VIRUSES

Virus-encoded proteins that directly modulate cytokine networks act by a variety of mechanisms. Table 1 defines several terms that have been adopted to illustrate the close relationship between immune antivirus responses and the collective viral anti-immune strategies. Although in this chapter we focus on virokines and viroceptors, particularly those that are secreted from infected cells or located at the surface of infected cells, it should be noted that many other viral strategies exist to subvert immune function. Table 2

Table 1 Virus-Encoded Modulators of Cytokines or Their Receptors

Term	Definition	Localization
Virokine	Mimic of host ligand or soluble immune regulator	Secreted
Viroceptor	Mimic of host receptor or related surface immune molecules	Type I: secreted Type II: cell surface Type III: intracellular
Cytokine response modifier (CRM)	Modulator of host cytokine induction, secretion, activation, or ligand function	Cell associated or secreted

Table 2 Strategies of Immune Subversion by Virus Proteins

Localization of viral protein	Mode of action	Virus example
I. Secreted	Virokine	See Table 3
	Viroceptor (Type I)	See Table 4
II. Cell surface	Viroceptor (Type II)	See Table 5
	Membrane channel	Influenza (M2)
	Superantigen	MMTV (orf1)
III. Intracellular	Viroceptor (Type III)	See Table 4
	Inhibition of apoptosis	Epstein-Barr virus (BHRF1)
	Block cytokine signal transduction	Adenovirus (E3-14.7K)
	Activate or repress cellular immune genes	Bovine papillomavirus (E2)
	Block cytokine secretion and/or activation	Cowpox virus (crmA)
	Inhibit viral antigen presentation	Adenovirus (E3-gp19K)
	Regulate host surface immune molecules	HIV-1 (nef)
	Regulate cell receptor function	Bovine papillomavirus (E5)
IV. Virion component	Modulate cell surface immune molecules	HIV-1 (gp120)

lists examples of some of the many intracellular pathways also known to be modified by viral proteins, and the reader is referred to recent reviews on this subject for further information (1,4,7–10).

Cytokines such as interferon (IFN)-α, β, and γ, and tumor necrosis factor (TNF)-α, and β can contribute to the antiviral response directly by inducing an antiviral state or causing a cytotoxic response in infected cells. Also, cytokines such as IFN-γ, interleukin-1β (IL-1β), and chemotactic cytokines can indirectly contribute to the antiviral responses via the upregulation of antiviral activities carried out by critical immune effector cells. Additionally, certain cytokines or the ligands for TNF, FAS, and CD40 can stimulate apoptotic responses to the viral infection via signaling through receptors that trigger programmmed cell death pathways (e.g., TNF, FAS, or CD40). Thus, the classification of which cytokines can be considered to exert "antiviral" effects can be a matter of definition and in its broadest scope could include any cytokine that becomes upregulated in response to viral challenge. It is noteworthy that, to date, analysis of the various anticytokine strategies exercised by viruses has consistently identified certain specific cytokines already shown, or suspected, to be mobilized during virus infections. The interferons are well documented as antiviral cytokines that serve to illustrate this point. A large variety of experimental systems have been used to document the importance of IFN in the antiviral response: (1) induction of the antiviral state in cultured cells, (2) antibody-mediated ligand depletion studies in animals, (3) the use of transgenic animals with knockouts in the ligand or receptor genes, and (4) infections with virus recombinants that either express IFN or else are deleted for the virus genes responsible for anti-interferon strategies. Such studies have also revealed that multiple intracellular and extracellular mechanisms are utilized by viruses against IFN-α, β, and γ (7,8). What follows, then, is a "viral perspective" of the extracellular immune network as revealed by the collective spectrum of virus-encoded proteins whose apparent function is to compromise the immune pathways critical for virus recognition and clearance.

VIROKINES

The term virokine was proposed in 1988 when vaccinia virus–encoded growth factor and complement inhibitor were first shown to be secreted from infected cells as soluble regulator molecules (11). The current list of virokines secreted from virus-infected cells include members as diverse as the cytokines, growth factors, and immune regulators they mimic (Table 3). It is likely that viruses originally acquired at least some of these genes, or their cDNA counterparts, from the hosts they infect. The larger DNA viruses, particularly the poxviruses and herpesviruses, as well as the oncogenic retroviruses, have been particularly successful at capturing host genes that aid in altering the microenvironment at infected sites to favor virus survival.

Mimics of cellular growth factors are good examples of secreted virus-encoded proteins that confer a selective advantage to the virus. To date, viruses have been found to express secreted proteins, which directly mimic the ligand activities of epidermal growth factor (EGF)/transforming growth factor-α (TGF-α), vascular endothelial growth factor (VEGF), and platelet-derived growth factor (PDGF). Sequence analysis of vaccinia virus, an orthopoxvirus,

Table 3 Secreted Virokines

Virus	Gene	Host homolog	Function
I. Poxvirus			
vaccinia	VCP (C21L)	C3b/C4b-binding proteins	Inhibit complement cascade
	VGF[a]	EGF/TGF-α	Paracrine stimulation of cellular EGF-R
rabbitpox	35KDa[b]	?	Retard host clinical symptoms
orf	A2R	Endothelial growth factor	Vascular remodeling?
myxoma	SERP-1/SPI-4	Serpin superfamily	Anti-inflammatory
II. Herpesvirus			
H. saimiri	CCPH/ORF-4	C4b-binding protein	Inhibit complement cascade
	ORF-13	CTLA-8	Unknown
Epstein-Barr virus	BCRF1/vIL-10[c]	IL-10	Pleiotropic immunoregulator?
III. Retrovirus			
simian sarcoma virus	v-sis	PDGF-B chain	Paracrine and autocrine stimulation of cellular PDGF-R

[a]Homologous growth factors also found in other poxviruses (e.g., myxoma, SFV, variola, etc.).
[b]Homologous to 35 kDa protein encoded by vaccinia strain Lister.
[c]Homolog also found in equine herpesvirus-2.

revealed the first example of a virus-encoded growth factor mimic, designated vaccinia growth factor (VGF) (12–14). VGF is expressed from vaccinia virus–infected cells as a 19 kDa glycoprotein. Alignment analysis with cellular proteins indicates that the VGF primary sequence includes a region that is homologous to the EGF and TGF-α superfamily, including the six cysteine residues important for the folding and biological activity of the ligand (reviewed in Ref. 15). VGF is first synthesized as a glycosylated transmembrane precursor that is subsequently cleaved and released as an active growth factor mimic, a scenario similar to the secretion of EGF and TGF-α (16,17). EGF or TGF-α normally bind to the cell surface EGF receptor (EGF-R) with high affinity, triggering an autophosphorylation event followed by downstream events that culminate in a mitogenic response (18). Virus-encoded growth factors such as VGF exploit this system by binding to the EGF-R and inappropriately stimulate cell growth of quiescent cells in a juxtacrine or paracrine fashion, so that subsequent virus replication in adjacent cells is enhanced. In vitro studies have shown that VGF is not required for virus growth in tissue culture, but deletion of the VGF gene from the vaccinia virus genome results in an attenuated disease in infected animals, indicating that the presence of VGF confers a selective advantage for virus growth in vivo (19). Similar secreted EGF mimics are encoded by other poxviruses, including variola virus (20), *Molluscum contagiosum* (21), Shope fibroma virus (22), malignant rabbit fibroma virus (23), and myxoma virus (24). In addition to EGF-like virokines, orf virus, a member of the *Parapoxvirus* genus, expresses a 16 kDa protein, with amino acid sequence homology to mammalian VEGF (25). This is of significance because infection of sheep and goats with orf virus results in extensive capillary proliferation and swelling of the dermis, a feature suspected to be due to the virus-encoded VEGF since mammalian VEGF is an important regulator of angiogenesis (25,26). In support of this, crude supernatants from cell cultures infected with orf virus have mitogenic activity for endothelial cells in vitro (25). Another example of a virus-encoded growth factor mimic is the v-*sis* oncogene of simian sarcoma virus, an acute transforming retrovirus, which encodes a protein that is closely related to the B chain of platelet-derived growth factor (PDGF) (27).

In addition to the presence of growth factor–like genes, some viruses express soluble proteins that are involved in modulation of the immune or inflammatory responses. The BCRF1 gene of Epstein-Barr virus (EBV) encodes a functional and structural homolog of human IL-10 (28). In general, IL-10 is an immune response suppressor and inhibits the synthesis of several potent antiviral cytokines (29). The EBV-encoded IL-10 (vIL-10) is 78% identical in amino acid sequence to human IL-10, suggesting that the vIL-10 was originally derived from the host gene (28). In addition, vIL-10 has retained many of the functions of the cellular IL-10 (reviewed in Ref. 30). For example, vIL-10 inhibits the synthesis and secretion of cytokines such as IFN-γ, IL-1, IL-12, and TNF-α (31). Thus, by directly antagonizing multiple potent antiviral cytokines, vIL-10 presumably dampens the overall antiviral response. In addition, both cellular IL-10 and vIL-10 stimulate B-cell growth and differentiation (32), which may directly contribute to the establishment or maintenance of latency in B cells.

Myxoma virus and malignant rabbit fibroma virus are related poxviruses that express a secreted 50–55 kDa protein designated SERP1 (also called SPI-4). SERP1 possesses structural and functional homology to the serpin family of serine proteinase inhibitors (33,34). Disruption of the two copies of the SERP1 gene in myxoma virus results in significant attenuation of the virus in infected rabbits due to a more effective inflammatory response, indicating that SERP1 may contribute to viral pathogenesis by inhibiting serine proteinases involved in the regulation of inflammation (34). SERP1 inhibits the activity of a number of extracellular human serine proteinases, including plasmin, urokinase, plasminogen activator, and C1s, but the in vivo substrate for SERP1 in infected tissues is unknown (35).

The complement pathway is an important mechanism for eliminating invading pathogens, and regulation of the pathway is tightly controlled by a regulatory network that includes specific complement-binding proteins. Both poxviruses and herpesviruses express proteins with significant homology to host complement regulatory proteins that function to inhibit complement-mediated clearance of infectious virus. Vaccinia virus complement-control protein (VCP) is homologous to the human complement regulatory protein C4-BP and, like C4-BP, binds both C4b and C3b, subsequently inhibiting progression of the complement pathway (4,11,36,37). The fourth open reading frame of herpesvirus saimiri (HVS), a lymphotropic tumor virus that infects New World monkeys, expresses a protein-designated complement-control protein homolog (CCPH) (38). CCPH is homologous to the membrane-associated complement control proteins such as decay accelerating factor and membrane cofactor protein (reviewed in Ref. 39). Alternative splicing of the mRNA accounts for the presence of two forms of CCPH, a soluble form (sCCPH) and a membrane-bound form (mCCPH), and transfected cells expressing mCCPH become resistant to complement-mediated lysis (38,39).

HVS also encodes proteins with potential relevance to the viral tropism for T cells. The ORF13 gene is 57% identical to a cellular T-cell–specific protein referred to as cytotoxic T-lymphocyte–associated antigen (CTLA-8) (40,41). CTLA-8 is expressed by activated T cells, but no biological function has yet been associated with the protein (40). The 5′-untranslated region of CTLA-8, however, exhibits characteristics of tightly regulated proteins such as cytokines, growth factors, and oncogenes, suggesting an important function for this protein (40). Transformation of human T-cell clones with HVS results in an alteration in the normal cytokine secretion profile of these cells (42), but the question of potential ORF13 involvement in this function remains to be answered.

Thus, virus-encoded proteins secreted from infected cells can have diverse effects on the microenvironment, particularly the innate immune system and early inflammatory pathways. While some of these viral proteins possess demonstrable homology to proteins important for immune regulation, a large number of viral proteins have yet to be assigned specific functions. An example is the 35 kDa major secreted protein of rabbitpox virus and vaccinia virus (strain Lister) (43). The rabbitpox protein seems to play a role in retarding the appearance of clinical symptoms in infected animals; however, no significant homology to any cellular protein is currently apparent (43). Systematic analysis of host proteins that potentially interact with this 35 kDa viral protein might well reveal novel elements of the host immune repertoire.

SECRETED (TYPE I) VIROCEPTORS

As indicated in Table 4, secreted viral homologs of cellular cytokine receptors have to date only been described in the poxvirus system (4–6,44,45). The first member of this novel class of cytokine regulators to be discovered is encoded by the T2 gene of Shope fibroma virus (SFV). The SFV T2 open reading frame was sequenced in 1987 and shown to express an early viral gene product (46). Analysis for potential homologs of T2 at that time revealed a significant relationship with the ligand-binding domain of what was then called the low-affinity nerve growth factor receptor, but it was not until the cellular type I and II TNF receptors were cloned and sequenced that it was realized T2 belonged to a larger TNF-receptor superfamily that includes a variety of important immunomodulatory receptors (47; reviewed in Ref. 48). Like both cellular TNF receptors (p55 and p75), the T2 viral protein includes a fourfold repeated cysteine-rich region, but unlike the cellular counterparts, it possesses a signal sequence and contains no obvious transmembrane domain. This observation provided the first clue that the T2 gene might express a truncated and secreted receptor capable of binding and inhibiting a cellular cytokine important for the normal antiviral response by the host. Consistent with this hypothesis, the expressed T2 protein of SFV was shown to be secreted as an extracellular protein that could bind TNF-α and TNF-β (47). A close homolog of T2 was later discovered in myxoma virus, and insertional mutagenesis of both copies of the myxoma T2 gene demonstrated that T2 was required for full virus pathogenesis in infected rabbits (49). The term "viroceptor" was then proposed to describe viral homologs of cellular receptors that function by intercepting the ligand-dependent signaling of cellular receptors

Table 4 Secreted (Type I) and Intracellular (Type III) Viroceptors

Viroceptor	Virus	Gene	Host homolog	Function
Type I	Poxvirus			
	vaccinia	B15R[a]	IL-1R (type II)	Inhibit IL-1β
	myxoma/SFV	T2[b]	TNF-R	Inhibit TNF-α/β
	myxoma/SFV	T7[c]	Interferon-γR	Inhibit interferon-γ
	vaccinia	B18R[d]	Ig superfamily	Inhibit interferon-α/β
	tanapox	Unmapped	Unknown	Inhibit interferon-γ, IL-2, and IL-5
Type III	Retrovirus	v-erbA	Thyroid hormone-R (intracellular)	Transforming oncogene
	avian erythroblastosis			

[a]Homologs also expressed by cowpox and myxoma.
[b]Homologs also expressed by cowpox, variola, and certain strains of vaccinia.
[c]Homologs in most poxviruses.
[d]Homolog encoded by variola.

(49). More recently, the T2 protein of myxoma was shown to inhibit the biological activity of rabbit TNF-α, but not the human or murine ligand, indicating that evolutionary pressures have resulted in strict species specificity in cases where the virus has co-evolved with only one specific host species (50). In the case of myxoma virus, for example, it has been estimated that this virus has been established in new world rabbit populations for at least 20 million years (51,52). Interestingly, the vaccinia virus genome includes two discontinuous and presumably nonfunctional fragments of TNF receptor genes, but this virus has been highly attenuated by two centuries of passage outside a true vertebrate host (53). In contrast, cowpox virus possesses two intact copies of apparently full-length TNF viroceptors denoted crmB and crmC (54). Since many other poxviruses also possess related viroceptors, it has been suggested that each might have a species specificity that reflects their particular and unique evolutionary history in vertebrate hosts (50,55).

In a systematic search for other secreted viroceptors expressed by poxviruses, it was discovered that the major secreted protein from myxoma virus–infected cells, a 37 kDa protein encoded by the T7 gene of the virus, exhibited significant sequence homology to the ligand-binding domain of the human and mouse IFN-γ receptor α chains (56). Binding studies indicated that the T7 protein formed 1:1 inhibitory complexes with rabbit IFN-γ, but, like the case with the myxoma TNF viroceptor, the myxoma IFN-γ viroceptor was specific for the rabbit ligand and did not inhibit human or mouse IFN-γ (57). Similarly, T7-like IFN-γ receptors have been reported in other poxviruses, such as SFV (56), vaccinia (58), ectromelia (55), cowpox (55), variola virus (20,59), and rabbitpox (55). It is likely that most, if not all, poxviruses encode and express similar viroceptors specific for IFN-γ, which probably reflects the importance of this ligand as a critical effector molecule of activated T cells and NK cells in the antiviral response (reviewed in Ref. 60).

The third cytokine known to be targeted by a poxvirus viroceptor is IL-1β, which synergizes with TNF in the early inflammatory responses to virus infection (reviewed in Ref. 61). The B15R gene of vaccinia and a closely related gene in cowpox both express a 50–60 kDa secreted protein that binds and inactivates IL-1β, but not IL-1α (62,63). The B15R sequence exhibits homology with the extracellular ligand-binding domain of the type II IL-1 receptor, and there is evidence that comparable genes exist in other poxvirus genomes as well, although the homolog in variola is apparently fragmented and nonfunctional (20). The role of B15R in virus pathogenesis is controversial. One group has reported that virus lacking B15R was less virulent than wild-type virus following intracranial inoculation of mice (63), while a second group showed an increase in disease symptoms following intranasal inoculation of mice (62). These discrepancies may be related to the fact that IL-1β production in response to virus infection can itself exacerbate fever and thus increase disease symptoms. Therefore, the route of virus inoculation and the host animal itself may profoundly influence how the IL-1 viroceptor alters the overall cytokine balance in response to the infection. Future studies with IL-1 viroceptors of poxviruses that have a defined pathogenic spectrum in a native host may shed light on this issue.

Very recently, a fourth class of secreted viroceptors expressed by poxviruses has been described exhibiting specificity for IFN-α/β. The B18R gene of vaccinia virus, originally defined as a member of the immunoglobulin superfamily, was shown to express a protein that could be detected on the surface of infected cells and as a soluble secreted protein capable of binding and inhibiting IFN-α/β, but not IFN-γ (64). Since poxviruses also express intracellular proteins that inhibit intermediates of the IFN-induced antiviral state (7), the existence of both intra- and extracellular mechanisms of subverting IFN function provides powerful testament to the importance of IFN as a major defense strategy against viral infections.

Finally, there is reason to suspect that more examples of secreted viroceptors remain to be discovered. For example, the major secreted protein expressed from tanapox-infected cells has been shown to bind and inhibit three distinct and independent human cytokines: IFN-γ, IL-2, and IL-5 (65). Although the gene encoding this protein has not yet been identified, it is possible that this is an example of a viroceptor with broader specificity for more than one ligand class. When considered collectively, the virus-encoded viroceptor "network" may yet rival the complexity of the host cytokine network activated by pathogenic challenge. The recent discovery that CD40 ligand has antiviral activity (by a still-unknown mechanism) gives further credence to the speculation that more viral anticytokine strategies targeted by viroceptors remain to be uncovered (66).

MEMBRANE-BOUND (TYPE II) VIROCEPTORS

In addition to type I secreted viroceptors, many viruses also express viroceptors that are expressed on the surface of infected cells. Many of these type II viroceptors show significant homology to host proteins normally expressed on the cell surface, and their localization suggests that the activities of this class of immunomodulator proteins are largely restricted to the infected cells themselves, most likely to confer some competitive advantage for the virus in the face of ongoing immune surveillance. To date, type II viroceptors have been described only for herpesviruses, poxviruses, and certain transforming retroviruses (Table 5).

The UL18 open reading frame of human cytomegaloviruses (HCMV) encodes a protein with homology to the heavy chain of the major histocompatibility complex (MHC) class I molecule (67). Originally it was speculated that the function of UL18 was to bind and sequester β_2-microglobulin (β2m), the light chain of the MHC class I molecules, thus preventing expression of viral peptide–

Table 5 Membrane-Associated (Type II) Viroceptors

Virus	Gene	Host homolog	Function
I. Poxvirus			
swinepox	K2R	Chemokine-R	Unknown
swinepox	C1L	IgE-c chain	Unknown
myxoma	M11L	?	Anti-inflammatory
vaccinia	B5R	Complement receptor	? (EEV formation)
ASFV	EP402R	CD2	Hemadsorption
II. Herpesvirus			
HSV	gC	C3 receptor	Binds C3b, iC3b, iC3b, and factor X
HSV	gE/gI[a]	FcR	Binds IgG
CMV	US28	Chemokine-R	Chemotactic modulation?
CMV	UL18	MHC-I (H chain)	Unknown
H. saimiri	ORF15/HVSCD59	CD59	Complement inhibition
H. saimiri	ECRF3/ORF74	Chemokine-R	Chemotactic modulation?
EBV	gp340	C3d	B cell adsorption
III. Retrovirus			
avian erythroblastosis virus	v-erbB	EGF-R	Transforming oncogene
McDonough feline sarcoma virus	v-fms	M-CSF-R	Transforming oncogene
feline sarcoma virus	v-kit	Stem cell factor-R	Transforming oncogene
URII avian sarcoma virus	v-ros	Insulin-R	Transforming oncogene
S13 avian erythroblastosis virus	v-sea	HGF-R family	Transforming oncogene
RPL30 avian retrovirus	v-eyk/v-ryk	Transmembrane tyrosine kinase superfamily[b]	Transforming oncogene

[a]Homologs encoded by CMV and several other herpesviruses.
[b]Includes eyk, ufo, axl, ark, mer, sky.

displaying class I molecules on the surface of HCMV-infected cells. Evidence now suggests, however, that UL18 is not involved in the downregulation of MHC class I from the surface of infected cells (68), since disruption of the UL18 gene resulted in no obvious effect on class I surface expression in vitro. Thus, a role for UL18 in viral replication or pathogenesis remains to be defined.

The ECRF3 open reading frame of HVS encodes a seven-transmembrane (serpentine) receptor that is related to the mammalian IL-8 receptor type B (69). Despite the overall low sequence homology (30%), studies have demonstrated that ECRF1 encodes a functional receptor for α-chemokines such as IL-8, neutrophil-activating peptide-2 (NAP-2), and GRO/melanoma growth-stimulatory activity (GRO/MGSA) as monitored by intracellular calcium mobilization (69). Like the host receptor that is its presumed progenitor, ECRF3 responds poorly to members of the β-chemokine family (69). In contrast to ECRF3, the open reading frame US28 of HCMV encodes a chemokine-like receptor that binds β-chemokines such as RANTES, MIP-1α, MIP-1β, and MCP-1, but not α-chemokines (70,71). Human erythroleukemic cells transfected with US28 were capable of initiating calcium mobilization in response to β-chemokine binding similarly to ECRF3 in response to α-chemokines (71). The subject of chemokine receptor mimicry by herpesviruses is reviewed elsewhere (72–74). The only known example of a chemokine-like receptor in the poxvirus family was found by sequence analysis of the genome of a swinepox virus (75). The open reading frame K2R bears significant homology to the chemokine receptor family and the US28 gene of HCMV, suggesting that it may also have chemokine-binding specificity, but this prediction remains to be confirmed experimentally. The function of virus-encoded chemokine receptor homologs is unknown, but since two receptor homologs studied to date are able to mediate a ligand-induced calcium signal, it is likely that chemokine-dependent modification of the host cell environment may somehow be beneficial, at least for the replication of HVS and HCMV. Additionally, the binding of chemokines to virus-infected cells might be a necessary component for immune evasion, for example, by inappropriate chemotaxis that disfavors virus clearance.

The M11L open reading frame of myxoma virus encodes a 166-amino-acid membrane-bound protein with no homology to any known cellular protein (76,77). Disruption of the M11L gene, however, results in the dramatic influx of leukocytes into myxoma-induced lesions, unlike infection with the wild-type virus, indicating the M11L is an important virulence factor involved in inhibiting leukocyte infiltration or preventing the elaboration of proinflammatory signals (76). Cell-surface expression of M11L is necessary for M11L function since a mutant of M11L that is not transported to the surface behaves similarly to the disrupted M11L virus (77), however, the biochemical mechanism of M11L is unknown.

The B5R gene of vaccinia virus encodes a 42 kDa membrane-bound protein with significant homology to host complement control proteins (78,79). To date, no complement inhibitory function has been demonstrated, but disruption of the B5R gene results in a decrease in the production of extracellular enveloped virus and attenuation of the virus in vivo (80,81). However, it is also possible that B5R could function as a membrane-bound complement receptor, since both herpes simplex virus (HSV-1) and HVS express membrane-bound complement regulatory proteins (82–87). Glycoprotein C (gC-1) is expressed on the surface of HSV-1–infected cells and binds C3 cleavage molecules of the complement pathway (82,83). Studies have shown that gC-1 and its homolog, gC-2, can protect infected cells against complement-mediated lysis (84,85). Herpesvirus saimiri expresses a protein (HVSCD59) with 48% identity to human CD59, an 18 kDa cell-surface glycoprotein that is involved in the regulation of complement-mediated lysis (86,87). CD59 regulates the complement cascade by preventing activation of C9 (88), and also plays a role in T-cell adhesion via interaction with CD2 on the surface of T cells (89). In addition to being a structural homolog of CD59, HVSCD59 is also a functional homolog since studies have shown that upon expression of HVSCD59, 3T3 cells become resistant to complement-mediated lysis (87).

The EBV envelope glycoprotein gp350 is necessary for virus attachment to the C3d complement receptor (CR2, also called CD21) on the surface of B lymphocytes (90,91). Two regions of gp350 primary amino acid sequence with signifant homology to the complement protein C3d have been identified (90). Interestingly, CR2/CD21 is a 155 kDa transmembrane glycoprotein that not only binds complement, but is also involved in B-cell activation (92).

In addition to the expression of viral-encoded complement receptors, HSV-infected cells also express receptors with specificity for the Fc domain of IgG (reviewed in Ref. 93). Fc binding is mediated by the interaction of two virus-encoded proteins, gE and gI, which are encoded by the US8 and US7 open reading frames, respectively (94,95). Both glycoproteins are required for optimal Fc-binding activity (95), but gE has been reported to bind low levels of IgG (96,97). The HSV Fc receptor is important for immune evasion, since the expression of gE/gI has been shown to protect cells from complement-mediated lysis and antibody-dependent cellular cytotoxicity (98,99). Similar virus-encoded Fc receptors have been reported for other herpesviruses such as varicella-zoster virus (100), human CMV (101), and EBV (102).

Amino acid sequence analysis of the genome of African swine fever virus (ASFV), the causative agent of African swine fever in domestic pigs, has revealed an open reading frame, EP402R, with significant homology to the membrane-bound T-cell adhesion molecule CD2 (103). CD2 is a member of the immunoglobulin supergene family present on the surface of T cells and NK cells that is important for adhesion of erythrocytes to lymphocytes and T-cell activation (104). In fact, disruption of EP40R2 results in decreased adhesion of erythrocytes to the surface of infected cells (103). EP40R2 is not essential for replication of the virus in vitro, but no data is currently available in regards to its contribution to pathogenesis in vivo. In addition, swinepox contains a gene (C1L) with significant homology to another member of the immunoglobulin supergene family, the constant chain of IgE (75), however, no function has currently been described for this protein.

In contrast to the membrane-bound viroceptors subverted from the host and encoded in the genomes of the larger DNA viruses (poxviruses and herpesviruses), the percent identity is often much higher (> 80%) for viroceptors encoded by retroviruses, indicating that these genes were more recently acquired from the host. Many retroviral oncogenes are functional viral encoded receptors that mediate signal transduction in a ligand-independent fashion. The oncogenes v-*kit* and v-*fms* are the only two oncogenes currently known to encode receptor homologs specific for cytokines known to be critical for immune regulation (105,106). v-*kit* is homologous to the stem cell factor receptor and is a member of the immunoglobulin supergene family, while v-*fms* is homologous to the receptor for colony-stimulating factor (CSF). To date, only one retroviral oncogene (v-erbA) expresses an intracellular receptor homolog, and this is the only known example of a type III viroceptor (Table 4).

PERSPECTIVES

Currently the DNA sequence database includes in excess of 1000 open reading frames encoded by the larger DNA viruses, most of which have unknown biological function. Since only a minority of these exhibit significant homology with known cellular genes (2), the potential exists for some of these virus genes to express immunomodulatory proteins that mimic or interact with new and as-yet-undiscovered elements of the vertebrate immune system. Although virokines and viroceptors represent only a subset of the known viral strategies to counteract the immune response to viral infections, the dissection of the biological roles of viral proteins that interact with the host cytokine/receptor network should continue to provide new insights into how the immune system recognizes and clears pathogens in general.

REFERENCES

1. Gooding LR. Virus proteins that counteract host immune defenses. Cell 1992; 61:5–7.
2. Murphy PM. Molecular mimicry and the generation of host defense protein diversity. Cell 1993; 72:823–826.
3. Marrack P, Kappler J. Subversion of the immune system by pathogens. Cell 1994; 76:323–332.
4. McFadden G. DNA viruses that affect cytokine networks. In: Aggarwal BB, Puri RK, eds. Human Cytokines: Their Role in Health and Disease. Blackwell Press, 1994:403–422.
5. Pickup DJ. Poxviral modifiers of cytokine responses to infection. Infect Agents Dis 1994; 3:116–127.
6. Smith GL. Virus strategies for evasion of the host response to infection. Trends Microbiol 1994; 2:80–88.
7. McNair ANB, Kerr IM. Viral inhibition of the interferon system. Pharmacol Ther 1993; 56:79–95.
8. Mathews MB. Viral evasion of cellular defense mechanisms: regulation of the protein kinase DAI by RNA effectors. Semin Virol 1993; 4:247–257.
9. Baserga R. Oncogenes and the strategy of growth factors. Cell 1994; 79:927–930.
10. Oltvai ZN, Korsmeyer SJ. Checkpoints of dueling dimers foil death wishes. Cell 1994; 79:189–192.
11. Kotwal GJ, Moss B. Vaccinia virus encodes a secretory polypeptide structurally related to complement control proteins. Nature 1988; 335:176–178.
12. Bloomquist MC, Hunt LT, Barker WC. Vaccinia virus 19-kilodalton protein: relationship to several mammalian proteins, including two growth factors. Proc Natl Acad Sci 1984; 81:7363–7367.
13. Brown JP, Twardsik DR, Marquardt H, Todaro GJ. Vaccinia virus encodes a polypeptide homologous to epidermal growth factor and transforming growth factor. Nature 1985; 313:491–492.
14. Reisner AH. Similarity between the vaccinia virus 19K early protein and epidermal growth factor. Nature 1985; 313:801–803.
15. McFadden G, Graham K, Opgenorth A. Poxvirus growth factors. In: McFadden G, ed. Viroceptors, Virokines and Related Immune Modulators Encoded by DNA Viruses. Austin, TX: R. G. Landes Company, 1995:1–15.
16. Stroobant P, Rice AP, Gullick WJ, Cheng DJ, Kerr IM, Waterfield MD. Purification and characterization of vaccinia virus growth factor. Cell 1985; 42:383–393.
17. Chang W, Lim JG, Hellstrom I, Gentry LE. Characterization of vaccinia virus growth factor biosynthetic pathway with an antipeptide antiserum. J Virol 1988; 62:1080–1083.
18. Hernandez-Sotomayor SMT, Carpenter G. Epidermal growth factor receptor: elements of intracellular communication. J Membrane Biol 1992; 128:31–89.
19. Buller RM, Chakrabarti S, Cooper JA, Twardzik DR, Moss B. Deletion of the vaccinia virus growth factor gene reduces virus virulence. J Virol 1988; 62:866–874.
20. Massung RF, Esposito JJ, Liu LI, Qi J, Utterback TR, Knight JC, Aubin L, Yuran TE, Parsons JM, Loparev VN, Selevanov NA, Cavallaro KF, Kerlavage AR, Mahy BWJ, Venter JC. Potential virulence determinants in terminal regions of variola smallpox virus genome. Nature 1993; 366:748–751.
21. Porter CR, Archard LC. Characterization and physical map-

ping of Molluscum contagiosum virus DNA and location of a sequence capable of encoding a conserved domain of epidermal growth factor. J Gen Virol 1987; 68:673–682.
22. Chang W, Upton C, Hu S, Purchio AF, McFadden G. The genome of Shope fibroma virus, a tumorigenic poxvirus, contains a growth factor gene with sequence similarity to those encoding epidermal growth factor and transforming growth factor alpha. Mol Cell Biol 1987; 7:535–540.
23. Upton C, McFadden G. Tumorigenic poxviruses: analysis of viral DNA sequences implicated in the tumorigenicity of Shope fibroma virus and malignant rabbit virus. Virology 1986; 152:308–321.
24. Upton C, Macen JL, McFadden G. Mapping and sequencing of a gene from myxoma virus that is related to those encoding epidermal growth factor and transforming growth factor alpha. J Virol 1987; 61:1271–1275.
25. Lyttle DJ, Fraser KM, Fleming SB, Mercer AA, Robinson AJ. Homologs of vascular endothelial growth factor are encoded by the poxvirus orf virus. J Virol 1994; 68:84–92.
26. Ferrara N, Houck K, Jakeman L, Leung DW. Molecular and biological properties of the vascular endothelial growth factor family of proteins. Endocrine Rev 1992; 13:18–32.
27. Doolittle RF, Hood LE, Devare SG, Robbins KC, Aaronson SA, Antoniades HN. Simian sarcoma virus onc gene, v-sis, is derived from the gene (or genes) encoding a platelet-derived growth factor. Science 1983; 221:275–277.
28. Moore KW, Vieira P, Fiorentino D, Trounstine ML, Khan TA, Mosmann T. Homology of cytokine synthesis inhibitory factor (IL-10) to the Epstein Barr virus gene BCRF1. Science 1990; 248:1230–1234.
29. Moore KW, O'Garra A, de Waal Malefyt R, Vieira P, Mosmann T. Interleukin-10. Annu Rev Immunol 1993; 11:165–190.
30. Swaminathan S, Kieff E. The role of BCRF1/vIL-10 in the life cycle of Epstein-Barr virus. In: McFadden G. ed. Viroceptors, Virokines and Related Immune Modulators Encoded by DNA Viruses. Austin, TX: R. G. Landes Company, 1995:111–125.
31. Hsu DH, de Waal Malefyt R, Fiorentino DF, Dang MN, Vieira P, de Vries J, Mosmann TR, Moore KW. Expression of interleukin-10 activity by Epstein-Barr virus protein BCRF1. Science 1990; 250:830–832.
32. Rousset F, Garcia E, DeFrance T, Peronne C, Vezzio N, Hzu DH, Kastelein R, Moore KW. Interleukin 10 is a potent growth and differentiation factor for activated human B lymphocytes. Proc Natl Acad Sci USA 1992; 89:1890–1893.
33. Upton C, Macen JL, Wishart DS, McFadden G. Myxoma virus and malignant rabbit fibroma virus encode a serpin-like protein important for virus virulence. Virology 1990; 179:618–631.
34. Macen JL, Upton C, Nation N, McFadden G. SERP1, a serine proteinase inhibitor encoded by myxoma virus, is a secreted glycoprotein that interferes with inflammation. Virology 1993; 195:348–363.
35. Lomas DA, Evans DL, Upton C, McFadden G, Carrell RW. Inhibition of plasmin, urokinase, tissue plasminogen activator, and C1s by a myxoma virus serine proteinase inhibitor. J Biol Chem 1993; 268:516–521.
36. Kotwal GJ, Isaacs SN, McKenzie R, Frank MM, Moss B. Inhibition of the complement cascade by the major secretory protein of vaccinia virus. Science 1990; 250:827–830.
37. Isaacs SN, Kotwal GJ, Moss B. Vaccinia virus complement-control protein prevents antibody-dependent complement-enhanced neutralization of infectivity and contributes to virulence. Proc Natl Acad Sci USA 1992; 89:628–632.
38. Albrecht JC, Fleckenstein B. New members of the multigene family of complement control proteins in Herpesvirus saimiri. J Virol 1992; 66:3937–3940.
39. Albrecht JC, Fleckenstein B. Complement regulatory proteins of Herpesvirus saimiri. In: McFadden G, ed. Viroceptors, Virokines and Related Immune Modulators Encoded by DNA Viruses. Austin, TX: R. G. Landes Company, 1995:127–145.
40. Rouvier E, Luciani MF, Mattei MG, Denizot F, Golstein P. CTLA-8, cloned from an activated T cell, bearing AU-rich messenger RNA instability sequences, and homologous to a Herpesvirus saimiri gene. J Immunol 1993; 150:5445–5456.
41. Albrecht JC, Nicholas J, Biller D, Cameron KR, Biesinger B, Newman C, Wittmann S, Craxton MA, Coleman H, Fleckenstein B, Honess RW. Primary structure of the Herpesvirus saimiri genome. J Virol 1992; 66:5047–5058.
42. De Carli M, Berthold S, Fickenscher H, Muller Fleckenstein I, D'Elios MM, Gao Q, Biagiotti R, Grazia Giudizi M, Kalden JR, Fleckenstein B, Romagnani S, Del Prete G. Immortalization with Herpesvirus saimiri modulates the cytokine secretion profile of established Th1 and Th2 human T cell clones. J Immunol 1993; 151:5022–5030.
43. Martinez-Pomares L, Thompson JP, Moyer RW. Mapping and investigation of the role in pathogenesis of the major unique secreted 35-kDa protein of rabbitpox virus. Virology 1995; 206:591–600.
44. Spriggs MK. Cytokine and cytokine receptor genes 'captured' by viruses. Curr Opin Immunol 1994; 6:526–529.
45. Spriggs MK. Poxvirus-encoded soluble cytokine receptors. Virus Res 1994; 33:1–10.
46. Upton C, DeLange AM, McFadden G. Tumorigenic poxviruses: genomic organization and DNA sequence of the telomeric region of the Shope fibroma virus genome. Virology 1987; 160:20–30.
47. Smith CA, Davis T, Wignall JM, Din WS, Farrah T, Upton C, McFadden G, Goodwin RG. T2 open reading frame from the Shope fibroma virus encodes a soluble form of the TNF receptor. Biochem Biophys Res Comm 1991; 176:335–342.
48. Smith CA, Goodwin RG. Tumor necrosis factor receptors in the poxvirus family: biological and genetic implications. In: McFadden G. ed. Viroceptors, Virokines and Related Immune Modulators Encoded by DNA Viruses. Austin, TX: R. G. Landes Company, 1995:29–40.
49. Upton C, Macen JL, Schreiber M, McFadden G. Myxoma virus expresses a secreted protein with homology to the tumor necrosis factor receptor gene family that contributes to viral virulence. Virology 1991; 184:370–382.
50. Schreiber M, McFadden G. The myxoma virus TNF-receptor homologue (T2) inhibits tumor necrosis factor α in a species-specific fashion. Virology 1994; 204:692–705.
51. Fenner F, Ratcliffe FN. Myxomatosis. London: Cambridge University Press, 1965.
52. Fenner F, Ross J. Myxomatosis. In: Thompson HV, King CM, eds. The European Rabbit. Oxford: Oxford University Press, 1994:205–239.
53. Howard ST, Chan YS, Smith GL. Vaccinia virus homologues of the Shope fibroma virus inverted terminal repeat proteins and a discontinuous ORF related to the tumor necrosis factor receptor family. Virology 1991; 180:633–647.
54. Hu F, Smith CA, Pickup DJ. Cowpox virus contains two copies of an early gene encoding a soluble secreted form of the type II TNF receptor. Virology 1994; 204:343–356.
55. Mossman K, Upton C, Buller RML, McFadden G. Species specificity of ectromelia virus and vaccinia virus interferon-γ binding proteins. Virology 1995; 208:762–769.
56. Upton C, Mossman K, McFadden G. Encoding of a homolog of the IFN-γ receptor by myxoma virus. Science 1992; 258:1369–1372.
57. Mossman K, Upton C, McFadden G. The myxoma virus-

soluble interferon-γ receptor homolog, M-T7, inhibits interferon-γ in a species-specific manner. J Biol Chem 1995; 270:3031–3038.
58. Goebel SJ, Johnson GP, Perkus ME, Davis SW, Winslow JP, Paoletti E. The complete DNA sequence of vaccinia virus. Virology 1990; 179:247–266.
59. Shchelkunov SN, Blinov VM, Sandakhchiev LS. Genes of variola virus and vaccinia viruses necessary to overcome the host protective mechanism. FEBS Lett 1993; 319:80–83.
60. Mossman K, Barry M, McFadden G. Interferon-γ receptors encoded by poxviruses. In: McFadden G, ed. Viroceptors, Virokines and Related Immune Modulators Encoded by DNA Viruses. Austin, TX: R. G. Landes Company, 1995:41–54.
61. Alcami A, Smith GL. Interleukin-1 receptors encoded by poxviruses. In: McFadden G, ed. Viroceptors, Virokines and Related Immune Modulators Encoded by DNA Viruses. Austin, TX: R. G. Landes Company, 1995:17–27.
62. Alcami A, Smith GL. A soluble receptor of interleukin-1β encoded by vaccinia virus: a novel mechanism of virus modulation of the host response to infection. Cell 1992; 71:153–167.
63. Spriggs MK, Hruby DE, Maliszewski CR, Pickup DJ, Sims JE, Buller RM. Vaccinia and cowpox viruses encode a novel secreted interleukin-1-binding protein. Cell 1992; 71:145–152.
64. Symons JA, Alcami A, Smith GL. Vaccinia virus encodes a soluble type I interferon receptor of novel structure and broad species specificity. Cell 1995; 81:1–20.
65. Essani K, Chalasani S, Eversole R, Beuving L, Birmingham L. Multiple anti-cytokine activities secreted from tanapox virus infected cells. Micro Path 1994; 17:347–353.
66. Ruby J, Bluethmann H, Aguet M, Ramshaw I. CD40 ligand has potent antiviral activity. Nature Med 1995; 1:437–441.
67. Browne H, Smith G, Beck S, Minson T. A complex between the MHC class I homologue encoded by human cytomegalovirus and β2 microglobulin. Nature 1990; 347:770–772.
68. Browne H, Churcher M, Minson T. Construction and characterization of a human cytomegalovirus mutant with the UL18 (class I homolog) gene deleted. J Virol 1992; 66:6784–6787.
69. Ahuja SK, Murphy PM. Molecular piracy of mammalian interleukin-8 receptor type B by herpesvirus saimiri. J Biol chem 1993; 268:20691–20694.
70. Neote K, DiGregorio D, Mak JY, Horuk R, Schall TJ. Molecular cloning, functional expression, and signaling characteristics of a C-C chemokine receptor. Cell 1993; 72:415–425.
71. Gao J, Murphy PM. Human cytomegalovirus open reading frame US28 encodes a functional β chemokine receptor. J Biol Chem 1994; 269:28539–28542.
72. Murphy PM. Molecular piracy of chemokine receptors by herpesviruses. Infect Agents Dis 1994; 3:137–154.
73. Ahuja SK, Gao JL, Murphy PM. Chemokine receptors and molecular mimicry. Immunol Today 1994; 15:281–287.
74. Schall TJ, Bacon KB. Chemokines, leukocyte trafficking, and inflammation. Curr Opin Immunol 1994; 6:865–873.
75. Massung RF, Jayarama V, Moyer RW. DNA sequence analysis of conserved and unique regions of swinepox virus: identification of genetic elements supporting phenotypic observations including a novel G protein-coupled receptor homologue. Virology 1993; 197:511–528.
76. Opgenorth A, Graham K, Nation N, Strayer D, McFadden G. Deletion analysis of two tandemly arranged virulence genes in myxoma virus, M11L and myxoma growth factor. J Virol 1992; 66:4720–4731.
77. Graham KA, Opgenorth A, Upton C, McFadden G. Myxoma virus M11L ORF encodes a protein for which cell surface localization is critical in manifestation of viral virulence. Virology 1992; 191:112–124.
78. Engelstad M, Howard ST, Smith GL. A constitutively expressed vaccinia gene encodes a 42-kDa glycoprotein related to complement control factors that forms part of the extracellular virus envelope. Virology 1992; 188:801–810.
79. Isaacs SN, Wolffe EJ, Payne LG, Moss B. Characterization of a vaccinia virus-encoded 42-kilodalton class I membrane glycoprotein component of the extracellular virus envelope. J Virol 1992; 66:7217–7224.
80. Engelstad M, Smith GL. The vaccinia virus 42-kDa envelope protein is required for the envelopment and egress of extracellular virus and for virus virulence. Virology 1993; 194:627–637.
81. Wolffe EF, Isaacs SN, Moss B. Deletion of the vaccinia virus B5R gene encoding a 42 kilodalton membrane glycoprotein inhibits extracellular virus envelope formation and dissemination. J Virol 1993; 67:4732–4741.
82. Clines DB, Lyss AP, Bina M. Fc and C3 receptors induced by herpes simplex virus on cultured human endothelial cells. J Clin Invest 1983; 69:123–128.
83. Smiley ML, Friedman HM. Binding of complement component C3b to glycoprotein C is modulated by sialic acid on herpes simplex virus type-1 infected cells. J Virol 1985; 55:857–861.
84. McNearney TA, Odell C, Holers VM, Spear PG, Atkinson JP. Herpes simplex virus glycoproteins gC-1 and gC-2 bind to the third component of complement and provide protection against complement mediated neuralization of viral infectivity. J Exp Med 1987; 166:1525–1535.
85. Harris SL, Frank I, Yee A, Cohen GH, Eisenberg RJ, Friedman HM. Glycoprotein C of herpes simplex virus type 1 prevents complement-mediated lysis and virus neutralization. J Infect Dis 1990; 162:331–337.
86. Albrecht JC, Nicholas J, Cameron KR, Newman C, Fleckenstein B, Honess RW. Herpesvirus saimiri has a gene specifying a homologue of the cellular membrane glycoprotein CD59. Virology 1992; 190:527–530.
87. Rother RP, Rollins SA, Fodor WL, Albrecht JC, Setter E, Fleckenstein B, Squinto SP. Inhibition of complement-mediated cytolysis by the terminal complement inhibitor of herpesvirus saimiri. J Virol 1994; 68:730–737.
88. Rollins SA, Sims PJ. The complement-inhibitory activity of CD59 resides in its capacity to block incorporation of C9 into membrane C5b-9. J Immunol 1990; 144:3478–3483.
89. Hahn WC, Menu E, Bothwell ALM, Sims PJ, Bierer BE. Overlapping but nonidentical binding sites on CD2 for CD58 and a second ligand CD59. Science 1992; 256:1805–1807.
90. Nemerow GR, Mold C, Keivens Schwend V, Tollefson V, Cooper NR. Identification of gp350 as the viral glycoprotein mediating attachment of Epstein-Barr virus (EBV) to the EBV/C3d receptor of B cells: sequence homology of gp350 and C3 complement fragment C3d. J Virol 1987; 61:1416–1420.
91. Nemerow GR, Houghten RA, Moore MD, Cooper NR. Identification of an epitope in the major envelope protein of Epstein-Barr virus that mediates viral binding to the B lymphocyte EBV receptor (CR2). Cell 1989; 56:369–377.
92. Tedder TF, Zhou LJ, Engel P. The CD19/CD21 signal transduction complex of B lymphocytes. Immunol Today 1994; 15:437–442.
93. York IA, Johnson DC. Inhibition of humoral and cellular immune recognition by herpes simplex viruses. In: McFadden G, ed. Viroceptors, Virokines and Related Immune Modulators Encoded by DNA Viruses. Austin, TX: R. G. Landes Company, 1995:89–110.
94. McGeoch DJ, Dolan A, Donald S, Rixon FJ. Sequence deter-

mination and genetic content of the short unique region in the genome of herpes simplex virus type 1. J Mol Biol 1985; 181:1–13.
95. Johnson DC, Frame MC, Ligas MW, Cross AM, Stow ND. Herpes simplex virus immunoglobulin G Fc receptor activity depends on a complex of two viral glycoproteins, gE and gI. J Virol 1988; 62:1347–1354.
96. Bell S, Cranage M, Borysiewicz L, Minson T. Induction of immunoglobulin G Fc receptors by recombinant vaccinia viruses expressing glycoproteins E and I of herpes simplex virus type I. J Virol 1990; 64:2181–2186.
97. Henke T, Graham FL, Lulitanond V, Johnson DC. Herpes simplex virus IgG Fc receptors induced using recombinant adenovirus vectors expressing glycoproteins E and I. Virology 1990; 177:437–444.
98. Frank I, Friedman HM. A novel function of the herpes simplex virus type 1 Fc receptor: participation in bipolar bridging of antiviral immunoglobulin G. J Virol 1989; 63:4479–4488.
99. Dubin G, Socolof E, Frank I, Friedman HM. Herpes simplex type 1 Fc receptor protects infected cells from antibody-dependent cellular cytotoxicity. J Virol 1991; 65:7046–7050.
100. Ogata M, Sigeta S. Appearance of immunoglobulin G Fc receptor in cultured human cells infected with varicella-zoster virus. Infect Immunol 1979; 26:770–774.
101. Furukawa T, Hornberger E, Sakuma S, Plotkin SA. Demonstration of immunoglobulin G receptors induced by human cytomegalovirus. J Clin Microb 1975; 2:332–336.
102. Yee C, Costa J, Hamilton V. Changes in the expression of Fc receptor produced by induction of Epstein-Barr virus in lymphoma cell lines. Virology 1982; 120:376–382.
103. Rodriguez JM, Yanez RJ, Almazan F, Vinuela E, Rodriguez JF. African swine fever virus encodes a CD2 homolog responsible for the adhesion of erythrocytes to infected cells. J Virol 1993; 67:5312–5320.
104. Bierer BE, Burakoff SJ. T-lymphocyte activation: the biology and function of CD2 and CD4. Immunol Rev 1989; 111:267–294.
105. Besmer P, Murphy JE, George PC, Qiu F, Bergold PJ, Lederman L, Snyder HW, Brodeur D, Zuckerman EE, Hardy WD. A new acute transforming feline retrovirus and relationship of its oncogene v-kit with the protein kinase gene family. Nature 1986; 320:415–421.
106. Sherr CJ, Rettenmier CW, Sacca R, Roussel MF, Look AT, Stanley ER. The c-fms proto-oncogene product is related to the receptor for the mononuclear phagocyte growth factor, CSF-1. Cell 1985; 41:665–676.

19
Cytokine Receptors

Virginia L. Shepherd and Rasul Abdolrasulnia
Vanderbilt University and VA Medical Center, Nashville, Tennessee

INTRODUCTION AND HISTORICAL BACKGROUND

The term cytokine defines a specific set of cell-derived protein mediators that have a wide range of activities on a variety of target cells. In 1979, the term interleukin (IL) was first used to define two specific proteins, one of macrophage origin (IL-1) and one of T-cell origin (IL-2). Both were able to induce a spectrum of T-cell activation events from proliferation to the generation of helper and cytolytic T cells. Since 1979, 15 interleukins, related colony-stimulating factors (CSF), the interferon family, and tumor necrosis factors (TNF) α and β have been described in molecular detail. These proteins comprise the group of mediators called cytokines. An area of intense research involves an understanding of how these cytokines act on their target cells to elicit specific responses. Central to this understanding is the hypothesis that cytokines act through a cell surface receptor and that this interaction is absolutely required for generation of the specific cytokine signal. Receptors for IL-2 were first described in 1981. Since that time, receptors have been identified for IL-1 through IL-13 and IL-15, the interferons, TNF-α and TNF-β, and related colony-stimulating factors. Binding studies from many laboratories using radiolabeled ligand have demonstrated that receptor number on the cell surface is low and ligand binds with relatively high affinity. Receptor characteristics were first identified by affinity cross-linking studies, offering some information about molecular size of the receptors involved and, later, supporting the idea that many, if not all, cytokines interact with a membrane-associated multimolecular complex. To date, the cDNAs encoding the binding and signaling subunits of most of the receptors for these cytokines have been cloned and sequenced.

The cytokine receptors fall into a unique class of endocytic, signal-transducing membrane proteins. Similar to the growth factor and hormone receptors, they bind a specific ligand at the cell surface, promote internalization, and generate a signal to the target cell. Unlike their counterparts, the cytokine receptors do not appear to function as protein kinases. This chapter will examine the current knowledge about the receptors for interleukin-1 through interleukin-13, interleukin-15, IFN-γ, and TNF-α, and how the characteristics of receptor structure and ligand-receptor interaction relate to the mechanism of cytokine action. In the past several years a wealth of information describing receptor function at the cellular and gene level has appeared, giving us a more complete understanding of how these protein mediators use cell-surface receptors to affect the wide range of activities on the variety of cells with which they interact.

CYTOKINE RECEPTOR CLASSES: STRUCTURE AND BIOCHEMISTRY

Accompanying chapters in this volume deal in depth with the structural and functional properties of interleukin-1 through interleukin-13, IFN-γ, and TNF-α. Cell-surface receptors have now been identified for all of these cytokines, as well as for the recently described IL-15. Binding affinities, cell distribution of receptors, receptor number per cell, and at least some biochemical characteristics have been reported for all of these cytokine receptors. The cDNAs for most of the receptors have been cloned and sequenced, and the genes for many of the receptor subunits

(a)

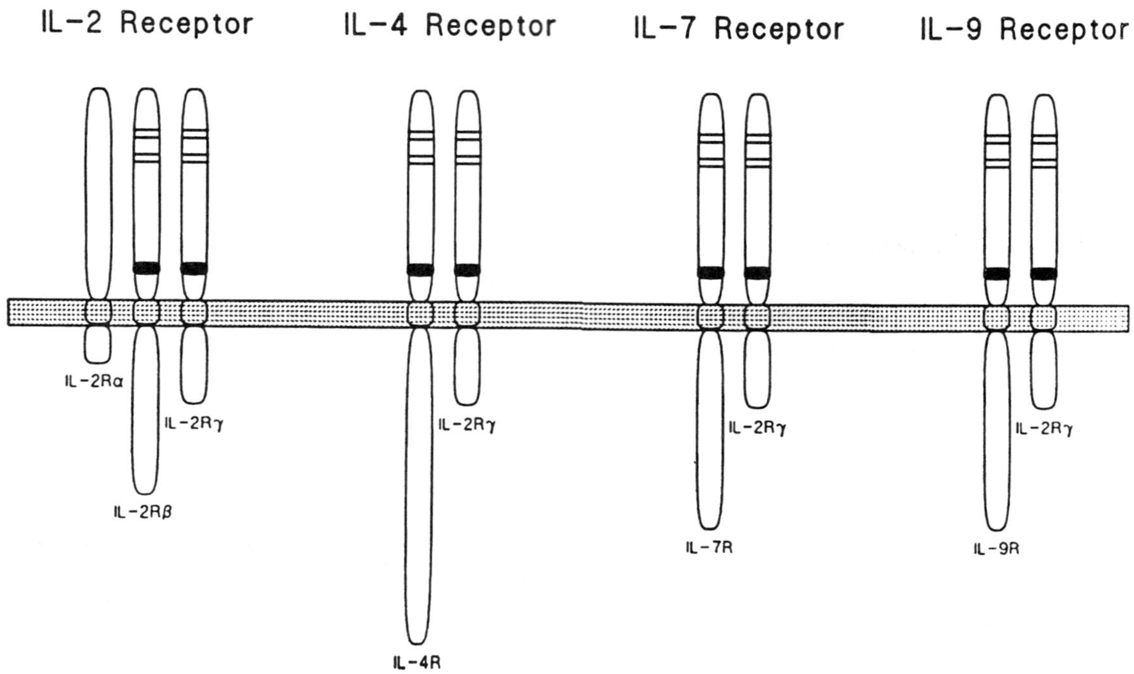

Figure 1 Structures of the class I cytokine receptor subfamilies. Schematic representations of the (a) IL-2 receptor, (b) IL-3 receptor, and (c) IL-6 receptor subfamilies of the class I cytokine receptor family are shown. Receptor domains are drawn to reflect approximate size of both intracellular and extracellular domains. Receptor-binding subunits are grouped with proposed signaling components. Motifs highlighted in each domain are as follows: conserved cysteine pairs, horizontal lines; conserved WSXWS motifs, solid box. The proposed IL-12–binding subunit (IL-12Rα) is marked with an asterisk.

have been isolated. This section will provide an overview of what is known today concerning the structures of these receptors and a discussion of the similarities and differences among the various receptors. Exciting new information has revealed that all of these receptors fall within definable subfamilies or classes and that the members of these classes share many structural and functional properties. Figure 1 is simplified representation of the structures of the receptors in class I, emphasizing orientation within the membrane, superfamily classification, lengths of intracellular and extracellular domains, and common structural features. Data from molecular cloning studies for the receptor subunits in classes I and II and for the IL-1 and TNF receptors are summarized in Table 1. Receptor classes, subfamilies, and conserved motifs are shown in Table 2. Following is a discussion of each of these classes and their respective member cytokines.

Class I Receptors

Class I receptors contain three distinct subfamilies: IL-2R, IL-3R, and IL-6R (27–29). Members of these subfamilies contain several conserved domains: a 200-amino-acid N-terminal domain that contains 14 antiparallel β strands, conserved N-terminal cysteines involved in dimerization, and a WSXWS motif just proximal to the transmembrane domain on the extracellular side that is postulated to be essential for ligand binding. Each member of class I is a multisubunit receptor, and receptors within subfamilies share common signaling subunits. None of these receptor components contains any motif in its cytoplasmic domain that is characteristic of a potential tyrosine or serine/threonine kinase, although all cytokines in class I induce rapid tyrosine phosphorylation of receptor components and intracellular substrates. Receptors in class I as well as class II require at least a portion of the cytoplasmic domain for signaling and contain one or more conserved motifs in their cytoplasmic portions that are necessary for signal transduction.

Receptors in class I are characterized by interaction of ligand with a specific α or binding subunit, followed by receptor association with a nonbinding signaling component. In each case, binding of ligand induces the formation

(b)

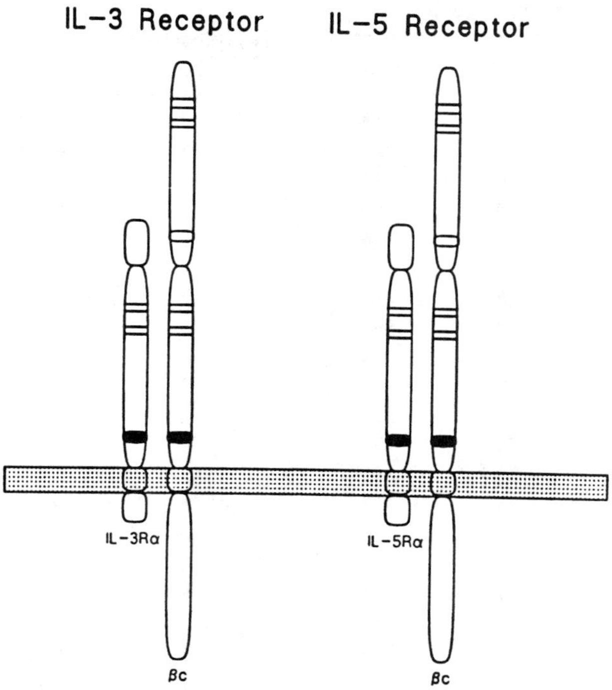
IL-3 Receptor
IL-5 Receptor
IL-3Rα
IL-5Rα
βc
βc

(c)

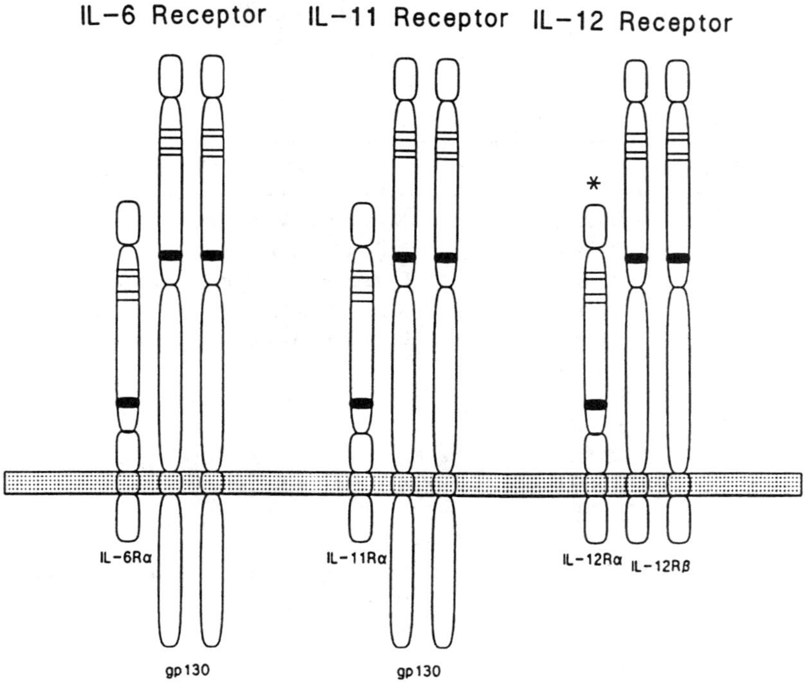
IL-6 Receptor
IL-11 Receptor
IL-12 Receptor
*
IL-6Rα
IL-11Rα
IL-12Rα
IL-12Rβ
gp130
gp130

Table 1 Summary of Cytokine Receptor Structural Characteristics

Receptor	Distribution of amino acids: Total	Extracellular	TM	Cytoplasmic	Extracellular N-glycosylation sites	Class	Subfamily	Mature mol. weight	Ref.
IL-1R-I	552	319	20	213	7	IL-1R	Ig	80	1
IL-1R-II				29	5	IL-1R	Ig	60	2
IL-2Rα	251	219	19	13	2			55	3–5
IL-2Rβ	525	214	25	286	4	I	IL-2R	75	6
IL-2Rγ	347	232	29	86	6	I	IL-2R	64	7
IL-3Rα	369	287	20	53	6	I	IL-3R	70	8
IL-3Rβ (βc)	856	417	26	413	2	I	IL-3R	120	9
IL-4R	800	207	24	569	6	I	IL-2R		10
IL-5Rα*	398	322	22	54	4	I	IL-3R	60	11
IL-6Rα	449	338	29	82	5	I	IL-6R	80	12
gp130	896	597	22	277	14	I	IL-6R	130	13
IL-7Rα	439	219	25	195	6	I	IL-2R	75	14
IL-9Rα	483	231	21	231	2	I	IL-2R	64	15
IL-10Rα	557	215	25	317	6	II	IFNR	90–100	16,17
IL-11Rα*	409	344	26	39	2	I	IL-6R		18
IL-12Rα	638	516	31	91	6	I	IL-6R	70	19
IFNγRα	489	245	21	223	5	II	IFNR	90	20,21
IFNγRβ	310	220	26	64	6	II	IFNR		22,23
TNFαRI	415	172	23	220	3	TNFR		55	24,25
TNFαRII	439	235	30	174	2	TNFR		75	26

All data are for human receptors unless indicated (* = murine).

Table 2 Cytokine Receptor Classes

Class	Subfamily	Cytokine receptor	Common features
I	IL-2R	IL-2Rβ	No intrinsic kinase activity
		IL-2Rγ	200 aa N-terminal sequence with 14 antiparallel β strands
		IL-4R	Conserved N-terminal cysteines
		IL-7R	Multiple subunits
		IL-9R	Conserved WSXWS C-terminal motif
		IL-13R	
		IL-15R	
	IL-3R	IL-3R	
		IL-5R	
	IL-6R	IL-6	
		IL-11	
		IL-12	
II	IFNR	IFN-γR	Conserved N- and C-terminal cysteines
		IL-10R	No WSXWS motif
			No kinase activity
Ig		IL-1R type I	
		IL-1R type II	
TNF R		TNF-αR type I	Conserved Cys repeats
		TNF-αR type II	
Chemokine R		IL-8R	Seven transmembrane-spanning domains

of a multisubunit complex that represents a signaling receptor. Class I contains three separate subfamilies that have been identified on the basis of overall similarities in receptor structure. The IL-2 receptor (IL-2R) subfamily contains the binding subunits for IL-2 (IL-2Rβ), IL-4, IL-7, IL-9, IL-13, and IL-15, and the common signaling subunit IL-2Rγ. The binding subunits for IL-3 and IL-5 and the common signaling chain βc constitute the IL-3R subfamily. The IL-6 subfamily contains receptors for IL-6, IL-11, and IL-12, and the common signal transduction molecule gp130. The binding subunit and the associated signaling subunit for each class I cytokine is shown in Table 3.

The IL-2 Receptor Subfamily

IL-2 has been called the major regulatory cytokine of the immune system (30). IL-2 is a 15 kDa protein produced primarily by CD4+ T cells. Its principal function is to amplify and regulate the immune response of a variety of cells. IL-2 function is mediated through high-affinity receptors expressed on these cells. The IL-2R complex is composed of three components: an α chain of 55 kDa (originally termed the Tac antigen) (3–5), a β chain of 75 kDa (6), and a γ chain of 64 kDa (7). The IL-2Rα subunit contains 219 amino acids in its extracellular domain. This subunit is unique among class I receptors in that it contains no extracellular motifs characteristic of this class of receptors, its primary structure is not similar to any other known cytokine receptor, and its cytoplasmic domain contains only 13 amino acids. Expression of the α chain increases IL-2 binding, but there is no evidence that this subunit contributes to IL-2 signaling (31). IL-2Rβ is required for high-affinity binding of IL-2 on activated T cells and intermediate binding on cells that do not express IL-2Rα (32). The IL-2Rβ component contains a total of 525 amino acids with a 214-amino-acid extracellular region (6). The conserved cysteine residues and the WSXWS motif place this receptor component in the class I family of receptors. The cytoplasmic domain of IL-2Rβ (286 amino acids) is significantly larger than the IL-2Rα intracellular portion. IL-2Rβ mutants lacking the cytoplasmic domain still associate with IL-2Rα to reconstitute high-affinity binding complexes (33), but deletion of the entire cytoplasmic domain blocks all IL-2 signaling. This domain is a rich in proline and contains two distinct subdomains that are high in serine and acidic amino acids. Specific deletion of the serine-rich region blocks IL-2–dependent proliferation and deletion of the acidic region blocks IL-2–dependent phosphorylation (6,33–35). In 1991 Murakami et al. reported that these specific regions are conserved in other receptors in class I, and they termed them the box 1 and box 2 cytoplasmic motifs (36). The role of these domains in signaling will be discussed below.

Table 3 Binding and Signaling Subunits of the Class I Family of Cytokines

Cytokine	Binding subunit(s)	Signaling subunit
IL-2	IL-2Rα + IL-2Rβ	IL-2Rγ
IL-3	IL-3	βc
IL-4	IL-4R	IL-2Rγ
IL-5	IL-5R	βc
IL-6	IL-6R	gp130
IL-7	IL-7R	IL-2Rγ
IL-9	IL-9R	IL-2Rγ
IL-11	IL-11R	gp130
IL-13	IL-4R + IL-13R	IL-13β
IL-15	IL-15R + IL-2Rβ	IL-2Rγ

An additional accessory component to the αβ complex was identified by Takeshita et al. (7), who detected a 64 kDa protein associated with IL-2Rβ in immunoprecipitates using anti–IL-2Rβ antibodies. These same workers and Hatakayama et al. (6) cloned the third IL-2R component, the γ chain. Based on the presence of the conserved cysteines and the WSXWS motif, this component is also a member of the class I family. The IL-2Rγ contains only 86 amino acids in its cytoplasmic domain, but appears to contain the conserved box 1 and box 2 motifs (36). Using cells transfected with various combinations of the IL-2R cDNAs, Takeshita et al. demonstrated that IL-2Rα and IL-2Rβ each bind IL-2 with low affinity (10^{-8} and 10^{-7} M), while no IL-2 binding to the γ chain is detectable (7). The αβ and βγ heterodimers bind IL-2 with intermediate affinity, and the trimeric structure αβγ binds IL-2 with high (10^{-11} M) affinity. Both the IL-2Rβγ and IL-2Rαβγ complexes are the biologically active forms of the receptor, whereas no physiological function for the IL-2Rαβ complex has yet been described.

IL-4 is a T-cell product originally described as a B-cell growth factor (37). IL-4 effects a wide range of hematopoietic cells as well as nonhematopoietic cells such as fibroblasts, epithelial cells, and tumor lines. Cross-linking studies in human cells demonstrated that the IL-4 binding protein was 140 kDa (38). Cloning and sequencing of both the murine and human 140 kDa IL-4 receptor have been reported (10,11). The human IL-4 receptor contains a total of 785 amino acids with 208 amino acids in the extracellular domain. This domain contains five potential N-linked glycosylation sites and seven cysteine residues. The cystoplasmic domain is high in proline and serine and high in acidic residues, similar to the IL-2Rβ. Transfection of the receptor cDNA into COS cells results in expression of an IL-4–binding protein with characteristics similar to the natural IL-4 receptor. Results from cross-linking studies in both murine and human cells suggest that additional proteins are involved in the IL-4 receptor complex. Recent studies have demonstrated that the γ chain of the IL-2R associates with IL-4R following binding of IL-4. Kondo et al. reported that antibodies to the IL-2Rγ chain inhibited IL-4-induced growth of CTLL-2 cells (40). In addition, Russel et al. reported that γ-chain monoclonal antibodies detect γ cross-linked to the IL-4R (41). However, recent studies by He et al. have raised the possibility that IL-4 may not require the presence of γ for its biological function in certain cell types (42). The finding that the IL-4–responsive B9 plasmacytoma does not express γ suggests that IL-4 does not necessarily require the IL-2Rγ chain for signaling. A recent hypothesis by He and Malek (43) suggests that two forms of the IL-4R might exist: a type I receptor that is expressed only on T cells that do not express IL-13R, and which might then use the γc chain, and a type II receptor expressed on IL-13–responsive cells that would use the IL-13Rβ component.

IL-7 is a 25 kDa protein involved in the regulation of lymphopoiesis (44). Park et al. demonstrated that murine cell lines and primary cells including cells of lymphoid and myeloid origin (45) express IL-7 receptors. Goodwin et al. (14) reported the cloning of the human and murine IL-7 receptors. Transfection of these cDNAs into COS cells results in expression of receptors that display identical binding properties to the native receptor. Both the murine and human IL-7 receptors contain 439 amino acids with an extracellular domain of 219 amino acids and a cytoplasmic domain of 195 amino acids. Cross-linking studies demonstrate a receptor molecular weight of 75 kDa for the transfected murine receptor, while cross-linking of IL-7 to its receptor on 70Z/3 cells showed a second protein at approximately 150 kDa. Kondo et al. demonstrated that the IL-2Rγ chain is the signaling component of the IL-7R complex (46). Anti–γ-chain antibodies inhibit proliferation of IL-7–dependent cells and specifically precipitate an IL-7–γ-chain cross-linked complex from a pre–B-cell line. The γ chain alone does not bind IL-7.

IL-9 is a multifunctional cytokine that is produced by activated T cells (47). Specific high-affinity receptors for IL-9 have been identified on a variety of mouse hematopoietic cells including T cells, mast cells, and macrophages. The murine cDNA encodes a protein of 468 amino acids with a 23-amino-acid signal peptide (15). The extracellular domain contains 233 amino acids, two potential N-linked glycosylation sites, and six cysteines. The position of the cysteine residues and the presence of a WSEWS sequence located on the extracellular side of the transmembrane domain indicate that this receptor belongs to the class I receptor family. The human IL-9R contains 522 amino acids with 53% homology to the murine receptor (15). The cytoplasmic domain of the human IL-9R is significantly larger than the murine counterpart (231 amino acids vs. 177 residues). Transfection of the human cDNA into IL-9R–deficient cells results in IL-9–responsive cells. The cytoplasmic domain of the IL-9R contains the high percentage of serines and prolines characteristic of this class of receptors. In addition, the IL-9R contains a Pro-X-Pro sequence preceded by a cluster of hydrophobic residues that is shared by the IL-2Rβ, IL-4R, and IL-7R (box 1). There is a 40% identity in the first 33 residues of the cytoplasmic domain between the IL-9R and the IL-2Rβ.

IL-13 is secreted by activated T cells and regulates inflammatory and immune responses (48). The action of IL-13 on peripheral blood monocytes includes inhibition of the production of inflammatory cytokines induced by LPS, stimulation of the production of IL-1R antagonist, and modulation of expression of cell surface molecules such as CD14, class II MHC antigens, and the mannose receptor (48–50). Other cells targets include B cells, fibroblasts, and endothelial cells. The observation that IL-13 and IL-4 share many of the same biological activities suggests that these two cytokines might bind to similar receptors, although the biochemical nature of the IL-13 receptor is unknown at this point. Zurawski et al. recently reported that the IL-4R and the IL-13R are structurally related (51). These same workers demonstrated that a mutated form of IL-4 that blocks IL-4 function also antagonizes IL-13 activity and that IL-13 and IL-4 cross-compete for binding to several cell types. These results suggest that IL-13 might bind to the IL-4R as well as a unique IL-13R, then complex with IL-2Rγ for signal transduction. However, recent studies by He et al. demonstrated that a γ-minus plasmacytoma cell line responds to low doses of IL-13 suggesting that this cytokine might not use the γ chain for signaling (42). In addition, IL-13–induced proliferation is not inhibited by anti-γc antibodies; B cells from X-linked SCID patients who lack a functional γc respond to IL-2 and IL-15 but not to IL-4 or IL-13, and expression of a mutant γc in B9 cells affects IL-4 but not IL-13 action. He and Malek have proposed that the IL-13 receptor complex does not contain γc, but does contain an as-yet-unidentified signaling component (IL-13Rβ) (43). Since it has been shown that IL-4 and IL-13 share a common receptor component in certain cells (52,53), this component may be the IL-13Rβ subunit. In support of these findings, Obiri et al. (54) and Lin et al. (55) have recently shown that γc is not a subunit of the human IL-13R.

IL-15 is a recently described 15 kDa glycoprotein that has biological functions similar to IL-2 (56). IL-15 stimulates proliferation of CTLL.2 cells and PHA-stimulated peripheral blood mononuclear cells and induces the generation of lymphokine-activated killer cell activity. The overlap with IL-2 functions suggests that IL-15 might share components of the IL-2R complex. In addition, IL-15 belongs to the family of four helix bundle–type cytokines that includes cytokines that bind to the IL-2R family. Giri and colleagues have demonstrated that cells transfected with IL-2Rβ and IL-2Rγ bind IL-15 and transduce IL-15 biological responses (57). However, the differential response of cells to IL-2 and IL-15 suggests that an additional IL-15–specific binding component exists. Their results demonstrate that IL-15, unlike the other IL-2 receptor family cytokines, shares the β subunit of the IL-2 receptor complex.

In summary, IL-2, IL-4, IL-7, and IL-9 all appear to interact with an α ligand–binding subunit, then interact with a common IL-2Rγ chain that mediates signaling. IL-13 may be unique in that it appears that γc may not be involved in IL-13 function. In addition, IL-15 is the only other cytokine besides IL-2 that interacts with the IL-2Rβ component.

IL-3 Receptor Subfamily

The IL-3 receptor (IL-3R) subfamily contains IL-3 and IL-5 and the related cytokine GM-CSF. The binding components of the IL-3R subfamily are 60–80 kDa glycoproteins with the common extracellular motifs found in class I cytokine receptors and short cytoplasmic domains. They share the same β-signaling subunit (βc) of 120 kDa that contains two repeats of the common motif of cytokine receptors in its extracellular portion and a large cytoplasmic domain that lacks intrinsic kinase activity.

IL-3 is a glycoprotein of 133 amino acids that is produced primarily by activated T cells (58). IL-3 is a potent hemopoietic growth factor that stimulates the proliferation and differentiation of a variety of hemopoietic cells including multipotent hemopoietic stem cells and lineage-committed cells. From cross-linking studies, human IL-3R was shown to consist of two proteins of 70 and 120 kDa (59). The larger protein was later shown to be identical to the β chain of the GM-CSF receptor (now referred to as the common β chain or βc) (8). Kitamura et al. reported the cloning of the human IL-3Rα subunit (70 kDa) (8). The extracellular domain (287 amino acids) contains the conserved WSXWS motif and cysteine pairs of the class I family. The cytoplasmic domain contains only 53 residues and does not contain the serine-rich domain found in the other class I receptors. It does, however, contain a box 1–like motif just after the transmembrane domain. This eight-residue motif is almost identical to the same domain in both the murine IL-5Rα (11) and the human GM-CSFRα subunits (60).

IL-5 regulates the production and function of B cells, eosinophils, and basophils (61). Binding studies and cross-linking experiments on murine B cell lines have revealed that high-affinity IL-5 receptors consist of at least two different polypeptide chains: an α chain (60 kDa) corresponding to a low-affinity binding protein, and the βc chain that associates with the α chain to form the high-affinity receptor (62,63). The βc component is required for high-affinity binding of IL-5 to the α chain as well as for signal transduction (64). Both the human and murine cDNAs for the IL-5Rα component have been cloned and sequenced (11). The IL-5Rα is a membrane-associated glycoprotein. The mature protein contains an extracellular domain of 322 amino acids containing three tandemly repeated sets of fibronectin type 3 domains, four cysteine residues in the amino-terminal half, and the WSXWS motif characteristic of class I receptors. The cytoplasmic portion contains 54 amino acids. This domain contains no consensus sequences for either a tyrosine kinase domain or a catalytic domain of a protein kinase. COS7 cells transfected with the murine IL-5Rα subunit bind IL-5 with low affinity but do not respond to IL-5. When FDC-P1 cells that are IL-3 dependent but do not bind IL-5 are transfected with the IL-5Rα cDNA, they bind IL-5 with high and low affinity and proliferate in response to IL-5, indicating that these cells constitutively express the βc chain. The βc chain by itself does not show any specific binding of IL-5. COS7 cells transfected with the human IL-5Rα bind IL-5 with high affinity, suggesting that the human IL-5 receptor is not as dependent on the β chain for high-affinity binding. To investigate the role of the cytoplasmic domain in signal transduction, FDC-P1 cells were transfected with a mutant IL-5Rα chain that lacked the total cytoplasmic region. These cells bound IL-5 with high affinity, but did not respond to IL-5. Using deletion mutants, regions rich in proline residues of the cytoplasmic domain were found to play a crucial role in mediating IL-5 signaling.

The βc chain is also a member of the IL-3R subfamily. It contains a total of 417 amino acids in its extracellular domain and 413 amino acids in the cytoplasmic region. The extracellular domain contains two repeats of the class I domains. The cytoplasmic domain contains the PXP box 1 motif and a box 2 motif similar to that found in IL-2Rβ. Both the α and βc components are necessary for responses to IL-3 and IL-5, but they serve distinct functions (65). The α chains are ligand specific and bind their respective ligands with low affinity. The βc subunit is used by both IL-3 and IL-5, and although it does not bind ligand by itself, it converts the binding of the cytokine to high affinity. Interestingly, the βc chain differentially enhances the affinity of the α chains: 500- to 1000-fold for IL-3 and only 2- to 5-fold for IL-5. Binding of ligand to the αβc complex results in a signaling receptor.

IL-6 Receptor Subfamily

The IL-6R family contains the binding subunits (α subunits) for IL-6, IL-11, and IL-12 and the related cytokines leukemia inhibitory factor (LIF), oncostatin M, and ciliary neurotrophic factor (CNTF), as well as the common signal transduction subunit gp130 (66). In each instance, ligand binding to the α subunit triggers association with gp130. Structural analyses revealed that these factors conform to the four–α-helix bundle structure first described for growth hormone (67). Based on similar effects on a variety of cell types and the shared structural similarities, it was suggested that these cytokines constitute a functionally related subfamily.

IL-6 is a multifunctional cytokine, which acts on a variety of cells (68). IL-6 is involved in the stimulation of immunoglobulin production by activated B cells, differentiation of T cells, regulation of the acute phase response, and induction of neural differentiation. It is produced by a number of cells including monocytes, endothelial cells, and fibroblasts in response to injury, inflammation, neoplasia, and autoimmune disease. Results from cross-linking studies demonstrate that the IL-6R is an 80 kDa protein (69). The nucleotide sequence of the cloned cDNA reveals that the receptor consists of an extracellular domain of 357 amino acids and a relatively short cytoplasmic domain of 82 residues (12). Taga et al. first reported that IL-6 binding to the IL-6R triggers the association of IL-6R with the signal transducer gp130 (70). Hibi et al. reported the cloning and expression of gp130 (13). Analysis of the cDNA shows an extracellular domain of 597 amino acids and a cytoplasmic domain of 277 amino acids. The extracellular domain contains six fibronectin-type III repeats, the second and third of which contain the four conserved cysteine residues and the WSXWS motif. It was previously shown that IL-6–induced IL-6R association with gp130 occurred through the extracellular domains of both receptor components since the soluble IL-6R bound to IL-6 can stimulate cells lacking IL-6R but expressing gp130 (71). Horsten et al. showed that the distal half of gp130 is sufficient for the interaction with IL-6 bound to the IL-6R (72). In addition, Grube and Cochran reported that residues 249–258 in this domain are involved in the regulation of IL-6 signal transmission and might represent the gp130–IL-6R recognition site (73). Ward et al. recently reported that the functional high-affinity human IL-6R complex consists of two molecules each of IL-6, IL-6R, and gp130 (74). The gp130 homodimer is formed by a disulfide linkage. The interactions involved in IL-6R or IL-6 homodimer formation are not known.

IL-11 was first isolated based on its ability to trigger the proliferation of a murine IL-6–dependent cell line (75). IL-11 shares many of the biological activities of IL-6. A receptor for IL-11 was identified on mouse plasmacytoma cells (76). Although this cell line was responsive to both IL-6 and IL-11, the two cytokines did not compete for receptor binding. Biochemical characterization of the IL-11 receptor demonstrated that the cytokine binds to a 150 kDa binding (α) subunit (77). Overlapping biological functions of IL-11 and IL-6 led Fourcin et al. (78) and Yin et al. (79) to study the possible involvement of gp130 in IL-11–mediated signaling. These studies showed that anti-gp130 monoclonal antibodies that block IL-6 function are also capable of blocking IL-11 function. A murine IL-11Rα chain was cloned by Hilton et al. (18) using degenerate oligonucleotides designed from the sequence encoding the WSXWS motif to screen a murine liver cDNA library. Using this strategy, these authors cloned the cDNAs for four separate proteins containing the WSXWS motif including gp130, the LIF α receptor, the IL-7 receptor, and the α-binding subunit for IL-11. The extracellular domain of the IL-11Rα contains 344 amino acids and includes the four conserved cysteine residues and the WSXWS motif found in class I receptors. IL-11R is most similar to the IL-6Rα chain with 24% amino acid identity. The cytoplasmic domain of the IL-11R contains only 39 amino acids with no motifs found in other receptor-binding subunits. Transfection of Ba/F3 cells with the IL-11Rα cDNA clone results in binding of ^{125}I–IL-11. However, a proliferative response of these cells to IL-11 requires coexpression of gp130, indicating that IL-11 binding to the α subunit requires gp130 for signal transduction.

IL-12 is a heterodimeric cytokine consisting of two disulfide-linked subunits: p35 and p40 (80). There is no sequence homology between p40 and p35. However, the p35 subunit is related to IL-6, while the p40 subunit is related to the extracellular domain of the IL-6 receptor. It has been postulated that IL-12 may have evolved from a cytokine/cytokine receptor complex that became covalently linked (81). A unique IL-12 receptor has been identified on PHA-stimulated lymphoblasts (82). Cross-linking experiments demonstrated that the IL-12R has a size of approximately 110 kDa. Chua et al. recently reported the cloning of the human IL-12R (19). The mature IL-12R contains a total of 638 amino acids. The extracellular domain is 516 residues long and contains six potential N-linked glycosylation sites. The presence of the two pairs of conserved cysteine residues and the WSXWS motif place this receptor in the class I family. The extracellular domain lacks the N-terminal immunoglobulin domain found in gp130 and therefore has only five fibronectin type III repeats. The IL-12R is most closely related over its entire length to gp130. However, its cytoplasmic domain is significantly shorter than the corresponding domain in gp130: 91 residues versus 276. This domain contains the sequences termed boxes 1 and 2 found in the first 100 residues of the intracellular portion of gp130 that have been shown to be sufficient for signal transduction. It has been postulated that IL-12 is a complex between a cytokine (p35) and a soluble receptor (p40) that closely parallels the IL-6 signaling pattern. It is possible that the IL-12R consists of only a single binding and signaling component. However, Chizzonite et al. have recently identified a receptor-associated protein of approximately 85 kDa that may represent an additional IL-12R component (83).

Class II Receptors

Class II receptors contain the IFN receptor family. The IFN-γR and IL-10R complexes are members of this family. These receptors contain conserved cysteines in both the N-terminal and C-terminal domains, but do not contain the WSXWS motif found in class I receptors. Both the IFN-γR and the IL-10R contain cytoplasmic domains that are similar to box 1 and box 2 previously described for the class I receptors. Similar to the class I receptors, both of these receptors appear to consist of a specific binding subunit and an associated signaling component.

IFN-γ is produced primarily by T cells and NK cells and plays an important role in host defense and modulation of immune responses (84). The IFN-γRα component of the high-affinity receptor has been cloned and sequenced (20,21). The intracellular domain of this receptor contains two functionally important regions: a 48-amino-acid sequence proximal to the transmembrane domain required for receptor-mediated internalization and biological response induction, and a second domain near the C-terminus containing residues Y440, D441, and H444, which are required for function. When the α chain is expressed in heterologous cells, the cells are unresponsive to IFN-γ, suggesting the existence of at least one additional component for signal transduction. Soh et al. (22) and Hemmi et al. (23) reported the cloning of an IFN-γR accessory molecule, which they called IFN-γRβ. The mature protein contains 224 amino acids in the extracellular domain and 66 residues in its cytoplasmic domain. The presence of two cysteine pairs and conserved proline, tryptophan, and tyrosine residues identifies it as a member of the IFN receptor family together with IFN-γRα and IL-10R. The relatively short cytoplasmic domain contains an LEVLD motif similar to the box 2 motif in IL-2Rβ and other class I receptors. Mutations in this domain have demonstrated that it is crucial for mitogenic responses to IFN-γ (85).

IL-10 is produced by B cells, T-helper cells, monocytes and macrophages, and keratinocytes (86). IL-10 has multiple effects on cells of the immune system, including inhibition of activation of T cells, NK cells, and macrophages. It has been proposed that these activities of IL-10 are mediated by specific cell-surface receptors. Both the murine and human IL-10R cDNAs have been cloned and sequenced (16,17). The human receptor contains a total of 557 amino acids, with an extracellular domain of 215 residues and a cytoplasmic portion of 317 amino acids. Although IL-10 is a member of the four–α-helix bundle cytokine family, which includes cytokines in the class I family, the structure of the IL-10R is most like the class II IFN-γRα, thus placing it in the IFNR family. The extracellular portion of the human IL-10R contains the conserved tryptophans and one of the cysteine pairs typical of class II receptors. IL-10 is known to inhibit a number of functions that are mediated by IFN-γ, suggesting several possible scenarios for relationships between the receptors for these two cytokines. First, the IFN-γR complex consists of a binding component and a signaling component. Similar to the class I receptors, IL-10 might share and compete for a common signaling subunit with IFN-γ. Alternatively, IL-10 might compete for similar intracellular downstream components of the class II signaling pathway. The latter hypothesis is supported by the recent finding that both IL-10 and IFN-γ activate a transcription factor that interacts with IFN-response elements in the FcγR1 gene (87).

IL-1 Receptor Family

The IL-1 subfamily contains the type I and II IL-1 receptors. IL-1 was originally described as a product of activated macrophages that stimulated proliferation of thymocytes in the presence of lectins or mitogens (reviewed in Refs. 88,89). It is now known that IL-1 has multiple biological effects on cells involved in immune or inflammatory responses including stimulation of lymphocyte differentiation and activation and stimulation of hematopoietic cell growth. In addition, IL-1 has profound proinflammatory effects on a variety of cell types. IL-1 exists in two forms, designated α and β. The primary structures of IL-1α and IL-1β show little amino acid similarity, although both interact with the same cell-surface receptors. IL-1 receptors have been identified on all IL-1–responsive cells. Two types of IL-1 receptors have been identified: a type I (80 kDa) receptor found predominantly on T cells and fibroblasts, and a type II (60 kDa) receptor expressed primarily on B cells and myeloid cells (90). It is now known that all IL-1–responsive cells express low levels of the type I receptor. Analyses of the deduced amino acid sequences for these receptors revealed that both contain three immunoglobulinlike motifs in their extracellular domains, placing these receptors in the immunoglobulin gene superfamily (1,2). These motifs share a common three-dimensional structure and are suggested to be involved in ligand binding. The cysteines involved in disulfide bonds are highly conserved. The two receptors differ considerably in their cytoplasmic domains: the type I receptor (IL-1RI) contains 215 amino acids in the cytoplasmic region, while IL-1RII contains only 29 amino acids. In general, IL-1RI binds IL-1α with the highest affinity, while IL-1RII binds IL-1β. However, each ligand competes completely with the other for binding to both receptor types (2). Recent work has demonstrated that type I IL-1 receptors mediate the biological function of IL-1 (91). Cells that are nonresponsive to IL-1 such as Jurkat cells gain responsiveness when they are transfected with the IL-1RI cDNA. It was originally suggested that IL-1RII functioned by forming a heterodimer with IL-1RI. However, work from several studies has not demonstrated any interaction of I and II following IL-1 binding (92). The findings that all IL-1 biological responses can be blocked with anti–IL-1RI antibodies while no signaling is blocked by anti–IL-1RII antibodies suggests that IL-1RII has no signaling role. In cells such as monocytes and neutrophils, which express greater than 90% type II receptors, biological responses appear to be mediated by the very low levels of type I receptors present: none of the biological activity is blocked by antibodies that inhibit IL-1 binding to type II receptors (91). The finding that some cells express predominantly type II receptors and that IL-2RII is increased during T-cell activation and monocyte differentiation (2,93) suggests that this receptor plays an important physiological role. The current hypothesis is that IL-1RII is a "decoy" receptor and functions to modulate responses of cells to IL-1 by binding IL-1 and preventing interaction with the type I signaling receptor (94). IL-1RII is released from cells as a soluble receptor (95) and specifically binds extracellular IL-1β with high affinity, thereby preventing interaction with IL-1R1. Thus, the IL-1RII may not function as a conventional cell-surface signaling receptor, but as an IL-1β–binding protein reservoir to modulate IL-1 activity during infection or inflammation. Recent exciting findings that viruses carry a gene for a soluble IL-1β–binding protein that is homologous to IL-1RII and blocks the inflammatory response of the host supports this unique role for IL-1RII (96).

TNF Receptor Family

Tumor necrosis factor-α (TNF) is a pleiotropic cytokine that is secreted predominantly by macrophages in response to LPS (97). TNF was originally defined by its ability to induce cytolytic or cytotoxic effects in some tumor cells. It is now known that virtually all nucleated cells studied express receptors for TNF. Two distinct receptors have been characterized: the type II 75 kDa receptor (TNF-R75) and the type I 55 kDa receptor (TNF-R55) (24–26). In the past several years, 12 related receptors have been identified (98). Based on their similar structures, they have been assigned to the newly described TNFR family. All of the sequence homology is in the extracellular domains, with conserved cysteine-rich repeats.

Both TNF receptors have been cloned and sequenced (24–26). TNF-R55 contains a total of 415 amino acids. The extracellular domain (172 amino acids) contains 22 cysteine residues and three potential N-glycosylation sites. There is no similarity to any of the previously described cytokine receptors, nor is there any immunoglobulinlike domain. Based on the presence of four cysteine-rich regions in the extracellular domain, this receptor belongs to the TNF-R family. The intracellular domain (220 amino acids) contains no kinase motif. TNF-R75 contains a total of 439 amino acids. The extracellular domain of this receptor (235 residues) also contains the conserved cysteine domains, placing it in the TNF-R family. Signaling via the TNF receptors appears to require the aggregation of receptor monomers initiated by binding of the trimeric TNF molecule. Although the extracellular domains of the two TNF receptors are 28% identical, the intracellular portions are completely unrelated, suggesting separate biological functions (99). TNF-R55 is known to be important in mediating a variety of functions including induction of cytolytic activity (100), increased antiviral activity (101), expression of manganous superoxide dismutase (102) and intercellular adhesion molecule (103), increased IL-6 mRNA expression (104), enhancement of the respiratory burst in neutrophils (105), and induction of NF-κB expression (100,103,106). In recent studies with TNF-R55–deficient mice, this receptor has been shown to play a key role in LPS-induced septic shock and in antimicrobial responses to *Listeria monocytogenes* (107,108). The role of the TNF-R75 is less defined. It has recently been suggested that both R55 and R75 are necessary for induction of apoptosis (109,110), and TNF-R75 has been shown to be involved in the inhibition of early hematopoiesis (111). A novel role for TNF-R75 has recently been suggested by Tartaglia et al., who suggested that it might play a role in TNF signaling by presenting TNF to neighboring TNF-R55 molecules (112). In this model, it appears that TNF-R75 regulates the rate of TNF association with TNF-R55 by increasing the local concentration of TNF and the cell surface through rapid ligand association and dissociation.

Chemokine Receptor Family

Monocyte-derived neutrophil chemotactic factor (MDNCF or IL-8) is an 8000 dalton protein produced by stimulated monocytes (113). IL-8 is a member of the α-subfamily of chemoattractant cytokines (chemokines) that contain two essential cysteine residues separated by a third intervening amino acid (C-X-C). Members of the α-subfamily interact primarily with neutrophils. Two IL-8 receptors, IL-8RA and IL-8RB, have been cloned from human neutrophils (114,115). The deduced amino acid sequences of the receptors indicate that they belong to the superfamily of seven transmembrane receptors that couple to G nucleotide–binding proteins. When expressed in mammalian cells, IL-8 appears to bind to both receptors with high affinity, while the GROα/MGSA homolog binds to IL-8RA with low affinity and IL-8RB with high affinity. Differences in the N-terminal amino acids of these two receptors appear to play an important role in ligand specificity (116). IL-8RA is expressed exclusively on neutrophils, while IL-8RB is also expressed on myelomonocytic cell lines (117). The IL-8RA cDNA predicts a 350-amino-acid protein with five potential glycosylation sites and seven hydrophobic membrane domains. IL-8RB is 355 amino acids long with one potential glycosylation site. Sequence alignment of the IL-8 receptors with other G protein–associated receptors demonstrates that these receptors contain conserved specific regions within the transmembrane and cytoplasmic domains. For example, replacement of the conserved aspartate at position 85 in the second transmembrane domain with an alanine residue abolishes ligand binding. At the end of the third transmembrane domain is a conserved 9-amino-acid motif that has been shown to be involved in coupling to G proteins for other receptors in this family. In addition, the IL-8 receptors contain a serine/threonine-rich C-terminal domain that is homologous to domains in other G protein–coupled receptors that is a target for phosphorylation. This phosphorylation appears to be important in switching off receptor signaling.

BIOLOGICAL ACTIVITIES: CYTOKINE SIGNALING AND JAK KINASES

As described above, the ligand-binding subunits for the class I and II cytokine receptors form signaling complexes with signal-transducing subunits that are structurally related to the cytokine receptors, but are themselves unable to bind ligand. IL-3 and IL-5 bind to α-receptor subunits; the α molecules then interact with a common βc molecule. IL-6, IL-11, and IL-12 share a common signaling component, gp130. Binding of these cytokines to their α-receptor subunits induces the formation of homodimers of gp130. Signaling by members of the third subfamily (IL-2, IL-4, IL-7, IL-9, IL-13, and IL-15) involves the formation of heterodimers between the binding subunits and a common signaling component (IL-2Rγ in most cases). In the case of IL-2, the ligand-binding affinity is increased by a third component (IL-2Rα), which has a structure unrelated to any of the other receptor components. The class I and II cytokine receptors do not have kinase domains and only limited similarities in their intracellular portions. However, in each case, rapid tyrosine phosphorylation of cellular substrates as well as the receptors themselves occurs following ligand-induced oligomerization. In most cases, the receptors require at least a portion of their cytoplasmic domain for function. This region contains two motifs termed box 1 and box 2 that have some similarity among these receptors (Table 4). These sequences appear to be involved in association of both signaling and binding components with intracellular kinases.

A variety of kinases have been identified that are activated upon ligand binding. Recent evidence suggests that the Janus family of kinases (Jaks) are associated with these receptors and are the kinases involved in the early initiation of cytokine signaling (118,119). Four members of the Jak family have been identified: Jak1, Jak2, Jak3, and tyk2. These kinases appear to bind to specific motifs in the cytoplasmic domains of the receptor subunits and phosphorylate specific substrates. A summary of the Jaks known to be associated with members of the class I and class II cytokine receptors is shown in Table 5. Although other kinases have been implicated in signaling of these cytokines, the Jak family of tyrosine kinases appears to become rapidly associated with binding and signaling subunits of these receptor complexes and might represent the kinases involved in the initiation of signal transduction. Following is a discussion of the motifs in the cytoplasmic

Table 4 Conserved Cytoplasmic Box 1 and Box 2 Motifs

Receptor	Box 1	Box 2
IL-2Rβ	LKCNTPDP	ISPLEVLERDK-
IL-4Rα	WWDQIPNP	
IL-7R	VWPSLPDH	
IL-9R	FYQNVPSP	
IL-2Rγ	ERTMPRIP	LEDLV
IL-3R	QRLFPRIP	
IL-5R	IRLFPPVP	
βc	WKEKIPNP	VSPLTIEPPN—
gp130	IWPNVPDP	VSVVEIEANDKK
IL-12R	LCPPLPTP	VVEMSWDKGERT
IL-10R	LLFKKPSP	PLDEEAEFLKVS
IFN-γRα		LPKS
IFN-γRβ		QFILEVLDKDG-

Table 5 Cytokine Receptors and Associated Jak Family Kinases

Receptor	Kinase
IL-2	Jak1, Jak2, Jak3
IL-3	Jak2
IL-4	Jak1, Jak3, Tyk2
IL-5	Jak2
IL-6	Jak1, Jak2, Tyk2
IL-7	Jak3
IL-11	Jak1, Jak2, Tyk2
IL-12	Jak2, Tyk2
IL-13	Jak1, Tyk2
IFN-γ	Jak1, Jak2

domains of these receptors involved in the interaction with Jak kinases.

Murakami et al. in 1991 described two regions within the cytoplasmic domain of the gp130 signaling molecule that are absolutely required for transduction of IL-6 biological activities (36). The 61 amino acids proximal to the transmembrane domain are sufficient for generating a growth signal. Mutations within either of these domains (box 1 and box 2) block IL-6 signaling and coincide with disappearance of IL-6–induced tyrosine phosphorylation of the gp130 molecule. Additionally, mutations in box 1 block phosphorylation of cellular substrates, suggesting that a tyrosine kinase might associate with gp130 via this conserved sequence. A third cytoplasmic sequence in the C-terminal region is required for IL-6–induced synthesis of acute-phase proteins and appears to be specific to the gp130 molecule (box 3). Tanner et al. further investigated the specific sequence in gp130 that interacts with the Jak kinases (120). The sequence in box 1 that is absolutely required for Jak2 association is PXXPXP. These data suggest that the function of this box 1 sequence is to allow association with Jak kinases and support previous findings that mutations in this box prevent class I cytokine receptor signaling. Other members of the IL-6R family also associate with and activate Jak kinases. Berger et al. and Yin et al. recently reported that IL-11 induces similar profiles of Jak phosphorylation as IL-6 (121,122). IL-12 induces the activation and phosphorylation of Jak2 and tyk2, but not Jak 1 (123).

Recent studies have demonstrated that the box 1 and box 2 motifs found in other receptors in class I and class II are necessary and sufficient for cytokine signaling. In the IL-2R subfamily, several laboratories have studied the role of the cytoplasmic motifs in both the binding subunits and the common γc chain involved in cytokine signaling and phosphorylation events. Goldsmith et al. reported that dimeric IL-2Rβγ complexes and trimeric IL-2Rαβγ complexes can bind IL-2 and initiate signal transduction (124). Internal deletion of box 1 or box 2 motifs within IL-2Rβ significantly impairs receptor function. Point mutations within these boxes or deletion of the C-terminal 94 amino acids also block function. Internal deletion of the 119 amino acids immediately downstream of box 2 or replacement of the transmembrane domain with the transmembrane region of the CD4 molecule has no effect. Kawahara et al. reported that overexpression of an IL-2Rγ mutant lacking the cytoplasmic domain blocks IL-2–mediated signaling in cells that express endogenous IL-2Rα and IL-2Rγ or all three IL-2R components (125). This domain is also critical for IL-4– and IL-7–induced signaling in these cells. Howard et al. reported that box 1 and 2 plus a C-terminal PQPLXP motif in IL-2Rβ are required for the proliferative response to IL-2 (126). The major tyrosine-phosphorylated component is a molecule of 116 kDa, which was identified as Jak3 kinase. Jak3 appears to interact with the IL-2Rβ through box 1 and 2 and the PQPLXP domain. Miyazaki et al. reported that IL-2 induces tyrosine phosphorylation of both Jak1 and Jak3 (127). These kinases were found to be selectively associated with the serine-rich region of IL-2Rβ and the carboxy-terminal domain of IL-2Rγ. Both regions are necessary for IL-2 signaling. Jak3-negative fibroblasts expressing the IL-2R complex became responsive to IL-2 after transfection with Jak3 cDNA.

Evidence now exists that the Jak family of kinases is also involved in signaling by other members of the IL-2R subfamily. IL-4 causes rapid tyrosine phosphorylation of the IL-4Rα chain and of a substrate of approximately 170 kDa in hematopoietic cell lines and in L cells that have been transfected with the γc cDNA (128). However, unlike other members of the IL-2R family, the IL-4R appears to mediate signaling independently of box 1 and 2 via a unique element located carboxyl-terminal to box 2. Truncation mutants of the IL-4Rα indicate that the region between amino acids 437 and 557 is important for phosphorylation of cellular substrates. This domain contains a sequence that has been designated the IL-4 receptor (I4R) motif. Replacement of the tyrosine at position 497 in this motif with a phenylalanine residue blocks phosphorylation in response to IL-4. From immunoprecipitation experiments it was found that a tyrosine kinase interacts with the I4R motif. Yin et al. reported that Jak1 kinase forms a complex with IL-4R (129). Jak3 has been shown to be functionally linked to the IL-4 receptor in T cells (130), and recent work reported by Musso et al. demonstrates that Jak3 is phosphorylated in monocytes in response to IL-4

(130). Zeng et al. reported that IL-7 also induces phosphorylation of Jak3 (131).

Welham et al. demonstrated that IL-13 and IL-4 induce the phosphorylation of tyk2 and Jak1 (132). IL-4 also stimulates phosphorylation of Jak3, while IL-13 has no effect on this kinase. Since Jak3 associates with IL-2Rγ, this supports previous findings that IL-3 does not use the γc component for signaling. IL-13 does induce phosphorylation of the IL-4R (133), suggesting that IL-4R is a component of both the IL-13R and IL-4R complexes.

In the IL-3R subfamily, the IL-3 and IL-5 α chains and the βc signaling component all contain the PXP box 1 motif, although the human IL-3Rα has an additional amino acid in this sequence (PRIP). All three contain a motif of acidic amino acids that might represent the box 2 domain. The role of the α chains of the IL-3 and IL-5 receptor complexes is unknown, although both have been shown to be required for IL-3 and IL-5 signaling. They may play a role in βc-chain dimerization or oligomerization or may bind a Jak kinase through the box 1 motif. The βc chain contains a cytoplasmic domain of 413 amino acids. This domain can be truncated at amino acid 517 leaving a membrane proximal 67 amino acids without loss of signaling in hematopoietic cells (134). Region 520–562 is homologous to the cytoplasmic region in IL-2Rβ that binds the kinase p561ck, suggesting that this region in βc might be involved in binding tyrosine kinases. Quelle et al. (135) demonstrated that the box 1 motif in the βc chain is essential for proliferation and c-myc induction as well as Jak2 kinase activation. Silvennoinen et al. demonstrated that Jak2 kinase is activated in IL-3–stimulated cells (136). Rao and Mufson (137) reported that the sequence 520–562 binds Jak2. Takaki et al. have recently suggested two possible models for IL-5R assembly and signaling (138): IL-5 might induce multimerization of IL-5R and dimerization of βc, bringing into close proximity two Jak2 molecules associated with βc, which initiates the phosphorylation. Second, the proline-rich region of IL-5Rα binds Jak2 following ligand binding, and this kinase and the βc-associated Jak2 initiate phosphorylation.

The role of Jak kinases was originally identified through the use of mutants that were selected based on their inability to respond to IFN-α/β or IFN-γ. From these studies it was determined that IFN-γ function required Jak1 and Jak2, and that IFN-γ induces the phosphorylation of both of these kinases (139). Recent work has concentrated on defining the sequences in IFN-γRα and IFN-γRβ that bind and activate Jak1 and 2. Farrar et al. reported that the IFN-γRα chain contains 48 amino acids in its cytoplasmic domain proximal to the transmembrane domain that are required for phosphorylation and biological activity (140). Substitution of an LKPS sequence within this domain with alanine residues blocks IFN-γ–induced phosphorylation, suggesting that this region is involved in tyrosine kinase activation or kinase-receptor association. An additional C-terminal YDH motif is necessary for biological response to IFN-γ (141). Another motif (LEKEV) at residues 312–316 is similar to box 2 in the class I receptors, although there is no evidence that this region is required for activity. The relatively short cytoplasmic domain of IFN-γRβ contains a motif (LEVLD) that is similar to the box 2 domain found in IL-2Rβ, IL-2Rγ, and IL-9R (23). Mutations within this domain suggest that it is crucial to mitogenic responses to IFN-γ. The current picture of IFN-γ signaling is as follows: IFN-γ, a homodimeric ligand, binds to the IFN-γRα subunit and causes rapid receptor dimerization. IFN-γRβ then associates with this complex and brings it into close proximity with two tyrosine kinases: one that interacts with the α chain through the box 1 motif and a second that interacts with the β chain through box 2. The kinases phosphorylate each other and the α chain. This phosphorylated complex then phosphorylates cellular substrates, which continues the signaling cascade.

CONCLUSIONS AND FUTURE DIRECTIONS

This chapter reviewed the current knowledge concerning cytokine receptor structure and how specific domains within the receptor sequence contribute to transduction of cytokine signals. It is now known that each of the cytokines interacts with a multicomponent receptor complex, and signaling is mediated via associated tyrosine kinases. Distinct classes of receptors have been defined based on structural features determined from the cDNA sequences. Conserved motifs within the extracellular and cytoplasmic domains are required for interaction with ligand and subsequent association with intracellular kinases. The recent identification of the Janus family of kinases has provided a more complete view of the steps involved in cell surface–to–nucleus signaling initiated by cytokine binding. Elucidation of the spatial and temporal interactions among the receptor subunits, associated kinases, and phosphorylated substrates will undoubtedly be an important focus of future research.

REFERENCES

1. Sims JE, Acres RB, Grubin CE, McMahan CJ, Wignall JM, March CJ, Dower SK. Cloning of the interleukin-1 receptor from human T cells. Proc Natl Acad Sci USA 1989; 86:8946–8950.
2. McMahan CJ, Slack JL, Mosley B, Cosman D, Lupton SD, Brunton LL, Grubin CE, Wignall JM, Jenkins NA, Cannizzaro LA, Benjamin D, Dower SK, Spriggs MK, Sims JE. A novel interleukin-1 receptor, cloned from B cells by mammalian ex-

pression, is expressed in many cell types. EMBO J 1991; 10:2821–2832.

3. Leonard WJ, Depper MJ, Crabtree CR, Rudikoff S, Pumphrey J, Robb RJ, Kronke M, Svetlik PB, Pfeffer NJ, Waldmann TA, Greene WC. Molecular cloning and expression of cDNAs for the human interleukin-2 receptor. Nature 1984; 311:626–631.
4. Nikaido T, Shimizu A, Ishida N, Sake H, Teshigawara K, Maeda M, Uchiyama T, Yodoi J, Honjo T. Molecular cloning of cDNA encoding human interleukin-2 receptor. Nature 1984; 311:631–635.
5. Cosman D, Cerretti DP, Larsen A, Park L, March C, Dower S, Gillis S, Urdal D. Cloning, sequence and expression of human interleukin-2 receptor. Nature 1984; 312:768–772.
6. Hatakeyama M, Tsudo M, Minamoto S, Kono T, Doi T, Miyata T, Miyasaka M, Taniguchi T. Interleukin-2 receptor chain gene: generation of three receptor forms by cloned human α and β chain cDNA's. Science 1989; 244:551–556.
7. Takeshita T, Asao H, Ohtani K, Ishii N, Kumaki S, Tanaka N, Munakata H, Nakamura M, Sugamura K. Cloning of the γ chain of the human IL-2 receptor. Science 1992; 257:379–382.
8. Kitamura T, Sato N, Arai K, Miyajima A. Expression cloning of the human IL-3 receptor cDNA reveals a shared β subunit for the human IL-3 and GM-CSF receptors. Cell 1991; 66:1165–1174.
9. Hayashida K, Kitamura T, Gorman DM, Arai K, Yokota T, Miyajima A. Molecular cloning of a second subunit of the receptor for human granulocyte-macrophage colony stimulating factor (GM-CSF): reconstitution of a high-affinity GM-CSF receptor. Proc Natl Acad Sci USA 1990; 87:9655–9659.
10. Idzerda R, March CJ, Mosley B, Lyman SD, Vanden Bos T, Gimpel SD, Din WS, Grabstein KH, Widmer MB, Park LS, Cosman D, Beckmann P. Human interleukin-4 receptor confers biologic responsiveness and defines a novel receptor superfamily. J Exp Med 1990; 171:861–868.
11. Murata Y, Takaki S, Migita M, Kikuchi Y, Tominaga A, Takatsu K. Molecular cloning and expression of the human interleukin-5 receptor. J Exp Med 1992; 175:341–351.
12. Yamasaki K, Taga T, Hirata Y, Yawata H, Kawanishi Y, Seed B, Taniguchi T, Hirano T, Kishimoto T. Cloning and expression of the human interleukin-6 (BSF-2/IFNβ2) receptor. Science 1988; 241:825–828.
13. Hibi M, Murakami M, Saito M, Hirano T, Taga T, Kishimoto T. Molecular cloning and expression of an IL-6 signal transducer, gp130. Cell 1990; 63:1149–1157.
14. Goodwin RG, Friend D, Ziegler SF, Jerzy R, Falk BA, Gimpel S, Cosman D, Dower SK, March CJ, Namen AE, Park LS. Cloning of the human and murine interleukin-7 receptor: demonstration of a soluble form and homology to a new receptor superfamily. Cell 1990; 60:941–951.
15. Renauld J-C, Druez C, Kermouni A, Houssiau F, Uyttenhove C, Van Roost E, Van Snick J. Expression cloning of the murine and human interleukin 9 receptor cDNAs. Proc Natl Acad Sci USA 1992; 89:5690–5694.
16. Ho ASY, Liu Y, Khan TA, Hsu D-H, Bazan JF, Moore KW. A receptor for interleukin 10 is related to interferon receptors. Proc Natl Acad Sci USA 1993; 90:11267–11272.
17. Liu Y, Wei SH-Y, Ho ASY, de Waal Malefyt R, Moore KW. Expression cloning and characterization of a human IL-10 receptor. J Immunol 1994; 152:1821–1829.
18. Hilton DJ, Hilton AA, Raicevic A, Rakar S, Harrison-Smith N, Begley CG, Metcalf D, Nicola NA, Willson TA. Cloning of a murine IL-11 receptor α-chain; requirement for gp130 for high affinity binding and signal transduction. EMBO J 1994; 13:4765–4775.
19. Chua AO, Chizzonite R, Desai BB, Truitt TP, Nunes P, Minetti LJ, Warrier RR, Presky DH, Levine JF, Gately MK, Gubler U. Expression cloning of a human IL-12 receptor component. J Immunol 1994; 153:128–136.
20. Aguet M, Dembic Z, Merlin G. Molecular cloning and expression of the human interferon-γ receptor. Cell 1988; 55:273–280.
21. Gray PW, Leong S, Fennie EH, Farrar MA, Pingel JT, Fernandez-Luna J, Schreiber RD. Cloning and expression of the cDNA for the murine interferon γ receptor. Proc Natl Acad Sci USA 86:8497–8501.
22. Soh J, Donnelly RJ, Kotenko JR, Wang N, Emanuel S, Schwartz B, Miki T, Pestka S. Identification and sequence of an accessory factor required for activation of the human interferon γ receptor. Cell 1994; 76:793–802.
23. Hemmi S, Bohni R, Stark G, Di Marco F, Aguet M. A novel member of the interferon receptor family complements functionality of the murine interferon γ receptor in human cells. Cell 1994; 76:803–810.
24. Loetscher H, Pan Y-CE, Lahm H-W, Gentz R, Brockhaus M, Tabuchi H, Lesslauer W. Molecular cloning and expression of the human 55 kDa tumor necrosis factor receptor. Cell 1990; 61:351–359.
25. Schall TJ, Lewis M, Koller KJ, Lee A, Rice GC, Wong GHW, Gatanaga T, Granger GA, Lentz R, Raab H, Kohr WJ, Goeddel DV. Molecular cloning and expression of a receptor for human tumor necrosis factor. Cell 1990; 61:361–370.
26. Smith CA, Davis T, Anderson D, Solam L, Beckmann MP, Jerzy R, Dower SK, Cosman D, Goodwin RG. A receptor for tumor necrosis factor defines an unusual family of cellular and viral proteins. Science 1990; 248:1019–1023.
27. Sato N, Miyajima A. Multimeric cytokine receptors: common versus specific functions. Curr Biol 1994; 6:174–179.
28. Bazan JF. Structural design and molecular evolution of a cytokine receptor superfamily. Proc Natl Acad Sci USA 1990; 87:15842–15848.
29. Davies DR, Wlodawer A. Cytokines and their receptor complexes. FASEB J 1995; 9:50–56.
30. Gaulton GN, Williamson P. Interleukin-2 and the interleukin-2 receptor complex. Lymph Activ 1994; 59:91–114.
31. Minami YT, Kono T, Miyazaki T, Taniguchi T. The IL-2 receptor complex: its structure, function, and target genes. Annu Rev Immunol 1993; 11:245–272.
32. Ringheim GE, Freimark BD, Robb RJ. Quantitative characterization of the intrinsic ligand-binding affinity of the interleukin-2 receptor beta chain and its modulation by the alpha chain and a second affinity-modulating element. Lymph Cytokine Res 1991; 10:219–226.
33. Goldsmith MA, Amaral MC, Greene WC. Ligand binding by the IL-2 receptor is modulated by intracellular determinants of the IL-2 receptor β-chain. J Immunol 1995; 154:2033–2040.
34. Hatakeyama M, Mori H, Doi T, Taniguchi T. A restricted cytoplasmic region of IL-2 receptor β-chain is essential for growth signal transduction but not for ligand binding and internalization. Cell 1989; 59:837–845.
35. Merida I, Williamson P, Gaulton G. The serine rich cytoplasmic domain of the IL-2 receptor β chain is essential for IL-2 dependent tyrosine protein kinase and phosphatidylinositol-3-kinase activation. J Biol Chem 1993; 268:6765–6770.
36. Murakami M, Hibi M, Nakagawa N, Nakagawa T, Yasukawa K, Yaminishi K, Taga T, Kishimoto T. IL-6-induced homodimerization of gp130 and associated activation of a tyrosine kinase. Science 1993; 260:1808–1810.
37. Sonoda Y. Interleukin-4—a dual regulatory factor in hematopoiesis. Leuk Lymph 1994; 14:231–240.
38. Park LS, Friend D, Sassenfeld HM, Urdal DL. Characteriza-

tion of the human B cell stimulatory factor 1 receptor. J Exp Med 1987; 166:476–488.
39. Mosley B, Beckmann MP, March CJ, Idzerda RL, Gimpel SD, VandenBos T, Friend D, Alpert A, Anderson D, Jackson J, Wignall JM, Smith C, Gallis B, Sims JE, Urdal D, Widmer MB, Cosman D, Park LS. The murine interleukin-4 receptor: molecular cloning and characterization of secreted and membrane bound forms. Cell 1989; 59:335–348.
40. Kondo M, Takeshita T, Ishii N, Nakamura M, Watanabe S, Arai K-I, Sugamura K. Sharing of the interleukin-2 (IL-2) receptor γ chain between receptors for IL-2 and IL-4. Science 262:1874–1879.
41. Russell SM, Keegan AD, Harada N, Nakamura Y, Noguchi M, Leland P, Friedmann MC, Miyajima A, Puri RK, Paul WE, Leonard WJ. Interleukin-2 receptor γ chain: a functional component of the interleukin-4 receptor. Science 262:1880–1884.
42. He Y-W, Adkins B, Furse RK, Malek TR. Expression and function of the γc subunit of the IL-2, IL-4, and IL-7 receptors. J Immunol 1995; 154:1596–1605.
43. He Y-W, Malek TR. The IL-2 receptor γc chain does not function as a subunit shared by the IL-4 and IL-13 receptors. Implications for the structure of the IL-4 receptor. J Immunol 1995; 155:9–12.
44. Ohara J, Paul WE. Receptors for B-cell stimulatory factor-1 expressed on cells of hematopoietic lineage. Nature 1987; 325:5337–5340.
45. Park LS, Friend D, Sassenfeld HM, Urdal DL. Characterization of the human B cell stimulatory factor 1 receptor. J Exp Med 1987; 166:476–488.
46. Kondo M, Takeshita T, Higuchi M, Nakamura M, Sudo T, Nishidawa S-I, Sugamura K. Functional participation of the IL-2 receptor γ chain in IL-7 receptor complexes. Science 263:1453–1457.
47. Renauld J-C, Kermouni A, Vink A, Louahed J, Van Snick J. Interleukin-9 and its receptor: involvement in mast cell differentiation and T cell oncogenesis. J Leuk Biol 1995; 57:353–360.
48. Zurawski G, de Vries JE. Interleukin 13, an interleukin 4-like cytokine that acts on monocytes and B cells, but not on T cells. Immunol Today 1994; 15:19–26.
49. de Waal Malefyt R, Figdor CG, Huijbens R, Mohan-Peterson S, Bennett B, Culpepper J, Dang W, Zurawski G, de Vries JE. Effect of IL-13 on phenotype, cytokine production, and cytotoxic function of human monocytes. Comparison with IL-4 and modulation by IFN-gamma or IL-10. J Immunol 1993; 151:6370–6381.
50. Doyle A, Herbein G, Montaner LJ, Minty AJ, Caput D, Ferrara P, Gordon S. Interleukin-13 alters the activation state of murine macrophage in vitro: comparison with interleukin-4 and interferon-gamma. Eur J Immunol 1994; 24:1441–1445.
51. Zurawski SM, Vega F Jr, Huyghe B, Zurawski G. Receptors for interleukin-13 and interleukin-4 are complex and share a novel component that functions in signal transduction. EMBO J 1993; 12:2663–2670.
52. Welham MJ, Learmonth L, Bone H, Schrader JW. Interleukin-13 signal transduction in lymphohemopoietic cells. J Biol Chem 1995; 270:12286–12296.
53. Vita N, Lefort S, Laurent P, Caput D, Ferrara P. Characterization and comparison of the interleukin 13 receptor with the interleukin 4 receptor on several cell types. J Biol Chem 1995; 270:3512–3517.
54. Obiri NI, Debinski W, Leonard WJ, Puri RK. Receptor for interleukin 13. J Biol Chem 270: 8797–8804.
55. Lin JX, Migon TS, Tsang M, Friedmann M, Weatherbee JA, Zhon L, Yamauchi A, Bloom ET, Mietz J, John S. The role of shared receptor motifs and common STAT proteins in the generation of cytokine pleiotropy and redundancy by IL-2, IL-4, IL-7, IL-13, and IL-15. Immunity 1995; 2:331–339.
56. Carson WE, Giri JG, Lindemann MJ, Linett ML, Ahdieh M, Paxton R, Anderson D, Eisenmann J, Grabstein K, Caligiuri MA. Interleukin (IL) 15 is a novel cytokine that activates human natural killer cells via components of the IL-2 receptor. J Exp Med 1994; 180:1395–1403.
57. Giri JG, Ahdieh M, Eisenmann J, Shanebeck K, Grabstein K, Kumaki S, Namen A, Park LS, Cosman D, Anderson D. Utilization of the β and γ chains of the IL-2 receptor by the novel cytokine IL-15. EMBO J 1994; 13:2822–2830.
58. Metcalf D. Control of granulocytes and macrophages: Molecular, cellular, and clinical aspects. Science 1991; 254:529–533.
59. Kuwaki T, Kitamura T, Tojo A, Matsuki S, Tamai Y, Miyazono K, Takaku F. Characterization of human interleukin-3 receptors on a multi-factor-dependent cell line. Biochem Biophys Res Commun 1989; 161:16–22.
60. Gearing DP, King JA, Gough NM, Nicola NA. Expression cloning of a receptor for human granulocyte-macrophage colony-stimulating factor. EMBO J 1989; 8:3367–3376.
61. Takatsu K, Takaki S, Hitoshi Y. Interleukin-5 and its receptor system: implications in the immune system and inflammation. Adv Immunol 1994; 57:145–190.
62. Tavernier J, Devos R, Cornelis S, Tuypens T, Van der Heyden J, Fiers W, Plaetinck G. A human high affinity interleukin-5 receptor α chain and a β chain shared with the receptor for GM-CSF. Cell 1991; 66:1175–1184.
63. Devos R, Plaetinck G, Van der Heyden J, Cornelis S, Vandekerckhove J, Fiers W, Tavernier J. Molecular basis of a high affinity murine interleukin-5 receptor. EMBO J 1991; 10:2133–2137.
64. Takaki S, Murata Y, Kitamura T, Miyajima A, Tominaga A, Takatsu K. Reconstitution of the functional receptors for murine and human interleukin 5. J Exp Med 1993; 177:1523–1529.
65. Mui A, Muto A, Sakamaki K, Sato N, Kinoshita T, Watanabe S, Yokota T, Arai K, Miyajima A. Function of the common β subunit of the GM-CSF/IL-3/IL-5 receptors. In: Gupta S, ed. Mechanisms of Lymphocyte Activation and Immune Regulation V. New York: Plenum Press, 1994:217–223.
66. Bazan JF. Hematopoietic receptors and helical cytokines. Immunol Today 1990; 11:350–353.
67. Bazan JF. A novel family of growth factor receptors: a common binding domain in the growth hormone, prolactin, erythropoietin, and IL-6 receptors, and the p75 IL-2 receptor β chain. Biochem Biophys Res Commun 1989; 164:788–795.
68. Hirano T, Akira S, Taga T, Kishimoto T. Biological and clinical aspects of interleukin 6. Immunol Today 1990; 11:443–449.
69. Hirano T, Matsuda T, Nakajima K. Signal transduction through gp130 that is shared among the receptors for the interleukin 6 related cytokine subfamily. Stem Cells 1994; 12:262–277.
70. Taga T, Hibi M, Hirata Y, Yamasaki K, Yasukawa K, Matsuda T, Hirano T, Kishimoto T. Interleukin-6 triggers the association of its receptor with a possible signal transducer, gp130. Cell 1989; 58:573–581.
71. Yasukawa K, Saito T, Tukunaga T, Sekimori Y, Koishihara Y, Fukui H, Ohsugi Y, Matsuda T, Yawata H, Hirano T, Taga T, Kishimoto T. Purification and characterization of soluble human IL-6 receptor expressed in CHO cells. J Biochem 1990; 108:673–676.
72. Horsten U, Schmitz-Van de Leur H, Mullberg J, Heinrich PC, Rose-John S. The membrane distal half of gp130 is responsible for the formation of a ternary complex with IL-6 and the IL-6 receptor. FEBS Lett 1995; 360:43–46

73. Grube BJ, Cochran CG. Identification of a regulatory domain of the interleukin-6 receptor. J Biol Chem 1994; 269:20791–20797.
74. Ward LD, Howlett GJ, Discolo G, Yasukawa K, Hammacher A, Moritz RL, Simpson RJ. High affinity interleukin-6 receptor is a hexameric complex consisting of two molecules each of interleukin-6, interleukin-6 receptor, and gp130. J Biol Chem 1994; 269:23286–23289.
75. Quesniaux VFJ. Interleukin 11. Leuk Lymph 1994; 14:241–249.
76. Paul SR, Bennett F, Calvetti JA, Kelleher K, Wood CR, O'Hara RM, Leary AC, Sibley B, Clark SC, Williams DA, Yang YC. Molecular cloning of a cDNA encoding interleukin-11, a stromal cell-derived lymphopoietic and hematopoietic cytokine. Proc Natl Acad Sci USA 1990;87:7512–7516.
77. Yin T, Taga T, Tsang MLS, Yasukawa K, Kishimoto T, Yang YC. Interleukin (IL)-6 signal transducer, gp130, is involved in IL-11 mediated signal transduction. Blood 1992; 80:593.
78. Fourcin M, Chevalier S, Lebrun J-J, Kelly P, Pouplard A, Wijdenes J, Gascan H. Involvement of gp130/interleukin-6 receptor transducing component in interleukin-11 receptor. Eur J Immunol 1994; 24:277–280.
79. Yin T, Taga T, Tsang ML, Yasukawa K, Kishimoto T, Yang YC. Involvement of IL-6 transducer SP130 in IL-11-mediated signal transduction. J Immunol 1993; 151:2555–2561.
80. Brunda MJ. Interleukin-12. J Leuk Biol 1994; 55:280–288.
81. Gearing DP, Cosman D. Homology of the p40 subunit of natural killer cell stimulatory factor (NKSF) with the extracellular domain of the interleukin-6 receptor. Cell 1991; 66:9–10.
82. Chizzonite R, Truitt T, Podlaski FJ, Wolitzky AG, Quinn PM, Nunes P, Stern AS, Gately MK. IL-12: monoclonal antibodies specific for the 40-kDa subunit block receptor binding and biologic activity on activated human lymphoblasts. J Immunol 1991; 147:1548–1556.
83. Chizzonite R, Truitt T, Desai BB, Nunes P, Podlaski FJ, Stern AS, Gately MK. IL-12 receptor. I. Characterization of the receptor on phytohemagglutinin-activated human lymphoblasts. J Immunol 1992; 148:3117–3124.
84. Pestka S, Langer JA, Zoon KC, Samuel CE. Interferons and their actions. Annu Rev Biochem 1987; 56:727–777.
85. Miura O, Cleveland JL, Ihle JN. Inactivation of erythropoietin receptor function by point mutations in a region having homology with other cytokine receptors. Mol Cell Biol 1993; 13:1788–1795.
86. Mosmann TR. Properties and functions of interleukin-10. Adv Immunol 1994; 56:1–26.
87. Larner AC, David M, Feldman GM, Igarashi K-I, Hackett RH, Webb DSA, Sweitzer SM, Petricoin EF III, Finbloom DS. Tyrosine phosphorylation of DNA-binding proteins by multiple cytokines. Science 1993; 261:1730–1732.
88. Mizel SB. The interleukins. FASEB J 1989; 3:2379–2388.
89. Mizel SB. Interleukin 1. Biology and molecular biology. In: Poste G, Crooke ST, eds. Cellular and Molecular Aspects of Inflammation. New York: Plenum, 1988:75–96.
90. Kuno K, Matsushima K. The IL-1 signaling pathway. J Leuk Biol 1994; 56:542–547.
91. Sims JE, Gayle MA, Slack JL, Alderson MR, Bird TA, Giri JG, Colotta R, Re F, Montovani A, Shanebeck K. Interleukin 1 signaling occurs exclusively via the type I receptor. Proc Natl Acad Sci USA 1993; 90:6155–6159.
92. Slack J, McMahan CJ, Waugh S, Schooley K, Spriggs MK, Sims JE, Dower SK. Independent binding of interleukin-1α and interleukin-1β to type I and type II interleukin-1 receptors. J Biol Chem 1993; 268:2513–2524.
93. Dower SK, Sims JE, Cerretti DP, Bird TA. The interleukin-1 system: receptors, ligands, and signals. In: Ishizaka K, Lachman PJ, Lerner R, Waksman BH, eds. Interleukins: Molecular Biology and Immunology. Basel: Karger, 1992:33.
94. Colotta F, Dower SK, Sims JE, Mantovani A. The type II 'decoy' receptor: a novel regulatory pathway for interleukin 1. Immunol Today 1994; 15:562–566.
95. Sims JE, Giri JG, Dower SK. The two interleukin-1 receptors play different roles in IL-1 actions. Clin Immunol Immunopath 1994; 72:9–14.
96. Alcami A, Smith GL. A soluble receptor for interleukin-1β encoded by vaccinia virus: a novel mechanism of virus modulation of the host response to infection. Cell 1992; 71:153–159.
97. Old LJ. Tumor necrosis factor. In: Bonavida B, Granger G, eds. Tumor Necrosis Factor: Structure, Mechanism of Action, Role in Disease and Therapy. Basel: Karger, 1990:1–30.
98. Smith CA, Farrah T, Goodwin RG. The TNF receptor superfamily of cellular and viral proteins: Activation, costimulation, and death. Cell 1994; 76:959–962.
99. Dembic Z, Loetscher H, Gubler U, Pan YE, Lahm H, Gentz R, Brockhaus M, Lesslauer W. Two human TNF receptors have similar extracellular, but distinct intracellular, domain sequences. Cytokine 1990; 2:231–237.
100. Tartaglia LA, Rothe M, Hu Y-F, Goeddel DV. Tumor necrosis factor's cytotoxic activity is signaled by the p55 TNF receptor. Cell 1993; 73:213–216.
101. Wong GWH, Tartaglia LA, Lee MS, Goeddel DV. Antiviral activity of tumor necrosis factor (TNF) is signaled through the 55-kDa receptor, type I TNF. J Immunol 1992; 149:3350–3353.
102. Tartaglia LA, Goeddel DV. Two TNF receptors. Immunol Today 1992; 13:151–153.
103. Mackay F, Loetscher H, Stueber D, Gehr G, Lesslauer W. Tumor necrosis factor α (TNF-α)-induced cell adhesion to human endothelial cells is under dominant control of one TNF receptor type, TNF-R55. J Exp Med 1993; 177:1277–1286.
104. Brakebusch C, Nophar Y, Kemper O, Engelmann H, Wallach D. Cytoplasmic truncation of the p55 tumour necrosis factor (TNF) receptor abolishes signaling, but not induced shedding of the receptor. EMBO J 1992; 11:943–950.
105. Menegazzi R, Cramer R, Patriarca P, Scheurich P, Dri P. Evidence that tumor necrosis factor α (TNF)-induced activation of neutrophil respiratory burst on biologic surfaces is mediated by the p55 TNF receptor. Blood 1994; 84:287–293.
106. Kruppa G, Thoma B, Machleidt T, Wiegmann K, Kronke M. Inhibition of tumor necrosis factor (TNF)-mediated NF-kappa B activation by selective blockade of the human 55-kDa TNF receptor. J Immunol 1992; 148:3152–3157.
107. Pfeffer K, Matsuyama T, Kundig TM, Wakeham A, Kishihara K, Shahinian A, Wiegmann K, Ohashi PS, Kronke M, Mak TW. Mice deficient for the p55 tumor necrosis factor receptor are resistant to endotoxic shock, yet succumb to *L. monocytogenes* infection. Cell 1993; 73:457–467.
108. Rothe J, Lesslauer W, Loetscher H, Lang Y, Koebel P, Kontgen F, Althage A, Zindernagel R, Steinmetz M, Bluethmann H. Mice lacking the tumour necrosis factor receptor 1 are resistant to TNF-mediated toxicity but highly susceptible to infection by *Listeria monocytogenes*. Nature 1993; 364:798–802.
109. Heller RA, Song K, Fan N. Cytotoxicity by tumor necrosis factor is mediated by both p55 and p70 receptors. Cell 1993; 73:216.
110. Vadenabeele P, Declercq W, Banhaesebroeck B, Grooten J, Fiers W. Both TNF receptors are required for TNF-mediated induction of apoptosis in PC60 cells. J Immunol 1995; 154:2904–2913.
111. Jacobsen FW, Rothe M, Rusten L, Goeddel DV, Smeland EB, Veiby OP, Slordal L, Jacobsen SEW. Role of the 75-kDa tumor

necrosis factor receptor: Inhibition of early hematopoiesis. Proc Natl Acad Sci USA 1994; 91:10695–10699.
112. Tartaglia LA, Pennica D, Goeddel DV. Ligand passing: the 75-kDa tumor necrosis factor (TNF) receptor recruits TNF for signaling by the 55-kDa TNF receptor. J Immunol 1993; 268:18542–18548.
113. Baggiolini M, Dewald B, Moser B. Interleukin-8 and related chemotactic cytokines—CXC and CC chemokines. Adv Immunol 1994; 55:97–179.
114. Murphy PM, Tiffany HL. Cloning of complementary DNA encoding a functional human interleukin-8 receptor. Science 1991; 253:1280–1283.
115. Holmes WE, Lee J, Kuang WJ, Rice GC, Wood WI. Structure and functional expression of a human interleukin-8 receptor. Science 1991; 253:1278–1280.
116. Suzuki H, Prado GN, Wilkinson N, Navarro J. The N terminus of interleukin-8 (IL-8) receptor confers high affinity binding to human IL-8. J Biol Chem 1994; 269:18263–18266.
117. Morohashi H, Miyawaki T, Nomura H, Kuno K, Murakami S, Matsushima K, Mukaida N. Expression of both types of human interleukin-8 receptors on mature neutrophils, monocytes, and natural killer cells. J Leuk Biol 1995; 57:180–187.
118. Ihle JN. The Janus kinase family and signaling through members of the cytokine receptor superfamily. Proc Soc Exp Biol Med 1994; 206:268–272.
119. Ihle JN, Witthuhn BA, Quelle FW, Yamamoto K, Thierfelder WE, Kreider B, Silvennoinen O. Signaling by the cytokine receptor superfamily: JAKs and STATs. Trends Biochem Sci 1994; 19:222–227.
120. Tanner JW, Chen W, Young RL, Longmore GD, Shaw AS. The conserved box 1 motif of cytokine receptors is required for association with JAK kinases. J Biol Chem 1995; 270:6523–6530.
121. Berger LC, Hawley TS, Lust JA, Goldman SJ, Hawley RG. Tyrosine phosphorylation of JAK-TYK kinases in malignant plasma cell lines growth-stimulated by interleukins 6 and 11. Biochem Biophys Res Commun 1994; 202:595–605.
122. Yin T, Yasukawa K, Taga T, Kishimoto T, Yang YC. Identification of a 130 kDa tyrosine-phosphorylated protein induced by interleukin 11 as JAK2 tyrosine kinase, which associates with gp130 signal transducer. Exp Hematol 1994; 22:467–472.
123. Bacon CM, McVicar DW, Ortaldo JR, Rees RC, O'Shea JJ, Johnston JA. Interleukin 12 (IL-12) induces tyrosine phosphorylation of JAK2 and TYK2: Differential use of Janus family tyrosine kinases by IL-2 and IL-12. J Exp Med 1995; 181:399–404.
124. Goldsmith MA, Xu W, Amaral MC, Kuczek ES, Greene WC. The cytoplasmic domain of the interleukin-2 receptor β chain contains both unique and functionally redundant signal transduction elements. J Biol Chem 1994; 269:14698–14704.
125. Kawahara A, Minami Y, Taniguchi T. Evidence for a critical role for the cytoplasmic region of the interleukin 2 (IL-2) receptor γ chain in IL-2, IL-4, and IL-7 signalling. Mol Cell Biol 1994; 14:5433–5440.
126. Howard OMZ, Kirken RA, Garcia GG, Hackett RH, Farrar WL. Structural domains of interleukin-2 receptor β critical for signal transduction: kinase association and nuclear complex-formation. Biochem J 1995; 306:217–224.
127. Miyazaki T, Kawahara A, Fujii H, Nakagawa Y, Minami Y, Liu Z-J, Oishi I, Silvennoinen O, Witthuhn BA, Ihle JN, Taniguchi T. Functional activation of Jak1 and Jak3 by selective association with IL-2 receptor subunits. Science 1994; 266:1045–1047.
128. Keegan A, Nelms K, Paul WE. The IL-4 receptor-signaling mechanisms. Mech Lymph Activ Immune Regul V (S. Gupta, ed.) 1994; 211–215.
129. Yin T, Tsang ML, Yang YC. JAK1 kinase forms complexes with interleukin-4 receptor and 4PX/insulin receptor substrate-1-like protein and is activated by interleukin 4 and interleukin 9 in lymphocytes. J Biol Chem 1994; 269:26614–26617.
130. Musso T, Johnston JA, Linnekin D, Varesio L, Rowe TK, O'Shea JJ, McVicar DW. Regulation of JAK3 expression in human monocytes: Phosphorylation in response to interleukins 2, 4 and 7. J Exp Med 1995; 181:1425–1431.
131. Zeng Y-X, Takahashi H, Shibata M, Hirokawa K. JAK3 Janus kinase is involved in interleukin 7 signal pathway. FEBS Lett 1994; 353:289–293.
132. Welham MJ, Learmonth L, Bone H, Schrader JW. Interleukin-13 signal transduction in lymphohemopoietic cells. Similarities and differences in signal transduction with interleukin-4 and insulin. J Biol Chem 1995; 270:12286–12296.
133. Smerz-Bertling C, Duschl A. Both interleukin 4 and interleukin 13 induce tyrosine phosphorylation of the 140-kDa subunit of the interleukin 4 receptor. J Biol Chem 1995; 270:966–970.
134. Sakamaki K, Miyajima I, Kitamura T, Miyajima A. Critical cytoplasmic domains of the common β subunit of the human GM-CSF, IL-3 and IL-5 receptors for growth signal transduction and tyrosine phosphorylation. EMBO J 1992; 11:3541–3549.
135. Quelle FW, Sato N, Witthuhn BA, Inhorn RC, Eder M, Miyajima A, Griffin JD, Ihle JN. JAK2 associates with the beta c chain of the receptor for granulocyte-macrophage colony-stimulating factor, and its activation requires the membrane-proximal region. Mol Cell Biol 1994; 4335–4341.
136. Silvennoinen O, Witthuhn B, Zuella FW, Cleveland JL, Yi T, Ihle JN. Structure of the murine JAK2 protein-tyrosine kinase and its role in interleukin 3 signal transduction. Proc Natl Acad Sci USA 1993; 90:8429–8433.
137. Rao P, Mufson RA. A membrane proximal domain of the human interleukin-3 receptor βc subunit that signals DNA synthesis in NIH 3T3 cells specifically binds a complex of src and Janus family tyrosine kinases and phosphatidylinositol 3-kinase. J Biol Chem 1995; 270:6886–6893.
138. Takaki S, Kanazawa H, Shiiba M, Takatsu K. A critical cytoplasmic domain of the interleukin-5 (IL-5) receptor α chain and its function in IL-5-mediated growth signal transduction. Mol Cell Biol 1994; 14:7404–7413.
139. Igarashi K, Garotta G, Ozmen L, Ziemiecki A, Wilks AF, Harpur AG, Larner AC, Finbloom DS. Interferon-γ induces tyrosine phosphorylation of interferon-γ receptor and regulated association of protein tyrosine kinases, Jak1 and Jak2, with its receptor. J Biol Chem 1994; 269:14333–14336.
140. Farrar MA, Fernandez-Luna J, Schreiber RD. Identification of two regions within the cytoplasmic domain of the human interferon-γ receptor required for function. J Biol Chem 1991; 266:19626–19635.
141. Greenlund AC, Farrar MA, Viviano BL, Schreiber RD. Ligand-induced IFNγ receptor tyrosine phosphorylation couples the receptor to its signal transduction system (p91). EMBO J 1994; 13:1591–1600.

20
Quantitation of Cytokines

Daniel G. Remick
University of Michigan, Ann Arbor, Michigan

IMMUNOASSAY PRINCIPLES

This chapter briefly discusses the quantitation of the biological activity of cytokines and the basic principles of immunoassays. It is not intended to represent a complete description of the methodologies involved but to review the nature of the assays and potential pitfalls in interpreting the data. Documenting the presence of cytokines increases our understanding of the mediators of disease that may be present. If the presence of a cytokine or an elevation of the cytokine is associated with a particular disease, then this cytokine may represent a potential target for inhibition.

Cytokines may also be assessed in immunological assays, either radioimmunoassays (RIAs) or enzyme-linked immunoabsorbent assays (ELISAs). Both of these techniques are predicated on the ability of a specific antibody to recognize the cytokine in question. With RIAs, the radioactivity (usually ^{125}I) is detected in a gamma counter, and the radioactivity is then related to a standard curve used to calculate the amount of cytokine. The RIAs are rapid and easily quantifiable and may have exquisite sensitivity. They have the disadvantage of working with radioactivity, especially the gamma emitter ^{125}I. ELISAs are routinely performed in plates with 96 wells, and at the end of the reaction a color is developed in each well. The intensity of the color may be quantitated spectrophotometrically in each well with an ELISA reader and the degree of activity directly related to a standard curve. ELISAs are often easy to perform, and several of the steps may be automated. The assays generally take longer to run than RIAs and may not offer the same degree of sensitivity.

ELISAs may be run in either a direct or indirect manner. Figure 1 depicts a direct ELISA. For this assay, the cytokine to be tested is directly bound, or captured, to the plastic of a 96-well plate. Unbound substances are washed away, and an antibody recognizing the cytokine is added to each well. This antibody is conjugated to an enzyme. After washing away unbound antibody, a substrate solution is added. The enzyme conjugated to the antibody will reduce/oxidize the substrate to yield a colored reaction, which may be quantitated in an ELISA reader. This method is convenient, rapid, and very specific if the antibody directed against the cytokine has specificity. However, it relies on the ability of the cytokine to bind to the plate before it is detected. This may be difficult, either because the cytokine does not bind well, or because the cytokine is present in a complex protein mixture where several other proteins will compete for binding, such as plasma or serum.

In an indirect ELISA (Fig. 2) these difficulties are overcome. In the first step, an antibody directed against the cytokine is bound to the ELISA plate. Antibodies, which are large proteins, bind extremely well to ELISA plates. Purified antibodies are used, and since this is the only protein present during the binding step, there is no competition for sites. Excess, unbound antibody is removed, and the antigen (cytokine) is added. After incubation and washing, the enzyme-conjugated antibody is added next. After washing, the antigen (cytokine) is sandwiched between two antibodies. For this reason this type of ELISA is often referred to as a sandwich ELISA. The color is developed by the addition of the appropriate substrate.

Since both RIAs and ELISAs depend on an antibody recognizing a specific antigen, they both have the same inherent shortcomings. Clearly, the antibody must be spe-

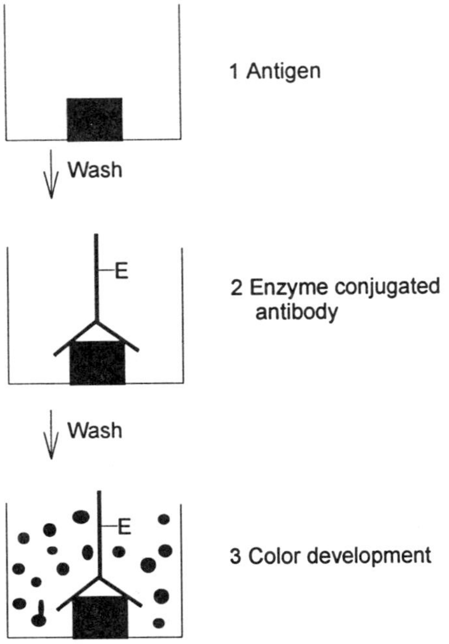

Figure 1 Direct ELISA. In the first step the antigen, which is the cytokine, is bound to the plate. After washing away unbound material, an enzyme-conjugated antibody specific for the cytokine is added in the second step. This will bind to the cytokine in a specific manner. In step three, unbound antibody is washed away, and a substrate solution is added. The enzyme on the antibody reacts with the solution to produce a color reaction. The intensity of the color is directly related to the amount of enzyme conjugated antibody present, which is related to the amount of antigen.

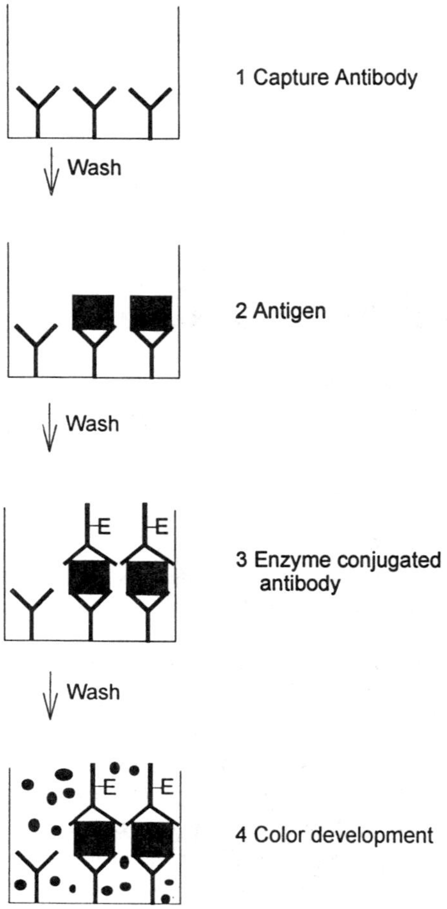

Figure 2 Indirect or sandwich ELISA. In the first step the capture antibody is bound to the plate. Excess antibody is removed and the antigen is added in the second step. The capture antibody (captured on the plate) will bind to the antigen. An enzyme-conjugated antibody is added in the third step, which binds to the antigen, after washing substrate solution is added and the color developed.

cific for the cytokine in question, and it should be a high-titer antibody. Monoclonal antibodies usually provide unmatched specificity, but their sensitivity may be lacking. Polyclonal antibodies may react with other factors present in the biological samples to yield a false positive. This may happen especially in situations in which there is some structural homology between two cytokines, such as lymphotoxin and tumor necrosis factor (1). Finally, the immunoassay may be detecting degraded peptide fragments that are immunoreactive but not biologically active. Thus, in certain disease settings, it is possible to have a positive immunoassay but a negative biological assay.

One method of differentiating between cytokines with structural similarity (and thus possible cross-reactivity of antibodies) is to perform a Western blot analysis. For this procedure, the samples are first subjected to polyacrylamide gel electrophoresis, which separates the proteins based primarily on their molecular weight. The proteins are then transferred to a support matrix, usually nitrocellulose. The individual bands are detected with an antibody, often the same antibody that would be used in the ELISA or RIA. With this method, the molecular size of the cytokine is known, which may give important information about which cytokine is being recognized by the antibody. Also, if multiple smaller bands are detected, it may indicate that there has been some breakdown.

With this brief description of the advantages and disadvantages of immunoassay, we now describe the biological assays. Limitations in interpretation of the data are discussed, although this is by no means a complete presentation of the potential pitfalls one may encounter when performing these complex assays.

GENERAL CONSIDERATIONS FOR BIOASSAYS

Bioassays require knowledge of tissue culture techniques. These are not difficult to learn, but they do require a considerable investment in equipment. It is absolutely critical to maintain sterility in the equipment, solutions, and cells. Bacterial overgrowth may make it impossible to interpret the data.

To set up a tissue culture facility requires purchase of or access to a sterile, laminar flow hood. All cell work should be performed in this hood to minimize the potential of contamination, and all solutions should be prepared and aliquoted in the laminar flow hood. A CO_2 incubator is used to grow the cells and is also used for conducting the bioassay. All material that comes into contact with the cells (tissue culture flasks, pipettes, and pipette tips) must be sterile. These items may be purchased as sterile or may be sterilized if an autoclave is available. Tissue culture media also may be prepared and then filter sterilized, or they may be purchased in an already sterile condition. Some cells require multiple, special additives for optimal growth. There may be great variation in quality, and the best supplier for each reagent must be determined. Some investigators assay for the best reagents and then buy a large quantity of the particular lot.

Bioassays are often performed using cell lines, that is, cells that may grow indefinitely in vitro. Cell lines are advantageous since they provide a uniform population that should respond to the cytokine in a uniform manner. Many cell lines have an absolute requirement for a particular cytokine (i.e., they do not continue to grow in culture unless the cytokine is present). This forms the basis for the assay, since the cells die out if the sample does not contain the cytokine. Tumor necrosis factor (TNF) is measured by the opposite property: it kills the target cells, and cell death indicates the presence of the cytokine.

Cell lines may sometimes be obtained from other investigators. Another excellent source is the American Type Culture Collection (ATCC WEB address, www.atcc.org), a nonprofit organization in Rockville, Maryland. Similar repositories are available in Europe. The ATCC serves as a central repository for cell lines, and frozen cells may be purchased for a minimal price. The internet address of ATCC is listed in Table 1, and the ATCC homepage includes a catalog, which may be searched.

CALCULATION OF ACTIVITY

The amount of activity in the fluid that was assayed is often expressed in terms of half-maximal units. These units are defined as the dilution that produces a half-maximal re-

Table 1 List of Commercial Sources of Cytokines and Anticytokine Antibodies

Company Name	Web address	Cytokines	Ab	Kits
Accurate Chemical	http://biores.com/mfg/Accurate.html	Yes	Yes	Yes
Amersham Life Science Inc.	http://www.amersham.co.uk/	Yes	—	Yes
ATTC	http://www.atcc.org/	—	Yes	—
Biosource International	http://biosource.com/	Yes	Yes	Yes
Boehringer Mannheim	http://biochem.boehringer-mannheim.com/	Yes	Yes	Yes
Calbiochem	http://www.calbiochem.com/	Yes	Yes	—
Cayman Chemical	http://www.caymanchem.com/index.htm	—	—	Yes
Chemicon	http://www.chemicon.com/	Yes	Yes	Yes
DuPont NEN	http://www.dupont.com/index.html	Yes	—	—
Endogen	http://www.endogen.com/	Yes	Yes	Yes
Genzyme Diagnostics	http://www.genzyme.com/	Yes	Yes	Yes
ICN Pharmaceuticals Inc.	http://www.icnpharm.com/	Yes	Yes	—
INCSTAR	http://www.incstar.com./	—	—	Yes
Life Technologies	http://www.lifetech.com/	Yes	Yes	—
PeproTech, Inc.	http://biores.com/mfg/PeproTech.html	Yes	Yes	—
Pharmacia Biotech	http://www.biotech.pharmacia.se	—	—	—
PharMingen	http://www.pharmingen.com/	Yes	Yes	—
Promega	http://www.promega.com/	Yes	—	—
R & D Systems	http://www.rndsystems.com/	Yes	Yes	Yes
SIGMA	http://www.sigma.sial.com/	Yes	Yes	—
Upstate Biotechnology Inc.	http://www.biosignals.com/	Yes	Yes	—

Ab indicates that they sell antibodies to cytokines.

sponse (2). When using the ^{3}H-thymidine method, the counts per minute (cpm) are plotted against the dilutions, with both axes a logarithmic scale. Figure 3 shows a typical plot of a standard and unknown samples. In this example, the standard's maximum response was 16,000 cpm, the half-maximal was 8000 cpm, so the log of the half-maximal was 3.9. As can be seen on the graph, the standard (middle curve) plateaus at the maximum and then falls off in a linear fashion. Regression analysis of the curve discloses an R^2 value of 0.9999. The number of units is calculated by determining the equation of the line and then solving for the x value that corresponds to the y value, giving the half-maximal response. In this example, the standard has 40 U/ml. The sample on the left is subjected to a similar analysis and has 5 U/ml; the sample on the right has 300 U/ml. This figure also shows the importance of calculating half-maximal units, since both the standard and the sample on the right produce the same maximal response but have markedly different activities.

Instead of using the radioactive ^{3}H-thymidine, the cells may be pulsed with MTT-tetrazolium (3-[4,5-dimethylthiazol-2-yl-]2,5-diphenyltetrazolium bromide). This is an extremely valuable method for quantitating cell proliferation or cytotoxicity (3,4). With this method, MTT-tetrazolium is added to the wells and the cells are incubated an additional 4–6 h. As the cells undergo oxidative respiration, the dye is metabolized to form insoluble dark purple crystals. The crystals are dissolved by removing some of the supernatant and then adding an appropriate solvent, such as a detergent or acidified isopropanol (4). The

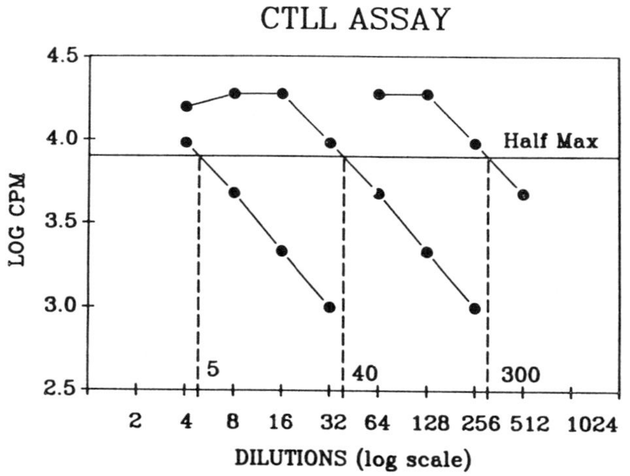

Figure 3 Determination of half-maximal units. This calculation was performed using the CTLL assay as an example. Similar data analysis may be performed for the other cytokine assays. Half-maximal units are defined as the reciprocal of the dilution that results in half-maximal response of the assay.

96-well plates are then conveniently read in a micro-ELISA reader. If the plate reader is interfaced with a computer, the data may then be easily analyzed. Figure 4 shows an example of a plate with samples of varying activity. As dilutions are continued across the plate, there is less biological activity and less color. Samples A, B, and D have less activity than samples E and F, since the color is less intense as one moves from left to right on the plate.

In most bioassays, a recombinant standard is also included in the assay. This recombinant standard may be used to calculate the concentration of the unknown sample. For these types of calculations the activity of the unknown sample is directly compared to the standard. This is a different approach than the half-maximal units, where the activity of the sample is compared only to the maximal response of the cells. Several computer programs exist that will calculate the equation for a curve generated by the recombinant standard. Figure 5 shows the results from a typical assay, in this case a proliferation assay using the B9 cell line responding to recombinant human IL-6. Most biological assays have a sigmoid-shaped curve and not a straight line, as shown in Figure 5. In a typical assay, the curve for the standard is generated based on the absorbance of the wells. The absorbance of the samples is then used in the equation to calculate the concentration of the cytokine in the sample. We generally do six separate dilutions and average the results where the absorbance falls on the standard curve.

BIOASSAYS FOR INDIVIDUAL CYTOKINES

Interleukin-1

The original assay for IL-1 was based on its ability to conduce the proliferation of murine thymocytes (5–8). In this system, IL-1 was able to augment proliferation in conjunction with a suboptimal concentration of a lectin, either phytohemagglutinin (PHA) or concanavalin A. Other lectins may also be used to improve the sensitivity of the assay (9). Since two mitogens were involved, this assay has also been called the comitogen or coproliferation assay. The assay is done by performing dilutions of the samples to be tested and adding the samples to 96-well microtiter plates. Murine thymocytes are then prepared as a single-cell suspension, mixed with the appropriate, suboptimal concentration of lectin, and then added to the wells. Incubation of the thymocytes, IL-1, and lectin causes the production of IL-2, and IL-2 then induces proliferation of IL-2–responsive thymocytes. After 2 days of incubation at 37°C, the proliferation of the thymocytes is detected by pulsing the cells with ^{3}H-thymidine. As the cells proliferate, the radioactive thymidine is incorporated into the DNA

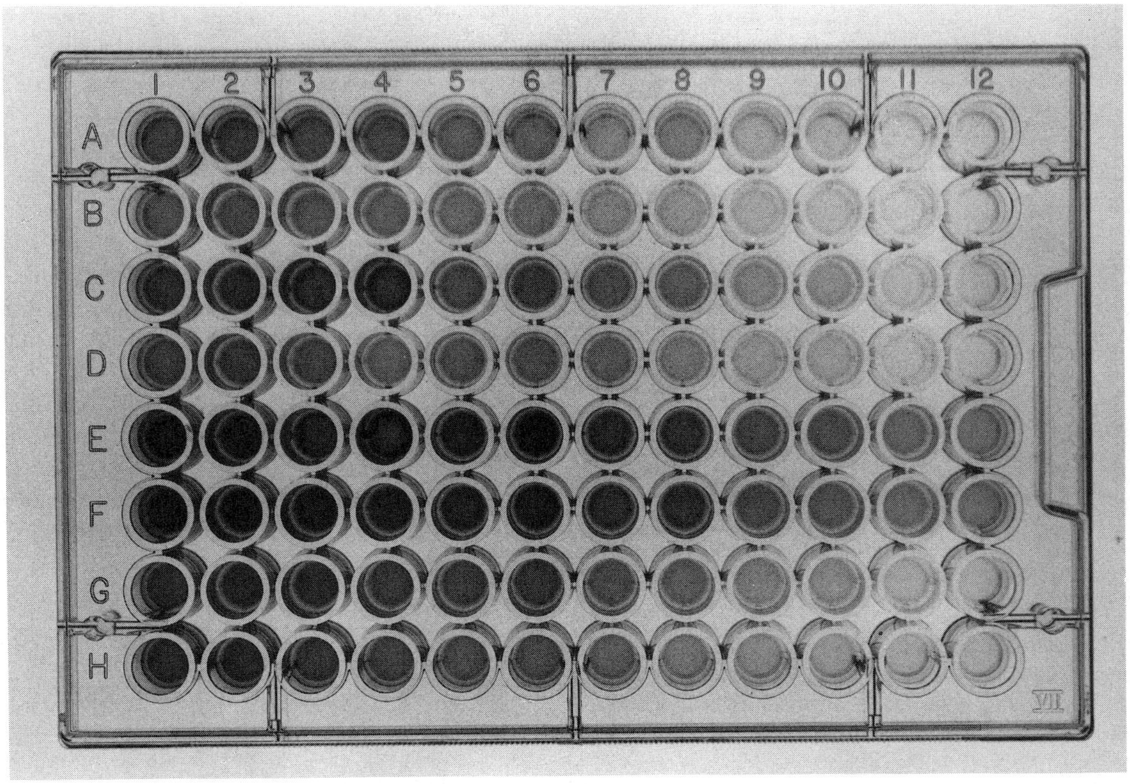

Figure 4 MTT-tetrazolium dye reduction. Serial dilutions of cytokines were placed in each well of this 96-well plate, in duplicate. Each row (A, B, C, etc.) is used for one sample, so there are six dilutions for each sample. This plate is from a cell proliferation assay, so the activity decreases as one moves from left to right. In this plate, samples E and F have the most activity, since the dye is still intense even in wells 7 and 8. Samples A and B have the least activity, since the dye is almost completely gone by wells 7 and 8.

of the cell. The amount of incorporated thymidine is determined by harvesting the cells on glass fiber filters. The DNA adheres to the filters, and the radioactivity may be counted in a scintillation counter. The activity in the sample is then ascertained by determining the dilution that induces a half-maximal response. The thymocyte coproliferation assay has several factors that may cause a false-positive reaction. IL-6 also induces proliferation of thymocytes (10,11), and murine TNF, but not human TNF, also causes proliferation (12). Since the assay actually measures the response to IL-2, this cytokine also results in a false-positive reaction (13). IL-7 will also have an effect in the assay (14). To determine if the thymocyte proliferation is truly due to IL-1, it is necessary either to use specific assays for IL-2, IL-6, and TNF to rule out their presence or to inhibit the proliferative activity with specific antibodies. Since either IL-1α or IL-1β will drive the assay, the use of antibodies has the advantage of being able to distinguish between these forms.

Another assay for IL-1 activity that is similar in principle to the thymocyte assay utilizes cell lines instead of the heterogeneous thymocyte populations (15–17). For this assay, the lymphoma cell line LBRM is stimulated with IL-1 and PHA. This stimulation results in the production of IL-2. The supernatants are transferred to another plate, leaving the LBRM cells behind, and the IL-2 detected in a standard IL-2 assay using an IL-2–dependent cell line. The NOB-1 assay is also a two-step assay with good specificity for IL-1 (18). Obviously, samples that contain IL-2 cause false-positive results: however, IL-6 does not cause a positive reaction in this assay (11,19).

An additional cell line has been described that responds to IL-1 directly: the D10.G4.1 assay (20). In the presence of IL-1, these cells proliferate. A subclone of the D10 cell line has recently been described that is extremely sensitive to the effects of IL-1 and is not significantly affected by other cytokines, such as TNF, IL-2, or IL-6 (11,21). The proliferation may be monitored by either incorporation of ^{3}H-thymidine (as in the thymocyte assay), or the degree of proliferation may be assessed by their

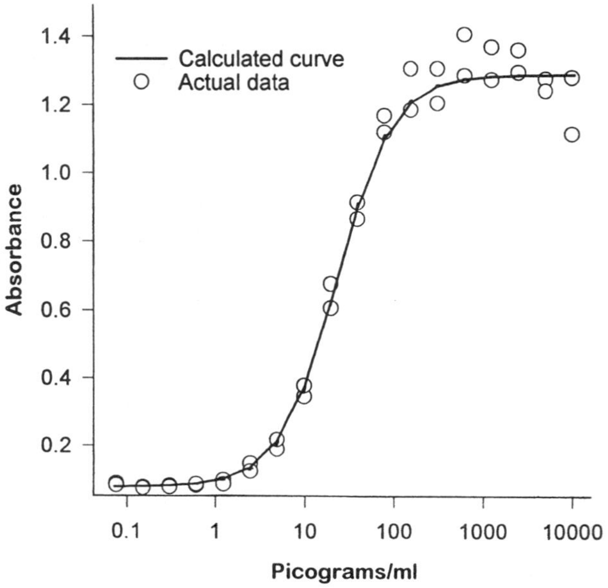

Figure 5 Typical bioassay standard curve. A standard curve of human recombinant IL-6 was assayed in the B9 proliferation assay. The circles are the actual absorbances from the assay, and the line is the calculated line using a four-parameter fit. While there is variation when the cells are maximally stimulated (1000 pg/ml), the curve follows the standard very closely in the middle and lower portions. The correlation coefficient for this curve was 0.999.

ability to metabolize a tetrazolium dye, MTT-tetrazolium, described in greater detail in the section on IL-2. Other cell lines have also recently been described that will react to IL-1, including the A375 (22) or RPMI 1788 (23) cells.

IL-1 in biological samples may also be quantitated by determining its ability to inhibit the binding of radiolabeled IL-1 to the IL-1 receptor present on fibroblasts. This is a very sensitive assay that measures both IL-1α and IL-1β. It is similar in principle to a radioimmunoassay, but instead of evaluating binding to antibody, one measures binding to receptors.

No discussion of IL-1 measurement would be complete without mentioning the presence of IL-1 inhibitors in biological samples (24). This is particularly a problem in blood products and has caused great difficulties in data interpretation. Various methods have been tried to remove the inhibitors, including performing serial dilutions, extraction with chloroform (25), and treatment with polyethylene glycol (26). The biological relevance of IL-1 values obtained from samples subjected to the treatments is not known with certainty. There has been controversy about the ELISAs for human IL-1β. As described in another chapter, this cytokine exists in two forms, a larger intracellular form and the secreted, 17 kDa cytokine. Some of the commercial ELISAs will recognize both forms, while others only recognize the smaller secreted form (27).

Interleukin-2

IL-2 was originally described as a factor that would support the long-term growth of T-cell clones. The most widely used biological assay today to test for the presence of IL-2 measures the ability of a sample to stimulate the proliferation of a cloned line of T cells. Two commonly employed lines are the HT-2 and the CTLL-2 (28). Both cell lines are available from the ATCC (CTLL, TIB-214; HT-2 CRL 1841) and may be easily maintained in culture or frozen for long-term storage. Both the HT-2 and CTLL cell lines have been derived from mice. A human T-cell line that responds to IL-2 has also been described (29), however, it should be noted that HT-2 and CTLL respond to human IL-2 very well.

The proliferation of the cells may be measured either by the amount of ^{3}H-thymidine incorporated or by the ability of the cells to metabolize MTT (3,4). The IL-2 assay is based on the knowledge that the cell lines have a requirement for IL-2 to proliferate or remain viable for longer than 20 h. Furthermore, it has been demonstrated that many of the prostaglandins (PGE_2 and $PGF_{2\alpha}$) and leukotrienes (LTB_4, LTC_4, LTD_4, and LTE_4) do not block the ability of IL-2 to induce CTLL proliferation (30). However, high-dose dexamethasone or cyclosporin A inhibits the ability of the assay to detect IL-2 (30). For either assay, samples are serially diluted in 96-well flat-bottomed microtiter plates in 0.1-ml volumes of tissue culture media. Since both cell lines are lymphocytes, it is critical to include 2β-mercaptoethanol in the media for optimal cell proliferation. Both cell lines are sufficiently sensitive to the effects of IL-2 that only 4,000–10,000 cells per well are required. The samples are allowed to incubate overnight at 37°C, and the next morning they are pulsed for 4 h with either MTT-tetrazolium or ^{3}H-thymidine.

Some lines of CTLL-2 and HT-2 cells also respond to IL-4 (31,32), and a positive assay may thus indicate the presence of either IL-2 or IL-4. To differentiate between these two possibilities, three approaches may be used. The first is to test the cell line with recombinant IL-4 and ascertain whether the cell line being used is sensitive to IL-4. The second approach is to inhibit any potential IL-4 using the 11B11 antibody, which blocks murine IL-4 (ATCC# HB188) (33). Last, anti–IL-2 receptor antibodies may be included to specifically inhibit the ability of IL-2 to induce proliferation. Additional information regarding measurement of IL-4 may be found in the section on interleukin-4.

Interleukin-3 and Colony-Stimulating Factors

Interleukin-3 has also been named multi-CSF to indicate that it has colony-stimulating activity for several different cell lineages. A number of assays exist for the quantitation of IL-3 (34). These assays have included the induction of the enzyme 20α-hydroxysteroid dehydrogenase (SDH) in splenic lymphocytes from nude mice, induction of proliferation of IL-3–dependent T-cell lines from long-term bone marrow cultures, induction of proliferation of mast like cells, and colony formation in soft agar cultures. The least complex assay involves the use of IL-3–dependent T-cell lines and measurement of proliferation induced by IL-3. The other assays may lack specificity, that is, induction of colonies in soft agar cultures may be due to other colony-stimulating factors or involve the expense of nude mice and the complexity of measuring the induction of 20α-SDH (35). Different cell lines have maintained their dependence on IL-3 for proliferation, such as prebasophil and prelymphoid cell lines, and for this reason they have been used to assay this cytokine (36).

Some cell lines respond only to murine IL-3, not to human IL-3. An alternative is to enrich a population of bone marrow cells for MY-10–positive cells, which proliferate in response to IL-3. This may be done by adhering the antibody to plastic dishes, adding the bone marrow cells in suspension, and gently removing the nonadherent T cells (37). The MY-10 cells remain adherent to the dishes and may be removed by more vigorous pipetting and scraping. These cells are used in colony-forming assays as an indicator for the presence of IL-3. When long-term cell lines are used to assay IL-3, the degree of proliferation is assessed by ^{3}H-thymidine incorporation in the same fashion as described for the IL-2 assay, except that an IL-3–dependent cell line is used, such as the FDC-P cell line (38,39), the FC.C/1 (40), or the 32D cells (41). The assay is then performed similarly to that described for IL-2.

It has recently been shown that interferon-γ (IFN-γ) inhibits the ability of the FDC-P cell line to respond to IL-3 (42). Since the same T cells that produce IL-3 also make IFN-γ, it is easy to mask the IL-3 activity unless specific antibodies are used to inhibit the IFN-γ material. Another potential problem with the IL-3–dependent cell lines is the use of antibodies to block IL-3 to confirm that the biological activity is really due to IL-3. The IL-3 cells are very responsive to immunoglobulin aggregates. One way to prevent immune complex stimulation of the cells is to block the Fc receptors on the cells with a monoclonal antibody (2.4G2, ATCC HB 197).

IL-3–dependent cell lines must be maintained in long-term cultures by adding an exogenous source of IL-3. This may be obtained by culturing an IL-3–producing cell line, such as the WEHI-3 (ATCC #TIB 68) for 48 h and collecting the supernatant. This supernatant is then used as the source of IL-3 for growth of the IL-3–dependent cell line. The optimal concentration of supernatant must be determined for each batch of conditioned media.

Cells lines have also been developed that respond to other colony-stimulating factors in a relatively specific manner. International standards for GM-CSF and M-CSF have also been established, which will help to validate the bioassays (43,44). The anemic mouse spleen cell assay has been used to detect erythropoietin, although several cytokines will act as inhibitors in this assay (45). The Rauscher cells, a murine erythroleukemia cell line, will proliferate in response to erythropoietin (46), and M-TAT cells will secrete hemoglobin in response to erythropoietin (47). Some cell lines will respond to both GM-CSF and IL-3 (48,49), and neutralizing antibodies are required in order to differentiate between the two cytokines (50).

All of the hematopoietic stimulating factors or colony-stimulating factors may be detected by determining their ability to stimulate colony formation in soft agar cultures. This is a complex assay procedure that requires not only expertise in tissue culture work but also the ability to differentiate the morphology of the cells growing in colonies. The basic protocol consists of collecting bone marrow cells from the femurs of mice. The cells are resuspended in tissue culture media containing 0.4% agar. Serial dilutions of the samples to be tested are added to the cells in agar, and the samples are incubated for 3–7 days at 37°C. At the end of the assay, the number of colonies are counted and the types of cells in the colonies are determined.

Interleukin-4

One assay for the quantitation of IL-4 utilizes a costimulation assay, since IL-4 stimulates B-cell proliferation in the presence of the polyclonal activator IgM (51). Briefly, B cells are purified from T-cell–depleted mouse spleens, their purity confirmed, and the cells resuspended in tissue culture media and placed into 96-well plates. The cells are then stimulated with anti-μ chain, which serves as one stimulus.

The samples containing IL-4 are serially diluted and also added to the well, which serves as the second stimulus. After 54 h of incubation, the wells are pulsed with ^{3}H-thymidine for 18 h and the degree of proliferation determined. A unit of IL-4 is expressed in terms of half-maximal units. IL-4 is species specific, and a different assay is used for quantitating human material. The assay is similar to that described for murine IL-4, but the B cells are isolated from blood, tonsils, or spleen and then reactivated with either anti-μ antibodies or *Staphylococcus aureus* Cowan strain (52).

As mentioned in the section of IL-2, some CTLL-2 cell lines respond to IL-4 (31,32). Using an IL-4–responsive cell line allows detection of IL-4 if appropriate controls are included to ensure that one is not detecting IL-2. A cell line has been cloned that responds to IL-4 and not IL-2—the CT.4S (53). This variant of the CTLL-2 cells is very sensitive to IL-4 and was originally reported not to respond to other cytokines (53); unfortunately, this cell line responds only to murine IL-4, not human IL-4. Since CT.4S cells do not respond to murine IL-2, these cells may be used to quantitate the presence of murine IL-4 directly. The other alternative is to use a cell line responsive to both IL-2 and IL-4 and use an antibody to inhibit the activity of either IL-2 or IL-4 (54). If anti–IL-2 is used (a monoclonal rat antimurine IL-2 is available from ATCC, S4B6, #HB 8794), then the remaining CTLL-2 activity is due to IL-4.

The CT.4S cells are generally unresponsive to several other cytokines, although they will proliferate in response to high levels of IL-2. Additionally, CT.4S cells proliferate in response to simian IL-15, presumably by reacting with the β chain of the IL-2 receptor (55). The cells will also respond to murine but not human IL-12 (56).

The CT.4S cell line must be maintained by culturing in the presence of 100 U/ml of murine IL-4. This can be an expensive procedure since it requires the continued purchase of recombinant material. A much more economical approach is to obtain IL-4–rich supernatants and use these as a source of exogenous IL-4. Such supernatants may be obtained by stimulating D10.G4.1 cells (ATCC, #TIB 224) with concanavalin A (2.5 μg/ml) and collecting the overnight supernatant. The optimal concentration of the supernatant to maintain long-term growth of the CT.4S cells needs to be determined empirically.

A recently described clone of the CT.4S cells has been prepared where the cells have been transfected with the receptor for human IL-4 (57). These are CT.h4S cells, and since they have the human IL-4 receptor, they will respond to human IL-4 in the same manner that they react to murine IL-4. Other derivatives of the CTLL cells have been described that are hyporesponsive to IL-2 but proliferate in response to IL-4 (58). Other cell lines have also been described that respond to IL-4 (59,60).

Interleukin-5

One of the first biological activities described for IL-5 was its ability to promote B-cell proliferation, which gave rise to one of its older names—B-cell growth factor II. Several assays may be used to detect the presence of IL-5 (61). In the first, purified B cells are distributed into 96-well plates, and the cells are stimulated with dextran sulfate along with serial dilutions of test supernatant. After 66 h of incubation, the wells are pulsed with ^{3}H-thymidine and harvested after 6 h. Other assays for IL-5 are based on its ability to induce antibody secretion. There are generally indirect assays, with the IL-5 inducing antibody secretion and the actual assay detecting the presence of antibody. Antibody production is usually measured by plaque formation (62). IL-5 may also be detected by examining its ability to induce eosinophil colonies from bone marrow in soft agar cultures. However, other cytokines also have this property, and this assay will not discriminate between IL-5 and IL-3 or GM-CSF (63).

The most widely used assay for IL-5 uses mouse B-cell lymphoma cells (BCL_1 #TIB 197) as the indicator cell line. These cells are prepared by injecting the cells intravenously into Balb\c mice. The mice are sacrificed 4–8 weeks later and the spleens dispersed in a sterile manner. T cells are removed from the suspension by incubation with anti-Thy-1.2 (ATCC #TIB 90) and anti-Lyt-2 (ATCC #TIB 105) antibodies and complement. Macrophages are removed by passage over Sephadex G-10 columns, and the dead cells and debris are removed by density centrifugation on Ficol-Hypaque gradients. These purified BCL_1 cells are then aliquoted and frozen for future use. The actual assay is performed in a similar fashion to most of the other cytokine assays: serial dilutions are prepared, the BCL_1 cells are added, and the plates are incubated overnight. Routinely, the samples are then pulsed with ^{3}H-thymidine for 6 h, the cells harvested, and the radioactivity counted.

The BCL_1 assay is not affected by IL-2 or granulocyte-macrophage colony-stimulating factor (64). However, it is augmented by IL-4 and inhibited by interferon (65). Human IL-5 has considerably less activity in the BCL_1 assay compared to murine IL-5. IL-5 quickly loses activity if stored at 4°C.

Interleukin-6

IL-6 has previously been called B-cell stimulatory factor 2 (66), hybridoma growth factor (67,68), hepatocyte-stimulating factor (69), and interferon-β2 (70). An international standard for biologically active IL-6 has been established (71). IL-6 is responsible for a number of biological activities, and thus several different assays exist to quantitate its presence. These assays have been investigated, and the authors determined that the most sensitive and specific test for IL-6 is its hybridoma growth factor activity (72). Standard bioassays for the detection of IL-6 hybridoma growth-promoting activity involve the use of IL-6–dependent cell lines in a fashion similar to that already described for the other interleukins. Two such cell lines are the T1165 (62) and B9 (68), both of which are murine plasmacytomas. These cells have a requirement for IL-6 for growth, and in the absence of IL-6 they die over a 3-day period. Several interleukins have been tested in B9 cell line, and some clones will respond to IL-13. Also, high concentrations of lipopolysaccharide (10 μg/ml) result in a small degree of augmentation (73). A more recent report indicates that even lower levels of LPS (100 ng/ml) will drive the B9 assay (74). The B9 cell line responds to IL-6 from multiple sources, including rat, mouse, and human. However, it does not respond to rabbit IL-6 (unpublished data from author's laboratory). To run the assay, samples are serially diluted in tissue culture media (either RPMI-1640 or Iscoves, although we have had considerably better success with B9 cells in Iscoves) containing fetal calf serum. Inclusion of fetal calf serum in the diluting media enhances the sensitivity of the assay since it provides a source of protein to prevent nonspecific adherence of the IL-6 to the plasticware. The diluted samples are placed in a 96-well plate and the cells are added on top. As with the CTLL-2 cells, it is important to include 2β-mercaptoethanol in the assay media since the cell lines are lymphocyte derived. After about 60–70 h of incubation at 37°C, the cells are pulsed with either ^{3}H-thymidine or MTT-tetrazolium and incubated an additional 6 h. The ^{3}H-thymidine is then counted or the color developed, and the units are calculated from the standard curve, which is always run on the same assay.

The B9 cells have an absolute requirement for the presence of IL-6 for continued growth. Several sources of IL-6 exist, the most expensive involving purchase of recombinant material. We have taken conditioned media from fibroblasts as a source of IL-6. Fibroblasts are grown to near confluence in tissue culture flasks. The medium is removed, fresh medium containing 10% fetal calf serum is added, and the cells are carried in culture without stimulus for an additional 1 or 2 days. This medium is then removed and used as a rich source of IL-6. The B9 cells proliferate quickly, and they will easily exhaust the medium of nutrients and IL-6 if not subcultured. We routinely split our cells three times a week.

Interleukin-7

IL-7 is a growth factor for early B cells, pre-B cells, and immature T cells and is a co-mitogen for peripheral T cells (75). Activated T cells will also proliferate in response to IL-7 alone (76). One of the first assays for detecting IL-7 used progenitor B cells from bone marrow cells as the indicator cell line (77,78). While this assay had specificity, it was very difficult to maintain the cells because of the unusual culture conditions. As a result, IL-7–dependent cell lines were established that have specificity for IL-7.

One widely used cell line for detecting IL-7 is the IxN\2bx, which is a pre–B-cell line. These cells will proliferate in the presence of IL-7, but not other cytokines. The cells have been tested with IL-1 through IL-12, several colony-stimulating factors, and the interferons and was found to be specific for IL-7 (79). Other cell lines have also been described for the measurement of IL-7, including the 2E8 cell line (ATCC #TIB-239) (80) and the DW34 cell line.

The IL-7 assay is a proliferation assay and is run in the same manner as the other proliferation assays. The samples to be tested are serially diluted and added to 96 well plates. The IxN/2bx cells are added and the plate is allowed to incubate for 2 days. At this time, proliferation is determined by either ^{3}H-thymidine incorporation or MTT tetrazolium. Units are calculated from a standard curve run in the same assay.

Interleukin-8/Chemokines

The chemokines are a family of cytokines whose primary activities are directed toward leukocytes. This family is divided into C-X-C and C-C chemokines based on the positioning of cysteine residues with (C-X-C) or without (C-C) an intervening amino acid. IL-8 is the best described of the chemokines and functions primarily as an activator of neutrophils; it has been described as neutrophil activating peptide or monocyte-derived neutrophil chemotactic factor (81,82). Since it functions as a neutrophil activator, most assays quantitate its presence by its ability to alter specific neutrophil functions. IL-8 has been shown to induce in neutrophils the following functions: chemotaxis, calcium fluxes, respiratory burst, chemokinesis, and release of several enzymes (83). Other chemokines will induce migration of T cells (84–87) or monocytes (88,89). Many of the chemokines have overlapping biological activities so that the assays lack specificity. To precisely determine if a particular chemokine is present, it is necessary to remove the biological activity by blocking with a specific antibody. Unlike most of the other cytokines, which use cloned cell lines, these assays use normal neutrophils, T cells, or monocytes.

One of the more sensitive assays for quantitating IL-8 activity is the chemotaxis assay. There are several variations in the performance of the chemotaxis assay, but all essentially measure the ability of neutrophils to migrate toward a positive gradient of the chemotactic factor. The chemotaxis may be measured by the cell's ability to migrate in soft agar or, more commonly, to migrate through the pores of a membrane. Many investigators use a multiwell chemotaxis chamber, which permits analysis of several samples and controls at the same time (90). For this assay, the samples to be tested are placed in the bottom of the chambers and a filter with micropores placed over the top and clamped into place. Purified neutrophils are pipetted into the wells of the upper chamber and allowed to migrate towards the chemotactic agent for 1 h. Nonmigrated cells are removed from the filter, and the migrated cells are stained and counted. The cells may be counted under microscopic examination or using an image analyzer. A more rapid method, which works for larger numbers of samples, is to label the neutrophils with a fluorescent dye, perform the chemotaxis in a 96-well format, and then read the membrane in a fluorescent plate reader (91). The critical parameter in either assay is to prevent the neutrophils from clumping prior to placing them in the chemotaxis chamber.

The chemokines will also induce chemotaxis of T cells, which may be measured by a modification of the neutrophil chemotaxis assay. For this assay the membrane is coated with extracellular matrix proteins, such as collagen, fibronectin, or lamin. The assay is then carried in a manner similar to that described above, although the incubation time is increased to 4 h. For this assay, it is imperative to freshly isolate the T cells under conditions that cause minimal activation (92). Monocyte chemotaxis may also be performed, but the membrane is not coated with extracellular matrix proteins, and the incubation time is only 1.5 h.

The biological activity of the chemokines may also be quantitated by measuring T-cell adherence to either cytokine-activated endothelial cells (89) or microtiter plates coated with matrix proteins (92). For both of these assays, the T cells are freshly isolated using the same precautions as for the T-cell chemotaxis assays. The T cells are then radiolabeled, mixed with the chemokines, and added to either the matrix proteins or endothelial cells. After incubation, the plates are washed and the number of bound T cells are determined by measuring the radioactivity.

The chemokines are also capable of inducing calcium fluxes in peripheral blood monocytes or monocyte cell lines. For these studies, the cells are loaded with fluorescent dyes, which can be used to monitor intracellular calcium levels. Examples of these dyes include Quin-3, Fura-2, and Indo-1. Upon addition of the chemokines, there is a rapid rise of intracellular calcium, which can be detected in a spectrofluorimeter. Recently, a very sensitive and specific ELISA was described for IL-8 (93).

Interleukin-9

The major biological activities of IL-9 include proliferation of a megakaryocyte cell line and mast cells. The most commonly used assay for the detection of IL-9 uses the human megakaryoblastic leukemia cell line M-07e as the indicator cell. Serial dilutions of the samples are mixed with the cell line, incubated for 3 days, and then cell proliferation determined by either MTT-tetrazolium or ^{3}H-thymidine incorporation.

The M-07e cell line is very sensitive to the effects of IL-9, but will also proliferate very well in response to IL-3 or GM-CSF (49). Additionally, the cells will weakly proliferate in response to IL-2, IL-4, and IL-6. To demonstrate that the biological activity is due to IL-9, it is necessary to use neutralizing antibodies to the other factors. (The M-07e cell line is available from the Genetics Institute, Biological Sample Request Co-ordinator, 87 Cambridge Park Drive, Cambridge, MA 02140-2387.)

Interleukin-10

IL-10 was originally called cytokine synthesis inhibitory factor (95). It was identified as a factor from T_H2 cells that inhibits the synthesis of cytokines from T_H1 cells. The original assay evaluated the reduction in interferon-γ in the presence of this factor. The bioassay actually measures the decrease in IFN-γ production of antigen-stimulated T cells. For the assay, cloned T_H1 cells are placed together with irradiated mouse spleen cells in 96-well plates. These cells are mixed with the appropriate antigen, plus the samples containing IL-10. After 24 h of incubation, the supernatants from each well are tested for IFN-γ (described below). If IL-10 is not present, the cells will secrete high levels of IFN-γ, whereas the presence of IL-10 will result in a marked reduction. This is an indirect assay, since the final determinant of IL-10 activity is actually measuring IFN-γ. There are cytokines that may affect the assay in either a positive or a negative manner. IL-2 will increase IFN-γ and may result in a false-positive assay. Transforming growth factor-β or IL-4 both decrease IFN-γ synthesis, and their presence may result in a false-positive assay. As with several of the other cytokine assays, it is necessary to include neutralizing antibodies to ensure specificity.

A mouse mast cell line has been described that is sensitive to IL-10 and does not respond to several other cytokines (96). This cell line, D36, requires the presence of murine IL-4 in order to be responsive to IL-10. Additionally, human serum inhibits the cells, so the assay cannot be used to measure IL-10 in serum or plasma samples.

Interleukin-11

IL-11 has several biological properties, including promoting the growth of hematopoietic and lymphoid cells (97,98). It shares several biological activities with IL-6, and one of the standard bioassays for IL-11 requires that IL-6 bioactivity be blocked with an antibody to ensure that the activity is due to IL-11. This cell line is called T10 and is derived from the T1165 cell line, which is highly responsive to IL-6. This is a proliferation assay, so the cells are mixed with serial dilutions of the samples and allowed to incubate for 3 days. At the end of the assay, cell proliferation is determined either by MTT-tetrazolium dye reduction or ^{3}H-thymidine uptake.

Another cell line has recently been described that is extremely sensitive to the proliferative effects of IL-11 (99). This cell line was derived from the B9 cells, and is called B9-11. Similar to the T10 cell line, it will respond to IL-6, and it is necessary to include IL-6–neutralizing antibodies in order to ensure the specificity of the assay. The cell line does not respond to several other cytokines or hematopoietic growth factors.

Interleukin-12

IL-12 was originally described as natural killer cell stimulatory factor (100) and is composed of two distinct subunits, which form a heterodimer (101). IL-12 also has a number of biological activities that it shares with other cytokines, including augmenting the proliferation of T lymphoblasts. Because of the large overlap in activities, a different approach to measuring IL-12 in complex media is used. For other cytokines, samples were tested and, if found to be positive, retested in the presence of neutralizing antibodies. For IL-12, an ELISA plate is first coated with an anti–IL-12 antibody, and then the samples are added to the ELISA plate. After incubation, all of the IL-12 will remain bound to the antibody present on the plate. The T lymphoblasts are prepared and added and proliferation determined after 2 days by either MTT-tetrazolium or ^{3}H-thymidine incorporation. If the sample is felt to only contain IL-12, such as a preparation of purified recombinant protein, the antibody capture step is not required, and the samples may be directly tested by incubating the samples with the T lymphoblasts.

This assay uses T lymphoblasts prepared from normal peripheral blood mononuclear cells and is in contrast to many of the cytokine assays that use cell lines. In this regard, it is similar to the IL-8 assays, which also use normal blood cells rather than the cell lines. The T lymphoblasts are prepared by stimulating the mononuclear cells with PHA for 2 days.

As mentioned above, IL-12 is a heterodimer composed of p40 and p35 subunits. The heterodimer of these subunits binds to the receptor to transduce a signal. However, the p40 subunits may combine to produce a homodimer. In the normal stimulation of IL-12 production, substantial amounts of p40 homodimer are produced. This moiety will bind to the receptor, but will not transduce a signal. Thus, production of the p40 homodimer will decrease the activity of the IL-12 in the sample and potentially lead to a false-negative result. Production of the p40 homodimer occurs in both mouse (102) and humans (103).

An alternative to the use of T-lymphoblast proliferation is to measure the ability of IL-12 to induce IFN-γ production (104). This assay has the advantage that T lymphoblasts do not need to be prepared. The first portion of this assay is similar to the T-lymphoblast assay, where antibodies to IL-12 are used to coat the plate, followed by the samples containing IL-12. After washing to remove substances not bound to the antibody, peripheral blood mononuclear cells are added to the plate in the presence of IL-2. This combination of IL-12 captured on the plate and the IL-2 will induce strong production of IFN-γ, which is measured in the supernatant. The assay is very sensitive and will measure IL-12 concentrations in the picogram/ml range. IFN-γ production has also been used to measure mouse IL-12 (105).

While the standard bioassay for IL-12 uses normal cells, there are two cell lines that will respond to IL-12 and may be used. The first cell line is Kit225/K6, which is a human T-cell leukemia cell line (106). This will respond to IL-12 and has been used in the antibody capture/bioassay in place of the T lymphoblasts. It was recently discovered that the murine CT.4S will proliferate in response to IL-12 and IL-4 (56). As mentioned in the section on IL-4, these cells will also respond to IL-15, presumably by interactions with the receptor.

Interleukin-13

IL-13 has several properties, many of which are immunosuppressive (107,108). Its activity has primarily been measured by evaluating the proliferation of B9 cells, the same cells used for the detection of IL-6. This assay is run in the same manner as described for the IL-6 assay, and any activity attributable to IL-13 is blocked with antibodies. A neutralizing antihuman IL-13 antibody is available from ATCC with the permission of DNAX. A subclone of the B9 cells has been described that is highly dependent on the presence of IL-13 for its growth (109). However, this cell line is still responsive to IL-6, and even with this cell line, neutralizing antibodies to IL-6 must be included to ensure specificity.

Interleukin-15

IL-15 induces proliferation of T cells primarily by acting through components of the IL-2 receptor complex (55,110). As such, many of the bioassays used for the detection of IL-2 will also measure IL-15, and the relative contribution of each cytokine to the proliferation needs to be determined by antibody inhibition studies. Cloned cell lines may be used to measure IL-15 or T lymphoblasts.

Tumor Necrosis Factor

The classic biological activity of TNF is its ability to induce the lysis of target cells. There exist several assays for the detection of tumor necrosis factor, and they vary in sensitivity and specificity. Given the popularity of measuring TNF, it is not surprising that several different cell lines have been described that are capable of being killed by TNF. The original assay involved the fibrosarcoma cell line L929 or LM (111). Samples are serially diluted and added to plates that already contain the adherent cells. A standard of recombinant TNF is also prepared and added to wells in the same assay. To enhance the sensitivity and increase the speed of the assay, an inhibitor of either RNA synthesis (actinomycin D) or protein synthesis (cycloheximide) is added. The cells are incubated overnight at 37°C, the supernatants and dead, nonadherent cells discarded, and the remaining viable cells are stained with crystal violet. The plates are washed with tap water, and the absorbance is measured at 620 nm. Units are calculated based on the recombinant standard. There exist conflicting reports regarding the ability of IL-1 to lyse TNF-sensitive target cells (112). The L929 cell line is also lysed by TNF-β (lymphotoxin), and to discriminate between these two structurally related cytokines it is necessary to use specific, neutralizing antibodies (113). Investigators should check the cell line they are using to determine if their targets are susceptible to lysis by IL-1, and if they are, appropriate controls should be used. It has also been reported that lipopolysaccharide will synergize with TNF to cause cytotoxicity (114). Therefore, LPS contamination, which could frequently be present in samples from stimulated cells or septic patients, could cause a sample to appear to have more activity.

Samples that contain low levels of TNF activity, such as serum or plasma, may be measured with the highly sensitive cell line WEHI 164 subclone 13 (115). These cells are exquisitely sensitive to the lyric effects of TNF and can detect as little as 1 pg/ml (111,115). This assay is also run in the presence of actinomycin D to enhance its sensitivity. To perform this assay, the samples are serially diluted in the plates and the cells are added on top. After overnight incubation, MTT-tetrazolium is added, the plates placed back in the incubator for an additional 4 h, and the color developed with isopropanol. Units are calculated based on a recombinant standard run in the same assay. The WEHI 164 cells have been shown to respond both TNF and lymphotoxin and are not lysed by several other cytokines. Furthermore, there is no synergism with IFN-γ (111), which has been shown to enhance the effects of TNF in some systems.

Even more rapid or sensitive methods are being described for the measurement of TNF. TNF induces fragmentation of DNA into nucleosomes, and it has been used as a rapid method for the detection of TNF (116). Cells from the human leukemia cell line ML-1a are first labeled with ^{3}H-thymidine and then mixed with TNF in the presence of cycloheximide. Release of DNA into the supernatant is detectable within 90 min. Ethidium homodimer-1 is a DNA stain taken up through altered cell membranes and may be used to detect the presence of TNF. Samples are mixed with the cells in the presence of the dye and the fluorescence determined (117). A sensitive new bioassay was described using a subclone of the L929 cells, L929-8 (118). These cells could detect TNF at a concentration of 130–200 fg/ml, and also did not react with LPS or IFN. An even further improvement pushed the limit of sensitivity to 68 fg/ml using a fluorescence-based assay and another subclone, the NCTC clone 929 cell line (119). In some situations there is species specificity; for example, it has been reported that one cell line is substantially more sensitive to porcine TNF than other cell lines (120). The addition of sodium fluoride has been reported to increase the sensitivity of the L929 assay to TNF (121).

Many of the bioassays for TNF are subject to nonspecific effects from plasma or serum, which causes reduction of the MTT-tetrazolium and a high background. Briefly heating the samples will help to overcome these obstacles (122,123). Another important problem is the lack of correlation between the bioassays and immunoassays. Part of this problem is that virtually all of the bioassays are inhibited by soluble TNF receptors (124), and these receptors are frequently released during the normal immune response. The soluble receptors bind to and inactivate TNF by preventing the TNF from binding to the receptors on the target cell surface. This then raises the important question of which is more important, the presence of immunoreactive TNF or bioactive TNF.

Interferon

The most widely used assay to measure the biological activity of IFN is the capacity of this cytokine to confer resistance to the cytopathic effects of a virus. All of the interferons (α, β, and γ) exhibit this antiviral effect. For this assay, it is necessary to have live virus, which is potentially pathogenic. Thus, all laboratory personnel who come into contact with the virus must be properly trained about adequate precautions. One of the commonly used viruses is the vesicular stomatitis virus (VSV), which exerts a cytopathic effect toward the L929 fibroblasts. Frozen stocks of virus may be kept for several years.

The assay is performed by adding the samples to be tested with the L929 cells and incubating overnight. The next morning the supernatants are aspirated and the VSV stock is added to the wells. The ratio of virus to cells should be about 0.1. The L929 cells are incubated overnight and the supernatant removed. After washing dead cells and debris from the plates, the cells are fixed with formalin. After fixation, the virus is no longer hazardous. The remaining cells may be stained with crystal violet, which then may be scored by hand or the dye eluted with methanol and read on an ELISA plate reader. Alternatively, the cellular viability may be measured with MTT-tetrazolium (125) or ^{3}H-thymidine (126). Different cell lines and different types of viruses may be used for the biological assay, since in all cases the final determinant is the ability of the IFN to enhance the antiviral capacity of the cell (127).

IFN-γ and LPS will synergize to enhance production of nitric oxide by macrophages. This biological activity has been exploited in a bioassay. For this assay, the murine macrophage cell line RAW 264.7 (ATCC #TIB 71) is mixed with the samples and LPS. The next morning the production of nitrate is directly correlated to the IFN-γ concentration (128). Type I interferons (INF-α and IFN-β) will exert antiproliferative effects on lymphocyte cells lines, which may also be used as a bioassay (129,130).

SUMMARY

There are obviously many ways to quantitate the bioactivity of the cytokines. ELISA assays are specific, but they require antibodies for each cytokine, and there may be species specificity to the antibodies. Additionally, the reagents or kits may be extremely expensive. The cell line assays are less costly and generally easier to perform but require maintenance of the cell cultures between assays. One must be constantly vigilant with the cell lines since they may be responding to yet-to-be-discovered cytokines. This chapter lists some of the more useful biological cytokine assays, and newer methodologies and cell lines are constantly being discovered.

ACKNOWLEDGMENTS

This work was supported in part by National Institutes of Health grants GM 44918 and GM50401. The technical assistance of Gerald Bolgos and David Newcomb is gratefully acknowledged.

REFERENCES

1. Pennica D, Nedwin GE, Hayflick JS, Seeburg PH, Derynck R, Palladino MA, Kohr WJ, Aggarwal BB, Goeddel DV. Human tumour necrosis factor: precursor structure, expression and homology to lymphotoxin. Nature 1984; 312:724–729.
2. Remick DG, Scales WE, Chensue SW, Kunkel SL. The pathophysiology of interleukins and tumor necrosis factors. In: Sayeed M, ed. Cellular Pathophysiology. Boca Raton, FL: CRC Press, 1989:41–89.
3. Mosmann T. Rapid colorimetric assay for cellular growth and survival: application to proliferation and cytotoxicity assays. J Immunol Methods 1983; 65:55–63.
4. Tada H, Shiho O, Kuroshima K, Koyama M, Tsukamoto K. An improved colorimetric assay for interleukin 2. J Immunol Methods 1986; 93:157–165.
5. Mizel SB, Oppenheim JJ, Rosenstreich DL. Characterization of lymphocyte-activating factor (LAF) produced by the macrophage cell line, P388D1. I. enhancement of LAF production by activated T lymphocytes. J Immunol 1978; 120:1497–1503.
6. Van Damme J, Van Beeumen J, Decock B, Van Snick J, De Ley M, Billiau A. Separation and comparison of two monokines with lymphocyte-activating factor activity: IL-1 beta and hybridoma growth factor (HGF). Identification of leukocyte-derived HGF as IL-6. J Immunol 1988; 140:1534–1541.
7. Koopman WJ, Farrar JJ, Fuller-Bonar J. Evidence for the identification of lymphocyte activating factor as the adherent cell-derived mediator responsible for enhanced antibody synthesis by nude mouse spleen cells. Cell Immunol 1978; 35:92–98.
8. Lotz M, Jirik F, Kabouridis P, Tsoukas C, Hirano T, Kishimoto T, Carson DA. B cell stimulating factor 2/interleukin 6 is a costimulant for human thymocytes and T lymphocytes. J Exp Med 1988; 167:1253–1258.
9. Sakai Y, Kimura S, Yamashita T, Umetsu K, Suzuki T, Sendo F. A sensitive IL-1 thymocyte co-stimulator assay using a novel beta-D-galactoside specific lectin from the beetle, Allomyrina dichotoma. Microbiol Immunol 1989; 33:569–577.
10. Helle M, Brakenhoff JP, De Groot ER, Aarden LA. Interleukin 6 is involved in interleukin 1-induced activities. Eur J Immunol 1988; 18:957–959.
11. Suda T, Hodgkin P, Lee F, Zlotnik A. Biological activity of recombinant murine interleukin-6 in interleukin-1 T cell assays. J Immunol Methods 1989; 120:173–178.
12. Ranges GE, Zlotnik A, Espevik T, Dinarello CA, Cerami A, Palladino MA. Tumor necrosis factor alpha/cachectin is a growth factor for thymocytes. Synergistic interactions with other cytokines. J Exp Med 1988; 167:1472–1478.
13. Lowenthal JW, Cerottini JC, MacDonald HR. Interleukin 1-dependent induction of both interleukin 2 secretion and interleukin 2 receptor expression by thymoma cells. J Immunol 1986; 137:1226–1231.
14. Everson MP, Eldridge JH, Koopman WJ. Synergism of interleukin 7 with the thymocyte growth factors interleukin 2, interleukin 6, and tumor necrosis factor alpha in the induction of thymocyte proliferation. Cell Immunol 1990; 127:470–482.
15. Conlon PJ. A rapid biologic assay for the detection of interleukin 1. J Immunol 1983; 131:1280–1282.
16. Larrick JW, Brindley L, Doyle MV. An improved assay for the detection of interleukin 1. J Immunol Methods 1985; 79:39–45.
17. Gillis S, Mizel SB. T-Cell lymphoma model for the analysis of interleukin 1-mediated T-cell activation. Proc Natl Acad Sci USA 1981; 78:1133–1137.
18. Remvig L, Vibe-Petersen J, Svenson M, Enk C, Bendtzen K. Biological assays for interleukin 1 detection. Comparison of

human T lymphocyte, murine thymocyte and NOB-1 assays. Allergy 1991; 46:59–67.

19. Suda T, Hodgkin P, Lee F, Zlotnik A. Biological activity of recombinant murine interleukin-6 in interleukin-1 T cell assays. J Immunol Methods 1989; 120:173–178.
20. Kaye J, Gillis S, Mizel SB, Shevach EM, Malek TR, Dinarello CA, Lachman LB, Janeway CA, Jr. Growth of a cloned helper T cell line induced by a monoclonal antibody specific for the antigen receptor: interleukin 1 is required for the expression of receptors for interleukin 2. J Immunol 1984; 133:1339–1345.
21. Orencole SF, Dinarello CA. Characterization of a subclone (D10S) of the D10.G4.1 helper T-cell line which proliferates to attomolar concentrations of interleukin-1 in the absence of mitogens. Cytokine 1989; 1:14–22.
22. Nakai S, Mizuno K, Kaneta M, Hirai YA. A simple, sensitive bioassay for the detection of interleukin-1 using human melanoma A375 cell line. Biochem Biophys Res Commun 1988; 154:1189–1196.
23. Vandenabeele P, Declercq W, Libert C, Fiers W. Development of a simple, sensitive and specific bioassay for interleukin-1 based on the proliferation of RPMI 1788 cells. Comparison with other bioassays for IL-1. J Immunol Methods 1990; 135:25–32.
24. Larrick JW. Native interleukin 1 inhibitors. Immunol Today 1989; 10:61–66.
25. Cannon JG, van der Me, Kwiatkowski D, Endres S, Lonnemann G, Burke JF, Dinarello CA. Interleukin-1 beta in human plasma: optimization of blood collection, plasma extraction, and radioimmunoassay methods. Lymphokine Res 1988; 7:457–467.
26. Hopkins SJ, Humphreys M. Bioassay of interleukin-1 in serum and plasma following removal of inhibitory activity with polyethylene glycol. J Immunol Methods 1990; 133:127–131.
27. Herzyk DJ, Berger AE, Allen JN, Wewers MD. Sandwich ELISA formats designed to detect 17 kDa IL-1 beta significantly underestimate 35 kDa IL-1 beta. J Immunol Methods 1992; 148:243–254.
28. Gillis S, Ferm MM, Ou W, Smith KA. T cell growth factor: parameters of production and a quantitative microassay for activity. J Immunol 1989; 120:2027–2032.
29. Antonen J, Mitsuya H, Krohn K. The use of an HTLV-I-infected human T-cell line (ATH8) in an interleukin-2 bioassay. J Immunol Methods 1987; 99:271–275.
30. Eskandari MK, Kunkel SL, Remick DG. Effects of arachidonic acid metabolites and other compounds on the CTLL assay for interleukin-2. J Immunol Methods 1989; 118:85–89.
31. Powers CD, Abbas AK, Miller RA. Frequencies of IL-2- and IL-4-secreting T cells in naive and antigen-stimulated lymphocyte populations. J Immunol 1988; 140:3352–3357.
32. Grabstein K, Eisenman J, Mochizuki D, Shanebeck K, Conlon P, Hopp T, March C, Gillis S. Purification to homogeneity of B cell stimulating factor. A molecule that stimulates proliferation of multiple lymphokine-dependent cell lines. J Exp Med 1986; 163:1405–1414.
33. Ohara J, Paul WE. Production of a monoclonal antibody to and molecular characterization of B-cell stimulatory factor-1. Nature 1985; 315:333–336.
34. Ihle JN. Biochemical and biological properties of interleukin-3: a lymphokine mediating the differentiation of a lineage of cells that includes prothymocytes and mastlike cells. Contemp Top Mol Immunol 1985; 10:93–119.
35. Gearing AJ, Johnstone AP, Thorpe R. Production and assay of the interleukins. J Immunol Methods 1985; 83:1–27.
36. Rennick DM, Lee FD, Yokota T, Arai KI, Cantor H, Nabel GJ. A cloned MCGF cDNA encodes a multilineage hematopoietic growth factor: multiple activities of interleukin 3. J Immunol 1985; 134:910–914.
37. Otsuka T, Miyajima A, Brown N, Otsu K, Abrams J, Saeland S, Caux C, de Waal Malefijt R, de Vries J, Meyerson P, et al. Isolation and characterization of an expressible cDNA encoding human IL-3. Induction of IL-3 mRNA in human T cell clones. J Immunol 1988; 140:2288–2295.
38. Dexter TM, Garland J, Scott D, Scolnick E, Metcalf D. Growth of factor-dependent hemopoietic precursor cell lines. J Exp Med 1980; 152:1036–1047.
39. Ihle JN, Keller J, Greenberger JS, Henderson L, Yetter RA, Morse HC, 3d. Phenotypic characteristics of cell lines requiring interleukin 3 for growth. J Immunol 1982; 129:1377–1383.
40. Le Gros GS, Gillis S, Watson JD. Induction of IL 2 responsiveness in a murine IL 3-dependent cell line. J Immunol 1985; 135:4009–4014.
41. Greenberger JS, Sakakeeny MA, Humphries RK, Eaves CJ, Eckner RJ. Demonstration of permanent factor-dependent multipotential (erythroid/neutrophil/basophil) hematopoietic progenitor cell lines. Proc Natl Acad Sci USA 1983; 80:2931–2935.
42. Gajewski TF, Fitch FW. Anti-proliferative effect of IFN-gamma in immune regulation. IV. Murine CTL clones produce IL-3 and GM-CSF, the activity of which is masked by the inhibitory action of secreted IFN-gamma. J Immunol 1990; 144:548–556.
43. Mire-Sluis AR, Das RG, Thorpe R. The international standard for macrophage colony stimulating factor (M-CSF). Evaluation in an international collaborative study. J Immunol Methods 1995; 179:141–151.
44. Mire-Sluis AR, Das RG, Thorpe R. The international standard for granulocyte-macrophage colony stimulating factor (GM-CSF). Evaluation in an international collaborative study. Participants of the Collaborative Study. J Immunol Methods 1995; 179:127–135.
45. Kendall RG, Ingham E, Norfolk DR. The effect of cytokines on the in-vitro erythropoietin bioassay. Clin Lab Haematol 1994; 16:273–279.
46. Feldman L, Davis KL, Feeley DM, Sytkowski AJ. A sensitive new bioassay for erythroid colony-stimulating factor. Exp Hematol 1993; 21:1657–1662.
47. Minegishi N, Minegishi M, Tsuchiya S, Fujie H, Nagai T, Hayashi N, Yamamoto M, Konno T. Erythropoietin-dependent induction of hemoglobin synthesis in a cytokine-dependent cell line M-TAT. J Biol Chem 1994; 269:27700–27704.
48. Hintz-Obertreis P, Krumwieh D, Seiler FR. Development of a rapid, highly sensitive, non-radioactive assay system for hematopoietic growth factors. Behring Inst Mitt 1991; 99–103.
49. Avanzi GC, Brizzi MF, Giannotti J, Ciarletta A, Yang YC, Pegoraro L, Clark SC. M-07e human leukemic factor-dependent cell line provides a rapid and sensitive bioassay for the human cytokines GM-CSF and IL-3. J Cell Physiol 1990; 145:458–464.
50. Nakoinz I, Lee MT, Weaver JF, Ralph P. Differentiation of the IL-3-dependent NFS-60 cell line and adaption to growth in macrophage colony-stimulating factor. J Immunol 1990; 145:860–864.
51. Howard M, Farrar J, Hilfiker M, Johnson B, Takatsu K, Hamaoka T, Paul WE. Identification of a T cell-derived B cell growth factor distinct from interleukin 2. J Exp Med 1982; 155:914–923.
52. Muraguchi A, Fauci AS. Proliferative responses of normal human B lymphocytes. Development of an assay system for human B cell growth factor (BCGF). J Immunol 1982; 129:1104–1108.
53. Hu-Li J, Ohara J, Watson C, Tsang W, Paul WE. Derivation of

a T cell line that is highly responsive to IL-4 and IL-2 (CT.4R) and of an IL-2 hyporesponsive mutant of that line (CT.4S). J Immunol 1989; 142:800–807.
54. Ho SN, Abraham RT, Gillis S, McKean DJ. Differential bioassay of interleukin 2 and interleukin 4. J Immunol Methods 1987; 98:99–104.
55. Grabstein KH, Eisenman J, Shanebeck K, Rauch C, Srinivasan S, Fung V, Beers C, Richardson J, Schoenborn MA, Ahdieh M, et al. Cloning of a T cell growth factor that interacts with the beta chain of the interleukin-2 receptor. Science 1994; 264:965–968.
56. Xu H, Rizzo LV, Caspi RR. IL-12 induces growth of the IL-4-dependent CT4S line and has a synergistic effect on IL-4-induced CT4S proliferation. J Immunol Methods 1995; 181:245–251.
57. Kinzer CA, Keegan AD, Beckmann MP, Widmer M, Idelson GH, Ben-Sasson SZ, Paul WE. CT.h4S cells, a murine cell line expressing human IL-4 receptors and capable of assaying human IL-4. FASEB J 1994; 8:506.
58. Favre N, Erb P. Use of the CTL44 cell line, a derivative of CTL/L cells, to identify and quantify mouse interleukin-4 by bioassay. J Immunol Methods 1993; 164:213–220.
59. Siegel JP, Mostowski HS. A bioassay for the measurement of human interleukin-4. J Immunol methods 1990; 132:287–295.
60. Maeda M, Noma T, Hama K, Honjo T. Application of a human T cell line derived from a Sezary syndrome patient for human interleukin 4 assay. Immunol Lett 1988; 18:247–253.
61. Dutton RW, Wetzel GD, Swain SL. Partial purification and characterization of a BCGFII from EL4 culture supernatants. J Immunol 1984; 132:2451–2456.
62. Takatsu K, Harada N, Hara Y, Takahama Y, Yamada G, Dobashi K, Hamaoka T. Purification and physicochemical characterization of murine T cell replacing factor (TRF). J Immunol 1985; 134:382–389.
63. Sanderson CJ. Eosinophil differentiation factor (interleukin-5). Immunol Ser 1990; 49:231–256.
64. McKenzie DT, Filutowicz HI, Swain SL, Dutton RW. Purification and partial sequence analysis of murine B cell growth factor II (interleukin 5). J Immunol 1987; 139:2661–2668.
65. Lohoff M, Sommer F, Rollinghoff M. Suppressive effect of interferon-gamma on the BCL1 cell-dependent interleukin 5 bioassay. Eur J Immunol 1989; 19:1327–1329.
66. Hirano T, Yasukawa K, Harada H, Taga T, Watanabe Y, Matsuda T, Kashiwamura S, Nakajima K, Koyama K, Iwamatsu A, et al. Complementary DNA for a novel human interleukin (BSF-2) that induces B lymphocytes to produce immunoglobulin. Nature 1986; 324:73–76.
67. Van Snick J, Cayphas S, Vink A, Uyttenhove C, Coulie PG, Rubira MR, Simpson RJ. Purification and NH2-terminal amino acid sequence of a T-cell-derived lymphokine with growth factor activity for B-cell hybridomas. Proc Natl Acad Sci USA 1986; 83:9679–9683.
68. Aarden LA, De Groot ER, Schaap OL, Landsdorp PM. Production of hybridoma growth factor by human monocytes. Eur J Immunol 1987; 17:1411–1416.
69. Gauldie J, Richards C, Harnish D, Lansdorp P, Baumann H. Interferon beta 2/B-cell stimulatory factor type 2 shares identity with monocyte-derived hepatocyte-stimulating factor and regulates the major acute phase protein response in liver cells. Proc Natl Acad Sci USA 1987; 84:7251–7255.
70. Zilberstein A, Ruggieri R, Korn JH, Revel M. Structure and expression of cDNA and genes for human interferon-beta-2, a distinct species inducible by growth-stimulatory cytokines. EMBO J 1986; 5:2529–2537.
71. Gaines Das RE, Poole S. The international standard for interleukin-6. Evaluation in an international collaborative study. J Immunol Methods 1993; 160:147–153.
72. Van Damme J, Van Snick J. Induction of hybridoma growth factor (HGF), identical to IL-6, in human fibroblasts by IL-1: use of HGF activity in specific and sensitive biological assays for IL-1 and IL-6. Dev Biol Stand 1988; 69:31–38.
73. Ulich TR, Guo K, Remick D, del Castillo J, Yin S. Endotoxin-induced cytokine gene expression in vivo III. IL-6 mRNA and serum protein expression and the in vivo hematologic effects of IL-6. J Immunol 1991; 146:2316–2323.
74. Pedersen MR, Jensen S, Christensen JD, Hansen EW. Lipopolysaccharide in concentrations above 40 ng/ml stimulates proliferation of the IL-6-dependent B9 cell line. J Immunol Methods 1995; 180:159–163.
75. Okaza'i H, Ito M, Sudo T, Hattori M, Kano S, Katsura Y, Mina'o N. IL-7 promotes thymocyte proliferation and maintains immunocompetent thymocytes bearing alpha beta or gamma delta T-cell receptors in vitro: synergism with IL-2. J Immunol 1989; 143:2917–2922.
76. Grabstein KH, Namen AE, Shanebeck K, Voice RF, Reed SG, Widmer MB. Regulation of T cell proliferation by IL-7. J Immunol 1990; 144:3015–3020.
77. Whitlock CA, Robertson D, Witte ON. Murine B cell lymphopoiesis in long term culture. J Immunol Methods 1984; 67:353–369.
78. Whitlock CA, Witte ON. Long-term culture of B lymphocytes and their precursors from murine bone marrow. Proc Natl Acad Sci USA 1982; 79:3608–3612.
79. Park LS, Friend DJ, Schmierer AE, Dower SK, Namen AE. Murine interleukin 7 (IL-7) receptor. Characterization on an IL-7-dependent cell line. J Exp Med 1990; 171:1073–1089.
80. Ishihara K, Medina K, Hayashi S, Pietrangeli C, Namen AE, Miyake K, Kincade PW. Stromal-cell and cytokine-dependent lymphocyte clones which span the pre-B- to B-cell transition. Dev Immunol 1991; 1:149–161.
81. Baggiolini M, Dewald B, Moser B. Interleukin-8 and related chemotactic cytokines—CXC and CC chemokines. Adv Immunol 1994; 55:97–179.
82. Baggiolini M, Walz A, Kunkel SL. Neutrophil-activating peptide-1/interleukin 8, a novel cytokine that activates neutrophils. J Clin Invest 1989; 84:1045–1049.
83. Peveri P, Walz A, Dewald B, Baggiolini M. A novel neutrophil-activating factor produced by human mononuclear phagocytes. J Exp Med 1988; 167:1547–1559.
84. Larsen CG, Anderson AO, Appella E, Oppenheim JJ, Matsushima K. The neutrophil-activating protein (NAP-1) is also chemotactic for T lymphocytes. Science 1989; 243:1464–1466.
85. Taub DD, Lloyd AR, Wang JM, Oppenheim JJ, Kelvin DJ. The effects of human recombinant MIP-1 alpha, MIP-1 beta, and RANTES on the chemotaxis and adhesion of T cell subsets. Adv Exp Med Biol 1993; 351:139–146.
86. Murphy WJ, Taub DD, Anver M, Conlon K, Oppenheim JJ, Kelvin DJ, Longo DL. Human RANTES induces the migration of human T lymphocytes into the peripheral tissues of mice with severe combined immune deficiency. Eur J Immunol 1994; 24:1823–1827.
87. Taub DD, Proost P, Murphy WJ, Anver M, Longo DL, Van Damme J, Oppenheim JJ. Monocyte chemotactic protein-1 (MCP-1), -2, and -3 are chemotactic for human T lymphocytes. J Clin Invest 1995; 95:1370–1376.
88. Brieland JK, Flory CM, Jones ML, Miller GR, Remick DG, Warren JS, Fantone JC. Regulation of monocyte chemoattractant protein-1 gene expression and secretion in rat pulmonary alveolar macrophages by lipopolysaccharide, tumor necrosis

factor-alpha, and interleukin-1 beta. Am J Respir Cell Mol Biol 1995; 12:104–109.
89. Taub DD, Lloyd AR, Conlon K, Wang JM, Ortaldo JR, Harada A, Matsushima K, Kelvin DJ, Oppenheim JJ. Recombinant human interferon-inducible protein 10 is a chemoattractant for human monocytes and T lymphocytes and promotes T cell adhesion to endothelial cells. J Exp Med 1993; 177:1809–1814.
90. Harvath L, Falk W, Leonard EJ. Rapid quantitation of neutrophil chemotaxis: use of a polyvinylpyrrolidone-free polycarbonate membrane in a multiwell assembly. J Immunol Methods 1980; 37:39–45.
91. DeForge LE, Kenney JS, Jones ML, Warren JS, Remick DG. Biphasic production of IL-8 in lipopolysaccharide (LPS)-stimulated human whole blood. Separation of LPS- and cytokine-stimulated components using anti-tumor necrosis factor and anti-IL-1 antibodies. J Immunol 1992; 148:2133–2141.
92. Taub DD, Key ML, Clark D, Turcovski-Corrales SM. Chemotaxis of T lymphocytes on extracellular matrix proteins. Analysis of the in vitro method to quantitate chemotaxis of human T cells. J Immunol Methods 1995; 184:187–198.
93. DeForge LE, Remick DG. Sandwich ELISA for detection of picogram quantities of interleukin-8. Immunol Invest 1991; 20:89–97.
94. Sylvester I, Rankin JA, Yoshimura T, Tanaka S, Leonard EJ. Secretion of neutrophil attractant/activation protein by lipopolysaccharide-stimulated lung macrophages determined by both enzyme-linked immunosorbent assay and N-terminal sequence analysis. Am Rev Respir Dis 1990; 141:683–688.
95. Fiorentino DF, Bond MW, Mosmann TR. Two types of mouse T helper cell. IV. Th2 clones secrete a factor that inhibits cytokine production by Th1 clones. J Exp Med 1989; 170:2081–2095.
96. Schlaak JF, Schmitt E, Huls C, Meyer zum Buschenfelde KH, Fleischer B. A sensitive and specific bioassay for the detection of human interleukin-10. J Immunol Methods 1994; 168:49–54.
97. Quesniaux VF, Mayer P, Liehl E, Turner K, Goldman SJ, Fagg B. Review of a novel hematopoietic cytokine, interleukin-11. Int Rev Exp Pathol 1993; 34 Pt A:205–214.
98. Yang YC. Interleukin 11: an overview. Stem Cells (Dayt) 1993; 11:474–486.
99. Lu ZY, Zhang XG, Gu ZJ, Yasukawa K, Amiot M, Etrillard M, Bataille R, Klein B. A highly sensitive quantitative bioassay for human interleukin-11. J Immunol Methods 1994; 173:19–26.
100. Kobayashi M, Fitz L, Ryan M, Hewick RM, Clark SC, Chan S, Loudon R, Sherman F, Perussia B, Trinchieri G. Identification and purification of natural killer cell stimulatory factor (NKSF), a cytokine with multiple biologic effects on human lymphocytes. J Exp Med 1989; 170:827–845.
101. Brunda MJ. Interleukin-12. J Leukoc Biol 1994; 55:280–288.
102. Gillessen S, Carvajal D, Ling P, Podlaski FJ, Stremlo DL, Familletti PC, Gubler U, Presky DH, Stern AS, Gately MK. Mouse interleukin-12 (IL-12) p40 homodimer: a potent IL-12 antagonist. Eur J Immunol 1995; 25:200–206.
103. Ling P, Gately MK, Gubler U, Stern AS, Lin P, Hollfelder K, Su C, Pan YC, Hakimi J. Human IL-12 p40 homodimer binds to the IL-12 receptor but does not mediate biologic activity. J Immunol 1995; 154:116–127.
104. D'Andrea A, Aste-Amezaga M, Valiante NM, Ma X, Kubin M, Trinchieri G. Interleukin 10 (IL-10) inhibits human lymphocyte interferon gamma-production by suppressing natural killer cell stimulatory factor/IL-12 synthesis in accessory cells. J Exp Med 1993; 178:1041–1048.
105. Skeen MJ, Ziegler HK. Activation of gamma delta T cells for production of IFN-gamma is mediated by bacteria via macrophage-derived cytokines IL-1 and IL-12. J Immunol 1995; 154:5832–5841.
106. Zhang M, Gately MK, Wang E, Gong J, Wolf SF, Lu S, Modlin RL, Barnes PF. Interleukin 12 at the site of disease in tuberculosis. J Clin Invest 1994; 93:1733–1739.
107. McKenzie AN, Culpepper JA, de Waal Malefyt R, Briere F, Punnonen J, Aversa G, Sato A, Dang W, Cocks BG, Menon S, et al. Interleukin 13, a T-cell-derived cytokine that regulates human monocyte and B-cell function. Proc Natl Acad Sci USA 1993; 90:3735–3739.
108. de Waal Malefyt R, Figdor CG, Huijbens R, Mohan-Peterson S, Bennett B, Culpepper J, Dang W, Zurawski G, de Vries JE. Effects of IL-13 on phenotype, cytokine production, and cytotoxic function of human monocytes. Comparison with IL-4 and modulation by IFN-gamma or IL-10. J Immunol 1993; 151:6370–6381.
109. Bouteiller CL, Astruc R, Minty A, Ferrara P, Lupker JH. Isolation of an IL-13-dependent subclone of the B9 cell line useful for the estimation of human IL-13 bioactivity. J Immunol Methods 1995; 181:29–36.
110. Carson WE, Giri JG, Lindemann MJ, Linett ML, Ahdieh M, Paxton R, Anderson D, Eisenmann J, Grabstein K, Caligiuri MA. Interleukin (IL) 15 is a novel cytokine that activates human natural killer cells via components of the IL-2 receptor. J Exp Med 1994; 180:1395–1403.
111. Eskandari MK, Nguyen DT, Kunkel SL, Remick DG. WEHI 164 subclone 13 assay for TNF: sensitivity, specificity and reliability. Immun Inves 1990; 19:69–79.
112. Onozaki K, Matsushima K, Aggarwal BB, Oppenheim JJ. Human Interleukin 11 is a cytocidal factor for several tumor cell lines. J Immunol 1985; 135:3962–3968.
113. Bringman TS, Aggarwal BB. Monoclonal antibodies to human tumor necrosis factors alpha and beta: application for affinity purification, immunoassays, and as structural probes. Hybridoma 1987; 6:489–507.
114. Pfister H, Hennet T, Jungi TW. Lipopolysaccharide synergizes with tumour necrosis factor-alpha in cytotoxicity assays. Immunology 1992; 77:473–476.
115. Espevik T, Nissen Meyer J. A highly sensitive cell line, WEHI 164 clone 13, for measuring cytotoxic factor/tumor necrosis factor from human monocytes. J Immunol Methods 1986; 95:99–105.
116. Higuchi M, Singh S, Aggarwal BB. Characterization of the apoptotic effects of human tumor necrosis factor: development of highly rapid and specific bioassay for human tumor necrosis factor and lymphotoxin using human target cells. J Immunol Methods 1995; 178:173–181.
117. Levesque A, Paquet A, Page M. Improved fluorescent bioassay for the detection of tumor necrosis factor activity. J Immunol Methods 1995; 178:71–76.
118. Branch DR, Shah A, Guilbert LJ. A specific and reliable bioassay for the detection of femtomolar levels of human and murine tumor necrosis factors. J Immunol Methods 1991; 143:251–261.
119. Shahan TA, Siegel PD, Sorenson WG, Kuschner WG, Lewis DM. A sensitive new bioassay for tumor necrosis factor. J Immunol Methods 1994; 175:181–187.
120. Bertoni G, Kuhnert P, Peterhans E, Pauli U. Improved bioassay for the detection of porcine tumor necrosis factor using a homologous cell line: PK(15). J Immunol Methods 1993; 160:267–271.
121. Jeffes EW, 3d, Schmitz K, Yamamoto R, Tomich JM, Beckman M, Nep R, Knauer M, Jeffes EW. A simple nonisotopic in vitro bioassay for LT and TNF employing sodium fluoride-treated L-929 target cells that detects picogram quantities of LT and

TNF and is as sensitive as TNF assays done with ELISA methodology. Lymphokine Cytokine Res 1991; 10:147–151.
122. Nargi FE, Yang TJ. Optimization of the L-M cell bioassay for quantitating tumor necrosis factor alpha in serum and plasma. J Immunol Methods 1993; 159:81–91.
123. Hay H, Cohen J. Studies on the specificity of the L929 cell bioassay for the measurement of tumour necrosis factor. J Clin Lab Immunol 1989; 29:151–155.
124. Engelberts I, Stephens S, Francot GJ, van der Li, Buurman WA. Evidence for different effects of soluble TNF-receptors on various TNF measurements in human biological fluids [letter]. Lancet 1991; 338:515–516.
125. Berg K, Hansen MB, Nielsen SE. A new sensitive bioassay for precise quantification of interferon activity as measured via the mitochondrial dehydrogenase function in cells (MTT-method). APMIS 1990; 98:156–162.
126. Ishitsuka H, Nomura Y, Takano K. A simple and efficient microassay method for titration of interferon. Microbiol Immunol 1977; 21:583–591.
127. Yousefi S, Escobar MR, Gouldin CW. A practical cytopathic effect/dye-uptake interferon assay for routine use in the clinical laboratory. Am J Clin Pathol 1985; 83:735–740.
128. Migliorini P, Corradin G, Corradin SB. macrophage NO2-production as a sensitive and rapid assay for the quantitation of murine IFN-gamma. J Immunol Methods 1991; 139:107–114.
129. Nederman T, Karlstrom E, Sjodin L. An in vitro bioassay for quantitation of human interferons by measurements of antiproliferative activity on a continuous human lymphoma cell line. Biologicals 1990; 18:29–34.
130. Hirai N, Hattori N, Hill NO, Osther K. Antiproliferative assay of human leukocyte interferon. Tohoku J Exp Med 1987; 151:309–316.

21
Fibroblasts and Cytokines

Soverin Karmiol and Sem Hin Phan
University of Michigan Medical School, Ann Arbor, Michigan

INTRODUCTION

The fibroblast, along with the myofibroblast and pericyte, belongs to a group of cells located primarily in the interstitial space of tissues. These cells arise from mesenchymal tissue and are not always easy to distinguish from each other in culture, despite differences in morphology, contractile protein content and composition, cell surface antigen expression, and extracellular matrix production. Furthermore, fibroblasts from one tissue may show distinguishing features from those in another organ or tissue (1), and even within the same tissue there may be heterogeneity in fibroblast populations (2). However, in many in vitro studies scant attention has been paid to these differences between subpopulations or to differences between fibroblast (and even fibroblastlike) cell lines and primary cell cultures. Since the topic of this chapter is derived primarily from such in vitro studies, it would be a futile endeavor to attempt to differentiate between these cell populations. Consequently, except where clear differences have been unequivocally established, fibroblast in this chapter refers to any one of these cell populations in a generic manner and all findings may not be applicable to all fibroblast subpopulations or types. Hence, any cell-specific application of the following discussion and review of fibroblast interaction with cytokines must be considered with this caveat in mind.

During the past decade, a considerable change has occurred in how the fibroblast is viewed in terms of its in vivo role under normal and diseased conditions. Prior to recent advances, this cell was viewed primarily as an important structural component of tissues, responsible for the production of interstitial collagens and other extracellular matrix components. Replacement of normal functional organs by fibrotic lesions has long been known to be stereotypically associated with replacement of parenchymal differentiated cells by fibroblasts. By virtue of their inability to express the differentiated and often highly specialized functions of the original parenchymal cells, these fibroblasts have been viewed as passive structural cells devoid of vital functions, whose unchecked proliferation is considered to be inimical to successful repair and regeneration of normal functional parenchyma. Early studies focused on fibroblast production of extracellular matrix and the enzymes involved in its metabolism, with little attention paid to its possible active role in paracrine or autocrine regulatory mechanisms. With the realization that these cells are a primary source of the connective tissue found in fibrotic lesions, increasing emphasis has been placed on understanding the mechanisms underlying regulation of fibroblast growth, chemotaxis, and extracellular matrix synthesis. Studies have concentrated on the role of exogenous factors in regulating these cellular functions, primarily treating the fibroblast as a passive target for regulatory signals from other cells. This view gradually began to change with the discovery that certain fibroblast products, including the extracellular matrix components or their degradation products, have regulatory effects on other cells and the fibroblast itself. Since this cell and cell lines derived from it are readily available due to relative ease in isolation and maintaining in tissue culture, they have been used in many in vitro model systems to study regulation of mediator production in other cells that are more difficult or impossible to culture. These studies serendipitously provided important clues that the fibroblast, not just the cells it is supposed to mimic or be a model for, is in its own right capable of responding to paracrine as well as autocrine signals by the production of mediators of regulatory and pathogenetic consequence on surrounding cells and itself. A wealth of information is now available which convincingly demonstrates that the fibroblast is an important cell in vivo and a key participant in normal and pathological processes driven by cytokines. Since the fibroblast is ubiquitous, the production of cytokines by this cell would be expected to play an important role in most if not all tissues, and, considering the myriad of activities of the cytokines produced by this cell, it has been postulated as playing a role in an ever-expanding repertoire of physiological as well as pathophysiological processes, such as the regulation of tissue differentiation, repair and fibrosis, hematopoiesis, bone metabolism, inflammation, and immune response.

This chapter has been organized into subsections discussing the known classes of cytokines that either are produced by fibroblasts or have an effect on fibroblast function. This organization is certainly not an accurate reflection of the pleiotropic effects of these cytokines, nor does it reflect their diverse multiple cellular sources. Hence this attempt at providing some structural framework merely represents a starting point by which to review the available literature on fibroblasts and cytokines. In each instance where sufficient physicochemical and biochemical information on each cytokine, its receptor, and associated signal transduction pathways is available in other chapters, it will not be repeated here.

INTERLEUKINS

The interleukin family of cytokines currently consists of some fifteen members (3). The nomenclature reflects the original concept that these molecules represent mediators by which leukocytes communicate regulatory signals with each other. Discoveries during the last decade however have gradually make this restriction to leukocytes untenable. The discussion that follows will be subdivided into interleukins with known effects on fibroblasts or which may be produced by them. The others either have not been reported to affect, or be produced by, fibroblasts, or their effects on fibroblasts have not been sufficiently well-characterized to warrant discussion as separate sections at this time.

Interleukin-1

In accordance with its numbering system, interleukin-1 (IL-1) was the first of this group of mediators to be discovered. Its role in regulating the immune and inflammatory response is well known (4) and is discussed elsewhere. Although there are two distinct forms of IL-1, IL-1α and IL-1β, their functional effects are very similar (4,5) and are mediated by the same receptor (6). The discussion that follows will generically refer to IL-1, unless a distinction is necessary to emphasize differences in biological activity.

The effects of IL-1 on fibroblasts were first examined using crude macrophage or monocyte supernates before it was identified as such (7). These early studies focused on the destructive process in arthritic joints, hence the interest in the synovial fibroblast (8). Studies using purified IL-1 reproduced some of these effects (9). Arthritis is a destructive process, and IL-1 inhibition of fibroblast growth and matrix synthesis and stimulation of collagenase production were among the first effect to be reported (10). The mechanism of inhibition was subsequently found to be partly due to the stimulation of prostaglandin E_2 (PGE_2) production by IL-1, which in autocrine fashion causes downregulation of matrix synthesis and cell proliferation (11). This inhibitory effect is blunted by the addition of cyclooxygenase inhibitors to the target fibroblast during IL-1 treatment (12). Stimulation of prostaglandin production by IL-1 appears to be mediated by upregulation in phospholipase A and cyclooxygenase activities due at least in part to increased synthesis of these enzymes. Additionally, IL-1 amplifies stimulation of PGE_2 production by bradykinin and other agonists by augmenting the bradykinin-induced increase in GTPase activity without any significant alteration in agonist receptor expression (13). IL-1 promotes the increased transcription of proteases such as plasminogen activator and matrix degradative enzymes such as collagenase, stromelysin, and other matrix metalloproteinases (14,15). With respect to collagenase, the increased transcription does not appear to be mediated by increased PGE_2 production (16). Increased collagenase transcription is preceded by a rapid but brief burst of c-*jun* transcription, whose gene product is a component of activator protein-1 (AP-1), important for binding to DNA (17). Since binding of AP-1 is enhanced by the formation of a heterodimer between the c-*jun* and c-*fos* gene products and IL-1 also induces c-*fos* expression (18), it has been hypothesized that IL-1 promotes procollagenase gene transcription by increasing AP-1 levels and binding to the responsive element in the 5′ flanking region of the procollagenase gene, which then results in increased collagenase production (19). IL-1 also causes rapid and transient expression of the c-*myc* proto-oncogene, which may be related to its growth factor activity (18). Since the prostromelysin (transin) gene has similar regulatory DNA sequences (20), it is likely that similar mechanisms may be operative in IL-1–induced stimulation of stromelysin production. A recent study, however, indicates that stromelysin-1 gene expression does not require the AP-1 site (21).

The effect of IL-1 on fibroblast collagen gene expression is complicated by the inhibitory effects of PGE_2 on collagen synthesis (22). Thus, the conditions of assay are critical in determining whether a stimulatory or inhibitory effect is seen (12). PGE_2 is known to stimulate fibroblast collagenase secretion and elevates intracellular cAMP, which inhibits net collagen production (23). In contrast to this variable inhibitory effect on net collagen synthesis, there is more consistent evidence of IL-1–induced increase in steady-state levels of procollagen mRNA (24). These studies underline the discrepancy between mRNA concentration and net production of collagen, suggesting that regulation is occurring at multiple levels of gene expression (12). Complex effects of IL-1 on glycosaminoglycan synthesis have also been reported (25).

Similar unmasking of the growth-promoting properties of IL-1 was also observed when the PGE_2 inhibitory effect was blocked with cyclooxygenase inhibitors or minimized by allowing its metabolism to inactive product prior to testing for cell proliferation (11). This activity, however, was relatively weak in comparison to other well-defined fibroblast growth promoting cytokines (26), thus making it uncertain if IL-1 could play a significant role in vivo in this regard. In view of these opposing effects, it is also difficult to pinpoint the actual in vivo role of IL-1 vis-à-vis fibroblast function in the pathogenesis of fibrosis. In pulmonary fibrosis there is evidence of an early and relatively brief increase in IL-1 production (27). This in conjunction with the in vitro data would suggest that an early IL-1–induced inhibitory effect on fibroblast proliferation and net matrix production and even the promotion of matrix destruction may be followed by promotion of fibrogenesis upon metabolism of PGE_2 and removal of its inhibitory effects. On the other hand, a persistent IL-1 effect may result in a sustained destructive process by the continued heightened production of matrix degradative enzymes and inhibition of matrix production by PGE_2, such as that seen in rheumatoid arthritis (28). Adding to this destructive role is the recent demonstration of the ability of IL-1 and other cytokines to upregulate NO production in fibroblasts (29). Further destructive potential is expected from the ability of IL-1 to promote inflammation and/or the immune response by increasing fibroblast expression of adhesion molecules, such as ICAM-1 (30), and by increasing IL-1 receptor expression in this same fibroblast (31). Paradoxically, however, blocking of IL-1 action using the IL-1 receptor antagonist inhibits pulmonary fibrosis (32). The complex regulation of fibroblast proliferation and extracellular matrix production by IL-1 and other cytokines is summarized in Figures 1 and 2.

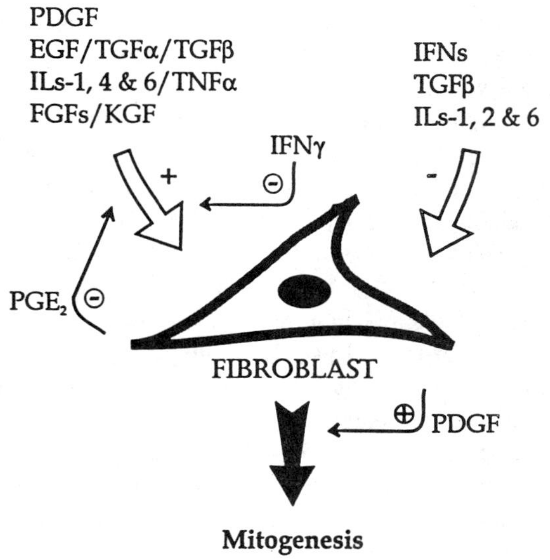

Figure 1 Cytokine regulation of fibroblast proliferation. The effects (positive sign indicates stimulation; negative sign indicates inhibition) of cytokines on fibroblast proliferation is shown. The thin lines with half arrow heads and originating from the fibroblast itself represent the effects of agents induced by the initial cytokine stimulation, and indicate autocrine regulation. The modulatory effect of exogenous IFN-γ is also shown.

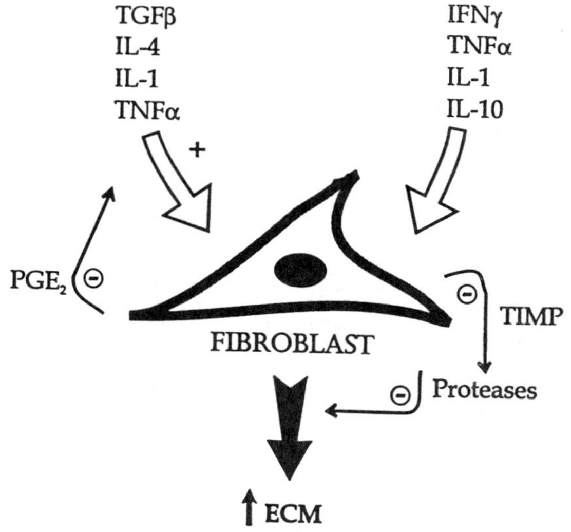

Figure 2 Cytokine regulation of extracellular matrix production. The complex effects of cytokines on fibroblast matrix production are shown. The thin and thick arrows have the same significance as in Figure 1. The figure shows multilevel regulation of net matrix production, as exemplified by the paradoxical effects of induction of both matrix-degrading enzymes ("proteases") and their inhibitors (e.g., TIMP). ECM refers to extracellular matrix production.

Alternatively, or in addition to these direct effects, IL-1 may play a role by regulating the production of other cytokines with similar and additional effects.

In addition to these effects on fibroblasts, IL-1 regulates the production of other cytokines by these cells (Table 1). In addition to stimulating the production of IL-1 itself and the IL-1 receptor antagonist (4,33), it upregulates fibroblast production of a growing number of other cytokines and mediators, including IL-6 (28), IL-8 (34,38), IL-11 (35), leukemia inhibitory factor (LIF) (36), tumor necrosis factor-α (TNF-α) (37), monocyte chemotactic proteins-1 (MCP-1) and -2 (MCP-2) (39), other C-C and C-X-C chemokines (40,41), a number of colony-stimulating factors (CSFs) (42), and serum complement components (43). IL-1–dependent regulation of specific cytokines will be discussed below. Fibroblasts appear to produce only IL-1β (44), with virtually none of it being secreted (45). This lack of secretion may be related to the lack of a signal sequence in this molecule (46), and recent data suggest that secretion by monocytes occurs via a reverse endocytotic process since most of the IL-1 appears to be located inside vesicles. In contrast, U937 cells do not package their IL-1β inside vesicles and do not secrete this cytokine (47). It is possible that the situation in fibroblasts may be similar to that in U937 cells, explaining why IL-1β is not secreted even though high levels are encountered intracellularly.

In addition to self-regulation, IL-1 production by fibroblasts is upregulated by other cytokines, such as TNF-α (48) and interferon-γ (IFN-γ) (49), and noncytokines such as phorbol esters, teleocidin, calcium ionophores, ouabain (50), and lipopolysaccharide (LPS) (49),

Table 1 Interleukins and Fibroblasts

IL	Made by fibroblast?	Effects on fibroblast
IL-1	Yes (Self-induced and induced by TNF-α and IFN-γ)	Inhibits/stimulates cell proliferation Inhibits/stimulates ECM production Induces collagenase and stromelysin Induces various cytokines Stimulates PGE_2 production Induces complement factor B and C3 Inhibits IL-6 induction of antiviral state Stimulates KGF, NGF, and LIF production Decreases TNF-α receptor expression
IL-2	No	Inhibits cell division
IL-3	No	Stimulates megakaryocyte-induced cell division
IL-4	No (?)	Stimulates cell proliferation Inhibits TNF-α–induced factor B production Stimulates collagen production Decreases TNF-α receptor expression Stimulates IL-6 expression Stimulates/inhibits ICAM-1 expression
IL-6	Yes (Self-induced and induced by IL-1 and TNF-α)	Induces antiviral state Induces complement factor B and C3 Regulates class I histocompatibility antigens Upregulates TIMP production and PAI-1 Inhibits growth-promoting properties of TNF-α Stimulates KGF production
IL-8	Yes (induced by TNF-α and IL-1)	Unknown
IL-10	No	Inhibits collagen synthesis Inhibits IL-1–induced MCP-1 expression Stimulates collagenase and stromelysin expression Stimulates release of soluble p55 TNF-α receptor
IL-11	Yes (induced by IL-1 and TGF-β)	Inhibits megakaryocyte-induced cell division
IL-13	No	Stimulates IL-6 expression Inhibits IFN-γ–induced ICAM-1 expression Stimulates C3 but inhibits factor B expression

but downregulated by steroids (51). The diversity of agents suggests multiple regulatory steps, with the evidence suggesting that regulation occurs at both transcriptional and posttranscriptional levels. Although increased intracellular Ca^{2+} and activation of protein kinase C (PKC) appear to upregulate IL-1β production, the exact mechanism remains to be identified. The gene structure for IL-1β is described in detail in a separate chapter, but suffice it to note here that the presence of identifiable regulatory sequences suggests possible sites where these agents could modulate IL-1β gene expression, although definitie data are lacking (52). Increased stability of IL-1β mRNA has been reported in fibroblasts treated by a number of agents, although this may not always be mediated by the AU-rich sequences present in the 3′ untranslated region (53). Inhibition of IL-1 production by steroids may be mediated via the glucocorticoid responsive element (GRE), which has been described in the IL-1β gene (54). Stimulation of IL-1β production by LPS, unlike that by the other agents, only occurs if the fibroblast has been primed by IFN-γ. The priming stimulus alone is insufficient to induce IL-1β gene expression. This study using synovial fibroblasts also shows significant secretion of IL-1β upon sequential IFN-γ and LPS treatments, which seems to contradict other studies showing no or minimal secretion using other agonists (50). This apparent discrepancy may be due to the cytotoxicity of combined IFN-γ and LPS treatment (55), which has been suggested as a means by which IL-1β is released to the extracellular environment (45). Thus, although it is clear that a number of cytokine and noncytokine agonists are capable of regulating IL-1β expression, the lack of significant secretion makes it doubtful that the fibroblast is a significant in vivo source of extracellular IL-1β. However, IL-1β expressed on the cell membrane surface (56) is of consequence since intimate contact with cells in the interstitium may transmit the regulatory signals associated with this cytokine (48), and release of IL-1β following lysis of fibroblasts is potentially of pathogenetic importance.

In addition to regulating the production of other cytokines, IL-1 also stimulates the production of components of the alternative pathway of complement, namely, factors B and C3 (Table 1) (57). This effect is enhanced by pretreatment with IFN-γ but is inhibited by dexamethasone and platelet-derived growth factor (PDGF), epidermal growth factor (EGF), and basic fibroblast growth factor (bFGF). Inhibition by the growth factors is specific for factor B only and appears to occur at a pretranslational level (43). Thus, IL-1 could also influence complement-dependent pathophysiological events in this manner.

Many of the effects of IL-1 on fibroblasts described above are similar to those induced by TNF-α (58), and synergistic effects have been reported when these two cytokines are used together (37). This occurs despite the fact that they are dissimilar in structure and bind to two different receptors (59). Despite these synergistic effects, IL-1 along with IL-4 paradoxically causes decreased TNF-α receptor expression in rheumatoid synovial fibroblasts by stimulating the shedding of the p55 component of the TNF-α receptor (60).

Interleukin-4

IL-4 has inhibitory effects on TNF-α–stimulated fibroblast factor B production (61) and MHC class II ligand stimulation of IL-8, RANTES, and MCP-1 expression by fibroblastlike synoviocytes (62). But recently IL-4 has been shown to have potentially more important stimulatory effects on fibroblasts. These studies suggest that this cytokine has significant fibroblast growth-promoting properties (63) and is capable of stimulating extracellular matrix, IL-6, and ICAM-1 expression by fibroblasts presumably via the IL-4 receptors present on these cells (64–67), although IFN-γ–induced ICAM-1 expression is apparently reduced by IL-4 (68). The stimulatory effect of IL-4 may be mediated partly by IL-4's ability to inhibit PGE_2 production by these cells (60,69), since this prostaglandin is a known inhibitor of fibroblast proliferation and matrix production (discussed above). Inhibition of IL-1–induced PGE_2 production appears to occur some time after the induction of cyclooxygenase-2 (69). As mentioned above, IL-4 also causes shedding of the p55 subunit of the TNF-α receptor, which can potentially result in diminished responsiveness to this cytokine (60). In view of these IL-4 fibroblast stimulatory properties and its production by TH2 lymphocytes, the role of this interleukin in fibrotic diseases with a significant TH2 component merits further exploration.

Interleukin-6

IL-6 is another example of a multifunctional cytokine, with important roles in the acute-phase response, regulation of T-cell function, terminal B-cell differentiation, and thrombopoiesis (70). Its role in the regulation of fibroblast function is as yet incompletely defined (summarized in Table 1). As with IL-1 and TNF-α, IL-6 is capable of self-induction but does not induce IL-1 or TNF-α in fibroblasts (71). It inhibits the fibroblast growth-promoting properties of TNF-α (72) and consistent with this inhibitory activity is reported to be important in downregulating fibrosis (73). Furthermore, wound fluid is found to contain fibroblast growth suppressor activity, which is neutralized by antibody to IL-6 (74). Recent studies, however, suggest that IL-6 has autocrine growth-promoting activity in certain fibroblast subpopulations (75), making it unclear what the actual role of this cytokine is in fibrotic diseases. IL-6 induces an antiviral state in fibroblasts and is thought to be the endogenous or autocrine mediator for induction of this state by TNF-α. On the other hand, IL-1 and PDGF inhibit IL-6 as well as IFN-γ induction of the antiviral state (76). IL-6 plays a role in the expression of class I histocompatibility antigens (77) and in the induction of the acute-phase proteins, complement factor B and C3 (78), in human fibroblasts. A recent report suggests that IL-6 may play a role in fibroblast extracellular matrix metabolism via its ability to enhance the production of tissue inhibitor of metalloproteinase (TIMP) (79). It may also regulate protease activity via its ability to induce plasminogen activator inhibitor type 1 gene expression (80). Specific receptors for IL-6 have been reported in a variety of cells (81) and presumably mediate the observed biological functions of this cytokine. Details on this receptor are available in an earlier chapter.

Much more information is available concerning the fibroblast as a source of IL-6. In the absence of any added agonists, IL-6 is produced at very low levels by fibroblasts (82). Upon stimulation by irradiation and an increasing array of substances, ranging from LPS to cytokines and phorbol esters, increased IL-6 mRNA accumulation and secretion is detected. Among the cytokines, the effects of IL-1 and TNF-α have been studied extensively (83), and more recently IL-4 and IL-13 have been found to also stimulate fibroblast IL-6 expression (68). Other agonists have also been used as a means of deciphering the stimulus-response coupling mechanisms involved in upregulation of IL-6 production induced by these cytokines. Studies using IL-1 and TNF-α as agonists implicate adenyl cyclase and its attendant effects on intracellular cAMP levels as being important in this regard. This conclusion is supported by the observation of the inhibitory effects on IL-6 induction by selective inhibitors of cAMP-dependent protein kinase A (84). PKC appears not to be important in IL-6 induction by IL-1, TNF-α, or TNF-β (70). Other studies, however, show that activators of PKC may stimulate IL-6 production (85), and certain agonists such as LPS may work via this pathway (86). There is also some suggestion of a role for G-binding proteins. In addition, recent studies suggest mediation by a novel pathway involving sphingomyelin in IL-1 induction of IL-6 (87). Increases in IL-6 production are associated with increases in IL-6 mRNA levels, which appear to be due to a combination of increased transcription and stabilization of the mRNA. Addition of cycloheximide causes superinduction (70). Stimulation of transcription appears to be mediated by multiple transcriptional enhancer elements in the 5′ flanking region of the IL-6 gene (88), which could account for the variety of substances capable of regulating IL-6 gene expression. The region from -225 to -111 base pairs of the IL-6 gene appears to be important in induction by stimulants such as cytokines and viruses (89).

Other Interleukins and Miscellaneous Cytokines

Much less is known about the other interleukins with regard to their effects on and production by fibroblasts (see Figs 1 and 2, Table 1). IL-2 has been shown to inhibit proliferation of bone marrow fibroblasts by preventing entry into S phase, an effect not mediated by IFN-γ (90). The latter is known to have inhibitory effects on fibroblast collagen production and proliferation (91). IFN-α and IFN-β are also known to inhibit fibroblast proliferation (92). Fibroblasts are an important cellular source for C-X-C (alpha) members of the chemokine family, including gro/CINC, macrophage inflammatory protein-2 (MIP-2), KC, and IL-8, when appropriately stimulated by agonists including other cytokines such as IL-1 and TNF-α (34,38–41). Whether IL-8 or other members of this family have any significant biological effects on the fibroblasts themselves is unclear or undetermined at this time.

More recently, fibroblasts have been reported to also secrete monocyte chemotactic activity in response to TNF-α and IL-1 (38). This activity has now been determined to be similar to that isolated from activated monocyte-conditioned media and endothelial cells (93) and designated variously as monocyte chemotactic and activating factor (MCAF) or MCP-1. It has been identified as the human homolog of the murine JE gene product (94). This chemotactic protein has an apparent molecular weight of approximately 15 kDa and has 76 amino acid residues (95). The chemotactic activity appears to be specific for monocytes and is not active against neutrophils (93). Its effects on fibroblasts have not been evaluated. In the later stages of pulmonary fibrosis, the lung myofibroblast appear to be the predominant source of MCP-1 (96), thus implicating this cell type in the pathogenetic events associated with this chemokine. Macrophage inflammatory protein-1α (MIP-1α) and RANTES are other members of the C-C chemokine family known to be expressed by fibroblasts upon appropriate stimulation (40,97).

With respect to the other interleukins, IL-3 is known to stimulate megakaryocyte-dependent bone marrow fibroblast proliferation, which appears to be inhibited by IL-11 (98). These two interleukins are known to be important in hematopoiesis, and the effects on marrow fibroblasts may be related to this overall hematopoietic activity. Additionally, IL-11 is known to be expressed by lung fibroblasts upon stimulation by cytokines such as IL-1 and TGF-β (35). IL-10 is important in regulation of the immune response and is known to have inhibitory properties on cytokine gene expression (99). Consistent with these inhibitory properties, recent studies have reported suppressive effects of IL-10 on IL-1–induced MCP-1 expression by fibroblasts, while stimulating release of soluble p55 TNF-α receptor (100). Additionally, IL-10 is reported to inhibit type I collagen gene expression in human dermal fibroblasts while stimulating collagenase and stromelysin mRNA expression (101). These effects appear to be mediated via IL-10 receptors expressed by fibroblasts (102). These effects taken together support IL-10's potential role as an anti-inflammatory and antifibrogenic cytokine. Finally, IL-13 has effects on fibroblasts similar to IL-4. It is capable of stimulating IL-6 expression by inhibiting IFN-γ–induced ICAM-1 expression (68). It also has effects on complement expression in fibroblasts by stimulating IFN-α, IL-1, or IFN-γ–induced C3 expression, but inhibiting Factor B production (103). Neither IL-4 nor IL-10 is known to be produced by fibroblasts. Little information on the other interleukins is available with regard to production by and effects on fibroblasts.

TUMOR NECROSIS FACTORS

TNF-α–cachectin and TNF-β–lymphotoxin are primarily products of mononuclear phagocytes (104) and lymphoid cells (105), respectively, and share many biological activities (106). However, TNF-α has been observed in human peripheral blood lymphocytes carefully depleted of monocytes. The message for the TNF-β gene is also detected in these cells, but the regulation of TNF-β gene expression appears to be independent of the TNF-α gene. Nevertheless, in both the human and mouse, the two genes are closely linked on the same chromosome as are their corresponding histocompatibility (HLA) complexes (107). It is, thus, interesting to note that TNF-α activates dermal fibroblasts to express class 1 HLA antigens (108), suggesting an immunoregulatory role that would enhance the participation of cytolytic T lymphocytes in the vascular and interstitial spaces. The proteins are 30% homologous in their primary structure (109), cause necrosis of Meth A sarcoma in vivo, are toxic to selected tumor cells in vitro, and bind to the same receptor on cells (110).

Mature human TNF-α and TNF-β are polypeptides of 157 amino acids and 171 amino acid residues, respectively. Both cytokines are highly conserved across species. TNF-β is N-glycosylated but TNF-α is not, and the biologically active native forms of both are noncovalently linked trimers (111). The molecular masses of TNF-α and TNF-β are 17 and 25 kDa, respectively (112). Two distinct receptor types have been identified that bind both cytokines. These receptors are designated TNFRI (also referred to as Type B or Type β) and TNFRII (also known as Type A or Type α) with molecular masses of 55 and 75 kDa, respectively (113). The two receptor types are distinct immunologically, but their binding domains show similarities in the pattern of cysteine residues. Signaling occurs when two or three identical receptor subunits are brought close together by binding to the TNF trimer (114). There is evidence to suggest that the 55 kDa receptor is responsible for mediating the lytic activity of TNF-α, while the 75 kDa receptor is responsible for transducing the cell proliferation signal, but this observation requires further elucidation (115).

As with other cytokines, the influence of TNF-α on fibroblasts is varied and inseparable from the influences of other cytokines (Figs. 1 and 2; Table 2). TNF-α stimulates the release of PGE_2, IL-6, and IL-8, as well as DNA synthesis by fibroblasts from various tissues, although at different levels. Monoclonal antibodies to either receptor block these stimulatory effects, although neutralization of the 55 kDa receptor had a more potent effect (116). TNF-α inhibits fibroblast collagen synthesis (117), although there is a recent conflicting report showing that it stimulates human lung fibroblasts in the absence of serum to synthesize more type I and III collagens. The basis for this discrepancy is unknown, but may be attributable to differences in the conditions of assay. INF-γ attenuated the stimulatory effect of TNF-α or IL-1. In the presence of 1% serum, TNF-α has a minor effect on collagen synthesis; however, the combination of IL-1 and TNF-α inhibits collagen production, while IFN-γ inhibits collagen production by itself. The inhibition of collagen synthesis by TNF-α and IL-1 has been attributed to the production of PGE_2 stimulated by these two cytokines. It has been demonstrated that TNF-α and IL-1 inhibit the expression of the α_1(I) procollagen gene in human lung fibroblasts at the transcriptional level by a PGE_2-independent mechanism and as well as through the effect of endogenous PGE_2 released by their stimulation (118). The amount of serum could influence this observation by supplying arachidonic acid to the cells for subsequent PGE_2 production. It is interesting that under conditions where one cytokine is ineffective the addition of another produces an effect, e.g., the collagen synthesis inhibition mentioned above, perhaps working through the PGE_2-independent mechanism. Generally these effects correlate with the mRNA levels for these collagens (119).

TNF-α in concert with IL-1 may play an essential role in regulating the regeneration of peripheral nerves following injury by stimulating nerve growth factor (NGF) production in fibroblasts. In Swiss 3T3 cells TNF-α stimulates NGF expression, which is synergistically amplified by IL-1 (120).

In cultured human synovial fibroblasts, TNF-α stimu-

Table 2 Cytokine Interactions with Fibroblasts

Cytokine	Made by fibroblast?	Fibroblast interactions
IFNs-α/β	Yes (β only, induced by IL-6, TNF-α, and IL-1)	Inhibits cell proliferation Induces antiviral state
IFN-γ	No	Inhibits cell proliferation Inhibits collagen synthesis Induces IL-1 and ICAM-1 expression
MCP-1	Yes (induced by TNF-α and IL-1)	Unknown
TNF-α	No?	Stimulates cell proliferation Inhibits (stimulates) collagen synthesis Causes class I HLA expression Induces ILs-1, 6, and 8; MCP-1; factor B Stimulates PGE_2 production Stimulates KGF and LIF production
EGF/TGF-α		Stimulates cell proliferation Induces PDGF Stimulates collagen synthesis Is chemotactic (EGF) Stimulates HGF production
TGF-β	Yes (self-induced)	Inhibits/stimulates cell proliferation Stimulates extracellular matrix synthesis Stimulates synthesis of protease inhibitors Induces TGF-β and PDGF Is chemotactic Stimulates LIF production
PDGF	Yes (self-induced and induced by EGF, TGF-β, and IL-1)	Stimulates cell proliferation Is chemotactic Stimulates HGF production Stimulates IL-1 receptor expression Stimulates α_2-integrin expression
FGFs/KGF	Yes (self-induced; stimulated by laser-light. KGF induced by IL-1, IL-6, TNF-α, and serum)	Stimulates proliferation Inhibits collagen and elastin production Increases TNF-α–stimulated PGE_2 production Increases TNF-α receptor expression Induces HGF and NGF production Increases PDGF chemotactic response

lates intercellular adhesion molecule 1 (ICAM-1) gene expression as assessed functionally by measuring the adhesion of monocytes to the fibroblasts. Both synovial fibroblasts from patients with rheumatoid and nonrheumatoid arthropathies responded similarly to TNF-α. The accumulation of ICAM-1 mRNA was inhibited by glucocoricoids (121). IFN-β can significantly upregulate TNF-α–induced ICAM-1 expression in dermal fibroblasts (122). In human dermal fibroblasts, TNF-α preferentially induces $\alpha_2\beta_1$ integrins, receptors for collagen, and laminin. TNF-α is, thereby, a potentially important promoter of adhesion of dermal fibroblasts to type 1 collagen during an inflammatory response (123).

Leukemia inhibitory factor (LIF) is a pleiotropic cytokine with biological activities that overlap those of IL-6 and IL-11 (124). TNF-α in conjunction with IL-1 and TGF-β stimulates the expression of LIF in synovial (125), lung (126), and dermal fibroblasts (127).

Human dermal fibroblasts express a membrane-associated IL-1 activity when stimulated with either TNF-α or TNF-β, while LPS-stimulated fibroblasts do not. IFN-γ has no effect on the membrane IL-1 expression (128). IFN-γ alone can be cytolytic for normal mouse fibroblasts. The dose required for the cytotoxicity is higher than that required for its antiviral effect or for the activation of macrophages. However, the dose required for cytotoxicity can be reduced 10- to 100-fold by concomitant treatment with TNF-α or IL-1, or both, although neither one alone is cytotoxic for these cells. Also, LPS, which alone has no effect on the viability of mouse fibroblasts, induces cytotoxicity in the presence of IFN-γ. Antibodies to TNF-α or IL-1 do not block the cytotoxic effect of LPS and IFN-γ. This toxicity has been shown to be transcription and translation dependent. It is noteworthy that preexposure of fibroblasts to IFN-γ enhances the binding of TNF-α to these cells (129).

Both TNFs have potent mitogenic activity for fibroblasts, which is downregulated by IFN-γ, synergized by EGF, PDGF, basic fibroblast growth factor (bFGF), and TGF-β (see relevant sections of this chapter). Stimulation of fibroblasts with TNF-α is accompanied by the phosphorylation of cytosolic proteins of 41 and 43 kDa at tyrosine, threonine, and/or serine residues. The mitogenic response correlates well with the increased stimulation of tyrosine phosphorylation (130). TNF-α treatment of human fibroblasts is associated with transcriptional activation of c-*jun*, resulting in the accumulation of c-*jun* mRNA and protein expression (131). TNF-α also induces a small and transient increase in the extracellular signal–regulated kinase (ERK) activity with no measurable increase in the Raf-1 kinase activity (132).

TNF-α or IL-1 causes a transient downregulation of EGF binding to quiescent fibroblast monolayers. This results from a reduction in EGF receptor affinity and appears to be mediated by a PKC-independent mechanism, since cytokine-mediated EGF receptor phosphorylation is resistant to staurosporine, a potent inhibitor of PKC (133,134). TNF-α enhances the transcription and translation of IL-6 in human fibroblasts. Human embryonic lung fibroblasts (WI-38) respond to both TNF-α and TNF-β by increasing the accumulation of IL-6 mRNA, with the former being about threefold more potent. Stimulation is independent of PKC, does not require new protein synthesis, and is at least in part a result of increased stabilization of the IL-6 mRNA (70). IL-1α or TNF-α, but not LPS, induces the transcription and translation of NAP/IL-8 and MCP-1 in fibroblasts (135).

TNF-α is chemotactic for fibroblasts (136), and whereas fibroblasts normally do not invade type I collagen gels, preincubation of the cells with TNF-α greatly enhances their invasiveness into the gel. This seems to be associated with specific degradation of type I collagen, because TNF-α does not promote the invasion of skin fibroblasts through a reconstituted basement membrane. TNF-α has been observed to induce the synthesis of collagenase in dermal fibroblasts, but only at a relatively high dose (137,138).

TNF-α in combination with exogenous arachidonic acid is toxic to confluent fibroblasts in serum-free culture medium. Indomethacin, a cyclooxygenase inhibitor, inhibits this cytotoxic effect. This suggests that TNF-α becomes cytotoxic when arachidonic acid, present in the culture medium, is converted to prostaglandins (139), consistent with the observation that TNF-α stimulates the production of PGE_2 in target cells (140).

EPIDERMAL GROWTH FACTOR/ TRANSFORMING GROWTH FACTOR-α

This family of EGF-like molecules are encoded by distinct genes and consists of EFG, TGF-α, amphiregulin (keratinocyte autocrine factor), heparin-binding EGF, and betacellulin (BTC). These molecules are characterized by high-affinity binding to the EGF receptor, but do not have identical biological effects. In a nontransformed intestinal epithelial cell line, each EGF-related peptide induces expression of the mRNA for the remaining family members, indicating existence of an extensive auto- and cross-induction regulatory network within the EFG-related family (141). EGF is folded into three loops held in place by three intrachain disulfide bonds. This triple-loop structure is required for biological activity and distinguishes all members of the EGF family from other families of growth factors (142).

In this section only EGF and TGF-α will be discussed in view of the more abundant data for these two cytokines with regard to fibroblasts. The similarity between EGF and TGF-α is mainly in the cystines comprising the intrachain disulfide bonds and consequently the triple-loop structure with about 30% identity in the other amino acids (143). Both EGF and TGF-α are processed from larger transmembrane protein precursors. The EGF molecule is synthesized as an active prepro-EGF glycoprotein containing approximately 1200 residues that is proteolytically cleaved to release a small biologically active 53-amino-acid fragment with a molecular weight of approximately 6 kDa (144,145). TGF-α is synthesized as part of a larger precursor, pro–TGF-α, which is an integral membrane glycoprotein. A 50-amino-acid sequence, with a molecular weight of approximately 5.6 kDa, in pro–TGF-α is exposed to the extracellular fluid phase where it can be cleaved by an appropriate enzyme (143). TGF-α was initially thought to be a fetal form of EGF in that its expression is detected only in fetal/embryonic tissue or ectopically in tumors. However, the expression of TGF-α has subsequently been observed in normal adult tissues such as the pituitary, ovary, keratinocytes, and macrophages. Yet, oncogenic transformation frequently results in enhanced expression of TGF-α and not EGF (146). TGF-α and EGF exert similar biological actions in many systems. But there are exceptions, which appear to stem from the ability of various cells to process the EGF receptor differently depending on whether EGF or TGF-α is bound. In NIH-3T3 cells the data suggest that (1) EGF routes the EGF receptor directly to a degradative pathway, whereas TGF-α allows receptor recycling prior to degradation, and (2) tyrosine phosphorylation could play a role in distinguishing these two modes of receptor processing (147). Using monoclonal antibodies to the EGF receptor, it is possible to selectively inhibit TGF-α but not EGF binding to this receptor. This implies that both cytokines bind to the receptor at different epitopes and may thereby elicit similar but not identical responses (148).

The EGF receptor is a single 170 kDa polypeptide chain of 1186 residues with a substantial amount of N-linked oligosaccharide (149). It is similar to receptors for several other growth factors (PDGF, insulin, IGF-1, CSF-1, and FGF). The cytoplasmic tyrosine kinase domain of these receptors is similar to a number of oncogene products (150). With respect to the EGF receptor, the tyrosine kinase domain has a high homology to the product of *erb B* (151). Near the carboxyl terminus of the receptor are four sites of EGF-dependent tyrosine autophosphorylation (152) perhaps defining an autoinhibitory region that can be relieved by phosphorylation or truncation (153). Interaction of either TGF-α or EGF with the EGF receptor leads to a change in the receptor conformation that promotes its dimerization and ultimately an increase in the formation of phosphoserine, phosphothreonine, and phosphotyrosine on the EGF receptor. The nontyrosine phosphorylations are attributable to nonreceptor kinases, such as PKC, indicating that the activation of these kinases is a consequence of EGF binding (154). The PKC phosphorylation of the EGF receptor inhibits tyrosine phosphorylation and, in some cells, ligand binding (142). Tyrosine kinase–deficient receptors are unable to transduce any biological responses to EGF, although ligand binding is not compromised. Also, it has been demonstrated that agents that interfere with the degradation and processing of the EGF and EGF-receptor complex inhibit the EGF-induced growth response (155). The EGF receptor tyrosine kinase can phosphorylate other proteins, such as phospholipase C-γ1 (PLC-γ1), phosphoinositol 3-kinase, and guanine nucleoside triphosphate–activating protein (GAP). Activation of these proteins leads to activation of other kinases: casein kinase II, microtubule-associated protein (MAP) kinase, *raf*-1 kinase (a serine kinase), ribosomal protein kinase, and PKC (156). These events cause cellular responses, such as intracellular increase in calcium, increase in glucose and amino acid transport, and transcription of early response genes such as c-*jun*, c-*fos*, and c-*myc* (150,157), in cells that ultimately participate in the mitogenic and chemotactic responses.

The role of phosphatidylinositol turnover in EGF-induced signal transduction has not been fully defined in fibroblasts. Upon addition of EGF to cells, approximately 60% of PLC-γ1 is rapidly phosphorylated on tyrosine residues. PLC-γ1 hydrolyzes phosphoinositol 4,5-bisphosphate, yielding inositol 1,4,5-triphosphate and diacylglycerol. The former acts as a second messenger to liberate stored calcium from the endoplasmic reticulum, while the latter is an activator of PKC (158). Most phosphoinositides do not contain a phosphate at the 3-position of the inositol ring, but a small pool phosphorylated at this site has been identified recently in growth factor–stimulated cells (159). All these phosphorylations occur within an hour of EGF binding, but EGF must remain in the extracellular environment for nearly 8 h before the cells become committed to increased DNA synthesis (160). Both EGF and PDGF elicit a mitogenic response. However, during PDGF stimulation, phospholipase C (PLC) activity remains high, while during EGF stimulation phosphoinositide breakdown is not observed (161). This suggests that polyphosphoinositide hydrolysis is not an exclusive signaling pathway for commitment to DNA replication and cell division. Treatment with α-thrombin following chymotrypsin, EGF, or PDGF pretreatment results in the increase of diacylglycerols primarily through the stimulation of phosphatidylcholine hydrolysis. But EGF does not stimulate the hydrolysis of other phospholipids, including the phosphoinositides. It would thus appear that EGF stimulation is associated with the activation of a PLC specific for phosphatidylcholine, reflecting cytokine-sensitive phosphatidylcholine pools or selective diacylglycerol metabolism (162).

Preexposure of quiescent fibroblasts to PDGF increases the sensitivity to EGF by 10-fold, independent of whether EGF alone is found to be mitogenic (163). IFN-γ inhibits the mitogenic activity of PDGF, EGF, and FGF on fibroblasts. This inhibition is accomplished by blocking the transition of the G_0/G_1 to S phase of the cell cycle (164). Bradykinin, which activates PKC, blocks the IGF-1– and EGF-dependent mitogenic signals without affecting the response to serum or PDGF (165).

The extracellular matrix influences fibroblast responses to cytokines. The addition of EGF to cultures of human skin fibroblasts grown on plastic coated with or without various macromolecules results in repression of collagen synthesis while cell multiplication is slightly stimulated. However, the addition of EGF to fibroblasts grown in a collagen matrix results in stimulation of both noncollagenous protein and collagen synthesis as has been observed in vivo (Fig. 2, Table 2) (166). Fibroblast chemotaxis is elicited by EGF in vitro, and PDGF and TGF-β can downregulate this activity (Fig. 3) (167).

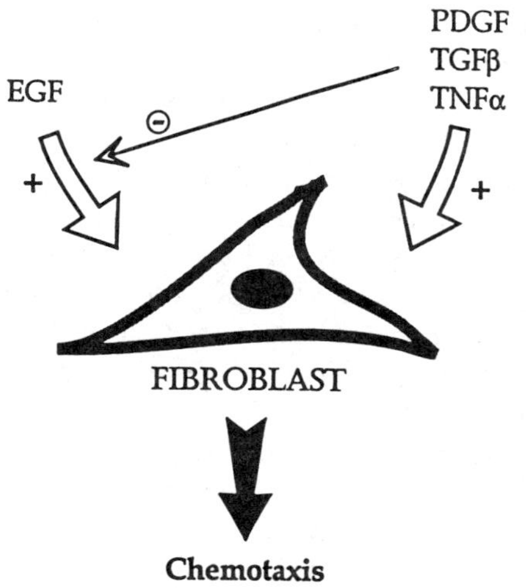

Figure 3 Cytokine regulation of fibroblast chemotaxis. All cytokines shown here are chemotactic for fibroblasts. However, as shown by the thin line, PDGF and TGF-β inhibit the chemotactic activity of EGF.

TRANSFORMING GROWTH FACTOR-β

TGF-β belongs to a large superfamily of closely related polypeptides. Over two dozen TGF-β–related factors have been identified including the activins and bone morphogenetic proteins (168,169). There are currently five distinct forms of TGF-β with significant amino acid sequence homology to the first member of this family, TGF-β1. The first three of these, TGF-β1, TGF-β2, and TGF-β3, have been found in mammals, while TGF-β4 has been reported in chickens, and TGF-β5 in *Xenopus* species (170,171). Discussion here will focus on the first three forms in view of the topic of this chapter. These forms of TGF-β are synthesized as larger precursor polypeptides with an amino-terminal signal sequence, a pro-domain of varying size, and a mature domain. The precursor is cleaved at a dibasic site to yield a C-terminal 110- to 140-amino-acid monomer of approximately 12.5 kDa. The biologically active form is a 25 kDa dimer composed of identical disulfide-linked monomers; however, heterodimer molecules also have biological activity (172). Functionally TGF-β1 exerts activities similar to the other two TGF-β species. The distinct differences in the potencies that have been observed between the different TGF-β forms depend on the cell type and assay used. These forms of TGF-β arise from different genes, and the promoter structures and regulatory elements for them are quite different (173). Also, there would appear to be a very complicated regulatory network whereby each TFGβ type either upregulates or downregulates the expression of the other (174).

The precursor form, which constitutes >98% of secreted TGF-β1, is unable to bind to its cellular receptor without prior activation (175). Normal fibroblasts secrete TGF-β in latent form, activation of which occurs after treatment with acid or proteases such as plasmin (176). A comparison of the human TGF-β1, TGF-β2, and TGF-β3 precursor cDNAs reveals an amino acid sequence identity of about 75–80% within the mature peptide sequences, while the rest of the precursor sequence, the pro-domain, is only 25–30% conserved (177,178). Among mammals, the sequence for TGF-β1 is highly conserved. In humans, pigs, cows, and monkeys the sequence is identical, and it differs by only one amino acid in mice (179). The most conserved feature of this family of proteins is the spacing of seven cysteines. Six of these form a rigid central structure known as the cysteine knot. This knot locks the base of several β-sheet strands together and probably accounts for the strong resistance of many TGF-β family members to heat, denaturants, and extremes of pH. The remaining cysteine residue in each monomer forms an additional disulfide bond that links two monomers into a dimer. Many hydrophobic contacts exist between the two monomer subunits that promote dimer formation even in the absence of this disulfide bond (180).

A similarity exists between members of the TGF-β superfamily, nerve growth factor (NGF) and PDGF. Though the monomers of NGF and PDGF show only <10% amino acid homology with the TGF-β monomer, they all contain the cysteine knot motif with the β-sheet strand arrangement. The three-dimensional structures of the cores of the monomers are nearly superimposable. However, the interactions between the monomers forming active dimers are completely different, resulting in different receptor interactions (181). This implies that selection pressure favored the cysteine knot anchoring of the β-sheet arrangement of the monomers but subsequent dimerization and interaction with their respective receptors evolved separately.

TGF-β receptors have been identified on the surface of almost all cell types examined. Nine membrane proteins that bind TGF-β have been identified (182). Two of these receptors, types I and II, are glycoproteins of 53 and 73–95 kDa, respectively, and display higher affinity for TGF-β1 than for TGF-β2 (183). The third receptor type (type III, or betaglycan) is a 200–400 kDa membrane proteoglycan with a core protein of 110 kDa that binds TGF-β1 and TGF-β2 with similar affinity (184). All three receptors bind all three TGF-β forms with high affinity (Kd = 10–200 pM) (185). In the human, rodent, and avian fibroblasts, the type III receptor is a larger disulfide-linked multisubunit complex of 565–615 kDa (186). The types I and II TGF-β receptors are involved in signaling, whereas the type III receptors regulate the access of TGF-β to the signaling receptors. Apparently, type I and II TGF-β receptors form a complex that is essential for the transduction of the TGF-β signal to the nucleus; TGF-β cannot bind to the type I receptor in the absence of the type II receptor. Ligand binding to the type II receptor activates a serine/threonine protein kinase, which then phosphorylates the type I receptor, itself containing serine/threonine protein kinase activity. The latter serine/threonine kinase phosphorylates an as yet unidentified protein (187). Each type II receptor can interact with a repertoire of type I receptors, and some type I receptors can interact with several type II receptors (168). This situation can then result in a wide range of receptor combinations for a given factor, which may vary from cell to cell depending on the relative amounts of receptor subunits. The diverse responses elicited by TGF-β may in part be explained by this receptor heterogeneity.

An inverse relationship exists between receptor affinity and receptor density, such that at low TGF-β concentrations, all cells bind a similar number of TGF-β molecules. There is little or no ligand-induced receptor downregulation so that the receptor appears to be essentially unmodulated (188). Increased cell density and differentiation appear to downregulate TGF-β binding due to a decrease in the number of high-affinity receptors and not from a change in receptor affinity (189).

Signal transduction after TGF-β binding is not well understood; however, some information is available. The growth stimulatory effect of TGF-β1 on AKR-2B fibroblastic cells is coupled to a pertussis toxin–sensitive guanine nucleotide regulatory protein (G-protein) (190). However, pertussis toxin does not affect TGF-β1–stimulated fibronectin or collagen mRNA expression, nor does it have any effect on TGF-β1–induced morphological alterations. These observations imply that TGF-β1–stimulated gene expression is coupled to multiple pathways distinguished by their sensitivity to pertussis toxin (191). In fibroblastic cells that are growth inhibited by TGF-β, FGF- and thrombin-induced alterations in intracellular pH, PKC activity, protein phosphorylation, and phosphoinositide metabolism are not affected by TGF-β (192).

TGF-β is a growth inhibitor for a wide variety of cell types, including selected cell types of mesenchymal origin, although it is originally discovered because of its ability to stimulate anchorage-independent growth of nontransformed fibroblastic cells (193). Growth arrest by TGF-β has been suggested to be mediated by the activation of the retinoblastoma protein (RP). Since the RP is regulated by its phosphorylation state (when underphosphorylated it acts as a growth suppressor), TGF-β may influence RP phosphorylation by inhibiting the induction of cycline-dependent kinase 4, resulting in the underphosphorylation of RB (194). However, for cells of mesenchymal origin, whether TGF-β stimulates or inhibits proliferation depends on the presence of other growth factors (185). The cell-cycle kinetics of TGF-β stimulation of fibroblast proliferation results in an approximate 12-h prolongation of G_1 and appears to be mediated by an indirect mitogenic response. For example, in ARK-2B cells TGF-β induces c-*sis* mRNA associated with an increase in PDGF activity in the conditioned media. Thus, TGF-β indirectly stimulates proliferation by inducing the synthesis of the c-*sis* gene product, PDGF B chain (195).

TGF-β is a potent upregulator of extracellular matrix production, stimulating synthesis of collagen types I, III, and V, fibronecton (196), the fibronectin-receptor complex, and glycosaminoglycans (197). It inhibits matrix degradation by increasing synthesis of inhibitors of proteases involved in matrix degradation, such as metalloproteinase inhibitor and plasminogen activator inhibitor 1 (198), and by inhibiting gene expression and synthesis of the proteases, transin/stromelysin, and collagenase (Fig. 2) (199). Regulation of matrix gene expression appears to be at both transcriptional and posttranscriptional levels (200,201). Despite the net collagen stimulatory activity, TGF-β also paradoxically stimulates the production of PGE_2 (202), which is inhibitory to collagen synthesis (203). It may be that the inhibitory effect of endogenous PGE_2 can be overcome by the stimulatory effect of TGF-β or that the kinetics of the two effects differ to the extent that the matrix-stimulatory effects predominate at the time of assay.

The effect of TGF-β on matrix production appears to be longlasting, if not irreversible, since an elevated expression of TGF-β, procollagen α_2(I), and fibronectin mRNAs persists for at least 96 h after the removal of the original TGF-β stimulus (204). Administration of TGF-β subcutaneously to rats results in granulation tissue in which myofibroblasts expressing α-smooth muscle actin are abundantly found. Since myofibroblasts are the key source of collagen in lung fibrosis (205), this effect of TGF-β on actin expression may be important in the pathogenesis of fibrotic lesions. Other cytokines like PDGF, basic fibroblast growth factor (bFGF), and TNF-α do not induce α-smooth muscle actin in the myofibroblasts and in fact may inhibit actin expression (206). In contrast, TGF-β upregulates α-smooth muscle actin expression in fibroblasts and concomitantly stimulates contraction of fibroblast populated collagen lattices or gels (207–209). This appears to be a possible mechanism by which fibrotic lesions in lung assume their increased contractility (210).

Another key biological activity of TGF-β in fibroblasts and macrophage is chemotaxis (Fig. 3) (211,212). Histological analysis of treated wounds demonstrates an in vivo chemotactic response of macrophages and fibroblasts to both PDGF-BB and TGF-β1, but the response to TGF-β1 is significantly less than that observed with PDGF-BB (213). Intraarticular injection of TGF-β results in a marked synovitis and synovial hyperplasia consistent with its parallel monocyte and fibroblast chemotactic properties and ability to induce macrophage-derived growth factors (214).

Finally, TGF-β autoinduces itself in fibroblasts (215). Although many cell types are capable of expressing TGF-β, studies of fibrotic lesions indicate that the fibroblast is a significant source of this cytokine, especially in the later stages of fibrosis (216). It is spontaneously secreted by synovial and other fibroblastlike cells. For example, fibroblasts from affected skin of scleroderma patients and normal control donors produce comparable amounts of TGF-β and possess comparable number of TGF-β receptors (217). This would suggest that an alteration in the activation of latent TGF-β or in postreceptor responsiveness of the scleroderma fibroblasts may be the critical factor in promoting fibrosis. Since TGF-β activity is modulated by other cytokines, it is possible that the conditions present in vivo in scleroderma and perhaps other fibrogenic processes may influence the action of TGF-β so that a fibrotic lesion results. In contrast, fibroblasts from fibrotic lungs express significantly greater amounts of TGF-β than those from control or unaffected lungs (209,218). Hence, given all these properties (summarized in Figs. 1–3 and Table 2), TGF-β has been strongly implicated in the pathogenesis of fibrotic lesions.

PLATELET-DERIVED GROWTH FACTOR

PDGF is a major mitogen in serum for connective tissue and other cells (Fig. 1). A closely related protein is vascular endothelial growth factor (VEGF). PDGF and VEGF have similar structures but bind different receptors and have different actions. VEGF appears to be mitogenic exclusively for endothelial cells (156). The PDGF from human platelets is a heparin-binding cationic glycoprotein of approximately 30 kDa (219). PDGF has been implicated, directly or indirectly, in neoplasia, arthritis, arteriosclerosis, and bone marrow fibrosis. In human platelets, PDGF, which is stored in α-granules, is a disulfide-bonded dimer of two similar but not identical polypeptides, denoted A and B. Both A and B chains share 60% amino acid homology and are synthesized as larger precursor molecules (211 and 109 amino acid residues, respectively). All three combinations of polypeptide chains (PDGF-AA, PDGF-AB, and PDGF-BB) have been identified and purified from platelets or other cells and have been shown to be mitogenic (220). The gene for the B chain is the normal cellular counterpart for the v-*sis* oncogene of the simian virus (221). The PDGF isoforms differ in their functional effects on human fibroblasts; whereas PDGF-AA induces transmodulation of the receptor for EGF to the same extent as the other isoforms, it is a less potent mitogen and has no stimulatory effect on actin reorganization or chemotaxis (222). However, it is found with macrophage-derived PDGF-like protein in human wound fluid for several days after surgery (223). PDGF-AA is a potent mitogen, however, for Swiss 3T3 cells (224).

In experimental wounds, PDGF treatment accelerates granulation tissue formation in a dose-dependent manner (225). In chronic pressure ulcers recombinant PDGF-BB accelerates wound healing by induction of fibroblast proliferation and differentiation (226). In experimental corneal wounds, PDGF significantly increases wound strength, which correlates with increased numbers of fibroblasts at the wound junction and elevated type III and type IV collagen expression (227).

The functional differences and binding characteristics of the isoforms have led to the identification of two classes of receptors. The PDGF B–type receptor (β-receptor, 190 kDa) binds PDGF-BB with high affinity and PDGF-AB with about 10 times lower affinity, but does not bind PDGF-AA with any appreciable affinity. The PDGF A–type receptor (α-receptor, 170 kDa) binds all three isoforms with high affinity. There is a close structural similarity between the two receptor types. Both have extracellular components, which can be divided into five immunoglobulinlike domains, and cytoplasmic elements, which possess split protein tyrosine domains (228). Ligand binding results in dimerization of both α and β receptors. The gene product of *raf*-1 is a serine kinase, which is a tyrosine phosphorylation substrate of the PDGF receptor kinase (229).

With human fibroblasts, which contain three to four times more B-type than A-type PDGF receptors, PDGF-AB and PDGF-BB were found to be potent mitogens, whereas PDGF-AA induced only a limited mitogenic stimulation (230). Whereas both receptors can transduce a mitogenic signal, only the B-type receptor appears to be able to mediate chemotaxis or the formation of circular membrane ruffles in human foreskin fibroblasts (156). PDGF-BB stimulates the synthesis of the integrin α_2 subunit in human diploid fibroblasts but not the integrin α_1 or α_3 subunit. It is possible that the chemotactic response stimulated by PDGF-BB may be mediated by the synthesis of this type of integrin subtype (231).

PDGF is a potent chemoattractant as well for cells that respond to PDGF as a mitogen. The chemotactic response of human diploid fibroblasts from primary cultures to PDGF is related to the proliferative state of the cells. Proliferating cell of low density has a greater migratory response than cells whose growth or proliferation have been density arrested. Treatment of these cells with TGF-β and bFGF results in a greater chemotactic response to PDGF (232).

In addition to the above effects on fibroblasts, PDGF upregulates IL-1 receptor mRNA levels and expression (233). In the presence of exogenous IL-1, this effect on IL-1 receptor expression may in turn modulate the PDGF mitogenic effect, in view of the regulatory effects of IL-1 on fibroblast growth (11,12). Cell-cycle regulation of human diploid fibroblasts is located in the proximal half of G_1, and in order to traverse this subphase, cells require certain factors such as PDGF. Human foreskin fibroblasts exposed to PDGF respond by increasing the mRNA levels of the proto-oncogenes c-*fos* and c-*myc*. The effect of c-*fos* mRNA level peaks at 30 min after exposure, while c-*myc* levels peak after 8 h (234). One strain of Hutchinson-Gilford (progeria) syndrome fibroblasts, AG3513, responds to PDGF by a reduced c-*fos* mRNA induction. The diminished response of AG3513 fibroblasts to PDGF is apparently not due to a reduced PDGF receptor number, affinity, or autophosphorylation (235). PDGF is known to activate the PLC-PKC pathway (236).

The response of fibroblasts to PDGF and/or EGF depends on the type of fibroblast used. With respect to 3T3 fibroblasts, PDGF and EGF act as a competence and progression factor, respectively. However, in EL2 rat fibroblasts, EGF appears to be able to induce the synthesis of putative intracellular regulatory proteins that govern the PDGF-induced competence in 3T3 cells. As a consequence, EGF in EL2 cells may act as both a competence and a progression factor via induction of the mechanisms, regulated in other cell lines by cooperation between different growth factors (237).

Exposure to PDGF inhibits high-affinity binding of EGF by several types of fibroblasts. In human fetal lung fibroblasts (WI-38), PDGF-dependent inhibition of EGF binding correlates with PDGF-induced phosphorylation of threonine 654 (a residue phosphorylated when PKC is activated) of the EGF receptor and with inhibition of EGF receptor autophosphorylation on tyrosine residues (238). However, in human foreskin fibroblasts (FS4) PDGF do not stimulate phosphorylation of threonine 654, nor was there an effect on EGF-dependent, tyrosine-specific autophosphorylation of the receptor (239). The differences in the amount of PDGF A and B chains and the α and β receptors result in a large range of possibilities for biological responses by PDGF.

FIBROBLAST GROWTH FACTORS

The FGF family currently consists of nine members showing 30–50% amino acid sequence homology with similar exon-intron structures in their genes. They are designated FGF-1 through FGF-9, although they also have common names (e.g., FGF-1 and FGF-2 are known as acidic FGF (aFGF) and basic FGF (bFGF), respectively). These two members of the FGF family are prototypical and most easily studied. As their names imply, they have different isoelectric points: pH 5.6 and pH 9.6 for aFGF and bFGF, respectively (240). The three-dimensional structure of human bFGF has been shown to closely resemble interleukin-1, implying membership in a larger cytokine family (241). Both aFGF and bFGF are single-chain proteins (140 and 146 amino acids, respectively) derived proteolytically from different precursor molecules of 155 amino acids each. When isolated from natural sources, bFGF has a mass of 18 kDa; however, larger forms up to 24 kDa are also found resulting from amino-terminal extensions of the protein produced by initiation of translation starting from CUG codons upstream from the normal initiation codon. These forms are found in the nucleus, and their role in that location remains obscure (242). During purification, aFGF of various sizes (16–18 kDa) have been observed. These different sizes are generated by proteolysis.

Neither aFGF, bFGF, nor glia-activating factor (FGF-9) has a leader sequence that allows the peptides to be secreted via known pathways, even though receptors for FGFs suggest extracellular modes of action exist. There is evidence, however, for the existence of alternative transport systems that do not involve the classical protein secretory pathway. Several candidate mechanisms have been suggested (243,244).

An interesting property of FGF activity is the association with heparin and heparan sulfate (245). The FGFs are associated with the extracellular matrix and in basement membranes attached to heparan sulfate (246). Heparan sulfate protects bFGF from proteolytic degradation, and the binding of a FGF to heparin or heparan sulfate in the membranes of cells causes the affinity of FGF binding to its receptor to increase two- to threefold (247). There is a great deal of specificity in FGF binding to proteoglycans since closely related compounds such as chondroitin sulfate do not bind FGF. Other growth factors, such as granulocyte-macrophage colony-stimulating factor (GM-CSF), amphiregulin (keratinocyte autocrine factor, KAF), IL-3, heparin-binding epidermal growth factor (HB-EGF), VEGF, hepatocyte growth factor (HGF) and PDGF, also interact with heparin.

The association of FGF with the extracellular matrix provides a means by which FGF can be regulated in times of tissue injury. The release of heparinase, cathepsin D, or collagenase following injury would release FGF to bind its receptors on target cells, implying that the extracellular matrix behaves as a storage site for FGFs (156). Aside from local release of FGF during wound healing and inflammatory episodes, other cell types and stimuli can increase the activity of FGF. The infiltration of T lymphocytes with their subsequent synthesis and release of FGF can supply this growth factor during wound healing (248). Low-level laser irradiation has been observed to be associated with the autocrine production of bFGF from fibroblasts, resulting in their proliferation (249).

As of now, five distinct FGF receptors have been discovered, designated FGFR1 through FGFR5. Although their genes are located on different chromosomes, they all have similar structures characterized by an extracellular domain with three immunoglobulinlike regions that bind FGFs, a transmembrane domain, and a cytoplasmic domain that contains a tyrosine-specific kinase (250). The second and third immunoglobulin loops, which have been implicated in FGF binding (the first loop appears to be dispensable for binding), and the cytoplasmic tyrosine kinase domains are highly conserved between the different receptors. The molecular weights of these glycoprotein receptors vary from 95,000 to 150,000 kDa. FGFR1 and FGFR2 bind aFGF and bFGF with similar affinity (Kd = 20 pM to 600 pM), whereas FGFR4 only binds aFGF (Kd = 200–600 pM) (156). Two classes of FGF-binding sites have been characterized: the so-called high-affinity, low-capacity receptors (Kd = 200–600 pM) and low-affinity, high-capacity receptors containing heparan sulfate side chains (Kd = 2–20 nM). The high-affinity receptors are signal transducing, while the low-affinity receptors play a role in storage and stabilization (251).

The cDNA sequences of both FGFR1 and FGFR2 differ in their extracellular regions. The FGFR1 and FGFR2 forms with two or three immunoglobulin loops are created by differential splicing. Differential splicing and polyadenylation also result in structural variation in the region of the third immunoglobulinlike loop (252,253). Some of the receptors are activated by multiple forms of FGFs, while others demonstrate greater specificity. With nine distinct ligands and five receptors and their variants, a great deal of versatility exists in the action of the FGFs.

Like other tyrosine kinase receptors, the first step in signal transduction involves receptor dimerization and trans- or autophosphorylation. Heparin plays a critical role in FGF receptor dimerization by dimerizing the ligand (254). It has also been demonstrated that ligand-induced dimerization occurs between different FGF receptors (255). Tyrosine residue 766 at the carboxy terminus of the FGFR1 is one of the autophosphorylation sites. Tyrosine residues corresponding to the 766 residue of FGFR1 have been found in the other members of the FGFR family. Phospholipase C-γ-1, and raf kinase are activated by FGFR1, presumably as substrates for the receptor tyrosine kinase (256,257).

The FGF family acts on cells of mesodermal, ectodermal, neuroectodermal, and endodermal origin with respect to survival, proliferation, cell migration, and neovascularization (250). These cell types are involved in wound healing, and members of the FGF family are considered to be important regulators of wound healing. As their name implies, these proteins are mitogens for fibroblasts but are by no means restricted to fibroblasts for mitogenic or other functions. Many of the cells responsive to FGF also synthesize the peptide: fibroblasts, astrocytes, endothelial cells, smooth muscle cells, chondrocytes, and osteoblasts (258).

Although the FGFs are mitogenic for the cells found in wounds, they tend to inhibit matrix protein production. FGF has been shown to increase TNF-α–mediated PGE_2 production and to increase the expression of TNF-α receptors on human fibroblasts (259). Interestingly, aFGF and bFGF have been effective in downregulating the excess collagen production of keloid fibroblasts. This inhibitory effect appears to be at the pretranslational level (260). In human dermal fibroblasts FGF inhibits collagen synthesis in monolayer culture and when embedded in collagen gels (261). In neonatal rat lung fibroblasts, bFGF is observed to inhibit elastin production and exerts its inhibitory effect at the transcriptional level (262). Human fetal skin fibroblasts will autonomously migrate into a denuded area in a cell monolayer, with both collagen and bFGF mediating this process. The fetal fibroblasts produce and secrete both these substances (263).

aFGF, bFGF, and other growth factors, such as EGF, PDGF, and TGF-α, induce the production of hepatocyte growth factor (HGF, scatter factor) in human skin fibroblasts (264,265). HFG has marked effects on epithelial and endothelial cells. The production of HFG by pro-fibrotic cytokines in skin fibroblasts implies that HGF could participate in wound healing and other inflammatory episodes. FGF has also been observed to induce the production of nerve growth factor in fibroblasts and astrocytes (266,267). This activity points to the fact that peripheral nerve regeneration is a part of the wound-healing process.

Though many of the FGFs can activate many different cell types, various degrees of specificity have been observed. Keratinocyte growth factor (KGF, FGF7) is an example of an FGF that is specific for epithelial cells. KGF is predominantly synthesized in mesenchymal cells with fibroblasts known to specifically produce it (268). The KGF receptor binds KGF and aFGF with high affinity and bFGF with low affinity. This receptor is almost exclusively found on the epidermal keratinocyte. During wound healing, an increase in both KGF protein and receptor is observed with an increase in KGF mRNA expression in the dermis preceding epithelial cell proliferation and subsequent reepithelialization (269). IL-1β, IL-6, TNF-α, and serum induce the expression of KGF in cultured murine and human dermal fibroblasts (270).

CONCLUSION

This chapter has underlined the increasing importance of the fibroblast as both a source of cytokines as well as a responder to cytokines in a manner that may have potentially significant impact on diverse disease processes. Increasingly it appears that the fibroblast has, in many instances, equal or greater potential as a source of inflammatory and other cytokines than do other cells, including the "professional" inflammatory or immune cells. The fact that many of these cytokines are autoinduced in fibroblasts further argues for the importance of these cells in prolonging and/or enhancing the activity of these cytokines. The effects of cytokines on fibroblasts are also so wide-ranging as to further suggest that these cells have a central role in a number of disease processes. Fibroblasts can no longer be looked upon as passive structural cells, standing by quietly while critical events are happening in their environment, but must be considered as active participants in these events.

REFERENCES

1. Conrad GW, Hart GW, Chen Y. Differences IN VITRO between fibroblast-like cells from cornea, heart, and skin of embryonic chicks. J Cell Sci 1977; 26:119–137.
2. Korn JH, Brinckerhoff CE, Edwards RL. Synthesis of PGE collagenase and tissue factor by fibroblast substrains: substrains are differentially activated for different metabolic products. Collagen Rel Res 1985; 5:437–447.
3. Anderson DM, Johnson L, Glaccum MB, Copeland NG, Gilbert DJ, Jenkins NA, Valentine V, Kirstein MN, Shapiro DN, Morris SW, et al. Chromosomal assignment and genomic structure of IL-15. Genomics 1995; 25:701–706.
4. Dinarello CA. The interleukin-1 family: 10 years of discovery. FASEB J 1994; 8:1314–1325.
5. Wood DD, Bayne EK, Goldring MB. The four biochemically distinct species of human interleukin-1 all exhibit similar biological activities. J Immunol 1985; 134:895–903.
6. Kilian PL, Kaffka KL, Stern AS. Interleukin-1-alpha and interleukin-1-beta bind to the same receptor on T cells. J Immunol 1986; 136:4509–4514.
7. Hamilton JA, Slywka J. Stimulation of human synovial fibroblast plasminogen activator production by mononuclear cell supernatants. J Immunol 1981; 126:851–855.
8. Dayer J-M, Russell RGG, Krane SM. Collagenase production by rheumatoid synovial cells: stimulation by a human lymphocyte factor. Science 1977; 195:181–183.
9. Hamerman D, Wood DD. Interleukin-1 enhances synovial cell hyaluronate synthesis. Proc Soc Exp Biol Med 1984; 177:205–210.
10. Dayer J-M, Bierard J, Chess L, Krane SM. Participation of monocyte/macrophages and lymphocytes in the production of a factor that stimulates collagenase and prostaglandin release by rheumatoid synovial cells. J Clin Invest 1979; 64:1386–1392.
11. Elias JA. Tumor necrosis factor interacts with interleukin-1 and interferons to inhibit fibroblast proliferation via fibroblast prostaglandin-dependent and -independent mechanisms. Am Rev Respir Dis 1988; 138:652–658.
12. Mauviel A, Teyton L, Bhatnagar R, Penfornis H, Laurent M, Hartmann D, Bonaventure J, Loyau G, Saklatvala J, Pujol J-P. Interleukin-1α modulates collagen gene expression in cultured synovial cells. Biochem J 1988; 252:247–255.
13. Burch RM, Connor JR, Axelrod J. Interleukin 1 amplifies receptor-mediated activation of phospholipase A2 in 3T3 fibroblasts. Proc Natl Acad Sci USA 1988; 85:6306–6309.
14. Frisch SM, Ruley HE. Transcription from the stromelysin promoter is induced by interleukin-1 and repressed by dexamethasone. J Biol Chem 1987; 262:16300–16304.
15. Tewari DS, Qian Y, Tewari M, Pieringer J, Thornton RD, Taub R, Mohan EO. Mechanistic features associated with induction of metalloproteinases in human gingival fibroblasts by interleukin-1. Arch Oral Biol 1994; 39:657–664.
16. Dayer J-M, de Rochemonteix B, Burrus B, Demczuk S, Dinarello C. Human recombinant interleukin-1 stimulates collagenase and prostaglandin E2 production by synovial cells. J Clin Invest 1986; 77:645–648.
17. Conca W, Kaplan PB, Krane SM. Increases in levels of procollagenase messenger RNA in cultured fibroblasts induced by human recombinant interleukin-1β or serum follow *c-jun* expression and are dependent on new protein synthesis. J Clin Invest 1989; 83:1753–1757.
18. Lin J-X, Vilcek J. Tumor necrosis factor and interleukin-1 cause a rapid and transient stimulation of *c-fos* and *c-myc* mRNA levels in human fibroblasts. J Biol Chem 1987; 262:11908–11911.
19. Angel P, Baumann I, Stein B, Delius H, Rahmsdorf HJ, Herrlich P. 12-O-tetradecanoyl-phorbol-13-acetate induction of the human collagenase gene is mediated by an inducible enhancer element located in the 5′-flanking region. Mol Cell Biol 1987; 7:2256–2266.
20. Matrisian LM, Leroy P, Ruhlmann C, Gesnel M-C, Breathnach R. Isolation of the oncogene and epidermal growth factor-induced transin gene: complex control in rat fibroblasts. Mol Cell Biol 1986; 6:1679–1686.
21. Quinones S, Buttice G, Kurkinen M. Promoter elements in the transcriptional activation of the human stromelysin-1 gene by the inflammatory cytokine, interleukin-1. Biochem J 1994; 302(pt 2):471–477.
22. Bhatnagar R, Penfornis H, Mauviel A, Loyau G, Saklatvala J, Pujol J-P. Interleukin-1 inhibits the synthesis of collagen by fibroblasts. Biochem Int 1986; 13:709–720.
23. Clark JG, Kostal KM, Marino BA. Modulation of collagen production following bleomycin-induced pulmonary fibrosis in hamsters. Presence of a factor in lung that increases fibroblast prostaglandin E2 and cAMP and suppresses fibroblast proliferation and collagen production. J Biol Chem 1982; 257:8098–8105.
24. Postlethwaite AE, Raghow R, Stricklin GP, Poppleton J, Seyer JM, Kang AH. Modulation of fibroblast functions by interleukin 1: increased steady-state accumulation of type I procollagen messenger RNAs and stimulation of other functions

but not chemotaxis by human recombinant interleukin 1α and β. J Cell Biol 1988; 106:311–318.
25. Bronson RE, Argenta JG, Bertolami C. Interleukin-1-induced changes in extracellular glycosaminoglycan composition of cutaneous scar-derived fibroblasts in culture. Collagen Rel Res 1988; 8:199–208.
26. Raines EW, Dower SK, Ross R. Interleukin-1 mitogenic activity for fibroblasts and smooth muscle cells is due to PDGF-AA. Science 1989; 343:393–396.
27. Jordana M, Richards C, Irving LB, Gauldie J. Spontaneous in vitro release of alveolar-macrophage cytokines after the intratracheal instillation of bleomycin in rats. Characterization and kinetic studies. Am Rev Respir Dis 1988; 137:1135–1140.
28. Dinarello CA. Biology of interleukin 1. FASEB J 1988; 2:108–115.
29. Jorens PG, Van Overveld FJ, Vermeire PA, Bult H, Herman AG. Synergism between interleukin-1 beta and interferon-gamma, an inducer of nitric oxide synthase, in rat lung fibroblasts. Eur J Pharmacol 1992; 224:7–12.
30. Chin JE, Winterrowd GE, Krzesicki RF, Sanders ME. Role of cytokines in inflammatory synovitis. The coordinate regulation of intercellular adhesion molecule 1 and HLA class I and class II antigens in rheumatoid synovial fibroblasts. Arthritis Rheum 1990; 33:1776–1786.
31. Takii T, Hayashi H, Marunouchi T, Onozaki K. Interleukin-1 down-regulates type I interleukin 1 receptor mRNA expression in a human fibroblast cell line TIG-1 in the absence of prostaglandin E2 synthesis. Lymphokine Cytokine Res 1994; 13:213–219.
32. Piguet PF, Vesin C, Grau GE, Thompson RC. Interleukin 1 receptor antagonist (IL-1ra) prevents or cures pulmonary fibrosis elicited in mice by bleomycin or silica. Cytokine 1993; 5:57–61.
33. Martel-Pelletier J, McCollum R, Pelletier JP. The synthesis of IL-1 receptor antagonist (IL-1ra) by synovial fibroblasts is markedly increased by the cytokines TNF-alpha and IL-1. Biochim Biophys Acta 1993; 1175:302–305.
34. Strieter RM, Phan SH, Showell HJ, Remick DG, Lynch JP III, Genord M, Raiford C, Eskandari M, Marks RM, Kunkel SL. Monokine-induced neutrophil chemotactic factor gene expression in human fibroblasts. J Biol Chem 1989; 264:10621–10626.
35. Suen Y, Chang M, Lee SM, Buzby JS, Cairo MS. Regulation of interleukin-11 protein and mRNA expression in neonatal and adult fibroblasts and endothelial cells. Blood 1994; 84:4125–4134.
36. Hamilton JA, Waring PM, Filonzi EL. Induction of leukemia inhibitory factor in human synovial fibroblasts by IL-1 and tumor necrosis factor-alpha. J Immunol 1993; 150:1496–1502.
37. Elias JA, Freundlich B, Kern JA, Rosenbloom J. Cytokine networks in the regulation of inflammation and fibrosis in the lung. Chest 1990; 97:1439–1445.
38. Larsen CG, Anderson AO, Oppenheim JJ, Matsushima K. Production of interleukin 8 by human dermal fibroblasts and keratinocytes in response to interleukin 1 or tumor necrosis factor. Immunology 1989; 68:31–36.
39. VanDamme J, Proost P, Put W, Arens S, Lenaerts JP, Conings R, Opdenakker G, Heremans H, Billiau A. Induction of monocyte chemotactic proteins MCP-1 and MCP-2 in human fibroblasts and leukocytes by cytokines and cytokine inducers. J Immunol 1994; 152:5495–5502.
40. Lukacs N, Chensue SW, Smith RE, Strieter RM, Warmington K, Wilke C, Kunkel SL. Production of monocyte chemoattractant protein-1 and macrophage inflammatory protein-1 alpha by inflammatory granuloma fibroblasts. Am J Pathol 1994; 144:711–718.
41. Ohmori Y, Hamilton TA. Cell type and stimulus specific regulation of chemokine gene expression. Biochem Biophys Res Commun 1994; 198:590–596.
42. Zucali JR, Dinarello CA, Oblon DJ, Gross MA, Anderson L, Weiner RS. Interleukin 1 stimulates fibroblasts to produce granulocyte-macrophage colony-stimulating activity and prostaglandin 2. J Clin Invest 1986; 77:1857–1863.
43. Circolo A, Pierce GF, Katz Y, Strunk RC. Antiinflammatory effects of polypeptide growth factors. J Biol Chem 1990; 265:5066–5071.
44. Mauviel A, Temime N, Charron D, Loyau G, Pujol J-P. Interleukin-1α and β-induced gene expression in human dermal fibroblasts. Biochem Biophys Res Commun 1988; 156:1209–1214.
45. Young PR, Hazuda DJ, Simon PL. Human interleukin 1β is not secreted from hamster fibroblasts when expressed constitutively from a transfected cDNA. J Cell Biol 1988; 107:447–456.
46. Gray PW, Glaister D, Chen E, Goeddel DV, Pennica D. Two interleukin 1 genes in the mouse: cloning and expression of the cDNA for murine interleukin 1 beta. J Immunol 1986; 137:3644–3648.
47. Rubartelli A, Cozzolino F, Talio M, Sitia R. A novel pathway for interleukin-1β, a protein lacking a signal sequence. EMBO J 1990; 9:1503–1510.
48. Le J, Weinstein D, Gubler U, Vilcek J. Induction of membrane-associated interleukin 1 by tumor necrosis factor in human fibroblasts. J Immunol 1987; 138:2137–2142.
49. Johnson WJ, Breton J, Newman-Tarr T, Connor JR, Meunier PC, Dalton BJ. Interleukin-1 release by rat synovial cells is dependent on the sequential treatment with γ-interferon and lipopolysaccharide. Arthritis Rheum 1990; 33:261–267.
50. Yamato K, El-Hajjaoui Z, Koeffler HP. Regulation of levels of IL-1 mRNA in human fibroblasts. J Cell Physiol 1989; 139:610–616.
51. Larsson E-L. Cyclosporin A and dexamethasone suppress T cell responses by selectively acting at distinct sites of the triggering process. J Immunol 1980; 124:2828–2833.
52. Kovacs EJ, Radzioch D, Young HA, Varesio L. Differential inhibition of IL-1 and TNF-α mRNA expression by agents which block second messenger pathways in murine macrophages. J Immunol 1988; 141:3101–3105.
53. Caput D, Beutler B, Hartog K, Thayer R, Brown-Shimer S, Cerami A. Identification of a common nucleotide sequence in the 3′-untranslated region of mRNA molecules specifying inflammatory mediators. Proc Natl Acad Sci USA 1986; 83:1670–1674.
54. Lee SW, Tsou A, Chan H, Thomas J, Petrie K, Eugui EM, Allison AC. Glucocorticoids selectively inhibit the translation of the interleukin 1β gene and decrease the stability of interleukin 1β mRNA. Proc Natl Acad Sci USA 1988; 85:1204–1208.
55. Dijkmans R, Van Dame J, Cornette F, Heremans H, Billiau A. Bacterial lipopolysaccharide potentiates gamma interferon-induced cytotoxicity for normal mouse and rat fibroblasts. Infect Immun 1990; 58:32–36.
56. Fuhlbrigge RC, Fine SM, Unane ER, Chaplin DD. Expression of membrane interleukin 1 by fibroblasts transfected with murine pro-interleukin 1 cDNA. Proc Natl Acad Sci USA 1988; 85:5649–5653.
57. Katz Y, Strunk RC. IL-1 and tumor necrosis factor: similarities and differences in stimulation of expression of alternative

pathway of complement and IFN-β/IL-6 genes in human fibroblasts. J Immunol 1989; 142:3862–3867.

58. Dinarello CA. Interleukin-1: amino acid sequences, multiple biological activities and comparison with tumor necrosis factor (cachectin). Year Immunol 1986; 2:68–89.

59. Beutler BA, Cerami A. Recombinant interleukin 1 suppressed lipoprotein lipase activity in 3T3-L1 cells. J Immunol 1985; 135:3969–3975.

60. Taylor DJ. Interleukin-4 (IL-4) induces down-modulation and shedding of the p55 tumour necrosis factor receptor and inhibits TNF alpha's effect on rheumatoid synovial fibroblasts. Rheumatol Int 1994; 14:21–25.

61. Katz Y, Strunk RC. Enhanced synthesis of factor B induced by tumor necrosis factor in human skin fibroblasts is decreased by IL-4. J Immunol 1990; 144:4675–4680.

62. Mehindate K, al-Daccak R, Schall TJ, Mourad W. Induction of chemokine gene expression by major histocompatibility complex class II ligands in human fibroblast-like synoviocytes. Differential regulation by interleukin-4 and dexamethasone. J Biol chem 1994; 269:32063–32069.

63. Monroe JG, Haldar H, Prystowsky MB, Lammie P. Lymphokine regulation of inflammatory processes; interleukin-4 stimulates fibroblast proliferation. Clin Immunol Immunopathol 1988; 49:292–298.

64. Postlethwaite AE, Holness MA, Katai H, Raghow R. Human fibroblasts synthesize elevated levels of extracellular matrix proteins in response to interleukin-4. J Clin Invest 1992; 90:1479–1485.

65. Feghali CA, Bost KL, Boulware DW, Levy LS. Human recombinant interleukin-4 induces proliferation and interleukin-6 production by cultured human skin fibroblasts. Clin Immunol Immunopathol 1992; 63:182–187.

66. Piela-Smith TH, Broketa G, Hand A, Korn JH. Regulation of ICAM-1 expression and function in human dermal fibroblasts by IL-4. J Immunol 1992; 148:1375–1381.

67. Sempowski GD, Beckmann MP, Derdak S, Phipps RP. Subsets of murine lungs fibroblasts express membrane-bound and soluble IL-4 receptors. Role of IL-4 in enhancing fibroblast proliferation and collagen synthesis. J Immunol 1994; 152:3606–3614.

68. Schlaak JF, Schwarting A, Knolle P, Meyer zum Buschenfelde KH, Mayet W. Effects of Th1 and Th2 cytokines on cytokine production and ICAM-1 expression on synovial fibroblasts. Ann Rheum Dis 1995; 54:560–565.

69. Endo T, Ogushi F, Sone S, Ogura T, Taketani Y, Hayashi Y, Ueda N, Yamamoto S. Induction of cyclooxygenase-2 is responsible for interleukin-1 beta-dependent prostaglandin E2 synthesis by human lung fibroblasts. Am J Respir Cell Mol Biol 1995; 12:358–365.

70. Akashi M, Loussararian AH, Adelman DC, Salto M, Koeffler HP. Role of lymphotoxin in expression of interleukin 6 in human fibroblasts: stimulation and regulation. J Clin Invest 1990; 85:121–129.

71. Zhang YH, Lin JX, Yip YK, Vilcek J. Enhancement of cAMP levels of protein kinase activity by tumor necrosis factor and interleukin 1 in human fibroblasts: role in the induction of interleukin 6. Proc Natl Acad Sci USA 1989; 85:6802–6805.

72. Kohase M, Henriksen-DeStefano D, May LT, Vilcek J, Sehgal PB. Induction of β2 interferon by tumor necrosis factor: a homeostatic mechanism in the control of cell proliferation. Cell 1986; 45:659–666.

73. Denis M. Interleukin-6 in mouse hypersensitivity pneumonitis: changes in lung free cells following depletion of endogenous IL-6 or direct administration of IL-6. J Leuk Biol 1992; 52:197–201.

74. Mateo RB, Reichner JS, Albina JE. Interleukin-6 activity in wounds. Am J Physiol 1994; 266:R1840–R1844.

75. Fries KM, Felch ME, Phipps RP. Interleukin-6 is an autocrine growth factor for murine lung fibroblast subsets. Am J Respir Cell Mol Biol 1994; 11:552–560.

76. Kohase M, May LT, Tamm I, Vilcek J, Sehgal PB. A cytokine network in human diploid fibroblasts: interactions of β-interferons, tumor necrosis factor, platelet-derived growth factor, and interleukin-1. Mol Cell Biol 1987; 7:273–280.

77. May LT, Helfgott DC, Sehgal PB. Anti-beta-interferon antibodies inhibit the increased expression of HLA-B7 mRNA in tumor necrosis factor-treated human fibroblasts: structural studies of the beta 2 interferon involved. Proc Natl Acad Sci USA 1986; 83:8957–8961.

78. Katz Y, Revel M, Strunk RC. Interleukin 6 stimulates synthesis of complement proteins factor B and C3 in human fibroblasts. Eur J Immunol 1989; 19:983–988.

79. Sato T, Ito A, Mori Y. Interleukin 6 enhances the production of tissue inhibitor of metalloproteinases (TIMP) but not that of matrix metalloproteinases of human fibroblasts. Biochem Biophys Res Commun 1990; 170:824–829.

80. Samad F, Berftrom G, Amrani DL. Regulation of plasminogen activation by interleukin-6 in human lung fibroblasts. Biochim Biophys Acta 1994; 1221:307–314.

81. Taga T, Kawanishi K, Hardy RR, Hirano T, Kishimoto T. Receptor for B cell stimulatory factor 2 (BSF2): quantitative, specificity, distribution and regulation of the expression. J Exp Med 1987; 166:967–981.

82. Heinrich PC, Castell JV, Andus T. Interleukin-6 and the acute phase response. Biochem J 1990; 265:621–636.

83. Akira S, Hirano T, Taga T, Kishimoto T. Biology of multifunctional cytokines: IL 6 and related molecules (IL 1 and TNF). FASEB J 1990; 4:2860–2867.

84. Zhang Y, Lin J-X, Vilcek J. Synthesis of interleukin 6 (interferon-β2/B-cell stimulatory factor 2) in human fibroblasts is triggered by an increase in intracellular cyclic AMP. J Biol Chem 1988; 263:6177–6182.

85. Sehgal PB, Walther Z, Tamm I. Rapid enhancement of β2 interferon/B-cell differentiation factor BSF-2 gene expression in human fibroblasts by diacylglycerols and the calcium ionophore A23187. Proc Natl Acad Sci USA 1987; 84:3663–3667.

86. Dennis GJ, Mond JJ. Corticosteroid-induced suppression of murine B cell immune response antigens. J Immunol 1986; 136:1600–1604.

87. Laulederkind SJ, Bielawska A, Raghow R, Hannun YA, Bollou LR. Ceramide induces interleukin 6 gene expression in human fibroblasts. J Exp Med 1995; 182:599–604.

88. Ray A, Tatter SB, May LT, Sehgal PB. Activation of human beta2-interferon/hepatocyte-stimulating factor/interleukin 6 promoter by cytokines, viruses, and second messenger agonists. Proc Natl Acad Sci USA 1988; 85:6701–6705.

89. Ray A, Sassone-Corsi P, Sehgal PB. A multiple cytokine- and second messenger-responsive element in the enhancer of the human interleukin-6 gene: similarities with c-fos gene regulation. Mol Cell Biol 1989; 9:5537–5547.

90. MacDonald D, Adams JA, McCarthy D, Barrett AJ. Interleukin-2 inhibits growth of fibroblasts derived from human bone marrow. Acta Haemat 1990; 83:26–30.

91. Elias JA, Freundlich B, Kern JA, Rosenbloom J. Cytokine networks in the regulation of inflammation and fibrosis in the lung. Chest 1990; 97:1439–1445.

92. Elias JA, Jimenez SA, Freundlich B. Recombinant gamma, alpha, and beta interferon regulation of human lung fibroblast proliferation. Am Rev Respir Dis 1987; 135:62–65.

93. Matsushima K, Oppenheim JJ. Interleukin 8 and MCAF: novel

inflammatory cytokines inducible by IL 1 and TNF. Cytokine 1989; 1:2–13.
94. Rollins BJ, Morrison ED, Stiles CD. Cloning and expression of JE, a gene inducible by platelet-derived growth factor and whose product has cytokine-like properties. Proc Natl Acad Sci USA 1988; 85:3738–3742.
95. Furutani Y, Nomura H, Notake M, Oyamada Y, Fukui T, Yamada M, Larsen CG, Oppenheim JJ, Matsushima K. Cloning and sequencing of the cDNA for human monocyte chemotactic and activating factor (MCAF). Biochem Biophys Res Commun 1989; 159:249–255.
96. Zhang K, Gharaee-Kermani M, Jones ML, Warren JS, Phan SH. Monocyte chemoattractant protein-1 gene expression in bleomycin-induced pulmonary fibrosis. J Immunol 1994; 153:4733–4741.
97. Rathanaswami P, Hachicha M, Sadick M, Schall TJ, McColl SR. Expression of the cytokine RANTES in human rheumatoid synovial fibroblasts. Differential regulation of RANTES and interleukin-8 genes by inflammatory cytokines. J Biol Chem 1993; 268:5834–5839.
98. Schmitz B, Thiele J, Witte O, Kaufmann R, Wickenhauser C, Fischer R. Influence of cytokines (IL-1 alpha, IL-3, IL-11, GM-CSF) on megakaryocyte-fibroblast interactions in normal human bone marrow. Eur J Haematol 1995; 5524–5532.
99. Moore KW, O'Garra A, de Waal Malefyt R, Vieira P, Mosmann TR. Interleukin-10. Annu Rev Immunol 1993; 11:165–190.
100. Seitz M, Loetscher P, Dewald B, Towbin H, Gallati H, Baggiolini M. Interleukin-10 differentially regulates cytokine inhibitor and chemokine release from blood mononuclear cells and fibroblasts. Eur J Immunol 1995; 25:1129–1132.
101. Reitamo S, Remitz A, Tamai K, Uitto J. Interleukin-10 modulates type I collagen and matrix metalloprotease gene expression in cultured human skin fibroblasts. J Clin Invest 1994; 94:2489–2492.
102. Weber-Nordt RM, Meraz MA, Schreiber RD. Lipopolysaccharide-dependent induction of IL-10 receptor expression on murine fibroblasts. J Immunol 1994; 153:3734–3744.
103. Katz Y, Stav D, Barr J, Passwell JH. IL-13 results in differential regulation of the complement proteins C3 and factor B in tumour necrosis factor (TNF)-stimulated fibroblasts. Clin Exp Immunol 1995; 101:150–156.
104. Carswell EA, Old LJ, Kassel RL, Green S, Fiore N, Williamson B. An endotoxin-induced serum factor that causes necrosis of tumors. Proc Natl Acad Sci USA 1975; 72:3666–3670.
105. Old LJ. Tumor necrosis factor (TNF). Science 1985; 230:630–632.
106. Goeddel DV, Aggarwal BB, Gray PW, Leung DW, Nedwin GE, Palladino MA, Patton JS, Pennica D, Shepard HM, Sugarman BJ, Wong GHW. Tumor necrosis factors: gene structure and biological activities. Cold Spring Harbor Symp Quant Biol 1986; 51:597–609.
107. Nedospasov SA, Turetskaya RL, Shakhov AN. Cloning and characterization of the rabbit lymphotoxin (TNF-β) gene and its tight linkage to the gene coding for tumor necrosis factor. Nucleic Acids Res 1988; 16:3112.
108. Collins T, Lapierre LA, Fiers W, Strominger JL, Pober JS. Recombinant human tumor necrosis factor increases mRNA levels and surface expression of HLA-A, B antigens in vascular endothelial cells and dermal fibroblasts in vitro. Proc Natl Acad Sci USA 1986; 83:446–450.
109. Rosenblum MG, Donato NJ, Gutterman JU. Characterization of human recombinant tumor necrosis factor-α antiproliferative effects on human cells in culture. Lymphokine Res 1988; 7:107–117.
110. Stauber GB, Aggarwal BB. Characterization and affinity cross-linking of receptors for human recombinant lymphotoxin (tumor necrosis factor-β) on a human histiocytic lymphoma cell line, U-937. J Biol Chem 1989; 264:3573–3576.
111. Vilcek J, Lee TH. Tumor necrosis factor: new insights into the molecular mechanisms of its multiple actions. J Biol Chem 1991; 266:7313–7316.
112. Aggarwal BB, Moffat B, Harkins RN. Human lymphotoxins: production by a lymphoblastoid cell line, purification and initial characterization. J Biol Chem 1984; 259:686–691.
113. Tartaglia LA, Goeddel DV. Two TNF receptors. Immunol Today 1992; 13:151–153.
114. Banner DW, D'Arcy A, Janes W, Gentz R, Schoenfeld H-J, Broger C, Loetscher H, Lesslauer W. Crystal structure of soluble human 55 kD TNF receptors-human TNF complex: implications for TNF receptor activation. Cell 1993; 73:431–445.
115. Tartaglia LA, Goeddel DV, Reynolds C, Figari IS, Weber RF, Fendly BM, Palladino MA Jr. Stimulation of human T-cell proliferation by specific activation of the 75-kDa tumor necrosis factor receptor. J Immunol 1993; 151:4637–4641.
116. Butler DM, Feldman M, Di Padova F, Brennan FM. p55 and p75 tumor necrosis receptors are expressed and mediate common functions in synovial fibroblasts and other fibroblasts. Eur Cytokine Netw 1994; 5:441–448.
117. Scharffetter K, Heckman M, Hatamochi A, Mauch C, Stein B, Reithmuller G, Zeigler-Heitbrock HW, Krieg T. Synergistic effect of tumor necrosis factor-alpha and interferon-gamma on collagen synthesis of human skin fibroblasts in vitro. Exp Cell Res 1989; 181:409–419.
118. Diaz A, Munoz E, Johnston R, Korn JH, Jimenez SA. Regulation of human lung fibroblast 1 (I) procollagen gene expression by tumor necrosis factor alpha, interleukin-1 beta and prostaglandin E2. J Biol Chem 1993; 268:10364–10371.
119. Elias JA, Freunlich B, Adams S, Rosenbloom J. Regulation of human lung fibroblast collagen production by recombinant interleukin-1, tumor necrosis factor, and interferon-gamma. Ann NY Acad Sci 1990; 580:233–244.
120. Hattori A, Iwasaki S, Murase K, Tsujimoto M, Sato M, Hayashi K, Kohno M. Tumor necrosis factor is markedly synergistic with interleukin 1 and interferon-γ in stimulating the production of nerve growth factor in fibroblasts. FEBS Lett 1994; 340:177–180.
121. Tessier P, Audette M, Cattaruzzi P, McColl SR. Up-regulation by tumor necrosis factor alpha of intercellular adhesion molecule 1 expression and function in synovial fibroblasts and its inhibition by glucocorticoids. Arthritis Rheum 1993; 36:1528–1539.
122. Horikoshi T, Ezoe K, Nakagawa H, Eguchi H, Hanada N, Hamaoka S. Up-regulation of ICAM-1 expression on human dermal fibroblasts by IFN-γ in the presence of TNF-α. FEBS Lett 1995; 363:141–144.
123. Ezoe K, Horikoshi T. Tumor necrosis factor-alpha increased the integrin alpha 2 beta expression and cell attachment to type I collagen in human dermal fibroblasts. Biochem Biophys Res Commun 1993; 192:281–287.
124. Gearing DP. The leukemia inhibitory factor and its receptor. Adv Immunol 1993; 53:31–58.
125. Hamilton JA, Waring PM, Filonzi EL. Induction of leukemia inhibitory factor in human synovial fibroblasts by IL-1 and tumor necrosis factor-alpha. J Immunol 1993; 150:1496–1502.
126. Elias JA, Zheng T, Whiting NL, Marcovici A, Trow TK. Cytokine-cytokine synergy and protein kinase C in the regulation of lung fibroblast leukemia inhibitory factor. Am J Physiol 1994; 266:L426–L435.
127. Lorenzo JA, Jastrzebski SL, Kalinowski JF, Downie E, Korn JH. Tumor necrosis factor alpha stimulates production of

leukemia inhibitory factor in human dermal fibroblast culture. Clin Immunol Immunopathol 1994; 70:260–265.
128. Kurt-Jones EA, Fiers W, Pober JS. Membrane interleukin-1 induction on human endothelial cells and human dermal fibroblasts. J Immunol 1987; 139:2317–2324.
129. Dijkmans R, Van Damme J, Cornette F, Heremans H, Billiau A. Bacterial lipopolysaccharide potentiates gamma interferon-induced cytotoxicity for normal mouse and rat fibroblasts. Infect Immun 1990; 58:32–36.
130. Kohno M, Nishizawa N, Tsujimoto M, Nomoto H. Mitogenic signalling pathway of tumor necrosis factor involves the rapid tyrosine phosphorylation of 41,000 and 43,000-Mr cytosol proteins. Biochem J 1990; 267:91–98.
131. Brach MA, Gruss HJ, Scott C, Herrmann F. The mitogenic response to tumor nencrosis factor alpha requires c-Jun/AP-1. Mol Cell Biol 1993; 13:4284–4290.
132. Westwick JK, Weitzel C, Minden A, Karin M, Brenner DA. Tumor necrosis factor alpha stimulates AP-1 activity through prolonged activation of the c-Jun kinase. J Biol Chem 1994; 269:26396–26401.
133. Bird TA, Saklatvala J. Down-modulation of epidermal growth factor affinity in fibroblasts treated with interleukin 1 or tumor necrosis factor is associated with phosphorylation at a site other than threonine 654. J Biol Chem 1990; 265:235–240.
134. Schutze S, Nottrott S, Pfizenmaier K, Kronke M. Tumor necrosis factor signal transduction. Cell-type-specific activation and translocation of protein kinase C. J Immunol 1990; 144:2604–2608.
135. Schroder JM, Stichering M, Henneicke HH, Preissner WC, Christophers E. IL-1 alpha or tumor necrosis factor-alpha stimulate release of three NAP-1/IL-8-related neutrophil chemotactic proteins in human dermal fibroblasts. J Immunol 1990; 144:2223–2232.
136. Postlethwaite AE, Seyer JM. Stimulation of fibroblast chemotaxis by human recombinant tumor necrosis factor alpha (TNF-alpha) and a synthetic TNF-alpha 31-68 peptide. J Exp Med 1990; 172:1749–1756.
137. Schirren CG, Scharffetter K, Hein R, Braun-Falco O, Krieg T. Tumor necrosis factor alpha induces invasiveness of human skin fibroblasts in vitro. J Invest Dermatol 1990; 94:706–710.
138. Lee TH, Lee GW, Ziff EB, Vilcek J. Isolation and characterization of eight tumor necrosis factor-induced gene sequences from human fibroblasts. Mol Cell Biol 1990; 10:1982–1988.
139. Hori T, Kashiyama S, Hayakawa M, Shibamoto S, Tsujimoto M, Oku N, Ito F. Tumor necrosis factor is cytotoxic to human fibroblasts in the presence of exogenous arachidonic acid. Exp Cell Res 1989; 185:41–49.
140. Elias JA, Gustilo K, Baeder W, Freundlich B. Synergistic stimulation of fibroblast prostaglandin production by recombinant interleukin-1 and tumor necrosis factor. J Immunol 1987; 138:3812–3816.
141. Barnard JA, Graves-Deal R, Pittelkow MR, DuBois R, Cook P, Ramsey GW, Bishop PR, Damstrup L, Coffey RJ. Auto- and cross-induction within the mammalian epidermal growth factor-related peptide family. J Biol Chem 1994; 269:22817–22822.
142. Carpenter G, Wahl MI. Epidermal growth factor. Handbook Exp Pharmacol 1990; 951:69–171.
143. Derynck R. Transforming growth factor-α. Mol Reprod Devel 1990; 27:3–9.
144. Gray A, Dull TJ, Ullrich A. Nucleotide sequence of epidermal growth factor cDNA predicts a 128,000-molecular weight protein precursor. Nature 1983; 303:722–725.
145. Mroczkowski B, Reich M, Chen K, Bell GI, Cohen S. Recombinant human epidermal growth factor precursor is a glycosylated membrane protein with biological activity. Mol Cell Biol 1989; 9:2771–2778.
146. Wong DTW, Weller PF, Galli SJ, Elovic A, Rand TH, Gallagher GT, Chiang T, Chou MY, Matossian K, McBride J, Todd R. Human eosinophils express transforming growth factor α. J Exp Med 1990; 172:673–681.
147. Decker SJ. Epidermal growth factor and transforming growth factor-alpha induce differential processing of the epidermal growth factor receptor. Biochem Biophys Res Commun 1990; 166:615–621.
148. Winkler ME, O'Conner L, Winget M, Fendly B. Epidermal growth factor and transforming growth factor α bind differently to the epidermal growth factor receptor. Biochemistry 1989; 28:6373–6378.
149. Ushiro H, Cohen S. Identification of phosphotyrosine as a product of epidermal growth factor activated protein kinase in A-431 cell membranes. J Biol Chem 1980; 255:8363–8365.
150. Carpenter G, Cohen S. Epidermal growth factor. J Biol Chem 1990; 265:7709–7712.
151. Ullrich A, Coussens L, Hayflick JS, Dull TJ, Gray A, Tam AW, Lee J, Yarden Y, Libermann TA, Schlessinger J, Downward J, Mayes ELV, Whittle N, Waterfield MD, Seeburg PH. Human epidermal growth factor receptor cDNA sequence and aberrant expression of the amplified gene in A431 epidermoid carcinoma cells. Nature 1984; 309:418–424.
152. Margolis BL, Lax I, Dombalagian M, Honegger AM, Howk R, Givol D, Ullrich A, Schlessinger J. All autophosphorylation sites of epidermal growth factor (EGF) receptor and HER2/neu are located in their carboxyl-terminal tails. Identification of a novel site in EGF receptor. J Biol Chem 1989; 264:10667–10671.
153. Walton GM, Chen WS, Rosenfeld MG, Gill GN. Analysis of deletions of the epidermal growth factor receptor reveals self-phosphorylation at tyrosine 992 and enhanced in vivo tyrosine phosphorylation of cell substrates. J Biol Chem 1990; 265:1750–1754.
154. Heisermann GJ, Gill GN. Epidermal growth factor receptor threonine and serine residues phosphorylated in vivo. J Biol Chem 1988; 264:13152–13158.
155. Chen WS, Lazar CS, Poenie M, Tsien Y, Gill GN, Rosenfeld MG. Requirement of intrinsic protein tyrosine kinase in the immediate and late actions of the EGF receptor. Nature 1987; 328:820–823.
156. Bennett NT, Schultz GS. Growth factors and wound healing: biochemical properties of growth factors and their receptors. Am J Surg 1993; 165:728–737.
157. Carpenter G, Wahl MI. The epidermal growth factor family. In: Sporn MB, Roberts AB, eds. Handbook of Experimental Pharmacology: Peptide Growth Factors and Their Receptors. Vol. 95. Berlin: Springer-Verlag, 1990:70–173.
158. Rhee SG, Pann-Ghill S, Sung-Ho R, Lee SY. Studies of inositol phospholipid-specific phospholipase C. Science 1989; 244:546–550.
159. Whitman M, Cantley L. Phosphoinositide metabolism and the control of cell proliferation. Biochim Biophys Acta 1988; 948:327–344.
160. Carpenter G, Cohen S. Human epidermal growth factor and the proliferation of human fibroblasts. J Cell Physiol 1976; 88:227–238.
161. Besterman JM, Watson SP, Cuatrecasas P. Lack, of association of epidermal growth factor-, insulin-, and serum-induced mitogenesis with stimulation of phosphoinositide degradation in BALB/c 3T3 fibroblasts. J Biol Chem 1986; 261:723–727.
162. Pessin MS, Baldassare JJ, Raben DM. Molecular species analysis of mitogen-stimulated 1,2-diglycerides in fibroblasts.

Comparison of alpha-thrombin, epidermal growth factor, and platelet-derived growth factor. J Biol Chem 1990; 265:7959–7966.
163. Wharton W, Leof E, Olashaw N, O'Keefe EJ, Pledger WJ. Mitogenic response to epidermal growth factor (EGF) modulated by platelet-derived growth factor in cultured fibroblasts. Exp Cell Res 1983; 147:443–448.
164. Oleszak E. Inhibition of mitogenic activity of PDGF, EGF, and FGF by interferon-gamma. Exp Cell Res 1988; 179:575–580.
165. Newman EL, Hyldahl L, Larsson O, Engstrom W, Rees AR. Bradykinin blocks the action of EGF, but not PDGF, on fibroblast division. FEBS Lett 1989; 251:225–229.
166. Colige A, Nusgens B, Lapiere CM. Effect of EGF on the skin fibroblasts is modulated by the extracellular matrix. Arch Dermatol Res 1988; 280 (suppl):S42–S46.
167. Adelmann-Grill BC, Wach F, Cully Z, Hein R, Krieg T. Chemotactic migration of normal dermal fibroblasts toward epidermal growth factor and its modulation by platelet-derived growth factor and transforming growth factor-beta. Eur J Cell Biol 1990; 51:322–326.
168. Massague J, Attisano L, Wrana JL. The TGF-β family and its composite receptors. Trends Cell Biol 1994; 4:172–178.
169. Derynck R. TGF-β-receptor-mediated signaling. Trends Biochem Sci 1994; 19:548–553.
170. Jakowlew SB, Dillard PJ, Sporn MB, Roberts AB. Complementary deoxyribonucleic acid cloning of a messenger ribonucleic acid encoding transforming growth factor beta 4 from chicken embryo chondrocytes. Mol Endocrinol 1988; 2:1186–1195.
171. Kondaiah P, Sands MJ, Smith JM, Fields A, Roberts AB, Sporn MB, Melton DA. Identification of a novel transforming growth factor-beta (TGF-beta 5) mRNA in Xenopus laevis. J Biol Chem 1990; 265:1089–1093.
172. Derynck R, Jarrett JA, Chen EY, Eaton DH, Bell JR, Assoian RK, Roberts AB, Sporn MB, Goeddel DV. Human transforming growth factor-beta complementary DNA sequence and expression in normal and transformed cells. Nature 1985; 316:701–705.
173. Barton DE, Foellmer BE, Du J, Tamm J, Derynck R, Francke U. Chromosomal mapping of genes for transforming growth factor beta 2 and beta 3 in man and mouse: dispersion of TGF-beta gene family. Oncogene Res 1988; 3:323–331.
174. Bascom CC, Wolfshohl RJ, Coffey J, Madisen L, Webb NR, Purchio AR, Derynck R, Moses HL. Complex regulation of transforming growth factor β1, β2, and β3 mRNA expression in mouse fibroblasts and keratinocytes by transforming growth factors β1 and β2. Mol Cell Biol 1989; 9:5508–5515.
175. Lawrence DA, Pircher R, Krycene-Martinerie C, Jullien P. Normal embryo fibroblasts release transforming growth factors in a latent form. J Cell Physiol 1984; 121:184–188.
176. Sato Y, Rifkin DB. Inhibition of endothelial cell movement by pericytes and smooth muscle cells; activation of a latent transforming growth factor-beta-like molecule by plasmin during co-culture. J Cell Biol 1989; 109:309–315.
177. Derynck R, Lindquist PB, Lee A, Wen D, Tamm J, Graycar JL, Rhee L, Mason AJ, Miller DA, Coffey RJ, Moses HL, Chen EY. A new type of transforming growth factor-beta, TGF-beta 3. EMBO J 1988; 7:3737–3743.
178. Madisen L, Lioubin MN, Farrand AL, Brunner AM, Purchio AF. Analysis of proteolytic cleavage of recombinant TGF-β1. Production of hybrid molecules with increased processing efficiency. Ann NY Acad Sci 1990; 593:7–25.
179. Roberts AB, Kim S-J, Kondaiah P, Jakowlew SB, Denhez F, Glick AB, Geiser AG, Watanabe S, Noma T, Lechleider R, Sporn MB. Transcriptional control of expression of the TGF-s. Ann NY Acad Sci 1990; 593:43–50.
180. Kingsley D. The TGF-β superfamily: new members, new receptors, and new genetic tests of function in different organisms. Genes Dev 1994; 8:133–146.
181. McDonald NQ, Hendrickson WA. A structural superfamily of growth factors containing a cystine knot motif. Cell 1993; 73:421–424.
182. Massague J. Receptors for the TGF-β family. Cell 1992; 69:1067–1070.
183. Cheifetz S, Weatherbee JA, Tsang ML-S, Anderson JK, Mole JE, Lucas R, Massague J. The transforming growth factor-beta system, a complex pattern of cross-reactive ligands and receptors. Cell 1987; 48:409–415.
184. Segarini PR, Seyedin SM. The high molecular weight receptor to transforming growth factor-β contains glycosaminoglycan chains. J Biol Chem 1988; 263:8366–8370.
185. Barnard JA, Russette ML, Moses HL. The biology of transforming growth factor β. Biochim Biophys Acta 1990; 1032:79–87.
186. Massague J. Subunit structure of a high-affinity receptor for type β-transforming growth factor: evidence for a disulfide-linked glycosylated receptor complex. J Biol Chem 1985; 260:7059–7066.
187. Wrana JL, Attisano L, Carcamo J, Zentella A, Doody J, Laiho M, Wang X-F, Massague J. TGFβ signals through a heteromeric protein kinase receptor complex. Cell 1992; 71:1003–1014.
188. Wakefield LM, Smith DM, Masui T, Harris CC, Sporn MB. Distribution and modulation of the cellular receptor for transforming growth factor-beta. J Cell Biol 1987; 105:965–975.
189. Rizzino A, Kazakoff P, Ruff E, Kuszynski C, Nebelsick J. Regulatory effects of cell density on the binding of transforming growth factor β, epidermal growth factor, platelet-derived growth factor, and fibroblast growth factor. Can Res 1988; 48:4266–4271.
190. Murthy US, Anzano MA, Stadel JM, Grieg R. Coupling of TGF-beta-induced mitogenesis to G-protein activation in AKR-2B cells. Biochem Biophys Res Commun 1988; 152:1228–1235.
191. Howe PH, Cunningham MR, Leof EB. Distinct pathways regulate transforming growth factor beta 1-stimulated proto-oncogene and extracellular matrix gene expression. J Cell Physiol 1990; 142:39–45.
192. Chambard JC, Pouyssegur J. TGF-beta inhibits growth factor-induced DNA synthesis in hamster fibroblasts without affecting the early mitogenic events. J Cell Physiol 1988; 135:101–107.
193. Moses HL, Branum EB, Proper JA, Robinson RA. Transforming growth factor production by chemically transformed cells. Can Res 1981; 41:2842–2848.
194. Roberts AB, Kim SJ, Sporn M. Is there a common pathway mediating growth inhibition by TGF-β and the retinoblastoma gene product? Cancer Cells 1991; 3:19–21.
195. Loef EB, Proper JA, Goustin AS, Shipley GP, Dicorleto PE, Moses HL. Induction of c-*sis* mRNA and activity similar to platelet-derived growth factor by transforming growth factor β; a proposed model for indirect mitogenesis involving autocrine activity. Proc Natl Acad Sci USA 1986; 83:2453–2457.
196. Sporn MB, Roberts AB, Wakefield LM, de Crombrugghe B. Some recent advances in the chemistry and biology of transforming growth factor-beta. J Cell Biol 1987; 105:1039–1045.
197. Ignotz RA, Massague J. Cell adhesion protein receptors as targets for transforming growth factor-beta action. Cell 1987; 51:189–197.
198. Overall CM, Wrana JL, Sodek J. Transforming growth factor-

beta regulation of collagenase, 72 kDA-progelatinase, TIMP and PAI-1 expression in rat bone cell populations and human fibroblasts. Connect Tissue Res 1989; 20:289–294.

199. Edwards DR, Murphy G, Reynolds JJ, et al. Transforming growth factor-beta modulates the expression of collagenase and metalloproteinase inhibitor. EMBO J 1987; 6:1899–1904.
200. Rossi P, Karsenty G, Roberts AB, Roche NS, Sporn MB, de Crombrugghe B. A nuclear factor 1 binding site mediates the transcriptional activation of a type I collagen promoter by transforming growth factor-beta. Cell 1988; 52:405–414.
201. Raghow R, Postlethwaite AE, Keski-Oja J, Moses HL, Kang AH. Transforming growth factor-beta increases steady state levels of type I procollagen and fibronectin messenger RNAs posttranscriptionally in cultured human dermal fibroblasts. J Clin Invest 1987; 79:1285–1288.
202. Diaz A, Varga J, Jimenez SA. Transforming growth factor-beta stimulation of lung fibroblast prostaglandin E2 production. J Biol Chem 1989; 264:11554–11557.
203. Fine A, Poliks CF, Donahue LP, Smith BD, Goldstein RH. The differential effect of prostaglandin E2 on transforming growth factor-beta and insulin-induced collagen formation in lung fibroblasts. J Biol Chem 1989; 264:16988–16991.
204. Ishikawa O, Yamakage A, LeRoy EC, Trojanowska M. Persistent effect of TGF-beta 1 on extracellular matrix gene expression in human dermal fibroblasts. Biochem Biophys Res Commun 1990; 169:232–238.
205. Zhang K, Rekhter MD, Gordon D, Phan SH. Co-Expression of α-smooth muscle actin and type I collagen in fibroblast-like cells of rat lungs with bleomycin-induced pulmonary fibrosis: a combined immuno-histochemical and in situ hybridization study. Am J Pathol 1994; 145:114–125.
206. Schmitt-Graff A, Desmouliere A, Gabbiani G. Heterogeneity of myofibroblast phenotypic features: an example of fibroblastic cell plasticity. Virchows Arch 1994; 425:3–24.
207. Montesano R, Orci L. Transforming growth factor-β stimulates collagen-matrix contraction by fibroblasts: implications for wound healing. Proc Natl Acad Sci USA 1988; 85:4894–4897.
208. Desmouliere A, Genoz A, Gabbiani F, Gabbiani G. Transforming growth factor-β1 induces α-smooth muscle actin expression in granulation tissue myofibroblasts and in quiescent and growing cultured fibroblasts. J Cell Biol 1993; 122:103–111.
209. Zhang H, Gharaee-Kermani M, Zhang K, Phan SH. Lung fibroblast contractile and α-smooth muscle actin phenotypic alterations in bleomycin-induced pulmonary fibrosis. Am J Pathol 1996; 148:527–537.
210. Evans JN, Kelley J, Low RB, Adler KB. Increased contractility of isolated lung parenchyma in an animal model of pulmonary fibrosis induced by bleomycin. Am Rev Respir Dis 1982; 125:89–94.
211. Postlethwaite AE, Keski-Oja J, Moses HL, Kang AH. Stimulation of the chemotactic migration of human fibroblasts by transforming growth factor beta. J Exp Med 1987; 165:251–256.
212. Wahl SM, Hunt DA, Wakefield LM, McCartney-Francis N, Wahl LM, Roberts AB, Sporn MB. Transforming growth factor beta induces monocyte chemotaxis and growth factor production. Proc Natl Acad Sci USA 1987; 84:5788–5792.
213. Pierce GF, Mustoe TA, Lingelbach J, Masakowski VR, Griffin GL, Senior RM, Deuel TF. Platelet-derived growth factor and transforming growth factor-beta enhance tissue repair activities by unique mechanisms. J Cell Biol 1989; 109:429–440.
214. Allen JB, Manthey CL, Hand AR, Ohura K, Ellingsworth L, Wahl SM. Rapid onset synovial inflammation and hyperplasia induced by transforming growth factor beta. J Exp Med 1990; 171:231–247.
215. Kelley J, Shull S, Walsh JJ, Cutroneo KR, Absher M, Baldor L. Auto-induction of transforming growth factor-β in human lung fibroblasts. Am J Respir Cell Mol Biol 1993; 8:417–424.
216. Zhang K, Flanders KC, Phan SH. Cellular localization of transforming growth factor β expression in bleomycin-induced pulmonary fibrosis. Am J Pathol 1995; 147:352–361.
217. Needleman BW, Choi J, Burrows-Mezu A, Fontana JA. Secretion and binding of transforming growth factor beta by scleroderma and normal dermal fibroblasts. Arthritis Rheum 1990; 33:650–656.
218. Breen E, Shull S, Burne S, Absher M, Kelley J, Phan SH, Cutroneo K. Bleomycin regulation of TGF-β mRNA in rat lung fibroblasts. Am J Resp Cell Mol Biol 1992; 6:146–152.
219. Heldin C-H, Wasteson A, Westermark B. Platelet-derived growth factor. Mol Cell Endocrinol 1985; 39:169–187.
220. Ross R, Raines EW, Bowen-Pope DF. The biology of platelet-derived growth factor. Cell 1986; 46:155–169.
221. Waterfield MD, Scrace GT, Whittle N, Stroobant P, Johnsson H, Wasteson A, Westermark B, Heldin C-H, Huang S, Deuel TF. Platelet-derived growth factor is structurally related to the putative transforming protein p28 sis of simian sarcoma virus. Nature 1983; 304:35–39.
222. Nister M, Hammacher A, Mellstrom K, Siegbahn A, Ronnstrand L, Westermark B, Heldin C-H. A glioma-derived PDGF A chain homodimer has different functional activities from a PDGF AB heterodimer purified from human platelets. Cell 1988; 52:791–799.
223. Dvonch VM, Murphey RJ, Matsuoka J, Grotendorst GF. Changes in growth factor levels in human wound fluid. Proc Natl Acad Sci USA 1989; 86:4416–4420.
224. Kazlauskas A, Bowen-Pope D, Siefert R, Hart CE, Cooper JE. Different effects of homo- and heterodimers of platelet-derived growth factor A and B chains on human and mouse fibroblasts. EMBO J 1988; 12:3727–3735.
225. Lepisto J, Kujari H, Niinikoski J, Laato M. Effects of heterodimeric isoforms of platelet-derived growth factor PDGF-AB on wound healing in the rat. Eur Surg Res 1994; 26:267–272.
226. Pierce GF, Tarpley JE, Allman RM, Goode PS, Serdar CM, Morris B, Mustoe TA, Berg JV. Tissue repair processes in healing chronic pressure ulcers treated with recombinant platelet-derived growth factor BB. Am J Pathol 1994; 145:1399–1410.
227. Murali S, Hardten DR, Demartelaere S, Olevsky OM, Mindrup EA, Hecht ML, Karlstad R, Chan CC, Holland EJ. Effect of topically administered platellet-derived growth factor on corneal wound strength. Curr Eye Res 1994; 13:857–862.
228. Hammacher A, Nister M, Heldin C-H. The A-type receptor for platelet-derived growth factor mediates protein tyrosine phosphorylation, receptor transmodulation and mitogenic response. Biochem J 1989; 264:15–20.
229. Morrison DK, Kaplan DR, Escobedo JA, Rapp UR, Roberts TM, Williams LT. Direct activation of the serine/threonine kinase activity of Raf-1 through tyrosine phosphorylation by the PDGF beta-receptor. Cell 1989; 58:649–657.
230. Ostman A, Backstrom G, Fong N, Betsholtz C, Wernstedt C, Hellman U, Westermark B, Valenzuela P, Heldin CH. Expression of three recombinant homodimeric isoforms of PDGF in Saccharomyces cerevisiae: evidence for difference in receptor binding and functional activities. Growth Factors 1989; 1:271–281.
231. Ahlen K, Rubin K. Platelet-derived growth factor-BB stimulates synthesis of the integrin alpha (2)-subunit in human diploid fibroblasts. Exp Cell Res 1994; 215:237–353.
232. Soma Y, Takehara K, Ishibashi Y. Alteration of the chemotactic

response of human skin fibroblasts to PDGF by growth factors. Exp Cell Res 1994; 212:274–277.
233. Chiou WJ, Bonin PD, Harris PK, Carter DB, Singh JP. Platelet-derived growth factor induces interleukin-1 receptor gene expression in Balb/c 3T3 fibroblasts. J Biol Chem 1989; 264:21442–21445.
234. Paulsson Y, Bywater M, Heldin C-H, Westermark B. Effects of epidermal growth factor and platelet-derived growth factor on c-fos and c-myc mRNA levels in normal human fibroblasts. Exp Cell Res 1987; 171:186–194.
235. Stachniak JB, Sharp ZD, Pool TB, Cameron IL. Impaired proliferation response after PDGF induction in fibroblasts from Hutchinson-Guilford progeria syndrome. Cell Biol Int Rep 1988; 12:621–626.
236. Hesketh TR, Morris JD, Moore JP, Metcalfe JC. Ca^{2+} and pH responses to sequential additions of mitogens in single 3T3 fibroblasts: Correlation with DNA synthesis. J Biol Chem 1988; 263:11879–11886.
237. DiFrancesco P, Testa EP, Testa U, Liboi E. Altered growth factor sensitivity in EL2 rat fibroblasts: influence of this biological characteristic on cell growth. Eur J Cell Biol 1989; 49:196–201.
238. Davis RJ, Czech MP. Stimulation of epidermal growth factor receptor threonine 654 phosphorylation by platelet-derived growth factor in protein kinase C-deficient human fibroblasts. J Biol Chem 1987; 262:6832–6841.
239. Decker SJ, Harris P. Effects of platelet-derived growth factor on phosphorylation of the epidermal growth factor receptor in human skin fibroblasts. J Biol Chem 1989; 264:9204–9209.
240. Baird A. Fibroblast growth factors: activities and significance of non-neurotrophin neurotropic growth factors. Curr Opinion Neurobiol 1994; 4:78–86.
241. Zhang J, Cousens LS, Barr PJ, Sprang SR. Three-dimensional structure of human basic fibroblast growth factor, a structural homolog of interleukin 1. Proc Natl Acad Sci USA 1991; 88:3446–3450.
242. Bugler B, Amalric F, Prats H. Alternative initiation of translation determines cytoplasmic or nuclear localization of basic fibroblast growth factor. Mol Cell Biol 1991; 11:573–577.
243. Wiedmann B, Sakai H, Davis TA, Wiedmann A. A protein complex required for signal-sequence-specific sorting and translation. Nature 1994; 370:434–440.
244. Yoeman LC. An autocrine model for cell-and-matrix-associated fibroblast growth factor. Oncol Res 1993; 5:489–499.
245. Rapraeger AC, Krufka A, Olwin BB. Requirement of heparan sulfate for bFGF-mediated fibroblast growth and myoblast differentiation. Science 1991; 252:1705–1708.
246. Saksela O, Moscatelli D, Sommer A, Rifkin DB. Endothelial cell-derived heparan sulfate bind basic fibroblast growth factor and protects it from proteolytic degradation. J Cell Biol 1988; 107:743–751.
247. Kiefer MC, Stephans JC, Crawford K, Okino K, Barr PJ. Ligand-affinity cloning and structure of a cell surface heparan sulfate proteoglycan that binds basic fibroblast growth factor. Proc Natl Acad Sci USA 1990; 87:6985–6989.
248. Blotnick S, Peoples GE, Freeman MR, Eberlein TJ, Klagsburn M. T lymphocytes synthesis and export heparin-binding epidermal growth factor-like growth factor and basic fibroblast growth factor, mitogens for vascular cells and fibroblasts: differential production and release by CD4+ and CD8+ T cells. Proc Natl Acad Sci USA 1994; 91:2890–2894.
249. Yu W, Naim JO, Lanzafame RJ. The effects of laser irradiation on the release of bFGF from 3T3 fibroblasts. Photochem Photobiol 1994; 59:167–170.
250. Partanen J, Vainikka S, Korhonen J, Armstrong E, Alita K. Diverse receptors for fibroblast growth factors. Prog Growth Factor Res 1992; 4:69–83.
251. Givol D, Yayon A. Complexity of FGF receptors: genetic basis for structural diversity and functional specificity. FASEB J 1992; 6:1347–1354.
252. Werner S, Duan D-SR, de Vries C, Peters KG, Johnson DE, Williams LT. Differential splicing in the extracellular region of fibroblast growth factor receptor 1 generates receptor variants with different ligand-binding specificities. Mol Cell Biol 1992; 12:82–88.
253. Miki T, Bottaro DP, Fleming TP, Smith CL, Burgess WH, Chan AM-L, Aaronson SA. Determination of ligand-binding specificity by alternative slicing: two distinct growth factor receptors encoded at a single gene. Proc Natl Acad Sci USA 1992; 89:246–250.
254. Orniz DM, Yayon A, Flanafgan JG, Svahan CM, Levi E, Leder P. Heparin is required for cell-free binding of basic fibroblast growth factor to a soluble receptor and for mitogenesis in whole cells. Mol Cell Biol 1992; 12:240–247.
255. Bellot F, Crumley G, Kaplow JM, Schlessinger L, Jaye M, Dionne CA. Ligand-induced transphosphorylation between different FGF receptors. EMBO J 1991; 10:2849–2854.
256. Mohammadi M, Hanegger AM, Rotin D, Fischer R, Bellot F, Li W, Dionne CA, Rubistein M, Schlessinger JA. A tyrosine-phosphorylated carboxy-terminal peptide of the fibroblast growth factor receptor (Flg) is a binding site for the SH2 domain of phospholipase C-gamma 1. Mol Cell Biol 1991; 11:5068–5078.
257. Burgess WH, Dionne CA, Kaplow J, Mudd R, Friesel R, Zilberstein A, Schlessinger J, Jaye M. Characterization and cDNA cloning of phospholipase C-gamma, a major substrate for heparin-binding growth factor 1 (acidic fibroblast growth factor)-activated tyrosine kinase. Mol Cell Biol 1990; 10:4770–4777.
258. Cordon-Cardo C, Vlodavsky I, Haimovitz-Friedman A, Hicklin D, Fuks Z. Expression of basic fibroblast growth factor in normal human tissue. Lab Invest 1990; 63:832–840.
259. Candela M, Barker SC, Ballou LR. Fibroblast growth factor increases TNF alpha-mediated prostaglandin E2 production and TNF alpha receptor expression in human fibroblasts. Mol Cell Biochem 1993; 120:43–50.
260. Tan EM, Rouda S, Greenbaum SS, Moore JH Jr, Fox JW, Sollberg S. Acidic and basic fibroblast growth factors down-regulate collagen gene expression in keloid fibroblasts. Am J Pathol 1993; 142:463–470.
261. Gessin JC, Brown LJ, Gordon JS, Berg RA. Regulation of collagen synthesis in human dermal fibroblasts in contracted collagen gels by ascorbic acid, growth factors and inhibitors of lipid peroxidation. Exp Cell Res 1993; 206:283–290.
262. Brettell LM, McGowan SE. Basic fibroblast growth factor decreases elastin production by neonatal rat lung fibroblasts. Am J Respir Cell Mol Biol 1994; 10:306–315.
263. Kondo H, Matsuda R, Yonezawa Y. Autonomous migration of human fetal skin fibroblasts into a denuded area in a cell monolayer is mediated by basic fibroblast growth factor and collagen. In Vitro Cell Dev Biol Anim 1993; 29A:929–935.
264. Gohda E, Matsunaga T, Kataoka H, Takebe T, Yamamoto I. Induction of hepatocyte growth factor in human skin fibroblasts by epidermal growth factor, platelet-derived growth factor and fibroblast growth factor. Cytokine 1994; 6:633–640.
265. Tamura M, Arakaki N, Tsubouchi H, Takada H, Daikuhara Y. Enhancement of human hepatocyte growth factor production by interleukin-1 alpha and -1 beta and tumor necrosis

factor-alpha by fibroblasts in culture. J Biol chem 1993; 268:8140–8145.
266. Yoshida K, Gage FH. Cooperative regulation of nerve growth factor synthesis and secretion in fibroblasts and astrocytes by fibroblast growth factor and other cytokines. Brain Res 1992; 569:14–25.
267. Yoshida K, Kakihana M, Chen LS, Ong M, Baurd A, Gage FH. Cytokine regulation of nerve growth-mediated cholinergic neurotropic activity synthesized by astrocytes and fibroblasts. J Neurochem 1992; 59:919–931.
268. Aaronson SA, Bottaro DP, Miki T, Ron D, Finch PW, Fleming TP, Ahn J, Taylor WG, Rubin JS. Keratinocyte growth factor: a fibroblast growth factor family member with unusual target specificity. Ann NY Acad Sci 199; 638:62–77.
269. Werner S, Peters KG, Lomgaker MT, Fuller-Pace F, Banda M, Williams LT. Large induction of keratinocyte growth factor expression in the dermis during wound healing. Proc Natl Acad Sci USA 1992; 89:6896–6900.
270. Brauche M, Augermeyer K, Hubner G, Werner S. Large induction of keratinocyte growth factor expression by serum factors and pro-inflammatory cytokines in cultured fibroblasts. Oncogene 1994; 9:3199–3204.

22

Cytokines and Endothelial Cells

Alberto Mantovani, Elisabetta Dejana, and Martino Introna
Istituto di Ricerche Farmacologiche "Mario Negri," Milan, Italy

Federico Bussolino
Università di Torino, Torino, Italy

INTRODUCTION

Endothelial cells (EC) were long thought to form a "passive" lining of blood vessels, endowed with negative properties, the most important being a nonthrombogenic substrate for blood. As such, EC were believed to participate in tissue reactions essentially as targets for injurious agents. The possibility of isolating and culturing EC from various tissues provided the tools for studying their complex reactions to a variety of activating stimuli. EC have, in this way, emerged as active participants in many physiological and pathological processes. It is now evident that hemostasis, inflammatory reactions, and immunity involve close interactions between immunocompetent cells and vascular endothelium. In particular, the ontogeny and function of white blood cells require an intimate relationship with vascular EC. Cytokines are mediators of these complex bidirectional interactions between leukocytes and vascular elements (for review see Refs. 1,2).

EC are strategically located at the interface between blood and tissues. In addition to responding rapidly (seconds to minutes) to agonists such as histamine and thrombin, vascular cells, upon exposure to cytokines, undergo profound alterations in function that involve gene expression and protein synthesis, require hours to develop, and are relatively long-lasting. This functional reprogramming of endothelial cells has been referred to as activation, analogous to the terminology long in use for the monocyte-macrophage lineage.

Vascular cells are both a target for and a source of cytokines, and these soluble polypeptide mediators serve to communicate with leukocytes (the primary producers of these molecules) as well as with diverse tissues and organs. Here we will review the response elicited by cytokines in vascular cells and analyze how vessel wall elements participate in cytokine cascades by producing copious amounts of some of these polypeptide mediators. Previous reviews of this rapidly expanding area of research (1,3,4) have provided the background and framework for this contribution and the reader is referred to these for more comprehensive referencing.

EC RESPONSES TO CYTOKINES

The spectrum of responses elicited in endothelial cells by different cytokines is extremely vast and varied. Different cytokines activate distinct, largely nonoverlapping, sets of functions, which can be grouped into programs of activation/differentiation (Fig. 1). The phenomenology of endothelial cell activation by cytokines has been reviewed in detail (2). Here we will update these previous descriptions and emphasize production and action of relatively novel mediators (e.g., chemokines) as well as the molecular basis for EC activation.

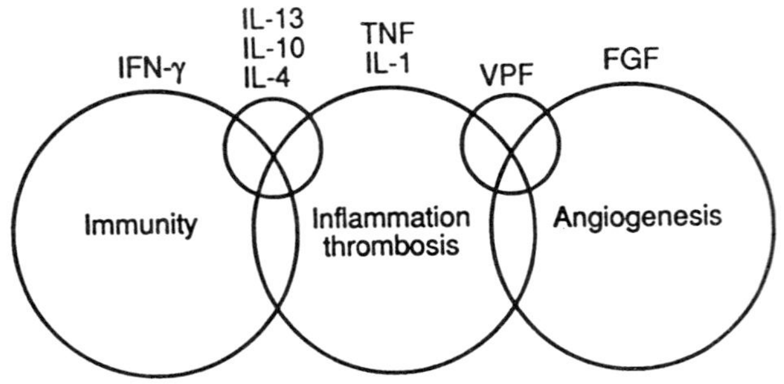

Figure 1 Cytokines activate distinct, but overlapping, sets of functions in EC. (Modified from Ref. 2.)

IL-1 and TNF

Activation

As previously described in detail, interleukin-1 (IL-1) and tumor necrosis factor (TNF) induce the expression of thrombosis and inflammation (Fig. 1). Briefly, these cytokines facilitate thrombus formation by inducing procoagulant activity, inhibiting the thrombomodulin/protein C anticoagulation pathway and blocking fibrin dissolution via stimulation of the type I inhibitor of plasminogen activator.

IL-1 and TNF induce augmented production of the vasodilatory mediators PGI_2 and NO (2,5,6). It has been shown that IL-1 induces enzymes in EC crucial for arachidonate metabolism, including phospholipase A_2 and cyclooxygenase 2 (7,8). Aspirin inhibits induction of prostaglandin H synthase in EC (9). IL-1 and TNF induce NO synthase by augmenting tetrahydropterin levels (5,10), an action counteracted by the anti-inflammatory cytokines IL-10 and TGF-β (10). Induction of NO and PGI_2 accounts for hemodynamic and permeability alterations in the systemic inflammatory response syndrome (SIRS). Vasodilation, by reducing shear stress, also facilitates the action of adhesion molecules and chemoattractants in leukocyte recruitment (Fig. 2) (11).

Inflammatory cytokines promote leukocyte extravasation by altering the rheology of microcirculation by induction of adhesion molecules [E-selectin, vascular cell adhesion molecule-1 (VCAM-1) and intercellular adhesion molecule-1 (ICAM-1)] and of chemotactic cytokines. The regulated expression of adhesion molecules in the multistep process of recruitment has been the object of recent extensive reviews, and chemokine production is discussed below. IL-1 and TNF cause production of platelet-activating factor (PAF) in EC (2). A serine protease is involved in the synthesis of PAF in activated EC (12,13). PAF can act on leukocytes as well as on EC themselves (14). PAF is not important for leukocyte adhesion, but primes leukocytes and facilitates transmigration (15).

IL-1 and TNF induce production in EC of copious amounts of IL-6, which acts locally and systemically (16). IL-6 does not activate EC, but was recently reported to induce c-*sis* expression in this cell type (17).

Gene Induction During Endothelial Activation

Endothelial cells have been the object of several systematic studies aimed at cloning genes induced by exposure to different signals, generally by differential screening approaches. The general aim of such studies was to clone members of the "immediate early inducible genes" (i.e., in the absence of protein synthesis) in order to elucidate the crucial steps that occur in the setting of different metabolic processes in the EC.

In particular, two groups have explored the "proinflammatory, prothrombotic" process. Dixit and coworkers cloned several genes induced in EC during TNF exposure (18), while we concentrated on the IL-1 effect

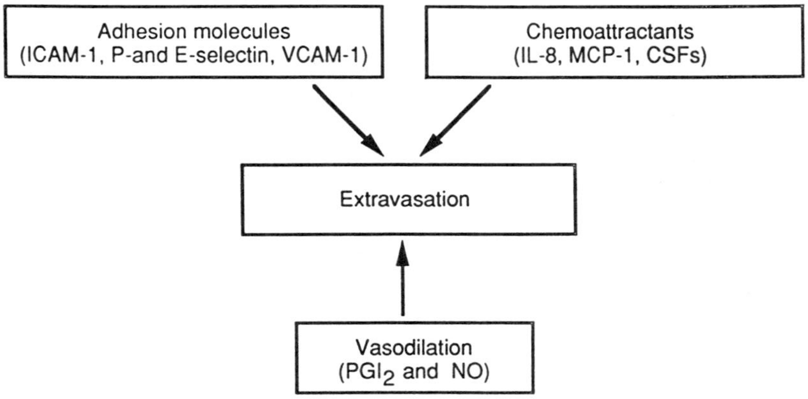

Figure 2 Pathways leading to leukocyte recruitment.

(19). As expected, several genes have been cloned by both groups, which had already been characterized as part of the IL-1 response, including E-selectin, IL-8, MCP-1 PA-I-1, gro α, and gro β (18,19).

More interestingly, new genes have been characterized in both systems, which include (1) a novel zinc finger transcription factor, A20, subsequently shown to be protective against TNF-induced cytotoxicity in a variety of cell types (20), (2) B12, a novel gene differentially expressed in the heart and liver during mouse embryogenesis and rapidly degraded (21), and (3) B94, which after its initial cloning (22) was further characterized as being expressed in a complex, modulated way during vasculogenesis and other processes including hematopoietic differentiation and spermatogenesis (23). Very recently, yet another clone, the secreted B61 protein (24), was shown to be a chemoattractant in vitro for endothelial cells and an angiogenetic factor in vivo (25). Its proposed function as the ligand for the Eck receptor may provide a clue for one circuit of angiogenesis during inflammation through the production of TNF and subsequently the production and release of B61 (25).

We cloned a gene (ptx3) from both human and mouse cells (26) and found that a partial sequence, including several sequencing errors, was already in the EMBL data base library and identified as TSG-14, cloned from another group of "gene searchers," as with the cloning of TNF inducible in fibroblasts (27). ptx3 is a secreted protein of 381 amino acids composed of two halves: the NH_2 terminal domain, showing no precise sequence similarity, and the COOH domain, being perfectly alignable with all known members of the acute-phase "pentraxin" gene family, including C-reactive protein (CRP) and serum amyloid P component (SAP), from many different species (28). While CRP and SAP are expressed mainly in the liver in vitro and in vivo following IL-6 stimulation (28), ptx3 shows a much wider range of responder cells in vitro, i.e., endothelial cells, fibroblasts, hepatocytes, and monocytes (26,29), but only in response to IL-1/TNF stimulation (26,29). In vivo ptx3 is inducible in several organs, particularly the heart and the muscle, upon i.v. injection of lipopolysaccharides (LPS) and may well represent, therefore, a new acute-phase marker with novel tissue specificity. The secondary structure prediction, based on the SAP crystal (30), indicates the presence of many α-helices in the NH_2 domain. Recently, two other pentraxins have been cloned from guinea pig testis (31,32) and brain (33), which also have "long" NH_2 domains, although the alignment among them is poor; ptx3 may represent the first cloned member of a "long-pentraxin" family, whose expression is mainly confined to extrahepatic sites. When a similar cloning strategy was applied to long-term IL-1– and TNF-stimulated endothelial cells, the isolation of a nuclear protein, C-193, was reported, which still behaves as a rapidly induced and protein synthesis–independent gene (34).

With a different perspective, another group searched for immediate early genes induced by exposure to PMA (which is able to induce a phenotyopic transition from a nonpolar cobblestone monolayer into a polar elongated fibroblastlike phenotype). One gene was isolated and called edg-1, which is a novel member of the G-protein coupled receptor family (35).

Bacterial Products and Toxins

Bacterial LPS elicit a spectrum of EC responses similar to that of IL-1 and TNF (2). It has recently been shown that EC, which are CD14 negative, interact with soluble CD14 and LPS-binding protein, both present in plasma (36–40). While direct activation of EC by LPS does occur, the indirect pathway involving mononuclear phagocyte activation and production of inflammatory cytokines is more effective in stimulation of EC by this bacterial product (41).

The *Lexosceles* venom toxin causes selective expression of IL-8, granulocyte-macrophage colony-stimulating factor (GM-CSF), and E-selectin, but not of IL-6 or ICAM-1 (42).

Infection with *E. coli* producing verocytotoxin (also known as Shiga toxin) is causally related to subsequent development of hemolytic uremic syndrome. Verocytotoxin directly activates EC in a prothrombotic-proinflammatory sense and synergizes with IL-1 and TNF (43,44). These cytokines have recently been shown to augment the expression of receptors for verocytotoxin, providing a molecular basis for synergism (44).

Hematopoietic Growth Factors

The hematopoietic growth factors GM-CSF, G-CSF, IL-3, and erythropoietin were shown to affect EC (for a recent review, see Ref. 45). By and large, these factors are relatively weak agonists, which affect migration and proliferation and amplify EC responsiveness to other signals (45). In the same studies, M-CSF, IL-5, and IL-6 were inactive. However, in two studies M-CSF induced expression of monocyte chemotactic protein-1 (MCP-1) in endothelial cells (46) and IL-6 induced c-*sis* expression (17). We failed to detect c-*fms* mRNA, encoding the M-CSF receptor, in human umbilical vein EC. Analysis of receptor expression revealed that EC express c-*kit* and the β chain common to the GM-CSF, IL-3, and IL-5 receptors. Moreover, they also have detectable mRNA for αIL-3 and αGM, but not for αIL-5 (F. Bussolino, unpublished). It has been speculated that the ability of various hemopoietic cytokines to affect EC is a remnant of common ancestors, i.e., a reflection of the common ontogenetic origin of hematopoietic and endothelial elements in blood islands (45).

Interferon-γ

Interferon-γ (IFN-γ) was the first molecularly defined immunological cytokine demonstrated to act on EC. IFN-γ induced the expression of class II MCH antigens and of the invariant chain in EC (see Ref. 3 for review). Moreover, it augments the levels of class I MHC antigens. Production of IL-1 by LPS-stimulated EC is augmented by IFN-γ. IFN-γ also amplified the action of TNF in EC, as with other cell types. MHC class II expression, as well as IL-1 and IL-6 production, underlies the ability of EC to act as antigen-presenting cells. IFN-γ causes a slow increase in the expression of ICAM-1 and renders EC more adhesive for T lymphocytes. It was recently reported (47) that $CD4^+$ T cells that produce IFN-γ inhibit class II (but not ICAM-1) expression. This would indicate a novel, as yet undefined, pathway of negative regulation of EC function. IFN-γ also downregulates von Willebrand's factor modulation induced by IL-1 (48). The invasion and proliferation of *Toxoplasma gondii* in EC is blocked by IFN-γ (49).

IL-4 and IL-13

IL-4 has growth factor activity and induces uPA in micro- but not in macrovascular EC (for review, see Refs. 2,50). It is intriguing that IL-4 and IL-13, but not IL-10, inhibited induction of RANTES expression in EC stimulated by IFN-γ and TNF (51). IL-4 selectively induces VCAM-1 and inhibits ICAM-1 and E-selectin expression. IL-4 is a weak inducer of IL-6 and MCP-1 in EC and amplifies production of these mediators in concert with other stimuli (see Ref. 2 for review). Recently it was shown that IL-13 has similar activities in EC (52). Sharing of receptor components may underlie the similarity of action of IL-4 and IL-13 in EC. It is intriguing that these cytokines have divergent effects on monocytes and EC (where they inhibit and amplify production of certain cytokines, respectively). Stimulation of certain EC function may be important in the induction and local expression of TH_2 type responses by, for instance, inducing recruitment of eosinophils and basophils and favoring the transition to the late phase reaction.

IL-10

Information on the interaction of IL-10 with vascular endothelium is scarce and fragmentary. This cytokine was a weak stimulus for expression of chemokines and IL-6 in mouse polyoma middle T (PmT) immortalized EC lines and amplified the action of IL-1 and TNF (53). This effect was associated with prolongation of mRNA half-life. Stimulation of HUVEC was variable and not reproducible. Under the same conditions, it inhibited cytokine production in human and murine mononuclear phagocytes. This dissociation of effects (inhibition vs. costimulation) is reminiscent of the activity of IL-4 and IL-13 (see above). IL-10 was reported to inhibit antigen presentation by human dermal microvascular EC (54) and induction of tetrahydrobiopterine in HUVEC (10).

Chemokines

By and large, the spectrum of action of chemokines is restricted to leukocytes. However, recent reports have suggested that some chemokines may affect EC function. IL-8 has recently been shown to induce migration and proliferation of EC (55,56). It has also been shown to cause angiogenesis in in vivo models in rodents (55,56). IL-8 was also reported to be chemotactic for SMC (57). Culture conditions critically affect the action of IL-8 on EC (58). Expression of IL-8R was recently demonstrated in HUVEC. In the same vein, gro α affected EC proliferation in one study (59), but others failed to observe binding and activity of this chemokine in EC (60). We have been unable to detect activities of various chemokines, in particular of MCP-1, in EC. However, the mouse C-C chemokine FIC has been shown to bind EC (61). The expression of IL-8 receptors and responsiveness of EC have been the object of conflicting reports (62,63).

Platelet factor 4 (PF4), a C-C chemokine contained in the α granules of platelets and of largely unknown function, has been reported to inhibit growth factor–induced proliferation of EC and angiogenesis in vivo (64).

The above results point to the possibility that some chemokines may regulate the proliferation and migration of vessel wall elements.

THE CYTOKINE REPERTOIRE OF ENDOTHELIAL CELLS

EC are important producers of various polypeptide mediators that regulate the hematopoietic system, the differentiation of T and B cells, and the extravasation of leukocytes at inflammatory sites. Here we will briefly summarize the available information on the production by EC of cytokines that affect leukocytes. TGF-β, a product and a regulator of EC (e.g., Refs. 1,65,66), is not considered in this chapter.

IL-1, TNF, and IL-6

IL-1, a cytokine that profoundly affects vascular cells (see above), is also a product of vessel wall elements (for review, see Refs. 1,3,4). Inducers of IL-1 production by vascular cells include LPS, TNF, and IL-1 itself. In vitro passage of EC results in spontaneous expression of IL-1α message and refractoriness to activation by exogeneous IL-1 (see above). IL-1 gene expression and protein were also detected in vivo after inoculation of LPS and feeding an atherogenic diet (67,68). Endothelial cells produce IL-1, but efforts to demonstrate expression of the IL-1 receptor antagonist (IL-1ra) in umbilical vein cells yielded negative results (69). IL-1ra blocks the action of IL-1 on EC, and these cells may represent a major target for the therapeutic activity of this molecule in vivo.

Upon exposure to inflammatory signals and superinduction by protein synthesis inhibitors, vascular SMC, but not EC, can express TNF mRNA and protein (70).

SMC and EC produce high levels of IL-6 (e.g., Ref. 16). Variable levels of spontaneous IL-6 production can usually be detected, and these are markedly augmented by exposure to IL-1, TNF, and LPS. IL-1 acts on T cells, at least in part, by inducing IL-6. Because IL-6 does not affect the proinflammatory and prothrombotic functions of IL-1 on EC and anti–IL-6 antibodies do not affect the activity of IL-1, it can be concluded that IL-6 is not the ultimate mediator of the effects of IL-1 on vascular cells (16). Production of IL-6 is amplified by costimulation with IL-4 or IL-13 (71–73). IL-6 is a pleiotropic cytokine that affects T and B lymphocytes and the production of acute-phase proteins by liver. The production of IL-6 by EC integrates these cells in immunological circuits and in the regulation of the acute-phase response.

CSFs

EC are an important source of CSFs. Hematopoietic growth factors produced by EC include G-CSF, GM-CSF, and M-CSF. CSF production is induced or augmented by a variety of stimuli, including LPS, IL-1, and TNF (see Refs. 1,3,4 for review), and minimally modified low-density lipoproteins (MM-LDL) (74). By producing CSFs as well as IL-1 and IL-6, which have CSF activity, EC participate in the regulation of hematopoiesis. Because EC respond to certain CSF molecules (G-CSF, GM-CSF, and erythropoietin; see above), autocrine and paracrine circuits involving EC and CSFs could be important in the generation and maintenance of an appropriate bone marrow mileu.

Chemokines: In Vitro Studies

Endothelial cells

EC produce various chemokines in response to signals representative of inflammatory reactions, immunity, and thrombosis (2). Inflammatory cytokines (IL-1 and TNF) and bacterial LPS induce the expression and release of IL-8 and gro α (18,59,75–80). Monocyte-associated IL-1α induces IL-8 expression in EC via a juxatacrine mechanism (81). Induction of IL-8 expression is associated with and depends on gene transcription (77). IL-4 and IL-13 are weak inducers of IL-8 expression and amplify induction by inflammatory cytokines (52,71–73,82). Histamine induces IL-8 production in EC (83). Hypoxia has recently been shown to induce IL-8 and MCP-1 expression in EC, a finding potentially relevant for pathological conditions in which activation and recruitment of leukocytes may amplify tissue damage (84,85). Platelets contain IL-1 and, when they interact with vascular EC, induce IL-8 gene expression (86).

As a result of proteolytic cleavage, IL-8 forms with a different NH_2 termini and lengths can be produced (87). It has been suggested that EC release predominantly a 77-amino-acid form of IL-8, which is less effective at activating leukocytes than the most common 73-residue form (80,88). The proteolytic conversion to smaller versions of the molecule can be catalyzed by thrombin (88).

The influence of IL-8 on the interaction of polymorphonuclear cells with vascular EC has been the object of conflicting observations, which seem now to reflect different experimental protocols and, most interestingly, different functions exerted by this cytokine under different pathophysiological conditions. IL-8 increased the adhesiveness of normal PMN for normal EC (89). In apparent contrast with these findings, EC-derived IL-8 was reported to inhibit binding of the PMN to activated EC (80). Although it elicits PMN extravasation when given locally, IL-8 inhibits recruitment if administered systemically by the i.v. route (90,91). The seemingly paradoxic anti-inflammatory effects of high levels of systemic IL-8, possibly dependent upon the action of a reverse chemotactic gradient and leukocyte deactivation, may represent a feedback mechanism to control tissue damage. Chemoattractants cause rapid release from leukocytes of the p75 TNF receptor and of the type II decoy IL-1 receptor (92). These molecules are likely to play an important role in inhibiting distant inflammatory reactions by chemokines.

The role of IL-8 produced locally by vascular cells was reexamined using reconstructed vessel wall models (93). Unequivocal evidence was obtained for the importance of IL-8 in transendothelial migration induced by inflammatory cytokines (93,94).

EC of postcapillary venules bind IL-8, possibly via heparinlike molecules (60). EC of postcapillary venules in kidney and other tissues express the promiscuous chemokine receptor present also on erythrocytes known as Duffy antigen (95,96). This receptor is present on EC of both Duffy-positive and -negative individuals and may serve to present chemokines to circulating leukocytes. Solid-phase IL-8 elicits haptotactic migration (97). Thus, locally produced IL-8 may be retained on the surface of EC and activate adhesive interactions and migration (60).

EC activated in vitro by inflammatory cytokines express gro α, which according to one report could in turn act on EC (59). It has been suggested that EC-bound gro α may promote monocyte adhesion (98).

IP10, a member of the C-X-C family but unique in that it attracts monocytes, is expressed in certain endothelia of mice exposed in vivo to IFN-γ or to LPS (99,100). There are no reports of in vitro expression of this chemokine in EC.

EC produce substantial amounts of the C-C chemokine MCP-1 (78,79). The proinflammatory signals IL-1, TNF, and, to a lesser extent, LPS are potent stimuli for MCP-1 production (78,79). Under the same conditions, MCP-2 was undetectable in EC supernatants (J. Van Damme, personal communication). IL-4 and IL-13 are active, though less potent inducers of MCP-1 expression (71–73). IFN-γ was recently shown to induce MCP-1 in human microvascular EC (101). M-CSF was reported to induce MCP-1, though we did not detect the M-CSF receptor c-*fms* in HUVEC by Northern analysis (46). Given the role that lipids and monocytes play in the natural history of atherosclerosis, it is of interest that MM-LDL induce MCP-1 production in EC and smooth muscle cells (102). Thrombin was recently found to induce expression of MCP-1 in monocytes and, less prominently, in EC (103). The C-C chemokine RANTES was produced by EC exposed to TNF and IFN-γ (51).

The molecular basis of stimulation of chemokine expression in EC has been studied to a limited extent. Induction by inflammatory signals and thrombin is protein synthesis independent in EC but, interestingly, not in monocytes (103). Direct demonstration of enhanced gene transcription was obtained for MCP-1 and IL-8 by nuclear run-off analysis (77,78).

SMC and Mesangial Cells

Early studies on MCP-1 found that vascular SMC secreted a monocyte chemoattractant (104,105), which resembled a similar factor produced by tumor lines (106). The SMC-derived factor was subsequently identified as MCP-1. Indeed IL-1, TNF, and MM-LDL induce expression of MCP-1 in SMC with mechanisms and kinetics resembling those observed in EC (102,107). IL-1 and TNF also stimulate expression and release of IL-8 in SMC (107).

Kidney mesangial cells may have an ontogenetic relationship with SMC. They express and release MCP-1 and IL-8 in response to IL-1 and TNF (108–111). Of special importance in relation to pathology is induction of MCP-1 and IP10 by aggregated immunoglobulins (110,111). Reactive oxygen intermediates act as second messengers in induction of MCP-1 by aggregated Ig (111). These results may have considerable relevance in kidney pathology involving mononuclear phagocyte infiltration.

In Vivo Studies

EC at sites of delayed-type hypersensitivity reactions and kidney allograft rejection express the C-C chemokine RANTES (112,113). In vivo studies on chemokines in vessel wall pathology have been largely restricted to atherosclerosis. MCP-1 expression has been detected in atheromatous lesions of rabbits, primates, and humans (114–117). IL-8 and MCP-1 mRNA have been detected in increased amounts in aortic aneurisms (118). Chemokine gene expression has been detected in various cellular elements, including SMC, EC, and mononuclear phagocytes, with somewhat different results in different studies. In the only study with mAb (114), cell populations positive for MCP-1 were different in lesions representative of different stages of the natural history of atherosclerosis. EC staining was prominent in diffuse intimal thickening and in fatty streaks, whereas it was weak in atheromatous lesions. Subendothelial macrophages were strongly positive for MCP-1 in fatty streak lesions and in atherosclerotic plaques. In plaques, a few intimal SMC stained for MCP-1. These results suggest that EC and macrophages are the major source of MCP-1 in early atherosclerotic lesions (114). Monocyte adhesion and infiltration is an early event in the natural history of atherosclerosis (4,119). Mononuclear phagocyte infiltration is also a prominent feature of vasculitis (120). Locally produced MCP-1 may play an important role in regulating extravasation of leukocytes, of monocytes in particular, in vessel wall pathology.

EC HETEROGENEITY

Vascular endothelia are remarkably heterogeneous in terms of morphology, marker expression (e.g., von Willebrand's factor), and function. Heterogeneity is evident at multiple levels and emerges from analysis of macro- vs. microvascular EC, of the vascular bed from different organs, and of EC obtained from different compartments in the same organ. However, studies of EC heterogeneity in response to and production of cytokines are scanty and fragmentary. It has been shown that, unlike large vessel EC, microvascular EC express PDGF receptors (121) and respond with growth and uPA production to IL-4 (50,75). When EC obtained from different levels of the lung (pulmonary artery and microvessels) were investigated for responsiveness to TNF, only the former were damaged by this cytokine, which increased PGI_2 production in both (122).

In various studies human microvascular EC of dermal or brain origin have been studied. These differed from HUVEC in terms of expression of adhesion molecules in response to inflammatory cytokines or endothelins (123–125), antigen presentation regulated by IFN-γ and IL-10 (54), tetrahydrobiopterine production regulated by various mediators (10). As discussed above, production of chemokines can be differentially regulated in different EC populations (101). All in all, the phenomenology of EC heterogeneity is growingly complex. However, its molecular and cellular basis, as well as the formulation of a conceptual framework for this descriptive phenomenology, remains elusive.

MOLECULAR BASIS FOR ACTIVATION OF THE PROINFLAMMATORY PROGRAM

In this section, the molecular basis for endothelial cell activation by IL-1 and TNF will be reviewed in some detail. Attention will be focused on these molecules because, among cytokines active on EC, IL-1 and TNF have been the most extensively studied and elicit the widest spectrum of responses. These include regulation of genes whose expression is largely (e.g., VCAM-1) or absolutely (e.g., E-selectin) restricted to EC, thus raising the largely unsolved issue of the molecular basis of endothelial specificity.

Receptors and Second Messengers

EC express both p75 and p55 TNF receptors. Following binding to the receptors, the importance of a subsequent step of internalization has been recently shown for the induction of several genes, showing that most of the internalized receptor is p75 (the most abundant in the membrane), which is then transported to the lysosomes via coated pits and coated vesicles (126,127).

On the other hand, p55 receptor is expressed at much lower levels in the membrane but is more abundant overall and is detectable mainly in the Golgi apparatus and in the cytoplasmic vacuoles (126).

TNF muteins (128) and agonistic or antagonistic antibodies (129) have been used to define the relative importance of p55 versus p75 in EC activation. By and large these studies indicate that TNF activates EC predominantly via p55 receptors (130,131). However, there is also evidence for a contribution of the p75 receptor (43), best observed at low TNF concentrations, consistent with the "ligand-passing" model of function of this molecule (132).

EC express only type I IL-1 receptor (133). Under resting conditions or upon exposure to inducers (e.g., IL-4) active on myelomonocytic cells, EC do not express the type II decoy receptor (133), a molecule without a demonstrable role in signaling, capable of blocking IL-1 (134).

Blocking of type I receptor inhibited the activation of EC by IL-1 (133). IL-1α has been suggested to act intracellularly in the regulation of EC life span (135) and to localize to the nucleus (136). The role, if any, of intracellular receptors in these putative intracellular activities of IL-1 remains to be elucidated.

Second messengers involved in EC activation by IL-1 or TNF have not been systematically investigated. The role of PKA-mediated phosphorylation was studied by forskolin-mediated activation of the adenylate cyclase in HUVEC (137), and, surprisingly, a reduction in cytokine-induced E-selectin and VCAM-1 expression was observed concomitantly with a stimulation in ICAM-1 expression (137). The data on E-selectin were further studied at a molecular level, showing a direct reduction in the transcriptional activity of the promoter, perhaps the result of phosphorylation of some component of the protein complexes, but without affecting the binding to NF-kB (see below) or CREB/ATF elements in the E-selectin promoter (see below).

Somewhat similar conclusions were also reached by others (138), who affirm that cAMP is neither an intracellular mediator nor a physiological regulator of TNF-induced E-selectin and VCAM-1 expression, but pharmacological elevations of cAMP in EC could have a role because of their inhibitory effects on the TNF-induced synthesis of E-selectin and VCAM-1 (138). Interestingly, IL-4 has been hypothesized to induce a yet uncharacterized adhesive molecule on the EC for PBL (different from VCAM-1) through a cAMP-dependent pathway (139).

Apart from still controversial results on the distribution of the various PKC isoforms in the different endothelial cell population (140), several studies indicate that PKC does not play a major role in the TNF-inducible expression of several genes, such as E-selectin (43,141,142), although PMA can indeed recapitulate the adhesive phenotype; recently, one interesting exception suggested a role for PKC in uPA (but not uPAR) reported expression (142).

Finally, few reports have explored the role of tyrosine phosphorylation in cytokine activation of EC. Van Hinsbergh et al. reported that the TNF-induced signal transduction pathway of PAI-1 transcription involves a genistein-sensitive step that is not involved in the induction of u-PA by TNF (143). Using a vast array of tyrosine protein kinase inhibitors and phosphatase inhibitors, they showed that the effect of genistein was quite selective, irrespective of the signal used (IL-4, thrombin, and TNF mutants (143). VCAM-1 expression induced by IL-4 seems to require a tyrosine phosphorylation step, including the phosphorylation of its own receptor (144), like that observed with oncostatin M, which phosphorylates its own receptor and requires tyrosine phosphorylation to induce IL-6 secretion (145). All in all, second messenger studies implying some phosphorylation event have to be considered in conjunction with the following observations on gene expression by NF-kB molecules, because the final activation step for NF-kB activity is phosphorylation of the I-kB inhibitor (see below).

Interestingly, many data suggest a role for reactive oxygen intermediates (ROI) as a common and critical denominator for various activating signals leading to NF-kB activation (see below). In particular, many different antioxidants [(such as *N*-acetyl-L-cysteine (NAC), dithiocarbamates, vitamin E derivatives, metal chelators, and pyrolidinedithiocarbamate (PDTC)] have been shown to effectively block NF-kB activation and subsequent gene induction. This issue has been little explored so far in endothelial cells: α-tocopherol was shown to inhibit the cytokine-induced expression of E-selectin, but the authors failed to detect any change in NF-kB activity on the promoter (146). Other agents like PDTC and NAC were shown, on the contrary, to inhibit the expression of VCAM-1, possibly through a reduction in the binding of NF-kB proteins to promoter elements, but interestingly in the same study no effect was reported in respect to E-selectin or ICAM-1 (147).

Transcription Factors

IL-1 and TNF regulate expression of a large and complex set of genes in EC, as revealed by investigations aimed at defining the molecular basis of functional responses and by "fishing expeditions" (see above). An analysis of regulatory elements of this vast array of genes is beyond the scope of this review. Here we will focus on regulation of genes whose expression is, at least to some extent, restricted to EC, such as E-selectin, VCAM-1, ICAM-1, and TM. In most cases, both from studies with deletion mutants of the different promoters and from gel retardation assays, molecules of the NF-kB family have been implicated.

The best-studied gene so far is E-selectin, whose promoter has been extensively characterized in EC. A fragment of 233 base pairs was found essential but not sufficient to confer cytokine inducibility and defined one NF-kB–binding site (148). With transfection assays in EC or nuclear extracts from cells of endothelial origin, at least six distinct elements exist in the minimal promoter, all of which are required for maximal inducibility, but none of which is sufficient (Fig. 3) (149,150). Canonical NF-kB proteins (dimers p50/p65) seem to be able to bind at least three such elements, which share the canonical consensus (5′-GGGPuNNPyPyCC-3′), although the possible presence of alternative NF-kB dimers has not yet been completely clarified. Many reports suggest that the dimers sitting on the promoter in resting conditions are in dynamic equilibrium with other, and possibly different, dimers that are mobilized upon activation (151). A fourth element,

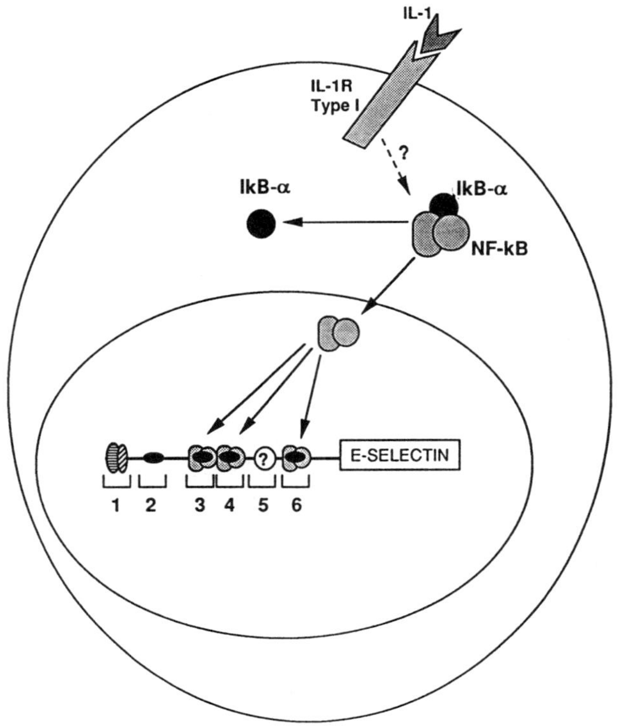

Figure 3 Model for the NF-kB–mediated activation of E-selectin promoter. (1) Protected site previously defined as NF ELAM-1 bound by ATF-2 (-153/-146 nucleotides upstream the transcriptional start site); (2) site bound by HMG I (high mobility group protein I) (-140/-135); (3) protected site previously defined as NF ELAM-3 or NF-kB or as site 1 bound by NF-kB (-126/-117); (4) protected site previously identified as site 2 bound by NF-kB (-116/-106) this site includes also a binding site for HMG-1 (-113/-110); (5) protected site identified as site 2 NF-ELAM-2 the bound protein has not been yet characterized (around -100); (6) protected site previously identified as NF-kB I or site 3 bound by NF-kB (-94/-83). The site also includes a binding site for HMG-1 (-90/-86).

NF-ELAM-1, is a CRE/ATF-like binding site, recognized by ATFa, ATF2, and ATF3 homodimers as well as ATFa/ATF2/c-*jun* heterodimers (152,153), while the fourth element, NF-ELAM-2, binds a still unknown protein. In most of these binding sites, there is yet another protein member of the high-mobility group protein I (Y), HMG I (Y) (154), which facilitates the binding of the NF-kB (and possibly also of the ATF) proteins to the DNA and also has a binding site on its own (Fig. 3). Finally, a role for the relative helix orientation of ATF and one NF-kB element, more than for the distance between them, has been reported as crucial in determining a correctly oriented transcription factor–induced DNA binding, which is pivotal for transcription to occur in endothelial cells (155). Cooperativity between different families of transcription factors is increasingly recognized as a crucial element in gene expression. Only accurate and consistent studies in that same cellular context and with few selected stimuli will eventually lead to the understanding of how a relatively small number of transcription factors can be used to achieve the high level of specificity required to regulate the complex pattern of gene expression in higher eukaryotes (156). Recent findings of interaction between NF-kB and other transcription factors, as in the case of the E-selectin and VCAM promoter, greatly expand the possibilities for combinatorial control of promoters (157,158).

A further recently recognized element of complexity for the regulation of the NF-kB activity in endothelial cells is the family of the NF-kB inhibitors known as IkB (159). In resting cells IkB proteins bind and sequester NF-kB in the cytoplasm. Stimulation rapidly degrades IkB (160) and releases NF-kB, which translocates to the nucleus. Recently, it was suggested that TNF-induced phosphorylation of Ik-Bα is a signal for its degradation, which occurs while it is still complexed with NF-kB (161). In any case, phosphorylation is a key regulatory element upstream of NF-kB activation (161).

It has been observed that immediately after TNF stimulation, protein levels of IkBα fall rapidly (162,163), while in parallel RNA levels are dramatically upregulated (164). Interestingly enough, the IkB α clone was one of the IL-1 immediate early-inducible genes cloned from HUVEC and described to be rapidly upregulated even in the absence of protein synthesis (19,164). It then became clear that the transcriptional activation of the IkB α gene was itself under the control of NF-kB activity (162,163). This circuit may explain the transient induction in the endothelial cells of many cytokine-inducible genes (e.g., ELAM-1).

VCAM-1 promoter was found to contain two canonical NF-kB–binding sites (165,166), which activate the gene when bound by the heterodimers p65/p50 or by the p65 homodimers (167). In addition, two GATA sites were described, which were found necessary for the induction of the full responsiveness to the cytokines (165). Very recently a functional interferon regulatory factor 1 (IRF-1) sequence motif was demonstrated in the VCAM-1 promoter and evidence was produced of slow cooperative DNA binding and transactivation properties between NF-kB and IRF-1 (158). Interestingly, like for the E-selectin promoter, the HMGI (Y) protein appears to facilitate NF-kB and IRF-1 activation (158).

The ICAM-1 promoter has recently been studied in endothelial cells, and it was shown that transcription induced by TNF-α critically depends on a single atypical NF-kB site, which possibly also mediates IL-1 and LPS inducibility (168). Even if p65 homodimers and p65/p50 heterodimers can bind in vitro to this element, functional studies implicate the p65 homodimer as the key transactivator on this site (168). This information may have more general implications when compared with a previous report on in vitro studies with nuclear extracts from IL-1–stimulated HUVEC with several fragments derived from many IL-1–inducible genes, which showed that while one canonical NF-kB sequence from E-selectin promoter could bind the canonical p65/p50 heterodimer, the other genes all contain variant NF-kB sites, including IL-8, GM-CSF, ICAM-1, and TF, and specifically bound c-Rel/p65 heterodimers (169). One possible conclusion is that all of the NF-kB sequences described so far, in spite of different decameric sequences, if they possess the triplet of guanine residues at the 5′ end, may bind and be activated by the p65/p50 heterodimer (like ELAM-1 and VCAM-1), while the variant may be activated by different dimers (like ICAM-1). A different mechanism seems to be involved in the transcriptional activation of the LDL receptor gene, which in endothelial cells shows a binding of a SP-1 transcription factor and does not have any NF-kB sites (170). Finally, the promoter of TM gene is repressed rather than induced by TNF. Both the basal activity and TNF-mediated repression can be reconducted to the same element, which binds an ets-like factor, not further characterized (171).

The results discussed here provide a basis for a new level of comprehension of EC activation by IL-1 and TNF by identifying transcription factors and regulatory elements involved in the induction or inhibition of gene expression and its time course. However, the central issue of specificity to endothelium of expression of certain genes remains unsolved. For instance, E-selectin is exclusively expressed in cytokine-activated EC, and considerable progress has been made in defining the molecular basis of induction and ensuing inhibition (see above), yet no genetic clues as to the endothelial specificity of this molecule is available. The difficulty in transfecting primary EC and the lack of appropriate EC lines may have hampered solving this critical question, which may ultimately require in vivo studies in transgenic mice.

CONCLUDING REMARKS

EC, long considered to form a layer of nucleated cellophane, play an important, active role in the onset and regulation of inflammatory and immune reactions. The varied spectrum of responses elicited by cytokines is a dramatic demonstration of EC plasticity. The vast phenomenology of EC responses can be rationalized and represented in terms of cytokine-induced activation/differentiation programs. Recent work summarized here has begun to unravel the molecular basis of EC activation by cytokines. However, central questions related to the mechanisms underlying expression restricted to EC and the phenomenology of EC heterogeneity remain without answer.

REFERENCES

1. Mantovani A, Dejana E. Cytokines as communication signals between leukocytes and endothelial cells. Immunol Today 1989; 10:370–375.
2. Mantovani A, Bussolino F, Dejana E. Cytokine regulation of endothelial cell function. FASEB J 1992; 6:2591–2599.
3. Pober J, Cotran RS. Cytokines and endothelial cell biology. Physiol Rev 1990; 70:427–451.
4. Libby P, Hansson GK. Biology of disease. Involvement of the immune system in human atherogenesis: current knowledge and unanswered questions. Lab Invest 1991; 64:5–15.
5. Rosenkranz Weiss P, Sessa WC, Milstien S, Kaufman S, Watson CA, Pober JS. Regulation of nitric oxide synthesis by proinflammatory cytokines in human umbilical vein endothelial cells. Elevations in tetrahydrobiopterin levels enhance endothelial nitric oxide synthase specific activity (see comments). J Clin Invest 1994; 93:2236–2243.
6. Rossi V, Breviario F, Ghezzi P, Dejana E, Mantovani A. Prostacyclin synthesis induced in vascular cells by interleukin-1. Science 1985; 229:174–176.
7. Jackson BA, Goldstein RH, Roy R, Cozzani M, Taylor L, Polgar P. Effects of transforming growth factor beta and interleukin-1 beta on expression of cyclooxygenase 1 and 2 and phospholipase A2 mRNA in lung fibroblasts and endothelial cells in culture. Biochem Biophys Res Commun 1993; 197:1465–1474.
8. Habib A, Creminon C, Frobert Y, Grassi J, Pradelles P, Maclouf J. Demonstration of an inducible cyclooxygenase in human endothelial cells using antibodies raised against the carboxyl-terminal region of the cyclooxygenase-2. J Biol Chem 1993; 268:23448–23454.
9. Wu KK, Sanduja R, Tsai AL, Ferhanoglu B, Mitchell DS. Aspirin inhibits interleukin 1-induced prostaglandin H synthase expression in cultured endothelial cells. Proc Natl Acad Sci USA 1991; 88:2384–2387.
10. Schoedon G, Schneemann M, Blau N, Edgell CJ, Schaffner A. Modulation of human endothelial cell tetrahydrobiopterin synthesis by activating and deactivating cytokines: new perspectives on endothelium-derived relaxing factor. Biochem Biophys Res Commun 1993; 196:1343–1348.
11. Mantovani A, Colotta F, Sozzani S, Allavena P, Dejana E. Vasodilation in the multistep paradigm of leukocyte extravasation. Lancet 1994; 343:1499–1500.
12. Bussolino F, Arese M, Silvestro L, et al. Involvement of a serine protease in the synthesis of platelet-activating factor by endothelial cells stimulated by tumor necrosis factor-alpha or interleukin-1 alpha. Eur J Immunol 1994; 24:3131–3139.
13. Heller R, Bussolino F, Ghigo D, et al. Human endothelial cells are target for platelet-activating factor. II. Platelet-activating factor induces platelet-activating factor synthesis in human umbilical vein endothelial cells. J Immunol 1992; 149:3682–3688.
14. Bussolino F, Silvagno F, Garbarino G, et al. Human endothelial cells are targets for platelet-activating factor (PAF). Activation of alpha and beta protein kinase C isozymes in endothelial cells stimulated by PAF. J Biol Chem 1994; 269:2877–2886.
15. Hill ME, Bird IN, Daniels RH, Elmore MA, Finnen MJ. Endothelial cell-associated platelet-activating factor primes neutrophils for enhanced superoxide production and arachidonic acid release during adhesion to but not transmigration across IL-1 beta-treated endothelial monolayers. J Immunol 1994; 153:3673–3683.
16. Sironi M, Breviario F, Proserpio P, et al. IL-1 stimulates IL-6 production in endothelial cells. J Immunol 1989; 142:549–553.
17. Calderon TM, Sherman J, Wilkerson H, Hatcher VB, Berman JW. Interleukin 6 modulates c-sis gene expression in cultured human endothelial cells. Cell Immunol 1992; 143:118–126.
18. Dixit VM, Green S, Sarma V, et al. Tumor necrosis factor-alpha induction of novel gene products in human endothelial cells including a macrophage-specific chemotaxin. J Biol Chem 1990; 265:2973–2978.
19. Introna M, Breviario F, D'Aniello E, Golay J, Dejana E, Mantovani A. IL-1 inducible genes in human umbilical vein endothelial cells. Eur Heart J 1993; 14 (suppl K):78–81.
20. Opipari AWJ, Hu HM, Yabkowitz R, Dixit VM. The A20 zinc finger protein protects cells from tumor necrosis factor cytotoxicity. J Biol Chem 1992; 267:12424–12427.
21. Wolf FW, Marks RM, Sarma V, et al. Characterization of a novel tumor necrosis factor-alpha-induced endothelial primary response gene. J Biol Chem 1992; 267:1317–1326.
22. Sarma V, Wolf FW, Marks RM, Shows TB, Dixit VM. Cloning of a novel tumor necrosis factor-alpha-inducible primary response gene that is differentially expressed in development and capillary tube-like formation in vitro. J Immunol 1992; 148:3302–3312.
23. Wolf FW, Sarma V, Seldin M, et al. B94, a primary response gene inducible by tumor necrosis factor-alpha, is expressed in developing hematopoietic tissues and the sperm acrosome. J Biol Chem 1994; 269:3633–3640.
24. Holzman LB, Marks RM, Dixit VM. A novel immediate-early response gene of endothelium is induced by cytokines and encodes a secreted protein. Mol Cell Biol 1990; 10:5830–5838.
25. Pandey A, Shao H, Marks RM, Polverini PJ, Dixit VM. Role of B61, the ligand for the Eck receptor tyrosine kinase, in TNF-alpha-induced angiogenesis. Science 1995; 268:567–569.
26. Breviario F, d'Aniello EM, Golay J, et al. Interleukin-1-inducible genes in endothelial cells. Cloning of a new gene related to C-reactive protein and serum amyloid P component. J Biol Chem 1992; 267:22190–22197.
27. Lee TH, Lee GW, Ziff EB, Vilcek J. Isolation and characterization of eight tumor necrosis factor-induced gene sequences from human fibroblasts. Mol Cell Biol 1990; 10:1982–1988.
28. Pepys MB, Baltz ML. Acute phase proteins with special

reference to C-reactive protein and related proteins (pentaxis) and serum amyloid A protein. Adv Immunol 1983; 34:141–160.
29. Vidal Alles V, Bottazzi B, Peri G, Golay J, Introna M, Mantovani A. Inducible expression of PTX3, a new member of the pentraxin family, in human mononuclear phagocytes. Blood 1994; 84:3483–3493.
30. Emsley J, White HE, O'Hara BP, et al. Structure of pentameric human serum amyloid P component. Nature 1994; 367:338–345.
31. Reid MS, Blobel CP. Apexin, an acrosomal pentaxin. J Biol Chem 1994; 269:32615–32620.
32. Noland TD, Friday BB, Maulit MT, Gerton GL. The sperm acrosomal matrix contains a novel member of the pentaxin family of calcium-dependent binding proteins. J Biol Chem 1994; 269:32607–32614.
33. Schlimgen AK, Helms JA, Vogel H, Perin MS. Neuronal pentraxin, a secreted protein with homology to acute phase proteins of the immune system. Neuron 1995; 14:519–526.
34. Chu W, Burns DK, Swerlick RA, Presky DH. Identification and characterization of a novel cytokine-inducible nuclear protein from human endothelial cells. J Biol Chem 1995; 270:10236–10245.
35. Hla T, Maciag T. An abundant transcript induced in differentiating human endothelial cells encodes a polypeptide with structural similarities to G-protein-coupled receptors. J Biol Chem 1990; 265:9308–9313.
36. von Asmuth EJ, Dentener MA, Bazil V, Bouma MG, Leeuwenberg JF, Buurman WA. Anti-CD14 antibodies reduce responses of cultured human endothelial cells to endotoxin. Immunology 1993; 80:78–83.
37. Pugin J, Schurer Maly CC, Leturcq D, Moriarty A, Ulevitch RJ, Tobias PS. Lipopolysaccharide activation of human endothelial and epithelial cells is mediated by lipopolysaccharide-binding protein and soluble CD14. Proc Natl Acad Sci USA 1993; 90:2744–2748.
38. Haziot A, Rong GW, Silver J, Goyert SM. Recombinant soluble CD14 mediates the activation of endothelial cells by lipopolysaccharide. J Immunol 1993; 151:1500–1507.
39. Read MA, Cordle SR, Veach RA, Carlisle CD, Hawiger J. Cell-free pool of CD14 mediates activation of transcription factor NF-kappa B by lipopolysaccharide in human endothelial cells. Proc Natl Acad Sci USA 1993; 90:9887–9891.
40. Goldblum SE, Brann TW, Ding X, Pugin J, Tobias PS. Lipopolysaccharide (LPS)-binding protein and soluble CD14 function as accessory molecules for LPS-induced changes in endothelial barrier function, in vitro. J Clin Invest 1994; 93:692–702.
41. Pugin J, Ulevitch RJ, Tobias PS. A critical role for monocytes and CD14 in endotoxin-induced endothelial cell activation. J Exp Med 1993; 178:2193–2200.
42. Patel KD, Modur V, Zimmerman GA, Prescott SM, McIntyre TM. The necrotic venom of the brown recluse spider induces dysregulated endothelial cell-dependent neutrophil activation—differential induction of GM-CSF, IL-8, and E-selectin expression. J Clin Invest 1994; 94:631–642.
43. van de Kar NC, Kooistra T, Vermeer M, Lesslauer W, Monnens LA, van Hinsbergh VW. Tumor necrosis factor alpha induces endothelial galactosyl transferase activity and verocytotoxin receptors. Role of specific tumor necrosis factor receptors and protein kinase C. Blood 1995; 85:734–743.
44. van de Kar NC, Monnens LA, Karmali MA, van Hinsbergh VW. Tumor necrosis factor and interleukin-1 induce expression of the verocytotoxin receptor globotriaosylceramide on human endothelial cells: implications for the pathogenesis of the hemolytic uremic syndrome. Blood 1992; 80:2755–2764.
45. Bussolino F, Bocchietto E, Silvagno F, Soldi R, Arese M, Mantovani A. Actions of molecules which regulate hemopoiesis on endothelial cells: memoirs of common ancestors? Path Res Pract 1994; 190:834–839.
46. Shyy YJ, Wickham LL, Hagan JP, et al. Human monocyte colony-stimulating factor stimulates the gene expression of monocyte chemotactic protein-1 and increases the adhesion of monocytes to endothelial monolayers. J Clin Invest 1993; 92:1745–1751.
47. Doukas J, Pober S. Lymphocyte-mediated activation of cultured endothelial cells (EC). CD4+ T cells inhibit EC class II MHC expression despite secreting IFN-gamma and increasing EC class I MHC and intercellular adhesion molecule-1 expression. J Immunol 1990; 145:1088–1098.
48. Tannenbaum SH, Gralnick HR. Gamma-interferon modulates von Willebrand factor release by cultured human endothelial cells. Blood 1990; 75:2177–2184.
49. Woodman JP, Dimier IH, Bout DT. Human endothelial cells are activated by IFN-gamma to inhibit Toxoplasma gondii replication. Inhibition is due to a different mechanism from that existing in mouse macrophages and human fibroblasts. J Immunol 1991; 147:2019–2023.
50. Wojta J, Gallicchio M, Zoellner H, Filonzi EL, Hamilton JA, Mcgrath K. Interleukin-4 stimulates expression of urokinase-type-plasminogen activator in cultured human foreskin microvascular endothelial cells. Blood 1993; 81:3285–3292.
51. Marfaingkoka A, Devergne O, Gorgone G, et al. Regulation of the production of the RANTES chemokine by endothelial cells—Synergistic induction by IFN-gamma plus TNF-alpha and inhibition by IL-4 and IL-13. J Immunol 1995; 154:1870–1878.
52. Sironi M, Sciacca FL, Matteucci C, et al. Regulation of endothelial and mesothelial cell function by interleukin-13: selective induction of vascular cell adhesion molecule-1 and amplification of interleukin-6 production. Blood 1994; 84:1913–1921.
53. Sironi M, Munoz C, Pollicino T, et al. Divergent effects of interleukin-10 on cytokine production by mononuclear phagocytes and endothelial cells. Eur J Immunol 1993; 23:2692–2695.
54. Vora M, Yssel H, de Vries JE, Karasek MA. Antigen presentation by human dermal microvascular endothelial cells. Immunoregulatory effect of IFN-gamma and IL-10. J Immunol 1994; 152:5734–5741.
55. Koch AE, Polverini PJ, Kunkel SL, et al. Interleukin-8 as a macrophage-derived mediator of angiogenesis. Science 1992; 258:1798–1801.
56. Strieter RM, Kunkel SL, Elner VM, et al. Interleukin-8. A corneal factor that induces neovascularization. Am J Pathol 1992; 141:1279–1284.
57. Yue TL, Mckenna PJ, Gu JL, Feuerstein GZ. Interleukin-8 is chemotactic for vascular smooth muscle cells. Eur J Pharmacol 1993; 240:81–84.
58. Watson CA, Camera-Benson L, Palmer-Crocker R, Pober JS. Variability among human umbilical vein endothelial cultures. Science 1995; 268:447–448.
59. Wen D, Rowland A, Derynck R. Expression and secretion of *gro*/MGSA by stimulated human endothelial cells. EMBO J 1989; 8:1761–1766.
60. Rot A. Endothelial cell binding of NAP-1/IL-8: role in neutrophil emigration. Immunol Today 1992; 13:291–294.
61. Heinrich JN, Ryseck RP, Macdonaldbravo H, Bravo R. The product of a novel growth factor-activated gene, fic, is a

biological active C-C-type cytokine. Mol Cell Biol 1993; 13:2020–2030.

62. Schonbeck U, Brandt E, Petersen F, Flad HD, Loppnow H. IL-8 specifically binds to endothelial but not to smooth muscle cells. J Immunol 1995; 154:2375–2383.
63. Petzelbauer P, Watson CA, Pfau SE, Pober JS. IL-8 and angiogenesis: evidence that human endothelial cells lack receptors and do not respond to IL-8 in vitro. Cytokine 1995; 7:267–272.
64. Maione TE, Gray GS, Petro J, et al. Inhibition of angiogenesis by recombinant human platelet factor-4 and related peptides. Science 1990; 247:77–79.
65. Sporn MB, Roberts AB, Wakefield LM, Crombrugghe B. Some recent advances in the chemistry and biology of transforming growth factor-beta. J Cell Biol 1987; 105:1039–1045.
66. Gamble JR, Vadas MA. Endothelial cell adhesiveness for human T lymphocytes is inhibited by transforming growth factor beta-1. J Immunol 1991; 146:1149–1154.
67. Moyer CF, Sajuthi D, Tulli H, Williams JK. Synthesis of IL-1 alpha and IL-1 beta by arterial cells in atherosclerosis. Am J Pathol 1991; 138:951–960.
68. Clinton SK, Fleet JC, Loppnow H, et al. Interleukin-1 gene expression in rabbit vascular tissue in vivo. Am J Pathol 1991; 138:1005–1014.
69. Bertini R, Sironi M, Martin Padura I, et al. Inhibitory effect of recombinant intracellular interleukin 1 receptor antagonist on endothelial cell activation. Cytokine 1992; 4:44–47.
70. Warner SJ, Libby P. Human vascular smooth muscle cells. Target for and source of tumor necrosis factor. J Immunol 1989; 142:100–109.
71. Rollins BJ, Pober JS. Interleukin-4 induces the synthesis and secretion of MCP-1JE by human endothelial cells. Am J Pathol 1991; 138:1315–1319.
72. Howell G, Pham P, Taylor D, Foxwell B, Feldmann M. Interleukin 4 induces interleukin 6 production by endothelial cells: synergy with interferon-gamma. Eur J Immunol 1991; 21:97–101.
73. Colotta F, Sironi M, Borre A, Luini W, Maddalena F, Mantovani A. Interleukin 4 amplifies monocyte chemotactic protein and interleukin 6 production by endothelial cells. Cytokine 1992; 4:24–28.
74. Rajavashisth TB, Andalibi A, Territo MC, et al. Induction of endothelial cell expression of granulocyte and macrophage colony-stimulating factors by modified low-density lipoproteins. Nature 1990; 344:254–257.
75. Strieter RM, Kunkel SL, Showell HJ, et al. Endothelial cell gene expression of a neutrophil chemotactic factor by TNF-alpha, LPS, and IL-1 beta. Science 1989; 243:1467–1469.
76. Schroder JM, Christophers E. Secretion of novel and homologous neutrophil-activating peptides by LPS-stimulated human endothelial cells. J Immunol 1989; 142:244–251.
77. Sica A, Matsushima K, Van Damme J, et al. IL-1 transcriptionally activates the neutrophil chemotactic factor/IL-8 gene in endothelial cells. Immunology 1990; 69:548–553.
78. Sica A, Wang JM, Colotta F, et al. Monocyte chemotactic and activating factor gene expression induced in endothelial cells by IL-1 and tumor necrosis factor. J Immunol 1990; 144:3034–3038.
79. Rollins BJ, Yoshimura T, Leonard EJ, Pober JS. Cytokine-activated human endothelial cells synthesize and secrete a monocyte chemoattractant, MCP-1/JE. Am J Pathol 1990; 136:1229–1233.
80. Gimbrone MAJ, Obin MS, Brock AF, et al. Endothelial interleukin-8: a novel inhibitor of leukocyte-endothelial interactions. Science 1989; 246:1601–1603.
81. Kaplanski G, Farnarier C, Kaplanski S, et al. Interleukin-1 induces interleukin-8 secretion from endothelial cells by a juxtacrine mechanism. Blood 1994; 84:4242–4248.
82. Korpelainen EI, Gamble JR, Smith WB, et al. The receptor for interleukin 3 is selectively induced in human endothelial cells by tumor necrosis factor alpha and potentiates interleukin 8 secretion and neutrophil transmigration. Proc Natl Acad Sci USA 1993; 90:11137–11141.
83. Jeannin P, Delneste Y, Gosset P, et al. Histamine induces interleukin-8 secretion by endothelial cells. Blood 1994; 84:2229–2233.
84. Karakurum M, Shreeniwas R, Chen J, et al. Hypoxic induction of interleukin-8 gene expression in human endothelial cells. Blood 1993; 81:2492–2495.
85. Karakurum M, Shreeniwas R, Chen J, et al. Hypoxic induction of interleukin-8 gene expression in human endothelial cells. J Clin Invest 1994; 93:1564–1570.
86. Kaplanski G, Porat R, Aiura K, Erban JK, Gelfand JA, Dinarello CA. Activated platelets induce endothelial secretion of interleukin-8 in vitro via an interleukin-1-mediated event. Blood 1993; 81:2492–2495.
87. Baggiolini M, Dewald B, Moser B. Interleukin-8 and related chemotactic cytokines—CXC and CC chemokines. Adv Immunol 1994; 55:99–179.
88. Hebert CA, Luscinskas FW, Kiely JM, et al. Endothelial and leukocyte forms of IL-8. Conversion by thrombin and interactions with neutrophils. J Immunol 1990; 145:3033–3040.
89. Carveth HJ, Bohnsack JF, McIntyre TM, Baggiolini M, Prescott SM, Zimmerman GA. Neutrophil activating factor (NAF) induces polymorphonuclear leukocyte adherence to endothelial cells and to subendothelial matrix proteins. Biochem Biophys Res Commun 1989; 162:387–393.
90. Hechtman DH, Cybulsky MI, Fuchs HJ, Baker JB, Gimbrone MAJ. Intravascular IL-8. Inhibitor of polymorphonuclear leukocyte accumulation at sites of acute inflammation. J Immunol 1991; 147:883–892.
91. Ley K, Baker JB, Cybulsky MI, Gimbrone MA, Luscinskas FW. Intravenous interleukin-8 inhibits granulocyte emigration from rabbit mesenteric venules without altering L-selectin expression or leukocyte rolling. J Immunol 1993; 151:6347–6357.
92. Colotta F, Orlando S, Fadlon EJ, Sozzani S, Matteucci C, Mantovani A. Chemoattractants induce rapid release of the interleukin 1 type II decoy receptor in human polymorphonuclear cells. J Exp Med (in press).
93. Huber AR, Kunkel SL, Todd RF, Weiss SJ. Regulation of transendothelial neutrophil migration by endogenous interleukin-8. Science 1991; 254:99–102.
94. Smith WB, Gamble JR, Clarklewis I, Vadas MA. Chemotactic desensitization of neutrophils demonstrates interleukin-8 (IL-8)-dependent and IL-8-independent mechanisms of transmigration through cytokine-activated endothelium. Immunology 1993; 78:491–497.
95. Hadley TJ, Lu ZH, Wasniowska K, et al. Postcapillary venule endothelial cells in kidney express a multispecific chemokine receptor that is structurally and functionally identical to the erythroid isoform, which is the Duffy blood group antigen. J Clin Invest 1994; 94:985–991.
96. Peiper SC, Wang ZX, Neote K, et al. The Duffy antigen receptor for chemokines (DARC) is expressed in endothelial cells of Duffy negative individuals who lack the erythrocyte receptor. J Exp Med 1995; 181:1311–1317.
97. Wang JM, Taraboletti G, Matsushima K, Van Damme J, Mantovani A. Induction of haptotactic migration of melanoma

cells by neutrophil activating protein/interleukin-8. Biochem Biophys Res Commun 1990; 169:165–170.
98. Schwartz D, Andalibi A, Chaverrialmada L, et al. Role of the GRO family of chemokines in monocyte adhesion to MM-LDL-stimulated endothelium. J Clin Invest 1994; 94:1968–1973.
99. Narumi S, Wyner LM, Stoler MH, Tannenbaum CS, Hamilton TA. Tissue-specific expression of murine IP-10 mRNA following systemic treatment with interferon-gamma. J Leukoc Biol 1992; 52:27–33.
100. Gómez-Chiarri M, Hamilton TA, Egido J, Emancipator SN. Expression of IP-10, a lipopolysaccharide- and interferon-gamma-inducible protein, in murine mesangial cells in culture. Am J Pathol 1993; 142:433–439.
101. Brown Z, Gerritsen ME, Carley WW, Strieter RM, Kunkel SL, Westwick J. Chemokine gene expression and secretion by cytokine-activated human microvascular endothelial cells—Differential regulation of monocyte chemoattractant protein-1 and interleukin-8 in response to interferon-gamma. Am J Pathol 1994; 145:913–921.
102. Cushing SD, Berliner JA, Valente AJ, et al. Minimally modified low density lipoprotein induces monocyte chemotactic protein 1 in human endothelial cells and smooth muscle cells. Proc Natl Acad Sci USA 1990; 87:5134–5138.
103. Colotta F, Sciacca FL, Sironi M, Luini W, Rabiet MJ, Mantovani A. Expression of monocyte chemotactic protein-1 by monocytes and endothelial cells exposed to thrombin. Am J Pathol 1994; 144:975–985.
104. Valente AJ, Fowler SR, Sprague EA, Kelley JL, Suenram CA, Schwartz CJ. Initial characterization of a peripheral blood mononuclear cell chemoattractant derived from cultured arterial smooth muscle cells. Am J Pathol 1984; 117:409–417.
105. Valente AJ, Graves DA, Vialle-Valentin CE, Delgado R, Schwartz CJ. Purification of a monocyte chemotactic factor secreted by nonhuman primate vascular cells in culture. Biochemistry 1988; 27:4162–4168.
106. Bottazzi B, Polentarutti N, Acero R, et al. Regulation of the macrophage content of neoplasms by chemoattractants. Science 1983; 220:210–212.
107. Wang JM, Sica A, Peri G, et al. Expression of monocyte chemotactic protein and interleukin-8 by cytokine-activated human vascular smooth muscle cells. Arterioscler Thromb 1991; 11:1166–1174.
108. Zoja C, Wang JM, Bettoni S, et al. Interleukin-1 beta and tumor necrosis factor-alpha induce gene expression and production of leukocyte chemotactic factors, colony-stimulating factors, and interleukin-6 in human mesangial cells. Am J Pathol 1991; 138:991–1003.
109. Rovin BH, Yoshimura T, Tan L. Cytokine-induced production of monocyte chemoattractant protein-1 by cultured human mesangial cells. J Immunol 1992; 148:2148–2153.
110. Satriano JA, Shuldiner M, Hora K, Xing Y, Shan Z, Schlondorff D. Oxygen radicals as second messengers for expression of the monocyte chemoattractant protein, JE/MCP-1, and the monocyte colony-stimulating factor, CSF-1, in response to tumor necrosis factor-alpha and immunoglobulin-G—evidence for involvement of reduced nicotinamide adenine dinucleotide phosphate (NADPH)-dependent oxidase. J Clin Invest 1993; 92:1564–1571.
111. Satriano JA, Hora K, Shan Z, Stanley ER, Mori T, Schlondorff D. Regulation of monocyte chemoattractant protein-1 and macrophage colony-stimulating factor-1 by IFN-gamma, tumor necrosis factor-alpha, IgG aggregates, and cAMP in mouse mesangial cells. J Immunol 1993; 150:1971–1978.
112. Devergne O, Marfaingkoka A, Schall TJ, et al. Production of the RANTES chemokine in delayed-type hypersensitivity reactions: Involvement of macrophages and endothelial cells. J Exp Med 1994; 180:775.
113. Pattison J, Nelson PJ, Huie P, et al. RANTES chemokine expression in cell-mediated transplant rejection of the kidney. Lancet 1994; 343:209–211.
114. Takeya M, Yoshimura T, Leonard EJ, Takahashi K. Detection of monocyte chemoattractant protein-1 in human atherosclerotic lesions by an anti-monocyte chemoattractant protein-1 monoclonal antibody. Hum Pathol 1993; 24:534–539.
115. Yu X, Dluz S, Graves DT, et al. Elevated expression of monocyte chemoattractant protein 1 by vascular smooth muscle cells in hypercholesterolemic primates. Proc Natl Acad Sci USA 1992; 89:6953–6957.
116. Yla Herttuala S, Lipton BA, Rosenfeld ME, et al. Expression of monocyte chemoattractant protein 1 in macrophage-rich areas of human and rabbit atherosclerotic lesions. Proc Natl Acad Sci USA 1991; 88:5252–5256.
117. Nelken NA, Coughlin SR, Gordon D, Wilcox JN. Monocyte chemoattractant protein-1 in human atheromatous plaques. J Clin Invest 1991; 88:1121–1127.
118. Koch AE, Kunkel SL, Pearce WH, et al. Enhanced production of the chemotactic cytokines interleukin-8 and monocyte chemoattractant protein-1 in human abdominal aortic aneurysms. Am J Pathol 1993; 142:1423–1431.
119. Ross R. The pathogenesis of atherosclerosis: a perspective for the 1990s. Nature 1993; 362:801–809.
120. Fauci AS, Haynes FB, Katz P. The spectrum of vasculitis. Ann Intern Med 1978; 89 (Part 1):660–676.
121. Beitz JG, Kim IS, Calabresi P, Frackelton AR Jr. Human microvascular endothelial cells express receptors for platelet-derived growth factor. Proc Natl Acad Sci USA 1991; 88:2021–2025.
122. Meyrick B, Christman B, Jesmok G. Effects of recombinant tumor necrosis factor-alpha on cultured pulmonary artery and lung microvascular endothelial monolayers. Am J Pathol 1991; 138:93–101.
123. Rott O, Tontsch U, Fleischer B, Cash E. Interleukin-6 production in "normal" and HTLV-1 tax-expressing brain-specific endothelial cells. Eur J Immunol 1993; 23:1987–1991.
124. McCarron RM, Wang L, Stanimirovic DB, Spatz M. Differential regulation of adhesion molecule expression by human cerebrovascular and umbilical vein entothelial cells. Endothelium 1995; 2:339–346.
125. Petzelbauer P, Bender JR, Wilson J, Pober JS. Heterogeneity of dermal microvascular endothelial cell antigen expression and cytokine responsiveness in situ and in cell culture. J Immunol 1993; 151:5062–5072.
126. Bradley JR, Thiru S, Pober JS. Disparate localization of 55-kd and 75-kd tumor necrosis factor receptors in human endothelial cells. Am J Pathol 1995; 146:27–32.
127. Bradley JR, Johnson DR, Pober JS. Four different classes of inhibitors of receptor-mediated endocytosis decrease tumor necrosis factor-induced gene expression in human endothelial cells. J Immunol 1993; 150:5544–5555.
128. Barbara JA, Smith WB, Gamble JR, et al. Dissociation of TNF-alpha cytotoxic and proinflammatory activities by p55 receptor- and p75 receptor-selective TNF-alpha mutants. EMBO J 1994; 13:843–850.
129. Mackay F, Loetscher H, Stueber D, Gehr G, Lesslauer W. Tumor necrosis factor alpha (TNF-alpha)-induced cell adhesion to human endothelial cells is under dominant control of one TNF receptor type, TNF-R55. J Exp Med 1993; 177:1277–1286.
130. Paleolog EM, Delasalle SA, Buurman WA, Feldmann M.

Functional activities of receptors for tumor necrosis factor-alpha on human vascular endothelial cells. Blood 1994; 84:2578–2590.

131. Slowik MR, De Luca LG, Fiers W, Pober JS. Tumor necrosis factor activates human endothelial cells through the p55 tumor necrosis factor receptor but the p75 receptor contributes to activation at low tumor necrosis factor concentration. Am J Pathol 1993; 143:1724–1730.
132. Tartaglia LA, Pennica D, Goeddel DV. Ligand passing: the 75-kDa tumor necrosis factor (TNF) receptor recruits TNF for signaling by the 55-kDa TNF receptor. J Biol Chem 1993; 268:18542–18548.
133. Colotta F, Sironi M, Borre A, et al. Type II interleukin-1 receptor is not expressed in cultured endothelial cells and is not involved in endothelial cell activation. Blood 1993; 81:1347–1351.
134. Colotta F, Re F, Muzio M, et al. Interleukin-1 type II receptor: a decoy target for IL-1 that is regulated by IL-4. Science 1993; 261:472–475.
135. Maier JA, Voulalas P, Roeder D, Maciag T. Extension of the life-span of human endothelial cells by an interleukin-1 alpha antisense oligomer. Science 1990; 249:1570–1574.
136. Maier JA, Statuto M, Ragnotti G. Endogenous interleukin 1 alpha must be transported to the nucleus to exert its activity in human endothelial cells. Mol Cell Biol 1994; 14:1845–1851.
137. Ghersa P, Hooft van Huijsduijnen R, Whelan J, Cambet Y, Pescini R, DeLamarter JF. Inhibition of E-selectin gene transcription through a cAMP-dependent protein kinase pathway. J Biol Chem 1994; 269:29129–29137.
138. Pober JS, Slowik MR, De Luca LG, Ritchie AJ. Elevated cyclic AMP inhibits endothelial cell synthesis and expression of TNF-induced endothelial leukocyte adhesion molecule-1, and vascular cell adhesion molecule-1, but not intercellular adhesion molecule-1. J Immunol 1993; 150:5114–5123.
139. Galea P, Thibault G, Lacord M, Bardos P, Lebranchu Y. IL-4, but not tumor necrosis factor-alpha, increases endothelial cell adhesiveness for lymphocytes by activating a cAMP-dependent pathway. J Immunol 1993; 151:588–596.
140. Mattila P, Majuri ML, Tiisala S, Renkonen R. Expression of six protein kinase C isotypes in endothelial cells. Life Sci 1994; 55:1253–1260.
141. Ritchie AJ, Johnson DR, Ewenstein BM, Pober JS. Tumor necrosis factor induction of endothelial cell surface antigens is independent of protein kinase C activation or inactivation. Studies with phorbol myristate acetate and staurosporine. J Immunol 1991; 146:3056–3062.
142. Niedbala MJ, Stein Picarella M. Role of protein kinase C in tumor necrosis factor induction of endothelial cell urokinase-type plasminogen activator. Blood 1993; 81:2608–2617.
143. van Hinsbergh VW, Vermeer M, Koolwijk P, Grimbergen J, Kooistra T. Genistein reduces tumor necrosis factor alpha-induced plasminogen activator inhibitor-1 transcription but not urokinase expression in human endothelial cells. Blood 1994; 84:2984–2991.
144. Palmercrocker RL, Pober JS. IL-4 induction of VCAM-1 on endothelial cells involves activation of a protein tyrosine kinase. J Immunol 1995; 154:2838–2845.
145. Schieven GL, Kallestad JC, Brown TJ, Ledbetter JA, Linsley PS. Oncostatin M induces tyrosine phosphorylation in endothelial cells and activation of p62yes tyrosine kinase. J Immunol 1992; 149:1676–1682.
146. Faruqi R, de la Motte C, DiCorleto PE. Alpha-tocopherol inhibits agonist-induced monocytic cell adhesion to cultured human endothelial cells. J Clin Invest 1994; 94:592–600.
147. Marui N, Offermann MK, Swerlick R, et al. Vascular cell adhesion molecule-1 (VCAM-1) gene transcription and expression are regulated through an antioxidant-sensitive mechanism in human vascular endothelial cells. J Clin Invest 1993; 92:1866–1874.
148. Whelan J, Ghersa P, Hooft van Huijsduijnen R, et al. An NF kappa B-like factor is essential but not sufficient for cytokine induction of endothelial leukocyte adhesion molecule 1 (ELAM-1) gene transcription. Nucleic Acids Res 1991; 19:2645–2653.
149. Whitley MZ, Thanos D, Read MA, Maniatis T, Collins T. A striking similarity in the organization of the E-selectin and beta interferon gene promoters. Mol Cell Biol 1994; 14:6464–6475.
150. Schindler U, Baichwal VR. Three NF-kappa B binding sites in the human E-selectin gene required for maximal tumor necrosis factor alpha-induced expression. Mol Cell Biol 1994; 14:5820–5831.
151. Hooft van Huijsduijnen R, Pescini R, DeLamarter JR. Two distinct NF-kappa B complexes differing in their larger subunit bind the E-selectin promoter kappa B element. Nucleic Acids Res 1993; 21:3711–3717.
152. Pescini R, Kaszubska W, Whelan J, DeLamarter JF, Hooft van Huijsduijnen R. ATF-a0, a novel variant of the ATF/CREB transcription factor family, forms a dominant transcription inhibitor in ATF-a heterodimers. J Biol Chem 1994; 269:1159–1165.
153. Kaszubska W, van Huijsduijnen RH, Ghersa P, et al. Cyclic AMP-independent ATF family members interact with NF-kappa B and function in the activation of the E-selectin promoter in response to cytokines. Mol Cell Biol 1993; 13:7180–7190.
154. Lewis H, Kaszubska W, DeLamarter JF, Whelan J. Cooperativity between two NF-kappa B complexes, mediated by high-mobility-group protein I(Y), is essential for cytokine-induced expression of the E-selectin promoter. Mol Cell Biol 1994; 14:5701–5709.
155. Meacock S, Pescini Gobert R, DeLamarter JF, Hooft van Huijsduijnen R. Transcription factor-induced, phased bending of the E-selectin promoter. J Biol Chem 1994; 269:31756–31762.
156. Tjian R, Maniatis T. Transcriptional activation: a complex puzzle with few easy pieces. Cell 1994; 77:5–8.
157. Siebenlist U, Franzoso G, Brown K. Structure, regulation and function of NF-kB. Annu Rev Cell Biol 1994; 10:405–455.
158. Neish AS, Read MA, Thanos D, Pine R, Maniatis T, Collins T. Endothelial interferon regulatory factor 1 cooperates with NF-kB as a transcriptional activator of vascular cell adhesion molecule 1. Mol Cell Biol 1995; 15:2558–2569.
159. Thanos D, Maniatis T. NF-kappa B: a lesson in family values. Cell 1995; 80:529–532.
160. Grimm S, Baeuerle PA. The inducible transcription factor NF-kappa B: structure-function relationship of its protein subunits. Biochem J 1993; 290:297–308.
161. Miyaloto S, Maki M, Schmitt MJ, Hatanaka M, Verna IM. Tumor necrosis factor alpha-induced phosphorylation of IkB alpha is signal for its degradation but not dissociation from NF-kB. Proc Natl Acad Sci USA 1994; 91:12740–12744.
162. Read MA, Whitley MZ, Williams AJ, Collins T. NF-kappa B and I kappa B alpha: an inducible regulatory system in endothelial activation. J Exp Med 1994; 179:503–512.
163. de Martin R, Vanhove B, Cheng Q, et al. Cytokine-inducible expression in endothelial cells of an I kappa B alpha-like gene is regulated by NF kappa B. EMBO J 1993; 12:2773–2779.
164. d'Aniello EM, Breviario F, Martin Padura I, et al. Interleukin-1

and tumor necrosis factor induce transient expression of an inhibitor of nuclear factor kB in endothelial cells. Endothelium 1993; 1:161–165.

165. Neish AS, Williams AJ, Palmer HJ, Whitley MZ, Collins T. Functional analysis of the human vascular cell adhesion molecule 1 promoter. J Exp Med 1992; 176:1583–1593.
166. Iademarco MF, McQuillan JJ, Rosen GD, Dean DC. Characterization of the promoter for vascular cell adhesion molecule-1 (VCAM-1). J Biol Chem 1992; 267:16323–16329.
167. Shu HB, Agranoff AB, Nabel EG, et al. Differential regulation of vascular cell adhesion molecule 1 gene expression by specific NF-kappa B subunits in endothelial and epithelial cells. Mol Cell Biol 1993; 13:6283–6289.
168. Ledebur HC, Parks TP. Transcriptional regulation of the intercellular adhesion molecule-1 gene by inflammatory cytokines in human endothelial cells—essential roles of a variant NF-kappa B site and p65 homodimers. J Biol Chem 1995; 270:933–943.
169. Parry GC, Mackman N. A set of inducible genes expressed by activated human monocytic and endothelial cells contain kappa B-like sites that specifically bind c-Rel-p65 heterodimers. J Biol Chem 1994; 269:20823–20825.
170. Hamanaka R, Kohno K, Seguchi T, et al. Induction of low density lipoprotein receptor and a transcription factor SP-1 by tumor nencrosis factor in human microvascular endothelial cells. J Biol Chem 1992; 267:13160–13165.
171. von der Ahe D, Nischan C, Kunz C, et al. Ets transcription factor binding site is required for positive and TNF alpha-induced negative promoter regulation. Nucleic Acids Res 1993; 21:5636–5643.

23
Cytokines and Macrophages

Mark D. Wewers
The Ohio State University, Columbus, Ohio

INTRODUCTION

In the day-to-day surveillance and regulation of the body's interactions with foreign substances, it is difficult to find a more central player than the mononuclear phagocyte. Monocytes and macrophages stationed at sentinel posts throughout the body function as critical regulators of immune response to foreign invasion. This chapter will focus on the role that cytokines play in providing the communication network that puts this array of macrophage defenders in direct contact with each other and with the vascular, mesenchymal, and inflammatory cells important in the defense of self.

Macrophage Life Cycle

The ability to produce cytokines is the forté of the macrophage and will be the focus of this chapter. However, afferent cytokines also have direct effect mononuclear phagocytes, e.g., controlling the numbers of macrophages, their state of differentiation, and their state of activation. It is now recognized that macrophages are derived from bone marrow precursors, which are induced by colony-stimulating factors to differentiate from a myelocytic precursor to the monocyte lineage (1–3). These bone marrow progenitors develop into monocytes within a week and are released as blood monocytes, which circulate in the blood for 1–3 days. In response to poorly characterized inducers, these mononuclear phagocytes then migrate into the tissues of the body to differentiate into tissue macrophages (3). Proof that alveolar macrophages are of bone marrow origin comes from observing the phenotypic and genotypic changes that occur within the lung population several months after either bone marrow transplantation or lung transplantation (4,5). In both instances, the lung becomes populated by macrophages autologous with the cell type of the functioning bone marrow.

The movement of macrophages from blood to the tissues represents an important crossroads in the functional role of the macrophage. Macrophages that have differentiated perform very different functions. In this context cytokine differences occur in accordance with differentiation. Monocytes are characterized by peroxidase-positive granules that are rich in leukocyte elastase and cathepsin G (6), which are lost with maturation into macrophages. Conversely, monocytes are potent releasers of IL-1 in response to endotoxin, however, this potency wanes with maturation in vitro and in vivo (7,8). In contrast, TNF and platelet-derived growth factor release by mononuclear phagocytes is enhanced with maturation into macrophages (9–12). These differentiation-induced, cytokine regulatory changes broaden our understanding of macrophage activation. They suggest that part of the activities previously attributed to macrophage activation may also be a reflection of relative stages of differentiation in the macrophage life cycle.

Role of Apoptosis in Macrophage Differentiation

The factors that promote monocyte migration to tissues such as the lung are beginning to be better understood. For example, local MCP-1 production likely provides a signal to enter the tissue compartment (13). Additional signals are needed to maintain macrophage viability. Environmental factors are also undoubtedly important in preventing monocyte apoptosis. For example, it has been shown that monocytes cultured in vitro undergo classic changes of apoptosis with the first 24 h unless they receive additional signals. In this regard, IL-1, TNF, and lipopolysaccharide provide antiapoptotic signals (14,15) (Fig. 1).

The apoptosis connection to macrophage viability is of particular interest in light of the recent discovery that ced-3, a critical regulator of programmed cell death in *C. elegans*, is homologous with the mammalian cysteine proteinase IL-1 converting enzyme (ICE) (16,17). ICE is the enzyme that converts the inactive IL-1β precursor to the mature functional molecule (18,19). When transfected into neuronal cells, ICE expression promotes apoptosis of these cells. This death can be inhibited by co-expression of the ICE inhibitory protein Crm A or by treatment with the synthetic inhibitor Ac-YVAD chloromethyl ketone (20). These observations suggest that there is an intimate connection between cytokine production and the differentiation of the monocyte to the macrophage.

Macrophage Functions

After setting up tissue residence, macrophages have at least three classic functions (2). First, they operate as housekeepers by removing apoptotic cells (21) and both endogenous and exogenous debris. In the process of this "garbage collection," macrophages are scrutinizing the environment for microorganisms. Their second function is as the body's primary defense against microorganisms, i.e.,

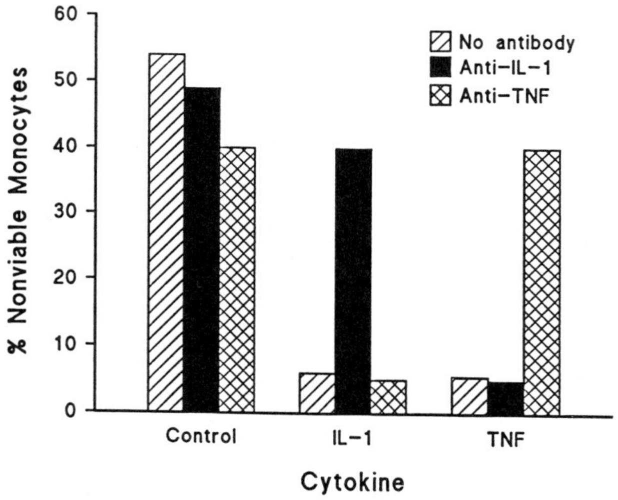

Figure 1 Cytokine regulation of macrophage apoptosis. Programmed cell death occurs spontaneously in monocytes after overnight culture, as shown in control. This apoptosis is prevented by both IL-1 and TNF. This cytokine prevention of apoptosis is blocked by antibody to IL-1 (solid bar) and antibody to TNF (cross-hatched bar). (From Ref. 15.)

the phagocytosis and killing of these invaders. The destruction of microorganisms first requires recognition of the invaders, which is aided by cell surface receptors for opsonins such as immunoglobulins (Fc receptors) and components of C3. Interaction with these receptors not only enhances phagocytosis but also "activates" the macrophages to heightened oxidant production and, as will be discussed later, cytokine production. The third function, related to phagocytic digestion of foreign material, involves processing and presenting foreign antigen to lymphocytes, inducing the proliferation of specific T lymphocytes as well as triggering the effector cells of the humoral system. It is now recognized that cytokine production by macrophages is intimately a part of these classic functions and that in addition to these "classic" functions macrophages may have long-range effects on virtually any cell type by the hormonal modulation of macrophage-derived cytokines.

HISTORY

Macrophages have been linked to cell-derived messenger proteins since the first descriptions of cytokines in the early 1960s. The first well-characterized cytokine was a lymphokine (lymphocyte product), which inhibited the in vitro migration of macrophages. Termed macrophage migration inhibitory factor (MIF), this cytokine, produced in response to a sensitized antigen, demonstrated that soluble nonimmunoglobulin cell proteins could have significant effects upon cell function (22,23). (MIF activity is now known to be due, at least in part, to interferon-γ.)

Later it was recognized that these "kines" had effects on other functions besides kinetic (cell movement) phenomena. From the perspective of macrophage-produced cytokines, the discovery of interleukin-1 (IL-1) represents a typical history of the recognition and then refinement of understanding of particular cytokines. IL-1 was first recognized as a soluble factor in rabbit peritoneal exudates, which could cause fever when injected into other animals (24). It was called endogenous pyrogen (EP) to distinguish it from exogenous pyrogens such as endotoxin. EP was determined to be a macrophage product, and an independent body of literature was generated in the EP field for the next 30 years. It was not appreciated until the late 1970s that EP and IL-1 were the same molecule. IL-1 was first identified in the early 1970s as the soluble macrophage protein [lymphocyte activating factor (LAF)] that substituted for the macrophage requirement in lymphocyte proliferation (25,26). IL-1 became the first interleukin in 1979 at the Second International Lymphokine Workshop (27). The term interleukin suggested its function as a message "between leukocytes" and, in this specific case, as a signal to lymphocytes to produce IL-2, the growth factor for T lymphocytes. This nomenclature unintentionally ignored the multifunctional components of IL-1, particularly its effects upon nonleukocyte cells (such as its function as an endogenous pyrogen), but set the precedent for the current practice of giving new cytokines an interleukin nomenclature once the amino acid sequence has been determined.

MACROPHAGE INTERACTIONS AND CYTOKINE PRODUCTION

Cytokine production by macrophages generally requires stimulation or activation, as resident "resting" macrophages are typically not constitutive cytokine producers. Stimulation can come from various particulate and cell interactions with macrophages. Endotoxin provides one of the most potent "activators" of macrophages. In addition, IFN-γ is a lymphocyte-derived macrophage-activating cytokine (responsible at least in part for MIF activity), and interactions with macrophage Fc receptors and complement receptors (process linked to their classic phagocytic function) have been extensively studied (Table 1).

Macrophage/Endotoxin Interactions

Endotoxin represents the classic in vitro stimulator of macrophages. It is likely an important in vivo stimulator as well, since it is the key moiety of the bacterial cell responsible for activating the immune response. Current concepts hold that macrophages regulate host response to bacterial lipopolysaccharides, linking macrophage-generated cytokines with sepsis (28). In this context, the predominant cell responsible for endotoxin handling is the macrophage. When radiolabeled endotoxin is injected into animals, it is cleared almost exclusively by tissue macrophages (29). The highest concentrations occur in the liver, spleen, and lung (29). Thus, the mononuclear phagocyte defense system (formerly known as the reticuloendothelial system) represents the first-line defense against bacteria.

Understanding the macrophage's role in defense against gram-negative infection has been aided greatly by the model provided by the endotoxin-nonresponsive mouse strain. C3H/HeJ mice are resistant to endotoxin shock at LPS concentrations that are lethal to endotoxin-responsive strains (C3H/HeN) (30). At least part of this resistance can be attributed to the inability of the macrophage from C3H/HeJ strain to release TNF in response to endotoxin (31). However, the endotoxin nonresponder strain is more susceptible to infection with *Salmonella* (32). These observations suggest that the ability to produce cytokines like TNF may be beneficial in fighting infection, but that an overwhelming release of TNF may, on the other hand, be lethal.

Table 1 Examples of Macrophage Cytokine Inducers[a]

Inducer	Cytokine
Infectious material	
Endotoxin	IL-1(9–11), TNF-α(9–11), IL-6(200–202), IL-8(201–204), IL-10(187)
Virus	
RSV	IL-1(97), TNF-α, IL-6, IL-8(205)
influenza	IL-1(97), TNF (206)
S. aureus (toxic shock toxin)	IL-1(96)
Immune complexes	
Fc receptor triggering	TNF(65), IL-1(69), IL-1ra(72,74,75,113), PDGF(189)
Fibrogenic material	
Silica	IL-1(78,79,86), TNF(86)
Asbestos	TNF(85,207), IL-1(207,208), IL-8(209)
Complement	
C5a	IL-1(104,105), IL-6(210)
Cytokines	
IL-1	PDGF(190), IL-8 (211)
TNF	IL-8(212), IL-1(213)
GM-CSF + IF	TNF(214)
Neuropeptides	
Substance P	IL-1(215)

[a]Cytokines released by macrophages directly in response to the interacting agent. References documenting these effects are indicated in parentheses after each cytokine.

Endotoxin Receptors

The exact mechanism governing how lipopolysaccharide interacts with macrophages is under active investigation. Studies are complicated by the heterogeneity and complexity of endotoxin molecules. Although lipophilic endotoxin may interact with cells directly via the plasma membrane, several candidate receptors have been identified. For example, the adhesion-promoting integrins of the CD11/CD18 family interact with endotoxin (33), and using radiolabeled endotoxin, specific macrophage binding to a surface receptor has been reported (34). Recently, CD14, a surface marker specific for mononuclear phagocytes, has been demonstrated to complex with endotoxin. This endotoxin complex, however, requires a second protein, lipopolysaccharide-binding protein (LBP). CD14 will complex with LBP, which is an acute-phase protein, but it does so only after LBP has first bound endotoxin (lipopolysaccharide) (33,35,36). Endotoxin incorporated into red blood cell membranes will rosette macrophages only if the red cell–endotoxin complex has been preincubated with LBP (37). CD14 antibodies can block this interaction, and LBP is absolutely required for endotoxin to be recognized by CD14. This interaction is of interest first because it demonstrates the complexity of the in vivo response to endotoxin. Second, this model is intriguing because it completes a cycle by linking endotoxin-induced cytokines, which have potent effects on acute-phase proteins (e.g., IL-6, IL-1, and TNF), to the ability of the macrophage to interact with endotoxin.

Endotoxin Sensitivity

Endotoxin stimulation of macrophages produces a burst of proinflammatory cytokines including IL-1, TNF, and IL-6 (38–40). Its interaction with the circulating mononuclear phagocyte, the blood monocyte, is astonishingly potent. The monocyte's sensitivity to endotoxin is in the pg/ml range for IL-1 production, rivaling the Limulus amebocyte lysate for sensitivity in this regard (41–43). The reason for the marked endotoxin sensitivity can be partly explained by new information from lipopolysaccharide receptor studies, but this work continues.

Since host response to sepsis is to a large extent linked to the macrophage's ability to interact with endotoxin, it is also notable that macrophages downregulate their responsiveness to endotoxin in response to corticosteroids and to endotoxin itself (31,44,45). Most of the endotoxin-induced cytokines share a 3′ untranslated repetitive TTATTTAT consensus sequence (46). Recent data argue that this region of the genetic message may be responsible for post-transcriptional control of cytokine expression; this has been demonstrated convincingly for TNF biosynthesis using a CAT gene construct (47).

Inflammatory cells recruited by macrophages may also affect the sensitivity of macrophages to endotoxin. For example, tissue toxicity caused by lipopolysaccharides can be dramatically reduced by neutrophil acyloxyacyl hydrolase, which removes acyl groups from endotoxin (48,49). Deacylated endotoxin was markedly depressed activity. Thus, endotoxin-induced release of neutrophil chemoattractants such as IL-8 may provide a means to detoxify endotoxin by the recruitment of neutrophil acyloxyacyl hydrolase. Ongoing investigations into the regulation of endotoxin receptor density and into the regulation of cytokine and endotoxin stability will provide new insights into therapeutic strategies in management of diseases characterized by endotoxin-induced cytokine excess.

α_2-Macroglobulin and Macrophage Cytokines

Macrophages interact with cytokines via specific cytokine receptors but likely also interact in various indirect ways, such as with α_2-macroglobulin (α2-MG). Interaction with the serum antiprotease α2-MG represents a mechanism that has been evolutionarily preserved for more than 550 million years, dating back to arthropodia (50). Macrophages express a 420 kDa cell surface glycoprotein, which binds the activated conformation of α2-MG (51). The activated conformation ("fast" α2-MG) is the protease-complexed form of α2-MG, which is rapidly cleared from the circulation ($T_{1/2}$ of 6 min) (52). A number of investigations now suggest that this "fast" α2-MG may also noncovalently complex various cytokines such as IL-1 (53), TGF-β (54), PDGF (55), IL-6 (56), and others such as MAF, NGF, and FGF (57). The significance of this interaction is not known at present. However, macrophage receptors for α2-MG may rapidly clear cytokine–α2-MG complexes, preventing natural cytokine receptor interactions. It is clear that in vitro systems that use sera must take into account the possible interference of α2-MG in the functional measurements (54). Furthermore, since macrophages are capable of α2-MG synthesis, autocrine effects on macrophage cytokine release must be considered as well.

Fc Receptors and Cytokines

The interaction of antigen-antibody complexes with macrophages is a key activation signal for macrophages. Bacteria coated with antibody are opsonized or more readily phagocytosed by macrophages. This event is mediated by the expression of receptors for the Fc component of immunoglobulin G. It is now recognized that macrophages express three distinct types of Fcγ receptors (FcγR I, FcγR II, and FcγR III) (58) and that these receptors have important links to cytokines. Certain cytokines effect the expression of the Fcγ receptors, and conversely, Fcγ receptor triggering may regulate the expression of cytokines.

In vitro, recombinant interferon-γ (IFN-γ) induces monocytes to increase their Fcγ receptor density 5- to 10-fold, thus sensitizing the macrophage to opsonized particles and immune complexes (59). This may be one of the mechanisms whereby IFN-γ acts as macrophage-activating factor (MAF/MIF). Interestingly, other cytokines have potentially relevant functions in regard to Fcγ receptor expression. For example, IL-4, which itself may be induced by Fcγ receptor cross-linking of non-B non-T cells from spleen and certain mast cell lines (60), can inhibit the IFN-γ–induced upregulation of Fcγ receptor expression (61). Endotoxin, IL-1, and, to a lesser extent, TNF have also been shown to counteract the IFN-γ effects on Fcγ receptor density (62). Paradoxically, IL-10, which in most instances works to suppress IFN-γ–mediated events, also upregulates the expression of FcγR I (63). Thus, cytokines may regulate the density of Fcγ receptors on macrophages.

From the perspective of function, Fcγ receptors may direct the production and release of various cytokines. The best studied examples are with TNF. Cross-linking of FcγRI on monocytes includes TNF-α release within 2 h (64,65). Of note, monocyte subsets that are Fcγ receptor positive are much better producers of TNF than are Fcγ receptor–negative subsets (66). IL-6 is induced by aggregated IgG and the Fcγ fragment of IgG in monocytes (67). The case of IL-1 is more complicated. Although several studies have implicated Fcγ receptor cross-linking in IL-1 release by monocytes (68–71), this finding is not universal and is complicated by concerns of endotoxin contamination of antibody preparations (72). Recent studies demonstrate that endotoxin primes monocytes for response to Fcγ receptor cross-linking (73); furthermore, the combination of FcγR cross-linking and endotoxin stimulation profoundly affects the cytokine profiles (74) (Fig. 2).

The connection of Fcγ receptors to IL-1 is further supported by the recent cloning of a specific IL-1 inhibitory protein, the IL-1 receptor antagonist (IL-1ra) (75). IL-1ra was purified and subsequently characterized from the inhibitory supernatant molecules produced by monocytes grown on adherent immune complexes or IgG (76). It is a unique cytokine that functions by acting as a competitive inhibitor for both IL-1α and IL-1β (76). IL-1ra is similar in structure to IL-1β and IL-1α (26 and 19% homology at gene level, respectively) (75) and functions by competing with IL-1 for its receptor. The full significance of this molecule has yet to be elucidated.

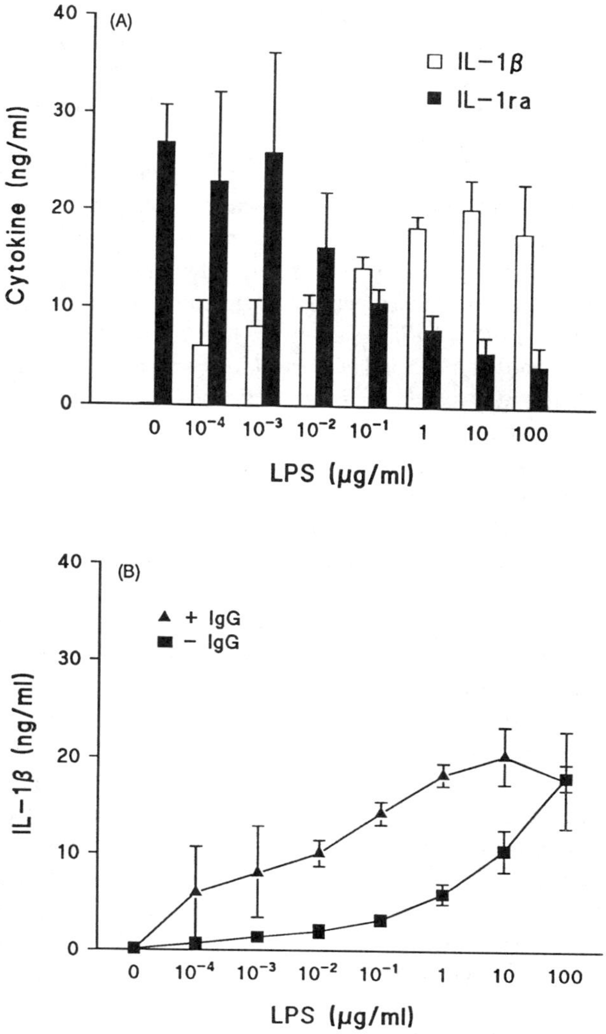

Figure 2 Endotoxin and Fc receptor cross-linking have diverse effects on IL-1β and IL-1ra release by monocytes. (A) Mononuclear phagocytes cultured overnight with increasing concentrations of lipopolysaccharide (LPS) on immobilized IgG to cross-link Fc receptors. Shown are the amounts of IL-1β (open bar) and IL-1ra (solid bar) released into the supernatant. (B) Effect of Fc receptor cross-linking on mononuclear responsiveness to LPS. Cells were cultured on either plastic (box) or immobilized human pooled IgG (triangle) while varying the LPS concentration. Note the Fc receptor cross-linking enhances the sensitivity of the IL-1β response to LPS but inhibits the IL-1ra response to LPS. (Adapted from Ref. 74.)

Interactions with Fibrogenic Dust

It has long been believed that certain types of dusts are fibrogenic in humans and animals and that the fibrogenicity is related to the interaction of the dust particles with macrophages. Of the fibrogenic dusts, silica and asbestos are prime examples. These particles have long been known to have a profound effect upon macrophages. The induction of cytokine production by fibrogenic dusts is currently an area of active investigation.

Silica represents a powerful membrane perturbation to macrophages and produces a selective toxicity to macrophages. Silicosis has been linked to diseases of macrophage dysfunction such as alveolar proteinosis (inability to clear surfactant) and tuberculosis (impaired killing of the tubercle bacillus) (77). Therefore, it is reasonable to hypothesize that silica-induced cytokine production by macrophages may also be involved in disease states. Silica is a potent inducer of IL-1 from monocytes and macrophages (78). It appears to have a greater effect upon inducing release of IL-1 precursor stores than does endotoxin, an effect possibly related to its cytotoxicity (78–81). Furthermore, silica sensitizes macrophages to endotoxin with respect to IL-1 release (82).

Silicotic lung disease is primarily a granulomatous and fibrotic disease, which suggests the release of fibroblast growth factors. IL-1 may have growth-promoting activity (83). In vitro, silica induces fibroblast growth factor activity but without the concomitant PGE_2 release that blocks this activity in endotoxin-induced macrophages (84). Silica also releases TNF in both in vitro and in vivo models (85,86). An exciting study has linked the lung fibrosis of silica to TNF (87). Rats given intratracheal silica developed pulmonary fibrosis, which was associated with the continued expression of TNF mRNA for more than 70 days without a demonstrable increase in IL-1 or TGF-β. Furthermore, an antibody to TNF prevented increased collagen deposition in this model.

Asbestos is also linked to fibrotic changes in the lung, and like silica, asbestos is cytotoxic for macrophages. Although less is known about asbestos's ability to induce cytokines, it has been linked to TNF release (86) and to alveolar macrophage-derived growth factor (IGF-1) (88). Of note, the mononuclear phagocytes harvested by bronchoalveolar lavage of asbestos workers demonstrate an increase in the population of younger macrophages (89). This change in the population has implications regarding the sensitivity of this lung population to cytokine release, since monocytes and macrophages differ with respect to IL-1 and TNF release (10,11,90,91)

Other Interactions

As information about cytokines expands, the types of macrophage interactions that may play a physiological role in macrophage cytokine production is growing exponentially. The largest body of new data is centering around deciphering the messages that cytokines produce when they are presented in concert. However, changes induced by viruses, diet, and bacterial exotoxins have been recognized as important macrophage modulators as well.

Diet

A fish oil diet (high in N-3 polyunsaturated fatty acids) has surprisingly potent effects upon macrophage cytokine release (92). Within 6 weeks of supplementation, inducible levels of IL-1α, IL-1β, and TNF-α are decreased in normal individuals. These changes are accompanied by changes in PGE_2 inducibility and a decrease in the arachidonic acid to eicosapentaenoic acid ratio in mononuclear cell membranes. The mechanism of this effect on cytokines is not known but may be at least partly explained by known prostanoid effects on cytokine production (93–95).

Infectious Agents

As mentioned above, bacterial endotoxins are the classic inducers of macrophage cytokines, but other infectious agents have marked effects also. The toxin produced by the strain of *Staphylococcus aureus* responsible for toxic shock syndrome is an inducer of IL-1 from monocytes (96). Viral particles from influenza and respiratory syncytial virus both can induce IL-1, although RSV induces a high molecular weight IL-1 inhibitor concomitantly (97). HIV-1 has been examined extensively with conflicting results regarding cytokines. As with RSV, part of the conflict has to do with the concomitant production of inhibitors that has been demonstrated for IL-1 (98,99). In addition, the stage of infection with HIV has an effect. For example, endotoxin-induced TNF release may be high in asymptomatic HIV infection but drops progressively with advancement through the stages of AIDS-related complex and full-blown AIDS (100). Of note, however, normal macrophages are not induced to produce IL-1, IL-6, or TNF with HIV or with purified envelope protein (101,102). Furthermore, when hyperplastic lymph nodes of HIV-infected individuals were studied, although IL-1 and IL-6 were detected in sinus cells, they did not differ from non–HIV-infected lymph nodes in this regard (103). In contrast, there was a marked enhancement of IFN-γ expression in the HIV lymph node cells that are in direct contact with HIV-infected cells, suggesting ongoing anti-HIV T-cell immune response (103).

Complement

Complement has been linked to macrophage cytokine release. C5a may act upon macrophages to induce cytokines. It has been shown that C5a, endotoxin-free by limulus amebocyte lysate assay, can induce IL-1 release. It also synergizes with endotoxin or with IFN-γ (104) and is believed to work by acting upon macrophage receptors for C5a (105). As another example, monocytes that are enriched for CR3 are better IL-1 producers than CR3- monocytes, implying a complement connection (106).

Surfactant

Although alveolar macrophages normally live in a surfactant-rich environment, until recently very little was known about the effect of surfactant upon macrophage cytokine production. Recent data, however, suggest that surfactant may serve a largely inhibitory function in macrophage proinflammatory cytokine release (107). Bovine surfactant preparations inhibit both TNF-α and IL-1β release from alveolar macrophages in a dose-dependent fashion (Fig. 3).

C-Reactive Protein

Acute-phase proteins may provide important mechanisms to modulate macrophage function. In humans, the chief acute-phase reactant is C-reactive protein (CRP), a pentameric hepatocyte protein that may increase 1000-fold in plasma in response to stress or inflammation (108). CRP has recently been shown to have important effects on macrophages. Macrophages possess CRP receptors (109,110), and CRP can potently upregulate proinflammatory cytokine production (111). Interestingly, in our own recent investigations it appears that acute-phase concentrations of

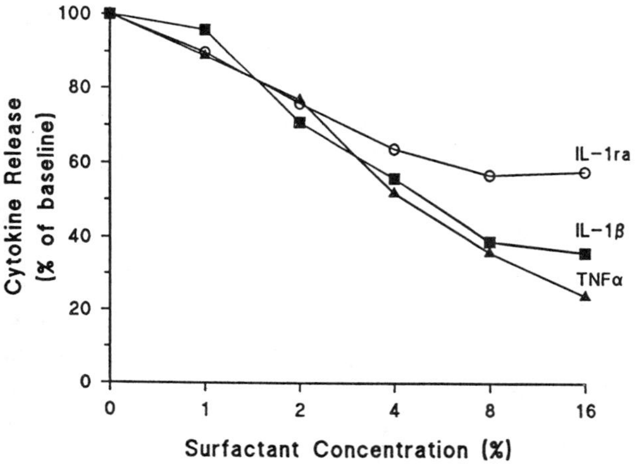

Figure 3 Effect of surfactant upon cytokine production by alveolar macrophages. Normal human alveolar macrophages cultured overnight in the presence of lipopolysaccharide (1 μg/ml) released significant amounts of IL-1ra, IL-1-β, and TNF-α as measured by specific ELISAs. The addition of increasing concentrations of bovine surfactant (Survanta) caused a dose-dependent decrease in cytokine release, which was more pronounced for the proinflammatory cytokines IL-1β and TNF-α. (Adapted from Ref. 107.)

CRP enhance blood monocyte but inhibit alveolar macrophage cytokine production (111a).

MACROPHAGE-DERIVED CYTOKINES

The macrophage-produced cytokines fall largely into the group of cytokines termed proinflammatory. These cytokines are the first messengers that macrophages release in response to perturbations in their environment and represent one of the more potent ways that macrophages serve to coordinate and direct immune reactions. These proinflammatory cytokines are reviewed briefly here and in more detail in specific chapters.

The IL-1 Family

The IL-1 family is now known to contain three members that arose from a common ancestral gene. All three members, IL-1α, IL-1β, and IL-1 receptor antagonist (IL-1ra), exist in close proximity on chromosome 2 (112). The two proinflammatory members, IL-1α and IL-1β, are produced as 31 kDa precursor proteins, which are produced in large quantities by macrophages in response to many stimuli, most notably endotoxin. IL-1ra is also induced by endotoxin but is an anti-inflammatory molecule by virtue of its ability to act as a pure receptor antagonist for the type I IL-1 receptor (113). IL-1β may represent as much as 1–2% of the mRNA produced after endotoxin challenge (114). IL-1α is a pI 5 molecule that is functional in the propeptide state; IL-1β is a pI 7 protein that requires processing to the 17 kDa protein for activation. These molecules share 50% homology at the gene level and 26% homology at the protein level.

IL-1α, IL-1β, and IL-1ra react with the same receptor molecule (115), and the α and β members have almost, if not, complete overlap in functional characteristics (114,116). However, when stimulated, monocytes produce about 10 times more IL-1β than IL-1α (114). Since IL-1α is functional in the precursor state (117) and since it is highly myristylated, which implies membrane targeting (118), it may be important in reactions within the localized milieu of the macrophage, whereas IL-1β may be the hormonal form of IL-1.

IL-1 acts upon many cell types, including lymphocytes, hepatocytes, endothelial cells, fibroblasts, chrondrocytes, and neutrophils (119). With lymphocytes the effects include augmentation of T-cell proliferation and B-cell activation in response to antigen challenge (119–121). With hepatocytes IL-1 is an inducer of the acute-phase response (122). In response to IL-1, endothelial cells present new neutrophil adherence molecules such as ELAMs and become more permeable (123,124). Fibroblast proliferation is enhanced with IL-1; collagenase, PGE_2, and IL-8 are released (83,125–127). Neutrophils experience heightened oxidant production, and smooth muscle cells are induced to relaxation in response to IL-1 by inducible nitric oxide synthase production (128–130). In addition to these direct effects, IL-1 synergizes with other cytokines for much of its activity, the best studied examples having involved TNF (8,131,132).

IL-1 is regulated at a number of levels, both during its induction and after its release. For example, IL-1 release decreases with maturation of monocytes into macrophages (7,10). IL-1 mRNA translation is inhibited by corticosteroids (45). Its production can be superinduced with cycloheximide, suggesting that there is a repressor protein. The processing from the inactive IL-1β precursor requires the coexpression of a convertase activity provided by IL-1β–converting enzyme (ICE) (133–137). ICE activation itself also appears to be tightly controlled (138). The cowpox virus expresses one such molecule, crmA, which is an inhibitor of ICE activity and appears to regulate the intensity of inflammation in response to cowpox infection (139). Newly described ICE analog, such as ICH-1, may also regulate ICE and hence control IL-1β activation (140).

Another level of IL-1 regulation occurs after its release. Two molecules directly function to inhibit IL-1. The first, IL-1ra, can block the ability of either IL-1α or IL-1β to interact with its signaling receptor, the type I receptor (76). This inhibition is relatively inefficient and requires a significant excess of IL-1ra over IL-1 in order to function. Interestingly, in experimental human endotoxemia an excess of IL-1ra/IL-1 (100-fold) has been documented (141). This suggests that IL-1ra may provide a mechanism to fine-tune functional IL-1 levels. The second inhibitory molecule is the soluble type II IL-1 receptor (IL-1RII). This is a 45 kDa molecule released by neutrophils in response to dexamethasone and IL-4 and has been likened to a "decoy" receptor (142). It selectively complexes IL-1β over IL-1α, but importantly, its effects are not inhibited by IL-1ra (143). This suggests that the soluble IL-1RII is a specific IL-1β inhibitor and functions in concert with IL-1ra to inhibit IL-1.

The TNF Family

One cytokine that is completely distinct in structure from IL-1 but that has enormous functional overlap is TNF-α. TNF-α is a 17 kDa protein that circulates as a trimer. Like IL-1, it is largely a macrophage product that is produced in large quantities within 90 min of endotoxin challenge (144). It exists on chromosome 6 within the major histocompatibility complex in tandem with lymphotoxin-α and lymphotoxin-β, the lymphocyte representatives of the TNF family (145,146). Lymphotoxin-α (also called TNF-β) and lymphotoxin-β also exist as trimers. Like IL-1α and IL-1β, these TNF family proteins have significant homology at both the gene and protein levels and have almost identical functional characteristics. As homotrimers, TNF-α and lymphotoxin-α both can interchangeably cross-link the p75 and p55 TNF receptors and can exist in either a membrane-bound form or as soluble molecules. TNF-α is released from the membrane by a metalloproteinase (147). In contrast, lymphotoxin-β appears to exist as a heterotrimer with lymphotoxin-α and functions through the TNFR-rp (148). Lymphotoxin-β function is not clear but appears to play a role in lymph node development (148).

The classic functions of tumor necrosis factor are its ability to cause necrosis of tumors transplanted into mice and its ability to cause cachexia (149). TNF also has important effects on many cells that overlap almost completely with IL-1, e.g., lymphocyte proliferation (150), fibroblast PGE_2 and collagenase release (151), fever (152), endothelial expression of ELAM (153), induction of acute-phase proteins (154), induction of secondary cytokines such as IL-8 (155), INOS synthesis (130), and activation of neutrophils (129,156). TNF has been singled out as a critical if not central player in the cascade of events that occurs with the syndrome of septic shock (157,158).

In contrast to IL-1, TNF release is enhanced in macrophages compared to monocytes, but like IL-1 its production is also enhanced by IFN-γ (159,160) and inhibited by corticosteroids (161). As mentioned above, TNF (and most proinflammatory peptides) contains a 3′ untranslated repeating sequence, TTATTTAT, which for TNF has been shown to be important in regulating translation (46,47). TNF is produced within 90 min of endotoxin challenge in humans (144), but its production is downregulated after endotoxin challenge as well (162).

IL-6

IL-6 is a macrophage-derived cytokine (also produced by many other cell types) 21 kDa in size with multiple functions, some of which overlap with IL-1 and TNF (163,164). IL-6 classically is a B-cell growth factor but has important effects as the major hepatic acute-phase protein inducer, as a T-cell activator, and as a thrombopoietin (154). As mentioned earlier, IL-1 and TNF are potent inducers of IL-6 production, and many of the functions previously attributed to IL-1 or TNF may be IL-6 effects [e.g., acute-phase protein production, thymocyte proliferation, and T-cell activation (165,166)]. IL-6 does not induce IL-1 or TNF but feeds back to inhibit endotoxin-induced TNF and IL-1 production.

The control of IL-6 production is intimately linked to

that of IL-1 and TNF. An NF-κB–factor produced in response to IL-1 or TNF interacts with the NF-κB site, and other regulatory sequences in the IL-6 promoter are also responsive to IL-1 and TNF (154). Cycloheximide superinduces IL-6, as it does for IL-1 and TNF. Interestingly, IL-6 does not appear to contain the TATTTATT sequence in the 3′ untranslated region, which is standard for most of the proinflammatory cytokines (46).

IL-6 is induced by endotoxin (167) and is present in the serum of septic patients (168). Similar to IL-1ra, it is induced by Fc receptor cross-linking on monocytes (169). IL-6 is upregulated in the lung in sarcoidosis and berylliosis (170) and also HIV infection (171). Interestingly, a specific IL-6 inhibitory activity has recently been identified in alveolar macrophage supernatants (172).

The receptor system for IL-6 is quite unique (173). The IL-6 receptor is an 80 kDa glycoprotein that has structural homology to both the immunoglobulin superfamily and the hematopoietic cytokine receptor family. The IL-6 receptor does not induce signal transduction but instead causes homodimerization of the gp130 transmembrane protein (173).

IL-8

IL-8 (formerly monocyte-derived chemotactic factor or neutrophil-activating peptide-1) is an 8 kDa peptide member of the C-X-C chemokine family, which likely exists as a dimer. It is produced by mononuclear phagocytes (and other cell types), which is a potent neutrophil attractant and stimulant (174–176). Like IL-6, IL-8 can be induced by endotoxin, IL-1, and TNF (177,178), and like IL-1, TNF and IL-6 are superinducible by cycloheximide. IL-8 is responsible for many of the neutrophil chemotactic properties that have been previously attributed to IL-1 and TNF in animal experiments (179). Importantly, IL-8 is a member of a large family of C-X-C members, which includes GROα, GROβ, GROγ, NAP-2, ENA-78, PF4, IP-10, and GCP-2 (180).

In human endotoxemia, circulating IL-8 levels peak 2 h after endotoxin and are augmented with the addition of ibuprofen (181). By itself, circulating IL-8 is not toxic since infusions of rIL-8 had no effect in nonhuman primates (182). It has been suggested that IL-8–induced neutrophil migration requires haptotaxis, i.e., binding to localized proteins such as matrix that produces a gradient (183). In this regard, circulating red blood cells may serve as a sink for IL-8 since the Duffy antigen appears to act as a receptor for IL-8 (184).

IL-8 is avidly induced in the lung by macrophages and by lung parenchymal cells. IL-8 is a major constituent of lavage fluid in patients with cystic fibrosis (185), and it is clear that IL-1 and TNF have potent capacity to induce IL-8 from lung epithelial cells and fibroblasts (127,155).

There are two known functional IL-8 receptors that share significant homology. These are abundantly expressed on neutrophils and to a lesser extent on macrophages and lymphocytes (180). Both receptors are typical G-protein coupled receptors with seven transmembrane regions (180).

IL-10

IL-10 is an 18 kDa protein with striking inhibitory effect on macrophage cytokine production (186). IL-10 was initially cloned from a mouse T-cell cDNA library as a product of TH_2 lymphocytes (63). The function of IL-10 closely mimics the function of TH_2 lymphocytes since it appears to direct the immune response toward a humoral response and away from cell-mediated immunity. In fact, it behaves as a counterregulator for IFN-γ. IFN-γ promotes cell-mediated immunity, activates macrophage cytokine production, and inhibits IL-10 production. In contrast, IL-10 inhibits macrophage cytokine production and inhibits IFN-γ production (63).

Although initially cloned from T cells, IL-10 is readily produced by macrophages (187). It is released from macrophages late after endotoxin, with levels peaking at 24–48 h, whereas TNF-α levels peak at 8 hours (187) (Fig. 4). At low levels it powerfully inhibits TNF-α, IL-1, and IL-6 release (186), and antibodies to IL-10 demonstrate that macrophage IL-10 production has an important autocrine inhibitory effect (187). When given in vivo to mice, IL-10 prevents the lethality and hypothermia induced by an endotoxin challenge (188). Finally, the EB viral genome contains an open reading frame that expresses BCRF1, which is highly homologous to IL-10, meaning that it may enhance infectivity by preventing proinflammatory cytokine induced antiviral inflammation.

Growth Factors

Platelet-Derived Growth Factor

Platelet-derived growth factor (PDGF) is a 31 kDa glycoprotein product of platelets and macrophages, which functions as a potent mesenchymal cell chemoattractant and as a fibroblast growth factor (189). It acts as a "competence" factor for fibroblast growth by inducing entrance into the early G_1 phase of the cell cycle, allowing the cell to respond to subsequent "progression" factors for proliferation. PDGF is released in exaggerated amounts in idiopathic pulmonary fibrosis (189) and may play a pathogenetic role in the fibrotic reaction.

PDGF does complex with α_2-macroglobulin, and in

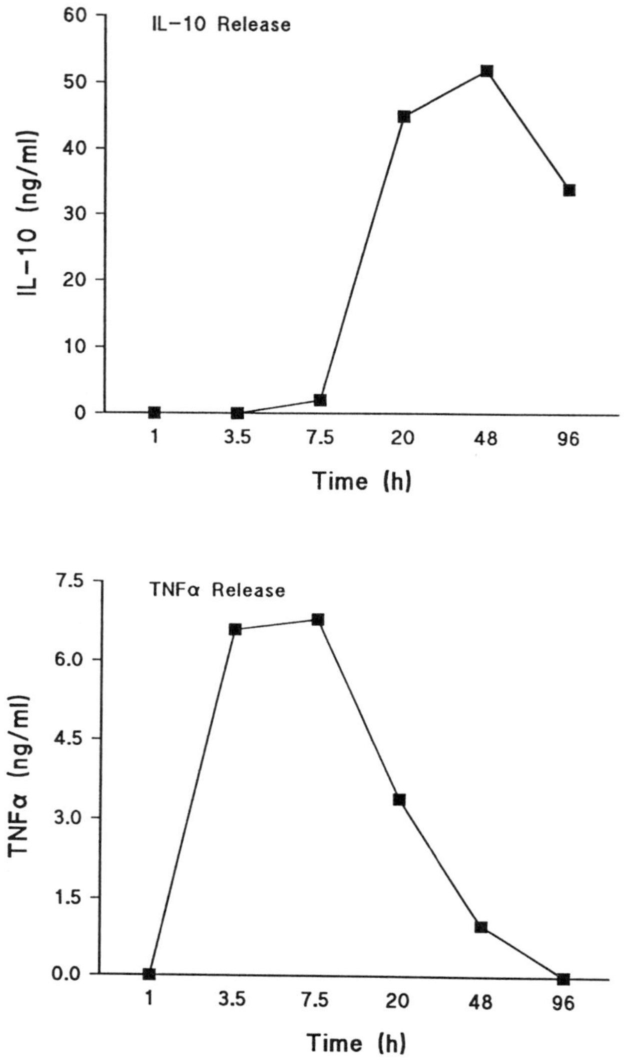

Figure 4 The kinetics of IL-10 and TNF-α release from purified human monocytes cultured in vitro with lipopolysaccharide (1 μg/ml) as assayed by ELISA. This delayed IL-10 synthesis has been linked to the downregulation of proinflammatory cytokine synthesis. (Adapted from Ref. 187.)

this regard its function often requires a pretreatment step to allow its expression (55,57). For example, supernatants from macrophages are dialyzed against 1 M acetic acid to allow full expression of chemotactic activity (189); this may remove the α_2-macroglobulin.

In the context of cytokine interactions, it is of note that IL-1 can induce PDGF from fibroblasts and smooth muscle cells, which may explain in part the ability of IL-1 to induce fibroblast proliferation (83,190). Furthermore, it has been shown that PDGF can induce TGF-β expression in wound macrophages (191).

Insulinlike Growth Factors

Insulinlike growth factors (IGFs) are 7.5 kDa peptides that have considerable homology with each other and with proinsulin, and which function as stimulators of cellular proliferation and differentiation (192). Virtually all cells have IGF receptors, with IGF-2 being primarily important in prenatal growth and development and IGF-1 being the postnatal growth factor (192). IGF-1 is known to be produced by alveolar macrophages, and it may represent most if not all of the activity formerly known as alveolar macrophage-derived growth factor (AMDGF) (193). Like AMDGF, IGF-1 acts as a progression factor, i.e., it can take a cell made "competent" by PDGF or fibronectin and push it through the G_1 phase of the cell cycle (193). It has been speculated that IGF-1 may be a key cytokine in the induction and maintenance of the fibrosis of certain interstitial lung disorders (193).

Transforming Growth Factor-β

Transforming growth factor-β (TGF-β) is a 25 kDa dimer that is multifunctional regulator of cell growth and differentiation produced and stored by platelets (194) but also by macrophages (195). Its function overlaps with that of epidermal growth factor and PDGF in their effects upon fibroblasts. TGF-β is chemotactic for fibroblasts (196). TGF-β mRNA appears to be constitutively expressed in monocytes and macrophages, but translation and protein release require mononuclear phagocyte activation (195).

The bleomycin model of lung fibrosis has been demonstrated to be linked to TGF-β production by macrophages (197). TGF-β, like IL-6, induces the production of acute-phase proteins, an effect that is both direct and due to modulation of the IL-6 effect (198). It may be linked to PDGF in that PDGF can induce its expression in experimental wounds (191). In this context TGF-β belongs with other macrophage cytokines, like PDGF and IGF-1, that may be critical regulators of fibrotic disorders of the lung (197). Distinct functions have been noted as well. For example, an interesting new twist to the repertoire of cytokine functions is provided by the recent observation that TGF-β may also have anti-inflammatory functions. TGF-β can produce burst suppression for H_2O_2 production by macrophages (199).

SUMMARY

Macrophages are key early responders to invaders and bodily insults. These first-line defenders are armed with a host of tools to recognize and destroy invaders. For example, immunoglobulin, endotoxin, complement, and C-reactive protein receptors enhance their ability to recognize foreign invaders. In addition, macrophages are equipped with an array of monitoring devices (receptors) that interpret the host's state of alertness in order to direct a response appropriate to the particular insult. For example, elevated IL-10 levels promote a humoral response and inhibit a delayed-type hypersensitivity, whereas IFN-γ encourages an aggressive cell-mediated response. Macrophages send out signals to neighboring parenchymal cells and distant organs via messengers like IL-1, TNF, and IL-8 to recruit support for the inflammatory response. Much of this essential cytokine communication is beginning to be understood, but not all of the signals have been identified, and the complexity of the networks will undoubtedly continue to challenge our understanding of the role of macrophages and cytokines in inflammation.

REFERENCES

1. van Furth R, Cohn ZA. The origin and kinetics of mononuclear phagocytes. J Exp Med 1968; 128:415.
2. Cline MJ, Lehrer RI, Territo C, Golde DW. Monocytes and macrophages: functions and diseases. Ann Intern Med 1978; 88:78.
3. Johnston RB Jr. Monocytes and macrophages. N Engl J Med 1988; 318:747.
4. Thomas ED, Ramberg RE, Sale GE, Sparkes RS, Golde DW. Direct evidence for a bone marrow origin of the alveolar macrophage in man. Science 1976; 192:1016.
5. Fung JJ, Zeevi A, Kaufman C, Paradis IL, Dauber JH, Hardesty RL, Griffith B, Duquesnoy RJ. Interactions between bronchoalveolar lymphocytes and macrophages in heart-lung transplant recipients. Hum Immunol 1985; 14:287.
6. Kargi HA, Campbell EJ, Kuhn C III. Elastase and cathepsin G of human monocytes: heterogeneity and subcellular localization to peroxidase-positive granules. J Histochem Cytochem 1990; 38:1179.
7. Wewers MD, Saltini C, Sellers S, Tocci MJ, Bayne EK, Schmidt JA, Crystal RG. Evaluation of alveolar macrophages in normals and individuals with active pulmonary sarcoidosis for the spontaneous expression of the interleukin-1 beta gene. Cell Immunol 1987; 107:479.
8. Elias JA, Gustilo K, Freundlich B. Human alveolar macrophage and blood monocyte inhibition of fibroblast proliferation: evidence for synergy between interleukin-1 and tumor necrosis factor. Am Rev Respir Dis 1988; 138:1595.
9. Martinet Y, Yamauchi K, Crystal RG. Differential expression of the tumor necrosis factor/cachectin gene by blood and lung mononuclear phagocytes. Am Rev Respir Dis 1988; 138:659.
10. Wewers MD, Herzyk DJ. Alveolar macrophages differ from blood monocytes in human interleukin 1 beta release: quantitation by enzyme linked immunoassay. J Immunol 1989; 143:1635.
11. Rich EA, Panuska JR, Wallis RS, Wolf CB, Leonard ML, Ellner JJ. Dyscoordinate expression of tumor necrosis factor-alpha by human blood monocytes and alveolar macrophages. Am Rev Respir Dis 1989; 139:1010.
12. Mornex J-F, Martinet Y, Yamauchi K, Bitterman PB, Grotendorst GR, Chytil-Weir A, Martin GR, Crystal RG. Spontaneous expression of the c-sis gene and release of a platelet-derived growth factorlike molecule by human alveolar macrophages. J Clin Invest 1986; 78:61.
13. Strieter RM, Koch AE, Antony VB, Fick Jr. RB, Standiford TJ, Kunkel SL. The immunopathology of chemotactic cytokines: The role of interleukin-8 and monocyte chemoattractant protein-1. J Lab Clin Med 1994; 123:183.
14. Mangan DF, Welch GR, Wahl SM. Lipopolysaccharide, tumor necrosis factor-α, and IL-1β prevent programmed cell death (apoptosis) in human peripheral blood monocytes. J Immunol 1991; 146:1541.
15. Mangan DF, Wahl SM. Differential regulation of human monocyte programmed cell death (apoptosis) by chemotactic factors and pro-inflammatory cytokines. J Immunol 1991; 147:3408.
16. Yuan J, Shaham S, Ledoux S, Ellis HM, Horvitz HR. The C. elegans cell death gene ced-3 encodes a protein similar to mammalian interleukin-1B-converting enzyme. Cell 1993; 75:641.
17. Miura M, Zhu H, Rotello R, Hartwieg EA, Yuan J. Induction of apoptosis in fibroblasts by IL-1B-converting enzyme, a mammalian homolog of the C. elegans cell death gene ced-3. Cell 1993; 75:653.
18. Thornberry NA, Bull HG, Calaycay JR, Chapman KT, Howard AD, Kostura MJ, Miller DK, Molineaux SM, Weidner JR, Aunins J, Elliston KO, Ayala JM, Casano FJ, Chin J, Ding GJF, Egger LA, Gaffney EP, Limjuco G, Palyha OC, Raju SM, Rolando AM, Salley JP, Yamin TT, Lee TD, Shively JE, MacCoss M, Mumford RA, Schmidt JA, Tocci MJ. A novel heterodimeric cysteine protease is required for interleukin-1beta processing in monocytes. Nature 1992; 356:768.
19. Cerretti DP, Kozlosky CJ, Mosley B, Nelson N, Van Ness K, Greenstreet TA, March CJ, Kronheim SR, Druck T, Cannizzaro LA, Huebner K, Black RA. Molecular cloning of the interleukin-1beta converting enzyme. Science 1992; 256:97.
20. Boudreau N, Sympson CJ, Werb Z, Bissell MJ. Suppression of ICE and apoptosis in mammary epithelial cells by extracellular matrix. Science 1995; 267:891.
21. Savill J, Hogg N, Ren Y, Haslett C. Thrombospondin cooperates with CD36 and the vitronectin receptor in macrophage recognition of neutrophils undergoing apoptosis. J Clin Invest 1992; 90:1513.
22. Bloom BR, Bennett B. Mechanism of a reaction in vitro associated with delayed-type hypersensitivity. Science 1966; 153:180.
23. David JR. Delayed hypersensitivity in vitro: its mediation by cell-free substances formed by lymphoid-cell-antigen interactions. Proc Natl Acad Sci USA 1966; 56:73.
24. Beeson PB. Temperature-elevating effect of a substance obtained from polymorphonuclear leukocytes. J Clin Invest 1948; 27:524.
25. Bach FH, Alter BJ, Solliday S, Zoschke DC, Janis M. Lymphocyte reactivity in vitro. II. Soluble reconstituting factor permitting response of purified lymphocytes. Cell Immunol 1970; 1:219.
26. Gery L, Waksman BH. Potentiation of the T-lymphocyte response to mitogens: II. The cellular source of potentiating mediator(s). J Exp Med 1972; 136:143.
27. Aarden LA, Burnner TK, Cerottini JD, Dayer J-M, de Weck AL, Dinarello CA, et al. Letter to the Editor. Revised nomenclature for antigen-nonspecific T cell proliferation and helper factors. J Immunol 1979; 123:2928.
28. Ulevitch RJ, Wolfson N, Virca GD, Kim S, Kline L, Mathison JC. Macrophages regulate the host response to bacterial lipopolysaccharides. In: Perspectives in Shock Research: Metabo-

lism, Immunology, Mediators, and Models. Alan R. Liss, Inc., 1989:193.

29. Mathison JC, Ulevitch RJ. The clearance, tissue distribution, and cellular localization of intravenously injected lipopolysaccharide in rabbits. J Immunol 1979; 123:2133.
30. Michalek SM, Moore RN, McGhee JR, Rosenstreich DL, Mergenhagen SE. The primary role of lymphoreticular cells in the mediation of host responses to bacterial endotoxin. J Infect Dis 1980; 141:55.
31. Beutler B, Krochin N, Milsark IW, Leudke C, Cerami A. Control of cachectin (tumor necrosis factor) synthesis: mechanisms of endotoxin resistance. Science 1986; 232:977.
32. von Jeney N, Gunther E, Jann K. Mitogenic stimulation of murine spleen cells: relation to susceptibility to Salmonella infection. Infect Immun 1977; 15:26.
33. Wright SD, Jong MTC. Adhesion-promoting receptors on human macrophages recognize E. coli by binding to lipopolysaccharide. J Exp Med 1986; 164:1876.
34. Haeffner-Cavaillon N, Chaby R, Cavaillon J-M, Szabo L. Lipopolysaccharide receptor on rabbit peritoneal macrophages: binding characteristics. J Immunol 1982; 128:1950.
35. Tobias PS, Mathison JC, Ulevitch RJ. A family of lipopolysaccharide binding proteins involved in responses to gram-negative sepsis. J Biol Chem 1988; 263:13479.
36. Ferrero E, Hsieh C-L, Francke U, Goyert SM. CD14 is a member of the family of leucine-rich proteins and is encoded by a gene syntenic with multiple receptor genes. J Immunol 1990; 145:331.
37. Wright SD, Tobias PS, Ulevitch RJ, Ramos RA. Lippolysaccharide (LPS) binding protein opsonizes LPS-bearing particles for recognition by a novel receptor on macrophages. J Exp Med 1989; 170:1231.
38. Durum SK, Schmidt JA, Oppenheim JJ. Interleukin 1: an immunological perspective. Ann Rev Immunol 1985; 3:263.
39. Dinarello CA. Human pyrogen: a monocyte product mediating host defenses. In: Gallin JI, Fauci AS, eds. Advances in Host Defense Mechanisms. New York: Raven Press, 1982:57.
40. May LT, Ghrayeb J, Santhanam U, Tatter SB, Sthoeger Z, Helfgott DC, Chiorazzi N, Grieninger G, Sehgal PB. Synthesis and secretion of multiple forms of beta 2-interferon/B-cell differentiation factor 2/hepatocyte-stimulating factor by human fibroblasts and monocytes. J Biol Chem 1988; 263:7760.
41. Duff GW, Atkins E. The detection of endotoxin by in vitro production of endogenous pyrogen: comparison with Limilus amebocyte lysate gelation. J Immunol Methods 1982; 52:323.
42. Wood DD, Cameron PM. The relationship between bacterial endotoxin and human B cell-activating factor. J Immunol 1978; 121:53.
43. Northoff H, Gluck D, Wolpl A, Kubanek B, Galanos C. Lipopolysaccharide-induced elaboration of interleukin 1 by human monocytes: use for detection of lipopolysaccharide in serum and the influence of serum-lipopolysaccharide interactions. Rev Inf Dis 1987; 9:S599.
44. Virca GD, Kim SY, Glaser KB, Ulevitch RJ. Lipopolysaccharide induces hyporesponsiveness to its own action in RAW 264. 7 cells. J Biol Chem 1990; 264:21951.
45. Kern JA, Lamb RJ, Reed JC, Daniele RP, Nowell PC. Dexamethasone inhibition of interleukin 1 beta production by human monocytes: posttranscriptional mechanisms. J Clin Invest 1988; 81:237.
46. Caput D, Beutler B, Hartog K, Thayer R, Brown-Shimer S, Cerami A. Identification of a common nucleotide sequence in the 3′-untranslated region of mRNA molecules specifying inflammatory mediators. Proc Natl Acad Sci USA 1986; 83:1670.
47. Han J, Brown T, Beutler B. Endotoxin-responsive sequences control cachectin/tumor necrosis factor biosynthesis at the translational level. J Exp Med 1990; 171:465.
48. Munford RS, Hall CL. Detoxification of bacterial lipopolysaccharides (endotoxins) by a human neutrophil enzyme. Science 1986; 234:203.
49. Kovach NL, Yee E, Munford RS, Raetz CRH, Harlan JM. Lipid IVa inhibits synthesis and release of tumor necrosis factor induced by lipopolysaccharide in human whole blood ex vivo. J Exp Med 1990; 172:77.
50. Sottrup-Jensen L, Borth W, Hall M, Quigley JP, Armstrong PB. Sequence similarity between α_2-macroglobulin from the horseshoe crab, Limulus polyphemus, and proteins of the α_2-macroglobulin family from mammals. Comp Biochem Physiol 1990; 96B:621.
51. Ashcom JD, Tiller SE, Dickerson K, Cravens JL, Argraves WS, Strickland DK. The human α_2-macroglobulin receptor: Identification of a 420-kD cell surface glycoprotein specific for the activated conformation of α_2-macroglobulin. J Cell Biol 1990; 110:1041.
52. Starkey PM, Barrett AJ. Alpha 2-macroglobulin, a physiological regulator of proteinase activity. In: Proteinases in Mammalian Cells and Tissues. Barrett AJ, ed. Elsevier/North-Holland Biomedical Press, 1977:663.
53. Borth W, Urbanski A, Prohaska R, Susani M, Luger TA. Binding of recombinant interleukin-1β to the third complement component and α_2-macroglobulin after activation of serum by immune complexes. Blood 1990; 75:2388.
54. Danielpour D, Sporn MB. Differential inhibition of transforming growth factor β1 and β2 activity by α_2-macroglobulin. J Biol Chem 1990; 265:6973.
55. Bonner JC, Hoffman M, Brody AR. Alpha-macroglobulin secreted by alveolar macrophages serves as a binding protein for a macrophage-derived homologue of platelet-derived growth factor. Am J Respir Cell Mol Biol 1989; 1:171.
56. Matsuda T, Hirano T, Nagasawa S, Kishimoto T. Identification of alpha-2-macroglobulin as a carrier protein for IL-6. J Immunol 1989; 142:148.
57. James K. Interactions between cytokines and α_2-macroglobulin. Immunol Today 1990; 11:163.
58. Anderson CL, Shen L, Eicher DM, Wewers MD, Gill JK. Phagocytosis mediated by three distinct Fcgamma receptor classes on human leukocytes. J Exp Med 1990; 171:1333.
59. Guyre PM, Morganelli PM, Miller R. Recombinant immune interferon increases immunoglobulin G Fc receptors on cultured human mononuclear phagocytes. J Clin Invest 1983; 72:393.
60. Conrad DH, Ben-Sasson SZ, Le Gros G, Finkelman FD, Paul WE. Infection with Nippostrongylus brasiliensis or injection of anti-IgD antibodies markedly enhances Fc-receptor-mediated interleukin 4 production by non-B, non-T cells. J Exp Med 1990; 171:1497.
61. Te Velde AA, Huijbens RJF, De Vries J, Figdor CG. IL-4 decreases FcgammaR membrane expression and FcgammaR-mediated cytotoxic activity of human monocytes. J Immunol 1990; 144:3046.
62. Arend WP, Ammons JT, Kotzin BL. Lipopolysaccharide and interleukin 1 inhibit interferon-gamma induced Fc receptor expression on human monocytes. J Immunol 1987; 139:1873.
63. Mosmann TR. Properties and functions of interleukin-10. [Review]. Adv Immunol 1994; 56:1.
64. Debets JMH, Van der Linden CJ, Dieteren EI, Seeuwenberg JFM, Buurman WA. Fc-receptor cross-linking induces rapid secretion of tumor necrosis factor (cachectin) by human peripheral blood monocytes. J Immunol 1988; 141:1197.

65. Debets JMH, Van de Winkel JGJ, Ceuppens JL, Dieteren IE, Buurman WA. Cross-linking of both Fc gamma RI and Fc gamma RII induces secretion of tumor necrosis factor by human monocytes, requiring high affinity Fc-Fc gamma R interactions: functional activation of Fc gamma RII by treatment with proteases or neuraminidase. J Immunol 1990; 144:1304.
66. Szabo G, Miller-Graziano CL, Wu J-Y, Takayama T, Kodys K. Differential tumor necrosis factor production by human monocyte subsets. J Leukocyte Biol 1990; 47:206.
67. Ling Z-D, Ziltener HJ, Webb BT, Matheson DS. Aggregated immunoglobulin and Fc fragment of IgG induce IL-6 release from human monocytes. Cell Immunol 1990; 129:95.
68. Blyden G, Handschumacher RE. Purification and properties of human lymphocyte activating factor (LAF). J Immunol 1977; 118:1631.
69. Morgan EL, Hobbs MV, Noonan DJ, Weigle WO. Induction of IL-1 secretion from human monocytes by Fc region subfragments of human IgG1. J Immunol 1988; 140:3014.
70. Chou YK, Sherwood T, Virella G. Erythrocyte-bound immune complexes trigger the release of interleukin-1 from human monocytes. Cell Immunol 1985; 91:308.
71. Dayer J-M, Passwell JH, Schneeberger EE, Krane SM. Interactions among rheumatoid synovial cells and monocyte-macrophages: production of collagenase-stimulatory factor by human monocytes exposed to concanavalin A or immunoglobulin Fc fragments. J Immunol 1980; 124:1712.
72. Arend WP, Joslin FG, Massoni RJ. Effects of immune complexes on production by human monocytes of interleukin 1 or an interleukin 1 inhibitor. J Immunol 1985; 134:3868.
73. Kindt GC, Moore SA, She Z-W, Wewers MD. Endotoxin priming of monocytes augments Fcgamma receptor cross-linking-induced TNF-α and IL-1β release. Am J Physiol Lung Cell Mol Physiol 1993; 265:L178.
74. Marsh CB, Pope HA, Wewers MD. Fcgamma receptor crosslinking down-regulates IL-1 receptor antagonist and induces IL-1beta in mononuclear phagocytes stimulated with endotoxin or Staphylococcus aureus. J Immunol 1994; 152:4604.
75. Eisenberg SP, Evans RJ, Arend WP, Verderber E, Brewer MT, Hannum CH, Thompson RC. Primary structure and functional expression from complementary DNA of a human interleukin-1 receptor antagonist. Nature 1990; 343:341.
76. Hannum CH, Wilcox CJ, Arend WP, Joslin FG, Dripps DJ, Heimdal PL, Armes LG, Sommer A, Eisenberg SP, Thompson RC. Interleukin-1 receptor antagonist activity of a human interleukin-1 inhibitor. Nature 1990; 343:336.
77. Ziskind M, Jones RN, Weill H. Silicosis. Am Rev Respir Dis 1976; 113:643.
78. Schmidt JA, Oliver CN, Lepe-Zuniga JL, Green I, Gery I. Silica-stimulated monocytes release fibroblast proliferation factors identical to interleukin 1: a potential role for interleukin 1 in the pathogenesis of silicosis. J Clin Invest 1984; 73:1462.
79. Gery I, Davies P, Derr J, Krett N, Barranger JA. Relationship between production and release of lymphocyte-activating factor (interleukin 1) by murine macrophages: effects of various agents. Cell Immunol 1981; 64:293.
80. Apte RN, Durum SK, Oppenheim JJ. Opioids modulate interleukin-1 production and secretion by bone-marrow macrophages. Immunol Lett 1990; 24:141.
81. Lepe-Zuniga JL, Gery I. Production of intra- and extracellular interleukin-1 (IL-1) by human monocytes. Clin Immunol Immunopath 1984; 31:222.
82. Vogel SN, English KE, O'Brien AD. Silica enhancement of murine endotoxin sensitivity. Infect Immun 1982; 38:681.
83. Schmidt JA, Mizel SB, Cohen D, Green I. Interleukin 1, a potential regulator of fibroblast proliferation. J Immunol 1982; 128:2177.
84. Brown GP, Monick M, Hunninghake GW. Fibroblast proliferation induced by silica-exposed humna alveolar macrophages. Am Rev Respis Dis 1988; 138:85.
85. Dubois CM, Bissonnete E, Rola-Pleszczynski M. Asbestos fibers and silica particles stimulate rat alveolar macrophages to release tumor necrosis factor: autoregulatory role of leukotriene B4. Am Rev Respir Dis 1989; 139:1257.
86. Driscoll KE, Sindenschmidt RC, Maurer JK, Higgins JM, Ridder G. Pulmonary response to silica or titanium dioxide: inflammatory cells, alveolar macrophage-derived cytokines, and histopathology. Am J Respir Cell Mol Biol 1990; 2:381.
87. Piguet PF, Collart MA, Grau GE, Sappino A-P, Vassalli P. Requirement of tumour necrosis factor for development of silica-induced pulmonary fibrosis. Nature 1990; 344:245.
88. Rom WN, Bitterman PB, Rennard SI, Cantin A, Crystal RG. Characterization of the lower respiratory tract inflammation of nonsmoking individuals with interstitial lung disease associated with chronic inhalation of inorganic dusts. Am Rev Respir Dis 1987; 136:1429.
89. Spurzem JR, Saltini C, Rom W, Winchester RJ, Crystal RG. Mechanisms of macrophage accumulation in the lungs of asbestos-exposed subjects. Am Rev Respir Dis 1987; 136:276.
90. Wewers MD, Rennard SI, Hance AJ, Bitterman PB, Crystal RG. Normal human alveolar macrophages obtained by bronchoalveolar lavage have a limited capacity to release interleukin-1. J Clin Invest 1984; 74:2208.
91. Elias JA, Schreiber AD, Gustilo K, Chien P, Rossman MD, Lammie PJ, Daniele RP. Differential interleukin 1 elaboration by unfractionated and density fractionated human alveolar macrophages and blood monocytes: relationship to cell maturity. J Immunol 1985; 135:3198.
92. Endres S, Ghorbani R, Kelley VE, Georgilis K, Lonnemann G, van der Meer JWM, Cannon JG, Rogers TS, Klempner MS, Weber PC, Shaefer EJ, Wolff SM, Dinarello CA. The effect of dietary supplementation with n-3 polyunsaturated fatty acids on the synthesis of interleukin-1 and tumor necrosis factor by mononuclear cells. N Engl J Med 1989; 320:265.
93. Kunkel SL, Spengler M, May MA, Spengler R, Larrick J, Remick D. Prostaglandin E2 regulates macrophage-derived tumor necrosis factor gene expression. J Biol Chem 1988; 263:5358.
94. Rola-Pleszczynski M, Lemaire I. Leukotrienes augment interleukin 1 production by human monocytes. J Immunol 1985; 135:3958.
95. Kunkel SL, Chensue SW, Phan SH. Prostaglandins as endogenous mediators of interleukin 1 production. J Immunol 1986; 136:186.
96. Ikejima T, Dinarello CA, Gill DM, Wolff SM. Induction of human interleukin-1 by a product of Staphylococcus aureus asociated with toxic shock syndrome. J Clin Invest 1984; 73:1312.
97. Roberts NJ, Prill AH, Mann TN. Interleukin 1 and interleukin 1 inhibitor production by human macrophages exposed to influenza virus or respiratory syncitial virus. J Exp Med 1986; 163:511.
98. Berman MA, Sandborg CI, Calabia BS, Andrews BS, Friou GJ. Interleukin 1 inhibitor masks high interleukin 1 production in acquired immunodeficiency syndrome (AIDS). Clin Immunol Immunopath 1987; 42:133.
99. Pennica D, Kohr W, Kuang W-J, Glaister D, Aggarwal BB, Chen EY, Goeddel DV. Identification of human uromodulin as the Tamm-Horsfall urinary glycoprotein. Science 1987; 236:83.

100. Cox RA, Anders GT, Cappell PJ, Johnson JE, Blanton HM, Seaworth BJ, Treasure RL. Production of tumor necrosis factor-alpha and interleukin-1 by alveolar macrophages from HIV-1-infected persons. AIDS Res Hum Retroviruses 1990; 6:431.
101. Molina J-M, Scadden DT, Amirault C, Woon A, Vannier E, Dinarello CA, Groopman JE. Human immunodeficiency virus does not induce interleukin-1, interleukin-6, or tumor necrosis factor in mononuclear cells. J Virol 1990; 64:2901.
102. Molina J-M, Schindler R, Ferriani R, Sakaguchi M, Vannier E, Dinarello CA, Groopman JE. Production of cytokines by peripheral blood monocytes/macrophages infected with human immunodeficiency virus type I (HIV-1). J Infect Dis 1990; 161:888.
103. Emilie D, Peuchmaur M, Maillot MC, Crevon MC, Brousse N, Delfraissy JF, Dormont J, Galanaud P. Production of interleukins in human immunodeficiency virus-1-replicating lymph nodes. J Clin Invest 1990; 86:148.
104. Okusawa S, Dinarello CA, Yancey KB, Endres S, Lawley TJ, Frank MM, Burke JF, Gelfand JA. C5a induction of human interleukin 1: synergistic effect with endotoxin or interferon gamma. J Immunol 1987; 139:2635.
105. Goodman MG, Chenoweth DE, Weigle WO. Induction of interleukin 1 secretion and enhancement of humoral immunity by binding of human C5a to macrophage surface C5a receptors. J Exp Med 1982; 1156:912.
106. Whisler RL, Newhouse YG, Lachman LB. Heterogeneity of human monocyte subsets in the promotion of B cell colonies and the role of interleukin 1. J Immunol 1982; 129:455.
107. Allen JN, Moore SA, Pope-Harman AL, Marsh CB, Wewers MD. Immunosuppressive properties of surfactant and plasma on alveolar macrophages. J Lab Clin Med 1995; 125:356.
108. Dowton SB, Colten HR. Acute phase reactants in inflammation and infection. Sem Hematol 1988; 25:84.
109. Zahedi K, Tebo JM, Siripont J, Klimo GF, Mortensen RF. Binding of human C-reactive protein to mouse macrophages is mediated by distinct receptors. J Immunol 1989; 142:2384.
110. Zeller J, Kubak B, Gewurz H. Binding sites for C-reactive protein on human monocytes are distinct from IgG Fc receptors. Immunology 1989; 67:51.
111. Rochemonteix BG, Wiktorowicz K, Kushner I, Dayer J-M. C-reactive protein increases production of IL-1α, IL-1β, and TNF-α, and expression of mRNA by human alveolar macrophages. J Leukocyte Biol 1993; 53:439.
111a. Pue CA, Mortensen RF, Marsh CB, Pope HA, Wewers MD. Acute phase levels of c-reactive protein enhance IL-1b and IL-1ra production by human blood monocytes but inhibit IL-1b and IL-1ra production by alveolar macrophages. J Immunol 1996; 156:1594.
112. D'Eustachio P, Jadidi S, Fuhlbrigge RC, Gray PW, Chaplin DD. Interleukin-1 alpha and beta genes: linkage on chromosome 2 in the mouse. Immunogenetics 1987; 26:339.
113. Arend WP, Welgus HG, Thompson RC, Eisenberg SP. Biological properties of recombinant human monocyte-derived interleukin 1 receptor antagonist. J Clin Invest 1990; 85:1694.
114. March CJ, Mosley B, Larsen A, Cerretti DP, Braedt G, Price V, Gillis S, Henney CS, Kronheim SR, Grabstein K, Conlon PJ, Hopp TP, Cosman D. Cloning, sequence and expression of two distinct human interleukin-1 complementary DNAs. Nature 1985; 315:641.
115. Dower SK, Kronheim SR, Hopp TP, Cantrell M, Deeley M, Gillis S, Henney CS, Urdal DL. The cell surface receptors for interleukin-1 alpha and interleukin-1 beta are identical. Nature 1986; 324:266.
116. Rupp EA, Cameron PM, Ranawat CS, Schmidt JA, Bayne EK. Specific bioactivities of monocyte-derived interleukin 1 alpha and interleukin beta are similar to each other on cultured murine thymocytes and on cultured human connective tissue cells. J Clin Invest 1986; 78:836.
117. Minnich-Carruth LL, Suttles J, Mizel SB. Evidence against the existence of a membrane form of murine IL-1 alpha. J Immunol 1989; 142:526.
118. Stevenson FT, Bursten SL, Fanton C, Locksley RM, Lovett DH. The 31-kDa precursor of interleukin 1α is myristoylated on specific lysines within the 16-kDa N-terminal propiece. Proc Natl Acad Sci USA 1983; 90:7245.
119. Dinarello CA, Cannon JG, Mier JW, Bernheim HA, LoPreste G, Lynn DL, Love RN, Webb AC, Auron PE, Reuben RC, Rich A, Wolff SM, Putney SC. Multiple biological activities of human recombinant interleukin 1. J Clin Invest 1986; 77:1734.
120. Bhardwaj N, Lau LL, Friedman SM, Crow MK, Steinman RM. Interleukin 1 production during accessory cell-dependent mitogenesis of T lymphocytes. J Exp Med 1989; 169:1121.
121. Durum SK, Gershon RK. Interleukin 1 can replace the requirement for I-A-positive cells in the proliferation of antigen-primed T cells. Proc Natl Acad Sci USA 1982; 79:4747.
122. Dinarello CA. Interleukin-1 and the pathogenesis of the acute-phase response. N Engl J Med 1984; 311:1413.
123. Pober JS, Bevilacqua MP, Mendrick DL, Lapierre LA, Fiers W, Gimbrone MA. Two distinct monokines, interleukin 1 and tumor necrosis factor, each independently induce biosynthesis and transient expression of the same antigen on the surface of cultured human vascular endothelial cells. J Immunol 1986; 135:1680.
124. Goldblum SE, Yoneda K, Cohen DA, McClain CJ. Provocation of pulmonary vascular endothelial injury in rabbits by human recombinant interleukin-1 beta. Infect Immun 1988; 56:2255.
125. Dayer J-M, de Rochemonteix B, Burrus B, Demczuk S, Dinarello CA. Human recombinant interleukin 1 stimulates collagenase and prostaglandin E2 production by human synovial cells. J Clin Invest 1986; 77:645.
126. Mielke V, Bauman JGJ, Sticherling M, Ibs T, Zomershoe AG, Seligmann K, Henneicke H-H, Schröder J-M, Sterry W, Christophers E. Detection of neutrophil-activating peptide NAP/IL-8 and NAP/IL-8 mRNa in human recombinant IL-1α- and human recombinant tumor necrosis factor-α-stimulated human dermal fibroblasts: an immunocytochemical and fluorescent in situ hybridization study. J Immunol 1990; 144:153.
127. Rolfe MW, Kunkel SL, Standiford TJ, Chensue SW, Allen RM, Evanoff HL, Phan SH, Strieter RM. Pulmonary fibroblast expression of interleukin-8: a model for alveolar macrophage-derived cytokine networking. Am J Respir Cell Mol Biol 1991; 5:493.
128. Beasley D, Cohen RA, Levinsky NG. Interleukin 1 inhibits contraction of vascular smooth muscle. J Clin Invest 1989; 83:331.
129. She Z-W, Wewers MD, Herzyk DJ, Sagone AL, Davis WB. Tumor necrosis factor primes neutrophils for hypochlorous acid production. Am J Physiol 1989; 257:L338.
130. Robbins RA, Springall DR, Warren JB, Kwon OJ, Buttery LD, Wilson AJ, Adcock IM, Riveros-Moreno V, Moncada S, Polak J, et al. Inducible nitric oxide synthase is increased in murine lung epithelial cells by cytokine stimulation. Biochem Biophys Res Commun 1994; 198:835.
131. Elias JA, Lentz V. IL-1 and tumor necrosis factor synergistically stimulate fibroblast IL-6 production and stabilize IL-6 messenger RNA. J Immunol 1990; 145:161.
132. Oksusawa S, Gelfand JA, Ikejima T, Connolly RJ, Dinarello CA. Interleukin 1 induces a shock-like state in rabbits: syner-

gism with tumor necrosis factor and the effect of cyclo-oxygenase inhibition. J Clin Invest 1988; 81:1162.
133. Dalboge H, Bayne S, Christensen T, Hejnaes KR. Cloning and expression of an interleukin-1 beta precursor and its conversion to interleukin-1 beta. FEBS Lett 1989; 246:89.
134. Kostura MJ, Tocci MJ, Limjuco G, Chin J, Cameron P, Hillman AG, Chartrain NA, Schmidt JA. Identification of a monocyte specific pre-interleukin 1 beta convertase activity. Proc Natl Acad Sci USA 1989; 86:5227.
135. Black RA, Kronheim SR, Sleath PR. Activation of interleukin-1 beta by a co-induced protease. FEBS Lett 1989; 247:386.
136. Black RA, Kronheim SR, Merriam JE, March CJ, Hopp TP. A pre-aspartate-specific protease from human leukocytes that cleaves pro-interleukin-1 beta. J Biol Chem 1989; 264:5323.
137. Kobayashi Y, Yamamoto K, Saido T, Kawasaki H, Oppenheim JJ, Matsushima K. Identification of calcium-activated neutral protease as a processing enzyme of human interleukin 1α. Proc Natl Acad Sci USA 1990; 87:5548.
138. Ayala JM, Yamin T-T, Egger LA, Chin J, Kostura MJ, Miller DK. IL-1beta-converting enzyme is present in monocytic cells as an inactive 45-kDa precursor. J Immunol 1994; 153:2592.
139. Ray CA, Black RA, Kronheim SR, Greenstreet TA, Sleath PR, Salvesen GS, Pickup DJ. Viral inhibition of inflammation: cowpox virus encodes an inhibitor of the interleukin-1 beta converting enzyme. Cell 1992; 69:597.
140. Wang L, Miura M, Bergeron L, Zhu H, Yuan J. Ich-1, and Ice/ced-3-related gene, encodes both positive and negative regulators of programmed cell death. Cell 1994; 78:739.
141. Granowitz EV, Santos AA, Poutsiaka DD, Cannon JG, Wilmore DW, Wolff SM, Dinarello CA. Production of interleukin-1-receptor antagonist during experimental endotoxaemia. Lancet 1991; 338:1423.
142. Colotta F, Re F, Muzio M, Bertini R, Polentarutti N, Sironi M, Giri JG, Dower SK, Sims JE, Mantovani A. Interleukin-1 Type II receptor: decoy target for IL-1 that is regulated by IL-4. Science 1993; 261:472.
143. Arend WP, Malyak M, Smith MF Jr, Whisenand TD, Slack JL, Sims JE, Giri JG, Dower SK. Binding of IL-1α, IL-1β, and IL-1 receptor antagonist by soluble IL-1 receptors and levels of soluble IL-1 receptors in synovial fluids. J Immunol 1994; 153:4766.
144. Michie HR, Manogue KR, Spriggs DR, Revhaug A, O'Dwyer S, Dinarello CA, Cerami A, Wolff SM, Wilmore DW. Detection of circulating tumor necrosis factor after endotoxin administration. N Engl J Med 1988; 318:1481.
145. Carroll MC, Katzman P, Alicott EM, Koller BH, Geraghty DE, Orr HT, Strominger JL, Spies T. Linkage map of the human major histocompatibility complex including the tumor necrosis factor genes. Proc Natl Acad Sci USA 1987; 84:8535.
146. Spies T, Blanck G, Bresnahan M, Sands J, Strominger JL. A new cluster of genes within the human major histocompatibility complex. Science 1989; 243:214.
147. Gearing AJH, Beckett P, Christodoulou M, Churchill M, Clements J, Davidson AH, Drummond AH, Galloway WA, Gilbert R, Gordon JL, Leber TM, Mangan M, Miller K, Nayee P, Owen K, Patel S, Thomas W, Wells G, Wood LM, Woolley K. Processing of tumour necrosis factor-α precursor by metalloproteinases. Nature 1994; 370:555.
148. Beutler B, van Huffel C. Unraveling function in the TNF ligand and receptor families. Science 1994; 264:667.
149. Beutler B, Cerami A. Cachectin: more than a tumor necrosis factor. N Engl J Med 1987; 316:379.
150. Ranges G, Zlotnik A, Espevik T, Dinarello CA, Cerami A, Palladino MA. Tumor necrosis factor alpha/cachectin is a growth factor for thymocytes: synergistic interactions with other cytokines. J Exp Med 1988; 167:1472.
151. Dayer J-M, Beutler B, Cerami A. Cachectin/tumor necrosis factor stimulates collagenase and prostaglandin E2 production by human synovial cells and dermal fibroblasts. J Exp Med 1985; 162:2163.
152. Dinarello CA, Cannon JG, Wolff SM, Bernheim HA, Beutler B, Cerami A, Figari IS, Palladino MA Jr, O'Connor JV. Tumor necrosis factor (cachectin) is an endogenous pyrogen and induces production of interleukin 1. J Exp Med 1986; 163:1433.
153. Pober JS Jr, Gimbrone MA, Lapierre LA, Mendrick DL, Fiers W, Rothlein R, Springer TA. Overlapping patterns of activation of human endothelial cells by interleukin 1, tumor necrosis factor, and immune interferon. J Immunol 1986; 137:1893.
154. Akira S, Hirano T, Taga T, Kishimoto T. Biology of multifunctional cytokines: IL 6 and related molecules (IL 1 and TNF). FASEB J 1990; 4:2860.
155. Standiford TJ, Kunkel SL, Basha MA, Chensue SW, Lynch JP III, Toews GB, Westwick J, Strieter RM. Interleukin-8 gene expression by a pulmonary epithelial cell line. A model for cytokine networks in the lung. J Clin Invest 1990; 86:1945.
156. Berkow RL, Wang D, Larrick JW, Dodson RW, Howard TH. Enhancement of neutrophil superoxide production by incubation with recombinant human tumor necrosis factor. J Immunol 1987; 139:3783.
157. Tracey KJ, Beutler B, Lowry SF, Merryweather J, Wolpe S, Milsark IW, Hariri RJ, Fahey TJ III, Zentella A, Albert JD, Shires GT, Cerami A. Shock and tissue injury induced by recombinant human cachectin. Science 1986; 234:470.
158. Waage A, Halstensen A, Espevik T. Association between tumour necrosis factor in serum and fatal outcome in patients with meningococcal disease. Lancet 1987; 1:355.
159. Beutler B, Tkacenko V, Milsark I, Krochin N, Cerami A. Effect of gamma interferon on cachectin expression by mononuclear phagocytes: reversal of the lps d (endotoxin resistance) phenotype. J Exp Med 1986; 164:1791.
160. Burchett SK, Weaver WM, Westall JA, Larsen A, Kronheim S, Wilson CB. Regulation of tumor necrosis factor/cachectin and IL-1 secretion in human mononuclear phagocytes. J Immunol 1988; 140:3473.
161. Han J, Thompson P, Beutler B. Dexamethasone and pentoxifylline inhibit endotoxin-induced cachectin/tumor necrosis factor synthesis at separate points in the signaling pathway. J Exp Med 1990; 172:391.
162. Mathison JC, Virca GD, Wolfson E, Tobias PS, Glaser K, Ulevitch RJ. Adaptation to bacterial lipopolysaccharide controls lipopolysaccharide-induced tumor necrosis factor production in rabbit macrophages. J Clin Invest 1990; 85:1108.
163. Sehgal PB, May LT, Tamm I, Vilcek J. Human B2 interferon and B-cell differentiation factor BSF-2 are identical. Science 1987; 235:731.
164. Van Snick J. Interleukin 6: an overview. Ann Rev Immunol 1990; 8:253.
165. Elias JA, Trinchieri G, Beck JM, Simon PL, Sehgal PB, May LT, Kern JA. A synergistic interaction of IL-6 and IL-1 mediates the thymocyte-stimulating activity produced by recombinant IL-1-stimulated fibroblasts. J Immunol 1989; 142:509.
166. Gauldie J, Richards C, Harnish D, Lansdorp P, Baumann H. Interferon beta 2/B-cell stimulatory factor type 2 shares identity with monocyte-derived hepatocyte-stimulating factor and regulates the major acute phase protein response in liver cells. Proc Natl Acad Sci USA 1987; 84:7251.
167. Dentener MA, Bazil V, Von Asmuth EJ, Ceska M, Buurman

WA. Involvement of CD14 in lipopolysaccharide-induced tumor necrosis factor-alpha, IL-6 and IL-8 release by human monocytes and alveolar macrophages. J Immunol 1993; 150:2885.
168. Waage A, Brandtzaeg P, Halstensen A, Kierulf P, Espevik T. The complex pattern of cytokines in serum from patients with meningococcal septic shock. J Exp Med 1989; 169:333.
169. Krutmann J, Kirnbauer R, Kock A, Schwarz T, Schopf E, May LT, Sehgal PB, Luger TA. Cross-linking Fc receptors on monocytes triggers IL-6 production: role in anti-CD3-induced T cell activation. J Immunol 1990; 145:1337.
170. Bost TW, Riches DW, Schumacher B, Carre PC, Khan TZ, Martinez JA, Newman LS. Alveolar macrophages from patients with beryllium disease and sarcoidosis express increased levels of mRNA for tumor necrosis factor-alpha and interleukin-6 but not interleukin-1 beta. Am J Respir Cell Mol Biol 1994; 10:506.
171. Twigg HL, Iwamoto GK, Soliman DM. Role of cytokines in alveolar macrophage accessory cell function in HIV-infected individuals. J Immunol 1992; 149:1462.
172. Twigg HL, Wilkes DS, Soliman DM. Measurement of IL-6 inhibitory activity in cultured cell supernatants and lysates. J Lab Clin Med 1994; 124:283.
173. Kishimoto T, Taga T, Akira S. Cytokine signal transduction. Cell 1994; 76:253.
174. Baggiolini M, Walz A, Kunkel SL. Neutrophil-activating peptide-1/interleukin 8, a novel cytokine that activates neutrophils. J Clin Invest 1989; 84:1045.
175. Djeu JY, Matsushima K, Oppenheim JJ, Shiotsuki K, Blanchard DK. Functional activation of human neutrophils by recombinant monocyte-derived neutrophil chemotactic factor/IL-8. J Immunol 1990; 144:2205.
176. Detmers PA, Lo SK, Olsen-Egbert E, Walz A, Baggiolini M, Cohn ZA. Neutrophil-activating protein 1/interleukin 8 stimulates the binding activity of the leukocyte adhesion receptor CD11b/CD18 on human neutrophils. J Exp Med 1990; 171:1155.
177. Matsushima K, Morishita K, Yoshimura T, Lavu S, Kobayashi Y, Lew W, Appella E, Kung HF, Leonard EJ, Oppenheim JJ. Molecular cloning of a human monocyte-derived neutrophil chemotactic factor (MDNCF) and the induction of MDNCF mRNA by interleukin and tumor necrosis factor. J Exp Med 1988; 167:1883.
178. Strieter RM, Kunkel SL, Showell HJ, Remick DG, Phan SH, Ward PA, Marks RM. Endothelial cell gene expression of a neutrophil chemotactic factor by TNF alpha, LPS, and IL-1 beta. Science 1989; 243:1467.
179. Maestrelli P, Tsai J-J, Cromwell O, Kay AB. The identification and partial characterization of a human mononuclear cell-derived neutrophil chemotactic factor apparently distinct from IL-1, IL-2, GM-CSF, TNF and IFN-gamma. Immunology 1988; 64:219.
180. Baggiolini M, Deward B, Moser B. Interleukin-8 and related chemotactic cytokines—CXC and CC chemokines. [Review]. Adv Immunol 1994; 55:97.
181. Martich GD, Danner RL, Ceska M, Suffredini AF. Detection of interleukin 8 and tumor necrosis factor in normal humans after intravenous endotoxin: the effect of antiinflammatory agents. J Exp Med 1991; 173:1021.
182. Van Zee KJ, Fischer E, Hawes AS, Hebert CA, Terrell TG, Baker JB, Lowry SF, Moldawer LL. Effects of intravenous IL-8 administration in nonhuman primates. J Immunol 1992; 148:1746.
183. Huber AR, Kunkel SL, Todd RF III, Weiss SJ. Regulation of transendothelial neutrophil migration by endogenous interleukin-8. Science 1991; 254:99.
184. Horuk R, Chitnis CE, Darbonne WC, Colby TJ, Rybicki A, Hadley TJ, Miller LH. A receptor for the malarial parasite Plasmodium vivax: the erythrocyte chemokine receptor. Science 1993; 261:1182.
185. McElvaney NG, Nakamura H, Birrer P, Hebert CA, Wong WL, Alphonso M, Baker JB, Catalano MA, Crystal RG. Modulation of airway inflammation in cystic fibrosis. In vivo suppression of interleukin-8 levels on the respiratory epithelial surface by aerosolization of recombinant secretory leukoprotease inhibitor. J Clin Invest 1992; 90:1296.
186. Fiorentino DF, Zlotnik A, Mosmann TR, Howard M, O'Garra A. IL-10 inhibits cytokine production by activated macrophages. J Immunol 1991; 147:3815.
187. De Waal Malefyt R, Abrams J, Bennett B, Figdor CG, de Vries JE. Interleukin 10 (IL-10) inhibits cytokine synthesis by human monocytes: an autoregulatory role of IL-10 produced by monocytes. J Exp Med 1991; 174:1209.
188. Gerard C, Bruyns C, Marchant A, Abramowicz D, Vandenabeele P, Delvaux A, Fiers W, Goldman M, Velu T. Interleukin 10 reduces the release of tumor necrosis factor and prevents lethality in experimental endotoxemia. J Exp Med 1993; 177:547.
189. Martinet Y, Rom WN, Grotendorst GR, Martin GR, Crystal RG. Exaggerated spontaneous release of platelet-derived growth factor by alveolar macrophages from patients with idiopathic pulmonary fibrosis. N Engl J Med 1987; 317:202.
190. Raines EW, Dower SK, Ross R. Interleukin-1 mitogenic activity for fibroblasts and smooth muscle cells is due to PDGF-AA. Science 1989; 243:393.
191. Pierce GF, Mustoe TA, Lingelbach J, Masakowski VR, Gramates P, Deuel TF. Transforming growth factor beta reverses the glucocorticoid-induced wound-healing deficit in rats: possible regulation in microphages by platelet-derived growth factor. Proc Natl Acad Sci USA 1989; 86:2229.
192. Stiles AD, D'Ercole AJ. The insulin-like growth factors and the lung. Am J Respir Cell Mol Biol 1990; 3:93.
193. Rom WN, Basset P, Fells GA, Nukiwa T, Trapnell BC, Crystal RG. Alveolar macrophages release an insulin-like growth factor I-type molecule. J Clin Invest 1988; 82:1685.
194. Assoian RK, Komoriya A, Meyers CA, Miller DM, Sporn MB. Transforming growth factor-beta in human platelets: identification of a major storage site, purification, and characterization. J Biol Chem 1983; 258:7155.
195. Assoian RK, Fleurdelys BE, Stevenson HC, Miller PJ, Madtes DK, Raines EW, Ross R, Sporn MB. Expression and secretion of type beta transforming growth factor by activated human macrophages. Proc Natl Acad Sci USA 1987; 84:6020.
196. Postlethwaite AE, Keski-Oja J, Moses HL, Kang AH. Stimulation of the chemotactic migration of human fibroblasts by transforming growth factor beta. J Exp Med 1987; 165:251.
197. Khalil N, Bereznay O, Sporn M, Greenberg AH. Macrophage production of transforming growth factor beta and fibroblast collagen synthesis in chronic pulmonary inflammation. J Exp Med 1989; 170:727.
198. Mackiewicz A, Ganapathi MK, Schultz D, Brabenec A, Weinstein J, Kelley MF, Kushner I. Transforming growth factor β_1 regulates production of acute-phase proteins. Proc Natl Acad Sci USA 1990; 87:1491.
199. Tsunawaki S, Sporn M, Nathan C. Comparison of transforming growth factor-beta and a macrophage-deactivating polypeptide from tumor cells: differences in antigenicity and mechanism of action. J Immunol 1989; 142:3462.
200. Baumgartner JD, Heumann D, Gerain J, Weinbreck P, Grau GE, Glauser MP. Association between protective efficacy of anti-lipopolysaccharide (LPS) antibodies and suppression of

LPS-induced tumor necrosis factor α and interleukin 6. Comparison of O side chain-specific antibodies with core LPS antibodies. J Exp Med 1990; 171:889.

201. DeForge LE, Remick DG. Kinetics of TNF, IL-6, and IL-8 gene expression in LPS-stimulated human whole blood. Biochem Biophys Res Commun 1991; 174:18.
202. Dentener MA, Bazil V, Von Asmuth EJU, Ceska M, Buurman WA. Involvement of CD14 in lipopolysaccharide-induced tumor necrosis factor-Alpha, IL-6 and IL-8 release by human monocytes and alveolar macrophages. J Immunol 1993; 150:2885.
203. Schroder J-M, Persoon NLM, Christophers E. Lipopolysaccharide-stimulated human monocytes secrete, apart from neutrophil-activating peptide 1/interleukin 8, a second neutrophil-activating protein: NH2-terminal amino acid sequence identity with melanoma growth stimulating activity. J Exp Med 1990; 171:1091.
204. Standiford TJ, Kunkel SL, Kasahara K, Milia MJ, Rolfe MW, Strieter RM. Interleukin-8 gene expression from human alveolar macrophages: the role of adherence. Am J Respir Cell Mol Biol 1991; 5:579.
205. Becker SJ, Quay J, Soukup J. Cytokine (tumor necrosis factor, IL-6, and IL-8) production by respiratory syncytial virus-infected human alveolar macrophages. J Immunol 1991; 147:4307.
206. House M, Arora DJS. Stimulation of tumor necrosis factor secretion by purified influenza virus neuraminidase. Cell Immunol 1990; 129:104.
207. Zhang Y, Lee TC, Guillemin B, Yu M-C, Rom WN. Enhanced IL-1β and tumor necrosis factor-α release and messenger RNA expression in macrophages from idiopathic pulmonary fibrosis or after asbestos exposure. J Immunol 1993; 150:4188.
208. Kline JN, Schwartz DA, Monick MM, Floerchinger CS, Hunninghake GW. Relative release of interleukin-1β and interleukin-1 receptor antagonist by alveolar macrophages: a study in asbestos-induced lung disease, sarcoidosis, and idiopathic pulmonary fibrosis. Chest 1993; 104:47.
209. Rosenthal GJ, Germolec DR, Blazka ME, Corsini E, Simeonova P, Pollock P, Kong LY, Kwon J, Luster MI. Asbestos stimulates IL-8 production from human lung epithelial cells. J Immunol 1994; 153:3237.
210. Scholz W, McClurg MR, Cardenas GJ, Smith M, Noonan DJ, Hugli TE, Morgan EL. C5a-mediated release of interleukin 6 by human monocytes. Clin Immunol Immunopathol 1990; 57:297.
211. Van Zee KJ, DeForge LE, Fischer E, Marano MA, Kenney JS, Remick DG, Lowry SF, Moldawer LL. IL-8 in septic shock, endotoxemia, and after IL-1 administration. J Immunol 1991; 146:3478.
212. Strieter RM, Chensue SW, Basha MA, Standiford TJ, Lynch JP III, Baggiolini M, Kunkel SL. Human alveolar macrophage gene expression of interleukin-8 by tumor necrosis factor-alpha, lipopolysaccharide, and interleukin-1-beta. Am J Respir Cell Mol Biol 1990; 2:321.
213. Bachwich PR, Chensue SW, Larrick JW, Kunkel SL. Tumor necrosis factor stimulates interleukin-1 and prostaglandin E2 production in resting macrophages. Bioch Biophys Res Comm 1986; 136:94.
214. Hart PH, Whitty GA, Piccoli DS, Hamilton JA. Synergistic activation of human monocytes by granulocyte-macrophage colony-stimulating factor and IFN-gamma: increased TNF-alpha but not IL-1 activity. J Immunol 1988; 141:1516.
215. Laurenzi MA, Persson MAA, Dalsgaard C-J, Haegerstrand A. The neutropeptide substance P stimulates production of interleukin 1 in human blood monocytes: activated cells are preferentially influenced by the neuropeptide. Scand J Immunol 1990; 31:529.

24
Cytokines and Lymphocytes

Stanley C. Jordan and Ragini Fredrich
Cedars-Sinai Medical Center, Los Angeles, California

INTRODUCTION AND HISTORICAL PERSPECTIVE

Cytokines are peptide molecules that are responsible for intercellular signaling during immune-activation events. They are responsible, in large part, for the regulation of all local and systemic immune inflammatory and immunoregulatory events.

Cytokines are produced by a variety of immune and nonimmune cells. Cytokines produced by activated lymphocytes are referred to as "lymphokines," while those produced by macrophages and monocytes are referred to as "monokines." The application of advanced molecular and immunological techniques has helped in defining the diverse functional properties of cytokines. In addition, the development of cytokine gene knock-out mice has provided a more precise understanding of the role of specific cytokines in immunological events such as allograft rejection and viral infections (1–3). In general, these models have shown that the cytokine network is highly redundant as interleukin-2 (IL-2) and interferon γ (IFN-γ) gene knock-out mice can still reject allografts while IL-4 gene knock-out mice can still develop tolerance (2–4). Knowledge of cytokine structure and function holds great potential benefit since induction or abrogation of cytokine effector functions should be of value in controlling inflammatory diseases and aberrant or dysfunctional immune cell activation as in autoimmune diseases or allograft rejection.

David et al. (5) and Bloom and Bennett (6) were the first to demonstrate that soluble immune mediators distinct from immunoglobulins had distinct effects on immune cells. These investigators described the existence of a soluble mediator present in the culture supernatants of activated lymphocytes, which was inhibitory to the normal migration of macrophages. This factor was subsequently designated macrophage inhibitory factor (MIF). Although the mechanism(s) of action of MIF were unclear, these investigators fundamentally established that lymphocyte responses to antigen are not exclusively limited to the production of antibody.

The discovery of MIF resulted in an intense search for other cellular mediators. Among the early cytokines discovered were lymphotoxin, macrophage-activating factor (MAF), and interferon (7–9). Dumonde et al. (10) first applied the term "lymphokine" to these soluble factors derived from activated lymphocytes. The discovery of lymphokines and their role in regulation of cell functions in an autocrine and paracrine manner brought new understanding to the traditional concepts of cell-mediated immunity and lymphocyte functions. Lymphocytes were no longer the mere appendages of red blood cells. The discovery of lymphokines and their functions were fundamental accomplishments, which helped usher in the era of modern immunology.

The lymphokine concept provided an explanation for immunologically mediated inflammatory events including clinical manifestations of autoimmune diseases and allograft rejection. In addition, acquired resistance to intracellular infections and delayed-type hypersensitivity could also be explained. The initial observations were limited to proinflammatory effector functions of lymphokines, but investigators soon turned their attention to the possibility that these soluble factors might also regulate the functions of T and B lymphocytes. Waksman (11) observed that adherent nonlymphoid cells released a soluble mediator, which augmented the growth of lymphocytes. This factor was initially referred to as lymphocyte activating factor (LAF) but is now known to be the potent monokine interleukin-1 (IL-1). Cohen et al. (12) subsequently proposed the term "cytokine" to account for a broader array of soluble factors produced by lymphoid and nonlymphoid cells having the capacity to interact with and regulate the functions of other immune and nonimmune cells.

The goals of this chapter will be to define the functions of a growing number of cytokines and how they interact with and regulate lymphocyte functions.

CYTOKINES AND T CELLS

T-Cell Activation and Cytokine Production

T cells recognize, respond to, and remember antigenic challenges in the context of HLA-DR (class II) or HLA-A,B,C (class I) antigen presentation. These HLA molecules "restrict" the subset of T cell—CD4+ T cells interact with HLA-DR while CD8+ T cells interact with HLA-A,B,C—with which the antigen-presenting cell (APC) interacts. The classic concept of CD4+ T-cell activation involves antigen presentation via HLA-DR molecules on an APC. This presentation to the T-cell receptor (TCR/CD3) complex then activates the T cell through calcium-dependent pathways and induces the expression of the IL-2 receptor (IL-2Rα) on the cell's surface. The T cells subsequently produce IL-2, which is consumed in an autocrine and paracrine fashion. Macrophages were initially felt to be the only cell types capable of presenting antigen to T cells, but any HLA-DR–expressing cell can present antigen to CD4+ T cells. These include dendritic cells, endothelial cells, and B cells (13). B-cell presentation of antigen recognized by surface Ig molecules and presented to CD4+ T cells via HLA-DR can result in rapid activation of the CD4+ T cell with subsequent production of IL-2 and other cytokines that results in the activation, proliferation, and differentiation of B cells into antigen-specific antibody–producing plasma cells (Fig. 1).

The Th1/Th2 Paradigm in T-Cell Activation

After antigen presentation to T cells, a number of cytokines are produced, which regulate subsequent T-cell activation and differentiation. Mosmann et al. (14) demonstrated that

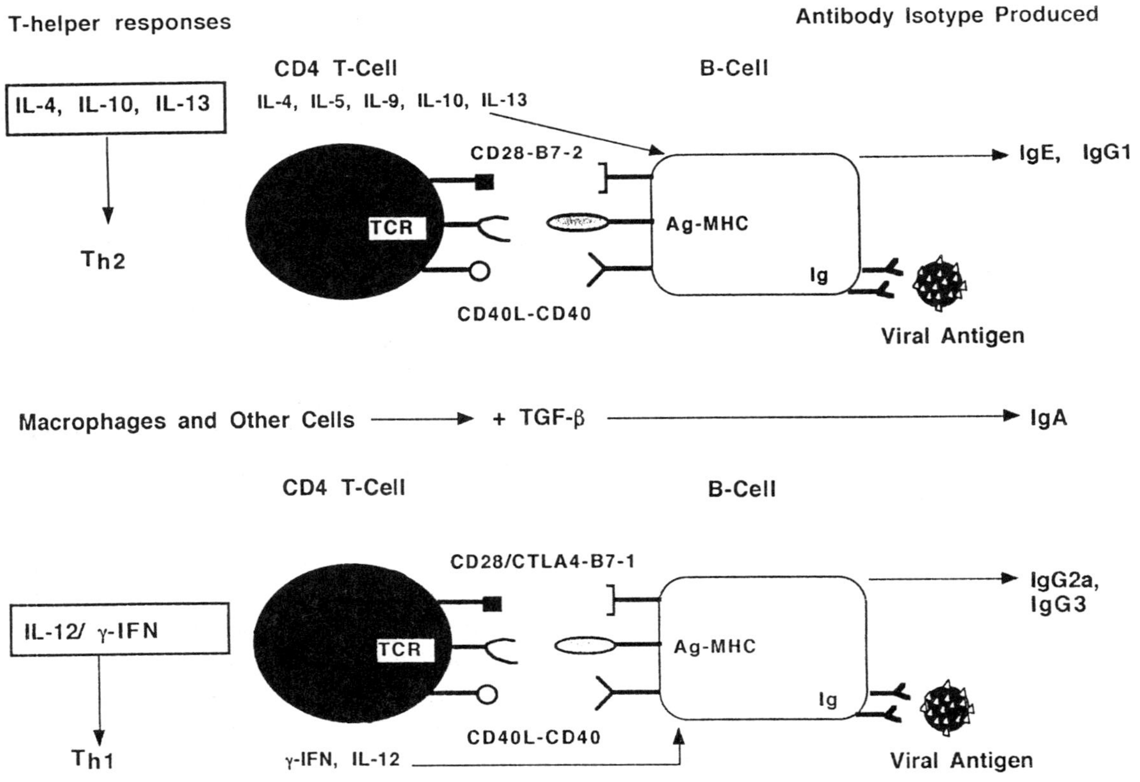

Figure 1 Cytokines elaborated by activated T cells (Th1, Th2) regulate the isotype of Ig produced by antigen-reactive B cells. Exposure to Th2 cytokines induces IgE and IgG1 responses, while exposure to Th1 cytokines induces the more phlogistic, complement-activating IgG2a antibody. TGF-β induces IgA.

cloned CD4+ T cells from mice and humans can be further divided into two functional subpopulations based on patterns of differential cytokine secretion. T-helper 1 (Th1) cells are defined as those that produce the pro-inflammatory cytokines IL-2, IFN-γ, TNF-β, and TNF-α in response to an antigenic challenge. Th1 cells promote the differentiation of cytotoxic T-cell precursors into fully functional cytotoxic effectors and enhance the production of phlogistic, complement-fixing IgG antibodies (15). Th1 cells are important mediators of delayed-type hypersensitivity (16). In general, Th1 cells are critical mediators of immune effector functions that include immunogenic inflammation and cytodestruction. Th1 cells are important in immune-destructive events such as allograft rejection and immunogenic inflammation in autoimmune diseases. In contrast, Th2 cells selectively secrete a distinct pattern of cytokines (IL-4, IL-5, IL-6, and IL-10) in response to antigenic confrontation (17,18).

Th2 cells provide cytokines that direct the synthesis of non–complement-fixing IgG antibodies, IgE and IgA synthesis. Th2 cell activation results in little immunogenic inflammation or cytodestructive events.

Current thinking suggests that both Th1 and Th2 CD4+ cells are derived from a common precursor and that the unique circumstances of the antigenic challenge to the Th0 cell (see Fig. 2) dictate deviation of the immune response toward a Th1 or Th2 pattern. This deviation is aided by specific cytokines. For example, antigenic encounters that stimulate IL-12 in the APCs (macrophages) result in IFN-γ production by T cells and deviation of the immune response to a Th1 profile, while antigenic encounters that induce IL-4/IL-13 transform the Th0 cells to a Th2 profile.

What has also become apparent is the reciprocal effects that Th1 and Th2 cells can have on each other. In this regard, IFN-γ produced by Th1 cells can inhibit the proliferation of Th2 clones in vitro. Conversely, IL-10 produced by Th2 clones can inhibit the proliferation and secretion of IFN-γ by Th1 clones (18,20). A limited discussion of the cytokines produced by and/or acting on

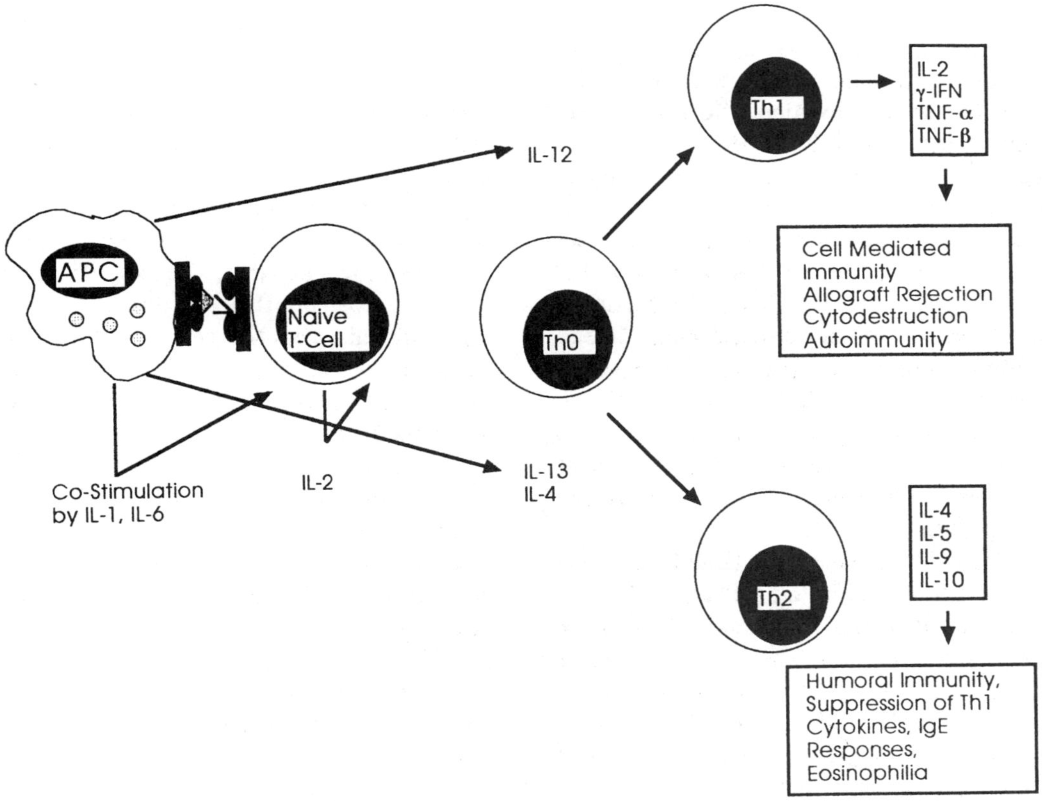

Figure 2 The Th1/Th2 paradigm of T-cell regulation of the immune response. Th0 cells deviate to a Th1 profile if IL-12 predominates during the antigenic encounter, with subsequent mediation of cell-mediated immunity and inflammatory reactions. If IL-4/IL-13 predominate, the Th0 cells are changed to a Th2 profile with a proclivity to produce immunoregulatory cytokines, induce IgE and eosinophilia, etc.

activated T cells is presented below, and more exhaustive review appears in other chapters.

Interleukin-2

IL-2 is a lymphokine produced by activated Th1 cells. IL-2 has a broad range of biological activities in vitro, which results primarily from its ability to expand and differentiate different subpopulations of T cells and B cells (21,22).

The biological activities of IL-2 are mediated by the binding of IL-2 to its multimolecular cellular receptor. Initially the receptor was thought to consist of two glycoprotein chains, an α chain (IL-2Rα) and a β chain (IL-2Rβ), which acted in concert to form a high-affinity receptor for IL-2 with subsequent transduction of the activation signal to the cell. The IL-2Rα is a 55 kDa transmembrane glycoprotein that is induced by T-cell activation. The IL-2Rβ is a transmembrane glycoprotein (75 kDa) that is constitutively expressed. Eventually a third component to the IL-2 receptor complex was found (IL-2Rγ) (23). This was found to be necessary for high-affinity binding. A current model of the IL-2R would describe the high-affinity receptor as an αβγ trimer in which all three chains are in contact with the ligand.

The importance of the γ chain was shown shortly after its discovery, when it was found that the human gene for the γ chain mapped to the same location as X-linked severe combined immune deficiency (XSCIDS) (23). In this condition, T cells are absent and B cells are dysfunctional. It is of further interest since a complete lack of IL-2 in humans (24) and mice (25) does not result in such a severe immunodeficiency. This led to the demonstration that several other cytokines (IL-4, IL-7, IL-9) (26–28) make use of the IL-2Rγ receptor for signal transduction. Due to this commonality of transduction pathway, the chain is now called γ-common or γ_c. Of further interest is IL-15, which utilizes the IL-2Rβ and γ_c but not the IL-2Rα chain. IL-15 shares many of the same activities of IL-2, explaining why ablation of the IL-2 gene does not perturb the immune system more in vivo (22,29).

The effector functions mediated by IL-2/IL-2R interactions occupies a central role in the mediation of cellular and humoral immunity. Interaction of IL-2 with the IL-2R-αβγ complex appears to activate T cells through a translocation of protein kinase C (PKC) from cytosol to plasma membrane. Hydrolysis of phosphatidylinositol to form inositol triphosphate (IP_3) and diacylglycerol (DAG) subsequently results in an intracellular rise in Ca^{2+} with subsequent increase in PKC activation. Thus, the IL-2/IL-2R-αβγ interaction results in the activation of intracellular signaling events that result in cytokine gene activation, cell proliferation, and differentiation. In addition to its ability to activate T cells, IL-2 can also induce the release of other lymphokines from activated T cells. These include IFN-γ, IL-4, IL-5, and TNF-β (30–33). This has significant diagnostic and therapeutic implications, as IL-2 has anticancer activities and the immunophilin drugs (cyclosporin A and FK506), which have made modern-day transplantation possible, exert their immunosuppressive actions through inhibition of IL-2 gene transcription.

Interferon-γ

IFN-γ, a pleiotropic cytokine, was first described by Wheelock (34). IFN-γ is produced by both CD4+ and CD8+ T cells and by natural killer (NK) cells. It functions primarily as an inhibitor of viral replication and performs numerous other immunological functions. The human IFN-γ gene has been cloned and located on chromosome 12 (35). A receptor for IFN-γ has been identified and located on chromosome 6 (36). The receptor is a single-chain 90 kDa glycoprotein that shows a very high degree of species specificity for IFN-γ (37). The engagement of IFN-γ with its cognate receptor results in dimerization of the receptor and activation of the kinases Jak1 and Jak2. This results in the phosphorylation of the IFN-γ receptor and an additional protein STAT 1. Through a combination of events, phosphorylated STAT 1 enters the cell nucleus, where the complex binds to control sites in the IFN-γ response gene. This interaction results in the induction of at least 20 distinct proteins, 12 of which are distinct to IFN-γ stimulation (38). Most of these genes have potent immunological actions, which include the induction of HLA-B, HLA-DR, IP-10, and P1-kinase activity (22,38).

The functional consequences of IFN-γ/IFN-γR engagement are many. IFN-γ produced by CD8+ T cells, Th1 CD4+ T cells, and NK cells has documented antiviral, antiprotozoan, immunomodulatory, and inflammatory activities (22,39). IFN-γ has extensive and pleiotropic immunomodulatory effects. In monocytes, IFN-γ increases the expression of HLA class I and class II molecules and increases the production of IL-1, NO, H_2O_2, and pterin. IFN-γ also transforms monocytes to a cytotoxic profile by downregulating TGF-β receptors and upregulating IL-2Rγ (γ_c) expression (40–42). It is also chemotactic for monocytes and neutrophils (43). IFN-γ has profound effects on B cells by inducing the production of the phlogistic complement-activating IgG_{2a} and IgG_3 subclass antibodies (44). IFN-γ also induces the expression of ICAM-1 on endothelial cells (45).

Tumor Necrosis Factor-β

TNF-β is a cytolytic cytokine produced by activated Th1 T cells. Human TNF-β appears to be coded by a single gene (46,47). It is an extremely pleiotropic peptide capable of producing a wide variety of biological effects on virtually all cell types. This is attributable to the ubiquity of the TNF receptors, which then may activate multiple signal transduction pathways and induce or suppress the expression of a wide variety of cytokine, growth factor, cytokine receptor, inflammatory mediator, and acute-phase reactant genes (47). Mice with the TNF-β gene knock-out fail to develop lymph nodes but have normal numbers of circulating lymphocytes. This is not the case for TNF-α, suggesting that the interaction of TNF-β with the lymphotoxin-β (LTβ) receptors produces complexes, which are essential for normal lymphocyte development (48). TNF-β has been implicated in the mediation of cytotoxicity, antibody-dependent cell-mediated cytotoxicity (ADCC), and NK cell cytotoxic reactions. In addition, it has been shown that TNF-β can synergize with other cytokines (i.e., IFN-γ) to increase growth-inhibitory effects (47). The TNF family of cytokine/receptor pairs include CD40/ CD40 ligand, which controls B-cell activation, Fas/Fas ligand, important in apoptotic signaling, and other signal/ligand pairs important in regulating growth and differentiation of a number of cell types (47,49). Elevated levels of soluble TNF receptors have been identified in a number of pathological disease states including active systemic lupus erythematosus (50), endotoxemia, and HIV infections (22,51,52). We have shown that a polyclonal antibody specific for TNF-β can significantly inhibit the cell-mediated tissue injury seen in a rat model of lung allograft rejection (53). We have also shown that TNF-β levels are often elevated in patients with active SLE compared to normal controls and those with inactive disease (54). Therefore, TNF-β appears to be a potent cytolytic and immunoregulatory cytokine produced after T-cell activation (Th1) and may be an important effector cytokine in autoimmune disease and allograft rejection.

Interleukin-3

IL-3 is a pleiotropic factor, derived primarily from activated T cells, which stimulates the proliferation and differentiation of pluripotential hematopoietic stem cells (55). This categorizes IL-3 as a multicolony-stimulating factor (multi-CSF). IL-3 stimulates the growth of multilineage colonies from bone marrow stem cells including erythroid, granulocyte-macrophage, and megakaryocyte precursors. In vivo use of IL-3 has shown a positive effect on hematopoiesis and has been used in limited trials of bone marrow failure states (56,57). IL-3 also has proinflammatory and immunoregulatory effects. It can act on monocytes to induce IL-8 (chemotactic factor) secretion and endothelial cell adhesion and on eosinophils by promoting chemotaxis and phagocytosis (58,59).

Granulocyte-Macrophage Colony-Stimulating Factor

Granulocyte-macrophage colony-stimulating factor (GM-CSF) is another pleiotropic cytokine with the ability to stimulate the growth and differentiation of hematopoietic stem cells along the granulocyte and monocyte pathway. GM-CSF has the capacity to activate tumoricidal activity in macrophages. The genes for GM-CSF have been mapped to human chromosome 5 and are closely linked to the genes for IL-3, IL-4, IL-5, and M-CSF. GM-CSF is produced by T cells, macrophages, and endothelial cells and mediates its effects via interaction with specific cell surface receptors (60,61). Although GM-CSF was initially characterized as a growth factor for granulocytes, monocytes, and macrophages, recent data have established additional functions (61). GM-CSF can increase the ADCC of neutrophils, eosinophils, and macrophages toward tumor cells and induce the synthesis and release of a number of cytokines, including IL-1, IL-8, and TNF-α (62). GM-CSF can also induce human endothelial cells to proliferate and migrate (63).

Interleukin-4

IL-4 was initially characterized as a B-cell stimulatory and differentiation factor (64). As is the case with most cytokines, IL-4 is a pleiotropic cytokine that regulates multiple immune functions in diverse cell types. IL-4 is produced by CD4 Th0 and Th2 cells and by a variety of other immune cell types (65,66). IL-4 shares approximately 25% amino acid homology with IL-13, a recently discovered cytokine with similar activities (67). IL-4 mediates its biological activities through interaction with cell surface receptors. A common type of binding is IL-4/IL-4R-γ_c. The γ_c or IL-2Rγ is a receptor shared by several other cytokines including IL-2, IL-7, IL-9, and IL-15.

IL-4 is a prototypic Th2 cytokine that has potent anti-inflammatory properties and strong immunomodulatory properties on a variety of cell types, including T cells, B cells, monocytes, neutrophils, fibroblast, endothelial cells, and epithelial cells. IL-4 (derived from mast cells and other Th2-like cells) transforms Th0 cells to a Th2 profile. These T cells are important in humoral immunity and antibody production (65,66). IL-4 also serves as an inhibitor of Th1 cytokine production, resulting in an inhibition of proinflammatory immune events and cell-mediated immune events (3,68).

IL-4 has a strong effect on B cells by increasing viability, size, and upregulation of the expression of several important activation markers (CD23, CD40, sIgM, B7-1/B7-2, Thy-1, IL-4R, and MHC class II antigens). The ability of IL-4 to enhance B-cell proliferation is dependent on B-cell/T-cell interactions (presumably through the

CD40/CD40L) since IL-4 acts to suppress non–T-cell–regulated B-cell proliferation that could occur with Epstein-Barr infection and result in tumor formation (65,66).

In contrast to the proinflammatory effects of IFN-γ on macrophages, IL-4 acts to exert an anti-inflammatory effect. IL-4 suppresses the expression of FcγR and inhibits the expression of inflammatory cytokines (IL-1, IL-6, IL-8, and TNF-α) by macrophages. IL-4 also upregulates the expression of the anti-inflammatory decoy peptide IL-1 receptor antagonist (IL-1RA). IL-4 also inhibits superoxide and collagenase release by macrophages (65,66).

IL-4 induces VCAM-1 expression on endothelial cells and allows T cells, eosinophils, and basophils to adhere. IL-4 also has potent effects on fibroblast, resulting in proliferation and collagen secretion by these cells. This may have important implications for scarring after immunological injury to tissues (69,70). IL-4 also had potent antitumor effects in an animal model using IL-4–transfected tumor cells in athymic mice. IL-4 transfectants did not allow for the growth of other tumors in these mice, suggesting an antitumor effect of IL-4 independent of T-cell activation (71). Clearly, the diverse effects exhibited by IL-4 in vivo suggest that this cytokine is a central player in the regulation of the human immune and inflammatory response systems (22).

Interleukin-5

IL-5 is a Th2 cytokine with pleiotropic effects on both T and B cells. IL-5 also has both early and late activating effects on B cells. Although IL-5 has profound effects on murine B cells, analogous effects of IL-5 on human B cells have been difficult to demonstrate (72). For example, IL-5 is the most critical factor for Ig class switch in murine B cells, but IL-6 is much more important in this role in human B cells (73). To date, the most profound effects of IL-5 appear to be on cells of the eosinophilic lineage. IL-5 is chemotactic for eosinophils and appears to stimulate eosinophil functions. Studies in humans have shown that IL-5 is selectively elevated in helminth-infected human patients with eosinophilia (74). Mice transgenic for IL-5 develop severe eosinophilia, and serum IL-5 levels are elevated in patients with idiopathic hypereosinophilic syndromes and in patients with the eosinophilia-myalgia syndrome that results from excessive consumption of L-tryptophan. IL-5 is also important in basophil development (75–77).

Based on these observations, IL-5 appears to be an excellent candidate for mediation of allergic and atopic disorders where eosinophilia is present. Targeting the IL-5/IL-5R complex may prove a beneficial strategy in treating these disorders.

Interleukin-6

IL-6 is a pleiotropic cytokine that plays an important role in the regulation of host defenses, acute-phase reactions, immune reactivity and hematopoiesis. IL-6 has been described as BSF-2, IFN-β2, hybridoma/plasmacytoma growth factor, hepatocyte-stimulating factor, cytotoxic T-cell differentiation factor, and macrophage-granulocyte inducing factor (78,79). IL-6 is expressed by a number of cell types, including T cells, B cells, macrophages, hepatocytes, fibroblasts, mesangial cells, osteoblasts, and vascular endothelial cells as well as numerous malignant cell types (80).

IL-6 exerts its biological functions through interaction with a high-affinity receptor complex, which consists of two membrane glycoproteins [80 kDa binding receptor and 130 kDa transduction protein (gp 130)] (81). The interaction of IL-6 with its receptor complex initiates multiple biological functions on numerous cell types. IL-6 plays an important role in the regulation of immune functions. For example, IL-6 is important for B-cell differentiation and antibody production and cytotoxic T-cell differentiation. IL-6 is a critical cytokine for the induction of IL-2Rα gene expression and IL-2 secretion by antigen-activated T cells (82). Clearly, IL-6 appears to have a very important role in the mediation of immunoinflammatory events. In vivo effects of IL-6 include a possible role in the mediation of autoimmune diseases such as systemic lupus erythematosus. In fact, data from Linker-Israeli et al. working in our lab showed that patients with active SLE have increased gene expression of IL-6 compared with negative controls and with inactive SLE patients (83). Other investigators have shown that IL-6 levels are elevated in membranoproliferative glomerulonephritis, psoriasis, and malignancies (82,84). The role of IL-6 in perpetuating dysfunctional autoimmunity or regulating malignant cell growth remains to be defined.

Interleukin-7

IL-7 is a cytokine that was originally identified as a factor important in the growth of pre–B cells in long-term marrow cultures (85). The bioactivity of IL-7 is mediated through interaction with its specific receptors. The high-affinity receptor consists of at least two components: IL-7R and the ubiquitous accessory receptor γ_c (IL-2Rγ). Subsequently, IL-7 has been shown to be a growth and differentiation factor for murine and human T and B cells (86). IL-7 also plays a role in thymic development and can induce the proliferation of thymocytes (87). IL-7 can stimulate the development of cytolytic T cells from CD8+ T-cell precursors and from CD4+ T cells as well (88). IL-7 also stimulates the growth of lymphokine-activated killer

(LAK) cells from human PBMCs. Therefore, IL-7 may be useful as a cytokine in cancer immunotherapy to induce LAK cells and possibly in thymic immunodeficient states (89).

Interleukin-9

IL-9 is a recently discovered cytokine that is derived from T cells. IL-9 was first discovered in a HTLV-1 transformed T-cell line (90). IL-9 expression has subsequently been shown in mitogen-activated human T cells. IL-9 also appears to be preferentially secreted by Th2 CD4+ T cells. IL-9 has been mapped to chromosome 5 in humans (91), a region that is rich in cytokine and growth factor genes (IL-3, IL-4, IL-5, M-CSF, GM-CSF, c-fms, and PDGF Rb). Deletion of this region of chromosome 5 often results in profound hematological disturbances, such as myelodysplasia and nonlymphocytic leukemia (92).

IL-9 appears to act as a cofactor for IL-2, IL-4, and IL-7 in regulating T-cell activation. IL-9 appears to synergize with IL-4 in stimulating IgG, IgE, and IgM production (93). The finding that IL-9 has Th2-like bioactivity could lead to development of strategies to block IL-9 gene activation and/or activity in the treatment of allergic diseases (22,93).

Interleukin-10

IL-10 was originally described as cytokine synthesis inhibitory factor because it is a Th2 cytokine that inhibits the production of Th1 cytokines (94). Human and mouse IL-10 are highly homologus to the Epstein-Barr virus (EBV) gene BCRF-1 (B-cell replacement factor-1), now designated viral-IL-10 or vIL-10 (95). In humans, IL-10 is produced by Th0, Th1, and Th2 CD4+ T cells. IL-10 is also produced by B cells and numerous other cell types.

IL-10 is a pleiotropic cytokine that exerts both immunosuppressive and immunostimulatory effects on various cell types. The predominant effect of IL-10 production is suppression of Th1 (IFN-γ and IL-2) cytokine production. In rat models of lung transplantation, we have shown that overexpression of IL-10 is seen in a model that develops spontaneous tolerance of the allograft. This is also associated with suppressed expression of Th1 cytokines. Conversely, in rat lung allograft models where rejection is seen, Th1 cytokines predominate and IL-10 expression is reduced (96).

IL-10 also has potent effects on macrophages, including downregulation of MHC class II expression, which would reduce the ability of macrophages to interact with T cells and NK cells (97). IL-10 also inhibits a number of potent cytokines involved in inflammatory responses including prostaglandin E_2, TNF-α, IL-1, IL-6, and IL-8 (98,99). In addition, IL-10 induces the release of soluble TNF receptors and inhibits the induction of ICAM and B7 gene expression (98,99). IL-10 also can induce the expression of IL-1 receptor antagonist (IL-1RA), which has potent anti-inflammatory properties. IL-10 induces FcRγI, which is important in mediating ADCC (100). The potent anti-inflammatory actions of IL-10 may account for the EBV capturing this mammalian gene in order to suppress any Th1-mediated inflammation that would jeopardize EBV survival in the host (95). We have also seen vIL-10 expression in fine needle aspiration biopsies of renal allografts from patients with posttransplant lymphoproliferative disorder associated with EBV infection. After cessation of immunosuppressive medication, vIL-10 expression continued and was associated with good allograft survival without medications. This suggests that overexpression of vIL-10 may suppress Th1 responses to the allograft as well.

IL-10 has potent effects on B cells. IL-10 induces IgA synthesis in CD40/CD40L-activated B cells (101). IL-10 also has a number of other activating effects on tonsilar and spleenic B cells (100,101). IL-10 knock-out mice appear to have normal B-cell and T-cell development and function but suffer from chronic bowel inflammation due to uncontrolled immune reactivity to bowel antigens (22,102).

Transforming Growth Factor-β

TGF-β is a multifunctional peptide, and T cells are known to produce it. TGF-β is a known modulator of immune cell growth and differentiation. It is known to have regulatory effects on T cells, B cells, NK cells, LAK cells, and monocytes/macrophages (103,104). TGF-β is a potent inhibitor of T- and B-cell activation, and when antigen is presented in the context of high concentrations of TGF-β, a transformation to a Th2-type immune response is seen (68). In conjunction with IL-2 and IL-5, TGF-β acts as a stimulus for IgA production by activated lymphocytes (105) (Fig. 1). TGF-β1 knock-out mice are born without obvious deformities, but they die at 3–4 weeks of age due to massive mononuclear infiltration of all organs. TGF-β's potent immunoregulatory capabilities suggests that it may find a use in the regulation of dysfunctional immune responses as are seen in autoimmune diseases of the kidney and other organs. In addition, TGF-β may find a role in the prevention and possible treatment of acute allograft rejection, since recent experiments in rat heart transplant models have shown that allografts transfected with the human TGF-β gene show significantly longer survival times (107). It is also of interest that the potent immunosuppressive drug cyclosporin A is known to induce TGF-β gene expression in tissues while suppressing the expression of IL-2 (108). Unfortunately, TGF-β is now known to play a very important role in promoting fibrosis of the kidney and lung in chronic transplant rejection and in autoimmune diseases (109). This would limit the broad application of TGF-β gene therapy in these circumstances.

Interleukin-15

IL-15 is a recently discovered cytokine that shares many of the biological activities of IL-2 but is structurally distinct (110,111). IL-15 utilizes the α and γ chains of the IL-2R-αβγ complex to activate IL-2R–bearing cells (110,112). The IL-2Rγ or γ_c is critical for signal transduction by IL-15 as is true for IL-2, IL-4, IL-7, IL-9, and IL-13 (112). IL-15 mRNA is expressed by a number of tissues including placenta and skeletal muscle. It is also found in adherent peripheral blood mononuclear cells (110). IL-15 is not produced by activated T cells. Again, early reports suggest that IL-15 shares many of the same properties as IL-2. For example, in the presence of CD40L on T cells, IL-15 stimulates the production of IgG, IgM, and IgA, similar to the pattern seen for IL-2 (110).

Cytokine Regulation of T-Cell Functions

T cells produce a wide variety of cytokines that regulate the functions of other cells, and they are susceptible to regulation by cytokines produced endogenously (acting in an autocrine manner) or exogenously (acting in a paracrine manner). The purpose of this section is to describe the cytokines that seem most important in regulating T-cell functions. Some of this information was briefly mentioned in the section on Th1/Th2 T-cell deviation.

Interleukin-1

IL-1 is an important coactivator of T cells during antigen-stimulation events. This was shown in experiments where recombinant IL-1 increased T-cell IL-2 production and IL-2R expression. IL-1 synergized with both IL-4 and IL-6 in augmenting T-cell activation. In addition, some activated T cells synthesize IL-1, suggesting a role for IL-1 in T- and B-cell activation (113–116). In human PBMCs activated in the absence of macrophages, IL-1 serves as an important cofactor for activation (113). A competitive inhibitor of IL-1 was recently identified (IL-1RA). This cytokine acts through competitive binding to the IL-1R and does not transduce the activation signal to the T cells (117). IL-1RA can inhibit T-cell activation, again suggesting an important role for IL-1 in T-cell activation. Interestingly, pooled human gammaglobulin preparations (IVIG), which are known to have immunoregulatory and anti-inflammatory effects in multiple autoimmune and inflammatory disorders, are known to be a potent stimulus for IL-1RA release (118).

Interleukin-12

The response to immunizations and other immunogenic stimuli is something we take for granted but, until recently, did not understand well. It was not appreciated until recently that cytokines were important in deviation of T-helper-cell differentiation toward CMI or antibody production (20) as discussed previously. Thus, the immune system has the selective ability to choose which pathway(s) will be utilized for specific antigenic confrontations. For example, many bacterial, protozoan, and viral pathogens stimulate CMI responses, while other pathogens such as helminths stimulate antibody responses (20). This paradigm is shown in Figure 2. The preferential deviation of an immune response toward CMI is now felt to be controlled by IL-12. IL-12 was originally called NK cell stimulatory factor and is produced primarily by macrophages and B lymphocytes. IL-12 is also known to stimulate the production of IFN-γ in T cells and NK cells. IL-12 also enhances the expansion of Th1 cells in vitro (20). IL-12 drives the differentiation of Th0 cells to the Th1 phenotype, while IL-4 transforms Th0 cells to a Th2 phenotype. Taken together, these data suggest that IL-12 production by APCs is a critical factor in changing the immune response to a Th1 profile. The implications of this information are great, since IL-12 could be used as an adjuvant to enhance responses to vaccines and to treat diseases where CMI is important. Adjuvants that incorporate bacterial components and are used to enhance responses to specific antigens probably already act by inducing IL-12 synthesis in the host (20).

Interleukin-13

IL-13 is a pleiotropic cytokine that is known to share amino acid homology and biological activities with the Th2 cytokine IL-4 (119). The biological effects of IL-13 are mediated by its binding to shared components of the IL-4R (possibly IL-2γ or γ_c) (120).

IL-13 is a potent regulator of immunoinflammatory events. IL-13 upregulates the monocyte/macrophage expression of CD23 and MHC class I and class II and downregulates the expression of FcγR (119,120). IL-13, like IL-4, is a potent inhibitor of the proinflammatory cytokines (IL-1, IL-6, IL-8, IL-10, and IL-12) and the chemokines (MIP-1 and MCP). IL-13 enhances the production of the T-cell–suppressive cytokine IL-1RA (120). In B cells, IL-13 acts independently of IL-4 to produce IgE and IgG_4 synthesis. These data suggest that IL-13 may be as important as IL-4 in the induction of atopic diseases in humans. Unlike IL-4, IL-13 has no discernible effects on human T cells (120). Thus, IL-13 could be characterized as a Th2-inducing cytokine. Data to support this contention include those mentioned above and recent data showing that IL-13 inhibits IL-2–induced IFN-γ production (120). It would appear that IL-13, TGF-β, IL-10, and IL-4 are all Th2-like cytokines that share redundant activities and are important inhibitors of CMI (Fig. 2).

Tumor Necrosis Factor-α

TNF-α is a cytokine produced primarily by lipopolysaccharide-stimulated macrophages, which mediates necrosis of transplanted tumors (121). TNF-α also has immunoregulatory functions in human and experimental animal systems (121). Bonavida et al. (121) showed that the addition of TNF-α (1-10 μg/ml) resulted in a dose-dependent inhibition of pokeweed mitogen–induced antibody production by PBMCs. Satoh et al. (122) showed that recombinant TNF-α is a potent inhibitor of the development of autoimmune diabetes mellitus in the nonobese diabetic (NOD) mice. These investigators felt that the inability to produce TNF in various mouse strains correlated with their propensity to develop autoimmune diseases. NOD mice were poor in vitro producers of TNF-α; it was therefore felt that the lack of TNF production allowed autoimmune antibody production to proceed and autoimmunity to develop. Exogenous administration of TNF-α inhibited the dysfunctional T-cell activation and prevented the development of autoimmunity by induction of suppressor T cells (123). Other investigators have shown that exogenous administration of TNF-α to (NZB × NZW) F_1 hybrid mice suppressed the progression of murine systemic lupus erythematosus (SLE). This was again postulated to be due to the induction of suppressor cell networks by TNF-α (124). It is unclear whether these observations will have clinical relevance.

CYTOKINES AND B CELLS

Currently there is no consensus regarding the mechanisms responsible for B-cell activation with subsequent differentiation to antibody-producing plasma cells. As previously mentioned, the Th2-inducing cytokines (IL-4 and IL-13) appear to play an important role in driving the immune response toward antibody production, but there are multiple other cytokines in the B-cell activation derby that vie for a role in activating and deviating B cells to a specific antibody production profile (125). A better understanding of the regulation of B-cell function has emerged with the discovery of Th1/Th2-driven cytokine patterns. Despite these advances, much effort is still needed to define how the panoply of cytokines impinge upon B-cell activation and antibody production. Understanding how cytokines regulate B-cell functions should lead to a better understanding of inflammatory processes, where dysfunctional or overactive B-cell responses are seen. This includes autoimmune diseases, allergic and atopic disorders, and allograft rejection. In this section, we will review the cytokines that appear most relevant to controlling B-cell activation and antibody production.

Interleukin-2

IL-2 is known to be a potent stimulator of B-cell activity. Initial studies suggested that the IL-2R was present only on activated T cells, but subsequent studies have demonstrated IL-2Rα on activated B cells as well (126). Data from our laboratory support a direct role for IL-2 in the activation of human B cells. (127). Our data show that lymphocytes isolated from patients with active SLE express high levels of IL-2R. Cell-sorting experiments showed that the majority of IL-2Rα–expressing cells were B cells and that addition of recombinant IL-2 to cultures of SLE lymphocytes significantly increased the spontaneous IgG production by these cells (128). Zubler et al. (126) suggest that IL-2 acts to clonally expand activated B cells. In addition, IL-2 increases IgM production and has been shown to significantly increase mRNA for J chain and μ chain in normal B cells (128).

Thus it appears that IL-2 has an important role in B-cell clonal expansion and possibly in stimulating IgG and IgM antibody production. An additional piece of evidence is relevant in this regard. Tiggers et al. (129) showed that recombinant IL-4 added to BCL-1 cells inhibited the J-chain mRNA accumulation stimulated by IL-2. This information suggests a possible mechanism for isotype switch mediated by IL-4, since other investigators have shown that IL-4 downregulates IgM production and increases IgG_1 production in B-cell lines (128).

Interleukin-4

IL-4 is a potent activator of B cells. IL-4 induces IgG production in LPS-activated B cells and acts synergistically with anti-IgM antibody to induce proliferation of B cells (130). IL-4 also induces the expression of MHC class II antigens on resting B cells, thus turning the resting B cell into an APC with the capability of activating T cells (128) (Fig. 1). IL-4 also stimulates the expression of low-affinity Fc receptors for IgE (FcεRII) (128). IL-4 stimulates IgG_1 synthesis by activated B cells and enhances the production of IgE by LPS-activated B cells (128). IFN-γ (Th1 cytokine) is a potent inhibitor of most IL-4 activities (3).

Interferon-γ

IFN-γ has a profound effect on B-cell functions. IFN-γ is primarily a late-acting factor of B-cell function, and studies with recombinant cytokine and anti–IFN-γ antibodies have established its importance as a regulator of B-cell activity (128). IFN-γ is important for the proliferation of the phlogistic IgG_{2a} subclass-producing B-cell clones, while suppressing the production of IgG_3 and IgG_{2b} (131). IL-4 is a potent inhibitor of IFN-γ–induced antibody production, thus a regulatory circuit exists where a predominant Th1 response could be controlled by an emerging Th2 response (3,128). Wilkes et al. (132) showed that the IgG_{2o}-to-IgG_1 ratio in bronchoalveolar lavage fluid from lung transplant recipients was significantly elevated in patients undergoing rejection compared with patients who were experiencing a stable course. These investigators felt that this was indirect evidence for a Th1 (IFN-γ) predominance in the patients with allograft rejection and could be responsible for some of the inflammatory injury to the lung allograft. Since effector functions are mediated by different immunoglobulin classes (i.e., IgE mediates hypersensitivity and atopic reactions while IgG_{2a} mediates complement-dependent cytotoxicity), one can postulate that T cells have developed a unique capacity to adapt their immune response profile to fit the characteristics of the invading pathogen (3).

Interleukin-6

Resting B cells do not express IL-6R; therefore, one would not expect IL-6 alone to have the capacity to induce B-cell proliferation and differentiation (133). The primary function of IL-6 on activated B cells appears to be the promotion of B-cell differentiation into antibody-producing plasma cells. IL-6 appears to have a critical role in enhancing antibody production in EBV-transformed B cells and in PBMCs stimulated by PWM. Muraguchi et al. showed that Ig production in PBMCs stimulated with PWM was substantially inhibited when anti–IL-6 monoclonal antibodies were added to the cultures (73).

We have also shown that IVIG added to PWM-stimulated PBMCs inhibits Ig production and IL-6 mRNA expression (134). This finding suggests that the inhibition of specific autoimmune disorders by IVIG where spontaneous Ig production and overexpression of IL-6 mRNA is seen could be due to an effect on IL-6 mRNA expression. IL-6 appears to augment Ig secretion from B cells independent of an action on proliferation. This is consistent with the role of IL-6 as a late-acting factor in B-cell proliferation.

Interleukin-1

IL-1 appears to be an important cofactor for the stimulation of murine B cells (128,135). Other studies have shown that IL-1 is an important coactivator of suboptimally activated B cells, suggesting that IL-1 is an important early activating factor for B-cell proliferation (128). IL-1 also enhances Ig secretion by preactivated B cells in response to IL-2, demonstrating that the specific effect of IL-1 is to enhance IL-2–directed B-cell differentiation.

THE TREATMENT OF HUMAN DISEASES

Robust immune responses are desirable in order to maintain the integrity of the individual's health. However, undesirable or dysfunctional immune responses can often occur and result in diseases such as systemic autoimmunity, uncontrolled allergic responses, allograft rejection, and nephritogenic immune responses (136).

Rapid advances in our understanding of the immune system and how T-cell activation and cytokine production occur resulted from investigations into the mechanisms of action of the powerful and specific immunosuppressive drug cyclosporin A (CSA). In fact, another drug, FK506, has emerged as an even more powerful inhibitor of T-cell activation, and investigations show that it has a similar mode of action as CSA (137). CSA, FK506, and rapamycin (which has potent immunosuppressive actions, but a distinct mechanism of action from CSA and FK506) fit into a new category of drug referred to as immunophilins. CSA and FK506 act primarily through inhibition of the enzyme calcineurin A (Fig. 3). During normal T-cell activation, engagement of the TCR results in an influx of Ca^{2+}. Ca^{2+} then binds to several intracellular Ca^{2+}-binding proteins including calmodulin and calcineurin A. The Ca^{2+}/calcineurin A/calmodulin complex acts as an enzyme to dephosphorylate the cytosolic component of the nuclear factor for activating transcription (N-FAT). Once dephosphorylated, N-FAT_c moves to the nucleus to interact with the nuclear N-FAT (N-FAT_n) in the upstream regulatory region of the IL-2 gene (and possibly other cytokine genes). The combination of N-FATc and N-FATn initiates transcription of the IL-2 gene with subsequent IL-2 production. Since CSA and FK506 are specific inhibitors of the dephosphorylation of N-FAT_c (an activation step), no transcription of IL-2 gene occurs, thus T-cell activation and proliferation is blocked (Fig. 3). Clearly the remarkable advances that have occurred in the treatment and prevention of allograft rejection and autoimmune diseases are a direct result of the clinical implementation of CSA and, more recently, FK506. The ability of CSA and FK506 to control the expression of Th1 cytokines such as IFN-γ and IL-2, which are important for allograft rejection, is critical to maintaining the integrity of transplanted organs and has allowed other solid organ transplants such as heart, liver, and lung to become commonplace. The clinical recognition of the powerful effects the immunophilin drugs have on

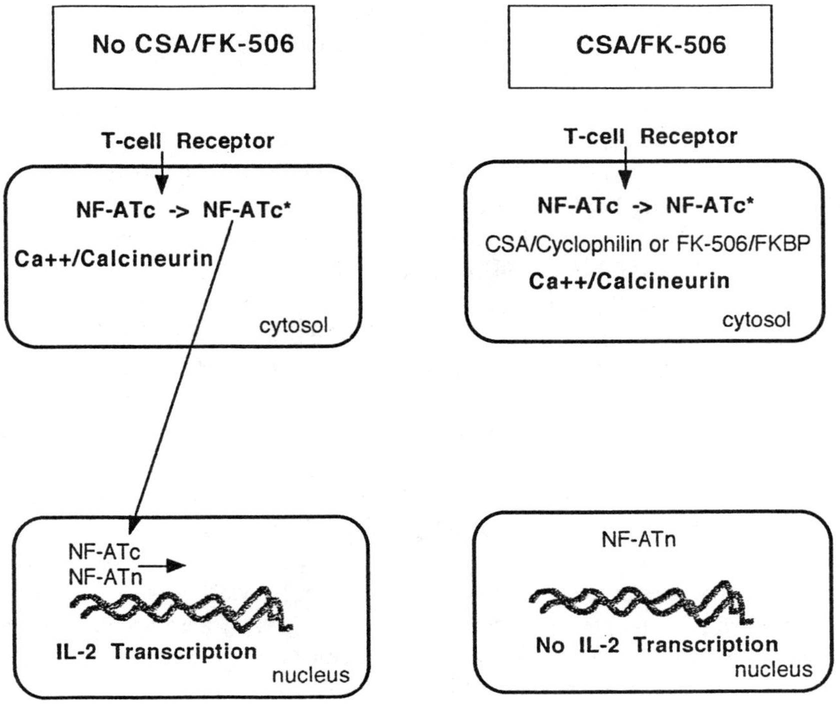

Figure 3 Immunophilin drugs cyclosporin A and FK506 inhibit T-cell activation by blocking Ca^{2+}-induced dephosphorylation of IL-2 transcriptional initiation factors. See text for details.

T-cell activation and cytokine production allowed basic scientists to use these drugs as probes to better understand the physiology of these events. In my opinion, this represents one of the most important advances in transplant immunology in this century.

Monoclonal antibodies have played an important role in defining the functional characteristics of specific T-cell molecules. The T-cell receptor complex (TCR/CD3) serves a critical role in transmitting signals initiated by antigen-presenting cell/T-cell interaction. The CD3 component of the TCR/CD3 complex is critical for transmission of signals to the interior of the T cell (138). OKT3, a monoclonal antibody that recognizes components of the CD3 complex on human T cells, is recognized as a potent immunosuppressive agent. The immunosuppression rendered by OKT3 appears to be mediated by effacement of the TCR/CD3 complex from all T cells, thus rendering them incapable of recognizing alloantigen presented in the context of a MHC-restricted fashion. An interesting complication of OKT3 therapy is the rapid development of what is now called the "cytokine-release syndrome." These patients often develop fever, chills, myalgias, capillary leak, and occasionally pulmonary edema. This syndrome results from the perturbation on the CD3 molecules on T cells with instant signaling for cytokine production to occur. The symptoms result from massive release of proinflammatory cytokines, including TNF-α, IL-2, and IFN-γ (139). Monoclonal antibodies to other T-cell receptor complex antigens such as CD4 are undergoing clinical trials, but none to date have the powerful immunosuppressive efficacy (or side effects) of OKT3.

A more ideal immunosuppressive agent would be one that could selectively block phlogistic cytokine production and change T cells to a tolerant or anergic state. Complete T-cell activation involves recognition of antigen through the T-cell receptor/CD3 complex and costimulatory signals delivered through other receptor/ligand pathways (140). This subsequently results in the release of proinflammatory cytokines such as IL-2 and IFN-γ. However, recognition of alloantigen via the T-cell receptor (TCR) alone is not sufficient to induce robust T-cell activation; T cells must receive one or more costimulatory signals from APCs to drive the high-level IL-2 production and cell proliferation necessary for T-cell effector functions.

In fact, antigen recognition in the absence of costimulation can lead to T-cell anergy or apoptosis (140,141). A central costimulatory pathway is activated by stimulation of the CD28 receptor on T cells by binding to ligands on APCs, notably B7-1/B7-2 and possibly others, resulting in a full immunizing signal to T cells. This is characterized by increased production and stability of IL-2 transcript, IL-2 secretion, and cell proliferation. This pathway, which is cyclosporine resistant, can be specifically and effectively blocked by CTLA4Ig (142). CTLA4Ig is a soluble recombinant fusion molecule, consisting of the extracellular domain of the human T-cell surface receptor CTLA-4 fused to a human Ig Cγ chain. CTLA4Ig binds to B7 and other ligands on APCs with high affinity and blocks their interaction with CD28 (142). CTLA4Ig has a 20-fold higher affinity for B7 than does CD28, allowing it to act as a competitive inhibitor of CD28 binding to B7-1/B7-2 with subsequent inhibition of coactivation signals and Th1 cytokine production. Engagement of CTLA4Ig with B7-1 and B7-2 also results in the expression of IL-4 with depressed IL-2 (Th2 >> Th1) (141). This is shown in Figure 4.

CTLA4Ig, although derived from human T cells, also binds to murine and rat B7. Thus, mixed lymphocyte cultures treated with CTLA4Ig generate far lower levels of IL-2 transcript than control, and T-cell proliferation is strongly suppressed. In vivo, CTLA4Ig is known to block T-cell–dependent, B-cell antibody synthesis and autoimmune glomerulonephritis and prevents cardiac allograft rejection in rats and mice (143,144).

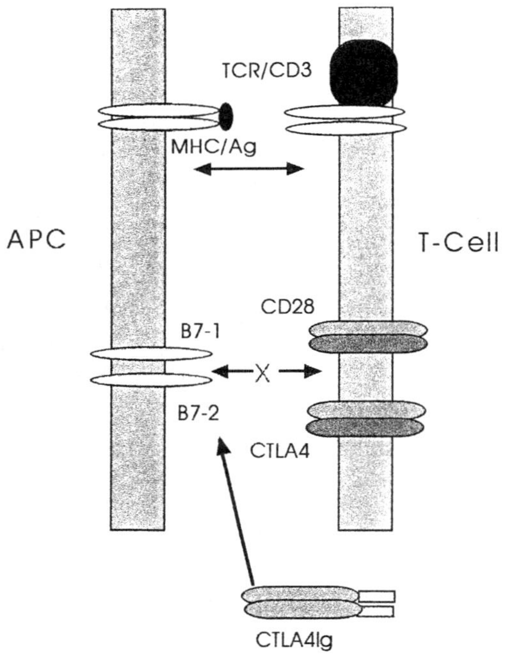

Figure 4 The synthetic peptide CTLA4Ig inhibits the costimulatory signals necessary for complete T-cell activation upon antigen/antigen-presenting cell (APC) encounter. CTLA4Ig blocks B7-1 and B7-2 interaction with CD28 and CTLA4 on T cells. This incomplete activation signal results in T-cell anergy (instead of immunization) and transformation of cytokine immune responses to a Th2 profile.

Since CTLA4Ig also binds to the costimulatory molecules in the rat, we were able to test the acute effect of CTLA4Ig treatment on the rejection process in rat lung allografts. Our data showed that CTLA4Ig inhibited inflammatory aspects of rat lung AR but did not result in significant inhibition of cytokine gene activation (145). More recently, Lin et al. (144) showed that a protocol employing donor-specific spleen cell transfusions and CTLA4Ig resulted in long-term acceptance of cardiac allografts in the BN->Lew model. These animals appeared to develop tolerance to their allografts while retaining the ability to reject third-party allografts. Data from other investigators suggest that CTLA4Ig not only results in blocking of the Th1 cytokine profile (141), but actually alters the immune response to Th2. This may account for the favorable anergic or tolerant response to allografts seen in these models and may pave the way for relevant applications of these agents to clinical transplantation and autoimmune diseases in the future.

ACKNOWLEDGMENTS

This work was supported by a grant from the NIH (1U01-AI37313-01) and the Rebecca Sakai Fund. We would like to express our gratitude to Ms. Bree Shelton and Ms. Rosilyn Clayton for their help in preparing this manuscript.

REFERENCES

1. Neta R, Oppenheim JJ, Durum S. The cytokine concept: Historical perspectives and current status of cloned cytokines. In: Cohen S, ed. Lymphokines and the Immune Response. Boca Raton, FL: CRC Press, 1990:30–40.
2. Steiger J, Nickerson PW, Steurer W, et al. IL-2 knockout recipient mice reject islet cell allografts. J Immunol 1995; 155:489–498.
3. Nickerson PW, Steurer W, Zheng X, et al. Cytokines and the Th1/Th2 paradigm in transplantation. Curr Opin Immunol 1994; 6:757–764.
4. Dalton DK, Pitts-Meek S, Kehr S, Figari IS, et al. Multiple defects of immune cell function in mice with disrupted interferon gamma genes. Science 1993; 259:1739–1742.
5. David JR, Al-Askari S, Lawrence HS, Thomas L. Delayed hypersensitivity in vitro. I. The specificity of inhibition of cell migration by antigens. J Immunol 1964; 93:264.
6. Bloom BR, Bennett B. Mechanism of a reaction in vitro associated with delayed type hypersensitivity. Science 1966:153:80.
7. Ruddle NH, Waksman BH. Cytotoxicity mediated by soluble antigen and lymphocytes in delayed hypersensitivity. III. Analysis of mechanisms. J Exp Med 1968; 128:1267.
8. Kolb WP, Granger GA. Lymphocyte in vitro cytotoxicity: Characterization of human lymphotoxin. Proc Natl Acad Sci USA 1968; 61:1250.
9. Green JA, Cooperband SR, Kibnick S. Immune specific induc-

tion of interferon production in cultures of human blood lymphocytes. Science 1969; 164:1415.
10. Dumonde DG, Wolstencroft RA, Panayi G, Matthew M, Morley J, Howson WT. "Lymphokines": non-antibody mediators of cellular immunity generated by lymphocyte activation. Nature 1969; 224:38.
11. Gery I, Waksman BH. Potentiation of the T-lymphocyte response to mitogens. II. The cellular source of the potentiating mediators. J Exp Med 1972; 136:143.
12. Cohen S, Bigazzi PE, Yoshida T. Similarities of T-cell function in cell-mediated immunity and antibody production. Cell Immunol 1974; 12:150.
13. Inaba K, Steinman RM. Resting and sensitized T-lymphocytes exhibit distinct stimulatory (antigen-presenting cell) requirements for growth and lymphokine release. J Exp Med 1984; 160:1717.
14. Mosmann TR, Coffman RL. Th1 and Th2 cells: Different patterns of lymphokine secretion to different functional properties. Ann Rev Immunol 1989; 7:145.
15. DeKruyff RH, Ju S-T, Hunt AJ, et al. Induction of antigen-specific antibody responses in primed and unprimed B cells: Functional heterogeneity among Th1 and Th2 T-cell clones. J Immunol 1989; 142:2575.
16. Cher DJ, Mosmann TR. Two types of murine helper T-cells clones: II. Delayed-type hypersensitivity is mediated by Th1 clones. J Immunol 1987; 138:3688.
17. Mosmann TR, Schumacher JH, Fiorentino DF, et al. Isolation of monoclonal antibodies specific for IL-4, IL-5, IL-6 and a new Th-2 specific cytokine (IL-10), cytokine synthesis inhibitory factor, by using a solid phase radioimmunoabsorbent assay. J Immunol 1990; 145:2928.
18. Mosmann TR, Moore KW. The role of IL-10 in cross regulation of Th1 and Th2 responses. Immunol Today 1991; 12:A49.
19. Maggi E, Parronchi P, Manetti R, et al. Reciprocal regulatory effects of IFN-gamma and IL-4 in the in vitro development of Th1 and Th2 clones. J Immunol 1992; 148:2142.
20. Scott P. IL-12: initiation cytokine for cell mediated immunity. Science 1993; 260:496–497.
21. Hatakeyama M, Taniguchi T. Interleukin-2. In: Sporn MB, Roberts AB, eds. Peptide Growth Factor and Their Receptors. New York: Springer-Verlag, 1990:523.
22. Cytokine Literature Review. R&D System 1995, 1:1–200.
23. Leonard WJ, Naguchi M, Russell SM, McBride OW. The molecular basis of x-linked severe-combined immunodeficiency: the role of the interleukin-2 receptor gamma chain as a common gamma chain. Immunol Rev 1994; 138:61–86.
24. Weinberg K, Parkman R. Severe combined immunodeficiency due to a specific defect in the production of IL-2. N Engl J Med 1990; 322:1718–1723.
25. Kundig T, Schorle H, Bachmann MF, et al. Immune responses in IL-2 deficient mice. Science 1993; 262:1059–1061.
26. Kondo M, Takeshita T, Ishii N, et al. Sharing of the IL-2 receptor gamma chain between receptors for IL-2 and IL-4. Science 1993; 262:1874–1877.
27. Kondo M, Takeshita T, Higuchi M, et al. Functional participation of the IL-2 receptor gamma chain in IL-7 receptor complexes. Science 1994; 263:1453–1454.
28. Russell SM, Johnston JA, Noguchi M, et al. Interaction of IL-2R beta and gamma c chains with Jak-1 and Jak-3: implications for XSCID and XCID. Science, 1994; 266:1042–1045.
29. Grabstein KH, Eisenman J, Shanebeck K, et al. Cloning of a T-cell growth factor that interacts with the β chain of the IL-2R. Science 1994; 264:965–968.
30. Strotter H, Rude E, Wagner H. T-cell factor (interleukin-2) allows in vivo induction of T help cells against heterologous erythrocytes in athymic (nu/nu) mice. Eur J Immunol 1980; 10:719.
31. Wagner H, Hardt C, Heeg K, Rollinghoff M, Pfizenmaier K. T-cell derived helper factors allows in vivo induction of cytotoxic T cells in nu/nu mice. Nature 1980; 284:278.
32. Yamamoto JK, Farrar WL, Johnson HM. Interleukin 2 regulation of mitogen induction of immune interferon (IFN) in spleen cells and thymocytes. Cell Immunol 1982; 66:333.
33. Howard M, Matis L, Malek TR, et al. Interleukin-2 induces antigen-reactive T cell lines to secrete BCGF-1. J Exp Med 1983; 158:202.
34. Wheelock EF. Interferon-like virus-inhibitor induced in human leukocytes by phytohemagglutinin. Science 1965; 149:310.
35. Naylor SL, Sakaguchi AY, Shows TB, Law ML, Goeddel DV, Gray PW. Human immune interferon gene located on chromosome T2. J Exp Med 1983; 157:1020.
36. Pfizenmaier K, Weigmann K, Scheurich P, et al. High-affinity human IFN-gamma-binding capacity is encoded by a single receptor gene located in proximity to c-ros on human chromosome region 6 q 22. J Immunol 1988; 141:856–860.
37. Aguet M, Dembic Z, Merlin G. Molecular cloning and expression of the human gamma interferon receptor. Cell 1988; 55:273–280.
38. Harris CA, Hunte B, Krauss MR, et al. Induction of leucine aminopeptidase by gamma IFN. Identification protein microsequencing after purification by preparative two-dimensional gel electrophoresis. J Biol Chem 1992; 267:6865–6869.
39. Paliard X, de Wall Malefijt R, Yssel H. Simultaneous production of IL-2, IL-4 and gamma interferon by activated human CD4+ and CD8+ T-cell clones. J Immunol 1988; 141:849–855.
40. Sen GC, Lengyel P. The interferon system, a birds eye view of its biochemistry. J Biol Chem 1992; 267:5017–5020.
41. Gusella GL, Musso T, Bosco MC, Espinoza-Delgado I, Matsushima K, Varesio L. IL-2 up-regulates but IFN-gamma suppresses IL-8 expression in human monocytes. J Immunol 1993; 151(5):2725–2732.
42. Bosco MC, Espinoza-Delgado I, Schwabe M, Russell SM, Leonard WJ, Longo DL, Varesio L. The gamma subunit of the interleukin-2 receptor is expressed in human monocytes and modulated by interleukin-2, interferon gamma, and transforming growth factor beta 1. Blood 1994; 83:3462–3467.
43. Issekutz AC, Issekkutz TB. Quantitation and kinetics of blood monocyte migration to acute inflammatory reactions, and IL-1 alpha, tumor necrosis factor-alpha, and IFN-gamma. J Immunol 1993; 151:2105–2115.
44. Snapper CM, McIntyre TM, Mandler R, Pecanha LM, Finkelman FD, Lees A, Mond JJ. Induction of IgG3 secretion by interferon gamma: a model for T cell-independent class switching in response to T cell-independent type 2 antigens. J Exp Med 1992; 175:1367–1371.
45. Renkonen R, Mattila P, Majuri ML, Paavonen T, Silvennoinen O. IL-4 decreases IFN-gamma-induced endothelial ICAM-1 expression by a transcriptional mechanism. Scand J Immunol 1992; 35:525–530.
46. Vilcek J, Lee TH. TNF: new insights into the molecular mechanisms of its multiple actions. J Biol Chem 1991; 266:7313.
47. Kronke M, et al. In: Aggarwal BB, Bilcek J, eds. Tumor Necrosis Factor: Structure, Function and Mechanism of Action. New York: Marcel Dekker, 1991.
48. De-Togni P, Ruddle NH, Streeter PR, Fick A, Mariathasan S, Smith SC, Carlson R, Shornick LP, Strauss-Schoenberger J, et al. Abnormal development of peripheral lymphoid organs in mice deficient in lymphotoxin. Science 1994; 264(5159):703.
49. Smith CA, Farrah T, Goodwin RG. The TNF receptor super-

family of cellular and viral proteins: activation, costimulation and death. Cell 1994; 76:959–962.
50. Aderka D, Wysenbeek A, Engelmann H, Cope AP, Brennan F, et al. Correlation between serum levels of soluble tumor necrosis factor receptor and disease activity in systemic lupus erythematosus. Arthritis Rheum 1993; 36:1111–1120.
51. Van Zee KJ, Kohno T, Fischer E, Rock CS, Moldawer LL, Lowry SF. Tumor necrosis factor soluble receptors circulate during experimental and clinical inflammation and can protect against excessive tumor necrosis factor alpha in vitro and in vivo. Proc Natl Acad Sci USA 1992; 89:4845–4849.
52. Kalinkovich A, Engelmann H, Harpaz N, Burstein R, Barak V, et al. Elevated serum levels of soluble tumor necrosis factor receptors (sTNF-R) in patients with HIV infection. Clin Exp Immunol 1992; 89:351–355.
53. Saito R, Prehn J, Zuo X-J, et al. Tumor necrosis factor participates in the pathogenesis of lung allograft rejection in the rat. Transplantation 1993; 55:967–972.
54. Yap H-K, Ng S-C, Lee B-W, et al. Modulation of MHC expression on human endothelial cells by sera from patients with SLE. Clin Immunol Immunopathol 1993; 68:321–326.
55. Ihle JN, Weinstein Y, Keller J, Henderson L, Palaszynski E. Interleukin-3. Methods Enzymol 1985; 116:540–552.
56. Oster W, Frisch J, Nicolay U, Schulz G. Interleukin-3. Biologic effects and clinical impact. Cancer 1991; 67:2712–2717.
57. Aglietta M, Sanavio F, Stacchini A, Morelli S, Fubini L, Severino A, Pasquino P, Volta C, Bretti S, Tafutp S, et al. Interleukin-3 in vivo: kinetic response of target cells. Blood 1993; 82:2054–2061.
58. Blom M, Tool AT, Kok PT, Koenderman L, Roos D, Verhoeven AJ. Granulocyte-macrophage colony-stimulating factor, interleukin-3 (IL-3), and IL-5 greatly enhance the interaction of human eosinophils with opsonized particles by changing the affinity of complement receptor type 3. Blood 1994; 83:2978–2984.
59. Warringa RA, Koenderman L, Kok PT, Kreukniet J, Bruijnzeel PL. Modulation and induction of eosinophil chemotaxis by granulocyte-macrophage colony-stimulating factor and interleukin-3. Blood 1991; 77:2694–2700.
60. Burgess AW. In: Spom MB, Roberts AB, eds. Peptide Growth Factors and Their Receptors I. New York: Springer-Verlag, 1990:190.
61. Moore WA. The clinical use of colony stimulating factors. Annu Rev Immunol 1991; 9:159–191.
62. Takahashi GW, Andrews DF 3d, Lilly MB, Singer JW, Alderson MR. Effect of granulocyte-macrophage colony-stimulating factor and interleukin-3 on interleukin-8 production by human neutrophils and monocytes. Blood 1993; 81:357–364.
63. Ebisawa M, Liu MC, Yamada T, Kato M, Lichtenstein LM, Bochner BS, Schleimer RP. Eosinophil transendothelial migration induced by cytokines. II. Potentiation of eosinophil transendothelial migration by eosinophil-active cytokines. J Immunol 1994; 152(9):4590–4596.
64. Noelle RJ, Kuziel WA, Maliszewski CR, McAdams E, Vitetta ES, Tucker PW. Regulation of expression of multiple class II genes in murine B cells by B cell stimulatory factor-1 (BSF-1). J Immunol 1986; 137:1718–1723.
65. Paul WE. Interleukin-4: a prototypic immunoregulatory lymphokine. Blood 1991; 77:1859–1870.
66. Banchereau J, Rybak ME. Interleukin 4. In: Thomson A, ed. The Cytokine Handbook. 2nd ed. New York: Academic Press, 1994:99.
67. Aversa G, Punnonen J, Cocks BG, de Waal Malefyt R, Vega F Jr, Zurawski SM, Zurawski G, de Vries JE. An interleukin 4 (IL-4) mutant protein inhibits both IL-4 or IL-13-induced human immunoglobulin G4 (IgG4) and IgE synthesis and B cell proliferation: support for a common component shared by IL-4 and IL-13 receptors. J Exp Med 1993; 178:2213–2218.
68. Matriano J, Streilein J. Induction of immunological tolerance. In: Sher L, Makowska L, eds. Intra-Abdominal Organ Transplantation 2000. R.G. Landes Co., 1994:249–280.
69. Thornhill MH, Kyan-Aung U, Haskard DO. IL-4 increases human endothelial cell adhesiveness for T cells but not for neutrophils. J Immunol 1990 144(8):3060–3065.
70. Verdegaal EM, Beekhuizen H, Blokland I, van Furth R. Increased adhesion of human monocytes to IL-4-stimulated human venous endothelial cells via CD11/CD18, and very late antigen-4 (VLA-4)/vascular cell adhesion molecule-1 (VCAM-1)-dependent mechanisms. Clin Exp Immunol 1993; 93:292–298.
71. Tepper RI, Pattengale PK, Leder P. Murine interleukin-4 displays potent anti-tumor activity in vivo. Cell 1989; 57:503–512.
72. Mandler R, Chu CC, Paul WE, Max EE, Snapper CM. Interleukin 5 induces S mu-S gamma 1 DNA rearrangement in B cells activated with dextran-anti-IgD antibodies and interleukin 4: a three component model for Ig class switching. J Exp Med 1993; 178:1577–1586.
73. Muraguchi A, Hirano T, Tang B, Matsuda T, Horii Y, Nakajima K, Kishimoto T. The essential role of B cell stimulatory factor 2 (BSF-2/IL-6) for the terminal differentiation of B cells. J Exp Med 1988; 167(2):332–344.
74. Steel C, Nutman TB. Regulation of IL-5 in onchocerciasis. A critical role for IL-2. J Immunol 1993; 150:5511–5518.
75. Dent LA, Strath M, Mellor AL, Sanderson CJ. Eosinophilia in transgenic mice expressing IL-5. J Exp Med 1990; 172:1425–1431.
76. Owen WF Jr, Petersen J, Sheff DM, et al. Hypodense eosinophils and IL-5 activity in the blood of patients with eosinophilia-myalgia syndrome. Proc Nat Acad Sci USA 1990; 87:8647–8651.
77. Denburg JA, Silver JE, Abrams JS. Interleukin-5 is a human basophilopoietin: induction of histamine content and basophilic differentiation of HL-60 cells and of peripheral blood basophil-eosinophil progenitors. Blood 1991; 77:1462–1468.
78. Tosato G, Pike SE. Interferon-beta 2/interleukin 6 is a co-stimulant for human T lymphocytes. J Immunol 1988; 141:1556–1562.
79. Uyttenhove C, Coulie PG, Van Snick J. T cell growth and differentiation induced by interleukin-HP1/IL-6, the murine hybridoma/plasmacytoma growth factor. J Exp Med 1988; 167(4):1417–1427.
80. Tosato G, Seamon KB, Goldman ND, Sehgal PB, May LT, Washington GC, Jones KD, Pike SE. Monocyte-derived human B-cell growth factor identified as interferon-beta 2. Science 1988; 239(4839):502–504.
81. Saito M, Yoshida K, Hibi M, Taga T, Kishimoto T. Molecular cloning of a murine IL-6 receptor-associated signal transducer, gp130 and its regulated expression in vivo. J Immunol 1992; 148:4066–4071.
82. Kishimoto T. The biology of IL-6. Blood 1989; 74:10.
83. Linker-Israeli M, Deans RJ, Wallace D, et al. Elevated levels of endogenous IL-6 in SLE. J Immunol 147:117–123.
84. Yashiro H, Atsush M, Masaynki I, et al. Involvement of IL-6 in mesangial proliferative glomerulonephritis. J Immunol 1989; 143:3949.
85. Namen AE, Schmierer AE, March CJ, Overell RW, Park LS, Urdal DL, Mochizuki DY. B cell precursor growth-promoting activity. Purification and characterization of growth

factor active on lymphocyte precursors. J Exp Med 1988; 167(3):988–1002.
86. Schlossman SF, Boumsell L, Gilks W, Harlan JM, Kishimoto T, Morimoto C, Ritz, Shaw S, Silverstein RL, Springer TA, et al. CD antigens. Blood 1994; (4):879–880.
87. Watson JD, Morrissey PJ, Namen AE, Conlon PJ, Widmer MB. Effect of IL-7 on the growth of fetal thymocytes in culture. J Immunol 1989; 143(4):1215–1222.
88. Kos FJ, Muellbacher A. IL-2 independent activity of IL-7 in the generation of secondary antigen-specific cytotoxic T-cell responses in vitro. J Immunol 1993; 150:387.
89. Lynch DH, Miller RE. Interleukin-7 promotes long term in vitro growth anti-tumor cytotoxic lymphocytes with immunotherapeutic efficacy in vivo. J Exp Med 1994; 179:31.
90. Yang YC, Ricciardi S, Ciarletta A, Calvetti J, Kelleher K, Clark SC. Expression cloning of cDNA encoding a novel human hematopoietic growth factor: human homologue of murine T-cell growth factor P4O. Blood 1989; 74(6):1880–1884.
91. Kelleher K, Bean K, Clark SC, Leung WY, Yang-Feng TL, Chen JW, Lin PF, Luo W, Yang YC. Human interleukin-9: genomic sequence, chromosomal location, and sequences essential for its expression in human T-cell leukemia virus (HTLV)-I-transformed human T cells. Blood 1991; 77(7):1436–1441.
92. Yang YC. Human interleukin-9: a new cytokine in hematopoiesis. Leuk Lymph 1992; 6:441–447.
93. Dugas B, Renauld JC, Pene J, Bonnefoy JY, Peti-Frere C, Braquet P, Bousquet J, Van Snick J, Mencia-Huerta JM. Interleukin-9 potentiates the interleukin-4-induced immunoglobulin (IgG, IgM and IgE) production by normal human B lymphocytes. Eur J Immunol 1993; (7):1687–1692.
94. Powrie F, Menon S, Coffman RL. Interleukin-4 and interleukin-10 synergize to inhibit cell-mediated immunity in vivo. Eur J Immunol 1993; 11:3043–3049.
95. Hsu D-I, Malefyt R, Fiorentino D, et al. Expression of IL-10 activity by Epstein-Barr virus protein BCRF-1. Science 1990; 250:830–832.
96. Zuo X-J, Matsumura Y, Prehn J, et al. Cytokine gene expression in rejecting and tolerant rat lung allograft models. Transplant Immunol 1995; 3:151–161.
97. de Waal Malefyt R, Abrams J, Bennett B, Figdor CG, de Vries JE. Interleukin 10 (IL-10) inhibits cytokine synthesis by human monocytes: an autoregulatory role of IL-10 produced by monocytes. J Exp Med 1991; 174(5):1209–1220.
98. Leeuwenberg JF, Jeunhomme TM, Buurman WA. Slow release of soluble TNF receptors by monocytes in vitro. J Immunol 1994; 152(8):4036–4043.
99. Willems F, Marchant A, Delville JP, Gerard C, Delvaux A, Velu T, de Boer M, Goldman M. Interleukin-10 inhibits B7 and intercellular adhesion molecule-1 expression on human monocytes. Eur J Immunol 1994; 24(4):1007–1009.
100. de Waal Malefyt R, Yssel H, Roncarolo MG, Spits H, de Vries JE. Interleukin-10. Curr Opin Immunol 1992; (3):314–320.
101. Briere F, Bridon JM, Chevet D, Souillet G, Bienvenu F, Guret C, Martinez-Valdez H, Banchereau J. Interleukin 10 induces B lymphocytes from IgA-deficient patients to secrete IgA. J Clin Invest 1994; (1):97–104.
102. Kuhn R, Lohler J, Rennick D, Rajewsky K, Muller W. Interleukin-10-deficient mice develop chronic enterocolitis. Cell 1993; 75:263–274.
103. Roberts AB, Sporn MB. Physiological actions and clinical applications of transforming growth factor-beta (TGF-beta). Growth Factors 1993; 8:1–9.
104. Hooper WC. The role of transforming growth factor-beta in hematopoiesis. A review. Leukemia Res 1991; 15:179–184.
105. Lebman DA, Lee FD, Coffman RL. Mechanism for transforming growth factor beta and IL-2 enhancement of IgA expression in lipopolysaccharide-stimulated B cell cultures. J Immunol 1990; 144:952–995.
106. Kulkarni AB, Karlsson S. Transforming growth factor-beta 1 knockout mice. A mutation in one cytokine gene causes a dramatic inflammatory disease. Am J Pathol 1993; 143:3–9.
107. Qin L, Chavin KD, Ding Y, Favaro JP, et al. Multiple vectors effectively achieve gene transfer in a murine cardiac transplant model. Immunosuppression with TGF β-1 or v-IL-10. Transplantation 1995; 59:809–816.
108. Wolf G, Thaiss F, Stahl RA. Cyclosporine stimulates expression of TGF-β in renal cells. Possible mechanism of CSA anti-proliferative effect. Transplantation 1995; 60:237–242.
109. Border W, Noble N. Targeting TGF-β for treatment of disease. Nature Med 1995; 1:1000.
110. Grabstein KH, Eisenman J, Shanebeck K, Rauch C, Srinivasan S, Fung V, Beers C, Richardson J, Schoenborn MA, Ahdieh M, et al. Cloning of a T cell growth factor that interacts with the beta chain of the interleukin-2 receptor. Science 1994; 264(5161):965–968.
111. Anderson DM, Johnson L, Glaccum MB, Copeland NG, Gilbert DJ, Jenkins NA, Valentine V, Kirstein MN, Shapiro DN, Morris SW, et al. Chromosomal assignment and genomic structure of IL-15. Genomics 1995; 25(3):701–706.
112. Giri JG, Kumaki S, Ahdieh M, Friend DJ, Loomis A, Shanebeck K, DuBose R, Cosman D, Park LS, Anderson DM. Identification and cloning of a novel IL-15 binding protein that is structurally related to the alpha chain of the IL-2 receptor. EMBO J 1995; 14(15):3654–3663.
113. Dinarello CA. Interleukin-1 and biologically related cytokines. In: Cohen S, ed. Lymphokines and the Immune Response. Boca Raton, FL: CRC Press, 1990:146–172.
114. Acres RB, Larsen A, Conlon PJ. IL-1 expression in a clone of human T-cells. J Immunol 1987; 138:2132.
115. Ho S, Abraham RT, Nilson A, Handwerger B, McKean D. Interleukin 1-mediated activation of interleukin-4 (IL-4) producing T. Lymphocytes proliferation by IL-4 dependent and IL-4 independent mechanisms. J Immunol 1987; 139:1532.
116. Tartakovsky B, Kovacs E, Takacs L, Durum SK. T-cell clone producing an IL-1 like activity after stimulation by antigen-presenting B-cells. J Immunol 1986; 137:160.
117. Hannum CH, Wilcox CJ, Arend WP, et al. Interleukin-1 receptor antagonist activity of a human interleukin-1 inhibitor. Nature 1990; 343:336–340.
118. Dinarello C. Is there a role for IL-1 blockade in IVIG therapy? Immunol Rev 1994; 139:174–188.
119. Minty A, Chalon P, Derocq JM, Dumont X, Guillemot JC, Kaghad M, Labit C, Leplatois P, Liauzun P, Miloux B, et al. Interleukin-13 is a new human lymphokine regulating inflammatory and immune responses. Nature 1993; 362(6417):248–250.
120. Zurawski G, de Vries JE. Interleukin 13, an interleukin 4-like cytokine that acts on monocytes and B cells, but not on T cells. Immunol Today, 1994; 15(1):19–26.
121. Bonavida B, Wright S, Kashiwa H, et al. Immunoregulation and cytotoxicity by tumor necrosis factor. In: Bonavida B, Gifford GE, Kirchner H, Old LJ, eds. Tumor Necrosis Factor/Cachectin and Related Cytokines. Basel: Karger, 1988:53–58.
122. Satoh J, Seino H, Abo T, et al. Recombinant human tumor necrosis factor-alpha suppresses autoimmune diabetes in non obese diabetic mice. J Clin Invest 1989; 84:1345–1348.
123. Serreze DV, Leiter E. Defective activation of T-suppressor function in non obese diabetic mice. Potential relation to cytokine deficiencies. J Immunol 1988; 140:3801–3807.

124. Jacob CO, McDevitt HO. Tumor necrosis factor alpha in murine autoimmune "lupus" nephritis. Nature 1988; 331:356–358.
125. Roehm N. Lymphokine regulation of B-cell function. In: Cambier J, ed. Ligands, Receptors and Signal Transduction in Regulation of Lymphocyte Function. Washington, DC: American Society for Microbiology, 1990:97–116.
126. Zubler RH, Lowenthal JW, Erard F, Hashimoto N, Devos R, MacDonald HR. Activated B cells express receptors for and proliferate in response to pure interleukin 2. J Exp Med 1984; 160:1170–1183.
127. Wigfall DR, Sakai R, Wallace DJ, Jordan SC. IL-2 receptor expression in peripheral blood lymphocytes from SLE patients: relationship to clinical activity. Clin Immunol Immunopathol 1988; 354–362.
128. Roehm N. Lymphokine regulation of B-cell function. In: Cambier J, ed. Ligands, Receptors and Signal Transduction in Regulation of Lymphocyte Function. Washington, DC: American Society for Microbiology, 1990:97–116.
129. Tiggers MA, Casey LS, Koshland ME. Mechanism of interleukin 2 signaling: mediation of different outcomes by a single receptor and transduction pathway. Science 1989; 243:781–786.
130. Howard MJ, Farrar J, Hilfiker M, et al. Identification of a T cell derived B cell growth factor distinct from interleukin 2. J Exp Med 1982; 155:914–923.
131. Kitari A, Strober W. Regulation of Cγ subclass germ-line transcripts in human peripheral blood B-cells. J Immunol 1993; 151:3478–3488.
132. Wilkes D, Heidler K, Niemeier M, et al. Increased broncho-alveolar IgG2/IgG1 ratio is a marker for human lung allograft rejection. J Invest Med 1994; 42:652–658.
133. Splanski J, Lipsky P. Human B-cell regulation by growth and differentiation factors. In: Cambier J, ed. Ligands, Receptors and Signal Transduction in Regulation of Lymphocyte Function. Washington, DC: American Society of Microbiology, 1990:149–195.
134. Toyoda M, Zhang X-M, Petrosian A, et al. Modulation of immunoglobulin production and cytokine mRNA expression in PBMCs by IVIG. J Clin Immunol 1994; 14:178–189.
135. Lipsky P, Thompson P, Rosenwasser L, Dinarello C. The role of interleukin 1 in human B cell activation: inhibition of B cell proliferation and the generation of immunoglobulin-secreting cells by an antibody against human leukocyte pyrogen. J Immunol 1983; 130:2708–2714.
136. Strom TB, Kelley V. Toward more selective therapies to block undesired immune responses. Kidney Int 1989; 31:1026–1033.
137. St. Georgiev V. Immunomodulating drugs: Major advances in research and development. Ann NY Acad Sci 1993; 685:1–12.
138. Strominger JL. Developmental biology of T-cell receptors. Science 1989; 244:943–950.
139. Reiss R, Makoff D, Rodriguez H, et al. Encephalopathy and cerebral infarction in OKT3-treated patients with concomitant elevation of CSF TNF-α levels. Nephrol Dial Transplant 1993; 8:464–468.
140. Harding FA, McArthur J, Gross JA, et al. CD28-mediated signaling co-stimulate murine T-cells and prevents induction of anergy in T-cell clones. Nature 1992; 356:607–609.
141. Tan P, Anasetti C, Hansen JA, et al. Induction of alloantigen-specific hypo-responsiveness in human T-lymphocytes by blocking interaction of CD28 with its natural ligand B7/BB1. J Exp Med 1993; 177:165–173.
142. Linsley PS, Wallace PM, Johnson J, Gibson MG, et al. Immunosuppression in vivo by a soluble form of the CTLA-4 T-cell activation molecule. Science 1992; 257:792–795.
143. Nishikawa K, Linsley P, Collins AB, Stanenkovic I, et al. Effect of CTLA4 chimeric protein on rat autoimmune anti-glomerular basement membrane glomerulonephritis. Eur J Immunol 1994; 24:1249–1254.
144. Lin H, Bolling S, Linsley P, et al. Long-term acceptance of major histocompatibility complex mismatched cardiac allografts induced by CTLA4Ig plus donor specific transfusions. J Exp Med 1993; 178:1801–1806.
145. Matsumura Y, Zuo X-J, Prehn J, et al. Soluble CTLA4-Ig modifies parameters of acute inflammation in rat lung allograft rejection without altering lymphocytic infiltration or transcription of key cytokines. Transplantation 1995; 59:551–558.

25

Cytokines and Neutrophils: Neutrophil-Derived Cytokines and the Inflammatory Response

Joseph C. Fantone
University of Michigan Medical School, Ann Arbor, Michigan

INTRODUCTION

Neutrophils have long been recognized as playing a sentinel role in the protection of the host against foreign pathogens. In addition, they are the principal cells initially recruited to sites of acute tissue injury (e.g., myocardial infarction) functioning to clear damaged cells and extracellular matrix in preparation for tissue repair. One of the important mechanisms by which neutrophils clear foreign pathogens and damaged tissues involves phagocytosis, degranulation of lysosomal contents, and activation of a cell membrane–associated NADPH oxidase with the generation of reactive oxygen metabolites. Further neutrophils have also been identified as an important source of inflammatory mediators including metabolites of arachidonic acid (e.g., leukotriene B_4). While generally functioning in a host-protective role, neutrophils may promote tissue injury if their activation is not controlled, the inflammatory stimulus is fixed to a host cell or structural component (e.g., basement membrane), or when stimulated in association with autoimmune diseases.

Results of studies from a variety of laboratories over the past 15 years have demonstrated that the development, activation, recruitment to sites of tissue injury, and function of neutrophils are tightly regulated by a variety of soluble and cell-associated inflammatory mediators (reviewed in Refs. 1,2). These include direct effects of a variety of cytokines (e.g., TNF-α, IFN-γ, IL-1, IL-8, GM-CSF) on neutrophil function. Until recently, neutrophils were thought to be primary targets of cytokine action with limited capacity to synthesize cytokines. The primary focus of this chapter will be to review the most recent work, predominantly generated in the past several years, that has implicated the neutrophil as a potential source of inflammatory cytokines and discuss the potential pathophysiological significance of this newly recognized function.

A central premise in understanding the inflammatory response has been that the tissue macrophage and recruited circulating monocytes function in a central role to orchestrate the development and evolution of the inflammatory response as a result of their ability to secrete a variety of cytokines. Beginning in the mid-1980s and through the past several years, it has become evident that endothelial cells, mesenchymal cells, and tissue parenchymal cells including epithelium also play critical roles in regulating the inflammatory response. This occurs as a result of their ability to not only respond to selected cytokines altering their phenotype (e.g., expression of cell surface adhesion molecules) but also to function as sources of regulatory cytokines. The reader is referred to other chapters in this book for information regarding the specific role of individual cells and their secretory products in regulating the host response to injury. However, in each of the paradigms developed to understand the regulatory mechanism involved in the inflammatory response, the neutrophil was viewed as a terminally differentiated cell whose capacity for gene transcription and protein synthesis was extremely limited, especially when compared to mononuclear phagocytic cells. Most investigators viewed the neutrophil's function to be as the first line of defense, ultimately to be cleared by metabolically active tissue macrophages. If the neutrophil did play a regulatory role, it was primarily indirect as a secondary effect of phagocytosis, protease release, and reactive oxygen metabolite action. One class of soluble products released by neutrophils that have been implicated in a regulatory role in the inflammatory process are low molecular weight lipid mediators, primarily products of phospholipase A_2 activation and arachidonic acid metabolism (3). The best characterized of these products is leukotriene B_4, a potent chemoattractant for neutrophils, which is thought to function to amplify the neutrophil recruitment to sites of tissue injury.

However, in the late 1980s it was reported that neutrophils were in fact capable of significant protein synthesis and postulated to play active roles in regulating not only their own function but also other cell activity at inflammatory sites. Initial reports indicated that neutrophils were capable of synthesis of both mRNA and their respective proteins including complement receptor 1 (CR1), complement receptor 3α chain (CR3α), MHC class I molecules, and actin (4,5). Neutrophil mRNA transcription and protein synthesis was upregulated within several hours following treatment with granulocyte-macrophage colony-stimulating factor (GM-CSF). These observations indicated that neutrophils did indeed have the capacity to synthesize both express proteins on the cell surface and in selected cases actively secrete them. Based on these initial observations and with the widespread application of molecular biology techniques, a number of laboratories during the past 5 years have attempted to determine whether neutrophils could be a significant source of inflammatory cytokines and under what conditions their expression and secretion are regulated. The results from several laboratories indicate that the neutrophil may be an important source of inflammatory cytokines and play a critical regulatory role in the development of the inflammatory response.

NEUTROPHIL RECRUITMENT AND THE INFLAMMATORY RESPONSE

To appreciate the potential implications of neutrophil synthesis and secretion of cytokines on the regulation of the inflammatory response, it is critical to understand the basic mechanisms involved in neutrophil recruitment to sites of inflammation. Derived from myeloid precursor stem cells, neutrophils differentiate within the bone marrow under the control of specific growth factors including granulocyte colony-stimulating factor (G-CSF) and GM-CSF. Neutrophils are subsequently released into the peripheral circulation and exist for 4–12 h in the circulation. Within the circulation, the cells localize into two pools: either marginated adjacent to the endothelium-lined vascular wall or moving freely along with blood flow. The cells can exchange between the two pools, and several studies suggest that there is heterogeneity within circulating neutrophil populations.

At sites of tissue injury such as may occur during gram-negative bacterial infection, a variety of mediators are initially generated within the tissue by activation of both soluble mediator systems and parenchymal cells including tissue macrophages. Activation of the coagulation and complement systems generates vasoactive mediators both directly and indirectly including stimulating mast cell degranulation and histamine release. The effects include increases in vascular permeability, localized fibrin deposition within tissues, and vascular thrombosis. In addition, neutrophil chemoattractants including bacterial formyl peptides and the complement product C5a are generated within the tissue. Bacteria-derived lipopolysaccharides stimulate tissue macrophages and recruited monocytes to generate a variety of inflammatory cytokines, including TNF-α, IL-1α and -β, IL-6, and the neutrophil chemoattractant IL-8. The coordinate expression of these inflammatory mediators is important in the recruitment of neutrophils into tissues.

Recent studies have shown that the recruitment of neutrophils into sites of tissue injury initially involves complex interactions between a specific class of cell adhesion molecules present on neutrophil and endothelial cell surfaces and their carbohydrate-rich counterreceptors (reviewed in Refs. 6–8). This class of receptors is known as the selectins, with L-selectin molecules expressed on neutrophils and E- and P- selectin expressed on endothelium. While L-selectin is constitutively expressed on neutrophils at sites of tissue injury, P-selectin is rapidly upregulated in conjunction with the neutrophil-activating molecule platelet-activating factor (PAF) on endothelial cells by thrombin and histamine (9,10). E-selectin, along with other adhesion molecules including intercellular adhesion molecule-1 (ICAM-1) expression, is increased over several hours following exposure of endothelium to the cytokine TNF-α or IL-1. The interaction of the selectin molecules with their cell-associated ligands results in neutrophil localization to and rolling at sites of inflammation. The binding of L-selectin to its endothelial cell ligand or exposure of neutrophils to chemoattractants such as IL-8, C5a, or bacteria formylated peptides results in enhanced expression of β_2-integrins (e.g., CD11b/CD18) on the neutrophil surface (reviewed in Refs. 10,11). These adhesion molecules interact with intercellular adhesion molecules (including ICAM-1, -2, and -3) on the endothelial cell to promote tight adhesion at the site of tissue injury, which in the presence of a chemoattractant gradient leads to migration from the vascular compartment into the tissue. Subsequent interaction of specific neutrophil β_1-integrin adhesion molecules with defined binding epitopes on extracellular matrix (ECM) and receptors for chemoattractants that may be bound to the ECM regulate the subsequent migration and localization of neutrophils to sites of tissue injury. Figure 1 is a schematic representation of these events.

Once the neutrophil encounters the injurious agent (e.g., bacteria), it will attempt to phagocytize it and clear the agent within the phagolysosome secondary to the action of lysosomal enzymes and NADPH oxidase–derived reactive oxygen species (1,2,12). Recent studies have also suggested that within neutrophils, specific stimuli can increase levels and activity of an inducible nitric oxide synthase (iNOS) and production of reactive nitrogen intermediates (13), although this remains controversial (14). Neutrophil phagocytosis and pathogen clearance may be enhanced by several processes. Opsonization of the bacteria by antibody molecule or complement proteins will enhance neutrophil phagocytosis and clearance secondary to cell activation via specific Fc (e.g., FcRγII, FcRγIII) or complement receptors. Further, macrophage-derived cytokines, especially TNF-α, can function to activate or "prime" neutrophils for enhanced phagocytic and bactericidal activity (1,2). While the interactions between multiple cell types and inflammatory mediators at sites of tissue injury are highly complex, this scenario of neutrophil recruitment to a site of bacterial injury has been purposely simplified as a model to highlight several points where the potential production of neutrophil-derived cytokines may function to modify the inflammatory response (Fig. 1).

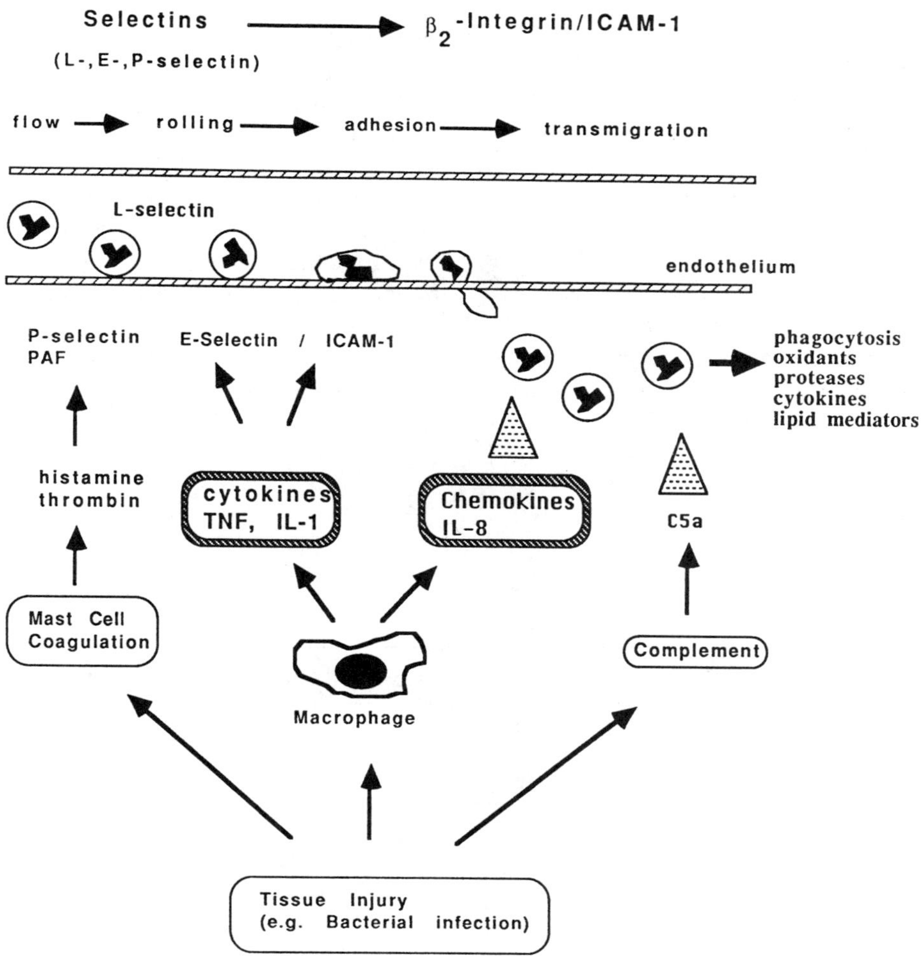

Figure 1 Regulation of neutrophil recruitment by macrophage-derived cytokines.

LPS AND NEUTROPHIL CYTOKINE PRODUCTION

In Vitro Studies

Studies from several laboratories have demonstrated that bacteria-derived LPS can cause a significant increase in the production and secretion of several cytokines by neutrophils (Table 1), including IL-8 (15–17), TNF-α (18), IL-1β (19,20), IL-1 ra (20,21), TGF-β (22), and MIP-1α and -1β (23,24). In general, the kinetics of the response to LPS results in a progressive increase in cytokine secretion over a 20- to 24-h period and is associated with increased intracellular levels of cytokine-specific mRNA. These increases occur at concentrations of LPS (100 ng/ml to 1 μg/ml) that potentially occur at sites of bacterial infection. While increases in selected cytokines are detectable in LPS-stimulated neutrophils, the levels of LPS required for maximal stimulation are frequently more than 10-fold those required for mononuclear phagocytic cells and the amount of cytokines produced range from less than 1% to 10% of that observed with LPS stimulation of comparative numbers of macrophages or monocytes. This raises the question as to the biological significance of neutrophil cytokine production. However, the presence of large numbers of neutrophils at sites of bacteria infection and the potential for these cytokines to have autocrine and paracrine effects within the local environment of tissue injury suggest that in specific inflammatory conditions (e.g., bronchopneumonia, bacterial meningitis, acute inflammatory colitis), neutrophil-derived cytokines may play critical roles in determining the evolution of the inflammatory response.

In Vivo Studies

Since the inception of experimental studies of the inflammatory response, investigators have established in vivo experimental models of local tissue and systemic inflammation in an attempt to better understand the pathophysiological mechanisms involved in localized bacterial infection and sepsis. Results of recent studies, depending on the specific experimental model and species studied, have indicated a role of phagocytic cells (including neutrophils) and cytokine production as important regulators of parenchymal and inflammatory cell function. Until recently, the neutrophil was relegated principally to an effector cell role in the host response. In these experimental models the neutrophil was thought to be a target of cytokine influences rather than a potential source. Recently, Gauldie and colleagues, using a rat model of intratracheal administration of LPS to induce acute lung inflammation, demonstrated production of TNF-α by both alveolar macrophages and neutrophils purified from bronchoalveolar lavage (25,26). Both cell types expressed measurable levels of TNF-α mRNA. In addition, neutrophil expression of TNF-α mRNA relative to actin mRNA was significantly increased compared to alveolar macrophages. Subsequent studies using the same model showed selective expression of TNF-α, MIP-2, IL-1β, and IL-6 in total lung homogenates. At 6–12 h after LPS instillation, significant levels of all cytokines were detected in a neutrophil-enriched fraction. The selectivity of the response was noted by the absence of LPS induction of either TGF-β or the chemokine RANTES.

Additional in vivo studies have also reported increased neutrophil levels of IL-1β and TNF-α mRNA. Increased levels of IL-1β mRNA were detected in neutro-

Table 1 Cytokine Production by Neutrophils

	In vitro stimulus						
Cytokine	LPS	TNF-α	IL-1β	G-CSF/GM-CSF	C5a/fmlp	L-selectin	In vivo
TNF-α	x					x	x
IL-1β	x					x	x
IL-1ra	x						
IL-8	x	x	x	x	x		
IL-6							x
MIP-1α	x						
MIP-1β	x						
MIP-2							x
TGF-β	x						
IFN-γ				x			

phils harvested from the pulmonary vasculature following intravenous infusion of LPS (27). Further, increased levels of neutrophil TNF-α have been identified in cells harvested from tissues following induction of acute immune complex inflammation (28). Using in situ hybridization techniques, TNF-α mRNA was detected in neutrophils associated with a skin wound (29). Levels of TNF-α mRNA were detectable 12 h after wounding, peaking at 72 h, and remaining detectable at 5 days postinjury. In another study, neutrophils isolated from human gingival crevicular fluid (GCF) from adult patients with periodontitis showed increased expression of mRNA for IL-1α, -1β, and TNF-α (30). No mRNA for IL-6 was detected in the GCF-derived neutrophils. Thus there is evidence not only in experimental animal models but within human inflamed tissues that neutrophils have the potential to upregulate expression of specific cytokines.

NEUTROPHIL-DERIVED IL-8

One of the potentially important and more extensively studied cytokines produced by activated neutrophils is the C-X-C chemokine IL-8 (15–17). Stimulation of neutrophils with LPS induces a biphasic increase in IL-8 production. Significant increases in IL-8 secretion and mRNA synthesis is detectable approximately 1 h after LPS stimulation. The direct effects of LPS resulted in an initial production of approximately 25% of the total IL-8 over the first 4–8 h of incubation. A second increase in IL-8 production was noted over the next 16 h. The second phase of IL-8 production, which accounts for approximately 70% of the total IL-8 produced, is inhibited by treatment of cells with neutralizing antibodies that bind TNF-α or IL-1β (Fig. 2). These observations suggest that LPS indirectly increases neutrophil IL-8 production secondary to induction of TNF-α and IL-1β. These findings are supported by studies indicating that TNF-α and IL-1β can directly induce neutrophil IL-8 production (16,17,31). This network model of cytokine induction and associated secondary effects is similar to that observed with LPS stimulation of mononuclear phagocytic cells (32) and indicates a potentially important autocrine/paracrine function of neutrophil-derived cytokines.

Additional studies have also demonstrated the production of IL-8 by neutrophils following stimulation with chemotactic stimuli, including *N*-formyl-methionyl-leucyl phenylalanine, C5a, and PAF (15,33). In the presence of chemotactic stimuli, peak production of IL-8 occurs between 2 and 6 h with no secondary phase suggesting a minimal role for TNF-α and IL-1β regulation. The significance of these findings is the potential for neutrophils to further amplify their recruitment and activation at inflammatory sites secondary to the production of at least one of the C-X-C chemokines (i.e., IL-8).

Interestingly, LPS has also been shown to enhance neutrophil production of IL-1 receptor antagonist (IL-1ra). Similar to other cytokines, the levels of IL-1ra produced by neutrophils are approximately 10% or less than mononuclear phagocytic cells and require higher doses of LPS stimulation (20,31,34–36). IL-1ra would potentially function to inhibit IL-8 production by blocking LPS-induced IL-1β effects. This is similar to the regulatory pathways effecting cytokine production by mononuclear phagocytic cells. However, in contrast to MPCs, which have the capacity to produce significant amounts of PGE_2 in response to stimulation and downmodulate cytokine production, neutrophils preferentially synthesize LTB_4 from arachidonic acid with minimal PGE_2 production. Thus, except for the generation of IL-1ra, LPS stimulation of IL-8 production is only minimally downregulated by endogenous neutrophil mediators, perhaps reflecting the terminal nature of neutrophil differentiation. These results suggested that there may be several mechanisms by which IL-8 production may be regulated.

TNF-α ACTIVATION OF NEUTROPHILS

It has become increasingly evident that one of the central roles of TNF-α in inflammation is the regulation of neutrophil function (37). This occurs through both direct and indirect mechanisms. For instance, TNF-α indirectly effects neutrophil recruitment by upregulating the adhesion molecules E-selectin and ICAM-1 on endothelium. In addition TNF-α can directly enhance neutrophil responses to secondary stimuli through a mechanism referred to as cell "priming" and, as noted above, induce IL-8 production. Multiple studies have demonstrated the in vivo suppression of neutrophil-dependent inflammatory processes utilizing neutralizing antibodies to TNF-α or soluble forms or derivatives of TNF receptors. These findings underscore the importance of defining the functional role of TNF-α in the regulation of neutrophil function.

The neutrophil expresses two distinct receptors for TNF-α, designated TR55 and TR75. Recent studies using blocking monoclonal antibodies to the specific receptors suggest that TR55 functions as the signaling receptor initiating intracellular signaling pathways (38,39). While the mechanism by which TNF-α activates neutrophils is not clear, phosphorylation/dephosphorylation of regulatory proteins appears to play a critical role. Recent studies suggest that the pathways involved in TNF-α activation of neutrophils are distinct from chemoattractant, FcRγ, and GM-CSF stimulation and likely involve mitogen activated protein kinase–independent mechanisms (40). Un-

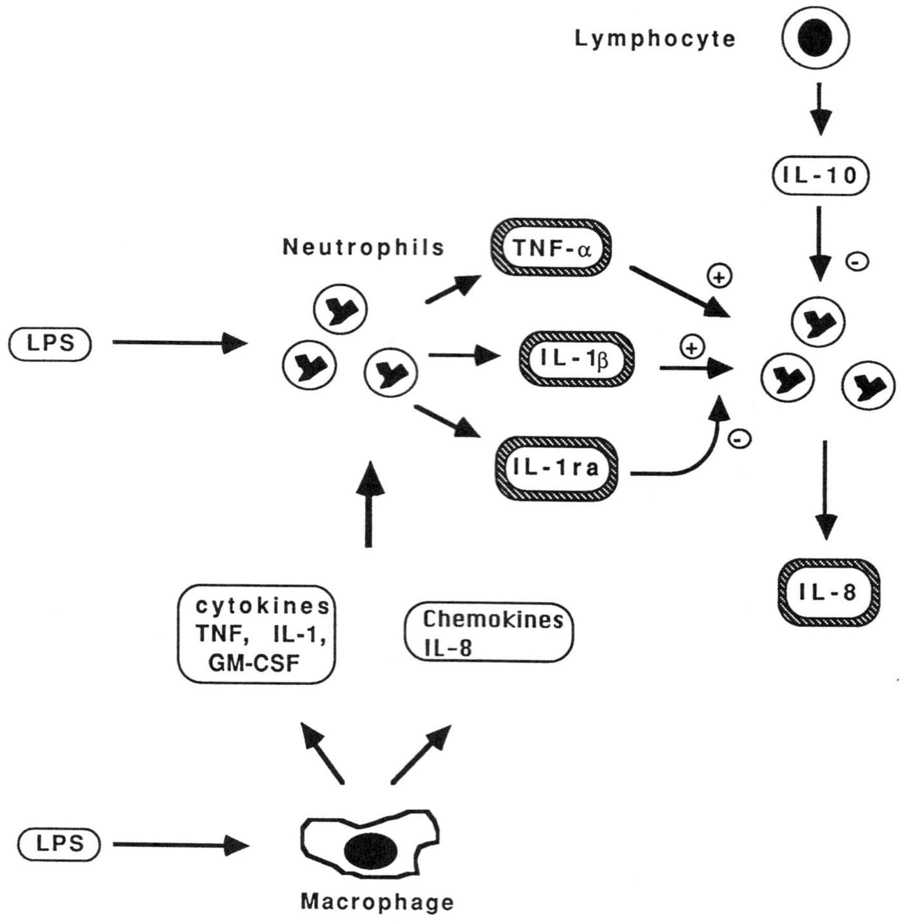

Figure 2 Regulation of neutrophil IL-8 production.

like TR55, the role for the neutrophil TR75 receptor is less clear. It has been suggested to play a co-regulatory role in modulating TNF-α binding to the TR55 receptor, although this hypothesis requires further validation. Several laboratories have reported that stimulation of neutrophils with chemoattractants such as f-MLP or cell adhesion induces proteolytic cleavage and release of both TR55 and TR75 from the neutrophil surface (44,45). This finding suggests an additional regulatory mechanism to control the local effects of endogenous TNF-α production by neutrophils.

REGULATION OF NEUTROPHIL CYTOKINE PRODUCTION

We have noted above that the endogenous production of TNF-α and IL-1β can significantly upregulate IL-8 production in LPS stimulated neutrophils. The stimulatory effect of LPS on neutrophil TNF-α production can be enhanced by addition of nitric oxide–generating compounds (43). These results suggest that NO• potentially produced by endothelial cells, smooth muscle cells, or macrophages could regulate neutrophil functional responses secondary to enhancing cytokine production. In contrast, the endogenous production or exogenous administration of IL-1ra can attenuate neutrophil IL-8 production. Further, treatment of LPS-stimulated neutrophils with soluble forms of TR75 can also inhibit cytokine production.

Additional studies have demonstrated a potent effect of IL-10 in inhibiting LPS-induced human neutrophil TNF-α, IL-1b, and IL-8 production (24,31,44). This effect was similar to that observed in monocytes. IL-10 appeared to function by inhibiting both gene transcription of the cytokines as well as mRNA stability. The inhibitory effect of IL-10 on neutrophil cytokine production was significantly greater than the inhibition observed with either IL-4 or TGF-β. Subsequent studies showed that IL-10 increased neutrophil production of IL-1ra, further supporting the role of IL-10 as an anti-inflammatory cytokine and its inhibitory function in the regulation of neutrophil activation (45). These observations are consistent with in vivo animal studies demonstrating the ability of IL-10 to inhibit LPS toxicity (46,47).

One of the important functional roles of interferon-γ (IFN-γ) identified in the past several years has been its ability to potentiate neutrophil function in addition to its proinflammatory effects on macrophages and immune stimulatory actions (48). IFN-γ is used clinically to enhance neutrophil function in patients with chronic granulomatous disease of childhood (CGD) (49). These patients represent a heterogeneous population exhibiting several different genetic deficiencies that result in defective function of phagocytic cell NADPH oxidase and the ability to produce reactive oxygen species. While several reports have suggested enhanced NADPH oxidase activity in different forms of the disease, this effect alone cannot explain the extent of benefit seen in the patients. Recent in vitro studies indicate that IFN-γ treatment of neutrophils inhibits the initial LPS-induced production of IL-8 but in contrast significantly enhances TNF-α, IL-1β, and IL-8 production during the last 18 h of incubation (50–52). The increases in TNF-α, IL-1β, and IL-8 protein production paralleled increases in their respective mRNA levels. Further, IL-8 production was attenuated by neutralizing antibodies to TNF-α and IL-1β, further supporting the autocrine function of both cytokines on IL-8 synthesis.

In another recent study, it was reported that exposure of human neutrophils to sulfatides, a ligand for L-selectin, stimulates an increase in both TNF-α and IL-8 mRNA levels over several hours (53). Whether neutralizing antibodies to TNF-α block IL-8 production in this system was not determined. However, the results raise a potentially important novel function for this class of adhesion molecules and suggest that the circulating neutrophil may be encountering its initial signal for increased cytokine production at the earliest stage in its recruitment to a site of inflammation.

CONCLUSION

Until recently, the functional role of the neutrophil in the inflammatory response was believed to be as a terminal cell functioning to phagocytize and clear pathogens and tissue debris. It is now becoming evident that the neutrophil may play an important active role in regulating the development of the inflammatory response. While it is unlikely to play a central role equivalent to that of the tissue macrophage in orchestrating inflammatory and immune responses, recent experimental evidence suggests that in the early phases of acute inflammation, the regulated production of neutrophil-derived cytokines may be critical in determining the progression of injury and ultimate restoration of tissue function resulting from a pathogenic insult.

REFERENCES

1. Gallin JI, Goldstein IM, Snyderman R, eds. Inflammation—Basic Principles and Clinical Correlates. 2nd ed. New York: Raven Press, 1992.
2. Cochrane CG, Gimbrone MA, eds. Cellular and Molecular Mechanisms of Inflammation. San Diego: Academic Press, 1992.
3. Zurier RB. Prostaglandins, leukotrienes and related compounds. In: Kelley WN, et al., eds. Textbook of Rheumatology. 4th ed. Philadelphia: W. B. Saunders Co., 1992:201–212.
4. Jack R, Fearon DT. Selective synthesis of mRNA and proteins by human peripheral blood neutrophils. J Immunol 1988; 140:4286–4293.
5. Neuman E, Huleatt JW, Jack RM. Granulocyte-macrophage colony stimulating factor increases synthesis and expression of CR1 and CR3 by human neutrophils. J Immunol 1990; 145:3325–3332.
6. Bevilacqua MP, Nelson RM. Selectins. J Clin Invest 1993; 91:379–387.
7. McEver RP. Selectins. Curr Opin Immunol 1994; 6:75–84.
8. Zimmerman GA, Prescott SM, McIntyre TM. Endothelial cell interactions with granulocytes: tethering and signaling molecules. Immunol Today 1992; 13:93–100.
9. Lorant DE, et al. Inflammatory roles of P-selectin. J Clin Invest 1993; 92:559–570.
10. Springer TA. Traffic signals for lymphocyte circulation and leukocyte migration: The multi-step paradigm. Cell 1994; 76:301–314.
11. Butcher EC. Leukocyte—endothelial cell recognition: three or more steps to diversity and sensitivity. Cell 1991; 67:1033–1036.
12. Ward PA, Mulligan MS. Leukocyte oxygen products and tissue damage. In: Jesaitis AJ, Dratz EA, eds. Molecular Basis of Oxidation Damage by Leukocytes. Boca Raton, FL: CRC Press, 1992:139–149.
13. Nussler AK, Billar TR. Inflammation, immunoregulation, and inducible nitric oxide synthase. J Leuk Biol 1993; 54:171–178.
14. Yan L, Vandivier RW, Suffredini AF, Danner RL. Human polymorphonuclear leukocytes lack detectable nitric oxide synthase activity. J Immunol 1994; 153:1825–1834.
15. Cassatella MA, Bazzoni F, Ceska M, Ferro L, Baggiolini M, Berton G. J Immunol 1992; 148:3216–3320.
16. Streiter RM, Kashahara K, Standiford TJ, Rolfe MW, Chensue SW, Kunkel SL. Cytokine induced neutrophil derived interleukin-8. Am J Pathol 1992; 141:397–407.
17. Fujishima S, Hoffman AR, Vu T, et al. Regulation of neutrophil interleukin 8 gene expression and protein secretion by LPS, TNF-alpha, and IL-1 beta. J Cell Physiol 1993; 154:478–485.
18. Dubravec DB, Spriggs DR, Mannick JA, Rodrick ML. Circulating human peripheral blood granulocytes synthesize and secrete tumor necrosis factor alpha. Proc Natl Acad Sci USA 1990; 87:6758–6761.
19. Lord PC, Wilmoth LM, Mizel SB, McColl CE. Expression of IL-1 alpha and beta genes by human blood polymorphonuclear leukocytes. J Clin Invest 1991; 87:1312–1321.
20. Jenkins JK, Malyak M, Arend WP. The effects of interleukin-10 on interleukin-1 receptor antagonist and interleukin-1 beta production in human monocytes and neutrophils. Lymphokine Cytokine Res 1994; 13:47–54.

21. Muzio M, Sironi M, Polentarutti N, Mantovani A, Colotta F. Induction by transforming growth factor-beta 1 of the interleukin-1 receptor antagonist and of its intracellular form in human polymorphonuclear cells. Eur J Immunol 1994; 24:3194–3198.
22. Grotendorst GR, Smale G, Pencev D. Production of transforming growth factor beta by human peripheral blood monocytes and neutrophils. J Cell Physiol 1989; 140:396–402.
23. Kasama T, Strieter RM, Standiford TJ, Burdick MD, Kunkel SL. Expression and regulation of human neutrophil-derived macrophage inflammatory protein 1 alpha. J Exp Med 1993; 178:63–72.
24. Kasama T, Strieter RM, Lukacs N, Burdick MD, Kunkel SL. Regulation of neutrophil-derived chemokine expression by IL-10. J Immunol 1994; 152:3559–3569.
25. Xing Z, Kirpalani H, Torry D, Jordana M, Gauldie J. Polymorphonuclear leukocytes as a significant source of tumor necrosis factor alpha in endotoxin challenged lung tissue. Am J Pathol 1993; 143:1009–1015.
26. Xing Z, Jordana M, Kirpalani H, Driscoll KE, Schall TJ, Gauldie J. Cytokine expression by neutrophils and macrophages in vivo: endotoxin induces tumor necrosis factor alpha, macrophage inflammatory protein-2, interleukin 1 beta, and interleukin-6, but not RANTES or transforming growth factor-beta mRNA expression in acute lung inflammation. Am J Respir Cell Mol Biol 1994; 10:148–153.
27. Williams JH, Patel SK, Hatakeyama D, et al. Activated pulmonary vascular neutrophils as early mediators of endotoxin-induced lung inflammation. Am J Respir Cell Mol Biol 1993; 8:134–144.
28. Zhang Y, Ramos BF, Jakschick BA. Neutrophil recruitment by tumor necrosis factor from mast cells in immune complex peritonitis. Science 1992; 258:1957–1959.
29. Feiken E, Romer J, Ericksen J, Lund LR, Romer J. Neutrophils express tumor necrosis factor-alpha during mouse skin wound healing. J Invest Dermatol 1995; 105:120–123.
30. Takeichi O, Saito I, Tsurumachi T, Saito T, Moro I. Human polymorphonuclear leukocytes derived from chronically inflamed tissue expresses inflammatory cytokines in vivo. Cell Immunol 1994; 156:296–309.
31. Cassatella MA, Meda L, Bonora S, Ceska M, Constantin G. Interleukin 10 inhibits the release of proinflammatory cytokines from human polymorphonuclear leukocytes. Evidence for an autocrine role of tumor necrosis factor and IL-1 beta in mediating the production of IL-8 triggered by lipopolysaccharide. J Exp Med 1993; 178:2207–2211.
32. Streiter R, Kunkel S. Acute lung injury; the role of cytokines in the elicitation of neutrophils. J Invest Med 1994; 42:640–651.
33. Ember JA, Sanderson SD, Hugli TE, Morgan EL. Induction of interleukin-8 synthesis from monocytes by human C5a anaphylatoxin. Am J Pathol 1994; 144:393–403.
34. Malyak M, Smith MF, Abel AA, Arend WP. Peripheral blood neutrophil production of interleukin-1 receptor antagonist and interleukin-1 beta. J Clin Immunol 1994; 14:20–30.
35. Ulich TR, Guo K, Yin S, et al. Endotoxin-induced cytokine gene expression in vivo. IV. Expression of interleukin-1 alpha/beta and interleukin-1 receptor antagonist mRNA during endotoxemia and during endotoxin-initiated local acute inflammation. Am J Pathol 1992; 141:61–68.
36. Re F, Mengozzi M, Muzio M, Dinarello CA, Mantovani A, Colotta A. Expression of interleukin-1 receptor antagonist (IL-1ra) by human circulating polymorphonuclear cells. Eur J Immunol 1993; 23:570–573.
37. Gamble JR, Smith WB, Vadas MA. TNF modulation of endothelial and neutrophil adhesion. In: Beutler B, ed. Tumor Necrosis Factors: The Molecules and Their Emerging Role in Medicine. New York: Raven Press, 1992; 65–86.
38. Della Bianca V, Dusi S, Nadalini KA, Donini M, Rossi F. Role of 55- and 75-kDA TNF receptors in the potentiation of Fc mediated phagocytosis in human neutrophils. Biochem Biophys Res Commun 1995; 214:44–50.
39. Abe Y, Osuka Y, Nakata T, Kashu Y, Kimua S. The functional role of 55- and 75-kDa tumor necrosis factor receptors in human polymorphonuclear leukocyte. Cytokine 1995; 7:39–49.
40. Waterman WH, Sha'afi RI. A mitogen activated protein kinase independent pathway involved in the phosphorylation and activation of cytosolic phospholipase A2 in human neutrophils stimulated with tumor necrosis factor alpha. Biochem Biophys Res Commun 1995; 209:271–278.
41. Menegazzi R, Cramer R, Patriarca P, Scheurich P, Dri P. Evidence that tumor necrosis factor alpha induced activation of neutrophil respiratory burst on biologic surfaces is mediated by the p55 TNF receptor. Blood 1994; 84:287–293.
42. Lantz M, Bjornberg F, Olsson I, Richter J. Adherence of neutrophils induces release of soluble tumor necrosis factor receptor isoforms. J Immunol 1994; 152:1362–1369.
43. Van Dervort AL, Yan L, Madara PJ, Cobb JP, Wesley RA, Corriveau CC, Tropea MM, Danner RL. Nitric oxide regulates endotoxin induced TNF-alpha production by neutrophils. J Immunol 1994; 152:4102–4109.
44. Wang P, Wu P, Anthes JC, Siegel MI, Eagan RW, Billah MM. Interleukin-10 inhibits interleukin-8 production in human neutrophils. Blood 1994; 83:2678–2683.
45. Cassatella M, Meda L, Gasperini S, Calzetti F, Bonora S. Interleukin-10 (IL-10) upregulates IL-1 receptor antagonist production from lipopolysaccharide-stimulated human polymorphonuclear leukocytes by delaying mRNA degradation. J Exp Med 1994; 179:1695–1699.
46. Gerard C, Bruyns C, Marchant A, et al. Interleukin 10 reduces the release of tumor necrosis factor and prevents lethality in experimental endotoxemia. J Exp Med 1993; 177:547–550.
47. Howard M, Muchamuel T, Andrade S, Menon S. Interleukin 10 protects mice from lethal endotoxemia. J Exp Med 1993; 176:1205–1208.
48. Todd PA, Goa KL. Interferon gamma-1β. A review of its pharmacology and therapeutic potential in chronic granulomatous disease. Drugs 1992; 43:111–122.
49. Bolinger AM, Taeubel MA. Recombinant interferon gamma for treatment of chronic granulomatous disease and other disorders. Clin Pharm 1992; 11:834–850.
50. Meda L, Gasperini S, Ceska M, Cassatella M. Modulation of proinflammatory cytokine release from human polymorphonuclear leukocytes by gamma interferon. Cell Immunol 1994; 157:448–461.
51. Cassatella MA, Gasparri I, Ceska M, Bazzoni F, Rossi F. Interferon-gamma inhibits interleukin-8 production by polymorphonuclear leukocytes. Immunology 1993; 78:177–184.
52. Van Dervort AL, Yan L, Madara PJ, et al. Nitric oxide regulates endotoxin induced tumor necrosis factor-alpha production by human neutrophils. J Immunol 1994; 152:4102–4109.
53. Laudanna C, Constantin G, Baron P, Scarpini E, Scarlato G, Cabrini G, Dechecchi C, Rossi F, Cassatelle MA, Berton G. Sulfatides trigger increase in cytosolic free calcium and enhanced expression of tumor necrosis factor alpha and interleukin-8 mRNA in human neutrophils. Evidence for a role of L-selectin as a signaling molecule. J Biol Chem 1994; 269:4021–4026.

26

Cytokines and Intermediary Metabolism

Riaz A. Memon, Kenneth R. Feingold, and Carl Grunfeld
University of California, San Francisco, California

INTRODUCTION

The host response to infection, inflammation, and cancer is frequently accompanied by profound alterations in lipid, protein, and carbohydrate metabolism. These metabolic changes occur in a variety of acute as well as chronic infections and may be a part of the host's acute-phase response, which can be beneficial. On the other hand, several of these metabolic disturbances are associated with chronic low-grade infections and malignancies and may contribute to wasting or cachexia. The frequently observed metabolic changes in lipid metabolism include hypertriglyceridemia, impaired triglyceride (TG) clearance, increased lipolysis, increased fatty acid (FA) and/or cholesterol synthesis, and decreased FA oxidation and ketogenesis (1–10). Alterations seen in protein metabolism during infections and cancer include increased hepatic amino acid uptake, increased synthesis of acute-phase proteins in the liver, enhanced skeletal muscle protein breakdown, and decreased body muscle mass (11–15). Changes in carbohydrate metabolism are variable and may include initial hyperglycemia followed by hypoglycemia, increased glucose utilization, altered gluconeogenesis, substrate cycling of glucose, and insulin resistance (16–20).

The metabolic disturbances that are associated with infections are not directly due to the growth of invading microorganisms. In fact, it has been observed that many of these metabolic changes can be reproduced by administration of bacterial products such as endotoxin [lipopolysaccharide (LPS)], lipoteichoic acid (LTA), or killed microorganisms (21–23). There is no correlation between metabolic disturbances observed in cancer to the magnitude of tumor burden; furthermore, cachexia is not produced by competition between the host and the tumor (24,25). There are also many similarities among the metabolic changes that occur in cancer, infections, and inflammatory disorders, suggesting that it is the host response to infection, inflammation, or cancer that induces these metabolic alterations.

Infection or LPS administration stimulates the production of multiple cytokines by many different cell types (26–28). Similarly, several cytokines are produced in different types of malignancies in humans as well as in tumor-bearing animal models; in malignancy cytokines may originate from the tumor or invading host cells (29,30). These cytokines are key mediators of several physiological and pathological responses that occur during the course of infection, inflammation, and cancer (31,32). Recent studies have shown that several cytokines such as tumor necrosis factor (TNF-α), interleukin-1 (IL-1), interleukin-6 (IL-6), interferons (INF-α, IFN-β, and IFN-γ), lymphotoxin (TNF-β), and leukemia inhibitory factor (LIF) have substantial effects on lipid, protein, and carbohydrate metabolism (33–35). From these studies it has emerged that several cytokines have multiple effects on intermediary metabolism and that more than one cytokine may exert or mediate the same or a very similar metabolic effect. Moreover, the metabolic effects of cytokines can be additive, synergistic, or antagonistic. Thus, the effect of cytokines on intermediary metabolism is analogous to their effects on the immune system.

In this chapter we will describe the effects of various cytokines on lipid and lipoprotein metabolism and the cellular and molecular mechanisms involved in producing these changes. We will also summarize the effects of cytokines on protein and carbohydrate metabolism. Finally, we will discuss the role of cytokine-induced metabolic alterations in the pathophysiology of cachexia as well as in the host acute-phase response.

EFFECTS OF CYTOKINES ON LIPID AND LIPOPROTEIN METABOLISM

Hypertriglyceridemia characterized by an increase in very-low-density lipoprotein (VLDL) levels is a consistent finding in a variety of bacterial, viral, and protozoal infections in humans as well as animals (1–8,21–23). An increase in VLDL levels could be due to increased VLDL production and/or decreased VLDL clearance. An increase in VLDL production could be driven by stimulation of hepatic fatty acid (FA) synthesis, inhibition of FA oxidation (ketogenesis), and/or an increase in adipose tissue lipolysis. All of these mechanisms could provide excess fatty acid substrate in the liver, which can be esterified and secreted in the circulation as VLDL. On the other hand, a decrease in VLDL clearance is usually due to the inhibition of the activity of lipoprotein lipase (LPL), the enzyme responsible for the catabolism of TG-rich lipoproteins. Studies have shown that all of these mechanisms are operative in producing hypertriglyceridemia in animal models of infections and that multiple cytokines are capable of producing these metabolic effects (see below).

In marked contrast to the effects of infection on TG metabolism, which are similar in all species, the changes in cholesterol metabolism during infection differ between nonprimates and primates. In most animals, infection or LPS administration increases serum cholesterol levels, which is primarily due to an increase in LDL levels (36–38). In nonhuman primates and humans, there is a decrease in serum cholesterol levels during infection (4,39). Like infection and LPS administration, several cytokines produce similar changes in cholesterol metabolism (see below).

In addition to the alterations in TG and cholesterol metabolism, ketone body (KB) metabolism is modulated during the course of infection (40,41). Infection and LPS administration decrease hepatic KB production and lower plasma KB levels through multiple mechanisms, which in turn are dependent upon the nutritional status of the host (42,43). Likewise, studies have now shown that several cytokines also have significant effects on KB metabolism (see below).

Cytokines and Triglyceride Metabolism

The administration of TNF to rats rapidly increases serum TG levels (44–46). This effect of TNF on serum TG levels is observed within 1 h of administration, peaks by 2 h, and persists for at least 17 h (44). TNF increases serum TG levels under a variety of dietary conditions ranging from a high-sucrose diet to a high-fat diet (45). TNF increases serum TG levels in insulinopenic diabetic animals (47,48) and in adrenalectomized rats (49), suggesting that its effects on TG metabolism are direct rather than being mediated through insulin, cortisol, or catecholamines. TNF also raises serum TG levels in mice (50), nonhuman primates (51), and humans (52).

The increase in serum TG levels after TNF administration is due to an increase in VLDL of normal composition (46). This increase in VLDL levels could be due to increased production or decreased clearance. Early in vitro studies in cultured mouse adipocytes demonstrated that TNF decreased the activity of LPL, inhibited de novo FA synthesis, and stimulated lipolysis (53). On the basis of these observations it was proposed that TNF produces hypertriglyceridemia and cachexia by decreasing the clearance of TG-rich lipoproteins and enhancing adipose tissue catabolism, respectively (53). However, several in vivo studies in rodents by our laboratory and others indicate that the acute increase in serum TG levels induced by TNF may not be due to the suppression of LPL activity (54–57). First, although TNF reduces LPL activity in epididymal fat pads in vivo, this decrease requires several hours, whereas the TNF-induced increase in serum TG levels occurs very rapidly and precedes the effect of TNF on LPL activity (54). Second, TNF administration does not decrease LPL activity in multiple other adipose tissue sites or in muscle (54–56). Third, while TNF does not decrease LPL activity in human adipocytes (55), TNF treatment increases serum TG levels in humans (52). In fact, TNF increases hepatic lipase and total postheparin plasma lipase activity (56). Finally, TNF administration does not decrease the clearance of chylomicrons or VLDL from the circulation (44,48,57).

There is now substantial experimental evidence that supports a role of increased VLDL production in TNF-induced acute hypertriglyceridemia. Several studies have demonstrated that TNF acutely stimulates de novo hepatic FA synthesis and total TG synthesis, with newly synthesized TGs appearing in the plasma (44,57,58,59). TNF has been shown to increase hepatic TG secretion, which is an index of VLDL production rate (57,58). TNF also has an acute stimulatory effect on adipose tissue lipolysis (60), and these peripherally derived FA contribute to increased hepatic VLDL production. The time course of the effects of TNF on hepatic FA and TG synthesis, adipose tissue lipolysis, and VLDL secretion parallels the time course for TNF-induced increase in serum TG levels (48,58–60). Thus, these studies demonstrate that TNF-induced acute increase in serum TG levels is primarily due to increased hepatic VLDL production. Of course, it is likely that under other experimental conditions such as different time points, higher doses of TNF, or different animal species, decreased VLDL clearance could play a role in TNF-induced increases in serum TG levels. Recent data indicate that low doses of LPS increase plasma TGs by increasing hepatic VLDL production, whereas high doses of LPS decrease LPL activity and TG clearance (21).

The stimulatory effect of TNF on adipose tissue lipolysis is observed only under certain dietary conditions. In chow-fed animals, TNF raises serum TGs by increasing hepatic FA synthesis and by stimulating lipolysis. In contrast, in sucrose-fed animals, TNF has no effect on lipolysis and increases serum TG levels primarily by stimulating hepatic FA synthesis (60). TNF has no effect on the activity of hepatic enzymes involved in the esterification of FA with glycerol (60), suggesting that the TNF-induced increase in hepatic TG synthesis is primarily driven by the availability of the FA substrate, which is provided either by hepatic synthesis or by adipose tissue lipolysis. This is further supported by studies demonstrating that fasting abolishes the effect of TNF on hepatic FA synthesis and serum TG levels (45), suggesting that the nutritional status of the host may be an important factor in modulating the mechanisms by which TNF increases serum TG levels.

Most of the early studies that implicated cytokines in altering TG metabolism were conducted with TNF in cultured adipocytes. However, multiple studies indicate that the ability to perturb lipid metabolism in cultured fat cells is not limited to TNF. In cultured adipocytes, IL-1, IL-6, lymphotoxin, IFN-α, IFN-γ, and LIF decrease LPL activity (61–66). While TNF, IFN-γ and LIF produce a parallel decrease in LPL mRNA (67,68), no such effect is observed for IL-1 and IFN-α (67). The effect on IL-6 on LPL mRNA have not been determined. TNF, lymphotoxin, IFN-α, and IFN-γ suppressed de novo FA synthesis in cultured fat cells (61,67). In contrast to TNF, both IL-1 and LIF increase FA synthesis in adipocytes (67,68). TNF significantly decreased the mRNA levels of both acetyl CoA carboxylase (ACC) and FA synthase, the regulatory enzymes in FA synthesis (67). IFN-α and IFN-γ decreased FA synthase mRNA levels without significantly altering ACC mRNA (67). In parallel with their stimulatory effect on FA synthesis in cultured fat cells, IL-1 and LIF increase ACC mRNA levels but had no effect on FA synthase mRNA (67,68). TNF, IL-1, IFN-α, IFN-γ and LIF stimulated lipolysis in cultured adipocytes (61–66); however, all of these cytokines produce a paradoxical decrease in the mRNA levels of hormone sensitive lipase (67–69), a key enzyme in intracellular lipolysis. Lymphotoxin also stimulates lipolysis, whereas IL-6 had little effect on lipolysis. The in vitro effects of cytokines on TG metabolism are summarized in Table 1. The data from these in vitro studies suggest that several cytokines exert effects on lipid metabolism in cultured adipocytes. It is likely that there are different modes of regulation of key enzymes involved in fat cell metabolism by cytokines as there is no coordinate decrease or increase in relevant mRNA levels.

It takes at least 8 h to see the effects of cytokines on cultured adipocytes; in contrast, the in vivo effects of most cytokines on TG metabolism are acute in onset. Like TNF, administration of IL-1, IL-6, lymphotoxin, and IFN-α rapidly increase hepatic FA synthesis in rodents within 1–2 h (59,70–72). The stimulatory effects of IL-1 and IFN-α on hepatic FA synthesis persist for several hours (59), whereas the effect of IL-6 on FA synthesis is transient (70). Several other cytokines such as IL-2, IFN-β, and IFN-γ had no effect on hepatic FA synthesis under similar conditions, indicating that stimulation of hepatic fatty acid synthesis is a distinct effect produced only by specific cytokines (59).

Other cytokines including IL-1, IL-6, ciliary neurotropic factor (CNTF), and keratinocyte growth factor (KGF) increase serum TG levels in rodents (71–75). As seen with TNF, each of these cytokines acutely increases serum TG levels by increasing VLDL production and not by decreasing the clearance of TG-rich lipoproteins (71,72). IL-6, CNTF, and LIF, which work via the same receptor trans-

Table 1 Effects of Cytokines on Lipid Metabolism in Cultured Adipocytes

Effect	TNF	Lymphotoxin	IL-1	IL-6	LIF	IFN-α	IFN-γ
LPL activity	↓	↓	↓	↓	↓	↓	↓
LPL mRNA	↓	nd	↔	nd	↓	↔	↓
FA synthesis	↓	↓	↑	nd	↑	↓	↓
ACC mRNA	↓	nd	↑	nd	↑	↔	↔
FAS mRNA	↓	nd	↔	nd	↔	↓	↓
Lipolysis	↑	↑	↑	↔ or ↑	↑	↑	↑
HSL mRNA	↓	nd	↔	nd	nd	↓	↓

LPL: lipoprotein lipase; FA: fatty acid; ACC: acetyl CoA carboxylase; FAS: fatty acid synthase; HSL: hormone-sensitive lipase; ↓: decreased; ↑: increased; ↔: no change; nd: not determined.

ducer (gp130), stimulate both hepatic FA synthesis and peripheral lipolysis, which provide FA for hepatic TG synthesis and secretion (72,73). Unlike TNF, IL-1 does not have a stimulatory effect on lipolysis in intact animals, and hence its effect on VLDL production is entirely due to enhanced hepatic FA synthesis and TG secretion (71). On the other hand, KGF does not stimulate hepatic FA synthesis, but increases hepatic TG synthesis and secretion solely through FA provided by increased peripheral lipolysis (75). Like TNF, both IL-1 and IL-6 have a delayed inhibitory effect on adipose tissue LPL activity in intact mice (76).

The effects of IFNs on TG metabolism are more complex, and there are substantial differences between their effects in rodents and humans. While treatment with IFN-α and IFN-γ increase serum TG levels in humans (77–80), they do not produce this effect in rodents (76). Studies have shown a strong correlation between hypertriglyceridemia and circulating IFN-α levels in AIDS (81). IFN-α levels also correlate with decreased LPL activity, impaired TG clearance, and increased basal hepatic FA synthesis in patients with HIV infection and AIDS (82,83), suggesting that multiple mechanisms could account for the effects of IFN-α on TG metabolism in humans. In contrast to these observations, while IFN-α acutely increases adipose tissue lipolysis (81) and hepatic FA synthesis (59) and has a delayed inhibitory effect on adipose tissue LPL activity (76) in mice, it does not increase serum TG levels (76). Similarly, IFN-γ treatment in cancer patients is associated with an inhibition of postheparin plasma lipase activity and an increase in serum TG levels (79), suggesting that IFN-γ may impair the clearance of lipoproteins. However, in spite of increasing lipolysis (84) and decreasing adipose tissue and skeletal muscle LPL activity (76). IFN-γ failed to increase serum TG levels in mice (76).

Recently, the effects of LIF have also been studied in mice. LIF produces a delayed increase in serum TG levels (76). LIF also produced a corresponding decrease in adipose tissue LPL activity, whereas it had no effect on skeletal muscle or heart LPL activity (76), suggesting that it may increase serum TG levels by decreasing the clearance of TG-rich lipoproteins. The effects of lymphotoxin on several of these parameters of TG metabolism have not been studied. Table 2 summarizes the in vivo effects of cytokines on TG metabolism in rodents.

Studies of the mechanisms of cytokine stimulation of hepatic FA synthesis suggest that these cytokines can be divided into two classes (85). Class I cytokines include TNF, IL-1, and IL-6, and these cytokines stimulate hepatic FA synthesis by acutely raising hepatic citrate levels. Citrate is an allosteric activator of acetyl-CoA carboxylase, which is the rate-limiting enzyme in fatty acid synthesis (85). Class II includes IFN-α, which has no effect on hepatic citrate levels. The mechanism by which IFN-α stimulates FA synthesis has not yet been determined. The stimulatory effects of cytokines from different classes on hepatic FA synthesis are either additive or synergistic, whereas there is no additive or synergistic effect among cytokines from the same class in stimulating hepatic FA synthesis (85).

IL-4 is a cytokine with marked inhibitory effects on the regulation of immune response (86). While IL-4 has no direct effect on hepatic FA synthesis, it inhibits the stimulatory effects of TNF, IL-1, and IL-6 on hepatic FA synthesis by preventing the increase in hepatic citrate levels (86). In contrast, IL-4 has no effect on the ability of IFN-α to stimulate hepatic FA synthesis (86). IFN-α does not work through an increase in hepatic citrate. These results indicate that cytokines use multiple mechanisms to regulate FA and TG metabolism. They also suggest that, as occurs with cytokine regulation of the immune response, there are complex metabolic interactions between cytokines which may be additive, synergistic, or antagonistic.

The administration of TNF or IL-1 alone has no effect on hepatic apo E mRNA levels, however, a combination of both cytokines decreases hepatic apo E mRNA levels (87).

Table 2 Effects of Cytokines on Triglyceride Metabolism in Intact Animals

Effect	TNF	IL-1	IL-6	CNTF	LIF	KGF	IFN-α	IFN-γ
Serum TG	↑	↑	↑	↑	↑	↑	↔	↔
Hepatic FAS	↑	↑	↑	↑	↑	↔	↑	↔
TG secretion	↑	↑	↑	↑	↑	↑	nd	nd
LPL activity	↓	↓	↓	nd	↓	nd	↓	↓
TG clearance	↔	↔	↔	↔	nd	↔	nd	nd
Lipolysis	↑	↔	↑	↑	↑	↑	↑	↑

TG: triglycerides; FAS: fatty acid synthesis; LPL: lipoprotein lipase; ↓: decreased; ↑: increased; ↔: no change; nd: not determined.

The effect of TNF and IL-1 combination on hepatic apo E mRNA levels is similar to that of endotoxin (88). Since apo E serves as a ligand for VLDL remnants and chylomicron remnants to bind to the LDL and LDL-related protein receptors, a decrease in apo E may slow the clearance of TG-containing lipoproteins and thereby contribute to cytokine-induced hyperlipoproteinemia.

Cytokines and Cholesterol Metabolism

There are marked differences in the effects of cytokines on cholesterol metabolism between primates and other animals. Like LPS, cytokines increase serum cholesterol levels in rodents, whereas in primates they produce hypocholesterolemia. The underlying mechanisms responsible for these interspecies differences have not been characterized.

TNF administration has been shown to increase serum total cholesterol levels in rats, mice, and hamsters (44,46,50,87). Unlike the change in serum TG levels, the increase in serum cholesterol levels is delayed in onset and is primarily due to an increase in LDL cholesterol levels (46). TNF also produces a decrease in serum HDL levels (87).

There are multiple cellular and molecular mechanisms by which TNF increases serum cholesterol levels in rodents. TNF has a delayed stimulatory effect on de novo hepatic cholesterol synthesis (44,50) via increases in the activity of hepatic β-hydroxy-β-methylglutaryl CoA (HMG-CoA) reductase (44,50), the rate-limiting enzyme for endogenous cholesterol synthesis. TNF increases hepatic HMG-CoA reductase protein and mRNA levels, accounting for the increase in HMG-CoA reductase activity (87). In contrast, TNF has no effect on hepatic LDL receptor protein or mRNA levels (87). The time course of TNF-induced increase in serum cholesterol levels corresponds with its effects on cholesterol synthesis and HMG-CoA reductase activity, protein, and mRNA levels, suggesting that increased cholesterol synthesis is possibly involved in increasing serum cholesterol levels. TNF also decreases the activity and mRNA levels of cholesterol 7α-hydroxylase (CAH), the rate-limiting enzyme in the conversion of cholesterol to bile acids in the liver (89). A decrease in CAH activity will result in increased availability of cholesterol for lipoprotein production. Thus, TNF uses multiple mechanisms to raise serum total cholesterol levels in rodents.

In addition to TNF, several other cytokines may influence cholesterol metabolism in rodents. Like TNF, IL-1 and IL-6 produce a delayed increase in serum cholesterol levels (50,72). IL-1, lymphotoxin, and IFN-γ increase hepatic cholesterol synthesis in mice, whereas IFN-α and IL-2 have no such effect (59). Like TNF, IL-1 increases the activity, protein, and mRNA levels for HMG-CoA reductase in liver, which may account for its ability to stimulate hepatic cholesterol synthesis and increase total serum cholesterol levels. The effects of TNF and IL-1 on hepatic HMG-CoA reductase mRNA levels and serum cholesterol levels are additive (87). IL-1 also decreases CAH activity and mRNA levels (89) which may also contribute increased lipoprotein production.

Both TNF and IL-1 decrease HDL cholesterol levels in hamsters (87). The HDL-lowering effect of TNF and IL-1 is additive, and the magnitude of this decrease is similar to that seen with LPS (88). The mechanism(s) by which cytokines lower HDL are less clear. TNF or IL-1 alone have a minimal effect on hepatic expression of apo A-I, the major apolipoprotein associated with HDL. However, a combination of TNF and IL-1 produces a significant decrease in apo A-I mRNA levels (87). LPS, TNF, and IL-1 increase the hepatic expression as well as serum levels of two other HDL-associated apoproteins: serum amyloid A (SAA) and apo J (90,91). Since SAA-rich HDL are rapidly cleared from the plasma, a cytokine-induced increase in SAA might contribute to the decrease in HDL levels observed following LPS, TNF, or IL-1 administration. SAA-rich HDL has a lower affinity for hepatocytes and a higher affinity for macrophages (92), suggesting that cytokine-induced changes in HDL structure may cause HDL to be preferentially directed toward macrophages. The effect of the cytokine-induced increase in apo J is not known, but, like SAA, apo J (90) may alter HDL metabolism.

LPS and TNF (but not IL-1) have also been shown to decrease lecithin cholesterol acyltransferase (LCAT) activity (51,93–95), which may also contribute to the decreased HDL cholesterol. This decrease in LCAT activity is caused by a decrease in hepatic levels of LCAT mRNA (95). Thus, cytokines can use multiple mechanisms to lower HDL cholesterol levels in rodents. However, it is of note that the decrease in HDL cholesterol appears to precede these changes in HDL structure (95). More recently, a mechanism by which cytokines might maintain HDL levels has been found. LPS, TNF, and IL-1 decrease cholesterol ester transfer protein (CETP) levels in plasma through decreases in CETP mRNA in nonhepatic tissues (96). Decreased CETP would prevent the transfer of cholesterol ester out of HDL. The effects of cytokines on cholesterol metabolism in rodents are summarized in Table 3.

In contrast to rodents, several cytokines including TNF, IL-2, IFN-α, IFN-β, IFN-γ, granulocyte-macrophage colony-stimulating factor (GM-CSF), and M-CSF decrease serum total cholesterol in nonhuman primates and humans (51,97–101). The decrease in serum cholesterol levels induced by TNF, IL-2, IFN-α, and IFN-γ is due to a decrease in both HDL and LDL fractions, whereas IFN-β,

Table 3 Effects of Cytokines on Cholesterol Metabolism in Rodents

Effect	TNF	IL-1	IL-6	IFN-α	IFN-γ
Serum cholesterol levels	↑	↑	↑	↔	↔
Hepatic cholesterol synthesis	↑	↑	nd	↔	↑
HMG-CoA reductase activity	↑	↑	nd	nd	↔
HMG-CoA reductase mRNA	↑	↑	nd	nd	nd
CAH activity	↓	↓	nd	nd	nd
CAH mRNA	↓	↓	nd	nd	nd
LCAT activity	↑	nd	nd	nd	nd
LCAT mRNA	↑	nd	nd	nd	nd
LDL receptor protein	↔	↔	nd	nd	nd
LDL receptor mRNA	↔	↔	nd	nd	nd
CETP protein	↓	↓	nd	nd	nd
CETP mRNA	↓	↓	nd	nd	nd

HMG-CoA: hydroxymethylglutaryl coenzyme A; CAH: cholesterol 7α-hydroxylase; LCAT: lecithin cholesterol acyltransferase; LDL: low-density lipoprotein; ↓: decreased; ↑: increased; ↔: no change; nd: not determined.

GM-CSF, and M-CSF primarily decrease LDL cholesterol levels. The decrease in LDL and HDL is usually accompanied by a respective decrease in serum apo B and apo A-I levels (51,97–101).

Only a few studies have investigated the mechanism(s) by which cytokines lower serum cholesterol levels in intact primates, however, several in vitro studies have been conducted to examine the effects of cytokines on cholesterol metabolism in cultured hepatocytes. In intact primates, TNF-induced decrease in serum cholesterol levels is associated with a decrease in cholesterol ester content in HDL and LDL fractions (51). TNF also decreases the activity and mass of LCAT in primates, which may contribute to the decrease in lipoprotein cholesterol ester content (51). In contrast to TNF, IL-1 has no effect on cholesterol metabolism in primates (94).

Recombinant IFNs have been shown to lower serum cholesterol levels in humans (97,101). A recent study demonstrated that IFN-β decreases the synthesis of apo B in normal subjects (102), suggesting that it may lower serum cholesterol levels by decreasing LDL production. It is not known whether the IFN-α– and IFN-γ–induced decrease in serum cholesterol levels is due to decreased lipoprotein synthesis or to increased lipoprotein clearance.

HepG2 cells, a hepatoblastoma-derived cell line, which has many functional similarities with human hepatocytes, have provided some clues to the mechanisms of cytokine-induced hypocholesterolemia in primates. TNF, IL-1, and IL-6 decrease apo A-1 and apo B secretion by HepG2 (103) cells, suggesting that these cytokines may lower HDL and LDL levels by decreasing their production. TNF, IL-1, and IL-6 decrease LCAT activity in the medium of HepG2 cells (103). Others have shown that TNF, IL-1, and Oncostatin M increase LDL binding to HepG2 cells in a dose-dependent manner, which is accompanied by a parallel increase in LDL receptor mRNA in HepG2 cells (104,105). An increase in LDL receptors could increase LDL clearance and contribute to decreased serum cholesterol levels.

Systemic administration of GM-CSF and M-CSF lowers serum cholesterol levels in humans (98,99). While studies in HepG2 cells demonstrate that GM-CSF and M-CSF have no significant effect on LDL binding (104), M-CSF has been shown to increase the number of LDL receptors in spleen and bone marrow cells (106), which may result in enhanced clearance of lipoproteins. Alternatively, it is possible that GM-CSF and M-CSF induce their cholesterol-lowering effect through the release of TNF and/or IL-1. This hypothesis is supported by the observations that treatment of human mononuclear cells with GM-CSF and M-CSF stimulates TNF and IL-1 release (107).

Cytokines and Ketone Body Metabolism

There are marked differences among the in vivo effects of cytokines on KB metabolism due to the complex interactions between the peripheral and hepatic effects of cytokines, the dose and timing of administration of cytokines, and the nutritional status of the host. Like infection or LPS administration, low doses of TNF and IL-1 rapidly decrease serum KB levels in mice in the fed state (108), however, only IL-1 is capable of lowering serum KB levels in the fasted state, as is seen with LPS (41). In contrast to the acute effect of TNF, the KB-lowering effect of IL-1 is sustained and persists for at least 16 h (108). IL-1 also lowers hepatic KB production, whereas TNF has no effect on hepatic KB levels (108). IL-6 does not have any effect on either serum or hepatic KB levels (109).

Serum KB levels are regulated by their rate of synthesis in the liver and also by their rate of peripheral utilization. On the other hand, the KB production rate in the liver depends upon the flux of FA substrate from the adipose tissue and the hepatic ketogenic capacity, which is regulated by malonyl-CoA levels (110). Malonyl-CoA is the first committed intermediate in FA synthesis and is an allosteric inhibitor of carnitine palmitoyl transferase I (CPT-I), the rate-limiting enzyme in FA oxidation and ketogenesis. Several studies have shown that infection or LPS administration could inhibit ketogenesis by suppressing adipose tissue lipolysis, raising hepatic malonyl-CoA levels, or increasing the peripheral KB utilization.

IL-1 uses multiple mechanisms to exert its antiketogenic effect. In the fed state, IL-1 increases the levels of hepatic malonyl-CoA (108), which is a potent inhibitor of CPT-I. As a result of CPT-I inhibition, the entry of FA into the mitochondria is prevented, resulting in decreased FA oxidation and KB production. In the fasting state, in addition to increasing hepatic malonyl-CoA levels, IL-1 also inhibits lipolysis and thus reduces the influx of FA substrate to the liver for KB synthesis (108). The effect of IL-1 on the peripheral utilization of KB has not been studied.

The mechanisms involved in the antiketogenic effect of TNF are more complex. Like IL-1, TNF also raises hepatic malonyl-CoA levels, but unlike IL-1, it does not seem to influence hepatic KB levels (108). The reason for this contradiction between the effects of TNF and IL-1 on hepatic KB levels is the fact that TNF also stimulates lipolysis. This results in an increased flux of FA substrate to the liver that counterbalances the inhibitory effect of malonyl-CoA on hepatic KB production by providing additional FA substrate. This conclusion is supported by experiments that demonstrate that when the lipolytic effect of TNF is blocked by pretreatment with phenylisopropyladenosine (PIA), a potent inhibitor of lipolysis (111), TNF decreases hepatic KB levels (108). Moreover, a lower dose of TNF, which does not stimulate lipolysis but does increases FA synthesis and malonyl-CoA levels, has also been shown to lower hepatic KB levels (108). Finally, kinetic studies demonstrate that TNF decreases the hepatic KB-production rate in rats (112). It is likely that TNF may also increase the peripheral utilization of KB, and this may decrease serum KB levels in the presence of apparently normal hepatic KB levels.

The inhibitory effect of IL-1 and TNF on hepatic ketogenesis suggests that in the presence of these cytokines, FA substrate is directed away from oxidation and KB formation and these FA are then available for reesterification, resulting in increased TG synthesis and VLDL production. Hence, the antiketogenic effect of IL-1 and TNF is probably another mechanism by which these cytokines induce hypertriglyceridemia.

The effects of IFNs on KB metabolism are in marked contrast to those of TNF or IL-1. IFN-γ exerts dose-dependent stimulatory effects on adipose tissue lipolysis, hepatic KB production, and serum KB levels (84). IFN-γ has no effect on hepatic FA synthesis or malonyl-CoA levels. When IFN-γ–induced lipolysis is blocked by prior treatment with PIA, the stimulatory effect of IFN-γ on hepatic KB levels is abolished, indicating that the ketogenic effect of IFN-γ is primarily due to increased flux of FA substrate from the periphery (84).

Like TNF, the effects of IFN-α on ketogenesis are more complex. IFN-α stimulates lipolysis in a dose-dependent manner but has a biphasic effects on serum and hepatic KB levels. Low doses of IFN-α increase serum and hepatic KB levels, whereas higher doses have no effect on KB levels (84). At low doses, IFN-α has no effect on hepatic FA synthesis or malonyl-CoA levels, and hence it raises KB levels primarily by mobilization of FA substrate. Thus, the effect of low doses of IFN-α on ketogenesis is very similar to that of IFN-γ. In contrast, at higher doses IFN-α stimulates hepatic FA synthesis and raises malonyl-CoA levels, which counteracts its lipolytic effects by inhibiting CPT I and preventing FA entry into the mitochondria for oxidation and KB formation. When IFN-α–induced lipolysis is blocked by prior treatment with PIA, IFN-α produces a decrease in hepatic KB levels (84), suggesting that in the absence of enhanced lipolysis IFN-α can be antiketogenic. Thus, the effect of high doses of IFN-α on hepatic ketogenesis is similar to the effect of TNF on hepatic KB levels. Finally, in the fasting state, when IFN-α has no effect on lipolysis, it lowers hepatic KB levels by raising hepatic malonyl-CoA levels. Thus, in the fasting state IFN-α may act like IL-1 on KB metabolism.

EFFECTS OF CYTOKINES ON PROTEIN METABOLISM

The host response to infection, sepsis, trauma, or cancer includes marked changes in protein metabolism. Increased peripheral release of amino acids and their enhanced uptake by the liver is typically observed in these conditions (11–15). Increased protein breakdown in skeletal muscle is the most important mechanism for the peripheral release of amino acids. Sepsis increases muscle proteolysis via increases in ubiquitin energy-dependent pathways (113). Some of the enhanced proteolysis of sepsis and burns is mediated by the induction of glucocorticoids (114,115). A large portion of the released amino acids is taken up by the liver to synthesize acute-phase proteins and to provide carbon skeleton for gluconeogenesis. In addition, some muscles release excess glutamine during sepsis, which may be utilized in the gastrointestinal tract (116). Thus, these metabolic changes are related to each other and may represent an integrated host response to infection and inflammation. However, in chronic infections or neoplastic diseases, protracted proteolysis may cause depletion of vital host proteins, resulting in muscle wasting and visceral organ atrophy.

Several in vivo and in vitro studies have been conducted to determine the role of cytokines in mediating the changes in protein metabolism (Table 4). In an early report, Goldberg et al. (117) studied the in vitro effect of different concentrations of several cytokines, including TNF, IL-1, and IFNs, on protein breakdown rates in rat or mouse muscle preparations. However, they were unable to detect a catabolic effect of these cytokines on skeletal muscle. They also reported that the doses of TNF that induce fever did not affect protein-turnover rates in skeletal muscle in rats (118). Similarly, Moldawer et al. (119) found that administration of single doses of TNF or IL-1 has no effect on muscle protein synthesis or degradation in intact mice. On the other hand, Flores et al. (120) showed that infusion of TNF alone or in combination with IL-1 over a period of 6 h increased muscle protein breakdown. Similarly, the group of Fischer and Hasselgren (121) have shown that repeated administration of TNF or IL-1 results in increased total and myofibrillar protein breakdown. It appears that the effects of TNF and IL-1 on muscle proteolysis are dose and time dependent, as high doses and prolonged periods of administration are needed to increase muscle protein breakdown. A recent study showed that IL-6, a major mediator of acute-phase protein synthesis, has no effect on protein breakdown in rats (122).

The administration of TNF and IL-1 increases the plasma levels of epinephrine, glucagon, ACTH, and corticosterone, whereas their effects on plasma insulin levels are variable (123). Since glucocorticoids stimulate skeletal muscle protein breakdown in both humans and experimental animals (124,125) and glucocorticoids have been shown to be a mediator of the increased proteolysis of sepsis and burns (114,115), it was proposed that TNF and IL-1 may exert their proteolytic effect through glucocorticoids. However, studies on the mechanisms of TNF- and IL-1–induced proteolysis reveal some interesting differences. TNF stimulates total and myofibrillar protein breakdown only when its administration results in increased plasma corticosterone levels (126). When the same protocol is used in adrenalectomized rats, TNF does not increase plasma corticosterone levels and muscle protein breakdown rates are not altered by TNF (126). In another study it was shown that prior administration of glucocorticoid receptor antagonist RU 38486 prevents TNF-induced proteolysis, suggesting that the effect of TNF on muscle protein breakdown is mediated by glucocorticoids (127). On the other hand, in vivo administration of IL-1 increases both total and myofibrillar protein breakdown rates (128) in control as well as adrenalectomized rats, suggesting that the proteolytic effect of IL-1 is independent of glucocorticoids. Similarly, prior treatment with glucocorticoid antagonist RU 38486 does not inhibit IL-1–induced proteolysis (129). These results suggest that while TNF-stimulated muscle proteolysis is mediated by glucocorticoids, IL-1 may affect muscle protein catabolism either directly or through some other mediator(s). Thus, TNF and IL-1 use different pathways to produce similar catabolic effects on muscle protein metabolism.

TNF and IL-1 have been shown to increase hepatic amino acid uptake in vivo (130,131). While in vitro addition of TNF or IL-1 has no direct effect on amino acid uptake in hepatocytes, both TNF and IL-1 enhance glucagon-induced amino acid uptake (130,131). Thus, TNF and IL-1 may stimulate amino acid uptake in vivo by increasing the circulating glucagon concentration as well as by potentiating the action of glucagon on amino acid metabolism in hepatocytes. Transforming growth factor-β inhibits amino acid uptake by cultured rat hepatocytes

Table 4 Effects of Cytokines on Protein Metabolism In Vivo

Effect	TNF	IL-1	IL-6	LIF	CNTF	TGF-β	IFN-γ
Muscle proteolysis	↑	↑	↔	nd	nd	nd	nd
Hepatic AA uptake	↑	↑	↑	nd	nd	↓	nd
AP protein synthesis	↑	↑	↑	↑	↑	↑	↑

AA: amino acid; AP: acute phase; ↓: decreased; ↑: increased: ↔: no change; nd: not determined.

(132), whereas IL-6 directly enhances amino acid uptake in hepatocytes (132). Since TNF and IL-1 induce IL-6, it is likely that part of their in vivo effect on amino acid uptake is mediated by IL-6.

Several cytokines stimulate the synthesis and secretion of a variety of acute-phase proteins by the liver (133). These proteins are considerd a part of the protective host response, as they have a primary role in homeostasis during infection and inflammation. The effects of cytokines on acute-phase proteins have been extensively reviewed (134). IL-1, IL-6, and TNF induce hepatic acute-phase protein synthesis in vivo as well as in vitro. While TNF and IL-1 induce the synthesis of a limited number of acute-phase proteins in cultured hepatocytes, IL-6 has a stimulatory effect on the synthesis of a broad range of acute-phase proteins (134). Several other cytokines including IFN-γ, LIF, IL-11, CNTF, TGF-β, and oncostatin M also stimulate the synthesis of several acute-phase proteins in cultured hepatocytes (134). The effects of these cytokines on muscle protein breakdown or hepatic amino acid uptake have not been determined. Although it is generally believed that muscle protein breakdown provides amino acids for the synthesis of acute-phase proteins, the extent to which proteolysis contributes to acute-phase response or the relationship between cytokine-induced proteolysis and acute-phase response has not been determined.

EFFECTS OF CYTOKINES ON CARBOHYDRATE METABOLISM

Infection or LPS administration may influence glucose homeostasis by altering the rates of glucose production and/or glucose utilization (16–19). Either hyperglycemia or hypoglycemia may occur during the course of sepsis depending upon the stage and/or severity of infection. Studies indicate that initially there is increased hepatic glucose production as well as increased peripheral glucose utilization (16–20). However, at early stages, the rate of glucose production is greater than the rate of peripheral glucose utilization, resulting in hyperglycemia. In later stages of sepsis, the glycogen reserves are depleted and there is also an inhibition of gluconeogenesis (16–20); thus, the rate of glucose production is lower than the rate of glucose utilization by peripheral tissues resulting in hypoglycemia.

In contrast with feeding, the enhanced glucose utilization induced by high doses of LPS occurs in macrophage-rich tissues. LPS stimulates glucose uptake in spleen, lung, intestine, Kupffer cells, and hepatic endothelial cells (135–137). The increased glucose uptake is mediated by increased expression of the GLUT1 glucose transporter (137). In Kupffer cells, the glucose is utilized heavily in the pentose cycle, which supports enhanced superoxide production and protein synthesis, which aid in host defense (138).

There is now substantial evidence in the literature to suggest that several cytokines can influence glucose metabolism (33,139,141). Cytokines can alter glucose metabolism either directly or through the stimulation of classical glucoregulatory hormones, and, similar to infection, the effects of these cytokines on glucose metabolism are time and dose dependent (Table 5). Doses of TNF that produce shock and death (140–142) usually produce initial hyperglycemia followed by a hypoglycemic phase. The effects of lower doses of TNF on glucose metabolism are more variable and dependent upon the dietary status of the animals and mode of TNF administration as well as the time after TNF administration.

The studies determining the mechanisms of TNF-induced hyperglycemia suggest that TNF increases serum glucose levels by stimulating hepatic glucose production (143). Although TNF decreases hepatic glycogen con-

Table 5 Effects of Cytokines on Glucose Metabolism

Effect	TNF	IL-1	IL-6	IFN-α	GM-CSF
Serum glucose levels	↑/↓	↓	↔	↑	↑
Glycogenolysis	↑	nd	↑	nd	nd
Gluconeogenesis	↑/↓	↓	↓	nd	↑
Glucose utilization	↑	↑	↔	↓	↑
Glucose absorption	↓	nd	nd	nd	nd
Glucoregulatory hormones	↑	↑	nd	↑	nd

GM-CSF: granulocyte-macrophage colony-stimulating factor; ↓: decreased; ↑: increased; ↔: no change; nd: not determined.

tent, glycogenolysis may play a minor role, as TNF enhances glucose production in fasted animals where hepatic glucose reserves are already low. Thus, TNF increases hepatic glucose output by increasing gluconeogenesis. Since TNF does not increase the activity of key gluconeogenic enzymes (144), it appears that the increase in de novo glucose synthesis is due to increased precursor availability. TNF administration is usually associated with enhanced lactate, alanine, and glutamine efflux from skeletal muscle (145), that are taken up by the liver to support gluconeogenesis. Additionally, TNF also produces a dose-dependent increase in key glucoregulatory hormones such as catecholamines, glucagon, and glucocorticoids (33,124,145–147), which, either alone or in combination, stimulate hepatic glucose production.

The role of TNF induction of insulin resistance in TNF-induced hyperglycemia has not been fully evaluated. One early study noted that TNF induced a rapid increase in plasma glucose that was accompanied by no change in plasma insulin, suggesting insulin resistance (147). Catecholamines, glucagon, and glucocorticoids, which are induced by TNF, are known to induce insulin resistance (33,124,145–148). Adrenergic blockade has been shown to ameliorate some of the insulin resistance secondary to TNF infusion (148).

Endogenous production of TNF in adipose tissue has recently been linked to the insulin resistance of obesity (148–156). Expression of adipose tissue TNF protein and mRNA is increased in genetic and acquired obesity in rodents and humans (149–153). There is a strong correlation between circulating insulin levels and TNF mRNA concentrations in adipose tissue (153). Diet-induced weight reduction leads to decreased adipose tissue TNF (151–153). Most strikingly, infusion of anti-TNF antibodies into genetically obese mice leads to reversal of insulin resistance (149,155). Some data indicate that TNF may induce insulin resistance by changes in phosphorylation of the insulin receptor (154–156). Other data indicate that TNF-induced decreases in the GLUT4 glucose transporter in fat cells may mediate some of TNF-induced insulin resistance (157). However, TNF increases glucose transport and the transporter mRNA in cultured muscle cells (158,159). The role of other secondary mediators of insulin resistance in the actions of TNF needs to be more carefully defined in obesity.

Several in vivo studies indicate that TNF enhances the rate of whole body glucose uptake (146,160–162), which could contribute to TNF-induced hypoglycemia. Meszaros et al. (160) reported that TNF administration to rats increases glucose utilization in tissues that are rich in mononuclear phagocytes such as spleen, liver, lungs, and intestine, and skin. In the liver, Kupffer cells account for most of the increase in hepatic glucose utilization (161). Other have also shown that TNF increases glucose uptake by macrophage-rich tissues in mice (162). Thus, TNF stimulation of glucose uptake in macrophages is consistent with the findings after high-dose LPS (135–137). The effect of TNF on glucose uptake by muscle is very variable (146,159,161). It appears that enhanced glucose uptake by skeletal muscles is a delayed effect and may require prolonged tissue exposure to TNF. This is supported by the in vitro finding that TNF stimulates glucose uptake in myotubes of the L6 muscle cell line after prolonged exposure (158), an effect that is dependent upon the synthesis of new glucose transporters (159).

TNF-induced increase in glucose uptake may initially be beneficial to the host as it provides more fuel for increased demands of the peripheral tissues. However, if hepatic glucose production is unable to meet the increased demands of peripheral tissues, hypoglycemia may result. This hypothesis is supported by the experiments that show that in fasted mice, TNF treatment inhibits gluconeogenesis (163) by decreasing the activity as well as transcription rates of hepatic phosphoenolpyruvate carboxykinase (PEPCK), the rate-limiting enzyme in gluconeogenesis. High doses of TNF also decrease glucose absorption by the gastrointestinal tract (164). Thus, TNF may produce hypoglycemia by decreasing glucose absorption, inhibiting gluconeogenesis, and increasing peripheral glucose utilization.

Like TNF, IL-1 also has a variable effect on plasma glucose levels. In the fed state, a low dose of LPS that does not produce shock increases glucose flux and leads to insulin resistance; blockade of IL-1 with IL-1ra attenuates both changes in glucose flux and insulin resistance (168). With regard to the hypoglycemia of sepsis, it is of note that the hypoglycemic effect of IL-1 is even more pronounced than that of TNF. IL-1 administration to mice acutely lowers serum glucose levels and increases serum insulin levels (168). Based on these observations, it was suggested that insulin may mediate the hypoglycemic effect of IL-1. However, IL-1 also decreases serum glucose levels in insulinopenic diabetic mice as well as in insulin-resistant Db/Db and Ob/Ob mice (167). Thus, it appears that the hypoglycemic effect of IL-1 is unlikely to be due to an increase in serum insulin levels alone; rather, multiple mechanisms may be involved. It is interesting to note that IL-1 induces hypoglycemia in spite of increasing the plasma levels of glucagon and corticosterone (166,168), which are known to have stimulatory effects on hepatic glucose production.

Therefore, studies were conducted to determine the mechanisms by which IL-1 produces hypoglycemia. Hill et al. (169) demonstrated that IL-1 inhibits glucocorticoid induction of PEPCK in the fasting state, which may decrease gluconeogenesis and lower plasma glucose levels. Moreover, IL-1 administration to rats also increases glucose uptake by macrophage-rich tissues such as spleen, liver, lungs, and skin as well as by skeletal muscle and diaphragm (170). In these studies IL-1 also increased plasma insulin levels. Somatostatin, given prior to IL-1, prevented both IL-1–induced hyperinsulinemia and increased glucose uptake by peripheral tissues (171). IL-1 also has no effect on tissue glucose uptake in streptozotocin-induced insulinopenic diabetic rats (139). Taken together, these studies suggest that IL-1–induced increases in peripheral glucose utilization may be mediated via insulin; however, other mechanisms must also be involved in decreasing plasma glucose levels.

In addition to these in vivo studies, several in vitro studies have also been conducted to determine the effect of IL-1 on glucose metabolism, and the results of these studies are complementary to the data obtained from studies in whole animals. IL-1β, but not IL-1α, produces a dose- and time-dependent increase in glucose transport in isolated rat adipocytes (172). The maximal effect of IL-1β is observed after 3 h of incubation and is partially blocked by cycloheximide, suggesting that this effect requires the synthesis of new proteins. Similarly, incubation of rheumatoid or nonrheumatoid synovial cells with IL-1 increases glucose uptake in a time- and dose-dependent manner (173). Finally, IL-1 has been shown to increase glucose uptake in fibroblasts, and this effect is accompanied by a proportional increase in the expression of glucose transporters (174).

The effects of IL-6 on glucose metabolism have not been studied in detail in vivo. IL-6 enhances glycogenolysis in cultured hepatocytes (175), which might increase plasma glucose levels. However, IL-6 also decreases the mRNA for hormone-induced PEPCK in Reuber hepatoma cells (176), an event that might inhibit gluconeogenesis and lower plasma glucose. IL-6 has no effect on the rates of glucose utilization in vivo (139). As a consequence, IL-6 administration does not influence glucose levels (141). However, increases in plasma IL-6 combined with other cytokines may influence glucose homeostasis.

Administration of IFN-α to healthy volunteers produces a mild increase in plasma glucose levels (177). IFN-α increases circulating levels of glucagon, cortisol, and growth hormone and also increases the clearance of insulin (177). Both of these factors may contribute to a state of insulin resistance, resulting in hyperglycemia. This is supported by experiments that show that IFN-α markedly impairs insulin-mediated glucose uptake by peripheral tissue under euglycemic hyperinsulinemic conditions (177).

GM-CSF has also been shown to modulate glucose metabolism. It produces a transient hyperglycemia due to an increase in hepatic glucose production (178). GM-CSF acutely increases glucose uptake by lung, liver, spleen, and skin (178). In the liver, GM-CSF stimulates glucose uptake in both parenchymal as well as nonparenchymal cells (178). It also increases glucose uptake severalfold in circulating neutrophils and mononuclear cells (139). GM-CSF has no effect on serum insulin levels, suggesting that the increased glucose uptake seen with GM-CSF is not insulin dependent. Similarly, prior treatment with indomethacin does not block GM-CSF–induced increase in glucose uptake by various tissues, indicating that this effect is independent of prostanoid production (139).

THE ROLE OF CYTOKINE-INDUCED METABOLIC DISTURBANCES IN THE PATHOPHYSIOLOGY OF CACHEXIA

Cachexia or wasting syndrome is commonly seen in chronic infections and malignancies. Its primary components include marked anorexia, progressive weight loss, and disturbances in protein, lipid, and carbohydrate metabolism (179,180). Since these abnormalities can be replicated by administration of cytokines, either alone or in combination, and the production of several cytokines is upregulated in different animal models of infections and cancer, a role of different cytokines in the pathophysiology of cachexia has been postulated by several investigators. As discussed above, cytokines exert a multitude of effects on protein, lipid, and carbohydrate metabolism. Whether these metabolic effects are beneficial or deleterious to the host depends upon several factors such as the nature of cytokine-inducing stimuli, which can be either transient or persistent, concentration of cytokine(s), duration of cytokine release, extent of the host exposure to the cytokine(s), presence or absence of other cytokines and inflammatory mediators, interactions among different cytokines, and finally the nutritional status of the host.

Several comprehensive reviews have been published on the role of cytokines in the pathophysiology of cachexia (30,33,35,181,182). While initial studies focused on the role of TNF as a mediator of cachexia (183,184), recent studies indicate that several other cytokines including IL-1, IL-6, IFN-γ, LIF, TGF-β, and CNTF are also able to produce the syndrome of cachexia (185–193). Similarly, passive immunization with antibodies directed against individual cytokines in different tumor models only results in a partial reversal of one or more effects (191–196). From these studies it has become clear that no single cytokine is the sole cause of cachexia. Rather, progressive weight loss and metabolic disturbances, which are the hallmarks of cachexia, are produced by the inappropriate release of a variety of cytokine mediators.

It has been shown by several investigators that treatment with TNF induces acute anorexia, weight loss, and hypertriglyceridemia in rodents (197–200). While repeated administration of the same dose of TNF for several days results in a state of tolerance to its anorectic and cachectic effects (197–200), the hypertriglyceridemic effect of TNF persists with prolonged treatment (200), suggesting that the hypertriglyceridemia is not necessarily linked to cachexia. Under these experimental conditions there is little evidence of decreased muscle mass, which is the main finding in true cachexia. Similarly, there is no correlation between hypertriglyceridemia and wasting in patients with AIDS (201). Patients with AIDS who develop hypertriglyceridemia often maintain stable weight and body composition for prolonged periods (201).

In contrast to the tolerance to the anorectic effect of TNF, repeated administration of IL-1 to rodents produces more sustained anorexia (202). Similarly, administration of murine IFN-γ to experimental animals produces a prolonged decline in food intake and induces weight loss (203). IL-6 has also been proposed to mediate tumor-induced anorexia in rodents (204). Clinical studies have shown that treatment of cancer patients with IFN-α and IFN-γ produces anorexia and weight loss (205,206). Thus, in addition to TNF, several other cytokines are able to produce anorexia and weight loss, which may contribute to the pathophysiology of cachexia.

In recent years, spontaneous or genetically engineered tumor cell lines that continuously secrete specific cytokine have been used by various investigators to study the role of cytokines in the pathogenesis of cachexia. Implantation of tumor cells that continuously secrete TNF, IL-6, IFN-γ, LIF, or CNTF (184,186,187,189,191,192) result in severe anorexia and profound weight loss, which is due to wasting of skeletal muscle and adipose tissue. However, the interactions between the primary cytokine secreted by tumor cells with other tumor products or with other cytokines produced by host cells in response to the implanted tumor have not been explored in genetically engineered tumor models. It is possible that there are synergistic or regulatory interactions between different cytokines in tumor-bearing animals, which could mediate the syndrome of cachexia. Such an interaction between IL-1 and IL-6 has recently been reported by Strassmann et al. (207) in a mouse tumor model which develops profound wasting of muscle and adipose tissues. IL-1, secreted by host macrophages that infiltrate the tumor, enhances tumor secretion of IL-6. When the host IL-1 is blocked, IL-6 secretion decreases and weight loss is reduced (207).

It is likely that the cachexia associated with chronic infections and cancer is the result of additive or synergistic effects of multiple cytokines acting on diverse metabolic pathways in a variety of different tissues. Sustained production of proteolytic cytokines such as TNF, IL-1, and other yet unidentified mediators may result in muscle wasting by inducing skeletal muscle protein breakdown (120,121,126–129). Catabolic effects of several cytokines such as TNF and IFNs on adipose tissue could deplete lipid stores (60,61,66,67,84), particularly in the absence of adequate food intake, resulting in reduced fat cell mass. The induction of hepatic fatty acid synthesis, triglyceride production and VLDL secretion by several cytokines (44,50,58,59,70,61,85) uses energy and may be an undesirable effect at times when the body needs to conserve energy. The simultaneous lipolytic and lipogenic effects of TNF and IFN-α (59–61,84,85) result in substrate (formerly called futile) cycling of substrate between adipose tissue and liver and may be a wasteful process. TNF also induces a substrate cycle in glucose metabolism in cultured myocytes (208). It is likely that other cytokines may induce similar substrate cycles in carbohydrate and protein metabolism. Cytokine-induced anorexia (197–206) may play a major role in the weight loss as the body loses its mechanisms to compensate for decreased food intake when cytokine-induced metabolic disturbances are present. Thus, a combination of cytokine-induced metabolic disturbances and anorexia can produce the syndrome of cachexia during chronic infections and cancer.

METABOLIC EFFECTS OF CYTOKINES AS A PART OF THE ACUTE-PHASE RESPONSE

Administration of low doses of cytokines that are not detrimental to the host exert several beneficial effects. Animals that genetically lack the ability to produce TNF and IL-1 are very sensitive to the lethal effects of infection (209–212), whereas pretreatment of rodents with very low doses of TNF, IL-1, and LIF protects against subsequent lethal endotoxemia or bacterial sepsis (213–216), suggesting that these cytokines may play a physiological role in host defense. Several cytokines such as TNF, IL-1, IL-6, LIF, and IFN-γ induce the synthesis of a variety of acute-phase proteins in the liver (134). Acute-phase proteins are involved in maintaining homeostasis during infection and inflammation. For example, C-reactive protein and complement 3 help in the opsonization of bacteria, immune complexes, and foreign particles (134). More recently, α-1 acid glycoprotein has been shown to protect against TNF-induced lethality (217). Cytokine-induced increase in glucose utilization by immunologically competent tissues may help to support the function of immune cells in counteracting the harmful effects of invading microorganisms (139). There is now substantial evidence to suggest that many of the changes produced in the lipid and lipoprotein metabolism during the course of infection or by cytokine administration may be beneficial to the host. The increase in lipoproteins may result in enhanced delivery of lipids to the cells activated during the immune response and to the cells involved in tissue repair. The dose of LPS that produces a half-maximal increase in serum triglyceride levels in rats is 1/500,000 the lethal dose of LPS (21). Similarly, the doses of TNF and IL-1 that stimulate hepatic FA synthesis and increase VLDL secretion (44,50,58,71) are comparable to those that produce fever and induce acute-phase protein synthesis, suggesting that LPS- and cytokine-induced changes in lipoprotein metabolism may also be a part of physiological response to immune stimulation.

Several experimental studies have now shown that lipoproteins can bind LPS and in fact LPS is found in VLDL from the circulation of normal individuals (218–221). Preincubation of LPS with chylomicrons, VLDL, LDL, or HDL prior to administration prevents the ability of LPS to induce fever, hypotension, or death in rodents (218–221). Moreover, the infusion of chylomicrons prior to LPS administration also prevents LPS-induced death in rodents (222). Indeed, when lipoprotein infusions were given after LPS injection, they were still able to neutralize the toxic effects of LPS in rodents, indicating that lipoproteins may have a therapeutic role during acute infection and endotoxemia (222). Recent studies demonstrate that infusion of chylomicrons or a synthetic triglyceride-rich lipid emulsion can also improve survival in a lethal polymicrobial gram-negative sepsis model (223). Finally, we have recently shown that hypolipidemic animals are more susceptible to the lethal effects of LPS (224), an effect that can be reversed by infusion of exogenous lipoproteins. These results suggest that lipoproteins are part of the host defense against infection.

A potential mechanism for this protective effect is suggested by in vitro studies, which reveal that lipoproteins can decrease the ability of LPS to induce cytokine secretion by macrophages (225,226). Recent in vivo studies indicate that lipoproteins can enhance the clearance rate of LPS by the liver (227,228). The binding of LPS to chylomicrons results in its decreased uptake by tissue macrophages, such as Kupffer cells in the liver (227), accompanied by lower serum TNF levels, which may decrease toxicity. On the other hand, the uptake of chylomicron-bound LPS by hepatocytes is increased with the LPS excreted through the bile (227,228). These studies suggest that lipoproteins can neutralize the lethal effects of LPS by redirecting its metabolism and by decreasing the release of cytokines responsible for LPS-induced toxicity.

In addition to detoxifying LPS, lipoproteins can reduce the toxic effects of several viruses (229–232). Human lipoproteins have also been shown to induce the lysis of the parasite *Trypanosoma brucei* (232,233). Finally, lipoproteins can bind to urate crystals (235), which reduces the inflammatory response induced by these crystals.

Thus, it can be proposed that LPS- and cytokine-induced increase in lipoprotein levels represents a nonspecific immune response that can decrease the toxicity of a variety of harmful biological and chemical agents (236). These studies also indicate that in addition to their role in lipid transport, lipoproteins may also serve as a part of the host defense mechanisms to combat the toxic effects of circulating LPS. Hence, LPS- and cytokine-induced changes in lipid metabolism may represent another facet of host acute-phase response.

REFERENCES

1. Gallin JI, Kaye D, O'Leary WM. Serum lipids in infection. N Engl J Med 1969; 281:1081–1086.
2. Guckian JC. Role of the metabolism in pathogenesis of bacteremia due to diplococcus pneumoniae in rabbits. J Infect Dis 1973; 127:1–8.
3. Kaufmann RL, Matson CF, Beisel WR. Hypertriglyceridemia produced by endotoxin: role of impaired triglyceride disposal mechanisms. J Infect Dis 1976; 133:548–555.
4. Sammalkorpi K, Valtonen V, Kerttula Y, Nikkila E, Taskinen MR. Changes in serum lipoprotein pattern induced by acute infections. Metabolism 1988; 37:859–865.
5. Wolfe RR, Shaw JH, Durkot MJ. Effect of sepsis on VLDL kinetics: responses in basal state and during glucose infusion. Am J Physiol 1985; 248:E732–E740.
6. Lanza-Jacoby S, Lansey SC, Cleary MP, Rosato FE. Alterations in lipogenic enzymes and lipoprotein lipase activity during gram negative sepsis in the rat. Arch Surg 1982; 117:144–147.
7. Bagby GJ, Corll CB, Martinez RR. Triacylglycerol kinetics in endotoxic rats with suppressed lipoprotein lipase activity. Am J Physiol 1987; 253:E59–64.
8. Lanza-Jacoby S, Tabares A. Triglyceride kinetics, tissue lipoprotein lipase, and liver lipogenesis in septic rats. Am J Physiol 1990; 258:E678–685.
9. Spiegel RJ, Schaefer EJ, Magrath IT, Edwards BK. Plasma lipid alterations in leukemia and lymphoma. Am J Med 1982; 72:775–782.
10. Vlassara H, Spiegel RJ, Doval DS, Cerami A. Reduced plasma lipoprotein lipase activity in patients with malignancy associated weight loss. Horm Metab Res 1983; 18:698–703.
11. Lusk G. Effect of infection on protein and ribonucleic acid synthesis in mammalian organs and tissues. Fed Proc 1966; 25:1688–1694.
12. Clowes GHA, George BC, Villee CA, Saravis CA. Muscle proteolysis induced by a circulating peptide in patients with trauma and sepsis. N Engl J Med 1983; 308:545–552.
13. Jepson MM, Pell JM, Bates PC, Millward DJ. The effect of endotoxaemia on protein metabolism in skeletal muscle and liver of fed and fasted rats. Biochem J 1986; 234:329–336.

14. Hasselgren PO, Pedersen P, Sax HC, Warner BW, Fischer JE. Current concepts of protein turnover and amino acid transport in liver and skeletal muscle during sepsis. Arch Surg 1988; 123:992–998.
15. Emery PW, Lovell L, Rennie MJ. Protein synthesis measured in vivo muscle and liver of cachectic tumor bearing mice. Cancer Res 1984; 44:2279–2784.
16. Lang CH, Bagby GJ, Spitzer JJ. Glucose kinetics and body temperature after lethal and non lethal doses of endotoxin. Am J Physiol 1985; 248:R471–R478.
17. Casteleijn E, Kuipers J, Koster JF, Van Berkel JC. Endotoxin stimulates glycogenolysis in the liver by means of intercellular communication. J Biol Chem 1988; 263:6953–6955.
18. Hill M, McCallum R. Altered transcriptional regulation of posphoenolpyruvate carboxykinase in rats following endotoxin treatment. J Clin Invest 1991; 88:811–816.
19. Meszaros K, Lang CH, Bagby GJ, Spitzer JJ. Contribution of different organs to increased glucose consumption after endotoxin administration. J Biol Chem 1987; 262:10965–10970.
20. Roh MS, Ekman L, Jeevanandam M, Brennan MF. Gluconeogenesis in tumor influenced hepatocytes. Surgery 1984; 96:427–433.
21. Feingold KR, Staprans I, Memon RA, et al. Endotoxin rapidly induces changes in lipid metabolism that produce hypertriglyceridemia: low doses stimulate hepatic triglyceride production while high doses inhibit clearance. J Lipid Res 1992; 33:1765–1776.
22. Nonogaki K, Moser AH, Pan XM, Staprans I, Grunfeld C, Feingold KR. Lipoteichoic acid stimulates lipolysis and hepatic triglyceride secretion in rats. J Lipid Res 1995; 36:1987–1995.
23. Fiser RH, Denniston JC, Beisel WR. Infection with diplococcus pneumoniae and salmonella typhimurium in monkeys: changes in plasma lipids and lipoproteins. J Infect Dis 1972; 125:54–60.
24. Young VR. Energy metabolism and requirements in the cancer patient. Cancer Res 1977; 37:2336–2347.
25. Argiles JM, Azcon-Bieto J. The metabolic environment of cancer. Mol Cell Biochem 1988; 81:3–17.
26. Ulich TR, Guo K, Castillo JD. Endotoxin induced cytokine gene expression in vivo. Expression of tumor necrosis factor mRNA in visceral organs under physiologic conditions and during endotoxemia. Am J Pathol 1988; 134:11–14.
27. Ulich TR, Guo K, Remick D, Castillo J, Yin S. Endotoxin induced cytokine gene expression in vivo. IL-6 mRNA and serum protein expression and the in vivo hematologic effects of IL-6. J Immunol 1991; 146:2316–2323.
28. Ulich TR, Guo K, Yin S, et al. Endotoxin induced cytokine gene expression in vivo. Expression of interleukin-1α/β and interleukin-1 receptor antagonist mRNA during endotoxemia and during endotoxin initiated local acute inflammation. Am J Pathol 1992; 141:61–68.
29. Pekarek L, Weichselbaum RR, Beckett MA, et al. Footprinting of individual tumors and their variants by constitutive cytokine expression patterns. Cancer Res 1993; 53:1978–1981.
30. Beutler B. Cytokines and cancer cachexia. Hosp Prac 1993; 28:45–52.
31. Tracey KJ, Cerami A. Tumor necrosis factor, other cytokines and disease. Ann Rev Cell Biol 1993; 9:317–343.
32. Dinarello CA, Wolef SM. The role of interleukin-1 in disease. N Engl J Med 1993; 328:106–113.
33. Grunfeld C, Feingold KR. The metabolic effects of tumor necrosis factor and other cytokines. Biotherapy 1991; 3:143–158.
34. Memon RA, Feingold KR, Grunfeld C. The effects of cytokines on intermediary metabolism. Endocrinologist 1994; 4:56–63.
35. Memon RA, Feingold KR, Grunfeld C. Role of cytokines in lipid metabolism and cachexia. In: Aggarwal BB, Puri PK, eds. Human Cytokines: Their Role in Disease and Therapy. Cambridge: Blackwell Science, 1995:239–251.
36. Vasconcelos PRL, Kettlewell MGW, Gibbons CF, Williamson DH. Increased rates of hepatic cholesterogenesis and fatty acid synthesis in septic rats in vivo: Evidence for the possible involvement of insulin. Clin Sci 1989; 76:205–211.
37. Lanza-Jacoby S, Wong SH, Tabares A, Baer D, Schneider T. Disturbances in the composition of plasma lipoproteins during gram negative sepsis in the rat. Biochim Biophys Acta 1992; 1124:233–240.
38. Hardardottir I, Grunfeld C, Feingold KR. Effects of endotoxin and cytokines on lipid metabolism. Curr Opin Lipidol 1994; 5:207–215.
39. Auerbach BJ, Parks JS. Lipoprotein abnormalities associated with lipopolysaccharide induced lecithin-cholesteryl acyltransferase and lipase deficiency. J Biol Chem 1989; 264:10264–10270.
40. Neufeld HA, Pace JA, White FE. The effect of bacterial infections on ketone concentrations in rat liver and blood and on free fatty acid concentrations in rat blood. Metabolism 1976; 25:877–884.
41. Lanza-Jacoby S, Rosato E, Braccia G, et al. Altered ketone body metabolism during gram negative sepsis in the rat. Metabolism 1990; 39:1151–1157.
42. Neufeld HA, Pace JG, Kaminski MV, et al. A probable endocrine basis for the depression of ketone bodies during infectious or inflammatory state in rats. Endocrinology 1980; 107:596–601.
43. Vary TC, Siegel JH, Nakatani T, et al. A biochemical basis for depressed ketogenesis. J Trauma 1986; 26:419–425.
44. Feingold KR, Grunfeld C. Tumor necrosis factor-alpha stimulates hepatic lipogenesis in the rat in vivo. J Clin Invest 1987; 80:184–190.
45. Feingold KR, Soued M, Serio MK, Adi S, et al. The effect of diet on tumor necrosis factor stimulation of hepatic lipogenesis. Metabolism 1990; 39:623–632.
46. Krauss RM, Grunfeld C, Doerrler WT, Feingold KR. Tumor necrosis factor acutely increases plasma levels of very low density lipoproteins of normal size and composition. Endocrinology 1990; 127:1016–1021.
47. Feingold KR, Soued M, Staprans I, et al. Effect of tumor necrosis factor on lipid metabolism in the diabetic rat: evidence that inhibition of adipose tissue lipoprotein lipase activity is not required for TNF induced hyperlipidemia. J Clin Invest 1989; 83:1116–1121.
48. Feingold KR, Soued M, Adi S, et al. Tumor necrosis factor increased hepatic very low density lipoprotein production and increased serum triglyceride levels in diabetic rats. Diabetes 1990; 39:1569–1574.
49. Evans RD, Williamson DH. Comparison of effects of platelet activating factor and tumor necrosis factor-α on lipid metabolism in adrenalectomized rats in vivo. Biochim Biophys Acta 1991; 1086:191–196.
50. Memon RA, Grunfeld C, Moser AH, Feingold KR. Tumor necrosis factor mediates the effects of endotoxin on cholesterol and triglyceride metabolism in mice. Endocrinology 1993; 132:2246–2253.
51. Ettinger WH, Miller LD, Albers JJ, Smith TK, Parks JS. Lipopolysaccharide and tumor necrosis factor cause a fall in plasma concentrations of lecithin cholesterol acyltransferase in cynomolgus monkeys. J Lipid Res 1990; 31:1099–1107.
52. Sherman ML, Spriggs DR, Arthur KA, Imamura K, Frei E, Kufe DW. Recombinant human tumor necrosis factor administered as a five day continuous infusion in cancer patients: Phase

I toxicity and effects on lipid metabolism. J Clin Oncol 1988; 6:344–350.
53. Beutler B, Cerami A. Cachectin and tumor necrosis factor as two sides of the same biological coin. Nature 1986; 320:584–588.
54. Grunfeld C, Gulli R, Moser AH, Gavin LA, Feingold KR. The effect of tumor necrosis administration in vivo on lipoprotein lipase activity in various tissues of the rat. J Lipid Res 1989; 30:579–585.
55. Kern PA. Recombinant human tumor necrosis factor does not inhibit lipoprotein lipase in primary cultures of isolated human adipocytes. J Lipid Res 1988; 29:909–914.
56. Semb H, Peterson J, Tavernier J, Olivecrona T. Multiple effects of tumor necrosis factor on lipoprotein lipase in vivo. J Biol Chem 1987; 262:8390–8394.
57. Chajek-Shaul T, Friedman G, Stein O, Shiloni E, Etienne J, Stein Y. Mechanisms of the hypertriglyceridemia induced by tumor necrosis factor administration to rats. Biochim Biophys Acta 1989; 1001:316–324.
58. Feingold KR, Serio MK, Adi S, et al. Tumor necrosis factor stimulates hepatic lipid synthesis and secretion. Endocrinology 1989; 124:2336–2342.
59. Feingold KR, Soued M, Serio MK, et al. Multiple cytokines stimulate hepatic lipid synthesis in vivo. Endocrinology 1989; 125:267–274.
60. Feingold KR, Adi S, Staprans I, et al. Diet affects the mechanisms by which TNF stimulates hepatic triglyceride production. Am J Physiol 1990; 259:E177–E184.
61. Patton JS, Shepard HM, Wilking H, et al. Interferons and tumor necrosis factor have similar catabolic effects on 3T3-L1 cells. Proc Natl Acad Sci USA 1986; 83:8313–8317.
62. Beutler B, Cerami A. Recombinant interleukin-1 suppresses lipoprotein lipase activity in 3T3-L1 cells. J Immunol 1985; 135:3969–3971.
63. Price SR, Mizel SB, Pekala PH. Regulation of lipoprotein lipase synthesis and 3T3-L1 adipocyte metabolism by recombinant interleukin-1. Biochim Biophys Acta 1986; 889:374–381.
64. Greenberg AS, Nordon RP, McIntosh J, Calvo JC, Scow RO, Jablons D. Interleukin-6 reduces lipoprotein lipase activity in adipose tissue of mice in vivo and in 3T3-L1 adipocytes. A possible role for interleukin-6 in cancer cachexia. Cancer Res 1992; 52:4113–4116.
65. Mori M, Yamaguchi K, Abe K. Purification of a lipoprotein lipase inhibiting protein produced by a melanoma cell line associated with cancer cachexia. Biochem Biophys Res Commun 1989; 160:1085–1092.
66. Feingold KR, Doerrler W, Dinarello CA, Fiers W, Grunfeld C. Stimulation of lipolysis in cultured fat cells by tumor necrosis factor, interleukin-1 and interferons is blocked by inhibition of prostaglandin synthesis. Endocrinology 1992; 130:10–16.
67. Doerrler W, Feingold KR, Grunfeld C. Cytokines induce catabolic effects in cultured adipocytes by multiple mechanisms. Cytokine 1994; 6:478–484.
68. Marshall MK, Doerrler W, Feingold KR, Grunfeld C. Leukemia inhibitory factor induces changes in lipid metabolism in cultured adipocytes. Endocrinology 1994; 135:141–147.
69. Sumida M, Sekiya K, Okuda H, et al. Inhibitory effect of tumor necrosis factor on gene expression of hormone sensitive lipase in 3T3-L1 adipocytes. J Biochem 1990; 107:1–2.
70. Grunfeld C, Adi S, Soued M, Moser AH, Fiers W, Feingold KR. Search for mediators of the lipogenic effects of tumor necrosis factor: possible role for interleukin-6. Cancer Res 1990; 50:4233–4238.
71. Feingold KR, Soued M, Adi S, et al. Effect of Interleukin-1 on lipid metabolism in the rat. Similarities to and differences from tumor necrosis factor. Arteriosclerosis Thromb 1991; 11:495–500.
72. Nonogaki K, Fuller GM, Fuentes NL, et al. Interleukin-6 stimulates hepatic triglyceride secretion in rats. Endocrinology 1995; 136:2143–2149.
73. Nonogaki K, Pan M, Moser AH, Shigenaga J, Staprans I, Sakamoto N, Grunfeld C, Feingold KR. LIF and CNTF, which share the gp130 transduction system, stimulate hepatic lipid metabolism in rats. Am J Physiol 1996 (in press).
74. Housley RM, Morris CF, Boyle W, Ring B, Blitz R, Tarpley JE, Aukerman SL, Devine PL, Whitehead RH, Pierce GF. Keratinocyte growth factor induces proliferation of hepatocytes and epithelial cells throughout the rat gastrointestinal tract. J Clin Invest 1994; 94:1764–1777.
75. Nonogaki K, Pan X-M, Moser AH, Staprans I, Feingold KR, Grunfeld C. Keratinocyte growth factor increases fatty acid mobilization and hepatic triglyceride secretion in rats. Endocrinology 1995; 136:4278–4284.
76. Feingold KR, Marshall M, Gulli R, Moser AH, Grunfeld C. Effect of endotoxin and cytokines on lipoprotein lipase activity in mice. Arteriosclerosis Thromb 1994; 14:1866–1872.
77. Olsen EA, Lichenstein GR, Wilkinson WE. Changes in serum lipids in patients with condylomata acuminata treated with interferon alpha. J Am Acad Dermatol 1988; 19:286–289.
78. Graessle D, Bonacini M, Chen S. Alpha-interferon and reversible hypertriglyceridemia. Ann Intern Med 1993; 118:316–317.
79. Kurzrock R, Rohde MF, Quesada JR, et al. Recombinant gamma interferon induces hypertriglyceridemia and inhibits post heparin lipase activity in cancer patients. J Exp Med 1986; 164:1093–1101.
80. Kurzrock R, Feinberg B, Talpaz M, Saks S, Gutterman JU. Phase I study of a combination of recombinant tumor necrosis factor and recombinant interferon-γ in cancer patients. J Interferon Res 1989; 9:435–444.
81. Grunfeld C, Kotler DP, Shigenaga JK, et al. Circulating interferon-α levels and hypertriglyceridemia in the acquired immunodeficiency syndrome. Am J Med 1991; 90:154–162.
82. Grunfeld C, Pang M, Doerrler W, et al. Lipids, lipoproteins, triglyceride clearance and cytokines in human immunodeficiency virus infection and the acquired immunodeficiency syndrome. J Clin Endo Metabol 1992; 74:1045–1052.
83. Hellerstein MK, Grunfeld C, Wu K, et al. Increased de novo hepatic lipogenesis in human immunodeficiency virus infection. J Clin Endo Metabol 1993; 76:559–565.
84. Memon RA, Feingold KR, Moser AH, Doerrler W, Grunfeld C. In vivo effects of interferon-α and interferon-γ on lipolysis and ketogenesis. Endocrinology 1992; 131:1695–1702.
85. Grunfeld C, Soued M, Adi S, et al. Evidence for two classes of cytokines that stimulate hepatic lipogenesis: relationships among tumor necrosis factor, interleukin-1 and interferon-alpha. Endocrinology 1990; 127:46–54.
86. Grunfeld C, Soued M, Adi S, et al. Interleukin-4 inhibits stimulation of hepatic lipogenesis by tumor necrosis factor, interleukin-1, and interleukin-6 but not interferon-α. Cancer Res 1991; 51:2803–2807.
87. Hardardottir I, Moser AH, Memon RA, Grunfeld C, Feingold KR. Effects of TNF, IL-1, and the combination of both cytokines on cholesterol metabolism in syrian hamsters. Lympho Cyto Res 1994; 13:161–166.
88. Feingold KR, Hardardottir I, Memon RA, Krul EJT, Moser AH, Taylor JM, Grunfeld C. Effect of endotoxin on cholesterol biosynthesis and distribution in serum lipoproteins in syrian hamsters. J Lipid Res 1993; 34:2147–2158.
89. Feingold KR, Spady DK, Pollock AS, Moser AH, Grunfeld C.

Endotoxin, TNF and IL-1 decrease cholesterol 7-alpha hydroxylase levels and activity. J Lipid Res 1996; 37:223–228.
90. Hardardottir I, Kunitake ST, Moser AR, et al. Endotoxin and cytokines increase hepatic messenger RNA levels and serum concentrations of apolipoprotein J (Clusterin) in syrian hamsters. J Clin Invest 1994; 94:1304–1309.
91. Dowton SB, Peters CN, Jestus JJ. Regulation of serum amyloid A gene expression in syrian hamsters by cytokines. Inflammation 1991; 15:391–397.
92. Kisilevsky R, Subrahamanyan L. Serum amyloid A changes high density lipoprotein's affinity. A clue to serum amyloid A's principal function. Lab Invest 1992; 66:778–785.
93. Auerbach BJ, Parks JS. Lipoprotein abnormalities associated with lipopolysaccharide-induced lecithin: cholesterol acyltransferase and lipase deficiency. J Biol Chem 1989; 264:10264–10270.
94. Ettinger WH, Miller LA, Smith TK, Parks JS. Effect of interleukin-1 alpha on lipoprotein lipids in cynomolgus monkeys: comparison to tumor necrosis factor. Biochim Biophys Acta 1992; 1128:186–192.
95. Ly H, Francone OL, Fielding CJ, et al. Endotoxin and TNF lead to reduced plasma LCAT activity and decreased hepatic LCAT mRNA levels in Syrian hamsters. J Lipid Res 1995; 36:1254–1263.
96. Hardardottir I, Moser AH, Fuller J, Fielding C, Feingold KR, Grunfeld C. Endotoxin and cytokines decrease serum levels and extra hepatic protein and mRNA levels of cholesterol ester transfer protein in Syrian hamsters. J Clin Invest 1996; 97:2585–2592.
97. Borden EC, Rosenzweig IB, Byrne GI. Interferons: from virus inhibitor to modulator of amino acid and lipid metabolism. J Interferon Res 1987; 7:591–596.
98. Nimer SD, Champlin RE, Golde DW. Serum cholesterol lowering activity of granulocyte macrophage colony stimulating factor. JAMA 1988; 260:3297–3300.
99. Stoudemire JB, Garnick MB. Effect of recombinant human macrophage colony stimulating factor on plasma cholesterol levels. Blood 1991; 77:750–755.
100. Rosenzweig IB, Wiebe DA, Hanks JA, et al. Effects of interleukin-2 on human plasma lipid, lipoprotein, and C-reactive protein. Biotherapy 1990; 2:193–198.
101. Rosenzweig IB, Wiebe DA, Borden EC, et al. Plasma lipoprotein changes in humans induced by beta interferon. Atherosclerosis 1987; 67:261–267.
102. Schectman G, Kaul S, Mueller RA, et al. The effect of interferon on the metabolism of LDLs. Arteriosclerosis Thromb 1992; 12:1053–1062.
103. Ettinger WH, Varma VK, Sorci-Thomas M, et al. Cytokines decrease apolipoprotein accumulation in medium from HepG2 cells. Arteriosclerosis Thromb 1994; 14:8–13.
104. Stopeck AT, Nicholson AC, Mancini FP, Hajjar DP. Cytokine regulation of low density lipoprotein receptor gene transcription in HepG2 cells. J Biol Chem 1993; 268:17489–17494.
105. Grove RI, Mazzucco CE, Radka SF, et al. Oncostatin M up-regulates low density lipoprotein receptors in HepG2 cells by a novel mechanism. J Biol Chem 1991; 266:18194–18199.
106. Shimano H, Yamada N, Ishibashi S, et al. Human monocyte colony stimulating factor enhances the clearance of lipoproteins containing apolipoprotein B-100 via both low density lipoprotein receptor dependent and independent pathways in rabbits. J Biol Chem 1990; 265:12869–12875.
107. Nelson S, Mason C, Bagby G, et al. Effect of murine recombinant granulocyto-macrophage colony stimulating factor on lipopolysaccharide induced tumor necrosis factor. Am Rev Respir Dis 1990; 141:677–652.
108. Memon RA, Feingold KR, Moser AH, et al. Differential effects of interleukin-1 and tumor necrosis factor on ketogenesis. Am J Physiol 1992; 263:E301–E309.
109. Memon RA, Feingold KR, Adi S, Grunfeld C. Which cytokines mediate the metabolic effects of endotoxin? In: Fiers W, Burman W, eds. Tumor Necrosis Factor. Vol. 4: Basel: Karger, 1993:107–112.
110. McGarry JD, Woeltje KF, Kuwajima M, Foster DW. Regulation of ketogenesis and renaissance of carnitine palmitoyltransferase. Diabetes Metab Rev 1989; 5:271–284.
111. Reaven GM, Chang H, Ho H, Jeng C, Hoffman BB. Lowering of plasma glucose in diabetic rats by antilipolytic agents. Am J Physiol 1988; 254:E23–E30.
112. Beylot M, Vidal H, Mithieux G, Odeon M, Martin C. Inhibition of hepatic ketogenesis by tumor necrosis factor-α in rats. Am J Physiol 1992; 263:E897–E902.
113. Tiao G, Fagan JM, Samuels N, James JH, Hudson K, Lieberman M, Fischer JE, Hasselgren PO. Sepsis stimulates nonlysosomal, energy-dependent proteolysis and increases ubiquitin mRNA levels in rat skeletal muscle. J Clin Invest 1994; 94:2255–2264.
114. Hall-Angeras M, Angeras U, Zamir O, Hasselgren PO, Fischer JE. Effect of the glucocorticoid receptor antagonist RU 38486 on muscle protein breakdown in sepsis. Surgery 1991; 109:468–473.
115. Fang CH, James HJ, Ogle C, Fischer JE, Hasselgren PO. Influence on burn injury on protein metabolism in different types of skeletal muscle and the role of glucocorticoids. J Am Coll Surg 1995; 180:33–42.
116. James JH, Hasselgren PO, King JK, James LE, Fischer JE. Intracellular glutamine concentration does not decrease in all muscles during sepsis. J Surg Res 1993; 54:558–564.
117. Goldberg A, Kettelhut K, Fagan J, Baracos V. Activation of protein breakdown and postaglandin E_2 production in rat skeletal muscle in fever in signaled by a macrophage product distinct from interleukin-1 or other known monokines. J Clin Invest 1988; 81:1378–1383.
118. Kettelhut IC, Goldberg AL. Tumor necrosis factor can induce fever in rats without activating protein breakdown in muscle or lipolysis in adipose tissue. J Clin Invest 1988; 81:1384–1389.
119. Moldawer LL, Svaninger G, Gelin J, Lundholm KG. Interleukin-1 and tumor necrosis factor do not regulate protein balance in skeletal muscle. Am J Physiol 1987; 253:C766–C773.
120. Flores EA, Bistrian BR, Pomposelli JJ, et al. Infusion of tumor necrosis factor/cachectin promotes muscle catabolism in the rat: a synegistic effect with interleukin-1. J Clin Invest 1989; 83:1614–1622.
121. Fischer JE, Hasselgren PO. Cytokines and glucocorticoids in the regulation of the "hepato-skeletal muscle axis" in sepsis. Am J Surg 1991; 161:266–271.
122. Garcia-Martinez C, Lopez-Soriano FJ, Argiles JM. Interleukin-6 does not activate protein breakdown in rat skeletal muscle. Cancer Lett 1994; 76:1–4.
123. Hiroo I, Jun-ichi F. Endocrine paracrine interaction in communication between the immune and endocrine systems. Activation of the hypothalamic pituitary adrenal axis in inflammation. Eur Endocrinol 1994; 130:32–37.
124. Odedra BR, Bates PC, Millward DJ. Time course of the effect of catabolic doses of corticosterone on protein turnover in rat skeletal muscle and liver. Biochem J 1983; 214:617–627.
125. Legaspi A, Albert JD, Calvano SE, et al. Proteolysis of skeletal muscle in response to acute elevation of plasma cortisol in man. Surg Forum 1985; 36:16–18.
126. Hall-Angeras M, Angeras Ulf, Zamir O, et al. Interaction between corticosterone and tumor necrosis factor stimulated

protein breakdown in rat skeletal muscle, similar to sepsis. Surgery 1990; 108:460–466.

127. Zamir O, Hasselgren PO, Kunkel SL, et al. Evidence that tumor necrosis factor participates in the regulation of muscle proteolysis during sepsis. Arch Surg 1992; 127:170–174.

128. Zamir O, Hasselgren PO, Allmen D, Fischer JE. In vivo administration of interleukin-1α induces muscle proteolysis in normal and adrenalectomized rats. Metabolism 1993; 42:204–208.

129. Zamir O, Hasselgren PO, Allmen DV, Fischer JE. The effect of interleukin-1α and the glucocorticoid receptor blocker RU 38486 on total and myofibrillar protein breakdown in skeletal muscle. J Surg Res 1991; 50:579–583.

130. Warren RS, Donner DB, Starnes HF, Brennan MF. Modulation of endogenous hormone action by recombinant human tumor necrosis factor. Proc Natl Acad Sci USA 1987; 84:8619–8622.

131. De Bandt JP, Lim SK, Plassart F, et al. Independent and combined actions of interleukin-1β, tumor necrosis factor-α, and glucagon on amino acid metabolism in the isolated perfused rat liver. Metabolism 1994; 43:822–829.

132. Andus T, Bauer J, Gerok W. Effects of cytokines on the liver. Hepatology 1991; 13:364–375.

133. Richards C, Gauldie J, Baumann H. Cytokine control of acute phase protein expression. Eur Cytokine Net 1991; 2:89–98.

134. Richards CD, Gauldie J. Role of cytokines in acute phase response. In: Aggarwal BB, Puri RK, eds. Human Cytokines: Their Role in Disease and Therapy. Cambridge: Blackwell Science, 1995:253–269.

135. Orlinska U, Newton RC. Role of glucose in interleukin-1 beta production by lipopolysaccharide-activated human monocytes. J Cell Physiol 1993; 157:201–208.

136. Spolarics Z, Spitzer JJ. Acute endotoxin tolerance is accompanied by stimulated glucose use in macrophage rich tissues. Biochem Biophys Res Commun 1995; 211:340–346.

137. Spolarics Z, Pekala PH, Bagby GJ, Spitzer JJ. Brief endotoxemia markedly increases expression of GLUT1 glucose transporter in Kupffer, hepatic endothelial and parenchymal cells. Biochem Biophys Res Commun 1993; 193:1211–1215.

138. Spolarics Z, Bautista AP, Spitzer JJ. Primed pentose cycle activity supports production and elimination of superoxide anion in Kupffer cells from rats treated with endotoxin in vivo. Biochim Biophys Acta 1993; 1179:134–140.

139. Lang CH. Role of cytokines in glucose metabolism. In: Aggarwal BB, Puri RK, eds. Human Cytokines: Their Role in Disease and Therapy. Cambridge: Blackwell Science, 1995:271–284.

140. Tracey KJ, Beutler B, Lowry SF, et al. Shock and tissue injury by recombinant human cachectin. Science 1986; 234:470–474.

141. Del Rey A. Metabolic and neuroendocrine effects of pro-inflammatory cytokines. Eur J Clin Invest 1992; 1:10–15.

142. Kettelhut IC, Fiers W, Goldberg AL. The toxic effects of tumor necrosis factor in vivo and their prevention by cyclooxygenase inhibitors. Proc Natl Acad Sci USA 1987; 84:4273–4277.

143. Flores EA, Istfan N, Pomposelli J, et al. Effect of interleukin-1 and tumor necrosis factor/cachectin on glucose turnover in the rat. Metabolism 1990; 39:738–743.

144. Yasmineh WG, Theologides A. Effect of tumor necrosis factor on enzymes of gluconeogenesis in the rat. Proc Soc Exp Biol Med 1992; 255:E206–E212.

145. Warren RS, Starnes F, Gabrilove JL, et al. The acute metabolic effects of tumor necrosis factor administration in humans. Arch Surg 1987; 122:1396–1400.

146. Evans DH, Jacob DO, Wilmore DW. Tumor necrosis factor enhances glucose uptake by peripheral tissues. Am J Physiol 1989; 257:R1182–R1189.

147. Van der Poll T, Romijn JA, Endert E, Borm JJ, Buller HR, Sauerwein HP. Tumor necrosis factor mimics the metabolic response to acute infection in healthy humans. Am J Physiol 1991; 261:E457–E465.

148. Lang CH. Beta-adrenergic blockade attenuates insulin resistance induced by tumor necrosis factor. Am J Physiol 1993; 264:R984–991.

149. Hotamisligil GS, Shargill NS, Spiegelman BM. Adipose expression of tumor necrosis factor alpha: direct role in obesity-linked insulin resistance. Science 1993; 259:87–91.

150. Hotamisligil GS, Spiegelman BM. TNF-α: a key component of obesity-diabetes link. Diabetes 1994; 43:1271–1291.

151. Hoffmann C, Lorenz K, Braithwaite SS, Palazuk BJ, Hotamisligil GS, Spiegelman BM. Altered gene expression for tumor necrosis factor-α and its receptors during drug and dietary modulation of insulin resistance. Endocrinology 1994; 134:264–270.

152. Kern PA, Saghizadeh M, Ong JM, Bosch RJ, Deem R, Simsolo RB. The expression of tumor necrosis factor in human adipose tissue. Regulation by obesity, weight loss, and relationship to lipoprotein lipase. J Clin Invest 1995; 95:2111–2119.

153. Hotamisligil GS, Arner P, Caro JF, Atkinson RL, Spiegelman BM. Increased adipose tissue expression of tumor necrosis factor-α in human obesity and insulin resistance. J Clin Invest 1995; 95:2409–2415.

154. Feinstein R, Kanety H, Papa MZ, Lunenfeld B, Karasik A. Tumor necrosis factor-alpha suppresses insulin-induced tyrosine phosphorylation of insulin receptor and its substrates. J Biol Chem 1993; 268:26055–26058.

155. Hotamisligil GS, Budavari A, Murray DL, Spiegelman BM. Reduced tyrosine kinase activity of the insulin receptor in obesity-diabetes: central role of tumor necrosis factor-α. J Clin Invest 1994; 94:1543–1549.

156. Hotamisligil GS, Murray DL, Choy LN, Spiegelman BM. TNF-α inhibits signaling from insulin receptor. Proc Natl Acad Sci USA 1994; 91:4854–4858.

157. Stephens JM, Pekala PH. Transcriptional repression of the GLUT4 and C/EBP genes in 3T3-L1 adipocytes by tumor necrosis factor-alpha. J Biol Chem 1991; 266:21839–21845.

158. Lee MD, Zentella A, Pekala PH, Cerami A. Effect of endotoxin induced monokines on glucose metabolism in the muscle cell line L-6. Proc Natl Acad Sci USA 1987; 84:2590–2594.

159. Cornelius P, Lee MD, Marlowe M, Pekala PH. Monokine regulation of glucose transporter mRNA in 16 myotubes. Biochem Biophys Res Commun 1989; 165:429–436.

160. Meszaros K, Lang CH, Bagby GJ, Spitzer JJ. Tumor necrosis factor increases in vivo glucose utilization of macrophage rich tissues. Biochem Biophys Res Commun 1987; 149:1–6.

161. Spolarics Z, Schuler A, Bagby GJ, et al. Tumor necrosis factor increases in vivo glucose uptake in hepatic nonparenchymal cells. J Leukocyte Biol 1991; 49:309–312.

162. Mahony SM, Tisdale MJ. Metabolic effects of TNF-α in NMRI mice. Br J Cancer 1990; 61:514–519.

163. Hill MR, McCallum RE. Identification of tumor necrosis factor as a transcriptional regulator of the phosphoenolpyruvate carboxykinase gene following endotoxin treatment of mice. Infect Immun 1992; 60:4040–4050.

164. Arbos J, Lopez-Soriano FJ, Carbo N, Argiles JM. Effect of tumor necrosis factor-α (cachectin) on glucose metabolism in rat: intestinal absorption and isolated enterocyte metabolism. Mol Cell Biochem 1992; 112:53–59.

165. Ling PR, Istfan NW, Colon E, Bistrian BR. Differential effects of interleukin-1 receptor antagonist in cytokine- and endotoxin-treated rats. Am J Physiol 1995; 268:E255–E261.

166. Del Ray A, Besedovsky H. Interleukin-1 affects glucose homeostasis. Am J Physiol 1987; 253:R794–R798.
167. Del Rey A, Besedovsky H. Antidiabetic effects of interleukin-1. Proc Natl Acad Sci USA 1989; 86:5943–5947.
168. Berkenbosch F, De Goeij DEC, Del Ray A, Besedovsky H. Neuroendocrine, sympathetic and metabolic responses induced by interleukin-1. Neuroendocrinol 1989; 50:570–576.
169. Hill MR, Stith RD, McCallum RE. Interleukin-1: a regulatory role in glucocorticoid regulated hepatic metabolism. J Immunol 1986; 137:858–862.
170. Lang CH, Dobrescu C. Interleukin-1 induced increases in glucose utilization are insulin mediated. Life Sci 1989; 45:2127–2134.
171. Jarrous A, Lang CH. Central administration of IL-1α increases glucose uptake by muscle. Cytokine 1995; 7:57–63.
172. Garcia-Welsh A, Schneiderman JS, Baly DL. Interleukin-1 stimulates glucose transport in adipose cells. Evidence for receptor discrimination between IL-1β and IL-1α. FEBS Lett 1990; 269:421–424.
173. Hernvann A, Aussel C, Cynober L, et al. IL-1β, a strong mediator for glucose uptake by rheumatoid and non-rheumatoid cultured human synoviocytes. FEBS Lett 1992; 303:77–80.
174. Bird TA, Daview A, Baldwin SA, Saklatvala J. Interleukin 1 stimulates hexose transport in fibroblasts by increasing the expression of glucose transporters. J Biol Chem 1990; 265:13578–13583.
175. Ritchie DG. Interleukin 6 stimulates hepatic glucose release from prelabeled glycogen pools. Am J Physiol 1990; 258:E57–E64.
176. Hill MR, Stith RD, McCallum RE. Mechanism of action of interferon-β_2/interleukin-6 on induction of hepatic liver enzymes. Ann NY Acad Sci 1989; 57:502–504.
177. Koivisto VA, Pelkonen R, Cantell K. Effect of interferon on glucose tolerance and insulin sensitivity. Diabetes 1989; 38:641–647.
178. Schuler A, Spolarics Z, Lang CH, et al. Upregulation of glucose metabolism by granulocyte-macrophage colony stimulating factor. Life Sci 1991; 49:899–906.
179. Brennan MF. Uncomplicated starvation versus cancer cachexia. Cancer Res 1977; 37:2359–2364.
180. Kern KA, Norton JA. Cancer cachexia. J Paren Enteral Nutr 1988; 12:286–298.
181. Moldawer LL, Rogy MA, Lowry SF. The role of cytokines in cancer cachexia. J Paren Enteral Nutr 1992; 16:43–49.
182. McNamara MJ, Alexander HR, Norton JA. Cytokines and their role in the pathophysiology of cancer cachexia. J Paren Enteral Nutr 1992; 16:50–55.
183. Tracey KJ, Lowry SF, Cerami A. Cachectin: a hormone that triggers acute shock and chronic cachexia. J Inf Dis 1988; 157:413–420.
184. Oliff A, Defco-Jones D, Boyer M, et al. Tumors secreting human TNF/cachectin induce cachexia in mice. Cell 1987; 50:555–563.
185. Gelin J, Moldawer LL, Lonnroth C, et al. Role of endogenous tumor necrosis factor-α and interleukin-1 for experimental tumor growth and the development of cancer cachexia. Cancer Res 1991; 51:415–421.
186. Black K, Garrett R, Mundy GR. Chinese hamster ovarian cells transfected with the murine interleukin-6 gene cause hypercalcemia as well as cachexia, leukocytosis and thrombocytosis in tumor bearing nude mice. Endocrinology 1991; 128:2657–2659.
187. Matthys P, Dijkmans R, Proost P, et al. Severe cachexia in mice innoculated with interferon gamma producing tumor cells. Int J Cancer 1991; 49:77–82.
188. Zugamaier G, Paik S, Wilding G, et al. Transforming growth factor β1 induces cachexia and systemic fibrosis without an antitumor effect in nude mice. Cancer Res 1991; 51:3590–3594.
189. Mori M, Yamaguchi K, Honda S, et al. Cancer cachexia syndrome developed in nude mice bearing melanoma cells producing leukemia inhibitory factor. Cancer Res 1991; 51:6656–6659.
190. Henderson JT, Seniuk NA, Richardson PM, Gauldie J, Roder JC. Systemic administration of ciliary neurotrophic factor induces cachexia in rodents. J Clin Invest 1994; 93:2632–2638.
191. Langstein HN, Doherty GM, Fraker DL, Buresh CM, Norton JA. The roles of interferon gamma and tumor necrosis alpha in an experimental rat model of cancer cachexia. Cancer Res 1991; 51:2302–2306.
192. Matthys P, Heremans H, Opdenakker G, Billiau A. Anti-interferon gamma antibody treatment, growth of Lewis lung tumours in mice and tumour-associated cachexia. Eur J Cancer 1991; 27:182–187.
193. Strassmann G, Fong M, Kenney JS, Jacob CO. Evidence for the involvement of interleukin-6 in experimental cancer cachexia. J Clin Invest 1992; 89:1681–1684.
194. Sherry BA, Gelin J, Fong Y, et al. Anticachectin/tumor necrosis factor alpha antibodies attenuate development of cachexia in tumor models. FASEB J 1989; 3:1956–1962.
195. Costelli P, Carbo N, Tessitore L, et al. Tumor necrosis factor-α mediates changes in tissue protein turnover in a rat cancer cachexia model. J Clin Invest 1993; 92:2783–2789.
196. Tisdale MJ. Newly identified factors that alter host metabolism in cancer cachexia. Trends Pharmacol Sci 1990; 11:473–475.
197. Socher SH, Friedman A, Martinez D. Recombinant human tumor necrosis factor induces acute reductions in food intake and body weight in mice. J Exp Med 1988; 167:1957–1962.
198. Stovroff MC, Fraker DL, Swedenberg JA, Norton JA. Cachectin/tumor necrosis factor, a possible mediator of cancer anorexia in the rat. Cancer Res 1988; 48:920–925.
199. Patton JS, Peters PM, McCabe J, et al. Development of partial tolerance to the gastrointestinal effects of high doses of recombinant tumor necrosis factor alpha in rodents. J Clin Invest 1987; 80:1587–1596.
200. Grunfeld C, Wilking H, Neese R, et al. Persistence of the hypertriglyceridemic effect of tumor necrosis factor despite development of tachyphylaxis to its anorectic/cachectic effect in rats. Cancer Res 1989; 49:2554–2560.
201. Grunfeld C, Feingold KR. Metabolic disturbances and wasting in acquired immunodeficiency syndrome. N Engl J Med 1992; 327:329–337.
202. Hellerstein MK, Meydani SN, Meydani M, Wu K, Dinarello CA. Interleukin-1 induced anorexia in the rat. J Clin Invest 1989; 84:228–235.
203. Langstein HN, Doherty GM, Fraker DL, Buresh CM, Norton JA. The roles of interferon gamma and tumor necrosis factor alpha in an experimental rat model of cancer cachexia. Cancer Res 1991; 51:2302–2306.
204. Sehgal PB. Interleukin-6 in infection and cancer. Proc Soc Exp Biol Med 1990; 193:183–191.
205. Sherwin SA, Knost JA, Fein S, et al. A multiple dose phase I trial of recombinant leukocyte A interferon in cancer patients. JAMA 1982; 248:2461–2466.
206. Vadhan-Raj S, Al-Katib A, Bhalla R, et al. Phase I trial of recombinant interferon gamma in cancer patients. J Clin Oncol 1986; 4:137–146.
207. Strassmann G, Jacob CO, Evans R, Beall D, Fong M. Mecha-

nisms of experimental cancer cachexia: interaction between mononuclear phagocytes and colon-26 carcinoma and its relevance to IL-6 mediated cancer cachexia. J Immunol 1992; 148:3674–3678.
208. Zentella A, Manogue K, Cerami A. Cachectin/TNF mediated lactate production in cultured myocytes is linked to activation of a futile substrate cycle. Cytokine 1993; 5:436–447.
209. Robson HG, Vas SI. Resistance of in-bred mice to *Salmonella typhimurium*. J Infect Dis 1972; 126:378–386.
210. von Jeney N, Gunther E, Jann K. Mitogenic stimulation of murine spleen cells: relation to susceptibility to Salmonella infection. Infect Immun 1977; 15:26–33.
211. O'Brien AD, Rosenstreich DL, Scher I, Campbell G, MacDermott RP, Formal SB. Genetic control of susceptibility to *Salmonella typhimurium* in mice: role of the Lps gene. J Immunol 1980; 124:20–24.
212. Beutler B, Krochin N, Milsark IW, Luedke C, Cerami A. Control of cachectin (tumor necrosis factor) synthesis: mechanisms of endotoxin resistance. Science 1986; 232:977–979.
213. Cross AS, Sadoff JC, Kelly N, Bernton E, Gemski P. Pretreatment with recombinant murine tumor necrosis factor-α/cachectin and murine interleukin-1α protects mice from lethal bacterial infection. J Exp Med 1989; 169:2021–2027.
214. Sheppard BC, Fraker DL, Norton JA. Prevention and treatment of endotoxin and sepsis lethality with recombinant human tumor necrosis factor. Surgery 1989; 106:156–162.
215. Alexander HR, Doherty GM, Fraker DL, et al. Human recombinant interleukin-1α protection against the lethality of endotoxin and experimental sepsis in mice. J Surg Res 1991; 50:421–424.
216. Alexander HR, Wong GH, Doherty GM, et al. Differentiation factor/leukemia inhibitory factor protection against lethal endotoxemia in mice: synergistic effect with interleukin-1 and tumor necrosis factor. J Exp Med 1992; 175:1139–1142.
217. Libert C, Brouckaert P, Fiers W. Protection by α1-acid glycoprotein against tumor necrosis factor-induced lethality. J Exp Med 1994; 180:1571–1575.
218. Ulevitch RJ, Johnston AR, Weinstein DB. New function for high density lipoproteins. Their participation in intravascular reactions of bacterial lipopolysaccharides. J Clin Invest 1979; 64:1516–1524.
219. VanLenten BJ, Fogelman AM, Haberland ME, Edwards PA. The role of lipoproteins and receptor mediated endocytosis in the transport of bacterial lipopolysaccharide. Proc Natl Acad Sci USA 1986; 83:2704–2708.
220. Navab M, Hough GP, VanLenten BJ, Berliner JA, Fogelman AM. Low density lipoproteins transfer bacterial lipopolysaccharides across endothelial monolayers in a biologically active form. J Clin Invest 1988; 81:601–605.
221. Harris HW, Grunfeld C, Feingold KR, Rapp JH. Human very low density lipoproteins and chylomicrons can protect against endotoxin induced death. J Clin Invest 1990; 86:696–702.
222. Read TE, Grunfeld C, Kumwenda ZL, et al. Triglyceride rich lipoproteins improve survival when given after endotoxin in rats. Surgery 1995; 117:62–67.
223. Read TE, Grunfeld C, Kumwenda ZL, et al. Triglyceride rich lipoproteins prevent septic death in rats. J Exp Med 1995; 182:267–272.
224. Feingold KR, Funk JL, Moser AH, et al. Role for circulating lipoproteins in the protection from endotoxin toxicity. Infect Immun 1995; 63:2041–2046.
225. Flegel WA, Wolpl A, Mannel DN, Northoff H. Inhibition of endotoxin induced activation of human monocytes by human lipoproteins. Infect Immun 1989; 57:2237–2245.
226. Cavaillon JM, Fitting C, Cavaillon NH, et al. Cytokine response by monocyte and macrophage to free and lipoprotein bound lipopolysaccharide. Infect Immun 1990; 58:2375–2382.
227. Harris HW, Grunfeld C, Feingold KR, et al. Chylomicrons alter the fate of endotoxin, decreasing tumor necrosis factor release and preventing death. J Clin Invest 1993; 91:1028–1034.
228. Read TE, Harris HW, Grunfeld C, et al. Chylomicrons enhance endotoxin excretion in bile. Infect Immun 1993; 61:3496–3502.
229. Leong JC, Kane JP, Oleszko O, Levy JA. Antigen specific nonimmunoglobulin factor that neutralizes xenotropic virus is associated with mouse serum lipoproteins. Proc Natl Acad Sci USA 1977; 74:276–280.
230. Seganti L, Grassi M, Matromarino P, Pana A, Superti F, Orsi N. Activity of human serum lipoproteins on the infectivity of rhabdoviruses. Microbiology 1983; 6:91–99.
231. Sernatinger J, Hoffman A, Harmon D, Kane JP, Levy JA. Neutralization of mouse xenotropic virus by lipoproteins involves binding to virons. J Gen Virol 1988; 69:2651–2661.
232. Heumer HP, Menzel HJ, Potratz D, et al. Herpes simplex virus binds to human serum lipoproteins. Intervirology 1988; 29:68–76.
233. Rifkin MR. Identification of the trypanocidal factor in normal human serum: high density lipoproteins. Proc Natl Acad Sci USA 1978; 75:3450–3454.
234. Hajduk SL, Moore DR, Vasudevacharya J, et al. Lysis of trypanosoma bruceii by a toxic subspecies of human high density lipoprotein. J Biol Chem 1989; 264:5210–5217.
235. Terkettaub R, Curtiss LK, Tenner AJ, Ginsberg MH. Lipoproteins containing apoprotein B are a major regulator of neutrophil responses to monosodium urate crystals. J Clin Invest 1984; 73:1719–1730.
236. Grunfeld C, Feingold KR. Tumor necrosis factor, interleukin and interferon induced changes in lipid metabolism as part of host defense. Proc Soc Exp Biol Med 1992; 200:224–227.

27
Cytokines and Liver Disease

Daniel Hill, Steven Shedlofsky, and Craig McClain
University of Kentucky Medical Center, Lexington, Kentucky

Anna Mae Diehl
Johns Hopkins University School of Medicine, Baltimore, Maryland

Hidekazu Tsukamoto
USC Center for Liver Diseases, University of Southern California, Los Angeles, California

INTRODUCTION

The liver plays a central role in the body's defense against external microbial and toxic agents, and it plays a pivotal role in the acute phase response following trauma, stress, or inflammation (1–7). Over the past decade, a wealth of information has emerged concerning the role of cytokines in both experimental and clinical liver disease (1–5). It is not surprising that many types of liver disease are associated with dysregulated cytokine metabolism, as the liver is both a major producer and a scavenger of cytokines (1–3,5). For example, Kupffer cells, the largest reservoir of fixed macrophages in the body, produce inflammatory/acute phase cytokines such as IL-1, TNF, and IL-6 (1–3,5,8,9). Stellate cells produce TGF-β vital for collagen production and fibrosis, hepatocytes produce the neutrophil chemoattractant IL-8, and endothelial cells express adhesion molecules such as ICAM-1 as a component of liver injury (10–14). This complex interaction of paracrine/autocrine functions of cytokines, differing liver cell types, and extracellular matrix plays a vital role in hepatic/total body defenses as well as in liver injury and repair. This chapter will deal with the role of cytokines in experimental and clinical liver disease and the potential role of cytokines and/or anticytokines in therapy for liver disease.

MODELS OF LIVER DISEASE

Experimental Liver Injury

Endotoxin or lipopolysaccharide (LPS) has been implicated in liver damage in a variety of types of experimental liver injury including that induced by toxins such as carbon tetrachloride, nutritional alterations such as choline deficiency, and virus-induced hepatotoxicity (1,2,15,16). Considerable evidence suggests that the hepatotoxic effects of LPS are mediated through cytokines (especially TNF), and it is well documented that LPS is a major stimulus for the production of many cytokines thought to be involved in liver injury (e.g., TNF, IL-6, IL-8, TGF-β) (1,2). The preponderance of research has investigated TNF as a mediator of hepatotoxicity; thus, models of TNF cytotoxicity will be the focus of this review (Table 1). Important in the understanding of TNF-mediated liver injury are the concepts of "sensitization" and "priming" (1,2). Liver cells are normally resistant to the cytolytic effects of TNF. However, they may become susceptible to TNF injury through at least these two mechanisms. "Priming" occurs when TNF-producing cells such as Kupffer cells are conditioned, or primed, to release massive amounts of TNF in response to stimuli that normally cause only minimal TNF release (e.g. small amounts of LPS) (1,2). A widely used in vivo model of priming is that of *Corynebacterium*

Table 1 TNF-Mediated Liver Injury

Agent	Increased serum/hepatic TNF	"Anti-TNF" decreases injury
Galactosamine	X	X
C. parvum	X	X
Ischemia reperfusion	X	X
Lead/LPS	X	X
Carbon tetrachloride	X	X
Polyhalogenated hydrocarbons	X	X
Ethanol	X	X
Acetaminophen	X	X
Cadmium	X	X
Liver transplantation	X	X
Sepsis	X	X
Choline deficiency	X	X
Bile duct ligation	X	
In vitro injury	X	X

parvum liver injury (17). A second important mechanism of TNF-induced hepatic injury involves the concept of sensitization. In this setting, the target cell (hepatocyte) is made exquisitely sensitive to normally innocuous amounts of TNF. Typical methods of sensitization employ the use of RNA or protein synthesis inhibitors such as actinomycin-D or cycloheximide (18). The classic liver injury model of sensitization is that of galactosamine-induced hepatotoxicity, with galactosamine selectively depleting hepatic uridine and UTP stores (19,20). Galactosamine-treated animals are sensitive to normally innocuous amounts of endotoxin or TNF. Liver injury in both of these models can be blocked or attenuated by agents that inhibit cytokine production or by drugs that specifically block TNF activity, e.g., TNF antibody or soluble receptor.

D-Galactosamine

The galactosamine model of liver injury is possibly the most frequently used model of hepatotoxicity (19–29). This agent selectively depletes hepatocytes of uridine nucleotides, with subsequent transcriptional blockade, and this hepatotoxicity can be reversed by administration of uridine. Grun et al. suggested a role for LPS in this model of liver injury when they demonstrated that colectomized rats had significantly less liver necrosis than control rats even though they had equal amounts of uracil nucleotide depletion (30,31). Subsequently, Galanos demonstrated that galactosamine markedly augmented the lethal effects of LPS more than 1000-fold (32). This galactosamine-induced sensitivity to endotoxin was dependent both on the state of the macrophage and the susceptibility of the hepatocyte to injury. Animals that were made LPS tolerant by previous low-dose exposure to LPS (or IL-1 or TNF) were rendered resistant to subsequent toxic effects of LPS/galactosamine (33,34). The mechanism(s) for this tolerance are probably multifactorial and include a relative insensitivity of monocytes/Kupffer cells to cytokine-inducing stimuli as well as hepatic production of acute phase proteins/enzymes such as α_1-acid glycoprotein and manganous superoxide dismutase (MN SOD), which protect against TNF cytotoxicity (24,35–38). Supporting the concept that some Kupffer cell–derived product plays a role in this liver injury, early studies by Diluzio's laboratory demonstrated that inhibition of Kupffer cell function by administration of methyl palmitate protected against galactosamine hepatotoxicity (39). A role for macrophage/Kupffer cell–derived cytokines was further demonstrated when it was shown that endotoxin-resistant mice were relatively insensitive to galactosamine liver injury (40). These resistant mice could then be rendered sensitive by transfer of macrophages or bone marrow from LPS-sensitive animals. Lehman et al. were the first to clearly define the involvement of TNF as a mediator of hepatotoxicity/lethality in the LPS/galactosamine model (41). They showed that high doses of recombinant human TNF caused lethality in mice in a time course similar to that of LPS, and TNF-induced lethality occurred in both endotoxin-sensitive and endotoxin-resistant mice. They then showed that galactosamine pretreatment markedly sensitized to TNF toxicity, with TNF lethality similar to endotoxin lethality (41). Teigs et al. demonstrated that TNF caused a dose-dependent hepatotoxicity in galactosamine-treated mice, that this could be prevented by uridine salvage, and that TNF was a "terminal mediator" of this hepatotoxicity (20). Studies from this same laboratory also showed in the isolated perfused liver system that mice initially treated with galactosamine/endotoxin or galactosamine/TNF in vivo developed massive liver injury and that this could be prevented by infusion of uridine into the perfused liver (26). Hishinuma et al. demonstrated unequivocally the role of TNF and LPS/galactosamine liver injury (42). They showed that LPS/galactosamine-treated mice were not protected against hepatotoxicity by nonspecific antiserum, but that anti-TNF antibody protected against hepatotoxicity in a dose-dependent fashion, and this has been confirmed by other investigators (42,43). Apoptosis precedes liver failure in the galactosamine model of hepatotoxicity, and this is one likely mechanism of liver injury (43). In summary, this is a model of hepatotoxicity in which there are only modest increases in hepatic TNF mRNA, but the hepatocyte is made exquisitely sensitive to TNF cytotoxicity by uridine depletion, with hepatotoxicity being reversed in vivo by uridine repletion or anti-TNF antibody (20,28).

Corynebacterium parvum

Administration of this bacteria creates a model of fulminant liver injury in which monocytes/macrophages/Kupffer cells are primed to release massive amounts of TNF and other acute phase/inflammatory cytokines such as IL-1, IL-6, and interferon-gamma (IFN-γ) in response to minimal amounts of stimuli such as LPS (17,44,45). In this model, heat-killed *C. parvum* are injected into animals. Approximately 1 week later small amounts of LPS are injected, which results in high serum TNF concentrations, severe hepatotoxicity, and possible lethality. Lethality has been blocked and hepatotoxicity attenuated by administering gadolinium chloride (inhibits Kupffer cell activity) by administering agents that downregulate TNF production such as prostaglandin analogs or glucocorticoids or by treatment with anti-TNF antibody (17,45,46). In contrast, treatment with neutralizing antibodies to IL-1α, IL-1β, and IL-6 rendered no protection against this liver injury (45). Moreover, as opposed to the galactosamine model of hepatotoxicity in which pretreatment with IL-1β or TNF protects against liver injury, neither of these agents attenuated liver injury in this model. Indeed, TNF "pretreatment" prior to LPS administration induced rapid mortality, presumably because macrophages were primed to release TNF in response to not only LPS, but also exogenous TNF (45). TNF appears to be a proximal mediator not only of hepatocyte injury, but also of early sinusoidal endothelial cell injury in this model (46). TNF has been shown to modulate nitric oxide synthetase activity in this model of liver injury, and studies from Billiars' group suggested that NO induction protects against hepatotoxicity in this model (47–49).

Carbon Tetrachloride

Carbon tetrachloride is a classic hepatotoxin that is dependent on bioactivation via cytochrome P450 with subsequent generation of free radicals and injury (50). Endotoxin has been long recognized to play a role in carbon tetrachloride liver injury. Endotoxin administration enhances this hepatotoxicity, and multiple agents known to block endotoxin were shown by Nolan et al. in the 1970s to attenuate carbon tetrachloride liver injury (51). Moreover, this animal model is regularly associated with endotoxemia (52). Several investigators recently reported that this model is associated with increased production of cytotoxic cytokines such as TNF-α (28,53,54). Mason's laboratory showed that gadolinium chloride, which blocks Kupffer cell function, attenuated liver injury (and TNF production) in rats treated with carbon tetrachloride plus LPS, while inhibition of nitric oxide generation with L-NMA increased both serum TNF level and liver injury (54). Studies from Czaja's laboratory reported that the endotoxin-neutralizing antibody E5 (a monoclonal IgM antibody directed against the core lipid A region of bacterial endotoxin) attenuated carbon tetrachloride liver injury (53). Moreover, these same investigators subsequently demonstrated that TNF-soluble receptor significantly reduced hepatotoxicity and lethality in carbon tetrachloride–treated rats (55). At the same time, liver regeneration was not inhibited. Finally, studies from Zern's laboratory showed that carbon tetrachloride liver injury was associated with activation of the oxidative stress sensitive transcription factor NFκB (56). This transcription factor has been shown to enhance the expression of cytotoxic cytokines such as TNF. Pretreating mice with water-soluble vitamin E markedly attenuated both liver injury and NFκB activation. They speculated that vitamin E may be a treatment option for inhibiting inflammatory cytokine production in this and other forms of cytokine-mediated liver injury.

LPS/Lead

Lead acetate, similar to galactosamine, causes markedly enhanced LPS lethality in rats (57). This finding has been confirmed in other animal species including mice, chicks, and baboons (58–60). The liver appears to be a major target for this lead-enhanced LPS toxicity. Honchel et al. demonstrated that lead markedly augmented both LPS- and TNF-mediated hepatotoxicity in rats (61). Lead not only "sensitized" the liver to TNF cytotoxicity, but "primed" the liver for enhanced TNF production (61). Thus, isolated Kupffer cells from lead-treated rats induced much more TNF in response to LPS than did control Kupffer cells. Similar to the *C. parvum* model of priming, this LPS/lead-induced liver injury was attenuated by pretreatment with a prostaglandin analog, dexamethasone, and anti-TNF antibody (62). This system appears to be a simple reproducible model of priming and sensitization related to TNF hepatotoxicity. How lead sensitizes to hepatotoxicity is unclear. It does not appear to impair clearance of LPS or inhibit RNA or protein synthesis, but it does interfere with several hepatic biochemical pathways and could enhance TNF-induced oxidative stress (63–66).

Polyhalogenated Hydrocarbons

A role for TNF in the toxicity of polyhalogenated aromatic compounds such as 2,3,7,8-tetrachlorodibenzo-*p*-dioxin (TCDD) (67,68) and related polyhalogenated biphenyls (69) has been demonstrated in animal models. The idea that TNF might be important in this toxicity originated from observations that polyhalogenated aromatic intoxicated animals developed a "wasting" syndrome with profound anorexia similar to TNF "cachexia." Although basal circulating TNF concentrations are not elevated in these models, the animals have a markedly enhanced serum TNF response to LPS administration with subsequent liver injury. In rats, severe liver injury only occurs if the LPS is given shortly after the polyhalogenated aromatic exposure (69), but the increase in LPS-induced TNF persists. In mice, the increase in LPS-induced TNF is associated with activation of the aryl-hydrocarbon (Ah) receptor and induction of CYP1A1 (67). TCDD administration to mice causes a mild nonfatal liver injury, along with eventual wasting and a 70% mortality at 24 days. By giving anti-TNF antibody every other day, these changes were partially reversed (mortality decreased to 30%), while dexamethasone treatment completely blocked mortality (68).

Ischemia Reperfusion

The critical role of the Kupffer cell in ischemia reperfusion was evidenced by the fact that gadolinium chloride suppression of Kupffer cell function attenuated this injury, while activation of Kupffer cells with latex particles aggravated it (70,71). This injury is associated with markedly increased serum TNF concentrations, as shown originally by Colletti and coworkers (72). TNF antiserum abrogated this liver injury (72). Other agents such as FK 506 and antioxidants that downregulate TNF production also decreased liver injury and mortality in this model (71,73). It has been suggested that there may be two phases of ischemia reperfusion injury. An early phase is neutrophil independent but is attenuated by inhibition of TNF. A second phase is felt to be due to recruitment of neutrophils with subsequent injury (74,75). Research from Remick's laboratory suggested that a combination of anti-TNF and antioxidants may be optimal therapy for this combined injury (76).

Ethanol

TNF and TNF-inducible cytokines such as IL-6 and IL-8 have been linked to several alcohol-related models with liver injury. Bhagwendeen and coworkers reported that rats chronically fed the liquid Lieber DeCarli ethanol diet were more susceptible to LPS hepatotoxicity than isocaloric controls, and this has been confirmed by several investigators (77,78). Studies from our laboratory demonstrated that rats chronically fed alcohol had increased serum TNF concentrations and liver injury in response to an LPS stimulus compared to isocaloric controls (79). Moreover, pretreatment with a prostaglandin E_1 analog attenuated both the liver injury and serum TNF response (79). Similarly, Kupffer cell TNF and IL-6 mRNA were increased in LPS-injected animals fed the Leiber DeCarli ethanol diet compared to isocaloric control rats (80).

Research from Thurman's laboratory showed that liver enzyme abnormalities in rats fed an ethanol diet were abrogated by gut sterilization (81). These same investigators demonstrated that treatment with gadolinium chloride also reduced liver injury induced by ethanol administration (82). Thus, liver injury was inhibited by the decreasing sources of endotoxin and cytokines.

The animal model that best simulates human alcoholic liver disease is the intragastric feeding model developed by Tsukamoto and French (83). This model overcomes the rats' natural aversion to ethanol, and these animals develop endotoxemia associated with their liver disease (84,85). To augment the liver injury that occurs, the diet can be enriched in polyunsaturated fatty acids and iron (83,86–88). Preceding the development of prominent histological liver injury is a depletion of hepatic mitochondrial glutathione (GSH) (89,90). GSH is not made in the mitochondria. Thus, it is presumed that chronic ethanol intake impairs mitochondrial GSH transport from the cytosol (89,90). These animals have increased hepatic levels of P450 2E1, which potentially increases hepatic oxidative stress (91,92). Nanji and coworkers demonstrated that the development of liver injury coincided with increased hepatic TNF mRNA (93). His group was able to inhibit liver injury in this model by feeding lactobacillus in the diet, thus altering gut flora and suppressing endotoxemia (94). Nanji also showed that rats that developed liver injury had depressed manganous SOD mRNA (95). The depression of this mitochondrial form of SOD is paradoxic in that TNF markedly induces this enzyme (96). This represents another mitochondrial defect in ethanol-fed animals, which may render them more susceptible to oxidative stress, an important mechanism of TNF cytotoxicity. Tsukamoto et al. demonstrated that Kupffer cells taken from rats that were intragastrically fed alcohol had not only increased

TNF mRNA, but also spontaneous release of TNF, which was not observed in isocaloric control rats (97). These same rats had increased TGF-β, which presumably played a role in the observed fibrosis, as discussed subsequently.

In summary, rats chronically fed ethanol appear more susceptible to LPS hepatotoxicity, and liver injury is attenuated by agents that inhibit cytokine production. In the intragastric alcohol feeding model, the development of injury coincides with the increased expression of hepatic TNF and is preceded by a defect in mitochondrial function and diminished mitochondrial GSH concentrations.

Other Forms of In Vivo Experimental Liver Injury

Several other nutritional/metabolic/medicinal/toxic forms of liver injury are associated with dysregulated TNF metabolism. Bile duct ligation in experimental animals creates a reproducible form of acute liver injury with subsequent fibrosis, and this model is associated with increased serum/macrophage TNF concentrations (98). Hepatic transplant rejection/failure is associated with elevated TNF levels, and liver dysfunction can be attenuated by anti-TNF antibody or drugs that downregulate TNF production (99–101). Acetaminophen liver injury is associated with Kupffer cell activation and increased TNF production (102,103). This liver injury can be markedly attenuated by gadolidium chloride treatment and can be delayed by anti-TNF antibody administration (103,104). Administration of the toxic metal cadmium to experimental animals induces increased hepatic TNF mRNA and secreted protein (105). Anti-TNF antibody attenuated the acute phase response and hepatic inflammation in this cadmium model of hepatotoxicity. Rats placed on a choline deficient diet rapidly develop a fatty liver (106). Research from our group indicated that these rats are highly susceptible to LPS hepatotoxicity, develop very elevated serum TNF concentrations, and that the liver injury is abrogated with anti-TNF antibody. Lastly, sepsis-related liver dysfunction is associated with increased serum/hepatic TNF, and lethality and liver injury are reduced with TNF inhibitors (107–113).

In Vitro Hepatocyte Injury

Cytokine cytotoxicity has been demonstrated in isolated rodent hepatocytes, in co-cultures of rodent hepatocytes with Kupffer cells, and in Hep G2 cells (18,27,114,115). Billings' laboratory demonstrated in mouse hepatocytes that TNF caused liver injury with GSH depletion, that IFN-γ acted synergistically to augment this toxicity, and that the antioxidant mannitol attenuated toxicity (114). Studies using rat Kupffer cells and hepatocytes from Wendel's laboratory showed that the co-cultures secreted TNF following LPS stimulation with resultant hepatocyte injury (27). TNF antiserum attenuated this cytotoxicity, but did not completely block hepatocyte killing. Thus, TNF appeared to be an obligatory, but not totally sufficient condition for hepatocyte killing. Further studies from this laboratory showed that wild-type mouse hepatocytes exposed to TNF developed direct hepatotoxicity that was preceded by DNA fragmentation (115). However, in mouse hepatocytes lacking the gene for the 55 kDA TNF receptor, injury could not be elicited. This study suggested that the TNF-R1 was both necessary and sufficient for TNF signaling/cytotoxicity in murine hepatocytes. Studies from our laboratory using isolated Hep G2 cells showed that this human hepatoma cell line could be sensitized by both actinomycin D or cycloheximide to dose-response TNF-mediated cytotoxicity (18). This cytotoxicity was attenuated by pretreatment with low doses of TNF (inducing in vitro tolerance) and by vitamin E. Injury was markedly augmented by antimycin A, an inhibitor of mitochondrial respiration, which increased oxidative stress. Thus, both rodent and human systems strongly suggest a role for in vitro TNF cytotoxicity.

In summary, the vast majority of models of in vivo experimental liver injury currently used are associated with increased serum/hepatic TNF concentrations and, in most models studied, anti-TNF antibody or agents that downregulate TNF production attenuated liver injury. Likewise, in vitro studies in both rodent and human hepatocytes support a role for TNF hepatocyte injury.

Mechanisms of TNF Cytotoxicity

Cytokines may cause hepatotoxicity indirectly through the actions of other cells or directly after binding to specific cytokine receptors on the hepatocyte to generate toxic intracellular signals. As an example of an indirect mechanism of cytotoxicity, cytokines (TNF and IL-1) from macrophages/Kupffer cells can stimulate neutrophils to release toxic metabolites, such as $OH^{-\cdot}$, $O_2^{\cdot}$, nitric oxide, and proteases. The killing that results from this release is exacerbated by the fact that TNF also can stimulate hepatocytes to produce IL-8, an attractant for neutrophils (12,116). The resident macrophage/Kupffer cell also can release substantial amounts of superoxide in response to factors such as LPS or alcohol, which could cause liver injury (117,118). Thus, neutrophil infiltration and/or activation of monocyte/Kupffer cells with release of toxic substances is one likely mechanism of liver cell injury.

Oxidant injury (especially mitochondrial injury) is one putative direct mechanism of TNF-mediated cytotoxicity (119–123). TNF causes morphological abnormalities in the mitochondria and inhibits hepatic mitochondrial respiration (122). Inhibition of normal mitochondrial respiration appears to sensitize to TNF cytotoxicity, and disrupted mitochondrial function is well documented in certain forms of liver injury such as ALD and acetaminophen hepatotoxicity (123–126). TNF normally induces a variety of protective proteins, which prevent further cellular destruction by TNF (e.g., mitochondrial MnSOD) (119). Transfected cells that overexpress MnSOD are protected against TNF toxicity, while cells with antisense manganous SOD are sensitive to TNF toxicity (119). Thus, there is an emerging tight interrelation between reactive oxygen intermediates and both activation of the oxidative stress sensitive transcription factor NFκB (with subsequent TNF production) and TNF cytotoxicity including mitochondrial oxidant injury. Apoptosis appears to be one important pathway for TNF-mediated cell death (43,127). TNF in nonhepatic cell types has been shown to act through several classic signaling pathways, with the ceramide pathway appearing to be a likely route of cytotoxicity (128–130). Dissecting the multiple potential pathways of cytokine cytotoxicity represents an important area of future investigation both for improved understanding of cell injury and for development of therapeutic intervention.

Other Cytokines

Neutrophil chemotactic peptides (IL-8) or factors that influence neutrophil function are probably the most frequently cited "other cytokines" thought to be important in the pathogenesis of experimental liver injury. In alcohol-related injury, Perez et al. initially described a chemotactic factor released from cultured rat hepatocytes exposed to ethanol (131). Shiratori et al. described a different chemotactic factor derived from alcohol-exposed rat hepatocytes subsequently shown to be a rodent counterpart to IL-8 (132). Antibody to this cytokine attenuated neutrophil chemotaxis. Overexpression of rodent IL-8 (CINC) recently has been shown to cause a neutrophilic hepatitis (Maher, personal communication). CINC levels increase in ischemia reperfusion injury, and levels decrease with gadolinium chloride treatment (133). Moreover, antibody to CD11b/CD18 on the surface of neutrophils significantly attenuated hepatic neutrophil infiltration and liver injury induced by endotoxin or by ischemia reperfusion (71,134,135). Similarly, monoclonal anti–ICAM-1 antibody was reported to attenuate significantly ischemia reperfusion liver injury (136). Thus, there is considerable evidence that agents that block neutrophil activation, adherence, or infiltration reduce liver injury, especially a second or later phase of liver injury.

IFN-γ has been shown in several animal models to act synergistically with TNF in inducing hepatocyte hepatotoxicity (45,114). Platelet-activating factor (PAF) is increased in some experimental forms of liver injury such as bile duct ligation, but data are much more limited for PAF than TNF (137). Extensive data suggest a role for TGF-β in hepatic fibrosis, but studies on liver injury are limited. IL-1 and IL-6, while frequently elevated in experimental forms of liver injury, generally have not been shown to induce hepatocyte injury (45). Indeed, most studies using antibodies to these cytokines have not shown beneficial effects in preventing liver injury.

Cytokines and Cholestasis

Jaundice is a well-recognized complication of infection/trauma/liver disease (138), and LPS has been shown to inhibit bile flow in isolated perfused livers (139). Following the recognition that cytokines mediated many of the effects of LPS, studies using conditioned medium from LPS-stimulated monocytes (containing multiple cytokines) also demonstrated inhibition of bile flow in isolated perfused livers (140). TNF was implicated as an etiological factor in LPS-induced cholestasis in experiments using rats with biliary drainage tubes. In this animal model, anti-TNF antibody prevented the LPS-induced decreases in bile flow and bile salt excretion (141). In the same study, isolated hepatocytes exposed either to TNF or LPS had decreased taurocholate uptake (the inhibition was noncompetitive). IL-6 also inhibited taurocholate uptake and depressed Na^+/K^+-ATPase activity (142), which drives biliary excretion. Further studies by these investigators have shown that expression of the gene for the Na^+-dependent taurocholate cotransporter at the mRNA level in mice was depressed after in vivo TNF and IL-1 administration, but not IL-6. In these same studies, LPS given to rats decreased expression of the taurocholate cotransporter at both the mRNA and protein level and depressed activity of the ATPase. However, this latter effect was posttranslational because there were no decreases in mRNA or protein concentration for the ATPase. In summary, considerable experimental data suggest a role for cytokines in the cholestasis relating to liver disease, trauma, and inflammation.

Cytokines and Liver Regeneration

The healthy liver has tremendous regenerative capacity. This is surprising since hepatocytes rarely proliferate in the normal adult liver. Labeling studies with tritiated thymidine demonstrate that fewer than one in 1000 hepatocytes is replicating at any point in time and suggest that the average life span of an adult hepatocyte is approximately a year. Although mature hepatocytes normally spend most of their life in a growth-arrested (G-0) state, they retain the capacity to proliferate. Indeed, after significant liver injury, virtually all the surviving hepatocytes will enter prereplicative stages of the cell cycle, synthesize DNA, and replicate (143,144). The amazing proliferative capacity of mature hepatocytes is well illustrated by the regenerative response evoked by 70% partial hepatectomy (PH). In the rat, waves of hepatocyte proliferation occur, which restore the original liver mass within 3 days (145). Similar responses to hepatic resection have been documented in humans (143,144).

Studies in partially hepatectomized rats indicate that liver regeneration requires a "reprogramming" of hepatocyte gene expression to upregulate both proliferation-specific genes and genes that perform liver-specific functions. This occurs because extracellular factors interact with specific receptors on the surface of hepatocytes and initiate signals that lead to increased transcription, translation, or posttranslational activation of these genes (Fig. 1). Many classes of hepatocyte receptors appear capable of modulating proliferation (146). Growth stimulatory actions clearly follow activation of receptor tyrosine kinases by classical mitogenic cytokines, such as epidermal growth factor (EGF), transforming growth factor alpha (TGF-α), and hepatocyte growth factor (HGF). Activation of receptors that lack endogenous tyrosine kinase activity can also influence proliferation. For example, activation of various transmembrane-spanning receptors (e.g., adrenergic receptors), which couple to heterotrimeric G proteins, can either stimulate or inhibit hepatocyte proliferation. Recently, it has also become apparent that activated cytokine receptors that recruit cytosolic tyrosine kinases to the membrane (e.g., growth hormone and prolactin receptors) can also promote proliferation. Given emerging evidence that a large number of extracellular signals can initiate proliferative signals, it is puzzling that hepatocytes seldom proliferate in vivo. This emphasizes the importance of factors that regulate the exposure and/or sensitivity of hepatocytes to mitogens.

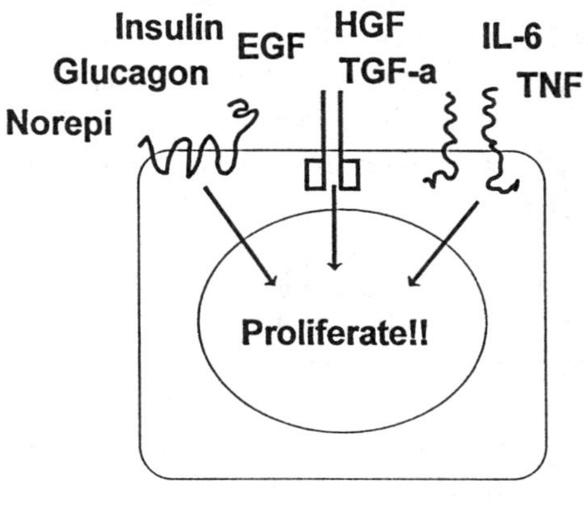

Figure 1 A number of different extracellular factors (EGF, TGF-α, hepatocyte growth factor, insulin, glucagon, norepinephrine, IL-6, TNF-α) can interact with specific receptors on the surface of hepatocytes to initiate signals that promote hepatocyte proliferation.

A growing body of information suggests that one of these factors is the proinflammatory cytokine tumor necrosis factor alpha (TNF-α). Reports that treatments that prevent gut-derived endotoxemia inhibit liver regeneration after PH (147) provided the first clue that TNF might be hepatotrophic. Subsequent evidence that hepatocyte proliferation and liver mass increase in rats after intravenous administration of human recombinant TNF added support for the theory (148,149). Recently, increases in hepatic TNF mRNA levels and circulating concentrations of TNF have been documented almost immediately after PH. TNF-α transcripts begin to increase in the liver within 5 minutes and peak within 30–60 minutes (150); serum TNF-α concentrations increase 4- to 5-fold within an hour of PH (151).

The most compelling evidence that TNF-α can be hepatotrophic is provided by studies with neutralizing anti-TNF antibodies. Treatment of rats with antibodies to TNF an hour before PH inhibits hepatocyte DNA synthesis for 72 h after PH (152). Furthermore, these antiproliferative effects are preceded by inhibition of several events, which normally occur during the prereplicative period that immediately follows PH. For example, induction of Jun nuclear kinase (JNK), a growth-related kinase that activates the c-Jun proto-oncogene, normally peaks within 15–30 minutes of PH (153,154). PH induction of JNK is virtually abolished in rats that have been pretreated with anti-TNF antibodies (154). Similarly, pretreatment with anti-TNF antibodies prevents accumulation of c-Jun protein in the nucleus and inhibits the induction of c-Jun DNA-binding activity which normally occurs during the first 3 hours after PH (154). Since recombinant TNF-α is known to induce JNK and c-Jun in cultured cells (155), it is likely that TNF interacts directly with hepatocytes to induce this proto-oncogene after PH. TNF-α also interacts with its receptors to decrease hepatocyte expression of C/EBPα (156), a constitutively expressed growth-arrest gene (157). Thus, by inducing normally silent growth-permissive genes and downregulating constitutively expressed growth-arrest genes, TNF receptor activation appears to help hepatocytes escape growth arrest and enter the cell cycle (Fig. 2).

TNF may also influence liver cell proliferation indirectly by inducing other factors that regulate the hepatocyte phenotype. For example, mRNA levels of the TNF-inducible cytokine interleukin 6 (IL-6) increase in the liver within 3 hours of PH (151). This is followed by increases in serum IL-6 concentrations (152) and accompanied by increases in hepatic expression of IL-6–inducible genes, such as the transcription factor C/EBPβ (158). Pretreat-

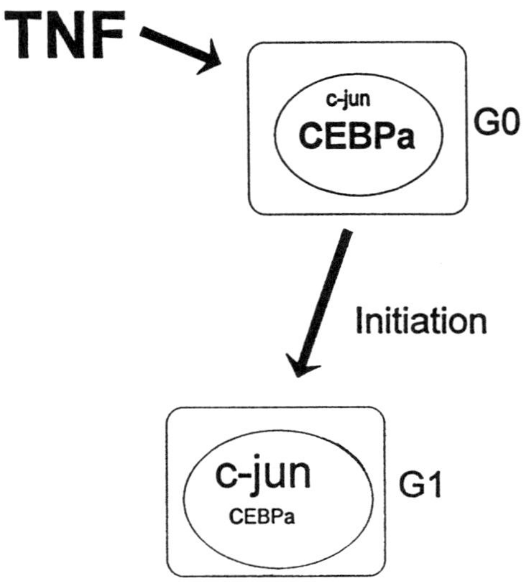

Figure 2 By activating its receptors on the surface of hepatocytes, TNF-α helps hepatocytes enter the cell cycle both by inducing proliferative genes (e.g., c-jun) that are not normally expressed and by downregulating constitutively expressed growth arrest genes (e.g., C/EBPα).

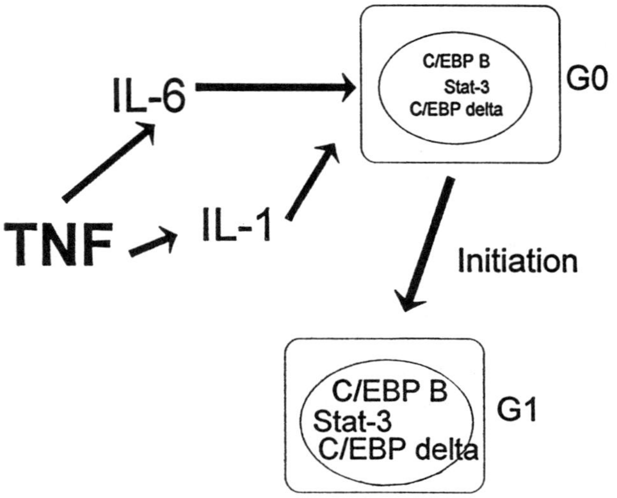

Figure 3 TNF-α–inducible cytokines, including IL-1 and IL-6, activate transcription factors, such as C/EBPβ, C/EBPδ, and Stat 3, which regulate the transcription of many hepatocyte genes. By this mechanism, TNF-α indirectly affects global changes in gene expression, which enable hepatocytes to adapt to new regenerative demands.

ment with anti-TNF antibodies inhibits induction of IL-6 (159) and at least one IL-6–responsive transcription factor (i.e., C/EBPβ) (159). Since IL-6 can also induce STAT-3 in liver (160) and the DNA-binding activity of this transcription factor increases after PH (161), it is likely that TNF indirectly modulates the expression of multiple hepatic genes by activating several IL-6–inducible transcription factors. TNF also stimulates production of other proinflammatory cytokines, such as interleukin-1 (IL-1) (162). This probably explains the induction of IL-1–inducible transcription factors, such as C/EBPδ, which follows PH (158), because pretreatment with anti-TNF antibodies also inhibits regenerative increases in C/EBPδ protein (159). Since the regulatory regions of many hepatocyte genes contain IL-6– and IL-1–responsive elements (162,163), TNF-dependent induction of these cytokines can affect global changes, which reprogram hepatocyte gene expression, permitting the cell to adapt to new regenerative demands (Fig. 3).

Given emerging evidence that IL-1– and IL-6–responsive elements also regulate the transcription of genes expressed by liver nonparenchymal cells, TNF may indirectly influence the phenotypes of these cells to promote other aspects of the regenerative response. In this regard, it is interesting that transcription of hepatocyte growth factor (HGF), a putative product of stellate cells, is regulated by IL-1 and IL-6 (164). We have reported that human recombinant TNF increases the expression of IL-6 and C/EBPβ mRNA in cultured rat stellate cells (165). Messenger RNA levels of HGF increase in the liver a few hours after PH (143,144). Preliminary experiments suggest that treatments that increase hepatic TNF enhance HGF expression, while those that inactivate TNF decrease induction of HGF after PH (Diehl, unpublished observations). If confirmed, these results suggest that TNF may act in a paracrine fashion to promote the local accumulation of hepatocyte mitogens. Since TNF also increases hepatocyte expression of the HGF receptor, c-met (166), this could contribute to its hepatotrophic actions after PH (Fig. 4). Thus, accumulating evidence suggests that TNF may initiate a cascade of events after PH, which both enhances hepatocyte sensitivity to mitogens and increases local production of these mitogens by liver nonparenchymal cells.

While PH has been a particularly useful model to characterize regenerative mechanisms, there has been some debate about the wisdom of extrapolating from results in this model to understand the regenerative response evoked by toxic or viral liver injury. Indeed, there is evidence that the timing of maximal liver DNA synthesis, pattern of proto-oncogene induction, and relative expression of differentiated and fetal hepatocyte genes in the latter models do not match those noted after PH (167). The molecular basis for these discrepancies is uncertain. Some may be explained by inherent differences in the model systems. Unlike toxins, PH does not injure the remnant liver. Thus,

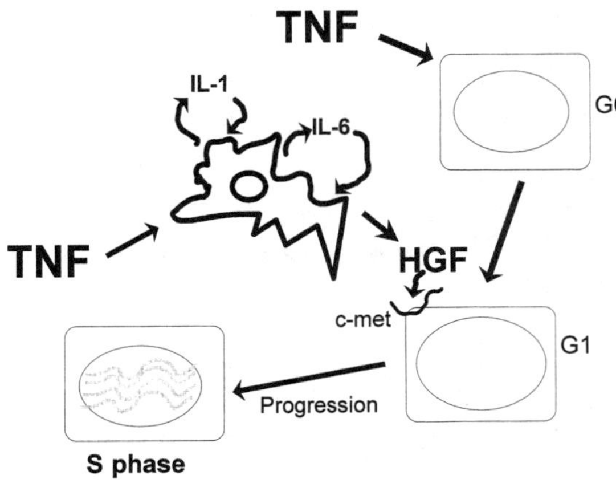

Figure 4 The TNF-α–inducible cytokines IL-1 and IL-6 increase stellate cell transcription of the hepatocyte mitogen HGF. TNF-α also increases the expression of c-met, the HGF receptor, on hepatocytes. Thus, TNF-α increases local production of hepatocyte mitogens and enhances hepatocyte.

after PH, regeneration can be studied in the absence of ongoing hepatocellular necrosis. In addition, since the injury signal (i.e., PH) is extremely circumscribed, it provokes a synchronous proliferative response in the surviving hepatocytes, which peaks 24 h after PH and antedates proliferation in other liver cell populations by at least a day. Finally, this in vivo model of hepatocyte proliferation is the only system available for evaluating the events that regulate synchronous entry of hepatocytes into the cell cycle. Given that mature hepatocytes spend most of their lives in G-0, transition into the cell cycle is a critically important step in liver regeneration. As detailed above, proinflammatory cytokines, including TNF-α, play a major role in facilitating G-0 to G-1 transition in hepatocytes after PH. Additional work will be required to determine if TNF-α performs a similar function during liver regeneration after toxic liver injury. Given that TNF has also been incriminated in the pathogenesis of several types of liver injury, it is critically important to understand which mechanisms drive trophic, as opposed to toxic, responses to this cytokine. These insights should empower us to manipulate the hepatocyte's response to obtain therapeutic benefits in patients with liver disease.

Cytokines, Hepatic Stellate Cells, and Liver Fibrosis

The critical role of the hepatic stellate cell in the development of liver fibrogenesis has unequivocally been demonstrated by numerous experimental (168–173) and clinical (174,175) studies over the past two decades. Several physiological functions have been described for the perisinusoidal cell. The most obvious function is the storage of vitamin A, as illustrated by the fact that as much as 80% of total body vitamin A is stored in the cells (176,177). Second, stellate cells appear to regulate sinusoidal blood flow via their contractile property and receptors for vasoactive agonists (178,179). Third, stellate cells produce collagen IV, laminin, and proteoglycans, which are all integral components of the basement membrane–like matrix in the perisinusoidal space of the normal liver. Through cell-matrix interactions, this matrix supports the maintenance of differentiated phenotypes for hepatocytes, sinusoidal endothelial cells, and stellate cells (180). It is believed that this matrix facilitates a quiescent, nonfibrogenic state of stellate cells (180). Thus, disruption of this normal liver matrix (as can occur in liver injury) is considered as a critical initial event in liver fibrogenesis. This process appears to be mediated by upregulation of gelatinase A (181,182) and stromelysin-1 (183) expression by stellate cells. The former enzyme specifically degrades type IV collagen (181), and the latter possesses degradative activity against all of the components of the basement membrane–like matrix (184). The resultant disruption of the normal matrix is associated with enhanced mobility and proliferation of stellate cells. This is followed by myofibroblastic activation of stellate cells characterized by the following key alterations: (1) upregulation of matrix gene expression (168–171), (2) upregulation of autocrine cytokines (TGF-α, TGF-β, PDGF) and their receptors (180,185–187), (3) acquisition of a myofibroblastic phenotype with smooth muscle α-actin gene expression (187,188), and (4) loss of vitamin A (189–191).

Among these changes, induction of an autocrine loop of TGF-β is considered to be a pivotal and proximal event leading to expression of other activation markers. TGF-β is the most potent cytokine for induction of matrix genes by mesenchymal cells including hepatic stellate cells (192). Upregulation of TGF-β expression by activated stellate cells has been demonstrated in vitro (187,193) and in vivo (186,190,194). The fibrogenic role of this cytokine is evidenced by induction of liver fibrosis in TGF-β transgenic mice (195). TGF-β also induces expression of smooth muscle α-actin (196) and may suppress retinol esterification, a step required for storage of vitamin A by stellate cells (197). This fibrogenic cytokine may also arise via a paracrine route. Kupffer cells have constitutive expression for TGF-β (194) and appear to be a source of this cytokine in experimental alcoholic liver fibrosis (97). Because TGF-β can be autoinduced (193), an autocrine-paracrine interaction may have a potentiating effect on stellate cell activation.

IL-6 is another cytokine with a potential fibrogenic

activity toward hepatic stellate cells. IL-6 stimulates collagen and glycosaminoglycan synthesis by cultured human dermal fibroblasts (198) and upregulates collagen fibronectin mRNA expression by hepatic stellate cell lines that exhibit minimal IL-6 expression (199). This cytokine may also have both autocrine (199) and paracrine routes (97) to exert its fibrogenic effects on stellate cells. Supporting a role for IL-6 in fibrosis is research from Rojkind's laboratory showing that turpentine-induced inflammation induced IL-6 and α1 (I) procollagen mRNA, and this was inhibited by IL-6 antibody (200).

Stellate cell activation is also characterized as a proliferative response, a change that appears to be induced by mitogenic cytokines such as PDGF (201,202), EGF (201), bFGF (201), TGF-α (201,202), IGF-1 (202), IL-1 (192,201), and TNF-α (192,201). The paracrine sources of these cytokines include Kupffer cells (97), platelets (203), hepatocytes (204), lymphocytes, and endothelial cells. Recent evidence suggests that activated stellate cells may express some of these mitogenic cytokines such as TGF-α (187), IGF-1 (205), and PDGF (206). Furthermore, a host of proinflammatory mediators may be expressed by the activated cells including MCP-1 (207), M-CSF (208), and ICAM (209). This suggests that hepatic stellate cells, upon activation, may recruit monocytes and neutrophils to promote inflammatory and consequently scarring processes.

An increasing body of evidence incriminates oxidative stress as a critical signal for liver fibrogenesis. Lipid peroxidation aldehydes may directly upregulate collagen gene expression by stellate cells (210) as shown for fibroblasts (211) even though their effects have not been reproduced consistently (212). A recent study, however, demonstrates that treatment of quiescent hepatic stellate cells with malondialdehyde or ascorbate and iron (a condition that results in enhanced oxidative stress) caused increased cell proliferation and expression of smooth muscle α-actin, and these changes were blocked by antioxidants (10). Interestingly, TGF-α treatment or culturing the cells on type I collagen also resulted in the same effects. Stimulation of hepatic stellate cells by these conditions was accompanied by enhanced binding activities to a smooth muscle α-actin promoter by NFκB and c-myb, redox-sensitive nuclear factors. Oxidative stress may also have an indirect effect on hepatic stellate cell activation via induction of NFκB-responsive genes such as TNF-α, IL-6, and IL-8. Expression of the former two cytokines by Kupffer cells is induced in liver injury and can be suppressed by administration of antioxidants (213,214) (Fig. 5). IL-8 is expressed by hepatocytes (132) and Kupffer cells (215), and it can promote fibrogenesis via recruitment of neutrophils with their subsequent degranulation and respiratory burst.

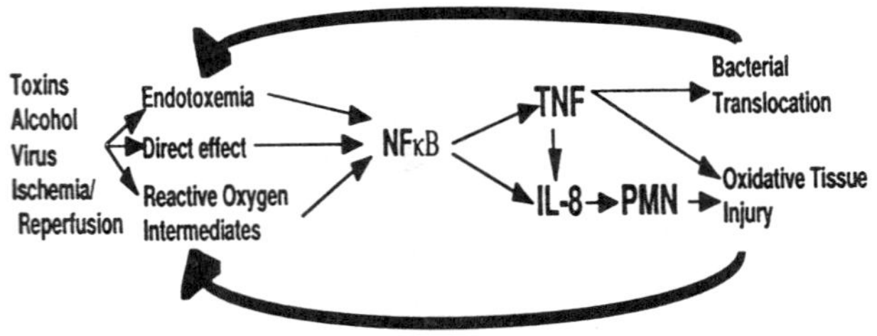

Figure 5 Multiple factors such as toxins or ischemia/reperfusion activate reactive oxygen-sensitive transcription factors such as NFκB with subsequent proinflammatory cytokine production. Antioxidants may decrease cytokine production and attenuate liver injury.

In summary, substantial evidence implicates cytokines as the major mediators for activation of hepatic stellate cells in liver fibrogenesis. Both autocrine and paracrine modes of action are present and may create an intricate interplay of signaling networks to achieve specific cellular and molecular changes, which characterize myofibroblastic activation. Oxidative stress seems a pivotal signal for either direct stimulation of stellate cells or indirect activation through upregulation of cytokines and proinflammatory mediators.

CYTOKINES IN CLINICAL LIVER DISEASE

Alcoholic Liver Disease

Alcoholic liver disease (ALD) is more prevalent in the developed world than all other forms of chronic liver disease combined (216). There also have been more reports of abnormal cytokine metabolism in ALD than in any other type of liver disease (4). Investigators originally became intersted in the potential role of cytokines in ALD when it became apparent that many of the metabolic actions of cytokines such as fever, neutrophilia, anorexia, hypoalbuminemia, and muscle wasting were frequent clinical complications of ALD (Table 2) (2). The initial reports on cytokine activity in patients with alcoholic hepatitis (AH) were of serum IL-1 (the only cytokine defined at the time) utilizing the thymocyte proliferation assay (217). It is now known that this bioassay detects both IL-1α and IL-1β as well as IL-6. Since that time, several groups have reported increased plasma IL-1 concentrations in patients with AH using immunoassays (218,219). Most groups reported increased IL-1β (218,219). It has also been reported that monocytes for patients with ALD secrete more IL-1 in response to LPS than monocytes from healthy controls (220).

Because IL-1 by itself did not appear to cause liver injury and TNF appeared to be hepatotoxic in experimental

Table 2 Biological Activities of Cytokines

Cytokine effects	Metabolic complications of alcoholic hepatitis
Fever	Fever
Anorexia	Anorexia
Neutrophilia	Neutrophilia
Altered amino acids, decreased glutathione, catabolism with muscle wasting	Altered amino acids, decreased glutathione, catabolism with muscle wasting
Hypermetabolism	Hypermetabolism
Decreased serum zinc	Decreased serum zinc
Increased acute phase reactants	Increased acute phase reactants
Decreased bile flow	Cholestasis
Decreased albumin	Decreased albumin
Bone loss	Bone loss
Collagen deposition	Hepatic fibrosis
Increased triglycerides	Increased triglycerides (Zieve syndrome)
Increased endothelial permeability	Ascites/peripheral edema
Slow wave sleep	Encephalopathy

models of liver injury, investigators turned to the possible role of TNF in ALD (1). We initially demonstrated that monocytes from AH patients had increased basal and LPS-stimulated TNF production (221), and this has been confirmed by others. Several investigators reported increased plasma TNF concentrations that correlated with increased mortality and/or liver dysfunction (218–220,222). Felver et al. measured plasma TNF concentrations in 23 patients with severe AH (219). The patients who had elevated TNF, either on admission or at the time of discharge 30 days later, had an 82% mortality. Those patients who did not have elevated TNF had 100% survival. Similarly, Bird et al. found that almost all patients with severe AH had elevated TNF plasma levels, that patients who subsequently died had higher plasma TNF levels than those who survived, and that TNF levels significantly correlated with serum bilirubin and creatinine values (222).

There is increased synthesis of hepatic acute phase reactants in ALD, and thus it seemed important to investigate the role of IL-6 in ALD (223,224). Serum IL-6 was initially shown by Deviere et al. to be elevated in ALD (225). The IL-6 concentrations correlated with factors known to be stimulated by IL-6 such as immunoglobulin A (IgA). Not only does IL-6 stimulate IgA production, but IgA also stimulates both IL-6 and TNF production by human monocytes (226,227). In ALD, the liver sinusoid is lined with IgA, and plasma IgA concentrations are increased (228). Thus, IgA may provide a local stimulus for TNF and IL-6 production in a positive feedback loop (226). This group also showed that monocytes from patients with ALD produced more IL-6 than monocytes from controls and that this increased IL-6 production was reversed by allowing the cells to rest in vitro for several days. They suggested that the monocytes were "primed" to release cytokines and that the priming "factor" may be some substance such as endotoxin found in the serum of ALD patients. We reported increased IL-6 bioactivity in patients with AH followed over a 6-month period (224). The IL-6 levels correlated with the clinical and biochemical course of the disease including serum bilirubin and markers of the hepatic acute phase response such as C-reactive protein. Thus, plasma IL-6 concentrations appear to correlate well with the clinical severity and course of alcoholic hepatitis. IL-6 also increases the expression of TNF receptors on hepatocytes and thus may enhance the effects of TNF (229).

In addition to the increased plasma and monocyte TNF, IL-1, and IL-6, there is increased hepatic immunohistochemical staining for these cytokines as well. We observed increased staining for all three cytokines in liver biopsies from 20 patients with moderate to severe AH (230). There was increased staining on bile duct epithelium, Kupffer cells, vascular structures, and hepatocytes. Ohlinger et al. also noted hepatocyte staining for TNF as well as increased ICAM-1 and CD11b staining from AH biopsies (231).

Hepatic neutrophil infiltration and neutrophilia are principal manifestations of alcoholic hepatitis (232). IL-8 mediates neutrophilia, neutrophil chemotaxis, and infiltration into tissues (233). Kunkel's group showed that hepatocytes secrete IL-8 in response to TNF (12,116). Together, these factors suggest an important role for IL-8 in the pathogenesis of AH. Sheron et al. reported that plasma IL-8 levels were highly elevated in patients with AH and correlated with the severity of disease (234). In contrast, only moderate elevations were found in patients with alcoholic cirrhosis or alcoholic patients undergoing alcohol withdrawal. They also found that liver tissue levels of IL-8 correlated with the degree of neutrophil infiltration. We reported serial plasma IL-8 concentrations in 40 patients followed over 6 months (235). We also found extremely elevated initial plasma IL-8 levels in AH patients, which remained moderately elevated throughout the 6-month period. We found only moderately elevated plasma IL-8 levels in alcohol-dependent patients without liver disease. These studies suggest that IL-8 plays an early role in the pathogenesis of hepatic neutrophil infiltration in ALD and that perhaps neutrophils are responsible, at least in part, for the sustained liver injury that can persist for months despite cessation of alcohol intake.

The mechanism(s) for dysregulated cytokine production in ALD are not well defined and are probably multifactorial (2). Alcohol, per se, does not appear to be a stimulus for monocyte production of proinflammatory cytokines (236). However, acute ethanol consumption does increase intestinal permeability, which may lead to increased endotoxemia (237), and patients with chronic liver disease frequently have endotoxemia (238). In addition, chronic alcohol intake may enhance cytokine production by increasing oxidative stress. Alcohol metabolism generates reactive oxygen intermediates (239). Moreover, acetaldehyde, a product of ethanol metabolism, is very reactive and can interact with proteins to produce neoantigens, which induce an immunological response (240). Polymorphoneutrophilic leukocytes which produce reactive oxygen species are attracted to the liver by the acetaldehyde-induced neoantigens (241). Acetaldehyde also promotes the attack of polyunsaturated acids in cell membranes by free radicals through lipid peroxidation (240). Deficiencies of nutritionally derived free radical scavengers such as reduced glutathione (GSH), vitamin E and selenium are associated with chronic alcohol abuse (242–244). Data from Ghezzi's laboratory suggest that GSH may be an important regulator of TNF production, with GSH inhibiting TNF production (111). Recently, the importance of oxidative stress–sensitive transcription factors for cytokine production has been emphasized. NFκB is an oxidative stress transcription factor for cytokines such as TNF, IL-1, IL-6, and IL-8 (127,245). We have preliminary data showing increased NFκB activation in monocytes from patients with AH. Other factors may be important as well, such as decreased prostaglandin (PGE_2) production by monocytes from AH patients, which may be partially responsible for ineffective endogenous downregulation of cytokine production in AH (246). Also important may be inadequate production of inhibitory cytokines such as IL-10 (247). Deviere's group studied monocytes from 13 patients with alcoholic cirrhosis (247). They found that LPS-stimulated monocytes from alcoholic cirrhotics had decreased IL-10 production and increased TNF production when compared with monocytes from normal controls. In addition, they found that the monocytes from the cirrhotic patients responded normally to recombinant IL-10, which induced a dose-dependent decrease in TNF secretion by the monocytes. Moreover, preincubating control monocytes with monoclonal anti–IL-10 antibody significantly increased TNF production, but the monoclonal anti–IL-10 antibody did not increase TNF production by monocytes from the patients who had alcoholic cirrhosis. Thus, inadequate amounts of inhibitors (e.g., IL-10) of proinflammatory cytokines may play a role in the increased TNF activity of ALD.

Viral Hepatitis

Current data do not suggest major differences in cytokine metabolism due to differing species of infecting virus in viral hepatitis. In this section, abnormal cytokine metabolism resulting from hepatitis A, B, and C viral infection will be discussed. The term non-A non-B hepatitis (NANB) is used to denote presumed hepatitis C in studies done before serologies to detect this virus were widely available. Acute viral infection (A, B, or C) can range from subclinical seroconversion to fulminant hepatic failure. NK cells under the influence of IL-1 and IL-2 become lymphokine-activated killer (LAK) cells and are thought to eliminate cells with active viral replication. In addition, it has been suggested that TNF-α contributes to cytolysis of infected hepatocytes. While this may be an effective means of eliminating the virus from the organism, if the response is great enough and destroys a major portion of hepatocytes, it can induce liver failure. Conversely, if the cytokine/immune response is not potent enough, infection with the hepatitis B or C virus can become chronic.

In an early report in children with acute hepatitis A or B, Muller et al. found decreased IL-1 production by monocytes (248). This same group later repeated these studies in adults with acute hepatitis A, B, or C and obtained similar results using an ELISA for IL-1 (249). However, elevated serum concentrations of IL-1β were reported in acute hepatitis B and NANB infection (250), and elevated serum levels of IL-1α and IL-1β were also reported in patients with acute hepatitis A, B, or C (251).

Peripheral blood mononuclear cells (PBMC—containing both monocytes and lymphocytes)—from patients with chronic hepatitis B have increased production of IL-1, which correlated with severity of fibrosis on liver biopsy

(252). Szkaradkiewicz found increased IL-1 activity in monocytes from patients with chronic persistent but not chronic active hepatitis B (253). Civeiera et al. studied patients with chronic hepatitis B and found a tendency to higher IL-1 production by PBMC in e antigen-positive patients (254). In contrast, Muller et al. found decreased IL-1 production by monocytes from patients with chronic hepatitis B or chronic hepatitis C (249). Serum IL-1β levels generally are elevated in patients with chronic hepatitis B or C (4).

Studies by Yoshioka et al. and Williams et al. reported reduced IL-2 activity in supernatants of PBMC from patients with acute and chronic active hepatitis (255,256). Later studies from several groups confirmed the decrease in IL-2 activity in patients with chronic viral hepatitis using an IL-2–dependent cell line rather than the PHA-induced lymphoblast proliferation assay employed in earlier studies (252,254,257).

Muller and Zielinski reported decreased TNF-α production by monocytes from patients with acute hepatitis B but normal production by monocytes from patients with acute hepatitis A or NANB (258). On the other hand, al-Wabel's group reported that TNF-α production by PBMC was higher in patients with acute hepatitis B (257). Patients with acute viral hepatitis had detectable serum TNF levels, which correlated with serum ALT levels, and TNF became undetectable during convalescence (251,259).

Initial studies of Yoshioka et al. reported that PBMC from patients with chronic viral liver disease had increased TNF production that correlated with the histological activity index (260). Sheron et al. reported elevated plasma TNF levels as well as increased TNF production by PBMC from patients with chronic hepatitis B that correlated with serum hepatitis B viral DNA levels if the patients were e antigen seropositive (261). Studies by al-Wabel confirmed that PBMC from patients with chronic active hepatitis B have increased TNF-α production (257). In contrast, Muller and Zielinski reported decreased TNF production by monocytes from patients with chronic viral hepatitis (258). Several groups have reported increased serum TNF in chronic viral liver disease (4,262). In liver biopsy samples from patients with chronic hepatitis B or NANB there is TNF staining of mononuclear cells infiltrating the liver that increases with the severity of disease (263,264). In situ hybridization also showed increased TNF-α mRNA activity in hepatocytes, Kupffer cells, and infiltrating mononuclear cells in chronic viral hepatitis (265).

Monocytes and PBMC cultures from patients with acute or chronic viral hepatitis have either normal or decreased IL-6 production (257,266). However, serum IL-6 levels are elevated in patients with acute hepatitis A, B, or NANB, and the serum IL-6 levels correlate with the patient's clinical condition and prothrombin time (251,267). Serum IL-6 levels also are elevated in chronic viral hepatitis and correlate with aminotransferase levels (4,268).

In summary, monocytes from patients with acute and chronic viral hepatitis generally have decreased IL-1, TNF, and IL-6 production. This may be a result of viral infection of the monocytes themselves, which has been shown in vitro to decrease monocyte cytokine production (269). The decreased IL-1 and IL-2 activities may allow the viral infections to become chronic due to LAK activation being insufficient to clear the infecting virus. However, PBMC from presumably similar patients with acute and chronic viral hepatitis generally have higher IL-1 and TNF production, and serum levels of IL-1, TNF, and IL-6 are elevated. PBMC cultures contain lymphocytes as well as monocytes. This and/or subtle differences in the patient groups between studies may explain the differences between monocyte and PBMC culture data. In many cases the cytokine activity correlates with the severity of liver disease. Thus, cytokines such as TNF may have some role in hepatocyte injury, or an insufficient TNF response may play a permissive role in the development of chronicity of viral infection. Impaired clearance of cytokines and/or endotoxins by the liver due to the viral infection may also contribute to the elevated plasma cytokine activity, especially in chronic viral hepatitis (251). Finally, cytokines such as TGF-β, whose mRNA was increased in liver biopsy specimens from patients with chronic viral hepatitis (270,271) and correlated with the histological activity index, may play a role in fibrosis and eventual cirrhosis of chronic viral liver disease.

Autoimmune/Metabolic Liver Disease

Primary biliary cirrhosis (PBC) and primary sclerosing cholangitis (PSC) are progressive diseases with biliary fibrosis. While the etiologies of these diseases are not known, their progressive fibrotic nature suggests a possible role for a cytokine-mediated inflammatory process. Patients with PSC generally have normal plasma TNF levels and normal PBMC cytokine (TNF and IL-1) production (261,272). Monocytes and PBMCs from patients with PBC have normal or low IL-1 and TNF production (249,252,258,272). However, IL-6 production by monocytes and PBMC from PBC patients is increased (266,273). PBC patients also have increased serum levels of IL-1β, IL-6, TNF-α, and IFN-γ, which are likely to be due to impaired hepatic clearance of cytokines (4). Moreover, elevated serum IL-2 and TNF levels in PBC patients decreased during treatment with colchicine (274). There is increased TGF-β mRNA in liver specimens of patients undergoing liver transplants because of end-stage PBC, suggesting that TGF-β may play a role in the fibrotic process (271).

The pathogenesis of idiopathic autoimmune hepatitis (AIH) remains unclear. However, corticosteroids are effective in treating AIH, and corticosteroids are known to decrease inflammatory cytokine production. PBMC from patients with chronic AIH produced highly elevated levels of IL-4 and IL-6 but normal TNF and decreased IL-2 (257). Patients with AIH have elevated serum IL-1β, IL-4, IL-6, TNF-α, and IFN-γ (4,275). TGF-β mRNA is increased in liver specimens of patients undergoing liver transplants because of end-stage autoimmune liver disease (271).

Tilg et al. (4) studied a large number of patients with chronic liver disease and found that all groups had elevated serum IL-1β, IL-6, TNF-α, and IFN-γ levels unrelated to indicators of disease/inflammatory activity such as C-reactive protein. Among the patients studied were 14 with hemochromatosis, 9 with Wilson's disease, and 16 with α_1-antitrypsin deficiency (4). This suggests that liver dysfunction per se is responsible, at least in part, for the observed increased cytokine activity, perhaps due to decreased hepatic cytokine clearance.

Fulminant Hepatic Failure/ Liver Transplantation

Cytokines such as TNF are important mediators of liver injury in animal models. This suggests a potential role for cytokines in fulminant hepatic failure (FHF). PBMC from patients with FHF have increased TNF and IL-1 production, with nonsurvivors having higher TNF and IL-1 activity (276). Serum TNF was detectable in all patients with fulminant hepatitis B but in only one of 25 controls (259). Patients with FHF have higher serum levels of IL-1β, TNF-α, and IL-6, and the increased levels generally correlate with the severity of liver disease, the development of complications, and mortality (259,262,267,277,278).

There is accumulating evidence that cytokines play a role in liver graft rejection in orthotopic liver transplantation (OLT) and that cytokine levels immediately post-transplant may be helpful in predicting rejection or other complications such as infection. During acute rejection, mononuclear cells infiltrate the liver. Immunohistochemical staining is positive for TNF-α in these cells (and for IL-1 if the episode is very severe) (279). Patients with acute liver graft rejection have elevated plasma TNF-α levels (279), with peak plasma TNF-α levels usually occurring at the time rejection is diagnosed clinically (280). In chronic graft dysfunction, increased numbers of mature hepatic macrophages are positive for TNF-α as well as IL-1-β (279). Some investigators have suggested that blood cytokine patterns can predict the development liver graft dysfunction. Tilg et al. reported elevated serum levels of TNF-α and IL-1, with the peak levels occurring 4 days before rejection was diagnosed (281). Fugger et al. found that intraoperative elevations of arterial TNF-α levels predicted rejection (282). Steininger et al. measured portal vein and hepatic vein cytokine levels intraoperatively immediately after liver graft reperfusion and found that there was hepatic release of TNF-α in patients with primary nonfunction or dysfunction (100). Kita et al. reported significantly higher blood levels of IL-6 up to 4 days prior to the histopathological diagnosis of rejection, and the pattern of IL-6 blood levels after transplant predicted long-term graft outcome (283).

A major complication of OLT is infection. Serum TNF is elevated during both active CMV infection and bacterial infection complicating OLT (281). Plasma IL-6 levels also are significantly increased if infection develops (284), and marked increases in IL-6 may predict onset of infection (282,285). Sautner et al. observed a sixfold increase in blood TNF-α and IL-6 concentrations between the anhepatic phase and the end of the operation in patients who subsequently developed infections (286). Serum IL-8 levels are elevated in patients with CMV hepatitis (characterized by neutrophil infiltration of the liver), but the serum IL-8 levels are more pronounced in the patients who have bacterial infections (287). In contrast, patients with acute rejection (which does not have hepatic neutrophil infiltration as a prominent feature) have only minor serum IL-8 elevations, and serum IL-8 is not detected in patients with an uneventful recovery after OLT (287). Plasma IL-10 concentrations increase dramatically within 2 h after liver allograft reperfusion, but IL-10 levels return to low levels by 48 h. Alterations in IL-10 metabolism may be important in decreased immunogenicity of liver grafts compared to other organ transplants (288).

THERAPY: CURRENT AND FUTURE

Successful treatment of liver disease has been hindered by our incomplete understanding of mechanisms/pathogenesis of many hepatic diseases. This lack of understanding of pathogenetic factors also impairs the monitoring of efficacy of specific therapeutic interventions. When new forms of therapy for liver diseases are assessed, mortality or imprecise measures of therapeutic efficacy such as liver enzymes or bilirubin are frequently used. Thus, even if a specific therapy has positive effects on the liver disease, this may not be detected as a significant improvement in mortality or "liver function tests" in an inherently nonhomogeneous group of patients treated at various stages of their disease with complicating metabolic features. As our understanding of the role of cytokines in liver disease evolves, we will likely use specific cytokines (e.g., interferons, IL-10) or anticytokines (e.g., anti-TNF antibody, inhibitors of TGF-β) to treat liver disease, and we will likely use serum/tissue cytokine determinations as markers of therapeutic efficacy.

As new cytokine drugs become available for use in humans, it is vital that physicians take into account the patient's total well-being and not just therapeutic intervention related to one organ system. It is becoming increasingly apparent that "anticytokine" therapeutic agents may be beneficial for one organ system but potentially harmful to another. For example, if an anticytokine antibody attenuates tissue injury in one organ but increases infection in another organ, little or no benefit is derived. Similarly, several cytokines with theoretic therapeutic efficacy have been unsuccessful in clinical trials because of major side effects. Thus, cytokines such as TNF (for tumor therapy) or ciliary neurotropic factor (for therapy of amyotrophic lateral sclerosis) have had important toxicities in clinical trials due to known metabolic effects of cytokines such as fever, anorexia, or organ injury (e.g., liver, lung injury) (289,290). It also is vital to remember that many cytokines may be either harmful or beneficial depending on cytokine concentration and local environmental influences (291). As already discussed, TNF can cause severe liver injury on the one hand and enhanced liver regeneration on the other. In a similar vein, marked overproduction of proinflammatory cytokines such as TNF by alveolar macrophages causes the adult respiratory distress syndrome (292,293). On the other hand, studies by Nelson's group and others have shown that inadequate alveolar macrophage TNF generation can predispose to pneumonia (294–296). Thus, future human cytokine studies will require careful attention to issues such as dosing, specific toxicities, targeted delivery, specific organ system sensitivity, and, most importantly, the well-being of the whole patient.

One strategy for treatment of liver disease is cytokine administration. In chronic viral hepatitis, cytokines postulated to be deficient are given as a therapy. PBMC from patients who have chronic hepatitis B have decreased production of IFN-α and IFN-γ (297). Interferons have antiviral effects, and deficient interferon responses may be a factor in the chronicity of hepatitis B viral infections. The use of the antiviral cytokine IFN-α has become widespread for the treatment of chronic hepatitis B and C. IFN-α has been shown to be more efficacious than IFN-γ and to have fewer side effects, although its exact mechanism(s) of action are not well defined (298). Unfortunately, IFN-α has limited efficacy in inhibiting viral proliferation and eradicating chronic viral infections; therefore its use has been the subject of critical discussion (299,300). For patients with chronic hepatitis B infection, long-term response is defined as normalization of liver injury with loss of HBV-DNA. Long-term response is achieved in about 35% of patients with chronic hepatitis B treated with 10 million units of IFN-α subcutaneously three times a week for 16 weeks. Most of these "responders" remain infected with the virus, and they continue to have hepatitis B surface antigen detected in their blood. They may remain at risk for hepatocellular carcinoma. For patients with chronic hepatitis C infection, long-term response is defined as normalization of liver injury with loss of the HCV-RNA. Long-term response is achieved in 5–20% of patients with chronic hepatitis C treated with 3 million units of INF-α subcutaneously three times a week for 26 weeks. A better long-term response may be achieved with longer therapy. More potent antiviral therapies are being developed for the treatment of chronic hepatitis B and C. Recently, it has been shown that increased hepatic iron is associated with poor response to interferon therapy in chronic hepatitis C (301), and phlebotomy may improve outcome (302). Iron reduction may decrease oxidative tissue injury and may influence oxidative stress transcription factors, such as NFκB, necessary for the production of important inflammatory cytokines such as TNF. Indeed, preliminary data (M. Kugelmas, personal communication) showed that phlebotomy in hepatitis C patients improved transaminase levels and decreased Th1/Th2 ratios.

A second important strategy to treat liver disease is to administer "anticytokines." Indeed, many of the agents currently used in clinical practice to treat liver disease are known to modulate cytokine metabolism (but obviously not developed for this purpose) (2). Some of these drugs and their effects are presented in Table 3. For example, abstinence and enteral nutrition support should enhance gastrointestinal integrity, decrease LPS translocation, and improve glutathione stores, which may inhibit TNF production. Another example is prednisone, which inhibits

Table 3 Current Therapy for Liver Disease

Therapy/Drug	Action
Abstinence	Improved gut integrity
Prednisone	Lowers TNF production
Colchicine	Lowers TNF production
Prostaglandin analogs	Lowers TNF production
Enteral nutrition	Improved gut integrity
Bile acids	Lowers TNF production
Interferon	Lowers TNF production
Pentoxifylline	Lowers TNF production
Propylthiouracil (PTU)	Lowers hepatic hypermetabolism
N-Acetylcysteine	Lowers TNF production

TNF and IL-8 production at the level of transcription and NFκB inhibition (303–306). In severe acute alcoholic hepatitis, clinical studies suggested that prednisone decreased the short-term mortality of some patients who did not have evidence of infection or GI bleeding (307). Recent studies from France reported that hepatic neutrophil infiltration predicted a good response to prednisone therapy in AH (308). We postulate that prednisone inhibits TNF and IL-8 production, thus diminishing neutrophil infiltration/activation and decreasing liver injury in this AH population with more readily reversible injury. Colchicine decreases TNF production, TNF receptor density, and TNF toxicity in mice (309,310). Although colchicine has not been shown to be beneficial in acute AH, it may benefit patients with ALD in the long term by slowing the development/progression of cirrhosis (311–313). Colchicine may have beneficial effects in treating PBC (313–316). Pentoxifylline decreases TNF production and in a preliminary report patients with AH who were treated with pentoxifylline had a decreased incidence of hepatorenal syndrome (2,317,318). Ursodeoxycholic acid decreases TNF production and may benefit patients who have PBC (319,320). Prostaglandin analogs decrease TNF activation and may be beneficial in patients with viral FHF (321). In acetaminophen toxicity, the liver's glutathione stores are depleted by a toxic metabolite of acetaminophen. *N*-Acetylcysteine (NAC) is given as an antedote for acetaminophen toxicity. NAC acts as a glutathione precursor to replace hepatic glutathione stores and decrease TNF production.

Unfortunately, for many major forms of liver disease, such as ALD and chronic viral liver disease, we have no optimal therapy. Moreover, many of our most widely used forms of therapy such as glucocorticoids and interferons have clinically important or unpleasant side effects. There are now in experimental animal/human trials many novel forms of anticytokine therapy that have great potential for various types of liver disease (Table 4). Understanding the appropriate timing of these interventions, possible combinations of interventions, and potential limitations/side effects of these agents is vital. One long-term intervention that may inhibit both injury and fibrosis is antioxidant therapy directed at oxidative stress sensitive transcription factors such as NFκB. NFκB is a multisubunit transcription factor, which rapidly activates expression of certain genes involved in the acute phase response, inflammatory cytokines, and certain viral replication (127). In unstimulated cells, NFκB resides in the cytoplasm in an inactive complex bound to the inhibitor IκB. Multiple stimuli, including endotoxin and reactive oxygen intermediates, cause release of IκB and permit NFκB to enter the nucleus, bind to DNA control elements, and induce synthesis of mRNA. One important intracellular signal to activate NFκB appears to be reactive oxygen intermediates, and there is extensive evidence that NFκB is an oxidative stress responsive transcription factor (127). Multiple antioxidants have been shown to inhibit NFκB activation (127). Thus, one could envision combination therapy for a patient hospitalized with acute AH. The patient may initially be given a drug such as anti-TNF antibody or soluble TNF receptor to acutely neutralize newly synthesized circulating and hepatic TNF levels. The patient could also be given a chronic form of antioxidant therapy to attenuate further production of proinflammatory cytokines such as TNF and IL-8 and/or fibrotic cytokines and collagen formation.

Table 4 Potential Future Therapeutic Agents for Cytokine-Induced Liver Injury

Antibodies against LPS
Antibodies against cytokines
Soluble cytokine receptors
Cytokine-receptor antagonists
Drugs that block cytokine production
Drugs that block a specific cytokine-induced response
Adhesion molecule antagonists
Drugs and foods that maintain or enhance gut integrity

In summary, many forms of experimental and human liver disease are associated with dysregulated cytokine metabolism. Strong evidence supports a role for specific cytokines in liver injury, cholestasis, liver regeneration, and hepatic fibrosis. Many current drugs used to treat liver disease are either cytokines or have "anticytokine" effects, and a host of new "anticytokine" agents have great therapeutic potential for several forms of liver disease.

REFERENCES

1. Shedlofsky SI, McClain CJ. Hepatic dysfunction due to cytokines. In: Kimball ES, ed. Cytokines and Inflammation. Boca Raton, FL: CRC Press, Inc., 1991:235–273.
2. McClain CJ, Hill DB, Schmidt J, Diehl AM. Cytokines and alcoholic liver disease. Semin Liver Dis 1993; 13:170–182.
3. McClain CJ, Hill DB, Marsano L, Cohen D, Shedlofsky S. A role for cytokines in alcoholic hepatitis. Proceedings of the 1992 conference on Alcohol, Drugs of Abuse, and Immunomodulators (AIDS). Vol. 86. New York: Pergamon Press, Inc., 1993:133–141.
4. Tilg H, Wilmer A, Vogel W, et al. Serum levels of cytokines in chronic liver diseases. Gastroenterology 1992; 103:264–274
5. Andus T, Bauer J, Gerok W. Effects of cytokines on the liver. Hepatology 1991; 13:364–375.
6. Trautwein C, Boker K, Manns MP. Hepatocyte and immune system: acute phase reaction as a contribution to early defense mechanisms. Gut 1994; 35:1163–1166.
7. Tiegs G. Immunotoxicology of host-response-mediated experimintal liver injury. J Hepatol 1994; 21:890–903.
8. Grewe M, Gausling R, Gyufko K, Hoffmann R, Decker K.

Regulation of the mRNA expression for tumor necrosis factor-alpha in rat liver macrophages. J Hepatol 1994; 20:811–818.
9. Luster MI, Germolec DR, Yoshida T, Kayama F, Thompson M. Endotoxin-induced cytokine gene expression and excretion in the liver. Hepatology 1994; 19:480–488.
10. Lee KS, Buck M, Houglum K, Chojkier M. Activation of hepatic atellate cells by TGFalpha and collagen type I is mediated by oxidative stress through c-myb expression. J Clin Invest 1995; 96:2461–2468.
11. Thornton AJ, Ham J, Kunkel SL. Kupffer cell-derived cytokines induce the synthesis of a leukocyte chemotactic peptide, interleukin-8, in human hepatoma and primary hepatocyte cultures. Hepatology 1992; 15:1112–1122.
12. Thornton AJ, Streiter RM, Lindley I, Baggiolini M, Kunkel SL. Cytokine-induced gene expression of a neutrophil chemotactic factor/IL-8 in human hepatocytes. J Immunol 1990; 144:2609–2613.
13. Ohira H, Ueno T, Shakado S, et al. Cultured rat hepatic sinusoidal endothelial cells express intercellular adhesion molecule-1 (ICAM-1) by tumor necrosis factor-alpha or interleukin-1 alpha stimulation. J Hepatol 1994; 20:729–734.
14. Essani NA, Fisher MA, Farhood A, Manning AM, Smith CW, Jaeschke H. Cytokine-induced upregulation of hepatic intercellular adhesion molecule-1 messenger RNA expression and its role in the pathophysiology of murine endotoxin shock and acute liver failure. Hepatology 1995; 21:1632–1639.
15. Nolan JP. Endotoxin, reticuloendothelial function, and liver injury. Hepatology 1981; 1:458–465.
16. Nolan JP. Intestinal endotoxins as mediators of hepatic injury—an idea whose time has come again. Hepatology 1989; 8:232–236.
17. Mizoguchi Y, Tsutsui H, Miyajima K, et al. The protective effects of prostaglandin E1 in an experimental massive hepatic cell necrosis model. Hepatology 1987; 7:1184–1188.
18. Hill DB, Schmidt J, Shedlofsky SI, Cohen DA, McClain CJ. In vitro tumor necrosis factor cytotoxicity in Hep G2 liver cells. Hepatology 1995; 21:1114–1119.
19. Keppler D, Lesch R, Reutter W, Decker K. Experimental hepatitis induced by d-galactosamine. Exp Mol Pathol 1968; 9:279–290.
20. Tiegs G, Wolter M, Wendel A. Tumor necrosis factor is a terminal mediator in galactosamine/endotoxin-induced hepatitis in mice. Biochem Pharmacol 1989; 38:627–631.
21. Blitzer B, Waggoner J, Jones E, et al. A model of fulminant hepatic failure in the rabbit. Gastroenterology 1978; 74:664–671.
22. Leist M, Wendel A. Tunicamycin potently inhibits tumor necrosis factor-induced hepatocyte apoptosis. Eur J Pharmacol 1995; 292:201–204.
23. Camara D, Caruana J, Schwartz K, Montes M, Nolan J. D-galactosamine liver injury: absorption of endotoxin and protective effect of small bowel and colon resection in rabbits. Proc Soc Exp Biol Med 1983; 172:255–259.
24. Libert C, Brouckaert P, Fiers W. Protection by alpha1-acid glycoprotein against tumor necrosis tumor-induced lethality. J Exp Med 1994; 180:1571–1575.
25. MacDonald J, Thayer K, White C. Inhibition of galactosamine cytotoxicity in an in vivo/in vitro hepatocellular toxicity model. Toxicol Appl Pharmacol 1987; 89:269–277.
26. Wang JF, Wendel A. Studies on the hepatotoxicity of galactosamine/endotoxin or galactosamine/TNF in the perfused mouse liver. Biochem Pharmacol 1990; 39:267–270.
27. Hartung T, Wendel A. Endotoxin-inducible cytotoxicity in liver cell cultures—I. Biochem Pharmacol 1991; 42:1129–1135.
28. Czaja MJ, Flanders KC, Biempica L, Klein C, Zern MA, Weiner FR. Expression of tumor necrosis factor-alpha and transforming growth factor-beta 1 in acute liver injury. Growth Factors 1989; 1:219–226.
29. Leist M, Gantner F, Bohlinger I, Germann PG, Tiegs G, Wendel A. Murine hepatocyte apoptosis induced in vitro and in vivo by TNF-alpha requires transcriptional arrest. J Immunol 1994; 153:1778–1788.
30. Grun M, Liehr H, Rasenack U. Significance of endotoxemia in experimental "galactosamine-hepatitis" in rats. Acta Hepat Gastroenterol 1988; 24:64.
31. Liehr H, Grun M, Seelig H-P, Seelig R, Reutter W, Heine W. On the pathogenesis of galactosamine hepatitis: indications of extrahepatocellular mechanisms responsible for liver cell death. Arch B Cell Pathol 1978; 26:331.
32. Galanos C, Freudenberg M, Reutter W. Galactosamine-induced sensitization to the lethal effects of endotoxin. Proc Natl Acad Sci USA 1979; 76:5939.
33. Freudenberg M, Galanos C. Induction of tolerance to lipopolysaccharide (LPS)-D-galactosamine lethality by pretreatment with LPS is mediated by macrophages. Infect Immun 1988; 56:1352.
34. Bohlinger I, Leist M, Barsig J, Uhlig S, Tiegs G, Wendel A. Interleukin-1 and nitric oxide protect against tumor necrosis factor alpha-induced liver injury through distinct pathways. Hepatology 1995; 22:1829–1837.
35. Alcorn JM, Fierer J, Chojkier M. The acute-phase response protects mice from D-galactosamine sensitization to endotoxin and tumor necrosis factor-alpha. Hepatology 1992; 15:122–129.
36. Wong G, Goeddel D. Induction of manganous superoxide dismutase by tumor necrosis factor: possible protective mechanism. Science 1989; 242:941–944.
37. Shaffer J, Treanor P, Del Vecchio P. Expression of bovine and mouse endothelial cell antioxidant enzymes following TNFalpha exposure. Free Rad Biol Med 1990; 8:497–502.
38. Opipari A, Hu H, Yabkowitz R, Dixit V. The A20 zinc finger protein protects cells from tumor necrosis factor cytotoxicity. J Biol Chem 1992; 267:12424–12427.
39. Al-Tuwaijri A, Akdamar K, Di Luzio NR. Modification of galactosamine-induced liver injury in rats by reticuloendothelial system stimulation or depression. Hepatology 1981; 1:107–113.
40. Chojkier M, Fierer J. D-galactosamine hepatotoxicity is associated with endotoxin sensitivity and mediated by lymphorecticular cells in mice. Gastroenterol 1985; 88:115–121.
41. Lehman V, Freudenberg MA, Galanos C. Lethal toxicity of lipopolysaccharide and tumor necrosis factor in normal and D-galactosamine-treated mice. J Exp Med 1987; 165:657.
42. Hishinuma I, Nagakawa J, Hirota K, et al. Tumor necrosis factor-α in galactosamine-induced hepatitis. Hepatology 1990; 12:1187–1191.
43. Leist M, Gantner F, Bohlinger I, Tiegs G, Germann PG, Wendel A. Tumor necrosis factor-induced hepatocyte apoptosis precedes liver failure in experimental murine shock models. Am J Pathol 1995; 146:1220–1234.
44. Arthur MJ. Corynebacterium parvum-elecited hepatic macrophages demonstrate enhanced respiratory burst activity compared with resident Kupffer cells in the rat. Gastroenterology 1986; 91:174–181.
45. Smith S, Calzetta A, Bankowski J, Kenworthy-Bott L, Terminelli C. Lipopolysaccharide-induced cytokine production and mortality in mice treated with corynebacterium parvum. J Leukoc Biol 1993; 54:23–29.
46. Arai M, Mochida S, Ohno A, Ogata I, Fujiwara K. Sinusoidal endothelial cell damage by activated macrophages in rat liver necrosis. Gastroenterology 1993; 104:1466–1471.

47. Rees D, Cunha F, Assreuy J, Herman A, Moncada S. Sequential induction of nitric oxide synthase by corynebacterium parvum in different organs of the mouse. Br J Pharmacol 1995; 114:689–693.
48. Harbrecht BG, Di Silvio M, Demetris AJ, Simmons RL, Billiar TR. Tumor necrosis factor-alpha regulates in vivo nitric oxide synthesis and induces liver injury during endotoxemia. Hepatology 1994; 20:1055–1060.
49. Harbrecht B, Stadler J, Demetris A, Simmons R, Billiar T. Nitric oxide and prostaglandins interact to prevent hepatic damage during murine endotoxemia. Am J Physiol 1994; 266:G1004–G1010.
50. Recknagel R, Glende E, Dolak J, Waller R. Mechanisms of carbon tetrachloride toxicity. Pharmacol Ther 1989; 43:139–154.
51. Nolan J, Liebowitz A. Endotoxin and the liver III. Modification of acute carbon tetrachloride injury by polymyxin B, an antiendotoxin. Gastroenterology 1978; 75:445–449.
52. Cornell RP. Endotoxin-induced hyperinsulinemia and hyperglucagonemia after experimental liver injury. Am J Physiol 1981; 241:E428–435.
53. Czaja MJ, Xu J, Ju Y, Alt E, Schmiedeberg P. Lipopolysaccharide-neutralizing antibody reduces hepatocyte injury from acute hepatotoxin administration. Hepatology 1994; 19:1282–1289.
54. Chamulitrat W, Blazka M, Jordan S, Luster M, Mason R. Tumor necrosis factor-alpha and nitric oxide production in endotoxin-primed rats administered carbon tetrachloride. Life Sci 1995; 57:2273–2280.
55. Czaja MJ, Xu J, Alt E. Prevention of carbon tetrachloride-induced rat liver injury by soluble tumor necrosis factor receptor. Gastroenterology 1995; 108:1849–1854.
56. Liu S-L, Esposti S, Yao T, Diehl AM, Zern MA. Vitamin E therapy of acute carbon tetrachloride-induced hepatic injury in mice is associated with inhibition of nuclear factor kappa b binding. Hepatology 1995; 22:1474–1481.
57. Selye H, Tuchweber B, Bertok L. Effect of lead acetate on the susceptibility of rats to bacterial endotoxins. J Bacteriol 1966; 91:884–890.
58. Rippe D, Berry L. Metabolic manifestations of lead acetate sensitization to endotoxin in mice. J Reticuloendothel Soc 1973; 13:527–535.
59. Truscott R. Endotoxin studies in chicks: effect of lead acetate. Can J Comp Med 1970; 34:134–137.
60. Holper K, Trejo R, Brettschneider L, Di Luzio N. Enhancement of endotoxic shock in lead sensitized sub-human primates. Surg Gynecol Obstet 1973; 136:593–601.
61. Honchel R, Marsano L, Cohen D, Shedlofsky S, McClain C. Lead enhances lipopolysaccharide and tumor necrosis factor liver injury. J Lab Clin Med 1991; 117:202–208.
62. Honchel R, Shedlofsky S, Cohen D, Marsano K, Lee E, McClain C. Is hepatic cytoprotection really cytokines protection? In: Dinarello C, Kluger M, Powanda M, eds. The Physiological and Pathological Effects of Cytokines. New York: Wiley-Liss, Inc., 1990:275–280.
63. Taki Y, Shimahara Y, Isselhard W. Derangement of hepatic energy metabolism in lead-sensitized endotoxicosis. Eur Surg Res 1985; 17:140–149.
64. Kato Y, Takimoto S, Ogura H. Mechanisms of induction of hypercalcemia and hyperphosphatemia by lead acetate in the rat. Calcif Tiss Res 1977; 24:41–46.
65. Kuttner R, Ebata T. The mediated effect of endotoxin and liver upon hepatic metabolism. Surg Gynecol Obstet 1984; 159:319–324.
66. Filkins J, Buchanan B. Effects of lead acetate on sensitivity to shock, intravascular carbon and endotoxin clearances, and hepatic endotoxin detoxification. Proc Soc Exp Biol Med 1973; 142:471–475.
67. Clark GC, Taylor MJ, Tritscher AM, Lucier GW. Tumor necrosis factor involvement in TCDD-mediated endotoxin hypersensitivity in C57BL/6L mice congenic at the Ah locus. Toxicol Appl Pharmacol 1991; 111:422–431.
68. Taylor MJ, Lucier GW, Mahler JF, Thompson M, Lockhart AC, Clark GC. Inhibition of acute TCDD toxicity by treatment with anti-TNF antibody or dexamethasone. Toxicol Appl Pharmacol 1992; 117:126–132.
69. Shedlofsky SI, Hoglen NC, Rodman LE, et al. 3,3′,4,4′-Tetrabromobiphenyl sensitizes rats to the hepatotoxic effects of endotoxin by a mechanism that involves more than tumor necrosis factor. Hepatology 1991; 14:1201–1208.
70. Shiratori Y, Kiriyama H, Fukushi Y, et al. Modulation of ischemia-reperfusion-induced hepatic injury by Kupffer cells. Dig Dis Sci 1994; 39:1265–1272.
71. Liu P, McGuire G, Fisher M, Farhood A, Smith C, Jaeschke H. Activation of kupffer cells and neutrophils for reactive oxygen formation is responsible for endotoxin-enhanced liver injury after hepatic ischemia. Shock 1995; 3:56–62.
72. Colletti LM, Remick DG, Burtch GD, Kunkel SL, Strieter RM, Campbell DA, Jr. Role of tumor necrosis factor-alpha in the pathophysiologic alterations after hepatic ischemia/reperfusion injury in the rat. J Clin Invest 1990; 85:1936–1943.
73. Sakr MF, McClain CJ, Gavaler JS, Zetti GM, Starzl TE, Van Thiel DH. FK 506 pretreatment is associated with reduced levels of tumor necrosis factor and interleukin 6 following hepatic ischemia/reperfusion. J Hepatol 1993; 17:301–307.
74. Jaeschke H, Farhood A, Smith C. Neutrophils contribute to ischemia/reperfusion injury in rat liver in vivo. FASEB J 1990; 4:3355–3359.
75. Jaeschke H. The inflammation response after hepatic ischemia/reperfusion. In: Lemasters J, Oliver C, eds. Cell Biology of Trauma. Boca Raton, FL: CRC Press, Inc., 1995:127.
76. Scales WE, Campbell DA, Jr., Green ME, Remick DG. Hepatic ischemia/reperfusion injury: importance of oxidant/tumor necrosis factor interactions. Am J Physiol 1994; 267:G1122–1127.
77. Bhagwandeen SB, Apte M, Manwarring L. Endotoxin induced hepatic necrosis in rats on an alcohol diet. J Pathol 1987; 152:47–53.
78. Arai M, Nakano S, Okuno F. Endotoxin-induced hypercoagulability: a possible aggravating factor of alcoholic liver disease. Hepatology 1989; 9:846–854.
79. Honchel R, Ray MB, Marsano L, et al. Tumor necrosis factor in alcohol enhanced endotoxin liver injury. Alcohol Clin Exp Res 1992; 16:665–669.
80. Hansen J, Cherwitz DL, Allen JI. The role of tumor necrosis factor-alpha in acute endotoxin-induced hepatotoxicity in ethanol-fed rats. Hepatology 1994; 20:461–474.
81. Adachi Y, Moore L, Bradford B, Gao W, Thurman R. Antibiotics prevent liver injury in rats following long-term exposure to ethanol. Gastroenterol 1995; 108:218–224.
82. Adachi Y, Bradford B, Gao W, Bojes H, Thurman R. Inactivation of Kupffer cells prevents early alcohol-induced liver injury. Gastroenterol 1994; 20:453–460.
83. Tsukamoto H, Towner S, Ciofalo L, French S. Ethanol-induced liver fibrosis in rats fed high fat diet. Hepatology 1986; 6:814–822.
84. French S. Nutrition factors in the pathogenesis of alcoholic liver disease. In: Watson R, Watzl B, eds. Nutrition and Alcohol. Boca Raton, FL: CRC Press, Inc., 1992; 337–362.
85. Nanji AA, Khettry U, Sadrzadeh SMH, Yamanaka T. Severity

of liver injury in experimental alcoholic liver disease. Am J Pathol 1993; 142:367–373.
86. Nanji A, French S. Dietary linoleic acid is required for development of experimentally induced alcoholic liver disease. Life Sci 1989; 44:223–227.
87. Stal P, Hultcrantz R. Iron increases ethanol toxicity in rat liver. J Hepatology 1993; 17:108–115.
88. Tsukamoto H, Horne W, Kamimura S, et al. Experimental liver cirrhosis induced by alcohol and iron. J Clin Invest 1995; 96:620–630.
89. Hirano T, Kaplowitz N, Tsukamoto H, Kamimura S, Fernandez-Checa J. Hepatic mitochondrial glutathione depletion and progression of experimental alcoholic liver disease in rats. Hepatology 1992; 16:1423–1427.
90. Garcia-Ruiz C, Morales A, Ballesta A, Rodes J, Kaplowitz N, Fernandez-Checa JC. Effect of chronic ethanol feeding on glutathione and functional integrity of mitochondria in periportal and perivenous rat hepatocytes. J Clin Invest 1994; 94:193–201.
91. Nanji A, Zhao S, Lamb R, Dannenberg A, Sadrzadeh S, Waxman D. Changes in cytochromes P450, 2E1, 2B1, and 4A, and phospholipases A and C in the intragastric feeding rat model for alcoholic liver disease: relationship to dietary fats and pathologic liver injury. Alcoholism 1994; 18:902–908.
92. Nanji A, Zhao S, Sadrzadeh S, Dannenberg A, Tahan S, Waxman D. Markedly enhanced cytochrome P450 2E1 induction and lipid peroxidation is associated with severe liver injury in fish oil-ethanol-fed rats. Alcohol Clin Exp Res 1994; 18:1280–1285.
93. Nanji AA, Zhao S, Sadrzadeh SM, Waxman DJ. Use of reverse transcription-polymerase chain reaction to evaluate in vivo cytokine gene expression in rats fed ethanol for long periods. Hepatology 1994; 19:1483–1487.
94. Nanji A, Khettry U, Sadrzadeh S. Lactobacillus feeding reduces endotoxemia and severity of experimental alcoholic liver disease. Proc Soc Exp Biol Med 1994; 205:243–247.
95. Nanji A, Griniuviene B, Sadrzadeh S, Levitsky S, McCully J. Effect of type of dietary fat and ethanol on antioxidant enzyme mRNA induction in rat liver. J Lipid Res 1995; 36:736–744.
96. Perara C, St. Clair D, McClain C. Differential regulation of manganese superoxide dismutase activity by alcohol and TNF in human hepatoma cells. Arch Biochem Biophys 1995; 323.
97. Kamimura S, Tsukamoto H. Cytokine gene expression by Kupffer cells in experimental alcoholic liver disease. Hepatology 1995; 22:1304–1309.
98. Bemelmans MHA, Gouma DJ, Greve JW, Burrman WA. Cytokines tumor necrosis factor and interleukin-6 in experimental biliary obstruction in mice. Hepatology 1992; 15:1132–1136.
99. Goto M, Takei Y, Kawano S, et al. Tumor necrosis factor and endotoxin in the pathogenesis of liver and pulmonary injuries after orthotopic liver transplantation in the rat. Hepatology 1992; 16:487–493.
100. Steininger R, Roth E, Fugger R, et al. Transhepatic metabolism of TNF-alpha, IL-6, and endotoxin in the early hepatic reperfusion period after human liver transplantation. Transplantation 1994; 58:179–183.
101. Yokoi Y, Nakamura S, Serizawa A, Suzuki S, Sakaguchi T, Baba S. Enhanced sensitivity to endotoxin shock following liver transplantation. Transplantation Proc 1995; 27:1560–1562.
102. Mathew J, Hines J, James O, Burt A. Non-parenchymal cell responses in paracetamol (acetaminophen)-induced liver injury. J Hepatology 1994; 20:537–541.
103. Blazka ME, Wilmer JL, Holladay SD, Wilson RE, Luster MI. Role of proinflammatory cytokines in acetaminophen hepatotoxicity. Toxicol Appl Pharmacol 1995; 133:43–52.
104. Laskin D, Gardner C, Price V, Jollow D. Modulation of macrophage functioning abrogates the acute hepatotoxicity of acetaminophen. Hepatology 1995; 21:1045–1050.
105. Kayama F, Yoshida T, Elwell MR, Luster MI. Role of tumor necrosis factor-alpha in cadmium-induced hepatotoxicity. Toxicol Appl Pharmacol 1995; 131:224–234.
106. Broitman S, Gottlieb L, Zamcheck N. Influence of neomycin and infected endotoxin in the pathogenesis of choline deficiency cirrhosis in the adult rat. J Exp Med 1963; 119:633–642.
107. Jiang J, Bahrami S, Leichtfried G, Redl H, Ohlinger W, Schlag G. Kinetics of endotoxin and tumor necrosis factor appearance in portal and systemic circulation after hemorrhagic shock in rats. Ann Surg 1995; 221:100–106.
108. Mohler K, Sleath P, Fitzner J, et al. Protection against a lethal dose of endotoxin by an inhibitor of tumour necrosis factor processing. Nature 1994; 370:218–220.
109. Iimuro Y, Yamamoto M, Kohno H, Itakura J, Fujii H, Matsumoto Y. Blockade of liver macrophages by gadolinium chloride reduces lethality in endotoxemic rats-analysis of mechanisms of lethality in endotoxemia. J Leukocyte Biol 1994; 55:723–728.
110. Rose-John S, Heinrich P. Soluble receptors for cytokines and growth factors: generation and biological function. Biochem J 1994; 300:281–290.
111. Peristeris P, Clark BD, Gatti S, et al. N-acetylcysteine and glutathione as inhibitors of tumor necrosis factor production. Cell Immunol 1992; 140:390–399.
112. Tracey K, Fong J, Hesse D, et al. Nature 1987; 330:662–664.
113. Liu P, Vonderfecht SL, McGuire GM, Fisher MA, Farhood A, Jaeschke H. The 21-aminosteroid tirilazad mesylate protects against endotoxin shock and acute liver failure in rats. J Pharmacol Exp Ther 1994; 271:438–445.
114. Adamson GM, Billings RE. Cytokine toxicity and induction of NO synthase activity in cultured mouse hepatocytes. Toxicol Appl Pharmacol 1993; 119:100–107.
115. Leist M, Gantner F, Jilg S, Wendel A. Activation of the 55 kDa TNF receptor is necessary and sufficient for TNF-induced liver failure, hepatocyte apoptosis, and nitrite release. J Immunol 1995; 154:1307–1316.
116. Thornton AJ, Ham J, Kunkel SL. Kupffer cell-derived cytokines induce the synthesis of a leukocyte chemotactic peptide, interleukin-8, in human hepatoma and primary hepatocyte cultures. Hepatology 1991; 14:1112–1122.
117. Bautista A, Spitzer J. Superoxide anion generation by in situ perfused rat liver: effect of in vivo endotoxin. Am J Physiol 1990; 259:G907–G912.
118. Bautista A, Meszaros K, Bojta J, Spitzer J. Superoxide anion generation in the liver during early stage of endotoxemia in rats. J Leukocyte Biol 1990; 48:123.
119. Wong GHW, Elwell JH, Oberley LW, Goeddel DV. Manganous superoxide dismutase is essential for cellular resistance to cytotoxicity of tumor necrosis factor. Cell 1989; 58:923–931.
120. Larrick JW, Wright SC. Cytotoxic mechanism of tumor necrosis factor-α. FASEB J 1990; 4:3215–3223.
121. Adamson G, Billings R. Tumor necrosis factor induced oxidative stress in isolated mouse hepatocytes. Arch Biochem Biophys 1992; 294:223–229.
122. Stadler J, Bentz B, Harbrecht B, et al. Tumor necrosis factor alpha inhibits hepatocyte mitochondrial respiration. Ann Surg 1992; 216:539–546.
123. Schulze-Osthoff K, Bakker A, Vanhaesebroeck B, Beyaert R, Jacob W, Fiers W. Cytotoxic activity of TNF is mediated by

early damage of mitochondrial functions. J Biol Chem 1992; 267:5317–5323.
124. Schulze-Osthoff K, Beyaert R, Vandevoorde V, Haegeman G, Fiers W. Depletion of the mitochondrial electron transport abrogates the cytotoxic and gene-inductive effects of TNF. EMBO J 1993; 12:3095–3104.
125. Cunningham CC, Coleman WB, Spach PI. The effects of chronic ethanol consumption on hepatic mitochondrial energy metabolism. Alcohol Alcoholism 1990; 25:127–136.
126. Lauderberg B, Liang D, Schwarzenbach F, Breen K. Mitochondrial dysfunction in alcoholic patients as assessed by breath analysis. Hepatology 1993; 17:418–422.
127. Schreck R, Albermann KAJ, Baeuerle PA. Nuclear factor κB: an oxidative stress-responsive transcription factor of eukaryotic cells (a review). Free Rad Res Commun 1992; 17:221–237.
128. Muller G, Ayoub M, Storz P, Rennecke J, Fabbro D, Pfizenmaier K. PKC-zeta is a molecular switch in signal transduction of TNF-alpha, bifunctionally regulated by ceramide and arachidonic acid. EMBO J 1995; 14:1961–1969.
129. Schutze S, Machleidt T, Kronke M. Mechanisms of tumor necrosis factor action. Semin Oncol 1992; 19:16–24.
130. Ginsburg I, Kohen R. Cell damage in inflammatory and infectious sites might involve a coordinated "cross-talk" among oxidants, microbial haemolysins and ampiphiles, cationic proteins, phospholipases, fatty acids, proteinases and cytokines (an overview). Free Rad Res 1995; 22:489–517.
131. Perez H, Roll F, Bissell D, Shak S, Goldstein I. Production of chemotactic activity for polymorphonuclear leukocytes by cultured rat hepatocytes exposed to ethanol. J Clin Invest 1984; 74:1350–1357.
132. Shiratori Y, Takada H, Hikiba Y, et al. Production of chemotactic factor, interleukin-8, from hepatocytes exposed to ethanol. Hepatology 1993; 18:1477–1482.
133. Hisama N, Yamaguchi Y, Miyanari N, et al. Ischemia-reperfusion injury: the role of Kupffer cells in the production of cytokine-induced neutrophil chemoattractant, a member of the interleukin-8 family. Transplant Proc 1995; 27:1604–1606.
134. Jaeschke H, Farhood A, Bautista A, Spolarics Z, Spitzer J, Smith CW. Functional inactivation of neutrophils with a mac-1 (CD11b/CD18) monoclonal antibody protects against ischemia-reperfusion injury in rat liver. Hepatology 1993; 17:915–923.
135. Fabian T, Croce M, Stewart R, Dockter M, Proctor K. Neutrophil CD18 expression and blockade after traumatic shock and endotoxin challenge. Ann Surg 1994; 220:552–563.
136. Farhood A, McGuire G, Manning A, Miyasaka M, Smith C, Jaeschke H. Intercellular adhesion molecule 1 (ICAM-1) expression and its role in neutrophil-induced ischemia reperfusion injury in rat liver. J Leukoc Biol 1995; 57:368–374.
137. Zhou W, Chao W, Levine B, Olson M. Role of platelet-activating factor in hepatic responses after bile duct ligation in rats. Am J Physiol 1992; 263:G587–G592.
138. Miller D, Keeton G, Webber S, Saunders S. Jaundice in severe bacterial infections. Gastroenterol 1976; 71:1799–1805.
139. Utili R, Abernathy C, Zimmerman H. Studies on the effects of E. coli endotoxin on canalicular bile formation in the isolated perfused rat liver. J Lab Clin Med 1977; 89:248–253.
140. Ott MT, Vore M, Barker DE, Strodel WE, McClain CJ. Monokine depression of bile flow in the isolated perfused rat liver. J Surg Res 1989; 47:248–250.
141. Whiting JF, Green RM, Rosenbluth AB, Gollan JL. Tumor necrosis factor-alpha decreases hepatocyte bile salt uptake and mediates endotoxin-induced cholestasis. Hepatology 1995; 22:1273–1278.
142. Green RM, Whiting JF, Rosenbluth AB, Beier D, Gollan JL. Interleukin-6 inhibits hepatocyte taurocholate uptake and sodium-potassium-adenosinetriphosphatase activity. Am J Physiol 1994; 267:G1094–1100.
143. Michalopoulos GD. Liver regeneration: molecular mechanisms of growth control. FASEB J 1991; 4:176–187.
144. Bucher NLR. Liver regeneration: an overview. J Gastroenterol Hepatol 1991; 6:615–624.
145. Higgins GM, Anderson RM. Experimental pathology of the liver: I. Regeneration of the liver by the white rat following partial surgical removal. Arch Pathol 1931; 12:186–202.
146. Diehl AM, Rai RM. Regulation of signal transduction during liver regeneration. FASEB J (in press).
147. Cornell R, Liljequist B, Bartizal K. Depressed liver regeneration after partial hepatectomy of germ-free, athymic and lipopolysaccharide-resistant mice. Hepatology 1990; 11:915–922.
148. Beyer HS, Stanley M, Theologides A. Tumor necrosis factor-alpha increases hepatic DNA and RNA and hepatocyte mitosis. Biochem Int 1990; 22:405–410.
149. Mealy K, Wilmore EW. Tumor necrosis factor increases hepatic cell mass. Br J Surg 1991; 78:331–333.
150. Rai R, Yang SQ, Karp C, Klein A, McClain C, Diehl AM. Kupffer cell depletion enhances liver regeneration after partial hepatectomy in rats. Am J Physiol (in press).
151. Rai R, Karp C, Lafredda S, Diehl AM. Kupffer cell depletion shifts the balance of growth-modulatory cytokines to promote hepatocyte proliferation after partial hepatectomy. Gastroenterology (in press).
152. Akerman PA, Cote P, Yang SQ, et al. Antibodies to tumor necrosis factor alpha inhibit liver regeneration after partial hepatectomy. Am J Physiol 1992; 263:G579–G585.
153. Westwick JK, Weitzel C, Leffert HL, Brenner DA. Activation of jun kinase is an early event in hepatic regeneration. J Clin Invest 1995; 95:803–810.
154. Diehl AM, Yin M, Fleckenstein J, et al. Tumor necrosis factor-alpha induces c-jun during the regenerative response to liver injury. Am J Physiol 1994; 267:G552–561.
155. Westwick JK, Weitzel C, Minden A, Karin M, Brenner DA. Tumor necrosis factor alpha stimulates AP-1 activity through prolonged activation of the c-jun kinase. J Biol Chem 1994; 269:26396–26401.
156. Diehl AM, Johns DA, Yang SQ, et al. Adenovirus-mediated transfer of C/EBP alpha identifies a dominant anti-proliferative role for this isoform in hepatocytes. J Biol Chem (in press).
157. Yin M, Yang SQ, Lin HZ, Chaterjee S, Diehl AM. Tumor necrosis factor alpha induces nuclear localization of cytokine-inducible C/EBP isoforms in cultured hepatocytes. Gastroenterology (in press).
158. Diehl AM, Yang SQ. Regenerative changes in C/EBP alpha and C/EBP beta expression modulate binding to the C/EBP site in the c-fos promoter. Hepatology 1994; 19:447–456.
159. Diehl AM, Yang SQ, Yin M, Lin HZ, Nelson S, Bagby G. Tumor necrosis factor-alpha modulates CCAAT/enhancer binding proteins-DNA binding activities and promotes hepatocyte-specific gene expression during liver regeneration. Hepatology 1995; 22:252–261.
160. Ruff-Jamison S, Zhong Z, Wen Z, Chen K, Cohen S. Epidermal growth factor and lipopolysaccharide activate Stat 3 transcription factor in mouse liver. J Biol Chem 1994; 269:21933–21935.
161. Cressman DE, Diamond RS, Taub R. Rapid activation of the Stat 3 transcription complex in liver regeneration. Hepatology 1995; 21:1443–1449.
162. Akira S, Hirano T, Taga T, Kishimoto T. Biology of multifunctional cytokines: IL-6 and its related molecules (IL-1 and TNF). FASEB J 1990; 4:2860–2867.

163. Hill CS, Treisman R. Transcriptional regulation by extracellular signals: mechanisms and specificity. Cell 1995; 80:199–211.
164. Liu Y, Michalopoulos GK, Zarnegar R. Structural and functional characterization of the mouse hepatocyte growth factor gene promoter. J Biol Chem 1994; 269:4152–4160.
165. Rai R, Diehl AM. Regulation of liver regeneration by proinflammatory cytokines. Eur J Hepatol (in press).
166. Zarnegar R. Regulation of HGF and HGFR gene expression. Exs 1995; 74:33-49.
167. Schmiedeberg P, Biempica L, Czaja MJ. Timing of protooncogene expression varies in toxin-induced liver regeneration. J Cell Physiol 1993; 154:294–300.
168. Shiratori Y, Ichida Y, Geerts A, Wisse A. Modulation of collagen synthesis by fat storing cells isolated from carbon tetrachloride or vitamin A-treated rats. Dig Dis Sci 1987; 32:1281–1289.
169. Matsuoka M, Zhang M, Tsukamoto H. Sensitization of hepatic lipocytes by high fat diet to stimulatory effects of Kupffer cell-derived factors: implication in alcoholic liver fibrogenesis. Hepatology 1990; 11:173–182.
170. Takahara T, Kojima T, Kojima T, et al. Collagen production in fat-storing cells after carbon tetrachloride intoxication in the rat. Immunoelectron microscopic observation of type I, type III collagens and prolyl hydroxylase. Lab Invest 1988; 59:509–521.
171. Milani S, Herbst H, Schuppan D, Kim K, Riecken E, Stein H. Procollagen expression by non-parenchymal rat liver cells in experimental biliary fibrosis. Gastroenterol 1990; 98:175–184.
172. Nakatsukasa H, Nagy P, Evarts R, Hsia C, Marsden E, Thorgerisson S. Cellular distribution of transforming growth factor-beta1 and pro-collagen type I, III, and IV transcripts in carbon tetrachloride-induced rat liver fibrosis. J Clin Invest 1990; 85:1833–1843.
173. Maher J, McGuire R. Extracellular matrix gene expression increases preferentially in rat lipocytes and sinusoidal endothelial cells during hepatic fibrosis in vivo. J Clin Invest 1990; 86:1641–1648.
174. Minato Y, Hasumura Y, Takeuchi J. The role of fat storing cells in Disse space fibrogenesis in alcoholic liver disease. Hepatology 1983; 3:559–566.
175. Okanoue T, Burbige E, French S. The role of Ito cell in perivenular and intralobular fibrosis in alcoholic hepatitis. Arch Pathol Lab Med 1982; 107:459–463.
176. Hendriks H, Verhoofstad W, Brouwer A, DeLeeuw A, Knook D. Perisinusoidal fat-storing cells are the main vitamin A storage sites in rat liver. Exp Cell Res 1985; 160:138–149.
177. Wake K. "Sternzellen" in the liver; perisinusoidal cells with special reference to storage of vitamin A. Am J Anat 1971; 132:429–462.
178. Kawada N, Tran-Thi T, Klein H, Decker K. The contraction of hepatic (Ito) stellate cells stimulated with vasoactive substances. Eur J Biochem 1993; 213:815–823.
179. Rockey DC, Housset CN, Friedman SL. Activation-dependent contractility of rat hepatic lipocytes in culture and in vivo. J Clin Invest 1993; 92:1795–1804.
180. Friedman S. The cellular basis of hepatic fibrosis. Mechanisms and treatment strategies. N Engl J Med 1993; 328:1828–1835.
181. Arthur MJ, Friedman SL, Roll FJ, Bissell DM. Lipocytes from normal rat liver release a neutral metalloproteinase that degrades basement membrane (type IV) collagen. J Clin Invest 1989; 84:1076–1085.
182. Arthur M, Stanley A, Iredale, J, Rafferty J, Hembry R, Friedman S. Secretion of 72 kDa type IV collagenase/gelatinase by cultured human lipocytes: analysis of gene expression, protein synthesis and proteinase activity. Biochem J 1992; 287:701–707.
183. Herbst H, Heinrichs O, Schuppan D, Milani S, Stein H. Temporal and spatial patterns of transin/stromelysin RNA expression following toxic injury in rat liver. Virchows Arch B Cell Pathol Incl Mol Pathol 1991; 60:295–300.
184. Arthur M. Matrix degradation in the liver. In: Surrenti C, Casini A, Milani S, Pinzain M, eds. Fat-Storing Cells and Liver fibrosis. Dordrecht: Kluwer Academic Publishers, 1994:110–129.
185. Weiner F, Giambrone M, Czaja M, et al. Ito-cell gene expression and collagen regulation. Hepatology 1990; 11:111–117.
186. Weiner F, Shah A, Biempica L, Zern M, Czaja M. The effects of hepatic fibrosis on Ito cell gene expression. Matrix 1992; 12:36–43.
187. Bachem M, Meyer D, Melchior R, Sell M, Gressner A. Activation of rat liver perisinusoidal lipocytes by transforming growth factors derived from myofibroblast-like cells. Potential mechanism of self perpetuation in liver fibrogenesis. J Clin Invest 1992; 89:193–203.
188. Rockey D, Boyles J, Gabbinai G, Friedman S. Rat hepatic lipocytes express smooth muscle actin upon activation in vivo and in culture. J Submicrosc Cytol Pathol 1992; 24:193–203.
189. Mak K, Leo A, Lieber C. Alcoholic liver injury in baboons: transformation of lipocytes to transitional cells. Gastroenterol 1984; 87:188–200.
190. Tsukamoto H, Cheng S, Blaner W. Effects of dietary polyunsaturated fat on ethanol-induced Ito cell activation. Am J Physiol (in press).
191. Friedman S, Wei S, Blaner W. Retinol release by activated rat hepatic lipocytes: regulation by Kupffer cell-conditioned medium and PDGF. Am J Physiol 1993; 264:G947–G952.
192. Matsuoka M, Pham NT, Tsukamoto H. Differential effects of interleukin-1 alpha, tumor necrosis factor alpha, and transforming growth factor beta 1 on cell proliferation and collagen formation by cultured fat-storing cells. Liver 1989; 9:71–78.
193. Weiner FR, Giambrone MA, Czaja MJ, et al. Ito-cell gene expression and collagen regulation. Hepatology 1990; 11:111–117.
194. Bissell DM, Wang SS, Jarnagin WR, Roll FJ. Cell-specific expression of transforming growth factor-beta in rat liver. Evidence for autocrine regulation of hepatocyte proliferation. J Clin Invest 1995; 96:447–455.
195. Sanderson N, Factor V, Nagy P, et al. Hepatic expression of mature transforming growth factor beta 1 in transgenic mice results in multiple tissue lesions. Proc Natl Acad Sci USA 1995; 92:2572–2576.
196. Desmouliere A, Geinoz A, Gabbiani F, Gabbiani G. Transforming growth factor-beta 1 includes alpha-smooth muscle actin expression in granulation tissue myofibroblasts and in quiescent and growing cultured fibroblasts. J Cell Biol 1993; 122:103–111.
197. Davis BH. Transforming growth factor beta responsiveness is modulated by the extracellular collagen matrix during hepatic ito cell culture. J Cell Physiol 1988; 136:547–553.
198. Duncan M, Berman B. Stimulation of collagen and glycosaminoglycan production in cultured human adult dermal fibroblasts by recombinant human interleukin-6. J Invest Dermatol 1992; 97:686–692.
199. Greenwel P, Rubin J, Schwartz M, Hertzberg EL, Rojkind M. Liver fat-storing cell clones obtained from a CCI4-cirrhotic rat are heterogeneous with regard to proliferation, expression of extracellular matrix components, interleukin-6, and connexin 43. Lab Invest 1993; 69:210–216.
200. Greenwel P, Iraburu MJ, Reyes-Romero M, et al. Induction

of an acute phase response in rats stimulates the expression of alpha 1(I) procollagen messenger ribonucleic acid in their livers. Possible role of interleukin-6. Lab Invest 1995; 72:83–91.
201. Pinzani M, Gesualdo L, Sababah G, Abboud H. Effects of platelet-derived growth factor and other polypeptide mitogens on DNA synthesis and growth of cultured rat liver fat storing cells. J Clin Invest 1989; 1786–1793.
202. Bachem M, Meyer D, Schafer W, et al. The response of rat liver perisinusoidal lipocytes to polypeptide growth regulator changes with their transdifferentiation into myofibroblast-like cells in culture. J Hepatol 1993; 18:40–52.
203. Bachem M, Melchior R, Gressner A. The role of thrombocytes in liver fibrogenesis: effects of platelet lysate and thrombocyte-derived growth factors on the mitogenic activity and glycosaminoglycan synthesis of cultured rat liver fat storing cells. J Clin Chem Clin Biochem 1989; 27:555–565.
204. Gressner A, Lotfi S, Gressner G, Lahme B. Identification and partial characterization of a hepatocyte-derived factor promoting proliferation of cultured fat-storing cells (parasinusoidal lipocytes). Hepatology 1992; 16:1250–1266.
205. Pinzani M, Abboud H, Aron D. Secretion of insulin-like growth factor-1 and binding proteins by rat liver fat-storing cells: regulatory role of platelet-derived growth factor. Endocrinology 1990; 127:2343–2349.
206. Caligiuri A, Sali C. Inflammatory mediators and fat-storing cell proliferation. In: Surrenti C, Casini A, Milani S, Pinzani M, eds. Fat-Storing Cells and Liver Fibrosis. Dordrecht: Kluwer Academic Publishers, 1994:222–235.
207. Marra F, Valente AJ, Pinzani M, Abboud HE. Cultured human liver fat-storing cells produce monocyte chemotactic protein-1. Regulation by proinflammatory cytokines. J Clin Invest 1993; 92:1674–1680.
208. Pinzani M, Abboud H, Gesualdo L, et al. Regulation of macrophage colony-stimulating factor in liver fat-storing cells by peptide growth factors. Am J Physiol Cell Physiol 1992; 262:C876–C891.
209. Hellerbrand C, Brenner D, Rippe R. Intracellular adhesion molecules 1 (ICAM-1) gene expression is induced following hepatic stellate cell activation. Hepatology 1995; 22:368A.
210. Parola M, Pinzani M, Casini A, et al. Stimulation of lipid peroxidation or 4-hydroxynonenal treatment increases procollagen alpha 1 (I) gene expression in human liver fat-storing cells. Biochem Biophys Res Commun 1993; 194:1044–1050.
211. Chojkier M, Houglum K, Solis-Herruzo J, Brenner D. Stimulation of collagen gene expression by ascorbic acid in cultured human fibroblasts. J Biol Chem 1989; 264:16957–16962.
212. Maher JJ, Tzagarakis C, Gimenez A. Malondialdehyde stimulates collagen production by hepatic lipocytes only upon activation in primary culture. Alcohol Alcohol 1994; 29:605–610.
213. Hennig B, Diana JN, Toborek M, McClain CJ. Influence of nutrients and cytokines on endothelial cell metabolism. J Am Coll Nutr 1994; 13:224–231.
214. Lin M, Pham T, Tsukamoto H. In vivo suppression of Kupffer cell NFκB activation and cytokine gene expression by iron chelator. Hepatology 1995; 22:365A.
215. Maher JJ. Rat hepatocytes and Kupffer cells interact to produce interleukin-8 (CINC) in the setting of ethanol. Am J Physiol 1995; 269:G518–523.
216. Senior J, Redeker A, Maddrey W, et al. Alcoholic hepatitis: a practical guide for physicians and other health care professionals. Vol. DHHS pub. No. (ADM) 88-1547. Washington, DC: National Institute on Alcohol Abuse and Alcoholism, 1988.
217. McClain CJ, Cohen DA, Dinarello CA, Cannon JG, Shedlofsky SI, Kaplan AM. Serum interleukin-1 (IL-1) activity in alcoholic hepatitis. Life Sc 1986; 39:1479–1485.
218. Khoruts A, Stahnke L, McClain CJ, Logan G, Allen JI. Circulating tumor necrosis factor, interleukin-1 and interleukin-6 concentrations in chronic alcoholic patients. Hepatology 1991; 13:267–276.
219. Felver ME, Mezey E, McGuire M, et al. Plasma tumor necrosis factor alpha predicts decreased long-term survival in severe alcoholic hepatitis. Alcohol Clin Exp Res 1990; 14:255–259.
220. Deviere J, Content J, Denys C, et al. Excessive in vitro bacterial lipopolysaccharide-induced production of monokines in cirrhosis. Hepatology 1990; 11:628–634.
221. McClain CJ, Cohen DA. Increased tumor necrosis factor production by monocytes in alcoholic hepatitis. Hepatology 1989; 9:349–351.
222. Bird GL, Sheron N, Goka AK, Alexander GJ, Williams RS. Increased plasma tumor necrosis factor in severe alcoholic hepatitis. Ann Intern Med 1990; 112:917–920.
223. Thiele DL. Tumor necrosis factor, the acute phase response and the pathogenesis of alcoholic liver disease. Hepatology 1989; 9:497–499.
224. Hill DB, Marsano L, Cohen D, Allen J, Shedlofsky S, McClain CJ. Increased plasma interleukin-6 activity in alcoholic hepatitis. J Lab Clin MEd 1992; 119:547–552.
225. Deviere J, Content J, Denys C, et al. High interleukin-6 serum levels and increased production by leucocytes in alcoholic liver cirrhosis. Correlation with IgA serum levels and lymphokines production. Clin Exp Immunol 1989; 77:221–225.
226. Deviere J, Content J, Denys C, et al. Immunoglobulin A and interleukin 6 form a positive secretory feedback loop: a study of normal subjects and alcoholic cirrhotics. Gastroenterology 1992; 103:1296–1301.
227. Deviere J, Vaerman JP, Content J, et al. IgA triggers TNFα secretion by monocytes. A study in normal subjects and patients with alcoholic liver cirrhosis. Hepatology 1991; 13:670–675.
228. Van de Wiel A, Delecroix D, Van Hattum J, et al. Characteristics of serum IgA and liver IgA deposits in alcoholic liver disease. Hepatology 1987; 7:95–99.
229. Van Bladel S, Libert C, Fiers W. Interleukin-6 enhances the expression of tumor necrosis factor receptors on hepatoma cells and hepatocytes. Cytokine 1991; 3:149–154.
230. Lee E, Marsano L, McClain CJ. Liver cytokines in alcoholic hepatitis as assessed by immunohistochemistry. Hepatology 1991; 14:A141.
231. Ohlinger W, Dinges HP, Zatloukal K, Mair S, Gollowitsch F, Denk H. Immunohistochemical detection of tumor necrosis factor-alpha, other cytokines and adhesion molecules in human livers with alcoholic hepatitis. Virchows Arch A Pathol Anat Histopathol 1993; 423:169–176.
232. Paronetto F. Cell mediated immunity in liver disease. Hum Pathol 1986; 17:168–178.
233. Baggiolini M, Imboden P, Detmers P. Neutrophil activation and the effects of interleukin-8/neutrophil-activating peptide 1 (IL-8/NAP-1). In: Baggiolini M, Sorg C, eds. Interleukin-8 (NAP-1) and Related Chemotactic Cytokines. Basel: Karger, 1992:1–17.
234. Sheron N, Bird G, Koskinas J, et al. Circulating and tissue levels of the neutrophil chemotaxin interleukin-8 are elevated in severe acute alcoholic hepatitis, and tissue levels correlate with neutrophil infiltration. Hepatology 1993; 18:41–46.
235. Hill DB, Marsano LS, McClain CJ. Increased plasma interleukin-8 concentrations in alcoholic hepatitis. Hepatology 1993; 18:576–580.
236. Verma B, Fogarasi M, Szabo G. Down-regulation of tumor

necrosis factor alpha activity by acute ethanol treatment in human peripheral blood monocytes. J Clin Immunol 1993; 13:8–22.
237. Bode C, Kugler V, Bode JC. Endotoxemia in patients with alcoholic and non-alcoholic cirrhosis and in subjects with no evidence of chronic liver disease following acute alcohol excess. J Hepatol 1987; 4:8–14.
238. Bigatello L, Broitman S, Fattori L, et al. Endotoxemia, encephalopathy, and mortality in cirrhotic patients. Am J Gastroenterol 1987; 82:11–15.
239. Lieber C. Hepatic, metabolic and toxic effects of ethanol: 1991 update. Alcohol Clin Exp Res 1991; 15:573–592.
240. Fridovich I. Oxygen radicals from acetaldehyde. Free Rad Biol Med 1989; 7:557–558.
241. Williams A, Barry R. Free radical generation by neutrophils: a potential mechanism of cellular injury in acute alcoholic hepatitis. Gut 1987; 28:1157–1161.
242. Situnayake R, Crump B, Thurnham D, Davies J, Gearty J, Davis M. Lipid peroxidation and hepatic antioxidants in alcoholic liver disease. Gut 1990; 31:1311–1317.
243. Marsano L, McClain C. Nutrition and alcoholic liver disease. JPEN 1991; 15:337–344.
244. Mitchell MC, Raiford DS, Mallat A. Effects of ethanol on glutathione metabolism. In: Watson RR, ed. Drug and Alcohol Abuse Reviews. Vol. 2: Liver Pathology and Alcohol. The Humana Press, Inc., 1991.
245. Beg A, Finco T, Nantermet P, Baldwin A. Tumor necrosis factor and interleukin-1 lead to phosphorylation and loss of I-kappa-B-alpha: a mechanism for NFkappaB activation. Mol Cell Biol 1993; 13:3301–3310.
246. Maxwell WJ, Keating JJ, Hogan FP, Kennedy NP, Keeling PWN. Prostaglandin E_2 and leukotriene B_4 synthesis by peripheral leucocytes in alcoholics. Gut 1989; 30:1270–1274.
247. Le Moine O, Marchant A, De Groote D, Azar C, Goldman M, Deviere J. Role of defensive monocyte interleukin-10 release in tumor necrosis factor-alpha overproduction in alcoholic cirrhosis. Hepatology 1995; 22:1436–1439.
248. Muller C, Godl I, Ahmad R, Wolf HM, Mannhalter JW, Eibl MM. Interleukin-1 production in acute viral hepatitis. Arch Dis Child 1989; 64:205–210.
249. Muller C, Knoflach P, Zielinski CC. Reduced production of immunoreactive interleukin-1 by peripheral blood monocytes of patients with acute and chronic viral hepatitis. Dig Dis Sci 1993; 38:477–481.
250. Li GY, Wang T, Huggins EM, Jr., et al. Cholylglycine measured in serum by RIA and interleukin-1 beta determined by ELISA in differentiating viral hepatitis from chemical liver injury. J Occup Med 1992; 34:930–933.
251. Torre D, Zeroli C, Giola M, et al. Serum levels of interleukin-1 alpha, interleukin-1 beta, interleukin-6, and tumor necrosis factor in patients with acute viral hepatitis. Clin Infect Dis 1994; 18:194–198.
252. Anastassakos C, Alexander GJ, Wolstencroft RA, et al. Interleukin-1 and interleukin-2 activity in chronic hepatitis B virus infection. Gastroenterology 1988; 94:999–1005.
253. Szkaradkiewicz A. IL-1 production by human monocytes in viral hepatitis type B [letter]. J Hepatol 1991; 13:381–382.
254. Civeira M, Prieto J, Morte S, Rinon M, Serrano M. Interleukins in chronic active hepatitis B. Relationship with viral markers. J Hepatol 1987; 5:37–44.
255. Yoshioka K, Kakumu S, Murakami H, Fukui K. Interleukin-2 activity in chronic active liver diseases: response by T cells and in the autologous mixed lymphocyte reaction. Clin Exp Immunol 1984; 56:668–676.
256. Saxena S, Nouri-Aria K, Anderson M, Williams R, Eddleston A. In vitro alpha-interferon treatment of peripheral blood mononuclear cells improves interleukin-2 activity in HBV-related chronic liver disease. J Hepatol 1985; 1:385–393.
257. Al-Wabel A, al-Janadi M, Raziuddin S. Cytokine profile of viral and autoimmune chronic active hepatitis. J Allergy Clin Immunol 1993; 92:902–908.
258. Muller C, Zielinski CC. Impaired lipopolysaccharide-inducible tumor necrosis factor production in vitro by peripheral blood monocytes of patients with viral hepatitis. Hepatology 1990; 12:1118–1124.
259. Cacopardo B, Fatuzzo F, Russo R, et al. Serum tumor necrosis factor in acute and fulminant hepatitis B [letter]. J Infect Dis 1992; 166:448–449.
260. Yoshioka K, Kakumu S, Arao M, Tsutsumi Y, Inoue M. Tumor necrosis factor alpha production by peripheral blood mononuclear cells of patients with chronic liver disease. Hepatology 1989; 10:769–773.
261. Sheron N, Lau J, Daniels H, et al. Increased production of tumour necrosis factor alpha in chronic hepatitis B virus infection. J Hepatol 1991; 12:241–245.
262. Fan X, Zhang Z. Increased tumour necrosis factor alpha production by neutrophils in patients with hepatitis B. J Clin Pathol 1994; 47:616–618.
263. Yoshioka K, Kakumu S, Arao M, et al. Immunohistochemical studies of intrahepatic tumour necrosis factor alpha in chronic liver disease. J Clin Pathol 1990; 43:298–302.
264. Hussain MJ, Lau JY, Williams R, Vergani D. Hepatic expression of tumour necrosis factor-alpha in chronic hepatitis B virus infection. J Clin Pathol 1994; 47:1112–1115.
265. Gonzalez-Amaro R, Garcia-Monzon C, Garcia-Buey L, et al. Induction of tumor necrosis factor alpha production by human hepatocytes in chronic viral hepatitis. J Exp Med 1994; 179:841–848.
266. Muller C, Zielinski CC. Interleukin-6 production by peripheral blood monocytes in patients with chronic liver disease and acute viral hepatitis. J Hepatol 1992; 15:372–377.
267. Sun Y, Tokushige K, Isono E. Yamauchi K, Obata H. Elevated serum interleukin-6 levels in patients with acute hepatitis. J Clin Immunol 1992; 12:197–200.
268. Kakumu S, Shinagawa T, Ishikawa T, et al. Serum interleukin 6 levels in patients with chronic hepatitis B. Am J Gastroenterol 1991; 86:1804–1808.
269. Knudsen P, Strom T, Zeldis J. Hepatitis B virus blocks monocyte release of interleukin 1 in vitro. Gastroenterology 1986; 90:1738.
270. Roulot D, Durand H, Coste T, et al. Quantitative analysis of transforming growth factor beta 1 messenger RNA in the liver of patients with chronic hepatitis C: absence of correlation between high levels and severity of disease. Hepatology 1095; 21:298–304.
271. Malizia G, Brunt EM, Peters MG, Rizzo A, Broekelmann TJ, McDonald JA. Growth factor and procollagen type I gene expression in human liver disease. Gastroenterology 1995; 108:145–156.
272. Broome U, Eriksson LS, Sundin U, Sundqvist KG. Decreased in vitro production of tumor necrosis factor in primary biliary cirrhosis patients. Scand J Gastroenterol 1992; 27:124–128.
273. Kakumu S, Shinagawa T, Ishikawa T, Yoshioka K, Wakita T, Ida N. Interleukin 6 production by peripheral blood mononuclear cells in patients with chronic hepatitis B virus infection and primary biliary cirrhosis. Gastroenterol Jpn 1993; 28:18–24.
274. Miller LC, Kaplam MM. Serum interleukin-2 and tumor necrosis factor-alpha in primary biliary cirrhosis: decrease by

colchicine and relationship to HLA-DR4. Am J Gastroenterol 1992; 87:465–470.
275. al-Janadi M, al-Wabel A, Raziuddin S. Soluble CD23 and interleukin-4 levels in autoimmune chronic active hepatitis and systemic lupus erythematosus. Clin Immunol Immunopathol 1994; 71:33–37.
276. Muto Y, Nouri-Aria KT, Meager A, Alexander GJ, Eddleston AL, Williams R. Enhanced tumour necrosis factor and interleukin-1 in fulminant hepatic failure. Lancet 1988; 2:72–74.
277. Sekiyama KD, Yoshiba M, Thomson AW. Circulating proinflammatory cytokines (IL-1 beta, TNF-alpha, and IL-6) and IL-1 receptor antagonist (IL-1Ra) in fulminant hepatic failure and acute hepatitis. Clin Exp Immunol 1994; 98:71–77.
278. Izumi S, Hughes RD, Langley PG, Pernambuco JR, Williams R. Extent of the acute phase response in fulminant hepatic failure. Gut 1994; 35:982–986.
279. Hoffmann MW, Wonigeit K, Steinhoff G, Herzbeck H, Flad HD, Pichlmayr R. Production of cytokines (TNF-alpha, IL-1-beta) and endothelial cell activation in human liver allograft rejection. Transplantation 1993; 55:329–335.
280. Imagawa D, Millis J, Olthoff K, et al. The role of tumor necrosis factor in allograft rejection. Transplantation 1990; 50:219–225.
281. Tilg H, Vogel W, Aulitzky W, et al. Evaluation of cytokines and cytokine-induced secondary messages in sera of patients after liver transplantation. Transplantation 1990; 49:1074–1080.
282. Fugger R, Hamilton G, Steininger R, Mirza D, Schulz F, Muhlbacher F. Intraoperative estimation of endotoxin, TNF-alpha, and IL-6 in orthotopic liver transplantation and their relation to rejection and operative infection. Transplantation 1991; 52:302.
283. Kita Y, Iwaki Y, Demetris AJ, Starzl TE. Evaluation of sequential serum interleukin-6 levels in allograft recipients. Transplantation 1994; 57:1037–1041.
284. Izumi S, Hughes RD, Langley PG, Pernambuco JR, Williams R. Acute phase response after liver transplantation for fulminant hepatic failure and cirrhosis. Transpl Int 1995; 8:340–345.
285. Tilg H, Nordberg J, Vogel W, et al. Circulating serum levels of interleukin 6 and C-reactive protein after liver transplantation. Transplantation 1992; 54:142.
286. Sautner T, Fugger R, Gotzinger P, et al. Tumour necrosis factor-alpha and interleukin-6: early indicators of bacterial infection after human orthotopic liver transplantation. Eur J Surg 1995; 161:97–101.
287. Tilg H, Ceska M, Vogel W, Herold M, Margreiter R, Huber C. Interleukin-8 serum concentrations after liver transplantation. Transplantation 1992; 53:800–803.
288. Le Moine O, Marchant LA, Durand F, et al. Systemic release of interleukin-10 during orthotopic liver transplantation. Hepatology 1994; 20:889–892.
289. Schilling PJ, Murray JL, Markowitz AB. Novel tumor necrosis factor toxic effects. Pulmonary hemorrhage and severe hepatic dysfunction. Cancer 1992; 69:256–260.
290. Richardson P. Ciliary neurotrophic factor: a review. Pharmac Ther 1994; 63:187–198.
291. Tracey K, Cerami A. Tumor necrosis factor and regulation of metabolism in infection: role of systemic versus tissue levels. PSEBM 1992; 200:233–239.
292. Goldblum S, Hennig B, Jay M, Yoneda K, McClain C. Tumor necrosis factor-induced pulmonary vascular endothelial injury. Infect Immunol 1989; 57:1218–1226.
293. Ferrari-Baliviera E, Mealy K, Smith R, Wilmore D. Tumor necrosis factor induces adult respiratory distress syndrome in rats. Arch Surg 1989; 124:1400–1405.
294. Nelson S, Summer W, Bagby G, et al. Granulocyte colony-stimulating factor enhances pulmonary host defenses in normal and ethanol-treated rats. J Infect Dis 1991; 164:901–906.
295. Nelson S, Bagby G, Summer W. Alcohol suppresses lipopolysaccharide-induced tumor necrosis factor activity in serum and lung. Life Sci 1989; 44:673–676.
296. Nelson S, Bagby G, Andresen J, Nakamura C, Shellito J, Summer W. The effects of ethanol, tumor necrosis factor, and granulocyte colony-stimulating factor on lung antibacterial defenses. Drugs Abuse, Immun Immunodef 1991:245–253.
297. Ikeda T, Lever A, Thomas H. Evidence for a deficiency of interferon production in patients with chronic hepatitis B virus infection acquired in adult life. Hepatology 1986; 6:962–965.
298. Di Bisceglie A, Rustgi V, Kassianides C, et al. Therapy of chronic hepatitis B with recombinant human alpha and gamma interferon. Hepatology 1990; 11:266–270.
299. Dusheiko GM, Roberts JA. Treatment of chronic type B and C hepatitis with interferon alfa: an economic appraisal [see comments]. Hepatology 1995; 22:1863–1873.
300. Koff R, Seeff L. Economic modeling of treatment in chronic hepatitis B and C: promises and limitations. Hepatology 1995; 22:1880–1882.
301. Olynyk JK, Reddy KR, Di Bisceglie AM, et al. Hepatic iron concentration as a predictor of response to interferon alfa therapy in chronic hepatitis C. Gastroenterology 1995; 108:1104–1109.
302. Hayashi H, Takikawa T, Nisamura N, al e. Improvement in serum amino transferase levels after phlebotomy in patients with chronic active hepatitis C and excessive hepatic iron. Am J Gastroenterol 1994; 89:1986–1988.
303. Sheron N, Williams R. IL-8 as a circulating cytokine: induction by recombinant tumour necrosis factor-alpha. Clin Exp Immunol 1992; 89:100–103.
304. Mukaida N, Matsushima K. Regulation of IL-8 production and the characteristics of the receptors for IL-8. In: Baggiolini M, Sorg C, eds. Interleukin 8 (NAP-1) and Related Chemotactic Cytokines. Basel: Karger, 1992:41–53.
305. Scheinberg R, Cogswell P, Lofquist A, Baldwin A. Role of transcriptional activation of I-kappa-B-alpha in mediation of immunosuppression by glucocorticoids. Science 1995; 270:283–290.
306. DeForge LE, Nguyen DT, Kunkel SL, Remick DG. Regulation of the pathophysiology of tumor necrosis factor. J Lab Clin Med 1990; 116:429–438.
307. Imperiale T, McCullough A. Do corticosteroids reduce mortality from alcoholic hepatitis? Ann Intern Med 1990; 113:299–307.
308. Bedossa P, Degott C, Duchatella V, et al. A polymorphonuclear cell infiltrate in liver biopsy of patients treated with prednisone for severe alcoholic hepatitis is associated with a favorable outcome. Gastroenterology 1992; 102:A780.
309. Ding A, Porteu F, Sanchez E, Nathan C. Downregulation of tumor necrosis factor receptors on macrophages and endothelial cells by microtubule depolymerizing agents. J Exp Med 1990; 171:715–727.
310. Tiegs G, Freudenberg MA, Galanos C, Wendel A. Colchicine prevents tumor necrosis factor-induced toxicity in vivo. Infect Immun 1992; 60:1941–1945.
311. Akriviadis E, Steindel H, Pinto P, Fong T, Reynolds T, Gupta S. Failure of colchicine to improve short term survival in patients with severe alcoholic hepatitis. Hepatology 1988; 8:1239.
312. Trinchet J, Meargrand M, Callard P, et al. Treatment of alcoholic hepatitis with colchicine: results of a randomized double blind trial. Gastroenterol Clin Biol 1989; 13:551–555.

313. Kershenobich D, Vargan F, Garcia-Tsao A, Tamayo R, Gent M, Rojkind M. Colchicine in the treatment of cirrhosis of the liver. N Engl J Med 1988; 318:1709–1713.
314. Kaplan M, Alling D, Zimmerman H, et al. A prospective trial of colchicine for primary biliary cirrhosis. N Engl J Med 1986; 315:1448–1454.
315. Warnes T, Smith A, Lee F, Haboubi N, Johnson P, Hunt L. A controlled trial of colchicine in primary biliary cirrhosis. Trial design and preliminary report. J Hepatol 1987; 5:1–7.
316. Bodenheimer H, Schaffner F, Pezzullo J. Evaluation of colchicine therapy in primary biliary cirrhosis. Gastroenterology 1988; 95:124–129.
317. McHutchinson J, Runyon B, Draguesku J, Cominelli F, Person J, Castracane J. Pentoxifylline may prevent renal impairment (hepatorenal syndrome) in severe acute alcoholic hepatitis. Hepatology 1991; 14:96A.
318. Zabel P, Wolter D, Schonharting M, Schade U. Oxpentifylline in endotoxaemia. Lancet 1989; 2:1476–1477.
319. Leuschner U, Fischer H, Kurtz W, al. e. Ursodeoxycholic acid in primary biliary cirrhosis: results of a controlled double-blind trial. Gastroenterol 1989; 97:1268–1274.
320. Heathcote E, Lindor K, Poupon R, et al. Combined analysis of french, american and canadian randomized controlled trials of ursodeoxycholic acid therapy in primary biliary cirrhosis. Gastroenterol 1995; 108:A1082.
321. Sinclair S, Greig P, Blendis L, et al. Biochemical and clinical response of fulminant viral hepatitis to administration of prostaglandin E. J Clin Invest 1989; 84:1063–1069.

28

Interaction of Inflammatory Cytokines, Bacterial Products, and Lipid Mediators in Intestinal Injury

Wei Hsueh, Xiaodi Tan, Xiaoming Sun, and Frank Gonzalez-Crussi
Children's Memorial Hospital, Northwestern University Medical School, Chicago, Illinois

INTRODUCTION

Inflammation is a complex pathophysiological process, which, on the one hand, defends the body from invading organisms or noxious agents and, on the other, may cause tissue damage in the host. Both defensive and injurious effects involve the release of inflammatory peptides and lipid mediators. Many of these mediators synergize with each other and induce one another's production, thereby forming a positive feedback control and perpetuating the inflammatory response. This chapter will discuss some interactions between pro-inflammatory cytokines and lipid mediators, specifically, TNF-α and platelet-activating factor (PAF), in the pathogenesis of intestinal injury. It should be kept in mind that, contrary to animal experiments in which simplicity of design dictates that a single cytokine be administered to induce an inflammatory response, in clinical patients multiple agents, including many pro-inflammatory cytokines and lipid mediators, are produced. The phenomena discussed in this chapter may represent only a small segment of a highly complex sequence of interrelated events whose net result is the production of irreversible tissue damage identified as necrosis. Nonetheless, the available evidence suggests that the biological phenomena here outlined occupy a central role in the pathophysiology of intestinal necrosis and may provide an opportunity for therapeutic intervention, with the consequent lessening of the heretofore serious morbidity and mortality attending intestinal diseases characterized by inflammation and necrosis.

TUMOR NECROSIS FACTOR-α, ENDOTOXIN SHOCK, AND INTESTINAL INJURY

Tumor necrosis factor-α (cachectin, TNF) is a cytokine produced mainly by activated macrophages (1,2). It has many potent proinflammatory actions (reviewed in Refs. 3–5), such as enhancement of chemotaxis (6–8) and phagocytosis (9) by polymorphonuclear leukocytes (PMNs) and macrophages (10); release of toxic oxygen intermediates by inflammatory cells (7,9,11–14); activation of endothelial cells (reviewed in Ref. 15); induction of expression of intercellular adhesion molecules (ICAMs) and endothelial leukocyte adhesion molecules (ELAM, or E-selectin) (16); increase in endothelial permeability (17) and facilitation of PMN migration across endothelial cells (18,19); induction of procoagulant activity (20,21); induction of other cytokines (5,22–27); and increasing the production of lipid mediators, such as thromboxane, prostaglandins (28–31), leukotrienes (32,33), and PAF (33–36). When administered in vivo, it produces shock and gastrointestinal injury (37,38). TNF has been proposed as a pivotal mediator of endotoxin shock because:

1. TNF induces shock (reviewed in Refs. 1–5, 39, and 40).
2. Administration of endotoxin (lipopolysaccharide, LPS) to animals (41–43) and human volunteers (44,45) increases the circulating TNF level.
3. Pretreatment of the animal with anti-TNF prevents LPS-induced shock (46).

The mechanism of TNF-induced bowel injury is unknown. It may simply be a consequence of severe splanchnic ischemia secondary to systemic hypotension, or TNF may act alone or in concert with other mediators to directly damage the intestinal epithelial cells. Ischemic bowel necrosis is a complication of septic shock in human patients. It is possible that during sepsis, TNF, a mediator of endotoxin shock, is released systemically or locally in the gut and causes the bowel necrosis. The lamina propria of the intestine contains abundant macrophages, which are believed to be the endogenous source of TNF during endotoxin shock. Recent investigation has shown that Paneth cells in normal murine small intestine contain TNF mRNA that is readily detectable by in situ hybridization (47). However, TNF protein is undetectable in these cells in normal mice. Our studies show that Paneth cells in normal human small intestine also contain TNF mRNA (48). Although it is still unclear whether these cells also participate in the production of TNF in endotoxin shock, we have observed a marked increase in TNF mRNA in Paneth cells in patients with necrotizing enterocolitis (48) (see below).

PAF, ENDOTOXIN SHOCK, AND INTESTINAL NECROSIS

Platelet-activating factor (PAF-acether) is an endogenous phospholipid mediator synthesized and secreted by many cell types, including monocytes, macrophages, PMNs, eosinophils, bone marrow–derived mast cells, endothelial cells, and platelets (for review, see Refs. 49–51). Recently, bacteria such as *Escherichia coli* have also been shown to be able to synthesize PAF (52,53). In vitro, PAF aggregates human and rabbit platelets (54,55), is chemotactic for eosinophils (56,57), and activates PMNs to release oxygen radicals (58,59) and lysosomal enzymes (58–61). When injected into animals, PAF has many potent pathophysiological actions (reviewed in Refs. 49, 50, 62, and 63), such as systemic hypotension, pulmonary hypertension, bronchoconstriction, and increased vascular permeability. PAF is proposed as an endogenous mediator of endotoxin shock based on the following evidence: (1) administration of PAF induces shock in animals (reviewed in Refs. 49, 50, 64, and 65), (2) treatment with PAF antagonists prevents endotoxin-induced shock (reviewed in Refs. 65 and 66), and (3) injection of endotoxin stimulates PAF production in vivo (67–69).

Injecting PAF either into the splanchnic circulation or systemically causes necrosis of the small intestine (70). The pathological features are similar to those of adult patients with ischemic bowel necrosis or neonates with necrotizing enterocolitis (NEC), both in gross and histological appearance (70) (Fig. 1). Mild or early bowel injury is found at the tips of the mucosal villi: only superficial loss of epithelium is observed at this stage. When a high dose of PAF is used, which produces prolonged shock, the necrotic change most typically extends throughout the mucosal layer and, in some instances, the entire thickness of the intestine. Priming the animal with a small intravenous (IV) dose of lipopolysaccharide (LPS) (70 μg/kg) substantially reduces the dose of PAF required to induce bowel necrosis. Thus, LPS and PAF act synergistically to induce bowel necrosis, as well as other systemic symptoms, such as hypotension, hemoconcentration, and leukopenia. LPS alone at this dose is without pathological effect (70).

PAF is probably the key endogenous mediator for bowel injury in endotoxemia (69). This is suggested by the following observations: (1) LPS (2 mg/kg, IV) injection increases intestinal PAF production (from control value of 0.13 ± 0.02 ng/g tissue to 2.04 ± 0.44 ng/g) in the rat, (2) LPS at this dose causes mild intestinal injury, and (3) pretreatment with a PAF antagonist, SRI 63–119 (3 mg/kg, Sandoz Research Institute. East Hanover, NJ), prevents LPS-induced bowel injury (69).

INTERACTION OF LPS, TNF, AND PAF IN THE DEVELOPMENT OF INTESTINAL INJURY

LPS and/or TNF Induce PAF Formation

Injection of TNF alone induces intestinal injury (37,38), but in our experience the extent and severity of such injury is milder than in lesions resulting from administration of PAF or PAF with LPS. Often, a near lethal dose of TNF must be used to cause severe intestinal damage. However, TNF acts in synergy with either LPS or PAF to induce severe and extensive bowel necrosis (71). To investigate the interaction of LPS and TNF in vivo, we injected recombinant TNF (0.5 mg/kg IV; Genentech, Inc., South San Francisco, CA) together with a low dose of LPS (200 μg/kg) into young male Sprague-Dawley rats and examined the systemic and local effects. We found that all animals treated with LPS and TNF developed severe shock, hemoconcentration, leukopenia, and segmental bowel necrosis, but animals treated with either agent alone at these doses maintained normal blood pressure (Fig. 2) and showed no intestinal injury (71). The leukopenia was also less severe. We also observed a synergy of TNF and LPS on the murine hypotensive response (Fig. 3). Results indicative of synergy between LPS and TNF have also been reported by other investigators (72,73).

One interesting observation of these studies was that

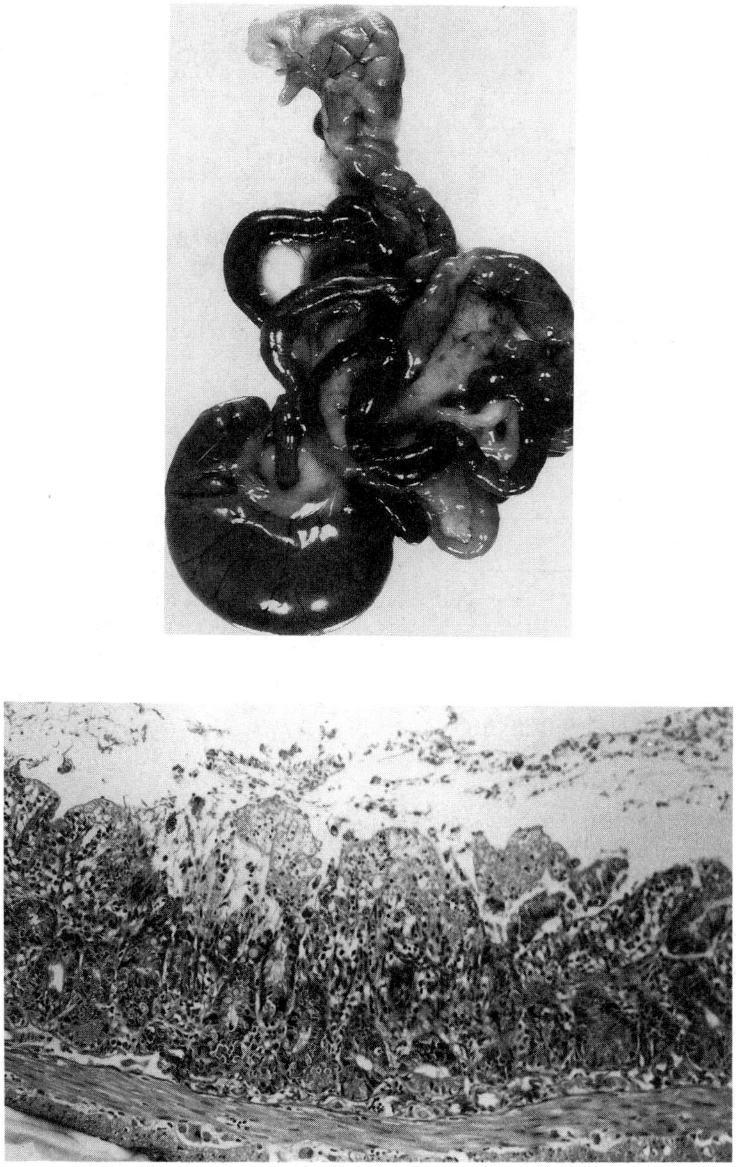

Figure 1 Gross and microscopic appearance of small intestine 2 h following intravenous injections of PAF (1-*O*-hexadecyl-2-acetyl-*sn*-glycero-3-phosphocholine; Sigma, St. Louis, MO; 3 μg/kg) into an adult Sprague-Dawley rat.

both LPS and TNF are able to induce PAF production in intestinal tissue (71). The effects of these two agents on intestinal PAF production appear to be additive. The bowel injury induced by LPS and TNF can be prevented by pretreatment with a PAF receptor antagonist, SRI 63-119, suggesting that LPS- and TNF-induced bowel injury is mediated by PAF (Table 1) (71). Moreover, systemic effects induced by LPS and TNF, such as hypotension and leukopenia, are also significantly ameliorated by the PAF antagonist, although not entirely reversed (Fig. 2). The increased hematocrit, however, is not affected by this PAF antagonist. A similar observation, that pretreatment with PAF antagonist reduced LPS-induced TNF elevation and mortality, has also been made in other models of endotoxin shock (74).

The cellular source of PAF in the intestine is unclear. In addition to the well-known origin of PAF from inflammatory cells, intestinal epithelial cells have also been shown to form PAF in vitro (75,76). TNF causes PAF production in endothelial cells (34,35) and macrophages

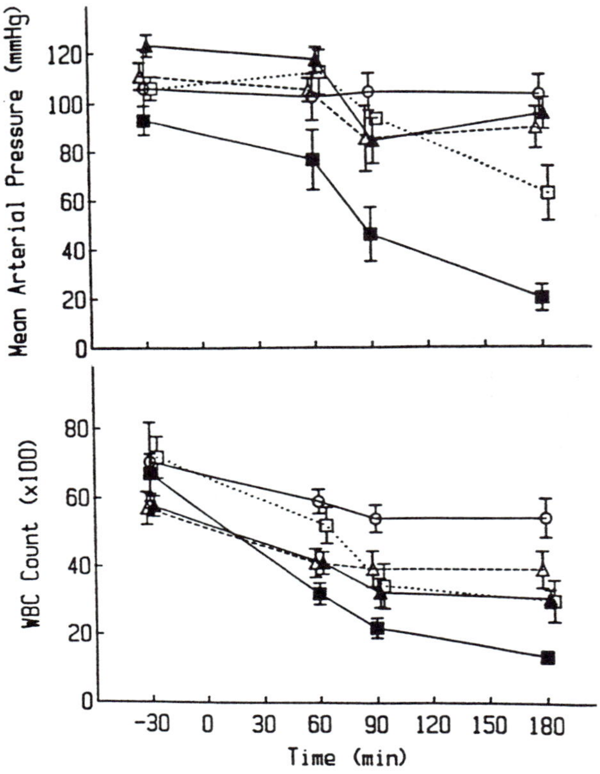

Figure 2 TNF- and LPS-induced systemic hypotension and leukopenia in young Sprague-Dawley rats (60–80 g): (open circles) control ($n = 5$, sham operated); (open triangles) 0.5 mg/kg of TNF injected at time 0 ($n = 6$); (solid triangles) 200 μg/kg of LPS injected at time 0 ($n = 7$); (solid squares) 0.5 mg/kg of TNF injected at time 0 + LPS injected at time 60 min (n = 5); and (open squares) 5 mg/kg of SRI 63-119 (divided into two doses, injected 20 min before TNF and 10 min before LPS) + 0.5 mg/kg of TNF, injected at time 0 + LPS injected at 60 min ($n = 3$).

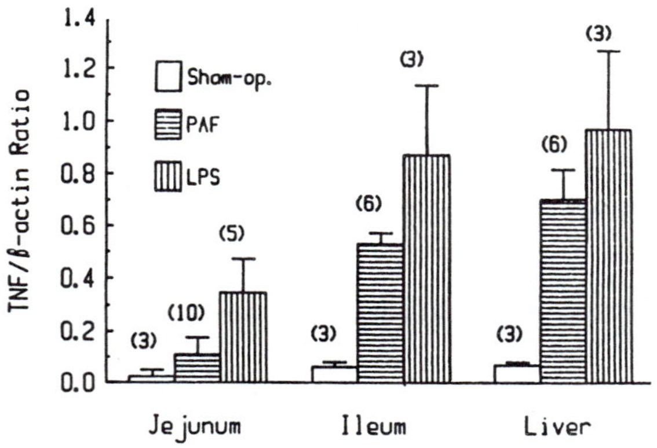

Figure 3 Increase of TNF mRNA in the liver and small intestine 30 min following injection of PAF (1 μg/kg) or LPS (2 mg/kg). TNF transcripts were analyzed by Northern blots and quantitated by densitometry.

Table 1 Synergistic Effects of TNF + LPS and TNF + PAF on Intestinal Injury in the Rat and Its Prevention by PAF Antagonist SRI 63-119

Treatment	No. of animals	Gross lesions[a] (%)	Microscopic necrosis[b]
None	5	0	0
TNF[c], 0.5 mg/kg	6	0	0
LPS, 200 μg/kg	7	0	0
TNF + LPS	5	35 ± 7	Mild (1), moderate (4)
PAF, 1 μg/kg	3	0	0
TNF + PAF	4	35 ± 4	Mild (1), moderate (3)
SRI 63119[d] + TNF + LPS	3	0	0

[a]Gross lesions: violaceous, dark, and lusterless areas as percentage of the entire length of small intestine.
[b]Microscopic lesions are graded as previously reported (70). Mild: necrosis is confined to the tips of mucosal villi. Moderate: partial or total loss of villi but necrosis confined to the mucosa.
[c]TNF (5.02×10^7 U/mg, 0.5 mg/kg) was injected IV into young Sprague-Dawley rats (60–80 g) 1 h before the intravenous injection of LPS or PAF.
[d]SRI 63-119 was divided into two doses injected 20 min before TNF and 10 min before LPS.

(35,36) in vitro. Moreover, TNF also acts synergistically with PAF in inducing bowel injury (Table 1) (71).

PAF and TNF Induce TNF Formation: Possible Role of Transcription Factor NF-κB

PAF induces TNF production in monocytes (77,78) and enhances TNF production by stimulated alveolar macrophages (79). Moreover, a rapid elevation of serum TNF was detected following PAF injection (80), suggesting that TNF may be an endogenous cytokine responsible for PAF-mediated injury. A related observation has been made, namely, that anti-TNF antibody prevented the lung injury induced by LPS/PAF (81).

The splanchnic bed has been suggested to be a major contributor to TNF production (82). Normal animals have very low levels of TNF mRNA in the liver and small intestine (Fig. 3). LPS (2 mg/kg) or PAF (1 μg/kg) induced a marked increase in TNF mRNA in the liver and distal ileum within 30 min (Fig. 3), which gradually diminished at 60 min. Interestingly, this response was much less pronounced in the jejunum and proximal ileum (Fig. 3). The doses of LPS and PAF used in these experiment were low, lower than those inducing prolonged shock and gross bowel injury, suggesting that TNF gene expression is not a nonspecific response to systemic multiorgan failure.

How does a lipid mediator induce transcription of a cytokine gene? Although we do not know the exact mechanism, our recent studies (83) showed that PAF, at a low dose (1 µg/kg) not inducing gut injury, rapidly activates NF-κB, a transcription factor involved in the induction of many pro-inflammatory cytokines (84,85). Furthermore, PAF, at the same dose, induces transcription of p50 mRNA, the precursor of NF-κB, in mouse intestine (83) (Fig. 4). The response was rapid and sustained, peaked at 30 minutes, and remained high at 90 min. In comparison, the response to TNF was much slower (Fig. 4), barely changing at 30 minutes, but much elevated at 90 min.

It is well established that TNF can activate NF-κB, probably via the signal transduction pathway PI-specific phospholipase C and sphingomyelinase (86). TNF has been shown to induce gene expression of p50/p105, a precursor of NF-κB, in vitro (87). This response may be mediated by NF-κB itself (88). TNF is also known to induce the production of other cytokines, probably also involving NF-κB (89,90).

Although both PAF and TNF induce p50/p105 expression, the regulatory pathways of these two mediators may be different. This is suggested by the observations that the action of PAF is much more rapid than TNF (Fig. 4), and that the effect of TNF is not blocked by PAF antagonists. The effect of PAF and TNF on the rapid activation and gene expression of these transcription factors may account, at least in part, for the pluripotent nature of these pro-inflammatory mediators.

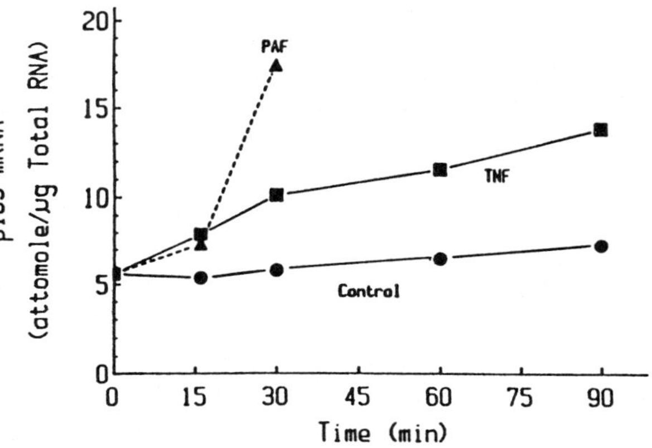

Figure 4 PAF and TNF induce the production of p50/p105 mRNA, the precursor of NF-κB, in the small intestine. (●): Sham operated. (■): TNF (1 mg/kg). (▲): PAF (1 µg/kg).

PAF Induces Endogenous PAF Production: Role of Phospholipase A_2

That exogenous PAF induces the production of endogenous PAF is a hypothesis supported by our observation that treatment with PAF antagonists decreases endogenous intestinal PAF formation in response to administration of LPS (69) or combined TNF with LPS (91). This self-perpetuating phenomenon could be partly accounted for by the formation of cytokines such as TNF, which in turn induce PAF production. In addition, PAF itself may activate the PAF-forming enzymes, i.e., phospholipase A_2 (PLA_2) and acetyltransferase. PAF has been shown to induce the activation of the latter enzyme in neutrophils (92). In the small intestine, PAF also activates PAF-forming PLA_2 (91). Combined TNF and LPS also activate this enzyme; this effect is blocked by PAF antagonists (91), suggesting that PAF is the endogenous mediator for PLA_2 activation. Although it is well established that the cytotoxic effect of TNF on tumor cells largely depends on PLA_2 activation (93), the effect of TNF on this enzyme in normal cells remains undefined. Several investigators have claimed that TNF cannot activate PLA_2 in PMNs (94). However, TNF, as a priming agent, could synergistically enhance PLA_2 activity induced by another agent, such as fMLP (94). Our studies also demonstrated a priming effect of TNF on PLA_2 activation in the small intestine (91).

Group II PLA_2 (PLA_2-II) is a 14 kDa secretory protein which has been detected in many inflammatory states (95,96). Recent reports demonstrated the preferential localization of this enzyme in the intestinal crypts, especially in the Paneth cells (97), and thus its physiological function may be similar to other Paneth cell defensins, namely, bacteriocidal (98). Although the role of this PLA_2 in PAF formation still remains to be explored, several reports have strongly suggested its participation in prostaglandin formation (99–102). PLA_2-II is known to be upregulated by TNF and LPS (103). TNF has been shown to activate PLA_2-II in endothelial cells (104) and liver cell lines (105) to release prostaglandins. We recently observed that PAF (1.5 µg/kg, IV) induced the gene expression as well as enzyme activation of PLA_2-II in the small intestine (106).

The Role of Endogenous LPS

Since at low doses LPS markedly enhances the injurious effect of TNF and PAF, and since the gut is a rich source of bacteria and their products, it is possible that endogenous LPS may participate in the deleterious actions of PAF and TNF. Analysis of the in vivo role of LPS is complicated by the fact that the experimental animals used in most published models of shock and intestinal injury were not germ-free. It is thus possible that LPS enters the circulation from the bowel lumen once the initial injury (loss of epithelium) develops. Furthermore, intestinal bacteria have been shown to be able to synthesize PAF (52,53). In healthy animals, bacteria-derived PAF is probably of little consequence. However, in animals with impaired gut mucosal barrier, bacteria-derived PAF may enter the intestinal tissue and aggravate tissue injury.

To examine the role of endogenous bacterial products, we injected PAF (3 μg/kg) into normally colonized rats, germ-free rats, and normal rats pretreated with combined antibiotics. We found that PAF did not cause prolonged shock and bowel necrosis in germ-free rats (107). However, when these rats were primed with a low dose (0.5 mg/kg) of LPS, the protection was lost. Furthermore, normally colonized rats pretreated with a combination of neomycin, polymyxin B, and metronidazole for 7 days showed an attenuated response of hypotension and intestinal necrosis to PAF (107). (Polymyxin B alone had no effect.) PAF probably induces shock, hemoconcentration, persistent leukopenia, and gross intestinal necrosis via a synergism with endogenous bacterial toxins, presumably from gut flora.

It should be noted that, contrary to animal experiments in which a single agent is usually administered to induce an inflammatory response, in clinical patients multiple bioactive agents are produced. Thus, only minute quantities of LPS, TNF, and PAF are needed to initiate the inflammatory response, but the synergistic actions of these agents and the positive feedback loops may potentiate and perpetuate the response, culminating in tissue injury.

ROLES OF EICOSANOIDS, NOREPINEPHRINE, REACTIVE OXYGEN SPECIES, AND THE COMPLEMENT SYSTEM

Platelet aggregation, from which PAF originally derived its name, is not a uniform response to PAF in all animal species. Rat platelets are refractory to PAF (70,108), and disseminated intravascular thrombi are not a feature of the pathohistology in our model. Thus, a thromboembolic phenomenon is not the cause of the intestinal ischemia and subsequent tissue necrosis seen in LPS-, TNF-, and PAF-induced bowel injury. Furthermore, neither LPS nor TNF has direct vasoconstricting actions. PAF at low doses elicits a vasodilating response in various beds (109,110). At high dose, however, PAF injection results in a severe and prolonged drop in mesenteric blood flow (110,111). Therefore, it seemed necessary to propose that vasoconstrictors are released following PAF production and that this, in turn, leads to intestinal ischemia. A close relationship of the metabolism of arachidonic acid and PAF in PMNs is well established (reviewed in Ref. 112). It is thus possible that leukotrienes and prostaglandins are released following PAF production. We have shown that leukotriene C4 (113,114) and norepinephrine (115) are released in vivo as well as from isolated, perfused rat small intestine after PAF injection. Thus, the prolonged splanchnic vasoconstriction in our model is at least in part mediated by these vasoconstricting substances.

Both TNF (7) and PAF (58–60) are known to stimulate neutrophils and monocytes to release reactive oxygen species (ROS). Furthermore, TNF induces conversion of xanthine dehydrogenase (XD) to xanthine oxidase (XO) in endothelial cells within 10 min (116). (XO is the enzyme responsible for superoxide (O_2^-) generation when catalyzing purine oxidation.) A free radical chain reaction is triggered and other ROS, including H_2O_2 and hydroxyl radicals (OH^-), are generated, the latter by an iron-catalyzed Fenton reduction. We investigated the role of ROS in PAF-induced intestinal injury by infusing the oxygen radical scavenger, superoxide dismutase (10 mg/kg), or pretreating the animal with the xanthine oxidase inhibitor allopurinol (5 mg/kg). We found that these treatments significantly ameliorated the intestinal injury. When PAF was administered alone, only 50% or less of the small bowel was perfused at the end of the experiment. Pretreatment with allopurinol or superoxide dismutase or catalase significantly improved the extent of perfusion to 65–70% (117). The histological grading of necrosis was also ameliorated. Most animals receiving PAF alone showed necrosis involving the entire mucosa; pretreatment, either with superoxide dismutase or catalase or allopurinol, suppressed the PAF-induced injury or resulted in injury limited to the villous tips (117). Although ROS seem important in the pathogenesis of experimental bowel necrosis, it remains unclear whether the main cellular source of ROS is inflammatory cells, endothelial cells, or the intestinal epithelium, which is an extremely rich source of XD/XO (118).

LPS and TNF also stimulate the production of NO and induce iNOS mRNA in macrophages (119,120), endothelial cells (121,122), and pulmonary epithelial cells (123,124). In contrast to NO produced by constitutive NOS, which helps to maintain gut mucosal integrity (125,126), cytokine-induced NO production is usually detrimental to the tissue. The mechanism of this injurious action is still poorly understood, but it is thought that the reaction between superoxide and NO generates peroxynitrite and subsequently hydroxy radicals (127,128) and that iron is released from ferritin by NO (129). We have shown that administration of NOS inhibitors aggravates PAF-induced intestinal injury, indicating a protective action of contitutive NO (130). The role of iNOS in the development of gut injury awaits to be explored.

The complement system plays an important role in inflammation, tissue injury, and endotoxin shock. LPS is known to activate complement through the classic and alternative pathways, and activation of the alternative pathway has been demonstrated in patients with gram-negative septicemia (131). Conceivably, the potentiating effect of LPS in our model could be related to activation of the complement system. To investigate the role of in vivo complement activation, we assayed serum CH100 activity in mice in response to injection of various agents (a decrease in CH100 indicates in vivo complement activation). We found that TNF (0.2 μg/g, 2.5 h) does not significantly activate the complement system, LPS (3 μg/g, 2.5 h) mildly activates it, and a combination of the two results in marked complement activation (Table 2) (132). Neither of these two agents by itself causes intestinal necrosis (Table 2). TNF and LPS at these doses have no systemic effect, but combined treatment of TNF followed by LPS resulted in severe shock (Fig. 5), reduction of intestinal perfusion, and bowel injury (Table 2) (132). These observations suggest an association of complement activation with LPS- and TNF-induced bowel injury.

To further investigate the role of the complement system, congenitally C5-deficient mice (B 10.D2/OSNJ strain) were injected with TNF (0.2 μg/g) followed at 30 minutes by LPS (3 μg/g), and the effects were compared with those observed in C5-sufficient mice treated in the same way (132). Several control mouse strains—B10.D2/NSNJ (having low C3 level), C57BL/10SNJ, and C57BL/6J—were used. The doses mentioned are lethal for normal control mice: 10 of 11 mice died, most of them within 15 minutes, and more than half of the B10.D2/NSNJ mice died. In contrast, all 14 B10.D2/OSNJ (C5-deficient) mice survived (Table 2). C5-deficient mice also show much less severe hypotension (Fig. 5) and less

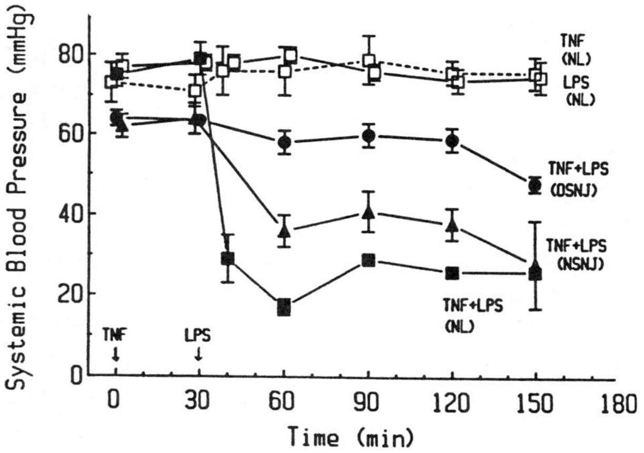

Figure 5 Change in systemic blood pressure following injection of TNF (0.2 μg/g), LPS (3 μg/g), or both in normal (C57BL/6J) and complement-deficient (B10.D2/OSNJ) mice: (□) C57BL/6J, TNF only ($n = 4$); (□) C57BL/6J, LPS only ($n = 4$); (■) C57BL/6J with TNF + LPS ($n = 8$); values from one mouse are presented at 90 min and on because of the death of all other mice); (▲) B10.D2/NSNJ, TNF + LPS ($n = 11$); and (●) B10.D2/OSNJ, TNF + LPS ($n = 14$).

Table 2 Lethality and Bowel Injury Induced by TNF and LPS in Normal Mice and Complement-Deficient Mice[a]

		Inducing agent				
Mouse strain	Plasma C5	TNF (μg/g)	LPS (μg/g)	CH100 (U/mL)	Mortality at 2.5 h	Perfused bowel (%)
C57BL/6J	Normal	0	0	18 ± 2	0/6	100
		0	3	16 ± 2	0/4	100
		0.2	0	12 ± 4	0/4	100
		0.2	3	1 ± 1	7/8	13 ± 5
B10.D2/NSNJ	Normal	0	3		0/3	100
		0.2	0		0/3	100
		0.2	3	2 ± 0	6/11	31 ± 7
B10.D2/OSNJ	0	0	3		0/3	100
		0.2	0		0/3	100
		0.2	3	0	0/14	80 ± 9

[a]B10.D2/OSNJ mice have no detectable C5.
TNF was injected (IV or intraarterially) 30 min before LPS injection (IV).

hemoconcentration than the control mice (132). This indicates that for shock and bowel injury to develop, an intact complement system, more specifically C5, is required. Although it is not known if C5a or the membrane attack complex, or both, are involved in the mediation of tissue injury, the following evidence suggests that C5a plays a significant role in our model: (1) C5a increases microvascular permeability (133), (2) C5a stimulates TNF production by mononuclear cells (134), (3) pretreatment of rats with F(ab′) fragments of rabbit antirat C5a significantly improves LPS-induced hypotension and hemoconcentration (135), and (4) C5a is required for the development of TNF-induced hemorrhagic necrosis of the skin in mice exposed to bacterial agents (136).

Although the complement system, especially C5, plays an essential role in TNF- and LPS-induced gut injury, activation of the complement system alone is insufficient to induce bowel necrosis (137). Administration of cobra venom factor, which activates the complement system in vivo, aggravated the hypotension and bowel necrosis induced by PAF and LPS but did not produce intestinal damage by itself, even when the complement level was depleted to 15% of control (137). Therefore, it is likely that other mediators, such as PAF, TNF, or LPS, must act in synergy with activated complement fragments to induce bowel necrosis. In our investigation of the role of complement, we determined that in complement-sufficient mice (132), TNF acts as a priming agent, since TNF alone did not cause systemic changes, such as blood pressure, temperature, CH100, or hematocrit. Following LPS administration, the primed system (inflammatory cells and endothelial cells) showed an exaggerated response to LPS. PAF may be the essential mediator in this response, since pretreatment with PAF antagonist abrogates both the systemic and local (intestinal) responses to TNF and LPS (138). The activated complement fragments, in synergy with TNF, LPS, and PAF, mediate shock and bowel necrosis.

The observation that congenitally C5-deficient mice are protected from TNF (0.2 μg/g)– and LPS (3 μg/g)–induced shock and tissue injury seems to indicate that C5 is absolutely required for their action. When 0.5 μg/g of TNF was used to prime the animal (with the same LPS dose), however, all animals, C5 deficient or sufficient, developed shock and severe hemoconcentration and died (132). This observation indicates that TNF and LPS induce shock and tissue injury through complement-dependent as well as complement-independent pathways. A high dose of TNF (with LPS) may have a direct effect on the vascular system and inflammatory cells or may release other mediators, independent of the complement system.

NECROTIZING ENTEROCOLITIS: A CLINICAL DISEASE THAT MAY INVOLVE TNF AND PAF

Necrotizing enterocolitis is an acute gastrointestinal disease, primarily of premature infants, with high morbidity and mortality (139,140). In some epidemiological surveys, NEC ranks as the third leading cause of death in children less than 1 year of age, next only to respiratory distress syndrome (RDS) and congenital malformations. This pattern may change in the next decade following widespread use of surfactant therapy for RDS, which dramatically decreases the mortality and morbidity of RDS. NEC occurs predominantly in the first 3 months of life and is characterized by the rapid onset of abdominal distention and tenderness, bloody stools, and the pathognomonic radiographic evidence of pneumatosis intestinalis. Frequently, NEC progresses and the affected patients develop shock, disseminated intravascular coagulation, neutropenia, thrombocytopenia, peritonitis, and pneumoperitoneum. Supportive therapy, consisting of fluids, bicarbonate, blood and platelet transfusions, pressors and antibiotics, is often inadequate, leading to surgical exploration, bowel resection, and, in 30–50% of cases, death (141).

NEC occurs either endemically or as an epidemic in the neonatal intensive care unit. It is generally accepted that an infectious agent may initiate the disease process in epidemic cases. However, the exact pathogenesis of bowel necrosis remains unclear. Even less is known about the endemic form of NEC. Several risk factors have been implicated. These include prematurity (142), feeding practice (143), hypoxia and intestinal ischemia (144), and infection (145). More than 90% of NEC patients develop the disease following enteral feedings. However, the cause of the correlation between enteral feeding and development of NEC is not understood. It has been thought to be due to massive proliferation of the intestinal flora following feeding (146) or to the speed or volume of feeding (147). Some investigators also believe that maternal milk protects against the development of NEC (148). Very little attention has been paid, heretofore, to the role of inflammatory mediators in this disease. Since the animal model we produced resembles clinical NEC in both gross and microscopic appearance, we hypothesize that LPS, PAF, and TNF may also play important roles in the pathogenesis of NEC in human patients (149).

To test our hypothesis and to investigate the role of these inflammatory mediators in NEC, we conducted a pilot study to measure plasma TNF, PAF, and the PAF-specific degrading enzyme acetylhydrolase in 12 patients with NEC and compared the results to 8 age-matched control infants (149). We found that plasma TNF concentrations

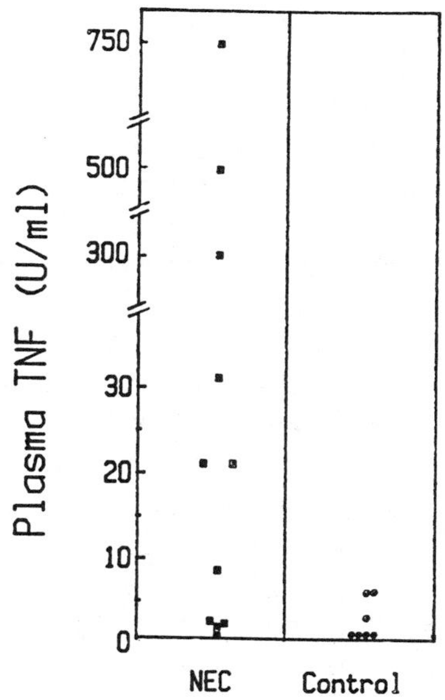

Figure 6 Individual plasma TNF values in NEC patients and controls; $p < 0.05$ with Mann-Whitney U test to compare nonparametric group means. (From Ref. 149.)

were significantly higher in NEC patients than in controls (136 ± 75 vs. 1.5 ± 0.8 U/ml, $p < 0.05$; Fig. 6). Plasma PAF levels were also higher in NEC patients compared to controls (18.1 ± 3.6 vs. 3.1 ± 0.9 ng/ml, $p < 0.01$; Fig. 7), sometimes markedly so. Plasma acetylhydrolase activity (the enzyme critical to PAF regulation) was lower in NEC patients compared to controls (10.6 ± 0.7 vs. 23.0 ± 1.4 nmol/ml/min, $p < 0.01$; Fig. 7). However, there was no correlation between PAF level and the severity of outcome of the disease. Interestingly, although plasma TNF concentrations were higher in NEC patients, there was wide individual variation and the levels did not correlate with the plasma PAF levels or the severity of the disease.

PAF has been shown to be produced by enterocytes (75,76). The cellular source of TNF was assumed to be macrophages and mast cells in the lamina propria; both cells have been shown to produce TNF in vitro (1–5,150). To our surprise, in situ hybridization study of human NEC specimens in the acute stage showed that TNF transcripts were localized in the Paneth cells and infiltrating eosinophils, and not in mucosal mast cells or resident macrophages (48). In one case which showed infiltrating macrophages in the submucosa and muscularis, TNF mRNA was also detected in these cells (48).

These observations, of course, do not prove that TNF and/or PAF are the primary mediators in the development of NEC. However, demonstration of abnormally high levels of these potent inflammatory mediators, as well as abnormally low activity of the PAF-degrading enzyme in the circulation of NEC patients, compellingly suggest that these mediators participate in the pathogenesis of NEC and may play an important role in its pathophysiology.

PAF, TNF, AND INTESTINAL ISCHEMIA/REPERFUSION

Ischemia/reperfusion injury has been implicated in many devastating intestinal diseases, including NEC (151) and hypovolemic shock (152). The ischemia-induced gut damage is aggravated during reperfusion mediated by reactive oxygen species (ROS) (153–156) and activated neutrophils (154–156,157). PAF may also contribute to the isch-

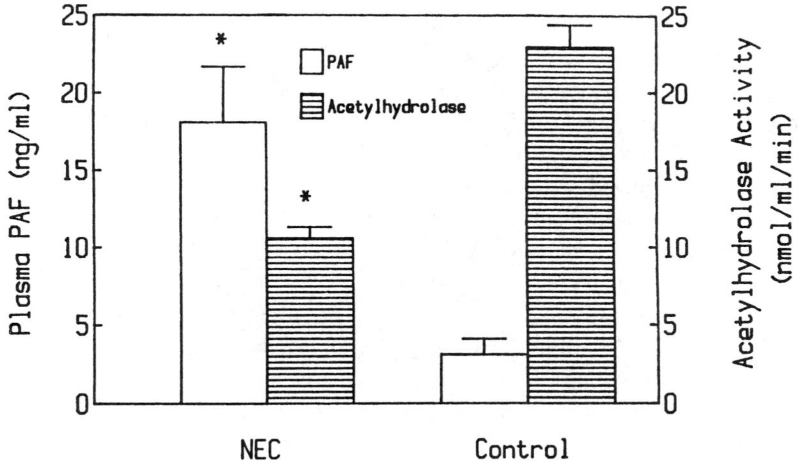

Figure 7 Plasma PAF and plasma acetylhydrolase activity in NEC patients and controls. $^*p < 0.01$ compared with control values with two-tailed unpaired Student's t-test. (From Ref. 149.)

emia/reperfusion injury. PAF is released by intestinal tissue during reperfusion, and PAF antagonists attenuate leukocyte adherence and emigration in postcapillary venules in ischemia/reperfusion injury (158). Besides local tissue necrosis, systemic and pulmonary injury may also develop after intestinal ischemia/reperfusion. The mediators for extraintestinal injury include PAF (159,160) and TNF (161). It has been shown that pretreatment of the animal with PAF antagonists prevents mesenteric ischemia/reperfusion–induced shock (159). TNF may be formed by hepatic macrophages probably induced by gut-derived bacterial toxins following reperfusion, and anti-TNF antibody attenuated the pulmonary microvascular injury (161).

THE ROLE OF CYTOKINES AND LIPID MEDIATORS IN INFLAMMATORY BOWEL DISEASE

Although the etiology and initiating events of ulcerative colitis (UC) and Crohn's disease (CD) are unknown, both of these diseases are associated with an inflammatory reaction in their active state. UC is characterized by neutrophil infiltration of the colonic mucosa with cryptitis and/or crypt abscesses, and CD is characterized by segmental, transmural, acute, and chronic inflammation of the small and/or large intestines with or without granuloma formation. The lamina propria contains many immune cells capable of producing various proinflammatory cytokines that may participate in the pathogenesis of inflammatory bowel disease (IBD) (reviewed in Ref. 162).

The importance of TNF-α in IBD remains controversial. Levels of mucosal TNF in CD and UC vary significantly. In some reports, TNF levels have been comparable to controls, in others high (163–167). Reportedly, TNF production in CD is higher than in UC (167). A recent study described increased numbers of TNF-secreting cells in biopsy specimens from children with nonspecific inflammation and IBD; the frequency was higher in CD (but not UC) compared with nonspecific inflammation (167). Immunohistochemical study has localized the TNF immunoreactivity in cells in the lamina propria of both UC and CD specimens (168). Serum TNF levels during relapse of CD and UC patients are also elevated (169).

The production of IL-1 by mucosal mononuclear cells from patients with active UC and CD is often increased (166,170). High mucosal IL-8 production was observed in IBD, especially in UC (163,171,172). Colonic IL-8 levels correlated well with the degree of inflammation and neutrophil numbers in the mucosa (171). Increased IL-6 gene expression and secretion have also been reported in the colonic mucosa of IBD patients (163,166,173). Recent reports suggest a possible role of another chemotactic agent, MCP-1, in IBD (174). MCP-1 was found to be constitutively expressed in normal intestinal epithelium, and its expression in epithelial cells, as well as lamina propria and endothelial cells, is increased in UC and CD patients. There are also changes in lymphocyte-derived cytokines. Lamina propria mononuclear cells from UC and CD produce less IL-2 than control patients (162); and UC cells also display diminished cytotoxicity (162). Whether this is due to downregulation or to binding by excess receptor is presently unclear. IFN-γ production of the mucosal biopsies in response to mitogen or IL-2 is also lower in patients with UC and CD than in controls (162).

In addition to lamina propria cells, intestinal epithelial cells also produce various inflammatory cytokines, including IL-6 (175), IL-8 (176), and MCP-1 (174) (Table 3). Whether the epithelial cell is also a source of IL-1 remains controversial (177,178). Many of these inflammatory cytokines have potent physiological effects on epithelial cell function (Table 3). Due to the pleiotropy and functional redundancy of the inflammatory cytokines and the lack of an ideal animal model for IBD, it has been almost impossible to fully delineate the role of each cytokine in the pathophysiology of IBD. However, the mutual interactions of these cytokines and their interaction with lipid mediators are likely to play an important role in the inflammatory process. Therefore, blockade of one or more of these mediators may benefit patients with IBD.

Levels of lipid mediators such as prostaglandins (179), thromboxane (180), and 5-lipoxygenase products (181,182), have long been known to be elevated in mucosa specimens from IBD patients. LTB4 is largely responsible for the lipid-extractable chemotactic activity in active UC (182) and has been a target for therapeutic intervention in IBD (183). LTB_4 activates macrophages and may promote the synthesis of proinflammatory cytokines such IL-1, TNF, and IL-8 (184). The role of PGs in intestinal inflammation may be more complex, since PGE_2 is immunosuppressive (185) and inhibits IL-1 production and function in vitro (186,187). Basal and stimulated PAF levels of epithelial and lamina propria cells have been found to be elevated in colonic biopsy specimens of UC (188), but not in CD. The increase in mucosal PAF in UC may not simply reflect the infiltrating PMNs, since only colonocytes (not lamina propria cells) from the biopsy specimens showed an increase in PAF production following ionophore stimulation (189). In contrast, cells isolated from biopsies of CD patients did not exhibit an upregulation of PAF production when compared to controls. There is also some evidence supporting a pathogenetic role of PAF in CD. Recent reports have described increased PAF concentrations in

Table 3 Proinflammatory Cytokines[a] Produced by Intestinal Epithelial Cells That Affect Epithelial Function

TNF
Made by Paneth cells:
Cytotoxic and growth inhibition for epithelial cells, in synergy with IFN-γ
Affects intestinal ion transport
Upregulates IL-8 production
May synergize with IFN-γ to increase intestinal permeability
IL-1
May be made by epithelial cells:
Increases mucus secretion
Regulates intestinal ion and water transport via prostaglandin release
Upregulates IL-8 production
IFN-γ
Detected in occasional epithelial cells
Increases epithelial permeability by altering intestinal barrier function
Inhibits epithelial cell proliferation probably by suppressing glutaminase
Upregulates polymeric Ig receptors in vitro
Upregulates ICAM-1 on epithelial cells in vitro
IL-6
Made by enterocytes
Secretion stimulated by TGF-β, enhanced by IL-6
IL-8
Secreted by intestinal epithelial cells upon stimulation
MCP-1
Expressed constitutively in the epithelium

[a]All produced by immune cells.

ileal or colonic biopsies (190) and in stools (191) of patients with active CD.

CONCLUSIONS

The pathogenetic mechanism of intestinal injury is highly complex. Evidence has emerged to suggest a central role of the intricate interaction and synergy between the various cytokines, lipid mediators, and endogenous bacteria. The resulting inflammatory response affects the functions of endothelial, epithelial, and lamina propria cells, as well as the circulating inflammatory cells that are recruited into the inflamed site. The mere presence of various cytokines and lipid mediators does not constitute proof of the primary pathogenetic role of any. However, since these mediators contribute significantly to the perpetuation and progress of inflammation, therapeutic targeting of a key mediator may hinder the inflammatory reaction, reduce tissue injury, and decrease the morbidity and mortality of the disease.

ACKNOWLEDGMENT

This work was supported by NIH Grants DK34574 and HD31840.

REFERENCES

1. Beutler B, Cerami A. Cacheain: more than a tumor necrosis fractor. N Engl J Med 1987; 316:379–385.
2. Old LJ. Tumor necrosis factor. Science 1985; 230:630–632.
3. Beutler B, Cerami A. Tumor necrosis, cachexia, shock, and inflammation: a common mediator. Annu Rev Biochem 1988; 57:505–518.
4. Tracey KJ, Vlassara H, Cerami A. Cachectin/tumour necrosis factor. Lancet 1989; 1:1122–1126.
5. Beutler B, Cerami A. The biology of cachectin/TNF-α primary mediator of the host response. Annu Rev Immunol 1989; 7:625–655.
6. Yonemaru M, Stephens KE, Ishizaka A, et al. Effects of tumor necrosis factor on PMN chemotaxis, chemiluminescence, and elastase activity. J Lab Clin Med 1989; 114:675–681.
7. Figari I, Mori N, Palladino MJ. Regulating of neutrophil chemotaxis and superoxide production by recombinant tumor necrosis factor-alpha and beta comparison to recombinant interferon-gamma and interleukin-1 alpha. Blood 1987; 70:979–984.
8. Ming WJ, Bersani L, Mantovani A. Tumor necrosis factor is chemotactic for monocytes and polymorphonuclear leukocytes. J Immunol 1987; 138:1469–1474.
9. Ferrante A. Tumor necrosis factor alpha potentiates neutrophil antimicrobial activity: increased fungicidal activity against *Torulopsis glabrata* and *Candida albicans* and associated increases in oxygen radical production and lysosomal enzyme release. Infect Immun 1989; 57:2115–2122.
10. Hoffman M, Weinberg JB. Tumor necrosis factor-alpha in-

duces increased hydrogen peroxide production and Fc receptor expression, but not increased la antigen expression by peritoneal macrophages. J Leuk Biol 1987; 42:704–707.
11. Klebanoff SJ, Vadas MA, Harlan JM, et al. Stimulation of neutrophils by tumor necrosis factor. J Immunol 1986; 136:4220–4225.
12. Berkow RL, Wang D, Larrick JW, Dodson RW, Howard TH. Enhancement of neutrophil superoxide production by preincubation with recombinant human tumor necrosis factor. J Immunol 1987; 139:3783–3791.
13. Nathan CF. Neutrophil activation on biological surfaces: massive secretion of hydrogen peroxide in response to product of macrophages and lymphocytes. J Clin Invest 1987; 80:1550–1560.
14. Laudanna C, Miron S, Berton G, Rossi F. Tumor necrosis factor-alpha/cachectin activates the O_2-generating system of human neutrophils independently of the hydrolysis of phosphoinositides and the release of arachidonic acid. Biochem Biophys Res Commun 1990; 166:308–315.
15. Pober JS. Cytokine-mediated activation of vascular endothelium. Physiology and pathology. Am J Pathol 1988; 133:426–433.
16. Luscinskas FW, Brock AF, Arnaout MA, Gimbrone MA Jr. Endothelial-leukocyte adhesion molecule-1-dependent and leukocyte (CDII/CD18)-dependent mechanisms contribute to polymorphonuclear leukocyte adhesion to cytokine-activated human vascular endothelium. J Immunol 1989; 142:2257–2263.
17. Brett J, Gerlach H, Nawroth P, Steinberg S, Godman G, Stern D. Tumor necrosis factor/cachectin increases permeability of endothelial cell monolayers by a mechanism involving regulatory G proteins. J Exp Med 1989; 169:1977–1991.
18. Furie MB, McHugh DD. Migration of neutrophils across endothelial monolayers is stimulated by treatment of the monolayers with interleukin-1 or tumor necrosis factor alpha. J Immunol 1989; 143:3309–3317.
19. Moser R, Schleiffenbaum B, Groscurth P, Fehr J. Interleukin I and tumor necrosis factor stimulate human vascular endothelial cells to promote transendothelial neutrophil passage. J Clin Invest 1989; 83:444–455.
20. McKenna TM, Titius WA. Role of monokines in altering receptor and non-receptor mediated vascular contraction in sepsis. Prog Clin Biol Res 1989; 286:279–303.
21. Bauer KA, ten Cate H, Barzegar S, Spriggs DR, Sherman ML, Rosenberg RD. Tumor necrosis factor infusions have a procoagulant effect on the hemostatic mechanism of humans. Blood 1989; 74:165–172.
22. Brouchaert PG, Libert C, Everaerdt B, Takahashi N, Fiers W. A role for interleukin 1 in the in vivo actions of tumor necrosis factor. Lymphokine Res 1989; 8(3):269–274.
23. Sica A, Wang JM, Colotta F, et al. Monocyte chemotactic and activating factor gene expression induced in endothelial cells by IL-I and tumor necrosis factor. J Immunol 1990; 144:3034–3038.
24. Shalaby MR, Waage A, Aarden L, Espevik T. Endotoxin, tumor necrosis factor-alpha and interleukin I induce interleukin 6 production in vivo. Clin Immunol Immunopathol 1989; 53:488–498.
25. Locksley RM, Heinzel FP, Shepard HM, et al. Tumor necrosis factors alpha and beta differ in their capacities to generate interleukin I release from human endothelial cells. J Immunol 1987; 139:1891–1895.
26. Dinarello CA, Cannon JG, Wolff SM, et al. Tumor necrosis factor (cachectin) is an endogenous pyrogen and induces production of interleukin 1. J Exp Med 1986; 163:1433–1450.
27. Kaushansky K, Broudy VC, Harlan JM, Adamson JW. Tumor necrosis factor-alpha and tumor necrosis factor-beta (lymphotoxin) stimulate the production of granulocyte-macrophage colony-stimulating factor, macrophage colony-stimulating factor, and IL-1 in vivo. J Immunol 1988; 141:3410–3415.
28. Dayer J, Beutler B, Cerami A. Cachectin/tumor necrosis factor stimulates collagenase and prostaglandin E2 production by human synovial cells and dermal fibroblasts. J Exp Med 1985; 162:2163–2168.
29. Endo H, Akahoshi T, Kashiwazaki S. Additive effects of IL-1 and TNF on induction of prostacyclin synthesis in human vascular endothelial cells. Biochem Biophys Res Commun 1988; 156:1007–1014.
30. Zavoico GB, Ewenstein BM, Schafer AI, Pober JS. IL-1 and related cytokines enhance thrombin-stimulated PGI2 production in cultured endothelial cells without affecting thrombin-stimulated von Willebrand factor secretion or platelet-activating factor biosynthesis. J Immunol 1989; 142:3993–3999.
31. Kettelhut IC, Fiers W, Goldberg AL. The toxic effects of tumor necrosis factor in vivo and their prevention by cyclooxygenase inhibitors. Proc Natl Acad Sci USA 1987; 84:4273–4277.
32. Huber M, Beutler B, Keppler D. Tumor necrosis factor alpha stimulates leukotriene production in vivo. Eur J Immunol 1988; 18:2085–2088.
33. Camussi G, Tetta C, Bussolino F, Baglioni C. Tumor necrosis factor stimulates human neutrophils to release leukotriene B4 and platelet-activating factor. Induction of phospholipase A2 and acetyl-CoA: I-alkyl-sn-glycero-3-phosphocholine 2-acetyltransferase activity and inhibition by antiproteinase. Eur J Biochem 1989; 182:661–666.
34. Bussolino F, Camussi G, Baglioni C. Synthesis and release of platelet-activating factor by human vascular endothelial cells treated with tumor necrosis factor or interleukin 1 alpha. J Biol Chem 1988; 263:11856–11861.
35. Camussi G, Bussolino F, Salvidio G, Baglioni C. Tumor necrosis factor/cachectin stimulates peritoneal macrophages, polymorphonuclear neutrophils, and vascular endothelial cells to synthesize and release platelet-activating factor. J Exp Med 1987; 166:1390–1404.
36. Valone FH, Epstein LB. Biphasic platelet-activating factor synthesis by human monocytes stimulated with IL-I-beta, tumor necrosis factor, or IFN-gamma. J Immunol 1988; 141:3945–3950.
37. Tracey KJ, Beutler B, Lowry SF, et al. Shock and tissue injury induced by recombinant human cachectin. Science 1986; 234:470–474.
38. Remick DG, Kunkel RG, Larrick JW, Kunkel SL. Acute in vivo effects of human recombinant tumor necrosis factor. Lab Invest 1987; 56:583–590.
39. Tracey KJ, Lowry SF, Cerami A. Cachectin/TNF-alpha in septic shock and septic adult respiratory distress syndrome. Am Rev Respir Dis 1988; 138:1377–1379.
40. Remick DG, Kunkel SL. Toxic effects of cytokines in vivo. Lab Invest 1989; 60:317–319.
41. Waage A. Production and clearance of tumor necrosis factor in rats exposed to endotoxin and dexamethasone. Clin Immunol Immunopathol 1987; 45:348–355.
42. Beutler BA, Milsark IW, Cerami A. Cachectin/tumor necrosis factor: production, distribution, and metabolic fate in vivo. J Immunol 1985; 135:3972–3977.
43. Mathison JC, Wolfson E, Ulevitch RJ. Participation of tumor necrosis factor in the mediation of Gram negative bacterial

lipopolysaccharide-induced injury in rabbits. J Clin Invest 1988; 81:1925–1937.

44. Cannon JG, Tompkins RG, Gelfand JA, et al. Circulating interleukin-1 and tumor necrosis factor in septic shock and experimental endotoxin fever. J Infect Dis 1990; 161:79–84.
45. Michie HR, Manogue KR, Spriggs DR, et al. Detection of circulating tumor necrosis factor after endotoxin administration. N Engl J Med 1988; 318:1481–1486.
46. Tracey KJ, Fong Y, Hesse DG, et al. Anti-cachectin/TNF monoclonal antibodies prevent septic shock during lethal bacteraemia. Nature 1987; 330:6624.
47. Keshav S, Lawson L, Chung LP, Stein M, Perry VH, Gordon S. Tumor necrosis factor mRNA localized to Paneth cells of normal murine intestinal epithelium by in situ hybridization. J Exp Med 1990; 171:327–332.
48. Tan XD, Hsueh W, Gonzalez-Crussi F. Cellular localization of TNF-alpha transcripts in normal bowel and in necrotizing enterocolitis. TNF gene expression in Paneth cells, intestinal eosinophils and macrophages. Am J Pathol 1993; 142:1858–1865.
49. Benveniste J. Paf-acether, an ether phospholipid with biological activity. Prog Clin Biol Res 1988; 282:73–85.
50. Hanahan DJ. Platelet activating factor: a biologically active phosphoglyceride. Annu Rev Biochem 1986; 55:483–509.
51. Snyder F. Biochemistry of platelet-activating factor: a unique class of biologically active phospholipids. Proc Soc Exp Biol Med 1989; 190:125–135.
52. Denizot Y, Dassa E, Kim HY, et al. Synthesis of PAF-acether from exogenous precursors by the prokaryote *Escherichia coli*. FEBS Lett 1989; 243:13–16.
53. Denizot Y, Dassa E, Benveniste J, Thomas Y. Paf-acether production by *Escherichia coli*. Biochem Biophys Res Commun 1989; 161:939–943.
54. McManus LM, Hanahan DJ, Pinckard RN. Human platelet stimulation by acetyl glyceryl ether phosphoryline. J Clin Invest 1981; 67:903–906.
55. Sugatani J, Miwa M, Hanahan DJ. Platelet-activating factor stimulation of rabbit platelets is blocked by serine protease inhibitor (chymotryptic protease inhibitor). J Biol Chem 1987; 262:5740–5747.
56. Wardlaw AJ, Moqbel R, Cromwell O, Kav AB. Platelet-activating factor. A potent chemotactic and chemokinetic factor for human eosinophils. J Clin Invest 1986; 78:1701–1706.
57. Sigal CE, Valone FH, Holtzman MJ, Geotzl EJ. Preferential human eosinophil chemotactic activity of the platelet-activating factor (PAF) I-0-hexadecyl-2-acetyl-sn-glyceryl-3-phosphocholine (AGEPC). J Clin Immunol 1987; 7:179–184.
58. Shaw JO, Pinckard RN, Ferrigni KS, McManus LM, Hanahan DJ. Activation of human neutrophils with I-0-hexadecyl/octadecyl-2-acetyl-sn-glyceryl-3-phosphorylcholine (platelet activating factor). J Immumol 1981; 127:1250–1255.
59. Gay JC, Beckman JK, Zaboy KA, Lukens JN. Modulation of neutrophil oxidative responses to soluble stimuli by platelet-activating factor. Blood 1986; 67:931–936.
60. O'Flaherty JT, Wykle RL, Miller CH, et al. I-0-Alkyl-sn-glyceryl-3-phosphorylcholines. A novel class of neutrophil stimulants. Am J Pathol 1981; 103:70–79.
61. Dewald B, Baggiolini M. Platelet-activating factor as a stimulus of exocytosis in human neutrophils. Biochim Biophys Acta 1981; 888:42–48.
62. Feuerstein G, Goldstein RE. Effect of PAF on the cardiovascular system. In: Snyder F, ed. Platelet-Activating Factor and Related Lipid Mediators. New York: Plenum Press, 1987:403–424.
63. Vargaftig BB. Bronchopulmonary pharmacology of PAF-acether. In: Synder F, ed. Platelet-Activating Factor and Related Lipid Mediators. New York: Plenum Press, 1987:341–353.
64. Feuerstein G, Siren AL. Platelet-activating factor and shock. Prog Biochem Pharmacol 1988; 22:181–190.
65. Handley DA. Platelet-activating factor as a mediator of endotoxin-related diseases. In: Handley DA, Saunders RN, Houlihan WJ, Tomesch JC, eds. Platelet-Activating Factor in Endotoxin and Immune Diseases. New York: Marcel Dekker, 1990:451–495.
66. Toth PD. The biological effects of PAF antagonists on endotoxemia. In: Handley DA, Saunders RN, Houlihan WJ, Tomesch JD, eds. Platelet-Activating Factor in Endotoxin and Immune Diseases. New York: Marcel Dekker, 1990:589–608.
67. Inarrea P, Gomez-Cambronero J, Pascual J, del Carmen Ponte M, Hernando L, Sanchez-Crespo M. Synthesis of PAF-acether and blood volume changes in gram-negative sepsis. Immunopharmacology 1985; 9:45–52.
68. Chang WS, Feddersen CO, Henson PM, Voelkel NF. Platelet-activating factor mediates hemodynamic changes and lung injury in endotoxin treated rat. J Clin Invest 1987; 79:1498–1509.
69. Hsueh W, Gonzalez Crussi F, Arroyave JL. Platelet-activating factor: an endogenous mediator for bowel necrosis in endotoxemia. FASEB J 1987; 1:403–405.
70. Gonzalez-Crussi F, Hsueh W. Experimental model of ischemic bowel necrosis. The role of platelet-activating factor and endotoxin. Am J Pathol 1983; 112:127–135.
71. Sun XM, Hsueh W. Bowel necrosis induced by tumor necrosis factor in rats is mediated by platelet-activating factor. J Clin Invest 1988; 81:1328–1331.
72. Neilson IR, Neilson KA, Yunis EJ, Rowe ML. Failure of tumor necrosis factor to produce hypotensive shock in the absence of endotoxin. Surgery 1989; 106:439–443.
73. Rothstein JL, Schreiber H. Synergy between tumor necrosis factor and bacterial products causes hemorrhagic necrosis and lethal shock in normal mice. Proc Natl Acad Sci USA 1988; 85:607–611.
74. Kuipers B, van der Poll T, Levi M, van Deventer SJ, ten Cate H, Imai Y, Hack CE, ten Cate JW. Platelet-activating factor antagonist TCV-309 attenuates the induction of the cytokine network in experimental endotoxemia in chimpanzees. J Immunol 1994; 152:2438–2446.
75. Gustafson C, Kald B, Sjodahl R, Tagesson C. Phospholipase C from *Clostridium perfringens* stimulates formation and release of platelet-activating factor (PAF-acether) in cultured intestinal epithelial cells (INT 407). Scand J Gastroenterol 1991; 26:1000–1006.
76. Ferraris L, Karmeli F, Eliakim R, Klein J, Fiocchi C, Rachmilewitz D. Intestinal epithelial cells contribute to the enhanced generation of platelet activating factor in ulcerative colitis. Gut 1993; 34:665–668.
77. Rose JK, Debs RA, Philip R, Ruis NM, Valone FH. Selective activation of human monocytes by the platelet-activating factor analog 1-0-hexadecyl-2-0-methyl-sn-glycero-3-phosphorylcholine. J Immunol 1990; 144:3513–3517.
78. Bonavida B, Mencia-Huerta J-M, Braquet P. Effect of platelet activating factor on monocyte activation and production of tumor necrosis factor. Int Arch Allergy Apply Immunol 1989; 88:157–160.
79. Dubois C, Bissonnette E, Rola-Pleszczynski M. Platelet-activating (PAF) enhances tumor necrosis factor production by alveolar macrophages. Prevention by PAF receptor antagonists and lipoxygenase inhibitors. J Immunol 1989; 143:964–970.

80. Huang L, Tan X, Crawford SE, Hsueh W. Platelet-activating factor and endotoxin induce tumour necrosis factor gene expression in rat intestine and liver. Immunology 1994; 83:65–69.
81. Rabinovici R, Bugelski PJ, Esser KM, Hillegass LM, Griswold DE, Vernick J, Feuerstein G. Tumor necrosis factor-alpha mediates endotoxin-induced lung injury in platelet activating factor-primed rats. J Pharmacol Exp Ther 1993; 267:1550–1557.
82. Fong YM, Marano MA, Moldawer LL, Wei H, Calvano SE, Kenney JS, Allison AC, Cerami A, Shires GT, Lowry SF. The acute splanchnic and peripheral tissue metabolic response to endotoxin in humans. J Clin Invest 1990; 85:1896–1904.
83. Tan X, Sun X, Gonzalez-Crussi FX, Ganzalez-Crussi F, Hsueh W. PAF and TNF increase the precursor of NF-kappa B p50 mRNA in mouse intestine: quantitative analysis by competitive PCR. Biochim Biophys Acta 1994; 1215:157–162.
84. Lenardo MJ, Baltimore D. NF-kB: a pleiotropic mediator of inducible and tissue-specific gene control. Cell 1990; 58:227–229.
85. Baeuerle PA. The inducible transcription activator NF-kB: regulation by distinct protein subunits. Biochim Biophys Acta 1991; 1072:63–80.
86. Schutze S, Potthoff K, Machleidt T, Berkovic D, Wiegmann K, Kronke M. TNF activates NF-kappa B by phosphatidylcholine-specific phospholipase C-induced "acidic" sphingomyelin breakdown. Cell 1992; 71:765–776.
87. Meyer R, Hatada EN, Hohmann HP, Haiker M, Bartsch C, et al. Cloning of the DNA-binding subunit of human nuclear factor kappa B: the level of its mRNA is strongly regulated by phorbol ester or tumor necrosis factor alpha. Proc Natl Acad Sci USA 1991; 88:966–970.
88. Ten RM, Paya CV, Isreal N, Bail OL, Mattei M-G, Virelizier JL, Kourilsky P, Isreal A. The characterization of the promoter of the gene encoding the p50 subunit of NF-kappa B indicates that it participates in its own regulation. EMBO J 1992; 11:195–203.
89. Zhang YH, Lin JX, Vilcek J. Interleukin-6 induction by tumor necrosis factor and interleukin-1 in human fibroblasts involves activation of a nuclear factor binding to a kappa B-like sequence. Mol Cell Biol 1990; 10:3818–3823.
90. Anisowicz A, Messineo M, Lew SW, Sager R. An NF-kB-like transcription factor mediates IL-1/TNF-alpha induction of gro in human fibroblasts. J Immunol 1991; 147:520–527.
91. Sun X, Caplan MS, Hsueh W. Tumour necrosis factor and endotoxin synergistically activate intestinal phospholipase A2 in mice. Role of endogenous platelet activating factor and effect of exogenous platelet activating factor. Gut 1994; 35:215–219.
92. Doebber TW, Wu MS. Platelet-activating factor (PAF) stimulates the PAF-synthesizing enzyme acetyl-CoA:1-alkyl-sn-glycero-3-phosphocholine O_2-acetyltransferase and PAF synthesis in neutrophils. Proc Natl Acad Sci USA 1987; 84:7557–7561.
93. Suffys P, Beyaert R, Devalck D, Vanhaesebroeck B, Van Roy F, Fiers W. Tumour-necrosis-factor-mediated cytotoxicity is correlated with phospholipase-A2 activity, but not with arachidonic cid release per se. Eur J Biochem 1991; 195:465–475.
94. Bauldry SA, McCall CE, Cousart SL, Bass DA. Tumor necrosis factor-alpha priming of phospholipase A2 activation in human neutrophils. An alternative mechanism of priming. J Immunol 1991; 146:1277–1285.
95. van den Bosch H, Aarsman AJ, van Schaik RH, Schalkwijk CG, Neijs FW, Sturk A. Structural and enzymological properties of cellular phospholipases A2. Biochem Soc Transact 1990; 18:781–785.
96. Mayer RJ, Marshall LA. New insights on mammalian phospholipase A2(s); comparison of arachidonoyl-selective and -nonselective enzymes. FASEB J 1993; 7:339–348.
97. Nevalainen TJ, Gronroos JM, Kallajoki M. Expression of group II phospholipase A2 in the human gastrointestinal tract. Lab Invest 1995; 72:201–208.
98. Harwig SS, Tan L, Qu XD, Cho Y, Eisenhauer PB, Lehrer RI. Bactericidal properties of murine intestinal phospholipase A2. J Clin Invest 1995; 95:603–610.
99. Kurihara H, Nakano T, Takasu N, Arita H. Intracellular localization of group II phospholipase A2 in rat vascular smooth muscle cells and its possible relationship to eicosanoid formation. Biochim Biophys Acta 1991; 1082:285–292.
100. Murakami M, Kudo I, Inoue K. Eicosanoid generation from antigen-primed mast cells by extracellular mammalian 14-kDa group II phospholipase A2. FEBS Lett 1991; 294:247–251.
101. Nakazato Y, Simonson MS, Herman WH, et al. Interleukin-1 alpha stimulates prostaglandin biosynthesis in serum-activated mesangial cells by induction of a non-pancreatic (type II) phospholipase A2. J Biol Chem 1991; 266:14119–14127.
102. Barbour SE, Dennis EA. Antisense inhibition of group II phospholipase A2 expression blocks the production of prostaglandin E2 by P388D1 cells. J Biol Chem 1993; 268:21875–21882.
103. Santos AA, Browning JL, Scheltinga MR, Lynch EA, Brown EF, Lawton P, Chambers E, Douglas I, Benjamin CD, Dinarello CA, et al. Are events after endotoxemia related to circulating phospholipasee A2?. Ann Surg 1994; 219:183–192.
104. Murakami M, Kudo I, Inoue K. Molecular nature of phospholipases A2 involved in prostaglandin I2 synthesis in human umbilical vein endothelial cells. Possible participation of cytosolic and extracellular type II phospholipase A2. J Biol Chem 1993; 268:839–844.
105. Suga H, Murakami M, Kudo I, Inoue K. Participation in cellular prostaglandin synthesis of type-II phospholipase A2 secreted and anchored on cell-surface heparan sulfate proteoglycan. Eur J Biochem 1993; 218:807–813.
106. Tan X-d, Wang H, Gonzalez-Crussi FX, Chang H, Gonzalez-Crussi F, Hsueh W. PAF and endotoxin increase the enzyme activity and gene expression of type II phospholipase A2 in the rat intestine: Role of polymorphonuclear leukocytes. J Immunol 1996; 156:2985–2990.
107. Sun X-M, MacKendrick W, Tien J, Huang W, Caplan MS, Hsueh W. Endogenous bacterial toxin is required for the injurious effect of PAF. Gastroenterology 1995; 109:83–88.
108. Sanchez-Crespo M, Alonso F, Inarrea P, Alvarez V, Egido J. Vascular actions of synthetic PAF-acether (a synthetic platelet-activating factor) in the rat: evidence for a platelet independent mechanism. Immunopharmacology 1982; 4:173–85.
109. Chu KM, Gerber JG, Nies AS. Local vasodilator effect of platelet activating factor in the gastric, mesenteric and femoral arteries of the dog. J Pharmacol Exp Ther 1988; 246:996–1000.
110. Siren AL, Feuerstein G. Effects of PAF and BN 52021 on cardiac function and regional blood flow in conscious rats. Am J Physiol 1989; 257:H25–32.
111. Zhang C, Hsueh W. PAF-induced bowel injury: effects of vasodilators. Dig Dis Sci 1991; 36:634–640.
112. Wykle R. Interrelationships in the metabolism of platelet-activating factor and arachidonate in neutrophils. In: Snyder F, ed. Platelet-activating factor and related lipid mediators. New York: Plenum Press, 1987:273–282.
113. Hsueh W, Gonzalez Crussi F, Arroyave JL. Release of leuko-

triene C4 by isolated, perfused rat small intestine in response to platelet-activating factor. J Clin Invest 1986; 78:108–114.
114. Hsueh W, Gonzalez-Crussi F, Arroyave JL. Platelet activating factor-induced ischemic bowel necrosis: an investigation of secondary mediators in its pathogenesis. Am J Pathol 1986; 122:231–239.
115. Hsueh W, Gonzalez Crussi F, Arroyave JL. Sequential release of leukotrienes and norepinephrine in rat bowel after platelet-activating factor. A mechanistic study of platelet-activating factor-induced bowel necrosis. Gastroenterology 1988; 94:1412–1418.
116. Friedl HP, Till GO, Ryan US, Ward PA. Mediator-induced activation of xanthine oxidase in endothelial cells. FASEB J 1989; 3:2512–2518.
117. Cueva JP, Hsueh W. Role of oxygen derived free radicals in platelet activating factor induced bowel necrosis. Gut 1988; 29:1207–1212.
118. Parks DA, Williams TK, Beckman JS. Conversion of xanthine dehydrogenase to oxidase in ischemic rat intestine: a reevaluation. Am J Physiol 1988; 254:G768–G774.
119. Corradin SB, Fasel N, Buchmuller-Rouiller Y, Ransijn A, Smith J, Mauel J. Induction of macrophage nitric oxide production by interferon-gamma and tumor necrosis factor-alpha is enhanced by interleukin-10. Eur J Immunol 1993; 23:2045–2048.
120. Thiemermann C, Wu CC, Szabo C, Perretti M, Vane JR. Role of tumour necrosis factor in the induction of nitric oxide synthase in a rat model of endotoxin shock. Br J Pharmacol 1993; 110:177–182.
121. Lamas S, Michel T, Brenner BM, Marsden PA. Nitric oxide synthesis in endothelial cells: evidence for a pathway inducible by TNF-alpha. Am J Physiol 1991; 261:C634–C641.
122. Estrada C, Gomez C, Martin C, Moncada S, Gonzalez C. Nitric oxide mediates tumor necrosis factor-alpha cytotoxicity in endothelial cells. Biochem Biophys Res Commun 1992; 186:475–482.
123. Robbins RA, Springall DR, Warren JB, Kwon OJ, Buttery LD, Wilson AJ, Adcock IM, Riveros-Moreno V, Moncada S, Polak J, et al. Inducible nitric oxide synthase is increased in murine lung epithelial cells by cytokine stimulation. Biochem Biophys Res Commun 1994; 198:835–843.
124. Robbins RA, Barnes PJ, Springall DR, Warren JB, Kwon OJ, Buttery LD, Wilson AJ, Geller DA, Polar JM. Expression of inducible nitric oxide in human lung epithelial cells. Biochem Biophys Res Commun 1994; 203:209–218.
125. Kubes P. Nitric oxide modulates epithelial permeability in the feline small intestine. Am J Physiol 1992; 262:G1138–G1142.
126. Kanwar S, Wallace JL, Befus D, Kubes P. Nitric oxide synthesis inhibition increases epithelial permeability via mast cells. J Immunol 1993; 151:6616.
127. Beckman JS, Beckman TW, Chen J, Marshall PA, Freeman BA. Apparent hydroxyl radical production by peroxynitrite: implications for endothelial injury from nitric oxide and superoxide. Proc Natl Acad Sci USA 1990; 87:1620–1624.
128. Carreras MC, Pargament GA, Catz SD, Poderoso JJ, Boveris A. Kinetics of nitric oxide and hydrogen peroxide production and formation of peroxynitrite during the respiratory burst of human neutrophils. FEBS Lett 1994; 341:65–68.
129. Reif DW, Simmons RD. Nitric oxide mediates iron release from ferritin. Arch Biochem Biophys 1990; 283:537–541.
130. MacKendrick W, Caplan M, Hsueh W. Endogenous nitric oxide protects against platelet-activating factor-induced bowel injury in the rat. Ped Res 1993; 34:222–228.
131. Fearon DT, Ruddy S, Schur PH, McCabe WR. Activation of the properdin pathway of complement in patients with Gram negative bacteremia. N Engl J Med 1975; 292:937–940.
132. Hsueh W, Sun X, Rioja LN, Gonzalez-Crussi F. The role of the complement system in shock and tissue injury induced by tumour necrosis factor and endotoxin. Immunology 1990; 70:309–314.
133. Bjork J, Hugli TE, Smedegard G. Microvascular effects of anaphylatoxins C3a and C5a. J Immunol 1974; 134:1115–1119.
134. Okusawa S, Yancey KB, van der Meer JWM, et al. C5a stimulates secretion of tumor necrosis factor from human mononuclear cells in vitro. Comparison with secretion of interleukin I beta and interleukin I alpha. J Exp Med 1988; 168:443–448.
135. Smedegard G, Cui L, Hugli TE. Endotoxin-induced shock in the rat: a role for C5a. Am J Pathol 1989; 135:489–497.
136. Rothstein TL, Lint TF, Schreiber H. Tumor necrosis factor/cachectin. Induction of hemorrhagic necrosis in normal tissue requires the fifth component of complement (C5). J Exp Med 1988; 168:2007–2021.
137. Llausas-Magana E, Hsueh W, Arroyave CM, Arroyave JL, Torre-Amione G, Gonzalez-Crussi F. Cobra venom factor, an activator of the complement system, enhances the bowel necrosis induced by platelet-activating factor. Immunopharmacology 1988; 15:31–38.
138. Sun XM, Hsueh W. PAF produces shock, in vivo complement activation and tissue injury in mice. J Immunol 1991; 147:509–514.
139. Kliegman RM, Fanaroff AA. Necrotizing enterocolitis. N Engl J Med 1984; 310:1093–1103.
140. Walsh MC, Kliegman RM. Necrotizing enterocolitis: treatment based on staging criteria. Pediatr Clin North Am 1986; 33:179–201.
141. Kliegman RM, Fanaroff AA. Neonatal necrotizing enterocolitis: a nine year experience. Am J Dis Child 1981; 135:608–611.
142. Wilson R, Kanto WP Jr, McCarthy BJ, et al. Epidemiologic characteristics of necrotizing enterocolitis: a population based study. Am J Epidemiol 1981; 114:880–887.
143. Frantz ID III, L'Heureux P, Engel RR, Hunt CE. Necrotizing enterocolitis. J Pediatr 1975; 86:259–263.
144. Touloukian RJ, Posch JN, Spencer R. The pathogenesis of ischemic gastroenterocolitis of the neonate: selective gut mucosal ischemia in asphyxiated neonatal piglets. J Pediatr Surg 1972; 7:194–205.
145. Kosloske AM, Ulrich JA. A bacteriologic basis for the clinical presentations of necrotizing enterocolitis. J Pediatr Surg 1980; 15:558–564.
146. Brown EG, Ainbender E, Henley WL, Hodes HL. Etiologic role of bacteria and intestinal function. In: Brown EG, Sweet AY, eds. Neonatal Necrotizing Enterocolitis. New York: Grune & Stratton. 1980:69–100.
147. LeGamma EF, Ostertag SG, Birenbaum H. Failure of delayed oral feedings to prevent necrotizing enterocolitis. Am J Dis Child 1985; 139:385–389.
148. Pitt J, Barlow B, Heird WC. Protection against experimental necrotizing enterocolitis by maternal milk: role of milk leukocytes. Pediatr Res 1971; 11:906–909.
149. Caplan MS, Sun XM, Hsueh H, Hageman JR. Role of platelet activating factor and tumor necrosis factor-alpha in neonatal necrotizing enterocolitis. J Pediatr 1990; 116:960–964.
150. Gordon JR, Galli SJ. Mast cells as a source of both preformed and immunologically inducible TNF-alpha achectin. Nature 1990; 346:274–276.
151. Crissinger KD, Granger DN. Mucosal injury induced by

ischemia and reperfusion in the piglet intestine: influences of age and feeding. Gastroenterology 1989; 97:920–926.
152. von Ritter C, Hinder RA, Oosthuizen MM, Svensson LG, Hunter SJ, Lambrecht H. Gastric mucosal lesions induced by hemorrhagic shock in baboons. Role of oxygen-derived free radicals. Dig Dis Sci 1988; 33:857–864.
153. McCord J. Oxygen-derived free radicals in postischemic tissue injury. N Engl J Med 1985; 312:159–163.
154. Granger DN. Role of xanthine oxidase and granulocytes in ischemia-reperfusion injury. Am J Physiol 1988; 255:H1269–1275.
155. Grisham MB, Granger DN. Neutrophil-mediated mucosal injury. Role of reactive oxygen metabolites. Dig Dis Sci (suppl) 1988; 33:6S–15S.
156. Parks DA, Gringer DN. Ischemia-induced vascular changes: role of xanthine oxidase and hydroxyl radicals. Am J Physiol 1983; 245:G285–289.
157. Kurtel H, Fujimoto K, Zimmerman BJ, Granger DN, Tso P. Ischemia-reperfusion-induced mucosal dysfunction: role of neutrophils. Am J Physiol 1991; 261:G490–G496.
158. Kubes P, Ibbotson G, Russell J, Wallace JL, Granger DN. Role of platelet-activating factor in ischemia/reperfusion-induced leukocyte adherence. Am J Physiol 1990; 259:G300–305.
159. Mozes T, Braquet P, Filep J. Platelet-activating factor: an endogenous mediator of mesenteric ischemia-reperfusion-induced shock. Am J Physiol 1989; 257:R872–R877.
160. Squadrito F, Sturniolo R, Altavilla D, Santoro G, Campo GM, Arena A, Caputi AP. Platelet activating factor involvement in splanchnic artery occlusion shock in rats. Eur J Pharmacol 1991; 192:47–53.
161. Caty MG, Guice KS, Oldham KT, Remick DG, Kunkel SI. Evidence for tumor necrosis factor-induced pulmonary microvascular injury after intestinal ischemia-reperfusion injury. Ann Surg 1990; 212:694–700.
162. Fiocchi C. Production of inflammatory cytokines in the intestinal lamina propria. Immunol Res 1991; 10:239–246.
163. Stevens C, Walz G, Singaram C, Lipman ML, Zanker B, Muggia A, Antolioni D, Peppercorn MA, Strom TB. Tumor necrosis factor-alpha, interleukin-8, and interleukin-6 expression in inflammatory bowel diseases. Dig Dis Sci 1992; 37:818–826.
164. Hyams JS, Treem WR, Eddy E, Wyzga N, Moore RE. Tumor necrosis factor-α is not elevated in children with inflammatory bowel disease. J Pediatr Gastroenterol Nutr 1991; 12:233–236.
165. MacDonald TT, Hutchings P, Choy MY, Murch S, Cooke A. Tumor necrosis factor-alpha and interferon gamma production measured at the single cell level in normal and inflamed human intestine. Clin Exp Immunol 1990; 81:301–305.
166. Steffen M, Witthoeft T, Pflueger I, Schreiber S, MacDermott RP, Raedler A. Enhanced secretion of tumour necrosis factor-alpha, IL-6, and IL-1 beta by isolated lamina propria mononuclear cells from patients with ulcerative colitis and Crohn's disease. Clin Exp Immunol 1993; 94:174–181.
167. Breese EJ, Michie CA, Nicholls SW, Murch SH, Williams CB, Domizio P, Walker-Smith JA, MacDonald TT. Tumor necrosis factor alpha-producing cells in the intestinal mucosal of children with inflammatory bowel disease. Gastroenterology 1994; 106:1455–1466.
168. Murch SH, Braegger CP, Walker-Smith JA, MacDonald TT. Location of tumour necrosis factor alpha by immunohistochemistry in chronic inflammatory bowel disease. Gut 1993; 34:1705–1709.
169. Murch SH, Lamkin VA, Savage MO, Walker-Smith JA, MacDonald TT. Serum concentrations of tumour necrosis factor alpha in childhood chronic inflammatory bowel disease. Gut 1991; 32:913–917.
170. Mahida YR, Wu K, Jewell D. Enhanced production of interleukin 1-β by mononuclear cells isolated from mucosa with active ulcerative colitis or Crohn's disease. Gut 1989; 30:835–838.
171. Raab Y, Gerdin B, Ahlstedt S, Hallgren R. Neutrophil mucosal involvement is accompanied by enhanced local production of interleukin-8 in ulcerative colitis. Gut 1993; 34:1203–1206.
172. Mitsuyama K, Toyonaga A, Sasaki E, Watanabe K, Tateishi H, Nishiyama T, Saiki T, Ikeda H, Tsuruta O, Tanikawa K. IL-8 as an important chemoattractant for neutrophils in ulcerative colitis and Crohn's disease. Clin Exp Immunol 1994; 96:432–436.
173. Isaacs KL, Sartor RB, Haskill S. Cytokine messenger RNA profiles in inflammatory bowel disease mucosa detected by polymerase chain reaction amplification. Gastroenterology 1992; 103:1587–1595.
174. Reinecker HC, Loh EY, Ringler DJ, Mehta A, Rombeau JL, MacDermott RP. Monocyte-chemoattractant protein 1 gene expression in intestinal epithelial cells and inflammatory bowel disease mucosa. Gastroenterology 1995; 108:40–50.
175. McGee DW, Beagley KW, Aicher WK, McGhee JR. Transforming growth factor-beta and IL-1 beta act in synergy to enhance IL-6 secretion by the intestinal epithelial cell line, IEC-6. J Immunol 1993; 151:970–978.
176. Schuerer-Maly CC, Eckmann L, Kagnoff MF, Falco MT, Maly FE. Colonic epithelial cell lines as a source of interleukin-8: stimulation by inflammatory cytokines and bacterial lipopolysaccharide. Immunology 1994; 81:85–91.
177. Radema SA, Van Deventer SJH, Cerami A. Interleukin 1β is expressed predominantly by enterocytes in experimental colitis. Gastroenterology 1991; 100:1180–1186.
178. Youongman KR, Simon PL, West GA, Cominelli F, Rachmilewitz D, Klein JS, Fiocchi C. Localization of intestinal interleukin 1 activity and protein and gene expression to lamina propria cells. Gastroenterology 1993; 104:749–758.
179. Sharon P, Ligmsky M, Rachmilewitz D, Zot U. Role of prostaglandins in ulcerative colitis. Enhanced production during active disease and inhibition by sulphasalazine. Gastroenterology 1978; 75:638–640.
180. Hawkey CJ, Karmeli F, Rachmilewitz D. Imbalance of prostacylcin and thromboxane synthesis in Crohn's disease. Gut 1983; 24:881–885.
181. Boughton-Smith NK, Hawkey CJ, Whittle BJR. The bioxynthesis of lipoxygenase and cyclooxygenase products from [^{14}C] arachidonic acid in human colonic mucosa. Gut 1983; 24:1176–1182.
182. Lobos EA, Sharon P, Stenson WF. Chemotactic activity in inflammatory bowel disease. Role of leukotriene B4. Dig Dis Sci 1987; 32:1380–1388.
183. Stenson WF, Lauritsen K, Laursene J, et al. A clinical trial of zileuton, a specific inhibitor of 5-lipoxygenase inhibitor, in ulcerative colitis. Gastroenterology 1991; 100:A253.
184. Rola-Pleszczynski M. Immunoregulation by leukotrienes and other lipoxygenase metabolites. Immunol Today 1985; 10:302–307.
185. Goodwin JS, Webb DR. Regulation of the immune response by prostaglandins. Clin Immunol Immunopathol 1980; 15:106.
186. Knudsen PJ, Dinarello CA, Stom TB. Prostaglandins post-transcriptionally inhibit monocyte expression of interleukin-1 activity by increasing intracellular cyclic adenosine monophosphate. J Immunol 1986; 137:3189–3194.
187. Fieren MW, van den Bend GJ, Ben-Efraim S, Bonta IL.

Prostaglandin E2 inhibits the release of tumor necrosis factor-alpha, rather than interleukin 1 beta, forem human macrophages. Immunol Lett 1992; 31:85–90.
188. Eliakim R, Karmeli F, Razin E, Rachmilewitz D. Role of platelet-activating factor in ulcerative colitis. Enhanced production in disease and inhibition by sulfasalazine and proednisolone. Gastroenterology 1988; 95:1167–1172.
189. Ferraris L, Karmeli F, Eliakim R, et al. Intestinal epithelial cells contribute to the enhanced generation of platelet-activating factor in ulcerative colitis. Gut 1993; 34:665–668.
190. Sobhani I, Hochlaf S, Denizot Y, et al. Raised concentrations of platelet-activating factor in colonic mucosa of Crohn's disease patients. Gut 1992; 33:1220–1225.
191. Denizot Y, Chaussade S, Nathan N, et al. PAF-acether and acetylhydrolase in stool of patients with Crohn's disease. Dig Dis Sci 1992; 37:432–437.

29
Cytokine Networks in Skin Disease

David P. Fivenson
Henry Ford Health System, Detroit, Michigan

Robert M. Strieter
University of Michigan Medical School, Ann Arbor, Michigan

Brian J. Nickoloff
Loyola University Medical Center, Maywood, Illinois

INTRODUCTION

The molecular basis for inflammation and immunological reactions in the skin involves cytokines produced by a wide variety of both indigenous and recruited cell types (1). The skin as an organ system has already proven to be characterized by a number of specific cytokine networks in vivo. Investigative dermatologists have succeeded in mapping a plethora of adhesion molecules, chemotactic factors, and the cytokines capable of modulating their expression in several different skin diseases (2). Indeed, it was the early recognition that keratinocytes themselves could produce interleukin-1 (originally termed epidermal cell–derived thymocyte activating factor, ETAF) (3) that helped launch the field of immunodermatology into the forefront of investigative dermatology. Keratinocytes have been shown to be veritable factories of cytokine production, including IL-1, IL-3, IL-6, IL-7, IL-8, IL-10, IL-12, IL-13, IL-15, G-CSF, M-CSF, MCAF/MCP-1, GM-CSF, TNF-α, TGF-α, KLIF, ELDIF, KGF/FGF-7, IGF-I, TGF-β, PDGF, IRAP, bFGF, IFN-α, IFN-β, and prostaglandins (4–14). This chapter summarizes these recent discoveries and emphasizes the distinctive contribution of keratinocytes, T cells, macrophages, dendritic cells, and endothelial cells to this exciting and rapidly advancing area of cutaneous immunobiology.

The onset and progress of inflammatory and immune-based processes in the skin requires the collaborative interaction between indigenous cell populations and recruited leukocytes. The notion of the skin as a specialized organ, immunologically distinguished from other organs because of the presence of skin-seeking T-cell subsets, was introduced in the term "skin-associated lymphoid tissue" by Streilein in 1978 (15) and subsequently expanded by many others (16–20). The remarkable similarity in the clinical and histological patterns of cutaneous inflammation for each skin disease (21), in the face of tremendous genetic diversity among dermatological patients, implies the existence of distinctive molecular mediators (22). These soluble molecular mediators (i.e., cytokines) must be capable of directing cell-cell communication and are responsible for maintaining the fidelity of cellular trafficking in each disease process by orchestrating the precise temporal expression of cellular adhesion molecules, chemotaxins, and growth factors. Since each skin disease possesses a unique constellation of cell-to-cell associations, it should not be surprising to learn that the profile of cytokines and participating cell types, or cytokine network, is relatively disease specific. In the following sections of this chapter, we compare and contrast the different cytokine networks in five relatively common skin conditions, with particular emphasis on the pathophysiological significance of these important molecules in allergic contact dermatitis, psoriasis, atopic dermatitis, wound healing, and cutaneous T-cell lymphoma (CTCL).

ALLERGIC CONTACT DERMATITIS

When human skin comes in contact with the poison ivy plant, the waxy catechol-bearing molecules on the leaf surface are transferred to the epidermis, and approximately 1–2 days later, a T-cell–mediated inflammatory reaction occurs at such contact points in previously sensitized individuals. Allergic contact dermatitis reactions in the skin are composed of two distinct functional phases: the first is referred to as the inductive or afferent limb, which occurs during the first exposure to the proinflammatory agent, and the elicitation or efferent limb occurs during subsequent exposures (23). This immunological reaction is known appropriately as allergic contact dermatitis and has been used to investigate the cellular basis of inflammation in the skin for several decades, following the pioneering work of Landsteiner and Chase in the early 1940s (24). Once an individual is sensitized to the poison ivy antigen (pentadecylcatechol or urushiol), subsequent exposures produce a rapid, dramatic, and transient series of cellular and molecular alterations in the skin referred to as rhus dermatitis. The urushiol molecule is just one of many exogenous contactants believed to act as haptens (i.e., dinitrocholorobenzene, nickel, *p*-phenylenediamine, and others). These haptens have the ability to penetrate the stratum corneum of the epidermis and conjugate to extracellular epidermal proteins before being recognized as "foreign" by the immune system (25).

Conventional dogma has portrayed this immunological delayed-type hypersensitivity phenomenon as a result of initial hapten-specific T-cell–Langerhans-cell interaction (26), with lymphokine production (27) and subsequent epidermal changes, including epidermal injury, edema, and reparative hyperplasia (28). In this schema, the epidermal keratinocytes are viewed wholly as passive bystander targets of the immunological reaction, having no significant participatory or immunomodulatory capability. This is in keeping with the previously held myth that the only role for keratinocytes was to function to maintain the structural integrity of the skin, providing a protective coat and an environmental barrier not unlike polystyrene wrap (e.g., Saran Wrap®). With the modern characterization of the ability of keratinocytes, both in vitro and in vivo, to produce a spectrum of multifunctional protein cytokine molecules including interleukins, interferons, adhesion molecules, chemotactic peptides, and growth factors, the keratinocyte must now be added to the growing list of nonhematopoietic cells that can function as an immunocompetent and immunoregulatory cell. Skin biologists have reversed this old dogma completely by proposing that, instead of the keratinocyte being the least important and passive cell in the skin, it is probably the most important initiating cell type in allergic contact dermatitis. In the poison ivy reaction, direct activation of keratinocytes by urushiol sets into motion a cascade of cytokines ultimately responsible for cellular infiltration typical of allergic contact dermatitis (29).

Our approach to the problem of characterizing the earliest detectable molecular events following topical application of urushiol has been to perform serial biopsies of skin before clinical changes are evident and before trafficking of leukocytes begins. Sequential biopsies were performed over a 3-week period on subjects after application of urushiol for use in both immunostaining and reverse transcriptase/PCR, and these studies have allowed us to divide the reaction into three phases. Phase I biopsies are obtained during the first 24 hours and precede clinical or histological changes. Phase II biopsies are obtained between 48 and 72 hours, when the clinical and histological changes are at their maximal levels. Phase III biopsies are obtained over 2–3 weeks following the single application of urushiol and represent the resolution of this reaction (29–32).

In phase I biopsies during the first 4–8 hours following topical application of urushiol, even though no significant light microscopic changes were detectable, immunohistochemical staining analysis revealed a remarkable series of reproducible changes (30). The antibodies employed included detection of the following molecules: TNF-α, TGF-α, TGF-β, IL-8, intercellular adhesion molecule-1 (ICAM-1), vascular cell adhesion molecule-1 (VCAM-1), and endothelial leukocyte adhesion molecule-1 (ELAM-1). At the mRNA level, there was upregulation of IL-4 and IL-10 on RNAs in excess of IFN-γ (31). In normal skin from these same subjects, there was no significant epidermal staining with any of these antibodies, and only weak dermal staining with ICAM-1 on endothelial cells and low levels of IL-10 and IFN-γ mRNAs were detected (31–33). The earliest immunostaining changes occurred at 4 hours, when the lower and midepidermal keratinocytes became positive for TNF-α. By 6–8 hours, the superficial and deep vascular plexus contained endothelial cells diffusely ELAM-1 + and VCAM-1, and the keratinocytes were focally ICAM-1 +. It should be emphasized that these adhesion molecule inductions occurred *subsequent* to the keratinocytes expressing detectable TNF-α but *before* significant movement of leukocytes into the dermis or epidermis.

Keratinocytes were judged to be actually producing TNF-α because the TNF-α immunoreactivity and mRNA expression were largely confined to the epidermis (30–34). TNF-α is the only known cytokine that can simultaneously induce endothelial cells to express ELAM-1 and VCAM-1, as well as stimulating keratinocyte ICAM-1. Furthermore, when purified, multipassage cultured keratinocytes were exposed to urushiol in vitro, TNF-α mRNA and protein were induced (32).

Recent studies have used serial tape-stripping of the epidermis to perturb barrier function and initiate inflammation without presentation of haptens such as urushiol (35). Using a rapid heat-saline technique to separate the epidermis from the dermis, we have shown specific epidermal upregulation of mRNAs for TNF-α, IL-4, IL-8, IL-10, IFN-γ, ICAM-1 TGF-α, and TGF-β after 4–6 hours. These changes were accompanied by immunohistochemical evidence for the expression of endothelial VCAM-1 and ELAM-1/E-selectin, as well as keratinocyte ICAM-1 also at 6 hours analogous to the changes seen after urushiol exposure (22,24). This cytokine profile was also confirmed in situ by immunolocalization of IL-10. Thus, both tape stripping and allergic contact dermatitis are characterized by what is referred to as a Th2 cytokine profile, while classic delayed-type hypersensitivity reactions (e.g., tuberculin reaction) have been shown by us to express predominately IL-2 and IFN-γ (so-called Th1 cytokine profile) (36).

We previously drew attention to the close spatial proximity of basal epidermal keratinocytes and the dermal capillary loop, with the distance separating epithelial cells and endothelial cells approximately 10–20 μm. Thus, we believe it is entirely possible that epidermal keratinocytes are directly activated by mechanical irritation (by barrier perturbation) or by haptens to produce TNF-α. Paracrine augmentation of TNF-α stimulates endothelial cells to produce ELAM-1/E-selectin and VCAM-1. Endothelial cell expression of these leukocyte adhesion molecules would then facilitate the attachment of T cells and monocytes bearing the appropriate complementary ligands to the vessel wall and thereby promoting the dermal influx of mononuclear cells, which occurrs in the second phase of allergic contact dermatitis (27,28,37).

The infiltrating T cells in rhus dermatitis include both CD4+ and CD8+ subsets (24,25). Their multifunctionality is also evidenced by detection of mRNAs for IL-2,4,5,8,10, and IFN-γ in phase II and phase III specimens (31–33). However, all the T cells belong to the memory subset (16), which means that they express extremely high levels of various adhesion molecules on their surface, which is probably how they gain a selective advantage over their naive T-cell counterparts in the initial binding to the endothelial cells (38,39). The potentially important adhesion molecule-ligand interactions in this model include HECA452/Sialyl-Lewis X–ELAM-1/E-selectin (40–42), VLA-4–VCAM-1 (43), and LFA-1–ICAM-1 (44), which are thought to contribute to lymphocyte migration (45).

Upregulation in the magnitude of delayed-type hypersensitivity reactions by anti–IL-10 and anti–IL-4 in murine ear-swelling model is further evidence for a complex hierarchy of cytokine-mediated regulation in cutaneous immune reactions (46,47). This observation implies that downregulatory signals are also expressed during phase II helping, to "blunt" the intensity of the local cutaneous eruption. This also suggests that more exaggerated reactions may be, in part, related to aberrant counterregulation as well as to higher levels of antigen exposure.

Once the T lymphocytes enter the dermis, they have a strong propensity for moving toward the epidermis, which may also require utilization of cell surface adhesion molecules and extracellular matrix receptors (48,49), as well as contact with dermal dendritic cells (50). This directional migration implies the existence of keratinocyte-derived T-cell chemotaxins (51,52). IL-8, along with several other members of the chemokine family, are clearly upregulated during this 24- to 48-h interval and spatially coincided with epidermal T-cell influx (53,54) (see below for a more complete discussion of chemokines in cutaneous disease).

Infiltrating dermal and epidermal T lymphocytes belonging to the memory subset (versus naive-type cells) produce IFN-γ mRNA and protein inducing the surrounding keratinocytes to express the class II major histocompatibility antigen HLA-DR (55,56). The IFN-γ produced by these activated T cells has important consequences for this cytokine network, because it can synergize with TNF-α (57) in the stimulation of keratinocytes to produce a variety of important immunoregulatory molecules, including IL-8 and monocyte chemoattractant-1, a CC chemokine previously known as MCAF (58). By recruiting more monocytes-macrophages and activating the indigenous population of other potential antigen-presenting cells, IFN-γ and MCP-1 can bring about the multicellular complex of "antigen" (i.e., urushiol) in and around keratinocytes, with antigen-presenting cells (i.e., Langerhans cells and dermal dendritic cells), to amplify the ongoing reaction. Upregulation of costimulation pathways such as the B7.2/CD28 interaction has been recently shown to be vital to this propagation. Induction of these molecules on indigenous LC and DDC serves to continue the ongoing reaction by providing the second signal required for T-cell activation after the antigen-class II T-cell receptor interaction (26,60–66).

During the resolution phase (phase III) of the rhus-allergic contact dermatitis reaction, biopsies have revealed a concomitant diminution in the expression of epidermal TNF-α, IL-8, and ICAM-1, despite continued endothelial cell expression of ELAM-1 and VCAM-1 (29,30,32,55). By the end of the third week, there was no significant epidermal population of T cells or monocytes/macrophages, and only a sparse perivascular lymphohistiocytic infiltrate remained. This transient expression of ICAM-1 by keratinocytes in vivo is in agreement with our in vitro results in which ICAM-1 expression tended to diminish by day 3 after cytokine stimulation (67). The endothelial cell ELAM-1/E-selectin expression in our model was somewhat unusual in that it was never associated with any neutrophil efflux (originally ELAM-1 was recognized for its ability to specifically bind neutrophils) (68). Its in vivo persistence contrasts with transient expression of ELAM-1/E-selection (< 24 h) in vitro on cultured endothelial cells. These differences may be related to diversity in the biological properties of endothelial cells when one compares cultured human umbilical vein (large vessel) endothelial cells (HUVEC) to human dermal microvascular endothelial cells (HDMEC) present in the skin. HDMEC have been recently shown to also have ELAM-1/E-selectin inducible in vitro (69,70). This expression persists at high levels long past the 24 h that ELAM-1/E-selectin remains detectable on similarly treated HUVEC (71).

These recent findings indicate that keratinocytes are important signal transducers as they are directly stimulated by haptens, mechanical irritation, croton oil, retinoids, and ultraviolet (UV) light to produce cytokines, including TNF-α, IL-12, IL-8, IL-10, TGF-α, and TGF-β (14,32,35). These early, proinflammatory cytokines upregulate Langerhans cells and help to nonspecifically recruit dermal TH cells into the epidermis. Once in the epidermis, these T cells cycle out of G_0 and are stimulated to produce T-cell–specific cytokines, including IFN-γ, IL-4, IL-2, and IL-10 (35,72). This epidermal cytokine production is then followed by endothelial cell expression of adhesion molecules critical for initiation and propagation of inflammatory dermatitis.

PSORIASIS

Psoriasis is a common (involving approximately 1–2% of the population) and chronic inflammatory skin disease (73) that features altered keratinocyte maturation (74) and marked keratinocyte hyperplasia (75,76). As in allergic contact dermatitis, previous investigators of psoriasis have suggested that the primary pathophysiological important cell type involved was the T cell, with the epidermal keratinocytes serving only as targets for T-cell–monocyte/macrophage–derived factors (77). The importance of the immune system was significantly advanced by the ability of cyclosporin A and other immunodulatory therapies to dramatically improve psoriasis (78), but we raised some doubt about this conclusion because cyclosporin A can also directly influence keratinocytes (79). Also, when keratinocytes are isolated from psoriatic lesions and separated from their microenvironment, they continue to exhibit an aberrant phenotype for multiple generations, indicating an inherent abnormality that can be distinguished from the immune cells (80,81). There is therefore still somewhat of a "chicken or the egg" controversy as to which specific cell type (keratinocyte vs. T cell) is primarily responsible for the cutaneous alterations of psoriasis (82).

When we began the dissection of the cytokine network in psoriasis (82), we were particularly interested in attempting to elucidate the possible molecular link between the overlying epidermal keratinocyte hyperplasia and the infiltrate of T cells and monocyte/macrophages. Psoriatic plaques were biopsied and immunostained with an antibody panel similar to the one used in the poison ivy model. mRNA expression was likewise studied simultaneously with a broad panel of molecular biological probes. In contrast to rhus dermatitis, in which the TNF-α was confined almost exclusively to the keratinocyte, in psoriasis dermal dendritic cells (macrophages) in the papillary dermis and occasional perivascular T cells, epidermal Langerhans cells, but no keratinocytes were TNF-α positive. This macrophage-derived TNF-α was thought to be actively produced and secreted because the immediately surrounding keratinocytes were strongly positive for the TNF-α–inducible proteins IL-8 and ICAM-1 (82) and the dermal microvascular endothelial cells were expressing ELAM-1 and VCAM-1 (72). Once again there was excellent spatial co-localization of keratinocytes expressing ICAM-1 and IL-8 with epidermotropic T-cell–bearing LFA-1.

The importance of LFA-1/ICAM-1 in the epidermal trafficking of T cells in psoriasis is supported by our previous experimental results. In these studies, cyclosporin A was injected intralesionally into psoriatic plaques and the first detectable immunological alteration was a reduction in keratinocyte ICAM-1 followed by loss of epidermal T-cell influx and eventually resolution of the lesion (83). In this disease, cyclosporin A may be clinically efficacious because it can both directly inhibit keratinocyte growth as well as inhibit IFN-γ (84) and TNF-α production by mononuclear leukocytes (85). The presence of potential targets for cyclosporin modulation including IL-2 and IFN-γ in psoriatic lesions was confirmed by PCR.

These observations of altered levels of T-cell, keratinocyte, and mononuclear cell–derived cytokines combined with the importance of T-cell–dependent immunity (as highlighted by the response of psoriasis to cyclosporin A) led to a concerted effort to fully characterize the cytokine network of psoriasis. We and others have shown that psoriatic lesional or psoriatic prone skin expresses a predominance of the T-helper cytokine IL-2, IFN-γ, as well as GM-CSF (33,34,86,87). Functional proteins for these have also been shown recently to be elevated within psoriatic lesions. This profile fits the so-called Th1 profile, which has been associated with augmented cell-mediated immunity (CMI) in vitro (33,36,86). Utilizing both Northern blots and RT/PCR, we have also identified elevated mRNA transcripts in psoriatic epidermis for IL-8, ICAM-1, and TGF-α, but not TNF-α, which nicely complements our earlier protein immunolocalization results reviewed above (33,34,86,87). The detection of IFN-γ in these studies was particularly important for pathophysiological considerations in psoriasis, because intradermal IFN-γ injection can induce lesions in the uninvolved skin of psoriatic patients (88,89).

Thus a dichotomy is beginning to become evident within inflammatory, T-cell–mediated dermatoses. These diseases seem to nicely fit into one group, which are largely mediated by Th1 type cytokines including psoriasis, lichen planus, tuberculoid leprosy (see Refs. 86 and 90 for a discussion of lichen planus and leprosy cytokine networks), and mycosis fungoides (see below). All of these diseases have been previously shown to be characterized by highly effective cellular immune reactions. Poison ivy dermatitis (see above), lepromatous leprosy (see Ref. 86), Sézary syndrome (see below), and tape-stripped epidermis (see above) all demonstrate excess humoral immunity (e.g., polyclonal hypergammaglobulinemia, eosinophilia) along with varying deficiencies in cellular immunity and share the Th2 cytokine profile (33–35,90).

We have focused on characterizing the role of antigen-presenting cells in both the epidermis and dermis of psoriatic lesions, as their function has been recently linked to the T-cell cytokines they are exposed to both in vitro and in vivo (64–66,92,93,97). Epidermal Langerhans cells as well as several subsets of dermal macrophages have been characterized as having accessory cell functions. We have been able to show that psoriatic dermal dendritic cells as well as epidermal Langerhans cells preferentially stimulate Th cells bearing this Th1 cytokine profile. As mentioned above, psoriatic epidermal keratinocytes also appear to have constitutively elevated expression of GM-CSF, IL-1, IL-6, IL-8, TNF-α, TGF-α, IGF-I, and ICAM-1 (11,91). The resulting cytokine milieu confers a significant level of autoreactivity when psoriatic dermal dendritic cells (DDCs) are combined in vitro with autologous resting T cells (compared to normal skin) (64,65,92).

The mechanism of this upregulated autostimulatory capacity appears to relate to endogenous high levels of DDC expression of the CD28 ligands B7.1 (CD80) and B7.2 (CD86). RT/PCR and flow cytometry have shown that there is direct correlation with the level of endogenous autostimulatory function and the levels of CD80 and CD86 expression on DDC in vitro (92,93). These levels can be augmented by GM-CSF and/or IFN-γ treatment (epidermal cytokines known to be upregulated in psoriasis) as well as downregulated by IL-10 (a Th2 cytokine) (92–94). Heightened APC ability to stimulate resting autologous T cells has also been shown to result from exposure to staphylococcal-derived superantigens in vitro. This antigen-independent pathway also results in selective Th1 cytokine expression in psoriasis patients (64,93,95).

Another critical observation has been made suggesting that there is substantially upregulated IL-12 expression by dermal dendritic cells in psoriasis (96,97). This cytokine stimulates selective expansion of Th1-type T cells and IFN-γ production and may be one of the important triggers initiating this self-propagating cycle of autoreactive, Th1 cytokine production, and T-cell proliferation, which seems to stimulate epidermal proliferation as well as combining with analogous positive feedback loops among epidermal cytokines.

Another part of the histology of psoriasis is the enlargement of postcapillary venules within dermal papillae (21,76). The growth of new blood vessels (or angiogenesis) is also an intricately controlled process that is impacted by many of the aforementioned psoriatic cytokines. We have recently been able to elucidate that there is abnormal upregulation of multiple angiogenic factors in psoriatic skin as compared to normal, including TNF-α and IL-8 (as above), NAP2, GRO-α, and ENA78, as well as the presence of abnormally high levels of a thrombospondin inhibitor (87). These all result in increased production of blood vessels in psoriatic skin. Many of these cytokines are also T-cell and monocyte chemoattractants, as well as mediators of PMN chemotaxis (87). The net effect of this complicated cytokine milieu is an intricate and dynamic series of interactions within the psoriatic epidermis and dermis, which result in angiogenesis, increased autoreactivity, Th1-type T-cell infiltration, and PMN efflux, all factors contributing to the characteristic hyperproliferative epidermal disease recognized as psoriasis.

These recent observations have also shed light (no pun intended) on the potential mechanisms of action of phototherapy, as well as other therapeutic modalities effective in psoriasis. Ultraviolet light has been shown to upregulate IL-10 expression in the epidermis as well as downregulate IL-12 expression by dendritic cells, resulting in decreases in Th1 cytokine and in the levels of CD80- and CD86-mediated antigen presentation (47,98,99). Likewise, tar preparations, topical and systemic steroids, cyclosporine, and antimetabolites (e.g., methotrexate) can now be potentially pinpointed by their effects on relevant parts of this cytokine network in psoriasis. Future therapies will no doubt be directed at specific modulation of these cytokine networks by such things as antisense oligonucleotides (99a) and the use of receptor specific cytotoxic cytokine therapies. This last concept was recently proven by Gottlieb et al. at the Rockefeller Institute (100). These investigators successfully treated a group of psoriasis patients using a targeted toxin created by linking the CD25-binding portion of the IL-2 molecule to diphtheria toxin. This resulted in selecting of the binding and cytotoxicity of activated CD25+ T cells and significant clearing of psoriasis in the majority of patients treated.

ATOPIC DERMATITIS

Atopic dermatitis is a pruritic skin disease often associated with other clinical abnormalities, including allergic rhinitis, asthma, and elevated circulating IgE antibodies (101). None of these other systemic abnormalities are associated with psoriasis, but both atopic dermatitis and psoriasis are characterized as chronic, recurring papulosquamous disorders associated with CD4+ T cell infiltration of the epidermis (102). There are increased numbers of activated mast cells in atopic dermatitis lesions, and IgE deposits are conspicuously found on dermal dendritic cells (103). It has been shown that this allergen-specific IgE present on the surface of cutaneous dendritic cells may be bound via IL-4–inducible Fc IgE receptors, which are specific to atopic dermatitis (103,104).

A number of T-cell abnormalities are reported in the blood of patients with atopic dermatitis, including impaired CD8 T-cell function and decreased capacity to produce IFN-γ (102,105,106). When T-cell clones have been isolated from the blood of atopic dermatitis patients, there has been a disproportionate number of IL-4–producing cells compared to normal skin (107,108). Molecular studies of relative levels of mRNA expressions combined with sensitive ELISA techniques have built upon these observations to show that atopic dermatitis is best characterized as a classic Th2-type cytokine-mediated dermatosis. Abnormally high levels of expression of IL-4, IL-5, and IL-10 have been associated with the inherent depressed cell-mediated immunity and the elevated IgE levels characteristic of active atopic dermatitis (109,110). The mechanism for this imbalance in T-helper cytokines has been recently suggested to involve a specific hyporesponsiveness to IL-12 in the atopic prepheral blood mononuclear cell population. Lester et al. (111) have shown that atopic patients produce adequate IL-12 and upregulate the IL-12 p40 subunit appropriately in response to superantigen stimulus, yet this does not result in production of IFN-γ as occurs in normal donors. These (allergen-primed) atopic T cells "default" to increased production of IL-4 and IL-10 in these studies. Naive T cells from atopic patients, however, appropriately produce IFN-γ after IL-12 stimulation (112). Thus, atopic dermatitis represents the antithesis of what we describe above as the role IL-2 and IFN-γ in psoriasis. In atopic dermatitis, elevated levels of IL-4 and IL-10 result in selective downregulation of the expression of CD28 ligands CD80 and/or CD86, resulting in depression of antigen presentation as well as polyclonal B-cell proliferation and immunoglobin synthesis (104). The increased levels of IL-4 potentiate the elevated IgE levels because IL-4 is a potent inducer of B-cell IgE antibody production as well as the specific Fc receptor mentioned above (113–115).

Another important consideration in atopic dermatitis is that, like psoriasis, cutaneous lesions can be exacerbated by stress (101,115,116). Several neuropeptides have been identified in nerve fibers of normal human skin (117). In psoriasis there are increased numbers of substance P–containing intraepidermal free nerve endings (118), and in atopic dermatitis there was also prominent immunohistochemical staining of substance P nerve fibers, neuropeptide Y–positive dendritic epidermal cells, and an absence of somatostatin nerves (119). These neuropeptides may be important in generating an acute inflammatory release, since substance P is a potent mast cell degranulator—which could produce histamine, accounting for the intense pruritus of atopic dermatitis (116)—and because several different neuropeptides are now recognized as potent immunomodulatory agents (120). Also, substance P can induce keratinocytes to release IL-1 and GM-CSF, which contribute to the local inflammatory response (121).

Taken together, the current pathophysiological pathway operative in atopic dermatitis includes a schema in which these patients display a hyperresponsiveness to a variety of ingested foods, contact with dust mites, colonization by staphylococci, or inhaled allergens (101). Any and all of these can be exacerbated by stress and the pool of inflammatory mediators, including the active participation of Th2-type T cells, whose cytokines may be responsible for many of the observed clinical findings in AD.

Investigators have taken advantage of the ability of IFN-γ to suppress IL-4 production by administering recombinant human IFN-γ systemically to atopic dermatitis patients (115). In this study, 22 patients with severe atopic dermatitis with elevated IgE levels were given daily subcutaneous injections of IFN-γ and, although there was no significant reduction in serum IgE levels during the 6-week treatment period, total clinical severity showed significant improvement ($p < 0.01$). Paradoxically, UV light [known to induce IL-10 expression in the epidermis (122)] is also an important modality in AD, as is cyclosporin A. Thus, although the Th2 cytokine pathway appears to be convincingly involved in the pathophysiology of atopic dermatitis, therapeutic agents in this disease may act via cytokine-dependent or -independent mechanisms.

WOUND HEALING

Normal wound healing involves the coordinated interplay of cells, extracellular matrix, and an ever-expanding host of soluble mediators. The cellular physiology of wound repair includes the coordinated stimulation of blood vessel growth (angiogenesis), keratinocyte and fibroblast proliferation, production of epithelial and endothelial basement membrane zones, and connective tissues (e.g., collagens, ground substance, etc). No other area in cutaneous biology is characterized by as diverse a cytokine network as has been described for wound healing (123–125). Cytokines families reported to be critically involved in wound healing include growth factors (bFGF, EGF, IGF, PDGF, VEGF, TGF-α, TGF-β), interleukins (IL-1, IL-2, IL-4, IL-6, IL-8), colony-stimulating factors (GM-CSF, G-CSF), interferons (α, β, and γ), TNF-α, TNF-β, as well as a host of other proteases (MMP-1, TIMP), clotting factors, and miscellaneous mediators of inflammation/cellular proliferation/cellular migration (123–126).

In skin, there are three well-recognized phases of acute wound repair: the inflammatory phase, the tissue formation or proliferative phase, and the tissue-remodeling phase. While these phases are not necessarily mutually exclusive, they are somewhat distinct and help to provide the investigator with a reasonable approach to understanding normal versus abnormal wound repair (123–125). It is becoming increasing more evident that the regulation of the different aspects of wound repair is affected by many of the above-mentioned families of cytokines. Rather than an exhaustive review of the mechanisms of action of dozens of these cytokines, which has been done by others as well as throughout this book, it is probably best to look at the groups of factors that impact on the key cellular constituents of each of these phases of wound repair.

To allow for effective keratinocyte migration and coverage in the acute or chronic wound, the wound bed must be hospitable. This is principally a result of fibroblast remodeling, neovascularization or angiogenesis, and resolution of acute inflammatory cells at the site of injury. Bacteria, necrotic debris, and fibrin/platelet clots are cleared from an acute wound by PMNs and macrophages during the inflammatory phase of wound repair. Which steps lead to leukocyte migration into a wound is dependent on a combination of endothelial cell and leukocyte activation, expression of adhesion molecules, leukocyte-endothelial cell adhesion, diapedesis, and migration via chemotactic gradients (127). Migration from the vessel requires not only adhesion between leukocytes (L-selectin, β_1-, and β_2-integrins) and endothelial cells (P-selectin, E-selectin, ICAM-1, and VCAM-1), but is also dependent upon both the continued expression of β_1- and β_2-integrins and the movement along a specific chemokine gradient (127).

Platelets and PMNs are usually the first leukocytes to arrive at an acute wound, initiating the inflammatory phases of healing. PMNs produce a number of cytokines that are important in wound repair, including IFN-α, M-CSF, G-CSF, TNF-α, IL-1, IRAP, IL-6, IL-8, MIP-α, and TGF-β (128–130), while platelets contribute TGF-α, TGF-β, PDGF, PF4, NAP-2, and PD-ECGF. The second wave of leukocytes consists of mononuclear cells producing IL-1, IL-6, IFN-α, G-CSF, GM-CSF, M-CSF, PDGF, TNF-α and TGF-β. Also contributing to the cytokine milieu are products from endothelial cells, mast cells, fibroblasts, and lymphocytes, which initiate transition to the proliferative phase of wound repair (127,131,132).

The proliferative phase of wound repair is dominated by two principal processes: angiogenesis and epithelialization. The role of the keratinocyte has been emphasized throughout this chapter in the context of psoriasis, atopic dermatitis, and contact dermatitis, and many are keratinocyte-derived and mediate parts of the wound-healing response. Wound healing is ultimately determined by wound closure, which involves both keratinocyte migration and proliferation. Many of these cytokines are multifunctional in their effects on the keratinocyte (123–126). For example, the following are all both keratinocyte mitogens and chemoattractants: IL-1, EGF, KGF, IGF, bFGF, PDGF, IL-8, and the related CXC and CC chemokine subfamilies. TGF-β has been extensively studied for its role in wound healing, yet it is an inhibitor of both keratinocyte growth and migration. Other inhibitors of keratinocyte growth and differentiation of importance in wound repair include the interferons, TNF-α and TNF-β.

Angiogenesis is critical for the development of lush granulation tissue, which serves to receive and nurture migrating keratinocytes in the healing wound or in the form of a split thickness skin graft. This is in and of itself an intricate process, regulated by a number of the redundant pathways (126). Cytokines reported as angiogenic include bFGF, aFGF, angiogenin, angiotropin, IL-8, ENA 78, GRO-α, VEGF, TNF-α, and TGF-β, while those reported as angiostatic include thrombospondin, the interferons, IP-10, PF4, MIG, and angiostatin. Similar inhibitors of angiogenesis have been reported for other cell and tissue extracts (132–135) and for a variety of natural and artificial agents, including inhibitors of basement membrane biosynthesis (136–139) placental RNase (140), lymphotoxin (141), interferons (142), prostaglandin synthetase inhibitors (143), heparin-binding fragments of fibronectin (144), protamine (145), angiostatic steroids (146), several antineoplastic and anti-inflammatory agents (147,148) and PF4 (149). Insufficient angiogenesis has been proposed as part of the mechanism for chronic venous insufficiency ulcers by us and others. Alternatively, the clinical lesion known as pyogenic granuloma has been described as extensive granulation tissue and is characterized by an exaggerated wound-healing phenotype with an overexpression of angiogenic cytokines as its proposed cause. The role of angiogenesis in the pathophysiology of psoriasis has already been reviewed in this chapter (87,150,153). Aberrant angiogenic activity is a feature common to most chronic inflammatory disorders. In rheumatoid arthritis, psoriasis, or tumorigenesis, neovascularization may be overinduced, whereas in chronic nonhealing wounds (ulcers), angiogenesis has been inhibited, resulting in a failure to heal. Although the complement of positive and negative regulators of angiogenesis may vary among different physiological and pathological settings, the recognition of this dual mechanism of control is necessary to gain a more thorough understanding of this complex process and its significance in wound repair (126).

The maturation phase of wound repair is characterized by increasing tensile strength of the wound, down regulation of angiogenesis (angiostatic agents reviewed above), increasing matrix molecules, and remodeling of the scar. During the early phases of granulation tissue formation, the immature connective tissue resembles undifferentiated mesenchyme with the presence of persistent fibrin, an embryonic form of fibronectin, a predominance of collagen type III, as compared to collagen type I, and a highly vascularized capillary bed (154–156). This phase is followed by transformation to granulation tissue, which is associated with increased deposition of collagen type I, fibronectin, and protease-dependent remodeling of the extracellular matrix (157). This granulation tissue provides the foundation for the initiation of reepithelialization of the wound surface. As basal keratinocytes migrate from the wound edge over the surface of the mature granulation tissue, keratinocytes at the wound margin begin to proliferate. This response is followed by epidermal regeneration and production of basement membrane extracellular constituents (fibronectin, type IV and VII collagen, heparin sulfate proteoglycans, and laminin), which provides the integrity of the epidermal to dermal structures (132). While the sequential yet overlapping interplay of coagulation, inflammation, formation of granulation tissue, and reepithelialization are necessary for restoration of tissue function, certain reparative events are central for the complete evolution of wound healing. Cytokines important to fibroblast and collagen remodeling include IL-1, IL-2, bFGF, aFGF, TGF-β, PDGF, EGF, IGF, PD-ECGF, K-FGF, and IL-10, while the interferons may help in counterregulation of these process (125).

Chronic venous insufficiency ulcers (CVIUs) have been postulated to be, at least in part, mediated by dysregulation of many of these cytokines. It has been suggested that longstanding venous hypertension results in the development of a perivascular fibrin cuff due to leakage of plasma proteins. These cuffs, although discontinuous, have been hypothesized to interfere with normal vascular and connective tissue remodeling after wounding, which inhibits its normal healing (123). Some investigators have also suggested that the nonhealing state is due to a disruption of the precise coordination of the temporal expression of these mediators and not necessarily an absolute absence of any particular one.

We have begun to investigate the etiology of chronic venous insufficiency ulcers from the perspectives of both what cytokines are relatively or absolutely out of balance in the chronic ulcer and what changes take place when appropriate medical and/or surgical therapy of these lesions induces wound healing. In addition to the above families of mediators involved in orchestrating effective wound healing are the recently described family of chemotactic cytokines collectively referred to chemokines, which may also be targets for dysregulation in the pathogenesis of chronic wounds.

The chemokines consistent of number of distinct 8–10 kDa cytokines that exhibit 20–55% homology in amino acid sequence. Members of the CXC chemokine subfamily are located on human chromosome number 4 (54). All chemokines have four cysteine residues that form two disulfide bridges. In the CXC subfamily, the first two conserved cysteines are separated by one unconserved amino acid, hence the designation "CXC." This family includes IL-8; growth-regulated gene product (GRO-α); platelet basic protein and its cleavage products (connective tissue activating peptide-III (CTAP-III), β-thromboglobulin (β-TG), and neutrophil-activating peptide-2 (NAP-2); interferon-γ–inducible protein (IP-10); epithelial neutrophil-activating peptide-78 (ENA-78); granulocyte chemotactic protein-2 (GCP-2); and monokine–γ-interferon-inducible gene (MIG). Many members of the C-X-C chemokine subfamily have chemotactic activity for neutrophils as well as T cells (54,158).

The CC chemokine subfamily genes are located on human chromosome number 17. The members of this subfamily have adjacent conserved cysteine amino acid residues. These proteins include MCP-1-3, macrophage inflammatory protein-1α and -1β (MIP-1α and MIP-1β), RANTES, and I-309. Many of the members of this subfamily have chemotactic activity for monocytes as well as T cells and NK cells (158–162).

These two subfamilies have been associated with a variety of important functions related to wound healing. The CC chemokine subfamily has its main effects on macrophage function and wound be expected to be important in the middle and later phases of wound healing, whereas the CXC chemokine subfamily has been associated with effects on keratinocyte, fibroblast, and endothelial cell growth. In particular, among the CXC chemokine family, IL-8, EN8-78, NAP2, and GRO-α all share a distinct ELR motif preceding the first cysteine in the N-terminus (158,163,164). These CXC chemokines have been associated with significant angiogenesis function in vitro as well as sharing abilities to stimulate both keratinocyte proliferation and migration. CXC chemokines including IP-10 and PF4 lack this ELR motif and are angiostatic (128,164,165).

As there is a necessary requirement for regulated angiogenesis in normal wound healing, chemokines are potentially meaningful targets for the investigation of the pathophysiology of CVIUs. We investigated tissue and wound fluid levels of CXC and CC chemokines in a series of venous insufficiency ulcer patients at baseline and throughout an 8-week clinical trial in which all patients were treated with a triple-layer compression dressing overlying a nonadherent hydrofoam wound dressing (Allevyn®, Smith and Nephew United). We showed that the angiostatic, ELR-CXC chemokines (PF4 and IP-10) are abnormally elevated in both tissues and wound fluids of these patients at baseline. ELR^- chemokines remained elevated throughout the 8-week observation period but were found to decrease in relation to the amounts of ELR^+ chemokines in association with wound healing. Our results suggest that the relative ratios of the CXC chemokines may be important in the pathogenesis of CVUs and that a shift in these ratios may be important part of the change that occurs when a CVU begins to heal (166).

CUTANEOUS T-CELL LYMPHOMA

Cutaneous T-cell lymphoma is a malignancy of CD4+ Th cells, which is typically slowly progressive but eventually involves extracutaneous sites (166,167). Two distinct clinicopathological subtypes of cutaneous T-cell lymphoma are mycosis fungoides and Sézary syndrome. Mycosis fungoides is the more indolent disease, characterized initially by localized Th infiltrates in the skin, with four typical stages: large plaque parapsoriasis or premycosis fungoides, patch, plaque, and tumor. Sézary syndrome is a more rapidly progressive leukemia, which is characterized by circulating malignant Th cells and diffuse erythroderma (166,167). Both mycosis fungoides and Sézary syndrome have characteristic cutaneous findings and histopathology, but diagnosis can be very difficult in the early phases of the disease. Malignant cells in cutaneous T-cell lymphoma arise from the expansion of a single clone of T lymphocytes that exhibit a propensity for invasion of the upper dermis and epidermis (i.e., epidermotropism). During advancing stages of cutaneous T-cell lymphoma, several immunological abnormalities are observed, including decreased T-cell responses to antigens and mitogens, decreased natural killer cell activity, decreased lymphokine-activated killer cell activity, eosinophilia, and increased levels of IgE and IgA, as well as loss of epidermotropism by the malignant cells with advancing disease (168–174).

As reviewed early in this chapter, the cytokine production profiles of CD4+ Th cells allows us to divide them into three distinct functional subsets: Th0, Th1, and Th2. Th1 clones synthesize IL-2 and IFN-γ, which result in augmented CMI, T-cell proliferation, and macrophage activation. Th2 clones preferentially synthesize IL-4, IL-5, and IL-10 and preferentially augment humoral responses while suppressing CMI (175–177). Th0 clones are uncommitted and can make both Th1 and Th2 cytokines. There have also been elegant reciprocal inhibition pathways described between these Th1 and Th2 cytokines (36). Leprosy was the first cutaneous disease to be shown to preferentially utilize Th1 versus Th2 cytokines in different stages of the disease. As reviewed elsewhere in this chapter, a dichotomy exists in many dermatoses, with Th2 cytokine patterns prominent in atopic dermatitis, late-phase allergic reactions, hyper-IgE syndrome, and chronic helminthic infection and lepromatous leprosy (90,178,179), while Th1 cytokines predominate in psoriasis and lichen planus (11,86). Th cytokine profiles appear to be important in the clinical progression of cutaneous carcinomas as basal and squamous cell carcinomas are characterized by local Th2 cytokines, while benign epidermal tumors (e.g., seborheic keratosis) have a Th1 local cytokine predominance (180,181).

We have used sensitive ELISA and RT/PCR techniques to generate substantial data indicating that the malignant T cell in cutaneous T-cell lymphoma is a Th2-type cell producing predominantly IL-4, IL-5, and IL-10 (183). We also see a dichotomy associated with disease stage as in leprosy with patch- and plaque-stage mycosis fungoides expressing a overall Th1 phenotype, while tumor-stage mycosis fungoides and Sézary syndrome and Th2 in profile. We, along with those studying cutaneous carcinomas, have hypothesized that it is the tumor-associated (benign or reactive) lymphocytes in these skin malignancies that generate Th1 cytokines, while the tumor cells are the source of Th2 cytokines. This scenario fits the clinicopathological findings of a greater percentage of reactive cells in the infiltrates of early cutaneous T-cell lymphoma (Th1 < Th2). This would be predicted to result in effective cellular immunity, leading to the relatively indolent disease observed clinically. In late-stage cutaneous T-cell lymphoma samples, tumor cells make up the majority of infiltrating T cells, resulting in defective cellular immunity and more rapidly progressive disease.

In both patch- and plaque-stage CTCL, there is excellent co-localization between LFA-1+, CD4+ T cells and epidermal keratinocyte ICAM-1 expression (183–185). With more advanced disease, especially Sézary syndrome, keratinocyte ICAM-1 expression becomes markedly diminished (185,186). CTCL has not been associated with loss of T-cell LFA-1 expression, in contrast with noncutaneous, higher grade non-Hodgkin's lymphomas, in which LFA-1 is frequently not expressed (187). When malignant T cells were enriched from the blood for a Sézary syndrome patient with > 90% atypical circulating T cells, no IFN-γ production was detectable either constitutively or after mitogenic stimulation (186). Thus, in Sézary syndrome, it appears that a lack of IFN-γ results in a loss of epidermotropism secondary to decreased ICAM-1 expression. This further suggests that induction of IFN-γ may be important in the treatment of cutaneous T-cell lymphoma.

In the sections above discussing the cytokine networks of psoriasis and atopic dermatitis, we described how the regulation of Th1 versus Th2 cytokine responses in dependent on the nature of the antigen-presenting cells, which stimulate the T cell during the afferent phase of the immune response (95). Professional antigen-presenting cells produce IL-12, which augments Th1 responses, while at least some nonprofessional antigen-presenting cells make IL-10, which favors Th2 responses. The costimulatory signals used by these antigen-presenting cells is important in dictating the Th cytokine response profile. Thus, to better understand the immunobiology of cutaneous T-cell lymphoma and the interplay between Th1 and Th2 T cells, we have begun to investigate the role(s) of antigen-presenting cells in cutaneous T-cell lymphoma. Antigen-presenting cells have been implicated in the presentation of immunogenic tumor antigens in cutaneous T-cell lymphoma, resulting in antitumoral cell–mediated immunity, which keeps early cutaneous T-cell lymphoma indolent. Antigen-presenting cells may also be involved in the chronic stimulation of T cells, which results in malignant transformation and/or generation of immunosuppressive signals that could lead to tumor progression (66,189–191).

Activity within cutaneous T-cell lymphoma skin lesions probably involves not only interaction between the cytokines of the tumor-infiltrating lymphocytes (Th1) and the tumor cells (Th2), but also how the cytokine milieu effects cutaneous antigen-presenting cell function. This concept is graphically represented in Figure 1. In early cutaneous T-cell lymphoma (Fig. 1A) tumor infiltrating lymphocytes outnumber tumor cells, thus Th1 cytokines and CD8+ cytotoxic cells predominate and Th2 cytokines from both malignant cells and IL-10–producing dermal macrophages are inhibited. Our data show that there is an expansion of antigen-presenting cells per unit area in early cutaneous T-cell lymphoma, along with parallel increases in B7.2 expression and relatively high levels of in situ GM-CSF production (66,188,191). The effect of all these is to augment antigen-presenting cell delivery of both signals 1 (tumor-associated antigen + Class II) and 2 (costimulatory molecule interaction).

With tumor progression (Fig. 1B) tumor cells outnumber tumor-infiltrating lymphocytes and the Th2 > Th1 cytokine effects are suppression of antigen-presenting cell function. This results in fewer functional dermal dendritic cells (DDCs) and LC and relative increases in the immunosuppressive effects of dermal macrophages. As these macrophages may produce IL-10 and constituently express CD86, their immunosuppressive effects may counterbalance the effects of DDC and LC and become accentuated in advanced-stage cutaneous T-cell lymphoma. This dynamic interaction is also supported by evidence that suggests that IFN-γ can augment antigen-presenting cell function through increased expression of the costimulatory B7.2/ CD86 antigen (95,191). Keratinocytes are included in Figure 1 as they have multiple roles in the biology of cutaneous T-cell lymphoma, including the expression of mediators of epidermotropism (TNF-α, ICAM-1, IL-1, IL-6, IL-7, IL-8, and other CXC chemokines), GM-CSF (which augments antigen-presenting cell function), and IL-10 (4–14,95,192).

IFN-γ and IL-12 treatment have been shown to inhibit the production of Th2 cytokines produced by the peripheral blood lymphocytes of patients with advanced Sézary syndrome (171,193). Other studies have revealed significant clinical responses among some patients with cutaneous T-cell lymphoma treated with recombinant human IFN-γ (194). IFN-α and photophoresis have been demonstrated to both have therapeutic activity in cutaneous T-cell lymphoma (195,196). IFN-α inhibits both IL-4 and IL-10 production in vitro by normal T cells as well as Sézary syndrome cells (36,197). Kim et al. suggest that IFN-α treatment of basal cell carcinoma functions via its ability to restore IL-2 mRNA production in the tumor-infiltrating lymphocytes in basal cell carcinomas (181).

Photophoresis is a potent activator of peripheral blood monocytes resulting in a marked release of the IL-6, TNF-α, and IL-1β (198). IFN-α may synergize with photophoresis by combining with TNF-α released from photoactivated macrophages to exert a more profound negative influence on proliferation of the malignant T-cell clone than either agent alone (196). In support of this line of reasoning, we have recently reported the augmentation of IL-2, IL-12, and IFN-γ in the Sézary cell line known as

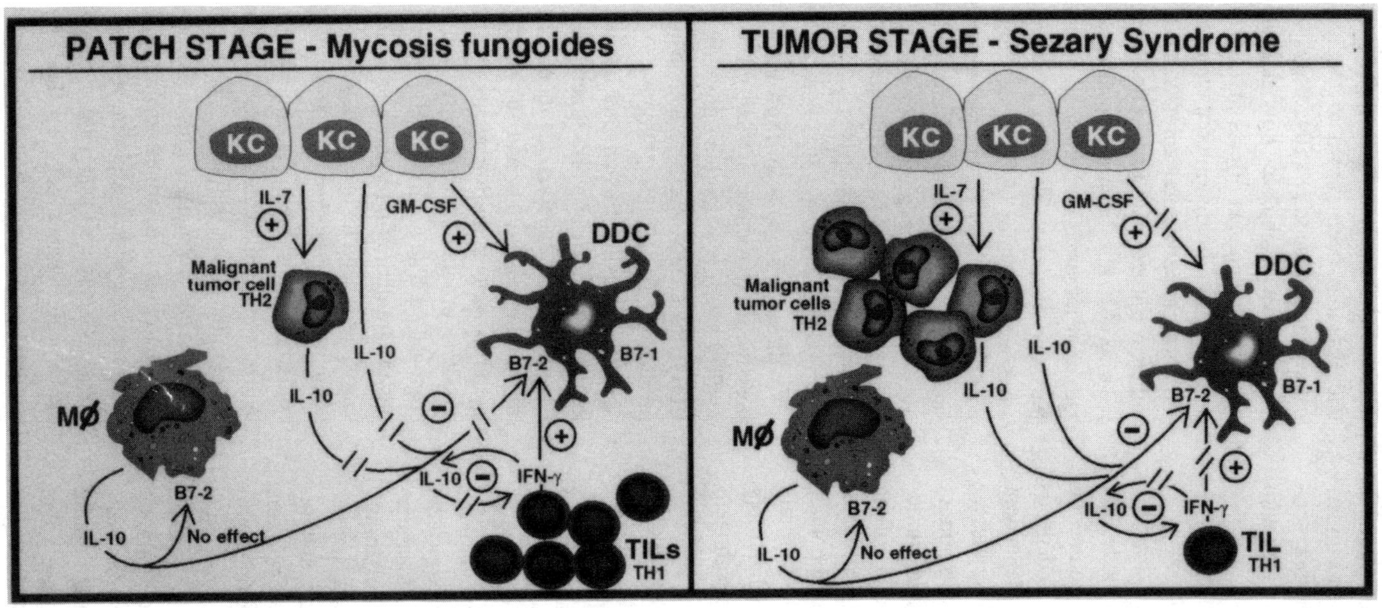

Figure 1 Cytokine interactions between antigen-presenting cell, tumor-infiltrating lymphocyte, keratinocytes, and malignant T cells in cutaneous T-cell lymphoma. (A) Early stage of cutaneous T-cell lymphoma. CD4+ tumor-infiltrating lymphocytes secreting Th1 cytokines (IFN-γ) predominate, resulting in inhibition of Th2 cytokines. Effective antigen-presenting cell function via B7.2 expression (signal 2) on dermal dendritic cells (DDCs) allows for immunogenic tumor antigens+ Class II (signal 1) to favor generation of CD8+ effector cells. Tumor-infiltrating lymphocyte cytokines are part of a positive feedback loop, also inhibiting IL-10 from dermal macrophages resulting in indolent disease. (B) Late stages of cutaneous T-cell lymphoma. Tumor-infiltrating lymphocyte secretion of Th1 cytokines is inhibited by expanding malignant (Th2) cells and by IL-10 from dermal macrophages. Antigen-presenting cell function is inhibited by IL-10 downregulation of DDC B7.2, resulting in loss of signal 2 and negative feedback to both Th1 and CD8+ tumor-infiltrating lymphocytes despite increasing tumor-associated antigen for signal 1. Constituent B7.2 on dermal macrophages favors their suppression of local cellular immunity.

HUT-78 cells treated with in vitro PUVA. We also have shown that the PBMC in Sézary syndrome, patients undergoing photophoresis can be documented to generate serially higher levels of mRNAs for these Th1 cytokines with each successive treatment cycle (199).

SUMMARY

The skin—not only the largest organ of the body, but also uniquely positioned at the external environmental interface—must be able to produce both antigen-independent inflammatory reactions as well as antigen-dependent immunological responses. Cytokines produced by indigenous and recruited cells within the skin are important contributors to both inflammatory reactions and immunological responses. In the past, chemical mediators of the acute inflammatory reaction included amines (i.e., histamine and epinephrine), kinins, complement components, and a variety of still emerging and poorly defined protein and tissue extract "factors" (200,201). Somewhat more recently, experimental and clinical studies of the pathogenesis of inflammation have been focused on a different group of nonprotein mediators, including prostaglandins, leukotrienes, and oxygen radicals and ions. Currently, the inflammatory cascade of mediators, which has largely replaced the aforementioned mediators, are the result of molecular studies principally involving adhesion molecules, glycoproteins, and cytokines. In contrast to previous tissue extracts and "factors," recombinant biotechnology has provided investigators with a host of purified protein molecules and animal models that can be used to understand and explain the classic manifestations of inflammation. Moreover, many immunological phenomena, such as autoimmunity, memory, organ-specific trafficking, and tolerance, can also be understood in terms of these cytokine networks and how they effect T lymphocytes, antigen-presenting cells, thymic epithelial cells, endothelial cells, keratinocytes, and other cells.

Finally, it should be emphasized that not only do cytokine networks provide an important molecular basis for understanding inflammatory reactions and immune responses with the skin (121), but the skin itself can produce cytokines that have a systemic impact on the immune system and neuroendocrine system as well as on the central nervous system (202). This is perhaps best underscored by the example of a severe sunburn. In this scenario, not only does the skin become intensely erythematous, but the sunbather subsequently experiences fever, malaise, and several other untoward systemic acute-phase responses. These noncutaneous effects are mediated in part by keratinocyte-derived IL-1, increased IL-1 levels can be detected in the serum following acute phototoxicity (203), and IL-1 has been shown to mediate many acute-phase pathogenetic responses (204,205).

Another example of how skin-derived cytokines can have extracutaneous effects involves ultraviolet radiation and systemic immunomodulation. Following ultraviolet radiation, experimental animals have a depressed ability either to initiate or to propagate various types of immune reactions. It has been proposed that the postultraviolet radiation increases in serum IL-1 induces the hypothalamus and pituitary gland to produce elevated levels of adrenocorticotropin (ACTH), leading to increased glucocorticosteroids and other immunomodulatory hormones, such as dehydroepiandrosterone and vitamin D_3 derivatives (206). These glucocorticosteroids have differential effects on Th1 versus Th2 subsets such that the Th1 cell cytokines (IL-2 and IFN-γ) are reduced, whereas the Th2 cytokines IL4 and IL-10 are increased. Keratinocytes also directly upregulate expression of the immunosuppressive IL-10 molecule (98,99). UV light also increases a particular subset of antigen-presenting cells referred to as CD1-DR+ CD36+ macrophages, which also constitutively produce IL-10 (190,207,208).

In conclusion, this chapter has reviewed the currently known cytokine networks operative within the skin of five different cutaneous disorders. The microenvironmental cellular costituents and their respective ability to respond to various inciting agents, their anatomical position, cytokine production capability, and diversity have been emphasized. The ability of keratinocytes to initiate cutaneous inflammation in response to a diversity of low molecular weight haptens by producing larger molecular weight cytokine polypeptides indicates the importance of this epidermal constituent. A role for cutaneously derived cytokines in both acute inflammatory reactions, as well as in immune-based responses both within the skin and beyond, has been highlighted. Clearly, many links exist between cytokines, adhesion molecules, chemotactic factors, neuropeptides, and immunomodulatory hormones. The skin is a readily accessible organ to investigate further the delicate balance between these positive and negative communication networks and how it mediates homeostasis in response to diverse array of environmental stimuli. Several lines of investigation support that epidermis can serve as the ultimate signal transducer as keratinocytes directly respond to these diverse stimuli by producing proinflammatory cytokines. These cytokines and the subsequent cytokine network that results from such stimulation will continue to intrigue investigative skin biologists for many years to come (11,200).

REFERENCES

1. Balkwill FR, Burke F. The cytokine network. Immunol Today 1989; 10:299–304.
2. Nickoloff BM, Griffiths CEM, Barker JNWN. The role of adhesion molecules, chemotatic factors, and cytokines in inflammatory and neoplastic skin disease—1990 update. J Invest Dermatol (suppl) 1990; 94:151–157.
3. Sauder DN, Carter C, Katz SI, Oppenheim JJ. Epidermal cell production of thymocyte activating factor (ETAF). J Invest Dermatol 1982; 79:34–39.
4. Csata M, Rosenbach T, Grabbe J, Czarnetzki BM. Epidermal scales. Are they a wastebasket of inflammatory mediators or active participants in epidermal inflammation? Int J Dermatol 1989; 28(2):86–89.
5. Barker JNWN, Mitra RS, Griffiths CEM, Dixit VM, Nickoloff BJ. Keratinocytes as initiators of inflammation. Lancet 1991; 337:211–214.
6. Luger TA, Schwarz T. Therapeutic use of cytokines in dermatology. J Am Acad Dermatol 1991; 24:915–926.
7. Yates RA, Nanney LB, Gates RE, King Jr LE. Epidermal growth factor and related growth factors. Int J Dermatol 1991; 30(10):687–694.
8. Baadsgaard O, Wang T. Immune regulation in allergic and irritant skin reactions. Int J Dermatol 1991; 30:161–172.
9. Staiano-Coico L, Krueger JG, Rubin JS, D'limi S, Vallat VP, Valentino L, Fahey III T, Hawes A, Kingston G, Madden MR, Mathwich M, Gottlieb AB, Aaronson SA. Human keratinocyte growth factor effects in a porcine model of epidermal wound healing. J Exp Med 1993; 178:865–878.
10. Krane JF, Gottlieb AB, Carter DM, Krueger JG. The insulin-like growth factor I receptor is overexpressed in psoriatic epidermis, but is differentially regulated from the epidermal growth factor receptor. J Exp Med 1992; 175:1081–1090.
11. Nickoloff BJ. The cytokine network in psoriasis. Arch Dermatol 1991; 127:871–884.
12. Barker JNWN, Jones ML, Swenson CL, Sarma V, Mitra RS, Ward PA, Johnson KJ, Fantone JC, Dixit VM, Nickoloff BJ. Monocyte chemotaxis and activating factor production by keratinocytes in response to IFN-γ. J Immunol 1991; 146:1192–1197.
13. Haan P de, Sampat S, Raaij EMH van der, Bruynzeel I, Boorsma DM, Willemze R, Stoof TJ. Effect of calcitriol on growth, differentiation, chemokine mRNA expression of cultured keratinocytes and on keratinocyte-T cell binding. Acta Derm Venereol (Stockh) 1994; 186:52–54.
14. Aragane Y, Riemann H, Bhardwaj RS, Schwarz A, Sawada Y, Yamada H, Luger TA, Kubin M, Trinchieri G, Schwarz T. J Immunol 1994; 153:5366–5372.
15. Streilein JW. Lymphocyte traffic T-cell malignancies and the skin. J Invest Dermatol 1978; 71:167–171.
16. Chin W, Hay JB. A comparison of lymphocyte migration through intestinal lymph nodes, subcutaneous lymph nodes, and chronic inflammatory sites in sheep. Gastroenterology 1980; 79:1231–1238.
17. Daynes RA, Spangrude GJ, Roberts LK, Krueger GG. Regulation by the skin of lymphoid cell recirculation and localization properties. J Invest Dermatol 1985; 85:145–205.
18. Tigelaar RE. Lymphocyte traffic and the skin. Dermatol Clin 1985; 3:569–585.
19. Knobler RM, Edelson RL. Cutaneous T cell lymphoma. Med Clin North Am 1986; 70:109–138.
20. Nickoloff BJ. Adhesion molecules and inflammatory cell migration pathways in the skin. In: Bos J, ed. Skin Immune System. Boca Raton, FL: CRC PRess, 1990:49–71.
21. Ackerman AB. Histologic Diagnosis of Inflammatory Skin Diseases. Philadelphia: Lea and Febiger, 1987.
22. Nickoloff BJ, Griffiths CEM. Abnormal cutaneous topobiology: the molecular basis for dermatopathological mononuclear cell patterns in inflammatory skin disease. J Invest Dermatol (suppl) 1990; 95:128–131.
23. Baer RL. The mechanisms of allergic contact dermatitis. In: Fisher AA, ed. Contact Dermatitis, 3rd ed. Philadelphia: Lea and Febiger, 1986:1–8.
24. Landsteiner R, Chase MW. Experiments on transfer of cutaneous sensitivity to simple chemicals. Proc Soc Exp Biol Med 1942; 49:688–690.
25. Eisen HN. Hypersensitivity to simple chemicals. In: Lawrence HS, ed. Cellular and Humoral Aspects of the Hypersensitive States. New York: Hoeber-Harper Inc., 1959:89–122.
26. Silberberg-Sinakin I, Gigli I, Baer RL, Thorbecke GJ. Langerhans cells: role in contact hypersensitivity, and relationships to lymphoid dendritic cells and to macrophages. Immunol Res 1980; 53:203–252.
27. Thestrup-Pedersen K, Larsen CG, Ronnevig J. The immunology of contact dermatitis. Contact Dermatitis 1989; 20:81–92.
28. Hartman A, Hoedemaker J, Nater JP. Histological aspects of DNCB sensitization and challenge tests. Br J Dermatol 1976; 94:407–416.
29. Griffiths CEM, Nickoloff BJ. Keratinocyte intercellular adhesion molecule-1 (ICAM-1) expression precedes dermal T lymphocyte infiltration in allergic contact dermatitis (rhus dermatitis). Am J Pathol 1989; 135:1045–1053.
30. Griffiths CEM, Barker JNWN, Kunkel S, Nickoloff BJ. Induction, distribution and dimunition of leukocyte adhesion molecules (ELAM-1, VCAM-1, ICAM-1), a T cell chemotaxin (IL-8) and modulatory cytokine (TNF-α) in allergic contact dermatitis (Rhus dermatitis). Br J Dermatol 1991; 124:519–526.
31. Ohmen JD, Hanifin JM, Nickoloff BJ, Rea TH, Wyzykowski R, Kim J, Jullien D, McHugh T, Nassif AS, Chan SC, Modlin RL. Overexpression of IL-10 in atopic dermatitis. Contrasting cytokine patterns with delayed-type hypersensitivity reactions. J Immunol 1995; 154:1956–1963.
32. Barker JNWN, Mitra RS, Griffiths CEM, Dixit VM, Nickoloff BJ. Keratinocytes as initiators of inflammation: a unifying explanation for the diverse array of environmental stimuli to produce cutaneous inflammation. Lancet 1991; 337:211–214.
33. Saed G, Fivenson DP, Naidu Y, Nickoloff BJ. Mycosis fungoides exhibits a Th1-type cell-mediated cytokine profile whereas Sezary syndrome expresses a Th2-type profile. J Invest Dermatol 1994; 103:29–33.
34. Nickoloff BJ, Fivenson DP, Kunkel SL, Strieter RM, Turka LA. Keratinocyte interleukin-10 expression is upregulated in tape-stripped skin, poison ivy dermatitis, and Sezary syndrome, but not in psoriatic plaques. Clin Immunol Imunopathol 1994; 73(1):63–68.
35. Nickoloff BJ, Naidu Y. Perturbation of epidermal barrier function correlates with initiation of cytokine cascade in human skin. J Am Acad Dermatol 1994; 30:535–546.
36. Parronchi P, DeCarli M, Manetti R, Simonelli C, Sampognaro S, Piccinni MP, Macchia D, Maggi E, Del Prete G, Romangnani S. IL-4 and IFN (α and γ) exert opposite regulatory effects on the development of cytolytic potential by Th1 or Th2 human T cell clones. J Immunol 1992; 149(8):2977–2983.
37. Gawkrodger DJ, McVittie E, Ross JA, Hunter JA. Phenotypic characterization of the early cellular responses in allergic and

irritant contact dermatitis. Clin Exp Immunol 1986; 66:590–598.
38. Picker LJ, Kishimoto TK, Smith CW, Warnock RA, Butcher EC. ELAM-1 is an adhesion molecule for skin-homing T cells. Nature 1991; 349:796–799.
39. Pitzalis C, Kingsley G, Haskard D, Panayi G. The preferential accumulation of helper-induce T lymphocytes in inflammatory lesions: evidence for regulation by selective endothelial and homotypic adhesion. Eur J Immunol 1988; 19:1397–1404.
40. Lowe JB, Stoolman LM, Nair RP, Larsen RD, Berhend TL, Marks RM. ELAM-1 dependent cell adhesion to vascular endothelium determined by a transfected human fucosyltransferase cDNA. Cell 1990; 63:475–484.
41. Phillips ML, Nudelman E, Gaeta FCA et al. ELAM-1 mediates cell adhesion by recognition of a carbohydrate ligand, Sialyl-Le X. Science 1990; 250:1130–1132.
42. Walz G, Aruffo A, Kolanus W, Bevilacqua M, Seed B. Recognition by ELAM-1 of the Sialy-Le X determinant on myeloid and tumor cells. Science 1990; 250:1132–1135.
43. Osborn L, Hession C, Tizzard R, et al. Direct expression cloning of vascular cell adhesion molecule-1, a cytokine-induced endothelial protein that binds to lymphocytes. Cell 1989; 59:1203–1211.
44. Dustin ML, Springer TA. Lymphocyte function-associated antigen-1 (LFA-1) in teraction with intercellular adhesion molecule-1 (ICAM-1) is one of at least three mechanisms for lymphocyte adhesion to cultured endothelial cells. J Cell Biol 1988; 107:321–331.
45. Stoolman LM. Adhesion molecules controlling lymphocyte migration. Cell 1989; 56:907–910.
46. Gautam SC, Chikkala F, Hamilton TA. Anti-inflammatory action of IL-4. Negative regulation of contact sensitivity to trinitrochlorobenzene. J Immunol 1992; 148:1411–1415.
47. Rivas JM, Ullrich SE. Systemic suppression of delayed-type hypersensitivity by supernatants from UV-irradiated keratinocytes. J Immunol 1992; 149(12):3868–3871.
48. Shiohara T, Moriya N, Gotoh C, et al. Differential expression of lymphocyte function associated antigen 1 (LFA-1) on epidermotropic and non-epidermotropic T-cell clones. J Invest Dermatol 1989; 93:804–808.
49. Hemler ME. Adhesive protein receptors on hematopoietic cells. Immunol Today 1988; 9:109–113.
50. Wilkinson PC, Higgins A. OKT3-activated locomotion of human blood lymphocytes:a phenomenon requiring contact of T cells with Fc receptor-bearing cells. Immunology 1987; 60:445–451.
51. Ternowitz T, Thestrup-Pedersen K. Epidermis and lymphocyte interactions during a tuberculin skin reaction. II. Epidermis contains specific lymphocyte chemotactic factors. J Invest Dermatol 1986; 87:613–616.
52. Larsen CG, Anderson AO, Appella E, Oppenheim JJ, Matsushima K. The neutrophil-activating protein (NAP-1) is also chemotactic for T lymphocytes. Science 1989; 243:1464–1467.
53. Enk A, Katz SI. Early molecular events in the induction phase of contact sensitivity. Proc Natl Acad Sci USA 1992; 89:1398–1402.
54. Streiter RM, Kunkel SL. Chemokines and the lung. In: Crystal R, West J, Weibel E, Barnes, eds. The Lung: Scientific Foundations. 2nd ed. New York: Raven Press, 1996.
55. Wood GS, Volterra AS, Abel EA, Nickoloff BJ, Adams RM. Allergic contact dermatitis: novel immunohistologic features. J Invest Dermatol 1986; 87:688–692.
56. Barker JNWN, Ophir J, MacDonald DM. Products of class II major histocompatibility complex gene regions are differentially expressed on keratinocytes in cutaneous disease. J Am Dermatol 1988; 19:667–672.
57. Barker JNWN, Sarma V, Mitra RS, Dixit VM, Nickoloff BJ. Marked synergism between tumor necrosis factor-alpha and interferon-gamma in regulation of keratinocyte-derived adhesion molecules and chemotactic factors. J Clin Invest 1990; 85:605–608.
58. Barker JNWN, Jones ML, Swenson CL, et al. Monocyte chemotaxis and activating factor production by keratinocytes in response to IFN gamma. J Immunol 1991; 4:1192–1197.
59. Stingl G, Katz S, Clement L, Green I, Shevach E. Immunologic function of Ia-bearing Langerhans cells. J Immunol 1978; 121:2005–2012.
60. Streilein JW, Bergstresser PR. Langerhans cell function dictates induction of contact hypersensitivity or unresponsiveness to DNFB in Syrian hamsters. J Invest Dermatol 1981; 77:272–277.
61. Res P, Kapsenberg ML, Bos JD, Steikema F. The crucial role of human dendritic antigen presenting cell subsets in nickel-specific T-cell proliferation. J Invest Dermatol 1987; 88:550–554.
62. Steinman R, Inaba K. Immunogenicity: role of dendritic cells. Bioessays 1989; 10:145–152.
63. Streilein JW. Antigen presenting cells in the induction of contact hypersensitivity in mice: evidence that Langerhans cells are sufficient but not required. J Invest Dermatol 1989; 93:443–448.
64. Nestle FO, Zheng XG, Thompson CB, Turka LA, Nickoloff BJ. Characterization of dermal dendritic cells obtained from normal human skin reveals phenotypic and functionally distinctive subsets. J Immunol 1993; 151:6535–6545.
65. Nestle FO, Nickoloff BJ. Role of dendritic cells in benign and malignant lymphocytic infiltrates of the skin. Clin Dermatol 1994; 12(2):271–278.
66. Fivenson DP. Dermal dendrocytes: dynamic determinants in the development of dermatologic disease. In: Dermal Immune System. Nickoloff BJ, ed. CRC Press, Boca Raton, FL, 1992:277–290.
67. Griffiths CEM, Esmann J, Fisher GJ, Voorhees JJ, Nickoloff BJ. Gamma interferon differentially modulates keratinocyte expression of HLA-DR and intercellular adhesion molecule-1 (ICAM-1): evidence for a protein kinase C-dependent and independent pathway. Br J Dermatol 1990; 122:333–342.
68. Bevilacqua MP, Pober JS, Wheeler ME, Cotran RS, Gimbrone MA. Interleukin-1 acts on cultured vascular endothelium to increase the adhesion of polymorphonuclear leukocytes, monocytes and related cell lines. J Clin Invest 1985; 76:2003–2011.
69. Xu Y, Swerlick RA, Sepp N, Bosse D, Ades EW, Lawley TJ. Characterization of expression and modulation of cell adhesion molecules on an immortalized human dermal microvascular endothelial cell line (HMEC-1). J Invest Dermatol 1994; 102:833–837.
70. Cornelius LA, Sepp N, Li LJ, Degitz K, Swerlick RA, Lawley TJ, Caughman SW. Selective upregulation of intercellular adhesion molecule (ICAM-1) by ultraviolet B in human dermal microvascular endothelial cells. J Invest Dermatol 1994; 103:23–28.
71. Petzelbauer P, Bender JR, Wilson J, Pober JS. Heterogeneity of dermal microvascular endothelial cell antigen expression and cytokine responsiveness in situ and in cell culture. J Immunol 1993; 151:5062–5072.
72. Nickoloff BJ, Griffiths CEM. Lymphocyte trafficking in psoriasis: a new perspective emphasizing the dermal dendrocyte with active dermal recruitment mediated via endothelial cells

followed by intraepidermal activation. J Invest Dermatol 1990; 95:35–37.

73. Krueger GG, Bergstresser PR, Lowe NJ, Voorhees JJ, Weinstein GD. Psoriasis. J Am Acad Dermatol 1984; 11:937–947.
74. Bernard BA, Robinson SM, Vandaele S, Mansbridge JN, Darmon M. Abnormal maturation pathway of keratinocytes in psoriatic skin. Br J Dermatol 1985; 112:647–653.
75. Weinstein GD, Frost P. Abnormal cell proliferation in psoriasis. J Invest Dermatol 1968; 50:254–258.
76. Cox AJ, Watson W. Histological variation in lesions of psoriasis. Arch Dermatol 1972; 106:503–506.
77. Valdimarsson H, Barker BS, Jonsdottir I, Fry L. Psoriasis: a disease of abnormal keratinocyte proliferation induced by T lymphocytes. Immunol Today 1986; 7:256–259.
78. Fry L. Immunointervention in skin disorders. J Autoimmun 1988; 1:593–601.
79. Nickoloff BJ, Fisher GS, Mitra RS, Voorhees JJ. Additive and synergistic antiproliferative effects of cyclosporine A and gamma interferon on cultured human keratinocytes. Am J Pathol 1988; 131:12–18.
80. Baker BS, Powles AV, Valdimarsson H, Fry L. An altered response by psoriatic keratinocytes to gamma interferon. Scand J Immunol 1988; 28:735–740.
81. Nickoloff BJ, Mitra RS, Elder JT, Fisher GJ, Voorhees JJ. Decreased growth inhibition by recombinant gamma interferon is associated with increased production of transforming growth factor-alpha in keratinocytes cultured from psoriatic lesions. Br J Dermatol 1989; 121:161–174.
82. Nickoloff BJ, Karabin GD, Barker JNWN, Griffiths CEM, Sarma V, Mitra RS, Elder JT, Kunkel SL, Dixit V. Cellular localization of interleukin-8 and its inducer-tumor necrosis factor-alpha in psoriasis. Am J Pathol 1991; 138:129–140.
83. Ho V, Griffiths CEM, Gupta A, et al. Intralesional cyclosporine A. The treatment of psoriasis: a clinical, immunologic and pharmacokinetic study. J Am Acad Dermatol 1990; 22:94–110.
84. Abb J, Abb H. Effect of cyclosporin on human keukocyte interferon production. Selective inhibition of gamma interferon synthesis. Transplant Proc 1983; 15:2380–2382.
85. Remick DG, Ngugyen DT, Eskanderl ME, Streiter RM, Kunkel SL. Cyclosporin A inhibits TNF production without decreasing TNF mRNA levels. Biochem Biophys Res Commun 1989; 161:551–555.
86. Uyermura K, Yamamura M, Fivenson DP, Modlin RL, Nickoloff BJ. The cytokine network in lesional and lesion-free psoriatic skin is characterized by a T-helper type 1 cell mediated response. J Invest Dermatol 1993; 101:701–705.
87. Nickoloff BJ, Mitra RS, Varani J, Dixit VM, Polverini PJ. Aberrant production of interleukin-8 and thrombospondin-1 by psoriatic keratinocytes mediates angiogensis. Am J Pathol 1994; 144:820–828.
88. Barker JNWN, Karabin GD, Stoof T, et al. Detection of gamma interferon mRNA in psoriatic epidermis by polymerase chain reaction. J Dermatol Sci 1991; 2:106–11.
89. Fierlbeck G, Rasner G, Muler G. Psoriasis induced at injection site by recombinant gamma interferon. Arch Dermatol 1991; 126:351–335.
90. Yamamura M, Uyermura K, Deans RJ, Weinberh K, Rea T, Bloom BR, Modlin RL. Defining protective responses to pathogens: cytokine profiles in leprosy lesions. Science 1991; 254:277–279.
91. Krueger JG, Krane JF, Carter DM, Gottlieb AB. Role of growth factors, cytokines and their receptors in the pathogenesis of psoriasis. J Invest Dermatol (suppl) 1990; 94:135–140.
92. Nestle FO, Turka LA, Nickoloff BJ. Characterization of dermal dendritic cells in psoriasis. Autostimulation of T lymphocytes and induction of Th1 type cytokines. J Clin Invest 1994; 94:202–209.
93. Mitra RS, Judge TA, Nestle FO, Turka LA, Nickoloff BJ. Psoriatic skin-derived dendritic cell function is inhibited by exogenous IL-10. Differential modulation of B7-1 (CD80) and B7-2 (CD86) expression. J Immunol 1995; 154:2668–2677.
94. Bonifati C, Carducci M, Fei PC, Trento E, Sacerdoti G, Fazio M, Ameglio F. Correlated increases of tumour necrosis factor-α, interleukin-6 and granulocyte monocyte-colony stimulating factor levels in suction blister fluids and sera of psoriatic patients-relationships with disease severity. Clin Exp Dermatol 1994; 19:383–387.
95. Turka LA, Nickoloff BJ. Immunologic functions of non-professional antigen presenting cells: New insights from studies of keratinocyte: T cell interactions. Immunol Today 1994; 15:464–468.
96. Macatonia SE, Hosken NA, Litton M, Vieira P, Hsieh CS, Culpepper JA, Wysocka M, Trinchieri G, Murphy KM, O'Garra A. Dendritic cells produce IL-12 and direct the development of Th1 cells from naive CD4+ T cells. J Immunol 1995; 154:5071–5079.
97. Macatonia SE, Hosken NA, Litton M, Vieira P, Hsieh CS, Culpepper JA, Wysocka M, Trinchieri G, Murphy KM, O'Garra A. Dendritic cells produce IL-12 and direct the development of Th1 cells from naive CD4+ T cells. J Immunol 1995; 154:5071–5079.
98. Schmitt DA, Owen-Schaub L, Ullrich SE. Effect of IL-12 on immune suppression and suppressor cell induction by ultraviolet radiation. J Immunol 1995; 154:5114–5120.
99. Ullrich SE. Mechanism involved in the systemic suppression of antigen-presenting cell function by UV irradiation. J Immunol 1994; 152:3410–3416.
99a. Nestle FO, Mitra RS, Bennett CF, Chan H, Nickoloff BJ. Cationic lipid is not required for uptake and selective inhibitory activity of ICAM-1 phosphorothioate antisense oligonucleotides in keratinocytes. J Invest Dermatol 1994; 103:569–575.
100. Gottlieb SL, Gilleaudeau P, Johnson R, Estes L, Woodworth TG, Gottlieb AB, Krueger JG. Response of psoriasis to a lymphocyte-selective toxin (DAB $_{389}$IL-2) suggests a primary immune, but not keratinocyte, pathogenic basis. Natl Med 1995; 1(5):442–447.
101. Hanifin JM. Atopic dermatitis. J Am Acad Dermatol 1982; 6:1–13.
102. Lever R, Turbitt M, Sanderson A, Mackie R. Immunophenotyping of the cutaneous infiltrate of the mononuclear cells in the peripheral blood in patients with atopic dermatitis. J Invest Dermatol 1987; 89:4–7.
103. Barker JNWN, Allern VA, MacDonald DM. Surface bound immunoglobulin E on antigen presenting cells in cutaneous tissue of atopic dermatitis. J Invest Dermatol 1988; 90:177–221.
104. Bieber T, de la Salle H, Wollenber A, Haimi J, Chizzonite R, Ring J, Hanau D, de la Sale C. Human epidermal Langerhans cells express the high affinity receptor for immunoglobulin E (FcεRI). J Exp Med 1992; 175:1285–1290.
105. Leung DYM, Wood N, Dubey D, Rhodes AR, Geha RS. Cellular basis of defective cell-mediated lympholysis in atopic dermatitis. J Immunol 1983; 130:1678–1682.
106. Rheinhold U, Pawelec G, Wehrmann W, Herold M, Kreggsel HW. Immunoglobulin E and immunoglobulin G subclass distribution in-vivo and relationship to in-vitro generation of interferon-gamma and neopterin in patients with severe atopic dermatitis. Clin Res Allergy 1986; 4:67–86.
107. Brown MA, Li SH, Chan SC, Hanifin J. Interleukin-4 mRNA

expression by normal and atopic T lymphocytes (abstract). Clin Res 1989; 37:406.
108. Wiererger EA, Sneuk M, Grout C, et al. Evidence for compartmentalization of functional subsets of CD4+ lymphocytes in atopic dermatitis. J Immunol 1990; 144:4651–4660.
109. Hamid Q, Boguniewicz M, Leung DYM. Differential in situ cytokine gene expression in acute versus chronic atopic dermatitis. J Clin Invest 1994; 94:870–876.
110. Holden CA. Atopic dermatitis-messengers, second messengers and cytokines. Clin Exp Dermatol 1993; 18:201–207.
111. Lester MR, Hofer MF, Gately M, Trumble A, Leung DYM. Down-regulating effects of IL-4 and IL-10 on the IFN-γ response in atopic dermatitis. J Immunol 1995; 154:6174–6181.
112. Ohmen JD, Hanifin JM, Nickoloff BJ, Rea TH, Wyzykowski R, Kim J, Jullien D, McHugh T, Nassif AS, Chan SC, Modlin RL. Overexpression of IL-10 in atopic dermatitis. Contrasting cytokine patterns with delayed-type hypersensitivity reactions. J Immunol 1995; 154:1956–1963.
113. Snapper CM, Paul WE. Interferon-γ and B cell stimulatory factor-1 reciprocally regulate Ig isotype production. Science 1987; 236:944–946.
114. Mosmann TR, Coffman RL. TH1 and TH2 cells: different patterns of lymphokine secretion lead to different functional properties. Am Rev Immunol 1989; 7:146–162.
115. Bieber T, Rieger A, Neuchrist C, et al. Induction of FceR2/CD23 on human epidermal Langerhans cells by recombinant interleukin 4 and γ-interferon. J Exp Med 1989; 170:309–314.
116. Hagermark O, Hokfelt T, Pernow B. Flare and itch induced by substance P in human skin. J Invest Dermatol 1978; 71:233–237.
117. Wallengren J, Ekman R, Sundler F. Occurrence and distribution of neuropeptides in the human skin. Acta Derm Venereol (Stockh) 1987; 67:185–192.
118. Naukkarinen A, Farber EM, Nickoloff BJ. Quantification of cutaneous nerves and their substance P content in psoriasis. J Invest Dermatol 1989; 92:126–129.
119. Pincelli C, Fantini F, Massimi P, Girolomoni G, Seidenari S, Giannetti A. Neuropeptides in skin from patients with atopic dermatitis: an immunohistochemical study. Br J Dermatol 1990; 122:745–750.
120. Goetzl EJ, Chernov T, Renold F, Pagan D. Neuropeptide regulation of the expression of immediate hypersensitivity. J Immunol 1985; 135:8025–8055.
121. Ansel J, Perry P, Brown J, et al. Cytokine modulation of keratinocyte cytokines. J Invest Dermatol (suppl) 1990; 94:101–107.
122. Marshall JD, Secrist H, DeKruyff RH, Wolf SF, Umetsu DT. IL-12 inhibits the production of IL-4 and IL-10 in allergen-specific human CD4+ T lymphocytes. J Immunol 1995; 155:111–117.
123. Falanga V. Chronic wounds: pathophysiologic and experimental considerations. J Invest Dermatol 1993; 100:721–725.
124. Jyung RW, Mustoe TA. Role of cytokines in wound repair. In: Gearting A, Rossio J, Oppenheim J, eds. Clinical Applications of Cytokines. New York, Oxford University Press, 1993:307–329.
125. Mauviel A, Uitto J. The extracellular matrix in wound healing: role of the cytokine network. Wounds 1993; 5:137–152.
126. Li V, Barnhill R. Angiogenesis in dermatology: pathophysiology and clinical implications. Prog Dermatol 1995; 29:1–12.
127. Strieter RM, Lukacs NW, Standiford TJ, Kunkel SL. Cytokines and lung inflammation. Thorax 1993; 48:765–769.
128. Wertheim WA, Kunkel SL, Standiford TJ, Burdick MD, Becker FS, Wilke CA, Gilbert AR, Strieter RM. Regulation of neutrophil-derived interleukin-8: the role of prostaglandin E_2, dexamethasone, and interleukin-4. J Immunol 1993; 151:2166–2175.
129. Grotendorst GR, Smale G, Pencev D. Production of transforming growth factor beta by human peripheral blood monocytes and neutrophils. J Cell Physiol 1989; 140:396–402.
130. Kasama T, Strieter RM, Standiford TJ, Burdick MD, Kunkel SL. Expression and regulation of human neutrophil-derived macrophage inflammatory protein-1α. J Exp Med 1993; 178:63–72.
131. Sibille Y, Reynolds HY. Macrophages and polymorphonuclear neutrophils in lung defense and injury. Am Rev Respir Dis 1990; 141:471–501.
132. Brem H, Folkman J. Inhibition of tumor angiogenesis mediated by cartilage. J Exp Med 1975; 141:427–439.
133. Langer R, Conn H, Vacanti J, Haudenschild CC, Folkman J. Control of tumor growth in animals by infusion of an antiangiogenesis inhibitor. Proc Natl Acad Sci USA 1980; 77:4331–4335.
134. Brem S, Preis I, Langer R, Brem H, Folkman J, Patz A. Inhibition of neovascularization by an extract derived from vitreous. Am J Ophthalmol 1977; 84:323–328.
135. Lutty GA, Thompson DC, Gallup JY, Mello RJ, Fenselau A. Vitreous: an inhibitor of retinal extract-induced neo-vascularization. Inv Ophthalmol Vis Sci 1983; 24:52–56.
136. Madri JA, Pratt BM, Tucker AM. Phenotypic modulation of endothelial cells by transforming growth factor-beta depends upon the composition and organization of the extracellular matrix. J Cell Biol 1988; 106:1375–1384.
137. Ingber DE, Madri JA, Folkman J. A possible mechanism for inhibition of angiogenesis by angiostatic steroids: induction of capillary basement membrane dissolution. Endocrinology 1986; 119:1768–1775.
138. Ingber DE, Folkman J. Inhibition of angiogenesis through modulation of collagen metabolism. Lab Invest 1988; 59:44–51.
139. Maragoudakis ME, Sarmonika M, Papakonstantinou E. Inhibition of basement membrane biosynthesis prevents angiogenesis. J Pharmacol Exp Ther 1988; 244:729–733.
140. Shapiro R, Vallee BL. Human placental ribonuclease inhibitor abolishes both angiogenic and ribonucleolytic activities of angiogenin. Proc Natl Acad Sci USA 1987; 84:2238.
141. Sato N, Fukuda K, Nariuchi H, Sagara N. Tumor necrosis factor inhibits angiogenesis in vitro. JNCI 1987; 79:1383.
142. Sidky YA, Borden EC. Inhibition of angiogenesis by interferons: effects on tumor- and lymphocyte-induced vascular responses. Cancer Res 1987; 47:5155–5161.
143. Peterson H-I. Tumor angiogenesis inhibition by prostaglandin synthetase inhibitors. Anticancer Res 1986; 6:251–254.
144. Homandberg GA, Kramer-Bjerke J, Grant D, Christianson G, Eisenstein R. Heparin-binding fragments of fibronectin are potent inhibitors of endothelial cell growth: structure and function correlates. Biochem Biophys Acta 1986; 874:61.
145. Taylor S, Folkman J. Protamine is an inhibitor of angiogenesis. Nature 1982; 297:307–312.
146. Crum R, Szabo S, Folkman J. A new class of steroids inhibits angiogenesis in the presence of heparin or a heparin fragment. Science 1985; 230:1375–1378.
147. Polverini PJ, Novak RF. Inhibition of angiogenesis by the antineoplastic agents mitoxantrone and bisantrene. Biochem Biophys Res Comm 1986; 140:901–907.
148. Lee K, Erturk E, Mayer R, Cockett ATK. Efficacy of antitumor chemotherapy in C3H mice enhanced by the antiangiogenesis steroid, cortisone acetate. Cancer Res 1987; 47:5021.
149. Maione TE, Gray GS, Petro J, Hunt AJ, Donner AL, Bauer SI,

Carson HF, Sharpe RJ. Inhibition of angiogenesis by recombinant human platelet factor-4. Science 1994; 247:77–79.
150. Rastinejad F, Polverini PJ, Bouck NP. Regulation of the activity of a new inhibitor of angiogenesis by a cancer suppresser gene. Cell 1989; 56:345–355.
151. Good DJ, Polverini PJ, Rastinejad F, LeBeau MM, Lemons RS, Frazier WA, Bouck NP. A tumor suppresser-dependent inhibitor of angiogenesis is immunologically and functionally indistinguishable from a fragment of thrombospondin. Proc Natl Acad Sci USA 1990; 87:6624–6628.
152. Tolsma SS, Volpert OV, Good DJ, Frazier WA, Polverini PJ, Bouck N. Peptides derives from two separate domains of the matrix protein thrombospondin-1 have antiangiogenic activity. J Cell Biol 1993; 122:497–511.
153. DiPietro LA, Polverini PJ. Angiogenic macrophages produce the angiogenic inhibitor thrombospondin 1. Am J Pathol 1993; 143:678–684.
154. French-Constant C, Van DWL, Dvorak HF, Hynes RO. Reappearance of an embryonic pattern of fibronectin splicing during wound healing in the adult rat. J Cell Biol 1989; 109:903–914.
155. Kurkinen M, Vaheri A, Roberts PJ, Stenan S. Sequential appearance of fibronectin and collagen in experimental granulation tissue. Lab Invest 1980; 43:47–51.
156. Epstein EHJ. [al(III)3] human skin collagen: release by pepsin digestion and preponderance in fetal life. J Biol Chem 1974; 249:3225–3231.
157. Donoff RB, McLennan JE, Grillo HC. Preparation and properties of collagenases from epithelium and mesenchyme of healing mammalian wounds. Biochem Biophys Acta 1971; 227:639–653.
158. Baggiolini M, Dewald B, Moser B. Interleukin-8 and related chemotactic cytokines-CXC and CC chemokines. Adv Immunol 1994; 55:97–179.
159. Taubb DD, Conlon K, Lloyd AR, Oppenheim JJ, Kelvin DJ. Preferential migration of activated CD4+ and CD8+ T cells in response to MIP-1alpha and MIP-1beta. Science 1993; 260:355–358.
160. Carr MW, Roth SJ, Luther E, Rose SS, Springer TA. Monocyte chemoattractant protein 1 acts as a T-lymphocyte chemoattractant. Proc Natl Acad Sci USA 1994; 91:3652–3656.
161. Taub DD, Proost P, Murphy WJ, Anver M, Longo DL, Van Damme J, Oppenheim JJ. Monocyte chemotactic protein-1 (MCP-1), -2, and -3 are chemotactic for human T lymphocytes. J Clin Invest 1995; 95:1370–1376.
162. Maghazachi A, Al-Aoukaty A, Schall TJ. CC chemokines induce the chemotaxis of NK and IL-2-activated NK cells. J Immunol 1994; 153:4969–4977.
163. Koch AE, Polverini PJ, Kunkel SL, Harlow LA, DiPietro LA, Elner VM, Elner SG, Strieter RM. Interleukin-8 (IL-8) as a macrophage-derived mediator of angiogenesis. Science 1992; 258:1798–1801.
164. Strieter RM, Kunkel SL, Arenberg DA, Burdick MD, Polverini PJ. Interferon γ-inducible protein 10 (IP-10) a member of the CXC chemokine family is an inhibitor of angiogenesis. Biochem Biophys Res Comm 1995; 210:51–57.
165. Maione TE, Gray GS, Petro J, Hunt AJ, Donner AL, Bauer SI, Carson HF, Sharpe FJ. Inhibition of angiogenesis by recombinant human platelet factor-4. Science 1990; 247:77–79.
166. Fivenson D, Faria D, Nickoloff B, Poverini P, Kunkel S, Burdick M, Strieter R. CXC chemokine efflux during chronic wound healing: critical role of the ELR motif in angiogenesis. J Invest Dermatol 1995; 104:625.
167. Winkler CF, Bunn PA Jr. Cutaneous T cell lymphoma. CRC Crit Rev Oncol Hematol 1983; 1:49–92.
168. Wood NL, Kitces EN, Blaylock K. Depressed lymphokine activated killer cell activity in mycosis fungoides. Arch Dermatol 1990; 126:907–913.
169. Laroche L, Kaiserlian D. Decreased natural-killer cell activity in cutaneous T-cell lymphomas. N Engl J Med 1983; 308:101–102.
170. Zachariae H, Ellegaard J, Grunnet E, Sogaard H, Thulin H. T- and B-cells and IgE in mycosis fungoides. Acta Dermatol Venereol 1975; 55:466–468.
171. Vowles BR, Cassin M, Vonderheid E, Rook AH. Aberrant cytokine production by Sezary syndrome patients: cytokine secretion pattern resembles murine Th 2 cells. J Invest Dermatol 1992; 99:90–94.
172. Nickoloff BJ, Griffiths CEM, Baadsgaard O, et al. Markedly diminished epidermal keratinocyte expression of intercellular adhesion molecule-1 (ICAM-1) in Sezary syndrome. J Am Med Assoc 1989; 261:2217–2221.
173. Edelson R, Berger C, Gasparro F, et al. Treatment of cutaneous T-cell lymphoma by extracorporial photochemotherapy: preliminary results. N Engl J Med 1987; 316:297–303.
174. Claudy AL, Rouchouse B, Boucherons S, LePetit JC. Treatment of cutaneous lymphoma with etretinate. Br J Dermatol 1983; 109:49–56.
175. Mosmann TR, Cherwinski H, Bond MW, Giedlin MA, Coffman RL. Two types of murine T helper cell clones. Definition according to profile of lymphokines activities and secreted proteins. J Immunol 1986; 136:2348–2357.
176. Rotteveel FTM, Kokkelink I, vanLier RAW, et al. Clonal analysis of functionally distinct human $CD4^+$ T cell subsets. J Exp Med 1988; 168:1659–1673.
177. Umetsu DT, Jabara HH, DeKruff RH, Abbas AK, Abrams JS, Geha RS. Functional heterogeneity among human inducer T cell clones. J Immunol 1988; 140:4211–4216.
178. Bos JD, Wierenga EA, Sillevis Smitt JH, et al. Immune dysregulation in atopic eczema. Arch Dermatol 1992; 128:1509–1512.
179. Del Prete G, Tiri A, Maggi E, et al. Defective in vitro production of gamma interferon and tumor necrosis factor alpha by circulating T cells from patients with the hyperimmunoglobulin E syndrome. J Clin Invest 1989; 84:1830–1835.
180. Yamamura M, Modlin RL, Ohmen JD, Moy RL. Local expression of antiinflammatory cytokines in cancer. J Clin Invest 1993; 91:1005–1010.
181. Kim J, Modlin RL, Moy RL, Dubinett SM, McHugh T, Nickoloff BJ, Uyemura K. IL-10 production in cutaneous basal and squamous cell carcinomas: a mechanism for evading the local T cell immune response. J Immunol 1995; 155:2240–2247.
182. Saed GM, Fivenson DP, Naidu Y, Nickoloff BJ. Mycosis fungoides exhibit a Th1 type cell mediated cytokine profile while Sezary syndrome expresses a Th2 type profile. J Invest Dermatol 1994; 103:29–33.
183. Nickoloff BJ, Lewinsohn DM, Butcher EC, Krensky AM, Clayberger C. Recombinant gamma interferon increases the binding of peripheral blood mononuclear leukocytes and a Leu 3+ T lymphocyte clone to cultured keratinocytes and to a malignant squamous carcinoma cell line that is blocked by antibody against the LFA-1 molecule. J Invest Dermatol 1988; 9:17–22.
184. Griffiths CEM, Voorhees JJ, Nickoloff BJ. Characterization of intercellular adhesion molecule-1 and HLA-DR expression in normal and inflamed skin: modulation by recombinant gamma interferon and tumor necrosis factor. J Am Acad Dermatol 1989; 20:617–629.
185. Veijlsgaard GL, Ralfkiaer E, Aunstorp C, Czajkowski M,

Marlin SD, Rothlein R. Kinetics and characterization of intercellular adhesion molecule-1 (ICAM-1) on keratinocytes in various inflammatory skin lesions and malignant cutaneous lymphomas. J Am Acad Dermatol 1989; 20:782–790.
186. Nickoloff BJ, Griffiths CEM, Baadsgaard O, Voorhees JJ, Hanson CA, Cooper KD. Markedly diminished epidermal keratinocyte expression of intercellular adhesion molecule-1 (ICAM-1) in Sezary syndrome. JAMA 1989; 261:2217–2221.
187. Clayberger C, Medeiros LJ, Link MP, Krensky A. Absence of cell surface LFA-1 as a mechanism of escape from immunosurveillance. Lancet 1987; 2:533–536.
188. Fivenson DP, Douglass MC, Nickoloff BJ. Cutaneous expression of Thy-1 in mycosis fungoides. Am J Pathol 1992; 141:1373–1380.
189. Enk AH, Angeloni VL, Udey MC, Katz SI. Inhibition of Langerhans cell antigen-presenting function by IL-10. A role for IL-10 in induction of tolerance. J Immunol 1993; 151:2390–2398.
190. Meunier L, Gonzalez-Ramos A, Cooper KD. Heterogenous populations of class II MHC+ cells in human dermal cell suspensions. Identification of a small subset responsible for potent dermal antigen-presenting cell activity with features analogous to Langerhans cells. J Immunol 1993; 151:4067–4080.
191. Nickoloff BJ, Nestle FO, Zheng XG, Turka LA. T lymphocytes in skin lesions of psoriasis and mycosis fungoides express B7-1: a ligand for CD28. Blood 1994; 83:2580–2586.
192. Nickoloff B, Turka L. Keratinocytes: key immunocytes of the integument. Am J Pathol 1993; 143:325–331.
193. Rook AH, Kubin M, Cassin M, Vonderheid EC, Vowels BR, Wolfe JJ, Wolf SF, Singh A, Trincheri G, Lessin SR. Interleukin 12 (IL-12) reverses cytokine and immune abnormalities of Sezary syndrome. J Immunol 1995; 154:1491–1498.
194. Kaplan EH, Rosen ST, Norris DB, Roenigk HH, Saks SR, Bunn PA. Phase II study of recombinant human interferon gamma for treatment of cutaneous T-cell lymphoma. J Natl Cancer Inst 1990; 82:208–212.
195. Olson EA, Rosen ST, Vollmer RT, et al. Interferon alfa-2a in the treatment of cutaneous T-cell lymphoma. J Am Acad Dermatol 1989; 20:395–407.
196. Edelson RL. Photopheresis: a clinically relevant immunobiologic response modifier. Ann NY Acad Sci 1991; 636:154–164.
197. Romagnani S. Induction of Th 1 and Th2 responses: a key role for the natural immune response? Immunol Today 1992; 13:379–381.
198. Vowels BR, Cassin M, Boufal M, Walsh LJ, Rook AH. Extracorporeal photochemotherapy induces the production of tumor necrosis factor or by monocytes: implications for the treatment of cutaneous T-cell lymphoma and systemic sclerosis. J Invest Dermatol 1992; 98:686–692.
199. Saed GM, Fivenson DP. Augmentation of Th-1 cytokines in the peripheral blood of Sezary syndrome patients after treatment with extracorporeal photopheresis. J Invest Dermatol 1994; 1102:586.
200. Schroder J-M. Cytokine networks in the skin. J Invest Dermatol 1995; 105:20s–24s.
201. Wilhelm DL. The mediation of increased vascular permeability in inflammation. Pharmacol Rev 1962; 14:251–269.
202. Ansel JC, Luger TA, Green I. Fever and increased serum IL-1 activity as a systemic manifestation of acute photoxicity in New Zealand white rabbits. J Invest Dermatol 1987; 89:32–37.
203. Dinarello CA. Interleukin-1 and the pathogenesis of acute phase response. N Engl J Med 1984; 311:1413–1417.
204. Akira S, Hirano T, Taga T, Kishimoto T. Biology of multifunctional cytokines: IL-6 and related molecules (IL-1 and TNF). FASEB J 1990; 4:2860–2867.
205. Araneo BA, Dowell T, Moon HB, Daynes RA. Regulation of murine lymphokine production in-vivo. J Immunol 1985; 143:1737–1744.
206. Daynes RA, Araneo BA, Dowel TA, Huang K, Dudley D. Regulation of murine lymphokine production in-vivo. III. The lymphoid tissue microenvironment exerts regulatory influences over T-helper cell functions. J Exp Med 1990; 171:979–996.
207. Baadsgaard O, Salvo B, Mannie A, Dass B, Fox DA, Cooper KD. In vivo ultraviolet-exposed human epidermal cells activate T suppressor cell pathways that involve CD4+ CD45RA+ suppressor-inducer T cells. J Immunol 1990; 145:2854–2861.
208. Baadsgaard O, Lisby S, Cooper KD, Vejlsgaard L. Decreased number and function of antigen-presenting cells in the skin following application of irritant agents: relevance for skin cancer? J Invest Dermatol 1989; 92:842–847.

30
Cytokines and Rheumatology

Deborah M. DeMarco and Robert B. Zurier
University of Massachusetts Medical Center, Worcester, Massachusetts

Joint inflammation in rheumatoid arthritis (RA) and other diseases that involve chronic inflammatory arthritis is characterized by the presence of lymphocytes and monocyte-derived macrophages in a hyperplastic synovium. The reasons why cells invade the synovium and why synovial cells proliferate in RA patients have not been defined. Experimental evidence does indicate that products exported by monocytes, macrophages, lymphocytes, and platelets help regulate synovial cell proliferation and take part in cartilage degradation and bone resorption. Factors specifically chemotactic for particular cells (monocytes, lymphocytes, neutrophils) have also been identified in several cell types. These cytokines interact in a complex and poorly understood manner. For example, interleukin-1 (IL-1) and tumor necrosis factor (TNF) are each capable of inducing the production of their respective molecules as well as each other, and IL-1 and IL-6 act in synergy to induce T-cell proliferation. This self-amplification network may play an important role in sustaining a pathological process. Normally, inhibitors of these cytokines serve to modify their activities. The delicate balance is not preserved in the rheumatic diseases. In this chapter we review the association of a variety of cytokines with rheumatic disease in general and inflammatory arthritis in particular.

INTERLEUKIN-1

Long before the diverse biological activities of IL-1 were recognized, IL-1–like factors (i.e., pyrogens) were reported to be present in synovial fluids from patients with inflammatory arthritis (1). It was subsequently shown that IL-1–like activity stimulated bone resportion in vitro by a mechanism independent of prostaglandin synthesis by the bone cells (2). In addition, IL-1–like activity induces proteoglycan resorption in cartilage explants (3).

Substantial experimental evidence now exists to support the view that IL-1 is an important mediator in the development of synovitis, although incontrovertible evidence is lacking (4). Injection of IL-1 into rabbit knee joints initiates both a transient acute arthritis and a chronic synovitis (5–7). Increased IL-1 activity has been demonstrated in synovial fluid (SF) from patients with RA, osteoarthritis (OA), and other forms of synovitis (8–12). Plasma concentrations of IL-1β, usually below the limit of detection in normal subjects, are detectable in patients with exacerbation of RA (1). Duff et al. (13) also found increased expression of the pro–IL-1β gene in synovial cells from OA and RA patients, observations that also suggest that IL-1 is important in the initiation and/or progression of arthritis. DiGiovine et al. (14) and Duff et al. (15) showed that monosodium urate (MSU) crystals stimulate both blood monocytes and adherent synovial cells to release IL-1 activity. The MSU-induced IL-1–containing supernatant was then shown to stimulate the long-term growth of human synovial fibroblast lines in vitro.

Mean plasma levels of IL-1β were increased significantly in a group of 51 RA patients when compared to 21 healthy aged-matched controls (16). The increased plasma IL-1β levels in these patients correlated with disease activity as measured by the Ritchie articular index (17), pain score, and erythrocyte sedimentation rate (ESR).

The role of IL-1 in the destructive arthritis seen in many connective tissue diseases appears due in large part to its effects on bone and cartilage. For example, osteoclast-activating factor (OAF), a soluble factor produced by peripheral blood mononuclear cells that leads to bone resorption, has amino acid sequence homology with human IL-1β (18). IL-1 is now considered to be one of the most potent OAF known. It increases bone resorption by stimulation of mediators from the synovium and by its systemic effects. Pettipher et al. (6) showed that injection of IL-1 into rabbit knees induced the accumulation of polymorphonuclear leukocytes and mononuclear cells in the joint space and loss of proteoglycan from the articular cartilage. Treatment of rheumatoid synovial fibroblasts in vitro with a combination of IL-1β and TNF-α activates IL-8 expression and protein secretion, which enhances the inflammatory response by recruiting neutrophils to the area (19). Injections of endotoxin, which also stimulate leukocyte accumulation, have no effect on proteoglycan loss, suggesting a direct effect for IL-1 in erosive arthritis. Dayer et al. (20–23) have shown that IL-1–stimulated adherent synovial cells, dermal fibroblasts, chondrocytes, and bone-derived cells produce increased amounts of collagenase and prostaglandin E_2 (PGE_2). IL-1 also stimulates osteoblasts to release macrophage colony-stimulating factor (M-CSF) in a dose-dependent manner (24). Thus IL-1 might potentiate bone resorption by increasing the number of available osteoclast precursors.

Cartilage destruction, associated with irreversible degradation of proteoglycan and collagen in joints of RA patients, is mediated by the combined activities of proteases and collagenase derived from the proliferative synovial lining (pannus). Plasminogen activator (PA), a serine protease that converts plasminogen into the active protease plasmin, may also be involved in cartilage destruction in RA. Increased amounts of PA are found in synovial fluid from RA patients (25); cartilage proteoglycan is degraded by plasmin (26), and plasmin activates collagenase (27), neutral proteoglycanase (28), complement, and kininogen (29), all of which are produced by the synovium in inactive forms. Mochan and colleagues (25) showed that addition of human IL-1 to cultured human synovial cells caused a 30- to 40-fold increase in cell-associated and extracellular PA activity. Further, the IL-1 present in mononuclear cell–conditioned medium stimulated synovial cells to produce PA. Bathon et al. (30) have shown that IL-1 can downregulate the production of type VI collagen from cultured RA and OA human synovial fibroblasts. Type VI collagen is a recently discovered protein that may play a role in cellular adhesion. Thus, the catabolic activity of IL-1 in cartilage and bone very likely contributes to joint tissue injury leading to deforming arthritis in RA and perhaps in other connective tissue diseases as well.

In addition to the role it plays in stimulating synovial cells to produce inflammatory mediators, the interaction of IL-1, with endothelial cells (EC) also contributes to the acute and chronic inflammation seen in rheumatic disease. IL-1 activates human endothelial cells in vitro to synthesize and release PGI_2 and PGE_2, both potent vasodilators (31). In RA, the synovial tissue—the development of which is dependent on certain functions of synovial endothelial cells (32)—is infiltrated with a variety of mononuclear cells. For example, IL-1 alters endothelial cell plasma membranes so that binding of blood lymphocytes and monocytes to endothelial cells is increased. IL-1, which is also a potent lymphocyte chemotactic factor, perpetuates the lymphocytic infiltrate in synovium by causing increased lymphocyte movement through postcapillary venules.

IL-1 also stimulates endothelial cell procoagulant activity and increases the production of plasminogen activator inhibitor (33). These factors might contribute to development of vasculitis in connective tissue disease by activating the clotting cascade, decreasing blood flow, and increasing the numbers of leukocytes and platelets in affected tissue. IL-1 also induces endothelial cell proliferation and neovascularization. Production of type IV collagen, the type found in vascular basement membrane (34), is also increased by IL-1. Basement membrane thickening in RA synovial vessels is heterogeneous: transition areas in which lymphocytes, macrophages, and plasma cells are widely distributed are the most dense; the lining is minimal in lymphocyte-rich areas containing aggregates of mainly T-helper cells (34).

Neutralizing antibodies to IL-1β have potential clinical value since they can reduce endotoxin-induced fever in animals (35). A naturally occurring IL-1 inhibitor has been identified in rheumatoid synovial fluids and consists of the extracellular portion of the type II IL-1 receptor (36,37). However, more attention has been focused on the development of soluble IL-1 receptors to inhibit the action of IL-1. Early studies in humans suggest that soluble type I receptors can be administered safely and that they block inflammation. Recombinant soluble IL-1 receptors have been generated and are being evaluated for therapeutic efficacy in RA (38). Results of a phase I double-blind, randomized trial in 16 RA patients indicate that intraarticular administration of soluble type 1 IL-1 receptors is well tolerated and that this agent appears to have some early beneficial effects (39).

Considerable information is now available about blockade of IL-1 receptors with the naturally occurring IL-1 receptor antagonist (40,41). Recombinant human IL-1 receptor antagonist protein (IL-1ra) has been shown to suppress the incidence of collagen-induced arthritis in mice. IL-1ra has been localized to perivascular areas of the sublining layer of rheumatoid synovial tissue (42). Recombinant IL-1ra is also under investigation as a therapeutic agent in RA. In a phase I clinical trial, 15 RA patients received IL-1ra by subcutaneous injection daily for 28 days (43). After 7 days the patients showed a significant reduction in the number of tender joints and a decrease in serological indices of inflammation. A phase II trial of IL-1ra has been completed in 175 RA patients, and significant improvement in four clinical measures (number of painful joints, number of swollen joints, and assessment of disease severity by patient and physician) was observed (44).

Another approach to long-term therapy in RA is to upregulate endogenous production of IL-1ra via administration of other cytokines. In addition, gene therapy has potential application in RA due to the ability of synovium to sequester materials that are injected intraarticularly. Extraneous genes can be incorporated into the synovial lining cells and their products secreted directly into the joint space. This is a particularly attractive approach in a chronic illness like RA because it promises a permanent therapeutic effect. Candidate genes code for immunosuppressive, anti-inflammatory, chondroprotective, or chondroreparative proteins (45). Both ex vivo and in vivo approaches have been used successfully to transfer genes to the synovia of rabbit knee joints (46–48). Retroviruses have been used to transfer various genes into rabbit synoviocytes. Transfer of the IL-1ra protects the joint from the deleterious effect of a single intraarticular injection of human, recombinant IL-1β (48).

Elevated levels of IL-1 receptor antagonist have also been found in patients with active polymyositis. These levels decreased in response to treatment with corticosteroids (42).

It is not clear whether the IL-1 inhibitors distributed in rheumatic diseases are the same as or closely related to IL-1ra protein. It will be important to determine whether the imbalances between IL-1 and its endogenous inhibitors contribute to the pathology of diseases such as RA.

Several drugs used in the treatment of arthritis may exert their beneficial effects via modulation of IL-1 activity. Corticosteroids and cyclosporin A, both of which generally downregulate immune responses, suppress IL-1 secretion (49–51).

INTERLEUKIN-2

Interleukin-2 (IL-2), first described as a T-cell growth factor (52), is a 15,000 kDa glycoprotein produced by T cells of helper phenotype in response to antigenic and mitogenic stimuli (53). IL-2 represents an important lymphokine in the initiation and regulation of cell-mediated immune responses; its production is stimulated by accessory cells via the secretion of IL-1. By binding to high-affinity receptors, IL-2 supports the proliferation of stimulated T cells (54), B lymphocytes (55), and natural killer (NK) cells (56). In addition, IL-2 stimulates the production of interferon-γ (IFN-γ) (57), augments the reactivity of cytolytic T cells, NK cells (58), and monocytes (59), and is essential for the induction of lymphokine-activated killer cells (60). Human T lymphocytes can be activated by interaction with different stimuli, such as antigens, mitogenic lectins, and antibodies directed to cell surface membrane structures, the best characterized of which are the CD3-TCR complex and the CD2 (T11, sheep E receptor) protein (61,62). During activation, T cells acquire a number of cell surface glycoproteins, some of which (IL-2, transferrin and insulin receptors, 4F2 and EA.1 antigens) appear early, even before DNA synthesis; others (HLA-DR, CB.1, Ta-1, TliSA1, T10 and VLA-1) appear later (63–70). The activation process also stimulates the production and secretion of IL-2 (71), which is accompanied by the upregulation of the Tac (CD25) antigen (72) and the ultimate expansion of sensitized T lymphocytes (73). The progressive loss of high-affinity IL-2 receptors leads to termination of T-cell growth in vitro.

Activated T cells appear to have a central role in the pathogenesis of many autoimmune and inflammatory diseases. Cell surface molecules can be shed by activated T lymphocytes and measured in serum to assess in vivo T-cell activation. Thus, many investigators have evaluated the relationship between these serum markers and disease activity in rheumatoid arthritis and have found that patients with RA have significantly elevated levels of soluble IL-2 receptors (sIL-2R) at baseline compared with age-matched healthy controls. Although measurements of IL-2 in RA patients have produced conflicting results (74), excessive production of IL-2, as well as B-cell activation and high levels of soluble IL-2 receptors (sIL-2R), have been observed in these patients (75,76) and in patients with multiple sclerosis (MS) (77) and systemic lupus erythematosus (SLE) (78). IL-2–responsive activated T-cell clones have been described at the site of tissue injury in patients with RA (79,80), MS (81), and pulmonary sarcoidosis (82). Synovial T lymphocytes from patients with RA exhibit spontaneous in vivo gene transcription of IL-2. The amounts of message for IL-2 were comparable to those produced by normal peripheral blood T lymphocytes activated in vitro with PHA; thus, synovial T lymphocytes appear to be activated in vivo in RA patients (83). In three male patients treated with IL-2 for metastatic cancer, inflammatory arthritis occurred, suggesting that IL-2 may facilitate synovitis; in two of these patients the clinical picture was consistent with rheumatoid arthritis (84). Thus, downregulation of IL-2 and IL-2–dependent T-cell proliferation may be a useful therapeutic maneuver in these patients. Indeed, gold sodium thiomalate, in concentrations attainable during chryso-therapy for RA, significantly reduces the proliferative responses of cultured human T cells stimulated by IL-2 (85). Gamma-linolenic acid (GLA), which suppresses joint pain and swelling in RA patients with active synovitis (86), also reduces IL-2–driven proliferation of circulating and synovial compartment T cells (87,88).

On the other hand, the immunostimulatory functions of IL-2 may allow its use as therapy in immunocompromised patients. In patients with psoriatic arthritis, for example, activation of peripheral blood mononuclear cells leads to reduced IL-2 production and T-cell proliferation compared to normal controls, an impairment that is not corrected by addition to cultures of indomethacin (89).

Levels of pleural fluid sIL-2R in patients with rheumatoid arthritis are much higher than in pleural fluid obtained from patients with SLE, cancer, pneumonia, and congestive heart failure; it is of interest that pleural fluid sIL-2R levels in patients with tuberculous pneumonia are equivalent to pleural fluid levels in patients with RA. Ratios of pleural fluid to serum sIL-2R are high, suggesting that receptors are shed locally (90). It is possible that the increased levels of circulating sIL-2R in patients with RA and perhaps in SLE is due in part to impaired urinary excretion of these molecules. Such a phenomenon could be due to altered structural features of the IL-2R or perhaps due to the fact that IL-2R binds serum proteins in patients with these rheumatic diseases (91). Circulating levels of sIL-2R have been measured serially in efforts to correlate these with disease activity. In one study, no correlation was found between circulating levels of receptor and disease activity (92). On the other hand, sIL-2R levels paralleled the course of disease in patients with juvenile RA and receptor levels were reduced after methotrexate therapy. Thus, sIL-2R levels may be useful for monitoring therapeutic responses in children with JR (93). During treatment with NSAID, sIL-2R levels among responders decreased whereas levels of sIL-2R increased in nonresponders (94). Serum levels of sIL-2R receptor did not predict clinical responses to methotrexate in patients with severe RA; whether these levels would correlate better in patients with early RA is not clear (95). In addition, intravenous pulse therapy with corticosteroid did reduce sIL-2R levels in 25 patients with active RA (96).

A major mode of action of cyclosporin is reduction of IL-2 synthesis (97). GLA also reduces IL-2 production (98). Sulfasalazine and 5-aminosalicylic acid, but not sulfapyridine, inhibit IL-2 production and steady-state IL-2 mRNA levels, as well as T-cell proliferation (99). It has been suggested that the antirheumatic effects of gold salts are due in part to suppression of IL-2 and IL-2R biosynthesis (100).

Attempts have been made to treat RA with antibody to the IL-2R and to IL-2 itself (101), but toxicity has not been trivial (102). Because IL-2 is immunostimulatory, antibodies directed against IL-2 receptors have been proposed as potential immunosuppressive therapy. Antibodies specific for the IL-2 receptor may prove to be protective in conditions of aberrant immune function, such as systemic lupus erythematosus (103). Treatment of RA patients refractory to other therapy was evaluated in an open trial of intravenous DAB_{486}IL-2 given for 5–7 days. Results of the open trial were encouraging, but subsequent studies indicated that the disadvantages of toxicity outweighed the benefit to patients. DAB_{486}IL-2 is a fusion toxin and the product of a synthetic gene engineered by replacing the codons for the receptor-binding domain of the diphtheria toxin with the codons for human IL-2. DAB_{486}IL-2 targets cells expressing the high-affinity form of the IL-2 receptor and achieves selective diphtheria toxin–mediated cytotoxicity of activated T cells.

INTERLEUKIN-3

Interleukin-3 (IL-3) is one of a group of cytokines known as colony-stimulating factors (CSF). These cytokines facilitate granulocyte and macrophage colony formation in bone marrow cultures (104,105) and stimulate growth of mast cells and T and B lymphocytes. IL-3 is produced predominantly by activated T cells; thus IL-1, which is important in T-cell proliferation and differentiation, may play a role in production of IL-3.

Factors with IL-2–like activity that induce proliferation of IL-2–dependent murine cell lines have been isolated from inflamed joints (106). It has been determined, however, that there is a high degree of species specificity for IL-3, and message for human IL-3 is not detectable in RA synovial tissue (107). It is therefore unlikely that human IL-3 is responsible for the proliferation observed in the murine cells. Nonetheless, IL-3 produced during immune reactions may serve to recruit additional hematopoietic cells to amplify these responses.

IL-3 also has nonhematopoietic effects. Although a direct role for IL-3 in rheumatic diseases has not been found, several possibilities exist. IL-3 induces proliferation of gut and skin epithelium $CD4^-$ $CD8^-$ αB T cells and mast cells/basophils (108). It also induces IL-8 secretion from monocytes, stimulates endothelial cell adhesion, and protects mast cells from apoptosis (109). These activities of IL-3 may be important in development of autoimmunity and to regulation of inflammation.

IL-3 activity has been well studied in two murine models of SLE. Sakihama et al. have shown that serum from MRL/lpr mice contain a factor with IL-3 activity that supports growth of IL-3–dependent cell lines (110). The activity was subsequently shown to be an IgG antibody to the IL-3 receptor. It has been suggested that autoantibodies directed toward the IL-3 receptor are involved in the lymphoid proliferation seen in MRL/lpr mice. Fishman and colleagues have shown increased levels of serum IL-3 in patients with both autoimmune diseases such as SLE or Sjögren's syndrome and with lymphoma (111). Preliminary data from animal studies suggest that IL-3 may play a role in pregnancy loss associated with antiphospholipid antibodies (112).

INTERLEUKIN-4

Interleukin-4 (IL-4), initially known as B-cell stimulatory factor-1, is produced by some activated T cells (113). It acts on resting B cells to increase their expression of class II major histocompatibility complex molecules (114), augments proliferation of activated B cells (115) and is involved in isotype differentiation in lipopolysaccharide (LPS)-stimulated murine B cells (116,117). These B cells exhibit markedly increased production of IgG1 and IgE when stimulated with LPS in the presence of IL-4.

Although originally thought to influence B cells only, IL-4 affects T cells as well. IL-4 enhances T-cell proliferation directly (118) and promotes proliferation induced by exposure to co-stimuli (119). The effects of IL-4 on T cells occurs by several different mechanisms, including direct stimulation of resting cells into the cell cycle, enhancement of IL-2 production, and induction of high-affinity IL-2 receptors (120). The capacity of IL-4 to stimulate proliferation of CD4+ and CD8+ cells may be enhanced by IL-1 (121). Whereas induction of IL-4 by expression of class II major histocompatibility complex antigens on B cells and B-cell synthesis of IgG1 and IgE are both blocked by exposure of cells to IFN-γ (121), the stimulatory effect of IL-4 on T cells is not influenced by IFN-γ (122).

In a study by Fay (123), IL-4 activity was found in all synovial fluids from 37 RA patients but in none of the fluids from patients with other types of synovitis such as Reiter's disease, trauma, or psoriasis. Other authors, however, have been unable to detect IL-4 in rheumatoid synovium (124,125). Although its precise role in autoimmune disease has not been defined, IL-4 may be important in T-cell–B-cell interactions such as antigen presentation (126), local immunoglobulin secretion, and recruitment of new T cells to the ongoing inflammatory process (118). Also, there is some evidence that IL-4 may act to restrain inflammation under certain conditions. For example, macrophages, which normally produce IL-1, tumor necrosis factor-α (TNF-α), and PGE_2 when exposed to LPS and IFN-γ, do not produce these factors if IL-4 is also present (127). Miossec et al. have shown in an ex vivo model of rheumatoid synovitis that IL-4 is capable of reducing spontaneous production of inflammatory cytokines such as Il-1β, IL-6, and TNF-α (125). IL-4 also reduces bone resorption in an ex vivo model of bone resorption (128). These anti-inflammatory properties of IL-4 provide the rationale for the clinical use of IL-4 in RA.

INTERLEUKIN-5

Interleukin-5 (IL-5) in humans is the major cytokine involved in production of eosinophilia. IL-5 is also a B-cell-stimulatory factor that may play a role in the induction and maintenance of B-cell hyperactivity and autoantibody production. Increased IL-5 production has been demonstrated in several models of SLE (129). IL-5 was identified initially in supernatants from concanavalin A–stimulated murine lymphocytes as a factor capable of causing B cells to either differentiate into antibody-producing cells or to undergo proliferation in the absence of T cells (130,131). It was therefore called T-cell-replacing factor (TRF) or B-cell growth factor (BCGF)-II. It was later determined that IL-2 and IL-4 were components of TRF (132), but IL-5 was found to be distinct both biochemically and functionally from other lymphokines (133–135). Whereas IL-4 acts on resting cells, IL-5 influences cells only after they have been activated. Available evidence suggests that IL-5 receptors are expressed mainly on activated B cells, eosinophils, and IL-5–dependent cell lines (136).

While murine IL-5 is clearly capable of inducing the expression of high-affinity IL-2 receptors on thymocytes (137) and of inducing B-cell differentiation with resultant antibody secretion (138–141), functional studies of human recombinant IL-5 (rhIL-5) are few in number and are controversial. Purified rhIL-5 exhibits potent eosinophil colony-stimulating activity (142,143). However, although initial cloning studies showed that IL-5 had weak but definite TRF activity (144) and increased the activity of IgG and IgM producing cells, subsequent studies (143) did not demonstrate human B-cell proliferation or differentiation, even after B cells were activated by a variety of stimulants. Other laboratories (145) have shown B-cell-stimulatory effects, but less pure preparations of rhIL-5 were used. Sera from patients with the eosinophilia myalgia syndrome (EMS) have increased IL-5 activity, which may be responsible for the eosinophilia that is a characteristic component of the illness (146). Serum from patients with EMS but not normal human serum can increase markedly the viability of normal eosinophils ex vivo, an effect that is attenuated with neutralizing antibody to IL-5.

Although the effects of IL-5 in humans remain controversial, the strong B-cell proliferative and differentiative functions exhibited in murine cell lines suggest a role for IL-5 in the pathogenesis of autoimmune diseases characterized by autoantibody production (147). It is possible that the activity of IL-5 as well as the panel of BCGFs may be different in humans and mice and raises the possibility that the mouse may not be the best model for study of the human antibody response.

INTERLEUKIN-6

Interleukin-6 (IL-6), also known as interferon-β_2, B-cell-stimulating factor 2, and hepatocyte-stimulating factor, is a glycoprotein derived from a wide variety of cell types, including T and B cells, fibroblasts, macrophages, monocytes, endothelial cells, and certain tumor cells (121). Normal cells do not produce IL-6 unless stimulated (148,149). Gene transcription of IL-6 is enhanced by IL-1 and to a lesser extent by TNF, platelet-derived growth factor (PDGF), and IFN-β_1, LPS also increases IL-6 mRNA synthesis in fibroblasts. The multiple effects of IL-6 are usually seen only in the presence of stimuli (such as IL-1), and IL-6 mediates its effects via the same receptor as type 1 IFN (i.e., IFN-α and IFN-β_1).

IL-6 is produced during immune responses and has marked effects on the growth and differentiation of B and T lymphocytes. The stimulation of T cells by IL-6 involves the induction of IL-2 receptors, which converts the T cells to an IL-2–responsive state. IL-1 and IL-6 are strongly synergistic in the induction of T-cell proliferation (150,151).

IL-6 probably plays a role in the induction and propagation of the inflammatory response, since the hepatocyte (a major source of acute-phase proteins) is a major target for IL-6 (152). Stimulation by IL-6 of acute-phase protein release from liver, coupled with its effects as a pyrogen, suggest that IL-6, together with IL-1, is responsible for many of the systemic signs and symptoms observed in connective tissue diseases, including fever, weight loss, leukocytosis, and elevated ESR.

Another important function of IL-6 is to induce activated B lymphocytes to terminally differentiate into immunoglobulin-producing plasma cells (153,154). IL-6 may therefore be involved in rheumatoid factor (RF) production. Increased levels of IL-6 have been observed in supernatants from tumor cells associated with autoimmunity, suggesting an important role for IL-6 in autoimmune diseases (155,156). Tumors often associated with autoantibody production, such as cardiac myxoma, cervical carcinoma, and bladder cell carcinoma, produce large amounts of IL-6 (156).

Waage and colleagues (157) found concentrations of IL-6 in serum and synovial fluid to correlate with the activity of inflammatory arthritis. This group studied paired samples of serum and synovial fluid from seropositive and seronegative RA patients as well as synovial fluid only from patients with psoriatic arthritis, Reiter's syndrome, and traumatic synovitis. They found that the level of IL-6 in synovial fluid correlated with the ESR in patients with inflammatory disease and that the median level of IL-6 was higher in patients with inflammatory arthritis than in those with a traumatic etiology, although this difference was not statistically significant.

Guerne and colleagues (158) showed that IL-6 is produced spontaneously by synovial cells in culture and that IL-1, TNF-α, and lymphotoxin (TNF-β) augment its synthesis. They found very high levels of IL-6 in synovial fluid obtained from patients with inflammatory synovitis. This synovial cell–derived IL-6 was able to induce hybridoma-plasmacytoma proliferation as well as immunoglobulin and acute-phase protein synthesis. They also showed that IL-6 is present in rheumatoid synovium in much higher concentrations than either IL-1 or TNF-α, suggesting an important role for this cytokine in the hypergammaglobulinemia and RF production seen in RA. However, the levels of IL-6 found in seropositive RA are comparable to those found in seronegative RA and psoriatic arthritis, both of which exhibit hypergammaglobulinemia but not autoantibodies. IL-6 may therefore act as a general stimulus for antibody synthesis rather than on inducer of specific autoantibodies.

Spontaneous production of IL-6 by rheumatoid synovial cells and culture is suppressed by IL-4, again reinforcing the notion of complicated cytokine-cytokine interactions in the rheumatic diseases (159). These findings that IL-6 is increased in synovial fluid from rheumatoid arthritis patients and that synovial cells and cells from synovial fluid from RA patients produce greater amounts of IL-6 than cells from patients with other forms of arthritis have been documented many times over (160,161). Concentrations of IL-6 are also increased in bone marrow serum from patients with RA (162). It is of interest that serum IL-6 levels exhibit a circadian rhythm, in that IL-6 levels are high in the morning and low in the evening. It has been suggested that the high morning levels of IL-6 may relate to the increased morning pain and stiffness associated with RA (163).

Allison and colleagues (164) have shown that glucocorticoids can suppress IL-6 formation by virtue of reducing stability of message for IL-6 in human monocytes.

After 16 weeks of cyclosporin A therapy, serum IL-6 was reduced significantly in 24 RA patients with active synovitis (165). Tenidap, a new drug indicated for the treatment of RA, which can be given once daily, also reduces circulating IL-6 levels in patients with RA (166). On the other hand, Lindsley's group (167) found that neither naproxen nor prinomide therapy for 16 weeks reduced serum IL-6 levels in RA patients; nor does methotrexate appear to reduce serum IL-6 levels in RA patients (168). Also, azathioprine did not reduce IL-6 levels in RA patients (169). Thus it appears that no therapy for RA reduces IL-6 serum levels despite the fact that the therapy may be clinically useful. It may be that if IL-6 production by circulating cells or especially by cells from the rheumatoid joint space were measured, then correlation of IL-6 production would be seen.

Given the apparent correlation of IL-6 with active synovitis in RA patients, it is no surprise that RA patients have been treated with monoclonal antibody to IL-6; five RA patients were given 10 mg/day anti–IL-6 monoclonal antibody intravenously for 10 consecutive days. No side effects were noted, clinical improvement appeared during the treatment, and C-reactive protein levels fell. However, serum IL-6 levels increased in four patients, in whom the monoclonal antibody appeared to suppress inflammation (70).

INTERLEUKIN-7

Interleukin-7 (IL-7) promotes growth and possibly differentiation of pre–B cells and immature thymocytes. IL-7 also enhances proliferation of some mature T-cell populations, such as CD4+ CD8–, CD4– CD8+, cytotoxic T cells, and lymphokine-activated killer cells (171). IL-7 probably acts synergistically with other cytokines to stimulate B-lineage development.

The role of IL-7 in autoimmune diseases is not known, but it may prove useful for therapy of human cancers. Due to its effects on T cells, it is likely to be involved in some way in the pathophysiology of rheumatic diseases.

INTERLEUKIN-8

Interleukin-8 (IL-8), also known as neutrophil chemotactic factor and neutrophil-activating peptide-1, is a tissue-derived cytokine secreted by several different cell types including human peripheral blood mononuclear cells, fibroblasts, epithelial cells, and hepatocytes stimulated by IL-1α and TNF-α. Monocytes, alveolar macrophages, and endothelial cells can also be induced by LPS to secrete IL-8. Watson et al. (172) have shown that supernatants from rhIL-1–stimulated human synovial cells express an activity that stimulates neutrophils, as assessed by an increase in neutrophil free calcium concentration and chemokinesis. They termed this activity polymorphonuclear leukocyte–stimulating factor (PSF). In addition, human synovial cells stimulated by IL-1α and IL-1β express mRNA for IL-8 (173).

The activity of IL-8 is mediated by its binding to high-affinity receptors present on the neutrophil surface (174). IL-8 rapidly downregulates > 90% of its receptors. GM-CSF and TNF-α also downregulate expression of IL-8 receptors (175).

IL-8 has been implicated in several rheumatic disorders. Several groups (176,177) have demonstrated IL-8 in psoriatic plaques. These lesions also contain high levels of IL-1, which could be the stimulus for IL-8 production. IL-8 has also been recovered from synovial fluid of patients with RA (178–180) and gout (181). These conditions are characterized by inflammation accompanied by neutrophil infiltration and tissue damage. The exact role of IL-8 in arthritis and other inflammatory diseases is not known, but since neutrophils are a major source of cartilage-degrading enzymes, at least one role for IL-8 may be in the cartilage degradation commonly seen in these disorders. IL-1 and TNF are considered to be the major cytokines responsible for destruction of articular cartilage by virtue of their ability to liberate proteoglycans and metalloproteinases from the synovial pannus. IL-1 stimulation of synovial cell IL-8 may provide an additional mechanism by which IL-1 causes cartilage destruction in inflammatory arthritis.

An important focus of IL-8 research in the future will be on the effect of neutralizing IL-8 antibodies in various models of disease and whether IL-8 antagonists may provide effective therapy.

INTERLEUKIN-9

Interleukin-9 (IL-9) is a T-cell–derived cytokine with pleiotropic activities. It was originally described as a cytokine produced by murine Th2 lymphocytes that allows for the antigen-independent proliferation of murine T-helper cell clones (182). However, freshly isolated normal T cells do not proliferate in response to this cytokine even when co-stimulated by lectins or anti-CD3 antibodies (183).

Independent investigations by Hültner and colleagues (184,185) revealed that a factor produced by activated splenocytes could enhance proliferation of mast cell lines induced by IL-3 or IL-4. This factor was subsequently shown by molecular cloning to be IL-9 (186), and the proliferative activity of IL-9 for mast cells, either alone or in synergy with IL-3, has been confirmed in other permanent mast cell lines (187). In addition to its proliferative activity, IL-9 also induces functional changes such as IL-6 secretion by mast cell lines (186), and it is a potent regulator of the expression of protease genes like those of the granzyme family (188).

Although most of the work to date with IL-9 has

involved oncogenesis and certain parasitic infections, it may play a role in the pathogenesis of at least some of the connective tissue diseases. Mast cells may be important in the pathogenesis of RA since they have been identified in biopsies from early lesions. In addition, granzyme one may play a role in the endothelial injury seen in scleroderma. This leads to release of vasoactive substances, end organ ischemia, and fibrosis.

Much is yet to be learned about IL-9. It appears to be a multifunctional cytokine with diverse effects on a variety of cell types. The possibility exists that IL-9 antagonists could be used to downregulate mast cell activation. In addition, there is evidence that IL-9 may be an "anti-apoptotic" factor (188), which may have relevance to the pathogenesis of autoimmune diseases.

INTERLEUKIN-10

Interleukin-10 (IL-10) produced by macrophages, T cells, and other cell types affects several functions of immune cells (189). IL-10 inhibits macrophage functions such as antigen presentation to THP-1 cells (human monocyte line) and subsequent cytokine secretion. In contrast, IL-10 is a potent stimulator of B cells and mast cells.

The anti-inflammatory properties of IL-10 have been evaluated in vivo in mice. IL-10 pretreatment reduces the amounts of circulating TNF induced by LPS and completely prevents mortality after challenge by a dose of LPS normally toxic for 50% of the mice (190), effects consistent with the ability of IL-10 to block secretion of cytokines by activated macrophages.

IL-10 gene regulation was evaluated in patients with RA, primary Sjögren's, and SLE (191). These rheumatic diseases and characterized by prominent B-lymphocyte hyperactivity that results in an increased production of immunoglobulins and the emergence of autoantibodies. Increased IL-10 gene expression by unstimulated PBMC from untreated patients was observed, leading the authors to postulate that it reflects a spontaneous inbalance of the cytokine network. Increased IL-10 levels did not correlate with clinical or biological markers of disease activity. It is likely that IL-10 gene dysregulation plays a major role in the polyclonal B-cell hyperactivity associated with these diseases.

Cush et al. also found elevated levels of IL-10 in both serum and synovial fluid from RA patients compared to normal subjects and to patients with other types of arthritis (192). They postulate that IL-10 may contribute to the decreased T-cell function and increased rheumatoid factor production observed in these patients.

Administration of IL-10 to lupus mice accelerates onset of disease, whereas administration of neutralizing antibodies to IL-10 delays emergence of autoimmune manifestations (193). It is not clear whether IL-10 influences human autoimmunity in the same fashion. However, Katsikis et al. suggest that IL-10 is an important human immunoregulatory cytokine, regulating monocyte and in some cases T-cell cytokine production (194). This group found expression of both mRNA and protein for IL-10 in rheumatoid and osteoarthritis joints. IL-10 protein was produced spontaneously in all synovial cultures studied. Neutralization of endogenously produced IL-10 by synovial cells in culture derived from rheumatoid arthritis patients resulted in a marked increase in protein levels of two inflammatory cytokines (TNF-α and IL-1β).

TRANSFORMING GROWTH FACTOR-β

Transforming growth factor-β (TGF-β), first isolated from human platelets, is made by almost all cell types. It has three structurally and functionally similar isoforms and is related to a large superfamily of homologous proteins, whose members mediate key events in growth and development (195). Several lines of evidence suggest that TGF-β is an important cytokine in initiation and termination of tissue repair. Sustained production of TGF-β appears to facilitate fibrosis (196). There is a dichotomy between effects of local and systemically disseminated TGF-β. In an experimental arthritis model, local secretion or injection of TGF-β drives the inflammatory response, whereas systemic administration inhibits the same response (197). The terms "multifunctional" or "pleiotropic" have been applied to TGF-β because it stimulates proliferation of some cells, especially cells from connective tissue, but inhibits proliferation of lymphocytes and epithelial cells (198). TGF-β also influences collagen synthesis and other molecules of the extracellular matrix. TGF-β is especially important during formation of normal tissues in embryogenesis and during response to injury with inflammation and repair. The relationship of TGF-β to the rheumatic diseases has focused on its effects on bone and cartilage, effects in RA, and its potential role in the pathogenesis of scleroderma and EMS.

TGF-β enhances proliferation of many cells of mesenchymal origin including the osteoblast, a cell type important to formation of new bone during fracture healing and to maintenance of existing bone. TGF-β also stimulates osteoblasts to produce extracellular matrix proteins. Although little is known about the role of TGF-β in the pathogenesis of bone disease, there is evidence that the estrogen receptor can activate the gene for TGF-β, thus promoting formation of new bone matrix (199). Therefore, further studies on the role of TGF-β in modulation of bone metabolism and osteoporosis need to be done.

TGF-β has direct actions on fibroblasts and is a strong chemotactic factor for these cells (200). The TGF-β1 gene is upregulated in response to tissue injury, and it is the TGF-β isoform most implicated in fibrosis (201). TGF-β has therefore been the subject of much research relating to scleroderma, a disease characterized by collagen overproduction. Accumulation of extracellular matrix (ECM) produced by fibroblasts and composed of self-aggregating macromolecules, including fibronectin, collagens, and proteoglycans (202), is the chief pathological feature of fibrotic diseases. Cells attach to the ECM by means of surface receptors called integrins (203). TGF-β1 causes deposition of ECM by stimulating cells to increase synthesis of matrix proteins and decrease production of matrix-degrading proteases such as plasminogen-activating factor (204), a thiol protease (205), and a metalloprotease (206); TGF-β also increases production of protease inhibitors such as plasminogen activator inhibitor (204). TGF-β also modulates expression of integrins so that cellular adhesion to the ECM is increased (201).

Skin fibroblasts from scleroderma patients produce more type I and type III collagens, glycosaminoglycans, and fibronectin than do cells from matched controls (207,208). Levels of messenger RNA (mRNA) for matrix proteins are increased in scleroderma (209,210), but fibroblasts from unaffected skin of these patients do not exhibit exaggerated matrix synthesis. It has been shown that TGF-β binds to one or more membrane receptors and enhances fibroblast collagen mRNA expression 5- to 10-fold (210,211). This response is now thought to involve the induction of collagen gene promoter activity and is associated with stimulation of DNA binding by nuclear proteins in the target fibroblasts (212,213). Because TGF-β causes both sustained stimulation of collagen production and autoinduction of its own synthesis, a brief exposure of fibroblasts to TGF-β may resist in persistent alteration of their biosynthetic phenotype (214). Similarly, fibroblasts from patients with EMS also display elevated expression of the gene for TGF-β (215). It also enhances formation of fibronectin by increasing transcription of the gene coding for this protein (211).

The effects of TGF-β on collagen and extracellular matrix formation and fibroblast chemotaxis provide evidence for the probable involvement of TGF-β in systemic sclerosis. The finding that TGF-β activates a collagen gene promoter provides a framework for potential intervention in collagen synthesis at the transcriptional level (211) and may provide an exciting therapeutic advance in the fibrosing diseases, which have thus far been resistant to all therapy. Injection of an antiserum capable of neutralizing the activity of TGF-β1 in mice inhibits TGF-β action in the lung (216) and joints (197). In each case, excessive deposition of ECM was prevented and tissue repair was normal.

Finally, TGF-β probably plays a role in development of synovial pannus and maintenance of inflammation in RA. Several cellular responses known to be mediated by TGF-β may be responsible: leukocyte recruitment and activation (217), fibroblast chemotaxis (200), increased production of extracellular matrix proteins by fibroblasts which may relate to the increased levels of atypical isoforms of fibronectin that have been found in RA synovial fluid (218,219), chemotaxis of monocytes and antiproliferative or immunosuppressive effects on T cells (220–222). Fava et al. (223) found TGF-β–like activity in inflammatory synovial fluid, with RA patients having the highest levels. They postulate that activation of latent TGF-β by proteases in the synovial pannus could cause a local increase in active TGF-β leading to perpetuation of the destructive, inflammatory process. Research with TGF-β n RA RA may provide clues to the pathogenesis of inflammatory arthritis.

INTERFERON-γ

Interferons, the first cytokines discovered, have a wide range of biological activity. Three major groups of interferons are known: IFN-α, produced by leukocytes, and IFN-β, produced by fibroblasts, were formerly known as type I interferons. These cytokines are secreted in response to stimulation by virus or double-stranded RNA. IFN-γ (type II) produced primarily by mitogen- or antigen-stimulated T lymphocytes (224), specifically TH_1 cells (225), shares only 10% structural homology with IFN-α and IFN-β and has a different cell surface receptor. The receptor for IFN-γ is believed to be expressed on most cells, is encoded on chromosome 6q, and displays little homology with any other receptor. This group of cytokines has antiviral and antiproliferative activities, which is why they have been used as anticancer agents. All three are immunoregulatory, but IFN-γ is the most active in this regard. Although IFN-γ will be the focus of this discussion, it must be realized that there may be synergy between the various interferons. Most of the work relating to the rheumatic diseases has focused on IFN-γ.

IFN-γ has diverse effects on immune responses, and it interacts with other lymphokines in either a synergistic or an antagonistic manner: IFN-γ augments B-cell–stimulating lymphokines and increases antibody production but counteracts the effects of IL-1 on connective tissue by inhibiting collagen synthesis (226–228). The main functions of IFN-γ appear to be stimulatory. It induces the appearance of new surface markers associated with differentiation, and it enhances proliferation of T lymphocytes and NK cells (229); IFN-γ also enhances the expression of class I and class II MHC antigens (230) and mediates antigen presentation by its effects on a monocyte-macrophage cell line (231).

IFN-γ affects various functions of vascular epithelium which are pivotal in the immune response. IFN-γ receptors are present on vascular endothelial cells, and IFN-γ increases endothelial class I and II MHC antigen expression (232). IFN-γ upregulates expression of ICAM-1, an endothelial cell surface molecule that mediates endothelial cell-leukocyte adherence (233).

IFN-γ–primed macrophages are crucial for host defense, but they are also dangerous cells because they produce free radicals, lytic enzymes, and inflammatory cytokines. These mediators can cause extensive local damage and are responsible for many of the systemic symptoms associated with acute and chronic inflammation. Apoptosis is one mechanism known to regulate potentially dangerous cells in the immune system. Munn et al. (234) have shown that even transient exposure to IFN-γ causes upregulation of activation-induced apoptosis in macrophages and that this is a potential mechanism to restrict the destructive potential of inflammatory macrophages.

Natural antibodies to IFN-γ have been detected in the serum of virus-infected patients and in healthy subjects, although at lower titers (235). These antibodies were shown to interfere with the immunomodulating activity but not with the antiviral and antiproliferative activity of this cytokine. These antibodies might therefore lead to new therapy for diseases in which there is evidence of activated cell-mediated immunity.

In summary, IFN-γ plays a pivotal role in host defense. After antigenic stimulation, T cells activated by IL-2 produce IFN-γ, which then influences their response as well as their interactions with B cells. Although much work has been done with IFN-γ in rheumatic diseases, especially RA and scleroderma, controversy remains about its role in the pathogenesis of these disorders.

Several investigators have assessed IFN-γ in peripheral blood mononuclear cells from patients with RA and other rheumatic diseases (236–239). In 1979, Hooks et al. (239) found "immune" interferon (now known as IFN-γ) in serum of patients with SLE, RA, scleroderma, and Sjögren's syndrome. A positive correlation between serial IFN titers, clinical disease activity, and serological activity (complement levels and anti-ds-DNA antibodies) was observed in the lupus patients. Cesario et al. (240) detected IFN from 11 of 84 synovial fluid samples and 10 of 84 simultaneous serum samples of patients with RA, SLE, seronegative spondyloarthritis, pseudogout, and unclassified arthritis; the highest titers were identified in RA patients. IFN activity was not identified in synovial fluids from patients with gout, disseminated gonococcal infection, or degenerative arthritis. No relationship was found between serum and synovial fluid levels; in fact, it was uncommon for IFN to be found in both body fluids simultaneously, suggesting local production of IFN in the joint space. More recently, several groups have found decreased or defective production of γ-IFN from peripheral blood mononuclear cells in patients with SLE and RA (237,241). Reuter's group found (241) that patients with active disease produced significantly less IFN-γ than those who were in remission. Firestein and Zvaifler (242,243) postulated that IFN-γ was the primary macrophage-activating factor in synovial tissue and synovial fluid because the phenotype of synovial fluid monocytes resembled that of peripheral blood monocytes exposed to IFN-γ. However, when they measured synovial fluid monocytes by sensitive radioimmunoassay (RIA), very little IFN-γ was identified, suggesting that activation of synovial fluid monocytes occurs by other mechanisms.

Despite these controversies, several observations suggest the possibility that IFN-γ might be beneficial in RA. Although RA is a disease of unknown etiology, polyclonal B-lymphocyte activation has been identified in synovial tissue, synovial fluid, and peripheral blood from RA patients (244,245), and IFN-γ inhibits activation of B lymphocytes by the Epstein-Barr virus (EBV) (246). Also, IFN-γ reduces IgM synthesis by EBV-sensitized B lymphocytes from patients with RA (247). Additionally, many investigators feel that despite an increase in the number of activated T lymphocytes in peripheral blood and in synovium of RA patients, at least some of these lymphocytes may be defective in that they exhibit decreased production of IL-2 and IFN-γ (248,249). Finally, decreased NK activity has been reported in patients with active RA (250), and IFN-γ has been shown to enhance NK activity (229). These activities of IFN-γ prompted several clinical trials in RA patients. Firestein et al. (251) evaluated the effects of IFN-γ in 12 patients with active RA during a 12-week open trial. Nine patients were unchanged clinically, and only 2 had 50% improvement. Surprisingly, parenteral IFN-γ failed to increase DR on peripheral blood monocytes from the RA patients. Since this was the first time parenteral IFN-γ had been evaluated, the dose given may not have been sufficient to produce the anticipated results.

Veys et al. (252), in a double-masked trial of 6 months' duration, evaluated the effects of subcutaneous IFN-γ compared to placebo in 22 RA patients. Although all clinical parameters tested showed a trend in favor of the treatment group, only the tender joint score improved significantly. Serious adverse effects were not observed.

The authors concluded that the optimal regimen for using IFN-γ had not been established and that further double-blind studies using this agent in a larger number of patients were needed.

In 1989, Cannon et al. (253) completed a 12-week, double-blind randomized, prospective, placebo-controlled trial of recombinant human IFN-γ for the treatment of RA; 54 patients received recombinant human IFN-γ and 51 were given placebo. Forty-two patients in each group completed the study, and some clinical improvement occurred in both groups of patients. Although the improvement with IFN-γ was greater than that observed in the placebo group, the differences were not statistically significant.

A multicenter clinical trial of IFN-γ in 249 patients with RA was conducted by the German Lymphokine Study Group (254). The joint pain index after 3 months of study was significantly improved in the interferon group compared to controls. The treated group was also able to reduce their intake of corticosteroids.

IFN-γ has also been evaluated as therapy for systemic juvenile rheumatoid arthritis (JRA) in a small, open study involving nine severely ill children who did not respond to conventional therapy (255). Seven of nine patients improved, but as in the adult studies, changes in clinical parameters were not statistically significant.

In addition to the trials in RA, the use of IFN-γ has been studied in scleroderma. The pathogenesis of scleroderma remains uncertain, but histopathological examination reveals excessive fibroblast activity with collagen deposition in many organs (256). Satisfactory treatment for scleroderma is not available. IFN-γ, because of its ability to inhibit collagen production by normal human dermal fibroblasts (257) and by scleroderma dermal fibroblasts (226), was evaluated in an open trial of 6 months duration involving 10 patients (258); 9 patients completed the study. After the 6-month treatment period, significant improvement (compared to baseline) was observed in total skin score, maximal mouth opening, range of motion of wrists and elbows, grip strength, functional index, dysphagia, and creatinine clearance. Serious side effects were not observed, but a significant decrease in blood and leukocyte counts and peripheral blood lymphocytes was noted. The study had several limitations, including the small sample size and the open, uncontrolled design. Also, some degree of spontaneous regression in dermal sclerosis may occur in some scleroderma patients, making evaluation of any drug difficult in this disease. Despite these reservations, IFN-γ may be useful in the treatment of scleroderma. However, a large double-blind, controlled trial must be done.

GRANULOCYTE-MACROPHAGE COLONY-STIMULATING FACTOR

Granulocyte-macrophage colony-stimulating factor (GM-CSF) is a glycoprotein with growth-enhancing effects on granulocyte, macrophage, and eosinophil colonies in the bone marrow. It was purified from conditioned medium of a T-lymphoblast cell line (259), and a cDNA has been cloned (260). Recombinant GM-CSF has been produced from three cell types: mammalian, yeast, and bacterial. The gene for GM-CSF is located on the long arm of chromosome five (260).

GM-CSF is produced in large amounts in response to inflammatory stimuli after activation of the producer cells. T lymphocytes produce GM-CSF when stimulated by antigens, lectins, or IL-2 (260), and monocytes release GM-CSF after stimulation with endotoxin (261). Fibroblasts and endothelial cells are also sources of GM-CSF (262–264). This cytokine is undetectable in the serum with current assays (265). It appears to be produced locally and to become compartmentalized in the extracellular matrix of bone marrow (266).

GM-CSF is considered a multipotential hematopoietic growth factor. In addition to its effect on myeloid precursors in the bone marrow, GM-CSF also causes neutrophil and eosinophil activation (267), induces neutrophil phagocytosis (268), and stimulates the proliferation and activation of monocyte-macrophages. Production of cytokines by monocyte-macrophages is also induced by GM-CSF (269).

As already discussed, cytokines play a critical role in establishing and maintaining the inflammatory response seen in many types of chronic synovitis such as rheumatoid arthritis. In addition to its role as a hematopoietic growth factor, GM-CSF has been found in chronic inflammatory joint effusions from RA patients and in the supernatants of cultured RA synovial tissue (270). GM-CSF may have a dual role in the pathogenesis of RA. It is likely that GM-CSF augments the inflammatory response in RA (and possibly other types of inflammatory synovitis) by inhibiting neutrophil migration out of the joint space (271) and by potentiating neutrophil activation (272,273). GM-CSF also increases the cell surface expression of HLA class II histocompatibility antigens (HLA-DR or Ia molecules) on blood monocytes from normal donors (274). Ia molecules are essential elements in antigen presentation. The synovial lining cells from RA patients express large amounts of Ia on their surface, and monocytes derived from rheumatoid synovial effusions exhibit an activated phenotype with increased expression of HLA-DR and low levels of CD14 (275).

Identification of GM-CSF at the site of a chronic inflammatory disease such as RA marks the first time that this cytokine has been implicated in a human disease state. Variation in Ia expression plays a major role in immune regulation (276). Parental administration of GM-CSF to mice augments the primary antibody response due to enhanced function of antigen-presenting cells, including increased Ia expression (277). Monocytes and macrophages isolated from inflammatory foci resemble GM-CSF–treated monocytes, which suggests that local production of GM-CSF may play a role in macrophage-activation disease states (274).

PLATELET-DERIVED GROWTH FACTOR

Platelet-derived growth factor (PDGF) plays a role in the regulation of cell proliferation during embryonal development and in wound healing. Originally described as a platelet protein with mitogenic activity for connective tissue cells, PDGF is now known to be produced by several cell types including monocytes/macrophages, endothelial cells, smooth muscle cells, mesangial cells, and many transformed or tumor cells. It is chemotactic for fibroblasts, smooth muscle cells, monocytes, and neutrophils and stimulates production of matrix proteins. These activities of PDGF have led to speculation that it may function as a "wound hormone" in the adult, but it is likely that it exerts its effects in vivo locally in an autocrine or paracrine fashion (278).

PDGF is a potent growth factor for synoviocytes in vitro (279). In rheumatoid arthritis, synoviocytes proliferate in a "tumorlike" manner, and the resulting pannus is highly invasive. These RA synovial cells exhibit a variety of phenotypic abnormalities similar to malignantly transformed cells, but the transformed phenotype is only displayed on cells locally (280).

Normal connective tissue cells do not express PDGF receptors or express them at very low levels, but their expression can be induced in inflammatory lesions (281). PDGF receptors contain tyrosine kinase in their cytoplasm domains (282). Binding of PDGF to its receptor stimulates a series of events including free arachidonic acid release, increased cyclooxygenase and prostaglandin synthesis, and calcium mobilization (280).

Data implicating PDGF as at least one of the cytokines driving synovial hyperplasia in RA are abundant. PDGF-like factors are present and synthesized locally in rheumatoid synovium. Remmers et al. have shown that synovial fluid from RA patients contains PDGF-like activity (279). They have also shown that explanted rheumatoid synovial tissue secretes PDGF-like activity that immunoprecipitates with PDGF antibody (283). RNA transcripts for PDGF are found in RNA extracted from rheumatoid synovial tissue, and rheumatoid synovial tissue stains intensely for PDGF (280,283).

PDGF appears to contribute directly to the synovial abnormalities found in RA: PDGF immunostaining is more extensive in RA synovium than in OA or normal synovium. Moreover, the intensity of staining correlates with the degree of inflammation as assessed by light microscopy (283). PDGF receptor expression is increased in RA synovium compared to tissue from OA patients and normal controls (284). PDGF is a potent mitogen for cultured synovial fibroblasts, which proliferate in rheumatoid pannus (284), and it is chemotactic for monocytes/macrophages, a prominent cell type in rheumatoid synovium (285).

The known effects of PDGF in normal cell growth coupled with the findings in RA suggest that unregulated or dysregulated PDGF expression can contribute to disease processes. A role for PDGF has been shown in oncogenesis (286) as well as in the fibroproliferative response, which characterizes diseases such as inflammatory kidney disease (287), chronic organ rejection (288), and scleroderma (289). In all of these diseases (including RA), abnormal cell proliferation occurs in response to an inflammatory infiltrate. The accumulating evidence for involvement of PDGF in several diseases would make specific PDGF antagonists desirable. A soluble form of the PDGF receptor that binds and inactivates PDGF has been produced (290). Low molecular weight compounds that disturb the interaction between PDGF and its receptors have also been described, but further development regarding specificity and potency is needed before they can be used in vivo. Attempts have been made to design specific inhibitors of the tyrosine kinases of growth factor receptors (291) such as PDGF, but further development is also needed in this area.

TUMOR NECROSIS FACTOR

Tumor necrosis factor, a polypeptide product of stimulated macrophages, and the closely related TNF-β (lymphotoxin) produced by activated T lymphocytes, were named for their ability to lyse tumor targets (292–295). These cytokines cause hemorrhagic necrosis of certain tumors by destruction of neovasculature. An historical perspective as well as a description of the isolation, characterization, and cloning of TNF are provided by Kunkel et al. (296).

TNF-α and -β share a common receptor (297) and are coded for by genes located on chromosome 6 near the MHC locus (298). They share 50% amino acid sequence homology and stimulate many of the same functional activities. Administration of TNF in experimental models leads to a syndrome indistinguishable from that of endotoxic shock, suggesting that TNF is the direct mediator of organ dysfunction seen in this disorder (299,300).

TNF has several clearly defined activities related to patients with arthritis and chronic rheumatic diseases, including development of constitutional symptoms, increased bone resorption, stimulation of enzyme release from synovial cells, stimulation of fibroblast growth, and effects on cell-cell interactions important in the immune response.

Several groups have found TNF activity in serum of patients with inflammatory arthritis, such as RA, reactive arthritis, ankylosing spondylitis, and gout (301–303). TNF activity has also been found in children with Kawasaki disease (304) and in adults with scleroderma (305). At least one group (303) showed that RA patients with TNF-α present in serum and synovial fluid had higher sedimentation rates and synovial fluid leukocyte counts than RA patients without elevated TNF.

TNF, like IL-1, acts as a potent endogenous pyrogen (306), inducing a biphasic fever response. The initial temperature elevation is mediated by direct effect on the hypothalamus; the secondary response is caused by augmentation of IL-1 production (306). Release of acute-phase proteins is also stimulated by TNF. TNF is likely responsible in part for another systemic reaction seen in chronic inflammatory diseases—cachexia with muscle wasting. TNF suppresses the expression of mRNA that codes for lipogenic enzymes, thereby inducing lipolysis (307). TNF administration to animals results in depletion of protein stores (308) with subsequent muscle wasting.

In RA, bone resorption occurs in area adjacent to the inflamed synovial pannus. Although there are probably several mechanisms of bone destruction, it is likely that TNF produced locally by macrophages in pannus plays a major role. TNF is a potent stimulus to bone resorption in vitro (309–313). It stimulates osteoclasts directly by a mechanism that is not prostaglandin mediated, since indomethacin has no effect on the process (313). TNF-α stimulates synoviocyte PGE_2 and collagenase production (314,315); it also stimulates osteoblasts to release PGE_2 and M-CSF (316). TNF suppression of bone formation by osteoblasts leads to decreased intracellular alkaline phosphatase and reduced synthesis of collagen and noncollagen matrix proteins (309). These activities all contribute to loss of bone and cartilage in RA.

TNF-α and TNF-β, as well as IL-1α and IL-1β, increase the amount of hyaluronic acid (HA) secreted by human synovial fibroblasts (317). The effect is inhibited by indomethacin, suggesting a prostaglandin-mediated mechanism. Increased levels of HA, a glycosaminoglycan that has structural and regulatory functions in connective tissues, have been found in synovial fluid and serum from patients with RA (318). Also, rheumatoid synovial cells in culture produce greater amounts of HA than synovial cells form normal controls (319). Although the significance of increase HA in RA patients is not clear, it has been postulated that this is one alteration in synovial cell properties that help to explain the invasive nature of the pannus in RA (320). There also may be a positive feedback (amplification) loop established by the TNF-induced increase in HA levels. HA stimulation of IL-1 production by human monocytes (321) may help explain the increased amounts of IL-1 found in rheumatoid pannus. Synergy between IL-1 and TNF-α, known to be important in rheumatoid synovium, has been shown in the induction of prostaglandin synthesis by lung fibroblasts (322) and in osteoclast-mediated bone resorption (323). Firestein and Zvaifler (324) noted that IL-1 and TNF-α together are more potent stimulators of synovial cell proliferation than either alone.

Finally, in such diseases as OA and RA, TNF may also cause destruction of articular cartilage. Shinmei et al. (325) studied the effects of IL-1 and TNF on collagenase and caseinase production by and proteoglycan release from human articular cartilage. They showed that IL-1 and TNF caused increased proteoglycan release and increased production of metalloproteinases leading to destruction of articular cartilage. TNF-α–containing cells have been localized to the cartilage-pannus junction in tissue from RA patients (326), suggesting production near the site of tissue destruction. Higher levels of soluble TNF receptors in synovial fluid than in serum from RA patients (327) suggests local production of these potentially protective molecules.

Some activities of TNF itself may protect the host. In addition to inducing fever, TNF stimulates human neutrophils to produce oxygen free radicals and potentiates their ability to kill pathogens (328). Acting synergistically with IFN-γ, TNF enhances leukocyte phagocytosis and antibody-dependent cellular cytotoxicity (329). Thus, whereas TNF contributes to organ dysfunction, is also augments the host immune status and promotes the inflammatory response. This heightened immune response may be beneficial or harmful depending on the location of the TNF produced. For example, the interaction of TNF with endothelium results in suppression of proteins important in maintaining a patient vasculature. This may be useful in tumor lysis, or it may be harmful, leading to vasculitis associated with some of the rheumatic diseases.

The notion that TNF-α plays a central role in propagation of tissue injury in RA patients led to experiments designed to suppress TNF-α production and action in animals and in patients. Success is not guaranteed: modification of TNF-α mRNA in transgenic mice resulted in spontaneous development of destructive arthritis (330).

Administration of TNF-α accelerates onset and enhances severity of collagen-induced arthritis in mice and rats (331,332), and administration of monoclonal antibodies against TNF-α (333) and of soluble TNF receptors (334) attenuates inflammation and joint tissue injury in the animal mode. Administration of specific TNF monoclonal antibodies to RA patients has engendered cautious optimism. Initial studies employed single infusions of antibodies (335–337). In a more recent study (338), seven RA patients completed two to four cycles of therapy. Median improvements in swollen joint counts and in C-reactive protein exceeded 80%. TNF impairs T-cell function in vitro. Administration of the chimeric monoclonal antibody to TNF-α in RA patients corrects abnormalities in T-cell proliferative responses to mitogens and antigens.

REFERENCES

1. Bodel PT, Hollingsworth JW. Pyrogen release from human synovial cell exudates. Br J Exp Path 1968; 49:11–19.
2. Gowen M, Wood DD, Ihrie EJ, McGuire MKB, Russell GG. An IL-1 like factor stimulates bone resorption in vitro. Nature 1983; 306:378–380.
3. Saklatvala J, Curry VA, Sarsfield SJ. Purification to homegeneity of pig monocyte catabolism, a protein that causes cartilage resorption in vitro. Biochem J 1983; 215:385–392.
4. Hollander AP. Criteria for identifying mediators of tissue damage in human autoimmune diseases. Autoimmunity 1991; 9:171–176.
5. Andreis M, Stastny P, Ziff M. Experimental arthritis produced by injection of mediators of delayed hypersensitivity. Arthritis Rheum 1974; 17:537–551.
6. Pettipher ER, Higgs GA, Henderson B. Interleukin 1 induces leukocyte infiltration and cartilage proteoglycan degradation in the synovial joint. Proc Natl Acad Sci 1986; 83:8749–8753.
7. Dingle JT, Page-Thomas DP, King B, Bard DR. In vivo studies of atricular tissue damage mediated by catabolin/interleukin 1. Ann Rheum Dis 1987; 46:527–533.
8. Ise K, Nakamura S, Ohkawara S, Yoshinaga M. DNA synthesis-potentiating activity on mouse thymocytes of synovial fluid of rheumatoid arthritis patients. Acta Pathol Jpn 1982; 32:491–503.
9. Fontana A, Hengartner H, Weber E, Fehr K, Grob PJ, Cohen G. Interleukin 1 activity in the synovial fluid of patients with rheumatoid arthritis. Rheumatol Int 1982; 2:49–53.
10. Wood DD, Ihrie EJ, Dinarello CA, Cohen PL. Isolation of an interleukin 1-like factor from human joint effusions. Arthritis Rheum 1983; 26:975–983.
11. Hopkins SJ, Humphreys M, Jayson MIV. Cytokines in synovial fluid. I. The presence of biologically active and immunoreactive IL-1. Clin Exp Immunol 1988; 72:422–427.
12. Di Giovine F, Situnavake D, Duff GW. Radioimmunoassay of interleukin 1β in acute and chronic rheumatic diseases. Br J Rheum 1987; 26(suppl 1):34.
13. Duff G, Di Giovine F, Dickens E, Symons J, Wood N, Manson J. Tumor necrosis factor and interleukin 1 in arthritis. Immunobiology 1987; 175:10–11.
14. DiGiovine FS, Malawista SE, Nuki G, Duff GW. Interleukin 1 as a mediator of crystal arthritis. J Immunol 1987; 138:3213–3218.
15. Duff GW, Atkins E, Malawista SE. The fever of gout: urate crystals activate endogenous pyrogen production from human and rabbit mononuclear phagocytes. Trans Assoc Am Phys 1983; 96:234–245.
16. Eastgate JA, Wood NC, Di Giovine FS, Symons JA, Grinlinton FM, Duff GW. Correlation of plasma interleukin 1 levels with disease activity in rheumatoid arthritis. Lancet 1988; 706–709.
17. Ritchie DM, Boyle JA, McInness JM, Jasani MK, Dalakos TG, Grieveson P, Buchanan WW. Clinical studies with an articular index for the assessment of joint tenderness in patients with rheumatoid arthritis. Quart J Med 1968; 37:393–406.
18. Dewhirst FE, Stashenko PE, Mole JE, Tsurumachi T. Purification of human osteoclast activating factor: identity with IL-1β. J Immunol 1985; 135:2562–2568.
19. Rathanswami P, Hachicha M, Wong WL, Schall TJ, McColl SR. Synergistic effect of IL-1β and TNFα on IL-8 gene expression in synovial fibroblasts. Arthritis Rheum 1993; 36:1295–1304.
20. Dayer JM, Breard J, Chess L, Krane SM. Participation of monocyte-macrophages and lymphocytes in the production of a factor that stimulates collagenase and prostaglandin release by rheumatoid synovial cells. J Clin Immunol 1979; 64:1386–1392.
21. Mizel SB, Dayer JM, Krane SM, Mergenhagen SE. Stimulation of rheumatoid synovial cell collagenase and prostaglandin production by partially purified lymphocyte activating factor (IL-1). Proc Natl Acad Sci 1981; 78:2473–2477.
22. Dayer JM, De Rochemonteix B, Burrus B, Demczuk S, Dinarello CA. Human recombinant interleukin 1 stimulates collagenase and prostaglandin E_2 production by human synovial cells. J Clin Invest 1986; 77:645–648.
23. Dayer JM, Demczuk S. Cytokines and other mediators in rheumatoid arthritis. Springer Sem Immunopathol 1984; 7:387–413.
24. Sato K, Fujii Y, Asano S, Ontsuki T, Kwakami M, Kasono K, Tsuchima T, Shizume K. Recombinant human IL-1 and β production stimulate mouse osteoblast-like cells (mc 3T3-E_1) to produce macrophage colony stimulating activity and PGE_2. Biochem Biophys Res Commun 1986; 141:285–291.
25. Mochan E, Uhl J, Newton R. Interleukin 1 stimulation of synovial cell plasminogen activator production. J Rheumatol 1986; 13:15–19.
26. Lack CH, Rogers HJ. Action of plasmin on cartilage. Nature 1958; 182:948–949.
27. Werb Z, Mainardi CL, Vater CA, Harris ED. Endogenous activation of latent collagenase by rheumatoid synovial cells. N Engl J Med 1977; 296:1017–1023.
28. Vaes G, Eickhout Y, Lenaers-Claeys G, Francois-Gillet C, Druetz J. The simultaneous release by bone explants in culture and the parallel activation of procollagenase and of a laent neutral proteinase that degrades cartilage proteoglycans and denatured collagen. Biochem J 1978; 172:261–274.
29. Wueppes KK. In: Lepore IH, Wand PA, eds. Inflammation Mechanisms and Control. New York: Academic Press, 1974:93.
30. Bathon JM, Hwang JJ, Shin LH, Precht PA, Towns MC, Horton WE. Type VI collagen specific messenger RNA is expressed constitutively by cultured human synovial fibroblasts and is suppressed by IL-1. Arthritis Rheum 1994; 37:1350–1356.
31. Rossi V, Breviario F, Ghezzi P, Dejana E, Mantovani A. Prostacyclin synthesis induced in vascular cells by interleukin 1. Science 1985; 229:1174–1176.

32. Ziff M. Rheumatoid arthritis—its present and future. J Rheumatol 1990; 17:127–133.
33. Nachman RL, Hajjar KA, Silverstein RL, Dinarello CA. IL-1 induces endothelial cell synthesis of plasminogen activator inhibitor. J Exp Med 1986; 163:1595–1600.
34. Iguchi T, Ziff M. Electron microscopic study of rheumatoid synovial vasculature. J Clin Invest 1986; 77:355–361.
35. Dinarello CA, Woff SM. The role of IL-1 in disease. NEJM 1993; 328:106–113.
36. Symons JA, Eastgate JA, Duff GW. A soluble binding protein specific for interleukin 1β is produced by activated mononuclear cells. Cytokine 1990; 2:190–198.
37. Symons JA, Eastgate JA, Duff GW. Purification and characterization of a novel soluble receptor for IL-1. J Exp Med 1991; 174:1251–1254.
38. Arend WP, Dayer JM. Inhibition of the production and effects of IL-1 and TNFα in rheumatoid arthritis. Arthritis Rheum 1995; 38:151–160.
39. Dreylow B, Capizio J, Lovis R, Jacobs C, Landar A, Pope RM. Phase I study of recombinant human IL-1 receptor administered intra-articularly in active RA (abst). Arthritis Rheum 1993; 36(suppl 9):S39.
40. Arend WP. IL-2 receptor antagonist: a new member of the IL-1 family. JCI 1991; 88:1445–1451.
41. Dinarello CA, Thompson RC. Blocking IL-1: IL-1 receptor antagonist in vivo and in vitro. Immunol Today 1991; 12:404–410.
42. Gabay C, Gay-Crosier F, Roux-Lombard P, Meyer O, Maineti C, Guerne PA, Vischer T, Dayer JM. Elevated serum levels of IL-1ra in PM/DM. Arthritis Rheum 1994; 1744–1751.
43. Lebsack ME, Paul CC, Bloedow DC, Burch FX, Sack MA, Chase W, Catalano MA. Subcutaneous IL-1 receptor antagonist in patients with RA (abst). Arthritis Rheum 1991; 34(suppl 9):S45.
44. Lebsack ME, Paul CC, Martindale JJ, Catalano MA. A dose- and regimen-ranging study of IL-1ra in patients with RA (abst). Arthritis Rheum 1993; 36(suppl 9):S39.
45. Evans C, Robbins PD. Prospects of treating arthritis by gene therapy. J Rheum 1994; 21:779–782.
46. Bandara G, Robbins PD, Georgescu HI, Mueller GM, Glorioso JC, Evans CH. Gene transfer to synoviocytes: prospects for gene treatment of arthritis. DNA Cell Biol 1992; 11:227–231.
47. Bandara G, Mueller GM, Galea-Lauri J, et al. Intraarticular expression of biologically active interleukin-1 receptor antagonist protein by ex vivo gene transfer. Proc Natl Acad Sci USA 1993; 90:10764–10768.
48. Hung GL, Galea-Lauri J, Mueller GM, et al. Suppression of intraarticular response to IL-1 by transfer of the IL-1 receptor antagonist gene to synovium. Gen Ther 1994; 1:64–69.
49. Bunjes D, Hardt C, Rollingnoff M, Wagner H. Cyclosporin A mediates immunosuppression of primary cytotoxic T cell responses by impairing the release of IL1 and IL-2. Eur J Immunol 1981; 11:657–661.
50. Snyder DS, Unanue ER. Corticosteroids inhibit murine macrophage 1a expression and interleukin 1 production. J Immunol 1982; 129:1803–1805.
51. Vane J, Botting R. Inflammation and the mechanism of action of antiinflammatory drugs. FASEB J 1987; 1:89–196.
52. Morgan FS, Riscetto FW, Gallo RC. Selective in vitro growth of T lymphocytes from normal human bone marrow. Science 1976; 193:1007.
53. Schreier MH, Iscove NN, Tees R, Aarden L, von Boehmer H. Clones of killer and helper T cells: growth requirement, specificity and retention of function in long-term cultures. Immunol Rev 1980; 51:315.
54. Gillis S, Fern M, Ou W, Smith K. T cell growth factor parameters of production and a quantitative microassay for activity. J Immunol 1978; 120:2027.
55. Waldmann TA, Goldman CK, Robb RJ, Depper JM, Leonard WJ, Sharrow SO, Bongiovanni KF, Korsmeyer SJ, Greene WC. Expression of interleukin 2 receptors on activated human B cells. J Exp Med 1984; 160:1450–1458.
56. Ortaldo JR, Mason AT, Gerard JP, Henderson LE, Farrar W, Hopkins III RF, Herberman RB, Rabin H. Effects of natural and recombinant human IL-2 in regulation of IFN-γ production and natural killer activity: lack of involvement of the Tac antigen for these immunoregulatory effects. J Immunol 1984; 133:779.
57. Kasahara T, Hooks JJ, Dougherty SF, Oppenheim JJ. Interleukin-2 mediated immune interferon production by human T cells and T cell subsets. J Immunol 1983; 130:1784.
58. Timonen T, Ortaldo JR, Stadler BM, Bonnard GD, Sharrow SO, Herberman RB. Cultures of purified human natural killer cells growth in the presence of interleukin-2. Cell Immunol 1982; 72:178.
59. Malkovsky M, Loeland B, North M, Asherson GL, Gao L, Ward P, Fiers W. Recombinant interleukin 2 directly augments the cytotoxicity of human monocytes. Nature 1987; 325:262.
60. Grimm EA, Mazumder A, Rosenberg SA. Lymphokine-activated killer cell phenomenon. Lysis of natural killer-resistant fresh solid tumor cells by interleukin 2-activated autologous human peripheral blood lymphocytes. J Exp Med 1982; 155:1823.
61. Meuer SC, Hussey RE, Canatrell A, Hodgdon JC, Schlossman SF, Smith KA, Reinherz EL. Triggering of the T3-Ti antigen receptor complex results in clonal T-cell proliferation through an interleukin 2 dependent autocrine pathway. Proc Natl Acad Sci USA 1984; 81:1509–1513.
62. Meuer SC, Hussey RE, Fabbi M, Fox DA, Acuto O, Fitzgerald KA, Hodgdon JC, Protentis JP, Schlossman SF, Reinherz EL. An alternative pathway of T cell activation: a functional role for the 50 Kd T11 sheep erythrocyte receptor protein. Cell 1984; 36:897–906.
63. Fox DA, Hussey RE, Fitzgerald KA, Acuto O, Poole C, Palley L, Daley JF, Schlossman SF, Reingerz EL. Ta_1, a novel 105 Kd human T cell activation antigen defined by a monoclonal antibody. J Immunol 1984; 133:1250–1256.
64. Reinherz EL, Meuer S, Fitzgerald KA, Hussey RE, Levine H, Schlossman SF. Antigen recognition by human T lymphocytes is linked to surface expression of the T3 molecular complex. Cell 1972; 30:735–743.
65. Swack JA, Nakatsuji T, Ito K, Tsuji K. Preservation of membrane glycoprotein labile epitopes during Western blotting with monoclonal antibodies. Biotechniques 1987; 5:564–671.
66. Morimoto C, Letvin N, Distaso JA, Aldrich WR, Schlossman SF. The isolation and characterization of the human suppressor induced T cell subset. J Immunol 1985; 134:1508–1515.
67. Hara T, Fu SM, Hansen JA. Human T cell activation. II. A new activation pathway used by a major T cell population via a disulfide-bonded dimer of a 44 kilodalton polypeptide (9.3 antigen). J Exp Med 1985; 162:1513–1524.
68. Moretta A, Pantaleo G, Lopez-Botet M, Moretta L. Involvement of T44 molecules in an antigen-independent pathway of T cell activation. Analysis of the correlations to the T cell antigen-receptor complex. J Exp Med 1985; 162:823–838.
69. Fox DA, Schlossman SF, Reinherz EL. Regulation of the alternative pathway of T cell activation by anti-T3 monoclonal antibody. J Immunol 1986; 136:1945–1950.
70. Fleischer B. A novel pathway of human T cell activation via a

103 KD T cell activation antigen. J Immunol 1987; 138:1346–1350.
71. Ruscetti FW, Morgan DA, Gallo RC. Functional and morphologic characterization of human T cells continuously growth in vitro. J Immunol 1977; 119–131–138.
72. Welte KA, Androeff MA, Platzer E, Holloway K, Rubin B, Moore MAS, Mertelsmann R. Interleukin-2 regulates the expression of Tac antigen on peripheral blood lymphocytes. J Exp Med 1984; 160:1390–1403.
73. Smith KA. T-cell growth factor. Immunol Rev 1980; 51:337–357.
74. Emery P, Williamson DJ, MacKay IR. The role of cytokines in rheumatological inflammation. In: Cruse JM, Lewis RE, eds. Concepts in Immunology: Autoimmunoregulation and Autoimmune Disease. Vol 4. Basel: Karger, 1987:143–155.
75. Symons JA, Wood NC, DiGiovini FS, Duff GW. Soluble IL-2 receptors in rheumatoid arthritis: correlation with disease activity, IL-1 and IL-2 inhibition. J Immunol 1988; 141:2612–2618.
76. Warrington RJ. Interleukin 2 abnormalities in systemic lupus erythematosus and rheumatoid arthritis. A role for overproduction of interleukin-2 in human autoimmunity? J Rheumatol 1987; 15:616–620.
77. Merrill JE, Mohlstrom C, Uittenbogaart C, Kermani-Arab V, Ellison GW, Myers LW. Response to and production of interleukin 2 by peripheral blood and cerebrospinal fluid lymphocytes of patients with multiple sclerosis. J Immunol 1984; 133:1931–1937.
78. Wolf RE, Brelsford WG. Soluble interleukin-2 receptors in systemic lupus erythematosus. Arthritis Rheum 1988; 31:729–735.
79. Buchan G, Barrett K, Fujita T, Taniguchi T, Maini R, Feldmann M. Detection of activated T cell products in the rheumatoid joint using cDNA probes to interleukin-2 (IL-2) receptor and IFN-γ. Clin Exp Immunol 1988; 71:295–301.
80. Stamenkovic I, Stegagno M, Wright KA, Krane SM, Amento EP, Colvin RB, Duquesnoy RJ, Kurnick JT. Clonal dominance among T-lymphocyte infiltrates in arthritis. Proc Natl Acad Sci USA 1988; 85:1179–1183.
81. Konttinen YT, Bergroth V, Kinnunen E, Nordstrom D, Kouri T. Activated T lymphocytes in patients with multiple sclerosis in clinical remission. J Neurol Sci 1987; 81:133–139.
82. Konishi K, Moller DR, Saltini C, Kirby M, Crystal RG. Spontaneous expression of the interleukin 2 receptor gene and presence of functional interleukin 2 receptors on T lymphocytes in the blood of individuals with active pulmonary sarcoidosis. J Clin Invest 1988; 82:775–781.
83. Waalen K, Sioud M, Natvig JB, Forre O. Spontaneous in vivo gene transcription of interleukin-2, interleukin-3, interleukin-4, interleukin-6, interferon-γ, interleukin-2 receptor, and protoncogene c-myc by rheumatoid synovial T-lymphocytes. Scand J Immunol 1992; 36:865–873.
84. Massarotti EM, Liu NY, Mier J, Atkins MB. Chronic inflammatory arthritis after treatment with high dose interleukin-2 for malignancy. Am J Med 1992; 92:693–697.
85. Wolf RE, Hall VC. Inhibition of in vitro proliferative response of culture T lymphocytes to interleukin-2 by gold sodium thiomalate. Arthritis Rheum 1988; 31:176–181.
86. Leventhal LJ, Boyce EG, Zurier RB. Treatment of rheumatoid arthritis with gammalinolenic acid. Ann Intern Med 1993; 119:867–873.
87. Santoli D, Phillips PD, Colt TL, Zurier RB. Suppression of interleukin 2-dependent human T cell growth *in vitro* by prostaglandins E and their precursor fatty acids. Evidence for a PGE-independent mechanism of inhibition by fatty acids. J Clin Invest 1990; 85:424–432.
88. DeMarco DM, Santoli D, Zurier RB. Effects of fatty acids on proliferation and activation of human synovial compartment lymphocytes. J Leuk Biol 1994; 56:612–615.
89. Fort JG, Smith JD, Abruzzo JL. Abnormal T cell function in patients with psoriatic arthritis: evidence for decreased interleukin-2 production. Rheumatol Int 1993; 13:151–154.
90. Pettersson T, Soderblom T, Nyberg P, Riska H, Linko L, Klockars M. Pleural fluid soluble interleukin-2 receptor in rheumatoid arthritis and systemic lupus erythematosus. J Rheumatol 1994; 21:1020–1024.
91. Manoussakis MN, Germanidis GS, Drosos AA, Moutsopoulos HM. Impaired urinary excretion of soluble IL-2 receptors in patients with systemic lupus erythematosus and rheumatoid arthritis. Lupus 1992; 1:105–109.
92. Boiardi L, Macchioni P, Salvarani C, Crossi F, Akcasadei-Maldini M, Mancini R, Beltrandi E, Portioli I. Serum soluble interleukin-2 receptor levels in rheumatoid arthritis: correlation with clinical and immunological parameters and with the response to auranofin. Clin Exp Rheumatol 1994; 12:357–362.
93. Rose CB, Fawcett PP, Gibney K, Doughty RA, Singsen PH. Serial measurements of soluble interleukin-2 receptor levels in children with juvenile rheumatoid arthritis treated with oral methotrexate. Ann Rheum Dis 1994; 53:471–474.
94. Dooley MA, Cush JJ, Lipsky PE, Dawson DV, Pisetsky DS. Effects of non-steroidal anti-inflammatory therapy in early rheumatoid arthritis on serum levels of soluble interleukin-2 receptor, CD4, and CD8. J Rheumatol 1993; 20:1857–1862.
95. Polisson P, Dooley MA, Dawson DV, Pisetsky DS. Arthritis Rheum 1994; 37:50–56.
96. Van Deen, Brink HR, Wijk MJ, Geertzen RG, Bijlsma JW. Influence of corticosteroid pulse therapy on the serum levels of soluble interleukin-2 receptor, interleukin-6, and interleukin-8 in patients with rheumatoid arthritis. J Rheumatol 1994; 21:430–434.
97. Kronke M, Leonard WJ, Depper JM, Arya SK, Wong-Staal F, Gallo RC, Waldmann TA, Greene WC. Cyclosporin A inhibits T cell growth factor gene expression at the level of mRNA transcription. Proc Natl Acad Sci USA 1984; 81:5214–5218.
98. Santoli D, Zurier RB. Prostaglandin E precursor fatty acids inhibit human interleukin-2 production by a prostaglandin E-independent mechanism. J Immunol 1989; 143:1303–1309.
99. Steven C, Lipman M, Fabry S, Moscovitch-Lopatin M, Almawi W, Keresztes S, Peppercorn MA, Strom TB. 5-aminosalicylic acid appogate T cell proliferation by blocking interleukin-2 production in peripheral blood mononuclear cell. J Pharmacol Exp Ther 1995; 272:399–406.
100. Sfikakis PP, Souliotis VO, Panayiotidis PT. Suppression of interleukin-2 and interleukin-2 receptor biosynthesis by gold compounds in vitro activated human peripheral blood mononuclear cells. Arthritis Rheum 1993; 36:200–212.
101. Pitzalis C, Choy E, Kingsley G. Monoclonal antibody in rheumatic disease. Presse Med 1994; 23:532–539.
102. Elliot MW, Maini RN. New directions from new biological therapy in rheumatoid arthritis. Int Arch Allergy Immunol 1994; 104:112–125.
103. Kelley VE, Gaulton GN, Hattori M, Ikegami H, Eisenbarth G, Strom TB. Anti-interleukin 2 receptor antibody suppresses murine diabetic insulitis and lupus nephritis. J Immunol 1988; 140:59–61.
104. Nicola NA, Vadas M. Hemopoietic colony-stimulating factors. Immunol Today 1984; 5:76–80.
105. Sieff C. Hematopoietic growth factors. J Clin Invest 1987; 79:1549–1557.

106. Firestein GS, Zvaifler NJ. The pathogenesis of rheumatoid arthritis. A critical assessment of the role of autologous stimulation in the perpetuation of rheumatoid synovitis. Rheum Dis Clin NA 1987; 13:447–461.
107. Firestein GS, Xu WD, Townsend K, Broide D, Alvaro-Garcia J, Glasebrook A, Zvaifler NJ. Cytokines in chronic inflammatory arthritis. 1. Failure to detect T cell lymphokines (IL-2 and IL-3) and presence of macrophage SCF-1 and a novel mast cell growth factor in rheumatoid synovium. J Exp Med 1988; 168:1573–1586.
108. Murison JG, Quaratino S, Kahan M, Verhoef A, Landei M. Definition of unique traits of human CD4–CD8-α β T cells. Clin Exp Immunol 1993; 93:464–470.
109. Magnelli L, Cinelli M, Turchetti A, Chiarugi VP. Aptosis induction in 32D cells by IL-3 withdrawal is preceded by a drop in the intracellular calcium level. Biochem Biophys Res Comm 1993; 194:1394.
110. Sakihama T, Akasu F, Iwamoto M, et al. Polyclonal and monoclonal IgG's from MRL/Mp-Ipr/Ipr mice induce an IL-3-dependent cell line to produce IL-3 through all Fcγ-receptor-mediated mechanism. Cell Immunol 1990; 125:160–170.
111. Fishman P, Shoenfeld Y. Cytokines, IL-3 and autoimmunity. ISR J Med Sci 1993; 29:159–164.
112. Fishman P, Bakimer R, Blank M, et al. The putative role of cytokines in the induction of primary anti-phospholipid syndrome in mice. Clin Exp Immunol 1992; 90:266–270.
113. Howard M, Farrar J, Hilfiker M, Johnson B, Takatsu K, Hammaoka T, Paul WE. Identification of a T cell derived B cell growth factor distinct from interleukin-2. J Exp Med 1982; 255:914–923.
114. Noelle R, Krammer PH, O'Hara J, Uhr JW, Vitetta ES. Increased expression of 1a antigens on resting B cells: an additional role for B cell growth factor. Proc Natl Acad Sci USA 1984; 81:6159–6153.
115. Rabin RM, Mond JJ, O'Hara J, Paul WE. B cell stimulatory factor 1 (BSF-1) prepares resting B cells to enter S phase in response to anti-IgM and lipopolysaccharide. J Exp Med 1986; 164:517–531.
116. Vitetta ES, O'Hara J, Myers CD, Layton JE, Krammer PH, Paul WE. Serologic, biochemical and functional identity of B cell stimulatory factor-1 and B cell differentiation factor for IgG. J Exp Med 1985; 162:1726–1731.
117. Coffman RL, O'Hara J, Bond MW, Carty J, Zlotnik A, Paul WE. B cell stimulatory factor-1 enhances the IgE response of lipopolysaccharide-activated B cells. J Immunol 1986; 136:4538–4541.
118. Paul WE. Interleukin 4/B cell stimulatory factor 1: one lymphokine, many functions. FASEB J 1987; 1:456–461.
119. Hu-Li J, Shevach EM, Mizuuchi J, O'Hara H, Mosmann T, Paul WE. B cell stimulatory factor 1 (interleukin 4) is a potent costimulant for normal resting T lymphocytes. J Exp Med 1987; 165:157–172.
120. Mitchell LC, David LS, Lipsky PE. Promotion of human T lymphocyte proliferation by IL-4. J Immunol 1989; 142:1548–1557.
121. Mizel SB. The interleukins. FASEB J 1989; 3:2379–2388.
122. Paul WE, O'Hara J. B cell stimulatory factor-1/interleukin 4. Ann Rev Immunol 1987; 5:429–459.
123. Fay AC, Trudgett A, McCrea JD, Kirk F, Thompson JM, Mitchell ES, Boys MJ, Roberts SD, McNeill TH. Detection and partial characterization of human B cell colony stimulating activity in synovial fluids of patients with rheumatoid arthritis. Clin Exp Immunol 1985; 60:316–322.
124. Miossec P, Naviliat M, Dupuy d'Angeac A, Sany J, Banchereau J. Low levels of IL-4 and High levels of TGF-β in rheumatoid synovitis. Arthritis Rheum 1990; 33:1180–1187.
125. Miossec P, Briolay J, Dechanet J, Wijdenes J, Martinez-Valdez H, Banchereau J. Inhibition of the production of proinflammatory cytokines and immunoglobulins by IL-4 in an ex-vivo model of rheumatoid synovitis. Arthritis Rheum 1992; 24:874–883.
126. Matis LA, Glimcher LH, Paul WE, Schwartz RH. Magnitude of response of histocompatibility restricted T cell clones is a function of the product of the concentrations of antigen and 1a molecules. Proc Natl Acad Sci USA 1983; 80:6019–6023.
127. Hart PH, Vitti GF, Burgess DR, Whitt GA, Piccoli DS, Hamilton JA. Potential anti-inflammatory effects of IL-4: Suppression of human monocyte TNF-α, IL-1 and PGE_2. Proc Natl Acad Sci USA 1989; 86:3803–3807.
128. Miossec P, Chomarat P, Dechanet J, Moreau JF, Roux JP, Demas P, Banchereau J. IL-4 Inhibits bone resorption through an effect on osteoclasts and pro-inflammatory cytokines in an ex-vivo model of bone resorption in RA. Arthritis Rheum 1994; 37:1715–1722.
129. Handwerger BS. T-cell and B-cell functions in lupus. Curr Op Rheum 1991; 3:757–779.
130. Takatsu K, Tominaga A, Hamoaka T. Antigen induced T cell replacing factor (TRF). J Immunol 1980; 124:2414–2422.
131. Swain SC, Dennert G, Warner JF, Dutton RW. Culture supernatants of a stimulated T cell line have helper activity that acts synergistically with IL-2 in the response of B cells to antigen. Proc Natl Acad Sci USA 1981; 78:2517–2521.
132. Swain SL, McKenzie DT, Dutton RW, Tonkonogy SL, English M. The role of IL-4 and IL-5: characterization of a distinct helper T cell subset that makes IL-4 and IL-5 (Th2) and requires priming before induction of lymphokine secretion. Immunol Rev 1988; 102:77–105.
133. Takatsu K, Harada N, Hara Y, Takahama Y, Yamada G, Dobashi K, Hamaoka T. Purification and physiochemical characterization of murine T cell replacing factor (TRF). J Immunol 1985; 134:382–389.
134. Swain SL. Role of BCGFII in the differentiation to antibody secretion normal and tumor B cells. J Immunol 1985; 134:3934–3943.
135. Swain SL, Howard M, Kappler J, Marrack P, Watson J, Booth R, Wetzel GD, Dutton RW. Evidence for two distinct classes of murine B cell growth factors with activities in different functional assays. J Exp Med 1983; 158:822–835.
136. Mahanty S, Nutman T. The biology of IL-5 and its receptor. Cancer Invest 1993; 11:624–634.
137. Takatsu ?, Kikuch Y, Takahasui T, Honjo T, Matsumoto M, Harada M, Yagtanuchi N, Tominaga A. Il-5, a T cell derived B cell differentiation factor also induces cytotoxic T lymphocytes. Proc Natl Acad Sci USA 1987; 84:4234–4238.
138. Pure E, Isaksan PC, Kappler JW, Marrack P, Krammer PH, Vitetta ES. T cell derived B cell growth and differentiation factors. J Exp Med 1983; 157:600–612.
139. Nakanishi K, Malek TR, Smith KA, Hamaoka T, Shevach EM, Paul WE. Both IL-2 and second T cell-derived factor in IL-4 supernatant have activity as differentiation factors in IgM synthesis. J Exp Med 1984; 160:1605–1621.
140. Brooks K, Yuan D, Uhr JW, Krammer PH, Vitetta ES. Lymphokine induced IgM secretion by clones of neoplastic B cells. Nature (London) 1983; 302:825–826.
141. Pure E, Isakson C, Takatsu K, Hamaoka T, Swain SL, Button RW, Dennert G, Uhr JW, Vitetta ES. Induction of B cell differentiation by T cell factors. J Immunol 1981; 127:1953–1958.
142. Clutterbuck E, Shields JG, Gordon J, Smith SH, Boyd A,

Callard RE, Campbell HD, Young IG, Sanderson CJ. Recombinant human IL-5 is an eosinophil differentiation factor but has no activity in standard human B cell growth factor assays. Eur J Immunol 1987; 17:1743–1750.
143. Campbell HD, Tucker WQ, Hort Y, Martinson ME, Mayo G, Clutterbuck EJ, Sanderson CJ, Young IG. Molecular cloning nucleotide sequence, and expression of the gene encoding human eosinophil differentiation factor (IL-5). Proc Natl Acad Sci USA 1987; 84:6629–6633.
144. Azuma C, Tanabe T, Konishi M, Kinashi T, Noma T, Matsuda F, Yaiota Y, Takatsu K, Hammerstron L, Smith C, et al. Cloning of cDNA for human T cell replacing factor (IL-5) and comparison with the murine homologue. Nucl Acid Res 1986; 14:9149–9158.
145. Yokota T, Coffman RL, Hagiwara HH, Rennick DM, Takebe Y, Yokota K, et al. Isolation and characterization of lymphokine cDNA clones encoding mouse and human IgG enhancing factor and eosinophil colony-stimulating factor activities. Relationship to IL-5. Proc Natl Acad Sci USA 1987; 84:7388–7392.
146. Varga J, Uitto J, Jimenez SA. The cause and pathogenesis of the Eosinophilia Myalgia Syndrome. Ann Intern Med 1992; 116:140–147.
147. Koike M, Takatsu K. IL-5 and its receptor: Which role do they play in the immune response? Int Arch All Imm 1994; 104:1–9.
148. Billiau A, Van Damme J, Ceuppens J, Baroja M. Interleukin-6, a ubiquitous cytokine with paracrine as well as endocrine functions. In: Libby J, ed. Lymphokine Receptor Interactions. London: Eurotext Ltd, 1989:133–142.
149. Kishimoto T, Hirano T. Molecular regulation of B lymphocyte response. Ann Rev Immunol 1989; 6:485–512.
150. Revel M, Zuberstein A. Interferon-β_2 living up to its name. Nature 1987; 325:581–582.
151. Elias JA, Trinchieri G, Beck JM, Simon PL, Sehgal PB, May LT, Kern JA. A synergistic interaction of IL-6 and IL-1 mediates the thymocyte-stimulating activity produced by recombinant IL-1 stimulated thymocytes. J Immunol 1989; 142:509–514.
152. Gauldie J, Richard C, Harnish D, Landsdorp P, Baumann H. Interferon-β_2/B cell stimulatory factor type 2 shares identity with monocyte-derived hepatocyte-stimulating factor and regulates the major acute phase protein response in liver cells. Proc Natl Acad Sci USA 1987; 84:7251–7255.
153. Hirano T, Taga T, Nakano N, Yasukawa K, Kashiwamura S, Shimizu K, Nakajima K, Pyun KH, Kishimoto T. Purification to homogeneity and characterization of human B cell differentiation factor (BCDF or BSF P-2). Proc Natl Acad Sci USA 1985; 82:5490–5494.
154. Yasukawa K, Hirano T, Watanabe Y, Muratani K, Matsuda T, Nakai S, Kishimoto T. Structure and expression of human B cell stimulatory factor-2 (BSF-2/IL-6) gene. EMBO J 1987; 6:2939–2945.
155. Bataille R, Klein B. Autoimmune conditions induced by epithelial solid tumors. Rheumatol Int 1987; 7:231–233.
156. Hirano T, Taga T, Yasukawa K, Nakajima K, Nakano N, Takatsuki F, Shimizu M, Murashima A, Tsunasawa S, Sakiyama F, Kishimoto T. Human B cell differentiation factor defined by an anti-peptide antibody and its possible role in autoantibody production. Proc Natl Acad Sci USA 1987; 84:228–231.
157. Waage A, Kaufmann C, Espevik T, Husby G. Interleukin-6 in synovial fluid from patients with arthritis. Clin Immunol Immunopathol 1989; 50:394–398.
158. Guerne PA, Zuraw BL, Vaughan JH, Carson DA, Lotz M. Synovium as a source of interleukin-6 in vitro. J Clin Invest 1989; 83:585–592.
159. Suzuki H, Sugiyama E, Tunru IS, Yamashita N, Matsuno H, Hamazaki T, Kobayashi M. Suppressive effect of interleukin-4 on IL-6 production by inherent rheumatoid synovial cells. Clin Immunopathol 1993; 66:67–72.
160. van Leeuwen MA, Westra J, Limburg PC, van Riel PL, van Rijswijk MH. Interleukin-6 in relation to other pro-inflammatory cytokines, chemotactic activity and neutrophil activation in rheumatoid synovial fluid. Ann Rheum Dis 1995; 54:33–38.
161. Duff GW. Cytokines and acute phase proteins in rheumatoid arthritis. Scand J Rheumatol 1994; 100:9–19.
162. Tanabe M, Ochi T, Tomita F, Suzuki R, Sakata T, Shimaoka Y, Nakagawa S, Ono K. Remarkable elevation of interleukin-6 and interleukin-8 levels in the bone marrow of patients with rheumatoid arthritis. J Rheumatol 1994; 21:830–835.
163. Arvidson NG, Gudbjornsson B, Elfman L, Ryden AC, Totterman TH, Hallgren R. Circadian rhythm of serum interleukin-6 in rheumatoid arthritis. Ann Rheum Dis 1994; 53:521–524.
164. Amano Y, Lee SW, Allison AC. Inhibition by glucocorticoids of the formation of interleukin-1 alpha, interleukin-1 beta, and interleukin-6: mediation by decreased mRNA stability. Mol Pharmacol 1993; 43:176–182.
165. Crilly A, Kolta S, Dougados M, Turrock RD, Amor B, Capell HA, Madhok R. Effect of cyclosporin A on interleukin-6 and soluble interleukin 2 receptor in patients with rheumatoid arthritis. Ann Rheum Dis 1995; 54:137–139.
166. Breedveld F. Tenidap: a novel cytokine-modulating anti-rheumatic drug for the treatment of rheumatoid arthritis. Scand J Rheumatol 1944; 100:31–44.
167. Cohick CB, Furst DE, Quagliata S, Corcoran KA, Steere KJ, Yager JG, Lindsley HB. Analysis of elevated serum interleukin-6 levels in rheumatoid arthritis: Correlation with erythrocytes sedimentation rate or C reactive protein. J Lab Clin Med 1994; 123:721–727.
168. Wascher TC, Hermann J, Brezinschek R, Brezinschek HP, Wilders-truschnig M, Raner F, Krejs GJ. Serum levels of interleukin-6 and tumour necrosis factor alpha are not correlated to disease activity in patients with rheumatoid arthritis after treatment with low dose methotrexate. Eur J Clin Invest 1994; 24:73–75.
169. Crilly A, McInness IB, Capell HA, Madhok R. The effect of Azathioprine on serum levels of interleukin-6 and soluble interleukin-2 receptor. Scand J Rheumatol 1994; 23:87–91.
170. Wendling D, Racadot E, Wijdenes J. Treatment of severe rheumatoid arthritis by anti interleukin-6 monoclonal antibody. J Rheumatol 1993; 20:259–262.
171. Appasamy PM. IL-7 biology and potential clinical applications. Cancer Invest 1993; 11:487–499.
172. Watson ML, Westwick J, Fincham NJ, Camp RDR. Elevation of PMN cytosolic free calcium and locomotion stimulated by novel peptides from IL-1-treated human synovial cell cultures. Biochem Biophys Res Commun 1988; 155:1154–1160.
173. DeMarco D, Kunkel SL, Strieter RM, Basha M, Zurier RB. Interleukin-1 induced gene expression of neutrophil activating peptide (IL-8) and monocyte chemotactic peptide in human synovial cells.
174. Hebert CA, Baker JB. IL-8: a review. Cancer Invest 1993; 11:743–750.
175. Vachula M, Bornstein I, Van Epps D. Role of GM-CSF in regulating the IL-8 receptor. FASEB J 1991; 5:A1632.
176. Schroeder JM, Christophers E. Identification of C5a des arg and an anionic neutrophil-activating peptide (ANAP) in psoriatic scales. J Invest Dermatol 1985; 87:53–58.
177. Sticherling M, Bornscheuer E, Schroder JM. Localization of

neutrophil activating peptide-1/IL-8 immunoreactivity in normal and psoriatic skin. J Invest Dermatol 1991; 96:26–30.
178. Brennan F, Zachariae C, Chantry D. Detection of IL-8 biological activity in synovial fluid from patients with RA and production of IL-8 mRNA by isolated synovial cells. Eur J Immunol 1990; 20:2141–2144.
179. Peichl P, Ceska M, Effenberger F. Presence of NAP-1/IL-8 in synovial fluids indicates a possible pathogenic role in RA. Scand J Immunol 1991; 34:333–339.
180. Koch A, Kunkel S, Burrows J. Synovial tissue macrophage as a source of the chemotactic cytokine IL-8. J Immunol 1991; 147:2187–2195.
181. Terkeltaub R, Zachariae C, Santoro D. Monocyte-derived neutrophic chemotactic factor/IL-8 is a potential mediator of crystal induced inflammation. Arthritis Rheum 1991; 34:894–903.
182. Van Snick J, Goethals A, Renauld J-C, Van Roost E, Uyttenhove C, Rubira MR, Moritz RL, Simpson RJ. Cloning and characterization of a cDNA for a new mouse T cell growth factor. (p40) J Exp Med 1989; 169:363–368.
183. Schmitt E, Van Brandwijk R, Van Snick J, Siebold B, Rüde E. TCGF III/p40 is produced by naive murine CD4+ T cells but is not a general T cell growth factor. Eur J Immunol 1989; 19:2167–2170.
184. Hültner L, Moeller J, Schmitt E, Jägu G, Reisbach G, Ring I, Dörmer P. J Immunol 1989; 142:3440–3446.
185. Moeller J, Hültner L, Schmitt E, Dörmer P. J Immunol 1989; 142:3447–3451.
186. Hültner L, Druez C, Moeller J, Uyttenhove C, Schmitt E, Rüde E, Dörmer P, Van Snick J. Eur J Immunol 1990; 20:1413–1416.
187. Renauld J-C, Houssian F, Louahed J, Vink A, Van Snick J, Uyttenhove C. Interleukin-9. Adv Immunol 1993; 54:79–97.
188. Renauld J-C, Kermouni A, Vink A, Louahed J, Van Snick J. Interleukin-9 and its receptor: involvement in mast cell differentiation and T cell oncogenesis. J Leukocyte Biol 1995; 57:353–360.
189. Mosmann TR. Properties and functions of IL-10. Adv Immunol 1994; 56:1–26.
190. Girard C, Bryns C, Marchant A, Abramowicz D, Vandenabelle P, Delvaux A, Fiers W, Goldman M, Velu T. IL-10 reduces the release of TNF and prevents lethality in experimental endotoxemia. J Exp Med 1993; 177:547–550.
191. Llorente L, Richaud-Patin Y, Fior R, Alcocer-Varela J, Wigdenes J, Fournier BN, Galanaud P, Emilie D. In vivo production of IL-10 by non-T cells in RA, SS, and SLE. Arthritis Rheum 1994; 37:1647–1655.
192. Cush JJ, Splawski JB, Thomas R, McFarlin JE, Schulze-Koops H, Davis LS, Fujita K, Lipsky PE. Elevated IL-10 levels in patients with RA. Arthritis Rheum 1995; 1:96–104.
193. Ishida H, Muchamuel T, Sakaguchi S, Andrade S, Menon S, Howard M. Continuous administration of IL-10 antibodies delays onset of autoimmunity in NZB/WF1 mice. J Exp Med 1994; 179:305–310.
194. Katsikis PD, Chu C-Q, Brennan FM, Maini RN, Feldmann M. Immunoregulatory Role of interleukin 10 in rheumatoid arthritis. J Exp Med 1994; 1517–1527.
195. Kingsley DM. The TGF-β superfamily: new members, new receptors, and new genetic tests of function in different organisms. Genes Dev 1994; 8:133–146.
196. Border WA, Ruoslahti E. TGF-β in disease: the dark side of tissue repair. J Clin Invest 1992; 90:1–7.
197. Wahl SM, Allen JB, Costa GL, Wong HL, Dasch JR. Reversal of acute and chronic synovial inflammation by anti-TGF-β. J Exp Med 1993; 177:225.
198. McCartney-Francis NL, Wahl SM. TGF-β a matter of life and death. J Leuk Biol 1994; 55:401–409.
199. Komm BS, Terpening CM, Benz DJ, et al. Estrogen binding, receptor mRNA and biologic response in osteoblast-like osteosarcoma cells. Science 1988; 241:81–84.
200. LeRoy EC. Collagen deposition in autoimmune diseases: the expanding role of the fibroblast in human fibrotic disease. Fibrosis: Ciba Foundation Symposium 114. London: Pitman, 1985.
201. Border WA, Noble NA. TGF-β in tissue fibrosis. N Engl J Med 1994; 331:1286–1292.
202. LeRoy EC. Connective tissue synthesis by scleroderma skin fibroblasts in cell future. J Exp Med 1972; 135:1351–1362.
203. Ruoslahit E. Integrins. JCI 1991; 87:1–5.
204. Laiho M, Sakesela O, Andreasen PA, Keski-Oja J. Enhanced production and extracellular deposition of the endothelial type plasminogen activator inhibitor in cultured human lung fibroblasts by TGF-β. Cell Biol 1986; 103:2403–2410.
205. Chiang CP, Nilsen-Hamilton M. Opposite and selective effects of epidermal growth factor and human platelet TGF-β on the production of secreted proteins by murine 3T3 cells and human fibroblasts. J Biol Chem 1986; 261:10478–10481.
206. Matrisian LM, LeRoy P, Rahlmann C, Gesnel MC, Breathrach R. Isolation of the oncogene and epidermal growth factor induced transin gene: Complex control in rat fibroblasts. Mol Cell Biol 1986; 6:1679–1686.
207. Uitto J, Bauer EA, Elson AZ. Scleroderma: increased biosynthesis of triple-helical type I and type II procollagens associated with unaltered expression of collagenase by skin fibroblasts in culture. J Clin Invest 1979; 64:921–930.
208. Kahari VM, Multimaki P, Vuorio E. Elevated pro-alpha$_2$(1) collagen mRNA levels in cultured scleroderma fibroblasts result from an increased transcription rate of the corresponding gene. FEBS Lett 1987; 215:331–334.
209. Varga J, Rosenbloom J, Jimenez SA. Transforming growth factor-β causes a persistent increase in steady state amounts of type I and type III collagen and fibronectin mRNAs in normal human dermal fibroblasts. Biochem J 1987; 247:597–604.
210. Rossi P, Karsenty G, Roberts AB, Roche NS, Sporn MB, DeCrombrugghe B. A nuclear factor 1 binding site mediates the transcriptional activation of a type I collagen promoter by enhancing TGF-β cells. Cell 1988; 52:405–414.
211. Ignotz R, Massague J. TGF-β stimulates the expression of fibronectin and collagen and their incorporation into the extracellular matrix. J Biol Chem 1986; 261:4337–4345.
212. Jimenez SA, Varga J, Olsen A, Li L, Diaz A, Herhal J, et al. Functional analysis of human α1 (I) procollagen gene promoter. Differential activity of collagen-producing and on producing cells and response to TGF-β1. J Biol Chem 1994; 169:12684–12691.
213. Inagaki Y, Truter S, Ramirez F. TGF-β stimulates α2 (I) collagen gene expression through a cisacting element that contains an spl-binding site. J Biol Chem 1994; 269:14828–14834.
214. Varga J, Jimenez SA. Modulation of collagen gene expression: its relation to fibrosis in systemic sclerosis and other disorders. Ann Intern Med 1995; 122:60–62.
215. Peltonen J, Varga J, Sollberg S, Uitto J, Jimenez SA. Elevated expression of the genes for TGF-β1 and type VI collagen in diffuse fasciitis associated with the eosinophil myalgia syndrome. J Invest Dermatol 1991; 96:20–25.
216. Giri SN, Hyde DM, Hollinger MA. Effect of antibody to TGF-β on bleomycin induced accumulation of lung collagen in mice. Thorax 1993; 48:959–966.

217. Wahl S. TGF-β: The good, the bad and the ugly. J Exp Med 1994; 80:1587–1590.
218. Carnemolla B, Cutolo M, Castellanai P, Balza E, Raffanti S, Zardi L. Characterization of synovial fluid fibronectin from patients with rheumatic inflammatory diseases and healthy subjects. Arthritis Rheum 1984; 27:913–921.
219. Carsons SE, Clausen H, Berkowitz E. Expression of a developmentally regulated oncofetal fibronectin epitope in inflammatory synovium. Arthritis Rheum 1988; 31:S22.
220. Kehrl JH, Wakefield LM, Roberts AV, Jakowlew S, Alvarez-Man M, Perynck R, Sporn MB, Fauci AS. Production of TGF-β by human T lymphocytes and its potential role in the regulation of T cell growth. J Exp Med 1986; 163:1037–1050.
221. Kehrl JH, Roberts AB, Wakefield LM, Jacowlew S, Sporn MB, Fauci AS. TGF-β is an important immunomodulatory protein for human lymphocytes. J Immunol 1986; 137:3855–3860.
222. Wahl SM, Hunt DA, Wong HL, Dougherty S, McCartney-Francis N, Wahl LM, Ellingsworth L, Schmidt JA, Hall G, Roberts AB, Sporn MB. TGF-β is a potent immunosuppressive agent that inhibits IL-1 dependent lymphocyte proliferation. J Immunol 1988; 140:3026–3032.
223. Fava R, Olsen N, Keski-Oja J, Moses H, Pincus T. Active and latent forms of TGF-β activity in synovial effusions. J Exp Med 1989; 169:291–296.
224. Younger JS, Salvin SB. Production and properties of migration inhibitory factor and interferon in the circulation of mice with delayed hypersensitivity. J Immunol 1973; 111:1914–1922.
225. Halloran PF. IFN-γ, prototype of the proinflammatory cytokines—importance in activation, suppression, and maintenance of the immune response. Transplant Proc 1995; 25:10–15.
226. Rosenbloom J, Feldman G, Fruendlich B, Jimenez SA. Inhibition of excessive scleroderma fibroblast collagen production by recombinant γ-interferon. Arthritis Rheum 1986; 29:851–856.
227. Jurado A, Carballido J, Griffel H, Hochkeppel HK, Wetzel GD. The immunomodulatory effects of interferon-gamma on mature B-lymphocyte responses. Experientia 1989; 45:521–536.
228. Dinarello CA, Mier JW. Current Concepts. Lymphokines. NEJM 1987; 317:940–945.
229. Trincheri B, Perussia B. Immune interferon: a pleiotropic lymphokine with multiple effects. Imunol Today 1985; 6:131–136.
230. Seeg PS, Moore RN, Johnson HM, Oppenheim JJ. Regulation of murine macrophage 1a antigen expression by a lymphokine with immune interferon activity. J Exp Med 1982; 156:1780–1793.
231. Williams JG, Jurkovich GJ, Maier RV. Interferon γ: a key immunoregulatory lymphokine. Surg Res 1993; 54:79–93.
232. Cotran RS. New roles for the endothelium in inflammation and immunity. Am J Pathol 1987; 129:407–413.
233. Thornhill MH, Haskard DO. IL-4 regulates endothelial cell activation by IL-1, TNF or IFN-γ. J Immunol 1990; 145:865–872.
234. Munn DH, Beall AC, Song D, Wrenn RW, Throckmorton DC. Activation-induced apoptosis in human macrophages: developmental regulation of a novel cell death pathway by macrophage colony-stimulating factor and interferon-γ. J Exp Med 1995; 181:127–136.
235. Caruso A, Folghera S, Martinelli F, Turano A. Natural antibodies to interferon-γ in humans: Inhibition of the biological activity of IFN-γ by human anti-IFN-γ antibodies. J Interferon Res 1994; 14:161–164.
236. Franchimont P, Reuter A, Vrindts-Gevaert Y, Bastings M, Malaise M, Sondag C, Frere M-C, Gysen P. Production of tumour necrosis factor-α, interferon-γ and interleukin-2 by peripheral blood mononuclear cells in subjects suffering from rheumatoid arthritis. Scand J Rheumatol 1988; 17:203–212.
237. Stolzenburg T, Binz H, Fontana A, Felder M, Wagenhaeuser F-J. Impaired mitogen-induced interferon γ production in rheumatoid arthritis and related diseases. Scand J Immunol 1988; 27:73–81.
238. Hovdenes J, Gaudernack G, Kvien TK, Hovdenes AB, Egeland T. Mitogen-induced interleukin 2 and gamma interferon production by CD4+ and CD8+ cells of patients with inflammatory arthritides. A comparison between cells from synovial fluid and peripheral blood. Scand J Immunol 1989; 30:597–603.
239. Hooks JJ, Haralampos M, Moutsopoulos MD, Geis SA, Stahl NI, Decker JL, Notkins AL. Immune interferon in the circulation of patients with autoimmune disease. N Engl J Med 1979; 301:5–8.
240. Cesario TC, Andrews BS, Martin DA, Jason M, Treadwell T, Friou G, Tilles JG. Interferon in synovial fluid and serum of patients with rheumatic disease. J Rheumatol 1983; 10:647–650.
241. Reuter A, Bernier J, Vrindts-Gevaert Y, Meuleman-Gathy R, Malaise M, Fiers W, Franchimont P. Production of interferon gamma by peripheral blood mononuclear cells from normal subjects and from patients with rheumatoid arthritis. Clin Exp Rheumatol 1988; 6:347–354.
242. Firestein GS, Zvaifler NJ. Peripheral blood and synovial fluid monocyte activation in inflammatory arthritis. Arthritis Rheum 1987; 30:864–871.
243. Firestein GS, Zvaifler NJ. Peripheral blood and synovial fluid monocyte activation in inflammatory arthritis. I. A cytofluorographic study of monocyte differentiation antigens and class II antigens and their regulation by γ-interferon. Arthritis Rheum 1987; 30:857–863.
244. Janossy G, Duke O, Poulter LW, Panayi G, Bofill M, Goldstein G. Rheumatoid arthritis: a disease of T lymphocyte-macrophage immunoregulation. Lancet 1981; 11:839–842.
245. Veys EM, Hermanns P, Schindler J, et al. Evaluations of T cell subsets with monoclonal antibodies in patients with rheumatoid arthritis. J Rheumatol 1982; 9:25–29.
246. Lotz M, Tsoukas CD, Fong S, Dinarello CA, Carson DA, Vaughan JH. Release of lymphokines after infection with Epstein Barr virus in vitro. II. A monocyte-dependent inhibitor of interleukin 1 downregulates the production of interleukin 2 and interferon-γ in rheumatoid arthritis. J Immunol 1986; 136:3643–3648.
247. Lotz M, Tsoukas CD, Curd JG, Carson DA, Vaughan JH. Effects of recombinant human interferons on rheumatoid arthritis B lymphocytes activated by Epstein-Barr virus. J Rheumatol 1987; 14:42–45.
248. Kitas GD, Salmon M, Allan IM, Bacon PA. The T cell system in rheumatoid arthritis: activated or defective? Scand J Rheumatol 1988; 76:161–173.
249. Lotz M, Tsoukas CD, Robinson CA, Dinarello CA, Carson DA, Vaughan JH. Basis for defective responses of rheumatoid arthritis synovial fluid lymphocytes to anti-CD3 (T3) antibodies. J Clin Invest 1986; 78:713–721.
250. Barada FA, O'Brien W, Horwitz DA. Defective monocyte cytotoxicity in rheumatoid arthritis. A correlation with disease activity and reversal by levamisole. Arthritis Rheum 1982; 25:10–16.
251. Firestein GS, Weisman MH, Zvaifler NJ. Parenteral IFN-γ in rheumatoid arthritis (RA): effect on disease activity and peripheral blood monocytes (PM) (abstr). Arthritis Rheum 1987; 30:S115.
252. Veys EM, Mielanats H, Verbruggen G, Grosclaude J-P, Meyer

W, Galazka A, Schindler J. Interferon gamma in rheumatoid arthritis—a double blind study comparing human recombinant interferon gamma with placebo. J Rheumatol 1988; 15:570–574.
253. Cannon GW, Pincus SH, Emkey RD, Denes A, Cohen SA, Wolfe F, Saway PA, Jaffer AM, Weaver AL, Cogen L, Schindler JD. Double-blind trial of recombinant γ-interferon versus placebo in the treatment of rheumatoid arthritis. Arthritis Rheum 1989; 32:964–973.
254. Hofschneider PH, Winter U, Lemmel EM, Brölz B, et al. Double blind controlled phase III multicenter clinical trial with interferon gamma in RA. Rheum Int 1992; 12:175–185.
255. Pernice W, Schuchmann L, Dippell J, Suschke J, Vogel P, Truckenbrodt H, Schincera F, Humburg CH, Brzoska J. Therapy for systemic juvenile rheumatoid arthritis with γ-interferon: a pilot study of nine patients. Arthritis Rheum 1989; 32:643–646.
256. LeRoy EC. Scleroderma (systemic sclerosis). In: Kelley WN, Harris ED, Ruddy S, Sledge CB, eds. Textbook of Rheumatology. Philadelphia: WB Saunders, 1985:1183–1205.
257. Jimenez SA, Freundlich B, Rosenbloom J. Selective inhibition of human diploid fibroblast collagen synthesis by interferons. J Clin Invest 1984; 74:1112–1116.
258. Kahan A, Amor B, Menkes CJ, Strauch G. Recombinant interferon-γ in the treatment of systemic sclerosis. Am J Med 1989; 87:273–277.
259. Gasson JC, Weisbart RH, Kaufman SE, et al. Purified human GM-CSF: direct action on neutrophils. Science 1984; 226:1339–1342.
260. Wong GG, Witek JS, Temple PA, et al. Human GM-CSF molecular cloning of the complementary DNA and purification of the natural and recombinant proteins. Science 1985; 228:810–815.
261. Sieff CA, Niemeyer CM, Mentzer SJ, Faller DV. Interleukin-1, tumor necrosis factor and the production of colony-stimulating factors by cultured mesenchymal cells. Blood 1988; 72:1316–1323.
262. Koeffler HP, Gasson J, Ranyard J, Souza L, Shepard M, Meinker R. Recombinant human tumor necrosis factor-α stimulates production of granulocyte-CSF. Blood 1987; 70:55–59.
263. Bagby GC Jr, Dinarello CA, Wallace P, Wagner C, Hefeneiders B, McCall E. IL-1 stimulates GM-CSF release by vascular endothelial cells. J Clin Invest 1986; 78:1316–1323.
264. Karshanskyi K, Lin N, Adamson JW. Interleukin-1 stimulates fibroblasts to synthesize GM-G-CSF mechanism for the hematopoietic response to inflammation. J Clin Invest 1988; 81:92–97.
265. Groopman JE, Molina JM, Scadden DT. Hematopoietic growth factors. N Engl J Med 1989; 321:1449–1459.
266. Gordon MY, Riley GP, Watt SM, Greaves MF. Compartmentalization of a hematopoietic growth (GM-CSF) by glycosaminoglycans in the bone marrow microenvironment. Nature 1987; 326:4033–4405.
267. Vadas MA, Varigos G, Nicola N, Pincus S, Dessein A, Metcalf D, Battye FL. Eosinophil activation by colony stimulating factor in man. Metabolic effects and analysis by flow cytometry. Blood 1983; 61:1232–1241.
268. Metcalf D, Begley LG, Johnson GR, Nicola NA, Vadas MA, Lopex AF, Williamson DG, Wong GG, Clark SC, Wang EA. Biologic properties in vitro of a recombinant human GM-CSF. Blood 1986; 67:37–45.
269. Gabrilove JL. Introduction and overview of hematopoietic growth factors. Semin Hem 1989; 26:1–4.
270. Xu WD, Firestein GS, Teatle R, Kaushansky K, Zvaifler NJ. Cytokines in chronic inflammatory arthritis. II. GM-CSF in rheumatoid synovial effusions. J Clin Invest 1989; 83:876–882.
271. Gasson JC, Weisbart RH, Kaufman SE, et al. Purified human GM-CSF: direct action on neutrophils. Science 1984; 226:1339–1342.
272. Weisbart RH, Golde DW, Clark SC, Wong GG, Gasson JC. Human GM-CSF is a neutrophil activator. Nature 1985; 314:361–363.
273. Fleischmann J, Golde DW, Weisbart RH, Gasson JC. GM-CSF enhances phagocytosis of bacteria by human neutrophils. Blood 1986; 68:707–711.
274. Alvaro-Garcia JM, Zvaifler NJ, Firestein GS. Cytokines in chronic inflammatory arthritis. IV. GM-CSF-mediated induction of class II MHC antigen on human monocytes: a possible role in rheumatoid arthritis. J Exp Med 198; 170:865–875.
275. Firestein GS, Zvaifler NJ. Down regulation of human monocyte differentiation antigens by gamma interferon. Cell Immunol 1987; 104:343–354.
276. Janeway C, Bottomly K, Babich J, Conrad P, Conzen S, Jones B, Kaye J, Katz M, McVay L, Murphy D, Tite J. Quantitative variation in 1a antigen expression plays a central role in immune regulation. Immunol Today 1985; 5:99–104.
277. Morrissey PJ, Bressler L, Park LS, Alpert A, Gillis S. GM-CSF augments the primary antibody response by enhancing the functions of antigen-presenting cells. J Immunol 1987; 139:1113–1119.
278. Heldin CH, Westermark B. Possible in vivo effect and clinical utility of platelet derived growth factor and PDGF antagonists. Transplant Proc 1993; 25:2074–2076.
279. Remmers EF, Lafyatis R, Kumkumian GK, Case JP, Roberts AB, Sporn MB, Wilder RL. Cytokines and growth regulation of synoviocytes from patients with rheumatoid arthritis and rats with streptococcal cell wall arthritis. Growth Factors 1990; 2:179–188.
280. Remmers EF, Sano H, Wilder RL. Platelet-derived growth factors and heparin-binding (fibroblast) growth factors in the synovial tissue pathology of RA. Sem A&R 1991; 21:191–199.
281. Rubin K, Terracio L, Ronnstrand L, et al. Expression of Platelet-derived growth factor receptors is induced on connective tissue cells during chronic synovial inflammation. Scand J Immunol 1988; 27:285–294.
282. Hart CE, Bowen-Pope DF. PDGF receptor current views of the two-subunit mode. J Invest Dermatol 1990; 94(suppl):535–573.
283. Remmers EF, Sano H, Lafyatis R, et al. Production of PDGF B chain (PDGF-β/c-sis) mRNA and immuno reactive PDGF-β like polypeptide by rheumatoid synovium. Coexpression with heparin binding acidic fibroblast growth factor-1. J Rheumatol 1991; 18:7–13.
284. Lafyatis R, Remmers EF, Roberts AB, et al. Anchorage-independent growth of synoviocytes from arthritic and normal joints: stimulation by exogenous platelet-derived growth factor and inhibition by transforming growth factor-beta and retinoids. JCI 1989; 83:1267–1276.
285. Ross R, Bowen-Pope DF, Raines EW. Platelet-derived growth factor and its role in health and disease. Phil Trans R Soc Lond B 1990; 327:155–169.
286. Ross R, Raines EW, Bowen-Pope DF. The biology of platelet-derived growth factor. Cell 1986; 46:155–169.
287. Fellstrom B, Klareskog L, Heldin CH, et al. Platelet-derived growth factor receptors in the kidney-upregulated expression in inflammation. Kidney Int 1989; 36:1099–1102.
288. Fellstrom B, Dimeny E, Larsson E, et al. Importance of PDGF receptor expression in accelerated atherosclerosis—chronic rejection. Transplant Proc 1989; 21:3689–3691.

289. Gay S, Jones RE Jr, Huang G-Q, et al. Immunohistologic demonstration of platelet-derived growth factor (PDGF) and sis-oncogene expression in scleroderma. J Invest Dermatol 1989; 92:301–303.
290. Duan D-Sr, Pazin MJ, Fretto LJ. A functional soluble extracellular region of the platelet-derived growth factor (PDGF) beta-receptor antagonizes PDGF-stimulated responses. J Biol Chem 1991; 266:413–418.
291. Bryckaert MC, Eldor A, Fontenay M. Inhibition of platelet-derived growth factor-induced mitogenesis and tyrosine kinase activity in cultured bone marrow fibroblasts by tyrophosphotins exp. Cell Res 1992; 199:255–261.
292. Coley WB. Contribution to the knowledge of sarcoma. III. The treatment of sarcoma by innoculation with erysipelas, with a report of three recent cases. Ann Surg 1981; 14:199–220.
293. Coley WB. The treatment of malignant tumors by repeated innoculation of erysipelas: with a report of ten original cases. Am J Med Sci 1983; 105:487–511.
294. Coley WB. The therapeutic value of the mixed toxins of the streptococcus of erysipelas and bacillus prodigiosus in the treatment of inoperable malignant tumors with a report of one hundred sixty cases. Am J Med Sci 1986; 112:251.
295. Carswell EA, Old LJ, Kassel RL, Green S, Fiore N, Williamson B. An endotoxin-induced serum factor that causes necrosis of tumors. Proc Natl Acad Sci USA 1975; 72:3666–3670.
296. Kunkel SL, Remick DG, Strieter RM, Larrick JW. Mechanisms that regulate the production and effects of tumor necrosis factor-α. Crit Rev Immunol 1989; 9:93–117.
297. Aggarwal BB, Essalu TE, Hass PE. Characterization of receptors for human tumor necrosis factor and their regulation by interferon. Nature 1985; 318:665–667.
298. Beutler BA, Milsark IW, Cerami A. Cachectin/tumor necrosis factor: production, distribution and metabolic fate in vivo. J Immunol 1985; 135:3972–3977.
299. Beutler BA, Cerami A. Cachectin and tumor necrosis factor as two sides of the same biological coin. Nature 1986; 320:584–588.
300. Beutler BA, Cerami A. The endogenous mediators of endotoxic shock. Clin Res 1987; 35:192–197.
301. Duff G, DiGiovine F, Dickens E, Symons J, Wood D, Manson J. Tumor necrosis factor and interleukin-1 in arthritis. Immunobiology 1987; 175:10–11.
302. DiGiovine FS, Nuke G, Duff GW. Tumor necrosis factor in synovial exudates. Ann Rheum Dis 1988; 47:768–772.
303. Saxne T, Palladino MA, Heinegard D, Tala N, Wollheim FA. Detection of tumor necrosis factor-α but not tumor necrosis factor-β in rheumatoid arthritis synovial fluid and serum. Arthritis Rheum 1988; 31:1041–1045.
304. Yoke K, Suzuki J, Tsunekawa N, et al. Detection of TNF in blood from patients with Kawasaki syndrome. International Conference on TNF and related cytokines. Immunobiology 1987; 175:D24, 106.
305. Andrews BS, Mitrick G, Sarkozi J, Yamamoto RS, Friou G. Serum from patients with scleroderma contains lymphotoxin and tumor necrosis factor-α (abstr). Arthritis Rheum 1988; 59.
306. Dinarello CA, Cannon JG, Wolff SM, Bernheim HA, Beutler BA, Cerami A, Figari IS, Palladino MA, O'Connor JV. Tumor necrosis factor is an endogenous pyrogen and induces production of interleukin-1. J Exp Med 1986; 163:1433–1450.
307. Kawakami M, Pekala PH, Lane MD, Cerami A. Lipoprotein lipase suppression in 3T3-L1 cells by an endotoxin-induced mediator from exudate cells. Proc Natl Acad Sci USA 1982; 79:912–916.
308. Tracey KJ, Wei H, Manogue KR, Fong Y, Hesse DG, Nguyen HT, Kuo CG, Butler B, Cotran RS, Cerami A, et al. Cachectin/TNF induces cachexia, anemia and inflammation. J Exp Med 1988; 167:1211–1227.
309. Bertolini DR, Nedwin GE, Bringman TS, et al. Stimulation of bone resorption and inhibition of bone formation in vitro by human tumor necrosis factor. Nature 1986; 319:515–518.
310. Stashenko P, Dewhirst FE, Peros WJ, et al. Synergistic action between interleukin-1, tumor necrosis factor and lymphotoxin in bone resorption. J Immunol 1987; 138:1464–1468.
311. Tashjian AH, Voelkel EF, Lazzaro M, et al. Tumor necrosis factor-α (cachectin) stimulates bone resorption in mouse calvaria via a prostaglandin-mediated mechanism. Endocrinology 1987; 120:2029–2036.
312. Andrews BS, Mirick G, Friou GJ, et al. Effect of colchicine and d-penicillamine on interleukin-1 (IL-1) and IL-1 inhibitors production by human peripheral blood mononuclear cells (PBMC). Arthritis Rheum 1988; 31:S114.
313. Thomson BM, Mundy GR, Chambers TJ. Tumor necrosis factors-α and β induce osteoblastic cells to stimulate osteoclastic bone resorption. J Immunol 1987; 138:775–779.
314. Dayer JM, DeRochemonteix B, Burrus B, Demczuk S, Dinarello CA. Human recombinant IL-1 stimulates collagenase and prostaglandin E_2 production by human synovial cell. J Clin Invest 1986; 77:645–648.
315. Dayer JM, Beutler B, Cerami A. Cachectin/tumor necrosis factor stimulates collagenase and prostaglandin E_2 production by human synovial cells and dermal fibroblasts. J Exp Med 1985; 162:2163–2168.
316. Sato K, Kasono K, Fujii YK, Kawakami M, Tsushima T, Shizume K. Tumor necrosis factor-α (cachectin) stimulates mouse osteoblast-like cells (MC 3T3-E1) to produce macrophage colony stimulating activity and prostaglandin E_2. Biochem Biophys Res Commun 1987; 145:323–329.
317. Butler DM, Vitti GF, Leizer T, Hamilton JA. Stimulation of the hyaluronic acid levels of human synovial fibroblasts by recombinant human tumor necrosis factor-α, tumor necrosis factor-β (lymphotoxin), IL-1α and IL-1β. Arthritis Rheum 1988; 31:1281–1289.
318. Engstrom-Laurent A, Hallgren R. Circulating hyaluronate in rheumatoid arthritis: Relationship to inflammatory activity and the effect of corticosteroid therapy. Ann Rheum Dis 1985; 44:83–88.
319. Dahl IM, Husby G. Hyaluronic acid production in vitro by synovial lining cells from normal and rheumatoid joints. Ann Rheum Dis 1985; 44:647–657.
320. Hamilton JA. Hypothesis: in vitro evidence for the invasive and tumor-like properties of the rheumatoid pannus. J Rheumatol 1983; 10:845–861.
321. Hiro D, Ito A, Matsuto K, Mori V. Hyaluronic acid is an endogenous inducer of IL-1 production by human monocytes and rabbit macrophages. Biochem Biophys Res Commun 1986; 140:715–722.
322. Elias JA, Gustilo K, Baeder W, Freundlich B. Synergistic stimulation of fibroblast prostaglandin production by recombinant IL-1 and tumor necrosis factor. J Immunol 1987; 138:3812–3816.
323. Stashenko P, Dewhirst FE, Peros WJ, Kent RL, Ago JM. Synergistic interactions between IL-1, tumor necrosis factor and lymphotoxin in bone resorption. J Immunol 1987; 138:1464–1468.
324. Firestein GS, Zvaifler NJ. The pathogenesis of rheumatoid arthritis: a critical assessment of the role of autologous stimulation in the perpetuation of rheumatoid synovitis. Rheum Dis Clin North Am 1987; 13:447–461.
325. Shinmei M, Masuda K, Kikuchi T, Shimomura Y. Interleukin-

1, tumor necrosis factor and interleukin-6 as mediators of cartilage distruction. Semin Arthritis Rheum 1989; 18:27–32.
326. Chu CQ, Field M, Feldmann M, Maini RN. Localization of tumor necrosis factor γ in synovial tissues and at the cartilage-pannus junction in patients with rheumatoid arthritis. Arthritis Rheum 1991; 34:1125–1132.
327. Cope AP, Aderka D, Doherty M, Engelmann H, Gibbons D, Jones AC, Brennan FM, Maini RN, Wallach D, Feldmann M. Increased levels of soluble tumor necrosis factor receptors in the sera and synovial fluids of patients with rheumatic diseases. Arthritis Rheum 1992; 35:1160–1169.
328. Nawroth PP, Stern DM. Modulation of endothelial cell hemostatic properties by tumor necrosis factor. J Exp Med 1986; 163:740–745.
329. Shalaby MR, Aggarwal BB, Rinderknecht E, Svedersky LP, Finkle BS, Palladino MA. Activation of human polymorphonuclear neutrophil functions by interferon-γ and tumor necrosis factors. J Immunol 1985; 135:2069–2073.
330. Keffer J, Probert L, Cazlaris H, Georgopoulos S, Kaslaris E, Kioussis D, Kollias G. Transgenic mice expressing human tumor necrosis factor: a predictive genetic model of arthritis. EMBO J 1991; 10:4025–4031.
331. Brahn E, Peacock DJ, Banquerigo ML, Liu DY. Effects of tumor necrosis factor alpha (TNF-α) on collagen arthritis. Lymphokine Cytokine Res 1992; 11:253–256.
332. Cooper WO, Fava RA, Gates CA, Cremer MA, Townes AS. Acceleration onset of collagen induced arthritis by intraarticular injection of tumour necrosis factor of transforming growth factor-beta. Clin Exp Immunol 1992; 898:244–250.
333. Williams RO, Feldmann M, Maini RN. Anti-tumor necrosis factor ameliorates joint disease in murine collagen-induced arthritis. Proc Natl Acad Sci USA 1992; 89:9784–9788.
334. Wooley PH, Dutcher J, Widmer MB, Gillis S. Influence of a recombinant human soluble tumor necrosis factor receptor-Fc fusion protein on type II collagen-induced arthritis in mice. J Immunol 1993; 151:6602–6607.
335. Elliot MJ, Maini RN, Feldmann M, Long-Fox A, Charles P, Katsikis P, Brennan FM, Walker J, Bij LH, Ghrayeb J, Woody JN. Treatment of rheumatoid arthritis with chimeric monoclonal antibodies to tumor necrosis factor α. Arthritis Rheum 1993; 36:1681–1690.
336. Elliot MJ, Maini RN, Feldmann M, Kalden JR, Antoni C, Smolen J, Leeb B, Breedveld EC, Macfarlane JD, Bijl JA, Woody JN. TNFα blockade in rheumatoid arthritis: results of a multi-center, placebo-controlled, double-blind, trial of chimeric anti-TNF-α monoclonal antibody (cA2) in refractory disease (abstract). Arthritis Rheum 1994; 37:5283.
337. Rankin ECC, Choy EHS, Kassimos D, Sopwith M, Kingsley G, Isenberg DA, Panayi GS. A double-blind, placebo-controlled, ascending dose trial of the recombinant humanised anti-TNFα antibody CDP571 in patients with rheumatoid arthritis: a preliminary report (abstract). Arthritis Rheum 1994; 37:8295.
338. Elliot MJ, Maini RN, Feldmann M, Long-Fox A, Charles P, Bij LH, Woody JN. Repeated therapy with monoclonal antibody to tumour necrosis factor alpha in patients with rheumatoid arthritis. Lancet 1994; 344:1125–1127.

31
Hematological Effects of Cytokines

William P. Sheridan
Amgen, Inc., Thousand Oaks, and UCLA School of Medicine, Los Angeles, California

Pamela Hunt, Scott Simonet, and Thomas R. Ulich
Amgen, Inc., Thousand Oaks, California

INTRODUCTION

Cytokines exert a wide range of hematological effects on white and red blood cells in the bone marrow and in the peripheral circulation. The cytokines whose in vivo hematological effects will be reviewed in this chapter can be broadly classified as the classical colony-stimulating factors (multi-CSF, GM-CSF, G-CSF, M-CSF, erythropoietin, and thrombopoietin), the interleukins, the tumor necrosis factors, and the interferons. A complete understanding of the hematological effects of cytokines is only possible through the complementary study of both their in vitro and in vivo effects. The interpretation of an exogenously administered cytokine's hematological effects in vivo are unfortunately always clouded by the inability to discern between direct and indirect effects. Most cytokines are well known to induce the release of other cytokines so that the indirect hematological effects of exogenously administered cytokines will be of great significance. As monoclonal antisera to the endogenously released cytokines of experimental animals become more readily available, these indirect effects of the "cytokine cascade" will be deciphered. In the meantime, in vitro assays of cytokines on relatively purified bone marrow cell subpopulations provide an important means of dissecting the direct and indirect effects of cytokines on hematopoietic precursors and accessory cells.

Despite the limitations of in vivo experiments, the in vivo study of cytokines provides information that cannot be obtained in vitro. Margination or demargination of peripheral leukocytes, a process that may involve the cytokine-induced expression of leukocyte- or endothelial-adhesion molecules, can only be studied in vivo. In this regard, one must remember that the number of blood cells as enumerated by phlebotomy represents the circulating number of peripheral blood cells and, because of the existence of a substantial marginated pool, may in some situations (such as paradoxical leukopenias) not be representative of the total number of intravascular cells (1,2). The release of mature marrow cells such as neutrophils or reticulocytes from the marrow into the circulation can also not be appreciated in vitro. A cytokine's marrow cell–releasing activity and hemoproliferative activity may be functionally reciprocal in that the continued presence of mature cells in the microenvironment of stem cells may initiate negative feedback mechanisms that would disappear in vivo with the release of the mature cells into the circulation. The proliferative effects of cytokines are kinetically different in vivo and in vitro. Proliferative effects in vivo as judged by marrow smears are usually apparent within hours, whereas the results of in vitro colony-forming assays are assessed after days or weeks. The microenvironments of the actual marrow and of bone marrow culture systems are clearly quite different and thus the in vivo hematological effects of cytokines as determined by the sum total of all of a cytokine's direct and indirect effects (permissive, additive, synergistic, or antagonistic) can only be judged by direct examination of the blood and marrow. The catabolism and consequently the half-lives of growth factors in vivo and in vitro must be disparate due to the different mechanisms of clearance in vivo (e.g., renal clearance) with a presumably more prolonged exposure of hematopoietic cells to growth factors in vitro. Finally, the spatial limitations imposed on hematopoiesis by the marrow's microanatomy and by the confines of the osseous cortex are not reproduced in vitro, so that the hierarchical expansion of one marrow cell line in preference to another may not be appreciated in vitro. The study of the hematological effects of cytokines in vivo thus provides a complementary perspective to in vitro investigations.

This chapter, as previously mentioned and unlike others in this book, will be largely limited to the description of the hematological effects of the therapeutic administration of exogenous cytokines. The focus will be on nonmyelosuppressed hosts. The clinical applications of G-CSF, GM-CSF, and erythropoietin will be briefly mentioned. The in vivo role of hemopoietic negative regulators (3), such as transforming growth factor-β, must be mentioned here as a subject that will no doubt grow rapidly in the next decade, but is as of now still largely uncharted. The role of endogenous cytokine release under physiological and pathophysiological conditions is another important facet of the in vivo hematological effects of cytokines that can as of today only be touched upon sketchily. Cytokines no doubt play an important role in normal hematopoiesis as well as during the profound hematological changes noted during endotoxemia and during many infectious and inflammatory diseases. As the measurement of cytokines in serum and in inflammatory effusions becomes clinically more routine, the roles of cytokines in recruiting leukocytes from the marrow into the circulation and inflammatory sites will become more apparent. One of the insights that assays of cytokine levels in the clinic will provide is the knowledge of the pathophysiological (as opposed to the pharmacological) range of cytokine concentrations that are actually achieved in vivo in various diseases. The hematological effects noted after the administration of pharmacological amounts of cytokines may easily be misinterpreted as also being relevant to the effects of endogenously released cytokines during disease states. In this regard one should note that the hematological effects of many cytokines are very dose-dependent as well as being dependent upon the route and frequency of administration. Since the kinetics, magnitude, and sources of the endogenous release of many cytokines under both physiological and pathophysiological conditions are still largely unexplored, caution must be exerted in extrapolating the results of pharmacological experiments toward an understanding of physiological and clinical hematological observations.

THE CLASSICAL COLONY-STIMULATING FACTORS AND RELATED CYTOKINES

IL-3 or Multi–Colony-Stimulating Factor

Multi-CSF, IL-3, is a growth factor acting on proximal hematopoietic stem cells as evidenced by its in vitro direct and indirect effects on cells of erythroid, neutrophil, eosinophil, macrophage, mast cell, and megakaryocytic lineages (4,5). In vivo, multi-CSF has been reported by several groups of investigators (6–10) to induce a number of hematological effects, although the effects on circulating numbers of peripheral blood cells have been generally much less marked compared to the effects of the other colony-stimulating factors or many other cytokines. In our experience (6,7), human multi-CSF injected IV as a single injection in rats induces a mild neutrophilia and monocytosis, peaking at 8 h. The peripheral neutrophilia at 8 h is accompanied by a slight decrease in marrow neutrophils and by a left-shifted myeloid hyperplasia characterized by a striking increases in myeloblasts. The marrow also demonstrates a slight increase in pronormoblasts. The peripheral blood and marrow at 24 h after injection are unremarkable. The effects of IL-3 did not demonstrate any significant dose-response relationship over a large range of doses. In vitro culture of rat marrow with human IL-3 did not promote colony formation. On the other hand, culture of rat marrow with human G-CSF also did not result in colony formation, suggesting that the in vitro action of cytokines on marrow cultures is not an absolute predictor of a cytokine's in vivo effects since human G-CSF is known to induce a dose-dependent neutrophilia and myeloid hyperplasia in vivo in the rat.

Metcalf et al. (8) studied the effects of homologous multi-CSF in the mouse and reported the occurrence of two- to threefold increases in circulating neutrophils and monocytes, and 10-fold increases in peripheral eosinophil numbers after repeated daily injections for up to 6 days. Splenomegaly was observed due to increases in the numbers of nucleated erythroid forms, granulocytes, megakaryocytes, and up to 100-fold rises in mast cells. No significant changes were seen in the marrow. Kindler et al. (9) administered multi-CSF via continuous infusion for 3–7 days in a murine model and noted enhancement of the number of myeloid and erythroid precursors in the spleen concomitantly with a decrease in progenitors in the marrow as measured by in vitro assay. The authors postulated that progenitor cells may be stimulated by multi-CSF to leave the marrow and colonize the spleen. Kimoto et al. (10) examined the effect of IL-3 on lymphopoiesis after administration of the cytokine to mice by osmotic minipumps for 1–4 weeks. Although a marked splenomegaly was described due to the accumulation of hemopoietic precursors, no increase in either T or B cells was seen in lymphoid organs. Lord et al. (11) noted that the proliferation rate of CFU-S is rapidly increased after the administration of IL-3 to mice. A similar effect was observed by the same authors after the administration of endotoxin. Since IL-3 was not detectable in the blood of endotoxin-treated mice, the hypothesis was forwarded that lipopolysaccharide (LPS) might act indirectly on blood cell precursors by eliciting the local production of IL-3. In this context, our laboratory has observed a similarity in the effect of intravenously administered multi-CSF and LPS on the proliferation of myeloblasts in rat marrow smears (6).

Multi-CSF has been reported to synergize with several other growth factors in vitro in colony-forming assays (12–14). In vivo, Donahue et al. (15) showed that multi-CSF and GM-CSF acted synergistically to promote hematopoiesis in primates that had been primed with multi-CSF prior to receiving GM-CSF. The coinjection of multi-CSF and G-CSF in rats induced only an additive peripheral neutrophilia, an effect that was generally not statistically significant because of the much greater magnitude of the G-CSF– as opposed to the multi-CSF–induced neutrophilia (7). On the other hand, the combination of multi-CSF and G-CSF did induce a synergistic myeloid hyperplasia in the marrow after daily injections for 1 week. The combination of multi-CSF and IL-6, in addition to its reported synergistic effects on progenitors in vitro (12), has been reported to induce a synergistic increase in marrow mast cells in vivo (6), a finding that may be of relevance to the observation that IL-3 in the presence of pokeweed mitogen spleen cell–conditioned medium supports the expansion of mast cells in vitro (16).

Granulocyte-Macrophage Colony-Stimulating Factor

GM-CSF in vitro, in addition to supporting the growth of myeloid cells, can cause the growth of erythroid precursors and multipotential cells under appropriate assay conditions (17). GM-CSF in vivo induces peripheral neutrophilia, monocytosis, and eosinophilia in rodents (18,19), nonhuman primates (20,21), and in humans (22,23).

Metcalf et al. (18) reported that mice injected intraperitoneally with murine GM-CSF three times daily for six days developed a two-fold increase in blood neutrophils as the only significant change in the peripheral circulation. In the peritoneal cavity, up to 15-fold increase in macrophages, a 100-fold increase in neutrophils, and a 100-fold increase in eosinophils was noted. The peritoneal macrophages also exhibited increased phagocytic activity for antibody-coated erythrocytes. Increased numbers of neutrophils and monocytes were observed in the liver and lung.

A single IV injection of murine GM-CSF in rats (19) causes a brief initial neutropenia and monocytopenia reaching a nadir at 15 min and followed by neutrophilia and monocytosis peaking between 4 and 8 h and subsiding to baseline levels by 12 h. The marrow at 6 h shows a decrease in mature neutrophils and a slight left-shifted myeloid hyperplasia. The marrow differential at 24 h is unremarkable. GM-CSF also causes a mild peripheral lymphopenia. GM-CSF–induced lymphopenia may not reflect a direct effect on lymphocytes, but rather may represent GM-CSF–induced gene expression of IL-1 or other cytokines that are known to induce lymphopenia. Daily injection of GM-CSF in rats for 1 week induced a repetitive daily neutrophilia with no evidence of either the development of tolerance or of any cumulative increase in the number of circulating neutrophils.

Donahue et al. (20) reported that continuous infusion of GM-CSF in rhesus macaque monkeys induced an increase in circulating neutrophils, eosinophils, monocytes, and lymphocytes, as well as a modest increase in circulating reticulocytes. GM-CSF in rhesus monkeys was similarly reported by Mayer et al. (21) to steadily increase total circulating leukocyte counts with fivefold increases in granulocytes, two- to fourfold increases in lymphocytes, and three- to fourfold increases in monocytes.

GM-CSF was administered for 3 days by Aglietta et al. (23) to patients who had normal bone marrows. Peripheral neutrophilia, eosinophilia, and monocytosis was noted as well as an increase in morphologically recognizable granulomonopoietic precursors in posttreatment marrow smears. The left-shifted myeloid hyperplasia in the marrow resulted in a relative decrease in marrow lymphocytes, but no peripheral lymphopenia occurred. As an interesting correlate to the previously mentioned in vitro effect of GM-CSF on erythroid precursors, GM-CSF in vivo caused an increase in the percentage of marrow BFU-E in S phase, although no accompanying increase in marrow normoblasts was morphologically recognizable (23).

GM-CSF in vivo also activates circulating neutrophils as evidenced by their enhanced oxidative metabolism and killing of bacteria (21) and by the induction of IgA-mediated phagocytosis (25). GM-CSF has been observed to act synergistically with other growth factors such as IL-3, G-CSF, and M-CSF in vitro (26–28). In vivo, GM-CSF acted synergistically with IL-3 when primates were primed with IL-3 for several days before receiving GM-CSF (15). GM-CSF coinjected with G-CSF in rats as either a single injection or as daily injections for a week induced an approximately additive increase in peripheral neutrophils (19).

The range of biological effects of GM-CSF noted in animals is wider than for G-CSF, and this difference has been confirmed in human clinical trials. Administration of GM-CSF causes acute transient leukopenia affecting neutrophils, eosinophils, and monocytes, followed by leukocytosis with neutrophilia of mature and immature forms, monocytosis, lymphocytosis, and eosinophilia (29). A first-dose effect probably related to release of inflammatory mediators from monocytes/macrophages was noted in phase I studies, especially in patients with lung cancer (29,30). This syndrome can include fever, dyspnea, hypoxia, and hypotension and tends to recur with the first dose in each course of therapy in susceptible patients (30). In addition to bone pain, GM-CSF administration is associated in a proportion of patients with fever, injection site inflammation, maculopapular rash, and serositis (29). The dose range effective at elevating the neutrophil count appears to overlap that associated with significant adverse events (29). Other effects of GM-CSF include inhibition of transmigration of leukocytes from the vascular space to the interstitium (31) and an increase in the incidence of vascular catheter occlusion and associated thrombosis when given by continuous IV infusion (32).

Clinical trials of GM-CSF have demonstrated a reduction in the period of severe neutropenia after autologous BMT for lymphoid malignancies (33) and mobilization of progenitor cells into the blood, similar to but weaker than that seen with G-CSF (34,35).

The primary physiological roles of GM-CSF remain unclear despite intense investigation. In mice engineered to be totally deficient in GM-CSF, the most prominent phenotypical finding is development of alveolar proteinosis and chronic pulmonary infection, suggesting that an important role of GM-CSF in vivo is to stimulate tissue macrophage phagocytic activity (36).

Granulocyte Colony-Stimulating Factor

G-CSF supports the proliferation and differentiation of neutrophils in vitro (37) and induces neutrophilia in rodents (38,39), nonhuman primates (40), and humans (24) in vivo. A single injection of G-CSF causes an initial peripheral neutropenia at 15 min in rats (39). A similar transient early neutropenia occurs in humans. The mechanism of G-CSF–induced neutropenia is unknown, but may conceivably involve G-CSF–induced activation of circulating neutrophils with resultant expression of neutrophil adhesion molecules that would promote systemic margination of neutrophils. G-CSF–induced neutropenia is followed by neutrophilia beginning at 30 min and peaking at 12 h in rats or hamsters. The number of circulating neutrophils in rats subsides to baseline between 30 and 36 h, but the morphology of the neutrophils at these time points is remarkable for the hypersegmentation of their nuclei. Hypersegmentation of neutrophils has also been reported in humans receiving G-CSF (200). The marrow of G-CSF–treated rats at 4 and 12 h exhibits an increase in number, mitoses, size, and cytoplasmic granulation of myeloblasts and promyelocytes but a decrease in mature neutrophils, demonstrating that G-CSF acts not only as a mitogen and growth factor for early cells in the myeloid series but also as a releasing factor for mature marrow neutrophils. The bone marrow at 48 h contains slightly increased numbers of myelocytes and metamyelocytes and large numbers of hypersegmented neutrophils. Some of the hypersegmented neutrophils contain nuclear drumsticklike protrusions (39).

Chronic administration of G-CSF to cynomolgus monkeys for 2–4 weeks induces a dose-dependent increase in circulating neutrophils that plateaus after 1 week and is thereafter sustained (40). In rats, single daily IV injections of a constant dose of G-CSF induces an initial large peripheral neutrophilia that is followed on days 2–4 by a decrease in the number of circulating neutrophils and on days 5–7 with an increase in circulating neutrophils (41). The mechanism of the transient trough in neutrophil numbers during daily injection of G-CSF might reasonably be postulated to be due to the marrow-releasing effect of G-CSF and a resultant depleted marrow reserve of neutrophils. The marrow after 3 days of G-CSF injections, however, contains over twice as many mature neutrophils as control or time zero marrows. The trough in peripheral neutrophilia might therefore be postulated to be due to an as yet unidentified inhibitor of neutrophil release, although other mechanisms of tolerance are possible. A similar trough in peripheral neutrophil numbers after daily injections of G-CSF has not been reported in primates or humans. The marrow at 24 h after a week of daily injections of G-CSF shows a striking myeloid hyperplasia at the maturational level of mature neutrophils with little or no increase in numbers of myeloid precursors.

Many of the effects of exogenously administered G-CSF noted in animals have been confirmed in clinical trials (42,43). Injection of G-CSF, by either SC or IV infusion or SC injection, causes an acute transient fall in total white cell count and neutrophil count, followed within hours by a sustained leukocytosis due to neutrophilia (44,45). The acute fall in neutrophil count recurs after each injection, is not associated with clinical symptoms, and is probably due to margination and extravasation of neutrophils (44). The neutrophilia is dose-related (44,45). With continued injection of G-CSF, the peripheral blood neutrophil count stabilizes at a new, high level dependent on the dose used (44,45). In children with severe chronic neutropenia, chronic G-CSF administration can maintain normal neutrophil levels for years, apparently indefinitely (46). G-CSF directly stimulates granulocytic progenitor cell division at all stages and shortens the time required to produce mature neutrophils from precursor cells in vivo by accelerating maturation and release (47).

The bone marrow shows myeloid hyperplasia with an increase in the myeloid:erythroid ratio, the proportion of immature neutrophils in the blood increases, and in addition large forms (polymacrocytes) are produced (44). Effects of G-CSF on other white cell lineages are weak, with mild increases in lymphocytes and monocytes noted (45).

Peak serum concentrations of G-CSF are directly related to dose, and elimination after single doses follows first-order kinetics. Absorption from subcutaneous injection sites is rapid and levels of ≥ 10 ng/ml are sustained for 8–16 h after SC injection of 5 μg/kg (44). Following onset of neutrophil leukocytosis, G-CSF serum levels fall despite continued administration, probably due to receptor binding and cellular clearance by circulating neutrophils (48).

An important effect of G-CSF first noted in phase I clinical trials is mobilization of a wide range of hemopoietic progenitor cells from the marrow to the peripheral blood (45,49). This phenomenon has been exploited as a means of obtaining transplantable hemopoietic stem cells (50,51) and has been confirmed by numerous investigators (52). The mechanism of PBPC mobilization by G-CSF is poorly understood, but interestingly this general characteristic is one that is now known to be shared by many of the hemopoietic cytokines (52). G-CSF appears to be the most potent PBPC-mobilizing agent of the various cytokines tested.

G-CSF has found a major clinical role in the supportive care of patients undergoing cancer chemotherapy (43), for the primary prophylaxis of severe neutropenia. The period of severe neutropenia is shortened, the depth of neutropenia lessened, and both the onset of the neutrophil nadir and the recovery of normal neutrophil levels are accelerated by daily subcutaneous injection of G-CSF, beginning after the completion of chemotherapy (44,53,54). These effects translate into the clinical benefit of a reduction in the proportion of patients with complicating infections (55,56). An additional benefit observed in phase I and II studies (54,57) is a reduction in severity and duration of mucositis. There are also benefits from the administration of G-CSF, with antibiotics, once a neutropenic infection has developed (58), but the magnitude of the benefit is less and it is clear that the most appropriate clinical use in this setting is prophylactic.

Chemotherapy or chemo-radiotherapy regimens used prior to allogeneic or autologous bone marrow transplantation lead to more severe and prolonged periods of pancytopenia than conventional cancer chemotherapy. Nearly all patients undergoing these procedures will develop an infection during the neutropenic period, usually within a few days of the onset of severe neutropenia. Administration of G-CSF after autologous BMT shortens the duration of neutropenia and fever and reduces the use of IV antibiotic administration and the length of hospital stay (57,59,60). Similar benefits have been noted after allogeneic BMT (61). A reduction in the use of parenteral nutrition was noted in early trials in BMT (57) and was thought to be due to effects of early neutrophil recovery on the gastrointestinal mucosa. The normal migration to the mucosa of neutrophils produced after G-CSF stimulation following BMT is detectable a few days in advance of a rise in peripheral blood levels (62).

Patients with severe chronic neutropenia (SCN) suffer from recurrent mouth ulcerations, skin infections, and episodes of bacteremia, pneumonia, and other infections (63). Clinical trials of G-CSF in these disorders have demonstrated persistent increase in neutrophil counts and associated reduction in all of the infectious complications of the neutropenia, transforming the treatment of this disease (64). The pathophysiology of SCN is heterogeneous, and a minority of patients fail to respond to G-CSF or develop resistance during therapy. The cause of this resistance to G-CSF is unclear, but no patients have had a detectable antibody response (63). Other clinical trials of G-CSF have included studies of effects in the neutropenia associated with AIDS (65,66), hairy cell leukemia (67,68), drug-associated agranulocytosis (69), and nonneutropenic infection (70).

The physiological role of G-CSF in normal regulation of granulopoiesis has recently been underscored by studies in the G-CSF knock-out mouse. Homozygous G-CSF –/– mice have circulating neutrophil levels only 20–30% of normal and are less able to withstand infectious challenge with *Listeria monocytogenes* (71).

Adverse effects of G-CSF in cancer patients have been few, with the most frequent being bone pain (72). This is thought to be due to the potent stimulation of myeloid activity and is more frequent in the setting of G-CSF administration without chemotherapy compared to G-CSF administration after chemotherapy (51). In patients with severe chronic neutropenia treated chronically with G-CSF, other adverse events observed in a small proportion of patients were anemia, splenomegaly, thrombocytopenia, and injection site reactions. The incidence of allergic-type reactions is less than 1 in 4000, and rarely (less than 1 in 7000) G-CSF is associated with cutaneous vasculitis (72).

Macrophage Colony-Stimulating Factor

M-CSF in vitro stimulates the proliferation of macrophages and synergizes in regard to this function with multi-CSF, G-CSF, and TNF (73–75). M-CSF also activates cells of the monocyte-macrophage system as evidenced by the enhancement of macrophage tumoricidal activity (76) and by the induction of monocyte migration (77) in vitro. M-CSF in vivo causes peripheral monocytosis in mice (78–80), rats (81), and primates (82). Human M-CSF is active in vivo in all these species. The increased number of monocytes in the circulation of M-CSF–treated animals may contribute to an increase in reticuloendothelial cells or else M-CSF may cause proliferation of reticuloendothelial cells as evidenced by the increase in Kupffer cells in M-CSF–treated mice (79). In mice, Koren et al. (78) reported a slight monocytosis, neutrophilia, and lymphopenia after 4 days of M-CSF administration. Also in mice injected with M-CSF for 4 days, Hume et al. (79) reported a 10-fold increase in monocytes, as well as a lymphopenia. Broxmeyer et al. (80) also noted an increase in circulating monocytes in M-CSF–treated mice. In rats, a single injection of M-CSF was observed in our laboratory (81) to induce a mild peripheral neutrophilia and lymphopenia between 2 and 16 h after injection and a monocytosis beginning at 20–24 h, peaking at 28–32 h, and subsiding to baseline levels by 36–48 h. The M-CSF–induced monocytosis was postulated to represent a direct effect of M-CSF on marrow monocyte precursor proliferation, maturation, and release, whereas the neutrophilia and lymphopenia might reflect indirect effects mediated by the known ability of M-CSF to cause the release of other cytokines (83).

Unlike neutrophils, which are present in a large reserve pool in the marrow, monocytes are present in the marrow in relatively small numbers (approximately 1–2% of marrow cells) and must therefore first proliferate and mature before they can be released into the circulation in large numbers. Interestingly, the peripheral monocytosis in M-CSF–treated rats was characterized not only by mature monocytes, but also by promonocytes with nonlobated nuclei. Stoudemire and Garnick (82) have noted striking vacuolization and increased Ia antigen expresion in the circulating monocytes and macrophagelike cells of M-CSF–treated rats and primates, suggesting activation of these cells by M-CSF. In a toxicological study in cynomolgus monkeys by the same authors, a reversible dose-related thrombocytopenia as well as slight and reversible decreases in the red blood cell count, hemoglobin, and hematocrit were observed.

Erythropoietin

EPO in vitro stimulates erythroid colony formation and also stimulates megakaryocytopoiesis via a direct effect on megakaryocyte progenitors that have receptors for EPO (84).

EPO in vivo given as a single IV injection to rats causes peripheral reticulocytosis beginning within 24 h and reaching a maximum at 3 days (85,86). The reticulocytosis is left-shifted as defined by a relative and absolute increase in immature or so-called Type I and II reticulocytes (86). The marrow shows an increase in pronormoblasts and early normoblasts within the first 24 h after the injection of EPO followed by a right-shifted erythroid hyperplasia on the following days. Daily injection of EPO in rats for 1 week results in a sustained erythroid hyperplasia in the marrow, a striking peripheral reticulocytosis, increases in the hematocrit and hemoglobin, and massive splenomegaly (86). Chronic daily administration of EPO for 12–14 days results in steady-state erythrocyte levels 20% above controls (87). The steady state is associated with a return of reticulocyte levels to control levels despite the continued administration of EPO. EPO administered on the same daily basis also caused an increase in platelets on days 2 through 8, but not at later time points (87). Interestingly, a transient increase in circulating white blood cells was also reported during the administration of EPO (87).

Erythropoietin has long been known to be the primary physiological regulator of erythropoiesis, and its importance has been underscored by the devastating effects of total EPO deficiency in the transgenic EPO knock-out mouse (90) and in pathophysiological states of erythropoietin deficiency in humans. Mice genetically engineered to lack the EPO gene (or the EPO receptor) are nonviable, dying in utero at about day 13 owing to failure of liver erythropoiesis (90). The kidney is the most important site of endogenous erythropoietin production, and in the majority of conditions that underly chronic renal failure, erythropoietin deficiency is the principal cause of the chronic hypoproliferative anemia complicating this disease. Erythropoietin has now found major clinical roles in the treatment of the anemia of chronic renal failure (88, and reviewed in Ref. 91) and also in the anemia associated with cancer treatment and AIDS treatment (92,93). Treatment of the anemia of chronic renal failure with EPO has led to dose-related rises in hematocrit, reductions in the number of red blood cell transfusions, and improvement in the debilitating symptoms of anemia, with reductions in fatigue and lethargy and overall improvement in physical activity levels. Current clinical trials focus on determining the benefits and risks of achieving a normal hematocrit and measurement of the degree of cardiovascular benefit associated with correction of chronic anemia. Iron supplementation is needed to maintain the therapeutic efficacy of EPO as body iron stores decline as a result of the increase in erythropoiesis. The peripheral blood of EPO-treated patients is reported to contain an increased number not only of BFU-E, but also of CFU-GEMM (89). Circulating numbers of CFU-GM were not elevated (89).

EPO is well tolerated by patients. The adverse effects of erythropoietin include those of erythrocytosis and in patients with chronic renal failure, hypertension (91), possibly due to the increased vascular resistance associated with the rise in hematocrit.

Thrombopoietin

The regulation of platelet formation has been the object of intense study for nearly 40 years. The early growth and the latter differentiation phases of the process have traditionally been considered as separate events under distinct regulatory controls. The primary factor guiding early megakaryopoiesis, megakaryocyte colony-stimulating factor (Meg-CSF), has been thought to induce growth of megakaryocyte precursor cells (colony formation) but not megakaryocyte maturation. The later-acting factor, thrombopoietin (TPO), has been thought to induce terminal megakaryocyte differentiation and platelet shedding, but not progenitor cell colony formation (94,95). Both of these activities have been found in the urine, serum, or plasma from aplastic and/or thrombocytopenic humans (96–98), rats (99), rabbits (100), and dogs (101). However, the low concentrations of these activities as well as the complexity of these sources had prevented the unambiguous identification or characterization of either putative cytokine.

In 1993, a study from Methia et al. implicated the hematopoietic cytokine receptor c-*mpl* as critical for human megakaryopoiesis and strongly suggested that the unknown ligand for this receptor was present in thrombocytopenic plasma (102). This was later confirmed by several groups that isolated and characterized a ligand for c-*mpl* from thrombocytopenic pig (103), dog (104,105), rat (106), or sheep (107) plasma. Plasma-derived c-*mpl* ligands specifically stimulated human (105) and rodent (106,107) in vitro megakarypoiesis and in vivo murine platelet production (107). This new cytokine, encoded on mouse chromosome 16 (108) and human chromosome 3 (109,110), has been cloned, expressed, and named (1) TPO, invoking the historical name for a platelet-forming factor (106,111), (2) Mpl ligand (ML), to reflect the ligand/receptor relationship with c-*mpl* (103), or (3) megakaryocyte growth and development factor (MGDF), to reflect the multiple biological properties of the molecule (104,105). Studies evaluating the relationship of endogenous serum TPO to platelet counts in human (112) and rodent (105,113–115) models of platelet recovery and work with *mpl*-deficient mice (116) confirm the physiological relevance of this molecule to thrombopoiesis.

Recombinant human MGDF is now available and has been used in a number of in vitro and in vivo studies. In vitro, it induces colony formation within murine bone marrow (117) or fetal liver–derived stem cell populations (118) and on human CD34+ cells (112). The addition of stem cell factor (SCF) but not interleukin-3 (IL-3), to rHuMGDF results in increased numbers of megakaryocyte colonies, presumably through SCF-mediated expansion of precursor cell populations (112). Since *mpl* expression appears to be restricted to hematopoietic tissues, specifically CD34+ cells, megakaryocytes, and platelets (119), it is not surprising that megakaryocytes are apparently the only cell type stimulated with rHuMGDF. However, as is the case with the other lineage-dominant hematopoietic factors, such as granulocyte colony-stimulating factor (G-CSF) and erythropoietin (EPO), this lineage restriction may not be absolute and subtle relationships could emerge with more study. For example, in experiments that coincubate murine fetal liver stem cell populations with recombinant murine TPO and murine stromal cell lines, enhanced myelopoiesis results (118).

ML is not only a megakaryocyte growth factor, it is a megakaryocyte differentiation factor. Human megakaryocytes stimulated with ML undergo increased expression of platelet-specific antigens (103,104,106) and become increasingly polyploid (112). Maturation is complete and terminal in that morphologically normal, functional platelets are produced from rHuMGDF-stimulated megakaryocytes (120). This duality of function is reminiscent of other cytokines that influence both growth and differentiation of their target cells (121) and obviates the need for two distinct megakaryopoietic regulatory factors inferred from the early work in this field. These biological effects of ML are mediated through cell receptor/ligand interactions. Blocking these interactions with a truncated, soluble form of the receptor prevents the biological effects (104,114), which appear to be mediated through classic receptor-induced second-signal events (122).

The properties of ML have also been evaluated in several animal models and to date the in vivo data match the promise of the in vitro work. Daily administration of ML into normal mice increases de novo platelet synthesis as detected by incorporation of ^{35}S into nacsent platelets (103) and ultimately increases the number of platelets to three to four times baseline values without affecting red or white cell counts (113). Platelet counts start to rise after 3 days of administration and reach maximum levels after 5–9 days. rHuMGDF also has potent thrombopoietic effects when administered into normal, nonhuman primates. Platelet counts in rhesus monkeys rise to six times baseline levels after 10 days of daily administration. After cessation of dosing, platelet counts return to normal values with kinetics consistent with a platelet half-life of 5 days (123). The elevation in platelet numbers occuring after 5–7 days is presumably engineered through increases in the maturity and number of marrow megakaryocytes and megakaryocyte progenitors (123), and not through an immediate effect on platelet shedding.

Although it is important to establish the thrombopoietic effect of rHuMGDF on normal animals, it is critical to establish efficacy in models of experimentally induced thrombocytopenia. Mice treated with the chemotherapeutic agent carboplatin show a 90% reduction in platelet counts 11–14 days after drug and without therapy do not recover normal platelet counts until days 17–20. In contrast, daily dosing with rHuMGDF commencing the day after carboplatin administration can significantly alleviate, and in some cases completely abrogate, the resultant thrombocytopenia (113). In other murine models of semilethal myeloablation, daily administration of rHuMGDF can reduce mortality from 94% in excipient-treated animals to 14%. rHuMGDF administered in combination with recombinant murine G-CSF completely prevents mortality.

Significant clinical benefits may be obtained from the administration of pharmacological doses of recombinant hematopoietic growth factors, such as alleviation of anemia with EPO or reduction of neutropenia with G-CSF. The in vitro and in vivo activities of rHuMGDF indicate that this cytokine holds similar promise as a clinical tool for prevention of thrombocytopenia associated with chemotherapy, irradiation, and bone marrow transplantation.

INTERLEUKINS

Interleukin-1

IL-1 is a cytokine that no doubt contributed very significantly to the activity of the preparation described by Kampschmidt and his colleagues (124) as leukocyte endogenous mediator (LEM). LEM was shown to induce a neutrophilia at 60–90 min after injection into rats. The neutrophilia was accompanied by a decrease in marrow neutrophils, suggesting that the release of marrow neutrophils contributed to the increase in circulating numbers of neutrophils. Recombinant IL-1 α and β at low doses (25 U/rat) were reported by our laboratory (125) to induce a monophasic neutrophilia in rats peaking at 90 min after IV injection. IL-1α has been hypothesized to be the membrane form of IL-1 and IL-1 β the major cytosolic or secreted form, although the validity of this concept has come into question (126). At the dose studied in our experiments, IL-1β caused a greater peripheral neutrophilia than IL-1α and exhibited more potent neutrophil releasing activity than IL-1α. IL-1β caused the release of 70% of the marrow reserve of mature neutrophils, whereas IL-1α caused the release of approximately 30% of marrow neutrophils. The peripheral neutrophilia was characterized by an increase in circulating band forms as well as by the presence of occasional metamyelocytes, observations that also suggest marrow release of neutrophils. IL-1α and β also both induced a significant absolute lymphopenia accompanying the peripheral neutrophilia. At higher doses, IL-1 induces a rapid and profound granulocytopenia (127). The granulocytopenia is followed at later time points (such as 12 h) by granulocytosis. We have also observed that IL-1 induces neutropenia at high doses (unpublished data), although at the high doses studied in our laboratory the rats were clinically hypotensive. IL-1 is known to cause hypotension at high doses (128). Whereas the discrimination between the direct and indirect hematological effects of a cytokine in vivo are problematic in any case, the effects of a cytokine at doses that cause hypotension are even more difficult to interpret since hypotension itself causes neutrophilia (129). At low doses, IL-1 causes neutrophilia in the absence of any changes in blood pressure (125). Since the concentrations of IL-1 measured in septic patients have generally been only two to three times greater than background levels (130), the hematological observations made in vivo after the administration of high doses of IL-1 may not be relevant to an understanding of the pathophysiological role of IL-1, although they are certainly of great pharmacological importance. IL-1 at low doses also causes neutrophilia in adrenalectomized rats (125), an observation that is of interest since IL-1 has been reported to possess corticotropin-releasing activity (127) and because both adrenal cortical and medullary hormones affect circulating numbers of granulocytes and lymphocytes (2).

Daily injection of low doses of IL-1 induces a repetitive daily neutrophilia and a mild myeloid hyperplasia in the marrow at the end of a week (41). The myeloproliferative effect of IL-1 may in part reflect the release of endogenous serum colony-stimulating activity at 3–6 h and the increase in marrow CFU-GM at 48 h after the injection of IL-1 as reported by Stork et al. (131) and similarly by Vogel et al. (132) in mice.

Interleukin-2

IL-2, also known as T-cell growth factor, promotes the growth of T cells in vitro. Among the interleukins, IL-2's in vivo hematological effects in humans are at present perhaps the best studied because of the use of IL-2 in combination with lymphokine-activated killer (LAK) cells as an experimental therapy for cancer patients by Rosenberg and colleagues (133).

In rats, daily administration of IL-2 was reported by Anderson et al. (134,135) to cause a dose- and regimen-related lymphocytosis, neutrophilia, eosinophilia, thrombocytopenia, hemolytic anemia, lymphoid hyperplasia, and extramedullary erythropoiesis and eosinopoiesis in the spleen. Large numbers of mononuclear cells accumulated in the hepatic sinusoids. Pleural effusions occurred composed predominantly of large granular lymphocytes (135). In mice receiving repeated injections of IL-2, Ettinghausen et al. (136) observed prominent accumulations of activated lymphocytes in lungs, liver, spleen, kidneys, and mesenteric lymph nodes as confirmed both by an in vivo DNA labeling technique as well as by histological study. In sheep receiving a daily brief infusion of IL-2, Jesmok et al. (137) noted a lymphopenia and neutrophilia after the first day's infusion, and a more pronounced lymphopenia and neutropenia within 2–4 h after the subsequent days' infusions. Baseline leukocyte levels in the IL-2–treated sheep (i.e., 24 h after the previous day's infusion) increased progressively. As is clear from the foregoing experiments and from the following results in patients, the hematological effects of IL-2 are very dose and regimen dependent as well as dependent upon the time of sampling of the peripheral blood.

A single bolus of IL-2 in humans causes a dramatic decrease in peripheral blood T cells beginning as early as 15 min after injection with a progressively greater lymphopenia for the first 4 h followed by a rebound lymphocytosis at 48 h (138). Continuous infusion of IL-2 often resulted in a depressed lymphocyte count during the first week followed by a progressive increase in circulating lymphocytes ranging from 2- to 16-fold over baseline (138). In a review of the hematological effects of IL-2 in combination with LAK cells, Ettinghausen et al. (133) reported that the large majority of patients developed an anemia that could not be explained on the basis of phlebotomy, cytophereses, or hemodilution and that required red cell transfusions. Anemia was accompanied by reticulocytopenia. IL-2 administration also resulted in an often severe thrombocytopenia, a lymphopenia, and an eosinophilia. Mild neutropenia and a rebound lymphocytosis followed the discontinuation of IL-2 therapy. Colony assays on the peripheral blood of patients showed a decline in circulating erythroid and granulocyte/macrophage progenitors. Furthermore, IL-2 also induced the appearance of circulating interferon-γ, suggesting that IL-2 may also indirectly suppress hematopoiesis.

Interleukin-4

IL-4 in vitro, in addition to its many effects on B cells, downregulates LPS-induced IL-1 and TNF expression by monocytes (139), an action that may potentially implicate IL-4 as a downregulating influence on LPS-induced hematological effects in vivo. IL-4 apparently does not cause any changes in the total leukocyte count or differential of the peripheral blood after experimental administration of IL-4 (D. Rennick, personal communication).

Overexpression of IL-4 in transgenic mice under the control of an immunoglobulin promoter/enhancer cassette resulted in an increase in serum IgE levels (attributable to effects on B-cell function) and the appearance of inflammatory ocular lesions (140). The eyelid inflammation that develops in these mice was associated with mast cell degranulation and tissue fibrosis (141). Transgenic mice that express IL-4 under the direction of the lymphocyte-specific lck promoter exhibit decreased bone formation by osteoblasts and develop severe osteoporosis of cortical and trabecular bone by 2 months of age (142).

Interleukin-5

IL-5, also known as eosinophil differentiation factor, causes the proliferation of eosinophil colonies in vitro (143). IL-5 also acts as a B-cell growth factor in vitro in mice, but apparently not in human lymphocyte cultures (143). The concept that the pathogenesis of eosinophilia is regulated in vivo by a lymphokine is suggested by the T-cell dependence of parasite-induced eosinophilia in rats (144). Although administration of other cytokines such as GM-CSF and IL-2 causes peripheral and tissue eosinophilia, IL-5 is especially implicated in the pathogenesis of eosinophilia in vivo because antiserum to IL-5 inhibits the eosinophilia noted during experimental parasitemia in mice (145). Samoszuk et al. (146) have found a correlation between the expression of IL-5 mRNA as detected by in situ hybridization in the Reed-Sternberg cells of Hodgkin's disease and tissue eosinophilia in the lymph nodes of these patients.

Interleukin-6

IL-6 is an extraordinarily multifunctional peptide growth factor that has also been identified under such diverse names as interferon-B2, B-cell stimulating factor, hybridoma growth factor, hepatocyte-stimulating factor, 26 kDa protein, and myeloid blood cell differentiation–inducing protein (147,148). IL-6 in vitro has been reported to enhance IL-3–dependent proliferation of hematopoietic precursors (12) as well as to enhance colony formation of almost all the hematopoietic lineages when combined with appropriate lineage-restricted growth factors (149). The relative contributions of IL-6's direct and indirect effects on the observed in vitro results is still unclear (149). Continuous perfusion of mice with IL-6 for 7 days increases the number of splenic, but not marrow, hematopoietic stem cells (CFU-S) (150). IL-6 administered as a single IV injection to rats induces a mild peripheral neutrophilia (151). The marrow shows a mild left-shifted hyperplasia but does not show a significant decrease in mature marrow neutrophils, suggesting that IL-6 affects the proliferation of early myeloid precursors, but that the source of the increase in peripheral neutrophils may be from the marginated pool rather than from the marrow reserve. The most striking peripheral hematological change after an injection of IL-6 is the reticulocytosis that occurs at 12–24 h. The reticulocytosis is accompanied by an erythroid hyperplasia of late normoblasts in the marrow. The kinetics of the IL-6–induced reticulocytosis and erythroid hyperplasia are different from those of EPO (86). The number of circulating reticulocytes returns to baseline within 48 h after the injection of IL-6, whereas the reticulocyte count is still increasing at 48 h after the injection of EPO and does not reach a maximum until 72 h. Whereas the marrow at 24 h after the injection of IL-6 demonstrates a right-shifted erythroid hyperplasia, the marrow at the same time after injection of EPO shows a more left-shifted erythroid hyperplasia. At 48 h after the injection of IL-6, the marrow differential is unremarkable, whereas at the same time after the injection of EPO a right-shifted erythroid hyperplasia has developed.

The foregoing sequence of changes in the marrow is consistent with the hypothesis that EPO acts on an earlier erythroid precursor than IL-6. Furthermore, continued daily injections of IL-6 for 1 week do not result in a continued erythroid hyperplasia, whereas daily injections of EPO causes a progressive reticulocytosis and erythroid marrow hyperplasia (86). In regard to the erythropoietic effect of IL-6 in vivo, one may note that Rennick et al. (152) reported that IL-6 increased the number of erythrocyte colonies supported by IL-4 and EPO, although IL-6 had no erythroid growth-promoting activity of its own. The transient in vivo enhancement of erythropoiesis by IL-6 may thus be an indirect effect or the result of an in vivo interaction with endogenous IL-4, EPO, or other growth factor. Injections of IL-6 every 12 h for 5 days in mice caused a thrombocytosis as well as an increase in the size but not number of megakaryocytes in the marrow (153). The dual in vivo effect of IL-6 on erythropoiesis and thrombocytopoiesis is qualitatively somewhat reminiscent of the in vivo effect of EPO. IL-6 has been reported to synergize in vitro with several other growth factors such as IL-3, IL-4, M-CSF, G-CSF, and erythropoietin (reviewed in Ref. 149). The combination of IL-6 and IL-3 in vivo in rats caused an increase in marrow mast cells not seen after the administration of either factor alone (6). Of potential significance for an understanding of the interactions of the hematological effects of cytokines in vivo is the report of Schindler et al. (154) that IL-6 suppresses TNF and IL-1 production by LPS in human peripheral blood mononuclear cells and the report of Vilcek and coworkers (155) that IL-6 decreases LPS-induced serum TNF levels in vivo in mice. Our laboratory has recently observed that splenic IL-6 mRNA peaks 1–2 h after the expression of TNF and IL-1 mRNA in LPS-injected rats (unpublished observations), a time sequence that might be consistent with a negative feedback effect of IL-6.

The ability of IL-6 to stimulate differentiation of B lymphocytes to antibody-producing plasma cells is supported by transgenic mouse studies, where mice overexpressing IL-6 under the control of an immunoglobulin heavy chain enhancer develop severe hypergammaglobulinemia and exhibit a concomitant plasmacytosis (156). Overexpression of IL-6 in pancreatic islets (157) resulted in islet hyperplasia and inflammation in young mice, giving way to a predominant B-lymphocyte infiltration of the islets in older mice. These infiltrates contained many IgG-positive plasma cells, again supporting an important role for IL-6 in stimulating differentiation of B lymphocytes to antibody-producing cells.

Interleukin-7

IL-7 in vitro supports the growth of early B cells and is a comitogen for T cells (158). Human IL-7 injected in mice increases the number of lymphocytes, NK cells, and macrophages in the spleen and lymph nodes (159). Administration of IL-7 to irradiated mice is reported to accelerate T- and B-cell reconstitution (160).

Interleukin-8

IL-8, also known by such names as monocyte-derived neutrophil chemotactic factor and neutrophil activating factor, is an inflammatory cytokine inducible by IL-1 and TNF and is apparently one of the major neutrophil chemotactic factors in the supernatants of endotoxin-stimulated human peripheral blood mononuclear cells (161). IL-8 injected intraperitoneally was reported by Furuta et al. (162) to induce a rapid neutrophilia at low doses and an initial neutropenia followed by neutrophilia at higher doses.

In addition to its proinflammatory role in recruiting neutrophils to local sites of acute and chronic inflammation, IL-8 may have some anti-inflammatory properties. IL-8 has been reported to inhibit neutrophil adhesion to activated vascular endothelium (163,164), and short-term intravascular delivery of IL-8 to rabbits was found to block neutrophil migration to sites of cytokine-induced inflammation (165,166). In transgenic mice engineered to express human IL-8 in the liver (171), circulating blood levels of IL-8 ranged from 1 to > 100 ng/ml. Circulating neutrophils were elevated in proportion to the levels of IL-8 in the blood of different transgenic lineages. Peripheral blood neutrophils within the IL-8 transgenic mice exhibited decreased surface expression of L-selectin and a decreased ability to extravasate into the inflamed peritoneal cavity. Despite the sustained elevated levels of IL-8, and the chronic neutrophilia, within the blood of these mice, the IL-8 transgenics survived normally with no gross abnormalities, plasma exudation, or localized tissue damage. This phenotype is very similar to that reported in baboons infused with sufficient IL-8 to maintain circulating levels between 10 and 15 ng/ml for a period of 8 h (172). Thus, the biological effects associated with chronic overexpression of IL-8 in mice correlates well with the effects of IL-8 in primates.

Interleukin-9

IL-9 has pleiotropic in vitro activity on T cells, B cells, and hemopoietic progenitors (167). In vivo, IL-9 causes mast cell proliferation and an increase in serum in immunoglobulin titers in transgenic mice. The occurrence of lymphoblastic lymphomas has also been described in IL-9 transgenic mice (168).

Interleukin-12

IL-12 administered daily to mice at a toxicological dose of 1 μg/day causes anemia, lymphopenia, neutropenia, and a mild thrombocytopenia (169). Extramedullary splenic hematopoiesis occurs in these mice. IL-12 administered to squirrel monkeys is reported to cause mild to moderate anemia and leukocytosis with trilinear bone marrow hyperplasia and reactive hyperplasia of lymph nodes (170). Positive immunomodulatory activity was detectable at doses far less than toxic doses (170).

INTERFERONS

Interferons, originally discovered because of their antiviral effects, also display a wide range of antiproliferative and immunomodulatory effects. In vitro, the IFNs all inhibit myeloid colony formation, with IFN-γ being the more potent inhibitor and synergizing in this regard with IFN-α and -β (173). The in vitro inhibitory effects of the IFNs can be overcome by the addition of colony-stimulating factor in conditioned medium (173). The occurrence of granulocytopenia is one of the most well known in vivo hematological effects of interferons in humans (174). In mice, all three types of IFNs (α, β, and γ) were reported by Koren et al. (78) to cause a generalized reduction of granulocytes, monocytes, and lymphocytes in the circulation. IFN-γ was 50- to 100-fold more potent in its myelosuppressive activity. The IFN-induced peripheral lymphopenia in mice is accompanied by peripheral lymphadenopathy and a decrease in the number of thoracic duct lymphocytes as reported by Gresser et al. (175), who postulated that by inhibiting the egress of lymphocytes from lymphoid tissues IFN may contribute to the lymphadenopathy noted clinically in viral diseases. Interferon has been postulated to be a mediator of the hematopoietic suppression observed in aplastic anemia patients whose sera may contain elevated levels of IFN and whose marrow cells demonstrate an increase in colony-formation activity after incubation with antiserum to IFN in vitro (176).

TUMOR NECROSIS FACTORS

TNF-α and TNF-β are proteins that share a moderate amino acid homology and can bind to the same receptor (177). TNF in vitro exerts both stimulatory and inhibitory hemopoietic effects in varying colony-forming assays, and the significance of these results to the in vivo effects of the same molecules is sometimes difficult to ascertain (178). TNF-α and -β induce similar hematological effects in vivo (129,179) and will therefore be discussed together.

TNF in mice was reported by Remick et al. (180) to induce a brief neutropenia followed by a monophasic left-shifted neutrophilia and accompanied by a peripheral lymphopenia characterized by a preferential decrease in B cells. TNF in rats also causes neutropenia at 30 min followed at 1–2 h by neutrophilia (125). The neutropenia may be due to transient intravascular margination since epinephrine, a demarginating factor that does not cause the release of marrow neutrophils, reverses TNF-induced neutropenia (2). Further support for the hypothesis that the initial neutropenia is due to intravascular margination is found in the observation that TNF administration will cause a reduction in the circulating number of neutrophils during experimental neutrophilia (2), suggesting that TNF actively causes neutrophils to leave the circulation, perhaps by margination resulting from the expression of neutrophil-endothelial adherence molecules. In this regard, Gamble et al. (181) reported that TNF in vitro stimulates the adherence of neutrophils to endothelium by a rapid protein synthesis independent effect on neutrophils as well as by a slower de novo RNA and protein synthesis dependent effect on endothelium. TNF in vitro also activates neutrophils, promotes neutrophil aggregation, and is chemotactic for neutrophils and monocytes (182,183).

Neutrophil aggregation in vivo after the administration of TNF could hypothetically contribute to neutropenia as a result of microembolization.

At relatively low doses (doses not accompanied by hypotension, clinically observable toxicity, or histologically observable tissue injury), the TNF-induced neutrophilia in rats, in contrast to that observed in mice, is biphasic (125). The peripheral neutrophilia is accompanied by a decrease in marrow neutrophils and an increase in circulating band cells, suggesting that the marrow reserve of neutrophils contributes to the increase in circulating neutrophils. The second wave of TNF-induced neutrophilia is of lesser magnitude and is temporally broader (i.e., of longer duration) than the first peak, suggesting that the second wave is due to the release of an endogenous factor. The second wave of neutrophilia is augmented in *Corynebacterium parvum*–primed rats, which exhibit hyperplasia of the reticuloendothelial system (184), and in IFN-γ–primed rats (2). The second wave of neutrophilia has specifically been postulated to be due to the TNF-induced release of endogenous IL-1 (125) on the basis that (1) TNF has been reported to induce the release of IL-1 both in vitro (185) and in vivo (186), (2) indomethacin and dexamethasone, which inhibit IL-1–induced neutrophilia, also abrogate the second peak of TNF-induced neutrophilia (125), and (3) TNF induces a biphasic febrile response in rabbits that has similar kinetics to the biphasic neutrophilia in rats and in which the second wave of fever is accompanied by an increase in serum IL-1 (186). Our laboratory has recently also noted that TNF-injected rats demonstrate an increase in splenic IL-1 mRNA at the time of the second wave of neutrophilia (unpublished observation). TNF in vitro has been shown to induce the synthesis of G-CSF and GM-CSF (187), raising the possibility that these myeloid growth factors contribute to TNF-induced neutrophilia and to the striking myeloid marrow hyperplasia observed in rats treated chronically with high doses of TNF (188). TNF at higher doses in rats induces a more pronounced initial peripheral neutropenia followed by a greater and more sustained monophasic neutrophilia (188). The qualitative and quantitative changes in circulating numbers of neutrophils observed after high doses of TNF are thus indistinguishable from those observed after experimental administration of moderate to high doses of endotoxin (184), a correlation that takes on important pathophysiological significance since endotoxin is known to cause the release of TNF in vivo (189) and because Remick et al. (190) have reported that antiserum to TNF inhibits endotoxin-induced neutrophilia in mice. Daily injections of high doses of TNF for 1 week cause a striking myeloid hyperplasia (188), an effect that may in part be postulated to be due to the in vivo release of colony-stimulating factor after administration of TNF (132).

TNF, like endotoxin as well as a number of other previously mentioned cytokines, causes peripheral lymphopenia. Among the mechanisms that might be postulated to account for the lymphopenia are a cytotoxic effect on lymphocytes, intravascular margination, emigration of lymphocytes into lymphoid or nonlymphoid tissues, or a decrease in recirculation of lymphocytes from the thoracic duct into the circulation (2). The possibility that TNF might induce lympholysis was suggested by Remick et al. (180). However, no in vitro cytotoxicity of TNF for murine peripheral blood mononuclear cells could be demonstrated. A transient decrease in lymphocyte recirculation from the thoracic duct could also explain the observation of peripheral lymphopenia. The importance of thoracic duct recirculation in the maintenance of circulating lymphocyte levels is apparent in the rapid peripheral lymphopenia that occurs in rats in whom lymph from the thoracic duct is drained. In rats in whom the thoracic duct had been cannulated, however, no evidence of a decrease in either lymphocyte recirculation or lymph flow was noted after the administration of TNF (2). The hypothesis that TNF induces lymphopenia via a receptor-mediated interaction between lymphocytes and endothelium is attractive in view of in vitro evidence by Bevilacqua et al. (191) and by Cavender et al. (192) that TNF induces the synthesis of endothelial adhesion molecules that bind lymphocytes and in view of the immunohistochemical finding of Cotran et al. (193) that endothelial adhesion molecules are expressed in vivo within the high endothelium at the site of delayed hypersensitivity reactions. The in vitro binding of lymphocytes to endothelial monolayers reaches a maximum 4–6 h after TNF treatment (192), kinetics that are quite analagous to our observations on the in vivo kinetics of TNF-induced lymphopenia. TNF was reported by Michie et al. (194) to be elevated in the serum of healthy men after the experimental intravenous injection of *E. coli* endotoxin. Since endotoxin induces the release of TNF into serum, endotoxin and TNF are likely to be present together in the circulation of patients when endotoxemia occurs during gram-negative bacterial infections. Coinjection of endotoxin and TNF in rats markedly increases both the magnitude and duration of endotoxin-induced neutropenia, neutrophilia, and lymphopenia (184). TNF also interacts with IL-1 and adrenal hormones in a fashion that would suggest that these endogenous factors may all contribute to the changes in circulating numbers of leukocytes noted during gram-negative infections (2). In addition to effects on myeloid and lymphoid cells, TNF exerts profound and dose-dependent effects on erythropoiesis in vivo. Cerami and colleagues (195) noted that rats treated with high doses of TNF developed anemia characterized by a decrease in red blood cell synthesis and a reduced life span of circulating erythrocytes. An anemia of similar magnitude was described after the administration of endotoxin, but not after the administration of IL-1, suggesting that TNF may be implicated in the pathogenesis of the anemia of chronic inflammatory diseases. TNF has been reported by Broxmeyer et al. (196) to inhibit erythropoiesis in vitro. Our laboratory (188) noted a peripheral anemia and marrow erythroid hypoplasia in rats treated for a week with high doses of TNF. The erythroid hypoplasia was postulated to represent in part a myelophthisic anemia secondary to a striking myeloid hyperplasia. EPO has been reported to prevent the decrease in hematocrit observed after TNF treatment of mice for 5 days (197).

TNF has also, however, been reported to stimulate erythropoiesis in both ex vivo (198) and in vivo experiments (41,188). Johnson et al. (198) found that a single IV injection of a relatively high dose of TNF caused an increase in immature erythroid cells (BFU-E), although inhibiting later-stage erythropoiesis as measured by the number of CFU-E in the marrow. Our laboratory (41,188) observed that low doses of TNF administered to rats as a daily IV dose for one week causes an erythroid hyperplasia of late normoblasts. The normoblasts are, however, morphologically dysplastic in their sometimes striking nuclear fragmentation, and no increase in circulating red blood cells is observed, suggesting a state of ineffective erythropoiesis (188). Interestingly, TNF has been reported by Clark and Chandri to induce dyserythropoiesis and erythrophagocytosis in experimental malaria (199).

REFERENCES

1. van Furth R, Sluiter W. Distribution of blood monocytes between a marginating and circulating pool. J Exp Med 1986; 163:474–479.
2. Ulich TR, del Castillo J, Ni RX, Bikhazi N, Calvin L. Mechanisms of TNF-induced lymphopenia, neutropenia, and biphasic neutrophilia. J Leuk Biol 1989; 45:155–167.
3. Axelrad AA. Some hemopoietic negative regulators. Exp Hematol 1990; 18:143–150.
4. Bot F, Dorssers L, Wagemaker G, Lowenberg B. Stimulating spectrum of human recombinant multi-CSF (IL-3) on human marrow precursor: importance of accessory cells. Blood 1988; 71:1609.
5. Prystowsky MB, Otten G, Nanjokas MG, Vardiman J, Ihle JN, Goldwasser E, Fitch FW. Multiple hemopoietic lineages are found after stimulation of mouse bone marrow precursor cells with interleukin-3. Am J Pathol 1984; 117:171.
6. Ulich TR, del Castillo J, Busser K, Guo K, Yin S. Acute in vivo effects of IL-3 alone and in combination with IL-6 on the blood cells of the circulation and bone marrow. Am J Pathol 1989; 135:663–670.
7. Ulich TR, del Castillo J, McNiece IK, Yin S, Irwin B, Busser K, Guo K. Acute and subacute hematologic effects of multi-CSF in combination with G-CSF in vivo. Blood 1990; 75:48–53.
8. Metcalf D, Bagley CG, Johnson GR, Nicola NA, Lopez AF, Williamson DJ. Effects of purified bacterially synthesized murine multi-CSF (IL-3) on hematopoiesis in normal adult mice. Blood 1986; 68:46–57.
9. Kindler V, Thorens B, de Kossodo S, Allet B, Eliason JF, Thatcher D, Farber N, Vassali P. Stimulation of hematopoiesis in vivo by recombinant bacterial murine IL-3. Proc Natl Acad Sci USA 1986; 83:1001–1005.
10. Kimoto M, Kindler V, Higaki M, Ody C, Izui S, Vassalli P. Recombinant murine IL-3 fails to stimulate T or B lymphopoiesis in vivo, but enhances immune responses to T-cell dependent antigens. J Immunol 1988; 140:1889–1894.
11. Lord BI, Molineux G, Testa NG, Kelly M, Spooncer E, Dexter TM. The kinetic response of haematopoietic precursor cells, in vivo, to highly purified recombinant IL-3. Lymphokine Res 1986; 5:97–104.
12. Leary AG, Ikebuchi K, Clark SC, Ihle JN, Hirai Y, Ogawa M. Interleukin-6 enhancement of interleukin-3 dependent proliferation of multipotential hematopoietic progenitors. Proc Natl Acad Sci USA 1987; 84:9035–9039.
13. Paquette RL, Zhou JY, Yang YC, Clark SC, Koeffler HP. Recombinant gibbon IL-3 acts synergistically with recombinant human G-CSF and GM-CSF in vitro. Blood 1988; 71:1596–1600.
14. Warren DJ, Moore MAS. Synergism among interleukin-1, interleukin-3 and interleukin-5 in the production of eosinophils from primitive hemopoietic stem cells. J Immunol 1988; 140:94.
15. Donahue RE, Seehra J, Metzger M, Lefebvre D, Rock B, Carbone S, Nathan DG, Garnick M, Sehgal PK, Laston D, Lavallie E, McCoy J, Schendel PF, Norton C, Turner K, Yang YC, Clark SC. Human IL-3 and GM-CSF act synergistically in stimulating hematopoiesis in primates. Science 1988; 241:1820.
16. Mori M, Suda T, Suda J, Miura Y, Saito M. The role of IL-3 in the proliferation of murine mast cells in vitro. Acta Haematol Jpn 1987; 50:59–64.
17. Metcalf D, Johnson GR, Burgess AW. Direct stimulation by purified GM-CSF of the proliferation of multipotential and erythroid precursor cells. Blood 1980; 55:138–147.
18. Metcalf D, Begley CG, Williamson DJ, Nice EC, De Lamarter J, Mermod JJ, Thatchter D, Schmidt A. Hemopoietic responses in mice injected with purified recombinant GM-CSF. Exp Hematol 1987; 15:1–9.
19. Ulich TR, del Castillo J, McNiece I, Watson L, Yin S, Andresen J. Hematologic effects of recombinant murine GM-CSF on the peripheral blood and bone marrow. Am J Pathol (in press).
20. Donahue RE, Wang EA, Stone DK, Kamen R, Wong GG, Sehgal PK, Nathan DG, Clark SL. Stimulation of hematopoiesis in primates by continuous infusion of recombinant human GM-CSF. Nature 1986; 321:872–875.
21. Mayer P, Lam C, Obenans H, Liehl E, Besemer J. Recombinant human GM-CSF induces leukocytosis and activates peripheral blood polymorphonuclear neutrophils in non-human primates. Blood 1987; 70:206–213.
22. Brandt SJ, Peters WP, Atwater SK, Kurtzberg J, Borowitz MJ, Jones RB, Shpall EJ, Bast RC, Gilbert CJ, Oette DH. Effect of recombinant human granulocyte-macrophage colony stimulating factor on hematopoietic reconstitution after high dose chemotherapy and autologous bone marrow transplantation. N Engl J Med 1988; 318:869–876.
23. Aglietta M, Piacibello W, Sanario F, Stacchini A, Apra F,

Schena M, Mossetti C, Carnino F, Caligaris-Cappib F, Gavosto F. Kinetics of human hematopoietic cells after in vivo administration of granulocyte-macrophage colony stimulating factor. J Clin Invest 1989; 83:551–557.
24. Morstyn G, Lieschke GJ, Sheridan W, Layton J, Cebon J, Fox RM. Clinical experience with recombinant human granulocyte-colony stimulating factor and granulocyte-macrophage colony stimulating factor. Semin Hematol 1989; 26(suppl. 2):9.
25. Weisbart RH, Kacena A, Schuh A, Golde DW. GM-CSF induces human neutrophil IgA-mediated phagocytosis by an IgA Fc receptor activation mechanism. Nature 1988; 332:647–648.
26. McNiece IK, Robinson BE, Quesenberry PJ. Stimulation of murine colony-forming cells with high proliferative potential by the combination of GM-CSF and CSF-1. Blood 1988; 72:191–195.
27. McNiece IK, Stewart MF, Deacon DM, Quesenberry PJ. Synergistic interactions between hematopoietic growth factors as detected by in vitro mouse bone marrow colony formation. Exp Hematol 1988; 16:383–388.
28. McNiece IK, Andrews RG, Stewart MF, Clark S, Boone T, Quesenberry P. Action of interleukin-3, G-CSF and GM-CSF on highly enriched human hematopoietic progenitor cells: synergistic interaction of GM-CSF plus G-CSF. Blood 1989; 74:110.
29. Lieschke GJ, Maher D, Cebon J, et al. Effects of bacterially synthesized recombinant human granulocyte-macrophage colony-stimulating factor in patients with advanced malignancy. Ann Intern Med 1989; 110:357–364.
30. Lieschke GJ, Cebon J, Morstyn G. Characterization of the clinical effects after the first dose of bacterially synthesized recombinant human granulocyte-macrophage colony-stimulating factor. Blood 1989; 74:2634–2643.
31. Peters WP, Stuart A, Affronti ML, Kim CS, Coleman RE. Neutrophil migration is defective during recombinant human granulocyte-macrophage colony-stimulating factor infusion after autologous bone marrow transplantation in humans. Blood 1988; 72:1310–1315.
32. Stephens LC, Haire WD, Schmit-Pokorny K, Kessinger A, Kotaluk G. Granulocyte macrophage colony stimulating factor: high incidence of apheresis catheter thrombosis during peripheral stem cell collection. Bone Marrow Transplant 1993; 11:51–54.
33. Nemunaitis J, Rabinowe SN, Singer JW, et al. Recombinant granulocyte-macrophage colony-stimulating factor after autologous bone marrow transplantation for lymphoid cancer. N Engl J Med 1991; 324:1773–1778.
34. Socinski MA, Cannistra SA, Elias A, Antman KH, Schnipper L, Griffin JD. Granulocyte-macrophage colony-stimulating factor expands the circulating haemopoietic progenitor cell compartment in man. Lancet 1988; 1:1194–1198.
35. Gianni AM, Siena S, Bregni M, et al. Granulocyte-macrophage colony-stimulating factor to harvest circulating haemopoietic stem cells for autotransplantation. Lancet 1989; 2:580–585.
36. Stanley E, Lieschke GJ, Grail D, et al. Granulocyte/macrophage colony-stimulating factor-deficient mice show no major perturbation of hematopoiesis but develop a characteristic pulmonary pathology. Proc Natl Acad Sci USA 1994; 91:5592–5596.
37. Souza LM, Boone TC, Gabrilove J, Lai PH, Zsebo KM, Murdock DC, Chazin VR, Bruszewski J, Lu H, Chen KK, Barendt J, Platzer E, Malcolm AS, Moore ?, Mertelsmann R, Welte K. Recombinant human granulocyte colony stimulating factor: effects on normal and leukemic myeloid cells. Science 1986; 232:61–65.
38. Cohen AM, Zsebo K, Inoue H, Hines D, Boone TC, Chazin VR, Tsai L, Ritch T, Souza LM. In vivo granulocyte colony stimulating factor. Proc Natl Acad Sci USA 1987; 84:2484–2488.
39. Ulich TR, del Castillo J, Souza L. Kinetics and mechanisms of recombinant human granulocyte colony stimulating factor-induced neutrophilia. Am J Pathol 1988; 133:630–638.
40. Welte K, Bonilla MA, Gillio AAP, Boone TC, Potter GK, Gabrilove JL, Moore MAS, O'Reilly RJ, Souza LM. Recombinant human granulocyte colony stimulating factor: effects on hematopoiesis in normal and cyclophosphamide-treated primates. J Exp Med 1987; 165:941–948.
41. Ulich TR, del Castillo J, Guo K, Souza L. The hematologic effects of the chronic administration of the monokines tumor necrosis factor, interleukin-1, and granulocyte colony stimulating factor on bone marrow and circulation. Am J Pathol 1989; 134:149–159.
42. Lieschke GJ, Burgess AW. Granulocyte colony-stimulating factor and granulocyte-macrophage colony-stimulating factor (1). N Engl J Med 1992; 327:28–35.
43. Lieschke GJ, Burgess AW. Granulocyte colony-stimulating factor and granulocyte-macrophage colony-stimulating factor (2). N Engl J Med 1992; 327:99–106.
44. Morstyn G, Campbell L, Souza LM, et al. Effect of granulocyte colony stimulating factor on neutropenia induced by cytotoxic chemotherapy. Lancet 1988; 1:667–672.
45. Gabrilove JL, Jakubowski A, Fain K, et al. Phase I study of granulocyte colony-stimulating factor in patients with transitional cell carcinoma of the urothelium. J Clin Invest 1988; 82:1454–1461.
46. Bonilla MA, Dale D, Zeidler C, et al. Long-term safety of treatment with recombinant human granulocyte colony-stimulating factor (r-metHuG-CSF) in patients with severe congenital neutropenias. Br J Haematol 1994; 88:723–730.
47. Lord BI, Bronchud MH, Owens S, et al. The kinetics of human granulopoiesis following treatment with granulocyte colony-stimulating factor in vivo. Proc Natl Acad Sci USA 1989; 86:9499–9503.
48. Layton JE, Hockman H, Sheridan WP, Morstyn G. Evidence for a novel in vivo control mechanism of granulopoiesis: mature cell related control of a regulatory growth factor. Blood 1989; 74:1303–1307.
49. Duhrsen U, Villeval JL, Boyd J, Kannourakis G, Morstyn G, Metcalf D. Effects of recombinant granulocyte colony-stimulating factor on hematopoietic progenitor cells in cancer patients. Blood 1988; 72:2074–2081.
50. Sheridan WP, Begley CG, Juttner CA, et al. Effect of peripheral-blood progenitor cells mobilised by filgrastim (G-CSF) on platelet recovery after high-dose chemotherapy. Lancet 1992; 339:640–644.
51. Sheridan WP, Begley GC, To LB, et al. Phase II study of autologous Filgrastim (G-CSF)-mobilized peripheral blood progenitor cells to restore hemopoiesis after high dose chemotherapy for lymphoid malignancies. Bone Marrow Transplant 1994; 14:105–111.
52. Sheridan WP. Cytokine-only approaches to mobilization of progenitoe cells. In: Morstyn G, Sheridan WP, eds. Cell Therapy. Boston: Cambridge University Press, 1995.
53. Bronchud MH, Scarffe JH, Tatcher N, Crowther D, Souza LM, Alton NK. Phase I/II study of recombinant human granulocyte colony-stimulating factor in patients receiving intensive chemotherapy for small cell lung cancer. Br J Cancer 1987; 56:809–813.
54. Gabrilove JL, Jakubowski A, Scher H, et al. Effect of granulocyte colony-stimulating factor on neutropenia and associated

morbidity due to chemotherapy for transitional-cell carcinoma of the urothelium. N Engl J Med 1988; 318:1414–1422.
55. Crawford J, Ozer H, Stoller T, et al. Reduction by granulocyte colony-stimulating factor of fever and neutropenia induced by chemotherapy in patients with small-cell lung cancer. N Engl J Med 1991; 325:164–170.
56. Trillet Lenoir V, Green J, Manegold C, et al. Recombinant granulocyte colony stimulating factor reduces the infectious complications of cytotoxic chemotherapy. Eur J Cancer 1993; 29A:319–324.
57. Sheridan WP, Morstyn G, Wolf M, et al. Granulocyte colony-stimulating factor and neutrophil recovery after high-dose chemotherapy and autologous bone marrow transplantation. Lancet 1989; 2:891–895.
58. Maher DW, Lieschke GJ, Green M, et al. Filgrastim in patients with chemotherapy-induced febrile neutropenia: A double-blind, placebo-controlled trial. Ann Intern Med 1994; 121:492–501.
59. Taylor KM, Jagannath S, Spitzer G, et al. Recombinant human granulocyte colony-stimulating factor hastens granulocyte recovery after high-dose chemotherapy and autologous bone marrow transplantation in Hodgkin's disease. J Clin Oncol 1989; 7:1791–1799.
60. Schmitz N, Linch DC, Reger P, et al. A randomized phase III study of filgrastim-mobilized peripheral blood progenitor cell transplantation (PBPCT) in comparison with autologous bone marrow transplantation (ABMT) in patients with Hodgkin's disease (HD) and non-Hodgkin's lymphoma (NHL) (abstr). Blood 1994; 84 (suppl 1):204a.
61. Masaoka T, Moriyama Y, Kato S, et al. A randomized, placebo-controlled study of KRN8601 (recombinant human granulocyte colony-stimulating factor) in patients receiving allogeneic bone marrow transplantation. Jpn J Med 1990; 3:233–239.
62. Lieschke GJ, Ramenghi U, O'Connor MP, Sheridan W, Szer J, Morstyn G. Studies of oral neutrophil levels in patients receiving G-CSF after autologous marrow transplantation. Br J Haematol 1992; 82:589–595.
63. Dale DC, Davis M, Vincent ME. Use of filgrastim (r-metHuG-CSF) in severe chronic neutropenia. In: Morstyn G, Dexter TM, eds. Figrastim (r-metHuG-CSF) in Clinical Practice. New York: Marcel Dekker, 1994:83–101.
64. Dale DC, Bonilla MA, Davis MW, et al. A randomized controlled phase III trial of recombinant human granulocyte colony-stimulating factor (filgrastim) for treatment of severe chronic neutropenia. Blood 1993; 81:2496–2502.
65. Miles SA, Mitsuyasu RT, Moreno J, et al. Combined therapy with recombinant granulocyte colony-stimulating factor and erythropoietin decreases hematologic toxicity from zidovudine. Blood 1991; 77:2109–2117.
66. van der Wouw PA, van Leeuwen R, van Oers RH, Lange JM, Danner SA. Effects of recombinant human granulocyte colony-stimulating factor on leucopenia in zidovudine-treated patients with AIDS and AIDS related complex, a phase I/II study. Br J Haematol 1991; 78:319–324.
67. Glaspy JA, Baldwin GC, Robertson PA, et al. Therapy for neutropenia in hairy cell leukemia with recombinant human granulocyte colony-stimulating factor. Ann Intern Med 1988; 109:789–795.
68. Glaspy JA, Souza L, Scates S, et al. Treatment of hairy cell leukemia with granulocyte colony-stimulating factor and recombinant consensus interferon or recombinant interferon-alpha-2b. J Immunother 1992; 11:198–208.
69. Teitelbaum AH, Bell AJ, Brown SL. Filgrastim (r-metHuG-CSF) reversal of drug-induced agranulocytosis. Am J Med 1993; 95:245–246.
70. Nelson S. Role of granulocyte colony-stimulating factor in the immune response to acute bacterial infection in the non-neutropenic host: an overview. Clin Infect Dis 1994; 18(suppl 2):S197–S204.
71. Lieschke GJ, Grail D, Hodgson G, et al. Mice lacking granulocyte colony-stimulating factor have chronic neutropenia, granulocyte and macrophage progenitor cell deficiency, and impaired neutrophil mobilization. Blood 1994; 84:1737–1746.
72. Decoster G, Rich W, Brown SL. Safety profile of filgrastim (r-metHuG-CSF). In: Morstyn G, Dexter TM, eds. Filgrastim (r-metHuG-CSF) in Clinical Practice. New York: Marcel Dekker, 1994:267–290.
73. Zhan YQ, Stanley RE, Clark SC, Hatzfeld SC, Levesquo JP, Federici C, Watt SM, Hatzfeld A. Interleukin-3 and interleukin-1a allow early bone marrow progenitors to respond to human colony-stimulating factor-1. Blood 1988; 72:1870.
74. McNiece IK, Stewart FM, Deacon DM, Quesenberry PJ. Synergistic interactions between hematopoietic growth factors as detected by in vitro mouse bone marrow colony formation. Exp Hematol 1988; 16:383.
75. Branch DR, Turner AR, Guilbert LJ. Synergistic stimulation of macrophage proliferation by the monokines tumor necrosis factor-alpha and colony-stimulating factor-1. Blood 1989; 73:307.
76. Ralph P, Nakoinz I. Stimulation of macrophage tumoricidal activity by the growth and differentiation factor CSF-1. Cell Immunol 1987; 105:270.
77. Wang JM, Griffin JD, Rambaldi A, Chen ZG, Mantorvani A. Induction of monocyte migration of recombinant macrophage-colony stimulating factor. J Immunol 1988; 141:575.
78. Koren S, Klimpel GR, Fleishman WR. Treatment of mice with macrophage-colony stimulating factor (CSF-1) prevents the in vivo myelosuppression induced by murine alpha, beta and gamma interferons. J Biol Resp Mod 1986; 5:481.
79. Hume DA, Parli P, Donahue RE, Fidler IJ. The effect of human recombinant macrophage colony-stimulating factor (CSF-1) on the murine mononuclear phagocyte system in vivo. J Immunol 1988; 141:3405–3409.
80. Broxmeyer HE, Geissler K, Cooper S, Williams DE. Influence of purified recombinant human macrophage colony stimulating factor in mice. Biotechnol Ther 1989–1990; 1:43.
81. Ulich TR, del Castillo J, Watson L, Yin S, Garnick MB. In vivo hematologic effects of recombinant human macrophage-colony stimulating factor. Blood 1990; 75:846.
82. Garnick MB, Stoudemire J, Donahue RE, Metzger M, Bree A, Timony G, Wong G, Munn D, Cheung NKV. In vitro and in vivo biological effects of recombinant human macrophage colony stimulatory factor. Proc ASCO (in press).
83. Warren MK, Ralph P. Macrophage growth factor CSF-1 stimulates human monocyte production of interferon, tumor necrosis factor and colony stimulating activity. J Immunol 1986; 137:2281.
84. Burstein SA, Ishibashi T. Erythropoietin and megakaryocytopoiesis. Blood Cells 1989; 15:202–204.
85. Eder H, Rosslenbroich B, Failing K. A dose-dependent effect of recombinant erythropoietin in the reticulocyte population of rats. Blut 1989; 59:184–187.
86. Ulich TR, del Castillo J, Yin S, Egrie JS. The erythropoietic effects of IL-6 and erythropoietin in vivo. Exp Hematol 1991; 19:29–34.
87. Berridge M, Fraser JK, Carter JM, Lin FK. Effect of recombinant erythropoietin on megakaryocytes and platelet production in the rat. Blood 1988; 72:970–977.
88. Adamson JW. The promise of recombinant erythropoietin. Sem Hematol 1989; 26:5–8.

89. Ganser A, Bergman M, Volkers B, Gruetzmacher P, Scigalla P, Hoelzer D. In vivo effects of recombinant human erythropoietin on circulating human hemopoietic progenitor cells. Exp Hematol 1989; 17:433–435.
90. Wu H, et al. Generation of committed erythroid BFU-E and CFU-E progenitors does not require erythropoietin or the erythropoietin receptor. Cell 1995; 83:1–20.
91. Erslev AJ. Erythropoietin [see comments]. N Engl J Med 1991; 324:1339–1344.
92. Miles SA, Mitsuyasu RT, Moreno J, et al. Combined therapy with recombinant granulocyte colony stimulating factor and erythropoietin decreases hematologic toxicity from zidovudine. Blood 1991; 77:2109–2117.
93. Danna RP, Rudnick SA, Abels RI. Erythropoietin therapy for the anemia associated with AIDS and AIDS therapy and cancer. In: Garnick MA, ed. Erythropoietin in Clinical Applications. New York: Marcel Dekker, 1990:301–324.
94. Williams N. Stimulators of megakaryocyte development and platelet production. Prog Growth Factor Res 1990; 2:81–95.
95. Hoffman R. Regulation of megakaryopoiesis. Blood 1989; 74:1196–1212.
96. Ogata K, Zhang Z-G, Abe K, Murphy MJ. Partial purification and characterization of human megakaryocyte colony-stimulating factor (Meg-CSF). Int J Cell Cloning 1990; 8:103–120.
97. Mazur EM, Cohen JL, Newton J, Sohl P, Narendran A, Gesner TG, Mufson RA. Human serum megakaryocyte colony-stimulating activity appears to be distinct from interleukin-3, granulocyte-macrophage colony-stimulating factor, and lymphocyte conditioned medium. Blood 1990; 76:290–297.
98. Mazur EM, de Alarcon P, South K, Miceli L. Human serum megakaryocyte colony stimulating activity increases in response to intensive cytotoxic chemotherapy. Exp Hematol 1984; 12:624.
99. Odell TT, McDonald TP, Detwiler TC. Stimulation of platelet production by serum of platelet-depleted rats. Proc Soc Exp Biol Med 1961; 107:428–431.
100. Mazur EM, Cohen JL, Wong GG, Clark SC. Modest stimulatory effect of recombinant human GM-CSF on colony growth from peripheral blood human megakaryocyte progenitor cells. Exp Hematol 1987; 15:1128–1133.
101. Mazur EM, South K. Human megakaryocyte colony-stimulating factor in sera from aplastic dogs: Partial purification, characterization, and determination of hematopoietic cell lineage specificity. Exp Hematol 1985; 13:1164–1172.
102. Methia N, Louache F, Vainchenker W, Wendling F. Oligonucleotides antisense to the proto-oncogene c-mpl specifically inhibit in vitro megakaryopoiesis. Blood 1993; 5:1395–1401.
103. de Sauvage FJ, Hass PE, Spencer SD, Malloy BE, Gurney AL, Spencer SA, Darbonne WC, Henzel WJ, Wong SC, Kuang W-L, Oles KJ, Hultgren B, Solberg LA, Goeddel DV, Eaton DL. Stimulation of megakaryopoiesis and thrombopoiesis by the c-Mpl ligand. Nature 1994; 369:533–538.
104. Bartley TD, Bogenberger J, Hunt P, Li Y-S, Lu HS, Martin F, Chang M-S, Samal B, Nichol JL, Swift S, Johnson MJ, Hsu R-Y, Parker VP, Suggs S, Skrine JD, Merewether LA, Clogston C, Hsu E, Hokom MM, Hornkohl A, Choi E, Pangelinan M, Sun Y, Mar V, McNinch J, Simonet L, Jacobsen R, Xie C, Sutter J, Chute H, Basu R, Selander L, Trollinger D, Sieu L, Padilla D, Trail G, Elliot G, Izumi R, Covey T, Crouse J, Garcia A, Xu W, Del Castillo J, Biron J, Cole S, Hu MC-T, Pacifici R, Ponting I, Saris C, Wen D, Yung YP, Lin L, Bosselman RA. Identification and cloning of a megakaryocyte growth and development factor (MGDF) which is a ligand for Mpl. Cell 1994; 77:1117–1124.
105. Hunt P, Li YS, Nichol JL, Hokom MM, Bogenberger JM, Swift SE, Skrine JD, Hornkohl AC, Lu H, Clogston C, Merewether LA, Johnson MJ, Parker V, Knudten A, Farese AM, Hsu RY, Garcia A, Stead R, Bosselman RA, Bartley TD. Purification and biological characterization of plasma-derived megakaryocyte growth and development factor. Blood 1995; 86:540–547.
106. Kato T, Ogami K, Shimada Y, Iwamatsu A, Sohma Y, Akahori H, Horie K, Kokubo A, Kudo Y, Maeda E, Kobayashi K, Ohashi H, Ozawa T, Inoue H, Kawamura K, Miyasaki H. Purification, cloning and expression of human thrombopoietin. Proc Natl Acad Sci 1995; 118:229–236.
107. Kuter DJ, Beeler DL, Rosenberg RD. The purification of megapoietin: a physiological regulator of megakaryocyte growth and development production. Proc Natl Acad Sci USA 1994; 91:32768.
108. Chang MS, McNinch J, Hsu RY, Copeland NG, Jenkins NA. The gene for murine megakaryocyte growth and development factor (MGDF) is located on mouse chromosome 16. Genomics 1995; 26:636–637.
109. Gurney AL, Kuang WJ, Xie MH, Malloy BE, Eaton DL, de Sauvage FJ. Genomic structure, chromosomal localization, and conserved alternative splice forms of thrombopoietin. Blood 1995; 85:981–988.
110. Chang MS, McNinch J, Basu R, Shuter J, Hsu RY, Perkins C, Mar V, Suggs S, Welcher A, Li YS, Lu H, Bartley TD, Hunt P, Martin F, Samal B, Bogenberger J. Cloning and characterization of the human megakaryocyte growth and development factor (MGDF) gene. J Biol Chem 1995; 270:511–514.
111. Lok S, Kaushansky F, Holly RD, Kuijper JL, Lofton-Day CE, Oort PJ, Grant FJ, Heipel MD, Burkhead SK, Kramer JM, Bell LA, Sprecher CA, Blumberg H, Johnson R, Prinkard D, Ching AFT, Mathewes SL, Bailey MC, Forstrom JW, Muddle MM, Osborn SG, Evans SJ, Sheppard PO, Presnell SR, O'Hara PJ, Hagen FS, Roth GJ, Foster DC. Cloning and expression of murine thrombopoietin cDNA and stimulation of platelet production in vivo. Nature 1994; 369:565–568.
112. Nichol JL, Hokom MM, Hornkohl AC, Sheridan WP, Ohashi H, Kato T, Li YS, Bartley TD, Choi E, Bogenberger J, Skrine JD, Knudten A, Chen J, Trail G, Sleeman L, Cole S, Grampp G, Hunt P. Megakaryocyte Growth and Development Factor: Analyses of in vitro effects on human megakaryopoiesis and endogenous serum levels during chemotherapy-induced thrombocytopenia. J Clin Invest 1995; 95:2973–2978.
113. Ulich TR, Del Castillo J, Yin S, Swift S, Padilla D, Senaldi G, Bennet L, Bogenberger J, Sun D, Samal B, Shimamoto G, Lee R, Steinbrink R, Boone T, Sheridan W, Hunt P. Megakaryocyte growth and development factor ameliorates carboplatin-induced thrombocytopenia in mice. Blood 1995; 86:971–976.
114. Wendling F, Maraskovsky E, Debili N, Fiorindo C, Teepe M, Titeux M, Methia N, Breton-Gorius J, Cosman D, Vainchenker W. c-Mpl ligand is a humoral regulator of megakaryopoiesis. Nature 1994; 369:571–574.
115. Kuter D, Rosenberg RD. The reciprocal relationship of thrombopoietin (c-mpl ligand) to changes in the platelet mass during busulfan-induced thrombocytopenia in the rabbit. Blood 1995; 85:2720–2730.
116. Gurney AL, Carver-Moore K, de Sauvage FJ, Moore MW. Thrombocytopenia in c-mpl deficient mice. Science 1994; 265:1445–1447.
117. Broudy VC, Lin NL, Kaushansky K. Thrombopoietin (c-mpl ligand) acts synergistically with erythropoietin, stem cell factor, and interleukin 11 to enhance murine megakaryocyte colony growth and increases megakaryocyte ploidy in vitro. Blood 1995; 85:1719–1726.
118. Zeigler FC, de Sauvage F, Widmer HR, Keller GA, Donahue C, Schreiber RD, Malloy B, Hass P, Eaton D, Matthews W. In

vitro megakaryopoietic and thrombopoietic activity of c-mpl ligand (TPO) on purified murine hematopoietic stem cells. Blood 1994; 84:4045–4052.
119. Debili N, Wendling F, Cosman D, Titeux M, Florindo C, Dusanter-Fourt I, Schooley K, Charon M, Nador R, Bettaieb A, Vainchenker W. The mpl receptor is expressed in the megakaryocyte lineage from late progenitors to platelets. Blood 1995; 95:391–401.
120. Choi E, Hokom MM, Bartley TD, Li YS, Ohashi H, Kato T, Nichol JL, Skrine JD, Knudten A, Chen J, Hornkohl AC, Grampp G, Sleeman L, Cole S, Trail G, Hunt P. Recombinant human megakaryocyte growth and development factor produces functional human platelets in vitro. Stem Cells 1995; 13:317–322.
121. Mazur EM, Cohen JL. Basic concepts of hematopoiesis and the hematopoietic growth factors. Clin Pharmacol Ther 1989; 46:250–256.
122. Drachman JG, Griffin JD, Kaushansky K. The c-mpl ligand (Thrombopoietin) stimulates tyrosine phosphorylation of Jak2, Shc, and c-mpl. J Biol Chem 1995; 370:4979–4982.
123. Farese AM, Hunt P, Boone T, MacVittie TJ. Recombinant human megakaryocyte growth and development factor stimulates thrombopoiesis in normal primates. Blood (in press).
124. Kampschmidt RF, Long RD, Upchurch HF. Neutrophil releasing activity in rats injected with endogenous pyrogen. Proc Soc Exp Biol Med 1972; 139:1224.
125. Ulich TR, del Castillo J, Keys M, Granger GA, Ni RX. Kinetics and mechanisms of recombinant interleukin-1 and tumor necrosis factor alpha-induced changes in circulating numbers of neutrophils and lymphocytes. J Immunol 1987; 139:3406–3415.
126. Suttles J, Carruth LM, Mizel SB. Detection of IL-1a and IL-1b in the supernatants of paraformaldehyde-treated human monocytes. J Immunol 1990; 144:170–174.
127. Fibbe WE, Falkenburg JH, Schaafsma MR, Willemze R. The hematopoietic activities of interleukin-1. Biotherapy 1989; 1:263–271.
128. Okusawa S, Gelfand GA, Ikejima T, Connoly RJ, Dinarello C. IL-1 induces a shock-like state in rabbits: Synergism with TNF and the effect of cyclooxygenase inhibition. J Clin Invest 1988; 81:1162–1172.
129. Ulich TR, Keys M, Ni RX, del Castillo J, Dakay E. The contributions of adrenal hormones, hemodynamic factors, and the endotoxin related stress reaction to stable prostaglandin analog induced peripheral lymphopenia and neutrophilia. J Leuk Biol 1988; 43:5–10.
130. Dinarello CA, Neta R. An overview on interleukin-1 as a therapeutic agent. Biotherapy 1989; 1:245–254.
131. Stork LC, Peterson VM, Rundus CH, Robinson WA. Interleukin-1 enhances murine granulopoiesis in vivo. Exp Hematol 1988; 16:163–167.
132. Vogel SN, Douches SD, Kaufman EN, Neta R. Induction of colony stimulating factor in vivo by recombinant IL-1 and recombinant TNF. J Immunol 1987; 138:2143–2148.
133. Ettinghausen SE, Moore JG, White DE, Plataniass L, Young N, Rosenberg S. Hematologic effects of immunotherapy with lymphokine activated killer cells and recombinant interleukin-2 in cancer patients. Blood 1987; 1654–1660.
134. Anderson TD, Hayes TJ. Toxicity of human recombinant IL-2 in rats. Pathologic changes are characterized by marked lymphocytic and eosinophilic proliferation and multisystem involvement. Lab Invest 1989; 60:331–346.
135. Anderson TD, Hayes TJ, Gately MK, Bontempo JM, Stern LL, Truitt GA. Toxicity of human recombinant IL-2 in the mouse is mediated by interleukin activated lymphocytes. Lab Invest 1988; 59:598–612.
136. Ettinghausen SE, Lipford E, Mule J, Rosenberg SA. Systemic administration of recombinant IL-2 stimulates in vivo cell proliferation in tissues. J Immunol 1985; 135:1488–1497.
137. Jesmok GJ, Gunther R. In vivo biology of recombinant IL-2 infusion in sheep. Inflammation 1989; 3:267–283.
138. Lotze M, Matory Y, Ettinghausen S, Raynor A, Sharrow S, Seipp C, Custer M, Rosenberg SA. In vivo administration of interleukin-2. J Immunol 1985; 135:2865–2875.
139. Essner R, Rhoades K, McBride WH, Morton DL, Econom JS. IL-4 downregulates IL-1 and TNF gene expression in human monocytes. J Immunol 1989; 142:3857.
140. Tepper RI, Levinson DA, Stanger BZ, Campos-Torres J, Abbas AK, Leder P. IL-4 induces allergic-like inflammatory disease and alters T cell development in transgenic mice. Cell 1990; 62:457–467.
141. Dvorak AM, Tepper RI, Weller PF, Morgan ES, Estrella P, Monahan-Earley RA, Galli SJ. Piecemeal degranulation of mast cells in the inflammatory eyelids lesions of interleukin-4 transgenic mice. Blood 1994; 83:3600–3612.
142. Lewis DB, Liggitt HD, Effmann EL, Motley ST, Teitelbaum SL, Jepsen KJ, Goldstein SA, Bonadio J, Carpenter J, Perlmutter RM. Osteoporosis induced in mice by overproduction of interleukin 4. Proc Natl Acad Sci 1993; 90:11618–11622.
143. Sanderson CJ, Campbell HD, Young IG. Molecular and cellular biology of eosinophil differentiation factor (IL-5) and its effects on human and mouse B cells. Immunol Rev 1988; 102:29–50.
144. Basten A, Beeson PB. Mechanism of eosinophilia. II Role of the lymphocyte. J Exp Med 1970; 131:1288.
145. Coffman RL, Seymour BW, Hudak S, Jackson J, Rennick D. Antibody to interleukin-5 inhibitors helminth-induced eosinophilia in mice. Science 1989; 245:308–310.
146. Samoszuk M, Nansen L. Detection of IL-5 mRNA in Reed-Sternberg cells of Hodgkins disease with eosinophilia. Blood 1990; 75:13–16.
147. Sehgal PB, May LT. Human interferon-beta$_2$. J Interferon Res 1987; 7:521.
148. Shabo Y, Lotem J, Rubenstein M, Revel M, Clark S, Wolf S, Kamen R, Sachs L. The myeloid blood cell differentiation-inducing protein MGI-2A is interleukin-6. Blood 1988; 72:2070–2073.
149. Le J, Vilcek J. Interleukin-6: a multifunctional cytokine regulating immune reactions and the acute phase protein response. Lab Invest 1989; 61:588–602.
150. Suzuki C, Okano A, Takatsuki F, Miyasaki Y, Hirano T, Kishimoto T, Ejima D, Akiyama Y. Continuous perfusion with IL-6 enhances production of hematopoietic stem cells. Biochem Biophys Res Comm 1989; 159:933–938.
151. Ulich TR, del Castillo J, Guo K. In vivo hematologic effects of recombinant IL-6 on hematopoiesis and circulating numbers of RBCs and WBCs. Blood 1989; 73:108–110.
152. Rennick D, Jackson J, Yang G, Wideman J, Lee F, Hudak S. Interleukin-6 interacts with IL-4 and other hematopoietic growth factors to selectively enhance the growth of megakaryocytic, erythroid, myeloid and multipotential progenitor cells. Blood 1989; 73:1828–1835.
153. Ishibashi T, Kimura H, Shikama Y, Uchida T, Karyone Y, Hirano T, Kishimoto T, Takatsuki F, Akiyama Y. Interleukin-6 is a potent thrombopoietic factor in mice. Blood 1989; 74:1241–1244.
154. Schindler R, Mancilla J, Endres S, Ghorbani R, Clark S, Dinarello C. Correlations and interactions in the production of

IL-6, IL-1, and TNF in human blood mononuclear cells: IL-6 suppresses IL-1 and TNF. Blood 1990; 75:40–47.
155. Aderka D, Le J, Vilcek J. IL-6 inhibits LPS-induced TNF production in cultured human monocytes, U937 cells, and in mice. J Immunol 1989; 143:3517–3523.
156. Suematsu S, Matsuda T, Aozasa K, Akira S, Nakano N, Ohno S, Miyazaki J-I, Yamamura K-I, Hirano T, Kishimoto T. IgG1 plasmacytosis in interleukin 6 transgenic mice. Proc Natl Acad Sci 1989; 86:7547–7551.
157. Campbell IL, Hobbs MV, Dockter J, Oldstone MBA, Allison J. Islet inflammation and hyperplasia induced by the pancreatic islet-specific overexpression of interleukin-6 in transgenic mice. Am J Pathol 1994; 145:157–166.
158. Armitage RJ, Namen AE, Sassenfeld HM, Grabstein KH. Regulation of human T-cell proliferation by IL-7. J Immunol 1990; 144:938–941.
159. Komschlies KL, Gregorio TA, Gruys ME, Back TC, Falynek CR, Wiltront RH. Administration of recombinant human IL-7 to mice alters the composition of B-lineage cells and T-cell subsets, enhances T-cell function, and induces regression of established metastases. J Immunol 1994; 152:5776–5784.
160. Boerman OC, Gregorio TA, Crzegorzewski KJ, Faltynek CR, Kenney JJ, Wiltront RH, Kamschlies KL. Recombinant human IL-7 administration in mice affects colony forming units, spleen and lymphoid precursor cell localization and accelerates engraftment of bone marrow transplants. J Leuk Biol 1995; 58:151–158.
161. Matsushima K, Oppenheim JJ. Interleukin-8 and MCAF: novel inflammatory cytokines inducible by IL-1 and TNF. Cytokine 1989; 1:2–13.
162. Furuta R, Yamagishi J, Kotani H, Sakamoto F, Fukui T, Matsui Y, Sohmura Y, Yamada M, Yoshimura T, Larsen CG, Oppenheim JJ, Matsushima K. Production and characterization of recombinant human neutrophil chemotactic factor. J Biochem 1989; 106:436.
163. Gimbrone MA Jr, Obin MS, Brock AF, Luis EA, Hass PE, Hebert CA, Yip YK, Leung DW, Lowe DG, Kohn WJ. Endothelial interleukin-8: a novel inhibitor of leukocyte-endothelial interactions. Science 1989; 246:1601–1603.
164. Hebert CA, Luscinskas FW, Kiely JM, Luis EA, Darbonne WC, Bennett GL, Liu CC, Obin MS, Gimbrone MA, Baker JB. Endothelial and leukocyte forms of IL-8. Conversion by thrombin and interactions with neutrophils. J Immunol 1990; 145:3033–3040.
165. Hechtman DH, Cybulsky MI, Fuchs HJ, Baker JB, Gimbrone MA. Intravascular IL-8: inhibitor of polymorphonuclear leukocyte accumulation at sites of acute inflammation. J Immunol 1991; 147:883–892.
166. Ley K, Baker JB, Cybulsky MI, Gimbrone JA, Luscinskas FW. Intravenous Interleukin-8 inhibits granulocyte emigration from the rabbit mesenteric venules without altering L-selectin expression or leukocyte rolling. J Immunol 1993; 151:6347–6357.
167. Renauld J-C, Kermouni A, Vink A, Louahed J, Van Snick J. Interleukin-9 and its receptor: involvement in mast cell differentiation and T cell oncogenesis. J Leuk Biol 1995; 57:353–360.
168. Renauld J-C, van der Lugt N, Vink A, van Roon M, Godfraind C, Warnier G, Merz H, Feller A, Berns A, Van Snick J. Thymic lymphomas in interleukin 9 transgenic mice. Oncogene 1994; 9:1327–1332.
169. Hendrzak JA, Brunda MJ. Biology of disease: interleukin-12 biologic activity, therapeutic utility, and role in disease. Lab Invest 1995; 72:619–637.
170. Sarmiento UM, Riley JH, Knaack PA, Lipman JM, Becker JJ, Gately MK, Chizzonite R, Anderson TD. Biologic effects of recombinant human interleukin-12 in squirrel monkeys (Sciureus saimiri). Lab Invest 1994; 71:862–873.
171. Simonet WS, Hughes TM, Nguyen HQ, Trebasky LD, Danilenko DM, Medlock ES. Long-term impaired neutrophil migration in mice overexpressing human interleukin-8. J Clin Invest 1994; 94:1310–1319.
172. Van Zee KJ, Fischer E, Hawes AS, Hebert CA, Terrell TG, Baker JB, Lowry SF, Moldawer LL. Effects of intravenous IL-8 administration in nonhuman primates. J Immunol 1992; 148:1746–1752.
173. Klimpel GR, Fleishmann WR, Klimpel KD. Gamma IFN and IFN alpha/beta suppresses murine myeloid colony formation: magnitude of suppression is dependent upon level of colony stimulating factor. J Immunol 1982; 29:76–80.
174. Kurzrock R, Rosenblum M, Sherwin S, Rios A, Talpaz M, Quesada J, Gutterman J. Pharmacokinetics, single dose tolerance, and biological activity of recombinant gamma interferon in cancer patients. Canc Res 1985; 45:2866–2872.
175. Gresser I, Guy-Grand D, Maury C, Maunoury MT. Interferon induces peripheral lymphadenopathy in mice. J Immunol 1981; 127:1569–1575.
176. Zoumbos NC, Gascon P, Djeu JY, Young N. Interferon is a mediator of the hematopoietic suppression in aplastic anemia in vitro and possibly in vivo. Proc Natl Acad Sci 1985; 82:188–192.
177. Old LJ. Tumour necrosis factor—polypeptide mediator network. Nature 1987; 326:330–331.
178. Le J, Vilcek J. Tumor necrosis factor and interleukin-1: cytokines with multiple overlapping biological activities. Lab Invest 1987; 56:234–248.
179. Ulich TR, del Castillo J, Keys M, Granger GA. Recombinant human alpha lymphotoxin (TNF-b) induces peripheral neutrophilia and lymphopenia in the rat. Am J Pathol 1987; 128:5–12.
180. Remick DG, Kunkel RG, Larrick JW, Kunkel SL. Acute in vivo effects of recombinant TNF. Lab Invest 1987; 56:583–590.
181. Gamble JR, Harlan JM, Klebanoff SJ, Vadas MA. Stimulation of the adherence of neutrophils to umbilical vein endothelium by human recombinant TNF. Proc Natl Acad Sci 1985; 82:8667–8671.
182. Larrick JW, Graham D, Toy K, Lin L, Senyk G, Fendly B. Recombinant TNF causes activation of human granulocytes. Blood 1987; 69:640–644.
183. Ming WJ, Bersani L, Mantovani A. TNF is chemotactic for monocytes and polymorphonuclear leukocytes. J Immunol 1987; 138:1469–1474.
184. Ulich TR, del Castillo J, Ni RX, Bikhazi N. Hematologic interactions of endotoxin, TNF, IL-1, and adrenal hormones and the hematologic effects of TNF in C. parvum-primed rats. J Leuk Biol 1989; 45:546–557.
185. Bachwich PR, Chensue SW, Larrick JW, Kunkel SL. TNF stimulates IL-1 and PGE2 production in resting macrophages. Biochem Biophys Res Comm 1986; 136:94.
186. Dinarello CA, Cannon JG, Wolff SM, Bernheim HA, Beutler B, Cerami A, Figari IS, Palladino MS, O'Conner JV. TNF is an endogenous pyrogen and induces production of IL-1. J Exp Med 1986; 163:1433.
187. Koeffler HP, Gasson J, Ranyard J, Souza L, Shepard M, Munker R. Recombinant human TNF stimulates production of G-CSF. Blood 1987; 70:55–59.
188. Ulich TR, del Castillo J, Yin S. TNF exerts dose-dependent effects on erythropoiesis and myelopoiesis in vivo. Exp Hematol 1990; 18.

189. Waage A. Production and clearance of TNF in rats exposed to endotoxin and dexamethasone. Clin Immunol Immunopath 1987; 45:348–355.
190. Remick D, Strieter RM, Eskandari MK, Nguyen DT, Genord MA, Raiford CL, Kunkel SL. Role of TNF in LPS-induced pathologic alterations. Am J Pathol 1990; 136:49–60.
191. Bevilacqua M, Pober JS, Mendrick DL, Cotran R, Gimbrone MA. Identification of an inducible endothelial-leukocyte adhesion molecule. PNAS USA 1987; 84:9238.
192. Cavender D, Saegusa Y, Ziff M. Stimulation of endothelial cell binding by tumor necrosis factor. J Immunol 1987; 139:1855–1860.
193. Munro JM, Pober JS, Cotran RS. TNF and IFN gamma induce distinct patterns of endothelial activation and associated leukocyte accumulation in the skin of Papio Anubis. Am J Pathol 1989; 135:121–133.
194. Michie HR, Manogue KR, Spriggs Dr, et al. Detection of circulating TNF after endotoxin administration. N Engl J Med 1988; 318:1481–1486.
195. Moldawer LL, Marano M, Wei H, Fong Y, Silen M, Kuo G, Manogue K, Vlassaor H, Coheb H, Cerami A, Lowry SF. Cachectin/TNF-alpha alters red blood cell kinetics and induces anemia in vivo. FASEB J 1989; 3:1637–1643.
196. Broxmeyer HE, Williams DE, Lu L, Cooper S, Anderson SL, Beyer GS, Hoffman R, Rubin B. The suppressive influences of tumor necrosis factors on bone marrow hematopoietic progenitor cells from normal donors and patients with leukemia: synergism of TNF and IFN-gamma. J Immunol 1986; 136:4487.
197. Johnson CS, Cook CA, Furmanski P. In vivo suppression of erythropoiesis by TNF: reversal with exogenous EPO. Exp Hematol 1990; 18:109–113.
198. Johnson CS, Chang MJ, Furmanski P. In vivo hematopoietic effects of TNF in normal and erythroleukemic mice: characterization and therapeutic applications. Blood 1988; 72:1875.
199. Clark IA, Chandri G. TNF may contribute to the anemia of malaria by causing dyserythropoiesis and erythrophagocytosis. Br J Haematol 1988; 70:99.

32
Cytokines and the Cardiovascular System

Göran K. Hansson, Veronika Stemme, and Tasuku Yokota
King Gustaf V Research Institute, Karolinska Institute, Stockholm, Sweden

INTRODUCTION

The immune system is directly exposed to the blood vessels through which its components continuously circulate. Interactions between immune mediators and vascular cells regulate basic functions of the immune system such as lymphocyte migration, regional immunity, and inflammatory responses. Immune mediators also modulate cardiovascular functions controlling both systemic and regional hemodynamics. Finally, several cardiovascular diseases are caused or modulated by immune responses. Many of these events are mediated by cytokines; in fact, vascular cells both respond to and produce many different cytokines.

In this chapter, we will first describe how cytokines modulate the functions of vascular cells. In particular, we will focus on smooth muscle cells, the quantitatively dominating cell type in the vessel wall. We will then discuss how cytokines may influence two basic cardiovascular activities: vascular tone and hemostasis. Finally, the involvement of cytokines in thrombosis and atherosclerosis will be reviewed.

For obvious reasons, we will build upon several previous chapters in this volume and particularly the one on endothelial cells. Readers who wish to obtain a comprehensive view of cytokine regulation of the cardiovascular system should therefore read these two chapters together.

EFFECTS OF CYTOKINES ON VASCULAR CELLS

Antigen Presentation by the Vascular Endothelium

The vascular endothelium plays an important role in the cytokine-regulated recruitment of leukocytes from blood to tissues, as discussed in several previous chapters of this volume. Endothelial cells (EC) also perform a series of other functions relating to immunity, inflammation, hemostasis, and permeability, which can all be modulated by cytokines. This large repertoire of cytokine-controlled events puts the endothelium in a central position as a target for and source of proinflammatory and immune-regulating cytokines. Several of these phenomena are discussed in other chapters, and we will focus only on the role of the endothelium in antigen presentation.

One of the most important, cytokine-regulated endothelial functions is the capacity of these cells to activate immunocompetent T cells by presenting foreign antigens. Endothelial cells cannot present antigens under baseline conditions, but stimulation with the immune-regulating cytokine interferon-γ (IFN-γ) renders the endothelium capable of antigen presentation.

This activity was discovered by Hirschberg et al., who showed that cultures of umbilical vein endothelial cells can activate allogeneic T lymphocytes in mixed cultures (1). Pober et al. unveiled the cytokine-dependent mechanism that endows antigen-presenting capacity to the endothelium. IFN-γ induces transcriptional expression of major histocompatibility complex (MHC) class II molecules, which can bind fragments of internalized peptide antigens in an endosomal compartment. The MHC-oligopeptide fragments are transported to the cell surface, where they can be recognized by antigen-specific CD4+ T lymphocytes (2–5). Although only IFN-γ is necessary for induction of MHC class II expression by EC, several other cytokines modulate the IFN-γ–induced MHC expression, including TNF-α, IL-1, and IL-3 (5,6).

Antigen presentation to CD4+ T cells depends not only on MHC expression but also on the ability of the antigen-presenting cell to internalize and process antigen, secrete IL-1, and express costimulatory adhesion molecules such as ICAM-1, LFA-3, and B7/CD80. The IFN-γ–activated endothelial cell must therefore also be able to carry out these functions (5). Similar and in part identical molecular interactions operate in endothelial antigen presentation via the synthetic route to HLA class I–restricted CD8+ T lymphocytes (7).

An interesting functional difference exists between endothelial cells, which respond to IFN-γ by MHC class II expression and antigen-presenting capacity, on one hand, and fibroblasts and smooth muscle cells, which also respond to IFN-γ by MHC class II expression but cannot de novo activate resting T cells. This difference could be due to differences in the capacity to express costimulatory molecules and/or antigen-processing capacity (8–11).

In vivo, endothelial cells of the microvasculature are often seen to express MHC class II molecules (5) and such expression can be observed also in large vessel endothelium under pathological conditions such as chronic rejection of organ transplants (12). Due to its huge surface area, the endothelium should have a huge potential for antigen presentation and immune activation, which would be activated by IFN-γ secreted by activated T cells. This probably represents an important amplification loop for immune activation.

Smooth Muscle Cells

Vascular smooth muscle cells (SMC) constitute the bulk of tissue in arteries and veins. They are important both as regulators of hemodynamics (blood flow, regional perfusion) and as producers of the matrix of blood vessels. They must also be able to repair vascular defects by scar formation. In order to accomplish all these functions, SMC must be able to act both as myocytes and as fibroblasts.

This problem is solved by the pluripotency of vascular SMC, which are relatively primitive compared with other muscle cells. Thus, they have the capacity to differentiate into highly contractile muscle cells but also to dedifferentiate into rapidly dividing mesenchymal cells with a sparse contractile apparatus (13–17). The differentiation program is regulated by the microenvironment but also by several different soluble factors. Some of the most important of these are cytokines. In this section, we will discuss cytokine regulation of smooth muscle proliferation, differentiation, and contractility and the expression of immune-related surface molecules.

Regulation of SMC Proliferation

In the normal blood vessel, SMC are in a quiescent state with regard to their cell cycle (16,17). However, when they are dissociated from the matrix of the vascular media and put on a plastic surface, they start to divide vividly. Proliferation continues even after the formation of a monolayer on the bottom of the culture dish since the cells do not respect contact inhibition. They are, however, strictly dependent on serum factors for their growth.

Experiments with blocking antibodies and recombinant growth factors have shown that basic fibroblast growth factor (bFGF) and platelet-derived growth factor (PDGF) are the two most important serum factors that induce G_0/G_1 transition in the cell cycle of cultured SMC (16,18). In addition, IGF-1 is necessary for growth, and angiotensin II appears to play a modest, growth-stimulatory role (19–21).

Recent in vivo studies have dissected growth regulation in the intact artery using a model in which the endothelium, intima, and innermost media are removed from the carotid artery of rats by a balloon catheter. This results in a wave of replication that is elicited by bFGF, which is probably released from extracellular stores in the artery (22–24). The accompanying microthrombus formed on the deendothelialized arterial surface can lead to PDGF release from adhering platelets. This factor acts as a chemoattractant, stimulating SMC migration from the media to the intima (25,26), where SMC continue to proliferate under the influence of growth factors.

Cell culture studies have shown that proinflammatory and immune-regulatory cytokines modify these proliferative responses. IL-1 has a weak, growth-promoting effect on SMC by upregulating PDGF receptors and increasing autocrine PDGF production (27,28). This is, however, counteracted by its induction of the growth-inhibitory prostaglandin PGE_1 (27). In vivo, it is likely that cellular growth and differentiation is regulated in a complex fashion by integration of signals derived from cytokines, growth factors, and autacoids (29–32).

Interferons are potent growth inhibitors for cultured SMC (11,33–36). This is due to a direct inhibitory effect on growth factor–induced movement through the first part of the G_1 phase of the cell cycle (33,37). IFN-γ also inhibits collagen production by SMC (38), which could contribute to its drastic effects on arterial scar formation (see below). The inhibitory effect of IFN-γ on SMC replication is, however, dependent on the simultaneous presence of growth factors in the extracellular milieu. Therefore, IFN-γ upregulates PDGF receptors, leading to a paradoxically increased growth factor sensitivity in a serum-free environment (39). The divergent effects of IFN-γ in growth factor–poor versus growth factor–rich cell culture environment might reflect a differential regulation of vascular cell proliferation by IFN-γ–producing T and NK cells depending on the local availability of growth factors in different stages of inflammation in vivo.

Animal experiments using the rat carotid artery injury model have revealed that IFN-γ is an in vivo modulator of SMC proliferation and tissue accumulation in the arterial intima. Injection of recombinant IFN-γ inhibits SMC proliferation and reduces the size of intimal hyperplastic lesions (40,41). Administration of IFN-γ during the first week after injury causes persistent growth inhibition and reduces lesion size even 10 weeks later (41). This suggests that interferon-mediated growth modulation during the early phase of the response determines the final outcome of the lesion.

Removal of IFN-γ–producing T lymphocytes using cytolytic antibodies, in contrast, increases SMC proliferation and lesion formation in the injured artery (40). Similarly, lesions become significantly larger in T-cell–deficient, homozygous rnu/rnu rats compared to T-cell–competent, heterozygous rnu/+ littermates (40). Together, the results from genetically T-cell–defective, T-cell–depleted, and IFN-γ–treated rats suggest that the Th1 response inhibits vascular and other connective tissue repair processes. It is possible that recombinant IFN-γ might be useful as a therapeutic agent to reduce scar formation and restenosis after surgical and angioplastic procedures in arteries.

Smooth Muscle Differentiation and Matrix Production

SMC produce the extracellular matrix of the vessel wall and therefore have a high capacity to synthesize structural proteins such as collagens, elastins, basement membrane components, and core proteins of proteoglycans. This is reflected in the ultrastructure, which is dominated by endoplasmic profiles together with contractile filaments. In fact, the proportions of these two components can be used to determine the phenotypic state of the smooth muscle cell (17,42). Thus, contractile filaments dominate in contractile SMC of the media, which regulate vascular tone, while endoplasmic reticular structures are abundant in "synthetic" SMC found in the intima and in cell culture systems.

Agents such as PDGF, which induce SMC proliferation, also affect differentiation and matrix formation (16,17,43). In contrast, agents that degrade the extracellular matrix modulate SMC phenotype (44). This implies that these phenomena are linked and that SMC growth and differentiation could be regulated by cell-matrix interactions. Studies on cytokine effects on SMC support this notion.

The extracellular matrix of the vessel wall is controlled by inflammatory cytokines. Interferon-γ is a potent inhibitor of collagen synthesis (38) and also inhibits production of α-actin, the major component of contractile filaments, and DNA synthesis in SMC (34). IL-1 and TNF exert important control of the extracellular matrix by inducing metalloproteinases that degrade matrix components (45). Proteolytic degradation of the matrix reduces adhesive interactions between SMC and their microenvironment. This in turn causes the SMC to dedifferentiate and proliferate in response to growth factors (44,46,47).

The antiinflammatory cytokine TGF-β exerts effects on SMC opposite of those of the proinflammatory ones. Thus, TGF-β stimulates collagen (20,48,49) and α-actin production (50) and induces fibrotic hyperplasia when transfected into the arterial wall (51). Consequently, antibodies to TGF-β inhibit neointimal hyperplasia after mechanical injury (52). TGF-β also modulates growth of SMC, but its effects are complex, dependent on phenotypic state, and mediated via expression of other growth factors and their receptors (53,54).

It has recently been shown that M-CSF may regulate phenotypic modulation of SMC (55). Several growth factors, including PDGF, EGF, and FGF, increase SMC expression of M-CSF receptors, i.e., the c-*fms* protooncogene, in parallel with inducing phenotypic modulation and DNA synthesis. It is therefore possible that engagement of the c-fms/M-CSF receptor could be a critical eliciting step in phenotypic modulation of SMC (55).

The effects of immune-derived cytokines on vascular cells are not limited to changing the balance between gene products normally expressed by these cells. In addition, immune-regulating and proinflammatory cytokines (i.e., IFN-γ, TNF-α, and IL-1) also endow on vascular SMC and EC the capacity to interact directly with immunocompetent cells. In addition to the antigen-presenting function discussed above, these cytokines also induce expression of leukocyte adhesion molecules. Their expression on EC is well known and described at length elsewhere in this volume. However, SMC also respond to IFN-γ, TNF-α, and IL-1 by expressing leukocyte adhesion molecules (56,57). This renders SMC capable of binding T cells;

the functional consequences of such interactions are still largely unknown.

To conclude, vascular cells are important targets for immune-derived cytokines, which regulate growth and differentiated functions both in EC and SMC. However, vascular cells are not only targets for but also sources of cytokines.

Vascular Synthesis of Cytokines

The first hint that vascular cells could produce immune cytokines came from the studies of antigen presentation. The fact that EC can activate antigen-specific T lymphocytes suggests that they not only process and present antigens but also produce co-stimulatory molecules. One of these is the cytokine IL-1. Studies by Libby et al. have shown that both EC and SMC produce IL-1 upon stimulation with proinflammatory cytokines or endotoxin (27,58,59). In addition, both cell types also make IL-6, M-CSF, and several chemokines (60–64).

In addition to their production of proinflammatory cytokines, EC and SMC synthesize several growth factors. Expression of PDGF and its receptors has been demonstrated both in vitro and in vivo (65–75) and is likely to be important in wound repair and under several pathological conditions (see below).

REGULATION OF SMC CONTRACTILITY AND VASCULAR TONE

Vascular contractility is caused by interaction between actin and myosin filaments in SMC. Some of the filament-forming proteins such as α-SM-actin are specific for SMC, but the filaments themselves are not nearly as well developed as in striated muscle. The contractile process is dependent on energy provided by ATP derived from mitochondrial respiration. It is regulated by several different soluble factors, including α-adrenergic agonists from local nerve endings and from the blood, circulating angiotensin, and endothelial-derived factors such as endothelin and nitric oxide (NO).

Proinflammatory cytokines (TNF, IL-1, IFN-γ) modulate SMC contractility on several levels of this regulatory system. The most long-lasting effect on SMC contractility is caused by direct interference with the production of contractile filaments. IFN-γ downregulates expression of the α-SM-actin gene (34).

A more rapid effect on vascular contractility is accomplished by cytokine regulation of NO production (for reviews, see Refs. 76–79). As mentioned, NO is normally released by endothelial cells. It diffuses over to SMC, where it nitrosylates the heme group of guanylyl cyclase, resulting in an activation of the enzyme to produce cGMP. The elevated cGMP level activates the myosin kinase cascade, resulting in dissociation of myosin from actin and a relaxation of the SMC. The artery is normally under modest NO-dependent vasodilation, and systemic administration of NO synthase inhibitors results in increased vascular tone and elevated blood pressure (79).

Endothelial NO synthesis is accomplished by the enzyme, endothelial NO synthase (eNOS or NOS-3). It is constitutively expressed as a protein but requires activation by Ca^{2+}/calmodulin (79). This occurs after stimulation of the EC by bradykinin, acetylcholine, and other stimuli. Since NO synthase inhibitors increase vascular tone under baseline conditions, the normal vessel wall is probably in a state of NO-dependent, partial relaxation (79,80).

SMC do not normally produce NO but can be stimulated to do so by proinflammatory cytokines. TNF-α, IL-1, lipopolysaccharide (LPS), and IFN-γ induce the production of large amounts of NO in cultured SMC (81–84). The inducible NO synthase isoform expressed upon cytokine stimulation, iNOS or NOS-2, binds calmodulin with high affinity immediately after translation and its activity is therefore independent of Ca^{2+} (85,86). This results in a high capacity for NO synthesis, which lasts until the enzyme protein is degraded (85,86). The iNOS expressed by SMC appears to be identical to the one expressed by cytokine-activated macrophages and is highly conserved between species (87).

iNOS expression is induced synergistically by IFN-γ and either TNF-α, IL-1, or LPS. Its promoter contains 10 copies of the IFN-γ response element, 3 copies of the IFN-γ activated site, and 2 copies each of NF-κB, AP-1, and TNF-response element (88). Cytokine-induced transcription factors therefore act in concert to induce transcription of the iNOS gene.

NO produced by cytokine-stimulated SMC acts as an auto- and paracrine mediator to control contractility and metabolism of the cells. Moderate cytokine stimulation induces sufficient NO to nitrosylate SMC heme proteins, including guanylyl cyclase (89). At higher levels of cytokine stimulation, the output of NO is sufficient to react with iron atoms in the iron-thiol groups of many intracellular enzymes, which therefore lose their activity (89). This causes inhibition of mitochondrial respiration, ATP deficiency, and a switch to anaerobic glycolysis in the cell (83,89). The combined effects of guanylyl cyclase activation and reduced energy levels are likely to explain the vasodilation of inflammation (83).

In vivo, the iNOS gene is activated in SMC during the response to vascular injury. Deendothelializing balloon catheter injury to the rat carotid artery causes rapid expression of iNOS in the underlying SMC layer (90). iNOS expression is maintained in the SMC for more than a week, probably due to cytokine stimulation. Enzyme expression is, however, induced so rapidly after deendothelialization (hours) that it is unlikely to be caused by de novo production of IFN-γ, TNF-α, and/or IL-1 from infiltrating leukocytes. Instead, there may be release of extracellularly deposited cytokines or activation of transcription through cytokine-independent mechanisms. iNOS activity is likely to be important by dilating vessels during inflammation, modulating platelet deposition and thrombus formation on the injured vessel, and controlling SMC proliferation during the response to injury (90,91).

CYTOKINES IN CARDIOVASCULAR DISEASES

Many diseases of the cardiovascular system involve cytokines as mediators of pathogenesis and defense mechanisms. Most of these reactions are presently not completely known, although much new information has been obtained during the last 10 years. We will focus on four important cardiovascular diseases in which cytokines are known to play a role, namely atherosclerosis, thrombosis, vasculitis, and arteriosclerosis of transplanted organs.

Thrombosis and Hemostasis

Several diseases in which inflammatory mediators are present, e.g., sepsis, vasculitides and atherosclerosis, are often complicated by thrombosis. Under normal conditions, the endothelial cells constitute a nonthrombogenic surface due to their production of antithrombotic factors such as prostacyclin (92), heparinlike molecules (93), nitric oxide (94), and an activator of the fibrinolytic system, tissue-type plasminogen activator (t-PA) (95). However, endothelial cells are also able to produce procoagulant substances, e.g., von Willebrand factor (96), tissue factor (97), and the fibrinolytic system inhibitors plasminogen activator inhibitor-1 (PAI-1) (98) and plasminogen activator inhibitor-2 (PAI-2) (99).

TNF-α and IL-1 have been shown to increase the expression of PAI-1 (100,101). The effect of TNF-α on PAI-1 expression has been confirmed in vivo in healthy volunteers (102). TNF-α also increases the expression of urokinase plasminogen activator (u-PA) (103) and PAI-2 (104) in cultured endothelial cells. These effects are modulated by IFN-γ: the TNF-α–induced expression of u-PA is antagonized by IFN-γ (105), while the TNF-α–induced expression of PAI-2 is synergistically increased by IFN-γ (106). The net effect of TNF-α and IFN-γ is antifibrinolytic in cultured endothelial cells due to increased expression of PAI-1 and PAI-2.

TNF-α also induces a procoagulative state of endothelial cells by stimulating their production of tissue factor, an essential cofactor for activation of the extrinsic pathway of coagulation (107). Procoagulant activity is also induced by a T-cell–derived cytokine, macrophage procoagulant-inducing factor (MTIF) (108,109).

Increased expression of PAI-1 (110,111), t-PA, and u-PA (112,113) has been demonstrated in smooth muscle cells and macrophages of atherosclerotic arteries. The expression of PAI-1 exceeds that of the plasminogen activators, suggesting that fibrinolysis is inhibited (113). In addition, tissue factor has been detected in atherosclerotic plaques, which appears to be in a procoagulative state (114). Cytokines produced in plaques could be important by switching the hemostatic balance toward coagulation and thrombosis.

Cytokine-induced expression of plasminogen activators and their inhibitors may have a role in the development of the plaque as well as in the induction of thrombotic complications. t-PA and u-PA could induce degradation of the extracellular matrix by activating matrix metalloproteinases (MMP). These enzymes degrade extracellular matrix and are activated by plasmin. Their activation is inhibited by plasmin inhibitors, tissue inhibitors of metalloproteinases, and PA inhibitors (115,116). PAI-1 may limit MMP activation and thus protect a developing plaque from uncontrolled matrix degradation. In advanced atherosclerosis, increased expression of MMPs and matrix-degrading activity has been detected in vulnerable regions of the plaque (117,118), and it is possible that cytokine-induced MMPs may promote destabilization and rupture of the plaque.

PAI-1 could therefore exert effects in advanced atherosclerotic lesions that appear to be contradictory. By inhibiting MMP activation, it may protect the plaque against rupture. However, if the plaque is already destabilized, PAI-1 may promote thrombosis formation on the plaque surface. The role of proinflammatory cytokines in these processes is intriguing and deserves further clarification.

Atherosclerosis

Atherosclerosis, the quantitatively dominating lethal disease in Western societies, has been the subject of huge research efforts over many years. It is clear that cholesterol-bearing lipoproteins are important in its pathogenesis, where local cholesterol accumulation together with tissue repair processes appear to have gone awry in the arterial intima (119). Increasing evidence pinpoints inflammation as an additional, important pathogenetic mechanism in atherosclerosis (31,120). In this chapter, we will discuss the possible role of cytokines in the different phases of the atherosclerotic process.

Fatty Streak Formation

Atherosclerosis is a chronic pathological process in the intima of large arteries. It starts when cholesterol-bearing low-density lipoprotein (LDL) infiltrates the arterial intima in such amounts that the normal elimination mechanisms are exceeded (119,121). In the intima, LDL adheres to proteoglycans of the extracellular matrix (122) and is oxidized by free oxygen radicals and enzymes produced by vascular endothelial cells and macrophages (121).

Oxidized LDL is a ligand for the macrophage scavenger receptor (MSR), which binds proteins and particles with clustered negative charges (123). Receptor-ligand complexes are internalized and degraded intracellularly, and fragments of the ligands can associate with MHC molecules and be presented as T-cell antigens (124). LDL cholesterol entering the macrophage via MSR accumulates in cytoplasmic droplets; the increasing intracellular cholesterol accumulation gradually transforms the macrophage into a foam cell, which is the hallmark of atherosclerosis (125). Foam cells form fatty streaks, i.e., lesions of the arterial intima, which predispose to atherosclerosis. Together with macrophages and macrophage-derived foam cells, T lymphocytes are also observed in the fatty streak (126,127).

In contrast to the case of the "normal" receptor for native LDL, MSR expression is not regulated by the intracellular cholesterol content (125). This permits continuing cholesterol accumulation and foam cell formation. MSR expression is, however, controlled by cytokines. M-CSF increases MSR expression on the mRNA level and enhances foam cell formation (128–130). In addition to M-CSF, an S-100–like chemotactic cytokine has also been shown to control scavenger receptor expression (131).

As a differentiating factor for the macrophage lineage as well as a regulator of lipoprotein uptake, M-CSF can be expected to have profound effects on atherosclerosis. Its importance was recently confirmed by the observation that *op/op* mice, which carry a defective M-CSF gene, only form diminutive fatty streak lesions when fed an atherogenic diet, even when bred onto the highly atherosclerosis-prone, apolipoprotein E-knock-out background (132).

Proinflammatory cytokines and endotoxins exert effects on MSR opposite to those of M-CSF. Both IFN-γ and TNF-α downregulate MSR expression and inhibit foam cell formation (133–135). In the case of IFN-γ, modulation is accomplished by destabilization of cytoplasmic mRNA, whereas TNF-α inhibits transcription of the MSR gene (135,136). Endotoxin has similar effects on MSR expression as TNF-α and appears to act by inducing autocrine TNF-α secretion, which in turn inhibits MSR gene transcription (135,137). The importance of proinflammatory cytokines for MSR expression is supported by the in vivo observation that areas of atherosclerotic plaques with inflammatory activity and T-cell activation show reduced expression of MSR proteins (138). This suggests that macrophage–T-cell interactions may be decisive for the development of the fatty streak.

Transition into Atherosclerosis

Studies of cholesterol-fed primates and rabbits have shown that fatty streaks progress into atherosclerotic plaques (119). However, only some of the fatty streaks develop into plaques, while others disappear. Progression into plaques is most commonly observed at sites of hemodynamic strain, and it is likely that hemodynamic factors such as pressure, turbulence, and shear stress determine the localization of the plaques.

Plaque formation is characterized by SMC migration from the media into the intima and the formation of a fibrous, SMC-rich cap around the foam cells (119). Many of the latter die, and a plaque core of extracellular lipid is formed. It appears likely that growth factors govern the formation of the fibrous cap, but the precise molecular mechanisms are not fully known (31,119). It is, however, clear that plaque SMC express receptors for PDGF and FGF and that plaque macrophages and endothelial cells produce mRNA for PDGF (65). Since endothelial dysfunction and damage is often observed at this stage, it is possible that microthrombi formed on denuded plaque surfaces release PDGF that stimulate SMC immigration and the formation of the cap.

The atherosclerotic plaque invariably carries a local immune activation. In the cap, approximately 20% of the cells are T lymphocytes (139), and nearly half of these show signs of activation (140,141). RT-PCR and immunohistochemical analyses have shown that the T-cell cytokine IFN-γ is produced in the plaque (138,140). Furthermore, SMC, EC, and macrophages all express high levels of MHC class II proteins (HLA-DR etc.), suggesting cellular responses to IFN-γ secretion (142).

The T-cell–activating antigens of atherosclerotic plaques have been unclear, but recent data show that a substantial proportion of plaque $CD4^+$ T cells react immunospecifically to oxidized LDL (143). This could imply that an autoimmune response to oxidized LDL may play an important role in atherosclerosis (143,144). Other antigens such as heat-shock proteins can, however, also elicit local T-cell responses in the plaque (145).

In experimental animals, immunosuppressive treatment aggravates atherosclerosis (146,147). Furthermore, immunodeficient MHC class I knock-out mice of the

C57BL/6 background develop significantly larger fatty streak lesions than the immunocompetent C57BL/6 strain (148). It may appear surprising that the net effect of T-cell activation and the concomitant cytokine secretion in the plaque could be beneficial. However, the antiatherosclerotic effects of T cells could be explained by the growth-inhibiting, vasodilatory, and cellular cholesterol-reducing effects of T-cell–derived and proinflammatory cytokines. This notion is supported by the recent finding that leukemia inhibitory factor (LIF), a cytokine expressed by T cells and activated macrophages, inhibits fatty streak formation in a rabbit model (149).

In addition to the specific T-cell cytokines, proinflammatory and macrophage-stimulating cytokines as well as chemokines are also produced in the atherosclerotic plaque. TNF-α and IL-1 are both found in plaques and can be produced both by SMC, EC, and macrophages (150–152). As discussed, they may be important as regulators both of SMC growth and the activation of EC, macrophages, and T cells. M-CSF produced by plaque cells may stimulate macrophage differentiation and MSR expression and might explain the surprising finding of replicating macrophages in the plaque (153,154). Finally, chemokines such as monocyte chemoattractant protein-1 (MCP-1) (155) and interleukin-1 (156) could be important for the recruitment of inflammatory and immunocompetent cells into the plaque.

The roles of systemic cytokine responses in atherosclerosis are less well known. However, recent data suggest that the serum concentration of active TGF-β is depressed in advanced atherosclerosis (157). This may be a useful disease marker, but its importance awaits further confirmation.

Plaque Complications and the Precipitation of Acute Events

The final stage of plaque development is characterized by the appearance of fissures, thrombi, and actual plaque rupture (158,159). It has been suggested that macrophage activation elicits this process by secreting proteases that degrade the extracellular matrix (45,118,159–161). Mast cells, which are also present at sites of plaque rupture (162,163), can exert similar functions (164). Interestingly, leukocyte elastase and mast cell chymase activate TGF-β (165); this may counteract tissue destruction by inducing a fibrotic response.

Proinflammatory cytokines (TNF-α, IFN-γ) could play a role in this process by virtue of their macrophage-activating and matrix-inhibiting activities. In addition, such cytokines (TNF-α, IL-1) enhance PAI-1 (plasminogen activator inhibitor-1) and tissue factor expression by endothelial cells (see above); this would be expected to tip the balance toward thrombosis formation on the arterial surface.

Proinflammatory cytokines might, however, also activate protective mechanisms. In particular, the induction by IL-1, TNF, and IFN-γ of NO production in vascular SMC and EC could inhibit platelet deposition and thrombus formation (91,166). The degree of inflammation might determine whether the responses are protective or harmful, since large concentrations of proinflammatory cytokines are strongly macrophage-activating and can induce apoptosis in vascular cells (167). Experimental evidence is still insufficient, however, to permit more detailed predictions concerning the effects of proinflammatory cytokines on atherosclerotic complications.

CONCLUSIONS

Studies in many laboratories over the last decade have shown that the cardiovascular system is a major target as well as an important source of cytokines. Thus, proinflammatory, immune-regulatory, and migration-inducing cytokines contribute to the regulation of basic functions of the blood vessel under physiological conditions. They also participate in the development of atherosclerosis, thrombosis, and other cardiovascular diseases. Further research will clarify whether cytokines can be used to treat and/or monitor these important diseases.

ACKNOWLEDGMENTS

Our work is supported by the Swedish Medical Research Council (Proj. 6816) and the Swedish Heart-Lung Foundation.

REFERENCES

1. Hirschberg H, Bergh OJ, Thorsby E. Antigen-presenting properties of human vascular endothelial cells. J Exp Med 1980; 152:249–255.
2. Collins T, Korman AJ, Wake CT, et al. Immune interferon activates multiple class II major histocompatibility genes and the associated invariant chain gene in human endothelial cells and dermal fibroblasts. Proc Natl Acad Sci USA 1984; 81:4917–4921.
3. Pober JS, Gimbrone MA. Expression of Ia-like antigens by human vascular endothelial cells is inducible in vitro: demonstration by monoclonal antibody binding and immunoprecipitation. Proc Natl Acad Sci USA 1982; 79:6641–6645.
4. Pober JS, Gimbrone MA, Cotran RS, et al. Ia expression by vascular endothelium is inducible by activated T cells and by human gamma interferon. J Exp Med 1983; 157:1339–1353.
5. Pober JS, Cotran RS. Cytokines and endothelial cell biology. Physiol Rev 1990; 70:427–456.
6. Korpelainen EI, Gamble JR, Smith WB, Dottore M, Vadas MA, Lopez AF. Interferon-γ upregulates interleukin-3 (IL-3) receptor expression in human endothelial cells and synergizes

with IL-3 in stimulating major histocompatibility complex class II expression and cytokine secretion. Blood 1995; 86:176–182.
7. Kaneko K, Savage CO, Pottinger BE, Shah V, Pearson JD, Dillon MJ. Antiendothelial cell antibodies can be cytotoxic to endothelial cells without cytokine pre-stimulation and correlate with ELISA antibody measurement in Kawasaki disease. Clin Exp Immunol 1994; 98:264–269.
8. Umetsu DT, Pober JS, Jabara HH, et al. Human dermal fibroblasts present tetanus toxoid antigen to antigen-specific T cell clones. J Clin Invest 1985; 76:254–260.
9. Geppert TD, Lipsky PE. Dissection of defective antigen presentation by interferon-gamma-treated fibroblasts. J Immunol 1987; 138:385–392.
10. Fabry Z, Waldschmidt MM, Moore SA, Hart MN. Antigen presentation by brain microvessel smooth muscle and endothelium. J Neuroimmunol 1990; 28:63–71.
11. Stemme S, Fager G, Hansson GK. MHC class II antigen expression in human vascular smooth muscle cells is induced by interferon-gamma and modulated by tumour necrosis factor and lymphotoxin. Immunology 1990; 69:243–249.
12. Salomon RN, Hughes CC, Schoen FJ, Payne DD, Pober JS, Libby P. Human coronary transplantation-associated arteriosclerosis. Evidence for a chronic immune reaction to activated graft endothelial cells. Am J Pathol 1991; 138:791–798.
13. Kocher O, Gabbiani G. Expression of actin mRNAs in rat aortic smooth muscle cells during development, experimental intimal thickening, and culture. Differentiation 1986; 32:245–251.
14. Barja F, Coughlin C, Belin D, Gabbiani G. Actin isoform synthesis and mRNA levels in quiescent and proliferating rat aortic smooth muscle cells in vivo and in vitro. Lab Invest 1986; 55:226–233.
15. Clowes AW, Clowes MM, Kocher O, Ropraz P, Chaponnier C, Gabbiani G. Arterial smooth muscle cells in vivo: relationship between actin isoform expression and mitogenesis and their modulation by heparin. J Cell Biol 1988; 107:1939–1945.
16. Schwartz SM, Campbell GR, Campbell JH. Replication of smooth muscle cells in vascular disease. Circ Res 1986; 58:427–444.
17. Thyberg J, Hedin U, Sjîlund M, Palmberg L, Bottger BA. Regulation of differentiated properties and proliferation of arterial smooth muscle cells. Arteriosclerosis 1990; 10:966–990.
18. Ross R. Platelet-derived growth factor. Lancet 1989; 1:1179–1182.
19. Daemen MJ, Lombardi DM, Bosman FT, Schwartz SM. Angiotensin II induces smooth muscle cell proliferation in the normal and injured rat arterial wall. Circ Res 1991; 68:450–456.
20. Galis ZS, Muszynski M, Sukhova GK, et al. Cytokine-stimulated human vascular smooth muscle cells synthesize a complement of enzymes required for extracellular matrix digestion. Circ Res 1994; 75:181–189.
21. Gibbons GH, Pratt RE, Dzau VJ. Vascular smooth muscle cell hypertrophy vs. hyperplasia. Autocrine transforming growth factor-β1 expression determines growth response to angiotensin II. J Clin Invest 1992; 90:456–461.
22. Lindner V, Lappi DA, Baird A, Majack RA, Reidy MA. Role of basic fibroblast growth factor in vascular lesion formation. Circ Res 1991; 68:106–113.
23. Lindner V, Reidy MA. Expression of basic fibroblast growth factor and its receptor by smooth muscle cells and endothelium in injured rat arteries. An en face study. Circ Res 1993; 73:589–595.
24. Olson EN, Chao S, Lindner V, Reidy MA. Intimal smooth muscle cell proliferation after balloon catheter injury: The role of basic fibroblast growth factor. Am J Pathol 1992; 140:1017–1023.
25. Ferns GA, Raines EW, Sprugel KH, Motani AS, Reidy MA, Ross R. Inhibition of neointimal smooth muscle accumulation after angioplasty by an antibody to PDGF. Science 1991; 253:1129–1132.
26. Fingerle J, Johnson R, Clowes AW, Majesky MW, Reidy MA. Role of platelets in smooth muscle cell proliferation and migration after vascular injury in rat carotid artery. Proc Natl Acad Sci USA 1989; 86:8412–8416.
27. Libby P, Warner S, Friedman GB. Interleukin-1: a mitogen for human vascular smooth muscle cells that induces the release of growth-inhibitory prostanoids. J Clin Invest 1988; 81:487–498.
28. Raines EW, Dower SK, Ross R. Interleukin-1 mitogenic activity for fibroblasts and smooth muscle cells is due to PDGF-AA. Science 1989; 243:393–396.
29. Hansson GK, Holm J, Jonasson L, Stemme S. The immune system modulates cellular growth and differentiation in the vessel wall. Excerpta Med 1989; 1989:571–574.
30. Hajjar DP, Pomerantz KB. Signal transduction in atherosclerosis: integration of cytokines and the eicosanoid network. FASEB J 1992; 6:2933–2941.
31. Libby P, Hansson GK. Involvement of the immune system in human atherogenesis: current knowledge and unanswered questions. Lab Invest 1991; 64:5–15.
32. Hansson GK. Immune responses in atherosclerosis. In: Hansson GK, Libby P, eds. Immune Functions of the Vessel Wall. Sydney: Gordon & Breach, 1995:143–158.
33. Hansson GK, Jonasson L, Holm J, Clowes MM, Clowes AW. Gamma-interferon regulates vascular smooth muscle proliferation and Ia antigen expression in vivo and in vitro. Circ Res 1988; 63:712–719.
34. Hansson GK, Hellstrand M, Rymo L, Rubbia L, Gabbiani G. Interferon-γ inhibits both proliferation and expression of differentiation-specific α-smooth muscle actin in arterial smooth muscle cells. J Exp Med 1989; 170:1595–1608.
35. Warner S, Friedman GB, Libby P. Immune interferon inhibits proliferation and induces 2′,5′-oligoadenylate synthetase gene expression in human vascular smooth muscle cells. J Clin Invest 1989; 83:1174–1182.
36. Palmer H, Libby P. Interferon-beta. A potential autocrine regulator of human vascular smooth muscle cell growth. Lab Invest 1992; 66:715–721.
37. Bennett MR, Evan GI, Newby AC. Deregulated expression of the c-myc oncogene abolishes inhibition of proliferation of rat vascular smooth muscle cells by serum reduction, interferon-γ, heparin, and cyclic nucleotide analogues and induces apoptosis. Circ Res 1994; 74:525–536.
38. Amento EP, Ehsani N, Palmer H, Libby P. Cytokines and growth factors positively and negatively regulate interstitial collagen gene expression in human vascular smooth muscle cells. Arterioscl Thromb 1991; 11:1223–1230.
39. Yokota T, Shimokado K, Kosaka C, Sasaguri T, Masuda J, Ogata J. Mitogenic activity of interferon gamma on growth-arrested human vascular smooth muscle cells. Arterioscl Thromb 1992; 12:1393–1401.
40. Hansson GK, Holm J, Holm S, Fotev Z, Hedrich HJ, Fingerle J. T lymphocytes inhibit the vascular response to injury. Proc Natl Acad Sci USA 1991; 88:10530–10534.
41. Hansson GK, Holm J. Interferon-g inhibits arterial stenosis after injury. Circulation 1991; 84:1266–1272.
42. Thyberg J, Blomgren K, Hedin U, Dryjski M. Phenotypic modulation of smooth muscle cells during the formation of

neointimal thickenings in the rat carotid artery after balloon injury: an electron-microscopic and serological study. Cell Tissue Res 1995; 281:421–433.
43. Fager G, Hansson GK, Ottosson P, Dahllîf B, Bondjers G. Human arterial smooth muscle cells in culture. Effects of platelet-derived growth factor and heparin on growth in vitro. Exp Cell Res 1988; 176:319–335.
44. Campbell JH, Rennick RE, Kalevitch SG, Campbell GR. Heparan sulfate-degrading enzymes induce modulation of smooth muscle phenotype. Exp Cell Res 1992; 200:156–167.
45. Farber HW, Beer DJ. Restricted secretion of a T-lymphocyte chemotactic cytokine by serotonin-stimulated cultured aortic endothelial cells. Circ Res 1991; 69:257–265.
46. Snow AD, Bolender RP, Wight TN, Clowes AW. Heparin modulates the composition of the extracellular matrix domain surrounding arterial smooth muscle cells. Am J Pathol 1990; 137:313–330.
47. Kenagy RD, Nikkari ST, Welgus HG, Clowes AW. Heparin inhibits the induction of three matrix metalloproteinases (stromelysin, 92-kD gelatinase, and collagenase) in primate arterial smooth muscle cells. J Clin Invest 1994; 93:1987–1993.
48. Varga J, Olsen A, Herhal J, Constantine G, Rosenbloom J, Jimenez SA. Interferon-gamma reverses the stimulation of collagen but not fibronectin gene expression by transforming growth factor-beta in normal human fibroblasts. Eur J Clin Invest 1990; 20:487–493.
49. Kähäri VM, Chen YQ, Su MW, Ramirez F, Uitto J. Tumor necrosis factor-alpha and interferon-gamma suppress the activation of human type I collagen gene expression by transforming growth factor-beta I. J Clin Invest 1990; 86:1489–1495.
50. Desmoulière A, Geinoz A, Gabbiani F, Gabbiani G. Transforming growth factor-β1 induces α-smooth muscle actin expression in granulation tissue myofibroblasts and in quiescent and growing cultured fibroblasts. J Cell Biol 1993; 122:103–111.
51. Nabel EG, Shum L, Pompili VJ, et al. Direct transfer of transforming growth factor β1 gene into arteries stimulates fibrocellular hyperplasia. Proc Natl Acad Sci USA 1993; 90:10759–10763.
52. Wolf YG, Rasmussen LM, Ruoslahti E. Antibodies against transforming growth factor-β1 suppress intimal hyperplasia in a rat model. J Clin Invest 1995; 93:1172–1178.
53. Battegay EJ, Raines EW, Seifert RA, Bowen PD, Ross R. TGF-beta induces bimodal proliferation of connective tissue cells via complex control of an autocrine PDGF loop. Cell 1990; 63:515–524.
54. Stouffer GA, Owens GK. TGF-β promotes proliferation of cultured SMC via both PDGF-AA-dependent and PDGF-AA-independent mechanisms. J Clin Invest 1994; 93:2048–2055.
55. Zeiher AM, Fisslthaler B, Schray UB, Busse R. Nitric oxide modulates the expression of monocyte chemoattractant protein 1 in cultured human endothelial cells. Circ Res 1995; 76:980–986.
56. Stemme S, Patarroyo M, Hansson GK. Adhesion of activated T lymphocytes to vascular smooth muscle cells and dermal fibroblasts is mediated by beta-1- and beta-2-integrin receptors. Scand J Immunol 1992; 36:233–242.
57. Libby P, Li H. Vascular cell adhesion molecule-1 and smooth muscle cell activation during atherogenesis. J Clin Invest 1993; 92:538–539.
58. Clinton SK, Fleet JC, Loppnow H, et al. Interleukin-1 gene expression in rabbit vascular tissue in vivo. Am J Pathol 1991; 138:1005–1014.
59. Warner S, Auger KR, Libby P. Interleukin-1 induces interleukin-1. II Recombinant human interleukin-1 induces interleukin-1 production by adult human vascular endothelial cells. J Immunol 1987; 139:1911–1917.
60. Epperson DE, Pober JS. Antigen-presenting function of human endothelial cells. Direct activation of resting CD8 T cells. J Immunol 1994; 153:5402–5412.
61. Wang JM, Sica A, Peri G, et al. Expression of monocyte chemotactic protein and interleukin-8 by cytokine-activated human vascular smooth muscle cells. Arterioscl Thromb 1991; 11:1166–1174.
62. Loppnow H, Libby P. Proliferating or interleukin 1-activated human vascular smooth muscle cells secrete copious interleukin 6. J Clin Invest 1990; 85:731–738.
63. Loppnow H, Libby P. Adult human vascular endothelial cells express the IL6 gene differentially in response to LPS or IL. Cell Immunol 1989; 122:493–503.
64. De Caterina R, Tanaka H, Nakagawa T, Hauptman PJ, Libby P. The direct effect of injectable cyclosporine and its vehicle, cremophor, on endothelial vascular cell adhesion molecule-1 expression. Ricinoleic acid inhibits coronary artery endothelial activation. Transplantation 1995; 60:270–275.
65. Wilcox JN, Smith KM, Williams LT, Schwartz SM, Gordon D. Platelet derived growth factor mRNA detection in human atherosclerotic plaques by in situ hydribization. J Clin Invest 1988; 82:1134–1139.
66. Walker LN, Bowen PD, Ross R, Reidy MA. Production of platelet-derived growth factor-like molecules by cultured arterial smooth muscle cells accompanies proliferation after arterial injury. Proc Natl Acad Sci USA 1986; 83:7311–7315.
67. Sjölund M, Rahm M, Claesson WL, Sejersen T, Heldin CH, Thyberg J. Expression of PDGF alpha- and beta-receptors in rat arterial smooth muscle cells is phenotype and growth state dependent. Growth Factors 1990; 3:191–203.
68. Sjölund M, Hedin U, Sejersen T, Heldin CH, Thyberg J. Arterial smooth muscle cells express platelet-derived growth factor (PDGF) a chain mRNA, secrete a PDGF-like mitogen, and bind exogenous PDGF in a phenotype- and growth state-dependent manner. J Cell Biol 1988; 106:403–413.
69. Seifert RA, Schwartz SM, Bowen PD. Developmentally regulated production of platelet-derived growth factor-like molecules. Nature 1984; 311:669–671.
70. Rubin K, Tingstrîm A, Hansson CK, et al. Induction of B-type receptors for platelet-derived growth factor in vascular inflammation: possible implications for development of vascular proliferative lesions. Lancet 1988; 1:1353–1356.
71. Miano JM, Vlasic N, Tota RR, Stemerman MB. Smooth muscle cell immediate-early gene and growth factor activation follows vascular injury. A putative in vivo mechanism for autocrine growth. Arterioscl Thromb 1993; 13:211–219.
72. Majesky MW, Reidy MA, Bowen PD, Hart CE, Wilcox JN, Schwartz SM. PDGF ligand and receptor gene expression during repair of arterial injury. J Cell Biol 1990; 111:2149–2158.
73. Libby P, Warner S, Salomon RN, Birinyi LK. Production of platelet-derived growth factor-like mitogen by smooth-muscle cells from human atheroma. N Engl J Med 1988; 318:1493–1498.
74. DiCorleto PE, Gajdusek CM, Schwartz SM, Ross R. Biochemical properties of the endothelium-derived growth factor: comparison to other growth factors. J Cell Physiol 1983; 114:329–345.
75. Barrett TB, Gajdusek CM, Schwartz SM, McDougall JK, Benditt EP. Expression of the sis gene by endothelial cells in culture and in vivo. Proc Natl Acad Sci USA 1984; 81:6772–6774.
76. Lowenstein CJ, Snyder SH. Nitric oxide, a novel biologic messenger. Cell 1992; 70:705–707.

77. Bredt DS, Snyder SH. Nitric oxide: a physiologic messenger molecule. Annu Rev Biochem 1994; 63:175–195.
78. Fleming I, Busse R. Control and consequences of endothelial nitric oxide formation. Adv Pharmacol 1995; 34:187–206.
79. Moncada S, Higgs A. The L-arginine-nitric oxide pathway. N Engl J Med 1993; 329:2002–2012.
80. Chu A, Chambers DE, Lin CC, et al. Effects of inhibition of nitric oxide formation on basal vasomotion and endothelium-dependent responses of the coronary arteries in awake dogs. J Clin Invest 1991; 87:1964–1968.
81. Beasley D, Schwartz JH, Brenner BM. Interleukin 1 induces prolonged L-arginine-dependent cyclic guanosine monophosphate and nitrite production in rat vascular smooth muscle cells. J Clin Invest 1991; 87:602–608.
82. Busse R, Mülsch A. Induction of nitric oxide synthase by cytokines in vascular smooth muscle cells. FEBS Lett 1990; 275:87–90.
83. Geng YJ, Hansson GK, Holme E. Interferon-gamma and tumor necrosis factor synergize to induce nitric oxide production and inhibit mitochondrial respiration in vascular smooth muscle cells. Circ Res 1992; 71:1268–1276.
84. Wood KS, Buga GM, Byrns RE, Ignarro LJ. Vascular smooth muscle-derived relaxing factor (MDRF) and its close similarity to nitric oxide. Biochem Biophys Res Com 1990; 170:80–88.
85. Xie QW, Cho HJ, Calaycay J, et al. Cloning and characterization of inducible nitric oxide synthase from mouse macrophages. Science 1992; 256:225–228.
86. Cho HJ, Xie QW, Calaycay J, et al. Calmodulin is a subunit of nitric oxide synthase from macrophages. J Exp Med 1992; 176:599–604.
87. Geng YJ, Almqvist M, Hansson GK. cDNA cloning and expression of inducible nitric oxide synthase from rat vascular smooth muscle cells. Biochim Biophys Acta 1994; 1218:421–424.
88. Xie QW, Whisnant R, Nathan C. Promoter of the mouse gene encoding calcium-independent nitric oxide synthase confers inducibility by interferon gamma and bacterial lipopolysaccharide. J Exp Med 1993; 177:1779–1784.
89. Geng YJ, Petersson AS, Wennmalm Å, Hansson GK. Cytokine-induced expression of nitric oxide synthase results in nitrosylation of heme-and non-heme iron proteins in vascular smooth muscle cells. Exp Cell Res 1994; 214:418–428.
90. Hansson GK, Geng YJ, Holm J, Hårdhammar P, Wennmalm Å, Jennische E. Arterial smooth muscle cells express nitric oxide synthase in response to endothelial injury. J Exp Med 1994; 180:733–738.
91. Yan Z-q, Yokota T, Zhang W, Hansson GK. Expression of inducible nitric oxide synthase inhibits platelet adhesion and restores blood flow in the injured artery. Circ Res 1996; 79:38–44.
92. Moncada S, Vane JR. Pharmacology and endogenous roles of prostaglandin endoperoxides, thromboxane A_2 and prostacyclin. Pharmacol Rev 1979; 30:292–331.
93. Marcum JA, McKenney JB, Rosenberg RD. Acceleration of thrombin-antithrombin complex formation in rat hindquarters via heparinlike molecules bound to the endothelium. J Clin Invest 1984; 74:341–350.
94. Radomski MW, Vallance P, Whitley G, Foxwell N, Moncada S. Platelet adhesion to human vascular endothelium is modulated by constitutive and cytokine induced nitric oxide. Cardiovasc Res 1993; 27:1380–1382.
95. van Hinsbergh VWM. Regulation of the synthesis and secretion of plasminogen activators by endothelial cells. Haemostasis 1988; 18:307–327.
96. Jaffe EA, Hoyer LW, Nachman RL. Synthesis of von Willebrand factor by cultured human endothelial cells. Proc Natl Acad Sci USA 1974; 71:1906–1909.
97. Lyberg T, Galdal KS, Evensen SA, Prydz H. Cellular cooperation in endothelial cell thromboplastin synthesis. Br J Haematol 1983; 53:85–95.
98. Sprengers ED, Verheijen JH, van Hinsbergh VWM, Emeis JJ. Evidence for the presence of two different fibrinolytic inhibitors in human endothelial cell conditioned medium. Biochim Biophys Acta 1984; 801:163–170.
99. Schleef RR, Wagner NV, Loskutoff DJ. Detection of both type 1 and type 2 plasminogen activator inhibitors in human cells. J Cell Physiol 1988; 134:269–274.
100. Emeis JJ, Kooistra T. Interleukin 1 and lipopolysaccharide induce an inhibitor of tissue-type plasminogen activator in vivo and in cultured endothelial cells. J Exp Med 1986; 163:1260–1266.
101. van den Berg EA, Sprengers ED, Jaye M, Burgess W, Maciag T, van Hinsbergh VWM. Regulation of plasminogen activator inhibitor-1 mRNA in human endothelial cells. Thromb Haemost 1988; 60:63–67.
102. van der Poll T, Levi M, Büller HR, et al. Fibrinolytic response to tumor necrosis factor in healthy subjects. J Exp Med 1991; 174:729–732.
103. van Hinsbergh VWM, van den Berg EA, Fiers W, Dooijewaard G. Tumor necrosis factor induces the production of urokinase-type plasminogen activator by human endothelial cells. Blood 1990; 75:1991–1998.
104. Zoellner H, Wojta J, Gallicchio M, McGrath K, Hamilton JA. Cytokine regulation of the synthesis of plasminogen activator inhibitor-2 by human vascular endothelial cells. Comparison with plasminogen activator inhibitor-1 synthesis. Thromb Haemost 1993; 69:135–140.
105. Niedbala Mj, Stein Picarella M. Tumor necrosis factor induction of endothelial cell urokinase-type plasminogen activator mediated proteolysis of extracellular matrix and its antagonism by γ-interferon. Blood 1992; 79:678–687.
106. Arnman V, Stemme S, Rymo L, Risberg B. Interferon-gamma modulates the fibrinolytic response in cultured human endothelial cells. Thromb Res 1995; 77:431–440.
107. Conway EM, Bach R, Rosenberg RD, Konigsberg WH. Tumor necrosis factor enhances expression of tissue factor mRNA in endothelial cells. Thromb Res 1989; 53:231–241.
108. Ryan J, Geczy CL. Characterization and purification of mouse macrophage procoagulant-inducing factor. J Immunol 1986; 137:2864–2870.
109. Gregory SA, Kornbluth RS, Helin H, Remold HG, Edgington TS. Monocyte procoagulant inducing factor: a lymphokine involved in the T cell-instructed monocyte procoagulant response to antigen. J Immunol 1986; 137:3231–3239.
110. Schneiderman J, Sawdey MS, Keeton MR, et al. Increased type 1 plasminogen activator inhibitor gene expression in atherosclerotic human arteries. Proc Natl Acad Sci USA 1992; 89:6998–7002.
111. Lupu F, Bergonzelli GE, Heim DA, et al. Localization and production of plasminogen activator inhibitor-1 in human healthy and atherosclerotic arteries. Arterioscler Thromb 1993; 13:1090–1100.
112. Lupu F, Heim DA, Bachmann F, Hurni M, Kakkar VV, Kruithof EKO. Plasminogen activator expression in human atherosclerotic lesions. Arterioscl Thromb Vasc Biol 1995; 15:1444–1455.
113. Raghunath PN, Tomaszewski JE, Brady ST, Caron RJ, Okada SS, Barnathan ES. Plasminogen activator system in human coronary atherosclerosis. Arterioscl Thromb Vasc Biol 1995; 15:1432–1443.

114. Wilcox JN, Smith KM, Schwartz SM, Gordon D. Localization of tissue factor in the normal vessel wall and in the atherosclerotic plaque. Proc Natl Acad Sci USA 1989; 86:2839–2843.
115. Danø K, Andreasen PA, Grøndahl-Hansen J, Kristensen P, Nielsen LS, Skriver L. Plasminogen activators, tissue degradation, and cancer. Adv Cancer Res 1985; 44:139–266.
116. Mignatti P, Robbins E, Rifkin DB. Tumor invasion through the human amniotic membrane: requirement for a proteinase cascade. Cell 1986; 47:487–498.
117. Galis ZS, Sukhova GK, Lark MW, Libby P. Increased expression of matrix metalloproteinases and matrix degrading activity in vulnerable regions of human atherosclerotic plaques. J Clin Invest 1994; 94:2493–2503.
118. Henney AM, Wakeley PR, Davies MJ, et al. Localization of stromelysin gene expression in atherosclerotic plaques by in situ hybridization. Proc Natl Acad Sci USA 1991; 88:8154–8158.
119. Ross R. The pathogenesis of atherosclerosis: a perspective for the 1990s. Nature 1993; 362:801–809.
120. Nilsson J. Cytokines and smooth muscle cells in atherosclerosis. Cardiovasc Res 1993; 27:1184–1190.
121. Steinberg D, Parthasarathy S, Carew TE, Khoo JC, Witztum JL. Beyond cholesterol: modifications of low-density lipoprotein that increase its atherogenicity. N Engl J Med 1989; 320:915–924.
122. Camejo G, Hurt CE, Olsson U, Bondjers G. Proteoglycans and lipoproteins in atherosclerosis. Curr Opin Lipidol 1993; 4:385–391.
123. Doi T, Higashino KI, Kurinara Y, et al. Charged collagen structure mediates the recognition of negatively charged macromolecules by macrophage scavenger receptors. J Biol Chem 1993; 268:2126–2133.
124. Abraham R, Singh N, Mukhopadhyay A, Basu SK, Bal V, Rath S. Modulation of immunogenicity of proteins by maleylation to target scavenger receptors on macrophages. J Immunol 1995; 154:1–8.
125. Brown MS, Goldstein JL. Lipoprotein metabolism in the macrophage: implifications for cholesterol deposition in atherosclerosis. Annu Rev Biochem 1983; 52:223–261.
126. Hansson GK, Seifert PS, Olsson G, Bondjers G. Immunohistochemical detection of macrophages and T lymphocytes in atherosclerotic lesions of cholesterol-fed rabbits. Arterioscl Thromb 1991; 11:745–750.
127. van der Munro JM, et al. An immunohistochemical analysis of human aortic fatty streaks. Hum Pathol 1987; 18:375–380.
128. Ishibashi S, Inaba T, Shimano H, et al. Monocyte colony-stimulating factor enhances uptake and degradation of acetylated low density lipoproteins and cholesterol esterification in human monocyte-derived macrophages. J Biol Chem 1990; 265:14109–14117.
129. Clinton SK, Underwood R, Hayes L, Sherman ML, Kufe DW, Libby P. Macrophage colony-stimulating factor gene expression in vascular cells and in experimental and human atherosclerosis. Am J Pathol 1992; 140:301–316.
130. Li H, Freeman MW, Libby P. Regulation of smooth muscle cell scavenger receptor expression in vivo by atherogenic diets and in vitro by cytokines. J Clin Invest 1995; 95:122–133.
131. Lau W, Devery JM, Geczy CL. A chemotactic S100 peptide enhances scavenger receptor and Mac-1 expression and cholesteryl ester accumulation in murine peritoneal macrophages in vivo. J Clin Invest 1995; 95:1957–1965.
132. Smith JD, Trogan E, Ginsberg M, Grigaux C, Tian J, Miyata M. Decreased atherosclerosis in mice deficient in both macrophage colony-stimulating factor (op) and apolipoprotein E. Proc Natl Acad Sci USA 1995; 92:8264–8268.
133. Fong LG, Fong AT, Cooper AD. Inhibition of mouse macrophage degradation of acetyl-low density lipoprotein by interferon-gamma. J Biol Chem 1990; 265:11751–11760.
134. Geng YJ, Hansson GK. Interferon-gamma inhibits scavenger receptor expression and foam cell formation in human monocyte-derived macrophages. J Clin Invest 1992; 89:1322–1330.
135. Van Lenten BJ, Fogelman AM. Lipopolysaccharide-induced inhibition of scavenger receptor expression in human monocyte-macrophages is mediated through tumor necrosis factor-alpha. J Immunol 1992; 148:112–116.
136. Dufva M, Svenningsson A, Hansson GK. mRNA quantification of macrophage scavenger receptor isoforms by the polymerase chain reaction. J Lipid Res 1995; 36:2282–2290.
137. Van Lenten BJ, Fogelman AM, Seager J, Ribi E, Haberland ME, Edwards PA. Bacterial endotoxin selectively prevents the expression of scavenger-receptor activity on human monocyte-macrophages. J Immunol 1985; 134:3718–3721.
138. Geng YJ, Holm J, Nygren S, Bruzelius M, Stemme S, Hansson GK. Expression of macrophage scavenger receptor in atherosclerosis. Relationship between scavenger receptor isoforms and the T cell cytokine, interferon-γ. Arterioscler Thromb Vasc Biol 1995; 15:1995–1202.
139. Jonasson L, Holm J, Skalli O, Bondjers G, Hansson GK. Regional accumulations of T cells, macrophages, and smooth muscle cells in the human atherosclerotic plaque. Arteriosclerosis 1986; 6:131–138.
140. Hansson GK, Holm J, Jonasson L. Detection of activated T lymphocytes in the human atherosclerotic plaque. Am J Pathol 1989; 135:169–175.
141. Stemme S, Holm J, Hansson GK. T lymphocytes in human atherosclerotic plaques are memory cells expressing CD45RO and the integrin VLA-1. Arterioscl Thromb 1992; 12:206–211.
142. Jonasson L, Holm J, Skalli O, Gabbiani G, Hansson GK. Expression of class II transplantation antigen on vascular smooth muscle cells in human atherosclerosis. J Clin Invest 1985; 76:125–131.
143. Stemme S, Faber B, Holm J, Wiklund O, Witztum JL, Hansson GK. T lymphocytes from human atherosclerotic plaques recognize oxidized LDL. Proc Natl Acad Sci USA 1995; 92:3893–3897.
144. Frostegård J, Wu R, Giscombe R, Holm G, Lefvert AK, Nilsson J. Induction of T-cell activation by oxidized low density lipoprotein. Arterioscler Thromb 1992; 12:461–467.
145. Wick G, Schett G, Amberger A, Kleindienst R, Xu Q. Is atherosclerosis an immunologically mediated disease? Immunol Today 1995; 16:27–33.
146. Emeson EE, Shen ML. Accelerated atherosclerosis in hyperlipidemic C57B1/6J mice treated with cyclosporine. Am J Pathol 1993; 142:1906–1915.
147. Roselaar SE, Schonfeld G, Daugherty A. Enhanced development of atherosclerosis in cholesterol-fed rabbits by suppression of cell-mediated immunity. J Clin Invest 1995; 96:1389–1394.
148. Fyfe AI, Qiao JH, Lusis AJ. Immune-deficient mice develop typical atherosclerotic fatty streaks when fed an atherogenic diet. J Clin Invest 1994; 94:2516–2520.
149. Moran CS, Campbell JH, Simmons DL, Campbell GR. Human leukemia inhibitory factor inhibits development of experimental atherosclerosis. Arterioscler Thromb 1994; 14:1356–1363.
150. Barath P, Fishbein MC, Cao J, Berenson J, Helfant RH, Forrester JS. Detection and localization and tumor necrosis factor in human atheroma. Am J Cardiol 1990; 65:297–302.
152. Moyer CF, Sajuthi D, Tulli H, Williams JK. Synthesis of IL-1 alpha and IL-1 beta by arterial cells in atherosclerosis. Am J Pathol 1992; 138:951–960.

153. Gordon D, Reidy MA, Benditt EP, Schwartz SM. Cell proliferation in human coronary arteries. Proc Natl Acad Sci USA 1990; 87:4600–4604.
154. Rosenfeld ME, Ross R. Macrophage and smooth muscle cell proliferation in atherosclerotic lesions of WHHL and comparably hypercholesterolemic fat-fed rabbits. Arteriosclerosis 1990; 10:680–687.
155. Nelken NA, Coughlin SR, Gordon D, Wilcox JN. Monocyte chemoattractant protein-1 in human atheromatous plaques. J Clin Invest 1991; 88:1121–1127.
156. Koch AE, Kunkel SL, Pearce WH, et al. Enhanced production of the chemotactic cytokines interleukin-8 and monocyte chemoattractant protein-1 in human abdominal aortic aneurysms. Am J Pathol 1993; 142:1423–1431.
157. Grainger DJ, Kemp PR, Metcalfe JC, et al. The serum concentration of active transforming growth factor-β is severely depressed in advanced atherosclerosis. Nature Med 1995; 1:74–79.
158. Davies MJ, Thomas A. Thrombosis and acute coronary-artery lesions in sudden cardiac ischemic death. N Engl J Med 1984; 310:1137–1140.
159. Davies MJ, Woolf N, Rowles P, Richardson PD. Lipid and cellular constituents of unstable human aortic plaques. Basic Res Cardiol 1994; 1:33–40.
160. Moreno PR, Falk E, Palacios IF, Newell JB, Fuster V, Fallon JT. Macrophage infiltration in acute coronary syndromes. Implications for plaque rupture. Circulation 1994; 90:775–778.
161. Shah PK, Falk E, Badimon JJ, et al. Human monocyte-derived macrophages induce collagen breakdown in fibrous caps of atherosclerotic plaques. Potential role of matrix-degrading metalloproteinases and implications for plaque rupture. Circulation 1995; 92:1565–1569.
162. Kaartinen M, Penttilñ A, Kovanen PT. Accumulation of activated mast cells in the shoulder region of human coronary atheroma, the predilection site of atheromatous rupture. Circulation 1994; 90:1669–1678.
163. Kovanen PT, Kaartinen M, Paavonen T. Infiltrates of activated mast cells at the site of coronary atheromatous erosion or rupture in myocardial infarction [see comments]. Circulation 1995; 92:1084–1088.
164. Saarinen J, Kalkkinen N, Welgus HG, Kovanen PT. Activation of human interstitial procollagenase through direct cleavage of the Leu83-Thr84 bond by mast cell chymase. J Biol Chem 1994; 269:18134–18140.
165. Taipale J, Lohi J, Saarinen J, Kovanen PT, Keski-Oja J. Human mast cell chymase and leukocyte elastase release latent transforming growth factor-beta 1 from the extracellular matrix of cultured human epithelial and endothelial cells. J Biol Chem 1995; 270:4689–4696.
166. Groves PH, Penny WJ, Cheadle HA, Lewis MJ. Exogenous nitric oxide inhibits in vivo platelet adhesion following balloon angioplasty. Cardiovasc Res 1992; 26:615–619.
167. Geng YJ, Wu Q, Muszynski M, Hansson GK, Libby P. Apoptosis of vascular smooth muscle cells induced by in vitro stimulation with interferon-γ, tumor necrosis factor-α and interleukin-1β. Arterioscl Thromb Vasc Biol 1996; 16:19–27.

33
The Role of Cytokines in Lung Injury

Jeffrey S. Warren
University of Michigan Medical School, Ann Arbor, Michigan

INTRODUCTION

Perusal of recent biomedical literature calls attention to the explosive growth of knowledge in the biology of cytokines. Dozens of lymphokine-cytokine activities were described during the 1960s and 1970s, but really only during the last decade has it become possible to attribute specific biological activities to biochemically defined mediators. Numerous cytokines have been characterized and produced in recombinant form. As mediators of cell-to-cell communication, cytokines appear to participate, in at least some manner, in most fundamental pathophysiological processes. For example, tumor necrosis factor-α (TNF-α) and interleukins-1 and -6 (IL-1, IL-6) are important in the systemic septic and endotoxic shock responses (1). TNF-α, IL-1, and interferon-γ (IFN-γ) appear to be important local inducers of endothelial activation, a complex, highly regulated process that dictates the temporal characteristics, intensity, and cellular composition of the localized inflammatory response (2). Activated endothelial cells, as well as many other cell types, express an array of cell surface leukocyte adhesion-promoting molecules [e.g., P-selectin, E-selectin, and intercellular adhesion molecule-1 (ICAM-1)] and secrete a large number of soluble leukocyte chemotactic and activating proteins [e.g., chemokines such as interleukin-8 (IL-8) and monocyte chemoattractant protein-1 (MCP-1)] (2,3). Interleukin-2 is a pivotal mediator of T-lymphocyte proliferation in such cellular immune processes as the delayed-type hypersensitivity response (4). Interleukin-4, -5, and -6 (IL-4, IL-5, IL-6) play fundamental roles in B-lymphocyte differentiation and function (5). Cytokine-associated growth factor activities also appear to be important in chronic inflammation, tissue repair, and neoplasia (6). In view of the pleiotypic nature of cytokine activities, it is not surprising that these mediators participate in systemic as well as local processes, both physiological and pathological (5). Undoubtedly there are locally determined tissue-specific nuances in cytokine function. In some cases, local differences in the role of individual cytokines may be a function of anatomy. For example, within the lungs there is a readily accessible reservoir of alveolar macrophages that orchestrate numerous cytokine-driven responses. The lungs also contain a vast microvascular surface located in close proximity to the airspace. It should be kept in mind that the biologically relevant functions of cytokines are defined not only by their biochemical actions but by the microenvironment.

The focus of this chapter is the role of cytokines in experimental lung injury. A large number of animal models mimic human pulmonary diseases and have thus served as valuable tools for providing insights into understanding basic pathophysiological mechanisms. Attention is directed toward two well-characterized experimental models of lung injury. Immunoglobulin G (IgG) immune complex–induced alveolitis in the rat serves as a general model of acute inflammation and, more specifically, mimics processes thought to occur in human diseases such as lupus pneumonitis, rheumatoid pneumonitis, and some types of extrinsic allergic alveolitis. It should be noted that in addition to experimental models of acute lung inflammation, several animal models also bear relevance to chronic disorders. Glucan-induced granulomatous vasculitis in the rat serves as a model of chronic inflammation, specifically foreign body-induced pulmonary granulomatous vasculitis. This particular model mimics Wegener's granulomatosis, the cause of which is unknown, and a variety of occupational lung diseases (7). IgG immune complex–induced alveolitis and glucan-induced granulomatous vasculitis serve as paradigms for relatively common acute and chronic types of inflammatory lung disease. Particularly well-characterized models of inflammatory lung disease include murine *Schistosoma*-induced pulmonary granulomatosis (8), asbestos-induced pulmonary fibrosis (9), and bleomycin-induced pulmonary fibrosis (10). For all of these models, there is burgeoning evidence that various cytokines and growth factors are involved.

ACUTE IgG IMMUNE COMPLEX ALVEOLITIS IN THE RAT

Relevance to Human Lung Disease

Deposition of circulating immune complexes and in situ formation of immune complexes have been invoked as central pathophysiological events leading to injury in many organs, including the lungs (Table 1). In many cases, deposits of complement components and immunoglobulin can be demonstrated by immunofluorescence or electron microscopy (16,17,35,36). In some cases, evidence of systemic complement activation supports the premise that immune complex formation leads to complement activation and subsequent tissue injury. For example, pulmonary parenchymal involvement is present to some degree in most patients with systemic lupus erythematosus (17). The severity of pulmonary parenchymal lupus generally correlates with the intensity of systemic disease and ranges from an asymptomatic, clinically inapparent form to an acute fulminating hemorrhagic pneumonitis (37). Deposits containing DNA, IgG anti-DNA, and C3 can be demonstrated in the lungs of patients with acute lupus pneumonitis (35,36,38). Electron-dense immune deposits have also been identified along the alveolar capillary basement membrane in lung biopsies obtained from such patients (39). Despite clinical examples that seem clear-cut, there are many inconsistencies and unknowns. Effector components of complement (e.g., C3a, C3b, C5a, and the membrane attack complex, C5b-9) are thought to be critical mediators of phagocyte activation, inflammatory cell recruitment, and microvascular injury, yet complement components are not always found in lung tissue from patients with lupus pneumonitis (17,37). Immune deposits apparently do not always lead to inflammatory cell recruitment, and inflammatory cells are not always present in lung tissue obtained from patients with lupus pneumonitis (17,37). Histopathological patterns and distributions of inflammatory cells vary considerably among lung diseases considered to be mediated by immune complexes or other immunological mechanisms. It is clear that many factors influence the pathogenesis of immunologically mediated lung disease. Experimental studies of acute immune com-

Table 1 Human Lung Diseases Associated with Immune Complex Deposition

Disease state	Comments	Ref.
Idiopathic interstitial pneumonitis[a]	Interstitial and alveolar septal deposits of IgG, IgM, and C3 in some patients; elevated levels of circulating immune complexes in some patients	11–15
Systemic lupus erythematosus	DNA, IgG anti-DNA, and C3 deposits found in most patients, but not all; infiltrating phagocytes not always present	16,17
Rheumatoid pneumonitis	Evidence of local complement consumption and immune complex formation	18–20
Extrinsic allergic alveolitis	Evidence supports immune complex role during acute phase, but biopsies are often negative for immune deposits; thought to be a good example of type III hypersensitivity[b]	21–23
Goodpasture's syndrome	Prototype type II hypersensitivity; rare patients have circulating immune complexes in addition to anti–basement membrane antibodies	24
Wegener's granulomatosis	Occasional interstitial pneumonitis and circulating immune complexes	25
Sjögren's syndrome	Circulating immune complexes	26
Mixed connective tissue disease	Circulating immune complexes	27–29
Polymyositis-dermatomyositis	Interstitial lung disease in some patients	30
Sarcoidosis	Circulating immune complexes in some patients; rarely pulmonary immune deposits	31,32
Asthma	Unusual patients develop broncho-constriction later than in classic type I hypersensitivity; some of these exhibit stigmata suggestive of type II hypersensitivity reactions	

[a]This group includes patients with the following pathological descriptors: usual interstitial pneumonia (UIP), desquamative interstitial pneumonia (DIP), lymphocytic interstitial pneumonia (LIP), and diffuse fibrosis.
[b]This designation refers to the Gell and Coombs classification of hypersensitivity reactions (34).

plex–mediated alveolitis have provided valuable information regarding phagocyte-mediated tissue injury, inflammatory cell recruitment, and host defense mechanisms that counterregulate inflammatory tissue damage. In recent years, it has become clear that cytokines play a pivotal role in the pathogenesis of acute immune complex–mediated lung injury.

Rat Model of Immune Complex Alveolitis

The first reported example of experimental immune complex-mediated lung injury was observed by Kniker and Cochrane (40), who identified reversible acute inflammatory reactions in the walls of pulmonary arteries of rabbits with "one-shot" acute serum sickness. It was thought that circulating immune complexes, formed within the vascular space, deposited in the vessel walls. No inflammatory changes were present in the interstitial or alveolar compartments. Brentjens et al. (41) demonstrated irreversible changes in the alveolar interstitium and capillary basement membranes of rabbits with chronic serum sickness. Electron microscopic examination and immunofluorescence studies confirmed the presence of immune complex and complement deposition within the alveolar septa. In 1974, Johnson and Ward (42) characterized an IgG immune complex–mediated acute alveolitis model in rats. The remainder of this section outlines studies that have led to our current understanding of the pathogenesis of this model. The following section focuses on more recent work that addresses the pathophysiological roles of cytokines.

In this model, affinity-purified rabbit IgG anti–bovine serum albumin (BSA) is instilled into the alveolar compartment via a tracheal cannula and BSA is infused intravenously. The typical concentrations of anti-BSA (300–600 μg in 0.2–0.4 ml) and BSA (10 mg in 1.0 ml) used to elicit injury have been chosen empirically. Acute alveolitis develops over the ensuing 4–6 h. Lung injury can be quantitated by morphometric analysis by measuring the accumulation of isotope-labeled colloid in the injured lungs and by measuring increases in lung weight. Morphometric analysis allows quantitation of neutrophil influx, alveolar hemorrhage (expressed as red blood cells per alveolus), and endothelial or epithelial injury (expressed as lengths of exposed basement membrane per length of epithelial or endothelial surface). Lung weights and measurements of isotope-labeled colloid reflect pulmonary vascular permeability status and in this model correlate well with morphometric injury measurements (43). The total lung content of [^{125}I] colloid (usually nonspecific rabbit IgG) compared to the amount of [^{125}I] colloid in 1.0 ml of whole blood provides a reproducible and quantitative index (permeability index) of lung injury. After development of injury, the heart and lungs are removed en bloc and the lungs perfused via the pulmonary artery to flush out nonparenchymal blood. Comparison of [^{125}I] colloid in 1.0 ml of blood normalizes slight animal-to-animal variations in intravenous injection technique. Pulmonary neutrophil influx can also be quantitated by measurement of whole lung myeloperoxidase activity (44).

Development of lung damage in this model is neutrophil and complement dependent, thus resembling the cutaneous reversed passive Arthus reaction (45). The physical configuration of immune complexes is an important determinant of the intensity of pulmonary inflammation. There is considerable variation in the capacity for preformed IgG-BSA immune complexes to elicit injury when instilled into the alveolar space. Large, precipitating complexes (formed near antigen-antibody equivalence) are the most potent in terms of inducing lung injury (46). In vitro studies indicate that such complexes generally fix complement more efficiently than small, soluble complexes (47). Precipitating complexes are also potent agonists of phagocyte activation (as measured by respiratory burst or enzyme release) (47). Histological examination of lung tissue obtained from rats with IgG immune complex alveolitis reveals a massive influx of neutrophils (Fig. 1). Electron-dense immune complexes can be seen within neutrophils and alveolar macrophages. The importance of neutrophils in acute immune complex alveolitis has been demonstrated in neutrophil-depleted rats (42). Rats that have been 95% depleted of neutrophils by nitrogen mustard, cyclophosphamide, or antiserum to neutrophils exhibit an up to 70% reduction in pulmonary vascular permeability following development of alveolitis (42).

There is both direct and indirect evidence that complement is required for full development of acute IgG immune complex alveolitis. Indirect immunofluorescence staining of rat lung sections reveals C3 deposits along alveolar septa (42). Numerous in vitro studies have implicated complement proteins in acute inflammatory processes, for example, C5a-mediated neutrophil chemotaxis and activation, C5a-mediated phagocyte priming, C5a-mediated endothelial cell activation (upregulation of P-selectin), and C3b opsonization (48,49). Recent studies suggest that the membrane attack complex (MAC; C5b-9) may play a pathogenetic role in some complement-mediated inflammatory lesions (50,51). Hattori et al. (52) observed that assembly of the MAC on endothelial cells promotes the upregulation of P-selectin expression. We recently confirmed and extended this observation using purified C5-C9 to produce membrane active MAC in the absence of serum (K. S. Kilgore and J. S. Warren, unpublished results). In addition, assembly of the MAC in TNF-α–activated endothelial cells results in increased adhesiveness for neutro-

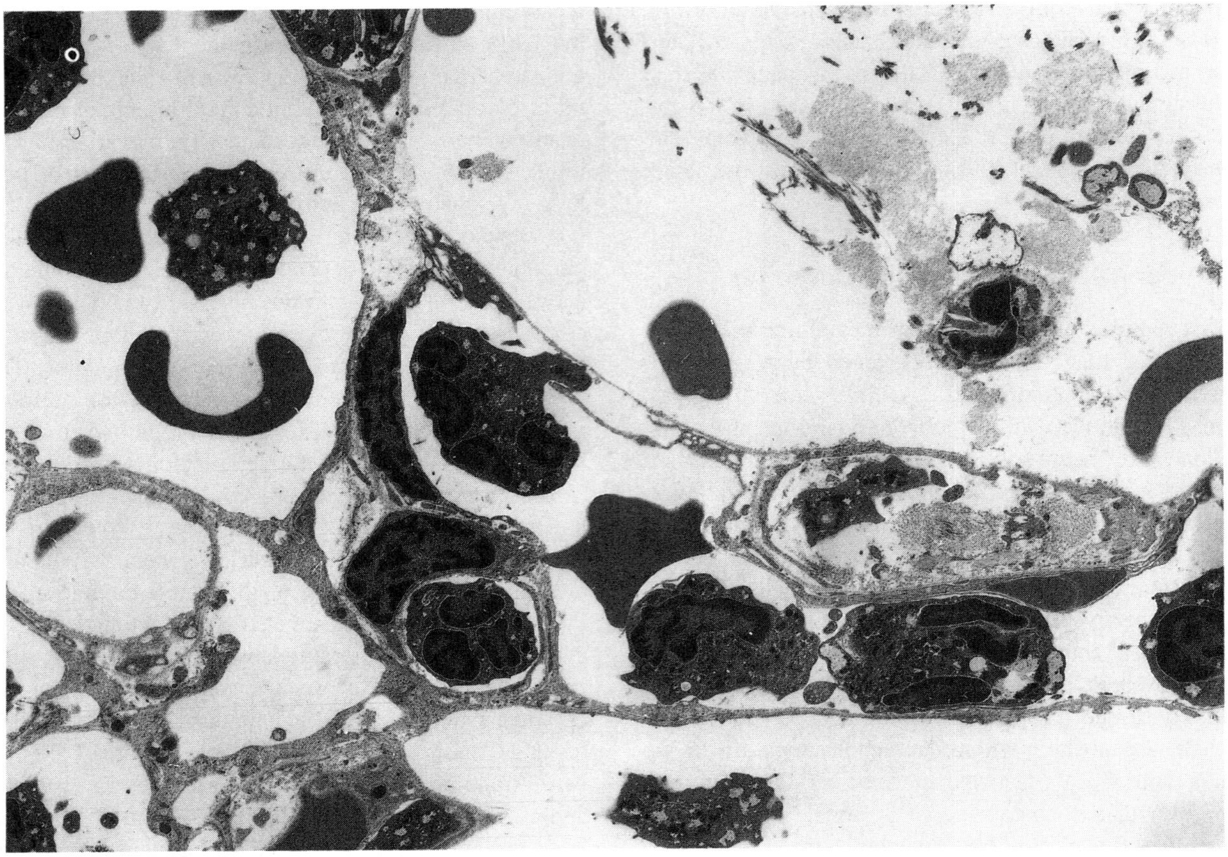

Figure 1 Acute IgG immune complex alveolitis in the rat. Extensive alveolar hemorrhage is accompanied by many neutrophils and fibrin strands. (Uranyl acetate and lead citrate stain, transmission electron micrograph; X4100.)

phils through both E-selectin and ICAM-1 upregulation (53). There is also evidence suggesting that endothelial surface-bound and processed complement components (iC3b) may directly participate in early neutrophil-endothelial adhesive interactions (54). In this circumstance, iC3b bound to the luminal surface of the endothelium links neutrophils via their CR3. Several in vivo studies suggest that complement is required for full development of lung injury. Initiation of IgG immune complex alveolitis in rats that have been complement-depleted (through repeated systemic injections of cobra venom factor) or systemically treated with soluble CR1 (sCR1), an inhibitor of C3 and C5 convertase activities, results in an attenuated form of lung injury (42,55). Similarly, deposition of immune complexes in the lungs of mice that are genetically C5 deficient results in negligible lung injury (56).

Pretreatment of rats with either superoxide dismutase, catalase, hydroxyl radical scavengers, or iron chelators also suppresses the development of alveolitis, supporting the hypothesis that oxygen-derived metabolites play a critical role in neutrophil-mediated tissue injury (43). Local generation of nitric oxide (NO·) by endogenous arginine-dependent pathways also appears to contribute to IgG immune complex–mediated lung injury (57), either through the toxic effects of NO· itself or derivatives, which include hydroxyl radical (HO ·) after the formation of short-lived peroxynitrite radicals from NO· and O_2– (58). Nitric oxide is synthesized from L-arginine via a biochemical pathway found in a variety of cell types (59). Suppression of immune complex alveolitis by superoxide dismutase is time dependent and early in the evolation of lung injury (first 60 min) is associated with reduced intrapulmonary neutrophil recruitment (44). This observation also supports the hypothesis that oxygen-derived metabolites influence neutrophil recruitment. It is unclear how O_2– influences neutrophil recruitment, but it may be linked to the formation of oxidant-generated chemotactic factors. Several potent oxidant-mediated or oxidant-amplified chemotactic mechanisms have been described. In vitro exposure of plasma to xanthine-xanthine oxidase (a cell-free superoxide anion-generating system) results in the generation of an arachidonate-derived neutrophil chemotactic factor (60,61). Shingu and Nobunaga reported that exposure of purified C5 to an oxygen radical-generating system results in the formation of C5a (62). von Zabern et al. demonstrated that exposure of C5 to H_2O_2 results in conformational changes in C5 and the formation of a membrane attack complex–like aggregate (63). Recent studies suggest that endogenously generated oxidants may directly upregulate the expression of genes that encode chemotactic peptides (IL-1, IL-8, MCP-1) recently shown to be important in endothelial activation and both neutrophil and monocyte recruitment in acute immune complex alveolitis and glucan-induced granulomatous vasculitis, respectively (64–68; see below). Similarly, in vitro studies also indicate that oxygen-derived metabolites can upregulate leukocyte adhesion molecule (E-selectin) expression in endothelial cells (69). Recently, Lewis et al. showed that exposure of endothelial cell monolayers to H_2O_2 results in platelet-activating factor (PAF) production (70). In vitro studies suggest that endothelium-associated PAF helps to mediate a rapid neutrophil-endothelial adhesive interaction (71). Platelet-activating factor is also a potent neutrophil chemotactic and activating factor (72). In vivo studies also suggest that PAF is an important mediator of IgG immune complex alveolitis (see below) (73).

Interactions between oxygen metabolites and lysosomal proteases may also be important in the pathogenesis of immune complex alveolitis. Weiss et al. have shown in vitro that exposure of latent (inactive) gelatinase or collagenase to hypochlorous acid results in enzyme activation (74,75). Hypochlorous acid is derived from myeloperoxidase-H_2O_2 and chloride ions and is produced by activated neutrophils. There is in vivo evidence that proteases and oxygen-derived species may exhibit a synergistic role in immune complex–associated lung injury in rats (76).

It has become clear that the acute inflammatory milieu is exceedingly complex. The capacity for endothelial cells to produce PAF, cytokines, leukocyte adhesion molecules, and even oxidant species bears witness to this premise. For instance, it now appears that conditions of ischemia, exposure to activated neutrophils, and some soluble inflammatory mediators (e.g., TNF-α, C5a, formyl peptides) can promote the conversion of endothelial cell xanthine dehydrogenase to xanthine oxidase (77,78). In turn, xanthine oxidase catalyzes the formation of O_2^- (79). The role of such pathways in vivo remains to be determined. Alveolar macrophages, once viewed as passive scavenger cells, clearly help to orchestrate many inflammatory processes. Alveolar macrophages not only produce oxidants, PAF, and complement components, but are also a major source of proinflammatory cytokines (73). As will become more clear in the ensuing portions of this chapter, such cytokines play a central role in the pathogenesis of both acute and chronic inflammatory processes.

Role of Cytokines in Acute Alveolitis

It has been recognized during the past decade that acute immune complex–mediated alveolitis is not simply the result of immune complex deposition, complement fixation, generation of C5a, and neutrophil recruitment and activation. In vitro and in vivo studies suggest that TNF-α, IL-1β, and IL-8 (or an IL-8–like homolog in the rat) are important in activation of blood vessel wall cells and neutrophil recruitment (Fig. 2). In addition, it appears that inflammatory cell recruitment is orchestrated through the regulated expression of leukocyte-endothelial cell adhesion molecules including P-selectin, E-selectin, ICAM-1, and CD11/CD18.

Perhaps the best studied cytokine in the context of acute immune complex–induced lung injury is TNF-α. Recent studies suggest that formation of IgG-BSA immune complexes within rat alveolar septa results in the elaboration of intrapulmonary TNF-α, which in turn plays a role in lung injury through a pathway involving neutrophil recruitment (80). Over the 4-h course of evolving lung injury, there is a 10-fold increase in TNF activity in retrieved bronchoalveolar lavage fluid compared to that present in fluid from negative controls. Immunohistochemical staining and in vitro studies suggest that resident alveolar macrophages are the chief source of TNF-α. Administration of rabbit polyclonal IgG that specifically neutralizes rat TNF-α activity results in a marked reduction in lung injury as measured by permeability indices. The reduction in pulmonary microvascular permeability is accompanied by a marked decrease in intrapulmonary neutrophil recruitment. Analogous studies indicate that IL-1β plays a similar role in the evolution of IgG immune complex alveolitis (81,82). The mechanisms responsible for TNF-α– and IL-1β–mediated pulmonary neutrophil recruitment are not completely clear, but several recent studies suggest possible mechanisms. The temporal patterns of TNF-α– and IL-1β–dependent neutrophil recruitment are in agreement with several in vitro studies, which indicate that TNF-α– and IL-1β–induced expression of E-selectin and ICAM-1 occurs over several hours through protein synthesis–dependent mechanisms (83–86). In turn, E-selectin and ICAM-1 play major roles in neutrophil-endothelial cell adhesive interactions (87,88). Recent in vivo studies have implicated these endothelial adhesion molecules and their complementary leukocytic counter-receptors in the development of acute immune complex alveolitis (89–92). A second TNF-α–mediated mechanism

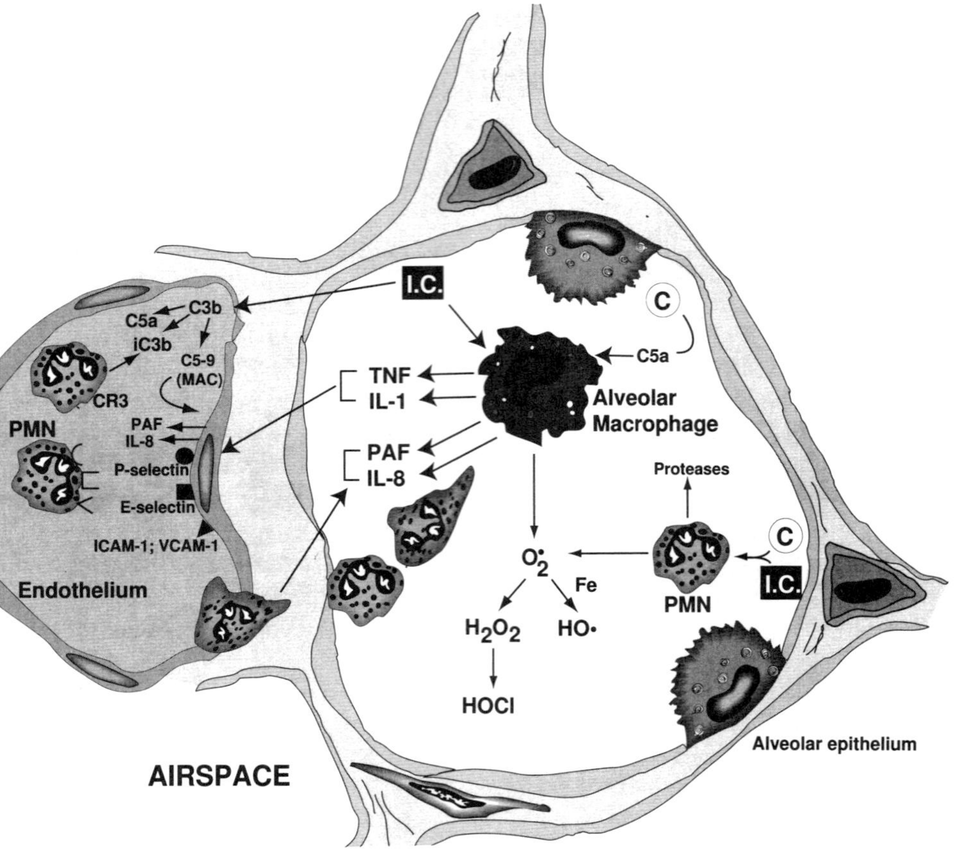

Figure 2 Inflammatory mediator pathways in the pathogenesis of acute immune complex alveolitis. Alveolar macrophages play a prominent role in orchestrating the inflammatory response. C, complement; I.C., immune complexes.

of neutrophil recruitment involves PAF. Intratracheal administration of specific PAF antagonists (L-652,731 or WEB-2086) results in decreased lung injury and decreased mobilization of neutrophils from the pulmonary vascular and/or interstitial compartment into the alveolar airspace (73,82). Exposure of isolated alveolar macrophages to TNF-α results in dose-dependent PAF secretion. These studies implicate PAF as a TNF-α–triggered mediator of lung injury. The precise role of PAF is unclear, but this lipid is a potent chemotactic factor, it can prime neutrophils, it is reported to mediate endothelial-neutrophil adhesive interactions, and it directly increases microvascular permeability (72). Finally, several studies indicate that TNF-α and IL-1β can trigger IL-8 production by several cell types, including endothelial cells (93,94). Strieter et al. (93) have demonstrated that exposure of cells indigenous to the lung (endothelial cells and alveolar macrophages) to TNF-α results in the expression of IL-8 mRNA and secretion of biologically active IL-8. In view of the potent chemotactic activity attributable to IL-8 (3), this may be an important neutrophil recruitment mechanism in acute alveolitis. In vivo studies employing neutralizing antihuman IL-8 antibodies that react with a partially characterized rat IL-8–like analog suggest that a locally produced IL-8–like mediator is involved in IgG immune complex–induced alveolitis in the rat (95).

The foregoing overviews of the roles of TNF-α, and to a lesser extent IL-1 and IL-8, in experimental acute immune complex alveolitis serve to emphasize the complexity of acute inflammatory lung injury. IgG immune complex alveolitis serves as a paradigm not only for acute inflammatory lung disease but for acute inflammation in general.

GLUCAN-INDUCED GRANULOMATOUS VASCULITIS

Relevance to Human Lung Disease

The spectrum of human granulomatous lung disease is broad, encompassing diseases caused by readily identifiable microbial agents (e.g., tuberculosis, schistosomiasis), foreign particulates (e.g., beryllium, talc), and diseases in which no causative agent has been identified (e.g., Wegener's granulomatosis, sarcoidosis). Granulomatous diseases are very common, contributing to vast worldwide morbidity and mortality. Several approaches to the classification of granulomatous diseases have evolved (96). Many granulomatous diseases can be classified according to etiology. Classification schemes based on the mechanism of granuloma formation often divide granulomas into "immunological" and "nonimmunological" groups depending upon whether or not development of lesions is thought to be antigen and T-lymphocyte dependent. Finally, morphological classifications depend upon the composition of lesional inflammatory cells and their anatomical relationships to each other and resident structures (e.g., angiocentric or bronchocentric granulomas). Much of what is known about mechanisms of granuloma formation is based on clinical data from humans, immunophenotypic analyses of granulomatous lesions, in vitro analyses of mediators and cells through to be important in granuloma formation, and experimental animal models.

Quantitative estimates of the cellular composition of human lung granulomas are baed on immunohistochemical analyses of lung biopsies and immunophenotypic analyses of cells retrieved by lung lavage. Such studies have allowed inferences to be made regarding cellular interactions in granuloma formation. Lung lesions typically contain 40–85% mononuclear phagocytes and 15–60% lymphocytes (97). The mononuclear phagocytes include resident alveolar macrophages, recruited blood monocytes, and macrophages produced by local cell proliferation (98,99). In many cases, retrieved mononuclear phagocytes express HLA-DR antigens (98,100). Granulomas are dynamic; in relatively recent formed lesions, the majority of lymphocytes are CD4+ (helper-inducer) T cells located in the center of the cellular aggregates; in "older" granulomas, CD8+ (cytotoxic-suppressor) T cells constitute a higher proportion of the cells and are located at the periphery of the lesion (98,101). Lung T cells from patients with sarcoidosis spontaneously produce IL-2 and IFN-γ (102). Retrieved T cells also express markers of cell activation: high-affinity IL-2 receptors, HLA-DR molecules, and the VLA-1 late activation marker (103–105). Recent studies indicate that sarcoid lung T lymphocytes express decreased numbers of T-cell antigen receptors (TCR) and increased quantities of TCR β-chain mRNA (106). These changes infer the existence of antigen-driven T-cell responses and suggest that immune granulomas are produced when antigen is presented to antigen-specific CD4+ T lymphocytes by mononuclear phagocytes in the context of histocompatible class II MHC molecules. In turn, granuloma formation is amplified via IL-2 and IFN-γ. Many of the putative mechanisms involved in the pathogenesis of human granulomatous lung diseases are further supported by data derived from animal models.

Rat Model of Granulomatous Vasculitis: Role of MCP-1, TNF-α, and IL-1β

Intravenous infusion of particulate yeast cell wall glucan (5–10 mg) into rats results in the rapid (48 h) and synchronous development of angiocentric monocyte/macrophage-rich pulmonary granulomas (Fig. 3). It should be noted (see below) that within 1–6 h after glucan infusion there is a transient accumulation of neutrophils in blood vessel walls at sites of particle embolization (7). Previous studies have shown that MCP-1 activity in bronchoalveolar lavage fluid and whole lung MCP-1 mRNA rise in association with pulmonary granuloma development (107). Time course studies of whole lung MCP-1 mRNA expression have shown that there is a biphasic rise in MCP-1 in the early

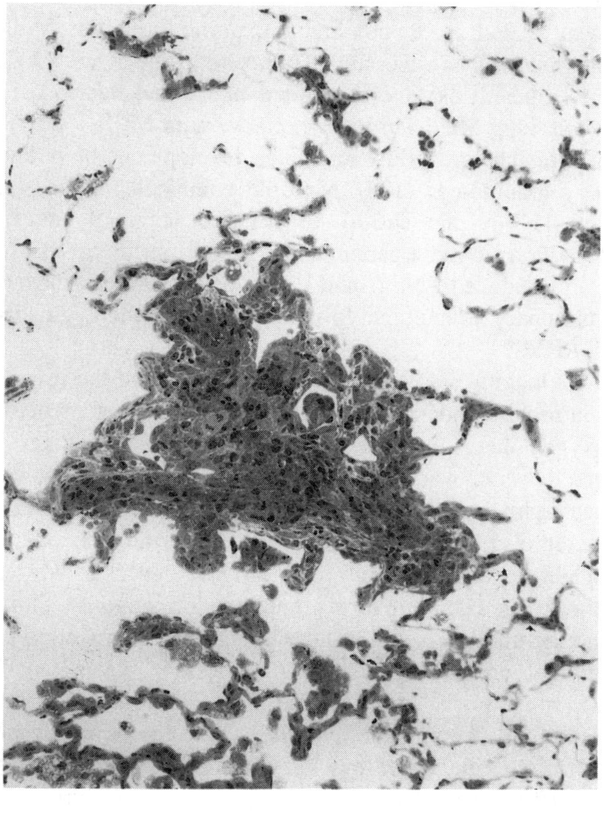

Figure 3 Glucan-induced granulomatous vasculitis. Angiocentric lesions, composed of mononuclear phagocytes, form within 48 h at sites of glucan embolization. The depicted lesion occurred 48 h after glucan (5 mg, i.v.) infusion. (Toluidine blue-stained 1 μm methacrylate-embedded sections.)

phase of granuloma development with discrete peaks of expression at 1 and 6–24 h after glucan infusion (107). In situ hybridization and immunohistochemical analyses have revealed that components of the bronchial and vascular walls are responsible for the early rise (1 h) in MCP-1 mRNA and protein expression while granuloma-associated macrophages are the predominant source of MCP-1 later (6–24 h) in the evolution of these lesions (107). We have also demonstrated that granuloma development is markedly attenuated in rats given neutralizing concentrations of anti–MCP-1 antibody (107). These studies suggest that MCP-1 plays a pivotal role in the pathogenesis of glucan-induced granulomatous vasculitis.

MCP-1 is secreted in vitro by a variety of cell types in response to endotoxin or proinflammatory cytokines such as TNF-α and IL-1 (3). However, few in vivo studies have addressed the mechanisms of MCP-1 induction in inflammatory lesions. We have observed that there are abrupt, transient rises in TNF-α and IL-1 activities in lung lavage fluid early in the evolution of glucan-induced pulmonary granulomatous vasculitis (7,107). We therefore examined the potential importance of TNF-α and IL-1 activities in local MCP-1 production and granuloma formation. Multiple infusions of either anti–TNF-α antibody or anti–IL-1β into glucan-treated rats resulted in marked decreases in whole lung MCP-1 mRNA expression, in MCP-1 activity in lung lavage fluid and in the development of pulmonary granulomas (108). Multiple combined infusions of anti–TNF-α and anti–IL-1β resulted in small additive reductions in these parameters. These data suggest that locally produced TNF-α and IL-1 modulate glucan-induced pulmonary granuloma formation through the regulation of MCP-1.

Ongoing studies are beginning to address the roles of neutrophils and/or their secreted products (e.g., reactive oxygen metabolites) in the induction of chemokines in blood vessel wall cells. Preliminary studies suggest that neutrophil depletion results in attenuation of the early (1–6 h) increase in MCP-1 expression in vessel wall cells at sites of glucan embolization (K. S. Kilgore and J. S. Warren, unpublished data). In turn, granuloma formation is retarded in neutrophil-depleted rats despite the fact that neutrophils are not present in the definitive lesions (48–72 h). In addition, the roles of intercellular adhesion molecules, neutrophil chemotactic cytokines, and monocyte chemotactic cytokines other than MCP-1 in glucan-induced granulomatous vasculitis are now beginning to be addressed.

PERSPECTIVES

It is particularly exciting to consider the role of cytokines in inflammatory processes in the context of the advances that have been made in understanding cell-cell adhesive interactions (e.g., neutrophils and endothelial cells). It appears that cytokine-activated endothelium exhibits a highly regulated set of responses that help dictate the intensity, temporal pattern, and composition of leukocytes that are recruited into an inflammatory site. For instance, TNF-α– or IL-1–activated endothelium expresses E-selectin and IL-8 within hours, a sequence consistent with their putative roles in acute inflammation (1). In contrast, upregulated ICAM-1 expression occurs more slowly and persists, a time course more compatible with its suspected role in chronic inflammation (1). Both in vitro and in vivo studies suggest that there is redundancy within the system, that is, these are probably several receptor-ligand pairs for each inflammatory cell type. The relative importance of such interactions may be tissue, time, and circumstance dependent. An example is CD18-dependent versus CD18-independent mechanisms of neutrophil-to-endothelial cell adhesion in which one adhesion interaction is stronger, allowing it to withstand greater shear stress than another (2).

The value of recombinant cytokines to the development of models of experimental lung injury is difficult to overestimate. In vitro systems, and to a greater extent in vivo systems, are exceedingly complex. The availability of pure cytokines or derived products (e.g., specific antibodies) have proven invaluable in defining the biological significance of various mediators. This is particularly true in cases in which cytokines exhibit many overlapping activities (e.g., TNF-α versus IL-1) (1). This chapter has focused on two models of lung injury. It must be kept in mind that numerous other model systems have yielded valuable insights into mechanisms of lung disease.

ACKNOWLEDGMENTS

The author thanks Charlene Fegan for manuscript preparation and Robin G. Kunkel for artwork and figure preparation. This work was supported in part by the American Heart Association of Michigan and National Institutes of Health Grants HL-40526 and HL-48287.

REFERENCES

1. Warren JS. Interleukins and tumor necrosis factor in inflammation. Crit Rev Clin Lab Sci 1990; 28:37–59.
2. Pober J, Cotran RS. Cytokines and endothelial cell biology. Physiol Rev 1990; 70:427–437.
3. Murphy PM. The chemokines. Annu Rev Immunol 1994; 12:593–622.
4. Smith K. Interleukin 2: inception, impact, and implication. Science 1988; 240:1169–1176.
5. Mizel SB. The interleukins. FASEB J 1989; 3:2379–2388.

6. Goustin AS, Leof EB, Shipley GD, Moses HL. Growth factors and cancer. Cancer Res 1986; 46:1015–1029.
7. Jones ML, Warren JS. Monocyte chemoattractant protein-1 in a rat model of pulmonary granulomatosis. Lab Invest 1992; 66:448–503.
8. Chensue SW, Warmington KS, Lukacs NW, Lincoln PM, Burdick MD, Strieter RM, Kunkel SL. Monocyte chemotactic protein expression during schistosome egg granuloma formation. Am J Pathol 1995; 146:130–138.
9. Brody AR, Hill LH, Adkins B, O'Connor RW. Chrysotile asbestos inhalation in rats: deposition pattern and reaction of alveolar epithelium and pulmonary macrophages. Am Rev Respir Dis 1981; 123:640–679.
10. Zhang K, Gharacee-Kermani M, Jones ML, Warren JS, and Phan SH. Lung monocyte chemoattractant protein-1 gene expression in bleomycin-induced pulmonary fibrosis. J Immunol 1994; 153:4733–4741.
11. Dreisin RB, Schwarz MI, Theofilopoulos AN, et al. Circulating immune complexes in the idiopathic interstitial pneumonias. N Engl J Med 1978; 298:353–357.
12. Haslam P, Thompson B, Mohammed I, et al. Circulating immune complexes in patients with cryptogenic fibrosing alveolitis. J Immunol 1979; 37:381–390.
13. Schwarz MI, Dreisin RB, Pratt DS, et al. Immunofluorescent patterns in the idiopathic interstitial pneumonias. J Lab Clin Med 1978; 91:929–938.
14. Eisenberg H, Barnett E, Simmons H. Diffuse pulmonary interstitial disease, an immune disease. Clin Res 1977; 25:132A.
15. Warwick MT, Haslam P, Weeks J. Antibodies in some chronic fibrosing lung diseases. Clin Allergy 1971; 1:209–219.
16. Silberstein SL, Barland P, Grayzel A, et al. Pulmonary dysfunction in systemic lupus erythematosus. Prevalence, classification, and correlation with other organ involvement. J Rheumatol 1980; 7:187–194.
17. Eagen JW, Memoli VA, Roberts JL, et al. Pulmonary hemorrhage in systemic lupus erythematosus. J Med 1978; 57:545–560.
18. Hunder GG, McDuffie FC, Huston KA, et al. Pleural fluid complement, complement conversions and immune complexes in immunologic and nonimmunologic diseases. J Lab Clin Med 1977; 90:971–979.
19. Wallser WV, Wright V. Pulmonary lesions and rheumatoid arthritis. Medicine (Baltimore) 1968; 47:501–520.
20. DeHoratius RJ, Abruzzo JL, Williams RC. Immunofluorescent and immunologic studies of rheumatoid lung. Arch Intern Med 1972; 129:441–446.
21. Wenzel FJ, Emanual DA, Gray RL. Immunofluorescent studies in patients with farmer's lung. J Allergy Clin Immunol 1971; 48:224–229.
22. Barrowcliff DF, Arblaster PG. Farmer's lung: a study of an early acute fatal case. Thorax 1968; 23:490–500.
23. Geha RS. Circulating immune complexes and activation of the complement sequence in acute allergic bronchopulmonary aspergillosis. J Allergy Clin Immunol 1977; 60:357–359.
24. Poskitt TR. Immunologic and electron microscopic studies in Goodpasture's syndrome. Am J Med 1970; 49:250–257.
25. Shasby DM, Schwarz MI, Forstot JZ, Theofilopoulous AN, Kassan SS. Pulmonary immune complex deposition in Wegener's granulomatosis. Chest 1982; 81:338–340.
26. Lawley TJ, Moutsopoulos HM, Ketz SL, et al. Demonstration of circulating immune complexes in Sjogren's syndrome. J Immunol 1979; 123:1382–1387.
27. Parker MD, Marion T. Circulating complement-fixing immune complexes in mixed connective tissue disease. Clin Res 1978; 26:374A.
28. Halla JT, Volankis JE, Schrohenloher RE. Circulating immune complexes in mixed connective tissue disease. Arthritis Rheum 1979; 22:484–489.
29. Cunningham PH, Andrews BS, Davis JS. Immune complexes in patients with progressive systemic sclerosis and mixed connective tissue disease. Clin Res 1978; 26:374A.
30. Schwarz MI, Matthay RA, Sahn SA, et al. Interstitial lung disease in polymyositis and dermatomyositis: analysis of six cases and review of the literature. Medicine (Baltimore) 1976; 55:89–104.
31. Daniele RP, McMillan LJ, Dauber JH, Rossman MD. Immune complexes in sarcoidosis: a correlation with activity and duration of disease. Chest 1978; 74:261–264.
32. Johnson NM, McNicol MW, Burton-Kee JE, Mowbray JF. Circulating immune complexes in sarcoidosis. Thorax 1980; 35:286–289.
33. Pepys J. Immunopathology of allergic lung disease. Clin Allergy 1972; 3:1.
34. Cotran RS. Pathogenesis of vasculitis. An update. Adv Pathol 1990; 3:301–320.
35. Inoue T, Kanayama Y, Ohe A, et al. Immunopathologic studies of pneumonitis in systemic lupus erythematosus. Ann Intern Med 1979; 91:30–34.
36. Rodriguez-Iturbe B, Garcia R, Rubio L, et al. Immunohistologic findings in the lung in systemic lupus erythematosus. Arch Pathol Lab Med 1977; 101:342–344.
37. Eagen JW, Roberts JL, Schwartz MM, et al. The composition of pulmonary immune deposits in systemic lupus erythematosus. Clin Immunol Immunopathol 1979; 12:204–219.
38. Gamsu G, Webb WR. Pulmonary hemorrhage systemic lupus erythematosus. J Can Assoc Radiol 1977; 29:66–68.
39. Kuhn C. Systemic lupus erythematosus in a patient with ultrastructural lesions of the pulmonary capillaries previously reported in the review as due to idiopathic pulmonary hemosiderosis. Am Rev Respir Dis 1972; 106:931–932.
40. Kniker WT, Cochrane CG. Pathogenic factors in vascular lesions of experimental serum sickness. J Exp Med 1965; 122:83–98.
41. Brentjens JR, O'Connell DW, Pawlowski B, Hsu KC, Andres GA. Experimental immune complex disease of the lung. The pathogenesis of a laboratory model resembling certain human interstitial diseases. J Exp Med 1974; 140:105–125.
42. Johnson KJ, Ward PA. Acute immunologic pulmonary alveolitis. J Clin Invest 1974; 54:49–57.
43. Johnson KJ, Ward PA. Role of oxygen metabolities in immune complex injury lung. J Immunol 1981; 126:1365–1368.
44. Warren JS, Yabroff KR, Mandel DM, Johnson KJ, Ward PA. Role of O2– in neutrophil recruitment into sites of dermal and pulmonary vasculitis. Free Rad Biol Med 1990; 8:163–172.
45. Cochran CG, Janoff A. The Arthus reaction: a model of neutrophil and complement-mediated injury. In: Zweifach B, Grant L, McCluskey R, eds. The Inflammatory Process. 2d ed. Vol. 3. New York: Academic Press, 1974:39–51.
46. Scherzer H, Ward PA. Lung injury produced by immune complexes of varying composition. J Immunol 1978; 121:948–952.
47. Ward PA, Duque RE, Sulavik MC, Johnson KJ. In vitro and in vivo stimulation of rat neutrophils and alveolar macrophages by immune complexes: production of O_2– and H_2O_2. Am J Pathol 1983; 110:297–311.
48. Schreiber RD. The chemistry and biology of complement receptors. Springer Semin Immunopathol 1984; 7:221–234.
49. Ross GD, Medof ME. Membrane complement receptor specific for bound fragments of C3. Adv Immunol 1985; 37:217–229.

50. Kunkel RG, Warren JS, Johnson KJ, Ward PA. Demonstration of the complement membrane attack complex (MAC) in IgA-immune complex induced acute lung injury. Fed Proc 1988; 2:A1176.
51. Asghar SS. Membrane regulations of complement activation and their aberrant expression in disease. Lab Invest 1995; 72:254–271.
52. Hattori R, Hamilton KK, McEver RP, Sims PJ. Complement proteins C5b-9 induce secretion of high molecular weight multimers of endothelial von Willebrand factor and translocation of granule membrane protein GMP-140 to the cell surface. J Biol Chem 1989; 264(15):9053–9060.
53. Kilgore KS, Shen J, Miller BF, Ward PA, Warren JS. Enhancement by the complement membrane attack complex of TNFα-induced endothelial cell expression of E-selectin and ICAM-1. J Immunol (in press).
54. Marks RM, Todd RJ, Ward PA. Rapid induction of neutrophil-endothelial adhesion by endothelial complement fixation. Nature 1989; 339:314–317.
55. Mulligan MS, Yeh CG, Rudolph AR, Ward PA. Protective effects of soluble CR1 in complement- and neutrophil-mediated tissue injury. J Immunol 1992; 148:1479–1485.
56. Tvedten HW, Till GO, Ward PA. Mediators of lung injury in mice following systemic activation of complement. Am J Pathol 1985; 119:92–100.
57. Mulligan MS, Hevel JM, Marletta MA, Ward PA. Tissue injury caused by deposition of immune complexes is L-arginine dependent. Proc Natl Acad Sci USA 1991; 88:6338–6342.
58. Beckman JS, Beckman TW, Chen J, Marshall PA, Freeman BA. Apparent hydroxyl radical production by peroxynitrite: implications for endothelial injury from nitric oxide and superoxide. Proc Natl Acad Sci USA 1990; 87:1620–1624.
59. Moncada S, Palmer RJM, Higgs EA. Nitric oxide: physiology, pathophysiology, and pharmacology. Pharmacol Rev 1991; 43:109–142.
60. Petrone WF, English DK, Wong K, McCord JM. Free radicals and inflammation: the superoxide dependent activation of a neutrophil chemotactic factor in plasma. Proc Natl Acad Sci USA 1980; 77:1159–1163.
61. Perez HD, Weksler BB, Goldstein IM. Generation of a chemotactic lipid from arachidonic acid by exposure to a superoxide-generating system. Inflammation 1980; 4:313–328.
62. Shingu M, Nobunaga M. Chemotactic activity generated in human serum from the fifth component of complement by hydrogen peroxide. Am J Pathol 1984; 117:201–206.
63. von Zabern WV, Hesse D, Nolte R, Haller Y. Generation of an activated form of human C5 (C5b-like C5) by oxygen radicals. Immunol Lett 1987; 14:209–215.
64. Koga S, Ogawa S, Kuwabara K, Brett J, Leavy JA, Ryan J, Koga Y, Plocinski J, Benjamin W, Burns DK, Stern DJ. Synthesis and release of interleukin-1 by reoxygenated human mononuclear phagocytes. Clin Invest 1992; 90:1007–1015.
65. Metinko AP, Kunkel SL, Standiford TJ, Strieter RM. Anoxia-hyperoxia induces monocyte-derived interleukin-8. J Clin Invest 1992; 90:791–798.
66. Israel N, Gougerot-Pocidalo MA, Aillet F, Virelizier JL. Redox status of cells influences constitutive or induced NF-κB translocation and HIV long terminal repeat activity in human T and monocytic cell lines. J Immunol 1992; 149:3386–3393.
67. DeForge LE, Preston AM, Takeuchi E, Kenney J, Boxer LA, Remick DG. Regulation of interleukin 8 gene expression by oxidant stress. J Biol Chem 1993; 268:25568–25576.
68. Satriano JA, Shuldiner M, Hora K, Xing Y, Shan Z, Schlondorff D. Oxygen radicals as second messengers for expression of the monocyte chemoattractant protein, JE/MCP-1, and the colony-stimulating factor, CSF-1, in response to tumor necrosis factor-α and immunoglobulin G. J Clin Invest 1993; 92:1564–1571.
69. Toledano MB, Leonard WJ. Modulation of transcription factor NF-κB binding activity by oxidation-reduction in vitro. Proc Natl Acad Sci USA 1991; 88:4328–4332.
70. Lewis MS, Whatley RE, Cain P, McIntyre TM, Prescott SM, Zimmerman GA. Hydrogen peroxide stimulates the synthesis of platelet-activating factor by endothelium and induces endothelial cell-dependent neutrophil adhesion. J Clin Invest 1988; 82:2045–2055.
71. Zimmerman GA, McIntyre TM, Prescott SM. Thrombin stimulates the adherence of neutrophils to human endothelial cells in vitro. J Clin Invest 1985; 76:2235–2246.
72. Benveniste J, Arnoux B. Platelet Activating Factor and Structurally Related Ether-Lipids. Amsterdam: Elsevier, 1983.
73. Warren JS, Barton PA, Mandel DM, Matrosic K. Intrapulmonary tumor necrosis factor triggers local platelet-activating factor production in rat immune complex alveolitis. Lab Invest 1990; 63:746–754.
74. Weiss SJ, Peppin G, Ortiz H, Ragsdale CM, Test SI. Oxidative autoactivation of latent collagenase by human neutrophils. Science 1985; 227:747–749.
75. Peppin GJ, Weiss SJ. Activation of the endogenous metalloproteinase, gelatinase, by triggered human neutrophils. Proc Natl Acad Sci USA 1986; 83:4322–4326.
76. Gannon DE, He X, Ward PA, Varani J, Johnson KJ. Time-dependent inhibition of oxygen radical induced lung injury. Inflammation 1990; 14:509–517.
77. Phan SH, Gannon DE, Varani J, Ryan US, Ward PA. Xanthine oxidase activity in rat pulmonary artery endothelial cells and its alteration by activated neutrophils. Am J Pathol 1989; 134:1201–1211.
78. Friedl HP, Till GO, Ryan US, Ward PA. Mediator-induced activation of xanthine oxidase in endothelial cells. FASEB J 1989; 3:2512–2518.
79. McCord JM, Fridovich I. The reduction of cytochrome c by milk xanthine oxidase. J Biol Chem 1968; 243:5753–5760.
80. Warren JS, Yabroff KR, Remick DG, et al. Tumor necrosis factor participates in the pathogenesis of acute immune complex alveolitis in the rat. J Clin Invest 1989; 84:1873–1882.
81. Warren JS. Intrapulmonary interleukin 1 mediates acute immune complex alveolitis in the rat. Biochem Biophys Res Commun 1991; 175:604–610.
82. Warren JS. Relationship between interleukin 1b and platelet-activating factor in the pathogenesis of acute immune complex alveolitis in the rat. Am J Pathol 1992; 141:551–560.
83. Gamble JR, Harlam JM, Klebanoff SJ, Vada MA. Stimulation of the adherence of neutrophil to umbilical vein endothelium by human recombinant tumor necrosis factor. Proc Natl Acad Sci USA 1985; 82:8667.
84. Pohlman TH, Stanness KA, Beaty PG, Ochs HD, Harlan JM. An endothelial cell surface factor(s) induced in vitro by lipopolysaccharide, interleukin 1 and tumor necrosis factor-α increases neutrophil adherence by a CDw18-dependent mechanism. J Immunol 1986; 136:4548.
85. Pohlman TH, Harlan JM. Human endothelial cell response to lipopolysaccharide, interleukin-1 and tumor necrosis factor is regulated by protein synthesis. Cell Immunol 1989; 119:41–52.
86. Luscinskas FW, Brock AF, Arnaout MA, Gimrone MA. Endothelial-leukocyte adhesion molecule-1-dependent and leukocyte (CD11/CD18)-dependent mechanisms contribute to polymorphonuclear leukocyte adhesion to cytokine-activated human vascular endothelium. J Immunol 1989; 142:2257–2263.
87. Bevilacqua MP, Pober JS, Mendrick DL, Cotran RS, Gimbrone

MA Jr. Identification of an inducible endothelial-leukocyte adhesion molecule. Proc Natl Acad Sci USA 1987; 84:9238.
88. Bevilacqua MP, Stengelin S, Gimbrone MA Jr, Seed B. Endothelial leukocyte adhesion molecule 1: an inducible receptor for neutrophil related to complement regulatory proteins and lectins. Science 1989; 243:1160.
89. Mulligan MS, Varani J, Dame MK, Lane CL, Smith CW, Anderson DC, Ward PA. Role of endothelial-leukocyte adhesion molecule 1 (ELAM-1) in neutrophil-mediated lung injury in rats. J Clin Invest 1991; 88:1396–1404.
90. Mulligan MS, Vaporciyan AA, Myasaka M, Tamatani T, Ward PA. Tumor necrosis factor α regulates in vivo pulmonary expression of ICAM-1. Am J Pathol 1993; 142:1739–1749.
91. Mulligan MS, Wilson GP, Todd RF, Smith CW, Anderson DC, Varani J, Issekutz TB, Myasaka M, Tamatani T, Rusche JR, Vaporciyan AA, Ward PA. Role of β, β 2 integrins and ICAM-1 in lung injury after deposition of IgG and IgA immune complexes. J Immunol 1993; 150:2407–2417.
92. Mulligan MS, Varani J, Warren JS, Till GO, Smith CW, Anderson DC, Todd RF, Ward PA. Roles of β2 integrins of rat neutrophils in complement- and oxygen radical-mediated acute inflammatory injury. J Immunol 1992; 148:1847–1857.
93. Strieter RM, Kunkel SL, Showell H, et al. Human endothelial cell gene expression of a neutrophil chemotactic factor by TNF-α, LPS, and IL-1β. Science 1989; 243:1468–1469.
94. Dixit VM, Green S, Sarma V, et al. Tumor necrosis factor-α induction of novel gene products in human endothelial cells including a macrophage-specific chemotaxin. J Biol Chem 1990; 165:2973–2978.
95. Mulligan MS, Jones ML, Bolanowski MA, Baganoff MP, Deppeler CL, Meyers DM, Ryan US, Ward PA. Inhibition of lung inflammatory reactions in rats with an anti-human IL-8 antibody. J Immunol 1993; 150:5585–5595.
96. Boros DL. Granulomatous inflammation. Prog Allergy 1978; 24:183–267.
97. Campbell DA, Poulter LW, du Bois RM. Immunocompetent cells in bronchoalveolar lavage reflect the cell populations in transbronchial biopsies in pulmonary sarcoidosis. Am Rev Respir Dis 1985; 132:1300–1306.
98. Hance AJ, Douches S, Winchester RJ, Ferrans VJ, Crystal RG. Characterization of mononuclear phagocyte subpopulations in the human lung by using monoclonal antibodies: changes in alveolar macrophage phenotype associated with pulmonary sarcoidosis. J Immunol 1985; 134:284–292.
99. Bitterman PB, Saltzman LE, Adelberg J, Ferrans VJ, Crystal RG. Alveolar macrophage replication. One mechanism for the expansion of the mononuclear phagocyte population in the chronically inflamed lung. J Clin Invest 1984; 72:460–469.
100. Unanue ER. Antigen-presenting function of the macrophage. Annu Rev Immunol 1984; 2:395–428.
101. Mishra BB, Poulter LW, Janossy G, James GJ. The distribution of lymphoid and macrophage like cell subsets of sarcoid and Kveim granulomata: possible mechanism of negative PPD reaction in sarcoidosis. Clin Exp Immunol 1983; 54:705–715.
102. Hunninghake GW, Hadek JE, Young RC, Kawanami O, Ferrans VJ, Crystal RG. Maintenance of granuloma formation in pulmonary sarcoidosis by T lymphocytes within the lung. N Engl J Med 1980; 302:594–598.
103. Semenzalo G, Agostini C, Trentin L. Evidence of cells bearing interleukin-2 receptor at sites of disease activity in sarcoid patients. Clin Exp Immunol 1984; 57:331–337.
104. Costabel U, Bross KJ, Ruhle KH, Lohr GW, Matthys H. Ia-like antigens on T-cells and their subpopulations in pulmonary sarcoidosis and in hypersensitivity pneumonitis. Analysis of bronchoalveolar and blood lymphocytes. Am Rev Respir Dis 1985; 131:337–342.
105. Saltini C, Hemler ME, Crystal RG. T-lymphocytes compartmentalized on the epithelial surface of the lower respiratory tract express the very late activation antigen complex VLA-1. Clin Immunol Immunopathol 1988; 46:221–233.
106. du Bois RM, Balbi B, Kirby M, Crystal RG. T-lymphocyte accumulation in pulmonary sarcoidosis: Evidence for persistent stimulation of sarcoid lung T-lymphocytes via the T-cell antigen receptor. Am Rev Respir Dis 1989; 139:A61.
107. Flory CM, Jones ML, Warren JS. Pulmonary granuloma formation in the rat is partially dependent on monocyte chemoattractant protein-1. Lab Invest 1993; 65:356–404.
108. Flory CM, Miller BF, Warren JS. Regulatory roles of tumor necrosis factor-α and interleukin 1β in monocyte chemoattractant protein-1-mediated pulmonary granuloma formation in the rat. Am J Pathol 1995; 146:450–462.

34
Cytokines and the Central Nervous System

Etty N. Benveniste
University of Alabama at Birmingham, Birmingham, Alabama

INTRODUCTION

This chapter will focus on the role of cytokines in causing and/or contributing to the development and progression of central nervous system (CNS) disease, as well as those involved in resolution of disease. Many of the cytokines to be discussed are products of activated lymphocytes and macrophages; however, recent work has conclusively demonstrated that glial cells of the CNS are also important sources of cytokines (for review see Ref. 1). There are two broad subgroups of glial cells: (1) macroglia, consisting of astrocytes and oligodendrocytes, and (2) microglia, the resident macrophage of the brain (for review see Ref. 2). Additionally, neurons have been shown to secrete selected cytokines (for review see Ref. 3). All four of these CNS cell types (astrocytes, oligodendrocytes, microglia, and neurons) can both respond to and produce many of the cytokines initially attributed to lymphocytes and macrophages (for review see Refs. 3,4). These cytokines include interleukin-1 (IL-1), interleukin-2 (IL-2), interleukin-4 (IL-4), interleukin-6 (IL-6), interleukin-10 (IL-10), interferon-alpha/beta (IFN-α/β), interferon-gamma (IFN-γ), tumor necrosis factor-alpha (TNF-α), lymphotoxin (LT), transforming growth factor-beta (TGF-β), and a number of chemokines. Thus, within the diseased CNS, there are local sources of cytokines, as well as cytokines produced by infiltrating cells of the peripheral immune system. The contribution of these cytokines to intracerebral immune responses and inflammation in neurological diseases that have an immunological or autoimmune component will be described, with a focus on the disease of multiple sclerosis (MS) and the animal model for MS, experimental allergic encephalomyelitis (EAE). Both in vitro as well as in vivo experiments will be discussed, and the limitations in interpretation of these findings will be alluded to. Additionally, there will be a discussion of the use of cytokines and/or cytokine inhibitors to modulate neurological disease and future directions for cytokine research.

MULTIPLE SCLEROSIS

MS is a chronic, inflammatory, demyelinating disease of the CNS that most commonly affects young adults during the second or third decade of life. The clinical course is quite variable, but the most common form is characterized by relapsing neurological deficits. Early MS lesions are characterized by the local accumulation of activated CD4+ and CD8+ T cells around small venules (5). Later, there is myelin degeneration associated with perivascular inflammation consisting of T cells, B cells, plasma cells, and macrophages (6,7). T-cells are also found at the leading edge of plaques and extend into the surrounding normal-appearing white matter. The T cells express activation molecules on their cell surface such as IL-2 receptors and class II MHC antigens (8). Additionally, class II MHC expression can also be detected on infiltrating macrophages and resident CNS cells including microglia, astrocytes, and brain capillary endothelial cells (8). Class II MHC expression in the CNS is presumably induced by IFN-γ secreted by activated T cells. These findings indicate that in early acute MS lesions, demyelination occurs in the face of an active immune response within the CNS. Gliosis is also a prominent feature of MS; this is a process characterized by astrocyte proliferation, hypertrophy, and increased synthesis of glial fibrillary acidic protein (GFAP), an astrocyte-specific protein (9). This reaction eventually leads to the formation of dense glial scars in the CNS, which can contribute to motor and sensory impairment. Different pathological findings are observed in chronic plaques: inflammatory infiltrates are less pronounced—there is evidence of abortive attempts at remyelination and T γ/δ cells in association with heat-shock proteins are detected (10). It has been suggested that different immunological processes take place as the disease enters a chronic phase. The primary demyelination observed in MS results from damage to the myelin sheath or the myelin-producing cells, the oligodendrocytes. Macrophages/microglia are active participants in myelin breakdown. Since myelin is critical for saltatory excitation along axons, demyelination leads to a loss of neurological function. Recent findings with magnetic resonance imaging (MRI) indicate that considerable subclinical disease occurs and that early in lesion development, there is a breakdown of the blood-brain barrier (BBB) (11,12), which may be a crucial event in the pathogenesis of new lesions in MS.

The etiology of MS has not been defined, although a large body of experimental evidence implicates immune-mediated processes in activation and progression of MS. Additionally, both genetic and environmental factors contribute to disease (for review see Ref. 13). Accumulating evidence supports the notion that MS is an autoimmune disorder mediated by T cells. This evidence includes (1) the pathology of the MS lesion, (2) immunological abnormalities in both the periphery and CNS of MS patients, (3) immunoglobulin synthesis within the CNS, (4) exacerbation of disease after treatment with IFN-γ, (5) putative autoantigens such as myelin basic protein (MBP) or proteolipid protein (PLP), (6) possible restricted T-cell receptor (TCR) usage, and (7) involvement of cytokine networks. The first six points have been extensively discussed elsewhere (for review see Ref. 13).

Although there is growing literature on the pattern of cytokine production in the CNS of MS patients and the cells responsible for cytokine expression within this site, the data is conflicting. More experimental data has been obtained examining cytokine levels in the serum of MS patients or from peripheral blood mononuclear cells from these patients, but the relationship between these findings and events in the CNS are not clear. Studies on cytokine expression in the cerebrospinal fluid (CSF) of MS patients have been inconclusive. At present, the number of studies examining cytokine gene expression in MS lesions has been limited, but those conducted have generated some interesting findings. T cells, particularly CD4+ T-helper cells, exist as two distinct subsets based on cytokine profiles: Th1 and Th2. Th1 cells predominantly produce IL-2, IFN-γ, LT, and TNF-α, while Th2 cells mainly produce IL-4, IL-5, IL-6, and IL-10 (for review see Ref. 14). Th1 cells are primarily involved in cellular immunity and inflammatory responses by activating cytotoxic T cells and macrophages through production of cytokines such as IL-2 and IFN-γ. In contrast, the cytokines produced by Th2 cells induce B-cell growth and differentiation and are involved in the development of humoral immunity. However, in the setting of the CNS, other cell types are capable of producing Th1- and Th2-type cytokines (i.e., macrophages, B cells, endothelial cells, glial cells), which complicates interpretation of data on cytokine expression. Given these caveats, the studies to be described have suggested that cytokines associated with a Th1 phenotype, such as TNF-α, LT, IL-2, and IFN-γ, may promote disease, while cytokines with a Th2 phenotype (IL-4, IL-10) may limit disease. In addition, IFN-α/β and TGF-β appear to contribute to downregulation of disease.

DETECTION OF CYTOKINES IN MULTIPLE SCLEROSIS

Tumor Necrosis Factor-α and Lymphotoxin

TNF-α is a proinflammatory cytokine recognized to be an important mediator of immunological and inflammatory responses in a variety of tissues, including the CNS (for review see Ref. 15). TNF-α is produced upon stimulation by a vast number of cell types including macrophages, T cells, B cells, astrocytes, and microglia (for review see Ref. 16). TNF-α has a diverse range of functions in the CNS because of its direct effects on oligodendrocytes, astrocytes, and brain endothelial cells. Most relevant to CNS disease is the ability of TNF-α to mediate myelin and oligodendrocyte damage in vitro (17,18). LT, the cytokine that is genetically and functionally related to TNF-α (19), exerts a more potent cytotoxic effect toward oligodendrocytes than does TNF-α, and mediates its effect via apoptosis (20). These aspects of TNF-α/LT activity may contribute directly to myelin damage and/or the demyelination process observed in MS. As mentioned previously, one of the pathological features of demyelinated plaques in MS is astrogliosis (21). In vitro, TNF-α induces astrocyte proliferation, which could contribute to the astrogliosis associated with MS (22).

TNF-α may also have an important role in the alteration of the BBB due to its effects on the adhesive properties of both astrocytes and brain endothelium. MS is characterized by the migration of inflammatory cells from blood into the brain and subsequent invasion of the extravascular tissue (23,24). Recent studies have shown that upregulation of adhesion molecules such as intercellular adhesion molecule-1 (ICAM-1) on brain endothelial cells by exposure to proinflammatory cytokines such as TNF-α,

IL-1β, and IFN-γ mediates leukocyte adhesion to endothelium (25). Similarly, astrocytes can be induced by TNF-α, as well as IL-1 and IFN-γ, to express ICAM-1 (26–28), and activated lymphocytes and monocytes bind to astrocytes in an ICAM-1–dependent manner (29). Studies on brain material from MS patients have shown aberrant expression of ICAM-1 on both brain endothelium and astrocytes (30). The presence of ICAM-1 and other adhesion molecules in the vessel walls, as well as on astrocytes, may guide inflammatory leukocytes into and through the brain, thereby contributing to impairment of the BBB and the neuropathology of MS.

TNF-α in Biological Fluids

The multiple effects of TNF-α on various cell populations in the CNS suggests that TNF-α has a central role in augmenting infiltration of inflammatory cells, gliosis, and demyelination, all pathogenic events involved in MS. Accordingly, there has been much interest in the pattern of TNF-α expression by MS patients. Studies to examine TNF-α expression in the CSF and/or serum of MS patients have yielded conflicting data. Two groups did not detect TNF-α in the CSF and/or serum of MS patients (31,32). Hauser et al. (33) detected higher levels of TNF-α in the CSF from patients with active MS than in MS patients in remission or patients with other neurological diseases (OND). They suggested that the TNF-α originated from cellular elements within the CNS, rather than from immune cells in the CSF, since there was no correlation between the degree of CSF pleocytosis and cytokine expression. Maimone et al. (34) detected TNF-α in the CSF and serum of ~ 20% of MS patients and ~ 30% of patients with OND, which agrees with another study in which TNF-α was detected in the CSF and serum of ~ 15% of patients with chronic progressive MS (35). The most compelling evidence for an association of TNF-α with MS comes from Sharief et al. (36), who documented that increased TNF-α expression in the CSF of MS patients correlated with disease progression. Another report demonstrated that patients with acute relapsing MS in exacerbation showed significantly higher CSF and serum levels of TNF-α compared with MS patients in remission (37). Sharief et al. (38,39) then demonstrated a strong correlation between increased CSF levels of TNF-α, disruption of the BBB, CSF pleocytosis, and increased levels of circulating ICAM-1 in both serum and CSF of patients with active MS.

A number of factors can account for differences in the ability of investigators to detect TNF-α in the CSF or serum of MS patients and correlate levels with disease progression. These include (1) differences in the clinical status of patients examined, (2) timing of CSF/serum collection, and (3) the short half-life of TNF-α in vivo. TNF-α secreted within the CNS or in the circulation is likely to interact with TNF-α receptors on a multitude of cells (T cells, B cells, endothelial cells, macrophages, microglia, astrocytes, and oligodendrocytes); thus little to any may filter to the lumbar CSF, where it may be undetectable. Additionally, many cell types express TNF-α in a functional membrane-bound form (40), making it unavailable for detection in the CSF or serum. TNF-α may also be bound by soluble TNF receptors, which are enhanced in the CSF and serum of MS patients (see below). These caveats also apply to the detection of all other cytokines discussed in this chapter.

TNF-α/LT Expression by Cells from MS Patients

Numerous studies have been performed to examine TNF-α expression by cells (T cells, macrophages) from MS patients during various stages of disease. Beck et al. (41) performed a longitudinal study on 20 MS patients and normal controls examining TNF-α production by peripheral blood mononuclear cells stimulated with phytohemagglutinin (PHA). An increase in TNF-α production was detected 2 weeks prior to disease exacerbation. In benign cases, or those that resolved fully, the increase disappeared rapidly, often before the appearance of symptoms. When the increase in TNF-α persisted, clinical sequalae were common. In MS patients with chronic progressive disease, frequently intervening increases in TNF-α were observed. The authors suggest that TNF-α may trigger exacerbations and may also play a role in maintaining disease in chronic progressive patients. Merrill et al. (32) examined patients with acute relapsing and chronic progressive MS. TNF-α was produced in greater quantities by monocytes isolated from the peripheral blood of MS patients and by cells isolated from the CSF, which were predominantly T cells, compared to healthy controls. Building on these findings, Benvenuto et al. (42) determined that CD4+ T-cell clones from the CSF of MS patients, upon stimulation with PHA, produced high levels of TNF-α, IFN-γ, and IL-2, but not IL-4, suggesting a Th1 phenotype for these cells. Peripheral blood monocytes/macrophages from MS patients with acute disease produced significantly higher amounts of TNF-α upon stimulation with lipopolysaccharide (LPS) or phorbol esters than did cells from patients with chronic stable MS or OND (43). A similar observation was made by Glabinski et al. (44), who determined that peripheral blood mononuclear cells from MS patients in relapse produced higher levels of TNF-α after stimulation with MBP than did patients with OND. When examining cytokine mRNA expression, TNF-α and LT mRNA levels were increased in peripheral blood mononuclear cells from MS patients with active disease compared to those with stable disease (45). Interestingly, TGF-β and IL-10 mRNA levels declined prior to a relapse, suggesting an inverse relationship of TNF-α/LT mRNA expression with TGF-β/IL-10 mRNA expression (45). In this regard, PLP-specific T-cell clones isolated from MS patients during different phases of disease produced different patterns of cytokines. During acute attacks, PLP-reactive T cells produced predominantly TNF-α, LT, and IFN-γ (i.e., Th1 phenotype) and no TGF-β, while PLP-reactive T cells isolated during remission produced primarily IL-10 and TGF-β, with little TNF-α expression (46). These studies collectively indicate that there is aberrant high expression of TNF-α and LT by immune cells from MS patients during acute phases of disease and that elevated expression of these cytokines may contribute to the disease progression of MS.

Expression of TNF-α/LT in MS Brain

The presence of TNF-α in MS brain lesions has been examined using double-labeling immunohistochemical techniques. Hofman et al. (47,48) demonstrated that TNF-α–positive astrocytes and macrophages were found in the plaque region of MS brain and were most numerous at the lesion edge. TNF-α–positive cells were not detected in normal brain specimens or in tissue from Alzheimer's patients. Selmaj et al. (49) determined that both TNF-α and LT were present in MS plaque regions and that TNF-α localized with astrocytes and macrophages, whereas LT was associated with microglia and T cells. Both TNF-α and LT were identified in acute and chronic active MS lesions, but were absent in chronic silent plaques. Further analysis of TNF-α expression in the CNS and association with lesion activity indicated that immunoreactivity for TNF-α was highest in chronic active MS lesions (50). Although TNF-α levels in MS lesions were significantly higher than in CNS tissue from noninflammatory controls, inflammatory OND controls also expressed high levels of TNF-α in the CNS. These results indicate that while elevated expression of TNF-α in the CNS is not unique to MS, it is associated with inflammation of the brain (51). These studies examining TNF-α and LT protein expression need to be verified by in situ hybridization to conclusively determine the cell types expressing TNF-α/LT mRNA. It is possible that the results observed with immunohistochemistry reflect the ability of astrocytes, microglia, T cells, and macrophages to bind TNF-α/LT. In vitro studies indicate that of the endogenous glial cells, both astrocytes and microglia are capable of producing TNF-α upon exposure to multiple stimuli, including the Th1 cytokine IFN-γ (52–58), while microglia can produce LT upon stimulation with LPS (59). Thus, it will be important to verify the identity of infiltrating cells as well as endogenous glial cells, which are sources of TNF-α/LT in patients with MS.

Soluble TNF Receptor Expression in MS

The presence of potential cytokine inhibitors has also been investigated in biological fluids. TNF-α and LT mediate their biological effects through binding to two distinct cell surface receptors of ~ 55 kDa (TNF-R1) and ~ 75 kDa (TNF-R2) (for review see Ref. 60). Soluble forms of the two receptor types (sTNF-Rs), which are released from the cell surface by proteolytic cleavage, have been detected in serum and urine from normal individuals (61–63). sTNF-Rs have been suggested to function as inhibitors by competing for TNF/LT with cell surface TNF receptors and thus blocking TNF activity (64,65); however, at low concentrations they can also stabilize the bioactivity of TNF/LT by attenuating the spontaneous denaturation of TNF/LT (66). With regard to MS, one study found that patients with acute relapsing MS and chronic progressive MS had significantly increased CSF levels of sTNF-R compared to OND (67). Interestingly, serum levels of sTNF-R1 were higher in MS patients with stable disease compared to those with active disease (68). These authors suggest that high serum levels of sTNF-R1 are associated with disease remission. At this time, data on sTNF-Rs in MS are sparse, and the utility of such measurements is unclear.

Interleukin-1

IL-1, a cytokine produced by a broad range of cells, is responsible for mediating a variety of processes in the host response to microbial and inflammatory diseases and has a number of effects on cells of the CNS which implicate it in contributing to the pathology of MS and other neurological disease states (for review see Ref. 15). IL-1 is mitogenic for astrocytes in vitro (69), while IL-1 directly injected into the brain can stimulate astrogliosis (70), suggesting that IL-1 may contribute to astroglial scarring. IL-1 also can modulate the expression of adhesion molecules on astrocytes and endothelium (27,71), thereby altering BBB permeability. Within the CNS, IL-1 can be produced by numerous cell types including infiltrating T cells, B cells, and macrophages, endothelial cells of the BBB, and glial cells (astrocytes and microglia).

IL-1 in Biological Fluids/Cells from MS Patients

In a study by Hauser et al. (33) in which higher levels of CSF TNF-α were detected in patients with active MS, comparable findings were observed for IL-1. However, in other studies, IL-1 could not be detected in the CSF or serum of MS patients with acute relapsing or chronic progressive forms of disease (32,34,37,72,73).

Merrill et al. (32) demonstrated that monocytes isolated from peripheral blood and cells isolated from CSF of

MS patients (acute relapsing and chronic progressive) produced more IL-1 than cells from healthy controls. Similar findings were reported by Imamura et al. (43), who determined that macrophages isolated from the peripheral blood of patients during the acute phase of MS produced higher levels of IL-1α and IL-1β upon stimulation with LPS or phorbol esters than cells from patients with chronic stable MS or OND. Spontaneous and IFN-γ–induced IL-1β secretion by monocytes was greater in chronic progressive MS patients than in normal controls (74). In contrast, Matsuda et al. (73) reported that relapsing MS patients in the acute phase had significantly higher levels of constitutive IL-1α production by peripheral blood mononuclear cells than normal subjects, while production of IL-1β was not enhanced. Examination of IL-1β mRNA from peripheral blood mononuclear cells from MS patients at various stages of disease did not reveal any correlation with disease activity; however, IL-1 protein levels were not examined (45).

Expression of IL-1 in MS Brain

Several studies have examined IL-1 expression in MS brain lesions. IL-1–positive cells have been detected by immunohistochemistry in acute, chronic active, and chronic silent plaques. IL-1 levels were maximal in acute MS lesions, and expression localized predominantly to microglial cells (47,48,50). IL-1 mRNA expression was examined in acute, subacute, and chronic lesions, as well as in normal-appearing white matter by reverse transcription–polymerase chain reaction (RT-PCR) and Southern blotting. IL-1 mRNA was detected in the majority of early acute, acute, and subacute plaques, but not in chronic lesions (75). Additionally, a very low level of IL-1 mRNA was observed in some of the control brains. No conclusions about the types of cells expressing IL-1 could be drawn from these studies. Thus, in the limited studies conducted, there is no correlation between IL-1 expression and type of MS plaque. The cell types capable of IL-1 expression within the CNS include infiltrating T cells, B cells, and macrophages, as well as astrocytes, oligodendrocytes, and microglia (54–56,76–78).

Interleukin-2

IL-2, a Th1 cytokine, is the principal cytokine responsible for proliferation and maintenance of T cells (for review see Ref. 79). IL-2 is produced primarily by CD4+ T cells (Th1) when they are activated and acts in both an autocrine and paracrine fashion on T cells. The principal actions of IL-2 are on lymphocytes and include functioning as the major autocrine growth factor for T cells. Binding of IL-2 by T cells results in proliferation of these cells, enhanced secretion of other cytokines such as IFN-γ and LT, and the enhanced expression of transferrin receptors. IL-2 can also stimulate the growth of NK cells, activated B cells, and macrophages.

IL-2 is present in the CNS only in disease states in which the BBB has been disrupted, and activated T cells have infiltrated into that site. Within the CNS, Il-2 has varied biological effects on oligodendrocytes, the myelin-producing cells. IL-2 has been shown to influence the proliferation and differentiation of rat oligodendrocytes (80). Oligodendrocyte numbers were increased approximately threefold in cultures containing IL-2, and IL-2 stimulated the maturation of oligodendrocytes, assessed as the enhanced expression of MBP mRNA and protein in IL-2–treated oligodendrocytes (81). These findings provide evidence that an important step in oligodendrocyte differentiation, MBP expression, can be amplified in part by IL-2. Interestingly, IL-2 appeared to have the opposite effect on rat oligodendrocyte progenitor cells since it inhibited proliferation of these cells (82). These results suggest that IL-2 can have varying biological effects on oligodendrocytes depending on the maturational stage of those cells. Several recent studies have identified a substance produced by regenerating optic nerves from fish which is cytotoxic to oligodendrocytes (83). The cytotoxic factor was identified as IL-2 and had the molecular weight of an IL-2 dimer. Apparently, only dimerized IL-2, not monomeric IL-2, is cytotoxic to oligodendrocytes (84). Since IL-2 affects the function of oligodendrocytes and T cells, two cell types important in the disease of MS (oligodendrocytes as a target cell, T cells as effector cells), there has been much interest in IL-2 expression in MS patients.

IL-2 in Biological Fluids/Cells from MS Patients

The literature on IL-2 levels in the CSF/serum of MS patients is conflicting. Gallo et al. (85) conducted a study to measure IL-2 levels in the CSF and serum from 30 patients with MS. Detectable levels of IL-2 were found in six serum and nine CSF samples of 21 patients with acute relapsing MS, while IL-2 was not detected in either CSF or serum from nine patients with chronic progressive MS. Another report documented increased levels of IL-2 in the serum from chronic progressive MS patients, which correlated with elevated serum levels of TNF-α (86). Sharief et al. (87) demonstrated that high levels of IL-2 were found in the CSF of acute relapsing-remitting MS patients, while levels in the serum were not significantly elevated. An extension of this study indicated that CSF concentrations of IL-2 were significantly higher in MS patients during relapse compared to those in remission, suggesting an association of CSF IL-2 levels with disease activity (88).

However, two studies found no correlation of CSF or serum levels of IL-2 with MS disease progression (72,89). It is unclear at this time whether IL-2 levels in the CSF or serum can be used as an objective marker of MS disease activity.

IL-2 production by peripheral blood cells has been reported to be elevated in chronic progressive MS (90); however, this could not be confirmed by others (91).

Expression of IL-2 in MS Brain

Two studies by Hofman et al. (8,48) demonstrated IL-2 protein expression in chronic active plaques of MS brain by immunohistochemistry, while normal brain specimens were negative. The source of IL-2 in MS brain was presumably activated T cells, although this was not formally demonstrated. In another study, expression of IL-2 was detected in acute, chronic active, and chronic silent MS lesions, with the highest level of IL-2 being in the acute MS lesions (50). Expression of IL-2 was widespread throughout the lesions, although microglia showed a tendency to stain, in keeping with the documented expression of IL-2 receptors on these cells (92). Wucherpfennig et al. (75) examined IL-2 mRNA expression in MS brain by RT-PCR and demonstrated occasional expression in subacute and acute lesions.

Soluble IL-2 Receptor Expression in MS

Soluble IL-2 receptor (sIL-2R) is released from the surface of activated lymphocytes, and levels are increased in various autoimmune diseases. This elevation of sIL-2R in body fluids can serve as an indicator of T-cell activation and may be useful for monitoring disease activity in autoimmune disorders. Some authors report that both serum and CSF levels of sIL-2R are increased in patients with MS compared to normal controls or patients with OND (93–95). The study by Adachi et al. (95) also demonstrated that sIL-2R CSF and serum levels were significantly higher in MS patients in relapse compared to those in remission and suggested that sIL-2R levels, particularly in CSF, closely reflected clinical disease activity. These findings were confirmed by Sharief et al. (87,88). In contrast, other studies indicate that CSF and serum sIL-2R levels do not correlate with MS disease activity (72,89).

Interleukin-6

IL-6, considered a Th2 cytokine, is a pleiotropic cytokine involved in the regulation of inflammatory and immunological responses (for review see Ref. 96). IL-6 serves as an important cytokine for inducing terminal differentiation of activated B cells into immunoglobulin-secreting plasma cells. IL-6 has been implicated in contributing to CNS inflammation and immune responsiveness due to its ability to promote astrogliosis, immunoglobulin production, and T-cell activation (22). However, IL-6 also has some immunosuppressive effects such as inhibition of TNF-α expression by macrophages and astrocytes (97,98), inhibition of ICAM-1 expression by astrocytes and microglia (90), and induction of expression of sTNF-Rs receptors and IL-1 receptor antagonist (IL-1Ra) (100). Recently, IL-6 was shown to suppress demyelination in a viral model of MS, although the mechanism of action is unknown (101).

IL-6 in Biological Fluids/Cells from MS Patients

One of the hallmarks of MS is intrathecal B-cell activation as evidenced by an elevation of the CSF IgG index and the presence of oligoclonal IgG bands in the CSF (102). Since IL-6 is involved in differentiation of B cells into immunoglobulin-secreting plasma cells, there has been interest in determining whether elevated IL-6 levels could be responsible for local B-cell responses within the CNS. Results have been conflicting: four groups reported that IL-6 was not detected reproducibly in MS CSF (33,103–105), while two more recent studies suggested that IL-6 was elevated in MS CSF (34) and MS plasma (106). These latter findings suggest that there is a heightened systemic B-cell response in MS. However, no relationship was observed between the incidence or the amount of intrathecal IgG synthesis or oligoclonal bands and IL-6 expression in the CSF.

Few studies have been performed to examine IL-6 production by MS patients. Monocytes/macrophages from MS patients with acute disease produced higher levels of IL-6 in response to LPS or phorbol esters than did cells from chronic stable MS or OND (43). This elevation in IL-6 production was accompanied by increases in TNF-α, IL-1α, and IL-1β. An additional study demonstrated that secretion of IL-6 was more frequent in cultures of unstimulated monocytes from chronic progressive MS patients than from normal controls (107). These two studies are suggestive that IL-6 production by monocytes from MS patients with active disease is heightened compared to production by patients with stable disease or OND.

Expression of IL-6 in MS Brain

In one study, IL-6 immunoreactivity was detected at the lesion edge of burnt-out plaques in MS brain, rather than in active plaques, suggesting an involvement of IL-6 in resolution of disease (108). Clearly, more studies need to be performed to determine the extent of IL-6 expression in the CNS and association with disease activity, as well as the cell type(s) expressing IL-6.

Interferon-γ

IFN-γ, a Th1 cytokine, is produced by a limited number of cell types that include T cells (both $CD4^+$ and $CD8^+$) and NK cells. Within the $CD4^+$ T-cell population, IFN-γ is produced by the Th1 subset. IFN-γ is a pleiotropic cytokine that has physiological importance in regulating immune and inflammatory events. Some of the immune effects of IFN-γ include the ability to enhance the functional activity of macrophages, promotion of T- and B-cell differentiation, and modulation of both class I and II MHC antigen expression on a variety of cells. In this regard, IFN-γ is considered the most potent inducer of class II MHC expression on most cell types, except B cells (for review see Ref. 109). IFN-γ also promotes inflammatory responses, largely through its ability to enhance TNF-α production. Two IFN-γ proteins self-associate to form a homodimer, and only the dimer displays biological activity due to its ability to mediate IFN-γ receptor dimerization (for review see Ref. 110).

IFN-γ has the capacity to modulate gene expression in astrocytes and microglia. IFN-γ induces expression of both class II MHC antigens and ICAM-1 on astrocytes and microglia (27,28,58,111), which can contribute to the ability of these cells to act as antigen-presenting cells within the CNS and stimulate aberrant immune responses within this site. Additionally, IFN-γ stimulates glial cells to produce a variety of cytokines, including TNF-α (for review see Ref. 4). IFN-γ has been shown to induce proliferation of astrocytes in vitro as well as enhance astrogliosis in vivo (112). Thus, many of the effects of IFN-γ on cells of the CNS appear to contribute to inflammatory events in this tissue. The source of IFN-γ within the CNS is likely to be infiltrating activated T cells since there are no endogenous cell sources of IFN-γ in the normal brain.

IFN-γ in Biological Fluids/Cells of MS Patients

The interest in IFN-γ with respect to neurological disease stems from the results of a pilot study to assess the effect of recombinant IFN-γ on patients with MS. Administration of IFN-γ to a group of 18 patients with relapsing-remitting MS resulted in worsening of disease in seven patients (113). This trial established IFN-γ as an important mediator of MS attacks. Examination of IFN-γ in the CSF or serum of MS patients indicated that IFN-γ can rarely be detected in these body fluids (114,115). However, when examining IFN-γ production by cells from MS patients, increased IFN-γ production was detected prior to exacerbations, suggesting that IFN-γ contributes to exacerbation of disease (41). More recent studies on the association of IFN-γ with MS demonstrated that peripheral blood lymphocytes from MS patients produced significantly more IFN-γ than those of normal control cells (41,116,117) and that there were significantly increased numbers of CSF cells secreting IFN-γ in MS patients (118). Peripheral blood cells that expressed IFN-γ mRNA and secreted IFN-γ protein were significantly higher in MS patients during relapse or patients with severe disability (46,119,120). The increase in IFN-γ production was associated with decreased expression of TGF-β and IL-10 (46,119), suggesting that IFN-γ has a role in promoting MS disease progression.

Expression of IFN-γ in MS Brain

IFN-γ–positive cells have been detected in the CNS of MS patients, and these cells have been identified as astrocytes (8,121). It is presumed that infiltrating activated T cells secrete IFN-γ, which then can bind to receptors on astrocytes. IFN-γ has been localized in acute, chronic active, and chronic silent lesions, and there was no correlation between the degree of immunoreactivity and disease stage (50). In this study, IFN-γ immunoreactivity was detected at high levels on microglia and leukocytes.

Interleukin-4

IL-4, a Th2 cytokine, is produced by activated CD4+ Th2 cells, mast cells, and B cells. IL-4 functions as a growth and differentiation factor for B cells, induces class II MHC expression on B cells, and is a switch factor for IgE production. IL-4 also induces differentiation of Th0 cells (uncommitted precursor cells) to those with a Th2 phenotype. IL-4 has a number of anti-inflammatory properties such as inhibition of cytokine production by macrophages and Th1 cells (for review see Ref. 122). IL-4 effects on cells from the CNS include inhibition of astrocyte proliferation (123), inhibition of superoxide production by microglia (124), and inhibition of TNF-α expression by microglia (125). All these in vitro studies suggest that IL-4 may provide protection during CNS inflammation.

IL-4 in Biological Fluids/Cells of MS Patients

It is not known whether IL-4 is involved in MS, however, there is speculation that IL-4 might have a beneficial influence on MS due to its anti-inflammatory actions, particularly inhibition of IFN-γ production. IL-4 was not detected in either the serum or CSF of MS patients (115), and no follow-up studies have been performed. The number of IL-4 secreting cells from the peripheral blood of MS patients was significantly elevated compared to normal controls (120). In support of this, Link et al. (119,126) demonstrated that mononuclear cells from the blood and CSF of MS patients expressed IL-4 mRNA at a higher level than OND and normal controls. IL-4 mRNA levels increased when such cells were stimulated with either MBP or PLP, but not in response to other antigens. Another study examined T-cell clones derived from the CSF of patients with MS and determined that these cells were unable to secrete IL-4, although they produced IFN-γ, TNF-α, and IL-2, suggestive of a Th1 phenotype (42).

Expression of IL-4 in MS Brain

Only one study has examined IL-4 expression in the CNS; IL-4 was detected in acute, chronic active, and chronic silent MS lesions, and was most pronounced at the lesion edge (50). Cells positive for IL-4 included microglia, macrophages, and oligodendrocytes.

Interleukin-10

IL-10, a Th2 cytokine, is produced by numerous cell types including activated CD4+ Th2 cells, B cells, monocytes, astrocytes, and microglia. IL-10 generally exerts anti-inflammatory effects such as inhibition of proinflammatory cytokine production (IFN-γ, TNF-α, IL-1, IL-8), inhibition of class II MHC expression and antigen presentation ability, inhibition of macrophage activation, and inhibition of adhesion molecule expression (for review see Ref. 127). Within the CNS, IL-10 has been shown to inhibit the antigen-presenting capabilities of both astrocytes and microglia (128), inhibit cytokine-induced ICAM-1 expression by astrocytes and microglia (99), and inhibit TNF-α expression by glial cells and encephalitogenic T cells (98,129).

IL-10 in Biological Fluids/Cells of MS Patients

Many of the biological effects of IL-10 are associated with inhibiting the function of T cells and macrophages, the major effector cells of MS. As such, recent studies have examined whether IL-10 expression is associated with MS. IL-10 protein was not detected in serum from MS patients (130). However, examination of IL-10 mRNA from mononuclear cells from patients with relapsing-remitting MS as well as stable MS revealed that IL-10 mRNA expression was higher in patients with stable disease (68). TGF-β mRNA expression was also associated with stable disease. Further longitudinal analysis revealed that while TNF-α and LT mRNA expression in peripheral blood mononuclear cells significantly increased prior to a relapse, IL-10 (as well as TGF-β) mRNA levels declined. In addition, IL-10 and TGF-β were expressed at high levels during relapse-free intervals (45). In support of these findings, PLP-reactive T-cell clones from MS patients isolated during disease remission secreted high levels of IL-10, while those isolated during acute attacks secreted lower levels (46). These results suggest that IL-10 expression correlates with disease activity; levels are low prior to relapse, increase during remission, and stay elevated during stable disease.

Expression of IL-10 in MS Brain

IL-10 protein was detected in lesions from MS patients with acute, chronic active, and chronic silent disease, with expression being highest in acute plaques. IL-10 expression was highest around vessels along the lesion edge, and cellular localization was almost exclusive to astrocytes (50).

Transforming Growth Factor-β

The TGF-β superfamily is a large group of cytokines that have diverse biological effects on cells derived from multiple cell lineages (for review see Ref. 131). TGF-β signaling involves both the type I (TβR-I) and type II (TβR-II) receptors. Wrana et al. (132) recently demonstrated that TGF-β binds directly to TβR-II, which is a constitutively active serine/threonine kinase. The bound TGF-β is then recognized by TβR-I, which is recruited into the complex and becomes phosphorylated by TβR-II. This phosphorylation event of TβR-I is required for subsequent propagation of signals to substrates downstream of the TGF-β receptor complex. Although most cells express type I and II receptors for TGF-β, the specificity of the biological response to TGF-β in a given cell type depends on the state of differentiation of the cell, as well as the concentration of TGF-β to which cells are exposed. TGF-β is generally inhibitory for cells of the immune system, suppressing proliferation of B and T cells, preventing induction of IL-1 and IL-2 receptors, inhibiting production of proinflammatory cytokines such as TNF-α and IL-1, and inhibiting expression of class II MHC antigens (for review see Ref. 133). Additionally, TGF-β promotes growth of new blood vessels, and serves as a chemotactic factor for macrophages.

TGF-β can modulate the activity of glial cells by inhibiting IFN-γ–induced class II MHC expression on astrocytes and microglia (57,134,135), inhibiting proliferation of astrocytes and microglia (57,136), inhibiting cytokine-induced C3 complement expression in astrocytes (137), and inhibiting TNF-α production by microglia and astrocytes (57,138). In addition, TGF-β is an important mediator of oligodendrocyte differentiation (139). Due to its immunosuppressive effects on astrocytes and microglia, and its contribution to oligodendrocyte differentiation, TGF-β may restrict inflammatory processes within the CNS, as well as promote potential remyelination events.

TGF-β in Biological Fluids/Cells of MS Patients

As will be discussed later, $TGF\text{-}\beta_1$ and $TGF\text{-}\beta_2$ have been shown to improve the clinical course of EAE (140–143). Additionally, TGF-β is detected in the CNS of animals that have recovered from EAE (144). In this regard, one paper has demonstrated that TGF-β–like activity could be detected in peripheral blood of MS patients with active disease, and within this group of patients, TGF-β activity was significantly correlated to the period of regression of symptoms (145). In studies to examine TGF-β production by cells of MS patients, a number of groups have demonstrated that TGF-β expression correlates with disease recovery and/or stable disease. Decreased TGF-β production by peripheral blood lymphocytes from MS patients correlated directly with disease activity, i.e., patients with active disease produced less TGF-β than patients with stable disease (117). As previously discussed, TGF-β mRNA from peripheral blood mononuclear cells from MS patients was significantly higher in patients with stable disease compared to those with relapsing-remitting MS (68,119). Additionally, TGF-β mRNA levels, as well as those for IL-10 mRNA, declined prior to relapses (45). Data from Correale et al. (46) demonstrated that PLP-reactive T-cell clones isolated from MS patients in remission secreted predominantly TGF-β, as well as IL-10 and IL-4, all anti-inflammatory cytokines that may have a protective function in MS.

Expression of TGF-β in MS Brain

TGF-β protein expression was examined in MS lesions by immunohistochemistry and was detected in acute, chronic active, and chronic silent lesions (50). There were no significant differences in the levels of immunoreactivity detected in lesions of different ages. The extracellular matrix of blood vessels stained heavily for TGF-β, and it appeared that TGF-β immunoreactivity localized to astrocytes. The isotype of TGF-β detected was not indicated.

Type I Interferons (IFN-α/β)

The IFNs are a family of proteins related by their ability to protect cells from viral infection. Based on several criteria, the IFNs have been divided into three distinct classes: α, β, and γ (for review see Ref. 110). IFN-α and IFN-β (known as type I IFNs) are the classical IFNs induced in response to viral infection. IFN-α and IFN-β display only 15–30% amino acid sequence homology, yet both forms bind to the same cell surface receptor. Twenty-six IFN-α genes have been identified, which encode at least 22 distinct proteins. Recent studies suggest that the different species of IFN-α induce a distinct array of biological responses in different cell types. IFN-β exists in a single form. IFN-α/β can be produced by a wide variety of cells in response to viral infection, bacteria, mycoplasma, and cytokines such as IL-2, TNF-α, and IL-1. IFN-α/β have primarily antiviral, antiproliferative, antitumor, and immunomodulatory activities (for review see Refs. 146,147).

Due to many of the antiviral and anti-inflammatory properties of IFN-α/β, there has been tremendous interest in testing their effectiveness in MS. As will be discussed later, IFN-β has been approved for treatment of relapsing-remitting MS (148,149). However, little has been done to address the issue of IFN-α/β production in MS. Those few studies indicated that IFN-α and IFN-β were not detected in the CSF of MS patients, although they were present in MS plaques (147,150).

Summary

This section has summarized studies examining expression of cytokines in MS. It is difficult to draw firm conclusions since some of the findings are conflicting, and some of the studies are of a preliminary nature. However, the data indicate that there is aberrant expression of cytokines in the CNS (TNF-α, LT, IL-1, IL-2, and IFN-γ), which are involved in mediating immune responsiveness and inflammation. Expression of three of these cytokines—TNF-α, LT, and IFN-γ—precedes exacerbations, suggesting a role in disease progression. Additionally, all of the above cytokines exert biological effects on glial cells, many of which contribute to demyelination, inflammation, gliosis, cytokine production, and immune reactivity. IL-2, IFN-γ, TNF-α, and LT expression likely originates from infiltrating lymphocytes, particularly activated Th1 cells, and macrophages. The subsequent activation of astrocytes and microglia to secrete cytokines such as IL-1, TNF-α, and LT may contribute to the propagation of intracerebral immune and inflammatory responses initiated by infiltrating T cells and macrophages.

Downregulation of cytokine cascades, immune responses, and inflammation may be mediated by the generation of cytokines with anti-inflammatory properties such as IFN-β, TGF-β, IL-10, and possibly IL-4. At least for IFN-β, there are encouraging data that this cytokine is beneficial in reducing MS disease severity. Whether immune and inflammatory responses within the CNS are propagated or suppressed depends on a number of parameters, including (1) the activation and differentiation status of cells, (2) cytokine receptor levels on glial and immune cells, (3) the concentration and location of cytokines in the CNS, (4) the timing of cytokine expression in the CNS, (5) synergistic and/or antagonistic interactions between different cytokines, and (6) the temporal sequence in which a particular cell is exposed to numerous cytokines. See Table 1 for a summary of the cytokines associated with MS.

Table 1 Cytokines Associated with Multiple Sclerosis

Cytokine	Peripheral blood cells	Serum	Cerebrospinal fluid	CNS	Association with disease progression	Association with disease remission
TNF-α	Increased	Increased	Increased	TNF-α–positive astrocytes, macrophages	+++	
LT	Increased	Not examined	Not examined	LT-positive T cells, microglia	+++	
IL-1	Increased	Conflicting results	Increased	IL-1–positive microglia	+++	
IL-2	Conflicting results	Increased	Increased	IL-2–positive microglia	+++	
IL-6	Increased	Conflicting results	Conflicting results	IL-6 present; cell type unknown	?	?
IFN-γ	Increased	Undetectable	Undetectable	IFN-γ–positive astrocytes, microglia, leukocytes	+++	
IL-4	Increased	Undetectable	Undetectable	IL-4–positive microglia macrophages, oligodendrocytes		+++
IL-10	Increased	Undetectable	Not examined	IL-10–positive astrocytes		+++
TGF-β	Increased	Not examined	Not examined	TGF-β–positive astrocytes		+++
IFN-α/β	Not examined	Not examined	Undetectable	IFN-α present; cell type unknown		+++

USE OF CYTOKINES TO MODULATE MULTIPLE SCLEROSIS

Clinical Use of Type I Interferons in MS

There has been much interest in the use of IFNs in clinical trials on MS patients. The rationale for treatment of MS with IFNs was based on the hypothesis that disease activity might be related to viral persistence or latency within the CNS (151). Clinical trials of natural and recombinant IFN-α all resulted in negative or equivocal clinical results (for review see Ref. 152). In 1981, an open trial of natural IFN-β administered intrathecally resulted in a decrease in frequency of exacerbations, with minimal side effects (153). A larger double-blind study followed, and MS patients with relapsing-remitting disease treated with IFN-β had a significantly lower exacerbation rate than the placebo group (154). A double-blind, placebo-controlled trial was recently completed at 11 centers in the United States and Canada in which a total of 372 patients with relapsing-remitting MS were enrolled. In this trial, IFN-β_{ser} (Betaseron, Berlex Laboratories, Richmond, CA) was used. IFN-β_{ser} is a nonglycosylated, recombinant form of IFN-β in which a serine residue replaces a cysteine at position 17, which greatly enhances its stability, making it suitable for clinical use. This clinical trial showed that IFN-β_{ser} significantly reduced the rate of exacerbations, the severity of exacerbations, and accumulation of MRI abnormalities, with minimal side effects (148,149). Based on these data, the U.S. Food and Drug Administration approved IFN-β_{ser} for use in relapsing-remitting MS early in 1994. Additional data from this trial as well as details relating to the entire 5-year blinded study were recently published. IFN-β_{ser} had a persistent beneficial effect on exacerbation rate and MRI burden of disease and was relatively free of long-term side effects (155). Of concern was the fact that neutralizing antibodies to IFN-β_{ser} were detectable in 38% of patients by the third year and were associated with a significant attenuation of treatment effect on exacerbation rate.

In October 1994, Biogen presented its findings on the treatment of relapsing-remitting MS patients with a glycosylated form of recombinant IFN-β—IFN-β-1a. Approximately 300 patients received an intramuscular injection, once weekly, for 2 years. The results from this study indicate that treatment with IFN-β-1a delayed sustained progression of disability, decreased exacerbation frequency, and significantly reduced the number and volume of active lesions (for review see Ref. 156).

The data from these two clinical trials indicate that IFN-β can favorably modulate relapsing-remitting MS in a safe manner. Future studies will need to address the influence of neutralizing antibodies, optimal dose and means of administration, and the concurrent use of other immunoregulatory agents to bolster the observed benefits of IFN-β. Whether IFN-β will prove to be beneficial in MS patients with progressive disease is unproven and also needs to be examined.

Much information has emerged recently to suggest the mechanism(s) of action by which IFN-β exerts its beneficial effect. Stone et al. (157) documented that 13 of 14 patients with relapsing-remitting MS demonstrated a reduction in the frequency of contrast-enhancing lesions (indicative of BBB breakdown) upon treatment with IFN-β_{ser} over a 6-month period, suggesting that IFN-β_{ser} treatment temporarily inhibited the opening of the BBB. As breakdown of the BBB is thought to be a crucial event in the pathogenesis of new lesions in MS (12), this effect of IFN-β_{ser} may prevent infiltration of activated immune cells into the CNS. A number of in vitro studies also shed light on how IFN-β may exert a protective effect in relapsing-remitting MS. IFN-β significantly decreased IFN-γ production and T-cell activation in MS patients (158), thus IFN-β may antagonize many of the biological effects mediated by IFN-γ that are related to immune activation. IFN-β has been shown to downregulate IFN-γ–induced class II MHC expression on astrocytes, endothelial cells, human astroglioma cells, and macrophages (159–163). This downregulation of class II MHC expression could result in a diminished antigen-presenting capacity of a number of cell types in the CNS, thereby reducing T-cell activation. Additionally, IFN-β downregulated both basal and IFN-γ–induced expression of VLA-4 antigen on monocytes (163). The VLA-4 antigen binds to the adhesion molecule vascular adhesion molecule-1 (VCAM-1), which can be induced by proinflammatory cytokines on a number of cell types including brain endothelial cells and astrocytes (164,165). VLA-4/VCAM-1 interactions are involved in mediating leukocyte adhesion to brain endothelial cells, and antibodies against VLA-4 inhibited leukocyte binding to inflamed brain vessels and blocked the development of EAE in Lewis rats (166). Downregulation of VLA-4 expression on monocytes by IFN-β may inhibit leukocyte recruitment into the CNS, thereby lessening disease severity. There are also several reports that IFN-β downregulates TNF production by peripheral blood cells (167,168), and we have data that IFN-β is capable of inhibiting TNF-α gene expression by human astrocytes (L. P. Tang and E. N. Benveniste, unpublished). Because of the deleterious effects of TNF-α in the CNS, this action of IFN-β would certainly be beneficial to MS patients. Suppressor cell function in MS patients is decreased in patients before and during relapses and in progressive disease, but appears normal in stable MS. Exposure of lymphocyte cultures from MS patients to recombinant IFN-β improved nonspecific suppressor cell activity (for review see Ref. 147), suggesting another beneficial effect of IFN-β.

Clinical Use of IFN-γ in MS

An entirely different outcome resulted from a dose-escalation pilot trial to assess the effect of recombinant IFN-γ on MS patients. It was recognized that IFN-γ induced class II MHC expression on a variety of cells, including cells of the CNS and activated macrophages, which might be potentially hazardous in MS. However, a number of studies had determined that IFN-γ production was deficient in cultures from stimulated peripheral blood cells from MS patients (169,170). Administration of IFN-γ to a group of 18 patients with relapsing-remitting MS produced a rapid worsening of disease in 7 patients, which resolved after cessation of treatment (113). This trial established IFN-γ as an important mediator of MS attacks, possibly because the IFN-γ or IFN-γ–activated cells (T cells, macrophages) had infiltrated the CNS at sites of previous BBB damage and reactivated cellular responses within this site.

EXPERIMENTAL ALLERGIC ENCEPHALOMYELITIS

The best characterized experimental model for CNS autoimmune disease is EAE. This disease is induced by injection of spinal cord components such as myelin basic protein (MBP) or proteolipid protein (PLP) with adjuvant, injection of MBP or PLP peptides, or transfer of encephalitogenic MBP- or PLP-specific T cells to naive recipients. Interestingly, recent studies have demonstrated that EAE can be induced by nonmyelin antigens such as the astrocyte-derived calcium-binding proteins S100β, suggesting that nonmyelin antigens may play an important role in the immunopathogenesis of inflammatory CNS diseases (171). EAE is characterized by an inflammatory infiltration of the CNS by activated T cells and macrophages, and acute, chronic, or chronic-relapsing paralysis. The mediators of this disease are, for the most part, myelin-reactive CD4+ T-helper cells that are class II MHC restricted (for review see Ref. 172). However, not all myelin antigen-reactive CD4+ T-cell lines induce EAE, thus, there are other factors that contribute to the encephalitogenicity of these cells. Most encephalitogenic T cells are of the Th1 subtype, which secrete IFN-γ, IL-2, TNF-α, and LT, but not IL-4 (173,174). It has been suggested that TNF-α/LT secretion by MBP-specific T-cell clones correlates with their encephalitogenic potential (175), although this has not been observed for PLP-reactive T-cell clones (176). It may be that TNF production is necessary, but not always sufficient, to induce EAE. Expression of adhesion molecules such as VLA-4 also influence the pathogenicity of encephalitogenic T cells (176,177). Cross et al. (23) have proposed that antigen-specific autoimmune T cells are responsible for initiation of disease and that perpetuation of disease may be the result of an influx of largely non–antigen-specific inflammatory cells of the recipient animal. Susceptibility to EAE appears to be linked to MHC alleles, although non-MHC genes may play a small role in contributing to EAE (178). The MHC-associated susceptibility to EAE has been related, in part, to the ability to mount a Th1-like response (174). Th1 cytokines have been implicated in contributing to EAE disease progression, while Th2 cytokines (IL-4, IL-10, IL-13) as well as TGF-β appear to mediate recovery from disease.

INVOLVEMENT OF CYTOKINES IN EAE

Tumor Necrosis Factor-α and Lymphotoxin

Use of Anti-TNF/LT Antibodies

The evidence implicating TNF-α/LT in contributing to the disease progression of EAE is very strong. A number of studies have been conducted to neutralize TNF-α/LT activity; these include the use of anti-TNF antibodies, soluble TNF receptors, and pharmacological inhibitors of TNF. Both polyclonal and monoclonal antibodies to TNF-α and LT could prevent the transfer of EAE by MBP-encephalitogenic T cells in SJL/J mice (179,180). CNS tissue from animals treated with anti-TNF antibodies were devoid of inflammation or demyelination compared to control animals. Preincubation of MBP-sensitized T-cells with anti–TNF-α antibody in vitro prior to injection did not diminish their ability to transfer EAE, suggesting that the effect of anti–TNF-α antibody is achieved by interfering with the effector phase of disease (180). Spleen cells from anti-TNF-α–treated animals did not secrete TNF-α upon appropriate stimulation; this inhibitory effect on TNF-α production may be one mechanism by which anti–TNF-α antibodies exert a beneficial effect (180,181). In addition, anti–TNF-α/LT antibody treatment prevented the upregulation of VCAM-1 on spinal cord vessels, which could account for the marked reduction in leukocytic infiltration into the CNS (182). These authors suggest that anti–TNF-α/LT treatment inhibits leukocyte recruitment, rather than the initial entry of MBP-specific T cells into the CNS. Additional evidence implicating TNF in the disease of EAE was provided by the study by Kuroda et al. (183) demonstrating that intraperitoneal injection of TNF-α resulted in a significant prolongation of clinical EAE in Lewis rats. Findings by Chung et al. (184) demonstrated that astrocytes from EAE-susceptible and resistant rat strains differed in their ability to express TNF-α protein. Astrocytes from Lewis rats (susceptible) produced TNF-α in response to IFN-γ/IL-1β, while astrocytes from brown Norway rats (resistant) did not. The capacity for TNF-α production by Lewis astrocytes, especially in response to disease-related cytokines such as IFN-γ and IL-1, may contribute to disease susceptibility and to the inflammation and demyelination associated with EAE.

In contrast, the disease course of EAE in SJL/J mice actively immunized with spinal cord homogenate was not affected by treatment with anti–TNF-α antibody (185). This result may reflect different immunological mechanisms responsible for induction of the active and passive transfer forms of EAE. This type of differential effect was also observed using anti–ICAM-1 antibody; anti–ICAM-1 treatment was very effective at inhibiting EAE induced by active immunization in the rat, while adoptively transferred EAE was not inhibited (186–188).

Use of sTNF-Receptors

Several studies have documented a beneficial effect of administration of sTNF-R1 and sTNF-R2 in EAE. Baker et al. (189) demonstrated that treatment with bivalent human sTNF-R1 and sTNF-R2 immunoglobulin (sTNFR-Ig) fusion proteins inhibited the development of actively induced EAE in Biozzi AB/H mice. The sTNFR-Ig fusion proteins have been shown to have a longer serum half-life, the ability to activate effector mechanisms via the Fc domain, and a higher TNF-neutralizing capacity than conventional monovalent sTNF-Rs. Intracranial injection of sTNFR-Ig was more potent in inhibiting disease than systemic administration, suggesting that the majority of TNF-α production and/or activity occurred within the CNS (189). Administration of monomeric sTNF-R1 completely blocked both the clinical signs of disease as well as demyelination and infiltration in SJL/J mice with EAE induced by adoptive transfer of MBP-specific encephalitogenic T cells (190). A dimeric form of sTNF-R1 was shown to be more effective in inhibiting EAE than the monovalent form (181). The mechanism by which sTNF-R1 works is not clear, but in vitro incubation of encephalitogenic T cells with sTNF-R1 prior to injection into recipient animals did not prevent the transfer of disease (181).

Collectively, the results from the above studies indicate that both anti-TNF antibodies and sTNF-Rs are potent inhibitors of EAE. It appears that these agents do not influence the function of MBP-reactive T cells, indicating an effect at steps subsequent to the generation and/or activation of reactive cells.

Use of Pharmacological Inhibitors

A number of pharmacological approaches have been taken to inhibit TNF production in EAE. The phosphodiesterase inhibitor pentoxifylline preferentially inhibited Th1 but not

Th2 cytokine production by MBP-activated T cells; i.e., TNF-α and IL-2 production was inhibited, but IL-4 and IL-6 production was unaffected (191). MBP-induced EAE in Lewis rats was almost completely abrogated by the oral administration of pentoxifylline during the induction phase of disease (191). These data suggest that a therapeutic agent like pentoxifylline, which has selectivity for Th1 responses, may have therapeutic benefit in a disease like MS. Another agent that inhibits TNF-α production, phosphatidylserine, reduced the severity of adoptively transferred EAE in SJL/J mice (192). Animals treated with phosphatidylserine had minimal CNS pathology, failed to relapse, and remained healthier than control animals. The authors speculate that the action of phosphatidylserine may involve a reversible inhibition of antigen-specific lymphocyte functions (including inhibition of TNF-α production).

Gelatinases, which belong to the family of matrix metalloproteases (MMPs), contribute to tissue destruction in diseases such as MS and EAE by their ability to degrade extracellular matrix components (193,194). A MMP inhibitor (GM 6001) induced a partial inhibition of gelatinase activity in the CSF of mice with actively induced EAE, suppressed the development of EAE, and reversed clinical signs of EAE (195). The beneficial action of GM 6001 appeared to be through restoration of the damaged BBB; i.e., animals treated with GM 6001 had normal permeability of the BBB compared to the enhanced permeability in nontreated animals. Metalloprotease inhibitors block TNF processing activity, thereby reducing the release of TNF both in vitro and in vivo (196). As TNF is a potent inducer of MMP expression (for review see Ref. 197), some of the benefit of GM 6001 may be due to inhibition of TNF processing, thereby leading to decreased gelatinase activity. Although speculative, it would be informative to examine TNF levels in animals treated with GM 6001.

Additional support for the involvement of TNF-α in EAE comes from a study by Karin et al. (198), in which a soluble peptide variant of a MBP epitope, used as a T-cell receptor antagonist, was used to inhibit EAE. Draining lymph node cells from rats treated with MBP 87-99 and the antagonist (p87-99[91K > A]) produced markedly reduced levels of TNF-α and IFN-γ compared with those immunized with MBP 87-99 alone. The authors speculate that inhibition of the Th1 cytokines TNF-α and IFN-γ is the means by which p87-99[91K > A] inhibits EAE.

Expression of TNF-α/LT in EAE Brain

Numerous studies have been conducted to examine TNF-α expression in EAE. TNF-α protein was detected by immunohistochemistry within inflammatory lesions during both the active and relapsing disease phases of chronic-relapsing EAE (199). TNF-α protein was also expressed in the CNS of Lewis rats at the peak of clinical disease, while in animals that had naturally recovered or in MBP orally tolerized animals that recovered, there was an absence of TNF-α expression (144). In SJL/J and B10.PL mice immunized with MBP peptides, cells staining positively for TNF-α and LT protein were present in the brain but not the spinal cord of these animals (200). Expression of TNF-α and LT occurred transiently (days 11 and 15 postinjection) and was not very pronounced. In SJL/J mice with MBP-induced EAE, TNF-α mRNA was barely detectable in the CNS of unstimulated mice, expression increased in the CNS during peak disease, and dropped to background levels during remission (201). Analysis of cells isolated from the CNS of mice with acute EAE showed that TNF-α was predominantly expressed by microglia and macrophages. This study did not address the involvement of astrocytes in TNF-α expression due to technical limitations. Thus, numerous studies demonstrate that TNF-α mRNA and protein expression in the CNS of animals with EAE correlate with disease activity; TNF-α is expressed during the peak of clinical disease, and declines upon remission.

Interleukin-1

Use of IL-1 and sIL-1 Receptor

Because of the known inflammatory effects of IL-1, there has been interest in the role of IL-1 in EAE. IL-1 has been shown to enhance the in vitro activation of encephalitogenic T cells, thereby enhancing the adoptive transfer of EAE (202), and guinea pigs with chronic EAE have elevated levels of IL-1 in their CSF (203). More recently, Jacobs et al. (204) demonstrated that in vivo administration of IL-1α enhanced the severity and chronicity of clinical paralysis associated with EAE, whereas treatment of animals with soluble IL-1 receptor (an IL-1 antagonist) significantly delayed the onset of EAE, reduced the severity of paralysis and weight loss, and reduced the duration of disease. These results suggest that IL-1 contributes to disease induction and/or progression of EAE.

Expression of IL-1 in EAE Brain

There have been conflicting reports as to whether IL-1 is present in the brains of animals with EAE. In a model of chronic relapsing EAE induced in Biozzi AB/H mice with autologous spinal cord homogenate, IL-1 was detected in the brains of these animals in both the acute and relapsing phases of disease (199), along with the cytokines TNF-α, IL-2, and IFN-γ. The cell type(s) expressing the cytokines was not determined. Similarly, a study by Kennedy et al. (205) examining EAE induced in SJL/J mice by adoptive transfer of MBP-sensitized cells demonstrated that IL-1 mRNA was detected in the acute phase of disease and remained elevated during the early chronic state. Khoury et al. (144) examined the cytokine profiles of Lewis rats with EAE induced by immunization with MBP, and the subsequent effect of oral administration of MBP, which has been shown to suppress EAE. In this study, animals were fed with ovalbumin (OVA) as a control or MBP. Brains from OVA-fed animals at the peak of clinical disease showed perivascular infiltration and expression of the cytokine IL-1, as well as IL-2, IL-6, IL-8, IFN-γ, and TNF-α. In contrast, in MBP orally tolerized animals, there was a marked reduction of the perivascular infiltrate, downregulation of all the inflammatory cytokines, and an increase in the expression of the inhibitory cytokines TGF-β and IL-4. In a separate study, IL-1 protein expression was detected in the CNS of Lewis rats 10 days after immunization with spinal cord homogenate. During full-blown EAE (day 13), the number of lesions with IL-1 positive cells had increased, and in the remission phase (day 25), IL-1 positive cells could still be detected but were reduced in number (206). Based on microscopic analysis, IL-1 immunoreactivity was detected in cells with ultrastructural features of macrophages and microglia. In contrast, Merrill et al. (200) did not detect IL-1 protein in the brains of SJL/J and B10.PL mice that were immunized with MBP peptides. This would suggest that the method of immunization by which EAE is induced may influence cytokine expression in the brain.

Interleukin-2

Inhibition of IL-2 Activity in EAE

As mentioned previously, many cloned MBP and PLP-specific encephalitogenic T cells are of the Th1-phenotype; i.e., they produce high levels of IL-2 (as well as IFN-γ, TNF-α, and LT) (174,176,207,208). As such, it has been assumed that IL-2 production contributes to the clinical signs and pathology of EAE. Additionally, an early event in EAE is infiltration into the CNS of IL-2 receptor positive T cells (209). To address the involvement of IL-2 in EAE, Rose et al. (210) used a chimeric protein (IL2-PE40), that is a cytotoxin capable of selectively killing IL-2 receptor–positive lymphocytes. Intraperitoneal administration of IL2-PE40 inhibited the clinical manifestations of adoptively transferred EAE and also reduced inflammation and demyelination (210). It was assumed that the beneficial effect of IL2-PE40 was due to the elimination of IL-2 receptor–positive T cells and a subsequent reduction in IL-2 production. Another study demonstrated that treatment of MBP-stimulated spleen cells from donor rats with an IL-2 receptor–specific monoclonal antibody dramatically suppressed the ability of these cells to induce EAE in Lewis rats (211).

Anti–IL-2 monoclonal antibody treatment dramatically reduced the manifestations of adoptively transferred EAE in SJL/J mice but had only a marginal effect in the active induction of disease by MBP (212).

Expression of IL-2 in EAE Brain

IL-2 mRNA and protein expression was detected at the preclinical stage of disease (200) and during acute disease (144,199,205) and diminished upon recovery (144,205). T cells isolated from the spinal cord and cells isolated from the CSF showed an upregulation of IL-2 mRNA expression, which correlated with disease progression (207,213).

Interleukin-6

The literature on IL-6 and EAE is sparse and inconclusive at the present time.

Use of IL-6 and Anti–IL-6 Antibodies

A study was recently conducted to assess the effect of IL-6 in a murine model of MS induced by Theiler's murine encephalomyelitis virus (TMEV). Administration of IL-6 both prior to infection and after infection dramatically reduced demyelination and inflammation in the spinal cord of SJL/J mice (101). Although the mechanism by which IL-6 treatment diminished the extent of myelin destruction is unknown, the authors speculate that IL-6 inhibition of viral replication and virus antigen production may have contributed to this beneficial effect.

A preliminary study of the effect of anti–IL-6 antibodies on active or adoptive transfer EAE indicated that the clinical severity of disease was reduced when anti–IL-6 antibody was given at the time of immunization or transfer, but disease worsened when antibody was administered at the start of clinical symptoms (214). Since anti–IL-6 antibody administration is documented to enhance circulating levels of IL-6 (215), it will be important to examine this parameter in EAE studies.

IL-6 in Biological Fluids and EAE Brain

Increased IL-6 levels have been found in the CSF of mice suffering acute EAE, and it was suggested that local production of IL-6 was responsible for the elevated expression since serum levels of IL-6 were not increased (216). IL-6 mRNA levels increased rapidly during acute EAE in SJL/J mice and declined when clinical symptoms resolved (205). A similar observation was made in Lewis rats with EAE (144). In contrast, IL-6 protein was not found in the CNS of SJL/J mice that had acquired EAE through MBP peptide immunization (200).

Interferon-γ

Use of IFN-γ and Anti–IFN-γ Antibodies

In the human disease of MS, it is widely believed that local production of IFN-γ in MS lesions contributes to disease and that treatments that inhibit production of IFN-γ or prevent its action may help to block progression of the disease. In EAE, however, treatment with IFN-γ appeared to exert a protective effect, while treatment with anti–IFN-γ antibody aggravated disease. Administration of neutralizing antibody against IFN-γ caused an increase in disease severity and mortality (212,217), while treatment with IFN-γ itself resulted in enhanced survival (217). Another study examined the effect of IFN-γ on EAE in Lewis rats. Systemically administered IFN-γ did not change the disease course of EAE, whereas IFN-γ applied locally into the ventricular system of the CNS resulted in complete suppression of clinical signs (218). Additionally, systemic administration of anti–IFN-γ antibody prior to the onset of clinical symptoms resulted in a more severe disease course. These results suggest that IFN-γ downregulates the development of EAE, and it has been speculated that low levels of IFN-γ may be crucial to healing at early stages of disease (for review see Ref. 219). The findings in both murine and rat models of EAE differ from those observed in human MS trials, in which administration of IFN-γ led to an exacerbation of disease (113). Further experimentation will be needed to determine the reasons for the seemingly contradictory actions of IFN-γ in EAE compared to MS.

Expression of IFN-γ in EAE Brain

A number of studies have been conducted to examine IFN-γ expression in the CNS of animals with EAE. IFN-γ mRNA and protein could be detected before the onset of clinical disease (200,205,213,220,221) and during the acute phase of disease (144,199,205,213,221) and diminished upon recovery (144,199,205,221). By immunohistochemistry, a subpopulation of IFN-γ positive cells were identified as T cells, the majority of which were macrophages (220).

Interleukin-4

Modulation of IL-4 Expression in EAE

IL-4 has been implicated in contributing to suppression and/or recovery in EAE (222). A recent study by Racke et al. (223) indicated that oral administration of retinoids inhibited EAE in Lewis rats induced by adoptive transfer of MBP-encephalitogenic cells. Associated with a decrease in clinical disease was a marked decrease in the amount of inflammation and demyelination in these animals. The action of retinoids appeared to be due to the development of T cells of a Th2 phenotype, which correlated with the subsequent loss in encephalitogenicity. Specifically, IL-4 production by T cells correlated with improved disease outcome. Another study has also documented a protective effect of IL-4 in EAE. Lewis rats were protected from MBP-induced EAE when they were treated with the encephalitogenic peptide of MBP covalently linked to mouse anti-rat IgD; this protective effect also correlated with a greatly reduced level of leukocyte infiltration into the CNS (224). The mechanism of action of the anti-IgD–peptide conjugate appeared to be by induction of a Th2-type cytokine response. Lymph node and spleen cells from protected rats expressed elevated levels of IL-4 and IL-13 mRNA compared to diseased rats, and neutralizing anti-rat IL-4 monoclonal antibody abrogated the protective effect of treatment with the anti-IgD–peptide conjugate. Collectively, these studies indicate that treatments such as retinoids and anti-IgD–peptide conjugates induce a Th2-type T-cell response (especially IL-4 production) that has a protective effect in the disease of EAE.

Expression of IL-4 in EAE Brain

In keeping with the presumed protective effect of IL-4, Khoury et al. (144) detected IL-4 in MBP orally tolerized Lewis rats, which correlated with recovery from disease, and also in the CNS of Lewis rats that had spontaneously recovered. Expression of IL-4 was not detectable during the peak of clinical disease. The kinetics of TGF-β expression, another anti-inflammatory cytokine, were comparable to that of IL-4 (144).

Interleukin-10

IL-10, another Th2 cytokine, has been suggested to contribute to recovery in EAE due to its immunosuppressive effects on T cells, macrophages, and glial cells. In particular, IL-10 production by PLP-specific Th2 cells inhibited the activity of encephalitogenic Th1 cells (225). In this regard, Rott et al. (129) demonstrated that IL-10 completely suppressed TNF-α production by encephalitogenic T cells in vitro, and systemic administration of IL-10 markedly suppressed the induction of EAE in Lewis rats. IL-10 mRNA expression in the CNS was associated with recovery from EAE (205). These studies suggest that IL-10 has a protective effect in EAE, which is most likely mediated by inhibiting Th1 cell development.

Interleukin-13

IL-13 is a recently identified cytokine produced by activated T cells, particularly cells belonging to the Th2 phenotype. The main role of IL-13 appears to be the suppression of cytotoxic and inflammatory functions of monocytes/macrophages and enhancement of B-cell immunoreactivity. Interestingly, no direct effect of IL-13 on T-cell function has been described (for review see Ref. 226). Because of its effects on inhibiting macrophage functions, IL-13 has been tested in EAE. IL-13 was able to significantly suppress the development of clinical and histological signs of EAE induced in Lewis rats by MBP immunization (227). There was no effect of IL-13 on MBP-directed T-cell autoreactivity, suggesting that the beneficial effect of IL-13 was due to inhibition of macrophage function, supporting the concept of a pivotal role of macrophages/microglia in the pathogenesis of EAE (228).

The study by Saoudi et al. (224) using anti-IgD peptide conjugates to prevent EAE was discussed earlier. Lymph node cells from protected rats produced markedly increased levels of IL-4 mRNA, as well as IL-13 mRNA, suggesting that IL-13 production in this system contributed to the protective effect of anti-IgD peptide conjugates.

Transforming Growth Factor-β

Use of TGF-β and Anti–TGF-β Antibodies

Numerous studies have demonstrated that TGF-β1 improved the clinical course of EAE in SJL/J mice (140–142,229). Injection of TGF-β1 could delay, but not suppress, development of EAE; however, TGF-β1 could prevent the incidence of relapse in these mice (142). Racke et al. (140) demonstrated that TGF-β1 inhibited the activation of MBP specific lymph node cells in vitro, which reduced the capacity of these cells to transfer EAE. Additionally, injection of TGF-β1 resulted in an improved clinical course, even when administered during ongoing disease. Johns et al. (141) showed that in vivo injection of TGF-β1 reduced the incidence of clinical disease, as well as the severity of inflammation and demyelination within the CNS. In addition, TGF-β1, β2, and β2 were present in inflammatory lesions within the brain. Comparable results have been obtained using TGF-β_2, which reduced the clinical severity of EAE and number of relapses and produced less inflammation and demyelination in the CNS (143,230). Of relevance to the above studies are the findings that CD4+ and CD8+ T-suppressor cells that mediate recovery from EAE produce TGF-β (231,232) and that TGF-β is capable of inhibiting IFN-γ and TNF-α production by the CD4+ effector T cells that transfer disease (229,331). The implication of these studies is that TGF-β expression by T-suppressor cells is critical for recovery from EAE, and TGF-β may mediate its action by inhibiting expression of Th1 cytokines by CD4+ effector T cells. Additionally, TGF-β can inhibit TNF-α production by endogenous glial cells (astrocytes and microglia) (57,138), thereby contributing to the control of EAE.

Santambrogio et al. (230) examined the mechanism by which TGF-β protects against EAE. TGF-β treatment on days 5–9 after immunization did not influence the appearance of myelin-sensitized cells in peripheral blood and lymph nodes, but did prevent the subsequent accumulation of T cells in brain and spinal cord. These findings indicated that the protective effect of TGF-β was exerted at the level of the target organ (the CNS) and/or its vascular endothelium. This has been confirmed by Fabry et al. (233), who documented that TGF-β inhibited the migration of lymphocytes into the CNS in animals with EAE.

In contrast to the beneficial effects of TGF-β, administration of anti–TGF-β antibody led to a worsening of EAE, both in incidence and severity (143,230,234).

Expression of TGF-β in EAE Brain

TGF-β has been detected in the CNS of animals with EAE, although there are reported differences in the timing of appearance. Khoury et al. (144) detected TGF-β in MBP orally tolerized Lewis rats, which correlated with recovery from disease, and also in the CNS of Lewis rats that had spontaneously recovered. Expression of TGF-β was not detectable during the peak of clinical disease. In Lewis rats immunized with MBP, expression of TGF-β mRNA in the spinal cord did not increase until day 13 postimmunization, shortly before recovery (221). TGF-β mRNA was expressed by cells resembling astrocytes, microglia, and neurons. In SJL/J mice with EAE, TGF-β was detected by immunohistochemistry in areas of CNS inflammation in both acute and chronic disease, and astrocytes showed positive staining for TGF-β (235). In at least some studies, expression of TGF-β within the CNS correlates with disease recovery.

Type I Interferons (IFNα/β)

As discussed previously, IFN-β has proven effective in lessening the frequency and severity of exacerbations in MS. Brod et al. (236,237) tested the effect of recombinant human IFN-α and rat IFN-α/β in actively induced and adoptive transfer models of EAE. Oral administration of type I IFNs decreased the clinical score and degree of inflammation in Lewis rats immunized with MBP (236). In addition, draining lymph node cells from IFN-treated animals secreted less IFN-γ compared to mock-fed animals. In another system EAE was induced in SJL/J mice, and following recovery from the initial attack the animals were fed with human or murine type I IFNs or mock IFN. The activated donor cells from animals fed type I IFN-α were less effective in transferring clinical disease compared to cells from mock-fed animals (237), which may be due to decreased IFN-γ production by these cells. These studies indicate that type I IFNs exert a protective effect in both actively induced and adoptive transfer models of EAE, and a possible mechanism of action is accomplished by reducing IFN-γ secretion.

Chemokines

Chemokines are small molecular weight (7–10 kDa) secreted proteins that mediate the recruitment and activation of leukocytes and other cells to sites of inflammation during an immune response. Chemokines are the products of three related gene families, members of which exhibit sequence homology and structural similarities (for review see Ref. 238). The two chemokine families involved in EAE include the CXC chemokines (or α chemokines), which are chemotactic for neutrophils and T cells, and CC chemokines (or β chemokines), which are chemotactic for monocyte/macrophage lineage cells. Members of the CXC family include IL-8, γIP-10, and GRO-α, β, γ; members of the CC family include macrophage inflammatory protein (MIP)-1α, MIP-1β, monocyte chemoattractant peptide-1 (MCP-1), and RANTES.

Chemokines are expressed locally in response to inflammatory stimuli and act to recruit inflammatory cells via their chemoattractant properties. A recent study by Ransohoff et al. (239) investigated chemokine production in the CNS during EAE to identify factors potentially governing inflammatory cell accumulation during immune-mediated demyelination. EAE was induced in SJL/J mice using PLP peptides, and expression of MCP-1 and γIP-10 mRNA was examined by RT-PCR and in situ hybridization. Astrocytes were the only cells in the CNS that expressed mRNA transcripts for MCP-1 and γIP-10, and expression correlated with the appearance of clinical and histological EAE. In the Lewis rat, levels of MCP-1 mRNA were elevated immediately before the onset of clinical signs, peaked with the height of disease, and declined with resolution of disease (240). The marked elevation of MCP-1 at the height of clinical disease also correlated with extensive perivascular accumulation of monocytes. The cell type(s) producing MCP-1 was not identified in this study. Also in the Lewis rat, IL-8 protein was expressed at the peak of disease and declined upon recovery (144). RT-PCR analysis of spinal cord from animals with both actively induced and adoptively transferred EAE demonstrated that mRNAs encoding RANTES, MIP-1α, MIP-1β, γIP-10, and MCP-1 were induced 1–2 days prior to clinical signs and achieved their highest levels at disease onset (241). These findings collectively indicate that expression of both CXC and CC chemokines that specifically target cells of the immune system such as T cells and macrophages are an important component of the disease process in EAE.

Summary

Data from EAE studies support, for the most part, the concept that Th1 cytokines (IL-2, TNF-α, LT) contribute to the initiation and/or disease progression of EAE, as do IL-1 and chemokines. As yet unresolved, the Th1 cytokine IFN-γ appears to impart a protective effect in EAE, which is in contrast to the detrimental effect of IFN-γ in MS. Th2 cytokines (IL-4, IL-10, IL-13), as well as TGF-β and type I IFNs (α/β) are associated with improving the clinical course of disease, and the temporal expression of TGF-β, IL-4, and IL-10 in the CNS correlates with disease recovery and remission. Many of the in vivo studies reported a protective effect in EAE when the pattern of cytokine expression was skewed to a Th2-like phenotype. These studies utilized TCR antagonists, anti–IgD-peptide conjugates that interacted with B cells, and retinoids, as well as pharmacological inhibitors that blocked Th1 cytokine production, leading to an amplification of Th2 responses. See Table 2 for a summary of the cytokines associated with EAE.

FUTURE DIRECTIONS FOR CYTOKINE RESEARCH

Antigen-Specific Therapies for MS

A number of therapeutic strategies for MS are based on the premise that an autoimmune T-cell response is a crucial element in the immunopathogenesis of MS. A large number of research groups and private companies are designing strategies aimed at selective inhibition of autoreactive T cells. Some of these approaches include (1) the use of class II MHC competitors and TCR antagonists to interfere with antigen recognition by autoreactive T cells, (2) vaccination against the autoreactive T cell itself by use of attenuated antigen-specific T cells or with peptides of selected antigen-specific TCR (T cell or TCR vaccination), and (3) oral tolerization with a relatively crude preparation of autoantigens, which appears to involve induction of regulatory T cells that release inhibitory cytokines such as TGF-β, IL-4, and IL-10 (for review see Refs. 156,219,242–244).

Cytokine-Specific Therapies for MS

Inhibitory Cytokines

Cytokine-based modulation of MS appears promising based on the beneficial effect of IFN-β on relapsing-remitting

Table 2 Cytokines Associated with Experimental Allergic Encephalomyelitis

Cytokine	Function	Expression in CNS during disease		Association with disease progression	Association with disease recovery
		Acute	Recovery		
TNF-α/LT	Enhances clinical course Antibody against TNF-α/LT inhibits disease sTNF-R inhibits disease	+++	—	+++	
IL-1	Enhances clinical course sIL-1R inhibits disease	+++	—	+++	
IL-2	Antibody against IL-2 or IL-2R inhibits disease	+++	—	+++	
IL-6	Inconclusive	+++	—	?	?
IFN-γ	Protects against disease Antibody against IFN-γ aggravates disease	+++	—		+++
IL-4	Protects against disease	—	+++		+++
IL-10	Protects against disease	—	+++		+++
IL-13	Protects against disease	Not examined			+++
TGF-β	Protects against disease	—	+++		+++
IFN-α/β	Protects against disease	Not examined			+++
Chemokines	Not examined	+++	—	+++	

MS. Additional cytokine therapies being designed or tested include the "inhibitory" cytokines TGF-β, IL-4, and IL-10. Some of the actions of these cytokines include inhibition of the biosynthesis of proinflammatory cytokines (IL-1, IFN-γ, TNF-α/LT, chemokines), thus, these reagents would provoke a deficiency in the cytokines implicated in MS disease progression. As discussed previously, TGF-β, IL-4, and IL-10 have been successfully used to inhibit EAE (see Table 2).

Neutralizing Antibodies Against Cytokines

Another method of inhibiting the activity of proinflammatory cytokines uses neutralizing antibodies against the cytokines of interest (for review see Ref. 245). Of concern with this approach is that such neutralizing antibodies may slow down the clearance of the offending cytokine from the circulation, thereby increasing the availability of the cytokine, as exemplified by anti–IL-6 neutralizing antibodies (215). For MS, the most likely approach would include the use of neutralizing antibodies against IL-1, TNF-α, LT, IFN-γ, and chemokines.

Soluble Cytokine Receptors

The use of modified soluble cytokine receptors may hold promise for treatment of MS. The best studied to date is the sTNF-R. Since sTNF-Rs display affinities for TNF similar to the membrane receptors, very large amounts of sTNF-R are required to inhibit TNF activity completely. To enhance the affinity of the soluble receptor for its ligand, chimeric proteins between sTNF-R and Ig domains have been generated. These fusion proteins, called immunoadhesins, are divalent and display a higher affinity (10- to 100-fold) for TNF than native monomeric sTNF-R. These immunoadhesins have been successfully used as therapeutic agents in a variety of animal models, including EAE (181,189,190). Candidates for MS treatment would include sTNF-Rs, sIL-1R, and sIFN-γR.

Receptor Antagonists

A number of naturally occurring secreted cytokine antagonists exist, the best studied being the IL-1 receptor antagonist (IL-1Ra). The IL-1Ra is structurally related to IL-1 and binds with high affinity to the IL-1R Type I, but not the IL-1R Type II. This aspect is important since only the IL-1R Type I is capable of transducing a signal following ligand binding; the IL-1R Type II is thought to serve as a decoy for IL-1 or act as a precursor for the soluble form of the receptor, thereby antagonizing the activity of IL-1 (for review see Ref. 246). Because the inhibitory activity of IL-1Ra is *diminished* by the simultaneous presence of sIL-1R Type I, the clinical use of sIL-1R Type I may not be advisable. It has been proposed that the simultaneous use of IL-1Ra and sIL-1R Type II may be beneficial since both molecules inhibit IL-1 biological activity without interfering with each other. Because of the documented involvement of IL-1 in MS disease progression, the use of IL-1Ra, alone or in conjunction with sIL-1R Type II, may be of benefit.

Immune Deviation

Specific cytokine patterns are associated with the cure or exacerbation of both infectious and immune-mediated diseases (for review see Ref. 247). As discussed previously, MS disease progression is associated with a Th1 response, while remission correlates with a Th2 response. A number of interventions have been successfully used in EAE to promote the development of a Th2 response, resulting in amelioration of disease. Some of these—retinoids, pentoxifylline, and TCR antagonists—may be of benefit in MS and warrant further investigation.

Summary

A number of methodologies have been described to obtain a reduction in proinflammatory cytokine expression, including inhibitory cytokines, neutralizing anticytokine antibodies, soluble cytokine receptors, receptor antagonists, and immune deviation to enhance Th2 responses. These procedures, as well as those directed toward selective inactivation of autoimmune T cells, may be beneficial in an autoimmune disease such as MS. Considerations for future therapy should encompass concurrent use of various methodologies, local rather than systemic application, and intermittent rather than prolonged therapy. Given the complexity of cytokine interactions, it will probably be necessary to interrupt the function of several cytokines at once to achieve the desired outcome.

ACKNOWLEDGMENTS

I thank Sue Wade for superb secretarial and editorial assistance in preparing this manuscript. This work was supported in part by National Multiple Sclerosis Society Grants 2269-A-4 and 2205-B-5, National Institutes of Health Grants NS-29719, NS-31096, and MH-50421, and a grant from Berlex Biosciences.

REFERENCES

1. Benveniste EN. Cytokines: influence on glial cell gene expression and function. In: Blalock JE, ed. Chemical Immunology: Neuroimmunoendocrinology. Vol. 52. Basel: Karger, 1992:106–153.
2. Raff MC. Glial cell diversification in the rat optic nerve. Science 1989; 243:1450–1455.
3. Patterson PH, Nawa H. Neuronal differentiation factors/cytokines and synaptic plasticity. Cell/Neuron 1993; 72/10:123–137.
4. Benveniste EN. Inflammatory cytokines within the central nervous system: sources, function, and mechanism of action. Am J Physiol Cell Physiol 1992; 263:32:C1–C16.
5. Hauser SL, Bhan AK, Gilles FH, et al. Immunohistochemical staining of human brain with monoclonal antibodies that identify lymphocytes, monocytes and the Ia antigen. J Neuroimmunol 1983; 5:197–205.
6. Prineas JW. Pathology of the early lesion in multiple sclerosis. Human Pathol 1975; 6:531–554.
7. Prineas JW, Wright RG. Macrophages, lymphocytes and plasma cells in the perivascular compartment in chronic multiple sclerosis. Lab Invest 1978; 38:409–421.
8. Hofman FM, VonHanwher R, Dinarello C, Mizel S, Hinton D, Merrill JE. Immunoregulatory molecules and IL-2 receptors identified in multiple sclerosis brain. J Immunol 1986; 136:3239–3245.
9. Bignami A, Eng LF, Dahl D, Uyeda CT. Localization of the glial fibrillary acidic protein in astrocytes by immunofluorescence. Brain Res 1972; 43:429–435.
10. Selmaj K, Brosnan CF, Raine CS. Colocalization of lymphocytes bearing $\gamma\delta$ T-cell receptor and heat shock protein hsp65$^+$ oligodendrocytes in multiple sclerosis. Proc Natl Acad Sci USA 1991; 88:6452–6456.
11. Thompson AJ, Miller D, Youl B, et al. Serial gadolinium-enhanced MRI in relapsing/remitting multiple sclerosis of varying disease duration. Neurology 1992; 42:60–63.
12. Gay D, Esiri M. Blood-brain barrier damage in acute multiple sclerosis plaques: an immunocytological study. Brain 1991; 144:557–572.
13. Martin R, McFarland HF, McFarlin DE. Immunological aspects of demyelinating diseases. Annu Rev Immunol 1992; 10:153–187.
14. Mosmann TR, Coffman RL. TH1 and TH2 cells: different patterns of lymphokine secretion lead to different functional properties. Annu Rev Immunol 1989; 7:145–173.
15. Benveniste EN. Role of cytokines in multiple sclerosis, autoimmune encephalitis, and other neurological disorders. In: Aggarawal B, Puri R, eds. Human Cytokines: Their Role in Research and Therapy. Boston: Blackwell Scientific Publications. 1995:195–216.
16. Tracey KJ, Cerami A. Tumor necrosis factor, other cytokines and disease. Annu Rev Cell Biol 1993; 9:317–343.
17. Selmaj KW, Raine CS. Tumor necrosis factor mediates myelin and oligodendrocyte damage in vitro. Ann Neurol 1988; 23:339–346.
18. Robbins DS, Shirazi Y, Drysdale BE, Lieberman A, Shin HS, Shin ML. Production of cytotoxic factor for oligodendrocytes by stimulated astrocytes. J Immunol 1987; 139:2593–2597.
19. Paul NL, Ruddle NH. Lymphotoxin. Annu Rev Immunol 1988; 6:407–438.
20. Selmaj K, Raine CS, Farooq M, Norton WT, Brosnan CF. Cytokine cytotoxicity against oligodendrocytes: apoptosis induced by lymphotoxin. J Immunol 1991; 147:1522–1529.
21. Raine CS. Biology of disease: analysis of autoimmune demyelination: its impact upon multiple sclerosis. Lab Invest 1984; 50:608–635.
22. Selmaj KW, Farooq M, Norton WT, Raine CS, Brosnan CF. Proliferation of astrocytes in vitro in response to cytokines. A primary role for tumor necrosis factor. J Immunol 1990; 144:129–135.
23. Cross AH, Cannella B, Brosnan CF, Raine CS. Homing to central nervous system vasculature by antigen-specific lymphocytes. Lab Invest 1990; 63:162–170.
24. Raine CS, Lee SC, Scheinberg LC, Duijvestijn AM, Cross AH. Adhesion molecules on endothelial cells in the central nervous

system: an emerging area in the neuroimmunology of multiple sclerosis. Clin Immunol Immunopathol 1990; 57:173–187.
25. McCarron RM, Wang L, Racke MK, McFarlin DE, Spatz M. Cytokine-regulated adhesion between encephalitogenic T lymphocytes and cerebrovascular endothelial cells. J Neuroimmunol 1993; 43:23–30.
26. Hurwitz AA, Lyman WD, Guida MP, Calderon TM, Berman JW. Tumor necrosis factor α induces adhesion molecule expression on human fetal astrocytes. J Exp Med 1992; 176:1631–1636.
27. Frohman EM, Frohman TC, Dustin ML, et al. The induction of intercellular adhesion molecule1 (ICAM-1) expression on human fetal astrocytes by interferon-γ, tumor necrosis factor–α, lymphotoxin, and interleukin-1: relevance to intracerebral antigen presentation. J Neuroimmunol 1989; 23:117–124.
28. Shrikant P, Chung IY, Ballestas M, Benveniste EN. Regulation of intercellular adhesion molecule-1 gene expression by tumor necrosis factor-α, interleukin-1β, and interferon-γ in astrocytes. J Neuroimmunol 1994; 51:209–220.
29. Héry C, Sébire G, Peudenier S, Tardieu M. Adhesion to human neurons and astrocytes of monocytes: the role of interaction of CR3 and ICAM-1 and modulation by cytokines. J Neuroimmunol 1995; 57:101–109.
30. Sobel RA, Mitchell ME, Fondren G. Intercellular adhesion molecule-1 (ICAM-1) in cellular immune reactions in the human central nervous system. Am J Pathol 1990; 136:1309–1316.
31. Gallo P, Piccinno MG, Krzalic L, Tavolato B. Tumor necrosis factor alpha (TNFα) and neurological diseases: failure in detecting TNFα in the cerebrospinal fluid from patients with multiple sclerosis, AIDS dementia complex, and brain tumors. J Neuroimmunol 1989; 23:41–44.
32. Merrill JE, Strom SR, Ellison GW, Myers LW. In vitro study of mediators of inflammation in multiple sclerosis. J Clin Immunol 1989; 9:84–96.
33. Hauser SL, Doolittle TH, Lincoln R, Brown RH, Dinarello CA. Cytokine accumulations in CSF of multiple sclerosis patients: frequent detection of interleukin-1 and tumor necrosis factor but not interleukin-6. Neurology 1990; 40:1735–1739.
34. Maimone D, Gregory S, Arnason BGW, Reder AT. Cytokine levels in the cerebrospinal fluid and serum of patients with multiple sclerosis. J Neuroimmunol 1991; 32:67–74.
35. Franciotta DM, Grimaldi LME, Martino GV, et al. Tumor necrosis factor in serum and cerebrospinal fluid of patients with multiple sclerosis. Ann Neurol 1989; 26:787–789.
36. Sharief MK, Phil M, Hentges R. Association between tumor necrosis factor-α and disease progression in patients with multiple sclerosis. N Engl J Med 1991; 325:467–472.
37. Tsukada N, Miyagi K, Matsuda M, Yanagisawa N, Yone K. Tumor necrosis factor and interleukin-1 in the CSF and sera of patients with multiple sclerosis. J Neurol Sci 1991; 102:230–234.
38. Sharief MK, Noori MA, Ciardi M, Cirelli A, Thompson EJ. Increased levels of circulating ICAM-1 in serum and cerebrospinal fluid of patients with active multiple sclerosis. Correlation with TNF-α and blood-brain barrier damage. J Neuroimmunol 1993; 43:15–22.
39. Sharief MK, Thompson EJ. In vivo relationship of tumor necrosis factor-α to blood-brain barrier damage in patients with active multiple sclerosis. J Neuroimmunol 1992; 38:27–34.
40. Kriegler M, Perez C, DeFay K, Albert I, Lu SD. A novel form of TNF/cachectin is a cell surface cytotoxic transmembrane protein: ramifications for the complex physiology of TNF. Cell 1988; 53:45–53.
41. Beck J, Rondot P, Catinot L, Falcoff E, Kirchner H, Wietzerbin J. Increased production of interferon gamma and tumor necrosis factor precedes clinical manifestation in multiple sclerosis: Do cytokines trigger off exacerbations? Acta Neurol Scand 1988; 78:318–323.
42. Benvenuto R, Paroli M, Buttinelli C, et al. Tumour necrosis factor-alpha synthesis by cerebrospinal-fluid-derived T cell clones from patients with multiple sclerosis. Clin Exp Immunol 1991; 84:97–102.
43. Imamura K, Suzumura A, Hayashi F, Marunouchi T. Cytokine production by peripheral blood monocytes/macrophages in multiple sclerosis patients. Acta Neurol Scand 1993; 8:281–285.
44. Glabinski A, Mirecka M, Pokoca L. Tumor necrosis factor alpha but not lymphotoxin is overproduced by blood mononuclear cells in multiple sclerosis. Acta Neurol Scand 1995; 91:276–279.
45. Rieckmann P, Albrecht M, Kitze B, et al. Tumor necrosis factor-α messenger RNA expression in patients with relapsing-remitting multiple sclerosis is associated with disease activity. Ann Neurol 1995; 37:82–88.
46. Correale J, Gilmore W, McMillan M, et al. Patterns of cytokine secretion by autoreactive proteolipid protein-specific T cell clones during the course of multiple sclerosis. J Immunol 1995; 154:2959–2968.
47. Hofman FM, Hinton DR, Johnson K, Merrill JE. Tumor necrosis factor identified in multiple sclerosis brain. J Exp Med 1989; 170:607–612.
48. Hofman FM, Hinton DR, Baemayr J, Weil M, Merrill JE. Lymphokines and immunoregulatory molecules in subacute sclerosing panencephalitis. Clin Immunol Immunopath 1991; 58:331–342.
49. Selmaj K, Raine CS, Cannella B, Brosnan CF. Identification of lymphotoxin and tumor necrosis factor in multiple sclerosis lesions. J Clin Invest 1991; 87:949–954.
50. Cannella B, Raine CS. The adhesion molecule and cytokine profile of multiple sclerosis lesions. Ann Neurol 1995; 37:424–435.
51. Brosnan CF, Cannella B, Battistini L, Raine CS. Cytokine localization in multiple sclerosis lesions: correlation with adhesion molecule expression and reactive nitrogen species. Neurology 1995; 45:S16–S21.
52. Chung IY, Benveniste EN. Tumor necrosis factor-α production by astrocytes: Induction by lipopolysaccharide, IFN-γ and IL-1β. J Immunol 1990; 144:2999–3007.
53. Lieberman AP, Pitha PM, Shin HS, Shin ML. Production of tumor necrosis factor and other cytokines by astrocytes stimulated with lipopolysaccharide or a neurotropic virus. Proc Natl Acad Sci USA 1989; 86:6348–6352.
54. Velasco S, Tarlow M, Olsen K, Shay JW, McCracken J, Nisen PD. Temperature-dependent modulation of lipopolysaccharide-induced interleukin-1β and tumor necrosis factor α expression in cultured human astroglial cells by dexamethasone and indomethacin. J Clin Invest 1991; 87:1674–1680.
55. Lee SC, Liu W, Dickson DW, Brosnan CF, Berman JW. Cytokine production by human fetal microglia and astrocytes: Differential induction by lipopolysaccharide and IL-1β. J Immunol 1993; 150:2659–2667.
56. Chao CC, Hu S, Close K, et al. Cytokine release from microglia: differential inhibition by pentoxifylline and dexamethasone. J Infect Dis 1992; 166:847–853.
57. Suzumura A, Sawada M, Yamamoto H, Marunouchi T. Transforming growth factor-β suppresses activation and proliferation of microglia in vitro. J Immunol 1993; 151:2150–2158.
58. Frei K, Siepl C, Groscurth P, Bodmer S, Schwerdel C, Fontana

A. Antigen presentation and tumor cytotoxicity by interferon-γ-treated microglial cells. Eur J Immunol 1987; 17:1271–1278.
59. Appel K, Honegger P, Gebicke-Haerter PJ. Expression of interleukin-3 and tumor necrosis factor-β mRNAs in cultured microglia. J Neuroimmunol 1995; 60:83–91.
60. Smith CA, Farrah T, Goodwin RG. The TNF receptor superfamily of cellular and viral proteins: activation, costimulation, and death. Cell 1994; 76:959–962.
61. Heller RA, Song K, Onasch MA, Fischer WH, Chang D, Ringold GM. Complementary DNA cloning of a receptor for tumor necrosis factor and demonstration of a shed form of the receptor. Proc Natl Acad Sci USA 1990; 87:6151–6155.
62. Lantz M, Gullberg U, Nilsson E, Olsson I. Characterization in vitro of a human tumor necrosis factor-binding protein. A soluble form of a tumor necrosis factor receptor. J Clin Invest 1990; 86:1396–1402.
63. Nophar Y, Kemper O, Brakebusch C, et al. Soluble forms of tumor necrosis factor receptors (TNF-Rs). The cDNA for the type 1 TNF-R, cloned using amino acid sequence data of its soluble form, encodes both the cell surface and a soluble form of the receptor. EMBO J 1990; 9:3269–3278.
64. Engelmann H, Aderka D, Rubinstein M, Rotman D, Wallach D. A tumor necrosis factor-binding protein purified to homogeneity from human urine protects cells from tumor necrosis factor toxicity. J Biol Chem 1989; 264:11974–11980.
65. Seckinger P, Isaaz S, Dayer J-M. Purification and biologic characterization of a specific tumor necrosis factor α inhibitor. J Biol Chem 1989; 264:11966–11973.
66. Aderka D, Engelmann H, Maor Y, Brakebusch C, Wallach D. Stabilization of the bioactivity of tumor necrosis factor by its soluble receptors. J Exp Med 1992; 175:323–329.
67. Tsukada N, Matsuda M, Miyagi K, Yanagisawa N. Increased levels of intercellular adhesion molecule-1 (ICAM-1) and tumor necrosis factor receptor in the cerebrospinal fluid of patients with multiple sclerosis. Neurology 1993; 43:2679–2682.
68. Rieckmann P, Martin S, Weichselbraun I, et al. Serial analysis of circulating adhesion molecules and TNF receptor in serum from patients with multiple sclerosis. Neurology 1994; 44:2367–2372.
69. Giulian D, Lachman LB. Interleukin-1 stimulation of astroglial proliferation after brain injury. Science 1985; 228:497–499.
70. Giulian D, Woodward J, Young DG, Krebs JF, Lachman LB. Interleukin-1 injected into mammalian brain stimulates astrogliosis and neovascularization. J Neurosci 1988; 8:2485–2490.
71. Aloisi F, Borsellino G, Samoggia P, et al. Astrocyte cultures from human embryonic brain: characterization and modulation of surface molecules by inflammatory cytokines. J Neurosci Res 1992; 32:494–506.
72. Peter JB, Boctor FN, Tourtellotte WW. Serum and CSF levels of IL-2, sIL-2R, TNF-α, and IL-1β in chronic progressive multiple sclerosis: expected lack of clinical utility. Neurology 1991; 41:121–123.
73. Matsuda M, Tsukada N, Miyagi K, Yanagisawa N. Increased interleukin-1 production by peripheral blood mononuclear cells in patients with multiple sclerosis. J Neurol Sci 1991; 102:100–104.
74. Porrini AM, Reder AT. IFN-γ, IFN-β, and PGE_1 affect monokine secretion: relevance to monocyte activation in multiple sclerosis. Cell Immunol 1994; 157:428–438.
75. Wucherpfennig KW, Newcombe J, Li H, Keddy C, Cuzner ML, Hafler DA. T cell receptor V_α-V_β repertoire and cytokine gene expression in active multiple sclerosis lesions. J Exp Med 1992; 175:993–1002.
76. Malipiero UV, Frei K, Fontana A. Production of hemopoietic colony-stimulating factors by astrocytes. J Immunol 1990; 144:3816–3821.
77. Giulian D, Baker TJ, Shih LC, Lachman LB. Interleukin-1 of the central nervous system is produced by ameboid microglia. J Exp Med 1986; 164:594–604.
78. Merrill JE, Matsushima K. Production of and response to interleukin-1 by cloned human oligodendroglioma cell lines. J Biol Reg Homeost Agents 1988; 2:77–86.
79. Smith KA. Interleukin-2: inception, impact and implications. Science 1988; 240:1169–1176.
80. Benveniste EN, Merrill JE. Stimulation of oligodendroglial proliferation and maturation by interleukin-2. Nature 1986; 321:610–613.
81. Benveniste EN, Herman PK, Whitaker JN. Myelin basic protein-specific RNA levels in interleukin-2-stimulated oligodendrocytes. J Neurochem 1987; 49:1274–1279.
82. Saneto RP, Altman A, Knobler R, Johnson HM, de Vellis J. Interleukin 2 mediates the inhibition of oligodendrocyte progenitor cell proliferation in vitro. Proc Natl Acad Sci USA 1986; 83:9221–9225.
83. Eitan S, Zisling R, Cohen A, et al. Identification of an interleukin 2-like substance as a factor cytotoxic to oligodendrocytes and associated with central nervous system regeneration. Proc Natl Acad Sci USA 1992; 89:5442–5446.
84. Eitan S, Schwartz M. A transglutaminase that converts interleukin-2 into a factor cytotoxic to oligodendrocytes. Science 1993; 261:106–108.
85. Gallo P, Piccinno M, Pagni S, Tavolato B. Interleukin-2 levels in serum and cerebrospinal fluid of multiple sclerosis patients. Ann Neurol 1988; 24:795–797.
86. Trotter JL, Collins KG, van der Veen RC. Serum cytokine levels in chronic progressive multiple sclerosis: interleukin-2 levels parallel tumor necrosis factor-α levels. J Neuroimmunol 1991; 33:29–36.
87. Sharief MK, Hentges R, Thompson EJ. The relationship of interleukin-2 and soluble interleukin-2 receptors to intrathecal immunoglobulin synthesis in patients with multiple sclerosis. J Neuroimmunol 1991; 32:43–51.
88. Sharief MK, Thompson EJ. Correlation of interleukin-2 and soluble interleukin-2 receptor with clinical activity of multiple sclerosis. J Neurol Neurosurg Psychiatry 1993; 56:169–174.
89. Freedman MS, Muth KL, Trotter JL, Yoshizawa CN, Antel JP. Prospective serial analysis of interleukin-2 and soluble interleukin-2 receptor in relapsing-remitting multiple sclerosis. Neurology 1992; 42:1596–1601.
90. Merrill JE, Mohlstrom C, Uittenbogaart C, Kermaniarab V, Ellison GW, Myers LW. Response to and production of interleukin-2 by peripheral blood and cerebrospinal fluid lymphocytes of patients with multiple sclerosis. J Immunol 1984; 133:1931–1937.
91. Selmaj K, Nowak Z, Tchorzewski H. Interleukin-1 and interleukin-2 production by peripheral blood mononuclear cells in multiple sclerosis patients. J Neurol Sci 1988; 85:67–76.
92. Sawada M, Suzumura A, Marunouchi T. Induction of functional interleukin-2 receptor in mouse microglia. J Neurochem 1995; 64:1973–1979.
93. Gallo P, Piccinno MG, Pagni S, et al. Immune activation in multiple sclerosis: study of IL-2, sIL-2R, and γ-IFN levels in serum and cerebrospinal fluid. J Neurol Sci 1989; 92:9–15.
94. Bansil S, Troiano R, Cook SD, Rohowsky-Kochan C. Serum soluble interleukin-2 receptor levels in chronic progressive, stable and steroid-treated multiple sclerosis. Acta Neurol Scand 1991; 84:282–285.
95. Adachi K, Kumamoto T, Araki S. Elevated soluble interleukin-

2 receptor levels in patients with active multiple sclerosis. Ann Neurol 1990; 28:687–691.
96. Van Snick JV. Interleukin-6: an overview. Annu Rev Immunol 1990; 8:253–278.
97. Aderka D, Le J, Vilcek J. IL-6 inhibits lipopolysaccharide-induced tumor necrosis factor production in cultured human monocytes, U937 cells, and in mice. J Immunol 1989; 143:3517–3523.
98. Benveniste EN, Tang LP, Law RM. Differential regulation of astrocyte TNF-α expression by the cytokines TGF-β, IL-6 and IL-10. Int J Dev Neurosci 1995; 13:341–349.
99. Shrikant P, Weber E, Jilling T, Benveniste EN. ICAM-1 gene expression by glial cells: differential mechanisms of inhibition by interleukin-10 and interleukin-6. J Immunol 1995; 155:1489–1501.
100. Tilg H, Trehu E, Atkins MB, Dinarello CA, Mier JW. Interleukin-6 (IL-6) as an anti-inflammatory cytokine: induction of circulating IL-1 receptor antagonist and soluble tumor necrosis factor receptor p55. Blood 1994; 83:113–118.
101. Rodriguez M, Pavelko KD, McKinney CW, Leibowitz JL. Recombinant human IL-6 suppresses demyelination in a viral model of multiple sclerosis. J Immunol 1994; 153:3811–3821.
102. Tourtellotte WW, Ma IB. Multiple sclerosis: the blood-brain barrier and the measurement of de novo central nervous system IgG synthesis. Neurology 1978; 28:76–83.
103. Frei K, Leist TP, Meager A, et al. Production of B cell stimulatory factor-2 and interferon-γ in the central nervous system during viral meningitis and encephalitis. Evaluation in a murine model infection and in patients. J Exp Med 1988; 168:449–453.
104. Houssiau FA, Bukasa K, Sindic CJM, Van Damme J, Van Snick J. Elevated levels of the 26K human hybridoma growth factor (interleukin 6) in cerebrospinal fluid of patients with acute infection of the central nervous system. Clin Exp Immunol 1988; 71:320–323.
105. Laurenzi MA, Sidén A, Persson MAA, Norkrans G, Hagberg L, Chiodi F. Cerebrospinal fluid interleukin-6 activity in HIV infection and inflammatory and noninflammatory diseases of the nervous system. Clin Immunol Immunopath 1990; 57:233–241.
106. Frei K, Fredrikson S, Fontana A, Link H. Interleukin-6 is elevated in plasma in multiple sclerosis. J Neuroimmunol 1991; 31:147–153.
107. Maimone D, Reder AT, Gregory S. T cell lymphokine-induced secretion of cytokines by monocytes from patients with multiple sclerosis. Cell Immunol 1993; 146:96–106.
108. Merrill JE. Proinflammatory and antiinflammatory cytokines in multiple sclerosis and central nervous system acquired immunodeficiency syndrome. J Immunother 1992; 12:167–170.
109. Glimcher LH, Kara CJ. Sequences and factors: a guide to MHC Class II transcription. Annu Rev Immunol 1992; 10:13–49.
110. Farrar MA, Schreiber RD. The molecular cell biology of interferon-γ and its receptor. Annu Rev Immunol 1993; 11:571–611.
111. Wong GHW, Clark-Lewis I, Harris AW, Schrader JW. Effect of cloned interferon-γ on expression of H-2 and Ia antigens on cell lines of hemopoietic, lymphoid, epithelial, fibroblastic and neuronal origin. Eur J Immunol 1984; 14:52–56.
112. Yong VW, Moumdjian R, Yong FP, et al. γ-Interferon promotes proliferation of adult human astrocytes in vitro and reactive gliosis in the adult mouse brain in vivo. Proc Natl Acad Sci USA 1991; 88:7016–7020.
113. Panitch HS, Hirsch RL, Schindler J, Johnson KP. Treatment of multiple sclerosis with gamma interferon: exacerbations associated with activation of the immune system. Neurology 1987; 37:1097–1102.
114. Abbott RJ, Bolderson I, Gruer PJK, Peatfield RC. Immunoreactive IFN-γ in CSF in neurological disorders. J Neurol Neurosurg Psychiatry 1987; 50:882–885.
115. Gallo P, Piccinno MG, Tavolato B, Sidén A. A longitudinal study on IL-2, sIL-2R, IL-4 and IFN-γ in multiple sclerosis CSF and serum. J Neurol Sci 1991; 101:227–232.
116. Hirsch RL, Panitch HS, Johnson KP. Lymphocytes from multiple sclerosis patients produce elevated levels of gamma interferon in vitro. J Clin Immunol 1985; 5:386–389.
117. Mokhtarian F, Shi Y, Shirazian D, Morgante L, Miller A, Grob D. Defective production of anti-inflammatory cytokine, TGF-β by T cell lines of patients with active multiple sclerosis. J Immunol 1994; 152:6003–6010.
118. Olsson T, Zhi WW, Hojeberg B, et al. Autoreactive T lymphocytes in multiple sclerosis determined by antigen-induced secretion of interferon-γ. J Clin Invest 1990; 86:981–985.
119. Link J, Söderström M, Olsson T, Höjeberg B, Ljungdahl A, Link H. Increased transforming growth factor-β, interleukin-4, and interferon-γ in multiple sclerosis. Ann Neurol 1994; 36:379–386.
120. Lu C-Z, Jensen MA, Arnason BGW. Interferon-γ- and interleukin-4-secreting cells in multiple sclerosis. J Neuroimmunol 1993; 46:123–128.
121. Traugott U, Lebon P. Multiple sclerosis: Involvement of interferons in lesion pathogenesis. Ann Neurol 1988; 24:243–251.
122. Paul WE, Seder RA. Lymphocyte responses and cytokines. Cell 1994; 76:241–251.
123. Barna BP, Estes ML, Pettay J, Iwasaki K, Zhou P, Barnett GH. Human astrocyte growth regulation: interleukin-4 sensitivity and receptor expression. J Neuroimmunol 1995; 60:75–81.
124. Chao CC, Hu S, Peterson PK. Modulation of human microglial cell superoxide production by cytokines. J Leuk Biol 1995; 85:65–70.
125. Chao CC, Molitor TW, Hu S. Neuroprotective role of IL-4 against activated microglia. J Immunol 1993; 151:1473–1481.
126. Link J, Söderström M, Ljungdahl A, et al. Organ-specific autoantigens induce interferon-γ and interleukin-4 mRNA expression in mononuclear cells in multiple sclerosis and myasthenia gravis. Neurology 1994; 44:728–734.
127. Moore KW, O'Garra A, de Waal Malefyt M, Vieira P, Mosmann TR. Interleukin-10. Annu Rev Immunol 1993; 11:165–190.
128. Frei K, Lins H, Schwerdel C, Fontana A. Antigen presentation in the central nervous system: the inhibitory effect of IL-10 on MHC class II expression and production of cytokines depends on the inducing signals and the type of cell analyzed. J Immunol 1994; 152:2720–2728.
129. Rott O, Fleischer B, Cash E. Interleukin-10 prevents experimental allergic encephalomyelitis in rats. Eur J Immunol 1994; 24:1434–1440.
130. Gallo O, Sivieri S, Rinaldi L, et al. Intrathecal synthesis of interleukin-10 (IL-10) in viral and inflammatory diseases of the central nervous system. J Neurol Sci 1994; 126:49–53.
131. Miyazono K, Hidenori I, Heldin CH. Transforming growth factor-β: latent forms, binding proteins and receptors. Growth Factors 1993; 8:11–22.
132. Wrana JL, Attisano L, Wieser R, Ventura F, Massagué J. Mechanism of activation of the TGF-β receptor. Nature 1994; 370:341–347.
133. Wahl SM. Transforming growth factor β: the good, the bad, the ugly. J Exp Med 1994; 180:1587–1590.
134. Panek RB, Lee Y-J, Benveniste EN. TGF-β suppression of

IFN-γ induced class II MHC gene expression does not involve inhibition of phosphorylation of JAK1, JAK2 or STAT1α or modification of IFNEX expression. J Immunol 1995; 154:610–619.
135. Schluesener HJ. Transforming growth factors type β1 and β2 suppress rat astrocyte autoantigen presentation and antagonize hyperinduction of class II major histocompatibility complex antigen expression by interferon-γ and tumor necrosis factor-α. J Neuroimmunol 1990; 27:41–47.
136. Lindholm D, Castren E, Kiefer R, Zafra F, Thoenen H. Transforming growth factor-β1 in the rat brain: increase after injury and inhibition of astrocyte proliferation. J Cell Biol 1992; 117:395–400.
137. Barnum SR, Jones JL. Transforming growth factor-β1 inhibits inflammatory cytokine-induced C3 gene expression in astrocytes. J Immunol 1994; 152:765–773.
138. Benveniste EN, Kwon JB, Chung WJ, Sampson J, Pandya K, Tang L-P. Differential modulation of astrocyte cytokine gene expression by TGF-β. J Immunol 1994; 153:5210–5221.
139. McKinnon RD, Piras G, Ida JA, Jr., Dubois-Dalcq M. A role for TGF-β in oligodendrocyte differentiation. J Cell Biol 1993; 121:1397–1407.
140. Racke MK, Jalbut SD, Cannella B, Albert PS, Raine CS, McFarlin DE. Prevention and treatment of chronic relapsing experimental allergic encephalomyelitis by transforming growth factor-β1. J Immunol 1991; 146:3012–3017.
141. Johns LD, Flanders KC, Ranges GE, Sriram S. Successful treatment of experimental allergic encephalomyelitis with transforming growth factor-β1. J Immunol 1991; 147:1793–1796.
142. Kuruvilla AP, Shah R, Hochwald GM, Liggitt HD, Palladino MA, Thorbecke GJ. Protective effect of transforming growth factor β_1 on experimental autoimmune diseases in mice. Proc Natl Acad Sci USA 1991; 88:2918–2921.
143. Racke MK, Sriram S, Carlino J, Cannella B, Raine CS, McFarlin DE. Long-term treatment of chronic relapsing experimental allergic encephalomyelitis by transforming growth factor-β2. J Neuroimmunol 1993; 46:175–184.
144. Khoury SJ, Hancock WW, Weiner HL. Oral tolerance to myelin basic protein and natural recovery from experimental autoimmune encephalomyelitis are associated with downregulation of inflammatory cytokines and differential upregulation of transforming growth factor β, interleukin 4, and prostaglandin E expression in the brain. J Exp Med 1992; 176:1355–1364.
145. Beck J, Rondot P, Jullien P, Wietzerbin J, Lawrence DA, TGF-β-like activity produced during regression of exacerbations in multiple sclerosis. Acta Neurol Scand 1991; 84:452–455.
146. Weinstock-Guttman B, Ransohoff RM, Kinkel RP, Rudick RA. The interferons: biological effects, mechanisms of action, and use in multiple sclerosis. Ann Neurol 1995; 37:7–15.
147. Arnason BGW, Reder AT. Interferons and multiple sclerosis. Clin Neuropharmacol 1994; 17:495–547.
148. Group TIMSS. Interferon beta-1β is effective in relapsing-remitting multiple sclerosis. I. Clinical results of a multicenter, randomized, double-blind, placebo-controlled trial. Neurology 1993; 43:655–661.
149. Paty DW, Li DKB, Group TIMSS. Interferon beta-1b is effective in relapsing-remitting multiple sclerosis. II. MRI analysis results of a multicenter, randomized, double-blind placebo-controlled trial. Neurology 1993; 43:662–667.
150. Brandt ER, Mackay IR, Hertzog PJ, Cheetham BF, Sherritt M, Bernard CC. Molecular detection of interferon-α expression in multiple sclerosis brain. J Neuroimmunol 1993; 44:1–6.
151. Johnson RT, Lazzarini RA, Waksman BH. Mechanisms of virus persistence. Ann Neurol 1981; 9:616–617.
152. Panitch HS, Bever CT, Jr. Clinical trials of interferons in multiple sclerosis: What have we learned? J Neuroimmunol 1993; 46:155–164.
153. Jacobs L, O'Malley J, Freeman A, Ekes R. Intrathecal interferon reduces exacerbations of multiple sclerosis. Science 1981; 214:1026–1028.
154. Jacobs L, Salazar AM, Herndon R, et al. Intrathecally administered natural human fibroblast interferon reduces exacerbations of multiple sclerosis. Arch Neurol 1987; 44:589–595.
155. The IFNB Multiple Sclerosis Study Group, University of British Columbia MS/MRI Analysis Group. Interferon beta-1b in the treatment of multiple sclerosis: final outcome of the randomized controlled trial. Neurology 1995; 45:1277–1285.
156. Steinman L, Conlon P. Designing rational therapies of multiple sclerosis. Bio/Technology 1995; 13:118–120.
157. Stone LA, Frank JA, Albert PS, et al. The effect of interferon-β on blood-brain barrier disruption demonstrated by contrast-enhanced magnetic resonance imaging in relapsing-remitting multiple sclerosis. Ann Neurol 1995; 37:611–619.
158. Noronha A, Toscas A, Jensen MA. Interferon β decreases T cell activation and interferon γ production in multiple sclerosis. J Neuroimmunol 1993; 46:145–154.
159. Inaba K, Kitaura M, Kato T. Contrasting effects of alpha/beta and gamma-interferons on expression of macrophage Ia antigens. J Exp Med 1986; 163:1030–1035.
160. Barna BP, Chou SM, Jacobs B, Lieberman BY, Ransohoff RM. Interferon-β impairs induction of HLA-DR antigen expression in cultured adult human astrocytes. J Neuroimmunol 1989; 23:45–53.
161. Satoh J-I, Paty DW, Kim SU. Differential effects of beta and gamma interferons on expression of major histocompatibility complex antigens and intercellular adhesion molecule-1 in cultured fetal human astrocytes. Neurology 1995; 45:367–373.
162. Huynh HK, Oger J, Dorovini-Zis K. Interferon-β downregulates interferon-γ-induced class II MHC molecule expression and morphological changes in primary cultures of human brain microvessel endothelial cells. J Neuroimmunol 1995; 60:63–73.
163. Soilu-Hanninen M, Salmi A, Salonen R. Interferon-β downregulates expression of VLA-4 antigen and antagonizes interferon-γ-induced expression of HLA-DQ on human peripheral blood monocytes. J Neuroimmunol 1995; 60:99–106.
164. Sasseville VG, Newman W, Brodie SJ, Hesterberg P, Pauley D, Ringler DJ. Monocyte adhesion to endothelium in simian immunodeficiency virus-induced AIDS encephalitis is mediated by vascular cell adhesion molecule-1/α4β1 integrin interactions. Am J Pathol 1994; 144:27–40.
165. Rosenman SJ, Shrikant P, Dubb L, Benveniste EN, Ransohoff RM. Cytokine-induced expression of vascular cell adhesion molecule-1 (VCAM-1) by astrocytes and astrocytoma cell lines J Immunol 1995; 154:1888–1899.
166. Yednock TA, Cannon C, Fritz LC, Sanchez-Madrid F, Steinman L, Karin N. Prevention of experimental autoimmune encephalomyelitis by antibodies against α4β1 integrin. Nature 1992; 356:63–66.
167. Abu-khabar KS, Armstrong JA, Ho M. Type I interferons (IFN-α and -β) suppress cytotoxin (tumor necrosis factor-α and lymphotoxin) production by mitogen-stimulated human peripheral blood mononuclear cells. J Leukocyte Biol 1992; 52:165–172.
168. Noronha A, Jensen MA, Toscas A, Sihag S. IFN-beta downregulates tumor necrosis factor release. Neurology 1992; 42:159.

169. Vervliet G, Claeys H, van Haver H, et al. Interferon production and natural killer (NK) activity in leukocyte cultures from multiple sclerosis patients. J Neurol Sci 1983; 60:137–150.
170. Vervliet G, Carton H, Meulepas E, Billiau A. Interferon production by cultured peripheral leucocytes of MS patients. Clin Exp Immunol 1984; 58:116–126.
171. Kojima K, Berger T, Lassmann H, et al. Experimental autoimmune panencephalitis and uveoretinitis transferred to the Lewis rat by T lymphocytes specific for the S100β molecule, a calcium binding protein of astroglia. J Exp Med 1994; 180:817–829.
172. Zamvil SS, Steinman L. The T lymphocyte in experimental allergic encephalomyelitis. Annu Rev Immunol 1990; 8:579–621.
173. Ando DG, Clayton J, Kono D, Urban JL, Sercarz EE. Encephalitogenic T cells in the B10.PL model of experimental allergic encephalomyelitis (EAE) are of the Th-1 lymphokine subtype. Cell Immunol 1989; 124:132–143.
174. Mustafa M, Vingsbo C, Olsson T, Ljungdahl A, Höjeberg B, Holmdahl R. The major histocompatibility complex influences myelin basic protein 63-88-induced T cell cytokine profile and experimental autoimmune. Eur J Immunol 1993; 23:3089–3095.
175. Powell MB, Mitchell D, Lederman J, et al. Lymphotoxin and tumor necrosis factor-alpha production by myelin basic protein-specific T cell clones correlates with encephalitogenicity. Int Immunol 1990; 2:539–544.
176. Kuchroo VK, Martin CA, Greer JM, Ju S-T, Sobel RA, Dorf ME. Cytokines and adhesion molecules contribute to the ability of myelin proteolipid protein-specific T cell clones to mediate experimental allergic encephalomyelitis. J Immunol 1993; 151:4371–4382.
177. Baron JL, Madri JA, Ruddle NH, Hashim G, Janeway CA, Jr. Surface expression of α4 integrin by CD4 T cells is required for their entry into brain parenchyma. J Exp Med 1993; 177:57–68.
178. Gasser DL, Newlin CM, Palm J, Gonatas NK. Genetic control of susceptibility to experimental allergic enchephalomyelitis in rats. Science 1973; 181:872–873.
179. Ruddle NH, Bergman CM, McGrath KM, et al. An antibody to lymphotoxin and tumor necrosis factor prevents transfer of experimental allergic encephalomyelitis. J Exp Med 1990; 172:1193–1200.
180. Selmaj K, Raine CS, Cross AH. Anti-tumor necrosis factor therapy abrogates autoimmune demyelination. Ann Neurol 1991; 30:694–700.
181. Selmaj KW, Raine CS. Experimental autoimmune encephalomyelitis: immunotherapy with anti-tumor necrosis factor antibodies and soluble tumor necrosis factor receptors. Neurology 1995; 45:S44–S49.
182. Barten DM, Ruddle NH. Vascular cell adhesion molecule-1 modulation by tumor necrosis factor in experimental allergic encephalomyelitis. J Neuroimmunol 1994; 51:123–133.
183. Kuroda Y, Shimamoto Y. Human tumor necrosis factor-α augments experimental allergic encephalomyelitis in rats. J Neuroimmunol 1991; 34:159–164.
184. Chung IY, Norris JG, Benveniste EN. Differential tumor necrosis factor α expression by astrocytes from experimental allergic encephalomyelitis-susceptible and -resistant rat strains. J Exp Med 1991; 173:801–811.
185. Teuschler C, Hickey WF, Korngold R. An analysis of the role of tumor necrosis factor in the phenotypic expression of actively induced experimental allergic orchitis and experimental allergic encephalomyelitis. Clin Immunol Immunopathol 1990; 54:442–453.
186. Archelos JJ, Jung S, Mäurer M, et al. Inhibition of experimental autoimmune encephalomyelitis by an antibody to the intercellular adhesion molecule ICAM-1. Ann Neurol 1993; 34:145–154.
187. Cannella B, Cross AH, Raine CS. Anti-adhesion molecule therapy in experimental autoimmune encephalomyelitis. J Neuroimmunol 1993; 46:43–56.
188. Willenborg DO, Simmons RD, Tamatani T, Miyasaka M. ICAM-1-dependent pathway is not critically involved in the inflammatory process of autoimmune encephalomyelitis or in cytokine-reduced inflammation of the central nervous system. J Neuroimmunol 1993; 45:147–154.
189. Baker D, Butler D, Scallon BJ, O'Neill JK, Turk JL, Feldmann M. Control of established experimental allergic encephalomyelitis by inhibition of tumor necrosis factor (TNF) activity within the central nervous system using monoclonal antibodies and TNF receptor-immunoglobulin fusion proteins. Eur J Immunol 1994; 24:2040–2048.
190. Selmaj K, Papierz W, Glabinski A, Kohno T. Prevention of chronic relapsing experimental autoimmune encephalomyelitis by soluble tumor necrosis factor receptor I. J Neuroimmunol 1995; 56:135–141.
191. Rott O, Cash E, Fleischer B. Phosphodiesterase inhibitor pentoxifylline, a selective suppressor of T helper type 1-but not type 2-associated lymphokine production, prevents induction of experimental autoimmune encephalomyelitis in Lewis rats. Eur J Immunol 1993; 23:1745–1751.
192. Monastra G, Cross AH, Bruni A, Raine CS. Phosphatidylserine, a putative inhibitor of tumor necrosis factor, prevents autoimmune demyelination. Neurology 1993; 43:153–163.
193. Gijbels K, Masure S, Carton H, Opdenakker G. Gelatinase in the cerebrospinal fluid of patients with multiple sclerosis and other inflammatory neurological disorders. J Neuroimmunol 1992; 41:29–34.
194. Gijbels K, Proost P, Masure S, Carton H, Billiau A, Opdenakker G. Gelatinase B is present in the cerebrospinal fluid during experimental autoimmune encephalomyelitis and cleaves myelin basic protein. J Neurosci Res 1993; 36:432–440.
195. Gijbels K, Galardy RE, Steinman L. Reversal of experimental autoimmune encephalomyelitis with a hydroxamate inhibitor of matrix metalloproteases. J Clin Invest 1994; 94:2177–2182.
196. Mohler KM, Sleath PR, Fitzner JN, et al. Protection against a lethal dose of endotoxin by an inhibitor of tumour necrosis factor processing. Nature 1994; 370:218–220.
197. Birkedal-Hansen H. Role of cytokines and inflammatory mediators in tissue destruction. J Periodontal Res 1993; 28:500–510.
198. Karin N, Mitchell DJ, Brocke S, Ling N, Steinman L. Reversal of experimental autoimmune encephalomyelitis by a soluble peptide variant of a myelin basic protein epitope: T cell receptor antagonism and reduction of interferon γ and tumor necrosis factor α production. J Exp Med 1994; 180:2227–2237.
199. Baker D, O'Neill JK, Turk JL. Cytokines in the central nervous system of mice during chronic relapsing experimental allergic encephalomyelitis. Cell Immunol 1991; 134:505–510.
200. Merrill JE, Kong DH, Clayton J, Ando DG, Hinton DR, Hofman FM. Inflammatory leukocytes and cytokines in the peptide-induced disease of experimental allergic encephalomyelitis in SJL and B10.PL mice. Proc Natl Acad Sci USA 1992; 89:574–578.
201. Renno T, Krakowski M, Piccirillo C, Lin J-Y, Owens T. TNF-α expression by resident microglia and infiltrating leukocytes in the central nervous system of mice with experimental allergic encephalomyelitis. J Immunol 1995; 154:944–953.
202. Mannie MD, Dinarello CA, Paterson PY. Interleukin 1 and

myelin basic protein synergistically augment adoptive transfer activity of lymphocytes mediating experimental autoimmune encephalomyelitis in Lewis rats. J Immunol 1987; 138:4229–4235.

203. Symons JA, Bundick RV, Suckling AJ, Rumsby MG. Cerebrospinal fluid interleukin 1-like activity during chronic relapsing experimental allergic encephalomyelitis. Clin Exp Immunol 1987; 68:648–654.

204. Jacobs CA, Baker PE, Roux ER, et al. Experimental autoimmune encephalomyelitis is exacerbated by IL-1α and suppressed by soluble IL-1 receptor. J Immunol 1991; 146:2983–2989.

205. Kennedy MK, Torrance DS, Picha KS, Mohler KM. Analysis of cytokine mRNA expression in the central nervous system of mice with experimental autoimmune encephalomyelitis reveals that IL-10 mRNA expression correlates with recovery. J Immunol 1992; 149:2496–2505.

206. Bauer J, Berkenbosch F, Van Dam A-M, Dijkstra CD. Demonstration of interleukin-1β in Lewis rat brain during experimental allergic encephalomyelitis by immunocytochemistry at the light and ultrastructural level. J Neuroimmunol 1993; 48:13–22.

207. Weinberg AD, Wallin JJ, Jones RE, et al. Target organ-specific up-regulation of the MRC OX-40 marker and selective production of Th1 lymphokine mRNA by encephalitogenic T helper cells isolated from the spinal cord of rats with experimental autoimmune encephalomyelitis. J Immunol 1994; 152:4712–4721.

208. Weinberg AD, Wyrick G, Celnik B, et al. Lymphokine mRNA expression in the spinal cords of Lewis rats with experimental autoimmune encephalomyelitis is associated with a host recruited CD45R hi/CD4$^+$ population during recovery. J Neuroimmunol 1993; 48:105–118.

209. Sedgwick J, Brostoff S, Mason D. Experimental allergic encephalomyelitis in the absence of a classical delayed-type hypersensitivity reaction. J Exp Med 1987; 165:1058–1075.

210. Rose JW, Lorberboum-Galski H, Fitzgerald D, et al. Chimeric cytotoxin IL2-PE40 inhibits relapsing experimental allergic encephalomyelitis. J Neuroimmunol 1991; 32:209–217.

211. Hayosh NS, Swanborg RH. Autoimmune effector cells: inhibition of adoptive transfer of autoimmune encephalomyelitis with a monoclonal antibody specific for interleukin 2 receptors. J Immunol 1987; 138:3771–3775.

212. Duong TT, St. Louis J, Gilbert JJ, Finkelman FD, Strejan GH. Effect of anti-interferon-γ and anti-interleukin-2 monoclonal antibody treatment on the development of actively and passively induced experimental allergic encephalomyelitis in the SJL/J mouse. J Neuroimmunol 1992; 36:105–115.

213. Renno T, Lin J-Y, Piccirillo C, Antel J, Owens T. Cytokine production by cells in cerebrospinal fluid during experimental allergic encephalomyelitis in SJL/J mice. J Neuroimmunol 1994; 49:1–7.

214. Gijbels K, Brocke S, Abrams PA, Steinman L. Effect of anti-interleukin-6 treatment on experimental autoimmune encephalomyelitis. Neurology 1993; 43:A412.

215. Martens E, Dillen C, Put W, Heremans H, van Damme J, Billiau A. Increased circulating interleukin-6 (IL-6) activity in endotoxin-challenged mice pre-treated with anti-IL-6 antibody is due to IL-6 accumulated in antigen-antibody complexes. Eur J Immunol 1993; 23:2026–2029.

216. Gijbels K, Van Damme J, Proost P, Put W, Carton H, Billiau A. Interleukin 6 production in the central nervous system during experimental autoimmune encephalomyelitis. Eur J Immunol 1990; 20:233–235.

217. Billiau A, Heremans H, Vandekerckhove F, et al. Enhancement of experimental allergic encephalomyelitis in mice by antibodies against IFN-γ. J Immunol 1988; 140:1506–1510.

218. Voorthuis JAC, Uitdehaag BMJ, De Groot CJA, Goede PH, van der Meide PH, Dijkstra CD. Suppression of experimental allergic encephalomyelitis by intraventricular administration of interferon-gamma in Lewis rats. Clin Exp Immunol 1990; 81:183–188.

219. van Noort JM, Nagelkerken L, Boog CJP. Immune intervention strategies in EAE. Int MSJ 1994; 1:42–50.

220. Stoll G, Müller S, Schmidt B, et al. Localization of interferon-γ and Ia-antigen in T cell line-mediated experimental autoimmune encephalomyelitis. Am J Pathol 1993; 142:1866–1875.

221. Issazadeh S, Mustafa M, Ljungdahl A, et al. Interferon γ, interleukin 4 and transforming growth factor β in experimental autoimmune encephalomyelitis in Lewis rats: dynamics of cellular mRNA expression in the central nervous system and lymphoid cells. J Neurosci Res 1995; 40:579–590.

222. Karpus WJ, Gould KE, Swanborg RH. CD4+ suppressor cells of autoimmune encephalomyelitis respond to T cell receptor-associated determinants on effector cells by interleukin-4 secretion. Eur J Immunol 1992; 22:1757–1763.

223. Racke MK, Burnett D, Pak S-H, et al. Retinoid treatment of experimental allergic encephalomyelitis. J Immunol 1995; 154:450–458.

224. Saoudi A, Simmonds S, Huitinga I, Mason D. Prevention of experimental allergic encephalomyelitis in rats by targeting autoantigen to B cells: evidence that the protective mechanism depends on changes in the cytokine response and migratory properties of the autoantigen-specific T cells. J Exp Med 1995; 182:335–344.

225. van der Veen RC, Stohlman SA. Encephalitogenic Th1 cells are inhibited by Th2 cells with related peptide specificity: relative roles of interleukin (IL)-4 and IL-10. J Neuroimmunol 1993; 48:213–220.

226. Zurawski G, de Vries JE. Interleukin 13, an interleukin 4-like cytokine that acts on monocytes and B cells, but not on T cells. Immunol Today 1994; 15:19–26.

227. Cash E, Minty A, Ferrara P, Caput D, Fradelizi D, Rott O. Macrophage-inactivating IL-13 suppresses experimental autoimmune encephalomyelitis in rats. J Immunol 1994; 153:4258–4267.

228. Huitinga I, van Rooijen N, de Groot CJA, Uitdehaag BMJ, Dijkstra CD. Suppression of experimental allergic encephalomyelitis in Lewis rats after eliminating of macrophages. J Exp Med 1990; 172:1025–1033.

229. Stevens DB, Gould KE, Swanborg RH. Transforming growth factor-β1 inhibits tumor necrosis factor-α/lymphotoxin production and adoptive transfer of disease by effector cells of autoimmune encephalomyelitis. J Neuroimmunol 1994; 51:77–83.

230. Santambrogio L, Hochwald GM, Saxena B, et al. Studies on the mechanisms by which transforming growth factor-β (TGF-β) protects against allergic encephalomyelitis: antagonism between TGF-β and tumor necrosis factor. J Immunol 1993; 151:1116–1127.

231. Karpus WJ, Swanborg RH. CD4$^+$ suppressor cells inhibit the function of effector cells of experimental autoimmune encephalomyelitis through a mechanism involving transforming growth factor-β1. J Immunol 1991; 146:1163–1168.

232. Miller A, Lider O, Roberts AB, Sporn MB, Weiner HL. Suppressor T cells generated by oral tolerization to myelin basic protein suppress both in vitro and in vivo immune responses by the release of transforming growth factor β after antigen-specific triggering. Proc Natl Acad Sci USA 1992; 89:421–425.

233. Fabry Z, Topham DJ, Fee D, et al. TGF-β2 decreases migration of lymphocytes in vitro and homing of cells into the central nervous system in viro. J Immunol 1995; 155:325–332.
234. Johns LD, Sriram S. Experimental allergic encephalomyelitis: neutralizing antibody to TGFβ1 enhances the clinical severity of the disease. J Neuroimmunol 1993; 47:1–8.
235. Racke MK, Cannella B, Albert P, Sporn M, Raine CS, McFarlin DE. Evidence of endogenous regulatory function of transforming growth factor-β1 in experimental allergic encephalomyelitis. Int Immunol 1992; 4:615–620.
236. Brod SA, Scott M, Burns DK, Phillips JT. Modification of acute experimental autoimmune encephalomyelitis in the Lewis rat by oral administration of Type 1 interferons. J Interferon Cytokine Res 1995; 15:115–122.
237. Brod SA, Khan M, Kerman RH, Pappolla M. Oral administration of human or murine interferon alpha suppresses relapses and modifies adoptive transfer in experimental autoimmune encephalomyelitis. J Neuroimmunol 1995; 58:61–69.
238. Baggiolini M, Dahinden CA. CC chemokines in allergic inflammation. Immunol Today 1994; 15:127–133.
239. Ransohoff RM, Hamilton TA, Tani M, et al. Astrocyte expression of mRNA encoding cytokines IP-10 and JE/MCP-1 in experimental autoimmune encephalomyelitis. FASEB J 1993; 7:592–600.
240. Hulkower K, Brosnan CF, Aquino DA, et al. Expression of CSF-1, c-fms, and MCP-1 in the central nervous system of rats with experimental allergic encephalomyelitis. J Immunol 1993; 150:2525–2533.
241. Godiska R, Chantry D, Dietsch GN, Gray PW. Chemokine expression in murine experimental allergic encephalomyelitis. J Neuroimmunol 1995; 58:167–176.
242. Hafler DA, Weiner HL. Antigen-specific therapies for the treatment of autoimmune diseases. Springer Semin Immunopathol 1995; 17:61–76.
243. Hohlfeld R, Meinl E, Weber F, et al. The role of autoimmune T lymphocytes in the pathogenesis of multiple sclerosis. Neurology 1995; 45:S33–S38.
244. Lassmann H, Vass K. Are current immunological concepts of multiple sclerosis reflected by the immunopathology of its lesions? Springer Semin Immunopathol 1995; 17:77–87.
245. Kroemer G, Moreno de Alborán I, Gonzalo JA, Martínez A-C. Immunoregulation by cytokines. Crit Rev Immunol 1993; 13:163–191.
246. Burger D, Dayer J-M. Inhibitory cytokines and cytokine inhibitors. Neurology 1995; 45:S39–S43.
247. Finkelman FD. Relationships among antigen presentation, cytokines, immune deviation, and autoimmune disease. J Exp Med 1995; 182:279–282.

35
Cytokines and Transplantation

Jeffrey D. Punch and Jonathan S. Bromberg
University of Michigan, Ann Arbor, Michigan

TH1/TH2 CYTOKINES

A major conceptual advance in cytokine research was the recognition that T cells can be broadly defined by their pattern of cytokine secretion and function (1). Thus TH1 cells secrete IL-2 and IFN-γ; respond to IL-12; and promote delayed-type hypersensitivity, cytotoxic T-lymphocyte generation, and IgG2a, IgG2b, and IgG3 antibody production. TH2 cells secrete IL-4, IL-5, IL-10, IL-13, and GM-CSF; respond to IL-4; inhibit antigen-presenting cell (APC) function and TH1 development; and promote IgG1 and IgE production. In general, it is considered that TH2 cells are suppressor cells and prevent the development of TH1 cell-mediated immunity. Many controversies and apparent contradictions have clouded this neat model of a compartmentalization and dichotomy in T-cell differentiation and function, nonetheless, the hypothesis has been successful in generating research into the roles of these cytokines and subsets in regulating immunopathological phenomena. In the setting of transplantation, the model predicts that IL-4 and IL-10 should promote immunosuppression, prolongation of graft survival, and tolerance, whereas IL-2, IL-12, and IFN-γ should promote active immunity and rejection.

In support of the model, Kreuger et al. (2) showed that TH2 cells not only fail to induce graft-versus-host disease (GVHD) but also inhibit the activity of TH1 cells in vivo and ameliorate GVHD after bone marrow transplantation. Donckier et al. (3) demonstrated that neonatal tolerance induced by injecting F_1 lymphoid cells into parental recipients results in the upregulation of IL-4, but not IL-2, production and that tolerance could be abrogated with anti-IL-4 monoclonal antibodies (mAb). Sayegh et al. (4) showed that CD28-B7 blockade with CTLA4Ig prolonged rat renal allograft survival and inhibited IL-2 and IFN-γ expression by graft infiltrating cells, while expression of IL-4 and IL-10 were increased. Supplementation with exogenous IL-2 reversed the immunosuppressive effects and promoted graft rejection. These reports support the TH1/TH2 model and suggest that interventions that bias responses toward TH2 cytokine profiles may be of utility in promoting graft survival.

However, the complexity of the alloantigen response reveals that a simple application of the TH1/TH2 model will not be a comprehensive enough approach. Chen et al. (5) attempted to prolong murine cardiac allograft survival with anti-CD8 mAb. This treatment did not enhance graft survival, and furthermore the cytokine profile of the graft infiltrating cells changed from exclusively TH1 to a mixed TH1 plus TH2. Steiger et al. (6) performed pancreatic islet allografts in IL-2 knock-out recipients; the grafts were rejected after only slight prolongation in survival and rejection was associated with IL-4 but not IL-2 production. Therefore there are likely to be factors other than a simple ratio of TH1 versus TH2 cytokines that determine responses to alloantigen.

Many studies have refined the TH1/TH2 model and focused on the effects of IL-10 in transplantation. In vitro IL-10 promotes TH2 and inhibits TH1 cells, cytokines, and related immune phenomena (7–13). However, there is not strict compartmentalization of these effects since human TH1 and TH2 clones can both produce and be inhibited by IL-10 (13). In vivo studies support the in vitro work and demonstrate that IL-10 can inhibit macrophage infiltration of tumors (14), contact hypersensitivity through inhibition of IL-1 and TNFα production (15), and delayed-type hypersensitivity to *Leishmania* (16). Studies of the influence of IL-10 on alloreactivity demonstrate that it can inhibit IL-2, IL-2R, and IFN-γ production in human mixed lymphocyte cultures (17), CD4+ and CD8+ proliferation initiated by Langerhans cells (18), allogeneic target cell lysis by an effect on antigen presentation (19), and lymph node enlargement in response to alloantigen (20). Taken together these studies suggest that IL-10 expression may correlate positively with graft tolerance in transplantation and that manipulation of IL-10 may provide therapeutic benefits.

In a cardiac transplant model in mice, Mottram et al. (21) showed that a tolerogenic regimen of anti-CD4 mAb resulted in intragraft expression of IL-4 and IL-10 protein, while untreated rejecting grafts expressed IL-2, IFN-γ, and IL-4 but not IL-10. Rats receiving cardiac transplants and sirolimus (rapamycin) immunosuppression had increased intragraft expression of IL-4 and IL-10 mRNA (22). Tolerated porcine proislet xenografts transplanted beneath the kidney capsule of mice treated with anti-CD4 mAb demonstrated intragraft expression of IL-2 and IL-10 mRNA (23). In a renal allograft model in rats, CTLA4Ig treatment resulted in decreased IL-2 and IFN-γ and increased IL-4 and IL-10 within the allograft (4). Taken together these results suggest, but do not prove, a role for IL-10 in the induction and maintenance of graft survival.

Further support for IL-10 in allograft acceptance comes from deliberate introduction or interference with IL-10 function. A study in SCID patients transplanted with allogeneic stem cells revealed that tolerant patients produced high levels of IL-10 (24). Neutralization of the IL-10 with specific mAb proliferation to alloantigen in vitro. Skin-grafted mice receiving donor lymphocytes via the portal vein plus cyclosporine lymphocytes via the portal vein plus cyclosporine showed decreased splenocyte IL-2 and enhanced IL-4 responses in vitro to specific alloantigen stimulation (25). Prolonged graft survival in these mice was abrogated by anti–IL-10 mAb treatment in vivo. Splenocytes from anti-IL-10–treated animals also had increased IL-2 and decreased IL-4 responses to alloantigen. In another murine skin allograft model, adoptive transfer of an alloantigen-specific, TH2 cell line producing IL-4 and IL-10 resulted in markedly prolonged graft survival and the inability to generate cytotoxic T lymphocytes (26). These reports provide stronger evidence that IL-10 is directly involved in the process of inducing tolerance and maintaining the tolerant state, and its presence is not merely an epiphenomenon.

Further circumstantial support for the role of IL-10–mediated immunosuppression in prolonged graft survival is provided by a study showing that the potent immunosuppressant tacrolimus markedly inhibited basal IL-4 mRNA and protein production by murine TH2 lines while only slightly inhibiting IL-10 mRNA and protein (27). Tacrolimus also blocked mitogen-induced IL-4 upregulation but did not block increases in IL-10.

Alternative investigations have examined rejecting allografts for the presence or absence of IL-10 and related that to immunopathology and the levels of other cytokines. Isolated graft infiltrating lymphocytes from irreversibly rejected human renal allografts had a high frequency of IL-10–secreting cells (28,29), suggesting that IL-10 did not inhibit rejection. However, these graft infiltrating cells also had a high frequency of IFN-γ production; IFN-γ production could be regulated by IL-2 or IL-4 in vitro, although frequency analysis of IL-2 and IL-4 was not performed, and there was no comparison to unrejected or reversibly rejecting controls. Studies on human liver transplants have shown significant levels of IL-10 mRNA in acute and chronic rejection (30) and high systemic levels of IL-10 at the time of transplantation in all patients (31). Comparable levels of IL-10 mRNA were noted in tolerant and rejecting hepatic grafts in rats (32). These studies all dispute the conclusion that IL-10 promotes tolerance. However, these reports were not controlled, showed significantly elevated TH1 cytokine mRNAs (30,32), and did not directly assess IL-10 function by blocking its activity with mAb. Likewise, significantly elevated levels of IL-10 mRNA were noted in murine GVHD models (33,34). However, whether IL-10 inhibited, promoted, or was irrelevant to the GVHD process was not investigated.

Another approach to investigating IL-10 function is the production of IL-10 transgenic mice. In an important series of experiments Sarvetnick and coworkers (35,36) targeted constitutive IL-10 transgene expression to pancreatic islet β-cells with the insulin promoter. Contrary to expectations, transgene expression resulted in significant lymphocytic infiltration of the islets. These islets were rejected in a normal fashion when used as donor cells to allogeneic recipients, and crossing the transgenic strain with an autoimmune strain resulted in accelerated rather that delayed autoimmune diabetes. These results suggest that rather than being immunosuppressive, IL-10 is immunostimulatory. Similar results were obtained by Strom and colleagues, who showed that an IL-10/Fc fusion protein accelerated islet allograft rejection (37). However, the relevance of artificially overexpressing IL-10 in a normal animal to the physiological function of IL-10 is uncertain. It is worth recalling that, as for all cytokines, IL-10 is produced by many cell types, many cells bear IL-10 receptors, and cytokine effects are pleiotropic, depending on the kinetics, dose, and context of secretion. IL-10 has known immunostimulatory effects (38–41). Therefore, while IL-10 is certainly immunosuppressive in many contexts, that is not always the case. The transgenic results (35,36) may mean that unregulated expression or high expression of IL-10 is preferentially immunostimulatory. This is supported by a report demonstrating enhancement of tumor immunity by IL-10–transfected carcinoma cells (42). What is not currently known is what variables determine immunosuppressive versus immunostimulatory effects of IL-10. Dose, kinetics, targets, and cytokine context are likely to be important, yet their effects have not been thoroughly investigated.

In the context of IL-10 immunosuppression, it is important to note that the Epstein-Barr virus *BCRF I* open reading frame encodes a gene with significant homology at both the nucleic acid and amino acid levels to human and murine IL-10 (43). The viral IL-10 (vIL-10) protein has IL-10 immunosuppressive activity (44). Significantly, vIL-10 seems also to lack the immunostimulatory properties of normal IL-10 (38–41). This suggests that vIL-10 is an effective immunosuppressant molecule. Using retroviral mediated transfer of vIL-10, we have been able to prolong cardiac allograft survival in mice (45). Significantly, retroviral transfer of cellular mIL-10 did not prolong graft survival (unpublished results). Lotze and colleagues have noted substantially similar results in a tumor model: vIL-10 inhibited and mIL-10 promoted tumor rejection (46). These results suggest that vIL-10 may be an effective therapeutic agent for immunosuppression and that studies of the structure-function relationship of cellular and viral IL-10 and their interaction with the IL-10 receptor may determine the choice between suppressive and stimulatory responses.

TGF-β does not fit cleanly into the TH1/TH2 model since it can be produced by, and suppress the immune function of, both cell subpopulations. Nonetheless, it is worth discussing the effects of TGF-β in this context because it can be exquisitely immunosuppressive, inhibiting the function of T, B, and APC. Studies have shown that TGF-β can inhibit the development of TH1 (47) and that certain TH1 cells can produce TGF-β and IL-10 and act in an immunosuppressive fashion (48). The administration of purified TGF-β to cardiac allograft recipients can prolong graft survival (49). There is strong evidence that one of the immunoregulatory consequences of cyclosporine is the upregulation of TGF-β production with subsequent immunosuppressive effects (50). Studies from our own laboratory (45,51) have used a gene transfer approach to introduce plasmid vectors that encode TGF-β1 into cardiac allografts. The results show that prolongation of graft survival can be achieved with this approach.

In sum, the TH1/TH2 model of interacting cells and cytokines has been successful in initiating investigation of the function of these entities in graft survival and rejection. While other influences such as antigen presentation, costimulator display, and ischemia-reperfusion undoubtedly play major roles, which have to be specifically addressed by other forms of clinical immunosuppression, the manipulation of specific TH1/TH2 cytokines may be of enormous therapeutic benefit. Importantly, IL-10 and TGF-β upregulation and IL-2, IL-12, and IFN-γ downregulation may prove to be of most utility.

THE ROLE OF CYTOKINE ASSAYS IN DIAGNOSING REJECTION

Despite major advances in immunosuppressive techniques in the last 20 years, allograft rejection remains the major clinical concern of the transplant surgeon. When diagnosed in a timely manner, rejection can be successfully treated most of the time. However, in practice there are often delays in diagnosis caused by the insensitivity of current allograft monitoring methods. The "gold standard" for diagnosis rejection has long been the allograft biopsy. However, even a biopsy may lack sensitivity since mild or early rejection may appear difficult to distinguish from normal tissue or virally infected tissue. In addition, biopsies are uncomfortable and entail potential risk of organ injury and hemorrhage. Finally, the patchy nature of allograft rejection can also lead to sampling errors with fine needle biopsy techniques. For these reasons, other methods of rapidly and accurately diagnosing rejection are continually being sought. IL-6 and members of the chemokine family including IL-8 and RANTES are prime candidates as markers of rejection since they all share proinflammatory properties.

IL-6, originally called B-cell stimulatory factor 2 (BCSF-2), was identified as a cytokine capable of inducing antibody production in B cells. It was later realized that this cytokine was identical to hybridoma growth factor (HGF), IFN-β2, and hepatocyte-stimulating factor. Early in vitro work suggesting that IL-6 may play a role in allograft rejection included the observation that IL-6 is required to move T cells from G_0 to G_1 of the cell cycle, where they become responsive to IL-2 (52). The induction of IL-2 responsiveness seems to occur via IL-6–mediated stimulation of IL-2 receptor expression (52). IL-6 also has been described as a necessary factor for the in vitro induction of cytotoxic T lymphocytes (53). These data, together with the understanding that IL-6 is important for B-cell proliferation and the production of alloantibody, suggested that IL-6 may play a prominent role in critical processes involved with rejection of allografts. Indeed, in a rat model of liver transplantation, Tono et al. showed that elevated biliary IL-6 levels correlated well with histological rejection (54).

Clinical studies have demonstrated that IL-6 levels are normally low in healthy volunteers and preoperative patients with end-stage renal failure (54). Early work by Van Oers et al. (55) suggested that IL-6 rises nonspecifically following major surgery but falls to normal levels within 2 or 3 weeks in uncomplicated cases. During episodes of rejection, IL-6 levels increased significantly beginning 2 days before the development of clinical signs or symptoms. This suggested that IL-6 may be useful as a marker of rejection. However, there was no evidence that increased IL-6 is in any way a specific marker for rejection. Subsequent work in renal transplantation confirmed elevated IL-6 levels in the serum during rejection compared to patients with stable graft function (56). However, IL-6 levels were also increased during episodes of infection, implying that IL-6 was a nonspecific marker of inflammation. Because the IL-6 levels were low in patients with cyclosporine nephrotoxicity, the IL-6 level was useful in distinguishing cyclosporine nephrotoxicity from acute rejection.

IL-6 levels have also been examined following liver transplantation in humans. Busuttil reported that IL-6 levels were elevated in patients experiencing acute rejection, but not in uncomplicated patients or patients with ischemia or cholangitis (57). However, IL-6 levels did not correlate with the histological grade of the rejection in this study. Tilg et al. found that serum IL-6 levels did not reliably predict rejection since only 2 of 6 patients with rejection had elevated serum IL-6 levels compared to 7 of 9 patients with bacterial infections and 5 of 7 patients with cytomegalovirus infections that had elevated IL-6 levels (58). Whether biliary IL-6 levels will become a useful clinical tool for the diagnosis of rejection in liver transplant recipients remains to be seen.

The role of IL-6 levels in predicting rejection has also been examined in heart and lung transplant. Kimball et al. showed that serum IL-6 levels rose dramatically in human heart transplant recipients 2 days prior to the diagnosis of rejection (59). Although encouraging, this study did not determine if an elevated IL-6 level was actually predictive of rejection. Weyand et al. were unable to find any correlation between serum IL-6 level and the clinical development of rejection in a population of 27 adult heart transplant recipients (60). Magnan et al. examined IL-6 levels in the sputum of human lung transplant recipients and found that elevated levels of IL-6 were present during acute rejection or acute cytomegalovirus infection (61). In this study, it was not possible to distinguish the two conditions based on the IL-6 level.

As it became apparent that elevated IL-6 levels are not specific enough to distinguish clinical rejection from other inflammatory conditions, interest in cytokine assays for this purpose has waned. Other potential candidates as cytokine markers of rejection have been the chemokine family, especially IL-8. This cytokine, formerly referred to as neutrophil chemotactic factor and TY cell chemotactic factor, causes T-cell migration to sites of inflammation. In a brief report Wada et al. described the appearance of IL-8 in the urine of renal transplant patients and founds that it predicted rejection (62). Similarly, in liver transplant patients elevated IL-8 levels were detected in the presence of acute rejection (59). Unfortunately, it was found that IL-8 levels were increased in the presence of bacterial and viral infection as well. In fact, the highest levels of IL-8 were seen in the presence of bacterial infection. Thus, it appears that IL-8, like IL-6, may be an important marker of rejection, but because it is not specific for acute rejection it will probably not be useful as a clinical tool. As more cytokines are identified and studied, their roles in graft rejection will undoubtedly be explored. Other candidates for potential markers for rejection include IL-12, IFN-γ, and the recently described IGIF (IFN-γ–reducing factor) (63). However, the complexity and redundancy of cytokine responses to both rejection and other inflammatory events described to date indicate that the search for a specific cytokine marker for rejection may be elusive.

REFERENCES

1. Mossman TR, Coffman RL. TH1 and TH2 cells: different patterns of lymphokine secretion lead to different functional properties. Annu Rev Immunol 1989; 7:145–173.
2. Krenger W, Snyder KM, Byon JCH, Falzarano G, Ferrara JLM. Polarized type 2 CD4+ CD8+ donor T cells fail to induce experimental acute graft-versus-host disease. J Immunol 1995; 155:585–593.
3. Donckier V, Wissing M, Bruyns C, Abramowicz D, Lybin M, Vanderhaeghen ML, Goldman M. Critical role of interleukin 4 in the induction of neonatal transplantation tolerance. Transplantation 1995; 59:1571–1576.
4. Sayegh MH, Akalin E, Hancock WW, Russell ME, Carpenter CB, Linsley PS, Turka LA. CD28-B7 blockade after alloantigenic challenge in vivo inhibits Th1 cytokines but spares Th2. J Exp Med 1995; 181:1869–1874.
5. Chan SY, DeBruyne LA, Goodman RE, Eichwald EJ, Bishop DK. In vivo depletion of CD8+ T cells results in Th2 cytokine production and alternate mechanisms of allograft rejection. Tranplantation 1995; 59:1155–1161.
6. Steiger J, Nickerson PW, Steurer W, Moscovitch-Lopatin M, Strom TB. IL-2 knockout recipient mice reject islet cell allografts. J Immunol 1995; 155:489–498.
7. Tripp CS, Wolf SF, Unanue ER. Interleukin 12 and tumor necrosis factor α are costimulators of interferon γ production by natural killer cells in severe combined immunodeficiency mice with listeriosis, and interleukin 10 is a physiologic antagonist. Proc Natl Acad Sci USA 1993; 90:3725–3729.
8. D'Andrea A, Aste-Amezaga M, Ma X, Kubin M, Trinchieri G. Interleukin 10 (IL-10) inhibits human lymphocyte interferon

γ-production by suppressing natural killer cell stimulatory factor/IL-12 synthesis in accessory cells. J Exp Med 1993; 178:1041–1048.
9. Ding L, Linsley PS, Huang LY, Germain RN, Shevach EM. IL-10 inhibits macrophage costimulatory activity by selectively inhibiting the up-regulation of B7 expression. J Immunol 1993; 151:1224–1234.
10. Chang CH, Furue M, Tamaki K. Selective regulation of ICAM-1 and major histocompatibility complex class I and II molecule expression on epidermal Langerhans cells by some of the cytokines released by keratinocytes and T cells. Eur J Immunol 1994; 24:2889–2895.
11. Enk AH, Angeloni VL, Udey MC, Katz SI. Inhibition of Langerhans cell antigen-presenting function by IL-10. J Immunol 1993; 151:2390–2398.
12. de Waal Malefyt R, Yssel H, de Vries JE. Direct effects of IL-10 on subsets of human CD4+ T cell clones and resting T cells. J Immunol 1993; 150:4754–4765.
13. Del Prete G, De Carli M, Almerigogna F, Giudizi MG, Biagiotti R, Romagnani S. Human IL-10 is produced by both type 1 helper (Th1) and type 2 helper (Th2) T cell clones and inhibits their antigen-specific proliferation and cytokine production. J Immunol 1993; 150:353–360.
14. Richter G, Kruger-Krasagakes S, Hein G, Huts C, Schmitt E, Diamantstein T, Blankenstein T. Interleukin 10 transfected into Chinese hamster ovary cells prevents tumor growth and macrophage infiltration. Cancer Res 1993; 53:4134–4137.
15. Enk AH, Saloga J, Becker D, Mohamadzadeh M, Knop J. Induction of hapten-specific tolerance by interleukin 10 in vivo. J Exp Med 1994; 179:1397–1402.
16. Powrie F, Menon S, Coffman RL. Interleukin-4 and interleukin-10 synergize to inhibit cell-mediated immunity in vivo. Eur J Immunol 1993; 23:2223–2229.
17. Danzer SG, Kirchner H, Rink L. Cytokine interactions in human mixed lymphocyte culture. Transplantation 1994; 57:1638–1642.
18. Peguet-Navarro J, Moulon C, Caux C, Dalbiez-Gauthier C, Banchereau J, Schmitt D. Interleukin-10 inhibits the primary allogeneic T cell response to human epidermal Langerhans cells. Eur J Immunol 1994; 24:884–891.
19. Matsuda M, Salazar F, Petersson M, Masucci G, Hansson J, Pisa P, Zhang QJ, Masucci MG, Kiessling R. Interleukin 10 pretreatment protects target cells from tumor- and allo-specific cytotoxic T cells and downregulates HLA class I expression. J Exp Med 1994; 180:2371–2376.
20. Delvaux A, Donckier V, Bruyns C, Florquin S, Gerard C, Amraout Z, Abramowicz D, Goldman M, Velu T. Effects of systemic administration of rIL-10 in an in vivo model of alloreactivity. Transplantation 1994; 58:972–974.
21. Mottram PL, Han WR, Purcell LJ, McKenzie IFC, Hancock WW. Increased expression of IL-4 and IL-10 and decreased expression of IL-2 and interferon-γ in long-surviving mouse heart allografts after brief CD4-monoclonal antibody therapy. Transplantation 1995; 59:559–565.
22. Ferraresso M, Tian L, Ghobrial R, Stepkowski SM, Kahan BD. Rapamycin inhibits production of cytotoxic but not noncytotoxic antibodies and preferentially activates T helper 2 cells that mediate long-term survival of heart allografts in rats. J Immunol 1994; 153:3307–3318.
23. Morris CF, Simeonovic CJ, Fung MC, Wilson JD, Hapel AJ. Intragraft expression of cytokine transcript during pig proislet xenograft rejection and tolerance in mice. J Immunol 1995; 154:2470–2482.
24. Bacchetta R, Biglet M, Touraine J, Parkman R, Tovo PA, Abrams J, De Waal Malefyt R, De Vries JE, Roncarto MG. High levels of interleukin 10 production in vivo are associated with tolerance in SCID patients transplanted with HLA mismatched hematopoietic stem cells. J Exp Med 1994; 179:493–502.
25. Gorczynski RM, Wojcik D. A role for nonspecific (cyclosporin A) or specific (monoclonal antibodies to ICAM-1, LFA-1, and IL-10) immunomodulation in the prolongation of skin allografts after antigen-specific pretransplant immunization or transfusion. J Immunol 1994; 152:2011–2019.
26. Maeda H, Takata M, Takahashi S, Ogoshi S, Fujimoto S. Adoptive transfer of Th2-like cell line prolongs MHC class II antigen disparate skin allograft survival in the mouse. Int Immunol 1994; 6:855–862.
27. Wang SC, Morel PA, Wang Q, Jordan ML, Simmons RL, Tweardy DJ. A dual mechanism of immunosuppression by FK-506. Transplantation 1993; 56:978–985.
28. Merville P, Pouteil-Noble, Wijdenes J, Potaux L, Touraine JL, Banchereau. Detection of single cells secreting IFN-gamma, IL-6, and IL-10 in irreversibly rejected human kidney allografts, and their modulation by IL-2 and IL-4. Transplantation 1993; 55:639–646.
29. Merville P, Lambert C, Durand I, Pouteil-Noble C, Touraine JL, Berthoux F, Banchereau J. High frequency of IL-10-secreting CD4+ graft-infiltrating T lymphocytes in promptly rejected kidney allografts. Transplantation 1995; 59:1113–1119.
30. Bishop GA, Rokahr KL, Napoli J, McCaughan GW. Intragraft cytokine mRNA levels in human liver allograft rejection analysed by reverse transcription and semiquantitative polymerase chain reaction amplification. Transplant Immunol 1993; 1:253–261.
31. Le Moine O, Marchant A, Durand F, Ickx B, Pradier O, Belghiti J, Abramowicz D, Gelin M, Goldman M, Deviere J. Systemic release of interleukin-10 during orthotopic liver transplantation. Hepatology 1994; 20:889–892.
32. Sun J, McCaughan GW, Matsumoto Y, Sheil AGR, Gallagher ND, Bishop GA. Tolerance to rat liver allografts. Transplantation 1994; 57:1349–1357.
33. Allen RD, Staley TA, Sidman CL. Differential cytokine expression in acute and chronic murine graft-versus-host-disease. Eur J Immunol 1993; 23:333–337.
34. Garlisi CG, Pennline KJ, Smith SR, Siegel MI, Umland SP. Cytokine gene expression in mice undergoing chronic graft-versus-host disease. Mol Immunol 1993; 30:669–677.
35. Wogensen L, Huang X, Sarvetnick N. Leukocyte extravasation into the pancreatic tissue in transgenic mice expressing interleukin 10 in the islets of Langerhans. J Exp Med 1993; 178:175–185.
36. Lee MS, Wogensen L, Shizuru J, Oldstone MBA, Sarvetnick N. Pancreatic islet production of murine interleukin-10 does not inhibit immune-mediated tissue destruction. J Clin Invest 1994; 93:1332–1338.
37. Zheng XX, Steele AW, Nickerson PW, Steurer W, Steiger J, Strom TB. Administration of noncytolic IL-10/Fc in murine models of lipopolysaccharide-induced septic shock and allogeneic islet transplantation. J Immunol 1995; 154:5590–5600.
38. MacNeil IA, Suda T, Moore KW, Mosmann TR, Zlotnik A. IL-10 a novel growth cofactor for mature and immature T cells. J Immunol 1990; 145:4167–4173.
39. Vieira P, de Waal Malefyt R, Dang MN, Johnson KE, Kastelein R, Fiorentino DF, DeVries JE, Roncarolo MG, Mosmann TR, Moore KW. Isolation and expression of human cytokine synthesis inhibitory factor (CSIF/IL-10) cDNA clones: homology to Epstein-Barr virus open reading frame BCRF1. Proc Natl Acad Sci USA 1991; 88:1172–1176.
40. Thompson-Snipes L, Dhar V, Bond MW, Mosmann TR, Moore

KW, Rennick DM. Interleukin 10: a novel stimulatory factor for mast cells and their progenitors. J Exp Med 1991; 173:507–510
41. Go NF, Castle BE, Barrett R, Kastelein R, Dang W, Mosmann TR, Moore KW, Howard M. Interleukin 10 a novel B cell stimulatory factor: unresponsiveness of X chromosome-linked immunodeficiency B cells. J Exp Med 1990; 172:1625–1631.
42. Allione A, Consalvo M, Nanni P, Lollini PL, Cavallo F, Giovarelli M, Forni M, Gulino A, Colombo MP, Dellabona P, Hock H, Blankenstein T, Rosenthal FM, Gansbacher B, Bosco MC, Musso T, Gusella L, Forni G. Immunizing and curative potential of replicating and nonreplicating murine mammary adenocarcinoma cells engineered with interleukin (IL)-2, IL-4, IL-6, IL-7, IL-10, tumor necrosis factor alpha, and gamma-interferon gene or admixed with conventional adjuvants. Cancer Res 1994; 54:6022–6026.
43. Moore KW, Vieira P, Fiorentino DF, Trounstine ML, Khan TA, Mosmann TR. Homology of cytokine synthesis inhibitory factor (IL-10) to the Epstein-Barr virus gene BCRFI. Science 1990; 248:1230–1234.
44. Hsu DH, de Waal Malefyt R, Fiorentino DF, Dang MN, Vieira P, DeVries J, Spits H, Mosmann TR, Moore KW. Expression of interleukin-10 activity by Epstein-Barr virus protein BCRF1. Science 1990; 250:830–832.
45. Qin LH, Chavin KD, Ding Y, Favaro JP, Woodward JE, Lin J, Tahara H, Robbins P, Shaked A, Ho DY, Sapolsky RM, Lotze MT, Bromberg JS. Multiple vectors effectively achieve gene transfer in a murine cardiac transplantation model: Immunosuppression with TGF-beta 1 or vIL-10. Transplant 1995; 59:809–816.
46. Suzuki T, Tahara H, Narula S, Moore KW, Robbins PD, Lotze MT. Viral interleukin 10 (vIL-10), the human herpes virus-4 cellular IL-10 (cIL-10) homologue, induces local anergy to allogeneic and syngeneic tumors. J Exp Med 1995;
47. Schmitt E, Hoehn P, Huels C, Goedert S, Palm N, Rude E, Germann T. T helper type 1 development of interleukin-12 and interferon-g and is inhibited by transforming growth factor β. Eur J Immunol 1994, 24:793–798.
48. Chen Y, Kuchroo VK, Inobe JI, Hafler DA, Weiner HL. Regulatory T cell clones induced by oral tolerance: suppression of autoimmune encephalomyelitis. Science 1994; 265:1237–1240.
49. Raju GP, Belland SE, Eisen HJ. Prolongation of cardiac allograft survival with transforming growth factor-b1 in rats. Transplantation 1994; 58:592–396.
50. Khanna A, Li B, Stenzel KH, Suthanthiran M. Regulation of new DNA synthesis in mammalian cells by cyclosporine. Transplantation 1994; 57:577–582.
51. Qin L, Ding Y, Bromberg JS. Gene therapy for transplantation: prolongation of allograft survival with TGFβ1. Ann Surg 1994; 220:508–519.
52. Van Snick J. Interleukin-6: an overview. Annu Rev Immunol 1990; 8:253–278.
53. Okada M, Kitahara M, Kishimoto S, Matsuda T, Hirano T, Kishimoto T. IL-6/BSF-2 functions as a killer helper factor in the in vitro induction of cytotoxic T cells. J Immunol 1988; 141:1543–1549.
54. Tono T, et al. Interleukin 6 levels in bile following liver transplantation. Transplant Proc 1991; 23:630–631.
55. Van Oers MHJ, van der Heyden AAPAM, Aarden LA. Interleukin-6 (IL-6) in serum and urine of renal transplantation recipients. Clin Exp Immunol 1988; 17:314–319.
56. Yoshimura N, Oka T, Kahan BD. Sequential determinations of serum interleukin-6 levels as an immunodiagnostic tool to differentiate rejection from nephrotoxicity in renal allograft recipients. Transplant 1991; 51:172–176.
57. Umeshita K, et al. Interleukin-6 in bile as a marker of allograft rejection following liver transplantation in man. Transplant Proc 1992; 24:1462–1464.
58. Tilg H, Ceska M, Vogel W, Herold M, Margreiter R, Huber C. Interleukin-8 serum concentrations after liver transplantation. Transplantation 1992; 54:142.
59. Kimball AU, Radovancevic B, Isom T, Frazier B, Kerman R. Cytokine panel predicts early rejection and therapeutic response after cardiac transplantation. Transplant Proc 1995; 27:1286–1287.
60. Weyland M, Krehmeier M, Fahrenkamp A, Middendorf C, Redmann M, Schwartz C, Konertz W, Scheld HH. Serial interleukin-6 blood levels early after cardiac transplantation. Transplant Proc 1992; 24:2546.
61. Magnan A, et al. Monitoring of alveolar macrophage production of tumor necrosis factor-alpha and interleukin-6 in lung transplant recipients. Am J Respir Crit Care Med 1994; 150:684–689.
62. Wada T, et al. Appearance of interleukin-8 in the urine predicts acute rejection of renal transplantation [letter]. Nephrol Dial Transplant 1994; 9:874–875.
63. Okamura H, et al. Cloning of a new cytokine that induces IFN-γ production by T cells. Nature 1995; 378:88–90.

36
Cytokines and Cancer

Frances Burke and Frances R. Balkwill
Biological Therapies Laboratory, Imperial Cancer Research Fund, London, England

Sylvie C. Kossodo
University Hospital, Geneva, Switzerland

INTRODUCTION

Whether used as therapy or produced locally during the development of a cancer, cytokines can act on tumor cells, tumor stroma, and the host (reviewed in Refs. 1,2). Cytokines influence many aspects of tumor cell biology, including proliferation, survival, and death, motility, surface antigen expression, and cell-cell or cell-matrix interactions. They may also control neovascularization, extracellular matrix synthesis, leukocyte infiltration, stromal cell proliferation, and local immune response. Finally, the cytokine context of a tumor can affect local nutritional balance, systemic metabolism, and host immune status.

Imbalances in the cytokine network are to be expected in tumors where there is genetic damage to cytokine ligands and/or receptors (e.g., Ref. 3). In addition, many components of cytokine pathways have been identified as proto-oncogenes or tumor suppressor genes (reviewed in Refs. 2,4).

Historically, the aim of cytokine therapy was to increase the immune response to tumor cell antigens, but antitumor effects can also be achieved by direct cytokine action on tumor cells, destruction of the tumor microvasculature, and/or alteration of the local nutritional balance (1,2,5). Other cytokines, particularly the colony-stimulating factors (CSFs), have been used to promote bone marrow recovery after ablative therapies (6). Clinical trials of systemically administered cytokines have met with some success: most notably interferon (IFN)-α in hematological malignancies (5,7) and as adjuvant therapy in melanoma (8); IL-2 in a small minority of patients with renal cell carcinoma and melanoma (9); and CSFs for myelosuppression (6). Aside from this, cytokine therapy has not fulfilled the predictions of preclinical studies.

The reasons for this are complex. They may be attributed, in part, to optimistic interpretation of animal model and tissue culture data, incomplete understanding of endogenous tumor cytokines, inappropriate methods of administration, and, because of the high doses required in systemic therapy, severe side effects and toxicity. Local administration of these locally acting agents may be more effective. Cytokine gene therapy and solid-phase protein delivery systems offer new possibilities in this respect.

In this chapter, we review the preclinical animal data of cytokine therapy, discuss the role of endogenous cytokines in human tumor biology, and describe clinical cancer trial data with cytokine therapies. We will concentrate on the cytokines for which there are the most data, both in experimental animal models and clinical trials. Detailed experience with these cytokines has revealed common themes and mechanisms, which are relevant to the more recently discovered cytokines.

CYTOKINE THERAPY OF EXPERIMENTAL ANIMAL CANCER

Systemic Delivery of Cytokines in Animal Models

Interferon-α

IFN-α and Nude Mice Xenografts Human xenografts from breast, colon, melanoma, osteosarcoma, and ovarian tumors have been used to study the direct activity of species-specific human IFN-α on human tumor cells in vivo (10–14). In general, IFN-α prolonged the survival of tumor-bearing mice and reduced tumor size, the effect being dose and schedule dependent. Continuous therapy was necessary for sustained growth inhibition. Tumor responses to IFN-α varied 10-fold in a series of human osteosarcoma xenografts, presumably reflecting a difference in the biology of individual tumors (14–16). Human IFN-α induced tumor differentiation of these xenografts and some form of tumor stroma communication, since there was mineralization with partial replacement of the tumor by normal heterotypic bone (17). In other models there was evidence that human IFNs inhibited tumor-inducd angiogenesis (18). IFN-α interacted positively with 5-fluorouracil (5-FU), cyclophosphamide, adriamycin, *cis*-platinum, DFMO, and mitomycin (19,20). IFN-α altered the activity of drug-metabolizing enzymes in the liver of mice, but this did not interfere with its ability to synergize with alkylating agents cyclophosphamide and adriamycin (20).

IFN-α in Syngeneic Systems There is well-documented evidence of the antitumor effects of IFN-αs in transplantable murine tumor models, where their effects have again been to increase survival rather than to cure mice (21). A mixture of cell source IFNα/βs have been commonly used in these experiments, rather than recombinant proteins. In a wide range of transplantable lines, IFN-α/βs were most effective when tumor burden was low (21). Repeated doses were more effective than one large single dose, but regression of established tumors was rarely observed (22). IFNs may act on tumor vasculature in some of these models (23), and responses in animals bearing tumor cells resistant to the in vitro actions of IFN-α (24) suggest that host-mediated actions can be important, especially those involving lymphocytes and macrophages.

IFN-α and Metastases The majority of metastasis models are based on the intravenous (i.v.) injection of transplantable tumor cells, which are then entrapped in organs such as the lungs and spleen. IFN-α/βs generally have activity in these models, and a host component is necessary for response. For instance, IFN-α/β treatment resulted in a 50- to 100-fold inhibition of Friend leukemia cells (FLCs) and multiplication in the liver and spleen after i.v. injection of metastatic tumor cells (25–27). A competent immune system was required for optimal effects, and mice were resistant to rechallenge with tumor cells. The host cell types mediating the antitumor effects varied between the different models. Antibodies to NK cells, CD4+ and CD8+ T cells, all reduced antimetastatic activity (28). However, in the colon 26 murine colon carcinoma model, treatment with IFN-α commencing 2 h after the injection of tumor cells had a significant antimetastatic effect that was also apparent in NK-deficient mice (29,30).

Oncogene Transgenic Mouse Models An oncogene transgenic mouse model of insulinoma was used to assess the potential of IFN-α/β as an antiangiogenic agent. The therapy consisted of IFN-α/β in combination with the antiangiogenic agents AGM-1470 and minocycline. Vessel density and volume were both decreased, and the rate of apoptosis was increased in mice receiving this therapy (31).

Interferon-γ

IFN-γ and Nude Mice Xenografts IFN-γ, like IFN-α, shows strict species specificity. Generally, IFN-γ was not as effective as IFN-α against human tumor xenografts, whichever injection route was chosen (32), although alterations of tumor cell surface antigen expression were described (33). IFN-γ was, however, active against xenografts growing intraperitoneally (i.p.) in nude mice (34). IFN-γ, administered i.p. at doses equivalent to those given in clinical trials, increased the survival of mice bearing human ovarian cancer xenografts growing as ascites or solid tumor (35). Mice bearing ascites were cured of peritoneal disease. In solid ovarian tumors, there was a gradual decline in the number of proliferating cells following IFN-γ treatment. End-labeling studies showed an increase in DNA fragmentation, which reached a maximum at 14 days. These changes were associated with induction of p53 and hypophosphorylation of pRb. Continued exposure to the cytokine was more effective than higher doses three times per week, and the animals only needed 14 days treatment to achieve a significant increase in survival.

IFN-γ in Syngeneic Systems and Metastatic Models Effects of IFN-γ in these models were of a similar nature to those seen for IFN-α. For instance, in murine osteogenic sarcomas and chemically induced fibrosarcomas, response to treatment was dependent on tumor burden and schedule, and tumor stasis was most commonly seen (36). IFN-γ was also active in experimental lung metastases models of colorectal cancer and melanoma (37). Conversely, pretreatment of metastatic colon 26 cells with IFN-γ in vitro significantly increased their metastatic potential in immunocompetent and nude mice (38), an effect that appeared to be dependent on the presence of NK cells (38,39). Results in some human clinical trials may relate to this effect.

Interleukin-2

The basis for the use of IL-2 as an anticancer treatment stems from its capacity to stimulate growth of lymphokine-activated killer (LAK) cells, natural killer (NK) cells, T cells, and tumor-infiltrating lymphocytes (TIL) (40).

IL-2 and Nude Mice Xenografts IL-2 has been used in combination with allogeneic human LAK cells in tumor xenograft models. Peritumoral injections of IL-2 and human LAK effector cells inhibited the growth of established tumors of human squamous cell carcinoma of the head and neck (41,42). Xenograft models of hematological malignancies are also responsive to IL-2 with or without LAK cells (43–46).

IL-2 and Syngeneic and Metastatic Models IL-2 alone or in combination with LAK cells reduced established pulmonary and hepatic metastases of a wide range of transplantable tumors such as sarcomas, melanomas, bladder carcinomas, and colon adenocarcinoma (47–49). Toxicity-related effects, including development of ascites, splenomegaly, lymphadenopathy, gastrointestinal effects, general weakness, and malaise, were noted in some of these studies (50–52). IL-2 induced regression of experimental tumors, primarily by enhancing the cytolytic activity of endogenous NK and LAK cells (53–58). Adoptive transfer of IL-2–activated lymphoid cells, together with IL-2, led to the regression of tumors, while either therapy alone was less effective (48,49). In mice with established metastasis, therapeutic efficacy was dependent on tumor burden (47) and the dose of both IL-2 and LAK (55). The beneficial effect of IL-2 treatment in the MCA-105 model was abolished by in vivo treatment with antibodies directed against different lymphocyte subsets (59). Transfection of IL-2–insensitive B16BL6 melanoma or AKR cells with class II MHC genes rendered them susceptible to rejection (60). The end point of the majority of reported experiments was a reduction in the number of tumor foci, which may not necessarily reflect a significant increase in the life span of the mice. There is in fact little information on the mean survival time (MST) of IL-2–treated mice. In the MCA-105 sarcoma model, for instance, MST was only doubled if mice received five separate LAK cell infusions with concomitant IL-2 (54).

IL-2 in Combination with Other Agents Combinations of IL-2 with other cytokines have been extensively studied (reviewed in Ref. 61). Synergistic interactions between IL-2 and IFN-α were first demonstrated in the M5076 reticulum cell sarcoma model (62). There was growth inhibition of subcutaneous (s.c.) tumors and a reduction in hepatic metastases. IL-2 plus IFN-γ was inhibitory in a s.c. tumor model after perilesional or systemic treatment (63). Antitumor activity was also demonstrated in s.c. and i.p. tumors following treatment with IL-2 in combination with tumor necrosis factor (TNF)-α (64–67).

Interleukin-12

Syngeneic Systems and Models of Metastases Of all the cytokines tested so far in animal model systems, IL-12 has shown the most promising preclinical results in a range of animal cancer models (68). IL-12 reduced the number of hepatic metastases of the M5076 tumor even when treatment was started as late as 28 days after s.c. injection. The number of metastases was also decreased in a model of B16F10 melanoma (68). MC-38 colon carinoma, and Lewis lung carcinoma (reviewed in Ref. 69). The antitumor effects were dose-dependent, with a maximum inhibition of tumor growth obtained with daily i.p. doses of 1 μg/5 times weekly. In all these models IL-12 was able to inhibit growth, sometimes of well-established tumors, occasionally induced complete tumor regression, and usually resulted in prolonged survival (69). Mice cured of tumor were resistant to subsequent rechallenge with the same tumor (69). The mechanism through which IL-12 mediates its antitumor effect is not altogether clear and may vary with the tumor type and location. IL-12 retained its antitumor actions in NK-deficient mice bearing the B16F10 melanoma (68) and in NK-depleted mice bearing either Renca or MCA-207 tumors (68), suggesting that NK cells do not play a critical role. While depletion of T cells resulted in a loss of the antitumor effect in the Renca and MCA-207 models, in mice bearing a B-cell lymphoma, IL-12 retained its antitumor capacity in T- and B-cell–deficient SCID mice (70). In some models, the IL-12 effect was abrogated by treating the mice with neutralizing antibodies to IFN-γ (71). While the induction of IFN-γ may be necessary for the antitumor effect of IL-12, it alone is not sufficient, since IL-12 therapy in tumor-bearing nude mice resulted in an 8-fold increase in serum IFN-γ as compared to euthymic controls (68). Moreover, murine IFN-γ was not as effective as IL-12 in producing an antitumor effect in Renca tumor–bearing mice. IL-12 was more effective than a combination of IL-2 and IFN-α in the B16F10 melanoma, Renca carcinoma, colon 38, and M5076 sarcoma models (68). While synergistic effects were observed with IL-12 and IL-2 in combination at low doses, the most effective therapy was still an optimal dose of IL-12 alone (68).

Tumor Necrosis Factor-α

TNF-α, a potent inflammatory cytokine mainly produced by monocytes/macrophages, has a wide range of activities, some of which overlap with IFN-γ and IL-1. The antitumor actions of TNF-α can be due to direct effect on tumor cells and/or indirect effects on host stroma, and many are potentiated by IFN-γ. Among its direct effects, TNF-α can induce both apoptotic and necrotic forms of cell death (72–74). Vascular damage is widely accepted as a mechanism of its tumoricidal actions. Under certain circumstances, TNF-α has direct cytotoxic effects on the vascular endothelium (72,74,75) and is known to modulate endothelial procoagulant properties (72,74,76).

TNF-α and Xenograft Models When TNF was used to treat human tumor xenografts growing in nude mice, the phenomenon of hemorrhagic necrosis was not commonly observed and growth retardation of tumor was seen rather than cure. Most investigators, using melanomas, colon and gastric cancer, lymphomas, and ovarian tumors growing s.c. in nude mice, have noted a dose-response relationship (34,77–83). The antitumor effect appeared to depend on the route of administration, with high continuous doses injected directly into the tumor being the most effective. With continual dosing, complete tumor regression could be observed (79). Intraperitoneal therapy of human ovarian xenografts with recombinant human TNF-α prolonged the survival of mice in two of three tumor models (34). TNF-α promoted adhesion of tumor cells to the peritoneal wall with the disappearance of free-floating ascites. The TNF-α–treated tumors developed an extensive stroma with the formation of new blood vessels, and their histological appearance resembled human ovarian cancer (84). TNF-α was also effective against established tumors of the ovarian cancer cell line OVCAR-3 if injected i.p. (83) rather than i.v. (80). Using a breast cancer xenograft model, de Kossodo et al. (72) looked at the mechanisms involved in the antitumor effects of a combination of TNF and IFN-γ. Locally injected human TNF-α resulted in growth inhibition of established tumors. However, macroscopic necrosis was observed in these mice when systemic rat IFN-γ, which has no activity alone, was also given. Within 4 h of treatment, platelet adherence to tumor cells was observed. This was followed by destruction of the tumor vasculature. Both necrosis and apoptosis of tumor cells was demonstrated and there was upregulation of mRNA for a range of stromal cytokines and adhesion molecules (72).

TNF-α and Syngeneic Models The antitumor actions of TNF-α have been most extensively studied in the murine Meth-A sarcoma model. Meth-A tumor cells are resistant to the cytotoxic effect of TNF in vitro, yet systemic administration of TNF-α consistently causes hemorrhagic necrosis of s.c. (vascular) but not i.p. (avascular, ascitic) tumors. Histologically, tumor blood vessels are occluded by thrombus, and selective toxicity to tumor blood vessels can be demonstrated (85). The vascular component of TNF-α action has been shown by monitoring red cell influx into s.c. implants of the semisyngeneic SA1 sarcoma in mice (86). TNF-α induced a rapid and prolonged extravasation of red cells into tumors, an effect that was partly T-cell dependent. Chemically induced murine tumors are often immunogenic to their hosts, and T-cell–mediated immunity is also an important component of TNF-α response in these models. This is suggested by the diminished antitumor effect of TNF-α in T-cell–deficient mice noted in several studies (78,86) as well as the development of T-cell–mediated immunity in animals cured of the Meth-A sarcoma tumor by TNF-α (87). It is important to note that many of these studies were conducted with the human recombinant cytokine prior to the discovery of species differences in TNF-α receptors (reviewed in Ref. 88). Human TNF-α, which was originally thought to be cross–species specific, does not recognize the murine p75 receptor, but does bind to the murine p55 receptor. Any activities of TNF-α mediated by the p75 receptors would therefore not have been recognized when treating murine tumors with human TNF. However, human TNF-α will induce endogenous murine TNF-α in xenograft models (72). Although lymphotoxin (TNF-β) binds to the same receptors as TNF-α and has similar properties, it has not been studied as extensively. Lymphotoxin has been shown, however, to induce necrosis of the Meth-A tumor (89).

Colony-Stimulating Factors

Purified CSFs have shown a dramatic ability to increase myelopoiesis both in normal animals and in those myelosuppressed by irradiation, chemotherapy, or infection (90). Unlike other cytokines, which often have transient effects on the bone marrow, the CSFs seem to have a sustained effect throughout therapy. Cells stimulated by CSFs appear to be functionally normal and may even show increased levels of activation. Administration of GM-CSF to mice, for instance, resulted in an increase in cycling rates and numbers of hematopoietic progenitor cells of all lineages in both marrow and spleen within 3–24 h of injection (91). Intraperitoneal injection of GM-CSF into three different strains of mice resulted in a 20-fold increase in peritoneal macrophages and granulocytes with prominent splenomegaly (92).

Human G-CSF is not species specific and was used in many preclinical experiments, where its most prominent activities were on granulocyte progenitors and mature circulating granulocytes (93,94). Both i.p. and s.c.

administration led to a 6- to 10-fold increase in total leukocyte count.

Local Delivery of Cytokines in Animal Models

Cytokine Gene Delivery

The rapid degradation or elimination of cytokines delivered systemically can decrease their effectiveness (95). An alternative means of cytokine delivery is the transfection of cytokine genes into tumor or carrier cells that will express the cytokine at the primary tumor site. This could resemble cytokine release in vivo and target the antitumor response with minimal side effects. As we understand more of the local cytokine environment of the tumors and the local actions of cytokines, methods for targeting cytokines to the tumor site become of prime importance. An increase in the local concentration of cytokines due to gene delivery seems to have three main effects. A marked local inflammatory response, with effector cells being dependent on the cytokine gene delivered, is followed by tumor growth inhibition and sometimes regression. In some instances, long-lasting immunity to tumor rechallenge develops.

IFN-α Gene Delivery A cytostatic effect was observed when NIH/3T3 fibroblasts, transfected with IFN-α, were implanted into nude mice bearing a chronic myeloid leukemia (CML) line (96). Cells from a spontaneous, invasive, and metastasizing mouse mammary adenocarcinoma were infected with a retroviral vector containing murine IFN-α1 (97). Tumors were rejected, or formed only small masses, following s.c. or i.v. injection into immunocompetent Balb/c mice. Rejection was CD8+ T-cell–mediated with some polymorphonuclear infiltrate. In addition, some of the mice developed an immune resistance to rechallenge. FLC cells transfected with an IFN-α–containing retroviral vector failed to grow when injected s.c. or i.p. (98).

IFN-γ Gene Delivery Expression of the IFN-γ gene in poorly immunogenic tumor cells resulted in tumor suppression in euthymic, but not athymic mice (99). Tumor MHC class I and class II molecules were generally upregulated (100). Antitumor effects in competent animals correlated with levels of IFN-γ produced and were abrogated with antibodies to IFN-γ (99,101).

IL-2 Gene Delivery In experimental models of IL-2 gene delivery, both antitumor specificity and immunological memory have been achieved (99,101–103). Mice rejecting IL-2–transfected tumor cells often mounted a response against further challenge (reviewed by Ref. 103). Both LAK cells and CD8+ T cells have been implicated in the antitumor response (102,104). The growth of parental tumors was inhibited if the IL-2–transfected cells were injected at the same site, but not at different sites (101). Compared to controls, IL-2 modified human and murine melanomas and sarcomas had much slower growth or failed to grow in nude mice (95,101,105,106). A mononuclear infiltrate was evident in the nude mice, while in immune-competent mice decreased tumorigenicity was associated with CD8+ T cells (102), NK cells (107), and a granulocytosis (108). IL-2 gene delivery has been compared with several other cytokines (IFN-γ, IFN-α, IL-7, GM-CSF). IL-2 was as good as, or superior to, other cytokines in mediating local antitumor effects, especially in nude mice (reviewed in Ref. 95).

IL-4 Gene Delivery Tumor cells engineered to produce IL-4 by transfection also had reduced or absent tumorigenicity in vivo, and they inhibited growth of untransfected cells when they were mixed with these before injection into animals (109,110). Histological analysis of five different transplantable tumors treated with IL-4–transfected cells showed a characteristic host inflammatory reaction with an early influx of eosinophils, followed by macrophage infiltration and fibroblast scarring (110). Treatment of mice with antibodies that specifically blocked the accumulation of neutrophils and eosinophils at the site of inflammation restored tumorigenicity of some of the transfected lines, but the IL-4–transfected cells were tumorigenic in T- and NK-cell–deficient mice (110).

IL-12 Gene Delivery The approach used here has been to transfect fibroblasts with IL-12 and then mix these with tumor cells (111). Treatment with IL-12–transfected fibroblasts reduced the size and growth rate of s.c. MCA-105 murine sarcoma (112) and BL-6 melanoma cells and induced antitumor immunity (111).

TNF-α Gene Delivery TNF cDNA has been successfully inserted into several tumor cell lines without modifying the morphology and phenotype of the tumor cells. Inoculation of TNF-α–secreting cells into syngeneic mice resulted in a decrease in tumorigenicity, which could be reversed by anti–TNF-α antibodies. Wild-type murine sarcomas were rejected by the gene-modified cells if the two cell types were injected at the same site (113). Liposomes containing the gene for TNF-α were injected s.c. into human gliomas growing in nude mice. Growth inhibition was observed, which was enhanced if IFN-γ was injected up to 12 h prior to the transfection. However, in other studies, transfection of TNF enhanced tumorigenicity and invasiveness of tumor cells (114).

Cytokine Gene Transfer by Gene Gun Genes for IL-6, TNF-α, IL-2, and IFN-γ have been directly transfected into tumors in vivo by particle-mediated gene transfer (115). A combination of murine TNF-α and IFN-γ genes inhibited the growth of the Renca tumor and murine IL-2 and IFN-γ genes eradicated tumor in 25% animals. This method of gene transfer resulted in significant serum levels of IL-6 and IFN-γ, and transgene espression was detected in treated tissues.

Protein Delivery

Gene therapy is not the only option for localizing cytokines. Slow-release delivery systems for cytokine protein may also be effective and provoke useful biological responses, as a recent study with GM-CSF microspheres showed (116).

Microsphere Delivery of GM-CSF High doses of GM-CSF were incorporated into gelatin/chondroitin sulfate microspheres and mixed with B16/ F10 tumor cells prior to s.c. injection (116). This mixture produced an intense host inflammatory infiltrate consisting of numerous eosinophils and macrophages. By 12 days postinjection neither microspheres nor tumor cells were detected, but a moderate infiltrate persisted. This was reflected in an increase in mouse survival. The combination of GM-CSF microspheres and tumor cells also produced systemic immunity to tumor cell rechallenge.

Microsphere Delivery of IFN-γ Protein In the study described above (116), a mixture of B16/F10 tumor cells and IFN-γ–containing microspheres also caused an inflammatory infiltrate, predominantly consisting of eosinophils. However, in contrast to the results with GM-CSF, microspheres were still present 12 days after injection. The tumor cells were fewer in number, but there was no significant increase in tumor-bearing mouse survival.

Antitumor Actions of Cytokines in Animal Tumor Models

Some general trends concerning experimental cytokine therapy emerge from the data reviewed above. Tumor burden was, not surprisingly, important, with the best responses seen with low tumor load. Chronic exposure is more successful than higher doses given as single injections. Responses to the IFNs, particularly of human tumor xenografts, are more dependent on individual tumors, while responses to IL-2 and IL-12 are consistent between tumor lines (although it is likely that negative results were not reported). The data would suggest that cytokines, such as the IFNs, IL-2, IL-12, and TNF-α, share some common mechanisms of action. In addition, mechanisms are likely to be complex with multiple actions on tumor and stroma contributing. Distinct mechanisms that have so far been identified from animal models include tumor cell cytostasis, necrosis, and apoptosis; inhibition of formation, or destruction of tumor vasculature; NK-cell–mediated killing, and the generation of specific T-cell immunity. These actions may contribute to a varying degree in the different model systems, and it is unlikely that one single action of a cytokine will be sufficient for an antitumor effect.

Cytokine Antagonists in Animal Models

Vascular permeability factor (VPF), also known as vascular endothelial growth factor (VEGF), is a pleiotropic cytokine produced by tumor cells, keratinocytes, activated macrophages, rheumatoid synovial cells, embryonic tissues, and epithelial and mesenchymal cell lines (117). This cytokine was first identified in the culture supernatant and ascites of rodent tumors (118). Upregulated by hypoxia (116), VEGF acts through its two known high-affinity receptors, KDR and flt, to selectively increase permeability of endothelial cells to macromolecules and promote their growth (for review see Ref. 117). When nude mice transplanted with human rhabdomyosarcoma, glioblastoma multiforme, or leiomyosarcoma cell lines were treated with a monoclonal antibody directed against VEGF, tumor growth was inhibited. These cell lines, however, did not respond to anti-VEGF treatment in vitro, supporting the notion that anti-VEGF antibody administration reduced vessel growth necessary for tumor development (119). Likewise, when athymic mice transplanted with human colon carcinoma cells were treated with an anti-VEGF antibody, tumor growth was reduced as was the number and size of liver metastases (120). Glioblastoma growth was also inhibited in nude mice infected with a retrovirus encoding a dominant-negative mutant of the Flk-1 receptor (mouse homolog of KDR) (121).

ENDOGENOUS CYTOKINES AND TUMOR BIOLOGY

Cytokine therapy is given against a background of a complex and highly regulated endogenous cytokine network, not only in the whole organism, but in the tumor microenvironment. As described above, endogenous cytokines are important mediators of tumor/stroma communication, as well as being autocrine and paracrine regulators of tumor cell growth and survival. However, the influence of these endogenous cytokines is generally not considered when designing therapies. Defining the cytokine context of tumors may not only enhance our understanding of tumor biology, but may lead to novel and more specific cytokine therapies.

Cytokine Networks in Cancers

One of the tumors in which this network is being examined is human epithelial ovarian cancer (1,122). This tumor microenvironment is rich in mRNA for growth factors, proinflammatory cytokines, and chemokines, but weak in lymphocyte-associated cytokines. The pattern of mRNA expression was unique to ovarian tissue but similar in normal, benign, and malignant biopsies, with > 80% of samples expressing 16 cytokines in common. Fourteen of these were also expressed by > 65% of cell lines, but fewer were detected in ovarian cancer xenografts. Potential autocrine loops existed for IL-1, IGF-1, M-CSF, GM-CSF, and TNF-α. IL-4 and IFN-γ receptors were expressed in the absence of ligand. Chemokines RANTES, MIP-1α, and MIP-1β were expressed in biopsies but were rarely detected in cell lines and were absent from xenografts. IGF-1 and its receptor was expressed in every sample, as was the IFN-γ receptor. Another 10 cytokine mRNAs and 6 receptors were expressed in > 80% of samples, including xenografts that had been through many passages in nude mice. These cytokines may contribute to key survival/growth loops and may be targets for therapy. Similarities between normal and malignant biopsies suggest that analogous processes of remodeling and repair occur.

The redundancy of the network detailed above would suggest that targeting a single endogenous cytokine may prove ineffective. However, a combination of cytokines and inhibitors, delivered systemically or locally, could provide a greater degree of specificity whether the desired result is apoptotic tumor cell death, destruction of the tumor vasculature, induction of a tumor-specific host immune response, or, optimally, a combination of all three.

Cytokines as Tumor Markers

In some advanced cancers, cytokines and/or their soluble receptors are found in blood or ascitic fluid. In general, high levels of cytokines or receptors are associated with a poor prognosis or advanced malignancy. For instance, serum levels of TNF-α were significantly higher in patients with head and neck cancer as compared to controls with benign disease (123). Serum levels of the α component of the IL-2 receptor were elevated in patients with non-Hodgkin's lymphoma compared to control subjects, and these levels were related to tumor stage (124). IL-6 levels were higher in cystic fluids of malignant ovarian tumors compared to benign tumors (125), but serum IL-6 concentrations were not altered. In another study, patients with advanced ovarian cancer had higher IL-6 levels in the serum and ascitic fluid as compared to patients with benign ovarian tumors (126). In patients with colorectal cancer, serum IL-6 levels were increased as compared to controls and highest in patients with liver or lung metastasis (123). Serum IL-8 levels were also elevated in colorectal cancer and showed significant differences according to histological type, being lower in well-differentiated adenocarcinomas as compared to other types (127). As with IL-6, serum IL-8 levels were significantly higher in patients with liver or lung metastasis (127).

The angiogenic cytokine VEGF is produced by tumor cells and stroma, and its expression often correlates with the degree of vascularization and grade of malignancy. Its receptors, KDR and flt, are mainly expressed by the tumor endothelium (128). Abundant levels of VEGF were found by immunoassay in the ascites of patients with epithelial ovarian cancer (129). Malignant epithelium was identified as one source of this VEGF (130). In breast cancer, in situ hybridization was used to show the presence of high levels of VEGF mRNA in tumor cells in comedo-type ductal carcinoma in situ, infiltrating ductal carcinoma, and metastatic ductal carcinoma. Flt and KDR were strongly expressed in endothelial cells of small capillaries adjacent to malignant tumor cells (131). In a study involving 103 primary breast cancers, VEGF expression, assessed by immunohistochemistry, was found to correlate with an increment in microvascular density. Moreover, the relapse-free survival rate of VEGF-rich tumors was significantly worse than the VEGF-poor tumors (132). Expression of VEGF and KDR is higher in metastatic than in nonmetastatic human colon cancers, as assessed by immunohistochemistry. The increased expression correlated with the extent of neovascularization and proliferative index of tumor cells (133). VEGF mRNA expression was significantly increased in 26 of 27 hypervascular renal cell carcinoma tissues when compared to adjacent normal kidney tissues. Even small tumors, when hypervascular, overexpressed VEGF mRNA (134).

CYTOKINE THERAPY OF HUMAN CANCERS

Systemic Cytokine Treatment

IFN-α

The antitumor activity of IFN-α seen in experimental animals has been confirmed in clinical trials over the last 15 years, although the types of tumor that respond to IFNs could not have been predicted from preclinical studies (7,135). In clinical trial so far, IFN-α has been most effective in the hematological malignancies. The high activity of several IFN-α preparations in hairy cell leukemia, a relatively uncommon leukemia, was confirmed, worldwide, although superior agents for treatment have now emerged (136). In a majority of hairy cell leukemia patients, IFN-α normalized peripheral blood cell counts, and cytogenetic improvement in the bone marrow was seen after several months of therapy.

Similar findings have been reported in chronic myelogenous leukemia, CML, a multilineage hematological malignancy characterized by a reciprocal translocation between chromosomes 9 and 22 to form the chimeric *bcr-abl* oncogene (7). In an analysis of 190 patients treated with partially purified IFN-αs or recombinant IFN-α, the complete hematological remission rate was 74%, with 53% of these remitting patients achieving cytogenetic remission in the bone marrow (7).

Approximately 30% of the total population achieved durable remission with a risk of relapse largely confined to the first 3 years. In an Italian study (137), over 300 patients were randomized to receive recombinant IFN-α or chemotherapy with hydroxyurea. IFN-α prolonged the time to disease progression compared to chemotherapy and improved survival. In one trial, the median survival of 293 IFN-α–treated patients was 61 months compared to 41 months for those receiving chemotherapy (138). Some data suggest that responses are as good with lower doses as they are with doses that cause significant toxicity, and further trials are underway to confirm this.

Other tumors of the myeloid system, which are partially differentiated, such as polycythemia vera and disorders involving high platelet counts, may respond to this cytokine. IFN-α caused a rapid and selective lowering of the platelet count in over 80% of patients with essential thrombocytemia (139). Normalization of bone marrow histology occurred in about half the patients who achieved a complete hematological response.

Multiple myeloma is another hematological malignancy in which IFN-α has activity during remission induction, although response rates are inferior to those seen with conventional therapy (mean response rate of 209 patients = 23.4%) (140,141,142). However, IFN-α may be useful as maintenance therapy, as it does not delay relapse. Following standard induction therapy, patients receiving IFN-α had an increased relapse-free survival (26 months vs. 14 months) (141) and the median duration of survival increased marginally, although this was lost on follow-up. In two other studies, maintenance IFN-α increased remission (143,144), but in a third it showed no effect (145). In many studies of IFN-α in myeloma, there is evidence that patients with IgA or Bence-Jones myeloma respond best (142,146).

Several studies have shown that IFN-αs have antitumor activity in low-grade "favorable histology" lymphoma, achieving a response rate of about 25% (142,147). IFN-α produced objective responses in patients with carcinoid and malignant endocrine pancreatic tumors (148), AIDS-related Kaposi's sarcoma (149), and a small minority of patients with renal cell carcinoma and melanoma (150). IFN-α had antiangiogenic activity in animal models, which may account for its action in Kaposi's sarcoma. Other evidence of this action comes from the dramatic regression of life-threatening pulmonary hemangiomas in children receiving IFN-α (151). However, results of IFN-α therapy in solid tumors are, in general, disappointing, although slowly proliferating variants of renal cell carcinoma and melanoma will undergo regression with response rates generally between 10 and 15% (152). There is some evidence that IFN-α works best against melanoma in an adjuvant/low tumor load setting (8). The trials so far suggest that very low doses over a long period, or very high doses over a short period, are less effective than moderately high doses for a period of approximately 1 year. The most persuasive results come from a randomized control study of 287 patients receiving IFN-α 20 mU/m^2/day i.v. for 1 month and 10 MU/m^2/day s.c. for 48 weeks for deep or regionally metastatic melanoma (153). Control patients received no therapy, only close observation. With a median follow-up time of 6.9 years, a significant prolongation of relapse-free survival (from 1 to 1.7 years, $p = 0.002$) and overall survival (from 2.8 to 3.8 years, $p = 0.02$) was observed in the treated compared to control group. The impact on relapse rate was most pronounced early during the treatment interval and of most benefit to node positive patients. IFN-α is, in fact, the first agent to show a significant benefit in a randomized trial with high-risk melanoma patients.

Mechanisms of IFN-α Action Against Human Cancer

IFN-αs have antitumor activity in a limited number of human cancers, mainly of those hematologic origin. Activity is generally seen in slow-growing tumors. As was found in animal models, continuous dosage is more effective than intermittent, and responses are slow, being maximal at 12 months or longer. It seems likely that IFN-α will act best at early stages of a cancer and that its activity in advanced cancer will be negligible. Dose-response relationships are still not entirely clear, but clinical efficacy is generally seen at doses of 2–5 MU (equivalent to 2–5 μg) per m^2 per day or 3 times weekly.

Current evidence would suggest that IFN-α acts as a negative regulator in responsive cancers, not as an immunomodulator of a host-cancer response. In this respect, it can be considered a prototype tumor suppressor protein, repressing the malignant phenotype in some cancers that are capable of differentiation (4). One way that it may act is by downregulating response to autocrine growth/survival factors. In a myeloma cell line, for instance, IFN-α downregulated expression of receptors, especially gp80, for the autocrine growth factor, IL-6 (154). IFN-α can also induce other tumor-suppressor proteins, such as the protein kinase that phosphorylates the eukaryotic peptide chain initiation factor, the IFN regulatory proteins IRF-1 and IRF-2, and a latent endoribonuclease Rnase L (4). It is of interest that some of the proteins encoded by oncogenic viruses can inhibit the action of these IFN-inducible regulators (4). There is evidence that some tumor cells may be deficient in IFN-α production. Cells from some patients with non-Hodgkin's lymphoma, chronic and acute lymphocytic leukemia are unable to make IFN-α, an ability that is restored at remission (142). This deficiency may lie in the loss of genes in p21-22 from chromosome 9 where the IFN gene cluster lies and may be secondary to the loss of the nearby retinoic acid receptor (155). However, in cell lines derived from some malignant melanomas, a deficiency in IFN-α secretion had a more specific cause (156). There was a failure of IFN-α gene activation at the promoter level because of disruption of a *trans*-acting IFN-α gene transcription factor.

Other potential targets for IFN-α action include tumor blood vessels, tumor cell signal transduction pathways, as well as components of the extracellular matrix and cellular adhesion molecules (2). It is likely that optimal activity will be achieved by combining IFN-α with other tumor suppressors, such as retinoids, or with specific antagonists of autocrine growth/survival factors.

IFN-γ

While sharing many antiproliferative and antiviral effects with IFN-α and IFN-β, IFN-γ has distinct immunomodulatory effects including macrophage activation, induction of MHC class II genes, and more pronounced synergistic interactions with cytokines such as TNF-α and IL-2. For these reasons, early clinical studies with IFN-γ concentrated on treatment doses and schedules that were capable of enhancing the immune function in patients with both malignant and infectious diseases (157). IFN-γ has now been tested in many advanced cancers; however, in solid tumors the results have been largely disappointing. IFN-γ has been used in such cancers as head and neck cancer (158), colorectal cancer (159), renal cell carcinoma (160,161), and ovarian cancer (162). In 100 patients in complete remission from SCLC, IFN-γ had no activity as adjuvant therapy (163), but it was effective in a very small trial (eight patients) of head and neck cancer patients receiving IFN-γ in 24-h infusions once per week for 4 weeks (158). Renal cell carcinoma has been the focus of a number of trials. Early indications were that a biologically defined low dose of IFN-γ once a week for a median period of 10 months resulted in 30% of patients responding with either partial or complete response (161). These results were not confirmed in a larger study of unselected patients (160), as those with poor prognosis features and large tumor burden did not respond to this nontoxic dose. In another study with 35 patients with renal cell carcinoma, a 15% response rate was demonstrated (164). In this latter study toxicity was minimal and treatment compared favorably with IL-2 or IFN-α.

IFN-γ has also been used in a number of studies in patients with ovarian cancer (162,165). Some responses were observed in patients with minimal residual disease. A large European multicenter phase II trial of i.p. IFN-γ in patients after chemotherapy with persistent residual disease at second-look laparotomy has recently been conducted. A dose of 20 MU/m^2 was administered twice per week for 3–4 months. Thirty-one percent of patients responded to therapy (23%, complete response; 8% partial response), with fever being the most common clinical adverse reaction. Both young age and tumor burden were predictive factors of response, with 52, 35, and 16% response in patients of less than 50 years, 50–59 years, and more than 60 years, respectively (162).

There is evidence that IFN-γ acts synergistically with other cytokines and more conventional therapeutic agents. IFN-γ in combination with TNF-α is discussed later. IFN-β and IFN-γ with cytotoxic drugs produced a remission rate of 47% in patients with metastatic colorectal cancer (166). In contrast, IFN-γ alone had no effect on a group of 50 patients with advanced colorectal cancer (159).

Perhaps reflecting the animal data described above and in contrast in the clinical experience with IFN-α, there is some indication that low-dose IFN-γ adjuvant therapy may have adverse effects. In a randomized phase III trial in melanoma that was halted prematurely, the number of relapses and deaths suggested that those receiving IFN-γ fared worse than untreated patients (167). Likewise, time to progression and survival were inferior in patients with SCLC treated in complete remission with IFN-γ compared to randomized control patients, although the differences were not statistically significant (163).

IL-2

Trials of systemic IL-2 therapy as a single agent have generally been carried out in advanced metastatic melanoma and renal cell carcinoma, tumor types thought from early studies to be most responsive to IL-2 (168). In a phase II setting, approximately 20% of patients will achieve a response, and the majority will receive two cycles of high-dose IL-2 administered i.v. over 3 weeks (169). Most of the responses were seen within 1 or 2 months, although maximal responses then took some time to evolve—some even after cessation of IL-2 therapy. In a review of many such trials, Jost and Stahel (9) concluded that complete remissions were almost exclusively achieved with doses approaching the MTD. Regression of advanced tumor is usually achieved at the expense of toxicity, especially hypotension. IL-2 therapy is accompanied by considerable side effects, including fever, chills, malaise, nausea, vomiting, disturbances of liver and renal function, respiratory failure, and hemodynamic changes resembling those seen in endotoxic shock, with the highest doses often requiring intensive care support (168,169). The most common side effect is the vascular leak syndrome, manifested by peripheral and/or pulmonary edema and weight gain. In a number of studies, the duration of IL-2 administration has been increased as the dose is reduced in attempt to reduce this toxicity. This lower dose treatment leads to a lymphocytosis with an increase of absolute numbers of NK cells (170).

The high toxicity of systemic IL-2 has encouraged regional administration. Patients with ovarian carcinoma, gastrointestinal tumors, and melanoma have been treated with intraperitoneal IL-2. With higher doses (~ 70 MIU/m^2/day), side effects were painful sterile or bacterial peritonitis, fibrosis, adhesion, and increased ascites formation (171). Systemic toxicity was also seen. With lower doses (approx 1 MIU/m^2/day) IL-2 persisted in the peritoneum and responses were noted. Complete and more commonly partial responses were achieved with peritumoral injections of IL-2 into locally metastatic head and neck tumors, and this treatment had low toxicity (172). Low-dose IL-2 given intrapleurally to treat mesothelioma was able to induce some partial responses with low toxicity (173).

Thus, the responses seen with IL-2 alone are probably no better than those achieved with high-dose IFN-α therapy, and long-term remissions are rare. In line with the significant benefit noted in animal experiments, IL-2 has been given to many patients in combination with ex vivo activated autologous lymphokine-activated cells (174). Once again the majority of trails were in renal cell carcinoma and melanoma, and response rates were in the region of 20–30%. However, after 10 years of clinical experimentation, this costly process has been dropped by most centers, as the additional benefit to patients has not been proved in randomized trials in the United States (175) or Europe (176). IL-2 has also been extensively used in combination with other cytokines, particularly IFN-α, but responses are not much improved. For instance, a phase II trial of low-dose s.c. IL-2 and IFN-α in metastatic renal cell carcinoma resulted in a 12% response rate and modest toxicity with a median survival of 14.5 months (177). More recent studies suggest that this combination may be synergistic with cisplatin-based regimes in malignant melanoma with response rates approaching 50% (178).

TNF

About 100 years ago, William Coley treated a series of cancer patients with filtrates of bacterial cultures (179). There were undoubtedly some dramatic and well-documented cures in a minority of patients, an effect now attributed to hyperthermia and the local or systemic induction of cytokines, particularly TNF-α. A further 70 years of laboratory experiments led to the isolation and subsequent cloning of this cytokine. The maximum tolerated dose (MTD) for recombinant TNF-α is about 200 μg/m^2/day (180), with hypotension being the principal limiting factor. Common acute side effects include fever, rigors, chills, headache, and, more rarely, pulmonary toxicity, chest pain, and intravascular coagulation. Responses to TNF therapy given at relatively nontoxic doses are, however, extremely rare.

TNF, IFN-γ Combination Therapy The ability of this cytokine to induce tumor necrosis has recently been confirmed in one specific clinical setting. High doses of locoregional TNF-α (4 × MTD), combined with systemic IFN-γ and local chemotherapy, mediated the specific destruction of tumor vasculature in a majority of patients with melanoma, squamous cell carcinoma, and soft tissue sarcoma whose tumors are accessible to isolated limb perfusion (ILP) (181,182). Complete response rates of 50–90% were achieved in these studies. The impressive antitumor effects of this treatment seem to be associated with vascular-mediated damage (183). This treatment, however, remains controversial: other groups using the same protocol have been unable to obtain similar results (184), and the impressive local antitumor effects are not accompanied by regression of distant metastasis.

G and GM-CSF

Animal and in vitro experiments suggested that colony-stimulating factors, particularly G-CSF and GM-CSF, would have clinical potential, not as antitumor agents, but as cytokines that could ameliorate the myelosuppression associated with chemotherapy. There are three different ways in which these cytokines could support myelotoxic therapies. First, they can be given during conventional therapy to shorten the depth and duration of neutropenia (185), reduce the incidence of infections (186), and facilitate delivery of the planned dose on time (187). Second, they can be used to increase cytotoxic dose-intensity above that of standard chemotherapy by supporting dose increases or accelerated administration (188). Third, they can be used alone or in combination with chemotherapy to mobilize hemopoietic progenitor cells into the circulation, where they can be harvested for transplantation after myeloablative chemotherapy or to support multicyclic chemotherapy (187).

In phase I and II trials, GM-CSF, given systemically at doses of 5–10 μg/kg, caused a reliable and significant reduction of chemotherapy-induced neutropenia, expressed as nadirs and/or duration of critical values (reviewed in Refs. 189,190). The cytokine was just as effective s.c. as i.v., and side effects were minimal. GM-CSF therapy was best given after chemotherapy and continued for 7–10 days. In phase II clinical, trials GM-CSF significantly reduced the incidence, duration, and extent of fever episodes. The first occurrence of infections was delayed and the incidence and severity of infectious complications reduced. This was translated into a reduction in hospitalization and antibiotic requirements. GM-CSF can be effective in severe neutropenia after repeated multiagent therapy and after myeloablative therapy and autologous bone marrow transplantation (189). Although the actions of GM-CSF have certainly contributed to patient well-being, they have not yet been shown to improve response to therapy and/or long-term survival, even when they permit increased dose intensity of chemotherapy regimes (189).

Results of G-CSF therapy have been much the same. For instance, in randomized trials, administration of G-CSF decreased the incidence of febrile neutropenia and allowed an increase in the number of courses of chemotherapy given (186). Results concerning dose intensity were more variable. In one study patients with SCLC randomized to G-CSF received higher doses of chemotherapy than controls and had a better 2-year survival rate (191), but in another G-CSF did not decrease dose reductions or treatment delays to a level that would allow an increase in received dose (192).

Cytokine Gene Delivery Trials

Prompted by encouraging preclinical studies, there are a range of clinical trials, planned or underway, using cytokine-directed gene therapy (95,103,193). Several approaches are being followed, all with the aim of provoking powerful local inflammation and generating systemic and specific immune responses to the tumor. The cytokines IL-2, IL-4, GM-CSF, and IL-12 are most frequently used in such protocols. Proposals include transfection of the relevant gene into autologous or allogeneic tumor cells or stromal components ex vivo. Other approaches use cell specific promoters to direct the gene in vivo, with direct injection, viral, or liposomal vectors being used for delivery (193).

CYTOKINE TREATMENT OF HUMAN CANCER: SUMMARY AND FUTURE DIRECTIONS

There appear to be two well-defined responses of human tumors to cytokine therapy. The first, typified by IFN-α–sensitive hematological tumors, is a slow differentiation or destruction of the malignant clone after several months or years of daily or thrice weekly therapy. This response may also achieved in a maintenance setting after chemotherapy and/or surgery to reduce the tumor burden. In successful protocols, there is little suggestion of a host immune response to tumor, although there is evidence for this in animal models of syngeneic transplantable tumors. It is more likely that IFNs cause a gradual decline in response of tumor cells to/or production of autocrine/paracrine growth factors. IFNs may also inhibit neovascularization of tumors preventing growth of primary tumors and micrometastases.

It is not clear whether this slow response will be seen with other tumors types or cytokine therapies. IFNs-α, -β, and -γ are unusual cytokines in that they are not known to be mitogens. Moreover, there is increasing molecular evidence that they, or the genes they induce, are tumor suppressors and that there are defects in the IFN system in several tumor types. IFNs, as growth-supressing cytokines, may be deregulated during the evolution of malignancy. Any novel cytokine that is characterized as a growth suppressor, especially in epithelial cells, may have similar potential. One obvious candidate is the TGF-βs, where mutations in the type II receptors are frequently found in colorectal and gastric carcinoma (3).

A second well-characterized response to a cytokine, most likely with IL-2, IL-12, and TNF-α/IFN-γ, is a more rapid regression of established tumor due to specific destruction of tumor vasculature and/or cytotoxic lymphocyte infiltration. This response may lead to specific immunity and long-term survival in a minority of patients, but is also often associated with severe side effects that resemble endotoxic shock.

Major advances in chemotherapy of cancer came from a combination of agents with different mechanisms of action. While combination of cytokines has been associated with increased and often unacceptable toxicity, sequential application of cytokines with different actions, combined with novel drugs that are likely to complement cytokine action, is an attractive option. Combination of IFN-α with other antiangiogenic drugs is a possibility, as is a sequential administration of IFN and agents such as matrix metalloprotease inhibitors that disrupt tumor/stroma communications. Cocktails of downregulatory cytokines and specific cytokine antagonists may have less toxicity. If designed with a knowledge of the growth/survival factors that typify a certain tumor, these may, over a short period of exposure, induce apoptosis of genetically damaged tumor cells without affecting their normal counterparts.

REFERENCES

1. Balkwill FR. Cytokine therapy of cancer. The importance of knowing the context. Eur Cytokine Netw 1994; 5:379–385.
2. Gutterman JU. Cytokine therapeutics: lessons from interferon α. PNAS 1994; 91:1198–1205.
3. Parsons R, Myeroff LL, Liu B, et al. Microsatellite instability and mutations of the transforming growth factor β type II receptor gene in colorectal cancer. Cancer Res 1995; 55:5548–5550.
4. Lengyel P. Tumor-suppressor genes: news about the interferon connection. Proc Natl Acad Sci 1993; 90:5893–5895.
5. Urabe A. Interferons for the treatment of hematological malignancies. Oncology 1994; 51:137–141.
6. Bronchud MH. Recombinant human granulocyte colony-stimulating factor in the management of cancer patients: five years on. Oncology 1994; 51:189–197.
7. Talpaz M. Use of interferon in the treatment of chronic myelogenous leukemia. Semin Oncol 1994; 21:3–7.
8. Balch CM, Buzaid AC. Finally, a successful adjuvant therapy for high-risk melanoma. J Clin Oncol 1996; 14:1–3.
9. Jost LM, Stahel RA. Interleukin-2 as single agent. In: Wagstaff J, ed. The Role of Interleukin-2 in the Treatment of Cancer Patients. Dordrecht: Kluwer Academic Publishers, 1993:7–28.
10. Balkwill FR, Taylor-Papadimitriou J, Fantes KH, Sebesteny A. Human lymphoblastoid interferon can inhibit the growth of human breast cancer xenografts in athymic (nude) mice. Eur J Cancer 1980; 16:569–573.
11. Balkwill FR, Moodie EM, Freedman V, Fantes KH. Human interferon inhibits the growth of established human breast tumours in the nude mouse. Int J Cancer 1982; 30:231–235.
12. Balkwill FR. Antitumour effects of interferons in animals. Interferon 4: in vivo and clinical studies, 1985:23–45.
13. Balkwill FR. The regulatory role of interferons in the human immune response. In: Interferons, Their Impact in Biology and Medicine. Oxford Medical Publications, 1985:61–80.
14. Brosjo O, Bauer HCF, Brostrom L-A, et al. Influence of human a-interferon on four human osteosarcoma xenografts in nude mice. Cancer Res 1985; 45:5598–5602.
15. Brosjo O, Bauer HCF, Brostrom L-A, Nilsson OS, Reinholt FP, Tribukait B. Growth inhibition of human osteosarcomas in nude mice by human interferon-alpha: significance of growth & tumour differation. Cancer Res 1987; 47:258–262.
16. Hofmann V, Groscurth P, Morant R, Cserhati M, Honegger HP, von Hochstetter A. Effects of leukocyte interferon (E. coli) on human bone sarcoma growth in vitro and in the nude mouse. Eur J Cancer Clin Oncol 1985; 21:859–863.
17. Forster S, Triffitt JT, Bauer HC, et al. Interferon-inhibited human osteosarcoma xenografts induce host bone in nude mice. J Bone Miner Res 1988; 3:199–202.
18. Sidky YA, Borden EC. Inhibition of angiogenesis by IFNs: effects on tumor and lymphocyte-induced vascular responses. Cancer Res 1987; 47:5155–5161.
19. Kase S, Kubota T, Watanabe M, Teramoto T, Kitajima M, Hoffman RM. Recombinant human interferon alpha-2a increases 5-fluorouracil efficacy by elevating fluorouridine concentration in tumor tissue. Anticancer Res 1994; 14:1155–1159.
20. Balkwill FR, Moodie EM. Positive interactions between human interferon and cyclophosphamide or adriamycin on a human model system. Cancer Res 1984; 44:904–908.
21. Gresser I. Antitumor effects of interferon. Acta Oncol 1989; 28:347–353.
22. Gresser I. How does interferon inhibit tumor growth? Interferon 1985; 6:93–126.
23. Dvorak HF, Gresser I. Microvascular injury in pathogenesis of interferon-induced necrosis of subcutaneous tumors in mice. JNCI 1989; 81:497–502.
24. Gresser I, Belardelli F, Tavernier J, et al. Anti-tumour effects of interferon in mice injected with interferon-sensitive and interferon-resistant friend leukemia cells. V. Comparisons with the action of tumor necrosis factor. Int J Cancer 1986; 38:771–778.
25. Gresser I, Maury C, Bandu M, Belardelli F. Importance of interferon alpha in the resistance of allogeneic c57b1/6 mice to the multiplication of Friend erythroleukemia cells in liver. Cancer 1990; 45:364–371.
26. Kaido T, Gresser I, Maury C, Maunoury M-T, Vignaux F, Belardelli F. Sensitized T lymphocytes render dba/2 beige mice responsive to IFN a/b therapy of friend erythroleukemia visceral metastases. Int J Cancer 1993; 54:475–481.
27. Yasui H, Proietti E, Vignaux F, Eid P, Gresser I. Inhibition of mouse a/b-interferon of the multiplication of a/b-interferon-resistant Friend erythroleukemia cells cocultured with mouse hepatocytes. Cancer Res 1990; 50:3533–3539.
28. Gresser I, Maury C, Carnaud C, De Maeyher E, Maunoury MT, Belardelli F. Anti-tumor effects of interferon in mice injected with interferon-sensitive and interferon-resistant Friend erythroleukemia cells. VIII. Role of the immune system in the inhibition of visceral metastases. Int J Cancer 1990; 46:468–474.
29. Ramani P, Hart IR, Balkwill FR. The effect of interferon on experimental metastases in immunocompetent and immunodeficient mice. Int J Cancer 1986; 37:563–568.
30. Ramani P, Balkwill FR. Action of recombinant alpha interferon against experimental and spontaneous metastases in a murine model. Int J Cancer 1989; 43:140–146.
31. Parangi S, O'Reilly MO, Holmgrem L, Folkman J, Hanahan D. Treatment of transgenic mice with a regimen of angiogenesis inhibitors impairs tumour development. Proc Am Assoc Can Res 1995; 36:104.
32. Balkwill FR, Griffin DB, Lee AE. Interferons alpha and gamma differ in their ability to cause tumour stasis and regression in vivo. Eur J Cancer Clin Oncol 1989; 25:1481–1486.
33. Balkwill FR, Stevens MH, Griffin DB, Thomas JA, Bodmer JG. Interferon gamma regulates hla-d expression on solid tumors in vivo. Eur J Cancer Clin Oncol 1987; 23:101–106.
34. Balkwill FR, Ward BG, Moodie E, Fiers W. Therapeutic potential of tumor necrosis factor alpha and gamma interferon in experimental human ovarian cancer. Cancer Res 1987; 47:4755–4758.
35. Malik STA, Knowles RG, East N, Lando D, Stamp G, Balkwill FR. Antitumour activity of gamma interferon in ascitic and solid tumour models of human ovarian cancer. Cancer Res 1991; 51:6643–6649.
36. Giovarelli M, Cofano F, Vecchi A, Forni M, Landolfo S, Forni G. Interferon-activated tumor inhibition in vivo. Small amounts of interferon-gamma inhibit tumor growth by eliciting host systemic immunoreactivity. Int J Cancer 1986; 37:141–148.
37. Kondo H, Tanaka N, Naomoto Y, Orita K. Antitumor effect of recombinant human interferon-beta and interferon-gamma in combination against human colon cancer cell line in vitro and in nude mice. Jpn J Cancer Res 1987; 78:1258–1265.
38. Ramani P, Balkwill FR. Enhanced metastases of a mouse

carcinoma after in vitro treatment with murine interferon gamma. Int J Cancer 1987; 40:830–834.
39. Kelly SA, Gschmeissner S, East N, Balkwill FR. Enhancement of metastatic poential by gamma-interferon. Cancer Res 1991; 51:4020–4027.
40. Lotze MT, Matory YL, Ettinghausen SE, et al. In vivo administration of purified human interleukin 2. J Immunol 1985; 135:2865–2875.
41. Sacchi M, Snyderman CH, Heo DS, et al. Local adoptive immunotherapy of human head and neck cancer xenografts in nude mice with lymphokine-activated killer cells and interleukin 2. Cancer Res 1990; 50:3113–3118.
42. Sacchi M, Vitolo D, Sedlmayr P, et al. Induction of tumor regression in experimental model of human head and neck cancer by human A-LAK cells and IL-2. Int J Cancer 1991; 47:784–791.
43. Greenberg PD. Therapy of murine leukemia with cyclophosphamide and immune Lyt-2+ cells: cytolytic T cells can mediate eradication of disseminated leukemia. J Immunol 1986; 136:1917–1922.
44. Mule JJ, Smith CA, Rosenberg SA. Interleukin 4 can mediate the induction of lymphokine-activated killer cell activity directed against fresh tumor cells. J Exp Med 1987; 166:792–797.
45. Talmadge JE, Phillips H, Schindler J, Tribble H, Pennington R. Systematic preclinical study on the therapeutic properties of recombinant human interleukin 2 for the treatment of metastatic disease. Cancer Res 1987; 47:5725–5732.
46. Foa R, Fierro MT, Raspadori D, et al. Lymphokine-activated killer (LAK) cell activity in B and T chronic lymphoid leukemia: defective LAK generation and reduced susceptibility of the leukemic cells to allogeneic and autologous LAK effectors. Blood 1990; 76:1349–1354.
47. Mule JJ, Shu S, Schwarz SL, Rosenberg SA. Adoptive immunotherapy of established pulmonary metastases with LAK cells and recombinant interleukin-2. Science 1984; 225:1487–1489.
48. Rosenberg SA, Spiess P, Lafreniere R. A new approach to the adoptive immunotherapy of cancer with tumor-infiltrating lymphocytes. Science 1986; 233:1318–1321.
49. Rosenberg SA, Mule JJ, Spiess J, Reichert CM, Schwarz SL. Regression of established pulmonary metastases and subcutaneous tumor mediated by systemic administration of high-dose recombinant interleukin. J Exp Med 1985; 161:1169–1188.
50. Marincola FM, Venzon D, White D, et al. HLA Association with response and toxicity in melanoma patients treated with interleukin 2-based immunotherapy. Cancer Res 1992; 52:6561–6566.
51. Moore AS, Theilen GH, Newell AD, Madewell BR, Rudolf AF. Preclinical study of sequential tumor necrosis factor and interleukin 2 in the treatment of spontaneous canine neoplasms. Cancer Res 1991; 51:233–238.
52. Papa MZ, Vetto JT, Ettinghausen SE, Mule JJ, Rosenberg SA. Effect of corticosteroid on the antitumor activity of lymphokine-activated killer cells and interleukin 2 in mice. Cancer Res 1986; 46:5618–5623.
53. Herberman RB, Nunn ME, Holden HT, Lavrin DH. Natural cytotoxic reactivity of mouse lymphoid cells against syngeneic and allogeneic tumors. II. Characterization of effector cells. Int J Cancer 1975; 16:230–239.
54. Lafreniere R, Rosenberg SA. Successful immunotherapy of murine experimental hepatic metastases with lymphokine-activated killer cells and recombinant interleukin 2. Cancer Res 1985; 45:3735–3741.
55. Mule JJ, Shu S, Rosenberg SA. The anti-tumor efficacy of lymphokine-activated killer cells and recombinant interleukin 2 in vivo. J Immunol 1985; 135:642–646.
56. Rosenberg S. Lymphokine-activated killer cells: a new approach to immunotherapy of cancer. JNCI 1985; 75:595–603.
57. Hirabayashi H, Yasumura S, Lin WC, et al. Production by human squamous cell carcinoma of a factor inducing activation and proliferation of immune cells. Arch Otolaryngol Head Neck Surg 1995; 121:285–292.
58. Henney CS, Kuribayashi K, Kern DE, Gillis S. Interleukin-2 augments natural killer cell activity. Nature 1981; 291:335–338.
59. Mule JJ, Yang JC, Afreniere RL, Shu SY, Rosenberg SA. Identification of cellular mechanisms operational in vivo during the regression of established pulmonary metastases by the systemic administration of high-dose recombinant interleukin 2. J Immunol 1987; 139:285–294.
60. Tanaka K, Hayashi H, Hamada C, Khoury G, Jay G. Expression of major histocompatibility complex class I antigens as a strategy for the potentiation of immune recognition of tumor cells. Proc Natl Acad Sci USA 1986; 83:8723–8727.
61. Brunda MJ. Antitumour activity of interleukin-2 combined with other cytokines. In: Waxman J, ed. Interleukin-2. London: Blackwell Scientific Publications, 1992:106–121.
62. Brunda MJ, Bellantoni D, Sulich V. In vivo anti-tumor activity of combinations of interferon alpha and interleukin-2 in a murine model. Correlation of efficacy with the induction of cytotoxic cells resembling natural killer cells. Int J Cancer 1987; 40:365–371.
63. Iigo M, Sakurai M, Tamura T, Saijo N, Hoshi A. In vivo antitumor activity of multiple injections of recombinant interleukin 2, alone and in combination with three different types of recombinant interferon, on various syngeneic murine tumors. Cancer Res 1988; 48:260–264.
64. Winkelhake JL, Stampfl S, Zimmerman RJ. Synergistic effects of combination therapy with human recombinant interleukin-2 and tumor necrosis factor in murine tumor models. Cancer Res 1987; 47:3948–3953.
65. Nishimura T, Ohta S, Sato N, Togashi Y, Goto M, Hashimoto Y. Combination tumor-immunotherapy with recombinant tumor necrosis factor and recombinant interleukin 2 in mice. Int J Cancer 1987; 40:255–261.
66. McIntosh JK, Mule JJ, Merino MJ, Rosenberg SA. Synergistic antitumor effects of immunotherapy with recombinant interleukin-2 and recombinant tumor necrosis factor alpha. Cancer Res 1988; 48:4011–4017.
67. Iigo M, Nishikata K, Hoshi A. Effect of recombinant interleukin-1 alpha, recombinant interleukin-2, recombinant interferon-beta, and recombinant tumor necrosis factor on subcutaneously implanted adenocarcinoma 755 and Lewis lung carcinoma. J Biol Response Mod 1990; 9:426–430.
68. Brunda MJ, Luistro L, Warrier RR, et al. Antitumor and antimetastatic activity of interleukin 12 against murine tumors. J Exp Med 1993; 178:1223–1220.
69. Gately MK, Gubler U, Brunda MJ, et al. Interleukin-12: a cytokine with therapeutic potential in oncology and infectious diseases. Ther Immunol (in press).
70. O'Toole M, Wolf S, O'Brien C, Hubbard B, Herrmann S. Effect of in vivo IL-12 administration on murine tumour cell growth. J Immunol 1993; 150:294A.
71. Nastala CL, Edington HD, McKinney TG, et al. Recombinant IL-12 administration induces tumour regression in association with IFN-gamma production. J Immunol 1994; 153:1697–1706.
72. de Kossodo S, Moore R, Gschmeissner S, East N, Upton C, Balkwill F. Changes in endogenous cytokines, adhesion mole-

cules and platelets during cytokine induced tumour necrosis. Br J Can 1995; 72:1165–1172.
73. Laster SN, Wood JG, Gooding LR. Tumour necrosis factor can induce both apoptic and necrotic forms of cell lysis. J Immunol 1988; 141:2629–2634.
74. Bellomo G, Perotti M, Taddei F, et al. Tumor necrosis factor alpha induces apoptosis in mammary adenocarcinoma cells by an increase in intranuclear free Ca^{2+} concentration and DNA fragmentation. Cancer Res 1992; 52:1342–1346.
75. Watanabe N, Niitsu Y, Umeno H, et al. Synergistic cytotoxic and antitumour effects of recombinant tumour necrosis factor and hyperthermia. Cancer Res 1988; 48:650–653.
76. Nawroth PP, Bank I, Handley D, Cassimeris J, Chess L, Stern D. Tumor necrosis factor/cachectin interacts with endothelial cell receptors to induce release of interleukin 1. J Exp Med 1986; 163:1363–1375.
77. Helson L, Helson C, Green S. Effects of murine tumor necrosis factor on heterotransplanted human tumors. Exp Cell Biol 1979; 47:53–60.
78. Haranaka K, Satomi N, Sakurai A. Antitumor activity of murine tumor necrosis factor (TNF) against transplanted murine tumors and heterotransplanted human tumors in nude mice. Int J Cancer 1984; 34:263–267.
79. Balkwill FR, Lee A, Aldam G, et al. Human tumour xenografts treated with recombinant human tumor necrosis factor alone or in combination with interferons. Cancer Res 1986; 46:3990–3993.
80. Creasey AA, Reynolds MT, Laird W. Cures and partial regression of murine and human tumors by recombinant human tumor necrosis factor. Cancer Res 1986; 46:5687–5690.
81. Nosoh Y, Toge T, Nishiyama M, et al. Antitumor effects of recombinant human tumor necrosis factor against human tumor xenografts transplanted into nude mice. Jpn J Surg 1987; 17:51–54.
82. Sohmura Y, Nakata K, Yoshida H, Kashimoto S, Matsui Y, Furuichi H. Recombinant human tumor necrosis factor—II. Antitumor effect on murine and human tumors transplanted in mice. Int J Immunopharmacol 1986; 8:357–368.
83. Manetta A, Podczaski E, Zaino RJ, Satyaswaroop PG. Therapeutic effect of recombinant human tumor necrosis factor in ovarian carcinoma xenograft in nude mice. Gynecol Oncol 1989; 34:360–364.
84. Malik STAA, Griffin DB, Fiers W, Balkwill FR. Paradoxical effects of tumour necrosis factor in experimental ovarian cancer. Int J Cancer 1989; 44:918–925.
85. Watanabe N, Niitsu Y, Umeno H, et al. Toxic effect of tumor necrosis factor on tumor vasculature in mice. Cancer Res 1988; 48:2179–2183.
86. Havell EA, Fiers W, North RJ. The antitumor function of tumor necrosis factor (TNF), I. Therapeutic action of TNF against an established murine sarcoma is indirect, immunologically dependent, and limited by severe toxicity. J Exp Med 1988; 167:1067–1085.
87. Palladino MA, Jr., Shalaby MR, Kramer SM, et al. Characterization of the antitumor activities of human tumor necrosis factor-alpha and the comparison with other cytokines: induction of tumor-specific immunity. J Immunol 1987; 138:4023–4032.
88. Malik S. The activity of TNF in experimental cancer models. In: Beutler B, ed. Tumor Necrosis Factors. New York: Raven Press, 1992.
89. Gray PW, Aggarwal BB, Benton CV, et al. Cloning and expressin of cdna for human lymphotoxin, a lymphokine with tumour necrosis activity. Nature 1984; 312:721–724.
90. Balkwill F. The colony stimulating factors. Cytokines in cancer therapy. Oxford: Oxford University Press, 1989:114–149.
91. Broxmeyer HE, Williams DE, Cooper S, et al. Comparative effects in vivo of recombinant murine interleukin 3, natural murine colony-stimulating factor-1, and recombinant murine granulocyte-macrophage. J Clin Invest 1987; 79:721–730.
92. Metcalf D, Begley CG, Williamson DJ, et al. Hemopoietic responses in mice injected with purified recombinant murine gm-csf. Exp Hematol 1987; 15:1–9.
93. Fujisawa M, Kobayashi Y, Okabe T, Takaku F, Komatsu Y, Itoh S. Recombinant human granulocyte colony-stimulating factor induces granulocytosis in vivo. Jpn J Cancer Res 1986; 77:866–869.
94. Tsuchiya M, Kaziro Y, Nagata S. The chromosomal gene structure for murine granulocyte colony-stimulating factor. Eur J Biochem 1987; 165:7–12.
95. Miller AR, McBride WH, Hunt K, Economou JS. Cytokine-mediated gene therapy for cancer. Ann Surg Oncol 1994; 1:436–450.
96. Tani K. [Implantation of genetically manipulated fibroblasts into mice as a model of gene therapy—supplementations of human granulocyte colony-stimulating factor (hG-CSF) and interferon-alpha (IFN-alpha)]. Hum Cell 1991; 4:25–32.
97. Ferrantini M, Giovarelli M, Modesti A, et al. IFN-alpha 1 gene expression into a metastatic murine adenocarcinoma (TS/A) results in CD8+ T cell-mediated tumor rejection and development of antitumor immunity. Comparative studies with IFN-gamma-producing TS/A cells. J Immunol 1994; 153:4604–4615.
98. Ferrantini M, Proietti E, Santodonato L, et al. A1-interferon gene transfer into metastatic friend leukemia cells abrogated tumorigenicity in immunocompetent mice: antitumor therapy by means of interferon-producing cells. Cancer Res 1993; 53:1107–1112.
99. Gansbacher B, Bannerji R, Daniels B, Zier K, Cronin K, Gilboa E. Retroviral vector-mediated gamma-interferon gene transfer into tumor cells generates potent and long lasting antitumor immunity. Cancer Res 1990; 50:7820–7825.
100. Esumi N, Hunt B, Itaya T, Frost P. Reduced tumorigenicity of murine tumor cells secreting gamma-interferon is due to nonspecific host responses and is unrelated to class I major histocompatibility complex expression. Cancer Res 1991; 51:1185–1189.
101. Gansbacher B, Zier K, Cronin K, et al. Retroviral gene transfer induced constitutive expression of interleukin-2 or interferon-gamma in irradiated human melanoma cells. Blood 1992; 80:2817–2825.
102. Fearon ER, Pardoll DM, Itaya T, et al. Interleukin-2 production by tumor cells bypasses T helper function in the generation of an antitumor response. Cell 1990; 60:397–403.
103. Foa R, Cignetti A, Riera L, Gillio Tos A, Guarini A. Cytokine gene therapy in oncology. Folia Biol Praha 1994; 40:37–48.
104. Bubenik J, Simova J, Jandlova T. Immunotherapy of cancer using local administration of lymphoid cells transformed by IL-2 cDNA and constitutively producing IL-2. Immunol Lett 1990; 23:287–292.
105. Russell SJ, Eccles SA, Flemming CL, Johnson CA, Collins MK. Decreased tumorigenicity of a transplantable rat sarcoma following transfer and expression of an IL-2 cDNA. Int J Cancer 1991; 47:244–251.
106. Abdel Wahab Z, Li WP, Osanto S, et al. Transduction of human melanoma cells with interleukin-2 gene reduces tumorigenicity and enhances host antitumor immunity: a nude mouse model. Cell Immunol 1994; 159:26–39.
107. Karp SE, Farber A, Salo JC, et al. Cytokine secretion by genetically modified nonimmunogenic murine fibrosarcoma. J Immunol 1993; 150:896–908.
108. Cavallo F, Giovarelli M, Gulino A, et al. Role of neutrophils

and CD4+ T lymphocytes in the primary and memory response to nonimmunogenic murine mammary adenocarcinoma made immunogenic by IL-2 gene. J Immunol 1992; 149:3627–3635.

109. Tepper RI, Pattengale PK, Leder P. Murine interleukin-4 displays potent anti-tumor activity in vivo. Cell 1989; 57:503–512.
110. Tepper RI, Coffman RL, Leder P. An eosinophil-dependent mechanism for the antitumor effect of interleukin-4. Science 1992; 257:548–551.
111. Tahara H, Zeh H Jr, Storkus WJ, et al. Fibroblasts genetically engineered to secrete interleukin 12 can suppress tumor growth and induce antitumor immunity to a murine melanoma in vivo. Cancer Res 1994; 54:182–189.
112. Pappo I, Tahara H, Robbins PD, et al. Administration of systemic or local interleukin-2 enhances the anti-tumor effects of interleukin-12 gene therapy. J Surg Res 1995; 58:218–226.
113. Asher AL, Mule JJ, Kasid A, et al. Murine tumor cells transduced with the gene for tumor necrosis factor-alpha. Evidence for paracrine immune effects of tumor necrosis factor against tumors. J Immunol 1991; 146:3227–3234.
114. Malik STA, Naylor S, East N, Oliff A, Balkwill FR. Cells secreting tumour necrosis factor show enhanced metastasis in nude mice. Eur J Cancer 1990; 26:1031–1034.
115. Sun WH, Burkholder JK, Sun J, et al. In vivo cytokine gene transfer by gene gun reduces tumor growth in mice. Proc Natl Acad Sci USA 1996; 92:2889–2893.
116. Golumbek PT, Ashari R, Jaffee EM, et al. Controlled release, biodegradable cytokine depots: a new approach in cancer vaccine design. Cancer Res 1993; 53:5841–5844.
117. Senger DR, Van de Water L, Brown LF, et al. Vascular permeability factor (VPF, VEGF) in tumor biology. Cancer Metastasis Rev 1993; 12:303–324.
118. Senger DR, Galli SJ, Dvorak AM, Peruzzi CA, Harvey VS, Dvorak HF. Tumor cells secrete a vascular permeability factor that promotes accumulation of ascites fluid. Science 1983; 219:983–985.
119. Kim KJ, Li B, Winer J, et al. Inhibition of vascular endothelial growth factor-induced angiogenesis suppresses tumour growth in vivo. Nature 1993; 362:841–844.
120. Warren RS, Yuan H, Matli MR, Gillett NA, Ferara N. Regulation of vascular endothelial growth factor of human colon cancer tumorigenesis in a mouse model of experimental liver metastasis. J Clin Invest 1995; 95:1789–1797.
121. Millauer B, Shawver LK, Plate KH, Risau W, Ullrich A. Glioblastoma growth inhibited in vivo by a dominant-negative flk-1 mutant. Nature 1994; 367:576–579.
122. Owens OJ, Stewart C, Leake RE. Growth factors in ovarian cancer. Br J Cancer 1991; 64:1177–1181.
123. Soylu L, Ozcan C, Cetik F, et al. Serum levels of tumor necrosis factor in squamous cell carcinoma of the head and neck. Am J Otolaryngol 1994; 15:281–285.
124. Stasi R, Zinzani L, Galieni P, et al. Detection of soluble interleukin-2 receptor and interleukin-10 in the serum of patients with aggressive non-Hodgkin's lymphoma. Identification of a subset at high risk of treatment failure. Cancer 1994; 74:1792–1800.
125. van der Zee AG, de Cuyper EM, Limburg PC, et al. Higher levels of interleukin-6 in cystic fluids from patients with malignant versus benign ovarian tumors correlate with decreased hemoglobin levels and increased platelet counts. Cancer 1995; 75:1004–1009.
126. Schroder W, Ruppert C, Bender H. Concomitant measurements of interleukin-6 (IL-6) in serum and peritoneal fluid of patients with benign and malignant ovarian tumours. Eur J Obstet Gynecol Reprod Biol 1994; 56:43–46.
127. Ueda T, Shimada E, Urakawa T. Serum levels of cytokines in patients with colorectal cancer: possible involvement of interleukin-6 and interleukin-8 in hematogenous metastasis. J Gastroenterol 1994; 29:423–429.
128. Boocock CA, Charnock-Jones DS, Sharkey AM, et al. Expression of vascular endothelial growth factor and its receptors flt and KDR in ovarian carcinoma. J Natl Cancer Inst 1995; 87:506–516.
129. Yeo K-T, Wang HH, Nagy JA, et al. Vascular permeability factor (vascular endothelial growth factor) in guinea pig and human tumor and inflammatory effusions. Cancer Res 1993; 53:2912–2918.
130. Olson TA, Mohanraj D, Carson LF, Ramakrishan S. Vascular permeability factor gene expression in normal and neoplastic human ovaries. Cancer Res 1994; 54:276–280.
131. Brown LF, Berse B, Jackman RW, et al. Expression of vascular permeability factor (vascular endothelial growth factor) and its receptors in breast cancer. Human Pathol 1995; 26:86–91.
132. Toi M, Hoshina S, Takayanagi T, Tominaga T. Association of vascular endothelial growth factor expression with tumor angiogenesis and with early relapse in primary breast cancer. Jpn J Cancer Res 1994; 85:1045–1049.
133. Takahashi Y, Kitadai Y, Bucana CD, Cleary KR, Ellis LM. Expression and its receptor, KDR, correlates with vascularity, metastasis and proliferation of human colon cancer. Cancer Res 1995; 55:3964–3968.
134. Takahashi A, Sasaki H, Kim SJ, et al. Markedly increased amounts of messenger RNAs for vascular endothelial growth factor and placenta growth factor in renal cell carcinoma associated with angiogenesis. Cancer Res 1994; 54:4233–4237.
135. Kumar L, Gulati SC. Alpha-interferon in chronic myelogenous leukaemia. Lancet 1995; 346:984–985.
136. Saven A, Piro LD. Treatment of hairy cell leukemia. Blood 1992; 79:1111–1120.
137. Interferon alfa-2a as compared with conventional chemotherapy for the treatment of chronic myeloid leukemia. The Italian Cooperative Study Group on Chronic Myeloid Leukemia [see comments]. N Engl J Med 1994; 330:820–825.
138. Allan NC, Richards SM, Shepherd PC. UK Medical Research Council randomised, multicentre trial of interferon-alpha n1 for chronic myeloid leukaemia: improved survival irrespective of cytogenetic response. The UK Medical Research Council's Working Parties for Therapeutic Trials in Adult Leukaemia. Lancet 1995; 345:1392–1397.
139. Giles FJ, Gray AG, Brozovic J, et al. Alpha-interferon therapy for essential thrombocythaemia, Lancet 1988;
140. Avvisati G, Mandelli F. Interferon treatment of multiple myeloma: an emerging dilemma for the 1990s. In: Talpaz M, Kurzrock R, Gutterman J, eds. Biological Response Modifiers in the Treatment of Hematopoietic Neoplasia. New York: Marcel Dekker, 1992:43–55.
141. Mandelli F, Avvisati G, Amadori S, et al. Maintenance treatment with recombinant interferon alfa-2b in patients with multiple myeloma responding to conventional induction chemotherapy. N Engl J Med 1990; 322:1430–1434.
142. Borden EC. Innovative treatment strategies for non-Hodgkin's lymphoma and multiple myeloma. Semin Oncol 1994; 21:14–22.
143. Westin J. Interferon therapy during the plateau phase of multiple myeloma: an update of a Swedish multicenter study. Semin Oncol 1991; 18:37–40.
144. Ludwig H, Kuhrer I. [The treatment of multiple myeloma]. Wien Klin Wochenschr 1994; 106:448–454.
145. Salmon S, Crowley J, Tucson A. Impact of glucocorticoids

(GC) and interferon (IFN) on outcome in multiple myeloma. Proc Am Soc Clin Oncol 1992; 11:316–316.
146. Osterborg A, Bjorkholm M, Bjoreman M, et al. Natural interferon-α in combination with melphalan/prednisone versus melphalan/prednisone in the treatment of multiple myeloma stages ii and iii: a randomized study from the myeloma group of central Sweden. Blood 1993; 81:1428–1434.
147. Goldstein D, Laszio J. Interferon therapy in cancer: from imaginon to interferon. Cancer Res 1986; 46:4315–4329.
148. Bajetta E, Zilembo N, Di Bartolomeo M, et al. Treatment of metastatic carcinoids and other neuroendocrine tumors with recombinant interferon-alpha-2a. A study by the Italian Trials in Medical Oncology Group. Cancer 1993; 72:3099–3105.
149. Volberding PA, Mitsuyasu RT, Golando JP, Spiegel RJ. Treatment of kaposi's sarcoma with interferon alfa-2b (intron a). Cancer 1987; 59:620–625.
150. Agarwala SS, Kirkwood JM. Interferons in the therapy of solid tumors. Oncology 1994; 51:129–136.
151. Folkman J, Shing Y. Angiogenesis. J Biol Chem 1992; 267:10931–10934.
152. Strander H. Interferon treatment of human neoplasia. Adv Cancer Res 1986; 46:1–265.
153. Kirkwood JM, Strawderman MH, Ernstoff MS, Smith TJ, Borden EC, Blum RH. Interferon Alfa-2b adjuvant therapy of high-risk resected cutaneous melanoma: the Eastern Cooperative Oncology Group Trial EST 1684. J Clin Oncol 1996; 14:7–17.
154. Schwabe M, Brini AT, Bosco MC, et al. Disruption by interferon-α of an autocrine interleukin-6 growth loop in IL-6-dependent U266 myeloma cells by homologous and heterologous down-regulation of the IL-6 receptor α- and β-chains. J Clin Invest 1994; 94:2317–2325.
155. Diaz MO, Rubin CM, Harden A, et al. Deletions of interferon genes in acute lymphoblastic leukemia. N Engl J Med 1990; 322:77–82.
156. Linge C, Gewert D, Rossmann C, Bishop JA, Crowe JS. Interferon system defects in human malignant melanoma. Cancer Res 1995; 55:4099–4104.
157. Jaffe HS, Sherwin SA. Early clinical trials of recombinant interferon gamma. In: Friedman RM, Sreevalsan T, eds. Interferons as Cell Growth Inhibitors and Antitumour Factors. Alan R. Liss, 1986:509–522.
158. Richtsmeier WJ, Koch WM, McGuire WP, Poole ME, Chang EH. Phase I-II study of advanced head and neck squamous cell carcinoma patients treated with recombinant human interferon gamma. Arch Otolaryngol Head Neck Surg 1990; 116:1271–1277.
159. Brown TD, Goodman PJ, Fleming T, et al. Phase II trial of recombinant DNA gamma-interferon in advanced colorectal cancer: a Southwest Oncology Group study. J Immunother 1991; 10:379–382.
160. Aulitzky WE, Lerche J, Thews A, et al. Low-dose γ-interferon therapy is ineffective in renal cell carcinoma patients with large tumour burden. Eur J Cancer 1994; 30A:940–945.
161. Aulitzky WE, Aulitzky W, Gastle G, et al. Acute effects of single doses of recombinant interferon-γ on blood cell counts and lymphocyte subsets in patients with advanced renal cell cancer. J Interferon Res 1989; 9:425–433.
162. Pujade-Lauraine E, Guastella N, Colombo N, et al. Intraperitoneal recombinant IFN Gamma in ovarian cancer patients with residual disease at second look laparotomy. In: Sharp F, Mason P, Blackett T, Berek J, eds. Ovarian Cancer 3. Cambridge: Chapman and Hall, 1995:231–242.
163. Jett JR, Maksymiuk AW, Su JQ, et al. Phase III trial of recombinant interferon gamma in complete responders with small-cell lung cancer. J Clin Oncol 1994; 12:2321–2326.
164. Ellerhorst JA, Kilbourn RG, Amato RJ, Zukiwski AA, Jones E, Logothetis CJ. Phase II trial of low dose gamma-interferon in metastatic renal cell carcinoma. J Urol 1994; 152:841–845.
165. D'Acquisto R, Markman M, Hakes T, Rubin S, Hoskins W, Lewis JL, Jr. A phase I trial of intraperitoneal recombinant gamma-interferon in advanced ovarian carcinoma. J Clin Oncol 1988; 6:689–695.
166. Klein HO, Golbach G, Voigt P, Coerper C, Bernhardt C. Combination of interferons and cytostatic drugs for treatment of advanced colorectal cancer. J Cancer Res Clin Oncol 1991; 117:S214–220.
167. Meyskens FL, Kopecky K, Samson M, et al. Recombinant human interferon g: adverse effects in high-risk stage i and ii cutaneous malignant melanoma. JNCI 1990; 82:1071.
168. Rosenberg SA, Lotze MT, Muul LM, et al. A progress report on the treatment of 157 patients with advanced cancer using lymphokine-activated killer cells and interleukin-2 or high-dose interleukin-2 alone. N Engl J Med 1987; 316:889–897.
169. Weiss GR, Margolin KA, Aronson FR, et al. A randomized phase ii trial of continuous infusion interleukin-2 or bolus injection interleukin-2 plus lymphokine-activated killer cells for advanced renal cell carcinoma. J Clin Oncol 1992; 10:275–281.
170. Caligiuri MA, Murray C, Robertson MJ, et al. Selective modulation of human natural killer cells in vivo after prolonged infusion of low dose recombinant interleukin 2. J Clin Invest 1993; 91:123–132.
171. Melioli G, Sertoli MR, Bruzzoni M, et al. A phase I study of recombinant interleukin-2 intraperitoneal infusion in patients with neoplastic ascites: toxic effects and immunologic results. Am J Clin Oncol 1991; 14:231–237.
172. Cortesina G, De Stefani A, Giovarelli M, et al. Treatment of recurrent squamous cell carcinoma of the head and neck with low doses of interleukin-2 injected perilymphatically. Cancer 1988; 62:2482–2485.
173. Thatcher N, Taylor P, Carroll KB, Loriaux EM, Palmer P, Franks CR. Interleukin-2 in malignant pleural mesothelioma (and adenocarcinoma of the lung). The use of intrapleural and continuous intravenous infusions: preliminary results. Cancer Treat Rev 1989; 16(suppl A):161–162.
174. Negrier S, Philip TO. Interleukin-2 and lymphokine activated killer (LAK) cells. In: Wagstaff J, ed. The role of interleukin-2 in the treatment of cancer patients. Dordrecht: Kluwer Academic Publishers, 1993:29–44.
175. Rosenberg SA. The development of new immunotherapies for the treatment of cancer using interleukin-2. A review. Ann Surg 1988; 208:121–135.
176. Negrier S, Philip T, Stoter G, et al. Interleukin-2 with or without LAK cells in metastatic renal cell carcinoma: a report of a European multicentre study. Eur J Cancer Clin Oncol 1989; 25:21–28.
177. Vogelzang NJ, Lipton A, Figlin RA. Subcutaneous interleukin-2 plus interferon alfa-2a in metastatic renal cancer: an outpatient multicenter trial. J Clin Oncol 1993; 11:1809–1816.
178. Buzaid AC, Legha SS. Combination of chemotherapy with interleukin-2 and interferon-alfa for the treatment of advanced melanoma. Semin Oncol 1994; 21:23–28.
179. Coley Nauts H, Fowler GA, Bogatko FH. A review of the influence of bacterial infection and of bacterial products (coley's toxins) on malignant tumors in man. Acta Med Scand 1953;
180. Spriggs DR, Sherman ML, Michie H, et al. Recombinant human tumor necrosis factor administered as a 24-hour intra-

venous infusion. A phase i and pharmacologic study. Blood 1988; 80:1039–1044.
181. Lienard D, Ewalenko P, Delmotte J-J, Renard N, Lejeune FJ. High-dose recombinant tumor necrosis factor alpha in combination with interferon gamma and melphalan in isolation perfusion of the limbs for melanoma and sarcoma. J Clin Oncol 1992; 10:52–60.
182. Eggermont AMM, Lienard D, Koops HS, Van Geel AN, Hoekstra AJ, Lejeune FJ. High dose tumor necrosis factor-alpha (tnf), gamma-interferon (ifn) and melphalan in isolation perfusion of the limb in patients with irresectable soft tissue sarcomas. 1992; 3:233.
183. Renard N, Lienard D, Lespagnard L, Eggermont A, Heimann R, Lejeune F. Early endothelium activation and polymorphonuclear cell invasion precede specific necrosis of human melanoma and sarcoma treated by intravascular high-dose tumour necrosis factor alpha (rTNFα). Int J Cancer 1994; 57:656–663.
184. Vaglini M, Belli F, Ammatuna M, et al. Treatment of primary or relapsing limb cancer by isolation perfusion with high-dose alpha-tumor necrosis factor, gamma-interferon, and melphalan. Cancer 1994; 73:483–492.
185. Bronchud MH, Scarffe JH, Thatcher N, et al. Phase i/ii study of recombinant human granulocyte colony-stimulating factor in patients receiving intensive chemotherapy for small cell lung cancer. Br J Cancer 1987; 56:809–813.
186. Crawford J, Ozer H, Stoller R, et al. Reduction by granulocyte colony-stimulating factor of fever and neutropenia induced by chemotherapy in patients with small-cell lung cancer. N Engl J Med 1991; 325:164–170.
187. Pettengell R, Gurney H, Radford JA, et al. Granulocyte colony-stimulating factor to prevent dose-limiting neutropenia in non-hodgkin's lymphoma: a randomized controlled trial. Blood 1992; 80:1430–1436.
188. Ardizzoni A, Venturini M, Crino L, et al. High dose-intensity chemotherapy, with accelerated cyclophosphamide-doxorubicin-etoposide and granulocyte-macrophage colony stimulating factor, in the treatment of small cell lung cancer. Eur J Cancer 1993; 5:687–692.
189. Engelhard M, Brittinger G. Clinical relevance of granulocyte-macrophage colony-stimulating factor. Semin Oncol 1994; 21:1–4.
190. Aglietta M, Placibello W, Pasquino P, et al. Rationale for the use of granulocyte-macrophage colony-stimulating factor in oncology. Semin Oncol 1994; 21:5–9.
191. Woll PJ, Hodgetts J, Lomax L, Bildet F, Cour Chabernaud V, Thatcher N. Can cytotoxic dose-intensity be increased by using granulocyte colony-stimulating factor? A randomized controlled trial of lenograstim in small-cell lung cancer. J Clin Oncol 1995; 13:652–659.
192. Miles DW, Fogarty O, Ash CM, et al. Received dose-intensity: a randomized trial of weekly chemotherapy with and without granulocyte colony-stimulating factor in small-cell lung cancer. J Clin Oncol 1994; 12:77–82.
193. Schmidt Wolf G, Schmidt Wolf IG. Cytokines and clinical gene therapy. Eur J Immunol 1995; 25:1137–1140.

37

Cytokines in Septic Shock

Anders Waage and Sigurd Steinshamn
University of Trondheim, Trondheim, Norway

INTRODUCTION

It is established that bacteria evoke a host response, including cytokines and other molecules, which mediate the cellular effects that eventually lead to the characteristic manifestations of septic shock. Since 1985 the role of cytokines in septic shock has been thoroughly investigated and the list of cytokines that act as mediator or somehow contribute to the pathogenesis has instantly increased. At present, TNF, IL-1, IL-6, IL-8, IL-10, IFN-γ, the two types of soluble TNF receptors (sTNFR), and IL-1 receptor antagonist (IL-1RA) comprise this list, in addition to other molecules, which are not included in the cytokine group. Our knowledge about the involvement of cytokines in septic shock is based on a multitude of studies, which can be divided into the following categories: (1) observation of pathophysiological and histopathological changes after injection of cytokines into experimental animals, (2) observation of the effects of specific cytokine inhibitors in experimental animals injected with endotoxin or bacteria, (3) pharmacokinetic studies in experimental animals and humans, (4) detection of cytokines in serum from septic shock patients, and (5) clinical trials with specific cytokine inhibitors. In general, two main goals guide most sepsis research. The first broad aim is to obtain as much precise information as possible about the mediators acting in septic shock in order to understand its pathogenesis; the second aim (which is dependent on the first) is to improve the treatment of septic shock based on the principle of anti-cytokine treatment.

TERMINOLOGY OF SEPSIS

Septic infections and related syndromes are heterogeneous with respect to etiology and clinical manifestations, and the terminology is confusing. The terms sepsis, septicemia, bacteremia, septic syndrome, and septic shock are often used interchangeably and without strict definition. A consensus conference held in 1991 defined some of these terms, which later were adopted by many studies (1). The main distinctions made were:

Systemic inflammatory response syndrome (SIRS) = the systemic inflammatory response to a variety of severe clinical insults
Sepsis = SIRS in response to infection
Severe sepsis = sepsis associated with organ dysfunction, hypoperfusion, or hypotension
Septic shock = sepsis induced with hypotension despite adequate fluid resuscitation
Multiple organ dysfunction syndrome (MODS) = presence of altered organ function in an acutely ill patient

In this chapter we will adhere to this terminology, however, many of the studies cited are based on other definitions, and too strict comparisons between studies are therefore not justified.

Figure 1 (1) demonstrates the underlying complexity by showing the overlapping pattern of these terms and how they are related to various infectious and noninfectious etiologies. Although it is clear that many etiologies have a final common pathway to the clinical manifestations, we

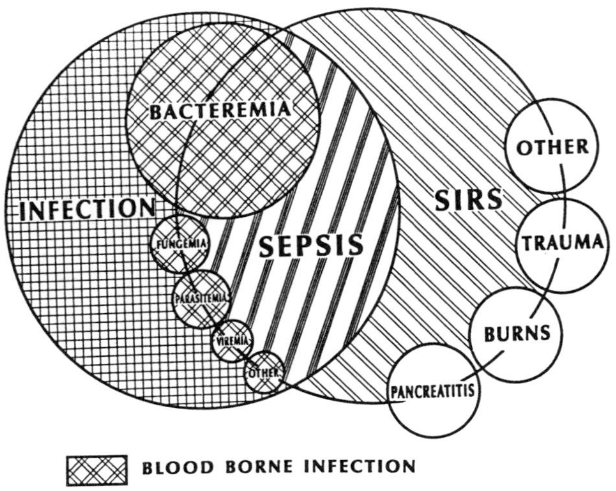

Figure 1 Etiology and terminology of sepsis and related syndromes. (From Ref. 1.)

do not know much about the variation in the involvement of cytokines in the various clinical settings.

ANIMAL STUDIES

Detection of Cytokines

One widely used model involves the injection of endotoxin or dead or live bacteria into experimental animals followed by blood sampling and examination for the presence of cytokines. In the various species examined there is a surprisingly uniform response pattern after bolus injection of endotoxin. TNF is the first cytokine to be released with a peak concentration 1–1½ h after injection of endotoxin. Within 2–3 h, the peak concentration of IL-1, IL-6, and IL-8 is attained (2,3). If bacteria are infused over a period of time (½–8 h) in baboons, the presence of IL-1, IL-6, and IL-8 in the circulation is stretched out to more than 8 h (4–6). Also, IFN-γ is released into the circulation later than TNF (7). It has been repeatedly demonstrated that IL-1, IL-6, and IL-8 are induced directly by endotoxin as well as indirectly by TNF, as antibodies to TNF attenuate the peak concentrations of these cytokines after injection of endotoxin (2,4,5)

Also, the kinetics of the inhibitory factors sTNFR p55 and sTNFR p75 and IL-1RA have been studied and are relevant to septic shock. In mice, endotoxin induces sTNFR p55 with a peak level after 30 min, and thereafter a gradual decrease, and sTNFR p75 with a peak after 4–8 h, after which both slowly diminish (8). Furthermore, TNF itself is also an inducer of release of both sTNFRs. Experiments in nephrectomized mice indicate an important role for sTNFR p75 in the inactivation of TNF and clearance of TNF-sTNFR complexes by the kidney (9).

In conclusion, animal experiments demonstrate that TNF is released 1–1½ h after endotoxin or bacteria and that the other cytokines are released a few hours later. Counterregulatory molecules like the sTNFRs and IL-1RA are released early, but later than the initial cytokine release.

Injection of Cytokines and Anticytokine Treatment

To assess the pathogenetic effect of cytokines in animal models, a common design has been to inject the factor into the animal and observe the histopathological organ changes and pathophysiological and circulatory consequences. Furthermore, the effect of specific cytokine inhibitory treatment has provided important information. Convincing evidence for TNF and IL-1 as mediators has been obtained as both cytokines induce changes that are very similar to those induced by endotoxin or live gram-negative bacteria (10–15). These changes include hemorrhagic and ischemic necrosis of the gastrointestinal tract, hemorrhages and leukocyte infiltration in the lungs, acute tubular necrosis of the kidneys, metabolic acidosis, hypermetabolism, stress hormone elaboration, hypotension, circulatory collapse, and death. When these changes are induced by live gram-negative bacteria shown to evoke the cytokine cascade, they can be completely inhibited by specific inhibition of TNF or IL-1 (16–18). The cytokine cascade is not triggered by endotoxin alone, as corresponding observations have been made with gram-positive bacteria (19). A similar effect on lethality has also been demonstrated with antibody to IFN-γ (20–22). High concentrations of IL-6 can be detected both in animals and in humans with shock, and, furthermore, antibodies to IL-6 have in some studies been shown to protect against lethal effects of *E. coli* and TNF (23) although this article was later retracted. When injected, IL-6 is rather nontoxic, and a definitive pathogenetic role for this molecule in sepsis has not yet been established. Injection of IL-8 induces leukocytosis and antibodies to this molecule do not interfere with lethality (24). Interleukin-10 has been shown to block the in vitro production of TNF as well as of other cytokines by LPS-activated monocytes and macrophages. In vivo, a corresponding effect was demonstrated as IL-10 administration before endotoxin challenge in mice reduced the amounts of TNF release in the circulation (25). It also efficiently prevented the hypothermia generated by endotoxin and completely prevented the mortality from a dose that was lethal in 50% of control mice. It has also been found that IL-10 protects mice against lethal shock induced by staphylococcal enterotoxin (26). Combinations of cytokines have been studied in animal experiments, and both IL-1 and IFN-γ have a strong synergistic effect in combination with TNF (13,27,28).

Altogether, TNF, IL-1, and IFN-γ (although not studied to the same extent) have been identified as important mediator molecules in the development of septic shock (Table 1). IL-10, sTNFRs and IL-1RA appear to have a counterregulatory effect. IL-6 and IL-8 probably contribute to manifestations like the acute phase response and leukocytosis, but are not involved in the essential pathogenetic mechanisms of septic shock.

Other Animal Models

Protective Effect of TNF and IL-1

Other animal models like cecal ligation and puncture (CLP) include ligation of the cecum below the ileal-cecal valve, which prevents obstruction of the intestines. The cecum is then punctured and the abdomen closed. Peritonitis develops and blood cultures become positive within a few hours. In this model, the kinetics of cytokine production is markedly different as TNF and IL-6 both peak 12 h after CLP (29–31). However, more important is the fact that anti-TNF increases lethality, indicating that TNF has a protective function in this model. This is in sharp contrast to its deleterious role in the previously described model.

Most studies have focused on gram-negative sepsis. However, a number of pathogens can induce septic shock, and it is of interest to examine the significance of the cytokines induced by different pathogens. *Candida albicans* induces weak release of bioactive TNF, which peaks after about 25 h and persists in the circulation for up to 40 h (32). In infections with *Candida* and other intracellular pathogens like *Listeria monocytogenes,* TNF and IL-1 protect against proliferation of the pathogen and reduce lethality in mice (32–35).

Of particular interest in relation to sepsis is the protective effect of endotoxin, TNF, and IL-1 demonstrated in galactosamine-treated or normal mice. Pretreatment with nonlethal doses of endotoxin, TNF, or IL-1 protects against lethal effect of a subsequent challenge (36–38). The basic mechanism of the protective effect is not known, although downregulation of receptors and induction of proteases or other soluble factors are suspected. Pretreatment with IL-6 and IL-8 have no protective effect (37).

In conclusion, the pathogenetic significance of the cytokines is strictly dependent on the model used, the actual pathogen, and the timing of the release of cytokines. Such experiments create a background for understanding the variation of the cytokine cascade in clinical sepsis.

Granulocytopenic Mice

Granulocytopenic mice are sources of information about the cellular sources of the cytokine cascade and the cellular interactions involved in cytokine effects. Furthermore, they are a realistic model for chemotherapy-induced granulocytopenic infections in humans. Somewhat surprisingly, levels of bioactive TNF were increased upon administration of LPS in mice with chemotherapy-induced granulocytopenia (39). One would expect the opposite effect to be the most likely, as monocytes—also being substantially reduced by cytostatic treatment—are a major source of TNF in vitro and experiments with mice deficient in monocytes/macrophages due to a genetic defect show attenuated response of bioactive TNF after endotoxin administration (40). However, the enhanced TNF response in the granulocytopenic situation is probably due to a reduction in sTNFR-p75 available for inactivation of TNF (39). TNF and IL-6 are also increased during granulocytopenia in mice infected with *C. albicans* (32), possibly through a parallel mechanism. The kinetics of release and

Table 1 Cytokine Effects in the Pathogenesis of Septic Shock

	Mediator[a]	Inhibitory effect	Other effects
TNF	+	–	
IL-1	+	–	Synergism with TNF
IFN-γ	+	–	Synergism with TNF
IL-6	+/–[b]	–	Acute phase response
IL-8	–	–	Leukocytosis
IL-10	–	+	
sTNFR p55	–	+	
sTNFR p75	–	–	Carrier for TNF
IL-1RA	–	+	

[a]Based on observations of circulatory collapse and death in experimental animals.
[b]Contradictory results have been published.

elimination of TNF and IL-6 are similar in the granulocytopenic and normal situations.

In the granulocytopenic situation, TNF has been shown to be a mediator of detrimental effects of infection, as an antibody to TNF protected rats from lethal infection with *Pseudomonas aeruginosa* (41). Furthermore, the protective effects of IL-1 in lethal bacterial infections have been reproduced in granulocytopenic mice (42). Despite striking similarities, some TNF effects are different in the normal and granulocytopenic situations. In rats and guinea pigs injected with TNF, granulocytopenia protects against lethality and multiorgan failure, indicating that the lethal effect of TNF in such models is mediated partially via granulocytes (43,44). In murine *C. albicans* infection, TNF has a protective role in the presence of granulocytes, but not in the granulocytopenic situation (32). The antifungal effect of TNF seems to be granulocyte-dependent.

In conclusion, granulocytopenia increases levels of TNF and IL-6 bioactivity upon injection of endotoxin and after infection with *C. albicans.* Data are supportive of a role for TNF as a mediator of septic infections during granulocytopenia. However, TNF effects may also be different from the normal situation, for instance, when effects are mediated by TNF-activated granulocytes.

Transgenic Mice

Another model of importance in determining the exact role of cytokines and their receptors is transgenic mice, which lack or overproduce a specific cytokine or receptor. Of particular interest are experiments conducted with mice lacking the p55 TNFR on the cell membrane and the corresponding soluble form (45,46). These used a model with D-galactosamine–treated mice dramatically sensitized to the lethal effect of LPS. There is agreement that death in LPS/D-galactosamine–challenged mice is due to TNF toxicity. Mice lacking p55 TNFR were completely insensitive to the lethal effect of LPS even at more than 100 times the lethal dose for wild-type littermates. The results were also similar if the mice were pretreated with IL-1, which has been shown to potentiate the lethal effect of TNF. The experiments demonstrate that TNF signaling through the p55 TNFR is necessary for the lethal pathology induced indirectly by LPS in galactosamine-sensitized mice or directly with TNF in IL-1–pretreated mice. The model was also used to test the pathway of the protective effect of TNF in infections with *Listeria monocytogenes*. It was shown that mice deficient in p55 TNFR had enhanced susceptibility to infection, with the intracellular bacterium demonstrating that p55 receptor is essential in mediating the protective effect of TNF. Corresponding experiments with p75 TNFR-deficient mice demonstrate that this receptor is also involved in TNF toxicity, but not to the same extent as p55 TNFR (47).

HUMAN MODELS OF SEPTIC SHOCK

A number of studies have investigated the kinetics of production and elimination of cytokines after injection of endotoxin in human volunteers. Briefly, the results are very similar to those in animals with peak levels of TNF, IL-1, IL-6, and IL-8 within a few hours and then a decline to undetectable levels (48–50). The two types of TNFRs have also been studied in the same way (51,52). The appearance of both receptors in the circulation was found to be monophasic, with peak levels occurring after that of TNF and detectable levels persisting several hours beyond the period in which circulating TNF was detectable. The IL-1RA peaks in the circulation about 3 h after injection of endotoxin, which is later than the peak of IL-1 (53,54). The concentration of IL-1RA is 100-fold higher than that of IL-1 on a molar basis.

The consequences of injection of TNF for a number of biological systems relevant to sepsis have been investigated in six human volunteers. TNF induced a 40-fold increase in IL-6 concentrations, whereas IL-1 remained undetectable. Peripheral blood leukocytes showed a biphasic change, which largely reflected a change in the number of circulating neutrophils (55). Initial profound decrease was followed by neutrophilia. Moreover, there were signs of neutrophil activation. TNF caused prolonged activation of the coagulation system as estimated by concentrations of the activation peptide of factor X and prothrombin fragment F1 + F2 (56). The effect on the fibrinolytic system was more complex, involving early activation and a following rapid inhibition (57).

Altogether, the results from experiments in human and animals on the kinetics of the cytokine cascade are in very good accordance. TNF has effects on neutrophil counts, the coagulation system, and the fibrinolytic system, which are relevant to sepsis.

DETECTION OF CYTOKINES IN HUMAN SEPSIS

There exist at least 30 reports on the cytokine levels in septic shock or related syndromes. Compared with the relatively uniform response pattern found in animals, the studies in human disease demonstrate much greater variation and the results are often difficult to interpret. There are several reasons for this variation, including the heterogeneity in etiologies and manifestations, which is obvious from Figure 1. In addition, it must be borne in mind that different assays have been used to quantify the cytokines, and these often produce different results. Of particular significance is the distinction between bioactive and immunoreactive TNF. Evidently many of the commercial assays for TNF which are based on immunological principles detect TNF degradation products or complexes that are not biologically active (58).

Meningococcal Disease

Meningococcal disease, caused by *Neisseria meningitidis,* is a worldwide endemic disease with the main manifestations of septic shock and meningitis. It comprises a relatively small proportion of the whole population of septic shock patients, however, the rationale to describe it as a separate entity is that it can be regarded as a model disease for septic shock. It strikes young, previously healthy individuals and has a sudden onset with very high levels of endotoxin, and the disease thus has similarities with the experimental situation of bolus injection of endotoxin. We have examined the cytokine response pattern including TNF, IL-1, IL-6, IL-8, and IL-10 in this disease (58–61). The serum levels of the five cytokines uniformly demonstrate several characteristics:

1. The serum levels are significantly higher in patients with septic shock than in other categories of meningococcal patients and in normal controls.
2. Fatal cases have with few exceptions very high concentrations in serum. There are significant associations between fatality and high levels of TNF, IL-1, IL-6, IL-8, and IL-10.
3. After the onset of shock, concentrations in the circulation gradually decline.

Other Categories of Septic Shock

The main group of septic shock patients have underlying conditions like surgery, polytrauma, burns, and abscessa or persisting infections. Based on the cited studies (59–75), it can be stated that:

1. TNF is elevated in serum in septic shock, and most, but not all, studies show association between severity of the shock and fatality. There is a discrepancy between immunoreactive and bioactive TNF, which rarely can be detected.
2. IL-1 is elevated in serum in most studies.
3. IL-6 is found to be elevated in all studies, and a high peak level or persisting elevation is better than the other cytokines as predictor of adverse outcome.
4. IL-8 and IL-10 are elevated during septic shock.
5. In longitudinal studies the concentration of cytokines fluctuates during sepsis. The highest level is not necessarily found at the onset of shock.

We conclude that cytokines are released in all categories of septic shock. However, the pattern of their presence is heterogeneous and not easily predictable from animal experiments. Meningococcal septic shock may be an exception, with a cytokine response similar to that observed in experimental studies.

PRINCIPLES OF ANTICYTOKINE TREATMENT

Inhibition of specific cytokines has been an extremely valuable tool for clarifying the pathogenetic significance of cytokines in animal models, and this approach has obvious potential in the treatment of clinical septic shock. We will therefore go into more detail about the principles for specific anticytokine treatments, which include antibodies to TNF, sTNFRs, fusion proteins with sTNFRs, and IL-1RAs (Table 2). Different types of antibodies have been used. Murine antibodies to TNF have been used in large clinical trials because of the unavailability of human tissue and difficulties of working with hybridomas derived from human cells. The drawback is that mouse antibodies may cause an immune response in humans, and to avoid this, variants of chimeric antibodies have been developed (designer antibodies, humanized murine antibodies). Another nonimmunogenic principle is to use naturally occurring inhibitors like the sTNFRs and the IL-1RA. One variant of sTNFR is a fusion molecule consisting of an IgG heavy chain and two molecules of the extracellular portion of either p55 or p75 TNFR. The advantage of using a dimeric fusion protein of TNFR and an IgG heavy chain is its greater affinity to TNF and its longer half-life in the circulation (76,77). With the sTNFR-p55 fusion protein a significant effect on survival in mice has been demonstrated, whereas the effect of sTNFR-p75 is incomplete (78). The difference in effects may be caused by a marked difference in dissociation rate between the complex and TNF (77,79,80). Also, for IL-1RA an effect on lethality from bacteria-induced shock has been clearly demonstrated (18,81). In addition to specific cytokine inhibition,

Table 2 Specific Cytokine Inhibition Used in Clinical Trials

Murine antibodies to TNF
Humanized murine antibodies to TNF
Dimeric sTNFR-IgG heavy-chain fusion protein
IL-1RA

there exists a number of drugs with a general anticytokine effect like glucocorticoids and pentoxiphylline, but these will not be discussed in this chapter.

ANTICYTOKINE TREATMENT OF HUMAN SEPTIC SHOCK

There is a strong scientific rationale for antimediator treatment of septic shock and related syndromes. Endotoxin, TNF, and IL-1 have been identified as proximal key mediators in cascade reactions and have been the most attractive targets for antimediator treatment. Endotoxin is outside the scope of this chapter and will not be discussed.

Anti-TNF treatment phase II studies have been performed with a murine TNF antibody (Celltech). There was no increase in 28-day survival, although a trend toward improved outcome was observed with the highest of four doses used (82). In another phase II study (Immunex) in 141 patients, no or even an adverse effect of IgG-TNFR was observed (83). This study used the p75 TNFR. Animal studies later showed a marginal effect for this receptor, whereas the p55 TNFR significantly reduces mortality (78). This may explain the somewhat worrisome finding of the clinical trial. Hoffmann-LaRoche has taken advantage of this and has launched a clinical trial with a fusion protein, including the p55 TNFR.

Two major phase III clinical trials conducted by Bayer-Miles with murine IgG antibody have been completed. The Norasept trial was conducted in the United States and Canada (84–86). Of 994 patients enrolled, 28-day mortality rates for shock patients were (1) placebo, 45.6%; (2) anti-TNF antibody 7.5 mg/kg, 37.7%; (3) anti-TNF antibody 15 mg/kg, 37.7%. The corresponding figures for all patients with or without shock were 33.1, 29.5, and 31.3%. The differences found were not significant, however, for shock patients there was a trend, and the study will be followed up with a new phase III trial including only shock patients.

The Intercept study was conducted in Europe and South Africa using the same antibody in doses of 3 and 15 mg/kg versus placebo (85,86). For the shock patients (n = 420) the overall 28-day mortality was (1) placebo, 42.9%; (2) 3 mg/kg, 36.7%; and (3) 15 mg/kg, 44.6%. The differences were not significant, and the correct interpretation is not obvious.

Phase II studies with IL-1RA indicated a dose-dependent effect of IL-1RA therapy in sepsis (87). A phase III study (Synergen) has been completed (88). The study evaluated 893 patients with sepsis or septic shock who were treated with IL-1RA 1 or 2 mg/kg versus placebo. The 28-day overall mortality was (1) placebo, 34%; (2) 1 mg/kg, 31%; and 2 mg/kg, 29%. For patients with shock, the corresponding figures were 36 and 31%. There was no significant effect of the treatment, however, a retrospective analysis could identify a subgroup of patients with sepsis and high risk with a reduction in mortality.

Altogether, the results of the completed phase III studies aiming to inhibit TNF or IL-1 (Table 3) are disappointing considering the expectations raised by animal studies. However, there may exist subgroups who will benefit from such treatment.

FUTURE DIRECTIONS

After years of investigations of antimediator treatment, the following paradox has become obvious. Antimediator treatment aiming to inhibit TNF and IL-1 as well as other principles like antiendotoxin and glucocorticoids have produced excellent results in animal experiments, including reversal of shock and reduction of mortality. However, when the same treatment is tested in phase III clinical studies, the effects are marginal if present at all.

What are the reasons for this discrepancy? One expla-

Table 3 Phase III Clinical Trials with Specific Inhibition of TNF and IL-1

			Mortality (%)	
	n	Treatment	Overall	Shock patients
Norasept	994	Murine anti-TNF	30 vs. 33[a]	38 vs. 46
Intersept	420	Murine anti-TNF		37 vs. 43
Synergen	893	IL-1RA	29 vs. 34	31 vs. 36

[a]Results are given for treatment group versus placebo group. When several doses were used, the best results are shown.

nation is the marked heterogeneity of the patients with septic shock and related syndromes. The clinical studies have included patients with a defined clinical syndrome irrespective of etiology and preexisting disease. From experiments we know that the significance of the cytokines may vary although the final clinical manifestation is septic shock. Examples mentioned in the present chapter are the opposite roles of TNF in fungal infection, listeriosis, and abdominal sepsis versus gram-negative sepsis. It is consequently a possibility that patients with adverse effects of anticytokine treatment have been included in the studies and may have hidden a positive effect for other subgroups. A major task for further clinical research is therefore to define criteria to identify subgroups that are likely to benefit from antimediator treatment. Another corollary is that the common animal model for septic shock on which the design of clinical trials has been based is too simplistic and is valid only for a small subgroup. Future research must develop new and more complex animal models to cover the spectrum of clinical situations.

REFERENCES

1. Bone RC, Balk RA, Cerra FB, et al. Definitions for sepsis and organ failure and guidelines for the use of innovative therapies in sepsis. Chest 1992; 101:1644–1655.
2. Shalaby MR, Waage A, Aarden L, et al. Endotoxin, tumor necrosis factor, and interleukin 1 induce interleukin 6 production in vivo. Clin Immunol Immunopathol 1989; 53:488.
3. van Zee KJ, DeForge LE, Fischer E, et al. IL-8 in septic shock, endotoxemia, and after IL-1 administration. J Immunol 1991; 146:3478–3482.
4. Fong Y, Tracey KJ, Moldawer LL, et al. Antibodies to cachectin/tumor necrosis factor reduce interleukin 1β and interleukin 6 appearance during lethal bacteremia. J Exp Med 1989; 170:1627–1633.
5. Redl H, Schlag G, Bahrami S, et al. The cytokine network in trauma and sepsis: TNF and IL-8. In: Schlag G, Redl H, eds. Pathophysiology of Shock, Sepsis and Organ Failure. Berlin: Springer, 1993:468–490.
6. Redl H, Schlag G, Bahrami S, et al. Plasma neutrophil-activating peptide-/interleukin-8 and neutrophil elastase in a primate bacteremia model. J Infect Dis 1991; 164:383–388.
7. Kiener PA, Marek, F, Rodgers G, et al. Induction of tumor necrosis factor, IFN-gamma, and acute lethality in mice by toxic and non-toxic forms of lipid A. J Immunol 1988; 141:870–874.
8. Bemelmans MHA, Gouma DJ, Buurman WA. LPS induced sTNF-receptor release in vivo in a murine model: investigation of the role of TNF, IL-1, LIF and IFN-gamma. J Immunol 1993; 151:5554–5562.
9. Bemelmans MHA, Gouma DJ, Buurman WA. Influence of nephrectomy on tumor necrosis factor clearance in a murine model. J Immunol 1993; 150:2007–2017.
10. Tracey KJ, Beutler B, Lowry SF, et al. Shock and tissue injury induced by recombinant human cachectin. Science 1986; 234:470–474.
11. Remick DG, Kunkel RG, Larrick JW, et al. Acute in vivo effects of human recombinant tumor necrosis factor. Lab Invest 1987; 56:583–590.
12. Hinshaw LB, Tekamp-Olson P, Chang ACK, et al. Survival of primates in LD_{100} septic shock following therapy with antibody to tumor necrosis factor (TNFα). Circ Shock 1990; 30:279–292.
13. Okusawa S, Gelfand JA, Ikejima T, et al. Interleukin 1 induces a shock-like state in rabbits. Synergism with tumor necrosis factor and the effect of cyclooxygenase inhibition. J Clin Invest 1988; 81:1162–1172.
14. Weinberg JR, Wright DJM, Guz A. Interleukin-1 and tumor necrosis factor cause hypotension in the conscious rabbit. Clin Sci 1988; 75:251–255.
15. Butler LD, Layman NK, Cain RL, et al. Interleukin 1-induced pathophysiology: induction of cytokines, development of histopathologic changes, and immunopharmacologic intervention. Clin Immunol Immunopathol 1989; 53:400–421.
16. Beutler B, Milsark IW, Cerami AC. Passive immunization against cachectin/tumor necrosis factor protects mice from lethal effect of endotoxin. Science 1985; 229:869–871.
17. Tracey KJ, Fong Y, Hesse DG, et al. Anti-cachectin/TNF monoclonal antibodies prevent septic shock during lethal bacteraemia. Nature 1987; 330:662–664.
18. Ohlsson K, Björk P, Bergenfeldt M, et al. Interleukin-1 receptor antagonist reduces mortlaity from endotoxin shock. Nature 1990; 348:550–552.
19. Hinshaw LB, Emerson TE, Taylor FB, et al. Lethal staphylococcus aureus-induced shock in primates: prevention of death with anti-TNF antibody. J Trauma 1992; 33:568–573.
20. Silva AT, Cohen J. Role of interferon-γ in experimental gram-negative sepsis. J Infect Dis 1992; 166:331–335.
21. Doherty GM, Lange JR, Langstein HN, et al. Evidence for IFN-γ as a mediator of the lethality of endotoxin and tumor necrosis factor-alpha. J Immunol 1992; 149:1666–1670.
22. Heinzel FP. The role of IFN-γ in the pathology of experimental endotoxemia. J Immunol 1990; 145:2920–2924.
23. Starnes HF, Parce MK, Tewari A, et al. Anti-IL-6 monoclonal antibodies protect against lethal Escherichia coli infection and lethal tumor necrosis factor-alpha challenge in mice. J Immunol 1990; 145:4185–4191.
24. van Zee KJ, Fischer E, Hawes AS, et al. Effects of intravenous IL-8 administration in nonhuman primates. J Immunol 1992; 148:1746–1752.
25. Gérard C, Bruyns C, Marchant A, et al. Interleukin 10 reduces the release of tumor necrosis factor and prevents lethality in experimental endotoxemia. J Exp Med 1993; 177:547–550.
26. Bean AGD, Freiberg RA, Andrade S, et al. Interleukin 10 protects mice against staphylococcal enterotoxin B-induced lethal shock. Infect Immun 1993; 61:4937–4939.
27. Waage A, Espevik T. Interleukin 1 potentiates the lethal effect of tumor necrosis factor alpha/cachectin in mice. J Exp Med 1988; 167:1987–1992.
28. Talmadge JE, Bowersox O, Tribble H, et al. Toxicity of tumor necrosis factor is synergistic with gamma-interferon and can be reduced with cyclooxygenase inhibitors. Am J Pathol 1987; 128:410–425.
29. Echtenacher B, Falk W, Männel DN, et al. Requirement of endogenous tumor necrosis factor/cachectin for recovery from experimental peritonitis. J Immunol 1990; 145:3762–3766.
30. Bagby GJ, Plessala KJ, Wilson LA, et al. Divergent efficacy of antibody to tumor necrosis factor-α in intravascular and peritonitis models of sepsis. J Infect Dis 1991; 163:83–88.
31. Eskandari MK, Bolgos G, Miller C, et al. Anti-tumor necrosis factor antibody therapy fails to prevent lethality after cecal ligation and puncture of endotoxemia. J Immunol 1992; 48:2724–2730.
32. Steinshamn S, Waage A. Tumor necrosis factor and interleukin-

6 in candida albicans infection in normal and granulocytopenic mice. Infect Immun 1992; 60:4003–4008.
33. Allendoerfer R, Magee DM, Smith JG, et al. Induction of tumor necrosis factor alpha in murine Candida albicans infection. J Infect Dis 1993; 167:1168–1172.
34. Nakane A, Minagawa T, Kato K. Endogenous tumor necrosis factor (cachectin) is essential to host resistance against Listeria monocytogenes infection. Infect Immun 1988;56:2563–2569.
35. Czuprynski CJ, Brown JF. Recombinant interleukin-1 alpha enhancement of nonspecific antibacterial resistance. Infect Immun 1987; 55:2061–2065.
36. Cross AS, Sadoff JC, Kelly N, et al. Pretreatment with recombinant murine tumor necrosis factor alpha/cachectin and murine interleukin 1 alpha protects mice from lethal bacterial infection. J Exp Med 1989; 169:2021–2027.
37. Shalaby R, Halgunset J, Haugen OA, et al. Cytokine associated tissue injury and lethality in mice: a comparative study. Clin Immunol Immunopathol 1991; 69:69–82.
38. Wallach D, Holtmann H, Engelmann H, et al. Sensitization and desensitization to lethal effects of tumor necrosis factor and IL-1. J Immunol 1988; 140:2994–2999.
39. Steinshamn S, Bemelmans MHA, Buurman WA, et al. Granulocytopenia reduces release of soluble TNF receptor p75 in-endotoxin-stimulted mice: a possible mechanism of enhanced TNF activity. Cytokine 1995; 7:50–56.
40. Wiktor-Jedrezejczak W, Ansari A, Szperl M, et al. Distinct in vivo functions of two macrophage subpopulations as evidenced by studies using macrophage deficient op/op mice. Eur J Immunol 1992; 22:1951–1954.
41. Opal SM, Cross AS, Kelly NM, et al. Efficacy of a monoclonal antibody directed against tumor necrosis factor in protecting rats from lethal infection with Pseudomonas aeruginosa. J Infect Dis 1990; 161:1148–1152.
42. Van der Meer JWM, Barza M, Wolff SM, et al. A low dose of recombinant interleukin 1 protects granulocytopenic mice from lethal gram-negative infection. Proc Natl Acad Sci 1988; 85:1620–1623.
43. Goto F, Watanabe E, Maruyama N, et al. Prevention of the toxic action of tumor necrosis factor by cyclooxygenase inhibitor and leukopenia. Circ shock 1989; 29:175–180.
44. Mallick AA, Ishizaka A, Stephens KE, et al. Multiple organ damage caused by tumor necrosis factor and prevented by prior neutrophil depletion. Chest 1989; 95:1114–1120.
45. Rothe J, Lesslauer W, Lötscher H, et al. Mice lacking the tumour necrosis factor receptor 1 are resistant to TNF-mediated toxicity but highly susceptible to infection by Listeria monocytogenes. Nature 1993; 364:798–802.
46. Pfeffer K, Matsuyama T, Kündig TM, et al. Mice deficient for the 55 kd tumor necrosis factor receptor are resistant to endotoxic shock, yet succumb to L. monocytogenes infection. Cell 1993; 73:457–467.
47. Erickson SL, de Sauvage FJ, Kikly K, et al. Decreased sensitivity to tumor-necrosis factor but normal T-cell development in TNF receptor-2-deficient mice. Nature 1994; 372:560–563.
48. Michie HR, Manogue KR, Spriggs DR, et al. Detection of circulating tumor necrosis factor after endotoxin administration. N Engl J Med 1988; 318:1481–1486.
49. van Deventer SJH, Büller HR, ten Cate JW, et al. Experimental endotoxemia in humans: analysis of cytokine release and coagulation, fibrinolytic, and complement pathways. Blood 1990; 76:2520–2526.
50. Martich GD, Danner RL, Ceska M, et al. Detection of interleukin 8 and tumor necrosis factor in normal humans after intravenous endotoxin: the effect of antiinflammatory agents. J Exp Med 1991; 173:1021–1024.
51. van Zee KJ, Kohno T, Fischer E, et al. Tumor necrosis factor soluble receptors circulate during experimental and clinical inflammation and can protect against excessive tumor necrosis factor α in vitro and in vivo. Proc Natl Acad Sci 1992; 89:4845–4849.
52. Spinas GA, Keller U, Brockhaus M. Release of soluble receptors for tumor necrosis factor (TNF) in relation to circulating TNF during experimental endotoxinemia. J Clin Invest 1992; 90:533–536.
53. Granowitz EV, Santos AA, Poutsiaka DD, et al. Production of interleukin-1-receptor antagonist during experimental endotoxaemia. Lancet 1991; 338:1423–1424.
54. Fischer E, van Zee KJ, Marano MA, et al. Interleukin-1 receptor antagonist circulates in experimental inflammation and in human disease. Blood 1992; 79:2196–2200.
55. van der Poll T, van Deventer SJH, Hack CE, et al. Effects on leukocytes following injection of tumor necrosis factor into healthy humans. Blood 1992; 79:693–698.
56. van der Poll T, Büller HR, ten Cate H, et al. Activation of coagulation after administration of tumor necrosis factor to normal subjects. N Engl J Med 1990; 322:1622–1627.
57. van der Poll T, Levi M, Büller HR, et al. Fibrinolytic response to tumor necrosis factor in healthy subjects. J Exp Med 1991; 174:729–732.
58. Engelberts I, Stephens S, Francot GJM, et al. Evidence for different effects of soluble TNF-receptors on various TNF measurements in human biological fluids. Lancet 1991; 338:515–516.
59. Waage A, Halstensen A, Espevik T. Association between tumour necrosis factor in serum and fatal outcome in patients with meningococcal disease. Lancet 1987; i:355–357.
60. Waage A, Brandtzæg P, Halstensen A, et al. The complex pattern of cytokines in serum from patients with menigococcal septic shock. Association between interleukin 6, interleukin 1, and fatal outcome. J Exp Med 1989; 169:333–338.
61. Halstensen A, Ceska M, Brandtzæg P, et al. Interleukin-8 in serum and cerebrospinal fluid from patients with meningococcal disease. J Infect Dis 1993; 167:471–475.
62. Lehmann AK, Halstensen A, Sørnes et al. High levels of interleukin 10 in serum is associated with case fatality in meningococcal disease. Infect Immun (in press).
63. Debets JM, Kampmeijer R, van der Linden MP, et al. Plasma tumor necrosis factor and mortality in critically ill septic patients. Crit Care Med 1989; 17:489–494.
64. Damas P, Reuter A, Gysen P, et al. Tumor necrosis factor and interleukin-1 serum levels during severe sepsis in humans. Crit Care Med 1989; 17:975–978.
65. Marks JD, Marks CB, Luce JM, et al. Plasma tumor necrosis factor in patients with septic shock. Mortality rate, incidence of adult respiratory distress syndrome, and effects of methyl-prednisolone administration. Am Rev Resp Dis 1990; 141:94–97.
66. Calandra T, Baumgartner JD, Grau GE, et al. Prognostic values of tumor necrosis factor/cachectin, interleukin-1, interferon-alpha, and interferon-gamma in the serum of patients with septic shock. Swiss-Dutch J5 Immunoglobulin Study Group. J Infect Dis 1990; 161:982–987.
67. Cannon JG, Tompkins RG, Gelfand JA, et al. Circulating interleukin-1 and tumor necrosis factor in septic shock and experimental endotoxin fever. J Infect Dis 1990; 161:79–94.
68. Damas P, Ledoux D, Nys M, et al. Cytokine serum level during severe sepsis in human: IL-6 as a marker of severity. Ann Surg 1992; 215:356–362.
69. Hack CE, De Groot ER, Felt-Bersma RJF, et al. Increased plasma levels of interleukin-6 in sepsis. Blood 1989; 74:1704–1710.

70. Endo S, Inada K, Inoue Y, et al. Two types of septic shock classified by the plasma levels of cytokines and endotoxin. Circ Shock 1992; 38:264–274.
71. Hack CE, Hart M, Strack RJM, et al. Interleukin-8 in sepsis: relation to shock and inflammatory mediators. Infect Immun 1992; 60:2835–2842.
72. Dofferhoff AM, Bom VJJ, de Vries-Hospers HG, et al. Patterns of cytokines, plasma endotoxin, plasminogen activator inhibitor, and acute-phase proteins during the treatment of severe sepsis in humans. Crit Care Med 1992; 20:185–192.
73. Pinsky MR, Vincent J-L, Deviere J, et al. Serum cytokine levels in human septic shock. Chest 1993; 103:565–575.
74. Moscovitz H, Shofer F, Mignott H, et al. Plasma cytokine determinations in emergency department patients as a predictor of bacteremia and infectious disease severity. Crit Care Med 1994; 22:1102–1107.
75. Marchant A, Devière J, Byl B, et al. Interleukin-10 production during septicaemia. Lancet 1994; 343:707–708.
76. Jacobs CA, Beckmann MP, Mohler K, et al. Pharmacokinetic parameters and biodistribution of soluble cytokine receptors. Int Rev Exp Pathol 1993; 34B:123–135.
77. Lesslauer W, Tabuchi H, Gentz R, et al. Recombinant soluble tumor necrosis factor receptor proteins protect mice from lipopolysaccharide-induced lethality. Eur J Immunol 1991; 21:2883–2886.
78. Evans TJ, Moyes D, Carpenter A, et al. Protective effect of 55- but not 75-kD soluble tumor necrosis factor receptor-immunoglobulin G fusion proteins in an animal model of gram-negative sepsis. J Exp Med 1994; 180:2173–2179.
79. Peppel K, Crawford D, Beutler B. A tumor necrosis factor (TNF) receptor-IgG heavy chain chimeric protein as a bivalent antagonist of TNF activity. J Exp Med 1991; 174:1483–1489.
80. Mohler KM, Torrance DS, Smith CA, et al. Soluble tumor necrosis factor (TNF) receptors are effective therapeutic agents in lethal endotoxemia and function simultaneously as both TNF carriers and TNF antagonists. J Immunol 1993; 151:1548–1561.
81. Alexander HR, Doherty GM, Buresh CM, et al. A recombinant human receptor antagonist to interleukin 1 improves survival after lethal endotoxemia in mice. J Exp Med 1991; 173:1029–1032.
82. Fisher CJ Jr., Opal SM, Dhainaut JF, et al. Influence of an anti-tumor necrosis factor monoclonal antibody on cytokine levels in patients with sepsis. The CB 0006 Sepsis Syndrome Study Group. Crit Care Med 1993; 21:318–327.
83. Sadoff J. Soluble TNF receptors. Presented at the Third International Congress in the Immune Consequences of Trauma, Shock, and Sepsis: Mechanisms and Therapeutic Approaches. Munich, FRG, March 5, 1994.
84. Wherry J, Wenzel R, Wunderink R, et al. Monoclonal antibody to human tumor necrosis factor (TNF MAb): multi-center efficacy and safety study in patients with the sepsis syndrome. 33rd Interscience Conference on Antimicrobial Agents and Chemotherapy, American Society for Microbiology, New Orleans, October 17–20, 1993, p. 246 (abstr. 696).
85. Abraham E, Wunderink R, Silverman H, Perl TM, Nasraway S, Levy H, Bone R, Wenzel RP, Balk R, Allred R, et al. Efficacy and safety of monoclonal antibody to human tumor necrosis factor alpha in patients with sepsis syndrome. A randomized, controlled, double-blind, multicenter clinical trial. TNF-alpha MAb Sepsis Study Group. JAMA 1995; 273:934–941.
86. Carlet J, Cohen J, Andersson J, et al. Intersept: An international efficacy and safety study of monoclonal antibody to human tumor necrosis factor in patients with the sepsis syndrome. 34th Interscience Conference on Antimicrobial Agents and Chemotherapy, American Society for Microbiology, Orlando, October 4–7, 1994, p. 7 (program addendum abstr. B/1).
87. Fisher CJ Jr, Slotman GJ, Opal SM, et al. Initial evaluation of human recombinant interleukin-1 receptor antagonist (IL-1ra) in the treatment of sepsis syndrome: A randomized open-label, placebo-controlled multicenter trial. Crit Care Med 1994; 22:12–21.
88. Fisher CJ Jr, Dhainaut JF, Opal SM, et al. Recombinant human interleukin-1 receptor antagonist in the treatment of patients with sepsis syndrome. JAMA 1994; 271:1836–1849.

38
Cytokines in Bacterial and Fungal Infections

Charles J. Czuprynski
School of Veterinary Medicine, University of Wisconsin-Madison, Madison, Wisconsin

Mary Haak-Frendscho
School of Veterinary Medicine, University of Wisconsin-Madison, and Promega Corporation, Madison, Wisconsin

INTRODUCTION

It has long been a goal of medicine to modulate host defense mechanisms in order to maximize resistance to infectious agents and minimize collateral damage to host tissues and organ systems. For decades these efforts relied on the use of crude microbial or plant products that enhanced resistance to infection in clinical or experimental studies. Although these biological response modifiers sometimes had potent protective effects, their crude nature and unacceptable side effects limited their use (1). The biotechnology revolution of the past decade, which resulted in cloning and production of recombinant cytokines in large quantities, fueled renewed enthusiasm for modulation of host defense mechanisms. The underlying rationale was that purified recombinant cytokines with known biological effects might be used to augment antimicrobial defense without causing other undesirable effects. Numerous investigations confirmed that cytokines can indeed be used to substantially increase resistance to a variety of pathogenic microbes (2,3). Practical implementation of this knowledge has been limited, however, by recognition that the beneficial effects of cytokines are critically dependent on dose, timing, and mode of administration. In particular, many of the beneficial effects require that the cytokine be given prior to or shortly after the onset of infection; therapeutic treatment of existing infection is rarely of substantial benefit (4). As a result, it is clear that recombinant cytokines are not a "magic bullet" for antimicrobial therapy. Nonetheless, experimental and clinical evaluation of cytokines is a dynamic area of research. New technologies, such as transgenic and knock-out mice, and novel reagents, including monoclonal antibodies, soluble receptors, and receptor chimeras, have provided powerful new tools for investigating the role of cytokines in both innate and acquired immunity to microbial infection. These investigations provided exciting new insights into the regulation of host defense mechanisms against microbial infection and underscore microbial strategies to circumvent these host defense mechanisms. Other evidence suggests that recombinant cytokines might have a place as an adjunct to conventional antibiotic therapy (5), as adjuvants for vaccination (6,7), or as part of a prophylactic regimen to reduce the risk of opportunistic infections in patients whose resistance is compromised by neoplasia, immunosuppressive drugs, or preexisting infection (8).

It is impossible to provide an encyclopedic review of all in vivo and in vitro studies relevant to the roles of cytokines in antibacterial resistance. The flood of new reports that emerge on a weekly basis would make even the most thorough effort incomplete within a short period after publication. As a result, in this review we will attempt the more modest goal of providing an overview of cytokines and cytokine interactions that protect against bacterial infection in vivo in experimental animals and in clinical trials. For a more thorough evaluation of the properties of individual cytokines, we refer the reader to the appropriate chapters in this book and to other reviews cited in this chapter. This review will be organized as follows: (1) role of proinflammatory cytokines (i.e., IL-1, TNF-α, IL-6, IL-8) in host defense; (2) regulation of T-cell cytokines (i.e., IL-2, IFN-γ, L-4, IL-10, IL-12) in protective cellular immunity; (3) prophylactic use of cytokines to reduce opportunistic infections; (4) potential use of cytokine inhibitors; and (5) prospective use of cytokines as adjuvants for vaccination or as adjuncts to antibiotic therapy.

CYTOKINES IN ANIMAL MODELS OF INFECTION AND IN HUMAN DISEASES

Inflammatory Cytokines

Interleukin-1

The interleukin-1 (IL-1) family contains three ligands [IL-1α, IL-1β, and the IL-1 receptor antagonist (IL-1ra)], and two receptors (type I and type II). The reader is referred to earlier chapters and elsewhere for thorough and excellent reviews of IL-1 and its receptors (9,10). The first two ligands (IL-1α and IL-1β) are powerful agonists, whereas the third (IL-1ra) is a potent naturally occurring antagonist of cell activation by either IL-1α or IL-1β. Expression of the genes for these ligands and receptors, which are all located on human chromosome 2, is tightly regulated (9). The principal secreted form of IL-1 is IL-1β; most IL-1α remains cell associated (11,12). In contrast, both secreted and intracellular forms of IL-1ra occur (13). Despite sharing a limited amino acid identity (~25%), IL-1α and IL-1β exhibit remarkably similar biological activities (9). Low numbers of type I receptors (IL-1RI) are found on many cell types and are the predominant type of IL-1R on T cells and fibroblasts (14). In contrast, the type 2 receptor (IL-1RII) is most common on neutrophils, monocytes, and B cells (15). Recent evidence suggests that it is the low numbers of type 1 receptors that are responsible for cell activation by IL-1 (16). Extrapolations from binding studies suggest that occupancy of less than 5% of IL-1RI with IL-1β is sufficient for activation of many cell types (17). Type 2 receptors appear to act as a "decoy" by binding IL-1β without mediating signal transduction (16). In addition, type 2 receptors can be shed to circulate as soluble receptors that can bind and neutralize IL-1β, but not IL-1ra (18).

Before IL-1 was cloned and its molecular structure delineated, there were reports of crude mediators produced by macrophages and other leukocytes (i.e., leukocyte endogenous mediator) that had a variety of proinflammatory effects, including enhancement of antibacterial resistance (19). The availability of recombinant IL-1 led to numerous demonstrations of its ability to enhance resistance to a variety of pathogens, including *Pseudomonas aeruginosa* (20,21), *Klebsiella pneumoniae* (20), *Listeria monocytogenes* (22), *Candida albicans* (23), *Staphylococcus aureus* (24), *Salmonella typhimurium* (24), *Mycobacterium lepraemurium* (25), *Streptococcus suis* (26), and *Brucella abortus* (27). Both IL-1α and IL-1β have been used in these studies, with similar results. In addition, both human and murine IL-1 are effective in mice. IL-1 is quite effective in enhancing microbial resistance, generally at doses of 0.1–1.0 μg per mouse (20,22). Typically IL-1 works best when given near the time of bacterial challenge; therapeutic administration of IL-1 is ineffective if delayed more than a few hours after infection challenge (4). Nor is IL-1 treatment beneficial for ameliorating chronic infection. The mechanisms by which IL-1 enhance antibacterial resistance have not been elucidated, although these probably relate to the various proinflammatory properties of this cytokine. IL-1 is a potent stimulus for neutrophil accumulation in vivo (28); however, protection has been observed in both normal and neutropenic mice (24). Although some investigators have identified an association with other cytokines, arachidonic acid metabolites, and acute phase proteins (29), a cause-and-effect relationship has not been firmly established. In addition to the benefits of administering IL-1 alone, there is evidence that combined administration of IL-1 with other cytokines, such as TNF-α, increases resistance in an additive or synergistic fashion (30). Furthermore, there is evidence that co-administration of IL-1 and antibiotics potentiates the antibacterial effects of both agents in mice and cattle (5,31). IL-1 also has been used successfully to augment immunization with inactivated bacterial and viral vaccines (32,33) and to potentiate the resistance transferred by adoptive immunization with immune splenocytes (34).

In addition to the information obtained from studies with recombinant IL-1, insights into the natural role of IL-1 were obtained from investigations that employed the IL-1 receptor antagonist, or antibodies against the type I IL-1 receptor, to impair resistance to *L. monocytogenes* (35), *Pneumocystis carinii* (36), and *Mycobacterium paratuberculosis* (37). Similar strategies have been used to blunt the adverse effects associated with sepsis caused by infection with gram-negative or gram-positive bacteria (38,39). The recent availability of soluble type 1 and type 2 IL-1R provides yet another potential tool for investigating the role of IL-1 in host defense to bacterial infection.

Tumor Necrosis Factor

Like IL-1, there is a family of tumor necrosis factor (TNF) ligands and receptors. In this case there are two ligands: TNF-α, which is produced by both mononuclear phagocytes and certain populations of activated T cells, and TNF-β, which is a product of cytotoxic T cells (40–42).

These two molecules have limited amino acid identity (approximately 34%) but mediate similar biological activities (40–42). There also are two known surface receptors, p55 and p75, both of which bind TNF-α and TNF-β (42,43). Some differences between p55 and p75 in tissue expression and biological activities have been reported, no doubt related to specific signal transduction pathways mediated by the two receptors (42,43). Both p55 and p75 TNF receptors can be shed from activated cells and neutralize the biological activities of TNF-α (44). For a thorough review of TNF and its receptor, the reader should consult the appropriate chapter in this book or other excellent reviews of the subject (40–43).

TNF-α is made in response to a variety of bacteria and bacterial products. Perhaps the best studied of these is bacterial lipopolysaccharide (LPS), which elicits a vigorous TNF-α response by macrophages in vitro and stimulates a brief burst of circulating TNF-α in the plasma of animals and humans experiencing septicemia or endotoxemia (45–47). Although a variety of mediators are released in response to sepsis, evidence suggests that TNF-α may be a key player in the pathogenesis of septic shock. A correlation between plasma TNF levels and clinical outcome of sepsis or infection has been noted (45). Neutralization of TNF-α activity, via administration of anti–TNF-α monoclonal antibodies (mAbs) or soluble TNF receptors, enhances survival in animal models of sepsis (48–50). Furthermore, there is evidence that plasma levels of TNF are elevated in some chronic infections, such as lepromatous leprosy, in which there is chronic inflammation without effective clearance of the offending microbe (51,52). However, there is also impressive evidence of the importance of TNF-α in host defense against a variety of pathogenic bacteria. Examples of bacterial or fungal infections against which recombinant TNF-α has been shown to be protective, or anti-TNF-α antibodies to be detrimental, include *Legionella pneumophila* (53), *L. monocytogenes* (30,49,50,54–56), *K. pneumoniae* (57), *Escherichia coli* (58), *Chlamydia trachomatis* (59), *P. carinii* (60), *Cryptococcus neoformans* (61), *Mycobacterium bovis* BCG (62), *M. lepraemurium* (63), *Neisseria meningitidis* (64), *C. albicans* (65), and *Histoplasma capsulatum* (66). In addition, the effects of TNF-α are enhanced in the presence of IL-1 or IFN-γ (30,56,58,60,61,67). Thus, TNF-α is a classic example of a biological double-edged sword, the effects of which may be either beneficial or detrimental, depending on circumstances such as timing, concentration, and presence of other cytokines.

Interleukin-6

Interleukin-6 (IL-6) was originally described as interferon-β2, and its biological activities initially were thought to chiefly involve B-cell maturation and immunoglobulin production. Later, it was shown that IL-6 was a potent hepatocyte-stimulating factor and a key player in the acute phase response (68). IL-6 is produced by a variety of cells, including mononuclear phagocytes, fibroblasts, epithelial cells, some T- and B-cell lines, mast cells, and granulocytes (68). Bacterial products, including LPS, are potent stimuli for IL-6 production, as are other cytokines such as IL-1 and TNF-α. The IL-6 receptor is a member of the hematopoietin receptor family, being composed of a 80 kDa binding protein and a 130 kDa signal transduction protein shared by other members of the receptor family (69). Plasma levels of IL-6 rise rapidly in response to bacterial infection and can remain elevated depending on the severity and duration of the infection (70–73). Unlike IL-1 and TNF-α, IL-6 does not directly stimulate an oxidative burst by phagocytic cells, nor does it stimulate expression of adhesion molecules on endothelial cells (68). Its role in host defense, therefore, may be more related to tissue repair resulting from bacterial infection, rather than direct activation of antibacterial defense mechanisms.

Several lines of evidence suggest an important role for IL-6 in the host response to bacterial infection. As indicated above, bacterial products are a potent stimulus for IL-6 expression by both mononuclear phagocytes and epithelial cells (74). Plasma levels of IL-6 can be quite high in human sepsis patients (68) and animals experimentally infected with both gram-negative (i.e., *E. coli* and *B. abortus*) and gram-positive bacterial pathogens (*L. monocytogenes, M. avium,* and *Streptococcus pneumoniae*) (70–73). Recombinant human IL-6 failed to protect mice against experimental *L. monocytogenes* infection (75), whereas higher doses of recombinant murine IL-6 elicited significant protection (76). Mice depleted of endogenous IL-6 by administration of neutralizing antibodies as well as IL-6 gene knock-out mice demonstrated reduced resistance to *L. monocytogenes* infection (77–79). This reduced resistance correlated with inadequate neutrophilia during the infection (79), an observation consistent with both the ability of IL-6 to act as a hematopoietic factor (68) and the importance of neutrophils in the early response to *L. monocytogenes* infection (80). A novel observation was made by Denis (81), who reported that human and murine IL-6 were potent growth factors for *M. avium,* and that these bacteria possessed an average of 15,000 binding sites per bacterium with a K_D of 50 nM. This observation has not been confirmed by other investigators, nor reported for other bacterial species.

Interleukin-8

Interleukin-8 (IL-8) is a member of the chemokine family and has potent effects on migration and activation of leukocytes and other cells. The chemokines are divided into three known subfamilies, based on the presence (CXC) or absence (CC) of an intervening amino acid between the first two cysteine residues of the repeat motifs (82). The third subfamily is the newly discovered C group (83). About 30 distinct chemokines have been described to date (84). Homologs of many of these have been described in mice and other species, although a murine counterpart of IL-8 has not yet been reported (82,85). Chemokines appear to play critical, and perhaps overlapping, roles in attracting specific cell types to sites of inflammation (82,85,86). For example, IL-8 and other CXC chemokines generally are chemoattractants for neutrophils and lymphocytes, whereas CC chemokines result in monocyte accumulation (82). IL-8 is perhaps the best studied of the chemokines in terms of its biological properties (Table 1). IL-8 is a chemoattractant for neutrophils in vitro and in vivo and elicits expression of adhesion molecules, an oxidative burst, and degranulation by neutrophils in vitro (82,85). In addition, IL-8 is a chemoattractant for a subpopulation of human T cells and a chemoattractant and activating agent for human basophils (82,85,86). Although the role of IL-8 and other chemokines has not been delineated in large numbers of infection models, there is sufficient evidence to infer that they are prominent players in the inflammatory response to microbial infection. Persuasive evidence has been obtained regarding the role of IL-8 in the adult respiratory disease syndrome (ARDS), which can be a devastating sequela of bacterial infection (87,88). Although plasma levels of IL-8 are not elevated in ARDS patients, bronchoalveolar lavage fluid levels are higher than in controls or patients who do not develop ARDS. IL-8 and other chemokines also appear to play important roles in the development of delayed hypersensitivity to mycobacterial products (85,89,90).

These findings suggest that IL-8 may be compartmentalized as part of the local inflammatory response, rather than being elevated in plasma as part of the systemic immune response to microbial infection (88). Another example is provided by the elevated urine levels, but not serum levels, of IL-8 in patients with *E. coli* urinary tract infection (91). These same authors observed epithelial cell production of IL-8 in response to some strains of *E. coli* in vitro, suggesting that epithelial cells were responsible for IL-8 production during urinary tract infection. Likewise, other investigators reported that invasive bacteria, but not noninvasive strains or bacterial products such as LPS, stimulated IL-8 expression by intestinal and cervical epithelial cell lines in vitro (92). Recently, a novel mechanism has been proposed wherein IL-8 also may act to protect against bacterial invaders by binding to carbohydrates, such as peptidoglycans and polysaccharides, on the surface of bacteria (93). This would form a multivalent array of solid-phase mediators that could efficiently activate leuko-

Table 1 Bioactivities of IL-8, a C-X-C Chemokine, Encoded on Human Chromosome 4 (q12 → q21)

Chemokine	Other names	Cell source	Cell target	Effects
IL-8, 8 kDa	NAP-1, NAF, MDNCF, MONAP, LYNAP, 3-10C, LAI, 9E3 (chicken), CEF (chicken)	Monocytes Macrophages Endothelial cells Fibroblasts Keratinocytes T cells Neutrophils Hepatocytes Chondrocytes	Neutrophils	Chemotaxis ↑ degranulation ↑ respiratory burst ↑ lysosomal enzyme release ↑ cytosolic free Ca^{2+} ↑ adherence to unstimulated endothelial monolayers, fibrinogen, and subendothelial matrix proteins ↑ granulocyte adhesion molecule expression Activates arachidonate-5-lipoxygenase and releases LTB_4 and 5-HETE (in the presence of exogenous AA)
			T cells	Chemotaxis
			Basophils	Chemotaxis, histamine, and leukotriene release from primed cells
			Monocytes	↑ adhesiveness to endothelial cells
			Keratinocytes	↑ proliferation
			Melanoma cells	↑ adhesiveness and haptotactic response

Source: Refs. 83, 85.

cyte cytotoxic mechanisms. And finally, IL-8 may play a role in homeostasis by helping to protect from bacterial infection through the epidermis. This is suggested by findings that human sweat, which is induced by physical or emotional stress, contains IL-8 (94).

T-Cell–Mediated Cytokines

Perhaps the most dynamic and influential area of cytokine biology during the past few years has been the delineation of the regulation of cytokine release by T-helper (TH) cells. This has resulted in general acceptance of the dichotomy between cellular and humoral immune responses on the basis of cytokine profiles. This paradigm evolved out of the seminal studies of Mosmann and Coffmann (95), who noted that some T-cell clones (TH1 cells) released cytokines (e.g., IL-2 and IFN-α) principally associated with macrophage activation and delayed-type hypersensitivity, while others (TH2 cells) released cytokines (e.g., IL-4 and IL-5) that promote production of IgG1 and IgE and events related to immediate-type hypersensitivity. Subsequent studies indicated that other cytokine patterns also were present, particularly a TH0 profile in which both TH1 and TH2 cytokines are produced (96). For in-depth analysis of this subject, the reader is referred to several excellent reviews, which offer both historical and contemporary analyses of the subject (97–101).

The TH1 versus TH2 dichotomy has been eagerly adopted as a working model to explain previously noted trends in immunoregulation and host defense. There are numerous reports that protective immunity to facultative intracellular pathogens requires a TH1-cell response (Table 2) and that resistance is impaired in the presence of TH2 cytokines (98–101). Particularly persuasive evidence was provided in reports that genetically determined innate resistance and immunologically acquired protective immunity to *L. major* infection in mice were associated with a TH1 cytokine profile (98). Although the TH1 versus TH2 paradigm possesses tremendous appeal because of its simple explanation of host defense, there is ample evidence that resistance to microbial infection is often a good deal more complex. For example, humans and cattle sometimes demonstrate a mixed cytokine response, similar to a TH0 cytokine profile, during microbial infection (101,102). Nonetheless, the TH1 versus TH2 paradigm provides a useful theoretical framework for considering immunoregulation of antibacterial resistance.

Considerable effort has focused on identifying the signals responsible for the progression of T cells to either a TH1- or TH2-type cytokine response. This is a complex and dynamic area of research. Current evidence suggests that release of IL-12 by mononuclear phagocytes (97,98) favors progression of TH0 cells to a TH1-type cytokine response, whereas IL-10 release by mononuclear phagocytes and IL-4 and IL-10 release by TH2 cells depresses

Table 2 Biological and Protective Effects of TH1 Cytokines

TH1 cytokine	Biological effects	Confers resistance against
IFN-γ	Macrophage activation	*Listeria monocytogenes*
	↑ NO production	*Mycobacterium tuberculosis*
	↑ LAK activity	*Salmonella typhimurium*
	↑ B-cell differentiation	*Chlamydia psittaci*
	↑ adhesion molecule expression	*Brucella abortus*
		Histoplasma capsulatum
IL-2	↑ NK activity	*Mycobacterium leprae*
	↑ LAK activity	*Listeria monocytogenes*
	↑ cytokine production	*Mycobacterium lepraemurium*
	↑ T-cell cytotoxicity	*Mycobacterium bovus* BCG
	↑ T-cell proliferation	*Escherichia coli*
		Klebsiella pneumoniae
		Coccidioides immitis
IL-12	↑ IFN-γ/cytokine production	*Leishmania donovani*
	↑ NK activity	*Mycobacterium tuberculosis*
	↑ LAK activity	*Mycobacterium avium*
	↑ T-cell cytotoxicity	*Mycobacterium bovis*
	↑ T-cell proliferation	*Histoplasma capsulatum*
		Listeria monocytogenes

Source: Refs. 98–101.

the TH1 cell response and predisposes TH0 cells to progress toward a TH2 cytokine profile (97,98,100,101,103–107). Perhaps the type of microbe or the background cytokine milieu helps to determine whether the relative timing and amounts of IL-12 and IL-10 favor development of a TH1- or TH2-like cytokine profile. More recently, there is evidence that the interactions of the co-stimulatory molecule B-7 on antigen-presenting cells with CD28 or CTLA-4 on TH0 cells substantially influences their progression along either the TH1 or the TH2 cytokine pathway (108).

TH1 Cytokines

Interferon-γ Arguably, the most studied and important of the TH1 cell cytokines is IFN-γ. Although evidence suggests that NK cells may be an important source of IFN-γ during the early stages of some bacterial infections (105,107), most investigators have interpreted their data as indicating that IFN-γ is pivotal in the TH1 cytokine response. Protection against a variety of pathogenic bacteria has been shown to be IFN-γ dependent. For example, infection with *L. monocytogenes* results in a significant upregulation of IFN-γ mRNA expression within 1 h (109) and elevated plasma levels of circulating IFN-γ concomitant with the peak bacterial burden in the spleen and liver (110). Neutralization of endogenous IFN-γ by injection of an anti–IFN-γ mAb significantly impairs resistance, an effect that is amplified when TNF-α is also neutralized (67). Likewise, anti-*Listeria* resistance is impaired in gene knock-out mice that lack the IFN-γ receptor (111). Conversely, administration of recombinant murine IFN-γ significantly enhances resistance to *L. monocytogenes* (56,104).

There is strong evidence for the importance of IFN-γ in both human and experimental murine mycobacterial infections. Accumulation of IFN-γ–positive cells in skin lesions of tuberculoid leprosy patients has been observed, whereas these are reduced or absent in lesions from lepromatous leprosy patients (112). Elevated levels of IFN-γ mRNA were detected in the blood and pleural fluid of human tuberculosis patients (113). However, mycobacterial infection in humans sometimes results in the expression of both TH1 and TH2 cytokines, similar to the TH0 cytokine profile in mice (101). Exciting results were obtained with the therapeutic use of IFN-γ in human lepromatous leprosy patients, in which multiple injections of IFN-γ into skin lesions resulted in enhanced emigration of CD4+ cells and reduced bacillary burden (114). Similar results were noted after injection of IL-2 or purified protein derivative of tuberculin (PPD), both of which were associated with an increased ability of the recipients' lymphocytes to produce IFN-γ in response to in vitro stimulation with mitogen or *M. leprae* antigens (115,116). Multiple intramuscular injections of IFN-γ increased the infiltration of mononuclear cells into skin lesions at sites distant to the IFN-γ injection (117). The beneficial use of IFN-γ in conjunction with antimycobacterial drug therapy in patients with atypical mycobacterial infection that was refractory to drug therapy alone also has been reported (118,119). Other studies have demonstrated the importance of IFN-γ in resistance of mice to experimental mycobacterial infection (101,104,120). The most compelling data were obtained in recent studies of gene knockout mice that lacked the ability to produce IFN-γ or the IFN-γ receptor. In both instances, resistance to *M. tuberculosis* infection plummeted, resulting in a devastating progressive infection that resembles tuberculosis in AIDS patients (121).

IFN-γ has also been shown to be important in host defense against experimental infection with *S. typhimurium* (122), *Chlamydia psittaci* (123) *H. capsulatum* (124), *Rickettsia conori* (125), and *Rhodococcus equi* (126). In addition, IFN-γ prolonged survival to *E. coli* and *K. pneumoniae* infection in mice subjected to surgical trauma (127). In general, these studies observed that host defense was impaired when endogenous IFN-γ was neutralized and enhanced when recombinant IFN-γ was administered. There are notable exceptions, however, which suggest that a TH1 cytokine response may not always be desirable for resistance to some bacterial or fungal infections. It has been suggested by some that resistance to *C. albicans* infection is impaired by injection of IFN-γ (128); however, other studies demonstrated the association of a TH1 cytokine response with resistance to candidiasis (129). IFN-γ may play a detrimental role in certain types of gram-negative infections that do not involve intracellular pathogens. Mortality was reduced when endogenous IFN-γ was neutralized in mice infected with *E. coli*; conversely, mortality increased when recombinant IFN-γ was injected (130). These findings may reflect the ability of IFN-γ to amplify the inflammatory cytokine response (e.g., IL-β, TNF-α) during septicemia. A recent report provided evidence of a detrimental role of IFN-γ in resistance to experimental infection with the Lyme disease spirochete (*Borrelia burgdorferi*). Susceptible strains of mice produced higher levels of IFN-γ and IL-2 and lower levels of IL-4 than did spleen and lymph node cells from a resistant mouse strain (131). Treatment with an anti–IL-4 mAb increased joint swelling and spirochete numbers, whereas administration of anti–IFN-γ mAb had the opposite effect. These findings correlate with previous reports that resistance to *B. burgdorferi* is associated with a borreliacidal antibody response (132).

Interleukin-2

It is somewhat paradoxical that expression of IL-2, which is the prototype T-cell–activating factor, is not always obviously increased in infections where IFN-γ is rapidly and dramatically elevated (109,133). Whether this reflects a lesser role for IL-2 in protective immunity, or an inability to accurately detect its expression is not clear. IL-2 knock-out mice are relatively normal, although they are prone to develop inflammatory bowel disease if not housed under gnotobiotic conditions (134). This suggests that in the complete absence of IL-2 there is an increased propensity for opportunistic infections. Human lepromatous leprosy patients exhibited a favorable response to intradermal IL-2 injection, resulting in increased mononuclear cell infiltration and a concomitant decrease in bacillary burden in most patients (116,135). This finding was consistent with an in vitro study, which showed that rIL-2-activated lymphocytes lyse mycobacterial infected human monocytes in vitro (136,137). Several studies have demonstrated a protective effect of recombinant IL-2 (rIL-2) in mice experimentally infected with *L. monocytogenes* (138), *M. lepraemurium* (139), *M. bovis* BCG (139), *E. coli* (140), *K. pneumoniae* (141), and *Coccidioides immitis* (142). Incubation of spleen cells from *L. monocytogenes* immunized donor mice with rIL-2 in vitro enhanced their ability to transfer resistance to naive recipient mice (143). This response appeared to be independent of CD4+ cells and only partially dependent on CD8+ cells. In contrast to the clinical success with leprosy patients and the beneficial effects of rIL-2 in animal model studies, administration of rIL-2 to AIDS patients resulted in an increased incidence of infection with *S. aureus* and other bacterial species (142). The mechanisms responsible for this decrease in antibacterial resistance were not determined.

TH2 Cytokines

Interleukin-4 Evidence suggests that IL-4 plays a detrimental role in resistance to infection with facultative intracellular bacteria. This probably reflects the ability of IL-4 to block macrophage activation by IFN-γ, resulting in reduced production of inflammatory cytokines (i.e., IL-1, TNF), increased production of IL-1ra, and diminished antibacterial activity (145). Human lepromatous leprosy lesions have increased expression of IL-4 mRNA as compared with skin biopsies of lesions from tuberculoid leprosy patients, suggesting an inverse correlation between IL-4 and resistance to *M. leprae* infection (112). Administration of anti–IL-4 mAb significantly increased resistance to *L. monocytogenes* infection in mice (133), an observation consistent with in vitro evidence that neutralization of IL-4 enhances the development of TH1 cells in response to in vitro stimulation with *L. monocytogenes* antigens (97). However, expression of IL-4 is barely detectable in *L. monocytogenes*–infected mice, and enhanced resistance in anti–IL-4 mAb–treated mice occurred without a concomitant shift to a TH1 cytokine profile (133). Furthermore, administration of rIL-4 does not impair resistance to *L. monocytogenes* (146). Administration of anti–IL-4 mAb has also been reported to increase the resistance of mice to infection with *Francisella tularensis* (147). In contrast, recently evidence indicated that anti–IL-4 mAb administration increased joint swelling and spirochetal burden in mice experimentally infected with *B. burgdorferi* (131). Thus, the role of IL-4 in resistance to bacterial infection appears to be more complex than the unequivocal negative effects of IL-4 in mice affected with *L. major* (98).

Interleukin-10 In vitro evidence indicates that IL-10 suppresses TH0 cell progression toward the TH1 cell phenotype and selects for the development of TH2 cells that produce IL-4 (97,105,148). The latter in turn further facilitates TH2 cell development. IL-10 may exert its effects both through downregulation of MHC class II expression and cytokine production by mononuclear phagocytes and by direct effects on T cells (149). Besides being a pivotal molecule in T-cell cytokine expression, IL-10 influences a broad range of other cell types, including granulocytes, mast cells, mononuclear phagocytes, and B cells (149). In general, IL-10 downregulates the production of inflammatory cytokines by mononuclear phagocytes and neutrophils and reduces or prevents macrophage activation by IFN-γ (150–153).

As one might predict, IL-10 appears to play a detrimental role in host defense against facultative intracellular bacterial pathogens. IL-10 mRNA expression is increased in skin lesions of lepromatous leprosy patients (112). Administration of anti–IL-10–neutralizing mAb enhanced the resistance of mice to *M. avium* infection (154,155). Similar observations were made regarding murine listeriosis, wherein expression of IL-10 mRNA is increased during infection (156). Administration of anti–IL-10 mAb before *L. monocytogenes* challenge improved resistance during the early stage of infection in both normal mice and mice whose resistance was compromised by injection of immune complexes (157,158). Cerebrospinal fluid from mice with *L. monocytogenes* meningitis contained IL-10 that decreased macrophage anti-*Listeria* activity in vitro (159). These authors also provided evidence that human meningitis patients have detectable levels of IL-10 in their cerebrospinal fluid.

Susceptible DBA/2 mice survived experimental *C. albicans* infection when treated with anti–IL-10 mAb at the time of infection, perhaps reflecting increased macrophage production of nitric oxide that results in increased candidacidal activity (160,161). Despite the findings described above, other evidence suggests that IL-10 may also have activities that contribute to rather than inhibit host defense. For example, IL-10 has been shown to enhance expression of certain Fc receptors (162). Curiously, the complete elimination of *L. monocytogenes* was impaired in mice that received anti–IL-10 mAb, despite the reduced bacterial burden during the early stage of infection (157). Likewise, anti-*Listeria* resistance of normal mice was seriously impaired if administration of the anti–IL-10 mAb was delayed until 48 h or later after *L. monocytogenes* challenge (157). The reason for this paradoxical effect of IL-10 is unclear, but it might reflect the ability of endogenous IL-10 to downregulate release of inflammatory mediators that enhance tissue damage, rather than promote repair, during the later stages of the infection (149). These findings suggest that the role of IL-10 in antibacterial resistance in vivo may be quite complex and is perhaps influenced by the timing of IL-10 expression and the presence of other mediators.

There are several reports that IL-10 may play a role in gram-negative infection and endotoxemia. High levels of IL-10 were present in the plasma and CSF of patients with meningococcal meningitis complicated with septic shock, whereas IL-10 was present in the CSF, but not plasma, of patients with uncomplicated meningitis (163). This finding may reflect a compartmentalization of IL-10 release and suggests that elevated plasma levels of IL-10 may indicate a poor prognosis in septic shock. Conflicting results regarding the role of IL-10 have been obtained in animal model studies. Anti–IL-10 administration protected mice against *K. pneumoniae* lung infection (164) but increased lethality in experimental endotoxemia (165). In other studies, recombinant IL-10 protected mice against experimental endotoxemia and lethal shock by staphylococcal enterotoxin B (152,166). The protective effects of rIL-10 were thought to result in part from a reduction in TNF release (152).

A randomized controlled trial of recombinant IL-10 administration to healthy human volunteers did not elicit any adverse clinical signs, had only a minor effect on T-lymphocyte numbers and proliferation, and reduced monocyte production of IL-1 and TNF-α, but not the p55 TNF receptor or IL-1ra (167). Gene knock-out mice for IL-10 develop normally, but displayed a propensity to develop inflammatory bowel disease when not housed under germ-free conditions (168). This observation suggests that IL-10 may play a role either in preventing translocation of the intestinal microflora or in downregulating the inflammatory response to microbial components in the intestinal epithelium.

Interleukin-12 A breakthrough in understanding the regulation of host defense occurred when the importance of IL-12 in initiation of a cellular immune response was elucidated. Once again, the reader is directed to the relevant chapter and other reviews for a detailed discussion of the subject (98,103). IL-12 is unique among the interleukins in being a heterodimer composed of 40 and 35 kDa proteins that are molecularly unrelated and whose genes reside on different chromosomes (103). It has been proposed that the biologically active IL-12 heterodimer may be analogous to a receptor-ligand complex, as the p 35 subunit is distantly related to IL-6 and G-CSF and the p40 subunit has sequence homology with the extracellular domain of the IL-6 receptor (103).

The role of IL-12 in protective cellular immunity was first demonstrated in vitro when macrophages that were fed killed *L. monocytogenes* released IL-12 that facilitated production of IFN-γ by both NK cells and TH1 cells (97,105). These studies also demonstrated the antagonistic role of IL-10 in this response, suggesting that a critical balance between IL-10 and IL-12 was pivotal in the initiation of cellular immunity. Other investigators confirmed the beneficial role of IL-12 by showing that resistance to murine listeriosis was enhanced by administration of recombinant IL-12 and diminished by injection of anti–IL-12 antibodies (169,170). These effects were IFN-γ dependent, in keeping with in vitro studies, which demonstrated the ability of IL-12 to promote IFN-γ production by TH1 and NK cells (97,105).

Increased expression of IL-12 was noted in the skin lesions of tuberculoid as compared to lepromatous leprosy patients (112). Similarly, IL-12 was increased in pleural fluid from patients with tuberculous pleuritis as compared with plasma levels from the same patients or in pleural fluid from patients with malignant pleural effusions (171). Additional animal model evidence suggests that IL-12 is important in host defense against mycobacterial infection. A flurry of publications demonstrated that IL-12 mRNA expression is increased during infection with *M. tuberculosis* (172), that endogenous IL-12 is important for resistance to *M. avium* (173), and that administration of recombinant IL-12 enhances resistance to infection with *M. tuberculosis* (174,175) and *M. avium* (176). Similar results have been noted for mice infected with the dimorphic fungal pathogen *H. capsulatum,* in which administration of mAbs against IL-12, TNF-α, or IFN-γ enhanced mortality, and injection of recombinant IL-12 or anti–IL-4 mAb reduced mortality (124). IL-12 is also required for TH1-cell development in mice infected with *C. albicans* (177).

Although IL-12 first appeared in the scientific literature in 1989, it already has entered human phase I clinical trials for the treatment of kidney cancer and AIDS. However, like other cytokines originally hailed as potential magic bullets, IL-12 is not without a dark side. The second clinical trial was halted for 4 months due to unforseen severe toxic effects (including two patients who died from the treatment), until it was understood that a unique treatment regimen was required to ensure patient safety at multiple high doses (178,179). Despite this tragic episode, IL-12 still is seen by many as a potential wonder drug. And, based on the aforementioned promising results in animal models of infection, it is likely that IL-12 is poised to enter clinical trials for the treatment of infectious diseases such as tuberculosis, leishmaniasis, schistosomiasis, and malaria, which infect hundreds of millions of people annually in developing countries (98,179).

USE OF CYTOKINES AND THEIR INHIBITORS TO MODULATE DISEASE

Prophylactic Use of Cytokines to Reduce Opportunistic Infections

Contemporary infectious diseases often involve opportunistic pathogens in humans whose immune response is compromised as a result of underlying infection (e.g., HIV) or treatment of some other medical condition (e.g., chemotherapy of cancer patients). These patients are at risk of contracting infections, which may be severe or life-threatening, with a variety of opportunistic bacterial and fungal pathogens that pose little risk to healthy adults. Many of these opportunists are facultative intracellular pathogens, and as such require an effective granulocyte and cellular immune response. Hospitalization and medical costs, as well as the obvious losses attributed to morbidity and mortality, make efforts to protect at-risk patients from these opportunistic infections a high priority. Cytokines have received considerable attention as a potential means of reducing the prevalence of opportunistic infections in this patient population. Table 3 lists cytokines being considered for the treatment of bacterial infections. Many of these attempts have not succeeded, which is not surprising since most animal model studies indicate that injection of recombinant cytokines alone augments host defense only if given at or before bacterial challenge. However, a notable exception is IFN-γ, which has been approved for use in the management of chronic granulomatous disease and thus for reducing the incidence of life-threatening infections in this immunocompromised population (181). Based on an increasing body of data, it also has been suggested that IFN-γ may be useful for treating infections characterized by intracellular persistence in macrophages, including toxoplasmosis, leishmaniasis, and mycobacteriosis (182).

Considerable success and practical benefit also have been realized by the prophylactic use of colony-stimulating factors (CSFs) in patients who are experiencing temporary leukopenia as a result of radiation treatment or chemotherapy (8). The CSFs are a series of growth factors that play critical roles in various stages of hematopoiesis (183). Some are relatively nonspecific and affect multipotential cells in the earlier stages of leukocyte maturation (i.e., IL-3 or multi-CSF and GM-CSF). Others are more specific and only affect cells that have proceeded down a single developmental pathway (i.e., G-CSF and CSF-1). Besides their effects on normal leukocyte maturation, these CSFs also often possess the ability to activate mature leukocytes of their respective lineage, or to prime these cells for enhanced responses to other activating agents. Animal studies suggested that administration of CSFs more rapidly re-

Table 3 Cytokines in Clinical Trials

Product	Company	Indication	U.S. Development Status
IFN-γ Actimmune	Genentech S. San Francisco, CA	Chronic granulomatous disease	Approved December 1990
GM-CSF Leukine	Immunex Seattle, WA	Reduction in postoperative infections Neonatal sepsis	Phase III
G-CSF Neupogen	Amgen Thousand Oaks, CA	Infectious disease	Phase II
TNF Tenefuse	Hoffmann-La Roche Nutley, NJ	Septic shock Severe sepsis	Phase II

Source: Adapted from Ref. 180.

stored normal numbers of leukocytes to animals that had been treated with radiation or radiomimetic agents (184). Clinical trials have confirmed the efficacy of CSFs in humans. For example, G-CSF more rapidly restores leukocyte numbers in leukopenic patients and reduced the incidence of fever and infection in some trials (8,185). Likewise, CSF-1 and GM-CSF have been shown to reduce the incidence of fungal infection in granulocytopenic patients (183,186). Beneficial effects also have been noted in autologous bone marrow transplantation patients, where G-CSF and GM-CSF have been used to restore normal leukocyte development and perhaps activate the antimicrobial capacity of the mature leukocytes (185).

Cytokine Inhibitors

Much of the efforts in cytokine biology have focused on identifying the beneficial effects of endogenous cytokines or the administration of recombinant cytokines in host defense. However, it must be remembered that cytokines are a double-edged sword that may both increase antimicrobial resistance and directly mediate or amplify tissue damage. As such, it would be useful to have tools to modulate cytokine production or activity. Listed in Table 4 are cytokine inhibitors currently being evaluated in U.S. human clinical trials. Several strategies have been developed to lower endogenous cytokine levels. These include (1) neutralizing monoclonal antibodies, (2) soluble receptors, (3) neutral inhibitors (i.e., UK-1ral), and (4) drugs that inhibit cytokine production. We will briefly examine each of these strategies. For a more detailed examination of the general principles and limitations of anticytokine therapy, the reader is referred elsewhere (187).

Neutralizing mAbs have been developed against many cytokines and used to inhibit cytokine activity in vitro and in vivo (187,188). Indeed, according to a 1995 survey by the Pharmaceutical Research and Manufacturers of America (Washington, DC), mAbs outnumber all other product categories among the 234 biotechnology drugs currently in clinical trials (180). Specificity is excellent, and the circulating half-life relatively long. However, these same problems can cause problems if the mAb is present in relatively low amounts in comparison with the ligand. The resulting soluble mAb-ligand complexes can dissociate and release the cytokine, which then binds to high-affinity receptors on the cell surface. In these circumstances, the mAb serves as a cytokine carrier rather than a neutralizing molecule. Such an effect has been shown for IL-3, IL-4, IL-6, and other cytokines (187). To reduce the time during which soluble cytokine-mAb complexes might remain in circulation, some investigators have utilized Fab fragments. Although the faster clearance of the smaller Fab-cytokine complexes is advantageous, the shorter half-life of free Fab fragments also limits the effectiveness of this strategy. One strategy to prevent anticytokine mAbs from acting as carriers is to use several mAbs that recognize different epitopes on the cytokine molecule (187–189). The multimeric cytokine-mAb complexes that form result in faster clearance of the cytokine, reducing the likelihood of its dissociating from the mAbs and binding to cell receptors.

Table 4 Cytokines Inhibitors in Clinical Trials

Product	Company	Indication	U.S. Development Status
Anti-TNF mAb	Chiron Emeryville, CA Miles, Inc. West Haven, CT	Sepsis syndrome	Phase III
CenTNF chimeric antibody	Centocor Malvern, PA	Inflammatory bowel disease	Phase II
Antril IL-1ra	Amgen Thousand Oaks, CA	Inflammatory bowel disease	Phase I/II
MAK195F F(ab′)2 TNF mAb	Knoll Pharmaceuticals Whippany, NJ	Severe sepsis	Phase I/II
Thalidomide		Tuberculosis	Phase I

Source: Adapted from Refs. 180, 198.

Soluble receptors for several cytokines are released by leukocytes and other cells in vitro and in vivo (18,44,187). These receptors are often upregulated during the natural production of the ligand, resulting in a coordinated system for downmodulating the biological responses mediated by that cytokine. Ligand specificity and binding affinity can differ somewhat between soluble and cell-surface cytokine receptors. For example, cell-surface and soluble-type 2 IL-1R demonstrate similar binding of IL-1β, but only the latter significantly binds the IL-1β precursor molecule (9). Unfortunately, these soluble receptors demonstrate relatively short circulating half-lives (in the order of hours) that limit their effectiveness in vivo (187,189). To circumvent this limitation, some investigators have engineered recombinant chimeric molecules that link the ligand-binding domain of the soluble receptor with the Fc region of a consensus human IgG molecule. Depending on the intended use, various isotypes of IgG can be employed to confer the desired effector function. Immunoadhesions demonstrate a longer circulating half-life (typically weeks) and have high specificity and improved affinity and avidity (190). In terms of utility for clinical use, immunoadhesions have another advantage over mAbs because they are entirely comprised of human molecules. The only potential antigenic sequence is that in the joining region. This was borne out in human clinical trials using CD4-IgG, in which the immunoadhesion was shown to be entirely nonimmunogenic (191,192). Results of in vitro animal studies have shown that TNF, IL-1, and IFN-γ receptor IgG chimeras all bind their respective ligands bivalently and with high affinity (190,193,194). In experimental infection models, these same immunoadhesions have been shown to block endogenous cytokines, thereby demonstrating their potential ability to act as a cytokine specific "sponge" to absorb excess cytokine in vivo.

At present there is only one known example of a naturally occurring receptor antagonist: IL-1ra (9,187). As indicated earlier, IL-1ra is related to IL-1, but does not transmit an activating signal when it binds to cell-surface IL-1R. IL-1ra binds cell surface type I IL-1R and less well to cell surface type 2 IL-1R, but it demonstrates little binding to soluble type 2 IL-1R (9). Thus, IL-1ra is an important part of an elegant regulatory system that coordinately links the potent inflammatory effects of IL-1β with two potent molecules (IL-1ra and type 2 IL-1R) that limit the extent and duration of action of IL-β. Although IL-1ra would seem to be a natural candidate for clinical use, it is severely limited by its very short circulating half-life (on the order of minutes), which is similar to that of IL-1β itself (9,187).

The use of drugs to prevent or reduce cytokine production also deserves attention. The ideal strategy would feature ease of administration, perhaps by oral ingestion, rather than require repeated injection of expensive bioengineered molecules. There are several drugs that inhibit expression of either cytokines or cytokine receptors. Perhaps the best recognized is cyclosporine, which inhibits IL-2 production and modulates resistance to various pathogens. Glucocorticoids, such as dexamethasone, have potent effects on immune cells that can be reflected in reduced production of various cytokines. In addition, dexamethasone also increases the production of certain cytokine inhibitors (i.e, soluble type 2 IL-1R) that dampen the effects of the cytokine cascade and probably contribute to the anti-inflammatory properties of this class of molecules (16,18). Other examples of cytokine inhibitors are provided by pentoxifylline, a methylxanthine derivative, and chlorpromazine, both of which have been shown to significantly reduce production of TNF-α following endotoxin challenge (195–198). Treatment with chlorpromazine also significantly reduced IL-1β and soluble TNF receptor in a mouse model of endotoxemia (197). Thalidomide, a drug that has caused devastating teratogenic effects in the human fetus, has a beneficial effect in human leprosy by reducing the release of TNF-α from infected macrophages (198). Other classes of small molecular weight compounds are being evaluated for their ability to prevent the production of TNF-α (199). Perhaps one or more of these may be used in the future to modulate the effects of TNF-α, thereby maximizing host defense and minimizing adverse effects.

Yet another strategy to inhibit cytokine function is the use of antisense oligonucleotides. They have the potential to specifically inhibit cytokine protein expression, rather than removing already expressed protein from the system. Significant research has been undertaken to develop methods for use of antisense technology to modulate cytokine expression (reviewed in Ref. 200). Clinical trials employing antisense oligonucleotides currently are under consideration for the treatment of viral infection and neoplasia, thus perhaps leading to use of this novel tool for immunomodulation in bacterial and fungal infection in the future.

Cytokines as Adjuvants and as Adjuncts to Antimicrobial Therapy

When recombinant cytokines became available in the mid-1980s, there was great enthusiasm that medicine's ability to harness the immune system would soon become possible. It was thought that administration of recombinant IFN-γ, or IL-1, or IL-2, or the succession of other cytokines, would soon provide humankind with new and potent weapons to make immunotherapy of infectious diseases a practical reality. Perhaps it reflects the maturation of cytokine biology as a discipline that such a blissfully optimistic scenario is no longer being painted. Instead, we have come to have a profound appreciation of the complexity and multiple levels of regulation of the immune system and the pivotal roles of cytokines in these processes. However, this does not mean that practical therapeutic use of cytokines is no longer a possibility.

One area in which cytokines have demonstrated some potential is as adjuvants for immunization. Several cytokines have been shown to augment the immune response to antigens. These include administration of IL-1 and IL-2, which have been shown to increase the antibody response to vaccination in cattle and sheep (32,33,201–202). IL-1 and IL-2 have also been used to augment adoptive transfer of resistance by spleen cells from *Listeria*-immunized mice (34,143). Exciting results were reported regarding the use of IL-12 as an adjuvant to immunize mice against *L. major* infection using an inactivated *Leishmania* vaccine, which by itself was unable to elicit substantial protection (203). Similar results were recently published regarding immunization of mice against *L. monocytogenes* using IL-12 and nonviable listerial antigens (204). These findings and others raise the hope that cytokines, or cytokine "cocktails," might be used in the future to fine-tune the immune response elicited by a given vaccine, to optimize the desired response and minimize untoward effects of the vaccination (6,7,98).

In addition, some successes have been reported in the use of cytokines as an adjunct to antibacterial therapy. Nakamura et al. (5) demonstrated significantly increased resistance to gram-negative bacterial infection in mice when they concurrently received recombinant human IL-1 and antibiotics (5). Other workers have demonstrated the benefits of combined antibiotic therapy with either IL-1 or IL-2 in cattle with naturally occurring or experimental *S. aureus* mastitis (31). A significant augmentation of the effectiveness of antibiotic therapy by co-administration of recombinant TNF-α also was reported for mice with experimental *S. aureus* mastitis (205). As described earlier, encouraging results have been obtained with IL-2 and IFN-γ in lepromatous leprosy patients (114–116). Promising results also were obtained with administration of IFN-γ to human patients with atypical mycobacterial infection that was refractory to treatment with antibiotics alone (118,119). These and other efforts should provide new insights into potential strategies to harness the beneficial effects of cytokines as an adjunct to conventional antimicrobial therapy.

FUTURE DIRECTIONS

A recurring theme in cytokine biology is the double-edged sword posed by their biological activities. Although essential for initiating and sustaining host defense against bacterial and fungal invaders, excess endogenous or exogenously administered cytokine can have grave toxic consequences for the host. Strategies to exploit the potential advantages of cytokine immunotherapy must prevent their deleterious biological effects without interfering with host defense and homeostasis. Like Odysseus passing between Scylla and Charybdis in the Strait of Messina, the hope and future success of cytokine immunotherapy will depend upon skillful navigation.

REFERENCES

1. Roth JA. Enhancement of nonspecific resistance to bacterial infection by biologic response modifiers. In: Roth JA, ed. Virulence Mechanisms of Bacterial Pathogens. Washington, DC: American Society for Microbiology, 1988:329–363.
2. Van Deuren M, Dofferhoff ASM, Van Der Meer JWM. Cytokines and the response to infection. J Pathol 1992; 168:349–356.
3. Lloyd AR, Johnston J. Cytokines and cytokine receptors in health and disease. Cytokine 1993; 5:399–406.
4. Czuprynski CJ, Brown JF. Recombinant murine interleukin-1α enhancement of nonspecific antibacterial resistance. Infect Immun 1987; 55:2061–2065.
5. Nakamura S, Minami A, Fujimato K, Kojima T. Combination effect of recombinant human interleukin-1α with antimicrobial agents. Antimicrob Agents Chemother 1989; 33:1804–1810.
6. Nohria A, Rubin RH. Cytokines as potential vaccine adjuvants. Biotherapy 1994; 7:261–269.
7. Taylor CE. Cytokines as adjuvants for vaccines: antigen-specific responses differ from polyclonal responses. Infect Immun 1995; 63:3241–3244.
8. Antman KS, Griffin JD, Elias A, Socinski MA, Ryan L, Cannistra SA, Oette D, Whitley M, Frei E, Schnipper LE. Effect of recombinant human granulocyte-macrophage colony-stimulatory factor on chemotherapy-induced myelosuppression. N Engl J Med 1988; 319:593–598.
9. Dinarello CA. The interleukin-1 family: 10 years of discovery. FASEB J 1994; 8:1314–1325.
10. Dinarello CA. Role of interleukin-1 in infectious diseases. Immunol Rev 1992; 127:119–146.
11. Kurt-Jones EA, Beller DI, Mizel SB, Unanue ER. Identification of a membrane-associated interleukin-1 in macrophages. Proc Natl Acad Sci USA 1985; 82:1204–1208.
12. Thornberry NA, Bull HG, Calaxcay JR, Champman KT, Howard AD, Kostora MJ, Miller DK, Molineaux SM, Weidner JR, Aunins J, Schmidt JA, Tocci M. A novel heterodimeric cystine protease is required for interleukin-1 beta processing in monocytes. Nature 1992; 356:768–774.
13. Haskill S, Martin M, VanLe L, Morris J, Peace A, Bigler CF, Jaffe GJ, Sporn SA, Fong S, Areid WP, Ralph P. cDNA cloning of a novel form of the interleukin-1 receptor antagonist associated with epithelium. Proc Natl Acad Sci USA 1991; 88:3681–3685.
14. Sims JE, March CJ, Cosman D, Widner MB, MacDonald HR,

McMahan CJ, Grubin CE, Wignall JM, Jackson JL, Cull SM, et al. cDNA expression cloning of the IL-1 receptor, a member of the immunoglobulin superfamily. Science 1988; 241:585–589.
15. McMahon CJ, Slack JL, Mosley B, Cosman D, Lupton SD, Bronton LL, Grubin CE, Wignall JM, Jenkins NA, Brannan CI, Copeland NG, Huebner K, Cruce CM, Cunnizano LA, Benjamin D, Dower S, Spriggs MK, Sims JE. A novel IL-1 receptor cloned from B cells by mammalian expression is expressed in many cell types. EMBO J 1991; 10:2821–2832.
16. Sims JE, Gayle MA, Slack JL, Alderson MR, Bard TA, Giri JG, Colotta F, Re F, Mantovani A, Shanebeck K, Grabstein KH, Dower SK. Interleukin-1 signalling occurs exclusively via the type I receptor. Proc Natl Acad Sci USA 1993; 90:6155–6159.
17. Gallis B, Prickett KS, Jackson J, Slack J, Schooley K, Sims JE, Dower SK. IL-1 induces rapid phosphorylation of the IL-1 receptor. J Immunol 1989; 143:3235–3240.
18. Symons JA, Eastgate JA, Duff GW. Purification and characterization of a novel soluble receptor for interleukin-1. J Exp Med 1991; 174:1251–1254.
19. Kampschmidt RF, Pullian LA. Stimulation of antimicrobial activity in the rat with leukocytic endogenous mediator. J Reticuloendothel Soc 1975; 17:162–169.
20. Ozaki Y, Ohashi T, Minami A, Nakamura S. Enhanced resistance to bacterial infection induced by recombinant human interleukin-1a. Infect Immun 1987; 55:1436–1440.
21. Van Der Meer JWM, Barza M, Wolff SM, Dinarello CA. Low dose recombinant interleukin-1 protects granulocytopenic mice from lethal gram-negative infection. Proc Natl Sci USA 1988; 85:1620–1623.
22. Czuprynski CJ, Brown JF, Young RM, Cooley AJ, Kurtz RS. Effects of murine recombinant interleukin-1α on the host response to bacterial infection. J Immunol 1988; 140:962–968.
23. Van't Wout JW, Van der Meer JWM, Barza M, Dinarello CA. Protection of neutropenic mice from lethal *Candida albicans* infection by recombinant interleukin-1. Eur J Immunol 1988; 18:1143–1146.
24. Minami A, Fujimoto K, Ozaki Y, Nakamura S. Augmentation of host resistance to microbial infections by recombinant human interleukin-1α. Infect Immun 1988; 56:3116–3120.
25. Denis M. Recombinant interleukin-1 infusion increases resistance of BALBC mice to murine leprosy. Int J Immunopharmacol 1991; 13:897–902.
26. Shi J, Goodbund RD, Chenagappa MM, Nelssen JL, Tokach MD, McVey DS, Blecha F. Influence of interleukin-1 or neutrophil function and resistance to *Streptococcus suis* in neonatal pigs. J Leukoc Biol 1994; 56:88–94.
27. Zhan YF, Stanley ER, Cheers C. Prophylaxis of experimental brucellosis with interleukin-1. Infect Immun 1991; 59:1790–1794.
28. Movat HZ, Burrowes CE, Cybulsky MI, Dinarello CA. Acute inflammation and a Shwartzman-like reaction induced by interleukin-1 and tumor necrosis factor. Am J Pathol 1987; 129:463–476.
29. Van Der Meer JWM, Vogels MTE, Kullberg BJ. Interleukin-1 and related pro-inflammatory cytokines in the treatment of bacterial infections in neutropenic and non-neutropenic animals. Biotherapy 1994; 7:161–167.
30. Roll JT, Young KM, Kurtz RS, Czuprynski CJ. Human rTNFα augments anti-bacterial resistance in mice: potentiation of its effects by recombinant human rIL-1α. Immunology 1990; 69:316–322.
31. Daley M, Williams T, Coyle P, Furda G, Dougherty R, Hayes P. Prevention and treatment of *Staphylococcus aureus* infections with recombinant cytokines. Cytokine 1993; 5:276–284.
32. Reddy DN, Reddy PG, Minocha HC, Fenwick BW, Baker PE, Davis WC, Blecha F. Adjuvanticity of recombinant bovine interleukin-1 beta: influence on immunity, infection, and latency in a bovine herpesvirus-1 infection. Lymphokine Res 1990; 9:295–307.
33. Reddy DN, Reddy PG, Zue W, Minocha HC, Daley MJ, Blecha F. Immunopotentiation of bovine respiratory disease virus vaccines by interleukin-1 β and interleukin-2. Vet Immunol Immunopathol 1993; 37:25–38.
34. Haak-Frendscho M, Kurtz RS, Czuprynski CJ. rIL-1α enhances adoptive transfer of resistance to *Listeria monocytogenes* infection. Microb Pathogen 1991; 10:385–392.
35. Havell EA, Moldawer LL, Helfgott D, Killian PL, Sehgal PB. Type I IL-1 receptor blockade exacerbates murine listeriosis. J Immunol 1992; 148:1486–1492.
36. Chen W, Havell EA, Moldawer LL, McIntyre KW, Chizzonite RA, Harmsen AG. Interleukin 1: an important mediator of host resistance against *Pneumocystis carinii.* J Exp Med 1992; 176:713–718.
37. Kenefick KB, Adams JL, Steinberg H, Czuprynski CJ. In vivo administration of a monoclonal antibody against the type I IL-1 receptor inhibits the ability of mice to eliminate *Mycobacterium paratuberculosis.* J Leukoc Biol 1994; 55:719–722.
38. Wakabayashi G, Gelfand JA, Bonke JF, Thompson RC, Dinarello CA. A specific receptor antagonist for interleukin-1 prevents *Escherichia coli* induced shock. FASEB J 1991; 5:338–343.
39. Aiura K, Gelfand JA, Burke JF, Thompson RC, Dinarello CA. Interleukin-1 (IL-1) receptor antagonist prevents *Staphylococcus epidermidis*-induced hypotension and reduces circulating levels of tumor necrosis factor and IL-1β in rabbits. Infect Immun 1993; 61:3342–3350.
40. Sherry B, Cerami A. Cachectin/tumor necrosis factor exerts endocrine, paracrine and autocrine control of inflammatory responses. J Cell Biol 1988; 107:1269–1277.
41. Ruddle NH. Tumor necrosis factor (TNF-α) and lymphotoxin (TNF-β). Curr Opin Immunol 1992; 4:327–332.
42. Tracey KJ, Cerami A. Tumor necrosis factor, other cytokines and disease. Annu Rev Cell Biol 1993; 9:317–343.
43. Beutler B, Van Huffel C. Unraveling function in the TNF ligand and receptor families. Science 1994; 264:667–668.
44. Spinas GA, Keller U, Brockhaus M. Release of soluble receptors for tumor necrosis factor (TNF) in relation to circulating TNF during experimental endotoxinemia. J Clin Invest 1992; 90:533–536.
45. Calandra T, Baumgartner J-D, Grau GE, Wu M-M, Lambert P-H, Schellekens J, Verhoef J, Glauser MP. Swiss-Dutch J5 Immunoglobulin Study Group. Prognostic values of tumor necrosis factor/cachectin, interleukin-1, interferon-α, and interferon-γ in the serum of patients with septic shock. J Infect Dis 1990; 161:982–987.
46. Michie HR, Manogue KR, Spriggs DR, Revhaug A, O'Dwyer S, Dinarello CA, Ceraui A, Wolff SM, Wilmore DW. Detection of circulating tumor necrosis factor after endotoxin administration. N Engl J Med 1988; 315:1481–1486.
47. Adams JL, Semrad SD, Czuprynski CJ. Administration of bacterial lipopolysaccharide elicits circulating tumor necrosis factor-alpha in neonatal calves. J Clin Microbiol 1990; 28:998–1001.
48. Beutler B, Milsark IW, Cerami A. Passive immunization against cachectin/tumor necrosis factor (TNF) protects mice

from the lethal effects of endotoxin. Science 1985; 229:869–871.
49. Pfeffer K, Matsuyama T, Kündig TM, Wakeham A, Kishihara K, Shahinian A, Wiegmann K, Ohashi PS, Krönke M, Mak TW. Mice deficient for the 55 kD tumor necrosis factor receptor are resistant to endotoxic shock, yet succumb to *L. monocytogenes* infection. Cell 1993; 73:457–467.
50. Rothe J, Lessiauer W, Lötscher H, Lang Y, Koebel P, Kontgen F, Althage A, Zinkernagel R, Steinmetz M, Bluethmann H. Mice lacking the tumour necrosis factor receptor 1 are resistant to TNF-mediated toxicity but highly susceptible to infection by *Listeria monocytogenes*. Nature 1993; 364:798–802.
51. Barnes PF, Chatterjee D, Brennan P, Rea TH, Modlin RL. Tumor necrosis factor production in patients with leprosy. Infect Immun 1992; 60:1441–1446.
52. Pisa P, Gennene M, Söder O, Ottenhoff T, Hansson M, Kiessling R. Serum tumor necrosis factor levels and disease dissemination in leprosy and leishmaniasis. J Infect Dis 1990; 161:988–991.
53. Blanchard DK, Djeu JY, Klein TW, Friedman H, Stewart WE II. Protective effects of tumor necrosis factor in experimental *Legionella pneumophila* infections of mice via activation of PMN function. J Leukoc Biol 1988; 43:429–435.
54. Havell EA. Evidence that tumor necrosis factor has an important role in antibacterial resistance. J Immunol 1989; 143:2894–2899.
55. Nakane A, Minagawa T, Kato K. Endogenous tumor necrosis factor (cachectin) is essential to host resistance against *Listeria monocytogenes* infection. Infect Immun 1988; 56:2563–2569.
56. Langermans JAM, Van Der Hulst MEB, Nibbering PH, Van Furth R. Endogenous tumor necrosis factor alpha is required for enhanced antimicrobial activity against *Toxoplasma gondii* and *Listeria monocytogenes* in recombinant gamma interferon-treated mice. Infect Immun 1992; 60:5107–5112.
57. Parant F, Tavernier J, Fiers W, Parant M. Comparative activity of human and murine tumor necrosis factor in toxicity and anti-infectious assays in mice. Microb Pathogen 1990; 8:143–149.
58. Cross AS, Sadoff JC, Kelly N, Bernton E, Gemski P. Pretreatment with recombinant murine tumor necrosis factor α/cachectin and murine interleukin 1 α protects mice from lethal bacterial infection. J Exp Med 1989; 169:2021–2027.
59. Williams DM, Magee DM, Bonewald LF, Smith JG, Bleicker CA, Byrne GI, Schachter J. A role in vivo for tumor necrosis factor alpha in host defense against *Chlamydia trachomatis*. Infect Immun 1990; 58:1572–1576.
60. Chen W, Havell EA, Harmsen AG. Importance of endogenous tumor necrosis factor alpha and gamma interferon in host resistance against *Pneumocystis carinii* infection. Infect Immun 1992; 60:1279–1284.
61. Aguirre K, Havell EA, Gibson GW, Johnson LL. Role of tumor necrosis factor and gamma interferon in acquired resistance to *Cryptococcus neoformans* in the central nervous system of mice. Infect Immun 1995; 63:1725–1731.
62. Kindler V, Sappino A-P, Grau GE, Piguet P-F, Vassalli P. The inducing role of tumor necrosis factor in the development of bactericidal granulomas during BCG infection. Cell 1989; 56:731–740.
63. Denis M. Cytokine modulation of *Mycobacterium lepraemurium* infection in mice; important involvement of tumor necrosis factor, interleukin-2 and dissociation from macrophage activation. Int J Immunopharmacol 1991; 13:889–895.
64. Nassif X, Mathison JC, Wolfson E, Koziol JA, Ulevitch RJ, So M. Tumour necrosis factor alpha antibody protects against lethal meningococcaemia. Mol Microbiol 1992; 6:591–597.
65. Steinshamn S, Waage A. Tumor necrosis factor and interleukin-6 in *Candida albicans* infection in normal and granulocytopenic mice. Infect Immun 1992; 60:4003–4008.
66. Wu-Hsieh BA, Lee G-S, Franco M, Hofman FM. Early activation of splenic macrophages by tumor necrosis factor alpha is important in determining the outcome of experimental histoplasmosis in mice. Infect Immun 1992; 60:4230–4238.
67. Nakane A, Minagawa T, Kohanawa M, Chen Y, Sato H, Moriyama M, Tsuruoka N. Interactions between endogenous gamma interferon and tumor necrosis factor in host resistance against primary and secondary *Listeria monocytogenes* infections. Infect Immun 1989; 57:3331–3337.
68. Cox G, Gauldie J. Structure and function of interleukin-6. In: Remick D, Kunkel S, eds. Cytokines in Health and Disease. New York: Marcel Dekker, 1992:97–120.
69. Kishimoto T, Taga T, Akira S. Cytokine signal transduction. Cell 1994; 76:253–262.
70. Starnes HF Jr, Pearce MK, Tewari A, Yim JH, Zou J-C, Abrams JS. Anti-IL-6 monoclonal antibodies protect against lethal *Escherichia coli* infection and lethal tumor necrosis factor-α challenge in mice. J Immunol 1990; 145:4185–4191.
71. Havell EA, Sehgal PB. Tumor necrosis factor-independent IL-6 production during murine listeriosis. J Immunol 1991; 146:756–761.
72. Saunders BM, Liu Z, Zhan Y, Cheers C. Interleukin-6 production during chronic experimental infection. Immunol Cell Biol 1993; 71:275–280.
73. Ziegler-Heitbrock HWL, Passlick B, Käfferlein E, Coulie PG, Izbicki JR. Protection against lethal pneumococcal septicemia in pigs is associated with decreased levels of interleukin-6 in blood. Infect Immun 1992; 60:1692–1694.
74. Hedges S, Svensson M, Svanborg C. Interleukin-6 response of epithelial cell lines to bacterial stimulation in vitro. Infect Immun 1992; 60:1295–1301.
75. Czuprynski CJ, Haak-Frendscho M, Maroushek N, Brown JF. Effects of human interleukin-6 alone and in combination with interleukin-1α and tumor necrosis factor α, on antibacterial resistance in mice. Antimicrob Agents Chemother 1992; 36:68–70.
76. Liu Z, Simpson RJ, Cheers C. Role of interleukin-6 in T-cell activation during primary and secondary infection with *Listeria monocytogenes*. Infect Immun 1995; 63:2790–2792.
77. Liu Z, Simpson RJ, Cheers C. Role of IL-6 in activation of T cells for acquired cellular resistance to *Listeria monocytogenes*. J Immunol 1994; 152:5375–5380.
78. Kopf M, Baumann H, Freers G, Freudenberg M, Lamers M, Kishimoto T, Zinkernagels R, Bluethmann H, Köhler. Impaired immune and acute-phase responses in interleukin-6-deficient mice. Nature 1994; 368:339–342.
79. Dalrymple SA, Lucian LA, Slattery R, McNeil T, Aud DM, Fuchino S, Lee F, Murray R. Interleukin-6-deficient mice are highly susceptible to *Listeria monocytogenes* infection: correlation with inefficient neutrophilia. Infect Immun 1995; 63:2262–2268.
80. Conlan JW, North RJ. Neutrophil-mediated dissolution of infected host cells as a defense strategy against a facultative intracellular bacterium. J Exp Med 1991; 174:741–744.
81. Denis M. Interleukin-6 is used as growth factor by virulent *Mycobacterium avium*: presence of specific receptors. Cell Immunol 1992; 141:182–188.
82. Miller MD, Krangel MS. Biology and biochemistry of the chemokines: a family of chemotactic and inflammatory cytokines. Crit Rev Immunol 1991; 12:17–46.
83. Kelner GS, Kennedy J, Bacon KB, Kleyensteuber S, Largaespada DA, Jenkins NA, Copeland NG, Bazan JF, Moore KW,

Schall TJ, Zlotnik A. Lymphotactin: a novel cytokine which represents a new class of chemokine. Science 1994; 266:1395–1399.

84. Schall TJ, Bacon KB. Chemokines, leukocyte trafficking, and inflammation. Curr Opin Immunol 1994; 6:865–873.
85. Baggiolini M, Dewald B, Moser B. Interleukin-8 and related chemotactic cytokines—CXC and CC chemokines. Adv Immunol Rev 1994; 55:97–179.
86. Schall TJ. The chemokines. In: Thomson A, ed. The Cytokine Handbook. 2d ed. New York: Academic Press, 1994:419–460.
87. Chollet-Martin S, Montravers P, Gibert C, Elbim C, Desmonts JM, Fagon JY, Gougerot-Pocidalo MA. High levels of interleukin-8 in the blood and alveolar spaces of patients with pneumonia and adult respiratory distress syndrome. Infect Immun 1993; 61:4553–4559.
88. Donnelly SC, Strieter RM, Kunkel SL, Walz A, Robertson CR, Carter DC, Grant IS, Pollok AJ, Haslett C. Interleukin-8 and development of adult respiratory distress syndrome in at-risk patient groups. Lancet 1993; 341:643–647.
89. Wilkinson PC, Newman I. Identification of IL-8 as a locomotor attractant for activated human lymphocytes in mononuclear cell cultures with anti-CD3 or purified protein derivative of *Mycobacterium tuberculosis.* J Immunol 1992; 149:2689–2694.
90. Larsen CG, Thomsen MK, Gesser B, Thomsen PD, Deleuran BW, Nowak J, Skødt, Thomsen HK, Deleuran M, Thestrup-Pedersen K, Harada A, Matsushima K, Menné T. The delayed-type hypersensitivity reaction is dependent on IL-8. J Immunol 1995; 155:2151–2157.
91. Agace WW, Hedges SR, Ceska M, Svanborg C. Interleukin-8 and the neutrophil response to mucosal gram-negative infection. J Clin Invest 1993; 92:780–785.
92. Eckmann L, Kagnoff MF, Fierer J. Epithelial cells secrete the chemokine interleukin-8 in response to bacterial entry. Infect Immun 1993; 61:4569–4574.
93. George AJT. Surface-bound cytokines—a possible effector mechanism in bacterial immunity? Immunol Today 1994; 15:89–90.
94. Jones AP, Webb LMC, Anderson AO, Leonard EJ, Rot A. Normal human sweat contains interleukin-8. J Leukoc Biol 1995; 57:434–439.
95. Mosmann TR, Cherwinski H, Bond MW, Giedlin MA, Coffman RL. Two types of murine helper T cell clones. I. Definition according to profiles of lymphokine activities and secreted proteins. J Immunol 1986; 136:2348–2357.
96. Kelso A. Th1 and Th2 subsets: paradigms lost? Immunol Today 1995; 16:374–379.
97. Shieh C-S, Macatonia SE, Tripp CS, Wolf SF, O'Gara A, Murphy KM. Development of T_H1 CD4+ T cells through IL-12 produced by *Listeria*-induced macrophages. Science 1993; 260:547–549.
98. Locksley RM. Interleukin 12 in host defense against microbial pathogens. Proc Natl Acad Sci USA 1993; 90:5879–5880.
99. Jacob CO. Cytokines and anti-cytokines. Curr Opin Immunol 1989; 2:249–257.
100. Modlin RL, Nutman TB. Type 2 cytokines and negative immune regulation in human infections. Curr Opin Immunol 1993; 5:511–517.
101. Wallis RS, Ellner JJ. Cytokines and tuberculosis. J Leukoc Biol 1994; 55:676–681.
102. Brown WC, Davis WC, Dobbelaere D, Rice-Ficht AC. CD4+ T-cell clones obtained from cattle chronically infected with *Fascida hepatica* and specific for adult worm antigen express both unrestricted and Th2 cytokine profiles. Infect Immun 1994; 62:818–827.
103. Hendrzak JA, Brunda MJ. Biology of disease. Interleukin-12 biologic activity, therapeutic utility, and role in disease. Lab Invest 1995; 72:619–637.
104. Kaufmann SHE. Immunity to intracellular bacteria. Annu Rev Immunol 1993; 11:129–163.
105. Tripp CS, Wolf SF, Unanue ER. Interleukin 12 and tumor necrosis factor α are costimulators of interferon γ production by natural killer cells in severe combined immunodeficiency mice with listeriosis, and interleukin 10 is a physiologic antagonist. Proc Natl Acad Sci USA 1993; 90:3725–3729.
106. Sieling PA, Wang X-H, Gately MK, Oliveros JL, McHugh T, Barnes PF, Wolf SF, Golkar L, Yamamura M, Yogi Y, Uyemura K, Rea TH, Modlin RL. IL-12 regulates T helper type 1 cytokine responses in human infectious disease. J Immunol 1994; 153:3639–3647.
107. De Maeyer E, De Maeyer-Guignard J. Interferon-γ. Curr Opin Immunol 1992; 4:321–326.
108. Thompson CB. Distinct roles for the costimulatory ligands B7-1 and B7-2 in T helper cell differentiation? Cell 1995; 81:979–982.
109. Iizawa Y, Brown JF, Czuprynski CJ. Early expression of cytokine mRNA in mice infected with *Listeria monocytogenes.* Infect Immun 1992; 60:4068–4073.
110. Liu W, Kurlander RJ. Analysis of the interrelationship between IL-12, TNF-α, and IFN-γ production during murine listeriosis. Cell Immunol 1995; 163:260–267.
111. Huang S, Hendriks W, Althage A, Hemmi S, Bluethmann H, Kamijo R, Vilcek J, Zinkernagel RM, Aguet M. Immune response in mice that lack the interferon-γ receptor. Science 1993; 259:1693–1694.
112. Yamamura M, Uyemura K, Deans RJ, Weinberg K, Rea TH, Bloom BR, Modlin RL. Defining protective responses to pathogens: cytokine profiles in leprosy lesions. Science 1991; 254:277–279.
113. Barnes PF, Lu S, Abrams JS, Wang E, Yamamura M, Modlin RL. Cytokine production at the site of disease in human tuberculosis. Infect Immun 1993; 61:3482–3489.
114. Kaplan G, Mathur NK, Job CK, Nath I, Cohn ZA. Effect of multiple interferon γ injections on the disposal of *Mycobacterium leprae.* Proc Natl Acad Sci USA 1989; 86:8073–8077.
115. Kaplan G. Recent advances in cytokine therapy in leprosy. J Infect Dis 1993; 167:S18–S22.
116. Kaplan G, Britton WJ, Hancock GE, Theuvenet WJ, Smith KA, Job CK, Roche PW, Molloy A, Burkhardt R, Barker J, Pradhan HM, Cohn ZA. The systemic influence of recombinant interleukin 2 on the manifestations of lepromatous leprosy. J Exp Med 1991; 173:993–1006.
117. Nathan C, Squires K, Griffo W, Levis W, Varghese M, Job CK, Nusrat AR, Sherwin S, Rapoport S, Sanchez E, Burkhardt RA, Kaplan G. Widespread intradermal accumulation of mononuclear leukocytes in lepromatous leprosy patients treated systemically with recombinant interferon γ. J Exp Med 1990; 172:1509–1512.
118. Holland SM, Eisenstein EM, Kuhns DB, Turner ML, Fleisher TA, Stroker W, Gallin JI. Treatment of refractory disseminated nontuberculous mycobacterial infection with interferon gamma. N Engl J Med 1994; 330:1348–1355.
119. Chatte G, Panteix G, Perrin-Fayolle M, Pacheco Y. Aerosolized interferon gamma for *Mycobacterium avium*-complex lung disease. Am J Respir Crit Care Med 1995; 152:1094–1096.
120. Appelberg R, Castro AG, Pedrosa J, Silva RA, Orme IM, Minóprio P. Role of gamma interferon and tumor necrosis factor alpha during T-cell-independent and -dependent phases of *Mycobacterium avium* infection. Infect Immun 1994; 62:3962–3971.

121. Flynn JL, Chan J, Triebold KJ, Dalton DK, Stewart TA, Bloom BR. An essential role for interferon γ in resistance to *Mycobacterium tuberculosis* infection. J Exp Med 1993; 178:2249–2254.
122. Matsumura H, Onozuka K, Terada Y, Nakano Y, Nakano M. Effect of murine recombinant interferon-γ in the protection of mice against *Salmonella*. Int J Immunopharmacol 1990; 12:49–56.
123. Zhong G, Peterson EM, Czarniecki CW, De La Maza LM. Recombinant murine gamma interferon inhibits *Chlamydia trachomatis* serovar L1 in vivo. Infect Immun 1988; 56:283–286.
124. Zhou P, Sieve MC, Bennett J, Kwon-Chung KJ, Tewari RP, Gazzinelli RT, Sher A, Seder RA. IL-12 prevents mortality in mice infected with *Histoplasma capsulatum* through induction of IFN-γ. J Immunol 1995; 155:785–795.
125. Li H, Jerrells TR, Spitalny GL, Walker DH. Gamma interferon as a crucial host defense against *Rickettsia conorii* in vivo. Infect Immun 1987; 55:1252–1255.
126. Kanaly ST, Hines SA, Palmer GH. Cytokine modulation alters pulmonary clearance of *Rhodococcus equi* and development of granulomatous pneumonia. Infect Immun 1995; 63:3037–3041.
127. Hershman MJ, Polk HC Jr, Ietsch JD, Shields RE, Wellhausen SR, Sonnenfeld G. Modulation of infection by gamma interferon treatment following trauma. Infect Immun 1988; 56:3412–2416.
128. Garner RE, Kuruganti U, Czarniecki CW, Chiu HH, Domer JE. In vivo immune responses to *Candida albicans* modified by treatment with recombinant murine gamma interferon. Infect Immun 1989; 57:1800–1808.
129. Romani L, Cenci E, Mencacci A, Spaccapelo R, Grohman U, Puccetti P, Bistoni F. Gamma interferon modifies CD4+ subset expression in murine candidiasis. Infect Immun 1992; 60:4950–4952.
130. Kohler J, Heumann D, Garotta G, LeRoy D, Bailat S, Barras C, Baumgartner J-D, Glauser MP. IFN-γ involvement in the severity of gram-negative infections in mice. J Immunol 1993; 151:916–921.
131. Keane-Myers A, Nickell SP. Role of IL-4 and IFN-γ in modulation of immunity to *Borrelia burgdorferi* in mice. J Immunol 1995; 155:2020–2028.
132. Schmitz JL, Schell RF, Callister SM, Lovrich SD, Day SP, Coe JE. Immunoglobulin GL confers protection against *Borrelia burgdorferi* infection in LSH hamsters. Infect Immun 1992; 60:2677–2682.
133. Haak-Frendscho M, Brown JF, Iizawa Y, Wagner RD, Czuprynski CJ. Administration of anti-IL-4 monoclonal antibody 11B11 increases the resistance of mice to *Listeria monocytogenes* infection. J Immunol 1992; 148:3978–3985.
134. Sadlack B, Merz H, Schorie H, Schimpl A, Feller AC, Horak I. Ulcerative colitis like disease in mice with a disrupted interleukin-2 gene. Cell 75:253–261.
135. Kaplan G, Sampaio EP, Walsh GP, Burkhardt RA, Fajardo IT, Gido LS, Machado ADM, Cellona RV, Abalos RM, Sarno EN, Cohn ZA. Influence of *Mycobacterium leprae* and its soluble products on the cutaneous responsiveness of leprosy patients to antigen and recombinant interleukin 2. Proc Natl Acad Sci USA 1989; 86:6269–6273.
136. Resnick M, Roguel N, Bercovier H, Enk C, Frankenburg S, Kedar E. Lysis of murine macrophages infected with intracellular pathogens by interleukin 2-activated killer (LAK) cells in vitro. Cell Immunol 1988; 113:214–219.
137. Blanchard DK, Michelini-Norris MB, Friedman H, Djeu JY. Lysis of mycobacteria-infected monocytes by IL-2 activated killer cells: role of LFA-1. Cell Immunol 1989; 119:402–411.
138. Haak-Frendscho M, Young KM, Czuprynski CJ. Treatment of mice with human recombinant interleukin-2 augments resistance to the facultative intracellular pathogen *Listeria monocytogenes*. Infect Immun 1989; 57:3014–3021.
139. Jeevan A, Asherson GL. Recombinant interleukin-2 limits the replication of *Mycobacterium lepraemurium* and *Mycobacterium bovis* BCG in mice. Infect Immun 1988; 56:660–664.
140. Chong KT. Prophylactic administration of interleukin-2 protects mice from lethal challenge with gram-negative bacteria. Infect Immun 1987; 55:668–672.
141. Iizawa Y, Nishi T, Kondo M, Tsuchiya K, Imada A. Effect of recombinant human interleukin-2 as the cause of experimental chronic respiratory tract infection caused by *Klebsiella pneumoniae* in mice. Infect Immun 1988; 56:45–50.
142. Hoeprich PD, Merry JM. Effect of recombinant human interleukin 2 in experimental murine coccidioidomycosis. Diagn Microbiol Infect Dis 1988; 9:115–118.
143. Haak-Frendscho M, Czuprynski CJ. Use of recombinant interleukin-2 to enhance adoptive transfer of resistance to *Listeria monocytogenes* infection. Infect Immun 1992; 60:1406–1414.
144. Murphy PM, Lane HC, Gallin JI, Fauci AS. Marked disparity in incidence of bacterial infections in patients with the acquired immunodeficiency syndrome receiving interleukin-2 or interferon-γ. Ann Intern Med 1988; 108:36–41.
145. Boulay J-L, Paul WE. The interleukin-4 family of lymphokines. Curr Opin Immunol 1992; 4:294–298.
146. Iizawa Y, Czuprynski C. Effects of administration of murine recombinant IL-4 on the resistance of mice to *Listeria monocytogenes* infection. Immunol Lett 1992; 32:185–190.
147. Leiby DA, Fortier AH, Crawford RM, Schreiber RD, Nacy CA. In vivo modulation of the murine immune response to *Francisella tularensis* LVS by administration of anticytokine antibodies. Infect Immun 1992; 60:84–89.
148. Sieling PA, Abrams JS, Yamamura M, Salgame P, Bloom BR, Rea TH, Modlin RL. Immunosuppressive roles for IL-10 and IL-4 in human infection. In vitro modulation of T cell responses in leprosy. J Immunol 1993; 150:5501–5510.
149. Howard M, O'Garra A. Biological properties of interleukin 10. Immunol Today 1992; 13:198–200.
150. Cassatella MA, Meda L, Bonora S, Ceska M, Constantin G. Interleukin 10 (IL-10) inhibits the release of proinflammatory cytokines from human polymorphonuclear leukocytes. Evidence for an autocrine role of tumor necrosis factor and IL-1β in mediating the production of IL-8 triggered by lipopolysaccharide. J Exp Med 1993; 178:2207–2211.
151. Kasama T, Strieter RM, Lukacs NW, Burdick MD, Kunkel SL. Regulation of neutrophil-derived chemokine expression by IL-10. J Immunol 1994; 152:3559–3569.
152. Gérard C, Bruyns C, Marchant A, Abramowicz D, Vandenabeele P, Delvaux A, Fiers W, Goldman M, Velu T. Interleukin 10 reduces the release of tumor necrosis factor and prevents lethality in experimental endotoxemia. J Exp Med 1993; 177:547–550.
153. Cassatella MA, Meda L, Gasperini S, Calzetti F, Bonora S. Interleukin 10 (IL-10) upregulates IL-1 receptor antagonist production from lipopolysaccharide-stimulated human polymorphonuclear leukocytes by delaying mRNA degradation. J Exp Med 1994; 179:1695–1699.
154. Bermudez LE, Champsi J. Infection with *Mycobacterium avium* induces production of interleukin-10 (IL-10), and administration of anti-IL-10 antibody is associated with enhanced resistance to infection in mice. Infect Immun 1993; 61:3093–3097.

155. Denis M, Chadirian E. IL-10 neutralization augments mouse resistance to systemic *Mycobacterium avium* infections. J Immunol 1993; 151:5425–5430.
156. Wagner RD, Czuprynski CJ. Cytokine mRNA expression in livers of mice infected with *Listeria monocytogenes*. J Leukoc Biol 1993; 53:525–531.
157. Wagner RD, Maroushek NM, Brown JF, Czuprynski CJ. Treatment with anti-interleukin-10 monoclonal antibody enhances early resistance to but impairs complete clearance of *Listeria monocytogenes* infection in mice. Infect Immun 1994; 62:2345–2353.
158. Tripp CS, Beckerman KP, Unanue ER. Immune complexes inhibit antimicrobial responses through interleukin-10 production. J Clin Invest 1995; 95:1628–1634.
159. Frei K, Nadal D, Pfister H-W, Fontana A. Listeria meningitis: identification of a cerebrospinal fluid inhibitor of macrophage listericidal function as interleukin 10. J Exp Med 1993; 178:1255–1261.
160. Romani L, Puccetti P, Mencacci A, Cenci E, Spaccapelo R, Tonnetti L, Grohmann U, Bistoni F. Neutralization of IL-10 up-regulates nitric oxide production and protects susceptible mice from challenge with *Candida albicans*. J Immunol 1994; 152:3514–3521.
161. Cenci E, Romani L, Mencacci A, Spaccapelo R, Schiaffella E, Puccetti P, Bistoni F. Interleukin-4 and interleukin-10 inhibit nitric oxide-dependent macrophage killing of *Candida albicans*. Eur J Immunol 1993; 23:1034–1038.
162. Te Velde AA, de Waal Malefijt R, Huijbens RJF, de Vries JE, Figdor CG. IL-10 stimulates monocyte FcγR surface expression and cytotoxic activity. J Immunol 1992; 149:4048–4052.
163. Lehmann AK, Halstensen A, Sornes S, Rokke O, Waage A. High levels of interleukin 10 in serum are associated with fatality in meningococcal disease. Infect Immun 1995; 63:2109–2112.
164. Greenberger MJ, Strieter RM, Kunkel SL, Danforth JM, Goodman RE, Standiford TJ. Neutralization of IL-10 increases survival in a murine model of *Klebsiella pneumonia*. J Immunol 1995; 155:722–729.
165. Standiford TJ, Strieter RM, Lukacs NW, Kunkel SL. Neutralization of IL-10 increases lethality in endotoxemia. Cooperative effects of macrophage inflammatory protein-2 and tumor necrosis factor. J Immunol 1995; 155:2222–2229.
166. Bean AGD, Freiberg RA, Andrade S, Menon S, Zlotnik A. Interleukin 10 protects mice against *Staphylococcal* enterotoxin B-induced lethal shock. Infect Immun 1993; 61:4937–4939.
167. Chernoff AK, Granowitz EV, Shapiro L, Vannier E, Lonnemann G, Angel JB, Kennedy JS, Rabson AR, Wolff SM, Dinarello CA. A randomized, controlled trial of IL-10 in humans. J Immunol 1995; 154:5492–5499.
168. Kuhn R, Lohler J, Rennick D, Rajewsky K, Muller W. Interleukin-10 deficient mice develop chronic enterocolitis. Cell 1993; 75:263–274.
169. Wagner RD, Steinberg H, Brown JF, Czuprynski CJ. Recombinant interleukin-12 enhances resistance of mice to *Listeria monocytogenes* infection. Microbial Pathogenesis 1994; 17:175–186.
170. Tripp CS, Gately MK, Hakimi J, Ling P, Unanue ER. Neutralization of IL-12 decreases resistance to *Listeria* in SCID and CB17 mice. J Immunol 1994; 152:1883–1887.
171. Zhang M, Gately MK, Wang E, Gong J, Wolf SF, Lu S, Modlin RL, Barnes PF. Interleukin 12 at the site of disease in tuberculosis. J Clin Invest 1994; 93:1733–1739.
172. Flelsch IEA, Hess JH, Huang S, Aguet M, Rothe J, Bluethmann H, Kaufmann SHE. Early interleukin 12 production by macrophages in response to mycobacterial infection depends on interferon γ and tumor necrosis factor α. J Exp Med 1995; 181:1615–1621.
173. Castro AG, Silva RA, Appelberg R. Endogenously produced IL-12 is required for the induction of protective T cells during *Mycobacterium avium* infections in mice. J Immunol 1995; 155:2013–2019.
174. Cooper AM, Roberts AD, Rhoades ER, Callahan JE, Getzy DM, Orme IM. The role of interleukin-12 in acquired immunity to *Mycobacterium tuberculosis* infection. Immunology 1995; 84:423–432.
175. Flynn JL, Goldstein MM, Triebold KJ, Sypek J, Wolf S, Bloom BR. IL-12 increases resistance of BALB/c mice to *Mycobacterium tuberculosis* infection. J Immunol 1995; 155:2515–2524.
176. Kobayashi K, Kasama T, Yamazaki J, Hosaka M, Katsura T, Mochizuki T, Soejima K, Nakamura RM. Protection of mice from *Mycobacterium avium* infection by recombinant interleukin-12. Antimicrob Agents Chemother 1995; 39:1369–1371.
177. Romani L, Mencacci A, Tonnetti L, Spaccapelo R, Cenci E, Puccetti P, Wolf SF, Bistoni F. IL-12 is both required and prognostic in vivo for T helper type 1 differentiation in murine candidiasis. J Immunol 1994; 152:5167–5176.
178. Hall SS. IL-12 at the crossroads. Science 1995; 268:1432–1434.
179. Cohen J. IL-12 deaths: explanation and a puzzle. Science 1995; 270:908.
180. Biotechnology medicines and vaccines approved and under development. Genetic Eng News 1995; (Aug):12–16.
181. Curnutte JT. Conventional versus interferon-γ therapy in chronic granulomatous disease. J Infect Dis 1993; 167:S8–S12.
182. Gallin JI, Farber JM, Holland SM, Nutman TB. Interferon-gamma in the management of infectious diseases (clinical conference). Ann Intern Med 1995; 123:216–224.
183. Groopman JE, Molina JM, Scadden JT. Hematopoietic growth factors. N Engl J Med 1989; 321:1449–1459.
184. Matsumoto M, Matsubara S, Matsumo T, Tamura M, Haffori K, Nomvia H, Ono M, Yokota T. Protective effect of human granulocyte colony-stimulating factor on microbial infection in neutropenic mice. Infect Immun 1987; 55:2715–2720.
185. Gorin NC, Coiffier B, Hayat M, Fouillard L, Kueutz M, Flesch M, Colombat P, Boivin P, Slavin S, Philip T. Recombinant human granulocyte-macrophage colony-stimulating factor after high dose chemotherapy and autologous bone marrow transplantation with unpurged and purged marrow in non-Hodgkins lymphoma: a double-blind placebo-controlled trial. Blood 1992; 80:1149–1157.
186. Nemunaitis J, Shannon-Dorcy K, Applebaum FR, Meyers J, Owens A, Day R, Ando D, O'Neil C, Buckner D, Suzer J. Long-term follow-up of patients with invasive fungal disease who received adjunctive therapy with recombinant human macrophage colony-stimulating factor. Blood 1993; 82:1422–1427.
187. Klein B, Brailly H. Cytokine-binding proteins: stimulating antagonists. Immunol Today 1995; 16:216–220.
188. Christman JW, Holden EPO, Blackwell TS. Strategies for blocking the systemic effects of cytokines in the sepsis syndrome. Crit Care Med 1995; 23:955–963.
189. Moldawer LL. Biology of proinflammatory cytokines and their antagonists. Crit Care Med 1994; 22:S3–S7.
190. Ashkenazi A, Capon DJ, Ward RHR. Immunoadhesins. Intern Rev Immunol 1993; 10:219–227.
191. Kahn JO, Allan JD, Hodges TL, Kaplan LD, Arri CJ, Fitch HF, Izu AK, Mordenti J, Sherwin SA, Groopman JE, Volgerding PA. The safety and pharmacokinetics of recombinant soluble

CD4 (rCD4) in subjects with the acquired immunodeficiency syndrome (AIDS) and AIDS-related complex. Ann Intern Med 1990; 112:254–261.

192. Hodges TL, Kahn JO, Kaplan LD, Groopman JE, Volgerding PA, Amman AJ, Arri CJ, Bouvier LM, Mordenti J, Izu AE, Allan JD. Phase I study of recombinant human CD4-immunoglobulin G therapy of patients with AIDS and AIDS-related complex. Antimicrob Agents Chemother 1991; 35:2580–2586.
193. Pitti RM, Marsters SA, Haak-Frendscho M, Osaka GC, Mordenti J, Chamow SM, Ashkenazi A. Molecular and biological properties of an interleukin-1 receptor immunoadhesin. Mol Immunol 1994; 31:1345–1351.
194. Haak-Frendscho M, Marsters SA, Chamow SM, Peers DH, Simpson NJ, Ashkenazi A. Inhibition of interferon-γ by an interferon-γ receptor immunoadhesin. Immunology 1993; 79:594–599.
195. Lee JC, Prabhakar U, Griswold DE, Dunnington D, Young PR, Badger A. Low-molecular-weight TNF biosynthesis inhibitors: strategies and prospectives. Circ Shock 1995; 44:97–103.
196. Dezube BJ, Lederman MM, Spritzler JG, Chapman B, Korvick JA, Flexner C, Dando S, Mattiacci MR, Ahlers CM, Zhang L, Novick WJ Jr, Kasdan P, Fahey JL, Pardee AB, Crumpacker CS, the National Institute of Allergy and Infectious Diseases AIDS Clinical Trials Group. High-dose pentoxifylline in patients with AIDS: inhibition of tumor necrosis factor production. J Infect Dis 1995; 171:1628–1632.
197. Netea MG, Blok WL, Kullberg BJ, Bemelmans M, Vogels MT, Buurman WA, van der Meer JW. Pharmacologic inhibitors of tumor necrosis factor production exert differential effects in lethal endotoxemia and in infection with live microorganisms in mice. J Infect Dis 1995; 171:393–399.
198. Kaplan G. Cytokine regulation of disease progression in leprosy and tuberculosis. Immunobiology 1994; 191:564–568.
199. Bradshaw M, Rutherford MS, Hoeper BJ, McWhinney CD, Borcherding DR, Schook LB, Edwards CK III. Specific transcriptional inhibition of bone marrow-derived macrophage tumor necrosis factor-α gene expression and protein production using novel enantiomeric carbocyclic nucleoside analogues. J Pharmacol Exp Ther 1995; 273:1506–1518.
200. Lefebvre-d'Hellencourt C, Diaw L, Guenounou M. Immunomodulation by cytokine antisense oligonucleotides. Eur Cytokine Network 1995; 6:7–19.
201. Hughes HPA, Campos M, Godson DL, Van Drunen Littel-Van Den Hurk S, McDougall L, Rapin N, Zamb T, Babiuk LA. Immunopotentiation of bovine herpes virus subunit vaccination by interleukin-2. Immunology 1991; 74:461–466.
202. Nash AD, Lofthouse SA, Barcham GJ, Jacobs HT, Ashman K, Meeusen ENT, Brandon MR, Andrews AE. Recombinant cytokines as immunological adjuvants. Immunol Cell Biol 1993; 71:367–379.
203. Afonso LCC, Scharton TM, Viera LQ, Mysocka M, Trinchieri G, Scott P. The adjuvant effect of interleukin-12 in a vaccine against *Leishmania major.* Science 1994; 263:235–237.
204. Miller MA, Skeen MJ, Ziegler HK. Nonviable bacterial antigens administered with IL-12 generate antigen-specific T cell responses and protective immunity against *Listeria monocytogenes.* J Immunol 1995; 155:4815–4828.
205. Sanchez MS, Ford CW, Yancey RJ. Effect of tumor necrosis factor α, interleukin-1β, and antibiotics on the killing of intracellular *Staphylococcus aureus.* J Dairy Sci 1994; 77:1251–1258.

39
Cytokines and Mycobacterial Infection

Jon S. Friedland
Royal Postgraduate Medical School, Hammersmith Hospital, London, England

INTRODUCTION

The clinical importance of the three principal mycobacterial infections, namely, tuberculosis, leprosy, and those due to the *Mycobacterium avium–intracellulare* complex (MAC), is undeniable. The World Health Organization estimates that approximately one third of the world's population has been infected with *Mycobacterium tuberculosis* and that there are approximately 3 million deaths each year from this pathogen. Leprosy is estimated to affect over 10 million people, although this number is beginning to decline with effective chemotherapy. In contrast, MAC infections are increasingly common in immunosuppressed patients, particularly those with concurrent human immunodeficiency virus (HIV) infection. In addition, there are numerous other mycobacterial pathogens such as *Mycobacterium kansasii* and *Mycobacterium marinum,* but these are less prevalent and will not be considered further in this chapter.

Two aspects of diseases caused by mycobacteria are immediately striking. First, there is the huge morbidity and mortality associated with such infection. Second, there is the fact that a large percentage of such infections are contained by the normal immune response. This has led to great interest in host defense mechanisms against these pathogens both for the general insight provided into immunity against intracellular organisms and because an understanding of the processes that lead to mycobacterial disease may aid in the development of immunotherapy. Current drug regimens are not entirely satisfactory, being of long duration, and compliance with therapy remains a major problem. In addition, the advent of multidrug-resistant tuberculosis has given such research even greater urgency.

The focus on the immune system has shown that cytokines have a central role in successful host defense against mycobacterial infection and in the pathophysiology of mycobacterial disease. Proinflammatory macrophage-derived cytokines such as tumor necrosis factor (TNF) and interferon (IFN)-γ initiate a cascade of immune responses to mycobacteria. In particular, TNF has a critical role in the formation of granuloma, the principal tissue immune response to mycobacteria. Chemokines are pivotal in cellular recruitment to areas of infection. The subsequent development of a T-helper (TH1) response (1) is necessary for successful control of infection. Finally, the downregulatory cytokines such as interleukin (IL)-10 and transforming growth factor (TGF)-β probably have an important role in controlling the proinflammatory response. This chapter will review in detail the involvement of such families of cytokines in the human immune response to pathogenic mycobacteria and important mycobacterial antigens.

CELLULAR MODELS OF MYCOBACTERIAL INFECTION

Mycobacterium tuberculosis

Wasting, fever, and a marked acute phase response are all clinical hallmarks of patients with tuberculosis. TNF was the first proinflammatory cytokine to be characterized and was originally described as cachectin, a weight-losing factor (2). In addition, TNF was found to be involved in generation of fever (3). Thus, it was not surprising when monocytes and macrophages phagocytosing *M. tuberculosis* were shown to upregulate TNF gene expression and secretion (4). TNF secretion was modulated by 1,25-

dihydroxyvitamin D_3, which has potentially important therapeutic implications in people with dietary deficiencies of vitamin D. In cellular models of tuberculosis, TNF mRNA accumulation was transient. However, mRNA accumulation of IL-1, the other principal acute proinflammatory and pyrogenic cytokine elevated in many disease states (5), persisted over a much longer period. More virulent strains of mycobacteria stimulated less TNF release from murine macrophages, and it has been postulated that this allows a virulent organism to multiply before immune responses are fully activated (6). However, this effect was not detected in human monocytic cells (7).

As in many infectious diseases, the secretion of proinflammatory cytokines such as TNF and IFN-γ heralds the onset of a cascade of cytokine secretion, in part due to the autologous actions of these cytokines. In particular, marked IL-6 secretion has been noted in both murine and human systems. IL-6 stimulates the acute phase protein response (8), and this is likely to be its major role in the immune response to mycobacterial disease. However, IL-6 secretion was similar in response to phagocytosis of strains of *M. tuberculosis* of differing virulence (9). TNF and IFN-γ secretion are also required for cellular secretion of IL-12 during infection of murine bone marrow–derived macrophages (10). IL-12 is involved in early activation of cytolytic activity of natural killer (NK) cells toward *M. tuberculosis*–infected monocytes (11). However, probably a more important action of this cytokine is stimulating development of a T-helper 1 response, critical in immunity to tuberculosis (see below). Phagocytosis of *M. tuberculosis* also results in gene expression and secretion of downregulatory cytokines such as TGF-β1 (12). TGF-β1 antagonizes the effects of TNF and IFN-γ, although pretreatment with TGF-β1 does decrease uptake of *M. tuberculosis* by human monocytes.

One major consequence of proinflammatory cytokine secretion is secondary chemokine release. However, phagocytosis of *M. tuberculosis* is also a direct stimulus to chemokine secretion. Chemokines are necessary for recruitment of monocytes and T lymphocytes to the site of infection. Such cellular recruitment is a prerequisite for granuloma formation. The process of phagocytosis stimulates gene expression and secretion of the monocyte chemoattractant MCP-1, although whether this is a nonspecific effect or additionally upregulated in tuberculosis remains controversial (13,14). In contrast, phagocytosis of *M. tuberculosis* is undoubtedly a potent stimulus to IL-8 gene expression and secretion (7). Such upregulation of IL-8 secretion did not occur when human monocytic cells phagocytosed *Toxoplasma gondii,* another intracellular pathogen (15). My current unpublished data indicates that 1,25-dihydroxyvitamin D_3 is a potent upregulator of IL-8 secretion following phagocytosis of *M. tuberculosis.* IL-8 is a T-cell chemoattractant (16) and appears selectively important in recruitment of T lymphocytes, which have been exposed to *M. tuberculosis* (17). An additional role of IL-8 may be to enhance nonoxidative killing mechanisms, although whether this occurs in the human response to tuberculosis has yet to be tested (18).

There has been considerable debate over whether the proinflammatory cytokines have direct antibacterial activity. In the murine system, it has been reported that both TNF and IFN-γ may inhibit replication of *M. tuberculosis* (19). This effect appeared to be dependent on the mycobacterial strain being investigated (20) and opposed by inhibitory macrophage-derived IL-10 (21). The generation of reactive nitrogen intermediates appeared to be involved in the actual killing process in murine bone marrow–derived macrophages (22,23). However, in human monocytes and macrophages, the role of reactive nitrogen intermediates in mycobacterial killing is more doubtful. The role of cytokines such as IFN-γ is more likely to be activation of T lymphocytes and other immune cells. Consistent with this is the fact that the majority of workers do not find that TNF has mycobacteriostatic or -cidal actions in human monocytes infected by *M. tuberculosis* (24–26). There are occasional dissenters to this view (27), although such atypical findings in human monocytes may be an artefact due to cell lysis (28). However, it is also possible that the state of differentiation and activation of the human macrophage may be important and hence TNF may have some inhibitory effect on growth of *M. tuberculosis* in particular cell types such as the alveolar macrophage (29). In addition, cell-cell contact may have a critical role in macrophage antimycobacterial mechanisms in tuberculosis (30,31). The balance of evidence suggests that this is not the main action of TNF in tuberculosis in humans and as will become apparent this cytokine has other important roles in host defense to granuloma.

Dissection of the specific elements of *M. tuberculosis* that stimulate monocyte and macrophage proinflammatory cytokine secretion has attracted considerable attention. Mycobacterial lipids and proteins may have important roles as adjuvants to novel vaccines and may be targets during development of new therapies. Lipoarabinomannan (LAM), a major cell wall component of tuberculosis, is a potent stimulus to TNF production by both human and murine macrophages (32,33). The specific structure of LAMs from different strains of *M. tuberculosis* may modulate TNF secretion from murine macrophages (34,35) and explain the differing TNF responses to whole mycobacteria observed in the murine cellular model (6). LAM can additionally increase the transcription rate of the IL-1 gene (36) as well as stimulating secretion of a wide variety of macrophage-derived cytokines (37). Lipomannans and phosphatidylinositol mannosides, which are derived from LAM but lack the arabinan and most mannan residues respectively, may also stimulate cytokine secretion, which suggests that the biological activity of LAM resides in the phosphatidylinositol portion of the molecule. In addition, it should be noted that LAM may subsequently decrease cellular IL-1 mRNA accumulation and modulate gene expression of other cytokines following exposure to a second stimulus such as LPS (38).

LAM may upregulate secretion of chemokines from both murine and human macrophages (39,40). In the murine system, accumulation of mRNA for the α chemokine KC was regulated by the Bcg locus, which codes for the natural resistance-associated macrophage protein (41). KC gene expression was dependent on intracellular calcium concentration as well as on activation of protein kinases A and C. Protein kinase C is known to be important in control of the transcriptional regulator NF-κB, and it has been shown that upregulation of proinflammatory cytokine gene expression by LAM may involve both NF-κB and NF-IL6 (42,43). Other lipids in the cell wall of *M. tuberculosis,* such as sulfatide, may be antagonistic and decrease protein kinase C activity (44). LAM itself can upregulate macrophage gene expression of downregulatory mediators such as IL-10 (45), which inhibits secretion of proinflammatory cytokines from monocytes (46). However, regulation of IL-10 secretion, which has the important function of limiting the inflammatory response in tuberculosis, is likely to be complex. For example, the degree of IL-10 gene expression itself may be regulated by the initial secretion of proinflammatory cytokines also stimulated by LAM.

Protein components of *M. tuberculosis* may also stimulate gene expression and secretion of proinflammatory cytokines (47–49). In addition, purified protein derivatives of *M. tuberculosis* increased gene expression and secretion of downregulatory TGF-β1 (50). However, it is the effect of specific proteins important in immune responses that is of most interest. The 65 kDa mycobacterial heat-shock protein (hsp) induced secretion of proinflammatory and chemotactic cytokines from human monocytic cells (51–53). The 65 kDa hsp is important because mice receiving monocytes transfected with this hsp are protected against infection by virulent *M. tuberculosis* (54). However, other proteins, such as the 30 kDa antigen of *M. tuberculosis,* have been shown to regulate TNF secretion and may have vital accessory functions in vivo (55). In addition, protein antigens may specifically modulate immune responses of certain cell types. For example, the *M. tuberculosis* 71 kDa hsp induces cytokine secretion from intraepithelial lymphocytes (56). There remains a great deal to be learned from such cellular studies, particularly in terms of signal transduction and molecular control of cytokine production in tuberculosis.

Mycobacterium avium–intracellular Complex

Cellular models of infection have also been extremely valuable in the investigation of the role of cytokines in MAC infections. There are interesting contrasts and comparisons with cytokine responses following *M. tuberculosis* infection. It should be pointed out at this stage that cellular models (but not stimulation of cells from infected patients) have no real place in the study of leprosy since it remains impossible to culture *M. leprae* in vitro. Such observations as the fact that IFN-γ inhibited phagocytosis of *M. leprae* by human macrophages (57) and enhanced growth of *M. lepraemurium,* a rather different organism, in murine macrophages (58) are of uncertain significance.

It is clear that MAC organisms may stimulate gene expression and secretion of proinflammatory cytokines such as TNF. In a manner similar to that observed in models of tuberculosis, transcription-dependent TNF secretion in response to phagocytosis of MAC is transient. It initially appears to involve activation of protein kinase C (59), although tyrosine kinases may be involved in subsequent pulses of TNF secretion that follow after a refractory period (60). Different colonial variants of *M. avium* stimulate secretion of proinflammatory cytokines from monocytes to varying extents (61). A subtype of *M. avium* that replicated faster within macrophages induced less cytokine secretion than one whose growth was relatively restricted (62). This suggests that the ability to downregulate a cytokine response could be important in establishing infection. However, others have found that survival of MAC within macrophages correlated with extent and not inhibition of TNF-α and IL-6 secretion (63). Such differences may in part reflect the extent to which cultured monocytes were allowed to develop a macrophage phenotype before infection with MAC. The interaction of MAC with the macrophage is complex, as was observed in *M. tuberculosis* infection, and phagocytosis of MAC stimulates both up- and downregulatory cytokine secretion. For example, phagocytosis of MAC by monocytes stimulated secretion of TGF-β (64).

There is considerable evidence in diseases due to atypical mycobacteria, such as MAC, that macrophages may be activated by cytokines to kill intracellular organisms. In murine cellular models of infection, TNF decreases MAC replication (65,66). The effects of IFN-γ appeared to be mediated by secondary activation of TNF and were blocked by antibodies to TNF. The effects of

IFN-γ can be upregulated by concomitant addition of IL-4 or by coculture with the prostaglandin inhibitor indomethacin (67). The mode of action of these cytokines remains uncertain. Although MAC strains may be sensitive to nitric oxide–mediated killing mechanisms (68), neither reactive oxygen nor reactive nitrogen intermediates appear critical in cytokine-mediated mycobacteriostasis (69,70). There is some evidence that acidification of the phagosome, normally maintained at about pH 6.5 (71), may be important in the killing of MAC organisms.

Studies in human monocytes and alveolar macrophages confirm that TNF has a central role in inhibiting growth of MAC organisms (72,73). These effects of TNF are once again dependent on local concentrations of 1,25-dihydroxyvitamin D_3 (74). Macrophage anti-MAC activity may be stimulated by NK cells, particularly when NK cells have been activated by IL-12 (75,76). Interestingly, the downregulatory cytokine TGF-β, which opposes many of the immunological actions of TNF, also antagonizes the effect of TNF on macrophages infected by MAC (94). Since both TNF and TGF-β are secreted by macrophages in response to MAC infection, the net effect is hard to predict. However, TNF is produced earlier in infection, so it is possible that if MAC infection is not controlled by the initial immune response, subsequent disease progression may be more rapid. Like TNF, IFN-γ may significantly inhibit MAC growth within human monocytes (77,78). IFN-γ has similar inhibitory effects on other atypical mycobacteria (79), although more mature culture-derived human macrophages did not respond to exogenous IFN-γ (80).

However, not all proinflammatory cytokines decrease replication of MAC. IL-6, produced by infected macrophages, antagonized the effects of TNF by downregulating TNF receptor expression on the cell surface (81) and also opposed the actions of IFN-γ (78). Other proinflammatory cytokines such as IL-1, IL-3, and M-CSF may decrease growth of MAC within monocytes. In addition, in concurrent studies, IL-1α and IL-6 had direct inhibitory effects on the growth of *M. avium* in the absence of monocytes (82). However, the induction of such cytokines within the human monocyte may be influenced by specific strain variants of MAC (33). Thus, although monocytes have the potential to inhibit growth of MAC organisms, the net effect of immune activation in cellular models is dependent on the interactions of a variety of cytokines. The exact signal transduction pathways that determine the monocyte response to intracellular MAC infection are not known. In addition, monocytes are unlikely to act alone in host defense to MAC, and interactions between monocytes and T lymphocytes are important in anti-MAC activity (84). In order to better characterize the involvement of cytokines in these and other responses in mycobacterial infection the next logical stage is to examine cytokine-mediated immunity in vivo.

ANIMAL MODELS OF MYCOBACTERIAL INFECTION

Animal models of tuberculosis have proved extremely helpful in dissecting the nature of cytokine involvement in immunity to this infection. Their advantage is that they allow in vivo manipulation of the immune system, but many cytokine responses show considerable divergence between humans and rodents, as do other mechanisms for control of infection. For example, while reactive nitrogen intermediates are important in the killing of intracellular pathogens in murine systems, this is not true in humans.

Mycobacterium tuberculosis

Murine models of tuberculosis often study infection with *M. bovis,* BCG strain, which is rarely a pathogen in humans. However, the organism is genetically and structurally very similar to *M. tuberculosis.* An early investigation established the importance of TNF in granuloma formation and control of infection. Mice treated with anti-TNF did not form granuloma, and growth of BCG was unrestricted in tissues (85). The importance of TNF has recently been confirmed in murine tuberculosis with administration of adenovirus vectors expressing the TNF 55 kDa receptor exacerbating disease (86).

As in cellular models, not only TNF but also IFN-γ has a central role in host defense against tuberculosis. Administration of either TNF or IFN-γ augmented resistance of mice to a lethal innoculum of *M. tuberculosis,* strain H37-Rv (87). The importance of IFN-γ in the immune response to tuberculosis was further demonstrated in murine models using knock-out animals. Mice with disruptions of either the IFN-γ gene (88,89) or the IFN-γ receptor gene (90) were very sensitive to *M. tuberculosis* infection. This appeared in part to be a consequence of impaired production of reactive nitrogen intermediates. This is consistent with the finding that BCG-susceptible mice have macrophages that exhibit impaired activation of the nitric oxide pathway (91). IFN-γ probably has its major effect on the production of reactive nitrogen intermediates by upregulating the IFN-γ regulatory factor-1 (IRF-1) gene, which in turns switches on the gene for inducible nitric oxide. Thus, it was entirely consistent that IRF-1 knock-out mice were demonstrated to have reduced resistance to infection by *M. bovis,* BCG strain (92). In addition, IFN-γ receptor knockout mice have aberrant regulation of expression of class II major histocompatibility antigens (93), and both class I and II MHC responses are required for immunity to tuberculosis (94). Third, IFN-γ together with TNF is required for in vivo generation of IL-12 (10), and administration of IL-12 has been shown to increase survival of *M. tuberculosis*–infected BALB/c mice (95).

Other proinflammatory cytokines have also been detected in murine models of tuberculosis. For example, IL-6 production, detected in cellular models and an activator of the acute phase protein response, has also been detected in murine tuberculosis (96). In addition, chemokine gene expression was found in the lungs of mice infected by aerosolized *M. tuberculosis* (97). Differences in chemokine mRNA accumulation were noted depending on the infected strain of mice, although no data on secreted cytokine or granuloma formation were collected in this study, so the significance of this observation is uncertain. Chemokines may interact with IFN-γ to modulate cellular recruitment to areas of mycobacterial infection as the immune response develops (98). It is certainly clear from in vivo experiments using neutralizing antibodies that IL-8 is critical during granuloma formation in the delayed-type hypersensitivity (DTH) reaction to mycobacterial antigens (99).

Murine systems have been particularly useful in the investigation of the role of T lymphocytes in immunity to tuberculosis. T cells are certainly required in the immune response to tuberculosis, and the disease is more common in humans with suppressed T-lymphocyte activity whether this is congenital due to immunosuppressive drugs or a result of another infection such as HIV. In addition to macrophages, CD4+ T cells are an important source of IFN-γ production (100). A TH1 response, characterized by secretion of IFN-γ and IL-2 (1), is critical in the control of tuberculosis. Specific mycobacterial antigens such as the *M. bovis* BCG strain 85 kDa protein are powerful activators of the TH1 response mediated by αβ T cells (101). However, both αβ and γδ T cells are involved in the successful in vivo immune response to tuberculosis in murine models (102,103). It is generally accepted that IL-4 and IL-10 (qv) are important in switching off such TH1 responses. However, in murine *M. bovis* infection the upregulation of IL-10, which may be important in controlling inflammatory responses, does not appear adequate to shift the T-cell response from a TH1 to a TH2 type (104). This indicates just how important the appropriate helper CD4+ T-lymphocyte response is. However, it should be remembered that such T cells act in conjunction with cytolytic T cells, which appear to secrete relatively low concentrations of cytokines in *M. tuberculosis* infection (105).

Mycobacterium avium–intracellulare Complex

Murine models of MAC infection are readily available, but they show one important difference when compared to MAC infection in humans: the majority of model systems do not involve immunosuppression. However, the resurgence of interest in MAC stems from the increased incidence of infection by *M. avium* strains in patients with advanced AIDS and not from the minority of patients with indolent *M. intracellulare* infection in chronically damaged lung tissue. As in models of tuberculosis infection and in cellular studies, production of upregulatory cytokines such as TNF and downregulatory cytokines such as TGF-β has been observed in murine MAC infection (106). An animal model system has indicated that TNF is involved in limiting growth of MAC in vivo, which confirms cellular studies (107). TNF appears to act via reactive nitrogen intermediates and is influenced by the presence of the BCG-resistance gene as well as by endogenous and exogenous steroids (108,109).

IL-1 is another proinflammatory cytokine which may have a beneficial role in murine resistance to MAC infection (110). In contrast, IL-10, which is also produced during murine MAC infection, appears to decrease resistance to this mycobacterium (111,112). Such information is derived from administration of anti–IL-10 allowing relatively unopposed action of proinflammatory cytokines. Although these studies may point the way toward novel therapeutic approaches, they tend to disregard the normal complex interactions between monocytes and T cells needed for elimination of intracellular infections. For example, in chronic MAC infection the TH1/TH2 balance, affected by IL-10 secretion, may be critical in determining the action of proinflammatory cytokines such as TNF-α, which is itself regulated by IL-10 (113).

The TH1 phenotype is associated with control of MAC as well as other mycobacterial infections in murine models (114). Similar data were derived from a leprosy model, where mice were injected with killed *M. leprae* and lymph node cytokine mRNA investigated (115). This is necessarily an artificial system since this organism only grows slowly in the footpad of nude mice and in the armadillo. In MAC infection, IL-12 appears central to both upregulation of the TH1 and switching off the TH2 response (116). Interestingly, IL-6 may have a role in the development of T-cell–mediated immunity in MAC, although the details of this effect require further investigation (117). In summary, cellular and animal studies have helped develop an overview of the interactions between monocytes and T cells and the up- and downregulatory cytokines that they secrete in order to control mycobacterial infection. Two questions now arise: Do these findings hold true in humans? If so, can modulation of cytokine cascades be used in the treatment of mycobacterial disease? The rest of this chapter will review the important data that address these questions, which remain to be fully answered.

DETECTION OF CYTOKINES IN HUMAN MYCOBACTERIAL INFECTION

There is no doubt that the study of cytokine responses in infection in humans is critically important. However, such studies present particular problems for the researcher. First, it is not always easy to access and biopsy an infected human tissue, which may be necessary for detailed study. Second, a patient presents during the course of his or her infection after initial events in the immune response have been initiated. Thus, clinical data must first be combined with human cellular information and then interpreted in the light of animal experiments. Unlike studies discussed earlier, most clinical data in this case originates from tuberculosis and leprosy patients and not MAC disease. The advantage of leprosy is that skin and nerves are readily biopsied for close scrutiny. In addition, leprosy patients may have a spectrum of disease from the tuberculoid form with good granuloma formation to the lepromatous form where granuloma formation is poor. Therefore, leprosy has the potential to provide insight into cytokines required for development of granuloma. In contrast, MAC infection is more difficult to investigate since the majority of infected patients have advanced HIV often with other concurrent secondary infections, which makes interpretation of data difficult.

Mycobacterium tuberculosis

Proinflammatory cytokines have been detected in tuberculosis infection in a variety of tissue sites confirming the findings of model systems. In a nonspecific manner, TNF and IL-6 secretion from peripheral blood mononuclear cells of *M. tuberculosis*–infected patients was increased (118). TNF production from ex vivo stimulated monocytes was greater if cells were harvested from patients with acute as opposed to chronic infection (119). In a similar manner, IL-1 secretion from monocytic cells was also elevated in patients with tuberculosis (120,121). In pulmonary tuberculosis, TNF appears to be concentrated in pleural effusions partly due to secretion from pleural tissue where TNF mRNA can be detected (122). IL-1 is also found in pleural disease (123), and both IL-1 and TNF together with the naturally occurring IL-1 receptor antagonist were present in cerebrospinal fluid (CSF) of patients with tuberculous meningitis (124,125). Concentrations of proinflammatory cytokines in CSF were lower than those in bacterial meningitis but higher than in viral meningitis. TNF production in tuberculosis complicating HIV infection has been reported to be increased, which would fit with the marked wasting seen in such patients (126).

Other proinflammatory cytokines such as IL-12 have also been detected in patients with tuberculosis (127). Considerable evidence in human studies supports data indicating that chemokines have a significant role in the immune response to tuberculosis. Both the α chemokine IL-8 and the β chemokine MCP-1 have been detected in the pleural fluid of patients with tuberculosis (128). In addition, IL-8 was also present in bronchoalveolar lavage fluid obtained from patients with tuberculosis (129). In addition, mRNA for RANTES, chemoattractant for memory T cells (130), can be found in lymph nodes from patients with tuberculosis (131). This supports the cellular and animal studies, which indicate that chemokines are important in the recruitment of cells required for antimycobacterial granuloma formation. Interestingly, patients with fatal tuberculosis had elevated plasma IL-8 concentrations compared to survivors, but their leukocytes were unable to secrete further IL-8 on LPS stimulation (Fig. 1) (132). This was partly due to the presence of an inhibitor of IL-8 secretion in plasma. In patients who die, an inability to secrete such cytokines may be important in determining outcome, even though IL-8 had been secreted earlier in the course of infection. Leukocytes from patients who recover from tuberculosis are surprisingly sensitive to LPS stimulation, and IL-8 secretion is increased (133). The reasons for this are uncertain. The proinflammatory response is certain to be controlled by downregulatory cytokines such as TGF-β, which has also been detected in patients (134).

In terms of the T-lymphocyte response, there can be no doubt that a TH1 phenotype is part of the normal immune response. Cultured human peripheral and alveolar T cells from subjects previously exposed to mycobacteria and then stimulated with mycobacterial antigens predominantly secrete the cytokines IL-2 and IFN-γ (123,135–138). γδ as well as αβ T lymphocytes contribute to the observed secretion of cytokines (139). Peripheral T cells from patients with active disease may have depressed cytokine production (137,140,141). In addition, at a later stage of infection, cells of the TH0 or TH2 phenotype particularly ones with marked IL-4 secretion have frequently been reported in human studies of tuberculosis (142–146). However, in pleural fluid T lymphocytes, IFN-γ and IL-2 gene expression was relatively increased and IL-4 mRNA decreased when compared to peripheral T cells (147). IL-10 mRNA was also present in pleural cells but

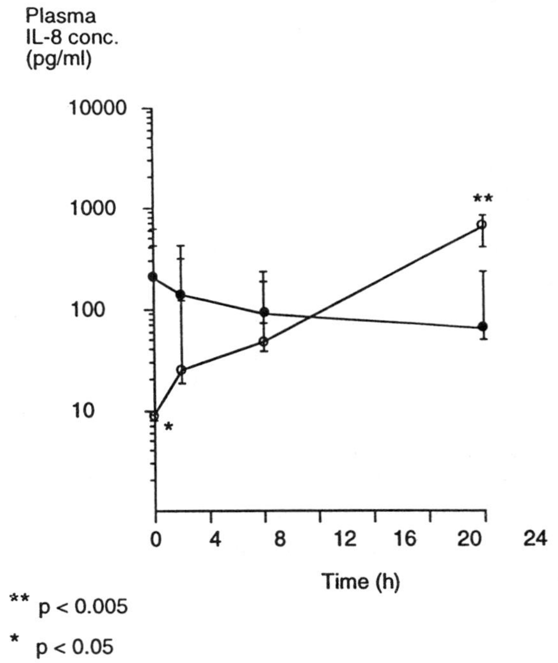

Figure 1 IL-8 plasma concentrations in 34 patients who survived (○) and 7 who subsequently died (●) from tuberculosis. All points show median values (note log scale on vertical axis) ± interquartile ranges. Value at time = 0 h is plasma IL-8 concentration on admission to study. Subsequent time points indicate IL-8 secretion after ex vivo stimulation of whole blood by LPS at a final concentration of 10 μg/ml. IL-8 secretion increases with time in patients who survive and control subjects (data not shown) but not in patients who died from acute tuberculosis infection. Sample sizes are identical at all time points. (Reproduced from Ref. 132 with permission.)

was mainly macrophage derived. This pattern of a TH1 response to *M. tuberculosis* modulated by macrophage-derived cytokines is most consistent with the previous experimental data. The functional importance of other T-cell types in vivo remains uncertain. It is interesting to note that HIV-infected patients with tuberculosis did not appear to have an augmented TH2 response but rather diminished TH1 responses (141). Potential alterations in proinflammatory and chemotactic cytokine secretion may be critical in these patients, since it is initial granuloma formation that is impaired (148).

Mycobacterium leprae

The immune response to tuberculosis and that to tuberculoid or paucibacillary leprosy have much in common. In both diseases, histological lesions are characterized by granuloma containing few mycobacteria. The focus of research has been on what determines the development of this histological picture as opposed to the one in lepromatous or multibacillary leprosy where granuloma formation is poor and organisms are in abundance. In addition, there is another interestingly clinical phenomenon in leprosy called a reversal reaction when there is natural upgrading from the lepromatous type of disease toward the tuberculoid pattern. There is also a similar but distinct upgrading in the immune response called erythema nodosum leprosum (ENL), which usually occurs with treatment. Thus, leprosy patients present an unusual opportunity to study a spectrum of immune responses to mycobacteria and their development.

Entirely consistent with the key role of TNF in experimental granuloma formation is the fact that mononuclear cells from tuberculoid leprosy patients stimulated by *M. leprae* or its LAM secrete considerably more TNF than such cells from lepromatous patients (149). Interestingly, the ENL reaction was associated with the greatest TNF secretion. IFN-γ was also shown to be important in the immunity to leprosy, and defective secretion was observed in patients with lepromatous disease (150). Contacts of patients with reduced IFN-γ responses to mycobacterial antigen were more at risk of developing disease (151).

One important observation was that tuberculoid lesions tend to express mRNA for TH1 cytokines and lepromatous lesions have a TH2 phenotype (152). The delayed cell-mediated response to *M. leprae* is critical in the human immune response (153). The importance of the TH1 phenotype upregulated by IL-12 and secreting IL-2 and IFN-γ has now been demonstrated repeatedly in investigations in humans (150,154,155). Thus, it was rather surprising that HIV infection did not affect IFN-γ production in patients coinfected with *M. leprae* (156). This was consistent with the fact that clinical leprosy does not seem to interact with HIV infection but is somewhat at odds with the apparent importance of CD4+ T cells in production of IFN-γ. However, monocytes are also a source of IFN-γ. In general, decreased IL-2 and IFN-γ secretion did appear to be involved in development of lepromatous disease. Early investigations then showed that intradermal injection of IL-2 into lepromatous lesions resulted in enhanced in vitro antigen-dependent cytolytic activity (157). It was as a consequence of such clinical experiments that the first cytokine-based antimycobacterial therapies emerged.

CYTOKINES AND THE TREATMENT OF MYCOBACTERIAL INFECTION

Leprosy patients have been treated with both IL-2 and IFN-γ. IFN-γ injected intradermally resulted in reduction of the bacillary load, and in some cases a granulomatous response was observed (158,159). The positive effect of this treatment was still evident 6 months later. When IFN-γ therapy was given by intramuscular injection, intradermal accumulation of mononuclear cells was a widespread phenomenon (160). However, prolonged therapy of lepromatous leprosy with IFN-γ was associated with the development of ENL (161). This appeared to be due in part to elevated TNF secretion from patient monocytes. This problem may be alleviated by treatment with thalidomide (162), which enhances breakdown of TNF-α mRNA (163). However, since this agent has potent teratogenic properties, it can only be given in a highly supervised environment. The concept of using both an anti-TNF agent and multiple cytokine therapies in this situation has been established. The alternative approach of using IL-2 also successfully reduced the bacillary load in lepromatous leprosy patients in association with a generalized enhancement of cell-mediated immunity (164,165).

The use of cytokines in other mycobacterial infections is still in its infancy. Steroids are established treatment for pericardial tuberculosis and CNS tuberculomas and have a tendency to decrease cytokine-mediated inflammatory responses at the gene level. This is a very nonspecific approach. No directed cytokine therapies are in routine use in the treatment of *M. tuberculosis* infection. However, there are known patients with a familial susceptibility to atypical mycobacterial infection who have been identified and treated with cytokines. Many of these patients are persistently infected with MAC organisms. Impaired IL-2 and IFN-γ generation led to therapy with exogenous IFN-γ in addition to antimycobacterial drugs, with consequent clinical and radiological improvement (166). However, the response was less dramatic in another series of six patients of whom three died and the remainder had persistent chronic infection (167). Interestingly, TNF production in response to LPS and IFN-γ was defective in both affected children and to a lesser extent in their parents. Hence restoration of TNF function may be important. It has been shown that macrophages from AIDS patients may have impaired ability to secrete TNF, which was exacerbated by pentoxyfylline treatment (168). The fact that this anti-TNF therapy was accompanied by increased MAC replication supports the central role of TNF in host defense against mycobacteria.

Cytokines may have therapeutic uses in mycobacterial disease other than stimulating immune responses. They may have a direct role in enhancing the effects of antibiotics by increasing their uptake into cells. For example, both TNF and IFN-γ increased cellular uptake of the antimycobacterial agent azithromycin (169). Similarly, GM-CSF appeared to augment the in vivo efficacy of anti-MAC therapy in model systems (170). This remains a relatively unexplored area.

CONCLUSIONS AND FUTURE DIRECTIONS

In the last 12 years considerable insight has been gained into the role of cytokines in the immune response to mycobacterial infection. The picture is complex, but the overall pattern has become clearer. In very broad terms, proinflammatory and chemotactic cytokines initiate recruitment of T cells and monocytes to areas of infection. TH1-mediated cytokines become involved in control of the spread of mycobacteria before the inflammatory response is downregulated. Cytokines have entered the therapeutic arena, although to a strictly limited extent. What can be expected next? More detailed information is required focusing on control of the cytokine response. Predicting the future is always hazardous, but there do seem to be promising areas of investigation. I feel that the focus must continue to be on three areas of human cellular models and human disease.

First, I think that cytokines have the potential to be employed as adjuvants to traditional antimycobacterial antibiotic therapy. This may improve outcome in immunosuppressed individuals, although this is a difficult area with HIV patients, for example, having high TNF levels but poor granuloma formation. More importantly, cytokine therapy may potentially reduce the current 6- to 12-month treatment regimens to a few weeks, which is vital for good patient compliance and disease control. It is possible to envisage both recruiting additional leukocytes to an area of infection by chemokines and then stimulating drug penetration to infected macrophages. The second area is the use of downregulatory cytokines. In disseminated disease, it seems that the inflammatory response may contribute to wasting and other symptoms. In cases where cure is not possible, palliation using IL-10 or a similar cytokine may be. The third potential role for cytokines is in combination with mycobacterial vaccines: not only vaccines against mycobacterial disease but also vaccines that use nonpathogenic mycobacteria, such as attenuated BCG strains, as carriers to deliver antigens protecting against other diseases. In these systems, expression of cytokines, particularly chemokines, may allow recruitment of the appropriate immune response to generate high-level immunity. It is certainly already possible to express cytokines in BCG vectors (171). Thus, the future looks exciting, and the next edition of this book will probably have a much larger section on therapeutics!

REFERENCES

1. Mossman TR, Coffman RL. TH1 and TH2 cells: different patterns of lymphokine secretion lead to different functional properties. Ann Rev Immunol 1989; 7:145–173.
2. Beutler B, Mahoney J, Le Trang N, Pekala P, Cerami A. Purification of cachectin, a lipoprotein lipase-suppressing hormone secreted by endotoxin-induced RAW 264.7 cells. J Exp Med 1985; 161:984–995.
3. Beutler B, Milsark IW, Cerami AC. Passive immunization against cachectin/tumour necrosis factor protects mice from lethal effect of endotoxin. Science 1985; 229:869–871.
4. Rook GAW, Taverne J, Leverton C, Steele J. The role of gamma-interferon, vitamin D3, metabolites and tumour necrosis factor in the pathogenesis of tuberculosis. Immunology 1987; 62:229–234.
5. Dinarello CA, Wolff SM. The role of interleukin-1 in disease. N Engl J Med 1993; 328:106–113.
6. Falcone V, Bassey EB, Toniolo A, Conaldi PG, Collins FM. Differential release of tumor necrosis factor-α from murine peritoneal macrophages stimulated with virulent and avirulent species of mycobacteria. FEMS Immunol Med Microbiol 1994; 8:225–232.
7. Friedland JS, Remick DG, Shattock R, Griffin GE. Secretion of Interleukin-8 following phagocytosis of *Mycobacterium tuberculosis* by human monocyte cell lines. Eur J Immunol 1992; 22:1373–1378.
8. Gauldie J, Richards C, Harnish D, Lansdorf P, Baumann H. Interferon β2/B-cell stimulatory factor type 2 shares identity with monocyte-derived hepatocyte-stimulating factor and regulates the major acute phase protein response in liver cells. Proc Natl Acad Sci USA 1990; 84:7251–7255.
9. Friedland JS. Phagocytosis, cytokines and *Mycobacterium tuberculosis.* Lymphokine Cytokine Res 1993; 12:127–133.
10. Flesch IEA, Hess JH, Huang S, et al. Early interleukin-12 production by macrophages in response to mycobacterial infection depends on interferon γ and tumor necrosis factor α. J Exp Med 1995; 181:1615–1621.
11. Denis M. Interleukin-12 (IL-12) augments cytolytic activity of natural killer cells towards *Mycobacterium tuberculosis*-infected human monocytes. Cell Immunol 1994; 156:529–536.
12. Hirsch CS, Yoneda T, Averill L, Ellner JJ, Toossi Z. Enhancement of intracellular growth of *Mycobacterium tuberculosis* in human monocytes by transforming growth factor-β1. J Infect Dis 1994; 170:1229–1237.
13. Friedland JS, Shattock R, Griffin GE. Phagocytosis of *M. tuberculosis* or particulate stimuli by human monocytic cells induces equivalent monocyte chemotactic protein 1 gene expression. Cytokine 1993; 5:150–156.
14. Kasahara K, Tobe T, Tomita M, et al. Selective expression of monocyte chemotactic and activating factor/monocyte chemoattractant protein 1 in human blood monocytes by *Mycobacterium tuberculosis.* J Infect Dis 1994; 170:1238–1247.
15. Friedland JS, Shattock R, Johnson J, Remick DG, Holliman RM, Griffin GE. Differential cytokine gene expression and secretion after phagocytosis by a human monocyte cell line of *Toxoplasma gondii* and *Mycobacterium tuberculosis.* Clin Exp Immunol 1993; 91:282–286.
16. Larsen CG, Anderson AO, Appella E, Oppenheim JJ, Matsushima K. The neutrophil-activating protein (NAP-1) is also chemotactic for lymphocytes. Science 1989; 243:1464–1466.
17. Wilkinson PC, Newman I. Identification of IL-8 as a locomotor attractant for activated human lymphocytes in mononuclear cell cultures with anti-CD3 or purified protein derivative of *Mycobacterium tuberculosis.* J Immunol 1992; 149:2689–2694.
18. Nibbering PH, Pos O, Stevenhagen A, van Furth R. Interleukin-8 enhances nonoxidative killing of *Mycobacterium fortuitum* by human granulocytes. Infect Immun 1994; 61:3111–3116.
19. Flesch IEA, Kaufmann SHE. Activation of tuberculostatic macrophage functions by gamma interferon, interleukin-4 and tumor necrosis factor. Infect Immun 1990; 58:2675–2677.
20. Flesch I, Kaufmann SHE. Mycobacterial growth inhibition by Interferon-γ-activated bone marrow macrophages and differential susceptibility among strains of *Mycobacterium tuberculosis.* J Immunol 1987; 1384408–4413.
21. Flesch IEA, Hess JH, Oswald IP, Kaufmann SHE. Growth inhibition of *Mycobacterium bovis* by IFN-γ stimulated macrophages: regulation by endogenous tumor necrosis factor-α, and IL-10. Int Immunol 1994; 6:693–700.
22. Flesch IEA, Kaufmann SHE. Mechanisms in mycobacterial growth inhibition by gamma interferon-activated bone marrow macrophages: role of reactive nitrogen intermediates. Infect Immun 1991; 59:3213–3218.
23. Yoshida A, Koide Y, Uchijima M, Yoshida TO. Dissection of strain difference in acquired protective immunity against *Mycobacterium bovis* Calmette-Guerin bacillus (BCG). J Immunol 1995; 155:2057–2066.
24. Douvas GS, Looker DL, Vatter AE, Crowle AJ. Gamma interferon activates human macrophages to become tumoricidal and leishmanicidal but enhances replication of macrophage-associated Mycobacteria. Infect Immun 1985; 50:1–8.
25. Robertson AK, Andrew PW. Interferon gamma fails to activate human monocyte-derived macrophages to kill or inhibit the replication of a non-pathogenic mycobacterial species. Microb Pathogen 1991; 11:283–288.
26. Meylan PRA, Richman DD, Kornbluth RS. Reduced intracellular growth of mycobacteria in human macrophages cultivated at physiologic oxygen pressure. Am Rev Respir Dis 1992; 145:947–953.
27. de Sousa JPC, Rastogi N. Comparative ability of human monocytes and macrophages to control the intracellular growth of *Mycobacterium avium* and *Mycobacterium tuberculosis:* effect of interferon-gamma and indomethacin. FEBS Microbiol Immunol 1992; 89:329–334.
28. Warwick-Davies J, Dhillon J, O'Brien L, Andrew PW, Lowrie DB. Apparent killing of *Mycobacterium tuberculosis* by cytokine-activated human monocytes can be an artefact of a cytotoxic effect on the monocytes. Clin Exp Immunol 1994; 96:214–217.
29. Hirsch CS, Ellner JJ, Russell DG, Rich EA. Complement receptor-mediated uptake and tumor necrosis factor-α-mediated-growth inhibition of *Mycobacterium tuberculosis* by human alveolar macrophages. J Immunol 1994; 152:743.
30. Filley EA, Bull HA, Dowd PM, Rook GAW. The effect of *Mycobacterium tuberculosis* on the susceptibility of human cells to the stimulatory and toxic effects of tumour necrosis factor. Immunology 1992; 77:505–509.
31. Sypek JP, Jacobson S, Vorys A, Wyler DJ. Comparison of gamma interferon, tumor necrosis factor and direct cell contact in activation of antimycobacterial defense in murine macrophages. Infect Immun 1993; 61:3901–3906.
32. Moreno C, Taverne J, Mehlert A, et al. Lipoarabinomannan from *Mycobacterium tuberculosis* induces the production of tumour necrosis factor from human and murine macrophages. Clin Exp Immunol 1989; 76:240–245.
33. Adams JL, Czuprynski CJ. Mycobacterial cell wall components induce the production of TNF-a, IL-1 and IL-6 by bovine

monocytes and the murine macrophage cell line RAW 264.7. Microb Pathogen 1994; 16:401–411.

34. Chatterjee D, Roberts AD, Lowell K, Brennan PJ, Orme IM. Structural basis of capacity of lipoarabinomannan to induce secretion of tumor necrosis factor. Infect Immun 1992; 60:1249–1253.
35. Adams LB, Fukutomi Y, Krahenbuhl JL. Regulation of murine macrophage effector functions by lipoarabinomannan from mycobacterial strains with different degrees of virulence. Infect Immun 1993; 61:4173–4181.
36. Zhang Y, Doerfler M, Lee TC, Guillemin B, Rom WN. Mechanisms of stimulation if interleukin-1β and tumor necrosis factor-α by *Mycobacterium tuberculosis* components. J Clin Invest 1993; 91:2076–2083.
37. Barnes PF, Chatterjee D, Abrams JS, et al. Cytokine production induced by *Mycobacterium tuberculosis* lipoarabinomannan. J Immunol 1992; 149:541–547.
38. Chujor CSN, Kuhn B, Schwerer B, Bernheimer H, Levis WR, Bevec D. Specific inhibition of mRNA accumulation for lymphokines in human T cell line Jurkat by mycobacterial lipoarabinomannan antigen. Clin Exp Immunol 1992; 87:398–403.
39. Roach TIA, Barton CH, Chatterjee D, Blackwell JM. Macrophage activation: lipoarabinomannan from avirulent strains of *Mycobacterium tuberculosis* differentially induces the early genes c-fos, KC, JE and tumor necrosis factor-α. J Immunol 1993; 150:1886–1896.
40. Zhang Y, Broser M, Cohen H, et al. Enhanced Interleukin-8 release and gene expression in macrophages after exposure to *Mycobacterium tuberculosis* and its components. J Clin Invest 1995; 95:586–952.
41. Roach TIA, Chatterjee D, Blackwell JM. Induction of early-response genes KC and JE by mycobacterial lipoarabinomannans: regulation of KC expression in murine macrophages by *Lsh/Ity/Bcg* (Candidate *Nramp*). Infect Immun 1994; 62:1176–1184.
42. Zhang Y, Rom WN. Regulation of the Interleukin-1β (IL-1β) gene by mycobacterial components and lipopolysaccharide is mediated by two nuclear factor-IL-6 motifs. Mol Cell Biol 1993; 13:3831–3837.
43. Zhang Y, Broser M, Rom WN. Activation of the interleukin 6 gene by *Mycobacterium tuberculosis* or lipopolysaccharide is mediated by nuclear factors NF-IL6 and NF-κB. Proc Natl Acad Sci USA 1994; 91:2225–2229.
44. Bronza JP, Horan M, Rademacher JM, Pabst KM, Pabst MJ. Monocyte responses to sulfatide from *Mycobacterium tuberculosis:* inhibition of priming for enhanced release of superoxide associated with increased secretion of interleukin-1 and tumor necrosis factor alpha and altered protein phosphorylation. Infect Immun 1991; 59:2542–2548.
45. Roach TIA, Barton CH, Chatterjee D, Liew FY, Blackwell JM. Opposing effects of interferon-γ on iNOS and interleukin-10 expression in lipopolysaccharide- and mycobacterial lipoarabinomannan-stimulated macrophages. Immunology 1995; 85:106–113.
46. de Waal R, Abrams J, Bennett B, Figdor CG, de Vries JE, Interleukin 10 (IL-10) inhibits cytokine synthesis by human monocytes: an autoregulatory role of IL-10 produced by monocytes. J Exp Med 1991; 174:1209–1220.
47. Wallis RS, Fujiwara H, Ellner JJ. Direct stimulation of monocyte release of Interleukin-1 by mycobacterial protein antigens. J Immunol 1986; 136:193–196.
48. Valone SE, Rich EA, Wallis RS, Ellner JJ. Expression of tumor necrosis factor in vitro by human mononuclear phagocytes stimulated with whole *Mycobacterium bovis* BCG and mycobacterial antigens. Infect Immun 1988; 56:3313–3315.
49. Wallis RS, Amir-Tahmasseb M, Ellner JJ. Induction of Interleukin 1 and tumor necrosis factor by mycobacterial proteins: the monocyte Western blot. Proc Natl Acad Sci 1990; 87:3348–3352.
50. Toossi Z, Young T, Averill LE, Hamilton BD, Shiratsuchi H, Ellner JJ. Induction of transforming growth factor β1 by purified protein derivative of *Mycobacterium tuberculosis.* Infect Immun 1995; 63:224–228.
51. Friedland JS, Shattock R, Remick DG, Griffin GE. Mycobacterial 65kD heat shock protein induces release of TNF, IL-6 and IL-8 from human monocytic cells. Clin Exp Immunol 1993; 91:58–62.
52. Petermans WE, Raats CJI, Langermans JAM, van Furth R. Mycobacterial heat-shock protein 65 induces proinflammatory cytokines but does not activate human mononuclear phagocytes. Scand J Immunol 1994; 39:613–617.
53. Retzlaff C, Yamamoto Y, Hoffman PS, Friedman H, Klein TW. Bacterial heat shock proteins directly induce cytokine mRNA and Interleukin-1 secretion in macrophage cultures. Infect Immun 1994; 62:5689–5693.
54. Silva CL, Lowrie DB. A single mycobacterial protein (hsp 65) expressed by a transgenic antigen-presenting cell vaccinates mice against tuberculosis. Immunology 1994; 82:244–248.
55. Averill L, Toossi Z, Aung H, Boom WH, Ellner JJ. Regulation of production of Tumor Necrosis Factor alpha in monocytes stimulated by the 30-kilodalton antigen of *Mycobacterium tuberculosis.* Infect Immun 1995; 63:3206–3208.
56. Beagley K, Fujihashi K, Black CA, et al. The *Mycobacterium tuberculosis* 71-kDa heat-shock protein induces proliferation and cytokine secretion by murine gut intraepithelial lymphocytes. Eur J Immunol 1993; 23:2049–2052.
57. Schlesinger LS, Horwitz MA. Phagocytosis of *Mycobacterium leprae* by human monocyte-derived macrophages is mediated by complement receptors CR1 (CD35), CR3 (CD11b/CD18) and IFN-γ activation inhibits complement receptor function and phagocytosis of the bacterium. J Immunol 1991; 147:1983–1994.
58. Denis M. Modulation of *Mycobacterium lepraemurium* growth in murine macrophages: benefical effect of tumor necrosis factor alpha and granulocyte-macrophage colony-stimulating factor. Infect Immun 1991; 59:705–707.
59. Gan H, Newman G, McCarthy PL, Remold HG. TNF-α response of human monocyte-derived macrophagers to *Mycobacterium avium,* serovar 4, is of brief duration and protein kinase C dependent. J Immunol 1993; 150:2892–2900.
60. Gan H, Newman GW, Remold HG. Human macrophages acquire a hyporesponsive state of tumor necrosis factor alpha production in response to successive *Mycobacterium avium* serovar 4 stimulation. Infect Immun 1995; 63:1921–1926.
61. Shiratsuchi H, Toossi Z, Mettler MA, Ellner JJ. Colonial morphotype as a determinant of cytokine expression by human monocytes infected with *Mycobacterium avium.* J Immunol 1993; 150:2943–2954.
62. Fattorini L, Xiao Y, Li B, Santoro C, Ippoliti F, Orefici G. Induction of IL-1β, IL-6, TNF-α, GM-CSF and G-CSF in human macrophages by smooth transparent and smooth opaque colonial variants of *Mycobacterium avium.* J Med Microbiol 1994; 40:129–133.
63. Newman GW, Gan HX, McCarthy PL, Remold HG. Survival of human macrophages infected with *Mycobacterium avium intracellulare* correlates with increased production of tumor necrosis factor-α and IL-6. J Immunol 1991; 147:3942–3948.
64. Bermudez LE. Production of transforming growth factor-β by

Mycobacterium avium-infected human macrophages is associated with unresponsiveness to IFN-γ. J Immunol 1993; 150:1838–1845.
65. Bermudez LE, Young LS. Tumor necrosis factor alone or in combination with IL-2 but not IFN-γ is associated with macrophage killing of *Mycobacterium avium* complex. J Immunol 1988; 140:3006–3013.
66. Furney SK, Skinner PS, Roberts AD, Appelberg R, Orme IM. Capacity of *Mycobacterium avium* isolated to grow well or poorly in murine macrophages resides in their ability to induce secretion of tumor necrosis factor. Infect Immun 1992; 60:4410–4413.
67. Denis M, Gregg EO. Modulation of *Mycobacterium avium* growth in murine macrophages: reversal of unresponsiveness to Interferon-gamma by indomethacin of Interleukin-4. J Leuk Biol 1991; 49:65–72.
68. Doi T, Ando M, Akaike T, Suga M, Sato K, Maeda H. Resistance to nitric oxide in *Mycobacterium avium* complex and its implication in pathogenesis. Infect Immun 1993; 61:1980–1989.
69. Appelberg R, Orme IM, De Sousa MJ, Silva MT. In vitro effects of interleukin-4 on interferon-γ-induced macrophage activation. Immunology 1992; 76:553–559.
70. Appelberg R, Orme IM. Effector mechanisms involved in cytokine-mediated bacteriostasis of *Mycobacterium avium* infections in murine macrophages. Immunology 1993; 80:352–359.
71. Sturgill-Koszycki S, Schlesinger PH, Chakraborty P, et al. Lack of acidification in *Mycobacterium* phagosomes produced by exclusion of the vesicular proton-ATPase. Science 1994; 263:678–681.
72. Bermudez LE, Young LS. Tumor necrosis factor α stimulates mycobactericidal/mycobacteriostatic activity in human macrophages by a protein kinase C independent pathway. Cell Immunol 1992 144:258–268.
73. Suzuki K, Lee WJ, Hashimoto T, et al. Recombinant granulocyte-macrophage colony-stimulating factor (GM-CSF) or tumour necrosis factor-alpha (TNF-α) activate human alveolar macrophages to inhibit growth of *Mycobacterium avium* complex. Clin Exp Immunol 1994; 98:169–173.
74. Bermudez LE, Young LS, Gupta S. 1,25 Dihydroxyvitamin D_3-dependent inhibition of growth or killing of *Mycobacterium avium* complex in human macrophages is mediated by TNF and GM-CSF. Cell Immunol 1990; 127:432–441.
75. Bermudez LE, Young LS. Natural killer cell-dependent mycobacteriostatic and mycobacteriocidal activity in human macrophages. J Immunol 1991; 146:265–270.
76. Bermudez LE, Wu M, Young LS. Interleukin-12-stimulated natural killer cells can activate human macrophages to inhibit growth of *Mycobacterium avium.* Infect Immun 1995; 63:4099–4104.
77. Shiratsuchi H, Johnson JL, Toba H, Ellner JJ. Strain- and donor-related differences in the interaction of *Mycobacterium avium* with human monocytes and its modulation by interferon-γ. J Infect Dis 1990; 162:932–938.
78. Denis M. Growth of *Mycobacterium avium* in human monocytes: identification of cytokines which reduce and enhance intracellular microbial growth. Eur J Immunol 1991; 21:391–395.
79. Geertsma MF, Nibbering PH, Pos O, van Furth R. Interferon-γ-activated human granulocytes kill ingested *Mycobacterium fortuitum* more efficiently than normal granulocytes. Eur J Immunol 1990; 20:869–873.
80. Blanchard DK, Michelini-Norris MB, Djeu JY. Interferon decreases the growth inhibition of *Mycobacterium avium-intracellulare* complex by fresh human monocytes but not by culture-derived macrophages. J Infect Dis 1991; 164:152–157.
81. Bermudez LE, Wu M, Petrofsky M, Young LS. Interleukin-6 antagonizes tumor necrosis factor-mediated mycobacteriostatic and mycobacteriocidal activities in macrophages. Infect Immun 1992; 60:4245–4252.
82. Shiratsuchi H, Johnson JL, Ellner JJ. Bidirectional effects of cytokines on the growth of *Mycobacterium avium* within human monocytes. J Immunol 1991; 146:3165–3170.
83. Michelini-Norris MB, Blanchard DK, Pearson CA, Djeu JY. Differential release of Interleukin (IL)-1α, IL-1β, and IL-6 from normal human monocytes stimulated with a virulent and an avirulent isogenic variant of *Mycobacterium avium-intracellulare* complex. J Infect Dis 1992; 165:702–709.
84. Hubbard RD, Collins FM. Immunomodulation of mouse macrophage killing of *Mycobacterium avium* in vitro. Infect Immun 1991; 59:570–574.
85. Kindler V, Sappino A, Grau GE, Piguet P, Vassalli P. The inducing role of tumour necrosis factor in the development of bactericidal granulomas during BCG infection. Cell 1989; 56:731–740.
86. Adams LB, Mason CM, Kolls JK, Scollard D, Krahenbuhl JL, Nelson S. Exacerbation of acute and chronic murine tuberculosis by administration of a tumor necrosis factor receptor-expressing adenovirus. J Infect Dis 1995; 171:400–405.
87. Denis M. Involvement of cytokines in determining resistance and acquired immunity in murine tuberculosis. J Leuk Biol 1991; 50:495–501.
88. Cooper AM, Dalton DK, Stewart TA, Griffin JP, Russell DG, Orme IM. Disseminated tuberculosis in Interferon γ gene-disrupted mice. J Exp Med 1993; 178:2243–2247.
89. Flynn JL, Chan J, Triebold KJ, Dalton DK, Stewart TA, Bloom BR. An essential role for interferon γ in resistance to *Mycobacterium tuberculosis* infection. J Exp Med 1993; 178:2249–2254.
90. Kamijo R, Le J, Shapiro D, et al. Mice that lack the Interferon-γ receptor have profoundly altered responses to infection with bacillus Calmette-Guerin and subsequent challenge with lipopolysaccharide. J Exp Med 1993; 178:1425–1440.
91. Barerra LF, Kramnik I, Skamene E, Radzioch D. Nitrite production by macrophages derived from BCG-resistant and -susceptible congenic mouse strains in response to IFN-γ and infection with BCG. Immunology 1994; 82:457–464.
92. Kamijo R, Rarada H, Matsuyama T, et al. Requirement for transcription factor IRF-1 in NO synthase induction in macrophages. Science 1994; 263:1612–1615.
93. Kamijo R, Shapiro D, Le J, Huang S, Aguet M, Vilcek J. Generation of nitric oxide and induction of major histocompatibility complex class II antigen in macrophages from mice lacking the interferon γ receptor. Proc Natl Acad Sci USA 1993; 90:6626–6630.
94. Ladel CH, Daugelat S, Kaufmann SHE. Immune response to *Mycobacterium bovis* bacille Calmette Guerin infection in major histocompatibility complex class I- and II-deficient knock-out mice: contribution of CD4 and CD8 T cells to acquired resistance. Eur J Immunol 1995; 25:377–384.
95. Flynn JL, Goldstein MM, Triebold KJ, Sypek J, Wolf S, Bloom BR. IL-12 increases resistance of BALB/c mice to *Mycobacterium tuberculosis* infection. J Immunol 1995; 155:2515–2524.
96. Huygen K, Vandenbussche P, Heremans H. Interleukin-6 production in *Mycobacterium bovis* BCG-infected mice. Cell Immunol 1991; 137:224–231.
97. Rhoades ER, Cooper AM, Orme IM. Chemokine responses in

mice infected with *Mycobacterium tuberculosis*. Infect Immun 1995; 63:3871–3877.
98. Appelberg R. Interferon-gamma (IFN-γ) and macrophage inflammatory proteins (MIP)-1 and -2 are involved in the regulation of the T cell-dependent chronic peritoneal neutrophilia of mice3 infected with mycobacteria. Clin Exp Immunol 1992; 89:269–273.
99. Larsen CG, Thomsen MK, Gesser B, et al. The delayed-type hypersensitivity reaction is dependent on IL-8. J Immunol 1995; 155:2151–2157.
100. Kawamura I, Tsukada H, Yoshikawa H, Fujita M, Nomoto K, Mitsuyama M. IFN-γ-producing ability as a possible marker for the protective T cells against *Mycobacterium bovis* BCG in mice. J Immunol 1992; 148:2887–2893.
101. Huygen K, Lozes E, Gilles B, et al. Mapping of TH1 helper T-cell epitopes on major secreted mycobacterial antigen 85A in mice infected with live *Mycobacterium bovis* BCG. Infect Immun 1994; 62:363–370.
102. Inoue T, Yoshikai Y, Matsuzaki G, Nomoto K. Early appearing γ/δ-bearing T cells during infection with Calmette Guerin bacillus. J Immunol 1991; 146:2754–2762.
103. Ladel CH, Hess J, Daugelat S, Mombaerts P, Tonegawa S, Kaufmann SHE. Contribution of α/β and γ/δ T lymphocytes to immunity against *Mycobacterium bovis* bacille Calmette Guerin: studies with T cell receptor-deficient mice. Eur J Immunol 1995; 25:838–846.
104. Flesch IEA, Kaufmann SHE. Role of macrophages and αβ T lymphocytes in early interleukin 10 production during Listeria monocytogenes infection. Int Immunol 1994; 6:463–468.
105. Orme IM, Miller ES, Roberts AD, et al. T lymphocytes mediating protection and cellular cytolysis during the course of *Mycobacterium tuberculosis* infection. Evidence for different kinetics and recognition of a wide spectrum of protein antigens. J Immunol 1992; 148:189–196.
106. Champsi J, Young LS, Bermudez LE. Production of TNF-α, IL-6 and TGF-β and expression of receptors for TNF-α and IL-6 during murine *Mycobacterium avium* infection. Immunology 1995; 84:549–554.
107. Sarmento AM, Appelberg R. Relationship between virulence of *Mycobacterium avium* strains and induction of tumor necrosis factor alpha production in infected mice and in in vitro-cultured mouse macrophages. Infect Immun 1995; 63:3759–3764.
108. Brown DH, Sheridan J, Pearl D, Zwilling BS. Regulation of mycobacterial growth by the hypothalamus-pituitary-adrenal axis: differential responses of *Mycobacterium bovis* BCG-resistant and -susceptible mice. Infect Immun 1993; 61:4793–4800.
109. Brown DH, LaFuse W, Zwilling BS. Cytokine-mediated activation of macrophages from *Mycobacterium bovis* BCG-resistant and susceptible mice: differential effects of corticosterone on mycobacterial activity and expression of the *Bcg* gene (candidate *Nramp*). Infect Immun 1995; 63:2983–2988.
110. Denis M, Ghadirian E. Interleukin-1 is involved in mouse resistance to *Mycobacterium avium*. Infect Immun 1994; 62:457–461.
111. Bermudez LE, Champsi J. Infection with *Mycobacterium avium* induces production of interleukin-10 (IL-10), and administration of anti-IL-10 antibody is associated with enhanced resistance to infection in mice. Infect Immun 1993; 61:3093–3097.
112. Denis M, Ghadirian E. IL-10 neutralization augments mouse resistance to systemic *Mycobacterium avium* infections. J Immunol 1993; 151:5425–5430.
113. Hernandez-Pando R, Rook GAW. The role of TNF-α in T-cell-mediated inflammation depends on the TH1/TH2 cytokine balance. Immunology 1994; 82:591–595.
114. Huygen K, Abramowicz D, Vandenbussche P, et al. Spleen cell cytokine secretion in *Mycobacterium bovis* BCG-infected mice. Infect Immun 1992; 60:2880–2886.
115. Walker KB, Butler R, Colston MJ. Role of Th-1 lymphocytes in the development of protective immunity against *Mycobacterium leprae*. J Immunol 1992; 148:1885–1889.
116. Castro AG, Silva RA, Appelberg R. Endogenously produced IL-12 is required for the induction of protective T cells during *Mycobacterium avium* infections in mice. J Immunol 1995; 155:2013–2019.
117. Appelberg R, Castro AG, Pedrosa J, Minoprio P. Role of interleukin-6 in the induction of protective T cells during mycobacterial infections in mice. Immunology 1994; 82:361–364.
118. Ogawa T, Uchida H, Kusumoto Y, Mori Y, Yamamura Y, Hamada S. Increase in tumor necrosis factor alpha- and interleukin-6-secreting cells in peripheral blood mononuclear cells from subjects infected with *Mycobacterium tuberculosis*. Infect Immun 1991; 59:3021–3025.
119. Takashima T, Ueta C, Tsuyuguchi I, Kishimoto S. Production of tumor necrosis factor alpha by monocytes from patients with pulmonary tuberculosis. Infect Immun 1990; 58:3286–3292.
120. Chensue SW, Davey MP, Remick DG, Kunkel SL. Release of Interleukin-1 by peripheral blood mononuclear cells in patients with tuberculosis and active inflammation. Infect Immun 1986; 52:341–343.
121. Fujiwara H, Kleinhenz ME, Wallis RS, Ellner JJ. Increased Interleukin-1 production and monocyte suppressor cell activity associated with human tuberculosis. Am Rev Respir Dis 1986; 133:73–77.
122. Barnes PF, Fong S, Brennan PJ, Twomey PE, Mazumder A, Moldin RL. Local production of tumor necrosis factor and IFN-γ in tuberculous pleuritis. J Immunol 1990; 145:149–154.
123. Shimokata K, Saka H, Murate T, Hasegawa Y, Hasegawa T. Cytokine content of pleural effusions: comparison between tuberculous and carcinomatous pleurisy. Chest 1991; 99:1103–1107.
124. Glimaker M, Kragsbjerg P, Forsgren M, Olcen P. Tumor necrosis factor-α (TNF-α) in cerebrospinal fluid from patients with meningitis of different etiologies: high levels of TNFα indicated bacterial meningitis. J Infect Dis 1993; 167:882–889.
125. Akalin H, Akdis AC, Mistik R, Helvaci S, Kilicturgay K. Cerebrospinal fluid interleukin-1β/interleukin-1 receptor antagonist balance and tumor necrosis factor-α concentrations in tuberculous, viral and acute bacterial meningitis. Scand J Infect Dis 1994; 26:667–674.
126. Wallis RS, Vjecha M, Amir-Tahmasseb M, et al. Influence of tuberculosis on human immunodeficiency virus (HIV-1): enhanced cytokine expression and elevated β2-microglobulin in HIV-1-associated tuberculosis. J Infect Dis 1993; 167:43–48.
127. Zhang M, Gately MK, Wang E, et al. Interleukin 12 at the site of disease in tuberculosis. J Clin Invest 1994; 93:1733–1739.
128. Antony VB, Godbey SW, Kunkel SL, et al. Recruitment of inflammatory cells to the pleural space. J Immunol 1993; 151:7216–7223.
129. Grunewwald T, Schuler-Mane W, Ruf B. Interleukin-8 and granulocyte colony-stimulating factor in bronchoalveolar lavage fluid and plasma of human immunodeficiency virus-infected patients with *Pneumocystis carinii* pneumonia, bacterial pneumonia or tuberculosis. J Infect Dis 1993; 168:1077–1078.
130. Schall TJ, Bacon K, Toy KJ, Goeddel DV. Selective attraction of monocytes and T lymphocytes of the memory phenotype by cytokine RANTES. Nature 1990; 347:669–671.

131. Devergne O, Marfaing-Koka A, Schall TT, et al. Production of the RANTES chemokine in delayed-type hypersensitivity reactions: involvement of macrophages and endothelial cells. J Exp Med 1994; 179:1689–1694.
132. Friedland JS, Hartley JC, Hartley CGC, Shattock RJ, Griffin GE. Inhibition of ex vivo proinflammatory cytokine secretion in fatal *Mycobacterium tuberculosis* infection. Clin Exp Immunol 1995; 100:233–238.
133. Friedland JS, Hartley JC, Hartley CGC, Shattock RJ, Griffin GE. Cytokine secretion in vivo and ex vivo following chemotherapy of *Mycobacterium tuberculosis* infection. Trans R Soc Trop Med Hyg 1996; 90:199–203.
134. Toossi Z, Gogate P, Shiratsuchi H, Young T, Ellner JJ. Enhanced production of TGF-β by blood monocytes from patients with active tuberculosis and presence of TGF-β in tuberculous granulomatous lung lesions. J Immunol 1995; 154:465–473.
135. Del Prete GF, De Carli M, Mastromauro C, et al. Purified protein derivative of *Mycobacterium tuberculosis* and excretor-secretory antigen(s) of *Toxocara canis* expand in vitro human T cells with stable and opposite (type 1 T helper or type 2 T helper) profile of cytokine production. J Clin Invest 1991; 88:346–350.
136. Follows GA, Munk ME, Gatrill AJ, Conradt P, Kaufmann SHE. Gamma interferon and Interleukin 2, but not interleukin 4, are detectable in γ/δ T-cell cultures after activation with bacteria. Infect Immun 1992; 60:1229–1231.
137. Johnson BJ, McMurray DN. Cytokine gene expression by cultures of human lymphocytes with autologous *Mycobacterium tuberculosis*-infected monocytes. Infect Immun 1994; 62:1444–1450.
138. Robinson DS, Ying S, Taylor IK, et al. Evidence for a Th1-like bronchoalveolar T-cell subset and predominance of interferon-gamma gene activation in pulmonary tuberculosis. Am J Respir Crit Care Med 1994; 149:989–993.
139. Barnes PF, Grisso CL, Abrams JS, Band H, Rea TH, Modlin RL. γδ T lymphocytes in human tuberculosis. J Infect Dis 1992; 165:506–512.
140. Schauf V, Rom WN, Smith KA, et al. Cytokine gene activation and modified responsiveness to Interleukin-2 in the blood of tuberculosis patients. J Infect Dis 1993; 168:1056–1059.
141. Zhang M, Lin Y, Iyer DV, Gong J, Abrams JS, Barnes PF. T-cell cytokine responses in human infection with *Mycobacterium tuberculosis*. Infect Immun 1995; 63:3231–3234.
142. Barnes PF, Abrams JS, Lu S, Sieling PA, Rea TH, Modlin RL. Patterns of cytokine production by mycobacterium-reactive human T cell clones. Infect Immn 1993; 61:197–203.
143. Mutis T, Cornelisse YE, Ottenhoff THM. Mycobacteria induce CD4+ T cells that are cytotoxic and display Th1-like cytokine secretion profile: heterogenicity in cytotoxicity and cytokine secretion levels. Eur J Immunol 1993; 23:2189–2195.
144. Orme IM, Roberts AD, Griffin JP, Abrams JS. Cytokine secretion by CD4 T lymphocytes acquired in response to *Mycobacterium tuberculosis* infection. J Immunol 1993; 151:518–525.
145. Surcell H-M, Troye-Blomberg M, Paulie S, et al. Th1/Th2 profiles in tuberculosis based on the proliferation and cytokine response of blood lymphocytes to mycobacterial antigens. Immunology 1994; 81:171–176.
146. Conradt P, Kaufmann SHE. Impact of antigen-presenting cells on cytokine profiles of human Th clones established after stimulation with *Mycobacterium tuberculosis* antigens. Infect Immun 1995; 63:2079–2081.
147. Barnes PF, Lu S, Abrams JS, Wang E, Yamamura Y, Modlin RL. Cytokine production at the site of disease in human tuberculosis. Infect Immun 1993; 61:3482–3489.
148. Nambuya A, Sewankambo N, Mugerwa J, Goodgame R, Lucas S. Tuberculous lymphadenitis associated with human immunodeficiency virus (HIV) in Uganda. J Clin Pathol 1988; 41:93–96.
149. Barnes PF, Chatterjee D, Brennan PJ, Rea TH, Modlin RL. Tumor necrosis factor production in patients with leprosy. Infect Immun 1992; 60:1441–1446.
150. Nogueira N, Kaplan G, Levy P, et al. Defective γ interferon production in leprosy. J Exp Med 1983; 158:2165–2170.
151. Sampaio EP, Moreira AL, Kaplan G, et al. *Mycobacterium leprae*-induced interferon-γ production by household contacts of leprosy patients: association with development of active disease. J Infect Dis 1991; 164:990–993.
152. Yamamura M, Uyemura K, Deans RJ, et al. Defining protective responses to pathogens: cytokine profiles in leprosy lesions. Science 1991; 254:277–279.
153. Kaplan G, Sheftel G, Job CK, Mathur NK, Nath I, Cohn ZA. Efficacy of a cell-mediated reaction to the purified protein derivative of tuberculin in the disposal of *Mycobacterium leprae* from human skin. Proc Natl Acad Sci USA 1988; 85:5210–5214.
154. Haanen JBAG, de Waal Malefijt R, Res PRP, et al. Selection of a human T helper type 1-like T cell subset by mycobacteria. J Exp Med 1991; 174:583–592.
155. Sieling PA, Wang X-H, Gately MK, et al. IL-12 regulates T helper type 1 cytokine responses in human infectious disease. J Immunol 1994; 153:3639–3647.
156. Sampaio EP, Caneshi JRT, Nery JAC, et al. Cellular immune response to *Mycobacterium leprae* infection in human immunodeficiency virus-infected individuals. Infect Immun 1995; 63:1848–1854.
157. Hancock GE, Cohn ZA, Kaplan G. The generation of antigen-specific, major histocompatibility complex-restricted cytotoxic T lymphocytes of the CD4+ phenotype. J Exp Med 1989; 169:909–919.
158. Nathan C, Kaplan G, Levis W, et al. Local and systemic effects of intradermal recombinant interferon-γ in patients with lepromatous leprosy. N Engl J Med 1986; 315:6–15.
159. Kaplan G, Mathur NK, Job CK, Nath I, Cohn ZA. Effect of multiple interferon γ injections on the disposal of *Mycobacterium leprae*. Proc Natl Acad Sci USA 1989; 86:8073–8077.
160. Nathan C, Squires K, Griffo W, et al. Widespread accumulation of mononuclear leukocytes in lepromatous leprosy patients treated systemically with recombinant interferon γ. J Exp Med 1990; 172:1509–1512.
161. Sampaio EP, Moreira AL, Sarno EN, Malta AM, Kaplan G. Prolonged treatment with recombinant interferon γ induces erythema nodosum leprosum in lepromatous leprosy patients. J Exp Med 1992; 175:1729–1737.
162. Sampaio EP, Kaplan G, Miranda A, et al. The influence of thalidomide on the clinical and immunological manifestation of erythema nodosum leprosum. J Infect Dis 1993; 168:408–414.
163. Moreira AL, Sampaio EP, Zmuidzinas A, Frindt P, Smith KA, Kaplan G. Thalidomide exerts its inhibitory action on tumor necrosis factor α by enhancing mRNA breakdown. J Exp Med 1993; 177:1675–1680.
164. Kaplan G. Kiessling R, Teklemariam S, et al. The reconstitution of cell-mediated immunity in the cutaneous lesions of lepromatous leprosy by recombinant interleukin 2. J Exp Med 1989; 169:893–907.
165. Kaplan G, Britton WJ, Hancock GE, et al. The systemic influence of recombinant interleukin-2 on the manifestations of lepromatous leprosy. J Exp Med 1991; 173:993–1006.
166. Holland SM, Eisenstein EM, Kuhns DB, et al. Treatment of refractory disseminated nontuberculous mycobacterial infec-

tion with inteferon gamma. N Engl J Med 1994; 330:1348–1355.
167. Levin M, Newport MJ, D'Souza S, et al. Familial disseminated atypical mycobacterial infection in childhood: a human mycobacterial susceptibility gene? Lancet 1995; 345:79–83.
168. Sathe SS, Sarai A, Tsingler D, Nedunchezian D. Pentoxifylline aggravates impairment in Tumor Necrosis Factor-α secretion and increases mycobacterial load in macrophages from AIDS patients with disseminated *Mycobacterium avium-intracellulare* complex infection. J Infect Dis 1994; 170:484–487.
169. Bermudez LE, Inderlied C, Young LS. Stimulation with cytokines enhances penetration of azithromycin into human macrophages. Antimicrob Agents Chemother 1991; 35:2625–2629.
170. Bermudez LE, Martinelli J, Petrofsky M, Kolonoski P, Young LS. Recombinant granulocyte-macrophage colony-stimulating-factor enhances the effects of antibiotics against *Mycobacterium avium* complex in the beige mouse model. J Infect Dis 1994; 169:575–580.
171. O'Donnell MA, Aldovani A, Duda RB, et al. Recombinant *Mycobacterium bovis* BCG secreting functional Interleukin-2 enhances gamma Interferon production by splenocytes. Infect Immun 1994; 62:2508–2514.

40
Cytokines in HIV-1 Infection

Linde Meyaard, Hanneke Schuitemaker, and Frank Miedema
Central Laboratory of the Netherlands Red Cross Blood Transfusion Service and the Laboratory of Experimental and Clinical Immunology of the University of Amsterdam, Amsterdam, The Netherlands

INTRODUCTION

Acquired immunodeficiency syndrome (AIDS) is the clinical stage of human immunodeficiency virus type 1 or 2 (HIV-1, HIV-2) infection. Acute infection with HIV, which may present with influenza- or mononucleosislike symptoms (1), is followed by an asymptomatic period of a few months to more than 13 years. This asymptomatic period is characterized by declining numbers of circulating CD4+ Th cells, eventually leading to immunodeficiency and clinical AIDS. The consequences of this include lymphadenopathy, infections with opportunistic pathogens, neoplasms like Kaposi's sarcoma and non-Hodgkin lymphoma, neurological symptoms, and dementia. The opportunistic infections are mainly caused by intracellular pathogens like viruses, mycobacteria, protozoa, and fungi, which like malignancies, are indicative of failing cellular immunity.

One feature of HIV-1 is its great variability with respect to biological properties such as replication rate and cytotropism (2–6). The HIV-1 biological phenotype is believed to be an important determinant in the variable clinical course of the infection. The asymptomatic phase of HIV-1 infection is characterized by low frequencies of infected cells in peripheral blood (7,8), and the predominant HIV-1 variants replicate both in primary T cells and macrophages and are non–syncytium inducing (NSI) (9). HIV-1–infected macrophages in the tissue compartment are believed to be the viral reservoir because in the asymptomatic phase of infection, peripheral blood T cells harbor macrophage-tropic viruses compatible with recent infection of these T cells by HIV-1 derived from macrophages (9,10). In the course of an infection, a gradual shift in the composition of virus populations toward T-cell tropic HIV-1 variants is observed (9). In 50–60% of infected individuals, this shift is accompanied by the emergence of SI variants (11). These SI variants in general are absent in the asymptomatic phase of infection. Macrophages are the main initial target cells for the virus (12,13), and in the first phase of infection only NSI variants are present. The period of time needed to generate specific mutations was correlated with the SI phenotype (14,15).

The ultimate event in AIDS pathogenesis is the loss of CD4+ cells. The risk for development of AIDS-associated symptoms increases with declining CD4+ T-cell numbers (16,17). However, even prior to the loss of CD4+ T cells in later stages of infection, HIV-1 infection is characterized by functional defects of T cells in vitro. These may be observed in early infection. Proliferation of T cells in response to ligation of the T-cell receptor (TCR) CD3 complex is impaired (18–21) and IL-2 production decreased (22,23). In addition, delayed-type hypersensitivity (DTH) reactions are decreased in vivo (24,25). The fact that both CD4+ and CD8+ T cells in HIV-1–infected individuals have disturbed function (20,22) even when the number of infected T cells is low (26), requires a systemic explanation. One attractive mechanism of immune dysregulation is a disturbance of cytokine networks induced by persistent HIV-1 infection.

When T-cell dysfunction in HIV-1–infected individuals is studied in more detail, it is clear that functional properties ascribed to Th1 cells are specifically affected. Clinical symptoms associated with a relative dominance of Th2-mediated responses have been described in HIV-1–

infected individuals and include eosinophilia (28), elevation of serum IgE levels (29), and allergic manifestations (30,31). Shearer and coworkers were the first to propose a shift to Th2 cytokine patterns to explain T-cell dysfunction and the failure of immune mechanisms to control HIV-1 replication (32). In this chapter, studies on cytokine networks in HIV-1 are reviewed, with an emphasis on the recent literature investigating disturbances in Th1:Th2 ratios.

CYTOKINES IN ANIMAL MODELS OF HIV-1 INFECTION

One problem with studies of the pathogenesis of HIV-1 infection is the lack of relevant animal models. Longitudinal studies in humans are bound to be descriptive and can only point to correlations but can never establish cause-effect relations. Therefore, especially for vaccination and intervention studies, animal models of HIV-1 infection are being explored, and pathogenesis studies done in these models are often consistent with previous findings in infected humans. In HIV, there are few studies on cytokines in animal models, in contrast to the many studies in other viral and parasitic infections.

Simian Immunodeficiency Virus

Simian immunodeficiency virus (SIV) infection of macaques is considered a good animal model for HIV-1 infection and AIDS (33). Certain SIV strains cause an AIDS-like syndrome in experimentally infected rhesus macaques (34). In contrast, sooty mangabey monkeys can be naturally infected with SIV but seem resistant to development of disease. Comparison of cloned CD4+ T-cell lines generated from cell samples obtained during the asymptomatic phase from SIV-infected macaques with T-cell lines obtained from mangabeys revealed a Th1-like profile in the first and a predominant Th2-like profile in the latter animals (35). There is, however, lack of good definition of primate Th1 and Th2 cells, and the relevance of these findings for human HIV-1 infection is not clear (36). The data contradict not only the Th1/Th2 paradigm in HIV-1 infection, which will be discussed later, but also findings in animal models of other infectious diseases, such as *L. major* infection, where a Th1 profile is associated with protection from and a Th2 profile with progression to disease (37). Furthermore, another study in simian AIDS showed increased levels of IFN-γ and increased IL-10 levels during the asymptomatic phase of infection in rhesus macaques, with a decrease of both these cytokines during progression to disease (38). Cytokine mRNA expression during acute SIV infection and up to 255 days after inoculation appeared to be related to viral load for all cytokines tested, indicative of chronic active infection (39).

HIV-1 Infection in Chimpanzees

Chimpanzees are the most commonly used nonhuman primates that can become persistently infected with HIV-1. Despite having many similarities with HIV-1 infection in humans, such as persistence of virus (40,41) and emergence of neutralization escape mutants (42), persistently HIV-1–infected chimpanzees do not develop clinical symptoms (43,44). Furthermore, no evidence for systemic immunological abnormalities, including T-cell nonresponsiveness, accessory cell function, and programmed cell death, can be found (45,46). To date, no studies on cytokine levels and release in HIV-1-infected chimpanzees have been published.

Chimpanzee monocytes are relatively resistant to infection with HIV-1, and no monocytotropic HIV-1 variants could be recovered from the peripheral blood of HIV-1 infected chimpanzees (45,47,48). These results suggest that in chimpanzees monocyte resistance to HIV-1 infection may contribute to control of virus replication and maintenance of the asymptomatic state (45). In any event, low replication rates of HIV-1 in chimpanzees do seem related to lack of pathology. In conclusion, given the absence of T-cell defects, HIV-1 infection of chimpanzees does not seem an appropriate model to study immune disturbances and intervention with immune modulators.

Murine Models of HIV-1 Infection

MAIDS

Infection of C57BL/6 mice with the LP-BM5 mixture of type C retroviruses leads to an AIDS-like syndrome characterized by B-cell lymphoproliferation and T-cell anergy, termed MAIDS (murine acquired immunodeficiency syndrome) (49). In this model, the cytokine network has been studied in detail. A preferential activation of Th2 cells is observed (50), and because of increased IL-10 production, LP-BM5–infected mice appear to be protected against the Th1-mediated pathology associated with experimental cerebral malaria (51). Furthermore, in vivo treatment with IL-12 protects LP-BM5–infected mice from immune deficiency (52). All these findings are compatible with the hypothesis of a switch to Th2-like T-cell responses during progression of human HIV-1 infection. The reported resistance to MAIDS of IL-4–deficient mice infected with LP-BM5 virus suggested a role of Th2 cytokines in MAIDS pathogenesis by downregulation of Th1 and cytotoxic T-cell responses (53). Others, however, showed that due to technical difficulties these conclusions might not be valid and resistance to MAIDS might not be due to absence of IL-4 (54,55). In addition, although there are gross similarities between HIV-1 infection and AIDS and the LP-BM5 model, there are many more differences (for instance, the target cell of the viruses) which make comparison of the pathogenic events in LP-BM5 infection in mice with those in HIV-1 infection rather difficult (see Ref. 56 for a recent review).

SCID-hu/PBL

One novel approach to provide an animal model for human HIV-1 infection involves using severe combined immune-deficiency syndrome (SCID) mice reconstituted with human peripheral blood lymphocytes (PBL) (57) or human hemato-lymphoid organs (58). In the SCID-PBL mice, only T cells and very few viable human cells from the monocyte/macrophage lineage are present, while the SCID-hu mouse has a lymphoid system more resembling the human immune system. Therefore, the latter model may be the best to provide useful information on HIV-1 pathogenesis. Studies reported thus far in this model involve acute HIV-1 infection, and its use in studying the asymptomatic phase of HIV-1 infection remains to be seen. So far, little data on cytokines in such models have emerged. In one recent paper, increased TNF-α, TNF-β, and IL-2 secretion in spleen, lymph nodes, and peripheral blood mononuclear cells from HIV-1–infected SCID-hu mice was observed (59), in good agreement with data reported in acute HIV-1 infection in humans.

DETECTION OF CYTOKINES IN HIV-1–INFECTED INDIVIDUALS

Th Cytokines

Shearer and Clerici in 1993 proposed that T-cell dysfunction in HIV-1 infection was associated with a change in cytokines secreted by Th cells from Th1 to a Th2 phenotype (32). One approach to studying the capacity of patient T cells to secrete a certain cytokine is to generate T-cell clones. In contrast to studies in mice, the distinction between Th1, Th2, and Th0 clones in humans is mainly based on quantitative rather than qualitative differences in cytokine secretion patterns. Maggi et al. reported in 1987 a reduced number of T-cell clones producing IL-2 and IFN-γ in AIDS patients (60). Recently, the same group generated a large panel of CD4+ T-cell clones specific for PPD or *T. gondii* and observed a significant increase in the production of Th2 cytokines by clones generated from HIV-1–infected individuals, with an overall increased percentage of Th0 type clones (61). Consistent with this, T-cell clones generated by random cloning procedures from CD4+ memory cells from asymptomatically HIV-1–infected individuals were more likely to be of Th0 type, mainly due to increased IL-4 production, as were cells from a single individual after HIV-1 infection compared to when uninfected (62). In general, the observed changes are relatively modest (61,62).

In addition to the increased percentage of Th0/Th2 CD4+ T-cell clones, CD8+ T-cell lines and clones have been generated from asymptomatic HIV-1–infected individuals that can provide helper activity for IgE synthesis (63) and show decreased cytolytic activity (64). These CD8+ T cells also have Th2-like cytokine secretion patterns (63,64). Romagnani's group reported that these cells preferentially express CD30 (65), which they suggest to be a marker specific for Th2 cells (66). However, primary CD30+ T cells were previously demonstrated to be able to produce both Th2 and Th1 cytokines, such as IFN-γ (67), the production of which can be enhanced by IL-12 (68). Furthermore, extensive study of peripheral T cells and T-cell clones provided little support for a division of Th1 and Th2 type T cells based on CD30 expression (69). Thus, a clear link between CD30 expression and Th2 cytokine secretion in general, as proposed for HIV-1 infection (66), is still controversial.

In addition to analyses of T-cell lines and T-cell clones, studies on cytokine production after stimulation of PBMC in vitro have been published. Decreased IL-2 production by T cells from HIV-1–infected individuals in vitro has been extensively documented (70–76). However, for other Th1 and Th2 cytokines, conflicting results have been reported. A clear-cut shift to Th2 responses in bulk cultures has only been reported by Clerici et al., who showed decreased IL-2 production induced by recall antigen but increased mitogen-induced IL-4 production by PBMC from HIV-1–infected individuals in certain phases of infection (71). Autran et al. (77) described a specific subset of CD4+ T cells, lacking CD7 expression, which have a Th2-like cytokine secretion pattern and are increased in number in HIV-1–infected individuals. However, increased IL-4 production as reported by Clerici et al. was only found by two other groups (75,78), while most reports show unchanged (51,73,79,80) or decreased (61,76,81–84) IL-4 production. It should be noted that most studies on PBMC cultures show a general decrease of all cytokines studied, probably due to the fact that T cells from HIV-1–infected individuals respond less well to polyclonal T-cell activation signals in vitro (61,72,76). In agreement with a decrease in Th1 cytokine secretion, decreased IFN-γ production was shown in very early (27,85) and in more recent studies (72–76). However, increased IFN-γ production, probably originating from mainly CD8+ T cells, and possibly reflecting activated cellular immunity, has been reported on more than one occasion (70,78,80,86,87). IL-10, a Th2 cytokine in mice, is produced by both Th1 and Th2-type T cells in humans (88), but might be considered Th2 because of its immunosuppressive effects (89). IL-10 can be produced by different cell types, which may explain conflicting data on IL-10 production in HIV-1 infection. In line with other monokines, the increase in IL-10 production in HIV-1–infected individuals (70,72,87,90,91) might be due to production by monocytes (80). Unresponsiveness of T cells might be the cause of the observation that IL-10 production is decreased or unaltered in other studies (73,76,82–84).

It is important to keep in mind that the varying results concerning cytokine secretion in HIV-1-infected individuals might be related to the stage of infection of the patients studied and to differences in cell preparations, mode of activation of cells, and cytokine assays. Furthermore, one must be careful when comparing data on cytokine responsiveness obtained by T-cell cloning with results from bulk PBMC cultures. By generating T-cell clones, the potential of T cells to secrete a certain cytokine is studied under optimal circumstances. By analyzing PBMC stimulated in bulk, the intrinsic decreased responses to the stimulus by cells from HIV-1–infected subjects might bias observations as to whether the cells have a commitment in vivo to secrete a certain cytokine. Furthermore, cross-regulation between different cell types in bulk cultures might influence the outcome. Single-cell analysis of cytokine secretion by T cells from HIV-1–infected individuals indicate the number of cells capable of producing either IFN-γ or IL-4, regardless of the amount of cytokines they are producing. Indeed, upon single-cell analysis by intracellular staining, we observed a moderate but significant decrease in the ratio of cells producing IFN-γ and IL-4 in HIV-1–infected individuals, in agreement with the findings in T-cell clones (76) (Table 1).

The conclusion from our data and those discussed here might be that there is a decrease in Th1-type cytokine production causing a disturbance in the balance between Th1 and Th2 cytokines, leading to a Th0-like cytokine profile. Thus, a change in cytokine secretion seems to be a *result* of disturbed T-cell function, and so far there seems no conclusive evidence to justify a shift to Th2-like responses as a *cause* for T-cell dysfunction in HIV infection.

Cytokines Produced by Antigen-Presenting Cells

Possibly due to persistent immune activation in vivo, production of monokines in HIV-1–infected individuals is increased. There are conflicting reports on TNF-α levels in serum, probably because detection of TNF-α in serum is not easy to reproduce although TNF-α production in vitro by PBMC from HIV-1–infected individuals reportedly is increased (70,86,92,93). Serum IL-6 levels are reportedly increased in HIV-1–infected individuals (94,95), as is IL-6 production ex vivo (93,96) and in situ in lymph nodes, where it is probably produced by cells of macrophage/endothelial origin (80). TGF-β, a cytokine produced by many different cell types, is also produced in increased amounts compared to controls by PBMC from HIV-infected individuals (97,98).

In recent years, attention has been focused on cytokines produced by antigen-presenting cells that are critical for polarization of Th-cell responses. The main cytokines in this context are IL-10 and IL-12 (see above). IL-12 plays a critical role in Th1 cell differentiation (99). Chehimi et al. were the first to publish that PBMC from HIV-1–infected individuals have impaired IL-12 p40 and p70 production upon stimulation with *S. aureus* antigen (83). A similar finding was obtained with stimulation with *T. gondii* antigen for IL-12 p40 (74). Alveolar macrophages from asymptomatically infected individuals, however, produce increased amounts of IL-12 p70 when stimulated with bacterial antigens, and a decreased IL-12 secretion was only found in patients with AIDS (91). These results, at this time, are suggestive of an underlying failure to produce the required amount of IL-12 to mount proper Th1 responses.

Table 1 Schematic Overview of Reports Concerning the Production of Th1 and Th2 Type Cytokines by Cells from HIV-Infected Individuals

Cell type analyzed	Th1 cytokines	Th2 cytokines	Ref.
T-cell clones	Decreased	Increased	60–62
PBMC			
bulk culture	Decreased	Increased	71
bulk culture	Decreased	Decreased	61,72,76
single-cell analysis	Decreased	Increased	76
Lymph nodes (constitutive expression)	Increased	Undetectable	80,87

Production of HIV-1–Suppressive Cytokines

An interesting component of host immunity that was ascribed an important role in AIDS pathogenesis is the noncytolytic inhibition of virus replication by CD8+ T lymphocytes in a MHC class I nonrestricted way (100–102). Inhibition has been demonstrated to be mediated at least in part by a soluble factor, although cell-cell contact appeared to be required for optimal inhibition of virus replication (103–105). This suppressive activity has been demonstrated in CD8+ T cells from both HIV-1–positive and HIV-1–negative individuals as well as from nonhuman primates (106–111). The extent of CD8+ suppressor activity correlates with progression to disease (111–113) and therefore may contribute to maintenance of the asymptomatic stage of the disease. In addition, CD8+ lymphocytes from long-term asymptomatic HIV-1–infected individuals show vigorous HIV-1–suppressive activity, and from most of these individuals infectious virus can only be isolated after CD8+ T-cell depletion (114). Two groups recently reported cloning of these factors, revealing its identities as IL-16 in one report (115) and as a combination of RANTES, MIP-1α, and MIP-1β in another (116). It is anticipated that the viral inhibitory factors are produced in response to persistent viral infection and thus are not specific for HIV-1 infection.

USE OF CYTOKINES OR CYTOKINE INHIBITORS TO MODULATE HIV-1 INFECTION

Cytokines and Regulation of HIV-1 Replication In Vitro

Since expression of cytokines in HIV-1–infected individuals is altered and some of these cytokines can influence HIV-1 replication, it was tempting to speculate that cytokines could influence AIDS pathogenesis by altering HIV-1 expression. The earliest studies on cytokine regulation of HIV-1 gene expression were performed in a CD4-expressing promonocytic cell line (U937) latently infected with HIV-1. Stimulation of these cells with phorbol esters, mitogens, or cytokines like TNF-α and TNF-β resulted in induction of virus production (117). Subsequently investigators employed primary T cells and macrophages in such cytokine regulation studies. TNF-α and TNF-β appeared to be the only cytokines to induce HIV-1 replication in both cells of the monocyte and the T-lymphocyte lineages. The effect of several other cytokines appears to be cell lineage restricted. Interleukin-3, IL-6, IL-10, IL-13, M-CSF, and GM-CSF influence viral replication only in cells of monocyte lineage, whereas IL-2, IL-7, and IL-12 stimulate HIV-1 replication specifically in T cells (118). For some cytokines this cell type–specific effect can simply be explained by the absence of the cytokine receptor on the target cell. Finally, IL-4 enhances viral replication in primary T cells and dose dependently inhibits or enhances viral replication in monocyte-derived macrophages (MDM) (119–121).

HIV-1 replication in macrophages is dependent on the activation and differentiation state of the cell and may thus be sensitive to several cytokines. Virus expression in primary macrophages when cultured in the presence of IL-1 (119), IL-3 (119,122,123), low-dose IL-4 (119), GM-CSF (119,122), M-CSF (124), or TNF-α and TNF-β (125) is enhanced. Exposure of MDM to TGF-β (126), IL-13 (127), IFN-γ (128), or high-dose IL-4 (119) results in inhibition of HIV-1 replication. The altered levels of some of these cytokines reported in HIV-1–infected individuals (86) may thus have important consequences for virus replication in vivo. The mechanism of action by which these cytokines regulate HIV-1 expression has not yet been elucidated. However, activation of nuclear factor kappa B (NF-κb) due to IL-1, TNF-α, or IL-3 treatment of macrophages, thus mediating enhancement of virus replication by triggering the viral promotor in the long terminal repeat (LTR), seems plausible (119,129).

High-dose IL-4 (50 ng/ml), probably by inducing terminal differentiation of monocytes (130), inhibits proliferation of MDM and concomitantly prevents productive HIV-1 infection (119). To establish productive HIV-1 infection in MDM, in addition to differentiation, cell proliferation seems to be required (131), although there is some controversy on this issue (132–134). Since HIV-1 is a retrovirus, it requires reverse transcription of its RNA genome into a proviral DNA species, which integrates in the host cell DNA. In quiescent MDM, the process of reverse transcription cannot be completed, resulting in incomplete pro-viral DNA species (119,131). Low-dose IL-4 treatment, on the other hand, does not affect proliferation of MDM but gives rise to enhanced HIV-1 replication, most likely a result of activation of NF-kB (119).

IL-4 and IL-10 in many respects have common actions on macrophages. They both inhibit the production of IL-1α, IL-1β, IL-6, IL-8, TNF-α, GM-CSF, and G-CSF at the transcriptional level (135–137). IL-10 also appeared to interfere with HIV-1 replication in primary macrophages (138–141). An inhibitory effect of IL-10 at the level of transcription has been reported (135), but in our experience, MDM cultured in the presence of IL-10 underwent abortive HIV-1 infection despite normal expression of HIV-1–specific RNA transcripts. In these cells, immunoblot analysis revealed the absence of properly processed core protein p24, suggesting disturbed processing of viral proteins. In already infected MDM, IL-10 treatment resulted in intracellular accumulation of viral proteins. IL-10 treatment thus seems to induce a posttranslational defect in macrophage protein synthesis.

Expression of TNF-α and IL-6, cytokines that may enhance HIV-1 replication in an autocrine fashion (125,142), is downregulated by IL-10 (135,143). Limiting TNF-α expression may thus be the cause of the observed IL-10–mediated inhibitory effects on HIV-1 replication in MDM. Exogenous TNF-α and IL-6 indeed restore HIV-1 replication in IL-10–treated MDM (143). Interestingly, TNF-α and IL-6 reportedly enhance HIV-1 replication at a replication at a posttranscriptional level (142). Interference with HIV-1 replication by IL-10 at a posttranscriptional level is in agreement with the hypothesis that inhibition may be mediated to a large extent by endogenous TNF-α (143).

Cytokines and Immune Modulation

In Vitro Modulation of T-Cell Function in HIV-1–Infected Individuals

Proliferative responses to recall antigen are disturbed early in HIV-1 infection but can be restored in vitro by IL-2 (18,144). IL-2 is also capable of restoring HIV-1–specific proliferative responses in vitro (145,146). In addition, the recently described IL-2–like cytokine IL-15 seems to have similar and synergistic effects to IL-2 on T-cell proliferation from HIV-1–infected individuals and controls (146).

In recent years, in view of the publications of Th1/Th2 cytokine disturbances, in vitro manipulation of proliferative responses by various cytokines or anticytokine antibodies has been widely studied. IL-12 is capable of enhancing proliferation of T cells from HIV-1–infected individuals in response to influenza, HIV-1 peptides (147), *Mycobacterium avium* (146,148), and polyclonal T-cell stimulators (146). Furthermore IFN-γ production in vitro in response to several T-cell stimulators is enhanced upon addition of IL-12 (73,146,147). When T cells from HIV-1–infected individuals are cloned in the presence of IL-12, an increase in the number of IFN-γ producing cells is observed (149). Thus, IL-12 seems capable of restoring T-cell defects in vitro. However, it should be noted that in most publications it was reported that T-cell proliferative responses and IFN-γ production in noninfected controls were also enhanced in the presence of IL-12 (146,148). The increased number of IFN-γ–producing T cells in the presence of IL-12 was also observed when T cells from noninfected controls were used (149–152). This suggests that IL-12 does not restore a disturbed balance, but in vitro enhances Th1-mediated responses, a phenomenon observable in normal conditions. In agreement with this, a general enhancing effect of IL-12 on NK cell function was observed in both HIV-1–infected individuals and controls (153).

Programmed cell death of T cells has been proposed as another mechanism of T-cell dysfunction in HIV-1 infection (154). Indeed CD4+ and CD8+ T cells from HIV-1–infected individuals die upon culture in vitro (155,156). Recently, this process of T-cell death was reported to be differentially influenced by Th1 and Th2 cytokines. Th1 cytokines and anti-Th2 cytokine antibodies were reported to inhibit programmed cell death of T cells from HIV-1–infected individuals in vitro (84,157). However, there is a caveat in that the same cytokines enhance T-cell proliferation, which might result in a decreased *percentage* of apoptotic cells, which could be achieved without actually inhibiting cell death. Furthermore, it should be noted that apoptosis in HIV-1 infection is often regarded as a result of immune activation rather than a principal cause of T-cell dysfunction (158–161). Therefore, modulation of the apoptotic process might not in itself be beneficial for the outcome of disease.

Experience with Cytokine Therapy In Vivo

Since T cells from HIV-1–infected individuals are impaired in their capacity to produce IL-2 needed for T-cell proliferative responses, treatment with IL-2 would be a logical form of immune therapy in HIV-1–infected individuals. Indeed, several studies of IL-2 therapy have been published in the last few years, all in individuals that were simultaneously treated with antiviral drugs, usually zidovudine. Low-dose IL-2 treatment was demonstrated in several studies to increase certain immune functions, especially in individuals with higher CD4+ T-cell counts, without increases in viral load (162–164). Prolonged treatment with IL-2 recently demonstrated once again benefical effects in individuals with CD4+ T-cell counts above 200/mm^3, while patients with fewer CD4+ T cells showed little immunological improvement but did show evidence of increased viral activation and suffered from toxic effects (165).

Another candidate for immune therapy is IFN-α, not primarily because of its immune modulatory potential but because of its antiviral capacities (166). IFN-α is produced in increased amounts during natural HIV-1 infection, and resistance to IFN-α of HIV-1 variants was demonstrated in the course of infection. Thus it was proposed that IFN-α therapy would only be useful in the early stages of HIV-1 infection (166). A phase I trial of IFN-α in combination with IL-2 showed few side effects and provided preliminary evidence for immune modulation and antiviral activity (167). IFN-α together with zidovudine had beneficial, though transient, effects on CD4+

T-cell numbers and T-cell function when compared with zidovudine alone (167).

Myelosuppression is a major symptom in AIDS patients. Moreover, zidovudine has myelosuppressive side effects (168). Therefore, combination therapy with zidovudine and multilineage haemopoietic growth factors has been considered (169). Treatment of AIDS patients with GM-CSF indeed resulted in a qualitative and quantitative increase in mature neutrophils, eosinophils, and monocytes (170). However, viral load was not monitored in this study. This is an even more important parameter when considering that all candidate growth factors like IL-3, GM-CSF, and M-CSF show enhancement of viral replication in primary macrophages in vitro (122,124). When used in combination with zidovudine in vitro, no enhancement of viral replication mediated by IL-3 and GM-CSF could be observed, but proliferative capacity of primary monocytes/macrophages was restored (123). Combination therapy with GM-CSF and zidovudine indeed suggested that the neutropenia associated with zidovudine therapy can be overcome by GM-CSF (169). GM-CSF has also been used to treat neutropenias associated with zidovudine.

POTENTIAL FUTURE DIRECTIONS FOR CYTOKINE RESEARCH

IL-12 as Immune Modulator in HIV-1 Infection

Recently, IL-12 has been greeted enthusiastically as a potential immune modulator to administer to HIV-1–infected individuals (171). Although single-dose treatment induced little or no adverse effects, in clinical trials in cancer patients, repetitive infusions caused severe toxicity, which suggests that caution is needed in using IL-12 as a therapeutic agent (172). Furthermore, although there are many reports of beneficial effects of IL-12 in animal models of parasitic and bacterial infections, relatively little is still known about the effects on IL-12 on viral infections in animal models. Whereas low-dose treatment with IL-12 is capable of enhancing the cellular immune response against lymphocytic choreomeningitis virus (LCMV) infection in mice, high-dose IL-12 treatment results in enhanced virus burden in these animals (173). IL-12–mediated cytotoxicity in these high-dose IL-12–treated animals might be explained by the increased production of TNF-α upon IL-12 treatment (174). In addition, in human T cells, IL-12 is capable of inducing the production of the immunosuppressive cytokine IL-10 in vitro (175). Thus prolonged IL-12 treatment might lead to immunosuppression due to increased production of cytokines as TNF-α and IL-10 and lead to increased viral burden and inability of the immune system to combat this.

Is Immunotherapy Useful in HIV-1 Infection?

The rationale behind immune modulation in HIV-1–infected individuals should not only be to correct secretion of cytokines that are over- or underproduced in the course of infection. To estimate the chance of success of immune modulation in vivo, a more basic question must first be answered: How important is the gradual loss of cellular immunity in immunopathogenesis of HIV-1 infection? One might expect that with a properly functioning immune system, HIV-1 will never get the chance to give rise to overt virus replication and severe CD4+ T-cell loss. Some findings argue strongly for an important role of the failing immune system in progression to AIDS. First, loss of T-cell function is predictive for rapid progression to AIDS independent of CD4+ T-cell numbers (176–178), and second, in individuals who remain asymptomatic for a prolonged period of time, immune function is preserved (179,180).

Many questions remain to be answered in the search for understanding of the role of the deteriorating immune system in AIDS pathogenesis. One of the these is the contribution of cytotoxic T lymphocytes to protection from disease. There is some controversy as to the role of HIV-1–specific CTL in controlling virus infection (182). Studies in long-term asymptomatic individuals have demonstrated preserved HIV-1 gag-specific CTL responses (183). However, despite the presence of strong CTL responses, viral load can increase in people who progress rapidly to AIDS (183). Based on these results, one may conclude that HIV-1–specific CTL do not play a critical role in determining the rate of progression to AIDS. However, the CTL precursor frequency in this study was determined by limiting dilution culture and propagation of cells under optimal in vitro conditions. One could argue that in vivo, although in these individuals CTL precursors are present, they are not capable of properly exerting their function. Defective presentation of HIV-1 antigens, lack of IL-12, or increased IL-10 production and decreased Th1 responses in vivo may all lead to failure of cellular immune responses. This would point to an important role for antigen-presenting and Th cells. Lack of cellular immunity would lead to lack of CTL responses and increasing viral loads. Cellular immunity has even been proposed to be correlated with protection from HIV-1 infection (184), but clear proof for this is lacking (181).

To be able to adequately and successfully intervene by immunotherapy in HIV-1 infection, we need more insight into why the host fails to successfully handle this virus infection. Given the complexity of virus-host interactions, integrated viroimmunological studies are required to determine what the critical event(s) are that eventually lead to loss of CD4+ T cells and progression to disease in HIV-1–infected individuals (185). Hopefully, this will lead to the development of successful therapeutic strategies.

REFERENCES

1. Tindall B, Cooper DA. Primary HIV infection: host responses and intervention strategies. AIDS 1991; 5:1–14.
2. Asjo B, Albert J, Karlsson A, et al. Replicative properties of human immunodeficiency virus from patients with varying severity of HIV infection. Lancet 1986; ii:660–662.
3. Von Briesen H, Becker WB, Henco K, et al. Isolation frequency and growth properties of HIV variants: multiple simultaneous variants in a patient demonstrated by molecular cloning. J Med Virol 1987; 23:51–66.
4. Evans LA, McHugh TM, Stites DP, et al. Differential ability of HIV isolates to productively infect human cells. J Immunol 1987; 138:3415–3418.
5. Cheng-Mayer C, Seto D, Tateno M, et al. Biologic features of HIV-1 that correlate with virulence in the host. Science 1988; 240:80–82.
6. Tersmette M, De Goede REY, Al BJM, et al. Differential syncytium-inducing capacity of human immunodeficiency virus isolates: frequent detection of syncytium-inducing isolates in patients with acquired immunodeficiency syndrome (AIDS) and AIDS-related complex. J Virol 1988; 62:2026–2032.
7. Koot M, Van't Wout AB, Kootstra NA, et al. Relation between changes in cellular load, evolution of viral phenotype, and the clonal composition of virus populations in the course of human immunodeficiency virus type 1 infection. J Infect Dis 1996; 173:
8. Connor RI, Mohri H, Cao Y, et al. Increased viral burden and cytopathicity correlate temporally with CD4+ T-lymphocyte decline and clinical progression in human immunodeficiency virus type 1 infected individuals. J Virol 1993; 67:1772–1777.
9. Schuitemaker H, Koot M, Kootstra NA, et al. Biological phenotype of human immunodeficiency virus type 1 clones at different stages of infection: progression of disease is associated with a shift from monocytotropic to T-cell-tropic virus populations. J Virol 1992; 66:1354–1360.
10. Schuitemaker H, Kootstra NA, De Goede REY, et al. Monocytotropic human immunodeficiency virus 1 (HIV-1) variants detectable in all stages of HIV infection lack T-cell line tropism and syncytium-inducing ability in primary T-cell culture. J Virol 1991; 65:356–363.
11. Tersmette M, Gruters RA, De Wolf F, et al. Evidence for a role of virulent human immunodeficiency virus (HIV) variants in the pathogenesis of acquired immunodeficiency syndrome: studies on sequential HIV isolates. J Virol 1989; 63:2118–2125.
12. Van't Wout AB, Kootstra NA, Mulder-Kampinga GA, et al. Macrophage-tropic variants initiate human immunodeficiency virus type 1 infection after sexual, parenteral and vertical transmission. J Clin Invest 1994; 94:2060–2067.
13. Palacio J, Souberbielle BE, Shattock RJ, et al. In vitro HIV-1 infection of human cervical tissue. Res Virol 1994; 145:155–161.
14. Fouchier RAM, Groenink M, Kootstra NA, et al. Phenotype-associated sequence variation in the third variable domain of the human immunodeficiency virus type 1 gp120 molecule. J Virol 1992; 66:3183–3187.
15. Fouchier RAM, Broersen SM, Brouwer M, et al. Temporal relationship between elongation of the HIV-1 gp120 V2 domain and the conversion towards a syncytium inducing phenotype. AIDS Res Hum Retroviruses 1995; 11:1473–1478.
16. Eyster ME, Gail MH, Ballard JO, et al. Natural history of human immunodeficiency virus infections in hemophiliacs: effects of T-cell subsets, platelet counts, and age. Ann Intern Med 1987; 107:1–6.
17. Moss AR, Bachetti P, Osmond D, et al. Seropositivity of HIV and the development of AIDS or AIDS-related conditions: three year follow-up of the San Francisco General Hospital cohort. Br Med J 1988; 269:745–750.
18. Shearer GM, Bernstein DC, Tung KSK, et al. A model for the selective loss of major histocompatibility complex self-restricted T cell immune responses during the development of acquired immune deficiency syndrome (AIDS). J Immunol 1986; 137:2514–2521.
19. Giorgi JV, Fahey JL, Smith DC, et al. Early effects of HIV on CD4 lymphocytes in vivo. J Immunol 1987; 138:3725–3730.
20. Miedema F, Petit AJC, Terpstra FG, et al. Immunological abnormalities in human immunodeficiency virus (HIV)-infected asymptomatic homosexual men. HIV affects the immune system before CD4+ T helper cell depletion occurs. J Clin Invest 1988; 82:1908–1914.
21. Ballet JJ, Couderc LJ, Rabian-Herzog C, et al. Impaired T-lymphocyte-dependent immune responses to microbial antigens in patients with HIV-1-associated persistent generalized lymphadenopathy. AIDS 1988; 2:291–297.
22. Gruters RA, Terpstra FG, De Jong R, et al. Selective loss of T-cell functions in different stages of HIV infection. Eur J Immunol 1990; 20:1039–1044.
23. Clerici M, Via CS, Lucey DR, et al. Functional dichotomy of CD4+ T helper lymphocytes in asymptomatic human immunodeficiency virus infection. Eur J Immunol 1991; 21:665–670.
24. Blatt SP, Hendrix CW, Butzin CA, et al. Delayed type hypersensitivity skin testing predicts progression to AIDS in HIV-infected patients. Ann Intern Med 1993; 119:177–184.
25. Markowitz N, Hansen NI, Wilcosky TC, et al. Tuberculin and anergy testing in HIV-seropositive and HIV-seronegative persons. Ann Intern Med 1993; 119:185–193.
26. Schnittman SM, Psallidopoulos MC, Lane HC, et al. The reservoir for HIV-1 in human peripheral blood is a T cell that maintains expression of CD4. Science 1989; 245:305–308.
27. Murray HW, Scavuzzo DA, Kelly CD, et al. T4+ cell production of interferon gamma and the clinical spectrum of patients at risk for and with acquired immunodeficiency syndrome. Arch Intern Med 1988; 148:1613–1616.
28. Fleury-Feith J, Van Nhieu JT, Picard C, et al. Bronchoalveolar lavage eosinophilia associated with Pneumocystis carinii pneumonitis in AIDS patients. Comparative study with non-AIDS patients. Chest 1992; 95:1198–1201.
29. Israël-Biet D, Labrousse F, Tourani JM, et al. Elevation of IgE in HIV-infected subjects: A marker of poor prognosis. J Allergy Clin Immunol 1992; 89:68–75.
30. Sample S, Chernoff DN, Lenahan GA, et al. Elevated serum concentrations of IgE antibodies to environmental antigens in HIV-seropositive male homosexuals. J Allergy Clin Immunol 1990; 86:876–880.
31. Butkus Small C, Kaufman A, Armenaka M, et al. Sinusitis and atopy in human immunodeficiency virus infection. J Infect Dis 1993; 167:283–290.
32. Clerici M, Shearer GM. A T_H1->T_H2 switch is a critical

step in the etiology of HIV infection. Immunol Today 1993; 14:107–111.
33. Letvin NL. Animal models for AIDS. Immunol Today 1990; 11:322–326.
34. Letvin NL, Daniel MD, Sehgal PK, et al. Induction of AIDS-like disease in macaque monkeys with T cell tropic retrovirus STLV-III. Science 1985; 230:71–73.
35. Ansari AA, Mayne A, Hunt D, et al. Th1/Th2 subset analysis.-I.Establishment of criteria for subset identification in PBMC samples from nonhuman primates. J Med Primatol 1994; 23:102–107.
36. Rosenberg YJ, Lewis MG, Villinger F, et al. Cytokines and simian immunodeficiency virus infection. Res Immunol 1994; 145:706–712.
37. Liew FY. Functional heterogeneity of CD4+ T cells in leishmaniasis. Immunol Today 1989; 10:40–45.
38. Brodie SJ, Sasseville VG, Reimann KA, et al. Macrophage function in simian AIDS. Killing defects in vivo are independent of macrophage infection, associated with alterations in Th phenotype, and reversible with IFN-gamma. J Immunol 1994; 153:5790–5801.
39. Khatissian E, Chakrabarti L, Hurtrel B. Cytokine patterns in lymph nodes during the early stages of simian immunodeficiency virus infection. Res Immunol 1994; 145:702–705.
40. Fultz PN, McClure HM, Swenson B, et al. Persistent infection of chimpanzees with human T-lymphotropic virus type III/lymphadenopathy associated virus: a potential model for acquired immunodeficiency syndrome. J Virol 1986; 58:116–124.
41. Arthur LO, Bess JW, Waters D, et al. Challenge of chimpanzees (pantroglodytes) immunized with human immunodeficiency virus envelop glycoprotein gp120. J Virol 1989; 63:5046–5053.
42. Nara PL, Smit L, Dunlop N, et al. Emergence of virus resistant to neutralization by V3-specific antibodies in experimental human immunodeficiency virus type 1 IIIB infection of chimpanzees. J Virol 1990; 64:3779–3791.
43. Nara P, Hatch W, Kessler J, et al. The biology of human immunodeficiency virus-1 IIIB infection in the chimpanzee: in vivo and in vitro correlations. J Med Primatol 1989; 18:343–355.
44. Schwartz RH. A cell culture model for T lymphocyte clonal anergy. Science 1990; 248:1349–1356.
45. Schuitemaker H, Meyaard L, Kootstra NA, et al. Lack of T-cell dysfunction and programmed cell death in human immunodeficiency type-1 infected chimpanzees correlates with absence of monocytotropic variants. J Infect Dis 1993; 168:1140–1147.
46. Gougeon M, Garcia S, Heeney J, et al. Programmed cell death in AIDS-related HIV and SIV infections. AIDS Res Hum Retroviruses 1993; 9:553–563.
47. Gendelman HE, Ehrlich GD, Baca LM, et al. The inability of human immunodeficiency virus to infect chimpanzee monocytes can be overcome by serial viral passage in vivo. J Virol 1991; 65:3853–3863.
48. Watanabe M, Ringler DJ, Fultz PN, et al. A chimpanzee-passaged human immunodeficiency virus isolate is cytopathic for chimpanzee cells but does not induce disease. J Virol 1991; 65:3344–3348.
49. Mosier DE, Yetter RA, Morse III HC. Retroviral induction of acute lymphoproliferative disease and profound immunosuppression in adult C57BL/6 mice. J Exp Med 1985; 161:766–784.
50. Gazzinelli RT, Makino M, Chattopadhyay SK, et al. CD4+ subset regulation in viral infection. Preferential activation of Th2 cells during progression of retrovirus-induced immunodeficiency in mice. J Immunol 1992; 148:182–188.
51. Eckwalanga M, Marussig M, Dias Tavares M, et al. Murine AIDS protects mice against experimental cerebral malaria: down-regulation by interleukin 10 of a T-helper type 1 CD4+ cell-mediated pathology. Proc Natl Acad Sci USA 1994; 91:8097–8101.
52. Gazzinelli RT, Giese NA, Morse HC III. In vivo treatment with interleukin 12 protects mice from immune abnormalities observed during murine acquired immunodeficiency syndrome (MAIDS). J Exp Med 1994; 180:2199–2208.
53. Kanagawa O, Vaupel BA, Gayama S, et al. Resistance of mice deficient in IL-4 to retrovirus-induced immunodeficiency syndrome (MAIDS). Science 1993; 262:240–242.
54. Gilmore GL. Resistance to murine acquired immunodeficiency syndrome (MAIDS) (Technical comment). Science 1994; 265:264.
55. Morawetz RA, Doherty TM, Giese NA, et al. Resistance to murine acquired immunodeficiency syndrome (MAIDS) (technical comment). Science 1994; 265:264–266.
56. Torbett BE, Mosier DE. Does preferential Th subset activation contribute to the murine acquired immunodeficiency disease (MAIDS)? Res Immunol 1994; 145:696–701.
57. Mosier DE, Gulizia RJ, Baird SM, et al. Human immunodeficiency virus infection of human-PBL-SCID mice. Science 1991; 251:791–794.
58. Aldrovandi GM, Feuer G, Gao L, et al. The SCID-hu mouse as a model for HIV-1 infection. Nature 1993; 363:732–736.
59. Kollmann TR, Pettoello-Mantovani M, Zhuang X, et al. Disseminated human immunodeficiency virus 1 (HIV-1) infection in SCID-hu mice after peripheral inoculation with HIV-1. J Exp Med 1994; 179:513–522.
60. Maggi E, Macchia D, Parronchi P, et al. Reduced production of interleukin 2 and interferon-gamma and enhanced helper activity for IgG synthesis by cloned CD4+ T cells from patients with AIDS. Eur J Immunol 1987; 17:1685–1690.
61. Maggi E, Mazzetti M, Ravina A, et al. Ability of HIV to promote a Th1 to Th0 shift and to replicate preferentially in Th2 and Th0 cells. Science 1994; 265:244–248.
62. Meyaard L, Otto SA, Keet IPM, et al. Changes in cytokine secretion patterns of CD4+ T cell clones in HIV-1 infection. Blood 1994; 84:4262–4268.
63. Paganelli R, Scala E, Ansotegui IJ, et al. CD8+ T lymphocytes provide helper activity for IgE synthesis in human immunodeficiency virus-infected patients with hyper-IgE. J Exp Med 1995; 181:423–428.
64. Maggi E, Giudizi MG, Biagiotti R, et al. Th2-like CD8+ T cells showing B cell helper function and reduced cytolytic activity in human immunodeficiency virus type 1 infection. J Exp Med 1994; 180:489–495.
65. Manetti R, Annunziato F, Biagiotti R, et al. CD30 expression by CD8+ T cells producing type 2 helper cytokines. Evidence for large numbers of CD8+CD30+ T cell clones in human immunodeficiency virus infection. J Exp Med 1994; 180:2407–2411.
66. Del Prete G, Maggi E, Pizzolo G, et al. CD30, Th2 cytokines and HIV infection: a complex and fascinating link. Immunol Today 1995; 16:76–80.
67. Alzona M, Jäck H-M, Fisher RI, et al. CD30 defines a subset of activated human T cells that produce IFN-gamma and IL-5 and exhibit enhanced B cell helper activity. J Immunol 1994; 153:2861.
68. Alzona M, Jäck H-M, Fisher RI, et al. IL-12 activates IFN-gamma production through the preferential activation of CD30+ T cells. J Immunol 1995; 154:9–16.

69. Hamann D, Hilkens CMU, Grogan JL, et al. CD30 expression does not discriminate between human Th1- and Th2-type T cells. J Immunol 1996 (in press).
70. Fan J, Bass HZ, Fahey JL. Elevated IFN-gamma and decreased IL-2 gene expression are associated with HIV infection. J Immunol 1993; 151:5031–5040.
71. Clerici M, Hakim FT, Venzon DJ, et al. Changes in interleukin-2 and interleukin-4 production in asymptomatic human immunodeficiency virus-seropositive individuals. J Clin Invest 1993; 91:759–765.
72. Barcellini W, Rizzardi GP, Borghi MO, et al. Th1 and Th2 cytokine production by peripheral blood mononuclear cells from HIV-infected patients. AIDS 1994; 8:757–762.
73. Zhang M, Gong J, Iyer DV, et al. T cell cytokine responses in persons with tuberculosis and human immunodeficiency virus infection. J Clin Invest 1994; 94:2435–2442.
74. Gazzinelli RT, Bala S, Stevens R, et al. HIV infection suppresses type 1 lymphokine and IL-12 responses to *Toxoplasma gondii* but fails to inhibit the synthesis of other parasite-induced monokines. J Immunol 1995; 155:1565–1574.
75. Hyjek E, Lischner HW, Hyslop T, et al. Cytokine patterns during progression to AIDS in children with perinatal HIV infection. J Immunol 1995; 155:4060–4071.
76. Meyaard L, Hovenkamp E, Keet IPM, et al. Th1 and Th2 secretion by T cells from HIV-infected individuals: single cell analysis of IL-4 and IFN-gamma production.
77. Autran B, Legac E, Blanc C, et al. A Th0/Th2-like function of CD4+CD7− helper cells from normal donors and HIV-infected patients. J Immunol 1995; 154:1408–1417.
78. Navikas V, Link J, Wahren B, et al. Increased levels of interferon-gamma (IFN-gamma), IL-4 and transforming growth factor-β (TGF-β) mRNA expressing blood mononuclear cells in human HIV infection. Clin Exp Immunol 1994; 96:59–63.
79. Nye KE, Knox KA, Pinching AJ. HIV infection of H9 lymphoblastoid cells chronically activates the inositol polyphosphate pathway. AIDS 1990; 4:41–45.
80. Emilie D, Fior R, Llorente L, et al. Cytokines from lymphoid organs of HIV-infected patients: production and role in the immune disequilibrium of the disease and in the development of B lymphomas. Immunol Rev 1994; 140:5–34.
81. Tough DF, Sprent J. Turnover of naive- and memory-phenotype T cells. J Exp Med 1994; 179:1127–1135.
82. Diaz-Mitoma F, Kumar A, Karimi S, et al. Expression of IL-10, IL-4 and interferon-gamma in unstimulated and mitogen-stimulated peripheral blood lymphocytes from HIV-seropositive patients. Clin Exp Immunol 1995; 102:31–39.
83. Chehimi J, Starr SE, Frank I, et al. Impaired interleukin 12 production in human immunodeficiency virus infected patients. J Exp Med 1994; 179:1361–1366.
84. Estaquier J, Idziorek T, Zou W, et al. T helper type 1/T helper type 2 cytokines and T cell death: preventive effect of interleukin 12 on activation-induced and CD95 (FAS/APO-1)-mediated apoptosis of CD4+ T cells from human immunodeficiency virus-infected persons. J Exp Med 1995; 182:1759–1767.
85. Murray HW, Hillman JK, Rubin BY, et al. Patients at risk for AIDS-related opportunistic infections. Clinical manifestations and impaired gamma interferon production. N Engl J Med 1985; 313:1504–1510.
86. Vyakarnam A, Matear P, Meager A, et al. Altered production of tumour necrosis factors alpha and beta and interferon gamma by HIV-infected individuals. Clin Exp Immunol 1991; 84:109–115.
87. Graziosi C, Pantaleo G, Gantt KR, et al. Lack of evidence for the dichotomy of Th1 and Th2 predominance in HIV-infected individuals. Science 1994; 265:248–252.
88. Yssel H, De Waal Malefyt R, Roncarolo MG, et al. IL-10 is produced by subsets of human CD4+ T cell clones and peripheral blood T cells. J Immunol 1992; 149:2378–2384.
89. Clerici M, Shearer GM. The Th1-Th2 hypothesis of HIV infection: new insights. Immunol Today 1994; 15:575–581.
90. Clerici M, Wynn TA, Berzofsky JA, et al. Role of interleukin-10 (IL-10) in T helper cell dysfunction in asymptomatic individuals infected with the human immunodeficiency virus (HIV-1). J Clin Invest 1994; 93:768–775.
91. Denis M, Ghadirian E. Dysregulation of interleukin 8, interleukin 10, and interleukin 12 release by alveolar macrophages from HIV type 1-infected subjects. AIDS Res Hum Retroviruses 1994; 10:1619–1627.
92. Wright SC, Jewett A, Mitsuyasu R, et al. Spontaneous cytotoxicity and tumor necrosis factor production by peripheral blood monocytes from AIDS patients. J Immunol 1988; 141:99–104.
93. Steffen M, Reinecker HC, Petersen J, et al. Differences in cytokine secretion by intestinal mononuclear cells, peripheral blood monocytes and alveolar macrophages from HIV-infected patients. Clin Exp Immunol 1993; 91:30–36.
94. Birx DL, Redfield RR, Tencer K, et al. Induction of interleukin-6 during human immunodeficiency virus infection. Blood 1990; 76:2303–2310.
95. Breen EC, Rezai AR, Nakajima K, et al. Infection with HIV is associated with elevated IL-6 levels and production. J Immunol 1990; 144:480–484.
96. Twigg HL III, Iwamoto GK, Soliman DM. Role of cytokines in alveolar macrophage accessory cell function in HIV-infected individuals. J Immunol 1992; 149:1462–1469.
97. Kekow J, Wachsman W, McCutchan JA, et al. Transforming growth factor-β and suppression of humoral immune responses in HIV infection. J Clin Invest 1991; 87:1010–1016.
98. Kekow J, Wachsman W, McCutchan JA, et al. Transforming growth factor beta and non-cytopathic mechanisms of immunodeficiency in human immunodeficiency virus infection. Proc Natl Acad Sci USA 1990; 87:8321–8325.
99. Hsieh C-S, Macatonia SE, Tripp CS, et al. Development of Th1 CD4+ T cells through IL-12 produced by Listeria-induced macrophages. Science 1993; 260:547–549.
100. Saag MS, Hahn BH, Gibbons J, et al. Extensive variation of human immunodeficiency virus type I in vivo. Nature 1988; 334:440–444.
101. Wiviott LD, Walker CM, Levy JA. CD8+ lymphocytes suppress HIV production by autologous CD4+ cells without eliminating the infected cells. Cell Immunol 1990; 128:628–634.
102. Brinchmann JE, Gaudernack G, Vartal F. CD8+ T cells inhibit HIV replication in naturally infected CD4+ T cells. J Immunol 1990; 144:2961–2966.
103. Walker CM, Levy JA. A diffusible lymphokine produced by CD8+ T lymphocytes suppresses HIV replication. Immunol 1989; 66:628–630.
104. Walker CM, Erickson AL, Hsueh FC, et al. Inhibition of human immunodeficiency virus replication in acutely infected CD4+ cells by CD8+ cells involves a noncytotoxic mechanism. J Virol 1991; 65:5921–5927.
105. Mackewicz CE, Yang LC, Lifson JD, et al. Non-cytolytic CD8 T-cell anti-HIV responses in primary HIV-1 infection. Lancet 1994; 344:1671–1673.
106. Castro BA, Walker CM, Eichberg JW, et al. Suppression of human immunodeficiency virus replication by CD8+ cells from infected and uninfected chimpanzees. Cell Immunol 1991; 132:246–255.
107. Castro BA, Homsy J, Lennette E, et al. HIV-1 expression in

chimpanzees can be activated by CD8+ cell depletion or CMV infection. Clin Immunol Immunopathol 1992; 65:227–233.

108. Ennen J, Findeklee H, Dittmar MT, et al. CD8+ T lymphocytes of African green monkeys secrete an immunodeficiency virus-suppressing lymphokine. Proc Natl Acad Sci USA 1994; 91:7207–7211.

109. Hsueh FW, Walker CM, Blackbourn DJ, et al. Suppression of HIV replication by CD8+ cell clones derived from HIV-infected and uninfected individuals. Cell Immunol 1994; 159:271–279.

110. Walker CM, Thomson-Honnebier GA, Hsueh FC, et al. CD8+ T cells from HIV-1 infected individuals inhibit acute infection by human and primate immunodeficiency viruses. Cell Immunol 1991; 137:420–428.

111. Brinchmann JE, Gaudernack G, Vartdal F. CD8+ T cells inhibit HIV replication in naturally infected CD4+ T cells. Evidence for a soluble inhibitor. J Immunol 1990; 146:2961–2966.

112. Mackewicz CE, Ortega HW, Levy JA. CD8+ cell anti-HIV activity correlates with the clinical state of the infected individual. J Clin Invest 1991; 87:1462–1466.

113. Landay AL, Mackewicz CE, Levy JA. An activated CD8+ T cell phenotype correlates with anti-HIV activity and asymptomatic clinical status. Clin Immunol Immunopathol 1993; 69:106–116.

114. Cao Y, Qin L, Zhang L, et al. Virologic and immunologic characterization of long-term survivors of human immunodeficiency virus type 1 infection. N Engl J Med 1995; 332:201–208

115. Baier M, Werner A, Bannert N, et al. HIV suppression by interleukin-16. Nature 1995; 378:563.

116. Cocchi F, DeVico AL, Garzino-Demo A, et al. Identification of RANTES, MIP-1α, and MIP-1β as the major HIV-suppressive factors produced by CD8+ T cells. Science 1995; 270:1811–1815.

117. Folks TM, Clouse KA, Justement J, et al. Tumor necrosis factor α induces expression of human immunodeficiency virus in a chronically infected T-cell clone. Proc Natl Acad Sci USA 1989; 86:2365–2368.

118. Foli A, Saville MW, Baseler MW, et al. Effects of the Th1 and Th2 stimulatory cytokines interleukin-12 and interleukin-4 on human immunodeficiency virus replication. Blood 1995; 85:2114–2123.

119. Schuitemaker H, Kootstra NA, Koppelman MHGM, et al. Proliferation dependent HIV-1 infection of monocytes occurs during differentiation into macrophages. J Clin Invest 1992; 89:1154–1160.

120. Kazazi F, Mathijs JM, Chang J, et al. Recombinant interleukin-4 stimulates human immunodeficiency virus production by infected monocytes and macrophages. J Gen Virol 1992; 73:941–948.

121. Novak RM, Holzer TJ, Kennedy MM, et al. The effect of interleukin-4 (BSF-1) on infection of peripheral blood monocyte-derived macrophages with HIV-1. AIDS Res Hum Retroviruses, 1990; 6:973–981.

122. Koyanagi Y, O'Brien WA, Zhao JQ, et al. Cytokines alter production of HIV-1 from primary mononuclear phagocytes. Science 1988; 241:1673–1675.

123. Schuitemaker H, Kootstra NA, Van Oers M, et al. Induction of monocyte proliferation and HIV expression by IL-3 does not interfere with anti-viral activity of zidovudine. Blood 1990; 76:1490–1493.

124. Gendelman HE, Orenstein JM, Martin MA, et al. Efficient isolation and propagation of human immunodeficiency virus on recombinant colony-stimulating factor 1-treated monocytes. J Exp Med 1988; 167:1428–1441.

125. Poli G, Kinter A, Justement JS, et al. Tumor necrosis factor alpha functions in an autocrine manner in the induction of human immunodeficiency virus expression. Proc Natl Acad Sci USA 1990; 87:782–785.

126. Poli G, Kinter AL, Justement JS, et al. Transforming growth factor β suppresses human immunodeficiency virus expression and replication in infected cells of the monocyte/macrophage lineage. J Exp Med 1991; 173:589–597.

127. Montaner LJ, Doyle AG, Collin M, et al. Interleukin 13 inhibits human immunodeficiency virus Type 1 production in primary blood-derived human macrophages in vitro. J Exp Med 1993; 178:743–747.

128. Kornbluth RS, Oh PS, Munis JR, et al. Interferons and bacterial lipopolysaccharide protect macrophages from productive infection by human immunodeficiency virus in vitro. J Exp Med 1989; 169:1137–1151.

129. Osborn L, Kunkel S, Nabel GJ. Tumor necrosis factor α and interleukin 1 stimulate the human immunodeficiency virus enhancer by activation of the nuclear factor kappa B. Proc Natl Acad Sci USA 1989; 86:2336–2340.

130. Te Velde AA, Klomp JPG, Yard BA, et al. Modulation of phenotypic and functional properties of human peripheral blood monocytes by IL-4. J Immunol 1988; 140:1548–1554.

131. Schuitemaker H, Kootstra NA, Fouchier RAM, et al. Productive HIV-1 infection of macrophages restricted to the cell fraction with proliferative capacity. EMBO J 1994; 13:5929–5936.

132. von Schwedler U, Kornbluth RS, Trono D. The nuclear localization signal of the matrix protein of human immunodeficiency virus type 1 allows the establishment of infection in macrophages and quiescent T lymphocytes. Proc Natl Acad Sci USA 1994; 91:6992–6996.

133. Gallay P, Swingler S, Aiken C, et al. HIV-1 infection of non-dividing cells: C-terminal tyrosine phosphorylation of the viral matrix protein is a key regulator. Cell 1995; 80:379–388.

134. Bukrinsky MI, Haggerty S, Dempsey MP, et al. A nuclear localization signal within HIV-1 matrix protein that governs infection of non-dividing cells. Nature 1993; 365:666–669.

135. De Waal Malefyt R, Abrams J, Bennett B, et al. Interleukin 10 (IL-10) inhibits cytokine synthesis by human monocytes: an autoregulatory role of IL-10 produced by monocytes. J Exp Med 1991; 174:1209–1220.

136. Te Velde AA, Huijbens RJF, Heije K, et al. Interleukin-4 (IL-4) inhibits secretion of IL-1β, tumor necrosis factor α, and IL-6 by human monocytes. Blood 1990; 76:1392–1397.

137. Peleman R, Wu J, Fargeas C, et al. Recombinant Interleukin-4 suppresses the production of interferon-gamma by human mononuclear cells. J Exp Med 1989; 170:1751–1756.

138. Kootstra NA, Van't Wout AB, Huisman JG, et al. Interference of IL-10 with HIV-1 replication in primary monocyte derived macrophages. J Virol 1994; 68:6967–6975.

139. Weissman D, Poli G, Fauci AS. Interleukin 10 blocks HIV replication in macrophages by inhibiting the autocrine loop of tumor necrosis factor alpha and interleukin 6 induction of virus. AIDS Res Hum Retroviruses 1994; 10:1199–1206.

140. Montaner LJ, Griffin P, Gordon S. Interleukin 10 inhibits initial reverse transcription of human immunodeficiency virus type 1 and mediates a virostatic latent state in primary blood-derived human macrophages in vitro. J Gen Virol 1994; 75:3393–3400.

141. Saville MW, Taga K, Foli A, et al. Interleukin-10 suppresses human immunodeficiency virus-1 replication in vitro in cells of the monocyte/macrophage lineage. Blood 1994; 83:3591–3599.

142. Poli G, Bressler P, Kinter A, et al. Interleukin 6 induces human immunodeficiency virus expression in infected monocytic cells alone and in synergy with tumor necrosis factor α by transcrip-

tional and post-transcriptional mechanisms. J Exp Med 1990; 172:151–158.
143. Poli G, Weissman D, Kinter AL, et al. Complex regulation of HIV replication by IL-4 and IL-10. J Cell Biochem 1994; Suppl.18B:Abstr J261.
144. Lifson JD, Benike CJ, Mark DF, et al. Human recombinant interleukin-2 partly reconstitutes deficient in-vitro immune responses of lymphocytes from patients with AIDS. Lancet 1984; 698–702.
145. Bell SJD, Cooper DA, Kemp BE, et al. Heterogenous effects of exogenous IL-2 on HIV-specific cell-mediated immunity (CMI). Clin Exp Immunol 1992; 90:6–12.
146. Seder RA, Grabstein KH, Berzofsky JA, et al. Cytokine interactions in human immunodeficiency virus-infected individuals: roles of interleukin (IL)-2, IL-12 and IL-15. J Exp Med 1995; 182:1067–1078.
147. Clerici M, Lucey DR, Berzofsky JA, et al. Restoration of HIV-specific cell-mediated immune responses by interleukin-12 in vitro. Science 1993; 262:1721–1724.
148. Newman GW, Guarnaccia JR, Vance III EA, et al. Interleukin-12 enhances antigen-specific proliferation of peripheral blood mononuclear cells from HIV-positive and negative donors in response to *Mycobacterium avium.* AIDS 1994; 8:1413–1419.
149. Paganin C, Frank I, Trinchieri G. Priming for high interferon-gamma production induced by interleukin-12 in both CD4+ and CD8+ T cell clones from HIV-infected patients. J Clin Invest 1995; 96:1677–1682.
150. Manetti R, Parronchi P, Giudizi MG, et al. Natural killer cell stimulatory factor (interleukin 12 [IL-12]) induces T helper type 1 (Th1)-specific immune responses and inhibits the development of IL-4-producing Th cells. J Exp Med 1993; 177:1199–1204.
151. Manetti R, Gerosa F, Giudizi M-G, et al. Interleukin 12 induces stable priming for interferon gamma (IFN-gamma) production during differentiation of human T helper (Th) cells and transient IFN-gamma production in established Th2 cell clones. J Exp Med 1994; 179:1273–1283.
152. Yssel H, Fasler S, De Vries JE, et al. IL-12 transiently induces IFN-gamma transcription and protein synthesis in human CD4+ allergen-specific Th2 T cell clones. Int Immunol 1994; 6:1091–1096.
153. Chehimi J, Starr SE, Frank I, et al. Natural killer (NK) cell stimulatory factor increases the cytotoxic activity of NK cells from both healthy donors and human immunodeficiency virus-infected patients. J Exp Med 1992; 175:789–796.
154. Ameisen JC, Capron A. Cell dysfunction and depletion in AIDS: the programmed cell death hypothesis. Immunol Today 1991; 12:102–105.
155. Meyaard L, Otto SA, Jonker RR, et al. Programmed death of T cells in HIV-1 infection. Science 1992; 257:217–219.
156. Groux H, Torpier G, Monté D, et al. Activation-induced death by apoptosis in CD4+ T cells from human immunodeficiency virus-infected asymptomatic individuals. J Exp Med 1992; 175:331–340.
157. Clerici M, Sarin A, Coffman RL, et al. Type 1/type 2 cytokine modulation of T-cell programmed cell death as a model for human immunodeficiency virus pathogenesis. Proc Natl Acad Sci USA 1994; 91:11811–11815.
158. Meyaard L, Otto SA, Keet IPM, et al. Programmed death of T cells in HIV-1 infection: no correlation with progression to disease. J Clin Invest 1994; 93:982–988.
159. Lewis DE, Ng Tang DS, Adu-Oppong A, et al. Anergy and apoptosis in CD8+ T cells from HIV-infected individuals. J Immunol 1994; 153:412–420.
160. Meyaard L, Miedema F. Programmed death of T cells in HIV infection: result of immune activation? In: Kroemer G, Martinez-AC, eds. Current Topics in Microbiology and Immunology: Apoptosis in Immunology. Berlin-Heidelberg: Springer-Verlag, 1995:213–221.
161. Muro-Cacho CA, Pantaleo G, Fauci AS. Analysis of apoptosis in lymph nodes of HIV-infected persons. Intensity of apoptosis correlates with the general state of activation of the lymphoid tissue and not with stage of disease or viral burden. J Immunol 1995; 154:5555–5566.
162. Teppler H, Kaplan G, Smith K, et al. Efficacy of low doses of the polyethylene glycol derivative of interleukin-2 in modulating the immune response of patients with human immunodeficiency virus type 1 infection. J Infect Dis 1993; 167:291–298.
163. Wood R, Montoya JG, Kundu SK, et al. Safety and efficacy of polyethylene glycol-modified interleukin-2 and zidovudine in human immunodeficiency virus type 1 infection: a phase I/II study. J Infect Dis 1993; 167:519–525.
164. Teppler H, Kaplan G, Smith KA, et al. Prolonged immunostimulatory effect of low-dose polyethylene glycol interleukin 2 in patients with human immunodeficiency virus type 1 infection. J Exp Med 1993; 177:483–492.
165. Kovacs JA, Baseler M, Dewar RJ, et al. Increases in CD4 T lymphocytes with intermittent courses of interleukin-2 in patients with human immunodeficiency virus infection. N Engl J Med 1995; 332:567–575.
166. Gendelman HE, Baldwin T, Baca-Regen L, et al. Regulation of HIV1 replication by interferon alpha: from laboratory bench to bedside. Res Immunol 1994; 145:679–684.
167. Frissen PHJ, Van der Ende ME, Ten Napel CHH, et al. Zidovudine and interferon-α combination therapy versus zidovudine monotherapy in subjects with symptomatic human immunodeficiency virus type 1 infection. J Infect Dis 1994; 169:1351–1355.
168. Dournon E, Matheron S, Rozenbaum W, et al. Effects of zidovudine in 365 consecutive patients with AIDS or AIDS-related complex. Lancet 1988; ii:1297–1302.
169. Scadden DT, Zon LI, Groopman JE. Pathophysiology and management of HIV-associated hematologic disorders. Blood 1989; 74:1455–1463.
170. Groopman JE, Mitsuyasu RT, DeLeo MJ, et al. Effect of recombinant human granulocyte-macrophage colony-stimulating factor on myelopoiesis in the acquired immunodeficiency syndrome. N Engl J Med 1987; 317:593.
171. Hall SS. IL-12 holds promise against cancer, glimmer of AIDS hope. Science 1994; 263:1685–1686.
172. Cohen J. Clinical trials. IL-12 deaths: explanation and a puzzle. Science 1995; 270:908.
173. Orange JS, Wolf SF, Biron CA. Effects of IL-12 on the response and susceptibility to experimental viral infections. J Immunol 1994; 152:1253–1264.
174. Orange JS, Salazar-Mather TP, Opal SM, et al. Mechanism of interleukin 12-mediated toxicities during experimental viral infections: role of tumor necrosis factor and glucocorticoids. J Exp Med 1995; 181:901–914.
175. Meyaard L, Hovenkamp E, Otto SA, et al. IL-12-induced IL-10 production by human T cells as a negative feedback for IL-12-induced immune responses.
176. Hofmann B, Orskov Lindhardt B, Gerstoft J, et al. Lymphocyte transformation response to pokeweed mitogen as a marker for the development of AIDS and AIDS related symptoms in homosexual men with HIV antibodies. Br Med J 1987; 295:293–296.
177. Schellekens PThA, Roos MThL, De Wolf F, et al. Low T-cell

responsiveness to activation via CD3/TCR is a prognostic marker for AIDS in HIV-1 infected men. J Clin Immunol 1990; 10:121–127.

178. Roos MTL, Miedema F, Koot M, et al. T-cell function in vitro is an independent progression marker for AIDS in human immunodeficiency virus (HIV)-infected asymptomatic individuals. J Infect Dis 1995; 171:531–536.
179. Sheppard HW, Lang W, Ascher MS, et al. The characterization of non-progressors: long-term HIV-1 infection with stable CD4+ T-cell levels. AIDS 1993; 7:1159–1166.
180. Keet IPM, Krol A, Klein MR, et al. Characteristics of long-term asymptomatic infection with human immunodeficiency virus type 1 with normal and low CD4+ cell counts. J Infect Dis 1944; 169:1236–1243.
181. Miedema F, Meyaard L, Klein MR. Protection from HIV-1 infection or AIDS? Science 1993; 262:1074–1075.
182. Zinkernagel RM, Hengartner H. T-cell mediated immunopathology versus direct cytolysis by virus: implications for HIV and AIDS. Immunol Today 1994; 15:262–268.
183. Klein MR, van Baalen CA, Holwerda AM, et al. Kinetics of Gag-specific CTL responses during the clinical course of HIV-1 infection: a longitudinal analysis of rapid progressors and long-term asymptomatics. J Exp Med 1995; 181:1365–1372.
184. Salk J, Bretcher PA, Salk PL, et al. A strategy for prophylactic vaccination against HIV. Science 1993; 260:1270–1272.
185. Miedema F, Meyaard L, Koot M, et al. Changing virus-host interactions in the course of HIV-1 infection. Immunol Rev 1994; 140:35–72.

41
Cytokines and Parasitic Diseases

Edward J. Pearce
College of Veterinary Medicine, Cornell University, Ithaca, New York

Alan Sher
National Institute of Allergy and Infectious Disease, National Institutes of Health, Bethesda, Maryland

INTRODUCTION

The word "parasite," though in the truest sense applicable to all infectious organisms, is generally used to describe only eukaryotic single-cell and metazoan pathogens. Consequently, what we think of as parasites are a heterogeneous group of pathogens inhabiting niches as disparate as the intestinal lumen and the cytoplasm of individual cells.

The seminal studies by Mosmann and Coffman in the mid-1980s showing that murine CD4+ T-helper (Th) cell clones could be segregated into at least two types, Th1 and Th2, capable of secreting mutually exclusive panels of cytokines (1,2) provided an invaluable conceptual framework within which to ask questions about the roles of cytokines in immune responses to parasites. Indeed, studies of parasite infections have played a major role in establishing the veracity of the Th subset paradigm. The elucidation of the function of IL-4 and IL-5, as control cytokines for IgE production (3) and eosinophilia, respectively (4), indicated that Th2 cells, which secreted these cytokines, were likely to be involved in allergies and responses to parasitic worms. Likewise, the production of IFN-γ by Th1 but not Th2 cells indicated that these cells would play a role in cell-mediated immunity (CMI) against parasitic protozoa (2). With notable exceptions, these correlations have proven remarkably robust. A growing appreciation of cytokine-based abilities of Th1- and Th2-like responses [Type 1 and 2 responses (5)] to promote their own expansion while suppressing the opposite type of response has gone a long way in explaining why so many parasitic infections have been found to induce responses skewed often dramatically in either a Type 1 or Type 2 direction (6).

CYTOKINES IN ANIMAL MODELS OF PARASITIC DISEASES

Protozoa

It should be borne in mind that in many cases the outcome (resistance or susceptibility) of the interaction of a parasite with an inbred laboratory host is a product of both the strain or clone of parasite and the strain of host. Although this makes the extrapolation of results obtained from such systems to human disease difficult, it has nevertheless provided an extremely powerful tool with which to dissect and understand protective and exacerbative antiparasite immune responses. Nowhere is this more clear than in studies of *Leishmania* in mice; many of the findings from this system are likely to be of broad applicability to other infectious diseases. Leishmaniasis, despite its low incidence compared to some other parasitic infections, has of all parasitic diseases been the subject of the most extensive and perhaps most revealing immunological research. This infection is initiated when infected sandfly vectors bite the mammalian host and transmit into the skin the infectious promastigote stage of the parasite, which is adapted to invade host macrophages. As alluded to above, certain strains of mice, exemplified by BALB/c, are highly susceptible to the *L. major* species of parasite and upon experimental infection develop progressing lesions, which eventually metastasize to the viscera and prove fatal. In contrast, other strains such as C57BL6 and C3H are quite resistant, exhibiting only small self-healing skin lesions following infection.

It was appreciated for some time that CD4+ cells could play a role in exacerbating disease or protecting against leishmaniasis in mice (7). The current understanding, based on the analysis of Th responses in infected resistant and susceptible mice (8) and of protective and exacerbative Th lines (9), is that Th1 cells are protective while Th2 cells promote infection. Based on in vitro studies, Th1 cells are assumed to be protective due to their ability to secrete IFN-γ, which is likely to function by activating macrophages to kill *Leishmania* parasites, apparently through a NO-dependent mechanism (10). Th2 cells exacerbate infection most probably due to the ability of IL-4, IL-10, and IL-13 (all Type 2 cytokines) to synergistically deactivate macrophage killing functions (11–13), making them more permissive hosts for the *Leishmania* parasites. Initial experiments to confirm these assumed roles of IFN-γ and IL-4 did show that neutralization of IFN-γ or IL-4 with repeated injections of specific mAbs throughout infection could negate resistance or susceptibility in C3H or BALB/c mice, respectively (14–16). Moreover, Th1 cells in mice lacking the IFN-γ R are not protective (17). Unexpectedly, the mAb anticytokine treatments were ineffective if started after the initiation of infection at times when inhibition of cytokine effector function might have been expected to be most effective, but were optimal if given only as one dose at the time of infection. This is now known to be due to the crucial roles played by IFN-γ and IL-4 in promoting the differentiation of naive precursor Th (ThP) cells, which secrete only IL-2 upon initial activation into Th1 or Th2 cells, respectively (18). Consequently, neutralization of IFN-γ at the time of priming prevents the differentiation of Thp cells into Th1 cells, and neutralization of IL-4 blocks the ThP to Th2 pathway. Additionally, early neutralization of IL-12, a potent inducer of IFN-γ production and itself a powerful promoter of Th1 development, abrogates resistance (19). Analysis of the cytokines produced by Th cells from mAb anti–IL-4–treated *L. major*–infected BALB/c mice revealed in addition to the absence of a Th2 response, the presence of a strong parasite specific Th1 response (15). Likewise, mAb anti–IFN-γ–treated C3H mice develop Th2 responses (14). It is these stable, long-term alterations in Th responses that account for the dramatic effects of the mAb anti–IFN-γ or anti–IL-4 treatments.

The importance of IFN-γ, IL-12, and IL-4 in determining the outcome of infection through influencing Th subset selection has led to a search for the cellular sources responsible for the early secretion of these cytokines and for the activation mechanisms responsible for their stimulation. Since macrophages are the major recognized source of IL-12 and *Leishmania* are obligate parasites of these cells, a logical hypothesis was that Th1-cell development in vivo in response to *L. major* infection would be initiated by IL-12 produced by macrophages upon invasion by the parasites (20). IL-12 would then act as a potent stimulus for IFN-γ production by NK cells and both the IL-12 and IFN-γ would play roles in promoting ThP to Th1 differentiation. Evidence supporting this pathway is that, in C3H mice, depletion of NK cells prior to infection leads to diminished Th1 and increased Th2 responses and susceptibility, and that C3H mice have greater NK activity than BALB/c animals (21). Moreover, neutralization of IL-12 at the time of infection prevents NK cell activation and early IFN-γ production (19). In this pathway, a major difference between susceptible and resistant mice lies not, as might have been predicted, in the failure of the susceptible animals to make IL-12, since at day 1 postinfection draining LN cells from BALB/c mice make as much IL-12 as do those from C3H animals (19). Rather, recent evidence indicates that TGF-β and to a lesser extent IL-4 and IL-10 in susceptible animals suppress protective antileishmanial responses, probably by inhibiting the NK-activating effects of IL-12 (19). Indeed, production of TGF-β has been implicated as a major factor responsible for increasing host susceptibility to *Leishmania* (22,23).

At present there is some debate over exactly how soon following exposure of resistant mice to *L. major* IL-12 is produced. In an analysis of cytokine gene transcripts in LN-draining sites of parasite inoculation, no IL-12 transcripts were observed in resistant or susceptible strains until day 7 of infection, even though IFN-γ transcription was upregulated at earlier times (24). Moreover, the promastigote stage of *Leishmania,* introduced into the mammal by the vector, has been found by several groups not to stimulate IL-12 production when it infected macrophages in vitro (24,25). This led to a proposal that amastigotes, the intracellular form of *Leishmania* into which the promastigote transforms and the stage that spreads between macrophages, are responsible for inducing IL-12 production (24). In vitro, this stage of the parasite, in contrast with the promastigote, promotes the transcription of the IL-12 gene by macrophages. That the induction of the latter response is ultimately ineffective in BALB/c mice could be due to the elevated levels of IL-4 and possibly other macrophage-deactivating cytokines attained in the susceptible but not resistant animals.

IFN-γ and Type 1 responses play a crucial role in resistance to another intracellular protozoan parasite, *Toxoplasma gondii.* This normally avirulent parasite, which is able to infect most nucleated cells, has been the subject of much recent interest due to its importance as an opportunistic infection during AIDS (26). In immunologically competent animals, the parasite is found encysted in host tissues and especially those of the central nervous system. Maintenance of the cysts, which are benign, is IFN-γ and TNF-α dependent, as neutralization of either of these cytokines by mAb in chronically infected mice leads to reactivation of the infection, the emergence of the rapidly dividing tachyzoite stage, the development of toxoplasmic encephalitis (TE), and death (27–29). Additionally, mAb anti–IFN-γ decreases, and exogenous IFN-γ increases, the length of time to death of mice infected with virulent strains of *T. gondii,* which are able to cause a lethal acute disease (27,28). IFN-γ appears to mediate its protective effects in two ways. The first is by activating macrophages to become cytostatic toward tachyzoites. This activation step is TNF-α dependent and is effected by the production of reactive nitrogen intermediates (RNI) and oxygen free radicals. The second mechanism (so far only demonstrated in in vitro human assays) is directed toward nonmacrophage cell types and functions through the depletion of intracellular tryptophan, an amino acid essential for tachyzoite survival (27).

A number of different cell types can serve as sources of the protective IFN-γ produced in response to *T. gondii* infection. As in the case of *L. major,* NK cells appear to be major contributors of IFN-γ and again their activation is dependent upon IL-12 and to a lesser extent TNF-α and IL-1β, produced by macrophages in response to the parasite (30,31). This IL-12–induced NK-derived IFN-γ can itself exert protective effects even in SCID mice and in intact animals is clearly important endogenously for early IFN-γ synthesis as well as the initial control of infection (32,33). The induction of IL-12 is also likely to play a role in promoting the development of the Type 1 response pattern found in *T. gondii* infection (32). The involvement of Type 1 cells in resistance to *T. gondii* is suggested by the observations that T lymphocytes (both CD4+ and CD8+) from mice protectively vaccinated against this parasite by prior exposure to an attenuated *T. gondii* strain produce large amounts of IFN-γ and little of the Type 2 cytokines in response to restimulation with antigen. Nevertheless, while clearly responsive in vitro, Th1 cells do not appear to play a direct effector role in resistance in vivo. Thus, in both vaccine and cyst reactivation animal models, in vivo depletion of CD4+ cells alone has little effect on established immunity. Instead, a partial loss of resistance is seen with anti-CD8+ treatment, while combined CD4+ and CD8+ depletion results in rapid infection-induced death of the animals (27,34,35). These and related observations suggest that CD8+ cells may be the primary effectors of immunity, with CD4+ lymphocytes playing a secondary role as helper cells for the protective response (27). The effector function of CD8+ cells in resistance to *T. gondii* is likely to result from their parasite-induced secretion of IFN-γ, although they are also able to mediate classical class I restricted cytotoxicity against parasite-infected host cells in vitro (27). Recent experiments suggest that the protective CD8+ response in *T. gondii*–infected hosts may be driven in part by a superantigen activity, which selectively expands IFN-γ producing Vβ5 T lymphocytes of this subset (36).

During chronic infection of certain strains of mice with avirulent *T. gondii* strains, spontaneous reactivation occurs resulting in severe TE and death. This loss in resistance is associated with the increased expression in brain tissue of IL-10 (37) and TGF-β (28), two cytokines that antagonize the macrophage-activating effects of IFN-γ and TNF-α as well as inhibit the production of IFN-γ by T cells and NK cells. The cellular source(s) of these downregulatory cytokines in the brain during TE is unclear, although macrophages and/or microglial cells are likely candidates. On the other hand, depletion studies indicate that the development of TE is CD4+ dependent (38). Thus, as in leishmaniasis, malaria, and other parasitic infections, the expression of immunopathology versus resistance may depend on a fine balance of Type 1 versus Type 2 responses. Indeed, recent experiments in *T. gondii*–infected IL-10 knock-out mice support the general concept that the production of downregulatory cytokines may be necessary to protect the host against the dangers of potentially excessive parasite induced Type 1 responses.

Trypanosoma cruzi is a vector-transmitted protozoan parasite, which can infect any nucleated cell in the mammalian host, certain strains of which show a preference for heart myocytes and thereby cause severe heart disease during the chronic stages of infection. During acute experimental *T. cruzi* infection in mice, when parasites can be found circulating in the blood, there is a profound suppression in T-cell proliferation that is at least in part due to reduced IL-2 production as a result of absence of AP-1 activation in activated T cells (39). This and other characteristics of T cells from acutely infected animals suggest that they are anergized, although what it is about the parasite that leads to this is unclear at present. As for each of the the protozoan parasites discussed here, IFN-γ has a host protective effect against *T. cruzi,* such that the administration to infected mice of exogenous IFN-γ, or of neutralizing mAb anti-IFN-γ, leads to diminished or exacerbated infections, respectively (40). However, the source of protective IFN-γ during infection is not clear, but the fact that class I–deficient (β_2-microglobulin -/-) mice are extremely susceptible to *T. cruzi* (41) suggests that CD8+

cells could be a physiologically important source. As with *L. major, T. cruzi* exhibits different pathogenicities in different mouse strains, although an additional complication here is that different virulent clones of the same species of parasite also exhibit markedly different properties (41). For example, C57BL6 mice are extremely susceptible to the SylvioX10 clone, but relatively resistant to the Tulahuen strain, which itself causes severe disease (higher parasitemia and greater mortality) in BALB/c mice (42). A kinetic analysis of the cytokines made by antigen restimulated lymphoid cells from Tulahuen-infected BALB/c and C57BL6 mice revealed a correlation between an early (weeks 1–2) increase in the ability to produce IFN-γ and the low early parasitemia in C57BL6 mice, and a late increase in IL-4 production, beginning at week 5, in the susceptible BALB/c animals (43). Consistent with the idea that Th2 responses are associated with susceptibility to *T. cruzi,* susceptible but not resistant strains also make IL-10 following infection; B cells, T cells, and macrophages contribute to the production of this cytokine (40). Interestingly, exogenous TGF-β can make resistant mice susceptible to acute infection, probably by inhibiting the IFN-γ–mediated activation of macrophages (44).

Entamoeba histolytica is a waterborne extracellular intestinal parasite, which in certain circumstances is aggressively invasive, eventually metastasizing to the liver, where in worst cases it causes severe abcess formation. Control of this parasite can be effected in vitro by IFN-γ plus TNF-α or CSF-1–activated macrophages, via a process which involves NO and to a lesser extent reactive oxygen intermediates (45). A role for PGE2 in downregulating TNF-α production by abcess macrophages, and therefore their antiparasite effectiveness, during the progression of amebic disease in gerbils has been reported (46).

Malaria, one of the most important infectious diseases, is transmitted from infected individuals by mosquitoes. When it takes a blood meal, the infected vector injects into the mammalian host malaria sporozoites, which rapidly invade hepatocytes, where they divide and give rise to merozoites. Merozoites reenter the circulation and invade erythrocytes, within which they multiply before rupturing the host cell and reinfecting other erythrocytes. Some merozoites give rise to gametes, which fuse and initiate the sexual stage of the life cycle when taken up by the mosquito. Nonprimate models of malaria use *Plasmodium* species, which are not infective for humans, but which mimic several aspects of the human malarial diseases. The ANKA strain of *P. berghei,* for example, causes in certain strains of mice experimental cerebral malaria (CM) with symptoms similar, though not identical, to those associated with CM due to *P. falciparum* in humans (47). In mice with experimental CM, brain capillaries and venules become plugged with leukocytes. This event, plus its life-threatening sequelae, including hemorrhagic necrosis, can be prevented by mAb anti–TNF-α, strongly implicating the latter cytokine as an important mediator of pathology in malaria (48). This hypothesis is supported by the observation that exogenous TNF-α can induce experimental CM in normally resistant mouse strains infected with *P. berghei* ANKA (49,50). TNF-α production in mice with experimental CM appears to be greatly elevated in the brain, but also in other tissues and lymphoid organs, indicating a generalized upregulation of production of this cytokine. However, it is the TNF-α produced by macrophages and eventually leukocytes and endothelial cells at the site of the cerebral blood vessel plug that is believed to be most responsible for experimental CM (47,49). The parasite molecule responsible for inducing TNF-α production will be discussed below. Other cytokines such as GM-CSF, IL-3, and IFN-γ are believed to play a role in recruiting and activating macrophages to produce TNF-α (47). IL-6 levels are also greatly elevated during experimental CM, as might be expected in a condition where macrophage and endothelial cell activation is so extensive; however, neutralization of this cytokine with mAb has not been found to have any effect on experimental CM (47).

Studies of the mechanisms underlying immunological resistance to rodent parasites have led to different conclusions, depending on which of several different *Plasmodium* species is examined. In mice infected with *P. chabaudi chabaudi,* a recognized model for *P. falciparum* in humans, resistance to the erythrocytic stage of the infection (which causes disease) is Th-cell dependent. Detailed analysis revealed that both Th2 and Th1 cells contribute to this protection, the former by promting IgG1 production, presumably by secretion of IL-4, and the latter via an NO-dependent mechanism in which IFN-γ is implicated by virtue of its ability to strongly promote NO production by macrophages (51). IFN-γ is also thought to mediate its protective effects against the intrahepatocyte stage of several species of the parasite by stimulating the liver cells to make NO (52). IL-12, given around the time of challenge infection, can protect BALBc mice against *P. yoelii* sporozoites in an IFN-γ–dependent manner in which both T cells and NK cell may be playing a role (53). In summary, studies in rodent models have indicated that IFN-γ plays a role in both the inherent immunological resistance to the intra- and extrahepatic stages of primary malaria infections and additionally by the immunity conferred by one vaccination with radiation-attenuated sporozoites. A caveat to the last statement is that multiple vaccination with irradiated sporozoites leads to a very effective antibody-based immunity, which functions even in IFN-γ knockout mice (54).

Helminths

Classically, helminth infections are associated with increased eosinophil and IgE levels, suggesting that these kinds of parasite induce Th2 responses. Work in mouse models has, with certain important exceptions (discussed below), shown this to be the case.

Most experimental immunology on parasitic helminths has focused on *Schistosoma mansoni,* which lives intravascularly or on intestinal nematodes. Infection with *S. mansoni* is initiated when the cercarial stage, which is free living in fresh water, penetrates the skin. The parasite remains in the skin for a few days before entering the vasculature and migrating over the following weeks to the hepatic portal system, where males and females mate and females begin producing eggs at about weeks 5–6. Prior to infection, the cytokines produced in vitro by antigen-stimulated spleen cells from infected mice are indicative of an ongoing Type 1 response (55,56). However, the parasite eggs induce a strong Type 2 response (55,57), which through the production of IL-10 by Th2 cells and macrophages (58,59) downregulates cytokine production by the preexisting parasite antigen specific Th1 cells and by decreasing B7 expression on antigen-presenting cells (60) results in the anergizing of further specific Th1 cells. Many eggs pass through the intestinal wall to continue the schistosome life cycle, but others get swept to the liver where they become trapped and induce granulomatous lesions. Ultimately, granulomas precipitate elevated hepatic portal blood pressure and fibrosis. Which cytokines are involved in the development of these lesions has been a focus of recent research on the pathogenesis of schistosomiasis, with the prevailing though not uncontested hypothesis being that Type 2 cytokines play a central role. This was originally based on the noted Th dependence of granuloma formation and the prevailing Th2 nature of the response at the time of granuloma development. The most convincing evidence supporting this claim has come from a model system in which eggs are injected intravenously into mice to become trapped in the lungs, where they induce granulomas, which mimic those seen in the livers of infected animals. In this system, mAb anti-IL-4 (61,62) or the lack of the IL-4 gene causes a reduction in pulmonary pathology, and IL-12, in an IFN-γ–dependent manner, causes a dramatic reduction in the size of granulomas (63), the latter was associated with a switch in the predominant egg-induced response from Th2 to Th1. Neutralization of IL-2 and MIP-1α have also been reported to reduce granuloma size and mAb anti–IL-5 prevents eosinophilia or eosinophil involvement in granulomas, but otherwise seems to have no effect on egg-induced pathology (64). Although Type 2 responses are clearly involved in the development of fibrosis during experimental schistosomiasis (discussed below), which if any cytokines are responsible for stimulating fibroblasts to proliferate and produce the Type 1 and Type 2 collagens that form the basis of fibrotic lesions, remains unknown. However, TGF-β (65) and, most interestingly, a novel cytokine Fibrosin (66), discovered through work on schistosomiasis-associated fibrosis, are implicated in this process.

The role of the immune system in mediating a protective response against a primary schistosome infection is paradoxical. The percentage of a given infectious dose that survives in a SCID mouse is no greater than that which matures in an immunocompetent mouse. However, liver-trapped eggs in the SCID mice fail to elicit granuloma formation but rather lead to severe acute hepatocyte damage, which can kill the infected animal (67). The granuloma, though damaging in the long run, therefore appears to play a crucial host-protective role during acute disease. Unexpectedly, TNF-α is able to restore the SCID mouse's capacity to isolate schistosome eggs within granulomalike lesions and protect the animals from the lethal hepatic damage that would otherwise ensue (67). The details of how TNF-α mediates this effect are unclear.

Efforts to develop a vaccine against schistosomiasis led to the discovery that exposure to radiation-attenuated schistosomes, which migrate to the lungs but die there, results in the development of a CD4+-dependent protective immune response, which acts against challenge parasites as they migrate through the lungs. Early analyses of immune responses in these mice indicated a correlation between CMI and protection (68), a prediction supported by the subsequent demonstrations that spleen cells from vaccinated mice make the Th1 pattern of cytokines (56) and that transcription of IFN-γ and to a lesser extent TNF-α, IL-2, and inducible NO synthase (an indicator of macrophage activation via IFN-γ) are upregulated in the lungs at the time challenge infection parasites are traversing this site (69). Studies of macrophages and endothelial cells as effector cells against larval schistosomes in vitro show that they can be activated by IFN-γ and TNF-α to kill the parasites via an NO-dependent mechanism (70). In support of the concept that IFN-γ and NO are crucial for protection, neutralization of either of these (with mAb or aminoguanidine, respectively) increases the susceptibility of vaccinated mice to challenge infection (69,71).

A role for Type 2 responses in immunopathology during a helminth infection has been illustrated most dramatically recently using a model of *Onchocerca volvulus*–induced ocular keratitis. *O. volvulus* is a highly host-specific filarial parasite of humans transmitted by the bite of an infected arthropod. It is pathogenic largely as the result of the fact that larval stages of the parasites, called microfilariae, migrate through the skin where they can be picked up by the vector. In addition to causing skin damage, microfilariae can traverse the conjunctiva, enter the cornea, and cause blindness ("river blindness"). Although these parasites cannot live in rodents, presensitization of mice by injection of onchocercal antigen followed by intracorneal injection of the same antigen some time later results in an intense Type 2 immune response in the eye, keratitis, and neovascularization. These pathological effects are markedly diminished in mice that lack the IL-4 gene (72) or that have received IL-12 (E. Pearlman, personal communication).

The murine gut parasites *Nippostrongylus brasiliensis, Heligmosomoides polygyrus,* and *Trichuris muris* in many ways serve as excellent models of human intestinal helminthiasises. In the cases of *H. polygyrus* and *T. muris,* infection is acquired by ingestion of eggs (execreted from infected mice), whereas for *N. brasiliensis,* eggs hatch outside the host to release invasive larval stages, which initiate infection percutaneously. In each case, the relative protective roles of Th1 and Th2 cytokines IFN-γ and IL-4 are reversed with respect to their function in controlling protozoan infections. For example, BALB/k mice develop a transient infection with *T. muris,* expelling the parasites before they mature. This is associated with the expression in the mesenteric LN of a Th2-like response and can be prevented by the administration of mAbs against IL-4 and IL-4R (73). In contrast, AKR mice allow *T. muris* to survive to maturity and begin laying eggs, express a Th1-like response in their MLN, and can be cured of infection by treatment with IL-4 (73). A similar situation exists for *H. polygyrus.* This parasite normally is able to survive to maturity and persist in BALB/c mice, despite the fact that these animals do mount a Th2-like response upon infection. Nevertheless, treatment of infected animals with exogenous IL-4 leads to a decrease in worm fecundity followed by expulsion of the parasites, and mAb anti–IL-4/IL-4R causes the opposite effect—an increase in egg output (74,75). In both cases, CD4+ cells appear to be the major contributors of IL-4. This cytokine also appears to play a role in the expulsion of *N. brasiliensis.* In immunocompetent animals, expulsion occurs shortly following the onset of egg production and cannot be prevented by mAb anti–IL-4. However, in nude or CD4+ cell–depleted mice, which develop chronic infections, IL-4 alone is sufficient to stimulate expulsion (74,75). The mode of action of IL-4 in these protective responses against gut helminths is unclear, but the view that it is working simply by stimulating the production of protective IgE is probably oversimplistic as in the *H. polygyrus* model a neutralizing anti-IgE mAb fails to have any deleterious effect on the IL-4–dependent resistance to reinfection evident in challenged drug-cured mice (76).

In keeping with the observation that the Th2 cytokine IL-4 is key for resistance to reinfection, IL-12 and IFN-γ, which often antagonize IL-4, increase the fecundity and longevity of *N. brasiliensis* in mice (77,78). The effects of IL-12 can be completely abrogated by mAb anti-IFN-γ, indicating that the latter is the effector cytokine. Moreover, IFN-α can mimic the effects of IFN-γ in this system (77). As has been previously pointed out, this would suggest that IFN-γ mediates its intestinal helminth infection–potentiating effects through a property it shares with IFN-α, such as suppressing Th2-cell proliferation, rather than through a unique effect such as stimulating NO production (74).

REGULATION OF CYTOKINE EXPRESSION IN HUMAN DISEASE STATES

Protozoa

The worst cases of falciparum malaria are found in children who develop cerebral complications (CM). This common cause of fatality among sub-Saharan children is caused when infected erythrocytes adhere to the endothelium of brain capillaries, probably via increased expression of ICAM-1 on the surface of endothelial cells (79), rupture and release a phosphatidylinositol-like molecule (the "malaria toxin"), which stimulates local TNF-α production by monocytes, endothelial cells, and macrophages (80). An eventual consequence of this is a local elevation in NO, which through its effects on neurotransmission could account for the coma associated with CM (81). Local TNF-α production could also lead to increased expression of ICAM-1 by the endothelium, thereby exacerbating the situation. IL-6 and soluble IL-1R levels are also raised in children with CM (82). Although local TNF-α production in the brain is considered to be of most importance from the point of view of the development of CM, blood levels of this cytokine at admission constitute an excellent indicator of whether or not a patient will survive a CM incident (83). CM occurs in only a small percentage of individuals with malaria, and recent work has been directed toward understanding the basis of this. One possibility is that there exist differences in the abilities of malaria toxins from different parasite isolates to activate TNF-α production (84), while another is that different individuals are more likely to produce TNF-α in response to the toxin. The latter is supported by the fascinating observation that variation in the TNF-α promoter region is associated with susceptibility to CM (85). In addition to being immunopathogenic, TNF-α may play a role in resistance to malaria (86).

PBMC from apparently immune, healthy people who live in areas endemic for malaria fail to proliferate or make IL-2 or IFN-γ in response to restimulation with malaria antigen in vitro (52). Removal of CD8+ cells in part reverses this "suppression" (52). To the best of our knowledge, there have been no reports of whether or not PBMC respond by making IL-4 and or IL-5, though by analogy with other systems, this would seem a distinct possibility. The reported immunosuppression of acute malaria could also be a manifestation of depressed Type 1 responses due to enhanced Type 2 responses.

PBMC from patients with kala azar, severe visceral leishmansias due to *L. donovani,* also fail to proliferate or produce IFN-γ upon activation with parasite antigen in vitro, and yet produce IL-4. This condition is correlated with the upregulation of production of the downregulatory cytokine IL-10, a cytokine that inhibits cytokine synthesis by Th cells and macrophage activation to kill *Leishmania* (87–89). CD8+ cells in some as yet poorly understood fashion play a role in stimulating IL-10 production and deactivating host antigen specific Th1 cells (87). As in mice, TGF-β production appears to be associated with persistent noncuring leishmaniasis in humans. Human monocytes infected with *Leishmania* parasites produce TGF-β and the active form of this cytokine is found at elevated levels in the plasma of *Leishmania*-infected individuals, in lesions from patients with cutaneous disease, and in the supernatant of antigen stimulated PBMC from patients (11).

As would be predicted from the data on murine *T. gondii* infection, humans asymptomatically infected with this protozoan show Type 1 dominated cytokine responses after in vitro stimulation with parasite antigen and in particular express elevated levels of IFN-γ (90,91), a crucial mediator of resistance revealed in the work on experimental models (27). Recent studies indicate that when *T. gondii*–infected individuals become co-infected with HIV, parasite-induced IFN-γ production by PBMC becomes suppressed even before CD4+ counts have begun to decline in the patients (91). HIV-positive individuals are known to be defective in IL-12 responses to mitogens (92), a deficiency also noted in PBMC of *T. gondii*–infected, HIV-positive individuals stimulated with parasite extracts (91). Thus, the depression in parasite-induced IFN-γ responses seen in co-infected patients may stem from their IL-12 production defect. Combined with the subsequent decline in CD4+ and CD8+ counts and function, these Type 1 cytokine production abnormalities may set the stage for the development of TE.

Helminths

Most work to decipher cytokine effects during human helminthiases has been directed towards filarial infections. The four most common filarial parasites of man are *Brugia malayi, O. volvulus, Wuchereria bancrofti,* and *Loa Loa.* Two of these, *B. malayi* and *W. bancrofti,* have been most investigated immunologically. They are found as adult worms in the lymphatics and are transmitted between humans by hematophagus arthropod vectors. The adults produce microfilariae, which circulate in the blood and are acquired by the vector at the time a blood meal is taken. Inside the vector, microfilariae differentiate to become third-stage larvae, which are infectious for humans. Several possible outcomes of infection can transpire: asymptomatic infection in which circulating microfilarial levels are high, severe symptomatic infection in which lymphatic obstruction occurs and microfilaremia is low, and tropical pulmonary eosinophilia, during which a severe allergic type response to microfilariae occurs (93). The latter syndrome clearly involves the production of IL-4 and IL-5, since it is characterized by extremely high specific IgE levels and eosinophilia. The asymptomatic microfilaremic state was long thought to be characterized by a state of anergy, as T-cell proliferation and IL-2 and IFN-γ production by PBMC in response to stimulation by parasite antigen in vitro are low (94). However, this is accompanied by an increase in the frequency of IL-4–producing CD4+ cells in the PBMC population, suggesting that the patients mount a parasite-specific Th2 response (95). This also appears to be the case in individuals with chronic pathology, although cells from these patients exhibit pronounced proliferative responses to parasite antigens. It has been suggested that a crucial difference between these two groups is that in the asymptomatic cases downregulatory cytokines such as IL-10 and possibly TGF-β are produced and dampen the tissue-damaging effects of the parasite-specific response (93,95). Indeed, mAb-mediated neutralization of IL-10 sometimes restores the ability of PBMC from asymptomatic patients to proliferate in response to antigen in vitro (95).

As in mice, *S. mansoni* infection in humans has been shown to result in an upregulation of the production of Type 2 cytokines, accompanied by a relative lack of schistosome antigen–specific Type 1 responses (96,97) The age-related development of resistance to schistosomes, which prevents many postpubescent people in endemic areas from becoming heavily reinfected following treatment, appears to be related to the Th2 response. In particular, antiworm IgE antibodies and IL-5 production in vitro in response to antigen restimulation are correlated with persistent low intensity of infection following treatment and reexposure to the parasite (98,99).

USE OF CYTOKINES OR CYTOKINE INHIBITORS TO MODULATE DISEASE

This is an extremely fertile area with several recent breakthroughs, which promise application to human medicine in the not too distant future. Clearly the emphasis here is on using what has been learned by intervention studies in animal models or correlative studies in human populations to develop intervention strategies to ameliorate disease due to parasitic infection.

While neutralizing anticytokine mAbs have proven to be extremely effective in rodent models, their application to human or domestic animal disease is likely to be limited due to the relatively large amounts of mAb needed. An exceptional situation in which mAb could be useful is CM, where a one-time administration of a neutralizing mAb against TNF-α to children with life-threateningly high levels of TNF-α might see them through a crisis and allow the initiation of chemotherapy. In general, the administrations of mAbs, or their more physiological alternatives soluble cytokine receptors, to neutralize cytokines and thereby alleviate acute pathological effects mediated by cytokines are likely to be relatively expensive and hospital based. An additional use of neutralizing anticytokine mAbs is to form soluble complexes with cytokine, a process that significantly prolongs the functional life of exogenous cytokine in vivo (100). This is a particularly promising development as most exogenous cytokines have half-lives in vivo that are so short as to render their use in a pharmacological fashion impractical. A notable exception to this generalization is IL-12, the recombinant form which has been found to be extremely effective in vivo in a number of settings. Most notably, repeated small doses of this cytokine (< 1 μg/mouse) can cure BALB/c mice of leishmaniasis (101,102).

A most promising application of cytokines to controlling parasitic diseases is their use as adjuvants. The greatest impact of parasitic diseases is in underdeveloped countries, where repeated administration of chemotherapy is logistically and economically impractical. Without doubt, the most appropriate and ultimately most practical way of controlling parasitic diseases is through vaccination, and it is toward this goal that much current immunological research on parasitic diseases is focused. In many cases, the key ingredient of a vaccine is the adjuvant, which pushes the immune response in the appropriate direction to allow the vaccinee to mount an immune response that will be beneficial upon subsequent encounter of the pathogen. In this regard, IL-12, combined in initial studies with a crude *L. major* extract (103) and most recently with a single recombinant *Leishmania* antigen (104), has proved to have potent adjuvant effects in promoting a protective *Leishmania*-specific Th1 response in mice. The effectiveness of this particular cytokine when used exogenously, combined with the greater understanding of the roles of cytokines and Th subset selection in immunopathogenesis in schistosomiasis, has led to the recent development of a novel experimental antipathology vaccine in which mice inoculated with a soluble egg antigen plus IL-12 develop egg antigen specific Th1 responses and significantly less fibrotic livers following subsequent infection (105). This approach establishes the plausibility of developing antipathology vaccines.

Chemotherapy often works most effectively in immunocompetent hosts. The application of the appropriate cytokines to promote this interaction between antiparasite drugs and the immune response has been effective in treating leishmansiasis. Here, IFN-γ has successfully been used as an adjunct with antimony in patients with visceral leishmaniasis who have not previously responded to antimony alone. Experimentally, both IL-12 (106) and mAb anti–IL-4 (107) have increased the efficacy of antimony in treating visceralized *L. major* infections in BALB/c mice. The increasing incidence of drug resistance in parasites combined with the limited expenditure on the development of new drugs for parasitic diseases means that this "immunochemotherapy" approach may become more important in the near future.

POTENTIAL FUTURE DIRECTIONS

An area that is currently receiving considerable attention is that of the effect of cytokines produced as a result of immune responses to one parasite on the type of immune response mounted toward a distinct pathogen. Since individuals exposed to one type of parasite tend to be exposed to several other distinct parasites, not to mention viral, bacterial, and fungal pathogens, this is a matter of considerable importance. Substantiating this concern are the findings that, relative to normal animals, *S. mansoni*–infected mice have been found to mount depressed Type 1 and/or enhanced Type 2 responses to *Vaccinia* (108) or to the test antigen myoglobin (109); in the case of the latter at least, IL-10 was shown to play a role in this altered responsiveness. On a slightly more positive note, AKR mice, which normally mount a Type 1 response to *T. muris* and are susceptible to this parasite, were rendered resistant by prior infection with *S. mansoni* (110). Presumably, this was due to the strong Type 2 response induced by the schistosome eggs predisposing the animals to mount a

Type 2 response to the intestinal helminth. While beneficial to the host in this case, there must be concern that, for example, administration of a vaccine intended to induce a Type 1 response to patients with schistosomiasis would lead to an ineffective and possibly even deleterious vaccine antigen-specific Type 2 response. The recent demonstration that, in contrast to individuals from a nonendemic area, patients with schistosomiasis failed to exhibit a tetanus toxoid specific Type 1 response following tetanus vaccination substantiates this concern.

There is a growing appreciation that certain parasite molecules interact in an antigen receptor–independent fashion with cells of the innate or specific arms of the host's defensive systems. In so doing, these molecules activate the host cells to produce cytokines that can have marked and prolonged effects on the nature of specific antiparasite immune responses that develop subsequently. In addition to the malaria toxin discussed above, molecules of this type have also been identified in other parasites. For example, an elongation factor from *L. brasiliensis* (111) and a carbohydrate containing molecule from *T. gondii* are both able to stimulate IL-12 production by host cells of the monocyte/macrophage lineage. Moreover, the sugar lacto-N-fucopentaose III, found on certain *S. mansoni* egg molecules, is able to activate CD5+ B cells to make IL-10 (112). The discovery of more molecules with these immunomodulatory characteristics, along with an increase in our understanding of how they are able to interact with host cells, is likely in the near future and may help to explain why so many parasites tend to induce immune responses highly skewed in Type 1 or 2 directions.

A remaining provocative though largely unresolved possibility is that parasites produce cytokines and/or possess receptors for host cytokines. Perhaps the most striking evidence suggesting that this is the case is provided by the observation that female schistosomes respond to recombinant TNF-α, in vitro, by producing more eggs (67). While this concept in general may seem unrealistic, it is as well to remember that homologs of many mammalian cytokines are present in lower order eukaryotes; much of our understanding of TGF-β and TGF-β receptors comes from work on their homologs in *Caenorhabditis elegans,* for example (113). The latter subject is likely to be a fertile ground for future research.

ACKNOWLEDGMENTS

The authors wish to thank Dr. Tanya Scharton-Kersten for helpful discussions and Dr. Eric Pearlman for allowing them to cite unpublished data. Edward Pearce is a recipient of a Burroughs Wellcome Fund New Investigator Award in Molecular Parasitology.

REFERENCES

1. Mosmann TR, Cherwinski H, Bond MA, Giedlin I, Coffman RL. Two types of murine helper T cell clone. I. Definition according to the profiles of lymphokine activities and secreted proteins. J Immunol 1986; 136:2348–2354.
2. Mosmann TR, Coffman RL. Th1 and Th2 cells: different patterns of lymphokine secretion lead to different functional properties. Annu Rev Immunol 1989; 7:145–173.
3. Finkelman FD, Katona IM, Urban JF, Snapper CM, Ohara J, Paul WE. Suppression of in vivo polyclonal IgE responses by monoclonal antibody to the lymphokine B-cell stimulatory factor. Proc Natl Acad Sci 1986; 83:9675–9678.
4. Coffman RL, Seymour BWP, Haduk S, Jackson J, Rennick D. Antibody to interleukin 5 inhibits helminth-induced eosinophilia in mice. Science 1989; 245:308–310.
5. Clerici M, Shearer G. The Th1-Th2 hypothesis. Immunol Today 1994; 15:575–581.
6. Sher A, Gazzinelli RT, Oswald IP, et al. Role of T-cell derived cytokines in the down regulation of immune responses in parasitic and retroviral infections. Immunol Rev 1992; 127:183.
7. Titus RG, Milon G, Marchal G, Vassali P, Cerotinni JC, Louis JA. Therapeutic effect of anti-L3T4 monoclonal antibody GK1.5 in cutaneous leishmaniasis in genetically susceptible BALB/c mice. J Immunol 1985; 135:2108–2114.
8. Heinzel FP, Sadick MD, Holaday BJ, Coffman RL, Locksley RM. Reciprocal expression of interferon-γ or interleukin-4 during the resolution or progression of murine leishmaniasis. J Exp Med 1989; 169:59–72.
9. Scott P, Natovitz P, Coffman RL, Pearce E, Sher A. Immunoregulation of cutaneous leishmaniasis. J Exp Med 1988; 168:1675–1684.
10. Wei X, Charles IG, Smith A, et al. Altered immune responses in mice lacking inducible nitric oxide synthase. Nature (Lond.) 1995; 375:408–411.
11. Reed SG. Cytokine control of the macrophage parasites *Leishmania* and *Trypanosoma cruzi.* In: Boothroyd JC, Komuniecki R, eds. Molecular Approaches to Parasitology. New York: Wiley-Liss, Inc., 1995:442–453.
12. Lehn M, Weiser WY, Engelhorn S, Gillis S, Remold RG. IL-4 inhibit H2O2 production and antileishmanial capacity of human cultured monocytes mediated by IFN-γ. J Immunol 1989; 143:3020–3024.
13. Oswald IP, Gazzinelli RG, Sher A, James SL. IL-10 synergizes with IL-4 and TGF-B to inhibit macrophage cytotoxic activity. J Immunol 1992; 148:3578–3582.
14. Scott P. IFN-γ modulates the early development of Th1 and Th2 responses in a murine model of cutaneous leishmaniasis. J Immunol 1991; 147:3149–3155.
15. Sadick MD, Heinzel FP, Holaday BJ, Pu RT, Dawkins RS, Locksley RM. Cure of murine leishmaniasis with anti-interleukin 4 monoclonal antibody. J Exp Med 1990; 171:115–127.
16. Chatelain R, Varkila K, Coffman RL. IL-4 induces a Th2 response in *Leishmania major*-infected mice. J Immunol 1992; 148:1182–1187.
17. Swihart K, Fruth U, Messmer N, et al. Mice from a genetically resistant background lacking the interferon gamma receptor are susceptible to infection with Leishmania major but mount a polarized T helper cell 1-type CD4+ response. J Exp Med 1995; 181:961–971.
18. Seder RA, Paul WE. Acquisition of lymphokine-producing phenotype by CD4+ T cells. Annu Rev Immunol 1994; 12:635–673.
19. Scharton-Kersten T, Afonso LC, Wysocka M, Trinchieri G, Scott P. IL-12 is required for natural killer cell activation and

subsequent T helper 1 cell development in experimental leishmaniasis. J Immunol 1995; 154:5320–5330.
20. Locksley RM. Interleukin 12 in host defense against microbial pathogens. Proc Natl Acad Sci USA, 1993; 90:5879–5880.
21. Scharton TM, Scott P. Natural killer cells are a source of interferon-γ that drives differentiation of CD4+ T cell subsets and induces early resistance to *Leishmania major* in mice. J Exp Med 1993; 178:567–577.
22. Barral-Netto M, Barral A, Brownell CE, et al. Transforming growth factor-beta in leishmanial infection: a parasite escape mechanism. Science 1992; 257:545–548.
23. Barral A, Barral-Netto M, Yong EC, Brownell CE, Twardzik DR, Reed SG. Transforming growth factor beta as a virulence mechanism for *Leishmania brasiliensis*. Proc Natl Acad Sci USA 1993; 90:3442–3446.
24. Reiner SL, Zheng S, Wang Z-E, Stowring L, Locksley RM. *Leishmania* promastigotes evade interleukin 12 (IL-12) induction by macrophages and stimulate a broad range of cytokines from CD4+ T cells during initiation of infection. J Exp Med 1994; 179:447–456.
25. Vieira LQ, Hondowicz BD, Afonso LCC, Wysocka M, Trinchieri G, Scott P. Infection with *Leishmania major* induces interleukin-12 production in vivo. Immunol Lett 1994; 40:157–161.
26. Luft BJ, Brooks RG, Conley FK, McCabe RE, Remington JS. Toxoplasmic encephalitis in patients with acquired immune deficiency syndrome. JAMA 1984; 252:913–917.
27. Gazzinelli RT, Denkers E, Sher A. Host resistance to *Toxoplasma gondii*: a model for studying the selective induction of cell-mediated immunity by intracellular parasites. Infect Agents Dis 1993; 2:139–149.
28. Hunter CA, Remington JS. Immunopathogenesis of toxoplasmic encephalitis. J Infect Dis 1994; 170:1057–1067.
29. Sher A. Regulation of cell-mediated immunity by parasites: the ups and downs of an important host adaptation. In: Boothroyd JC, Komuniecki R, eds. Molecular Approaches to Parasitology. New York: Wiley-Liss, Inc., 1995:431–442.
30. Gazzinelli RT, Hieny S, Wynn TA, Wolf S, Sher A. Interleukin 12 is required for the T-lymphocyte-independent induction of interferon gamma by an intracellular parasite and induces resistance in T-cell-deficient hosts [see comments]. Proc Natl Acad Sci USA; 1993; 90:6115–6119.
31. Hunter CA, Remington JS. Regulation of IL-12 and TNF-a induced NK production of IFN-γ by IL-1B and TGF-B. Eur Cytokine Netwk 1994; 5:145.
32. Gazzinelli RT, Wysocka M, Hayashi S, et al. Parasite-induced IL-12 stimulates early IFN-gamma synthesis and resistance during acute infection with *Toxoplasma gondii*. J Immunol 1994; 153:2533–2543.
33. Khan IA, Matsuura T, Kasper LH. Interleukin-12 enhances murine survival against acute toxoplasmosis. Infect Immun 1994; 62:1639–1642.
34. Gazzinelli R, Xu Y, Hieny S, Cheever A, Sher A. Simultaneous depletion of CD4+ and CD8+ T lymphocytes is required to reactivate chronic infection with *Toxoplasma gondii*. J Immunol 1992; 149:175–180.
35. Gazzinelli RT, Hakim FT, Hieny S, Shearer GM, Sher A. Synergistic role of CD4+ and CD8+ T lymphocytes in IFN-γ procution and protective immunity induced by an attenuated *Toxoplasma gondii* vaccine. J Immunol 1991; 146:286–292.
36. Denkers KY, Caspar P, Sher A. *Toxoplasma gondii* possesses a superantigen activity that selectively expands murine T cell receptor V beta 5-bearing CD8+ lymphocytes. J Exp Med 1994; 180:985–994.
37. Burke JM, Roberts CW, Hunter CA, Murray M, Alexander J. Temporal differences in the expression of mRNA for IL-10 and IFN-gamma in the brains and spleens of C57BL/10 mice infected with *Toxoplasma gondii*. Parasite Immunol 1994; 16:305–314.
38. Israelski DM, Araujo FG, Conley FK, Suzuki Y, Sharma S, Remington JS. Treatment with anti-L3T4 (CD4) monoclonal antibody reduces the inflammatory response in toxoplasmic encephalitis. J Immunol 1989; 142:954–960.
39. Soong L, Tarleton RL. *Trypanosoma cruzi* infection suppresses nuclear factors that bind to specific sites on the interleukin-2 enhancer. Eur J Immunol 1994; 24:16–23.
40. Silva JS, Morrissey PJ, Grabstein KH, Mohler KM, Anderson D, Reed SG. Interleukin 10 and interferon gamma regulation of experimental *Trypanosoma cruzi* infection. J Exp Med 1992; 175:169–174.
41. Tarleton RL, Koller BH, Latour A, Postan M. Susceptibility of beta 2-microglobulin-deficient mice to infection with *Trypanosoma cruzi* [see comments]. Nature (Lond.) 1992; 356:338–340.
42. Tarleton RL. The role of T cells in *Trypanosoma cruzi* infections. Parasitol Today 1995; 11:7–9.
43. Hoft DF, Lynch RG, Kirchhoff LV. Kinetic analysis of antigen-specific immune responses in resistant and susceptible mice during infection with *Trypanosoma cruzi*. J Immunol 1993; 151:7038–7047.
44. Silva JS, Twardzik DR, Reed SG. Regulation of *Trypanosoma cruzi* infections in vitro and in vivo by transforming growth factor-B. J Exp Med 1991; 174:539–549.
45. Lin JY, Seguin R, Keller K, Chadee K. Tumor necrosis factor alpha augments nitric oxide-dependent macrophage cytotoxicity against *Entamoeba histolytica* by enhanced expression of the nitric oxide synthase gene. Infect Immun 1994; 62:1534–1541.
46. Wang W, Keller K, Chadee K. Modulation of tumor necrosis factor production by macrophages in *Entamoeba histolytica* infection. Infect Immun 1992; 60:3169–3174.
47. Grau GE, Bieler G, Pointaire P, et al. Significance of cytokine production and adhesion molecules in malarial immunopathology. Immunol Lett 1990; 25:189–194.
48. Grau GE, Fajardo LF, Piguet PF, Allet B, Lambert PH, Vassalli P. Tumor necrosis factor (cachectin) as an essential mediator in murine cerebral malaria. Science 1987; 237:1210–1212.
49. Clark IA, Ilschner S, MacMicking JD, Cowden WB. TNF and *Plasmodium berghei* ANKA-induced cerebral malaria. Immunol Lett 1990; 25:195–198.
50. Grau GE, Piguet PF, Vassalli P, Lambert PH. Tumor-necrosis factor and other cytokines in cerebral malaria: experimental and clinical data. Immunol Rev 1989; 112:49–70.
51. Taylor-Robinson AW, Phillips RS, Severn A, Moncada S, Liew FY. The role of TH1 and TH2 cells in a rodent malaria infection. Science 1993; 260:1931–1934.
52. Riley EM, Hviid L, Theander TG. Malaria. In: Kierszenbaum F, ed. Parasitic Infections and the Immune System. New York: Academic Press, Inc., 1994:119–143.
53. Sedegah M, Finkelman F, Hoffman S. Interleukin 12 induction of interferon γ-dependent protection against malaria. Proc Natl Acad Sci USA 1994; 91:10700–10702.
54. Tsuji M, Miyahira Y, Nussenzweig RS, Aguet M, Reichel M, Zavala F. Development of antimalaria immunity in mice lacking IFN-gamma receptor. J Immunol 1995; 154:5338–5344.
55. Gryzch JM, Pearce EJ, Cheever A, et al. Egg deposition is the major stimulus for the production of Th2 cytokines in murine schistosomiasis mansoni. J Immunol 1991; 146:1322–1327.
56. Pearce EJ, Caspar P, Gryzch JM, Lewis FA, Sher A. Down-

regulation of Th1 cytokine production accompanies induction of Th2 responses by a helminth, *Schistosoma mansoni.* J Exp Med 1991; 173:159–165.
57. Vella AT, Pearce EJ. CD4+ Th2 response induced by *Schistosoma mansoni* eggs develops rapidly, through an early, transient, Th0-like stage. J Immunol 1992; 148:2283–2290.
58. Sher A, Fiorentino D, Casper P, Pearce EJ, Mosmann TR. Production of IL-10 by CD4+ T lymphocytes correlates with down regulation of Th1 cytokine synthesis. J Immunol 1991; 147:2713–2718.
59. Flores Villanueva PO, Harris TS, Ricklan DE, Stadecker MJ. Macrophages from schistosomal egg granulomas induce unresponsiveness in specific cloned Th-1 lymphocytes in vitro and down-regulate schistosomal granulomatous disease in vivo. J Immunol 1994; 152:1847–1855.
60. Flores Villanueva PO, Reiser H, Stadecker MJ. Regulation of T helper cell responses in experimental murine schistosomiasis by IL-10. Effect on expression of B7 and B7-2 costimulatory molecules by macrophages. J Immunol 1994; 153:5190–5199.
61. Wynn TA, Eltoum I, Cheever AW, Lewis FA, Gause WC, Sher A. Analysis of cytokine mRNA expression during primary granuloma formation induced by eggs of *Schistosoma mansoni.* J Immunol 1993; 151:1430–1440.
62. Chensue SW, Terebuh PD, Warming KS, Hershey SD, Evanoff HL, Higashi GI. Role of IL-4 and IFN-γ in *Schistosoma mansoni* egg-induced hypersensitivity granuloma formation. J Immunol 1992; 148:900–906.
63. Wynn TA, Eltoum I, Oswald IP, Cheever AW, Sher A. Endogenous interleukin 12 (IL-12) regulates granuloma formation induced by eggs of *Schistosoma mansoni* and exogenous IL-12 both inhibits and prophylactically immunizes against egg pathology. J Exp Med 1994; 179:1551–1561.
64. Sher A, Coffman RL, Hieny S, Scott P, Cheever AW. Interleukin 5 (IL-5) is required for the blood and tissue eosinophilia but not granuloma formation induced by infection with *Schistosoma mansoni.* Proc Natl Acad Sci 1990; 87:61–66.
65. Czaja MJ, Weiner FR, Flanders KC, et al. In vitro and in vivo association of transforming growth factor-B1 with hepatic fibrosis. J Cell Biol 1989; 108:2477–2482.
66. Prakash S, Robbins PW, Wyler DJ. Cloning and analysis of murine cDNA that encodes a fibrogenic lymphokine, fibrosin. Proc Natl Acad Sci USA 1995; 92:2154–2158.
67. Amiri P, Locksley RM, Parslow TG, et al. Tumour necrosis factor α restores granulomas and induces parasite egg-laying in schistosome-infected SCID mice. Nature (Lond.) 1992; 356:604–606.
68. James SL, Scott P. Induction of cell mediated immunity as a strategy for vaccination against parasites. In: Englund PT, Sher A, eds. The Biology of Parasitism. New York: Alan R. Liss, 1988:249–264.
69. Wynn TA, Oswald IP, Eltoum IA, et al. Elevated expression of Th1 cytokines and nitric oxide synthase in the lungs of vaccinated mice after challenge infection with *Schistosoma mansoni.* J Immunol 1994; 153:5200–5209.
70. Oswald IP, Eltoum I, Wynn TA, et al. Endothelial cells are activated by cytokine treatment to kill an intravascular parasite, *Schistosoma mansoni,* through the production of nitric oxide. Proc Natl Acad Sci USA 1994; 91:999–1003.
71. Smythies LE, Coulson PS, Wilson RA. Monoclonal antibody to IFN-gamma modifies pulmonary inflammatory responses and abrogates immunity to *Schistosoma mansoni* in mice vaccinated with attenuated cercariae. J Immunol 1992; 149:3654–3658.
72. Pearlman E, Lass JH, Bardenstein DS, Hazlett FE, Diaconu E, Kazura JW. IL-4 and Th2 cells are required for the development of experimental onchocercal keratitis (river blindness). J Exp Med 1995;
73. Else KJ, Finkelman FD, Maliszewski CR, Grencis RK. Cytokine-mediated regulation of chronic intestinal helminth infection. J Exp Med 1994; 179:347–351.
74. Finkelman FD, Gause WC, Urban JFJ. Cytokine control of protective immunity against nematode infections. In: Boothroyd JC, Komuniecki R, eds. Molecular Approaches to Parasitology. Vol. 12. New York: Wiley-Liss, Inc., 1995:467–476.
75. Urban JF, Jr., Maliszewski CR, Madden KB, Katona IM, Finkelman FD. IL-4 treatment can cure established gastrointestinal nematode infections in immunocompetent and immunodeficient mice. J Immunol 1995; 154:4675–4684.
76. Urban JF, Jr., Katona IM, Paul WE, Finkelman FD. Interleukin 4 is important in protective immunity to a gastrointestinal nematode infection in mice. Proc Natl Acad USA 1991; 88:5513–5517.
77. Urban JF, Jr., Madden KB, Cheever AW, Trotta PP, Katona IM, Finkelman FD. IFN inhibits inflammatory responses and protective immunity in mice infected with the nematode parasite, *Nippostrongylus brasiliensis.* J Immunol 1993; 151:7086–7094.
78. Finkelman FD, Madden KB, Cheever AW, et al. Effects of interleukin 12 on immune responses and host protection in mice infected with intestinal nematode parasites. J Exp Med 1994; 179:1563–1572.
79. Berendt AR, Simmons DL, Tansy J, Newbold CI, Marsh K. Intracellular adhesion molecule 1 is an endothelial cell adhesion receptor for *Plasmodium falciparum.* Nature (Lond.) 1989; 341:57–59.
80. Jakobsen PH, Bate CAW, Taverne J, Playfair JH. Malaria: toxins, cytokines and disease. Parasite Immunol 1995; 17:223–231.
81. Clark IA, Rockett KA, Cowden WB. Proposed central role of nitric oxide in conditions clinically similar to cerebral malaria. Lancet 1992; 340:894–895.
82. Jakobsen PH, McVay V, Morris-Jones D, et al. Increased concentrations of interleukin-6 and interleukin-1 receptor antagonist and decreased concentrations of beta-2-glycoprotein I in Gambian children with cerebral malaria. Infect Immun 1994; 62:4374–4379.
83. Kwiatkowski D, Hill AVS, Sambou I, et al. TNF concentration in fatal cerebral, non-fatal cerebral, and uncomplicated *Plasmodium falciparum* malaria. Lancet 1990; 336:1201–1204.
84. Allan RJ, Beattie P, Bate C, et al. Strain variation in tumor necrosis factor induction by parasites from children with acute *falciparum* malaria. Infect Immun 1995; 63:1173–1175.
85. McGuire W, Hill AV, Allsopp CE, Greenwood BM, Kwiatkowski D. Variation in the TNF-alpha promoter region associated with susceptibility to cerebral malaria. Nature (Lond.) 1994; 371:508–510.
86. Taverne J. Unravelling the cytokine network in malaria. Parasitol Today 1993; 9.
87. Holaday BJ, Pompeu MM, Jeronimo S, et al. Potential role for interleukin-10 in the immunosuppression associated with kala azar. J Clin Invest 1993; 92:2626–2632.
88. Karp CL, el-Safi SH, Wynn TA, et al. In vivo cytokine profiles in patients with kala-azar. Marked elevation of both interleukin-10 and interferon-gamma [see comments]. J Clin Invest 1993; 91:1644–1648.
89. Ghalib HW, Piuvezam MR, Skeiky YA, et al. Interleukin 10 production correlates with pathology in human *Leishmania donovani* infections. J Clin Invest 1993; 92:324–329.
90. Murray HW, Kilmann JK, Rubin BY, et al. Patients at risk for AIDS-related opportunistic infections. Clinical manifesta-

tion and impaired gamma interferon. N Engl J Med 1985; 313:1504–1508.

91. Gazzinelli RT, Bala S, Stevens S, et al. HIV infection suppresses Type 1 lymphokine and IL-12 responses to *Toxoplasma gondii* but fails to inhibit the synthesis of other parasite induced monokines. J Immunol 1995;
92. Chehimi J, Starr SE, Frank I, et al. Impaired interleukin 12 production in human immunodeficiency virus-infected patients. J Exp Med 1994; 179:1361–1366.
93. Nutman TB. Immune responses in lymphatic dwelling filarial infections. In: Boothroyd JC, Komuniecki R, eds. Molecular Approaches to Parasitology. New York: Wiley-Liss, Inc., 1995:511–523.
94. King CL, Kumaraswami V, Poindexter RW, et al. Immunologic tolerance in lymphatic filariasis. Diminished parasite-specific T and B lymphocyte precursor frequency in the microfilaremic state. J Clin Invest 1992; 89:1403–1410.
95. King CL, Mahanty S, Kumaraswami V, et al. Cytokine control of parasite-specific anergy in human lymphatic filariasis. Preferential induction of a regulatory T helper type 2 lymphocyte subset. J Clin Invest 1993; 92:1667–1673.
96. Williams ME, Montenegro S, Domingues AL, et al. Leukocytes of patients with *Schistosoma mansoni* respond with a Th2 pattern of cytokine production to mitogen or egg antigens but with a Th0 pattern to worm antigens. J Infect Dis 1994; 170:946–954.
97. Araujo MI, Bacellar O, Ribeiro de Jesus A, Carvalho EM. The absence of gamma-interferon production of *S. mansoni* antigens in patients with schistosomiasis. Braz J Med Biol Res 1994; 27:1619–1625.
98. Hagan P, Blumenthal UJ, Dunn D, Simpson AJG, Wilkins HA. Human IgE, IgG4 and resistance to reinfection with *Schistosoma haematobium*. Nature (Lond.) 1991; 349:243–246.
99. Pearce EJ. Immunology of schistosomiasis. In: Boothroyd JC, Komuniecki R, eds. Molecular Approaches to Parasitology. New York: Wiley-Liss, 1995.
100. Finkelman FD, Madden KB, Morris SC, et al. Anti-cytokine antibodies as carrier proteins. Prolongation of in vivo effects of exogenous cytokines by injection of cytokine-anti-cytokine antibody complexes. J Immunol 1993; 151:1235–1244.
101. Heinzel FP, Schoenhaut DS, Rerko RM, Rosser LE, Gately MK. Recombinant interleukin 12 cures mice infected with *Leishmania major*. J Exp Med 1993; 177:1505–1509.
102. Sypek JP, Chung CL, Mayor SEH, et al. Resolution of cutaneous leishmaniasis: interleukin 12 initiates a protective T helper 1 immune response. J Exp Med 1993; 177:1797–1802.
103. Afonso LCC, Scharton TM, Vieria LQ, Wysocka M, Trinchieri G, Scott P. The adjuvant effect of interleukin-12 in a vaccine against *Leishmania major*. Science 1994; 263:235–237.
104. Mougneau E, Altare F, Wakil AK, et al. Expression cloning of a protective *Leishmania* antigen. Science 1995; 268:563–566.
105. Wynn T, Cheever AW, Jankovic D, et al. An IL-12 based vaccination method for preventing fibrosis induced by schistosome infection. Nature (Lond.) 1995;
106. Nabors GS, Afonso LC, Farrell JP, Scott P. Switch from a type 2 to a type 1 T helper cell response and cure of established *Leishmania major* infection in mice is induced by combined therapy with interleukin 12 and Pentostam. Proc Natl Acad Sci USA 1995; 92:3142–3146.
107. Nabors GS, Farrell JP. Depletion of interleukin-4 in BALB/c mice with established *Leishmania major* infections increases the efficacy of antimony therapy and promotes Th1-like responses. Infect Immun 1994; 62:5498–5504.
108. Actor JK, Shirai M, Kullberg MC, Buller RML, Sher A, Berzofsky JA. Helminth infection results in decreased virus-specific CD8+ cytotoxic T-cell and Th1 cytokine responses as well as delayed viral clearance. Proc Natl Acad Sci USA 1993; 90:948–952.
109. Kullberg MC, Pearce EJ, Hieny SA, Sher A, Berzofsky JA. Infection with *Schistosoma mansoni* alters Th1/Th2 cytokine responses to a non-parasite antigen. J Immunol 1992; 148:3264.
110. Curry AJ, Else KJ, Jones F, Bancroft A, Grencis RK, Dunne DW. Evidence that cytokines-mediated immune interactions induced by *Schistosoma mansoni* alter disease outcome in mice concurrently infected with *Trichuris muris*. J Exp Med 1995; 181:769–774.
111. Skeiky YA, Guderian JA, Benson DR, et al. A recombinant *Leishmania* antigen that stimulates human peripheral blood mononuclear cells to express a Th1-type cytokine profile and to produce interleukin 12. J Exp Med 1995; 181:1527–1537.
112. Velupillai P, Harn DA. Oligosaccharide-specific induction of interleukin 10 production by B220+ cells from schistosome-infected mice: a mechanism for regulation of CD4+ T-cell subsets [see comments]. Proc Natl Acad Sci USA 1994; 91:18–22.
113. Derynck R. TGF-β-receptor mediated signalling. Trends Biochem Sci 1994; 19:548.

42
Cytokines and Malaria

Sylvie C. Kossodo and Georges E. Grau
University Hospital, Geneva, Switzerland

INTRODUCTION

Malaria is considered to be the most important parasitic disease, affecting more than 300 million people annually worldwide. People can be infected with four species of protozoan parasites of the genus *Plasmodium: P. vivax, P. malariae, P. ovale,* and *P. falciparum.* While the three first species cause a debilitating and incapacitating disease, the latter one is responsible for life-threatening complications such as anemia and cerebral malaria (CM), which account for 500,000 to 1.5 million deaths annually, principally among young African children.

Malaria is responsible for the activation of numerous pathways of the inflammatory and immune responses, only some of which appear to be critically important in the pathogenesis of the neurological complications associated with *P. falciparum* infection. In this chapter, we will summarize the current knowledge on the close relationship between cytokines and either protection against parasites or pathology caused by them. The relevance of certain animal models, notably in rodents and monkeys, to human disease can be arguable. However, certain data pertaining to animal models can be extrapolated to what is observed in human malaria. We have therefore chosen to intermingle studies performed in animal models and in humans. Intervention studies in animals have proven to be useful tools to decipher pathophysiological mechanisms, eventually leading to new therapeutic approaches.

CYTOKINE INDUCTION IN THE HOST

Th-1 Versus Th-2 T-Lymphocyte Subsets

There is now compelling evidence supporting the notion that both T-cell compartments, namely, Th-1 and Th-2 subsets (1), are involved in the various phases of malaria infection, where they may play either a protective or deleterious role.

The elegant studies by Taylor-Robinson and coworkers (2) showed that both Th-1 and Th-2 subsets of CD4+ T cells could protect mice against *P. chabaudi chabaudi* infection. Two distinct mechanisms were found to be involved: Th-1 cells protected by a nitric oxide–dependent pathway, while Th-2 cells protected by the enhancement and accelerated production of specific IgG1 antibody. As reviewed in Ref. 3, Th-1–related cytokines are predominantly produced during the acute phase of infection with *P. chabaudi chabaudi* and lead mainly to the production of macrophage-derived cytokines. This antibody-independent pathway is probably sufficient for parasite control early in infection via macrophage-associated inflammatory responses, but can also lead to the pathological consequences of infection. As the disease advances, the pattern of cytokine production shifts toward a Th-2–like response. Later in infection, a major consequence of this switch to a production of Th-2–related cytokines would be the down-regulation of interferon (IFN)-γ–induced macrophage activation and the promotion of antibody production by mature

B cells, suggesting that the mechanism of parasite control in the later stages of infection is predominantly antibody dependent.

The functional importance of Th-2 cells has been investigated in IL-4 knock-out mice. IL-4–deficient mice infected with *P. chabaudi chabaudi* are able to control and clear a primary infection, although recrudescent parasitemias are significantly higher in these mice compared with wild-type mice. In vitro Th-1 responses and ex vivo IL-12 mRNA levels are sustained in the IL-4–deficient mice. Th-2–associated cytokine mRNAs such as IL-5 and IL-6, but not IL-10, are reduced early in infection but expressed at comparable levels at the later stages of infection in normal and IL-4 knock-out mice. All of these impairments of Th-2–related activities are insufficient to affect parasite clearance in the deficient mice, probably because such activities are only delayed and can take place normally at the later stages of infection (4).

Parasite clearance and survival do not necessarily correlate. For example, treatment of susceptible *P. chabaudi* AS-infected A/J mice with various doses of recombinant IL-12 significantly decreases the peak parasitemia level, but only treatment with 0.1 μg results in increased survival (5). Treatment of resistant B6 mice with 0.1 μg of IL-12 using the same regimen also significantly decreases the peak parasitemia level, but 40% of the animals die. Treatment of these mice with anti–IL-12 monoclonal antibody results in a more severe course of infection, but survival is not significantly altered. Compared with spleen cells from untreated mice, cells from IL-12–treated mice produce significantly higher levels of IFN-γ spontaneously as well as in response to ConA or antigen stimulation on day 7 postinfection. There are significantly higher serum levels of IFN-γ and tumor necrosis factor (TNF)-γ in IL-12–treated mice, which correlated with high levels of the nitric oxide (NO) metabolite, NO_3^-. Furthermore, CD4+ T-cell depletion abrogates IL-12–induced resistance. Simultaneous depletion of both IFN-γ or TNF in vivo, by injection of neutralizing monoclonal antibody (mAb) in IL-12–treated mice, results in 100% mortality. Significantly increased mortality is observed following treatment with aminoguanidine, a selective inhibitor of cytokine-inducible nitric oxide synthase, without effect on parasitemia. In conclusion, these results demonstrate that IL-12 regulates the development of resistance to *P. chabaudi* AS via a CD4+ Th-1 response, which involves the cytokines IFN-γ and TNF and is in part NO dependent.

The levels of mRNA for diverse cytokines have been monitored over the course of malarial infection. Acute *P. yoelii* murine malaria is associated with a marked depression of splenic T-cell responses. Early in infection, no transcripts for IL-2R p55, IL-2, IL-6, IL-10, and IFN-γ are detected (6). In contrast, late in the infection, when mice have cleared the parasites and became resistant to reinfection, mRNAs for the above cytokines as well as for IL-4, IL-5, GM-CSF, and TNF are detectable. During this late phase of infection, lymphocytes proliferated vigorously in response to TCR- and T-cell mitogen-mediated stimulation. Surprisingly, during an early phase (as early as 3 days postinfection) with low parasitemia, before the establishment of T-cell unresponsiveness, a broad array of cytokine expression including IL-2 and IFN-γ expression as well as marked lymphoproliferative response upon T-cell mitogen- and TCR-mediated stimulation can be observed. A similar pattern of cytokine profile is seen in freshly isolated (ex vivo) splenocytes from *P. yoelii*–infected animals, suggesting that inactivation or/and ablation of reactive T cells may explain T-cell hyporesponsiveness during acute malaria.

Cytokine Induction by *Plasmodium* Antigens

Malaria infection in the vertebrate host occurs after sporozoite inoculation by infected mosquitoes. The sporozoites migrate into the liver, where they differentiate into merozoites. Merozoites invade red blood cells, where they either develop asexually or differentiate into male and female gametocytes (sexual forms). Transmission occurs when gametocytes are transferred back to mosquitoes. The parasite, at each stage of its cycle, expresses specific antigens. In addition to the molecules expressed on the infected erythrocyte surface, plasmodia are able to release soluble antigens. All these molecules elicit cytokine responses and thereby modulate the host response to infection.

Surface and Soluble Parasite Antigens

One of the first responses of the host to infection is the induction of fever, which is triggered by erythrocytic schizont rupture (7), as shown over a hundred years ago. More recently, in vitro studies have shown that this event stimulates human mononuclear cells to release TNF (8). The TNF-inducing capacity has been studied in various *P. falciparum* isolates originating from children with or without CM (9). The isolates from severe cases tend to induce more TNF by human mononuclear cells, but there is a considerable overlap with isolates from uncomplicated cases.

In mice malaria toxins stimulate a T-cell–independent antibody response, generating short-lived immunoglobulin M (IgM) antibodies that inhibit the TNF-inducing activity of the toxins. A similar antibody response is seen in humans. Serum from a European adult infected with *P. falciparum* inhibits the induction of TNF by malaria toxins derived from *P. falciparum*–infected erythrocytes. IgM antibodies are found to be responsible for the inhibitory activity. These inhibitory antibodies cannot be detected in convalescent-phase serum collected from the same patient 6 weeks later or in sera from healthy European and African controls. The antibodies appear to be malaria specific in that they inhibit TNF induction by a variety of *P. falciparum* isolates but fail to inhibit TNF induction by bacterial lipopolysaccharide or lipoteichoic acid. The inhibitory antibodies bind to liposomes containing phosphatidylinositol but not other phospholipids. Serum from a European adult infected with *P. vivax* also inhibits the activity of toxins derived from *P. falciparum*–infected erythrocytes, and this too is mediated by IgM antibodies which are malaria specific and bind to phosphatidylinositol liposomes (10).

T cells from most adult nonexposed donors, which express a memory phenotype (CD45RO+), can respond by proliferation to *P. falciparum* asexual stages in vitro (11). Such cells may have arisen from exposure to environmental organisms. Unfractionated PBMC from both malaria-exposed and nonexposed donors inhibit parasite growth by up to 62%. Purified T cells in the presence of adherent cells exert a similar effect, but purified T cells alone or adherent cells alone have a minimal effect. Antigens released at the time of schizont rupture are maximally effective in stimulating IFN-γ production. Neutralizing antibodies to IFN-γ show a partial reduction of growth inhibition in some individuals tested suggesting that different mechanisms may be operative. Neutralizing antibody to TNF has a partial effect in combination with anti–IFN-γ. Antibodies to IL-1 and IL-4 have no effect. Hence, T-cell fractionation experiments show that while purified CD4+ T cells from some donors produced IFN-γ and inhibit parasite growth, purified CD8+ T cells can inhibit parasite growth to a greater extent without production of detectable IFN-γ.

Characterized antigens have also been found to cause TNF release. For example, *P. falciparum* FCR3 ring-erythrocyte surface antigen (RESA)–defective strain induces lower levels of TNF after schizont rupture than the SGE1 strain of *P. falciparum* (12). Although previous reports have described TNF-inducing activity of phospholipid from *P. falciparum,* these findings strengthen the evidence for Pf155/RESA antigens also being involved in TNF production during malaria. T- and B-cell responses to the *P. falciparum* blood-stage antigen Pf155/RESA have been investigated in a cohort of Cameroonian women (13). A proliferative response is induced in 25% of the donors by Pf155/RESA and in 7–11% by any peptide. Cytokine release occurs in 23–30% of the Pf155/RESA-stimulated cultures and in 8–25% of the peptide-stimulated cultures. Overall, each peptide induces a cellular response (proliferation and/or cytokine release) in 44% of the donors. T cells from 23% of the donors fail to respond to any peptide. Responding cells do not usually respond in all readouts, and proliferation and release of any of the three cytokines do not correlate. Similarly, antibody and T-cell responses are not related. Selected epitopes of Pf155/RESA, an important vaccine candidate, are well recognized in naturally exposed individuals and are able to activate T cells to proliferate and to produce various lymphokines in numerous individuals from a malaria-endemic area.

Cytokine release can also be triggered by malarial antigens from nonerythrocytic stages. Cellular and humoral responses to the circumsporozoite (CS) protein of *P. falciparum,* a prime vaccine candidate antigen, have been studied in individuals naturally exposed to malaria in a highly endemic region of Papua New Guinea (14). A paucity of any CD4+ lympoproliferative response to this protein by Papua New Guineans parallels a paucity of CD8+ T-cell response and contrasts markedly with the responses of other endemic populations. This study also discloses an inverse relationship between IFN-γ production and IL-6 production and a positive correlation between IFN-γ production and CS peptide-specific lymphoproliferation. High levels of peptide-specific IL-6 production are associated with high levels of peptide-specific serum antibodies. These data provide evidence for the limited activation of distinct CD4+ T-cell subsets and for the existence of functionally distinct subpopulations of human CD4+ T cells with respect to cytokines known to be important in sporozoite immunity. In spite of this cytokine-inducing capacity, vaccination with *P. falciparum* sporozoites does not lead to fluctuations in serum levels of TNF, IFN-γ, or IL-6 (15). Three volunteers out of the four vaccinees were nonetheless protected against infection after subsequent mosquito bites.

A longitudinal study of cellular and serological responses to the major merozoite surface protein of *P. falciparum* (PfMSP1) has been conducted in a malaria immune population living in The Gambia, where malaria transmission is seasonally endemic (16). Recombinant or native proteins representing the sequence of PfMSP1 from the Wellcome strain of *P. falciparum* have been used in in vitro lymphocyte proliferation, cytokine, and antibody assays. Cellular responses of individual donors fluctuate over time, independent of seasonal changes in malaria transmission, whereas anti-PfMSP1 antibody levels are remarkably stable. At a population level, IFN-γ responses are both more prevalent and of greater magnitude at the end of the rainy (malaria transmission) season than during the dry season. Malaria infections are more likely to be symptomatic in urban dwellers than in inhabitants of rural villages, but there is no significant difference in the level or prevalence of cellular or serological responses between the two groups.

Parasitized erythrocytes, and in particular some exo-antigens, are potent cytokine inducers (8). Taverne, Playfair, and colleagues have extensively studied malarial exo-antigens and have shown their ability to stimulate the production of cytokines, mainly TNF, in vitro (17,18) and in vivo (19). These exo-antigens are highly immunogenic and induce the production of T-cell–independent antibodies (20). The capacity of exo-antigens to stimulate macrophages to produce TNF is associated with a phospholipid (21). It has also been noted that lysed erythrocytes release a specific TNF-inducing factor that may contribute to the fever response to noninfectious tissue injury (22). The TNF-inducing activity per cell is about 200 times higher in malaria-parasitized erythrocytes than in uninfected erythrocytes.

The development of antidisease immunity in children infected with *P. falciparum* is thought to be related to their immunological responses to certain soluble parasite-derived exo-antigens (23). Cellular and humoral responses to these antigens have been assessed in a cross-sectional study of a cohort of Gabonese schoolchildren who live in an area where malaria is holoendemic and perennially transmitted. Concurrent parasitemia has a significant influence on lymphoproliferative and antibody responses to the exo-antigens. Cytokine responses are unaffected by either the presence or the intensity of parasitemia and are dissociated from both proliferative and antibody response to the exo-antigens. However, the mitogen-stimulated production of TNF and IL-6 positively correlates with the corresponding lymphoproliferative responses. At the individual level, mitogen-stimulated TNF, IFN-γ, IL-2, and IL-6 responses correlate, as well as mitogen- and exo-antigen–induced TNF.

Recently it has been shown that glycosylphosphatidylinositols are a novel class of glycolipid toxin produced by the parasite, which substitute for the endogenous inositolglycan-based signal transduction pathways of the host (24). Glycosylphosphatidylinositol stimulates high levels of TNF and IL-1 production by macrophages and induces hypoglycemia through an insulin-mimetic activity, and may therefore contribute to the cerebral syndrome and other malarial pathophysiology. The authors also demonstrate that monoclonal antibodies to parasite-derived glycosylphosphatidylinositol can neutralize the toxic activities of whole parasite extracts. These findings suggest a central role for glycosylphosphatidylinositol of parasite origin in the etiology of severe malaria and suggest novel approaches for the immunotherapy or immunoprophylaxis of disease.

Malaria Parasite DNA

Interestingly, apart from the numerous immunogenic proteins or lipids, DNA from malaria parasites can also induce immunological responses. In a study of humoral immunity, the number, type, and location of cytokine- and Ab-secreting cells activated in mice immunized and boosted with plasmid DNA encoding the circumsporozoite protein of the malarial parasite *P. yoelii* (PyCSP) have been monitored (25). The initial humoral response, characterized by production of IgG1 anti-PyCSP Abs and the Th-2 cytokine IL-4, localizes to the draining lymph nodes. Conversely, the secondary response is dominated by IFN-γ production and secretion of IgG2a anti-PyCSP Abs in the spleen. PyCSP DNA and mRNA are detected only in the quadriceps muscles (sites of plasmid injection), yet these sites lack either cytokine- or Ab-secreting cells. These findings indicate that circulating lymphocytes encounter plasmid-encoded Ag in the muscle bed, initiate a humoral response in the draining lymph nodes, and then seed distal lymphoid organs. Profound differences observed between the primary and secondary immune responses induced by plasmid immunization may influence vaccine efficacy.

Utilization of Cytokine Receptor by the Parasite

Among the numerous types of relationship between infectious agents and host molecule receptors, an interesting example is the interaction recently described between plasmodia and the IL-8 receptor. *P. vivax* and *P. falciparum* are the major causes of human malaria, except in sub-Saharan Africa where people lack the Duffy blood group antigen. Duffy-negative human erythrocytes are resistant to invasion by *P. vivax* and the related monkey malaria, *P. knowlesi*. The Duffy blood group antigen appears to be the erythrocyte receptor for the chemokines IL-8 and melanoma growth stimulatory activity (MGSA). While IL-8 binds minimally to Duffy-negative erythrocytes, a monoclonal antibody to the Duffy blood group antigen blocks binding of IL-8 and other chemokines to Duffy-positive erythrocytes. In addition, both MGSA and IL-8 block the binding of the parasite ligand and the invasion of human erythrocytes by *P. knowlesi*, suggesting the possibility of receptor blockade for antimalarial therapy (26) (for review see Refs. 27, 28). Alanine scanning mutagenesis of the charged amino acids of MGSA was used to identify specific residues that are involved in binding to the human erythrocyte Duffy antigen/chemokine receptor (DARC) and to the type B interleukin-8 receptor (IL-8RB) on neutrophils (29). An alanine scan MGSA mutant, E6A, binds to human erythrocytes and is able to inhibit malaria invasion as efficiently as wild-type MGSA but has a severely reduced ability to bind to or signal through the IL-8RB. Mutant chemokines like E6A could prove to be useful therapeutically for the design of receptor blocking drugs that inhibit erythrocyte invasion by *P. vivax* malaria.

Cytokines as Effectors of Protection Against Liver and Blood Stages of Malaria Parasites

The protective effect of TNF on erythrocytic stages of malaria parasites has been previously described, notably by Taverne et al. (30) and Clark et al. (31), who documented reduced parasite counts in the blood of animals injected with exogenous TNF. Similarly, TNF infusion suppresses the rise in parasitemia in PbA-infected mice when given before it induces cerebral signs and acute death in CM-resistant mice (32). The killing effect of TNF on malaria parasites is indirect and involves neutrophil and macrophage activation (33). However, as discussed below, this beneficial effect of TNF during malaria is not restricted to the erythrocytic stages.

A number of cytokines interfere with the plasmodium during its hepatic course, both directly and/or in the context of complex cytokine interactions, with the participation of various cells including the hepatocyte itself. The role of IFN-γ during this stage is suggested by studies showing that several interferon inducers protect mice against *P. berghei* infection. This protection is far greater against sporozoite-induced infections than against blood form–induced infection (34). Administration of mouse or rat IFN-γ a few hours before sporozoite challenge inhibits, in a dose-dependent manner, the development of exo-erythrocytic forms in the liver of mice and rats infected with *P. berghei* sporozoites. These observations suggest that immunologically induced IFN-γ could be involved in controlling malaria infection under natural conditions. The mechanisms of action of IFN-γ have been analyzed in vitro: IFN-γ impairs the intrahepatocytic development of sporozoites in human hepatoma cell line HepG2 (35) or in fresh human hepatocytes (36). In contrast, experiments using fresh hepatocytes do not provide evidence for an effect of TNF at the hepatocyte level, whatever the dose and the schedule of administration and whatever the strain of mice. In vivo administration of recombinant mouse TNF nonetheless prevents the appearance of *P. yoelii* erythrocytic stages. The fact that TNF induces different acute phase proteins in tumor cell lines and in normal rat hepatocytes (37) might explain the differences observed in these two experimental models. Indeed, acute phase proteins secreted after cytokine stimulation have been shown to inhibit parasite development in vitro and in vivo (38,39). A crucial element in the effect of TNF is the presence of nonparenchymal cells (NPC), consisting mainly of pit cells, Kupffer cells, endothelial cells, T cells, epithelial cells, fibroblasts, and Ito cells. Some of these cells are known to produce one or several of the following cytokines: IL-1, IFN-γ, and IL-6 (40,41). The addition of NPCs to hepatocytes restores the capacity of TNF to modulate parasite hepatic stage development in vitro. Markedly elevated IL-6 concentrations are present in co-culture supernatants upon addition of TNF, whereas IL-1 and IFN-γ remain undetectable. Anti–IL-6 mAb dramatically decreases the TNF-induced inhibition when added 24 h before sporozoite inoculation. These results strongly suggest that IL-6 is a crucial mediator in the observed TNF effect (42).

Two mechanisms seem to be involved: first, the induction of an oxidative burst, because catalase, a scavenger of H_2O_2 or with superoxide dismutase a scavenger of O_2^-, reverses the IL-6–mediated inhibitory activity and, second, the induction of the L-arginine–dependent cytotoxic effector pathway. Kupffer and endothelial cells are able, when co-cultured with hepatocytes, to block protein synthesis through conversion of L-arginine into nitrogen oxides in response to IFN-γ and/or LPS (43). The total reversion of IL-6 activity on *P. yoelii* intrahepatic development by two specific inhibitors of the L-arginine–dependent cytotoxic effector pathway or O_2^--generating system indicates the intricate involvement of the two mechanisms (44). Another study in rats has shown that a fish oil–supplemented diet reduces the production capacity of IL-1 and IL-6 and inhibits schizont development after intravenous inoculation of *P. berghei* sporozoites without influencing nitric oxide production by peritoneal macrophages (45).

Recombinant human or mouse interleukin-1 alpha (IL-1) also protect against the preerythrocytic stages of *P. falciparum* malaria (36). It has been postulated that this effect of IL-1 may be mediated in part by the C-reactive protein (CRP), an acute phase protein synthesized by IL-1–stimulated hepatocytes. Evidence for this has been obtained with both in vitro and in vivo models of *P. yoelii* infection (38). It is clear, however, that CRP by itself cannot account for all the effects of IL-1 or more generally, for the pleiomorphic manifestations of the inflammatory syndrome. Indeed, upon analysis of the respective roles played by hepatocytes and/or rat serum with increased levels of acute phase proteins, it appears that inflammatory factors other than CRP are equally involved (39). It has been demonstrated that hepatic IL-1 activity is mediated by the secretion of IL-6 (46). Evidence has been obtained that the effect of IL-1 on the hepatic development of *P. falciparum* or *P. yoelii* sporozoites is in part mediated by IL-6, since anti–IL-6 mAbs partially neutralized the effect of IL-1 (47). Various types of experiments have been performed to analyze the mode of action of IL-6 (42). Addition of recombinant IL-6 to *P. yoelli*-infected hepatocyte cultures results in a specific dose-dependent inhibition of parasite development. Time course experiments show that, without any effect on free sporozoites, IL-6 exerts its action during both the early phase of infection and during the subsequent maturation of the schizonts.

The production of various cytokines, and particularly IFN-γ, TNF, and IL-6, is increased during blood-stage malaria and might thus modulate the preerythrocytic stage of the infection (48–52), since one should bear in mind that people in endemic areas are regularly exposed to mosquito bites and hence to the hepatic phases. Taken together, these data further illustrate the interdependence of different stages of development of malaria parasites.

Cytokines and Sexual Stages

The relationship between cytokine release and gametocyte infectivity was not studied until recently. Particular attention has been paid to the period of transient loss of infectivity of the parasite to the mosquitoes. Infectivity of malarial infections to mosquitoes is due to the presence of circulating gametocytes, the sexual stages that fertilize in the midgut of a blood-fed mosquito. In several malaria host-parasite systems, it has been noted that peaks of parasitemia are associated with "crisis," characterized by the appearance of morphologically abnormal intraerythrocytic parasites. This peculiar event is synchronous with the reduction or loss or infectivity of the parasite to the mosquito, which persists for 4–6 days. Neither of these phenomena has been explained.

In the model of *P. cynomolgi* infection in its natural host, the toque monkey, *Macaca sinica,* it has been shown that the loss of infectivity at crisis is caused by the death of circulating intraerythrocytic gametocytes mediated by crisis serum (53). In addition, this killing effect in the crisis serum is due to the presence of both TNF and IFN-γ. Indeed, simultaneous addition of rabbit antihuman TNF and antihuman IFN-γ Igs completely reverses the killing effect of the crisis serum observed in vitro. In contrast, the addition of antibodies to either cytokine alone or normal rabbit Igs has no effect. Neither cytokine when tested individually, nor a combination of both cytokines, affects the infectivity of the gametocytes to mosquitoes. However, when either cytokine is added to crisis serum previously depleted of TNF and IFN-γ, the killing effect of the crisis serum is totally restored. These results indicate that TNF and IFN-γ each mediate gametocyte killing independently of the other, but they demonstrate that each cytokine requires the presence of an unknown factor in the crisis serum in order to mediate this effect.

Recently, further evidence for a role of TNF in gametocyte killing has been obtained. When studying *P. vivax*–infected subjects, it has been found that soon after a paroxysm (an episode of fever with chills and rigors, which coincides with the rupture of erythrocytic schizonts) the serum of malaria nonimmune patients mediates killing of blood parasites (asexual erythrocytic stages and gametocytes). Sera taken 1 h after paroxysm contain elevated TNF concentrations. This in vitro killing effect of human paroxysm sera is completely abrogated by the addition of anti-TNF antibodies and is restored by the addition of recombinant human TNF. When added to normal serum, however, recombinant TNF had no parasite killing effect, indicating, as found for monkey crisis serum, that a complementary factor present in paroxysm serum, but not in normal serum, is necessary for TNF to kill blood-stage parasites. The elevated serum TNF levels and gametocyte-killing effects of human serum taken 1 h after paroxysm are transient: 4 h following a paroxysm, serum TNF levels have fallen in most patients and the loss of killing effect is dramatic in all patients. Even when recombinant TNF is added to serum taken 4 h after a paroxysm, it fails to kill gametocytes demonstrating the very transient presence of complementary factor(s) needed for cytokine-mediated killing of blood-stage parasites (54). The observation of parasite-killing effects in human paroxysm serum has been done in nonimmune individuals with no previous history of malaria having contracted the disease on a visit to an endemic region in Sri Lanka. These individuals have no clinical immunity and consequently suffer severe clinical symptoms. In contrast, semi-immune individuals resident in a malaria-endemic region of Sri Lanka, who experience relatively mild symptoms during an infection, have low serum TNF levels and hardly any parasite-killing effect in their serum.

Thus, in infection crisis in both monkey and human malaria, the presence of parasite-killing factors coordinates closely with the period of clinical symptoms. These observations corroborate the view that cytokines including the complementary factor(s) may mediate the clinical symptoms of malaria (55).

CEREBRAL MALARIA PATHOGENESIS: IMPORTANCE OF CYTOKINES AND MECHANISMS OF CYTOKINE-MEDIATED PATHOLOGY

Role of TNF in Cerebral Malaria

A role for endotoxin-induced, macrophage-derived mediators in the pathogenesis of severe malaria was suggested in

1948 by Maegraith (56). Since then, several lines of evidence have indicated that TNF plays a central part in the pathogenesis of murine CM. This contention is supported by the following evidence (32,57). First, serum TNF levels are markedly elevated only at the time of neurological syndrome. For instance, the depletion in CD4+ T cells, which prevents CM, also inhibits the rise in serum TNF levels. Second, passive immunization of *P. berghei*–infected mice of a CM-susceptible strain significantly protects them against CM and the associated early mortality. The anti-TNF antibody treatment prevents all pathological abnormalities including hemorrhages and the focal arrest of monocytes and other circulating leukocytes within brain capillaries and venules. In addition, the conspicuous macrophage accumulation in lymphoid organs is also suppressed by anti-TNF antibody treatment, suggesting that amplification loops of TNF on its own synthesis and on macrophage recruitment are interrupted by the antibody treatment, which, however, has no effect on the course of infection itself. In the same murine model, inhibiting TNF by pentoxifylline has also been reported to prevent CM (58). Third, administration of recombinant murine TNF to *P. berghei*–infected mice of a CM-resistant strain induces a lethal neurological syndrome. Interestingly, TNF infusion prevents the rise in parasitemia, suggesting a role for this cytokine in protection against malaria (see above). In addition, we have shown increased TNF mRNA accumulation in several organs during malaria infection. TNF mRNA accumulation appears to be significantly higher in brains from mice with CM than in infected mice without CM (59). More recently, it has been shown that transgenic mice overexpressing soluble TNF receptor are also protected against CM, strengthening the concept that TNF is a cardinal molecule in CM pathogenesis in the rodent *P. berghei* ANKA model (60).

There is now a growing amount of evidence suggesting that in human CM, TNF may also be playing a pathogenic role. As described in some detail below, TNF overproduction has been found to correlate with disease severity in *P. falciparum*–infected patients. Moreover, TNF production was localized to mononuclear cells trapped in brain capillaries, as shown by immunohistochemistry performed on brain autopsy specimens (61). Interestingly, a monoclonal antibody directed against TNF is able to reduce fever in *P. falciparum*–infected children (62). However, it is too early yet to know whether this treatment will have any significant effect on morbidity or mortality of children with severe malaria. A crucial issue will be the time of antibody administration: several hours with coma are likely to have elapsed before the treatment can be initiated. It is obvious that the later the anti-TNF antibody is injected, the lower the chances are that TNF-induced lesions can be prevented or reversed. Indeed, in the experimental model, when the anti-TNF antibody was injected in mice that had developed coma for several hours, there was no protection against CM-induced lethality.

Mechanisms of Excessive TNF Secretion: Importance of the Immune Response

The first indication that CM is the result of T-cell–dependent immunopathological reactions came from the observation that athymic nude mice do not develop this neurological syndrome, although their susceptibility to *P. berghei* ANKA infection is identical to that of euthymic mice (63). Upon further analysis of T cells in CM, we demonstrated that CD4+ T lymphocytes play a significant role in the development of murine cerebral malaria (64). Along this line, it has been shown that a murine leukemia virus, which induces murine AIDS, reduces the severity of neurological complications in *P. berghei* ANKA-infected susceptible mice (65). Protection against CM appears to involve IL-10 and correlates with the severity of immunodeficiency.

Since there is a need for T cells in the development of CM and since the main source of TNF is the monocyte/macrophage, the link between these two cell types has been investigated. In vivo experiments involving the injection of various anticytokine antibodies have led to the hypothesis that IL-3, GM-CSF, and IFN-γ act in a cascade leading to TNF overproduction and the ensuing neurovascular damage (66,67). On the one hand, IL-3 and GM-CSF act by enlarging the pool of mononuclear phagocytes, which is manifested by the accumulation of these cells in lymphoid organs, and, on the other hand, IFN-γ is responsible for their activation, further stimulating the synthesis and release of excessive amounts of TNF.

Consequences of TNF Overproduction: Endothelial Activation

While the cytotoxic effects of TNF on the vascular endothelium may be direct, they are more likely to be indirect via the upregulation of cell adhesion molecules such as intercellular adhesion molecule 1 (ICAM-1) (68) on endothelial cells, which in turn mediates cell trapping in brain microvessels and further endothelial cell damage (69). Pathogenic mechanisms of brain microvascular injury have been studied in vivo using the *P. berghei* ANKA model. Brain lesions, leading to perivascular microhemorrhages, are associated with the sequestration of macrophages and parasitized erythrocytes in cerebral venules, but appear to involve platelets as critical effectors (69). Electron microscopy indicates that during CM platelets adhere to and probably damage brain endothelial cells. Radiolabeled platelet distribution studies show that platelets sequester in the brain and lung vasculature during CM. Conversely, noncerebral malaria such as that induced by infection with *P. yoelii* is not associated with cerebral sequestration of platelets. In vivo treatment with a monoclonal antibody to LFA-1 (found expressed on platelets) selectively abrogates the cerebral sequestration of platelets, and this correlates with prevention of CM (70,71). Moreover, malaria-infected animals rendered thrombocytopenic are protected against CM. Thus, an interaction between LFA-1 expressed on platelets and its ligand, ICAM-1, upregulated on endothelial cells appears pivotal to microvascular damage. These data suggest a novel mechanism of action for anti–LFA-1 mAb in vivo and illustrate an unexpected role of platelets, in addition to monocytes, in vascular pathology. Platelets have also been reported to adhere to falciparum-infected erythrocytes and to hold them together (72). An indication that adhesion molecules could likewise be involved in human cerebral malaria was provided by Berendt and colleagues, who demonstrated that ICAM-1 is one of the molecules involved in cytoadherence of *P. falciparum*–infected erythrocytes, at least in vitro (73). Since TNF, among other cytokines, is known to upregulate the surface expression of ICAM-1 on endothelial cells, it is likely that the markedly high TNF levels observed in patients with CM (50,51) lead to sequestration of parasitized erythrocytes via this pathway. Upregulation of ICAM-1 and other cell adhesion molecules has indeed been found in brain microvessels from malaria-infected patients (61,74).

Ferrante and colleagues have recently described a peptide corresponding to residues 70–80 of the TNF molecule that can differentially affect host cell responses, including endothelial cell adhesiveness (75). Interestingly, their peptide, unlike TNF, does not increase the expression of adhesion molecules on endothelial cells and fails to promote binding of *P. falciparum*–infected erythrocytes to endothelial cells. TNF (70–80) also inhibits the TNF-induced increase in adhesion of *P. falciparum*–infected erythrocytes to endothelial cells. The peptide nonetheless enhances human PMN-mediated killing of *P. falciparum* in vitro and reduces the *P. chabaudi* parasitemia in mice. Thus, the host-protective effects of TNF can be retained while toxic effects are eliminated using a selected, characterized subunit of the cytokine.

Nitric Oxide

The pathogenic role of other intermediates has also been studied in the mouse CM model. Because TNF can enhance NO production, the generation of this mediator has been considered as one of the pathways potentially involved in TNF-induced cell damage. Cells able to release NO intermediates comprise macrophages and endothelial cells, two cell types central to the lesion of CM. While the pathogenic role of NO remains unclear, it has obvious protective roles. At least seven pieces of evidence support this contention:

1. NO acts as a relaxing factor.
2. NO is able to downregulate cell adhesion molecules.
3. NO can reduce leukocyte-endothelial adherence (76).
4. NO can scavenge free oxygen radicals (77).
5. NO inhibition can cause endothelial alterations, such as albumin leakage (78) and platelet deposition (79).
6. NO can reduce platelet aggregation and adhesion to endothelium (78,80,81).
7. Unlike TNF blockade, efficient in vivo NO blockade, either in experimental LPS shock (82; P. Brouckaert, personal communication) or in experimental cerebral malaria (83–85), is not able to protect against pathology; it rather aggravates it (86).

Thus it is conceivable that NO mediates early changes of cerebral malaria, such as neurotransmission disturbances, when the neurological syndrome is still reversible (although L-NMMA treatment in comatose mice fails to wake them up), but this mediator is not likely to be involved in the actual process of neurovascular damage. It has recently been shown that human red blood cells infected by *P. falciparum* synthesize NO. The synthesis of NO appears to be very high in comparison with human endothelial cells; no citrulline and nitrate production is detectable in noninfected red blood cells. The NO synthase activity was very high in the lysate of infected red blood cells (while not measurable in that of normal red blood cells), suggesting that increased NO synthesis in *P. falciparum* malaria can be directly elicited by soluble factor(s) released by the blood stages of the parasite, without necessarily requiring the intervention of host cytokines.

Cytokines might yet be involved in other aspects of malarial pathology, even if they do not exert direct toxic effects. IL-6 is one such example. Serum IL-6 levels are increased in mice with *P. berghei* ANKA infection, the highest concentrations being observed in mice with full-blown neurological syndrome (52). High IL-6 levels are also observed, however, in the absence of pathology in nonlethal malaria infection. Even though IL-6 is produced in large amounts during malaria infection, it does not play a major role in the pathogenesis of CM. Indeed, passive immunization against IL-6 does not prevent CM, but significantly reduces serum IgG levels in malaria-infected mice. Thus, IL-6 is possibly involved in malarial-induced hypergammaglobulinemia and the associated complications, such as immune-complex diseases and glomerulonephritis. In addition, overexpression of IL-6 might contribute to Burkitt's lymphomas, which have been associated with plasmodium infections in epidemiological studies. The significance of IL-6 production in malaria may thus go beyond protection during the liver stages (52).

CYTOKINES AS MARKERS OF DISEASE ACTIVITY OR GENETIC SUSCEPTIBILITY

Serum Cytokine Levels in Malaria Patients

The significance of TNF in the pathology of human CM has been evaluated by measuring serum concentrations of this cytokine in Malawian children with severe *P. falciparum* malaria. In spite of considerable overlap, the mean admission serum level of TNF is significantly higher in the patients who died than in survivors. The mortality rate increases with the concentration of TNF, and the highest TNF concentrations are found in the serum of patients with profound coma, the varying degrees of neurological involvement, being evaluated by a modified Glasgow coma scale (50). High serum TNF concentrations correlate with hyperparasitemia, young age, and severity of illness. In contrast, TNF levels are not significant elevated in cerebrospinal fluid of patients with CM. Thus, TNF may be an important component in the pathogenesis of severe falciparum malaria, and in particular in the cerebral syndrome and the hypoglycemia that can complicate this disease (50). An association between high serum TNF levels and severity of falciparum infection has also been noted by other groups (51,88–90).

Other biological parameters have been studied in Malawian patients with mild or complicated malaria. Mild is defined as normal neurological status (coma score = 5), whereas complicated includes all patients with a coma score of 4 or less. In patients with neurological involvement, plasma levels of TNF, IL-6, and lactate are significantly higher. Blood acidosis, the degree of hepatomegaly, splenomegaly, and the duration of fever are significantly more important in patients with complicated than in patients with mild disease. In contrast, there is no difference in the level of parasitemia, IL-1β, or IFN-γ. The high degree of overlap in admission levels of TNF, i.e., the discrepancy in many patients between TNF and prognosis, has led to measuring TNF inhibitors in these patients (91,92). During CM, plasma concentrations of soluble receptors (sTNF-R1 and STNF-R2), which act as binding proteins for TNF, are greatly increased when compared with levels at convalescence and in controls without malaria. Concentrations of both receptors correlate with plasma levels of TNF measured by immunoradiometric assay, but not with those of IL-6. The mean plasma concentrations of both immunoreactive TNF and soluble TNF receptors are greater in patients with CM than in those with uncomplicated malaria. Despite high levels of immunoreactive TNF in the plasma of patients acutely ill with malaria, no bioactive TNF is detectable in these patients. Soluble TNF receptors are present in greatly increased concentrations in the plasma of patients with malaria and may play a role in mediating or modulating the pathogenic effects of the cytokine.

The discrepancy between results obtained when different assays are used to test for cytokines has been highlighted by a collaborative study involving several international research groups, conducted in order to test the validity and reproducibility of TNF measurements in serum (93). Twenty serum samples obtained from cerebral malaria patients have been assayed for TNF using various immunoassays, radioimmunoassays, and bioassays and assessed by 11 different laboratories. Substantial differences were observed in terms of sensitivity and values obtained with the different commercial and in-house assays, emphasizing the necessity of establishing international standards for cytokine assays to render the interpretation of biological and medical data more reliable.

Other studies have found proinflammatory cytokines such as TNF or IFN-γ, GM-CSF, IL-2, and IL-8 in circulating blood during malaria (94–96), but also anti-inflammatory molecules, such as IL-10, a monocyte/lymphocyte-derived cytokine which inhibits certain cellular immune responses such as delayed-type hypersensitivity. In particular, the production of TNF, IL-1, and IL-6 is strongly inhibited by IL-10. Fourteen cerebral, 11 severe, and 20 mild malaria cases have been found to have mean IL-10 levels of 2812, 2882, and 913 pg/ml, respectively, while 98% of healthy individuals had undetectable (less than 100 pg/ml) circulating IL-10 (97). Thirteen of the 25 cerebral/severe cases have 2000 pg/ml. In 11 hospitalized patients, circulating IL-10 levels return to virtually normal levels 7 days after antimalarial chemotherapy when biological and clinical features disappear. In another study in Thailand, markedly elevated serum levels of IL-10 and IFN-γ have been found in patients with acute *P. falciparum* malaria prior to treatment (98). Serum levels for both cytokines decrease significantly during antimalarial chemotherapy, returning to normal values 2–3 weeks after treatment initiation. The relation between the immune response and the clinical features of severe falciparum malaria has been studied in Burundian adults with or without cerebral involvement (99). At the time of admission, none of the immunological variables investigated (TNF, IFN-γ, IL-10, soluble ICAM-1, and soluble ELAM-1) appears to be specific to cerebral malaria or to a fatal outcome. It remains to be determined, however, whether elevated levels of IL-10 during infection play a beneficial role by reducing the parasite-induced inflammatory response or a detrimental one by decreasing the cellular immune responses.

In another study, while both IL-10 and IFN-γ are produced regardless of the falciparum malaria patients' clinical severity, circulating IFN-γ and IL-10 are significantly higher in patients with severe disease. In vitro PBMC stimulation by malarial antigens during acute infection shows that although there is no lymphoproliferation, the cells can produce IL-10 and IFN-γ. Recombinant human IL-10 completely abolishes in vitro TNF production in response to malarial antigens, as well as the antigen-specific proliferative response of convalescent patients. However, anti–IL-10 mAb is insufficient to restore proliferation of PBMC from acutely infected patients. IL-10 may thus have an important negative feedback action on the production of inflammatory cytokines in acute falciparum malaria without contributing to the defect in antigen-specific proliferation (100).

Increased concentrations of IL-6, soluble IL-6 receptor, IL-1 receptor antagonist, and some IgM antibodies against phosphatidylinositol are found in Gambian children with cerebral malaria, but those who die have decreased concentrations of β_2-glycoprotein I (phosphatidylinositol-binding serum protein I) (101). The authors conclude that increased concentrations of cytokines and soluble cytokine receptors represent a normal host response to *P. falciparum* infections but that excessive secretion of cytokines like IL-6 may predispose to cerebral malaria and a fatal outcome while β_2-glycoprotein I may protect against a fatal outcome of cerebral malaria.

The chemokine macrophage inflammatory protein 1 alpha (MIP-1α) and IL-8 have been investigated in patients with falciparum malaria. Maximal concentrations of MIP-1α and IL-8 are found on day 14, at a time when parasites are not detected in blood smears (102).

Cytokine Production Capacity

Apart from the quantitation of cytokines in serum or plasma, the determination of cytokine-production capacity may be helpful in understanding the modulation of host responses in malaria patients. This can be achieved either by in vitro cultures or by direct whole-blood stimulation test. Cultures of peripheral blood mononuclear cells have shown a higher malaria-specific IFN-γ production capacity in nonimmune individuals compared to that of immune subjects (103). Since it is known that nonimmune individuals, such as children in endemic areas or adult visitors from nonendemic areas, are particularly prone to CM, these data suggest that a "virgin" immune system is associated with a higher susceptibility to develop CM. The whole-blood stimulation test (by either mitogens or specific antigens) is simpler, faster, and therefore more suitable to large studies than the culture of blood mononuclear cells. In Gabonese patients with severe or mild, but no cerebral, malaria, it has been observed that an elevated TNF production capacity predicts a more rapid cure (104), but the kinetics of cytokine production capacity remain to be studied in patients with CM.

Genetic Studies

In a large case-control study in Gambian children, homozygotes for the TNF2 allele, a variant of the TNF gene promoter region, are seven times more likely to die or develop severe neurological sequelae due to cerebral malaria (105). Although the TNF2 allele is in linkage disequilibrium with several neighboring HLA alleles, this disease association is independent of HLA class I and class II variation. These data suggest that regulatory polymorphisms of cytokine genes can affect the outcome of severe infection. The maintenance of the TNF2 allele at a gene frequency of 0.16 in The Gambia implies that the increased risk of cerebral malaria in homozygotes is counterbalanced by some biological advantage.

CONCLUSIONS, NOVEL THERAPIES, AND FUTURE DIRECTIONS

The exceedingly complex phenomenon of malaria is an illustration of the duality of effects of cytokines, and particularly of TNF. As outlined in this chapter, a balance in the cytokine network is needed to achieve protective immunity while avoiding eventual deleterious consequences. The final outcome of malaria depends on the amount of cytokines released, their rate, and the time and site of production (Fig. 1). Since cytokines are harmful only when produced in inappropriately high amounts and during limited periods of time (as observed in human CM), they might represent a target for therapeutic intervention. As stated almost 10 years ago (106), new approaches in the treatment of severe falciparum malaria, and particular CM, are still needed. The main cytokine that should be neutralized, in view of its potent cytotoxic effects and the correlation between serum concentrations and severity in human CM (50,51,88–90), is TNF. Neutralization of this cytokine, its receptor, or the molecules triggered by it, when technically feasible, is one of the possible therapeutic

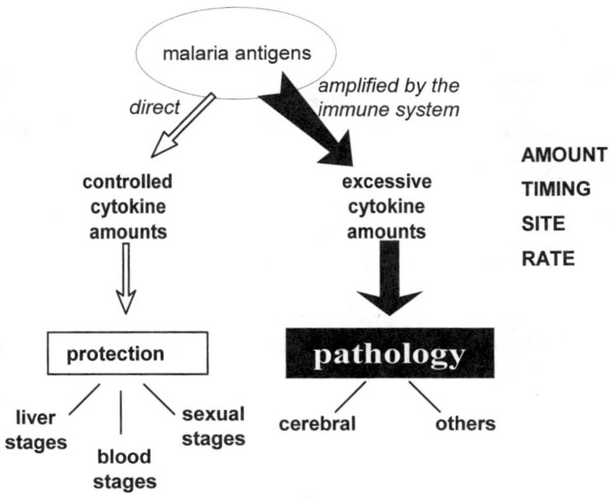

Figure 1 IL-8 plasma concentrations in 34 patients who survived (○) and 7 who subsequently died from tuberculosis (●). All points show median values (note log scale on vertical axis) ± interquartile ranges. Value at time = 0 h is plasma IL-8 concentration on admission to study. Subsequent time points indicate IL-8 secretion after ex vivo stimulation of whole blood by LPS at a final concentration of 10 μg/ml. IL-8 secretion increases with time in patients who survive and control subjects (data not shown), but not in patients who died from acute tuberculosis infection. Sample sizes are identical at all time points.

regimes to be envisaged not only in severe malaria but also in acute complications of various infectious diseases where TNF has been shown to play a pathogenic role. Administration of monoclonal anti-TNF antibodies is now considered in addition to conventional treatment of severe falciparum malaria. It is clear, however, that non–antibody inhibitors, such as IL-10 or soluble TNF receptors, should be tested in these patients.

The understanding of the mechanisms of cytokine-induced microvascular damage in CM will benefit from other models of CM: an elegant retinal whole mount system has been described (107,108), as well as cultures of brain microvascular endothelial cells (probably more relevant to neurovascular damage than umbilical vein endothelial cells (109)), which have been purified from mice (110), monkeys (111), and humans (112,113). Still in addition to these in vitro approaches, in vivo models in mice may still be useful to decipher new pathogenic mechanisms and test new therapies, as clearly stated recently (114).

ACKNOWLEDGMENTS

Our work was supported by grants from the Swiss National Research Foundation and from the UNDP/World Health Organization/TDR. GEG was supported by the Cloëtta Foundation, Zürich, Switzerland.

REFERENCES

1. Mosmann TR, Cherwinski H, Bond MW, Giedlin MA, Coffman RL. Two types of murine helper T cell clones. I. Definition according to profiles ot lymphokine activities and secreted proteins. J Immunol 1986; 136:2348–2357.
2. Taylor-Robinson AW, Phillips RS, Severn A, Moncada S, Liew FY. The role of Th1 and Th2 cells in a rodent malaria infection. Science 1993; 260:1931–1934.
3. Vonderweid T, Langhorne J. The roles of cytokines produced in the immune response to the erythrocytic stages of mouse malarias. Immunobiology 1993; 189:397–418.
4. Vonderweid T, Kopf M, Kohler G, Langhorne J. The immune response to *Plasmodium chabaudi* malaria in interleukin-4-deficient mice. Eur J Immunol 1994; 24:2285–2293.
5. Stevenson MM, Tam MF, Wolf SF, Sher A. IL-12-induced protection against blood-stage *Plasmodium chabaudi* AS requires IFN-gamma and TNF-alpha and occurs via a nitric oxide-dependent mechanism. J Immunol 1995; 155:2545–2556.
6. Lucas B, Smith K, Haque A. Plasmodium yoelii in mice: differential induction of cytokine gene expression during hyporesponsiveness induction and restimulation. Cell Immunol 1995; 160:79–90.
7. Golgi C. On the cycle of development of malarial parasites in tertian fever: differential diagnosis between the intracellular malarial parasites of tertian and quartan fever. Arch Sci Med 1889; 13:173–196.
8. Kwiatkowski D, Cannon JG, Manogue KR, Cerami A, Dinarello CA, Greenwood BM. Tumour necrosis factor production in falciparum malaria and its association with schizont rupture. Clin Exp Immunol 1989; 77:361–366.
9. Allan RJ, Beattie P, Bate C, Vanhensbroek MB, Morrisjones S, Greenwood BM, et al. Strain variation in tumor necrosis factor induction by parasites from children with acute falciparum malaria. Infect Immun 1995; 63:1173–1175.
10. Bate CA, Kwiatkowski D. Inhibitory immunoglobulin M antibodies to tumor necrosis factor-inducing toxins in patients with malaria. Infect Immun 1994; 62:3086–3091.
11. Fell AH, Currier J, Good MF. Inhibition of Plasmodium falciparum growth in vitro by CD4(+) and CD8(+) T cells from non-exposed donors. Parasite Immunol 1994; 16:579–586.
12. Picot S, Peyron F, Deloron P, Boudin C, Chumpitazi B, Barbe G, et al. Ring-infected erythrocyte surface antigen (Pf/155 RESA) induces tumour necrosis factor-alpha production. Clin Exp Immunol 1993; 93:184–188.
13. Fievet N, Maubert B, Cot M, Chougnet C, Dubois B, Bickii J, et al. Humoral and cellular immune responses to synthetic peptides from the *Plasmodium falciparum* blood-stage antigen, Pf155/RESA, in Cameroonian women. Clin Immunol Immunopathol 1995; 76:164–169.
14. Doolan DL, Beck HP, Good MF. Evidence for limited activation of distinct CD4+ T-cells subsets in response to the *Plasmodium falciparum* circumsporozoite protein in Papua New Guinea. Parasite Immunol 1994; 16:129–136.
15. Egan JE, Hoffman SL, Haynes JD, Sadoff JC, Schneider I, Grau GE, et al. Humoral immune responses in volunteers immunized with irradiated *Plasmodium falciparum* sporozoites. Am J Trop Med Hyg 1993; 49:166–173.
16. Riley EM, Morrisjones S, Blackman MJ, Greenwood BM, Holder AA. A longitudinal study of naturally acquired cellular

and humoral immune responses to a merozoite surface protein (MSP1) of *Plasmodium falciparum* in an area of seasonal malaria transmission. Parasite Immunol 1993; 15:513–524.
17. Bate CA, Taverne J, Playfair JH. Malarial parasites induce TNF production by macrophages. Immunology 1988; 64:227–231.
18. Taverne J, Bate CA, Sarkar DA, Meager A, Rook GAW, Playfair JH. Human and murine macrophages produce TNF in response to soluble antigens of *Plasmodium falciparum*. Parasite Immunol 1990; 12:33–43.
19. Bate CA, Taverne J, Playfair JH. Soluble material antigens are toxic and induce the production of tumour necrosis factor in vivo. Immunology 1989; 66:600–605.
20. Bate CA, Taverne J, Dave A, Playfair JH. Malaria exoantigens induce T-independent antibody that blocks their ability to induce TNF. Immunology 1990; 70:315–320.
21. Bate CAW, Taverne J, Roman E, Moreno C, Playfair JHL. Tumour necrosis factor induction by malaria exoantigens depends upon phospholipid. Immunology 1992; 75:129–135.
22. Bate CAW, Kwiatkowski DP. Stimulators of tumour necrosis factor production released by damaged erythrocytes. Immunology 1994; 83:256–261.
23. Luty AJF, Mayombo J, Lekoulou F, Mshana R. Immunologic responses to soluble exoantigens of *Plasmodium falciparum* in Gabonese children exposed to continuous intense infection. Am J Trop Med Hyg 1994; 51:720–729.
24. Schofield L, Vivas L, Hackett F, Gerold P, Schwarz RT, Tachado S. Neutralizing monoclonal antibodies to glycosylphosphatidylinositol, the dominant TNF-alpha-inducing toxin of *Plasmodium falciparum*—prospects for the immunotherapy of severe malaria. Ann Trop Med Parasitol 1993; 87:617–626.
25. Mor G, Klinman DM, Shapiro S, Hagiwara E, Sedegah M, Norman JA, et al. Complexity of the cytokine and antibody response elicited by immunizing mice with *Plasmodium yoelii* circumsporozoite protein plasmid DNA. J Immunol 1995; 155:2039–2046.
26. Horuk R, Chitnis CE, Darbonne WC, Colby TJ, Rybicki A, Hadley TJ, et al. A receptor for the malarial parasite *Plasmodium vivax*—the erythrocyte chemokine receptor. Science 1993; 261:1182–1184.
27. Ahuja SK, Gao JL, Murphy PM. Chemokine receptors and molecular mimicry. Immunol Today 1994; 15:281–287.
28. Horuk R. The interleukin-8-receptor family—from chemokines to malaria. Immunol Today 1994; 15:169–174.
29. Hesselgesser J, Chitnis CE, Miller LH, Yansura DG, Simmons LC, Fairbrother WJ, et al. A mutant of melanoma growth stimulating activity does not activate neutrophils but blocks erythrocyte invasion by malaria. J Biol Chem 1995; 270:11472–11476.
30. Taverne J, Tavernier J, Fiers W, Playfair JH. Recombinant tumor necrosis factor inhibits malaria parasites in vivo but not in vitro. Clin Exp Immunol 1987; 67:1–4.
31. Clark IA, Hunt NH, Butcher GA, Cowden WB. Inhibition of murine malaria (*Plasmodium chabaudi*) in vivo by recombinant interferon-gamma or tumor necrosis factor, and its enhancement by butylated hydroxyanisole. J Immunol 1987; 139:3493–3496.
32. Grau GE, Piguet PF, Vassalli P, Lambert PH. Tumor necrosis factor and other cytokines in cerebral malaria: experimental and clinical data. Immunol Rev 1989; 112:49–70.
33. Kumaratilake LM, Ferrante A. T-cell cytokines in malaria: their role in the regulation of neutrophil- and macrophage-mediated killing of *Plasmodium falciparum* asexual blood forms. Res Immunol 1994; 145:423–429.
34. Jahiel RI, Vilcek J, Nussenweig RS. Exogenous interferon protects mice against *Plasmodium berghei* malaria. Nature 1970; 227:1350–1351.
35. Schofield L, Ferreira A, Altszuler R, Nussenzweig V, Nussenzweig RS. Interferon-gamma inhibits the intrahepatocytic development of malaria parasites in vitro. J Immunol 1987; 139:2020–2025.
36. Mellouk S, Maheshwari RK, Rhodes-Feuillette A, Beaudoin RL, Berbiguier N, Matile H, et al. Inhibitory activity of interferons and interleukin 1 on the development of *Plasmodium falciparum* in human hepatocyte cultures. J Immunol 1987; 139:4192–4195.
37. Perlmutter DH, Dinarello CA, Punsel PI, Colten HR. Cachectin/Tumor necrosis factor regulates hepatic acute-phase gene expression. J Clin Invest 1986; 78:1378.
38. Pied S, Nussler A, Pontet M, Miltgen F, Matile H, Lambert PH, et al. C-reactive protein protects against pre-erythrocytic stages of malaria. Infect Immun 1989; 57:278–282.
39. Nussler A, Pied S, Pontet M, Miltgen F, Rénia L, Gentilini M, et al. Inflammatory status and the pre-erythrocytic stages of malaria. Role of the C-reactive protein. Exp Parasitol 1991; 72:1–7.
40. Lotze M, Jirik F, Kabouridis P, Tsoukas C, Hirano T, Kishimoto T, et al. B cell stimulating factor 2/interleukin 6 is a costimulant for human thymocytes and T lymphocytes. J Exp Med 1988; 167:1253.
41. Katz Y, Strunk RC. Similarities and differences in stimulation of expression of alternative pathway of complement and IFN-beta2/IL-6 genes in human fibroblasts. J Immunol 1989; 142:3862–3867.
42. Nussler A, Pied S, Goma J, Renia L, Miltgen F, Grau GE, et al. TNF inhibits malaria hepatic stages in vitro via synthesis of IL-6. Int Immunol 1991; 3:317–321.
43. Billiar TR, Curran RD, Stuehr DJ, West MA, Bentz BG, Simmons RL. An L-arginine-dependent mechanism mediate Kupffer cell inhibition of hepatocyte protein synthesis in vitro. J Exp Med 1989; 169:1467–1472.
44. Nussler A, Drapier JC, Renia L, Pied S, Miltgen F, Gentilini M, et al. L-Arginine-dependent destruction of intrahepatic malaria parasites in response to tumor necrosis factor and/or interleukin-6 stimulation. Eur J Immunol 1991; 21:227–230.
45. Vreden SG, Blok WL, Sauerwein RW, Oettinger MC, Verhave JP, Meuwissen JE, et al. Inhibition of *Plasmodium berghei* liver schizont development and reduction of cytokine production capacity in rats by dietary fish oil supplementation. Am J Trop Med Hyg 1995; 53:206–210.
46. Helle M, Brakenhoff JP, De Groot ER, Aarden LA. Interleukin 6 is involved in interleukin 1 induced activities. Eur J Immunol 1988; 18:957–959.
47. Pied S, Civas A, Berlotpicard F, Renia L, Miltgen F, Gentilini M, et al. IL-6 induced by IL-1 inhibits malaria pre-erythrocytic stages but its secretion is down-regulated by the parasite. J Immunol 1992; 148:197–201.
48. Rhodes Feuillette A, Jaureguiberry G, Ballet JJ, Andrieu B, Le Bras J, Galibert F, et al. The interferon compartment of the immune response in human malaria: I. Interferon inducers in *Plasmodium falciparum* cultures. J Interferon Res 1985; 5:159–168.
49. Lebbar S, Cavaillon JM, Caroff M, Ledur A, Brade H, Sarfati R, et al. Molecular requirement for interleukin 1 induction by lipoploysaccharide-stimulated human monocytes: involvement of the heptosyl-2-keto-3-deoxyoctulosonate region. Eur J Immunol 1986; 16:87–91.
50. Grau GE, Taylor TE, Molyneux ME, Wirima JJ, Vassalli P, Hommel M, et al. Tumor necrosis factor and disease severity

in children with falciparum malaria. N Engl J Med 1989; 320:1586–1591.

51. Kern P, Hemmer CJ, Vandamme J, Gruss HJ, Dietrich M. Elevated tumor necrosis factor-alpha and interleukin-6 serum levels as markers for complicated *Plasmodium falciparum* malaria. Am J Med 1989; 87:139–143.
52. Grau GE, Frei K, Piguet PF, Fontana A, Heremans H, Billiau A, et al. Interleukin 6 production in experimental cerebral malaria. Modulation by anti-cytokine antibodies and possible role in hypergammaglobulinemia. J Exp med 1990; 172:1505–1508.
53. Naotunne TD, Rathnayake KD, Jayasinghe A, Carter R, Mendis KN. Plasmodium-cynomolgi-serum-mediated blocking and enhancement of infectivity to mosquitoes during infections in the natural host, Macaca-Sinica. Exp Parasitol 1990; 71:305–313.
54. Mendis KN, Naotunne TD, Karunaweera ND, Del Giudice G, Grau GE, Carter R. Anti-parasite effects of cytokines in malaria. Immunol Lett 1990; 25:217–220.
55. Naotunne TD, Karunaweera ND, Del Giudice G, Kularatne MU, Grau GE, Carter R, et al. Cytokines kill malaria parasites during infection crisis—extracellular complementary factors are essential. J Exp Med 1991; 173:523–529.
56. Maegraith B. Pathological Processes in Malaria and Blackwater Fever. Oxford: Blackwell 1948.
57. Grau GE, Fajardo LF, Piguet PF, Allet B, Lambert PH, Vassalli P. Tumor necrosis factor (cachectin) as an essential mediator in murine cerebral malaria. Science 1987; 237:1210–1212.
58. Kremsner PG, Grundmann H, Neifer S, Sliwa K, Sahlmuller G, Hegenscheid B, et al. Pentoxifylline prevents murine cerebral malaria. J Infect Dis 1991; 164:605–608.
59. de Kossodo S, Grau GE. Profiles of cytokine production in relation with susceptibility to experimental cerebral malaria. J Immunol 1993; 151:4811–4820.
60. Garcia I, Araki K, Miyazaki Y, Araki M, Juillard P, Grau GE, et al. Transgenic mice expressing soluble TNF-R1-IgG3 molecules are protected from lethal septic shock and cerebral malaria, and are highly sensitive to *Listeria monocytogenes* and *Leishmania major* infections. Eur J Immunol 1995;
61. Porta J, Carota A, Pizzolata GP, Wildi E, Widmer MC, Margairaz C, et al. Immunopathological changes in human cerebral malaria. Clin Neuropathol 1993; 12:142–146.
62. Kwiatkowski D, Molyneux ME, Stephens S, Curtis N, Klein N, Pointaire P, et al. Response to monoclonal anti-TNF antibody in children with cerebral malaria. Q J Med 1993; 86:91–98.
63. Finley RW, Mackey LJ, Lambert PH. Virulent *P. berghei* malaria: prolonged survival and decreased cerebral pathology in T-cell-deficient nude mice. J Immunol 1982; 129:2213–2218.
64. Grau GE, Piguet PF, Engers HD, Louis JA, Vassalli P, Lambert PH. L3T4+ T lymphocytes play a major role in the pathogenesis of murine cerebral malaria. J Immunol 1986; 137:2348–2354.
65. Eckwalanga M, Marussig M, Tavares MD, Bouanga JC, Hulier E, Pavlovitch JH, et al. Murine AIDS protects mice against experimental cerebral malaria: downregulation by interleukin 10 of a T-helper type 1 CD4(+) cell-mediated pathology. Proc Natl Acad Sci USA 1994; 91:8097–8101.
66. Grau GE, Kindler V, Piguet PF, Lambert PH, Vassalli P. Prevention of experimental cerebral malaria by anticytokine antibodies. IL-3 and GM-CSF are intermediates in increased TNF production and macrophage accumulation. J Exp Med 1988; 168:1499–1504.
67. Grau GE, Heremans H, Piguet PF, Pointaire P, Lambert PH, Billiau A, et al. Monoclonal antibody against interferon-gamma can prevent experimental cerebral malaria and its associated overproduction of tumor necrosis factor. Proc Natl Acad Sci USA 1989; 86:5572–5574.
68. Rothlein R, Czajkowski M, O Neill MM, Marlin SD, Mainolfi E, Merluzzi VJ. Induction of intercellular adhesion molecule 1 on primary and continuous cell lines by pro-inflammatory cytokines. J Immunol 1988; 141:1665–1669.
69. Grau GE, Tacchini-Cottier F, Vesin C, Milon G, Lou J, Piguet PF, et al. TNF-induced microvascular pathology: active role for platelets and importance of the LFA-1/ICAM-1 interaction. Eur Cytokine Netw 1993; 4:415–419.
70. Grau GE, Pointaire P, Piguet PF, Vesin C, Rosen H, Stamenkovic I, et al. Late administration of monoclonal antibody to leukocyte function-antigen 1 abrogates incipient murine cerebral malaria. Eur J Immunol 1991; 21:2265–2267.
71. Falanga PB, Butcher EC. Late treatment with anti-LFA-1 (CD11a) antibody prevents cerebral malaria in a mouse model. Eur J Immunol 1991; 21:2259–2263.
72. Wahlgren M, Abrams JS, Fernandez V, Bejarano MT, Azuma M, Torii M, et al. Adhesion of *Plasmodium falciparum*-infected erythrocytes to human cells and secretion of cytokines (IL-1-beta, IL-1RA, IL-6, IL-8, IL-10, TGF beta, TNF alpha, G-CSF, GM-CSF). Scand J Immunol 1995; 42:626–636.
73. Berendt AR, Simmons DL, Tansey J, Newbold CI, Marsh K. Intercellular adhesion molecule-1 is an endothelial cell adhesion receptor for *Plasmodium falciparum*. Nature 1989; 341:57–59.
74. Turner GDH, Morrison H, Jones M, Davis TME, Looareesuwan S, Buley ID, et al. An immunohistochemical study of the pathology of fatal malaria—evidence for widespread endothelial activation and a potential role for intercellular adhesion molecule-1 in cerebral sequestration. Am J Pathol 1994; 145:1057–1069.
75. Kumaratilake LM, Rathjen DA, Mack P, Widmer F, Pasertsiriroj V, Ferrante A. A synthetic tumor necrosis factor agonist peptide enhances human polymorphonuclear leukocyte-mediated killing of *Plasmodium falciparum* in vitro and suppresses *Plasmodium chabaudi* infection in mice. J Clin Invest 1995; 95:2315–2323.
76. Kubes P, Suzuki M, Granger DN. Nitric oxide—an endogenous modulator of leukocyte adhesion. Proc Natl Acad Sci USA 1991; 88:4651–4655.
77. Gaboury J, Woodman RC, Granger DN, Reinhardt P, Kubes P. Nitric oxide prevents leukocyte adherence—role of superoxide. Am J Physiol 1993; 265:H862–H867.
78. Siney L, Lewis MJ. Endothelium-derived relaxing factor inhibits platelet adhesion to cultured porcine endocardial endothelium. Eur J Pharmacol 1992; 229:223–226.
79. Herbaczynskacedro K, Lembowicz K, Pytel B. NG-Monomethyl-L-arginine increases platelet deposition on damaged endothelium in vivo—a scanning electron microscopic study. Thromb Res 1991; 64:1–9.
80. Radomski MW, Palmer RMJ, Moncada S. Modulation of platelet aggregation by an L-arginine nitric oxide pathway. Trends Pharmacol Sci 1991; 12:87–88.
81. Radomski MW, Vallance P, Whitley G, Foxwell N, Moncada S. Platelet adhesion to human vascular endothelium is modulated by constitutive and cytokine induced nitric oxide. Cardiovasc Res 1993; 27:1380–1382.
82. Evans T, Carpenter A, Silva A, Cohen J. Inhibition of nitric oxide synthase in experimental gram-negative sepsis. J Infect Dis 1994; 169:343–349.
83. Kremsner PG, Nussler A, Neifer S, Chaves MF, Bienzle U, Senaldi G, et al. Malaria antigen and cytokine-induced produc-

tion of reactive nitrogen intermediates by murine macrophages: no relevance to the development of experimental cerebral malaria. Immunology 1993; 78:286–290.
84. Senaldi G, Kremsner PG, Grau GE. Nitric oxide and cerebral malaria. Lancet 1992; 340:1554–1555.
85. Asensio VC, Oshima H, Falanga PB. *Plasmodium berghei*—Is nitric oxide involved in the pathogenesis of mouse cerebral malaria? Exp Parasitol 1993; 77:111–117.
86. Robertson FM, Offner PJ, Ciceri DP, Becker WK, Pruitt BA, Billiar TR. Detrimental hemodynamic effects of nitric oxide synthase inhibition in septic shock. Arch Surg 1994; 129:149–156.
87. Ghigo D, Todde R, Ginsburg H, Costamagna C, Gautret P, Bussolino F, et al. Erythrocyte stages of *Plasmodium falciparum* exhibit a high nitric oxide synthase (NOS) activity and release an NOS-inducing soluble factor. J Exp Med 1995; 182:677–688.
88. Kwiatkowski D, Hill AV, Sambou I, Twumasi P, Castracane J, Manogue KR, et al. TNF concentration in fatal cerebral, non-fatal cerebral, and uncomplicated *Plasmodium falciparum* malaria. Lancet 1990; 336:1201–1204.
89. Shaffer N, Grau GE, Hedberg K, Davachi F, Lyamba B, Hightower AW, et al. Tumor necrosis factor and severe malaria. J Infect Dis 1991; 163:96–101.
90. Butcher GA, Garland T, Ajdukiewicz AB, Clark IA. Serum tumor necrosis factor associated with malaria in patients in the Solomon Islands. Trans R Soc Trop Med Hyg 1990; 84:658–661.
91. Kern P, Hemmer CJ, Gallati H, Neifer S, Kremsner P, Dietrich M, et al. Soluble tumor necrosis factor receptors correlate with parasitemia and disease severity in human malaria. J Infect Dis 1992; 166:930–934.
92. Molyneux ME, Engelmann H, Taylor TE, Wirima JJ, Aderka D, Wallach D, et al. Circulating plasma receptors for tumour necrosis factor in Malawian children with severe falciparum malaria. Cytokine 1993; 5:604–609.
93. de Kossodo S, Houba V, the WHO Study Group, Grau GE. Assaying tumour necrosis factor concentrations in human serum: a WHO international collaborative study. J Immunol Methods 1995; 11:107–114.
94. Harpaz R, Edelman R, Wasserman SS, Levine MM, Davis JR, Sztein MB. Serum cytokine profiles in experimental human malaria. J Clin Invest 1992; 90:505–523.
95. Mshana RN, Boulandi J, Mshana NM, Mayombo J, Mendome G. Cytokines in the pathogenesis of malaria: levels of IL-1beta, IL-4, IL-6, TNF-alpha and IFN-gamma in plasma of healthy individuals and malaria patients in a holoendemic area. J Clin Lab Immunol 1991; 34:131–139.
96. Friedland JS, Ho M, Remick DG, Bunnag D, White NJ, Griffin GE. Interleukin-8 and *Plasmodium falciparum* Malaria in Thailand. Trans R Soc Trop Med Hyg 1993; 87:54–55.
97. Peyron F, Burdin N, Ringwald P, Vuillez JP, Rousset F, Banchereau J. High levels of circulating IL-10 in human malaria. Clin Exp Immunol 1994; 95:300–303.
98. Wenisch C, Parschalk B, Narzt E, Looareesuwan S, Graninger W. Elevated serum levels of IL-10 and IFN-gamma in patients with acute *Plasmodium falciparum* malaria. Clin Immunol Immunopathol 1995; 74:115–117.
99. Deloron P, Dumont N, Nyongabo T, Aubry P, Astagneau P, Ndarugirire F, et al. Immunologic and biochemical alterations in severe falciparum malaria: relation to neurological symptoms and outcome. Clin Infect Dis 1994; 19:480–485.
100. Ho M, Sexton MM, Tongtawe P, Looareesuwan S, Suntharasamai P, Webster HK. Interleukin-10 inhibits tumor necrosis factor production but not antigen-specific lymphoproliferation in acute *Plasmodium falciparum* malaria. J Infect Dis 1995; 172:838–844.
101. Jakobsen PH, Mckay V, Morrisjones SD, Mcguire W, Vanhensbroek MB, Meisner S, et al. Increased concentrations of interleukin-6 and interleukin-1 receptor antagonist and decreased concentrations of beta- 2-glycoprotein I in Gambian children with cerebral malaria. Infect Immun 1994; 62:4374–4379.
102. Burgmann H, Hollenstein U, Wenisch C, Thalhammer F, Looareesuwan S, Graninger W. Serum concentrations of MIP-1 alpha and interleukin-8 in patients suffering from acute *Plasmodium falciparum* malaria. Clin Immunol Immunopathol 1995; 76:32–36.
103. Chizzolini C, Grau GE, Geinoz A, Schrijvers D. Lymphocyte-T interferon-gamma production induced by *Plasmodium falciparum* antigen is high in recently infected non-immune and low in immune subjects. Clin Exp Immunol 1990; 79:95–99.
104. Kremsner PG, Winkler S, Brandt C, Wildling E, Jenne L, Graninger W, et al. Elevated capacity of tumor necrosis factor production predicts accelerated cure in patients with Plasmodium falciparum malaria. Am J Trop Med Hyg 1995;
105. McGuire W, Hill AVS, Allsopp CEM, Greenwood BM, Kwiatkowski D. Variation in the TNF-alpha promoter region associated with susceptibility to cerebral malaria. Nature 1994; 371:508–511.
106. Wyler DJ. Steroids are out in the treatment of cerebral malaria: what's next. J Infect Dis 1988; 158:320–324.
107. Chanling T, Neill AL, Hunt NH. Early microvascular changes in murine cerebral malaria detected in retinal whole mounts. Am J Pathol 1992; 140:1121–1130.
108. Neill AL, Chanling T, Hunt NH. Comparisons between microvascular changes in cerebral and non-cerebral malaria in mice, using the retinal whole-mount technique. Parasitology 1993; 107:477–487.
109. Grau GE, Lou J. TNF in vascular pathology: importance of platelet-endothelium interactions. Res Immunol 1993; 144:355–363.
110. Lou J, Grau GE. Platelet fusion to endothelial cells: a critical mechanism of TNF alpha-induced vascular changes. Pathol Biol 1994; 42:810 Abstract.
111. Gay F, Robert C, Pouvelle B, Peyrol S, Scherf A, Gysin J. Isolation and characterization of brain microvascular endothelial cells from Saimiri monkeys—an in vitro model for sequestration of *Plasmodium falciparum*-infected erythrocytes. J Immunol Methods 1995; 184:15–28.
112. Tacchini-Cottier F, Lou J, Roberts DJ, Garcia AM, Grau GE. Detection of an LFA-1 epitope on the surface of erythrocytes infected with a strain of *P. falciparum*. Immunology 1995; 85:205–213.
113. Crandall I, Sherman IW. Antibodies to synthetic peptides based on band 3 motifs react specifically with *Plasmodium falciparum* (human malaria)-infected erythrocytes and block cytoadherence. Parasitology 1994; 108:389–396.
114. Taylor-Robinson AW. Murine models of cerebral malaria: a qualified defence. Parasitol Today 1995; 11:407–409.

Index

About the Editors

DANIEL G. REMICK is an Associate Professor of Pathology at the University of Michigan, and the Director of Autopsy Service as well as Electron Microscopy Service at the University of Michigan Health Systems, Ann Arbor. The author of numerous scientific publications, he is a member of the American Society of Investigative Pathologists, the Shock Society, the American Association of Immunologists, and the International Cytokine Society, among others. Dr. Remick received the M.D. degree (1982) from the Mayo Medical School, Rochester, Minnesota.

JON S. FRIEDLAND is a Senior Lecturer as well as an Honorary Consultant Physician in Infectious Diseases and Medicine at the Royal Postgraduate Medical School, Hammersmith Hospital, London, England. The author or coauthor of numerous scientific publications, he is a Fellow of the Royal Society of Tropical Medicine and Hygiene and a member of the International Cytokine Society and the Royal College of Physicians, among others. Dr. Friedland received the M.B., B.S. degree (1984) from the University of London, England, the M.A. degree (1985) from Cambridge University, England, and the Ph.D. degree (1993) from the University of London, England.